RAVEN

BIOLOGY OF PLANTS

Vincent van Gogh (1853–1890)
Field of Poppies, Auvers-sur-Oise, June 1890

In May 1890, Vincent van Gogh moved to Auvers-sur-Oise to be near his brother Theo, who supported him financially and emotionally throughout his life. Lacking the money to pay for models, van Gogh turned to still lifes and landscapes, and in the 70 days before his death, he feverishly painted 70 works of art. He had moved to Auvers in late spring, when poppies bloom, and he admired the contrasts of the brilliant red flowers growing in the fields of "vigorously" green alfalfa. Van Gogh wrote in a letter to his sister Wilhelmina that "there are colors which cause each other to shine brilliantly, which form a couple, which complete each other like man and woman."

EIGHTH EDITION

RAVEN BIOLOGY OF PLANTS

Ray F. Evert
University of Wisconsin, Madison

Susan E. Eichhorn
University of Wisconsin, Madison

 W. H. Freeman and Company Publishers

Publisher: Peter Marshall
Associate Director of Marketing: Debbie Clare
Developmental Editor: Sally Anderson
Supplements and Media Editor: Marni Rolfes
Project Editor: Vivien Weiss
Photo Editors: Bianca Moscatelli and Elyse Rieder
Cover and Text Designer: Blake Logan
Illustrations: Rhonda Nass
Senior Illustration Coordinator: Bill Page
Production Coordinator: Susan Wein
Composition: Sheridan Sellers
Printing and Binding: RR Donnelly

Cover image: Haags Gemeentemuseum, The Hague, Netherlands/
The Bridgeman Art Library International

Library of Congress Control Number: 2011945124
ISBN-13: 978-1-4292-1961-7
ISBN-10: 1-4292-1961-0

Printed in the United States of America
Second printing

W. H. Freeman and Company
41 Madison Avenue
New York, NY 10010
Houndmills, Basingstoke RG21 6XS, England
www.whfreeman.com

To Peter H. Raven, who has been a co-author of this book since its inception. It was he and Helena Curtis who saw the need for a new introductory plant biology textbook, with a fresh approach, for botany and plant science majors. The success of *Biology of Plants* has been due in large part to Peter's efforts through the first seven editions. We dedicate this eighth edition to him.

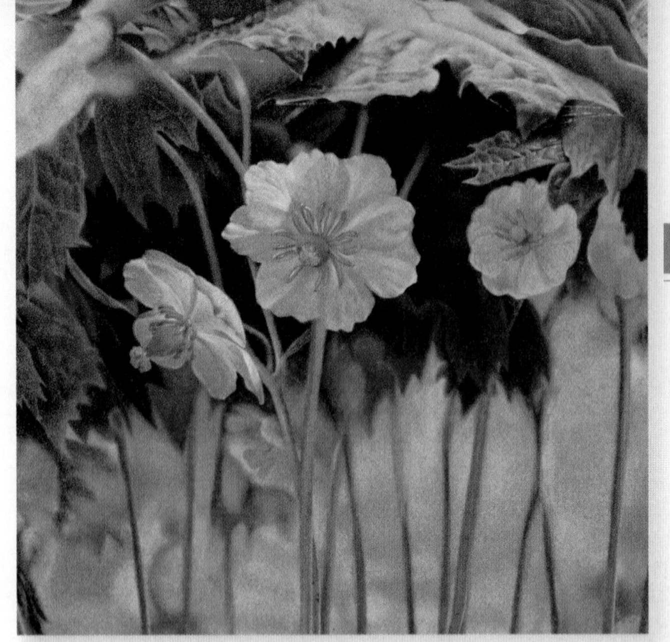

CONTENTS IN BRIEF

CONTENTS

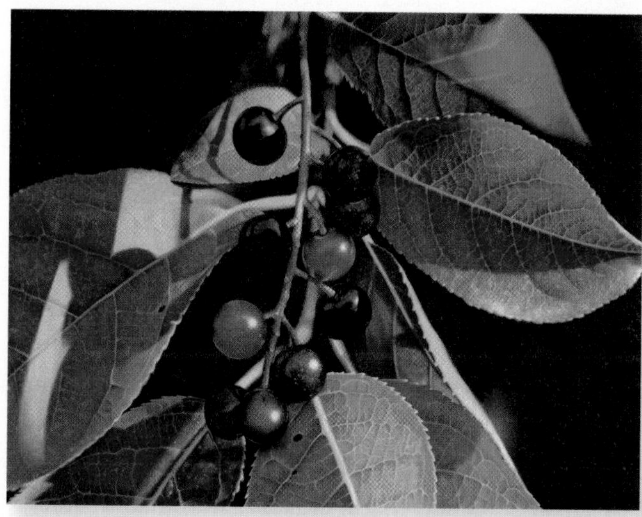

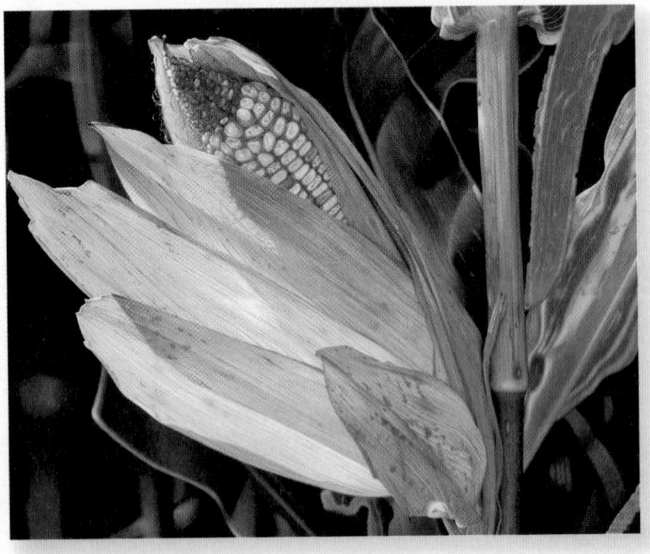

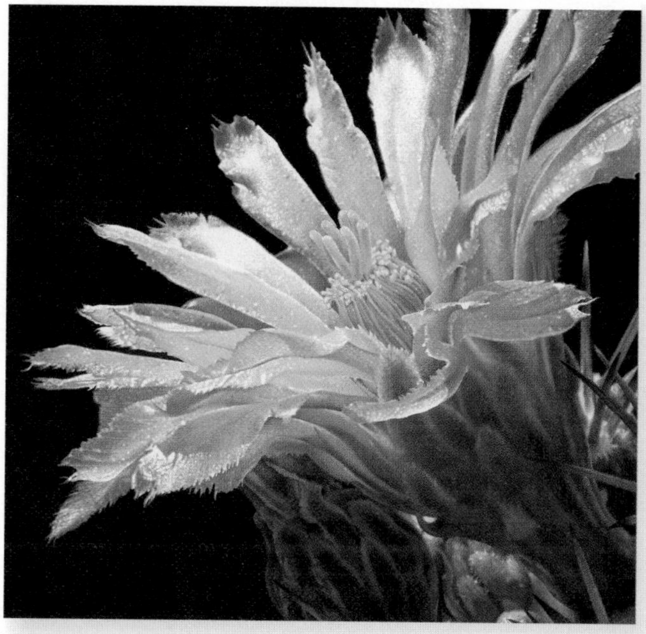

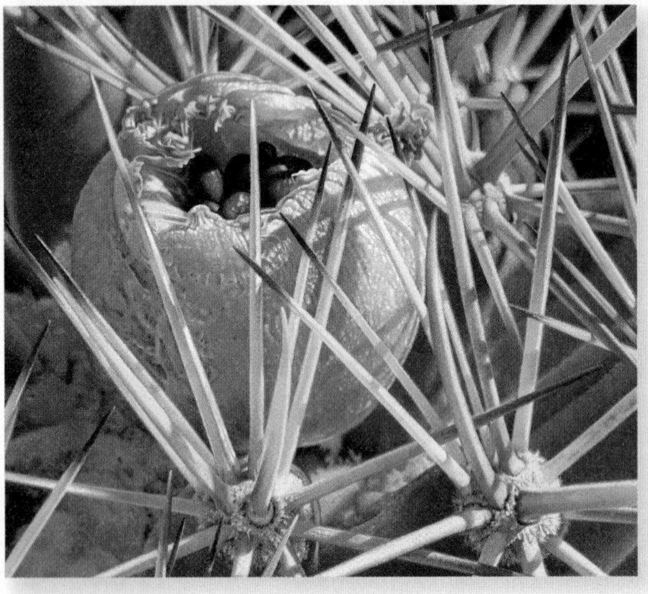

SECTION 7 ECOLOGY

PREFACE

As we approached this revision of *Biology of Plants,* we recognized that extensive work would be needed to address the advances that have been made in all areas of plant biology. From new molecular details about photosynthesis to the vast differences in taxonomic relationships that have been revealed by comparison of DNA and RNA sequences, to advances in genomics and genetic engineering to an enhanced understanding of the anatomy and physiology of plants, there have been exciting developments in the field. This current edition of *Biology of Plants* has undergone the most significant revision in its history, with every topic scrutinized and, where necessary, revised and updated.

While covering these advances, we have strengthened the narrative by expanding and clarifying discussions; carefully defining new terms; and adding new diagrams, photos, and electron micrographs. Each chapter now begins with an attractive photograph and informative caption that relates to the chapter content but in a tangential way that often touches on an environmental topic.

With each revision, we continue to pay special attention to the book's interlocking themes: (1) the functioning plant body as the dynamic result of processes mediated by biochemical interactions; (2) evolutionary relationships as valuable for understanding form and function in organisms; (3) ecology as an integrated theme that pervades the book and emphasizes our dependence on plants to sustain all life on Earth; and (4) molecular research as essential for revealing details about plant genetics, cellular function, and taxonomic relationships.

Changes Reflecting Major Recent Advances in Plant Science

Every chapter has been carefully revised and updated, most notably:

- **Chapter 7** (Photosynthesis, Light, and Life)—presents an expanded discussion of light reactions, including an updated diagram on the transfer of electrons and protons during photosynthesis; new essay on "Global Warming: The Future Is Now"
- **Chapter 9** (The Chemistry of Heredity and Gene Expression)—incorporates histone acetylation, DNA methylation, epigenesis, and noncoding RNAs
- **Chapter 10** (Recombinant DNA Technology, Plant Biotechnology, and Genomics)—updates material on the impact of new molecular methods for studying plants, resulting in the development of golden rice, as well as plants that are resistant to herbicides, pesticides, and diseases
- **Chapter 11** (The Process of Evolution)—covers recombination speciation (speciation not involving polyploidy) and includes two new essays on "Invasive Plants" and "Adaptive Radiation in Hawaiian Lobeliads"
- **Chapter 12** (Systematics: The Science of Biological Diversity)—presents an expanded discussion of the chloroplast as the main source of plant DNA sequence data and introduces DNA barcoding and supergroups; new essay on "Google Earth: A Tool for Discovering and Protecting Biodiversity"
- **Chapter 14** (Fungi)—reorganized and updated with the latest classifications; includes nucleariids and the phyla Microsporidia and Glomeromycota, as well as a new phylogenetic tree of the fungi

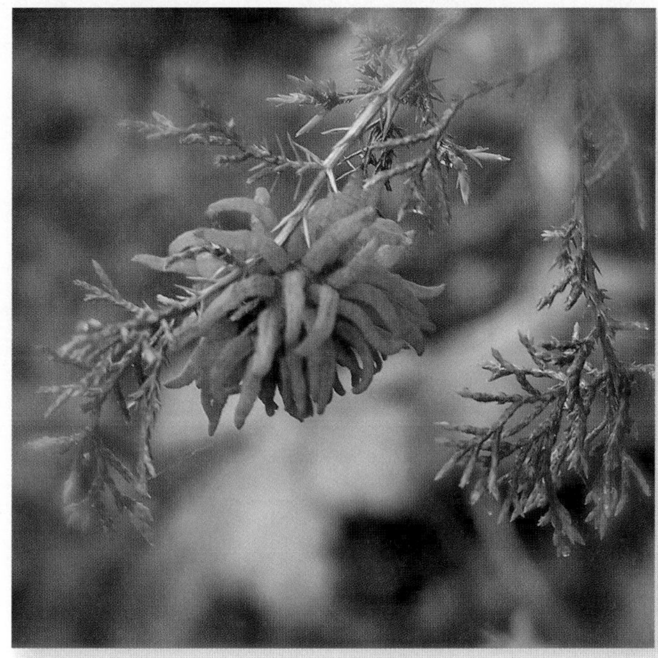

The cedar-apple rust fungus alternates between two hosts, cedar and apple trees, causing damage to apple harvests (page 278).

- **Chapter 15** (Protists: Algae and Heterotrophic Protists)—incorporates the latest classifications, including a phylogenetic tree showing the relationships of the algae; new discussion of the cultivation of algae for biofuel production and a new essay on "Coral Reefs and Global Warming"
- **Chapter 18** (Gymnosperms)—includes an expanded discussion of double fertilization in the gnetophytes, as well as a cladogram of the phylogenetic relationships among the major groups of embryophytes and a new figure depicting alternative hypotheses of relationships among the five major lineages of seed plants
- **Chapter 19** (Introduction to the Angiosperms)—follows the classification recommended by the Angiosperm Phylogeny Group and presents an expanded discussion of embryo sac types
- **Chapter 20** (Evolution of the Angiosperms)—presents an expanded discussion of angiosperm ancestors and includes new cladograms depicting the phylogenetic relationships of the angiosperms

Austrobaileya scandens is considered to have evolved separately from the main lineage of angiosperms (page 480).

The Chinese brake fern removes arsenic from contaminated soils (page 391).

- **Chapter 21** (Plants and People)—updated and revised to include a new figure depicting independent centers of plant domestication and discusses efforts to develop perennial versions of important annual grains; new essay on "Biofuels: Part of the Solution, or Another Problem?"
- **Chapter 22** (Early Development of the Plant Body)—the discussion of seed maturation and dormancy has been revised, and considerable fine-tuning and updating have occurred throughout this and the other anatomy chapters, with an emphasis on structure/function relationships
- **Chapter 23** (Cells and Tissues of the Plant Body)—the presence of forisomes in sieve-tube elements of some legumes has been added
- **Chapter 24** (The Root: Structure and Development)—the topic of border cells and their functions has been added
- **Chapter 25** (The Shoot: Primary Structure and Development)—includes a new discussion accompanied by micrographs on leaf vein development and the ABCDE model of flower development; new essay on "Strong, Versatile, Sustainable Bamboo"
- **Chapter 26** (Secondary Growth in Stems)—a new diagram depicting the relationship of the vascular cambium to secondary xylem and secondary phloem has been added
- **Chapter 27** (Regulating Growth and Development: The Plant Hormones)—expanded discussions of the role of auxin

in vascular differentiation and on hormone receptors and signaling pathways for the plant hormones; new discussions of brassinosteroids as a major class of plant hormones and of strigolactones, which interact with auxin in regulating apical dominance

- **Chapter 28** (External Factors and Plant Growth)—extensive revision of gravitropism, circadian rhythms, floral stimulus, and thigmonastic movements; new discussion of genes and vernalization, as well as hydrotropism, phytochrome-interaction factors (PIFs), and the shade-avoidance syndrome; new essay on "Doomsday Seed Vault: Securing Crop Diversity"

- **Chapter 29** (Plant Nutrition and Soils)—discusses strategies involving nitrogen uptake by plants, along with the new topics of beneficial elements, determinate and indeterminate nodules, and strategies by plants for the acquisition of phosphate; essay added on "The Water Cycle"

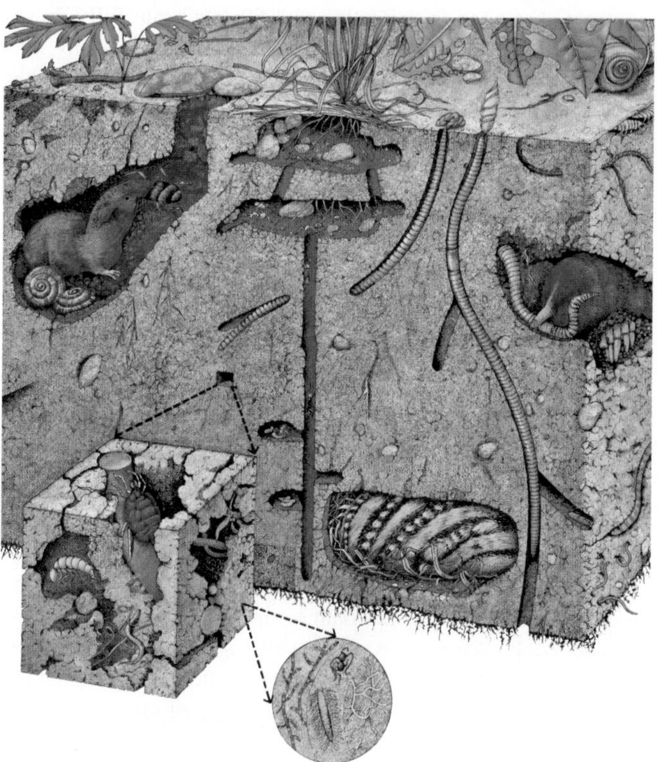

Living organisms of the A soil horizon, or "topsoil" (page 690).

- **Chapter 30** (The Movement of Water and Solutes in Plants)—expanded discussion on hydraulic redistribution and on the mechanisms of phloem loading, including the polymer trapping mechanism; new essay on "Green Roofs: A Cool Alternative"

- **Chapter 31** (The Dynamics of Communities and Ecosystems) and Chapter 32 (Global Ecology)—remain online at www.whfreeman.com/raven8e. Fully illustrated, these chapters have been thoroughly updated by Paul Zedler of the University of Wisconsin, Madison

Media and Supplements

Companion Web Site

www.whfreeman.com/raven8e

For students, this free interactive Web site includes:

- **Interactive study aids** to help foster understanding of important concepts from the text, including multiple-choice quizzes, flashcards, plus interactive figures and tables

- **Animations and live-action videos** illustrate several topics from the text for a deeper understanding of the more difficult concepts

- **Two ecology chapters** in easily downloadable and printable PDF format, fully illustrated and thoroughly updated with major contributions by Paul Zedler, University of Wisconsin, Madison.

The dynamic **eBook** is customizable and fully integrates the complete contents of the text and interactive media in a format that features a variety of helpful study tools, including full-text searching, note-taking, bookmarking, highlighting and more.

For instructors, the site includes **all figures** from the text in both PowerPoint and JPEG formats, optimized for excellent classroom projection. The site also includes the **Test Bank** by Robert C. Evans, Rutgers University, Camden, which has been thoroughly revised and organized by chapter in easy-to-edit Word files. The Test Bank features approximately 90 questions per chapter, structured as multiple-choice, true-false, and short-answer questions that can be used to test student comprehension of all major topics in the textbook.

Instructor's Resource DVD with Test Bank

Included here are all the resources from the site, with **all text images** in JPEG and PowerPoint formats, plus the Test Bank in easy-to-edit and -print Word format.

Overhead Transparency Set

The set includes full-color illustrations from the book, optimized for classroom projection.

Laboratory Topics in Botany

Laboratory Topics in Botany offers several exercises within each topic that can be selected for coverage that suits individual course needs. Questions and problems follow each topic; refinements and updating have been made throughout. Written by Ray F. Evert and Susan E. Eichhorn, University of Wisconsin, Madison; and Joy B. Perry, University of Wisconsin, Fox Valley.

Preparation Guide for Laboratory Topics in Botany

Preparation Guide for Laboratory Topics in Botany offers helpful information on how to cover each topic, the length of time needed to complete the material, the sources of supplies and ordering schedules, how to set up the laboratory, and ways to guide students in their laboratory work. Written by Susan E. Eichhorn, University of Wisconsin, Madison; Joy B. Perry, University of Wisconsin, Fox Valley; and Ray F. Evert, University of Wisconsin, Madison.

Acknowledgments

We are grateful for the enthusiastic response we have received from readers who have used previous editions of *Biology of Plants*, either in English or in one of the six foreign languages in which it has been published. As always, we have appreciated the support and recommendations made by teachers who used the last edition in their courses. We also wish to thank the following people who provided valuable critiques of chapters or portions of chapters for this edition:

Richard Amasino, *University of Wisconsin, Madison*
Paul Berry, *University of Michigan*
James Birchler, *University of Missouri*
Wayne Becker, *University of Wisconsin, Madison*
Clyde Calvin, *Portland State University*
Kenneth Cameron, *University of Wisconsin, Madison*
Nancy Dengler, *University of Toronto*
John Doebley, *University of Wisconsin, Madison*
Eve Emshwiller, *University of Wisconsin, Madison*
Thomas German, *University of Wisconsin, Madison*
Thomas Givnish, *University of Wisconsin, Madison*
Linda Graham, *University of Wisconsin, Madison*
Christopher Haufler, *University of Kansas*
David Hibbett, *Clark University*
Robin Kurtz, *University of Wisconsin, Madison*
Ben Pierce, *Southwestern University*
Scott Russell, *University of Oklahoma, Norman*
Dennis Stevenson, *New York Botanical Garden*
Joseph Williams, *University of Tennessee, Knoxville*
Paul Zedler, *University of Wisconsin, Madison*

The following people gave us helpful feedback as we planned this edition:

Richard Carter, *Valdosta State University*
Sara Cohen Christopherson, *University of Wisconsin, Madison*
Les C. Cwynar, *University of New Brunswick*
Brian Eisenback, *Bryan College*
Karl H. Hasenstein, *University of Louisiana at Lafayette*
Bernard A. Hauser, *University of Florida*
Jodie S. Holt, *University of California Riverside*
George Johnson, *Arkansas Tech University*
Carolyn Howes Keiffer, *Miami University*
Jeffrey M. Klopatek, *Arizona State University*
Rebecca S. Lamb, *Ohio State University*
Monica Macklin, *Northeastern State University*
Carol C. Mapes, *Kutztown University of Pennsylvania*
Shawna Martinez, *Sierra College*
Austin R. Mast, *Florida State University*
Wilf Nicholls, *Memorial University of Newfoundland*
Karen Renzaglia, *Southern Illinois University*
Frances M. Wren Rundlett, *Georgia State University*
A. L. Samuels, *University of British Columbia*
S. E. Strelkov, *University of Alberta*
Alexandru M. F. Tomescu, *Humboldt State University*
M. Lucia Vazquez, *University of Illinois at Springfield*
Justin K. Williams, *Sam Houston State University*
Michael J. Zanis, *Purdue University*

We are very grateful to our artist, Rhonda Nass, for the exquisite paintings that open each section and for her beautifully drawn artwork. She has worked closely with us through many editions, and we value her ability to interpret our pencil sketches and render them into drawings that are instructive and accurate, as well as attractive. We are grateful to Rick Nass, who contributed a number of expertly produced graphs. We also thank Sarah Friedrich and Kandis Elliot, Media Specialists, Department of Botany, University of Wisconsin, Madison, for the preparation of digital images of photomicrographs and herbarium specimens. Mark Allen Wetter, Collections Manager/Senior Academic Curator, and Theodore S. Cochrane, Senior Academic Curator, both of the Wisconsin State Herbarium, Department of Botany, University of Wisconsin, Madison, were very helpful in selecting and scanning herbarium specimens for our use.

We would especially like to thank Sally Anderson, our talented developmental editor, who has worked with us for five editions now. We are grateful for her many contributions at every stage of the process, from the early stages of planning the new edition through the manuscript and proof stages to the finished book. We have worked well together over the years, and we thank her for her dedication to all aspects of making this edition the most accurate and accessible of all.

We would also like to thank Richard Robinson, who has written the engaging ecology-oriented essays that have been added to this edition. The essays are highlighted in the text by a green leaf, and they cover such topics as green roofs, invasive plants, coral bleaching, the development of biofuels, and the use of Google Earth to map and study biodiversity.

The preparation of the eighth edition has involved the collaborative efforts of a large number of talented people at W. H. Freeman and Company. Particular thanks go to Peter Marshall, Publisher of Life Sciences, whose vision and support have made this new edition possible; to Vivien Weiss, who has skillfully managed the production process; to Elyse Rieder and Bianca Moscatelli, who enthusiastically tracked down photographs; and to Blake Logan, who applied her design talents to give this edition a new look. We especially want to thank Sheridan Sellers for the remarkable work she has done in making up the pages of this book—with her aesthetic and pedagogical talents, she has worked a miracle in placing the many large and complicated illustrations into a coherent layout. And we are grateful to Linda Strange, our long-time copyeditor who, with good humor and a steady hand, manages to hold us to a high standard in consistency and accuracy. We also thank Marni Rolfes, associate editor, who competently handled the day-to-day matters and kept us on track, and Bill Page, who coordinated the complicated illustration program. Our gratitude also goes to Debbie Clare, Associate Director of Marketing, who has been tireless in managing the sales and marketing efforts for this edition, and to Susan Wein, Production Coordinator, for her many contributions during the complex stages of production.

A great number of people, only some of whom are mentioned here, have contributed in many essential ways to this revision, and we extend to them our deepest appreciation and gratitude.

Ray F. Evert
Susan E. Eichhorn

INTRODUCTION

◀ Mayapple *(Podophyllum peltatum)*, which grows in open woods and pastures in the contiguous United States and Canada, blooms in early May. Utilizing the sun's energy, the plant rapidly produces stems, leaves, and flowers. Although called mayapple, the fruit, which is yellowish when ripe, is actually a berry. The fruits are edible and can be used in preserves and beverages, but the leaves and roots are poisonous.

CHAPTER 1

Botany: An Introduction

◀ **A change of habitat** Although plants are primarily adapted for life on land, some, such as the hardy water lily *(Nymphaea fabiola)*, have returned to an aquatic existence. Evidence of its ancestors' sojourn on land includes a water-resistant waxy outer layer, or cuticle, as well as stomata through which gases are exchanged, and a highly developed internal transport system.

CHAPTER OUTLINE

"What drives life is . . . a little current, kept up by the sunshine," wrote Nobel laureate Albert Szent-Györgyi. With this simple sentence, he summed up one of the greatest marvels of evolution—photosynthesis. During the photosynthetic process, radiant energy from the sun is captured and used to form the sugars on which all life, including our own, depends. Oxygen, also essential to our existence, is released as a by-product. The "little current" begins when a particle of light strikes a molecule of the green pigment chlorophyll, boosting one of the electrons in the chlorophyll to a higher energy level. The "excited" electron, in turn, initiates a flow of electrons that ultimately converts the radiant energy from the sun to the chemical energy of sugar molecules. Sunlight striking the leaves of the water lily shown above, for example, is the first step in the process leading to production of the molecules that make up the flowers, leaves, and stems, as well as all the molecular components that allow the plants to grow and develop.

Only a few types of organisms—plants, algae, and some bacteria—possess chlorophyll, which is essential for a living cell to carry out photosynthesis. Once light energy is trapped in chemical form, it becomes available as an energy source to all other organisms, including human beings. We are totally dependent on photosynthesis, a process for which plants are exquisitely adapted.

The word "botany" comes from the Greek *botanē,* meaning "plant," derived from the verb *boskein,* "to feed." Plants, however, enter our lives in innumerable ways other than as sources of food. They provide us with fiber for clothing; wood for furniture, shelter, and fuel; paper for books (such as the page you are reading at this moment); spices for flavor; drugs for medicines; and the oxygen we breathe. We are utterly dependent on plants. Plants also have enormous sensory appeal, and our lives are enhanced by the gardens, parks, and wilderness areas available to us. The study of plants has provided us with great insight into the nature of all life and will continue to do so in the years ahead. And, with genetic engineering and other forms of modern technology, we have entered the most exciting period in the history of botany, where plants can be transformed, for example, to resist disease, kill pests, produce vaccines, manufacture biodegradable plastic, tolerate high-salt soils, resist freezing, and provide higher levels of vitamins and minerals in food products, such as maize (corn) and rice.

CHECKPOINTS

After reading this chapter, you should be able to answer the following:

1. Why do biologists believe that all living things on Earth today share a common ancestor?
2. What is the principal difference between a heterotroph and an autotroph, and what role did each play on the early Earth?
3. Why is the evolution of photosynthesis thought to be such an important event in the evolution of life in general?
4. What were some of the problems encountered by plants as they made the transition from the sea to the land, and what structures in terrestrial plants evolved to solve those problems?
5. What are biomes, and what are the principal roles of plants in an ecosystem?

Evolution of Plants

Life Originated Early in Earth's Geologic History

Like all other living organisms, plants have had a long history during which they **evolved,** or changed, over time. The planet Earth itself—an accretion of dust and gases swirling in orbit around the star that is our sun—is some 4.6 billion years old (Figure 1–1). It is believed that Earth sustained a lethal meteor bombardment that ended about 3.8 to 3.9 billion years ago. Vast chunks of rubble slammed into the planet, helping to keep it hot. As the molten Earth began to cool, violent storms raged, accompanied by lightning and the release of electrical energy, and widespread volcanism spewed molten rock and boiling water from beneath the Earth's surface.

The earliest known fossils are found in rocks of Western Australia about 3.5 billion years old (Figure 1–2). These microfossils consist of several kinds of small, relatively simple filamentous microorganisms resembling bacteria. About the same age as these microfossils are ancient **stromatolites**—fossilized microbial mats consisting of layers of filamentous and other microorganisms and trapped sediment. Stromatolites continue to be formed today in a few places, such as in the warm, shallow oceans off the shores of Australia and the Bahamas (see Chapter 13). By comparing the ancient stromatolites with modern ones, which are formed by cyanobacteria (filamentous photosynthetic bacteria), scientists have concluded that the ancient stromatolites were formed by similar filamentous bacteria.

Whether life originated on Earth or reached Earth through space in the form of spores (resistant reproductive cells) or by some other means is problematic. Life may have formed on Mars, for example, whose early history paralleled that of Earth. Strong evidence, first discovered by the Opportunity rover in 2004, indicated that water once flowed across the planet, raising the possibility that, at one time, Mars could have supported life (Figure 1–3). In 2008, the Phoenix Mars

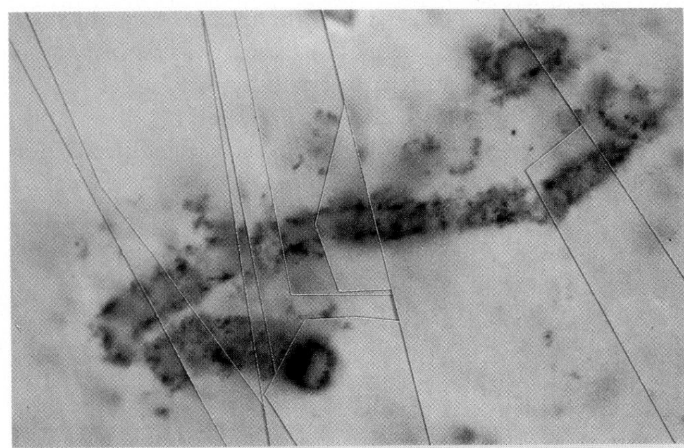

1–2 The earliest known fossils Obtained from ancient rocks in northwestern Western Australia, these fossilized prokaryotes are dated at 3.5 billion years of age. They are about a billion years younger than the Earth itself, but there are few suitable older rocks in which to look for earlier evidence of life. More complex organisms—those with eukaryotic cellular organization—did not evolve until about 2.1 billion years ago. For about 1.5 billion years, therefore, prokaryotes were the only forms of life on Earth. These so-called microfossils have been magnified 1000 times.

1–1 Life on Earth Of the nine planets in our solar system, only one, as far as we know, has life on it. This planet, Earth, is visibly different from the others. From a distance, it appears blue and green, and it shines a little. The blue is water, the green is chlorophyll, and the shine is sunlight reflected off the layer of gases surrounding the planet's surface. Life, at least as we know it, depends on these visible features of Earth.

1–3 Life on Mars? This color-enhanced image shows a portion of the Jezero Crater, a 25-mile-wide impact crater on northern Mars that once held a lake. Claylike minerals (indicated in green) were carried by ancient rivers into the lake, forming a delta. Because clays are able to trap and preserve organic matter, deltas and lakebeds are promising areas in which to search for signs of ancient life on Mars.

Lander found water ice in abundance near the surface. Moreover, its instruments monitored a diurnal water cycle: water vapor, originating from the shallow subsurface water ice and from water clinging to soil grains, is released into the Martian atmosphere during the morning hours, and at night it condenses and falls out by gravity. Most of the ice crystals evaporate as they fall through the atmospheric boundary layer, but snowfall on Mars has been observed.

No organic molecules or traces of previous or present biological activity were detected at the Phoenix landing site. However, one would expect organic molecules to be present in the Martian soil, given the steady influx of certain types of meteorites that contain considerable quantities of organic material. Meteorites that fall to Earth contain amino acids and organic carbon molecules such as formaldehyde. We will continue to assume, however, that life on Earth originated on Earth.

In 2011, NASA's Mars Reconnaissance Orbiter satellite found evidence of liquid water flowing down slopes and crater walls during the warm month on Mars. The liquid is thought to be highly salty and to occur just below the surface, where it is protected from freezing in the frigid temperatures found on Mars and from evaporating in the planet's low air pressures. These findings further raise the possibilities of finding life on Mars.

Most Likely, the Forerunners of the First Cells Were Simple Aggregations of Molecules

According to current theories, organic molecules, formed by the action of lightning, rain, and solar energy on gases in the environment or spewed out of hydrothermal vents, accumulated in the oceans. Some organic molecules have a tendency to aggregate in groups, and these groups probably took the form of droplets, similar to the droplets formed by oil in water. Such assemblages of organic molecules appear to have been the forerunners of primitive cells, the first forms of life. Sidney W. Fox and his coworkers at the University of Miami produced proteins that aggregated into cell-like bodies in water. Called "proteinoid microspheres," these bodies grow slowly by the accumulation of additional proteinoid material and eventually bud off smaller microspheres. Although Fox likened this process to a type of reproduction, the microspheres are not living cells. Some researchers have suggested that clay particles, or even bubbles, may have played a role in life's origin on Earth by collecting chemicals and concentrating them for synthesis into complex molecules.

According to current theories, these organic molecules may also have served as the source of energy for the earliest forms of life. The primitive cells or cell-like structures were able to use these abundant compounds to satisfy their energy requirements. As they evolved and became more complex, these cells were increasingly able to control their own destinies. With this increasing complexity, they acquired the *ability to grow, to reproduce,* and *to pass on their characteristics to subsequent generations (heredity).* Together with *cellular organization,* these properties characterize all living things on Earth.

Today, just about all organisms use an identical genetic code to translate DNA into proteins (see Chapter 9), whether they are fungi, plants, or animals. It seems quite clear, therefore, that life as we know it emerged on Earth only once and that all living things share a common ancestor: a DNA-based microbe that lived more that 3.5 billion years ago. Near the end of *On the Origin of Species,* Charles Darwin wrote: "Probably all the organic beings which have ever lived on this earth have descended from some one primordial form, into which life was first breathed."

Autotrophic Organisms Make Their Own Food, but Heterotrophic Organisms Must Obtain Their Food from External Sources

Cells that satisfy their energy requirements by consuming the organic compounds produced by external sources are known as **heterotrophs** (Gk. *heteros,* "other," and *trophos,* "feeder"). A heterotrophic organism is dependent on an outside source of organic molecules for its energy. Animals, fungi (Figure 1–4), and many of the one-celled organisms, such as certain bacteria and protists, are heterotrophs.

As the primitive heterotrophs increased in number, they began to use up the complex molecules on which their existence depended—and which had taken millions of years to accumulate. Organic molecules in free solution (that is, not inside a cell)

1–4 A modern heterotroph This fungus, an orange-cap boletus (*Leccinum* sp.) known as an Aspen mushroom, is growing on a forest floor in Colorado. Like other fungi, this boletus absorbs its food (often from other organisms).

1–5 Photosynthetic autotrophs Large-flowered trilliums *(Trillium grandiflorum)*, one of the first plants to flower in spring in the deciduous woods of eastern and midwestern North America, are seen here growing at the base of birch trees. Like most vascular plants, trilliums and birches are rooted in the soil; photosynthesis occurs chiefly in the leaves. Trillium produces flowers in well-lighted conditions, before leaves appear on surrounding trees. The underground portions (rhizomes) of the plant live for many years and spread to produce new plants vegetatively under the thick cover of decaying material on the forest floor. Trilliums also reproduce by producing seeds, which are dispersed by ants.

became more and more scarce, and competition began. Under the pressure of this competition, cells that could make efficient use of the limited energy sources now available were more likely to survive than cells that could not. In the course of time, by the long, slow process of elimination of the most poorly adapted, cells evolved that were able to make their own energy-rich molecules out of simple inorganic materials. Such organisms are called **autotrophs,** "self-feeders." Without the evolution of these early autotrophs, life on Earth would soon have come to an end.

The most successful of the autotrophs were those in which a system evolved for making direct use of the sun's energy—that is, the process of photosynthesis (Figure 1–5). The earliest photosynthetic organisms, although simple in comparison with plants, were much more complex than the primitive heterotrophs. Use of the sun's energy required a complex pigment system to capture the light energy and, linked to this system, a way to store the energy in an organic molecule.

Evidence of the activities of photosynthetic organisms has been found in rocks 3.4 billion years old, about 100 million years after the first fossil evidence of life on Earth. We can be almost certain, however, that both life and photosynthetic organisms evolved considerably earlier than the evidence suggests. In addition, there seems to be no doubt that heterotrophs evolved before autotrophs. With the arrival of autotrophs, the flow of energy in the **biosphere** (that is, the living world and its environment) came to assume its modern form: radiant energy from the sun channeled through the photosynthetic autotrophs to all other forms of life.

Photosynthesis Altered Earth's Atmosphere, Which in Turn Influenced the Evolution of Life

As photosynthetic organisms increased in number, they changed the face of the planet. This biological revolution came about because photosynthesis typically involves splitting the water molecule (H_2O) and releasing its oxygen as free oxygen molecules (O_2). Prior to 2.2 billion years ago, the oxygen released into the oceans and lakes reacted with dissolved iron and precipitated as iron oxides (Figure 1–6). From about 2.7 to 2.2 billion years ago, oxygen began gradually to accumulate in the atmosphere. By about 700 million years ago, atmospheric levels of oxygen increased markedly, and began to approach modern levels during the Cambrian period (570–510 million years ago).

This increase in oxygen level had two important consequences. First, some of the oxygen molecules in the outer layer of the atmosphere were converted to ozone (O_3) molecules. When there is a sufficient quantity of ozone in the atmosphere, it absorbs the ultraviolet rays—rays highly destructive to living organisms—from the sunlight that reaches the Earth. By about 450 million years ago, the ozone layer apparently protected organisms sufficiently so they could survive in the surface layers of water and on the shores, and life emerged on land for the first time.

1–6 Banded iron formations These 2-billion-year-old red bands of iron oxide (also known as rust), found at Jasper Knob in Michigan, are evidence of oxygen accumulation.

Second, the increase in free oxygen opened the way to a much more efficient utilization of the energy-rich carbon-containing molecules formed by photosynthesis. It enabled organisms to break down those molecules by the oxygen-utilizing process known as **respiration.** As discussed in Chapter 6, respiration yields far more energy than can be extracted by any **anaerobic,** or oxygenless, process.

Before the atmosphere accumulated oxygen and became **aerobic,** the only cells that existed were **prokaryotic**—simple cells that lacked a nuclear envelope and did not have their genetic material organized into complex chromosomes. It is likely that the first prokaryotes were heat-loving organisms called "archaea" (meaning "ancient ones"), the descendants of which are now known to be widespread, with many thriving at extremely high temperatures and in acid environments hostile to other forms of life. Bacteria are also prokaryotes. Some archaea and bacteria are heterotrophic, and others, such as the cyanobacteria, are autotrophic.

According to the fossil record, the increase of relatively abundant free oxygen was accompanied by the first appearance of **eukaryotic** cells—cells with nuclear envelopes, complex chromosomes, and organelles, such as mitochondria (sites of respiration) and chloroplasts (sites of photosynthesis), surrounded by membranes. Eukaryotic organisms, in which the individual cells are usually much larger than those of the bacteria, appeared about 2.1 billion years ago and were well established and diverse by 1.2 billion years ago. Except for archaea and bacteria, all organisms—from amoebas to dandelions to oak trees to human beings—are composed of one or more eukaryotic cells.

The Seashore Environment Was Important in the Evolution of Photosynthetic Organisms

Early in evolutionary history, the principal photosynthetic organisms were microscopic cells floating below the surface of the sunlit waters. Energy abounded, as did carbon, hydrogen, and oxygen, but as the cellular colonies multiplied, they quickly depleted the mineral resources of the open ocean. (It is this shortage of essential minerals that is the limiting factor in any modern plans to harvest the seas.) As a consequence, life began to develop more abundantly toward the shores, where the waters were rich in nitrates and minerals carried down from the mountains by rivers and streams and scraped from the coasts by the ceaseless waves.

The rocky coast presented a much more complicated environment than the open sea, and, in response to these evolutionary pressures, living organisms became increasingly complex in structure and more diversified. Not less than 650 million years ago, organisms evolved in which many cells were linked together to form an integrated, multicellular body. In these primitive organisms we see the early stages in the evolution of plants, fungi, and animals. Fossils of multicellular organisms are much easier to detect than those of simpler ones. The history of life on Earth, therefore, is much better documented from the time of their first appearance.

On the turbulent shore, multicellular photosynthetic organisms were better able to maintain their position against the action of the waves, and, in meeting the challenge of the rocky coast, new forms developed. Typically, these new forms developed relatively strong cell walls for support, as well as specialized structures to anchor their bodies to the rocky surfaces (Figure 1–7). As these organisms increased in size, they were confronted with the problem of how to supply food to the dimly lit, more deeply submerged portions of their bodies, where photosynthesis was not taking place. Eventually, specialized food-conducting tissues evolved that extended the length of the bodies of these organisms and connected the upper, photosynthesizing parts with the lower, nonphotosynthesizing structures.

1–7 Evolution of multicellular organisms Early in the course of their evolution, multicellular photosynthetic organisms anchored themselves to rocky shores. These kelp *(Durvillaea potatorum),* seen at low tide on the rocks along the coast of Victoria and Tasmania, Australia, are brown algae (class Phaeophyceae), a group in which multicellularity evolved independently of other groups of organisms.

Colonization of the Land Was Associated with the Evolution of Structures to Obtain Water and Minimize Water Loss

The body of a plant can best be understood in terms of its long history and, in particular, in terms of the evolutionary pressures involved in the transition to land. The requirements of a photosynthetic organism are relatively simple: light, water, carbon dioxide for photosynthesis, oxygen for respiration, and a few minerals. On land, light is abundant, as are oxygen and carbon dioxide, both of which circulate more freely in air than in water. Also, the soil is generally rich in minerals. The critical factor, then, for the transition to land—or as one investigator prefers to say, "to the air"—is water.

Land animals, generally speaking, are mobile and able to seek out water just as they seek out food. Fungi, though immobile, remain largely below the surface of the soil or within whatever damp organic material they feed on. Plants utilize an alternative evolutionary strategy. **Roots** anchor the plant in the ground and collect the water required for maintenance of the plant body and for photosynthesis, while the **stems** provide support for the principal photosynthetic organs, the **leaves.** A

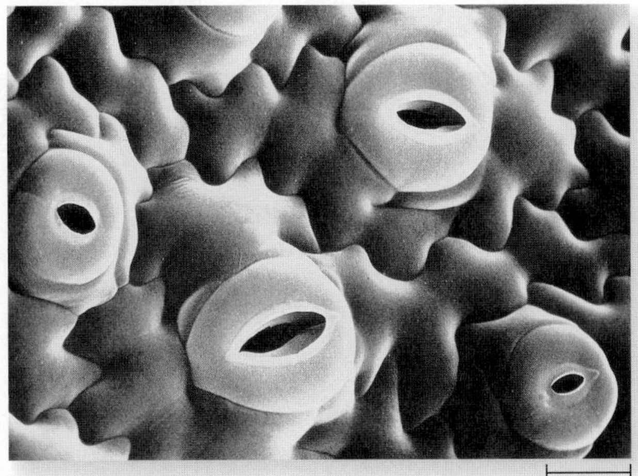

1–8 Stomata Open stomata on the surface of a tobacco *(Nicotiana tabacum)* leaf. Each stoma in the aerial parts of the plant is regulated by two guard cells.

continuous stream of water moves upward through the roots and stems, then out through the leaves. The outermost layer of cells, the **epidermis,** of all the aboveground portions of the plant that are ultimately involved in photosynthesis is covered with a waxy **cuticle,** which retards water loss. However, the cuticle also tends to prevent the exchange of gases between the plant and the surrounding air that is necessary for both photosynthesis and respiration. The solution to this dilemma is found in the **stomata** (singular: stoma), each consisting of a pair of specialized epidermal cells (the guard cells), with a small opening between them. The stomata open and close in response to environmental and physiological signals, thus helping the plant maintain a balance between its water losses and its oxygen and carbon dioxide requirements (Figure 1–8).

In younger plants and in **annuals**—plants with a life span of one year—the stem is also a photosynthetic organ. In longer-lived plants—**perennials**—the stem may become thickened and woody and covered with **cork,** which, like the cuticle-covered epidermis, retards water loss. In both annuals and perennials, the **vascular system,** or conducting system, of the stem conducts a variety of substances between the photosynthetic and nonphotosynthetic parts of the plant body. The vascular system has two major components: the **xylem,** through which water passes upward through the plant body, and the **phloem,** through which food manufactured in the leaves and other photosynthetic parts of the plant is transported throughout the plant body. It is this efficient conducting system that gives the main group of plants—the **vascular plants**—their name (Figure 1–9).

Plants, unlike animals, continue to grow throughout their lives. All plant growth originates in **meristems,** which are

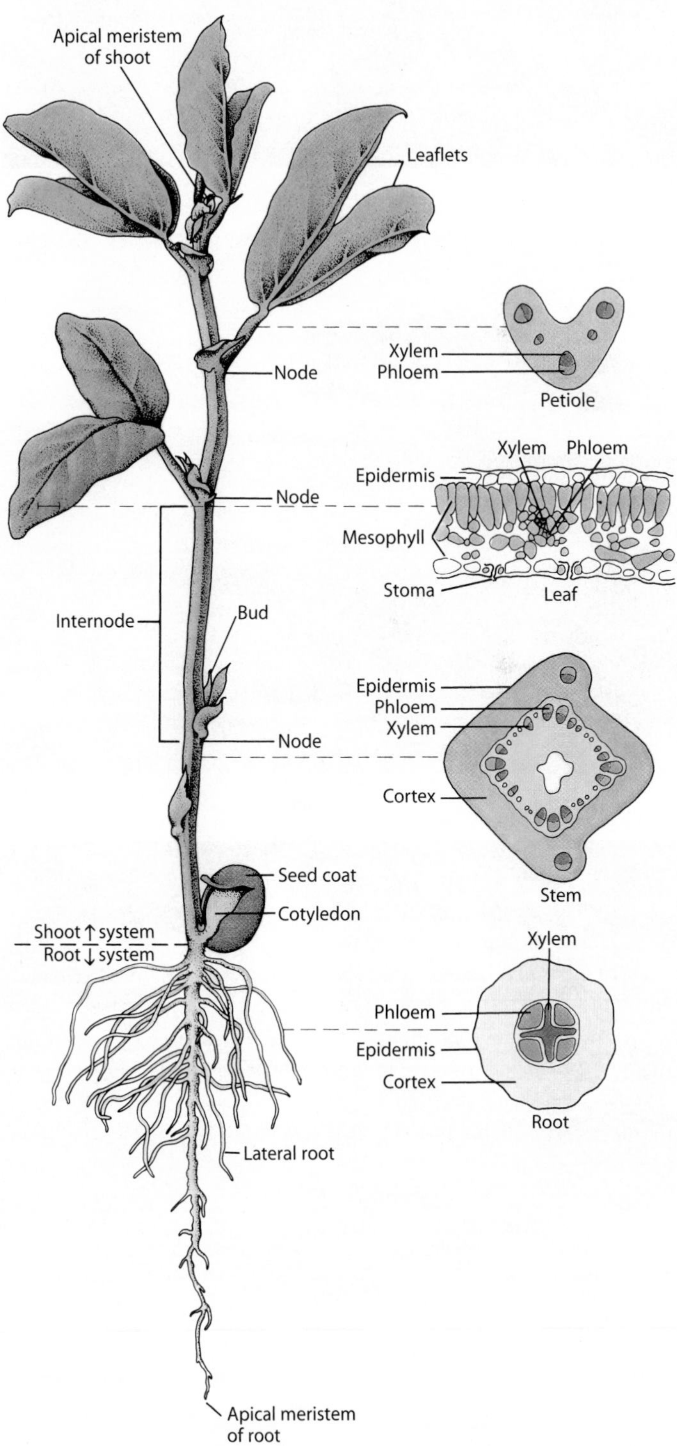

1–9 A modern vascular plant Diagram of a young broad bean *(Vicia faba)* plant, showing the principal organs and tissues of the modern vascular plant body. The organs—root, stem, and leaf—are composed of tissues, which are groups of cells with distinct structures and functions. Collectively, the roots make up the root system, and the stems and leaves together make up the shoot system of the plant. Unlike roots, stems are divided into nodes and internodes. The node is the part of the stem at which one or more leaves are attached, and the internode is the part of the stem between two successive nodes. In the broad bean, the first few foliage leaves are divided into two leaflets each. Buds, or embryonic shoots, commonly arise in the axils—the upper angle between leaf and stem—of the leaves. Lateral, or branch, roots arise from the inner tissues of the roots. The vascular tissues—xylem and phloem—occur together and form a continuous vascular system throughout the plant body. They lie just inside the cortex in root and stem. The mesophyll tissue of leaves is specialized for photosynthesis. In this diagram, a cotyledon, or seed leaf, can be seen through a tear in the seed coat.

(a)

(b)

(c)

(d)

1–10 Examples of the enormous diversity of biomes on Earth ***(a)*** The temperate deciduous forest, which covers most of the eastern United States and southeastern Canada, is dominated by trees that lose their leaves in the cold winters. Here are paper birches and a red maple photographed in early autumn in the Adirondack Mountains of New York State. ***(b)*** Underlain with permafrost, Arctic tundra is a treeless biome characterized by a short growing season. Shown here are tundra plants in full autumn color, photographed in Tombstone Valley, Yukon, Canada. ***(c)*** In Africa, savannas are inhabited by huge herds of grazing mammals, such as these zebras and wildebeests. The tree in the foreground is an acacia. ***(d)*** Moist tropical forests, shown here in Costa Rica, constitute the richest, most diverse biome on Earth, with perhaps half of all species of organisms on Earth found there. ***(e)*** Deserts typically receive less than 25 centimeters of rain per year. Here in the Sonoran desert in Arizona, the dominant plant is the giant saguaro cactus. Adapted for life in a dry climate, saguaro cacti have shallow, wide-spreading roots, as well as thick stems for storing water. ***(f)*** Mediterranean climates are rare on a world scale. Cool, moist winters, during which the plants grow, are followed by hot, dry summers, during which the plants become dormant. Shown here is an evergreen oak woodland on Mount Diablo in California.

(e)

(f)

embryonic tissue regions capable of adding cells indefinitely to the plant body. Meristems located at the tips of all roots and shoots—the **apical meristems**—are involved with the extension of the plant body. Thus the roots are continuously reaching new sources of water and minerals, and the photosynthetic regions are continuously extending toward the light. The type of growth that originates from apical meristems is known as **primary growth.** On the other hand, the type of growth that results in a thickening of stems and roots—**secondary growth**—originates from two **lateral meristems,** the vascular cambium and the cork cambium.

During the transition "to the air," plants also underwent further adaptations that made it possible for them to reproduce on land. The first of these adaptations was the production of drought-resistant spores. This was followed by the evolution of complex, multicellular structures in which the gametes, or reproductive cells, were held and protected from drying out by a layer of sterile cells. In the **seed plants,** which include almost all familiar plants except the ferns, mosses, and liverworts, the young plant, or embryo, is enclosed within a specialized covering (seed coat) provided by the parent. There the embryo is protected from both drought and predators and is provided with a supply of stored food. The embryo, the supply of stored food, and the seed coat are the components of the **seed.**

Thus, in summary, the vascular plant (Figure 1–9) is characterized by a root system that serves to anchor the plant in the ground and to collect water and minerals from the soil; a stem that raises the photosynthetic parts of the plant body toward its energy source, the sun; and leaves, which are highly specialized photosynthetic organs. Roots, stems, and leaves are interconnected by a complicated and efficient vascular system for the transport of food and water. The reproductive cells of plants are enclosed within multicellular protective structures, and in seed plants the embryos are protected by resistant coverings. All of these characteristics are adaptations to a photosynthetic existence on land.

Evolution of Communities

The invasion of the land by plants changed the face of the continents. Looking down from an airplane on one of Earth's great expanses of desert or on one of its mountain ranges, we can begin to imagine what the world looked like before the appearance of plants. Yet even in these regions, the traveler who goes by land will find an astonishing variety of plants punctuating the expanses of rock and sand. In those parts of the world where the climate is more temperate and the rains are more frequent, communities of plants dominate the land and determine its character. In fact, to a large extent, they *are* the land. Rainforest, savanna, woods, desert, tundra—each of these words brings to mind a portrait of a landscape (Figure 1–10). The main features of each landscape are its plants, enclosing us in a dark green cathedral in our imaginary rainforest, carpeting the ground beneath our feet with wildflowers in a meadow, moving in great golden waves as far as the eye can see across our imaginary prairie. Only when we have sketched these **biomes**—natural communities of wide extent, characterized by distinctive, climatically controlled groups of plants and animals—in terms of trees and shrubs and grasses can we fill in other features, such as deer, antelope, rabbits, or wolves.

How do vast plant communities, such as those seen on a continental scale, come into being? To some extent we can trace the evolution of the different kinds of plants and animals that populate these communities. Even with accumulating knowledge, however, we have only begun to glimpse the far more complex pattern of development, through time, of the whole system of organisms that make up these various communities.

Ecosystems Are Relatively Stable, Integrated Units That Are Dependent on Photosynthetic Organisms

Such communities, along with the nonliving environment of which they are a part, are known as ecological systems, or

ecosystems. An ecosystem is a kind of corporate entity made up of transient individuals. Some of these individuals, the larger trees, live as long as several thousand years; others, the microorganisms, live only a few hours or even minutes. Yet the ecosystem as a whole tends to be remarkably stable (although not static). Once in balance, it does not change for centuries. Our grandchildren may someday walk along a woodland path once followed by our great-grandparents, and where they saw a pine tree, a mulberry bush, a meadow mouse, wild blueberries, or a robin, these children, if this woodland still exists, will see roughly the same kinds of plants and animals in the same numbers.

An ecosystem functions as an integrated unit, although many of the organisms in the system compete for resources. Virtually every living thing, even the smallest bacterial cell or fungal spore, provides a food source for some other living organism. In this way, the energy captured by green plants is transferred in a highly regulated way through a number of different types of organisms before it is dissipated. Moreover, interactions among the organisms themselves, and between the organisms and the nonliving environment, produce an orderly cycling of elements such as nitrogen and phosphorus. Energy must be added to the ecosystem constantly, but the elements are cycled through the organisms, returned to the soil, decomposed by soil bacteria and fungi, and recycled. These transfers of energy and the cycling of elements involve complicated sequences of events, and in these sequences each group of organisms has a highly specific role. As a consequence, it is impossible to change a single component of an ecosystem without the risk of destroying the balance on which the stability of the ecosystem depends.

At the base of productivity in almost all ecosystems are the plants, algae, and photosynthetic bacteria. These organisms alone have the ability to capture energy from the sun and to manufacture organic molecules that they and all other kinds of organisms may require for life. There are roughly half a million kinds of organisms capable of photosynthesis, and at least 20 times that many heterotrophic organisms, which are completely dependent on the photosynthesizers. For animals, including human beings, many kinds of molecules—including essential amino acids, vitamins, and minerals—can be obtained only through plants or other photosynthetic organisms. Furthermore, the oxygen that is released into the atmosphere by photosynthetic organisms makes it possible for life to exist on the land and in the surface layers of the ocean. Oxygen is necessary for the energy-producing metabolic activities of the great majority of organisms, including photosynthetic organisms.

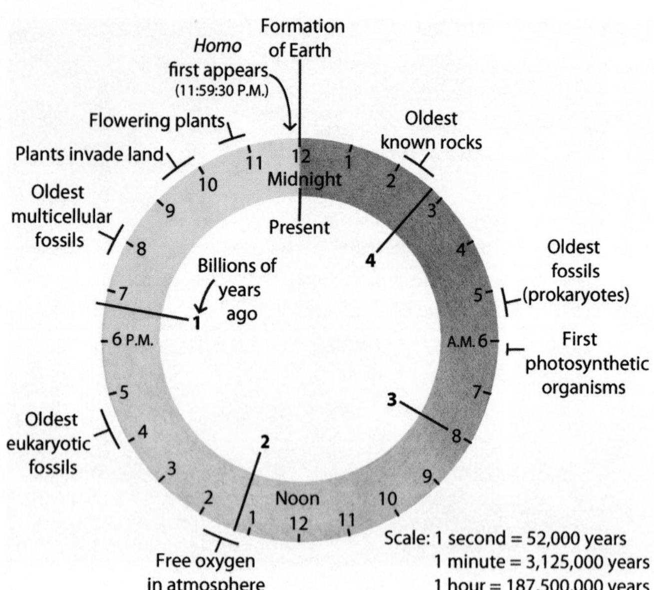

1–11 The clock face of biological time Life appears relatively early in the Earth's history, sometime before 6:00 A.M. on a 24-hour scale. The first multicellular organisms do not appear until the twilight of that 24-hour day, and the genus *Homo* is a very late arrival—less than a minute before midnight.

Appearance of Human Beings

Human beings are relative newcomers to the world of living organisms (Figure 1–11). If the entire history of the Earth were measured on a 24-hour time scale starting at midnight, cells would appear in the warm seas before dawn. The first multicellular organisms would not be present until well after sundown, and the earliest appearance of humans (about 2 million years ago) would be about half a minute before the day's end. Yet humans more than any other animal—and almost as much as the plants that invaded the land—have changed the surface of the planet, shaping the biosphere according to their own needs, ambitions, or follies.

With the cultivation of crops, starting about 10,500 years ago, it became possible to maintain growing populations of people that eventually built towns and cities. This development (reviewed in detail in Chapter 21) allowed specialization and the diversification of human culture. One characteristic of this culture is that it examines itself and the nature of other living things, including plants. Eventually, the science of biology developed within the human communities that had been made possible through the domestication of plants. The part of biology that deals with plants and, by tradition, with prokaryotes, fungi, and algae is called **botany,** or **plant biology.**

Plant Biology Includes Many Different Areas of Study

The study of plants has been pursued for thousands of years, but like all branches of science, it became diverse and specialized only during the twentieth century. Until the late 1800s, botany was a branch of medicine, pursued chiefly by physicians who used plants for medicinal purposes and who were interested in determining the similarities and differences between plants and animals for that purpose. Today, however, plant biology is an important scientific discipline that has many subdivisions: **plant physiology,** which is the study of how plants function, that is, how they capture and transform energy and how they grow and develop; **plant morphology,** the study of the form of plants; **plant anatomy,** the study of their internal structure; **plant taxonomy** and **systematics,** involving the naming and classifying of plants and the study of the relationships among them; **cytology,** the study of cell structure, function, and life histories; **genetics,** the study of heredity and variation; **genomics,** the study of the content, organization, and function of genetic information in whole genomes; **molecular biology,** the study of the structure and function of biological molecules; **economic botany,** the study of past, present, and future uses of plants by people; **ethnobotany,** the study of the uses of plants for medicinal and other purposes by indigenous peoples; **ecology,** the study of the relationships between organisms and their environment; and **paleobotany,** the study of the biology and evolution of fossil plants.

Included in this book are all organisms that have traditionally been studied by botanists: not only plants but also prokaryotes, viruses, fungi, and autotrophic protists (algae). Nonphotosynthetic eukaryotes and protists have traditionally been the province of zoologists. Although we do not regard algae, fungi, prokaryotes, or viruses as plants, and shall not refer to them as plants in this book, they are included here because of tradition and because they are normally considered part of the botanical portion of the curriculum, just as botany itself used to be considered a part of medicine. Moreover, both prokaryotes (e.g., nitrogen-fixing bacteria) and fungi (e.g., mycorrhizal fungi) form important, mutually beneficial symbiotic relationships with their plant hosts. Virology, bacteriology, phycology (the study of algae), and mycology (the study of fungi) are well-established fields in their own right, but they still fall loosely under the umbrella of botany.

A Knowledge of Botany Is Important for Dealing with Today's—and Tomorrow's—Problems

In this chapter, we have ranged from the beginnings of life on this planet to the evolution of plants and ecosystems to the development of agriculture and civilization. These broad topics are of interest to many people other than botanists, or plant biologists. The urgent efforts of botanists and agricultural scientists will be needed to feed the world's rapidly growing human population (Figure 1–12), as discussed in Chapter 21. Modern plants, algae, and bacteria offer the best hope of providing a renewable source of energy for human activities, just as extinct plants, algae, and bacteria have been responsible for the massive accumulations of gas, oil, and coal on which our modern industrial civilization depends. In an even more fundamental

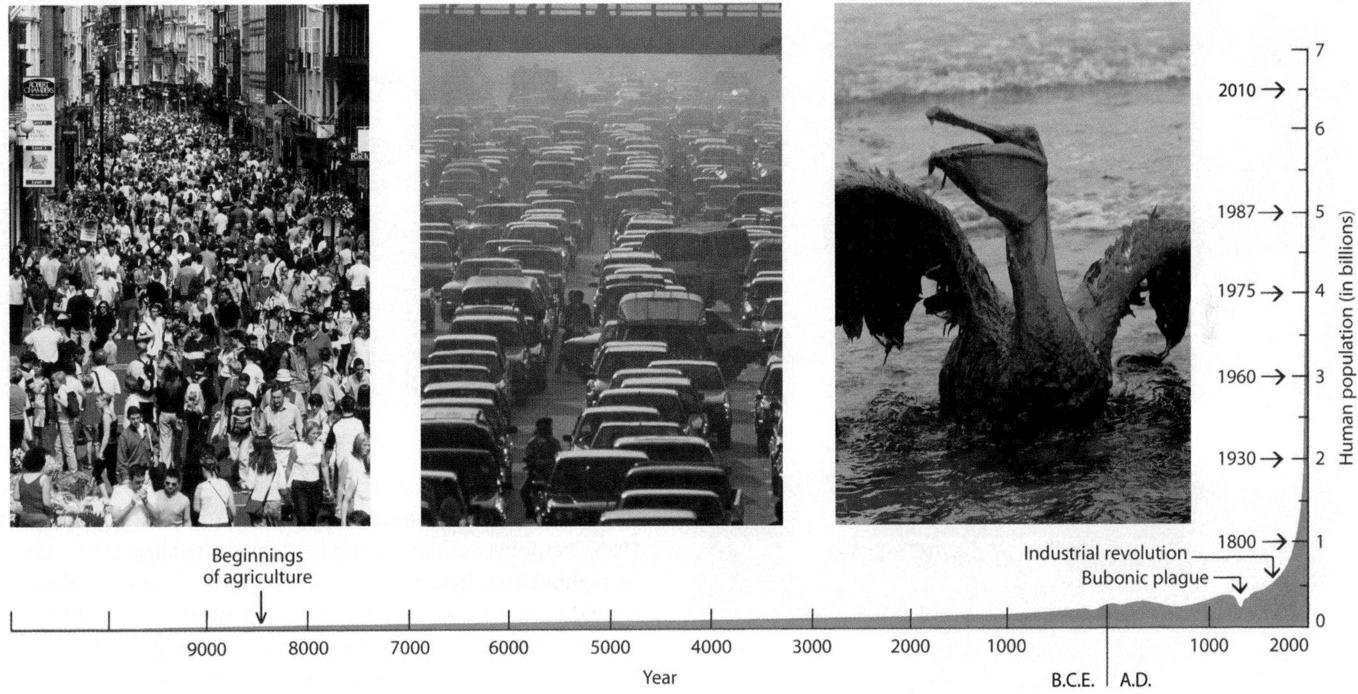

1–12 Growth of the human population Over the last 10,000 years, the human population has grown from several million to approximately 6.5 billion. A significant increase in the rate of population growth occurred as a result of the cultivation of plants as crops, and an even more dramatic increase began with the advent of the Industrial Revolution, which started in the middle of the eighteenth century and continues to the present.

The consequences of the rapid growth of the human population are many and varied. In the United States and other parts of the developed world, they include not only the sheer numbers of people but also heavy consumption of nonrenewable fossil fuels and the resulting pollution—both as the fuels are burned and as a result of accidents such as oil spills at drilling sites and during transport. In less developed parts of the world, the consequences include malnutrition and, all too often, starvation, coupled with a continuing vulnerability to infectious diseases. The consequences for other organisms include not only the direct effects of pollution but also—and most important—the loss of habitat.

sense, the role of plants, along with that of algae and photosynthetic bacteria, commands our attention. As the producers of energy-containing compounds in the global ecosystem, these photosynthetic organisms are the route by which all other living things, including ourselves, obtain energy, oxygen, and the many other materials necessary for their continued existence. As a student of botany, you will be in a better position to assess the important ecological and environmental issues of the day and, by understanding, help to build a healthier world.

In this second decade of the twenty-first century, it is clear that human beings, with a population of 6.5 billion in 2010 and a projected population of 9 billion by 2050, are managing the Earth with an intensity that would have been unimaginable a few decades ago. Every hour, manufactured chemicals fall on every square centimeter of the planet's surface. The protective stratospheric ozone layer formed 450 million years ago has been seriously depleted by the use of chlorofluorocarbons (CFCs), and damaging ultraviolet rays penetrating the depleted layer have increased the incidence of skin cancer in people all over the world. Moreover, it has been estimated that, by the middle of this century, the average temperature will have increased between 1.5° and 4.5°C due to the **greenhouse effect.** This **global-warming** phenomenon—the trapping of heat radiating from the Earth's surface out into space—is intensified through the increased amounts of carbon dioxide, nitrogen oxides, CFCs, and methane in the atmosphere resulting from human activities. And most seriously, a large portion of the total number of species of plants, animals, fungi, and microorganisms is disappearing during our lifetime—the victims of human exploitation of the Earth—resulting in a loss of biodiversity. All of these trends are alarming, and they demand our utmost attention.

(a)

(b)

(c)

1–13 Phytoremediation ***(a)*** Sunflowers growing on a lake contaminated with radioactive cesium and strontium following the 1986 Chernobyl nuclear disaster in Ukraine, then part of the USSR. Suspended from styrofoam rafts, the sunflowers' roots are able to remove up to 90 percent of the contaminants in 10 days. ***(b)*** Poplar (*Populus* spp.) and willow (*Salix* spp.) trees growing on a fuel-contaminated site in Elizabeth, North Carolina. The deep-rooted trees draw the contaminants up through their stems and leaves, reducing the need for mechanical pumping and treatment of contaminated groundwater. ***(c)*** Naturally occurring selenium accumulates in ditches containing run-off from irrigated croplands, creating bodies of standing water that are poisonous to wildlife, especially migrating birds. Furthermore, plants grown in the high-selenium soil that results from evaporation of the water are toxic. Pickleweed *(Salicornia bigelovii)*, a salt marsh plant, is highly efficient at removing selenium, which is absorbed by the plant and then released into the atmosphere to be dispersed by prevailing winds. For the endangered salt marsh harvest mouse, seen here, pickleweed is a staple food.

Marvelous new possibilities have been developed during the past few years for the better utilization of plants by people, and we discuss these developments throughout this book. It is now possible, for example, to clean up polluted environments through **phytoremediation** (Figure 1–13), to stimulate the growth of plants, to deter their pests, to control weeds in crops, and to form hybrids between plants with more precision than ever before.

The potential for exciting progress in plant biology grows with every passing year, as additional discoveries are made and new applications are developed. The methods of **genetic engineering,** discussed in Chapter 10, make it possible to accomplish the astonishing feat of transferring genes from a virus, a bacterium, an animal, or a particular plant into an entirely different species of plant in order to produce specific desirable characteristics in the recipient plant. These so-called **transgenic plants,** which contain genes from entirely different species, can be made to exhibit new and extraordinary properties. By inserting maize and bacterial genes into the rice nucleus, for example, a more nutritious rice, with higher levels of β-carotene, can be produced (see Figure 10–1). Another area of investigation is working toward increasing the iron content of rice. Both of these developments show promise for improving the health of the vast numbers of poorly nourished people with rice-dependent diets. In addition, pest-resistant varieties of maize and cotton have been developed by transferring genes from a soil bacterium that attacks the caterpillars that cause major crop losses. The transformed maize and cotton, with their ability to express the bacterial genes, kill the caterpillars that might attack them, allowing growers to reduce their use of pesticide.

1–14 Transgenic plants Citrus seedlings were transformed by insertion of flower-initiation genes from *Arabidopsis,* the small flowering plant of the mustard family that is widely used for genetic research. The six-month-old transgenic seedling on the right has developed flowers, whereas the control seedling on the left has not and will take years to flower and set fruit.

The Hawaiian papaya industry was saved by development of transgenic papaya trees able to resist the papaya ringspot virus (see Figure 10–13). Other improvements involve transgenic soybeans that can tolerate Roundup, an herbicide that kills both broad-leaved weeds and untransformed soybeans. In addition, citrus plants have been transformed to flower in 6 months rather than the usual 6 to 20 years, thereby reducing the time needed for a citrus tree to set fruit (Figure 1–14). Attempts are being made to increase the efficiency of photosynthesis and thereby increase crop yields, and to enhance the "glossiness" of crops by selectively breeding for plants with waxier leaves. This increased waxiness would benefit the plant by reducing water loss and, by increasing the reflectivity of the surfaces of the crop, could result in a slight cooling of summer temperatures in central North America and Eurasia.

Hopes for the future include, among numerous possibilities, biodegradable plastics, trees with higher fiber content for paper manufacture, plants with increased levels of healthy oils and anti-cancer proteins, and vaccines that can be produced in plants, holding the promise of someday being able to deliver hepatitis B vaccine, for example, in bananas. These methods, first applied in 1973, have already been the basis of billions of dollars in investments and increased hope for the future. Discoveries still to be made will undoubtedly exceed our wildest dreams and go far beyond the facts that are available to us now.

In addition, we have come to appreciate, even more, the importance of green spaces to our increasingly complex lives. In cities, former industrial sites are being skillfully developed into parks of various types. An abandoned elevated railroad track in New York City, originally slated for demolition, has been saved and now serves as a popular urban destination known as the High Line (Figure 1–15a). Running for roughly a mile, the park is planted primarily with the types of wildflowers, grasses, shrubs, and trees that grew along the track during the decades when it was no longer in use. A meandering pathway follows the route of the original track, and the setting, with its views of the Hudson River, as well as of the city life below, draws millions of visitors a year. Another example of the reclamation of an industrial site is a decommissioned military airbase at Magnuson Park in Seattle, Washington (Figure 1–15b). Cleared of asphalt and converted to thriving wetlands, the area with its newly created ponds and thickets of native plants has attracted a variety of wildlife. Winding trails invite visitors to experience the peaceful setting as they learn about the essential role played by wetlands habitats.

As we turn to Chapters 2 and 3, in which our attention narrows to a cell so small it cannot be seen by the unaided eye, it is important to keep these broader concerns in mind. A basic knowledge of plant biology is useful in its own right and is essential in many fields of endeavor. It is also increasingly relevant to some of society's most crucial problems and to the difficult decisions that will face us in choosing among the proposals for diminishing them. To quote from an editorial in the 19 November 2010 issue of *Science:* "Plants are essential to the survival of our planet—to its ecology, biodiversity, and climate." Our own future, the future of the world, and the future

(a)

(b)

1–15 Greening of abandoned industrial sites ***(a)*** The High Line in New York City is built on an abandoned railroad track elevated above a newly resurrected neighborhood of restaurants, galleries, and shops. Remnants of the original track can be seen among the shrubs, perennials, grasses, and trees planted along the popular promenade. ***(b)*** A recently created wetlands situated on a former airbase at Magnuson Park in Seattle, Washington, offers a rich habitat of native plants and various species of wildlife, including dragonflies, frogs, ducks, owls, hawks, shorebirds, and warblers.

of all kinds of plants—as individual species and as components of the life-support systems into which we all have evolved—depend on our knowledge and ability to critically assess the information we are given. Thus, this book is dedicated not only to the botanists of the future, whether teachers or researchers, but also to the informed citizens, scientists, and laypeople alike, in whose hands such decisions lie.

SUMMARY

Photosynthesis Is the Process by Which the Sun's Energy Is Captured to Form Organic Molecules

Only a few kinds of organisms—plants, algae, and some bacteria—have the capacity to capture energy from the sun and use it to form organic molecules by the process of photosynthesis. Almost all life on Earth depends, directly or indirectly, on the products of this process.

The Chemical Building Blocks of Life Accumulated in the Early Oceans

The planet Earth is some 4.6 billion years old. The oldest known fossils date back 3.5 billion years and resemble today's filamentous bacteria. Although the process by which living organisms arose is a matter of speculation, there is general agreement that life as we know it probably emerged on Earth only once—that is, all living things share a common ancestor.

Heterotrophic Organisms Evolved before Autotrophic Organisms, Prokaryotes before Eukaryotes, and Unicellular Organisms before Multicellular Organisms

Heterotrophs, organisms that feed on organic molecules or on other organisms, were the first life forms to appear on Earth. Autotrophic organisms, those that could produce their own food by photosynthesis, evolved no less than 3.4 billion years ago. Until about 2.1 billion years ago, the prokaryotes—archaea and bacteria—were the only organisms that existed. Eukaryotes, with larger, much more complex cells, evolved at that time. Multicellular eukaryotes began to evolve at least 650 million years ago, and they began to invade the land about 450 million years ago.

With the advent of oxygen-producing photosynthesis, in which water molecules are split and oxygen is released, oxygen began to accumulate in the atmosphere. The presence of this free oxygen enabled organisms to break down the energy-rich products of photosynthesis by aerobic respiration.

Colonization of the Land Was Associated with the Evolution of Structures to Obtain Water and Minimize Water Loss

Plants, which are basically a terrestrial group, have achieved a number of specialized characteristics that suit them for life on land. These characteristics are best developed among the members of the dominant group known as the vascular plants. Among these features are a waxy cuticle, penetrated by specialized openings known as stomata through which gas exchange takes place, and an efficient conducting system. This system consists of xylem, in which water and minerals pass from the roots to the stems and leaves, and phloem, which transports the products of photosynthesis to all parts of the plant. Plants increase in length by primary growth and expand in girth by secondary growth through the activity of meristems, which are embryonic tissue regions capable of adding cells indefinitely to the plant body.

Ecosystems Are Relatively Stable, Integrated Units That Are Dependent on Photosynthetic Organisms

As plants have evolved, they have come to constitute biomes, great terrestrial assemblages of plants and animals. The interacting systems made up of biomes and their nonliving environments are called ecosystems. Human beings, which appeared about 2 million years ago, developed agriculture about 10,500 years ago and thus provided a basis for the huge increase in their population levels. Subsequently, they have become the dominant ecological force on Earth. Humans have used their knowledge of plants to foster their own development and will continue to do so with increasingly greater importance in the future.

Genetic Engineering Allows Scientists to Transfer Genes between Entirely Different Species

With the advent of genetic engineering, it became possible for biologists to transfer genes from one species into an entirely different species. Genetic engineering has already resulted in the development of transgenic plants with desirable traits such as increased nutritive value and resistance to certain diseases and pests.

QUESTIONS

1. What was the likely source of the raw material incorporated into the first life forms?
2. What criteria would you use to determine whether an entity is a form of life?
3. What role did oxygen play in the evolution of life on Earth?
4. What advantages do terrestrial plants have over their aquatic ancestors? Can you think of any disadvantages to being a terrestrial plant?
5. Plants enter our lives in innumerable ways other than as sources of food. How many ways can you list? Have you thanked a green plant today?
6. A knowledge of botany—of plants, fungi, algae, and bacteria—is key to our understanding of how the world works. How is that knowledge important for dealing with today's and tomorrow's problems?

SECTION 1

BIOLOGY OF THE PLANT CELL

◀ Plants capture the sun's energy and use it to form the organic molecules essential to life. This process—photosynthesis—requires the green pigment chlorophyll, which is present in the leaves of this chokecherry *(Prunus virginiana)* plant. The organic molecules formed during photosynthesis provide both the energy and the larger structural molecules needed by the plant, including the anthocyanin pigments that are produced as the chokecherries ripen to a dark purple.

CHAPTER 2

The Molecular Composition of Plant Cells

◀ **The chemistry of chilis** As chili peppers ripen, their carotenoid pigments are produced and the colors of the peppers change from green to yellow to red. Capsaicin, the molecule that causes the burning sensation when we eat the chili peppers, deters browsing mammals but is not detected by birds, which eat the fruit and distribute the seeds in their droppings.

CHAPTER OUTLINE

Everything on Earth—including everything you can see at the moment, as well as the air surrounding it—is made up of chemical elements in various combinations. **Elements** are substances that cannot be broken down into other substances by ordinary means. Carbon is an element, as are hydrogen and oxygen. Of the 92 elements that occur naturally on Earth, only six were selected in the course of evolution to form the complex, highly organized material of living organisms. These six elements—carbon, hydrogen, nitrogen, oxygen, phosphorus, and sulfur (CHNOPS)—make up 99 percent of the weight of all living matter. The unique properties of each element depend on the structure of its atoms and on the way those atoms can interact and bond with other atoms to form molecules.

Water, a molecule consisting of two hydrogens and one oxygen (H_2O), makes up more than half of all living matter and more than 90 percent of the weight of most plant tissues. By contrast, electrically charged ions, such as potassium (K^+), magnesium (Mg^{2+}), and calcium (Ca^{2+}), important as they are, account for only about 1 percent. Almost all the rest of a living organism, chemically speaking, is composed of **organic molecules**—that is, of molecules that contain carbon.

In this chapter, we present some of the types of organic molecules that are found in living things. The molecular drama is a grand extravaganza with, literally, a cast of thousands. A single bacterial cell contains some 5000 different kinds of organic molecules, and an animal or plant cell has at least twice that many. As we noted, however, these thousands of molecules are composed of relatively few elements. Similarly, relatively few types of molecules play the major roles in living systems. Consider this chapter an introduction to the principal players in the drama. The plot begins to unfold in the next chapter.

CHECKPOINTS

After reading this chapter, you should be able to answer the following:

1. What are the four main types of organic molecules found in plant cells, and what are their basic structural subunits and their principal functions?
2. By what processes are all four types of organic molecules split into their subunits, and by what process can these subunits be joined together?
3. How do energy-storage polysaccharides and structural polysaccharides differ from one another? What are some examples of each?
4. What is an enzyme, and why are enzymes important to cells?
5. How is ATP different from ADP, and why is ATP important to cells?
6. What is the difference between primary and secondary metabolites?
7. What are the main types of secondary metabolites, and what are some examples of each?

Organic Molecules

The special bonding properties of carbon permit the formation of a great variety of organic molecules. Of the thousands of different organic molecules found in cells, just four different types make up most of the dry weight of living organisms. These four are **carbohydrates** (consisting of sugars and chains of sugars), **lipids** (most of which contain fatty acids), **proteins** (composed of amino acids), and **nucleic acids** (DNA and RNA, which are made up of complex molecules known as nucleotides). All of these molecules—carbohydrates, lipids, proteins, and nucleic acids—consist mainly of carbon and hydrogen, and most of them contain oxygen as well. In addition, proteins contain nitrogen and sulfur. Nucleic acids, as well as some lipids, contain nitrogen and phosphorus.

Carbohydrates

Carbohydrates are the most abundant organic molecules in nature and are the primary energy-storage molecules in most living organisms. In addition, they form a variety of structural components of living cells. The walls of young plant cells, for example, are made up of the universally important carbohydrate cellulose embedded in a matrix of other carbohydrates and proteins.

The simplest carbohydrates are small molecules known as **sugars;** larger carbohydrates are composed of sugars joined together. There are three principal kinds of carbohydrates, classified according to the number of sugar subunits they contain. **Monosaccharides** ("single, or simple, sugars"), such as ribose, glucose, and fructose, consist of only one sugar molecule. **Disaccharides** ("two sugars") contain two sugar subunits linked covalently. Familiar examples are sucrose (table sugar), maltose (malt sugar), and lactose (milk sugar). Cellulose and starch are **polysaccharides** ("many sugars"), which contain many sugar subunits linked together.

Macromolecules (large molecules), such as polysaccharides, that are made up of similar or identical small subunits are known as **polymers** ("many parts"). The individual subunits of polymers are called **monomers** ("single parts"); **polymerization** is the stepwise linking of monomers into polymers.

Monosaccharides Function as Building Blocks and Sources of Energy

Monosaccharides, or single sugars, are the simplest carbohydrates. They are made up of linked carbon atoms to which hydrogen atoms and oxygen atoms are attached in the proportion of one carbon atom to two hydrogen atoms to one oxygen atom. Monosaccharides can be described by the formula $(CH_2O)_n$, where n can be as small as 3, as in $C_3H_6O_3$, or as large as 7, as in $C_7H_{14}O_7$. These proportions gave rise to the term "carbohydrate" (meaning "carbon with water added") for sugars and for the larger molecules formed from linked sugar subunits. Examples of several common monosaccharides are shown in Figure 2–1. Notice that each monosaccharide has a carbon chain (the carbon "skeleton") with a hydroxyl group (—OH) attached to every carbon atom except one. The remaining carbon atom is in the form of a carbonyl group (—C=O). Both these groups are **hydrophilic** ("water-loving"), and thus

(a) **Glyceraldehyde** ($C_3H_6O_3$) *(b)* **Ribose** ($C_5H_{10}O_5$) *(c)* **Glucose** ($C_6H_{12}O_6$)

2–1 Some biologically important monosaccharides
(a) Glyceraldehyde, a three-carbon sugar, is an important energy source and provides the basic carbon skeleton for numerous organic molecules. ***(b)*** Ribose, a five-carbon sugar, is found in the nucleic acid RNA and in the energy-carrier molecule ATP. ***(c)*** The six-carbon sugar glucose serves important structural and transport roles in the cell. The terminal carbon atom nearest the double bond is designated carbon 1.

monosaccharides, as well as many other carbohydrates, readily dissolve in water. The five-carbon sugars (pentoses) and six-carbon sugars (hexoses) are the most common monosaccharides in nature. They occur in both chain and ring forms, and, in fact, they are normally found in a ring form when dissolved in water (Figure 2–2). As a ring is formed, the carbonyl group is converted to a hydroxyl group. Thus, the carbonyl group is a distinguishing feature of monosaccharides in the chain form but not in a ring form.

Monosaccharides are the building blocks—the monomers—from which living cells construct disaccharides, polysaccharides, and other essential carbohydrates. Moreover, the monosaccharide **glucose** is the form in which sugar is transported in the circulatory system of humans and other vertebrate animals. As we will see in Chapter 6, glucose and other monosaccharides are primary sources of chemical energy for plants as well as for animals.

The Disaccharide Sucrose Is a Transport Form of Sugar in Plants

Although glucose is the common transport sugar for many animals, sugars are often transported in plants and other organisms as disaccharides. **Sucrose,** a disaccharide composed of glucose and fructose, is the form in which sugar is transported in most plants from the photosynthetic cells (primarily in the leaves), where it is produced, to other parts of the plant body. The sucrose we consume as table sugar is commercially harvested from sugar beets (enlarged roots) and sugarcane (stems), where it accumulates as it is transported from the photosynthetic parts of the plant.

In the synthesis of a disaccharide from two monosaccharide molecules, a molecule of water is removed and a new bond is formed between the two monosaccharides. This type of chemical reaction, which occurs when sucrose is formed from

alpha-Glucose (ring form)

Glucose (chain form)

beta-Glucose (ring form)

2–2 Ring and chain forms of glucose In aqueous solution, the six-carbon sugar glucose exists in two different ring structures, alpha (α) and beta (β), which are in equilibrium with each other. The molecules pass through the chain form to get from one structure to the other. The sole difference in the two ring structures is the position of the hydroxyl (—OH) group attached to carbon 1; in the alpha form it is below the plane of the ring, and in the beta form it is above the plane.

glucose and fructose, is known as **dehydration synthesis,** or a condensation reaction (Figure 2–3). In fact, the formation of most organic polymers from their subunits occurs by dehydration synthesis.

When the reverse reaction occurs—for example, when a disaccharide is split into its monosaccharide subunits—a molecule of water is added. This splitting, which occurs when a disaccharide is used as an energy source, is known as **hydrolysis,** from *hydro,* meaning "water," and *lysis,* "breaking apart." Hydrolysis reactions are energy-yielding processes that are important in energy transfers in cells. Conversely, dehydration synthesis reactions—the reverse of hydrolysis reactions—require an input of energy.

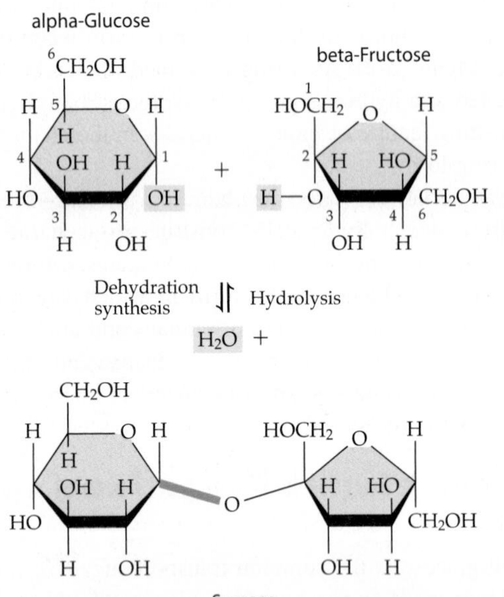

2–3 Sucrose formation and degradation Sugar is generally transported in plants as the disaccharide sucrose. Sucrose is made up of two monosaccharide subunits, one alpha-glucose and one beta-fructose, bonded in a 1,2 linkage (the carbon 1 of glucose is linked to the carbon 2 of fructose). The formation of sucrose involves the removal of a molecule of water (dehydration synthesis). The new chemical bond formed is shown in blue. The reverse reaction—splitting sucrose into its constituent monosaccharides—requires the addition of a water molecule (hydrolysis). Formation of sucrose from glucose and fructose requires an energy input of 5.5 kcal per mole by the cell. Hydrolysis releases the same amount of energy.

Polysaccharides Function as Storage Forms of Energy or as Structural Materials

Polysaccharides are polymers made up of monosaccharides linked together in long chains. Some polysaccharides function as storage forms of sugar and others serve a structural role.

Starch, the primary storage polysaccharide in plants, consists of chains of glucose molecules. There are two forms of starch: **amylose,** which is an unbranched molecule, and **amylopectin,** which is branched (Figure 2–4). Amylose and amylopectin are stored as starch grains within plant cells. **Glycogen,** the common storage polysaccharide in prokaryotes, fungi, and animals, is also made up of chains of glucose molecules. It resembles amylopectin but is more highly branched. In some plants—most notably grains, such as wheat, rye, and barley—the principal storage polysaccharides in leaves and stems are polymers of fructose called **fructans.** These polymers are water-soluble and can be stored in much higher concentrations than starch.

Polysaccharides must be hydrolyzed to monosaccharides and disaccharides before they can be used as energy sources or transported through living systems. The plant breaks down its starch reserves when monosaccharides and disaccharides are needed for growth and development. We hydrolyze these polysaccharides when our digestive systems break down the starch stored by plants in such foods as maize, or corn (a grain), and potatoes (tubers), making glucose available as a nutrient for our cells.

Polysaccharides are also important structural compounds. In plants, the principal component of the cell wall is the important polysaccharide known as **cellulose** (Figure 2–5). In fact, half of all the organic carbon in the living world is contained in cellulose, making it the most abundant organic compound known. Wood is about 50 percent cellulose, and cotton fibers are nearly pure cellulose.

Cellulose is a polymer composed of monomers of glucose, as are starch and glycogen, but there are important differences. Starch and glycogen can be readily used as fuels by almost all kinds of living organisms, but only some microorganisms—certain prokaryotes, protozoa, and fungi—and a very few animals, such as silverfish, can hydrolyze cellulose. Animals such as cows, termites, and cockroaches can use cellulose for energy only because it is broken down by microorganisms that live in their digestive tracts.

To understand the differences between structural polysaccharides, such as cellulose, and energy-storage polysaccharides,

(a) **Amylose—linear chain of repeated alpha-glucose monomers**

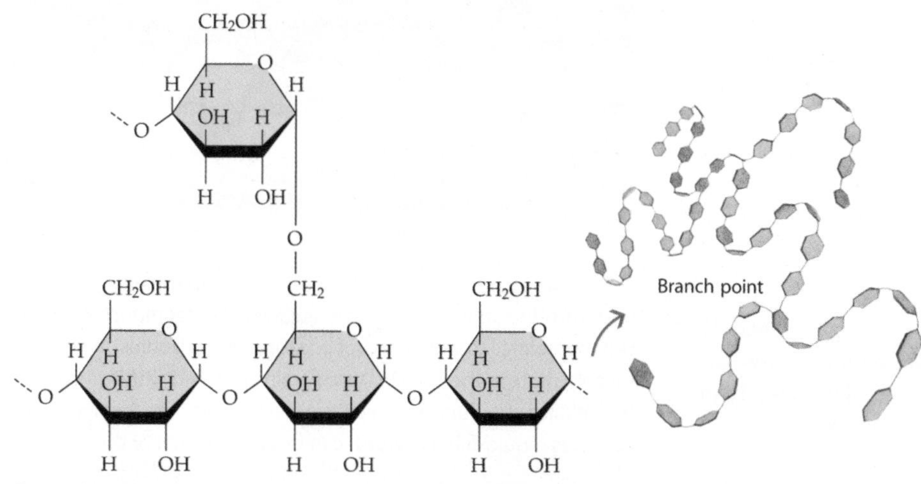

(b) **Amylopectin—branched chain of repeated alpha-glucose monomers**

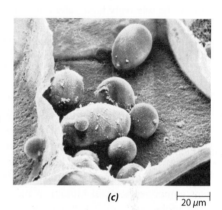

(c) 20 μm

2–4 Starch In most plants, accumulated sugars are stored as starch, which occurs in two forms: unbranched (amylose) and branched (amylopectin). ***(a)*** A single molecule of amylose may contain 1000 or more alpha-glucose monomers, with carbon 1 of one glucose ring linked to carbon 4 of the next (known as a 1,4 linkage) in a long, unbranched chain that winds to form a uniform coil. ***(b)*** A molecule of amylopectin may contain 1000 to 6000 or more alpha-glucose monomers; short chains of about 8 to 12 alpha-glucose monomers branch off the main chain at intervals of about every 12 to 25 alpha-glucose monomers. ***(c)*** Perhaps because of their coiled nature, starch molecules tend to cluster into grains. In this scanning electron micrograph of a single storage cell of potato *(Solanum tuberosum)*, the spherical structures are starch grains.

such as starch or glycogen, we must look again at the glucose molecule. You will recall that the molecule is basically a chain of six carbon atoms and that when it is in solution, as it is in the cell, it assumes a ring form. The ring may close in either of two ways (Figure 2–2): one ring form is known as alpha-glucose, the other as beta-glucose. The alpha (α) and beta (β) forms are in equilibrium, with a certain number of molecules changing from one form to the other all the time and with the chain form as the intermediate. Starch and glycogen are both made up entirely of alpha-glucose subunits (Figure 2–4), whereas cellulose consists entirely of beta-glucose subunits (Figure 2–5). This seemingly slight difference has a profound effect on the three-dimensional structure of the cellulose molecules, which are long and unbranched. As a result, cellulose is impervious to the enzymes that easily break down starch and glycogen. Once glucose molecules are incorporated into the plant cell wall in the form of cellulose, they are no longer available to the plant as an energy source.

Cellulose molecules form the fibrous part of the plant cell wall. The long, rigid cellulose molecules combine to form microfibrils, each consisting of hundreds of cellulose chains. In plant cell walls, the cellulose microfibrils are embedded in a

(a)

(b)

2–5 Cellulose ***(a)*** Cellulose resembles starch in that it consists of glucose monomers in 1,4 linkages. The cellulose polymer, however, consists of beta-glucose monomers, whereas starch is made up of alpha-glucose monomers. ***(b)*** Cellulose molecules, bundled together into microfibrils, are important structural components of plant cell walls. The —OH groups (blue) that project from both sides of the cellulose chain form hydrogen bonds (dashed lines) with —OH groups on neighboring chains, resulting in microfibrils made up of cross-linked parallel cellulose molecules. Compare the structure of cellulose with that of starch in Figure 2–4.

matrix containing two other complex, branched polysaccharides, namely, hemicelluloses and pectins (see Figure 3–29). Hemicelluloses stabilize the cell wall by hydrogen bonding to the cellulose microfibrils. Pectins make up most of the middle lamella, a layer of intercellular material that cements together the walls of adjacent plant cells. Pectins, which are especially plentiful in certain fruits, such as apples and cranberries, allow jams and jellies to thicken and gel.

Chitin is another important structural polysaccharide (Figure 14–5). It is the principal component of fungal cell walls and also of the relatively hard outer coverings, or exoskeletons, of insects and crustaceans, such as crabs and lobsters. The monomer of chitin is *N*-acetylglucosamine, which consists of a molecule of glucose to which a nitrogen-containing group has been added.

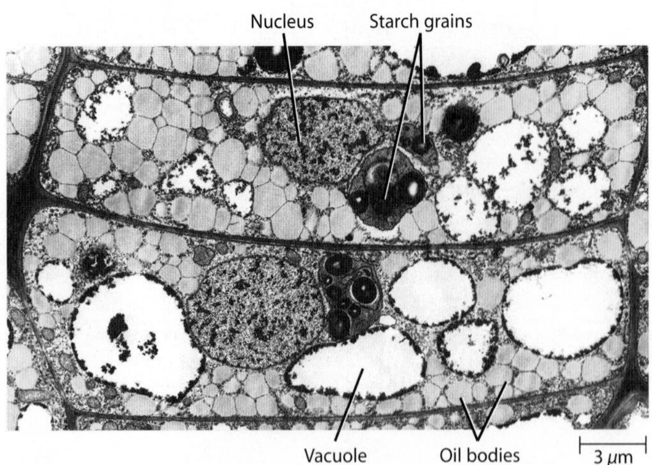

2–6 Storage of oil and starch Two cells from the fleshy, underground stem (corm) of the quillwort *Isoetes muricata*. These cells contain a large quantity of oil stored as oil bodies. In addition, carbohydrate in the form of starch grains is stored within amyloplasts, the cellular structures in which the starch grains are formed. Several vacuoles (liquid-filled cavities) can be seen in each of these cells.

Lipids

Lipids are fats and fatlike substances. They are generally **hydrophobic** ("water-fearing") and thus are insoluble in water. Typically, lipids serve as energy-storage molecules—usually in the form of fats or oils—and also for structural purposes, as in the case of phospholipids and waxes. Phospholipids are important components of all biological membranes. Although some lipid molecules are very large, they are not, strictly speaking, macromolecules, because they are not formed by the polymerization of monomers.

Fats and Oils Are Triglycerides That Store Energy

Plants, such as the potato, ordinarily store carbohydrates as starch. However, some plants also store food energy as oils (Figure 2–6), especially in seeds and fruits, as in the olives produced by olive trees. Animals, which have a limited capacity for storing carbohydrates (which they store as glycogen), readily convert excess sugar to fat. Fats and oils contain a higher proportion of energy-rich carbon-hydrogen bonds than do carbohydrates. Consequently, fats and oils contain more chemical energy. On average, fats yield about 9.1 kilocalories (kcal) per gram when oxidized to release energy, compared with 3.8 kcal per gram of carbohydrate or 3.1 kcal per gram of protein.

Fats and **oils** have similar chemical structures (Figure 2–7). Each consists of three fatty acid molecules bonded to one glycerol molecule. As with the formation of disaccharides and polysaccharides from their subunits, each of these bonds is formed by dehydration synthesis, which involves the removal of a molecule of water. Fat and oil molecules, also known as

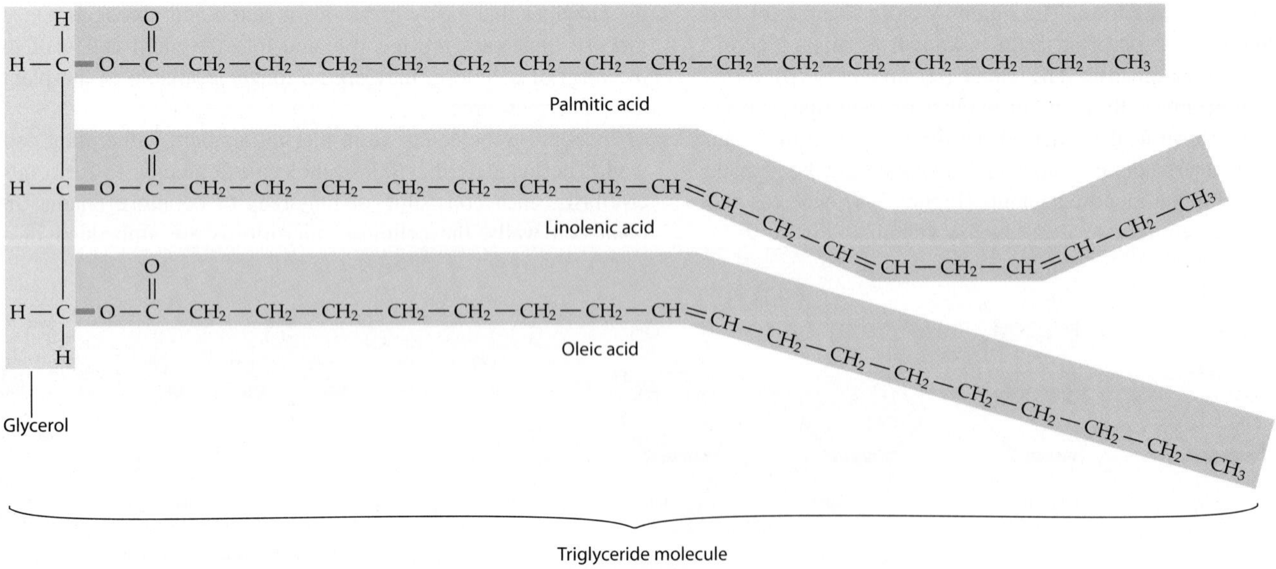

$H-C(H)-O-C(=O)-CH_2-CH_2-CH_2-CH_2-CH_2-CH_2-CH_2-CH_2-CH_2-CH_2-CH_2-CH_2-CH_2-CH_2-CH_3$

Palmitic acid

$H-C-O-C(=O)-CH_2-CH_2-CH_2-CH_2-CH_2-CH_2-CH_2-CH=CH-CH_2-CH=CH-CH_2-CH=CH-CH_2-CH_3$

Linolenic acid

$H-C(H)-O-C(=O)-CH_2-CH_2-CH_2-CH_2-CH_2-CH_2-CH_2-CH=CH-CH_2-CH_2-CH_2-CH_2-CH_2-CH_2-CH_2-CH_3$

Oleic acid

Glycerol

Triglyceride molecule

2–7 Triglyceride Fat and oil molecules consist of three fatty acid molecules bonded (blue) to a glycerol molecule (hence the term "triglyceride"). Three different fatty acids are shown here. Palmitic acid is saturated, and linolenic and oleic acids are unsaturated, as you can see by the double bonds in the hydrocarbon chains.

Polar head

Nonpolar tail

Phosphate group

Glycerol

Fatty acid

Phospholipid molecule

2–8 **Phospholipid** A phospholipid molecule consists of two fatty acid molecules linked to a glycerol molecule, as in a triglyceride, but the third carbon of glycerol is linked to the phosphate group of a phosphate-containing molecule. The letter "R" denotes the atom or group of atoms that makes up the "rest of the molecule." The phospholipid tail is nonpolar and uncharged and is therefore hydrophobic (insoluble in water); the polar head containing the phosphate and R groups is hydrophilic (soluble in water).

triglycerides (or triacylglycerols), contain no polar (hydrophilic) groups. Nonpolar molecules tend to cluster together in water, just as droplets of fats tend to coalesce, for example, on the surface of chicken soup. Nonpolar molecules are therefore hydrophobic, or insoluble in water.

You have undoubtedly heard a lot about "saturated" and "unsaturated" fats. A fatty acid that has no double bonds between carbon atoms is said to be **saturated.** Each carbon atom in the chain has formed covalent bonds to four other atoms, and its bonding possibilities are therefore complete. By contrast, a fatty acid that contains carbon atoms joined by double bonds is said to be **unsaturated.** The double-bonded carbon atoms have the potential to form additional bonds with other atoms.

The physical nature of a fat is determined by the length of the carbon chains in the fatty acids and by the extent to which the fatty acids are saturated or unsaturated. The presence of double bonds in unsaturated fats leads to kinks in the hydrocarbon chains, which prevents close packing among molecules. This tends to lower the melting point of the fat, such that unsaturated fats tend to be liquid (oily) at room temperature. Examples of unsaturated fats, which are primarily found in plants, are safflower oil, peanut oil, and corn oil, all obtained from oil-rich seeds. Animal fats and their derivatives, such as butter and lard, contain highly saturated fatty acids and are usually solid at room temperature. Thus, the terms *fat* and *oil* generally refer to the physical state of the triglyceride. Fats are triglycerides that are usually solid at room temperature, whereas oils are usually liquid.

Phospholipids Are Modified Triglycerides That Are Components of Cellular Membranes

Lipids, especially phospholipids, play very important structural roles, particularly in cellular membranes. Like triglycerides, **phospholipids** are composed of fatty acid molecules attached to a glycerol backbone. In the phospholipids, however, the third carbon of the glycerol molecule is linked not to a fatty acid but to a phosphate group to which another polar group is usually attached (Figure 2–8). Phosphate groups are negatively charged. As a result, the phosphate end of the molecule is hydrophilic and therefore soluble in water, whereas the fatty acid end is hydrophobic and insoluble. If phospholipids are added to water, they tend to form a film along its surface, with their hydrophilic "heads" under the water and their hydrophobic "tails" protruding above the surface. If phospholipids are surrounded by water, as in the watery interior of the cell, they tend to align themselves in a double layer—a **phospholipid bilayer**—with their phosphate heads directed outward and their fatty acid tails oriented toward one another (Figure 2–9). As we discuss further in Chapter 4, such configurations are important not only to the structure of cellular membranes but also to their functions.

Cutin, Suberin, and Waxes Are Lipids That Form Barriers to Water Loss

Cutin and **suberin** are unique lipids that are important structural components of many plant cell walls. The major function of these lipids is to form a matrix in which **waxes**—long-chain lipid compounds—are embedded. The waxes, in combination with cutin or suberin, form barrier layers that help prevent the loss of water and other molecules from plant surfaces.

A protective **cuticle,** which is characteristic of plant surfaces exposed to air, covers the outer walls of epidermal cells (the outermost cells) of leaves and stems. Composed of wax embedded in cutin **(cuticular wax),** the cuticle is frequently covered by a layer of **epicuticular wax** (Figure 2–10). When you polish a freshly picked apple on your sleeve, you are polishing this layer of epicuticular wax.

Suberin is a major component of the walls of cork cells, the cells that form the outermost layer of bark in woody stems and roots. As seen with the electron microscope, suberin-containing,

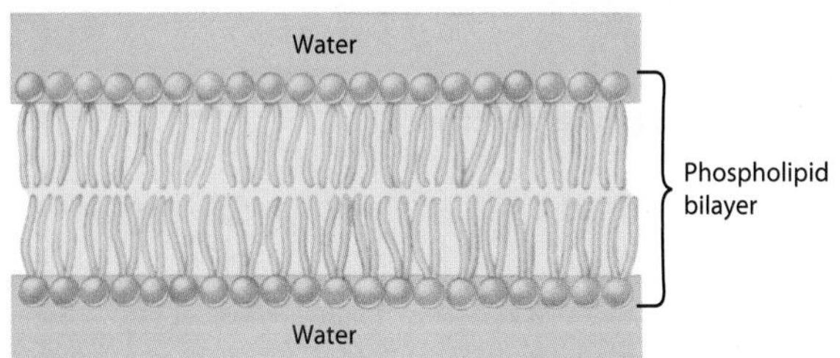

2–9 **Phospholipid bilayer** Surrounded by water, phospholipids spontaneously arrange themselves in two layers, with their hydrophilic heads extending outward into the water and their hydrophobic tails inward, away from the water. This arrangement—a phospholipid bilayer—forms the structural basis of cellular membranes.

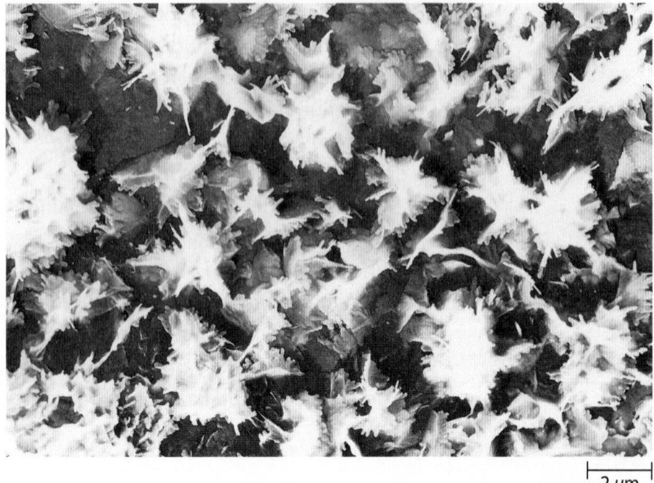

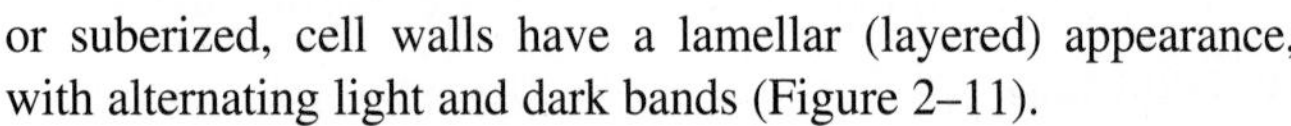

2–10 Epicuticular wax Scanning electron micrograph of the upper surface of a leaf of the eucalyptus tree *(Eucalyptus cloeziana)* showing deposits of epicuticular wax. Beneath these deposits is the cuticle, a wax-containing layer or layers covering the outer walls of the epidermal cells. Waxes help protect exposed plant surfaces from water loss.

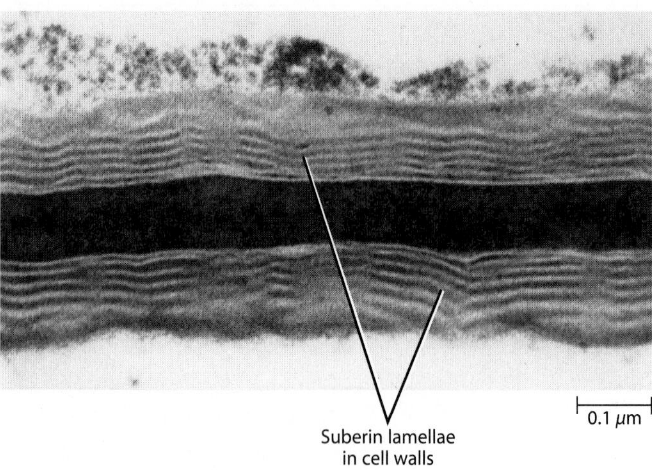

2–11 Suberin lamellae Electron micrograph showing suberin lamellae in the walls between two cork cells of a potato tuber. Note the alternating light and dark bands. Cork cells form the outermost layer of a protective covering in plant parts such as potato tubers and woody stems and roots.

or suberized, cell walls have a lamellar (layered) appearance, with alternating light and dark bands (Figure 2–11).

Waxes are the most water-repellent of the lipids. Carnauba wax used for car and floor polishes is harvested from the leaves of the carnauba wax palm *(Copernicia cerifera)* from Brazil.

Steroids Stabilize Cellular Membranes and Also Function as Hormones

Steroids can be easily distinguished from the other classes of lipids by the presence of four interconnected hydrocarbon rings. In living organisms, hydrocarbon chains of various lengths, as well as hydroxyl and/or carbonyl groups, may be attached to this skeleton, making possible a large variety of structural and other molecules. When a hydroxyl group is attached at the carbon-3 position, the steroid is called a **sterol** (Figure 2–12). Sitosterol is the most abundant sterol found in green algae and plants, and ergosterol occurs frequently in fungi. Cholesterol, so common in animal cells, is present in only trace amounts in plants. In all organisms except prokaryotes, sterols are important components of membranes, where they stabilize the phospholipid tails.

Steroids may also function as hormones. For example, the sterol antheridiol serves as a sex attractant in the water mold *Achlya bisexualis,* and a group of steroid derivatives called brassins promote the growth of certain stems. There is also evidence that some plants produce estrogen, one of the mammalian sex hormones, but its role in the plant is unknown.

2–12 Sterols ***(a)*** General structure of a sterol; ***(b)*** β-sitosterol; ***(c)*** ergosterol; ***(d)*** cholesterol.

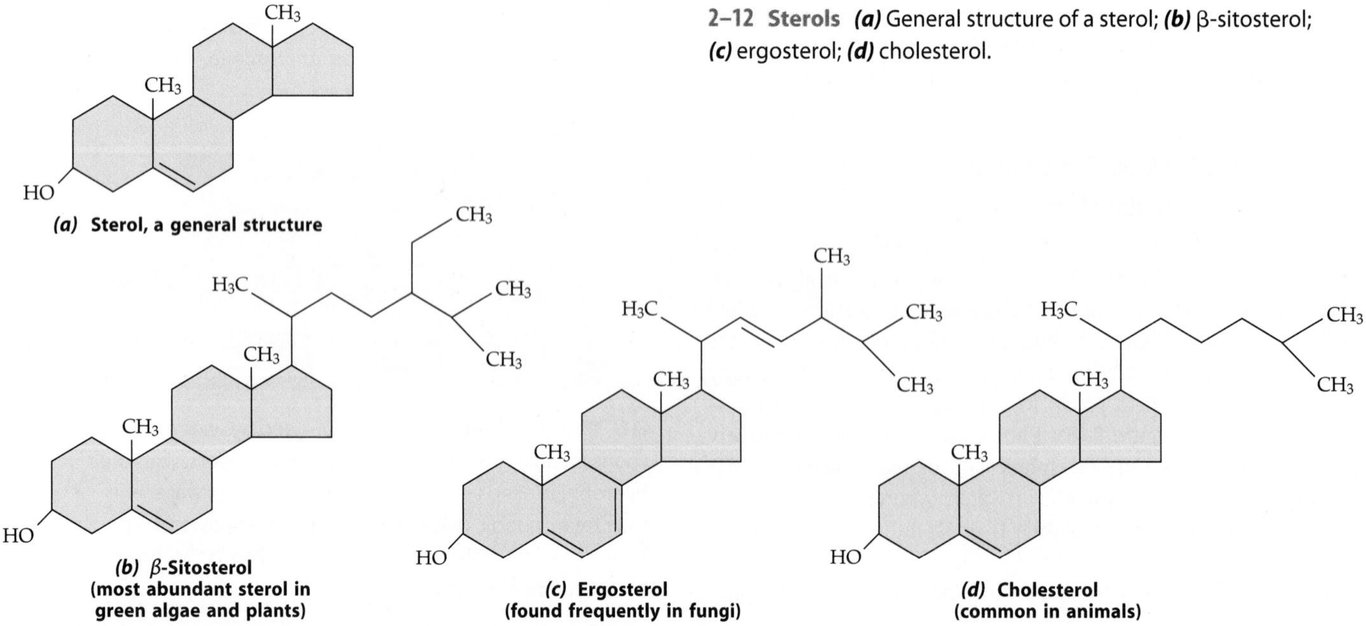

(a) Sterol, a general structure

(b) β-Sitosterol (most abundant sterol in green algae and plants)

(c) Ergosterol (found frequently in fungi)

(d) Cholesterol (common in animals)

Proteins

Proteins are among the most abundant organic molecules. In the majority of living organisms, proteins make up 50 percent or more of the dry weight. Only plants, with their high cellulose content, are less than half protein in their dry weight. Proteins perform an incredible diversity of functions in living organisms. In their structure, however, proteins all follow the same simple blueprint: they are all polymers of nitrogen-containing molecules known as **amino acids,** arranged in a linear sequence. Twenty different kinds of amino acids are used by living organisms to form proteins. (See "Vegetarians, Amino Acids, and Nitrogen" below.)

Protein molecules are large and complex, often containing several hundred or more amino acid monomers. Thus, the possible number of different amino acid sequences, and therefore the possible variety of protein molecules, is enormous—about as enormous as the number of different sentences that can be written with our own 26-letter alphabet. Organisms, however, synthesize only a very small fraction of the proteins that are theoretically possible. A single cell of the bacterium *Escherichia coli,* for example, contains 600 to 800 different kinds of proteins at any one time, and a plant or animal cell has several times that number. A complex organism has at least several thousand different kinds of proteins, each with a special function and each, by its unique chemical nature, specifically fitted for that function.

In plants, the largest concentration of proteins is found in certain seeds (for example, the seeds of cereals and legumes), in which as much as 40 percent of the dry weight may be protein. These specialized proteins function as storage forms of amino acids to be used by the embryo when it resumes growth upon germination of the seed.

Amino Acids Are the Building Blocks of Proteins

Each specific protein is made up of a precise arrangement of amino acids. All amino acids have the same basic structure,

VEGETARIANS, AMINO ACIDS, AND NITROGEN

Like fats, amino acids are formed within living cells using sugars as starting materials. Whereas fats contain only carbon, hydrogen, and oxygen atoms—all available in the sugar and water of the cell—amino acids also contain nitrogen. Most of the Earth's supply of nitrogen exists in the form of gas in the atmosphere. Only a few organisms, all of which are microorganisms, are able to incorporate nitrogen from the air into compounds—ammonia, nitrites, and nitrates—that can be used by living systems. Hence, only a small proportion of the Earth's nitrogen supply is available to the living world.

Plants incorporate the nitrogen from ammonia, nitrites, and nitrates into carbon-hydrogen compounds to form amino acids. Animals are able to synthesize some of their own amino acids, using ammonia derived from their diet as a nitrogen source. The amino acids they cannot synthesize, the so-called **essential amino acids,** must be obtained in the diet, either from plants or from the meat of other animals that have eaten plants. For adult human beings, the essential amino acids are lysine, tryptophan, threonine, methionine, histidine, phenylalanine, leucine, valine, and isoleucine. To take full advantage of the protein-building capabilities of these amino acids, it is important to have a diet that supplies them in the right ratio.

For many years, agricultural scientists concerned with the world's hungry people concentrated on developing plants with a high caloric yield. Recognition of the role of plants as a major source of amino acids for human populations, however, has led to an emphasis on the development of high-protein strains of food plants. Of particular importance has been the development of plants, such as "high-lysine" maize, with increased levels of one or more of the essential amino acids.

People who eat meat usually get enough protein and the correct balance of amino acids. People who are vegetarians, whether for philosophical, aesthetic, or economic reasons, have to be careful that they eat enough protein and, in particular, all of the essential amino acids.

Adequate protein is rarely a problem for vegetarians who eat milk, eggs, and other dairy products. These foods contain relatively high amounts of protein, with a good balance of essential amino acids. Vegans, who do not eat food from any animal source, may need to pay extra attention to obtaining enough protein from plants alone. High-protein plant foods include beans, nuts, and whole grains. Consuming a varied diet with adequate calories is usually sufficient to ensure adequate protein consumption. Vegans should also be sure to obtain enough calcium (dark green vegetables are a good source), iron (beans, seeds, and dried fruits), and especially vitamin B_{12} (from nutritional yeast or vitamin supplements).

One good approach to obtaining the right balance of amino acids from plant sources is to combine certain foods. Beans, for instance, are likely to be deficient in tryptophan and in the sulfur-containing amino acids cysteine and methionine, but they are a good-to-excellent source of isoleucine and lysine. Rice is deficient in isoleucine and lysine but provides an adequate amount of the other essential amino acids. Thus rice and beans in combination make just about as perfect a protein menu as eggs or steak, as some vegetarians have known for quite a long time.

Eating a variety of richly colored vegetables supplies our bodies with fiber and valuable nutrients, among them vitamins A, C, and E, as well as potassium, zinc, and selenium.

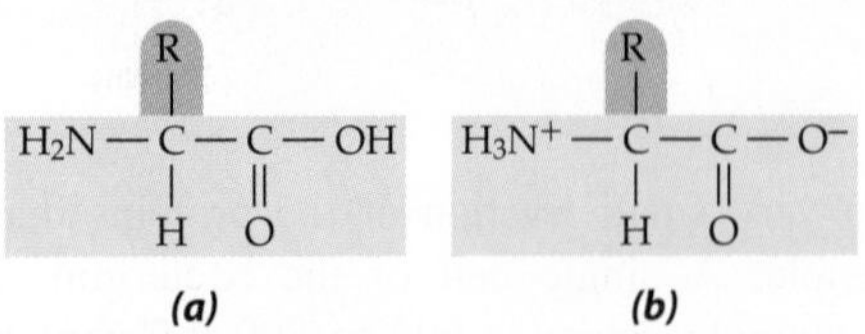

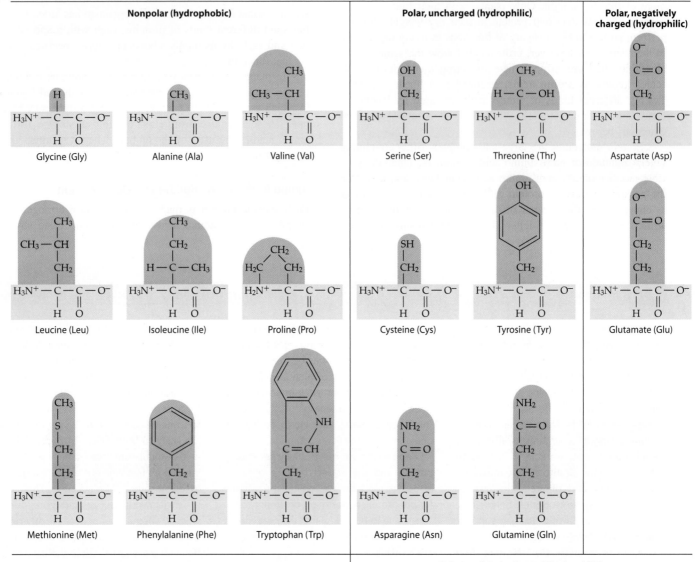

Nonpolar (hydrophobic)

Glycine (Gly) · Alanine (Ala) · Valine (Val)

Leucine (Leu) · Isoleucine (Ile) · Proline (Pro)

Methionine (Met) · Phenylalanine (Phe) · Tryptophan (Trp)

Polar, uncharged (hydrophilic)

Serine (Ser) · Threonine (Thr)

Cysteine (Cys) · Tyrosine (Tyr)

Asparagine (Asn) · Glutamine (Gln)

Polar, negatively charged (hydrophilic)

Aspartate (Asp)

Glutamate (Glu)

(c)

Polar, positively charged (hydrophilic)

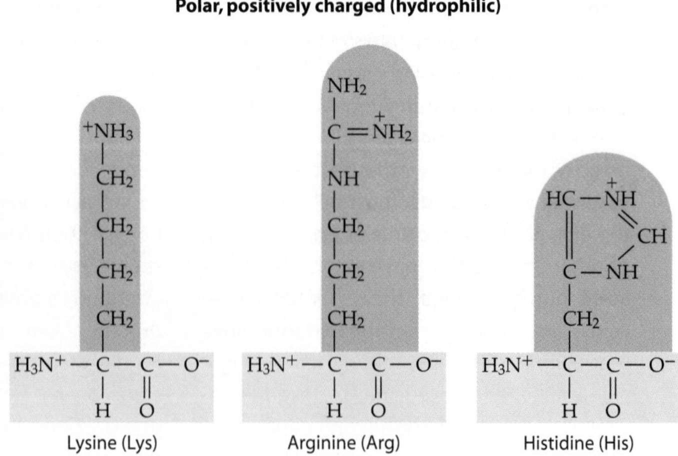

2–13 Amino acids ***(a)*** The general formula of an amino acid. Every amino acid contains an amino group (—NH_2) and a carboxyl group (—COOH) bonded to a central carbon atom. A hydrogen atom and a side group (R) are also bonded to the same carbon atom. This basic structure is the same in all amino acids, but the R side group is different in each kind of amino acid. ***(b)*** At pH 7, both the amino and the carboxyl groups are ionized. ***(c)*** The 20 amino acids present in proteins. As you can see, the essential structure is the same in all 20 molecules, but the R side groups differ. Amino acids with nonpolar R groups are hydrophobic and, as proteins fold into their three-dimensional form, these amino acids tend to aggregate on the inside of proteins. Amino acids with polar, uncharged R groups are relatively hydrophilic and are usually on the surface of proteins. Those amino acids with acidic (negatively charged) and basic (positively charged) R groups are very polar, and therefore hydrophilic; they are almost always found on the surface of protein molecules. All of the amino acids are shown in the ionized state predominating at pH 7. The letters in parentheses following the name of each amino acid form the standard abbreviation for that amino acid.

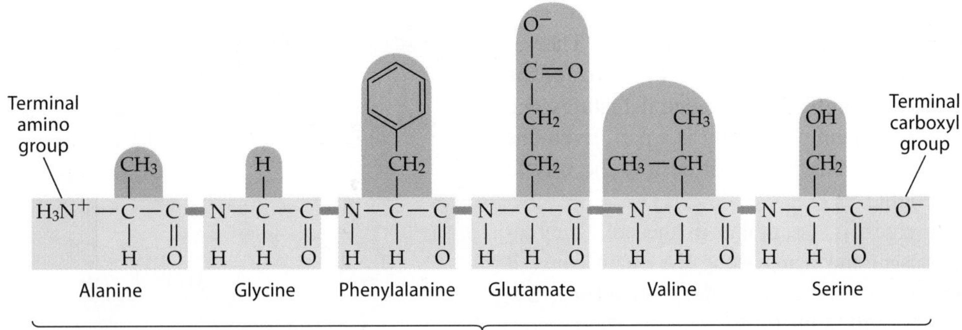

2–14 Polypeptide The links between amino acid residues are known as peptide bonds (blue). Peptide bonds are formed by the removal of a molecule of water (dehydration synthesis). The bonds always form between the carboxyl group (—COO^-) of one amino acid and the amino group (—NH_3^+) of the next. Consequently, the basic structure of a protein is a long, unbranched molecule. The short polypeptide chain shown here contains six different amino acids, but proteins consist of polypeptides with several hundred or even as many as 1000 linked amino acid monomers. This linear arrangement of amino acids is known as the primary structure of the protein.

consisting of an amino group (—NH_2), a carboxyl group (—COOH), and a hydrogen atom, all bonded to a central carbon atom. The differences arise from the fact that every amino acid has an "R" group—an atom or a group of atoms—also bonded to the central carbon (Figure 2–13a, b). It is the R group ("R" can be thought of as the "rest of the molecule") that determines the identity of each amino acid.

A large variety of amino acids is theoretically possible, but only 20 different kinds are used to build proteins. And it is always the same 20, whether in a bacterial cell, a plant cell, or a cell in your own body. Figure 2–13c shows the complete structure of the 20 amino acids found in proteins. The amino acids are grouped according to their polarity and electric charge, which determine not only the properties of the individual amino acids but also, more importantly, the properties of the proteins formed from them.

In yet another example of a dehydration synthesis, the amino group of one amino acid links to the carboxyl group of the adjacent amino acid by the removal of a molecule of water. Again, this is an energy-requiring process. The covalent bond formed is known as a **peptide bond,** and the molecule that results from the linking of many amino acids is known as a **polypeptide** (Figure 2–14). Proteins are large polypeptides, and, in some cases, they consist of several polypeptides. These macromolecules have molecular weights ranging from 10^4 (10,000) to more than 10^6 (1,000,000). In comparison, water has a molecular weight of 18, and glucose has a molecular weight of 180.

A Protein's Structure Can Be Described in Terms of Levels of Organization

In a living cell, a protein is assembled as one or more long polypeptide chains. The linear sequence of amino acids, which is dictated by the information stored in the cell for that particular protein, is known as the **primary structure** of the protein (Figures 2–14 and 2–15a). Each kind of polypeptide has a

2–15 The four levels of protein organization ***(a)*** The primary structure of a protein consists of a linear sequence of amino acids linked together by peptide bonds. ***(b)*** The polypeptide chain may coil into an alpha helix, one type of secondary structure. ***(c)*** The alpha helix may fold to form a three-dimensional, globular structure, the tertiary structure. ***(d)*** The combination of several polypeptide chains into a single functional molecule is the quaternary structure. The polypeptides may or may not be identical.

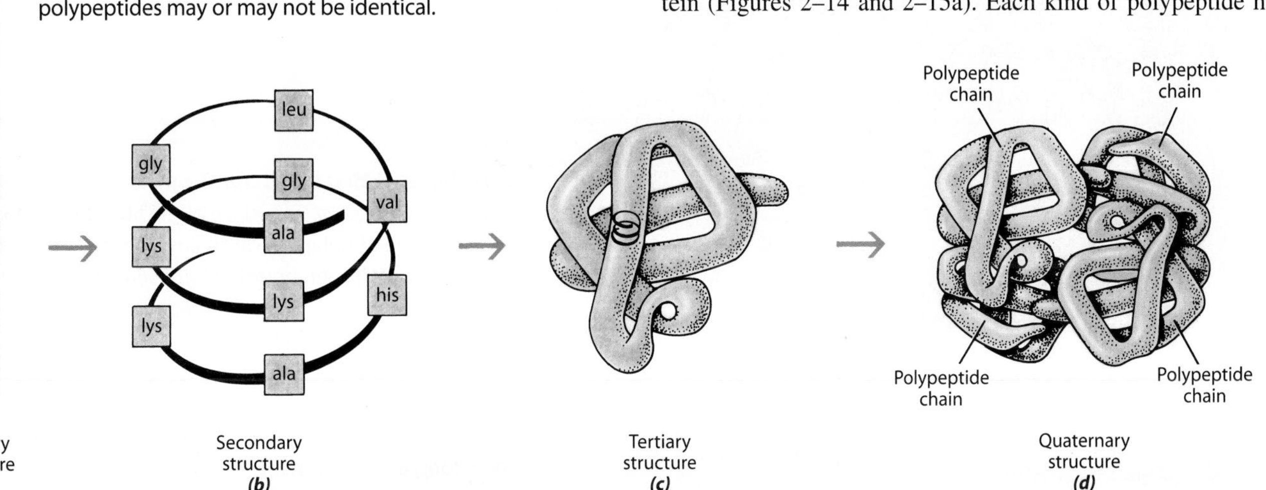

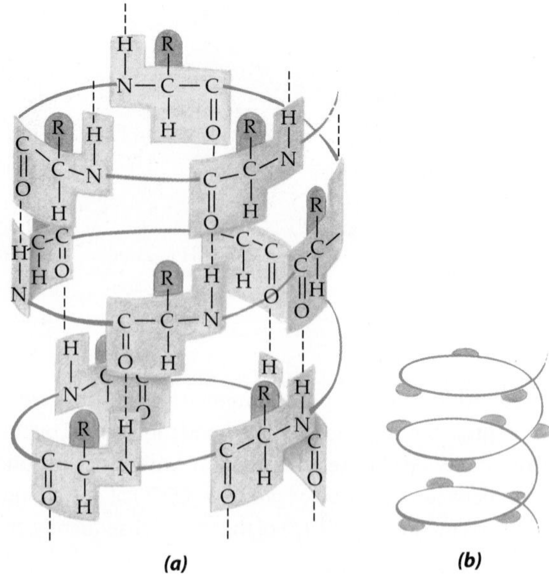

2–16 Alpha helix ***(a)*** The helix is held in shape by hydrogen bonds, indicated by the dashed lines. The hydrogen bonds form between the double-bonded oxygen atom in one amino acid and the hydrogen atom of the amino group in another amino acid situated four amino acids farther along the chain. The R groups, which appear flattened in this diagram, actually extend out from the helix, as shown in ***(b)***. In some proteins, virtually all of the molecule is in the form of an alpha helix. In other proteins, only certain regions of the molecule have this secondary structure.

different primary structure—a unique polypeptide "word" consisting of a unique sequence of amino acid "letters." The sequence of amino acids determines the structural features of the polypeptide molecule and therefore the structural features and biological function of the protein of which it is a part. Even one small variation in the sequence may alter or destroy the way in which the protein functions.

As a polypeptide chain is assembled in the cell, interactions among the various amino acids along the chain cause it to fold into a pattern known as its **secondary structure.** Because peptide bonds are rigid, a chain is limited in the number of shapes it can assume. One of the two most common secondary structures is the **alpha helix** (Figures 2–15b and 2–16), the shape of which is maintained by hydrogen bonds. Another common secondary structure is the **beta pleated sheet** (Figure 2–17). In the beta pleated sheet, polypeptide chains are lined up in parallel and are linked by hydrogen bonds, resulting in a zigzag shape rather than a helix.

Proteins that exist for most of their length in a helical or pleated-sheet secondary structure are known as **fibrous proteins.** Fibrous proteins play a variety of important structural roles, providing support and shape in organisms. In other proteins, known as **globular proteins,** the secondary structure folds to form a **tertiary structure** (Figure 2–15c). For some proteins, the folding occurs spontaneously, that is, by a self-assembly process. For others, certain proteins known as **molecular chaperones** facilitate the process by inhibiting incorrect folding. Globular proteins tend to be structurally complex, often having more than one type of secondary structure. Most biologically active proteins, such as enzymes, membrane proteins, and transport proteins, are globular, as are the subunits of some important structural proteins. The microtubules that occur within the cell, for example, are composed of a large number of spherical subunits, each of which is a globular protein (see Figure 3–25).

The tertiary structure forms as a result of complex interactions among the R groups of the individual amino acids. These interactions include attractions and repulsions among amino acids with polar R groups and repulsions between nonpolar R groups and the surrounding water molecules. In addition, the sulfur-containing R groups of two cysteine monomers can form a covalent bond with each other. These bonds, known as **disulfide bridges,** lock portions of the polypeptide molecule into a

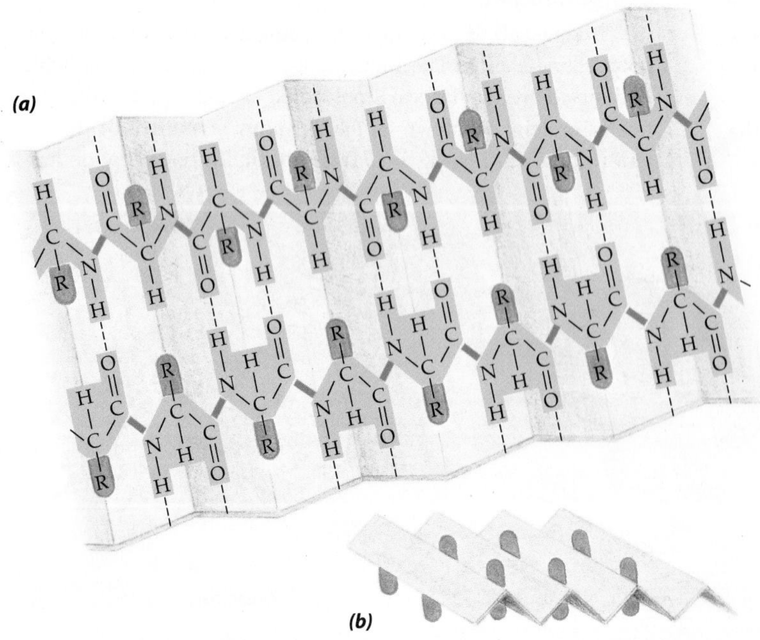

2–17 Beta pleated sheet ***(a)*** The pleats result from the alignment of the zigzag pattern of the atoms that form the backbone of polypeptide chains. The sheet is held together by hydrogen bonding between adjacent chains. The R groups extend above and below the pleats, as shown in ***(b)***. In some proteins, two or more polypeptide chains are aligned with one another to form a pleated sheet. In other proteins, a single polypeptide chain loops back and forth in such a way that adjacent portions of the chain form a pleated sheet.

particular position and sometimes link adjacent polypeptides together.

Most of the interactions that give a protein its tertiary structure are not covalent and are therefore relatively weak. They can be disrupted quite easily by physical or chemical changes in the environment, such as heat or increased acidity. This structural breakdown is called **denaturation.** The coagulation of egg white when an egg is cooked is a familiar example of protein denaturation. When proteins are denatured, the polypeptide chains unfold and the tertiary structure is disrupted, causing a loss of the biological activity of the protein. Most organisms cannot live at unusually high temperatures or outside a specific pH range because their enzymes and other proteins become unstable and nonfunctional due to denaturation.

Many proteins are composed of more than one polypeptide chain. These chains may be held to each other by hydrogen bonds, disulfide bridges, hydrophobic forces, attractions between positive and negative charges, or, most often, a combination of these types of interactions. This level of organization of proteins—the interaction of two or more polypeptides—is called a **quaternary structure** (Figure 2–15d).

Enzymes Are Proteins That Catalyze Chemical Reactions in Cells

Enzymes are large, complex globular proteins that act as catalysts. By definition, **catalysts** are substances that accelerate the rate of a chemical reaction by lowering the energy of activation, but remain unchanged in the process (see Figure 5–5). Because they remain unaltered, catalyst molecules can be used over and over again and so are typically effective at very low concentrations.

Enzymes are often named by adding the ending *-ase* to the root of the name of the substrate (the reacting molecule or molecules). Thus, amylase catalyzes the hydrolysis of amylose (starch) into glucose molecules, and sucrase catalyzes the hydrolysis of sucrose into glucose and fructose. Nearly 2000 different enzymes are now known, and each of them is capable of catalyzing some specific chemical reaction. The behavior of enzymes in biological reactions is further explained in Chapter 5.

Nucleic Acids

The information dictating the structures of the enormous variety of proteins found in living organisms is encoded in and translated by molecules known as nucleic acids. Just as proteins consist of long chains of amino acids, nucleic acids consist of long chains of molecules known as **nucleotides.** A nucleotide, however, is a more complex molecule than an amino acid.

As shown in Figure 2–18, a nucleotide consists of three components: a phosphate group, a five-carbon sugar, and a nitrogenous base—a molecule that has the properties of a base and contains nitrogen. The sugar subunit of a nucleotide may be either **ribose** or **deoxyribose,** which contains one fewer oxygen atom than ribose (Figure 2–19). Five different nitrogenous bases occur in the nucleotides that are the building blocks of nucleic acids: adenine, guanine, thiamine, cytosine, and uracil. In the nucleotide shown in Figure 2–18—adenosine monophosphate (AMP)—the nitrogenous base is adenine and the sugar is ribose.

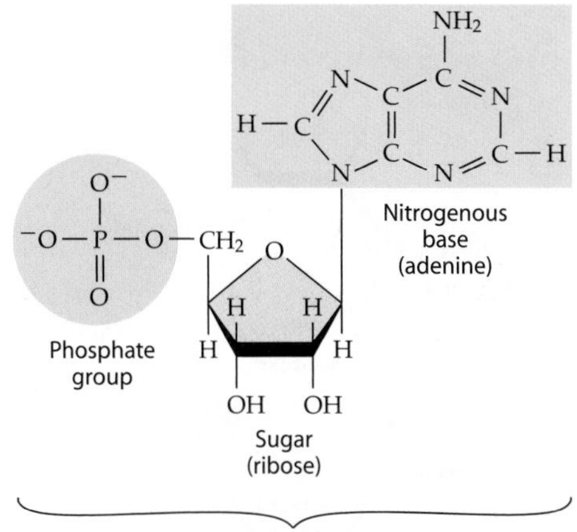

2–18 Nucleotide structure A nucleotide is made up of three different subunits: a phosphate group, a five-carbon sugar, and a nitrogenous base. The nitrogenous base in this nucleotide is adenine, and the sugar is ribose. Because there is a single phosphate group, this nucleotide is called adenosine monophosphate, abbreviated AMP.

Two types of nucleic acids are found in living organisms. In **ribonucleic acid (RNA),** the sugar subunit in the nucleotides is ribose. In **deoxyribonucleic acid (DNA),** it is deoxyribose. Like polysaccharides, lipids, and proteins, RNA and DNA are formed from their subunits in dehydration synthesis reactions. The result is a linear macromolecule consisting of one nucleotide after another (Figure 2–20). DNA molecules, in particular, are exceedingly long and, in fact, are the largest macromolecules in cells.

Although their chemical components are very similar, DNA and RNA generally play different biological roles. DNA is the carrier of genetic messages. It contains the information, organized in units known as **genes,** that we and other organisms inherit from our parents. RNA molecules are involved in the synthesis of proteins based on the genetic information provided by DNA. Some RNA molecules function as enzymelike catalysts, referred to as ribozymes.

The discovery of the structure and function of DNA and RNA is undoubtedly the greatest triumph thus far of the molecular approach to the study of biology. In Section 3, we will trace the events leading to the key discoveries and consider in some detail the marvelous processes—the details of which are

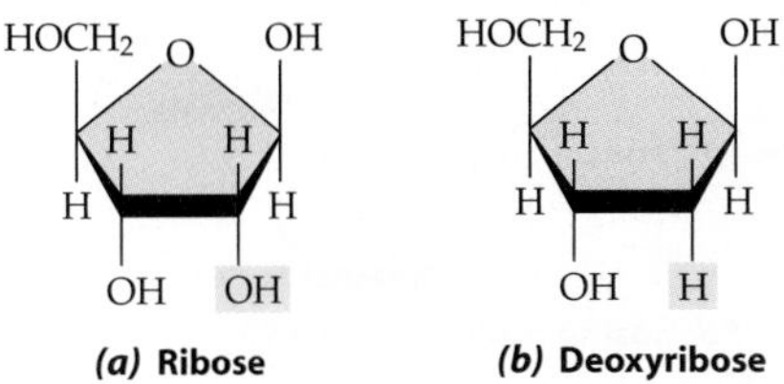

2–19 Ribose and deoxyribose The sugar subunit of a nucleotide may be either ***(a)*** ribose or ***(b)*** deoxyribose. RNA is formed from nucleotides that contain ribose, and DNA from nucleotides that contain deoxyribose.

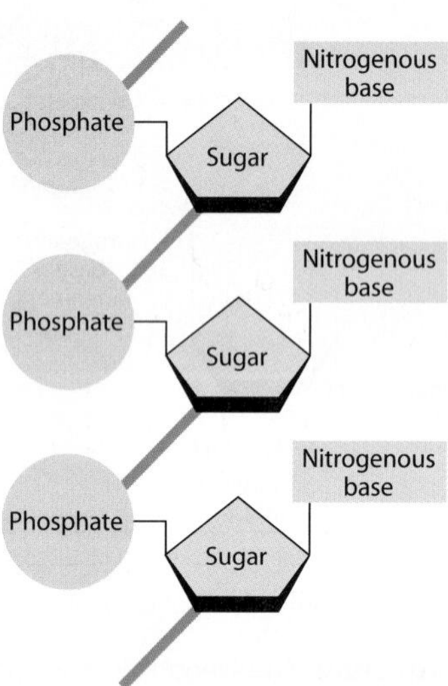

2–20 General structure of a nucleic acid Nucleic acid molecules are long chains of nucleotides in which the sugar subunit of one nucleotide is linked to the phosphate group of the next nucleotide. The covalent bond linking one nucleotide to the next—shown here in blue—is formed by a dehydration synthesis reaction. RNA molecules consist of a single chain of nucleotides, as shown here. DNA molecules, by contrast, consist of two chains of nucleotides, coiled around each other in a double helix.

still being worked out—by which the nucleic acids perform their functions.

ATP Is the Cell's Major Energy Currency

In addition to their role as the building blocks of nucleic acids, nucleotides have an independent and crucial function in living systems. When modified by the addition of two more phosphate groups, they are the carriers of the energy needed to power the numerous chemical reactions occurring within cells.

The principal energy carrier for most processes in living organisms is the molecule **adenosine triphosphate,** or **ATP,** shown schematically in Figure 2–21. Note the three phosphate groups on the left. The bonds linking these groups are relatively weak, and they can be broken quite readily by hydrolysis and release a large amount of energy in the process. The products of ATP hydrolysis are **ADP (adenosine diphosphate),** a free phosphate, and enough energy to drive many of the energy-requiring reactions in cells (Figure 2–21).

During respiration, ADP is "recharged" to ATP when glucose is oxidized to carbon dioxide and water, just as the money in your pocket is "recharged" when you cash a check or visit an automated teller machine. We consider this process in more detail in Chapter 6. For now, however, the important thing to remember is that ATP is the molecule that is directly involved in providing energy for much of the living cell.

Secondary Metabolites

Historically, the compounds produced by plants have been separated into primary and secondary metabolites, or products. **Primary metabolites,** by definition, are molecules that are found in all plant cells and are necessary for the life of the plant. Examples of primary metabolites are simple sugars, amino acids, proteins, and nucleic acids. **Secondary metabolites,** by contrast, are restricted in their distribution, both within the plant and among the different species of plants. Once considered waste products, secondary metabolites are now known to be important for the survival and propagation of the plants that produce them. Many serve as chemical signals that enable the plant to respond to environmental cues. Others function in the plant's defense against herbivores, pathogens (disease-causing organisms), or competitors. Some provide protection from radiation from the sun, while still others aid in pollen and seed dispersal.

As indicated above, secondary metabolites are not evenly distributed throughout the plant. Their production typically occurs in a specific organ, tissue, or cell type at specific stages of development (for example, during flower, fruit, seed, or seedling development). Some, the **phytoalexins,** are antimicrobial compounds produced only after wounding or after attack by bacteria or fungi (see page 57). Although secondary metabolites are produced at various sites within the cell, they are stored primarily within vacuoles. Moreover, their concentration in a plant often varies greatly during a 24-hour period. The three major classes of secondary plant compounds are alkaloids, terpenoids, and phenolics.

Alkaloids Are Alkaline Nitrogenous Compounds That Include Morphine, Cocaine, Caffeine, Nicotine, and Atropine

Alkaloids are among the most important compounds in terms of their pharmacological or medicinal effects. Interest in them has traditionally stemmed from their dramatic physiological or psychological effect on humans.

P — P — P — Ribose — Adenine + H_2O ⇌ P — P — Ribose — Adenine + P + Energy

Adenosine triphosphate (ATP) **Adenosine diphosphate (ADP)**

2–21 Hydrolysis of ATP With the addition of a molecule of water to ATP, one phosphate group is removed from the molecule. The products of this reaction are ADP, a free phosphate, and energy. A large amount of energy is released when an ATP molecule is hydrolyzed. With an equal energy input, the reaction can be reversed.

2–22 Some physiologically active alkaloids ***(a)*** Morphine is contained in the milky juice released by slitting the seed pods of the opium poppy *(Papaver somniferum)*; ***(b)*** cocaine is present in the leaves of the coca plant *(Erythroxylum coca)*; ***(c)*** coffee *(Coffea)* beans and tea *(Camellia)* leaves contain caffeine; and ***(d)*** cultivated tobacco *(Nicotiana tabacum)* plants contain nicotine.

The first alkaloid to be identified—in 1806—was **morphine,** from the opium poppy *(Papaver somniferum).* It is used today in medicine as an analgesic (pain reliever) and cough suppressant; however, excessive use of this drug can lead to strong addiction. Nearly 10,000 alkaloids have now been isolated and their structures identified, including cocaine, caffeine, nicotine, and atropine. The structures of some of these physiologically active alkaloids are shown in Figure 2–22.

Cocaine comes from coca *(Erythroxylum coca),* a shrub or small tree that is indigenous to the eastern slopes of the Andes Mountains of Bolivia and Peru. Many Incas living at the high elevations of these mountains chew coca leaves to lessen hunger pangs and fatigue while working in this harsh environment. Chewing the leaves, which contain small concentrations of cocaine, is relatively harmless compared with the smoking, snorting, or intravenous injection of cocaine. The habitual use of cocaine and its derivative "crack" can have devastating effects both physically and psychologically, and can lead to death. Cocaine has been used as an anesthetic in eye surgery and as a local anesthetic by dentists.

Caffeine, a stimulant found in such plants as coffee *(Coffea arabica),* tea *(Camellia sinensis),* and cocoa *(Theobroma cacao),* is a component of popular beverages. The high concentrations of caffeine present in the developing seedlings of the coffee plant have been shown to be highly toxic and lethal to both insects and fungi. In addition, caffeine released by the seedling apparently inhibits germination of other seeds in the vicinity of the seedling, preventing the growth of competitors. This process is called **allelopathy.**

Nicotine, another stimulant, is obtained from the leaves of the tobacco plant *(Nicotiana tabacum).* It is a highly toxic alkaloid that has received considerable attention because of concern over the harmful effects of cigarette smoking. Nicotine is synthesized in the roots and transported to the leaves, where it is sequestered in vacuoles. It is an effective deterrent to attack by grazing herbivores and insects. Nicotine is synthesized in response to wounding, apparently functioning as a phytoalexin.

Atropine-containing extracts from the Egyptian henbane *(Hyoscyamus muticus)* were used by Cleopatra in the first century B.C.E. to dilate her pupils, in the hope that she would appear more alluring. During the Medieval period, European women used atropine-containing extracts from the deadly nightshade *(Atropa belladonna)* for the same purpose. Atropine is used today as a cardiac stimulant, a pupil dilator for eye examinations, and an effective antidote for some nerve gas poisoning.

Terpenoids Are Composed of Isoprene Units and Include Essential Oils, Taxol, Rubber, and Cardiac Glycosides

Terpenoids, also called terpenes, occur in all plants and are by far the largest class of secondary metabolites, with over 22,000 terpenoid compounds described. The simplest of the terpenoids is the hydrocarbon isoprene (C_5H_8). All terpenoids can be classified according to their number of isoprene units (Figure 2–23a). Familiar categories of terpenoids are the monoterpenoids, which consist of two isoprene units; sesquiterpenoids (three isoprene units); and diterpenoids (four isoprene units). A single plant may synthesize many different terpenoids at

(b)

CH_3

H_2C C CH_2 CH

Isoprene (C_5H_8)

(a)

2–23 Isoprene, a terpenoid ***(a)*** A diverse group of compounds is formed from isoprene units. All sterols, for example, are built of six isoprene units. ***(b)*** Blue haze, composed largely of isoprene, hovering over the Blue Ridge Mountains in Virginia.

different locations within the plant for a variety of purposes and at different times during the course of its development.

Isoprene itself is a gas emitted in significant quantities by the leaves of many plant species and is largely responsible for the bluish haze that hovers over wooded hills and mountains in summer (Figure 2–23b). It is also a component of smog. Isoprene, which is emitted only in the light, is made in chloroplasts from carbon dioxide recently converted to organic compounds by photosynthesis. One may wonder why plants produce and discharge such large quantities of isoprene. Studies have shown that isoprene emissions are highest on hot days and, further, that the isoprene "blanket" may help the plant cope with heat by stabilizing photosynthetic membranes within the plant's cells.

Many of the monoterpenoids and sesquiterpenoids are called **essential oils** because, being highly volatile, they contribute to the fragrance, or essence, of the plants that produce them. In the mint *(Mentha),* large quantities of volatile monoterpenoids (menthol and menthone) are both synthesized and stored in glandular hairs (trichomes), which are outgrowths of the epidermis. The essential oils produced by the leaves of some plants deter herbivores, some protect against attack by fungi or bacteria, while others are known to be allelopathic. The terpenoids of flower fragrances attract insect pollinators to the flowers.

The diterpenoid **taxol** has attracted considerable attention because of its anti-cancer properties. It has been shown to shrink cancers of the ovary and the breast. At one time, the only source of taxol was the bark of the Pacific yew tree *(Taxus brevifolia).* Harvesting all of the bark from a tree yields only a very small amount of taxol (a scant 300-milligram dose from a 40-foot-tall, 100-year-old tree). Moreover, removal of the bark kills the tree. Fortunately, it has been found that extracts from the needles of the European yew tree *(Taxus baccata)* and *Taxus* bushes, as well as from a yew fungus, can yield taxol-like compounds. Needles can be harvested without destroying the yew trees and bushes. Taxol has now been synthesized in the laboratory, but the synthesis technique remains to be refined. Even then, it may be cheaper to make commercial forms of taxol from natural sources.

The largest known terpenoid compound is **rubber,** which consists of molecules containing 400 to more than 100,000 isoprene units. Rubber is obtained commercially from the milky fluid, called **latex,** of the tropical plant *Hevea brasiliensis,* a member of the Euphorbiaceae family (Figure 2–24). The latex is synthesized in cells or in a series of connected cells forming tubelike growths called laticifers. About 1800 species of plants have been reported to contain rubber, but only a few yield enough rubber to make them commercially valuable. In *Hevea,* rubber may constitute 40 to 50 percent of the latex. Latex is obtained from the rubber tree by making a V-shaped incision in the bark. A spout is inserted at the bottom of the incision, and

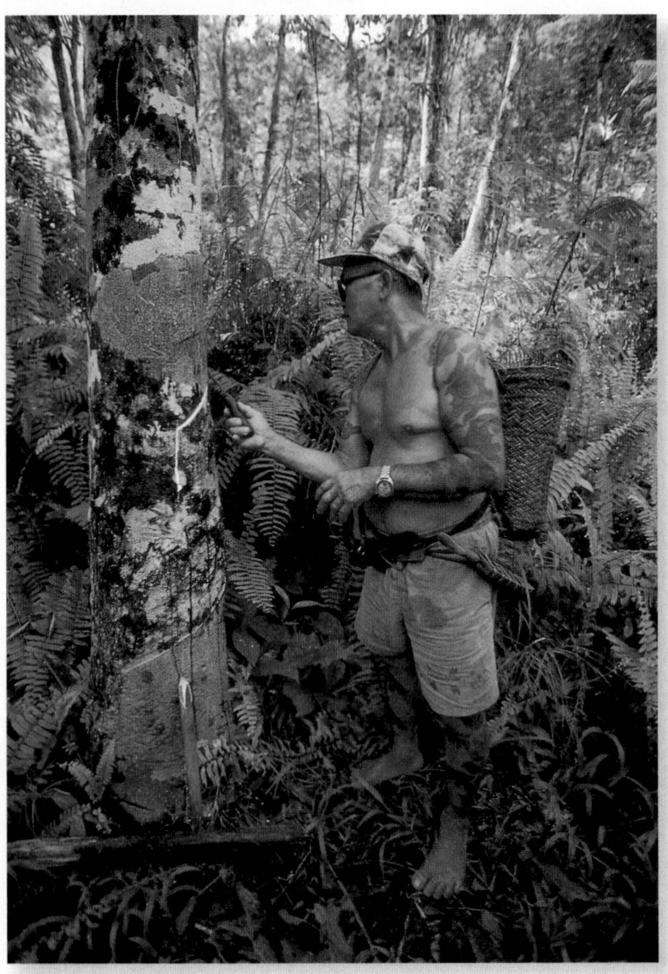

2–24 Tapping a rubber tree An incision is made in the trunk of the tropical rubber tree, *Hevea brasiliensis,* in order to obtain rubber, a terpenoid component of the milky latex. These cultivated trees are being tapped by an Iban village elder on the island of Borneo, Malaysia.

the latex flows down the incision and is collected in a cup attached to the tree. The latex is processed, and the rubber is removed and pressed into sheets for shipment to factories.

Many terpenoids are poisonous, among them the **cardiac glycosides,** which are sterol derivatives that can cause heart attacks. When used medicinally, cardiac glycosides can result in a slower and strengthened heartbeat. Foxglove plants *(Digitalis)* are the principal source of the most active cardiac glycosides, digitoxin and digoxin. Cardiac glycosides synthesized by members of the milkweed family of plants (Apocynaceae) provide an effective defense against herbivores. Interestingly, some insects have learned to adapt to these toxic substances. An example is the caterpillar of the monarch butterfly, which feeds preferentially on milkweed (*Asclepias* spp.) and stores the cardiac glycosides safely within its body (Figure 2–25). When the adult butterfly leaves the host plant, the bitter-tasting cardiac glycosides protect the butterfly against predator birds. Ingestion of the cardiac glycosides causes the birds to vomit, and they soon learn to recognize and avoid other brightly colored monarchs on sight alone.

Terpenoids play a multiplicity of roles in plants. In addition to the roles already mentioned, some are photosynthetic pigments (carotenoids) or hormones (gibberellins, abscisic acid), while others serve as structural components of membranes (sterols) or as electron carriers (ubiquinone, plastoquinone). All of these substances are discussed in subsequent chapters.

Phenolics Include Flavonoids, Tannins, Lignins, and Salicylic Acid

The term **phenolics** encompasses a broad range of compounds, all of which have a hydroxyl group (—OH) attached to an aromatic ring (a ring of six carbons containing three double bonds). They are almost universally present in plants and are known to accumulate in all plant parts (roots, stems, leaves, flowers, and fruits). Although they represent the most studied of secondary metabolites, the function of many phenolic compounds is still unknown.

The **flavonoids,** which are water-soluble pigments present in the vacuoles of plant cells, represent the largest group of plant phenolic compounds (see Chapter 20). Flavonoids found in red wines and grape juice have received considerable attention because of their reported lowering of cholesterol levels in the blood. Over 3000 different flavonoids have been described, and they are probably the most intensively studied phenolics of plants. Flavonoids are divided into several classes, including the widespread anthocyanins, flavones, and flavonols. The **anthocyanins** range in color from red through purple to blue. Most of the **flavones** and **flavonols** are yellowish or ivory-colored pigments, and some are colorless. The colorless flavones and flavonols can alter the color of a plant part through the formation of complexes with anthocyanins and metal ions. This phenomenon, called **co-pigmentation,** is responsible for some intensely blue flower colors (Figure 2–26).

Flower pigments act as visual signals to attract pollinating birds and bees, a role recognized by Charles Darwin and by naturalists before and after his time. Flavonoids also affect how plants interact with other organisms, such as symbiotic bacteria living within plant roots, as well as microbial pathogens.

(a)

(b)

2–25 Milkweed and monarch butterflies ***(a)*** A monarch butterfly caterpillar feeds on milkweed, ingesting and storing the toxic terpenoids (cardiac glycosides) produced by the plant. The monarch caterpillar and the monarch butterfly ***(b)*** thus become unpalatable and poisonous. The conspicuous coloration of caterpillar and butterfly warns would-be predators.

2–26 Co-pigmentation The intensely blue color of these day flowers, *Commelina communis*, is the result of co-pigmentation. In the day flower, anthocyanin and flavone molecules are joined to a magnesium ion to form the blue pigment commelinin.

For example, flavonoids released from the roots of legumes can stimulate or inhibit specific genetic responses in the different types of bacteria associated with them. Flavonoids may also provide protection against damage from ultraviolet radiation.

Probably the most important deterrents to herbivore feeding in angiosperms (flowering seed plants) are the **tannins,** phenolic compounds present in relatively high concentrations in the leaves of a wide array of woody plants. The bitter taste of the tannins is repellent to insects, reptiles, birds, and higher animals. Unripe fruits frequently have a high concentration of tannins in their outer cell layers. Humans have made use of tannins to tan leather, denaturing the protein of the leather and protecting it from bacterial attack. Tannins are sequestered in vacuoles within the plant cell in order to prevent damage to other cellular components (Figure 2–27).

Lignins, unlike other phenolic compounds, are deposited in the cell wall rather than in the vacuole. Second only to cellulose as the most abundant organic compounds on Earth, lignins are polymers formed from three types of monomers: *p*-coumaryl, coniferyl, and sinapyl alcohols. The relative amount of each monomer differs significantly depending on whether the lignin is from gymnosperms (nonflowering seed plants), woody angiosperms, or grasses. In addition, there is great variation in the monomeric composition of lignins from different species, organs, tissues, and even cell wall fractions.

The major importance of lignin is the compressive strength and stiffness it adds to the cell wall. **Lignification,** the process of lignin deposition, is believed to have played a major role in the evolution of terrestrial plants. Although unlignified cell walls can withstand substantial tensile (stretching) forces, they are weak against the compressive forces of gravity. With the addition of lignin to the walls, it was possible for terrestrial plants to increase in stature and to develop branch systems capable of supporting large photosynthetic surfaces.

Lignin also waterproofs the cell wall. It therefore facilitates upward transport of water in the conducting cells of the xylem by limiting the outward movement of water from these cells. In addition, lignin assists the water-conducting cells in resisting the tension generated by the stream of water (the transpiration stream) being pulled to the top of tall plants (see Chapter 30). A further role of lignin is indicated by its deposition in response to various types of injury and attack by fungi. This "wound lignin" protects the plant from fungal attack by increasing the resistance of walls to mechanical penetration, protecting them against fungal enzyme activity, and reducing the diffusion of enzymes and toxins from the fungi into the plant. It has been suggested that lignin may have first functioned as an antifungal and antibacterial agent and only later assumed a role in water transport and mechanical support in the evolution of terrestrial plants.

Salicylic acid, the active ingredient in aspirin, was first known for its analgesic properties. It was discovered by the ancient Greeks and by Native Americans, who obtained it for pain relief from tea brewed from willow *(Salix)* bark (Figure 2–28). Only recently, however, has the action of this phenolic acid in

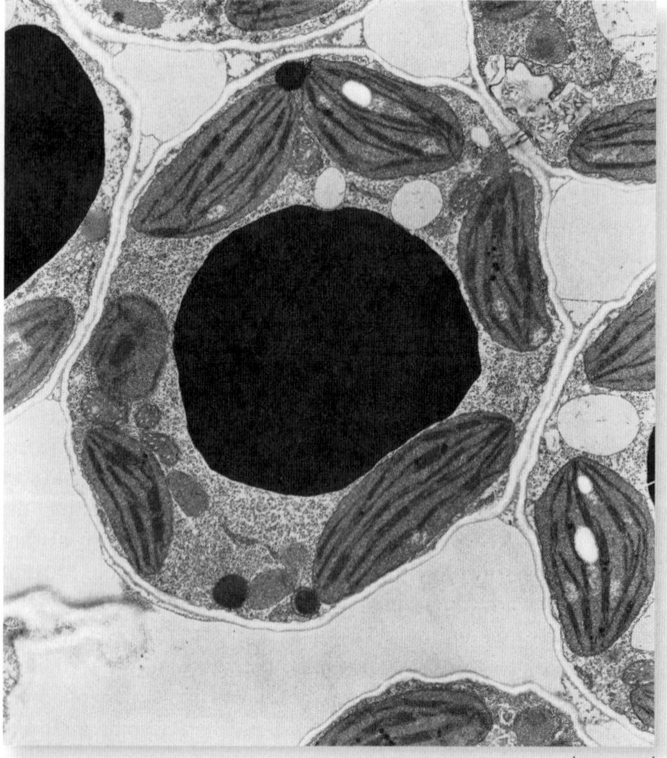

2–27 Tannin Tannin-containing vacuole in a leaf cell of the sensitive plant, *Mimosa pudica.* The electron-dense tannin, which makes the leaves unpalatable to eat, completely fills the central vacuole of this cell.

Salicylic acid

Acetylsalicylic acid
(aspirin)

(a)

(b)

2–28 Salicylic acid ***(a)*** Chemical structures of salicylic acid and aspirin. ***(b)*** Willow tree *(Salix)* growing along a stream bank.

plant tissues been discovered: it is essential for the development of **systemic acquired resistance,** commonly referred to as **SAR.** SAR develops in response to a localized attack by pathogenic bacteria, fungi, or viruses. As a result, other portions of the plant are provided with long-lasting protection against the same and unrelated pathogens. Salicylic acid also probably triggers the dramatic increases in temperature during flowering of the voodoo lily *(Sauromatum guttatum)* and other species of Araceae.

With this introduction to the molecular composition of plant cells, the stage has been set to turn our attention to the living cell—its structure and the activities by which it maintains itself as an entity distinct from the nonliving world surrounding it. We shall explore how the organic molecules introduced in this chapter perform their functions. It should come as no surprise that organic molecules rarely function in isolation but rather in combination with other organic molecules. The marvelous processes by which these molecules perform their functions still are not clearly understood and remain the object of intensive investigations.

SUMMARY

Living Matter Is Composed of Only a Few of the Naturally Occurring Elements

Living organisms are composed primarily of only six elements: carbon, hydrogen, nitrogen, oxygen, phosphorus, and sulfur. The bulk of living matter is water. Most of the rest of living material is composed of organic (carbon-containing) molecules—carbohydrates, lipids, proteins, and nucleic acids. Polysaccharides, proteins, and nucleic acids are examples of macromolecules, which are made up of similar monomers joined together by dehydration synthesis (removing H_2O) to form polymers. By the reverse process, called hydrolysis (adding H_2O), polymers can be split apart to their constituent monomers.

Carbohydrates Are Sugars and Their Polymers

Carbohydrates serve as a primary source of chemical energy for living systems and as important structural elements in cells. The simplest carbohydrates are the monosaccharides, such as glucose and fructose. Monosaccharides can be combined to form disaccharides, such as sucrose, and polysaccharides, such as starch and cellulose. Starch molecules are storage polysaccharides made of polymers of alpha-glucose that form into coils, whereas cellulose is a structural polysaccharide that forms linear microfibrils impervious to the enzymes that break down starch. Carbohydrates can usually be broken apart by the addition of a water molecule at each linkage, that is, by hydrolysis.

Lipids Are Hydrophobic Molecules That Play a Variety of Roles in the Cell

Lipids are another source of energy and structural material for cells. Compounds in this group—fats, oils, phospholipids, cutin, suberin, waxes, and steroids—are generally insoluble in water.

Fats and oils, also known as triglycerides, store energy. Phospholipids are modified triglycerides that are important components of cellular membranes. Cutin, suberin, and waxes are lipids that form barriers to water loss. The surface cells of stems and leaves are covered with a cuticle, composed of wax and cutin, that prevents water loss. Steroids are molecules having four interconnected hydrocarbon rings. Steroids are found in cellular membranes, and they may have other roles in the cell as well.

Proteins Are Versatile Polymers of Amino Acids

Amino acids have an amino group, a carboxyl group, a hydrogen, and a variable R group attached to the same carbon atom. Twenty different kinds of amino acids—differing in the size, charge, and polarity of the R group—are used to build proteins. By the process of dehydration synthesis, amino acids are joined by peptide bonds. A chain of amino acids is a polypeptide, and proteins consist of one or more long polypeptides.

A protein's structure can be described in terms of levels of organization. The primary structure is the linear sequence of amino acids joined by peptide bonds. The secondary structure, most commonly an alpha helix or a beta pleated sheet, is formed as a result of hydrogen bonding between amino and carboxyl groups. The tertiary structure is the folding that results from interactions between R groups. The quaternary structure results from specific interactions between two or more polypeptide chains.

Enzymes are globular proteins that catalyze chemical reactions in cells. Because of enzymes, cells are able to accelerate the rate of chemical reactions at moderate temperatures.

Nucleic Acids Are Polymers of Nucleotides

Nucleotides are complex molecules consisting of a phosphate group, a nitrogenous base, and a five-carbon sugar. They are the building blocks of the nucleic acids deoxyribonucleic acid (DNA) and ribonucleic acid (RNA), which transmit and translate the genetic information. Some RNA molecules function as catalysts.

Adenosine triphosphate (ATP) is the cell's major energy currency. ATP can be hydrolyzed, releasing adenosine diphosphate (ADP), phosphate, and considerable energy. This energy can be used to drive other reactions or physical processes in the cell. In the reverse reaction, ADP can be "recharged" to ATP with the addition of a phosphate group and an input of energy.

Secondary Metabolites Play a Variety of Roles Not Directly Related to the Basic Functioning of the Plant

Three main classes of secondary metabolites found in plants are alkaloids, terpenoids, and phenolics. Although the botanical functions of these substances are not clearly known, some are thought to deter predators and/or competitors. Examples of such compounds include caffeine and nicotine (alkaloids), as well as cardiac glycosides (terpenoids) and tannins (phenolics). Others, such as anthocyanins (phenolics) and es-

SUMMARY TABLE Biologically Important Organic Molecules

CLASS OF MOLECULE	TYPES	SUBUNITS	MAIN FUNCTIONS	OTHER FEATURES
Carbohydrates	Monosaccharides (e.g., glucose)	Monosaccharide	Ready energy source	Carbohydrates are sugars and polymers of sugars.
	Disaccharides (e.g., sucrose)	Two monosaccharides	Transport form in plants	To identify carbohydrates, look for compounds that consist of monomers with many hydroxyl groups (—OH) and usually one carbonyl (—C=O) group attached to the carbon skeleton. However, if the sugars are in the ring form, the carbonyl group is not evident.
	Polysaccharides	Many monosaccharides	Energy storage or structural components	
	Starch		Major energy storage in plants	
	Glycogen		Major energy storage in prokaryotes, fungi, and animals	
	Cellulose		Component of plant cell walls	
	Chitin		Component of fungal cell walls	
Lipids	Triglycerides	3 fatty acids + 1 glycerol	Energy storage	Lipids are nonpolar molecules that will not dissolve in polar solvents such as water. Thus lipids are the ideal molecule for long-term energy storage. They can be "put aside" in a cell and will not dissolve in the watery environment or "leak out" into the rest of the cell.
	Oils		Major energy storage in seeds and fruits	
	Fats		Major energy storage in animals	
	Phospholipids	2 fatty acids + 1 glycerol + 1 phosphate group	Major component of all cell membranes	Phospholipids and glycolipids are modified triglycerides with a polar group at one end. The polar "head" of the molecule is hydrophilic and thus dissolves in water; the nonpolar "tail" is hydrophobic and insoluble in water. This is the basis for their role in cell membranes, where they are arranged tail to tail in two layers.
	Cutin, suberin, and waxes	Vary; complex lipid structures	Protection	Act as waterproofing for stems, leaves, and fruits.
	Steroids	Four linked hydrocarbon rings	Component of cell membranes and hormones	A sterol is a steroid with a hydroxyl group at the carbon-3 position.
Proteins (polypeptides)	Many different types	Amino acids	Numerous functions, including structural and catalytic (enzymes)	Primary, secondary, tertiary, and quaternary structures.
Nucleic acids	DNA	Nucleotides	Carrier of genetic information	Each nucleotide is composed of a sugar, a nitrogenous base, and a phosphate group. ATP is a nucleotide that functions as the principal energy carrier for cells.
	RNA		Involved in protein synthesis	

sential oils (terpenoids), attract pollinators. Still others, such as phenolic lignins, are responsible for the compressive strength, stiffness, and waterproofing of the plant body. Some secondary metabolites, such as rubber (a terpenoid) and morphine and taxol (alkaloids), have important commercial or medicinal uses. Primary metabolites, in contrast to secondary metabolites, are found in all plant cells and are necessary for the plant to live.

QUESTIONS

1. Why must starch be hydrolyzed before it can be used as an energy source or transported?
2. Why is it an advantage for a plant to store food energy as fructans rather than as starch? As oils rather than as starch or fructans?
3. What is the principal difference between a saturated and an unsaturated fat or oil?
4. What aspect of their structure do all amino acids have in common? What part of an amino acid determines its identity?
5. What are the several levels of protein organization, and how do they differ from one another?
6. The coagulation of egg white when an egg is cooked is a common example of protein denaturation. What happens when a protein is denatured?
7. A number of insects, including the monarch butterfly, have adopted a strategy of utilizing certain secondary metabolites of plants for protection against predators. Explain.
8. Lignin, a cell wall constituent, is believed to have played a major role in the evolution of terrestrial plants. Explain in terms of all the presumed functions of lignin.

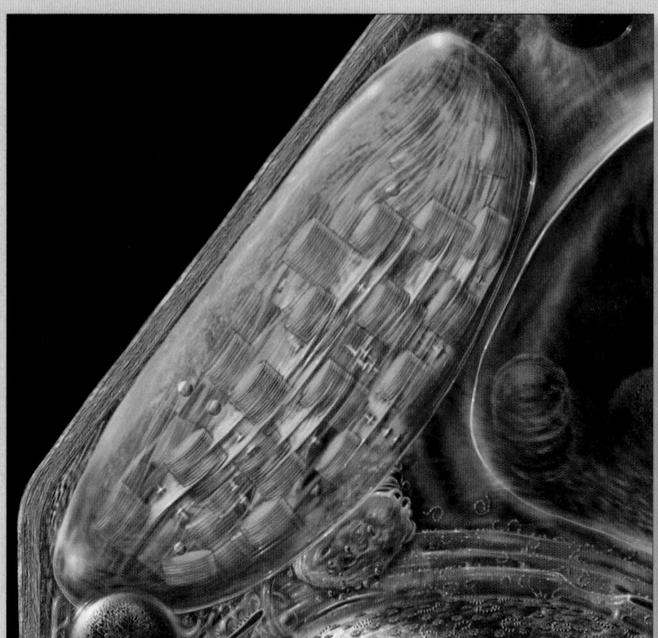

CHAPTER 3

The Plant Cell and the Cell Cycle

◀ **Powerhouse of the plant cell** The chloroplast is the site where light energy is used to produce the organic molecules required by the plant cell. In this image, the flattened and stacked membranes of the grana can be seen inside the chloroplast. Chlorophyll and other pigments embedded in the chloroplast membranes capture the sun's energy, the first step in the process essential to life—photosynthesis.

CHAPTER OUTLINE

In the last chapter, we progressed from atoms and small molecules to large, complex molecules, such as proteins and nucleic acids. At each level of organization, new properties appear. Water, we know, is not the sum of the properties of elemental hydrogen and oxygen, both of which are gases. Water is something more and also something different. In proteins, amino acids become organized into polypeptides, and polypeptide chains are arranged in new levels of organization—the secondary, tertiary, and, in some cases, quaternary structure of the complete protein molecule. Only at the higher levels of organization do the complex properties of the protein emerge, and only then can the molecule assume its function.

The characteristics of living organisms, like those of atoms or molecules, do not emerge gradually as the degree of

CHECKPOINTS

After reading this chapter, you should be able to answer the following:

1. How does the structure of a prokaryotic cell differ from that of a eukaryotic cell?
2. What are the various types of plastids, and what role(s) does each play in the cell?
3. What developmental and functional relationships exist between the endoplasmic reticulum and the Golgi bodies of the plant cell?
4. What is the "cytoskeleton" of the cell, and with what cellular processes is it involved?
5. How do primary cell walls differ from secondary cell walls?
6. What is the cell cycle, and what key events occur in the G_1, S, G_2, and M phases of the cell cycle?
7. What is the role of mitosis? What events occur during each of the four mitotic phases?
8. What is cytokinesis, and what roles do the phragmosome, the phragmoplast, and the cell plate play during the process?

(a)

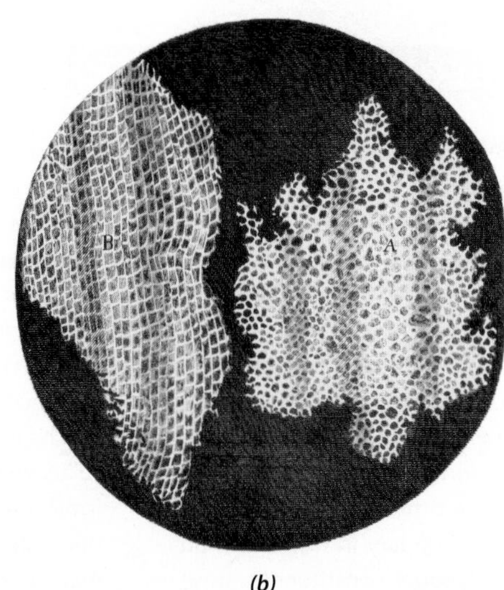

(b)

3–1 Hooke's microscope The English microscopist Robert Hooke first used the term "cell" to refer to the small chambers he saw in magnified slices of cork. ***(a)*** One of Hooke's microscopes, made for him around 1670. Light from an oil lamp (far left) was directed to the specimen through a water-filled glass globe that acted as a condenser. The specimen was mounted on a pin, just below the tip of the microscope. The microscope was focused by moving it up and down using a screw held to the stand by a clamp. ***(b)*** This drawing of two slices of cork appeared in Hooke's book, *Micrographia,* published in 1665.

organization increases. They appear quite suddenly and specifically, in the form of the living cell—something that is more than and different from the atoms and molecules of which it is composed. Life begins where the cell begins.

Cells are the structural and functional units of life (Figure 3–1). The smallest organisms are composed of single cells. The largest are made up of trillions of cells, each of which still lives a partly independent existence. The realization that all organisms are composed of cells was one of the most important conceptual advances in the history of biology, because it provided a unifying theme, encompassed in the **cell theory** (see essay on the next page), for the study of all living things. When studied at the cellular level, even the most diverse organisms are remarkably similar to one another, both in their physical organization and in their biochemical properties.

The word "cell" was first used in a biological sense some 340 years ago. In the seventeenth century, the English scientist Robert Hooke, using a microscope of his own construction, noticed that cork and other plant tissues are made up of what appeared to be small cavities separated by walls (Figure 3–1). He called these cavities "cells," meaning "little rooms." However, "cell" did not take on its present meaning—the basic unit of living matter—for more than 150 years.

In 1838, Matthias Schleiden, a German botanist, reported his observation that all plant tissues consist of organized masses of cells. In the following year, zoologist Theodor Schwann extended Schleiden's observation to animal tissues and proposed a cellular basis for all life. Formulation of the cell theory is usually connected with Schleiden and Schwann. In 1858, the idea that all living organisms are composed of one or more cells took on an even broader significance when the pathologist Rudolf Virchow generalized that cells can arise only from preexisting cells: "Where a cell exists, there must have been a preexisting cell, just as the animal arises only from an animal and the plant only from a plant."

From the perspective provided by Darwin's theory of evolution, published in the following year, Virchow's concept takes on an even larger significance. There is an unbroken continuity between modern cells—and the organisms they compose—and the first primitive cells that appeared on Earth at least 3.5 billion years ago.

Every living cell is a self-contained unit, and each is bound by an outer membrane—the plasma membrane, or plasmalemma (often simply called the cell membrane). The plasma membrane controls the passage of materials into and out of the cell and so makes it possible for the cell to differ biochemically and structurally from its surroundings. Enclosed within this membrane is the cytoplasm, which, in most cells, includes a variety of discrete structures and various dissolved or suspended molecules. In addition, every cell contains DNA (deoxyribonucleic acid), which encodes the genetic information (see pages 180 and 181), and this code, with rare exceptions, is the same for every organism, whether bacterium, oak tree, or human.

Prokaryotes and Eukaryotes

Two fundamentally distinct groups of organisms can be recognized: **prokaryotes** and **eukaryotes.** These terms are derived from the Greek word *karyon,* meaning "kernel" (nucleus). The

CELL THEORY VERSUS ORGANISMAL THEORY

In its classical form, the cell theory proposed that the bodies of plants and animals are aggregations of individual, differentiated cells. The proponents of this concept believed that the activities of the whole plant or animal might be considered the summation of the activities of the individual constituent cells, with the individual cells of prime importance. This concept has been compared to the Jeffersonian theory of democracy, which considered the nation to be dependent on and secondary in rights and privileges to its individual, constituent states.

By the latter half of the nineteenth century, an alternative to the cell theory was formulated. Known as the **organismal theory,** it considers the entire organism, rather than the individual cells, to be of prime importance. The many-celled plant or animal is regarded not merely as a group of independent units but as a more or less continuous mass of protoplasm, which, in the course of evolution, became subdivided into cells. The organismal theory arose in part from the results of physiological research that demonstrated the necessity for coordination of the activities of the various organs, tissues, and cells for normal growth and development of the organism. The organismal theory might be likened to the theory of government that holds that the unified nation is of prime importance, not the states of which it is formed.

The nineteenth-century German botanist Julius von Sachs concisely stated the organismal theory when he wrote, "Die Pflanze bildet Zelle, nicht die Zelle Pflanzen," which means "The plant forms cells, the cells do not form plants."

Indeed, the organismal theory is especially applicable to plants: their protoplasts do not pinch apart during cell division, as in animal cell division, but are partitioned initially by insertion of a cell plate. Moreover, the separation of plant cells is rarely complete—the protoplasts of contiguous cells remain interconnected by the cytoplasmic strands known as plasmodesmata. Plasmodesmata traverse the walls and unite the entire plant body into an organic whole called the symplast, which consists of the interconnected protoplasts and their plasmodesmata. As appropriately stated by Donald Kaplan and Wolfgang Hagemann, "Instead of higher plants being federal aggregations of independent cells, they are unified organisms whose protoplasts are incompletely subdivided by cell walls."

In its modern form, the cell theory states simply that (1) all living organisms are composed of one or more cells; (2) the chemical reactions of a living organism, including its energy-releasing processes and its biosynthetic reactions, take place within cells; (3) cells arise from other cells; and (4) cells contain the hereditary information of the organisms of which they are a part, and this information is passed from parent cell to daughter cell. The cell and organismal theories are not mutually exclusive. Together, they provide a meaningful view of structure and function at cellular and organismal levels.

name *prokaryote* means "before a nucleus," and *eukaryote,* "with a true nucleus."

Modern prokaryotes are represented by the archaea and the bacteria (see Chapter 13). Prokaryotic cells differ most notably from eukaryotic cells in that they lack nuclei; that is, their DNA is not surrounded by a membranous envelope (Figure 3–2). The DNA occurs in the form of a large, circular molecule, with which a variety of proteins are loosely associated. This molecule, known as the bacterial chromosome, is localized in a region known as the **nucleoid.** (Most prokaryotes have only one chromosome.) Prokaryotes also lack specialized membrane-bounded structures **(organelles)** that perform specific functions.

In eukaryotic cells, the chromosomes are surrounded by an envelope, made up of two membranes, which separates them from the other cell contents. The DNA in eukaryotic cells is linear and tightly bound to special proteins known as histones, forming a number of chromosomes that are structurally more complex than bacterial chromosomes. Eukaryotic cells are further divided into distinct compartments that perform different functions (Figure 3–3).

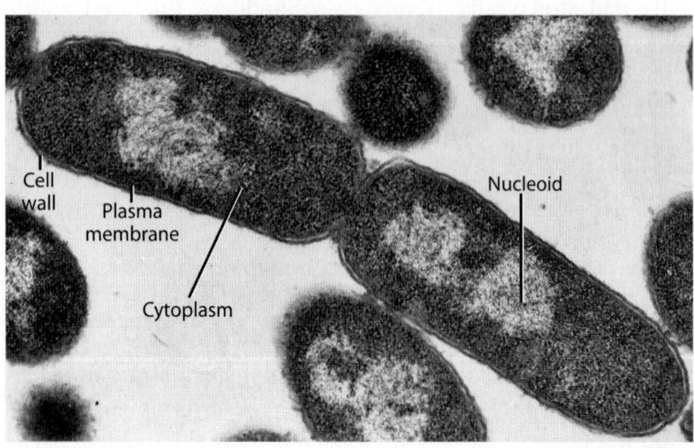

3–2 Prokaryote Electron micrograph of cells of *Escherichia coli,* a bacterium that is a common, usually harmless inhabitant of the human digestive tract. However, some *E. coli* strains, which generally are acquired by ingestion of contaminated food or water, produce toxins that cause massive secretion of fluids into the intestine, resulting in vomiting and diarrhea. This heterotrophic (nonphotosynthetic) prokaryote is the most thoroughly studied of all living organisms. Each rod-shaped cell has a cell wall, a plasma membrane, and cytoplasm. The genetic material (DNA) is found in the less granular area in the center of each cell. This region, known as the nucleoid, is not surrounded by a membrane. The densely granular appearance of the cytoplasm is largely due to the presence of numerous ribosomes, which are involved with protein synthesis. The two cells in the center have just divided but have not yet separated completely.

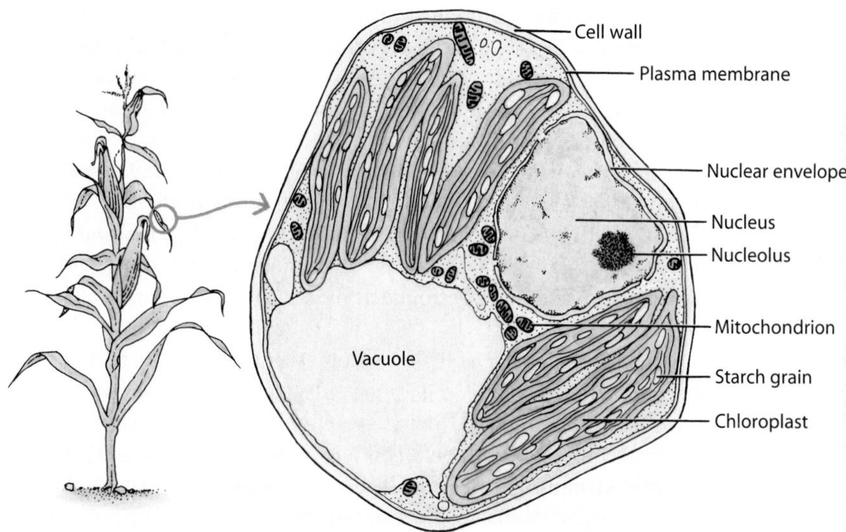

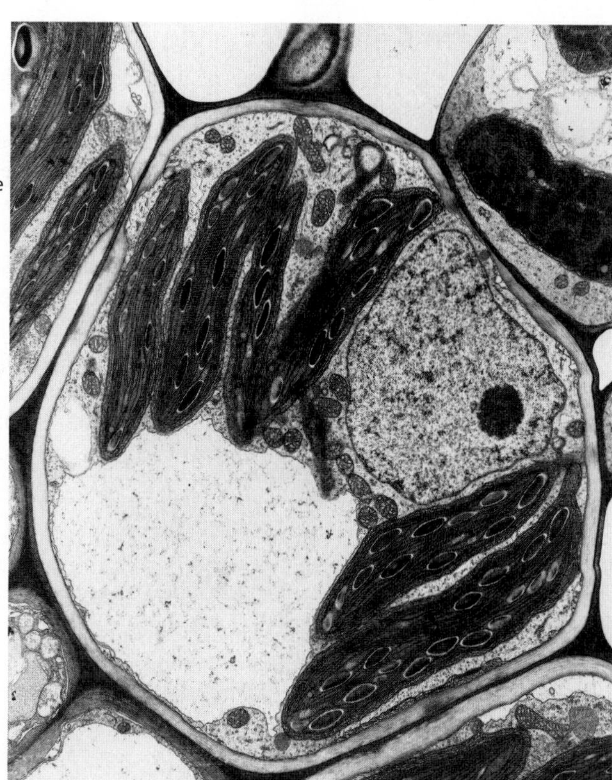

3–3 Photosynthetic eukaryote Electron micrograph of a cell from the leaf of a maize *(Zea mays)* plant. The granular material within the nucleus is chromatin. It contains DNA associated with histone proteins. The nucleolus is the region within the nucleus where the RNA components of ribosomes are synthesized. Note the many mitochondria and chloroplasts, all bounded by membranes. The vacuole, which is a fluid-filled region enclosed by a membrane, and the cell wall are characteristic of plant cells.

Some features that distinguish prokaryotic cells from eukaryotic cells are listed in Table 3–1.

Compartmentation in eukaryotic cells is accomplished by means of membranes, which, when seen with the aid of an electron microscope, look remarkably similar in various organisms. When suitably preserved and stained, these membranes have a three-layered appearance, consisting of two dark layers separated by a lighter layer (Figure 3–4). The term "unit membrane" has been used to designate membranes that have such an appearance.

Table 3-1 Comparison of Selected Features of Prokaryotic and Eukaryotic Cells

	Prokaryotic cells	Eukaryotic cells
Cell size (length)	Generally 1–10 micrometers	Generally 5–100 micrometers (animals at low end, plants at high end, of that range); many much longer than 100 micrometers
Nuclear envelope	Absent	Present
DNA	Circular, in nucleoid	Linear, in nucleus
Organelles (e.g., mitochondria and and chloroplasts)	Absent	Present
Cytoskeleton (microtubules and actin filaments)	Absent	Present

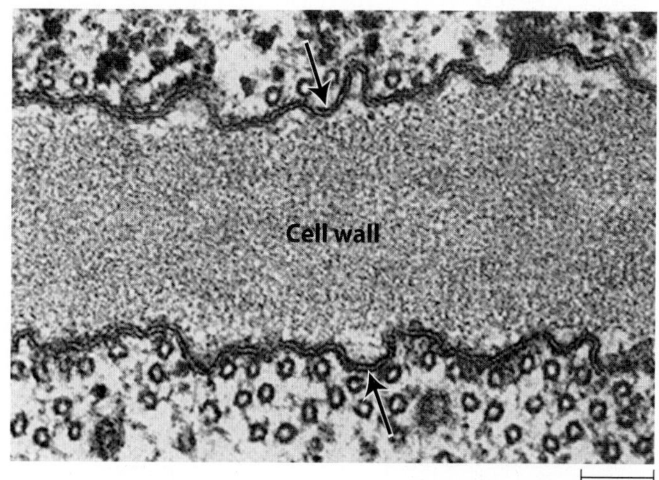

3–4 Plasma membrane Under high magnification, cellular membranes often have a three-layered (dark-light-dark) appearance, as seen in the plasma membranes on either side of the common cell wall between two cells from an onion *(Allium cepa)* root tip. The numerous circular structures bordering the wall are microtubules.

The Plant Cell: An Overview

The plant cell typically consists of a more or less rigid **cell wall** and a **protoplast.** The term "protoplast" is derived from the word **protoplasm,** which is used to refer to the contents of cells. A protoplast is the unit of protoplasm inside the cell wall.

A protoplast consists of **cytoplasm** and a **nucleus** (Table 3–2). The cytoplasm includes distinct, membrane-bounded entities (organelles such as plastids and mitochondria), systems of membranes (the endoplasmic reticulum and Golgi apparatus), and nonmembranous entities (such as ribosomes, actin filaments, and microtubules). The rest of the cytoplasm—the "cellular soup," or cytoplasmic matrix, in which the nucleus, various entities, and membrane systems are suspended—is called the **cytosol.** The cytoplasm is surrounded by a single membrane, the **plasma membrane.**

Table 3-2 **An Inventory of Plant Cell Component**

Cell wall	Middle lamella	
	Primary wall	
	Secondary wall	
	Plasmodesmata	
Protoplast	Nucleus	Nuclear envelope
		Nucleoplasm
		Chromatin
		Nucleolus
	Cytoplasm	Plasma membrane (outer boundary of cytoplasm)
		Cytosol
		Organelles surrounded by two membranes:
		Plastids
		Mitochondria
		Organelles surrounded by one membrane:
		Peroxisomes
		Vacuoles, surrounded by tonoplast
		Endomembrane system (major components):*
		Endoplasmic reticulum
		Golgi apparatus
		Vesicles
		Cytoskeleton
		Microtubules
		Actin filaments
		Ribosomes
		Oil bodies

*The endomembrane system also includes the plasma membrane, the nuclear envelope, the tonoplast, and all other internal membranes, with the exception of mitochondrial, plastid, and peroxisomal membranes.

The plasma membrane has several important functions: (1) it separates the protoplast from its external environment; (2) it mediates the transport of substances into and out of the protoplast (see Chapter 4); (3) it coordinates the synthesis and assembly of cell wall microfibrils (cellulose); and (4) it detects and facilitates responses to hormonal and environmental signals involved in the control of cell growth and differentiation.

In contrast to most animal cells, plant cells develop one or more, often liquid-filled cavities, or **vacuoles,** within their cytoplasm. The vacuole is surrounded by a single membrane called the **tonoplast.**

In a living plant cell, the cytoplasm is always in motion; the organelles, as well as various substances suspended in the cytosol, can be observed being swept along in an orderly fashion in the moving currents. This movement is known as **cytoplasmic streaming,** or **cyclosis,** and it continues as long as the cell is alive. Cytoplasmic streaming facilitates the transport of materials within the cell and between the cell and its environment (see essay on page 56).

Nucleus

The nucleus is often the most prominent structure within the protoplast of eukaryotic cells. The nucleus performs two important functions: (1) it controls the ongoing activities of the cell by determining which protein molecules are produced by the cell and when they are produced, as we shall see in Chapter 9; and (2) it stores the genetic information (DNA), passing it on to the daughter cells in the course of cell division. The total genetic information stored in the nucleus is referred to as the **nuclear genome.** Genetic information occurs also in the DNA of plastids (plastid genome) and the DNA of mitochondria (mitochondrial genome) of plant cells.

The nucleus is surrounded by a pair of membranes called the **nuclear envelope.** The nuclear envelope, as seen with the aid of an electron microscope, contains a large number of circular pores that are 30 to 100 nanometers (10^{-9} meter; see Metric Table at the back of the book) in diameter (Figure 3–5); the inner and outer membranes are joined around each pore to form the margin of its opening. The pores are not merely holes in the envelope; each has a complex structure—the largest supramolecular complex assembled in the eukaryotic cell. The pores provide a direct passageway through the nuclear envelope for the exchange of materials between the nucleus and the cytoplasm. In various places, the outer membrane of the nuclear envelope may be continuous with the **endoplasmic reticulum,** a complex system of membranes that plays a central role in cellular biosynthesis. The nuclear envelope may be considered a specialized, locally differentiated portion of the endoplasmic reticulum.

If the cell is treated by special staining techniques, thin threads and grains of **chromatin** can be distinguished from the **nucleoplasm,** or nuclear matrix (Figure 3–6). Chromatin is made up of DNA, which carries the hereditary information of the cell, combined with histone proteins. During the process of nuclear division, the chromatin becomes progressively more condensed until it is visible as distinct **chromosomes** (see Chapter 8). Chromosomes of nondividing nuclei are attached at

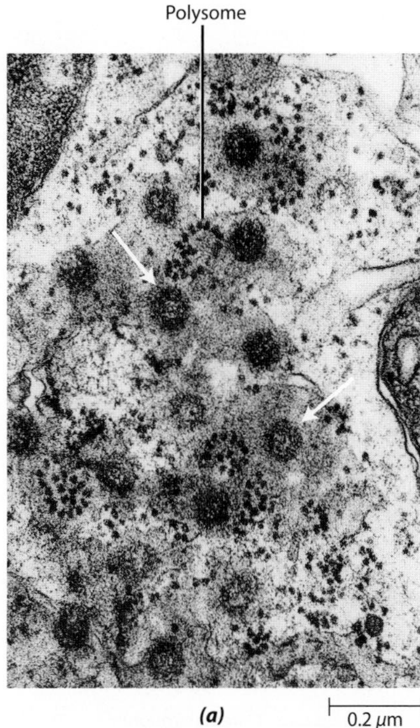

Endoplasmic reticulum

(b)

0.5 μm

3–5 Nuclear pores Electron micrographs of nuclei in parenchyma cells of the seedless vascular plant *Selaginella kraussiana* show the nuclear pores ***(a)*** in surface view and ***(b)*** in sectional view (see arrows). Note the polysomes (coils of ribosomes) on the surface of the nuclear envelope in *(a)*. The rough endoplasmic reticulum is seen paralleling the nuclear envelope in *(b);* the nucleus is to the left.

one or more sites to the nuclear envelope. The content of DNA per cell is much higher in eukaryotic organisms than in bacteria.

Different species vary in the number of chromosomes present in their somatic (body) cells. *Machaeranthera gracilis,* a desert annual, has 4 chromosomes per cell; *Arabidopsis thaliana,* the small flowering "weed" used widely for genetic research, 10; *Brassica oleraceae,* cabbage, 18; *Triticum vulgare,* bread wheat, 42; and one species of the fern *Ophioglossum,* about 1250. The gametes, or sex cells, however, have only half the number of chromosomes that is characteristic of the organism's somatic cells. The number of chromosomes in the gametes is referred to as the **haploid** ("single set") number and designated as n, and that in the somatic cells is called the **diploid** ("double set") number, which is designated as $2n$. Cells that have more than two sets of chromosomes are said to be **polyploid** ($3n$, $4n$, $5n$, or more).

Often the only structures within a nucleus that are discernible with the light microscope are the spherical structures

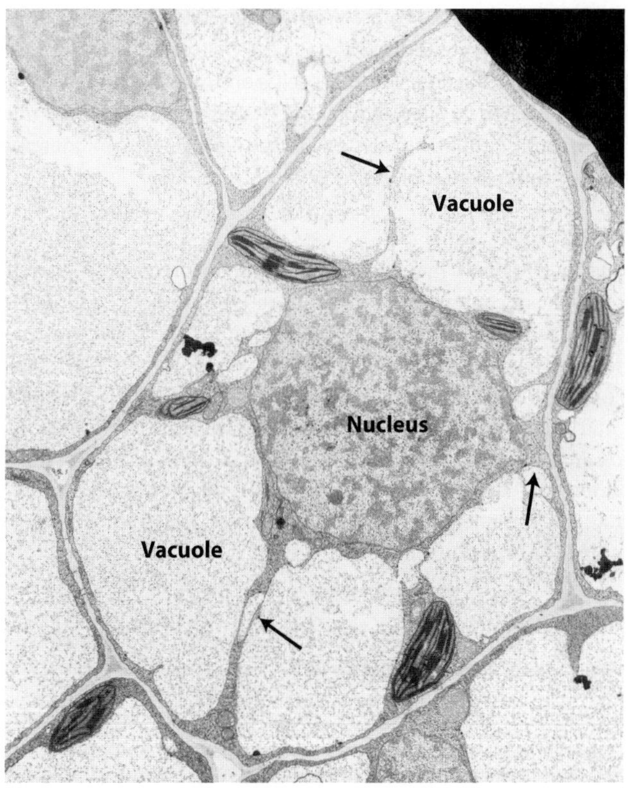

3–6 Chromatin Parenchyma cell from a tobacco *(Nicotiana tabacum)* leaf, with its nucleus "suspended" in the middle of the cell by strands of cytoplasm (arrows). The dense granular substance in the nucleus is chromatin, distinguishing it from the nucleoplasm. The less granular regions surrounding the nucleus are portions of a large central vacuole that merge beyond the plane of this cell section.

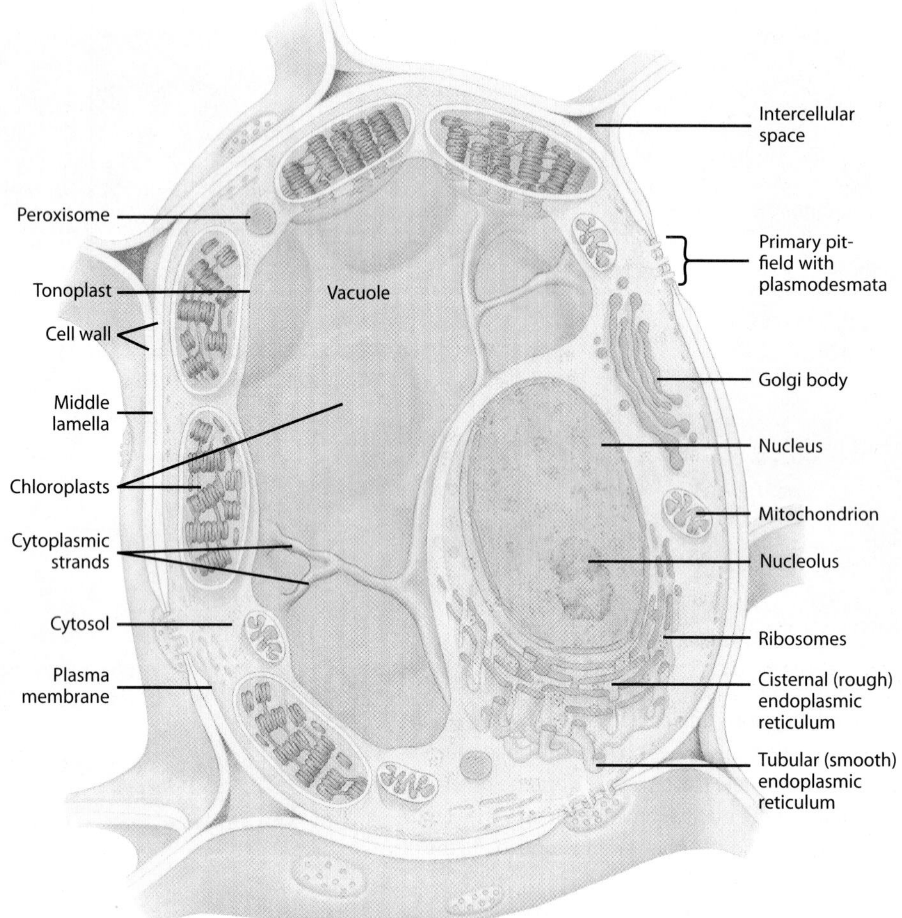

3–7 Diagram of a chloroplast-containing plant cell Typically, the disk-shaped chloroplasts are located in the cytoplasm along the cell wall, with their broad surfaces facing the surface of the wall. Most of the volume of this cell is occupied by a vacuole (surrounded by the tonoplast), which is traversed by a few strands of cytoplasm. In this cell, the nucleus lies in the cytoplasm along the wall, although in some cells (see Figure 3–6) it might appear suspended by strands of cytoplasm in the center of the vacuole.

known as **nucleoli** (single: nucleolus), one or more of which are present in each nondividing nucleus (Figures 3–3 and 3–7). Each nucleolus contains high concentrations of RNA (ribonucleic acid) and proteins, along with large loops of DNA emanating from several chromosomes. The loops of DNA, known as nucleolar organizer regions, are the sites of formation of ribosomal RNA. The nucleolus, in fact, is the site of formation of ribosomal subunits (large and small), which are then transferred, via the nuclear pores, to the cytosol, where they are assembled to form ribosomes.

Ribosomes Are Formed in the Cytosol and Serve as the Sites of Protein Synthesis

Ribosomes are small particles, only about 17 to 23 nanometers in diameter, consisting of protein and RNA. Although the number of protein molecules in ribosomes greatly exceeds the number of RNA molecules, RNA constitutes about 60 percent of the mass of a ribosome. Each ribosome consists of a large and a small subunit, which are produced in the nucleolus and exported to the cytoplasm, where they are assembled into a ribosome. As we discuss in Chapter 9, ribosomes are the sites at which amino acids are linked together to form proteins. Abundant in the cytoplasm of metabolically active cells, ribosomes are found both free in the cytosol and attached to the endoplasmic reticulum. Plastids and mitochondria contain smaller ribosomes, similar to those in prokaryotes.

Ribosomes actively involved in protein synthesis occur in clusters or aggregates called **polysomes,** or polyribosomes (Figure 3–8). Cells that are synthesizing proteins in large quantities often contain extensive systems of polysome-bearing endoplasmic reticulum. In addition, polysomes are often attached to the outer surface of the nuclear envelope (Figure 3–5a). All ribosomes of a particular organism are structurally and functionally identical, differing from one another only in the proteins they are making at any given time.

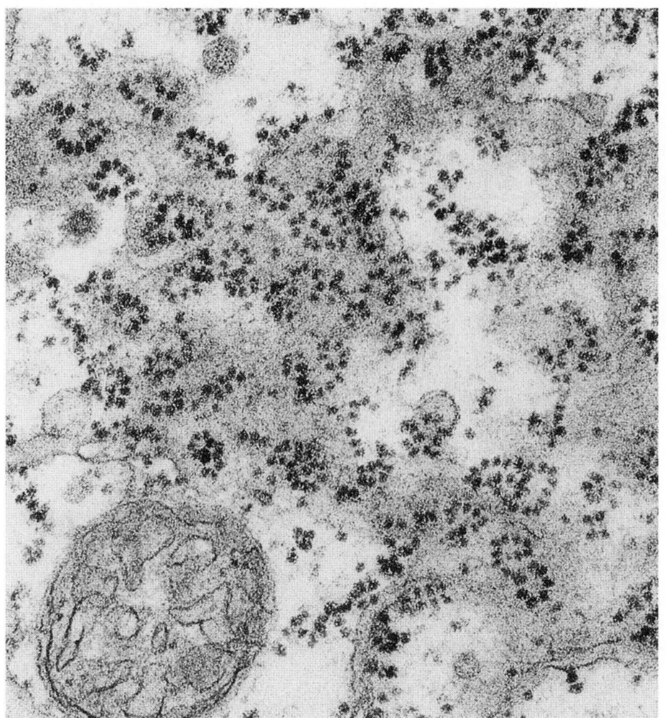

3–8 Polysomes Numerous polysomes (clusters of ribosomes) are seen here on the surface of rough endoplasmic reticulum. The endoplasmic reticulum is a network of membranes extending throughout the cytosol of the eukaryotic cell, dividing it into compartments and providing surfaces on which chemical reactions can take place. Polysomes are the sites at which amino acids are assembled into proteins. This electron micrograph shows a portion of a leaf cell from the fern *Regnellidium diphyllum.*

Chloroplasts and Other Plastids

Together with vacuoles and cell walls, **plastids** are characteristic components of plant cells, and they are concerned with such processes as photosynthesis and storage. The principal types of plastids are chloroplasts, chromoplasts, and leucoplasts. Each plastid is surrounded by an envelope consisting of two membranes. Internally, the plastid is differentiated into a membrane system consisting of flattened sacs called **thylakoids** and a more or less homogeneous matrix, called the **stroma.** The number of thylakoids present varies among plastid types.

Chloroplasts Are the Sites of Photosynthesis

Mature plastids are commonly classified, in part, on the basis of the kinds of pigments they contain. **Chloroplasts,** the sites of photosynthesis (see Chapter 7), contain chlorophylls and carotenoid pigments. The chlorophyll pigments are responsible for the green color of these plastids. The carotenoids are yellow and orange pigments that, in green leaves, are masked by the more numerous chlorophyll pigments. Chloroplasts are found in plants and green algae. In plants, they are usually disk-shaped and measure between 4 and 6 micrometers in diameter. A single mesophyll ("middle of the leaf") cell may contain 40 to 50 chloroplasts; a square millimeter of leaf contains some 500,000. The chloroplasts are usually found with their broad surfaces parallel to the cell wall (Figures 3–7 and 3–9). They can reorient in the cell under the influence of light (Figure 3–9).

The internal structure of the chloroplast is complex (Figure 3–10). The stroma is traversed by an elaborate system of thylakoids. The thylakoids are believed to constitute a single, interconnected system. Chloroplasts are generally characterized by the presence of **grana** (singular: granum)—stacks of disklike

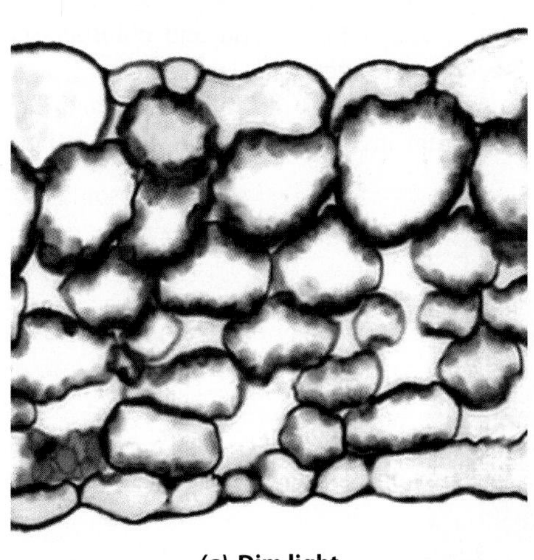

(a) **Dim light**

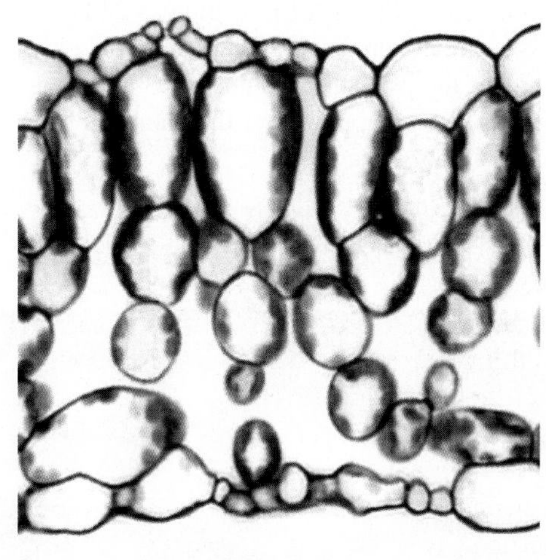

(b) **Bright light**

3–9 Orientation of chloroplasts As shown in cross sections through the leaf of *Arabidopsis thaliana*, chloroplasts move to maximize or minimize their absorption of light, which is entering from the top of the micrographs. ***(a)*** Under dim light, chloroplasts (dark objects along the edges of the cells) move to the cell walls parallel to the leaf surface, thus maximizing light absorption for photosynthesis. ***(b)*** In bright light, chloroplasts migrate to the cell walls perpendicular to the leaf surface, thus minimizing light absorption and photodamage.

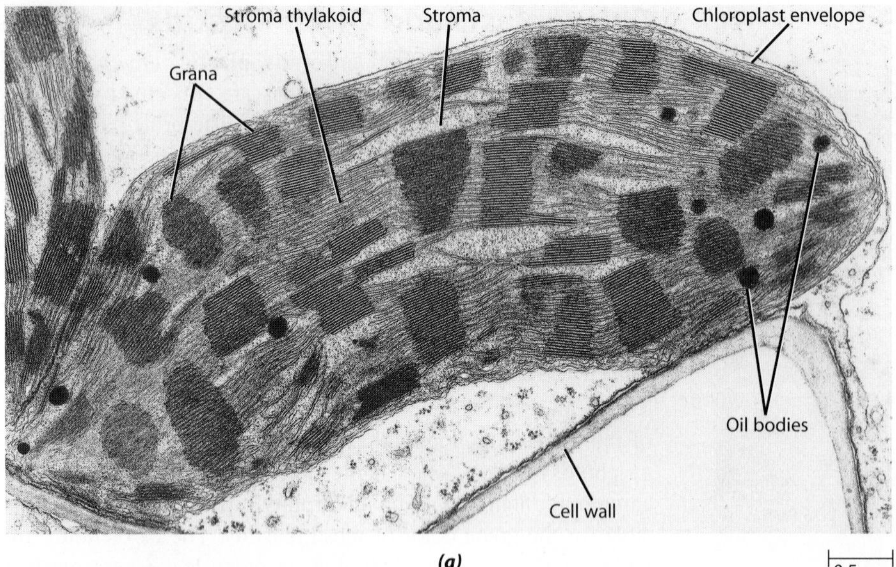

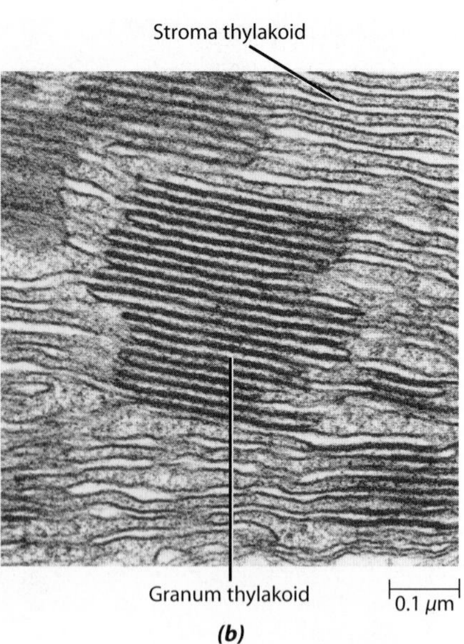

3–10 Internal structure of a chloroplast ***(a)*** Section of a chloroplast of a maize *(Zea mays)* leaf showing grana and stroma thylakoids. ***(b)*** Detail showing a granum, which is composed of stacks of disklike thylakoids. The thylakoids of the various grana are interconnected by the stroma thylakoids.

thylakoids that resemble a stack of coins. The thylakoids of the various grana—the **grana thylakoids**—are interconnected by the thylakoids that traverse the stroma—the **stroma thylakoids.** Chlorophylls and carotenoid pigments are found embedded in the thylakoid membranes.

The chloroplasts of green algae and plants often contain starch grains and small lipid (oil) droplets coated with proteins. The starch grains are temporary storage products and accumulate only when the alga or plant is actively photosynthesizing (Figure 3–3). Starch grains may be lacking in the chloroplasts of plants that have been kept in the dark for as little as 24 hours, because the starch has been broken down to sugar to supply carbon and energy to parts of the plant that are unable to photosynthesize. The starch grains often reappear after the plant has been in the light for only 3 or 4 hours.

Chloroplasts are semiautonomous organelles—that is, they contain the components necessary for the synthesis of some, but not all, of their own polypeptides. Chloroplasts resemble bacteria in several ways. For example, like bacterial DNA, the chloroplast DNA occurs in **nucleoids,** which are clear, grana-free regions containing DNA. However, unlike bacteria, with a single DNA molecule each, chloroplasts have multiple DNA copies (Figure 3–11). Further similarities are that the DNA of chloroplasts, like that of bacteria, is not associated with histones. And the ribosomes of both bacteria and plastids are about two-thirds the size of the eukaryotic cell's cytoplasmic ribosomes, and both bacteria and chloroplasts replicate by binary fission (see Chapter 13).

The formation of chloroplasts and the pigments associated with them involves contributions from both nuclear and plastid

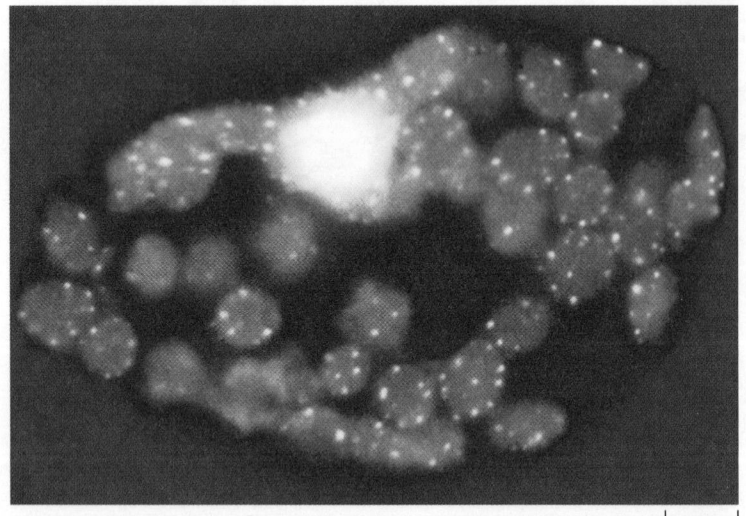

3–11 Nuclear and plastid DNA This photograph of a spinach *(Spinacea oleracea)* leaf mesophyll cell, stained with the fluorescent dye DAPI, shows the concentration of DNA (large white area) in the nucleus and the location of multiple copies of DNA (small white dots) in the chloroplasts. The chlorophyll molecules show red fluorescence.

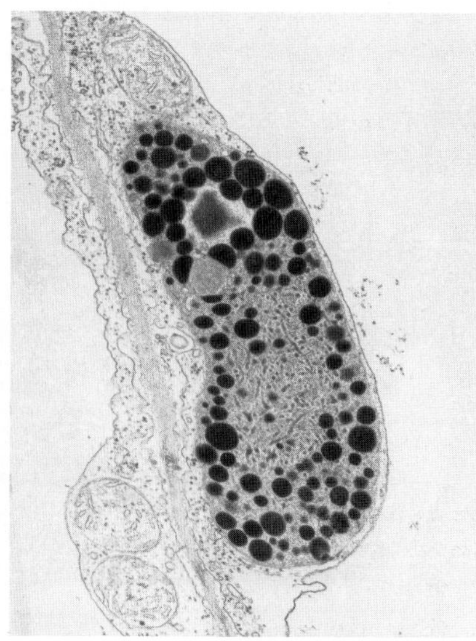

3–12 Chromoplast A chromoplast from a petal of *Forsythia,* a common garden shrub that is covered with yellow flowers in the early spring. The chromoplast contains numerous electron-dense oil bodies in which the yellow pigment is stored.

DNA. Overall control, however, clearly resides in the nucleus. Although some chloroplast proteins are synthesized within the chloroplast itself, most of these proteins are encoded by nuclear DNA, synthesized in the cytosol, and then imported into the chloroplast.

Chloroplasts—the workhorses of the plant world—are the ultimate source of virtually all of our food supplies and our fuel. Chloroplasts are not only the sites of photosynthesis, during which light energy and carbon dioxide are used to form carbohydrates, but they are also involved in the synthesis of amino acids, fatty acids, and a number of secondary metabolites. And, as we have noted, chloroplasts provide space for the temporary storage of starch.

Chromoplasts Contain Pigments Other Than Chlorophyll

Chromoplasts (Gk. *chroma,* "color"), like chloroplasts, are also pigmented plastids (Figure 3–12). Of variable shape, chromoplasts lack chlorophyll but synthesize and retain carotenoid pigments, which are often responsible for the yellow, orange, or red colors of many flowers, aging leaves, some fruits, and some roots, such as carrots. Chromoplasts may develop from previously existing green chloroplasts by a transformation in which the chlorophyll and internal membrane structure of the chloroplast disappear and masses of carotenoids accumulate, as occurs during the ripening of many fruits (tomatoes and chili peppers, for example). The precise functions of chromoplasts are not well understood, although at times they act as attractants to insects and other animals, with which they have coevolved, playing an essential role in the cross-pollination of flowering plants and the dispersal of fruits and seeds (see Chapter 20).

Leucoplasts Are Nonpigmented Plastids

Structurally the least differentiated of mature plastids, **leucoplasts** (Figure 3–13) lack pigments and an elaborate system of inner membranes. Some leucoplasts, known as *amyloplasts,* synthesize starch (Figure 3–14), whereas others are thought to be capable of forming a variety of substances, including oils and proteins.

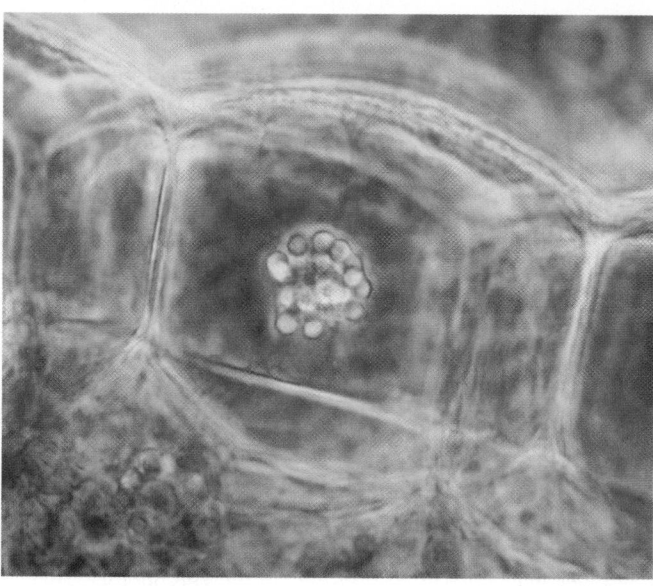

3–13 Leucoplasts Leucoplasts are small colorless plastids. Here they can be seen clustered around the nucleus in an epidermal cell of a leaf from the houseplant known as a wandering Jew *(Zebrina).* The purple color is due to anthocyanin pigments in vacuoles of the epidermal cells.

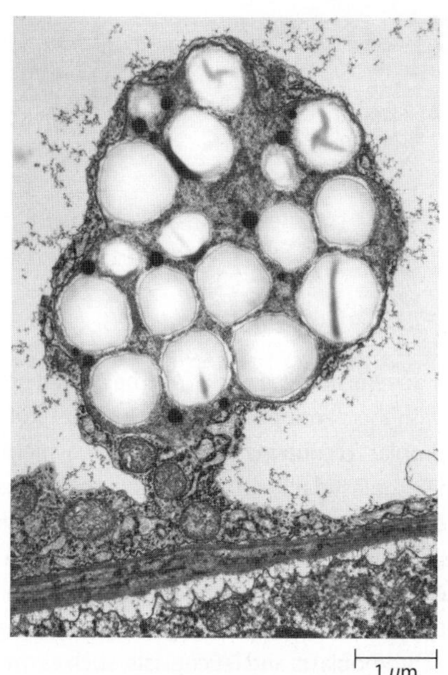

3–14 Amyloplast A type of leucoplast, this amyloplast is from the embryo sac of soybean *(Glycine max).* The round, clear bodies are starch grains. The smaller, dark bodies are oil bodies. Amyloplasts are involved in the synthesis and long-term storage of starch in seeds and storage organs, such as potato tubers.

Proplastids Are the Precursors of Other Plastids

Proplastids are small, colorless or pale green, undifferentiated plastids that occur in meristematic (dividing) cells of roots and shoots. They are the precursors of the other, more highly differentiated plastids such as chloroplasts, chromoplasts, or leucoplasts (Figure 3–15). If the development of a proplastid into a more highly differentiated form is arrested by the absence of light, it may form one or more **prolamellar bodies,** which are semicrystalline bodies composed of tubular membranes. Plastids containing prolamellar bodies are called **etioplasts** (Figure 3–16). Etioplasts form in leaf cells of plants grown in the dark. When exposed to light, the etioplasts develop into chloroplasts, and the membranes of the prolamellar bodies develop into thylakoids. In nature, the proplastids in the embryos of seeds first develop into etioplasts, and then, upon exposure to light, the etioplasts develop into chloroplasts. The various kinds of plastids are remarkable for the relative ease with which they can change from one type to another. These highly flexible organelles—the plastids—which are responsive to the environment and highly adaptable to many purposes, allow the plant to economize, in terms of energy.

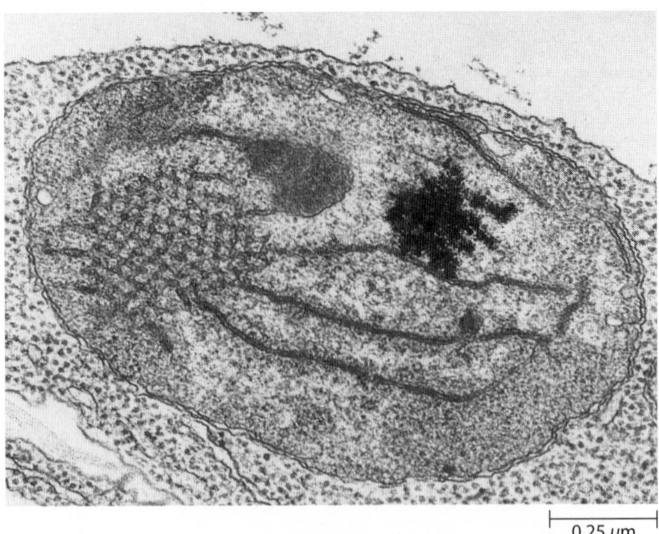

3–16 Etioplast An etioplast in a leaf cell of tobacco *(Nicotiana tabacum)* grown in the dark. Note the semicrystalline prolamellar body (checkerboard pattern at the left). When exposed to light, the tubular membranes of the prolamellar body develop into thylakoids.

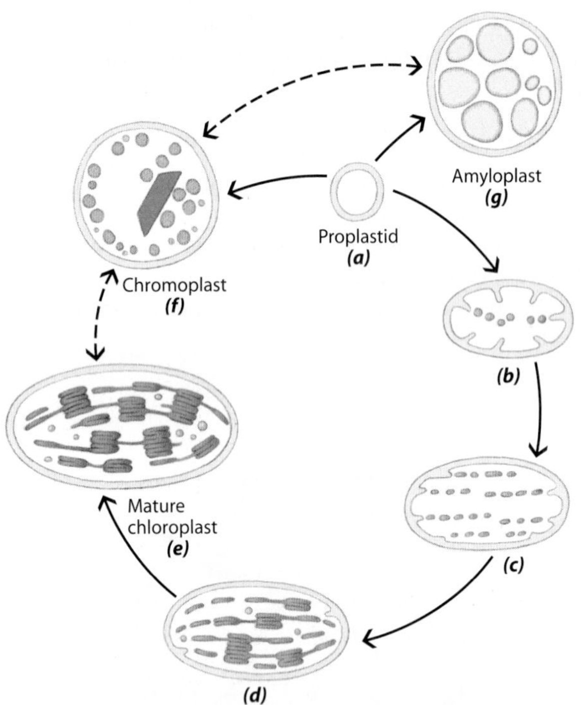

3–15 Plastid developmental cycle ***(a)*** The central process shown here begins with the development of a chloroplast from a proplastid. Initially, the proplastid contains few or no internal membranes. ***(b)–(d)*** As the proplastid differentiates into a chloroplast, flattened vesicles develop from the inner membrane of the plastid envelope and eventually align themselves into grana and stroma thylakoids. ***(e)*** The thylakoid system of the mature chloroplast appears discontinuous with the envelope. ***(f), (g)*** Proplastids may also develop into chromoplasts and leucoplasts, such as the starch-synthesizing amyloplast shown here. Note that chromoplasts may be formed from proplastids, chloroplasts, or leucoplasts (for example, an amyloplast). The various kinds of plastids can change from one type to another (dashed arrows).

Plastids reproduce by fission, the process of dividing into equal halves, which is characteristic of bacteria. In meristematic cells, the division of proplastids roughly keeps pace with cell division. In mature cells, however, most plastids are derived from the division of mature plastids.

Mitochondria

Mitochondria (singular: mitochondrion), like plastids, are surrounded by two membranes (Figures 3–17 and 3–18). The inner membrane has numerous invaginations called **cristae** (singular: crista). The cristae occur as folds or tubules, which greatly increase the surface area available to proteins and the reactions associated with them. Mitochondria are generally smaller than plastids, measuring about half a micrometer in diameter, and exhibit great variation in length and shape.

Mitochondria are the sites of respiration, a process involving the release of energy from organic molecules and its conversion to molecules of ATP (adenosine triphosphate), the chief immediate source of chemical energy for all eukaryotic cells. (These processes are discussed further in Chapter 6.) Most plant cells contain hundreds or thousands of mitochondria, the number of mitochondria per cell being related to the cell's demand for ATP.

Besides respiration, mitochondria are involved in numerous other metabolic processes, among them the biosynthesis of amino acids, vitamin cofactors, and fatty acids, and they play a pivotal role in **programmed cell death,** the genetically determined process that leads to the death of the cell. Programmed cell death is preceded by mitochondrial swelling and the release of cytochrome *c*, which is normally involved in electron

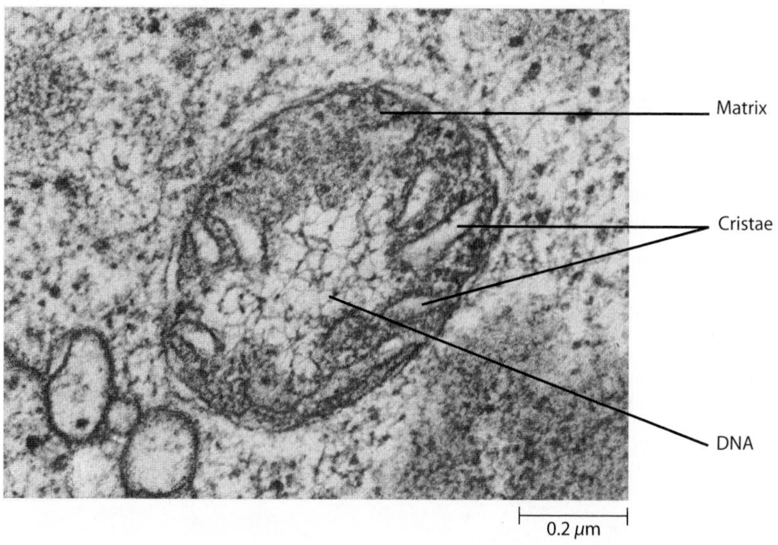

3–17 Mitochondrion A mitochondrion in a leaf cell of spinach *(Spinacia oleracea)*, in a plane of section revealing some strands of DNA in the nucleoid. The envelope of the mitochondrion consists of two separate membranes. The inner membrane folds inward to form cristae, which are embedded in a dense matrix. The small particles in the matrix are ribosomes.

transport (see Chapter 7). The release of cytochrome *c* appears to be critical for the activation of proteases and nucleases, enzymes that bring about degradation of the protoplast.

Mitochondria are in constant motion, turning and twisting and moving from one part of the cell to another; they also fuse and divide. Mitochondria tend to congregate where energy is required. In cells in which the plasma membrane is very active in transporting materials into or out of the cell, they often can be found arrayed along the membrane surface. In motile, single-celled algae, mitochondria are typically clustered at the bases of the flagella, presumably providing energy for flagellar movement.

Mitochondria, like plastids, are semiautonomous organelles. The inner membrane of the mitochondrion encloses a **matrix** that contains proteins, RNA, DNA, small ribosomes similar to those of bacteria, and various solutes (dissolved substances). The DNA of the mitochondrion, like that of the plastid, occurs in one or more clear areas, the nucleoids (Figure 3–17). Thus, as mentioned previously, in plant cells genetic information is found in three different compartments: nucleus, plastid, and mitochondrion. The nuclear genome, the total genetic information stored in the nucleus, is far larger than the genome of either the plastid or the mitochondrion and accounts for most of the genetic information of the cell. Both plastids and mitochondria can code for some, but by no means all, of their own polypeptides.

Mitochondria and Chloroplasts Evolved from Bacteria

On the basis of the close similarity between bacteria and the mitochondria and chloroplasts of eukaryotic cells, it is quite clear that mitochondria and chloroplasts originated as bacteria that were engulfed by larger heterotrophic cells (see Chapter 12). These larger cells were the forerunners of the eukaryotes. The smaller cells, which contained (and still contain) all the mechanisms necessary to trap and/or convert energy from their surroundings, donated these useful capacities to the larger cells. Cells with these respiratory and/or photosynthetic "assistants" had the advantage over their contemporaries of energetic self-sufficiency and undoubtedly soon multiplied at their contemporaries' expense. All but a very few modern eukaryotes contain mitochondria, and all autotrophic eukaryotes also contain chloroplasts; both seem to have been acquired by independent symbiotic events. (**Symbiosis** is the close association between two or more dissimilar organisms that may be, but is not necessarily, beneficial to each.) The smaller cells—now established as symbiotic organelles within the larger cells—obtained protection from environmental extremes. As a consequence, eukaryotes were able to invade the land and acidic waters, where the prokaryotic cyanobacteria are absent but the eukaryotic green algae abound.

Peroxisomes

Peroxisomes (also called microbodies) are spherical organelles that have a single bounding membrane and range in diameter from 0.5 to 1.5 micrometers. They have a granular interior, which may contain a body, sometimes crystalline, composed of protein (Figure 3–18). Peroxisomes contain no internal membranes, and, typically, they are closely associated with one or two segments of endoplasmic reticulum. Once thought to have their origin from the endoplasmic reticulum, peroxisomes are now known to be self-replicating organelles, like plastids and mitochondria. Unlike plastids and mitochondria, however, peroxisomes possess neither DNA nor ribosomes and must therefore import the materials required for their replication, as well as all of their proteins. Like plastids and mitochondria, peroxisomes have been shown to undergo movement within the cell.

Some peroxisomes play an important role in **photorespiration,** a process that consumes oxygen and releases carbon dioxide, exactly the reverse of what happens during photosynthesis (see Chapter 7). In green leaves, peroxisomes are closely associated with mitochondria and chloroplasts (Figure 3–18). Other peroxisomes, called **glyoxysomes,** contain the enzymes necessary for the conversion of stored fats to sucrose during germination in many seeds. The two types of peroxisome are interconvertible.

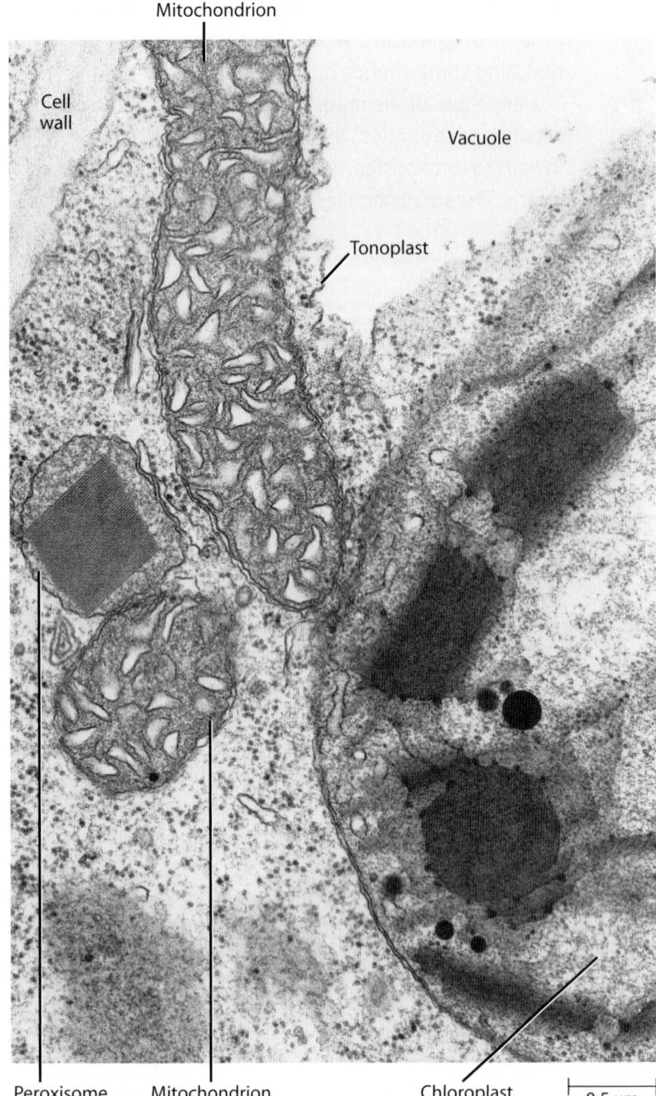

3–18 Organelles in a leaf cell A peroxisome with a large crystalline inclusion and a single surrounding membrane may be contrasted with the two mitochondria and the chloroplast, each of which has an envelope consisting of two membranes. Due to the plane of the section, the double nature of the chloroplast envelope is apparent only in the lower portion of this electron micrograph. A single membrane, the tonoplast, separates the vacuole from the rest of the cytoplasm in this leaf cell of tobacco *(Nicotiana tabacum)*.

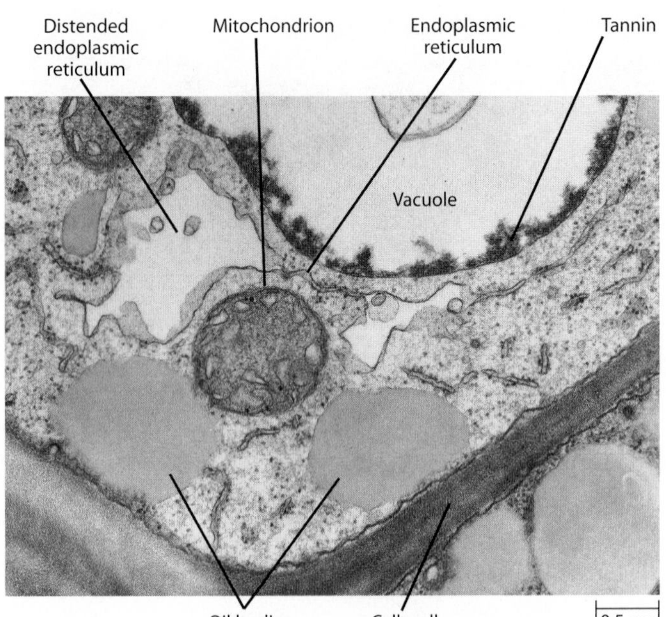

3–19 Cytoplasmic components of a plant cell Oil bodies are visible in this electron micrograph of a parenchyma cell from the thickened stem, or corm, of the quillwort *(Isoetes muricata)*. The dense material lining the vacuole is tannin. To the upper left of the mitochondrion, a cisterna of the endoplasmic reticulum is greatly distended. Some vacuoles are thought to arise from the endoplasmic reticulum in this way.

Vacuoles

Together with plastids and a cell wall, the **vacuole** is one of the three characteristic structures that distinguish plant cells from animal cells. As mentioned previously, vacuoles are organelles surrounded by a single membrane known as the **tonoplast,** or vacuolar membrane (Figures 3–7 and 3–18). The vacuole may originate directly from the endoplasmic reticulum (Figure 3–19), but most of the tonoplast and vacuolar proteins are derived directly from the Golgi apparatus, which is discussed on pages 52 to 54.

Many vacuoles are filled with a liquid commonly called **cell sap.** The principal component of the cell sap is water, with other components varying according to the type of plant, organ, and cell and to their developmental and physiological states. In addition to inorganic ions such as Ca^{2+}, K^+, Cl^-, Na^+, and HPO_4^{2-}, vacuoles commonly contain sugars, organic acids, and amino acids. Sometimes a particular substance is present in such a high concentration that it forms crystals. Calcium oxalate crystals, which can assume several different forms, are especially common (Figure 3–20). In most cases, vacuoles do not synthesize the molecules they accumulate, but instead receive them from other parts of the cytoplasm.

The immature plant cell typically contains numerous small vacuoles that increase in size and fuse into a single vacuole as the cell enlarges. In the mature cell, as much as 90 percent of the volume may be taken up by the vacuole, with the rest of the cytoplasm consisting of a thin peripheral layer closely pressed against the cell wall (Figure 3–7). By filling such a large proportion of the cell with "inexpensive" vacuolar contents, plants not only save "expensive" (in terms of energy) nitrogen-rich cytoplasmic material but also acquire a large surface between the thin layer of cytoplasm and the protoplast's external environment. Most of the increase in the size of the cell results from enlargement of the vacuole(s). A direct consequence of this strategy is the development of internal pressure and the maintenance of tissue rigidity, one of the principal roles of the vacuole and tonoplast (see Chapter 4).

Different kinds of vacuoles with distinct functions may be found in a single mature cell. Vacuoles are important storage compartments for primary metabolites, such as sugars and

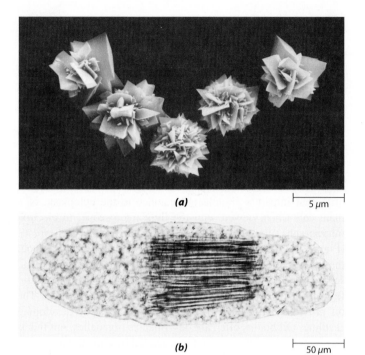

3–20 Calcium oxalate crystals Vacuoles can contain different forms of calcium oxalate crystals. ***(a)*** Druses, or aggregates of crystals composed of calcium oxalate, from the vacuoles of epidermal cells of redbud *(Cercis canadensis),* as seen with a scanning electron microscope. ***(b)*** A bundle of raphides, or needlelike crystals of calcium oxalate, in a vacuole of a leaf cell of the snake plant *(Sansevieria).* The tonoplast surrounding the vacuole is not discernible. The granular substance seen here is the cytoplasm.

organic acids and the reserve proteins in seeds. Vacuoles also remove toxic secondary metabolites, such as nicotine and tannin, from the rest of the cytoplasm (see Figure 2–27). Such substances are sequestered permanently in the vacuoles. As discussed in Chapter 2, the secondary metabolites contained in the vacuoles are toxic not only to the plant itself but also to pathogens, parasites, and/or herbivores, and they therefore play an important role in plant defense.

The vacuole is often a site of pigment deposition. The blue, violet, purple, dark red, and scarlet colors of plant cells are usually caused by a group of pigments known as the anthocyanins (page 33). Unlike most other plant pigments, the anthocyanins are readily soluble in water and are dissolved in the cell sap. They are responsible for the red and blue colors of many vegetables (radishes, turnips, cabbages), fruits (grapes, plums, cherries), and a host of flowers (cornflowers, geraniums, delphiniums, roses, peonies). Sometimes the pigments are so brilliant that they mask the chlorophyll in the leaves, as in the ornamental red maple.

Anthocyanins are also responsible for the brilliant red colors of some leaves in autumn. These pigments form in response to cold, sunny weather, when the leaves stop producing chlorophyll. As the chlorophyll that is present disintegrates, the newly formed anthocyanins in the vacuoles are unmasked. In leaves that do not form anthocyanin pigments, the breakdown of chlorophyll in autumn may unmask the more stable yellow-to-orange carotenoid pigments already present in the chloroplasts. The most spectacular autumnal coloration develops in years when cool, clear weather prevails in the fall.

Vacuoles are also involved in the breakdown of macromolecules and the recycling of their components within the cell. Entire cell organelles, such as mitochondria and plastids, may be deposited and degraded in vacuoles. Because of this digestive activity, vacuoles are comparable in function with the organelles known as lysosomes that occur in animal cells.

Endoplasmic Reticulum

The **endoplasmic reticulum (ER)** is a complex, three-dimensional membrane system that permeates the entire cytosol. In sectional view, the endoplasmic reticulum appears as two parallel membranes with a narrow space, or *lumen,* between them. The form and abundance of this system of membranes vary greatly from cell to cell, depending on the cell type, its metabolic activity, and its stage of development. For example, cells that store proteins have abundant **rough ER,** which consists of flattened sacs, or **cisternae** (singular: cisterna), with numerous polysomes on their outer surface (Figures 3–5b and 3–8). In contrast, cells that secrete lipids have extensive systems of **smooth ER,** which lacks ribosomes and is largely tubular in form. The tubular form is involved in lipid synthesis. Both rough and smooth forms occur within the same cell and have numerous connections between them. These membranes are in continuous motion and are constantly changing their shape and distribution.

In many cells, an extensive network of endoplasmic reticulum, consisting of interconnected cisternae and tubules, is located just inside the plasma membrane in the peripheral, or **cortical,** cytoplasm (Figure 3–21). The most likely function of this **cortical ER** appears to be in regulating the level of calcium ions (Ca^{2+}) in the cytosol. The cortical ER may therefore play a role in a host of developmental and physiological processes involving calcium. It has also been suggested that this cortical network serves as a structural element that stabilizes or anchors the cytoskeleton of the cell. The cortical ER is believed to be a general indicator of the metabolic and developmental status of a cell—quiescent cells have less and developing and physiologically active cells have more.

Some electron micrographs show the rough endoplasmic reticulum to be continuous with the outer membrane of the nuclear envelope. As we have mentioned, the nuclear envelope may be considered a specialized, locally differentiated portion of the endoplasmic reticulum. When the nuclear envelope breaks into fragments during prophase of nuclear division, it becomes indistinguishable from cisternae of the rough ER. When new nuclei are formed during telophase, vesicles of endoplasmic reticulum join to form the nuclear envelopes of the two daughter nuclei.

The endoplasmic reticulum functions as a communication system within the cell and as a system for channeling materials—such as proteins and lipids—to different parts of the cell. In addition, the cortical endoplasmic reticulum of adjacent plant cells is interconnected by cytoplasmic strands, called plasmodesmata, that traverse their common walls and play a role in cell-to-cell communication (see pages 61 and 62 and 88 to 90).

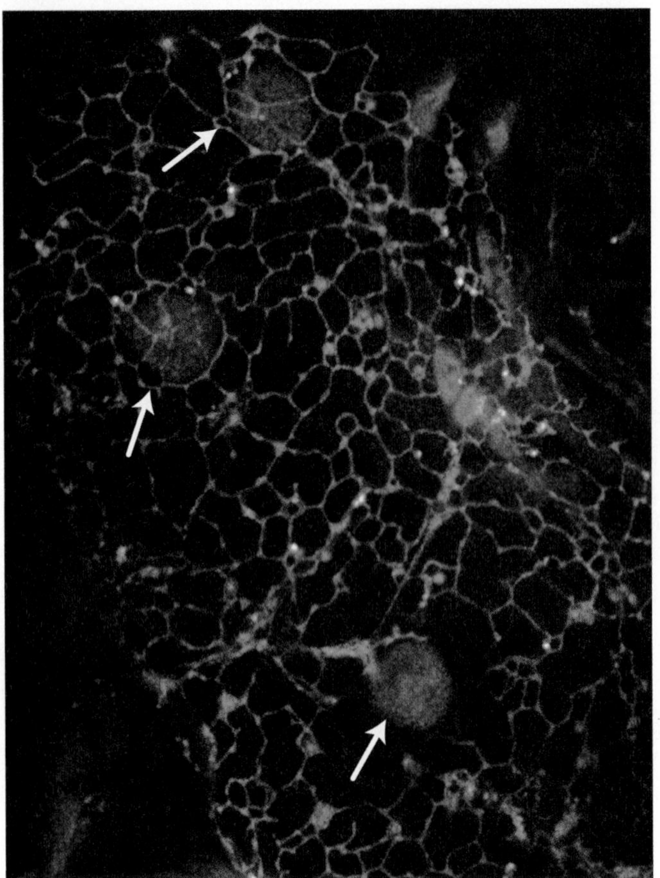

3–21 Cortical endoplasmic reticulum A single epidermal cell of *Nicotiana benthamiana* showing cortical endoplasmic reticulum. The cortical endoplasmic reticulum has been labeled with green fluorescent protein and photographed using confocal laser scanning microscopy. Tubular endoplasmic reticulum elements form a "cage" around each of the three chloroplasts (arrows) seen here.

The endoplasmic reticulum is one of the main sites of lipid synthesis in plants, the other being the plastid. **Oil bodies,** or lipid droplets, arise in the endoplasmic reticulum and then are released into the cytosol. They are more or less spherical structures that impart a granular appearance to the cytoplasm of a plant cell when viewed with the light microscope. In electron micrographs, the oil bodies have an amorphous appearance (Figure 3–19). Oil bodies are widely distributed throughout the cells of the plant body but are most abundant in fruits and seeds. In fact, approximately 45 percent of the weight of sunflower, peanut, flax, and sesame seeds is composed of oil. The oil provides energy and a source of carbon to the developing seedling. Oil bodies often are described as organelles, but this is incorrect because they are not surrounded by a membrane.

Golgi Apparatus

The term **Golgi apparatus,** or Golgi complex, is used to refer collectively to all of the **Golgi bodies** (also called Golgi stacks or dictyosomes) of a cell. Golgi bodies consist of five to eight stacks of flattened, disk-shaped sacs, or **cisternae,** which often

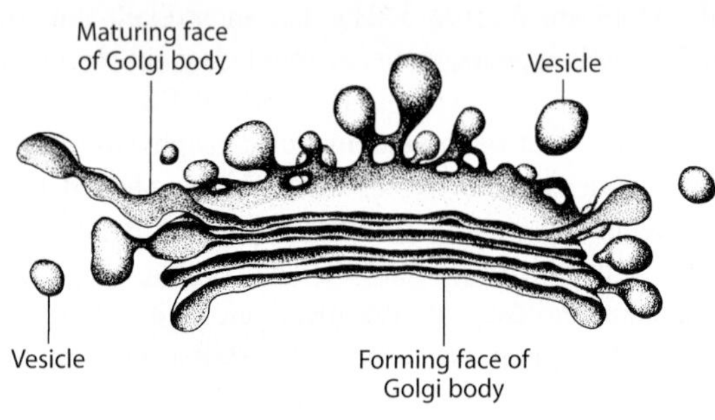

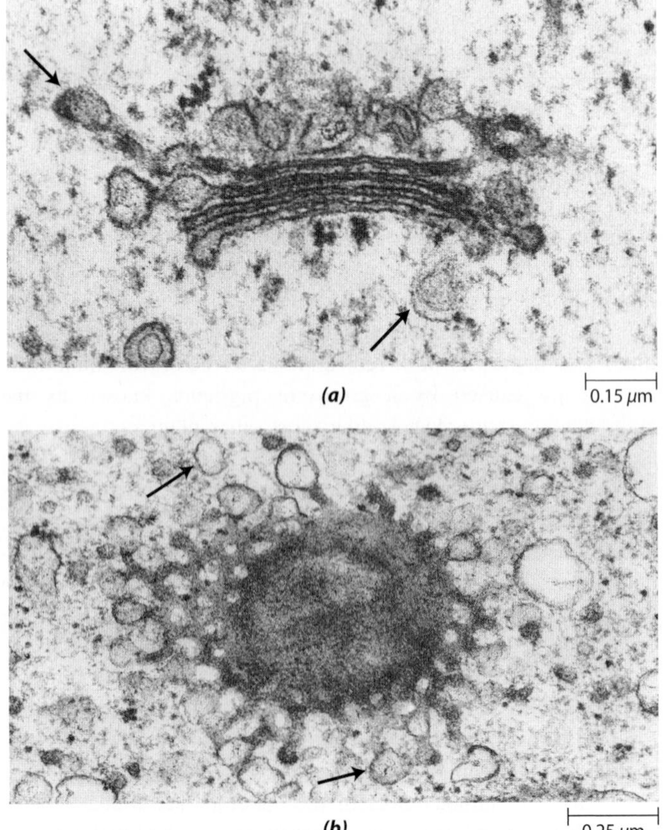

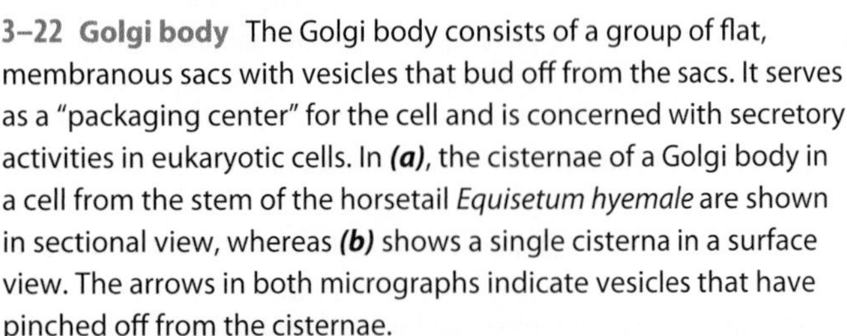

3–22 Golgi body The Golgi body consists of a group of flat, membranous sacs with vesicles that bud off from the sacs. It serves as a "packaging center" for the cell and is concerned with secretory activities in eukaryotic cells. In ***(a)***, the cisternae of a Golgi body in a cell from the stem of the horsetail *Equisetum hyemale* are shown in sectional view, whereas ***(b)*** shows a single cisterna in a surface view. The arrows in both micrographs indicate vesicles that have pinched off from the cisternae.

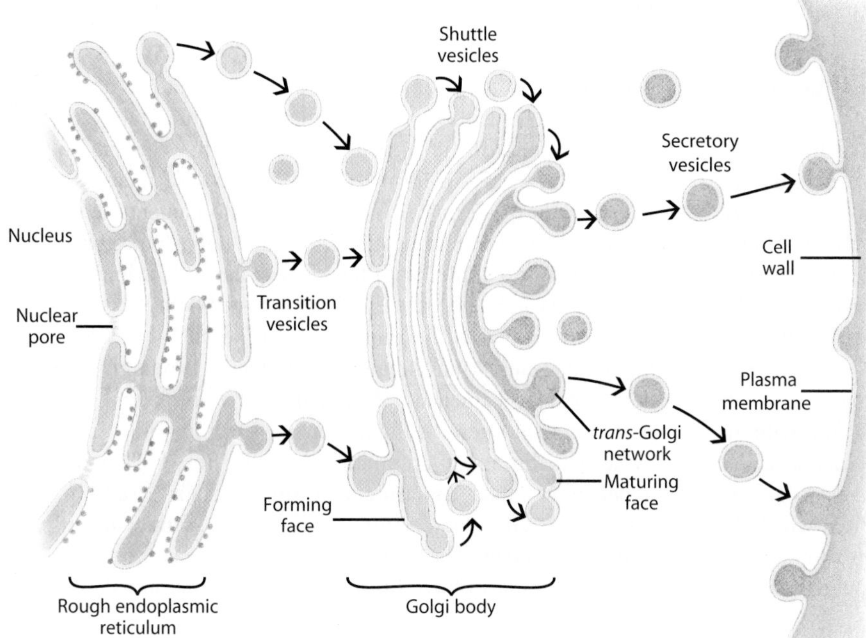

3–23 The endomembrane system The endomembrane system consists of an interconnected network of the endoplasmic reticulum, the nuclear envelope, Golgi bodies with their transport and secretory vesicles, the plasma membrane, and vacuolar membranes. This diagram depicts the origin of new membranes from the rough endoplasmic reticulum at the left. Transition vesicles pinch off a smooth-surfaced portion of the endoplasmic reticulum and carry membranes and enclosed substances to the forming *(cis)* face of the Golgi body. In this cell, cell wall substances are being transported stepwise across the Golgi stack to the *trans*-Golgi network by means of shuttle vesicles. Secretory vesicles derived from the *trans*-Golgi network then migrate to the plasma membrane and fuse with it, contributing new membrane to the plasma membrane and discharging their contents into the wall.

are branched into a complex series of tubules at their margins (Figure 3–22). Unlike the centralized Golgi of mammalian cells, the Golgi apparatus of plant cells consists of many separate stacks that remain active during mitosis and cytokinesis.

The Golgi apparatus is a dynamic, highly polarized membrane system. Customarily, the two opposite poles of a Golgi stack are referred to as forming (or *cis*) and maturing (or *trans*) faces. The stacks between the two faces comprise the middle (or *medial*) cisternae. An additional structurally and biochemically distinct compartment, the ***trans*-Golgi network,** occurs on the maturing face of the Golgi body (Figure 3–23).

Golgi bodies are involved in secretion. In plants, most of the Golgi bodies are involved in the synthesis and secretion of noncellulosic polysaccharides (hemicelluloses and pectins) destined for incorporation into the cell wall. Evidence indicates that different steps in the synthesis of these polysaccharides occur sequentially in different cisternae of the Golgi body. Golgi bodies also process and secrete glycoproteins (see page 58) that are transferred to them from the rough endoplasmic reticulum, via **transition vesicles.** These transition vesicles flow from the endoplasmic reticulum to the forming face of the Golgi body. The glycoproteins are transported stepwise across the stack to the maturing face by means of **shuttle vesicles.** They are then sorted in the *trans*-Golgi network for delivery to the vacuole or for secretion at the cell surface (Figure 3–23). Polysaccharides destined for secretion at the cell surface may also be sorted in the *trans*-Golgi network. A given Golgi body can process polysaccharides and glycoproteins simultaneously.

Newly formed vacuolar proteins are packaged at the *trans*-Golgi network into **coated vesicles,** so-called because these vesicles are coated with several proteins, including **clathrin,** which forms cages around the vesicles (Figure 3–24). Glycoproteins and complex polysaccharides destined for secretion at the cell surface are packaged in noncoated, or smooth-surfaced, vesicles. The movement of these vesicles from the *trans*-Golgi network to the plasma membrane appears to depend on the presence of actin filaments (see page 55). When the vesicles reach the plasma membrane, they fuse with it and discharge their contents into the cell wall. In enlarging cells the vesicle membranes become incorporated into the plasma membrane, contributing to its growth. The secretion of substances from cells in vesicles is called **exocytosis.**

Substances are also transferred from the cell wall to the *trans*-Golgi network or vacuoles through the formation of vesicles (called primary endocytic vesicles or endosomes) at the plasma membrane. The uptake of extracellular substances by infolding of the plasma membrane and pinching off of a vesicle is called **endocytosis** (see pages 86 and 87).

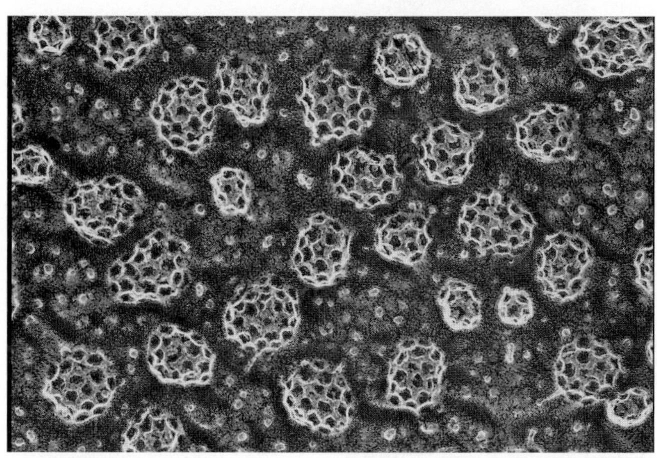

3–24 Coated vesicles These coated vesicles are surrounded by proteins, including clathrin. The three-pronged clathrin subunits are associated with one another to form cages around the vesicles.

The Endoplasmic Reticulum and Golgi Apparatus Are Components of the Endomembrane System

Thus far, we have considered the various components of the protoplast in isolation. With the exception of mitochondrial, plastid, and peroxisomal membranes, however, all cellular membranes—including plasma membrane, nuclear envelope, endoplasmic reticulum, Golgi apparatus, tonoplast, and various kinds of vesicles—constitute a continuous, interconnected system, known as the **endomembrane system** (Figure 3–23). The endoplasmic reticulum is the initial source of membranes. Transition vesicles from the endoplasmic reticulum transport new membrane material to the Golgi apparatus, and secretory vesicles derived from the *trans*-Golgi network contribute to the plasma membrane. The *trans*-Golgi network also supplies vesicles that fuse with the tonoplast and thus contribute to formation of the vacuoles. The endoplasmic reticulum, Golgi apparatus, and *trans*-Golgi network, therefore, may be considered a functional unit in which the Golgi bodies serve as the main vehicles for the transformation of endoplasmic-reticulum-like membranes into plasma-membrane-like and tonoplast-like membranes.

Cytoskeleton

All eukaryotic cells possess a **cytoskeleton,** a dynamic, three-dimensional network of protein filaments that extends throughout the cytosol and is intimately involved in many processes. These processes include cell division, growth, and differentiation, as well as the movement of organelles from one location to another within the cell. The cytoskeleton of plant cells consists of two types of protein filaments: microtubules and actin filaments. In addition, plant cells, like animal cells, may contain a third type of cytoskeletal filament, the intermediate filament. Little is known about the structure or role of intermediate filaments in plant cells.

Microtubules Are Cylindrical Structures Composed of Tubulin Subunits

Microtubules are cylindrical structures about 24 nanometers in diameter and of varying lengths. Each microtubule is built up of subunits of the protein called **tubulin.** The subunits are arranged in a helix to form 13 rows, or "protofilaments," around a core (Figure 3–25a, b). Within each protofilament the subunits are oriented in the same direction, and all the protofilaments are aligned in parallel with the same polarity. During assembly of the microtubules, one end grows faster than the other. The fast-growing end is the plus end, and the slow-growing end is the minus end. Microtubules are dynamic structures that undergo regular sequences of breakdown, re-formation, and rearrangement into new configurations at specific points in the cell cycle and during differentiation (see Figures 3–47 and 3–48), a behavior called **dynamic instability.** Their assembly takes place at specific "nucleating sites" known as **microtubule organizing centers.** The surface of the nucleus and portions of the cortical cytoplasm have been identified as microtubule organizing centers.

Microtubules have many functions. In enlarging and differentiating cells, microtubules just inside the plasma membrane (cortical microtubules) are involved in the orderly growth of the cell wall, especially through their control of the alignment of cellulose microfibrils as they are added to the cell wall (Figure 3–26; see Figure 3–47a). The direction of expansion of the cell is governed, in turn, by the alignment of cellulose microfibrils in the wall. Microtubules serve to direct secretory Golgi vesicles containing noncellulosic cell wall substances toward the developing wall. In addition, microtubules make up the spindle fibers that play a role in chromosome movement and in cell plate formation in dividing cells. Microtubules are also important components of flagella and cilia and are involved in the movement of these structures.

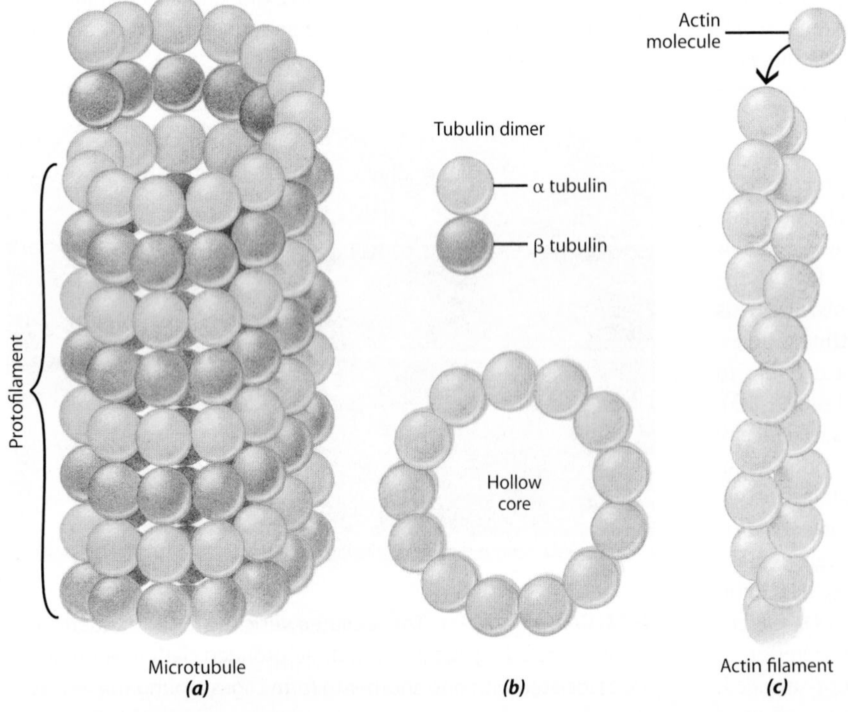

3–25 Microtubules and actin filaments Two components of the cytoskeleton—microtubules and actin filaments—are formed from globular protein subunits. ***(a)*** Longitudinal view and ***(b)*** transverse section (made at a right angle to the long axis) of a microtubule. Microtubules are hollow tubules composed of two different types of molecules, alpha (α) tubulin and beta (β) tubulin. These tubulin molecules first come together to form soluble dimers ("two parts"), which then self-assemble into insoluble hollow tubules. The arrangement results in 13 "protofilaments" around a hollow core. ***(c)*** Actin filaments consist of two linear chains of identical molecules coiled around one another to form a helix.

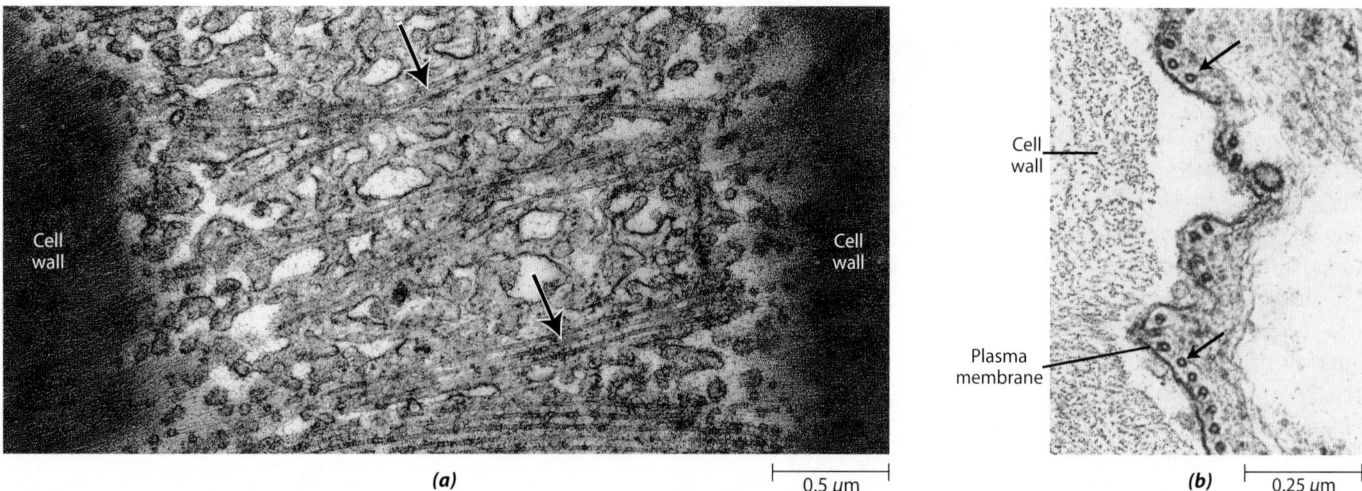

3–26 Cortical microtubules ***(a)*** A longitudinal view of cortical microtubules (indicated by arrows) in leaf cells of the fern *Botrychium virginianum.* The microtubules occur just inside the wall and plasma membrane. ***(b)*** A transverse view of cortical microtubules (arrows), which can be seen to be separated from the wall by the plasma membrane. Cortical microtubules play a role in the alignment of cellulose microfibrils in the cell wall.

Actin Filaments Consist of Two Linear Chains of Actin Molecules in the Form of a Helix

Actin filaments, or microfilaments, are also polar structures with distinct plus and minus ends. They are composed of a protein called **actin** and occur as long filaments 5 to 7 nanometers in diameter (Figure 3–25c). Some actin filaments are associated spatially with microtubules and, like microtubules, form new arrays, or configurations, at specific points in the cell cycle. In addition to single filaments, bundles of actin filaments occur in many plant cells (Figure 3–27).

Actin filaments are involved in a variety of activities in plant cells, including cell wall deposition, tip growth of pollen tubes, movement of the nucleus before and following cell division, organelle movement, vesicle-mediated secretion, organization of the endoplasmic reticulum, and cytoplasmic streaming (see the essay "Cytoplasmic Streaming in Giant Algal Cells" on the next page).

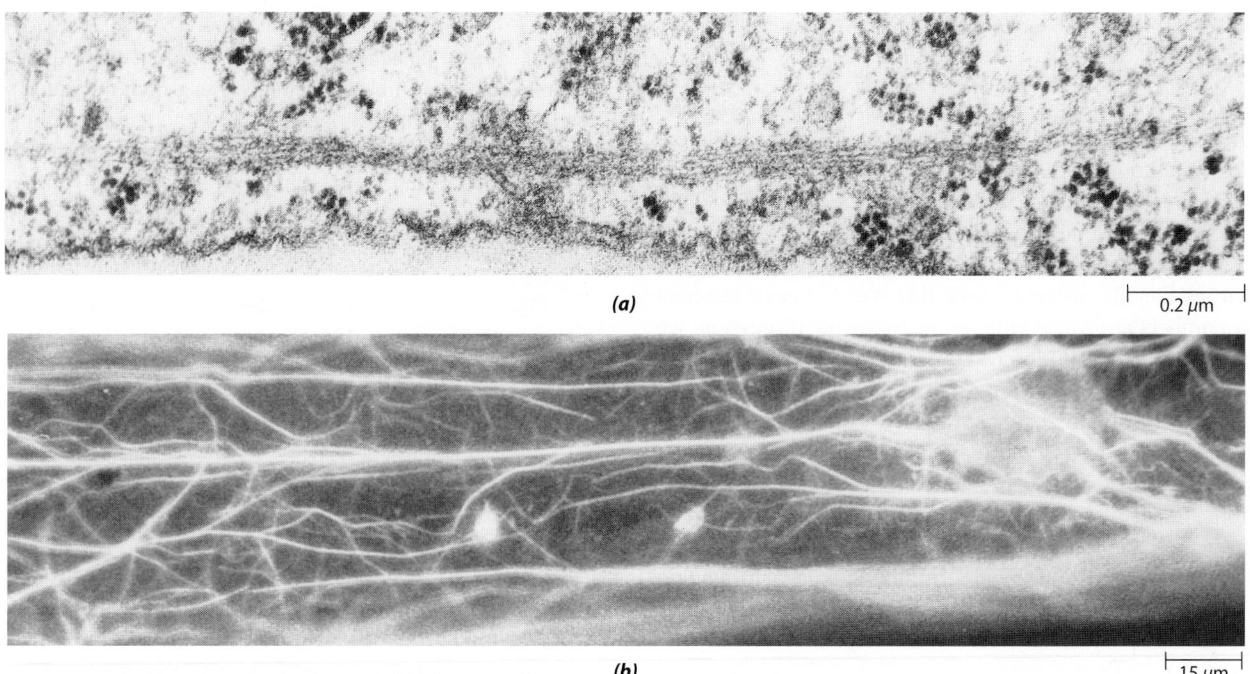

3–27 Actin filaments ***(a)*** A bundle of actin filaments as revealed in an electron micrograph of a leaf cell of maize *(Zea mays).* ***(b)*** Several bundles of actin filaments as revealed in a fluorescence micrograph of a stem hair of tomato *(Solanum lycopersicum).* Actin filaments are involved in a variety of activities, including cytoplasmic streaming.

CYTOPLASMIC STREAMING IN GIANT ALGAL CELLS

One cannot help but be astonished when witnessing for the first time the streaming of cytoplasm and the movement of plastids and mitochondria, which appear to be motoring freely within the cell. It's alive!

Much of our understanding of cytoplasmic streaming comes from work on the giant cells of green algae such as *Chara* and *Nitella*. In these cells, which are 2 to 5 centimeters long, the chloroplast-containing layer of cytoplasm bordering the wall is stationary. Spirally arranged bundles of actin filaments extend for several centimeters along the cells, forming distinct "tracks" that are firmly attached to the stationary chloroplasts. The moving layer of cytoplasm, which occurs between the bundles of actin filaments and the tonoplast, contains the nucleus, mitochondria, and other cytoplasmic components.

The generating force necessary for cytoplasmic streaming comes from an interaction between actin and myosin, a protein molecule with an ATPase-containing "head" that is activated by actin. ATPase is an enzyme that breaks down (hydrolyzes) ATP to release energy (page 30). Apparently, the organelles in the streaming cytoplasm are indirectly attached to the actin filaments by myosin molecules, which use the energy released by ATP hydrolysis to "walk" along the actin filaments, pulling the organelles with them. Streaming always occurs from the minus to the plus ends of the actin filaments, all of which are similarly oriented within a bundle.

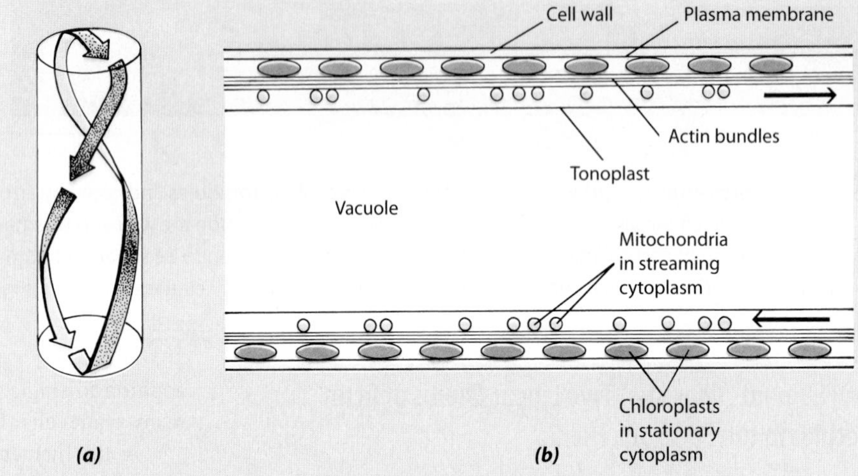

Cytoplasmic streaming ***(a)*** A track followed by the streaming cytoplasm in a giant algal cell. ***(b)*** A longitudinal section through part of the cell, showing the arrangement of stationary and streaming layers of cytoplasm. The proportions have been distorted in both diagrams for clarity.

Flagella and Cilia

Flagella and **cilia** (singular: flagellum and cilium) are hairlike structures that extend from the surface of many different types of eukaryotic cells. They are relatively thin and constant in diameter (about 0.2 micrometer), but they vary in length from about 2 to 150 micrometers. By convention, those that are longer or are present alone or in small numbers are usually referred to as flagella, whereas those that are shorter or occur in greater numbers are called cilia. In the following discussion, we use the term "flagella" to refer to both.

In some algae and other protists, flagella are locomotor structures, propelling the organisms through the water. In plants, flagella are found only in the reproductive cells (gametes), and then only in plants that have motile sperm, including the mosses, liverworts, ferns, cycads, and maidenhair tree *(Ginkgo biloba)*.

Each flagellum has a precise internal organization consisting of an outer ring of nine pairs of microtubules surrounding two additional microtubules (Figure 3–28). This basic 9-plus-2 pattern of organization is found in all flagella of eukaryotic organisms.

The movement of a flagellum originates from within the structure itself. Flagella are capable of movement even after they have been detached from cells. The movement is produced by a sliding microtubule mechanism in which the outer pairs of microtubules move past one another without contracting. As the pairs slide past one another, their movement causes localized bending of the flagellum. Sliding of the pairs of microtubules results from cycles of attachment and detachment of enzyme-containing "arms" between neighboring pairs in the outer ring (Figure 3–28).

Flagella grow out of cylinder-shaped structures in the cytoplasm known as **basal bodies,** which then form the basal portion of the flagellum. The internal structure of the basal body resembles that of the flagellum itself, except that the outer tubules in the basal body occur in triplets rather than in pairs, and the two central tubules are absent.

Cell Wall

The **cell wall**—above all other characteristics—distinguishes plant cells from animal cells. The cell wall constrains expansion of the protoplast and prevents rupture of the plasma membrane when the protoplast enlarges following the uptake of water by the cell. The cell wall largely determines the size and shape of the cell and the texture of the tissue, and it contributes to the

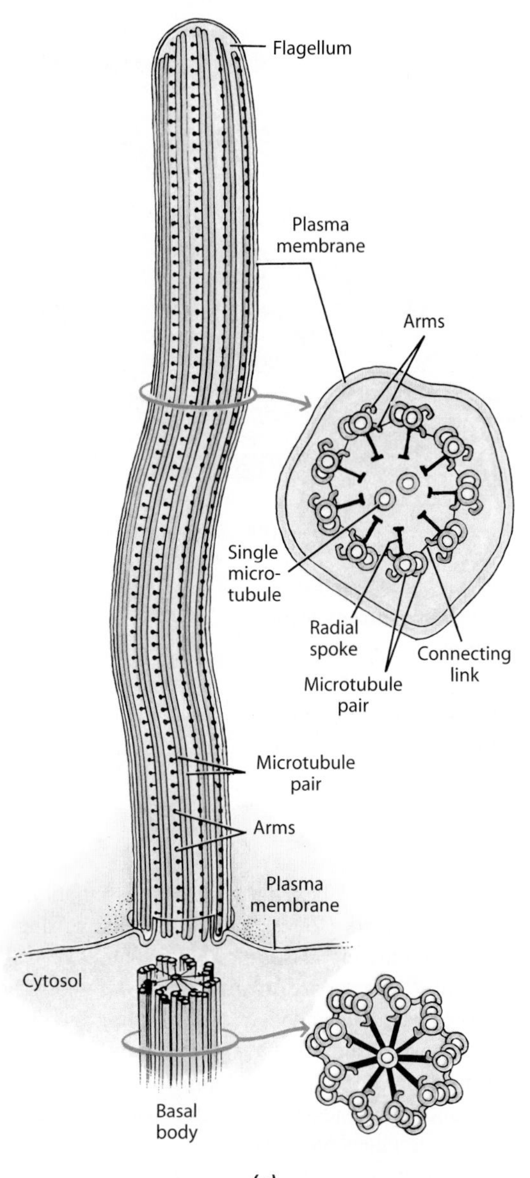

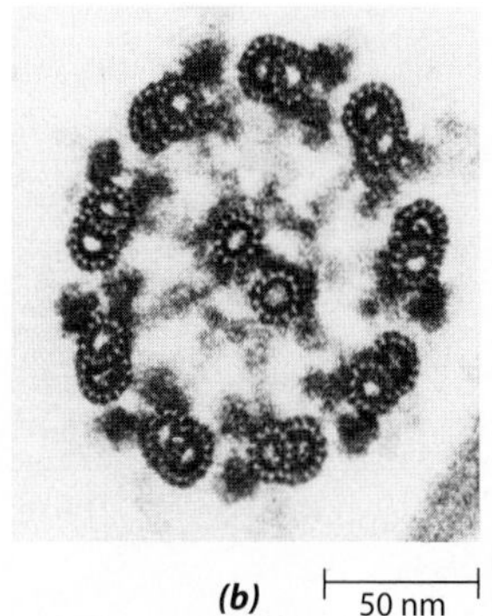

3–28 Structure of a flagellum ***(a)*** Diagram of a flagellum with its underlying basal body, and ***(b)*** an electron micrograph of the flagellum of *Chlamydomonas,* as seen in transverse section. Virtually all eukaryotic flagella have this same internal structure, which consists of an outer cylinder of nine pairs of microtubules surrounding two additional microtubules in the center. The "arms," the radial spokes, and the connecting links are formed from different types of protein. The basal bodies from which flagella arise have nine outer triplets, with no microtubules in the center. The "hub" of the wheel in the basal body is not a microtubule, although it has about the same diameter.

final form of the plant organ. Cell types are often identified by the structure of their walls, reflecting the close relationship between cell wall structure and cell function.

Once regarded as merely an outer, inactive product of the protoplast, the cell wall is now recognized as having specific and essential functions. Cell walls contain a variety of enzymes and play important roles in the absorption, transport, and secretion of substances in plants.

In addition, the cell wall may play an active role in defense against bacterial and fungal pathogens by receiving and processing information from the surface of the pathogen and transmitting this information to the plasma membrane of the plant cell. Through gene-activated processes (see Chapter 9), the plant cell may then become resistant through the production of phytoalexins, which are antimicrobial compounds toxic to the attacking pathogen (page 30). Resistance can also result from the synthesis and deposition of substances such as lignin (page 34), which act as barriers to invasion. Certain cell wall polysaccharides, dubbed "oligosaccharins," may even function as signal molecules.

Cellulose Is the Principal Component of Plant Cell Walls

The principal component of plant cell walls is **cellulose,** which largely determines their architecture. Cellulose is made up of repeating monomers of glucose attached end to end. The cellulose polymers are bundled into **microfibrils** about 10 to 25 nanometers in diameter (Figure 3–29). Cellulose has crystalline properties (Figure 3–30) because of the orderly arrangement of its molecules in certain parts, the **micelles,** of the microfibrils (Figure 3–31). Cellulose microfibrils wind together to form fine threads that may coil around one another like strands in a cable. Wound in this fashion, cellulose molecules have a strength exceeding that of an equivalent thickness of steel.

The cellulose framework of the wall is interlocked by a cross-linked matrix of noncellulosic molecules. These molecules are the polysaccharides known as hemicelluloses and pectins (Figure 3–29), as well as the structural proteins called glycoproteins.

The **hemicelluloses** vary greatly among different cell types and among different plant taxonomic groups. Being hydrogen-bonded to the cellulose microfibrils, the hemicelluloses limit the extensibility of the cell wall by tethering adjacent microfibrils; hence these polysaccharides play a significant role in regulating cell enlargement.

Pectins are characteristic of the first-formed (primary) cell wall layers and the intercellular substance (middle lamella) that cements together the walls of contiguous cells. They are highly hydrophilic polysaccharides, and the water they attract to the cell wall imparts plastic, or pliable, properties to the wall, a condition

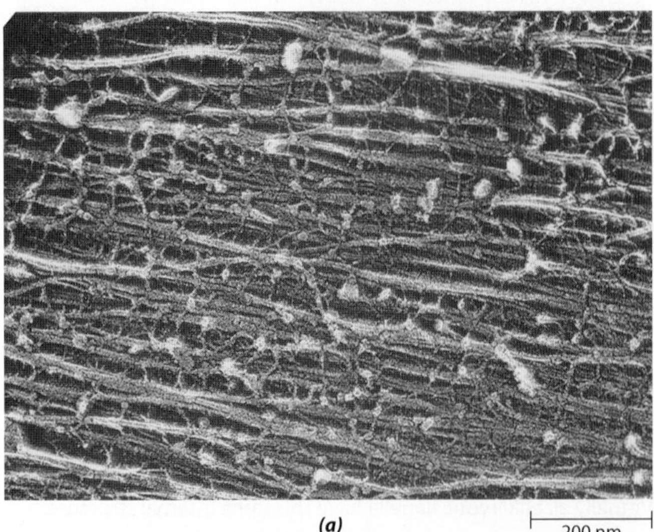

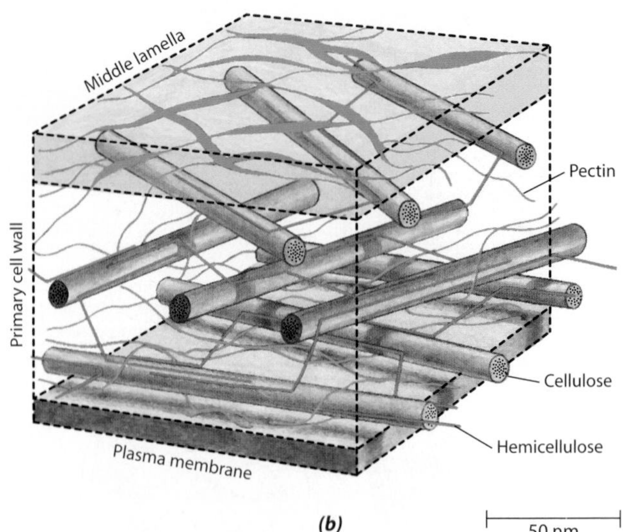

3–29 Primary walls ***(a)*** Surface view of the primary wall of a carrot *(Daucus carota)* cell, prepared by a fast-freeze, deep-etch technique, showing cellulose microfibrils cross-linked by an intricate web of matrix molecules. ***(b)*** Schematic diagram showing how the cellulose microfibrils are cross-linked into a complex network by hemicellulose molecules. The hemicellulose molecules are linked to the surface of the microfibrils by hydrogen bonds. The cellulose-hemicellulose network is permeated by a network of pectins, which are highly hydrophilic polysaccharides. Both hemicellulose and pectin are matrix substances. The middle lamella is a pectin-rich layer that cements together the primary walls of adjacent cells.

necessary for wall expansion. Growing primary walls are composed of about 65 percent water and are especially low in calcium (Ca^{2+}). Extensive Ca^{2+} cross-linking of pectins occurs after cell elongation is completed, preventing further stretching.

Callose, which is composed of spirally wound chains of glucose residues, is a widely distributed cell wall polysaccharide. It probably is best known for its association with the walls of sieve elements (food-conducting cells) of angiosperm phloem (food-conducting tissue) (see Chapter 23). Callose is deposited rapidly in response to mechanical wounding and environmental- or pathogen-induced stress, sealing the plasmodesmata (cytoplasmic strands) between contiguous cells. It occurs also in the normal course of development in pollen tubes (see Chapter 19) and is transiently associated with the cell plates of dividing cells (see pages 68 to 71).

Cell walls may also contain **glycoproteins—**structural proteins—as well as enzymes. The best characterized glycoproteins are the **extensins.** This family of hydroxyproline-rich proteins is so-named because they were originally presumed to be involved with the cell wall's extensibility. It appears, however, that the deposition of extensin may actually strengthen the wall, making it less extensible. A large number of enzymes have been reported in the primary wall layers. Such enzymes include peroxidases, phosphatases, cellulases, and pectinases.

Another important constituent of the walls of many kinds of cells is **lignin,** which adds compressive strength and bending stiffness (rigidity) to the cell wall (page 34). It is commonly found in the walls of plant cells that have a supporting or mechanical function.

Cutin, suberin, and **waxes** are fatty substances commonly found in the walls of the outer, protective tissues of the plant body. Cutin, for example, is found in the walls of the epidermis, and suberin is found in those of the secondary protective tissue, cork. Both substances occur in combination with waxes and function largely to reduce water loss from the plant (pages 23 and 24).

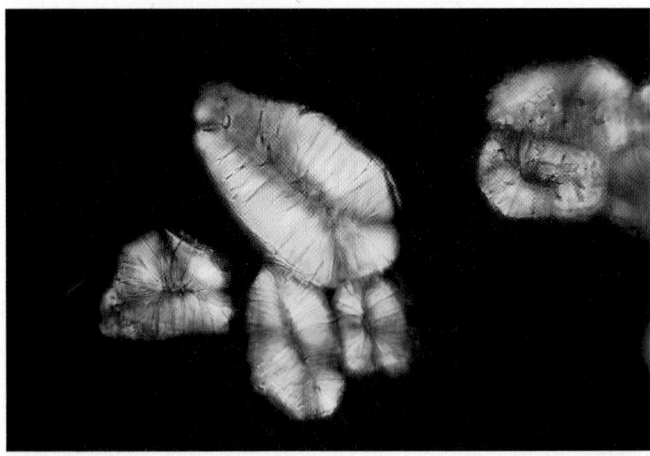

3–30 Stone cells Stone cells (sclereids) from the flesh of a pear *(Pyrus communis)* seen in polarized light. Clusters of such stone cells are responsible for the gritty texture of this fruit. The stone cells have very thick secondary walls traversed by numerous simple pits, which appear as lines in the walls. The walls appear bright in polarized light because of the crystalline properties of their principal component, cellulose.

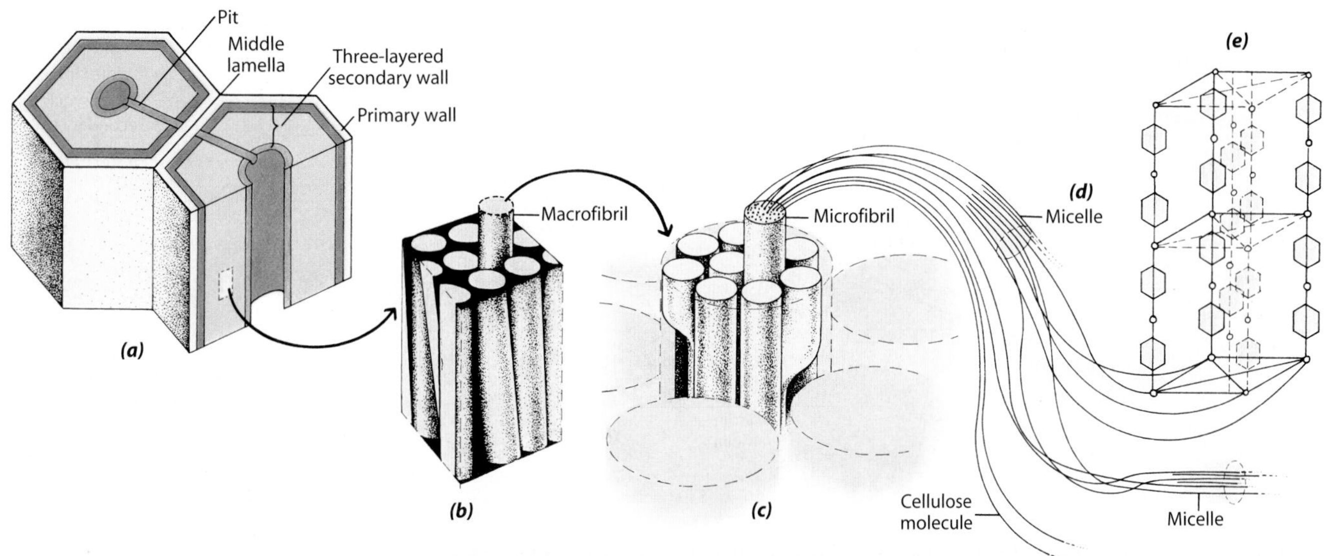

3–31 Detailed structure of a cell wall ***(a)*** Portion of wall showing, from the outside in, the middle lamella, primary wall, and three layers of secondary wall. Cellulose, the principal component of the cell wall, exists as a system of fibrils of different sizes. ***(b)*** The largest fibrils, macrofibrils, can be seen with the light microscope. ***(c)*** With the aid of an electron microscope, the macrofibrils can be resolved into microfibrils about 10 to 25 nanometers wide. ***(d)*** Parts of the microfibrils, the micelles, are arranged in an orderly fashion and impart crystalline properties to the wall. ***(e)*** A fragment of a micelle shows parts of the chainlike cellulose molecules in a lattice arrangement.

Many Plant Cells Have a Secondary Wall in Addition to a Primary Wall

Plant cell walls vary greatly in thickness, depending partly on the role the cells play in the structure of the plant and partly on the age of the individual cell. The cellulosic layers formed first make up the **primary wall.** The region of union of the primary walls of adjacent cells is the **middle lamella** (also called intercellular substance). Many cells deposit additional wall layers. These form the **secondary wall.** If present, the secondary wall is laid down by the protoplast on the inner surface of the primary wall (Figure 3–31a).

The Middle Lamella Joins Adjacent Cells The middle lamella is composed mainly of pectins. Frequently, it is difficult to distinguish the middle lamella from the primary wall, especially in cells that develop thick secondary walls. In such cases, the two adjacent primary walls and the middle lamella, and perhaps the first layer of the secondary wall of each cell, may be called a *compound middle lamella.*

The Primary Wall Is Deposited while the Cell Is Increasing in Size The primary wall is deposited before and during the growth of the plant cell. As described above, primary walls are composed of cellulose, hemicelluloses, pectins, proteins (both glycoproteins and enzymes), and water. Primary walls may also contain lignin, suberin, or cutin.

Actively dividing cells commonly have only primary walls, as do most mature cells involved in such metabolic processes as photosynthesis, respiration, and secretion. These cells—that is, living cells with only primary walls—typically are able to lose their specialized cellular form, divide, and differentiate into new types of cells. For this reason, it is principally cells with only primary walls that are involved in wound healing and regeneration in the plant.

Usually, primary walls are not of uniform thickness throughout but have thin areas called **primary pit-fields** (Figure 3–32a). Cytoplasmic strands, or plasmodesmata, which connect the living protoplasts of adjacent cells, are commonly aggregated in the primary pit-fields, but they are not restricted to such areas.

The Secondary Wall Is Deposited after the Primary Wall Has Stopped Increasing in Size Although many plant cells have only a primary wall, in others the protoplast deposits a secondary wall inside the primary wall. Secondary wall formation occurs mostly after the cell has stopped growing and the primary wall is no longer increasing in surface area. Secondary walls are particularly important in specialized cells that have a strengthening function and in those involved in the conduction of water. Many of these cells die after the secondary wall has been laid down.

Cellulose is more abundant in secondary walls than in primary walls, and pectins may be lacking; the secondary wall is therefore rigid and not readily stretched. The matrix of the secondary wall is composed of hemicellulose. Structural proteins and enzymes, which are relatively abundant in primary walls, are absent in secondary walls.

Frequently, three distinct layers—designated S_1, S_2, and S_3, for the outer, middle, and inner layer, respectively—can be distinguished in a secondary wall (Figure 3–33). The layers differ

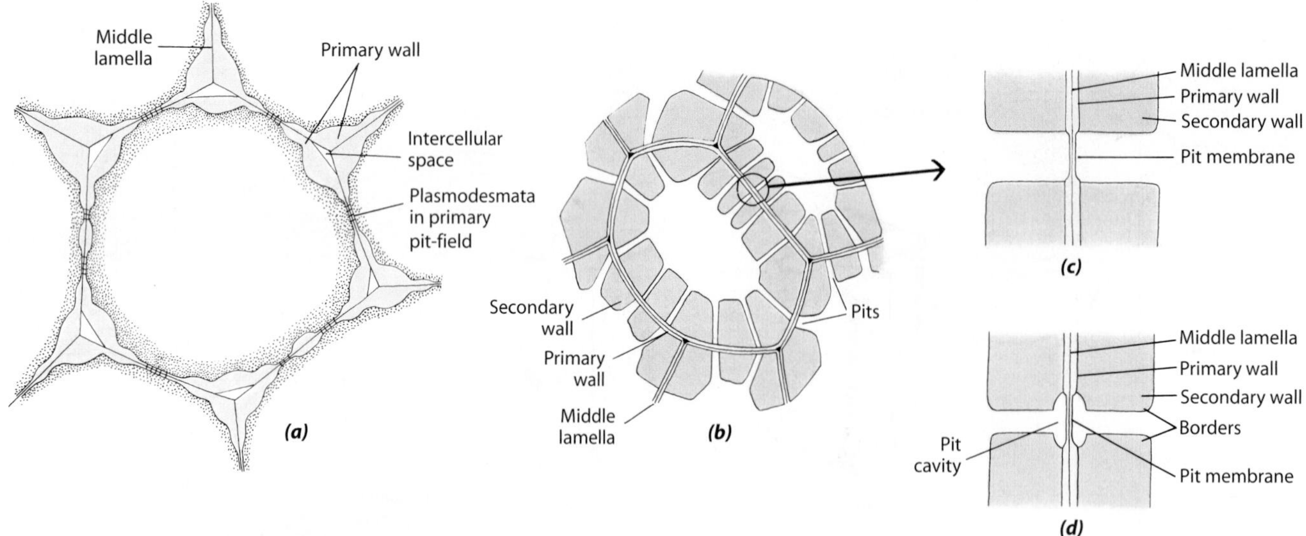

3–32 Primary pit-fields, pits, and plasmodesmata ***(a)*** Cells with primary walls and primary pit-fields, which are thin areas in the walls. As shown here, plasmodesmata commonly traverse the wall at the primary pit-fields. ***(b)*** Cells with secondary walls and numerous simple pits. ***(c)*** A simple pit-pair. ***(d)*** A bordered pit-pair.

from one another in the orientation of their cellulose microfibrils. Such multiple wall layers are found in certain cells of the secondary xylem, or wood. The laminated structure of these secondary walls greatly increases their strength, and the cellulose microfibrils are laid down in a denser pattern than in the primary wall. The secondary walls of cells found in wood commonly contain lignin.

Whereas the Primary Wall Has Pit-Fields, the Secondary Wall Has Pits When the secondary wall is deposited, it is not laid down over the primary pit-fields of the primary wall. Consequently, characteristic interruptions, or **pits,** are formed in the secondary wall (Figure 3–32b, c). In some instances, pits are also formed in areas where there are no primary pit-fields. Lignified secondary walls are not permeable to water, but, with the formation of pits, at least at those sites the adjacent cells are separated only by nonlignified primary walls.

A pit in a cell wall usually occurs opposite a pit in the wall of an adjoining cell. The middle lamella and two primary walls between the two pits are called the **pit membrane.** The two opposite pits plus the membrane constitute a **pit-pair.** Two principal types of pits are found in cells with secondary walls: **simple** and **bordered.** In bordered pits, the secondary wall arches over the **pit cavity.** In simple pits, there is no overarching. (See Chapter 23 for additional information on the properties of pit membranes in tracheary elements.)

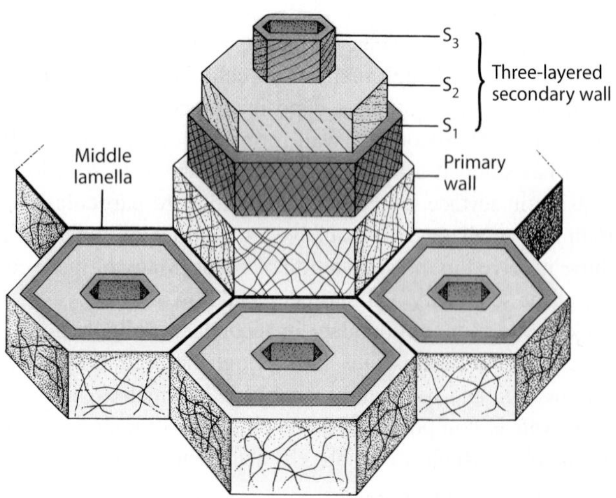

3–33 The layers of secondary cell walls Diagram showing the organization of the cellulose microfibrils and the three layers (S_1, S_2, S_3) of the secondary wall. The different orientations of the three layers strengthen the secondary wall.

Growth of the Cell Wall Involves Interactions among Plasma Membrane, Secretory Vesicles, and Microtubules

Cell walls grow both in thickness and in surface area. Extension of the wall is a complex process under the close biochemical control of the protoplast. During growth, the primary wall must yield enough to allow an appropriate extent of expansion, while at the same time remaining strong enough to constrain the protoplast. Growth of the primary wall requires loosening of wall structure, a phenomenon influenced by a novel class of wall proteins called **expansins** and by some hormones (see Chapter 27). There is also an increase in protein synthesis and in respiration to provide needed energy, as well as an increase in uptake of water by the cell. The new cellulose microfibrils are placed on top of those previously formed, layer upon layer.

In cells that enlarge more or less uniformly in all directions, the microfibrils are laid down in a random arrangement, forming an irregular network. In contrast, in elongating cells, the microfibrils of the side walls are deposited in a plane at right angles (perpendicular) to the axis of elongation (Figure 3–34).

Newly deposited cellulose microfibrils run parallel to the cortical microtubules lying just beneath the plasma membrane. It is generally accepted that the cellulose microfibrils are

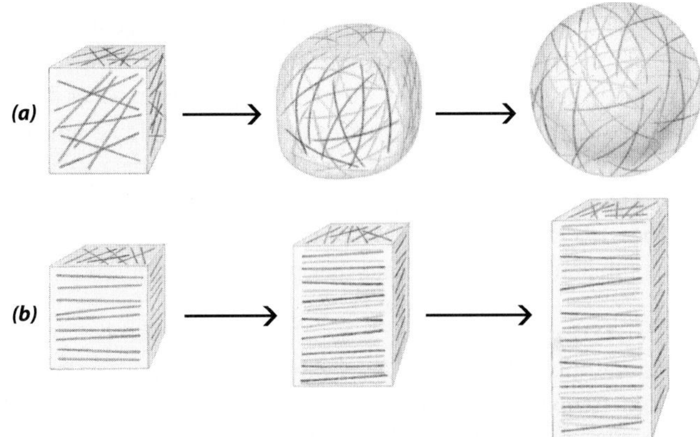

3–34 **Cell expansion** The orientation of cellulose microfibrils within the primary wall influences the direction of cell expansion. ***(a)*** If the cellulose microfibrils are randomly oriented in all walls, the cell will expand equally in all directions, tending to become spherical in shape. ***(b)*** If the microfibrils are oriented at right angles to the ultimate long axis of the cell, the cell will expand longitudinally along that axis.

synthesized by **cellulose synthase** complexes that occur in the plasma membrane (Figure 3–35). In seed plants, these enzyme complexes appear as rings, or rosettes, of six hexagonally arranged particles that span the membrane. During cellulose synthesis, the complexes, which move in the plane of the membrane, extrude the microfibrils onto the outer surface of the membrane. From there, the cellulose microfibrils are integrated into the cell wall. Movement of the enzyme complexes presumably is guided by the underlying cortical microtubules, although the mechanism by which the complexes are connected to the microtubules is unknown. The rosettes are inserted into the plasma membrane via secretory vesicles from the *trans*-Golgi network.

The matrix substances—hemicelluloses, pectins, and glycoproteins—are carried to the wall in secretory vesicles. The type of matrix substance synthesized and secreted by a cell at any given time depends on the stage of development. Pectins, for example, are more characteristic of enlarging cells, whereas hemicelluloses predominate in cells that are no longer enlarging. And, as mentioned previously, the best characterized glycoproteins, the extensins, are involved with the cell wall's extensibility.

Plasmodesmata Are Cytoplasmic Strands That Connect the Protoplasts of Adjacent Cells

As mentioned previously, the protoplasts of adjacent plant cells are connected with one another by **plasmodesmata** (singular: plasmodesma). Although such structures have long been visible with the light microscope (Figure 3–36), they were difficult to interpret. Not until they could be observed with an electron microscope was their nature as cytoplasmic strands confirmed.

Plasmodesmata may occur throughout the cell wall, or they may be aggregated in primary pit-fields or in the membranes between pit-pairs. When seen with the electron microscope, plasmodesmata appear as narrow, plasma-membrane-lined channels (about 30–60 nanometers in diameter) traversed by a modified tubule of endoplasmic reticulum known as the *desmotubule* (Figure 3–37). Many plasmodesmata are formed

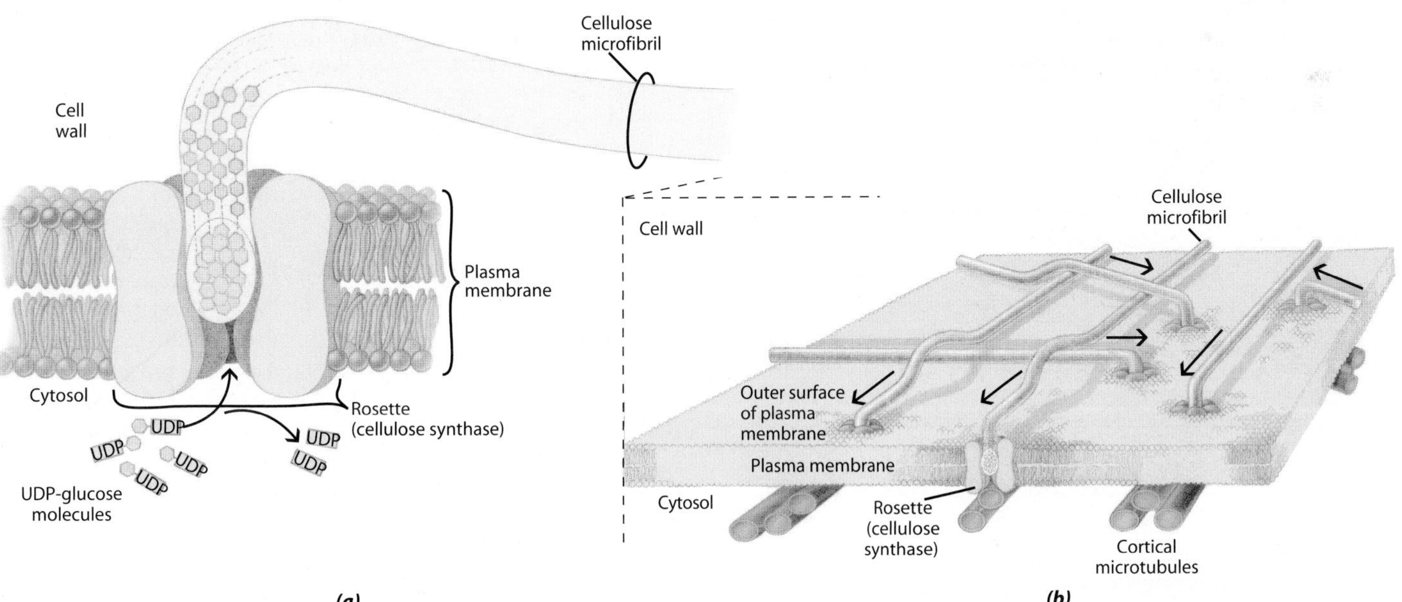

3–35 **Synthesis of cellulose microfibrils** Cellulose microfibrils are synthesized by enzyme complexes that move within the plane of the plasma membrane. ***(a)*** The enzymes are complexes of cellulose synthase that, in seed plants, form rosettes embedded in the plasma membrane. Each enzyme rosette, shown here in a longitudinal section, synthesizes cellulose from the glucose derivative UDP-glucose (uridine diphosphate glucose). The UDP-glucose molecules enter the rosette on the inner (cytoplasmic) face of the membrane, and a cellulose microfibril is extruded from the outer face of the membrane. ***(b)*** As the far ends of the newly formed microfibrils become integrated into the cell wall, the rosettes continue synthesizing cellulose, moving along a route (arrows) that parallels the cortical microtubules in the underlying cytoplasm.

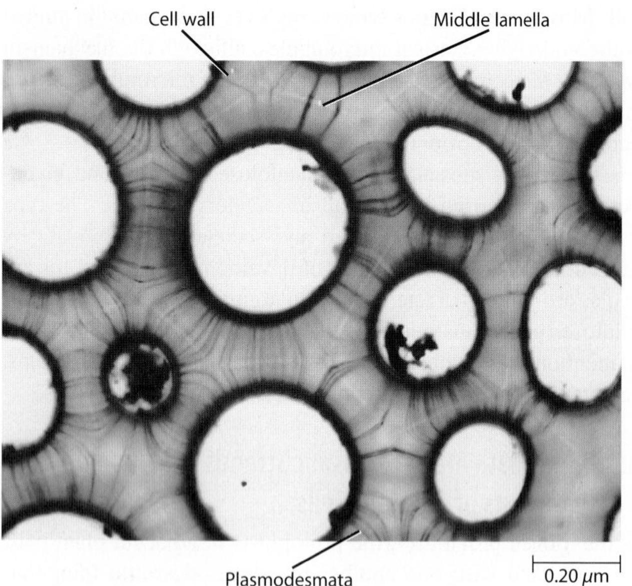

3–36 Plasmodesmata Light micrograph of plasmodesmata in the thick primary walls of persimmon *(Diospyros)* endosperm, the nutritive tissue within the seed. The plasmodesmata appear as fine lines extending from cell to cell across the walls. The middle lamella appears as a light line between these cells. Plasmodesmata generally are not discernible with the light microscope, but the extreme thickness of the persimmon endosperm cell walls greatly increases the length of the plasmodesmata, making them more visible.

during cell division as strands of tubular endoplasmic reticulum become trapped within the developing cell plate (see Figure 3–46). Plasmodesmata are also formed in the walls of nondividing cells. These structures provide a pathway for the transport of certain substances (sugars, amino acids, signaling molecules) between cells, and we discuss them in greater detail in Chapter 4 (see pages 88 to 90).

The Cell Cycle

Cells reproduce by a process known as **cell division,** in which the contents of a cell are divided between two **daughter cells.** In one-celled organisms, such as bacteria and many protists, cell

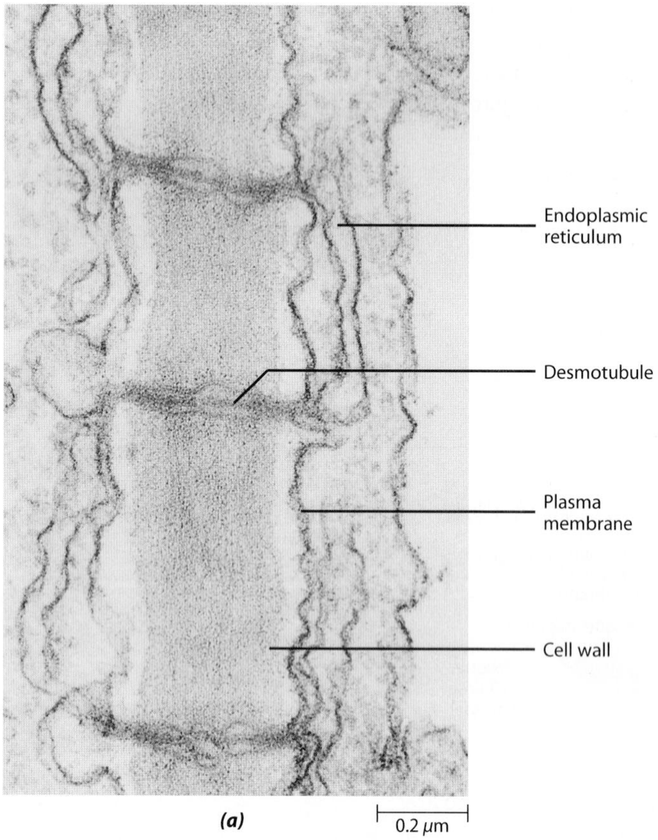

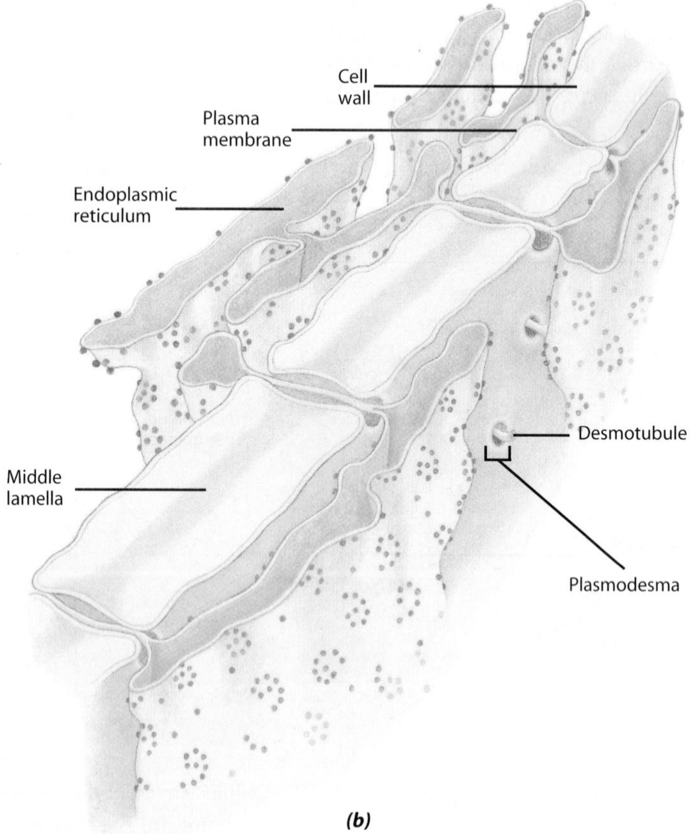

3–37 Plasmodesmata and desmotubules ***(a)*** Plasmodesmata connecting two leaf cells from the cottonwood *(Populus deltoides)* tree. ***(b)*** As seen with the electron microscope, plasmodesmata appear as narrow, plasma-membrane-lined channels in the walls, each traversed by a modified tubule of endoplasmic reticulum known as the desmotubule. Note the continuity of the endoplasmic reticulum on either side of the wall with the desmotubules of the plasmodesmata. The middle lamella between the adjacent primary walls is not discernible in the electron micrograph, as is common in such preparations.

division increases the number of individuals in a population. In many-celled organisms, such as plants and animals, cell division, together with cell enlargement, is the means by which the organism grows. It is also the means by which injured or worn-out tissues are repaired or replaced, most notably in animals.

The new cells are similar both in structure and function to the parent cell and to one another. They are similar, in part, because each new cell typically receives about half of each parent cell's cytoplasm. More important, in terms of structure and function, is that each new cell inherits an exact replica of the genetic, or hereditary, information of the parent cell. Therefore, before cell division can occur, all of the genetic information present in the parent cell nucleus must be faithfully replicated.

Cell division in eukaryotes consists of two overlapping, sequential stages: mitosis and cytokinesis. During **mitosis,** or nuclear division, a complete set of previously duplicated chromosomes is allocated to each of the two daughter cell nuclei. **Cytokinesis** is a process that divides the entire cell into two new cells. Each new cell contains not only a nucleus with a full chromosome complement but also approximately half of the cytoplasm of the parent cell.

Although mitosis and cytokinesis are the two events most commonly associated with the reproduction of eukaryotic cells, they represent the climax of a regular, repeated sequence of events known as the **cell cycle** (Figure 3–38). Commonly, the cell cycle is divided into interphase and mitosis. **Interphase** precedes and follows mitosis. It is a period of intense cellular activity, during which elaborate preparations are made for cell division, including chromosome duplication. Interphase can be divided into three phases, which are designated G_1, S, and G_2. Mitosis and cytokinesis together are referred to as the **M phase** of the cell cycle.

Some cell types pass through successive cell cycles during the life of the organism. This group includes the one-celled organisms and certain cell types in plants and animals. For example, the plant cells called **initials,** along with their immediate derivatives, or sister cells, constitute the **apical meristems** located at the tips of roots and shoots (see Figure 23–1). Meristems are embryonic tissue regions; most of the cell divisions in plants occur in and near the meristems. Initials may pause in their progress around the cell cycle in response to environmental factors, as during winter dormancy, and resume proliferation at a later time. This specialized resting, or dormant, state, during which the initials are arrested in the G_1 phase, is often called the **G_0 phase** (G-zero phase). Other fates include differentiation and programmed cell death.

Many, if not most, plant cells continue to replicate their DNA before they differentiate, a process called **endoreduplication,** or endoreplication. Depending on the cell type, one or more rounds of DNA synthesis may occur, sometimes resulting in gigantic nuclei with multiple copies of each gene. The multiple gene copies apparently provide a mechanism to increase the level of gene expression. The trichomes (epidermal hairs) of *Arabidopsis thaliana* undergo four endoreduplication cycles, increasing the DNA content sixteenfold beyond that of ordinary epidermal cells. The highly metabolic suspensor cells of developing common bean *(Phaseolus vulgaris)* embryos may have as many as 8192 copies of each gene.

In a multicellular organism, it is of critical importance that cells divide at an appropriate rate to produce the right number of cells needed for normal growth and development. In addition, the cell must have mechanisms that can sense whether certain conditions have been met before proceeding to the next phase. For example, replication of the DNA and synthesis of its associated

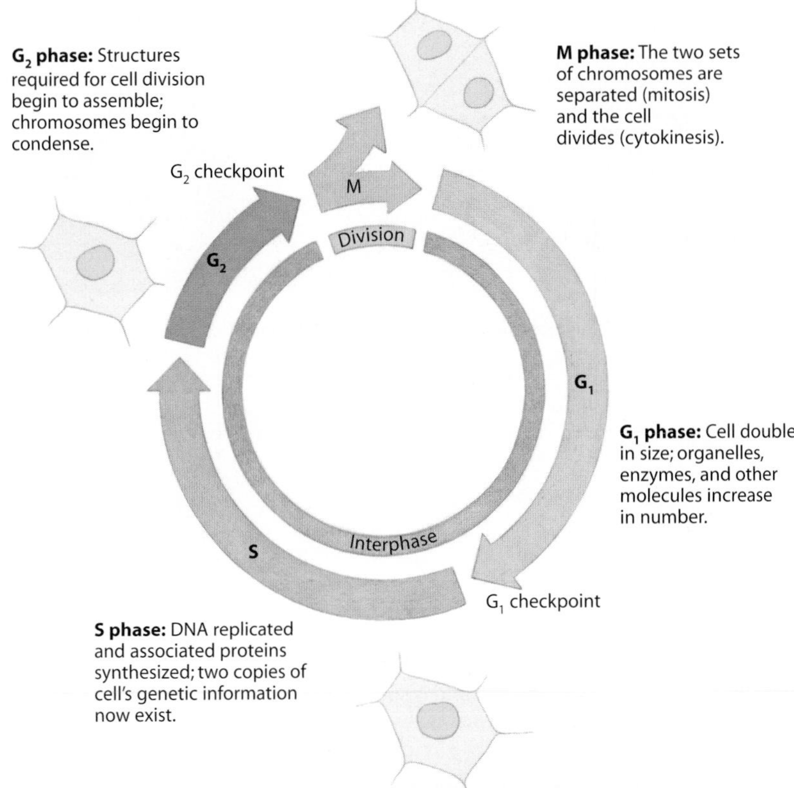
G_2 phase: Structures required for cell division begin to assemble; chromosomes begin to condense.
M phase: The two sets of chromosomes are separated (mitosis) and the cell divides (cytokinesis).
G_2 checkpoint
M
Division
G_2
G_1
G_1 phase: Cell doubles in size; organelles, enzymes, and other molecules increase in number.
Interphase
S
G_1 checkpoint
S phase: DNA replicated and associated proteins synthesized; two copies of cell's genetic information now exist.

3–38 The cell cycle Mitosis (the division of the nucleus) and cytokinesis (the division of the cytoplasm), which together constitute the M phase, take place after completion of the three preparatory phases (G_1, S, and G_2) of interphase. Progression of the cell cycle is mainly controlled at two checkpoints, one at the end of G_1 and the other at the end of G_2. In cells of different species or even of different tissues within the same organism, the various phases occupy different proportions of the total cycle.

proteins must be completed before the cell can proceed from the G_2 phase to mitosis and cytokinesis, or else daughter cells may receive incomplete sets of genetic information.

The nature of the control or controls that regulate the cell cycle is currently the subject of intense research; the control system is fundamentally conserved among all eukaryotic cells. In the typical cell cycle, progression is controlled at crucial transition points, called **checkpoints.** Two main checkpoints are near the end of G_1 and at the end of G_2 (Figure 3–38). It is at the G_1 checkpoint that the control system either arrests the cycle or initiates the S phase, which commits the cell to another round of division. At the G_2 checkpoint, the control system again either arrests the cycle or initiates mitosis. A third checkpoint, called the spindle assembly checkpoint, delays anaphase if some chromosomes are not properly attached to the mitotic spindle. Passage through the checkpoints depends on the successful activation and subsequent inactivation of protein kinases (enzymes that have the capacity to transfer phosphate groups from ATP to a specific amino acid) known as cyclin-dependent kinases. Activation of these kinases depends on their interaction with the regulatory protein known as cyclin and the subsequent phosphorylation of the kinase-cyclin complexes by other kinases.

Let us now examine some of the events that take place during each of the three phases of interphase.

Interphase

Before a cell can begin mitosis and actually divide, it must replicate its DNA and synthesize more of the proteins associated with the DNA in the chromosomes. Furthermore, it must produce sufficient quantities of organelles and other cytoplasmic components for two daughter cells, and assemble the structures needed to carry out mitosis and cytokinesis. These preparatory processes occur during interphase—that is, during the G_1, S, and G_2 phases of the cell cycle (Figure 3–38).

The key process of DNA replication occurs during the **S phase** (synthesis phase) of the cell cycle, at a time when many of the DNA-associated proteins, most notably histones, are also synthesized. G (gap) phases precede and follow the S phase.

The **G_1 phase,** which occurs after mitosis and precedes the S phase, is a period of intense biochemical activity. As the cell increases in size it synthesizes more enzymes, ribosomes, organelles, membrane systems, and other cytoplasmic molecules and structures.

In those cells that contain **centrioles**—that is, in most eukaryotic cells except those of fungi and plants—the centrioles begin to separate from one another and to duplicate. Centrioles, which are structures identical to the basal bodies of cilia and flagella (page 56), are surrounded by a cloud of amorphous material called the **centrosome.** The centrosome is also duplicated, so that each of the duplicated pairs of centrioles is surrounded by a centrosome.

The primary role of the **G_2 phase,** which follows the S phase and precedes mitosis, is to check that chromosome replication is complete and any DNA damage has been repaired. In centriole-containing cells, duplication of the centriole pair is completed, with the two mature centriole pairs lying just outside the nuclear envelope, somewhat separated from one another. By the end of interphase, the newly duplicated chromosomes, which are dispersed in the nucleus, begin to condense but are difficult to distinguish from the nucleoplasm.

Two Events That Occur in Interphase Are Unique to Plants

Before mitosis can begin, the nucleus must migrate to the center of the cell, if it is not already there. This migration ap-

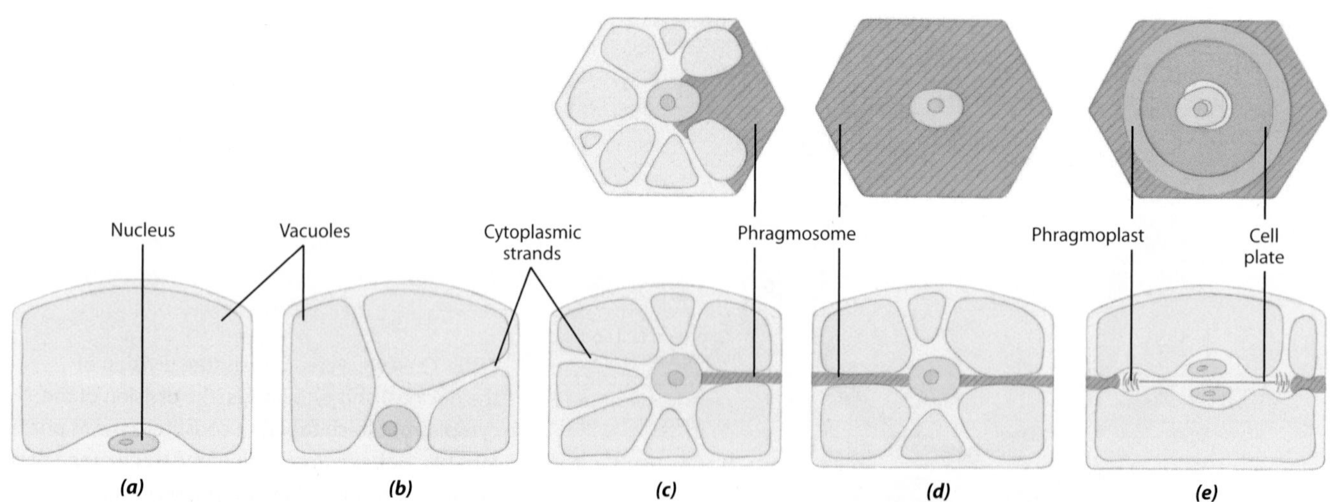

3–39 Cell division in a cell with a large vacuole ***(a)*** Initially, the nucleus lies along one wall of the cell, which contains a large central vacuole. ***(b)*** Strands of cytoplasm penetrate the vacuole, providing a pathway for the nucleus to migrate to the center of the cell. ***(c)*** The nucleus has reached the center of the cell and is suspended there by numerous cytoplasmic strands. Some of the strands have begun to merge to form the phragmosome through which cell division will take place. ***(d)*** The phragmosome, which forms a layer that bisects the cell, is fully formed. ***(e)*** When mitosis is completed, the cell will divide in the plane occupied by the phragmosome.

pears to start in the G_1 phase just before DNA replication and is especially discernible in plant cells with large vacuoles. In such cells, the nucleus initially becomes anchored in the center of the cell by cytoplasmic strands (Figure 3–39). Gradually, the strands merge to form a transverse sheet of cytoplasm that bisects the cell in the plane where it will ultimately divide. This sheet, called the **phragmosome,** contains both the microtubules and the actin filaments involved in its formation. Phragmosomes are clearly seen only in dividing cells containing large vacuoles.

In addition to the migration of the nucleus to the center of the cell, one of the earliest signs that a plant cell is about to divide is the appearance of a narrow, ringlike band of microtubules that lies just beneath the plasma membrane (see Figures 3–47b and 3–48b). This relatively dense band of microtubules encircles the nucleus in a plane corresponding to the equatorial plane of the future mitotic spindle. Because the band of microtubules appears during G_2 just before the first phase of mitosis (prophase), it is called the **preprophase band.** Actin filaments are aligned in parallel with the preprophase band microtubules. The preprophase band disappears after initiation of the mitotic spindle, long before initiation of the **cell plate.** The cell plate, the initial partition between the daughter cells, does not appear until telophase, the last phase of mitosis. Yet, as the cell plate forms, it grows outward to fuse with the cell wall of the parent cell precisely at the zone previously occupied by the preprophase band. The preprophase band therefore "predicts" the position of the future cell plate. Initially, callose is the principal cell wall polysaccharide of the developing cell plate; it gradually is replaced with cellulose and matrix components.

Mitosis and Cytokinesis

Mitosis is a continuous process but is conventionally divided into four major phases: prophase, metaphase, anaphase, and telophase (Figures 3–40 through 3–42). These four phases make up the process by which the genetic material synthesized during the S phase is divided equally between two daughter nuclei. Mitosis is followed by cytokinesis, during which the cytoplasm is divided and the two daughter cells separate.

During Prophase, the Chromosomes Shorten and Thicken

As viewed with a microscope, the transition from G_2 of interphase to **prophase,** the first phase of mitosis, is not a clearly defined event. During prophase, the chromatin, which is diffuse in the interphase nucleus, gradually condenses into well-defined chromosomes. Initially, however, the chromosomes appear as elongated threads scattered throughout the nucleus. (The threadlike appearance of the chromosomes when they first become visible is the source of the name "mitosis"; *mitos* is the Greek word for "thread.")

As prophase advances, the threads shorten and thicken, and as the chromosomes become more distinct, it becomes evident that each is composed of not one but two threads coiled about one another. During the preceding S phase, each chromosome duplicated itself; hence each chromosome now consists of two

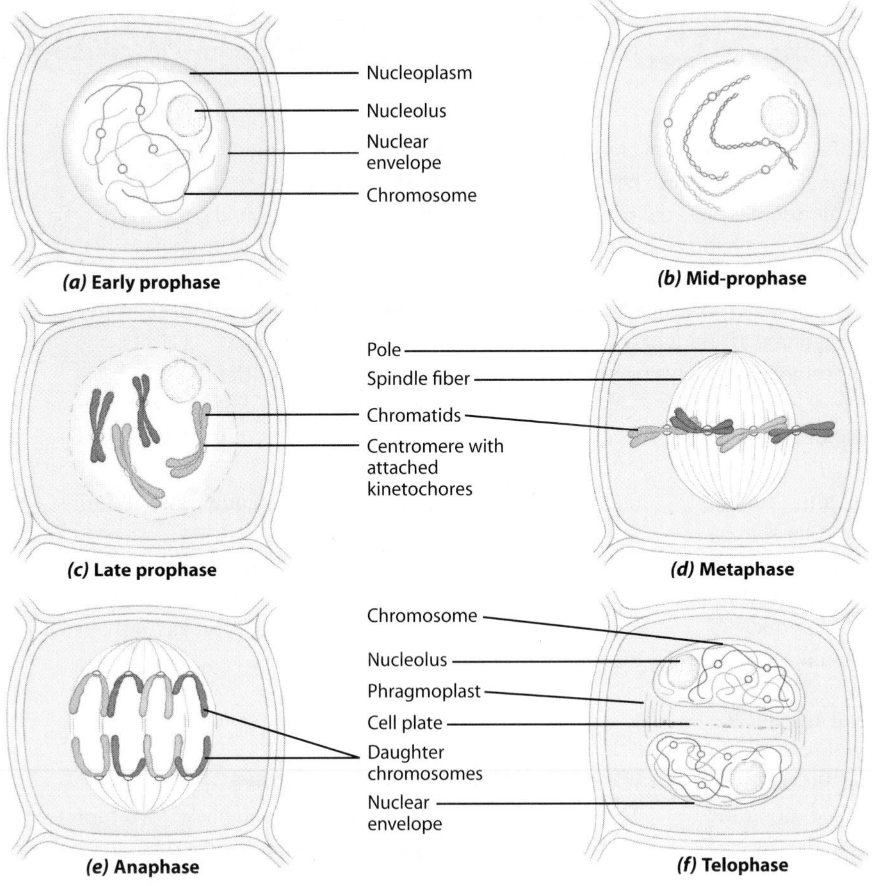

3–40 Mitosis, a diagrammatic representation ***(a)*** During early prophase, the four chromosomes shown here become visible as long threads scattered throughout the nucleus. ***(b)*** As prophase continues, the chromosomes shorten and thicken until each can be seen to consist of two threads (chromatids) attached to each other at their centromeres. ***(c)*** By late prophase, kinetochores develop on both sides of each chromosome at the centromere. Finally, the nucleolus and nuclear envelope disappear. ***(d)*** Metaphase begins with the appearance of the spindle in the area formerly occupied by the nucleus. During metaphase, the chromosomes migrate to the equatorial plane of the spindle. At full metaphase (shown here), the centromeres of the chromosomes lie on that plane.
(e) Anaphase begins as the centromeres of the sister chromatids separate. The sister chromatids, now called daughter chromosomes, then move to opposite poles of the spindle. ***(f)*** Telophase begins when the daughter chromosomes have completed their migration.

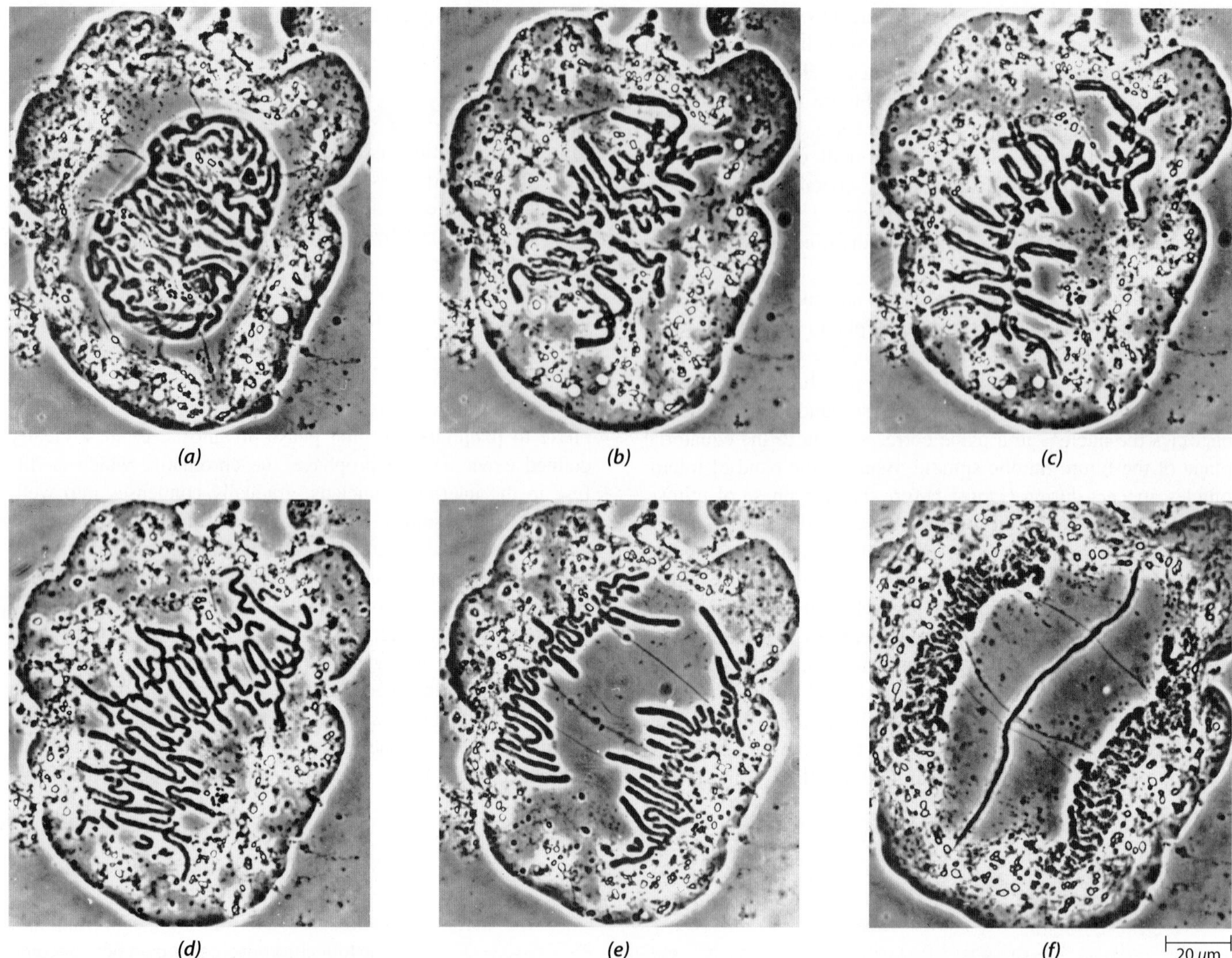

3–41 Mitosis in a living cell Phase-contrast optics of a cell of the African blood lily *(Haemanthus katherinae)* show the stages of mitosis. The spindle is barely discernible in these cells, which have been flattened to show all of the chromosomes more clearly. ***(a)*** Late prophase: the chromosomes have condensed. A clear zone has developed around the nucleus. ***(b)*** Late prophase–early metaphase: the nuclear envelope has disappeared, and the ends of some of the chromosomes are protruding into the cytoplasm. ***(c)*** Metaphase: the chromosomes are arranged with their centromeres on the equatorial plane. ***(d)*** Mid-anaphase: the sister chromatids (now called daughter chromosomes) have separated and are moving to opposite poles of the spindle. ***(e)*** Late anaphase. ***(f)*** Telophase–cytokinesis: the daughter chromosomes have reached the opposite poles, and the two chromosome masses have begun the formation of two daughter nuclei. Cell plate formation is nearly complete.

identical **sister chromatids.** By late prophase, after further shortening, the two chromatids of each chromosome lie side by side and nearly parallel, joined together along their length, with a constriction in a unique region called the **centromere** (Figure 3–43). The centromeres consist of specific DNA sequences needed to bind the chromosomes to the mitotic spindle, which forms during metaphase, the next phase of mitosis.

During prophase, a clear zone appears around the nuclear envelope (Figure 3–41a). Microtubules appear in this zone. At first the microtubules are randomly oriented, but by late prophase they are aligned parallel to the nuclear surface along the spindle axis. Called the **prophase spindle,** this is the earliest manifestation of mitotic spindle assembly, and it forms while the preprophase band is still present.

Toward the end of prophase, the nucleolus gradually becomes indistinct and disappears. Either simultaneously or shortly afterward, the nuclear envelope breaks down, marking the end of prophase.

During Metaphase, the Chromosomes Become Aligned on the Equatorial Plane of the Mitotic Spindle

The second phase of mitosis is metaphase. **Metaphase** begins as the **mitotic spindle,** a three-dimensional structure that

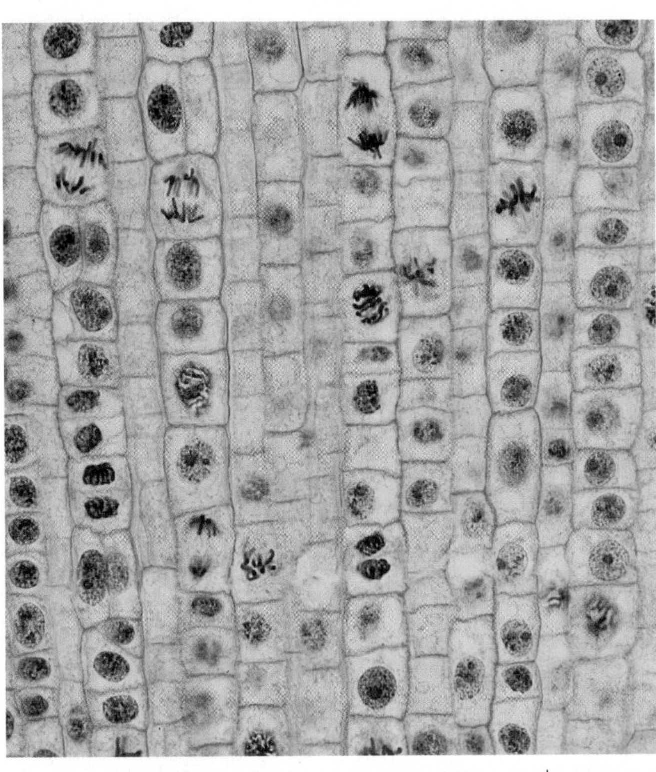

3–42 Dividing cells in a root tip By comparing these cells with the phases of mitosis illustrated in Figures 3–40 and 3–41, you should be able to identify the various mitotic phases shown in this photomicrograph of an onion *(Allium)* root tip.

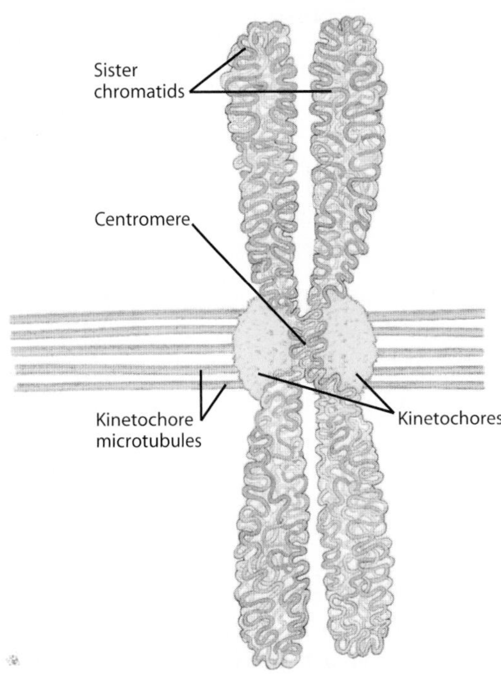

3–43 Fully condensed chromosome The chromosomal DNA was replicated during the S phase of the cell cycle. Each chromosome now consists of two identical parts, called sister chromatids, which are attached at the centromere, the constricted area in the center. The kinetochores are protein-containing structures, one on each chromatid, associated with the centromere. Attached to the kinetochores are microtubules that form part of the spindle.

is widest in the middle and tapers toward its poles, appears in the area formerly occupied by the nucleus (Figure 3–44). The spindle consists of **spindle fibers,** which are bundles of microtubules (see Figure 3–48c). With the sudden breakdown of the nuclear envelope, some of the spindle microtubules become attached to, or "captured" by, specialized protein complexes called **kinetochores** (Figures 3–43 and 3–44). These structures develop on both sides of each chromosome at the centromere, so that each chromatid has its own kinetochore. The attached spindle microtubules are called **kinetochore microtubules.** The remaining spindle microtubules, which extend from pole to pole, are called **polar microtubules.** The kinetochore microtubules extend to opposite poles from the sister chromatids of each chromosome.

Eventually, the kinetochore microtubules align the chromosomes midway between the spindle poles so that the kinetochores lie on the equatorial plane of the spindle. When all the chromosomes have moved to the equatorial plane, or metaphase plate, the cell has reached full metaphase. The chromatids are now in position to separate.

The Mitotic Spindle Consists of a Highly Organized Array of Kinetochore Microtubules and Polar Microtubules

As we described above, the mitotic spindle consists of two major classes of microtubules, the kinetochore microtubules and the polar microtubules, which are not attached to kinetochores (Figure 3–44). All spindle microtubules are oriented with one end (the minus end) at or near one of the poles and the other end (the plus end) pointing away from the pole. Some of the polar microtubules are relatively short, but most are long enough to overlap with polar microtubules from the opposite pole. Hence the mitotic spindle contains an array of microtubules and consists of two halves. Actin filaments are intermingled among the microtubules of the spindle and form an elastic "cage" around the spindle during mitosis.

During Anaphase, the Sister Chromatids Separate and, as Daughter Chromosomes, Move to Opposite Poles of the Spindle

The shortest phase of mitosis, **anaphase** (Figure 3–41d, e), begins abruptly with the simultaneous separation of all the sister chromatids at the centromeres. The sister chromatids are now called **daughter chromosomes** (Figures 3–40e and 3–41d). As the kinetochores of the daughter chromosomes move toward opposite poles, the arms of the chromosomes seem to drag behind. As anaphase continues, the two identical sets of chromosomes move rapidly toward the opposite poles of the spindle. By the end of anaphase, they have arrived at opposite poles (Figure 3–41e).

As the daughter chromosomes move apart, the kinetochore microtubules shorten by the loss of tubulin subunits, primarily

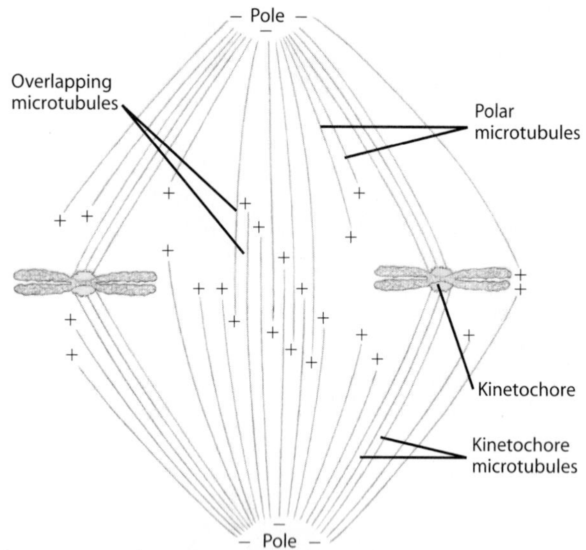

3–44 Mitotic spindle at metaphase The spindle consists of kinetochore microtubules and overlapping polar microtubules. Note that the minus ends of the microtubules are at or near the poles and the plus ends away from the poles. Following a tug-of-war, the chromosomes have come to lie on the equatorial plane.

at the kinetochore ends. It would seem, therefore, that movement of the chromosomes toward the poles results solely from shortening of the microtubules. Evidence indicates, however, that **motor proteins** (such as dynein or kinesin) use ATP to pull the chromosomes along the microtubules to the poles, while, at the same time, tubulin subunits are lost at the kinetochore.

During Telophase, the Chromosomes Lengthen and Become Indistinct

During **telophase** (Figure 3–41f), the separation of the two identical sets of chromosomes is completed as nuclear envelopes form around each set. The membranes of these nuclear envelopes are derived from vesicles of endoplasmic reticulum. The spindle apparatus disappears, and in the course of telophase, the chromosomes elongate and become slender threads again. Nucleoli re-form at this time. With telophase completed, the two daughter nuclei enter interphase.

The two daughter nuclei produced during mitosis are genetically identical to one another and to the nucleus that divided to produce them. This is important because the nucleus is the control center of the cell, as described in more detail in Chapter 9. The nucleus contains coded instructions that specify the production of proteins, many of which mediate cellular processes by acting as enzymes, whereas others serve directly as structural elements in the cell. This hereditary blueprint is faithfully transmitted to the daughter cells, and its precise distribution is ensured in eukaryotic organisms by the organization of chromosomes and their division during the process of mitosis.

The duration of mitosis varies with the tissue and the organism involved. However, prophase is always the longest phase and anaphase is always the shortest.

Cytokinesis in Plants Occurs by the Formation of a Phragmoplast and a Cell Plate

As we have noted, cytokinesis—the division of the cytoplasm—typically follows mitosis. In most organisms, cells divide by ingrowth of the cell wall, if present, and constriction of the plasma membrane, a process that cuts through the spindle fibers. In bryophytes and vascular plants and in a few algae, cell division occurs by formation of a cell plate, which starts in the middle of the cell and grows outward (Figures 3–45 through 3–48).

In early telophase, an initially barrel-shaped system of microtubules called the **phragmoplast** forms between the two daughter nuclei. The phragmoplast, like the mitotic spindle that preceded it, is composed of overlapping microtubules that form two opposing arrays on either side of the division plane. The phragmoplast also contains an extensive array of actin filaments, which parallel the microtubules but do not overlap. The cell plate is initiated as a disk suspended in the phragmoplast

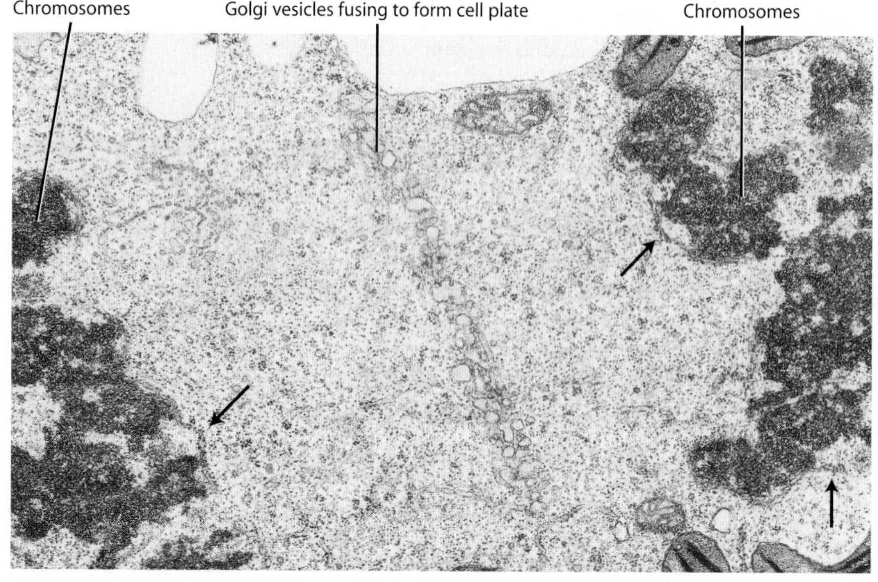

3–45 Cell plate formation In plant cells, separation of the daughter chromosomes is followed by formation of a cell plate, which completes the separation of the dividing cells. Here numerous Golgi vesicles can be seen fusing in an early stage of cell plate formation. The two groups of chromosomes on either side of the developing cell plate are at telophase. Arrows point to portions of the nuclear envelope reorganizing around the chromosomes.

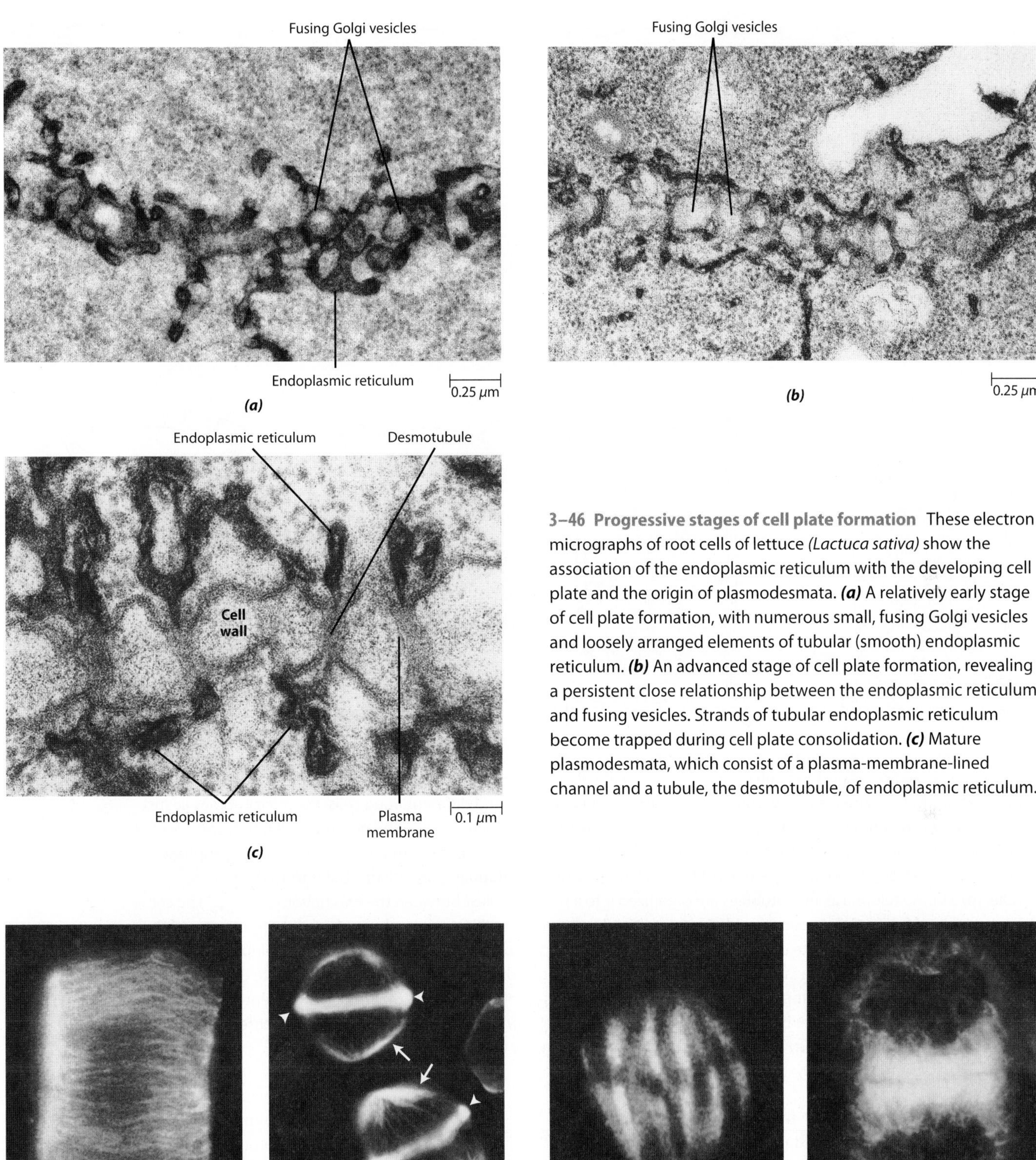

3–46 Progressive stages of cell plate formation These electron micrographs of root cells of lettuce *(Lactuca sativa)* show the association of the endoplasmic reticulum with the developing cell plate and the origin of plasmodesmata. ***(a)*** A relatively early stage of cell plate formation, with numerous small, fusing Golgi vesicles and loosely arranged elements of tubular (smooth) endoplasmic reticulum. ***(b)*** An advanced stage of cell plate formation, revealing a persistent close relationship between the endoplasmic reticulum and fusing vesicles. Strands of tubular endoplasmic reticulum become trapped during cell plate consolidation. ***(c)*** Mature plasmodesmata, which consist of a plasma-membrane-lined channel and a tubule, the desmotubule, of endoplasmic reticulum.

(a) **Interphase** *(b)* **Preprophase band and spindle** *(c)* **Mitotic spindle at metaphase** *(d)* **Phragmoplast at telophase**

3–47 Fluorescence micrographs of microtubular arrays in root tip cells of onion *(Allium cepa)*
(a) Prior to formation of the preprophase band of microtubules, most of the microtubules lie just beneath the plasma membrane. ***(b)*** A preprophase band of microtubules (arrowheads) encircles the nucleus at the site of the future cell plate. Other microtubules (arrows), forming the prophase spindle, outline the nuclear envelope, which itself is not visible. The lower right-hand cell is at a later stage than that above. ***(c)*** The mitotic spindle at metaphase. ***(d)*** During telophase, new microtubules form a phragmoplast, in which cell plate formation takes place.

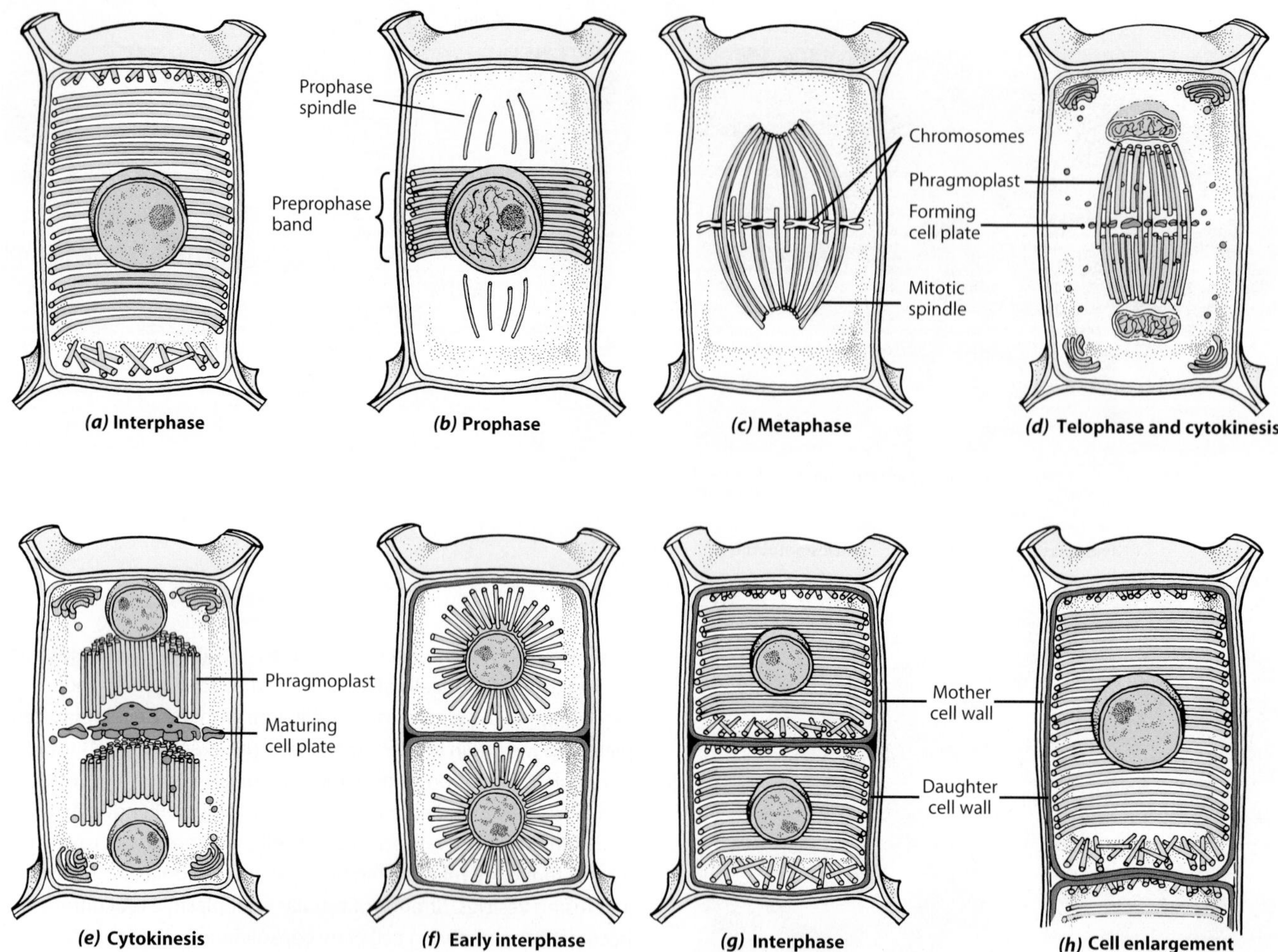

3–48 Microtubule arrays and the cell cycle Changes in the distribution of microtubules during the cell cycle and cell wall formation during cytokinesis. ***(a)*** During interphase, and in enlarging and differentiating cells, the microtubules lie just inside the plasma membrane. ***(b)*** Just before prophase, a ringlike band of microtubules, the preprophase band, encircles the nucleus in a plane corresponding to the equatorial plane of the future mitotic spindle, and microtubules of the prophase spindle begin to assemble on opposite sides of the nucleus. ***(c)*** During metaphase, the microtubules form the mitotic spindle. ***(d)*** During telophase, microtubules are organized into a phragmoplast between the two daughter nuclei. The cell plate, made up of fusing Golgi vesicles guided into position by the phragmoplast microtubules, forms at the equator of the phragmoplast. ***(e)*** As the cell plate matures in the center of the phragmoplast, the phragmoplast and developing cell plate grow outward until they reach the wall of the dividing cell. ***(f)*** During early interphase, microtubules radiate outward from the nuclear envelope into the cytoplasm. ***(g)*** Each sister cell forms its own primary wall. ***(h)*** With enlargement of the daughter cells (only the upper one is shown here), the mother cell wall is torn. In *(g)* and *(h)* the microtubules once more lie just inside the plasma membrane, where they play a role in the orientation of newly forming cellulose microfibrils.

(Figure 3–48d). At this stage, the phragmoplast does not extend to the walls of the dividing cell. Phragmoplast microtubules disappear where the cell plate has formed but are successively regenerated at the margins of the plate. The cell plate—preceded by the phragmoplast—grows outward until it reaches the walls of the dividing cell, completing the separation of the two daughter cells. In cells with large vacuoles, the phragmoplast and cell plate are formed within the phragmosome (Figure 3–39).

The cell plate is formed by a process involving the fusion of tubular outgrowths from secretory vesicles derived from the Golgi apparatus. Apparently, the vesicles are directed to the division plane by the phragmoplast microtubules and actin filaments, with the help of motor proteins. The vesicles deliver the hemicelluloses and pectins that form the cell plate. When the vesicles fuse, their membranes contribute to the formation of the plasma membrane on either side of the cell plate. Plasmodesmata are formed at this time as segments of smooth endoplasmic reticulum become entrapped between the fusing vesicles (Figure 3–46).

The developing cell plate fuses with the parent cell wall precisely at the position previously occupied by the preprophase band. Actin filaments have been found bridging the gap between the expanding perimeter of the phragmoplast and the cell wall, possibly providing an explanation as to how the expanding phragmoplast is directed to the site of the former preprophase band.

After the cell plate contacts the parent cell walls, a middle lamella develops within it. Each daughter cell then deposits a new layer of primary wall around the entire protoplast. The original wall of the parent cell stretches and ruptures as the daughter cells enlarge (Figure 3–48h).

SUMMARY

The Cell Is the Fundamental Unit of Life

All living matter is composed of cells. Although extremely varied in structure and function, all cells are remarkably similar in their basic structure. All cells possess an outer membrane, known as the plasma membrane, that isolates the cell's contents from the external environment. Enclosed within this membrane are the cytoplasm and the hereditary information in the form of DNA.

Cells Are of Two Fundamentally Different Types: Prokaryotic and Eukaryotic

Prokaryotic cells lack nuclei and membrane-surrounded organelles. Prokaryotic cells are represented today by the archaea and bacteria. The prokaryotic chromosome consists of a single circular molecule of DNA localized in the nucleoid. Eukaryotic cells have DNA contained within a true nucleus and have distinct organelles that perform different functions.

Plant Cells Typically Consist of a Cell Wall and a Protoplast

The protoplast consists of the cytoplasm and a nucleus. The plasma membrane is the outer boundary of the protoplast, next to the cell wall. The cytosol, or cytoplasmic matrix, of individual plant cells is always in motion, a phenomenon known as cytoplasmic streaming.

The Nucleus Is Surrounded by a Nuclear Envelope and Contains Nucleoplasm, Chromatin, and One or More Nucleoli

The nucleus is the control center of the cell and contains most, though not all, of the cell's genetic information; that contained in the nucleus is referred to as the nuclear genome. The nuclear envelope, which consists of a pair of membranes, may be considered a specialized, locally differentiated portion of the endoplasmic reticulum. The chromatin (chromosomes) consists of DNA and histone proteins. The nucleoli are the sites of ribosomal subunit formation.

Ribosomes Are the Sites of Protein Synthesis

Ribosomes, which are found both free in the cytosol and attached to the endoplasmic reticulum and the outer surface of the nuclear envelope, are the sites where amino acids are linked together to form proteins. During protein synthesis, the ribosomes occur in clusters called polysomes.

There Are Three Main Types of Plastids: Chloroplasts, Chromoplasts, and Leucoplasts

Plastids are characteristic components of plant cells. Each plastid is surrounded by an envelope consisting of two membranes. Mature plastids are classified, in part, on the basis of the kinds of pigments they contain: chloroplasts contain chlorophyll and carotenoid pigments; chromoplasts contain carotenoid pigments; and leucoplasts are nonpigmented. Proplastids are the precursors of plastids.

Mitochondria Are the Sites of Respiration

Mitochondria, like plastids, are organelles surrounded by two membranes. The inner membrane is folded to form an extensive inner membrane system, increasing the surface area available to enzymes and the reactions associated with them. Mitochondria are the principal sites of respiration in eukaryotic cells.

Plastids and Mitochondria Share Certain Features with Prokaryotic Cells

Both plastids and mitochondria are semiautonomous organelles. Their DNA is found in nucleoids, and they possess small ribosomes similar to those of bacteria. Plastids and mitochondria originated as bacteria that were engulfed within larger heterotrophic cells. Plant cells contain three genomes: those of the nucleus, the plastids, and the mitochondria.

Peroxisomes Are Surrounded by a Single Membrane

Unlike plastids and mitochondria, peroxisomes are organelles surrounded by a single membrane. Moreover, they possess neither DNA nor ribosomes. Some peroxisomes play an important role in photorespiration. Others are involved in the conversion of stored fats to sucrose during seed germination.

Vacuoles Perform a Variety of Functions

Vacuoles are organelles bounded by a single membrane called the tonoplast. Together with plastids and cell walls, they are characteristic components of plant cells. Many vacuoles are filled with cell sap, an aqueous solution containing a variety of salts, sugars, anthocyanin pigments, and other substances. Vacuoles play an important role in cell enlargement and the maintenance of tissue rigidity. Some vacuoles are important storage compartments for primary metabolites, whereas others sequester toxic secondary metabolites. In addition, many vacuoles are involved in the breakdown of macromolecules and the recycling of their components within the cell.

The Endoplasmic Reticulum Is an Extensive Three-Dimensional System of Membranes with a Variety of Roles

The endoplasmic reticulum exists in two forms: rough endoplasmic reticulum, which is studded with ribosomes, and smooth endoplasmic reticulum, which lacks ribosomes. The rough form is involved with membrane and protein synthesis, and the smooth form with lipid synthesis. Some of the lipids produced occur as oil bodies in the cytosol.

SUMMARY TABLE Plant Cell Components

MAJOR COMPONENTS	INDIVIDUAL CONSTITUENTS	KEY DESCRIPTIVE FEATURES	FUNCTION(S)
Cell wall		Consists of cellulose microfibrils embedded in a matrix of hemicelluloses, pectins, and glycoproteins. Lignin, cutin, suberin, and waxes may also be present.	Strengthens the cell; determines cell size and shape.
	Middle lamella	Pectin-rich layer between cells.	Cements adjacent cells together.
	Primary wall	First wall layers to form. Contains primary pit-fields.	Found in actively dividing and actively metabolizing cells.
	Secondary wall	Formed in some cells after the primary wall is laid down. Located interior to the primary wall. Contains pits.	Found in cells with strengthening and/or water-conducting functions. Is rigid and thus imparts added strength.
	Plasmodesmata	Cytoplasmic strands traversing the cell wall.	Interconnect protoplasts of adjacent cells, providing a pathway for the transport of substances between cells.
Nucleus		Surrounded by a pair of membranes, forming the nuclear envelope; contains nucleoplasm, nucleoli, and chromatin (chromosomes) consisting of DNA and histone proteins.	Controls cellular activities. Stores genetic information.
Plasma membrane		Single membrane, forming the outer boundary of the protoplast.	Mediates transport of substances into and out of the cell. Site of cellulose synthesis. Receives and transmits hormonal and environmental signals.
Cytoplasm	Cytosol cytoplasm.	The least differentiated part of the are suspended.	Matrix in which organelles and membrane systems
	Plastids	Surrounded by a double-membrane envelope. Semiautonomous organelles containing their own DNA and ribosomes.	Sites of food manufacture and storage.
	Chloroplasts	Contain chlorophyll and carotenoid pigments embedded in thylakoid membranes.	Sites of photosynthesis. Involved in amino acid synthesis and fatty acid synthesis. Temporary storage of starch.
	Chromoplasts	Contain carotenoid pigments.	May function in attracting insects and other animals essential for cross-pollination and fruit and seed dispersal.
	Leucoplasts	Lack pigments entirely.	Some (amyloplasts) store starch; others form oils.
	Proplastids	Undifferentiated plastids; may form prolamellar bodies.	Precursors of other plastids.

The Golgi Apparatus Is a Highly Polarized Membrane System Involved in Secretion

The Golgi apparatus consists of Golgi bodies, which are composed of stacks of flattened, disk-shaped sacs, or cisternae. Most plant Golgi bodies are involved in the synthesis and secretion of complex noncellulosic cell wall polysaccharides, which are transported to the surface of the cell in secretory vesicles derived from the *trans*-Golgi network. All cellular membranes, with the exception of mitochondrial, plastid, and peroxisomal membranes, constitute a continuous, interconnected system, known as the endomembrane system.

The Cytoskeleton Is Composed of Microtubules and Actin Filaments

The cytosol of eukaryotic cells is permeated by the cytoskeleton, a complex network of protein filaments, of which there are two well-characterized types in plant cells: microtubules and actin filaments. Microtubules are thin, cylindrical structures of variable length and are composed of subunits of the protein tubulin. They play a role in cell division, the growth of the cell wall, and the movement of flagella. Actin filaments are composed of the protein actin. These long filaments occur singly and in bundles, which play a role in cytoplasmic steaming.

The Cell Wall Is the Major Distinguishing Feature of the Plant Cell

The cell wall determines the structure of the cell, the texture of plant tissues, and many important characteristics that distinguish plants as organisms. Cellulose is the principal component of plant cell walls. All plant cells have a primary wall. In addition, many also have a secondary wall, which is laid down on the inner surface of the primary wall. The region of union of the primary walls of adjacent cells is a pectin-rich layer called the middle lamella. The cellulose microfibrils of primary walls occur in a cross-linked matrix of noncellulosic molecules, including hemicelluloses, pectins, and glycoproteins. Because

MAJOR COMPONENTS	INDIVIDUAL CONSTITUENTS	KEY DESCRIPTIVE FEATURES	FUNCTION(S)
Cytoplasm	Mitochondria	Surrounded by a double-membrane envelope. The inner membrane is folded into cristae. Semiautonomous organelles, containing their own DNA and ribosomes.	Sites of cellular respiration.
	Peroxisomes	Surrounded by a single membrane. Sometimes contain crystalline protein bodies.	Contain enzymes for a variety of processes such as photorespiration and conversion of fats to sucrose.
	Vacuoles	Surrounded by a single membrane (the tonoplast); may take up most of the cell volume.	Some filled with cell sap, which is mostly water. Often contain anthocyanin pigments; store primary and secondary metabolites; break down and recycle macromolecules.
	Ribosomes	Small, electron-dense particles, consisting of RNA and protein.	Sites of protein synthesis.
	Oil bodies	Have an amorphous appearance.	Sites of lipid storage, especially triglycerides.
	Endoplasmic reticulum	A continuous, three-dimensional membrane system that permeates the entire cytosol.	Multiple functions, including protein synthesis (rough endoplasmic reticulum) and lipid synthesis (smooth. endoplasmic reticulum.) Channels materials throughout the cell.
	Golgi apparatus	Collective term for the cell's Golgi bodies, stacks of flattened, membranous sacs.	Processes and packages substances for secretion and for use within the cell.
	Endomembrane system	The continuous, interconnected system of cellular membranes, consisting of the endoplasmic reticulum, Golgi apparatus, *trans*-Golgi network, plasma membrane, nuclear envelope, tonoplast, and various vesicles.	A dynamic network in which membranes and various substances are transported throughout the cell. See functions of various components.
	Cytoskeleton	Complex network of protein filaments, consisting of microtubules and actin filaments.	Involved in cell division, growth, and differentiation.
	Microtubules	Dynamic, cylindrical structures composed of tubulin.	Involved in many processes, such as cell plate formation, deposition of cellulose microfibrils, and directing the movement of Golgi vesicles and chromosomes.
	Actin filaments (microfilaments)	Dynamic, filamentous structures composed of actin.	Involved in many processes, including cytoplasmic streaming and the movement of the nucleus and organelles.

of the presence of pectins, primary walls are highly hydrated, making them more plastic. Actively dividing and elongating cells commonly have only primary walls. Secondary walls contain hemicelluloses but apparently lack pectins and glycoproteins. Lignin may also be present in primary walls but is especially characteristic of cells with secondary walls. Lignin adds compressive strength and rigidity to the wall.

The protoplasts of adjacent cells are connected to one another by cytoplasmic strands called plasmodesmata, which traverse the cell walls. Plasmodesmata provide pathways for the transport of certain substances between cells. Callose, a widely distributed cell wall polysaccharide, is deposited rapidly in response to wounding, sealing the plasmodesmata.

Cells Reproduce by Cell Division

During cell division, the cellular contents are apportioned between two new daughter cells. The new cells are structurally and functionally similar both to the parent cell and to one another, because each new cell inherits an exact replica of the genetic information of the parent cell. In eukaryotes, cell division consists of two overlapping stages: mitosis (the division of the nucleus) and cytokinesis (the division of the cytoplasm).

Dividing Eukaryotic Cells Pass through a Regular Sequence of Events Known as the Cell Cycle

Commonly, the cell cycle is divided into interphase and mitosis. Interphase consists of three phases (G_1, S, and G_2), which are the preparatory phases of the cycle. During the G_1 phase, the cell doubles in size. This size increase is accompanied by an increase in numbers of cytoplasmic molecules and structures. DNA replication occurs only during the S phase, resulting in duplication of the chromosomes. The primary role of the G_2 phase is to make sure that chromosome replication is complete and to allow for the repair of damaged DNA. Progression through the cycle is mainly controlled at two crucial checkpoints, one at the transition from G_1 to S and the other at the transition from G_2 to the initiation of mitosis. Mitosis and cytokinesis together are referred to as the M phase of the cell cycle.

During Prophase, the Duplicated Chromosomes Shorten and Thicken

When the cell is in interphase, the chromosomes are in an uncoiled state and are difficult to distinguish from the nucleoplasm. Mitosis in plant cells is preceded by migration of the nucleus to the center of the cell and the appearance of the preprophase band, a dense band of microtubules that marks the equatorial plane of the future mitotic spindle. As prophase of mitosis begins, the chromatin gradually condenses into well-defined chromosomes, each chromosome consisting of identical strands called sister chromatids, held together at the centromere. Simultaneously, the spindle begins to form.

Metaphase, Anaphase, and Telophase Followed by Cytokinesis Result in Two Daughter Cells

Prophase ends with the breakdown of the nuclear envelope and the disappearance of the nucleolus. During metaphase, the chromatid pairs, maneuvered by kinetochore microtubules of the mitotic spindle, come to lie in the center of the cell, with their centromeres on the equatorial plane. During anaphase, the sister chromatids separate and, as daughter chromosomes, move to opposite poles of the spindle. During telophase, the separation of the two identical sets of chromosomes is made final as nuclear envelopes are formed around each set. Nucleoli also re-form at this time.

Mitosis is generally followed by cytokinesis, the division of the cytoplasm. In plants and certain algae, the cytoplasm is divided by a cell plate that begins to form during mitotic telophase. The developing cell plate arises from fusing Golgi vesicles guided to the division plane by microtubules and actin filaments of the phragmoplast, an initially barrel-shaped system of microtubules that forms between the two daughter nuclei in early telophase.

QUESTIONS

1. Distinguish between the cell theory and organismal theory.
2. What three features of plant cells distinguish them from animal cells?
3. Both plastids and mitochondria are said to be “semiautonomous” organelles. Explain.
4. Once regarded as depositories for waste products in plant cells, vacuoles now are known to play many different roles. What are some of those roles?
5. Explain the phenomenon of autumn leaf coloration.
6. Distinguish between rough endoplasmic reticulum and smooth endoplasmic reticulum, both structurally and functionally.
7. Distinguish between microtubules and actin filaments. With what functions are each of these protein filaments associated?
8. Using the following terms, explain the process of cell wall growth and cellulose deposition in expanding cells: cellulose microfibrils, cellulose synthase complexes (rosettes), cortical microtubules, secretory vesicles, matrix substances, plasma membrane.
9. In what sense do plants differ from animals with respect to the location of most cell division activity?
10. In the typical cell cycle there are checkpoints. What are these checkpoints? What purpose do they serve?
11. Distinguish between a centromere and a kinetochore.
12. What is the preprophase band? What role does it play in plant cell division?

CHAPTER 4

The Movement of Substances into and out of Cells

Standing on water Water, with its high surface tension, supports two mating water striders. The surface tension is due to the attraction of the water molecules to each other, caused by hydrogen bonding, which produces a strong yet elastic "skin" at the water's surface. Specially adapted long, slender legs with tiny hairs allow the water striders to walk on the water's surface.

CHAPTER OUTLINE

All cells are separated from their surroundings by a surface membrane—the plasma membrane. Eukaryotic cells are further divided internally by a variety of membranes, including the endoplasmic reticulum, Golgi bodies, and the bounding membranes of organelles (Figure 4–1). These membranes are not impermeable barriers, for cells are able to regulate the amount, kind, and often the direction of movement of substances that pass across their membranes. This is an essential capability of living cells, because few metabolic processes could occur at reasonable rates if they depended on the concentrations of necessary substances found in the cell's surroundings.

As we shall see, membranes also permit differences in electrical potential, or voltage, to become established between the cell and its external environment and between adjacent compartments of the cell. Differences in chemical concentration (of various ions and molecules) and electrical potential across membranes are forms of potential, or stored, energy that are essential to many cellular processes. In fact, the existence of these differences is a criterion by which we may distinguish living organisms from the surrounding nonliving environment.

Of the many kinds of molecules surrounding and contained within the cell, by far the most common is water. Further, most of the other molecules and ions important to the life of the cell (Figure 4–2) are dissolved in water. Therefore, let us begin our consideration of transport across cellular membranes by looking at how water moves.

CHECKPOINTS

After reading this chapter, you should be able to answer the following:

1. What is water potential, and what value does the concept of water potential have for plant physiologists?
2. Distinguish between diffusion and osmosis. What kinds of substances enter and leave cells by each process?
3. What is the basic structure of cellular membranes?
4. What are transport proteins, and of what importance are they to plant cells?
5. What are the similarities and differences between facilitated diffusion and active transport?
6. What role is played by vesicle-mediated transport? Compare movement out of the cell with movement into the cell.
7. What are the roles of signal transduction and plasmodesmata in cell-to-cell communication?

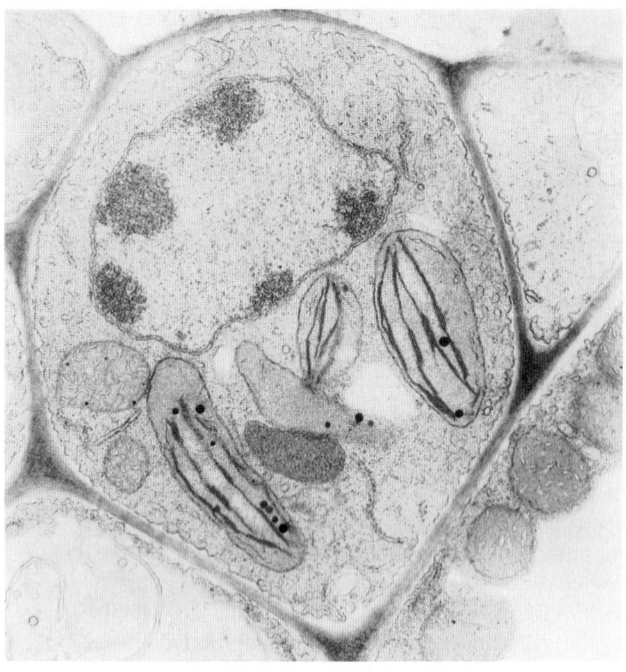

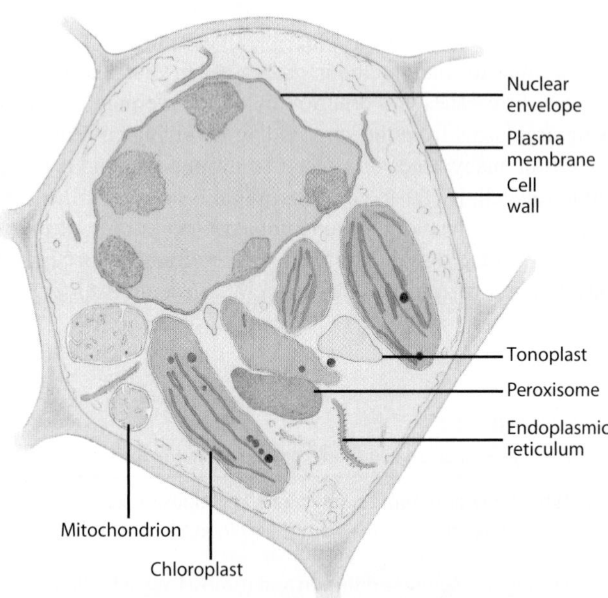

4–1 Cellular membranes In addition to the plasma membrane, which controls the movement of substances into and out of the cell, numerous internal membranes control the passage of substances within the cell. Membranes surround the nucleus, chloroplasts, and mitochondria (all three bounded by a pair of membranes), as well as peroxisomes and vacuoles (both bounded by a single membrane, which for vacuoles is known as the tonoplast). In addition, the endoplasmic reticulum (segments of which are seen here) is made up of membranes. This electron micrograph is of a leaf cell of *Moricandia arvensis,* a crucifer (cabbage family).

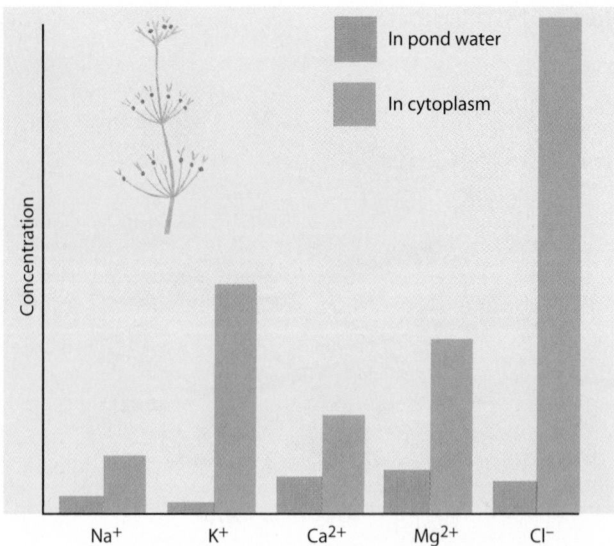

4–2 Relative concentrations of ions in cytoplasm and typical pond water Among the most important ions in many living cells are sodium (Na^+), potassium (K^+), calcium (Ca^{2+}), magnesium (Mg^{2+}), and chloride (Cl^-). The differences in concentrations of these ions in the cytosol of the green alga *Nitella* and in the surrounding pond water indicate that cells regulate their exchanges of materials with the surrounding environment. This regulation is accomplished by the plasma membrane.

Principles of Water Movement

The movement of water, whether in living organisms or in the nonliving world, is governed by three basic processes: bulk flow, diffusion, and osmosis.

Bulk Flow Is the Overall Movement of a Liquid

In **bulk flow** (or mass flow), the molecules of water (or some other liquid) move all together from one place to another because of differences in potential energy. **Potential energy** is the stored energy an object—or a collection of objects, such as a collection of water molecules—possesses because of its position. The potential energy of water is usually referred to as **water potential.**

Water moves from a region of higher water potential to one of lower water potential, regardless of the reason for the difference in potential. A simple example is water running downhill in response to gravity. Water at the top of a hill has more potential energy (that is, a higher water potential) than water at the bottom of a hill (Figure 4–3). As the water runs downhill, its potential energy can be converted to mechanical energy by a waterwheel or to mechanical and then electrical energy by a hydroelectric turbine.

Pressure is another source of water potential. If we put water into a rubber bulb and squeeze the bulb, this water, like the water at the top of a waterfall, has water potential, and it will move to an area of lower water potential. Can we make the water that is running downhill run uphill by means of pressure? Yes, we can, but only so long as the water potential produced by the pressure exceeds the water potential produced by gravity.

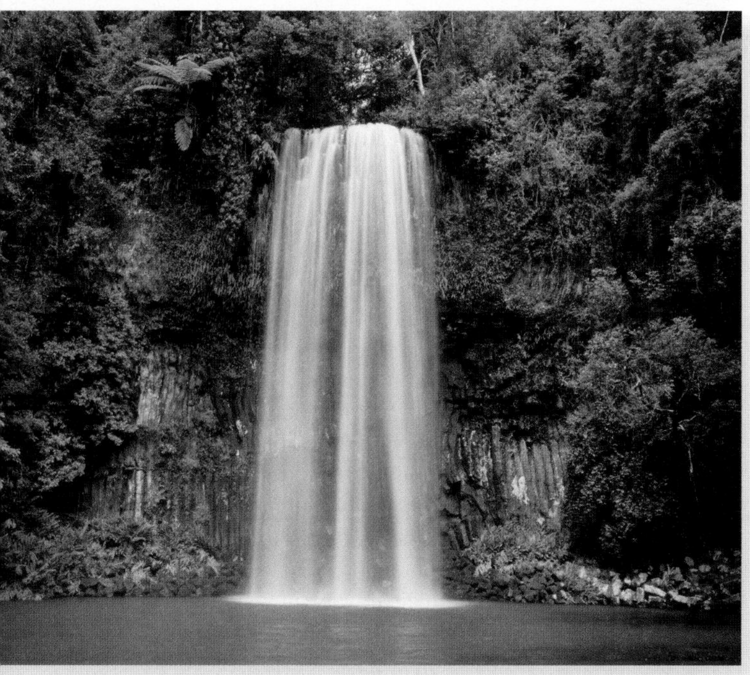

4–3 Potential energy of a waterfall Water at the top of a falls, like a boulder on a hilltop, has potential energy. The movement of water molecules as a group, as from the top of the falls (high water potential) to the bottom (low water potential), is referred to as bulk flow.

Pressure-driven bulk flow is the predominant mechanism responsible for the long-distance transport of sap, which is an aqueous solution of sugars and other solutes. This sap moves in the conduits of the food-conducting tissue—the phloem—of plants. (The **solutes** are the substances dissolved in a solution; the water is the **solvent** of the solution.) The sap moves by bulk flow from the leaves, where the sugars and other solutes are produced, to other parts of the plant body, where they are used for maintenance and growth.

The concept of water potential is useful because it enables plant physiologists to predict how water will move in the plant under various conditions. Water potential is usually measured in terms of the pressure required to stop the movement of water—that is, the **hydrostatic pressure**—under the particular circumstances involved. The units used to express this pressure are called pascals (Pa) or, more conveniently, megapascals (MPa). By convention, the water potential of pure water is set at zero. Under these conditions, the water potential of an aqueous solution of a substance will therefore have a negative value (less than zero), because a higher solute concentration results in lower water potential.

Diffusion Results in the Uniform Distribution of a Substance

Diffusion is a familiar phenomenon. If a few drops of perfume are sprinkled in one corner of a room, the scent will eventually permeate the entire room, even if the air is still. If a few drops of dye are put in one end of a tank of water, the dye molecules will slowly become evenly distributed throughout the tank (Figure 4–4). This process may take a day or more, depending on the size of the tank, the temperature, and the size of the dye molecules.

Why do the dye molecules move apart? If you could observe the individual dye molecules in the tank, you would

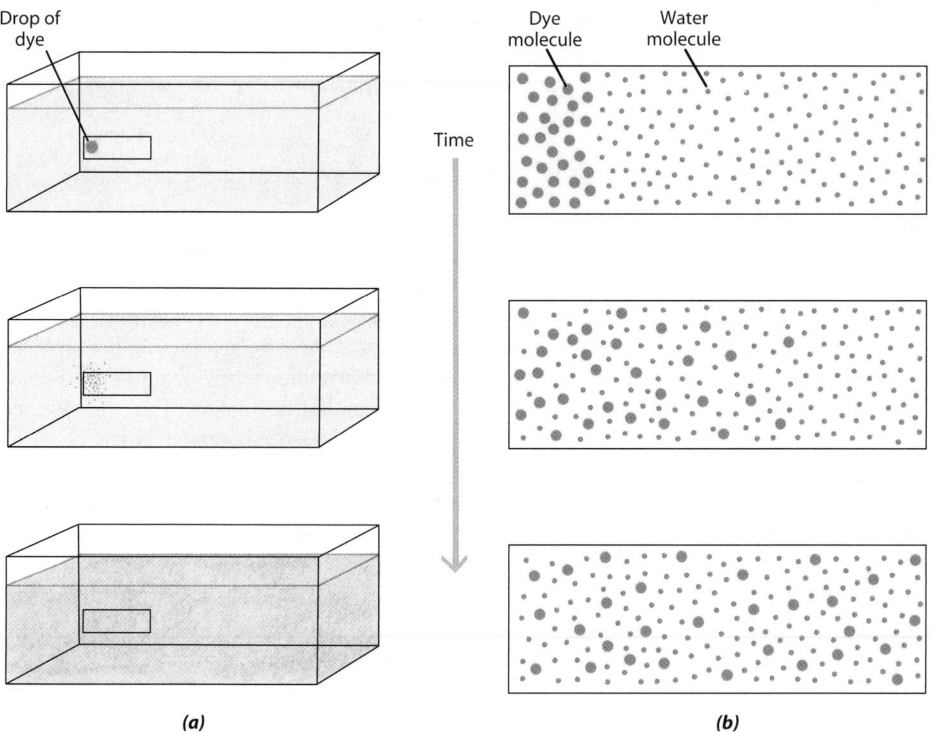

4–4 Diffusion Diffusion as revealed by the addition of a drop of dye to a tank of water (top left). The outlined section within each panel in ***(a)*** is shown in close-up in ***(b)***. The random movement of individual molecules (or ions) by diffusion produces a net movement from an area in which a substance has a high concentration to an area in which that substance has a lower concentration. Notice that as the dye molecules (indicated by red) diffuse to the right, the water molecules (indicated by blue) diffuse in the opposite direction. The result is an even distribution of the two types of molecules.

see that each one moves individually and at random. Imagine a thin section through the tank, running from top to bottom. Dye molecules will move into and out of the section, some moving in one direction, some moving in the other. But you would see more dye molecules moving from the side of greater dye concentration. Why? Simply because there are more dye molecules at that end of the tank. For example, if there are more dye molecules on the left, more of them will move randomly to the right, even though there is an equal probability that any one dye molecule will move from right to left. Consequently, the overall (net) movement of dye molecules is from left to right. Similarly, the net movement of water molecules is from right to left.

What happens when all the molecules are evenly distributed throughout the tank? Their even distribution does not affect the behavior of the molecules as individuals; they still move at random. But there are now as many molecules of dye and as many molecules of water on one side of the tank as on the other, and so there is no net direction of motion. There is, however, just as much individual motion (thermal agitation) as before, provided the temperature has not changed.

Substances that are moving from a region of higher concentration to a region of lower concentration are said to be moving *down a concentration gradient.* (The **concentration gradient** is the concentration difference of a substance per unit distance.) Diffusion occurs only down a concentration gradient. A substance moving in the opposite direction, toward a higher concentration of its own molecules, would be moving *against a concentration gradient,* which is analogous to being pushed uphill. The steeper the downhill gradient—that is, the larger the difference in concentration—the more rapid the net movement. Also, diffusion is more rapid in gases than in liquids and is more rapid at higher than at lower temperatures. Can you explain why?

Notice that there are two concentration gradients in our tank; the dye molecules are moving along one of them, and the water molecules are moving in the opposite direction along the other. The two kinds of molecules are moving independently of each other. In both cases, the movement is down a gradient. When the molecules have reached a state of equal distribution (that is, when there are no more gradients), they continue to move, but now there is no net movement in either direction. In other words, the net transfer of the molecules is zero; the system is said to be in a state of **equilibrium.**

The concept of water potential is also useful in understanding diffusion. A high concentration of solute in one region, such as dye in one corner of the tank, means a low concentration of water molecules there and, thus, low water potential. If the pressure is equal everywhere, water molecules, as they move down their concentration gradient, are moving from a region of higher water potential to a region of lower water potential. The region of the tank in which there is pure water has a greater water potential than the region containing water plus some dissolved substance. When equilibrium is reached, water potential is equal in all parts of the tank.

The essential characteristics of diffusion are: (1) each molecule moves independently of the others, and (2) these movements are at random. The net result of diffusion is that the diffusing substance eventually becomes evenly distributed. Briefly, **diffusion** may be defined as *the dispersion of substances by a movement of their ions or molecules, which tends to equalize their concentrations throughout the system.*

Cells and Diffusion

Water, oxygen, carbon dioxide, and a few other simple molecules diffuse freely across the plasma membrane. Carbon dioxide and oxygen, which are both nonpolar, are soluble in lipids and move easily through the lipid bilayer. Despite their polarity, water molecules also move through the membrane without hindrance, apparently through momentary openings created by spontaneous movements of the membrane lipids. Other uncharged polar molecules, provided they are small enough, also diffuse through these openings. The permeability of the membrane to these solutes varies inversely with the size of the molecules, indicating that the openings are small and that the membrane acts like a sieve in this respect.

Diffusion is also the principal way in which substances move within cells. One of the major factors limiting cell size is this dependence on diffusion, which is a slow process except over very short distances. Diffusion is not an effective way to move molecules over long distances at the rates required for cellular activity. In many cells, the transport of materials is speeded up by active streaming of the cytoplasm (pages 42 and 56).

Efficient diffusion requires a steep concentration gradient, that is, a short distance and a substantial concentration difference. Cells maintain such gradients by their metabolic activities. For example, in a nonphotosynthesizing cell, oxygen is used up within the cell almost as rapidly as it enters, thereby maintaining a steep gradient from outside to inside. Conversely, carbon dioxide is produced by the cell, and this maintains a gradient from inside to outside along which carbon dioxide can diffuse out of the cell. Similarly, within a cell, molecules or ions are often synthesized at one place and used at another. Thus a concentration gradient is established between the two regions of the cell, and the substance diffuses down the gradient from the site of production to the site of use.

Osmosis Is the Movement of Water across a Selectively Permeable Membrane

A membrane that permits the passage of some substances while inhibiting the passage of others is said to be **selectively permeable.** The movement of water molecules through such a membrane is known as **osmosis.** Osmosis involves a *net* flow of water from a solution that has higher water potential to a solution that has lower water potential (Figure 4–5). In the absence of other factors that influence water potential (such as pressure), the movement of water by osmosis is from a region of lower solute concentration (and therefore higher water concentration) into a region of higher solute concentration (and lower water concentration). The presence of solute decreases the water potential, creating a water potential gradient down which water moves. The water potential is not affected by *what* is dissolved in the water, only by *how much* is dissolved—that is, the concentration of particles of solute (molecules or ions) in the water.

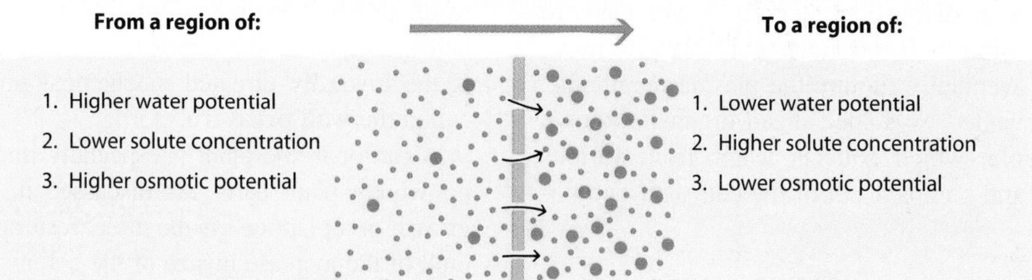

4–5 Osmosis In osmosis, the direction of water movement across a selectively permeable membrane is from a region of higher to a region of lower water potential.

A small solute particle, such as a sodium ion, counts just as much as a large solute particle, such as a sugar molecule.

Osmosis results in a buildup of pressure as water molecules continue to move across the membrane into a region of lower water concentration. If water is separated from a solution by a membrane that allows the ready passage of water but not of solute, in a system such as that shown in Figure 4–6, the water will move across the membrane and cause the solution to rise in the tube until an equilibrium is reached—that is, until the water potential is the same on both sides of the membrane. If enough pressure is applied on the solution in the tube by a piston, say, as in Figure 4–6c, it is possible to prevent the net movement of water into the tube. The pressure that would have to be applied to the solution to stop water movement is called the **osmotic pressure.** The tendency of water to move across a membrane because of the effect of solutes on water potential is called the **osmotic potential** (also called solute potential), which is negative.

Osmosis and Living Organisms

The movement of water across the plasma membrane in response to differences in water potential causes some crucial problems for living systems, particularly those in an aqueous environment. These problems vary according to whether the water potential of the cell or organism is higher than, similar to, or lower than the water potential of its environment. For example, the water potential of one-celled organisms that live in salt water is usually similar to the water potential of the medium they inhabit, which is one way of solving the problem.

Many types of cells live in environments with relatively high water potential. In freshwater single-celled organisms,

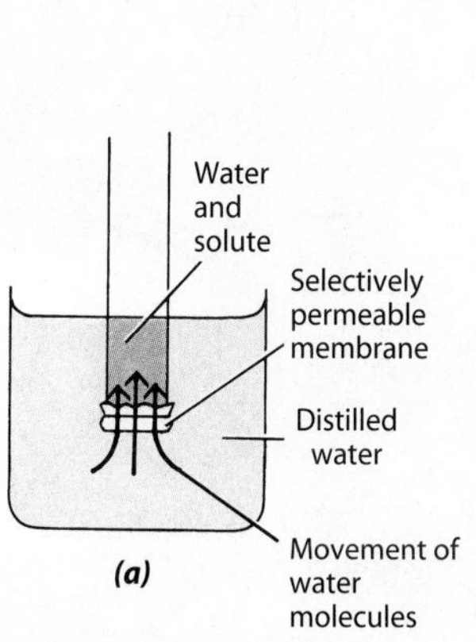

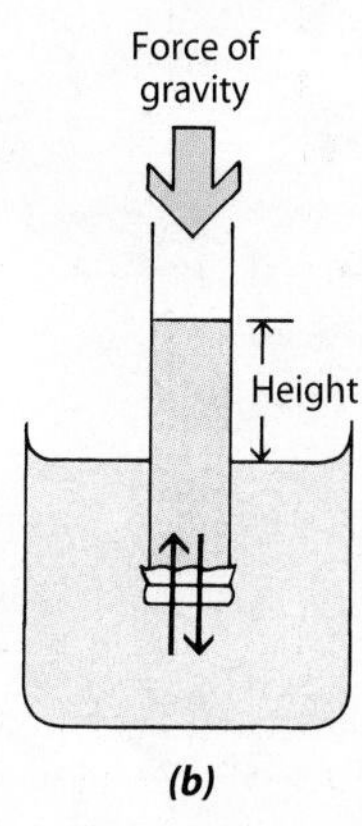

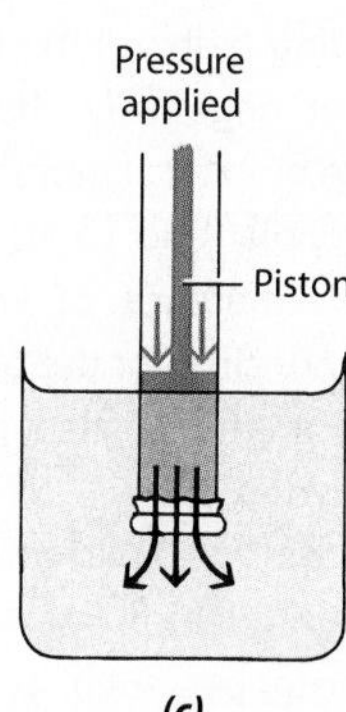

4–6 Osmosis and the measurement of osmotic potential ***(a)*** The tube contains a solution and the beaker contains distilled water. A selectively permeable membrane at the base of the tube permits the passage of water molecules but not of solute particles. ***(b)*** The diffusion of water into the solution causes its volume to increase, and thus the column of liquid rises in the tube. However, the downward pressure created by the force of gravity acting on the column of solution is proportional to the height of the column and the density of the solution. Thus, as the column of solution rises in the tube, the downward pressure gradually increases until it becomes so great that it counterbalances the tendency of water to move into the solution. In other words, the water potential on the two sides of the membrane becomes equal. At this point, there is no further net movement of water. ***(c)*** The pressure that must be applied to the piston to force the column of solution back to the level of the water in the beaker provides a quantitative measure of the osmotic potential of the solution—that is, of the tendency of water to diffuse across the membrane into the solution.

such as *Euglena,* the water potential of the cell is lower than that of the surrounding medium; consequently, water tends to move into the cell by osmosis. If too much water were to move into the cell, it could eventually rupture the plasma membrane. In *Euglena* this is prevented by a specialized organelle known as a contractile vacuole, which collects water from various parts of the cell body and pumps it out of the cell with a rhythmic contraction.

Turgor Pressure Contributes to the Stiffness of Plant Cells

If a plant cell is placed in a solution with a relatively high water potential, the protoplast expands and the plasma membrane stretches and exerts pressure against the cell wall. The plant cell does not rupture, however, because it is restrained by the relatively rigid cell wall.

Plant cells tend to concentrate relatively strong solutions of salts within their vacuoles, and they can also accumulate sugars, organic acids, and amino acids. As a result, plant cells absorb water by osmosis and build up their internal hydrostatic pressure. This pressure against the cell wall keeps the cell **turgid,** or stiff. Consequently, the hydrostatic pressure in plant cells is commonly referred to as turgor pressure. **Turgor pressure** is the pressure that develops in a plant cell as a result of osmosis and/or imbibition (see the essay "Imbibition" below). Equal to and opposing the turgor pressure at any instant is the inwardly directed mechanical pressure of the cell wall, called the **wall pressure.**

Turgor in the plant is especially important in the support of nonwoody plant parts. As discussed in Chapter 3, most of the growth of a plant cell is the direct result of water uptake, with the bulk of the increase in size of the cell resulting from enlargement of the vacuoles (page 50). Before the cell can increase in size, however, there must be a loosening of the wall structure in order to decrease the resistance of the wall to turgor pressure.

Turgor is maintained in most plant cells because they generally exist in a medium with a relatively high water potential. However, if a turgid plant cell is placed in a solution (for example, a sugar or salt solution) with a relatively low water potential, water will leave the cell by osmosis. As a result, the vacuole and rest of the protoplast shrink, thus causing the plasma membrane to pull away from the cell wall (Figure 4–7). This phenomenon is known as **plasmolysis.** The process can be reversed if the cell is then transferred to pure water. (Because of the high water permeability of membranes, the vacuole and

IMBIBITION

Water molecules exhibit a tremendous cohesiveness because of their polarity—that is, the difference in charge between one end of a water molecule and the other. Similarly, because of this difference in charge, water molecules can cling (adhere) to either positively charged or negatively charged surfaces. Many large biological molecules, such as cellulose, are polar and so attract water molecules. The adherence of water molecules is also responsible for the biologically important phenomenon called imbibition or, sometimes, hydration.

Imbibition (from the Latin *imbibere,* "to drink in") is the movement of water molecules into substances such as wood or gelatin, which swell as a result of the accumulation of water molecules. The pressures developed by imbibition can be astonishingly large. It is said that stone for the ancient Egyptian pyramids was quarried by driving wooden pegs into holes drilled in the rock face and then soaking the pegs with water. The swelling wood created a force that split the slab of stone. In living plants, imbibition occurs particularly in seeds, which may increase to many times their original size as a result. Imbibition is essential to the germination of the seed (see Chapter 22).

The germination of seeds begins with changes in the seed coat that permit a massive uptake of water by imbibition. The embryo and surrounding structures then swell, bursting the seed coat. In the acorn on the left, photographed on a forest floor, the embryonic root emerged after the tough layers of the fruit split open.

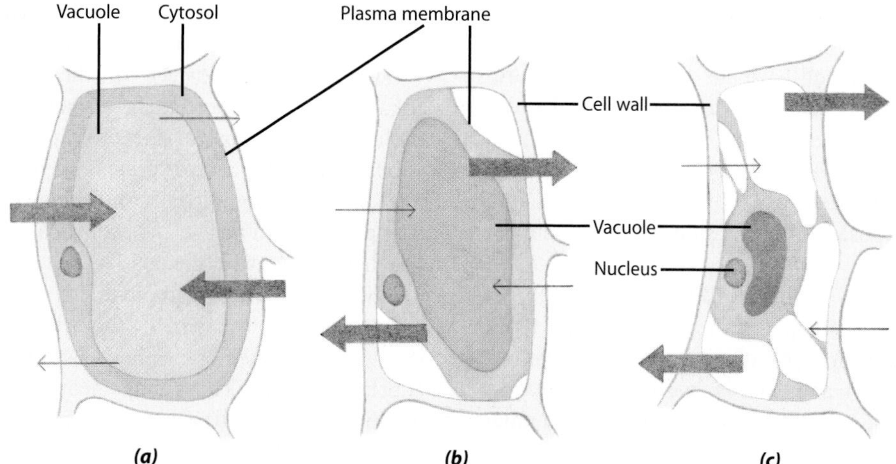

4–7 Plasmolysis in a leaf epidermal cell ***(a)*** Under normal conditions, the plasma membrane of the protoplast is in close contact with the cell wall. ***(b)*** When the cell is placed in a relatively concentrated sugar solution, water passes out of the cell, the protoplast contracts slightly, and the plasma membrane moves away from the wall. ***(c)*** When immersed in a more concentrated sugar solution, the cell loses even larger amounts of water and the protoplast contracts still further. As water is lost from the vacuole, its contents become more concentrated. The widths of the arrows indicate the relative amounts of water entering or leaving the cell.

the rest of the protoplast are at or very near equilibrium with respect to water potential.) Figure 4–8 shows *Elodea* leaf cells before and after plasmolysis. Although the plasma membrane and the tonoplast, or vacuolar membrane, are, with few exceptions, permeable only to water, the cell walls allow both solutes and water to pass freely through them. The loss of turgor by plant cells may result in **wilting,** or drooping, of leaves and stems, or of your dinner salad.

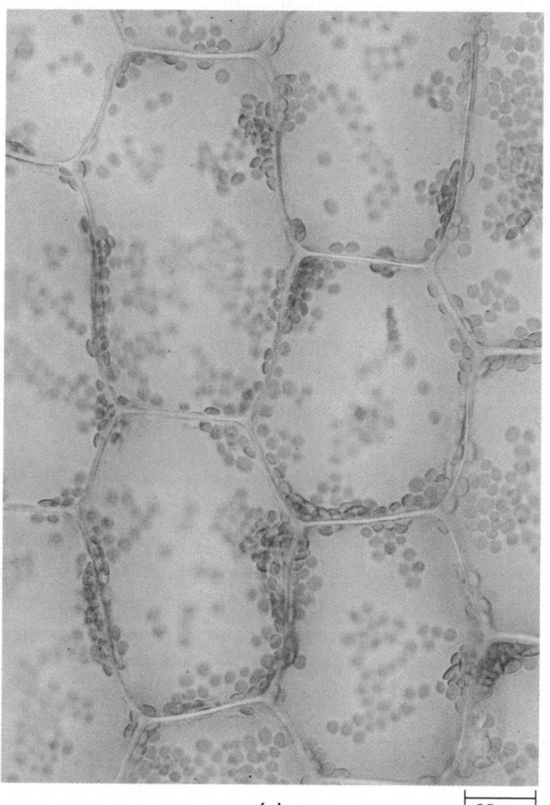

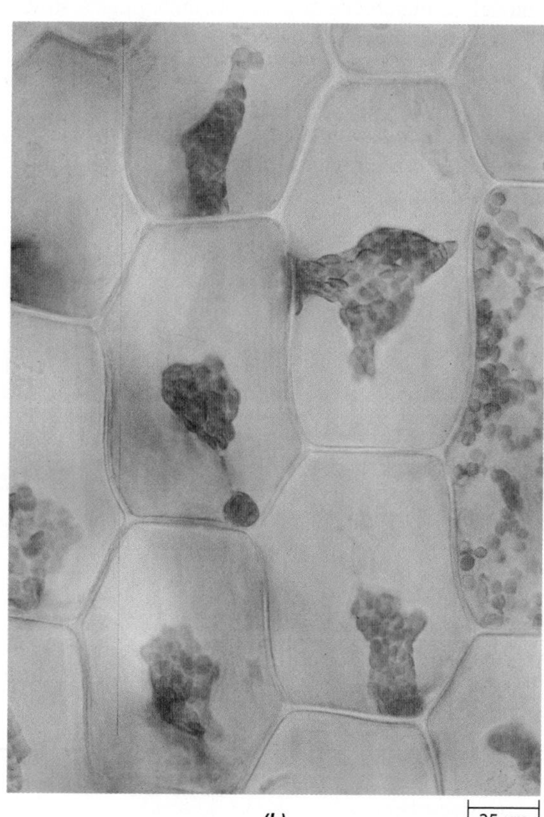

4–8 Plasmolysis in *Elodea* leaf cells ***(a)*** Turgid cells and ***(b)*** plasmolyzed cells after being placed in a relatively concentrated sucrose solution.

Structure of Cellular Membranes

All of the membranes of the cell have the same basic structure, consisting of a **lipid bilayer** in which are embedded globular proteins, many of which extend across the bilayer and protrude on either side (Figure 4–9). The portion of these **transmembrane proteins** embedded in the bilayer is hydrophobic, whereas the portions exposed on either side of the membrane are hydrophilic.

The two surfaces of a membrane differ considerably in chemical composition. For example, there are two major types of lipids in the plasma membrane of plants cells—**phospholipids** (the more abundant) and **sterols,** particularly stigmasterol (but not cholesterol, which is the major sterol in animal tissues)—and the two layers of the bilayer have different concentrations of each. Moreover, the transmembrane proteins have definite orientations within the bilayer, and the portions protruding on either side have different amino acid compositions and tertiary structures. Other proteins are also associated with membranes, including the so-called **peripheral proteins,** which, because they lack discrete hydrophobic sequences, do not penetrate into the lipid bilayer. Transmembrane proteins and other lipid-bound proteins tightly bound to the membrane are called **integral proteins.** Although some of the integral proteins appear to be anchored in place (perhaps to the cytoskeleton), the lipid bilayer is generally quite fluid. Some of the proteins float more or less freely in the bilayer, and they and the lipid molecules can move laterally within it, forming different patterns, or mosaics, that vary from time to time and place to place—hence the name **fluid-mosaic** for this model of membrane structure.

A new model of membrane structure is emerging: a structure that is less fluid, of variable thickness, and with a higher proportion of proteins. In this model, the proteins are organized in large functional complexes, some of which project over the membrane surface and thus occupy larger areas of the membrane than do their transmembrane regions. In addition, the lipids tend to group together, forming lipid-lipid and lipid-protein interactions and imparting a "patchiness" to the membrane.

On the outer surface of the plasma membrane of all eukaryotic cells, short-chain carbohydrates (oligosaccharides) are attached to most of the protruding proteins, forming **glycoproteins.** The carbohydrates are believed to play important roles in the recognition of molecules (such as hormones, the coat proteins of viruses, and molecules on the surfaces of bacteria) that interact with the cell.

Most membrane carbohydrates are present in the form of glycoproteins, but a small proportion are present as **glycolipids,** which are membrane lipids with short-chain carbohydrates attached to them. The arrangement of the carbohydrate groups on the external surface of the plasma membrane has been revealed largely by experiments using **lectins,** proteins that bind tenaciously to specific carbohydrate groups.

Two basic configurations have been identified among transmembrane proteins (Figure 4–10). One is a relatively simple rodlike structure consisting of a single alpha helix embedded in the hydrophobic interior of the membrane, with less regular, hydrophilic portions extending on either side. The other configuration is found in large globular proteins with complex three-dimensional structures that make repeated "passes" through the membrane. In such "multipass" membrane proteins, the polypeptide chain usually spans the lipid bilayer as a series of alpha helices.

Whereas the lipid bilayer provides the basic structure and impermeable nature of cellular membranes, the proteins are responsible for most membrane functions. Most membranes are composed of 40 to 50 percent lipid (by weight) and 60 to 50 percent protein, and the amounts and types of proteins in a membrane reflect its function. Membranes involved with energy transduction (the conversion of energy from one form to another), such as the internal membranes of mitochondria and chloroplasts, consist of about 75 percent protein. Some membrane proteins are enzymes that catalyze membrane-associated reactions, whereas

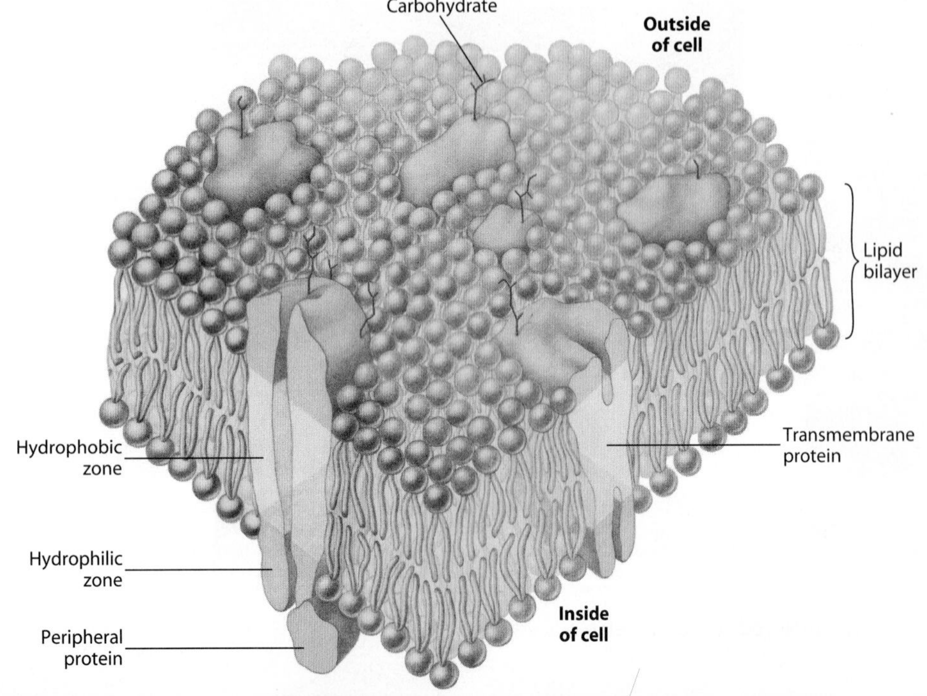

4–9 Fluid-mosaic model of membrane structure The membrane is composed of a bilayer (double layer) of lipid molecules—with their hydrophobic "tails" facing inward—and large protein molecules. The proteins traversing the bilayer are a type of integral protein known as transmembrane proteins. Other proteins, called peripheral proteins, are attached to some of the transmembrane proteins. The portion of a transmembrane protein molecule embedded in the lipid bilayer is hydrophobic; the portion or portions exposed on either side of the membrane are hydrophilic. Short carbohydrate chains are attached to most of the protruding transmembrane proteins on the outer surface of the plasma membrane. The whole structure is quite fluid, and hence the proteins can be thought of as floating in a lipid "sea."

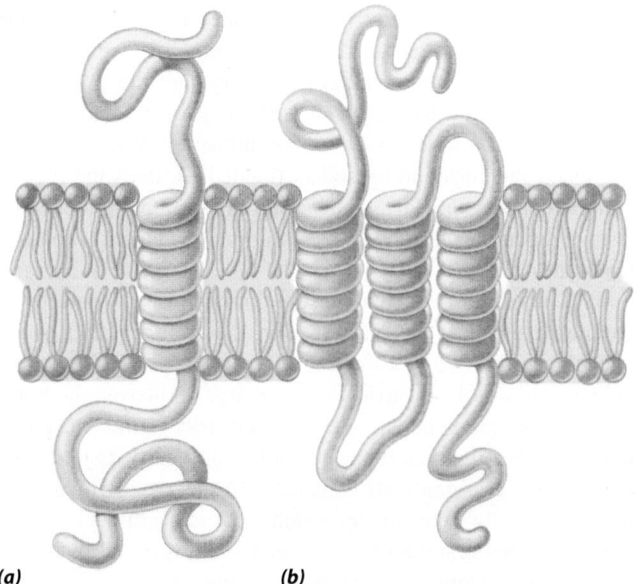

4–10 Two configurations of transmembrane proteins Some transmembrane proteins extend across the lipid bilayer as a single alpha helix ***(a)*** and others—multipass proteins—as multiple alpha helices ***(b).*** The portions of protein protruding on either side of the membrane are hydrophilic; the helical portions within the membrane are hydrophobic.

others are carriers involved in the transport of specific molecules or ions into and out of the cell or organelle. Still others act as receptors for receiving and transducing (converting) chemical signals from the cell's internal or external environment.

Transport of Solutes across Membranes

As mentioned previously, small nonpolar molecules, such as oxygen and carbon dioxide, and small uncharged polar molecules, such as water, can permeate cellular membranes freely by simple diffusion. The observation that hydrophobic molecules diffuse readily across plasma membranes provided the first evidence of the lipid nature of the membrane.

Most substances required by cells, however, are polar and require **transport proteins** to transfer them across membranes. Each transport protein is highly selective; it may accept one type of ion (such as Ca^{2+} or K^+) or molecule (such as a particular sugar or amino acid) and exclude a nearly identical one. These proteins provide the specific solutes they transport with a continuous pathway across the membrane, without the solutes coming into contact with the hydrophobic interior of the lipid bilayer.

Transport proteins can be grouped into three broad classes: pumps, carriers, and channels (Figure 4–11). Pumps are driven by either chemical energy (ATP) or light energy, and in plant and fungal cells they typically are proton pumps (the enzyme

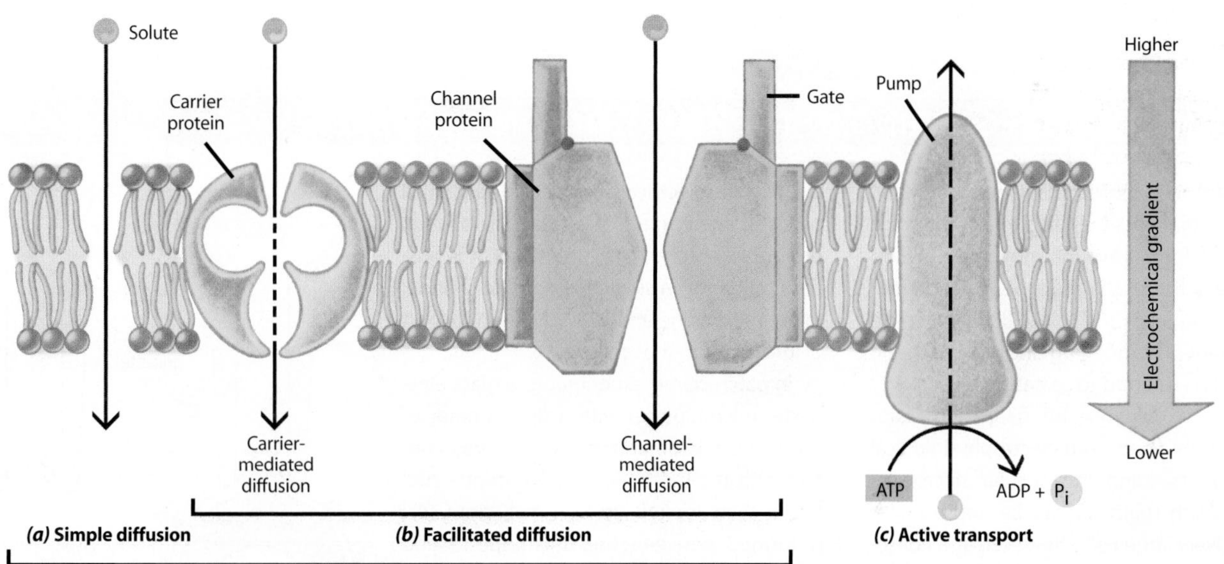

4–11 Modes of transport through the plasma membrane ***(a)*** In simple diffusion, small nonpolar molecules, such as oxygen and carbon dioxide, and small uncharged polar molecules, such as water, pass directly through the lipid bilayer, down their concentration gradient. ***(b)*** Facilitated diffusion occurs via either carrier proteins or channel proteins. Carrier proteins bind the specific solute and undergo conformational changes as the solute molecule is transported. Channel proteins allow selected solutes—commonly ions such as Na^+ and K^+—to pass directly through water-filled pores. Channel proteins are gated. When the gates are open, solutes pass through, but when they are closed, solute flow is blocked. As with simple diffusion, facilitated diffusion occurs down concentration or electrochemical gradients. Both diffusional processes are passive transport processes, which do not require energy. ***(c)*** Active transport, on the other hand, moves solutes against the concentration or electrochemical gradient and therefore requires an input of energy, usually supplied by the hydrolysis of ATP to ADP and P_i (see Figure 2–21). The transport proteins involved in active transport are known as pumps.

H^+-ATPase). Both carrier and channel proteins are driven by the energy of electrochemical gradients, as we discuss below. **Carriers** bind the specific solute being transported and undergo a conformational change to transport the solute across the membrane. **Channel proteins** form water-filled pores that extend across the membrane and, when open, allow specific solutes (usually inorganic ions such as Na^+, K^+, Ca^{2+}, and Cl^-) to pass through. The ion channels are not open continuously. Instead, they have "gates" that open briefly and then close again, a process referred to as **gating** (see the essay below).

Another basis for the three categories of transport protein is the speed of transport. The number of solute molecules transported per protein per second is relatively slow with pumps (fewer than 500 per second), intermediate with carriers (500–10,000 per second), and most rapid with channels (10,000 to many millions per second).

The plasma membrane and tonoplast also contain water channel proteins called **aquaporins,** which facilitate the movement of water and/or small neutral solutes (urea, boric acid, silicic acid) or gases (ammonia, carbon dioxide) across the membranes. Evidence indicates that water movement through aquaporins is increased in response to certain environmental stimuli that cause cell expansion and growth. Water passes relatively freely across the lipid bilayer of biological membranes, but aquaporins permit water to diffuse more rapidly into the cell and through the tonoplast into the vacuole. Why would cells need this increased water movement? One explanation is that the vacuole and cytosol must be in constant osmotic equilibrium, so rapid movement of water is required. It has also been suggested that aquaporins facilitate the rapid flow of water from the soil into root cells and through the xylem during periods of high transpiration (see Chapter 30). Aquaporins also have been shown to block the influx of water into root cells when the surrounding soil is flooded. Besides the plasma membrane and tonoplast, aquaporins are found in the endoplasmic reticulum and the inner chloroplast and mitochondrial membranes.

If a molecule is uncharged, the direction of its transport is determined only by the difference in the concentration of the molecule on the two sides of the membrane (the concentration gradient). If a solute carries a net charge, however, both the concentration gradient and the total electrical gradient across the membrane (the membrane potential) influence its transport. Together, the two gradients constitute the **electrochemical gradient.** Plant cells typically maintain electrical gradients across the plasma membrane and the tonoplast. The cytosol is electrically negative relative to both the aqueous medium outside the cell and the solution (cell sap) inside the vacuole. Transport down a concentration gradient or an electrochemical gradient is called **passive transport.** One example of passive transport is the **simple diffusion** of small polar molecules across a lipid bilayer (Figure 4–11a). Most passive transport, however, requires carrier proteins to facilitate the passage of ions and polar molecules across the hydrophobic interior of the membrane. Passive transport with the assistance of carrier proteins is called **facilitated diffusion** (Figure 4–11b).

PATCH-CLAMP RECORDING IN THE STUDY OF ION CHANNELS

The membranes of both plant and animal cells contain channel proteins that form pathways for passive ion movement. When activated, these channels become permeable to ions, allowing solute flow across the membrane. The opening and closing of ion channels is referred to as gating.

The first evidence for gated ion channels was obtained from electrophysiological experiments using intracellular microelectrodes. Such methods can be applied only to relatively large cells, however, and consequently the measured currents (that is, the ion movements) flow through many channels at a time. In addition, different types of channels may be opened simultaneously. Plant cells pose another problem: when a microelectrode is inserted into the protoplast, it usually penetrates both the plasma membrane and the tonoplast, so that the data reflect the collective behavior of both membranes.

The patch-clamp technique revolutionized the study of ion channels. This technique involves the electrical analysis of a very small patch of membrane of a naked (wall-less) protoplast or tonoplast. In this way, it is possible to identify a single ion-specific channel in the membrane and to study the transport of ions through that channel.

In patch-clamp experiments, a glass electrode (micropipette) with a tip diameter of about 1.0 micrometer is brought into contact with the membrane. When gentle suction is then applied, an extremely tight seal is formed between the micropipette and the membrane. With the patch still attached to the intact protoplast ***(a)***, the transport of ions can be recorded between the cytoplasm and the artificial solution that fills the micropipette. If the pipette is withdrawn from the attached protoplast, it is possible to separate only the small patch of membrane, which remains intact within the tip ***(b)***. With a detached patch it is easy to alter the composition of the solution on one or both sides of the membrane to test the effect of different solutes on channel behavior.

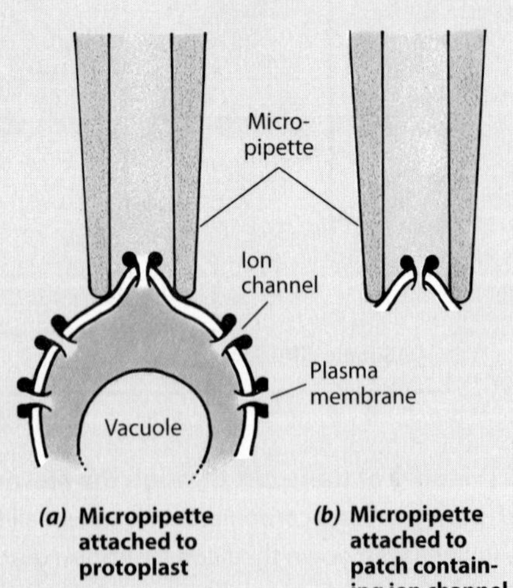

(a) **Micropipette attached to protoplast**

(b) **Micropipette attached to patch containing ion channel**

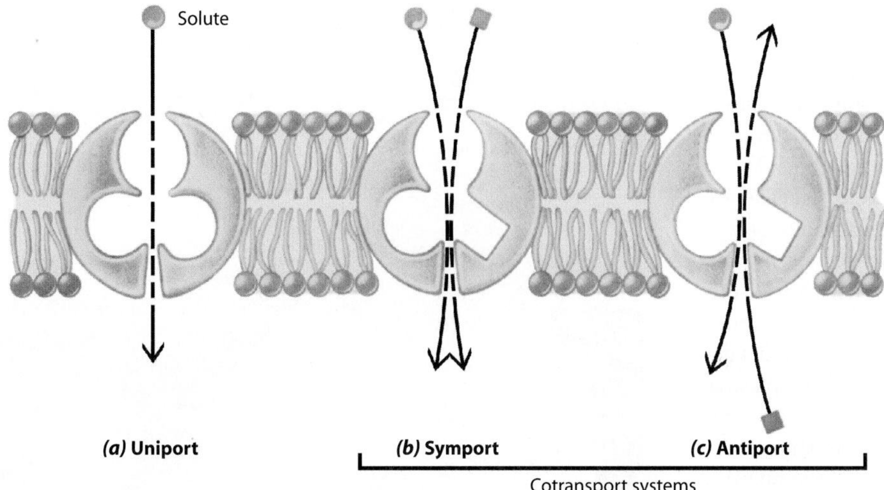

4–12 Passage of solutes through carrier proteins *(a)* In the simplest type, known as a uniport, one particular solute is moved directly across the membrane in one direction. Carrier proteins involved with facilitated diffusion function as uniporters, as do all channel proteins. *(b)* In the type of cotransport system known as a symport, two different solutes are moved across the membrane simultaneously and in the same direction. *(c)* In another type of cotransport system, known as an antiport, two different solutes are moved across the membrane either simultaneously or sequentially, but in opposite directions.

All channel proteins and some carrier proteins are **uniporters,** which are simply proteins that transport only one solute from one side of the membrane to the other. Other carrier proteins function as **cotransport systems,** in which the transfer of one solute depends on the simultaneous or sequential transfer of a second solute. The second solute may be transported in the same direction **(symport)** or in the opposite direction **(antiport)** (Figure 4–12). Neither simple diffusion nor passive transport is capable of moving solutes *against* a concentration gradient or an electrochemical gradient. The capacity to move solutes against a concentration or electrochemical gradient requires energy. This process is called **active transport** (Figure 4–11c), and it is always mediated by carrier proteins. As we have already seen, the proton pump in plant and fungal cells is energized by ATP and mediated by an H^+-ATPase located in the membrane. The enzyme generates a large electrical potential and a pH gradient—that is, a gradient of protons (hydrogen ions)—that provide the driving force for solute uptake by all the H^+-coupled cotransport systems. By this process, even neutral solutes can be accumulated to concentrations much higher than those outside the cell, simply by being cotransported with a charged molecule (for example, an H^+). The energy-yielding first process (the pump) is referred to as **primary active transport,** and the second process (the cotransporter) as **secondary active transport** (Figure 4–13).

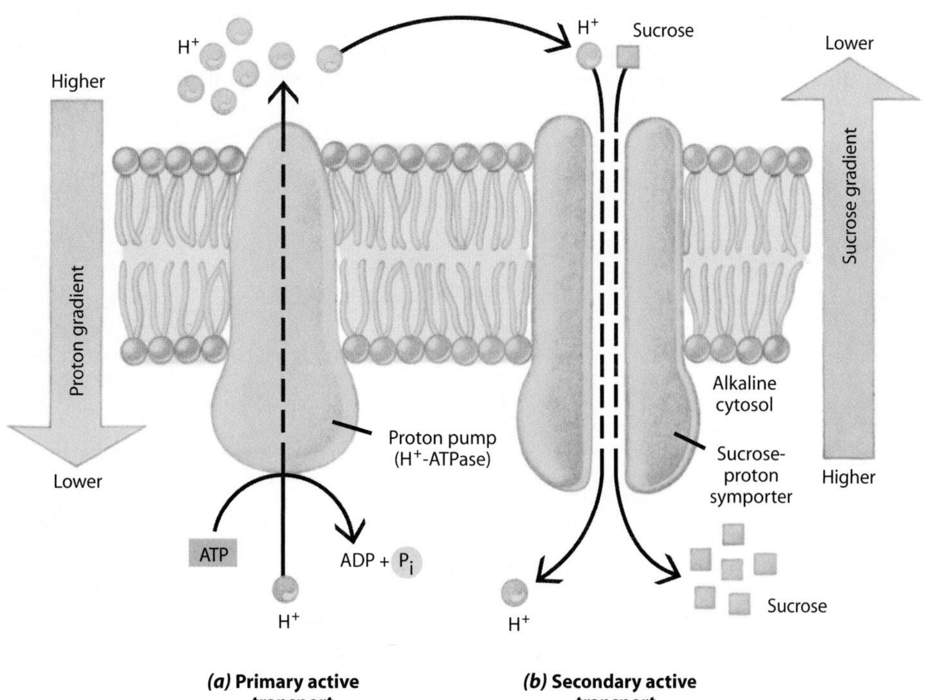

4–13 Primary and secondary active transport of sucrose *(a)* Primary active transport occurs when the proton pump (the enzyme H^+-ATPase) pumps protons (H^+) against their gradient. The result is a proton gradient across the membrane. *(b)* The proton gradient energizes secondary active transport. As the protons flow passively down their gradient, sucrose molecules are cotransported across the membrane against their gradient. The carrier protein is known as a sucrose-proton symporter.

Vesicle-Mediated Transport

The transport proteins that ferry ions and small polar molecules across the plasma membrane cannot accommodate large molecules, such as proteins and polysaccharides, or large particles, such as microorganisms or bits of cellular debris. These large molecules and particles are transported by means of vesicles that bud off from or fuse with the plasma membrane, a process called **vesicle-mediated transport.** As we saw in Figure 3–23, vesicles move from the *trans*-Golgi network to the surface of the cell. The hemicelluloses, pectins, and glycoproteins that form the matrix of the cell wall are carried to developing cell walls in secretory vesicles that fuse with the plasma membrane, thus releasing their contents into the wall. This process is known as **exocytosis** (Figure 4–14). The slimy substance that lubricates a growing root and helps it penetrate the soil is a polysaccharide that is transported by secretory vesicles and released into the walls of the rootcap cells by exocytosis (see pages 560 and 561). Exocytosis is not limited to the secretion of substances derived from Golgi bodies. For example, the digestive enzymes secreted by carnivorous plants, such as the Venus flytrap and butterwort (see Chapters 28 and 29), are carried to the plasma membrane in vesicles derived from the endoplasmic reticulum.

Transport by means of vesicles can also work in the opposite direction. In **endocytosis,** material to be taken into the cell induces the plasma membrane to bulge inward, producing a vesicle enclosing the substance. Three different forms of endocytosis are known: phagocytosis, pinocytosis, and receptor-mediated endocytosis.

Phagocytosis ("cell eating") involves the ingestion of relatively large, solid particles, such as bacteria or cellular debris, via large vesicles derived from the plasma membrane (Figure 4–15a). Many one-celled organisms, such as amoebas, feed in this way, as do plasmodial slime molds and cellular slime molds. A unique example of phagocytosis in plants is found in the nodule-forming roots of legumes during the release of *Rhizobium* bacteria from infection threads. The released bacteria

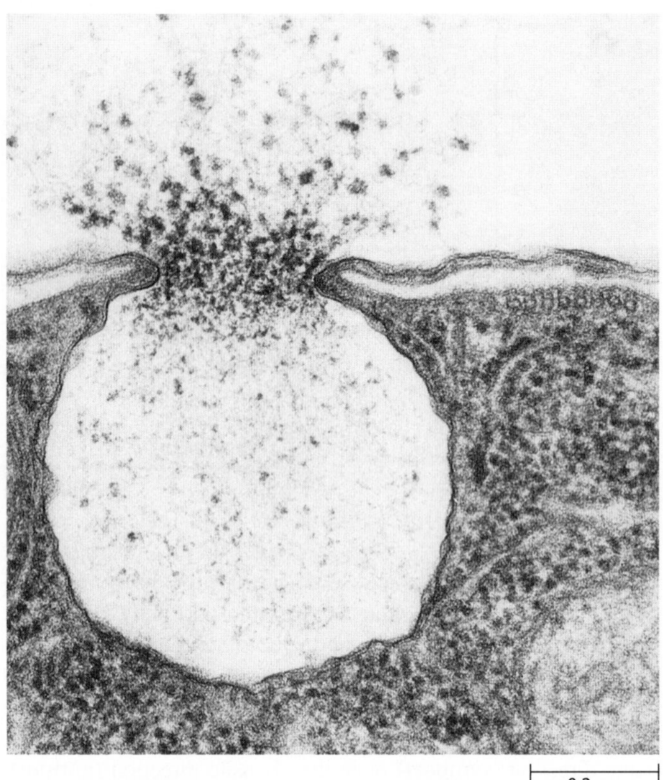

4–14 Exocytosis A secretory vesicle, formed by a Golgi apparatus of the protist *Tetrahymena furgasoni,* discharges mucus at the cell surface. Notice how the vesicle membrane has fused with the plasma membrane.

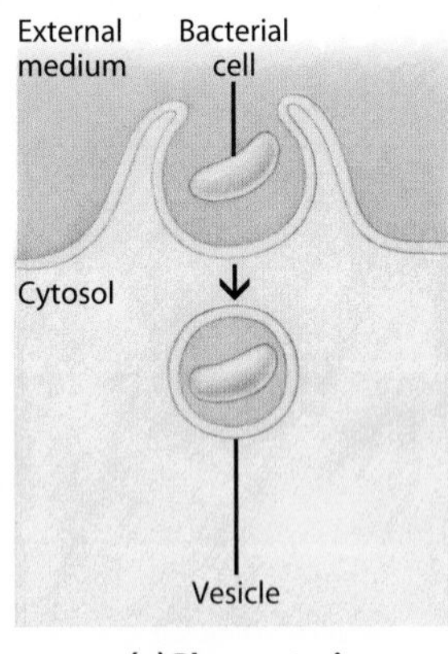

***(a)* Phagocytosis**

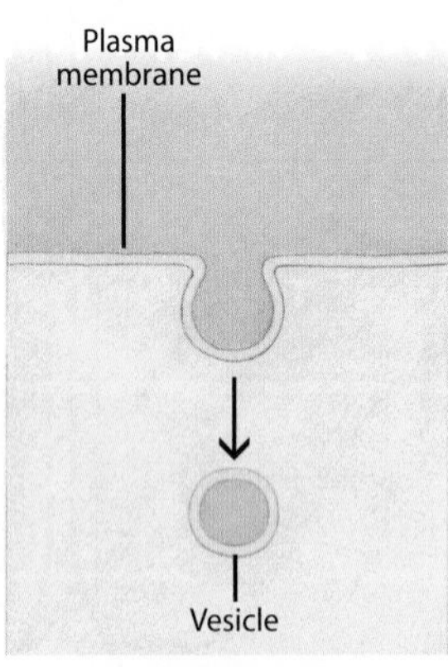

***(b)* Pinocytosis**

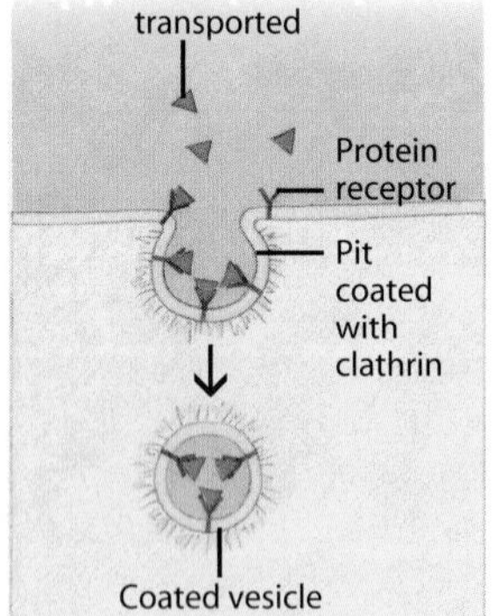

***(c)* Receptor-mediated endocytosis**

4–15 Three types of endocytosis ***(a)*** In phagocytosis, contact between the plasma membrane and particulate matter, such as a bacterial cell, causes the plasma membrane to extend around the particle, engulfing it in a vesicle. ***(b)*** In pinocytosis, the plasma membrane pouches inward, forming a vesicle around liquid from the external medium that is to be taken into the cell. ***(c)*** In receptor-mediated endocytosis, the molecules to be transported into the cell must first bind to specific receptor proteins. The receptors are either localized in indented areas of the plasma membrane known as coated pits or migrate to such areas after binding the molecules to be transported. When filled with receptors carrying their particular molecules, the pit buds off as a coated vesicle.

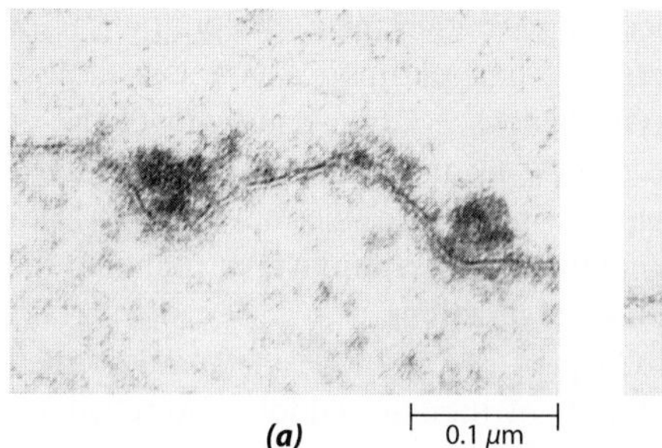
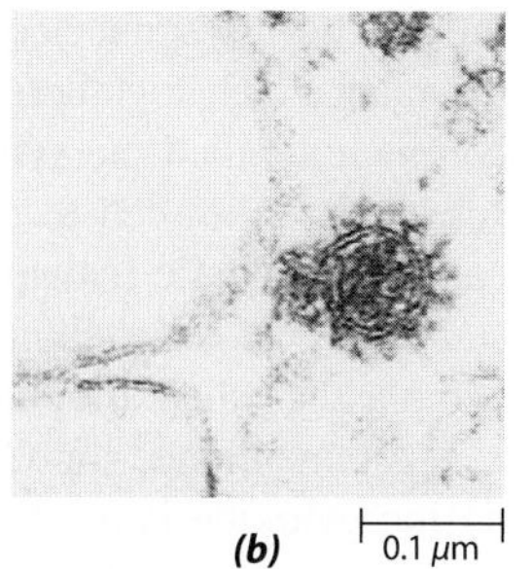
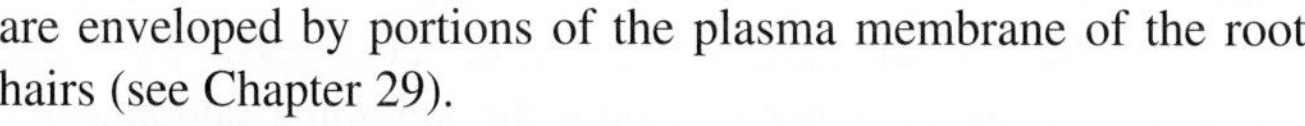

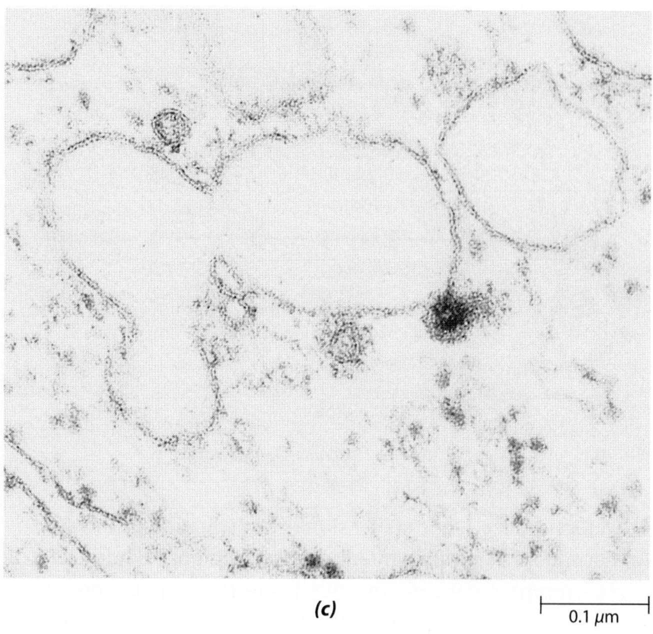

4–16 Receptor-mediated endocytosis Seen here are maize *(Zea mays)* rootcap cells that have been exposed to a solution containing lead nitrate. (The rootcap cells form a protective covering at the tip of the root.) ***(a)*** Granular deposits containing lead can be seen in two coated pits. ***(b)*** A coated vesicle with lead deposits. ***(c)*** One of two coated vesicles has fused with a large Golgi vesicle, into which it will release its contents. This coated vesicle (dark structure) still contains lead deposits, but it appears to have lost its coat, which is located just to the right of it. The coated vesicle to its left is clearly intact.

are enveloped by portions of the plasma membrane of the root hairs (see Chapter 29).

Pinocytosis ("cell drinking") involves the taking in of liquids, as distinct from solid matter (Figure 4–15b). It is the same in principle as phagocytosis. Unlike phagocytosis, however, which is carried out only by certain specialized cells, pinocytosis is believed to occur in all eukaryotic cells, as the cells continuously and indiscriminately "sip" small amounts of fluid from the surrounding medium.

In **receptor-mediated endocytosis,** particular membrane proteins serve as receptors for specific molecules that are to be transported into the cell (Figure 4–15c). This cycle begins at specialized regions of the plasma membrane called **coated pits** (Figure 4–16a). Coated pits are depressions of the plasma membrane coated on their inner, or cytoplasmic, surfaces with the peripheral protein clathrin (page 53). The substance being transported attaches to the receptors in a coated pit. Shortly thereafter (usually within minutes), the coated pit invaginates and pinches off to form a **coated vesicle** (Figure 4–16b). Vesicles thus formed contain not only the substance being transported but also the receptor molecules, and the vesicles have an external coating of clathrin (see Figure 3–24). Within the cell, the coated vesicles shed their coats (Figure 4–16c) and then fuse with some other membrane-bounded structure (for example, Golgi bodies or small vacuoles), releasing their contents in the process. As you can see in Figure 4–15c, the surface of the membrane facing the interior of a vesicle is equivalent to the surface of the plasma membrane facing the exterior of the cell. Similarly, the surface of the vesicle facing the cytoplasm is equivalent to the cytoplasmic surface of the plasma membrane.

As we noted in Chapter 3, new material needed for expansion of the plasma membrane in enlarging cells is transported, ready-made, in Golgi vesicles. During endocytosis, portions of the plasma membrane are returned to the Golgi bodies. And in exocytosis, portions of membranes used in forming endocytic vesicles are transported back to the plasma membrane. In the process, the membrane lipids and proteins, including the specific receptor molecules, are recycled.

Cell-to-Cell Communication

Thus far in our consideration of the transport of substances into and out of cells, we have assumed that individual cells exist in isolation, surrounded by a watery environment. In multicellular organisms, however, this is generally not the case. Cells are organized into **tissues,** which are groups of specialized cells with common functions. Tissues are further organized to form **organs,** each of which has a structure that suits it for specific functions.

Signal Transduction Is the Process by Which Cells Use Chemical Messengers to Communicate

As you might imagine, the successful existence of multicellular organisms depends on the ability of the individual cells to communicate with one another so that they can collaborate to create harmonious tissues and organs and, ultimately, a properly functioning organism. This communication is accomplished in large part by means of chemical **signals**—that is, by substances that are synthesized within and transported out of one cell and then travel to another cell. In plants, the chemical signals are represented largely by **hormones,** chemical messengers typically produced by one cell type or tissue in order to regulate the function of cells or tissues elsewhere in the plant body (see Chapter 27). These signal molecules must be small enough to get through the cell wall easily.

The plasma membrane plays a key role in signal recognition. When the signal molecules reach the plasma membrane of the **target cell,** they may be transported into the cell by any one of the endocytic processes we have considered. Alternatively, they may remain outside the cell, where they bind to specific **receptors** on the outer surface of the membrane. In most cases the receptors are transmembrane proteins that become

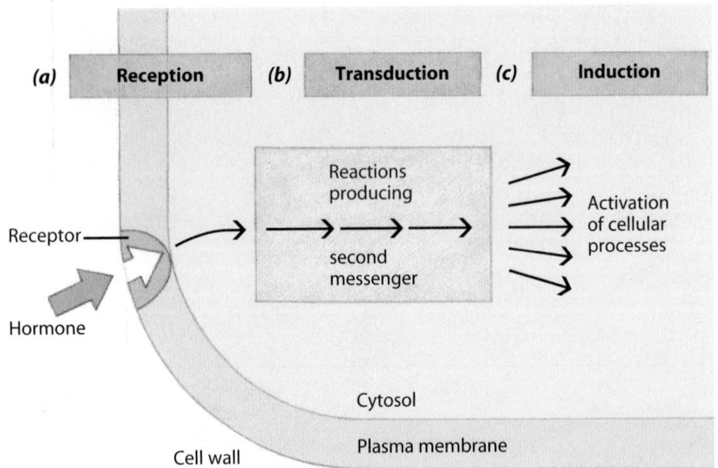

4–17 Generalized model of a signal-transduction pathway ***(a)*** Reception. A hormone (or other chemical signal) binds to a specific receptor in the plasma membrane. ***(b)*** Transduction. The receptor now stimulates the cell to produce a second messenger. ***(c)*** Induction. The second messenger enters the cytosol and activates cellular processes. In other cases, the chemical signals enter the cell and bind to specific receptors inside.

activated when they bind a signal molecule (the first messenger) and generate secondary signals, or **second messengers,** on the inside of the cell. The second messengers, which increase in concentration within the cell in response to the signal, pass the signal on by altering the behavior of selected cellular proteins, thereby triggering chemical changes within that cell. This process by which a cell converts an extracellular signal into a response is called **signal transduction.** Two of the most widely used second messengers are calcium ions and, in animals and fungi, cyclic AMP (cyclic adenosine monophosphate, a molecule formed from ATP).

A signal-transduction pathway may be divided into three steps: **reception, transduction,** and **induction** (Figure 4–17). The binding of the hormone (or of any chemical signal) to its specific receptor represents the reception step. During the transduction step, the second messenger, which is capable of amplifying the stimulus and initiating the cell's response, is formed in or released into the cytosol. The calcium ion, Ca^{2+}, has been identified as the second messenger in many plant responses. Binding of a hormone to its specific receptor triggers the release of Ca^{2+} ions stored in the vacuole into the cytosol. The Ca^{2+} ions enter the cytosol through specific Ca^{2+} channels in the tonoplast. In some plant cells, release of Ca^{2+} ions stored in the lumen of the endoplasmic reticulum may also be involved (page 51). The Ca^{2+} ions then combine with **calmodulin,** the principal calcium-binding protein in plant cells. The Ca^{2+}-calmodulin complex influences, or induces, numerous cellular processes, generally through activation of appropriate enzymes.

Plasmodesmata Enable Cells to Communicate

Plasmodesmata, the narrow strands of cytoplasm that interconnect the protoplasts of neighboring plant cells, are also important pathways in cell-to-cell communication. Because they are intimately interconnected by plasmodesmata, the protoplasts of the plant body, together with their plasmodesmata, constitute a continuum called the **symplast.** Accordingly, the movement of substances from cell to cell by means of the plasmodesmata is called **symplastic transport.** In contrast, the movement of substances in the cell wall continuum, or **apoplast,** surrounding the symplast is called **apoplastic transport.**

As mentioned in Chapter 3, plasmodesmata are formed during cytokinesis as strands of tubular endoplasmic reticulum become entrapped within the developing cell plate. Plasmodesmata can also be formed de novo across existing cell walls. Those plasmodesmata formed during cytokinesis are called **primary plasmodesmata,** and those formed after cytokinesis are referred to as **secondary plasmodesmata.** The formation of secondary plasmodesmata is essential in order to establish communication between neighboring cells not derived from the same cell lineage or precursor cell. Secondary plasmodesmata typically are branched, and many are interconnected by a cavity in the region of the middle lamella.

As seen with an electron microscope, a plasmodesma appears as a plasma-membrane-lined channel typically traversed by a tubular strand of tightly constricted endoplasmic reticulum, called a **desmotubule,** which is continuous with the endoplasmic reticulum of the adjacent cells. In most plasmodesmata, the desmotubule does not resemble the adjoining endoplasmic reticulum—it is much narrower in diameter and contains a central, rodlike structure. The central rod represents the merger of the inner portions of the bilayers of the tightly appressed endoplasmic reticulum forming the desmotubule (Figure 4–18). Although some molecules may pass through the desmotubule, nearly all transport via the plasmodesma is through the cytoplasmic channel surrounding the desmotubule. This channel, called the **cytoplasmic sleeve,** is subdivided into narrower channels by globular proteins, which are embedded in the inner part of the plasma membrane and the outer portion of the desmotubule and are interconnected by spoke-like structures. Plasmodesmata therefore consist of an outer plasma membrane, a middle cytoplasmic sleeve, and a central desmotubule.

Plasmodesmata apparently provide a more efficient pathway between neighboring cells than the alternative, less direct route through the plasma membrane of one cell, the cell wall, and plasma membrane of the second cell. Cells and tissues that are far removed from direct sources of nutrients are thought to be supplied with nutrients either by simple diffusion or by bulk flow through plasmodesmata. In addition, as discussed in Chapter 30, some substances move through plasmodesmata to and from the xylem and phloem, the tissues concerned with long-distance transport in the plant body.

Evidence for transport between cells via plasmodesmata comes from studies using fluorescent dyes or electric currents. The dyes, which do not easily cross the plasma membrane, can be observed moving from the injected cells into neighboring cells and beyond (Figure 4–19). Such studies have revealed that most plasmodesmata can allow the passage of molecules up to molecular weights of 800 to 1000. Thus, the effective pore sizes, or **size exclusion limits,** of such plasmodesmata are adequate for small solutes, such as sugars, amino acids, and signaling molecules, to move freely across these intercellular

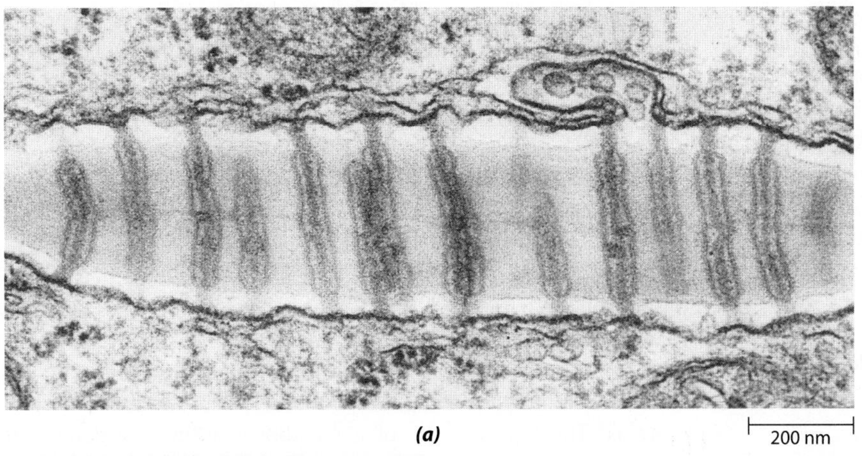

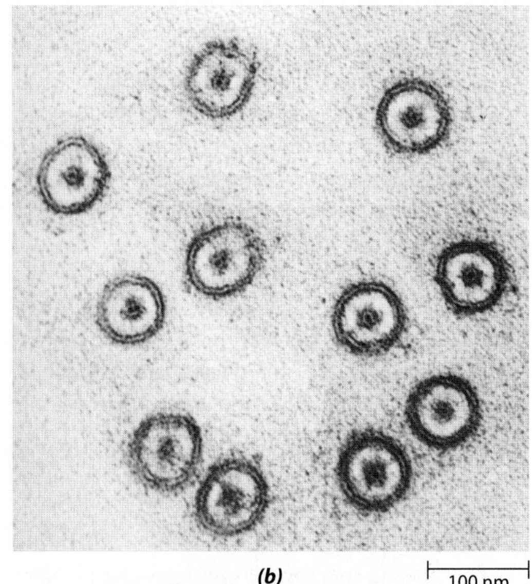

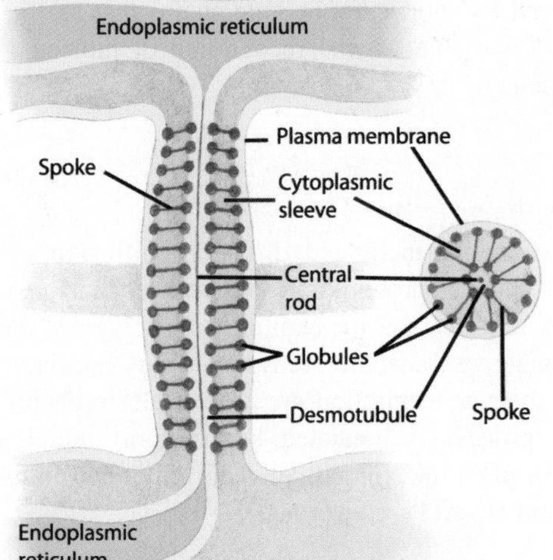

4–18 Plasmodesmata Electron micrographs of the cell walls of the sugarcane *(Saccharum)* leaf showing ***(a)*** a longitudinal view of plasmodesmata and ***(b)*** plasmodesmata in transverse view. Note that the endoplasmic reticulum is connected to the desmotubule, which contains a central rod, apparently formed by the merger of the inner portions of the endoplasmic reticulum bilayers. The cytoplasmic channel surrounding the desmotubule is called the cytoplasmic sleeve. The plasmodesma consists of the plasma membrane, the cytoplasmic sleeve, and the desmotubule. Globular proteins embedded in the inner and outer portions of the plasma membrane and desmotubule, respectively, are interconnected by spokelike extensions. Note that the globules, with their spokelike extensions, divide the cytoplasmic sleeve into a number of narrower channels.

connections. Studies involving fluorescent dyes, however, have revealed the presence of barriers at various regions in the symplast. These barriers are due to differences in the size exclusion limits of the plasmodesmata at the boundaries of the various regions. Such regions are referred to as **symplastic domains.**

The passage of pulses of electric current from one cell to another can be monitored by means of receiver electrodes placed in neighboring cells. The magnitude of the electrical force is found to vary with the density of plasmodesmata and with the number and length of cells between the injection and receiver electrode. This indicates that plasmodesmata can also serve as a path for electrical signaling between plant cells.

Plasmodesmata once were depicted as rather passive entities that exerted little direct influence over the substances that

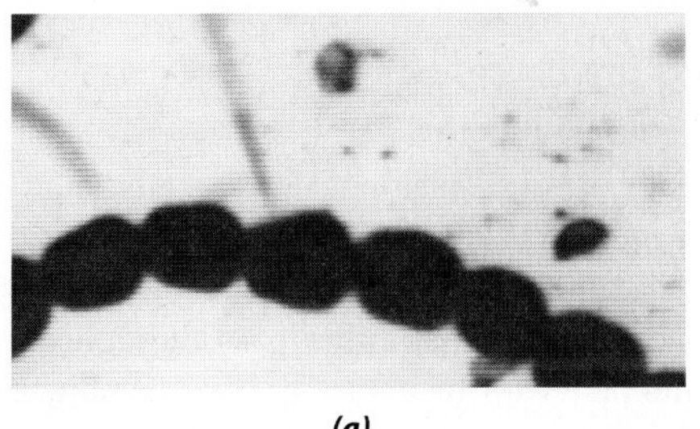

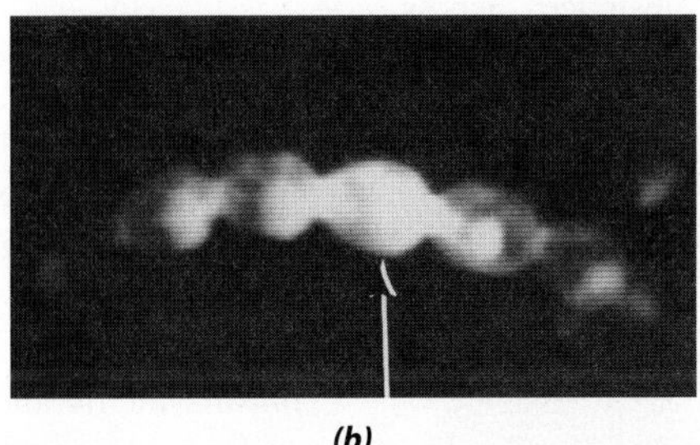

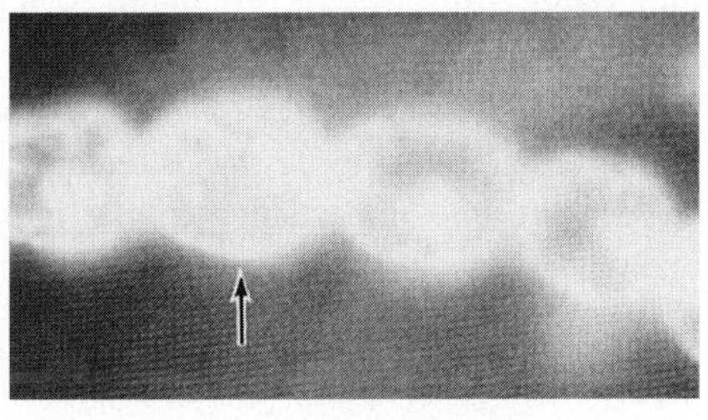

4–19 Movement through plasmodesmata ***(a)*** A hair from a stamen of *Setcreasea purpurea* before injection with the fluorescent dye disodium fluorescein. ***(b), (c)*** Two and five minutes after the dye was injected into the cytoplasm of the cell indicated by the arrow. Notice that the dye has passed in both directions into the cytoplasm of the neighboring cells. Because the plasma membrane is impermeable to the dye, the movement of dye from cell to cell must have occurred via plasmodesmata connecting the adjoining cells.

moved through them. We now know that plasmodesmata are dynamic structures capable of controlling, to varying degrees, the intercellular movement of molecules, and that they have the capacity to mediate the cell-to-cell transport of macromolecules, including proteins and a wide range of RNAs, thereby playing a primary role in the coordination of plant growth and development. Our understanding of the control mechanisms regulating plasmodesmatal permeability is still rudimentary. Clearly, the deposition and degradation of callose (page 58) at the ends of the plasmodesmata play a role in the control of plasmodesmatal permeability. Actin and myosin, which occur along the length of the plasmodesma, may provide a contractile mechanism to regulate the size of the openings.

SUMMARY

The plasma membrane regulates the passage of materials into and out of the cell, a function that makes it possible for the cell to maintain its structural and functional integrity. This regulation depends on interaction between the membrane and the materials that pass through it.

Water Moves down a Water Potential Gradient

Water is one of the principal substances passing into and out of cells. Water potential determines the direction in which the water moves; that is, the movement of water is from regions of higher water potential (lower solute concentration) to regions of lower water potential (higher solute concentration), provided that the pressure in the two regions is equal. The concept of water potential is useful because it enables plant physiologists to predict how water will move under various conditions.

Water Movement Takes Place by Bulk Flow and Diffusion

Bulk flow is the overall movement of water molecules, as when water flows downhill or moves in response to pressure. Sap moves by bulk flow from the leaves to other parts of the plant body. Diffusion involves the independent movement of molecules and results in net movement down a concentration gradient. Diffusion is most efficient when the distance is short and the gradient is steep. By their metabolic activities, cells maintain steep concentration gradients of many substances across the plasma membrane and between different compartments of the cytoplasm. The rate of movement of substances within cells is increased by cytoplasmic streaming. Carbon dioxide and oxygen are two important nonpolar molecules that move into and out of cells by diffusion across the plasma membrane.

Osmosis Is the Movement of Water across a Selectively Permeable Membrane

A selectively permeable membrane is a membrane that permits the movement of water but inhibits the passage of solutes. In the absence of other forces, the movement of water by osmosis is from a region of lower solute concentration, and therefore of higher water potential, to a region of higher solute concentration, and so of lower water potential. The turgor (rigidity) of plant cells is a consequence of osmosis, as well as of a strong but somewhat elastic cell wall.

Membranes Consist of a Lipid Bilayer and Proteins

The plasma membrane and other cellular membranes are composed of lipid bilayers in which proteins are embedded. The lipid bilayer provides the basic structure and impermeable nature of the membrane. In plant cells, the major types of lipids are phospholipids (the more abundant) and sterols. Different membrane proteins perform different functions; some are enzymes, others are receptors, and still others are transport proteins. The two surfaces of a membrane differ considerably in chemical composition. The exterior surface of the plasma membrane is characterized by short carbohydrate chains that are believed to play important roles in the recognition of molecules that interact with the cell.

Small Molecules Cross Membranes by Simple Diffusion, Facilitated Diffusion, or Active Transport

Both simple diffusion and facilitated diffusion (diffusion assisted by carrier or channel proteins) are passive transport processes. If transport requires the expenditure of energy by the cell, it is known as active transport. Active transport can move substances against their concentration gradient or electrochemical gradient. This process is mediated by transport proteins known as pumps. In plant and fungal cells, an important pump is a membrane-bound H^+-ATPase enzyme.

Large Molecules and Particles Cross Membranes by Vesicle-Mediated Transport

Controlled movement of large molecules into and out of a cell occurs by endocytosis or exocytosis, processes in which substances are transported within vesicles. Three forms of endocytosis are known: phagocytosis, in which solid particles are taken into the cell; pinocytosis, in which liquids are taken in; and receptor-mediated endocytosis, in which molecules and ions to be transported into the cell are bound to specific receptors in the plasma membrane. During exocytosis and endocytosis, portions of membranes are recycled between the Golgi bodies and the plasma membrane.

Signal Transduction Is the Process by Which Cells Use Chemical Messengers to Communicate

In multicellular organisms, communication among cells is essential for coordination of the different activities of the cells in the various tissues and organs. Much of this communication is accomplished by chemical signals that either pass through the plasma membrane or interact with receptors on the membrane surface. Most receptors are transmembrane proteins that become activated when they bind a signal molecule and generate second messengers inside the cell. The second messengers, in turn, amplify the stimulus and trigger the cell's response. This process is known as signal transduction.

SUMMARY TABLE The Movement of Substances across Membranes

MOVEMENT OF IONS AND SMALL MOLECULES

Name of Process	Movement Against or Down a Gradient	Transport Proteins Required?	Energy Source Such as ATP Required?	Substances Moved	Comments
Passive transport					
Simple diffusion	Down	No	No	Small nonpolar molecules (O_2, CO_2, and others)	Net movement of a substance down its concentration gradient.
Osmosis (a special case of diffusion)	Down	No	No	H_2O	Diffusion of water across a selectively permeable membrane.
Facilitated diffusion	Down	Yes	No	Polar molecules and ions	Carrier proteins undergo conformational changes to transport a specific solute. Channel proteins form water-filled pores for specific ions.
Active transport	Against	Yes	Yes	Polar molecules and ions	Often involves proton pumps. Enables cells to accumulate or expel solutes at high concentrations.

MOVEMENT OF LARGE MOLECULES AND PARTICLES (VESICLE-MEDIATED TRANSPORT)

Name of Process	Basic Function	Examples and Comments
Exocytosis	Releasing materials from the cell	Secretion of polysaccharides into the cell wall matrix; secretion of digestive enzymes by carnivorous plants.
Endocytosis	Taking materials into the cell	
Phagocytosis	Ingesting solids	Ingestion of bacteria or cellular debris.
Pinocytosis	Taking up liquids	Incorporation of fluid from the environment.
Receptor-mediated endocytosis	Taking up specific molecules	Binding of molecules by specific receptors in clathrin-coated pits, which then invaginate to form coated vesicles in the cell.

Plasmodesmata Enable Cells to Communicate

Plasmodesmata are also important pathways in cell-to-cell communication. All of the protoplasts of interconnected cells, together with their plasmodesmata, constitute a continuum called the symplast. The cell wall continuum surrounding the symplast is called the apoplast. Once thought of as rather passive entities through which ions and small molecules move by simple diffusion or bulk flow, plasmodesmata are now known to be dynamic structures capable of controlling the intercellular movement of various-sized molecules.

QUESTIONS

1. Distinguish between a substance moving *down* a concentration gradient and a substance moving *against* a concentration gradient.

2. At the close of the Punic Wars, when the Romans destroyed the city of Carthage (in 146 B.C.E.), it is said that they sowed the ground with salt and plowed it in. Explain, in terms of the physiological processes discussed in this chapter, why this action would make the soil barren to most plants for many years.

3. After the spring thaw and with April showers, homeowners often find water backing up in the basement drain because the sewer pipe is clogged with roots. Using the information you have obtained in this chapter, explain how the roots got into the sewer pipe.

4. Secondary active transport is significant to the plant because it enables a cell to accumulate even neutral solutes at concentrations much higher than those outside the cell. Using the terms proton pump (H^+-ATPase), proton gradient, proton-coupled cotransport, sucrose-proton cotransport, primary active transport, and secondary active transport, explain how this system works.

5. What is vesicle-mediated transport, and how does endocytosis differ from exocytosis?

6. What are the differences between phagocytosis and receptor-mediated endocytosis?

7. Explain, in general terms, what happens in each step—reception, transduction, and induction—of a signal-transduction pathway.

8. Diagram, label, and explain the structure of a plasmodesma.

SECTION 2

ENERGETICS

◀ The green, photosynthesizing leaves of the grapevine are its principal source of sugars, and its fruits—the grapes—are one of the important destinations for those sugars. The sugar-filled juices of the grapes can be extracted, stored under anaerobic conditions with yeast cells, and converted into wine by the metabolizing of glucose to ethanol.

CHAPTER 5

The Flow of Energy

Heating up Emerging in late winter, skunk cabbage *(Symplocarpus foetidus)* melts surrounding snow and ice by hydrolyzing ATP—which supplies the energy for most cellular activities—to ADP, releasing energy as heat. In spite of the cold, the plant maintains a constant internal temperature of 22°C (72°F) and emits a foul odor, luring the pollinating flies and bees that ensure its reproduction.

CHAPTER OUTLINE

Life here on Earth is solar-powered. Nearly every process vital to life depends on a steady flow of energy from the sun. An immense amount of solar energy—estimated to be 13 × 10^{23} calories per year—reaches the Earth. (A calorie is the amount of heat required to raise the temperature of one gram of water 1°C.) About 30 percent of this solar energy is immediately reflected back into space as light, just as light is reflected from the moon. About 20 percent is absorbed by the Earth's atmosphere. Much of the remaining 50 percent is absorbed by the Earth itself and converted to heat. Some of this absorbed heat energy serves to evaporate the waters of the oceans, producing the clouds that, in turn, produce rain and snow. Solar energy, in combination with other factors, is also responsible for the movements of air and water that help set patterns of climate over the surface of the Earth.

Less than one percent of the solar energy reaching the Earth is captured by the cells of plants and other photosynthetic organisms and converted by them into the energy that drives the processes of life. Living organisms change energy from one form to another, transforming the radiant energy of the sun into the chemical, electrical, and mechanical energy used by nearly all the living organisms on our planet (Figure 5–1).

This flow of energy is the essence of life. In fact, one way of viewing evolution is as a competition among organisms for the most efficient use of energy resources. A cell can be best understood as a complex system for transforming energy. At the other end of the biological scale, the structure of an ecosystem (that is, all the living organisms in a particular locale and the nonliving factors with which they interact), or of the biosphere (the zone of air, land, and water at the Earth's surface occupied by organisms) itself, is determined by the energy exchanges occurring among the groups of organisms within it.

In this chapter, we look first at the general principles governing all energy transformations. Then we turn our attention to the characteristic ways in which cells regulate the energy transformations that take place within living organisms. In the chapters that follow, we will examine the principal and com-

CHECKPOINTS

After reading this chapter, you should be able to answer the following:

1. What are the first and second laws of thermodynamics, and how do they relate to living organisms?
2. Why are oxidation-reduction reactions important in biology?
3. How do enzymes catalyze chemical reactions, and what are some of the factors that influence enzyme activity?
4. How does feedback inhibition regulate cellular activities?
5. What are coupled reactions, and how does ATP function as an intermediate between exergonic and endergonic reactions?

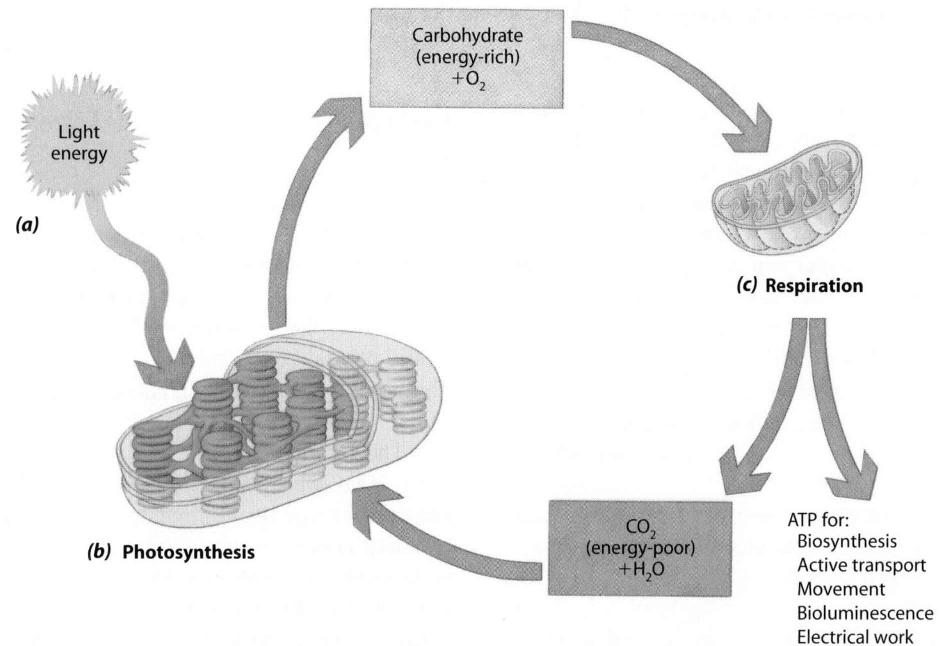

5–1 Energy flow in the biosphere ***(a)*** The radiant energy of sunlight is produced by nuclear fusion reactions taking place in the sun. ***(b)*** Chloroplasts, present in all photosynthetic eukaryotic cells, capture the radiant energy of sunlight and use it to convert water and carbon dioxide into carbohydrates, such as glucose, sucrose, or starch. Oxygen is released into the air as a product of the photosynthetic reactions. ***(c)*** Mitochondria, present in eukaryotic cells, carry out the final steps in the breakdown of these carbohydrates and capture their stored energy in ATP molecules. This process, cellular respiration, consumes oxygen and produces carbon dioxide and water, completing the cycling of the molecules. With each transformation, some energy is dissipated to the environment in the form of heat. Thus, the flow of energy through the biosphere occurs in one direction. It can continue only as long as there is an input of energy from the sun. Energy flow also occurs in prokaryotic cells, with photosynthesis limited to photosynthetic bacteria. Respiration occurs in the vast majority of photosynthetic organisms.

plementary processes of energy flow through the biosphere: respiration (in Chapter 6) and photosynthesis (in Chapter 7).

The Laws of Thermodynamics

Energy is an elusive concept. Today it is usually defined as the capacity to do work. Until about 200 years ago, heat—the form of energy most readily studied—was considered to be a separate, though weightless, substance called "caloric." An object was hot or cold depending on how much caloric it contained; when a cold object was placed next to a hot one, the caloric flowed from the hot object into the cold one; and when metal was pounded with a hammer, it became warm because the caloric was forced to the surface. Even though the idea of a caloric substance proved incorrect, the concept turned out to be surprisingly useful.

The development of the steam engine in the latter part of the eighteenth century, more than any other single chain of events, changed scientific thinking about the nature of energy. Energy became associated with work, and heat and motion came to be seen as forms of energy. This new understanding led to the study of **thermodynamics**—the science of energy transformations—and to formulation of its laws.

The First Law States That the Total Energy in the Universe Is Constant

The **first law of thermodynamics** states, quite simply: *Energy can be changed from one form to another but cannot be created or destroyed.* In engines, for instance, chemical energy (such as in coal or gasoline) is converted to heat, or thermal energy, which is then partially converted to mechanical movements (kinetic energy). Some of the energy is converted back to heat by the friction of these movements, and some leaves the engine in the form of exhaust products. Unlike the heat in the engine or the boiler, the heat produced by friction and lost in the exhaust cannot produce "work"—that is, it cannot move pistons or turn gears—because it is dissipated into the environment. But it is nevertheless part of the total equation. In fact, engineers calculate that most of the energy consumed by an engine is dissipated randomly as heat; most engines work with less than 25 percent efficiency.

The notion of **potential energy** developed in the course of such engine-efficiency studies. A barrel of oil or a ton of coal could be assigned a certain amount of potential energy, expressed in terms of the amount of heat it would liberate when burned. The efficiency of the conversion of the potential energy to usable energy depended on the design of the energy-conversion system.

Although these concepts were formulated in terms of engines running on heat energy, they apply to other systems as well. For example, a boulder pushed to the top of a hill contains potential energy. Given a little push (the energy of activation), it rolls down the hill again, converting the potential energy to energy of motion and to heat produced by friction. As mentioned in Chapter 4, water can also possess potential energy (page 76). As it moves by bulk flow from the top of a waterfall or over a dam, it can turn waterwheels that turn gears, for example, to grind maize. Thus, the potential energy of water, in this system, is converted to the kinetic energy of the wheels and gears and to heat, which is produced by the movement of the water itself as well as by the turning wheels and gears.

Light is another form of energy, as is electricity. Light can be changed to electrical energy, and electrical energy can be changed to light (by letting it flow through the tungsten wire in a light bulb, for example).

The first law of thermodynamics can be stated more completely as follows: *In all energy exchanges and conversions, the total energy of the system and its surroundings after the conversion is equal to the total energy before the conversion.* A "system" can be any clearly defined entity—for example, an exploding stick of dynamite, an idling automobile engine, a mitochondrion, a living cell, a tree, a forest, or the Earth itself. The "surroundings" consist of everything outside the system.

The first law of thermodynamics can be stated as a simple bookkeeping rule: when we add up the energy income and expenditures for any physical process or chemical reaction, the books must always balance. (Note, however, that the first law does not apply to nuclear reactions, in which energy is, in fact, created by the conversion of mass to energy.)

The Second Law States That the Entropy of the Universe Is Increasing

The energy released as heat in an energy conversion has not been destroyed—it is still present in the random motion of atoms and molecules—but it has been "lost" for all practical purposes. It is no longer available to do useful work. This brings us to the **second law of thermodynamics,** which is the more important one, biologically speaking. It predicts the direction of all events involving energy exchanges. Thus, it has been called "time's arrow."

The second law states: *In all energy exchanges and conversions, if no energy leaves or enters the system under study, the potential energy of the final state will always be less than the potential energy of the initial state.* The second law is entirely in keeping with everyday experience (Figure 5–2). A boulder will roll downhill but never uphill. A ball that is dropped will bounce—but not back to the height from which it was dropped. Heat will flow from a hot object to a cold one and never the other way.

A process in which the potential energy of the final state is less than that of the initial state is one that releases energy. Such a process is said to be **exergonic** ("energy-out"). Only exergonic processes can take place spontaneously. Although "spontaneously" has an explosive sound, the word says nothing about the speed of the process—just that it can take place without an input of energy from outside the system. By contrast, a process in which the potential energy of the final state is greater than that of the initial state is one that requires energy. Such processes are said to be **endergonic** ("energy-in"). For an

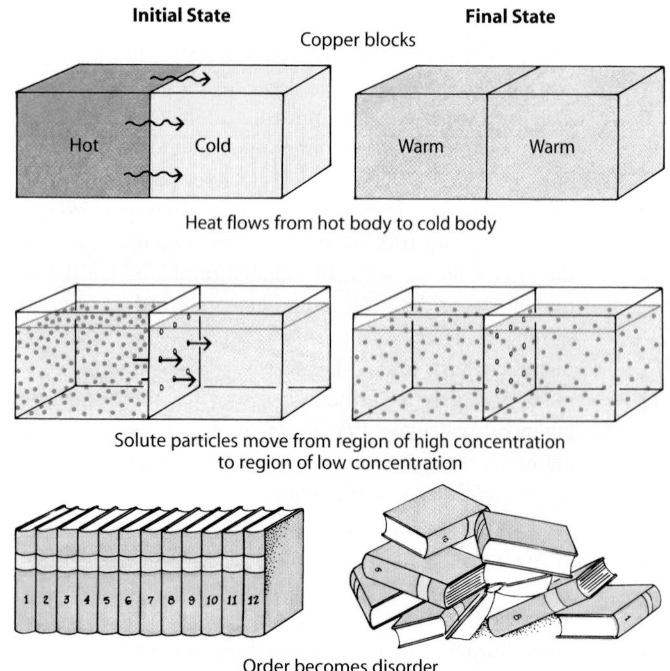

5–2 Some illustrations of the second law of thermodynamics In each case, a concentration of energy—in the hot copper block, in the solute particles on one side of a tank, and in the neatly organized books—is dissipated. In nature, processes tend toward randomness, or disorder. Only an input of energy can reverse this tendency and reconstruct the initial state from the final state. Ultimately, however, disorder will prevail, because the total amount of energy in the universe is finite.

endergonic process to proceed, there must be an input of energy to the system.

One important factor in determining whether or not a reaction is exergonic is ΔH, the change in heat content of the system, where Δ stands for change, and H for heat content (the formal term for H is "enthalpy"). In general, the change in heat content is approximately equal to the change in potential energy. The energy change that occurs when glucose, for instance, is oxidized can be measured in a calorimeter and expressed in terms of ΔH. The complete oxidation of a mole[1] of glucose to carbon dioxide and water yields 673 kilocalories:

$$\underset{\text{Glucose}}{C_6H_{12}O_6} + \underset{\text{Oxygen}}{6O_2} \longrightarrow \underset{\text{Carbon dioxide}}{6CO_2} + \underset{\text{Water}}{6H_2O} + \underset{\text{Heat released}}{673 \text{ kcal}}$$

$$\Delta H = -673 \text{ kcal/mol}$$

In many cases, an exergonic chemical reaction is also an exothermic reaction—that is, it gives off heat and thus has a negative ΔH. (In making metabolic measurements, the kilocalorie, or kcal, is generally used; 1 kilocalorie is the amount of heat required to raise the temperature of one kilogram of water 1°C.)

Another factor besides the gain or loss of heat determines the direction of a process. This factor, called **entropy** (symbolized as S), is a measurement of the disorder, or randomness, of a system. Let us return to water as an example. The change from ice to liquid water and the change from liquid water to water vapor are both endothermic processes—a considerable amount of heat is absorbed from the surroundings as they occur. Yet, under the appropriate conditions, they proceed spontaneously. The key factor in these processes is the increase in entropy. In the case of the change of ice into water, a solid is being turned into a liquid, and some of the hydrogen bonds that hold the water molecules together in a crystal (ice) are being broken. As the liquid water turns to vapor, the rest of the hydrogen bonds are ruptured as the individual water molecules separate, one by one. In each case, the disorder of the system has increased.

The notion that there is more disorder associated with more numerous and smaller objects than with fewer, larger ones is in keeping with our everyday experience. If there are 20 papers on a desk, the possibilities for disorder are greater than if there are 2 or even 10. If each of the 20 papers is cut in half, the entropy of the system—the capacity for randomness—increases. The relationship between entropy and energy is also a commonplace idea. If you were to find your room tidied up and your books in alphabetical order on the shelf, you would recognize that someone had been at work—that energy had been expended. To organize the papers on a desk similarly requires the expenditure of energy.

[1]A mole (abbreviated as mol) is the amount of a substance with a weight, in grams, that is numerically equal to its atomic weight or molecular weight. For example, carbon dioxide (CO_2) has a molecular weight of 44, so one mole of CO_2 is 44 grams of CO_2. Glucose ($C_6H_{12}O_6$) has a molecular weight of 180, so one mole of glucose is 180 grams of glucose.

Now let us return to the question of the energy changes that determine the course of chemical reactions. As discussed above, both the change in the heat content of the system (ΔH) and the change in entropy (ΔS) contribute to the overall change in energy. This total change—which takes into account both heat and entropy—is called the **free-energy change** and is symbolized as ΔG, after the American physicist Josiah Willard Gibbs (1839–1903), who was one of the first to integrate all of these ideas.

Keeping ΔG in mind, let us examine once again the oxidation of glucose. The ΔH of this reaction is –673 kcal/mol. The ΔG is –686 kcal/mol. (Note that these are values under *standard conditions,* with all reactants and products present at a concentration of one molar (1 mole per liter), as well as a temperature of 25°C and a pressure of 1 atmosphere. The real values under actual conditions are likely to be somewhat different.) Thus, the entropy factor has contributed 13 kcal/mol to the free-energy change of the process. The changes in heat content and in entropy both contribute to the lower energy state of the products of the reaction.

The relationship among ΔG, ΔH, and entropy is given in the following equation:

$$\Delta G = \Delta H - T\Delta S$$

This equation states that the free-energy change is equal to the change in heat content (a negative value in exothermic reactions, which give off heat) minus the change in entropy multiplied by the absolute temperature T. In exergonic reactions, ΔG is always negative, but ΔH may be zero or even positive. Because T is always positive, the greater the change in entropy, the more negative ΔG will be—that is, the more exergonic the reaction will be. Therefore it is possible to state the second law in another, simpler way: *All naturally occurring processes are exergonic.*

Living Organisms Require a Steady Input of Energy

The most interesting implication of the second law, as far as biology is concerned, is the relationship between entropy and order. Living systems are continuously expending large amounts of energy to maintain order. Stated in terms of chemical reactions, living systems continuously expend energy to maintain a position far from equilibrium. If equilibrium were to occur, the chemical reactions in a cell would, for all practical purposes, stop, and no further work could be done. At equilibrium, a cell would be dead.

The universe is a closed system—that is, neither matter nor energy enters or leaves the system. The matter and energy present in the universe at the time of the "big bang" are all the matter and energy it will ever have. Moreover, after each and every energy exchange and transformation, the universe as a whole has less potential energy and more entropy than it did before. In this view, the universe is running down. The stars will flicker out, one by one. Life—any form of life on any planet—will come to an end. Finally, even the motion of individual molecules will cease. Take heart, though. Even the most pessimistic among us do not believe this will occur for another 20 billion years or so.

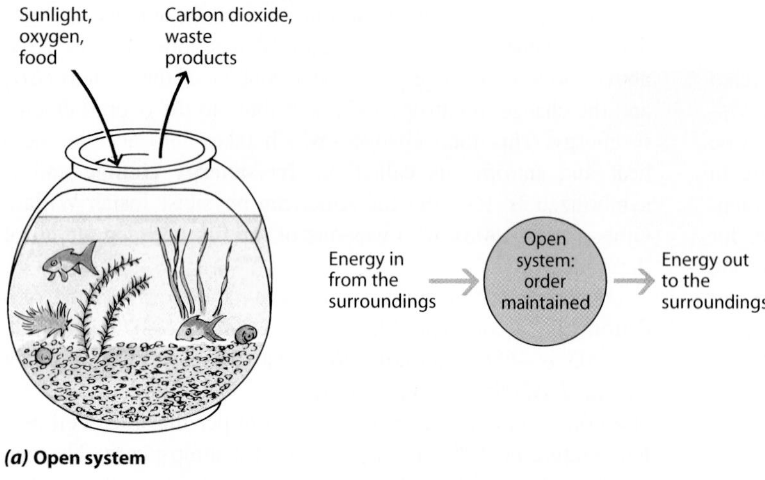

(a) **Open system**

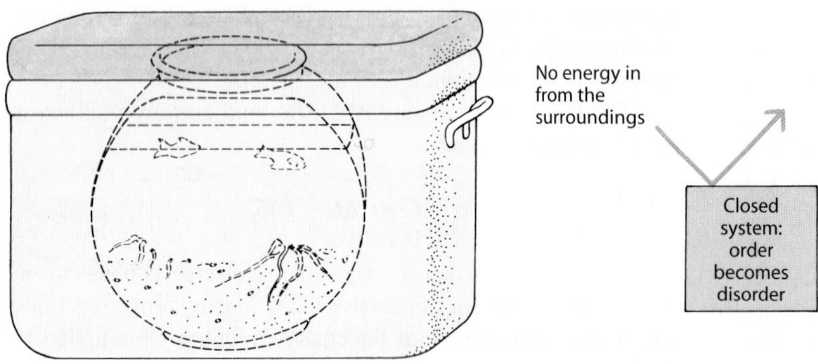

(b) **Closed system**

5–3 Open and closed systems ***(a)*** A fish bowl is, like the Earth, an open system—matter and energy enter and leave the system. Sunlight passes through the glass, oxygen diffuses into the water at its surface, and food is added by a human caretaker. Heat leaves the system through the glass and through the opening at the top, carbon dioxide diffuses out of the water at its surface, and the waste products of the animals are removed when the bowl is cleaned. Although energy is lost from the system with each and every energy exchange, a steady supply of energy—mainly food—from outside the system maintains its order. ***(b)*** If, however, the fish bowl is placed in an opaque container that is sealed and insulated, it becomes a closed system, consisting of the fish bowl, its contents, and the air within the container. Neither matter nor energy can enter or leave the system. For a period of time after the container is sealed, energy continues to be converted from one form to another by the organisms in the fish bowl. With each and every conversion, however, some of the energy is given off as heat and dissipated into the water, the glass, and the air within the container. In time, the system will run down—the organisms will die and their bodies will disintegrate. The order originally present in the system will have become the disorder of individual atoms and molecules moving at random.

In the meantime, life can exist *because* the universe is running down. Although the universe as a whole is a closed system, the Earth is not. It is an open system (Figure 5–3), receiving an energy input of about 13×10^{23} calories per year from the sun. Photosynthetic organisms are specialists at capturing the light energy released by the sun as it slowly burns itself out. They use this energy to organize small, simple molecules (water and carbon dioxide) into larger, more complex molecules (sugars). In the process, the captured light energy is stored as chemical energy in the bonds of sugars and other molecules.

Living cells—including photosynthetic cells—can convert this stored energy into motion, electricity, and light, and, by shifting the energy from one type of chemical bond to another, into more useful forms of chemical energy. At each transformation, energy is lost to the surroundings as heat. But before the energy captured from the sun is completely dissipated, organisms use it to create and maintain the complex organization of structures and activities that we know as life.

Oxidation-Reduction

Chemical reactions are essentially energy transformations in which energy stored in chemical bonds is transferred to other, newly formed chemical bonds. In such transfers, electrons shift from one energy level to another. In many reactions, electrons pass from one atom or molecule to another. These reactions, known as **oxidation-reduction** (or redox) **reactions,** are of great importance in living organisms.

The *loss* of an electron is known as **oxidation,** and the atom or molecule that loses the electron is said to be oxidized. The reason electron loss is called oxidation is that oxygen, which attracts electrons very strongly, is often the electron acceptor.

Reduction is, conversely, the *gain* of an electron. Oxidation and reduction reactions always take place simultaneously. The electron lost by the oxidized atom is accepted by another atom, which is thus reduced—hence the term "redox" (for reduction-oxidation) reactions (Figure 5–4).

Redox reactions may involve only a solitary electron, as when sodium loses an electron and becomes oxidized to Na^+, and chlorine gains an electron and is reduced to Cl^-. In biological reactions, however, the electron is often accompanied by a proton, that is, as a hydrogen atom. In such cases, oxidation involves the removal of hydrogen atoms, and reduction the gain of hydrogen atoms. For example, when glucose is oxidized in the process of cellular respiration, electrons and protons are lost by the glucose molecule and are gained by oxygen atoms, which are reduced to water:

$$\underset{\text{Glucose}}{C_6H_{12}O_6} + \underset{\text{Oxygen}}{6O_2} \longrightarrow \underset{\text{Carbon dioxide}}{6CO_2} + \underset{\text{Water}}{6H_2O} + \text{Energy}$$

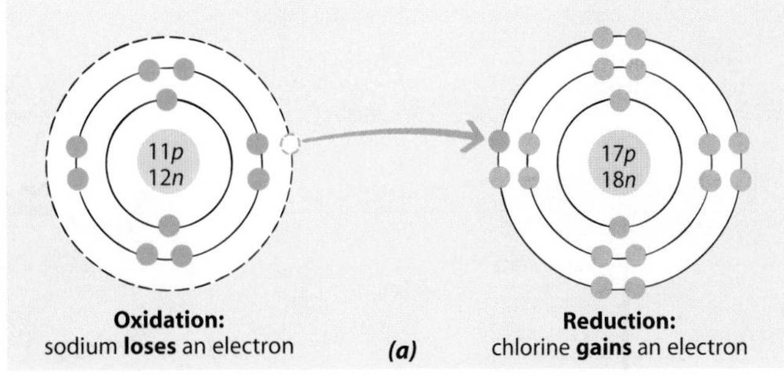

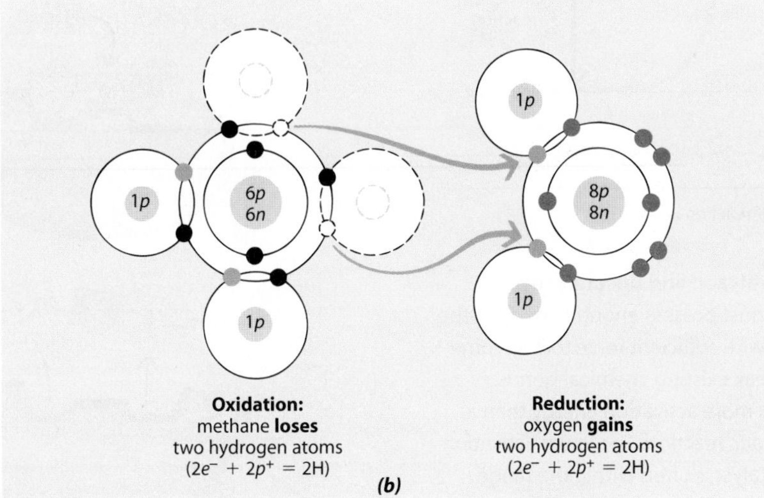

5–4 Redox reactions ***(a)*** In some oxidation-reduction reactions, such as the oxidation of sodium and the reduction of chlorine, a single electron is transferred from one atom to another. Such simple reactions typically involve elements or inorganic compounds. ***(b)*** In other oxidation-reduction reactions, such as the partial oxidation of methane (CH_4), the electrons are accompanied by protons. In these reactions, which often involve organic molecules, oxidation is the loss of hydrogen atoms, and reduction is the gain of hydrogen atoms. When an oxygen atom gains two hydrogen atoms, as shown here, the product is, of course, a water molecule.

The electrons are moved to a lower energy level, and energy is released. In other words, the oxidation of glucose is an exergonic process.

Conversely, during photosynthesis, electrons and hydrogen atoms are transferred from water to carbon dioxide, thereby oxidizing the water to oxygen and reducing the carbon dioxide to form three-carbon sugars:

$$\underset{\text{Carbon dioxide}}{6CO_2} + \underset{\text{Water}}{6H_2O} + \text{Energy} \longrightarrow \underset{\text{Three-carbon sugar}}{2C_3H_6O_3} + \underset{\text{Oxygen}}{6O_2}$$

In this case, the electrons are moved to a higher energy level, and an input of energy is required for the reaction to take place. The reduction of carbon dioxide to sugar, in other words, is an endergonic process.

The complete oxidation of a mole of glucose under standard conditions, as defined earlier, releases 686 kilocalories of energy (that is, $\Delta G = -686$ kcal/mol). Conversely, the reduction of carbon dioxide to form the equivalent of a mole of glucose stores 686 kilocalories of energy in the chemical bonds of glucose.

If the energy released during the oxidation of glucose by oxygen were to be released all at once, most of it would be dissipated as heat. Not only would it be of no use to the cell, but the resulting high temperature would destroy the cell. Mechanisms have evolved in living organisms, however, that regulate these chemical reactions—and a multitude of others—in such a way that, during the oxidation of glucose, the energy is released in a series of small increments. The energy released is, in turn, stored in certain chemical bonds from which it can later be released as the cell needs it. These mechanisms, which require only a few kinds of molecules, enable cells to use energy efficiently, without disrupting the delicate balances that characterize a living organism. To understand how they work, we must look more closely at the proteins known as enzymes and at the molecule known as ATP.

Enzymes

Most chemical reactions require an initial input of energy to get started. This is true even for exergonic reactions such as the oxidation of glucose or the burning of natural gas in a home furnace.

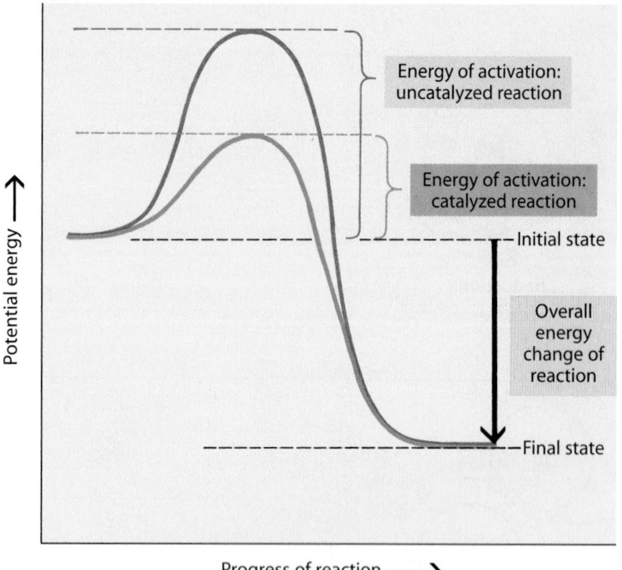

5–5 Energy of activation in catalyzed and uncatalyzed reactions To react, molecules must possess enough energy—the energy of activation—to collide with sufficient force to overcome their mutual repulsion and to break existing chemical bonds. An uncatalyzed reaction requires more activation energy than a catalyzed one, such as an enzymatic reaction. The lower activation energy in the presence of the catalyst is often within the range of energy possessed by the molecules within living cells, so the reaction can occur at a rapid rate with little or no added energy. Note, however, that the overall energy change from the initial state to the final state is the same with and without the catalyst.

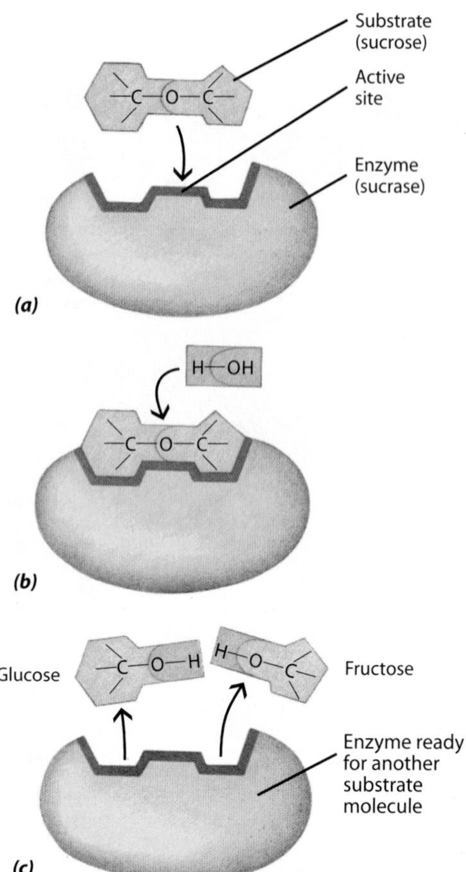

5–6 A model of enzyme action ***(a), (b)*** Sucrose, a disaccharide, is hydrolyzed to yield a molecule of glucose and a molecule of fructose ***(c)***. The enzyme involved in this reaction, sucrase, is specific for this process. As you can see, the active site of the enzyme fits the opposing surface of the sucrose molecule. The fit is so exact that a molecule composed of, for example, two subunits of glucose would not be affected by this enzyme.

The energy that must be possessed by the molecules in order to react is known as the **energy of activation** (Figure 5–5).

In the laboratory, the energy of activation is usually supplied as heat. In a cell, however, many different reactions are going on at the same time, and heat would affect all of these reactions indiscriminately. Moreover, excessive heat could break the hydrogen bonds that maintain the structure of many of the cell's molecules and would have other generally destructive effects. Cells avoid this problem by the use of **enzymes,** protein molecules that are specialized to serve as catalysts.

A **catalyst** is a substance that lowers the activation energy required for a reaction by forming a temporary association with the molecules that are reacting. This temporary association brings the reacting molecules close to one another and may also weaken existing chemical bonds, making it easier for new ones to form. As a result, little, if any, added energy is needed to start the reaction, and it then proceeds more rapidly than it would in the absence of the catalyst. The catalyst itself is not permanently altered in the process, so it can be used over and over again. (In Chinese, the word for "catalyst" is the same as the word for "marriage broker," and the functions are indeed analogous.)

Because of enzymes, cells are able to carry out chemical reactions at great speed and at comparatively low temperatures. A single enzyme molecule may catalyze the reaction of tens of thousands of identical molecules in a second. Thus enzymes are typically effective in very small amounts. The molecule on which an enzyme acts is known as its **substrate.** For example, in the reaction in Figure 5–6, sucrose is the substrate and sucrase is the enzyme. Many enzymes have multiple substrates.

An Enzyme Has an Active Site That Binds a Specific Substrate

A few enzymes are RNA molecules known as **ribozymes** (page 29). All other enzymes, however, are large, complex globular proteins consisting of one or more polypeptide chains (page 27). The polypeptide chains of an enzyme are folded in such a way that they form a groove or pocket on the protein surface (Figure 5–7). The substrate fits very precisely into this groove, which is the site of the reactions catalyzed by the enzyme. This portion of the enzyme is known as the **active site.**

The active site not only has a precise three-dimensional shape but also has exactly the correct array of charged and uncharged, or hydrophilic and hydrophobic, areas on its binding

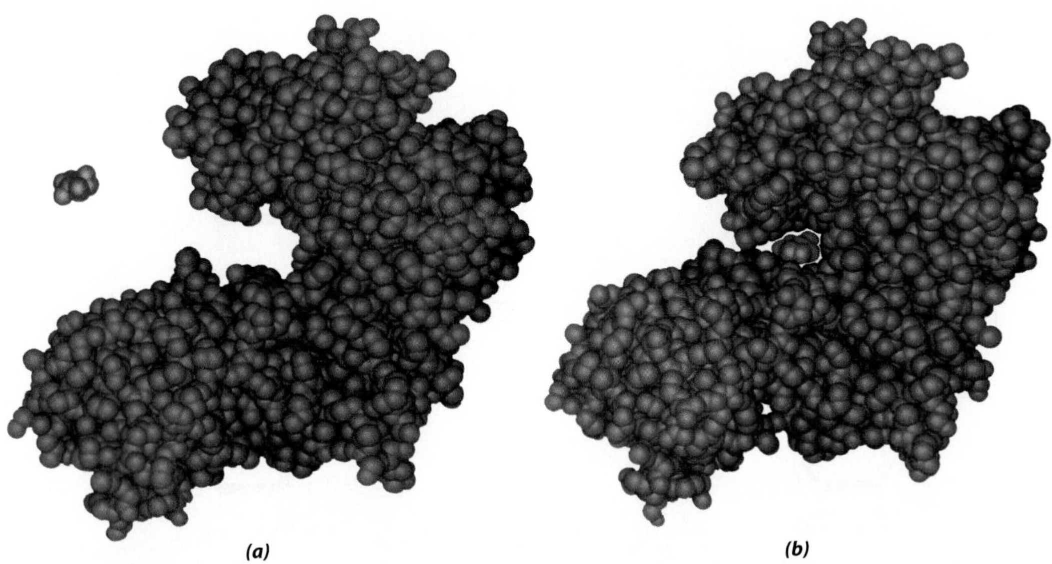

5–7 Hexokinase Space-filling models of the yeast enzyme hexokinase (blue) and one of its substrates, glucose (brown). Hexokinase catalyzes the first step in the breakdown of glucose in respiration. Such models, which are produced by computer techniques, show the three-dimensional shape of the molecules. Here, the glucose molecule is shown colliding with the enzyme and binding to the active site, which appears as a cleft in the side of the hexokinase molecule. ***(a)*** In the absence of glucose, hexokinase has an open cleft. ***(b)*** When glucose is bound by hexokinase, the cleft is partially closed.

surface. If a particular portion of the substrate has a negative charge, the corresponding feature on the active site has a positive charge, and so on. Thus, the active site both confines the substrate molecule and orients it in the correct manner.

The amino acids involved in the active site need not be adjacent to one another on the polypeptide chains. In fact, in an enzyme with quaternary structure (page 29), these amino acids may even be on different polypeptide chains. The amino acids are brought together at the active site by the precise folding of the polypeptide chains in the molecule.

Cofactors in Enzyme Action

The catalytic activity of some enzymes appears to depend only on their structure as proteins. Other enzymes, however, require one or more nonprotein components, known as **cofactors,** without which the enzymes cannot function.

Some Cofactors Are Metal Ions

Certain metal ions are cofactors for particular enzymes. For example, magnesium ion (Mg^{2+}) is required in most enzymatic reactions involving the transfer of a phosphate group from one molecule to another. The two positive charges on the magnesium ion hold the negatively charged phosphate group in position. Other ions, such as Ca^{2+} and K^+, play similar roles in other reactions. In some cases, ions serve to hold the enzyme protein in its proper three-dimensional shape.

Other Cofactors Are Organic Molecules Called Coenzymes

Nonprotein organic cofactors also play a crucial role in enzyme-catalyzed reactions. Such cofactors are called **coenzymes.** For example, in some oxidation-reduction reactions, electrons are passed along to a molecule that serves as an electron acceptor. There are several different electron acceptors in any given cell, and each is tailor-made to hold the electron at a slightly different energy level. As an example, let us look at just one, nicotinamide adenine dinucleotide (NAD^+), which is shown in Figure 5–8. At first glance, NAD^+ looks complex and unfamiliar, but if you look at it more closely, you will find that you recognize most of its component parts. The two ribose (five-carbon sugar) units are linked by a **pyrophosphate bridge.** One of the ribose units is attached to the nitrogenous base adenine. The other is attached to another nitrogenous base, nicotinamide. A nucleoside plus a phosphate is called a **nucleotide,** and a molecule that contains two of these combinations is called a dinucleotide.

The nicotinamide ring is the active end of NAD^+—that is, the part that accepts electrons. Nicotinamide is a derivative of niacin, which is one of the B vitamins. Vitamins are organic compounds that are required in small quantities by many living organisms. Whereas plants can synthesize all of their required vitamins, humans and other animals cannot synthesize most vitamins and so must obtain them in their diets. Many vitamins are precursors of coenzymes or parts of coenzymes.

When nicotinamide is present, our cells can use it to make NAD^+, which, like many other coenzymes, is recycled. That is, NAD^+ is regenerated when $NADH + H^+$ passes its electrons to another electron acceptor. Thus, although this coenzyme is involved in many cellular reactions, the actual number of NAD^+ molecules required is relatively small.

Some enzymes use nonprotein cofactors that are permanently attached to the enzyme protein. Such tightly bound cofactors (either ions or coenzymes) are referred to as

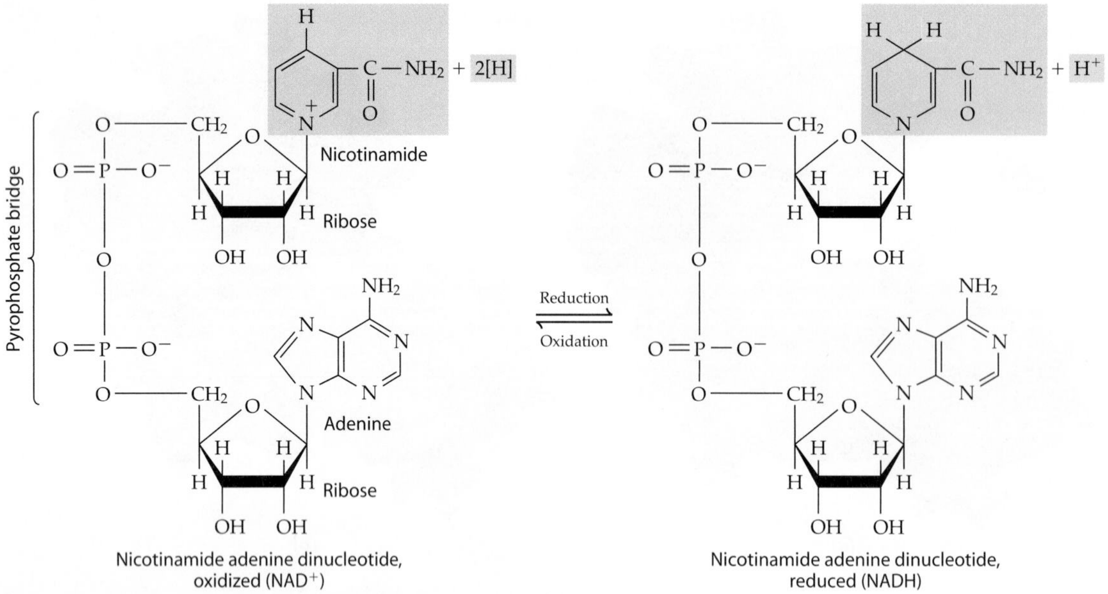

Pyrophosphate bridge

Nicotinamide

Ribose

Adenine

Ribose

Reduction

Oxidation

Nicotinamide adenine dinucleotide, oxidized (NAD^+)

Nicotinamide adenine dinucleotide, reduced (NADH)

(a)

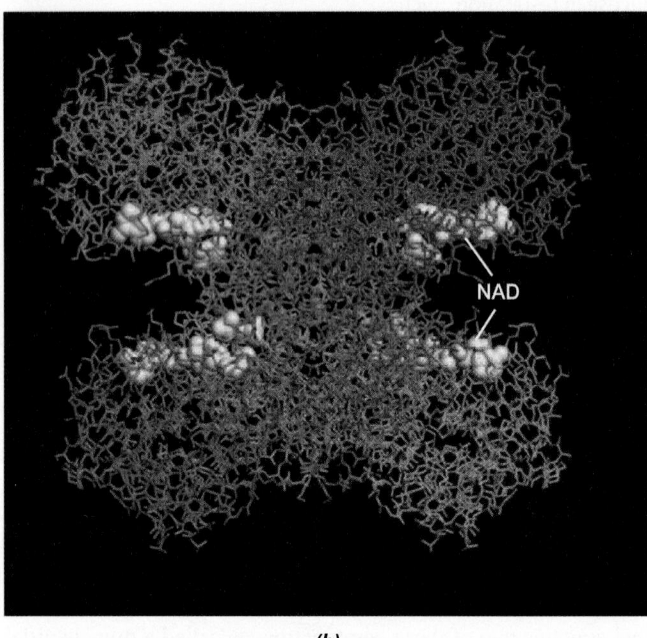

(b)

5–8 Nicotinamide adenine dinucleotide ***(a)*** This electron acceptor is shown here in its oxidized form, NAD^+, and in its reduced form, NADH. Nicotinamide is a derivative of niacin, one of the B vitamins. Notice how the bonding within the nicotinamide ring (shaded rectangle) shifts as the molecule changes from the oxidized to the reduced form, and vice versa. Reduction of NAD^+ to NADH requires two electrons and one hydrogen ion, or proton (H^+). The two electrons, however, generally travel as components of two hydrogen atoms; thus, one hydrogen ion is "left over" when NAD^+ is reduced. ***(b)*** The coenzyme NAD (white) and the four subunits (red, yellow, purple, and green) of the enzyme glyceraldehyde 3-phosphate dehydrogenase, which is involved in glycolysis, the metabolic breakdown of glucose (see Figure 6–3).

prosthetic groups. Examples of prosthetic groups are the iron-sulfur clusters of ferredoxins (see page 134).

Metabolic Pathways

Enzymes characteristically work in series, like workers on an assembly line. Each enzyme catalyzes one small step in an ordered series of reactions that together form a **metabolic,** or **biochemical, pathway** (Figure 5–9). Different metabolic pathways serve different functions in the life of the cell. For example, one pathway may be involved in the breakdown of the polysaccharides in bacterial cell walls, another in the breakdown of glucose, and another in the synthesis of a particular amino acid.

Cells derive several advantages from this sort of arrangement. First, the groups of enzymes making up a common pathway can be segregated within the cell. Some are found in solution, as in the vacuoles, whereas others are embedded in the membranes of specialized organelles, such as mitochondria and chloroplasts. A second advantage is that there is little accumulation of intermediate products, because each product tends to be used up in the next reaction along the pathway. A third advantage is that if any of the reactions along the pathway are highly exergonic (that is, energy-releasing), they will rapidly use up the products of the preceding reactions, pulling those reactions forward. Similarly, the products that accumulate from the exergonic reactions will push the next reactions forward by increasing the concentrations of the reactants for those next reactions.

Some reactions are common to two or more pathways in a cell. Frequently, however, identical reactions that occur in different pathways are catalyzed by different enzymes. Such

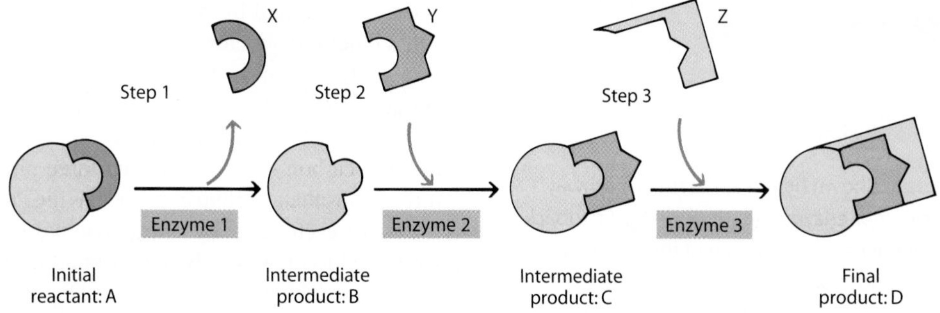

5–9 Schematic representation of a metabolic pathway To produce the final product (D) from the initial reactant (A), a series of reactions is required. Each reaction is catalyzed by a different enzyme, and each results in a small but significant modification in the substrate molecule. If any step of the pathway is inhibited—either because of a nonfunctioning enzyme or because a substrate is unavailable—the pathway will shut down, and the subsequent reactions of the series will not occur.

enzymes are referred to as **isozymes.** Commonly, each isozyme is coded for by a different set of genes. Isozymes are adapted to the specific pathway and cellular location where they are used.

Regulation of Enzyme Activity

Another remarkable feature of metabolism is the extent to which each cell regulates the synthesis of the products necessary for its well-being, making them in the appropriate amounts and at the rates required. At the same time, cells avoid overproduction, which would waste both energy and raw materials. The availability of reactant molecules or of cofactors is a principal factor in limiting enzyme action, and for this reason most enzymes probably work at a rate well below their maximum.

Temperature affects enzymatic reactions. An increase in temperature increases the rate of enzyme-catalyzed reactions, but only to a point. As can be seen in Figure 5–10, the rate of most enzymatic reactions approximately doubles for each 10°C rise in temperature in the range 10° to 40°C, but then drops off very quickly after about 40°C. The increase in reaction rate occurs because of the increased energy of the reactants. The decrease in reaction rate above about 40°C occurs as the structure of the enzyme molecule itself begins to unfold, due to disruption of the relatively weak forces that hold it in its specific, active shape. This unfolding of the enzyme molecule is known as **denaturation** (page 29).

The pH of the surrounding solution also affects enzyme activity. Among other factors, the three-dimensional shape of an enzyme depends on the attraction and repulsion between negatively charged (acidic) and positively charged (basic) amino acids. As the pH changes, these charges change, and so the shape of the enzyme changes, until it is so drastically altered that the enzyme is no longer functional. Probably more important, however, the charges of the active site and substrate are changed so that the binding capacity is affected. Some enzymes are frequently found at a pH that is not their optimum, suggesting that this discrepancy may not be an evolutionary oversight but a way of controlling enzyme activity. Several enzymes of the photosynthetic pathway are regulated by changes in pH.

Living organisms also have more precise ways of turning enzyme activity on and off. In each metabolic pathway there is at least one enzyme whose activity maintains considerable control over the rate of the overall metabolic pathway because it catalyzes the slowest, or rate-limiting, reaction. These **regulatory enzymes** exhibit increased or decreased catalytic activity in response to substrate levels and certain signals. Such regulatory enzymes constantly adjust the rate of each metabolic pathway to meet changes in the cell's demands for energy and for molecules required for growth and repair of the cell. In most metabolic pathways, the first enzyme of the sequence is a regulatory enzyme. By regulating a metabolic pathway at this early step, the cell ensures that minimal energy is spent and metabolites are diverted for more important processes.

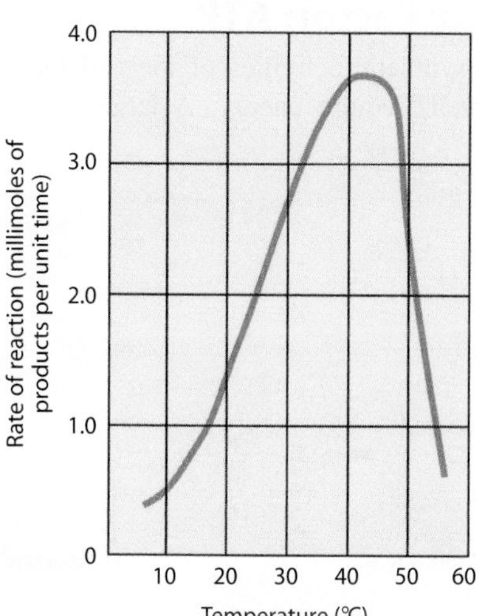

5–10 Effect of temperature on an enzyme-catalyzed reaction The concentrations of enzyme and reacting molecules (substrate) were kept constant. The rate of the reaction, as in most metabolic reactions, approximately doubles for every 10°C rise in temperature up to about 40°C. Above this temperature, the rate decreases as the temperature increases, and at about 60°C the reaction stops altogether, presumably because the enzyme is denatured.

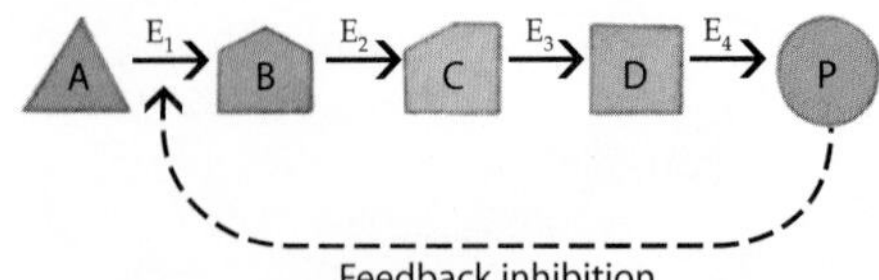

5–11 Feedback inhibition Shown here is a metabolic pathway catalyzed by a sequence of four enzymes (E_1 through E_4). Feedback inhibition typically involves allosteric inhibition of the first enzyme (E_1) in the sequence by the end product (P) of the pathway. Therefore, enzyme E_1 will be more active when amounts of P are low.

The most important kind of regulatory enzymes in metabolic pathways are **allosteric enzymes.** The term *allosteric* derives from the Greek *allos,* "other," and *stereos,* "shape" or "form." Allosteric enzymes have at least two sites: an **active site** that binds the substrate, and an **effector site** that binds the regulatory substance. When the regulatory substance is bound at the effector site, the enzyme is reversibly changed from one form to the other and the enzyme is said to be allosterically regulated.

In some metabolic pathways, the regulatory enzyme is specifically inhibited by the end product of the pathway, which accumulates when the amount of this product exceeds the cell's needs. By inhibiting the regulatory enzyme (the first enzyme in the sequence), all subsequent enzymes operate at reduced rates because their substrates become depleted. This type of regulation is called **feedback** (or **end-product**) **inhibition** (Figure 5–11). When more of this end product is needed by the cell, molecules of the end product dissociate from the regulatory enzyme and the activity of the enzyme increases again. In this way the cellular concentration of the pathway's final product is brought into balance with the cell's needs.

The Energy Factor: ATP

All of the biosynthetic activities of the cell (and many other activities as well) require energy. A large proportion of this energy is supplied by **ATP,** the nucleotide derivative that is the cell's chief energy currency.

At first glance, ATP (Figure 5–12) also appears to be a complex molecule. However, as with NAD^+, you will find that its component parts are familiar. ATP is made up of adenine, the five-carbon sugar ribose, and three phosphate groups. These three phosphate groups, each having negative charges, are linked to each other by phosphoanhydride bonds and, in turn, are linked to the ribose by a phosphoester bond.

To understand the role of ATP, let us briefly review the concept of **free energy,** the energy available to do work.

For any chemical reaction, the *direction* in which it spontaneously proceeds is determined by the free-energy difference between the reactants (or substrates, in the case of an enzyme-catalyzed reaction) and the resulting products. The free-energy change is symbolized as ΔG. Only if the reaction is exergonic (negative ΔG) will it proceed in the direction written. Yet many cellular reactions, including biosynthetic reactions—such as the formation of a disaccharide from two monosaccharide molecules—are endergonic (positive ΔG). In such reactions, the electrons forming the chemical bonds of the product are at a higher energy level than the electrons in the bonds of the starting materials. That is, the potential energy of the product is greater than the potential energy of the reactants, an apparent violation of the second law of thermodynamics. Cells circumvent this difficulty by using enzymes to catalyze **coupled reactions** in which otherwise endergonic reactions are linked to and driven by exergonic reactions that provide a surplus of energy. The result is that the net process is exergonic and thus able to proceed spontaneously (Figure 5–13). ATP is the molecule that most frequently serves as the intermediate between the exergonic and the otherwise endergonic reaction in such coupled reactions.

Enzymes catalyzing the hydrolysis of ATP are known as **ATPases.** A variety of different ATPases have been identified. The ATPase head of the myosin molecule catalyzes the release of energy that is used by the myosin to "walk" along actin filaments (see essay on page 56). Many of the proteins that move molecules and ions through cellular membranes against concentration gradi-

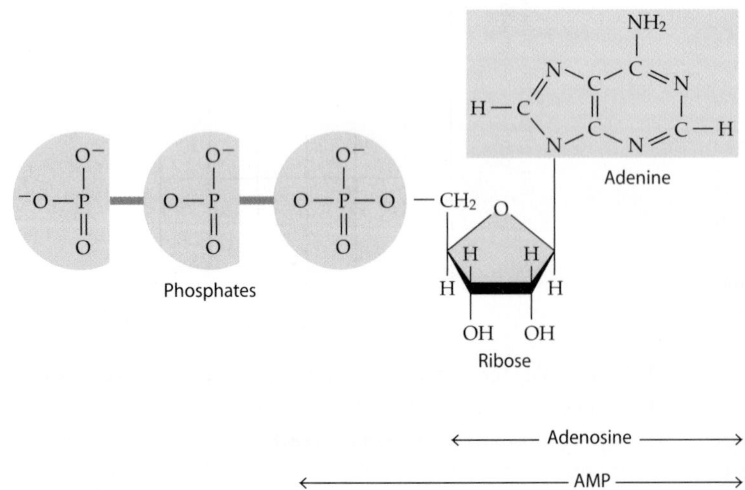

5–12 Adenosine triphosphate (ATP) The structures of ATP as well as adenosine diphosphate (ADP) and adenosine monophosphate (AMP) are shown. A phosphoester bond links the first phosphate group to the ribose of adenosine, and phosphoanhydride bonds (in blue) link the second and third phosphate groups. At pH 7, the phosphate groups are fully ionized.

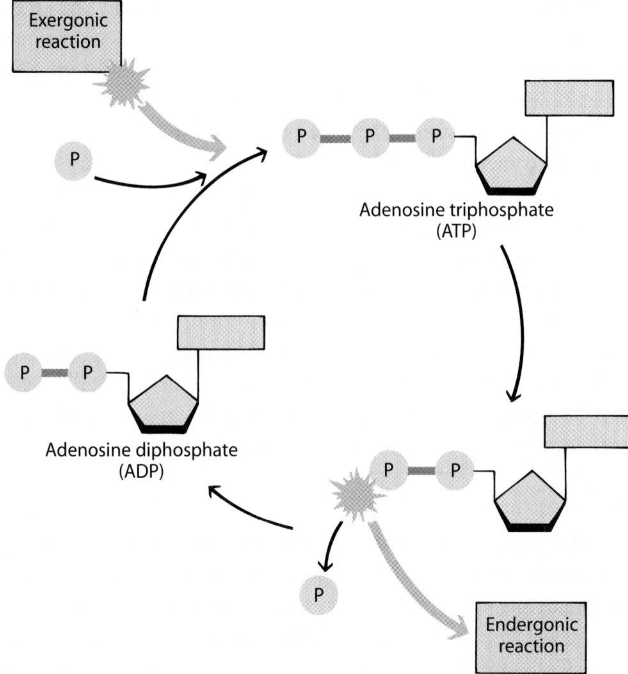

5–13 Endergonic and exergonic reactions In living organisms, endergonic reactions, such as biosynthetic reactions, are powered by the energy released in exergonic reactions to which they are coupled. In most coupled reactions, ATP is the intermediate that carries energy from one reaction to the other.

ents are not only transport proteins but also ATPases, releasing energy to power the transport process (page 85).

Because of its structure, the ATP molecule is well suited to this role in living organisms. Free energy is released from the ATP molecule when one phosphate group is removed by hydrolysis, producing a molecule of ADP (adenosine diphosphate) and a free phosphate ion:

$$\text{ATP} + \text{H}_2\text{O} \longrightarrow \text{ADP} + \text{Phosphate} + \text{Energy}$$

This reaction is highly exergonic. About 7.3 kilocalories of energy are released per mole of ATP hydrolyzed when the reaction is carried out under standard conditions. Under most cellular conditions, the energy yield is, in fact, significantly higher, often in the range of 12 to 15 kilocalories of energy released by the hydrolysis of 1 mole of ATP. Removal of a second phosphate group produces AMP (adenosine monophosphate) and releases an equivalent amount of free energy:

$$\text{ADP} + \text{H}_2\text{O} \longrightarrow \text{AMP} + \text{Phosphate} + \text{Energy}$$

The high negative free energy of hydrolysis of the terminal phosphoanhydride bond in ATP is due, in large part, to the fixed and closely localized negative charges on the three phosphates (Figure 5–12). Hydrolysis leaves the molecule with only two fixed, adjacent negative charges—a more stable arrangement. Hence, ADP + P_i has a lower overall free energy than ATP.

In many ATP-requiring reactions within a cell, the terminal phosphate group of ATP is not simply removed but is transferred to another molecule. This addition of a phosphate group to a molecule is known as **phosphorylation;** the enzymes that catalyze such transfers are known as **kinases.**

Let us look at a simple example of energy exchange involving ATP in the formation of sucrose in sugarcane. Sucrose is formed from the monosaccharides glucose and fructose (see Figure 2–3). Under standard thermodynamic conditions, sucrose synthesis is strongly endergonic, requiring an input of 5.5 kilocalories for each mole of sucrose formed:

$$\text{Glucose} + \text{Fructose} + \text{Energy} \longrightarrow \text{Sucrose} + \text{H}_2\text{O}$$

However, when coupled with the hydrolysis of ATP to ADP, the synthesis of the sucrose is actually exergonic. During the series of reactions involved in the formation of sucrose, two molecules of ATP are used to phosphorylate the glucose and fructose, thus energizing each of them:

$$\text{ATP} + \text{Glucose} \longrightarrow \text{Glucose phosphate} + \text{ADP}$$
$$\text{ATP} + \text{Fructose} \longrightarrow \text{Fructose phosphate} + \text{ADP}$$

They are then linked by hydrolyzing these phosphates. The overall equation for sucrose formation from the phosphorylated monosaccharides is:

$$\text{Glucose phosphate} + \text{Fructose phosphate} \longrightarrow \text{Sucrose} + 2\ \text{Phosphate}$$

The cell spends a total of 2 × 7.3 kilocalories = 14.6 kilocalories of ATP energy (assuming standard conditions), and uses 5.5 kilocalories of it to form a mole of sucrose. The other 9.1 kilocalories are used to drive the reaction forward irreversibly and are ultimately released as heat. Thus, the sugarcane plant is able to form sucrose by coupling the hydrolysis of two molecules of ATP to the synthesis of a covalent bond between glucose and fructose.

Where does the ATP originate? As we will see in the next chapter, energy released by the exergonic oxidation of molecules such as glucose is used to "recharge" the ADP molecule back to ATP. Of course, the energy released in these reactions is originally derived from the sun as radiant energy that is converted, during photosynthesis, into chemical energy. Some of this chemical energy is stored in the ATP molecule before being converted to the chemical bond energies of other organic molecules. Thus the ATP/ADP system serves as a universal energy-exchange system, shuttling between energy-releasing and energy-requiring reactions.

SUMMARY

Living Systems Operate According to the Laws of Thermodynamics

The first law of thermodynamics states that energy can be converted from one form to another but cannot be created or destroyed. The potential energy of the initial state (or reactants) is equal to the potential energy of the final state (or products) plus the energy released in the process or reaction. The second law of thermodynamics states that, in the course of energy conversions, if no energy enters or leaves a system, the potential energy of the final state will always be less than the potential energy of the initial state. Stated differently, all natural processes tend to proceed in such a direction that the disorder, or randomness, of the universe increases. This disorder, or randomness, is known as entropy. To maintain the organization on which life depends, living systems must have a constant supply of energy to overcome the tendency toward increasing disorder.

Life on Earth Is Dependent on the Flow of Energy from the Sun

A small fraction of the solar energy that reaches the Earth is captured in the process of photosynthesis and is converted to the energy that drives nearly all the processes of life. These processes include the many other metabolic reactions associated with living organisms and from which living organisms derive their order and organization.

In photosynthesis, the energy of the sun is used to forge high-energy carbon-carbon and carbon-hydrogen bonds of organic compounds. In respiration, these bonds are subsequently oxidized to carbon dioxide and water, and free energy is released. Some of this released free energy is used to synthesize ATP from ADP and P_i, but as in machines, some energy is lost as heat at each energy-conversion step. The resulting ATP is used to drive cellular processes.

Oxidation-Reduction Reactions Play an Important Role in the Flow of Energy

Energy transformations in cells involve the transfer of electrons from one energy level to another and, often, from one atom or molecule to another. Reactions involving the transfer of electrons from one molecule to another are known as oxidation-reduction (or redox) reactions. An atom or molecule that loses electrons is oxidized, and one that gains electrons is reduced; these two types of reactions always occur simultaneously. In photosynthesis, for example, electrons and protons are transferred from water to carbon dioxide, oxidizing the water to oxygen and reducing the carbon dioxide to form a sugar.

Enzymes Enable Chemical Reactions to Take Place at Temperatures Compatible with Life

Enzymes are the catalysts of biological reactions, lowering the free energy of activation and thus enormously increasing the rate at which reactions take place. With few exceptions, enzymes are large, complex, globular protein molecules folded in such a way that a particular group of amino acids forms an active site. The reacting molecule, known as the substrate, fits precisely into this active site, which is the site of the reactions catalyzed by the enzyme. Although the shape of an enzyme may change temporarily in the course of a reaction, the enzyme is not permanently altered.

Many enzymes require cofactors, which may be metal ions or nonprotein organic molecules known as coenzymes. Coenzymes often serve as electron carriers, with different coenzymes holding electrons at different energy levels.

Enzyme-catalyzed reactions take place in ordered series of steps called metabolic pathways. Each step in a pathway is catalyzed by a particular enzyme. The stepwise reactions of metabolic pathways enable cells to carry out their chemical activities with remarkable efficiency in terms of both energy and materials. Each metabolic pathway is under the control of one or more regulatory enzymes. In feedback inhibition, the regulatory enzyme, the first enzyme in the pathway, is inhibited by an excess of the end product, which accumulates when the cell's needs have been met. The entire pathway shuts down as the substrates for the intermediate enzymes are depleted.

ATP Supplies the Energy for Most Activities of the Cell

The ATP molecule consists of the nitrogenous base adenine, the five-carbon sugar ribose, and three phosphate groups. The three phosphate groups are linked by two phosphoanhydride bonds that release free energy, about 7.3 kcal/mol of ATP, when hydrolyzed. Cells are able to carry out endergonic (energy-requiring) reactions by coupling them with exergonic (energy-yielding) reactions that provide a surplus of free energy. Such coupled enzyme reactions usually involve ATP as the common intermediate that carries energy from one reaction to the other.

QUESTIONS

1. Distinguish between/among the following: active site and substrate; AMP, ADP, and ATP; ATPases and kinases.

2. At least four types of energy conversions take place in photosynthetic cells. Name them.

3. Enzymes characteristically work in series, called metabolic pathways, like workers on an assembly line. What are some advantages to the cell of this sort of arrangement?

4. The laws of thermodynamics apply only to closed systems, that is, to systems in which no energy is entering and leaving. Is an aquarium a closed system? If not, could you convert it to one? A space station may or may not be a closed system, depending on certain features of its design. What would these features be? Is the Earth a closed system? How about the universe?

5. The most interesting implication of the second law of thermodynamics, as far as living systems are concerned, is the relationship between entropy and order. Explain.

6. In some multienzyme systems, the end product of the metabolic pathway is brought into balance with the cell's needs by feedback inhibition. Explain.

CHAPTER 6

Respiration

◀ **From rice to wine** In this nineteenth-century Japanese woodblock print, a geisha holds a cup of sake, an alcoholic beverage made from fermented rice. The fungus *Aspergillus oryzae* converts the starch in rice to glucose, which yeast then convert to ethanol by alcoholic fermentation. The sake, after maturing for 9 to 12 months, is diluted with water to reduce its alcohol content from 20 to 15 percent.

ATP is the universal energy currency in living organisms. It takes part in a great variety of cellular events, including the biosynthesis of organic molecules, the flick of a flagellum, the streaming of cytoplasm, and the active transport of a molecule across the plasma membrane. In the following pages, we describe how a cell oxidizes carbohydrates and captures a portion of the released energy in the phosphoanhydride bonds of ATP. This process, which occurs largely in the mitochondrion (Figure 6–1), provides an excellent illustration both of the chemical principles described in the previous chapter and of the way in which cells carry out biochemical processes.

An Overview of Glucose Oxidation

As mentioned in Chapter 2, energy-yielding carbohydrate molecules are generally stored in plants as sucrose or starch. A necessary preliminary step to **respiration**—the complete oxidation of sugars or other organic molecules to carbon dioxide and water—is the hydrolysis of these storage molecules to the monosaccharides glucose and fructose. Respiration itself is generally considered to begin with glucose, an end product of the hydrolysis of both sucrose and starch.

The oxidation of glucose (and other carbohydrates) is complicated in detail but simple in its overall design. As we saw in the last chapter, **oxidation** is the loss of electrons, and **reduction** is the gain of electrons. In the oxidation of glucose, the glucose molecule is split apart, and the hydrogen atoms (that is, electrons and their accompanying protons) are removed from the carbon atoms and combined with oxygen, which is thereby reduced to water. As this happens, the electrons are transferred from higher energy levels to lower energy levels, and free energy is released.

CHECKPOINTS

After reading this chapter, you should be able to answer the following:

1. What is the overall reaction, or equation, for respiration, and what is the principal function of this process?
2. What are the main events that occur during glycolysis?
3. Where in the cell does the citric acid cycle occur, and what are the products formed?
4. How does the flow of electrons in the electron transport chain result in the formation of ATP?
5. How and why does the net energy yield under aerobic conditions differ from that obtained under anaerobic conditions?
6. What is the central role played by the citric acid cycle in the metabolism of the cell?

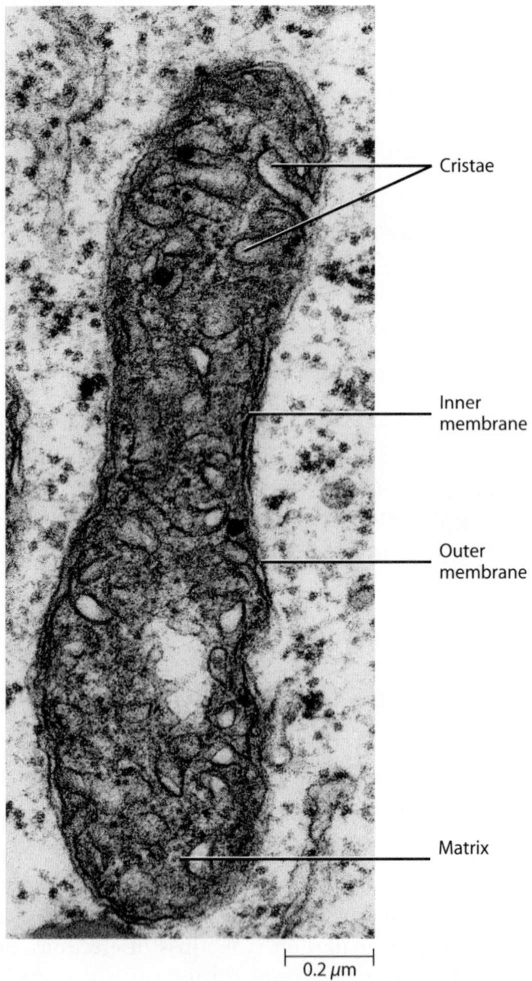

6–1 Mitochondrion in a leaf cell Mitochondria are the sites of respiration, the process by which chemical energy is transferred from carbon-containing compounds to ATP. Most of the ATP is produced on the surfaces of the cristae by enzymes embedded in these membranes. This electron micrograph is of a mitochondrion in a leaf cell of the fern *Regnellidium diphyllum.*

Glucose can be used as a source of energy under both aerobic (that is, in the presence of oxygen) and anaerobic (in the absence of oxygen) conditions. However, maximum energy yields for oxidizable organic compounds are achieved only under aerobic conditions. Consider, for example, the overall reaction for the complete oxidation of glucose:

$$\underset{\text{Glucose}}{C_6H_{12}O_6} + \underset{\text{Oxygen}}{6O_2} \longrightarrow \underset{\text{Carbon dioxide}}{6CO_2} + \underset{\text{Water}}{6H_2O} + \text{Energy}$$

With oxygen as the ultimate electron acceptor, this reaction is highly exergonic (energy-yielding), with the release of 686 kcal/mol under standard conditions. This reaction represents the overall process of respiration. When energy is extracted from organic compounds without the involvement of oxygen, the process is called fermentation, which yields much less energy, as we will see later in this chapter.

Respiration involves glycolysis, the formation of acetyl CoA from pyruvate, the citric acid cycle, and the electron transport chain, which produces a gradient that drives oxidative phosphorylation (Figure 6–2). In **glycolysis,** the six-carbon glucose molecule is broken down to a pair of three-carbon molecules of **pyruvate.** The pyruvate molecules are then oxidized to two molecules of acetyl CoA. In the **citric acid cycle,** the acetyl CoA molecules are completely oxidized to carbon dioxide, and the resulting electrons are transferred to the **electron transport chain.** In **oxidative phosphorylation,** the free energy that is released as electrons are moved through the electron transport chain (ultimately reducing oxygen to water) is used to form ATP from ADP and phosphate.

As the glucose molecule is oxidized, some of its energy is extracted in a series of small, discrete steps and is stored in the phosphoanhydride bonds of ATP. In accord with the second law of thermodynamics, however, much of its energy is dissipated as heat energy.

Glycolysis

As mentioned above, in glycolysis (from *glyco-,* meaning "sugar," and *lysis,* meaning "splitting"), the six-carbon glucose molecule is split into two molecules of pyruvate (Figure 6–3). Glycolysis occurs in a series of 10 steps, each catalyzed by a specific enzyme. This series of reactions is carried out by virtually all living cells, from bacteria to the eukaryotic cells of plants and animals. Glycolysis is an anaerobic process that occurs in the cytosol. Biologically, glycolysis may be considered a primitive process, in that it most likely arose before the appearance of atmospheric oxygen and before the origin of cellular organelles.

The glycolytic pathway is outlined in Figure 6–4. It illustrates the principle that biochemical processes in a living cell proceed in small, sequential steps. Each step is catalyzed by a specific enzyme. Note the formation of ATP from ADP and phosphate and the formation of NADH from NAD^+ (see Figure 5–8). **ATP** and **NADH** represent the cell's net energy harvest from glycolysis. In Chapter 7 we will see that reactions 4 through 6 also occur in the Calvin cycle, a part of the photosynthetic process. This repetition illustrates a principle of biochemical evolution: pathways do not arise entirely new; rather, a few new reactions are added to an existing set to make a "new" pathway.

Glycolysis begins with the preparatory phase, which requires an input of energy in the form of ATP (steps 1 and 3; Figure 6–4). In **step 1,** the first preparatory reaction, the terminal phosphate of an ATP molecule is transferred to the glucose molecule (a six-sided ring; Figure 6–3) to produce glucose 6-phosphate and ADP. In **step 2,** the glucose 6-phosphate is rearranged, becoming fructose 6-phosphate (a five-sided ring). In **step 3**—the second preparatory reaction—the fructose 6-phosphate gains a second phosphate to form fructose 1,6-bisphosphate ("bis" simply means two) as another ATP is converted to ADP. Note that thus far, two molecules of ATP have been converted to two molecules of ADP and no energy has been recovered. In other words, the energy yield thus far is –2ATP.

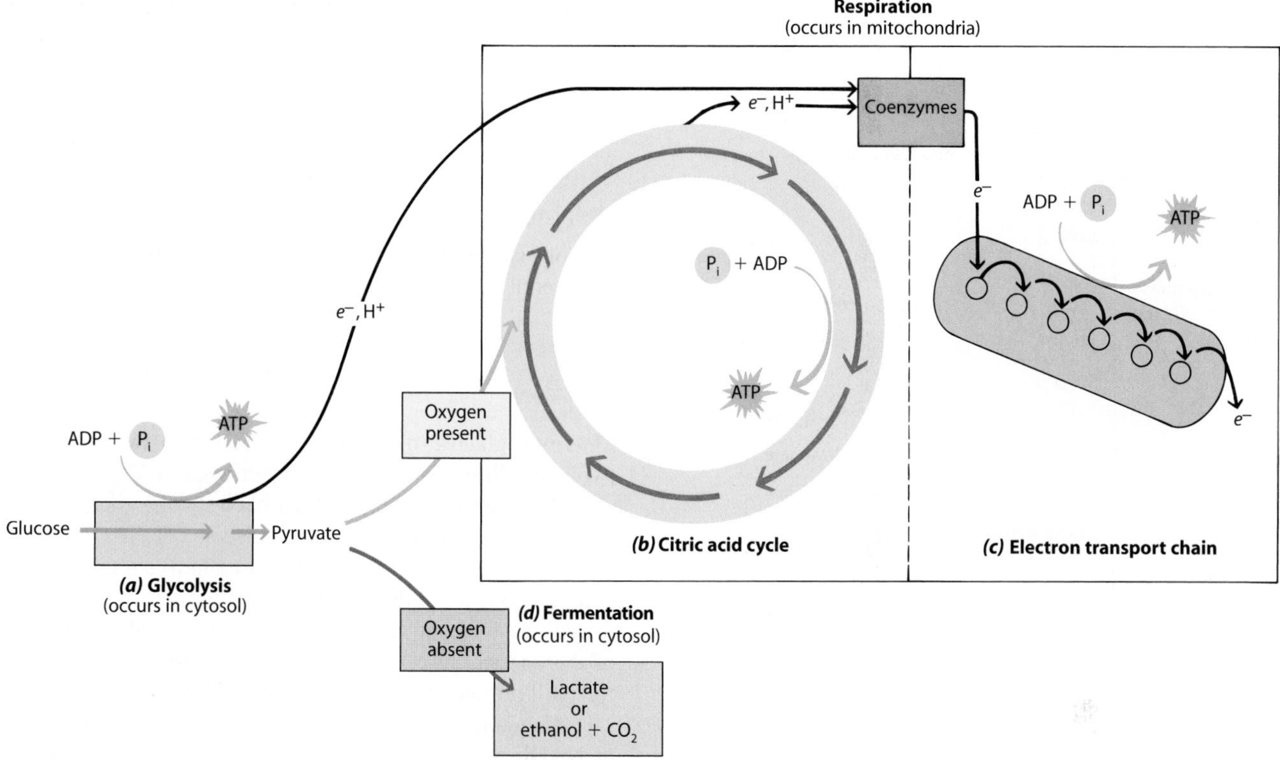

6–2 Aerobic and anaerobic breakdown of glucose ***(a)*** The complete oxidative breakdown of glucose, called respiration, consists overall of glycolysis, the citric acid cycle, and the electron transport chain. In glycolysis, glucose is split into pyruvate. A small amount of ATP is synthesized from ADP and phosphate, and a few electrons (e^-) and their accompanying protons (H^+) are transferred to coenzymes that function as electron carriers. ***(b)*** In the presence of oxygen (aerobic pathway), the pyruvate is converted to acetyl CoA, which is fed into the citric acid cycle. In the course of this cycle, additional ATP is synthesized and more electrons and protons are transferred to coenzymes. ***(c)*** The coenzymes then transfer the electrons to an electron transport chain in which the electrons drop, step by step, to lower energy levels, generating a proton gradient. The proton gradient is used to drive the formation of considerably more ATP. This process is called oxidative phosphorylation. At the end of the electron transport chain, the electrons reunite with protons and combine with oxygen to form water. ***(d)*** In the absence of oxygen (anaerobic pathway), the pyruvate is converted to either lactate or ethanol. This process, known as fermentation, produces no additional ATP, but it does regenerate the coenzymes that are necessary for glycolysis to continue.

Step 4 is the cleavage step from which glycolysis derives its name. The six-carbon sugar molecule is split in half, producing two three-carbon molecules (glyceraldehyde 3-phosphate and dihydroxyacetone phosphate). The dihydroxyacetone phosphate is converted to glyceraldehyde 3-phosphate, so that at the end of **step 5** there are two molecules of glyceraldehyde 3-phosphate. Therefore, *the products of subsequent steps must be counted twice to account for the fate of each glyceraldehyde 3-phosphate molecule.* With completion of step 5, the preparatory phase is complete.

Glucose —Glycolysis→ 2 Pyruvate + 4 Hydrogen atoms (4[H])

6–3 Overview of glycolysis In glycolysis, the six-carbon glucose molecule is split, via a series of 10 reactions, into two molecules of a three-carbon compound known as pyruvate. In the course of glycolysis, four hydrogen atoms are removed from the original glucose molecule.

The first point in the pathway that yields energy is at **step 6,** where two molecules of NAD^+ are reduced to two molecules of NADH. In this step some of the energy from the oxidation of glyceraldehyde 3-phosphate is stored as high-energy electrons in the NADH. At **steps 7** and **10,** two molecules of ADP take energy from the system, forming two molecules of ATP per molecule of glyceraldehyde 3-phosphate—or four molecules of ATP per molecule of glucose. Two of the four ATP are, in effect, replacements for the two ATP used in steps 1 and 3. The net ATP yield is only two molecules of ATP per molecule of glucose. The formation of ATP by the enzymatic

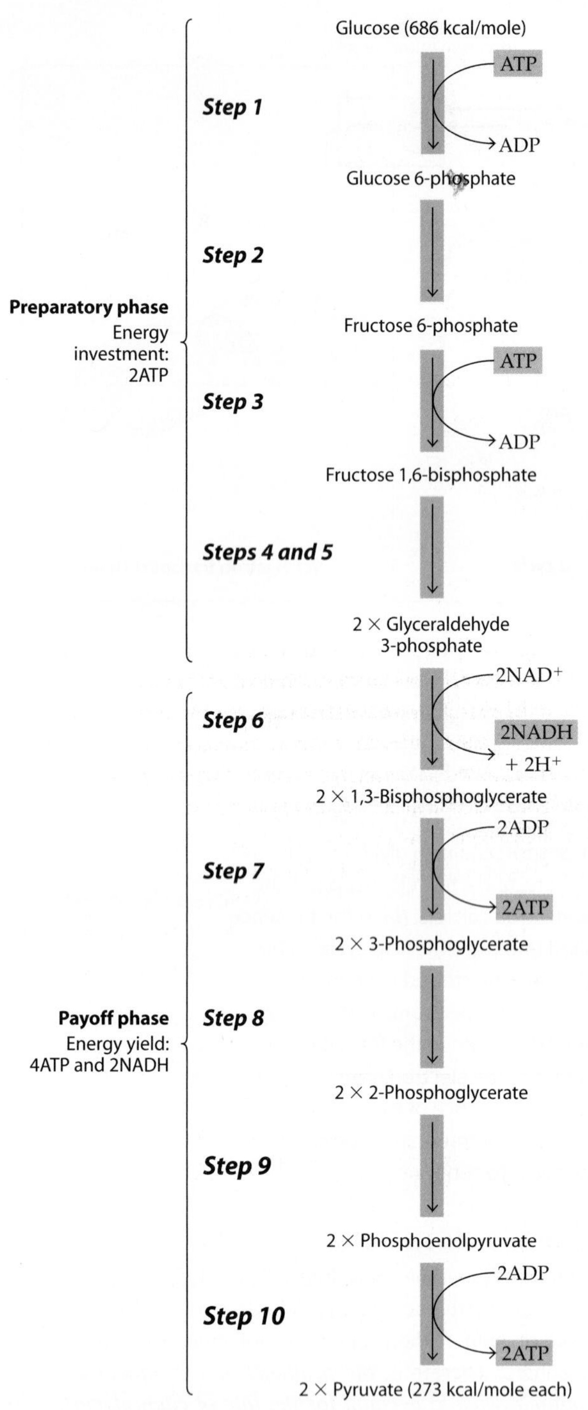

6–4 Summary of the two phases of glycolysis The preparatory phase requires an energy investment of 2 ATP per glucose molecule. This stage ends with the splitting of the six-carbon sugar molecule into two three-carbon molecules. The payoff phase produces an energy yield of 4 ATP and 2 NADH—a substantial return on the original investment. The net ATP yield is therefore 2 molecules of ATP per molecule of glucose. Carbohydrates other than glucose, including glycogen, starch, various disaccharides, and a number of monosaccharides, can undergo glycolysis once they have been converted to glucose 6-phosphate or fructose 6-phosphate.

transfer of a phosphate group from a metabolic intermediate to ADP, as occurs in steps 7 and 10, is referred to as **substrate-level phosphorylation.**

Glycolysis Ends with Most of the Energy of the Original Glucose Molecule Still Present in the Two Pyruvate Molecules

Glycolysis (from glucose to pyruvate) can be summarized by the overall equation:

$$\text{Glucose} + 2\text{NAD}^+ + 2\text{ADP} + 2\text{P}_i \longrightarrow 2\ \text{Pyruvate} + 2\text{NADH} + 2\text{H}^+ + 2\text{ATP} + 2\text{H}_2\text{O}$$

Thus one glucose molecule is converted to two molecules of pyruvate. (Water is the other product when ADP combines with phosphate to form ATP.) The *net* harvest—the energy yield—is two molecules of ATP and two molecules of NADH per molecule of glucose. Two moles of pyruvate have a total energy content of about 546 kilocalories, compared with 686 kilocalories stored in a mole of glucose. A large portion (about 80 percent) of the energy stored in the original glucose molecule is therefore still present in the two pyruvate molecules.

Also note that, under aerobic conditions, the two molecules of NADH are high-energy compounds in their own right; they can yield additional ATP molecules in the mitochondria when used as electron donors to the electron transport chain of the aerobic pathway.

The Aerobic Pathway

Pyruvate is a key intermediate in cellular energy metabolism because it can be utilized in one of several pathways. Which pathway it follows depends in part on the conditions under which metabolism takes place and in part on the specific organism involved—and, in some cases, the particular tissue in the organism. The principal environmental factor that determines which pathway is followed is the availability of oxygen.

In the presence of oxygen, pyruvate is oxidized completely to carbon dioxide, and glycolysis is only the initial phase of respiration. This entire aerobic pathway results in the complete oxidation of glucose and a much greater ATP yield than can be achieved by glycolysis alone. These reactions take place in two stages—the citric acid cycle and the electron transport chain—both occurring within mitochondria in eukaryotic cells.

Recall that mitochondria are surrounded by two membranes, with the inner one convoluted inward into folds called **cristae** (Figure 6–5). Within the inner compartment, surrounding the cristae, is the **matrix,** a dense solution containing enzymes, coenzymes, water, phosphates, and other molecules involved in respiration. Thus, a mitochondrion resembles a self-contained chemical factory. The outer membrane allows most small molecules to move in or out freely, but the inner one permits the passage of only certain molecules, such as pyruvate and ATP, while it prevents the passage of others. Most of the enzymes of the citric acid cycle are found in solution within the mitochondrial matrix. Other enzymes in the citric acid cycle and

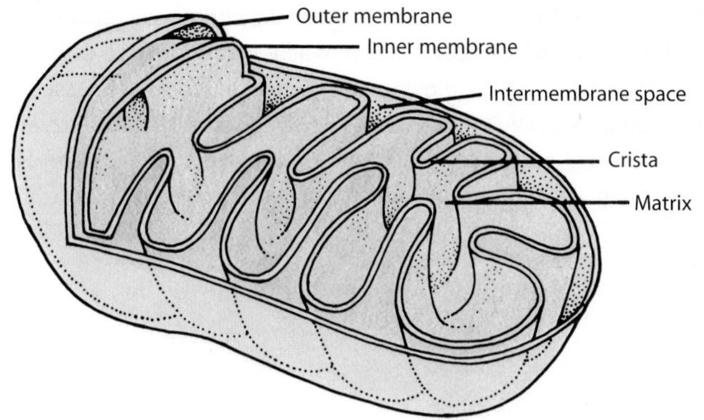

6–5 Structure of a mitochondrion A mitochondrion is surrounded by two membranes, as shown in this three-dimensional diagram. The inner membrane folds inward, forming the cristae. Many of the enzymes and electron carriers involved in respiration are present within the inner membrane.

the components of the electron transport chain are embedded in the inner membrane that makes up the cristae.

A Preliminary Step: Pyruvate Enters the Mitochondrion and Is Both Oxidized and Decarboxylated

Pyruvate passes from the cytosol, where it is produced by glycolysis, to the matrix of the mitochondrion, crossing the outer and inner membranes in the process. Pyruvate, however, is not used directly in the citric acid cycle. Within the mitochondrion, pyruvate is both oxidized and decarboxylated—that is, electrons are removed and CO_2 is split out of the molecule. In the course of this exergonic reaction, a molecule of NADH is produced from NAD^+ for each pyruvate molecule that is oxidized (Figure 6–6). The two pyruvate molecules derived from the original glucose molecule have now been oxidized to two acetyl ($—CH_3CO$) groups. In addition, two molecules of CO_2 have been liberated, and two molecules of NADH have been formed from NAD^+.

Each acetyl group is temporarily attached to **coenzyme A (CoA)**—a large molecule consisting of a nucleotide linked to pantothenic acid, one of the B-complex vitamins. The combination of the acetyl group and CoA is known as **acetyl CoA** and is the form in which carbon atoms from glucose enter the citric acid cycle.

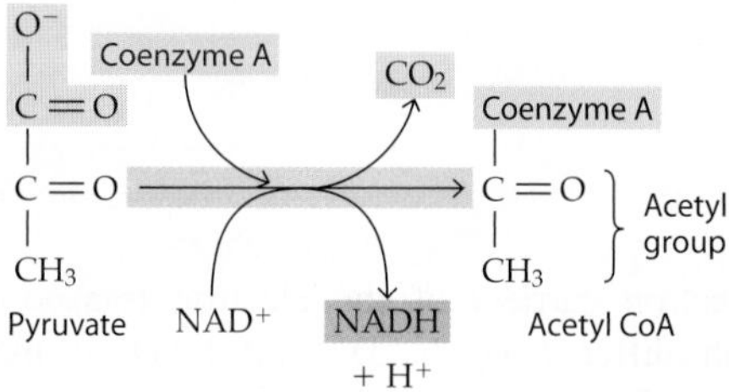

6–6 Formation of acetyl CoA from pyruvate The three-carbon pyruvate molecule is oxidized and decarboxylated to form the two-carbon acetyl group, which is attached to coenzyme A as acetyl CoA. The oxidation of the pyruvate molecule is coupled to the reduction of NAD^+ to NADH. Acetyl CoA is the form in which carbon atoms derived from glucose enter the citric acid cycle.

The Citric Acid Cycle Oxidizes the Acetyl Groups of the Acetyl CoA Molecules

The citric acid cycle was originally known as the Krebs cycle in honor of Sir Hans Krebs, whose research group was largely responsible for its elucidation. Krebs experimentally demonstrated this metabolic pathway in 1937 and later received a Nobel Prize in recognition of his brilliant work. The Krebs cycle is more commonly called the citric acid cycle or the TCA (tricarboxylic acid) cycle today, because it begins with the formation of citric acid, or citrate, which has three carboxylic acid ($—COO^-$) groups.

The citric acid cycle begins with acetyl CoA, its substrate. Upon entering the citric acid cycle (Figure 6–7), the two-carbon acetyl group is combined with a four-carbon compound (oxaloacetate) to produce a six-carbon compound (citrate). The coenzyme A is released to combine with a new acetyl group when another molecule of pyruvate is oxidized. In the course of the cycle, two of the six carbons are removed and oxidized to CO_2, and oxaloacetate is regenerated—thus literally making this series of reactions a cycle. Each turn around the cycle uses up one acetyl group and regenerates one molecule of oxaloacetate, which is then ready to enter the citric acid cycle again.

In the course of these steps, some of the energy released by the oxidation of the carbon atoms is used to convert ADP to ATP (one molecule per cycle; another case of substrate-level phosphorylation), but most is used to reduce NAD^+ to NADH (three molecules per cycle). In addition, some of the energy is used to reduce a second electron carrier—the coenzyme known as flavin adenine dinucleotide (FAD). One molecule of **$FADH_2$** is formed from FAD in each turn of the cycle. Oxygen is not directly involved in the citric acid cycle; the electrons and protons removed in the oxidation of carbon are all accepted by NAD^+ and FAD.

The overall equation for the citric acid cycle is therefore:

$$\text{Oxaloacetate} + \text{Acetyl CoA} + 3H_2O + \text{ADP} + P_i + 3NAD^+ + \text{FAD} \longrightarrow \text{Oxaloacetate} + 2CO_2 + \text{CoA} + \text{ATP} + 3\text{NADH} + 3H^+ + FADH_2$$

In the Electron Transport Chain, Electrons Removed from the Glucose Molecule Are Transferred to Oxygen

The glucose molecule is now completely oxidized to CO_2. Some of its energy has been used to produce ATP from ADP and P_i in substrate-level phosphorylation, both in glycolysis and in the citric acid cycle. Most of the energy, however, still remains in the electrons removed from the carbon atoms as they were oxidized. These electrons were passed to the electron carriers NAD^+ and FAD and are still at a high energy level in NADH and $FADH_2$.

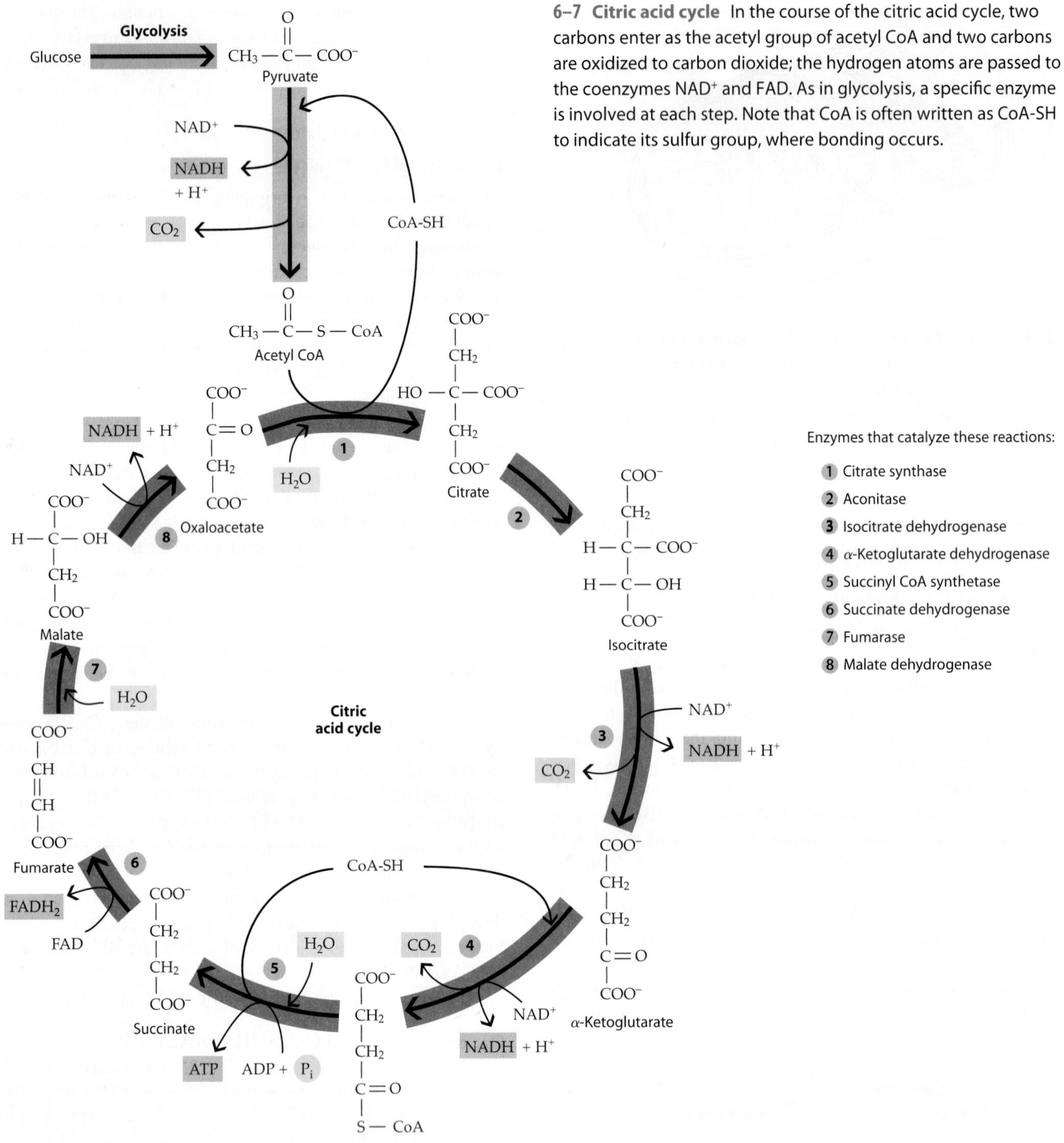

6–7 Citric acid cycle In the course of the citric acid cycle, two carbons enter as the acetyl group of acetyl CoA and two carbons are oxidized to carbon dioxide; the hydrogen atoms are passed to the coenzymes NAD^+ and FAD. As in glycolysis, a specific enzyme is involved at each step. Note that CoA is often written as CoA-SH to indicate its sulfur group, where bonding occurs.

In the next stage of respiration, these high-energy electrons of NADH and $FADH_2$ are passed, step by step, to form water (H_2O)—a compound in which the electrons from glucose are at a low energy level. This stepwise passage is made possible by the electron transport chain (Figure 6–8), a series of electron carriers, each of which holds the electrons at a slightly lower energy level than the previous carrier. Each carrier is capable of accepting or donating one or two electrons. Each component of the chain can accept electrons from the preceding carrier and transfer them to the following carrier in a specific sequence. With one exception, all of these carriers are embedded in the inner mitochondrial membrane.

The electron carriers of the electron transport chain of mitochondria differ from NAD^+ and FAD in their chemical structures. Some of them belong to a class of electron carriers known as **cytochromes**—protein molecules with an iron-containing porphyrin ring, or heme group, attached. Cytochromes pick up electrons on their iron atoms, which can be reversibly reduced from the ferric (Fe^{3+}) to the ferrous (Fe^{2+}) form. Each cytochrome differs in its protein structure and in the energy level at which it holds the electrons. In their reduced form, cytochromes carry a single electron without a proton.

Non-heme iron proteins—the **iron-sulfur proteins**—are additional components of the electron transport chain. The iron

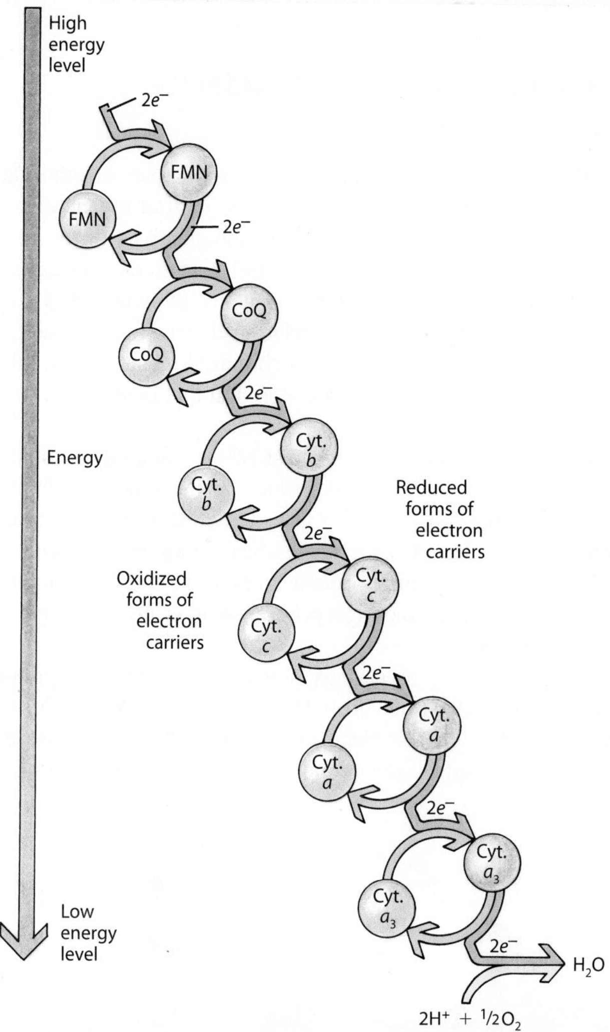

6–8 Schematic representation of the electron transport chain The molecules shown here—flavin mononucleotide (FMN), coenzyme Q (CoQ), and cytochromes *b*, *c*, *a*, and a_3—are the principal electron carriers of the chain. At least nine other molecules also function as intermediates between these electron carriers.

Electrons carried by NADH enter the chain where they are transferred to FMN, which is thus reduced (blue). Almost instantaneously, FMN passes on the electrons to CoQ. In the process, FMN returns to its oxidized form (gray), ready to receive another pair of electrons, and CoQ is reduced. CoQ then passes the electrons to the next carrier and returns to its oxidized form, and so on down the line. As the electrons move down the chain, they drop to successively lower energy levels. The electrons are ultimately accepted by oxygen, which combines with protons (hydrogen ions) to form water.

Electrons carried by $FADH_2$ are at a slightly lower energy level than those carried by NADH. They enter the electron transport chain farther down the line, at CoQ instead of at FMN, and generate only 2 ATP per coenzyme molecule versus the yield of 3 ATP per molecule for NADH.

of these proteins is attached not to a porphyrin ring but to sulfides and to the sulfur atoms of the sulfur-containing amino acids found in the protein chain. Like cytochromes, the iron-sulfur proteins pick up electrons on their iron atoms and therefore carry electrons but not protons.

The most abundant components of the electron transport chain are quinone molecules. In mitochondria the quinone is **ubiquinone,** also called **coenzyme Q (CoQ)** (Figure 6–9). Unlike cytochromes and iron-sulfur proteins, a quinone can accept or donate either one or two electrons. This permits CoQ to serve as an intermediate between two-electron carriers and one-electron carriers. In addition, CoQ picks up a proton along with each electron it carries—the equivalent of a hydrogen atom. By alternating electron transfer between carriers that carry electrons only and those that carry hydrogen atoms, CoQ can shuttle protons across the inner mitochondrial membrane. For example, every time a quinone molecule accepts an electron, it also picks up a proton (H^+) from the mitochondrial matrix. When the quinone gives up its electron to the next carrier, the proton is released into the intermembrane space. Because the electron carriers are oriented in the mitochondrial membrane so that protons are always picked up on the matrix side of the membrane and released into the intermembrane space, a proton gradient is generated across the inner mitochondrial membrane. Unlike the cytochromes and non-sulfur centers, CoQ is not tightly associated with any one protein complex but can freely move among protein complexes in the electron transport chain, receiving and donating electrons. Because it is small and hydrophobic, CoQ can move freely within the lipid bilayer of the membrane and thus can shuttle electrons between other, less mobile carriers.

At the "top," or beginning (most energetic end), of the electron transport chain are the electrons held by NADH and $FADH_2$. Recall that for every molecule of glucose oxidized, the yield of the citric acid cycle was two molecules of $FADH_2$ and six molecules of NADH (Figure 6–7), and the oxidation of pyruvate to acetyl CoA yielded two molecules of NADH

$$\text{CoQ (ring: } 2\,CH_3{-}O{-},\ CH_3,\ 2\,C{=}O\text{)}{-}(CH_2{-}CH{=}\overset{CH_3}{\overset{|}{C}}{-}CH_2)_nH + 2[H] \underset{\text{Oxidation}}{\overset{\text{Reduction}}{\rightleftharpoons}} \text{CoQH}_2\text{ (ring: } 2\,CH_3{-}O{-},\ CH_3,\ 2\,OH\text{)}{-}(CH_2{-}CH{=}\overset{CH_3}{\overset{|}{C}}{-}CH_2)_nH$$

Coenzyme Q, oxidized (CoQ)

Coenzyme Q, reduced ($CoQH_2$)

6–9 Oxidized and reduced forms of coenzyme Q The oxidized form of CoQ is reduced by accepting electrons from a donor in the electron transport chain. The protons picked up by CoQ come from the matrix side of the inner mitochondrial membrane. Coenzyme Q is also known as "ubiquinone," a reflection of the ubiquity of this compound—it occurs in virtually all eukaryotic cells.

(Figure 6–6). Recall also that an additional two molecules of NADH were produced in glycolysis (Figure 6–4); in the presence of oxygen, electrons from these NADH molecules are transported into the mitochondrion. Electrons from all NADH molecules are transferred to the electron acceptor flavin mononucleotide (FMN), the first component of the electron transport chain. Electrons from $FADH_2$ molecules are transferred to CoQ, which is farther down the electron transport chain than FMN.

As electrons flow along the electron transport chain from higher to lower energy levels, the released energy is harnessed and used to generate a proton gradient across the inner membrane by transferring protons from the matrix to the intermembrane space. The proton gradient, in turn, drives the formation of ATP from ADP and P_i by oxidative phosphorylation. At the end of the chain, the electrons are accepted by oxygen and combine with protons (hydrogen ions) to produce water. Each time one pair of electrons passes from NADH to oxygen, enough protons are "pumped" across the membrane to generate three molecules of ATP. Each time a pair of electrons passes from $FADH_2$, which holds them at a slightly lower energy level than NADH, enough protons are "pumped" to form about two molecules of ATP.

Oxidative Phosphorylation Is Achieved by the Chemiosmotic Coupling Mechanism

Until the early 1960s, the mechanism of oxidative phosphorylation was one of the most baffling puzzles in all of biochemistry. As a result of the insight and experimental creativity of the British biochemist Peter Mitchell (1920–1992)—and the subsequent work of many other investigators—much of the puzzle has now been solved. Oxidative phosphorylation depends on a gradient of protons (H^+ ions) across the mitochondrial membrane and the subsequent use of the free energy stored in that gradient to form ATP from ADP and phosphate.

As shown in Figure 6–10, most of the components of the electron transport chain are embedded in the inner membrane of the mitochondrion. In fact, most of the electron carriers are tightly associated with proteins embedded in the membrane, forming four distinct multiprotein complexes (I–IV). Complex II is not included in the figure because it is not involved in NADH oxidation.

The protein complexes, as noted above, are also proton pumps. As the electrons drop to lower energy levels during their transit through the electron transport chain, the released free energy is used by the protein complexes to pump protons

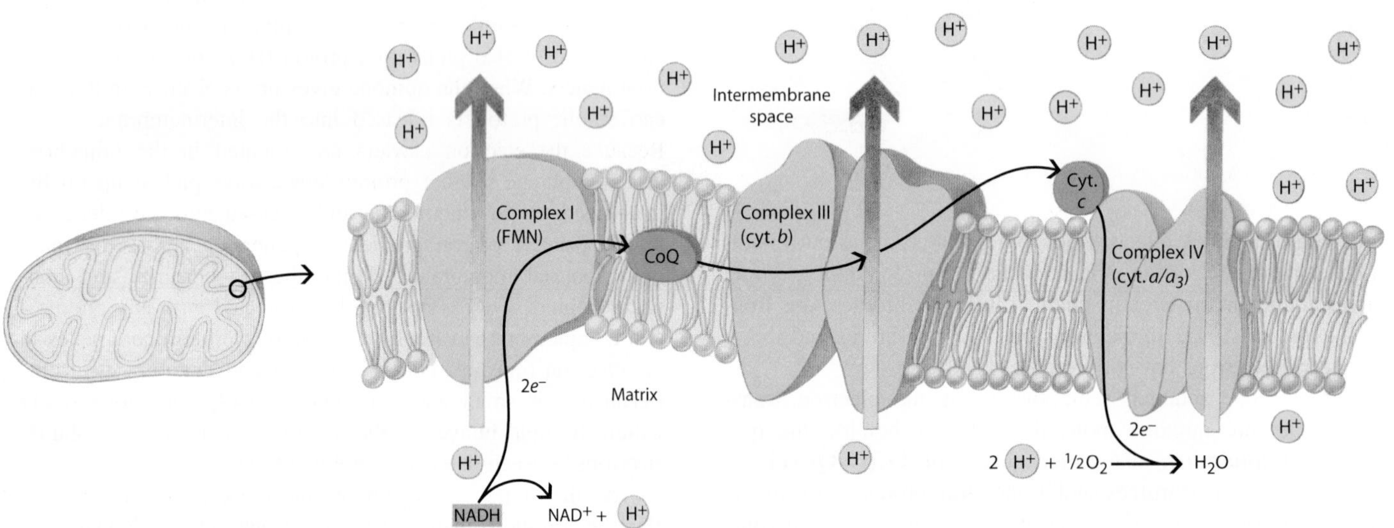

6–10 Electron transport chain The arrangement of the components of the electron transport chain in the inner membrane of the mitochondrion. Three complex protein structures (here labeled I, III, and IV) are embedded in the membrane. They contain electron carriers and the enzymes required to catalyze the transfer of electrons from one carrier to the next. Complex I contains the electron carrier FMN, which receives two electrons from NADH and passes them to CoQ. CoQ, located in the lipid interior of the membrane, ferries electrons from Complex I to Complex III, which contains cytochrome *b*. From Complex III, the electrons move to cytochrome *c*, a peripheral membrane protein on the intermembrane-space side that shuttles back and forth between Complexes III and IV. The electrons then move through cytochromes *a* and a_3, located in Complex IV, back into the matrix, where they combine with protons (H^+) and oxygen, forming water.

Complex II, another protein complex (not shown here) embedded in the inner mitochondrial membrane, contains FAD. Electrons are passed from succinate (in the citric acid cycle) to FAD, forming $FADH_2$, and then to CoQ. Complex II is not part of the transfer of electrons from NADH to O_2.

As the electrons make their way down the electron transport chain, protons are pumped through the three protein complexes from the matrix to the intermembrane space. This transfer of protons from the matrix side of the inner mitochondrial membrane to the intermembrane-space side establishes the electrochemical proton gradient that powers the synthesis of ATP.

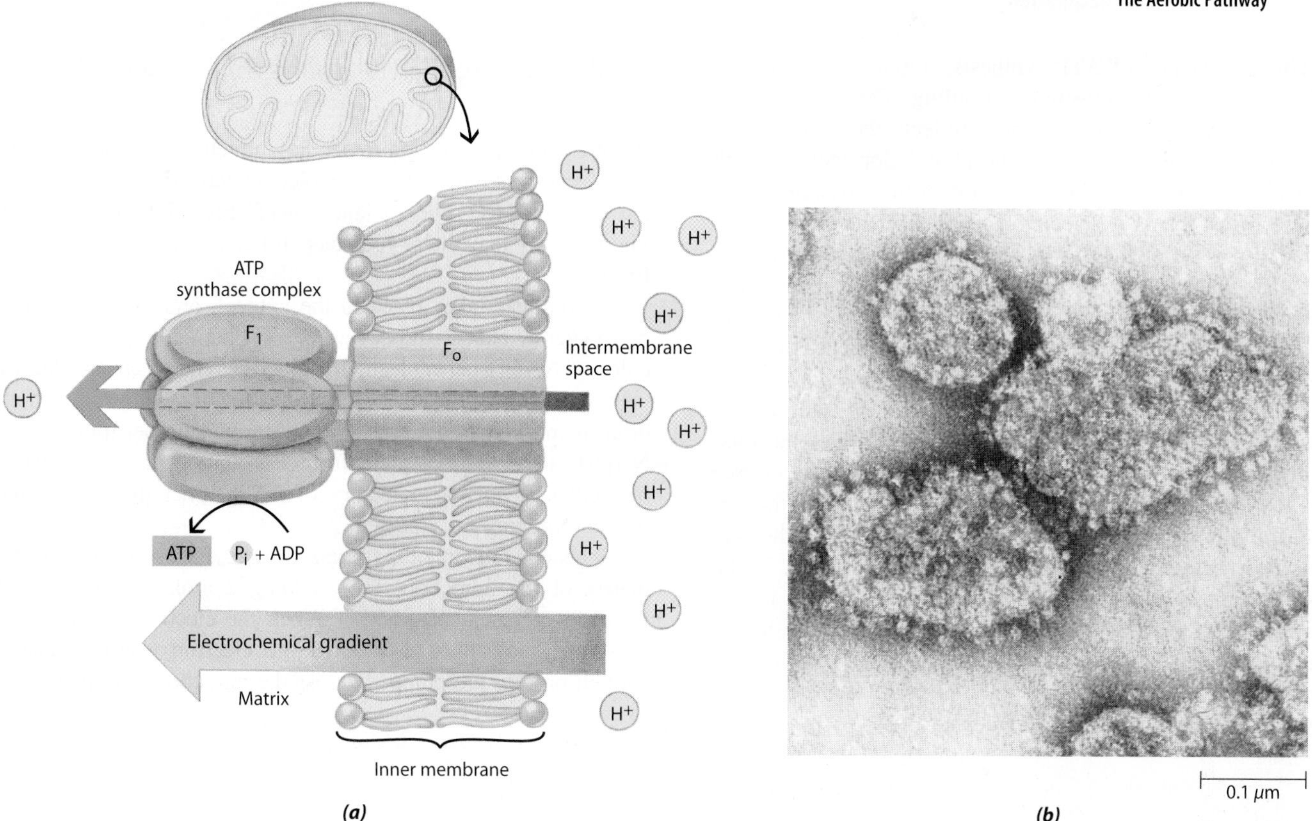

6–11 ATP synthase complex ***(a)*** This enzyme complex consists of two major portions, F_O, which is contained within the inner membrane of the mitochondrion, and F_1, which extends into the matrix. Binding sites for both ATP and ADP are located on the F_1 portion, which consists of nine separate protein subunits. A channel, or pore, connecting the intermembrane space to the mitochondrial matrix, passes through the entire complex. When protons flow through this channel, moving down the electrochemical gradient, ATP is synthesized from ADP and phosphate. ***(b)*** The knobs protruding from the vesicles in this electron micrograph are the F_1 portions of ATP synthase complexes. The F_O portions to which they are attached are embedded in the membrane and are not visible. The vesicles were prepared by disrupting the inner mitochondrial membrane with ultrasonic waves. When the membrane is disrupted in this way, the fragments immediately reseal, forming closed vesicles. The vesicles are, however, inside out. The outer surface here is the surface that faces the matrix in the intact mitochondrion.

across the inner membrane. In other words, as electrons flow from NADH to Complex I then to Complex III and finally to Complex IV, protons are transferred from the matrix side of the inner membrane to the intermembrane space. It is thought that for each pair of electrons moving down the electron transport chain from NADH to oxygen, about 10 protons are pumped out of the matrix.

As we noted earlier, the inner membrane of the mitochondrion is impermeable to protons. Thus, the protons that are pumped into the intermembrane space cannot easily move back across the membrane into the matrix. The result is a concentration gradient of protons across the inner membrane of the mitochondrion, with a much higher concentration of protons in the intermembrane space than in the matrix.

Like a boulder at the top of a hill or water at the top of a falls, the difference in the concentration of protons between the intermembrane space and the matrix represents potential energy. This potential energy results not only from the actual concentration difference (more hydrogen ions outside the matrix than inside) but also from the difference in electric charge (more positive charges outside than inside). The potential energy is thus in the form of an **electrochemical gradient.** It is available to power any process that provides a channel allowing the protons to flow down the gradient back into the matrix.

Such a channel is provided by a large enzyme complex known as **ATP synthase** (Figure 6–11). This enzyme complex, which is embedded in the inner membrane of the mitochondrion, has binding sites for ADP and phosphate. It also has an inner channel, or pore, through which protons can pass. When protons flow through this channel, moving down the electrochemical gradient from the intermembrane space back into the matrix, the energy that is released powers the synthesis of ATP from ADP and phosphate.

Note that ATP synthase functions in a manner that is opposite to that of the proton pump H^+-ATPase, described in Chapter 4. There, ATP was *used* as an energy source to pump protons against their electrochemical gradient. ATP synthase, on the other hand, uses the energy of protons moving down their gradient to *produce* ATP.

This mechanism of ATP synthesis, summarized in Figure 6–12, is known as **chemiosmotic coupling.** The term "chemiosmotic," coined by Peter Mitchell, reflects the fact that the production of ATP in oxidative phosphorylation includes both chemical processes (the "chemi" portion of the term) and transport processes across a selectively permeable membrane (the "osmotic" portion of the term). As we have seen, two distinct events take place in chemiosmotic coupling: (1) a proton gradient is established across the inner membrane of the mitochondrion, and (2) potential energy stored in the gradient is used to generate ATP from ADP and phosphate.

Chemiosmotic power also has other uses in living organisms. For example, it provides the power that drives the rotation of bacterial flagella. In photosynthetic cells, as we will see in the next chapter, chemiosmotic power is involved in the formation of ATP using energy supplied to electrons by the sun. It can also be used to power other transport processes. In the mitochondrion, for example, the energy stored in the proton gradient is also used to move, or transport, other substances through the inner membrane. Both phosphate and pyruvate are carried into the matrix by membrane proteins that simultaneously transport protons down the gradient.

The Overall Energy Harvest Involves NADH and $FADH_2$ as Well as ATP

We are now in a position to see how much of the energy originally present in the glucose molecule has been recovered in the form of ATP. The "balance sheet" for ATP yield given in Figure 6–13 may help you keep track of the discussion that follows.

Glycolysis takes place in the cytosol, and in the presence of oxygen yields 2 molecules of ATP directly plus 2 molecules of NADH per molecule of glucose. The electrons held by these 2 NADH molecules are transported across the mitochondrial membrane at a "cost" of 1 ATP molecule per molecule of NADH. Thus, the net yield from reoxidation of the 2 NADH molecules is only 4 molecules of ATP, rather than the 6 that would otherwise be expected.

The conversion of pyruvate to acetyl CoA occurs in the matrix of the mitochondrion, yielding 2 molecules of NADH for each molecule of glucose. When the electrons held by the 2 NADH molecules pass down the electron transport chain, enough protons are pumped across the mitochondrial membrane to synthesize 6 ATP.

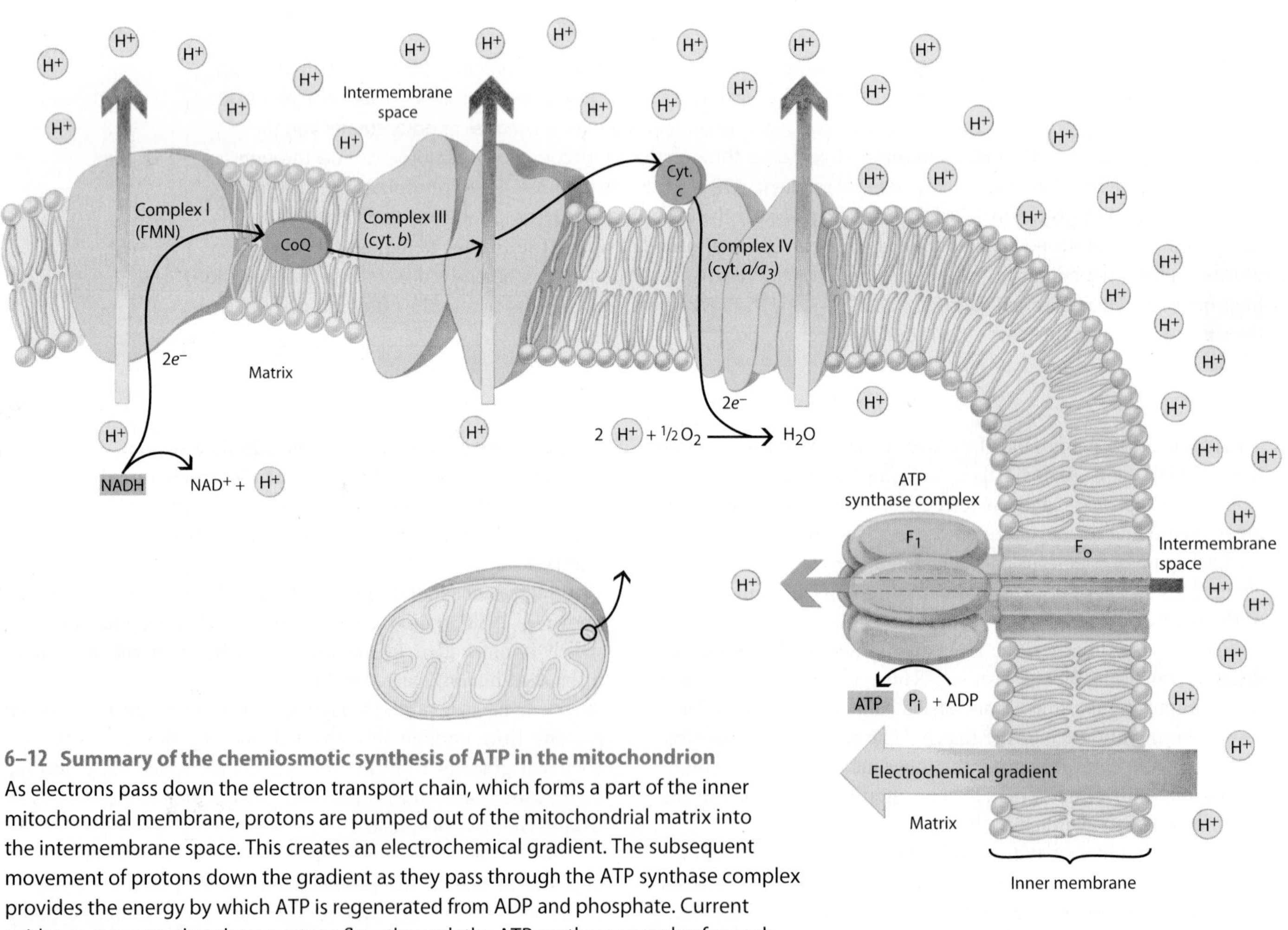

6–12 Summary of the chemiosmotic synthesis of ATP in the mitochondrion
As electrons pass down the electron transport chain, which forms a part of the inner mitochondrial membrane, protons are pumped out of the mitochondrial matrix into the intermembrane space. This creates an electrochemical gradient. The subsequent movement of protons down the gradient as they pass through the ATP synthase complex provides the energy by which ATP is regenerated from ADP and phosphate. Current evidence suggests that three protons flow through the ATP synthase complex for each molecule of ATP formed.

BIOLUMINESCENCE

In most organisms, the energy stored in ATP is used for cellular work, such as biosynthesis, transport, and movement. In some, however, a fraction of the chemical energy is reconverted to light energy. Bioluminescence is probably an accidental by-product of energy exchanges in most luminescent organisms, such as the fungus *Mycena lux-coeli* shown here, photographed by its own light. However, in some luminescent organisms, bioluminescence serves a useful function. For example, the firefly uses flashes of light as its mating signal.

The citric acid cycle also occurs in the matrix of the mitochondrion, yielding 2 molecules of ATP, 6 of NADH, and 2 of $FADH_2$. The passage down the electron transport chain of the electrons held by these NADH and $FADH_2$ molecules drives the pumping of enough protons across the membrane to yield 22 ATP—18 from the 6 NADH molecules plus 4 from the 2 $FADH_2$ molecules. Thus, for each molecule of glucose, the total yield of the citric acid cycle is 24 ATP.

As the balance sheet shows (Figure 6–13), the net yield from a single molecule of glucose is 36 molecules of ATP—6 from glycolysis, 6 from the conversion of pyruvate to acetyl CoA, and 24 from the citric acid cycle. All but 2 of the 36 molecules of ATP have come from reactions in the mitochondrion, and all but 4 involve the oxidation of NADH or $FADH_2$ down the electron transport chain, coupled with oxidative phosphorylation.

The total difference in free energy (ΔG) between the reactants (glucose and oxygen) and the products (carbon dioxide and water) is –686 kilocalories per mole under standard conditions. The terminal phosphoanhydride bonds of the 36 ATP molecules account for about 263 kilocalories (7.3×36) per mole of glucose (again, under standard conditions). In other words, about 38 percent of the energy is conserved in the ATP. The remainder is lost as heat.

Other Substrates for Respiration

Thus far we have regarded glucose as the main substrate for respiration. It is important to note, however, that fats and proteins can also be converted to acetyl CoA and enter the citric acid cycle (see Figure 6–15). For a fat, the triglyceride molecule is first hydrolyzed to glycerol and three fatty acids. Then, beginning at the carboxyl end of the fatty acids, two-carbon acetyl groups are successively removed as acetyl CoA by a process called beta oxidation. A molecule such as oleic acid (see Figure 2–7), which contains 18 carbon atoms, yields nine molecules of acetyl CoA that can be oxidized by the citric acid cycle. Proteins are similarly broken down into their constituent amino acids, and the amino groups are removed. Some of the residual carbon skeletons are converted to citric acid cycle intermediates such as α-ketoglutarate, oxaloacetate, and fumarate, and thereby enter the cycle.

Molecules produced:

	Cytosol	Matrix of mitochondrion	Electron transport and oxidative phosphorylation	
Glycolysis	2 ATP →	→	→	2 ATP
	2 NADH →	→	4 ATP (net yield) →	4 ATP
Pyruvate to acetyl CoA		2 × (1 NADH) →	2 × (3 ATP) →	6 ATP
Citric acid cycle		2 × (1 ATP) →	→	2 ATP
		2 × (3 NADH) →	2 × (9 ATP) →	18 ATP
		2 × (1 $FADH_2$) →	2 × (2 ATP) →	4 ATP
			Total:	36 ATP

6–13 Energy yield from glucose oxidation
A summary of the net energy yield of 36 ATP from the complete oxidation of one molecule of glucose.

Anaerobic Pathways

In most eukaryotic cells (as well as in most bacteria), pyruvate usually follows the aerobic pathway and is completely oxidized to carbon dioxide and water. However, in the absence or shortage of oxygen, pyruvate is not the end product of glycolysis. Under these conditions, the NADH produced during the oxidation of glyceraldehyde 3-phosphate cannot pass its electrons to O_2 via the electron transport chain, but must still be reoxidized to NAD^+. Without this reoxidation, glycolysis would soon stop because the cell would run out of NAD^+ as an electron acceptor.

In many bacteria, fungi, protists, and animal cells, this oxygenless, or anaerobic, process results in the formation of lactate, a three-carbon compound similar in structure to pyruvate. This process is therefore called **lactate fermentation.** In yeast and most plant cells, however, pyruvate is converted to ethanol (ethyl alcohol) and carbon dioxide. This anaerobic process is therefore called **alcohol fermentation.** In both cases, the two electrons (and one proton) from NADH are transferred to what was the middle carbon of pyruvate. In the case of alcohol fermentation, however, the reoxidation of NADH is preceded by the release of carbon dioxide (decarboxylation) (Figure 6–14).

Thermodynamically, lactate fermentation and alcohol fermentation are similar. In both, the NADH is reoxidized, and the energy yield for glucose breakdown is limited to the net gain of 2 ATP molecules produced during glycolysis. The complete, balanced equations for the two types of glucose fermentation can be written as follows:

$$\text{Glucose} + 2\text{ADP} + 2\text{P}_i \longrightarrow 2\ \text{Ethanol} + 2\text{CO}_2 + 2\text{ATP} + 2\text{H}_2\text{O}$$

and

$$\text{Glucose} + 2\text{ADP} + 2\text{P}_i \longrightarrow 2\ \text{Lactate} + 2\text{ATP} + 2\text{H}_2\text{O}$$

During alcohol fermentation, approximately 7 percent of the total available energy of the glucose molecule—about 52 kilocalories per mole—is released, with about 93 percent remaining in the two alcohol molecules. However, if one considers the efficiency with which the anaerobic cell conserves much of those 52 kilocalories as ATP (7.3 kcal/mol of ATP, or

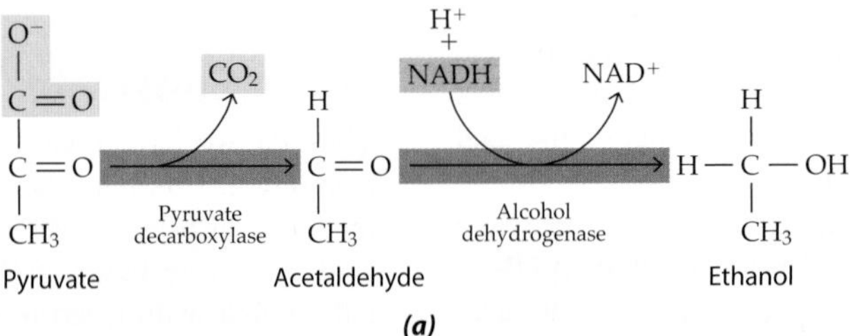

(a)

(b)

6–14 Alcohol fermentation ***(a)*** The two-step process by which pyruvate is converted anaerobically to ethanol. In the first step, carbon dioxide is released. In the second, NADH is oxidized and acetaldehyde is reduced. Most of the energy of the glucose remains in the alcohol, which is the principal end product of the sequence. However, by regenerating NAD^+, these steps allow glycolysis to continue, with its small but sometimes vital yield of ATP. ***(b)*** An example of anaerobic glycolysis. Ancient Egyptian wall paintings, such as the one shown here, are the earliest historical record of wine-making. They have been dated to about 5000 years ago. However, recently discovered pottery fragments, stained with wine, suggest that the Sumerians had mastered the art of wine-making at least 500 years before the Egyptians. The grapes were picked and then crushed by foot, and the juice was collected in jugs and allowed to ferment, producing wine. In modern wine-making, pure yeast cultures are added to relatively sterile grape juice for fermentation, rather than relying on the yeasts carried on the grapes.

THE BOTANY OF BEER

The bubbly and deliciously bitter beer we enjoy today represents centuries of the practical combination of botany and biochemistry. Humans have brewed beer for at least 5000 years, and probably much longer. Drinking beer "warms the liver and sates the heart," according to an ode written in 1800 B.C.E. to Ninkasi, the Sumerian goddess of beer. Providing a source of nutritious calories and clean water—the brewing process eliminates harmful microbes—beer has been a staple of human culture since civilization began, and was long considered far safer to drink than water.

Beer production begins with a source of carbohydrates. Nearly all modern beer is made from barley, but wheat, maize, rice, and other grains, and even starchy roots, will serve as well. The barley grains are sprouted to convert starches in the endosperm, the nutritious part of the grain, to sugars. This process is controlled by the enzyme amylase, which the barley seed contains in abundance. After a few days, the grain is rapidly heated and dried to stop germination, while preserving the amylase. Further roasting darkens the grain, now called malt, accounting for the final color of the beer. The malt is then crushed and mixed with water to complete the release of its sugars.

The resulting liquid is boiled with flower clusters of the hops plant, *Humulus lupulus,* which act as a preservative and add flavor. Bitter resins in the flowers balance the sweetness of the sugars and give beer its characteristic taste and aroma. The introduction of hops into the beer-making process occurred in the Middle Ages; prior to that, dates, raisins, and spices, such as cinnamon and cardamom, were added for flavor.

The liquid, now called bitter wort, is cooled and mixed with brewer's yeast (a species of the *Saccharomyces* genus), which ferments the sugars to alcohol and carbon dioxide. The choice of yeast and the temperature of the fermentation distinguish ales, which are fermented warm and have an assertive taste, from lagers, which are fermented cold and have a smoother, crisper taste. The fermentation process slows as sugar is consumed and the alcohol level increases. Most beers have an alcohol content of about 5 percent; higher levels of alcohol can be achieved by adding sugar or additional alcohol-tolerant yeast. Carbon dioxide, which gives beer its fizz, tends to bubble off, but it can be trapped by brewing in a closed container under pressure—or the carbon dioxide can be added back after brewing is complete.

Beer is now a multibillion-dollar global industry, with four companies producing over half of all the beer sold worldwide. Virtually the entire world production of hops is devoted to beer-making. While most barley is used as a food for animals and people, about 15 percent of the annual crop goes to beer production. Global beer consumption is almost 200 billion liters per year, or about 30 liters per person each year, but there are wide variations among countries.

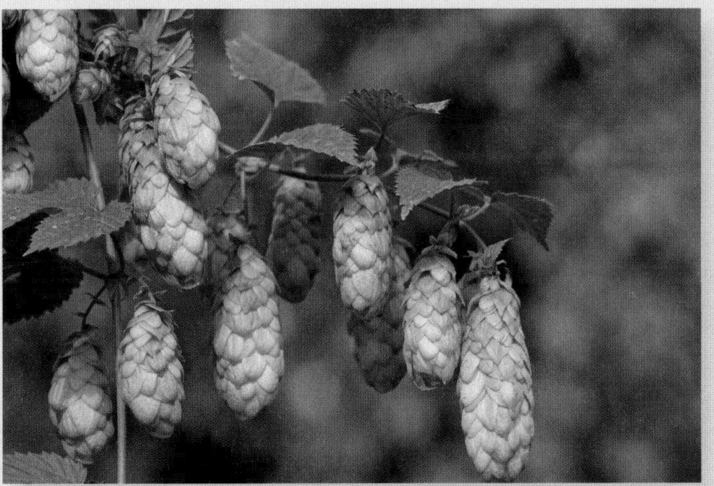

The female flower clusters, or inflorescences, of hops *(Humulus lupulus)* provide beer's distinctive flavor and act as an antibiotic, killing detrimental microorganisms while allowing growth of the yeast essential for fermentation.

14.6 kcal/mol of glucose), the efficiency of energy conservation is about 26 percent. The efficiency of human-made machines seldom exceeds 25 percent, and that of respiration, as we have seen in this chapter, is about 38 percent.

The fact that glycolysis does not require oxygen suggests that the glycolytic sequence evolved early, before free oxygen was present in the atmosphere. Presumably, primitive one-celled organisms used glycolysis (or something very much like it) to extract energy from organic molecules they absorbed from their surroundings. Although the anaerobic pathways generate only two molecules of ATP for each glucose processed, this low yield was and is adequate for the needs of many organisms or parts of organisms. The root systems of rice plants in flooded rice paddies, for example, often carry out extensive fermentation to provide energy for the growth and metabolism of the roots.

The Strategy of Energy Metabolism

The various pathways by which different organic molecules are broken down to yield energy are known collectively as **catabolism.** Catabolic intermediates, such as pyruvate or the acetyl group of acetyl CoA, are also central to the biosynthetic processes of life. These processes, known collectively as **anabolism,** are the pathways by which cells synthesize

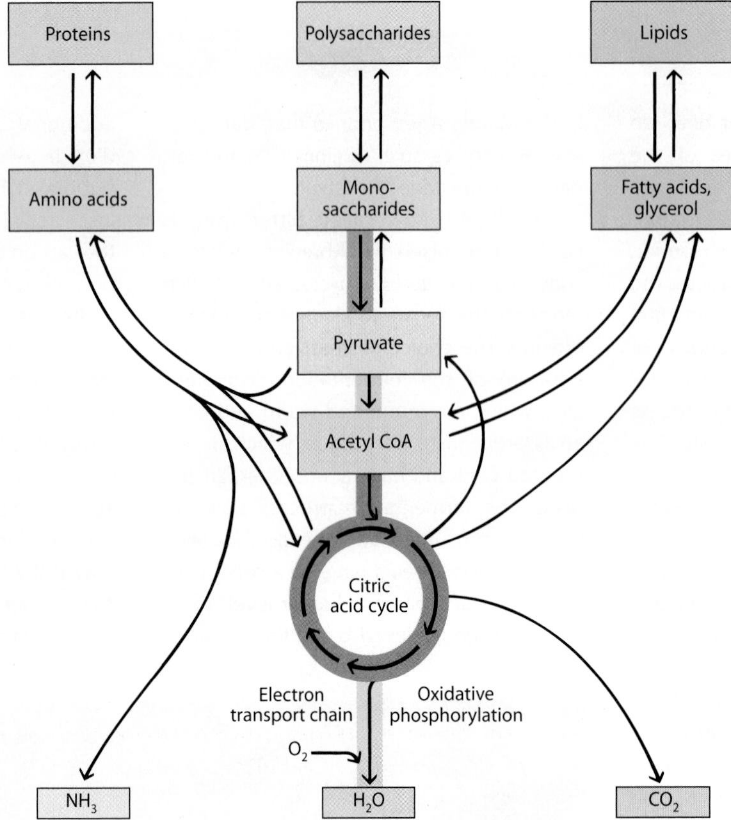

6–15 Some major pathways of catabolism and anabolism in the living cell Catabolic pathways (arrows pointing down) are exergonic. A significant portion of the energy released in these pathways is captured in the synthesis of ATP. Anabolic pathways (arrows pointing up) are endergonic. The energy that powers the reactions in these pathways is supplied primarily by ATP and NADH.

the diversity of molecules that constitute a living organism. Because many of these molecules, such as proteins and lipids, can be broken down and fed into the central pathway of glucose metabolism, you might guess that the reverse process can occur—namely, that the various intermediates of glycolysis and the citric acid cycle can serve as precursors for biosynthesis. This is indeed the case, as outlined in Figure 6–15. Hence, the citric acid cycle plays a critical role in both catabolic and anabolic processes and represents a major "hub" of metabolic activities in the cell.

For the reactions of the catabolic and anabolic pathways to occur, there must be a steady supply of organic molecules that can be broken down to yield not only energy but also building-block molecules. Without a supply of such molecules, the metabolic pathways cease to function and the organism dies. Heterotrophic cells (including the heterotrophic cells of plants, such as root cells) are dependent on external sources—specifically, autotrophic cells—for the organic molecules that are essential for life. Autotrophic cells, however, are able to synthesize their own energy-rich organic molecules from simple inorganic molecules and an external energy source. These molecules supply energy as well as building-block molecules.

By far the most important autotrophic cells are the photosynthetic cells of algae and plants. In the next chapter, we examine how these cells capture the energy of sunlight and use it to synthesize the monosaccharide molecules on which life on this planet depends.

SUMMARY

Respiration, or the Complete Oxidation of Glucose, Is the Chief Source of Energy in Most Cells

The overall reaction for respiration, which results in the complete oxidation of glucose, is:

$$\underset{\text{Glucose}}{C_6H_{12}O_6} + \underset{\text{Oxygen}}{6O_2} \longrightarrow \underset{\text{Carbon dioxide}}{6CO_2} + \underset{\text{Water}}{6H_2O} + \text{Energy}$$

As glucose is oxidized in a series of sequential enzyme-catalyzed reactions, some of the energy released is packaged in the form of terminal phosphoanhydride bonds in ATP and the rest is lost as heat.

In Glycolysis, Glucose Is Split into Pyruvate

The first phase in the oxidation of glucose is glycolysis, in which the six-carbon glucose molecule is split into two three-carbon molecules of pyruvate. This reaction occurs in the cytosol of eukaryotic cells and results in the formation of two molecules of ATP and two of NADH.

The Citric Acid Cycle Completes the Metabolic Breakdown of Glucose to Carbon Dioxide

In the course of respiration, the three-carbon pyruvate molecules are oxidized in the mitochondrial matrix to two-carbon acetyl groups, which then enter the citric acid cycle as acetyl CoA. In the citric acid cycle, each acetyl group is oxidized in a series of reactions to yield two additional molecules of carbon dioxide, one molecule of ATP, and four molecules of reduced electron carriers (three NADH and one $FADH_2$). With two turns of the cycle, the carbon atoms derived from the glucose molecule are completely oxidized and released as molecules of CO_2.

In the Electron Transport Chain, the Flow of Electrons Is Coupled to the Pumping of Protons across the Inner Mitochondrial Membrane and the Synthesis of ATP by Oxidative Phosphorylation

The next stage of respiration is the electron transport chain, which involves a series of electron carriers and enzymes embedded in the inner membrane of the mitochondrion. Along this series of electron carriers, the high-energy electrons carried by NADH and $FADH_2$ move "downhill" energetically, ultimately reducing oxygen to water. The large quantity of free energy released during the passage of electrons down the electron transport chain powers the pumping of protons (H^+ ions) out of the mitochondrial matrix. This creates an electrochemical gradient of potential energy across the inner membrane of the mitochondrion. When protons pass through the ATP synthase complex as they flow down the gradient back into the matrix, the free energy released is used to form ATP from ADP and phosphate. This process, known as chemiosmotic coupling, is the mechanism by which oxidative phosphorylation is accomplished.

In the course of the aerobic breakdown of a glucose molecule to CO_2 and H_2O, 36 molecules of ATP are generated, most of them in the mitochondrion in the final stage of respiration, oxidative phosphorylation.

Fermentation Reactions Occur under Anaerobic Conditions

In the absence or shortage of oxygen, pyruvate produced by glycolysis may be converted either to lactate (in many bacteria, fungi, protists, and animal cells) or to ethanol and carbon dioxide (in yeasts and most plant cells). These anaerobic processes—called fermentation—yield 2 ATP for each glucose molecule.

The Citric Acid Cycle Is the "Metabolic Hub" for the Breakdown and Synthesis of Many Different Types of Molecules

Although glucose is regarded as the main substrate for respiration in most cells, fats and proteins can also be converted to molecules that can enter the respiratory sequence at several different steps. The various pathways by which organic molecules are broken down to yield energy are known collectively as catabolism. The biosynthetic processes of life are known collectively as anabolism.

QUESTIONS

1. Distinguish between substrate-level phosphorylation and oxidative phosphorylation. Where do these processes occur in the cell in relation to respiration?
2. Sketch the structure of a mitochondrion. Describe where the various stages in the complete breakdown of glucose take place in relation to mitochondrial structure. What molecules and ions cross the mitochondrial membranes during these processes?
3. One might say that two distinct events take place in chemiosmotic coupling. What are those events? What are some uses of chemiosmotic power in living organisms?
4. Certain chemicals function as "uncoupling" agents when they are added to respiring mitochondria. The passage of electrons down the electron transport chain to oxygen continues, but no ATP is formed. One of these agents, the antibiotic valinomycin, is known to transport K^+ ions through the inner membrane into the matrix. Another agent, 2,4-dinitrophenol, transports H^+ ions through the membrane. How do these substances prevent the formation of ATP?
5. With some strains of yeast, fermentation stops before the sugar is exhausted, usually at an alcohol concentration in excess of 12 percent. What is a plausible explanation?

CHAPTER 7

Photosynthesis, Light, and Life

◀ **Our brilliant star** We, and all the life around us, owe our existence to the continued thermonuclear events taking place at the heart of a middle-aged, mid-sized star—our sun. Those thermonuclear events produce energy that arrives as sunlight on Earth, where photosynthetic systems have evolved to capture and convert the light energy to the chemical energy that flows through living organisms.

CHAPTER OUTLINE

In the previous chapter, we described the breakdown of carbohydrates to yield the energy required for the many kinds of activities carried out by living organisms. In the pages that follow, we complete the circle by describing how light energy from the sun is captured and converted to chemical energy (see Figure 5–1). This process—**photosynthesis**—is the route by which virtually all energy enters our biosphere.

Each year, more than 250 billion metric tons of sugar are produced worldwide by photosynthetic organisms. The importance of photosynthesis, however, extends far beyond the sheer weight of this product. Without this flow of energy from the sun, channeled largely through the chloroplasts of eukaryotic cells (Figure 7–1), the pace of life on this planet would swiftly diminish and then would virtually cease altogether due to entropy, as dictated by the inexorable second law of thermodynamics.

Photosynthesis: A Historical Perspective

The importance of photosynthesis was not recognized until comparatively recently. Aristotle and other Greeks, observing that the life processes of animals were dependent on the food they ate, thought that plants derived all of their food from the soil.

More than 350 years ago, in one of the first carefully designed biological experiments ever reported, the Belgian physician Jan Baptista van Helmont (ca. 1577–1644) offered the first experimental evidence that soil alone does not nourish the plant. Van Helmont grew a small willow tree in an earthenware pot, adding only water to the pot. At the end of five years, the willow had increased in weight by 74.4 kilograms, whereas the soil had decreased in weight by only 57 grams. On the basis of these results, van Helmont concluded that all the substance of the plant was produced from the water and none from the soil! Van Helmont's conclusions, however, were only partially correct; hydrogen from water does contribute to the mass of the

CHECKPOINTS

After reading this chapter, you should be able to answer the following:

1. What is the role of light in photosynthesis, and what are the properties of light that suggest it is both a wave and a particle?
2. What are the principal pigments involved in photosynthesis, and why are leaves green?
3. List the main products of the light reactions of photosynthesis.
4. List the main products of the carbon-fixation reactions of photosynthesis.
5. What are the main events associated with each of the two photosystems in the light reactions, and what is the difference between antenna pigments and reaction center pigments?
6. Describe the principal differences among the C_3, C_4, and CAM pathways for carbon fixation. What features do they have in common?

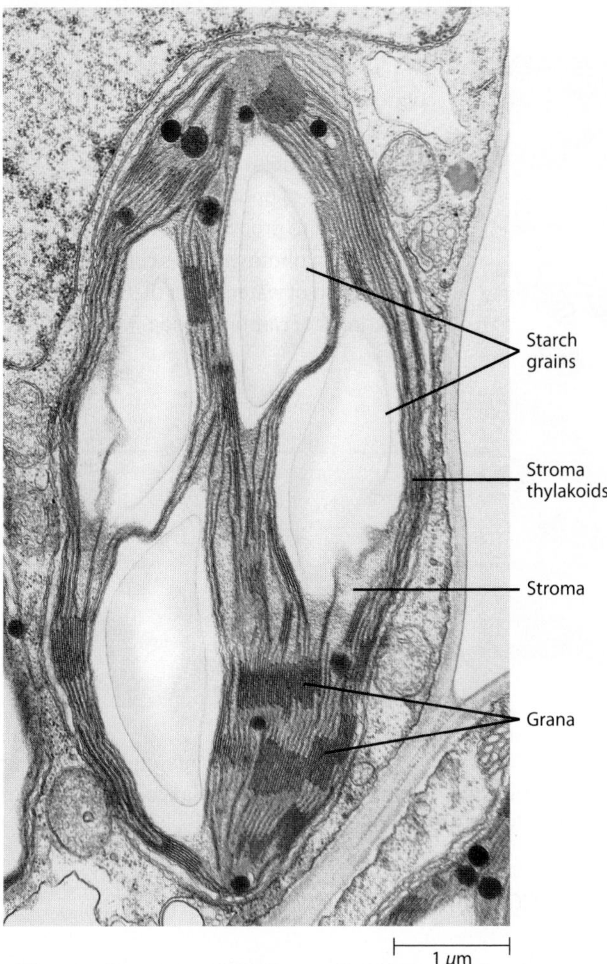

7–1 Chloroplast Shown here is a chloroplast from a mesophyll cell of the pigweed *(Amaranthus retroflexus)* leaf. The light-capturing reactions of photosynthesis occur in the internal membranes, or thylakoids, where the chlorophylls and other pigments are embedded. Many of the thylakoids occur in disklike stacks called grana. The thylakoids of different grana are interconnected by stroma thylakoids. The series of reactions by which the captured light energy is used to synthesize carbon-containing compounds occurs in the stroma, the material surrounding the thylakoids. During periods of intense photosynthesis, some of the carbohydrate is stored temporarily in the chloroplast as grains of starch. At night, sucrose is produced from the starch and exported from the leaf to other parts of the plant, where it is eventually used for the manufacture of other molecules needed by the plant.

plant, but the major contribution of atmospheric carbon dioxide was not yet appreciated.

Toward the end of the eighteenth century, the English clergyman-scientist Joseph Priestley (1733–1804) reported that he had "accidently hit upon a method of restoring air that had been injured by the burning of candles." On 17 August 1771, Priestley "put a [living] sprig of mint into air in which a wax candle had burned out and found that, on the 27th of the same month, another candle could be burned in this same air." The "restorative which nature employs for this purpose," he stated, was "vegetation." Priestley extended his observations and soon showed that air "restored" by vegetation was not "at all inconvenient to a mouse." Priestley's experiments offered the first logical explanation of how "injured" air was "restored" and was able to continue to support life despite the burning of countless fires and the breathing of many animals. When he was presented with a medal for his discovery, the citation read, in part: "For these discoveries we are assured that no vegetable grows in vain . . . but cleanses and purifies our atmosphere." Today, we would explain Priestley's experiments simply by saying that plants take up the CO_2 produced by combustion or exhaled by animals and that animals inhale the O_2 released by plants.

Shortly thereafter, the Dutch physician Jan Ingenhousz (1730–1799) confirmed Priestley's work and showed that the air was "restored" only in the presence of sunlight and only by the green parts of plants. In 1796, Ingenhousz suggested that carbon dioxide is split in photosynthesis to yield carbon and oxygen, with the oxygen then released as gas. Subsequently, the proportion of carbon, hydrogen, and oxygen atoms in sugars and starches was found to be about one atom of carbon per molecule of water (CH_2O), as the word "carbohydrate" indicates. Thus, in the overall reaction for photosynthesis,

$$CO_2 + H_2O + \text{Light energy} \longrightarrow (CH_2O) + O_2$$

it was generally assumed that the carbohydrate came from a combination of water molecules and the carbon atoms of carbon dioxide, and that the oxygen was released from the carbon dioxide. This entirely reasonable hypothesis was widely accepted; but, as it turned out, it was quite wrong.

The investigator who upset this long-held theory was C. B. van Niel of Stanford University. Van Niel, then a graduate student, was investigating the activities of different types of photosynthetic bacteria (Figure 7–2). One particular group of such

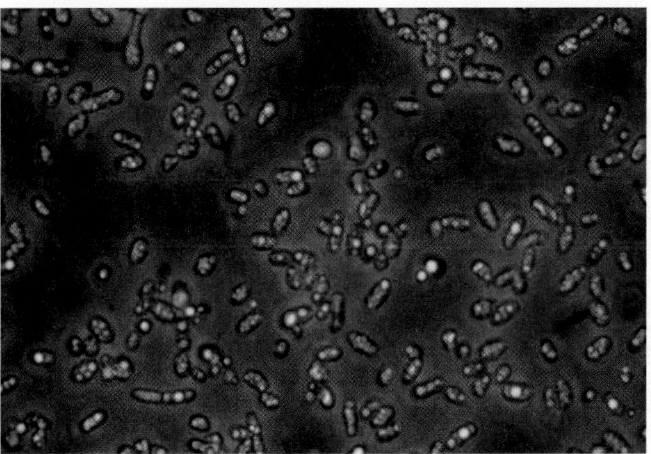

7–2 Purple sulfur bacteria These bacteria reduce carbon to carbohydrates during photosynthesis but do not release oxygen. In these cells, hydrogen sulfide (H_2S) plays the same role that water plays in the photosynthetic process of plants. The hydrogen sulfide is split, and the sulfur accumulates as globules, visible within these cells.

7–3 Oxygen production during photosynthesis The bubbles on the leaves of this submerged pondweed, *Elodea,* are bubbles of oxygen, one of the products of photosynthesis. Van Niel was the first to propose that the oxygen produced in photosynthesis comes from the splitting of water rather than the breakdown of carbon dioxide.

bacteria—the purple sulfur bacteria—reduces carbon to carbohydrates during photosynthesis but does not release oxygen. The purple sulfur bacteria require hydrogen sulfide for their photosynthetic activity. In the course of photosynthesis, globules of sulfur accumulate inside the bacterial cells (Figure 7–2). Van Niel found that the following reaction takes place during photosynthesis in these bacteria:

$$CO_2 + 2H_2S \xrightarrow{\text{Light}} (CH_2O) + H_2O + 2S$$

This finding was simple and did not attract much attention until van Niel made a bold extrapolation. He proposed the following generalized equation for photosynthesis:

$$CO_2 + 2H_2A \xrightarrow{\text{Light}} (CH_2O) + H_2O + 2A$$

In this equation, H_2A represents an oxidizable substance, such as hydrogen sulfide or free hydrogen. In algae and green plants, however, H_2A is water (Figure 7–3). In short, van Niel proposed that water, *not* carbon dioxide, was the source of the oxygen in photosynthesis.

In 1937, Robin Hill showed that isolated chloroplasts, when exposed to light, were able to produce O_2 in the absence of CO_2. This light-driven release of O_2 in the absence of CO_2—called the **Hill reaction**—occurred only when the chloroplasts were illuminated and provided with an artificial electron acceptor. This finding supported van Niel's proposal made six years earlier.

More convincing evidence that the O_2 released in photosynthesis is derived from H_2O came in 1941, when Samuel Ruben and Martin Kamen used a heavy isotope of oxygen (^{18}O) to trace the oxygen from water to oxygen gas:

$$CO_2 + 2H_2{}^{18}O \xrightarrow{\text{Light}} (CH_2O) + H_2O + {}^{18}O_2$$

Thus, in the case of algae and green plants, in which water serves as the electron donor, a complete, balanced equation for photosynthesis can be written as follows:

$$3CO_2 + 6H_2O \xrightarrow{\text{Light}} C_3H_6O_3 + 3O_2 + 3H_2O$$

Although glucose is commonly represented as the carbohydrate product of photosynthesis in summary equations, the more immediate carbohydrate products are trioses (three-carbon sugars), with the formula $C_3H_6O_3$.

When considering the equation above, you might wonder why H_2O is present on both sides of the equation. The reason is that the energy from splitting 6 molecules of H_2O is required to convert 3 molecules of CO_2 into a three-carbon sugar. And in the process of converting 3 molecules of CO_2 into a three-carbon sugar, some of the hydrogens that are removed from water are used to remove some of the oxygens from carbon by reactions that form H_2O.

As noted above, Ingenhousz was the first to deduce that light was required for the process we now call photosynthesis. We now know that photosynthesis occurs in two stages, only one of which actually requires light. Evidence for this two-stage process was first presented in 1905 by the English plant physiologist F. F. Blackman, as the result of experiments in which he measured the individual and combined effects of changes in light intensity and temperature on the rate of photosynthesis. These experiments showed that photosynthesis has both a light-dependent stage and a light-independent stage.

In Blackman's experiments, the light-independent reactions increased in rate as the temperature was increased, but only up to about 30°C, after which the rate began to decrease. From this evidence it was concluded that these reactions were controlled by enzymes, since this is the way enzymes are expected to respond to temperature (see Figure 5–10). This conclusion has since been proven to be correct.

The Nature of Light

Over 300 years ago, the English physicist Sir Isaac Newton (1642–1727) separated light into a spectrum of visible colors by passing it through a prism. In this way, Newton showed that white light actually consists of a number of different colors, ranging from violet at one end of the spectrum to red at the other. The separation of colors is possible because light of different colors is bent (refracted) at different angles in passing through a prism.

In the nineteenth century, the British physicist James Clerk Maxwell (1831–1879) demonstrated that light is but a small part of a vast continuous spectrum of radiation, the **electromagnetic spectrum** (Figure 7–4). All the radiations in this spectrum travel in waves. The **wavelengths**—that is, the distances from the crest of one wave to the crest of the next—range from those of gamma rays, which are measured in fractions of a nanometer, to those of low-frequency radio waves, which are measured in kilometers (1 kilometer = 10^3 meters = 0.6 mile). Radiation of each particular wavelength has a characteristic amount of energy associated with it. The shorter the wavelength, the greater the energy; conversely, the longer the wavelength, the lower the energy. Within the spectrum of visible light, violet light has the shortest wavelength and red light has the longest. The shortest rays of violet light have almost twice the energy of the longest rays of red light.

Light Has the Properties of Waves and Particles

By 1900 it had become clear that the **wave model** of light was not adequate. The key observation, a very simple one, was made in 1888: when a zinc plate is exposed to ultraviolet light, it acquires a positive charge. The metal becomes positively charged because the light energy dislodges electrons from the metal atoms. Subsequently it was discovered that this **photoelectric effect,** as it is known, can be produced in all metals. Every metal has a critical maximum wavelength for the effect; the light, or other radiation, must be of that particular wavelength or shorter (that is, more energetic) for the effect to occur.

For some metals, such as sodium, potassium, and selenium, the critical wavelength is within the spectrum of visible light, and as a consequence, visible light striking the metal can set up a moving stream of electrons (an electric current). Exposure meters, television cameras, and the electric eyes that open doors at supermarkets and airline terminals all operate on this principle of turning light energy into electrical energy.

What, then, is the problem with the wave model of light? Simply this: the wave model predicts that the brighter the light—that is, the stronger, or more intense, the beam—the greater the force with which electrons will be dislodged from a metal. Whether or not light can eject the electrons of a particular metal, however, depends only on the wavelength of the light, not on its intensity. A very weak beam of the critical

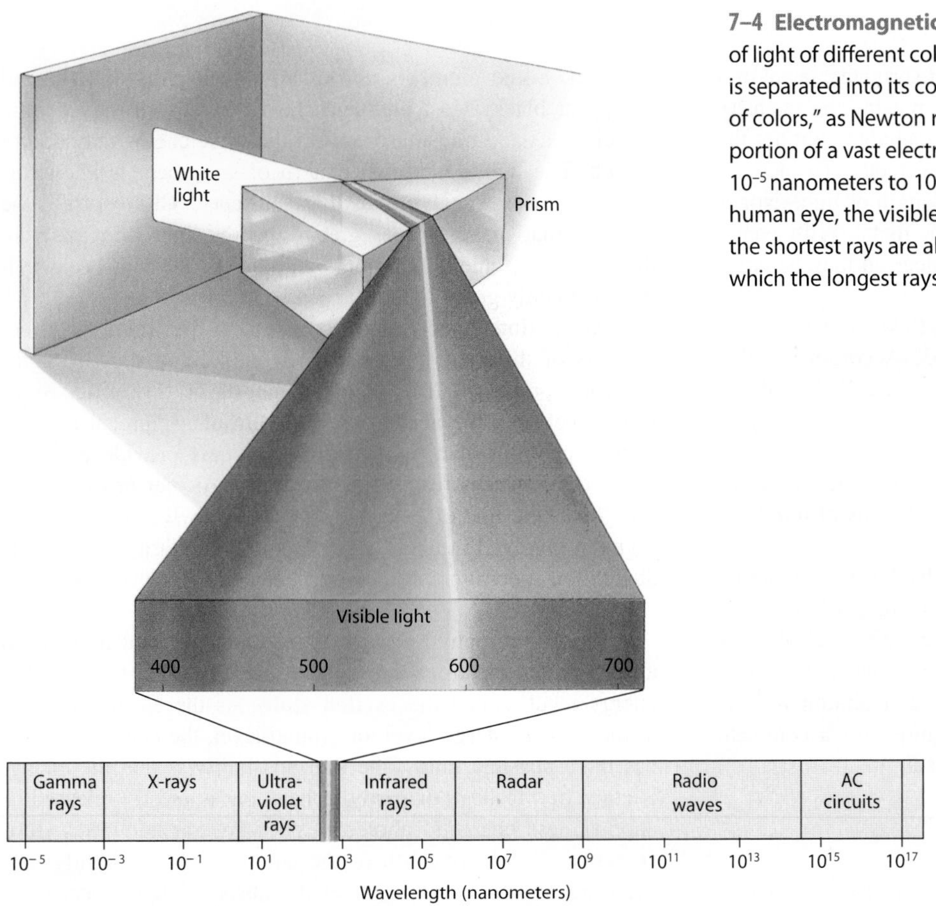

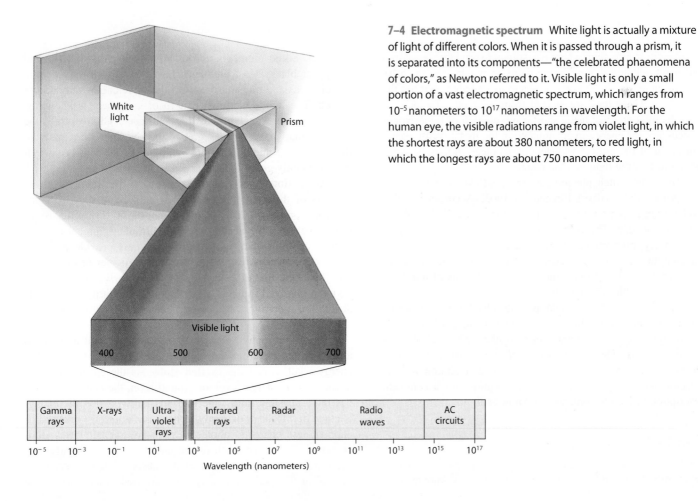

7–4 Electromagnetic spectrum White light is actually a mixture of light of different colors. When it is passed through a prism, it is separated into its components—"the celebrated phaenomena of colors," as Newton referred to it. Visible light is only a small portion of a vast electromagnetic spectrum, which ranges from 10^{-5} nanometers to 10^{17} nanometers in wavelength. For the human eye, the visible radiations range from violet light, in which the shortest rays are about 380 nanometers, to red light, in which the longest rays are about 750 nanometers.

THE FITNESS OF LIGHT

Light, as Maxwell showed, is only a tiny band in a continuous electromagnetic spectrum. From the physicist's point of view, the difference between light and darkness—so dramatic to the human eye—is only a few nanometers of wavelength, or, expressed differently, a small amount of energy. Why is it that this tiny portion of the spectrum is responsible for vision, for phototropism (the curving of an organism toward light), for photoperiodism (the seasonal changes that take place in an organism with the changing length of day and night), and for photosynthesis, on which all life depends? Is it an amazing coincidence that all these biological activities depend on these same wavelengths?

George Wald of Harvard University argued that the relationship between light and life was not a coincidence. He concluded that if life evolved elsewhere in the universe, it would probably be dependent on the same small portion of the spectrum as life on Earth. Wald based this conjecture on two points.

First, living things are composed of large, complicated molecules held in special configurations and relationships to one another by hydrogen bonds and other weak bonds. Radiation of even slightly higher energies (shorter wavelengths) than the energy of violet light breaks these bonds and so disrupts the structure and function of the molecules. DNA molecules, for example, are particularly vulnerable to such disruption. Radiations with wavelengths less than 200 nanometers—that is, with still higher energies—drive electrons out of atoms to create ions; hence, it is called ionizing radiation. On the other hand, radiations with wavelengths longer than those of the visible band—that is, with less energy than red light—are absorbed by water, which makes up the bulk of all living things on Earth. When such radiation is absorbed by organic molecules, it causes them to increase their motion (increasing heat), but it does not trigger changes in their electron configurations. Only those radiations within the range of visible light have the property of exciting molecules—that is, of moving electrons into higher energy levels—and so of producing chemical and, ultimately, biological change.

The second reason the visible band of the electromagnetic spectrum was "chosen" by living things is simply that it is what is available. Most of the radiation reaching the surface of the Earth from the sun is within this range. Most of the higher-energy wavelengths are absorbed by the oxygen and ozone high in the atmosphere. Much infrared radiation is screened out by water vapor and carbon dioxide before it reaches the Earth's surface.

This is an example of what has been termed "the fitness of the environment." The suitability of the environment for life and that of life for the physical world are exquisitely interrelated. If they were not, life could not exist.

wavelength is effective, whereas a stronger (brighter) beam of a longer wavelength is not. Furthermore, increasing the intensity of light increases the number of electrons dislodged but not the velocity at which they are ejected from the metal. To increase the velocity, one must use a shorter wavelength of light. Nor is it necessary for energy to accumulate in the metal. With even a dim beam of the critical wavelength, an electron may be emitted the instant the light hits the metal.

To explain such phenomena, the **particle model** of light was proposed by Albert Einstein in 1905. According to this model, light is composed of particles of energy called **photons,** or quanta of light. The energy of a photon (a quantum of light) is inversely proportional to its wavelength—the longer the wavelength, the lower the energy. Photons of violet light, for example, have almost twice the energy of photons of red light, the longest visible wavelength.

The wave model of light permits physicists to describe certain aspects of its behavior mathematically, whereas the photon model permits another set of mathematical calculations and predictions. These two models are no longer regarded as opposing one another. In fact, the wave and photon models are complementary, because both are required for a complete description of the phenomenon we know as light.

The Role of Pigments

For light energy to be used by living organisms, it must first be absorbed. A substance that absorbs light is known as a **pigment.** Some pigments absorb all wavelengths of light and so appear black. Most pigments, however, absorb only certain wavelengths and transmit or reflect the wavelengths they do not absorb. The light absorption pattern of a pigment is known as the **absorption spectrum** of that substance. **Chlorophyll,** the pigment that makes leaves green, absorbs light principally in the violet and blue wavelengths and also in the red; because it reflects mainly green light, it appears green.

An **action spectrum** demonstrates the relative effectiveness of different wavelengths of light for a specific light-requiring process, such as photosynthesis or flowering. Similarities between the absorption spectrum of a pigment and the action spectrum of a light-requiring process provide evidence that the pigment is responsible for that particular process (Figure 7–5). One line of evidence that chlorophyll is the principal pigment involved in photosynthesis is the similarity between its absorption spectrum and the action spectrum for photosynthesis (Figure 7–6).

When chlorophyll molecules (or other pigment molecules) absorb light, electrons are temporarily boosted to a higher energy level, called the **excited state.** As the electrons return to their lower energy level, or ground level, the energy released has three possible fates. The first possibility is that the energy is converted to heat, or partly to heat, but is mostly released as another, less energetic photon, a phenomenon known as **fluorescence.** The wavelength of the emitted light is slightly longer (and of lower energy) than the absorbed light, because a portion of the excitation energy is converted to heat before the

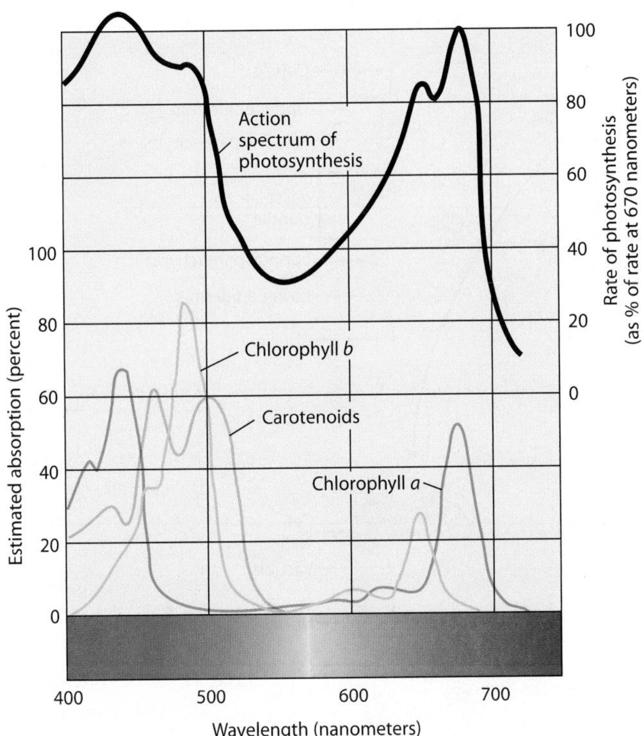

7–5 Comparison of action and absorption spectra The action spectrum for photosynthesis (upper curve) and the absorption spectra for chlorophyll *a*, chlorophyll *b*, and carotenoids (lower curves) in a plant chloroplast. Note the relationship between the action spectrum of photosynthesis and the absorption spectra for the three types of pigments, all of which absorb light at the wavelengths used in photosynthesis.

less energetic, or fluorescent, photon is emitted. A second possibility is that the energy—but not the electron—may be transferred from the excited chlorophyll molecule to a neighboring chlorophyll molecule, exciting the second molecule and allowing the first one to return to its ground (unexcited) state. This process is known as **resonance energy transfer,** and it may be repeated to a third, a fourth, or further chlorophyll molecules. The third possibility is that the high-energy electron itself may be transferred to a neighboring molecule (an electron acceptor) that is part of an electron transport chain, leaving an "electron hole" in the excited chlorophyll molecule. This possibility results in the oxidation of the chlorophyll molecule and the reduction of an electron acceptor.

During the process of photosynthesis in intact chloroplasts, both the second and third possibilities—namely, transfer of energy of the excited chlorophyll to a neighboring chlorophyll and transfer of the high-energy electron itself to a neighboring electron acceptor—are useful energy-releasing events, whereas the competing reaction, resulting in fluorescence, is not a productive reaction.

As you know from Chapter 3, photosynthesis in eukaryotic cells takes place in the chloroplast, and chloroplast structure plays a key role in these energy transfers (Figures 7–1 and 7–7). The chlorophyll molecules themselves, in association with hydrophobic proteins (pigment-protein complexes), are embedded in the thylakoids of the chloroplast.

The Main Photosynthetic Pigments Are the Chlorophylls, the Carotenoids, and the Phycobilins

There are several kinds of chlorophyll, which differ from one another both in the details of their molecular structure and in

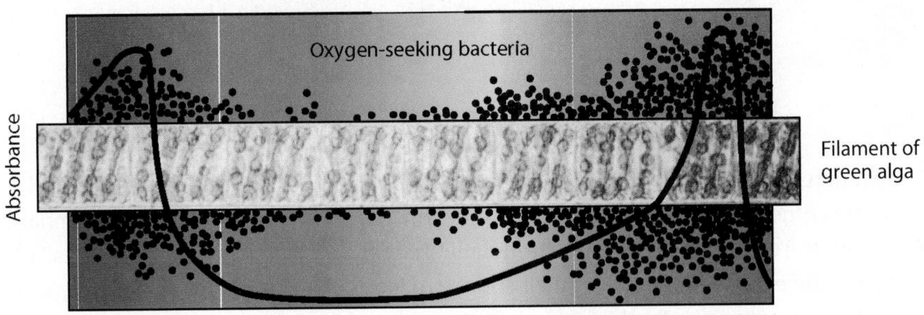

7–6 Correlating action and absorption spectra Results of an experiment performed in 1882 by T. W. Engelmann revealed the action spectrum of photosynthesis in the filamentous alga *Spirogyra.* Like investigators working more recently, Engelmann used the rate of oxygen production to measure the rate of photosynthesis. Unlike his successors, however, he lacked sensitive electronic devices for detecting oxygen. As his oxygen indicator, he chose motile bacteria that are attracted by oxygen. He replaced the mirror and diaphragm usually used to illuminate objects under view in his microscope with a "micro-spectral apparatus," which, as its name implies, projected a tiny spectrum of colors onto the slide under the microscope. Then he arranged a filament of algal cells parallel to the spread of the spectrum. The oxygen-seeking bacteria congregated mostly in the areas where the violet and red wavelengths fell upon the algal filament. As you can see, the action spectrum for photosynthesis paralleled the absorption spectrum of chlorophyll (as indicated by the heavy black line). Engelmann concluded that photosynthesis depends on the light absorbed by chlorophyll. This is an example of the sort of experiment that scientists refer to as "elegant," not only brilliant but also simple in design and conclusive in its results.

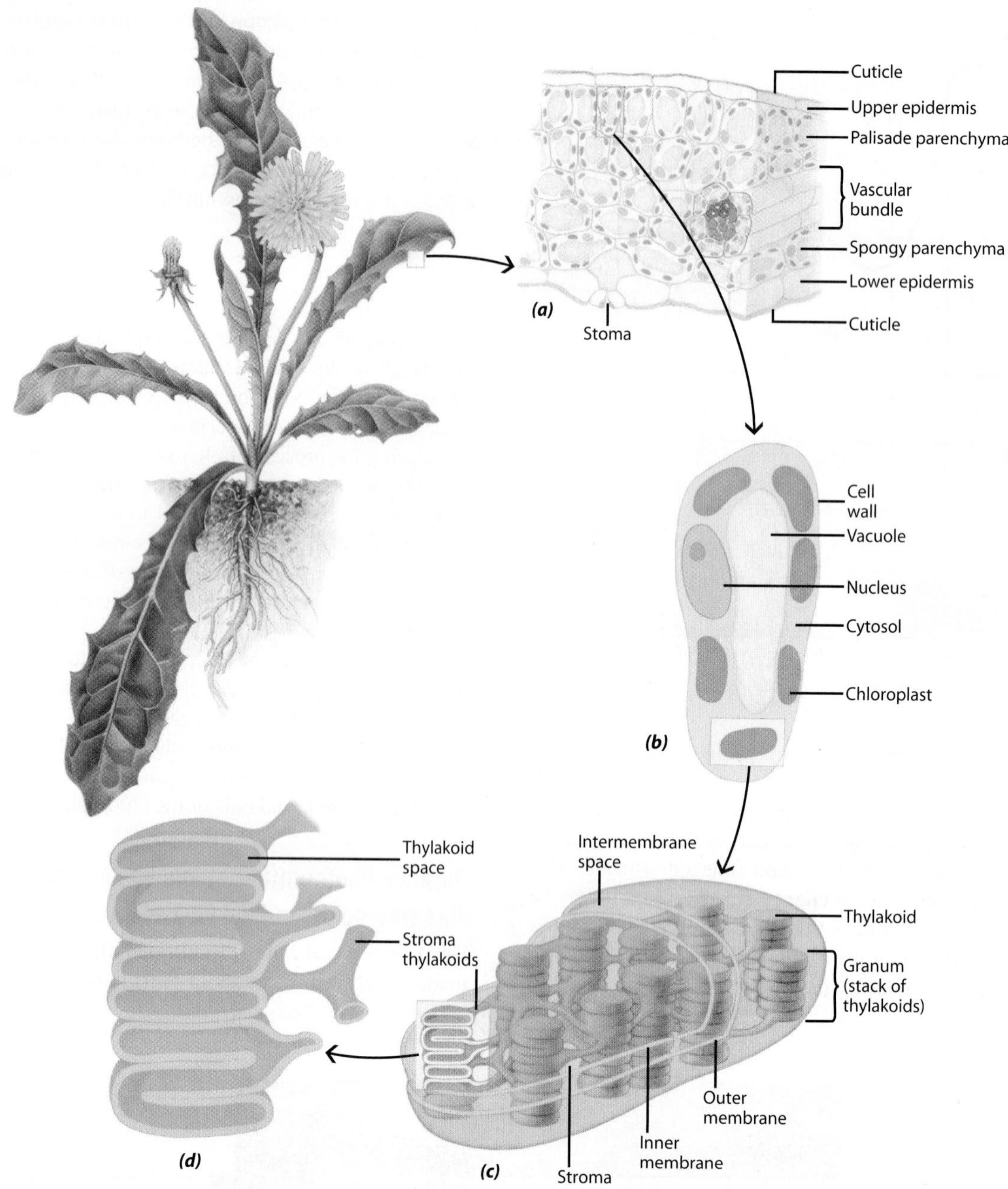

7–7 Journey into a leaf Shown here is a dandelion *(Taraxacum officinale)*. ***(a)*** The inner tissue of the leaf, called mesophyll, is specialized for photosynthesis. The mesophyll of the dandelion leaf consists of elongated, columnar-shaped cells (palisade parenchyma beneath the upper epidermis) and irregularly shaped cells (spongy parenchyma), all of which contain numerous chloroplasts and are exposed for much of their surface to intercellular (air) spaces. Oxygen, carbon dioxide, and other gases, including water vapor, enter the leaf largely through special openings, the stomata (singular: stoma). These gases fill the intercellular spaces, entering and leaving the leaf and cells by diffusion. Water and minerals taken up by the root enter the leaf in the water-conducting tissue (xylem, shown in red) of the vascular bundles, or veins, of the leaf. Sugars, the products of photosynthesis, move out of the leaf by way of the food-conducting tissue (phloem, shown in blue) of the vascular bundles, traveling to nonphotosynthetic parts of the plant. ***(b)*** The chloroplasts occur in the narrow layer of protein-rich cytoplasm that surrounds a large central vacuole. ***(c)*** The three-dimensional structure of a chloroplast and ***(d)*** the arrangement of the pigment-containing thylakoid membranes. The stacks of disklike thylakoids, known as grana, are interconnected by the thylakoids known as stroma thylakoids, which traverse the stroma.

their specific absorption properties. **Chlorophyll *a*** occurs in all photosynthetic eukaryotes and in the cyanobacteria. Not surprisingly, chlorophyll *a* is essential for the oxygen-generating photosynthesis carried out by organisms of these groups (Figure 7–8).

Plants, green algae, and euglenoid algae also contain the pigment **chlorophyll *b*,** which has a slightly different absorption spectrum than chlorophyll *a*. Chlorophyll *b* is an **accessory pigment**—a pigment that is not directly involved in photosynthetic energy transduction but serves to broaden the range of light that can be used in photosynthesis (Figure 7–5). When a molecule of chlorophyll *b* absorbs light, the energy is ultimately transferred to a molecule of chlorophyll *a*, which then transforms it into chemical energy during the course of photosynthesis. In the leaves of most green plants, chlorophyll *a* generally constitutes about three-fourths of the total chlorophyll content, with chlorophyll *b* accounting for the remainder.

Chlorophyll *c* takes the place of chlorophyll *b* in some groups of algae, most notably the brown algae and the diatoms (see Chapter 15). The photosynthetic bacteria (other than cyanobacteria) contain either **bacteriochlorophyll,** which is found in purple bacteria, or **chlorobium chlorophyll,** which occurs in green sulfur bacteria. These bacteria cannot extract electrons from water and thus do not evolve oxygen. Chlorophylls *b* and *c* and the photosynthetic pigments of purple bacteria and green sulfur bacteria are simply chemical variations of the basic structure shown in Figure 7–8.

Two other classes of pigments that are involved in the capture of light energy are the **carotenoids** and the **phycobilins.** The energy absorbed by these accessory pigments must be transferred to chlorophyll *a;* like chlorophylls *b* and *c,* these accessory pigments cannot substitute for chlorophyll *a* in photosynthesis. The phycobilins are found in the cyanobacteria and in chloroplasts of the red algae. They are water-soluble.

Carotenoids are red, orange, or yellow lipid-soluble pigments found in all chloroplasts and in cyanobacteria. Like the chlorophylls, the carotenoid pigments of chloroplasts are associated with hydrophobic proteins and are embedded in the thylakoid membranes. Two groups of carotenoids—**carotenes** and **xanthophylls**—are normally present in chloroplasts. The β-carotene found in plants is the principal source of the vitamin A required by humans and other animals. In green leaves, the color of the carotenoids is usually masked by that of the much more abundant chlorophylls, but in temperate regions of the world, carotenoids become visible when the chlorophylls break down in autumn. Although the carotenoid pigments can help collect light of different wavelengths, their principal function is that of an anti-oxidant, preventing oxidative damage to the chlorophyll molecules by light. Without the carotenoids, there would be no photosynthesis in the presence of oxygen.

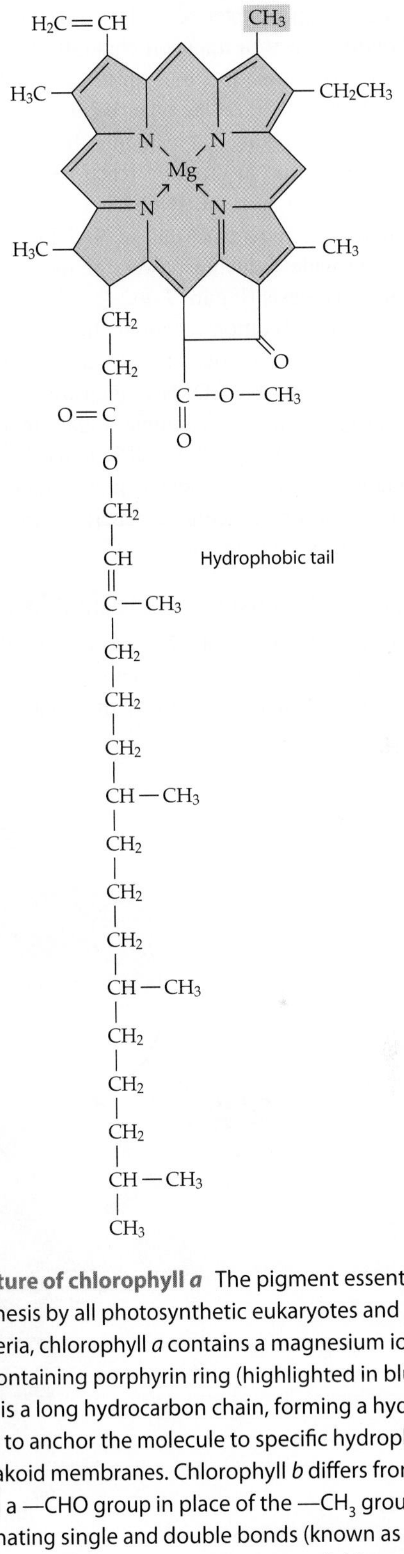

7–8 Structure of chlorophyll *a* The pigment essential for photosynthesis by all photosynthetic eukaryotes and the cyanobacteria, chlorophyll *a* contains a magnesium ion held in a nitrogen-containing porphyrin ring (highlighted in blue). Attached to the ring is a long hydrocarbon chain, forming a hydrophobic tail that serves to anchor the molecule to specific hydrophobic proteins of the thylakoid membranes. Chlorophyll *b* differs from chlorophyll *a* in having a —CHO group in place of the —CH_3 group indicated in gray. Alternating single and double bonds (known as conjugated bonds), such as those in the porphyrin ring of chlorophylls, are common among pigments.

The Reactions of Photosynthesis

The many reactions that occur during photosynthesis are divided into two major processes: **light reactions,** or energy-transduction reactions, which require light energy for photosynthesis to occur, and **carbon-fixation reactions,** in which carbon dioxide is converted into organic compounds.

In the light reactions, light energy is used to form ATP from ADP and inorganic phosphate and to reduce the electron-carrier molecules, notably the coenzyme $NADP^+$. $NADP^+$ is similar in structure to NAD^+ (see Figure 5–8)—it has an extra phosphate on one of the riboses—but its biological role is distinctly different. As we saw in Chapter 6, NADH transfers its electrons to the mitochondrial electron transport chain, thereby driving the pumping of protons across the inner mitochondrial membrane (see Figure 6–12). NADPH, on the other hand, is used to provide energy for biosynthetic pathways, including the synthesis of sugars during photosynthesis. In the light reactions of photosynthesis, water molecules are split, O_2 is liberated, and electrons that are released are used to reduce $NADP^+$ to NADPH. The NADPH is then used to provide reducing power for the carbon-fixation reactions of photosynthesis (Figure 7–9).

During the carbon-fixation reactions, the energy of ATP is used to link carbon dioxide covalently to an organic molecule, and the reducing power of NADPH is then used to reduce the newly fixed carbon atoms to a simple sugar. In the process, the chemical energy of ATP and NADPH is used to synthesize molecules suitable for transport (sugar) and storage (starch). At the same time, a carbon skeleton is generated from which all other organic molecules can be built.

Two Photosystems Are Involved in the Light Reactions

In the chloroplast (Figures 7–1, 7–7, and 7–9), the pigment molecules (chlorophylls *a* and *b* and carotenoids) are embedded in the thylakoids in discrete units of organization called **photosystems.** Each photosystem includes an assembly of about 250 to 400 pigment molecules and consists of two closely linked components: an **antenna complex** and a **reaction center.** The antenna complex consists of pigment molecules that gather light energy and "funnel" it to the reaction center. The reaction center is made up of a complex of proteins and chlorophyll molecules that enable light energy to be converted into chemical energy. Within the photosystems, the chlorophyll molecules are bound to specific chlorophyll-binding membrane proteins and held in place to allow efficient capture of light energy.

All of the pigments within a photosystem are capable of absorbing photons, but only one **special pair** of chlorophyll *a* molecules per reaction center can actually use the energy in the photochemical reaction. This special pair of chlorophyll *a* molecules is situated at the core of the reaction center of the photosystem. The other pigment molecules, called **antenna pigments** because they are part of the light-gathering network, are located in the antenna complex. In addition to chlorophyll, varying amounts of carotenoid pigments are also located in each antenna complex.

Each photosystem is generally associated with a **light-harvesting complex** composed of chlorophyll *a* and *b* molecules, along with carotenoids and pigment-binding proteins. Like the

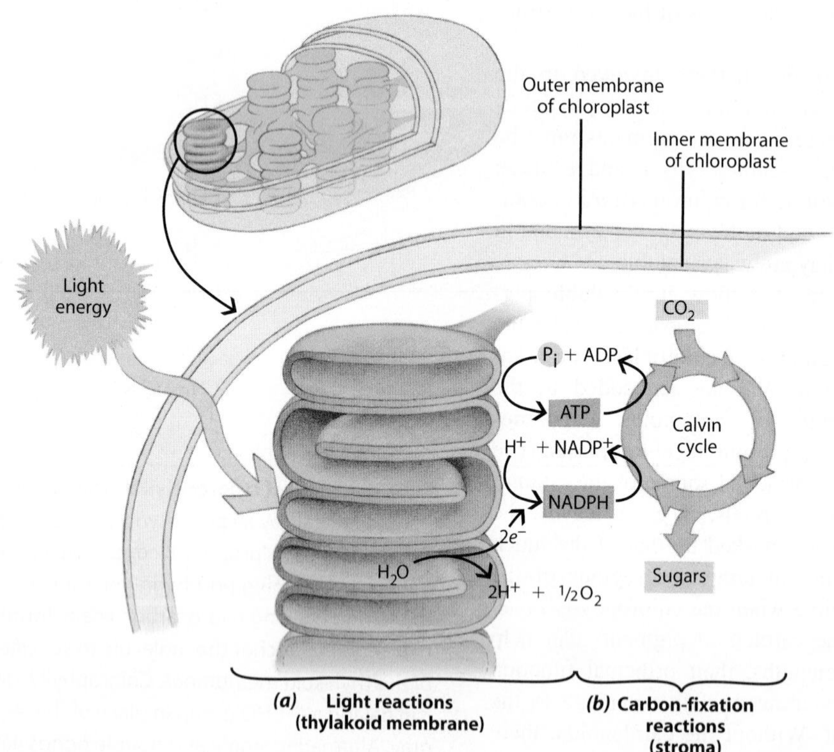

7–9 Overview of photosynthesis Photosynthesis takes place in two stages: the light reactions and the carbon-fixation reactions. ***(a)*** In the light reactions, light energy absorbed by chlorophyll *a* molecules in the thylakoid membrane is used indirectly to power the synthesis of ATP. Simultaneously, in the interior of the thylakoid, water is split into oxygen gas and hydrogen atoms (electrons and protons). The electrons are ultimately accepted by $NADP^+$ and H^+, producing NADPH. ***(b)*** In the carbon-fixation reactions, which occur in the stroma of the chloroplast, sugars are synthesized from carbon dioxide and the hydrogen carried by NADPH. This process is powered by the ATP and NADPH produced in the light reactions. As we shall see, it involves a series of reactions, known as the Calvin cycle, that are repeated over and over.

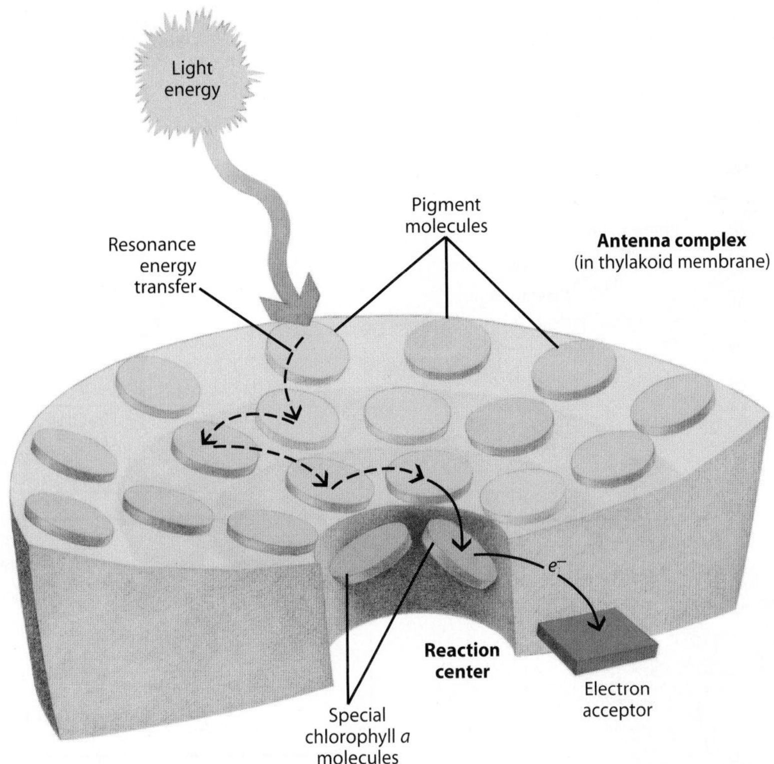

7–10 Energy transfer during photosynthesis The diagram shows a portion of an antenna complex, which occurs in the thylakoid membrane. Light energy absorbed by a pigment molecule anywhere in the antenna complex passes by resonance energy transfer from one pigment molecule to another until it reaches one of two special chlorophyll *a* molecules at the reaction center. When this chlorophyll *a* molecule absorbs the energy, one of its electrons is boosted to a higher energy level and is transferred to an electron-acceptor molecule.

antenna complex of a photosystem, the light-harvesting complex also collects light energy, but it does not contain a reaction center. A photosystem and its associated light-harvesting complexes are collectively referred to as a **photosystem complex.**

Light energy absorbed by a pigment molecule anywhere in an antenna complex or light-harvesting complex is transferred from one pigment molecule to the next by resonance energy transfer until it reaches the reaction center, with its special pair of chlorophyll *a* molecules (Figure 7–10). When either of the two chlorophyll *a* molecules of the reaction center absorbs energy, one of its electrons is boosted to a higher energy level and is transferred to an electron-acceptor molecule to initiate electron flow. The chlorophyll molecule is now in an oxidized state (electron deficient, or positively charged), and the electron acceptor molecule is reduced (electron rich, or negatively charged).

Two different kinds of photosystems, Photosystem I and Photosystem II, are linked together by an electron transport chain (Figure 7–11). The photosystems were numbered in order of their discovery. In **Photosystem I,** the special pair of chlorophyll *a* molecules of the reaction center is known as P_{700}. The "P" stands for pigment and the subscript "700" designates the optimal absorption peak in nanometers. The reaction center of **Photosystem II** also contains a special pair of chlorophyll *a* molecules. Its optimal absorption peak is at 680 nanometers, and accordingly it is called P_{680}.

In general, Photosystem I and Photosystem II work together simultaneously and continuously. Although, for convenience, the two photosystems generally are depicted as occurring more or less side by side in the same thylakoid membrane, as in Figure 7–12, Photosystems I and II are spatially separated. Photosystem II is located primarily in the grana thylakoids and Photosystem I almost entirely in the stroma thylakoids and at the margins, or outer portions of either side, of the grana thylakoids (Figure 7–13a). In addition, as we shall see, Photosystem I can operate independently. The structures of the photosystems and of other protein complexes of the photosynthetic apparatus are shown in Figure 7–13b.

Water Is Oxidized to Oxygen by Photosystem II

In Photosystem II, light energy is absorbed by molecules of P_{680} in the reaction center, either directly or indirectly by resonance energy transfer from one or more of the antenna molecules. When a P_{680} molecule is excited, its energized electron is transferred to a primary acceptor molecule, which transfers its extra electron to a secondary acceptor molecule (Figure 7–11). **Pheophytin,** a modified chlorophyll *a* molecule in which the central magnesium atom has been replaced by two protons, acts as an early electron acceptor. Pheophytin then passes the electron to **PQ_A,** a **plastoquinone,** which is tightly bound to the reaction center. Next, PQ_A passes two electrons to **PQ_B,** another

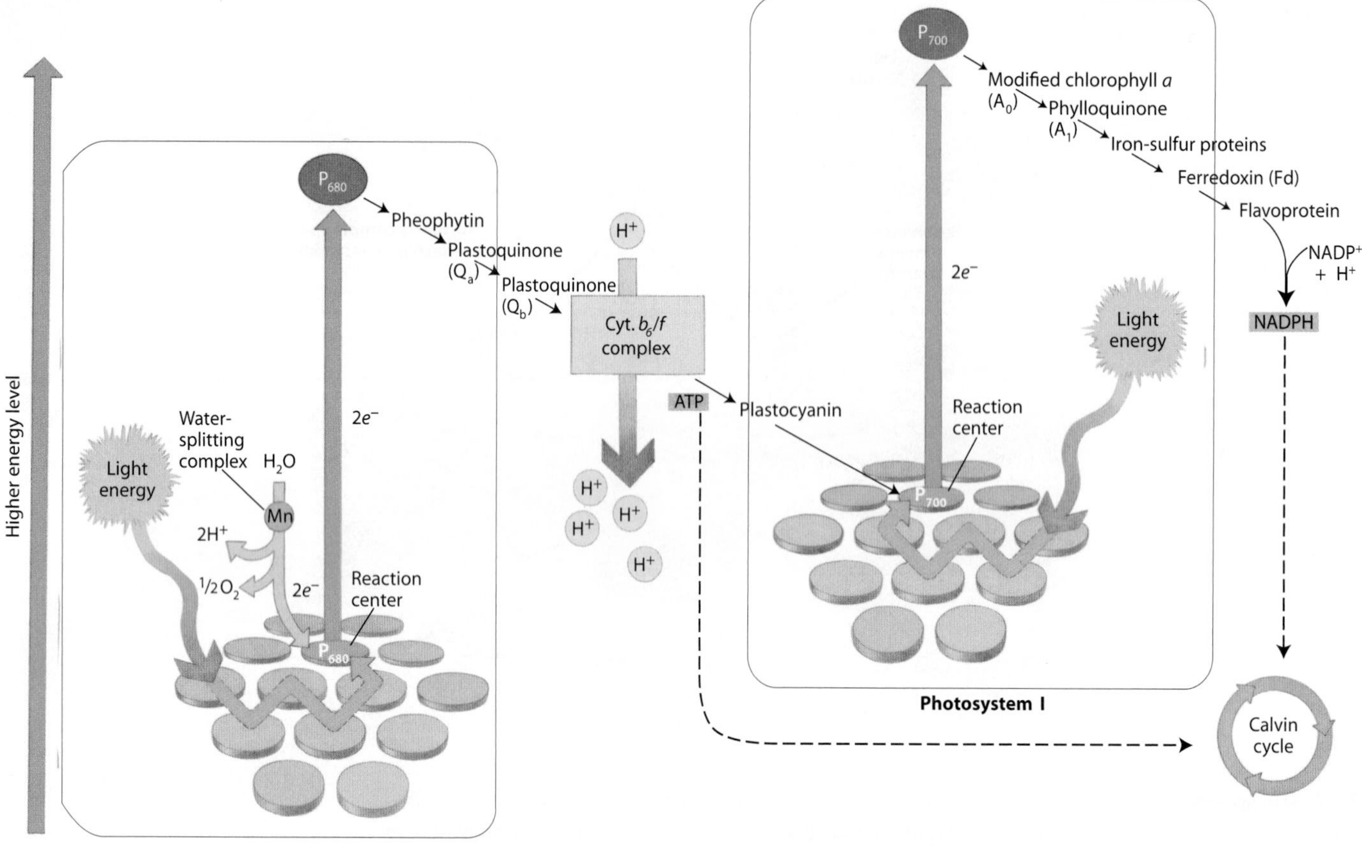

7–11 Noncyclic electron flow and photophosphorylation This zigzag scheme (called the Z scheme) shows the pathway of electron transfer from H_2O to $NADP^+$ that occurs during noncyclic electron flow, as well as the energy relationships. To raise the energy of electrons derived from H_2O by photolysis to the energy level required to reduce $NADP^+$ to NADPH, each electron must be energized twice (broad orange arrows) by photons absorbed in Photosystems I and II. After each excitation step, the high-energy electrons flow "downhill" via the electron-carrier chains shown (black arrows). Protons are pumped across the thylakoid membrane into the thylakoid lumen (interior) during the water-splitting reaction and during electron transfer through the cytochrome b_6/f complex, producing the proton gradient that is central to ATP formation (see Figure 7–12 for the details of this process). The formation of ATP by noncyclic electron flow is called noncyclic photophosphorylation.

plastoquinone, which simultaneously picks up two protons from the stroma, thereby becoming reduced to **plastoquinol, PQ_BH_2** (Figure 7–12). The plastoquinol then joins a pool of mobile plastoquinol molecules in the interior lipid portion of the thylakoid membrane. The plastoquinol can now pass two electrons and two protons (H^+) to the cytochrome b_6/f complex and is thus oxidized back to PQ_B.

Photosystem II has the unique ability to extract electrons from water and to use these electrons to replace those lost by P_{680} (now $P_{680}{}^+$) to plastoquinone. This is accomplished in the **oxygen-evolving complex,** an assembly of proteins and manganese ions (Mn^{2+}) that catalyze the splitting and oxidation of water (Figures 7–11 and 7–12). The oxygens of two water molecules bind to a cluster of four manganese atoms, which serve to gather the four electrons released as the two water molecules are oxidized. As soon as the four electrons are released from the two water molecules, oxygen is evolved. Thus, with the absorption of four photons, the two water molecules are split, yielding four electrons, four protons, and oxygen gas:

$$2H_2O \longrightarrow 4e^- + 4H^+ + O_2$$

This light-dependent oxidative splitting of water molecules is called **water photolysis.** The oxygen-evolving complex is located on the inside of the thylakoid membrane, and the protons are released into the lumen of the thylakoid, not directly into the stroma of the chloroplast. Thus, the photolysis of water molecules contributes to the generation of a proton gradient across the thylakoid membrane—the sole means by which ATP is generated during photosynthesis.

The Cytochrome b_6/f Complex Links Photosystems II and I

The mobile plastoquinol (PQ_BH_2), which is found in the interior lipid portion of the thylakoid membrane (Figure 7–12),

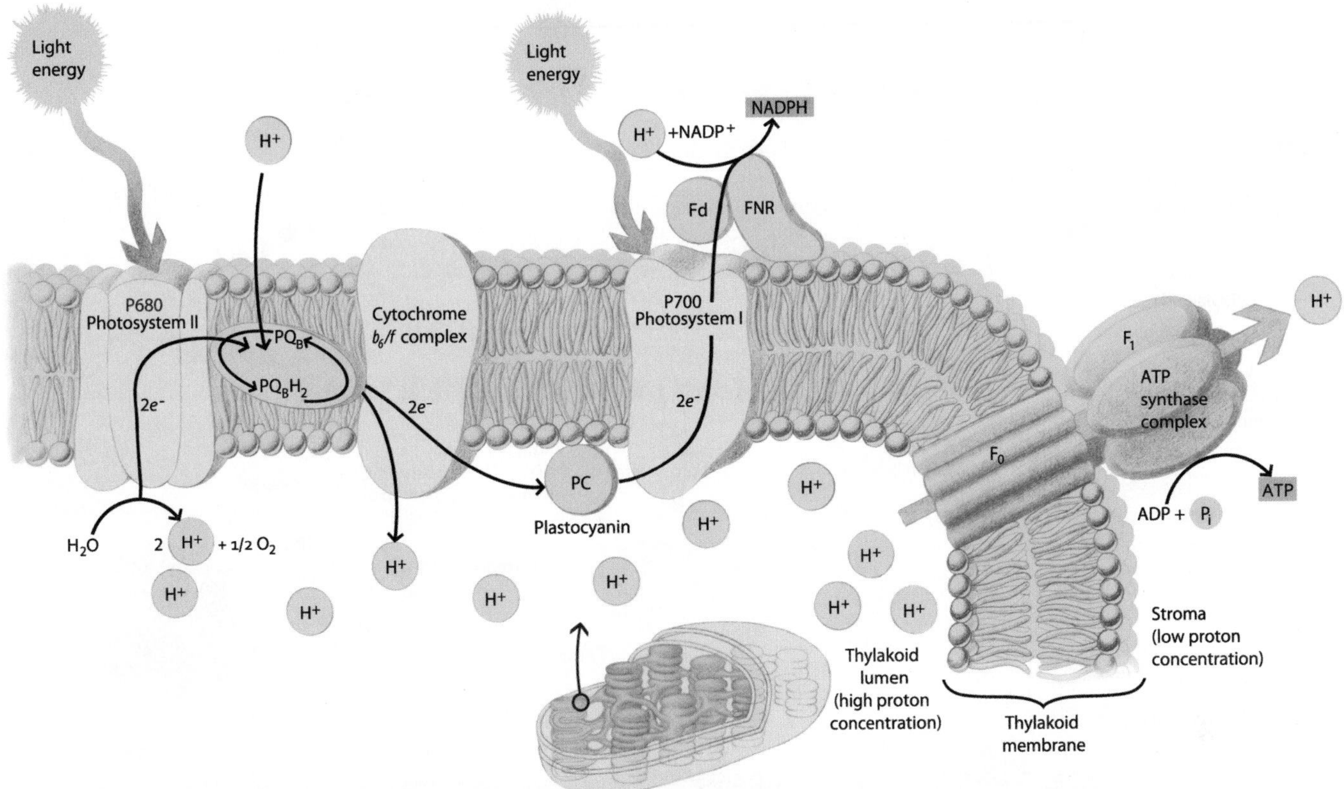

7–12 Transfer of electrons and protons during photosynthesis During the light reactions, electrons move from H_2O (lower left) through Photosystem II, the intermediate chain of electron carriers, Photosystem I, and finally to $NADP^+$. In Photosystem I, $NADP^+$ is reduced to NADPH in the stroma through the action of ferredoxin (Fd) and the flavoprotein ferredoxin-$NADP^+$ reductase (FNR). Plastoquinol (PQ_BH_2) transfers electrons from Photosystem II to the cytochrome b_6/f complex, and plastocyanin, in turn, transfers electrons from the cytochrome b_6/f complex to Photosystem I. Protons are released into the lumen during the oxidation of water and pumped into the lumen through the action of the cytochrome b_6/f complex. These protons, which contribute to an electrochemical proton gradient, must diffuse to the ATP synthase complex. There they diffuse down an electrochemical potential gradient, which is used to synthesize ATP.

donates two electrons received from Photosystem II, one at a time, to the **cytochrome b_6/f complex.** (This complex is analogous to Complex III of the electron transport chain of mitochondria, see Figure 6–10.) As each PQ_BH_2 is oxidized back to PQ_B, it releases two protons into the thylakoid lumen via the cytochrome b_6/f complex. The PQ_B now returns to the pool of plastoquinone, where it can accept more electrons from Photosystem II and more protons from the stroma. Next, reduced cytochrome *f* of the complex donates electrons to **plastocyanin,** a small, water-soluble, copper-containing protein found in the lumen. Plastocyanin, like plastoquinol, is a mobile electron carrier. It transfers one electron at a time between the cytochrome b_6/f complex and P_{700} of Photosystem I (Figure 7–12).

ATP Is Synthesized by an ATP Synthase Complex

The protons released into the thylakoid lumen during the oxidation of water and the pumping of protons across the thylakoid membrane and into the lumen via the cytochrome b_6/f complex together generate an electrochemical proton gradient that drives the synthesis of ATP. **ATP synthase complexes,** embedded in the thylakoid membrane, provide a channel through which the protons can flow down the gradient, back into the stroma (Figure 7–12). As they do so, the potential energy of the gradient drives the synthesis of ATP from ADP and P_i. This process is entirely analogous to proton-driven ATP synthesis in the mitochondrion but is called **photophosphorylation** here to emphasize that light provides the energy used to establish the proton gradient. Thus, chloroplasts and mitochondria generate ATP by the same basic mechanism: chemiosmotic coupling (Figure 7–12). Furthermore, bacteria also generate ATP by chemiosmotic coupling, which should not be surprising given that chloroplasts and mitochondria are derived from bacteria that were free-living.

$NADP^+$ Is Reduced to NADPH in Photosystem I

In Photosystem I, light energy excites antenna molecules, which pass the energy to the P_{700} molecules at the reaction center (Figure 7–11). When a P_{700} molecule is excited in this way, its energized electron is passed to a primary acceptor molecule called A_0, which is thought to be a special chlorophyll with

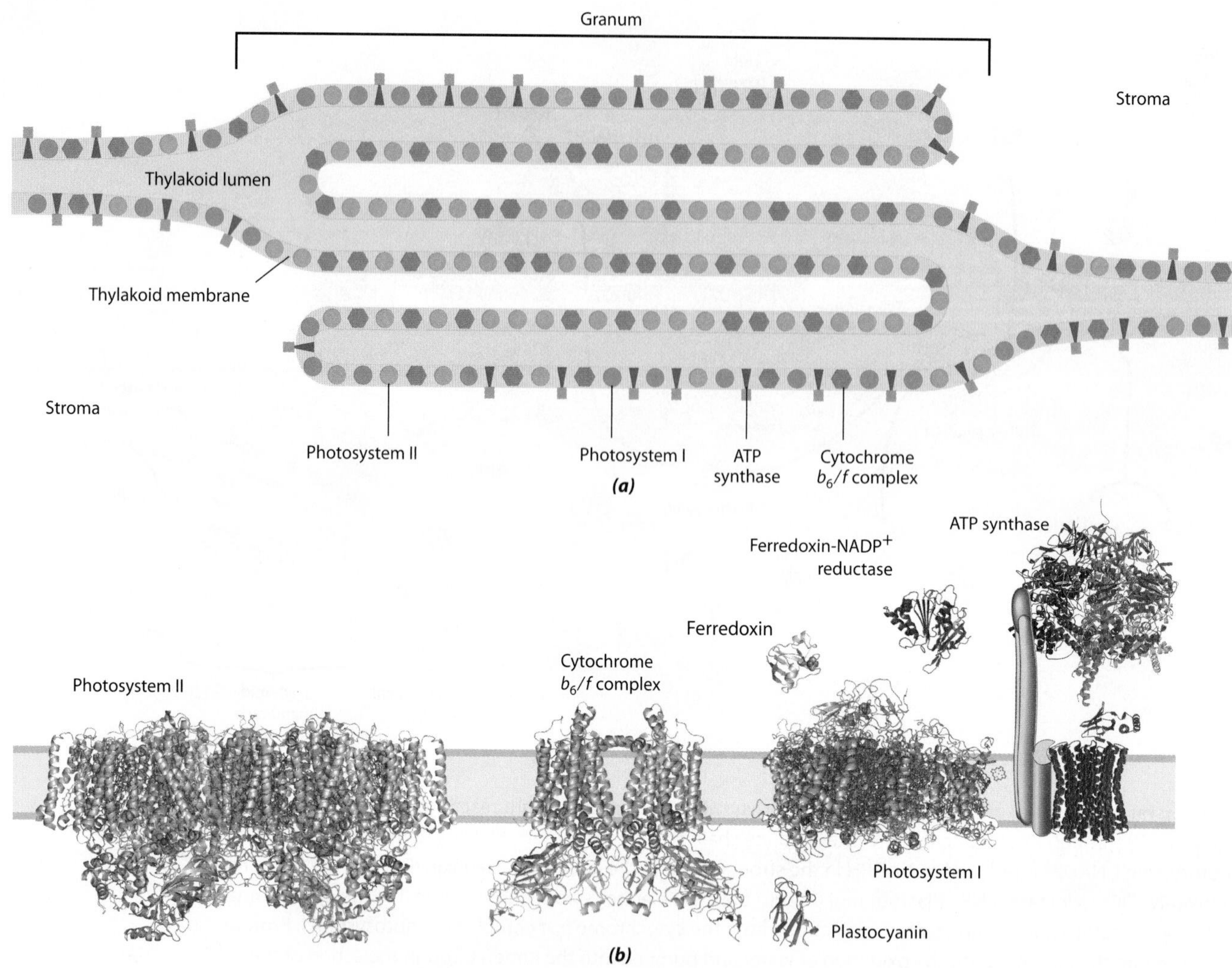

7–13 Organization and structure of the four major protein complexes of the thylakoid membranes ***(a)*** Photosystem II is located primarily in the grana thylakoids, and Photosystem I and ATP synthase almost entirely in the stroma thylakoids and the outer portions of the grana. The cytochrome b_6/f complexes are distributed evenly throughout the membranes. The spatial separation of the photosystems requires mobile electron carriers such as plastoquinol and plastocyanin to shuttle electrons between the separated membrane complexes. ***(b)*** The structure of the four major protein complexes and the soluble proteins of the photosynthetic apparatus.

a function similar to the pheophytin of Photosystem II. The electrons are then passed downhill through a chain of carriers, including phylloquinone (A_1), and iron-sulfur proteins, including ferredoxin. **Ferredoxin (Fd),** a mobile iron-sulfur protein, is found in the chloroplast stroma. It is the final electron acceptor of Photosystem I. Electrons are transferred from ferredoxin to $NADP^+$. This transfer is catalyzed by **ferredoxin-$NADP^+$ reductase (FNR),** a peripheral membrane protein found on the stromal side of the thylakoid membrane (Figure 7–12). This results in the reduction of the $NADP^+$ to NADPH and oxidation of the P_{700} molecule. The electrons removed from the P_{700} molecule are replaced by the electrons that have moved down the electron transport chain from Photosystem II and are carried to P_{700} by plastocyanin.

Thus, in the light, electrons flow continuously from water through Photosystems II and I to $NADP^+$, resulting in the oxidation of water (H_2O) to oxygen (O_2) and the reduction of $NADP^+$ to NADPH. This unidirectional flow of electrons from water to $NADP^+$ is called **noncyclic electron flow,** and the ATP production that occurs is called **noncyclic photophosphorylation.** Inasmuch as Photosystem II supplies electrons to Photosystem I, for photosynthetic efficiency, the rates of delivery of photons to the two reaction centers must be equal. When light conditions favor one or the other photosystem, the surplus energy is redistributed between them, resulting in an equal balance of energy at the two reaction centers.

The free energy change (ΔG) for the reaction

$$H_2O + NADP^+ \longrightarrow NADPH + H^+ + \tfrac{1}{2}O_2$$

is 51 kilocalories per mole. The energy of 700-nanometer light is about 40 kilocalories per mole of photons. Because

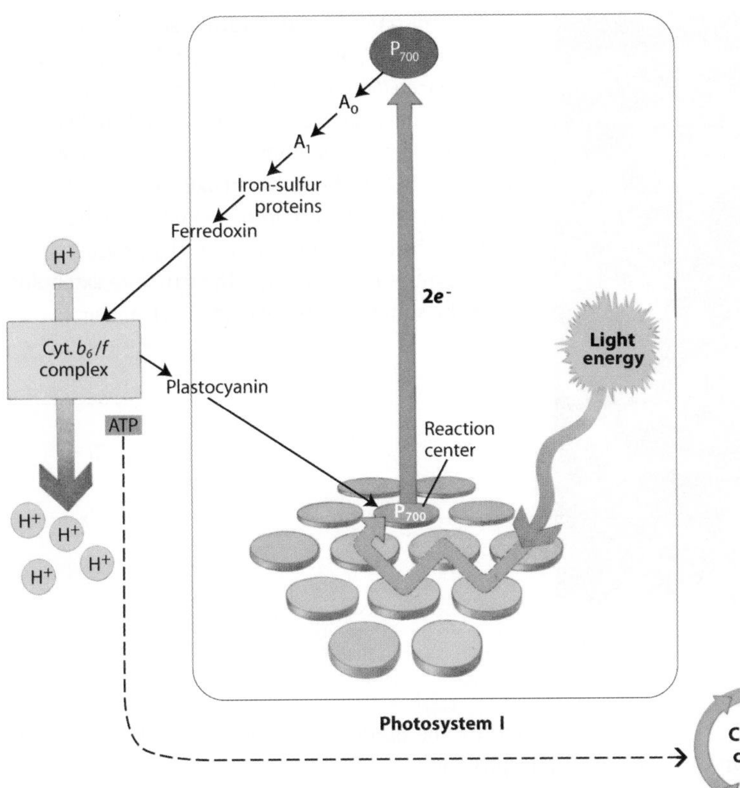

7–14 Cyclic electron flow Cyclic electron flow involves only Photosystem I. ATP is produced from ADP by the same chemiosmotic coupling mechanism of photophosphorylation shown in Figure 7–12, but oxygen is not released and $NADP^+$ is not reduced. The formation of ATP by cyclic electron flow is called cyclic photophosphorylation.

four photons are required to boost two electrons to the level of NADPH, about 160 kilocalories are available. Approximately one-third (51/160) of the available energy is captured as NADPH. The total energy harvest from noncyclic electron flow (based on the passage of 6 pairs of electrons from H_2O to $NADP^+$) is 6 ATP and 6 NADPH.

Cyclic Photophosphorylation Generates Only ATP

As mentioned previously, Photosystem I can work independently of Photosystem II. In this process, called **cyclic electron flow,** energized electrons are transferred from P_{700} to A_0, as before (Figure 7–14). Instead of being passed downhill to $NADP^+$, however, the electrons are shunted to an acceptor in the electron transport chain between Photosystems I and II. The electrons then pass downhill through that chain back into the reaction center of Photosystem I, driving the transport of protons across the thylakoid membrane and hence powering the generation of ATP. Because this process involves a cyclic flow of electrons, it is called **cyclic photophosphorylation.** It is believed that the earliest photosynthetic organisms had only a system like Photosystem I, and some extant photosynthetic bacteria have only a Photosystem I as well. With just Photosystem I, cyclic photophosphorylation is the only path of electron flow that is possible because, without Photosystem II, no water or hydrogen sulfide is split to provide the electrons that are necessary to produce NADPH. Thus, the only product of cyclic photophosphorylation is ATP. Organisms with just Photosystem I produce NADPH from other sources of high-energy electrons and do not use light for NADPH production.

As we have seen, the total energy harvest from noncyclic electron flow (based on the passage of 6 pairs of electrons from H_2O to $NADP^+$) is 6 ATP and 6 NADPH. Yet, the carbon-fixation reactions that we are about to encounter require more ATP than NADPH—at a ratio of about 3:2. Cyclic photophosphorylation, providing extra ATP, is therefore an ongoing necessity to meet the needs of the Calvin cycle, as well as to drive a host of other energy-requiring processes within the chloroplast.

The Carbon-Fixation Reactions

In the second series of photosynthetic reactions, the ATP and NADPH generated by the light reactions are used to fix and reduce carbon and to synthesize simple sugars. Carbon is available to photosynthetic cells in the form of carbon dioxide. For algae and cyanobacteria, this carbon dioxide is found dissolved in the surrounding water. In most plants, carbon dioxide reaches the photosynthetic cells through special openings, called stomata, in the leaves and green stems (Figure 7–15).

In the Calvin Cycle, CO_2 Is Fixed via a Three-Carbon Pathway

In many plant species, the reduction of carbon occurs exclusively in the stroma of the chloroplast by means of a series of reactions often called the **Calvin cycle** (named after its discoverer, Melvin Calvin, who received a Nobel Prize in 1961 for his work on the elucidation of this pathway). The Calvin cycle is analogous to other metabolic cycles (page 112) in that, by the end of each turn of the cycle, the starting compound is regenerated. The starting (and ending) compound in the Calvin cycle is a five-carbon sugar with two phosphate groups, known as **ribulose 1,5-bisphosphate (RuBP).**

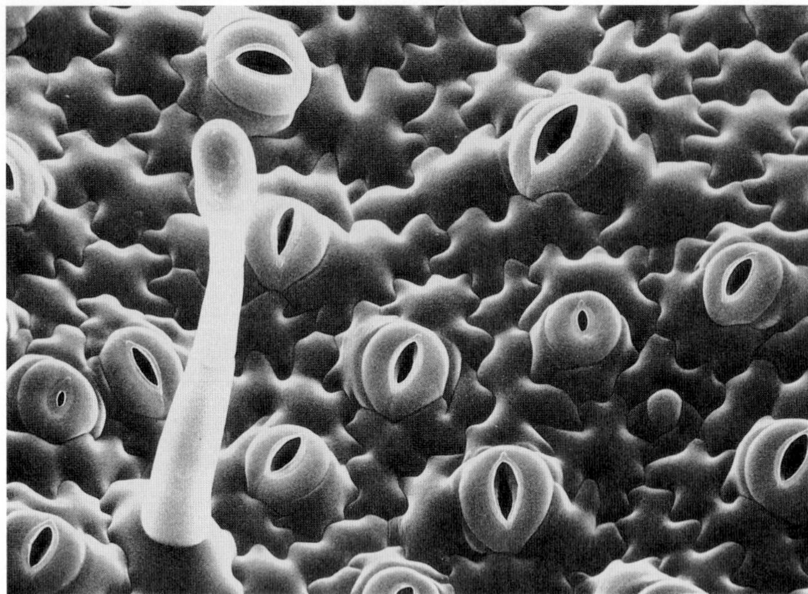

7–15 Stomata in a leaf Shown here is a scanning electron micrograph of stomata on the lower surface of a tobacco *(Nicotiana tabacum)* leaf. A plant can open or close its stomata as needed; in this micrograph, most of them are open. It is through the stomata that the carbon dioxide required for photosynthesis diffuses into the interior of the leaf and the oxygen produced as a by-product of photosynthesis diffuses out. The structure extending from the surface at the lower left is a leaf hair.

The Calvin cycle occurs in three stages. The first stage begins when carbon dioxide enters the cycle and is enzymatically combined with, or "fixed" (bonded covalently) to, RuBP. The resultant six-carbon compound, an unstable enzyme-bound intermediate, is rapidly hydrolyzed to generate two molecules of 3-phosphoglycerate, or 3-phosphoglyceric acid (PGA) (Figure 7–16). Each PGA molecule—the first detectable product of the Calvin cycle—contains three carbon atoms. Hence, the Calvin cycle is also known as the **C_3 pathway.**

RuBP carboxylase/oxygenase, often called **Rubisco** for short, is the enzyme that catalyzes this crucial initial reaction of carbon fixation. (The enzyme's oxygenase activity is discussed later in the chapter.) Rubisco is undoubtedly the world's most abundant enzyme; by some estimates this enzyme may account for as much as 40 percent of the total soluble protein of most leaves.

In the second stage of the cycle, 3-phosphoglycerate is reduced to **glyceraldehyde 3-phosphate,** or 3-phosphoglyceraldehyde (PGAL) (Figure 7–17). This occurs in two steps that are essentially the reversal of the corresponding steps in glycolysis, with one exception: the nucleotide cofactor for the reduction of 1,3-bisphosphoglycerate is NADPH, not NADH. Note that it takes the fixation of three molecules of CO_2 to three molecules of ribulose 1,5-bisphosphate to form six molecules of glyceraldehyde 3-phosphate.

In the third stage of the cycle, five of the six molecules of glyceraldehyde 3-phosphate are used to regenerate three molecules of ribulose 1,5-bisphosphate, the starting material.

The complete cycle is summarized in Figure 7–18. As in every metabolic pathway, each step in the Calvin cycle is catalyzed by a specific enzyme. At each full turn of the cycle, a molecule of carbon dioxide enters the cycle and is reduced, and a molecule of RuBP is regenerated. Three turns of the cycle, with the introduction of three atoms of carbon, are necessary to produce one molecule of glyceraldehyde 3-phosphate, the phosphorylated form of the $C_3H_6O_3$ in the last equation on page 124. The overall equation for the production of one molecule of glyceraldehyde 3-phosphate is:

$$3CO_2 + 9ATP + 6NADPH + 6H^+ \longrightarrow$$
$$\text{Glyceraldehyde 3-phosphate} + 9ADP + 8P_i + 6NADP^+ + 3H_2O$$

CH_2-O- P
$C=O$
HCOH
HCOH
CH_2-O- P
Ribulose 1,5-bisphosphate (RuBP)

CO_2 H_2O
RuBP carboxylase/oxygenase (Rubisco)

CH_2-O- P
HCOH
$C=O$
O^-
+
O^-
$C=O$
HCOH
CH_2-O- P
2 Molecules of 3-phosphoglycerate (PGA)

7–16 First stage of the Calvin cycle Melvin Calvin and his co-workers, Andrew A. Benson and James A. Bassham, briefly exposed photosynthesizing algae to radioactive carbon dioxide ($^{14}CO_2$), then killed the cells in boiling alcohol and separated the various ^{14}C-containing compounds by two-dimensional paper chromatography. They found that the various intermediates became radioactively labeled. They deduced that radioactive carbon is covalently linked to a molecule of ribulose 1,5-bisphosphate (RuBP). The resulting six-carbon compound then immediately splits to form two molecules of 3-phosphoglycerate (PGA). The radioactive carbon atom, indicated here in orange, appears in one of the two molecules of PGA.

$$\underset{\text{3-Phosphoglycerate (PGA)}}{\begin{array}{c} O^- \\ | \\ C{=}O \\ | \\ HCOH \\ | \\ CH_2{-}O{-}\text{P} \end{array}} \xrightarrow[\text{3-Phosphoglycerate kinase}]{\text{ATP} \;\; \text{ADP}} \underset{\text{1,3-Bisphosphoglycerate}}{\begin{array}{c} O \\ \| \\ C{-}O{-}\text{P} \\ | \\ HCOH \\ | \\ CH_2{-}O{-}\text{P} \end{array}} \xrightarrow[\text{Glyceraldehyde 3-phosphate dehydrogenase}]{H^+ + \text{NADPH} \;\; \text{NADP}^+ \;\; P_i} \underset{\text{Glyceraldehyde 3-phosphate (PGAL)}}{\begin{array}{c} O \\ \| \\ C{-}H \\ | \\ HCOH \\ | \\ CH_2{-}O{-}\text{P} \end{array}}$$

(a) *(b)*

7–17 Second stage of the Calvin cycle Overall, the second stage of the Calvin cycle involves the conversion of 3-phosphoglycerate (PGA) to glyceraldehyde 3-phosphate (PGAL) in two steps. ***(a)*** In the first step of the sequence, the enzyme 3-phosphoglycerate kinase in the stroma catalyzes the transfer of phosphate from ATP to PGA, yielding 1,3-bisphosphoglycerate. ***(b)*** In the second step, NADPH donates electrons in a reduction catalyzed by glyceraldehyde 3-phosphate dehydrogenase, producing PGAL. In addition to its role as an intermediate in CO_2 fixation, PGAL has several possible fates in the plant cell. It may be oxidized by glycolysis for energy production or used for hexose synthesis.

(Note again that the Calvin cycle requires more ATP than NADPH, hence the need for the ATP generated by cyclic photophosphorylation.)

The immediate product of the cycle, glyceraldehyde 3-phosphate, is the primary molecule transported from the chloroplast to the cytosol of the cell. This same triose phosphate is formed when the fructose 1,6-bisphosphate molecule is split at the fourth step in glycolysis, and it is interconvertible with another triose phosphate, dihydroxyacetone phosphate. Using the energy provided by hydrolysis of the phosphate

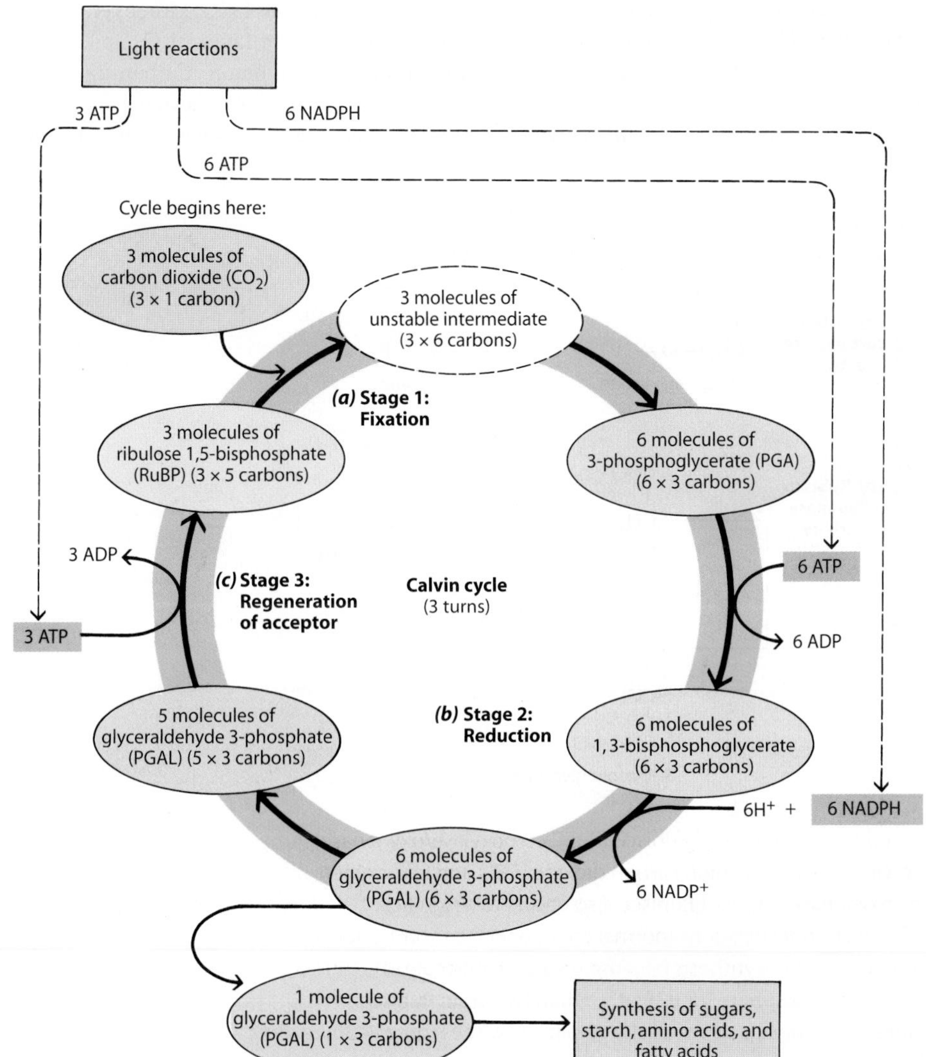

7–18 Summary of the Calvin cycle At each full "turn" of the cycle, one molecule of carbon dioxide (CO_2) enters the cycle. Three turns are summarized here—the number required to make one molecule of glyceraldehyde 3-phosphate (PGAL), the equivalent of one molecule of a three-carbon sugar. The energy that drives the Calvin cycle is provided in the form of ATP and NADPH, produced by the light reactions of photosynthesis. ***(a)*** Stage 1: Fixation. The cycle begins at the upper left when three molecules of ribulose 1,5-bisphosphate (RuBP), a five-carbon compound, are combined with three molecules of carbon dioxide. This produces three molecules of an unstable intermediate that splits apart immediately, yielding six molecules of 3-phosphoglycerate (PGA), a three-carbon compound. ***(b)*** Stage 2: Reduction. The six molecules of PGA are reduced to six molecules of glyceraldehyde 3-phosphate (PGAL). ***(c)*** Stage 3: Regeneration of acceptor. Five of the six PGAL molecules are combined and rearranged to form three five-carbon molecules of RuBP. The one "extra" molecule of PGAL represents the net gain from the Calvin cycle. PGAL serves as the starting point for the synthesis of sugars, starch, and other cellular components.

bonds, the first four steps of glycolysis can be reversed to form glucose from glyceraldehyde 3-phosphate.

Most Fixed Carbon Is Converted to Sucrose or Starch

As we have mentioned, although glucose commonly is represented as the carbohydrate product of photosynthesis in summary equations, in reality, very little free glucose is generated in photosynthesizing cells. Rather, most of the fixed carbon is converted either to **sucrose,** the major transport sugar in plants, or to **starch,** the major storage carbohydrate in plants (pages 19 and 20).

Much of the glyceraldehyde 3-phosphate produced by the Calvin cycle is exported to the cytosol, where, through a series of reactions, it is converted to sucrose. Most of the glyceraldehyde 3-phosphate that remains in the chloroplast is converted to starch, which is stored temporarily, during the light period, as starch grains in the stroma (Figure 7–1). At night, sucrose is produced from starch and exported from the leaf via the vascular bundles to other parts of the plant.

Photorespiration Occurs When Rubisco Binds O_2 Instead of CO_2

In the presence of ample CO_2, the enzyme Rubisco catalyzes the carboxylation of ribulose 1,5-bisphosphate with great efficiency. As mentioned previously, however, Rubisco is not absolutely specific for CO_2 as a substrate. Oxygen competes with CO_2 at the active site, and Rubisco catalyzes the condensation of O_2 with RuBP to form one molecule of 3-phosphoglycerate and one of **phosphoglycolate** (Figure 7–19). This is the enzyme's oxygenase activity, as reflected in its name: RuBP carboxylase/oxygenase. No carbon is fixed during this reaction, and energy must be expended to salvage the carbons from phosphoglycolate, which is not a useful metabolite (metabolic product).

The salvage pathway is long and employs three cellular organelles: chloroplast, peroxisome, and mitochondrion (Figure 7–20). The pathway involves, in part, the conversion of two molecules of phosphoglycolate into a molecule of the amino acid serine (which has three carbons) and a molecule of CO_2. The oxygenase activity of Rubisco combined with the salvage pathway *consumes* O_2 and *releases* CO_2, a process called **photorespiration.** Unlike mitochondrial respiration (in plants often referred to as "dark respiration" to distinguish it from photorespiration, which occurs only in the light), photorespiration is a wasteful process, yielding neither ATP nor NADH. In some plants, as much as 50 percent of the carbon fixed in photosynthesis may be reoxidized to carbon dioxide during photorespiration. Apparently, the evolution of Rubisco produced an active site that is not able to discriminate between CO_2 and O_2, perhaps because much of its evolution occurred before O_2 was an important component of the atmosphere.

The condensation of O_2 with RuBP occurs concurrently with CO_2 fixation under today's atmospheric conditions, with an atmosphere consisting of 21 percent O_2 and only 0.039 percent (390 parts per million, or ppm) CO_2. Moreover, conditions that can alter the CO_2/O_2 ratio in favor of O_2 and, hence, enhance photorespiration are quite common. Carbon dioxide is not continuously available to the photosynthesizing cells of a plant. As we have seen, it enters the leaf by way of the stomata,

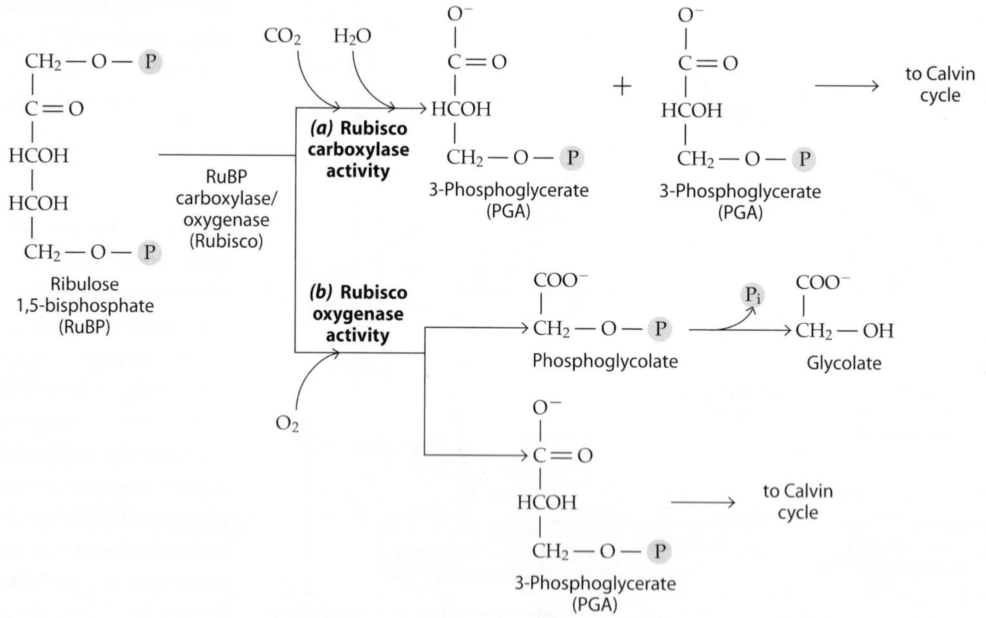

Ribulose 1,5-bisphosphate (RuBP)
RuBP carboxylase/oxygenase (Rubisco)
CO_2 H_2O
***(a)* Rubisco carboxylase activity**
3-Phosphoglycerate (PGA) + 3-Phosphoglycerate (PGA) → to Calvin cycle
O_2
***(b)* Rubisco oxygenase activity**
Phosphoglycolate → (P_i) Glycolate
3-Phosphoglycerate (PGA) → to Calvin cycle

7–19 Reactions catalyzed by Rubisco ***(a)*** The carboxylase activity of Rubisco (RuBP carboxylase/oxygenase), which results in the fixation of CO_2 in the Calvin cycle, is favored by high carbon dioxide (CO_2) and low oxygen (O_2) concentrations (see also Figure 7–16). ***(b)*** The oxygenase action of Rubisco also occurs to a significant extent, especially in the presence of low CO_2 and high O_2 concentrations (normal atmospheric concentrations). The oxygenase action of Rubisco lessens the efficiency of photosynthesis because only one molecule of 3-phosphoglycerate (PGA) is formed from ribulose 1,5-bisphosphate (RuBP) rather than two, as in Rubisco's carboxylase activity. The oxygenase activity of Rubisco combined with the salvage pathway (see Figure 7–20) consumes O_2 and releases CO_2, a process called photorespiration.

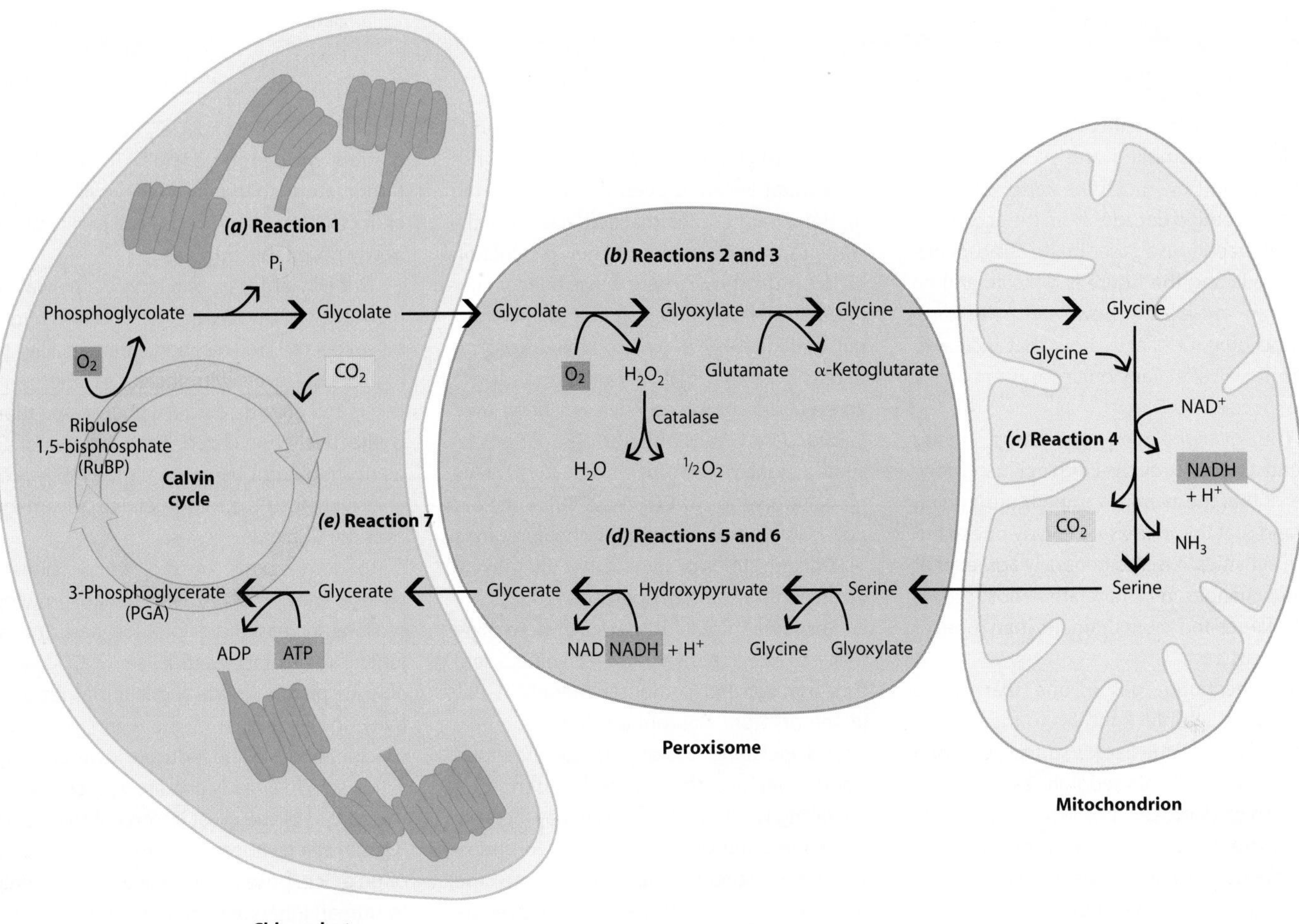

7–20 Salvage of phosphoglycolate Phosphoglycolate formed during photorespiration is salvaged by conversion into serine and thus to 3-phosphoglycerate (PGA), which is fed into the Calvin cycle. ***(a)*** Reaction 1: Phosphoglycolate is dephosphorylated in chloroplasts to form glycolate. ***(b)*** Reactions 2 and 3: In peroxisomes, the glycolate is oxidized to glyoxylate, which is then transaminated to glycine. ***(c)*** Reaction 4: In mitochondria, two glycine molecules condense to form serine and the CO_2 that is released during photorespiration. ***(d)*** Reactions 5 and 6: In peroxisomes, the serine is transaminated to hydroxypyruvate, which is then reduced to glycerate. The glycerate then enters the chloroplasts. ***(e)*** Reaction 7: The glycerate is phosphorylated to PGA, which rejoins the Calvin cycle. Oxygen is consumed at two points in the photorespiratory pathway, once in the chloroplast (the oxygenase activity of Rubisco) and once in the peroxisome (oxidation of glycolate to glyoxylate). Carbon dioxide is released at one point in the mitochondrion (condensation of two glycine molecules to form one molecule of serine).

the specialized pores that open and close depending on, among other factors, water stress. When a plant is subjected to hot, dry conditions, it must close its stomata to conserve water. This cuts off the supply of CO_2 and also allows the O_2 produced by photosynthesis to accumulate. The resulting low CO_2 and high O_2 concentrations favor photorespiration.

Also, when plants are growing in close proximity to one another, the air surrounding the leaves may be quite still, with little gas exchange between the immediate environment and the atmosphere as a whole. Under such conditions, the concentration of CO_2 in the air closest to the leaves may be rapidly reduced to low levels by the photosynthetic activities of the plant. Even if the stomata are open, the concentration gradient between the outside of the leaf and the inside may be so slight that little CO_2 diffuses into the leaf. Meanwhile, O_2 accumulates, favoring photorespiration and greatly reducing the plant's photosynthetic efficiency.

Although photorespiration may be viewed as wasteful, it does rescue 75 percent of the carbon in phosphoglycolate and, as some workers note, "makes the best of a bad situation caused by Rubisco's seemingly inevitable oxygenase activity." In addition to minimizing the loss of carbon, photorespiration acts to protect the photosynthetic apparatus from photoinhibition when leaves are exposed to more light than they can utilize and when NADPH production in the light reactions exceeds the demand of the Calvin cycle for reducing power. Moreover,

GLOBAL WARMING: THE FUTURE IS NOW

If it were easy to predict the future, anyone could win at gambling, and everyone would agree about how much the Earth will warm over the coming decades from the accumulation of greenhouse gases in the atmosphere. But predicting the future is difficult, and no one can say exactly how warm the Earth will become, or exactly how that increased warmth will affect any one ecosystem or any one species.

The overwhelming consensus among scientists who study the Earth's climate, however, is that the average temperature is going up and that this increase is largely due to human activities. It is also generally agreed that future changes in climate are almost certain to be larger and more dramatic than those of the recent past.

A greenhouse gas is one that allows short-wave, visible light from the sun to pass through to the Earth's surface, but then traps long-wave, infrared light as it radiates back toward space, thus warming the atmosphere. Carbon dioxide is the principal greenhouse gas; other greenhouse gases are methane, chlorofluorocarbons, and nitrogen oxides. The concentration of CO_2 in the atmosphere has varied over the long span of geologic time, but it remained relatively constant from about 10,000 years ago up until about 150 years ago.

At that time, the concentration of CO_2 began to climb, due mainly to our use of fossil fuels, such as coal, oil, and natural gas, as well as to the destruction and burning of forests, particularly in the tropics. In 1850, the concentration of CO_2 in the atmosphere was about 270 ppm. By 1960, it had climbed to 317 ppm, and by early 2011, it had increased further to 390 ppm. All forecasts indicate it will continue to rise for several decades, even if (and it is a big if) the world's industrialized nations take significant steps to reduce their production of CO_2 right now. There have been major international agreements to do so, most significantly in 1997 in Kyoto, Japan. Unfortunately, even the relatively mild goals those agreements contained have not been met. In 2007, the Group of Eight (G8) nations issued a nonbinding goal "to at least halve global emissions of CO_2 by 2030." They did not, however, set an ultimate goal in terms of concentrations or climate; nor did they indicate what should happen after 2050.

How warm will it get? To determine that—to predict the future—climate scientists use detailed computer models. A rise in temperature has hundreds of potential effects, including loss of polar ice, increased cloud cover, and changes in the strength of prevailing winds, each of which may increase or decrease the temperature rise. To understand the ultimate effect on climate, the models must account for all of them. Over the past several decades, a better understanding of these variables, and a vast increase in computing power, have greatly improved climate models, and as a result, different models now tell mostly the same story. Unless there is a significant drop in the level of CO_2 in the atmosphere, the average global temperature by the end of this century will be 1.5° to 4.5°C (2.5° to 7.5°F) warmer than it is now.

The consequences of this increase cannot be known with certainty. Wild plants adapted to a particular temperature range may move northward as the climate warms, but it is likely that for some species, their rate of "migration" will be too slow to keep up with the temperature change, and they will become extinct. Other plant species will proliferate in newly hospitable climates.

Not all plants respond in the same way to high levels of carbon dioxide. C_3 plants might be expected to respond dramatically to higher CO_2 levels with increased photosynthesis and growth, because photorespiration is effectively minimized. The response of C_4 plants might not be so dramatic as they progressively lose their competitive advantage over the C_3 plants.

In some parts of the world, climate change may result in lengthened growing seasons, increased precipitation, and, in conjunction with increased levels of CO_2 available to plants, greater agricultural productivity. In other parts of the world, however, precipitation may be reduced, lowering crop yields and, in already arid areas, accelerating the spread of the great deserts of the world. Rises in sea level, resulting from the melting of polar ice, pose a potential threat not only to human inhabitants of coastal regions but also to the multitude of marine organisms that live or reproduce in shallow waters at the continents' edges.

What is to be done? It is too late to avoid climate change altogether—it is already happening. The urgent question now is what must be done to reduce its extent and to manage its unavoidable effects. Many climate scientists think that reducing atmospheric CO_2 to 350 ppm is a critical goal. Some models predict that above that level, changes in climate will be faster, more far-reaching, and more chaotic than we can easily manage.

photorespiration is the only pathway in the plant for the removal of phosphoglycolate, which is a toxic compound.

The Four-Carbon Pathway Is a Solution to Photorespiration

The Calvin cycle is not the only pathway used in carbon-fixation reactions. In some plants, the first detectable product of CO_2 fixation is not the three-carbon molecule 3-phosphoglycerate but rather the four-carbon molecule **oxaloacetate,** which is also an intermediate in the citric acid cycle. Plants that employ this **C_4 pathway,** along with the Calvin cycle, are commonly called **C_4 plants** (for four-carbon), as distinct from the C_3 plants, which use *only* the Calvin cycle. The C_4 pathway is also referred to as the Hatch-Slack pathway after M. D. Hatch and C. R. Slack, two Australian plant physiologists who played key roles in its elucidation.

The oxaloacetate is formed when CO_2 is fixed to **phosphoenolpyruvate (PEP)** in a reaction catalyzed by the enzyme **PEP carboxylase,** which is found in the cytosol of **mesophyll cells** of C_4 plants (Figure 7–21). The CO_2 is obtained from air

Below that level, things will still change, but not as profoundly, and not as fast. To reach that goal, we will need to reduce our consumption of fossil fuels, develop new energy sources, and protect and redevelop carbon-storing forests. At the same time, we must begin to plan for the disruptions a changing climate will inevitably bring, including a rise in sea level, warmer oceans, and changing rainfall patterns.

Planning for these inevitabilities, while trying to head off the worst of them, will require research, ingenuity, and, most important, political will. The magnitude of the challenges can sometimes feel overwhelming. But when it comes to managing the future we want to live in, the real gamble lies in doing nothing.

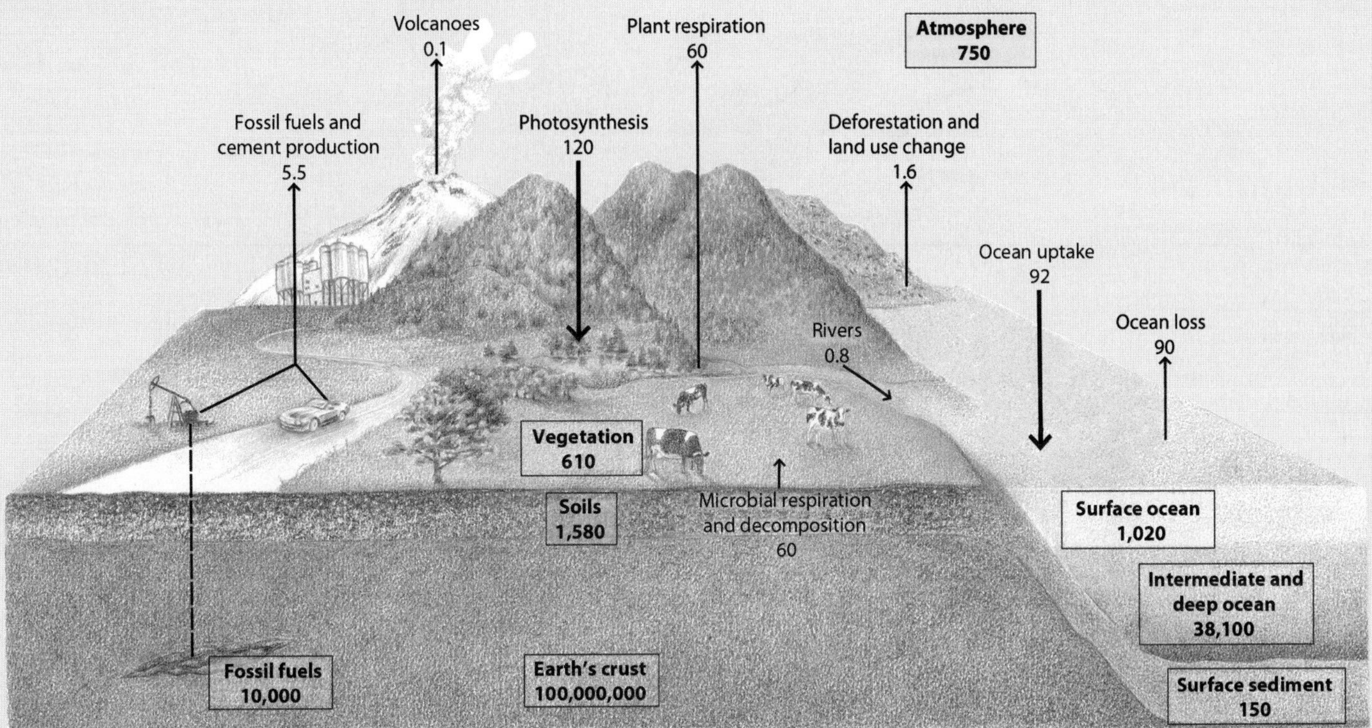

The carbon cycle In the diagram, carbon reservoirs (where carbon is stored) are represented by boxes, and carbon fluxes (transfers of carbon between reservoirs) by arrows. The numbers are all estimates of the amount of carbon, expressed in gigatons (1 gigaton = 1 billion metric tons). The amount of carbon present in the atmosphere is currently increasing by about 4 gigatons each year. The principal photosynthesizers in the carbon cycle are plants, phytoplankton, marine algae, and cyanobacteria.

Phosphoenolpyruvate (PEP) + HCO_3^- → (PEP carboxylase, − P_i) → Oxaloacetate → Malate / Aspartate

Malate: COO^-—HCOH—CH_2—COO^-

Phosphoenolpyruvate (PEP): O^-—C=O—C—O—Ⓟ, $\|CH_2$

Oxaloacetate: COO^-—C=O—CH_2—COO^-

Aspartate: COO^-—H_3N^+—CH—CH_2—COO^-

7–21 Carbon fixation by the C_4 pathway Carbon dioxide is "fixed" to phosphoenolpyruvate (PEP) by the enzyme PEP carboxylase. PEP carboxylase uses the hydrated form of CO_2, which is HCO_3^- (bicarbonate ion). Depending on the species, the resulting oxaloacetate is either reduced to malate or transaminated to aspartate through the addition of an amino group (—NH_2). The malate or aspartate moves into the bundle-sheath cells, where CO_2 is released for use in the Calvin cycle. Notice that PEP contains a phosphoanhydride bond. Like ATP, PEP is a high-energy compound.

spaces adjacent to the mesophyll cells. The oxaloacetate is then reduced to malate or converted, with the addition of an amino group, to the amino acid aspartate in the chloroplast of the same cell. The next step is a surprise: the malate or aspartate (depending on the species) moves from the mesophyll cells to **bundle-sheath cells** surrounding the vascular bundles of the leaf, where the malate or aspartate is decarboxylated to yield CO_2 and pyruvate. The CO_2 then enters the Calvin cycle by reacting with RuBP to form 3-phosphoglycerate. Meanwhile, the pyruvate returns to the mesophyll cells, where it reacts with ATP to regenerate PEP (Figure 7–22). Hence, the anatomy of the leaves of C_4 plants establishes a *spatial separation* between the C_4 pathway and the Calvin cycle, which occur in two different types of cells.

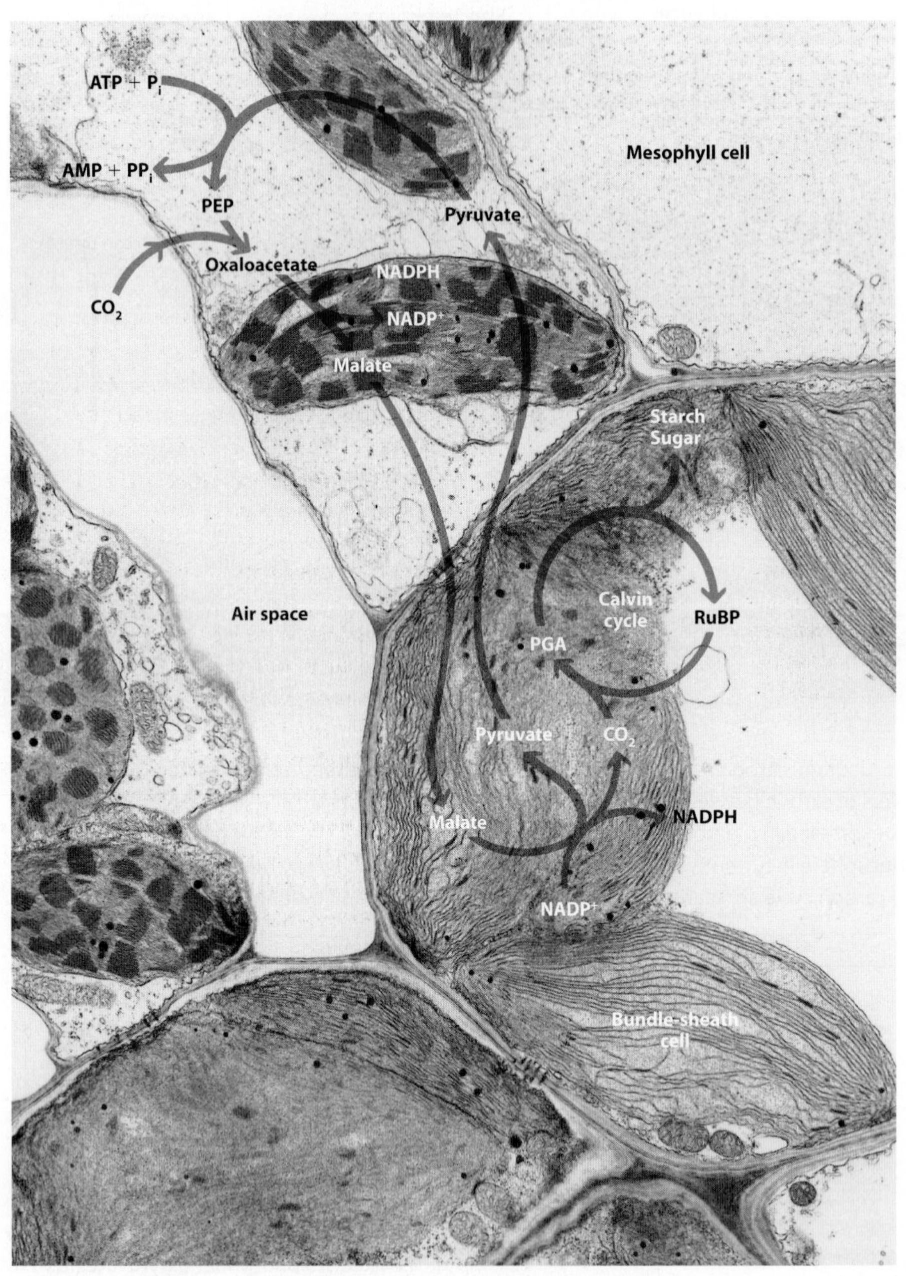

7–22 Carbon fixation in a C_4 plant Shown here is the pathway for carbon fixation in a leaf of maize *(Zea mays)*. Carbon dioxide is first fixed in mesophyll cells as oxaloacetate, which is rapidly converted to malate. The malate is then transported to bundle-sheath cells, where the CO_2 is released to enter the Calvin cycle, ultimately yielding sugars and starch. Pyruvate returns to the mesophyll cells for regeneration of phosphoenolpyruvate (PEP). Hence, there is a spatial separation between the C_4 pathway, which occurs in the mesophyll cells, and the Calvin cycle, which takes place in the bundle-sheath cells.

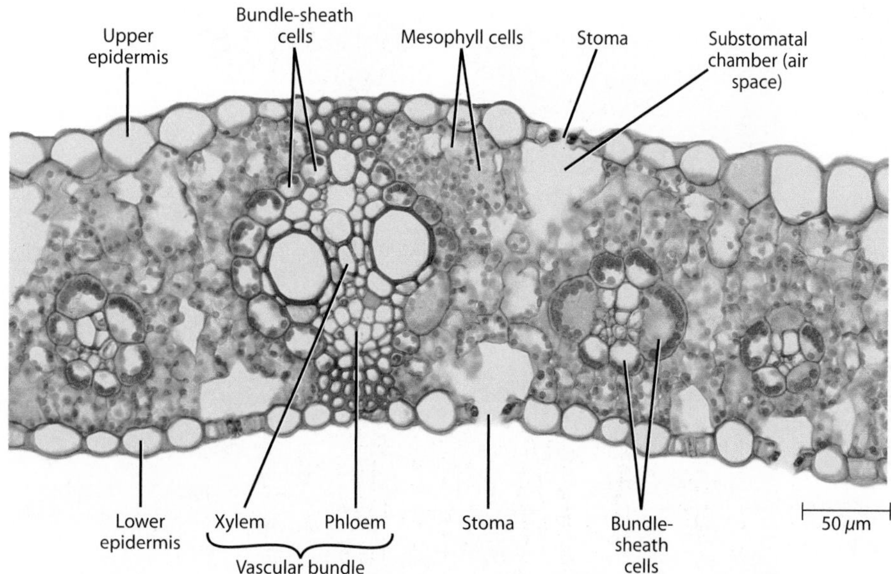

7–23 Vascular bundles in a C_4 plant As is typical of C_4 plants, the vascular bundles (composed of xylem and phloem), shown here in a transverse section of a portion of a maize *(Zea mays)* leaf, are surrounded by large, chloroplast-containing bundle-sheath cells. The bundle-sheath cells, where the Calvin cycle occurs, are, in turn, surrounded by a layer of mesophyll cells, where the C_4 pathway takes place. The bundle-sheath cells and surrounding mesophyll cells form concentric layers known as Kranz anatomy. Seen here are four vascular bundles—one large and three small. The uptake of sugar from the mesophyll occurs largely in the small bundles. The large bundles are involved primarily with export of the sugar from the leaf to other parts of the plant.

The two primary carboxylating enzymes of photosynthesis use different forms of the CO_2 molecule as a substrate. Rubisco uses CO_2, whereas PEP carboxylase uses the hydrated form of carbon dioxide, the bicarbonate ion (HCO_3^-), as its substrate. PEP carboxylase has a high affinity for bicarbonate and is not affected by the presence or concentration of O_2, in contrast to Rubisco. Hence, PEP carboxylase operates very efficiently even when the concentration of its substrate is quite low.

Typically, the leaves of C_4 plants are characterized by an orderly arrangement of the mesophyll cells around a layer of large bundle-sheath cells, so that together the two form concentric layers around the vascular bundle (Figure 7–23). This wreathlike arrangement has been termed **Kranz anatomy** (*Kranz* is the German word for "wreath"). In some C_4 plants, the chloroplasts of the mesophyll cells have well-developed grana, while those of the bundle-sheath cells have either poorly developed grana or none at all (Figure 7–24). In addition, when photosynthesis is occurring, the bundle-sheath chloroplasts commonly form larger and more numerous starch grains than the mesophyll chloroplasts.

C_4 Photosynthesis Occurs in Some Eudicots without Kranz Anatomy Kranz anatomy is not essential for C_4 photosynthesis, as has been shown in three species of the family Chenopodiaceae, the goosefoot family, which ranges from annuals to trees that grow in saline or alkaline desert soils. The three species are *Bienertia cycloptera* (Figure 7–25a), *Bienertia sinuspersici,* and *Suaeda aralocaspica.* In each of these species, C_4 photosynthesis occurs *within individual cells* in which dimorphic, or two forms of, chloroplasts and photosynthetic enzymes are compartmentalized in two distinct regions of the same cell. In *Bienertia,* the chloroplasts of one form are scattered throughout a thin outer layer of cytoplasm, which is connected by cytoplasmic strands that traverse the vacuole to a large, centrally located cytoplasmic compartment containing the second form of chloroplast (Figure 7–25b). In *Suaeda,* the two forms of chloroplasts are partitioned between the two ends of the cells (Figure 7–25c). Hence, in *Bienertia,* each cell has the equivalent of a bundle-sheath cell embedded within and surrounded by a mesophyll cell, whereas in *Suaeda,* each cell has the equivalent of a bundle-sheath cell at one end and a mesophyll cell at the other. There are no intervening walls between compartments.

Photosynthesis Is Generally More Efficient in C_4 Plants Than in C_3 Plants Fixation of CO_2 has a larger energy cost in C_4 plants than in C_3 plants. For each molecule of CO_2 fixed in the C_4 pathway, a molecule of PEP must be regenerated at the cost of two phosphate groups of ATP (Figure 7–22). Therefore, C_4 plants need five ATP molecules to fix one molecule of CO_2, whereas C_3 plants need only three. One might well ask why C_4 plants have evolved such an energetically expensive method of providing CO_2 to the Calvin cycle.

High CO_2 and low O_2 concentrations limit photorespiration. Consequently, C_4 plants have a distinct advantage over C_3 plants, because CO_2 fixed by the C_4 pathway is essentially "pumped" from the mesophyll cells into the bundle-sheath cells,

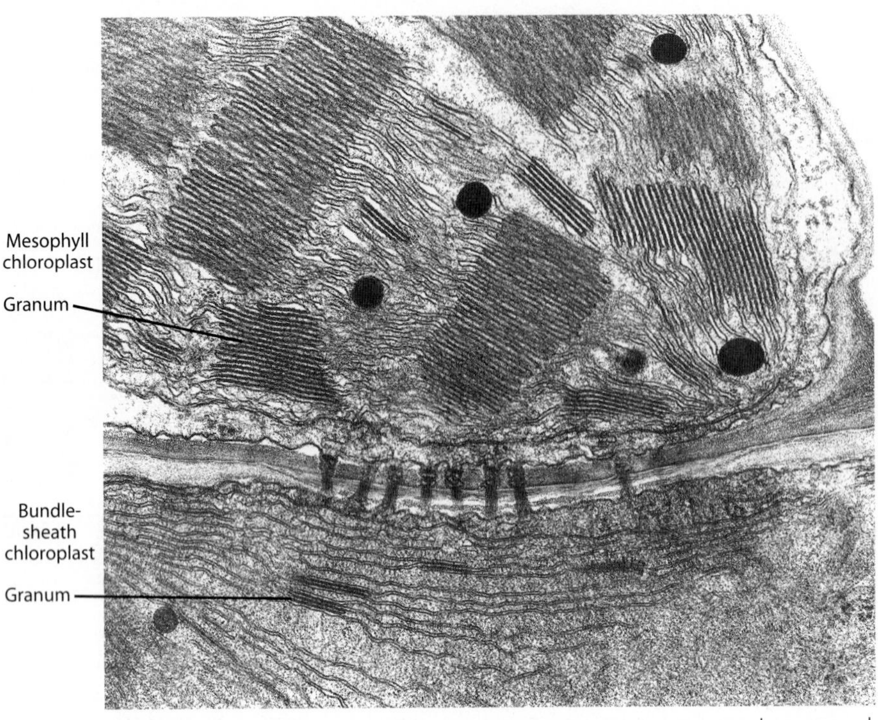

7–24 Comparison of plastids in mesophyll and bundle-sheath cells Electron micrograph showing portions of chloroplasts in a mesophyll cell (above) and a bundle-sheath cell (below) of a maize *(Zea mays)* leaf. Compare the well-developed grana of the mesophyll cell chloroplast with the poorly developed grana of the bundle-sheath cell chloroplast. Note the plasmodesmata in the wall between these two cells. The intermediates of photosynthesis move from one cell to the other via the plasmodesmata in this C_4 plant.

thus maintaining a high ratio of CO_2 to O_2 at the site of Rubisco activity. This high CO_2/O_2 ratio favors the carboxylation of RuBP. In addition, because both the Calvin cycle and photorespiration are localized in the inner, bundle-sheath layer of cells, any CO_2 liberated by photorespiration into the outer, mesophyll layer can be refixed by the C_4 pathway that operates there. The CO_2 liberated by photorespiration can thus be prevented from escaping from the leaf. Moreover, compared with C_3 plants, C_4 plants are superior utilizers of available CO_2. This is in part because PEP carboxylase activity is not inhibited by O_2. As a result, the net photosynthetic rates (that is, total photosynthetic rate minus photorespiratory loss) of C_4 grasses, for example, can be two to three times the net photosynthetic rates of C_3 grasses under the same environmental conditions. In short, the gain in efficiency from the elimination of photorespiration in C_4 plants more than compensates for the energetic cost of the C_4 pathway. Maize *(Zea mays)*, sugarcane *(Saccharum officinale)*, and sorghum *(Sorghum vulgare)* are examples of C_4 grasses. Wheat *(Triticum aestivum)*, rye *(Secale cereale)*, oats *(Avena sativa)*, and rice *(Oryza sativa)* are examples of C_3 grasses.

The C_4 plants evolved primarily in the tropics and are especially well adapted to high light intensities, high temperatures, and dryness. The optimal temperature range for C_4 photosynthesis is much higher than that for C_3 photosynthesis, and C_4 plants flourish even at temperatures that would eventually be lethal to many C_3 species. Because of their more efficient use of carbon dioxide, C_4 plants can attain the same photosynthetic rate as C_3 plants but with smaller stomatal openings and, hence, with considerably less water loss. The preponderance of C_4 plants in hotter, drier climates may be an expression of these advantages of C_4 photosynthesis at high temperatures. In addition, C_4 plants have only one-third to one-sixth the amount of Rubisco found in C_3 plants, and the overall leaf nitrogen content in C_4 plants is less than in C_3 plants. C_4 plants are therefore able to use nitrogen more efficiently than C_3 plants.

A familiar example of the competitive capacity of C_4 plants is seen in lawns in the summertime. In most parts of the United States, lawns consist mainly of C_3 grasses, such as Kentucky bluegrass *(Poa pratensis)* and creeping bent grass *(Agrostis tenuis)*. As the summer days become hotter and drier, these dark green, fine-leaved grasses are often overwhelmed by rapidly growing crabgrass *(Digitaria sanguinalis)*, which disfigures lawns as its yellowish-green, broader-leafed plants slowly take over. Crabgrass, you will not be surprised to learn, is a C_4 plant.

All of the plants now known to utilize C_4 photosynthesis are flowering plants (angiosperms), including at least 19 families, 3 of which are monocots and 16 eudicots (see Chapter 25). No family has been found, however, that contains only C_4

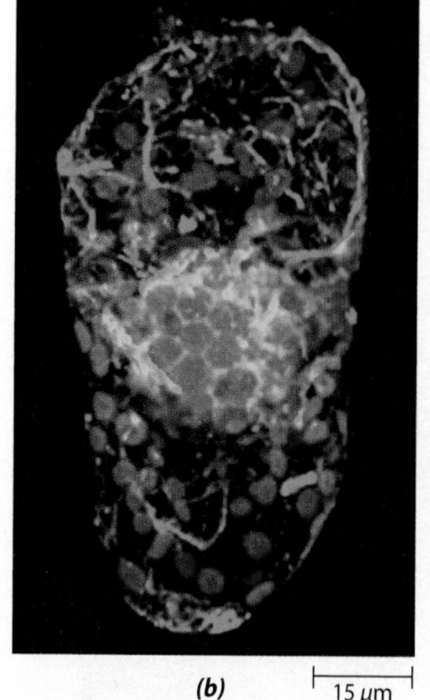

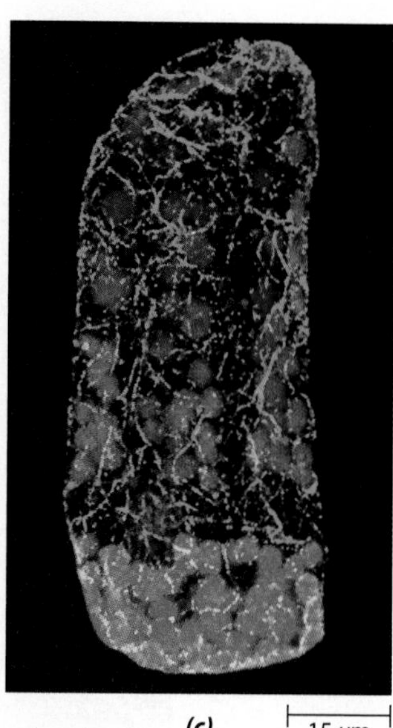

(a) *(b)* *(c)*

7–25 C_4 plants lacking Kranz anatomy ***(a)*** *Bienertia cycloptera* grows in salty depressions in semi-deserts in Central Asia. ***(b), (c)*** Immunofluorescent images of *Bienertia sinuspersici* and *Suaeda aralocaspica* showing positions of chloroplasts (red). In *B. sinuspersici (b),* one form of chloroplast is located in the outer layer of the cytoplasm and the other form is located centrally. In *S. aralocaspica (c),* the two forms of chloroplast are located in distinct regions of the cell; here one form is largely aggregated at the lower, or proximal, end of the cell.

species. This pathway has undoubtedly arisen independently many times in the course of evolution.

Species have been discovered in several genera that have photosynthetic characteristics intermediate between those of C_3 and C_4 species. These so-called **C_3-C_4 intermediates,** which are characterized by Kranz-like leaf anatomy, partially suppressed photorespiration, and a reduced sensitivity to O_2, are considered by some plant biologists as evidence of stepwise evolution of the C_4 pathway from C_3 ancestors.

Plants Having Crassulacean Acid Metabolism Can Fix CO_2 in the Dark

Another strategy for CO_2 fixation has evolved independently in many succulents (plants with fleshy, water-storing stems or leaves), including cacti and stonecrops such as *Bryophyllum, Kalanchoë,* and *Sedum.* Because it was first recognized in stonecrops (family Crassulaceae), it is called **crassulacean acid metabolism (CAM).** Plants that take advantage of CAM photosynthesis are called **CAM plants.** CAM plants, like C_4 plants, use both the C_4 pathway and the Calvin cycle. In CAM plants, however, there is a *temporal separation*—a separation in time—rather than a spatial separation of the two pathways (Figure 7–26).

Plants are considered to have CAM if their photosynthetic cells have the ability to fix CO_2 in the dark through the activity of PEP carboxylase in the cytosol. The initial carboxylation product is oxaloacetate, which is immediately reduced to malate. The malate so formed is stored as malic acid in the vacuole and is detectable as a sour taste. During the following light period, the malic acid is recovered from the vacuole and decarboxylated, and the CO_2 is transferred to RuBP of the Calvin cycle *within the same cell* (Figure 7–27). Thus, a structural precondition of all CAM plants is the presence of cells that have both large vacuoles, in which the malic acid can be temporarily stored in aqueous solution, and chloroplasts, where the CO_2 obtained from the malic acid can be transformed into carbohydrates.

The CAM plants are largely dependent on nighttime accumulation of CO_2 for their photosynthesis, because their stomata are closed during the day, retarding water loss. This is obviously advantageous in the conditions of high light intensity and water stress under which most CAM plants live. If all atmospheric CO_2 uptake in a CAM plant occurs at night, the water-use efficiency of that plant can be many times greater than that of a C_3 or C_4 plant. Typically, a CAM plant loses 50 to 100 grams of water for every gram of CO_2 gained, compared with 250 to 300 grams for C_4 plants and 400 to 500 grams for C_3 plants. During periods of prolonged drought, some CAM plants can keep their stomata closed both night and day, maintaining low metabolic rates by refixing the CO_2 produced by respiration.

Among the vascular plants, CAM is more widespread than C_4 photosynthesis. It has been reported in at least 23 families of flowering plants, mostly eudicots, including such familiar houseplants as the maternity plant and wax plant. Not all CAM

7–26 Comparison of C_4 and CAM photosynthesis C_4 and CAM plants utilize both C_4 and C_3 (Calvin cycle) pathways, with CO_2 initially being incorporated into four-carbon acids in the C_4 pathway. Subsequently, the CO_2 is transferred to the C_3 pathway, or Calvin cycle. ***(a)*** In C_4 plants, the two pathways occur at the same time but in different cells; hence, they are said to be spatially separated (see Figure 7–22). ***(b)*** In CAM plants, by contrast, they are temporally separated, functioning at different times (see Figure 7–27). The C_4 pathway, or initial fixation of CO_2, occurs at night, and the C_3 pathway functions during the day. The stomata of C_4 plants are open during the day and closed at night, whereas those of CAM plants are closed during the day and open at night.

plants are highly succulent; two examples of less succulent ones are the pineapple and Spanish "moss," both members of the family Bromeliaceae (monocots). Some nonflowering plants have also been reported to show CAM activity, including the bizarre gymnosperm *Welwitschia mirabilis* (see Figure 18–41), the aquatic species of quillwort *(Isoetes)* (see Figure 17–21),

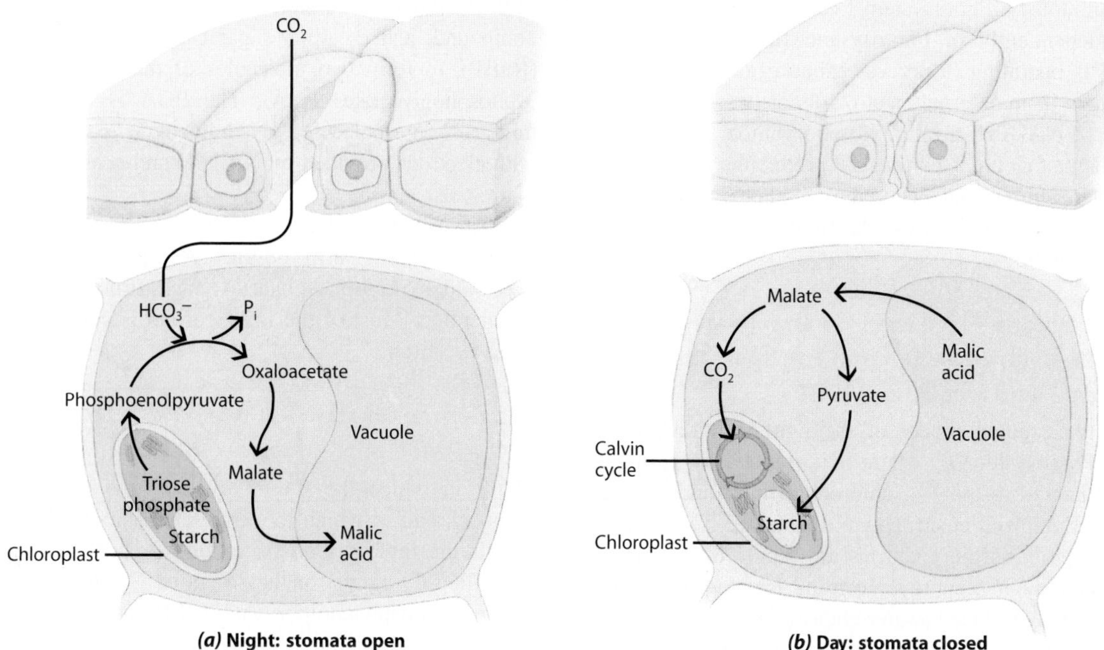

7–27 Crassulacean acid metabolism (CAM) Because CAM involves the formation of malic acid at night and its disappearance during daytime, CAM plants are known as plants that taste sour at night and sweet during the day. ***(a)*** CO_2 is first fixed at night when the stomata are open. At night, starch from the chloroplast is broken down as far as phosphoenolpyruvate (PEP). Carbon dioxide, hydrated to form HCO_3^- (bicarbonate ion), reacts with PEP to form oxaloacetate, which then is reduced to malate. Most of the malate is pumped into the vacuole and stored there as malic acid. ***(b)*** During the daytime, the malic acid is recovered from the vacuole and decarboxylated, producing CO_2 and pyruvate. The CO_2 enters the Calvin cycle, where it is refixed by Rubisco. Much of the pyruvate may be converted to sugars and starch by reverse glycolysis. Stomatal closure during the daytime prevents loss of water and of the CO_2 released by decarboxylation of malate.

and some ferns. *Welwitschia,* however, fixes CO_2 predominantly in the same way as a C_3 plant.

Each of the Carbon-Fixation Mechanisms Has Its Advantages and Disadvantages in Nature

The type of photosynthetic mechanism used by a plant is important, but it is not the only factor that determines where the plant lives. All three mechanisms—C_3, C_4, and CAM photosynthesis—have advantages and disadvantages, and a plant can compete successfully only when the benefits of its type of photosynthesis outweigh other factors. For example, although C_4 plants generally tolerate higher temperatures and drier conditions than C_3 species, C_4 plants may not compete successfully at temperatures below 25°C. This is, in part, because they are more sensitive to cold than C_3 species. In addition, as discussed above, CAM plants conserve water by closing their stomata during the day, a practice that severely reduces their ability to take in and fix CO_2. Hence, CAM plants grow slowly and compete poorly with C_3 and C_4 species under conditions other than extreme aridity. Thus, to some extent, each photosynthetic type of plant has limitations imposed by its own photosynthetic mechanism.

SUMMARY

In Photosynthesis, Light Energy Is Converted to Chemical Energy and Carbon Is "Fixed" into Organic Compounds

A complete, balanced equation for photosynthesis can be written as follows:

$$3CO_2 + 6H_2O \xrightarrow{\text{Light}} C_3H_6O_3 + 3O_2 + 3H_2O$$

The first step in photosynthesis is the absorption of light energy by pigment molecules. The pigments involved in eukaryotic photosynthesis include the chlorophylls and the carotenoids, which are packed in the thylakoids of chloroplasts as photosynthetic units called photosystems. Light absorbed by pigment molecules boosts their electrons to a higher energy level. Because of the way the pigment molecules are arranged in the photosystems, they are able to transfer this energy to a pair of special chlorophyll *a* molecules at the reaction centers. There are two different kinds of photosystems, Photosystem I and Photosystem II, which generally work together simultaneously and continuously. Photosystem I can also carry out photosynthesis independently of Photosystem II, but when only Photosystem I is operating (a process called cyclic photophosphorylation) there is no external electron donor (water) and thus no production of NADPH. Cyclic phosphorylation only results in proton gradients that are used for ATP production.

The many reactions that occur during photosynthesis are divided into two major processes: the light reactions and the carbon-fixation reactions.

In the Light Reactions, Electrons Flow from Water to Photosystem II, down an Electron Transport Chain to Photosystem I, and Finally to $NADP^+$

In the currently accepted model of the light reactions, light energy enters Photosystem II, where it is trapped by pigment molecules and passed to the P_{680} chlorophyll molecules of the reaction center. Energized electrons are transferred from P_{680} to an electron acceptor. As the electrons are removed from P_{680}, they are replaced by low-energy electrons from water molecules, and oxygen is produced (water photolysis).

Pairs of electrons then pass downhill to Photosystem I along an electron transport chain. This passage generates a proton gradient that drives the synthesis of ATP from ADP and phosphate (photophosphorylation). Meanwhile, light energy absorbed in Photosystem I is passed to the P_{700} chlorophyll molecules of the Photosystem I reaction center. The energized electrons are ultimately accepted by the coenzyme molecule $NADP^+$, and the electrons removed from P_{700} are replaced by the electrons from Photosystem II.

The energy yield from the light-dependent reactions is stored in the molecules of NADPH and in the ATP formed by photophosphorylation. Photophosphorylation also occurs in cyclic electron flow, a process that does not require Photosystem II. The only product of cyclic electron flow is ATP. This extra ATP is required by the Calvin cycle, which uses ATP and NADPH in a 3:2 ratio.

In the Electron Transport Chain, Electron Flow Is Coupled to Proton Pumping and ATP Synthesis by a Chemiosmotic Mechanism

Like oxidative phosphorylation in mitochondria, photophosphorylation in chloroplasts is a chemiosmotic process. As electrons flow down the electron transport chain from Photosystem II to Photosystem I, protons are pumped from the stroma into the thylakoid lumen, creating a gradient of potential energy. As protons flow down this gradient from the thylakoid lumen back into the stroma, they pass through an ATP synthase, generating ATP.

In the Calvin Cycle, CO_2 Is Fixed via a Three-Carbon Pathway

In the carbon-fixation reactions, which take place in the stroma of the chloroplast, the NADPH and ATP produced in the light reactions are used to reduce carbon dioxide to organic carbon. The Calvin cycle is responsible both for the initial fixation of CO_2 and for the subsequent reduction of the newly fixed carbon. In the Calvin cycle, a molecule of CO_2 combines with the starting compound, a five-carbon sugar called ribulose 1,5-bisphosphate (RuBP), to form two molecules of the three-carbon compound 3-phosphoglycerate (PGA). The PGA is then reduced to the three-carbon molecule glyceraldehyde 3-phosphate (PGAL), with electrons provided by NADPH and energy provided by ATP hydrolysis.

At each turn of the Calvin cycle, one carbon atom enters the cycle. Three turns of the cycle produce one molecule of glyceraldehyde 3-phosphate. At each turn of the cycle, RuBP is regenerated. Most of the fixed carbon is converted to either sucrose or starch.

The Carbon-Fixation Pathway in C_4 Plants Is a Solution to the Problem of Photorespiration

Plants in which the Calvin cycle is the only carbon-fixation pathway, and in which the first detectable product of CO_2 fixation is the three-carbon compound 3-phosphoglycerate (PGA), are called C_3 plants. In the so-called C_4 plants, CO_2 is initially fixed to phosphoenolpyruvate (PEP) to yield oxaloacetate, a four-carbon compound. This reaction occurs in the mesophyll cells of the leaf. The oxaloacetate is rapidly converted to malate (or to aspartate, depending on the species), which moves from the mesophyll cells to the bundle-sheath cells. There the malate is decarboxylated and the CO_2 enters the Calvin cycle by reacting with ribulose 1,5-bisphosphate (RuBP) to form PGA. Thus, the C_4 pathway takes place in the mesophyll cells, but the Calvin cycle occurs in bundle-sheath cells.

C_4 plants are more efficient utilizers of CO_2 than C_3 plants, in part because PEP carboxylase is not inhibited by O_2. Thus, C_4 plants can attain the same photosynthetic rate as C_3 plants, but with smaller stomatal openings and, hence, with less water loss. In addition, C_4 plants are more competitive than C_3 plants at high temperatures.

CAM Plants Can Fix CO_2 in the Dark

Crassulacean acid metabolism (CAM) occurs in many succulent plants. In CAM plants, the fixation of CO_2 to phosphoenolpyruvate (PEP) to form oxaloacetate occurs at night, when the stomata are open. The oxaloacetate is rapidly converted to malate, which is stored overnight in the vacuole as malic acid. During the daytime, when the stomata are closed, the malic acid is recovered from the vacuole and the fixed CO_2 is transferred to ribulose 1,5-bisphosphate (RuBP) of the Calvin cycle. The C_4 pathway and the Calvin cycle occur within the same cells in CAM plants; hence, these two pathways, which are spatially separated in C_4 plants, are temporally separated in CAM plants.

QUESTIONS

1. Explain how the Hill reaction and the use of ^{18}O provided evidence for van Niel's proposal that water, not carbon dioxide, is the source of the oxygen evolved in photosynthesis.

2. What is the relationship between the absorption spectrum of a pigment and the action spectrum of a process that depends on that same pigment?
3. As excited electrons return to ground level, the energy released has three possible fates. What are those fates, and which two are useful energy-releasing events in photosynthesis?
4. What is photophosphorylation, and what is the relationship between this process and the thylakoid membrane?
5. Distinguish between noncyclic and cyclic electron flow and photophosphorylation. What products are produced by each? Why is cyclic photophosphorylation essential to the Calvin cycle?
6. Briefly explain the role of each of the following protein complexes in photosynthesis: Photosystem II, cytochrome b_6/f, Photosystem I, and ATP synthase.
7. By means of a labeled diagram, explain the term "Kranz anatomy."
8. In what ways do C_4 plants have an advantage over C_3 plants?
9. Whereas the C_4 pathway and the Calvin cycle (C_3 pathway) are *spatially separated* in C_4 plants, in CAM plants these two pathways are *temporally separated.* Explain.
10. CAM plants are said to taste sweet during the day and sour at night. Explain why.

SECTION 3

GENETICS AND EVOLUTION

◀ Corn, or maize *(Zea mays)*, is one of the most important crop plants in the world. It has long been an important organism for genetic studies. The different color patterns of some of the kernels in this painting are the result of genetic transposition, the study of which brought American geneticist Barbara McClintock a Nobel Prize.

CHAPTER 8

Sexual Reproduction and Heredity

Exotic mutations Striped tulips were prized in seventeenth-century Holland, and vast fortunes were spent in pursuit of these treasured flowers. Unfortunately, the striped effect was due to a viral infection, so the tulips weakened and died, leaving investors bankrupt. Today, stable mutations in the tulips achieve the same dramatic effects.

CHAPTER OUTLINE

Ever since people first started to look at the world around them, they have puzzled and wondered about heredity. Why is it that the offspring of all living things—whether dandelions, dogs, aardvarks, or oak trees—always resemble their parents more than other members of their species, and never some other species? Although biological inheritance—**heredity**—has been the object of wonder since early in human history, only rather recently have we begun to understand how it works. In fact, the scientific study of heredity, known as **genetics,** did not really begin until the second half of the nineteenth century.

Much of the pioneering research on the molecular mechanisms of heredity was originally done on prokaryotes because of the ease of studying them, but in this chapter, we focus on the genetics of eukaryotes, primarily plants. The branch of genetics discussed here is generally referred to as **Mendelian genetics,** in recognition of the work of Gregor Mendel (Figure 8–1).

Mendel was born into a peasant family in 1822. In 1843, at 21, he entered the Augustinian Monastery in Brünn in the Austro-Hungarian Empire (now Brno in the Czech Republic), where he was able to receive an education. He attended the University of Vienna for two years, pursuing studies in

CHECKPOINTS

After reading this chapter, you should be able to answer the following:

1. What is the relationship between haploid and diploid chromosome numbers and meiosis and fertilization?
2. Describe the events that occur during crossing-over, and explain why this process is important.
3. What are the main events that occur during meiosis I? How is meiosis I different from meiosis II?
4. List the advantages and disadvantages of sexual and asexual reproduction.
5. What were the major findings of Gregor Mendel, and what were the unique aspects of his experimental method that contributed to his success?
6. How is it possible for a trait to be visible in the parents but not in the offspring? What type of test could you conduct to verify your answer?
7. What are linked genes? In what way is the concept of linkage at odds with the principle of independent assortment?
8. What are the different kinds of mutations, and how do mutations affect the evolution of a population of organisms?

8–1 Gregor Mendel, the founder of modern genetics Mendel discovered the principles of heredity by crossing different varieties of the garden pea *(Pisum sativum)* and analyzing the pattern of inheritance of traits in succeeding generations.

mathematics and botany. Mendel's experiments, carried out in the quiet monastery garden, were first reported in 1865 at a meeting of the Brünn Natural History Society. Although none of those in attendance seemed to understand the significance of Mendel's results, his paper was published the following year in the Proceedings of the Society, a journal that was circulated to libraries all over Europe. Unfortunately, his work, which marks the beginning of modern genetics, was ignored until after his death. It was not until 1900 that his paper was independently rediscovered by three scientists, each working in a different European country. And each found, in Mendel's brilliant analyses, that much of his own work had been anticipated.

Although Mendel never saw a chromosome, he developed principles that reflect the existence of genes, chromosomes, and the process of meiosis. In this chapter we examine the physical basis of heredity—that is, meiosis—and discuss Mendel's work, as well as some ways in which genes can be altered. We also consider why sexual reproduction is virtually universal among eukaryotes.

Sexual Reproduction

Sexual reproduction, one of the major characteristics of eukaryotes as a group, does not occur in prokaryotes. Although some eukaryotes do not reproduce sexually, it is evident that these asexual organisms have lost the capacity to do so during the course of their evolutionary history.

Sexual reproduction involves a regular alternation between meiosis and fertilization. **Meiosis** is the process of nuclear division in which the chromosome number is reduced from the diploid (2*n*) to the haploid (*n*) number. During meiosis, the nucleus of a diploid cell undergoes two divisions, the first of which is a reduction division. These divisions result in the production of four daughter nuclei, each containing one-half the number of chromosomes of the original nucleus (Figure 8–2). In plants, meiosis occurs during the production of spores (meiospores) in

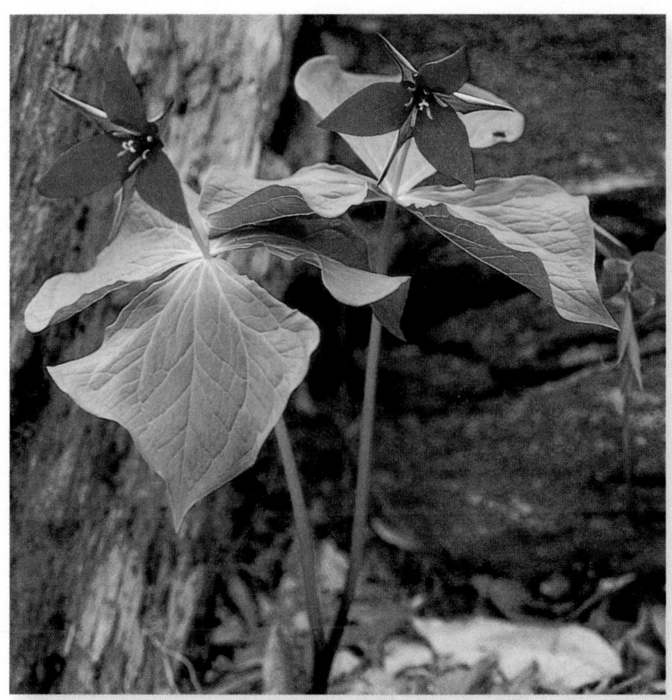

(a)

(b)

8–2 Meiosis in wake-robin ***(a)*** Wake-robin *(Trillium erectum)* flowers in the early spring. ***(b)*** Two groups of four developing microspores each at anaphase II of meiosis in *T. erectum*. Separation of the fully condensed, clearly visible chromosomes is nearly complete. When completed, each of the newly formed nuclei will possess a haploid set of chromosomes, containing a total of over 30 meters of DNA.

flowers, cones, or similar structures. **Fertilization,** or **syngamy,** is the process by which two haploid cells (gametes) fuse to form a diploid zygote, reestablishing the diploid chromosome number. Meiosis thus counterbalances the effects of fertilization, ensuring that the number of chromosomes remains constant from generation to generation.

In every diploid cell, each chromosome has a partner. The members of a pair of chromosomes are known as **homologous chromosomes,** or **homologs.** The two homologs resemble each other in size and shape and, also, as we shall see, in the kinds of hereditary information they contain. One homolog comes from the gamete of one parent and its partner from the gamete of the other parent. After fertilization, both homologs are present in the zygote. The interactions of the genes on each of these sets of chromosomes determine the genetic characteristics of the diploid organism.

The Eukaryotic Chromosome

As we noted in Chapter 3, chromosomes of eukaryotes are made of DNA and protein. The DNA content varies greatly among chromosomes, and this variation is reflected in great differences in chromosome length. Each eukaryotic chromosome contains only a single, threadlike molecule of double-stranded DNA, which is continuous throughout its entire length. This is certainly an impressive feat of packaging. DNA is an "exquisitely thin filament," in the words of E. J. DuPraw, who calculated that a length of DNA sufficient to reach from the Earth to the sun would weigh only half a gram. Each diploid cell of wake-robin *(Trillium),* a common spring wildflower (Figure 8–2), contains about 68 meters of DNA! We are only beginning to understand what *Trillium* does with this unusually large amount of DNA, much of which, it turns out, is repetitive and unexpressed.

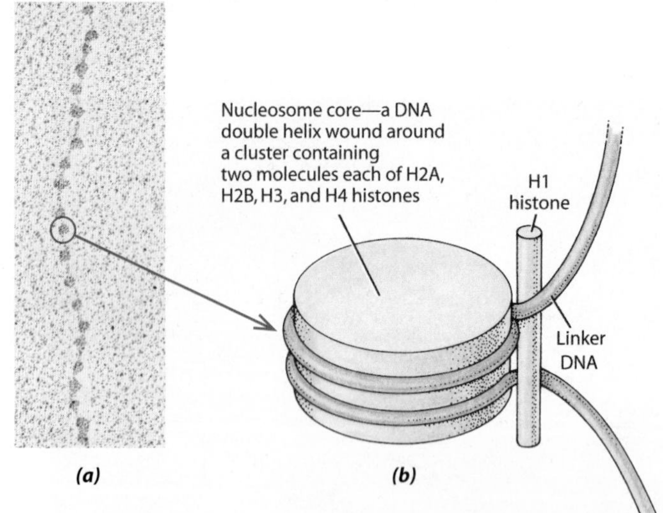

8–3 Nucleosome ***(a)*** Electron micrograph of nucleosomes and connecting threads of DNA from a chicken erythrocyte (red blood cell). The nucleosomes—the structures that look like beads on a string—are each approximately 10 nanometers in diameter. ***(b)*** Each nucleosome consists of eight histone molecules (two each of H2A, H2B, H3, and H4), around which the DNA helix is wound twice. The nucleosomes are separated from each other by linker DNA. Histone H1 binds to the outer surface of the nucleosomes.

Chromosomes Contain Histone Proteins

The combination of DNA and its associated proteins in eukaryotic chromosomes is known as **chromatin** ("colored threads"), because of its staining properties. Chromatin is more than half protein, and the most abundant proteins belong to a class of small polypeptides known as **histones.** Histones are positively charged (basic) and therefore are attracted to the negatively charged (acidic) DNA. They are always present in chromatin and are synthesized in large amounts during the S phase of the

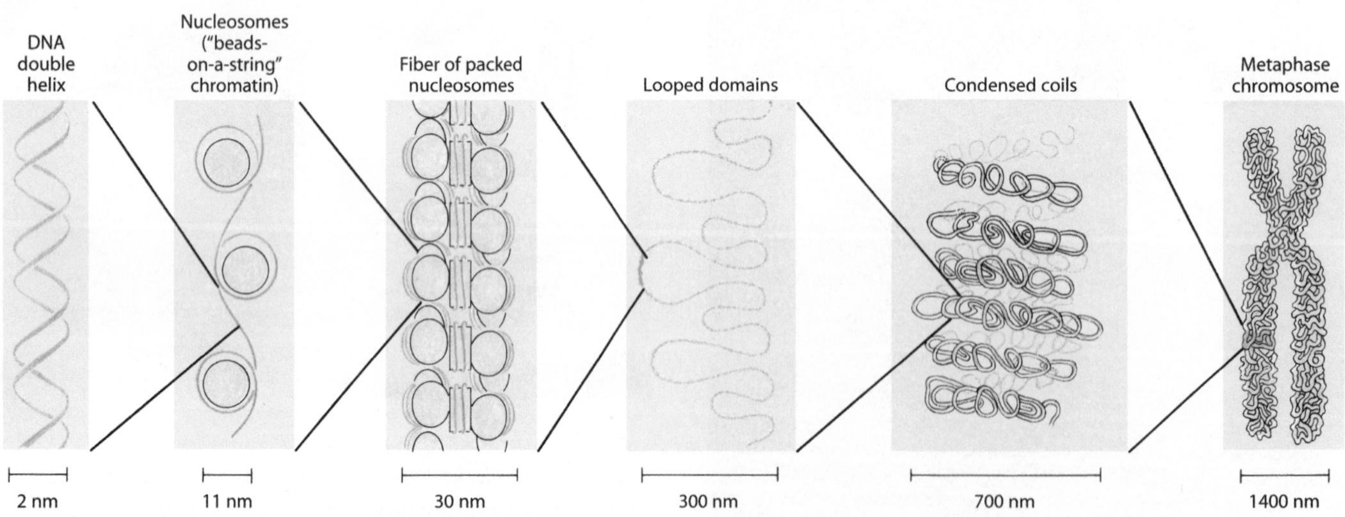

8–4 From DNA to a chromosome Stages in the folding of chromatin, culminating in a fully condensed metaphase chromosome. The model is derived from electron micrographs of chromatin at different degrees of condensation. According to present evidence, each chromatid of a replicated chromosome contains a single molecule of double-stranded DNA and is, by weight, about 60 percent protein.

cell cycle. The histones, of which there are five distinct types (H1, H2A, H2B, H3, and H4), are primarily responsible for the folding and packaging of DNA.

The fundamental packing units of chromatin are the **nucleosomes** (Figure 8–3), which resemble beads on a string. Each nucleosome consists of a core of eight histone molecules (two each of H2A, H2B, H3, and H4), around which the DNA filament is wrapped twice, like a thread around a spool. Histone H1 lies on the DNA outside the nucleosome core, that is, between the "beads." When a fragment of DNA is wrapped up in a nucleosome, it is about one-sixth the length it would be if fully extended.

As shown in Figure 8–4, additional packing of the nucleosomes produces a fiber that is about 30 nanometers in diameter. Further condensation of this fiber produces a series of loops, known as looped domains. The looped domains condense into the compact metaphase chromosomes that become visible with the light microscope during mitosis and meiosis.

Other proteins associated with the chromosomes are the enzymes involved with DNA and RNA synthesis, regulatory proteins, and a large number and variety of molecules that have not yet been isolated and identified. Unlike the histones, these molecules vary from one cell type to another and may be involved in the differential expression of genes in different cell types of the same organism.

The Process of Meiosis

Meiosis occurs only in specialized diploid cells and only at particular times in the life cycle of a given organism. Through meiosis and cytokinesis, a single diploid cell gives rise to four haploid cells—either gametes or meiospores (spores resulting from meiosis). A **gamete** is a cell that unites with another gamete to produce a diploid **zygote,** which may then divide, either meiotically or mitotically. A **spore** is a cell that can develop into an organism without uniting with another cell. Spores often divide mitotically, producing multicellular organisms that are entirely haploid and that eventually give rise to gametes by mitosis, as occurs in plants and many algae (see Figure 12–17c).

As we have already noted, meiosis consists of two successive nuclear divisions, which are designated **meiosis I** and **meiosis II.** As shown in Figure 8–5, homologous chromosomes pair and then separate from one another, and in meiosis II, the chromatids of each homolog separate.

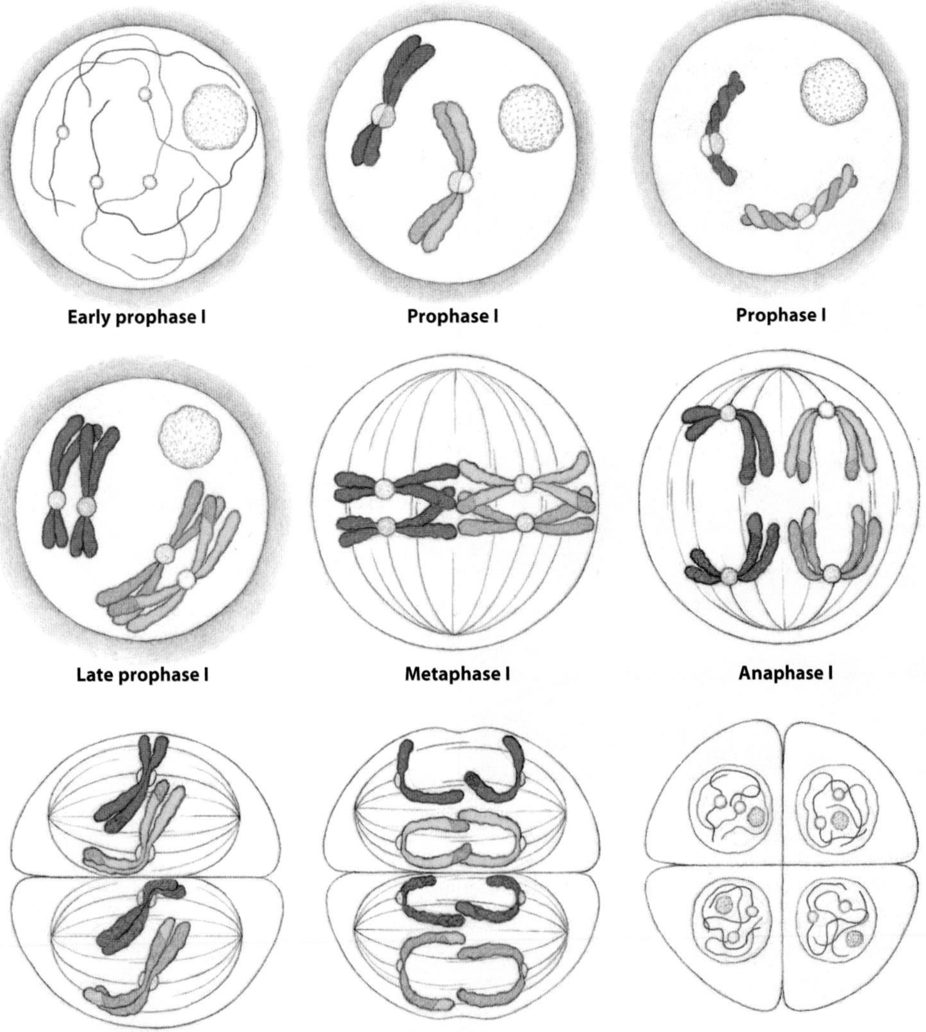

8–5 Meiosis A diagrammatic representation of meiosis I and II, with two pairs of chromosomes. Not all stages are shown.

Prophase I: The chromosomes become visible as elongated threads, homologous chromosomes come together in pairs, the pairs coil round each other, and the paired chromosomes become very short. Initially, each chromosome appears to be single rather than double; individual chromatids are not visible until late prophase. **Metaphase I:** The paired chromosomes move into position on the metaphase plate, with their centromeres evenly distributed on either side of the equatorial plane of the spindle. **Anaphase I:** The paired chromosomes separate and move to opposite poles.

Events in the second meiotic division are essentially the same as those in mitosis. **Metaphase II:** The chromosomes are lined up at the equatorial plane, with their centromeres lying on the plane. **Anaphase II:** The centromeres separate, and the chromatids separate and move toward opposite poles of the spindle. **Telophase II:** The chromosomes have completed their migration. Four new nuclei, each with the haploid number of chromosomes, are formed.

In Meiosis I, the Homologous Chromosomes Separate and Move to Opposite Poles

Following duplication of the chromosomes during the preceding interphase, the first of the two nuclear divisions of meiosis begins. It follows through the stages of prophase, metaphase, anaphase, and telophase.

In **prophase I** (prophase of the first meiotic division), the chromosomes—present in the diploid number—first become visible as long, slender threads. As in mitosis (see Chapter 3), the chromosomes have already duplicated during the preceding interphase. Consequently, at the beginning of prophase I, each chromosome consists of two identical chromatids attached at the centromere. At this early stage of meiosis, however, each chromosome appears to be single rather than double.

Before the individual chromatids become apparent, the homologous chromosomes pair with each other. The pairing is very precise, beginning at one or more sites along the length of the chromosomes and proceeding in a zipperlike fashion, such that the same portions of the homologous chromosomes lie next to one another. Each homolog is derived from a different parent and is made up of two identical chromatids. Thus, a homologous pair consists of four chromatids. The pairing of homologous chromosomes is a necessary part of meiosis; the process cannot occur in haploid cells because such homologs are not present. The pairing is called **synapsis,** and the associated pairs of homologous chromosomes are called **bivalents.**

During the course of prophase I, the paired threads become more and more condensed, and consequently the chromosomes shorten and thicken. With the aid of an electron microscope, it is possible to identify a densely staining axial core, consisting mainly of proteins, in each chromosome (Figure 8–6a). During mid-prophase, the axial cores of a pair of homologous chromosomes approach each other to within 0.1 micrometer, forming a **synaptonemal complex,** which unites the homologs (Figure 8–6b).

It can now be seen that each axial core is double; that is, each bivalent is made up of four chromatids, two per chromosome. During the time when the synaptonemal complex exists, portions of the chromatids break apart and rejoin with corresponding segments from their homologous chromatids (Figure 8–7). This **crossing-over** results in chromatids that are complete but have a different representation of genes than they had originally. Figure 8–8 shows the visible evidence that crossing-over has taken place—the X-like configuration called a **chiasma** (plural: chiasmata).

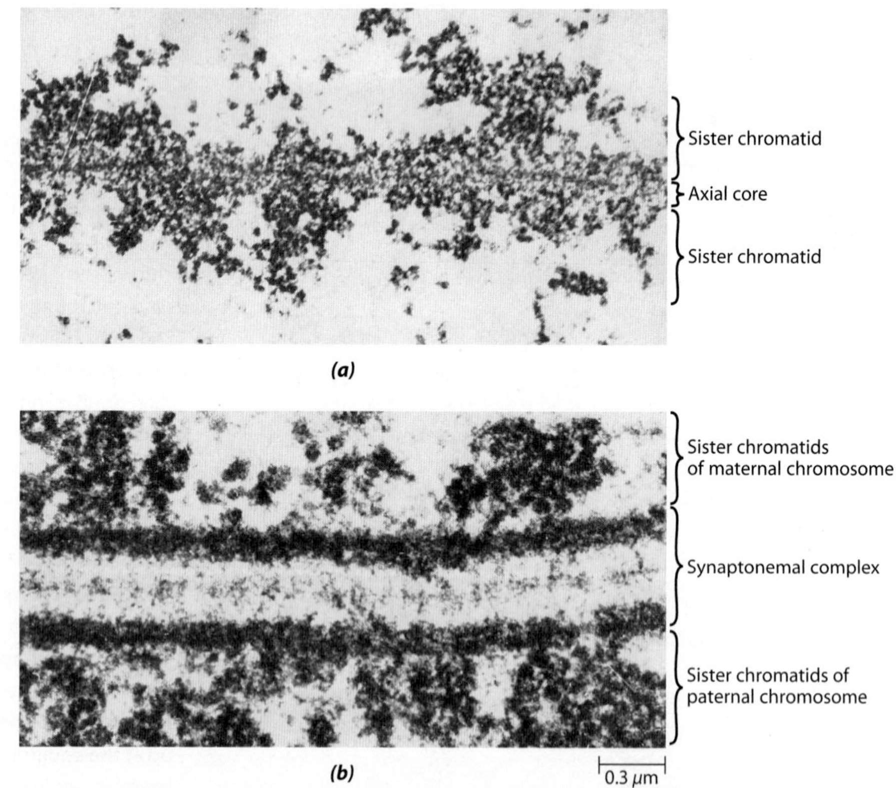

8–6 Synaptonemal complex ***(a)*** Portion of a chromosome of *Lilium* early in prophase I, prior to pairing with its homologous chromosome. Note the dense axial core, with sister chromatid material above and below it. This core, which consists mainly of proteins, may arrange the genetic material of the chromosome in preparation for pairing and genetic exchange. ***(b)*** Synaptonemal complex, consisting of a zipperlike protein structure connecting the two axial cores of the two homologous chromosomes, in a bivalent of *Lilium*.

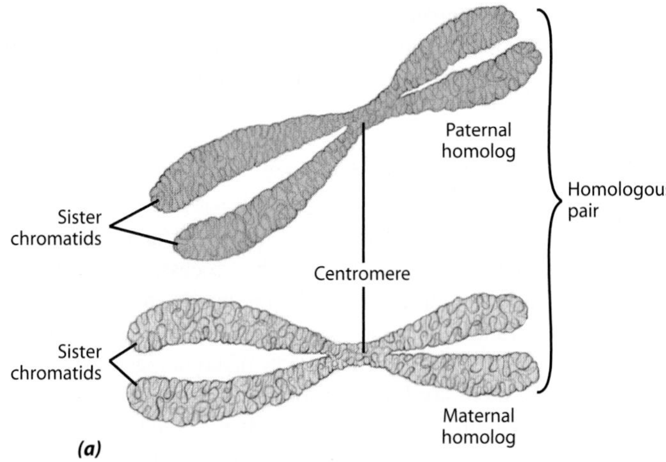

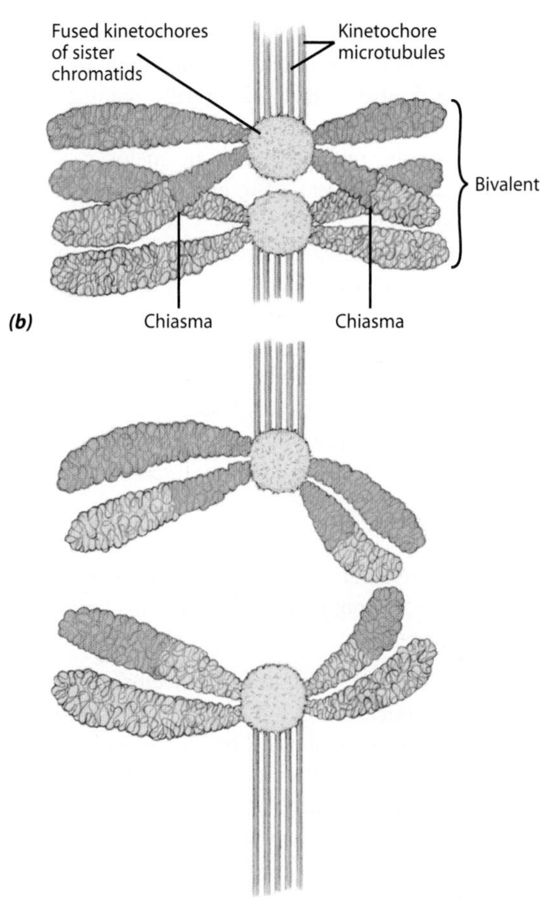

8–7 Crossing-over ***(a)*** A homologous pair of chromosomes, before meiosis. One member of the pair is of paternal origin, and the other of maternal origin. Each of these chromosomes has duplicated and consists of two sister chromatids connected at the centromere. ***(b)*** In prophase of the first meiotic division, the two homologs come together and become closely associated. The paired homologous chromosomes are called a bivalent. Within the bivalent, chromatids of the two homologs intersect at a number of points, making possible the exchange of chromatid segments. This phenomenon is known as crossing-over, and the locations at which it occurs are called chiasmata. ***(c)*** The result of crossing-over is a recombination of the genetic material of the two homologs. The sister chromatids of each homolog are no longer identical. Kinetochore microtubules attached to the fused kinetochores of sister chromatids separate the homologs at anaphase I.

As prophase I proceeds, the synaptonemal complex disappears. Eventually, the nuclear envelope breaks down. The nucleolus usually disappears as RNA synthesis is temporarily suspended. Finally, the homologous chromosomes appear to repulse one another. Their chromatids are held together at the chiasmata, however. These chromatids separate very slowly. One or more chiasmata may occur in each arm of the chromosome, or only one in the entire bivalent; the appearance of a particular bivalent can vary widely, depending on the number of chiasmata present (Figure 8–8).

In **metaphase I,** the spindle—an axis of microtubules similar to that which functions in mitosis—becomes conspicuous (Figure 8–9). As meiosis proceeds, individual microtubules become attached to the centromeres of the chromosomes of each bivalent. These paired chromosomes then move to the equatorial plane of the cell, where they line up randomly in a configuration characteristic of metaphase I. The centromeres of the paired chromosomes line up on opposite sides of the equatorial plane. In contrast, in mitotic metaphase, as we have seen, the centromeres of the individual chromosomes line up directly on the equatorial plane.

Anaphase I begins when the homologous chromosomes separate and begin to move toward the poles. Notice again the contrast with mitosis. In mitotic anaphase, the centromeres

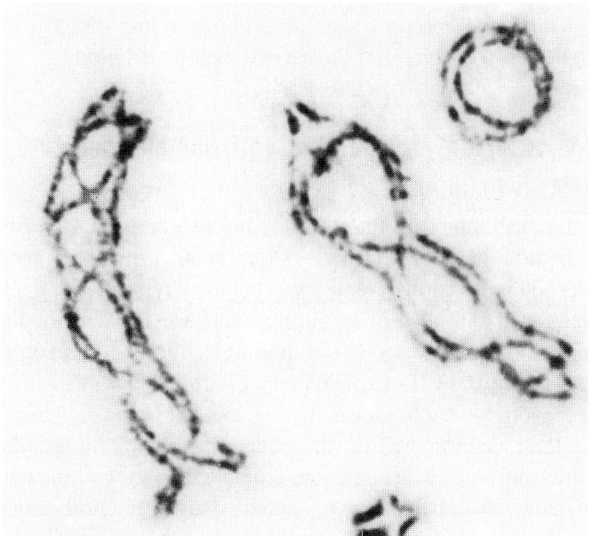

8–8 Chiasmata Variations in the number of chiasmata can be seen in the paired chromosomes of a grasshopper *(Chorthippus parallelus),* in which $n = 4$.

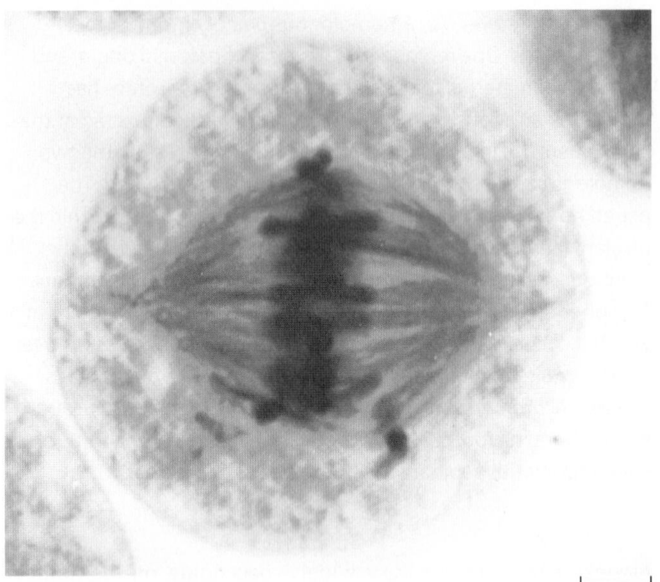

8–9 Spindle The spindle in a pollen mother cell of wheat *(Triticum aestivum)* during metaphase I of meiosis.

separate and the sister chromatids move toward the poles. In meiotic anaphase I, the centromeres do not separate and the sister chromatids remain together; it is the homologs that separate. Because of the exchanges of chromatid segments that resulted from crossing-over, however, the chromatids are not identical, as they were at the onset of meiosis.

In **telophase I,** the coiling of the chromosomes relaxes and the chromosomes become elongated and once again indistinct. New nuclear envelopes are formed from the endoplasmic reticulum, as telophase gradually gives way to interphase. Finally, the spindle disappears, the nucleolus is re-formed, and protein synthesis commences again. In many organisms, however, no interphase occurs between meiotic divisions I and II; in these organisms, the chromosomes pass more or less directly from telophase I to prophase of the second meiotic division.

In Meiosis II, the Chromatids of Each Homolog Separate and Move to Opposite Poles

At the beginning of the second meiotic division, the sister chromatids are still attached to one another by their centromeres. This division resembles a mitotic division: the nuclear envelope (if one re-formed during telophase I) becomes disorganized, and the nucleolus disappears by the end of **prophase II.** At **metaphase II,** a spindle again becomes obvious, and the chromosomes—each consisting of two chromatids—line up with their centromeres on the equatorial plane. At **anaphase II,** the centromeres separate and are pulled apart, and the newly separated chromatids, now called daughter chromosomes, move to opposite poles (Figures 8–2b and 8–7c). At **telophase II,** new nuclear envelopes and nucleoli are organized, and the contracted chromosomes relax as they fade into an interphase nucleus. Walls develop about each new cell. Thus, cells are formed with the haploid number of chromosomes. Note that although the mechanism of meiosis II resembles that of mitosis, meiosis II is not the same as mitosis because each cell formed in meiosis II has half the number of chromosomes of a cell formed in mitosis.

Meiosis Produces Genetic Variability

The end result of meiosis is that each cell has only half as many chromosomes as the original diploid nucleus. But more important are the genetic consequences of the process. At metaphase I, the orientation of the bivalents is random; that is, the chromosomes are randomly divided between the two new nuclei. If the original diploid cell had two pairs of homologous chromosomes, $n = 2$, there are four possible ways in which they could be distributed among the haploid cells. If $n = 3$, there are 8 possibilities; if $n = 4$, there are 16. The general formula is 2^n. In human beings, $n = 23$, and so the number of possible combinations is 2^{23}, which is equal to 8,388,608. Many organisms have much higher numbers of chromosomes than $n = 23$. In addition, because of crossing-over, the genes are present in new combinations not present in the gametes that gave rise to the diploid cell. Thus, crossing-over is another important mechanism for **genetic recombination**—that is, for recombining the genetic material from the two parents.

To summarize, two processes in meiosis contribute to genetic variation in the cells produced by meiosis. First, the pairs of homologous chromosomes can separate in different ways and, as the number of chromosome pairs increases, the number of different ways they can separate becomes larger and larger. Second, crossing-over mixes the genes present in homologous chromosomes into new combinations not present in the chromosomes of the original cells that fused to produce the diploid cells undergoing meiosis. Usually, at least one chiasma occurs between every pair of homologous chromosomes.

Meiosis differs from mitosis in three fundamental ways (Figure 8–10).

1. Two nuclear divisions are involved in meiosis and only one in mitosis, yet in both meiosis and mitosis the DNA is replicated only once.
2. Each of the four nuclei produced in meiosis is haploid, containing only one-half the number of chromosomes—that is, only one member of each pair of homologous chromosomes—present in the original diploid nucleus from which it was produced. By contrast, each of the two nuclei produced during mitosis has the same number of chromosomes as the original nucleus.
3. Each of the nuclei produced by meiosis contains different gene combinations from the others, whereas the nuclei produced by mitosis have identical gene combinations.

In meiosis, then, nuclei *different* from the original nucleus are produced, in contrast with mitosis, which produces nuclei with chromosome complements *identical* to those of the original nucleus. The genetic and evolutionary consequences of the behavior of chromosomes in meiosis are profound. Because of meiosis and fertilization, the populations of diploid organisms that occur in nature are far from uniform; instead, they

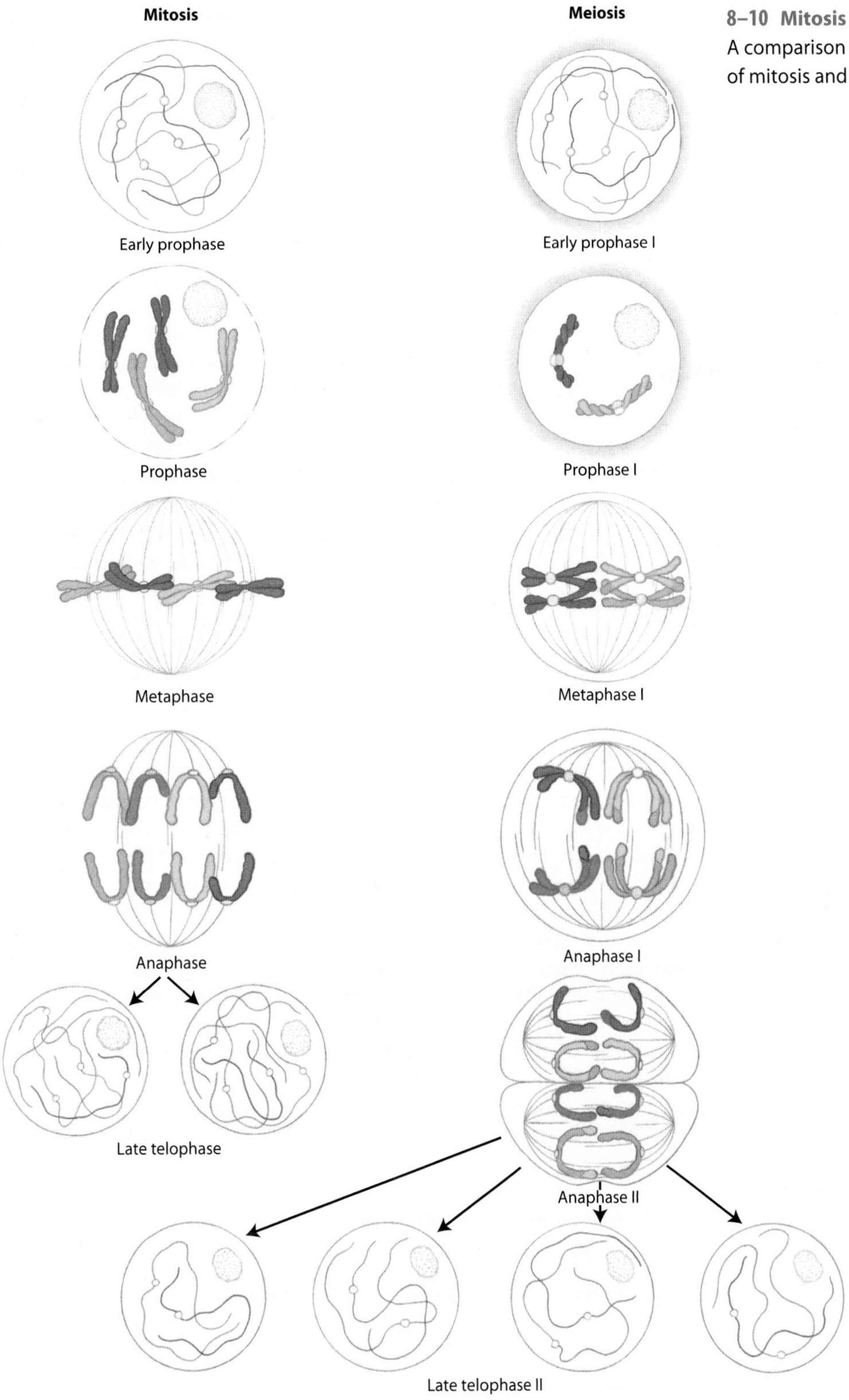

8–10 Mitosis and meiosis
A comparison of the main features of mitosis and meiosis.

consist of individuals that differ from one another in many characteristics.

How Characteristics Are Inherited

The characteristics of diploid organisms are determined by the interactions between alleles. An **allele** is one of two or more alternative forms of the same gene. Alleles occupy the same site, or **locus,** on homologous chromosomes. The ways in which alleles interact to produce particular characteristics were first revealed by Gregor Mendel. For his studies, Mendel chose cultivated varieties of the garden pea *(Pisum sativum)* and considered well-defined but contrasting traits, such as differences in flower color or seed shape. He then made large numbers of

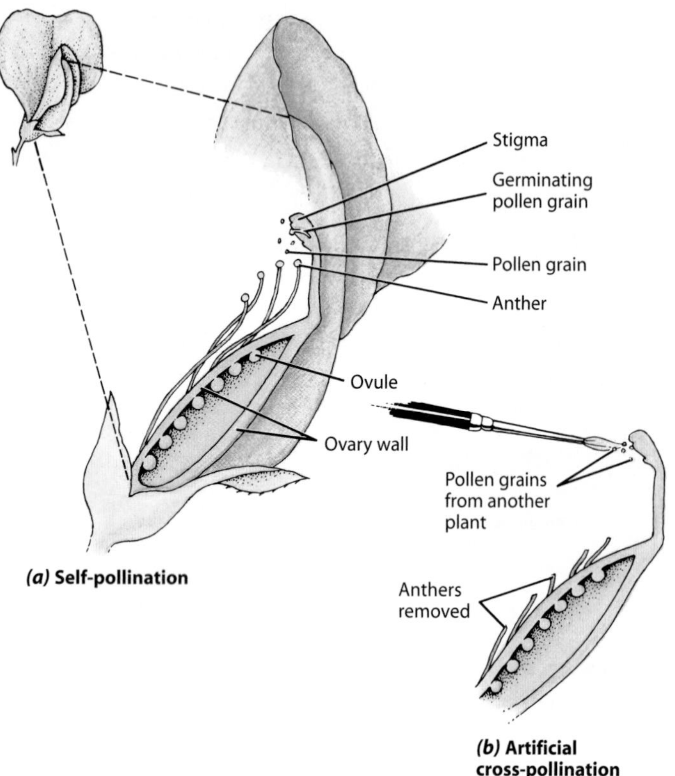

8–11 Pollination in the garden pea In a flower, the pollen develops in the anthers, and the egg cells develop in the ovules. Pollination occurs when the pollen grains are transferred from anther to stigma. The pollen grains then germinate, forming pollen tubes, which carry the sperm to the eggs. When the nuclei of sperm and egg unite, the fertilized egg, or zygote, develops into an embryo within the ovule. The ovule matures to form the seed and the ovary wall to form the fruit, which in the pea is the pod.

Pollination in most species of flowering plants involves the transfer (often by an insect) of pollen from one plant to the stigma of another plant. This is called cross-pollination. ***(a)*** In the pea flower, however, the stigma and anthers are completely enclosed by petals, and the flower, unlike most, does not open until after pollination and fertilization have taken place. Thus, self-pollination is the normal course of events in the pea flower—that is, pollen is deposited on the stigma of the same flower in which it is produced. Note that each pea in a pod represents an independent fertilization event. ***(b)*** In his cross-breeding experiments, Mendel pried open the flower bud before the pollen matured and removed the anthers with tweezers, preventing self-pollination. Then he artificially cross-pollinated the flower by dusting the stigma with pollen collected from another plant. The self-pollinating nature of peas was useful to Mendel, because he could grow the resulting hybrids without any manipulations to produce the next generation.

experimental crosses (Figure 8–11), and perhaps most important of all, he studied the offspring not only of the first generation but also of subsequent generations and their crosses.

Table 8–1 lists seven traits of the pea plants that Mendel used in his experiments. When Mendel crossed plants with these contrasting traits, he observed that, in every case, one of the alternative traits could not be seen in the first generation (now known as the F_1 generation, for "first filial generation"). For example, the seeds of all the progeny of the cross between yellow-seeded plants and green-seeded plants were as yellow as those of the yellow-seeded parent. Mendel called the trait for yellow seeds, as well as the other traits that were seen in the F_1 generation, **dominant.** He called the traits that did not appear in the first generation **recessive.** When plants of the F_1 generation were allowed to self-pollinate (Figure 8–12), the recessive trait reappeared in the F_2 generation in a ratio of approximately 3 dominant to 1 recessive (Table 8–1). Crosses between individuals that differ in a single trait, such as seed color, are called **monohybrid** crosses.

These results can be easily understood in terms of meiosis. Consider a cross between a white-flowered plant and a purple-flowered plant. The allele for white flower color, which is a recessive trait, is indicated by the lowercase letter *w*. The contrasting allele for purple flower color, which is a dominant trait, is indicated by the capital letter *W*. Diploid plants have two alleles at each locus, one allele inherited from each

Table 8-1 Results of Mendel's Experiments with Pea Plants

	Original Crosses			Second Filial Generation (F_2)	
Trait	**Dominant**	×	**Recessive**	**Dominant**	**Recessive**
Seed form	Round	×	Wrinkled	5474	1850
Seed color	Yellow	×	Green	6022	2001
Flower position	Axial	×	Terminal	651	207
Flower color	Purple	×	White	705	224
Pod form	Inflated	×	Constricted	882	299
Pod color	Green	×	Yellow	428	152
Stem length	Tall	×	Dwarf	787	277

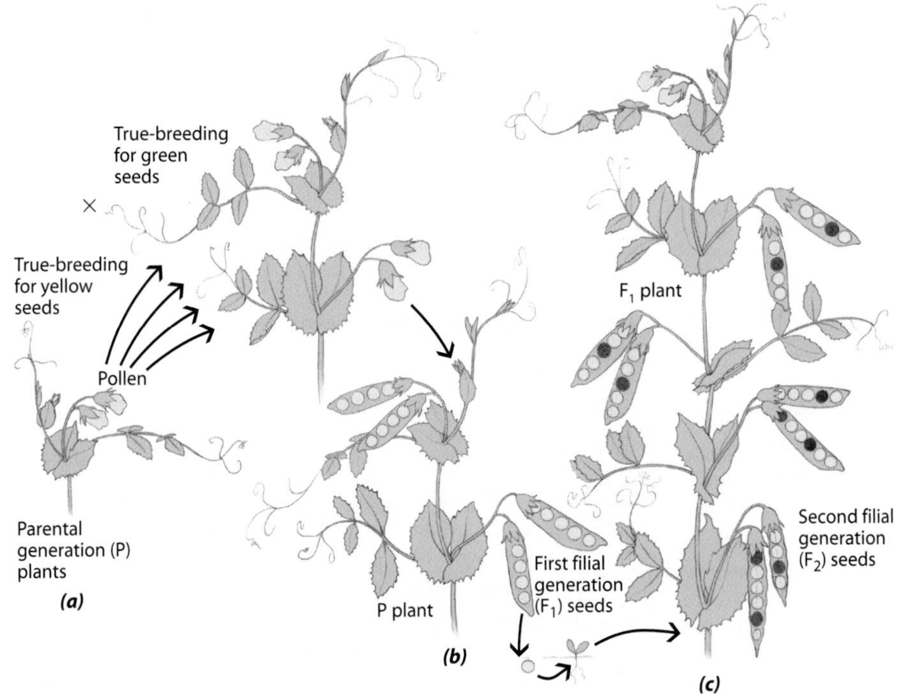

8–12 An outline of one of Mendel's experiments ***(a)*** A true-breeding yellow-seeded pea plant was crossed with a true-breeding green-seeded plant by removing pollen from the anthers of flowers on one plant and transferring it to the stigmas of flowers on the other plant. These plants are the parental (P) generation. ***(b)*** Pea pods containing only yellow seeds developed from the fertilized flowers. These peas (seeds) and the plants that grew from them when they were planted constitute the F_1 generation. When the F_1 plants flowered, they were not disturbed and were allowed to self-pollinate. ***(c)*** Pea pods developing from the self-pollinated flowers contained both yellow and green peas (the F_2 generation) in an approximate ratio of 3:1. That is, about 3/4 were yellow and 1/4 were green.

parent. The two alleles are found on different but homologous chromosomes. In the strains of garden pea with which Mendel worked, white-flowered individuals had the genetic constitution, or **genotype,** *ww*. Purple-flowered individuals had the genotype *WW*. Individuals such as these, which have two identical alleles at a particular site, or locus, on their homologous chromosomes, are said to be **homozygous.** Through the process of meiosis, the two alleles of a diploid genotype separate (because the two homologs on which they are located separate) and only one allele goes into each gamete. When plants with white and purple flowers are crossed, every individual in the F_1 generation receives a *W* allele from the purple-flowered parent and a *w* allele from the white-flowered parent, and thus they all have the genotype *Ww* (Figure 8–13). Such an individual is said to be **heterozygous** for the gene for flower color.

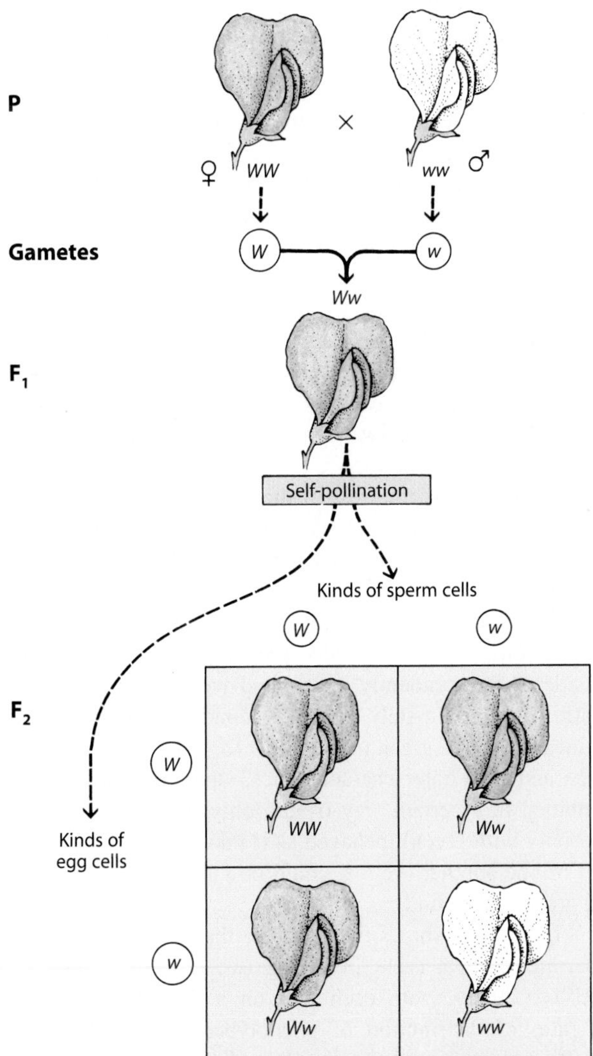

8–13 Crossing homozygotes Shown here are the F_1 and F_2 generations following a cross between two homozygous parent (P) pea plants, one with two dominant alleles for purple flowers (*WW*) and the other with two recessive alleles for white flowers (*ww*). The female symbol ♀ identifies the plant that contributes the egg cells (female gametes), and the male symbol ♂ indicates the plant that contributes the sperm cells (male gametes).

The phenotype of the offspring in the F_1 generation is purple, but note that the genotype is *Ww*. The F_1 heterozygote produces four gametes, two of each kind, ♀*W*, ♀*w*, ♂*W*, ♂*w*, in equal proportions. When this plant self-pollinates, the *W* and *w* egg and sperm cells combine randomly to form, on the average, 1/4 *WW* (purple), 2/4 (or 1/2) *Ww* (purple), and 1/4 *ww* (white) offspring. It is this underlying 1:2:1 genotypic ratio that accounts for the phenotypic ratio of 3 dominants (purple) to 1 recessive (white).

Because of meiosis, a heterozygous individual forms two kinds of gametes, *W* and *w,* which are present in equal proportions. As indicated in Figure 8–13, these gametes recombine to form one *WW* individual, one *ww* individual, and two *Ww* individuals, on average, for every four offspring produced. In terms of their appearance, or **phenotype,** the heterozygous *Ww* individuals are purple-flowered and are thus indistinguishable from the homozygous *WW* individuals. The action of the allele from the purple-flowered parent is sufficient to mask the action of the allele from the white-flowered parent. This, then, is the basis for the 3:1 phenotypic ratios that Mendel observed.

How can you tell whether the genotype of a plant with purple flowers is *WW* or *Ww*? As shown in Figure 8–14, you can tell by crossing such a plant with a white-flowered plant and counting the progeny of the cross. Mendel performed just this kind of experiment, which is known as a **testcross**—the crossing of an individual showing a dominant trait with a second individual that is homozygous recessive for that trait.

One of the simplest ways to predict the types of offspring that will be produced in a cross is to diagram the cross as shown in Figures 8–13 and 8–14. This sort of checkerboard diagram is known as a *Punnett square,* after the English geneticist who first used it for the analysis of genetically determined traits.

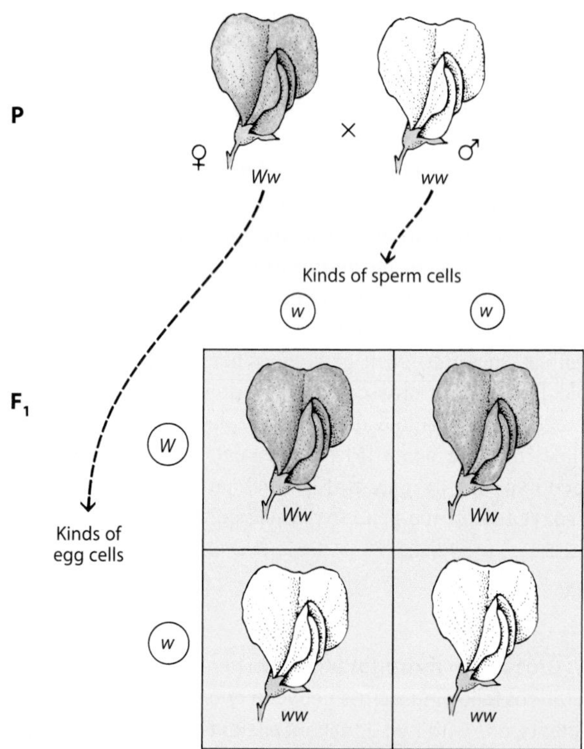

8–14 A testcross For a pea flower to be white, the plant must be homozygous for the recessive allele (*ww*). But a purple pea flower can be produced by a plant with either a *Ww* or a *WW* genotype. How can you determine the genotype of a purple-flowered plant? Geneticists solve this problem by crossing such plants with homozygous recessives. This sort of experiment is known as a testcross. As shown here, a 1:1 phenotypic ratio in the F_1 generation (that is, one purple to one white flower) indicates that the purple-flowering parent used in the testcross must have been heterozygous. What would the result have been if the test plant had been homozygous for the purple-flower allele?

Mendel's Two Principles

The Principle of Segregation: Individuals Carry Pairs of Genes for Each Trait and These Pairs Separate during Meiosis

The principle established by these experiments is the **principle of segregation,** which is sometimes known as Mendel's first law. According to this principle, hereditary characteristics are determined by discrete factors (now called **genes**) that appear in pairs, one of each pair being inherited from each parent. During meiosis, the pairs of factors are separated, or *segregated.* Hence, each gamete produced by an offspring at maturity contains only one member of the pair. This concept of a discrete factor explained how a characteristic could persist from generation to generation without blending with other characteristics, as well as how it could seemingly disappear and then reappear in a later generation.

The Principle of Independent Assortment: The Alleles of a Gene Segregate Independently of the Alleles of Other Genes

In a second series of experiments, Mendel studied hybrids that involved two traits—that is, he carried out **dihybrid** crosses. For example, he crossed a homozygous strain of garden pea that had round and yellow seeds with a homozygous strain that had wrinkled and green seeds.

The alleles for round seeds and yellow seeds are both dominant, and those for wrinkled and green seeds are recessive (Table 8–1). All the seeds of the F_1 generation were round and yellow. When the F_1 seeds were planted and the flowers allowed to self-pollinate, 556 F_2 seeds were produced. Of these, 315 seeds showed the two dominant characteristics—round and yellow—and 32 combined the recessive characteristics, wrinkled and green. All the rest of the seeds produced were unlike those of either parent: 101 were wrinkled and yellow, and 108 were round and green. Totally new combinations of characteristics had appeared.

This experiment did not, however, contradict Mendel's previous results. If the two traits, seed color and seed shape, are considered independently, round and wrinkled still appeared in a ratio of approximately 3:1 (423 round to 133 wrinkled), and so did yellow and green (416 yellow to 140 green). But the seed shape and seed color characteristics, which had originally been combined in a certain way (round only with yellow and wrinkled only with green), behaved as if they were entirely independent of one another (yellow could now be found with wrinkled, and green with round).

Figure 8–15 shows the basis for the results of the dihybrid experiments. In a cross involving two pairs of dominant and recessive alleles, with each pair on a different chromosome, the ratio of distribution of phenotypes is 9:3:3:1. The fraction 9/16 represents the proportion of the F_2 progeny expected

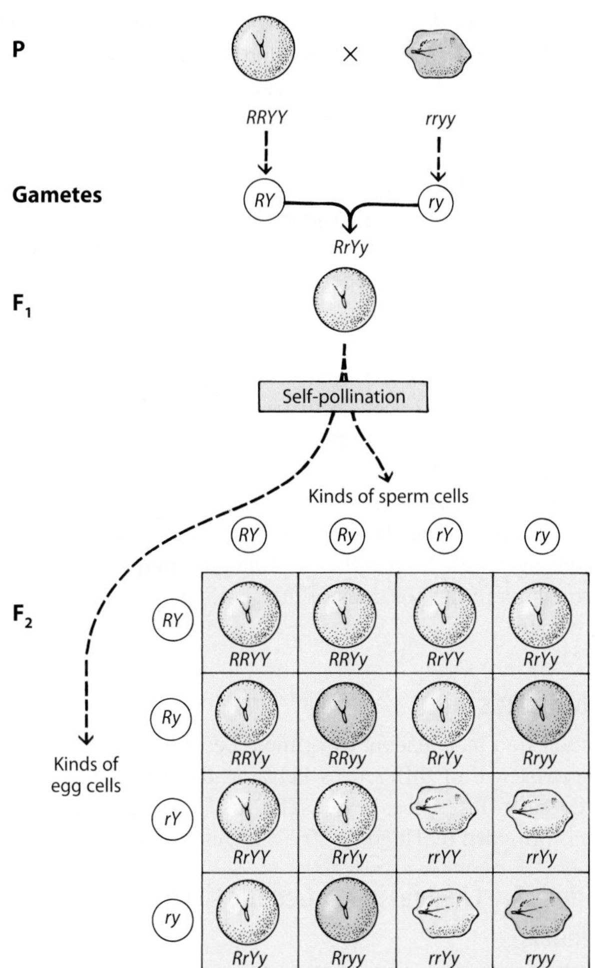

8–15 Independent assortment Shown here is one of the experiments from which Mendel derived his principle of independent assortment. A plant homozygous for round (*RR*) and yellow (*YY*) peas is crossed with a plant having wrinkled (*rr*) and green (*yy*) peas. The F_1 peas are all round and yellow, but notice how the characteristics appear, on the average, in the F_2 generation. Of the 16 possible combinations in the offspring, nine show the two dominant characteristics (round and yellow), three show one combination of dominant and recessive (round and green), three show the other combination (wrinkled and yellow), and one shows the two recessives (wrinkled and green). This 9:3:3:1 distribution of the phenotypes is always the expected result from a cross involving two independently assorting genes, each with one dominant and one recessive allele in each of the parents. Note that each single characteristic produces a dominant-to-recessive ratio of 3:1.

to show the two dominant characteristics, 1/16 the proportion expected to show the two recessive characteristics, and 3/16 and 3/16 the proportions expected to show the alternative combinations of dominant and recessive characteristics. In the example shown, one parent carries both dominant alleles and the other carries both recessive alleles. Suppose that each parent carried one recessive and one dominant allele. Would the results be the same? If you are not sure of the answer, try making a diagram of the possibilities using a Punnett square, as was done in Figure 8–15.

From these experiments, Mendel formulated his second law, the **principle of independent assortment.** This law states that the two alleles of a gene assort, or segregate, independently of the alleles of other genes. Note that the principle of segregation deals strictly with alleles of one gene, while the principle of independent assortment considers relationships between genes.

Linkage

Knowing that genes are located on chromosomes, one can readily guess that if two genes are located close together on the same chromosome, they generally will not segregate independently. Such genes, which are usually inherited together, are called **linked genes.**

The linkage of genes was first discovered by the English geneticist William Bateson and his co-workers in 1905, while they were studying the genetics of the sweet pea *(Lathyrus odoratus).* These scientists crossed a doubly homozygous recessive strain of sweet peas that had red petals and round pollen grains with a second strain that had purple petals and long (elongated) pollen grains. All the F_1 progeny had purple petals and long pollen grains, showing these traits to be dominant. When the F_1 was self-pollinated, they obtained the following characteristics in the F_2 generation:

4831	purple	long
390	purple	round
393	red	long
1338	red	round

If the genes for flower color and pollen shape were on the same chromosome, there should have been only two types of progeny. If the genes were on two different chromosomes, there should have been four progeny types in a ratio of 3910:1304:1304:434, or 9:3:3:1. Clearly, the experimental results were not in a 9:3:3:1 ratio.

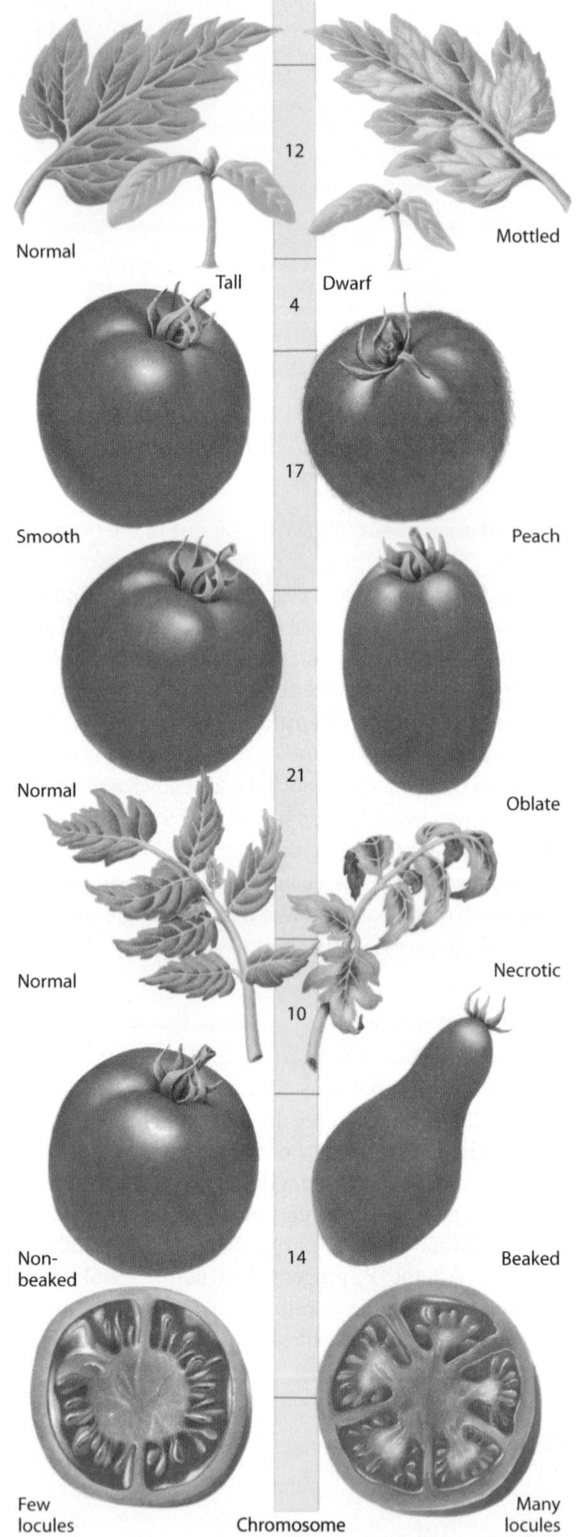

8–16 Linkage map A portion of a linkage map of chromosome 1 of a tomato *(Solanum lycopersicum)*, showing relative positions of genes on the chromosome. Each locus is flanked by drawings of the variant phenotype that first identified that genetic locus (shown on the right) and the normal phenotype (on the left). Distances between loci are shown in "map units," which are based on the frequency of crossing-over, or recombination, between genes. If two genes are more than 50 map units apart, the frequency of recombination is high enough that they appear to assort independently.

The explanation for the results obtained is that the two genes are "linked" on one chromosome but are sometimes exchanged between homologous chromosomes during crossing-over. It is now known that crossing-over—the breakage and rejoining of chromosomes that results in the appearance of chiasmata—occurs in prophase I of meiosis (Figure 8–7). The greater the distance between two genes on a chromosome, the greater the chance that crossing-over will occur between them. The closer together two genes are, the greater is their tendency to assort together in meiosis and, thus, the greater is their "linkage." Chromosome maps can be constructed based on the frequency of crossing-over between the genes and therefore on the extent to which alleles are exchanged. Such **genetic,** or **linkage, maps** provide an approximation of the positions of the genes on the chromosomes (Figure 8–16).

Mutations

The studies on independent assortment described above depend on the existence of differences between the alleles of a gene. How do such differences arise? The first answer to this question was provided by Hugo de Vries, a Dutch geneticist.

Mutations Are Changes in the Genetic Makeup of an Individual

In 1901, de Vries studied the inheritance of characteristics in an evening primrose *(Oenothera glazioviana)* that flourished along the coastal dunes of Holland. He found that, although the patterns of heredity in this plant were generally well ordered and predictable, a characteristic occasionally appeared that had not been observed previously in either parental line. De Vries hypothesized that this new characteristic was the phenotypic expression of a change in a gene. Moreover, according to his hypothesis, the new allele would then be passed along to subsequent generations just as the other alleles were. De Vries spoke of this hereditary change in one of the alleles of a gene as a "mutation" and of the organism carrying it as a "mutant."

Ironically, we now know that only 2 of roughly 2000 changes in the evening primrose that were observed by de Vries were due to a mutation, as defined by him. All of the rest were due to new genetic combinations or to the presence of extra chromosomes rather than actual abrupt changes in any particular gene.

Today, we call any change in the hereditary makeup of an organism a **mutation.** Such changes may occur at the level of the gene (de Vries's concept of a mutation, now called a point mutation) or at the level of the chromosome. In some **chromosome mutations** there are alterations in the structure of individual chromosomes; for example, a piece of chromosome may be deleted, duplicated, or inverted. In others, the number of chromosomes is altered, one or more individual chromosomes being added or deleted. In still others, whole sets of chromosomes are added.

A Point Mutation Occurs When One Nucleotide Is Substituted for Another

Point mutations involve only one or a few nucleotides of the DNA in a particular chromosome. They may occur spontaneously in nature or be induced by agents, called **mutagens,** that affect DNA. Mutagens, such as ionizing radiation, ultraviolet radiation, and various kinds of chemicals, generally cause point mutations. In human beings and other animals, this is often how cancer starts. Point mutations may also arise by the rare mispairings that can occur during the replication of DNA.

Deletions and Duplications Involve the Removal or Insertion of Nucleotides or Chromosome Segments

Chromosome mutations called **deletions** occur when segments of a chromosome are lost. Take, for example, the chromosome AB•CDEFG, in which • represents the centromere. A deletion of EF would generate the mutated chromosome AB•CDG. Large deletions can be readily detected because the chromosomes are noticeably shorter. A chromosome **duplication** is a mutation in which part of a chromosome has been doubled. For the chromosome AB•CDEFG, duplication of EF would give rise to the mutated chromosome AB•CDEFEFG.

Genes May Move from One Location to Another

Whereas genes usually occur in a fixed position on the chromosomes, they may on rare occasions move around. In bacteria, **plasmids**—small circular molecules of DNA separate from the main chromosome—may enter the chromosome at places where they share a common sequence of nucleotides. (The consequences of such a process for bacterial and viral evolution are discussed in Chapter 13, and the consequences for genetic engineering in Chapter 10.) In both bacteria and eukaryotes, genes may move as small segments of DNA from one location to another in the chromosomes. These movable genetic elements, known as **transposons** or "jumping genes," were first detected in maize by Barbara McClintock in the late 1940s (Figure 8–17). On the basis of these unexpected findings, McClintock was awarded a Nobel Prize in 1983. Whether occurring by plasmids or transposons, the changed position of the genes involved may disrupt the action of their new neighbors, or vice versa, and lead to effects—known as **position effects**—that we recognize as mutations.

Pieces of Chromosomes May Be Inverted or Moved to Another Chromosome

If two breaks occur in the same chromosome, the chromosomal segment between the breaks may rotate 180 degrees and reenter its chromosome with its sequence of genes oriented in the direction opposite to the original one. Such a change in chromosomal sequence results in a chromosome mutation called an **inversion.** For example, in chromosome AB•CDEFG, a 180-degree rotation of DEF would result in the mutated chromosome AB•CFEDG. Another sort of change that is common in certain groups of plants involves the exchange of parts between two nonhomologous chromosomes to produce a chromosome mutation called a **translocation.** Translocations are often reciprocal, meaning a segment from one chromosome is exchanged with a segment from another, nonhomologous one, so that the two translocation chromosomes are formed simultaneously.

Entire Chromosomes May Be Lost or Duplicated

Chromosome mutations may also be associated with changes in chromosome number, which occur spontaneously and rather commonly but are usually eliminated promptly. Whole chromosomes may, under certain circumstances, be added to or subtracted from an individual's or species' normal complement, a

(a)

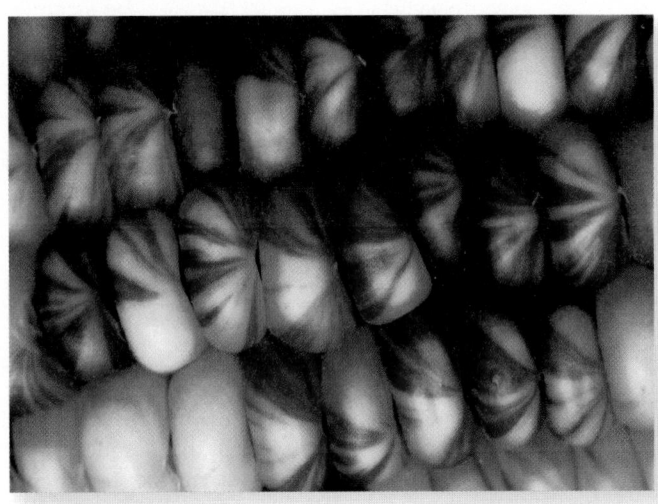

(b)

8–17 Barbara McClintock ***(a)*** Dr. McClintock is seen here holding an ear of corn, or maize, of the type she used in her landmark studies on transposons. Her work was largely unheralded until decades after her investigations were published. ***(b)*** The study of variegated (multicolored) kernels such as those shown here led to McClintock's discovery of transposons.

condition known as **aneuploidy. Polyploidy,** the duplication of whole sets of chromosomes, may also occur. (The evolutionary importance of polyploidy in plants is discussed in Chapter 11.) In any of these cases, associated alterations of the phenotype usually occur.

Mutations Provide the Raw Materials for Evolutionary Change

When a mutation occurs in a predominantly haploid organism, such as the fungus *Neurospora* or a bacterium, the phenotype associated with this mutation is immediately exposed to the environment. If favorable, the proportion of organisms with the mutation tends to increase in the population as a result of natural selection (see Chapter 11). If unfavorable, the mutant is quickly eliminated from the population. Some mutations may be more nearly neutral in their effects and then may persist by chance alone, but many have either negative or positive effects on the organism in which they occur. In a diploid organism, the situation is very different. Each chromosome and every gene is present in duplicate, and a mutation in one of the homologs, even if it would be unfavorable in a homozygous condition, may have much less effect or even be advantageous when present in a single dose. For this reason, such a mutation may persist in the population. The mutant gene could eventually alter its function, or the selective forces on the population could change in such a way that the effects of the mutant gene become advantageous.

Whether mutations are harmful or neutral, the capacity to mutate is extremely important, because it results in variation among the individuals in a species and may allow some individuals to adapt to changing conditions. Mutations thus provide the raw material for evolutionary change. Mutations in eukaryotes occur spontaneously at a rate of about 1 mutant gene at a given locus per 200,000 cell divisions. This, together with recombination, provides variation of the sort that is necessary for evolutionary change through natural selection.

Broadening the Concept of the Gene

Alleles Undergo Interactions That Affect the Phenotype

As the early studies in genetics proceeded, it soon became apparent that dominant and recessive characteristics are not always as clear-cut as in the seven traits studied by Mendel. Interactions can and do occur between the alleles of a gene that affect the phenotype.

Incomplete Dominance Produces Intermediate Phenotypes In cases of **incomplete dominance,** the phenotype of the heterozygote is intermediate between those of the parent homozygotes. For example, in snapdragons, a cross between a red-flowered plant and a white-flowered plant produces a plant that has pink flowers. As we shall see in Chapter 9, genes dictate the structure of proteins—in this case, proteins involved in pigment synthesis in the cells of the flower petals. For this heterozygote, the pigment produced in a flower petal cell by one allele is not completely masked by the action of the other allele. When the F_1 generation is allowed to self-pollinate, the characteristics segregate again, the result in the F_2 generation being a ratio of one red-flowered (homozygous) plant to two pink-flowered (heterozygous) plants to one white-flowered (homozygous) plant, 1:2:1 (Figure 8–18). Thus, the alleles themselves remain discrete and unaltered, conforming to Mendel's principle of segregation.

Some Genes Have Multiple Alleles Although any individual diploid organism can have only two alleles of any given gene, it is possible that more than two forms of a gene may be present in a population of organisms. When three or more alleles exist for a given gene, they are referred to as **multiple alleles.** The genes that control self-sterility (that is, genes that prevent self-pollination) in certain flowering plants can have large numbers of alleles. It is estimated, for example, that some populations of red clover have hundreds of alleles of the self-sterility gene. Fortunately for Mendel, the peas he chose to study are self-pollinators.

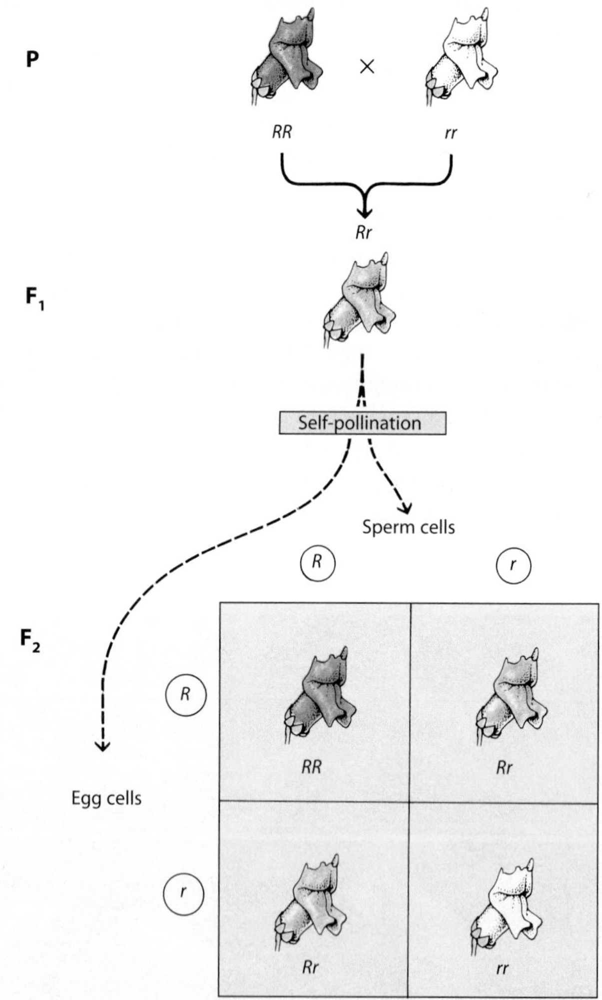

8–18 Incomplete dominance A cross between a red (*RR*) snapdragon and a white (*rr*) snapdragon. This looks very much like the cross between a purple- and a white-flowering pea plant, but there is a significant difference because, in this case, neither allele is dominant. The flower of the heterozygote is a blend of the two colors. The phenotypic ratio in this cross is also different: instead of the 3:1 ratio seen in the cross with a dominant trait, we see a 1:2:1 phenotypic ratio with incomplete dominance.

Gene Interactions Also Occur among Alleles of Different Genes

In addition to interactions between alleles of the same gene, interactions also occur among the alleles of different genes. Indeed, most of the characteristics (both structural and chemical) that constitute the phenotype of an organism are the result of the interaction of two or more distinct genes.

Epistasis Occurs When One Gene Interacts with Another In some cases, one gene may interfere with or mask the effect of another gene at a different locus. This type of interaction is called **epistasis** ("standing upon"). In foxglove *(Digitalis purpurea),* for example, two unlinked genes interact in the pathway that determines petal coloration. One gene affects the intensity of the red pigmentation in the corolla: allele *d* results in light red color, as occurs in the natural foxglove populations, whereas the mutant allele *D* produces dark red color. The second gene determines which cells are involved with pigment synthesis: allele *w* permits pigment synthesis throughout the corolla, as occurs in natural populations, but the mutant allele *W* restricts pigment synthesis to small spots on the throat of the flower (Figure 8–19). Self-pollination of the dihybrid *DdWw* results in the following F_2 ratio:

9 *D_W_*	white with spots
3 *ddW_*	white with spots
3 *D_ww*	dark red
1 *ddww*	light red

The dash represents a slot that may be filled with either another dominant allele or the recessive allele. Quite clearly, the dominant allele *W* is epistatic to *D* and *d.* It suppresses the synthesis of red pigment in all but the throat spots. When pigment synthesis is allowed, pigment can be produced in either high or low concentrations.

8–19 Epistasis in foxglove In foxglove *(Digitalis purpurea),* genes *D* and *d* produce dark red and light red pigments, respectively, and the epistatic gene *W* restricts pigment synthesis to spots in the throat of each flower.

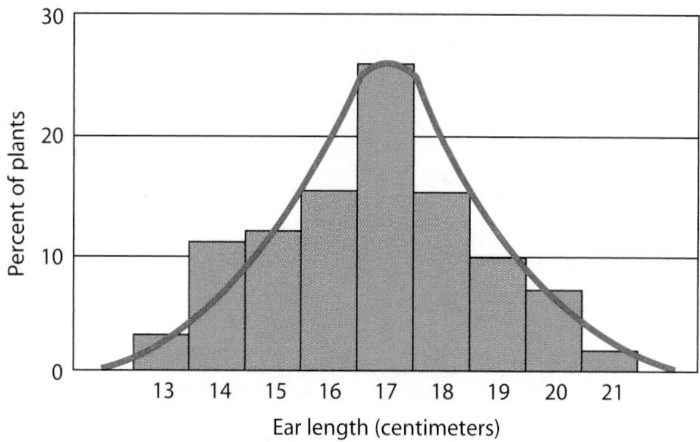

8–20 Continuous variation Distribution of ear length of the Black Mexican variety of maize *(Zea mays).* This is an example of a phenotypic characteristic that is determined by the interaction of a number of genes. Such a characteristic shows a gradation known as continuous variation. If the variation is plotted as a curve, the curve is bell-shaped, with the mean, or average, falling in the center of the curve.

Certain Traits Are Controlled by Several Genes Some traits, such as size, shape, weight, yield, and metabolic rate, are not the result of interactions between one, two, or even several genes. Instead, they are the cumulative result of the combined effects of many genes. This phenomenon is known as **polygenic inheritance.**

Many polygenic traits do not show a clear difference between groups of individuals—such as the differences tabulated by Mendel. Instead, they show a gradation known as **continuous variation.** If one makes a graph of differences among individuals for any trait affected by many genes, none of which is dominant over the other, the result is often a bell-shaped curve, with the mean, or average, usually falling in the center of the curve (Figure 8–20).

The first experiment that illustrated the way in which many genes can interact to produce a continuous pattern of variation in plants was carried out with wheat by the Swedish scientist H. Nilson-Ehle. Table 8–2 shows the phenotypic effects of various combinations of the alleles of two genes that all act semidominantly to control the intensity of color in wheat kernels.

A Single Gene Can Have Multiple Effects on the Phenotype

Although many genes have a single phenotypic effect, a single gene can often have multiple effects on the phenotype of an organism, affecting a number of apparently unrelated traits. This phenomenon is known as **pleiotropy** ("many turnings"). Mendel encountered pleiotropic genes during his breeding experiments with peas. When he crossed peas having purple flowers, brown seeds, and a dark spot on the axils of the leaves with a variety having white flowers, light seeds, and no spot on the axils, the same traits for flower, seeds, and leaves always stayed together as a unit. The inheritance of these traits can be accounted for by a single gene, which visibly influences several traits.

A case of particular interest to plant breeders is the relationship between yield in wheat and the presence or absence of awns (long, slender bristles) on the lemmas (bracts) associated

Table 8-2 The Genetic Control of Color in Wheat Kernels

Parents	$R_1R_1R_2R_2$ × $r_1r_1r_2r_2$ (Dark red) (White)		
F_1	$R_1r_1R_2r_2$ (Medium red)		
F_2	**Genotype**	**Phenotype**	
1	$R_1R_1R_2R_2$	Dark red	15 red to 1 white
2 } 4	$R_1R_1R_2r_2$	Medium-dark red	
2	$R_1r_1R_2R_2$	Medium-dark red	
4 } 6	$R_1r_1R_2r_2$	Medium red	
1	$R_1R_1r_2r_2$	Medium red	
1	$r_1r_1R_2R_2$	Medium red	
2 } 4	$R_1r_1r_2r_2$	Light red	
2	$r_1r_1R_2r_2$	Light red	
1	$r_1r_1r_2r_2$	White	

with wheat flowers. The same gene affects both. Wheat with awns has greater yields than wheat without awns, providing a method for judging potential yield without waiting for a mature crop.

The Inheritance of Some Characteristics Is under the Control of Genes Located in Plastids and Mitochondria

In Chapter 3, we noted that plastids and mitochondria contain their own DNA and encode some of their own proteins. Hence, not all characteristics of the cell are controlled exclusively by the DNA of the chromosomes located in the nucleus. The inheritance of characteristics under the control of genes located in the cytoplasm—strictly speaking, in plastids and mitochondria—is known as **cytoplasmic inheritance.**

In most organisms, including angiosperms (flowering plants), most cytoplasmically inherited characteristics are maternally inherited; that is, they are determined solely by the female parent. Any one of several phenomena may be responsible for **maternal inheritance.** In some angiosperms, for instance, the immediate precursor (the generative cell) of the sperm cells may receive no plastids when the microspore (immature pollen grain) divides to produce the larger tube cell and smaller generative cell (Figure 8–21a). In some orchids, the generative cell receives neither plastids nor mitochondria. In several species of angiosperms, the generative cells receive both plastids and mitochondria, but one or both types of organelles degenerate before the generative cell divides to produce the sperm cells. In yet other cases, plastids and mitochondria are excluded or removed from generative or sperm cells. Even though plastids and mitochondria may be present in sperm cells at the time of fertilization, it is still possible

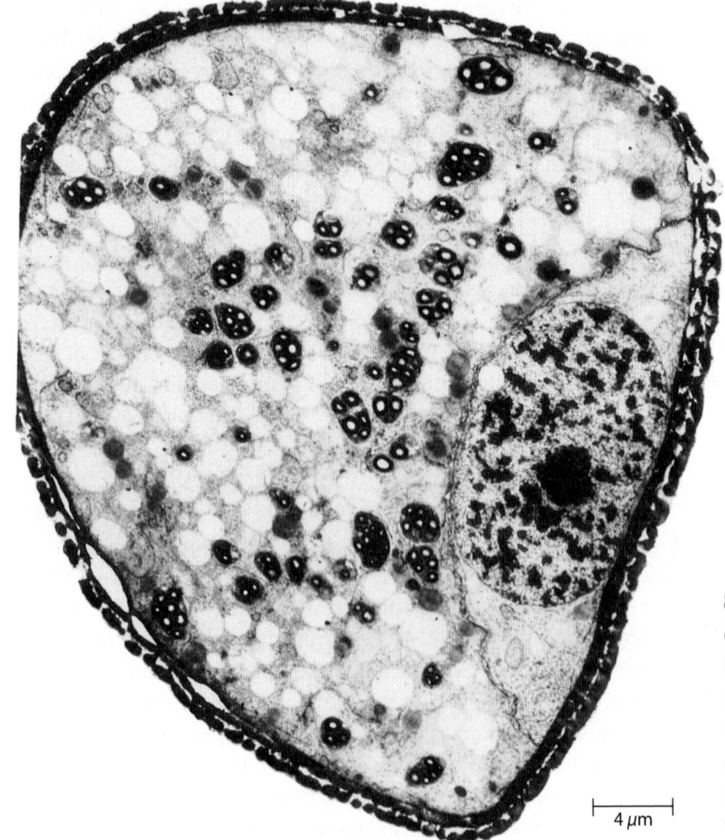

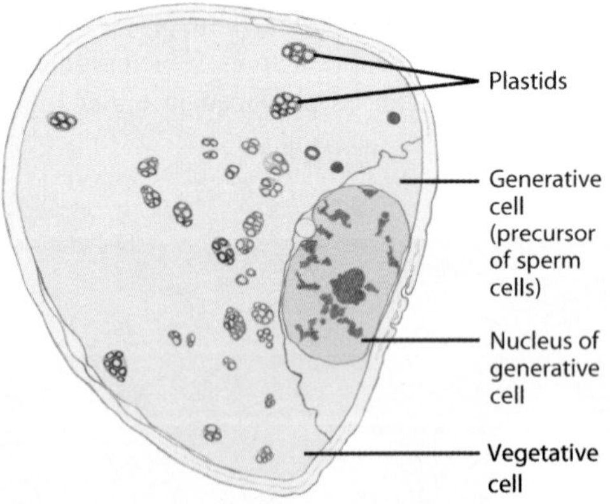

8–21 Maternal inheritance The sperm cells of most angiosperms do not contain plastids and mitochondria, so cytoplasmically inherited characteristics in the next generation are due to maternally inherited plastids and mitochondria. This electron micrograph of a pollen grain of wild garlic *(Tulbaghia violacea)* shows the newly formed generative cell, which will develop into sperm cells. Note that the generative cell lacks plastids. The larger vegetative cell, which will produce a pollen tube that carries the sperm cells to the egg apparatus, contains numerous plastids. Only the sperm cells or their nuclei will enter the egg cell.

8–22 Environmental effects on gene expression The common water-crowfoot *(Ranunculus peltatus)* grows with part of the plant body submerged in water. Leaves growing above the water are broad, flat, and lobed. The genetically identical underwater leaves at the right are thin and finely divided, appearing almost rootlike (arrows). These differences are thought to be related to differences in the turgor (page 80) of the immature leaf cells in the two environments. The degree of turgor affects the expansion of the cell walls and thus the ultimate size of the cells.

that they will not be transmitted to the egg. In such cases, only the sperm nucleus enters the egg cell, and the entire sperm cytoplasm is excluded from it.

Some conspicuous characteristics are due to cytoplasmic inheritance involving chloroplasts. Among these are the variegated or mottled leaves of certain plants that are prized as landscape plants and houseplants for their attractive foliage. In variegated coleus and hostas, for example, each of the lighter patches contains cells that developed from a cell containing only mutant (nongreen) plastids. Cytoplasmic inheritance involving mitochondria includes cytoplasmic male sterility, a maternally inherited trait that prevents the production of pollen but does not affect female fertility. The cytoplasmic male sterility phenotype has been widely used in the commercial production of F_1 hybrid seed (for example, in maize, onions, carrots, beets, and petunias), because it is unnecessary to remove the stamens to prevent self-pollination before making crosses.

The Phenotype Is the Result of the Interaction of the Genotype with the Environment

The expression of a gene is always the result of its interaction with the environment. To take a common example, a seedling may have the genetic capacity to be green, to flower, and to fruit, but it may never turn green if kept in the dark, and it may not flower and fruit unless certain precise environmental requirements are met.

The water buttercup is an especially striking example of environmental effects on gene expression. The common water-crowfoot grows with part of the plant body submerged in water and part floating on the surface of the water. Although the leaves are genetically identical, the broad, floating leaves differ markedly in both form and physiology from the finely divided leaves that develop under water (Figure 8–22).

Temperature often affects gene expression. Primrose plants that are red-flowered at room temperature are white-flowered above 30°C (86°F).

The expression of a gene may be altered not only by factors in the external environment but also by factors in the internal environment of the organism throughout plant development. These factors include temperature, pH, ion concentrations, hormones, and a multitude of other influences, including the action of other genes.

Asexual Reproduction: An Alternative Strategy

Asexual reproduction (also known as **vegetative reproduction**) results in progeny that are identical to their single parent. In this type of reproduction, the features of sexual reproduction discussed above, meiosis and fertilization, are absent. In eukaryotes, there is a wide variety of means of asexual reproduction, ranging from the development of an unfertilized egg cell to divisions of the parent organism into separate parts. In all such cases, however, the new organisms are the product of mitosis and are therefore genetically identical to the parent.

Vegetative reproduction is common in plants and is accomplished in many different ways (see "Vegetative Reproduction: Some Ways and Means" on page 171). Often, plants reproduce both sexually and asexually, thus hedging their evolutionary bets (Figure 8–23), but some species reproduce only asexually. Even among these, however, it is clear that their ancestors were capable of sexual reproduction and that vegetative reproduction represents a means to reproduce in the absence of mating. This choice, if rigidly adhered to, severely restricts the ability of the population to adapt to differing conditions. Lacking recombination and genetic variability, the

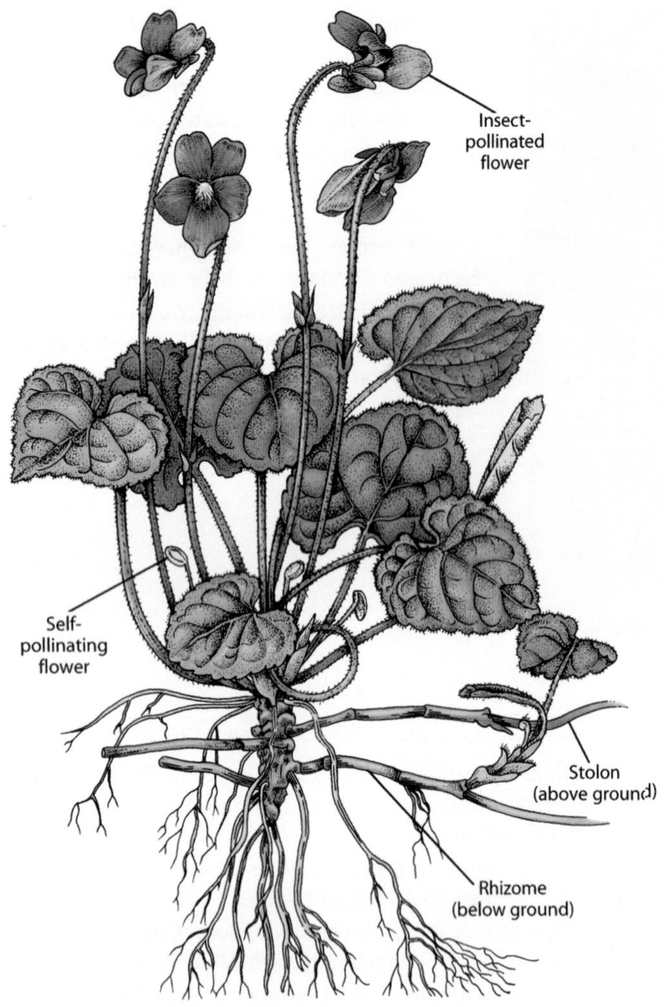

8–23 Sexual and asexual reproduction in violets The large and small flowers are involved in different forms of sexual reproduction. The large flowers are cross-pollinated by insects, and the seeds may be carried some distance from the parent plant by ants. The smaller flowers, closer to the ground, are self-pollinating and never open. Seeds from these flowers drop close to the parent plant and produce plants that are genetically similar to the parent. Presumably, such plants are better able (on the average) to grow successfully near the parent. Both of these sexual forms of reproduction involve genetic recombination. Asexual reproduction in violets occurs when horizontally creeping stems (known as stolons when they are above ground and rhizomes when they are below ground) produce new, genetically identical plants close to the parent.

population cannot adjust to changing conditions as readily as populations that can reproduce sexually.

Advantages and Disadvantages of Asexual and Sexual Reproduction

In general, asexual reproduction allows the exact replication of individuals that are particularly well suited to a certain environment or habitat. This suitability may include characteristics that are desirable or that facilitate survival under a particular set of environmental conditions. Lacking recombination and genetic variability, however, as noted above, an asexually reproducing population cannot adjust to changing conditions as readily as populations that can reproduce sexually. When a gene of an asexually reproducing individual is damaged or changed in a way that makes it deleterious (harmful), all of that individual's offspring inherit the gene. And, when a new strain of a disease-causing agent evolves, all of the asexually produced offspring are likely to be susceptible to that new strain.

As we have seen, because sexual reproduction involves a regular alternation between meiosis and fertilization, it produces a rich array of genetic diversity in natural populations and, to a certain extent, helps to maintain that diversity. Sexually reproducing organisms are less likely to produce offspring with the deleterious alleles present in the parent. And, inasmuch as the offspring of sexually reproducing organisms are genetically varied, it is likely that some of the offspring will have combinations of alleles that enable them to resist a new disease and to pass that resistance to their offspring.

In theory, sexual reproduction is unnecessary if an organism is particularly well adapted to its environment. What is needed in such a situation is the accurate reproduction of a particular "winning combination," generally speaking. In fact, however, natural populations have to keep adjusting to a constantly changing environment, and those that are able to invade new environments in competition with others will have an advantage.

Asexual reproduction is a simple and effective way to produce many new individuals. Sexual reproduction, by contrast, requires a great amount of energy and other resources. In angiosperms, sexual reproduction requires not only the production of gametes but also the development of flowers and various other devices that enhance the possibility that these gametes will be fertilized. The preponderance of sexual reproduction among eukaryotes today suggests that it has been favored over asexual reproduction, but the precise forces responsible for its being so widespread are still not completely understood.

SUMMARY

Sexual Reproduction Involves Meiosis and Fertilization

Sexual reproduction involves a special kind of nuclear division called meiosis. Meiosis is the process by which the chromosomes are reassorted and cells are produced that have the haploid number (n) of chromosomes. The other principal component of sexual reproduction is fertilization, the coming together of haploid cells to form a zygote. Fertilization restores the diploid number ($2n$) of chromosomes. There are characteristic differences among major groups of organisms as to when in the life cycle these events take place.

Eukaryotic Chromosomes Contain Histone Proteins

The eukaryotic chromosome differs in many ways from the chromosome of prokaryotes. Its DNA is always associated with proteins—primarily histones—that play a major role in

VEGETATIVE REPRODUCTION: SOME WAYS AND MEANS

The forms of vegetative reproduction in plants are many and varied. Some plants reproduce by means of runners, or stolons—long slender stems that grow along the surface of the soil. In the cultivated strawberry *(Fragaria × ananassa),* for example, leaves, flowers, and roots are produced at every other node on the runner. Just beyond the second node, the tip of the runner turns up and becomes thickened. This thickened portion first produces adventitious roots and then a new shoot, which continues the runner.

Underground stems, or rhizomes, are also important reproductive structures, particularly in grasses and sedges. Rhizomes invade areas near the parent plant, and each node can give rise to a new flowering shoot. The noxious character of many weeds results from this type of growth pattern, and many garden plants, such as irises, are propagated almost entirely from rhizomes. Corms, bulbs, and tubers are specialized for storage and reproduction. White potatoes are propagated artificially from tuber segments, each with one or more "eyes." It is the eyes of the "seed pieces" of potato that give rise to the new plant.

The roots of some plants—for example, cherry, apple, raspberry, and blackberry—produce "suckers," or sprouts, that give rise to new plants. Commercial varieties of banana do not produce seeds and are propagated by suckers that develop from buds on underground stems. When the root of a dandelion is broken, as it may be if one attempts to pull it from the ground, each root fragment may give rise to a new plant.

In a few species, even the leaves are reproductive. One example is the houseplant *Kalanchoë daigremontiana,* familiar to many people as the "maternity plant," or "mother of thousands." The common names of this plant are based on the fact that numerous plantlets arise from meristematic tissue located in notches along the margins of the leaves. The maternity plant is ordinarily propagated by means of these small plants, which, when they are mature enough, drop to the soil and take root. Another example of vegetative propagation is provided by the walking fern *(Asplenium rhizophyllum),* in which young plants form where the leaf tips touch the ground.

In certain plants, including some citrus trees, orchids, certain grasses (such as Kentucky bluegrass, *Poa pratensis*), and dandelions, the embryos in the seeds may be produced asexually from the parent plant. This is a kind of vegetative reproduction known as apomixis (see page 228). The seeds produced in this way give rise to individuals that are genetically identical to the parent, because meiosis is bypassed and fertilization is not required to produce an apomictic embryo—providing another instance of asexual reproduction.

(a)

(b)

(c)

Vegetative reproduction ***(a)*** The strawberry *(Fragaria × ananassa)* is propagated asexually by stolons. Strawberry plants also produce flowers and reproduce sexually. ***(b)*** *Kalanchoë daigremontiana,* showing small plants that have arisen in the notches along the margins of the leaves. ***(c)*** The walking fern *(Asplenium rhizophyllum),* showing how the leaves root at their tips and produce new plants (visible here on three of the leaves, one below and two at upper right). In this way, the fern is capable of forming large colonies of genetically identical plants.

SUMMARY TABLE Comparison of the Main Features of Mitosis and Meiosis*

MITOSIS (IN SOMATIC CELLS)	MEIOSIS (IN CELLS IN THE SEXUAL CYCLE)
One cell division, resulting in two daughter cells (Parental cell → Daughter cells)	Two cell divisions, resulting in four products of meiosis (Meiocyte → → Products of meiosis)
Chromosome number per nucleus maintained (e.g., for a diploid cell) (Parental cell 2n → Daughter cells 2n, 2n)	Chromosome number halved in the products of meiosis (2n → → n, n, n, n)
Normally, no pairing of homologs	Full synapsis of homologs at prophase I
Normally, no chiasmata	At least one chiasma per homologous pair
Centromeres divide at anaphase	Centromeres do not divide at anaphase I but do at anaphase II
Conservative process: daughter cells' genotypes identical with parental genotype	Promotes variation among the products of meiosis
Cell undergoing mitosis can be diploid or haploid	Cell undergoing meiosis is diploid

*After Anthony J. F. Griffiths, William M. Gelbart, Richard C. Lewontin, Jeffrey H. Miller, *Modern Genetic Analysis*, 2d ed. (New York: W. H. Freeman and Company, 2002), Figure 4-24.

chromosome structure. The DNA molecule wraps around histone cores to form nucleosomes, which are the basic packaging units of eukaryotic DNA.

Meiosis Involves Two Sequential Nuclear Divisions and Results in Four Nuclei (or Cells), Each of Which Has the Haploid Number of Chromosomes

In the first meiotic division (meiosis I), paired homologous chromosomes undergo crossing-over and eventually separate. Homologous chromosomes pair lengthwise to form bivalents. The chromosomes are double, each consisting of two chromatids. Chiasmata form between the chromatids of the homologs. These chiasmata are the visible evidence of crossing-over—the exchange of chromatid segments between homologous chromosomes. The bivalents line up at the equatorial plane in a random manner but with the centromeres of the paired chromosomes on either side of the plane. In this way, the chromosomes from the female parent and those from the male parent assort independently during anaphase I. This independent assortment, together with crossing-over, ensures that all the products of meiosis differ from the parental set of chromosomes and from each other. In this way, meiosis permits the expression of the variability that is stored in the diploid genotype. In the second meiotic division (meiosis II), the two chromatids of each chromosome separate as in mitosis.

Mendel's Experiments Provide the Foundation of Modern Genetics

Gregor Mendel's breeding experiments with pea plants revealed that hereditary characteristics are determined by discrete factors (now called genes). Different forms of the same gene are known as alleles. According to Mendel's principle of segregation, each individual possesses a pair of alleles, and the members of each pair of alleles segregate (that is, separate from one another) during gamete formation. When two gametes come together in fertilization, the offspring receives one allele from each parent. The members of a given pair of alleles may be the same (homozygous) or they may be different (heterozygous).

Alleles Can Be Dominant or Recessive

The genetic makeup of an organism is its genotype, and its observable characteristics constitute its phenotype. An allele that is expressed in the phenotype of a heterozygous individual to the exclusion of the other allele is a dominant allele. An allele whose effects are concealed in the phenotype of a heterozygous individual is a recessive allele. In crosses involving two individuals heterozygous for the same gene, the expected ratio of dominant to recessive phenotypes in the offspring is 3:1.

Unlinked Genes Assort Independently during Meiosis, but Genes on the Same Chromosome Are Linked

Mendel's second principle—independent assortment—applies to the behavior of two or more different genes. This principle states that, during gamete formation, the alleles of one gene segregate independently of the alleles of another gene. When organisms heterozygous for each of two independently assorting genes are crossed, the expected phenotypic ratio in the offspring is 9:3:3:1.

Studies by Bateson, however, showed that although some genes assort independently, others tend to remain together. Such genes are located relatively close together on a chromosome and are said to be linked. Linkage maps are based on the amount of crossing-over that occurs between genes and can give an approximation of the location of genes on chromosomes.

Mutations Are Changes in the Genetic Makeup of an Individual

Mutations are random changes in the genotype. Different mutations of a single gene increase the diversity of alleles of the gene in a population. As a consequence, mutation provides the variability among organisms that is the raw material for evolution. There are several different kinds of mutations: point mutations, deletions, duplications, inversions, translocations, and changes in chromosome number.

Alleles Can Be Incompletely Dominant and Can Exist in More Than Two Forms

Many traits are inherited according to the patterns revealed by Mendel. However, for other traits—perhaps the majority—the patterns are more complex. Although many alleles interact in a dominant-recessive manner, some show varying degrees of incomplete dominance, with intermediate phenotypes. In a population of organisms, multiple alleles of a single gene may exist, but only two alleles can be present in any given diploid individual.

Different Genes Can Also Interact with One Another

Novel phenotypes may result from gene interactions, or genes may affect one another in an epistatic manner, such that one hides the effect of the other. The phenotypic expression of many traits is influenced by multiple genes. This phenomenon is known as polygenic inheritance. Such traits often show continuous variation in a population, as represented by a bell-shaped curve. Conversely, a single gene can affect two or more superficially unrelated traits. This property of a gene is known as pleiotropy.

Some Genes Are Located in the Cytoplasm

Although most of the traits inherited by the cell are contributed by nuclear genes, some inherited traits are controlled by genes located in the cytoplasm—specifically, in plastids and mitochondria. This phenomenon is known as cytoplasmic inheritance.

The Phenotype Is the Result of the Genotype Interacting with the Environment

Gene expression is affected by factors in the external and internal environments. Variations in the expression of particular alleles may result from environmental influences, interactions with other genes, or both.

Sexual Reproduction Results in Diversity, Whereas Asexual Reproduction Does Not

Asexual reproduction results in progeny that are identical to their single parent, whereas sexual reproduction produces an infinite array of genetic diversity in natural populations. Lacking recombination and genetic variability, asexually produced plants are not able to adjust as readily to changing conditions as are plants of the same species produced sexually.

QUESTIONS

1. What are the two critical events in sexual reproduction in eukaryotes?
2. Distinguish between chiasma and crossing-over; between synaptonemal complex and synapsis.
3. What is the principal difference between meiotic anaphase I and meiotic anaphase II?
4. In what ways is meiosis different from mitosis?
5. Distinguish between the following terms: gene and allele; genotype and phenotype; epistasis and pleiotropy.
6. Why is a homozygous recessive always used in a testcross?
7. Explain how the movement of chromosomes during meiosis relates to Mendel's two principles, or laws.
8. A pea plant that breeds true for round, green seeds (*RRyy*) is crossed with a plant that breeds true for wrinkled, yellow seeds (*rrYY*). Each parent is homozygous for one dominant characteristic and for one recessive characteristic. (a) What is the genotype of the F_1 generation? (b) What is the phenotype? (c) The F_1 seeds are planted and their flowers are allowed to self-pollinate. Draw a Punnett square to determine the ratios of the phenotypes in the F_2 generation. How do the results compare with those of the experiment shown in Figure 8–15?
9. In Jimson weed, the allele for violet petals (*W*) is dominant over the allele for white petals (*w*), and the allele for prickly capsules (*S*) is dominant over the allele for smooth capsules (*s*). A plant with white petals and prickly capsules was crossed with one that had violet petals and smooth capsules. The F_1 generation was composed of 47 plants with white petals and prickly capsules, 45 plants with white petals and smooth capsules, 50 plants with violet petals and prickly capsules, and 46 plants with violet petals and smooth capsules. What were the genotypes of the parents?
10. Explain what is meant by cytoplasmic inheritance; maternal inheritance.

CHAPTER 9

The Chemistry of Heredity and Gene Expression

◀ **Fertilizing plants** Under the direction of nucleic acids, plants synthesize organic molecules, such as the proteins and oils present in pollen and the sugars contained in nectar. While searching for nectar, honey bees transport sperm-producing pollen grains from flower to flower. In the process, egg cells are ultimately fertilized and genetic information is recombined.

CHAPTER OUTLINE

When the existence of genes and the fact that they are carried in the chromosomes were no longer in doubt, scientists began to focus on the question of how chromosomes can carry such an enormous amount of very complex information.

Early chemical analyses revealed that the eukaryotic chromosome consists of both **deoxyribonucleic acid (DNA)** and protein, in about equal amounts. Once it became clear to researchers that chromosomes carried the genetic information, the problem became one of deciding whether the protein or the DNA plays this essential role. By the early 1950s, a great deal of evidence for the role of DNA as the genetic material had accumulated. It was not until the discovery of the structure of DNA, however, that scientists came to understand how DNA actually carries the genetic information.

The Structure of DNA

In the early 1950s, a young American scientist, James Watson, went to Cambridge, England, on a research fellowship to study problems of molecular structure. There, at the Cavendish Laboratory, he met physicist Francis Crick. Both were interested in DNA, and they soon began to work together to solve the problem of its molecular structure. They did not do experiments in the usual sense but, rather, undertook to examine all the data about DNA provided by others and attempt to solve the puzzle of the structure of DNA.

Each DNA Strand Consists of a Polymer of Four Nucleotides

By the time Watson and Crick began their studies of DNA, a lot of information on the subject had already accumulated. It was known that the DNA molecule is large, long and thin, and composed of four different kinds of molecules called nucleotides (page 29). Each nucleotide contains a phosphate group, the sugar deoxyribose, and one of four nitrogenous bases: **adenine, guanine, cytosine,** and **thymine.** Two of the bases, adenine and guanine, are similar in structure and are called

CHECKPOINTS

After reading this chapter, you should be able to answer the following:

1. How does DNA replication occur?
2. What is the nature of the genetic code, and in what sense is it universal?
3. Describe the main steps in the transcription of RNA from DNA.
4. Where does translation occur in a eukaryotic cell, and what are the major steps in the process?
5. What are some of the factors regulating gene expression in eukaryotes?

purines. The other two bases, cytosine and thymine, are also similar in structure and are called **pyrimidines.**

In 1950, Linus Pauling had shown that regions of proteins often take the form of a helix (page 28) and that the helical structure is maintained by hydrogen bonding between amino acids in successive turns of the helix. Pauling had even suggested that the structure of DNA might be similar. Subsequent X-ray studies by Rosalind Franklin and Maurice Wilkins at Kings College, London, provided strong evidence that the DNA molecule consisted entirely of a helix. Another critical clue came from data obtained by Erwin Chargaff indicating that the ratio of DNA nucleotides containing thymine to those containing adenine is approximately 1:1, and that the ratio of nucleotides containing guanine to those containing cytosine is also approximately 1:1.

DNA Exists in the Form of a Double Helix

By piecing together the various data, Watson and Crick were able to deduce that DNA is not a single-stranded helix, which is found in many proteins, but is instead an entwined double helix. If you were to take a ladder and twist it into the shape of a helix, keeping the rungs perpendicular to the sides, a crude model of a double helix would be formed (Figure 9–1). The two sides of the ladder are made up of alternating sugar molecules and phosphate groups. The rungs of the ladder are formed by pairs of adenine (A) with thymine (T) and guanine (G) with cytosine (C)—one base for each sugar-phosphate, with two bases forming each rung. The paired bases meet in the interior of the helix and are joined by hydrogen bonds, the relatively weak bonds that, as Pauling had demonstrated, contribute to protein structure.

As Watson and Crick worked their way through the data, they assembled tin-and-wire models of possible structures of the double helix, testing where each piece would fit into the three-dimensional puzzle (Figure 9–2a). They made the exciting discovery that to produce a model consistent with all of the data, an adenine could "pair" only with a thymine and a guanine could "pair" only with a cytosine in the double helix, because only these two combinations of bases form the correct hydrogen bonds. Adenine forms two hydrogen bonds with thymine, and guanine forms three hydrogen bonds with cytosine. Thus, the "rungs" of our "ladder" consist of A-T and G-C pairs only.

The double-stranded structure of a small portion of a DNA molecule is shown in Figure 9–3. In each strand, the phosphate group that joins two deoxyribose molecules is attached to one sugar at the 5′ position (carbon 5 of deoxyribose) and to the other sugar at the 3′ position (the third carbon in the ring of deoxyribose). Thus each strand has a 5′ end and a 3′ end. Moreover, the two strands run in opposite directions—that is, the direction from the 5′ end to the 3′ end of one strand is opposite to that of the other strand (that is, the strands are **antiparallel**).

The Watson-Crick model explains in a simple and logical way the ratios of bases that Chargaff had observed—that is, that the amount of A equals that of T, and the amount of C equals that of G. Perhaps the most important property of the model is that the two strands are complementary; that is, each strand contains a sequence of bases that will pair in a specific way with the bases in the other strand, and thus the sequence of bases in one strand can be used to determine the sequence of bases in the other, **complementary strand.**

In what could be considered one of the great understatements of all time, Watson and Crick wrote in their original brief publication: "It has not escaped our notice that the specific pairing we have postulated immediately suggests a

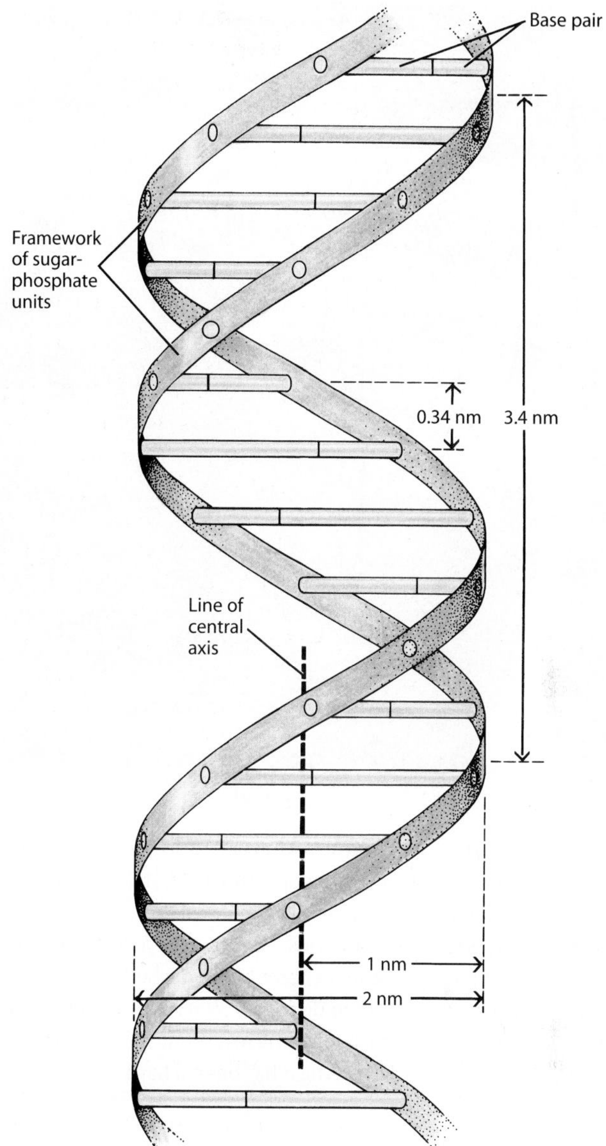

9–1 DNA helix The double-stranded helical structure of DNA, as first presented in 1953 by Watson and Crick. The framework of the helix is composed of the sugar-phosphate units of the nucleotides. The rungs are formed by the four nitrogenous bases adenine and guanine (the purines) and thymine and cytosine (the pyrimidines). Each rung consists of a pair of bases. Knowledge of the distances in nanometers (nm) shown here was crucial in establishing the detailed structure of the DNA molecule. These distances were determined from X-ray diffraction photographs of DNA taken by Rosalind Franklin.

(a)

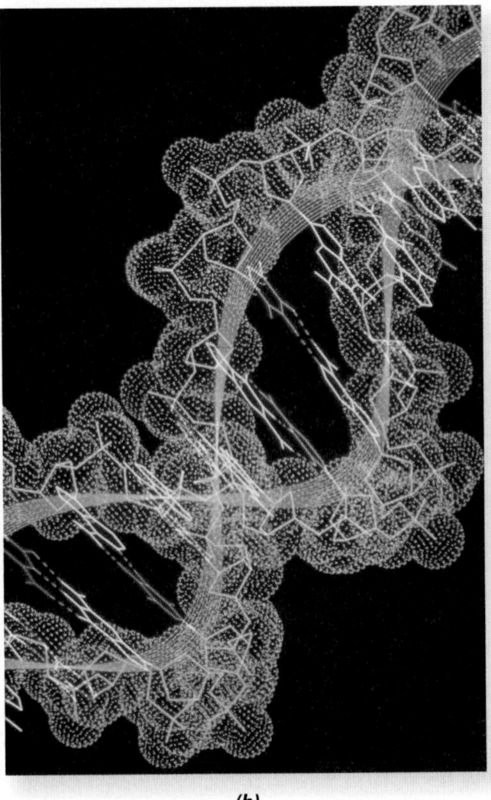

(b)

9–2 Models of DNA ***(a)*** James Watson (left) and Francis Crick in 1953, with one of their models of DNA. At the time they announced their deduction of the DNA structure, Watson was 23 and Crick was 34. ***(b)*** A computer-generated model of a portion of a DNA molecule. The sugar-phosphate backbone is indicated by the blue ribbons and green dots. The purines are shown in yellow, and the pyrimidines in red. The hydrogen bonds linking the base pairs are represented by blue dashed lines.

In 1993, 40 years after their discovery, Watson remarked: "The molecule is so beautiful. Its glory was reflected on Francis and me. I guess the rest of my life has been spent trying to prove that I was almost equal to being associated with DNA, which has been a hard task." Crick replied: "We were upstaged by a molecule."

possible copying mechanism for the genetic material." In 1962, nine years after their model for the structure of DNA was published, Watson, Crick, and Wilkins shared a Nobel Prize in recognition of their landmark studies. (Unfortunately, Rosalind Franklin had died and could not share the Nobel Prize posthumously.)

DNA Replication

An essential property of the genetic material is the ability to provide for exact copies of itself. As Watson and Crick initially observed, a mechanism by which DNA can replicate itself is implicit in the double-stranded and complementary structure of the DNA helix.

At the time of DNA replication, the molecule is "unzipped locally," with the paired bases separating as the hydrogen bonds are broken. As the two strands separate, they act as **templates,** or guides, for the synthesis of two new strands. Each strand provides the "blueprint" for the synthesis of the new complementary strand along its length (Figure 9–4), using free nucleotides in the cell. If a T is present on the original strand (the template), only an A can fit in the adjacent location of the new strand; a G will fit only with a C, and so on. In this way, each strand forms a copy of its original partner strand, and two exact replicas of the molecule are produced. The age-old question of how hereditary information is duplicated and passed on, generation after generation, has been answered.

Replication of DNA is a process that occurs only once in each cell generation, during the S phase of the cell cycle (page 64). It is the first event in the duplication of chromosomes. In most eukaryotic cells, DNA replication leads ultimately to mitosis, but in cells that give rise to meiospores or to gametes it leads to meiosis.

The principle of DNA replication, in which each strand of the double helix serves as a template for the formation of a new strand, is relatively simple and easy to understand. However, the process by which the cell accomplishes replication is actually very complex. Like other biochemical reactions of the cell, DNA replication requires a number of different enzymes, each catalyzing a particular step of the process.

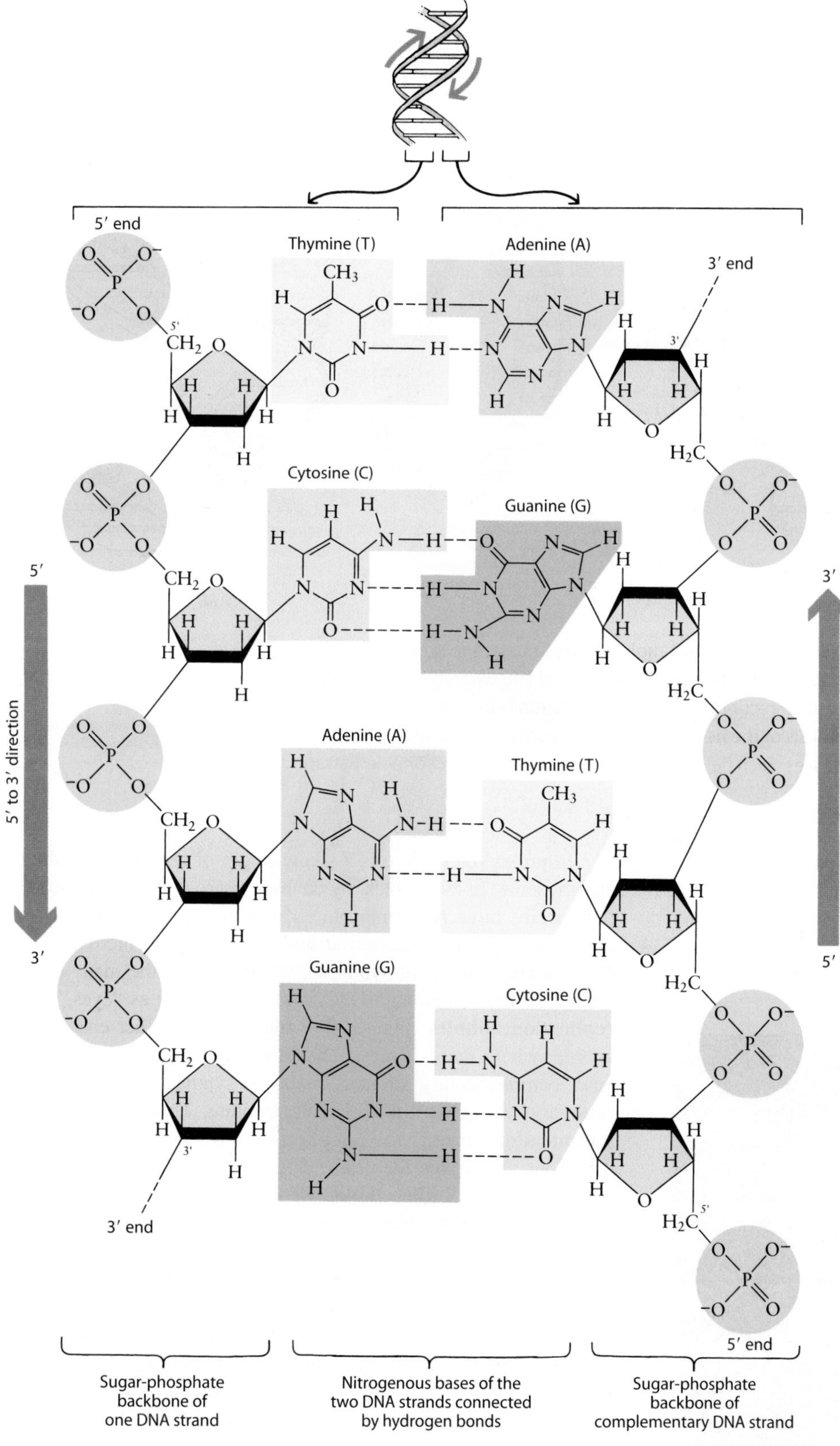

9–3 Molecular structure of DNA The double-stranded structure of a small portion of a DNA molecule. Each nucleotide consists of a phosphate group, a deoxyribose sugar, and a purine or pyrimidine base.

Note the repetitive sugar-phosphate-sugar-phosphate sequence that forms the backbone of each strand of the molecule. Each phosphate group is attached to the 5′ carbon of one sugar subunit and to the 3′ carbon of the sugar subunit in the adjacent nucleotide. Each strand of the DNA molecule thus has a 5′ end and a 3′ end, determined by these 5′ and 3′ carbons. The strands are antiparallel—that is, the direction from the 5′ to the 3′ end of one strand is opposite to that of the other strand.

The strands are held together by hydrogen bonds (represented here by dashed lines) between the bases. Notice that adenine and thymine form two hydrogen bonds, whereas guanine and cytosine form three. Because of these bonding requirements, adenine can pair only with thymine, and guanine can pair only with cytosine. Thus the order of bases along one strand determines the order of bases along the other strand.

The sequence of bases varies from one DNA molecule to another. It is customarily written as the sequence in the 5′ to 3′ direction of one of the strands. Here, using the strand on the left, the sequence is TCAG.

Initiation of DNA replication always begins at specific nucleotide sequences known as the **origins of replication.** Initiation requires special initiator proteins and enzymes known as **helicases** (see Figure 9–6), which break the hydrogen bonds linking the complementary bases at the origin of replication, opening up the helix so replication can occur. Single-strand binding proteins keep the two strands separated and allow the enzymes necessary for synthesis to bind to the strands.

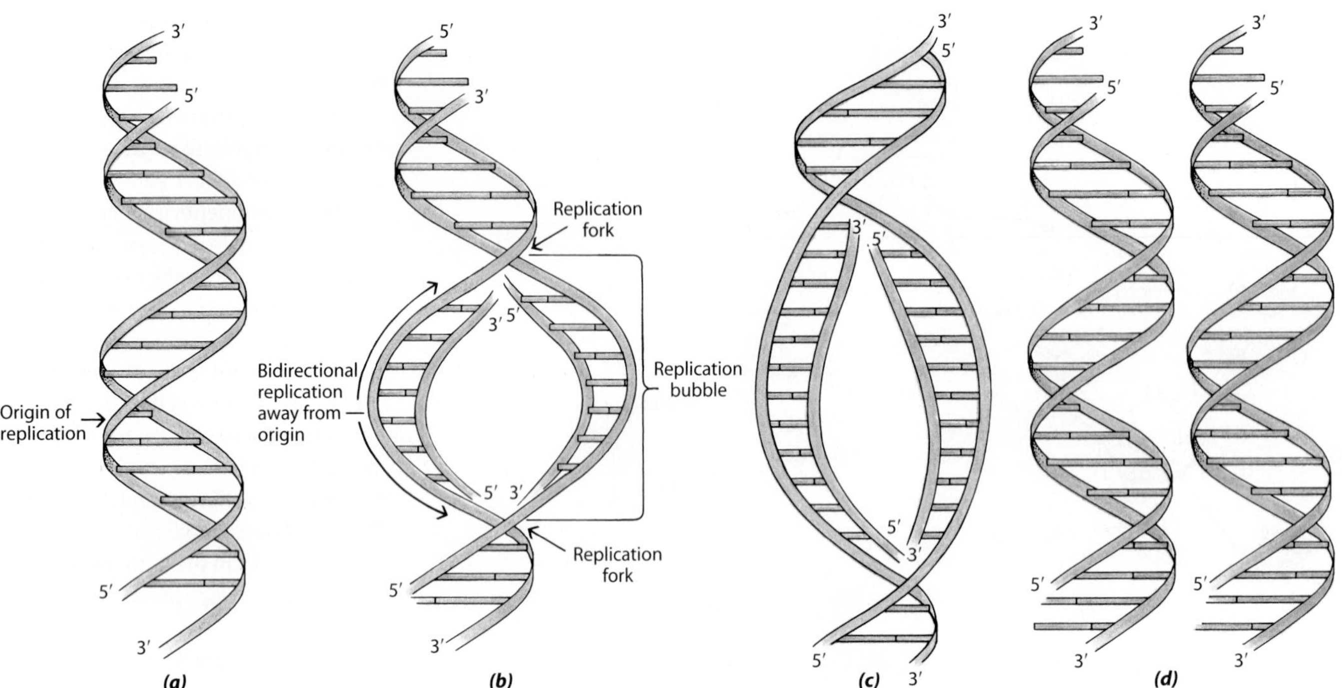

9–4 Overview of DNA replication ***(a)*** The two strands of the DNA molecule separate at the origin of replication as a result of the action of special initiator proteins and enzymes. ***(b), (c)*** The two replication forks move away from the origin of replication in opposite directions, forming a replication bubble that expands in both directions (bidirectionally). ***(d)*** When synthesis of the new DNA strands is complete, the two double-stranded chains separate into two new double helices. Each helix consists of one old strand and one new strand.

The synthesis of new strands is catalyzed by enzymes known as **DNA polymerases.** DNA replication requires a short RNA primer to initiate DNA synthesis; these primers are later removed and replaced with DNA nucleotides.

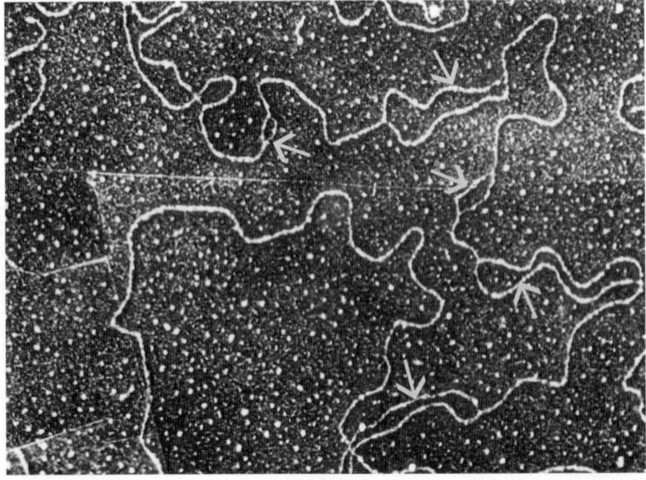

9–5 Replication bubbles Replication of eukaryotic chromosomes is initiated at multiple origins. The individual replication bubbles spread until, ultimately, they meet and join. In this electron micrograph of a replicating chromosome in an embryonic cell of the fruit fly *(Drosophila)*, the replication bubbles are indicated by arrows.

In prokaryotes, there is a single origin of replication in the chromosome. In eukaryotes, by contrast, there are many replication origins in each chromosome. If eukaryotic DNA caught in the act of replication is viewed under the electron microscope, the localized regions of synthesis, which form at the many origins of replication along a DNA molecule, appear as "eyes," or **replication bubbles** (Figures 9–4 and 9–5). At either end of a bubble, where the existing strands are being separated and the new, complementary strands are being synthesized, the molecule appears to form a Y-shaped structure. This is known as a **replication fork.** The two replication forks move in opposite directions away from the origin (Figure 9–4), and thus replication is said to be **bidirectional.** Replication proceeds along the linear chromosome as each bubble expands bidirectionally until it meets an adjacent bubble.

DNA polymerase synthesizes new DNA strands only in the 5′ to 3′ direction. Along one of the original strands, therefore, DNA can be synthesized continuously in the 5′ to 3′ direction as a single unit. This newly synthesized DNA is known as the **leading strand.** Because of DNA's antiparallel structure and the directionality of DNA polymerase, the other DNA strand is synthesized, again in the 5′ to 3′ direction, as a series of fragments. These fragments of what is known as the **lagging strand** are each individually synthesized in a direction *opposite* to the overall direction of replication. As the fragments, known as **Okazaki fragments,** lengthen, they are eventually joined together by an enzyme called **DNA ligase.** The complex process of DNA replication is summarized in Figure 9–6.

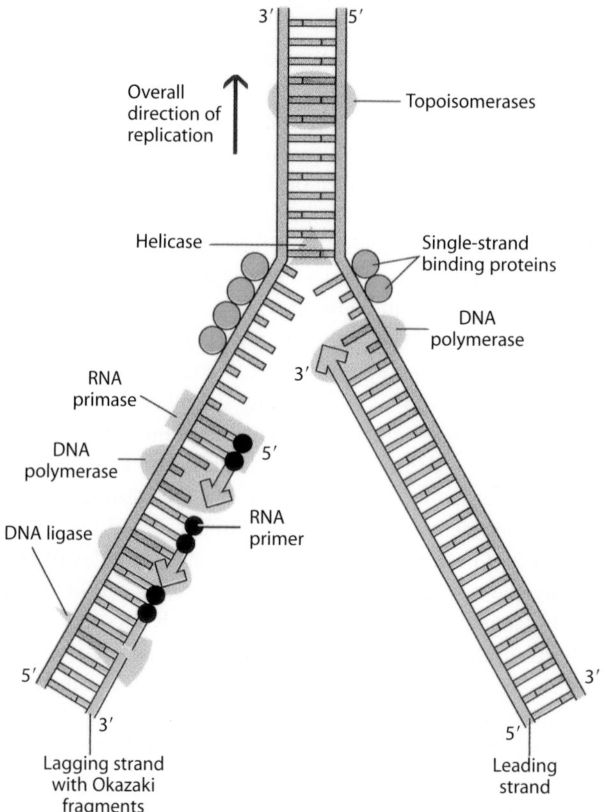

9–6 Summary of DNA replication The two strands of the DNA double helix separate, and new complementary strands are synthesized by DNA polymerase in the 5′ to 3′ direction, using the original strands as templates.

Each new strand begins with a short RNA primer formed by the enzyme RNA primase. Synthesis of the leading strand, which requires a single RNA primer (not shown in the diagram), is continuous. However, synthesis of the lagging strand, which occurs in a direction opposite to the overall direction of replication, is discontinuous. The short DNA segments of the lagging strand, each requiring individual RNA primers, are called Okazaki fragments. Synthesis of each Okazaki fragment ends when the DNA polymerase runs into the RNA primer attached to the 5′ end of the previous fragment. Following replacement of the RNA primer of the previous Okazaki fragment by DNA nucleotides, the fragment is joined to the growing strand by the enzyme DNA ligase.

Enzymes called topoisomerases prevent the DNA molecule from twisting upon itself as the two strands are separated and tension develops ahead of the replication fork.

From DNA to Protein: The Role of RNA

Watson and Crick disclosed the chemical nature of the gene and suggested how it duplicated itself; however, many questions remained unanswered, such as how the nucleotide sequence in DNA specifies the sequence of amino acids in a protein.

The search for the answer to this question led to **ribonucleic acid (RNA),** the sister molecule of DNA (discussed in Chapter 2). RNA's involvement was long suspected, because cells that are synthesizing large amounts of protein invariably contain large amounts of RNA. Moreover, unlike DNA, which is found mostly in the nucleus, RNA is found mostly in the cytoplasm, where protein synthesis takes place.

Like DNA, RNA is a long-chain macromolecule of nucleic acid, but it differs from DNA in two important properties (Figure 9–7).

1. In the nucleotides of RNA, the sugar component is ribose rather than deoxyribose.
2. The nitrogenous base thymine that is found in DNA does not occur in RNA. Instead, RNA contains a closely related pyrimidine, **uracil (U).** Uracil, like thymine, pairs only with adenine.

Three kinds of RNA play roles as intermediaries in the steps that lead from DNA to proteins: messenger RNA, transfer

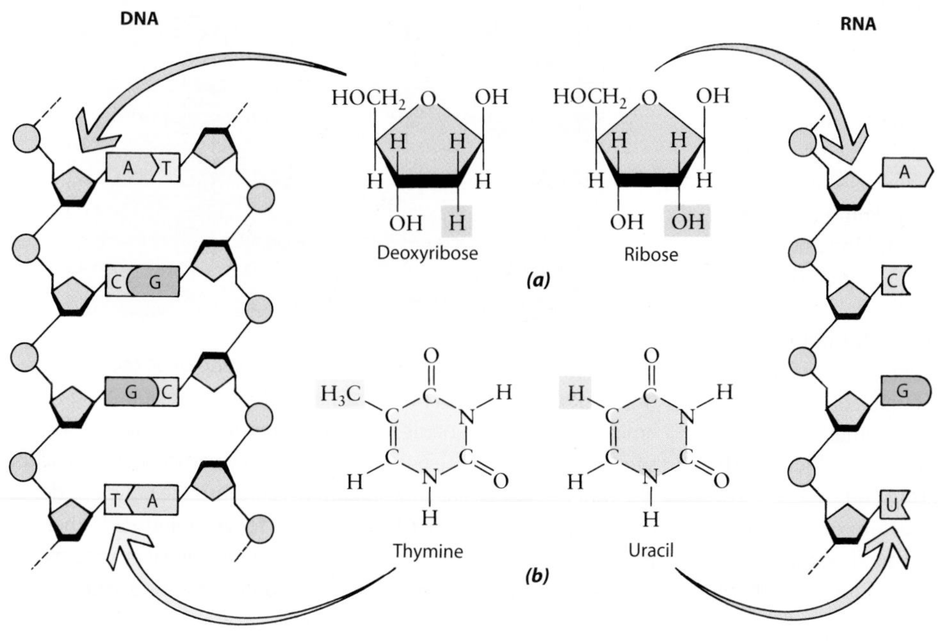

9–7 RNA structure Chemically, RNA is very similar to DNA, but there are two differences in its nucleotides. ***(a)*** One difference is in the sugar component. Instead of deoxyribose, RNA contains ribose, which has an additional oxygen atom. ***(b)*** The other difference is that instead of thymine, RNA contains the closely related pyrimidine uracil (U). Uracil, like thymine, pairs only with adenine. A third difference concerns the structure of RNA. Unlike DNA, RNA is usually single-stranded and does not form a regular helix.

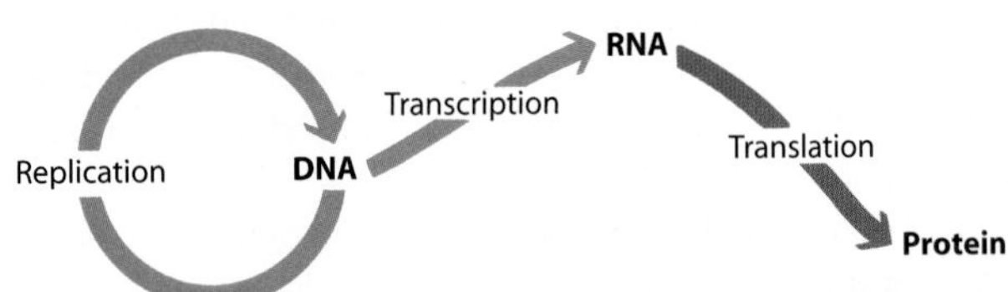

9–8 Information transfer in the cell The three processes of information transfer: replication, transcription, and translation. Replication of the DNA occurs only once in each cell cycle (pages 64 and 65), during the S phase prior to mitosis or meiosis. Transcription and translation, however, occur repeatedly throughout the interphase portions of the cell cycle.

RNA, and ribosomal RNA. DNA serves as a template for the synthesis of **messenger RNA (mRNA)** molecules, by a process called transcription. Transcription follows rules similar to those of replication and is catalyzed by the enzyme **RNA polymerase.** The role of mRNA is to carry the genetic message from the DNA to the ribosomes, which are composed of **ribosomal RNA (rRNA)** and proteins.

The ribosomes, which are located in the cytosol, are the actual sites of protein synthesis. The role of the **transfer RNA (tRNA)** molecules, each of which is specific for a particular amino acid, is to match the coded nucleotide sequence of mRNA with the proper amino acids to enable the ribosomes to join the amino acids together into a growing polypeptide. This ribosome-mediated synthesis of a polypeptide is called translation. It is the nucleotide sequence in the mRNA that determines the amino acid sequence in the protein.

Thus, RNA is synthesized from DNA, and RNA is used to synthesize protein. The three processes of information transfer are: **replication** (the synthesis of an entire DNA molecule), **transcription** (the synthesis of mRNA, which is a copy of a portion of one strand of the double-stranded DNA helix), and **translation** (the synthesis of a polypeptide directed by the nucleotide sequence of the mRNA) (Figure 9–8).

The Genetic Code

The identification of mRNA as the working copy of the genetic information that dictates the sequence of amino acids in proteins still left unresolved the big question: how is this accomplished? Proteins contain 20 different kinds of amino acids, but DNA and RNA each contain only four different kinds of nucleotides. Somehow, these nucleotides constitute a **genetic code** for the amino acids.

As it turned out, the idea of a code was useful not only as a dramatic metaphor but also as a working analogy. Scientists, seeking to understand how the sequence of nucleotides stored in the double helix of DNA could specify the quite dissimilar structures of protein molecules, approached the problem with methods used by cryptographers in deciphering codes.

As we saw in Chapter 2, the primary structure of a particular kind of protein molecule consists of a specific linear arrangement of the 20 different kinds of amino acids. Similarly, there are four different kinds of nucleotides, arranged in a specific linear sequence in a DNA molecule. If each nucleotide "coded" for one amino acid, only four amino acids could be specified by the four bases. If two nucleotides specified one amino acid, there could be a maximum number of 16, using all possible arrangements of the nucleotides ($4 \times 4 = 16$)—still not quite enough to code for all 20 amino acids. Therefore, following the code analogy, at least three nucleotides in sequence must specify each amino acid. This would provide for $4 \times 4 \times 4$, or 64, possible combinations, or **codons**—clearly more than enough.

The three-nucleotide, or triplet, codon was widely and immediately adopted as a working hypothesis. Its existence, however, was not actually demonstrated until the code was finally broken, a decade after Watson and Crick first presented their DNA structure. The scientists who performed the initial, crucial experiments were Marshall Nirenberg and his colleague Heinrich Matthaei, both of the U.S. National Institutes of Health. Utilizing the extracted contents of *Escherichia coli* cells, radioactively labeled amino acids, and synthetic RNAs that they had synthesized, they defined the first code word—UUU for phenylalanine—and provided a method for defining the others. As a result of these and similar experiments, subsequently performed in a number of laboratories, the mRNA codons for all of the amino acids were soon worked out. Of the 64 possible triplet combinations, 61 specify particular amino acids and 3 are stop signals. With 61 combinations coding only 20 amino acids, you can see that there must be more than one codon for many of the amino acids; hence, the genetic code is said to be *redundant.* As can be seen in Figure 9–9, codons specifying the same amino acid often differ only in the third nucleotide.

Second letter

First letter (5′ end)	U	C	A	G	Third letter (3′ end)
U	UUU, UUC Phe UUA, UUG Leu	UCU, UCC, UCA, UCG Ser	UAU, UAC Tyr UAA Stop UAG Stop	UGU, UGC Cys UGA Stop UGG Trp	U C A G
C	CUU, CUC, CUA, CUG Leu	CCU, CCC, CCA, CCG Pro	CAU, CAC His CAA, CAG Gln	CGU, CGC, CGA, CGG Arg	U C A G
A	AUU, AUC, AUA Ile AUG Met	ACU, ACC, ACA, ACG Thr	AAU, AAC Asn AAA, AAG Lys	AGU, AGC Ser AGA, AGG Arg	U C A G
G	GUU, GUC, GUA, GUG Val	GCU, GCC, GCA, GCG Ala	GAU, GAC Asp GAA, GAG Glu	GGU, GGC, GGA, GGG Gly	U C A G

9–9 Triplet codes The genetic code, consisting of 64 codons (triplet combinations of mRNA bases) and their corresponding amino acids. (For the names and structures of the 20 amino acids, see Figure 2–13.) Of the 64 codons, 61 specify particular amino acids. The other 3 codons are stop signals, which cause the polypeptide chain to terminate. Since 61 triplets code for 20 amino acids, there obviously must be "synonyms"; for example, leucine (Leu) has six codons. Each codon, however, specifies only one amino acid.

The Genetic Code Is Universal

One of the most remarkable discoveries of molecular biology is that the genetic code is nearly identical in all organisms, with only minor exceptions. Under proper circumstances, genes from bacteria, for example, can function perfectly in plant cells. Also, plant genes can be introduced into bacteria, where plant proteins can then be produced by the protein synthesis "machinery" of the bacterial cell. Not only does this observation provide a striking demonstration that all life on Earth is derived from a common ancestor, but it also provides the basis for the techniques of genetic engineering, which hold much promise for future development (see Chapter 10).

Protein Synthesis

The information encoded in the DNA and transcribed into mRNA is subsequently translated into a specific amino acid sequence of a polypeptide chain. The principles of protein synthesis are basically similar in prokaryotic and eukaryotic cells, although there are some differences in detail. First, we focus on the process as it takes place in prokaryotic cells, using *E. coli* as our model.

Messenger RNA Is Synthesized from a DNA Template

As we have seen, instructions for protein synthesis are encoded in nucleotide sequences in the DNA and are copied, or transcribed, into molecules of mRNA following the same base-pairing rules that govern DNA replication. Each new mRNA molecule is transcribed from one of the two strands of the DNA helix (Figure 9–10). Transcription is catalyzed by RNA polymerase. The single-stranded mRNA molecules produced in this process can be over 10,000 nucleotides long. Specific nucleotide sequences of the DNA, called **promoters,** are the binding sites for the RNA polymerase and thus determine where RNA synthesis starts and which DNA strand is used as a template. Once an RNA polymerase molecule attaches to a promoter and the DNA double helix is opened, the process of transcription, or RNA synthesis, can begin. Synthesis ends after the RNA polymerase transcribes a special sequence in the DNA called the **terminator.**

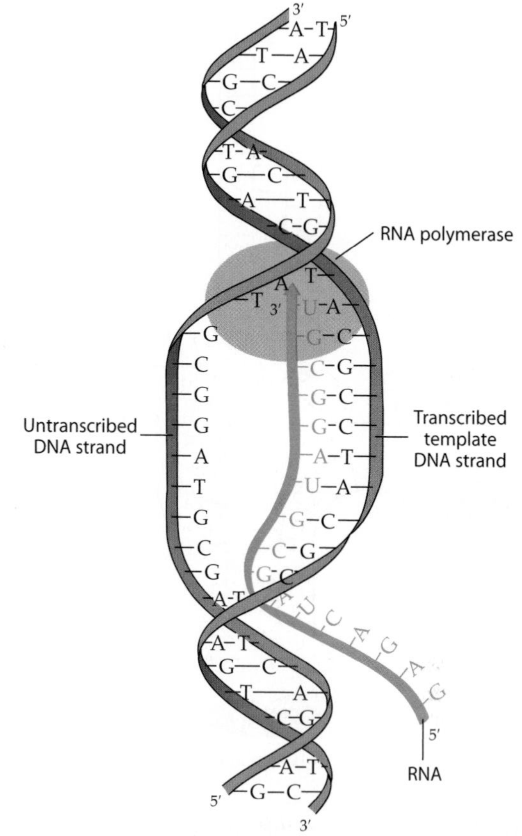

9–10 RNA transcription At the point of attachment of the enzyme RNA polymerase, the DNA opens up, and as the RNA polymerase moves along the DNA molecule, the two strands separate. Nucleotide building blocks are assembled in RNA in a 5′ to 3′ direction. Note that the RNA strand is complementary—not identical—to the 3′ to 5′ template strand from which it is transcribed. Its sequence is, however, identical to that of the untranscribed (5′ to 3′) DNA strand, except for the replacement of thymine (T) by uracil (U).

As we have mentioned, protein synthesis requires, in addition to mRNA molecules, the two other types of RNA: transfer RNA and ribosomal RNA. These molecules, which are transcribed from their own specific genes in the DNA of the cell, differ both structurally and functionally from mRNA.

Each Transfer RNA Carries an Amino Acid

Transfer RNA (tRNA) molecules are sometimes called "the dictionary of the language of life," because of the role they play in translating the nucleotide sequence of mRNA into the amino acid sequence of protein. Each tRNA molecule is relatively small, consisting of about 80 nucleotides with regions that "fold back" and base-pair into short stretches of double helix (Figure 9–11). There are over 60 different tRNA molecules in every cell, at least one for each of the 20 amino acids found in proteins.

Each tRNA molecule has two important attachment sites. One of these sites, the **anticodon,** consists of a sequence of three nucleotides that binds to the codon on an mRNA molecule. The other site, at the 3′ end of the tRNA molecule, attaches to a particular amino acid. The tRNA with its attached amino acid is called an **aminoacyl-tRNA.** Attachment of the tRNA molecules to their proper amino acids is brought about by enzymes known as **aminoacyl-tRNA synthetases.** There are at least 20 different aminoacyl-tRNA synthetases, one or more for each amino acid.

Ribosomal RNA Is Associated with Protein to Form a Ribosome

As the name implies, ribosomal RNA (rRNA) is an integral part of a ribosome, which is a large complex of RNA and protein molecules. Functionally, ribosomes are protein-synthesizing machines to which tRNA molecules bind in precise relationship to the mRNA molecules so as to read accurately the genetic message encoded in the mRNA. Ribosomes consist of two subunits, one small and the other large, each composed of specific rRNA molecules and proteins (Figure 9–12). The smaller

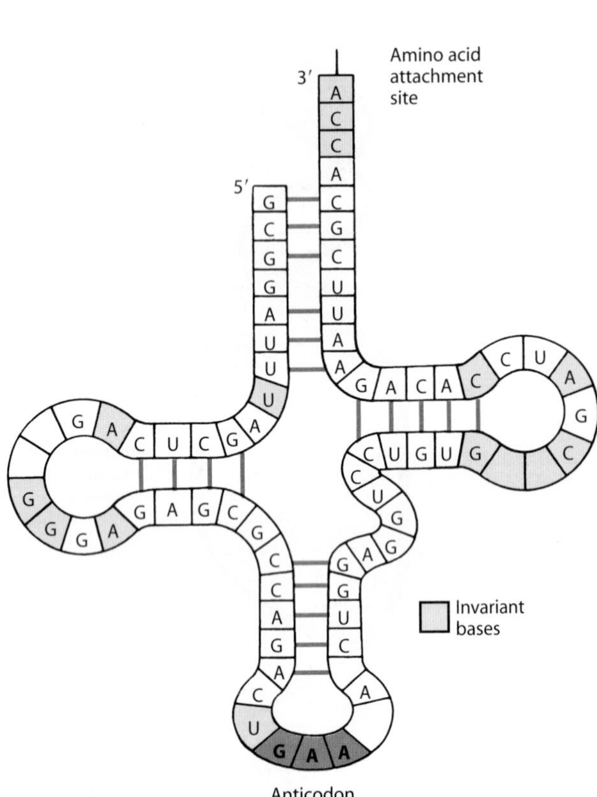

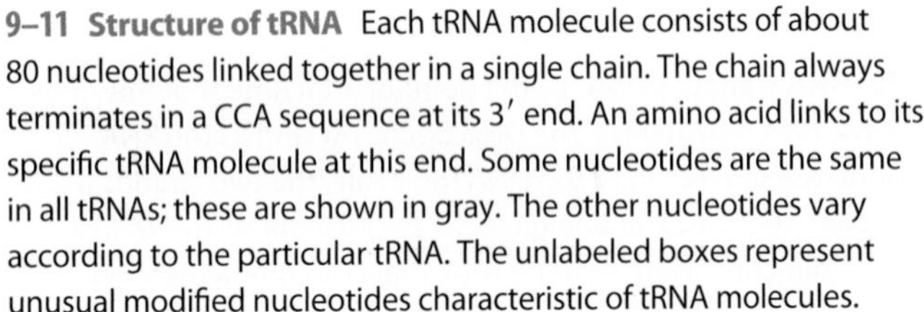

9–11 Structure of tRNA Each tRNA molecule consists of about 80 nucleotides linked together in a single chain. The chain always terminates in a CCA sequence at its 3′ end. An amino acid links to its specific tRNA molecule at this end. Some nucleotides are the same in all tRNAs; these are shown in gray. The other nucleotides vary according to the particular tRNA. The unlabeled boxes represent unusual modified nucleotides characteristic of tRNA molecules.

Some of the nucleotides are hydrogen-bonded to one another, as indicated by the red lines. In some regions, the unpaired nucleotides form loops. The loop on the right in this diagram is thought to play a role in binding the tRNA molecule to the surface of the ribosome. Three of the unpaired nucleotides (red) in the loop at the bottom of the diagram form the anticodon. They serve to "plug in" the tRNA molecule to an mRNA codon.

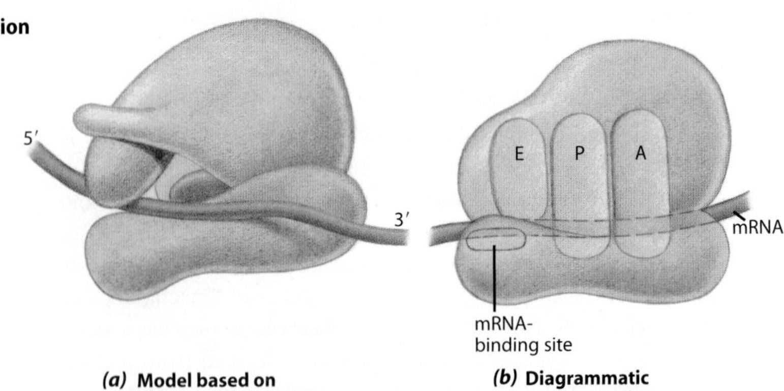

9–12 Ribosome structure In both prokaryotes and eukaryotes, ribosomes consist of two subunits, one large and one small. Each subunit is composed of specific rRNA and protein molecules. The size and density of both the subunits and the whole ribosome are greater in eukaryotes than prokaryotes. ***(a)*** One view of the three-dimensional structure of the *E. coli* ribosome, as revealed by electron micrographs. ***(b)*** A schematic diagram of another view of the *E. coli* ribosome. In protein synthesis, the ribosome moves along an mRNA molecule that is threaded between the two subunits. The A (aminoacyl) site, the P (peptidyl) site, and E (exit) site are binding sites for tRNA.

subunit has an **mRNA-binding site.** The larger subunit has three sites where tRNA can bind: an **A (aminoacyl) site,** where the incoming amino-acid-bearing tRNA usually binds; a **P (peptidyl) site,** where the tRNA bearing the growing polypeptide chain resides; and an **E (exit) site,** from which the tRNAs leave the ribosome after they have released their amino acids.

mRNA Is Translated into Protein

The synthesis of protein is known as "translation" because it involves the transfer of information from one language (a sequence of nucleotides) to another (a sequence of amino acids). In most cells, protein synthesis consumes more energy than any other biosynthetic process, because of the large numbers of proteins that are continually produced. The major stages in translation are initiation, elongation of the polypeptide chain, and chain termination (Figure 9–13).

9–13 Three stages in protein synthesis ***(a)*** Initiation. The smaller ribosomal subunit attaches to the 5′ end of the mRNA molecule. The first, or initiator, tRNA molecule, bearing the modified amino acid fMet, binds to the AUG initiation codon on the mRNA molecule. The larger ribosomal subunit locks into place, with the tRNA occupying the P (peptidyl) site. The A (aminoacyl) and E (exit) sites are vacant. ▶

(b) Elongation. A second tRNA with its attached amino acid moves into the A site, and its anticodon binds to the mRNA. A peptide bond is formed between the two amino acids brought together at the ribosome. At the same time, the bond between the first amino acid and its tRNA is broken. The ribosome moves along the mRNA chain in a 5′ to 3′ direction. The second tRNA, with the dipeptide attached, is moved from the A site to the P site as the first tRNA is released from the ribosome via the E site. A third tRNA moves into the A site, and another peptide bond is formed. The growing peptide chain is always attached to the tRNA that is moving from the A site to the P site, and the incoming tRNA bearing the next amino acid always occupies the A site. This step is repeated over and over until the polypeptide is complete.

(c) Termination. When the ribosome reaches a termination codon (in this example, UGA), a release factor occupies the A site. The polypeptide is cleaved from the last tRNA, the tRNA is released from the P site, and the two subunits of the ribosome separate.

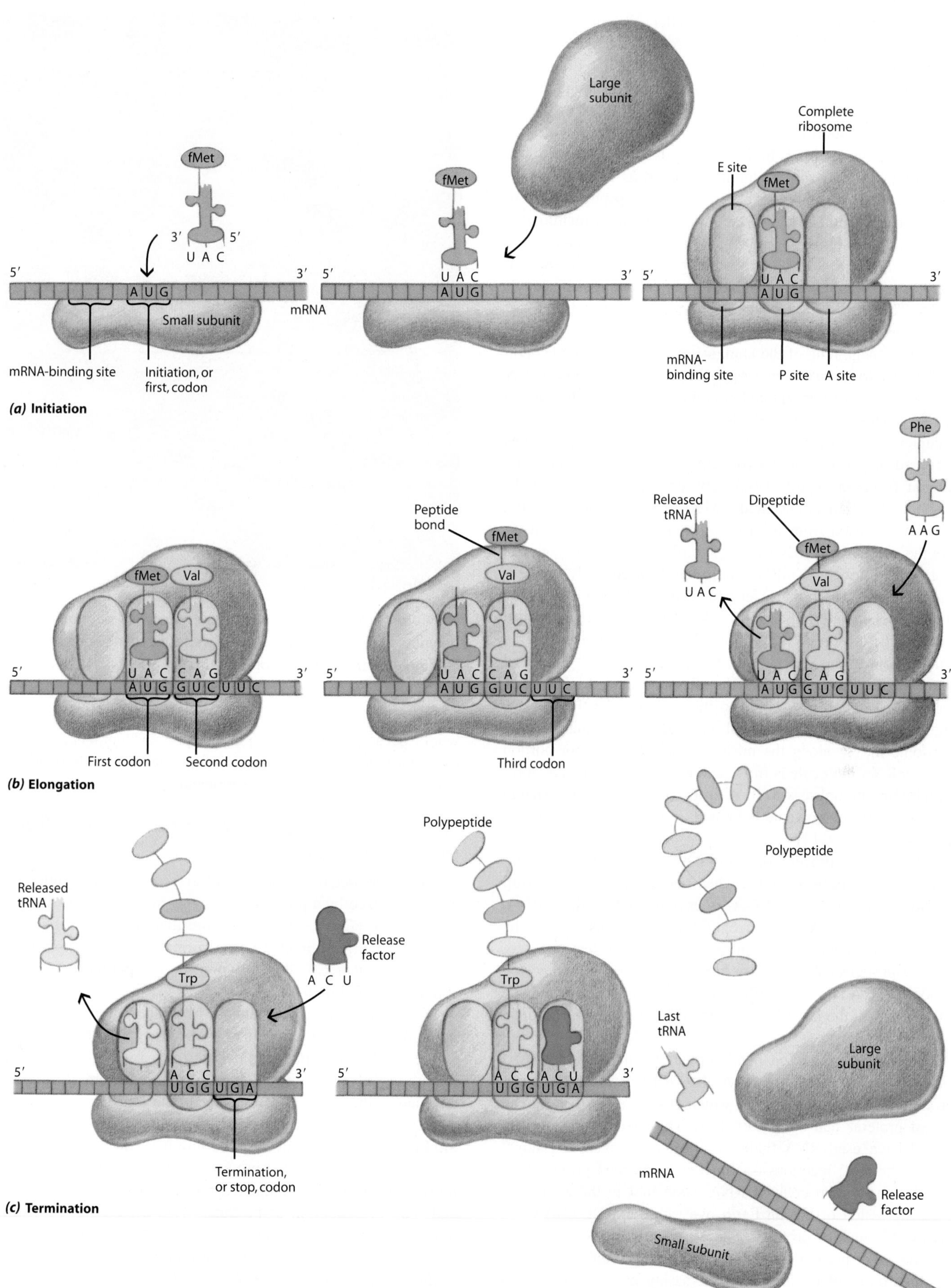
fMet
3′
5′
U A C
5′
3′
A U G
Small subunit
mRNA
mRNA-binding site
Initiation, or first, codon
fMet
U A C
A U G
5′
3′
Large subunit
Complete ribosome
E site
fMet
U A C
A U G
5′
3′
mRNA-binding site
P site
A site
(a) Initiation
fMet
Val
U A C C A G
A U G G U C U U C
5′
3′
First codon
Second codon
Peptide bond
fMet
Val
U A C C A G
A U G G U C U U C
5′
3′
Third codon
Released tRNA
U A C
Dipeptide
fMet
Val
Phe
A A G
C A G
A U G G U C U U C
5′
3′
(b) Elongation
Released tRNA
Release factor
A C U
Trp
A C C
U G G U G A
5′
3′
Termination, or stop, codon
Polypeptide
Trp
A C C A C U
U G G U G A
5′
3′
Polypeptide
Last tRNA
Large subunit
mRNA
Release factor
Small subunit
(c) Termination

Initiation begins when the smaller ribosomal subunit attaches to a strand of mRNA near its 5′ end, exposing its first codon, the *initiation codon.* The anticodon on the first tRNA then pairs with the initiation codon of the mRNA in an antiparallel fashion: the initiation codon is usually (5′)-AUG-(3′) and the tRNA anticodon is (3′)-UAC-(5′). In prokaryotes, the initiator tRNA, which binds to the AUG codon, carries a modified form of the amino acid methionine known as formyl methionine (fMet), the amino acid that starts the polypeptide chain. In eukaryotes, the initiator tRNA carries unmodified methionine. The larger ribosomal subunit then attaches to the smaller subunit, resulting in the fMet-tRNA being bound to the P site and the A site being made available for an incoming aminoacyl-tRNA. The energy for this step is provided by the hydrolysis of guanosine triphosphate (GTP).

At the beginning of the **elongation** stage, the second codon of the mRNA is positioned opposite the vacant ribosomal A site. A tRNA with an anticodon complementary to the second mRNA codon binds to the mRNA and, with its amino acid, occupies the A site of the ribosome. With both P and A sites occupied, **peptidyl transferase** activity of the larger ribosomal subunit forges a peptide bond between the two amino acids, attaching the first amino acid (fMet) to the second. The first tRNA is released from the ribosome via the E site and reenters the cytoplasmic tRNA pool. The ribosome then moves one more codon down the mRNA molecule. Consequently, the second tRNA, to which the fMet and second amino acid are now attached, is transferred from the A to the P position. A third aminoacyl-tRNA now moves into the A position opposite the third codon on the mRNA, and the step is repeated. Over and over, the P position accepts the tRNA bearing the growing polypeptide chain, and the A position accepts the tRNA bearing the new amino acid that will be added to the chain. As the ribosome moves along the mRNA strand, the initiator portion of the mRNA molecule is freed, and another ribosome can form an initiation complex with it. A group of ribosomes translating the same mRNA molecule is known as a **polysome,** or **polyribosome** (Figure 9–14).

The cyclic process of translation just described continues until **termination** occurs, when one of three possible **stop codons** (UAG, UAA, or UGA) is encountered on the mRNA. There are no tRNAs that recognize these codons, and so no tRNAs will enter the A site in response to them. Instead, cytoplasmic proteins called **release factors** bind directly to any stop codon at the A site on the ribosome, the completed polypeptide chain is freed, the last tRNA is released, and the two ribosomal subunits separate.

Remarkably, peptide bond formation on the ribosome is catalyzed by the ribosomal RNA. This is likely to be a remnant of an ancient "RNA world" in which DNA and the widespread use of proteins had not yet evolved. The ancient forms of life would have used RNA both for information and for catalyzing biochemical reactions—a function performed primarily by proteins in modern cells. If RNA came first in the evolutionary progression of life on Earth, we would expect RNA to be involved in making proteins. And we would also expect that, as the process of protein synthesis became more sophisticated during the course of evolution, an increasing number of proteins

9–14 Polysomes Groups of ribosomes are called polysomes, or polyribosomes. All the ribosomes in a specific cluster, or group, are reading the same molecule of mRNA.

would participate (for example, the protein components of the ribosome). Nevertheless, the core function of peptide bond formation still resides in an RNA molecule.

In Eukaryotes, Polypeptides Are Sorted According to Their Final Cellular Location

The polypeptides encoded by nuclear genes of eukaryotic cells are synthesized by a process that begins in the cytosol. A mechanism is therefore required to ensure that each of the polypeptides eventually is directed to the correct cellular compartment. This mechanism is called **polypeptide** (or **protein**) **targeting** and **sorting** (Figure 9–15).

There are several mechanisms by which polypeptide targeting and sorting take place. In one case, the ribosomes involved in the synthesis of polypeptides destined for the endoplasmic reticulum (ER), or for membranes derived from it, become attached to the ER membrane early in the translational process. As the polypeptides are synthesized, they are transferred across the ER membrane (or are inserted into it, in the case of integral membrane proteins). This process is therefore called **cotranslational import.** The completed polypeptide can remain in the endoplasmic reticulum or be transported across the membrane to the Golgi apparatus, which, in turn, produces various vesicles that will be transported to other destinations, such as a vacuole or the plasma membrane (see Chapter 3).

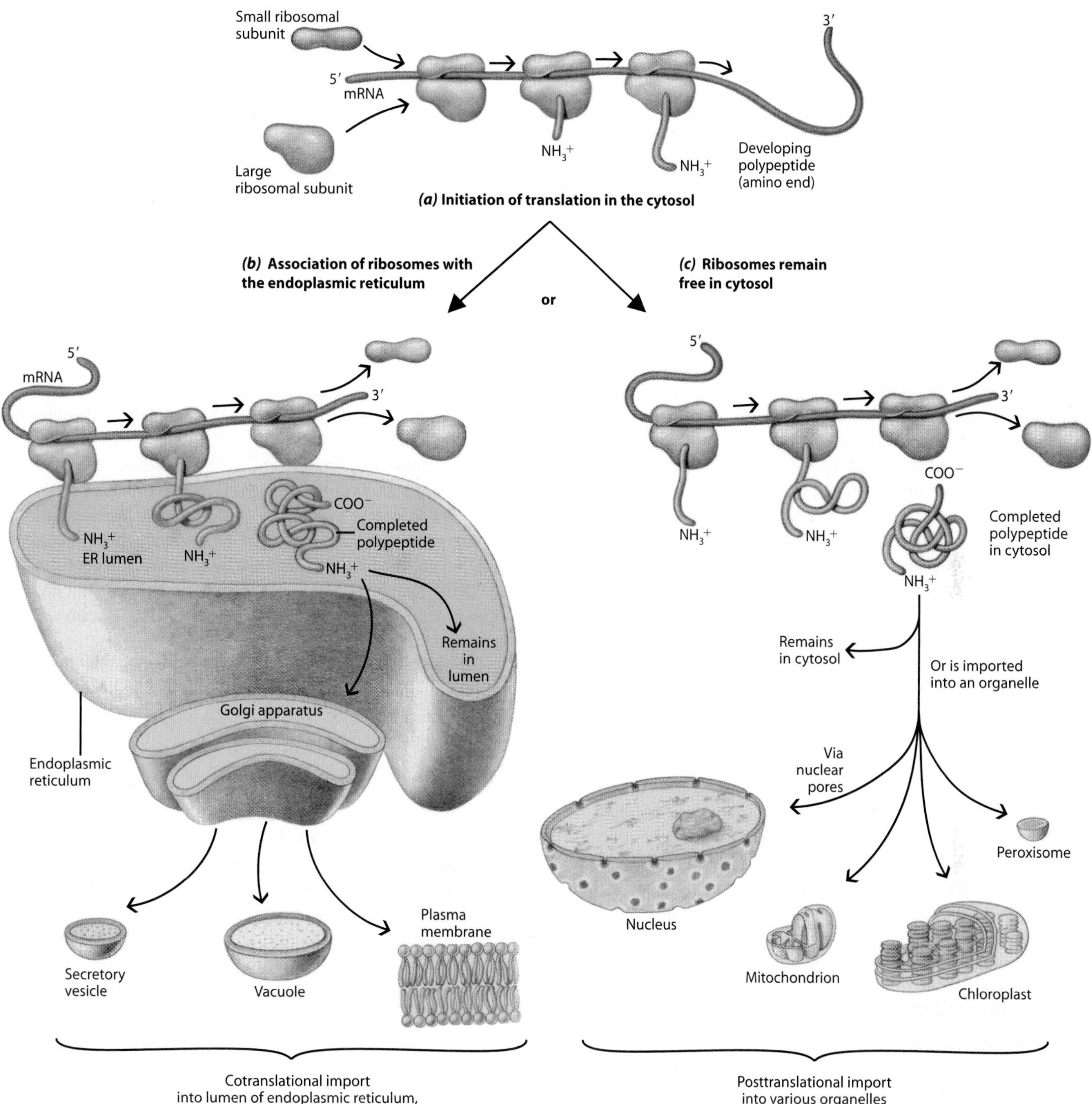

9–15 Polypeptide targeting and sorting ***(a)*** Synthesis of all polypeptides encoded by nuclear genes begins in the cytosol. The large and small ribosomal subunits associate with each other and with the 5′ end of an mRNA molecule, forming a functional ribosome that begins making the polypeptide. When the developing polypeptide is about 30 amino acids long, it enters one of two alternative pathways.

(b) If destined for any of the compartments of the endomembrane system (page 54), the polypeptide becomes attached to an endoplasmic reticulum (ER) membrane and is transferred into its lumen (the interior space) as synthesis continues. This process is called cotranslational import. The completed polypeptide then either remains in the endoplasmic reticulum or is transported via the Golgi apparatus and various vesicles to its final destination. Integral membrane proteins are inserted into the endoplasmic reticulum membrane as they are made.

(c) If the polypeptide is destined for the cytosol or for import into the nucleus, mitochondria, chloroplasts, or peroxisomes, its synthesis continues in the cytosol. When synthesis is complete, the polypeptide is released from the ribosome and either remains in the cytosol or is imported into an organelle. This process is called posttranslational import.

Regulation of Gene Expression in Eukaryotes

The differences between cell types are a result of selective, or differential, gene expression—that is, only certain genes are expressed in a specific cell type. In any cell, some genes are expressed continuously, others only as their products are needed, and still others not at all. The mechanisms that control gene expression—that turn genes "on" and "off"—are collectively called **gene regulation.**

Gene regulation in prokaryotes typically involves turning genes on and off in response to changes in the nutrients available in the environment. In eukaryotes, especially multicellular eukaryotes, the problems of regulation are very different. A multicellular organism usually starts life as a fertilized egg, the zygote. The zygote divides repeatedly by mitosis and cytokinesis, producing many cells. In flowering plants, for example, these cells begin to differentiate, becoming epidermal cells, photosynthetic cells, storage cells, water-conducting cells (tracheary elements), food-conducting cells (sieve elements), and so forth. As it differentiates, each cell type begins to produce characteristic proteins that distinguish it structurally and functionally from other types of cells.

Essentially all of the genetic information originally present in the zygote is also present in every diploid cell of the organism. This is especially clear for plants, in which a differentiated cell such as a mesophyll cell has the capacity to dedifferentiate and divide and to regenerate an entire plant genetically similar to the original plant. This phenomenon, known as **totipotency,** is discussed in Chapter 10. Thus, although the original differentiated cell produced only its characteristic proteins—and not proteins characteristic of other cell types—it may later produce other proteins that change its characteristics. (Of course, the various types of differentiated cells contain many proteins in common, consistent with the many functions that all cells carry out, such as the citric acid cycle and the processes of replication, transcription, and translation.) Clearly, the differentiation of the cells of a multicellular organism depends on the activation of certain groups of genes and the inactivation of others.

Chromatin Condensation Is an Important Factor in Gene Regulation

Many lines of evidence indicate that the degree of condensation of the chromatin, as shown by staining with certain dyes, plays a major role in the regulation of gene expression in eukaryotic cells. Staining reveals two types of chromatin: **euchromatin,** which undergoes the process of condensation and decondensation in the cell cycle and stains weakly, and **heterochromatin,** which remains highly condensed throughout the cell cycle, including interphase, and stains strongly. Transcription occurs in the euchromatin and takes place during interphase, when the euchromatin is less condensed. In the decondensed state, euchromatin is accessible to RNA polymerase and the other molecules required for transcription.

Most, but not all, of the highly condensed heterochromatin appears to be devoid of transcription. Some regions of heterochromatin are constant from cell to cell and are never expressed. An example is the highly condensed chromatin located in the centromere region of each chromosome. This region is believed to play a structural role in the movement of chromosomes during mitosis and meiosis. Similarly, no transcription takes place at the end sequences of the chromosomes, called **telomeres.**

You may recall that the basic packing units of chromatin are the nucleosomes, each consisting of a core of eight histone molecules around which a DNA filament is tightly wrapped about twice (see page 154). Overall, this chromatin structure represses gene expression. For a gene to be transcribed, transcription factors, activators, and RNA polymerase must bind to the DNA. How does the transcription machinery gain access to the tightly wrapped DNA? The answer is that, before transcription, the chromatin undergoes structural changes that make the DNA more accessible to the transcription machinery.

One such change involves **histone acetylation**—the addition of acetyl groups (CH_3CO) to histone proteins by acetyltransferase enzymes. Each histone in the core of the nucleosome consists of a globular portion, which associates with other histones and with the DNA, and a flexible, positively charged extension called the **histone tail.** The positively charged tails probably interact with the negatively charged phosphates of the DNA. When acetyl groups are added to the histone tails, the tails are neutralized and no longer bind to the DNA. This allows the DNA to separate from the histones, creating a conformation conducive to transcription. Other enzymes, called deacetylases, remove acetyl groups from histones and restore repression of the chromatin (Figure 9–16).

Another change in chromatin structure that is associated with the regulation of gene expression is **DNA methylation.** This involves the methylation of cytosine bases in the DNA, yielding 5-methylcytosine. In vertebrates, plants, and fungi, heavily methylated DNA is associated with the repression of transcription, whereas unmethylated DNA is usually transcriptionally active. In some species, including *Arabidopsis thaliana,* DNA methylation appears to be essential for the long-term inactivation of genes that occurs during normal cell differentiation in the embryo.

Note that the chromatin modifications resulting from histone acetylation and DNA methylation do not entail a change in the DNA sequence, yet they may be inherited by subsequent generations of cells. The inheritance of changes that do not directly involve the nucleotide sequence but can be reliably passed from one generation to the next is called **epigenetic inheritance.**

Specific Binding Proteins Regulate Gene Expression

In eukaryotes, as in prokaryotes, transcription is regulated by proteins that bind to specific sites on the DNA molecule. Working in conjunction with other proteins called **transcription factors** that directly or indirectly affect the initiation of transcription, these proteins rearrange the nucleosomes (page 155), allowing the transcription machinery access to the DNA. Many of these proteins and their binding sites have now been identified, and it is increasingly clear that this level of transcriptional control is far more complex in multicellular eukaryotes than in prokaryotes. Recent evidence indicates that histones may also play a role in the regulation of gene expression by selectively exposing genes to transcription.

A gene in a multicellular organism seems to respond to several, maybe even many, different regulatory proteins, some tending to turn the gene on and others to turn it off. The sites at which these proteins bind may be hundreds or even thousands

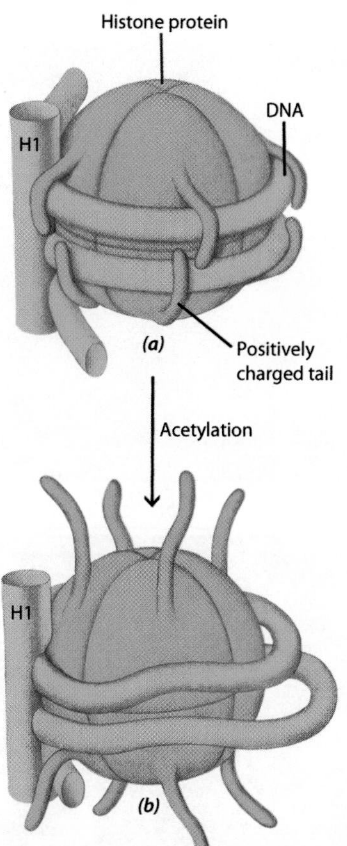

9–16 Histone acetylation ***(a)*** The positively charged tails of nucleosomal histone proteins probably interact with negatively charged phosphates of DNA, binding the DNA to the nucleosome. ***(b)*** Acetylation of the tails neutralizes them, weakening their interaction with the DNA and permitting a transcription factor to bind to the DNA.

of base pairs away from the promoter sequence at which RNA polymerase binds and transcription of that gene begins. This, as you might expect, adds to the difficulty of identifying the regions to which the regulatory molecules bind and also of understanding exactly how they exert their effects.

The DNA of the Eukaryotic Chromosome

Early studies of the DNA of eukaryotic cells revealed two surprising facts. First, with a few exceptions, the amount of DNA per cell—the **genome**—is the same for every diploid cell of any given species (which is not surprising), but the variations among different species are enormous (Figure 9–17 and Table 9–1). Second, every eukaryotic cell has what appears to be a great excess of DNA—that is, much more DNA than is required to encode cellular proteins. In fact, it is estimated that in an average eukaryotic cell, less than 10 percent of the DNA codes for proteins. In humans, it may even be as little as 1 percent. By contrast, prokaryotes, such as bacteria, use their DNA very thriftily. Except for regulatory or signal sequences, virtually all of their DNA encodes proteins and is expressed at some time in the organism's life.

Continuing research is revealing the organization of eukaryotic DNA, as well as some of the functions of the stretches of DNA that do not encode protein. For example, some of these stretches are transcribed into "noncoding RNAs," which are part of a system to regulate chromatin structure, as discussed later in the chapter.

In Eukaryotic DNA, Many Nucleotide Sequences Are Repeated

Eukaryotic genomes contain large numbers of copies of apparently nonessential nucleotide sequences that do not code for protein. The presence of a seemingly excess amount of DNA accounts, at least in part, for the large amounts of DNA in organisms such as *Paris japonica* and *Trillium hagae* (Figure 9–17).

Two major categories of repeated DNA can be recognized: tandemly repeated DNA and interspersed repeated DNA. In **tandemly repeated DNA,** repeated sequences are arranged one after another in a row—that is, in tandem series. The repeated unit may range from as few as 2 to as many as 2000 base pairs, and the units tend to be clustered at a few locations on the chromosomes. In **interspersed repeated DNA,** units are scattered throughout the genome rather than being arranged in tandem, and they may be hundreds or even thousands of base pairs long. The dispersed units, which may number in the hundreds of thousands, are similar to one another, rather than identical.

A subcategory of tandemly repeated DNA—**simple-sequence repeated DNA**—consists of sequences with fewer than 10 base pairs that are typically present in long repeats. Simple-sequence segments are believed to be vital to chromosome structure. They occur around the centromeres, which play an important role in chromosome movement, and at the telomeres, the natural ends of the chromosomes, which serve as caps and protect the chromosomes from degradation with each round of replication. In plants, the telomeric sequence is TTTAGGG.

The most abundant interspersed repeat sequences are the transposable elements, or **transposons** (page 165), which have a propensity to increase in number from one cell generation to the next by replicating themselves at higher rates than other genes. Characterized as "molecular parasites," they can be dangerous because they are inserted into and interrupt the function of host genes. To reduce the threat of transposons to the integrity of their genomes, all eukaryotes have a set of mechanisms that has evolved to epigenetically silence the transposons by modifying their chromatin to be in a heterochromatin state. It is believed that transposons have played a central role in the evolution of complex multicellular organisms.

Most Structural Genes Consist of Introns and Exons

One of the great surprises in the study of eukaryotic DNA was the discovery that the protein-coding sequences of genes, called **structural genes,** are usually not continuous but are instead interrupted by noncoding sequences—that is, by nucleotide sequences that are not translated into protein. These noncoding interruptions within a gene are known as intervening sequences, or **introns.** The coding sequences—the sequences that *are* translated into protein—are called **exons.**

The presence of introns in the genes of eukaryotes was reported almost simultaneously in 1977 by several groups of investigators. Electron micrographs of hybrids between mRNA molecules and the segments of DNA known to contain the genes coding for them showed that there was not a perfect match between the mRNA molecules and the genes from which

9–17 Largest and smallest angiosperm genomes ***(a)*** *Paris japonica,* with its 148,852 million base pairs, and ***(b)*** *Trillium hagae,* with its 129,505 million base pairs, have the largest angiosperm genomes studied so far. Much of the DNA is not expressed. ***(c)*** *Genlisea aurea* and ***(d)*** *Utricularia gibba* (bladderwort), both of which are carnivorous, have the smallest angiosperm genomes known to date, at 64 and 88 million base pairs, respectively.

they were transcribed (Figure 9–18). The nucleotide sequences of the genes were much longer than the complementary mRNA molecules found in the cytosol.

We now know that most, but not all, structural genes of multicellular eukaryotes contain introns. The introns are part of the newly transcribed RNA molecules, but they are snipped out before translation occurs. The number of introns per gene varies widely. Introns have also been found in genes coding for transfer RNAs and ribosomal RNAs. Prokaryotic genes, on the other hand, rarely contain introns. The presence of introns may have accelerated the rate of eukaryotic evolution by facilitating a process often referred to as "exon shuffling"—that is, the rearrangement of modules of proteins by DNA recombination such that new protein functions evolve.

Transcription and Processing of mRNA in Eukaryotes

Transcription in eukaryotes is the same, in principle, as in prokaryotes. It begins with the attachment of an RNA polymerase to a particular nucleotide sequence on the DNA molecule. The enzyme then moves along the molecule, using the 3′ to 5′ strand as a template for the synthesis of RNA molecules, as shown in Figure 9–10. The transcribed RNA molecules (rRNAs, tRNAs, and mRNAs) then play their various roles in the translation of the encoded genetic information into protein.

Despite these basic similarities, there are significant differences between prokaryotes and eukaryotes in transcription and translation, as well as the events that occur between these two processes. One important difference is that prokaryotic genes are grouped so that two or more structural genes are transcribed onto a single RNA molecule, but this is not usually the case in eukaryotes. In eukaryotes, each structural gene is typically transcribed separately, and its transcription is under separate controls.

Another important difference is that in eukaryotes, unlike prokaryotes, the DNA is separated from the sites of protein synthesis in the cytoplasm by the nuclear envelope. In eukaryotic cells, therefore, transcription and translation are separated in both time and space. After transcription is completed in the nucleus, the mRNA transcripts of eukaryotes are intensively

Table 9-1 Nuclear Genome Size of Selected Prokaryotes and Eukaryotes

Genome	Approximate Genome Size (millions of base pairs)*
Prokaryotes (circular chromosomes)	
Bacteria	
Escherichia coli (intestinal bacterium)	4.6
Haemophilus influenzae (rod-shaped bacterium)	1.8
Archaea	
Archaeoglobus fulgidus (sulfate reducer)	2.2
Methanosarcina barkeri (methanogen)	4.8
Eukaryotes (linear chromosomes)	
Fungi	
Saccharomyces cerevisiae (baking/brewing yeast)	12.1
Animals	
Drosophila melanogaster (fruit fly)	184
Homo sapiens (human)	2900
Algae	
Chlamydomonas reinhardtii (unicellular green alga)	136
Cyanidioschyzon merolae (unicellular red alga)	16.5
Moss	
Physcomitrella patens	480
Lycophyte	
Selaginella moellendorffii	212.6
Angiosperms	
Arabidopsis thaliana (thale cress)	125
Carica papaya (papaya)	372
Cucumis sativus (cucumber)	367
Hordeum vulgare (barley)	5179
Oryza sativa subsp. *indica* (rice)	466
Populus trichocarpa (black cottonwood)	485
Vitis vinifera (grapevine)	475
Zea mays (maize)	2300

*The number of base pairs for the "1C content"—the DNA content of a nucleus immediately after meiosis but before DNA replication.

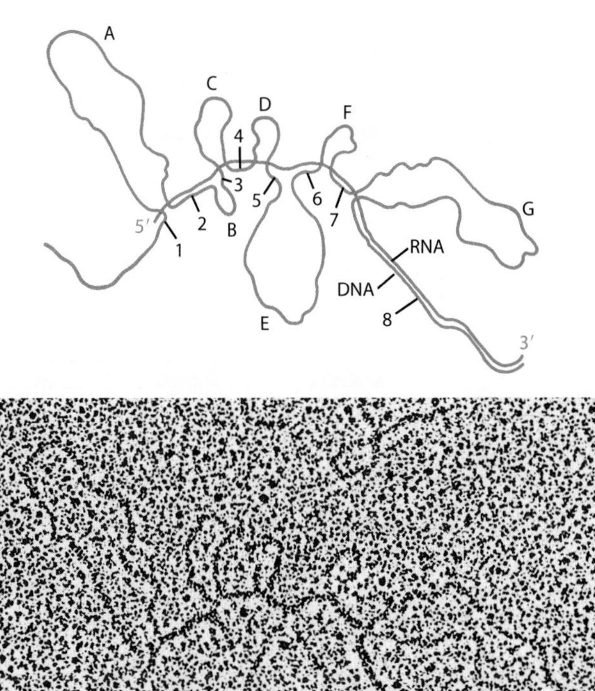

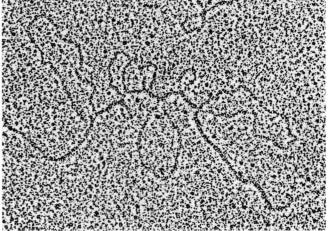

9–18 Exons and introns This electron micrograph reveals the results of an experiment in which a single strand of DNA containing the gene coding for ovalbumin (a protein of vertebrate animals) was hybridized with the messenger RNA for ovalbumin. The complementary sequences of the DNA and mRNA are held together by hydrogen bonds. There are eight such sequences (the exons labeled 1 through 8 in the accompanying diagram). Some segments of the DNA do not have corresponding mRNA segments and so loop out from the hybrid. These are the seven introns, labeled A through G. Only the exons are translated into protein.

modified before they are transported to the cytoplasm, the site of translation.

Even before transcription is completed, while the newly forming mRNA strand is only about 20 nucleotides long, a "cap" of an unusual nucleotide is added to its leading (5′) end. This cap is necessary for binding of the mRNA to the eukaryotic ribosome. After transcription is completed and the molecule released from the DNA template, special enzymes add a string of adenine nucleotides, known as the poly-A tail, to the trailing (3′) end of most eukaryotic mRNAs.

Before the modified mRNA molecules leave the nucleus, the introns are removed, and the exons are spliced together to form a single, continuous molecule (Figure 9–19). The splicing mechanism is exceedingly precise. As you can see, the addition or deletion of even a single nucleotide would shift the reading of the triplet codons so that completely different amino acids would be coded for. Consequently, an entirely new, probably nonfunctional, protein would be produced.

A number of instances have now been found in which identical mRNA transcripts are processed in more than one way. Such alternative splicing can result in the formation of different functional polypeptides from RNA molecules that were originally identical. In such cases, an intron may become an exon, or vice versa. Thus, the more that is learned about eukaryotic DNA and its expression, the more difficult it becomes to define "gene" or "intron" or "exon."

Molecular biologists once thought that the eukaryotic chromosome might be simply a large-scale version of the prokaryotic chromosome. That turned out not to be the case. The structure and organization of the chromosome, the regulation of gene expression, and the processing of mRNA molecules are all much more complex in eukaryotes than in prokaryotes. For many years it also seemed reasonable to believe that the chromosomes of eukaryotes were stable. Recombinations were produced by crossing-over, of course, but it was thought that—except for occasional point mutations and mistakes in

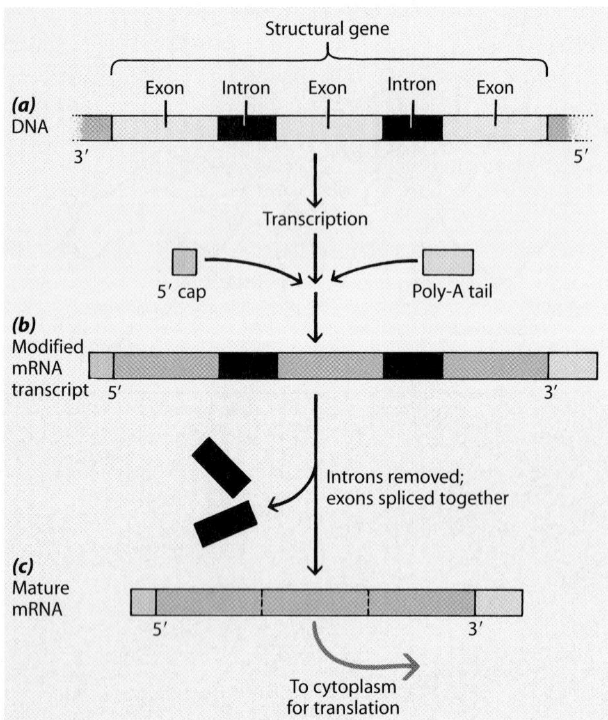

9–19 Processing of mRNA A summary of the stages in the processing of an mRNA transcribed from a structural gene of a eukaryote. ***(a)*** The genetic information encoded in the DNA is transcribed into an RNA copy. ***(b)*** This copy is then modified with the addition of a cap at the 5′ end and a poly-A tail at the 3′ end. ***(c)*** The introns are snipped out, and the exons are spliced together. The mature mRNA then passes to the cytoplasm, where it is translated into protein.

9–20 MicroRNAs and development In *Arabidopsis,* a particular microRNA (miR159) represses the expression of two related transcription factors in tissues such as leaves. Shown here are a wild-type *Arabidopsis* plant (left) and a mutant (right) in which miR159 is no longer expressed. This mutation permits the transcription factors to be expressed in leaves and results in developmental abnormalities, such as leaf curling.

meiosis—the structure of each chromosome was essentially fixed and unchanging. Perhaps the greatest surprise of all has been the discovery that this, too, is not the case. Segments of DNA can be moved through transposons and by viruses. And evidence suggests that many bacteria take up DNA from the environment and incorporate it into the bacterial chromosome, allowing the exchange of genetic information among bacteria.

Noncoding RNAs and Gene Regulation

As is becoming increasingly clear, there is a wide range of RNAs transcribed from most eukaryotic genomes that are not mRNAs, tRNAs, or rRNAs. A general term for such RNAs is **noncoding RNAs.**

Several types of noncoding RNAs have been shown to be involved in the regulation of gene expression. For example, a class of small (approximately 22-nucleotide-long) noncoding RNAs, known as microRNAs, decrease expression of target genes by a process known as **RNA interference,** in which the small RNA binds to a complementary region of an mRNA and either prevents mRNA translation into protein or promotes mRNA degradation. There are many examples in which loss of regulation by RNA interference leads to striking developmental abnormalities (Figure 9–20). Other classes of noncoding RNAs silence regions of the genome by causing the formation of heterochromatin.

SUMMARY

Watson and Crick Deduced That DNA Is a Double Helix

The Watson and Crick model of DNA is a double-stranded helix, shaped like a twisted ladder. The two sides of the ladder are composed of repeating subunits consisting of a phosphate group and the five-carbon sugar deoxyribose. The "rungs" are made up of paired nitrogenous bases. There are four bases in DNA—the purines adenine (A) and guanine (G) and the pyrimidines thymine (T) and cytosine (C). A can pair only with T, and G only with C, so a "rung" always consists of one purine and one pyrimidine. The four bases are the four "letters" used to spell out the genetic message. The paired bases are joined by hydrogen bonds.

When DNA Replicates, Each Strand Is Used as a Template to Synthesize a Complementary Strand

When the DNA molecule replicates, the two strands separate locally, breaking the hydrogen bonds that otherwise hold them together. Each strand acts as a template for the formation of a new, complementary strand from nucleotides available in the cell. The addition of nucleotides to the new strands is catalyzed by DNA polymerase. A variety of other enzymes also play key roles in the replication process.

DNA Replication Is Bidirectional

Replication begins at a particular nucleotide sequence on the chromosome, the origin of replication. It proceeds bidirectionally, by way of two replication forks that move in opposite directions.

The circular DNA molecules of prokaryotes usually have only one origin of replication; replication is completed when the leading and lagging strands complete the circle. By contrast, the linear DNA molecules of eukaryotes have many replication origins.

DNA Contains Encoded Hereditary Information

Genetic information is encoded in the sequence of nucleotides in molecules of DNA, and these, in turn, determine the sequence of amino acids in molecules of protein.

In the Process of Transcription, mRNA Is Synthesized Using DNA as a Template

The genetic information in DNA is not expressed directly but is transferred by messenger RNA (mRNA). The long molecules of mRNA are assembled by complementary base-pairing along one strand of the DNA helix. This process, called transcription, is catalyzed by the enzyme RNA polymerase.

The Genetic Code Is a Triplet Code

Each sequence of three nucleotides in the coding region of the mRNA molecule is the codon for a specific amino acid. Of the 64 possible triplet combinations of the four-lettered nucleotide code, 61 specify particular amino acids and 3 are termination codons. With 61 combinations coding for 20 amino acids, there is more than one codon for many amino acids. With only minor exceptions, the triplet code is nearly identical in all organisms.

In the Process of Translation, the Information Encoded in a Strand of mRNA Is Used to Synthesize a Specific Protein

Protein synthesis—translation—takes place on the ribosomes, which are located in the cytosol. A ribosome is formed from two subunits, one large and one small, each consisting of characteristic ribosomal RNAs (rRNAs) complexed with specific proteins. Also required for protein synthesis is another group of RNA molecules, known as transfer RNAs (tRNAs). These small molecules can carry an amino acid on one end, and they have a triplet of bases, the anticodon, on a central loop at the opposite end of the three-dimensional structure. The tRNA molecule is the adapter that pairs the correct amino acid with each mRNA codon during protein synthesis. There is at least one kind of tRNA molecule for each kind of amino acid found in proteins.

Regulation of Gene Expression Is More Complex in Eukaryotes Than in Prokaryotes

During development of multicellular eukaryotes, different groups of genes are activated or inactivated in different types of cells. Gene expression is correlated with the degree of condensation of the chromatin. Before transcription, chromatin structure changes, and the DNA becomes more accessible to the transcription machinery. A variety of specific regulatory proteins are also thought to play key roles in the regulation of gene expression.

In Eukaryotes, Most Structural Genes Contain Introns and Exons

Not all of the DNA in a eukaryotic gene specifies a protein sequence. In the process of transcription, some of the DNA sequences that are transcribed into RNA are introns that must be removed before the transcript can be used as an mRNA molecule. These mRNA segments are excised from the mRNA before it reaches the cytoplasm. The remaining mRNA segments, transcribed from portions of the DNA known as exons, are spliced together in the nucleus before the mRNA moves into the cytoplasm.

In Eukaryotic DNA, Many Nucleotide Sequences Are Repeated

In addition to introns, eukaryotic genomes contain large numbers of copies of other apparently excess DNA that does not code for proteins. There are two major categories of repeated DNA: tandemly repeated DNA and interspersed repeated DNA. A subcategory of tandemly repeated DNA, called simple-sequence repeated DNA, occurs at the centromeres and telomeres (the tips) of chromosomes. The most abundant interspersed repeat sequences are the transposons. Single-copy DNA makes up 50 to 70 percent of eukaryotic chromosomal DNA.

Transcription in Eukaryotes Involves Many Regulatory Proteins

The transcription of each gene in eukaryotes is regulated separately, and each gene produces an RNA transcript containing the encoded information for a single product. RNA transcripts are processed in the nucleus to produce the mature mRNA molecules that move from the nucleus to the cytoplasm. This processing includes the removal of introns and the splicing together of exons. Alternative splicing of identical RNA transcripts in different types of cells can produce different mRNA molecules and thus different polypeptides.

QUESTIONS

1. How are "complementary bases" and "antiparallel strands" involved in the structure of DNA?
2. Shown here is the sequence of bases, in the 5′ to 3′ direction, in one strand of a hypothetical DNA molecule. Identify the sequence of bases in the complementary strand.

   5′ – A – A – G – T – T – T – G – G – T – T – A – C – T – T – G – 3′

   3′ –   –   –   –   –   –   –   –   –   –   –   –   –   –   –   – 5′

3. What would be the sequence of an mRNA molecule transcribed from the above DNA molecule?
4. Distinguish among the following: origin of replication, replication bubble, and replication fork.
5. Distinguish among the following: replication, transcription, and translation; mRNA, tRNA, and rRNA; A site, P site, and E site.
6. Distinguish between codons and anticodons; euchromatin and heterochromatin; introns and exons.
7. Why is the genetic code said to be redundant?
8. Which amino acid is carried by the tRNA molecule shown in Figure 9–11? Consult Figure 9–9, and keep in mind the antiparallel nature of nucleic acid interactions.

CHAPTER 10

Recombinant DNA Technology, Plant Biotechnology, and Genomics

Defense against a virus Genetically modified to resist the devastating plum pox virus, the HoneySweet plum tree contains a coat-protein gene from the pox virus. The gene enables the tree to launch defenses against viral attack. Causing deformed fruit, the virus has led to the destruction of over a million stone-fruit trees, such as those bearing peaches, apricots, almonds, and cherries, as well as plums.

Ever since agriculture was developed, about 10,500 years ago, humans have selected the strains of crops that produce more and higher-quality foods. The seeds of plants with desirable traits were saved and used for the following year's crop, in the hope that those desirable traits would reappear. For some 200 years, humans have purposefully been crossing plants and selecting their progeny to produce improved strains. It was not until after the discovery of Mendel's work, however, in which he demonstrated how discrete factors—genes—behave during a cross, that scientifically based selective breeding could occur. But such breeding programs often require years to obtain desirable results. Furthermore, traditional breeding relies on the inherent genetic variability of a given species, that is, on the range of allelic variation that has resulted from that species' evolutionary path. Biotechnology has the potential to "go beyond" what is possible with traditional breeding, as illustrated in this chapter.

Recombinant DNA Technology

Since 1973 it has been possible to obtain DNA from virtually *any* organism, recombine it with the DNA of a carrier (a plasmid or a virus, most commonly), and insert it into the cells of any other organism. Using this **recombinant DNA technology,** geneticists are able to create novel genotypes, a feat impossible to achieve using traditional genetic techniques. This technology not only allows individual genes to be inserted very precisely into organisms, but also allows gene transfer between species that otherwise are incapable of hybridizing with one another.

As we shall see, recombinant DNA technology has had a major impact on agriculture, leading to such improvements as increased crop yields, improved nutritional value, and resistance to pests, diseases, and herbicides. First grown commercially in 1996, genetically modified plants covered 135 million hectares in 25 countries in 2009. With this technology, it is possible, for example, to insert genes from two mild viruses—watermelon

CHECKPOINTS

After reading this chapter, you should be able to answer the following:

1. How is recombinant DNA technology used to create novel genotypes?
2. What attributes of *Arabidopsis thaliana* make it an ideal model plant for molecular genetic research?
3. What are some of the techniques used in plant biotechnology to manipulate the genetic potential of plants?
4. What is the broad objective of genomics?
5. How does one go about determining the function of a newly discovered gene?

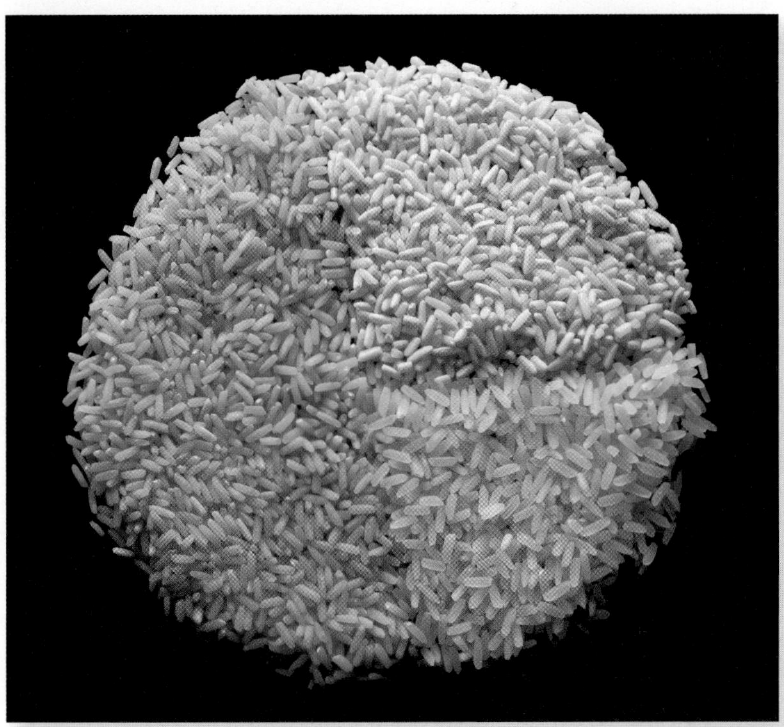

10–1 Genetically modified rice Two versions of genetically engineered rice called "Golden Rice" are shown here and compared with ordinary rice (lower right). The original version of Golden Rice (upper right) was transformed with a daffodil gene for the synthesis of β-carotene, which is present only in trace amounts in ordinary rice. A second version (left) was transformed with a maize gene, which produces 23 times more β-carotene. The presence of β-carotene imparts a yellow, or "golden," color to the transformed grain. The β-carotene is converted to vitamin A in the human body. If children receive insufficient amounts of β-carotene in their diet, they often suffer blindness, a condition that is widespread in southern Asia.

mosaic virus 2 and zucchini yellow mosaic virus—into cells of a squash, producing a genetically altered squash that is resistant to infection by more virulent viruses. Certain crops, such as maize (corn), tomato, potato, and cotton, have been genetically modified to resist specific insect pests. As a result, pesticide use can be reduced substantially. In addition, recombinant DNA has been used to make certain commercially important crops, such as tomatoes, soybeans, cotton, and oilseed rape (used for canola oil), resistant to herbicides. The result is that these herbicides can be used to increase yields by selectively killing the weeds that would otherwise compete with the crop plants for sunlight and nutrients. The economics involved are not trivial, because farmers currently spend about $40 billion annually on weed control in the United States alone. As discussed on page 205, recombinant DNA technology has also been used to increase the nutritional value of certain crops, such as "Golden Rice," which shows promise for reversing the vitamin A deficiencies suffered by millions of people with rice-based diets (Figure 10–1).

Restriction Enzymes Are Used to Make Recombinant DNA

Recombinant DNA technology is based in large part on the ability to precisely cut DNA molecules from different sources into specific pieces and to combine those pieces to produce new combinations. This procedure depends on the existence of **restriction enzymes,** which recognize specific sequences of double-stranded DNA known as **recognition sequences.** These sequences are typically four to six nucleotides long and are always palindromic (that is, one strand is identical to the other strand when read in the opposite direction).

Restriction enzymes cut DNA within or near their particular recognition sequences. Some restriction enzymes make straight cuts, while others cut through the two strands a few nucleotides apart, leaving what are called **sticky ends** (Figure 10–2). The single DNA strands at the two sticky ends are complementary, and they can therefore pair with one another and rejoin through

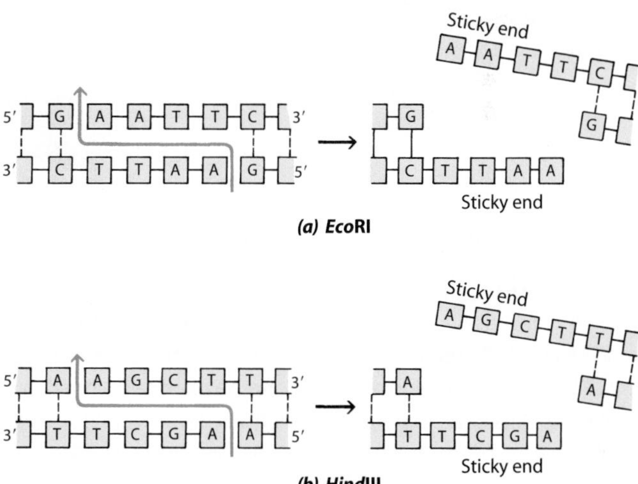

10–2 Restriction enzymes The DNA nucleotide sequences recognized by two widely used restriction enzymes, ***(a)*** *Eco*RI and ***(b)*** *Hind*III. These two enzymes cut the DNA so that sticky ends result. A sticky end can reattach to its complementary sequence at the end of a fragment of any DNA molecule, from any other organism, that has been cut by the same restriction enzyme. Restriction enzymes are named for the bacteria from which they are obtained by combining the first letter of the genus with the first two letters of the specific epithet. *Eco*RI is from *E. coli,* and *Hind*III is from *Haemophilus influenzae.*

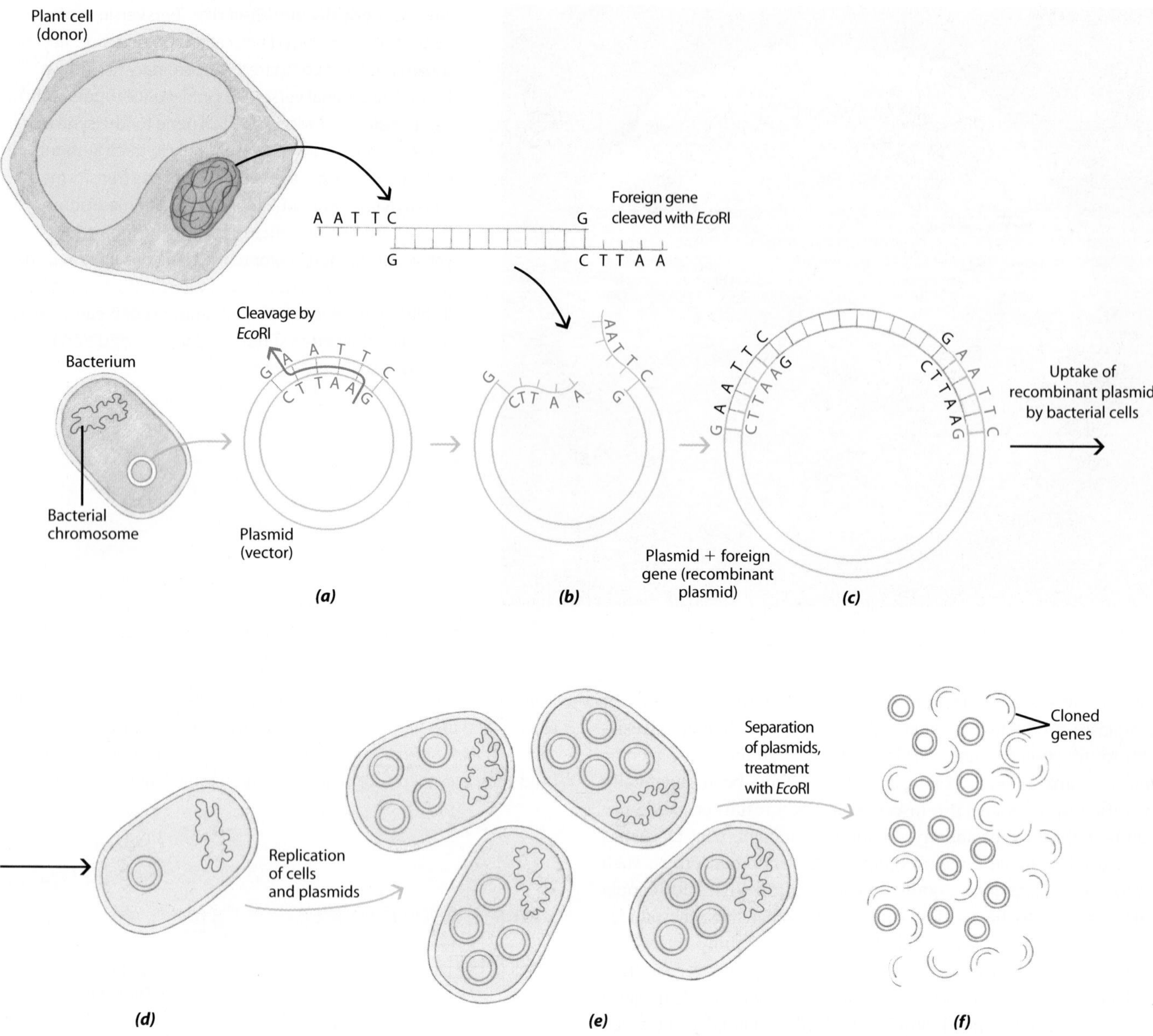

10–3 The use of plasmids in DNA cloning ***(a)*** The plasmid is cleaved by a restriction enzyme. In this example, the enzyme is *Eco*RI, which cleaves the plasmid at the sequence (5′)-GAATTC-(3′), leaving sticky ends exposed. ***(b)*** These ends, consisting of TTAA and AATT sequences, can join with any other segment of DNA that has been cleaved by the same enzyme. Thus it is possible to splice a foreign gene—say, from a plant cell—into the plasmid ***(c)***. (In this illustration, the length of the GAATTC sequences is exaggerated, and the lengths of the other portions of both the foreign gene and the plasmid are compressed.) ***(d)*** When plasmids incorporating a foreign gene are released into a medium in which bacteria are growing, they are taken up by some of the bacterial cells. ***(e)*** As these cells multiply, the recombinant plasmids replicate. The result is an increasing number of cells, all making copies of the same plasmid. ***(f)*** The recombinant plasmids can then be separated from the other cell contents and treated with *Eco*RI to release the copies of the cloned gene.

the action of DNA ligase (see Figure 9–6). More important, these sticky ends can join with any other segment of DNA—from any number of sources—that has been cut by the same restriction enzyme and therefore has complementary sticky ends. This property makes it possible to create virtually unlimited recombinations of genetic material, because the DNA fragments recombine regardless of the original sources of the fragments.

How is recombinant DNA made and manipulated? The basic procedure is to use restriction enzymes to cut DNA from a **donor organism.** Consisting of one to several genes each, the DNA fragments so formed are combined with small DNA molecules, such as bacterial **plasmids,** which can replicate autonomously (on their own) when introduced into bacterial cells. The plasmids act as carriers, or **vectors,** for the foreign DNA fragments. Together with their inserted foreign DNA, the plasmids are examples of recombinant DNA because they consist of DNA originating from two different sources—a donor and a bacterium. The recombinant DNA, now existing as recombinant plasmids, is then taken up by the bacterial **host cells,** which are said to be **transformed.** If a bacterial cell containing a recombinant plasmid is allowed to divide and grow into a colony with millions of cells, each cell will contain the same recombinant plasmid with the same DNA insert. This process of amplifying, or generating, many identical DNA fragments is called **DNA cloning,** or gene cloning. Figure 10–3 gives an overview of this procedure using a plasmid as the cloning vector. A virus can also be used as a cloning vector, and there are viruses that replicate in eukaryotic cells as well as in prokaryotic cells.

Selectable Marker Genes and Reporter Genes Are Used to Identify Host Cells That Contain Recombinant DNA

Even though a population of host cells may be exposed to recombinant plasmids, not all the cells will be transformed. It is essential, therefore, to have a method for identifying the host cells that contain the recombinant DNA.

A commonly used approach can be illustrated using *Escherichia coli* as host and a plasmid vector that carries a gene or genes for resistance to antibiotics, in addition to the gene of interest. One such plasmid carries the gene amp^R, which confers resistance to the antibiotic ampicillin. Only cells of *E. coli* carrying recombinant plasmids with this gene will be able to survive and grow in the presence of ampicillin; all nontransformed cells will die. The amp^R gene therefore provides selection for host cells containing the recombinant DNA with the gene of interest. Genes of this type are called **selectable marker genes.**

In addition to genes for antibiotic resistance, genes called **reporter genes** can be used to visually detect recombinant DNA in host cells. One such process involves the use of plasmid vectors that carry the gene *lacZ.* The *lacZ* gene encodes for β-galactosidase, the enzyme that hydrolyzes the sugar lactose. Such plasmids have a single recognition sequence for the restriction enzyme used, and the recognition sequence lies within the *lacZ* gene. Bacteria containing a plasmid with an intact (uninterrupted) *lacZ* gene, and thus producing β-galactosidase, will form blue colonies when grown on solid culture media that are supplemented with a modified sugar called X-gal. Colonies of cells that have had a foreign DNA fragment inserted into the recognition sequence of the plasmid will appear white, however, because the *lacZ* gene has been disrupted, preventing it from producing β-galactosidase. In some cases, both antibiotic resistance (selective screening) and color identification (nonselective screening) methods are used in tandem (Figure 10–4).

Other reporter genes have been developed to monitor where certain proteins reside in the cell or where gene promoters are active. Figure 10–5 shows an example of the use of the reporter gene encoding green fluorescent protein (GFP), from the jellyfish *Aequorea victoria,* to determine protein location. In this example, a chimeric gene ("chimeric" meaning with components from different sources) was created, consisting of the GFP coding sequences joined to the actin-binding protein gene. When this chimeric gene was introduced into plants, a chimeric protein, containing GFP fused to the end of the actin binding protein, was produced. The location of the actin-binding protein could be determined by monitoring the location of the fluorescing GFP portion of the chimeric protein. Figure 10–6 illustrates an example of fusing the reporter gene luciferase, which encodes the enzyme that causes fireflies to glow when the enzyme reacts with its substrates ATP and luciferin, to a virus promoter to determine in which plant cell types the promoter is expressed. The reporter gene assay showed that the promoter is active in almost all cell types of a tobacco plant.

DNA Libraries Can Be Either Genomic or Complementary

Creation of a genomic library involves cutting the entire genome of an organism into a large number of restriction fragments and then introducing them randomly into a large number of bacterial cells or viruses. Such a collection of DNA clones is called a **genomic library,** because it contains cloned fragments of most, if not all, of the genome.

Another kind of DNA library is the **complementary DNA (cDNA) library.** Complementary DNA is synthetic DNA made from messenger RNA, using a special enzyme called **reverse transcriptase.** This enzyme is obtained from animal viruses called retroviruses. Using mRNA as a template, reverse transcriptase catalyzes the synthesis of a single strand of cDNA by reverse transcription. The single-stranded DNA is then converted into double-stranded DNA, using DNA polymerase. Because it is made on an mRNA template, cDNA is devoid of introns and untranscribed DNA between genes. This means that, unlike genomic DNA, cDNA from eukaryotes contains only protein-coding sequences and can be translated into functional proteins in bacteria, which are unable to remove introns.

The Polymerase Chain Reaction Can Be Used to Amplify Segments of DNA

The **polymerase chain reaction (PCR)** is a technique in which any piece of DNA can be greatly amplified in a relatively short period of time. Literally millions of copies of a DNA segment can be made by the polymerase chain reaction in just a few hours. The procedure involves separating the two strands of the double helix by heating, and then adding a short primer to the selected DNA sequence on each strand by lowering the temperature in the presence of all the components necessary for polymerization. The separated DNA strands are exposed to a temperature-resistant DNA polymerase such as *Taq* polymerase, which catalyzes new strand synthesis beginning at each DNA primer and using nucleotides supplied in the solution. *Taq* polymerase is obtained from the bacterium *Thermus aquaticus,* which was discovered growing in the hot springs near Old Faithful in Yellowstone National Park. The enzyme is not destroyed by the temperatures used to separate DNA strands.

In the polymerase chain reaction, both strands of the DNA are copied simultaneously. After completion of the replication of the segment between the two primers, the two newly formed double-stranded DNA molecules are heated. This causes the strands to separate into four single strands, and a second cycle of replication is carried out by lowering the temperature in the presence of all the components necessary for polymerization. If this procedure is repeated for 20 cycles, amplifications of up to a millionfold can be achieved within a few hours.

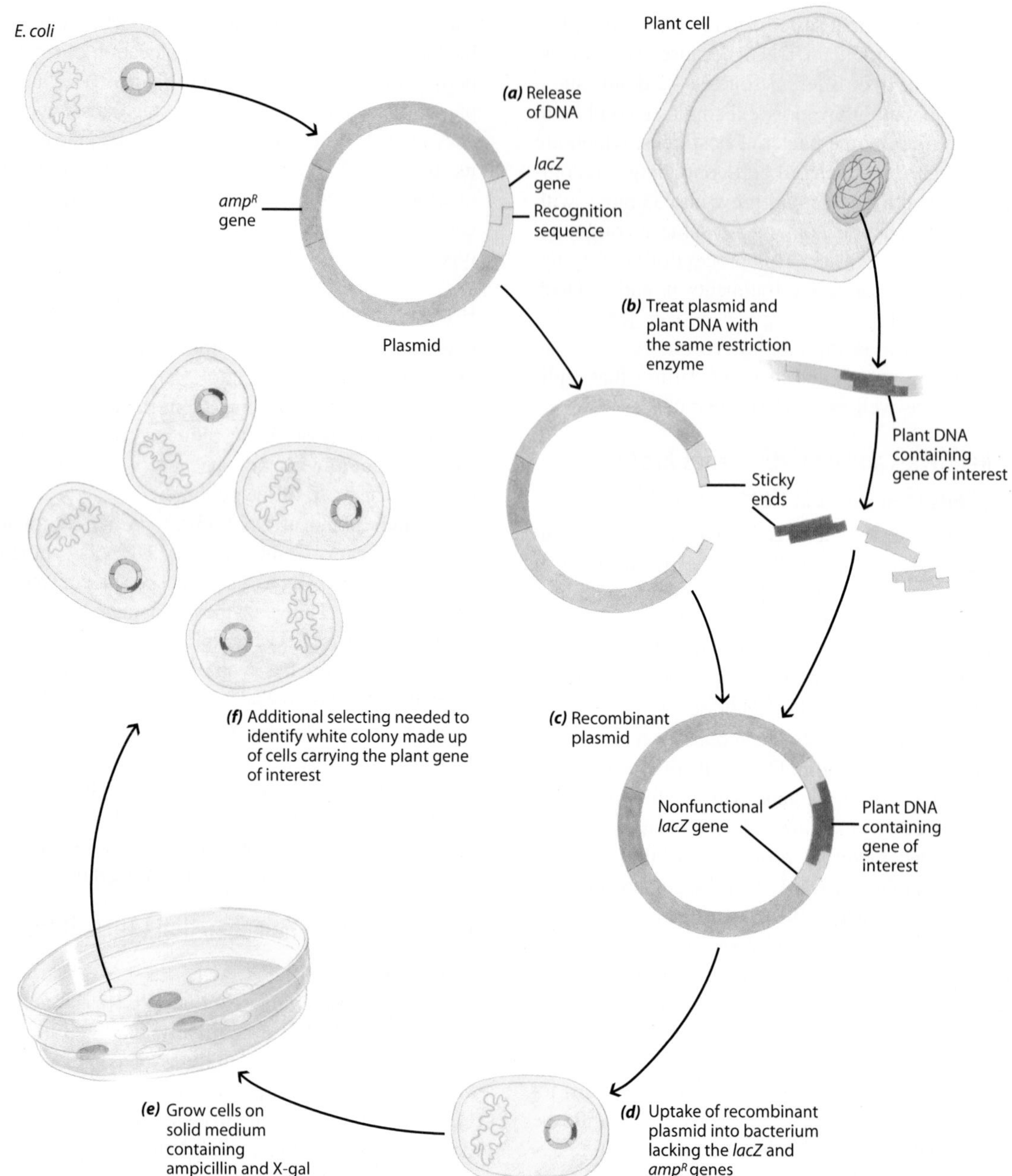

10–4 Using *lacZ* as a reporter gene ***(a)*** Cells of *E. coli* containing plasmids with the amp^R and *lacZ* genes are treated to release the plasmids. The amp^R gene confers resistance to the antibiotic ampicillin, and the *lacZ* gene produces β-galactosidase, an enzyme that hydrolyzes the sugar lactose. Both genes will help to identify transformed bacteria. In addition, plant cells containing the gene of interest are treated to release their DNA. ***(b)*** The plasmids and plant DNA are treated with the same restriction enzyme. The recognition sequence for the restriction enzyme is within the *lacZ* gene on the plasmid, so the gene is disrupted when the foreign DNA is inserted into the restriction site. For each plant cell, DNA is cut into many fragments, one of which contains the gene of interest. Treatment with the restriction enzyme results in sticky ends on both the plasmids and the plant DNA fragments. ***(c)*** Combining the treated plasmids and fragments results in recombinant plasmids containing the gene of interest, as well as recombinant plasmids containing other plant genes. Insertion occurs by base-pairing of the plasmid sticky ends with the complementary sticky ends of the plant DNA fragments. ***(d)*** The plasmid solution, consisting of recombinant plasmids and intact plasmids, is mixed with bacteria lacking the amp^R and *lacZ* genes. Some of the bacteria will take up plasmids. ***(e)*** The bacteria are grown on a solid culture medium with ampicillin and X-gal, a modified sugar that turns blue when digested by β-galactosidase. Bacteria containing plasmids are resistant to ampicillin and will therefore grow and form colonies on this medium. Bacterial colonies containing plasmids without a disrupted *lacZ* gene will appear blue because they can produce β-galactosidase and digest X-gal. Bacteria containing recombinant plasmids, however, will form white colonies because they have a nonfunctional *lacZ* gene. ***(f)*** Each white colony consists of a clone of bacteria carrying identical recombinant plasmids. Additional selection is needed to determine whether or not a white colony is made up of transformed bacteria carrying the plant gene of interest.

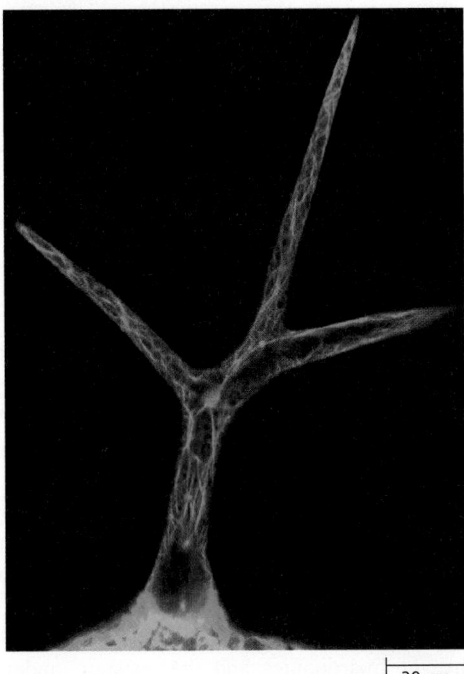

10–5 Green fluorescent protein A trichome (a leaf hair) of an *Arabidopsis* plant that was transformed with a fragment of the DNA sequence coding for an actin-binding protein (talin) fused to a DNA sequence coding for green fluorescent protein (GFP). The GFP-talin protein binds to actin filaments in all of the living cells of the transgenic plant. Confocal microscopy reveals the distribution of actin bundles, most of which run parallel to the long axis of the trichome.

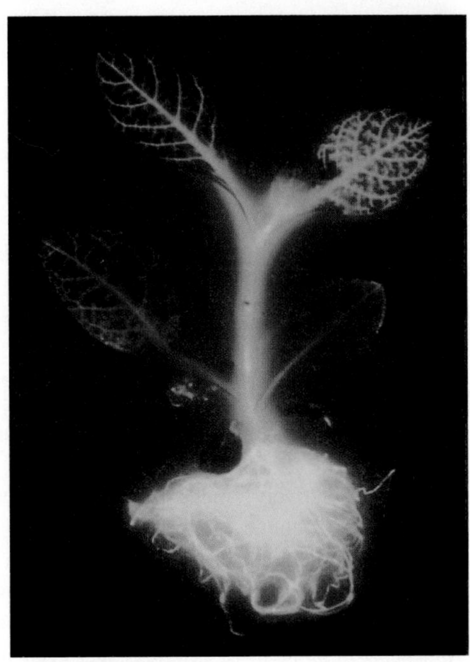

10–6 Glowing tobacco plant Firefly genes that code for the production of the enzyme luciferase were inserted into cells isolated from a normal tobacco plant *(Nicotiana tabacum),* using the Ti plasmid of *Agrobacterium tumefaciens* as vector. After the undifferentiated callus cells developed into a whole plant, the cells that incorporated the luciferase gene into their DNA became luminescent in the presence of luciferin, ATP, and oxygen.

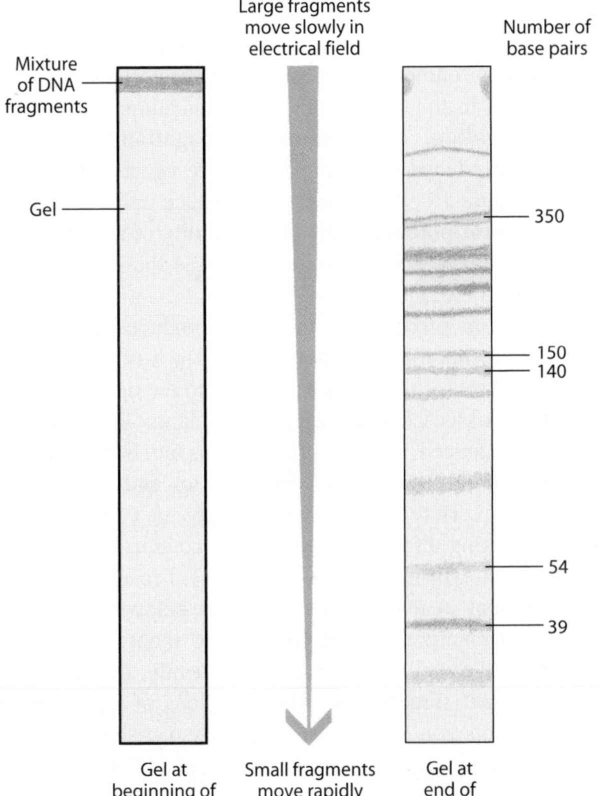

DNA Sequencing Has Revealed the Genomes of Organisms

With the development of techniques for cutting DNA molecules into smaller pieces and making multiple copies of those pieces, it became possible to determine the nucleotide sequence of any isolated gene. One of the most important features of restriction enzymes is that different enzymes cut DNA molecules at different sites. Cutting a DNA molecule with one restriction enzyme produces one particular set of short DNA fragments. Cutting an identical DNA molecule with a different restriction enzyme produces a different set of short DNA fragments. The fragments of each set can be separated from one another by electrophoresis on the basis of their length, or size (Figure 10–7), and cloned to make multiple copies.

10–7 Separating fragments of DNA by electrophoresis In electrophoresis, the electrical field separates molecules both by charge and by length, or size. Smaller molecules move faster than larger ones. A mixture of DNA fragments containing different numbers of base pairs can be cleanly separated according to size. The gel is then sliced into sections, and the separate, purified fragments are washed out of the gel unharmed. This separation procedure is important in many aspects of recombinant DNA work.

The DNA fragment copies can then be analyzed to determine the exact nucleotide sequence of each fragment. Because the sets of fragments produced by different restriction enzymes overlap, the information obtained from sequencing the different sets can be pieced together like a puzzle to reveal the entire sequence of a DNA molecule (Figure 10–8).

Sequencing is now rarely if ever done from specific restriction fragments. Rather, *random* clones are sequenced. In principle, the rest of the process (as illustrated in Figure 10–8b, c) has not changed: overlapping pieces are aligned to generate a sequence. Moreover, most DNA sequencing is now carried out automatically by DNA-sequencing machines, which greatly reduce the time required to determine the nucleotide sequences of short fragments and have made it possible to sequence entire genomes.

Because of their relatively small size, the first complete genome sequences were obtained for prokaryotes, that of the bacterium *Haemophilus influenzae* being the first, in 1995. Sequencing of the first eukaryotic genome, of the yeast *Saccharomyces cerevisiae,* was finished in 1996. Since then, the number of species with sequenced genomes has been increasing exponentially, due, in part, to advances in technology that have reduced the cost of sequencing and provided more efficient ways to assemble genome sequences. Also, it is becoming increasingly common to sequence many different individuals within a species. This permits an evaluation of genetic variability within the species and provides an opportunity to identify which genes are responsible for specific phenotypes. For example, sequencing the genomes of many rice varieties that are resistant to a particular disease and comparing them with the genomes of rice varieties that are susceptible will provide a list of candidate genes that may confer the disease resistance.

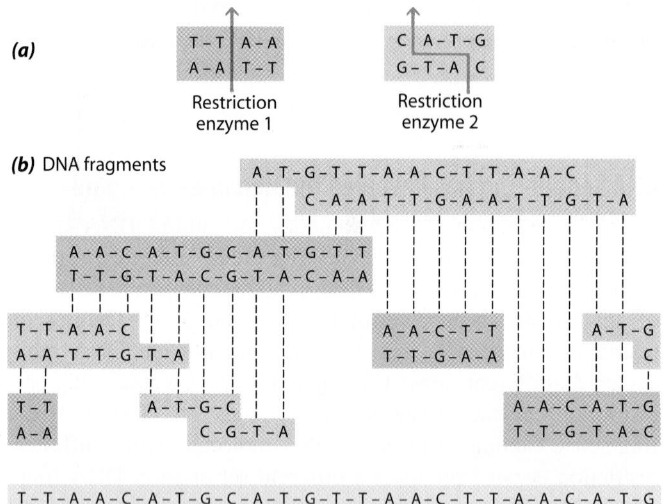

10–8 A simplified example of DNA sequencing Identical samples of the DNA molecule to be sequenced are treated with different restriction enzymes that cut the DNA at different sites. ***(a)*** One sample is treated with one enzyme (restriction enzyme 1), producing one set of fragments, and another sample is treated with another enzyme (restriction enzyme 2), producing a different set of fragments. The fragments of each set are then separated from one another, cloned, and analyzed, revealing the nucleotide sequence of each individual fragment ***(b)***. As you can see, the fragments produced by the two restriction enzymes overlap, making it possible to determine the nucleotide sequence of the molecule as a whole ***(c)***.

Plant Biotechnology

The origins of **plant biotechnology,** the application of an array of techniques to manipulate the genetic potential of plants, can be traced back to the late 1850s and 1860s and the work of the German plant physiologists Julius von Sachs and W. Knop. They demonstrated that many kinds of plants could be grown in water if provided with a few essential minerals. In other words, plants could be grown without placing their roots in soil, a technique we now call **hydroponics.** By the mid-1880s, it was known that at least 10 chemical elements found in plants are necessary for normal growth. Today, 17 elements are generally considered to be essential for most plants (see Table 29–1).

The use of hydroponics and the quest to understand the mineral nutrition of plants provided the impetus for studies on the growth of excised plant parts—isolated pieces of shoot tips, leaves, and embryos at various stages of development—in nutrient solutions. Gradually, substances such as sucrose, various vitamins, and other organic substances were added to the nutrient solutions in an attempt to sustain growth. It was not until the discovery of plant hormones and an understanding of their roles in the control of plant growth and development, however, that organ and tissue culture truly became feasible.

Plant Tissue Culture Can Be Used in Clonal Propagation

Plant **tissue culture** can be broadly defined as a collection of methods for growing large numbers of cells in a sterile and controlled environment. At present, the greatest impact of tissue culture is in the area of plant multiplication, referred to as **micropropagation,** or as **clonal propagation,** because the individuals produced from single cells are genetically identical **(clones)** (Figure 10–9). The goal is to induce individual cells to express their **totipotency,** meaning the ability of a single mature plant cell to grow into an entire plant (see the essay "Totipotency" on page 202).

Apart from providing a means to produce identical copies of a plant, micropropagation provides a way to circumvent many plant diseases. This is due, in part, to the decontamination of the plant tissues being used—the explants—and the sterile conditions practiced in micropropagation, but is primarily due to the use of meristem (embryonic tissue) and shoot-tip culture techniques. Here, only very small explants of meristem and shoot tips lacking differentiated vascular (conducting) tissues are cultured. Such explants are often virus-free, because viruses that may be present in mature vascular tissue below the meristems can reach the meristematic regions of the shoot tips only very slowly, by cell-to-cell movement. Consequently, the viruses may simply not enter some of the cells in regions of rapid cell division. The production of virus-free plants by meristem culture has greatly increased the yields of several crop plants, including potatoes and rhubarb.

MODEL PLANTS: *ARABIDOPSIS THALIANA* AND *ORYZA SATIVA*

Arabidopsis thaliana (thale cress), a small weed of the mustard family (Brassicaceae), was the first plant to be selected as a model experimental organism in the study of plant molecular genetics. There are several attributes that make it particularly suited for both classical and molecular genetic research.

1. Its short generation time. Only six weeks are required to go from seed to seed, and each plant has the potential to produce more than 10,000 seeds.
2. Its small size. *Arabidopsis* is such a small plant that literally dozens can be grown in a small pot, requiring only moist soil and fluorescent light for rapid growth.
3. Its adaptability. *Arabidopsis* plants grow well on sterile, biochemically defined media. In addition, *Arabidopsis* cells have been grown in culture, and plants have been regenerated from such cells.
4. Typically, it self-fertilizes. This allows new mutations to be made homozygous with minimal effort. Many mutations have been identified in *Arabidopsis,* including visible ones useful as markers in genetic mapping. Such maps provide an approximation of the positions of genes on the chromosomes.
5. Its susceptibility to infection by the bacterium *Agrobacterium tumefaciens,* which carries plasmids capable of transferring genes into plants (see pages 200 and 201).
6. Its relatively small genome (125 million base pairs, containing approximately 26,000 genes), which simplifies the task of identifying and isolating genes.

None of the crop plants share all of these traits with *Arabidopsis.* Typical crop plants have generation times of several months and require a great deal of space for growth in large numbers. In addition, those that have been used for recombinant DNA studies have large genomes and large amounts of repetitive DNA.

Rice *(Oryza sativa),* with the smallest genome of the major cereals (389 million base pairs, containing approximately 41,000 genes), was the second model system selected for plant science research. It is the most important food crop in the world, feeding over half of the global population. A high degree of **synteny** (many blocks of genes being the same across different species) exists among the genomes of all grasses (Poaceae). Therefore, the knowledge obtained from analysis of the rice genome can be applied quite directly to wheat, maize, barley, and other grasses.

The list of model plants for which complete genome sequences are now known includes *Carica papaya* (papaya), *Cucumis sativa* (cucumber), *Populus trichocarpa* (black cottonwood), *Sorghum bicolor* (sorghum), *Vitis vinifera* (grapevine), *Zea mays* (maize), and the moss *Physcomitrella patens.* Many more plants are in the sequencing pipeline.

(a) (b) (c) (d) (e)

The first two model plants ***(a)*** *Arabidopsis thaliana.* Its short generation time is one of the features that has made *A. thaliana* a model plant used by researchers worldwide. Seen here is the *Landsberg erecta strain* of *A. thaliana,* 14 days after planting. At this stage, the shoot contains a rosette of seven to eight leaves. ***(b)*** By 21 days, the plant has already undergone transition from a vegetative to a reproductive phase. ***(c)*** By 37 days, the plant has produced many flowers, and fruit and seed development is well under way. ***(d)*** By 53 days, the plant is undergoing senescence. ***(e)*** *Oryza sativa.* Rice was the first monocot selected as a model system.

(a) (b) (c) (d)

10–9 Plant tissue culture ***(a)*** Callus tissue derived from young sugarcane leaf tissue grown in a sterile culture medium containing relatively high levels of the hormones auxin and cytokinin. ***(b)*** When the concentration of cytokinin in the medium is reduced, shoots begin to regenerate from pieces of callus. ***(c)*** When the shoots have elongated, the callus giving rise to them is transferred to a medium with a high auxin concentration. ***(d)*** The higher auxin concentration induces the development of roots.

Genetic Engineering Allows the Manipulation of Genetic Material for Practical Purposes

Genetic engineering, the application of recombinant DNA technology, is providing one of the most important means by which crop plants will be improved in the future. It has two distinct advantages. First, genetic engineering allows individual genes to be inserted into organisms in a way that is both precise and simple. Second, the species involved in the gene transfer do not have to be capable of hybridizing with one another, and thus new genetic potential can be incorporated into an organism that could not be introduced with conventional plant breeding.

One way of transferring foreign genes into target plants is through the use of *Agrobacterium tumefaciens,* a soil-dwelling bacterium that infects a wide range of flowering plants, typically gaining entry through wounds. *A. tumefaciens* induces the formation of tumors, called crown-gall tumors (Figure 10–10), by transferring a specific region, the **transferred DNA,** or **T-DNA,** of a tumor-inducing (Ti) plasmid to the host plant's nuclear DNA.

Each **Ti plasmid** is a closed circle of DNA that consists of about 100 genes (Figure 10–11). The T-DNA consists of approximately 20,000 base pairs of DNA bounded by 25-base-pair repeats at each end (Figure 10–11b). The T-DNA carries

10–10 Crown-gall tumors *Agrobacterium tumefaciens,* a bacterium that is commonly used to transfer foreign genes into plants, induces the formation of crown galls, seen here growing on a tomato *(Solanum lycopersicum)* stem.

a number of genes, including one (gene *O*) that codes for an enzyme that synthesizes opines, which are unique amino acid derivatives. In addition, the *onc* region is a group of three genes, two of which code for enzymes involved in the synthesis of the hormone auxin, and one of which codes for an enzyme that catalyzes the synthesis of a cytokinin. The presence of these genes involved in hormone biosynthesis enables host cells to grow and divide uncontrollably into a tumor. The opines are used by the bacteria as sources of carbon and nitrogen. Another region of the Ti plasmid, the *vir* region, is essential for the transfer process but is not incorporated into the host's DNA. Thus, *Agrobacterium* is a natural genetic engineer. It reprograms plant cells by transferring new genetic information into the host's genome.

Agrobacterium tumefaciens, with its Ti plasmid, is a powerful tool for genetic engineering of eudicots and certain monocots. The tumor-promoting genes on the T-DNA can be removed and replaced by any genes that researchers wish to transfer to plants (Figure 10–12). Infection of a plant with *A. tumefaciens* containing these engineered plasmids will result in transfer of these genes into the plant's genome. **Transgenic plants** (plants containing foreign genes) obtained through the use of plasmids (see page 257) transmit the foreign genes to their progeny in a Mendelian fashion (see Chapter 8).

Other Methods Are Available for the Transfer of Genes

One method used for gene transfer is called **electroporation.** In this method, brief high-voltage electrical pulses are administered to a solution containing plant tissues or naked protoplasts and DNA. The electrical pulses cause the brief opening of pores in the plasma membrane, allowing the DNA to enter the protoplast, with or without a cell wall. The electroporated cells with this incorporated DNA can be regenerated into transgenic plants.

In a second method, called **particle bombardment,** or **biolistics,** high-velocity microprojectiles (small gold or tungsten beads about 1 micrometer in diameter) are used to deliver RNA or DNA into cells. The RNA or DNA coats the surface of the beads, which are shot from a modified firearm (known as a gene gun) at a plant cell target, such as a piece of callus or a leaf.

Particle bombardment was used to create a virus-resistant strain of a papaya grown in Hawaii. Genes for the protein coat of papaya ringspot virus (PRSV), a pathogen that severely damages papaya trees, were "shot" into papaya tissue. Some of the papaya cells incorporated the viral genes into their DNA, conferring an ability on the part of the plant cells to destroy viruses. Subsequent growth of the cells resulted in transgenic papaya trees resistant to infection by papaya ringspot virus, saving the papaya industry of Hawaii (Figure 10–13). Other countries are now using particle bombardment to develop their

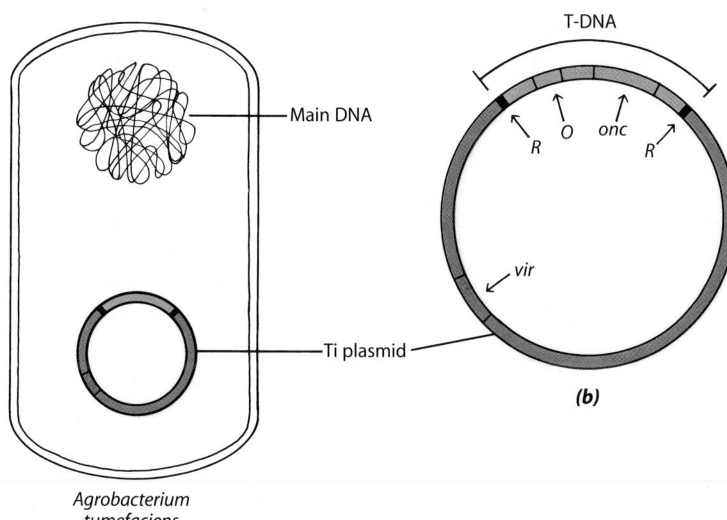

10–11 Ti plasmid ***(a)*** Diagrammatic representation of an *Agrobacterium tumefaciens* cell, showing the main DNA (the bacterial "chromosome") and the Ti plasmid. (The Ti plasmid and bacterial chromosome are not shown to scale.) ***(b)*** Detail of the Ti plasmid. *O* is the gene that codes for an opine-synthesizing enzyme; *onc* is a group of three genes coding for enzymes that are involved in the biosynthesis of plant hormones; and *R* represents sequences of 25 base pairs. Only the DNA between the two *R* regions is transferred into the plant's genome. The group of genes known as *vir* controls the transfer of the T-DNA to the host (plant) chromosome.

TOTIPOTENCY

The finding that either roots or shoots can be generated from the same undifferentiated callus cells has important implications for plant genetics (see Figure 10–9). As early as 1902, the German botanist Gottlieb Haberlandt suggested that all living plant cells are totipotent—that is, each cell possesses the potential to develop into an entire plant—but he was never able to demonstrate this. In fact, more than half a century passed before his hypothesis was proved to be essentially correct. Haberlandt did not know which substances to provide to the cells, because plant hormones had not yet been discovered.

In the late 1950s, F. C. Steward isolated small bits of phloem tissue from carrot *(Daucus carota)* root and placed them in liquid growth medium in a rotating flask. (Such pieces of tissue are called explants.) The medium contained sucrose and the inorganic nutrients necessary for plant growth, along with certain vitamins and coconut milk, which Steward knew to be rich in plant growth compounds—although the nature of these compounds was not then understood.

In the rotating flask, individual cells continuously broke away from the growing cell mass and floated free in the medium. These individual cells were able to grow and divide. Before long, Steward observed that roots had developed in many of these new cell clumps. If left in the swirling medium, the cell clumps did not continue to differentiate, but if they were transferred to a solid medium—which was agar in these experiments—some of the clumps developed shoots. If the clumps were then transplanted to soil, the little plants leafed out, flowered, and produced seed. Similar results were obtained a few years later by V. Vasil and A. C. Hildebrandt, who used explants of tobacco *(Nicotiana)* pith from a fresh stem of a hybrid. Rather than coconut milk, the medium used by Vasil and Hildebrandt contained the hormones IAA and kinetin, which are discussed in Chapter 27.

The results indicated that at least some of the cells of the mature carrot phloem and tobacco pith were totipotent—that is, they contained all the genetic potential for full plant development, although this potential was not expressed by these cells in the living plant. These experiments also showed that such differentiated cells can express portions of their previously unexpressed genetic potential in order to trigger particular developmental patterns. By achieving these results, Steward and Vasil and Hildebrandt confirmed Haberlandt's hypothesis concerning totipotency.

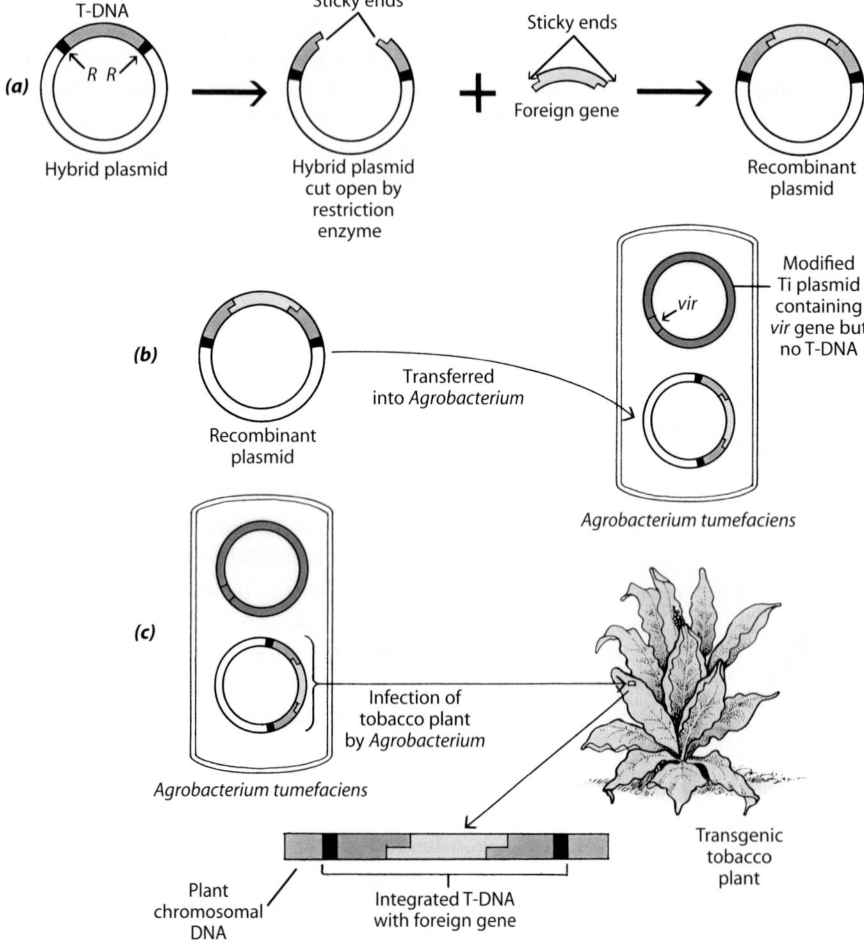

10–12 Plasmid vectors for DNA transfer Procedure for using plasmids of *Agrobacterium tumefaciens* as vectors in DNA, or gene, transfer. ***(a)*** A hybrid plasmid that carries only the T-DNA (blue) from a Ti plasmid is cut open with a restriction enzyme and a foreign gene (yellow) is inserted, creating a recombinant plasmid. ***(b)*** The recombinant plasmid is transferred into an *A. tumefaciens* cell that contains a Ti plasmid with its T-DNA removed (red), creating an engineered plasmid. ***(c)*** The *A. tumefaciens* containing the engineered plasmid is used to infect a plant. The *vir* region of the Ti plasmid without T-DNA controls the transfer of the foreign gene from the recombinant plasmid into the plant's chromosomes.

10–13 Transgenic virus-resistant papaya trees The insertion of protein-coat genes of papaya ringspot virus into papaya cells subsequently resulted in trees resistant to the virus. Seen here on the left are healthy transgenic papaya trees; on the right are virus-infected trees.

own transgenic papaya strains. Particle bombardment and *Agrobacterium*-mediated gene transfer are the two techniques most widely used today for the transfer of DNA to plants.

As noted by T. Erik Mirkov, genetic engineering will not completely replace traditional methods of plant breeding. He reminds us that "after gene manipulation and tissue culture have introduced a single gene into a crop, several years of plant breeding are always needed to make sure that the new plant has the right agronomic characteristics." The genes must also be shown to be transmitted in a stable way from generation to generation.

Genetic Engineering Is Being Used to Confer Resistance to Insects and Tolerance to Herbicides

To reduce the need for chemical insecticides, maize (corn), cotton, and other plants have been genetically engineered with genes from the bacterium *Bacillus thuringiensis (Bt).* These "transgenes" encode proteins (*Bt* toxins) that specifically kill the larvae of butterflies, moths, and beetles (Figure 10–14a), but do little or no harm to most other organisms, including people. Before the introduction of *Bt* corn, the European corn borer *(Ostrinia nubilalis)* caused losses of $1 billion per year in the United States alone. *Bt* cotton produces insecticidal toxins that resist the cotton bollworm *(Helicoverpa armigera),* which is also one of the most serious insect pests of wheat, corn, peanuts, soybeans, and vegetables. *Bt* cotton is grown extensively in China. Results of large-scale field monitoring of *H. armigera* in six provinces in northern China indicate that the deployment of *Bt* cotton there led to reduced populations of *H. armigera,* not only on cotton but also on other host crops. There are four generations of *H. armigera* per year in northern China. *Bt* cotton kills most of the larvae of the second generation and, apparently, works as a dead-end trap for much of the *H. armigera* population. Researchers have cautioned that prolonged use of *Bt* cotton for pest control increases the potential for insects to develop resistance to *Bt.* In 2009, *Bt*-resistant bollworms were reported in western India, the second largest cotton producer after China.

Plants have been engineered for tolerance to several herbicides, including glyphosate (marketed by Monsanto under the name Roundup), which works by blocking a single enzyme essential to plants for the production of aromatic amino acids. Although extremely effective and nontoxic to animals, glyphosate kills all plants, including crops. One successful approach to genetically engineering tolerance into crop plants resulted from the identification of a mutant form of the target enzyme from the bacterium *Salmonella* that is not blocked by glyphosate. Transfer of this mutant gene into crop plants, using the Ti plasmid, resulted in plants that were tolerant to the herbicide. The fields in which they grow can be treated with glyphosate, and the crops survive, with the weeds being controlled relatively inexpensively compared with more traditional methods.

Roundup Ready soybeans now make up more than 95 percent of the soybeans planted in the United States and about two-thirds of the soybeans planted globally. Some researchers have noted that the success of glyphosate-resistant soybean and other crop species has created an over-reliance on a single herbicide for weed management. In response to the resultant selection pressure, at least 13 weed species have developed glyphosate resistance.

Genetic Engineering Is Being Used to Reduce Postharvest Losses As genes involved in hormone biosynthesis and function are identified, genetic engineering is also being used to alter growth and development in plants. (Plant hormones are discussed in Chapter 27.) For example, genetically altered forms of the enzymes catalyzing ethylene biosynthesis have been transferred into tomato plants, resulting in diminished ethylene production and a significant delay in fruit ripening (Figure 10–14b). Ethylene regulates the ripening of many fleshy fruits. (Such fruits can be induced to ripen by application of ethylene after they have arrived at market.) Mutant forms of the ethylene receptor gene from *Arabidopsis* have been used to delay the senescence, or wilting, of flowers (Figure 10–14c, d). This achievement is of great interest to the cut-flower industry.

Genetic engineering has also been used to manipulate production of the hormone cytokinin. The coding sequence of the *Agrobacterium* gene for an enzyme that synthesizes a cytokinin was fused to the promoter sequence of an *Arabidopsis* gene that is expressed only in senescent leaves. When this chimeric gene was transferred into tobacco plants, the production of cytokinin in aging leaves dramatically delayed leaf senescence (Figure 10–14e). If this delayed-senescence effect could be made to operate in desirable crop plants, it could have a significant impact on yield.

Genetic Engineering Is Also Being Used to Improve Food Quality Alterations in fatty acid biosynthetic pathways have been used to reduce the proportion of saturated fats in soybean *(Glycine max)* and canola *(Brassica napus).* A soybean line with a high level of the unsaturated fatty acid oleic acid has been developed, resulting in a 20 percent overall reduction in saturated

(a) *(b)*

(c)

(d) *(e)*

10–14 Transgenic plants Gene transfer technology allows genetic engineers to alter growth, development, and disease resistance in plants. ***(a)*** The *BT* gene from the bacterium *Bacillus thuringiensis* codes for a protein that is toxic to butterflies and moths. When the *BT* gene is transferred into plants, the plant cells produce the toxin and become resistant to caterpillar damage, as seen here in transformed (left) and untreated (right) tomato *(Solanum lycopersicum)* plants four days after exposure to caterpillars. ***(b)*** Transfer of a mutant form of the ethylene receptor gene *(etr1-1)* from *Arabidopsis* into tomatoes renders the fruits insensitive to ethylene. On the left are transgenic tomatoes that remained the same golden color 100 days after picking, whereas untreated fruits, on the right, turned deep red and began to rot after the same length of time. ***(c), (d)*** Transfer of the *etr1-1* gene into petunia *(Petunia hybrida* cv. *Mitchell)* plants renders flowers insensitive to ethylene, resulting in their increased longevity. The flower from a transgenic plant in *(c)* is viable eight days after pollination, whereas the flower from an untreated plant in *(d)* is withered three days after pollination. ***(e)*** When tobacco *(Nicotiana tabacum)* plants are transformed with a gene that codes for a cytokinin biosynthetic gene that is expressed only in older leaves, a dramatic delay in leaf senescence results, as seen here 20 weeks after transplanting transgenic (left) and untreated (right) seedlings into soil.

fatty acids. Potatoes *(Solanum tuberosum)* are being developed with an increased starch content that would absorb less oil when fried, resulting in lower-fat french fries, and a project is under way in Australia to add vitamins A and E to bananas.

In 1999, an exciting new development was announced: production of a genetically modified line of rice, dubbed "Golden Rice," that has a significant β-carotene content (Figure 10–1); ordinary rice has essentially none. This development was accomplished by introducing into rice plants the genes for the last three enzymes in the biosynthetic pathway leading to the synthesis of β-carotene, which is converted in the human body into vitamin A. The genes came from a daffodil and a bacterium. Control sequences were also added to ensure that the β-carotene is synthesized in the rice endosperm, so that the vitamin will not be lost if milling removes the bran, as in the production of polished rice. The golden color is due to the presence of β-carotene in the endosperm. Vitamin A deficiency is widespread in south and southeast Asia, especially among children, and can cause blindness. The World Health Organization (WHO) estimates that between 250,000 and 500,000 children go blind every year as a result of this deficiency. Vitamin A deficiency is also responsible for childhood deaths, especially from diarrhea. Golden Rice holds the promise of eliminating these problems.

Almost immediately, however, opposition to genetically modified (GM) crops in general and Golden Rice in particular made it impossible to gain approval for Golden Rice. Scientists have since replaced the daffodil gene with a maize gene. This new version of Golden Rice, dubbed GR2, produces up to 23 times more β-carotene in its endosperm. In addition, GR2 and the original Golden Rice, now called GR1, were developed in subspecies *japonica,* which fares poorly in Asian fields. Thus it was necessary for researchers to backcross GR1 and GR2 lines with the long-grained nonsticky *indica* subspecies widely used by Asia's farmers. Sadly, there are no guarantees that Golden Rice will ever be approved in the targeted countries.

Gene transfer is being used in other plants to produce desirable products that the plant does not normally make. Current examples include the production of pharmaceutical proteins of mammalian origin, such as human growth hormone, which has been expressed in transgenic tobacco, and human serum albumin, expressed in tobacco and potato. Work is under way to produce polyhydroxybutyrate, a bioplastic that could replace petroleum-based products, in transgenic poplars.

The Introduction of Transgenes into Crops Has Both Benefits and Risks

Genetic engineering technology is providing biologists with an opportunity never before available—the transfer of genetic traits between very different organisms. For the plant biologist this means potentially increasing crop production by introducing genes that increase the crop's resistance to various pathogens or herbicides and enhance its tolerance to various stresses. Examples of the former were described earlier: the control of butterfly, moth, and beetle larvae and resistance to papaya ringspot virus and to the herbicide glyphosate. A major advantage of herbicide resistance is that it permits "no till" farming which, in turn, mitigates two very critical problems: (1) soil erosion (the rate of loss of good soils is alarming), and (2) carbon loss from soils (tilling greatly accelerates microbial conversion of soil carbon into CO_2). Insecticides, which can be highly toxic, are estimated to account for as many as 20,000 deaths per year, especially among farmworkers in developing countries. The use of *Bt* cotton, which provides resistance to the cotton bollworm, not only has resulted in yield increases over conventional cotton but also has led to huge reductions in insecticide use. Cultivation of *Bt* cotton has greatly reduced the exposure of farmworkers to broad-spectrum pesticides. In addition, a modified "ice-minus" strain of the bacterium *Pseudomonas syringae* is being used to reduce the susceptibility of certain crops to frost and thus allow earlier planting. Plants are being engineered for heat, drought, and salt tolerance, and attempts are under way to transform C_3 crops into C_4 crops (page 140) and to develop perennial cereal grains.

The benefits of genetic engineering do not come without some risks, which vary according to the traits introduced into the crop plants. There is no evidence that the process of transferring a gene from one plant or animal species to another poses a risk, but the properties of the new strain need to be carefully evaluated. There is also no evidence that the foods produced by genetically modified plants now on the market pose any risk to the health of humans or other animals. On the other hand, genes from a modified crop could reach its wild or weedy relatives through natural hybridization. The characteristics of the resulting hybrids or modified wild plants need to be considered in an environmental context, and the possibility of the formation of new weeds, although remote, should be taken into account.

Certain countries have been slow to accept genetically modified crops and the products derived from them, even though the use of pesticides in these countries is, in general, much heavier than in the United States. Genetically engineered crops may be regarded as one important element in promoting systems of agriculture that are sustainable. By no means, however, are genetically modified crops the only solution to this important problem. Biological controls and improved cultivation practices are also important elements in putting together such systems, which have a critical role to play in feeding a hungry world. How improved crop strains reach the needy poor worldwide, and what the role of large-scale agricultural corporations should be in an increasingly globalized economy, are also matters for careful consideration in a world that is changing rapidly.

Genomics

Genomics is the study of genomes in their entirety. Its goal is to understand the content, organization, function, and evolution of the genetic information of organisms. Genomics consists of three subfields: structural, functional, and comparative genomics.

Structural Genomics Is Concerned with the Organization and Sequence of the Genetic Information of Genomes

One of the first steps in characterizing a genome is to prepare genetic and physical maps of all of its chromosomes. **Genetic maps,** also called linkage maps (page 164), provide a *rough approximation* of the locations of genes relative to those of other known genes, as determined by rates of recombination. By contrast, **physical maps** are based on direct DNA sequencing information. Physical maps place genes in relation to the distances measured in numbers of base pairs. Figure 10–15

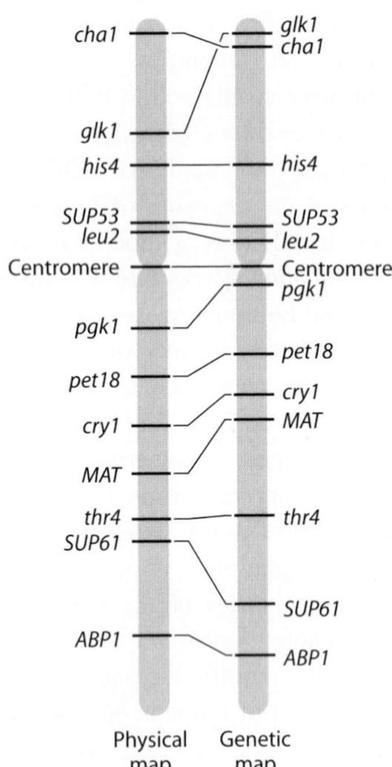

10–15 Maps of yeast chromosome III Physical (left) and genetic (right) maps may differ in the relative distances between genes and even in the order of genes on a chromosome. The physical map, which shows the actual locations of the genes, has greater resolution and accuracy than the genetic map.

compares genetic and physical maps of chromosome III of yeast. Although the two maps may differ in the distances between genes, because the rates of recombination are not constant across a chromosome, genetic maps have been critical to the development of physical maps and to the sequencing of whole genomes.

Functional Genomics Analyzes the Sequences Identified by Structural Genomics to Determine Their Function

A genomic sequence is, by itself, of limited use. The broad aim of functional genomics is to probe genome sequences for meaning—that is, to identify genes, to determine which genes are expressed (and under what conditions), and to determine their function or that of their protein products. The goals of functional genomics include identifying all the RNA molecules transcribed by a genome (the **transcriptome**) and all the proteins encoded by the genome (the **proteome**).

How does one go about determining the function of a newly discovered gene? A favored first approach is to conduct a search for **homology,** or relatedness of proteins' amino acid sequences, with other genes of known function. Public electronic databases containing millions of genes and proteins of known function, in a diverse group of organisms, are available for homology searches. If homology exists between a known gene and the newly discovered gene, a tentative biological function may be assigned to the newly discovered gene.

Collections of mutants, in which genes are inactivated by the insertion of a large piece of DNA, such as the T-DNA of *Agrobacterium,* have been developed for a large number of genes. These so-called **knockout mutants,** each with a different gene inactivated, are then screened for changes in their phenotype or for how they function in a defined environment. Any of the changes identified are traced back to the specific sequence mutated.

Considerable insight about gene function comes from knowing when and where the genes are expressed. With the development of DNA microarray analysis, scientists can monitor the expression of thousands of DNA fragments simultaneously. **Microarrays** consist of numerous closely packed DNA fragments attached to a glass slide, or "chip," in an orderly pattern or array, usually as dots. The DNA fragments correspond to all of the genes in a genome. The DNA chip is exposed to a sample of labeled RNA (referred to as a probe), formed from the RNA taken from a cell. Each RNA transcript will bind (hybridize) to its complementary DNA sequence. The spots on the chip to which it binds indicate the genes that were being actively transcribed in the cell under a given condition.

Microarray analysis enables researchers to study which genes are active in particular tissues and to learn how gene expression changes during specific developmental processes and under environmental stress, such as pathogen attack, water deficit, or lack of nutrients.

Comparative Genomics Provides Important Information about Evolutionary Relationships among Organisms

Comparative genomics compares the gene content, function, and organization of the genomes of different organisms. Comparisons of genome sequences are bringing about a greater understanding of the evolutionary relationships among organisms. The impact of such comparisons is evident in the diversity chapters of this book (see Chapters 12–20).

The first clear-cut result from comparative genomics was confirmation of the existence of three domains of living organisms: Bacteria, Archaea, and Eukarya (see Chapter 12). Greater sequence similarities exist among members of the same domain than between members of two different domains. In addition, many sequences are found only in a specific domain.

Prokaryotic Genomes Are Highly Diverse and Can Undergo Horizontal Gene Transfer

With few exceptions, prokaryotic genomes consist of a single circular chromosome (see Chapter 13). The total amount of DNA in prokaryotic genomes ranges from just 159,662 base pairs in *Carsonella ruddii,* an aphid endosymbiont, to more than 9.1 million base pairs in *Bradyrhizobium japonicum,* a root-nodule endosymbiont. *Escherichia coli,* a bacterium widely used in genetic studies, has 4.6 million base pairs.

Only about half of the genes identified in prokaryotic genomes can be assigned a function. About a quarter of the genes have no significant similarity to any other known genes in bacteria, indicating that considerable genetic diversity exists among bacteria. Indeed, sequencing of hundreds of strains of *E. coli* reveals that over 50 percent of the genes present in one strain are not present in any of the other strains. In addition, the results of genomic studies reveal that both closely and distantly

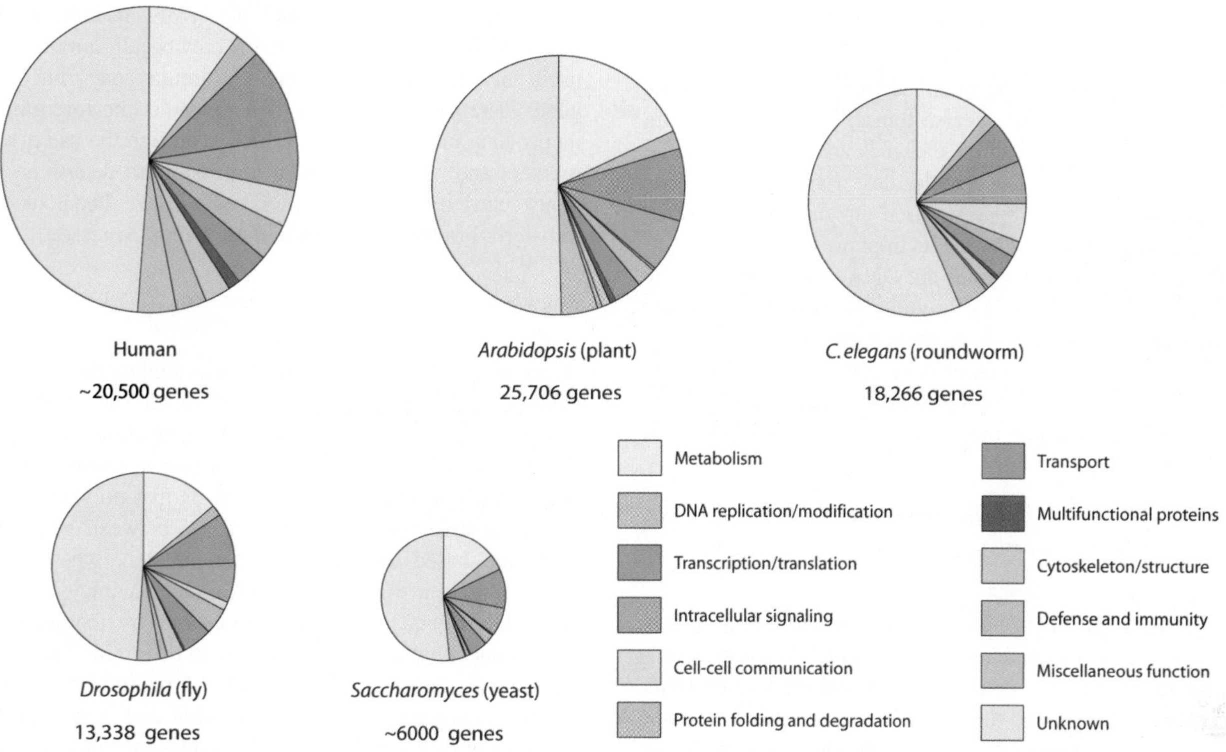

10–16 Comparison of the numbers and types of proteins encoded in the genomes of different eukaryotes For each eukaryote shown here, the area of the entire pie chart represents the total number of protein-coding genes, shown at roughly the same scale. In each case, the proteins encoded by about half of the genes are unknown. The functions of the rest of the genes are known or are predicted from their similarity to genes of known function.

related bacterial species periodically exchange genetic information over evolutionary time—a process called **horizontal,** or **lateral, gene exchange.** Gene transfers also have occurred between bacteria and eukaryotes, as in the case of the T-DNA of *Agrobacterium,* which is transferred into plant cell nuclei, and the migration into the nucleus of ancient mitochondrial and chloroplast genes (see Chapter 12).

Eukaryotic Genomes Vary Greatly in the Number of Protein-Coding Genes

The genomes of eukaryotic organisms are larger than those of prokaryotes, and, in general, multicellular eukaryotes have more DNA than do one-celled eukaryotes such as yeast (see Table 9–1). Figure 10–16 shows the total number of protein-coding genes in several multicellular eukaryotes that have been completely sequenced. The functions of about half of the proteins encoded by these genomes are known or have been tentatively assigned on the basis of sequence comparisons. Note that the number of protein-coding genes in different organisms does not seem proportional to biological complexity. For example, the roundworm *Caenorhabditis elegans* apparently has more genes than the fruit fly *Drosophila,* which has a much more complex body plan and more complex behavior. Moreover, both the roundworm and the fruit fly have fewer genes than the structurally less complex plant *Arabidopsis.* Plant genomes, however, have large numbers of duplicated genes, often from whole genome doublings (polyploidization), and animals such as *Drosophila* often produce multiple proteins from a single gene. Transposable elements are the largest contributors to genome size. Thus the significance of differences in gene number among organisms is not clear.

The genomes of many other organisms are in the process of being sequenced. These new findings, in addition to the large amounts of existing DNA sequence data, provide valuable information for applications in agriculture, human health, and biotechnology. Knowing the complete genome sequences of crop plants will help identify the genes that affect yield, disease resistance, and pest resistance, in addition to other agricultural traits. These genes can then be manipulated by genetic engineering or even by traditional breeding to increase yields and to produce more nutritious foods.

SUMMARY

Recombinant DNA Technology Is Used to Create Novel Genotypes

Recombinant DNA technology includes methods for (1) obtaining DNA segments short enough to be analyzed and manipulated, (2) obtaining large quantities of identical DNA segments, and (3) determining the exact sequence of nucleotides in a DNA segment.

Restriction Enzymes Are Used to Cut DNA into Fragments Having Sticky Ends

Short DNA segments can be obtained by transcribing mRNA into DNA with the enzyme reverse transcriptase or by cutting DNA molecules with restriction enzymes, which are bacterial enzymes that cut foreign DNA molecules at specific sites. The DNA segments produced by restriction enzymes can be separated on the basis of their size by electrophoresis.

Different restriction enzymes cut DNA at different specific nucleotide sequences. Instead of cutting the two strands of the molecule straight across, some restriction enzymes leave sticky ends. Any DNA cut by such an enzyme can readily be joined to another DNA molecule cut by the same enzyme. The discovery of restriction enzymes has made possible the development of recombinant DNA technology.

DNA Cloning and the Polymerase Chain Reaction Are Used to Produce Large Quantities of Identical DNA Segments

In cloning, the segments to be copied are introduced into bacterial cells by means of plasmids or viruses, which function as vectors. Once in the bacterial cell, the vector and the foreign DNA it carries are replicated, and the multiple copies can be harvested from the cells. The polymerase chain reaction is a much more rapid process by which millions of copies of a DNA segment can be made in just a few hours.

The availability of multiple copies makes possible, in turn, the determination of the exact order of nucleotides in a DNA segment. By combining sequencing information for sets of short segments produced by different restriction enzymes, molecular biologists can determine the complete sequence of a long DNA segment (such as an entire gene).

Plant Biotechnology Involves the Use of Tissue Culture

Advances in hormone research and DNA biochemistry have made it possible to manipulate the genetics of plants in specific ways. Among the most important procedures in biotechnology is tissue culture. In ideal cases, tissue culture is used to obtain complete plants from single, genetically altered cells. The potential for single cells to develop into entire plants is called totipotency.

Genetic Engineering Involves the Manipulation of Genes for Practical Purposes

Genetic engineering (recombinant DNA) technology is based on the ability to cut DNA molecules precisely into specific pieces and to combine those pieces to produce new combinations. The Ti plasmid of *Agrobacterium tumefaciens*, which induces the formation of crown-gall tumors, is being used as a vector for introducing foreign genes into a plant's genes. The transgenic plant then transmits the foreign genes to its progeny in a Mendelian fashion. Through the use of selective markers and reporter genes, it is possible to determine whether genes carried by the plasmid vector have been successfully transferred to the plant cells and are being expressed.

The Field of Genomics Includes Structural, Functional, and Comparative Genomics

Genomics is the field of genetics that studies the content, organization, and function of genetic information in whole genomes. Structural genomics focuses on the organization and sequence of the genetic information contained within a genome and is concerned with genetic and physical maps that provide information on the relative positions and distances between genes. Genetic maps are based on rates of recombination, whereas physical maps, which have greater resolution and accuracy, are based on direct DNA sequencing information and are measured in base pairs. Functional genomics attempts to determine the function of the genetic sequences elucidated by structural genomics. Comparative genomics compares the content and organization of the genomes of different species and provides information about evolutionary relationships.

Genomics has been characterized as the ultimate extension of recombinant DNA technology to the global analysis of the nucleic acids present in the nucleus, the cell, an organism, or a group of organisms. Information obtained by genomics has accelerated the process of identifying genes responsible for various biological phenomena and has provided an immensely powerful resource in the study of phylogenetic, or evolutionary, relationships among organisms.

QUESTIONS

1. What are the uses of restriction enzymes in recombinant DNA technology?
2. Describe the role of sticky ends in recombinant DNA technology. How are sticky ends produced? What enzyme is required to complete their recombination?
3. Explain how the Ti plasmid of *Agrobacterium tumefaciens* is used in the production of transgenic plants.
4. Distinguish between structural and functional genomics.
5. Of what value is the study of comparative genomics?

CHAPTER 11

The Process of Evolution

◀ **Carnivore adaptation** Growing in acidic bogs, where the bacteria that make nitrogen available to plants cannot live, carnivorous plants extract vital nitrogen from the insects they trap and digest. The sticky, shimmering filaments of sundews *(Drosera intermedia)* attract the unwary, such as this common blue damselfly. As Darwin noted, sundews "often grow in places where hardly any other plant can exist."

CHAPTER OUTLINE

In 1831, as a young man of 22, Charles Darwin (Figure 11–1) set forth on a five-year voyage as ship's naturalist on a British navy ship, HMS *Beagle.* The book he wrote about the journey, *The Voyage of the Beagle,* not only is a classic work of natural history but also provides us with insight into the experiences that led directly to Darwin's proposal of his theory of evolution by natural selection.

At the time of Darwin's historic voyage, most scientists—and nonscientists as well—still believed in the theory of "special creation." According to this idea, each of the many different kinds of living organisms was created (or otherwise came into existence) in its present form. Some scientists, such as Jean Baptiste de Lamarck (1744–1829), had proposed theories of evolution but could not convincingly explain the mechanism by which the process occurred.

Darwin's Theory

Darwin was able to bring about a great intellectual revolution simply because the mechanism for evolution he presented was so convincing that there was no longer room for reasonable scientific doubt. Particularly important in the genesis of Darwin's

CHECKPOINTS

After reading this chapter, you should be able to answer the following:

1. What is Darwin's theory of evolution?
2. How is the Hardy-Weinberg equilibrium important for studying evolution? What are the four agents, other than natural selection, that can change the composition of the gene pool in a population, and how do these changes occur?
3. What are some ways in which organisms adapt to their physical environments?
4. How does the biological species concept differ from the morphological species concept, and how do both of these concepts differ from the various phylogenetic species concepts?
5. How are new species formed? What mechanisms keep closely related species from interbreeding?

11–1 Charles Darwin In his book *The Voyage of the Beagle,* Darwin made the following comments about his selection as ship's naturalist for the voyage: "Afterwards, on becoming very intimate with FitzRoy [the captain of the *Beagle*], I heard that I had run a very narrow risk of being rejected on account of the shape of my nose! He . . . was convinced that he could judge of a man's character by the outline of his features; and he doubted whether anyone with my nose could possess sufficient energy and determination for the voyage. But I think he was afterwards well satisfied that my nose had spoken falsely."

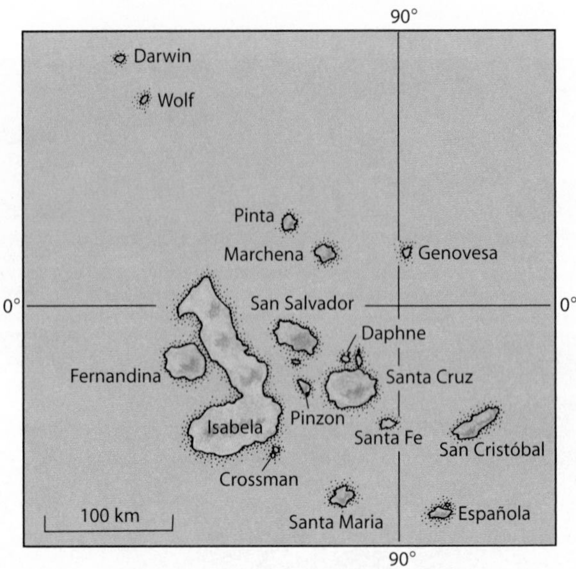

11–2 The Galápagos Archipelago Some 950 kilometers west of the coast of Ecuador, the Galápagos Islands consist of 13 principal volcanic islands and many smaller islets and rocks. These islands have been called "a living laboratory of evolution." "One is astonished," wrote Charles Darwin in 1837, "at the amount of creative force . . . displayed on these small, barren, and rocky islands."

ideas were his observations during a stay of some five weeks in the Galápagos Islands, an archipelago that lies in equatorial waters some 950 kilometers off the western coast of South America (Figure 11–2). There he made two particularly important observations. First, he noted that the plants and animals found on the islands, although distinctive, were similar to those on the nearby South American mainland. If each kind of plant and animal had been created separately and was unchangeable, as was then generally believed, why did the plants and animals of the Galápagos not resemble those of Africa, for example, rather than those of South America? Or indeed, why were they not utterly unique, unlike organisms anywhere else on Earth? Second, people familiar with the islands pointed out variations that occurred from island to island in such organisms as the giant tortoises. Sailors who took these tortoises on board and kept them as sources of fresh meat on their sea voyages were able to tell from which island a particular tortoise had come simply by noting its appearance. If the Galápagos tortoises had been specially created, why did they not all look alike?

Darwin began to wonder whether all the tortoises and other strange animals and plants of the Galápagos might not have been derived at different times from organisms that existed on the mainland of South America. Once they reached this remote archipelago, these organisms might have spread slowly from island to island, changing bit by bit in response to local conditions and eventually becoming distinct variants that could be differentiated easily by the human eye.

In 1838, after returning from his historic voyage, Darwin read a book entitled *Essay on the Principles of Population,* by Thomas Malthus, an English clergyman. Published in 1798, eleven years before Darwin's birth, the book sounded an early warning about the explosive growth of the human population: the human population was increasing so rapidly that it would soon be impossible to feed all of the Earth's inhabitants. Darwin saw that Malthus's reasoning was theoretically correct, not only for the human population but also for all other populations of organisms. Virtually any kind of animal or plant, if it were to reproduce unchecked, would cover the Earth's surface within a relatively short period of time. Darwin noted, however, that this does not occur. Instead, populations of species remain more or less constant in size year after year because death intervenes, limiting their numbers. To Darwin, it became obvious: there must be a struggle for existence, in which favorable variations tend to survive.

In 1842, when he was 33 years old, Charles Darwin wrote out his overall argument for evolution by natural selection and then continued to enlarge and refine his manuscript for many years. The independent discovery of these principles by another British naturalist, Alfred Russel Wallace (1823–1913), who

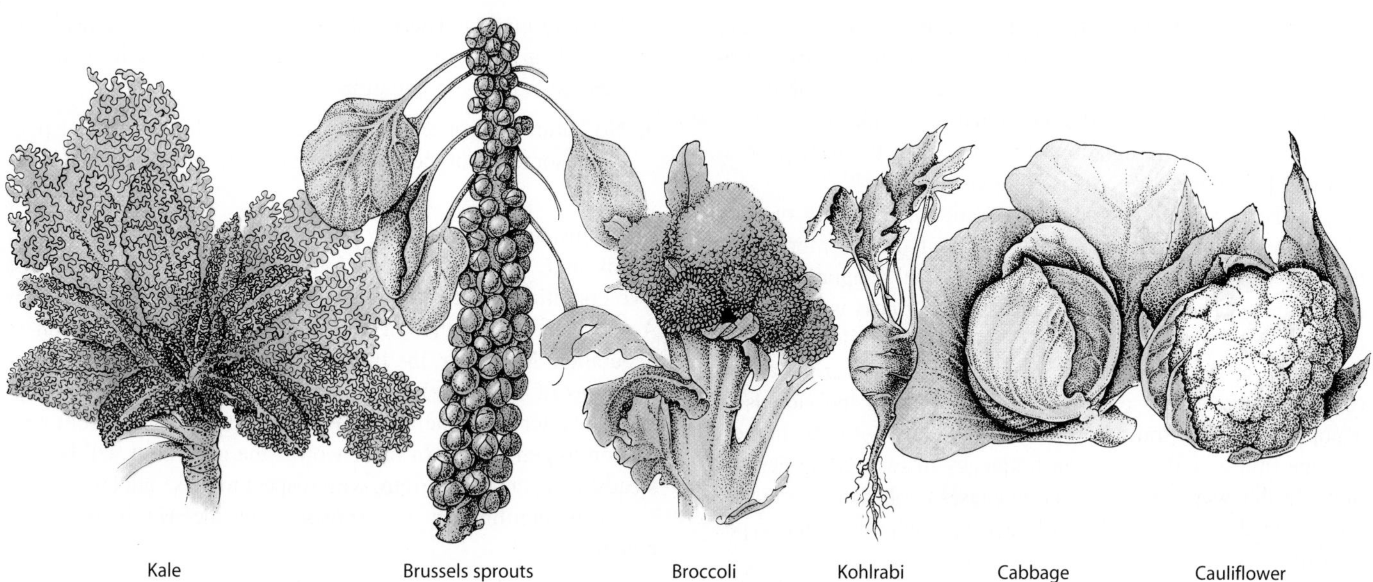

11–3 Artificial selection Shown here are six vegetables, all produced from a single species *(Brassica oleracea)*, a member of the mustard family. They are the result of selection for leaves (kale), lateral buds (brussels sprouts), flowers and stem (broccoli), stem (kohlrabi), enlarged terminal buds (cabbage), and flower clusters (cauliflower). Kale most resembles the ancestral wild plant. Artificial selection, as practiced by plant and animal breeders, gave Darwin the clue to the concept of natural selection.

sent Darwin an essay on the subject in 1858, induced Darwin finally to publish his book, *On the Origin of Species by Means of Natural Selection, or the Preservation of Favoured Races in the Struggle for Life.* It appeared in November 1859 and immediately was recognized as one of the most influential and important books of all time.

In his book, Darwin named the process by which surviving progeny are chosen **natural selection** and presented extensive evidence that this mechanism is the principal means by which evolution occurs. He likened natural selection to **artificial selection,** the process by which breeders of domesticated plants and animals deliberately change the characteristics of the strains or races in which they are interested (Figure 11–3). They do this by allowing only those individuals with desirable characteristics to breed. Darwin recognized that wild organisms are also variable. Some individuals have characteristics that enable them to produce more progeny than others under the prevailing environmental conditions. As a result of advantages in survival, fecundity, or mating success, the characteristics of the individuals with more progeny gradually become more common in the population than the characteristics of those with fewer progeny. Ultimately, this tendency can be counted on to produce slow but steady changes in the frequencies of different characteristics in populations. It is this process that leads to evolution.

In artificial selection, breeders can concentrate their efforts on one or a few characteristics of interest, such as fruit size or animal weight. In natural selection, however, the entire organism must be "fit" in terms of the total environment in which it lives. In other words, the entire phenotype is subject to natural selection. Such a process might reasonably be expected to require a long period of time, and it is no coincidence that the writings of the geologist Charles Lyell, who postulated that the Earth was much older than previously thought, had a profound influence on Darwin. Darwin needed an older Earth as a stage upon which to view the unfolding of the diversity of living things. The process of natural selection soon became known as the "survival of the fittest," which is an apt phrase but one, as we will see, that must be employed cautiously.

The Concept of the Gene Pool

A new branch of biology, **population genetics,** emerged from the synthesis of Mendelian principles with Darwinian evolution. A **population** can be defined as a local group of potentially interbreeding individuals belonging to the same species. For now, we can think of a **species** as a group of populations that have the potential to interbreed in nature.

A population is unified and defined by its **gene pool,** which is simply the sum total of all the alleles of all the genes of all the individuals in the population. From the viewpoint of the population geneticist, each individual organism is only a temporary vessel, holding a small sampling of the gene pool for a moment in time. Population geneticists are interested in gene pools, the changes in their composition over time, and the forces causing these changes.

In natural populations, some alleles increase in frequency from generation to generation, and others decrease. (The frequency of an allele is simply the proportion of that allele in a population in relation to all the alleles of the same gene.) If an individual has a favorable combination of alleles in its genotype, it is more likely to survive and reproduce. As a consequence, its alleles are likely to be present in an increased proportion in the next generation. Conversely, if the combination of alleles is not favorable, the individual is less likely to survive and reproduce. Representation of its alleles in the next generation will be reduced or perhaps eliminated.

In the context of population genetics, the **fitness** of each of the individuals that share possession of one or more traits does not mean physical well-being or optimal adaptation to the environment. The sole relevant measure of fitness is the rate of production of viable, surviving offspring per unit time, averaged over all individuals that share a particular heritable trait (or traits), divided by the average fitness for all individuals in a population. The traits borne by individuals with the highest average fitness will increase in frequency in succeeding generations. In this way, natural selection causes evolution to occur by increasing the representation of favored alleles and genotypes over time.

The Behavior of Genes in Populations: The Hardy-Weinberg Law

In the early 1900s, biologists raised an important question, alluded to earlier, about the maintenance of variability in populations. How, they asked, can both dominant and recessive alleles remain in populations? Why don't dominants simply drive out recessives? For example, if in a population of plants with contrasting flower color, the allele for purple color is dominant over the allele for white color, why are not all the flowers purple? The question is clearly important to understanding evolution, because it is the gene pool of populations that provides the "raw material" on which the process of natural selection operates. This question was answered in 1908 by G. H. Hardy, an English mathematician, and G. Weinberg, a German physician.

Working independently, Hardy and Weinberg showed that in a large population in which random mating occurs and in the absence of forces that change the proportions of alleles (discussed below), the original ratio of dominant alleles to recessive alleles is retained from generation to generation. In other words, the **Hardy-Weinberg law,** as it is now known, states that the proportions, or *frequencies,* of the alleles in a population's gene pool remain constant, or at equilibrium, from generation to generation unless acted upon by agents other than sexual recombination. Furthermore, the genotypic frequencies stabilize after one generation in proportions determined by the allelic frequencies. To demonstrate this, they examined the behavior of alleles in an idealized *nonevolving* population in which five conditions hold:

1. No mutations. Mutations alter the gene pool by changing one allele into another.
2. Isolation from other populations. The movement of individuals—with the transfer of their alleles—into or out of the population can change gene pools.
3. Large population size. If the population is large enough, the laws of probability apply—that is, it is highly unlikely that chance alone can alter the frequencies, or relative proportions, of alleles.
4. Random mating. The Hardy-Weinberg equilibrium will hold only if an individual of any genotype chooses its mates at random from the population.
5. No natural selection. Natural selection alters the gene pool when some genotypes produce more offspring than others.

Consider a single gene that has only two alleles, *A* and *a*. Hardy and Weinberg demonstrated mathematically that if the five conditions listed above are met, the frequencies of alleles *A* and *a* in the population will not change from generation to generation. Moreover, the frequencies of the three possible combinations of these alleles—the genotypes *AA, Aa,* and *aa*—will stabilize after one generation and then will not change from generation to generation. In other words, the gene pool will be in a steady state, an equilibrium, with respect to these alleles.

This equilibrium is expressed by the Hardy-Weinberg equation:

$$p^2 + 2pq + q^2 = 1$$

In this equation, the letter p designates the frequency of one allele at a specific locus (for example, *A*), and the letter q designates the frequency of the other allele *(a)*. The sum of p and q must always equal 1 (that is, 100 percent of the alleles of that particular gene in the gene pool). The expression p^2 designates the frequency of individuals homozygous for one allele *(AA),* q^2 the frequency of individuals homozygous for the other allele *(aa),* and $2pq$ the frequency of heterozygotes *(Aa).*

The Hardy-Weinberg Equilibrium Provides a Standard for Detecting Evolutionary Change

The Hardy-Weinberg equilibrium and its mathematical formulation have proved as valuable a foundation for population genetics as Mendel's principles have been for classical genetics. At first glance, this seems hard to understand, because the five conditions specified for a gene pool in equilibrium—that is, a nonevolving population—are seldom likely to be met in a natural population. How, then, can the Hardy-Weinberg equation be useful? An analogy from physics may be helpful. Newton's first law says that a body remains at rest or maintains a constant velocity when not acted upon by an external force. In the real world, bodies are always acted upon by external forces, but this first law is an essential premise for examining the nature of such forces. It provides a standard against which to measure.

Similarly, the Hardy-Weinberg equation provides a standard against which we can measure the changes in allele frequencies that are always occurring in natural populations. Without the Hardy-Weinberg equation, we would not be able to detect change, determine its magnitude and direction, or uncover the forces responsible for it.

The Agents of Change

For evolution to occur, the frequencies of alleles or genotypes in a population must deviate from the Hardy-Weinberg equilibrium. At the population level, evolution may be defined as a generation-to-generation change in a population's genetic structure. Such small-scale generation-to-generation change in the frequency of a population's alleles is referred to as **microevolution.**

According to modern evolutionary theory, natural selection is the major force in changing the composition of the gene pool. Let us first consider some other agents that can change the composition of a population's gene pool. There are four: mutations, gene flow, genetic drift, and nonrandom mating.

Mutations Provide the Variations on Which Evolutionary Forces Act

From the point of view of population genetics, **mutations** are heritable changes in the genotype (Figure 11–4). As we learned earlier, a mutation may involve the substitution of one or a few nucleotides in a DNA molecule or changes in whole chromosomes, segments of chromosomes, or even entire sets of chromosomes (pages 164 through 166). Most mutations occur "spontaneously"—meaning simply that we do not know the factors that triggered them. Mutations are generally said to occur at random, or by chance. This does not mean that mutations occur without cause but rather that the events triggering them are independent of their subsequent effects. Although the rate of mutation can be influenced by environmental factors, the specific mutations produced are independent of the environment—and independent of their potential for subsequent benefit or harm to the organism and its offspring.

Although the rate of spontaneous mutation is generally low, mutations provide the raw material for evolutionary change, because they provide the variation acted upon by evolutionary forces. However, because mutation rates are so low, the amount of change due to mutation is very small.

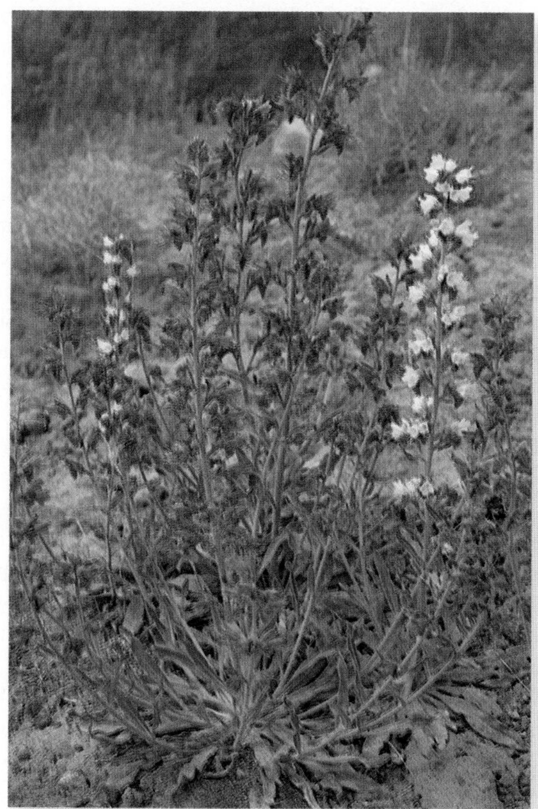

11–4 Mutation in viper's bugloss *(Echium vulgare)* A recessive mutation, *a*, occurred in cells of an *AA* (homozygous dominant) blue-flowered plant that were destined to develop into gametes, making those cells *Aa*. On self-pollination, the mutation was transmitted to progeny, some of which were *aa* (homozygous recessive) and expressed the mutant white-flowered phenotype, as seen here.

Gene Flow Is the Movement of Alleles into or out of a Population

The movement of alleles into or out of a population—**gene flow**—can occur as a result of immigration or emigration of individuals of reproductive age. In the case of plants, gene flow can also occur through the movement of gametes by transfer of pollen between populations.

Gene flow can introduce new alleles into a population or it can change existing allele frequencies. Its overall effect is to decrease the difference between populations. Natural selection, by contrast, can increase differences, producing populations more suited to local conditions. Thus, gene flow often counteracts natural selection. The overall effect of gene flow is the increased genetic variation within populations and decreased differences between populations.

The possibilities of gene flow between natural populations of most plant species diminish rapidly with distance. Although pollen sometimes can be dispersed over great distances, the chances of its landing on a receptive stigma at any great distance are slight, particularly if other plants are releasing pollen into the air or onto pollinators nearby. For several kinds of insect-pollinated plants that grow in temperate regions, a gap of only 300 meters may effectively isolate two populations. Rarely will more than 1 percent of the pollen that reaches a given individual come from this far away. In wind-pollinated plants, very little pollen falls more than 50 meters from the parent plant under normal circumstances.

Genetic Drift Refers to Changes That Occur Due to Chance

As we stated previously, the Hardy-Weinberg equilibrium holds true only if the population is large. This qualification is necessary because the equilibrium depends on the laws of probability. Consider, for example, an allele, say *a,* that has a frequency of 1 percent. In a population of 1 million individuals, 20,000 *a* alleles would be present in the gene pool. (Remember that each diploid individual carries two alleles for any given gene. In the gene pool of this population there are 2 million alleles for this particular gene, of which 1 percent, or 20,000, are allele *a.*) If a few individuals in this population were destroyed by chance before leaving offspring, the effect on the frequency of allele *a* would be negligible.

In a population of 50 individuals, however, the situation would be quite different. In this small population, it is likely that only one copy of allele *a* would be present. If the lone individual carrying this allele failed to reproduce or were destroyed by chance before leaving offspring, allele *a* would be completely lost. Similarly, if 10 of the 49 individuals homozygous for allele *A* were lost, the frequency of *a* would jump from 1 in 100 to 1 in 80.

This phenomenon, a change in the gene pool that takes place as a result of chance, is **genetic drift.** Population geneticists and other evolutionary biologists generally agree that genetic drift plays a role in determining the evolutionary course of small populations. Its relative importance compared with that of natural selection, however, is a matter of debate. There are at least two situations—the founder effect and the bottleneck effect—in which genetic drift has been shown to be important.

The Founder Effect Occurs When a Small Population Colonizes a New Area A small population that becomes separated from a larger one may or may not be genetically representative of the larger population from which it was derived (Figure 11–5). Some rare alleles may be overrepresented or, conversely, may be completely absent in the small population. An extreme case would be the initiation of a new population by a single plant seed. As a consequence, when and if the small population increased in size, it would continue to have a different genetic composition—a different gene pool—from that of the parent group. This phenomenon, a type of genetic drift, is known as the **founder effect.**

The Bottleneck Effect Occurs When Environmental Factors Suddenly Decrease Population Size The **bottleneck effect** is another type of situation that can lead to genetic drift. It occurs when a population is drastically reduced in numbers by an event, such as an earthquake, flood, or fire, that may have little or nothing to do with the usual forces of natural selection. A population bottleneck is likely not only to eliminate some alleles entirely but also to cause others to become overrepresented in the gene pool.

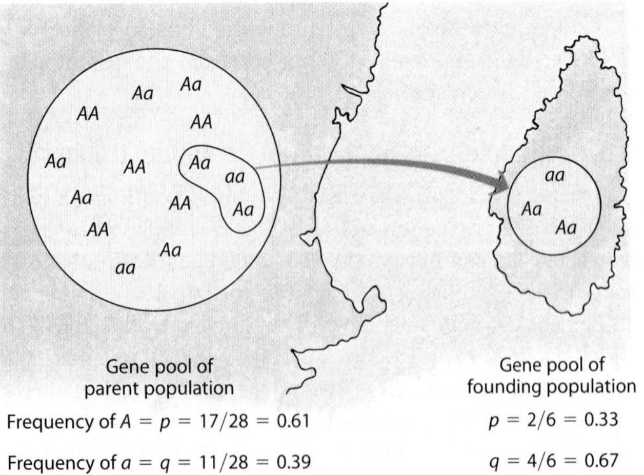

11–5 The founder effect When a small subset of a population founds a new colony (for example, on a previously uninhabited island), the allele frequencies within the founding group may be different from those within the parent population. Thus the gene pool of the new population will have a different composition from the gene pool of the parent population.

Nonrandom Mating Decreases the Frequency of Heterozygotes

Disruption of the Hardy-Weinberg equilibrium can also result from nonrandom mating. Typically, the members of a population mate more often with close neighbors than with more distant ones. Thus, within the large population, neighboring individuals tend to be closely related. Such nonrandom mating promotes **inbreeding,** the mating of closely related individuals. An extreme form of nonrandom mating that is particularly important in plants is self-pollination (as in the pea plants Mendel studied). Nonrandom mating does not change the frequencies of alleles within a population, but it does affect the genotypic frequencies.

Inbreeding and self-pollination tend to increase the frequencies of homozygotes in a population at the expense of the heterozygotes. Take, for example, Mendel's pea plants, in which only two alleles are involved in flower color, *W* (purple) and *w* (white). When *WW* plants and *ww* plants self-pollinate, all of their progeny are homozygous. When *Ww* plants self-pollinate, however, only half of their progeny are heterozygous. With succeeding generations, there will be a decrease in the frequency of heterozygotes, with a corresponding increase in the frequencies of the two homozygotes. Note, again, that although nonrandom mating, as exemplified by the pea plants, can change the ratio of genotypes and phenotypes in a population, the frequencies of the alleles in question remain the same.

Responses to Selection

In the course of the controversies that led to the synthesis of evolutionary theory with Mendelian genetics, some biologists argued that natural selection could serve only to eliminate the "less fit." As a consequence, it would tend to reduce the genetic variation in a population and thus reduce the potential for further evolution. Modern population genetics has demonstrated that this is not always the case. Natural selection can, in fact, be a critical factor in preserving and promoting genetic variability in a population.

The response of a population to selection is affected by many of the principles of genetics discussed in Chapter 8. In general, only the phenotype is being selected, and it is the relationship between the phenotype and the environment that determines the reproductive success of an organism. Because many characteristics in natural populations are determined by the interactions of many genes, phenotypically similar individuals can have many different genotypes.

When some feature, such as tallness, is strongly selected for, there is an accumulation of alleles that contribute to this feature and an elimination of those alleles that work in the opposite direction. But selection for a polygenic characteristic is not simply the accumulation of one set of alleles and the elimination of others. Gene interactions, such as epistasis (one gene affects the phenotypic expression of another gene at another locus) and pleiotropy (one gene affects a number of phenotypic characteristics), are of fundamental importance in determining the course of selection in a population.

Bear in mind that the phenotype is not determined solely by the interactions of the multitude of alleles making up the

(a)

(b)

11–6 Environmental effects on phenotype ***(a)*** Jeffrey pines *(Pinus jeffreyi)* usually grow tall and straight. ***(b)*** Environmental forces, however, can alter the normal growth patterns. This Jeffrey pine is growing on a mountaintop in Yosemite National Park, California, where it is exposed to strong, constant winds.

genotype. The phenotype is very much a product of the interaction of the genotype with the environment in the course of the individual's life (Figure 11–6).

Also keep in mind that the relative fitness associated with particular traits depends on the environment. That is, a trait that may be selectively favored in some environments may be selectively disadvantageous in others. For example, greater height is favored in crowded environments because it reduces the chance that a plant will be overtopped by competitors, thereby reducing its rate of energy capture. On the other hand, great height would be a disadvantage in sparsely populated microsites in which it is unlikely that a competing plant grows nearby. (A microsite is a small area with unique characteristics within an environment.) In the absence of an advantage of overtopping (or avoiding being overtopped), the energetic cost of producing taller stems favors the evolution of short plants in sparsely covered microsites.

Moreover, the relative fitness associated with particular traits depends on the range of other phenotypes present in a population. For example, many phenotypes (and associated genotypes) might be able to survive in a particular range of temperatures. But if some of those phenotypes (and genotypes) produce faster growth, the relative fitness of the others may fall to zero.

Evolutionary Changes in Natural Populations May Occur Rapidly

Under certain circumstances, the characteristics of populations may change rapidly, often in response to a rapidly changing environment. Particularly during the past few centuries, the influence of human beings in many areas has been so great that some populations of other organisms have had to adjust rapidly in order to survive. Evolutionary biologists have been particularly interested in examples of such rapid changes, because the principles involved are presumed to be the same as those that govern changes in populations generally.

In a Maryland study, plants growing in the ungrazed part of a pasture were found to be taller than those in the grazed part. Such differences clearly could have originated from the effects of grazing. In fact, when samples of the plant species involved were taken from grazed and ungrazed parts of the pasture and grown together in a garden, it was found that the differences were genetic. Plants of white clover *(Trifolium repens),* Kentucky bluegrass *(Poa pratensis),* and orchard grass *(Dactylis glomerata)* from the grazed part of the pasture remained shorter than those from the ungrazed part. This allowed the deduction that shorter plants of these species, perhaps because they were missed during intensive grazing, were at a selective advantage under conditions of grazing, and that the alleles responsible for short stature had increased in the grazed portion of the pasture. A similar example is illustrated in Figure 11–7.

In Wales, the tailings around a number of abandoned lead mines are rich in lead (up to 1 percent) and zinc (up to 0.03 percent)—substances that are toxic to most plants at these concentrations. Because of the presence of these metals, the tailings are often nearly devoid of plant life. Observing that one species of grass, *Agrostis tenuis,* was colonizing the mine soil, scientists took some *Agrostis* plants from these areas and others from nearby pastures and grew them together either in normal soil

(a)

(b)

11–7 Genetic versus environmental selection for phenotype *Prunella vulgaris* is a common herb of the mint family; it is widespread in woods, meadows, and lawns in temperate regions of the world. Most populations consist of erect plants, such as those shown in ***(a)***, which grow in open, often somewhat moist, grassy places throughout the cooler regions of the world. Populations found in lawns, however, always consist of prostrate plants, such as those shown in ***(b)***, growing in Berkeley, California. Erect plants of *P. vulgaris* cannot survive in lawns because they are damaged by mowing and do not have the capability for resprouting low branches from the base, which would be necessary for survival. When lawn plants are grown in an experimental garden, some remain prostrate whereas others grow erect. The prostrate habit is determined genetically in the first group and environmentally in the second group. Other than in lawns, the erect form of *P. vulgaris* most likely is favored by competition for light in crowded microsites where there is no counter-selection for shortness favored by herbivores eating relatively tall plants.

or in mine soil. In the normal soil, the mine plants were slower growing and smaller than the pasture plants. In the mine soil, however, the mine plants grew normally but the pasture plants did not grow at all. Half the pasture plants in mine soil were dead in three months and had misshapen roots that were rarely more than 2 millimeters long. But a few of the pasture plants (3 of 60) showed some tolerance to the effects of the metal-rich soil. They were doubtless genetically similar to the plants originally selected in the development of the lead-tolerant strain of *Agrostis*. The mines were no more than 100 years old, so the lead-tolerant strain had developed in a relatively short period of time. The tolerant plants had been selected from the genetically variable plants found in adjacent habitats and were then constituted, through the agency of natural selection, as a distinct strain (Figure 11–8).

11–8 Environmental selection for lead tolerance Bent grass *(Agrostis tenuis)* is growing (in the foreground) on the tailings of abandoned lead mines rich in lead, photographed in Wales. Such tailings are often devoid of plant life because the lead is present at levels toxic to plants. The lead-tolerant bent grass plants growing here are descendants of a population of plants that gradually adapted to the lead in the tailings.

INVASIVE PLANTS

In summer, marshes throughout the United States are swathed in a lush mantle of purple. The spiky flowering stalks of purple loosestrife *(Lythrum salicaria),* each over 3 meters tall, are abuzz with insect pollinators gorging on the abundant nectar. Their efforts will help each plant produce hundreds of thousands of seeds each summer, seeds that will be shed into the water to float to new marshes and waterways downstream.

Beautiful as it is, purple loosestrife is a major environmental pest. Today it is the predominant plant species in many sunny wetlands, where it forms a monoculture, crowding out native wetland species such as rushes and cattails. Brought from Europe in the early 1800s for use as an ornamental and medicinal plant, purple loosestrife quickly moved into the wild. Widely adaptable, prodigiously fertile, and lacking natural enemies, the plant spread quickly into every state of the contiguous United States except Florida and into every Canadian province.

Purple loosestrife is, by any measure, an evolutionary success. As an invasive species and a noxious weed, it has reduced both plant and animal diversity in millions of acres of wetlands. It shares these distinctions with several dozen other plant species that are reshaping the ecological landscape, including water hyacinth, curly-leaf pondweed, and Eurasian watermilfoil in aquatic environments, and kudzu, garlic mustard, and Japanese knotweed in terrestrial ecosystems.

There is nothing unnatural about plants spreading into new habitats. What distinguishes the spread of many invasive species is that they were introduced by humans, far away from the plants' established ranges. That leap—across oceans or mountain ranges or deserts—has meant, most crucially, that a plant arrives without the insects, birds, or fungi that have evolved to prey upon it. The absence of such natural checks is a key feature leading to the invasiveness of introduced plants.

Over evolutionary time, however, no niche goes unexploited. Some indigenous predator will one day develop a taste for purple loosestrife, but whether that will take 20 years or 20,000 is unknown. Not willing to wait, some groups are investigating whether it is safe and effective to introduce, as a biological control, various species of beetles and weevils that feed on the plant. Early results have been promising, with the insects controlling the loosestrife enough that native plant species have rebounded.

Making changes to any complex system, however, can lead to unexpected outcomes. Will the beetles also feed on native species that are too fragile to handle the predation? Will the weevil larvae provide food for a native insect that will then become a pest? Answering such questions, or at least thinking through their implications, before introducing a new species into an environment is crucial. Ignoring these issues led to the problems we now face with invasive species. Fully considering the consequences of such actions is the burden we must take on as we contemplate correcting the mistakes of the past.

(a)

(b)

Invasive terrestrial plants ***(a)*** Purple loosestrife and ***(b)*** kudzu *(Pueraria montana* var. *lobata),* known as "the plant that ate the South," are vigorous growers that outcompete and replace native plants, resulting in the destruction of habitat important to wildlife.

The Result of Natural Selection: Adaptation

Natural selection results in **adaptation,** a term with several meanings in biology. First, it can mean a state of being adjusted to the environment. Every living organism is adapted in this sense, just as Abraham Lincoln's legs were, as he remarked, "just long enough to reach the ground." Second, adaptation can refer to a particular characteristic that aids in the adjustment of an organism to its environment. Third, adaptation can mean the evolutionary process, occurring over the course of many generations, that produces organisms better suited to their environment. Fourth, and most precisely, adaptation is a variation in a trait that increases the fitness of the individual.

Natural selection involves interactions between individual organisms, their physical environment, and their biological environment—that is, other organisms. In many cases, the adaptations that result from natural selection can be clearly correlated with environmental factors or with the selective forces exerted by other organisms.

Clines and Ecotypes Are Reflections of Adaptation to the Physical Environment

The differences in appearance between individual plants may or may not have a genetic basis (Figures 11–6 and 11–7). When the observed differences arise directly because of environmental conditions, they are said to reflect **developmental plasticity.** Such plasticity is much greater in plants than in animals, because the open system of growth (indeterminate growth pattern) that is characteristic of plants can be more easily modified to produce striking differences in the expression of a particular genotype. Even the parts of individual plants that are formed in different environments may differ in appearance, as is known to every gardener who has observed how, on the same plants, leaves formed in shade may differ from those formed in the sun.

Most of the observed differences between plants of the same species growing in different habitats reflect genetic differences, as we will see. These differences may be correlated with such factors as precipitation, sun exposure, and soil moisture or type. If these environmental differences occur gradually, the features of the plant populations may do so as well. A gradual change in the characteristics of populations of an organism along an environmental gradient is called a **cline.** Many species exhibit north-south clines of various traits, such as requirements for flowering or for ending dormancy.

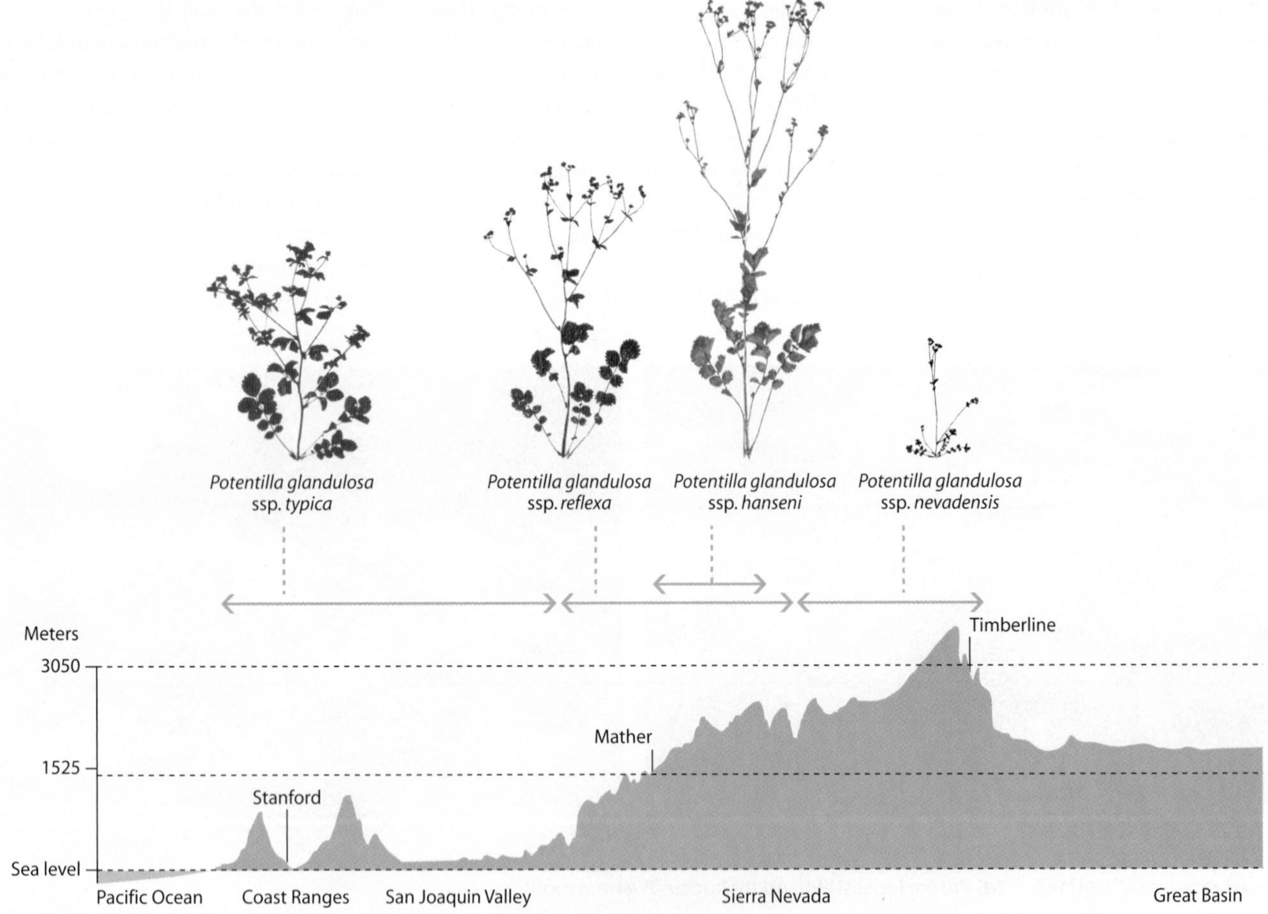

11–9 Ecotypes Plants from a number of populations of *Potentilla glandulosa,* a relative of the strawberry, were collected at 38° north latitude, from the Pacific Ocean to Timberline. They were then transplanted to experimental gardens at Stanford, Mather, and Timberline, all also at about 38° north latitude. The plants were propagated asexually so that genetically alike individuals could be grown at the three sites, which had very different climates. When grown side by side at each location, four distinct ecotypes became apparent. These four ecologically distinct ecotypes were correlated with differences in their morphology, especially flower and leaf characteristics. Because the differences between these ecotypes were maintained in the experimental gardens, they were shown to be genetically determined.

Each ecotype is distributed over a range of altitudes, and where the ranges overlap, the two ecotypes grow in different environments. The four ecotypes, which have been given subspecies names, are seen here at the flowering stage, and their approximate ranges are shown.

Clines are common in organisms that live in the sea, where the temperature often rises or falls very gradually with changes in latitude. Clines are also characteristic of organisms living in areas, such as the eastern United States, where rainfall gradients may extend over thousands of kilometers. When populations of plants are sampled along a cline, the differences are often proportional to the distance between the populations.

A species that occurs in different habitats may exhibit a different phenotype in each one. If the differences between the habitats are sharp, the features of the plant populations may also differ sharply. Each group of genetically distinct populations of the species is known as an **ecotype.** The differences in the features of ecotypes may be striking.

A clear demonstration of major ecotypic differences was presented by Jens Clausen, David Keck, and William Hiesey for the perennial herb *Potentilla glandulosa,* which ranges through a variety of climatic zones in the western United States. Experimental gardens were established at three sites in California where native populations of *P. glandulosa* occur: (1) Stanford, located between the inner and outer Coast Ranges, at 30 meters elevation, with warm temperate weather and predominant winter rainfall; (2) Mather, on the western slope of the Sierra Nevada, at 1400 meters elevation, with long, cold, snowy winters and hot, mostly dry summers; and (3) Timberline, east of the crest of the Sierra Nevada at roughly the same latitude as the two other stations but at 3050 meters elevation, with very long, cold, snowy winters and short, cool, rather dry summers (Figure 11–9).

When *P. glandulosa* plants from numerous locations were grown side by side in the gardens set up at the three sites, four distinct ecotypes became apparent. The morphological, or structural, characteristics of each ecotype were correlated with its physiological responses, which in turn were critical to the survival of each ecotype in its native environment.

For example, plants of the Coast Ranges ecotype grew actively in both winter and summer when cultivated at Stanford, which lies within their native range. These plants decreased in size but survived at Mather, outside their native range, even though they were subjected to about five months of cold winter weather. At Mather they became winter-dormant, but they stored enough food during their growing season to carry them through the long, unfavorable winter. At Timberline, plants of the Coast Ranges ecotype failed to survive, almost invariably dying during the first winter. The short growing season at this high elevation did not permit them to store enough food to survive the long winter. Other species that grow in California's Coast Ranges produce ecotypes that have physiological responses comparable to those of *P. glandulosa.* Indeed, strains of unrelated plant species that occur together naturally in a given location are often more similar to one another physiologically than they are to other populations of their own species.

The physiological and morphological characteristics of ecotypes, as in *P. glandulosa,* usually have a complex genetic basis, involving dozens (or, in some cases, perhaps hundreds) of genes. Sharply defined ecotypes are characteristic of regions, such as western North America, where the breaks between adjacent habitats are sharply defined. On the other hand, when the environment changes more gradually from one habitat to another, the characteristics of the plants growing in that region may do likewise.

Ecotypes Differ Physiologically

To understand why ecotypes flourish where they do, we must understand the physiological basis for their ecotypic differentiation. For example, Scandinavian strains of goldenrod *(Solidago virgaurea)* from shaded habitats and exposed habitats showed experimental differences in their photosynthetic response to light intensity during growth. The plants from shaded environments grew rapidly under low light intensities, whereas their growth rate was markedly retarded under high light intensities. In contrast, plants from exposed habitats grew rapidly under conditions of high light intensity, but much less well at low light levels.

In another experiment, Arctic and alpine populations of the widespread mountain sorrel *(Oxyria digyna)* were studied using strains from an enormous latitudinal gradient extending southward from Greenland and Alaska to the mountains of California and Colorado (Figure 11–10). Plants of northern populations had more chlorophyll in their leaves, as well as higher respiration rates at all temperatures, than plants from farther

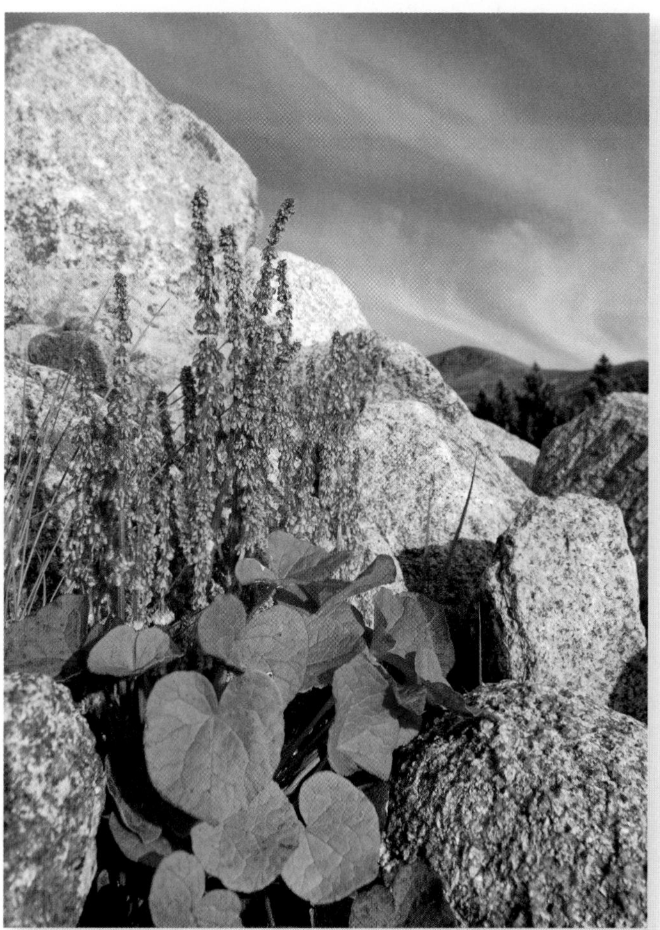

11–10 ***Oxyria digyna*** Mountain sorrel, a widespread Arctic and alpine eudicot, is shown here growing among granite rocks in the Highlands region of Scotland. The leaves and stems are edible. Physiological differences among its populations account for its adaptations to a wide variety of habitats.

south. High-elevation plants from near the southern limits of the species' range carried out photosynthesis more efficiently at high light intensities than did low-elevation plants from farther north. Thus each strain could function better in its own habitat, marked, for example, by high light intensities in high mountain habitats and lower intensities in the far north. The existence of *O. digyna* over such a wide area and such a wide range of ecological conditions is made possible, in part, by differences in metabolic potential among its various populations.

Coevolution Results from Adaptation to the Biological Environment

When populations of two or more species interact so closely that each exerts a strong selective force on the other, simultaneous adjustments occur that result in **coevolution.** One of the most important, in terms of sheer number of species and individuals involved, is the coevolution of flowers and their pollinators, which is described in Chapter 20. Another example of coevolution involves the monarch butterfly and milkweed plants (page 33).

The Origin of Species

Although Darwin titled his monumental book *On the Origin of Species,* he was never really able to explain how species might originate. However, an enormous body of work, mostly in the twentieth century, has provided many insights into the forming of separate species, or **speciation.** In addition, a great amount of time and discussion has gone into attempts to develop a clear definition of the term "species."

What Is a Species?

In Latin, **species** simply means "kind," and so species are, in the simplest sense, different kinds of organisms. By one definition—the **biological species concept**—a species is a group of natural populations whose members can interbreed with one another but cannot (or at least usually do not) interbreed with members of other such groups. The key criterion of this definition is **genetic isolation:** if members of one species freely exchanged genes with members of another species, they could no longer retain those unique characteristics that identify them as different kinds of organisms.

The biological species concept does not work in all situations, and it is difficult to apply to actual data in nature. Several alternative species concepts have therefore been proposed. In practice, species are generally identified purely on an assessment of their morphological, or structural, distinctness. The assumption is that reproductive isolation leads to genetic differences that are reflected in morphological differences. In fact, most species recognized by taxonomists have been designated as distinct species based on anatomical and morphological criteria; the **morphological species concept** is the name given to this practical approach.

The inability to form fertile hybrids has often been used as a basis for defining species. However, this criterion is not generally applicable. In some groups of plants—particularly long-lived plants such as trees and shrubs—species that are very distinct morphologically often can form fertile hybrids with one another. Take the case of the sycamores *Platanus orientalis* and *Platanus occidentalis,* which have been isolated from one another in nature for at least 50 million years. *Platanus orientalis* is native from the eastern Mediterranean region to the Himalayas, whereas *P. occidentalis* is native to eastern North America (Figure 11–11). Since Roman times, *P. orientalis* has been widely cultivated in southern Europe, but it cannot be grown in northern Europe away from the moderating influence of the sea. After the European discovery of the New World, *P. occidentalis* was brought into cultivation in the colder portions of northern Europe, where it flourished. In about 1670, these two very distinct trees produced intermediate and fully fertile hybrids when they were cultivated together in England. Called the London plane, the hybrid *(Platanus × hybrida)* is capable of growing in regions with cold winters and is now grown extensively as a street tree in New York City and throughout the temperate regions of the world.

Dissatisfaction with the prevailing species concepts has prompted plant systematists (see Chapter 12) to propose

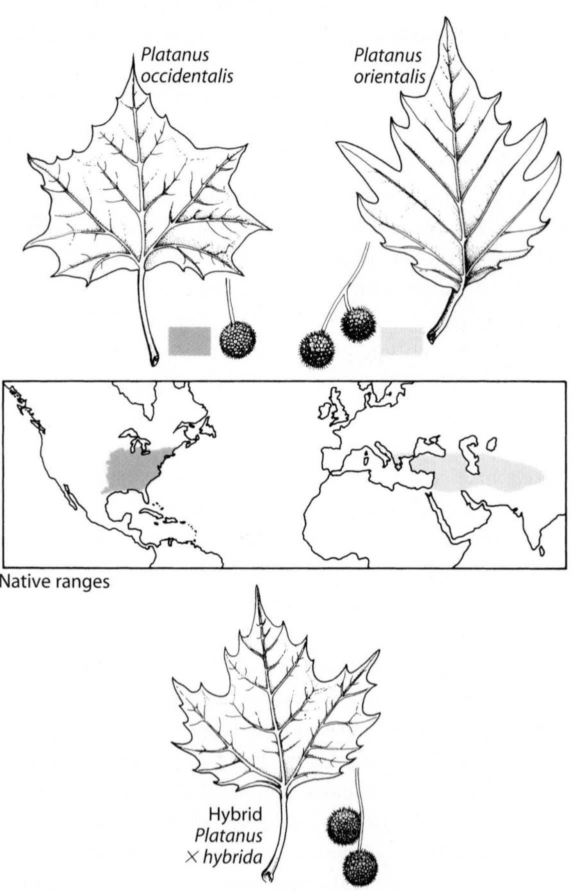

11–11 Hybrids The distribution of two species of sycamore, *Platanus orientalis,* which is native to areas from the eastern Mediterranean region to the Himalayas, and *Platanus occidentalis,* which is native to North America. The fully fertile hybrid, *Platanus × hybrida,* is the London plane, a sturdy tree suitable for growing along city streets.

several different **phylogenetic species concepts,** which are based on reconstructing the evolutionary history of populations. Although several phylogenetic species concepts have been proposed, David A. Baum and Michael J. Donoghue note that there appear to be two main approaches: character-based and history-based. In the character-based approach, an organism is considered to be a member of a given species if and only if it possesses certain characters or combination of characters. With a history-based approach, an organism is considered to be a member of a given species if and only if it is historically related to other organisms of the species. In short, one approach defines species on the basis of characters, whereas the other defines species in terms of historical relationships or ancestry.

How Does Speciation Occur?

By definition, the members of a species share a common gene pool that is effectively separated from the gene pools of other species. A central question, then, is, how does one pool of genes split off from another to begin a separate evolutionary journey? A subsidiary question is, how do two species, often very similar, inhabit the same place at the same time and yet remain reproductively isolated?

According to current thinking, speciation is most commonly the result of the geographic separation of a population of organisms; this process is known as **allopatric** ("other country") **speciation.** Under certain circumstances, speciation may also occur without geographic isolation, in which case it is known as **sympatric** ("same country") **speciation.**

Allopatric Speciation Involves the Geographic Separation of Populations

Every widespread species that has been carefully studied has been found to contain geographically representative populations that differ from each other to a greater or lesser extent. Examples are the ecotypes of *Potentilla glandulosa* and the strains of *Oxyria digyna.* A species composed of such geographic variants is particularly susceptible to speciation if geographic barriers arise, preventing gene flow from "blending away" any differences that might arise—through selection or drift—among nearby populations. Such barriers can arise as a consequence of rising mountains, climate change, or shifts in sea level, or as a result of long-distance dispersal.

Populations of a species that are characterized by gene flow over short distances (e.g., through seed dispersal or pollen flow) should be able to differentiate over those distances, leading to genetic divergence, or separation, and speciation. One pathway to speciation might be the accumulation of enough genetic divergence to create reproductive isolation, based on the inviability or infertility of crosses. Organisms with greatly restricted dispersal should therefore be able to form new species over small distances. Ultimately, these organisms produce species that are restricted to small areas, while those with greater dispersal ability should form species only at large distances. Flowering plants and snails, for example, are capable of differentiation and speciation over small distances, whereas more mobile groups, such as birds, tend to speciate over larger distances. Further studies may well show that speciation occurs over short distances in plants with very heavy seeds, or in tropical understory plants with fleshy fruits dispersed by notoriously sedentary understory birds.

Geographic barriers are of many different types. Islands, which are frequently sites for the development of new species, set the stage for the sudden (in geologic time) diversification of a group of organisms that share a common ancestor. This sudden diversification of such a group of organisms, forming new species with different ecological roles and adaptations, is called **adaptive radiation.** Adaptive radiation is one of the most important processes bridging ecology and evolution, and Darwin used the concept of adaptive radiation extensively in the *Origin of Species.*

Adaptive radiation is associated with the opening up of a new biological frontier that may be as vast as the land or the air or as small as an archipelago such as the Galápagos Islands. Adaptive radiation results in the almost simultaneous formation of many new species in a wide range of habitats. (See "Adaptive Radiation in Hawaiian Lobeliads" on pages 224 and 225.) Differentiation that occurs on islands is particularly striking because, in the absence of competition, organisms seem more likely to produce highly unusual forms than do related species on the continents. On islands, the characteristics of plants and animals may change more rapidly than on the mainland, and features that are never encountered elsewhere may arise. Similar clusters of species may also arise in mainland areas, of course, and may involve spectacular differentiation. The emergence of molecular systematics (see Chapter 12)—the use of DNA and RNA to infer relationships among organisms—is providing new, powerful tools for the study of adaptive radiation. Specifically, molecular data can provide insights into evolutionary relationships among species independent of the observable traits and ecological roles that appear to be undergoing radiation. This provides a direct means of inferring relationships by using genetic data and combining this information with morphological and ecological information.

Sympatric Speciation Occurs without Geographic Separation

A well-documented mechanism by which new species are produced through sympatric speciation—that is, when there is no geographic isolation—is polyploidy. By definition, **polyploids** are cells or individuals that have more than two sets of chromosomes (pages 165 and 166). Polyploids may arise as a result of **nondisjunction**—the failure of homologs to separate—during meiosis, or they may be generated when the chromosomes divide properly during mitosis or meiosis but cytokinesis does not follow. Polyploid individuals can be produced deliberately in the laboratory by the use of the drug colchicine, which disrupts microtubule formation and hence prevents the separation of chromosomes during mitosis.

Polyploidy that leads to the formation of new species due to a doubling of chromosome number within an individual organism is called **autopolyploidy** (Figure 11–12). Such individuals are called **autopolyploids.** Sympatric speciation by autopolyploidy was first discovered by Hugo de Vries, during

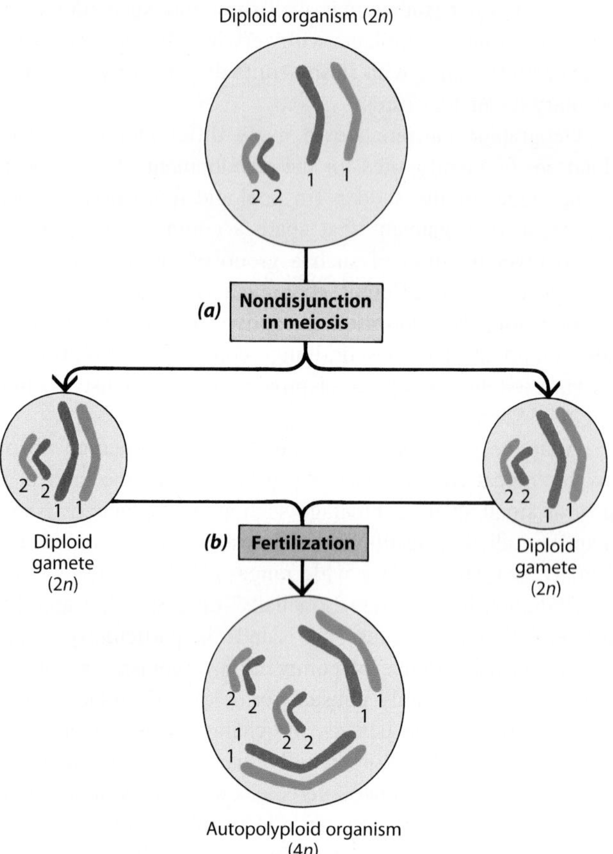

11–12 Autopolyploidy Polyploidy within individual organisms can lead to the formation of new species. ***(a)*** If the chromosomes of a diploid organism do not separate during meiosis (nondisjunction), diploid ($2n$) gametes may result. ***(b)*** Union of two such gametes, produced either by the same individual or by different individuals of the same species, produces an autopolyploid, or tetraploid ($4n$), individual. Although this individual may be capable of sexual reproduction, it will be reproductively isolated from the diploid parent species.

his investigation of the genetics of the evening primrose *(Oenothera glazioviana),* a diploid species with 14 chromosomes. Among his plants he found an unusual variant, which, upon microscopic examination, proved to be tetraploid (4 sets of chromosomes, or $4n$), with 28 chromosomes. De Vries was unable to breed the tetraploid primrose with the diploid primrose, because of problems with pairing of the chromosomes during meiosis. The tetraploid (an autopolyploid) was a new species, which de Vries named *Oenothera gigas* (*gigas* meaning "giant").

A much more common mode of polyploidy is **allopolyploidy,** which results from a cross between two different species, producing an **interspecific hybrid** (Figure 11–13). Such hybrids are usually sterile, because the chromosomes cannot pair at meiosis (having no homologs), a necessary step for producing viable gametes. If, however, autopolyploidy then occurs in the sterile hybrid, and the resulting cells divide by mitosis and cytokinesis, they eventually produce a new individual asexually. That individual—an **allopolyploid**—will have twice as many chromosomes as its parent. Consequently, it is reproductively isolated from its parental line. Moreover, its chromosomes—now duplicated—can pair, meiosis can occur normally, and fertility is restored. It is a new species capable of sexual reproduction.

Hybridization and sympatric speciation through polyploidy are important phenomena in plants, and clearly they have been important to the evolution of flowering plants. The extent of polyploidy in flowering plants ranges from 47 percent to over

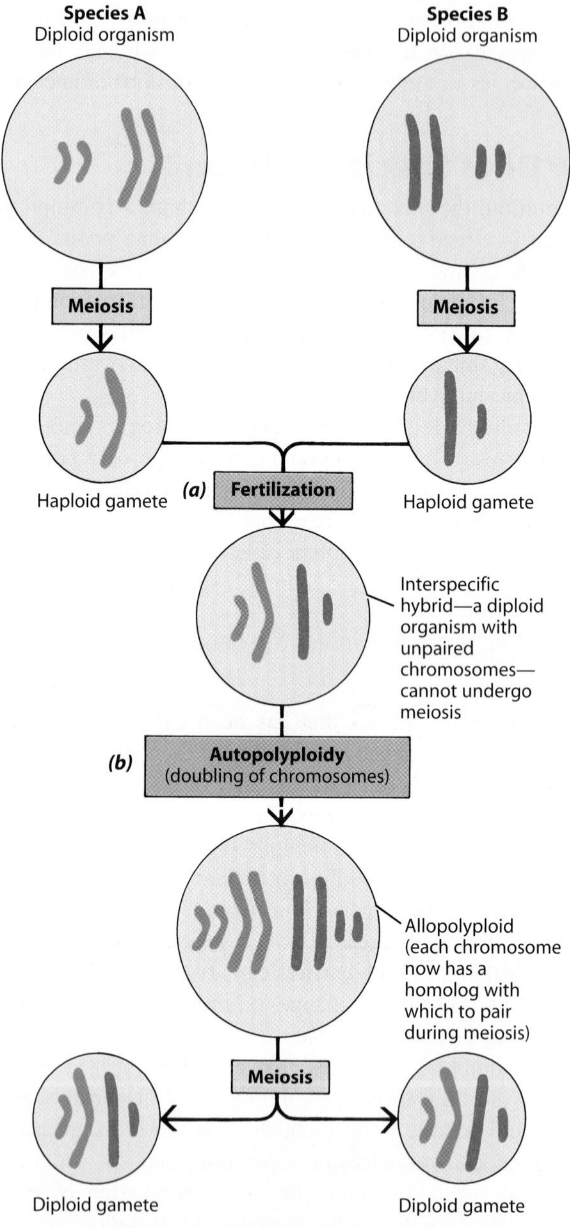

11–13 Allopolyploidy ***(a)*** An organism that is a hybrid between two different species—an interspecific hybrid—and that is produced from two haploid (n) gametes can grow normally because mitosis is normal. It cannot reproduce sexually, however, because the chromosomes cannot pair at meiosis. ***(b)*** If autopolyploidy subsequently occurs and the chromosome number doubles, the chromosomes can pair at meiosis. As a result, the hybrid—an allopolyploid—can produce viable diploid ($2n$) gametes and is now a new species capable of reproducing sexually.

70 percent, depending on the species, and in the grass family, 80 percent of the species are estimated to be polyploid. Furthermore, a large number of the major food crops are polyploid, including wheat, sugarcane, potatoes, sweet potatoes, and bananas.

Some polyploids that originated as weeds in habitats altered by human activity have been spectacularly successful. Probably the best-documented examples are two species of goat's beard, *Tragopogon mirus* and *Tragopogon miscellus*, which are products of allopolyploid speciation (Figure 11–14). Both arose during the last hundred years in the Palouse region of southeastern Washington and adjacent Idaho, following the introduction and naturalization of their Old World progenitors, *Tragopogon dubius*, *Tragopogon porrifolius*, and *Tragopogon pratensis*. All three of the Old World species are diploid ($2n = 12$), and each crosses readily with both of the others, forming F_1 hybrids that are highly sterile. In 1949, two tetraploid ($2n = 24$) hybrids were discovered that are fairly fertile, and these have increased substantially in the Palouse region since their discovery. Both *T. mirus* and *T. miscellus* have been reported in Arizona, and *T. miscellus* also occurs in Montana and Wyoming. Crosses of the Old World *T. dubius* and *T. porrifolius* produced the tetraploid *T. mirus*, and crosses of *T. dubius* and *T. pratensis* produced the tetraploid *T. miscellus*. The origin of these allopolyploids of *Tragopogon* has been confirmed by DNA sequencing and genomic technology.

11–14 Allopolyploidy in *Tragopogon* Three highly fertile diploid ($2n = 12$) species of *Tragopogon* (goat's beard) were introduced from Europe: ***(a)*** *Tragopogon dubius*, ***(b)*** *Tragopogon porrifolius*, and ***(c)*** *Tragopogon pratensis*. By 1930, all were well established in southeastern Washington and adjacent Idaho. These species hybridized easily, resulting in highly sterile diploid (F_1) interspecific hybrids: ***(d)*** *Tragopogon dubius* × *porrifolius*, ***(e)*** *Tragopogon porrifolius* × *pratensis*, and ***(f)*** *Tragopogon dubius* × *pratensis*. In 1949, four small populations of *Tragopogon* were discovered that were clearly different from the diploid hybrids and were immediately suspected of being two newly originated polyploid species. The suspected polyploids differed in none of their characters from the diploid hybrids *(d)*, *(f)*, except that they were much larger in every way and were obviously fertile, with flower heads containing many developing fruits. It was soon confirmed that these populations were tetraploid ($2n = 24$) species, and they were named ***(g)*** *Tragopogon mirus* and ***(h)*** *Tragopogon miscellus*. These tetraploid species are among the few allopolyploids whose time of origin is known with a high degree of certainty. ***(i)*** Note that this inflorescence of *Tragopogon porrifolius* × *pratensis* is a highly sterile diploid hybrid of a generation later than the F_1 seen here as *(e)*.

ADAPTIVE RADIATION IN HAWAIIAN LOBELIADS

Adaptive radiation—the evolution of species with a variety of ecological roles and adaptations that have descended from a single ancestor—is a key process shaping plant and animal groups of remote islands, lakes, and mountaintops. Few colonists from mainland sources ever reach such isolated areas. Descendants of each ancestor therefore encounter a wide range of ecologically "open" habitats and other resources, with few competitors other than their own relatives. Selection on closely related populations to diverge, avoid competition, and adapt to alternative conditions should then lead, almost inevitably, to adaptive radiation.

The Hawaiian lobeliads, with about 128 species in five genera, provide a striking example of adaptive radiation in plants. They differ so widely in habitat, growth form, leaf shape, flower morphology, and seed dispersal that they were first thought to represent five independent colonizations of the islands, but recent DNA analyses imply that the Hawaiian lobeliads all arose from one colonist.

Native to wet-forest interiors, *Cyanea* includes 76 species of unbranched small and large trees that produce bird-pollinated flowers and bird-dispersed fleshy fruits. Its tubular flowers are long and curved, like the bills of many Hawaiian honeycreepers, their primary pollinators. The closest relative of *Cyanea* is *Clermontia,* with 22 species of branched shrubs and epiphytes native to somewhat sunnier wet-forest edges, gaps, and canopies. Both genera range from near sea level to an elevation of 2000 meters (over 6000 feet). *Delissea,* with 10 species, grows in open moist scrub and has flowers and fruits similar to those of *Cyanea.*

Lobelia includes 5 species of giant rosette shrubs of mountain bogs and grasslands, with spectacular inflorescences of bird-pollinated flowers, as well as 9 species of smaller rosette shrubs from interior rock walls. The 2 species of *Brighamia* are some of the most bizarre

Cyanea floribunda, growing in cloud forest understory on the island of Hawaii, the youngest main island of the Hawaiian Islands.

Clermontia kakeana, growing in a gap in the cloud forest on the western side of the island of Maui.

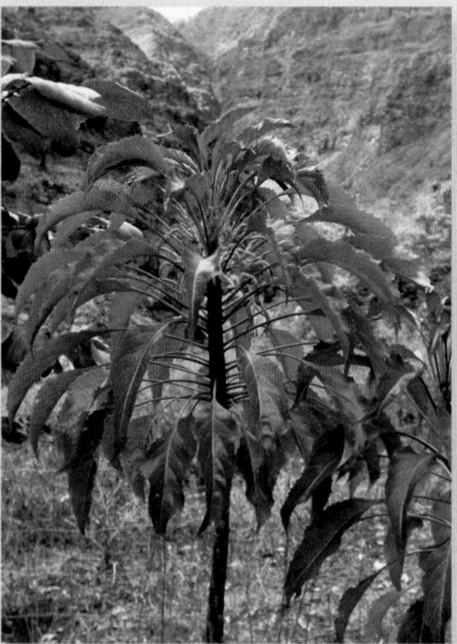

Delissea rhytidosperma, in moist scrub on Kauai, the oldest of the main islands of the Hawaiian Islands.

Lobelia gloria-montis, seen here in a mountain bog on the western side of Maui.

plants on Earth, growing mostly on tall sea cliffs and bearing a "cabbage head" of fleshy leaves on a succulent unbranched stem. Hawkmoths pollinate the flowers, which are fragrant at night. *Lobelia* and *Brighamia* produce capsular fruits that contain dustlike, wind-dispersed seeds.

The ancestor of the lobeliads appears to have been a woody plant adapted to open habitats at mid-elevations, with bird-pollinated flowers and wind-dispersed pollen. That ancestor reached what are now the Northwest Hawaiian Islands 13 million years ago, on Gardner or French Frigate Shoals, by long-distance seed dispersal from East African volcanoes or from other islands of the central Pacific. Within 3 million years, the lobeliad genera had radiated into distinctive habitats and growth forms.

Lobeliads occupy nearly the entire range of light availabilities found on moist sites in the Hawaiian Islands. Physiological studies show that species from sunnier habitats have higher rates of maximum photosynthesis and respiration, but require more light to reach peak photosynthesis. They therefore outperform shade species in bright light, while shade species have higher rates of photosynthesis in dimmer light.

Lobeliads vary tremendously in leaf size and shape. Groups from moist, shady habitats, such as *Cyanea,* have the largest leaves, associated with the low evaporation rates in such habitats. Some *Cyanea* species have divided leaves, which are almost always associated with thornlike prickles on stems and leaf veins and may have helped protect these lobeliads from the now extinct native herbivores, flightless Hawaiian ducks and geese.

Bird dispersal of fleshy fruits usually increases a plant's ability to disperse, but birds of the interiors of wet forests are notoriously reluctant to travel far. Forest-interior plant species with fleshy fruits should thus undergo genetic differentiation—and, ultimately, speciation—on smaller spatial scales than bird-dispersed groups from forest edges, or wind-dispersed groups with dustlike seeds in open habitats. Indeed, *Cyanea* species have narrower elevational ranges and occur on fewer islands than *Clermontia* species, resulting in over three times as many species of *Cyanea* overall. The evolution of *Cyanea* species numbers on each major island saturates surprisingly rapidly, within 1.2 million years of island emergence.

Tragically, many adaptive radiations—and the evidence they provide for evolution—are under threat. These species, many of which are approaching extinction, arose in the absence of mainland competitors, predators, and pathogens, but are now exposed to them because of increased global trade, species introductions, habitat destruction, and human population pressures.

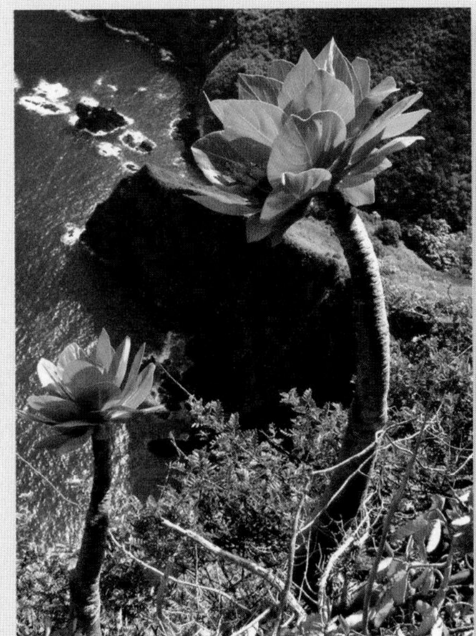

Brighamia rockii, growing on the steep Kapailoa cliffs on the island of Molokai.

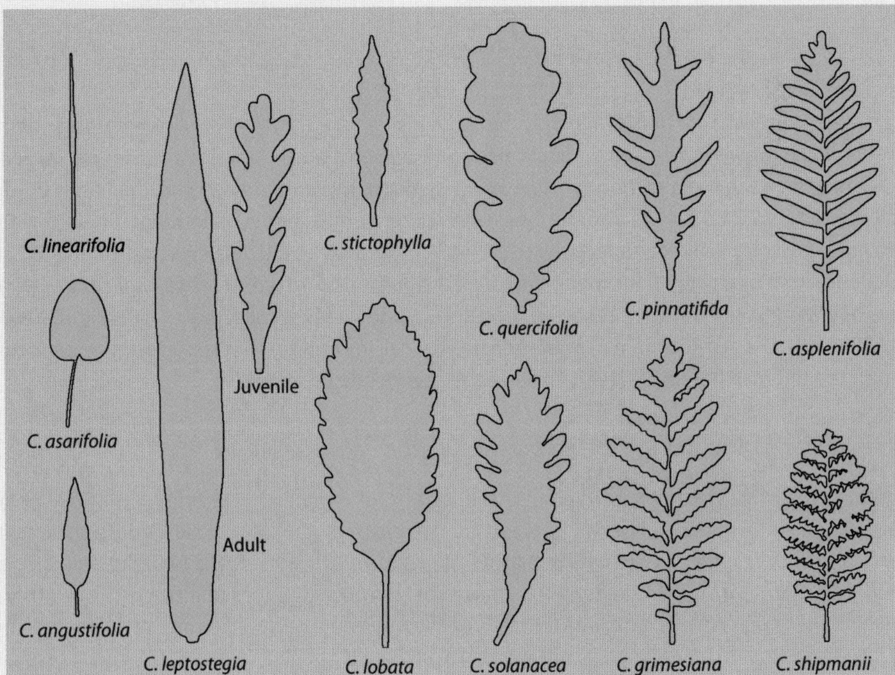

The leaves of *Cyanea* vary greatly in size and shape. All lobed or compound leaves, except those of juvenile *Cyanea leptostegia,* bear thornlike prickles restricted to their veins.

11–15 Polyploidy in *Spartina* Polyploidy has been investigated extensively among grasses of the genus *Spartina,* which grow in salt marsh habitats along the coasts of North America and Europe. ***(a)*** *Spartina* growing in a salt marsh on the coast of Virginia. ***(b)*** A *Spartina hybrid.* ***(c)*** *Spartina maritima,* the native European species of salt marsh grass, has $2n = 60$ chromosomes, shown here from a cell in meiotic anaphase I. ***(d)*** *Spartina alterniflora* is a North American species with $2n = 62$ chromosomes (30 bivalents and 2 unpaired chromosomes), shown here from a cell in meiotic metaphase I. ***(e)*** A vigorous polyploid, *Spartina anglica,* arose spontaneously from a hybrid between the species shown in *(c)* and *(d)* and was first collected in the early 1890s. This polyploid, which has $2n = 122$ chromosomes, shown here from a cell in meiotic anaphase I, is now extending its range through the salt marshes of Great Britain and other temperate countries.

One of the best-known polyploids, the origin of which is associated with human activity, is a salt-marsh grass of the genus *Spartina* (Figure 11–15). One native species, *Spartina maritima,* occurs in marshes along the coasts of Europe and Africa. A second species, *Spartina alterniflora,* was introduced into Great Britain from eastern North America in about 1800, and spread from where it was first planted to form large but local colonies.

In Britain, the native *S. maritima* is short in stature, whereas *S. alterniflora* is much taller, frequently growing to 0.5 meter and occasionally to 1 meter or even more. Near the harbor at Southampton, in southern England, both the native species and the introduced species existed side by side throughout the nineteenth century. In 1870, botanists discovered a sterile hybrid between these two species that reproduced vigorously by rhizomes. Of the two parental species, *S. maritima* has a somatic chromosome number of $2n = 60$ and *S. alterniflora* has $2n = 62$; the hybrid, owing perhaps to some minor error during meiosis, also has $2n = 62$ chromosomes. This sterile hybrid, which was named *Spartina* × *townsendii,* still persists. In about 1890, a vigorous seed-producing polyploid, named *Spartina anglica,* was derived naturally from the sterile hybrid. This fertile polyploid, which has a diploid chromosome number of $2n = 122$ (one chromosome pair was evidently lost), has spread rapidly along the coasts of Great Britain and northwestern France. It is often planted to bind mud flats, and such use has contributed to its further spread.

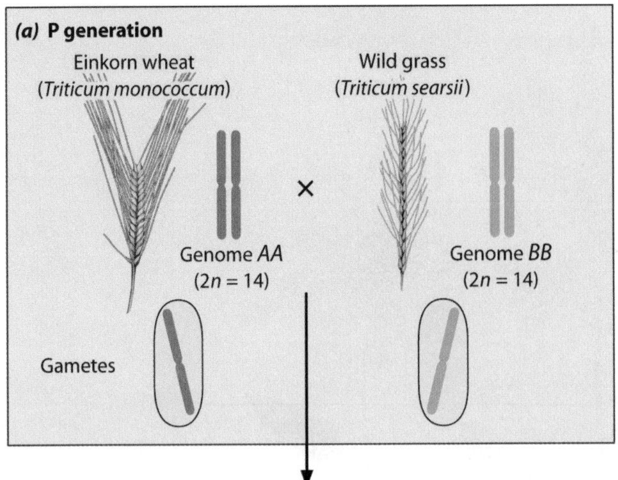

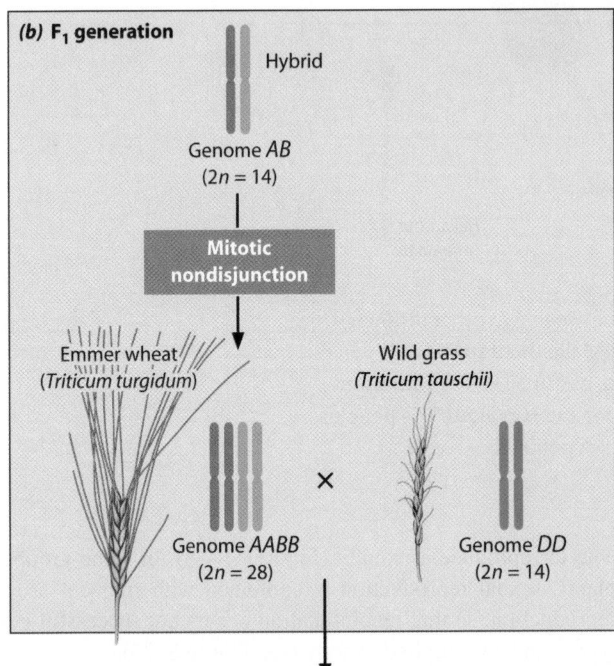

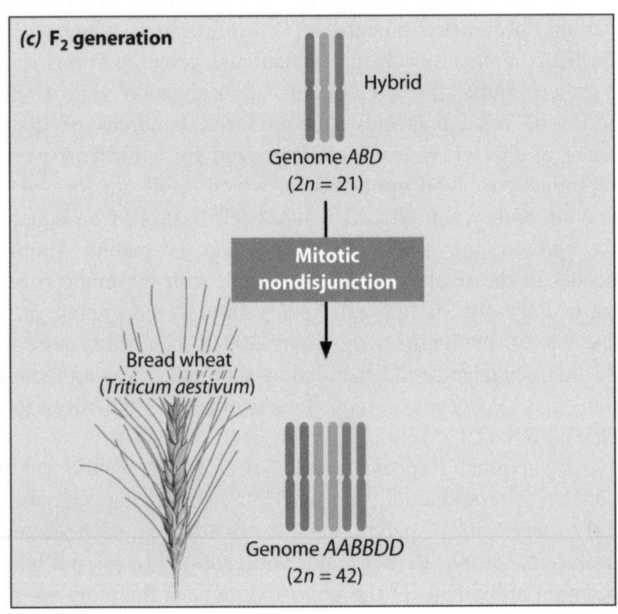

11–16 The origin of bread wheat *(Triticum aestivum)* The species *T. aestivum* ($2n = 42$) is a hexaploid with genes derived from three different species. ***(a)*** Two diploid species, *Triticum monococcum* ($2n = 14$) and a wild grass of the same genus ($2n = 14$), originally crossed to produce ***(b)*** a diploid hybrid ($2n = 14$), which underwent a mitotic nondisjunction to create the tetraploid *Triticum turgidum* ($2n = 28$), or emmer wheat. A cross between *T. turgidum* and *Triticum tauschii* ($2n = 14$) produced ***(c)*** a sterile triploid hybrid ($2n = 21$), which subsequently underwent mitotic nondisjunction to produce the hexaploid *T. aestivum* ($2n = 42$), or bread wheat.

One of the most important polyploid groups of plants is the genus *Triticum,* the wheats. The most commonly cultivated crop in the world, bread wheat *(Triticum aestivum)* is hexaploid (6 sets of chromosomes, or $6n$) and has $2n = 42$ chromosomes. Bread wheat originated at least 8000 years ago, probably in central Europe, following the natural hybridization of a cultivated tetraploid wheat, called emmer wheat *(Triticum turgidum),* with $2n = 28$ chromosomes, and a wild grass of the same genus *(Triticum tauschii)* with $2n = 14$ chromosomes (Figure 11–16). The wild grass probably occurred spontaneously as a weed in the fields where the emmer wheat was being cultivated. The hybridization that gave rise to bread wheat *(T. aestivum)* probably occurred between polyploids that arose from time to time within the populations of the two ancestral species.

It is likely that the desirable characteristics of the new, fertile, 42-chromosome wheat—such as larger, more separate kernels—were easily recognized, and the plant was selected for cultivation by the early farmers of Europe when it appeared in their fields. One of its parents, the 28-chromosome cultivated emmer wheat, had itself originated following hybridization between two wild 14-chromosome species in the Near East. Species of tetraploid wheat with $2n = 28$ chromosomes are still cultivated and are the chief grains used in pastas, such as spaghetti or macaroni. In contrast to tetraploid wheat, hexaploid bread wheat grains contain gluten, a "sticky" protein that traps the CO_2 produced by fermenting yeast, allowing bread to rise. Tetraploid wheat, on the other hand, is used for unleavened flat breads, such as pita bread.

An example of sympatric speciation not involving polyploidy is provided by the anomalous sunflower *(Helianthus anomalus),* the product of interbreeding between two other distinct species of sunflower, the common sunflower *(Helianthus annuus)* and the petioled sunflower *(Helianthus petiolaris)* (Figure 11–17). All three species occur widely in the western United States. Molecular evidence indicates that *H. anomalus* arose by **recombination speciation,** a process in which two distinct species hybridize, the mixed genome of the hybrid becoming a third species that is genetically (reproductively) isolated from its ancestors.

First-generation hybrids of *H. annuus* and *H. petiolaris* are semisterile, a condition apparently due to unfavorable interactions between the genomes of the parental species that result in difficulties during meiosis in the hybrid. Over several generations, however, full fertility is achieved in the hybrid as the

11–17 Sympatric speciation *Helianthus anomalus* is the result of a cross between two other distinct species of sunflower, *Helianthus annuus* and *Helianthus petiolaris*. Individually, the three species are easily distinguished from one another. The hybrid *H. anomalus,* for example, has small leaves with short petioles (leaf stalks) and few petals, which are wider than those in *H. petiolaris* and *H. annuus*. The petioles of *H. petiolaris* are long and thin, whereas *H. annuus* has large leaves with thick petioles.

gene combinations become rearranged. Individuals with the newly arranged genomes are compatible with one another—that is, *H. anomalus* with *H. anomalus*—but are incompatible with *H. annuus* and *H. petiolaris.*

In a study undertaken by Loren H. Rieseberg and co-workers, the genomic compositions of three experimentally produced hybrid lineages involving crosses between *H. annuus* and *H. petiolaris* were compared with that of the naturally occurring *H. anomalus*. Surprisingly, by the fifth generation, the genomes of all three lineages were remarkably similar to that of the naturally occurring *H. anomalus*. In addition, in all three experimentally produced hybrid lineages, the fertility was uniformly high (greater than 90 percent). It has been hypothesized that certain combinations of genes from *H. annuus* and *H. petiolaris* consistently work better together and so are always found together in surviving hybrids. This study has been characterized as "a first-ever re-creation of a new species," meaning that a hybrid produced experimentally matched a hybrid that occurred in nature.

Sterile Hybrids May Become Widespread If They Can Reproduce Asexually

Even if hybrids are sterile, as in the horsetail hybrid *Equisetum* × *ferrissii,* they may become widespread, providing they are able to reproduce asexually (Figure 11–18). In some groups of plants, sexual reproduction is combined with frequent asexual reproduction, so that recombination occurs but successful genotypes can be multiplied exactly (see Figure 8–23).

An outstanding example of such a system is the extremely variable Kentucky bluegrass *(Poa pratensis),* which in one form or another occurs throughout the cooler portions of the Northern Hemisphere. Occasional hybridization with a whole series of related species has produced hundreds of distinct races of this grass, each characterized by a form of asexual reproduction called **apomixis,** in which seeds are formed that contain embryos produced without fertilization. Consequently, the embryos are genetically identical to the parent. Apomixis occurs in the ovule or immature seed, with the embryo being formed by one of two different pathways, depending on the species. Of the angiosperms, over 300 species from more than 35 families have been described as apomictic. Among them are the grass (Poaceae), composite (Asteraceae), and rose (Rosaceae) families.

In apomictic species, or in species with well-developed vegetative reproduction, the individual strains may be particularly successful in specific habitats. In addition, such asexually propagated strains do not require **outcrossing** (cross-pollination between individuals of the same species) and therefore often do

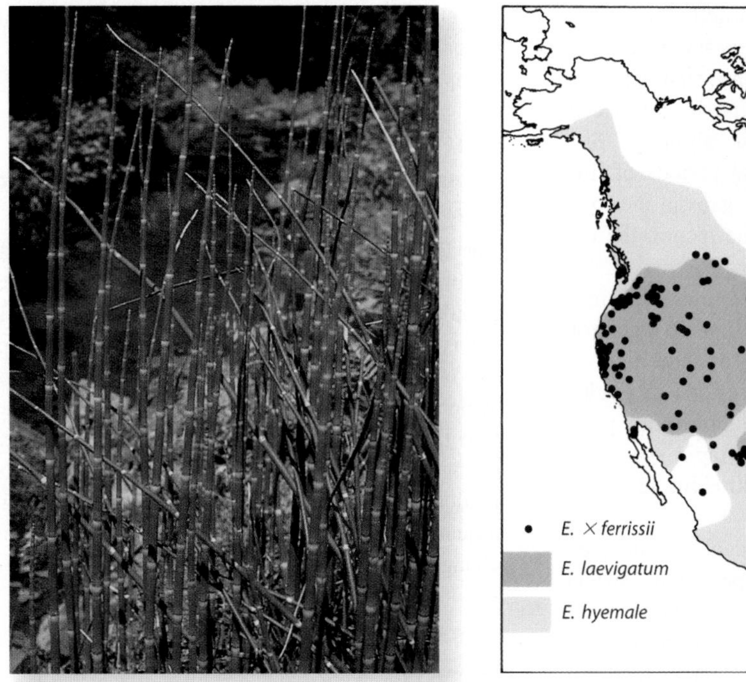

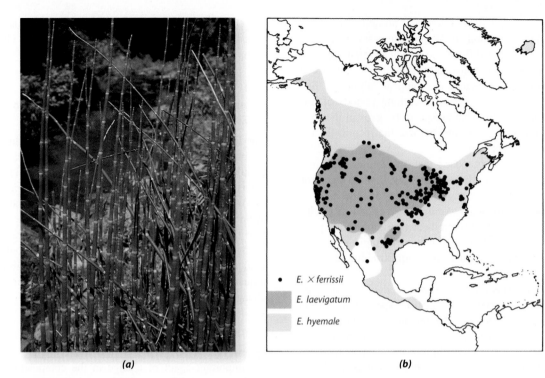

(a) (b)

11–18 A widespread sterile hybrid ***(a)*** One of the most abundant and vigorous of the horsetails (see Figure 17–37) found in North America is *Equisetum* × *ferrissii,* a completely sterile hybrid of *Equisetum hyemale* and *Equisetum laevigatum.* Horsetails propagate readily from small fragments of underground stems, and the hybrid maintains itself over its wide range through such vegetative propagation. ***(b)*** Ranges of *Equisetum* × *ferrissii* and its parental species.

well in environments such as high mountains, where pollination by insects may be uncertain.

The Origin of Major Groups of Organisms

As knowledge about the ways in which species may originate has become better developed, evolutionary biologists have turned their attention to the origin of genera and higher taxonomic groups of organisms. As suggested by the essay on adaptive radiation (pages 224 and 225), genera may originate through the same kinds of evolutionary processes that are responsible for the origin of species. If a particular species has a distinctive adaptation to a new habitat, the adapted species may become very different from its progenitor. It may gradually give rise to many new species, developing into a novel evolutionary line, a process called **phyletic change.** Gradually, it may become so distinctive that this line may be classified as a new genus, family, or even class of organisms. (The levels of classification, or taxonomic groups, are discussed in Chapter 12.) Accordingly, no special mechanisms would be necessary to account for the origin of taxonomic groups above the level of species—that is, for **macroevolution**—which is due only to discontinuity in habitat, range, or way of life and the accumulation of many small changes in the frequencies of alleles in gene pools. This is the **gradualism model** of evolution.

Although the fossil record documents many important stages in evolutionary history, there are numerous gaps, and gradual transitions of fossil forms are rarely found. Instead, new forms representing new species appear rather suddenly (in geologic terms) in strata, apparently persist unchanged for their tenure on Earth, and then disappear from the rocks as suddenly as they appeared. For many years, this discrepancy between the model of slow phyletic change and the poor documentation of such change in much of the fossil record was attributed to the imperfection of the record itself. Darwin, in the *Origin of Species,* noted that the geologic record is "a history of the world imperfectly kept, and written in a changing dialect; of this history we possess the last volume alone, relating only to two or three countries. Of this volume, only here and there a short chapter has been preserved; and of each page, only here and there a few lines."

In 1972, two young scientists, Niles Eldredge and Stephen Jay Gould, proposed that perhaps the fossil record is not so imperfect after all. Both Eldredge and Gould had backgrounds in geology and invertebrate paleontology, and both were impressed with the fact that there was very little evidence for a gradual phyletic change in the fossil species they studied.

Typically, a species would appear "abruptly" in fossil-bearing strata, last 5 million to 10 million years, and disappear, apparently not much different from when it first appeared. Another species, related but distinctly different, would take its place, persist with little change, and disappear "abruptly." Suppose, Eldredge and Gould argued, these long periods of little or no change, followed by what appear to be gaps in the fossil record, are not flaws in the record but *are* the record, the evidence of what really happens.

Eldredge and Gould proposed that species undergo most of their morphological modification as they first diverge from their progenitors, and then change little even as they give rise to additional species. In other words, long periods of gradual change or of no change at all (periods of equilibrium) are punctuated by periods of rapid change, that is, of rapid speciation. This theory is called the **punctuated equilibrium model** of evolution.

How could new species make such "sudden" appearances? Eldredge and Gould turned to allopatric speciation—speciation with geographic separation—for the answer. If new species formed principally in small populations isolated from parent populations and occurred rapidly (in thousands rather than millions of years), and if the new species then outcompeted the old species, taking over their geographic range, the resulting fossil pattern would be the one observed.

The punctuated equilibrium model has stimulated a vigorous and continuing debate among biologists, a reexamination of evolutionary mechanisms as currently understood, and a reappraisal of the evidence. Perhaps populations change more rapidly at some times than at others, particularly in periods of environmental stress. Regularly, new studies are published in support of gradualism or of punctuated equilibrium, but a consensus on either model has not been achieved. These activities have at times been misinterpreted as a sign that Darwin's theory is "in trouble." In fact, they indicate that evolutionary biology is alive and well and that scientists are doing what they are supposed to be doing—asking questions. Darwin, we think, would have been delighted.

SUMMARY

Darwin Proposed a Theory of Evolution by Natural Selection

Charles Darwin was not the first to propose a theory of evolution, but his theory differed from others in that it envisioned evolution as a two-part process, depending on (1) the existence in nature of heritable variations among organisms and (2) the process of natural selection by which some organisms, by virtue of their heritable variations, leave more surviving progeny than others. Darwin's theory is regarded as the greatest unifying principle in biology.

Population Genetics Is the Study of Gene Pools

Population genetics is a synthesis of the Darwinian theory of evolution with the principles of Mendelian genetics. For the population geneticist, a population is an interbreeding group of organisms, defined and united by its gene pool (the sum of all the alleles of all the genes of all the individuals in the population). Evolution is the result of accumulated changes in the composition of the gene pool.

The Hardy-Weinberg Law States That in an Ideal Population, the Frequency of Alleles Will Not Change over Time

The Hardy-Weinberg law describes the steady state in allele and genotype frequencies that would exist in an ideal, nonevolving population, in which five conditions are met: (1) no mutation, (2) isolation from other populations, (3) large population size, (4) random mating, and (5) no natural selection. The Hardy-Weinberg equilibrium demonstrates that the genetic recombination that results from meiosis and fertilization cannot, in itself, change the frequencies of alleles in the gene pool. The mathematical expression of the Hardy-Weinberg equilibrium provides a quantitative method for determining the extent and direction of change in allele and genotype frequencies.

Five Agents Cause Gene Frequencies in a Gene Pool to Change

An important agent of change in the composition of the gene pool is natural selection. Other agents of change include mutation, gene flow, genetic drift, and nonrandom mating. Mutations provide the raw material for change, but mutation rates are usually so low that mutations, in themselves, have little effect on allelic frequencies in a single generation. Gene flow, the movement of alleles from one population to another, may introduce new alleles or alter the proportions of alleles already present. It often has the effect of counteracting natural selection. In genetic drift, certain alleles increase or decrease in frequency, and sometimes even disappear, as a result of chance events. Circumstances that can lead to genetic drift, which is most likely to occur in small populations, include the founder effect and the bottleneck effect. Nonrandom mating causes changes in the proportions of genotypes but does not affect allele frequencies.

Natural Selection Acts on the Phenotype, Not the Genotype

Only the phenotype is accessible to selection. Similar phenotypes can result from very different combinations of alleles. Because of epistasis and pleiotropy, single alleles cannot be selected in isolation. Selection affects the entire genotype.

The Result of Natural Selection Is the Adaptation of Populations to Their Environment

Evidence of adaptation to the physical environment can be seen in gradual variations that follow a geographic distribution (cline) and in distinct groups of phenotypes (ecotypes) of the same species occupying different habitats. Adaptation to the biological environment results from the selective forces exerted by interacting species of organisms on each other (coevolution).

There Are Several Definitions of a Species

The biological species concept defines a species as a group of natural populations whose members can interbreed with one another but cannot (or at least usually do not) interbreed with members of other such groups. In practice, most species are generally identified purely on an assessment of their morphological, or structural, distinctness (the morphological species concept). Plant systematists have proposed several phylogenetic species concepts, which are based on reconstructing the evolutionary history of populations. For speciation—the formation of new species—to occur, populations that formerly shared a common gene pool must be reproductively isolated from one another and subsequently subjected to different selection pressures.

Allopatric Speciation Involves the Geographic Separation of Populations, Whereas Sympatric Speciation Occurs among Organisms Living Together

Two principal modes of speciation are recognized, allopatric ("other country") and sympatric ("same country"). Allopatric speciation occurs in geographically isolated populations. Islands are frequently sites for sudden diversification and the development of new species from a common ancestor, a pattern of speciation that is called adaptive radiation. Sympatric speciation, which does not require geographic isolation, occurs principally in plants through polyploidy, often coupled with hybridization. Hybrid populations derived from two species are common in plants, especially trees and shrubs. Even if hybrids are sterile, they may become widespread by asexual means of reproduction, including apomixis, in which seeds are formed but with embryos that are produced without fertilization.

The Gradualism and Punctuated Equilibrium Models Are Used to Explain Evolution of Major Groups of Organisms

Carried out over time, the same processes responsible for the evolution of species may give rise to genera and other major groups. This is the gradualism model of evolution. Paleontologists have presented evidence for an additional pattern of evolution known as punctuated equilibrium. They propose that new species are formed during bursts of rapid speciation among small isolated populations, that the new species outcompete many of the existing species (which become extinct), and that, in turn, many of the new species abruptly become extinct.

QUESTIONS

1. Explain the influence of Thomas Malthus and Charles Lyell on the development of Darwin's theory of evolution.
2. How did Darwin's concept of evolution differ primarily from that of his predecessors? What was the major weakness in Darwin's theory?
3. What is meant by developmental plasticity? Why is developmental plasticity much greater in plants than in animals?
4. Distinguish between each of the following: cline and ecotype; microevolution and macroevolution; allopatric speciation and sympatric speciation; autopolyploidy and allopolyploidy.
5. Define genetic isolation. Why is it such an important factor in speciation?

SECTION 4

DIVERSITY

◀ The pima pineapple cactus *(Coryphantha scheeri* var. *robustispina)* is a flowering plant, an angiosperm, that is native to the Sonoran Desert of southern Arizona and northern New Mexico. Soon after the rainy season begins in July, the silky yellow flowers burst into bloom. The flowers last one to three days and, because the cactus does not self-pollinate in the wild, bees are presumed to be the pollinators. The pima pineapple cactus is endangered, primarily due to loss of habitat, but conservation efforts are under way.

CHAPTER 12

Systematics: The Science of Biological Diversity

◀ **A large, diverse family** Bittersweet nightshade *(Solanum dulcamara),* shown here, is a widespread weed poisonous to humans but not as toxic as deadly nightshade *(Atropa belladonna),* which can be fatal. Both poisonous species belong to the Solanaceae, a family that includes important agricultural crops, such as potatoes, tomatoes, eggplants, and chili peppers.

CHAPTER OUTLINE

In the previous section, we considered the mechanisms by which evolutionary change occurs. Now we turn our attention to the products of evolution—that is, to the multitude of different kinds, or species, of living organisms that share our biosphere today. It is estimated that there are perhaps 10 million eukaryotic species and an unknown number of prokaryotic ones. The scientific study of this biological diversity and its evolutionary history is called **systematics.** The overall goal of systematists is to discover all the branches of the phylogenetic **tree of life,** the family tree depicting the genealogic relationships of organisms, with a single, ancestral species at its base.

Taxonomy: Nomenclature and Classification

An important aspect of systematics is **taxonomy,** which involves the identifying, naming, and classifying of species. The modern system of naming living things began with the eighteenth-century Swedish naturalist Carl Linnaeus (Figure 12–1), whose ambition was to name and describe all of the known kinds of plants, animals, and minerals. In 1753, Linnaeus published a two-volume work entitled *Species Plantarum* ("The Kinds of Plants") in which he described each species in Latin, in a sentence limited to 12 words. He regarded these descriptive Latin phrase names, or **polynomials,** as the proper names for the species, but in adding an important innovation devised earlier by Caspar Bauhin (1560–1624), Linnaeus made permanent the **binomial** ("two-term") **system** of nomenclature. In the margins of *Species Plantarum,* next to the "proper" polynomial name of each species, Linnaeus wrote a single word. This word, when

CHECKPOINTS

After reading this chapter, you should be able to answer the following:

1. Describe the binomial system of nomenclature.
2. Why is the term "hierarchical" used to describe taxonomic categories? Name the principal categories between the levels of species and kingdom.
3. What is cladistic analysis? What does a cladogram represent?
4. What evidence is there for the existence of the three major domains, or groups, of living organisms?
5. Name the three kingdoms of multicellular eukaryotes, and give the major identifying characteristics of each.

12–1 Carl Linnaeus (1707–1778) A professor, physician, and naturalist, Linnaeus devised the binomial system for naming species of organisms and established the major categories that are used in the hierarchical system of biological classification. When he was 25, Linnaeus spent five months exploring Lapland for the Swedish Academy of Sciences. He is shown here wearing a version of the traditional Lapp outfit and holding a sprig of twinflower *(Linnaea borealis)*, a species that is named after him.

combined with the first word of the polynomial—the **genus** (plural: genera)—formed a convenient "shorthand" designation for the species. For example, for catnip, which was formally named *Nepeta floribus interrupte spicatus pedunculatis* (meaning "*Nepeta* with flowers in an interrupted pedunculate spike"), he wrote the word "cataria" (meaning "cat-associated") in the margin of the text, thus calling attention to a familiar attribute of the plant. Linnaeus and his contemporaries soon began calling this species *Nepeta cataria,* and this Latin name is still used for this species today.

The convenience of this new system was obvious, and the cumbersome polynomial names were soon replaced by binomial names. The earliest binomial name given to a particular species has priority over other names applied to the same species later. The rules governing the scientific names of plants, photosynthetic protists, and fungi are embodied in the *International Code of Botanical Nomenclature.* Codes also exist for animals and nonphotosynthetic protists *(International Code of Zoological Nomenclature),* as well as for microbes *(International Code of Nomenclature of Bacteria).*

The Species Name Consists of the Genus Name Plus the Specific Epithet

A species name consists of two parts. The first is the name of the genus—also called the generic name—and the second is the **specific epithet.** For catnip, the generic name is *Nepeta,* the specific epithet is *cataria,* and the species name is *Nepeta cataria.*

A generic name may be written alone when one is referring to the entire group of species making up that genus; Figure 12–2 shows three species of the violet genus, *Viola.* A specific epithet is meaningless, however, when written alone. The specific epithet *biennis,* for example, is used in conjunction with dozens of different generic names. *Artemisia biennis,* a kind of wormwood, and *Lactuca biennis,* a species of wild lettuce, are two very different members of the sunflower family, and *Oenothera biennis,* an evening primrose, belongs to a different family altogether. Because of the danger of confusing names, a specific epithet is always preceded by the name or the initial letter of the genus that includes it: for example, *Oenothera biennis* or *O. biennis.* Names of genera and species are printed in italics or are underlined when written or typed.

If a species is discovered to have been placed in the wrong genus initially and must then be transferred to another genus, the specific epithet moves with it to the new genus. If there is already a species in that genus that has that particular specific epithet, however, an alternative name must be found.

Each species has a **type specimen,** usually a dried plant specimen housed in a museum or herbarium, which is designated either by the person who originally named that species or by a subsequent author if the original author failed to do so (Figure 12–3). The type specimen serves as a basis for comparison with other specimens in determining whether they are members of the same species.

The Members of a Species May Be Grouped into Subspecies or Varieties

Certain species consist of two or more subspecies or varieties (some botanists consider varieties to be subcategories of subspecies, and others regard them as equivalent). All of the members of a subspecies or variety of a given species resemble one another and share one or more features not present in other subspecies or varieties of that species. As a result of these subdivisions, although the binomial name is still the basis of classification, the names of some plants and animals may consist of three parts. The subspecies or variety that includes the type specimen of the species repeats the name of the species, and all the names are written in italics or underlined. Thus the peach tree is *Prunus persica* var. *persica,* whereas the nectarine is *Prunus persica* var. *nectarina.* The repeated *persica* in the name of the peach tree tells us that the type specimen of the species *P. persica* belongs to this variety, abbreviated "var." (see Figure 12–2b, c for other examples).

(a)

(b)

(c)

12–2 Three members of the violet genus ***(a)*** The common blue violet, *Viola sororia,* which grows in temperate regions of eastern North America as far west as the Great Lakes. ***(b)*** *Viola tricolor* var. *tricolor,* a yellow-flowered violet representing a mostly perennial species native to western Europe. ***(c)*** Pansy, *Viola tricolor* var. *hortensis,* an annual, cultivated strain of the wild species represented in *(b).* These taxa differ in flower color and size, leaf shape and margin, and other features that distinguish the species of this genus, even though there is an overall similarity among all of them. There are about 500 species of the genus *Viola.*

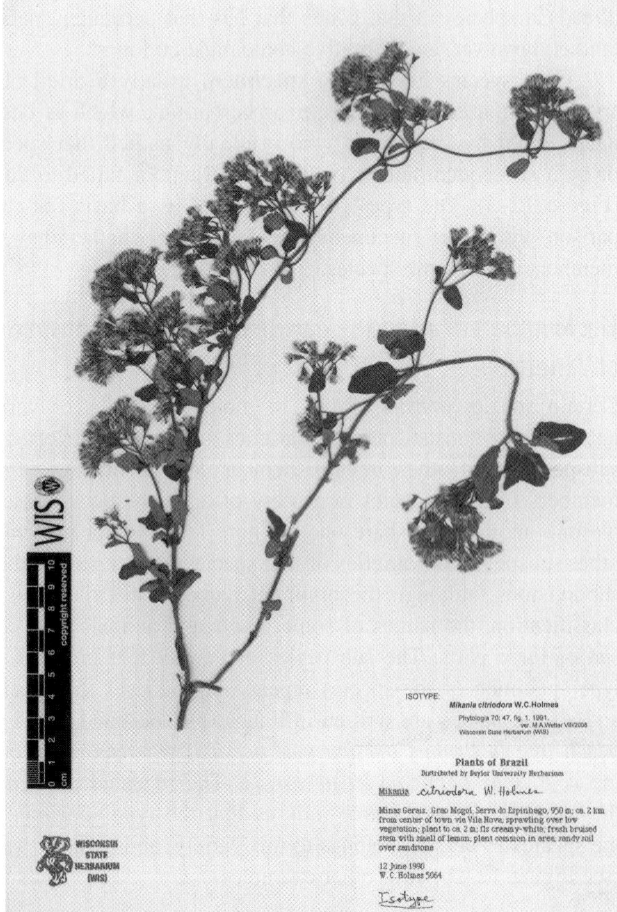

12–3 Type specimen The type specimen of the angiosperm *Mikania citriodora* (family Asteraceae), which is found in Brazil. This specimen was collected by W. C. Holmes and described by him in a paper published in the journal *Phytologia* (volume 70, pages 47–51, 1991).

Organisms Are Grouped into Broader Taxonomic Categories Arranged in a Hierarchy

Linnaeus (and earlier scientists) recognized three kingdoms—plant, animal, and mineral—and until recently, the kingdom was the most inclusive unit used in biological classification. In addition, several hierarchical taxonomic categories were added between the levels of genus and kingdom: genera were grouped into families, families into orders, and orders into classes. The Swiss-French botanist Augustin-Pyramus de Candolle (1778–1841), who invented the word "taxonomy," added another category—division—to designate groups of classes in the plant kingdom. Hence, the divisions became the largest inclusive groups of the plant kingdom. At the XV International Botanical Congress in 1993, however, the *International Code of Botanical Nomenclature* made the term **phylum** (plural: phyla) nomenclaturally equivalent to division. "Phylum" has long been used by zoologists for groups of classes and has been adopted for use in this book.

In this hierarchical system—that is, of groups within groups, with each group ranked at a particular level—the taxonomic group at any level is called a **taxon** (plural: taxa). The level at which it is ranked is called a **category.** For example, genus and species are categories, and *Prunus* and *Prunus persica* are taxa within those categories.

Regularities in the form of the names for the different taxa make it possible to recognize them as names at that level. For example, names of plant families end in *-aceae,* with a very few exceptions. Older names are allowed as alternatives for a few families, such as Fabaceae, the pea family, which may also be called by the older name, Leguminosae. Other examples are Apiaceae, the parsley family (also known as Umbelliferae), and Asteraceae, the sunflower family (also known as Compositae). Names of plant orders end in *-ales.*

Sample classifications of maize *(Zea mays)* and the commonly cultivated edible mushroom *(Agaricus bisporus)* are given in Table 12–1.

Many Different Classifications of Plants Have Been Proposed

The earliest classifications were based on the appearance, or the habit, of the plant. For instance, Theophrastus (370–285 B.C.E.), a student of Aristotle's and known as the Father of Botany, classified all plants on the basis of form: trees, shrubs,

Table 12-1 **Biological Classification. Notice how much you can tell about an organism when you know its place in the system. The descriptions here do not define the various categories but tell you something about their characteristics. The kingdoms Plantae and Fungi belong to the domain Eukarya.**

Category	Taxon	Description
Maize		
Kingdom	Plantae	Organisms that are primarily terrestrial, with chlorophylls *a* and *b* contained in chloroplasts, spores enclosed in sporopollenin (a tough wall substance), and nutritionally dependent multicellular embryos
Phylum	Anthophyta	Vascular plants with seeds and flowers; ovules enclosed in an ovary, pollination indirect; the angiosperms
Class	Monocotyledoneae	Embryo with one cotyledon; flower parts usually in threes; many scattered vascular bundles in the stem; the monocots
Order	Poales	Monocots with fibrous leaves; reduction and fusion in flower parts
Family	Poaceae	Hollow-stemmed monocots with reduced greenish flowers; fruit a specialized achene (caryopsis); the grasses
Genus	*Zea*	Robust grasses with separate staminate and carpellate flower clusters; caryopsis fleshy
Species	*Zea mays*	Maize, or corn
Edible Mushroom		
Kingdom	Fungi	Nonmotile, multinucleate, heterotrophic, absorptive organisms in which chitin predominates in the cell walls
Phylum	Basidiomycota	Dikaryotic fungi that form a basidium bearing four spores (basidiospores); subphyla Agaricomycotina, Pucciniomycotina, and Ustilaginomycotina
Class	Agaricomycetes	Fungi that produce basidiomata, or "fruiting bodies," and club-shaped, aseptate basidia that line gills or pores; the hymenomycetes
Order	Agaricales	Fleshy fungi with radiating gills or pores
Family	Agaricaceae	Agaricales with gills
Genus	*Agaricus*	Dark-spored soft fungi with a central stalk and gills free from the stalk
Species	*Agaricus bisporus*	The common edible mushroom

undershrubs, and herbs. Linnaeus used the "sexual system," by which plants were classified into 24 classes based on the number and arrangement of the stamens in each flower. Such systems of classification are referred to as **artificial systems,** because they classify organisms primarily as an aid to identification and generally by means of one or a few characters.

For Linnaeus and his immediate successors, the goal of taxonomy was the revelation of the grand, unchanging design of creation. After publication of Darwin's *On the Origin of Species* in 1859, however, differences and similarities among organisms came to be seen as products of their evolutionary history, or **phylogeny.** Biologists now wanted classifications to be not only informative and useful but also an accurate reflection of the evolutionary relationships among organisms. Such classifications are referred to as **natural classifications.** The evolutionary relationships among organisms have often been diagrammed as **phylogenetic trees,** which depict the genealogic relationships between taxa as *hypothesized* by a particular investigator or group of investigators.

Traditionally, the classification of a recently discovered organism and its phylogenetic relationship to other organisms was based on its outward similarities to other members of that taxon. Phylogenetic trees constructed by traditional methods rarely included detailed considerations of comparative information. Although this approach produced many useful results, it was based primarily on the investigator's opinion regarding which factors were most important in determining the classification. Therefore, it is not surprising that very different classifications were sometimes proposed for the same groups of organisms.

In a Classification Scheme That Accurately Reflects Phylogeny, Every Taxon Should Be Monophyletic

A **monophyletic group** (also called a **clade**) is composed of an ancestor and *all* its descendants; none of its descendants are excluded (Figure 12–4). Thus, a genus should consist of all species descended from the most recent common ancestor—and only of species descended from that ancestor. Similarly, a family should consist of all genera descended from a more distant common ancestor—and only of genera descended from that ancestor. Simply stated, a monophyletic group is one that can be removed from a phylogenetic tree by one "cut" of the tree. A phylogenetic classification attempts to give formal taxonomic names only to groups that are monophyletic, although not every monophyletic group may need a name.

As new information becomes available, researchers sometimes find that current taxonomic groups are not monophyletic. There are two such groups: paraphyletic and polyphyletic (Figure 12–4). A **paraphyletic group** is one consisting of a common ancestor, but *not all* descendants of that ancestor. In phylogenetic classification, paraphyletic groups are not given formal names. A **polyphyletic group** is a group with two or more ancestors, but not including the true common ancestor of its members.

Homologous Features Have a Common Origin, and Analogous Features Have a Common Function but Different Evolutionary Origins

Systematics is, to a great extent, a comparative science. It groups organisms into taxa from the categorical levels of genus through phylum, based on similarities in structure and other characters. From Aristotle on, however, biologists have recognized that superficial similarities are not useful criteria for taxonomic decisions. For example, birds and insects should not be grouped together simply because both have wings. A wingless insect (such as a silverfish) is still an insect, and a flightless bird (such as the kiwi) is still a bird.

A key question in systematics is the origin of a similarity or difference. Does the similarity of a particular feature reflect inheritance from a common ancestor, or does it reflect adaptation to similar environments by organisms that do not share a common ancestor? And a related question arises concerning differences between organisms. Does a difference reflect separate evolutionary histories, or does it reflect instead the adaptations of closely related organisms to very different environments? As we will see in later chapters, foliage leaves, cotyledons, bud scales, and floral parts have quite different functions and

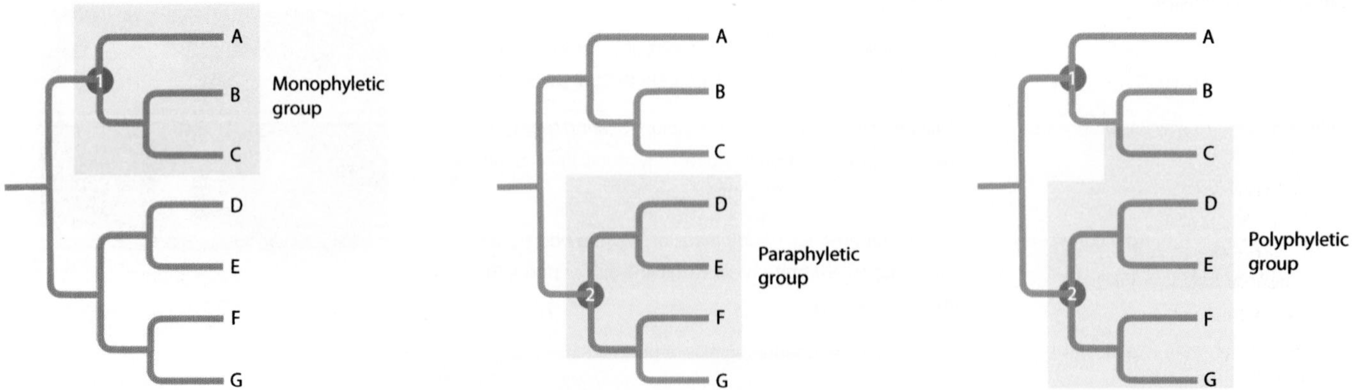

12–4 Monophyletic, paraphyletic, and polyphyletic groups A monophyletic group, or clade, includes the common ancestor 1 and all of its descendants (species A, B, and C). A paraphyletic group includes the common ancestor 2 of some (species D, E, and F), but not all, of its descendants (species G is not included). A polyphyletic group has two or more ancestors; species D, E, F, and G share common ancestor 2, but species C has a different one, ancestor 1.

CONVERGENT EVOLUTION

Comparable selective forces, acting on plants growing in similar habitats but different parts of the world, often cause totally unrelated species to assume a similar appearance. The process by which this happens is known as convergent evolution.

Let us consider some of the adaptive characteristics of plants growing in desert environments—fleshy, columnar stems (which provide the capacity for water storage), protective spines, and reduced leaves. Three fundamentally different families of flowering plants—the spurge family (Euphorbiaceae), the cactus family (Cactaceae), and the milkweed family (Apocynaceae)—have members with these features. The cactuslike representatives of the spurge and milkweed families shown here evolved from leafy plants that look quite different from one another.

Native cacti occur (with the exception of one species) exclusively in the New World. The comparably fleshy members of the spurge and milkweed families occur mainly in desert regions in Asia and especially Africa, where they play an ecological role similar to that of the New World cacti.

Although the plants shown here—***(a)*** *Euphorbia,* a member of the spurge family; ***(b)*** *Echinocereus,* a cactus; and ***(c)*** *Hoodia,* a fleshy milkweed—have CAM photosynthesis (page 145), all three are related to and derived from plants that have only C_3 photosynthesis. This indicates that the physiological adaptations involved in CAM photosynthesis also arose as a result of convergent evolution.

(a)

(b)

(c)

appearances, but all are evolutionary modifications of the same type of organ, namely, the leaf. Such structures, which have a common origin but not necessarily a common function, are said to be **homologous** (from the Greek *homologia,* meaning "agreement"). These are the features on which evolutionary classification systems, ideally, are constructed.

By contrast, other structures, which may have a similar function and superficial appearance, have an entirely different evolutionary background. Such structures are said to be **analogous** and are the result of **convergent evolution** (see the essay above). Thus, the wings of a bird and those of an insect are analogous, not homologous. Similarly, the spine of a cactus (a modified leaf) and the thorn of a hawthorn (a modified stem) are analogous, not homologous. Distinguishing between homology and analogy is seldom so simple, and it generally requires detailed comparison as well as evidence from other features of the organisms under study.

Cladistics

The most widely used method of classifying organisms today is known as **cladistics,** a form of **phylogenetic analysis** that explicitly seeks to understand phylogenetic relationships. The approach focuses on the branching of one lineage from another in the course of evolution. It recognizes a monophyletic group, or clade, by its *shared derived characters* (synapomorphies). **Synapomorphies** are character states that arose in the common ancestor of the group and are present in all of its members. Character states are two or more forms of a particular feature, such as the presence or absence of wood or flowers.

To develop an evolutionary tree, one must determine which changes are more recent and which occurred farther back in the past; that is, the tree must have direction—it must be **rooted.** By arranging the characters in a specific direction, rooting makes it possible to recognize shared derived character states that define monophyletic taxa.

Outgroups are used to root a tree. An **outgroup** is a taxon that is closely related to but not a member of the study group (the **ingroup**) under investigation. Character states possessed by the closest outgroups are considered to be ancestral, while those present in the ingroup, but absent in the nearest outgroups, are considered to be derived.

The result of cladistic analysis is a **cladogram,** which provides a graphical representation of a working model, or hypothesis, of the phylogenetic relationships among a group of organisms. These hypotheses can then be tested by attempting to incorporate additional species or characters that may or may not conform to the predictions of the model.

To see how a cladogram is constructed, let us consider four different groups of plants: hornworts (see Figure 16–29), ferns, pines, and oaks. For each of the plant groups, we have selected four homologous characters to be analyzed (Table 12–2). To

Table 12-2 Selected Characters Used in Analyzing the Phylogenetic Relationships of Four Plant Taxa

	Characters*			
Taxon	Xylem and Phloem	Wood	Seeds	Flowers
Hornworts	–	–	–	–
Ferns	+	–	–	–
Pines	+	+	+	–
Oaks	+	+	+	+

*The character state "present" (+) is the derived condition; the character state "absent" (–) is the ancestral condition.

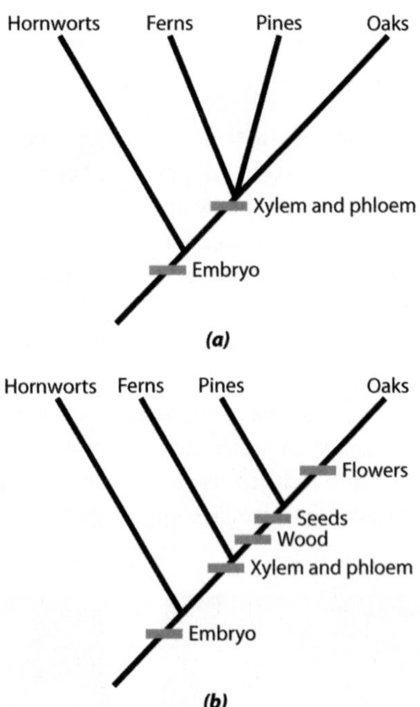

12–5 Cladograms These cladograms show the phylogenetic relationships among ferns, pines, and oaks, indicating the shared characters that support the patterns of relationships. ***(a)*** A cladogram based on the presence or absence of xylem and phloem. ***(b)*** Further resolution of the relationships, based on additional information on the presence or absence of wood, seeds, and flowers.

keep matters simple, the characters are considered to have only two different states: present (+) and absent (–).

Through their possession of embryos, hornworts are known to be related to the other three plant groups, which also have embryos. However, hornworts lack many features that the other three groups share—for example, xylem and phloem, and many characters not shown in the table. The hornworts can be used as the outgroup and can be considered to have diverged earlier than the other taxa from a common ancestor. Accordingly, the hornworts can be used to determine whether features shared among ferns, pines, and oaks can potentially be used to define a clade. For example, seeds are not present in hornworts, and therefore seeds can be hypothesized to be a potential shared derived feature that would support uniting pines and oaks as a monophyletic group. Applying this argument to our few characters results in the "absent" character state being consistently recognized as the ancestral condition and the "present" character state as the derived condition.

Figure 12–5a shows how one might sketch a cladogram based on the presence or absence of the vascular tissues xylem and phloem. Inasmuch as ferns, pines, and oaks all have xylem and phloem, they can be hypothesized to form a monophyletic group. Figure 12–5b shows how further resolution is obtained as information about other features is added.

How does one interpret the cladogram in Figure 12–5b? To begin, note that cladograms do not indicate that one group gave rise to another, as in many phylogenetic trees constructed by the traditional method. Rather, they imply that groups terminating in adjacent branches (the branch points are called nodes) shared a common ancestor. Such groups are said to be **sister groups,** or closest relatives. The cladogram of Figure 12–5b tells us that oaks shared a more recent common ancestor with pines than with ferns, and that they are more closely related to pines than to ferns. The relative positions of the various plants on the cladogram indicate their relative times of divergence.

A fundamental principle of cladistics is that a cladogram should be constructed in the simplest, least complicated, and most efficient way. This principle is called the **principle of parsimony.** When conflicting cladograms are constructed from the data at hand, the one with the greatest number of statements of homology and the fewest of analogy is preferred.

Molecular Systematics

Before the advent of molecular systematics, classification by any methodology was based largely on comparative morphology and anatomy, but plant systematics has been revolutionized by the application of molecular techniques. The techniques most widely used are those for determining the sequence of nucleotides in nucleic acids—sequences that are genetically determined (see Chapter 10). Molecular data are different from data obtained from traditional sources in several important ways: in particular, they are easier to quantify, they have the potential to provide many more characters for phylogenetic analysis, and they allow comparison of organisms that are morphologically very different. With the development of molecular techniques, it has become possible to compare organisms at the most basic level—the gene.

Analysis of nucleic acid sequences provides powerful data for understanding evolutionary relationships. Many different genes, with varying rates of change, can be used to study evolution in different lineages. Much of the variation in the homologous genes of different groups of organisms is due to neutral mutations that have accumulated at a nearly constant rate over evolutionary time. This variation is not the result of a selection process. Instead, it represents the differences in the number of nucleotide changes that have occurred in homologous

GOOGLE EARTH: A TOOL FOR DISCOVERING AND PROTECTING BIODIVERSITY

While examining satellite images, researchers at the Kew Royal Botanic Gardens in England came upon "an unexpected patch of green" in the mountainous areas of northern Mozambique. They were looking for a possible site for a conservation project, and three years later, in 2008, an expedition confirmed what the images suggested: the region, known as Mount Mabu, was home to the largest continuous expanse of midland rainforest in southern Africa. The forest, little known except to those living on its fringes, provides habitat to hundreds of animal species, among them pygmy chameleons, rare snakes, birds, and butterflies, and several species previously unknown to science. Over 500 plant specimens were collected, among them rare orchids and a new species of tropical mistletoe. The full extent of the region's biodiversity is currently being catalogued, and conservation efforts are underway with the government of Mozambique to protect the area.

In one sense, the discovery was not that unusual—government and academic researchers have been poring over satellite photos for decades, and announcements of previously unknown areas of wilderness have emerged before. What made the Mount Mabu discovery remarkable was that the images were not stored away in some government laboratory or available only to those with the right academic credentials—instead, they were freely available to anyone with a personal computer, courtesy of Google Earth.

Aiding in the discovery of new territories, however, may be the least of the program's strengths. Conservationists around the world are increasingly exploiting Google Earth to increase their understanding of the areas they already know, and to share their findings with others all around the globe.

To protect biodiversity, conservationists need to know what is out there. Satellite imagery, with its bird's-eye view of the entire landscape, allows them to survey millions of hectares without leaving their desks. That ability makes it practical to discover new habitats, such as Mount Mabu, or just an isolated desert stream, before mounting a field expedition.

Furthermore, the program allows the user to annotate any image, adding text, photos, videos, and website links. The San Diego Natural History Museum, for example, is mapping the plants of San Diego County (similar programs are underway in many other locales). A volunteer field botanist, after using a GPS device to pinpoint his or her location, can make a detailed list of all the plant species in the vicinity and upload the list to the museum's master map, along with photographs of the landscape and other natural features.

With hundreds of such annotations, the map may be used to track the spread of an invasive cactus, for instance, or to understand the barriers to pollination of a rare flowering plant. A map—technically known as an "overlay"—from one source can be superimposed on overlays from other sources. Combining overlays can show, for instance, whether a particular rare plant is near a proposed development, or whether an invasive cactus is encroaching on the habitat of an endangered songbird. Making it possible to look for patterns in the data collected may be the most powerful application of the program.

Biologists are just now thinking about how Google Earth can best be used to study the environment. As millions more users become involved in the coming years, this and similar programs will play a central role in understanding and protecting the world's biodiversity.

(a)

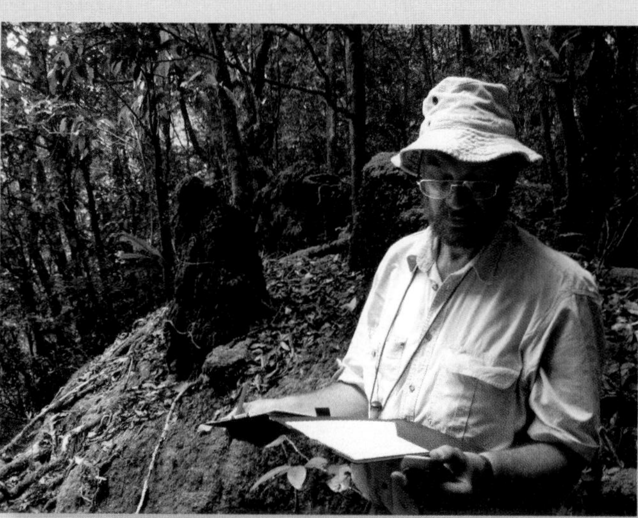

(b)

Mount Mabu ***(a)*** A satellite image showing Mount Mabu as a dark patch of green surrounded by lighter areas of cropland and human settlements. Scattered areas of dark green suggest the original extent of the forest, now degraded by logging and agricultural burning. ***(b)*** Botanist Jonathan Timberlake records his notes about the vegetation growing on the steep slopes of Mount Mabu.

genes since the lineages branched apart. Groups that diverged more recently tend to have fewer differences among them than groups that diverged from a common ancestor longer ago. Noncoding DNA sequences also provide nearly neutral markers that reflect past evolutionary events.

As sequences of nucleic acids from a variety of species have been determined, the information has been entered into computer banks, primarily in GenBank, which is supported by the National Institutes of Health as part of the National Center for Biotechnology Information. It is therefore possible to make detailed comparisons among large numbers of taxa.

The power of molecular systematics, when combined with the study of morphological character states, is exemplified by the Consensus Phylogeny of Flowering Plants, spearheaded by the group of plant systematists known as the Angiosperm Phylogeny Group. Virtually every family of flowering plants now has a well-supported phylogenetic position, and this community-wide endeavor is making solid progress at the genus level.

Many of the molecular systematic studies have produced surprises. For example, the parasitic plant *Rafflesia* (see Figure 19–5b), which has giant flowers, is placed in the order Malpighiales with such species as poinsettia *(Euphorbia pulcherrima),* which has tiny flowers. Also, the 10 families of angiosperms that form symbiotic associations with nitrogen-fixing bacteria in root nodules (see Chapter 29), which were long believed to have independently developed the capacity to fix nitrogen, belong to a single clade, along with a few families that do not fix nitrogen. Furthermore, the water lotus *(Nelumbo)* (see Figure 20-9b), thought to be related to water lilies or to some other aquatic flowering plants, is actually most

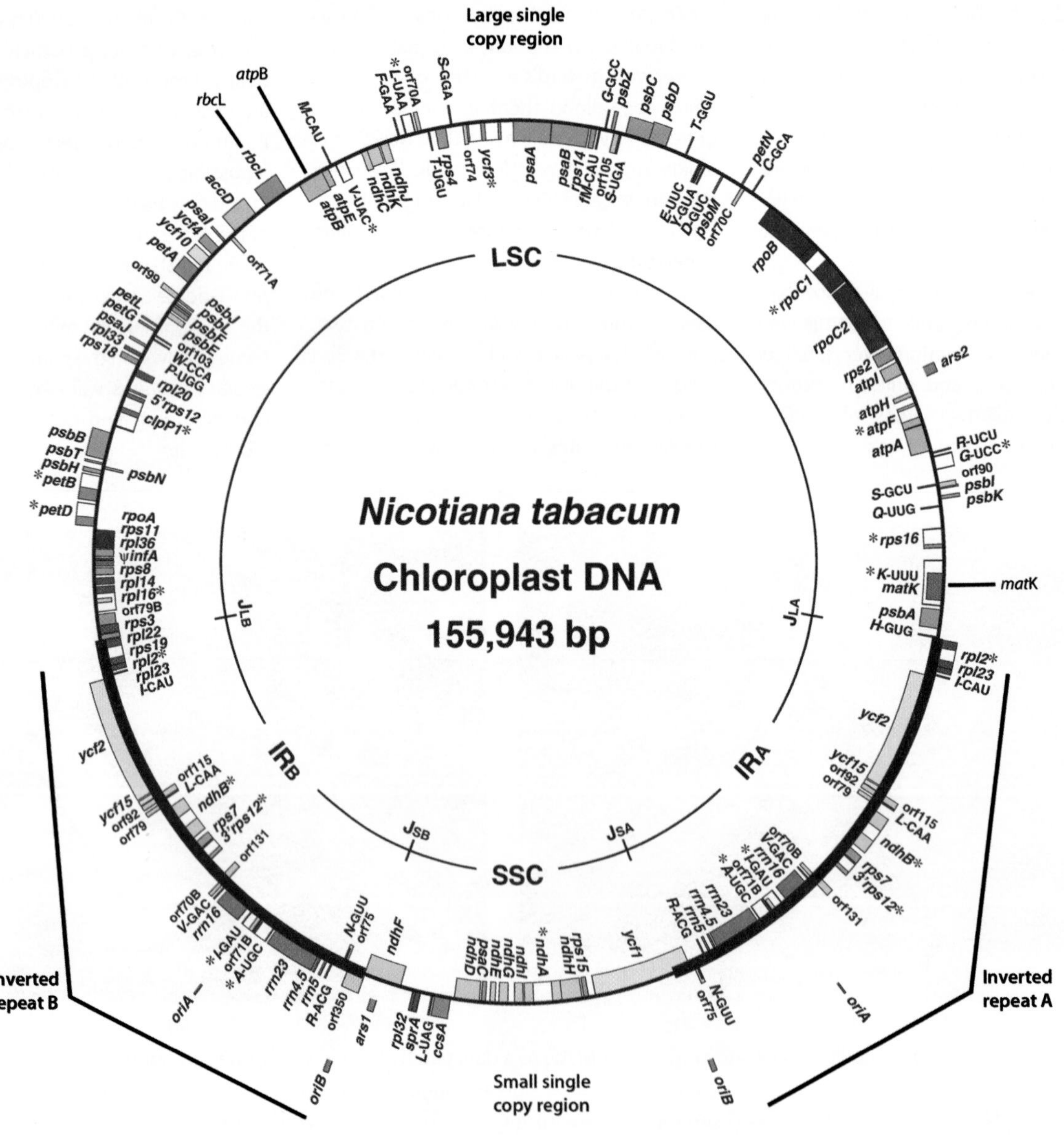

12–6 **Gene map of a chloroplast** This diagram of the chloroplast DNA of tobacco *(Nicotiana tabacum)* shows the locations of some major genes, the inverted repeats (IRA and IRB), and the small single-copy (SSC) and large single-copy (LSC) regions. Note the location of the *rbcL, atpB,* and *matK* genes.

closely related to the sycamore or plane tree *(Platanus),* along with trees and shrubs in the family Proteaceae, which includes the macadamia nut tree.

The Chloroplast Has Been the Main Source of Plant DNA Sequence Data

As noted in Chapter 3, the genome of the chloroplast exists as a circular molecule of DNA. In most plants it consists of 135 to 160 kilobase pairs (kbp), the smallest of the three plant genomes. The mitochondrial genome (also circular) consists of 200 to 2500 kbp, and the nuclear genome is much larger, with 1.1×10^6 to 1.1×10^{11} kbp.

The chloroplast genome is characterized by the presence of two regions that encode the same genes, but in opposite directions. These regions are known as **inverted repeats;** between them are a small single-copy region and a large single-copy region (Figure 12–6).

Plant systematists have been particularly interested in the generation of an extensive database of sequences of the chloroplast gene *rbcL,* which encodes the large subunit of the Rubisco enzyme of the Calvin cycle (page 136). The *rbcL* gene is present in all photosynthetic eukaryotes and cyanobacteria and is especially well suited for analysis of the relationships among large plant groups. Not only is it a slowly evolving, single-copy gene, but it lacks introns and is large enough (1428 base pairs) to preserve a significant number of phylogenetically informative characters. Because of its slow rate of change, the *rbcL* gene is not very useful for resolving relationships within or between closely related genera. Other chloroplast genes have been used for such purposes. Data for *atpB,* the gene that encodes a subunit of ATP synthase, in combination with data for *rbcL,* have been helpful in refining the picture of angiosperm relationships.

Relatively few studies have used mitochondrial or nuclear genes for studying plant systematics. In general, mitochondrial genes evolve too slowly to allow accurate discrimination between species. Some nuclear genes, such as the one that encodes alcohol dehydrogenase, are becoming more frequently used.

DNA Barcoding Is Providing a Means for Rapid Identification of Species

DNA barcoding, an invention of Paul Hebert, a geneticist at the University of Guelph, Ontario, is modeled after the Universal Product Code, the familiar barcode found on many consumer products (Figure 12–7). For the identification of animal species, Hebert suggested analyzing a small piece of the mitochondrial cytochrome oxidase 1 gene, called the *cox1,* or *CO1,* gene, which is present in the mitochondria of all animals, and using it as a universal DNA barcode. This small fragment of the *CO1* gene usually enables clear-cut identification of animal species; although it varies widely between species, it barely varies between individuals.

The standard region of *CO1* is not suitable as a DNA barcode for most plants, however, because their mitochondrial genes evolve too slowly for accurately distinguishing between species. Two plastid coding regions, *rbcL* and *matK,* have been recommended by the Plant Working Group of the Consortium for the Barcode of Life as a core barcode. These two regions can be supplemented with additional markers as required. Although the *rbcL* and *matK* barcode works better for some plants than for others, studies showed that it correctly identified 72 percent of all species, on average, and grouped 100 percent of plants into the correct genus.

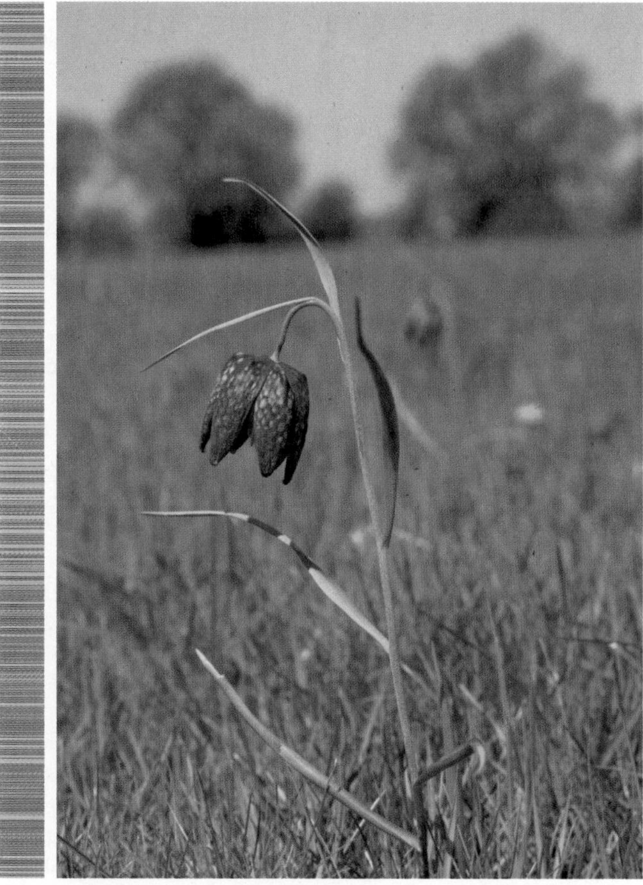

12–7 Barcode for *Fritillaria meleagris* Each of the four DNA bases—adenine, thymine, guanine, and cytosine—is represented by a different colored line in this barcode, which is based on the *rbcL* gene.

DNA barcoding should greatly aid in the identification and classification of organisms and in mapping the extent of biological diversity. Until recently, taxonomists needed a plant's flowers or fruits in order to identify it, but these are available only at certain times of the year. With DNA barcoding, any plant part at any stage of development can be used for that purpose.

The Major Groups of Organisms: Bacteria, Archaea, and Eukarya

In Linnaeus's time, as we mentioned earlier, three kingdoms were recognized—animals, plants, and minerals—and, until fairly recently, it was common to classify every living thing as either an animal or a plant. Kingdom Animalia included those organisms that moved and ate things and whose bodies grew to a certain size and then stopped growing. Kingdom Plantae comprised all living things that did not move or eat and that grew indefinitely. Thus the fungi, algae, and bacteria, or prokaryotes,

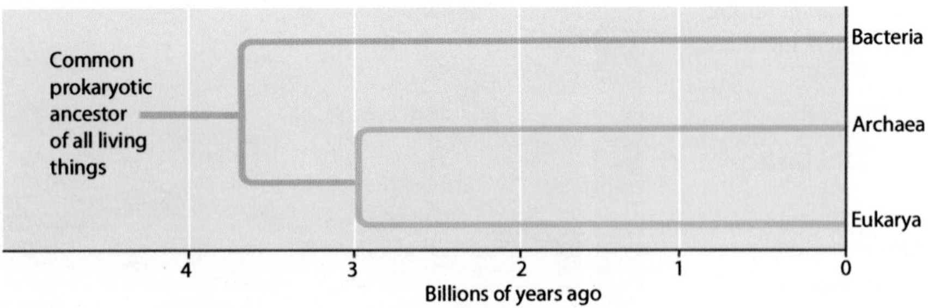

12–8 Evolutionary relationships of the three domains of life As indicated in this diagram, all living things share a common, very ancient prokaryotic ancestor, and the Archaea and Eukarya share a more recent ancestor with each other than with the Bacteria.

were grouped with the plants, and the protozoa—the one-celled organisms that ate and moved—were classified as animals. Jean Baptiste de Lamarck, Georges Cuvier, and most other eighteenth- and nineteenth-century biologists continued to place all organisms in one or the other of these kingdoms. This old division into plants and animals is still widely reflected in the organization of college textbooks, including this one. That is why, in addition to plants, we include algae, fungi, and prokaryotes in this text.

In the twentieth century, new data began to emerge. This was partly due to improvements in the light microscope and, subsequently, the development of the electron microscope. It was also due to the application of biochemical techniques to studies of differences and similarities among organisms. As a result, the number of groups recognized as constituting different kingdoms increased. The new techniques revealed, for example, the fundamental differences between prokaryotic and eukaryotic cells. These differences were sufficiently great to warrant placing the prokaryotic organisms in a separate kingdom, Monera. Then, in the 1970s, analysis of small-subunit ribosomal RNA by Carl Woese at the University of Illinois provided the first evidence that the world is divided into three major groups, or domains—**Bacteria, Archaea,** and **Eukarya** (Figures 12–8 and 12–9). The Bacteria and Archaea are prokaryotes. The Eukarya include all eukaryotes. Table 12–3 summarizes some of the major differences that distinguish the three domains.

Initially, the domain Eukarya consisted of four kingdoms: Protista, Fungi, Animalia, and Plantae. However, with the emergence of molecular systematics and the comparison of DNA sequences, in addition to comparison of cellular features, it became clear that the protists do not constitute a monophyletic group. More recently, researchers have hypothesized that the eukaryotes comprise seven groups, called supergroups (Figure 12–10). A **supergroup** lies between a domain and a kingdom. All of the supergroups include phyla of protists; most consist

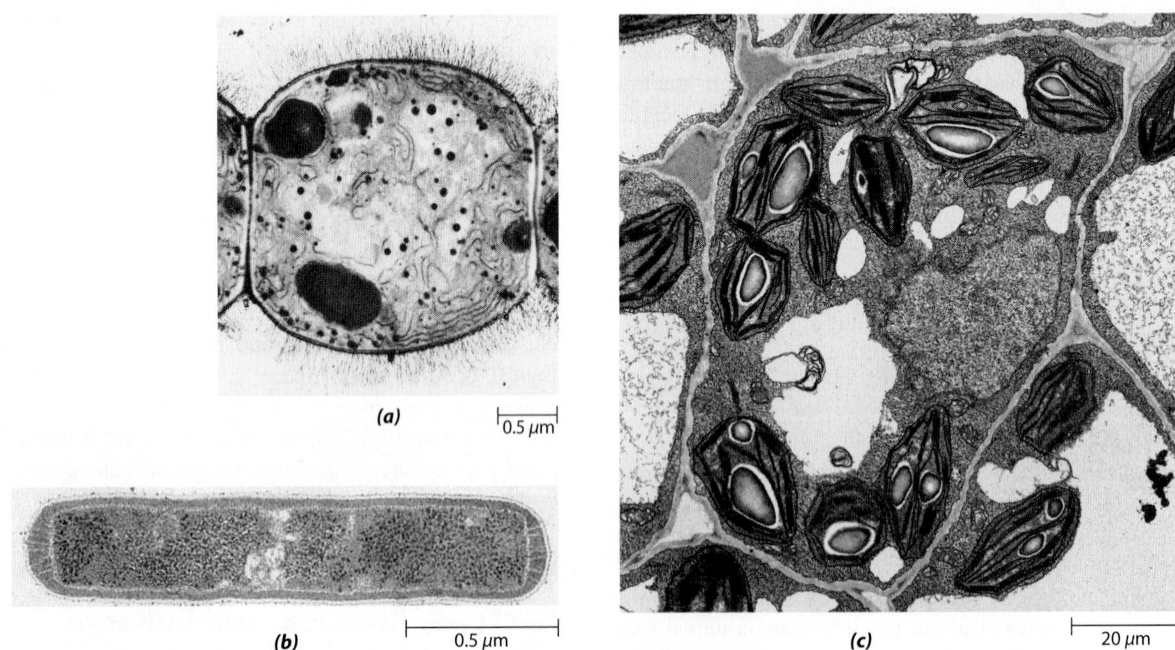

12–9 Representatives of the three domains Electron micrographs of ***(a)*** a prokaryote, the cyanobacterium *Anabaena* (domain Bacteria); ***(b)*** another prokaryote, the archaeon *Methanothermus fervidus* (domain Archaea); and ***(c)*** a eukaryotic cell, from the leaf of a sugar beet *(Beta vulgaris)* (domain Eukarya). The cyanobacterium is a common inhabitant of ponds, whereas *Methanothermus,* which is adapted to high temperatures, grows optimally at 83° to 88°C. Notice the greater complexity of the eukaryotic cell, with its conspicuous nucleus and chloroplasts, and its much larger size (note the scale markers).

Table 12-3 Some Major Distinguishing Features of the Three Domains of Life*

Characteristic	Bacteria	Archaea	Eukarya
Cell type	Prokaryotic	Prokaryotic	Eukaryotic
Nuclear envelope	Absent	Absent	Present
Number of chromosomes	1	1	More than 1
Chromosome configuration	Circular	Circular	Linear
Organelles (mitochondria and plastids)	Absent	Absent	Present
Cytoskeleton	Absent	Absent	Present
Chlorophyll-based photosynthesis	Present	Absent	Present

* Note that some features listed apply to only certain representatives of a particular domain.

entirely of protists. The kingdoms Fungi and Animalia and their unicellular relatives are within the supergroup Opisthokonta, and the kingdom Plantae (land plants), together with their algal relatives, belong to a supergroup for which we give no formal name. Names suggested for this supergroup have been challenged by several recent phylogenomic studies. The division of eukaryotes into supergroups is still being investigated and should be regarded as a work in progress.

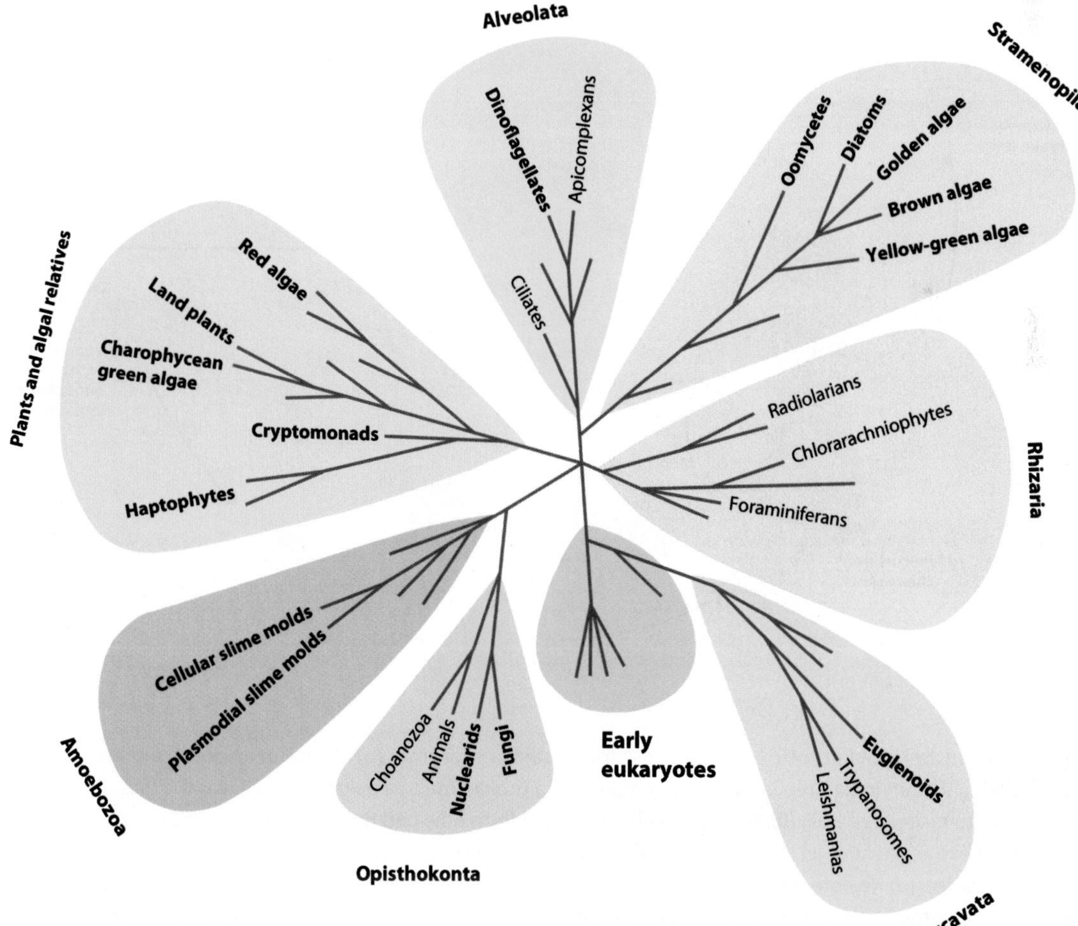

12–10 The major lineages of eukaryotes In the model depicted here, the eukaryotes are placed into seven supergroups (Opisthokonta, Amoebozoa, Plants and algal relatives, Alveolata, Stramenopila, Rhizaria, and Excavata). Selected subgroups are listed for each supergroup; those in bold are considered in this book. Only three kingdoms have survived from previous classifications: Plantae (land plants), Fungi, and Animalia. All other eukaryotes are protists. Recent evidence indicates that the stramenopiles, alveolates, and rhizarians can be grouped together, forming the SAR clade.

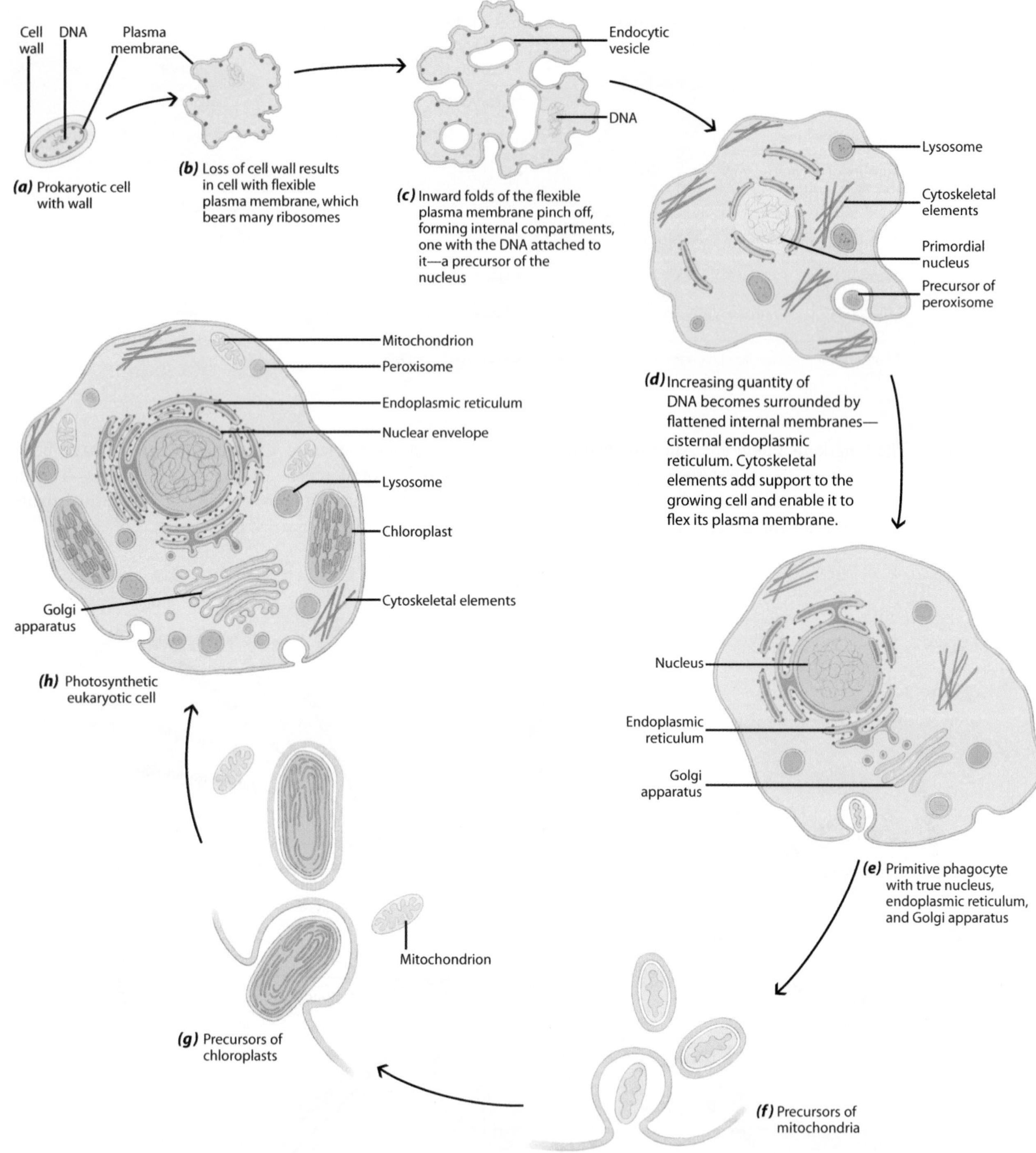

12–11 Origin of a photosynthetic eukaryotic cell from a heterotrophic prokaryote ***(a)*** Most prokaryotes contain a rigid cell wall, so it is likely that an initial step in the transformation of a prokaryote to a eukaryotic cell was the loss of the prokaryote's ability to form a cell wall. ***(b), (c)*** This free-living, naked form now had the ability to increase in size, change shape, and engulf extracellular objects by infolding of the plasma membrane (endocytosis), resulting in the formation of endocytic vesicles. ***(d), (e)*** Internalization of a patch of the plasma membrane to which DNA was attached was the probable precursor of the nucleus. The primitive phagocyte eventually acquired a true nucleus containing an increased quantity of DNA. A cytoskeleton must also have developed to provide inner support for the wall-less cell and to play a role in movement, of both the cell itself and its internal components. ***(f)*** The mitochondria of the eukaryotic cell had their origin as bacterial endosymbionts, which ultimately transferred most of their DNA to the host's nucleus. ***(g)*** Chloroplasts also are the descendants of bacteria. They, too, ultimately transferred most of their DNA to the host's nucleus. ***(h)*** The photosynthetic eukaryotic cell contains a complex endomembrane system and a variety of other internal structures, such as the peroxisomes, mitochondria, and chloroplasts depicted here.

Origin of the Eukaryotes

One of the most momentous series of events to take place in the evolution of life on Earth was the transformation of relatively simple prokaryotic cells into intricately organized eukaryotic cells. You will recall from Chapter 3 that eukaryotic cells are typically much larger than prokaryotes, and their DNA, which is more highly structured, is enclosed within a nuclear envelope. In addition to having an internal cytoskeleton, eukaryotic cells differ further from prokaryotic cells in their possession of mitochondria and, in plants and algae, chloroplasts that are about the size of a prokaryotic cell.

The Serial Endosymbiotic Theory Provides a Hypothesis for the Origin of Mitochondria and Chloroplasts

Both mitochondria and chloroplasts are believed to be the descendants of bacteria that were taken up and adopted by an ancient **host cell.** This concept for the origin of mitochondria and chloroplasts is known as the **serial endosymbiotic theory,** with the prokaryotic ancestors of mitochondria and chloroplasts as the endosymbionts. An **endosymbiont** is an organism that lives within another, dissimilar organism. The process by which eukaryotic cells originated is termed *serial* endosymbiosis because the events did not occur simultaneously—mitochondria definitely appeared before chloroplasts.

The Endomembrane System Is Thought to Have Evolved from Portions of the Plasma Membrane Endosymbiosis has had a profound influence on diversification of the eukaryotes. Most experts believe that the process of establishing an endosymbiotic relationship was preceded by the evolution of a prokaryotic host cell into a primitive **phagocyte** (meaning "eating cell")—a cell capable of engulfing large particles such as bacteria (Figure 12–11). It is likely that the ancestral host cell was a wall-less heterotroph living in an environment that provided it with food. Such cells would need a flexible plasma membrane capable of enveloping bulky food particles by folding inward. In this scenario, endocytosis was followed by the breakdown of the food particles within vesicles derived from the plasma membrane. The plasma membrane was rendered flexible by the incorporation of sterols, and the development of a cytoskeleton (especially of microtubules) provided the mechanism necessary for capturing food or prey and carrying it inward by endocytosis. The host cell's lysosomes (membrane-bounded vesicles that contain degradative enzymes) fused with the food vacuoles, breaking down their contents into usable organic compounds. The intracellular membranes derived from the plasma membrane gradually compartmentalized the host cells, forming what is known as the endomembrane system of the eukaryotic cell (page 54).

The genesis of the nucleus—the principal feature of eukaryotic cells—could also have begun by infolding of the plasma membrane. In prokaryotes, the circular DNA molecule, or prokaryotic chromosome, is attached to the plasma membrane. Infolding of this portion of the plasma membrane could have resulted in enclosure of the DNA within an intracellular sac, the primordial nucleus (Figure 12–11).

Mitochondria and Chloroplasts Are Thought to Have Evolved from Bacteria That Were Phagocytized The stage is set. A phagocyte now exists that can prey on bacteria, but the phagocyte still lacks mitochondria. The next step is for the phagocyte not to digest the bacterial precursors of the mitochondria (or chloroplasts) but to adopt them, establishing a symbiotic ("living together") relationship.

The green *Vorticella* shown in Figure 12–12 is an example of a modern protist that establishes endosymbioses with certain species of the green alga *Chlorella.* The algal cells remain intact within the host cells as endosymbionts, providing photosynthetic products useful to the heterotrophic host.

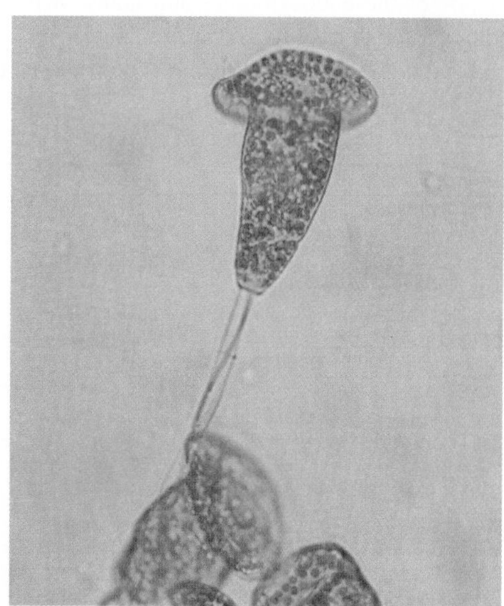

(a)

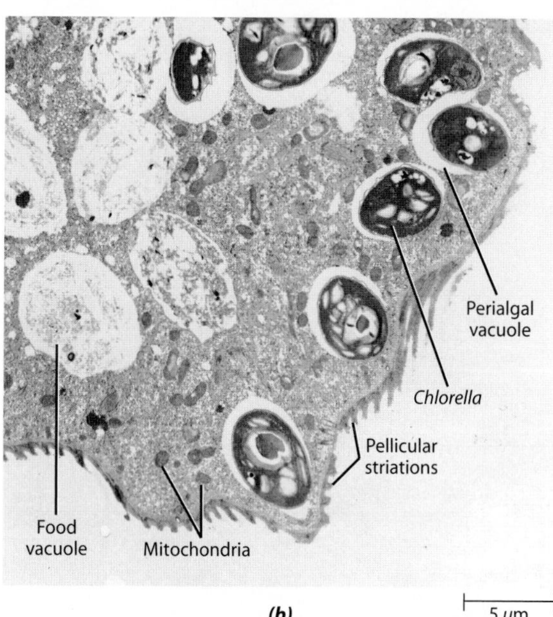

(b)

12–12 Endosymbiosis in *Vorticella* ***(a)*** Each bell-shaped cell of the protozoan *Vorticella* contains numerous cells of the autotrophic, endosymbiotic alga *Chlorella.* ***(b)*** Electron micrograph of a *Vorticella* containing cells of *Chlorella.* Each algal cell is found in a separate vacuole (perialgal vacuole) bounded by a single membrane. The protozoan provides the algae with protection and mineral nutrients, while the algae produce carbohydrates that serve as nourishment for their heterotrophic host cell.

In return, the algae receive essential mineral nutrients from the host. There are many examples of prokaryotic (bacterial) and eukaryotic endosymbionts in other protists, as well as in the cells of some 150 genera of freshwater and marine invertebrate animals. Algal endosymbionts, including those occurring in the polyps of reef-building corals, increase productivity and survival of the host (see Chapter 15).

Transformation of an endosymbiont into an organelle usually involved loss of the endosymbiont's cell wall (if any existed) and other unneeded structures. In the course of evolution, the DNA of the endosymbiont and many of its functions were gradually transferred to the host's nucleus. Hence, the genomes of modern mitochondria and chloroplasts are quite small compared with the nuclear genome. Although the mitochondrion or chloroplast cannot live outside a eukaryotic cell, both are self-replicating organelles that have retained many of the characteristics of their prokaryotic ancestors.

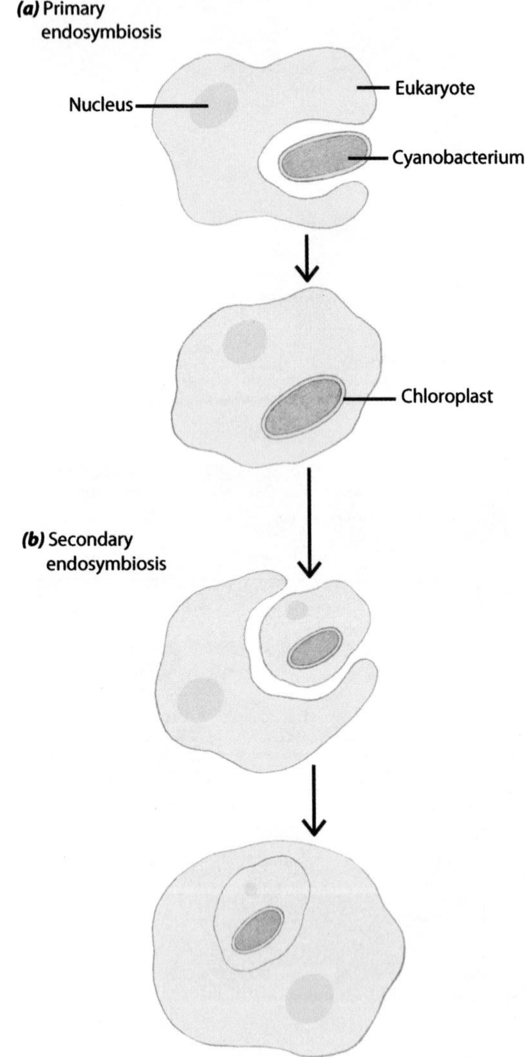

12–13 Endosymbiosis and the ancestry of chloroplasts ***(a)*** In primary endosymbiosis, a free-living cyanobacterium is engulfed by a eukaryotic cell but is not digested by it. Eventually, the cyanobacterium is transformed into a chloroplast. ***(b)*** In secondary endosymbiosis, a eukaryotic cell that has already acquired a chloroplast by primary endocytosis is itself taken up by a second eukaryote.

Mitochondria are widely accepted to have evolved from an alpha-proteobacterium in a common ancestor of all existing (extant) eukaryotes. By contrast, the chloroplasts of the algae (see Chapter 15) are universally believed to have evolved from cyanobacterial endosymbionts by three major types of endosymbiosis. The land plants, in turn, inherited their chloroplasts from green algae.

The three types of endosymbiosis involved in the evolutionary origin of algal plastids are designated primary, secondary, and tertiary endosymbiosis. In the process of **primary endosymbiosis,** cyanobacterial cells ingested by the eukaryotic host evolve into primary plastids, each of which is bounded by an envelope consisting of two membranes (Figure 12–13a). Primary plastids are found in red algae and green algae, and in the glaucophytes (a small group of freshwater algae that contain blue-green plastids resembling cyanobacteria). Whether primary plastids originated more than once is controversial, but many experts favor a common ancestor.

In the process of **secondary endosymbiosis,** eukaryotic cells containing plastids are themselves engulfed by another eukaryotic cell and evolve into secondary plastids (Figure 12–13b). Such plastids are characterized by the presence of three or four envelope membranes. Of the algal lineages considered in this book, secondary plastids are found in haptophytes, most cryptomonads, and many euglenoids, dinoflagellates, and stramenopiles.

In **tertiary endosymbiosis,** the eukaryotic cells have a plastid derived from a eukaryotic endosymbiont with a secondary plastid. The envelope of tertiary plastids consists of more than two membranes. Tertiary plastids derived independently from cryptomonad, haptophyte, or diatom endosymbionts with secondary plastids are found in several dinoflagellate species.

The Protists and Eukaryotic Kingdoms

The following is a synopsis of the protists and the three kingdoms included in the domain Eukarya (see Table 12–4, which does not include the kingdom Animalia).

The Kingdom Fungi Includes Eukaryotic, Multicellular Absorbers

Members of the kingdom Fungi, which are nonmotile, filamentous eukaryotes that lack plastids and photosynthetic pigments, absorb their nutrients from either dead or living organisms (Figure 12–14). The fungi have traditionally been grouped with plants, but there is no longer any doubt that the fungi are an independent evolutionary line. Moreover, comparisons of ribosomal RNA sequences indicate that the fungi are more closely related to animals than to plants. Apparently, animals and fungi diverged 1.5 billion years ago, the fungi arising from protists closely related to the modern genus *Nuclearia.* Aside from their filamentous growth habit, fungi have little in common with any of the protist groups that have been classified as algae.

Table 12-4 Classification of Living Organisms Included in This Book

Prokaryotic Domains
- Bacteria (bacteria)
- Archaea (archaea)

Eukaryotic Domain
- Eukarya
- Kingdom Fungi (fungi)
  - Phylum Microsporidia (microsporidians)
  - Chytrids
  - Zygomycetes
  - Phylum Glomeromycota (glomeromycetes)
  - Phylum Ascomycota (ascomycetes)
  - Phylum Basidiomycota (basidiomycetes)
- Protists
  - Algae
    - Euglenoids
    - Phylum Cryptophyta (cryptomonads)
    - Phylum Haptophyta (haptophytes)
    - Dinoflagellates
    - Class Bacillariophyceae (diatoms)*
    - Class Chrysophyceae (golden algae)*
    - Class Xanthophyceae (yellow-green algae)*
    - Class Phaeophyceae (brown algae)*
    - Phylum Rhodophyta (red algae)
    - Green algae
  - Heterotrophic protists
    - Phylum Oomycota (oomycetes)†
    - Phylum Myxomycota (plasmodial slime molds)
    - Phylum Dictyosteliomycota (cellular slime molds)
- Kingdom Plantae
  - Bryophytes
    - Phylum Marchantiophyta (liverworts)
    - Phylum Bryophyta (mosses)
    - Phylum Anthocerotophyta (hornworts)
  - Vascular plants
    - Seedless vascular plants
      - Phylum Lycopodiophyta (lycophytes)
      - Phylum Monilophyta (ferns and horsetails)
    - Seed plants
      - Phylum Coniferophyta (conifers)
      - Phylum Cycadophyta (cycads)
      - Phylum Ginkgophyta (ginkgo)
      - Phylum Gnetophyta (gnetophytes)
      - Phylum Anthophyta (angiosperms)

* These algae are known as photosynthetic stramenopiles.
† The oomycetes are plastidless, or heterotrophic, stramenopiles.

Typically, for example, the cell walls of fungi include a matrix of chitin. The structures in which fungi form their spores are often complex. Fungal reproductive cycles, which are also quite complex, typically involve both sexual and asexual processes. Fungi are discussed in Chapter 14.

The Kingdom Animalia Includes Eukaryotic, Multicellular Ingesters

The animals are multicellular organisms with eukaryotic cells lacking cell walls, plastids, and photosynthetic pigments. Nutrition is primarily ingestive—food is taken in through a mouth or other opening—with digestion occurring in an internal cavity. In some forms, however, nutrition is absorptive, and a number of groups lack an internal digestive cavity. The level of organization and tissue differentiation in complex animals far exceeds that of the other kingdoms, particularly with the evolution of sensory and neuromotor systems. The motility of the organism—or, in sessile forms, of its component parts—is based on contractile fibrils. Reproduction is predominantly sexual. Animals are not discussed in this book, except in relation to some of their interactions with plants and other organisms.

The Protists Include Unicellular, Colonial, and Simple Multicellular Eukaryotes

The protists (Figure 12–15) comprise all organisms traditionally regarded as protozoa (one-celled "animals"), which are heterotrophic, as well as all algae, which are autotrophic. Also included among the protists are some heterotrophic groups of organisms that have traditionally been placed with the fungi—including the water molds and their relatives (phylum Oomycota), the plasmodial slime molds (phylum Myxomycota), and the cellular slime molds (phylum Dictyosteliomycota).

The reproductive cycles of protists are varied but typically involve both cell division and sexual reproduction. Protists may be motile by means of 9-plus-2 flagella (see Figure 3–28) or cilia or by amoeboid movement, or they may be nonmotile. One group of protists, the green algae, are clearly very closely related to the bryophytes and vascular plants, and certainly the ancestral group from which bryophytes and vascular plants were derived. In this book, we consider the bryophytes and vascular plants, which are adapted for life on land, to constitute the kingdom Plantae. Some evolutionary plant biologists, however, place the green algae, bryophytes, and vascular plants together in a clade called **green plants,** or **viridophytes.** The bryophytes and vascular plants then are referred to as the "land plants."

In summary, the protists are paraphyletic and include a very heterogeneous assemblage of unicellular, colonial, and multicellular eukaryotes that do not have the distinctive characteristics of fungi, animals, or plants (bryophytes and vascular plants). The protists are discussed in Chapter 15.

The Kingdom Plantae Includes Eukaryotic, Multicellular Photosynthesizers

Plants—with three phyla of bryophytes (liverworts, mosses, and hornworts) and the seven extant phyla of vascular plants—constitute a kingdom of photosynthetic organisms adapted for

12–14 Fungi ***(a)*** Red blanket lichen *(Herpothallon sanguineum)* growing on a tree trunk in the Corkscrew Swamp Sanctuary in Florida. ***(b)*** A white coral fungus (family Clavariaceae). ***(c)*** Mushrooms (genus probably *Mycena*), with dew droplets, growing in a rainforest in Peru. ***(d)*** An earthball, *Scleroderma aurantium.*

life on the land (Figure 12–16). All plants are multicellular, and they are composed of eukaryotic cells that contain vacuoles and are surrounded by cell walls that contain cellulose. Their principal mode of nutrition is photosynthesis, although a few plants have become heterotrophic. Structural differentiation occurred during the evolution of plants on land, with trends toward the evolution of organs specialized for photosynthesis, anchorage, and support. In more complex plants, such organization has produced specialized photosynthetic, vascular, and covering tissues. Reproduction in plants is primarily sexual, with cycles of alternating haploid and diploid generations. In the more advanced members of the kingdom, the haploid generation (the gametophyte) has been reduced during the course of evolution. The unifying character of the Plantae is the presence of an embryo during the sporophytic phase of the life cycle. Consequently, the term "embryophyte" has become synonymous with "plant." The bryophytes are discussed in Chapter 16 and the vascular plants in Chapters 17 through 20.

Life Cycles and Diploidy

The first eukaryotic organisms were probably haploid and asexual, but once sexual reproduction was established among them, the stage was set for the evolution of diploidy. It seems likely that this condition first arose when two haploid cells combined to form a diploid zygote; such an event probably took place repeatedly. Presumably, the zygote then divided immediately by meiosis **(zygotic meiosis),** thus restoring the haploid

(a)  *(b)*

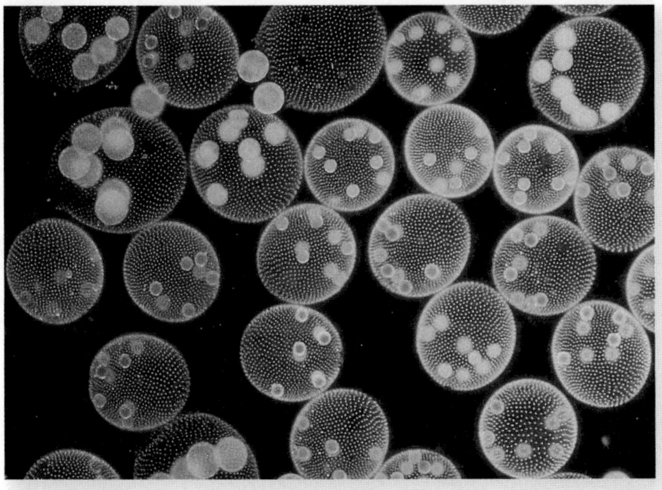

(c)  *(d)*

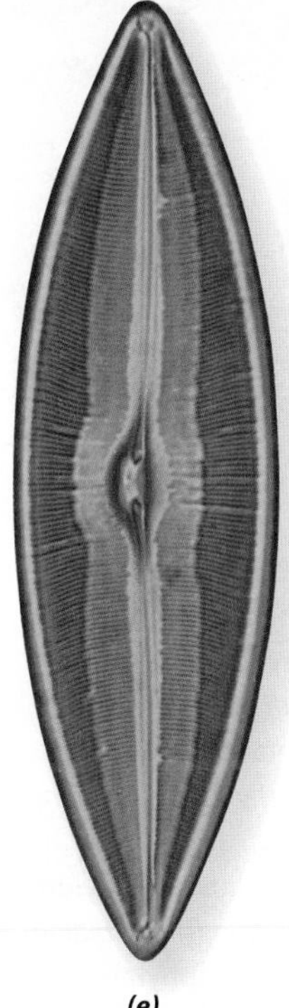

(e)

12–15 Protists ***(a)*** Plasmodium of a plasmodial slime mold, *Physarum* (phylum Myxomycota), growing on a leaf. ***(b)*** *Postelsia palmaeformis,* the "sea palm" (class Phaeophyceae), growing on exposed intertidal rocks off Vancouver Island, British Columbia. ***(c)*** *Volvox,* a motile colonial green alga (class Chlorophyceae). ***(d)*** *Callophyllis flabellulata,* a red alga (phylum Rhodophyta), photographed on exposed rocks at low tide, along the coast of central California. ***(e)*** A pennate diatom (class Bacillariophyceae), showing the intricately marked shell characteristic of this group.

condition (Figure 12–17a on page 254). In organisms with this simple kind of life cycle—the fungi and certain algae, including *Chlamydomonas*—the zygote is the only diploid cell.

By "accident"—an accident that occurred in a number of separate evolutionary lines—some of these zygotes divided mitotically instead of meiotically and, as a consequence, produced an organism that was composed of diploid cells, with meiosis occurring later. This delayed meiosis **(gametic meiosis)** results in the production of gametes and is characteristic of most animals and some protists (Oomycota, the water molds), as well as some green and brown algae (for example, *Fucus,* a brown alga). If they come together, these gametes fuse, an event that immediately restores the diploid state (Figure 12–17b). In animals, such as ourselves, gametes (eggs and sperm) are the only haploid cells. In fact, for all organisms undergoing gametic meiosis, gametes are the only haploid stage.

In plants, meiosis **(sporic meiosis)** results in the production of spores, not gametes. Spores are cells that can divide directly by mitosis to produce a multicellular haploid organism; this is in contrast to gametes, which can develop only following fusion with another gamete. Multicellular haploid organisms that appear in alternation with diploid forms are found

(a) (b) (c) (d) (e)

12–16 Plants ***(a)*** *Sphagnum,* the peat moss (phylum Bryophyta), forms extensive bogs in cold and temperate regions of the world. ***(b)*** *Marchantia* is by far the most familiar of the thallose liverworts (phylum Marchantiophyta). It is a widespread, terrestrial genus that grows on moist soil and rocks. ***(c)*** *Diphasiastrum digitatum,* a "club moss" (phylum Lycopodiophyta). ***(d)*** Wood horsetail, *Equisetum sylvaticum* (phylum Monilophyta). ***(e)*** Lady fern, *Athyrium filix-femina* (phylum Monilophyta). ***(f)*** The dandelion, *Taraxacum officinale,* and ***(g)*** strawberry cactus, *Echinocereus stramineus,* are eudicots (phylum Anthophyta). ***(h)*** Foxtail barley, *Hordeum jubatum,* and ***(i)*** *Cymbidium hamsey* × *musita* orchids are monocots (phylum Anthophyta). ***(j)*** Sugar pine, *Pinus lambertiana* (left), and incense cedar, *Calocedrus decurrens,* are both conifers (phylum Coniferophyta).

(f)

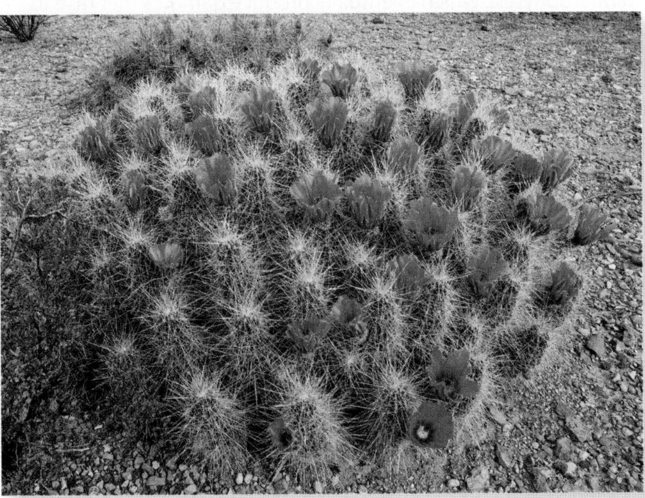

(g)

in plants, as well as in some brown, red, and green algae (and in two closely related genera of chytrids and one or more other groups of protists not discussed in this book). Such organisms exhibit the phenomenon known as **alternation of generations** (Figure 12–17c). Among the plants, the haploid, gamete-producing generation is called the **gametophyte** and the diploid, spore-producing generation is called the **sporophyte.** This same terminology is used for the algae and sometimes other groups as well.

In some algae—most of the red algae, many of the green algae, a few of the brown algae—the haploid and diploid forms are the same in external appearance. Such types of life cycles are said to exhibit an alternation of **isomorphic** generations. There are some life cycles, however, in which the haploid and diploid forms are not identical. During the history of these groups, mutations occurred that were expressed in only one generation, although the alleles were present, of course, in both the diploid and the haploid generations. In life cycles of this kind, the gametophyte and sporophyte became notably different from one another, and an alternation of **heteromorphic** generations originated. Such life cycles are characteristic of plants and some brown and red algae.

In the bryophytes (liverworts, mosses, and hornworts), the gametophyte is nutritionally independent from and usually larger than the sporophyte, which may be more complex structurally. In the vascular plants, on the other hand, the sporophyte is much larger and more complex than the gametophyte, which is nutritionally dependent on the sporophyte in nearly all groups.

Diploidy permits the storage of more genetic information and so perhaps allows a more subtle expression of the organism's genetic background in the course of development. This may be why the sporophyte is the large, complex, and nutritionally independent generation in vascular plants. One of the

(h)

(i)

(j)

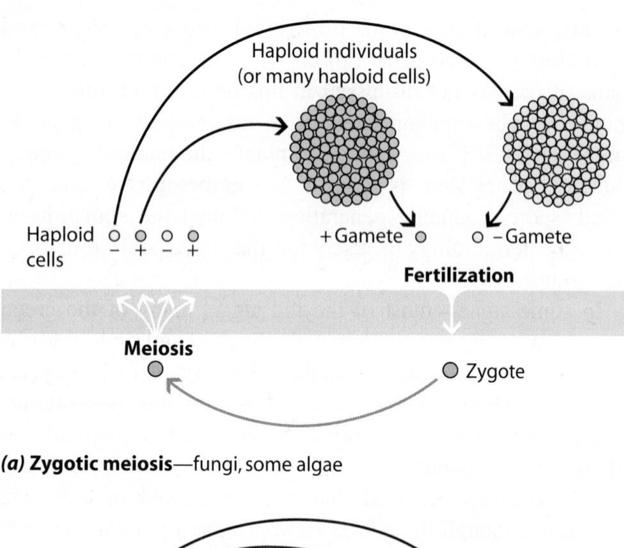

(a) Zygotic meiosis—fungi, some algae

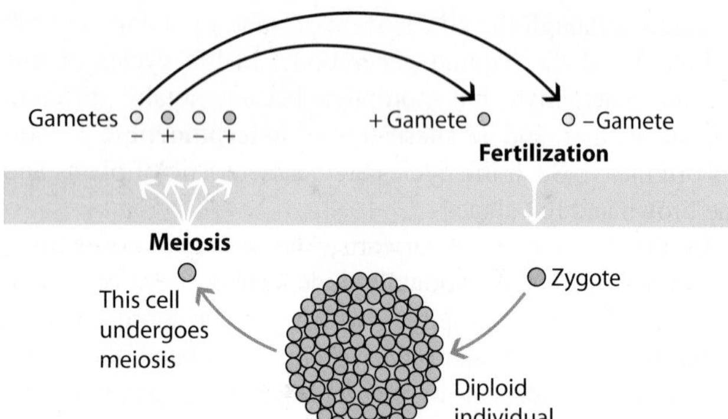

(b) Gametic meiosis—animals, some protists and algae

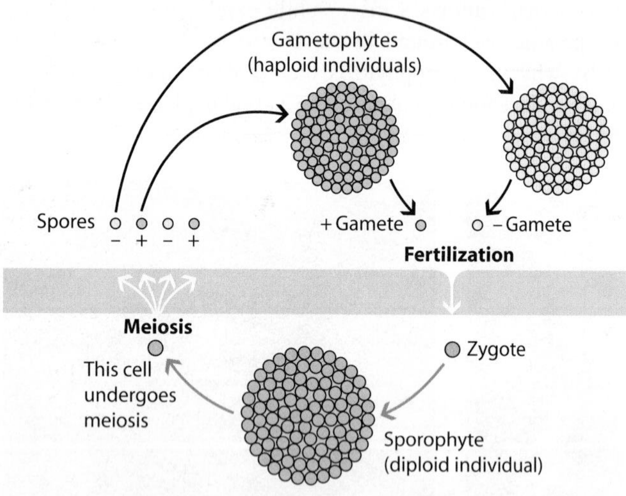

(c) Sporic meiosis, or alternation of generations—plants, many algae

12–17 The principal types of life cycles In these diagrams, the diploid phase of the cycle takes place below the broad bar, and the haploid phase occurs above it. The four white arrows signify the products of meiosis; the single white arrow represents the fertilized egg, or zygote. ***(a)*** In zygotic meiosis, the zygote divides by meiosis to form four haploid cells. Each of these cells divides by mitosis to produce either more haploid cells or a multicellular haploid individual that eventually gives rise to gametes by differentiation. This type of life cycle is found in a number of algae and in the fungi. ***(b)*** In gametic meiosis, the haploid gametes are formed by meiosis in a diploid individual and fuse to form a diploid zygote that divides to produce another diploid individual. This type of life cycle is characteristic of most animals and some protists, as well as some green and brown algae. ***(c)*** In sporic meiosis, the sporophyte, or diploid individual, produces haploid spores as a result of meiosis. These spores do not function as gametes but undergo mitotic division. This gives rise to multicellular haploid individuals (gametophytes), which eventually produce gametes that fuse to form diploid zygotes. These zygotes, in turn, differentiate into diploid individuals. This kind of life cycle, known as alternation of generations, is characteristic of plants and many algae.

clearest evolutionary trends in this group, which predominates in most terrestrial habitats, is the increasing "dominance" of the sporophyte and "suppression" of the gametophyte. Among the flowering plants, the female gametophyte is a microscopic body that typically consists of only seven cells, and the male gametophyte consists of only three cells. Both of these gametophytes are nutritionally dependent on the sporophyte.

SUMMARY

Systematics, the scientific study of biological diversity, encompasses both taxonomy—the identifying, naming, and classifying of species—and phylogenetics, which discerns the evolutionary interrelationships among organisms.

Organisms Are Named with a Binomial and Grouped into Taxonomic Categories Arranged in a Hierarchy

Organisms are designated scientifically by a name that consists of two words—a binomial. The first word in the binomial is the name of the genus, and the second word, the specific epithet, combined with the name of the genus, completes the name of the species. Species are sometimes subdivided into subspecies or varieties. Genera are grouped into families, families into orders, orders into classes, classes into phyla, phyla into kingdoms, and kingdoms into domains. It has been hypothesized that the eukaryotes comprise seven supergroups. A supergroup lies between a domain and a kingdom.

Organisms Are Classified Phylogenetically Using Homologous, Rather than Analogous, Features

In classifying organisms into the categories of genus through domain, systematists seek to group the organisms in ways that reflect their phylogeny (evolutionary history). In a phylogenetic system, every taxon should be monophyletic—that is, it should include every taxon of an ancestor and all its descendants. A major principle of such classification is that the similarities used in constructing the system should be homologous—that is, the result of common ancestry, rather than the result of convergent evolution.

The former traditional methods—essentially intuitive—of classifying organisms have been largely replaced by more explicit cladistic methods. Cladistic analysis attempts to understand branching sequences (genealogy) on the basis of shared derived character traits. This results in a graphical representation, or cladogram, that is a working model of the phylogenetic relationships of a group of organisms.

A Comparison of the Molecular Composition of Organisms Can Be Used to Predict Their Evolutionary Relationships

New techniques in molecular systematics are providing a relatively objective and explicit method of comparing organisms at the most basic level of all, the gene. These studies focus on nucleotide sequencing, in particular, for plants, of chloroplast DNA and the genes encoding the subunits of ribosomal RNA. As a result, valuable contributions have been made to more accurate classification schemes that reflect an improved understanding of biological diversity and its evolutionary history.

Organisms Are Classified into Two Prokaryotic Domains and One Eukaryotic Domain, Which Consists of the Protists and Three Kingdoms

In this text, living organisms are grouped—based largely on data obtained from the sequencing of the small-subunit ribosomal RNA—into the three domains Bacteria, Archaea, and Eukarya. The Bacteria and Archaea are two distinct lineages of prokaryotic organisms. The Archaea are more closely related to the Eukarya domain, which consists entirely of eukaryotes, than they are to the Bacteria. The protists and the Fungi, Plantae, and Animalia kingdoms occur within the Eukarya. The Fungi kingdom includes multicellular, nonmotile absorbers; the Animalia kingdom consists primarily of multicellular ingesters; and the Plantae kingdom includes multicellular photosynthesizers. The protists are a paraphyletic group that includes a very heterogeneous assemblage of unicellular, colonial, and simple multicellular organisms that lack the distinctive features of fungi, plants, or animals.

There Are Three Principal Types of Life Cycles Involving Sexual Reproduction

In primitive eukaryotes and all fungi, the zygote formed by fertilization divides immediately by meiosis (zygotic meiosis). In most animals and some groups of protists, meiosis results in the formation of gametes (gametic meiosis), which then fuse to give rise to a diploid individual. In plants and many algae, the diploid sporophyte produces haploid spores (sporic meiosis). The spores divide by mitosis and give rise to a haploid gametophyte, which eventually produces gametes. This type of life cycle is referred to as an alternation of generations. If the gametophyte and sporophyte in a particular life cycle are approximately equal in size and complexity, the alternation of generations is said to be isomorphic; if they differ widely in size and complexity, it is said to be heteromorphic.

QUESTIONS

1. Distinguish between or among the following: category and taxon; monophyletic, polyphyletic, and paraphyletic; host and endosymbiont.
2. Identify which of the following are categories and which are taxa: undergraduates; the faculty of The Pennsylvania State University; the Green Bay Packers; major league baseball teams; the U.S. Marine Corps; the family Robinson.
3. A key question in systematics is the origin of a similarity or difference. Explain.
4. Explain the advantages that molecular techniques have over comparative morphology and anatomy in assessing phylogenetic relationships.
5. Describe the role of endosymbiosis in the origin of eukaryotic cells.
6. The life cycle of organisms that undergo sporic meiosis is referred to as an alternation of generations. Explain.

CHAPTER 13

Prokaryotes and Viruses

◀ **Bananas under threat** When infected by the banana bunchy top virus, a banana plant produces yellowed, stiff leaves that grow in a tight bunch at the top of the stunted plant. Spread by the banana aphid, the virus causes one of the most serious diseases of bananas—a commercially valuable crop—and efforts are under way to produce a transgenic plant resistant to the virus.

CHAPTER OUTLINE

Of all organisms, the prokaryotes are the smallest, the simplest structurally, and the most abundant worldwide. Even though individuals are microscopically small, the total weight of prokaryotes in the world is estimated to exceed that of all other living organisms combined. In the sea, for example, prokaryotes make up an estimated 90 percent or more of the total weight of living organisms. In a single gram (about 1/28 of an ounce) of fertile agricultural soil, there may be 2.5 billion prokaryotic individuals (Figure 13–1). About 5000 species of prokaryotes are currently recognized, but thousands more await discovery, most likely with the use of DNA sequencing technologies.

The prokaryotes are, in evolutionary terms, the oldest organisms on Earth. The oldest known fossils are chainlike prokaryotes found in rocks from Western Australia, dated at about 3.5 billion years old (see Figure 1–2). Although some present-day prokaryotes resemble these ancient organisms in appearance, none of the prokaryotes living today are primitive. They are, instead, organisms that have succeeded extremely well in adapting to their particular environments.

The prokaryotes are, in fact, the most dominant and successful forms of life on Earth. Their success is undoubtedly due to their ability to metabolize a great variety of nutrients, as well as their rapid rate of cell division. Growing under optimal conditions, a population of the best-known prokaryote—*Escherichia coli*—can double in size and divide every 20 minutes. Prokaryotes can survive in many environments that support no

CHECKPOINTS

After reading this chapter, you should be able to answer the following:

1. Describe the basic structure of a prokaryotic cell.
2. How do prokaryotes reproduce, and in what ways does genetic recombination take place in prokaryotes?
3. In what ways are the cyanobacteria ecologically important?
4. Metabolically, what are the principal differences between the cyanobacteria and the purple and green bacteria?
5. How do the mycoplasmas and phytoplasmas differ from all other Bacteria?
6. Physiologically, what are the three large groups of Archaea?
7. Describe the basic structure of a virus. How do viruses reproduce?

other form of life. They live in the icy wastes of Antarctica, in the dark depths of the oceans, in the near-boiling waters of natural hot springs (Figure 13–2), and in the superheated water found near undersea vents. Some prokaryotes are among the very few extant organisms that can survive without free oxygen, obtaining their energy by anaerobic processes. Oxygen kills some types, whereas others can adapt and exist either with or without oxygen.

In Chapter 12, we emphasized that there are two distinct lineages of prokaryotes, the Bacteria and the Archaea. At the molecular level, these two domains, while both prokaryotic, are as evolutionarily distinct from one another as either is from all the rest of the living world—the Eukarya. We begin by considering some features largely shared by prokaryotes, while noting any differences that may exist between the two domains (see Table 12–3). We next turn our attention specifically to Bacteria, and then to Archaea. Finally, we look briefly at the viruses. Viruses are not cells and, hence, have no metabolism of their own. A virus consists primarily of a genome (either DNA or RNA) that replicates itself within a living host cell by directing the genetic machinery of that cell to synthesize viral nucleic acids and proteins.

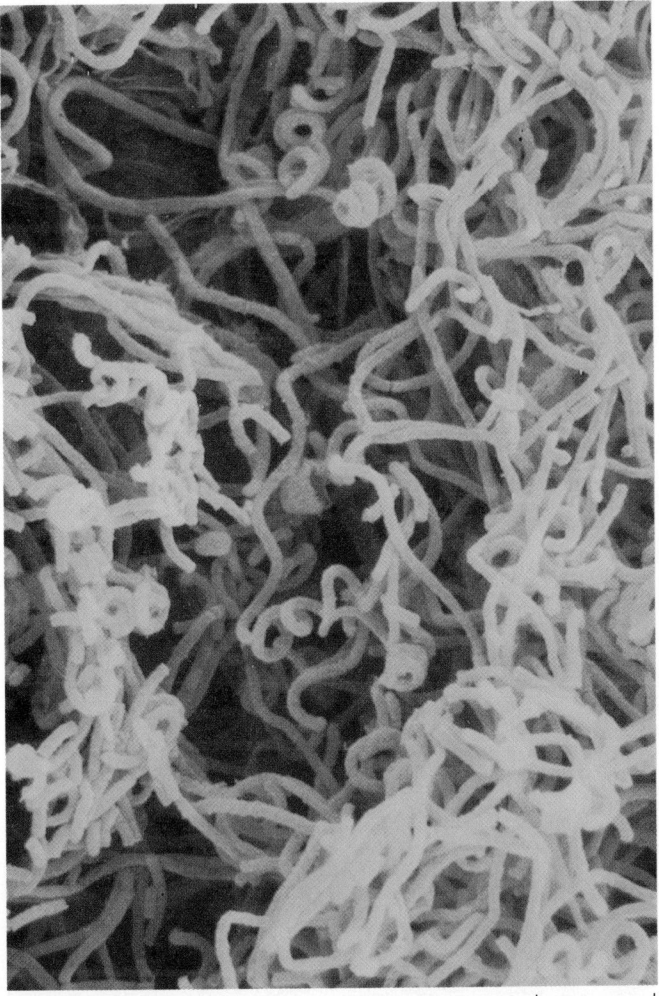

13–1 A filamentous actinomycete, *Streptomyces scabies* Actinomycetes are abundant in soil, where they are largely responsible for the "moldy" odor of damp soil and decaying material. *Streptomyces scabies* is the bacterium that causes potato scab disease.

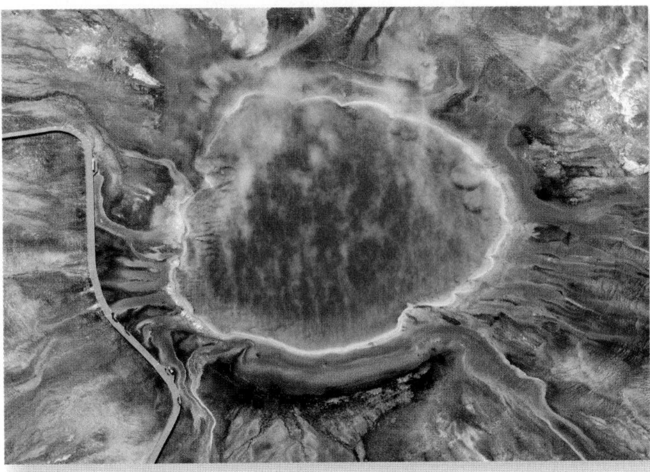

13–2 Thermophilic prokaryotes Aerial view of a very large hot spring, Grand Prismatic Spring, in Yellowstone National Park in Wyoming. Thermophilic ("heat-loving") prokaryotes thrive in such hot springs. Carotenoid pigments of thickly growing thermophiles, including cyanobacteria, color the runoff channels a brownish orange.

Characteristics of the Prokaryotic Cell

Prokaryotes lack a nucleus surrounded by a nuclear envelope (page 118). Instead, a single circular, or continuous, molecule of DNA, associated with non-histone proteins, is localized in a region of the cell called the **nucleoid.** In addition to its **chromosome,** a prokaryotic cell may also contain one or more smaller extrachromosomal pieces of circular DNA, called **plasmids,** that replicate independently of the cell's chromosome and carry important genetic traits.

The prokaryotic chromosome is highly organized within the nucleoid, with replication beginning and ending at points located at opposite sides of the circular chromosome. If stretched out, the chromosome would be much longer than the cell itself—in some cases, 1000 times longer—but twisting, or supercoiling, of the chromosome into a compact shape allows it to fit within the cell.

The notion that the cytoplasm of most prokaryotes is relatively unstructured is now known to be incorrect. Although prokaryotes lack membrane-bounded organelles, many possess numerous small, enzyme-containing microcompartments, each surrounded by a protein shell that may act as a semipermeable barrier. The cytoplasm often has a fine granular appearance owing to its many ribosomes—as many as 10,000 in a single cell. These prokaryotic ribosomes are smaller than the cytoplasmic ribosomes of eukaryotes. Prokaryotes occasionally contain **inclusions,** distinct granules consisting of storage material. The cyanobacteria and prochlorophytes contain extensive systems of membranes (thylakoids) bearing chlorophyll and other photosynthetic pigments (see Figures 13-11 and 13–16). Prokaryotes lack a cytoskeleton, but most prokaryotic cells possess both actinlike and tubulinlike polymers that function much like a cytoskeleton and play roles in chromosome segregation and cell division.

The Plasma Membrane Serves as a Site for the Attachment of Various Molecular Components

The plasma membrane of a prokaryotic cell is formed from a lipid bilayer and is similar in chemical composition to that of a eukaryotic cell. However, with rare exceptions, the plasma membranes of prokaryotes lack sterols. In prokaryotes capable of respiration (aerobic or anaerobic), the plasma membrane incorporates the electron transport chain that, in eukaryotic cells, is found in the mitochondrial inner membrane, lending further support to the serial endosymbiotic theory (page 247). In the photosynthetic purple bacteria, the sites of photosynthesis are found in the plasma membrane, which is often extensively convoluted, greatly increasing its working surface (see Figure 13–17). Also, the membrane contains specific attachment sites for the DNA molecule, ensuring proper placement of the chromosome within the cell.

The Cell Walls of Most Prokaryotes Contain Peptidoglycans

The protoplasts of almost all prokaryotes are surrounded by a cell wall, which gives the different types their characteristic shapes. Many prokaryotes have rigid walls, some have flexible walls, and only a few—the mycoplasmas, phytoplasmas, and species of the Thermoplasma group of Archaea—have no cell walls at all.

The cell walls of prokaryotes are complex and contain many kinds of molecules not present in eukaryotes. The walls of Bacteria contain complex polymers known as **peptidoglycans,** which are primarily responsible for the mechanical strength of the wall. Archaea do not contain these molecules, so peptidoglycan has been dubbed a "signature molecule" for differentiating species of Bacteria from species of Archaea.

Bacteria can be divided into two major groups on the basis of the capacity of their cells to retain the dye known as crystal violet. Those whose cells retain the dye are called **gram-positive,** whereas those that do not are called **gram-negative,** after Hans Christian Gram, the Danish microbiologist who discovered the distinction. Gram-positive and gram-negative bacteria differ markedly in the structure of their cell walls. In gram-positive bacteria, the wall, which ranges from 10 to 80 nanometers in thickness, has a homogeneous appearance and consists of as much as 90 percent peptidoglycan. In gram-negative bacteria, the wall consists of two layers: an inner peptidoglycan layer, only 2 to 3 nanometers thick, and an outer layer of lipopolysaccharides, phospholipids, and proteins. The molecules of the outer layer are arranged in a bilayer, about 7 to 8 nanometers in thickness, similar in structure to the plasma membrane. Gram staining is widely used to identify and classify bacteria, because it reflects a fundamental difference in the architecture of the cell wall.

Many prokaryotes secrete slimy or gummy substances on the outer surface of their walls. Most of these substances consist of polysaccharides; a few consist of proteins. Commonly known as a "capsule," the general term for these layers is **glycocalyx.** The glycocalyx plays an important role in infection, by allowing certain pathogenic bacteria to attach to specific host tissues. The glycocalyx may also protect the bacteria from desiccation and may be important in the ecology of microbes in natural environments.

Prokaryotes Store Various Compounds in Granules

A wide variety of prokaryotes—both Bacteria and Archaea—contain inclusion bodies, or storage granules, lipidlike compounds such as **poly-β-hydroxybutyric acid** and starchlike granules such as **glycogen,** that serve as a depository for carbon and energy. Inorganic compounds such as polyphosphates and sulfur granules are also important nutrient stores for some prokaryotes.

Prokaryotes Have Distinctive Flagella

Many prokaryotes are motile, and their ability to move independently is usually due to long, slender appendages known as **flagella** (singular: flagellum) (Figure 13–3). Lacking microtubules and a plasma membrane, these flagella differ greatly from those of eukaryotes (see Figure 3–28). Each prokaryotic flagellum is composed of subunits of a protein called flagellin, which are arranged into chains that are wound in a triple helix (three chains) with a hollow central core. Bacterial flagella grow at the tip. Flagellin molecules formed in the cell pass up through the hollow core and are added at the far end of the chains. In some species, the flagella are distributed over the entire cell surface; in others, they occur singly or in tufts at one or both ends of the cell.

Fimbriae and Pili Are Involved in Attachment

Fimbriae and pili—the two terms are often used interchangeably—are filamentous structures assembled from protein subunits in much the same way as the filaments of flagella. **Fimbriae**

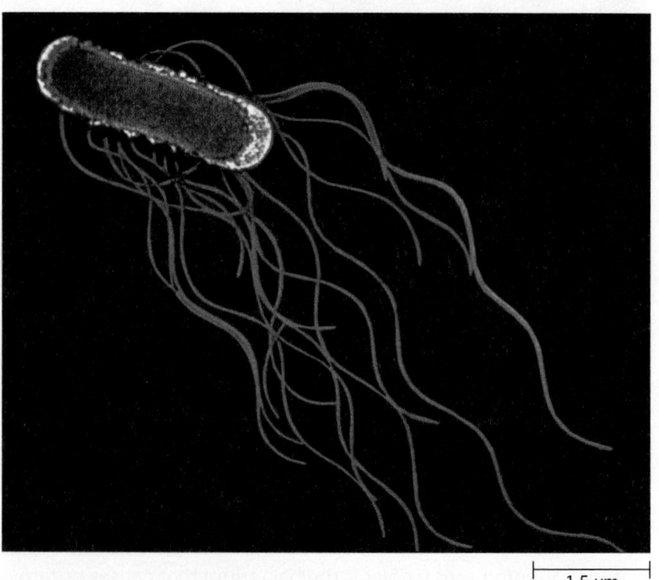

13–3 Flagella on *Salmonella* The *Salmonella* bacillus is a common cause of food-poisoning outbreaks.

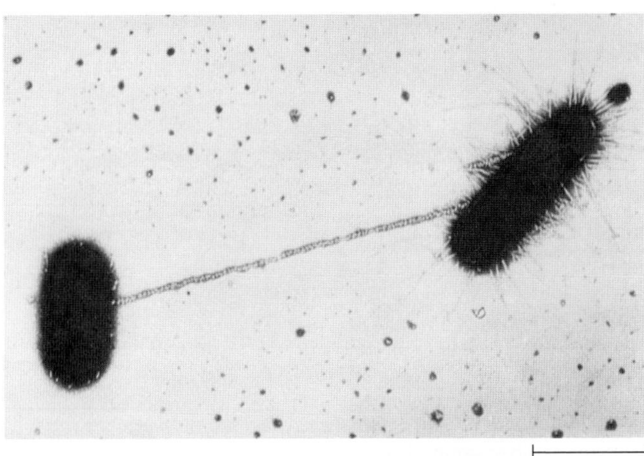

13–4 Conjugating *Escherichia coli* cells The elongated donor cell at the right in this electron micrograph is connected to the more rotund recipient cell by a long pilus, which is the first step in conjugation. Transfer of genetic material occurs through a cytoplasmic connection that forms once the two cells are in contact. Numerous short fimbriae are visible on the donor cell.

(singular: fimbria) are much shorter, more rigid, and typically more numerous than flagella (Figure 13–4). The fimbriae serve to attach the organism to a food source or other surfaces.

Pili (singular: pilus) are generally longer than fimbriae, and only one or a few are present on the surface of an individual cell. Some pili are involved in the process of conjugation between prokaryotes (Figure 13–4), serving first to connect the two cells and then, by retracting, to draw them together for the actual transfer of DNA. Some pili are also involved in the pathogenicity of bacteria for plants and animals.

Recently, variously sized tubules (up to 1 millimeter long and 30 to 130 nanometers wide) have been found connecting bacteria of both the same and different species (Figure 13–5). Called **nanotubes,** these tubules are composed of cell wall material, a plasma membrane, and cytoplasm, and are structurally distinguishable from pili involved in conjugation. Small cytoplasmic molecules and proteins have been shown to move between adjacent cells via the nanotubes. Thus, the nanotubes appear to provide a network for the exchange of cellular molecules within and between species of bacteria.

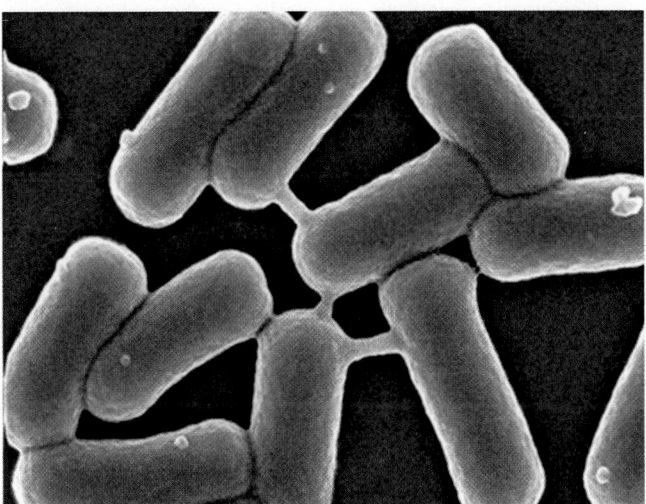

13–5 Nanotubes Neighboring *Bacillus subtilis* cells are seen here connected by nanotubes, which are tubules that provide a pathway for the exchange of cytoplasmic molecules between cells.

Diversity of Form

The oldest method of identifying prokaryotes is by their physical appearance. Prokaryotes exhibit considerable diversity of form, but many of the most familiar species fall into one of three categories (Figure 13–6). A prokaryote with a cylindrical shape is called a rod, or **bacillus** (plural: bacilli); spherical ones are called **cocci** (singular: coccus); and long curved, or spiral, rods are called **spirilla** (singular: spirillum). Cell shape is a relatively constant feature in most species of prokaryotes.

In many prokaryotes, after division, the cells remain together, producing filaments, clusters, or colonies that also have a distinctive shape. For example, cocci and rods may adhere to form chains, and such behavior is characteristic of particular genera. Bacilli usually separate after cell division. When they do remain together, they form long, thin chains of cells, as are found in the filamentous actinomycetes (Figure 13–1). The gram-negative rods of the myxobacteria aggregate and construct complex fruiting bodies, within which some cells become converted into resting cells called myxospores (Figure 13–7). The myxospores are more resistant to drying, UV radiation, and heat than the vegetative cells and are of survival value to the bacterium.

Nearly all prokaryotes growing on surfaces tend to form **biofilms,** assemblages of cells attached to the surface and enclosed in a matrix of polysaccharides, proteins, and DNA excreted by the prokaryotic cells. We all are familiar with biofilms. The film that develops on the surface of your unbrushed teeth is a biofilm, as is the film that forms in a pet's water bowl that has gone unattended for several days. Most biofilms are composed of several species of bacteria and archaea. Their formation requires intercellular communication by signaling molecules and coordinated gene expression of the various inhabitants. Biofilms increase the chances for survival of the cells in the biofilm and allow the cells to live in close association with each other, facilitating intercellular communication and opportunities for genetic exchange.

Reproduction and Gene Exchange

Most prokaryotes reproduce by a simple type of cell division called **binary fission,** which means "dividing in two" (Figure 13–8). According to one model, the segregation of replicated DNA molecules depends on the attachment of the origins of replication (see Chapter 9) of the replicated molecules to specific sites on the plasma membrane. When cell growth takes place between the two attachment sites, the replicated DNA molecules, or sister chromosomes, are separated passively as a by-product of cell elongation. After the sister chromosomes have moved apart, cytoskeleton-like elements form a contractile

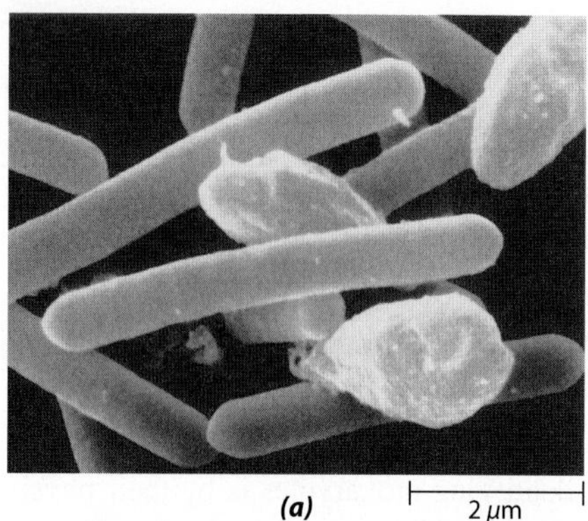

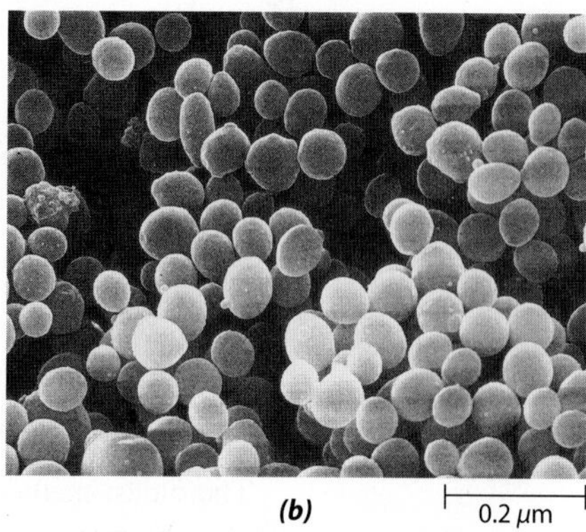

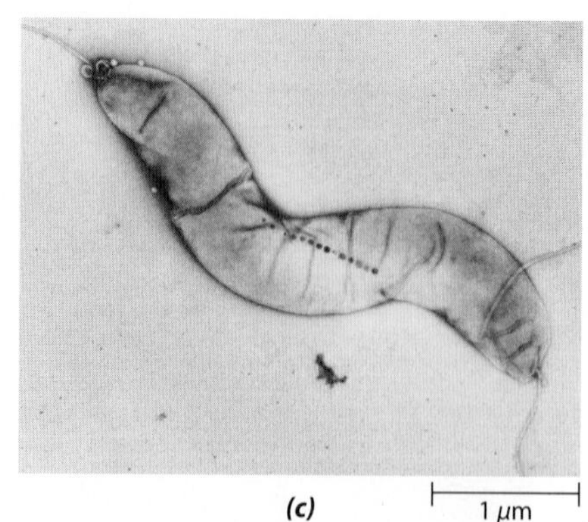

13–6 Three major forms of prokaryotes: bacilli, cocci, and spirilla ***(a)*** *Clostridium botulinum,* the source of the toxin that causes deadly food poisoning, or botulism, is a bacillus, or rod-shaped bacterium. The saclike structures are endospores, which are resistant to heat and cannot easily be destroyed. Bacilli are responsible for many plant diseases, including fire blight of apples and pears (caused by *Erwinia amylovora)* and bacterial wilt of tomatoes, potatoes, and bananas (caused by *Pseudomonas solanacearum).* ***(b)*** Many prokaryotes, such as *Micrococcus luteus,* shown here, take the shape of spheres. Among the cocci are *Streptococcus lactis,* a common milk-souring agent, and *Nitrosococcus nitrosus,* a soil bacterium that oxidizes ammonia to nitrites. ***(c)*** Spirilla, such as *Magnetospirillum magnetotacticum,* are less common than bacilli and cocci. Flagella can be seen at either end of this cell, which was isolated from a swamp. The string of dark magnetic particles orient the cell in the Earth's magnetic field.

ring in the middle of the cell. As the ring contracts, the cell is divided into two identical daughter cells.

In some prokaryotes, reproduction is by budding or by fragmentation of filaments of cells. As they multiply, prokaryotes, barring mutation, produce clones of genetically identical cells. Mutations do occur, however. It has been estimated that in a culture of *E. coli* that has divided 30 times, about 1.5 percent of the cells have mutations. Mutations, combined with a rapid generation time, are responsible for the extraordinary adaptability of prokaryotes. Further adaptability is provided by **horizontal,** or **lateral, gene transfer.** Three mechanisms of lateral gene transfer are known for prokaryotes: conjugation, transformation, and transduction. Such genetic recombinations are quite common in nature. Lateral gene transfer events occur within bacteria and archaea and also have been detected both between bacteria and archaea and between prokaryotes and eukaryotes.

Conjugation has been characterized as the prokaryotic version of sex. This form of mating takes place when a pilus produced by the donor cell comes in contact with the recipient cell (Figure 13–4). This "sex pilus" then retracts, pulling the two cells together so that they are in direct contact, held together by binding proteins. A portion of the donor chromosome then passes through this "conjugation junction" to the recipient cell. Conjugation is one mechanism used by plasmids to transfer copies of themselves to a new host. Conjugation can transfer genetic information between distantly related organisms; for example, plasmids can be transferred between bacteria and fungi, as well as between bacteria and plants. **Transformation**

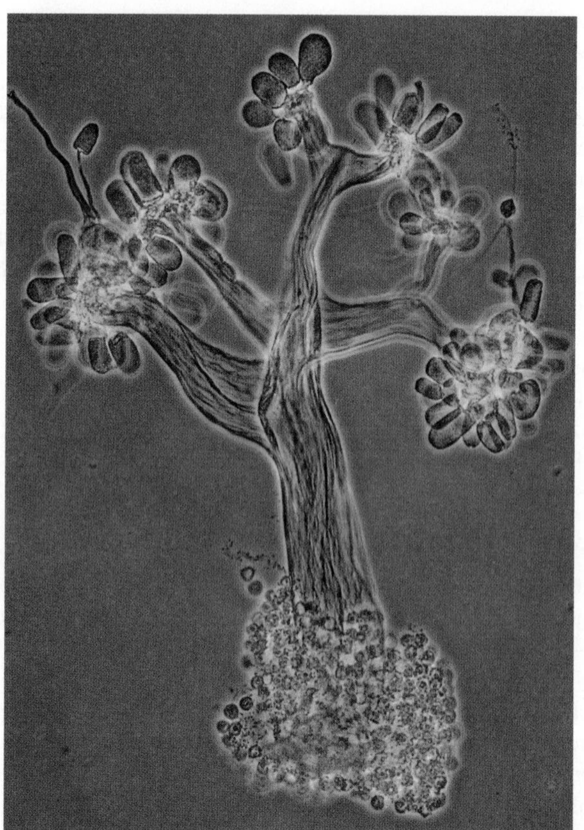

13–7 Fruiting body of a myxobacterium *Chondromyces crocatus,* a myxobacterium, or gliding bacterium, produces fruiting bodies, each of which may contain as many as 1 million cells. A fruiting body consists of a central stalk that branches to form clusters of myxospores. The myxobacteria spend most of their life as rods that glide together along slime tracks, but they eventually form fruiting bodies, as shown here.

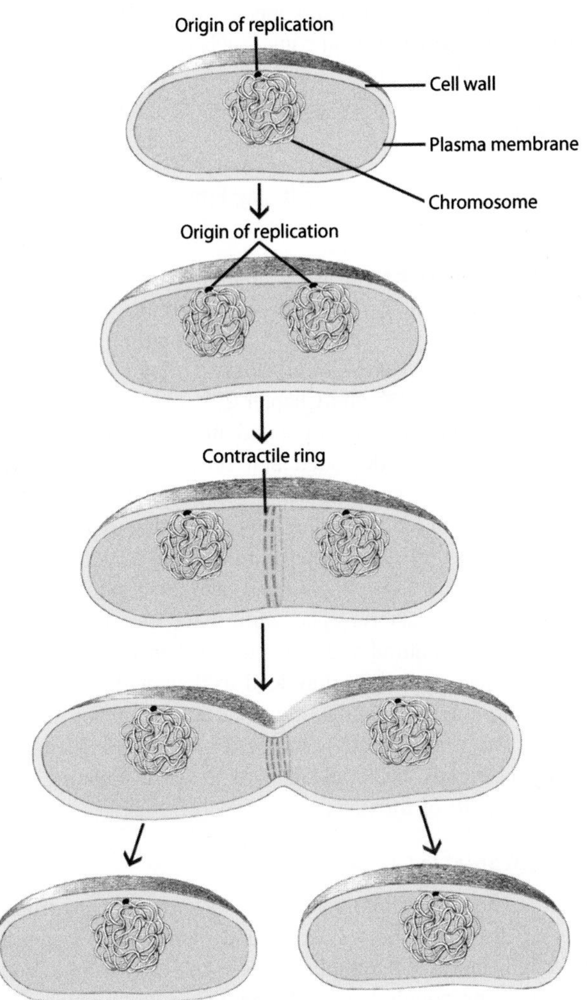

13–8 Cell division in a bacterium Attachment of the chromosome to the plasma membrane at the origin of replication ensures that, after duplication, a chromosome is distributed to each daughter cell as the plasma membrane elongates. After the sister chromosomes have moved apart, a ring of cytoskeleton-like elements contracts, dividing the cell in two.

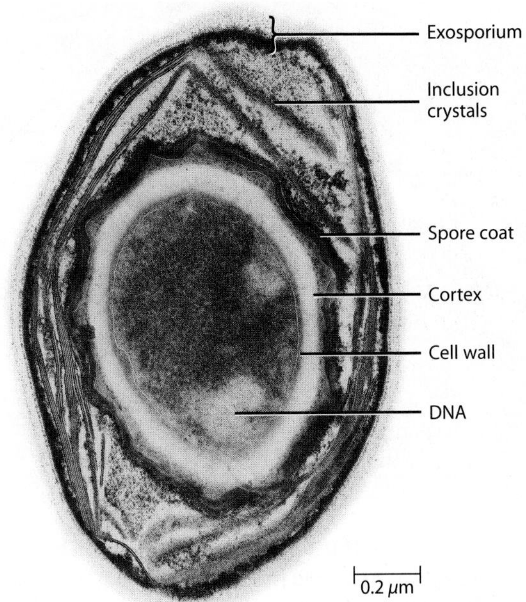

13–9 Mature endospore of *Bacillus megaterium* The outermost layer is the enveloping exosporium, made up of a faint peripheral layer and a dark basal layer. Underlying the exosporium are large inclusion crystals. The spore proper is covered with a protein spore coat. Beneath the spore coat is a thick peptidoglycan cortex, which is essential for the unique resistance properties of bacterial spores. Interior to the cortex is a thin cell wall, also of peptidoglycan, which covers a dehydrated spore protoplast containing DNA.

occurs when a prokaryote takes up free, or naked, DNA from the environment. The free DNA may have been released by an organism that died. Because DNA is not chemically stable outside cells, transformation is probably less important than conjugation. **Transduction** occurs when viruses that attack bacteria—viruses known as **bacteriophages**—bring with them DNA they have acquired from their previous host. Bacteriophages also attack archaea. The battle between bacteria and their phages has been described as, quantitatively, the dominant predator-prey relation in the biosphere. One researcher has estimated that every two days, half the bacteria on Earth are killed by bacteriophages.

Endospores

Certain species of bacteria have the capacity to form **endospores,** which are dormant resting cells (Figure 13–9). This process, called sporulation, has been studied extensively in the genera *Bacillus* and *Clostridium.* It characteristically occurs when a population of cells begins to use up its food supply.

Endospore formation greatly increases the capacity of the bacterial cell to survive. Endospores are exceedingly resistant to heat, radiation, and chemical disinfectants, primarily due to their dehydrated protoplasts. The endospores of *Clostridium botulinum,* the organism that causes often-fatal food poisoning, are not destroyed by boiling for several hours. In addition, endospores can remain viable (that is, they can germinate and develop into vegetative cells) for a very long time. For example, endospores recovered in 7000-year-old fractions of sediment cores from a Minnesota lake proved to be viable. More remarkably, ancient endospores preserved in the gut of an extinct bee trapped in amber were also reported to be revivable. The amber—and presumably the endospores—was estimated to be 25 to 40 million years old.

Metabolic Diversity

Prokaryotes Are Autotrophs or Heterotrophs

Prokaryotes exhibit tremendous metabolic diversity. Although some are autotrophs (meaning "self-feeding"), using carbon dioxide as their sole source of carbon, most prokaryotes are **heterotrophs,** requiring organic compounds as a carbon source. The vast majority of heterotrophs are **saprotrophs** (Gk. *sapros,* "rotten," or "putrid"), which obtain their carbon from dead

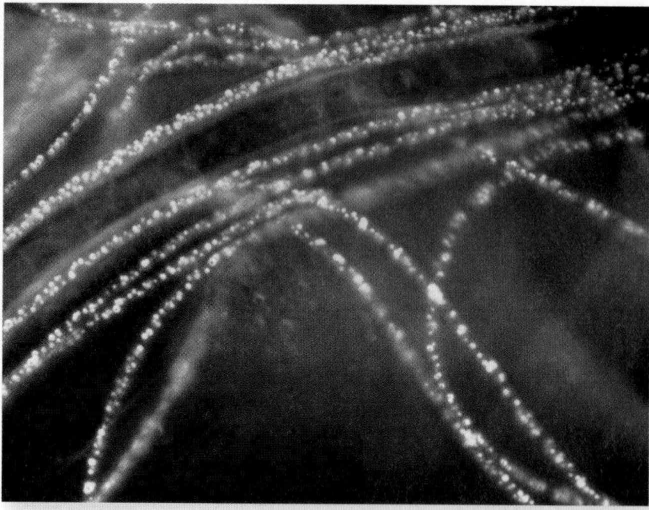

13–10 Filamentous sulfur-oxidizing bacteria Seen here are filaments of *Beggiatoa* isolated from a sewage treatment plant. The chains of cells, each filled with particles of sulfur, are found in areas rich in hydrogen sulfide, such as sulfur springs and polluted bodies of water.

organic matter. Saprotrophic bacteria and fungi are responsible for the decay and recycling of organic material in the soil; indeed, they are the recyclers of the biosphere.

Among the autotrophs are those that obtain their energy from light. These organisms are referred to as **photosynthetic autotrophs.** Some autotrophs, known as **chemosynthetic autotrophs,** are able to use inorganic compounds, rather than light, as an energy source (Figure 13–10). The energy is obtained from the oxidation of reduced inorganic compounds containing nitrogen, sulfur, or iron or from the oxidation of gaseous hydrogen.

Prokaryotes Vary in Their Tolerance of Oxygen and Temperature

Prokaryotes vary in their need for, or tolerance of, oxygen. Some species, called **aerobes,** require oxygen for respiration. Others, called **anaerobes,** lack an aerobic pathway and therefore cannot use oxygen as a terminal electron acceptor. Instead, they generate energy by anaerobic respiration—in which inorganic molecules such as sulfates are the terminal electron acceptor—or by fermentation. There are two kinds of anaerobes: **strict anaerobes,** which are killed by oxygen and thus can live only in the absence of oxygen, and **facultative anaerobes,** which can grow in the presence or absence of oxygen.

Prokaryotes also vary with regard to the range of temperatures under which they can grow. Some have a low optimum temperature (that is, a temperature at which growth is most rapid). These organisms, called **psychrophiles,** can grow at 0°C or lower and can survive indefinitely when it is much colder. At the other extreme are the **thermophiles** and **extreme thermophiles,** which have high and very high temperature optima, respectively. Thermophilic prokaryotes, with growth optima between 45° and 80°C, are common inhabitants of hot springs. Some extreme thermophiles have optimal growth temperatures greater than 100°C and have even been found growing in the 140°C water near deep-sea vents. Because their heat-stable enzymes are capable of catalyzing biochemical reactions at high temperatures, thermophiles and extreme thermophiles are being intensively investigated for use in industrial and biotechnological processes.

Prokaryotes Play a Vital Role in the Functioning of the World Ecosystem

The autotrophic bacteria make major contributions to the global carbon balance. The role of certain bacteria in fixing atmospheric nitrogen—that is, in incorporating nitrogen gas into nitrogen compounds—is likewise of major biological significance (see page 265 and Chapter 29). Through the action of decomposers, materials incorporated into the bodies of once-living organisms are degraded and released and made available for successive generations. More than 90 percent of the CO_2 production in the biosphere, other than that associated with human activities, results from the metabolic activity of bacteria and fungi. The CO_2 is converted back to organic matter by plants and some bacteria. The abilities of certain bacteria to decompose toxic natural and synthetic substances such as petroleum, pesticides, and dyes may lead to their widespread use in cleaning up dangerous spills and toxic dumps, when the techniques of using these bacteria are better developed. Meanwhile, naturally occurring bacteria are hard at work cleaning up oil spills in the Gulf of Mexico and elsewhere.

Some Prokaryotes Cause Disease

In addition to their ecological roles, bacteria are important as agents of disease in both animals and plants. Human diseases caused by bacteria include tuberculosis, cholera, anthrax, gonorrhea, whooping cough, bacterial pneumonia, Legionnaires' disease, typhoid fever, botulism, syphilis, diphtheria, and tetanus. In addition, there is a clear link between stomach ulcers and infection with *Helicobacter pylori.*

Biofilms of bacteria can dramatically affect humans. The major cause of death among people suffering from the genetic disease called cystic fibrosis is infections by *Pseudomonas aeruginosa,* which forms biofilms in the airways of the lungs. Biofilms formed by *Streptococcus* are a major problem with mechanical heart valves and surrounding tissues of the heart. Contamination by these bacteria may originate at the time of surgery, during dental work, or from indwelling devices such as central venous catheters. Gingivitis, a form of periodontal disease, is due to the long-term effects of plaque—biofilm—deposits. If the plaque is not removed, it turns into a hard deposit called tartar that becomes trapped at the base of the tooth, causing irritation and inflammation of the gums.

About 100 species of bacteria, including many strains that appear to be identical but differ in the species they infect, have been found to cause diseases of plants. Many of these diseases are highly destructive; some of them are described later in this chapter. There are no known plant or animal diseases caused by archaea.

Some Prokaryotes Are Used Commercially

Industrially, bacteria are sources of a number of important antibiotics: streptomycin, aureomycin, neomycin, and tetracycline, for example, are produced by actinomycetes. Bacteria are also widely used commercially for the production of drugs and other substances, such as vinegar, various amino acids, and enzymes. The production of almost all cheeses involves bacterial fermentation of the sugar lactose into lactic acid, which coagulates milk proteins. The same kinds of bacteria used in cheese making are also used in the production of yogurt and the lactic acid that preserves sauerkraut and pickles.

As noted in Chapter 10, maize, cotton, and other crops have been genetically engineered with genes from the bacterium *Bacillus thuringiensis,* conferring tolerance in the engineered plants to several herbicides, most notably glyphosate. Efforts are under way to reengineer *E. coli* and other easily grown microorganisms to produce biofuels essentially similar to existing fossil fuels.

Bacteria

Phylogenetic analysis, employing sequencing of ribosomal RNA, reveals that there are at least 17 major groups of Bacteria. The groups range from the most ancient lineage of extreme thermophilic chemosynthetic autotrophs that oxidize gaseous hydrogen or reduce sulfur compounds to the lineages of photosynthetic autotrophs, represented by the cyanobacteria and the purple and green bacteria. The bacteria selected for individual discussion here are those we consider to be of special evolutionary and ecological importance.

Cyanobacteria Are Important from Ecological and Evolutionary Perspectives

The cyanobacteria deserve special emphasis because of their great ecological importance, especially in the global carbon and nitrogen cycles, as well as their evolutionary significance. They represent one of the major evolutionary lines of Bacteria. Photosynthetic cyanobacteria have chlorophyll *a,* together with carotenoids and other, unusual accessory pigments known as **phycobilins.** There are two kinds of phycobilins: **phycocyanin,** a blue pigment, and **phycoerythrin,** a red one. Within the cells of cyanobacteria are numerous layers of membranes, often parallel to one another (Figure 13–11). These membranes are photosynthetic thylakoids that resemble those found in chloroplasts—and, in fact, chloroplasts correspond in size to an entire cyanobacterial cell. The main storage product of cyanobacteria is glycogen.

Many cyanobacteria produce a mucilaginous envelope, or sheath, that binds groups of cells or filaments together. The sheath is often deeply pigmented, particularly in species that sometimes occur in terrestrial habitats. The colors of the sheaths in different species include light gold, yellow, brown, red, emerald green, blue, violet, and blue-black. Despite their former name—"blue-green algae"—only about half of the species of cyanobacteria are blue-green in color, and they are definitely not algae.

Cyanobacteria often form filaments and may grow in large masses 1 meter or more in length. Some cyanobacteria are unicellular, a few form branched filaments, and a very few form plates or irregular colonies (Figure 13–12). Following division of the cyanobacterial cell, the resulting subunits may then separate to form new colonies. As in other filamentous or colonial bacteria, the cells of cyanobacteria are usually joined only by their walls or by mucilaginous sheaths, so that each cell leads an independent life.

Some filamentous cyanobacteria are motile, gliding and rotating around the longitudinal axis. Short segments called

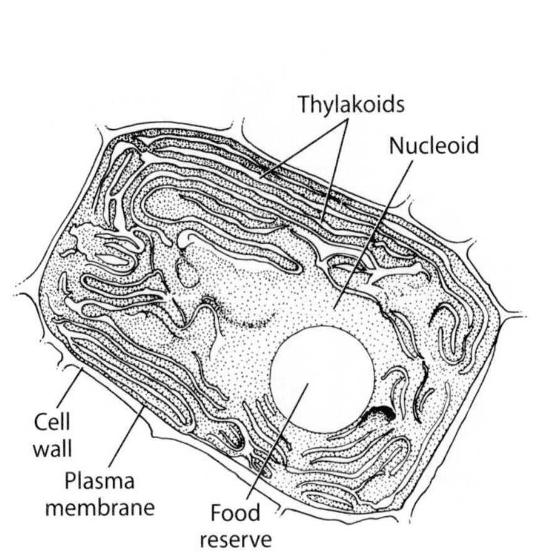

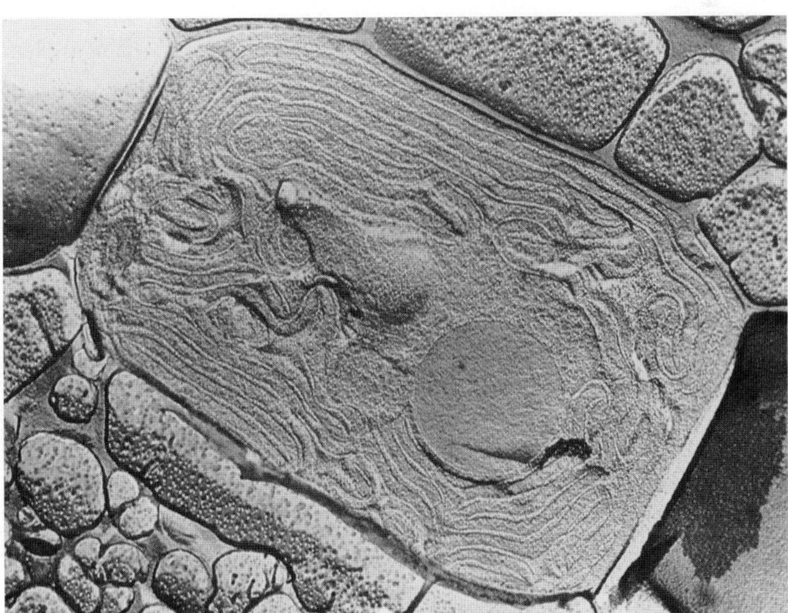

13–11 The cyanobacterium *Anabaena cylindrica* Photosynthesis takes place in the chlorophyll-containing membranes—the thylakoids—within the cell. The three-dimensional quality of this electron micrograph is due to the freeze-fracturing technique used in the preparation of the cells.

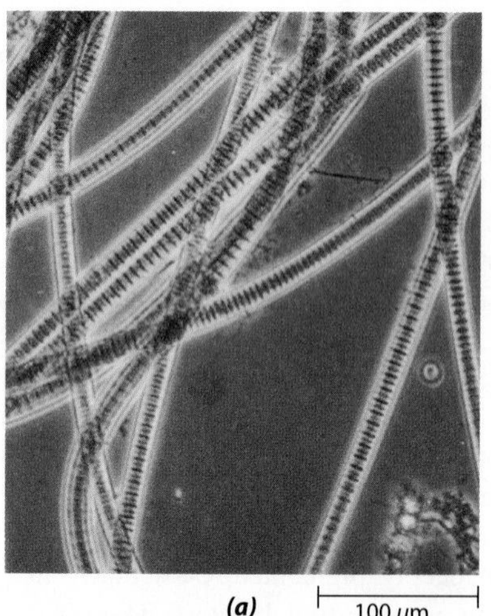

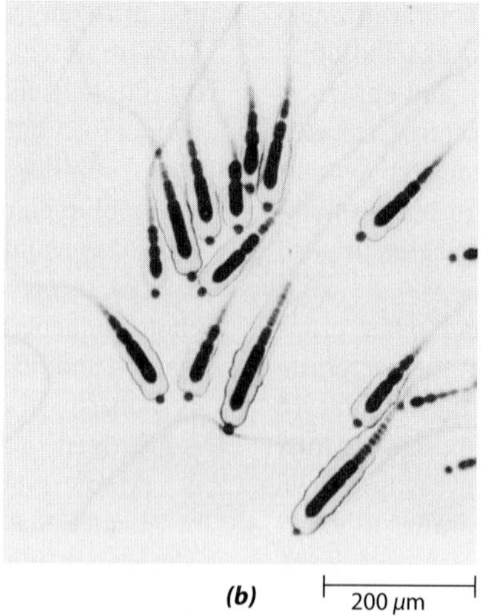

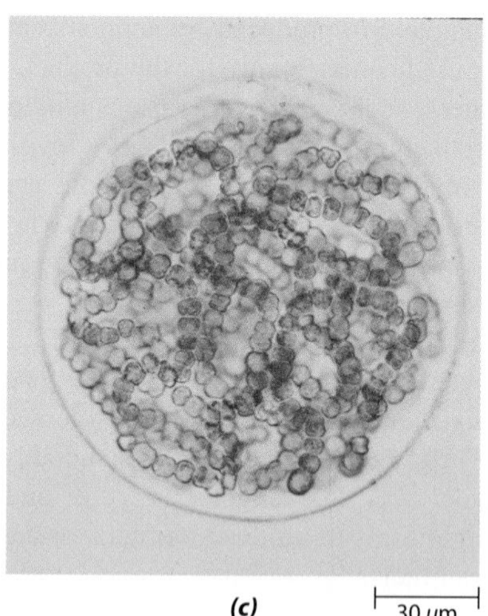

13–12 Three common genera of cyanobacteria ***(a)*** *Oscillatoria,* in which the only form of reproduction is fragmentation of the filament. ***(b)*** *Calothrix,* a filamentous form with a basal heterocyst (see Figure 13–14). *Calothrix* is capable of forming akinetes—enlarged cells that develop a resistant outer envelope—just above the heterocysts. ***(c)*** A gelatinous "ball" of *Nostoc commune,* containing numerous filaments. These cyanobacteria occur frequently in freshwater habitats.

hormogonia (singular: hormogonium) break off from a cyanobacterial colony and glide away from their parent colony at rates as rapid as 10 micrometers per second. This movement may be connected with the extrusion of mucilage through small pores in the cell wall, together with the production of contractile waves in one of the surface layers of the wall. Some cyanobacteria exhibit intermittent jerky movements.

Cyanobacteria Can Live in a Wide Variety of Environments Although more than 7500 species of cyanobacteria have been described and given names, there may actually be as few as 200 distinct, free-living, nonsymbiotic species. Like other bacteria, cyanobacteria sometimes grow under extremely inhospitable conditions, from the water of hot springs to the frigid lakes of Antarctica, where they sometimes form luxuriant mats 2 to 4 centimeters thick, in water beneath more than 5 meters of permanent ice. The greenish color of some polar bears in zoos is due to the presence of colonies of cyanobacteria within the hollow hairs of their fur. Cyanobacteria are absent in acidic waters, where eukaryotic algae are often abundant.

Layered chalk deposits called **stromatolites** (Figure 13–13), which have a continuous geologic record covering 2.7 billion years, are produced when colonies of cyanobacteria bind calcium-rich sediments. Today, stromatolites are formed in only a few places—particularly in shallow pools in hot, dry climates—such as Hamlin Bay and Shark Bay in Western Australia. Their

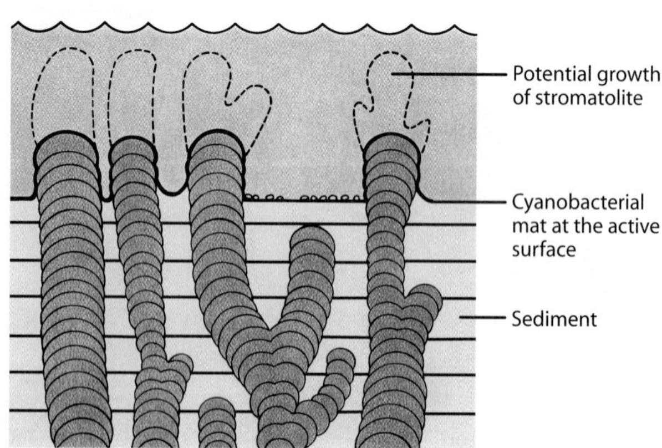

13–13 Stromatolites Stromatolites are produced when flourishing colonies of cyanobacteria bind calcium carbonate into domed structures, like the ones shown in the diagram and photograph, or into other, more intricate forms. Such structures are abundant in the fossil record but today are being formed in only a few, highly suitable environments, such as the tidal flats of Hamlin Pool in Western Australia in this photograph.

abundance in the fossil record is evidence that such environmental conditions were prevalent in the past, when cyanobacteria played the decisive role in elevating the level of free oxygen in the atmosphere of the early Earth. More ancient stromatolites (3 billion years or older), produced in an oxygen-free environment, probably were made by purple and green bacteria.

Many marine cyanobacteria occur in limestone (calcium carbonate) or lime-rich substrates, such as coralline algae (see page 340) and the shells of mollusks. Some freshwater species of cyanobacteria, particularly those that grow in hot springs, often deposit thick layers of lime in their colonies.

Cyanobacteria Form Gas Vesicles, Heterocysts, and Akinetes The cells of cyanobacteria living in freshwater or marine habitats—especially those that inhabit the surface layers of the water, in the community of microscopic organisms known as the **plankton**—commonly contain bright, irregularly shaped structures called **gas vesicles.** These vesicles provide and regulate the buoyancy of the organisms, thus allowing them to float at certain levels in the water. When numerous cyanobacteria become unable to regulate their gas vesicles properly—for example, because of extreme fluctuations of temperature or oxygen supply—they may float to the surface of the body of water and form visible masses called "blooms." Some cyanobacteria that form blooms secrete chemical substances that are toxic to other organisms, causing large numbers of deaths. The Red Sea apparently was given its name because of the blooms of planktonic species of *Trichodesmium,* a red cyanobacterium.

Many genera of cyanobacteria can fix nitrogen, converting nitrogen gas to ammonium, a form in which the nitrogen is available for biological reactions. In filamentous cyanobacteria, this **nitrogen fixation** often occurs within **heterocysts,** which are specialized, enlarged cells (Figure 13–14). Heterocysts are surrounded by thickened cell walls containing large amounts of glycolipid, which serves to impede the diffusion of oxygen into the cell. Within a heterocyst, the cell's internal membranes are reorganized into a concentric or reticulate pattern. Heterocysts are low in phycobilins, and they lack Photosystem II, so the cyclic photophosphorylation that occurs in these cells does not result in the evolution of oxygen (page 135). The oxygen that is present is either rapidly reduced by hydrogen, a by-product of nitrogen fixation, or expelled through the wall of the heterocyst. Nitrogenase, the enzyme that catalyzes the nitrogen-fixing reactions, is sensitive to the presence of oxygen, and nitrogen fixation therefore is an anaerobic process. Heterocysts have small plasmodesmatal connections—microplasmodesmata—with adjacent vegetative cells. The products of nitrogen fixation are transported through the microplasmodesmata from the heterocyst to the vegetative cells, and the products of photosynthesis move in the opposite direction through these same connections, from the vegetative cells to the heterocyst.

Among cyanobacteria that fix nitrogen are free-living species such as *Trichodesmium,* which lives in certain tropical oceans. *Trichodesmium* accounts for about a quarter of the total nitrogen fixed there, an enormous amount. Symbiotic cyanobacteria are likewise very important in nitrogen fixation. In the warmer parts of Asia, rice was often grown continuously on the same land, without the addition of fertilizers, because of the presence of nitrogen-fixing cyanobacteria in the rice paddies (Figure 13–15). Here, the cyanobacteria, especially members of the genus *Anabaena* (Figure 13–14), often occur in association with the small, floating water fern *Azolla,* which forms masses on the paddies.

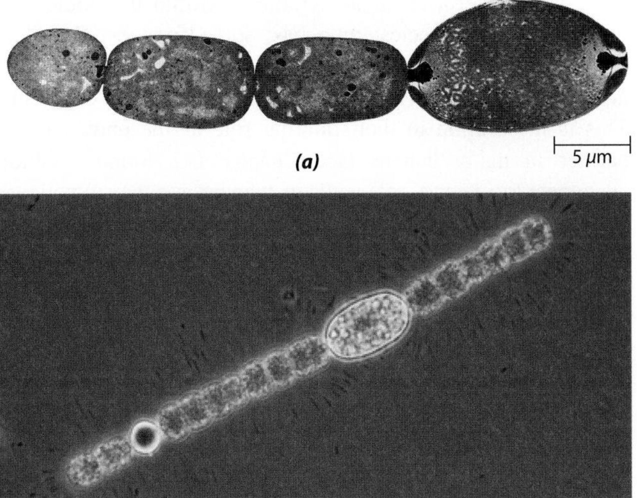

13–14 Filament of *Anabaena* ***(a)*** This electron micrograph shows a chain of cells held together along incompletely separated walls. The first cell at the right end of the chain is a heterocyst, in which nitrogen fixation takes place. The gelatinous matrix of this filament was destroyed during preparation of the specimen for electron microscopy. ***(b)*** In this preparation, the gelatinous matrix is barely discernible as striations extending outward from the cell surfaces. The third cell from the left is a heterocyst. *Anabaena,* like the *Calothrix* shown in Figure 13–12b, forms akinetes (large oval body toward the right).

13–15 Rice planting A farmer preparing his field for planting rice in the Dragon's Backbone rice terraces in Guangxi Province, China. In Southeast Asia, rice was often grown on the same land continuously, without the addition of fertilizers, because *Anabaena azollae,* which lives in the tissues of the water fern *Azolla* that grows in the rice paddies, can fix nitrogen.

Cyanobacteria occur as symbionts within the bodies of a vast number of species: amoebas, some sponges, flagellated protozoa, diatoms, green algae that lack chlorophyll, other cyanobacteria, mosses, liverworts, vascular plants, and oomycetes. This is in addition to their familiar role as the photosynthetic partner in many lichens (see Chapter 14). Some symbiotic cyanobacteria lack a cell wall, in which case they function as chloroplasts. The symbiotic cyanobacterium divides at the same time as its host cell by a process similar to chloroplast division.

In addition to the heterocysts, some cyanobacteria form resistant spores called **akinetes,** which are enlarged cells surrounded by thickened envelopes (Figures 13–12b and 13–14b). Similar to the endospores formed by other bacteria, akinetes are resistant to heat and drought and thus allow the cyanobacterium to survive during unfavorable periods.

Prochlorophytes Contain Chlorophylls *a* and *b* and Carotenoids

The **prochlorophytes** are a group of photosynthetic bacteria that contain chlorophylls *a* and *b,* as well as carotenoids, but do not contain phycobilins. Thus far, only three genera of prochlorophytes have been discovered. The first of these is *Prochloron,* which is found only along tropical seashores and lives as a symbiont within colonial sea squirts (ascidians). The cells of *Prochloron* are roughly spherical and contain an extensive system of thylakoids (Figure 13–16).

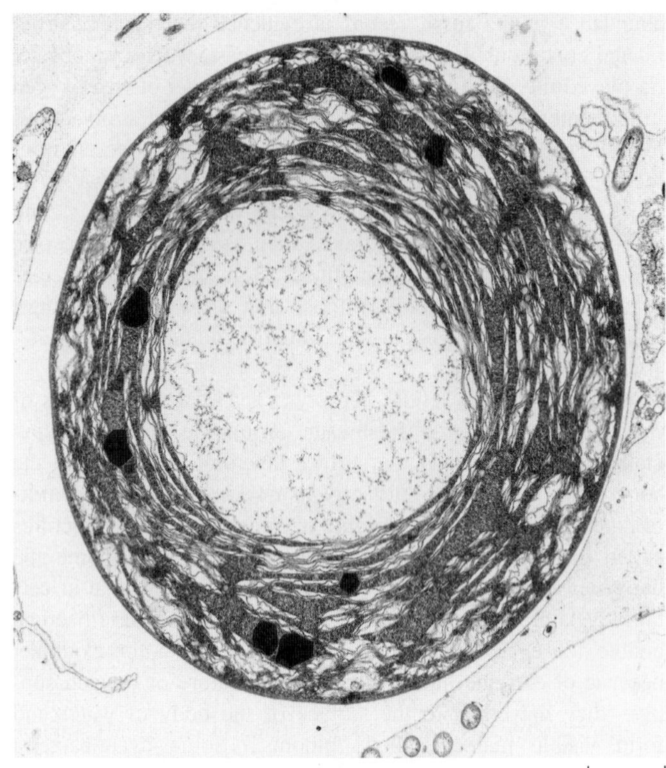

13–16 ***Prochloron*** A single cell of the bacterium *Prochloron,* showing the extensive system of thylakoids. *Prochloron* is a photosynthetic bacterium with chlorophylls *a* and *b* and carotenoids, the same pigments found in the green algae and plants. The prochlorophytes resemble both cyanobacteria (because they are prokaryotic and contain chlorophyll *a*) and the chloroplasts of green algae and plants (because they contain chlorophyll *b* instead of phycobilins).

The two other known prochlorophyte genera are *Prochlorothrix* and *Prochlorococcus. Prochlorothrix,* which is filamentous, has been found growing in several shallow lakes in the Netherlands. *Prochlorococcus* is the smallest known photosynthetic organism (about 0.6 micrometer in diameter), has the smallest genome of any photosynthetic cell, and is thought to be the most numerous photosynthetic organism on Earth. Prochlorococci are found in the mineral-poor oceans from 40° north to 40° south latitudes and from the surface down to the euphotic zone—the zone into which enough light penetrates for photosynthesis to occur. Prochlorococci represent 40 to 50 percent of the biomass of the phytoplankton, which produces half the oxygen on Earth, making *Prochlorococcus* of great ecological importance.

Purple and Green Bacteria Have Unique Types of Photosynthesis

The purple and green bacteria together represent the second major group of photosynthetic bacteria, after the cyanobacteria. The overall photosynthetic process and the photosynthetic pigments used by these bacteria differ from those used by the cyanobacteria and prochlorophytes. Whereas the cyanobacteria and prochlorophytes produce oxygen during photosynthesis, the purple and green bacteria do not. In fact, purple and green bacteria can grow in light only under anaerobic conditions, because pigment synthesis in these organisms is repressed by oxygen. Cyanobacteria employ chlorophyll *a* and two photosystems in their photosynthetic process. Prochlorophytes have chlorophylls *a* and *b* and two photosystems. By contrast, purple and green bacteria use several different types of bacteriochlorophyll, which differ in certain respects from chlorophyll, and have only one photosystem (Figure 13–17). The photosystems present in purple and green bacteria appear to be ancestors of individual photosystems—Photosystem II and Photosystem I, respectively. Unlike the purple and green bacteria, photosynthetic autotrophs such as plants and algae, as well as the cyanobacteria and prochlorophytes, have both photosystems.

The colors characteristic of photosynthetic bacteria are associated with the presence of several accessory pigments that function in photosynthesis. In two groups of purple bacteria, these pigments are yellow and red carotenoids. In the cyanobacteria, as we have seen, the pigments are the red and blue phycobilins, which are not found in purple and green bacteria.

The purple and green bacteria are subdivided into those that use mostly sulfur compounds as electron donors and those that do not. In the purple sulfur and green sulfur bacteria, sulfur compounds play the same role in photosynthesis that water plays in organisms containing chlorophyll *a* (page 124).

Purple sulfur or green sulfur bacterium:

$$\underset{\text{Carbon dioxide}}{CO_2} + \underset{\text{Hydrogen sulfide}}{2H_2S} \xrightarrow{\text{Light}} \underset{\text{Carbohydrate}}{(CH_2O)} + \underset{\text{Water}}{H_2O} + \underset{\text{Sulfur}}{2S}$$

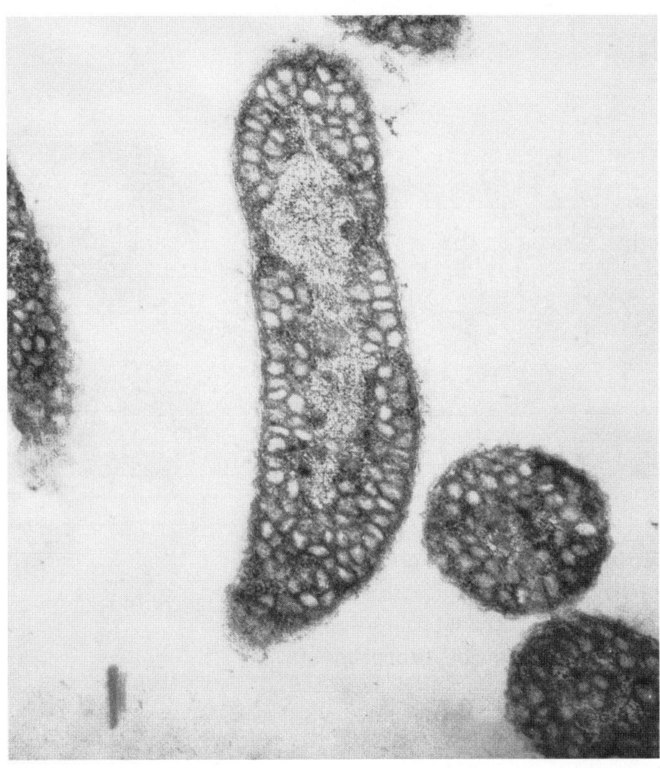

13–17 A purple nonsulfur bacterium, *Rhodospirillum rubrum* The structures that resemble vesicles are intrusions of the plasma membrane, which contains the photosynthetic pigments. This cell, with its many membrane intrusions, has a very high content of bacteriochlorophyll. It is from a culture that was grown in dim light. In cells grown in bright light, the membrane intrusions are less extensive because there is a decreased need for photosynthetic pigments.

Cyanobacterium, prochlorophyte, alga, or plant:

$$\underset{\text{Carbon dioxide}}{CO_2} + \underset{\text{Water}}{2H_2O} \xrightarrow{\text{Light}} \underset{\text{Carbohydrate}}{(CH_2O)} + \underset{\text{Water}}{H_2O} + \underset{\text{Oxygen}}{O_2}$$

The purple nonsulfur and green nonsulfur bacteria, which are able to use hydrogen sulfide (H_2S) at only low levels, also use organic compounds as electron donors. These compounds include alcohols, fatty acids, and a variety of other organic substances.

Because of their requirement for H_2S or a similar substrate, the purple sulfur and green sulfur bacteria can grow only in habitats that contain large amounts of decaying organic material, recognizable by the sulfurous odor. In these bacteria, as in the closely related colorless sulfur bacterium *Beggiatoa,* elemental sulfur may accumulate as deposits within the cell (Figure 13–10).

Mycoplasmas Are Wall-less Organisms That Live in a Variety of Environments

Mycoplasmas are bacteria that lack cell walls. Usually about 0.2 to 0.3 micrometer in diameter, they are probably the smallest organisms capable of independent growth. The genome of the mycoplasmas is likewise small, only one-fifth to one-fourth the size of the genome of *E. coli* and other common prokaryotes. Because they lack a cell wall and consequently lack rigidity, mycoplasmas can assume various forms. In a single culture, a mycoplasma may range from small rods to highly branched filamentous forms.

Mycoplasmas may be free-living in soil and sewage, or they may occur as parasites of the mouth or urinary tract of humans or as pathogens in animals and plants. Among the plant-pathogenic mycoplasmas are the **spiroplasmas,** long spiral or corkscrew-shaped cells less than 0.2 micrometer in diameter, which are motile even though they lack flagella (Figure 13–18). They move by means of a rotary motion or a slow undulation. Some spiroplasmas have been cultured on artificial media, including *Spiroplasma citri,* which causes citrus stubborn disease. Symptoms of this disease, such as bunchy upright growth of twigs and branches, are slow to develop and difficult to detect. Citrus stubborn disease is widespread and difficult to control, and in California and some Mediterranean countries, it is probably the greatest threat to the production of grapefruit and sweet oranges. *Spiroplasma citri* has also been isolated from corn (maize) plants suffering from corn stunt disease.

Phytoplasmas Cause Diseases in Plants

Like the mycoplasmas, **phytoplasmas** lack a cell wall and are very small. They have been identified in more than 200 distinct plant diseases affecting several hundred genera of plants. Some of these diseases are very destructive, such as X-disease of peach, which can render a tree commercially worthless in two to four years, and pear decline, so-called because it commonly

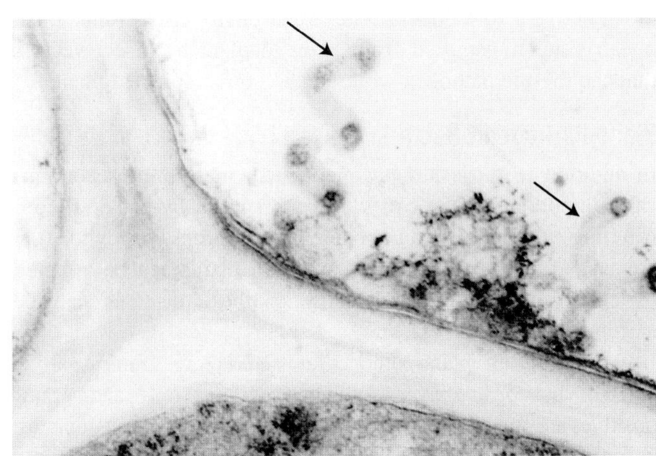

13–18 Spiroplasmas Arrows indicate two spiroplasmas in a sieve tube of a maize *(Zea mays)* plant with corn stunt disease. *Spiroplasma citri* causes both corn stunt disease and citrus stubborn disease.

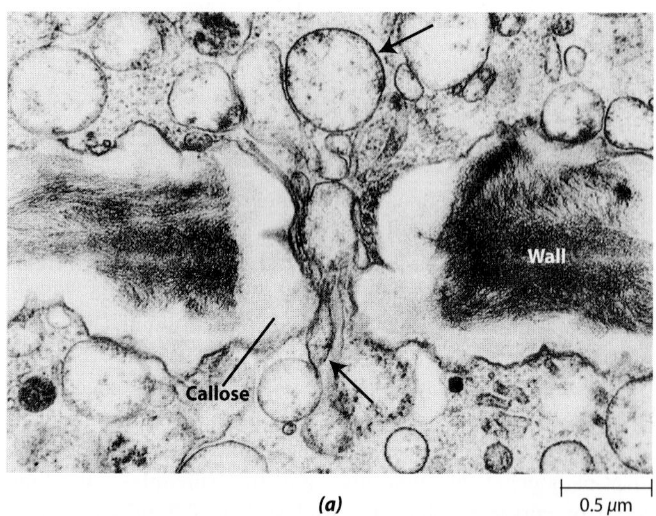

(a)

(b)

13–19 Phytoplasmas ***(a)*** Seen here are phytoplasmas (arrows) that seem to be traversing a sieve-plate pore in a young inflorescence of a coconut palm *(Cocos nucifera)* affected by lethal yellowing disease. The pore is partly occluded with callose, which lines the wall bordering the pore. ***(b)*** A devastated grove of coconut palms—now looking like telephone poles—in Ghana, Africa. Lethal yellowing has been responsible for the death of palms of many genera in southern Florida and elsewhere.

causes a slow, progressive weakening and ultimate death of pear trees. Aster yellows, another phytoplasma-caused disease, results in a general yellowing (chlorosis) of the foliage and infects a wide variety of host crops, ornamental plants, and weeds. Carrots are among the host crops suffering the greatest losses, commonly 10 to 25 percent and as high as 90 percent. Elm yellows, also known as elm phloem necrosis, and lethal yellowing of coconut are also caused by phytoplasmas (Figure 13–19).

In flowering plants, phytoplasmas are generally confined to the conducting elements of the phloem known as sieve tubes. Most phytoplasmas are believed to move passively from one sieve-tube element to another through the sieve-plate pores, as the sugar solution is transported in the phloem (Figure 13–19a). Motile spiroplasmas, also found in sieve tubes, may be able to move actively in phloem tissue. Most phytoplasmas and spiroplasmas are transmitted from plant to plant by insect vectors that acquire the pathogen while feeding on an infected plant.

Plant-Pathogenic Bacteria Cause a Wide Variety of Diseases

In addition to those already mentioned, many more economically important diseases of plants are caused by bacteria, contributing substantially to the one-eighth of crops worldwide that are lost to disease annually. Almost all plants can be affected by bacterial diseases, and many of these diseases can be extremely destructive.

Virtually all plant-pathogenic bacteria are gram-negative, and all but *Streptomyces,* which is filamentous and gram-positive, are rod-shaped. They are **parasites**—symbionts that are harmful to their hosts. The symptoms caused by plant-pathogenic bacteria are varied, with the most common appearing as spots of various sizes on stems, leaves, flowers, and fruits (Figure 13–20). Almost all such bacterial spots are caused by members of two closely related genera, *Pseudomonas* and *Xanthomonas.*

Some of the most destructive diseases of plants—such as blights, soft rots, and wilts—are also caused by bacteria. Blights are characterized by rapidly developing necroses (dead, discolored areas) on stems, leaves, and flowers. Fire blight in apples and pears, caused by *Erwinia amylovora,* is a widespread, economically important disease that can kill young trees within a single season. Bacterial soft rots occur most commonly in the fleshy storage tissues of vegetables (such as potatoes or carrots), as well as in fleshy fruits (examples are tomatoes and eggplants), and succulent stems or leaves (as in cabbage or lettuce). The most destructive soft rots are caused by bacteria of the genus *Erwinia,* with heavy losses occurring in the post-harvest period.

Bacterial vascular wilts affect mainly herbaceous plants. The bacteria invade the vessels of the xylem, where they multiply. They interfere with the movement of water and inorganic nutrients by producing high-molecular-weight polysaccharides, which results in the wilting and death of the plants. The bacteria commonly degrade portions of the vessel walls and can even cause the vessels to rupture. Once the walls have ruptured, the bacteria then spread to the adjacent parenchyma tissues, where they continue to multiply. Among the most important examples of wilts are bacterial wilt of alfalfa, tomato, and bean plants (each caused by different species of *Clavibacter*); bacterial wilt of cucurbits, such as squashes and watermelons (caused by *Erwinia tracheiphila*); and black vein of crucifers, such as cabbage (caused by *Xanthomonas campestris*). The most economically important wilt disease of plants, however, is caused by *Pseudomonas solanacearum.* It affects more than 40 different genera of plants, including such major crops as bananas, peanuts, tomatoes, potatoes, eggplants, and tobacco, to name a few. This disease occurs worldwide in tropical, subtropical, and warm temperate areas.

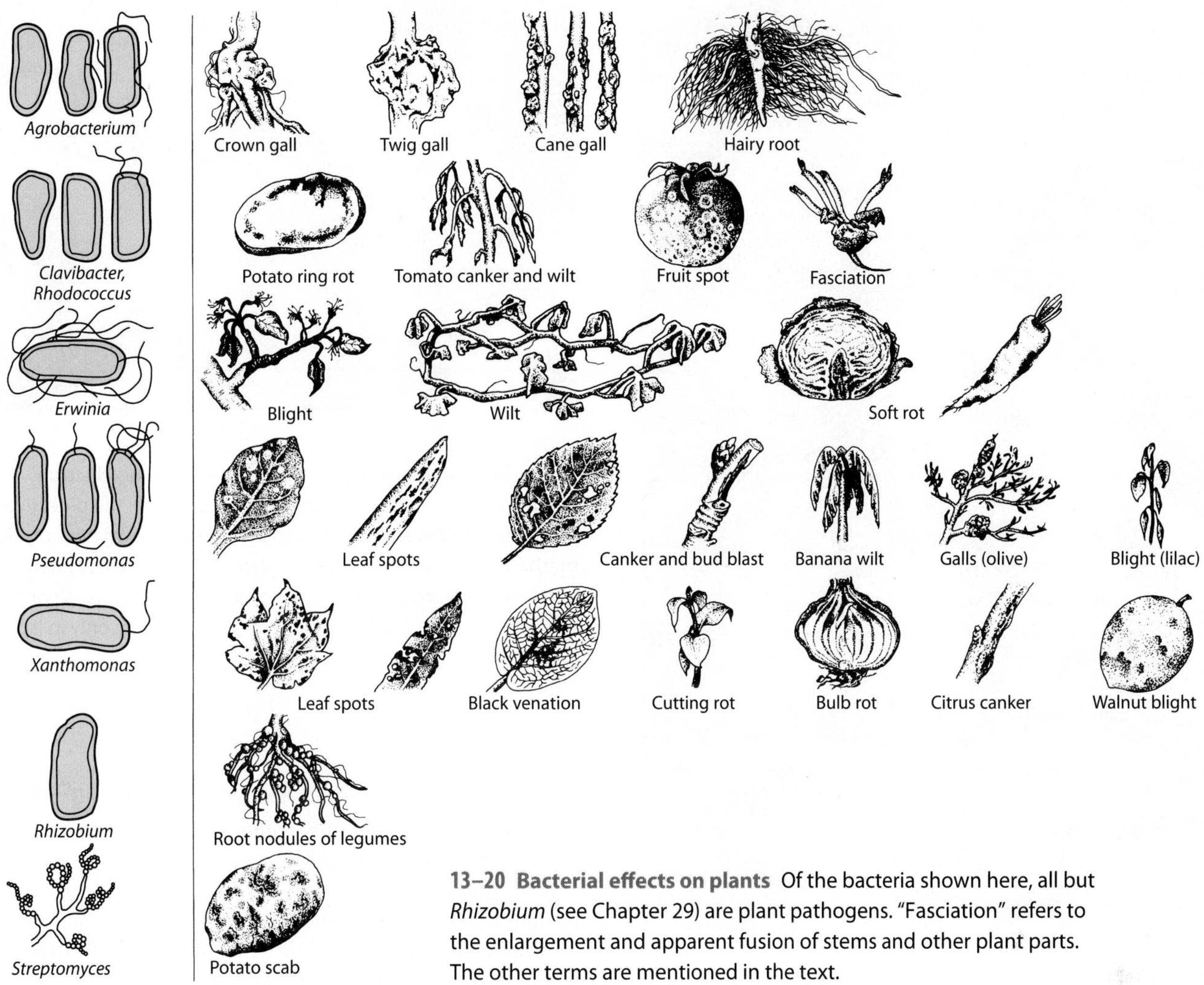

13–20 Bacterial effects on plants Of the bacteria shown here, all but *Rhizobium* (see Chapter 29) are plant pathogens. "Fasciation" refers to the enlargement and apparent fusion of stems and other plant parts. The other terms are mentioned in the text.

Archaea

The Archaea exhibit enormous physiological diversity. On the basis of that diversity, the Archaea that have been studied most thoroughly can be divided into three large groups—extreme halophiles, methanogens, and extreme thermophiles—and one small group represented by a thermophile that lacks a cell wall. Until very recently, the archaea were generally thought of as noncompetitive relics inhabiting hostile environments that were of little importance to global ecology. It is now known, however, that archaea are present in less hostile environments such as soil. Archaea also constitute a major component of the oceanic picoplankton (organisms smaller than 1 micrometer), possibly outnumbering all other oceanic organisms. There are no known pathogens in this domain of prokaryotes.

The Extreme Halophiles Are the "Salt-Loving" Archaea

The **extreme halophilic archaea** are a diverse group of prokaryotes that occur everywhere in nature where the salt concentration is very high—in places such as the Great Salt Lake and the Dead Sea, as well as in ponds where sea water is left to evaporate, yielding table salt (Figure 13–21). Extreme halophiles have a very high requirement for salt, with most of them requiring 12 to 23 percent salt (sodium chloride, NaCl) for optimal growth. Their cell walls, ribosomes, and enzymes are stabilized by the sodium ion, Na^+.

All extreme halophiles are chemoorganotrophs (heterotrophs that obtain their energy from the oxidation of organic compounds), and most species require oxygen. In addition, certain species of extreme halophiles exhibit a light-mediated synthesis of ATP that does not involve any chlorophyll pigments. Among them is *Halobacterium halobium,* the prevalent species of Archaea in the Great Salt Lake. Although the high concentration of salt in their environment limits the availability of oxygen for respiration, such extreme halophiles are able to supplement their ATP-producing capacity by using light energy to produce ATP, using a protein called bacteriorhodopsin, which is found in the plasma membrane.

The Methanogens Are the Methane-Producing Archaea

The **methanogens** are a unique group of prokaryotes—the only ones that produce methane gas, an important contributor

13–21 Extreme halophiles An aerial view of extreme halophilic archaea growing in evaporating ponds of sea water near San Francisco Bay, California. The ponds yield table salt, as well as other salts of commercial value. As the water evaporates and the salinity increases, the halophiles multiply harmlessly, forming massive growths, or "blooms," that produce bright colors in the sea water.

13–22 Methane-producing archaea A scanning electron micrograph of methane-producing archaea from the digestive tract of a cud-chewing animal. Cells such as those shown here produce methane and carbon dioxide. Methanogens are strict anaerobes and can therefore live only in the absence of oxygen—a condition prevailing on the young Earth but occurring today only in isolated environments.

to global warming (Figure 13–22). All methanogens are strictly anaerobic and will not tolerate even the slightest exposure to oxygen. Methanogens can produce methane (CH_4) from hydrogen (H_2) and carbon dioxide (CO_2); the needed electrons are derived from the H_2, and CO_2 serves as both carbon source and electron acceptor.

All methanogens use ammonium (NH_4^+) as a nitrogen source, and a few can fix nitrogen. Methanogens are common in sewage treatment plants, in bogs, and in the ocean depths. In fact, most of the natural gas reserves now used as fuel were produced by the activities of methane-producing prokaryotes in the past. Methanogens are also found in the digestive tracts of cattle and other ruminant (cud-chewing) animals, where they are important for the breakdown of cellulose. It is estimated that a cow belches about 50 liters of methane a day while chewing the cud. Some methanogens are endosymbionts of certain protozoa, and a subset of these are found in the gut of insects.

The Extreme Thermophiles Are the "Heat-Loving" Archaea

The **extreme thermophilic archaea** contain representatives of the most heat-loving of all known prokaryotes. The membranes and enzymes of these archaea are unusually stable at high temperatures: all have temperature optima above 80°C, and some grow at temperatures over 110°C. Most species of extreme thermophiles metabolize sulfur in some way, and, with only a few exceptions, they are strict anaerobes. These archaea are inhabitants of hot, sulfur-rich environments, such as the hot springs and geysers found in Iceland, Italy, New Zealand, and Yellowstone National Park (Figure 13–23). As noted, extreme thermophilic archaea also grow near deep-sea hydrothermal vents and cracks in the ocean floor from which geothermally superheated water is emitted.

Thermoplasma Is an Archaeon Lacking a Cell Wall

A fourth group of Archaea consists of a single known genus, *Thermoplasma,* containing a single species, *Thermoplasma acidophilum. Thermoplasma* resembles the mycoplasmas (page 267) in lacking a cell wall and being very small; individuals vary from spherical (0.3 to 2 micrometers in diameter) to filamentous. *Thermoplasma* has been found only in acidic, self-heating coal refuse piles in southern Indiana and western Pennsylvania, in sites within the piles where the temperatures range from 32° to 80°C—the sort of very unusual habitat where archaea seem to thrive.

Viruses

Viruses are simple submicroscopic parasites of plants, animals, archaea, and bacteria (including mycoplasmas). They also parasitize protists and fungi. In the extracellular state, the simplest

13–23 Extreme thermophiles A steaming hot spring in Yellowstone National Park, with hydrogen-sulfide-rich steam rising to the surface of the Earth. Unlike the large hot spring shown in Figure 13–2, this one, because of its high temperature and acidity, is almost entirely dominated by extreme thermophilic archaea growing in a habitat where other organisms, including most bacteria, cannot survive. The archaea form a mat around the spring.

viruses consist of a nucleic acid core, the **viral genome,** surrounded by a protein coat, which protects the genome in the external environment and helps the virus attach to the next cell or host. Outside its host cells, the entire infectious virus particle (genome plus coat)—also called a **virion**—is metabolically inert. The virion is the structure by which the viral genome is carried from one host to another. To multiply, viruses must enlist a host cell in which they can replicate (Figure 13–24).

Viruses Cause Terrible Diseases and Huge Economic Losses

Practically every kind of organism can be infected by distinct viruses, and it is clear that a tremendous diversity of viruses exists. A virus typically is associated with a specific type of host and is usually discovered and studied because it causes a disease in that host.

In humans, viruses are responsible for many infectious diseases, including chicken pox, measles, mumps, influenza, colds (often complicated by secondary bacterial infections), infectious hepatitis, polio, rabies, herpes, AIDS, and deadly hemorrhagic fevers (such as those caused by Ebola virus and Hantavirus).

In plants, well over 2000 diseases are known to be caused by the more than 600 different kinds of plant viruses identified. Viral diseases greatly reduce the productivity of many kinds of agricultural and horticultural crops, with worldwide losses estimated at about $15 billion annually.

Often the only symptom of viral infection in plants is reduced growth rate, resulting in various degrees of dwarfing or stunting. The most obvious symptoms are usually those that appear on leaves, where a virus interferes with chlorophyll production, thus affecting photosynthesis. Mosaics and ring spots are the most common symptoms produced by systemic viruses, which are those viruses that move throughout the plant. In mosaic diseases, light green, yellow, or white areas—ranging in size from small flecks to large stripes—appear intermingled

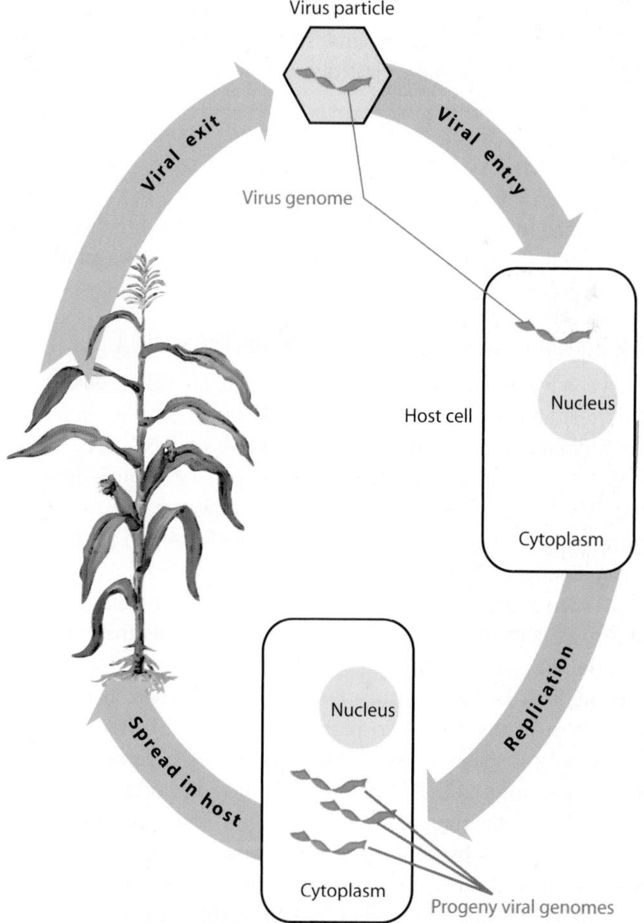

13–24 Generalized life cycle of a virus The life cycle, or infection cycle, of a virus consists of two stages: replication inside host cells and the spread to new hosts. For most plant viruses, the spread to new hosts is assisted by vector organisms.

13–25 Infection by tobacco mosaic virus A tobacco leaf infected with tobacco mosaic virus shows the typical splotches of pale color, indicating the breakdown of chlorophyll.

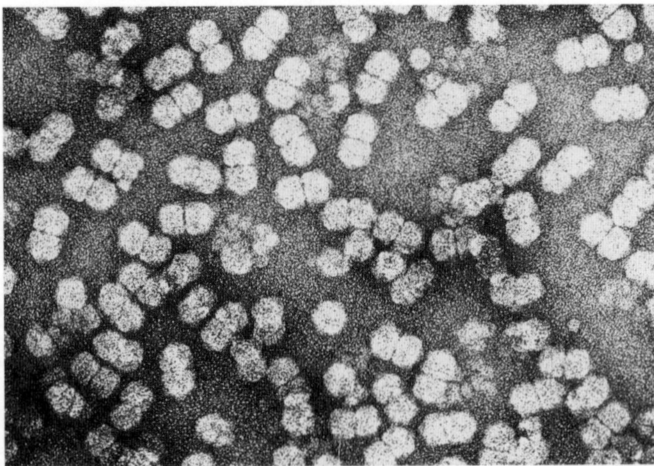

13–26 Geminivirus Purified geminivirus from the grass *Digitaria,* negatively stained in 2 percent aqueous uranyl acetate. Each geminivirus, which has DNA as its genetic material, typically appears as a two-part entity.

with the normal green of leaves and fruit (Figure 13–25). In ring spot diseases, chlorotic (yellow) or necrotic (dead tissue) rings appear on the leaves and sometimes also on the stems and fruits. Less common viral diseases include leaf roll (potato leaf roll), yellows (beet yellows), dwarf (barley yellow dwarf), canker (cherry black canker), and tumor (wound tumor). The yellow blotches or borders on the leaves of some prized horticultural varieties may be caused by viruses, and the variegated appearance of some flowers is the result of viral infections that are passed on from generation to generation in vegetatively propagated plants.

The Genomes of Viruses Can Be Either DNA or RNA

As we have seen, the genetic material in all cells consists of double-stranded DNA (see Chapter 9). In viruses, by contrast, the genomes are composed of either RNA or DNA. The RNA or DNA may be either single-stranded (ss) or double-stranded (ds). The great majority of plant viruses, however, are "positive-sense" single-stranded RNA viruses, whose RNA can act directly as a messenger RNA in infected host cells.

Three types of plant viruses—the geminiviruses, the badnaviruses, and the caulimoviruses—have DNA as their genetic material. A geminivirus is a small spherical particle that often looks as if it is made up of connected pairs, when in fact it is a single structure with two parts (Figure 13–26). Bean golden mosaic is a disease of bean plants caused by a geminivirus. It is spread from plant to plant by whiteflies and occurs in tropical climates. Another geminivirus causes maize (corn) streak, which is spread by leafhoppers and has the smallest known genome of any virus. Geminiviruses are responsible for other devastating diseases of cereal and vegetable crops worldwide as well, including wheat, sugarcane, beans, beets, cassava, cotton, peppers, and squash. The badnaviruses cause diseases of, for example, bananas, sugarcane, cacao, and raspberries. The caulimoviruses infect cauliflowers, blueberries, and carnations, among other plants.

Viruses Multiply by Redirecting the Host Cell's Biosynthetic Machinery

Unlike animal viruses, which enter cells through receptor-mediated endocytosis (see Chapter 4), plant viruses are unable to penetrate the cell wall, which is a formidable barrier. The transmission, or spread, of viruses from diseased to healthy plants most commonly involves insect vectors, such as aphids, leafhoppers, or whiteflies, with piercing and sucking mouthparts. In addition to insect vectors, plant viruses can enter the plant through wounds made mechanically by nematodes (roundworms) or during harvest operations, or by transmission into an ovule via a pollen tube of an infected pollen grain. Viruses are also spread by vegetative propagation of ornamental plants and crops such as potatoes.

Once inside a host cell, a virion sheds its coat, freeing its nucleic acid. Within the cell, the viral RNA or DNA then multiples by redirecting the cell's biosynthetic machinery, thus producing the nucleic acids and proteins to assemble additional viral particles.

On entry into a host cell, the single-stranded DNA of the geminivirus is transported to the nucleus, where it is converted to double-stranded DNA by synthesis of a complementary strand directed by the host cell. This double-stranded DNA then serves as a template for transcription of the viral replicase *(Rep)* gene, which is required for further replication.

Because they are capable of producing high numbers of genome copies in inoculated cells, geminiviruses are especially attractive as potential vector systems for the expression of specific gene products in plants. Engineered viral DNA, in which the gene for the coat protein has been replaced by a foreign gene of interest (see Chapter 10), can be transmitted to the plant by mechanical inoculation of leaves, where it will be amplified during viral DNA replication.

In positive-sense single-stranded RNA viruses—such as tobacco mosaic virus (TMV)—the virus-encoded replicase

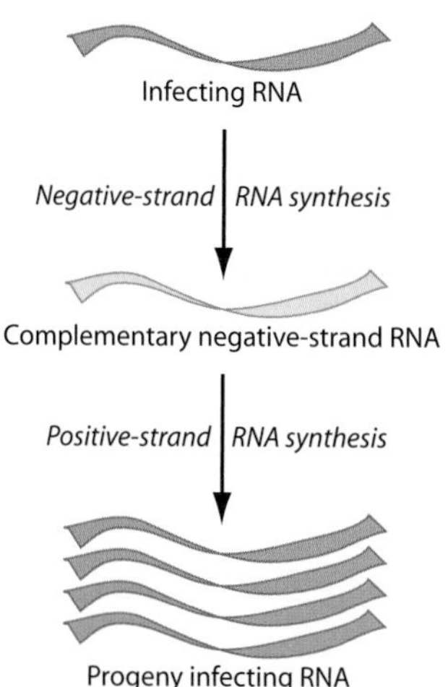

13–27 RNA virus replication Simplified diagram of the replication of a positive-sense single-stranded RNA virus genome.

synthesizes a complementary (negative) strand of RNA, using the positive strand as a template. New positive RNA strands then are synthesized from the negative-strand template (Figure 13–27). Replication of the positive strand is generally assumed to involve association with cellular membranes such as the endoplasmic reticulum, outer chloroplast membrane, and tonoplast (the membrane surrounding the vacuole). As mentioned previously, the positive-sense single-stranded RNA acts as messenger RNA. Utilizing the ribosomes of the host cell, it directs the synthesis of enzymes and of the subunits of the protein coat.

The Viral Capsid Is Composed of Protein Subunits

All viruses have one or more proteins, called capsid or coat proteins, that assemble in a precise symmetrical manner to form the **capsid,** a shell-like covering that protects the nucleic acid. Some viruses also have an envelope of lipid molecules interspersed with proteins on the outer surface of the capsid. The surface proteins and lipids help in recognizing potential host cells, and they also provide targets for the immune response of animals combating a viral infection.

Because viral capsids are important determinants in viral infection, scientists have been using electron microscopy and crystallography to study the structure of capsids. Knowledge about the structure of viruses leads to smarter designs of antivirals and antibodies to better control viral infections. The study of plant viruses has been instrumental in understanding viral diseases.

The tobacco mosaic virus—the first virus to be viewed under the microscope, in 1939—is a classic example of a **helical virion,** one of two major structural classes of viruses. TMV is a rodlike particle about 300 nanometers long and 15 nanometers in diameter. The RNA of more than 6000 nucleotide bases forms a single strand that fits into a groove in each of more than 2000 identical protein molecules arranged with helical symmetry, much like a coiled spring (Figure 13–28).

The other major, and most common, structural class of viruses is the **icosahedral viruses.** The icosahedron is a 20-sided structure, in which the capsid is assembled from 180 or more protein molecules arranged in a symmetry similar to a geodesic dome (Figure 13–29). Most of the icosahedral plant viruses are about 30 nanometers in diameter.

Viruses Move from Cell to Cell within the Plant via Plasmodesmata

As mentioned above, the cell wall is a formidable barrier, so most plant viruses require insect vectors to penetrate the protoplast. Some viruses remain confined to the initially infected cell, whereas others move throughout the plant body, resulting in a **systemic infection.**

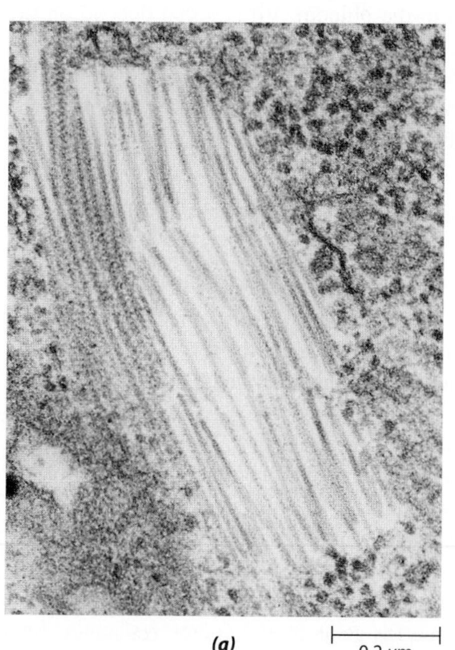

13–28 Tobacco mosaic virus (TMV) ***(a)*** Electron micrograph showing TMV particles in a mesophyll cell of a tobacco leaf. ***(b)*** A portion of the TMV particle, as determined by X-ray crystallography. The single-stranded RNA, shown here in red, fits into the grooves of the protein subunits that assemble into the helical arrangement of the capsid.

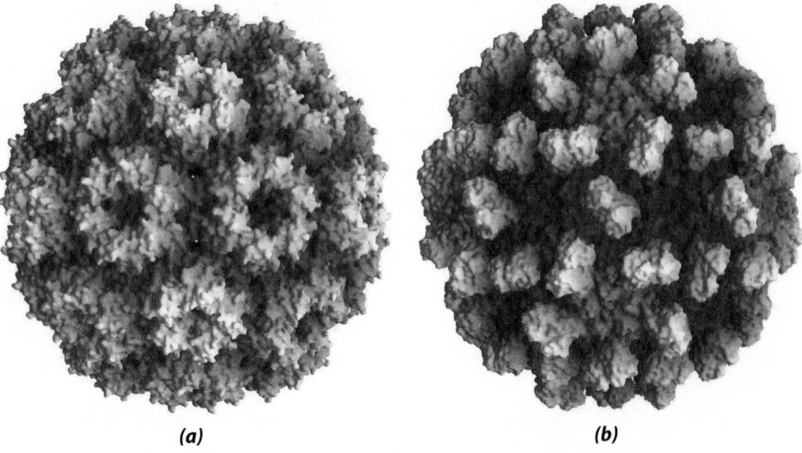

13–29 Icosahedral plant viruses Protein capsids, as determined by X-ray crystallography, for ***(a)*** cowpea chlorotic mottle virus and ***(b)*** tomato bushy stunt virus. Both of these viruses are icosahedral; that is, their subunits are arranged in a 20-sided structure or a variation of it.

The movement of many viruses throughout the plant can be divided into two phases: (1) cell-to-cell, or short-distance, movement between parenchymatous cells, and (2) long-distance movement in the conducting channels, or sieve tubes, of the phloem tissue.

Cell-to-cell movement of viruses in plants occurs through plasmodesmata (Figure 13–30). The effective pore sizes, or size exclusion limits (see Chapter 4), of plasmodesmata are too small under ordinary circumstances to allow the passage of virions or viral genomes (see Figure 4–18). Movement of the viruses is facilitated by virus-encoded proteins called **movement proteins** and involves two distinct mechanisms: tubule-guided movement and non-tubule-guided movement. In viruses such as Grapevine fanleaf virus, the plasmodesmata are structurally modified by insertion of a tubule assembled by the movement protein. The desmotubule is absent in these modified plasmodesmata. For other viruses, such as tobacco mosaic virus, there are no major changes in plasmodesmata structure. Instead, the movement protein increases the size exclusion limit of the plasmodesmata tenfold. Tobacco mosaic mutants deficient in the gene that specifies production of the movement protein are unable to move systemically throughout the plant body. After establishing that viral genomes move through plasmodesmata, researchers found that some plant messenger RNAs also move through plasmodesmata. In fact, the study of viruses has been instrumental in understanding the function of plasmodesmata and in uncovering a system by which plant cells can communicate.

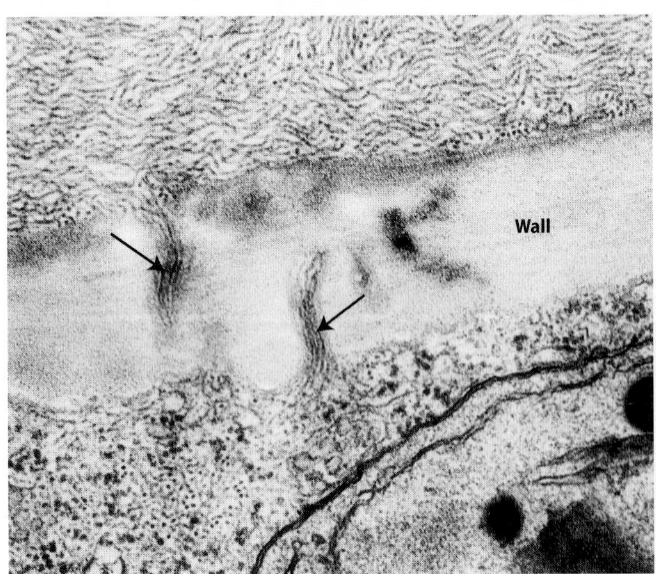

13–30 Viral movement through plasmodesmata Beet yellows virus particles in plasmodesmata (arrows), moving from a sieve-tube member of the phloem (above) to its sister cell, a companion cell (below).

Cell-to-cell movement via plasmodesmata is a slow process. In a leaf, for instance, a virus moves about 1 millimeter, or 8 to 10 parenchyma cells, per day. In the phloem, the movement of viruses can reach speeds of 1 centimeter per day. Once in the phloem, the viruses move systemically toward growing regions, such as shoot tips and root tips, and regions of storage, such as rhizomes and tubers, where the virus reenters the parenchyma cells adjacent to the phloem. Viruses that depend on the phloem for successful establishment of infections are introduced by the vector directly into the phloem. Some phloem-dependent viruses, such as the beet yellows virus, seem to be limited to the phloem and a few adjacent parenchyma cells.

Various Host Responses Confer Resistance to Plant Pathogens

A common response by a host to infection with different types of pathogens (fungi, bacteria, and viruses) is the **hypersensitive response (HR).** Activation of the hypersensitive response is dependent on recognition of the pathogen by a specific, dominant *resistance gene* (the host gene that determines resistance). Responses associated with the hypersensitive response include host cell death at the infection site accompanied by the accumulation of high concentrations of molecules with antimicrobial properties. These responses inhibit pathogen movement at the edge of the lesion.

Disease resistance is usually mediated by dominant genes. According to the gene-for-gene model, plant resistance to a pathogen occurs only when a plant possesses a dominant resistance gene and the pathogen expresses the complementary avirulence gene. The avirulence gene codes for a viral protein that interacts with the host's resistance gene and causes a failure at some stage of the infection cycle. Two examples of these dominant resistance genes are gene *N* of *Nicotiana glutinosa* and gene *Rx* of potato. The former recognizes the replicase protein of tobacco mosaic virus, resulting in a hypersensitive response to TMV; the latter confers immunity to potato virus X and is triggered by the viral coat protein.

Many pathogens can elicit a mechanism known as **systemic acquired resistance (SAR),** which develops in response to a localized attack by a pathogen. Induction of SAR requires formation of necrotic lesions in the plant, either as part of the hypersensitive response or as a disease symptom. The activation of SAR, which requires salicylic acid (page 34), provides other parts of the plant with long-lasting protection against the same and other, unrelated pathogens.

A specific defense response to plant-pathogenic viruses is **posttranscriptional gene silencing (PTGS).** Many viruses induce PTGS upon infection of the host plant. Genes silenced by PTGS continue to be transcribed, but the levels of messenger RNA are low to undetectable, because a sequence-specific mechanism is activated that degrades RNA molecules. The gene-silencing mechanism thus eliminates the protein product required for viral pathogenesis.

Viroids: Other Infectious Particles

Viroids are the smallest known agents of infectious disease. They consist of small, circular, single-stranded molecules of RNA and lack capsids of any kind (Figure 13–31). Ranging in size from 246 to 399 nucleotides, viroids are much smaller than the smallest viral genomes. Although the viroid RNA is a single-stranded circle, it can form a secondary structure that resembles a short double-stranded molecule with closed ends. The viroid RNA contains no protein-encoding genes, and hence it is totally dependent on the host for its replication. The viroid RNA molecule appears to be replicated in the nucleus of the host cell, where it apparently mimics DNA and allows the host cell RNA polymerase to replicate it. Viroids may cause their symptoms by interfering with gene regulation in the infected host cell.

The term "viroid" was first used by Theodor O. Diener of the U.S. Department of Agriculture in 1971, to describe the infectious agent causing spindle tuber disease of potato. Potatoes infected by the potato spindle tuber viroid are elongated (spindle-shaped) and gnarled. They sometimes have deep crevices in their surfaces. Two other well-studied viroids are the citrus exocortis viroid and the coconut cadang-cadang viroid. The coconut cadang-cadang viroid has resulted in the death of millions of coconut trees in the Philippines over the last half century.

SUMMARY

Bacteria and Archaea Are the Two Prokaryotic Domains

Prokaryotes are the smallest and structurally simplest organisms. In evolutionary terms, they are also the oldest organisms on Earth, and they consist of two distinct lineages, the domains Bacteria and Archaea. Prokaryotes lack a nucleus bounded by

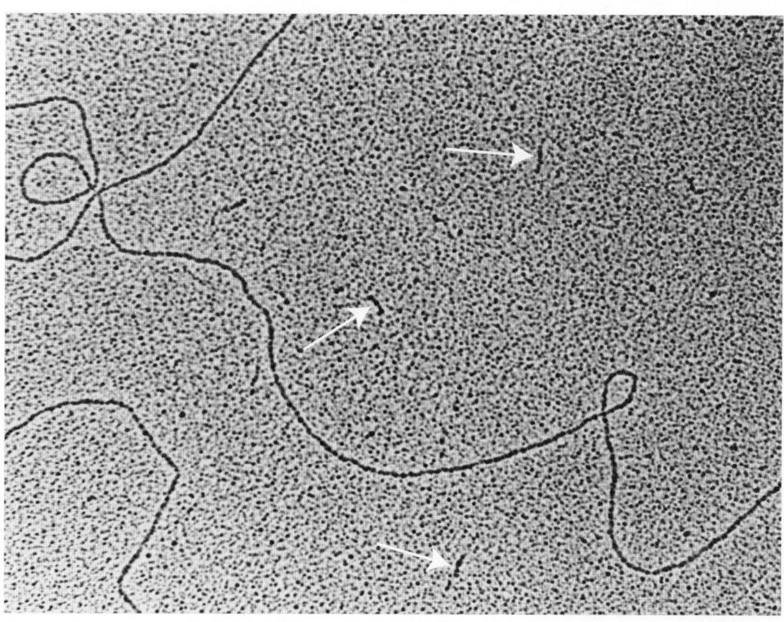

13–31 Viroids Electron micrograph of potato spindle tuber viroids (arrows) mixed with portions of the double-stranded DNA molecule of a bacteriophage. This micrograph illustrates the tremendous difference in size between the genetic material of a viroid and a virus.

a nuclear envelope, as well as membrane-bounded organelles. However, they possess numerous enzyme-containing microcompartments encapsulated by a protein shell. Prokaryotes lack a cytoskeleton, but they do possess homologs of actin and tubulin that play roles in cell division. Often, additional small pieces of circular DNA, known as plasmids, are also present. Almost all prokaryotes have rigid cell walls, notable exceptions being mycoplasmas and phytoplasmas. The bacterial cell wall is composed mainly of peptidoglycan. Gram-negative bacteria, which have walls that do not retain the dye known as crystal violet, have an outer layer of lipopolysaccharides, phospholipids, and proteins over the peptidoglycan layer. Many prokaryotes secrete slimy or gummy substances on the surface of their walls, forming a layer called the glycocalyx, or capsule. A wide variety of prokaryotes—both bacteria and archaea—contain granules of poly-β-hydroxybutyric acid and glycogen, which are food storage compounds.

Prokaryotic Cells Have Characteristic Shapes

Prokaryotic cells may be rod-shaped (bacilli), spherical (cocci), or spiral (spirilla). All prokaryotes are unicellular, but daughter cells may adhere in groups, in filaments, or in solid masses. Many prokaryotes have flagella and thus are motile; rotation of the flagella moves the cell through the medium. Lacking microtubules, the flagella of prokaryotes differ greatly from those of eukaryotes. Prokaryotes may also have fimbriae or pili. Variously sized tubules, called nanotubes, have recently been found connecting adjacent bacterial cells.

Prokaryotes Characteristically Reproduce by Binary Fission

Most prokaryotes reproduce by binary fission—or dividing in two. Mutations, combined with a rapid generation time, are responsible for the extraordinary adaptability of prokaryotes. Further adaptability is provided by lateral gene transfer that takes place as a result of conjugation, transformation, or transduction. Certain species of bacteria have the capacity to form endospores, dormant resting cells that can survive unfavorable conditions.

Prokaryotes Exhibit Tremendous Metabolic Diversity

Although some are autotrophs, most prokaryotes are heterotrophs. The vast majority of heterotrophs are saprotrophs and, together with the fungi, they are the recyclers of the biosphere. Some autotrophs, the photosynthetic autotrophs, obtain their energy from light. Other autotrophs obtain their energy from the reduction of inorganic compounds and are called chemosynthetic autotrophs. A number of genera play important roles in the cycling of nitrogen, sulfur, and carbon. Of all living organisms, only certain bacteria are capable of nitrogen fixation. Without bacteria, life on Earth as we know it would not be possible.

Some prokaryotes are aerobic, others are strict anaerobes, and still others are facultative anaerobes. Prokaryotes also vary with regard to the range of temperatures at which they grow, ranging from those that can grow at or below 0°C (psychrophiles) to those that can grow at temperatures higher than 100°C (extreme thermophiles).

The Bacteria Include Pathogenic and Photosynthetic Organisms

Many bacteria are important pathogens in plants and animals. One distinctive group of bacteria, the mycoplasmas and phytoplasmas, which lack a cell wall and are very small, include a number of disease-causing organisms.

Photosynthetic bacteria can be divided into three major groups: the cyanobacteria, the prochlorophytes, and the purple and green bacteria. The cyanobacteria and prochlorophytes contain chlorophyll *a,* the same molecule that occurs in all photosynthetic eukaryotes, and they produce oxygen during photosynthesis. Purple and green bacteria, on the other hand, contain several different types of bacteriochlorophyll and do not produce oxygen during photosynthesis. In addition, the prochlorophytes contain chlorophyll *b,* but lack phycobilins, which are present in cyanobacteria. Many genera of cyanobacteria can fix nitrogen.

The Archaea Are Physiologically Diverse Organisms That Occupy a Wide Variety of Habitats

The Archaea can be divided into three large groups: extreme halophiles, methanogens, and extreme thermophiles. A fourth group is represented by a single genus, *Thermoplasma,* which lacks a cell wall. Once believed to occupy primarily hostile environments, the Archaea are now known to constitute a major component of the oceanic picoplankton.

Viruses Are Submicroscopic Parasites Consisting of DNA or RNA Surrounded by a Protein Coat

Viruses possess genomes that replicate within a living host by directing the host cell's genetic machinery to synthesize viral nucleic acids and proteins. Viruses contain either RNA or DNA—single-stranded or double-stranded—surrounded by an outer protein coat, or capsid, and sometimes also by a lipid-containing envelope. Viruses are comparable in size to large macromolecules and come in a variety of shapes. Most are spherical, with icosahedral symmetry, and others are rod-shaped, with helical symmetry.

Viruses and Viroids Cause Diseases in Plants and Animals

Viruses are responsible for many diseases of humans and other animals, and also for more than 2000 known kinds of plant diseases. The transmission of viruses from diseased to healthy plants most commonly involves insect vectors. Once inside a host cell, the viral particle, or virion, sheds its capsid, freeing its nucleic acid. Most plant viruses are RNA viruses. In single-stranded RNA viruses, such as tobacco mosaic virus, the viral RNA directs formation of a complementary strand of RNA, which then serves as a template for the production of new viral RNA molecules. Utilizing the ribosomes of the host cell, the viral RNA directs the synthesis of capsid proteins. The new RNA strands and capsid proteins are then assembled into complete virions within the host cell.

Short-distance movement of viruses from cell to cell within host plants occurs through plasmodesmata. Such movement is facilitated by virus-encoded proteins called movement proteins. A large number of plant viruses move systemically throughout the plant in the phloem.

Viroids, the smallest known agents of infection, consist of small, circular, single-stranded molecules of RNA. Unlike viruses, viroids lack protein coats. They are thought to interfere with gene regulation in infected host cells, where they occur mainly in the nucleus.

QUESTIONS

1. Distinguish between the following: gram-positive bacterium and gram-negative bacterium; fimbria and pilus; endospore and akinete; virus and viroid.
2. Pear decline is so-called because it causes a slow, progressive weakening and ultimate death of the pear tree. It is a systemic disease caused by phytoplasmas. What is meant by a "systemic disease," and by what pathway do phytoplasmas move through the tree?
3. What genetic factors contribute to the extraordinary adaptability of prokaryotes to a broad range of environmental conditions?
4. What are some host responses that confer resistance to plant pathogens?
5. One might argue that viruses should be considered living organisms. By what criteria might viruses be considered alive?

CHAPTER 14

Fungi

◀ **Cedar-apple rust** The fungus causing cedar-apple rust alternates between two hosts, eastern red cedars and apple trees. Spectacularly colored galls on the cedars produce spores that are carried by wind to apple trees, which bear damaged fruit. On the apple trees, the fungus produces spores that are carried by wind back to the cedars. Fungicides applied to the economically important apple trees interrupt the cycle.

CHAPTER OUTLINE

The Fungi are heterotrophic organisms that were once considered to be primitive or degenerate plants lacking chlorophyll. It is now clear, however, that the only characteristics Fungi share with plants—other than those common to all eukaryotes—are their sessile nature and multicellular growth form. (A few fungi, including yeasts, are unicellular.) Molecular evidence strongly suggests that fungi are more closely related to animals than to plants. As we will see, the Fungi are a form of life so distinctive from all others that they have been assigned their own kingdom—the kingdom Fungi.

Over 100,000 species of Fungi have been identified thus far, with some 1200 new species discovered each year. Conservative estimates for the total number of species exceed 1.5 million, placing the Fungi second only to insects in this regard. The largest living organism on Earth today may be an individual of the tree root-rot fungus *Armillaria solidipes,* formerly known as *Armillaria ostoyae* (Figure 14–1), which encompasses nearly 900 hectares (2200 acres) of forest in the Blue Mountains in eastern Oregon. This fungus is estimated to be more than 2400 years old. A close relative of *A. solidipes, Armillaria gallica,* has been found occupying 15 hectares

CHECKPOINTS

After reading this chapter, you should be able to answer the following:

1. Describe the characteristics of the Fungi that differentiate them from all other life forms.
2. From what type of organism is it thought that Fungi evolved?
3. What are the distinguishing characteristics of the Microsporidia, chytrids, zygomycetes, Glomeromycota, Ascomycota, and Basidiomycota?
4. Describe yeasts and their relationship to filamentous fungi.
5. What are asexual fungi, and what is their relationship to other groups of Fungi?
6. Explain the kinds of symbiotic relationships that exist between fungi and other organisms.

(a)

(b)

14–1 Root-rot fungus ***(a)*** A white mycelial mat of root-rot fungus *(Armillaria solidipes)* is seen here growing under the bark of an infected tree. *Armillaria* causes disease in living trees, shrubs, vines, and other plants, and it acts as a decomposer when growing on dead trees and stumps. ***(b)*** *Armillaria* mushrooms emerge in the fall.

(37 acres) in northern Michigan. This "humongous fungus," as it has been dubbed, is estimated to be at least 1500 years old.

The Importance of Fungi

Fungi Are Important Ecologically as Decomposers

The ecological impact of the Fungi cannot be overestimated. Together with the heterotrophic bacteria, fungi are the principal decomposers of the biosphere. Decomposers are as necessary to the continued existence of the living world as are food producers. Decomposition breaks down the organic material incorporated into the bodies of organisms, releasing carbon dioxide into the atmosphere and returning nitrogenous compounds and other materials to the soil, where the molecules can be used again—recycled—by plants and eventually by animals. Estimates are that, on average, the top 20 centimeters of fertile soil contains nearly 5 metric tons of fungi and bacteria per hectare (2.47 acres). Some 500 known species of fungi, representing a number of distinct groups, are marine, breaking down organic material in the sea just as their relatives do on land. There are also many freshwater species.

As decomposers, fungi often come into direct conflict with human interests. A fungus makes no distinction between a rotting tree that has fallen in the forest and a fence post; the fungus is just as likely to attack one as the other. Equipped with a powerful array of enzymes that break down organic substances, including the lignin and cellulose of wood, fungi are often nuisances and are sometimes highly destructive. Fungi attack cloth, paint, leather, waxes, jet fuel, petroleum, wood, paper, insulation on cables and wires, photographic film, and even the coating of the lenses of optical equipment—in fact, almost any conceivable substance, including CDs and DVDs. Although individual species of fungi are highly specific to particular substrates, as a group they attack virtually anything. Everywhere, they are the scourge of food producers, distributors, and sellers alike, for they grow on bread, fresh fruits (Figure 14–2), vegetables, meats, and other products. Fungi reduce the nutritional value, as well as the palatability, of such foodstuffs. In addition, some produce very toxic substances known as mycotoxins on certain plant materials and on meats.

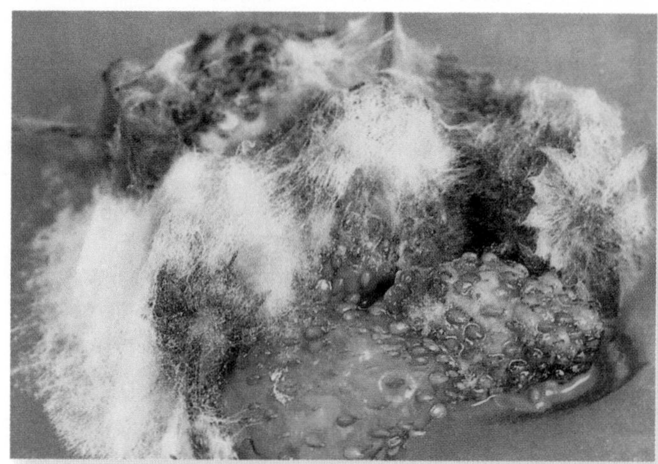

14–2 Decomposing fruit The common mold *Rhizopus* growing on strawberries.

Fungi Are Important Medically and Economically as Pests, Pathogens, and Producers of Certain Useful Chemicals

The importance of fungi as commercial pests is enhanced by their ability to grow under a wide range of conditions. Some strains of *Cladosporium herbarum* that attack meat in cold storage can grow at a temperature as low as –6°C. In contrast, one species of *Chaetomium* grows optimally at 50°C and survives even at 60°C.

Many fungi attack living rather than dead organisms and sometimes in surprising ways (see "Predaceous Fungi" on page 303). They are the most important causal agents of plant diseases. Well over 5000 species of fungi attack economically valuable crop and garden plants, as well as trees and many wild plants (Figure 14–1). A number of the most important pathogens are asexual fungi (see pages 294 and 295). Anthracnose diseases of plants, which cause lesions and blackening, are generally caused by asexual fungi. In addition, an often fatal disease of dogwoods *(Cornus florida)* that was detected over a wide area of the eastern United States in the late 1980s is caused by the asexual fungus *Discula destructiva.* Other fungi—more than 175 species have been identified—cause serious diseases in domestic animals and humans.

Although fungal infections of humans are most common in tropical regions, an alarming increase has occurred in the numbers of individuals infected with fungi in all regions of the world. This increase is due in part to the growing population of individuals with compromised immune systems, such as people with AIDS and cancer patients undergoing chemotherapy. Worldwide, the majority of AIDS deaths are due to pneumonia caused by *Pneumocystis carinii,* long thought to be a protozoan but now classified as a fungus. Other serious fungal pathogens of people with suppressed immune systems are species of *Candida* that cause thrush and other infections of mucous membranes, as well as *Cryptococcus neoformans,* a basidiomycete that causes cryptococcosis and has the growth habit of a yeast when it grows as a human pathogen.

The qualities that make fungi such important pests can also make them commercially valuable. Yeasts are utilized by vintners as a source of ethanol, by bakers as a source of carbon dioxide, and by brewers as a source of both substances. Many domestically useful strains of yeast have been developed by selection and breeding, and the techniques of genetic engineering are now being used to improve these strains further through the addition of useful genes from other organisms. Some of the flavors of wine come directly from the grape, but most arise from the action of the yeast. Most of the yeasts essential to the production of wine, cider, sake, and beer are strains of *Saccharomyces cerevisiae,* although other species also play a role. Most lager beer, for example, is made using *Saccharomyces carlsbergensis* (see essay on page 119). *Saccharomyces cerevisiae,* however, is now virtually the only species used in baking bread (see Figure 14–8a). Other fungi provide the distinctive flavors and aromas of specific kinds of cheese. For example, certain *Penicillium* species give some types of cheese the appearance, flavor, odor, and texture so highly prized by gourmets. Roquefort, Danish blue, Stilton, and Gorgonzola are all ripened by *Penicillium roqueforti.* Another species, *Penicillium camemberti,* gives Camembert and Brie cheeses their special qualities. Soy paste (miso) is produced by fermenting soybeans with *Aspergillus oryzae,* and soy sauce is made by fermenting soybeans with a mixture of *A. oryzae* and *Aspergillus sojae,* as well as lactic acid bacteria. *Aspergillus oryzae* is also important for the initial steps in brewing sake, the traditional alcoholic beverage of Japan; *S. cerevisiae* is important later in the process.

The commercial use of fungi in industry continues to grow, and many antibiotics—including penicillin, the first antibiotic to be used widely—are produced by fungi. Dozens of different kinds of fungi (mushrooms) are eaten regularly by humans, and many of them are cultivated commercially. The ability of fungi to break down substances is leading to investigations into the use of fungi in toxic waste cleanup programs. The white rot fungus *Phanerochaete chrysosporium,* which survives by degrading lignin as a necessary step to getting at the cellulose and hemicellulose in wood, has been very effective in degrading toxic organic compounds.

A striking example of the potential value of compounds derived from fungi is cyclosporin, a "wonder drug" isolated from the soil-inhabiting fungus *Tolypocladium inflatum.* Cyclosporin suppresses the immune reactions that cause rejection of organ transplants, but without the undesirable side effects of other drugs used for this purpose. This remarkable drug became available in 1979, making it possible to resume organ transplants, which had essentially been abandoned. Because of cyclosporin, successful organ transplantation has today become almost commonplace.

A number of yeasts, most notably *S. cerevisiae,* have become important laboratory organisms for genetic research. This yeast is the organism of choice for studies of the metabolism, molecular genetics, and development of eukaryotic cells, and for chromosome studies. Haploid cells of *S. cerevisiae* have 16 chromosomes, and the DNA sequences of all 16 have been determined. *Saccharomyces cerevisiae* was the first eukaryote to have its genome completely sequenced. Fungi have long been used as model systems for genetics and molecular biology. Notable are the experiments performed in the 1940s by George Beadle and Edward L. Tatum, who later shared a Nobel Prize for their work. Working with mutants of the red bread mold *Neurospora crassa,* Beadle and Tatum hypothesized that enzymes and other protein molecules are direct products of genes (the one gene–one enzyme hypothesis).

Fungi Form Important Symbiotic Relationships

The relationships between fungi and other organisms are extremely diverse. For example, at least 90 percent of all vascular plant species form mutually beneficial associations, called mycorrhizas, between their roots and fungi. These associations, which are discussed further beginning on page 312, play a critical role in plant nutrition. Lichens, many of which occupy extremely hostile habitats, are symbiotic associations between fungi and either algal or cyanobacterial cells (see page 306). Symbiotic relationships also exist between fungi and insects. In one such relationship, the fungi, which produce cellulase and other enzymes needed for digestion of plant material, are

14–3 Fungi Representatives of four of the six major groups of fungi. ***(a)*** The chytrid *Polyphagus euglenae* (cell on left) parasitizing a *Euglena* cell. The cytoplasm of the rounded-up *Euglena* cell is degraded. ***(b)*** A flower fly *(Syrphus)* that has been killed by the fungus *Entomophthora muscae,* a zygomycete. ***(c)*** Common morels, *Morchella esculenta,* a species of Ascomycota. The morels are among the most prized of the edible fungi. ***(d)*** A mushroom, *Hygrocybe aurantiosplendens,* a species of Agaricomycotina, a subphylum of Basidiomycota. A mushroom is made up of densely packed hyphae, collectively known as the mycelium.

cultivated by ants in fungus "gardens." The ants supply the fungus with leaf cuttings and anal droppings, and the ants eat nothing but the fungus. A similar association has evolved among some basidiomycetes and termites found in tropical Africa and Asia. Species of *Termitomyces* (termite fungus) are the fungi that most commonly form an association with termites. Other symbiotic relationships involve a great variety of fungi, known as **endophytes,** that live inside the leaves and stems of apparently healthy plants. Many of these fungi produce toxic secondary metabolites that protect their hosts against pathogenic fungi and attack by insects and grazing mammals (see "From Pathogen to Symbiont: Fungal Endophytes" on page 307).

Most mycologists now recognize six groups of fungi: Microsporidia, chytrids, zygomycetes, Glomeromycota, Ascomycota, and Basidiomycota (Figure 14–3; Table 14–1 on the next page).

Characteristics of Fungi

Most Fungi Are Composed of Hyphae

Fungi are primarily terrestrial. Most fungi are filamentous, and the fungi that produce structures such as mushrooms consist of a great many such filaments, packed tightly together (Figure 14–3c, d). Fungal filaments are known as **hyphae,** and a mass of hyphae from one organism is called a **mycelium** (Figure 14–1). (The words "mycelium" and **mycology**—the study of Fungi—are derived from the Greek *mykēs,* meaning "fungus.") Growth of hyphae occurs at their tips, but proteins are synthesized throughout the mycelium. Hyphae grow rapidly; an individual fungus may produce more than a kilometer of new hyphae in 24 hours.

Table 14-1 Important Characteristics of the Major Groups of Fungi

Group	Representatives	Nature of Hyphae	Method of Asexual Reproduction	Type of Sexual Spore
Microsporidia (1500 species)	*Bohuslavia, Microsporidium*	Unicellular	Nonmotile spores	Nonmotile spores
Chytrids (790 species)	*Allomyces, Coelomomyces*	Aseptate, coenocytic	Zoospores	None
Zygomycetes (1000 species)	*Rhizopus* (common bread mold)	Aseptate, coenocytic	Nonmotile spores (sporangiospores)	Zygospores (in zygosporangium)
Glomeromycota (200 species)	*Glomus* (endomycorrhizal fungus)	Aseptate, coenocytic	Nonmotile spores	None
Ascomycota (32,300 species)	*Neurospora,* powdery mildews, *Morchella* (edible morels), *Tuber* (truffles), yeasts	Septate or unicellular	Budding, conidia (nonmotile spores), fragmentation	Ascospores
Basidiomycota (22,300 species)	Mushrooms (*Amanita,* poisonous; *Agaricus,* edible), stinkhorns, puffballs, shelf fungi, rusts, smuts	Septate with dolipore in many species	Budding, conidia (nonmotile spores, including urediniospores), fragmentation	Basidiospores

The hyphae of most species of Fungi are divided by partitions, or cross walls, called **septa** (singular: septum). Such hyphae are said to be **septate.** In other species, septa typically occur only at the bases of reproductive structures (sporangia and gametangia) and in older, highly vacuolated portions of hyphae. Hyphae lacking septa are said to be **aseptate,** or **coenocytic,** which means "contained in a common cytoplasm" or multinucleate. In most septate fungi, the septa are perforated by a central pore so that the protoplasts of adjacent cells are essentially continuous from cell to cell. In members of the Ascomycota, the pores are usually unobstructed (Figure 14–4) and large enough to allow nuclei, which are quite small, to squeeze through. Such mycelia are therefore functionally coenocytic. The nuclei of fungal hyphae are haploid.

Not all fungi are filamentous. Some, the yeasts, are unicellular and reproduce by fission or, more frequently, by budding. Yeasts do not form a taxonomic group; they are merely a morphological growth form. The yeast growth form is exhibited by a broad range of unrelated fungi encompassing the zygomycetes, Ascomycota, and Basidiomycota. There are at least 80 genera of yeasts, with approximately 600 known species. Most yeasts are ascomycetes, but at least a quarter of the genera belong to the subphylum Agaricomycotina of the Basidiomycota.

Some fungi are **dimorphic,** that is, they exhibit both unicellular (yeast) and filamentous growth forms, shifting from one form to the other under changing environmental conditions. In many of these species, the fungus spends most of its life cycle in the filamentous form. Other fungi exist primarily as yeasts, including the most familiar yeast, *Saccharomyces cerevisiae.* Generally, most laboratory cultures provide all of the nutrients essential for *S. cerevisiae* to continue its existence as a yeast. Apparently, the filamentous phase is the form in which *S. cerevisiae* forages for food.

All fungi have cell walls. The cell walls of plants and many protists are built on a framework of cellulose microfibrils, interpenetrated by a matrix of noncellulosic molecules, such as hemicelluloses and pectic substances (page 57). In fungi, the

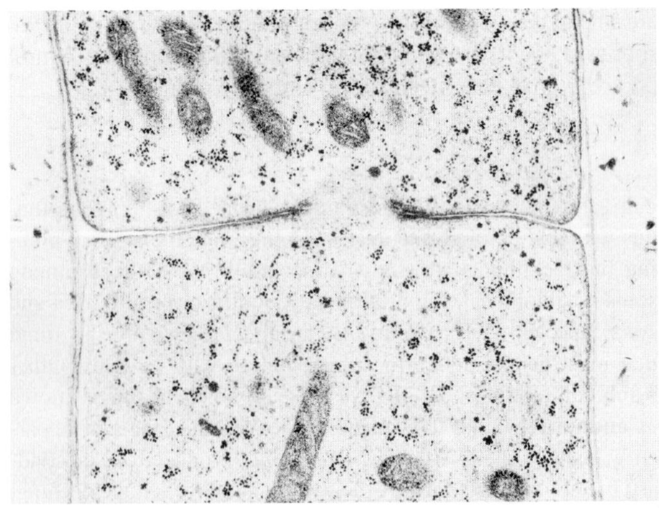

14–4 Septum with unobstructed central pore Electron micrograph of a septum between two cells in the ascomycete *Gibberella acuminata.* The large globular structures are mitochondria, and the tiny dark granules are ribosomes. This specimen was thin-sectioned through the central pore region of a septum.

14–5 **Chitin** The structure of chitin, which consists of β-1,4-linked *N*-acetylglucosamine units. A similar linkage is found in cellulose and in molecules of the cell walls of bacteria, suggesting that such a linkage provides a particularly strong polysaccharide. Chitin is characteristic of the cell walls of many fungi, as well as the exoskeletons of arthropods.

cell wall is composed primarily of another polysaccharide—**chitin**—the same material found in the hard shells, or exoskeletons, of arthropods, such as insects, arachnids, and crustaceans (Figure 14–5). Chitin is more resistant to microbial degradation than is cellulose.

With their rapid growth and filamentous form, fungi have a relationship to their environment that is very different from that of any other group of organisms. The surface-to-volume ratio of fungi is very high, so they are in as intimate a contact with the environment as are the bacteria. Usually, no somatic (body) part of a fungus is more than a few micrometers from its external environment, being separated from it only by a thin cell wall and the plasma membrane. With their extensive mycelia, fungi can have a profound effect on their surroundings—for example, in binding soil particles together. Hyphae of individuals of the same species often fuse, thus increasing the intricacy of the network.

Fungi Are Heterotrophic Absorbers

Because their cell walls are rigid, fungi are unable to engulf small microorganisms or other particles. Typically, a fungus secretes enzymes (called exoenzymes) onto a food source and then absorbs the smaller molecules that are released. Fungi absorb food mostly at or near the growing tips of their hyphae.

All fungi are heterotrophic. In obtaining their food, they function as saprotrophs (living on organic materials from dead organisms), or as parasites, or as mutualistic symbionts (see page 306). Some fungi, mainly yeasts, obtain their energy by fermentation, producing ethyl alcohol from glucose (page 118). Glycogen is the primary storage polysaccharide in fungi, as it is in animals and bacteria. Lipids serve an important storage function in some fungi.

Specialized hyphae known as **rhizoids** anchor some kinds of fungi to the substrate. Parasitic fungi often have similar specialized hyphae, called **haustoria** (singular: haustorium), that absorb nourishment directly from the cells of other organisms (Figure 14–6).

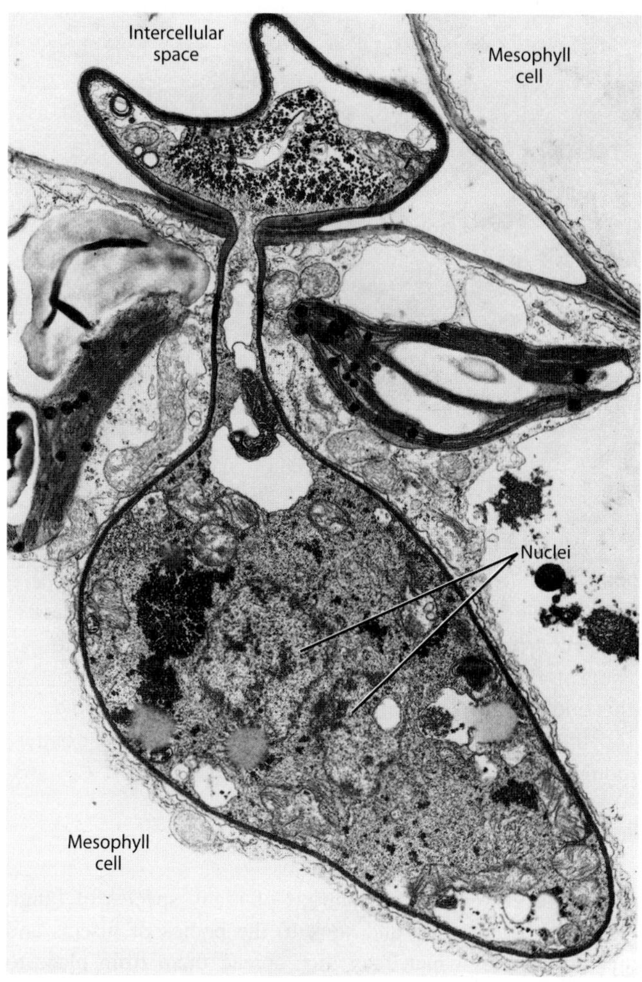

14–6 **Haustorium** Electron micrograph of a haustorium of *Melampsora lini,* a rust fungus, growing in a mesophyll cell of a flax *(Linum usitatissimum)* leaf. In the intercellular space at the top of the micrograph is the haustorial mother cell. The narrow penetration hypha leads to the large, bulbous haustorium within the lower mesophyll cell. The haustorium absorbs nutrients from the flax cell.

Fungi Have Unique Variations of Mitosis and Meiosis

One of the most characteristic features of the Fungi involves nuclear division. The processes of meiosis and mitosis are different from those that occur in plants, animals, and many protists. In most fungi, the nuclear envelope does not disintegrate and re-form but is constricted near the midpoint between the two daughter nuclei. In others, it breaks down near the mid-region. In most fungi, the spindle forms within the nuclear envelope, but in some Basidiomycota it appears to form within the cytoplasm and move into the nucleus. Except for the chytrids, all fungi lack centrioles, but they form unique structures called **spindle pole bodies,** which appear at the spindle poles (Figure 14–7). Like centrioles, spindle pole bodies function as microtubule organizing centers during mitosis and meiosis.

Fungi Reproduce Both Asexually and Sexually

Fungi reproduce through the formation of spores that are produced either sexually or asexually. Except in the chytrids, nonmotile spores are the characteristic means of reproduction in fungi. Some spores are dry and very small. They can remain suspended in air for long periods, and thus can be carried to great heights and for great distances. This property helps

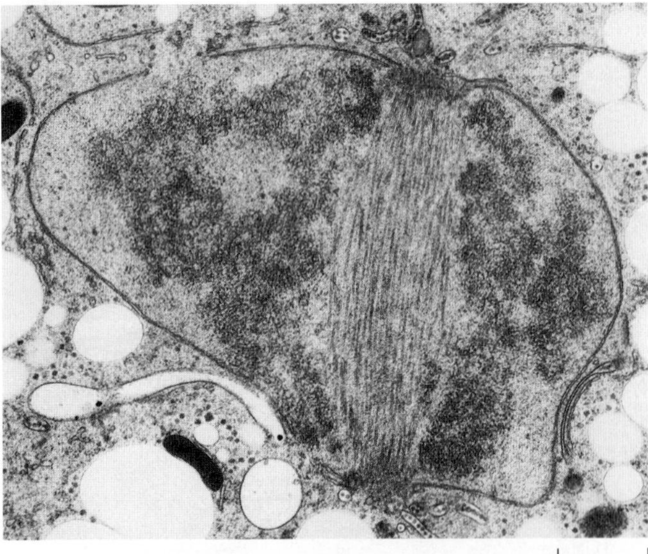

14–7 Spindle pole bodies Electron micrograph of a metaphase nucleus of *Arthuriomyces peckianus*, a rust fungus, showing the spindle inside the nucleus and the two spindle pole bodies at either end (top and bottom) of the spindle. Spindle pole bodies, which are microtubule organizing centers, are characteristic of the zygomycetes, Ascomycota, and Basidiomycota.

explain the very wide distributions of many species of fungi. Other spores are slimy and stick to the bodies of insects and other arthropods, which may then spread them from place to place. The spores of some fungi are propelled ballistically into the air (see "Phototropism in a Fungus" on the facing page). The bright colors and powdery textures of many types of molds are due to the spores. However, some fungi never produce spores.

Many fungi have been found (from molecular data and/or mating studies) to have geographically restricted distributions. Human activity seems to be a major force responsible for moving fungi around from place to place (on nursery stock or in agricultural commodities, for instance).

Sexual reproduction in fungi consists of three distinct phases: plasmogamy, karyogamy, and meiosis. The first two phases are phases of syngamy, or fertilization. **Plasmogamy** (the fusion of protoplasts) precedes **karyogamy** (the fusion of nuclei). In some species, karyogamy follows plasmogamy almost immediately, whereas in others, the two haploid nuclei do not fuse for some time, forming a **dikaryon** ("two nuclei"). Karyogamy may not take place for several months or even years. During that time the pairs of nuclei divide in tandem, producing a dikaryotic mycelium. Eventually, the nuclei fuse within a sporogenous cell to form a diploid nucleus, which quickly undergoes meiosis, reestablishing the haploid condition. This uniquely evolved life cycle characterizes the largest clade of fungi, the subkingdom Dikarya, which comprises the Ascomycota and Basidiomycota. Sexual reproduction in most fungi results in the formation of specialized spores such as zygospores, ascospores, and basidiospores.

It is important to emphasize that the diploid phase in the life cycle of a fungus is represented only by the zygote nucleus. Meiosis typically follows formation of the zygote nucleus; in other words, meiosis in fungi is zygotic (see Figure 12–17a). The general name of the gamete-producing structure of fungi is **gametangium** (plural: gametangia). The gametangia may form sex cells called **gametes** or simply contain nuclei that function as gametes.

The most common method of asexual reproduction in fungi is by means of spores, which are produced either in **sporangia** (singular: sporangium) or from hyphal cells called **conidiogenous cells.** The spores produced by conidiogenous cells occur singly or in chains and are called **conidia** (singular: conidium). The sporangium is a saclike structure, the entire contents of which are converted into one or more—usually many—spores. Some fungi also reproduce asexually by fragmentation of their hyphae.

A common form of reproduction exhibited by yeasts is **budding,** that is, the production of a small outgrowth, the bud,

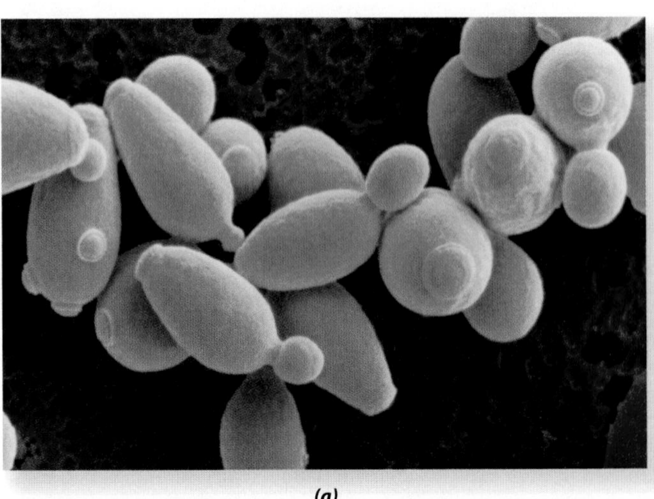

(a)

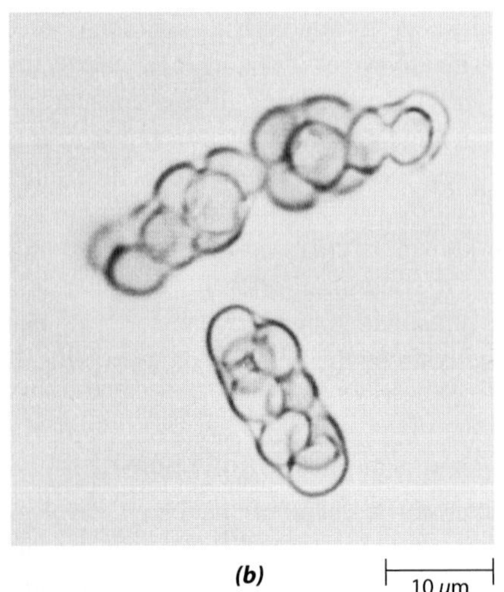

(b)

14–8 Yeasts ***(a)*** Budding cells of bread yeast, *Saccharomyces cerevisiae*. ***(b)*** Asci, with eight ascospores each, of *Schizosaccharomyces octosporus*.

PHOTOTROPISM IN A FUNGUS

Fungi have a variety of methods that ensure wide dispersal of their spores. One of the most ingenious is found in species of *Pilobolus,* a zygomycete that grows on dung. The sporangiophores of this fungus, which attain a height of 5 to 10 millimeters, are positively phototropic—that is, they grow toward the light. An expanded region of the sporangiophore located just below the sporangium (known appropriately as the subsporangial swelling) functions as a lens, focusing the sun's rays on a photoreceptive area at its base. Light focused elsewhere promotes maximum growth of the sporangiophore on the side away from the light, causing the sporangiophore to curve toward the light.

The vacuole in the subsporangial swelling contains a high concentration of solutes, which results in water moving into it by osmosis. Eventually, the turgor pressure becomes so great that the swelling splits, shooting the sporangium in the direction of the light. The initial velocity may approach 50 kilometers per hour, and the sporangium may travel a distance greater than 2 meters. Considering that the sporangia are only about 80 micrometers in diameter, this is an enormous distance. This mechanism is adapted to shoot spores away from the dung—where animals do not feed—and into the grass, where they can be eaten by herbivores and excreted in fresh dung to continue the cycle anew.

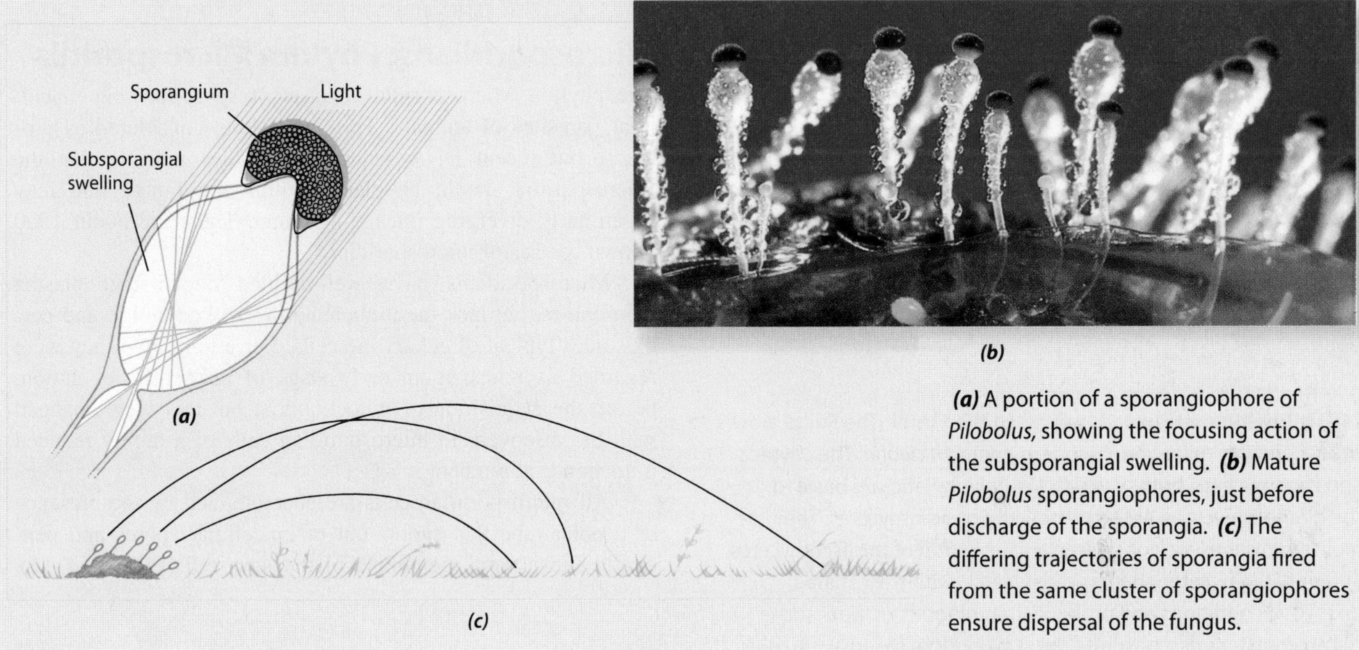

(a) A portion of a sporangiophore of *Pilobolus*, showing the focusing action of the subsporangial swelling. **(b)** Mature *Pilobolus* sporangiophores, just before discharge of the sporangia. **(c)** The differing trajectories of sporangia fired from the same cluster of sporangiophores ensure dispersal of the fungus.

from the parent cell (Figure 14–8a). Each yeast cell can thus be regarded as a conidiogenous cell. Budding is, of course, an asexual method of reproduction. Some yeasts multiply asexually only in the haploid condition. Each haploid cell is capable of serving as a gamete, and at times, two haploid cells may fuse to form a diploid cell, or zygote, which functions as an ascus (Figure 14–8b).

Fungi and Nucleariids Are Sister Groups

As noted earlier, there is considerable molecular evidence that fungi (kingdom Fungi) are more closely related to animals (kingdom Animalia) than they are to plants. The kingdoms Fungi and Animalia, together with a diverse collection of single-celled protists, form a eukaryotic supergroup known as the Opisthokonta. The two kingdoms are thought to have diverged about 1.5 billion years ago, the Fungi arising from a protist closely related to the modern genus *Nuclearia.* Members of this genus are multinucleate amoebas that use their fine pseudopodia to feed on algae

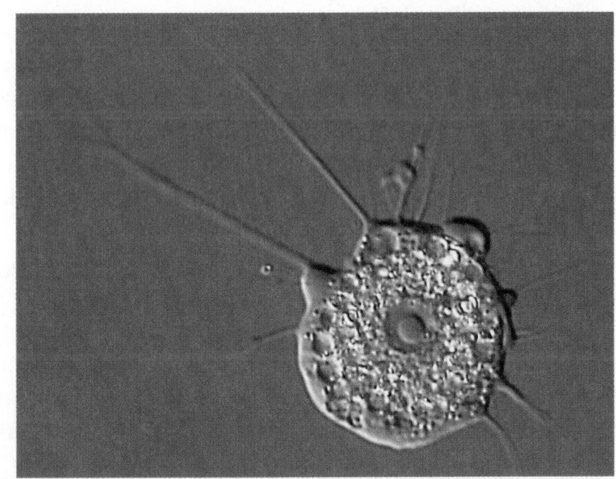

14–9 *Nuclearia* It is believed that fungi arose from protists closely related to members of the modern genus *Nuclearia,* which are multinucleate amoebas with slender pseudopodia.

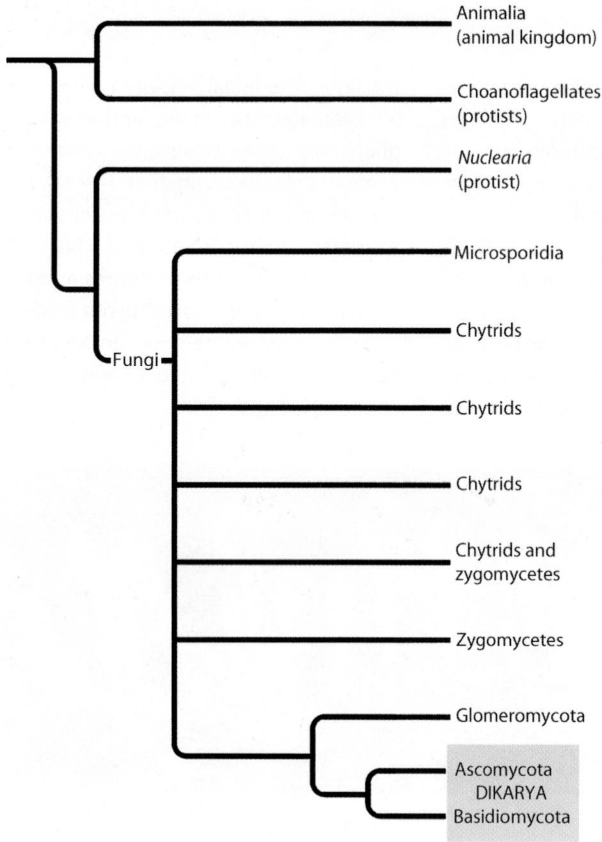

14–10 Phylogenetic relationships of the Fungi The Fungi arose from a protist similar to the modern genus *Nuclearia.* The chytrids and zygomycetes, both of which are polyphyletic, are basal to the glomeromycetes, ascomycetes, and basidiomycetes. The glomeromycetes are considered a sister group of the ascomycetes, and the ascomycetes and basidiomycetes, with their shared dikaryotic conditions and probably homologous croziers and clamp connections, constitute the subkingdom Dikarya.

and bacteria (Figure 14–9). It now seems that the chytrids are the earliest fungal lineage and that the flagellated condition—represented by flagellated zoospores—is a primitive character retained by the chytrids after they evolved from flagellated protists. *Nuclearia* and the fungal progenitors of the zygomycetes, Glomeromycota, Ascomycota, and Basidiomycota probably lost their flagella fairly early in their evolutionary history.

Although the Fungi appear to be a monophyletic lineage, relationships within the Fungi are far from certain. Figure 14–10 summarizes the current thinking about the relationships among the fungal groups. Whereas the chytrids and zygomycetes are phylogenetically basal to all other fungal groups, the ascomycetes and basidiomycetes are the most derived, or more recently evolved, groups. The glomeromycetes are accepted as a sister group to the ascomycetes. Neither the chytrids nor the zygomycetes are monophyletic groups.

Fungi have been around for a long time. Because of their soft structure, there are relatively few fossilized fungi. The oldest fossils resembling fungi are represented by aseptate filaments from the Lower Cambrian, about 544 million years ago. They are believed to have been reef saprotrophs. Fossil fungi morphologically similar to the chytrid *Allomyces* were found on the stems of *Aglaophyton major,* an Early Devonian plant more than 400 million years old (see Figure 17–13). Highly branched hyphae were also found in cortical cells of *A. major.* Such fungi, which belong to the Glomeromycota, form mycorrhizas, specifically endomycorrhizas, which penetrate plant root cells and increase the uptake of nutrients by both the plant and fungal cells (see page 312). These are one of the few symbiotic plant-fungus associations in the fossil record and are believed to have played a major role in moving plants onto the land. The oldest known Ascomycota fossils are found in rocks of Silurian age (438 million years ago), and the oldest known Basidiomycota are from the Middle Devonian (392 million years ago).

Microsporidians: Phylum Microsporidia

The phylum Microsporidia consists of spore-forming unicellular parasites of animals. They were long considered as protozoa, but recent DNA sequencing studies indicate that the Microsporidia should be placed within the Fungi, and may be an early diverging lineage of Fungi. There are about 1500 known species of microsporidians.

Microsporidians have well-defined nuclei and plasma membranes, but lack the mitochondria, stacked Golgi, and peroxisomes typical of eukaryotic cells. For a long time they were regarded as relics of an early stage of eukaryotic evolution, before the acquisition of mitochondria, but this view changed with the discovery in microsporidian cells of a highly reduced mitochondrial remnant.

All microsporidian cells are characterized by the presence of a polar tube that shoots out of unicellular spores and penetrates the plasma membrane of the host cell (Figure 14–11).

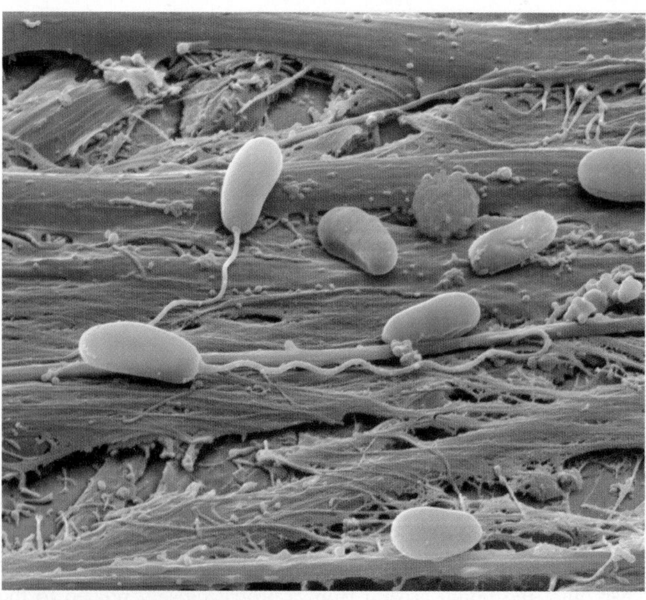

14–11 Microsporidia Spores of the microsporidium *Tubulinosema ratisbonensis,* two of which have everted (shot out) their polar tubes.

The contents of the microsporidian are then injected, via the polar tube, into the host cell. After infection, the microsporidian begins to multiply, relying on the host cell for energy. The cell wall of microsporidians is made of proteins and chitin. The chitin provides the spores with high resistance to unfavorable environmental conditions. Some species seem to reproduce only asexually, whereas others produce several different types of sexual or asexual spores.

All major groups of animals host microsporidians. Most infect insects, but they are also important parasites of fish and crustaceans. About 10 percent of the known species of microsporidians are parasites of vertebrates, including humans. Infection in humans, known as microsporidiosis, is primarily found in individuals with compromised immune systems, such as people with AIDS and organ transplant recipients.

Chytrids: A Polyphyletic Group of Fungi with Flagellated Cells

The polyphyletic chytrids are a predominantly aquatic group, consisting of about 790 species. Soils from ditches and the banks of ponds and streams are also inhabited by chytrids, and some chytrids are even found in desert soils and in the rumens of large herbivorous mammals such as cows. Chytrids are varied not only in form but also in the nature of their sexual interactions and in their life histories. The cell walls of chytrids contain chitin, and like other fungi, the chytrids store glycogen. Meiosis and mitosis resemble these processes in other fungi in that they are intranuclear; that is, the nuclear envelope remains intact until late telophase, when it breaks in a median plane and then re-forms around the daughter nuclei.

Almost all chytrids are coenocytic, with few septa at maturity. They are distinguished from other fungi primarily by their characteristic motile cells (zoospores and gametes), most of which have a single, posterior, smooth (whiplash) flagellum (Figure 14–12). Some chytrids are simple, unicellular organisms that do not develop a mycelium; the whole organism is transformed into a reproductive structure at the appropriate time. Other chytrids have slender rhizoids that extend into the substrate and serve as an anchor (Figure 14–13). Some species are parasites of algae, protozoa, and aquatic oomycetes and of the spores, pollen grains, or other parts of plants. The chytrid *Batrachochytrium dendrobatidis,* which causes a thickening of the skin of amphibians, has been implicated in the worldwide die-offs of frogs. Some other chytrid species are saprotrophic on such substrates as dead insects.

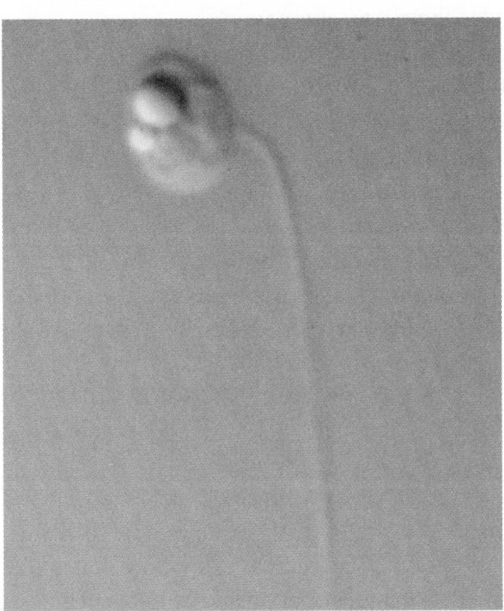

14–12 Chytrid zoospore A uniflagellated zoospore of the chytrid *Polyphagus euglenae.* The chytrids are distinguished from other fungi primarily by their characteristic motile zoospores and gametes.

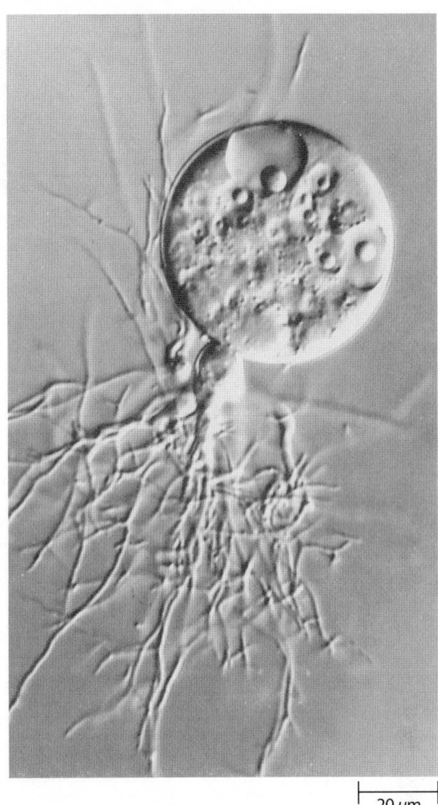

14–13 Chytrid with rhizoids *Chytridium confervae,* a common chytrid, as seen with the aid of differential interference contrast system optics. Note the slender rhizoids extending downward.

Several species of chytrids are plant pathogens, including *Physoderma maydis* and *Physoderma alfalfae,* which cause minor diseases known, respectively, as brown spot of corn and crown wart of alfalfa. *Synchytrium endobioticum* causes a disease of potatoes known as black wart disease, which is a serious problem in regions of Europe and Canada.

Chytrids exhibit a variety of modes of reproduction. Some species of *Allomyces,* for example, have an alternation of isomorphic generations like that shown in Figure 14–14, whereas in other species the alternating generations are heteromorphic—the haploid and diploid individuals do not closely resemble one another. Alternation of generations is characteristic of plants and of many algae but is otherwise found only in *Allomyces,* in one other closely related genus of chytrids, and in a very few heterotrophic protists not considered in this book. In terms of its life cycle, morphology, and physiology, *Allomyces* is the best-known chytrid.

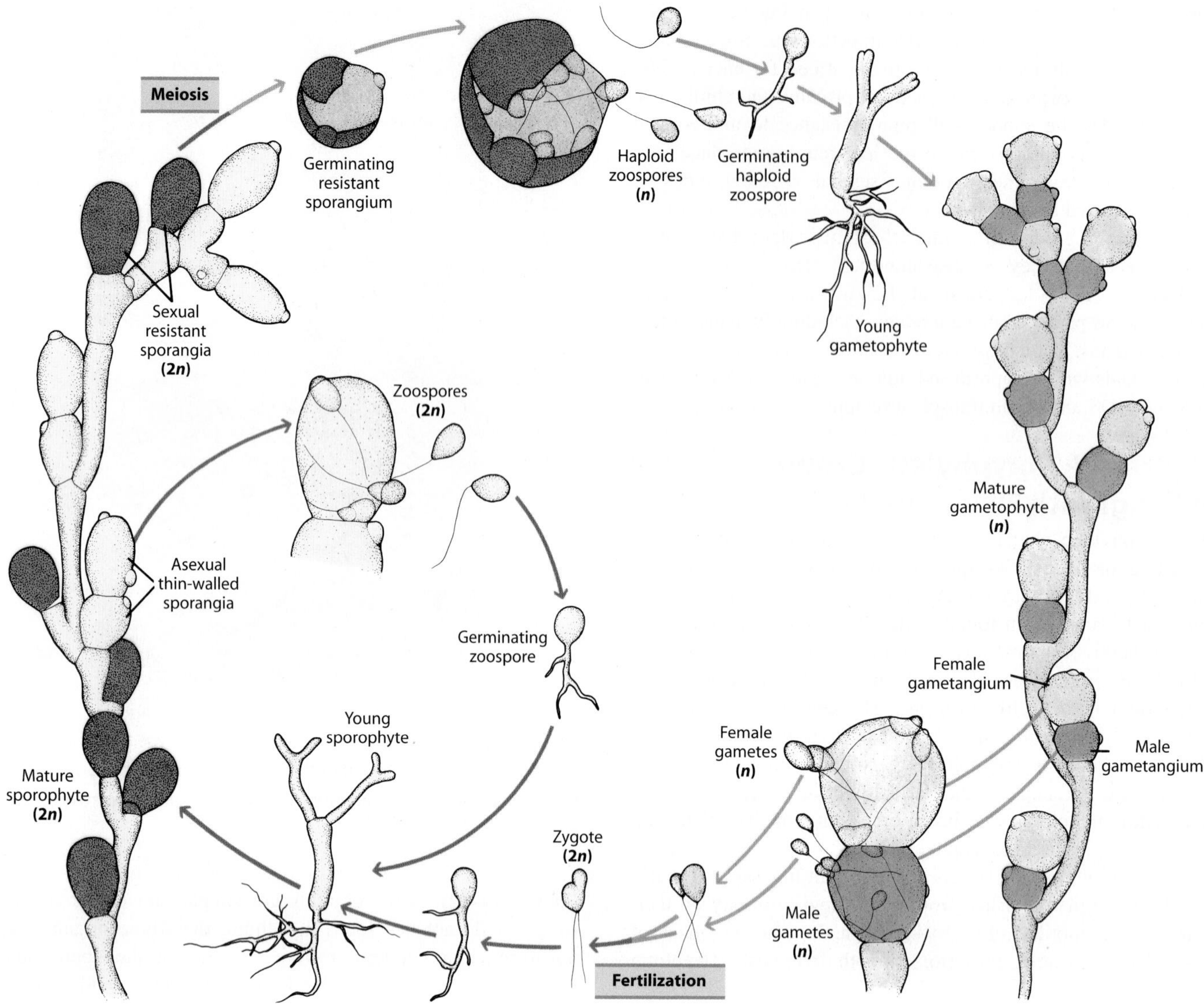

14–14 Life cycle of *Allomyces arbusculus* In the life cycle of the chytrid *A. arbusculus,* there is an alternation of isomorphic generations. The haploid and diploid individuals are indistinguishable until they begin to form reproductive organs. The haploid individuals (gametophytes) produce approximately equal numbers of colorless female gametangia and orange male gametangia (right). The gametes come in two sizes, a condition called anisogamy. The male gametes, which are about half the size of the female gametes, are attracted by sirenin, a hormone produced by the female gametes. The zygote loses its flagella and germinates to produce a diploid individual, the sporophyte, which forms two kinds of sporangia. The first are asexual sporangia, colorless, thin-walled structures that release diploid zoospores—which in turn germinate and repeat the diploid generation. The second kind are sexual sporangia, reddish-brown, thick-walled structures that are able to withstand severe environmental conditions. After a period of dormancy, meiosis occurs in these sexual, resistant sporangia, forming haploid zoospores. The zoospores develop into gametophytes, which produce gametangia at maturity.

Zygomycetes: A Polyphyletic Group of Filamentous Fungi

The zygomycetes, like the chytrids, are a polyphyletic group. At present, considerable controversy exists over the relationships within and between these two basal groups of Fungi. It is a work in progress.

Most species of zygomycetes live on decaying plant and animal matter in the soil, while some are parasites of plants, insects, or small soil animals. A few occasionally cause severe infections in humans and domestic animals. There are approximately 1000 described species of zygomycetes. Most of them have coenocytic hyphae, within which the cytoplasm can often be seen streaming rapidly. Zygomycetes can usually be recognized by their profuse, rapidly growing hyphae, but some also exhibit a unicellular, yeastlike form of growth under certain conditions. Asexual reproduction by means of haploid spores produced in specialized sporangia borne on the hyphae is almost universal in zygomycetes.

One of the best-known and most familiar zygomycete is *Rhizopus stolonifer,* a black mold that forms cottony masses on the surface of moist, carbohydrate-rich foods such as bread or similar substances exposed to air. This organism is also a serious pest of stored vegetables and fruits (Figure 14–2). The life cycle of *R. stolonifer* is illustrated in Figure 14–15. The mycelium is composed of several distinct kinds of haploid hyphae. Most of the mycelium consists of rapidly growing, coenocytic hyphae that grow through the substrate, absorbing nutrients. From these hyphae, arching hyphae called **stolons** are formed. The stolons form rhizoids wherever their tips come into contact with the substrate. From each of these points, a sturdy, erect branch arises, which is called a **sporangiophore** ("sporangium bearer") because it produces a spherical sporangium at its apex. Each sporangium begins as a swelling, into which a number of nuclei flow. The sporangium is eventually isolated by the formation of a septum. The protoplasm within is cleaved, and a cell wall forms around each of the asexually produced nuclei to form spores (sporangiospores). As the sporangium wall matures, it becomes black, giving the mold its characteristic color. With the breaking of the sporangium wall, the spores are liberated, and each spore can germinate to produce a new mycelium, completing the asexual cycle.

The zygomycetes are named for their chief characteristic—the formation of sexually produced resting spores called **zygospores,** which develop within thick-walled structures called **zygosporangia** (Figure 14–16) in species that reproduce sexually. The zygospores often remain dormant for long periods. Sexual reproduction in *R. stolonifer* requires the presence of two physiologically distinct mycelia, designated + and – strains. When two compatible individuals are in close

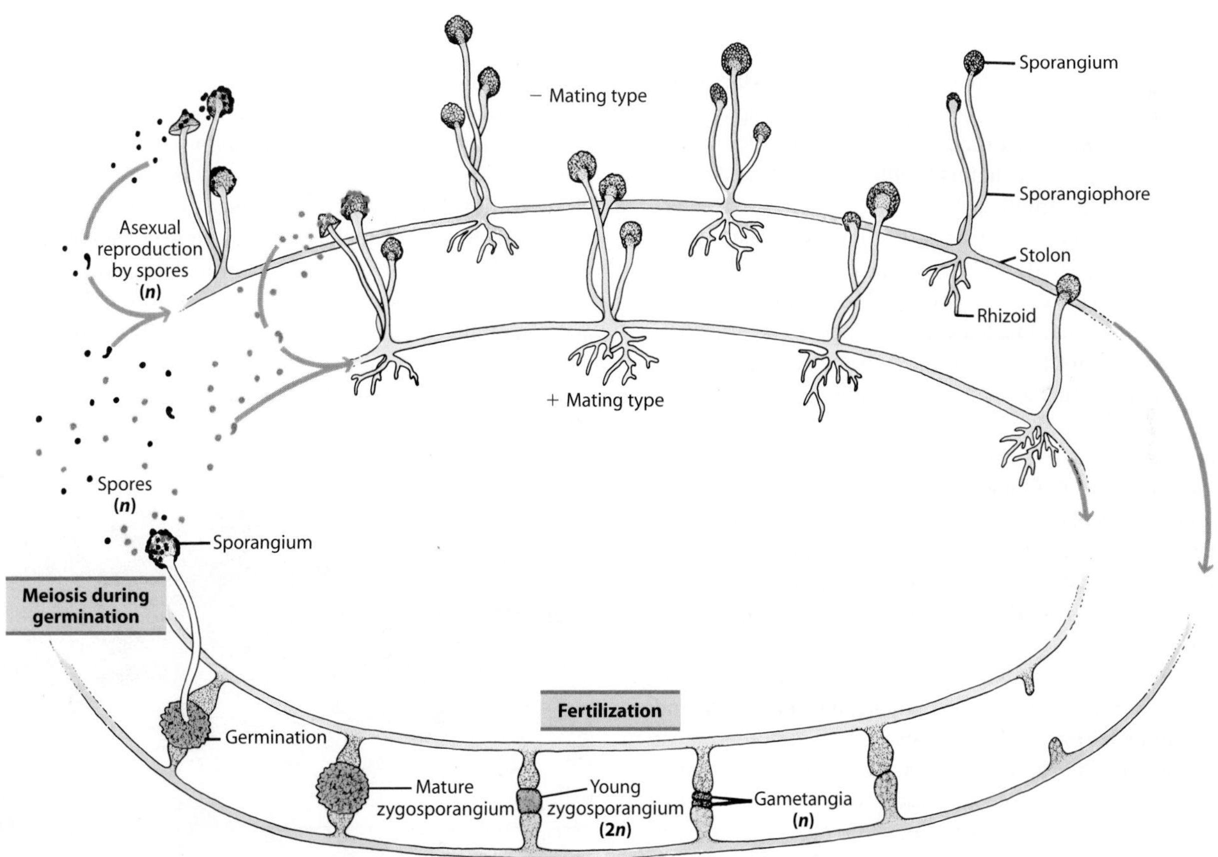

14–15 Life cycle of *Rhizopus stolonifer* In *R. stolonifer,* as in most other zygomycetes, asexual reproduction (upper left) by means of haploid spores is the chief mode of reproduction. Less frequently, sexual reproduction (lower middle) occurs. The spores are formed in sporangia, with black walls that give the mold its characteristic color. In this common species, sexual reproduction involves genetically differentiated mating strains, which have traditionally been labeled + and – types. (Although the two mating strains are morphologically indistinguishable, they are shown here in different colors.) Sexual reproduction results in the formation of a resting spore called a zygospore, which develops within a zygosporangium. The zygosporangia in species of *Rhizopus* develop a thick, rough, black coat, and the zygospore remains dormant, often for several months. The zygospore undergoes meiosis during germination from within the zygosporangium.

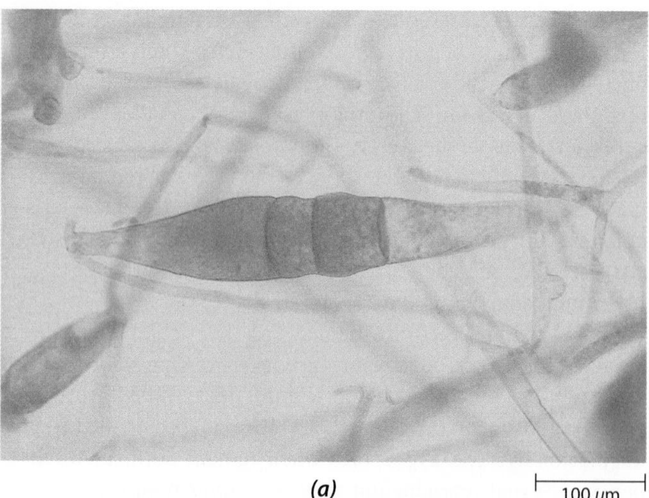

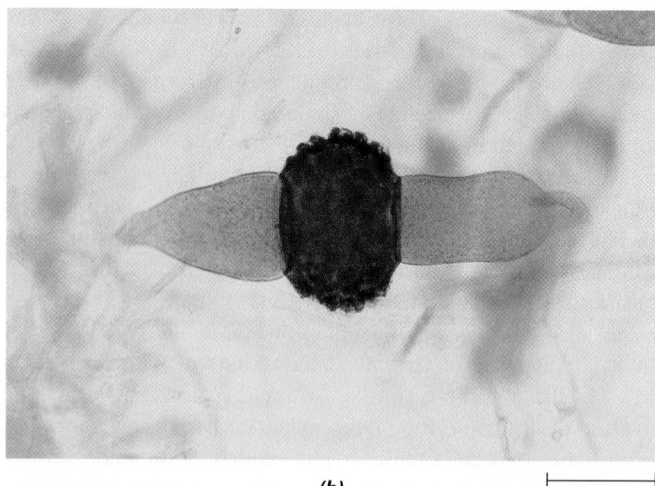

14–16 Reproduction in the zygomycete *Rhizopus stolonifer* ***(a)*** Gametangia, the gamete-producing structures, are in the process of fusing to produce a zygospore. ***(b)*** The zygospore develops within a thick-walled zygosporangium.

proximity, they produce hormones that cause outgrowths of the hyphae to come together and develop into gametangia. Species such as *R. stolonifer* that require + and – strains for sexual reproduction are said to be **heterothallic,** whereas self-fertile species are called **homothallic.**

In any event, the gametangia become separated from the rest of the fungal body by the formation of septa (Figure 14–15). The walls between the two touching gametangia dissolve, and the two multinucleate protoplasts come together. Following plasmogamy (fusion of the two multinucleate gametangia), the + and – nuclei pair, and a thick-walled zygosporangium is produced. Inside the zygosporangium, the paired + and – nuclei fuse (karyogamy) to form diploid nuclei, which develop into a single multinucleate zygospore. At the time of germination, the zygosporangium cracks open and a sporangiophore emerges from the zygospore. Meiosis occurs at the time of germination, so the spores produced asexually within the new sporangium are haploid. When these spores germinate, the cycle begins again.

A group of zygomycetes that has great ecological significance, the order Entomophthorales, is parasitic on insects and other small animals (Figure 14–3b). Species of this order, most of which reproduce by means of a terminal, asexual spore that is discharged at maturity, are increasingly being used in the biological control of insect pests of crops.

Only two genera of zygomycetes commonly cause disease in living plants and living plant tissue. One of them is *Rhizopus,* which causes soft rot of many flowers, fleshy fruits, seeds, bulbs, and corms. The other is *Choanephora,* which causes a soft rot of squash, pumpkin, okra, and pepper.

Glomeromycetes: Phylum Glomeromycota

Although only about 200 species of glomeromycetes have been described thus far, molecular sequences from environmental samples indicate that there is a huge, as yet undescribed diversity of these very important fungi. The glomeromycetes are of major ecological importance. All known species grow in association with roots, forming **mycorrhizas,** which literally means "fungus roots." In fact, glomeromycetes cannot be grown independently of their host plants, therefore they cannot be grown in culture. The type of mycorrhiza formed by glomeromycetes is called an arbuscular mycorrhiza (AM), and the glomeromycetes are commonly referred to as AM fungi. (Mycorrhizas are discussed further on pages 312 to 315.)

The glomeromycetes are widespread and occur in about 80 percent of vascular plants. They have mostly coenocytic

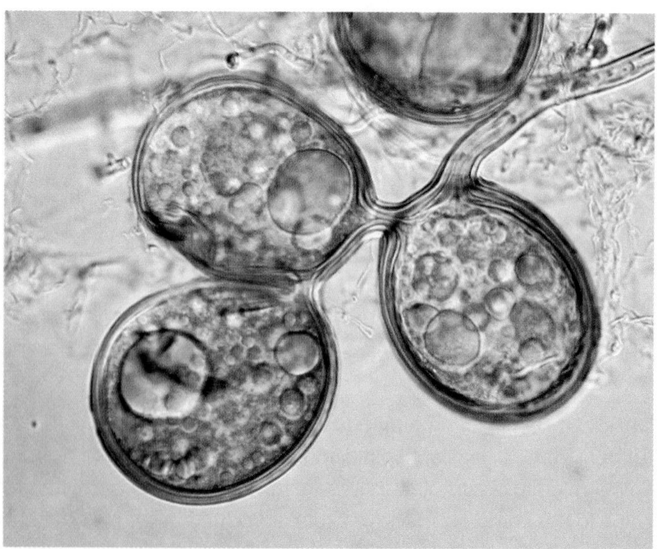

14–17 *Glomus* spores A glomeromycete of the genus *Glomus,* an arbuscular mycorrhizal (AM) fungus, reproduces asexually through the production of large multinucleate spores. Three spores and a portion of a fourth are seen in this micrograph.

hyphae and reproduce only asexually by means of unusually large multinucleate spores (Figure 14–17), which are produced underground.

Ascomycetes: Phylum Ascomycota

The ascomycetes, which comprise about 32,300 described species, include a number of familiar and economically important fungi. Most of the blue-green, red, and brown molds that cause food spoilage are ascomycetes. Ascomycetes are also the cause of some serious plant diseases, including the powdery mildews, which primarily attack leaves; brown rot of stone fruits, caused by *Monilinia fructicola;* Panama disease of bananas, caused by *Fusarium oxysporum,* which wiped out Gros Michel plantations in Central America in the 1960s and now, with a new strain, is threatening the Cavendish bananas in Asia; chestnut blight, caused by *Cryphonectria parasitica,* which was accidentally introduced into North America from Japan; and Dutch elm disease. The latter disease, which devastated entire populations of majestic elm trees in North America and Europe, is caused by *Ophiostoma ulmi* and by the more virulent *Ophiostoma novo-ulmi.* Many yeasts are ascomycetes, as are the edible morels and truffles (Figure 14–18). Many new families and thousands of additional species of ascomycetes—some undoubtedly of great economic importance—await discovery and scientific description.

There are three major groups of Ascomycota: the subphyla Taphrinomycotina, Saccharomycotina, and Pezizomycotina. The Taphrinomycotina and Saccharomycotina are dominated by yeasts. Pezizomycotina is the largest subphylum of Ascomycota; it includes all filamentous ascoma-producing species, with the exception of *Neolecta* of the Taphrinomycotina. Around 40 percent of the Pezizomycotina form lichens (see page 306 and pages 308 to 312).

Ascomycetes have either unicellular (yeasts) or filamentous growth forms. In general, their hyphae have perforated septa (Figure 14–4), which allow the cytoplasm and, rarely, the nuclei to move from one cell to the next. The hyphal cells of the vegetative mycelium may be either uninucleate or multinucleate. Some ascomycetes are homothallic, others heterothallic.

The life cycle of a filamentous ascomycete is shown in Figure 14–19. In most species of this phylum, asexual reproduction takes place through the formation of conidia that are usually multinucleate. The conidia are formed from conidiogenous cells (Figure 14–20), which are borne at the tips of modified hyphae called **conidiophores** ("conidia bearers"). Unlike zygomycetes, which produce spores internally within a sporangium, ascomycetes produce their asexual spores externally as conidia.

Sexual reproduction in ascomycetes always involves the formation of an **ascus** (plural: asci), a saclike structure within which haploid **ascospores** are formed following meiosis. Because the ascus resembles a sac, the ascomycetes commonly are referred to as the **sac fungi.** Both asci and ascospores are unique structures that distinguish the ascomycetes from all other fungi (Figure 14–21a). Ascus formation usually occurs within a complex structure composed of tightly interwoven hyphae—the **ascoma** (plural: ascomata), or ascocarp. Many ascomata are macroscopic. An ascoma may be open and more or less cup-shaped (an *apothecium;* Figure 14–18a), closed and spherical (a *cleistothecium;* Figure 14–21b), or spherical to flask-shaped with a small pore through which the ascospores escape (a *perithecium;* Figure 14–21c). Asci usually develop on an inner surface of the ascoma, a layer called the hymenium, or **hymenial layer** (Figure 14–22 on page 294).

In the life cycle of a filamentous ascomycete, the mycelium grows out from a germinating ascospore on a suitable substrate (Figure 14–19, top left). Soon after, the mycelium begins to reproduce asexually by forming conidia. Many generations

(a)

(b)

14–18 Ascomycetes ***(a)*** Eyelash cup, *Scutellinia scutellata.* ***(b)*** The highly prized, edible ascoma of a black truffle, *Tuber melanosporum.* In the truffles, this spore-bearing structure is produced below ground and remains closed, liberating its ascospores only when the ascoma decays or is broken open by digging animals. Truffles are mycorrhizal (see page 312), mainly on oaks and hazelnuts, and are searched for by specially trained dogs and pigs. The pigs used are sows, because truffles emit a chemical that mimics the pheromone in the saliva of a boar.

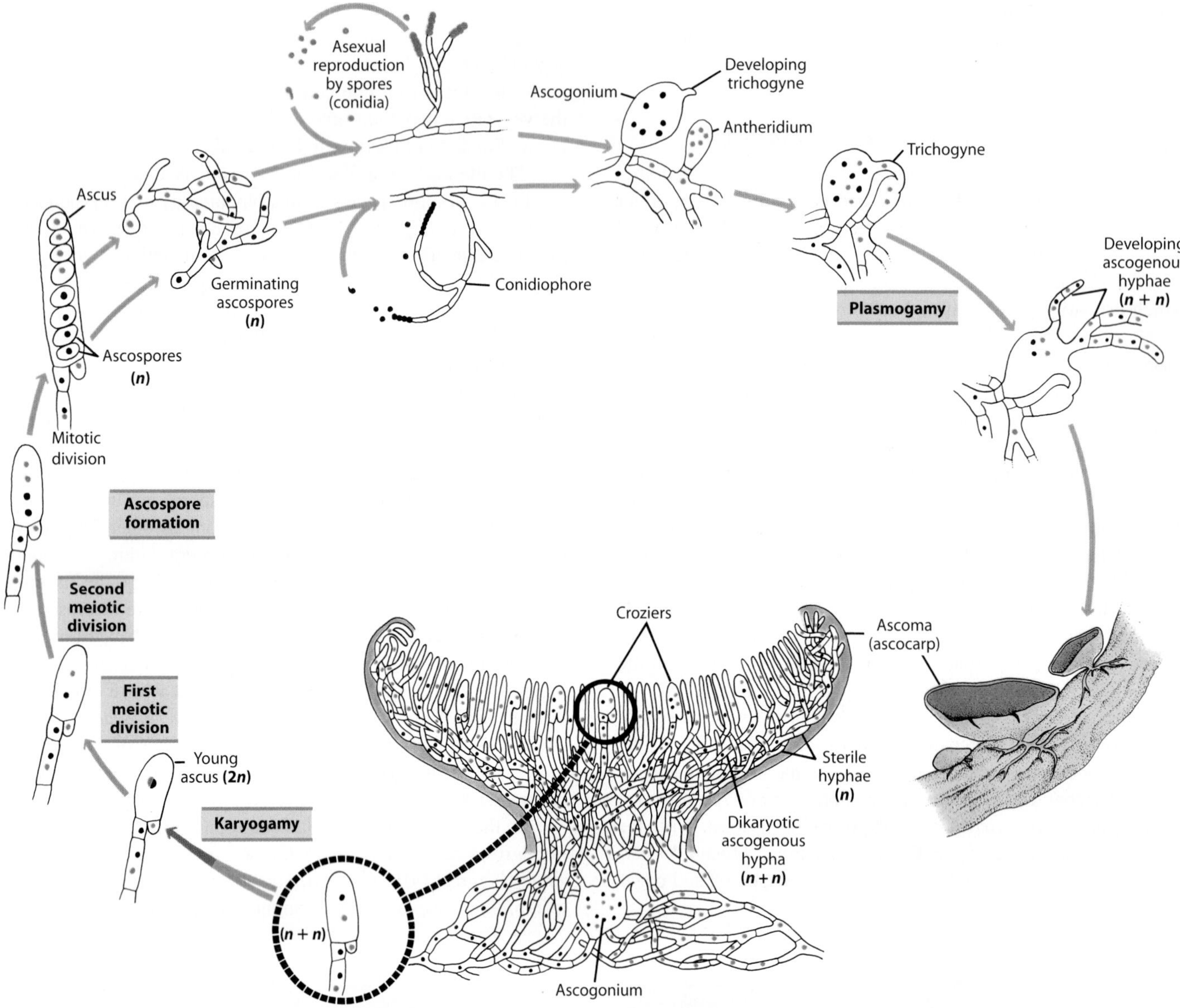

14–19 Typical life cycle of an ascomycete Asexual reproduction (upper left) occurs by way of specialized spores known as conidia, which are usually multinucleate. Sexual reproduction involves the formation of asci and ascospores. Plasmogamy produces fused protoplasts with as yet unfused nuclei, designated $n + n$. Fusion of the nuclei, karyogamy, is followed immediately by meiosis in the ascus, producing ascospores.

of conidia are produced during the growing season, and it is the conidia that are primarily responsible for propagating and disseminating the fungus.

Sexual reproduction, involving ascus formation, occurs on the same mycelium that produces conidia. The formation of multinucleate gametangia called **antheridia** (the "male" gametangia) and **ascogonia** (the "female" gametangia) precedes sexual reproduction. The male nuclei of the antheridium pass into the ascogonium via the **trichogyne,** which is an outgrowth of the ascogonium. Plasmogamy—the fusion of protoplasts—has now taken place. In the ascogonium, the male nuclei may pair with the genetically different female nuclei within the common cytoplasm, but they *do not yet fuse with them.* **Ascogenous hyphae** now begin to grow out of the ascogonium. As these hyphae continue to develop, compatible pairs of nuclei migrate into them, and cell division occurs in such a way that the resultant cells are invariably **dikaryotic,** which means they contain two compatible haploid nuclei. (Monokaryotic cells contain only one nucleus.)

The asci form near the tips of the dikaryotic, ascogenous hyphae. Commonly, the apical cell of the ascogenous hypha forms a hooked tip, or **crozier** (shepherd's crook), which allows the paired nuclei to divide simultaneously, one in the hypha and the other in the hook. (Similar, probably homologous, backwardly directed branches, called clamp connections, occur on many dikaryotic hyphae of basidiomycetes; see Figure 14–28.) Subsequent cell division occurs in such a way that the immature ascus contains a compatible pair of nuclei. These two nuclei

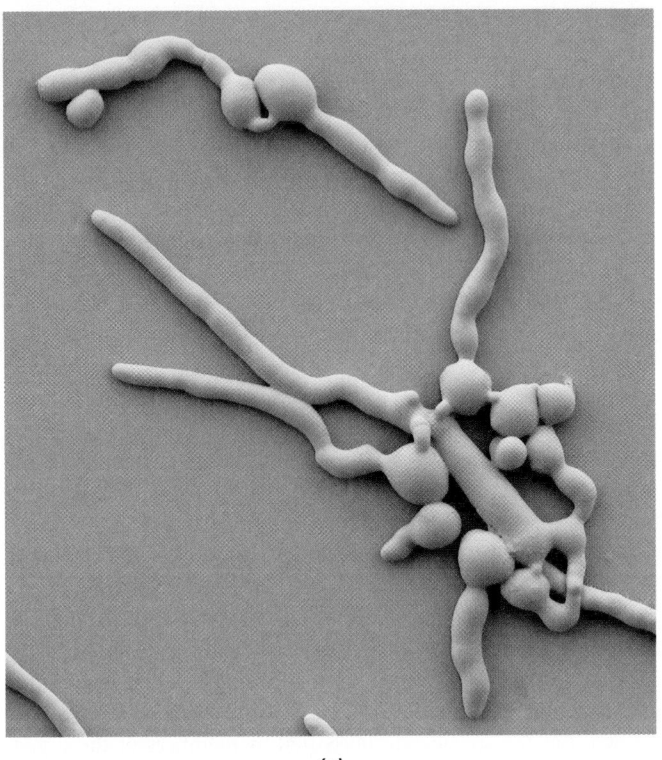

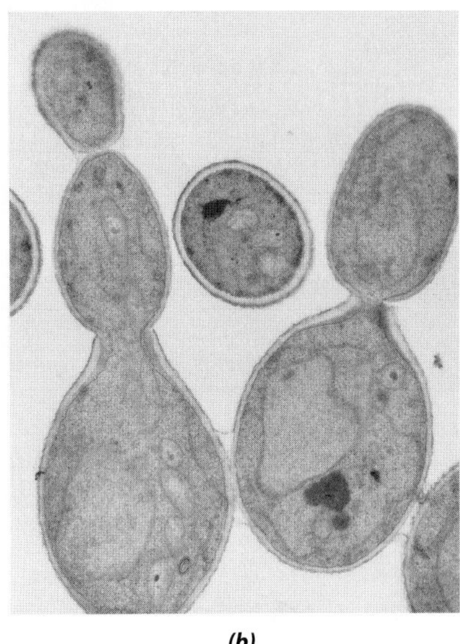

(a) (b)

14–20 Conidia of ascomycetes The characteristic asexual spores of ascomycetes—conidia—are usually multinucleate. These electron micrographs show stages in the formation of conidia. ***(a)*** Scanning electron micrograph of germinating conidia of *Neurospora crassa* at various stages of development. ***(b)*** Transmission electron micrograph of conidia in *Nomuraea rileyi,* which infects the velvetbean caterpillar.

then fuse (karyogamy) to form a diploid nucleus (zygote), the only diploid nucleus in the life cycle of the ascomycetes. Soon after karyogamy, the young ascus begins to elongate. The diploid nucleus then undergoes meiosis, which is generally followed by one mitotic division, producing an ascus with eight nuclei. These haploid nuclei are then cut off in segments of the cytoplasm to form ascospores. In many ascomycetes, the ascus becomes turgid at maturity and finally bursts, releasing its

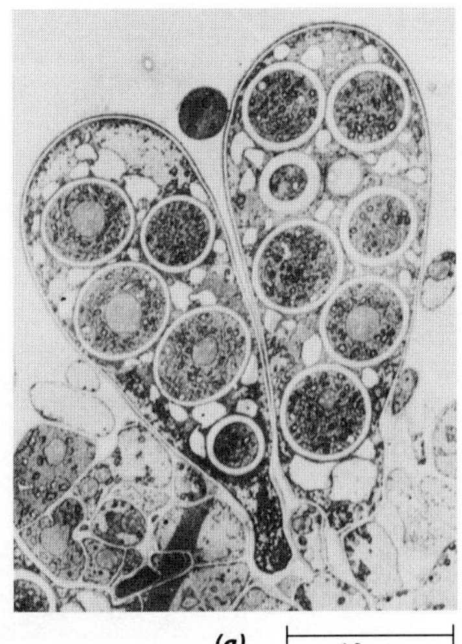

(a) 10 μm

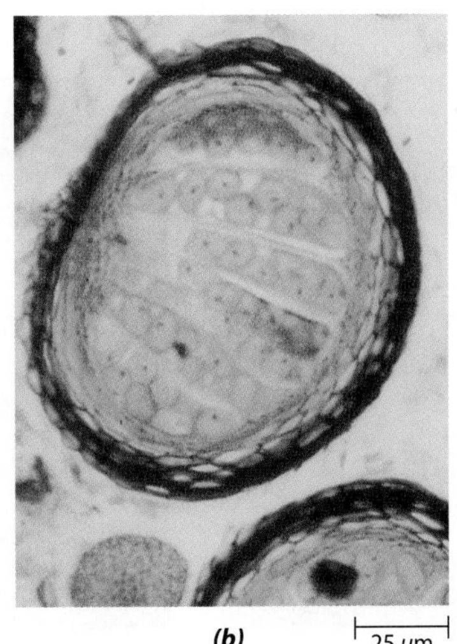

(b) 25 μm

(c) 100 μm

14–21 Asci and ascospores ***(a)*** An electron micrograph showing two asci of *Ascodesmis nigricans* in which ascospores are maturing. ***(b)*** Ascoma of *Erysiphe aggregata,* showing the enclosed asci and ascospores. This completely enclosed type of ascoma is called a cleistothecium. ***(c)*** An ascoma of *Coniochaeta,* showing the enclosed asci and ascospores. Note the small pore at the top. This sort of ascoma, with a small opening, is known as a perithecium.

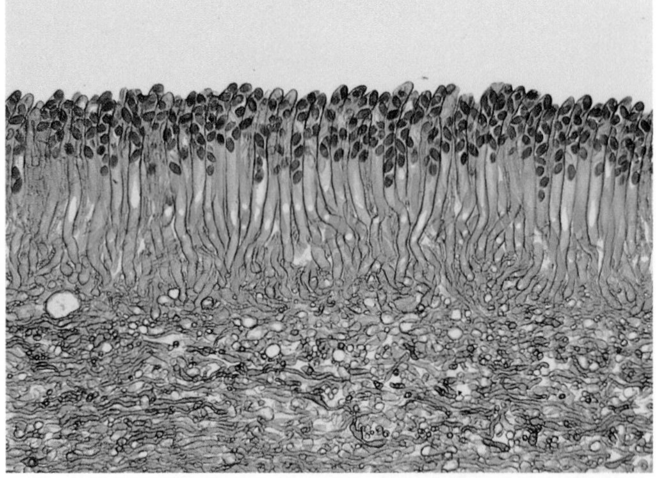

14–22 Hymenium of an ascomycete A stained thin section through the hymenial layer of a morel *(Morchella)*, showing asci with ascospores (dark structures).

ascospores explosively into the air in the cup fungi and some of the perithecium-forming species. The ascospores are generally propelled about 2 centimeters from the ascus, but some species propel them as far as 30 centimeters. This initiates airborne spore dispersal.

In yeasts, each haploid cell is capable of serving as a gamete, and two haploid cells sometimes fuse to form a diploid cell, or zygote, which functions as an ascus (Figure 14–8b). Meiosis occurs within the ascus. Usually, four ascospores are produced per ascus, although in some species meiosis is followed by one or more mitotic divisions, resulting in greater numbers of ascospores. In other yeasts, such as *Saccharomyces cerevisiae,* meiosis is sometimes delayed, and the zygote divides mitotically to form a population of diploid cells that reproduce asexually by budding. Thus, such yeasts have both haploid and diploid budding stages. Diploid cells may ultimately undergo meiosis and revert to the haploid condition. In still other yeasts, the ascospores fuse in pairs immediately after they are formed; in this case, the ascospore is the only haploid cell in the life cycle, which is predominantly diploid.

The Asexual Fungi Are Ascomycetes

Most fungi formerly classified as Deuteromycetes, or Fungi Imperfecti, are asexual forms, or anamorphs, of Ascomycota; a few have affinities with Basidiomycota or zygomycetes. Evidence for a relationship to Ascomycota comes from DNA sequencing data and from similarities in the structure of the mycelium, in the layering of the hyphal cell wall, and in the nature of nuclear division, as seen with an electron microscope.

Some genera contain species in which only the asexually reproducing state (anamorph) is known (the "imperfect" members) or in which features of the sexual reproductive state (teleomorph) are not used as a basis of classification. Thus, for some species of the well-known asexual fungi genera *Penicillium* and *Aspergillus* (Figures 14–23 and 14–24), the sexual state is known, but the species are classified as members of these genera because of their overall resemblance to the other asexual fungi species.

Many fungi exhibit the phenomenon of **heterokaryosis,** in which genetically different nuclei occur together in a common cytoplasm. The nuclei may differ because of mutation or because of the fusion of genetically distinct hyphae. Because genetically different nuclei may occur in different proportions in different parts of a mycelium, these sectors may have different properties.

Among the asexual fungi, as well as some other fungi, haploid nuclei that are genetically different occasionally fuse. Within the resulting diploid nuclei, the chromosomes may associate, recombination may follow, and genetically novel haploid nuclei may form. Restoration of the haploid condition

(a)

(b)

14–23 Asexual fungi *Penicillium* and *Aspergillus* are two of the common genera of asexual fungi. ***(a)*** A culture of *Penicillium notatum,* the original penicillin-producing fungus, showing the distinctive colors produced during growth and spore development. ***(b)*** A culture of *Aspergillus fumigatus,* a fungus that causes respiratory disease in humans. Notice the concentric growth pattern produced by successive "pulses" of spore production.

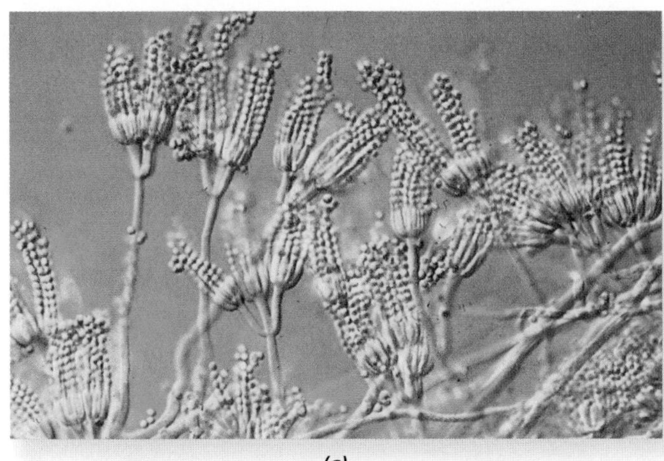
(a)

(b)

14–24 Conidia of asexual fungi The conidia and conidiophores—the specialized hyphae that bear the conidia—of asexual fungi are used in their classification. ***(a)*** *Penicillium* (brushlike conidia) and ***(b)*** *Aspergillus* (tightly clumped conidia arising from the swollen top of the conidiophore). Note the long chains of small, dry conidia in both organisms.

does not involve meiosis. Instead, it results from a gradual loss of chromosomes, a process called *haploidization.* This genetic phenomenon, in which plasmogamy, karyogamy, and haploidization occur in sequence, is known as **parasexuality;** it was discovered in *Aspergillus.* Within the hyphae of this common fungus, there is one diploid nucleus, on average, for every 1000 haploid nuclei. Parasexual cycles may add considerably to genetic and evolutionary flexibility in fungi that lack a true sexual cycle.

The commercial importance of several asexual fungi (for example, *Penicillium* and *Aspergillus)* has already been noted. The ubiquitous soil asexual fungus *Trichoderma* has many commercial applications. For example, cellulose-degrading enzymes produced by *Trichoderma* are used by clothing manufacturers to give jeans a "stone-washed" look. The same enzymes are added to some household laundry detergents to help remove fabric nubs. *Trichoderma* is also used by farmers in the biological control of other fungi that attack crops and forest trees.

Many important antibiotics are produced by asexual fungi. The first antibiotic was discovered by Sir Alexander Fleming, who noted in 1928 that a strain of *Penicillium* that had contaminated a culture of *Staphylococcus* growing on a nutrient agar plate had completely halted the growth of the bacteria. Ten years later, Howard Florey and his associates at Oxford University purified penicillin and later came to the United States to promote the large-scale production of the drug. Production of penicillin increased enormously with escalating demand during World War II. Penicillin is effective in curing a wide variety of bacterial diseases, including pneumonia, scarlet fever, syphilis, gonorrhea, diphtheria, and rheumatic fever.

Not all substances produced by asexual fungi are useful. For example, the **aflatoxins** are potent causative agents of liver cancer in humans. These highly carcinogenic mycotoxins, which show their effects at concentrations as low as a few parts per billion, are secondary metabolites produced by certain strains of *Aspergillus flavus* and *Aspergillus parasiticus.* Both of these fungi frequently grow in stored food products, especially peanuts, maize, and wheat. In tropical countries, aflatoxins have been estimated to contaminate at least 25 percent of the food. Aflatoxins have been detected occasionally in maize harvested in the United States, although strong efforts have been made to detect and destroy contaminated maize.

One group of asexual fungi, the dermatophytes (Gk. *derma,* "skin," and *phyton,* "plant"), are the cause of ringworm, athlete's foot, and other fungal skin diseases. Such diseases are especially prevalent in the tropics. The pathogenic stages of these fungi are asexual, but most of these organisms have now been correlated with species of ascomycetes. Nevertheless, they continue to be classified as asexual fungi on the basis of their disease-causing forms. During World War II, more soldiers had to be sent back from the South Pacific because of skin infections than because of wounds received in battle.

The Basidiomycetes: Phylum Basidiomycota

Phylum Basidiomycota, the last of the six groups of fungi to be discussed, includes some of the most familiar fungi. Among the 22,300 distinct species of this phylum are the mushrooms, toadstools, stinkhorns, puffballs, and shelf fungi, as well as two important groups of plant pathogens, the rusts and smuts. Members of the Basidiomycota play a central role in the decomposition of plant litter, often constituting two-thirds of the living biomass (not including animals) in the soil in temperate regions.

A diagram of the life cycle of a mushroom provides a convenient reference point as we proceed with our discussion

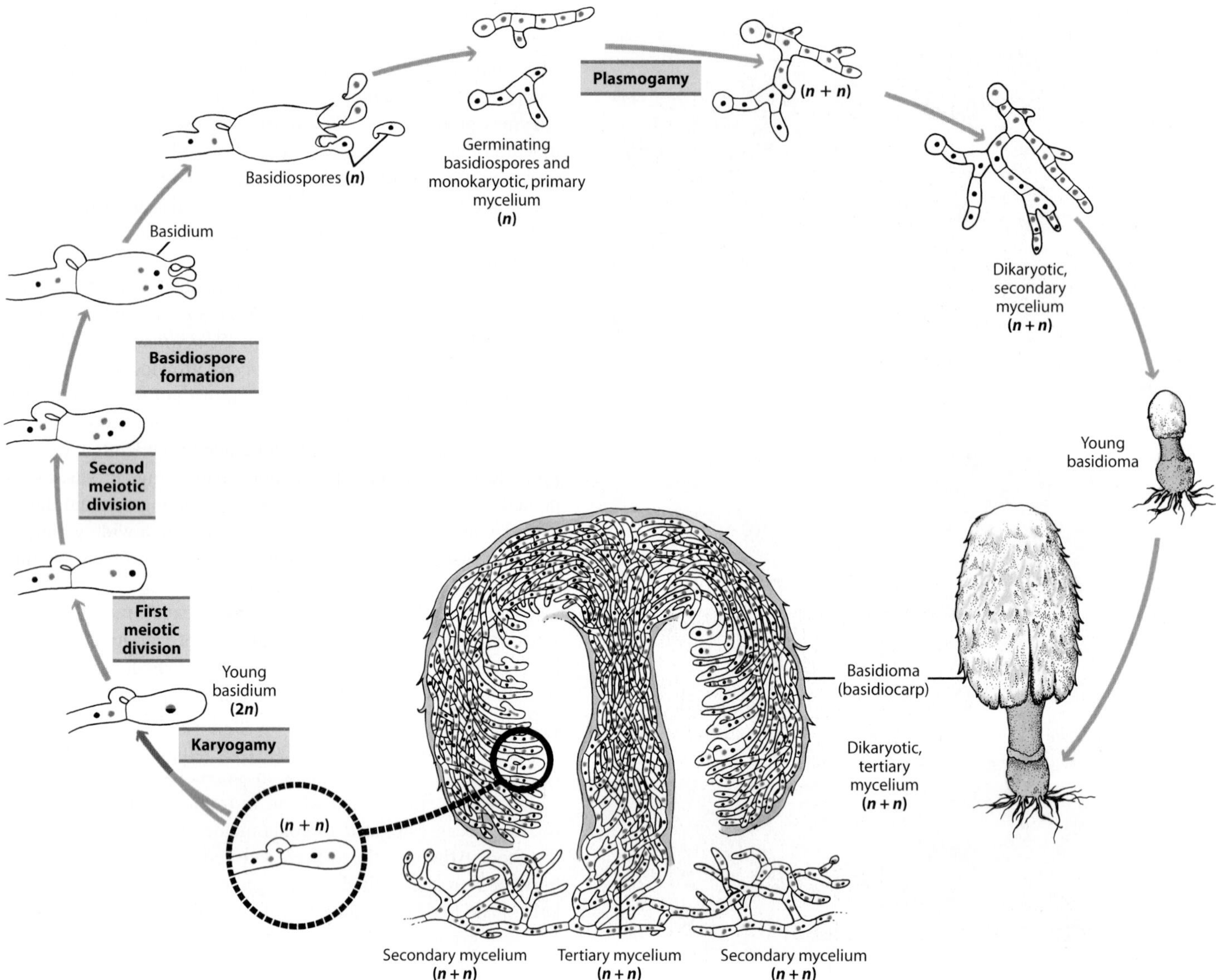

14–25 Life cycle of a mushroom Monokaryotic, primary mycelia are produced from basidiospores (upper left) of this hymenomycete (phylum Basidiomycota). These mycelia give rise to dikaryotic, secondary mycelia, often following the fusion of different mating types, in which case the mycelia are heterokaryotic. Dikaryotic, tertiary mycelia form the basidioma, within which basidia form on the hymenia that line the gills, ultimately releasing up to billions of basidiospores.

(Figure 14–25). The Basidiomycota are distinguished from other fungi by their production of **basidiospores,** which are borne outside a club-shaped spore-producing structure called the **basidium** (plural: basidia) (Figure 14–26). In nature, most basidiomycetes reproduce primarily through the formation of basidiospores.

The mycelium of the Basidiomycota is always septate, but the septa are perforated. In many species, the pore of the septum has an inflated doughnutlike or barrel-shaped margin called a **dolipore.** Any fungus with dolipore septa belongs to the Basidiomycota. On either side of the dolipore may be membranous caps called **parenthesomes,** so named because in profile they resemble a pair of parentheses (Figure 14–27). Many basidiomycetes, including the rusts and smuts, have septa that resemble those of the ascomycetes.

In most species of Basidiomycota, the mycelium passes through two distinct phases—monokaryotic and dikaryotic—during the life cycle. When it germinates, a basidiospore produces a mycelium that, initially, may be multinucleate. Septa soon form, however, and the mycelium is divided into **monokaryotic** (uninucleate) cells. This mycelium also is referred to as the **primary mycelium.** Commonly, the dikaryotic mycelium is produced by fusion of monokaryotic hyphae from different mating types (in which case it is heterokaryotic), resulting in formation of a **dikaryotic** (binucleate), or **secondary, mycelium,** since karyogamy does not immediately follow plasmogamy.

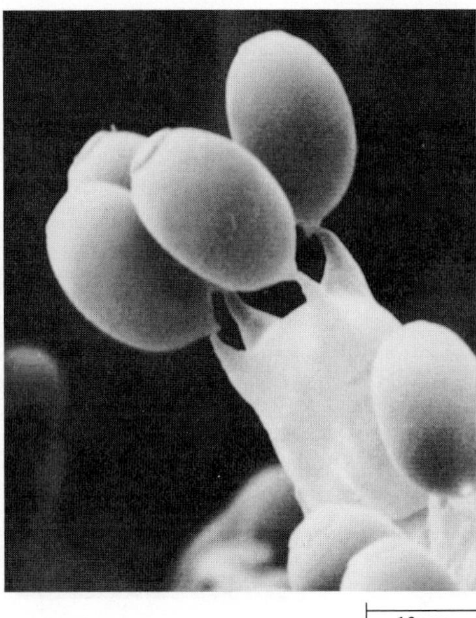

14–26 Basidiospores Scanning electron micrograph of basidiospores of the inky cap mushroom, *Coprinus cinereus.* The micrograph shows the top of a basidium, with four basidiospores, each attached to a stalklike sterigma.

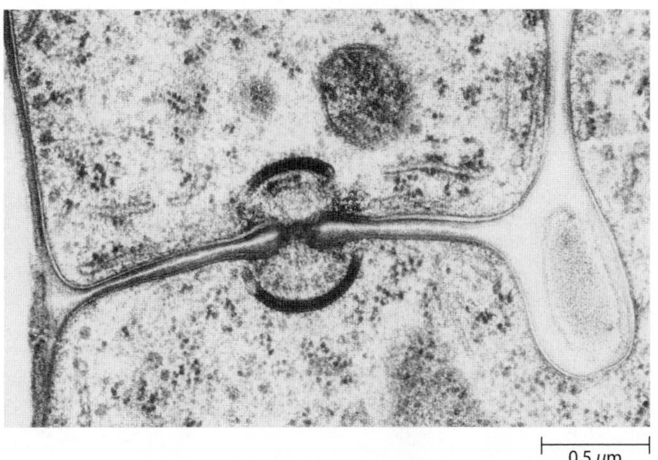

14–27 A dolipore septum Dolipore septa are common in Agaricomycotina, as shown here in *Auricularia auricula,* a common wood-decay species. Each dolipore septum is perforated by a pore. Parenthesomes are visible above and below the dolipore.

The apical cells of the dikaryotic mycelium usually divide by the formation of **clamp connections** (Figure 14–28). These clamp connections, which ensure the allocation of one nucleus of each type to the daughter cells, are found only in the Basidiomycota, although many of the species may not form them. As noted earlier, clamp connections and the croziers of the ascomycetes are probably homologous structures.

The mycelium that forms the **basidiomata** (singular: basidioma)—fleshy, basidiospore-producing bodies, such as mushrooms and puffballs—is also dikaryotic. It is called the **tertiary mycelium.** Formation of the basidiomata may require light and low CO_2 levels, both of which signal to the mycelium that it is "outside" its substrate. As it forms the basidiomata, the tertiary mycelium becomes differentiated into specialized hyphae that play different functions within the basidiomata.

The phylum Basidiomycota includes three subphyla: Agaricomycotina, Pucciniomycotina, and Ustilaginomycotina. The Agaricomycotina include all fungi that produce basidiomata,

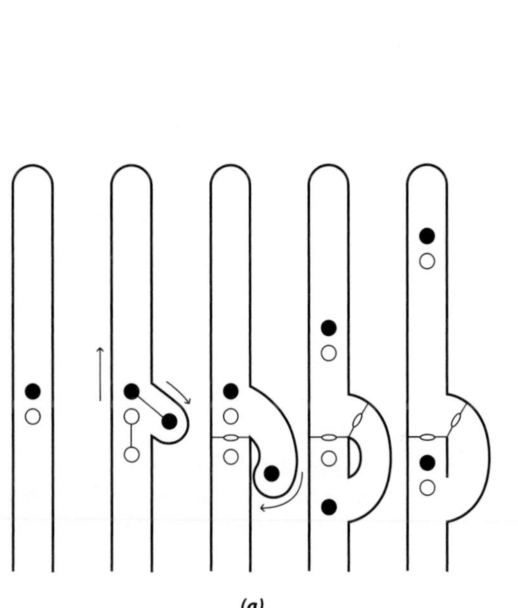

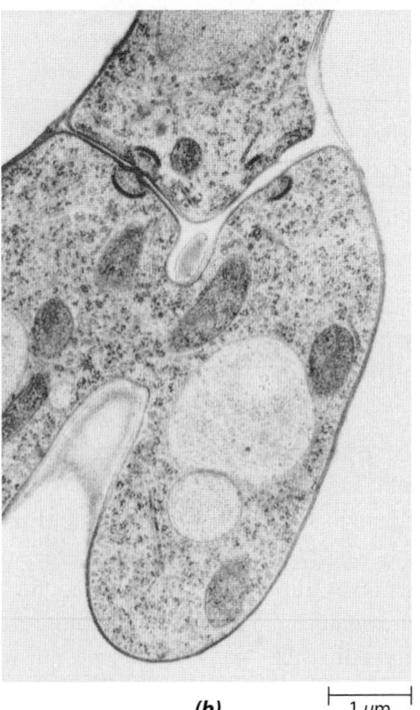

14–28 Clamp connections ***(a)*** In the Agaricomycotina, dikaryotic hyphae characteristically are distinguished by the formation of clamp connections during cell division in the tips of hyphae. Clamp connections presumably ensure the proper distribution of the two genetically distinct types of nuclei in the basidioma. Two septa form to divide the parent cell into two daughter cells. ***(b)*** Electron micrograph of a clamp connection and characteristic septa in a hypha of *Auricularia auricula.*

14–29 Hymenomycetes ***(a)*** The fly agaric, *Amanita muscaria.* The mushrooms are at various stages of growth. Among the characteristics of this genus of mushrooms, many members of which are poisonous, are the scales on the cap, the ring on the stalk, and the cup, or volva, around its base. ***(b)*** *Polyporus arcularius,* a polypore fungus. The polypores lack the gills found in most kinds of mushrooms. In *P. arcularius,* spores are shed through diamond-shaped pores. ***(c)*** Shelf fungi, such as *Ganoderma applanatum,* are wood-rotting fungi. ***(d)*** An edible tooth fungus, *Hericium coralloides.* The hymenium, an outer spore-bearing layer of basidia, is borne on the surface of the downwardly directed teeth.

such as mushrooms, shelf fungi, and puffballs. Neither the Pucciniomycotina (the rusts) nor the Ustilaginomycotina (the smuts) form basidiomata. Instead, these fungi produce their spores in **sori** (singular: sorus).

Subphylum Agaricomycotina Includes the Hymenomycetes and the Gasteromycetes

The subphylum Agaricomycotina includes the edible and poisonous mushrooms, coral fungi, tooth fungi, and shelf or bracket fungi (Figure 14–29). These Agaricomycotina commonly are referred to as "hymenomycetes," because they produce their basidiospores on a distinct fertile layer, the hymenium, which is exposed before the spores are mature (Figure 14–30). Another informal group of Agaricomycotina, the "gasteromycetes" (literally, the "stomach fungi"), includes forms in which no distinct hymenium is visible at the time the basidiospores are released. (Note that the hymenomycetes and gasteromycetes are not taxonomic groups.) Among the more familiar gasteromycetes are the stinkhorns, earthstars, false puffballs, bird's-nest fungi, and puffballs (see Figure 14–33). Most Agaricomycotina have club-shaped, aseptate (internally undivided) basidia, usually bearing four basidiospores, each on a minute projection called a **sterigma** (plural: sterigmata) (Figures 14-26 and 14-30).

The structure that one recognizes as a mushroom or toadstool is the basidioma (Figure 14–25). ("Mushroom" is sometimes popularly used to designate the edible forms of basidiomata, and "toadstool" is used to designate the inedible ones, but mycologists do not recognize such a distinction and use only the term "mushroom." In this book, all such forms are referred to as mushrooms; this does not mean that they are all edible.) The

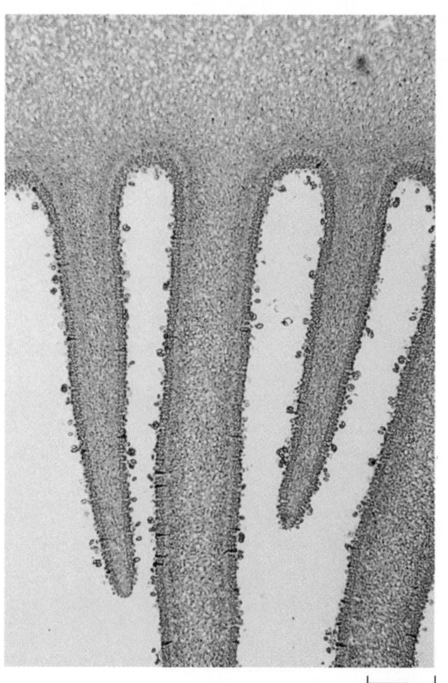

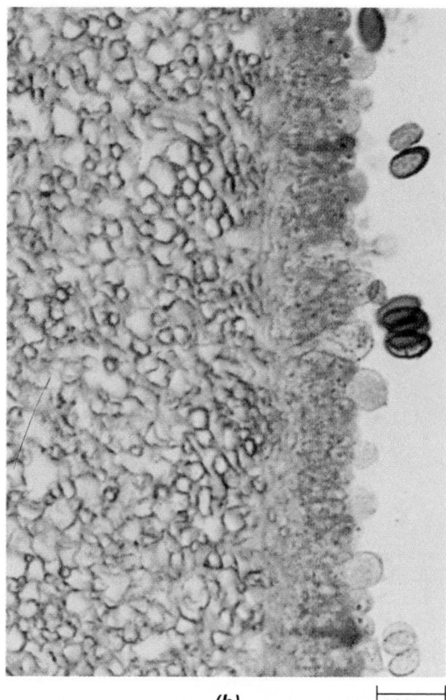

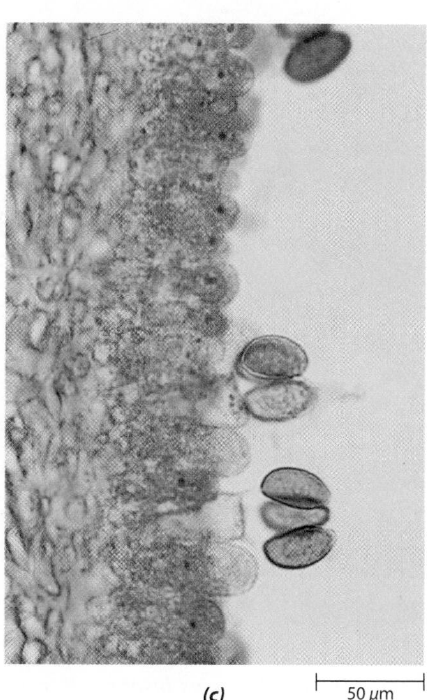

(a) (b) (c)

14–30 Hymenium of an Agaricomycotina Stained sections through the gills of *Coprinus*, a common mushroom, at progressively higher magnifications. The hymenial layer is stained darker in each of these preparations. ***(a)*** Outlines of some of the gills. ***(b)*** Developing basidia and basidiospores in a section through the hymenial layer. ***(c)*** Nearly mature basidiospores attached to basidia by sterigmata.

mushroom generally consists of a **pileus,** or **cap,** that sits atop a **stipe,** or **stalk.** The masses of hyphae in the basidiomata usually form distinct layers. Early in its development—the "button" stage—the mushroom may be covered by a membranous tissue that ruptures as the mushroom enlarges. In some genera, remnants of this tissue are visible as patches on the upper surface of the cap and as a cup, or volva, at the base of the stipe (Figure 14–29a). In many hymenomycetes, the lower surface of the cap consists of radiating strips of tissue called **gills** (Figure 14–30), which are lined with the hymenium. In other hymenomycetes, the hymenium is located elsewhere; for example, in the tooth fungi (Figure 14–29d) the hymenium covers downwardly directed spines. In the boletes and polypores (Figure 14–29b), the hymenium lines vertical tubes that open as pores.

As mentioned previously, in hymenomycetes, basidia form in well-defined hymenia that are exposed before the basidiospores are mature. Each basidium develops from a terminal cell of a dikaryotic hypha. Soon after the young basidium enlarges, karyogamy occurs. This is followed almost immediately by meiosis of each diploid nucleus, resulting in the formation of four haploid nuclei (Figure 14–25). Each of the four nuclei may then migrate into a sterigma, which enlarges at its tip to form a uninucleate, haploid basidiospore. In many basidiomycetes, meiosis is followed by mitosis, yielding eight haploid nuclei. The timing and location of the post-meiotic mitosis vary, as does the fate of the nuclei. Post-meiotic mitosis may occur in the basidium, the sterigmata, or the young spore. When post-meiotic mitosis occurs in the basidium or sterigmata, one nucleus enters each spore and the other four nuclei abort, yielding uninucleate spores. When post-meiotic mitosis occurs in the young spores, four of the daughter nuclei may migrate back into the basidium, where they abort, resulting in uninucleate spores, or all eight nuclei may remain in the spores, producing binucleate spores.

At maturity, the basidiospores are discharged forcibly from the basidioma but depend on the wind for dispersal. The reproductive capacity of a single mushroom is tremendous, with billions of spores produced by a single basidioma. This reproductive capacity is essential because each species occupies a narrow niche in the environment, and the chance that a given spore will land on a substrate suitable for germination and growth is slim.

In relatively uniform habitats, such as lawns and fields, the mycelium from which mushrooms are produced spreads underground, growing downward and outward and forming a ring of mushrooms on the edge of the colony. This ring may grow as large as 30 meters in diameter. In an open area, the mycelium expands evenly in all directions, dying at the center and producing basidiomata at the outer edges, where it grows most actively because this is the area where nutritive material in the soil is most abundant. As a consequence, the mushrooms appear in rings, and as the mycelium grows, the rings become larger. Such circles of mushrooms are known in European folk legends as "fairy rings" (Figure 14–31).

The best-known hymenomycetes are the gill fungi, including *Agaricus campestris,* the common field mushroom. The

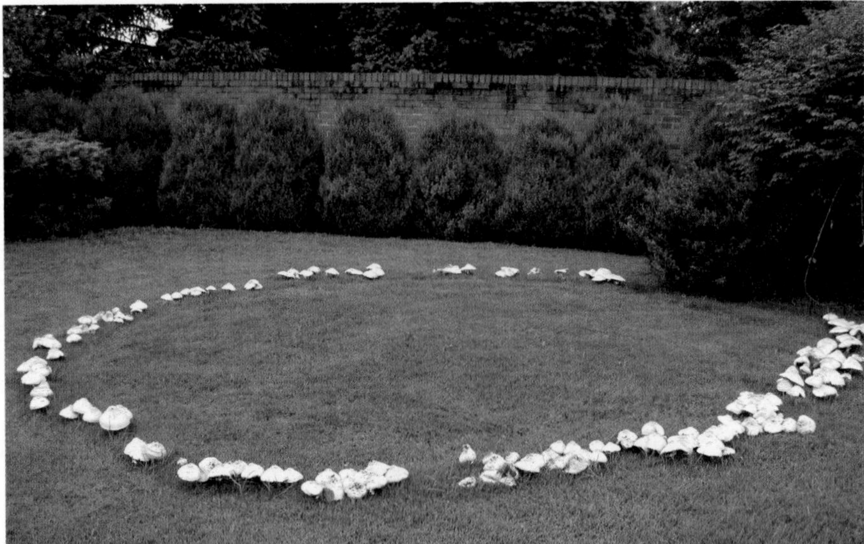

14–31 Fairy ring Seen here is a "fairy ring" formed by the mushroom *Marasmius oreades.* Some fairy rings are estimated to be up to 500 years old. Because of the exhaustion of key nutrients, the grass immediately inside such a ring is often stunted and lighter green than the grass outside the ring.

closely related *Agaricus bisporus* (see Table 12–1 on page 237) is one of many mushrooms now cultivated commercially. It is grown in more than 100 countries, and the world crop is valued at over $1 billion. *Agaricus bisporus,* together with the Asian shiitake mushroom, *Lentinula edodes,* makes up about 86 percent of the world's mushroom crop. Other mushrooms are also being cultivated, and some are gathered in large quantities in nature. An alarm has been sounded that mushrooms are declining, both in the total number of species and in the quantity of individual species, in forests of Europe and the Pacific Northwest of the United States. If this trend continues, it could result in a dramatic decline in the health of the trees that are dependent on mycorrhizal fungi for nutrient uptake, as well as in a disruption of the nutrient cycle in the ecosystem. The cause of the decline has not been identified, but pollutants such as nitrates are suspected.

The gill fungi also include many poisonous mushrooms. The genus *Amanita* includes the most poisonous of all mushrooms, as well as some that are edible. Just a few bites of the "destroying angel," *Amanita virosa,* can be fatal. Other Agaricomycotina contain chemicals that cause hallucinations in humans who eat them (Figure 14–32).

(a)

(b)

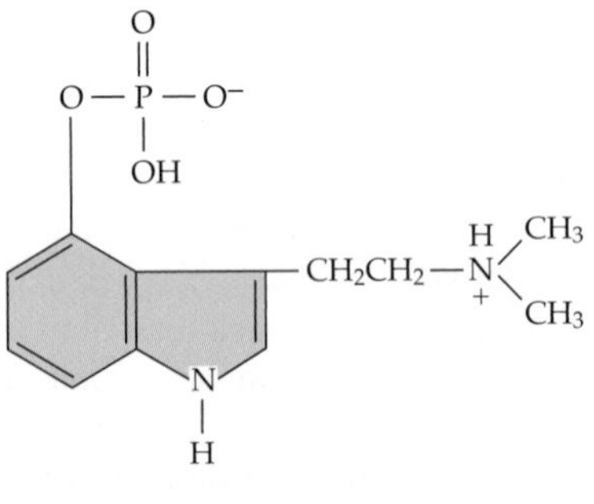

(c) **Psilocybin**

14–32 Hallucinogenic mushrooms Mushrooms figure prominently in the religious ceremonies of several groups of Indians in southern Mexico and Central America. The Indians eat certain hymenomycetes for their hallucinogenic qualities. ***(a)*** One of the most important of these mushrooms is *Psilocybe mexicana,* shown here growing in Guadalajara, Jalisco, Mexico. ***(b)*** The shaman María Sabina is shown eating *Psilocybe* in the course of a midnight religious ceremony. ***(c)*** Psilocybin, the chemical responsible for the colorful visions experienced by those who eat these "sacred" mushrooms, is a structural analog of LSD and mescaline (see Figure 20-31a).

(a)

(b)

14–33 Gasteromycetes ***(a)*** Puffballs, *Calostoma cinnabarina*. Raindrops cause the thin outer layer, or peridium, of the puffball to dimple, forcing out a puff of air, mixed with spores, through the opening. ***(b)*** The veiled, or netted, stinkhorn, *Dictyophora duplicata*. The basidiospores are released in a foul-smelling, slimy mass at the top of the fungus. Flies, attracted by the smell of rotting meat, visit the fungus, expecting a meal but getting only a spore mass. They fly off to richer sources of rotting material, spreading the spores, which adhere to their legs and bodies in great numbers. ***(c)*** White-egg bird's-nest fungus, *Crucibulum laeve*. The round structures ("eggs") of the basidiomata (the "nests") of these fungi contain the basidiospores, which are splashed out and dispersed by raindrops. ***(d)*** Earthstar, *Geastrum saccatum,* showing one fully opened individual and two others in earlier stages of development. The outer layers of the peridium fold back in this genus, raising the spore mass above the dead leaves.

(c)

(d)

The gasteromycetes (Figure 14–33) are characterized by the fact that their basidiospores mature inside the basidiomata and are not discharged forcibly from them. The basidiomata of the gasteromycetes possess a distinct outer covering, called the **peridium,** that varies from almost papery thin in some species to thick and rubbery or leathery in others. In some species, the peridium opens naturally when the spores are mature; in others, it remains permanently closed, with the spores being liberated only after it has been ruptured through the action of an external agent.

Stinkhorns (Figure 14–33b) have a remarkable morphology. These fruiting bodies develop underground as leathery, egg-shaped structures. At maturity, they differentiate into an elongating stalk and a pileus, or cap, bearing a **gleba,** which is the fertile portion of the basidioma. The gleba forms a foul-smelling, sticky mass of spores that attracts flies and beetles, which disperse the spores.

The puffballs are familiar gasteromycetes. At maturity, the interior of a puffball dries up, and it releases a cloud of spores when struck (Figure 14–33a). Some giant puffballs may grow to 1 meter in diameter and may produce several trillion basidiospores. The bird's-nest fungi (Figure 14–33c) begin their development like puffballs, but the disintegration of much of their internal structure leaves them looking like miniature bird's nests.

Subphylum Pucciniomycotina Consists Largely of Rust Fungi

Of the 8000 species of Pucciniomycotina that have been described, including yeasts, saprotrophs, and parasites of plants, animals, and fungi, the overwhelming majority (about 90 percent) are rust fungi. Unlike the Agaricomycotina, few rusts form basidiomata. As mentioned previously, their spores occur in masses called sori (Figure 14–34). They do, however, form dikaryotic hyphae and basidia, which are septate. As plant pathogens, the rusts are of tremendous economic importance, causing billions of dollars of damage to crops throughout the world each year. Among the more serious rust diseases are black stem rust of cereals, white pine blister rust, coffee rust, cedar-apple rust (page 278), peanut rust, wheat stripe rust, and soybean rust. The latter, which is caused by *Phakopsora pachyrhizi,* was detected in the United States in 2004. First reported in Japan in 1902, soybean rust has gradually been spreading throughout the world. In China and other Asian countries, it sometimes slashes soybean yields by as much as 80 percent.

The life cycles of many rusts are complex, and these pathogens are a constant challenge to **plant pathologists** whose task it is to keep them under control. An example of a rust life cycle is provided by *Puccinia graminis,* the cause of black stem rust of wheat, the most widely cultivated crop in the world. (Wheat provides about one-fifth of the calories consumed by humans worldwide.) Numerous strains of *P. graminis* exist, and, in addition to wheat, they parasitize other cereals such as barley, oats, and rye, as well as various species of wild grasses. As early as A.D. 100, Pliny described wheat rust as "the greatest pest of the crops." Plant pathologists combat black stem rust largely by breeding resistant wheat varieties, but mutation and recombination in the rust typically make any advantage short-lived. For four decades, wheat's resistance to *P. graminis* was provided primarily by a single gene, *Sr31,* which was discovered by Norman Borlaug, one of the architects of the Green Revolution (see Chapter 21). Then, in 1999, a new race of *P. graminis* capable of overcoming *Sr31* was discovered in Uganda. Known as Ug99, it has been characterized as "the most virulent" in 50 years. By 2010, Ug99 had spread to South Africa and as far as Iran. It is feared that the fungus will eventually appear in Punjab, an area of Pakistan and northern India that is one of the world's major suppliers of wheat. Plant pathologists and breeders are urgently working to identify wheat plants with genetic resistance to Ug99.

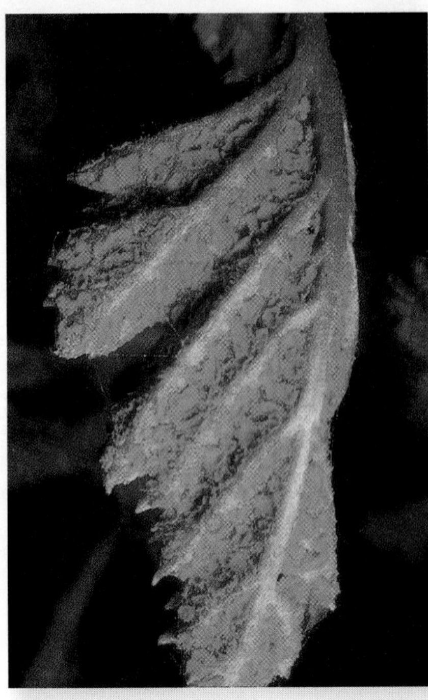

14–34 Rust sori Orange sori of the blister rust *Kuehneola uredinis* on a blackberry leaf, photographed in San Mateo County, California.

Puccinia graminis is **heteroecious,** that is, it requires two different hosts to complete its life cycle (Figure 14–35). **Autoecious** parasites, in contrast, require only one host. *Puccinia graminis* can grow indefinitely on its grass host, but there it reproduces only asexually. For sexual reproduction to take place, the rust must spend part of its life cycle on barberry *(Berberis)* and part on a grass. One method of attempting to eliminate this rust has been to eradicate barberry bushes. For example, the crown colony of Massachusetts passed a law ordering "whoever . . . hath any barberry bushes growing in his or their land . . . shall cause the same to be extirpated or destroyed on or before the thirteenth day of June, A.D. 1760."

Infection of the barberry occurs in the spring (Figure 14–35, upper left), when uninucleate basidiospores infect the plant by forming haploid mycelia, which first develop **spermogonia,** primarily on the upper surfaces of the leaves. The form of *P. graminis* that grows on barberry consists of separate + and – strains, so the basidiospores and the spermogonia derived from them are either + or –. Each spermogonium is a flask-shaped pustule lined by cells that form sticky, uninucleate cells called **spermatia.** The mouth of the spermogonium is surrounded by a brush of orange, stiff, unbranched, pointed hairs, the **periphyses,** which hold droplets of sugary, sweet-smelling nectar. The nectar, which is attractive to flies, contains the spermatia. Found among the periphyses are branched **receptive hyphae.** Flies visit the spermogonia and feed on the nectar. In moving from one spermogonium to another, on the barberry, the flies transfer spermatia. If a + spermatium of one spermogonium comes in contact with the – receptive hypha of another spermogonium, or vice versa, plasmogamy occurs and dikaryotic hyphae are produced. Aecial initials are produced from the dikaryotic hyphae that extend downward from the spermogonium. **Aecia** are then formed primarily on the lower surface of the leaf, where they produce chains of **aeciospores.** The dikaryotic aeciospores must then infect the wheat; they will not grow on barberry.

The first external manifestation of infection on the wheat is the appearance of rust-colored, linear streaks on the leaves and stems (the red stage). These streaks are **uredinia,** containing unicellular, dikaryotic **urediniospores.** Urediniospores are produced throughout the summer and reinfect the wheat; they are also the primary means by which the wheat rust has spread throughout the wheat-growing regions of the world. In late summer and early fall, the red-colored sori gradually darken and become **telia** with two-celled dikaryotic **teliospores** (the black stage). The teliospores are overwintering spores, which infect neither wheat nor barberries. Shortly after they are formed, karyogamy takes place, and the teliospores overwinter in the diploid state. Meiosis actually begins immediately but is arrested in prophase I. In

PREDACEOUS FUNGI

Among the most highly specialized of the fungi are the predaceous fungi, which have developed a number of mechanisms for capturing small animals they use as food. Some gilled fungi attack and consume the small roundworms known as nematodes. The oyster mushroom, *Pleurotus ostreatus*, for example, grows on decaying wood ***(a), (b).*** Its hyphae secrete a substance that anesthetizes nematodes, after which the hyphae envelop and penetrate these tiny worms. The fungus apparently uses them primarily as a source of nitrogen, thus supplementing the low levels of nitrogen that are present in wood.

Some of the microscopic asexual fungi secrete on the surface of their hyphae a sticky substance in which passing protozoa, rotifers, small insects, or other animals become stuck ***(c).*** More than 50 species of this group trap or snare nematodes. In the presence of these roundworms, the fungal hyphae produce loops that swell rapidly, closing the opening like a noose when a nematode rubs against its inner surface. Presumably, stimulation of the cell wall increases the amount of osmotically active material in the cell, causing water to enter the cells and increase their turgor pressure. The outer wall then splits, and a previously folded inner wall expands as the trap closes. Hyphal trapping rings, together with small nematodes, have been found in amber estimated to be about 100 million years old. Predaceous fungi have been around for a long time.

(a)

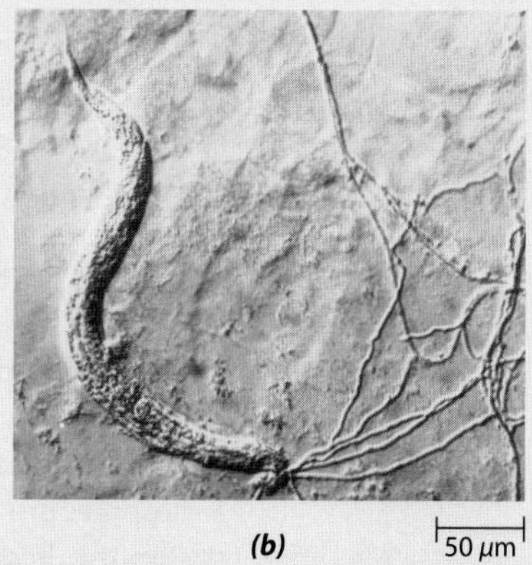

(b)

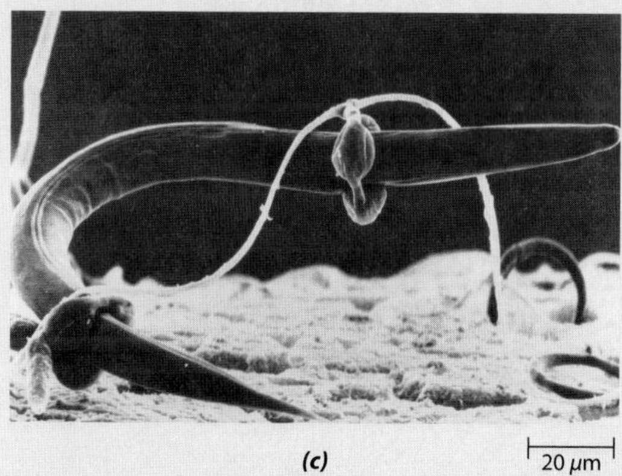

(c)

(a) The oyster mushroom, *Pleurotus ostreatus.* ***(b)*** Hyphae of the oyster mushroom, which produce a substance that anesthetizes prey, are seen here converging on the mouth of an immobilized nematode. ***(c)*** The predaceous conidial fungus *Arthrobotrys anchonia* has trapped a nematode. The trap consists of rings, each consisting of three cells, which, when triggered, swell to about three times their original size in just 0.1 second and strangle the nematode. Once the worm has been trapped, fungal hyphae grow into its body and digest it.

early spring, prior to germination, meiosis is completed in the short, curved basidia that emerge from the two cells of the teliospore. Septa are formed between the resultant nuclei, which then migrate into the sterigmata and develop into basidiospores. Thus, the year-long cycle is completed.

In certain regions, the life cycle of wheat rust can be short-cut through the persistence of the uredinial state when actively growing plant tissues are available throughout the year. In the North American plains, urediniospores from winter wheat in the southwestern states and Mexico drift northward to southern Manitoba. Later generations scatter westward to Alberta, and finally there is a southern drift at the end of summer, apparently moving along the eastern flank of the Rocky Mountains, and so back to the wintering grounds. Under such circumstances, wheat rust does not depend on barberry to persist, so barberry eradication was not an effective tool for controlling wheat rust in this area. In contrast, the spread of urediniospores from south to north is prevented in Eurasia, where there are extensive east-west mountain ranges, and barberry *is* necessary for survival of the pathogen. It should be noted that some species of barberry are resistant to rust and are safely marketed as ornamentals.

Subphylum Ustilaginomycotina Includes the Smuts

With few exception, the Ustilaginomycotina are parasites of flowering plants and are commonly referred to as smuts. The name "smut" refers to the sooty or smutty appearance of the black, dusty masses of teliospores, which are the characteristic resting spores of the smut fungi. Approximately 1070 species of Ustilaginomycotina have been described. Economically, the smuts are very important. They attack approximately 4000 species of flowering plants, including both food crops and ornamentals. Three of the better known smut fungi are *Ustilago*

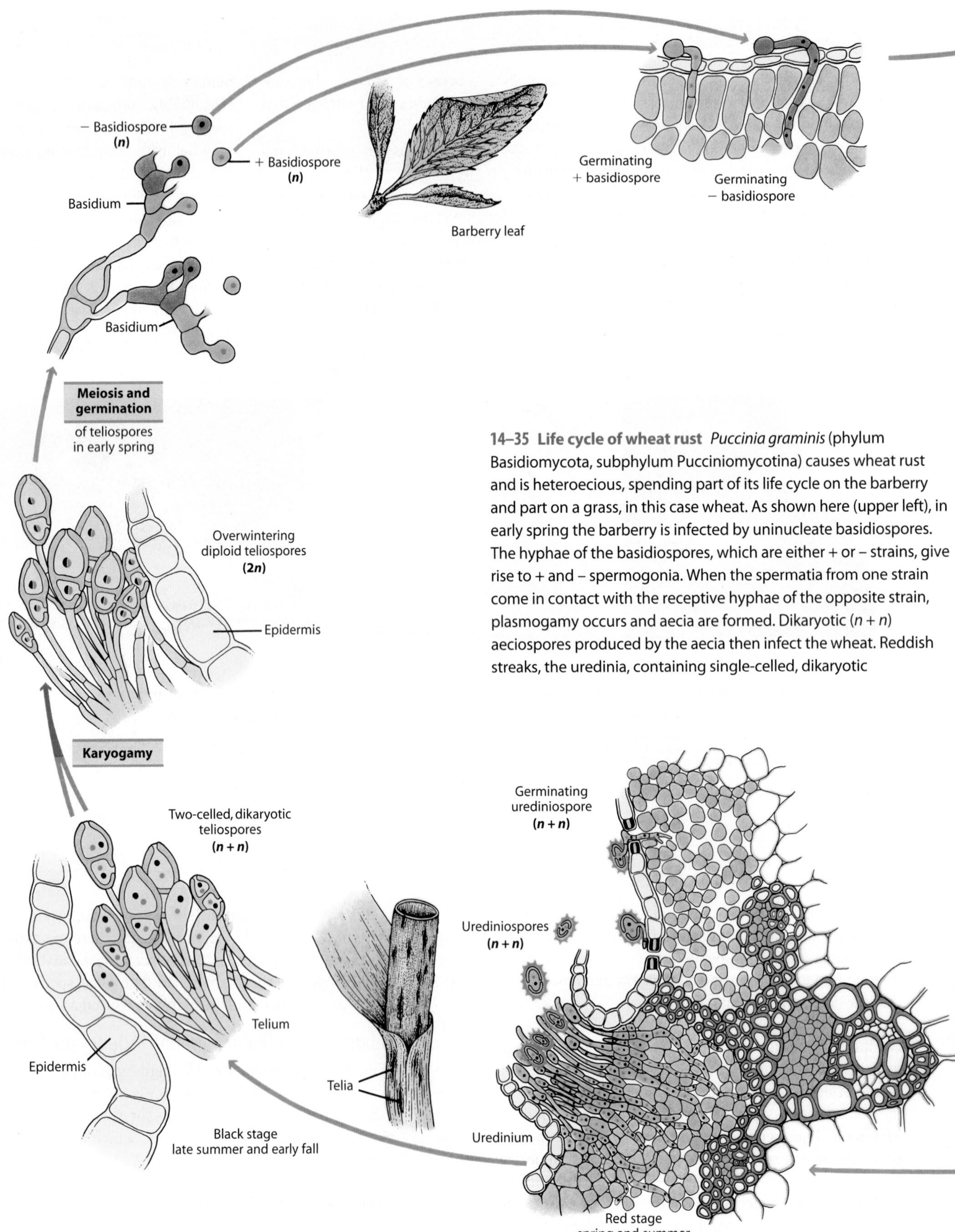

14–35 Life cycle of wheat rust *Puccinia graminis* (phylum Basidiomycota, subphylum Pucciniomycotina) causes wheat rust and is heteroecious, spending part of its life cycle on the barberry and part on a grass, in this case wheat. As shown here (upper left), in early spring the barberry is infected by uninucleate basidiospores. The hyphae of the basidiospores, which are either + or – strains, give rise to + and – spermogonia. When the spermatia from one strain come in contact with the receptive hyphae of the opposite strain, plasmogamy occurs and aecia are formed. Dikaryotic ($n + n$) aeciospores produced by the aecia then infect the wheat. Reddish streaks, the uredinia, containing single-celled, dikaryotic

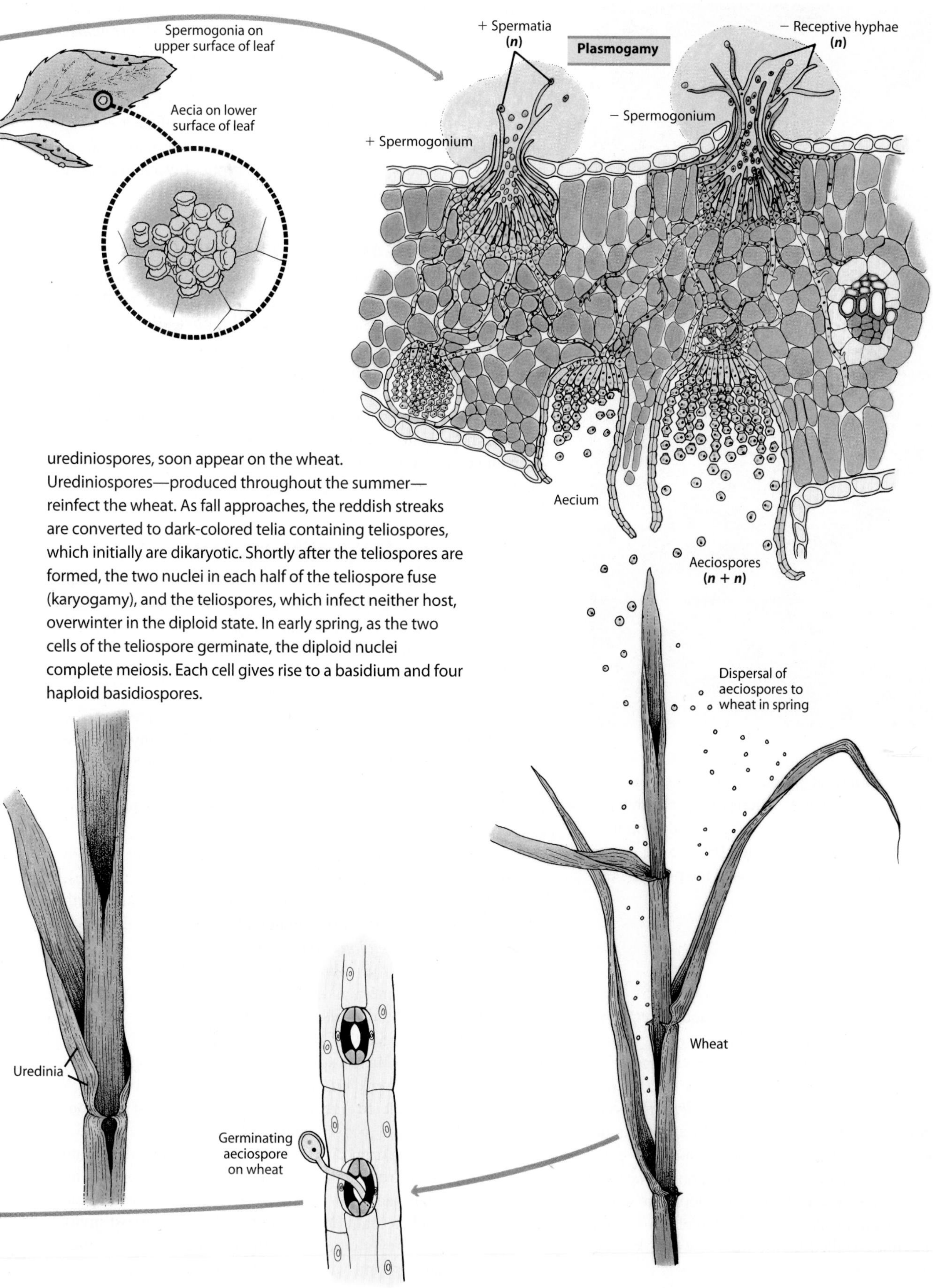

urediniospores, soon appear on the wheat. Urediniospores—produced throughout the summer—reinfect the wheat. As fall approaches, the reddish streaks are converted to dark-colored telia containing teliospores, which initially are dikaryotic. Shortly after the teliospores are formed, the two nuclei in each half of the teliospore fuse (karyogamy), and the teliospores, which infect neither host, overwinter in the diploid state. In early spring, as the two cells of the teliospore germinate, the diploid nuclei complete meiosis. Each cell gives rise to a basidium and four haploid basidiospores.

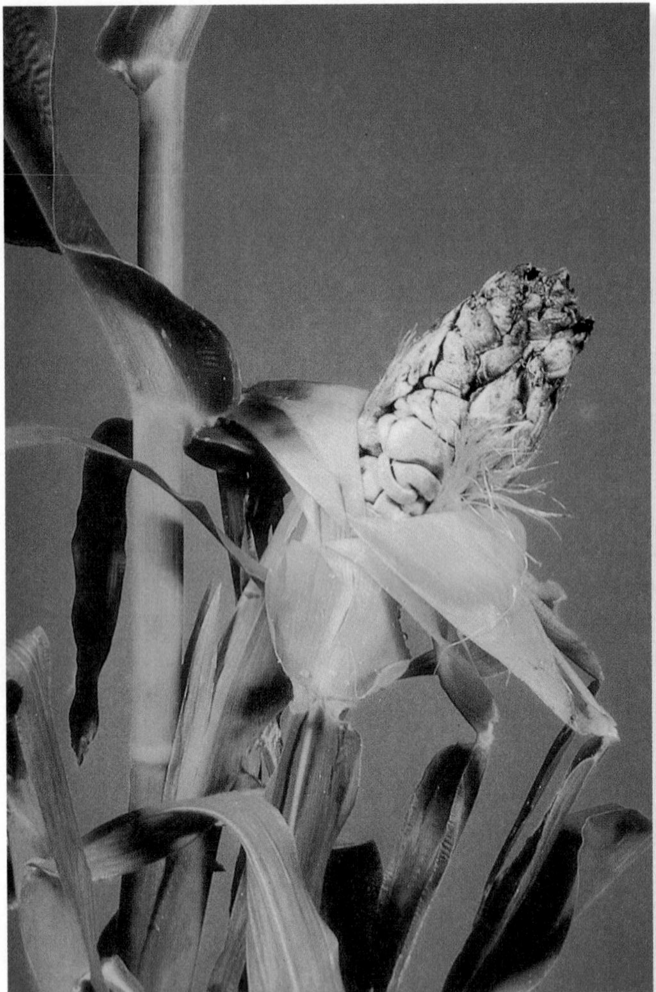

14–36 Corn smut The fungus *Ustilago maydis,* a smut that causes the familiar plant disease corn smut, produces black, dusty-looking masses of spores in ears of corn (maize). When young and white, these spore masses are cooked and eaten in Mexico and Central America, where they are regarded as a delicacy. *Ustilago* is a member of the subphylum Ustilaginomycotina of the phylum Basidiomycota.

maydis, the cause of common corn smut (Figure 14–36); *Ustilago avenae,* which causes loose smut of oats; and *Tilletia tritici,* the cause of bunt or stinking smut of wheat.

The life cycle of a smut, which is autoecious (requiring only one host), is considerably simpler than that of *Puccinia graminis.* Take for example the *Ustilago maydis* life cycle. Infections by spores of *U. maydis* remain localized, producing sori or large tumors. The most conspicuous tumors, or galls, occur on the ear of corn (maize), where the kernels become much enlarged and unsightly due to the development within them of a massive mycelium. A dikaryotic mycelium eventually gives rise to the thick-walled teliospores, in which karyogamy and meiosis take place.

On germination, the teliospore gives rise to a four-celled basidium (most smuts form septate basidia). Two + and two – haploid, uninucleate basidiospores are formed, one from each of the four cells of the basidium (*U. maydis,* like *P. graminis,* is heterothallic). The basidiospores may infect maize plants directly or give rise by budding to populations of uninucleate cells called **sporidia,** which can also infect maize plants. The basidiospores or sporidia germinate to produce either a + or – mycelium. When mycelia of opposite strains come into contact, plasmogamy occurs, producing a dikaryotic mycelium, most cells of which change into teliospores.

Symbiotic Relationships of Fungi

Symbiosis—"living together"—is a close and long-term association between organisms of different species. Some symbiotic relationships, typically disease-causing, are **parasitic.** One species (the parasite) benefits from the association and the other (the host) is harmed. Although many fungi are parasites, other fungi are involved in symbiotic relationships that are **mutualistic**—that is, the association is beneficial to both organisms. Two of these mutualistic symbioses—lichens and mycorrhizas—have been and continue to be of extraordinary importance in enabling photosynthetic organisms to become established in previously barren terrestrial environments.

A Lichen Consists of a Mycobiont and a Photobiont

A lichen is a mutualistic symbiotic association between a fungal partner and a population of unicellular or filamentous algal or cyanobacterial cells. The fungal component of the lichen is called the **mycobiont** (Gk. *mykēs,* "fungus," and *bios,* "life"), and the photosynthetic component is called the **photobiont** (Gk. *phōto-,* "light," and *bios,* "life"). The scientific name given to the lichen is the name of the fungus. About 98 percent of the lichen-forming fungal species belong to the Ascomycota, the remainder to the Basidiomycota. Lichens are polyphyletic. DNA evidence indicates that they have evolved independently on at least five occasions, and it is likely that they evolved independently many more times.

About 13,250 species of lichen-forming fungi have been described, representing almost half of all known ascomycetes. Some 40 genera of photobionts are found in combination with these ascomycetes. The most common photobionts are the green algae *Trebouxia, Pseudotrebouxia,* and *Trentepohlia,* and the cyanobacterium *Nostoc.* About 90 percent of all lichens have one of these four genera as a photobiont. A number of lichens incorporate two photobionts—a green alga and a cyanobacterium. Different species of the algal genus can serve as photobionts in a single lichen species. In addition, a single fungal species can form lichens with different algae or cyanobacteria.

Lichens are able to live in some of the harshest environments on Earth, and consequently they are extremely widespread. They occur from arid desert regions to the Arctic and grow on bare soil, tree trunks, sunbaked rocks, fence posts, and windswept alpine peaks all over the world (Figures 14-37 and 14-38). Some lichen thalli are so tiny that they are almost invisible to the unaided eye; others, like the reindeer "mosses," may cover kilometers of land with ankle-deep growth. One species, *Verrucaria serpuloides,* is a permanently submerged marine lichen. Lichens are often the first colonists of newly exposed

FROM PATHOGEN TO SYMBIONT: FUNGAL ENDOPHYTES

The leaves and stems of plants are often riddled with fungal hyphae. Such fungi are called **endophytes,** a general term used for a plant or fungus growing within a plant. Although some fungal endophytes cause disease symptoms in the plants they inhabit, others produce no such effects. Instead, they protect the host plants from insect and vertebrate herbivores and microbial pathogens, enhance drought tolerance and nutritional status, improve growth, and help some plants tolerate higher temperatures. The endophytes, in turn, gain nutrition and shelter and a means of dissemination via host propagules.

In many species of grasses, endophytic fungi infect the flowers of the host and proliferate in the seeds. Eventually, a substantial mass develops throughout the stems and leaves of the mature grass plant, with fungal hyphae growing between the host cells. A good example of such a relationship is provided by tall fescue, *Festuca arundinacea,* a grass that covers more than 15,000 square kilometers (35 million acres) of lawns, fields, and pastures in the United States, especially in the eastern states and the Midwest. Tall fescue plants that are free of fungus provide good forage for livestock, but cattle feeding on infected plants become lethargic and stop grazing, often panting and drooling excessively. If the animals are not moved to other forage, they develop gangrene, and eventually die. These symptoms are associated with the endophytic ascomycete *Sphacelia typhina.* The deterrent effects of endophytes on herbivores occur because the fungi produce alkaloids—bitter, nitrogen-rich compounds that are abundant in some plants. Alkaloids have physiological effects on humans and other animals (pages 30 to 31).

Some of the alkaloids produced by another fungus, *Claviceps purpurea,* which infects rye *(Secale cereale)* and other grasses, are identical to those produced by *Sphacelia.* The rye-infecting fungus, which replaces infected grain with a hard mycelial mat, or sclerotium, causes a plant disease called ergot. The sclerotium—also called an ergot—contains lysergic acid amide (LDA), a precursor of lysergic acid diethylamide (LSD). LSD was first discovered in studies of the alkaloids of *C. purpurea.* Domestic animals and people who eat the infected grain develop a disease called ergotism, which is often accompanied by gangrene, nervous spasms, psychotic delusions, and convulsions. It occurred frequently during the Middle Ages, when it was known as St. Anthony's fire. It has been suggested that the widespread accusations of witchcraft in Salem Village (now Danvers) and other communities in Massachusetts and Connecticut in 1692, which led to a number of executions, may have resulted from an outbreak of convulsive ergotism.

The endophytic fungi that infect hosts other than grasses are not usually transmitted through seeds of their host plants; their spores simply are blown from plant to plant or are carried by insects. In many of these relationships, the fungi may infect only the vegetative parts of the host and remain metabolically active for long periods. If the plant tissues are damaged by herbivores, the fungi may grow rapidly and produce toxins, protecting both the plants and new sites for fungal infection. Thousands of species of fungi may be involved in these relationships, almost all of them ascomycetes.

An interesting mutualistic association between an endophyte and plant host is that of the ascomycete *Curvularia protuberata* and the tropical panic grass, *Dichanthelium lanuginosum,* which grow at high soil temperatures in Yellowstone National Park. In symbiotic association, both partners can tolerate and survive temperatures up to 65°C; grown separately, neither the fungus nor the plant can grow at temperatures above 38°C. In 2007, Luis Márquez and co-workers reported that a third partner—a virus from the fungus—is involved in the mutualistic interaction. Moreover, they found that the ability of the fungus to confer heat tolerance to the host plant requires the presence of the virus, which was named *Curvularia* thermal tolerance virus (CThTV). When transferred to tomato *(Solanum lycopersicum),* the virus-infected fungus was able to confer heat tolerance to these plants. Thus, the fungus *C. protuberata,* when infected with CThTV, can confer heat tolerance to both a monocot (a grass) and a eudicot (tomato).

Sclerotia of *Claviceps purpurea,* which causes the plant disease known as ergot, are seen here growing on stalks of rye *(Secale cereale).*

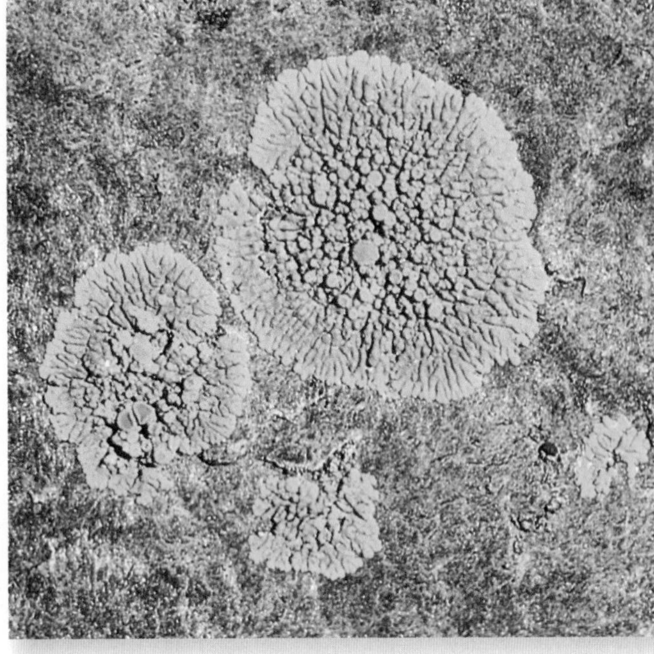

(a)

(c)

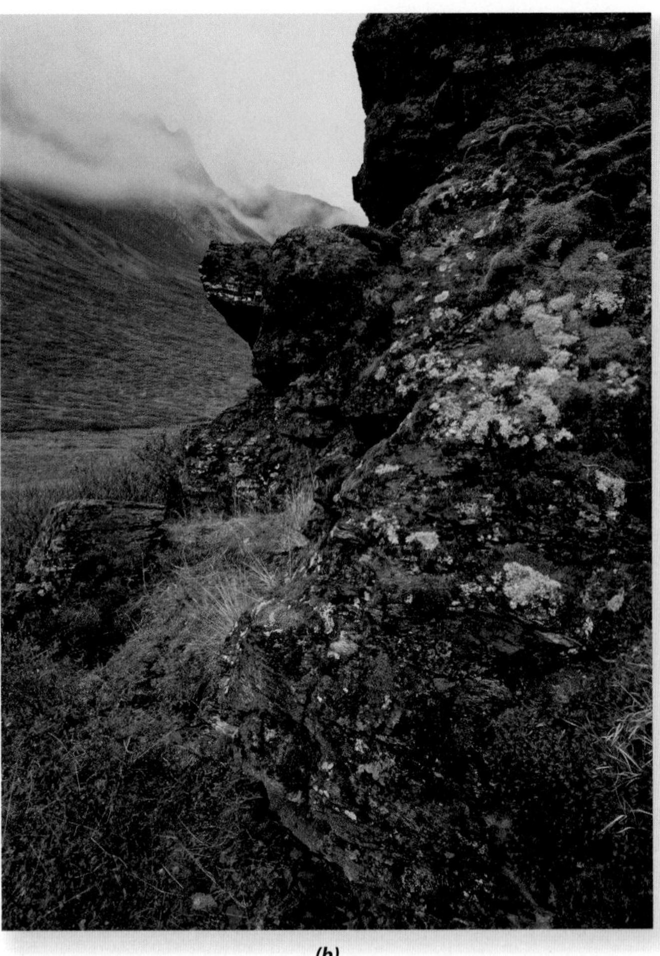

(b)

14–37 Crustose and foliose lichens ***(a)*** A crustose ("encrusting") lichen, *Caloplaca saxicola,* growing on a bare rock surface in central California. ***(b)*** Crustose and foliose ("leafy") lichens growing on a rocky outcrop in the Arctic National Wildlife Refuge in Alaska. ***(c)*** *Parmelia perforata,* a foliose lichen, has been gathered and incorporated into a hummingbird's nest on a dead tree branch in Mississippi.

rocky areas. In Antarctica, there are more than 350 species of lichens (Figure 14–39) but only two species of vascular plants; seven species of lichens actually occur within 4 degrees of the South Pole! Although widely distributed, individual lichen species usually occupy fairly specific substrates, such as the surfaces or interiors of rocks, soil, leaves, and bark. Some lichens provide the substrate for other lichens and parasitic fungi that may be closely related to the lichen being parasitized.

In almost all cases, the fungus makes up most of the thallus and plays the major role in determining the form of the lichen. There are two general types of lichen thalli. In one, the cells of the photobiont are more or less evenly distributed throughout the thallus; in the other, the cells of the photobiont form a distinct layer within the thallus (Figure 14–40a). Three major growth forms are recognized among the latter, stratified lichens: **crustose,** which is flattened and adheres firmly to the substrate, having a "crusty" (encrusting) appearance; **foliose,** which is leaflike; and **fruticose,** which is erect and often branched and "shrubby" (Figures 14-37 and 14-38).

The colors of lichens range from white to black, through shades of red, orange, brown, yellow, and green, and these organisms contain many unusual chemical compounds. Many lichens are used as sources of dyes; for example, the characteristic color of Harris tweed originally resulted from treatment of the wool with a lichen dye. Many lichens have also been used as medicines, components of perfumes, or minor sources of food. Some species are being investigated for their ability to secrete anti-tumor compounds.

Lichens commonly reproduce by simple fragmentation, by the production of special powdery propagules known as **soredia** (Figure 14–40b), or by small outgrowths known as **isidia.** Fragments, soredia, and isidia, which all contain both fungal

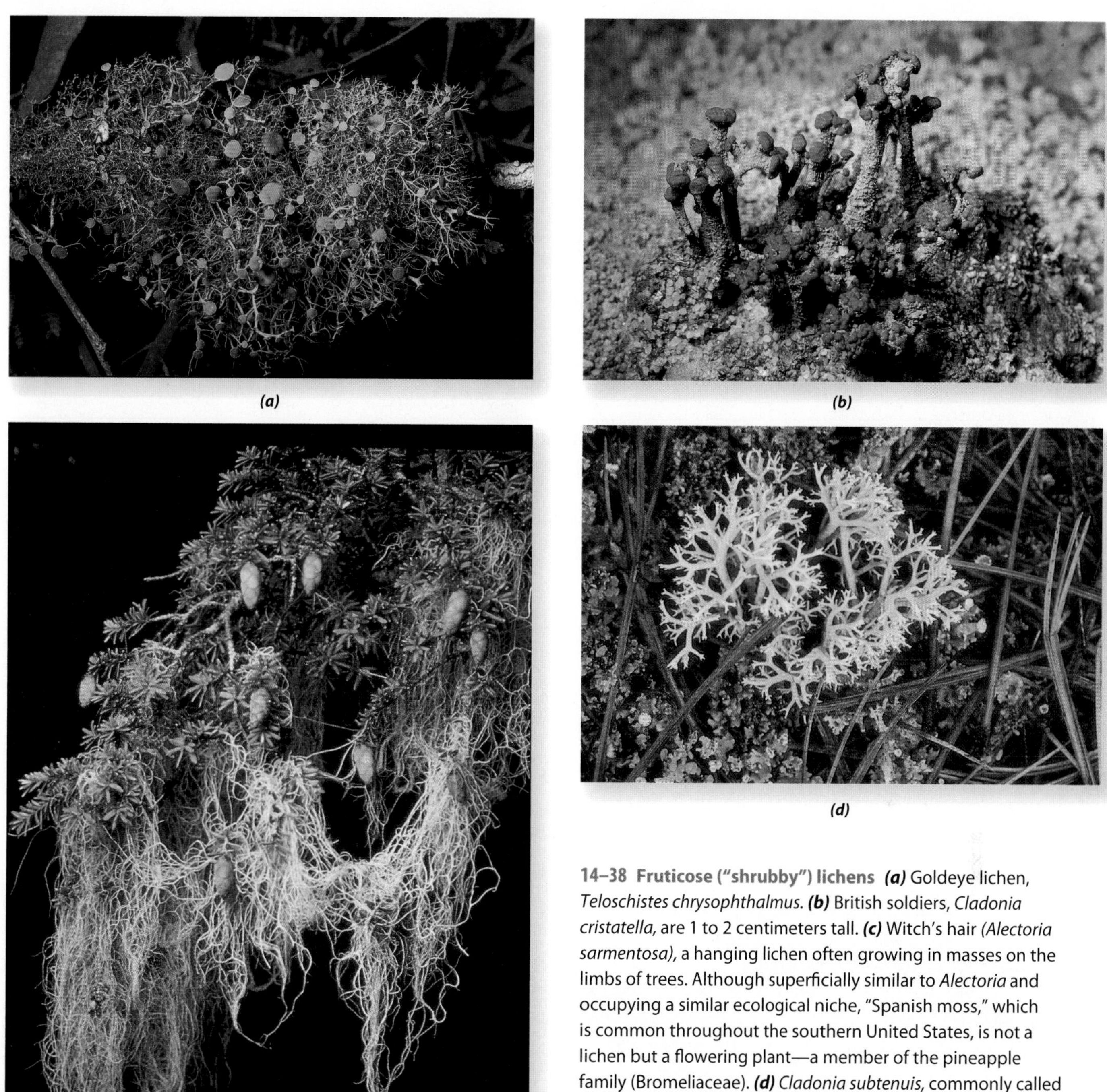

14–38 Fruticose ("shrubby") lichens ***(a)*** Goldeye lichen, *Teloschistes chrysophthalmus.* ***(b)*** British soldiers, *Cladonia cristatella,* are 1 to 2 centimeters tall. ***(c)*** Witch's hair *(Alectoria sarmentosa),* a hanging lichen often growing in masses on the limbs of trees. Although superficially similar to *Alectoria* and occupying a similar ecological niche, "Spanish moss," which is common throughout the southern United States, is not a lichen but a flowering plant—a member of the pineapple family (Bromeliaceae). ***(d)*** *Cladonia subtenuis,* commonly called "reindeer moss," is actually a lichen. Lichens of this group, which are abundant in the Arctic, concentrated radioactive substances following atmospheric nuclear tests and nuclear reactor accidents. Reindeer feeding on the lichens concentrated these radioactive substances still further and passed them on to the humans and other animals that consumed them or their products, especially milk and cheese.

hyphae and algae or cyanobacteria, act as small dispersal units to establish the lichen in new localities. The fungal component of the lichen produces ascospores, conidia, or basidiospores that are typical of its taxonomic group. If the fungus is an ascomycete, it may form ascomata, which are similar to those of other ascomycetes except that, in lichens, the ascomata may endure and produce spores slowly but continuously over a number of years. Regardless of the type of spore, any of them may form new lichens when they germinate and come into contact with the appropriate green algae or cyanobacteria.

(a)

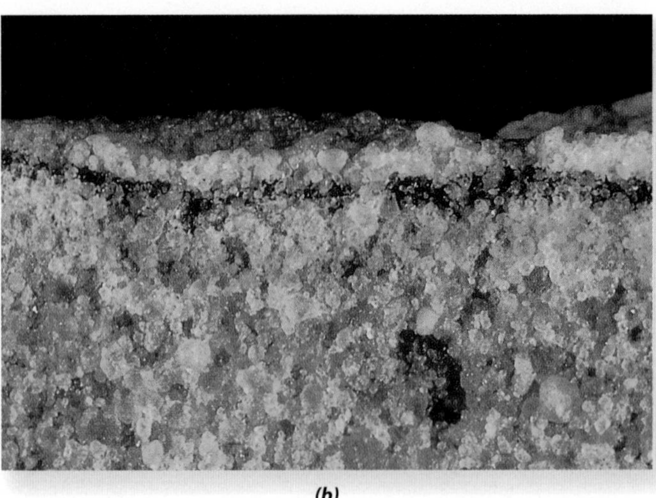

(b)

14–39 Lichens in Antarctica In this seemingly lifeless, dry region of Antarctica, lichens live just beneath the exposed surface of the sandstone ***(a)***. ***(b)*** In the fractured rock, the colored bands are distinct zones biologically. The white and black zones are formed by a lichen, and the lower green zone is produced by a nonlichenized unicellular green alga. The air temperatures in this part of Antarctica rise almost to freezing in summer and probably fall to –60°C in winter.

The Survival of Lichens Is Due to Their Ability to Dry Out Very Rapidly How can lichens survive under environmental conditions so severe that they exclude any other form of life? At one time, it was thought that the secret of a lichen's success was that the fungal tissue protected the alga or cyanobacterium from drying out. Actually, one of the chief factors in their survival seems to be that they dry out very rapidly. Lichens are frequently very desiccated, with a water content ranging from only 2 to 10 percent of their dry weight. When a lichen dries out, photosynthesis ceases. In this state of "suspended animation," some species of lichens can endure even blazing sunlight or great extremes of heat or cold. Cessation of photosynthesis depends, in large part, on the fact that the upper cortex of the lichen becomes thicker and more opaque when dry, cutting off the passage of light

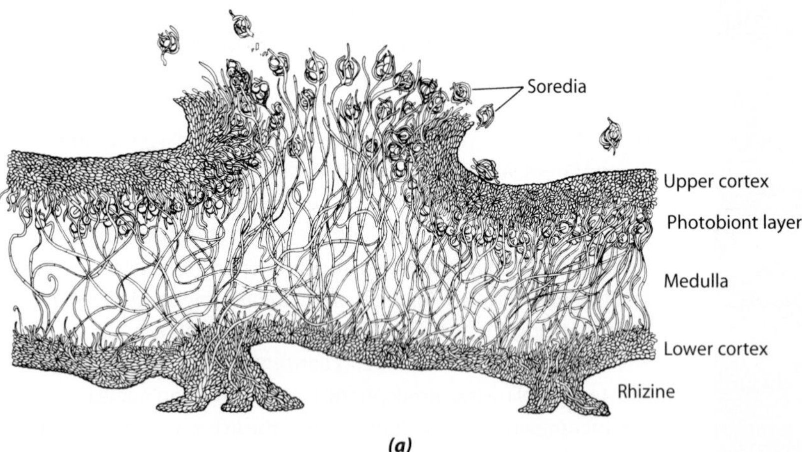

(a)

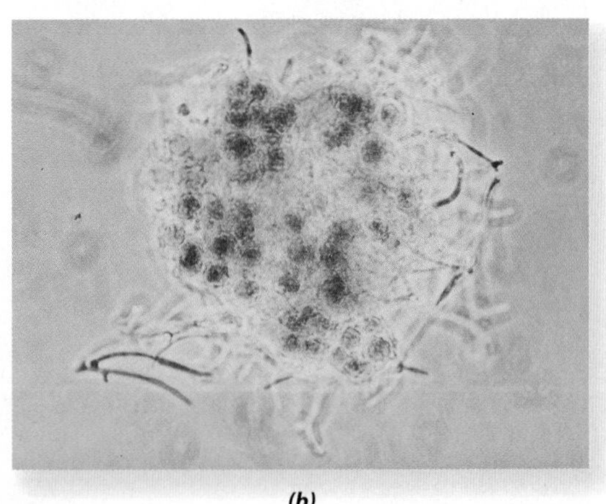

(b)

14–40 A stratified lichen ***(a)*** A cross section of the lichen *Lobaria verrucosa,* shown here releasing soredia consisting of hyphae wrapped around cyanobacteria. The simplest lichens consist of a crust of fungal hyphae entwining colonies of algae or cyanobacteria. More complex lichens, however, exhibit a definite growth form with a characteristic internal structure. The lichen shown here has four distinct layers: (1) the upper cortex, a protective surface of heavily gelatinized fungal hyphae; (2) the photobiont layer, which, in *Lobaria,* consists of cyanobacterial cells and loosely interwoven, thin-walled hyphae; (3) the medulla, consisting of loosely packed, weakly gelatinized hyphae, which makes up about two-thirds of the thickness of the thallus and appears to serve as a storage area; and (4) the lower cortex, which is thinner than the upper cortex and is covered with fine projections (rhizines) that attach the lichen to its substrate. ***(b)*** A soredium, composed of fungal hyphae and cells of the photobiont.

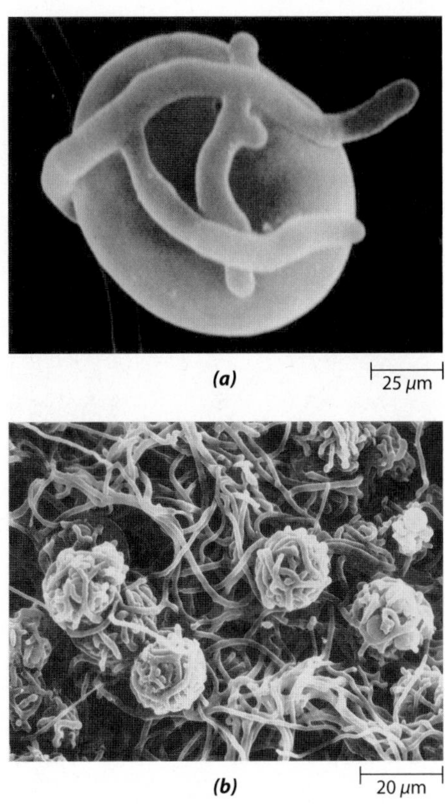

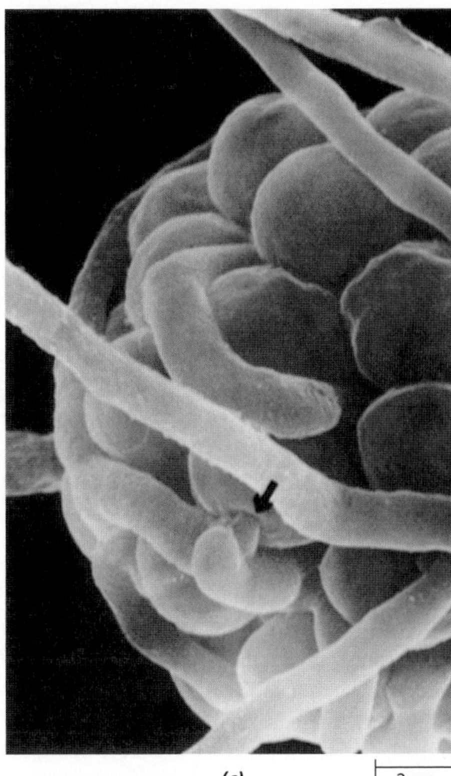

14–41 Initial development of a lichen Scanning electron micrographs of early stages in the interaction between fungal and algal components of British soldier lichens *(Cladonia cristatella)* in laboratory culture. The photosynthetic component in this lichen (see Figure 14–38b) is *Trebouxia,* a green alga. ***(a)*** An algal cell surrounded by fungal hyphae. ***(b)*** Mixed groups of fungal and algal components developing into the mature lichen. ***(c)*** Penetration of an algal cell by a fungal haustorium (arrow).

energy. A wet lichen may be damaged or destroyed by light intensities or temperatures that do not harm a dry lichen.

Lichens have an extremely slow rate of growth, their radius increasing at rates ranging from 0.1 to 10 millimeters a year. Calculated on this basis, some mature lichens are estimated to be as much as 4500 years old. They achieve their most luxuriant growth along seacoasts and on fog-shrouded mountains. The oldest known lichenlike fossils were found in marine phosphorite of the Doushantuo Formation (between 551 and 635 million years old) at Weng'an, South China.

The Relationship between Mycobiont and Photobiont Is Mutualistic Considerable debate has surrounded the nature of the relationship between mycobiont and photobiont—that is, whether the relationship is a parasitic or a mutualistic symbiosis. As noted by David L. Hawksworth, the issue is in reality one of scale. At the cellular level, individual photobiont cells can be considered parasitized by the mycobionts whose hyphae adhere closely to the phycobiont cell walls (Figure 14–41). These hyphae typically form haustoria, or **appressoria,** which are specialized hyphae that penetrate photobiont cells by means of pegs. These structures are involved with the transfer of carbohydrates and nitrogen compounds (in the case of nitrogen-fixing cyanobacteria such as *Nostoc*) from the photobiont to the fungus. In addition, the mycobiont controls the division rate of its photobiont. The mycobiont, in turn, provides the photobiont with a suitable physical environment in which to grow and absorbs needed minerals from the air in the form of dust or from rain. At the whole lichen level, the association clearly is mutualistic, as neither partner can flourish without the other in niches in which they occur in nature. Today, mutualism generally is judged on the basis of the dual functional unit, and therefore lichen symbiosis is considered to be mutualistic.

Lichens Are Important Ecologically Lichens clearly play an important role in ecosystems. Mycobionts produce large numbers of secondary metabolites called **lichen acids,** which sometimes amount to 40 percent or more of the dry weight of a lichen. These metabolites are known to play a role in the biogeochemical weathering of rock and soil formation. The lichens trap the newly formed soil, making it possible for a succession of plants to grow.

Lichens containing a cyanobacterium are of special importance, because they contribute fixed nitrogen to the soil. Such lichens are a crucial factor in supplying nitrogen in many ecosystems, including the old-growth forests of the Pacific Northwest of the United States, some tropical forests, and certain desert and tundra sites.

Because they have no means of excreting the elements that they absorb, some lichens are particularly sensitive to toxic compounds. The compounds cause the limited amount of chlorophyll in their algal or cyanobacterial cells to deteriorate. Lichens are very sensitive indicators of the toxic components—particularly sulfur dioxide—of polluted air, and they are increasingly being used to monitor atmospheric pollutants, especially around cities. Because lichens containing cyanobacteria are especially sensitive to sulfur dioxide, air pollution can substantially limit nitrogen fixation in natural communities, causing long-range changes in soil fertility. Both the state of health of the lichens and their chemical composition are used to monitor the environment. For example, analysis of lichens can detect the distribution of heavy metals and other pollutants around industrial sites. Fortunately, many lichens have the ability to bind heavy metals outside their cells and thus escape damage themselves.

There are many interactions between lichens and other groups of organisms. Lichens are eaten by a number of vertebrate and invertebrate animals. They are an important winter food source for reindeer and caribou in the far-north regions of North America and Europe, and are eaten by mites, insects, and slugs. Lichens are dispersed by birds that use lichens in their nests (Figure 14–37c). The nests of flying squirrels can contain up to 98 percent lichens. Some lichens have been found to have important functions as antibiotics.

Mycorrhizas Are Mutualistic Associations between Fungi and Roots

The most prevalent and possibly the most important mutualistic symbiosis in the plant kingdom is that of **mycorrhizas** (mycorrhizae), which, as noted earlier, literally means "fungus roots." Mycorrhizas are intimate and mutually beneficial symbiotic associations between fungi and roots, and they occur in the vast majority of vascular plants, both wild and cultivated. The few families of flowering plants that usually lack mycorrhizas include the mustard family (Brassicaceae) and the sedge family (Cyperaceae).

Mycorrhizal fungi benefit their host plants by increasing the plants' ability to capture water and essential elements (see Chapter 29), especially phosphorus. Increased absorption of zinc, manganese, and copper—three other essential nutrients—has likewise been demonstrated. For many forest trees, if seedlings are grown in a sterile nutrient solution and then transplanted to a prairie soil, they grow poorly, and many eventually die from malnutrition (Figure 14–42). Mycorrhizal fungi also provide protection against attack by pathogenic fungi and nematodes (small roundworms). In return for these benefits, the fungal partner receives from the host plant carbohydrates and vitamins essential for its growth. The roots of neighboring plants may be linked by the hyphal network of a shared mycorrhizal fungus and thereby provided with a pathway for the transfer of water and mineral nutrients from one plant to another.

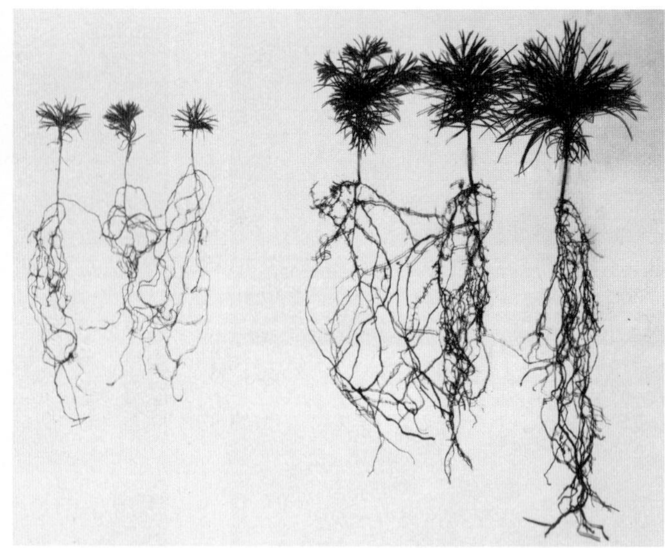

14–42 Mycorrhizas and tree nutrition Nine-month-old seedlings of white pine *(Pinus strobus)* were raised for two months in a sterile nutrient solution and then transplanted to prairie soil. The seedlings on the left were transplanted directly. The seedlings on the right were grown for two weeks in forest soil containing fungi before being transplanted to the prairie soil.

Endomycorrhizas Penetrate Root Cells

There are two major types of mycorrhizas: **endomycorrhizas,** which penetrate the root cells, and **ectomycorrhizas,** which surround the root cells. Of these two, endomycorrhizas are by far the more common, occurring in about 80 percent of all vascular plants. The fungal component is a glomeromycete. Thus, endomycorrhizal relationships are not highly specific. The fungal hyphae penetrate the cortical cells of the plant root, where they form highly branched structures called **arbuscules** (Figure 14–43a), hence these mycorrhizas commonly are called **arbuscular mycorrhizas** (AM). The arbuscules do not enter the protoplast but greatly invaginate the plasma membrane of the cortical cell, increasing its surface area and facilitating the transfer of metabolites and nutrients between the two mycorrhizal partners, the plant cells and the fungus. Most, or perhaps all, exchanges between plant and fungus take place at the arbuscules. In some cases, terminal swellings called **vesicles** (Figure 14–43b) may also occur between the host plant cells and are thought to function as storage compartments for the fungus. Such mycorrhizas are called vesicular-arbuscular (or V/A) mycorrhizas. The hyphae extend out into the surrounding soil for several centimeters and thus greatly increase the potential for the absorption of water and uptake of phosphates and other essential nutrients.

Ectomycorrhizas Surround but Do Not Penetrate Root Cells

Ectomycorrhizas (Figures 14-44 and 14-45) are characteristic of certain groups of trees and shrubs that are found primarily in temperate regions. Among these groups are the beech family (Fagaceae), which includes the oaks; the willow, poplar, and cottonwood family (Salicaceae); the birches (Betulaceae); the pine family (Pinaceae); and certain groups of tropical trees that form dense stands of only one or a few species. The trees that grow at timberline (the upper altitudes and latitudes of tree growth) in different parts of the world—such as pines in the northern mountains, *Eucalyptus* in Australia, and southern beech—almost always are ectomycorrhizal. The ectomycorrhizal association apparently makes the trees more resistant to the harsh, cold, dry conditions that occur at the limits of tree growth.

In ectomycorrhizas, the fungus surrounds but does not penetrate living cells in the roots. In the conifers, the hyphae grow between the cells of the root epidermis and cortex, forming a characteristic highly branched network, the **Hartig net** (Figure 14–46), which eventually surrounds many of the cortical and epidermal cells. In the roots of most angiosperms colonized by ectomycorrhizal fungi, the epidermal cells are triggered to enlarge primarily at right angles to the surface of the root,

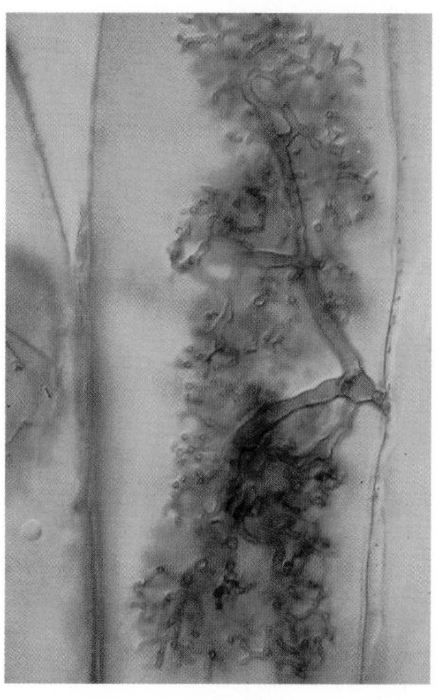

(a)

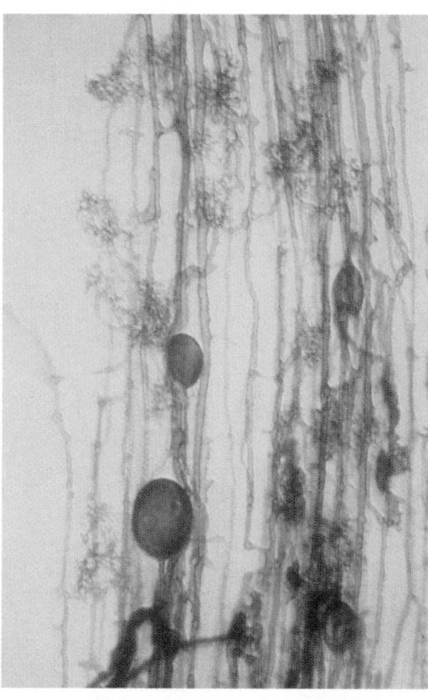

(b)

14–43 Endomycorrhizas *Glomus versiforme,* a glomeromycete, is shown here growing in association with the roots of leeks *(Allium porrum).* ***(a)*** Arbuscules growing inside a leek root cell. ***(b)*** Arbuscules (highly branched structures) and vesicles (dark, oval structures). Arbuscules predominate in young infections, with vesicles becoming common later.

thickening the root rather than extending it, and the Hartig net is confined to this layer (Figure 14–46b). The Hartig net functions as the interface between the fungus and the plant. In addition to the Hartig net, ectomycorrhizas are characterized by a **mantle,** or sheath, of hyphae that covers the root surface. Mycelial strands extend from the mantle into the surrounding soil (Figure 14–45). Typically, root hairs do not develop on ectomycorrhizas, and the roots are short and often branched. Ectomycorrhizas are mostly formed with Agaricomycotina, including many genera of mushrooms, but some ectomycorrhizas involve associations with ascomycetes, including truffles *(Tuber)* and the edible morels *(Morchella).* At least 6000 species of fungi are

14–44 Ectomycorrhizas A section through the extensive ectomycorrhizas of a lodgepole pine *(Pinus contorta)* seedling. The seedling extends about 4 centimeters above the soil surface.

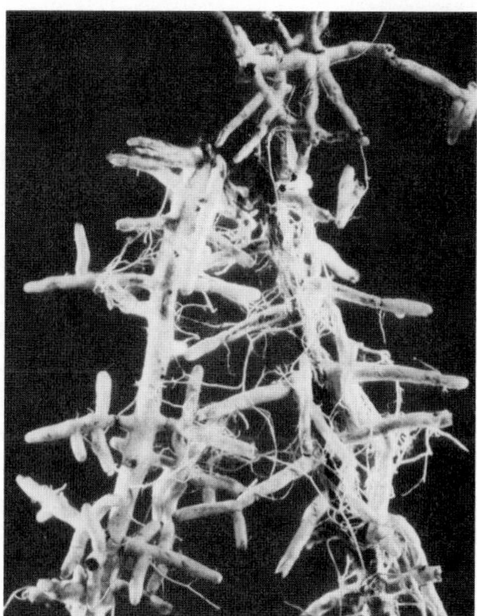

14–45 Fungal mantle Ectomycorrhizas from a western hemlock *(Tsuga heterophylla).* In such ectomycorrhizas, the fungus commonly forms a sheath of hyphae, called a fungal mantle, around the root. Hormones secreted by the fungus cause the root to branch. This growth pattern and the hyphal sheath impart a characteristic branched and swollen appearance to the ectomycorrhizas. The narrow mycelial strands, called rhizomorphs, extending from the the mycorrhizas function as extensions of the root system.

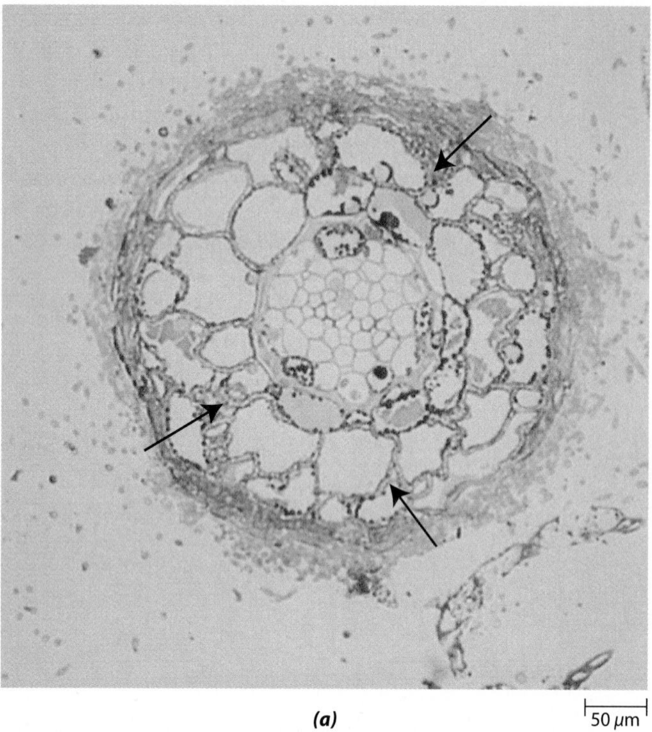

(a)

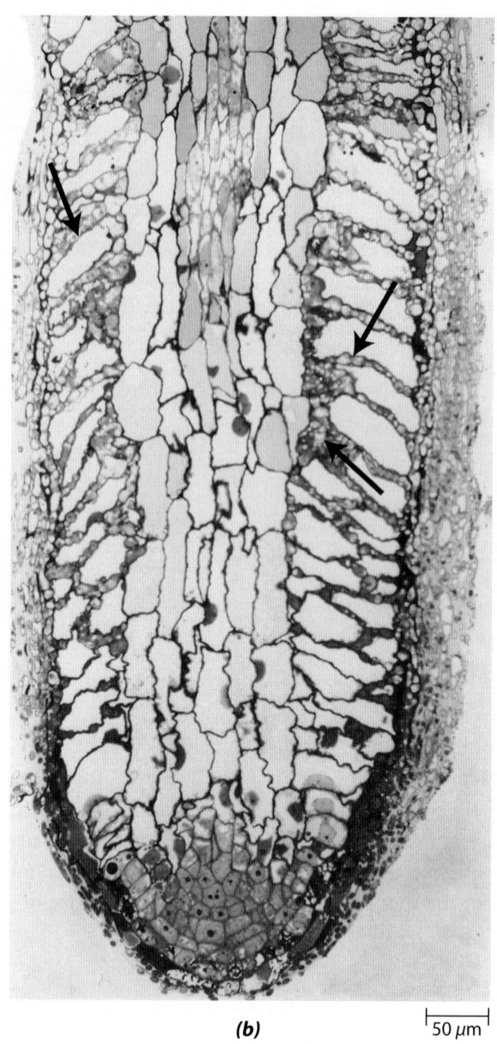

(b)

14–46 Ectomycorrhizal sections ***(a)*** Transverse section of an ectomycorrhiza of *Pinus*. The hyphae of the fungus form a mantle around the root and also penetrate between the epidermal and cortical cells, where they form the characteristic Hartig net (arrows). ***(b)*** Longitudinal section of ectomycorrhiza of the North American beech *(Fagus grandifolia)*. The fungus ensheathes the root, forming a mantle around it. The Hartig net (arrows) is confined to the layer of radially enlarged epidermal cells.

involved in ectomycorrhizal associations, often with a high degree of specificity.

The genome of *Laccaria bicolor,* a mushroom-forming fungus (Agaricomycotina), has shed some light on the characteristics of ectomycorrhizal fungi. For example, *L. bicolor* has an especially large number of genes associated with plasma membrane transport, apparently a reflection of the complexity of substance exchange between fungus and host root. Many genes have been found that encode secreted proteins suspected of playing a role in the establishment and maintenance of the symbiotic relationship. One protein is secreted only by hyphae located within the roots, indicating that it is the host not the fungus that influences expression of the pertinent gene. In addition, *L. bicolor* lacks the capacity to break down such cell wall components as cellulose and lignin, but then the fungus relies on the host to provide it with carbohydrates. It is likely that the genes required for the breakdown of such cell wall materials were lost over evolutionary time.

Other Kinds of Mycorrhizas Are Found in the Heather and Orchid Families Two other kinds of mycorrhizas are those characteristic of the heather family (Ericaceae) and a few closely related groups, and those associated with the orchids (Orchidaceae). In Ericaceae, the fungal hyphae form an extensive, loosely organized web over the root surface. Rather than extending the absorptive surface significantly, the principal role of the fungus is to release enzymes into the soil to break down certain compounds and make them available to the plant. The mycorrhizas of Ericaceae seem to function primarily in enhancing nitrogen, rather than phosphorus, uptake by the plants. This allows Ericaceae to colonize the kinds of infertile, acidic soils where they are especially frequent. The fungal components of ericoid mycorrhizas largely belong to a poorly defined group of ascomycetes.

In nature, orchid seeds, which usually lack endosperm and, consequently, have minimal storage reserves, germinate only in the presence of suitable fungi. The fungi, which colonize the cortical cells, form coils of hyphae called pelotons that are surrounded by the highly invaginated cortical cell plasma membrane. In addition to providing its host with mineral nutrients, the fungus also supplies it with carbon, at least when the host is a seedling. The fungi in such associations are mostly Agaricomycotina, with more than 100 species involved.

Mycorrhizal Associations Were Probably Important in Moving Plants onto the Land A study of the fossils of early plants has revealed that endomycorrhizal associations were as frequent then as they are

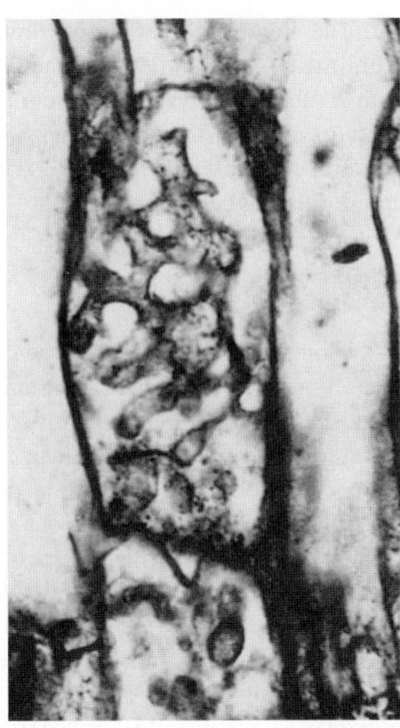

14–47 **Fossil endomycorrhiza** Silicified root of a gymnosperm from the Triassic period of Antarctica, showing well-developed arbuscules.

in their modern descendants (Figure 14–47). The first arbuscular mycorrhizas had their origin in the first bryophytelike land plants at least 400 million years ago. This finding led K. A. Pirozynski and D. W. Malloch to suggest that the evolution of mycorrhizal associations may have been a critical step in allowing colonization of the land by plants. Given the poorly developed soils that would have been available at the time of the first colonization, the role of mycorrhizal fungi (probably glomeromycetes) may have been of crucial significance, particularly in facilitating the uptake of phosphorus and other nutrients. A similar relationship has been demonstrated among contemporary plants colonizing extremely nutrient-poor soils, such as slag heaps: those individuals with endomycorrhizas have a much better chance of surviving. Thus it may have been not a single organism but rather a symbiotic association of organisms, comparable to a lichen, that initially invaded the land.

SUMMARY

Fungi Are Ecologically and Economically Important

Fungi, together with the heterotrophic bacteria, are the principal decomposers of the biosphere, breaking down organic products and recycling carbon, nitrogen, and other components to the soil and air. As decomposers, they often come into direct conflict with human interests, attacking almost any substance. Most fungi are saprotrophs; that is, they live on organic material from dead organisms. Many fungi attack living organisms, however, and cause diseases in plants, domestic and wild animals, and humans. A number of fungi are economically important to humans as destroyers of foodstuffs and other organic materials. The kingdom Fungi also includes yeasts, cheese molds, edible mushrooms, and *Penicillium* and other producers of antibiotics.

Most Fungi Are Composed of Hyphae

Fungi are rapidly growing organisms that characteristically form filaments called hyphae, which may be septate or aseptate. In most fungi, the hyphae are highly branched, forming a mycelium. Parasitic fungi often have specialized hyphae (haustoria) with which they extract organic carbon and other substances from the living cells of other organisms.

Fungi Are Absorbers That Reproduce by Means of Spores

Fungi, which are almost all terrestrial, reproduce by means of spores, usually wind-dispersed. Meiosis in fungi is zygotic—that is, the zygote, the only diploid phase in the fungal life cycle, divides by meiosis to form four haploid cells. No motile cells are formed at any stage of the fungal life cycle, except in chytrids. The primary component of the fungal cell wall is chitin. Typically, a fungus secretes enzymes onto a food source and then absorbs the small molecules that are released. Glycogen is the primary storage polysaccharide.

There Are Six Groups of Fungi

The kingdom Fungi includes Microsporidia, chytrids, zygomycetes, Glomeromycota, Ascomycota, and Basidiomycota. The Fungi are believed to have arisen from protists closely related to the modern genus *Nuclearia.*

Microsporidia May Be an Early Lineage of Fungi

The phylum Microsporidia consists of spore-forming unicellular parasites of animals. They are characterized by the presence of a polar tube that penetrates host cells and leads to infection.

The Chytrids Form Flagellated, Motile Cells

The polyphyletic chytrids are a predominantly aquatic group, and the only group of fungi with motile reproductive cells (zoospores and gametes). Most chytrids are coenocytic, with few septa at maturity. Some species of chytrids are parasitic and others are saprotrophic. Several species are plant pathogens causing minor diseases such as brown spot of corn and crown wart of alfalfa.

The Zygomycetes Form Zygospores in Zygosporangia

The polyphyletic zygomycetes have mostly coenocytic hyphae. Their asexual spores are generally formed in sporangia, saclike structures in which the entire contents are converted to spores. The zygomycetes are so named because they form resting spores called zygospores during sexual reproduction. The zygospores develop within thick-walled structures called zygosporangia.

The Glomeromycota Reproduce Only Asexually

The glomeromycetes (phylum Glomeromycota) have mostly coenocytic hyphae and reproduce by means of large multinucleate asexual spores. They are associated with roots as mycorrhizas and cannot be grown independently of the plant.

The Ascomycota Form Ascospores Internally in Asci

Phylum Ascomycota has some 32,300 named, distinct species, more than any other group of fungi. Ascomycetes have either unicellular (yeast) or filamentous growth forms. The distinguishing characteristic of ascomycetes is the ascus, a saclike structure in which the meiotic (sexual) spores, known as ascospores, are formed. In the ascomycete life cycle, the protoplasts of male and female gametangia fuse, and the female gametangia produce specialized hyphae that are dikaryotic (each compartment containing a pair of haploid nuclei). The ascus forms near the tip of a dikaryotic hypha. Ascospores are generally forcibly expelled. The asci are incorporated into complex, spore-producing bodies called ascomata. Typically, asexual reproduction occurs by formation of usually multinucleate spores known as conidia. There are three subphyla of Ascomycota. The Taphrinomycotina and Saccharomycotina are dominated by yeasts, which typically reproduce by budding, an asexual method of reproduction. With one exception, the Pezizomycotina are filamentous ascoma-producing species.

The Basidiomycota Form Basidiospores Borne Externally on Basidia

The phylum Basidiomycota includes many of the largest and most familiar fungi. They include mushrooms, puffballs, shelf fungi, stinkhorns, and others, as well as the rusts and smuts, which are important plant pathogens. The distinguishing characteristic of basidiomycetes is the production of basidia. The basidium is produced at the tip of a dikaryotic hypha and is the structure in which meiosis occurs. Each basidium typically produces four basidiospores, and these are the principal means of reproduction in the Basidiomycota.

The Mushrooms, Rusts, and Smuts Are Representatives of the Three Subphyla of Basidiomycota

The Basidiomycota can be divided into three subphyla: Agaricomycotina, Pucciniomycotina, and Ustilaginomycotina. In the Agaricomycotina, the basidia are incorporated into complex, spore-producing bodies called basidiomata. In the hymenomycetes, which include the mushrooms and shelf fungi, the basidiospores are produced on a distinct fertile layer, the hymenium. This layer often lines the gills or tubes of the hymenomycetes and is exposed before the forcibly discharged spores are mature. In the gasteromycetes, which include the stinkhorns and puffballs, the basidiospores mature inside the basidiomata and are not discharged forcibly from them. Members of the subphyla Pucciniomycotina and Ustilaginomycotina, the rusts and smuts, respectively, do not form basidiomata. Instead, they have septate basidia.

Lichens Consist of a Mycobiont and a Photobiont

Lichens are mutualistic symbiotic partnerships between a fungal partner (the mycobiont) and a population of either green algae or cyanobacteria (the photobiont). About 98 percent of the fungal partners belong to the Ascomycota, the rest to the Basidiomycota. The fungi receive carbohydrates and nitrogen compounds from their photosynthetic partners and provide the photobiont with a suitable physical environment in which to grow. The ability of a lichen to survive under adverse environmental conditions is related to its ability to withstand desiccation and remain dormant when dry.

Mycorrhizas Are Mutualistic Associations between Fungi and Roots

Mycorrhizas—symbiotic associations between plant roots and fungi—occur in all but a few families of vascular plants. Endomycorrhizas, also called arbuscular mycorrhizas, in which the fungal partners are glomeromycetes, occur in about 80 percent of vascular plants. In such associations, the fungus penetrates the cortical cells of the host but does not enter the protoplasts. In the second major type of mycorrhizal association, ectomycorrhizas, the fungus does not penetrate the host cells but forms a sheath, or mantle, that surrounds the roots and also a network (the Hartig net) that grows around the cortical cells. Mostly Agaricomycotina, but also some ascomycetes, are involved in ectomycorrhizal associations. Mycorrhizal associations are important in obtaining phosphorus and other nutrients for the plant and providing organic carbon for the fungus. Such associations were characteristic of the first plants to invade the land.

QUESTIONS

1. Distinguish between or among the following: hypha and mycelium; somatic and vegetative; rhizoids and haustoria; plasmogamy and karyogamy; sporangium and gametangium; heterothallic and homothallic; dikaryotic, monokaryotic, and diploid; parasitic and mutualistic; arbuscules and vesicles; endomycorrhizas and ectomycorrhizas.
2. "The fungi are of enormous importance both ecologically and economically." Elaborate on this statement both in general terms and with specific reference to each of the major groups of fungi, including the yeasts, lichens, and mycorrhizal fungi.
3. How might one determine, on the basis of hyphal structure alone, whether a given fungus is a member of the zygomycetes, Ascomycota, or Basidiomycota?
4. What do zygospores, ascospores, and basidiospores have in common? Zoospores, conidia, aeciospores, and urediniospores?
5. Many fungi produce antibiotics. What do you think the function of the antibiotics might be for the fungi that produce them?
6. In the life cycle of a mushroom, three types of hyphae, or mycelia, can be recognized: primary, secondary, and tertiary. How do these three types of mycelia relate to one another, and how do they fit into the life cycle?
7. "Both the state of health of the lichens and their chemical composition are used to monitor the environment." Explain.
8. "The most prevalent and probably the most important mutualistic symbiosis in the plant kingdom is the mycorrhiza." Explain.

CHAPTER 15

Protists: Algae and Heterotrophic Protists

◀ **Algae and a clam** Found in the shallow waters of Pacific coral reefs, the small giant clam *(Tridacna maxima)* opens its shell to expose the algae in its colorful mantle to sunlight. In a symbiotic relationship, the photosynthetic algae, known as zooxanthellae, supply most of the clam's nutritional requirements, while the algae feed off waste products and carbon dioxide produced by the clam.

CHAPTER OUTLINE

With approximately 70 percent of its surface covered by water, Earth is known as the "water planet." This abundance of water provided the watery habitat in which life began, at least 3.5 billion years ago, with the first appearance of the prokaryotes.

The oldest fossils interpreted as eukaryotic algae, because of their size and consistent form, are those of *Grypania,* estimated to be as much as 2.1 billion years old (Figure 15–1). These distinctive coiled fossils measure up to 0.5 meter in length and 2 millimeters in diameter. The oldest fossils that can be confidently assigned to a modern algal group are unbranched

CHECKPOINTS

After reading this chapter, you should be able to answer the following:

1. Of what ecological importance are algae?
2. In what ways are the euglenoids, cryptomonads, and dinoflagellates similar? Why is it difficult to classify these organisms on the basis of how they obtain their food?
3. What are the distinctive features of the phylum Haptophyta, and how are Haptophyta important in global climate control?
4. What are the basic characteristics of the brown algae? The red algae?
5. What characteristics of the green algae have led botanists to conclude that the charophycean green algae are the protist group from which the bryophytes and vascular plants have evolved?
6. How does the mode of cell division in the Chlorophyceae differ from that in the other classes of green algae?
7. How do the oomycetes differ from other stramenopiles, and what are some important plant diseases caused by oomycetes?
8. What features distinguish the plasmodial from the cellular slime molds? Why are these organisms not considered to be algae?

15–1 *Grypania spiralis* *Grypania spiralis* is the oldest known eukaryotic (multicellular) organism. These fossils date to about 2.1 billion years ago and are interpreted as eukaryotic algae.

15–2 A marine alga *Postelsia palmaeformis,* the sea palm, is a brown alga found along the Pacific coast, from British Columbia to California. It is well adapted to its habitat—the high intertidal to mid-intertidal zone of the shoreline in areas exposed to very heavy wave action. The flexible stipes (stalks) and tightly adhering holdfasts of *Postelsia* counter the drag forces exerted by the waves, which remove the mussels and seaweeds that would otherwise compete with *Postelsia* for space.

filaments known as *Bangiomorpha.* About 1.2 billion years old, these multicellular fossils are nearly indistinguishable from the modern red alga *Bangia.*

A diverse array of the descendants of the early eukaryotes—the protists—live today in the oceans and marine shorelines (Figure 15–2), as well as in freshwater lakes, ponds, and streams. Although some protists live in terrestrial habitats, their principal realm remains the water.

The protists include eukaryotes that do not have the distinctive characteristics of those of the kingdoms Plantae, Fungi, or Animalia. Most biologists would agree that plants, fungi, and animals are derived from ancient protists and that the study of modern protists sheds light on the origin of these important groups. In addition to the evolutionary significance of protists, some cause important diseases of plants or animals, whereas others are of great ecological significance.

Protist groups covered in this book include photosynthetic organisms that function ecologically like plants—that is, as primary producers, using light energy to manufacture their own food. These are the algae, the organisms studied by **phycologists.** Among the algae, the green algae are particularly significant because the plants are derived from an ancestor that, if still alive, would be classified as a green alga. In addition to these autotrophs, we will describe some colorless, heterotrophic protists that are not algae: the oomycetes and slime molds, which have traditionally been the concern of the mycologist, or student of fungi. Although these heterotrophic protists are not directly related to fungi, they continue to be described by terminology used for fungi. Similarities and differences among the protist groups discussed in this chapter are summarized in Table 15–1. A great many protists are not discussed in this chapter, including, for example, ciliates, radiolarians, and other large groups of heterotrophs.

Table 15-1 Comparative Summary of Characteristics of Protists

Group	Number of Species	Photosynthetic Pigments	Carbohydrate Food Reserve	Flagella	Cell Surface	Habitat
Euglenoids	800 to 1000	Most have none, or chlorophylls *a* and *b;* carotenoids	Paramylon	Usually 2; often unequal, with 1 forward and 1 behind or reduced to stub; extend apically	Flexible or rigid pellicle of proteinaceous strips beneath the plasma membrane	Mostly freshwater, some marine
Cryptomonads (Cryptophyta)	200	None, or chlorophylls *a* and *c;* phycobilins; carotenoids	Starch	2; unequal; subapical; hairy	Stiff layer of proteinaceous plates beneath plasma membrane	Marine and freshwater; cold waters
Haptophytes (Haptophyta)	300	Chlorophylls *a* and *c;* carotenoids, especially fucoxanthin	Chrysolaminarin	None or 2; equal or unequal; most have haptonema	Scales of cellulose; scales of calcified organic material in some	Great majority marine, a few freshwater

Group	Number of Species	Photosynthetic Pigments	Carbohydrate Food Reserve	Flagella	Cell Surface	Habitat
Dinoflagellates	4000	None in many, or chlorophylls *a* and *c*; carotenoids, mainly peridinin	Starch	None (except in gametes) or 2, dissimilar; lateral (1 transverse, 1 longitudinal)	Layer of vesicles beneath the plasma membrane, with or without cellulose plates	Mostly marine, many freshwater; some in symbiotic relationships
Diatoms* (Bacillariophyceae)	10,000 to 12,000 recognized	None, or chlorophylls *a* and *c*; carotenoids, mainly fucoxanthin	Chrysolaminarin	None or 1; only in male gametes of centric type; apical; hairy	Silica	Marine and freshwater
Golden algae* (Chrysophyceae)	1000	None, or chlorophylls *a* and *c*; carotenoids, mainly fucoxanthin	Chrysolaminarin	None or 2; apical; hairy forward, smooth behind	Silica scales; cellulose also in scales of some	Predominantly freshwater, a few marine
Yellow-green algae* (Xanthophyceae)	600	Chlorophylls *a* and *c*; carotenoids, but lack fucoxanthin	Oil	None or 2; hairy forward, smooth behind	Cellulose, sometimes with silica	Predominantly freshwater or soil, a few marine
Brown algae* (Phaeophyceae)	1500	Chlorophylls *a* and *c*; carotenoids, mainly fucoxanthin	Laminarin; mannitol (transported)	2; only in reproductive cells; lateral; hairy forward, smooth behind	Cellulose embedded in matrix of mucilaginous algin; plasmodesmata in some	Almost all marine; mostly temperate and polar, flourish in cold ocean waters
Red algae (Rhodophyta)	6000	Chlorophyll *a*; phycobilins; carotenoids	Floridean starch	None	Cellulose microfibrils embedded in matrix (usually galactans); deposits of calcium carbonate in many	Predominantly marine, about 100 freshwater species; many tropical species
Green algae	17,000	Chlorophylls *a* and *b*; carotenoids	Starch	None or 2 (or more); apical or subapical; equal or unequal; smooth	Glycoproteins, noncellulose polysaccharides or cellulose; plasmodesmata in some	Mostly aquatic, freshwater or marine; many in symbiotic relationships
Oomycetes* (Oomycota)	700	None	Glycogen	2; in zoospores and male gametes only; apical or lateral; hairy forward, smooth behind	Cellulose or cellulose-like	Marine, freshwater, and terrestrial (need water)
Plasmodial slime molds (Myxomycota)	700	None	Glycogen	Usually 2; apical; unequal; in reproductive cells only; smooth	Plasma membrane and slime sheath; cellulose-walled spores; plasmodium	Terrestrial
Cellular slime molds (Dictyosteliomycota)	50	None	Glycogen	None (amoeboid)	Plasma membrane and slime sheath on myxamoebas and slugs; thick cell wall rich in cellulose on mature macrocysts	Terrestrial

* Stramenopiles

Protists exhibit an amazing array of body types, including amoeba-like cells; single cells—with or without cell walls—that may or may not have flagella; colonies consisting of aggregations of cells that may be flagellated (bearing one or more flagella) or not; branched or unbranched filaments; one- or two-cell-thick sheets; tissue that resembles some tissues of plants and animals; and multinucleate masses of protoplasm with or without cell walls. Protists vary in size from microscopic single cells to 30-meter-long brown algae known as kelps. Both extremely small and very large sizes confer protection against being eaten by aquatic herbivores. Many protists reproduce sexually and have complex life histories, but some reproduce only by asexual means. All three types of sexual life cycles—zygotic, sporic, and gametic—occur among protists. It is common for different phases of the life history of a single protist species to be very different in size and appearance.

The phylogenetic relationships among the various protist groups are not yet firmly established. Considerable controversy exists, for instance, over the phylogenetic relationship between the green and the red algae. Whereas some evolutionary plant biologists contend that red algae are a sister group to green algae and thus should be grouped with them, others argue strongly against such a relationship. The apparent relatedness of the plasmodial and cellular slime molds is being questioned. Existing phylogenetic trees, however, such as that in Figure 12–10, show that the algae clearly belong to several lineages and that the oomycetes, once considered fungi, are much more closely related to brown and golden algae.

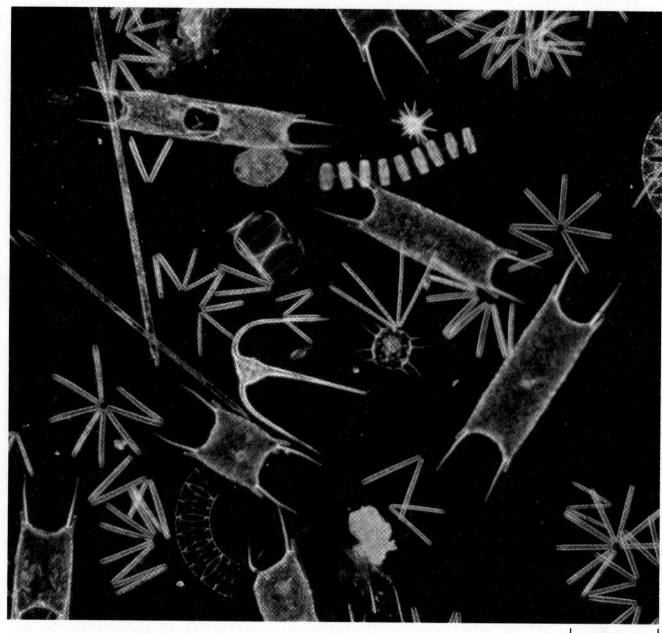

15–3 Marine phytoplankton The organisms shown here are dinoflagellates and filamentous and unicellular diatoms.

Ecology of the Algae

The open sea, the shore, and the land are the three life zones that make up our biosphere. Of these zones, the sea and the shore are more ancient. Here, algae play a role comparable to that of plants in the far younger terrestrial world. Algae often are also dominant in freshwater habitats—ponds, streams, and lakes—where they may be the most important contributors to the productivity of these ecosystems. Everywhere they grow, algae play an ecological role comparable to that of the plants in land habitats.

Along rocky shores can be found larger, more complex algae, the seaweeds, which are members of the red, brown, and green algae. At low tide, it is easy to observe fairly distinct banding patterns or layers that reflect the positions of seaweed species in relation to their ability to survive exposure. Seaweeds of this intertidal zone are subjected twice a day to large fluctuations of humidity, temperature, salinity, and light, in addition to the pounding action of the surf and forceful, abrasive water motions. Seaweeds found near the North and South Poles must endure months of darkness under the sea ice. Seaweeds are also prey to a host of herbivores and microbial pathogens. The complex biochemistries, structures, and life histories of seaweeds reflect adaptation to these physical and biological challenges.

Anchored offshore beyond the zone of waves, massive brown kelps form forests that provide shelter for a rich diversity of fish and invertebrate animals, some of which are valued human foods. Many large carnivores, including sea otters and tuna, find food and refuge in kelp beds, such as those off the coast of California. The kelps themselves, along with some red algal species, are harvested by humans for extraction of industrial products (see "Algae and Human Affairs" on the facing page).

In all bodies of water, minute photosynthetic cells and tiny animals occur as suspended **plankton** (from the Greek word for "wanderer"). The photosynthetic algae and cyanobacteria, which together constitute the **phytoplankton,** are the beginning of the food chain for the heterotrophic organisms that live in the ocean and in bodies of fresh water. The heterotrophic plankton includes the **zooplankton,** consisting mainly of tiny crustaceans and the larvae of many different phyla of animals, and many heterotrophic protists and bacteria **(bacterioplankton).**

In the sea, most small and some large fish, as well as most of the great whales, feed on the plankton, and still larger fish feed on the smaller fish. In this way, the "great meadow of the sea," as the phytoplankton is sometimes called, can be likened to the meadows of the land, serving as the source of nourishment for heterotrophic organisms. Floating or swimming single-celled or colonial chrysophytes, diatoms, green algae, and dinoflagellates are the most important organisms at the base of freshwater food chains. In marine waters, unicellular or colonial haptophytes, dinoflagellates, and diatoms are the most important eukaryotic members of the marine phytoplankton and therefore essential to the support of marine animal life (Figure 15–3).

Marine phytoplankton is increasingly being used as "fodder" to support the growth of farmed shrimp, shellfish, and other marketable seafood products. Seaweeds can be grown in aquatic farming operations to produce edible and industrially useful products. Both commercial uses of algae are examples of *mariculture,* by which marine organisms are cultivated in their natural environment, and are analogous to terrestrial agricultural systems.

Given the concern over the consequences of carbon release from fossil fuels and global warming, great effort has been made to find renewable sources of energy, especially liquid fuels for transportation. In the United States, corn (maize) has largely

ALGAE AND HUMAN AFFAIRS

People of various parts of the world, especially in the Far East, eat both red and brown algae. Kelps ("kombu") are eaten regularly as vegetables in China and Japan. They are sometimes cultivated but are mainly harvested from natural populations. *Porphyra* ("nori"), a red alga, is eaten by many inhabitants of the north Pacific Basin—and of the United States, as sushi has increased in popularity—and it has been cultivated in Japan, Korea, and China for centuries (page 342). Various other red algae are eaten on the islands of the Pacific and on the shores of the North Atlantic. Seaweeds are generally not of high nutritive value as a source of carbohydrates, because humans, like most other animals, lack the enzymes necessary to break down most of the materials in cell walls, such as cellulose and the protein-rich intercellular matrix. Seaweeds do, however, provide necessary salts, as well as a number of important vitamins and trace elements, and so are valuable supplementary foods. Some green algae, such as *Ulva,* or sea lettuce, are also eaten as greens.

In many northern temperate regions, kelp is harvested for its ash, which is rich in sodium and potassium salts and is therefore valued for industrial processes. Kelp is also harvested regionally and used directly for fertilizer.

Alginates, which are a group of substances derived from kelps, such as *Macrocystis,* are widely used as thickening agents and colloid stabilizers in the food, textile, cosmetic, pharmaceutical, paper, and welding industries. Off the west coast of the United States, *Macrocystis* kelp beds can be harvested several times a year by cropping them just below the water surface.

One of the most useful, direct commercial applications of any alga is the preparation of **agar,** which is made from a mucilaginous material extracted from the cell walls of several genera of red algae. Agar is used to make the capsules that contain vitamins and drugs, as well as for dental-impression material, as a base for cosmetics, and as a culture medium for bacteria and other microorganisms. Purified agarose is the gel often used in electrophoresis in biochemical experimentation. Agar is also employed as an anti-drying agent in bakery goods, in the preparation of rapid-setting jellies and desserts, and as a temporary preservative for meat and fish in tropical regions. Agar is produced in many parts of the world, but Japan is the principal manufacturer. A similar algal colloid called carrageenan is used in preference to agar for the stabilization of emulsions, such as paints, cosmetics, and dairy products. In the Philippines, the red alga *Eucheuma* is cultivated commercially as a source of carrageenan.

(a)

(b)

(c)

(a) A forest of giant kelp *(Macrocystis pyrifera)* growing off the coast of California. ***(b)*** Harvesting the seaweed *Nudaria* by hand from submerged ropes in Japan. ***(c)*** A kelp harvester operating in the near-shore waters of California. Cutting racks at the rear of the ship are lowered 3 meters below the water surface, and the ship moves backward, cutting through the kelp canopy. The harvested kelp is moved via conveyor belts to a collecting bin on board the ship.

15–4 Algal turf scrubbers A view of algal turf scrubbers at Falls City, Texas. Such scrubbers improve water quality by taking up nutrients from effluents or polluted waters and restoring oxygen. The biomass produced by the algae can be converted into fuel, and research efforts are under way to develop a cost-effective way of doing so.

been used to generate ethanol, but it is questionable whether this use of corn is economically viable. Moreover, the demand for corn ethanol competes with the need for corn as food. Cellulosic ethanol from switchgrass and wood chips is still in the research phase (see the essay "Biofuels: Part of the Solution, or Another Problem?" in Chapter 21).

One answer to the energy problem is the cultivation of algae for biofuel production. The cultivation of algae does not require the use of valuable agricultural land resources. In addition, algal biomass production can be 5 to 10 times greater than that of land-based agriculture. The two most common ways to produce biofuels from algae are (1) fermentation of the algal biomass and (2) industrial growth of algae for oil extraction. In the brown and green algae, much of the biomass consists of the cellulose-rich cell walls. The cellulose and other cellular carbohydrates can be fermented to ethanol. Other algae, such as diatoms, store relatively large amounts of oil and show promise for oil extraction in the future.

Algal turf scrubbers are providing a relatively inexpensive and highly productive system for the mass production of algae (Figure 15–4). Pioneered for wastewater treatment, the system consists of an attached algal community, the "turf," that grows on screens in a shallow trough or basin (called a raceway) through which water is pumped. As water flows down the raceway, the algae produce oxygen and remove nutrients through biological activity. The nutrients removed, or "scrubbed," from the water are stored in the algal biomass growing on the screens. The algae are harvested about once a week during the growing season.

In both marine and fresh waters relatively free of serious human disturbance, phytoplankton populations are usually held in check by seasonal climate changes, nutrient limitation, and predation. When humans disturb natural ecosystems, however, some algae may be released from these constraints, and their populations grow to undesirable "bloom" proportions. In the ocean, some of these blooms are known as "red tides" or "brown tides," because the water becomes colored by large numbers of algal cells containing red or brown accessory pigments. Algal blooms often correlate with the release of large quantities of toxic compounds into the water. These toxic compounds, which may have evolved as a defense against protist or animal predators, can cause human illness and massive die-offs of fish, birds, or aquatic mammals (see "Red Tides/Toxic Blooms" on the facing page). In recent years the frequency of toxic marine blooms has increased globally, although only a few dozen phytoplanktonic species are toxic. Some ecologists link this increase with the worldwide decline in water quality caused by the increase in human population.

The algae, so abundant on this watery planet, are important in carbon cycling (see "Global Warming: The Future Is Now" on pages 140 and 141). They are able to transform carbon dioxide (CO_2)—a so-called greenhouse gas that contributes to global warming—into carbohydrates by photosynthesis and into calcium carbonate by calcification. Large amounts of these carbohydrates and calcium carbonate are incorporated into algae and transported to the ocean bottom. Today, marine phytoplankton absorb about one-half of all the CO_2 that results from human activities, especially the burning of coal and other fossil fuels. The extent to which such carbon is transported to the ocean depths, and thus does not contribute to global temperature increases, is a matter of controversy.

The mechanism by which some phytoplanktonic organisms reduce the amount of CO_2 in the atmosphere is by forming calcium carbonate as they fix CO_2 during photosynthesis. The calcium carbonate is deposited as tiny scales covering the phytoplankton. The CO_2 thus removed from the water is replaced by atmospheric CO_2, creating a suction effect, also known as "CO_2 drawdown." Over the eons, phytoplankton covered with calcium carbonate settled to the ocean floor and ultimately gave rise to the famous White Cliffs of Dover and to the economically important oil deposits in the North Sea. Several kinds of red, green, and brown seaweeds can also become encrusted with calcium carbonate. The effect of calcification of these seaweeds on the global carbon cycle is not as well understood as the CO_2-drawdown effect of phytoplankton calcification.

Some marine phytoplankton, particularly haptophytes and dinoflagellates, produce significant amounts of a sulfur-containing organic compound that aids in regulating osmotic pressure within their cells. A volatile compound derived from this sulfur compound is excreted by the cells and is subsequently converted to sulfur oxides in the atmosphere. The sulfur oxides increase cloud cover and thus reflect sunlight away from the planet. Sulfur oxides, which are also generated by burning fossil fuels, contribute to acid rain and have a cooling effect on the climate. Scientists who work to predict future climates must take such cooling effects into account, along with the climate-warming effects of the greenhouse gases.

We begin our consideration of specific protist groups with the euglenoids, which are among the most ancient lineages of eukaryotic algae.

RED TIDES/TOXIC BLOOMS

Periodically, the west coast of Florida is ravaged by a major red tide incident. Commonly called "Florida red tides," these blooms are responsible for massive fish kills and the death of birds, sea turtles, dolphins, and Florida's endangered manatees. The organism responsible for Florida red tides is the dinoflagellate *Karenia brevis* (formerly called *Gymnodinium breve*) ***(a),*** which may be so abundant that the sea is colored reddish brown. Environmental factors that favor such blooms include warm surface temperatures, high nutrient content in the water, low salinity (which often occurs during rainy periods), and calm seas. Thus, rainy weather followed by sunny weather in the summer months is often associated with red tide outbreaks.

Karenia brevis produces dozens of toxins, some of which are airborne. Collectively known as brevetoxins, these airborne toxins cause respiratory problems in humans by constricting bronchioles in the lungs. This dinoflagellate also produces a compound called brevenal that in animals acts as a kind of antidote to brevetoxins, alleviating wheezing and shortness of breath and helping to clear mucus from the lungs. Brevenal is being evaluated as a potential treatment for cystic fibrosis, a debilitating lung disorder afflicting 30,000 people in the United States alone.

Red tides have been around for a long time. Records of their presence are found in the Old Testament and Homer's *Iliad.* There is considerable concern, however, that their incidence is increasing and spreading around the globe. Ecologists are not sure whether this increase is merely an upturn in a natural cycle or the beginning of a serious global epidemic. The increased frequency is correlated with increased nutrients in coastal waters originating from runoff from human development, the heavy use of fertilizers, and livestock farms.

Other dinoflagellates are responsible for the formation of red tides in different areas. *Gonyaulax tamarensis* is the organism involved in red tide blooms along the northeastern Atlantic coast, from the Canadian Maritime provinces to southern New England, *Gymnodinium catenella* sometimes causes red tides along the Pacific coast from Alaska to California, and *Protogonyaulax tamarensis* causes red tides in the North Sea off the coast of Northumberland in the United Kingdom. Over 40 marine species of dinoflagellates have been identified as producing toxic substances that kill birds and mammals, render shellfish toxic, or produce a widespread tropical fish-poisoning disease called ciguatera. The poisons produced by some dinoflagellates, such as *G. catenella* and several species of *Alexandrium,* which appear from the Gulf of Maine to the Gulf of Alaska, are extraordinarily powerful nerve toxins. The toxins bind to sodium or calcium channels in nerve cells, disrupting electrical conduction and uncoupling communication between nerve and muscle. Paralysis of breathing muscles may result.

When shellfish, such as mussels and clams, ingest toxic dinoflagellates or other organisms, they accumulate and concentrate the toxic substances. Depending on the species of toxic organisms the shellfish have consumed, they themselves may become dangerously toxic to the people who eat them. Along the Atlantic and Pacific coasts, fisheries are commonly closed in summer, and people regularly suffer from poisoning after consuming mussels, oysters, scallops, or clams taken from certain regions.

The term "red tide" is somewhat misleading. Many toxic events are called red tides even though the waters are not discolored. Conversely, a bloom of nontoxic, harmless algae can cause spectacular color changes in ocean waters, as seen in ***(b).*** In addition, some phytoplankton blooms neither discolor the water nor produce toxic compounds but kill marine animals in other ways—sometimes simply by depleting the oxygen in shallow waters.

(a)

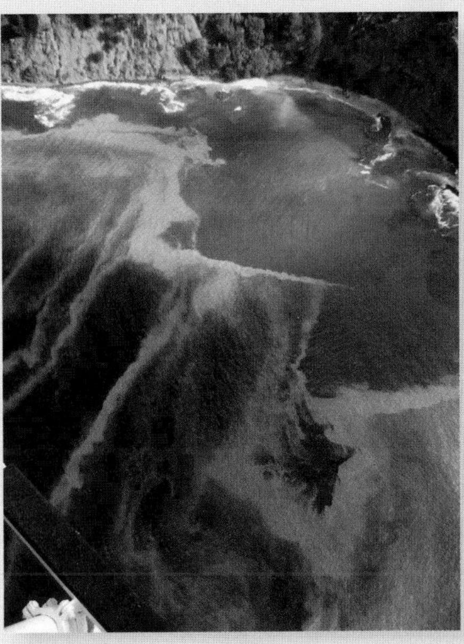

(b)

(a) *Karenia brevis,* the unarmored dinoflagellate responsible for the outbreaks of red tide along the west coast of Florida. Its curved, transverse flagellum lies in a groove encircling the organism. Its longitudinal flagellum, only a portion of which is visible, extends from the middle of the organism toward the lower left. The apical groove at the top is an identifying characteristic of *K. brevis.* ***(b)*** Spectacular red tide bloom caused by *Noctiluca scintillans* near Cape Rodney, New Zealand.

Euglenoids

Flagellates known as euglenoids comprise more than 80 genera and 800 to 1000 species. Molecular evidence suggests that the earliest euglenoids were phagocytes (particle eaters). About one-third of the genera, including the common *Euglena,* contain chloroplasts. The similarities between the chloroplasts of the euglenoids and those of the green algae—both possess chlorophylls *a* and *b* together with several carotenoids—suggest that the euglenoid chloroplasts were derived from endosymbiotic green algae. About two-thirds of the genera are colorless heterotrophs that rely on particle feeding or absorption of dissolved organic compounds. These feeding habits, along with a more general requirement for vitamins, explain why many of the euglenoids live in fresh waters that are rich in organic particles and compounds.

The structure of *Euglena* (Figure 15–5) illustrates many of the features typical of euglenoids in general. With one exception (the colonial genus *Colacium*), euglenoids are unicellular. *Euglena,* like most euglenoids, does not have a cell wall or any other rigid structure covering the plasma membrane. The genus *Trachelomonas,* however, has a wall-like covering made of iron and manganese minerals. The plasma membrane of euglenoids is supported by an array of helically arranged proteinaceous strips, which are in the cytosol, immediately beneath the plasma membrane. These strips form a structure called the **pellicle,** which may be flexible or rigid. The flexible pellicle of *Euglena* allows the cell to change its shape, facilitating movement in muddy habitats where flagellar swimming is difficult. Swimming *Euglena* cells have a single, long flagellum that emerges from the base of an anterior depression called the **reservoir,** or flagellar pocket, and a second, nonemergent flagellum (Figure 15–5). A swelling at the base of the emergent flagellum, together with a nearby, distinctive red **eyespot,** or **stigma,** in the cytosol, constitute a light-sensing system in euglenoids.

A **contractile vacuole** collects excess water from all parts of the euglenoid cell. The water is discharged from the cell via the reservoir. After each discharge, a new contractile vacuole is formed by coalescence of small vesicles. Contractile vacuoles are commonly observed in freshwater protists, which need to eliminate the excess water that accumulates as a result of osmosis. If the water is not removed, the cells will burst.

In contrast to the chloroplasts of green algae, euglenoid plastids do not store starch. Instead, granules of a unique polysaccharide known as **paramylon** form in the cytosol. Euglenoid plastids often resemble the plastids of many green and other algae in possessing a protein-rich region called a **pyrenoid,** which is the site of Rubisco and some other enzymes involved in photosynthesis.

Euglenoids reproduce by mitosis and lengthwise cytokinesis, continuing to swim while they divide. The nuclear envelope remains intact during mitosis, as it does in many other protists and in fungi. This suggests that an intact mitotic nuclear envelope is a primitive feature and that nuclear envelopes that break down during mitosis, as they do in plants, animals, and various protists, are a derived characteristic. Sexual reproduction and meiosis do not seem to occur in euglenoids, suggesting that these processes had not yet evolved when this group diverged from the main lineage of protists.

Cryptomonads: Phylum Cryptophyta

The cryptomonads are chocolate-brown, olive, blue-green, or red, fast-growing, single-celled flagellates that occur in marine

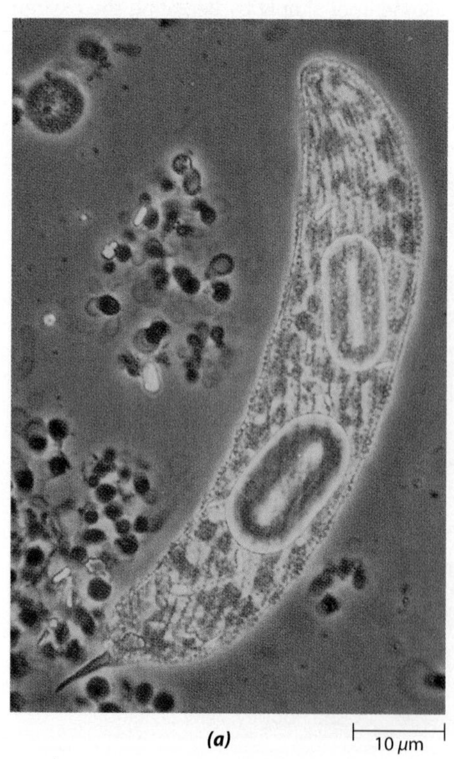

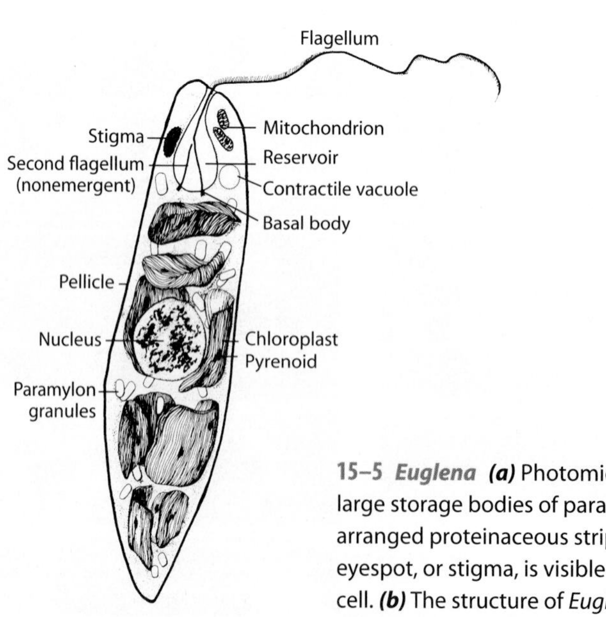

15–5 ***Euglena*** ***(a)*** Photomicrograph showing two large storage bodies of paramylon and the helically arranged proteinaceous strips of the pellicle. A red eyespot, or stigma, is visible at the upper end of the cell. ***(b)*** The structure of *Euglena,* as interpreted from electron micrographs.

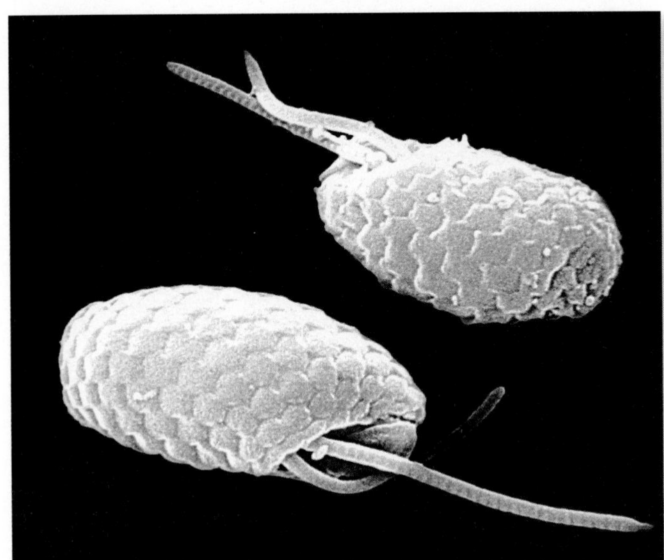

15–6 *Cryptomonas,* a cryptomonad This unicellular alga has two, slightly unequal flagella of about the same length as the cell. Both flagella emerge from the anterior end of the cell. Color has been added to this scanning electron micrograph to differentiate the flagella from the rectangular periplast plates that can be seen on the surface of the cells. The plates lie just under the plasma membrane.

and fresh waters (Figure 15–6). There are 200 known species of cryptomonads (Gk. *kryptos,* "hidden," and *monos,* "alone"). They are aptly named because their small size (3 to 50 micrometers) often makes them inconspicuous. They live primarily in cold or subsurface waters and are readily eaten by aquatic herbivores. Cryptomonad cells are particularly rich in polyunsaturated fatty acids, which are essential to the growth and development of zooplankton. Cryptomonads are ecologically important phytoplanktonic algae, both because of their edibility and because they are frequently the dominant algae in lakes and coastal waters when diatom and dinoflagellate populations undergo seasonal declines.

Like euglenoids, cryptomonads require certain vitamins and have both pigmented photosynthetic members and colorless phagocytic members that consume particles such as bacteria. Cryptomonads provide some of the best evidence that colorless eukaryotic hosts can obtain chloroplasts from endosymbiotic eukaryotes. The chloroplasts of cryptomonads and certain other algae have four bounding membranes. Evidence indicates that the cryptomonads arose through the fusion of two different eukaryotic cells, one heterotrophic and the other photosynthetic, establishing a **secondary endosymbiosis** (page 248). In addition to chlorophylls *a* and *c* and carotenoids, some cryptomonad chloroplasts contain a phycobilin, either phycocyanin or phycoerythrin. These water-soluble accessory pigments are otherwise known only in the cyanobacteria and red algae, providing evidence for the origin of the cryptomonad chloroplast.

The outer of the four membranes surrounding the cryptomonad chloroplast is continuous with the nuclear envelope and is called the **chloroplast endoplasmic reticulum** (Figure 15–7). The space between the second and third chloroplast membranes contains starch grains and the remains of a reduced nucleus, complete with three linear chromosomes and a nucleolus with typical eukaryotic RNA. The reduced nucleus, called a **nucleomorph** (meaning "looks like a nucleus"), is interpreted as the remains of the nucleus of a red algal cell that was ingested and retained for its photosynthetic capabilities by a heterotrophic

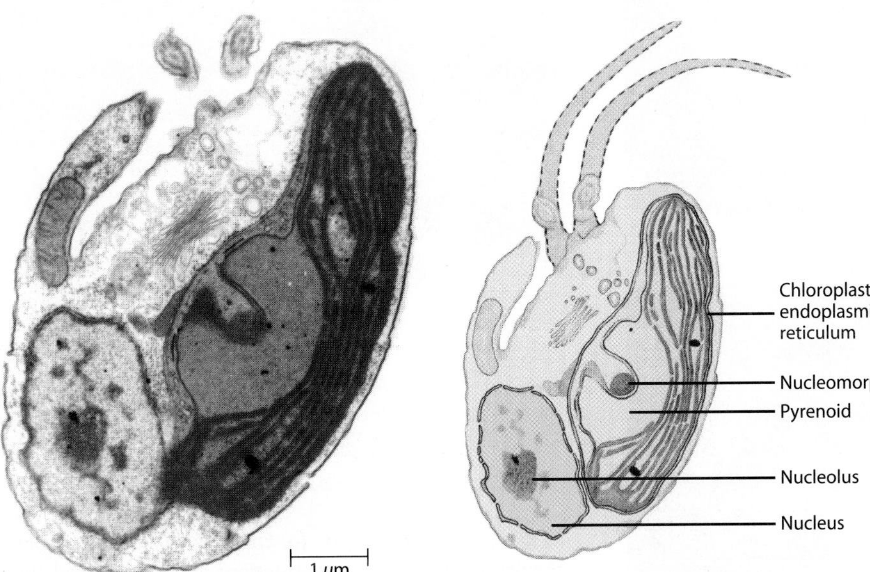

15–7 Cryptomonad structure This electron micrograph of a cryptomonad shows the nucleus and the nucleomorph, which is wedged in a cleft in the chloroplast. The nucleomorph is considered to be a vestigial nucleus of the endosymbiont—a red algal cell—engulfed by a heterotrophic host. The outer of the four membranes surrounding the chloroplast is the chloroplast endoplasmic reticulum. The connection between the nuclear membrane and the chloroplast membranes is not visible in the plane of this section.

host. The endosymbiont resembles the chloroplasts of other algae in that most of its genes have been transferred to the host's nucleus so that the endosymbiont is no longer capable of independent existence.

Haptophytes: Phylum Haptophyta

The phylum Haptophyta comprises a diverse array of primarily marine phytoplankton, though a few freshwater and terrestrial forms are known. The phylum consists of unicellular flagellates, colonial flagellates, and nonmotile single cells and colonies. There are about 300 known species in 80 genera, but new species are constantly being discovered. Haptophyte species diversity is highest in the tropics.

The most distinctive feature of the haptophyte algae is the **haptonema** (Gk. *haptein,* "to fasten"; relates to sense of touch), a threadlike structure that extends from the cell along with two flagella of equal length (Figure 15–8). It is structurally distinct from a flagellum. Although microtubules are present in the haptonema, they do not have the 9-plus-2 arrangement that is typical of eukaryotic flagella and cilia (see Figure 3–28). The haptonema can bend and coil but cannot beat like a flagellum. In some cases, it allows the haptophyte cell to catch prey food par-

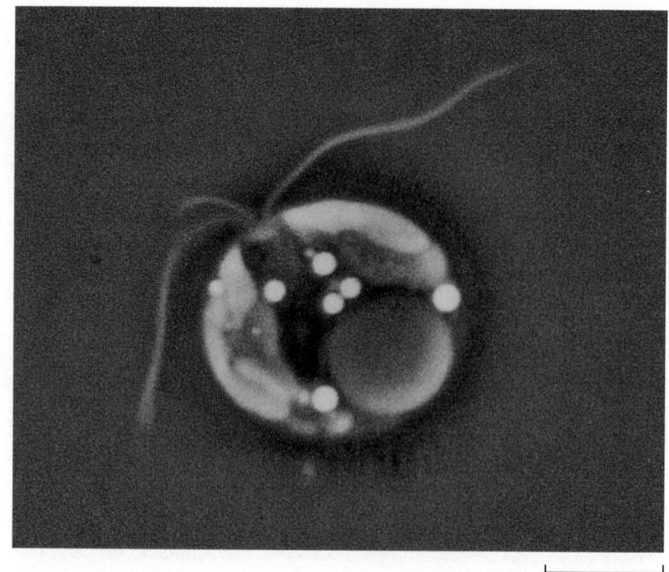

15–8 Haptonema The haptophyte *Prymnesium parvum.* Most haptophytes have two smooth and nearly equal flagella. Many also possess a haptonema (the smaller appendage shown here), which can bend or coil but cannot beat as do the flagella.

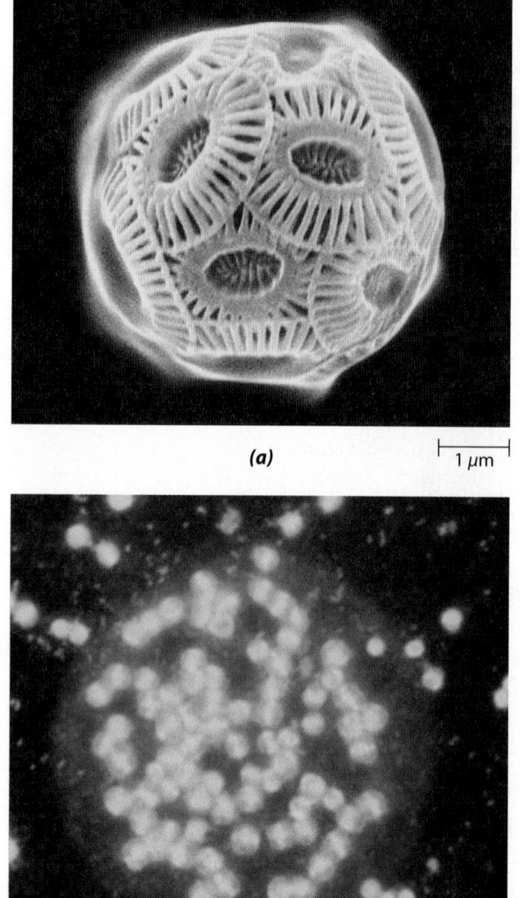

15–9 Haptophyte algae ***(a)*** *Emiliania huxleyi,* a coccolithophorid. This is the most widespread and abundant of the estimated 300 species of this group of extremely minute algae. The platelike scales covering cells of the coccolithophorids consist of calcium carbonate. ***(b)*** A fluorescence micrograph of a young colony of *Phaeocystis,* stained with acridine orange. The cells are embedded in a polysaccharide mucilage. ***(c)*** Massive *Phaeocystis* blooms clog fishing nets and wash up onto beaches, producing thick mounds of foam, as seen here on a beach in southern California.

ticles, functioning somewhat like a fishing rod. In other cases, it seems to help the cell sense and avoid obstacles.

Another characteristic of haptophyte algae is the presence of small, flat scales on the outer surface of the cell (Figure 15–9a). These scales are composed of organic material or calcified organic material. The calcified scales are known as **coccoliths,** and the 12 or more families of organisms that are decorated with coccoliths are known as the coccolithophorids. Coccoliths are of two types: those produced inside the cell—in Golgi vesicles—and transported to the cell's exterior, and those generated outside the cell. The two types of coccoliths are produced by alternate stages of the same life cycle. Coccoliths form the basis for a continuous fossil record back to their first appearance in the Late Triassic, some 230 million years ago.

Most haptophytes are photosynthetic, with chlorophyll *a,* some variation of chlorophyll *c,* and carotenoids, including **fucoxanthin,** a golden-brown carotenoid. In common with the cryptomonads, the plastids of haptophytes are surrounded by a chloroplast endoplasmic reticulum, which is continuous with the nuclear envelope. As in cryptomonads, the chloroplast endoplasmic reticulum is evidence that the plastids were acquired by secondary endosymbiosis. Sexual reproduction and alternation of heteromorphic generations occur in haptophytes, but the chromosome levels and life histories of many forms are as yet unknown.

Marine haptophytes are significant components of food webs, serving both as producers and, even though most are autotrophic, as consumers. As consumers, they graze on small particles such as cyanobacteria or absorb dissolved organic carbon. They are an important means by which organic carbon and two-thirds of the oceans' calcium carbonate are transported to the deep ocean. In addition, marine haptophytes are important producers of sulfur oxides connected with acid rain. The gelatinous, colonial stage of *Phaeocystis* (Figure 15–9b) dominates the phytoplankton of the marginal ice zone in polar regions and contributes some 10 percent of the atmospheric sulfur compounds that are generated by phytoplankton. In addition, its gelatin contributes significant organic carbon to the water. In all of the oceans, especially at mid-latitudes, *Emiliania huxleyi* can form blooms covering thousands of square kilometers. The two haptophyte genera *Chrysochromulina* and *Prymnesium* are notorious for forming marine toxic blooms that kill fish and other marine life.

Dinoflagellates

Molecular systematic data indicate that the dinoflagellates are closely related to ciliated protozoa such as *Paramecium* and *Vorticella* (see Figure 12–12) and to the apicomplexans, a phylum of protozoan parasites whose cells contain a nonpigmented plastid. The malaria parasite *Plasmodium* belongs to the apicomplexans. Collectively, these organisms form the **alveolates,** a supergroup characterized by the presence of small membrane-bounded cavities (alveoli) under their cell surfaces.

Most dinoflagellates are unicellular biflagellates—that is, single-celled organisms with two flagella (Figure 15–10). Some 4000 species are known, including both marine and freshwater forms. Dinoflagellates are unique in that their flagella beat within two grooves. One groove encircles the body like a belt, and the second groove is perpendicular to the first. The beating of the flagella in their respective grooves causes the dinoflagellate to spin like a top as it moves. The encircling flagellum is ribbon-like. There are also numerous nonmotile dinoflagellates, but they typically produce reproductive cells having flagella in grooves, from which their relationship to other dinoflagellates is deduced.

Dinoflagellates are unusual, though not unique, in having permanently condensed chromosomes. The chief method of reproduction of dinoflagellates is by longitudinal cell division, with each daughter cell receiving one of the flagella and a portion of the wall, or theca. Each daughter cell then reconstructs the missing parts in a very intricate sequence.

Many of the dinoflagellates are bizarre in appearance, with stiff cellulose plates forming the theca, which often looks like a strange helmet or part of an ancient coat of armor (Figures 15–10 and 15–11). The cellulose plates of the wall are within vesicles just inside the outermost membrane of the cell.

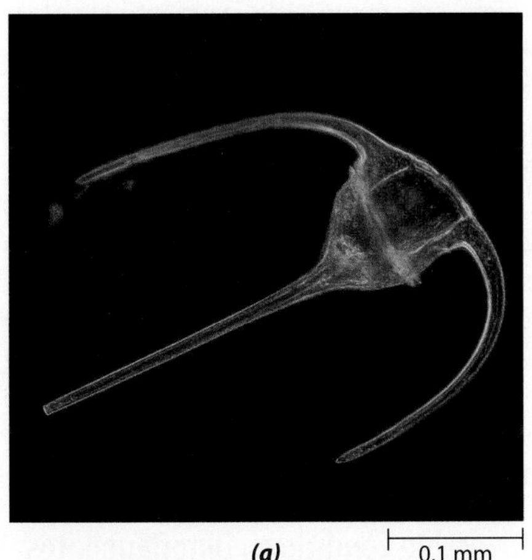

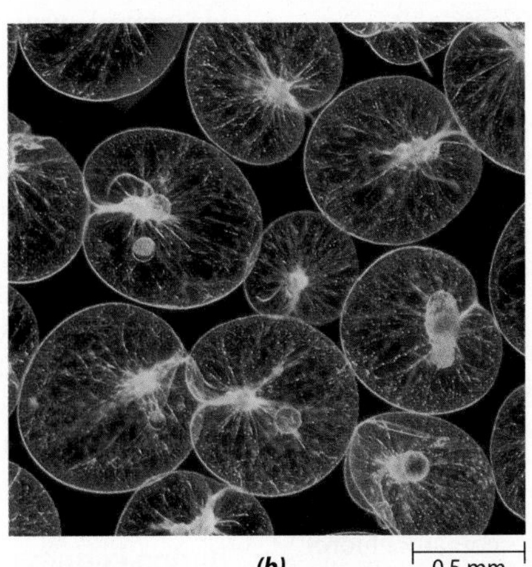

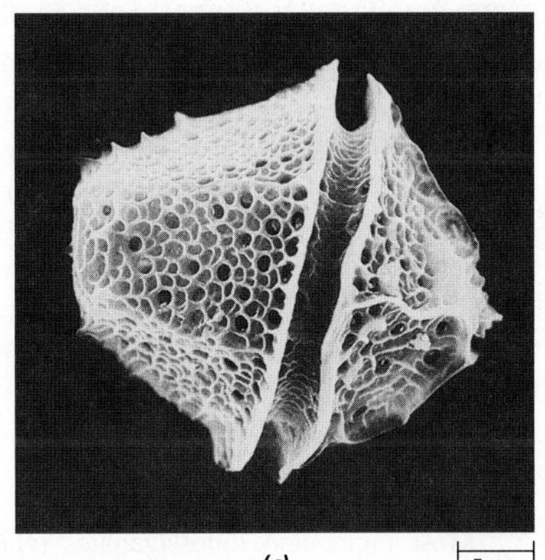

15–10 Dinoflagellates ***(a)*** *Ceratium tripos,* an armored dinoflagellate. ***(b)*** *Noctiluca scintillans,* a bioluminescent marine dinoflagellate. ***(c)*** *Gonyaulax polyedra,* the armored dinoflagellate responsible for spectacular red tides along the coast of southern California.

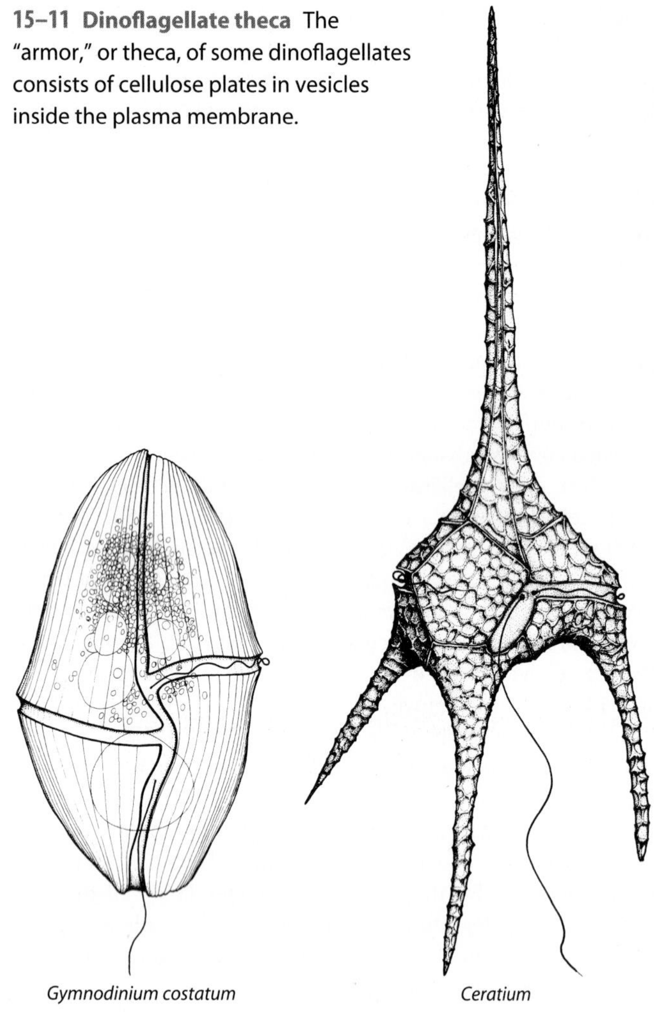

15–11 Dinoflagellate theca The "armor," or theca, of some dinoflagellates consists of cellulose plates in vesicles inside the plasma membrane.

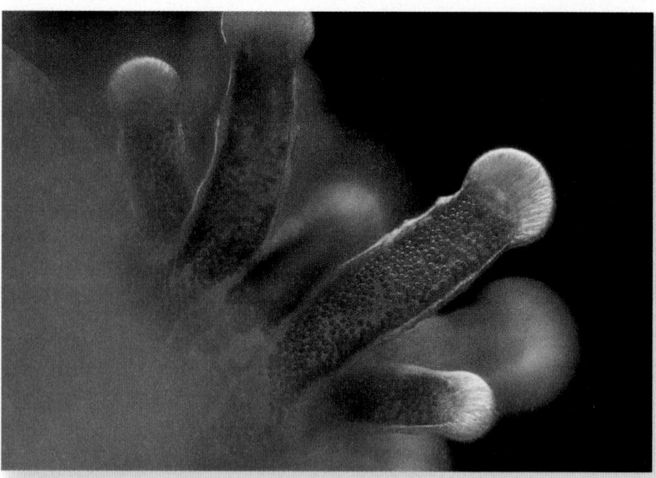

15–12 Zooxanthellae The symbiotic form of dinoflagellates, known as zooxanthellae, is seen here as golden spheres in tentacles of a coral animal. These symbionts are responsible for much of the productivity of coral reefs.

Dinoflagellates in the open ocean often have large, elaborate, sail-like thecal plates that aid in flotation. Other dinoflagellates have very thin or no cellulose plates and therefore do not appear to have a theca.

Many Dinoflagellates Ingest Solid Food Particles or Absorb Dissolved Organic Compounds

About half of all dinoflagellates lack a photosynthetic apparatus and hence obtain their nutrition either by ingesting solid food particles or by absorbing dissolved organic compounds. Even many pigmented—photosynthetic—and heavily armored dinoflagellates can feed in these ways. The ability of chlorophyll-containing algae such as these dinoflagellates to utilize both organic and inorganic carbon sources is referred to as **mixotrophy;** such algae are called mixotrophs. Some feeding dinoflagellates extrude a tubular process known as a peduncle, which can suction organic materials into the cell. The peduncle is retracted into the cell when feeding is finished.

Most pigmented dinoflagellates typically contain chlorophylls *a* and *c*, which are generally masked by carotenoid pigments, including **peridinin,** which is similar to fucoxanthin, an accessory pigment typical of chrysophytes (see page 333). The presence of peridinin supports the hypothesis that the chloroplasts of many dinoflagellates were derived from ingested diatoms and haptophytes by endosymbiosis, as described in Chapter 12. Other dinoflagellates have green or blue-green plastids obtained from ingested green algae or cryptomonads. The carbohydrate food reserve in dinoflagellates is starch, which is stored in the cytosol.

Pigmented dinoflagellates occur as symbionts in many other kinds of organisms, including sponges, jellyfish, sea anemones, tunicates, corals, octopuses and squids, snails, turbellarians, and certain other protists. In the giant clams of the family Tridacnidae, the dorsal surface of the inner lobes of the mantle may appear chocolate-brown as a result of the presence of symbiotic dinoflagellates. When they are symbionts, dinoflagellates lack armored plates and appear as golden spherical cells called **zooxanthellae** (Figure 15–12).

Zooxanthellae are primarily responsible for the photosynthetic productivity that makes possible the growth of coral reefs in tropical waters, which are notoriously nutrient-poor (see "Coral Reefs and Global Warming" on facing page). Coral tissues may contain as many as 30,000 symbiotic dinoflagellates per cubic millimeter, primarily in cells that line the gut of the coral polyps. Amino acids produced by the polyps stimulate the dinoflagellates to produce glycerol instead of starch. The glycerol is used directly for coral respiration. Because the dinoflagellates require light for photosynthesis, the corals containing them grow mainly in ocean waters less than 60 meters deep. Many of the variations in the shapes of coral are related to the light-gathering properties of different geometric arrangements. This relationship is somewhat similar to the ways in which various branching patterns of trees serve to maximally expose their leaves to sunlight.

During Periods of Unfavorable Conditions, Dinoflagellates Form Resting Cysts

Under conditions that do not allow continued population growth, such as low nutrient levels, dinoflagellates may produce

CORAL REEFS AND GLOBAL WARMING

Coral reefs, with their spectacular colors and convoluted surfaces, provide habitats for one-fourth of all ocean-dwelling species. As the rainforest is to the land, the coral reef is to the sea—a unique and irreplaceable ecosystem for fostering biodiversity. In addition, reef-dwelling fish are a major food source for humans, and, by absorbing much of the energy of large waves, coral reefs protect fragile coastlines from the worst of ocean storms.

In 1998, one-sixth of the corals in the world turned white and, within a few short months, died. This worldwide coral "bleaching" coincided with unusually high sea temperatures throughout the tropics, where most corals live. As global temperatures increase, what was unusual in 1998 is likely to become commonplace. With the warming of the oceans, coral reefs everywhere are becoming endangered ecosystems.

Corals are animals in the phylum Cnidaria, whose members include jellyfish and sea anemones. The individual organism, called a polyp, secretes a hard exoskeleton made of calcium carbonate, the same material found in chalk. A head or branch of coral, composed of many hundreds of polyps whose exoskeletons have fused together, creates, over time, the large structure we recognize as a reef.

Coral polyps obtain much of their nutrition by absorbing carbohydrates from symbiotic dinoflagellates called zooxanthellae, which live within the polyps and give color to the otherwise white corals. Bleaching occurs when the coral polyp loses its zooxanthellae. The primary trigger for this loss is environmental stress, especially high water temperature and acidification, both due to increased carbon dioxide in the atmosphere. A rise in water temperature of 1° to 2°C over several weeks is enough to provoke bleaching. If the temperature falls soon after, the polyp may regain its symbiont and survive. But if temperatures remain elevated for many weeks, the polyp is likely to die. Acidification of ocean waters occurs because carbon dioxide dissolved in water produces carbonic acid. The more CO_2, the more acidic the ocean becomes, and the more stress there is on corals.

The first known reports of coral bleaching date from the 1970s. With the steady rise in ocean temperatures, bleaching has increased. The 1998 event was the worst ever seen, but it is unlikely to hold that distinction for long. Sea surface temperatures fluctuate seasonally and over longer cycles, but on average they are rising in step with the increase in atmospheric CO_2. As the baseline temperature rises, fluctuations will raise the temperatures above the coral's limit of tolerance more often, and for longer. If current trends continue, bleaching conditions will occur annually throughout the tropics within the next 20 years. Within 40 years, average winter ocean temperatures could be too warm for corals. The concentration of CO_2 in the atmosphere is currently at 390 parts per million (ppm), and is rising by almost 2 ppm yearly. According to one of the world's experts on coral reefs, an atmospheric CO_2 concentration of 450 ppm is incompatible with the continued existence of the world's coral reefs. These essential ecosystems are indeed under significant threat.

Damaged reef A healthy Great Barrier reef in Australia (left), and a reef showing the effects of coral bleaching (right).

nonmotile resting cysts that drift to the lake or ocean bottom, where they remain viable for years. Ocean currents may transport these benthic (meaning "bottom of the water") cysts to other locations. When conditions become favorable, the cysts may germinate, reviving the population of swimming cells. Cyst production, movement, and germination explain many aspects of the ecology and geography of toxic dinoflagellate blooms. They explain why blooms do not necessarily occur at the same site every year, and why blooms are associated with nutrient pollution of the oceans by sewage and agricultural runoff. In addition, they explain why blooms appear to move from one location to another from year to year. Sexual reproduction has been found in a number of species of dinoflagellates.

Many Dinoflagellates Produce Toxic or Bioluminescent Compounds

About 20 percent of all known dinoflagellate species produce one or more highly toxic compounds that are economically and ecologically significant (see the essay on page 323). The toxic compounds produced by dinoflagellates may provide protection from predation.

Marine dinoflagellates are also famous for their bioluminescent capabilities (Figure 15–10b). These are responsible for the attractive sparkling of ocean waters that is commonly observed at night as boats or swimmers agitate the water. When dinoflagellate cells are disturbed, a series of well-understood biochemical events results in a reaction involving luciferin and the enzyme luciferase that, as in fireflies and other organisms, creates a brief flash of light (see Chapters 10 and 28). Bioluminescence is thought to serve as protection against predators, such as copepods, small crustaceans that are the most numerous members of the zooplankton. One hypothesis is that the dinoflagellates' light flashes directly disrupt feeding by startling the predators. Another hypothesis suggests a more indirect process: copepods that have fed on luminescent dinoflagellates become more visible to the fish that feed on them.

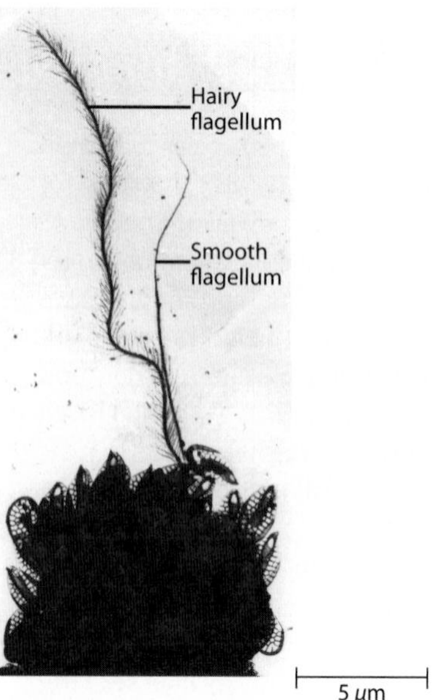

15–13 Stramenopile flagella The freshwater alga *Synura petersenii* has both hairy and smooth flagella, as is characteristic of stramenopiles. *Synura* is infamous for the unpleasant tastes and odors it imparts to the water in which it grows.

Photosynthetic Stramenopiles

Electron microscopists have long suspected that the diatoms, chrysophytes, xanthophytes, brown algae, and oomycetes, along with certain other groups not discussed in this book, are closely related, based on the presence of distinctive strawlike hairs on one of the two flagella of their swimming cells. The term **stramenopiles** is derived from the Greek meaning "straw hairs." In these organisms, also known as **heterokonts** (meaning "different flagella"), the flagellum ornamented with the distinctive hairs is long and the other is shorter and smooth, as shown in Figure 15–13. Molecular sequence analyses have now confirmed the suspicion, once based only on their unique flagella, that the groups listed above are indeed closely related. Moreover, studies have confirmed that these photosynthetic stramenopiles are also closely related to several clades of heterotrophic protists that lack plastids. Among the latter are the oomycetes, which are considered later in this chapter.

Diatoms: Class Bacillariophyceae

The diatoms are unicellular or colonial organisms that are exceedingly important components of the phytoplankton (Figure 15–14). It has been estimated that diatoms account for as much as 25 percent of global carbon fixation, equivalent to that of all terrestrial rainforests combined; they clearly are major players in the well-being of our planet. Diatoms, especially very small forms, account for the greatest biomass and species diversity of phytoplankton in polar waters. For aquatic animals in both marine and freshwater habitats, diatoms are a primary source of food. Species such as *Thalassiosira pseudonana* are commonly used as food in the commercial marine culture, or mariculture, of economically valuable bivalves, such as oysters. Diatoms provide the animals with essential carbohydrates, fatty acids, sterols, and vitamins.

There are about 285 genera and 10,000 to 12,000 recognized species of living diatoms; however, one expert believes that the actual number of living species may be in the millions. There are thousands of extinct species, known from the remains of their silica-containing cell walls. Fossil diatoms date back to about 180 million years ago. They first became abundant in the fossil record 100 million years ago, during the Cretaceous period. The fact that many of the fossil species are identical to those still living today indicates an unusual persistence through geologic time.

There are often tremendous numbers of individual diatoms in small areas. For example, over 30 to 50 million individuals of the freshwater genus *Achnanthes* may be found on 1 square centimeter of a submerged rock in the streams of North America. Large numbers of species likewise may occur together. In two small samples of mud from the ocean near Beaufort, North Carolina, for example, 369 species of diatoms were identified. Most species of diatoms occur in plankton, but some are bottom dwellers or grow on other algae or on submerged plants.

The Walls of Diatoms Consist of Two Halves A unique feature of diatoms is their two-part cell walls. Known as **frustules,** the walls, which are made of polymerized, opaline silica ($SiO_2 \cdot nH_2O$), consist of two overlapping halves. The two halves fit together like a laboratory Petri dish. Electron microscopy has shown that fine tracings

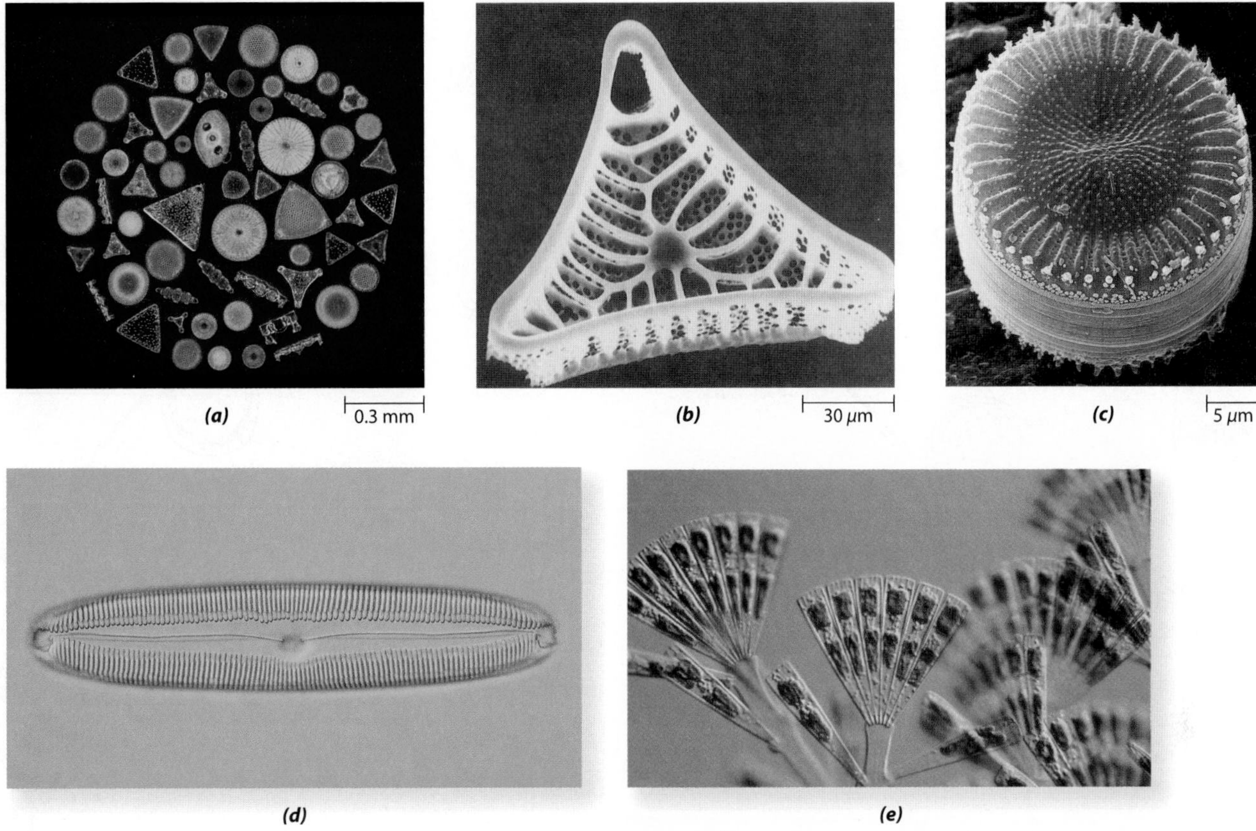

15–14 Diatoms ***(a)*** A selected array of marine diatoms, as seen with a light microscope. ***(b)*** Scanning electron micrograph of one half of an *Entogonia* frustule. ***(c)*** Scanning electron micrograph of *Cyclotella meneghiniana,* a centric diatom that occurs in both fresh water and brackish water. ***(d)*** Photomicrograph of *Pinnularia,* a pennate diatom, showing a clearly visible raphe. ***(e)*** *Licmophora flabellata,* a stalked pennate diatom, as seen with a light microscope.

on the diatom frustules are composed of a large number of minute, intricately shaped depressions, pores, or passageways, some of which connect the living protoplast within the frustule to the outside environment (Figure 15–14b, c). Species can be distinguished by differences in frustule ornamentation. In most cases, both halves of the frustule have exactly the same ornamentation, but in some cases, the ornamentation may differ.

On the basis of symmetry, two major types of diatoms are recognized: the **pennate** diatoms, which are bilaterally symmetrical (Figure 15–14d, e), and the **centric** diatoms, which are radially symmetrical (Figure 15–14c). Centric diatoms, which have a larger surface-to-volume ratio than the pennate forms and consequently float more easily, are more abundant than pennate diatoms in large lakes and marine habitats.

Diatoms lack flagella, except on some male gametes. Despite their lack of flagella or other locomotor organelles, many species of pennate diatoms and some species of centric diatoms are motile. Their locomotion results from a rigorously controlled secretion that occurs in response to a wide variety of physical and chemical stimuli. All pennate diatoms seem to possess, along one or both valves, a fine groove called the **raphe** (Figure 15–14d), which is basically a pair of pores connected by a complex slit in the siliceous wall. The diatoms secrete mucilage from the slit, allowing them to move.

Reproduction in Diatoms Is Mainly Asexual, Occurring by Cell Division When cell division takes place, each daughter cell receives one half of the frustule of its parental cell and constructs a new half (Figure 15–15). As a consequence, one of the two new cells is typically somewhat smaller than the parental cell, and after a long series of cell divisions, the size of the diatoms in the resulting population tends to decline. In some diatom populations, when the size decreases to a critical level, sexual reproduction occurs. The cells that develop from division of the zygote typically regain maximum size for the species. In some other cases, sexual reproduction is triggered by changes in the physical environment.

The sexual life history of diatoms is gametic (see Figure 12–17b), like that of animals and certain brown and green seaweeds. Sexual reproduction in the centric diatoms is **oogamous,** meaning that the female gamete is a relatively large, nonflagellated egg and the male gamete is noticeably smaller and flagellated. The male gametes, with a single flagellum, are the only

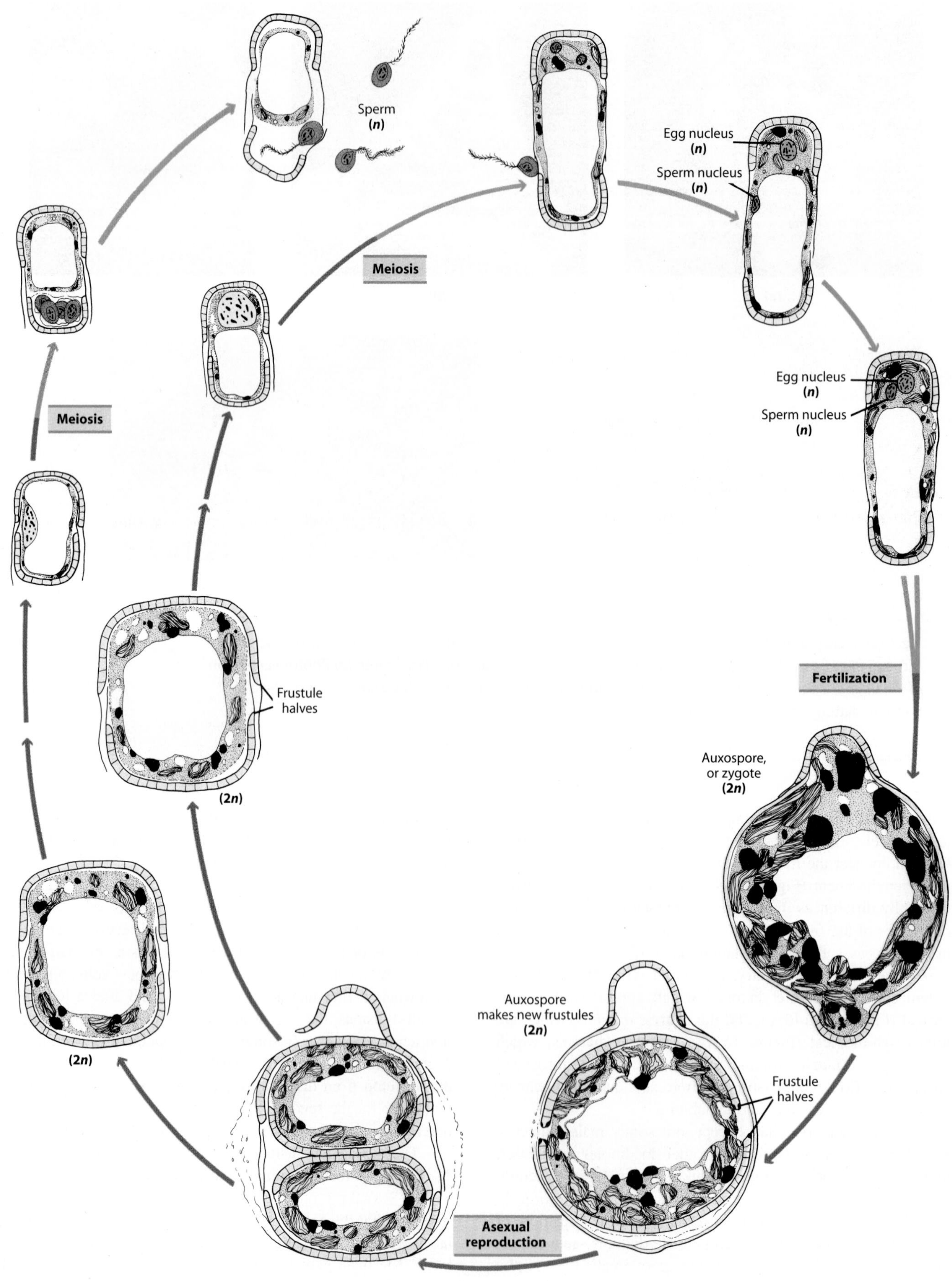
Sperm
(n)
Egg nucleus
(n)
Sperm nucleus
(n)
Meiosis
Meiosis
Egg nucleus
(n)
Sperm nucleus
(n)
Frustule
halves
(2n)
Fertilization
Auxospore,
or zygote
(2n)
(2n)
Auxospore
makes new frustules
(2n)
Frustule
halves
Asexual
reproduction

15–15 Generalized life cycle in a centric diatom
Reproduction in the diatoms is mainly asexual, occurring by cell division. The cell walls, or frustules, in all diatoms are composed of two overlapping portions, one of which encloses the other. When cell division takes place, each daughter cell (bottom left) receives one half of the parental frustule (bottom right) and constructs a new frustule half. The existing half is always the larger part of the silica wall, with the new half fitting inside it. Thus one daughter cell of each new pair tends to be smaller than the parental cell from which it is derived.

In some species, the frustules are expandable and are enlarged by the growing protoplast within them. In other species, however, the frustules are more rigid. Thus, in a population, the average cell size decreases through successive cell divisions. When the individuals of these species have decreased in size to about 30 percent of the maximum diameter, sexual reproduction may occur (top). Certain cells function as male gametangia, and each produces sperm through meiosis. Other cells function as female gametangia. In these female gametangia, two or three of the four products of meiosis are nonfunctional, so that one or two eggs are produced per cell. This is an example of gametic meiosis. After fertilization, the resulting auxospore, or zygote, expands to the full size characteristic of the species. The walls formed by the auxospore are often different from those of the asexually reproducing cells of the same species. Once the auxospore is mature, it divides and produces new frustule halves with intricate markings typical of the asexually reproducing cells.

flagellated cells found in diatoms at any stage of their life cycle. In the pennate diatoms, sexual reproduction is **isogamous,** meaning that the gametes are equal in size and shape; both male and female gametes are nonflagellated. Both types of sexual reproduction result in empty frustules that readily sediment (meaning "to sink down"). Researchers have observed that mass sexual reproduction of marine diatoms can result in the formation of layers of silica in southern ocean sediments.

Unfavorable conditions, such as low levels of mineral nutrients, can cause marine coastal or benthic diatoms to form resting stages. The resting cells have heavy frustules, which allow them to sink readily to the bottom (if they are not already there). These cells germinate when nutrient conditions improve. Diatoms are frequently most abundant in the spring and fall, when upwelling in the oceans or wind-driven turnover of stratified lakes occurs. These processes resuspend sufficient silica for diatom growth. When the silica becomes depleted, diatom blooms give way to dominance by other phytoplankton that do not require silica. Diatoms may bloom beneath the winter ice cover of lakes, because herbivorous animals do not actively feed during the cold season.

The silica frustules of diatoms have accumulated in ocean sediments over millions of years, forming the fine, crumbly substance known as diatomaceous earth. This substance is used as an abrasive in silver polish and as a filtering and insulating material. The Santa Maria, California, oil fields contain a subterranean deposit of diatomaceous earth that is 900 meters thick, and near Lompoc, California, more than 270,000 metric tons of diatomaceous earth are quarried annually for industrial use.

The most conspicuous features within the protoplast of diatoms are the brownish plastids that contain chlorophylls *a* and *c*, as well as fucoxanthin. There are usually two large plastids in the cells of pennate diatoms, whereas centric diatoms have numerous discoid plastids. Diatoms' reserve storage materials include lipids and the water-soluble polysaccharide **chrysolaminarin,** which is stored in vacuoles. Chrysolaminarin is similar to the laminarin found in brown algae.

Although most species of diatoms are autotrophic, some are heterotrophic, absorbing dissolved organic carbon. These heterotrophic species are primarily pennate diatoms that live on the bottom of the sea in relatively shallow habitats. A few diatoms are obligate heterotrophs. They lack chlorophyll and thus cannot produce their own food through photosynthesis. On the other hand, some diatoms, lacking their characteristic frustules, live symbiotically in large marine protozoa (order Foraminifera) and provide organic carbon to their hosts. Certain diatoms are associated with production of the neurotoxin domoic acid, which causes a condition called amnesic shellfish poisoning, which can result in permanent or short-term memory loss, or even death, in humans.

Golden Algae: Class Chrysophyceae

The golden algae, or chrysophytes, are primarily unicellular or colonial organisms that are abundant in freshwater habitats throughout the world; marine representatives are also known. There are a few plasmodial, filamentous, and tissuelike forms and about 1000 known species. Some chrysophytes are colorless, whereas others have chlorophylls *a* and *c*, the color of which is largely masked by an abundance of fucoxanthin. The golden color of this pigment gives rise to the name "chrysophyte" (Gk. *chrysos,* "gold," and *phyton,* "plant"). An individual pigmented cell usually contains one or two large chloroplasts. As in the diatoms, the carbohydrate food reserve is chrysolaminarin. It is stored in a vacuole usually found in the posterior of the cell.

Several chrysophytes are known to ingest bacteria and other organic particles. Individual cells of the freshwater, pigmented, motile colony *Dinobryon* can each consume about 36 bacteria per hour (Figure 15–16a). *Dinobryon* and its chrysophyte relatives are the major consumers of bacteria in some of the cooler lakes of North America. *Poterioochromonas* can ingest motile algal cells that are two to three times larger in diameter than itself. Its cell volume can expand as much as 30 times to accommodate the food it has eaten. *Uroglena americana,* another relative, seems to require prey food as a source of an essential phospholipid. Such particle-consuming chrysophytes can be dominant components of the phytoplankton in temperate lakes of low nutrient content (oligotrophic). The ability to consume particulate food provides an advantage under such low-nutrient conditions.

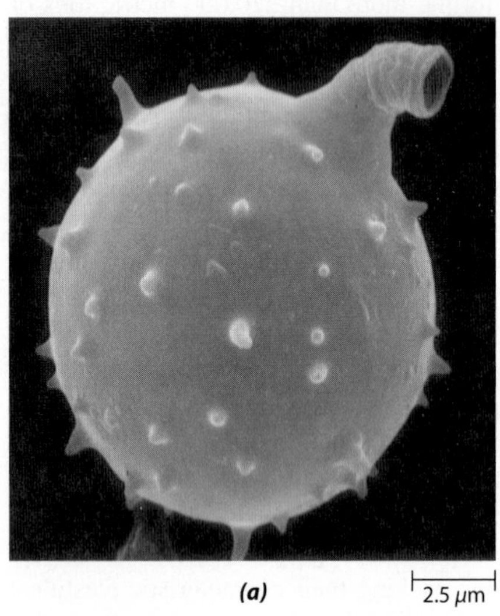

15–16 Representative chrysophytes ***(a)*** Scanning electron micrograph of a mature resting cyst of *Dinobryon cylindricum*. The hooked, cylindrical collar contains a pore through which the amoeboid cell emerges when it is ready to germinate. Surface spines represent ornamentation. ***(b)*** Scanning electron micrograph of silica scales and flagella of *Synura* cells in a colony. The scales are arranged in very regular, overlapping patterns in spiral rows.

Some chrysophytes have walls made of interwoven cellulosic fibrils that may be impregnated with minerals. Others lack walls, and some look very much like amoebas with plastids. The members of one group, known as the synurophytes after the motile, colonial genus *Synura* (Figure 15–16b), are covered with overlapping, ornamented silica scales. The scales are produced inside the cells in vesicles and are then transported to the outside. The presence of a scaly enclosure prevents these algae from feeding on particles. In cold, acidic lakes, their scales may persist in the sediments and provide useful information about past habitats.

Reproduction in most chrysophytes is asexual and, for some forms, involves zoospore formation. Sexual reproduction is known for some species. Characteristic resting cysts are commonly formed at the end of the growing season. Sometimes, but not always, cysts are the result of sexual reproduction. In some groups, the resting cysts contain silica, which, like silica-containing scales, can sediment and form valuable records of past ecological conditions.

Some freshwater chrysophytes can form blooms and are blamed for unpleasant tastes and odors that arise from their excretion of organic compounds into drinking water.

Yellow-Green Algae: Class Xanthophyceae

The yellow-green algae, or xanthophytes, are mostly nonmotile organisms, although some are amoeboid or flagellated, as are their gametes and zoospores. There are about 600 species of yellow-green algae, occurring primarily in freshwater or soil habitats. Their plastids contain chlorophylls *a* and *c* and carotenoids, but lack the accessory pigment fucoxanthin. The absence of fucoxanthin is responsible for the yellow-green appearance of the chloroplasts. The main storage product is oil, which exists as droplets in the cytosol. The cell walls are composed primarily of cellulose, with silica sometimes also present.

One of the better-known yellow-green algae is *Vaucheria*, the "water felt," a little-branched, filamentous alga that is coenocytic, with multiple nuclei not separated by cell walls. *Vaucheria* reproduces both asexually, by the formation of large, compound, multiflagellated zoospores, and sexually, by oogamy (Figure 15–17). *Vaucheria* is widespread in freshwater, brackish, and marine habitats, and often is found on mud that is alternately immersed in water and exposed to air.

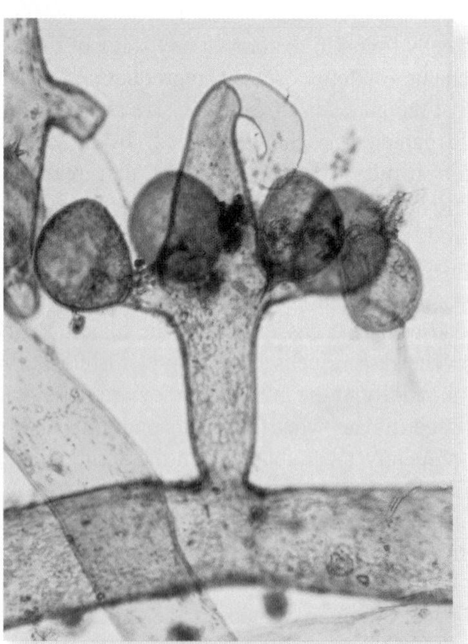

15–17 *Vaucheria*, the "water felt" This coenocytic, filamentous member of the class Xanthophyceae is oogamous, producing oogonia (blue spheres) and antheridia. The antheridium shown here is empty.

Brown Algae: Class Phaeophyceae

The brown algae, an almost entirely marine group, include the most conspicuous seaweeds of temperate, boreal (northern), and polar waters. Although there are only about 1500 species, the brown algae dominate rocky shores throughout the cooler regions of the world (Figure 15–18). Most people have observed seashore rocks covered with **rockweeds,** the common name for members of the brown algal order Fucales. The larger brown algae of the order Laminariales, a number of which form extensive beds offshore, are called **kelps.** In clear water, brown algae flourish from low-tide level to a depth of 20 to 30 meters. Surprisingly, tropical kelp beds have been discovered offshore at the Galápagos Islands at a depth of about 60 meters. On gently sloping shores, kelp beds may extend 5 to 10 kilometers from the coastline.

One member of the order Fucales, *Sargassum,* forms immense floating masses in the Sargasso Sea (so named because of the abundance of *Sargassum*) in the Atlantic Ocean, northeast of the Caribbean (Figure 15–19). *Sargassum muticum* and some other brown algae can form nuisance growths when introduced into nonnative areas and can seriously interfere with commercial mariculture operations. *Sargassum* can also outcompete and replace members of the Laminariales and Fucales, which are regarded as critical, or keystone, species in their communities.

The Basic Form of a Brown Alga Is a Thallus Though they are a monophyletic group, brown algae range in size from microscopic forms to the largest of all seaweeds, the kelps, which are as much as 60 meters long and weigh more than 300 kilograms. The basic form of a brown alga is a **thallus** (plural: thalli), a simple,

(a)

(b)

(c)

15–18 Brown algae ***(a)*** Bull kelp, *Durvillaea antarctica,* exposed at low tide on the rugged coastline on South Island, New Zealand. ***(b)*** Detail of the kelp *Laminaria,* showing holdfasts, stipes, and the bases of several fronds. ***(c)*** Rockweed, *Fucus vesiculosus,* densely covers many rocky shores that are exposed at low tide. When submerged, the air-filled bladders on the blades carry them up toward the light. Photosynthetic rates of frequently exposed marine algae are one to seven times as great in air as in water, whereas the rates are higher in water for those rarely exposed. This difference accounts, in part, for the vertical distribution of seaweeds in intertidal areas.

15–19 *Sargassum* ***(a)*** The brown alga *Sargassum* has a complex pattern of organization. A member of the order Fucales, *Sargassum* has a life cycle like that of *Fucus* (Figure 15–23). ***(b)*** Two species of this genus form immense free-floating masses in the Sargasso Sea.

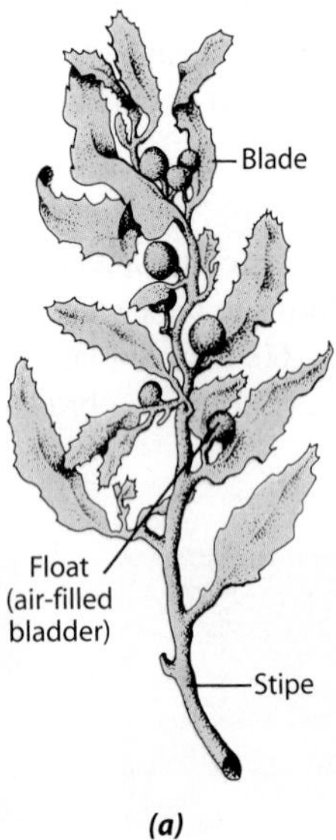

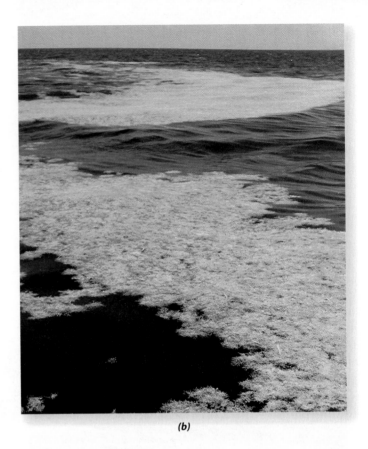

relatively undifferentiated vegetative body. The thalli range in complexity from simple branched filaments, as in *Ectocarpus* (Figure 15–20), to aggregations of branched filaments that are called pseudoparenchyma because they look tissuelike, to authentic tissues, such as are found in *Macrocystis* (Figure 15–21). As in some green algae and plants, adjacent cells are typically linked by plasmodesmata. Unlike plant plasmodesmata, those of brown algae do not seem to have desmotubules connecting the endoplasmic reticulum of adjacent cells (see Figure 4–18).

The Pigment Fucoxanthin Gives the Brown Algae Their Characteristic Color The cells of brown algae typically contain numerous disk-shaped, golden-brown plastids that are similar both biochemically and structurally to the plastids of chrysophytes and diatoms, with which they probably had a common origin. In addition to chlorophylls *a* and *c*, the chloroplasts of brown algae also contain various carotenoids, including an abundance of fucoxanthin, which gives the members of this class their characteristic dark-brown or olive-green color. The reserve storage material in brown algae is the carbohydrate **laminarin,** which is stored in vacuoles. Molecular analyses suggest that there are two major lineages of brown algae: those with starch-producing pyrenoids in their plastids, including *Ectocarpus* (Figure 15–20), and those without pyrenoids, such as *Laminaria* and its relatives. These two groups also consistently differ in the structure of their sperm.

The Kelps and Rockweeds Have Highly Differentiated Bodies Large kelps, such as *Laminaria,* are differentiated into regions known as the **holdfast, stipe,** and **blade,** with a meristematic region located between the blade and stipe (Figure 15–18b). The pattern of growth resulting from this type of meristematic activity is particularly important in the commercial use of *Macrocystis,* which is harvested along the California coast. When the older blades are harvested at the surface by kelp-cutting boats, *Macrocystis* is able to regenerate new blades. Giant kelps, such as *Macrocystis* and *Nereocystis,* may be more than 60 meters long. They grow very rapidly, so a considerable amount of material is available for harvest. One of the most important products derived from the kelps is a mucilaginous intercellular material called **algin,** which is important as a stabilizer and emulsifier for some foods and for paint and as a coating for paper. Algin, together with the cellulose in inner cell wall layers, provides the flexibility and toughness that allow seaweeds to withstand the mechanical stress imposed by waves and currents. Algin also helps to reduce drying when the seaweeds are exposed by low tides, and it increases buoyancy and helps to slough off organisms that attempt to colonize the algal blades.

The internal structure of kelps is complex. Some of them have, in the center of the stipe, elongated cells that are modified for food conduction. These cells resemble the food-conducting cells in the phloem of vascular plants, including the presence of sieve plates (Figure 15–21b). The cells can conduct food material rapidly—at rates as high as 60 centimeters per hour—from the blades at the water surface to the poorly illuminated stipe and holdfast regions far below. Lateral translocation from the outer photosynthetic layers to the inner cells takes place in many relatively thick kelps. **Mannitol** is the primary carbohydrate that is translocated, along with amino acids.

The rockweed *Fucus* (Figure 15–18c) is a dichotomously branching brown alga that has **air bladders** near the ends of its blades. The pattern of differentiation of *Fucus* otherwise resembles that of the kelps. *Sargassum* (Figure 15–19) is related to *Fucus.* Some species of *Sargassum* remain attached, whereas in others, the individuals form floating masses in which the hold-

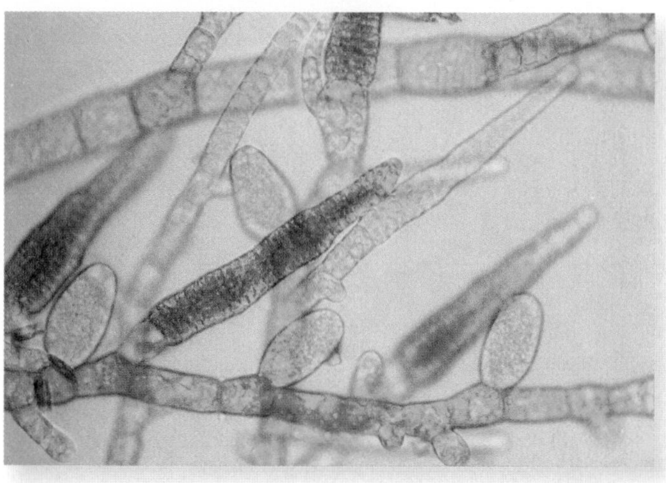

15–20 *Ectocarpus* The brown alga *Ectocarpus* has simple branched filaments. This micrograph of *Ectocarpus siliculosus* shows unilocular sporangia (the short, rounded, light-colored structures) and plurilocular sporangia (the longer, dark-colored structures), which are borne on sporophytes. Meiosis, resulting in haploid zoospores, takes place within the unilocular sporangia. Diploid zoospores are formed in the plurilocular sporangia. *Ectocarpus* occurs in shallow water and estuaries throughout the world, from cold Arctic and Antarctic waters to the tropics.

fasts have been lost. Both forms occur in some species. *Fucus* and *Sargassum* and some other brown algae grow by means of repeated divisions from a single apical cell, not from a meristem located within the body, as is characteristic of the kelps.

The Life Cycles of Most Brown Algae Involve Sporic Meiosis For most brown algae, the life cycles involve an alternation of generations and, therefore, sporic meiosis (see Figure 12-17c). The gametophytes of the more primitive brown algae, such as *Ectocarpus,* produce multicellular reproductive structures called **plurilocular gametangia.** These may function as male or female gametangia, or they may produce flagellated haploid spores that give rise to new gametophytes. The diploid sporophytes produce both **plurilocular** and **unilocular sporangia** (Figure 15–20). The plurilocular sporangia form diploid zoospores that produce new sporophytes. Meiosis takes place within the unilocular sporangia, producing haploid zoospores that germinate to produce gametophytes. Unilocular sporangia, along with algin and plasmodesmata, are defining features of the brown algae.

In *Ectocarpus,* the gametophyte and sporophyte are similar in size and appearance (isomorphic). Many of the larger brown algae, however, including the kelps, undergo an alternation of heteromorphic generations—a large sporophyte and a microscopic gametophyte, as in the common kelp *Laminaria* (Figure 15–22). In *Laminaria,* the unilocular sporangia are produced on the surface of the mature blades. Half of the zoospores produced in the sporangia have the potential to grow into male gametophytes, and half into female gametophytes. According to one hypothesis, the plurilocular gametangia borne on these gametophytes have become modified during the course of evolution into one-celled antheridia and one-celled oogonia. Each antheridium releases a single sperm, and each oogonium contains a single egg. The fertilized egg in *Laminaria* remains attached to the female gametophyte and develops into a new sporophyte. In several genera of brown algae, the female gametes attract the male gametes by organic compounds.

Fucus and its close relatives have a gametic life cycle (Figure 15–23), as do the diatoms and certain green seaweeds. Understanding the evolutionary pressures that stimulated the origin of gametic life cycles in these protists may help to explain

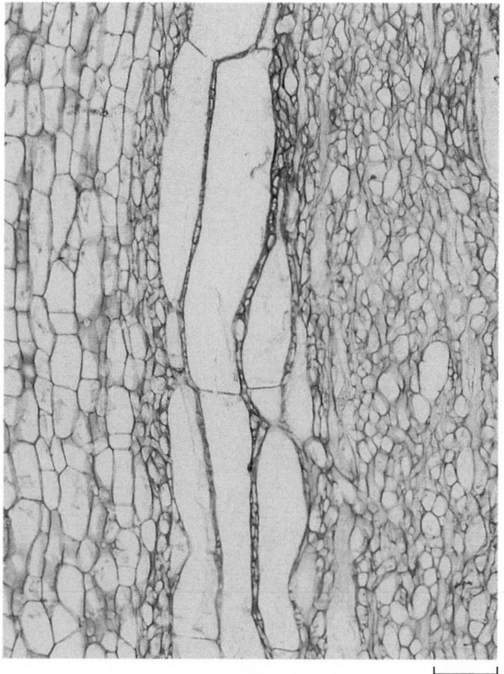

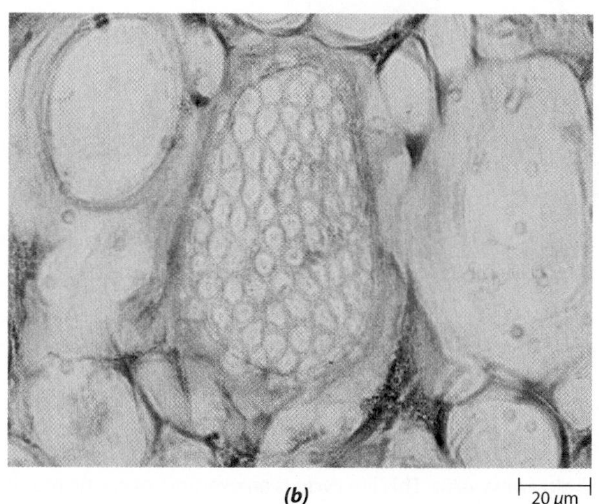

15–21 *Macrocystis* Some brown algae, such as the giant kelp *Macrocystis integrifolia,* have sieve tubes comparable to those found in food-conducting tissue of vascular plants. ***(a)*** Longitudinal section of part of a stipe, with sieve tubes, the relatively wide elements in the middle of the micrograph. The sieve-tube elements are joined end-on-end by the sieve plates, which appear as narrow cross walls here. ***(b)*** Cross section showing a sieve plate.

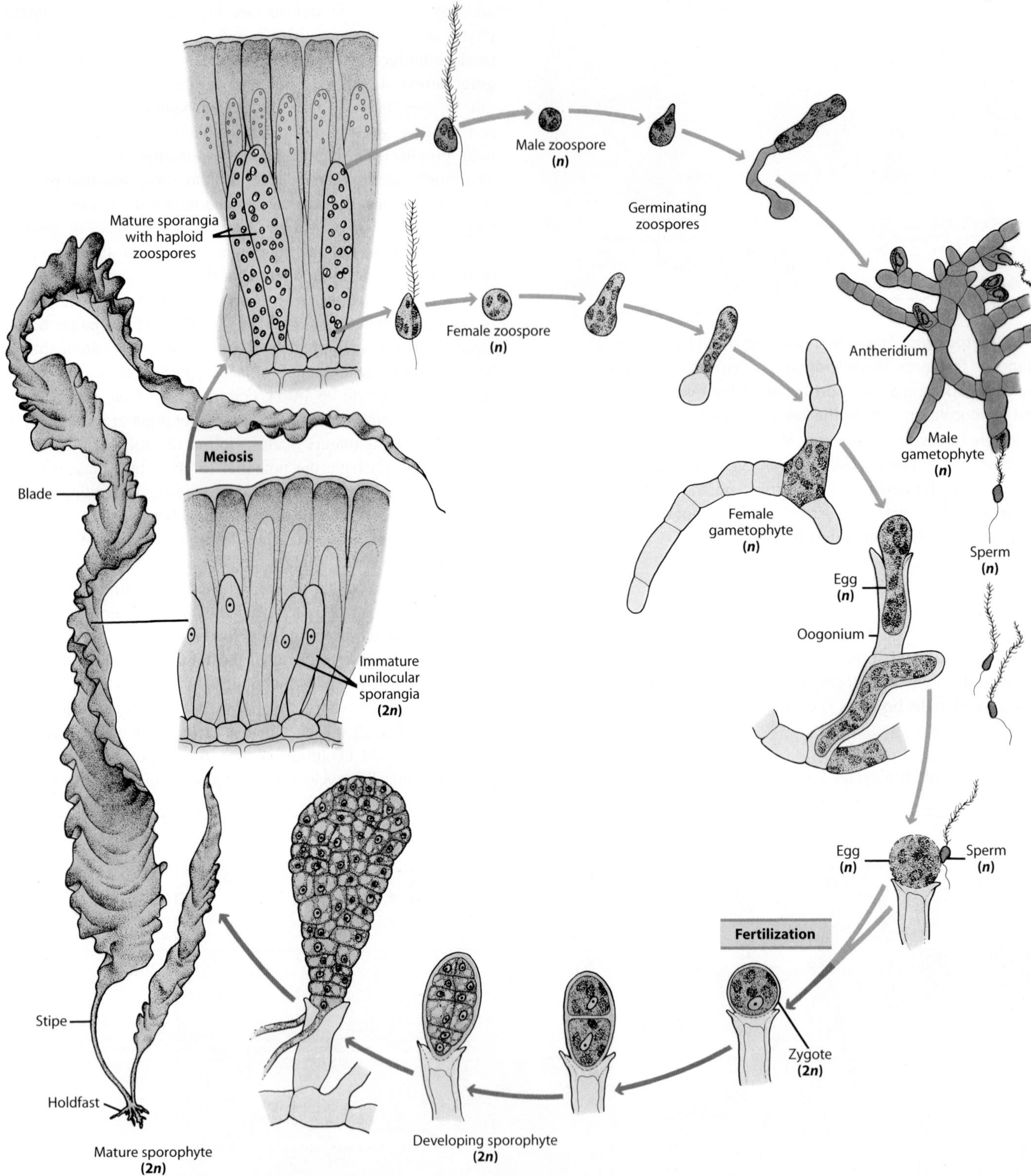

15–22 Life cycle of the kelp *Laminaria* This life cycle is an example of sporic meiosis, or alternation of generations. Like many of the brown algae, *Laminaria* has an alternation of heteromorphic generations in which the sporophyte is conspicuous. Motile haploid zoospores are produced in the sporangia following meiosis (upper left). From these zoospores grow the microscopic, filamentous gametophytes, which in turn produce the motile sperm and nonmotile eggs. In some other brown algae, the sporophyte and gametophyte are similar; they have an alternation of isomorphic generations.

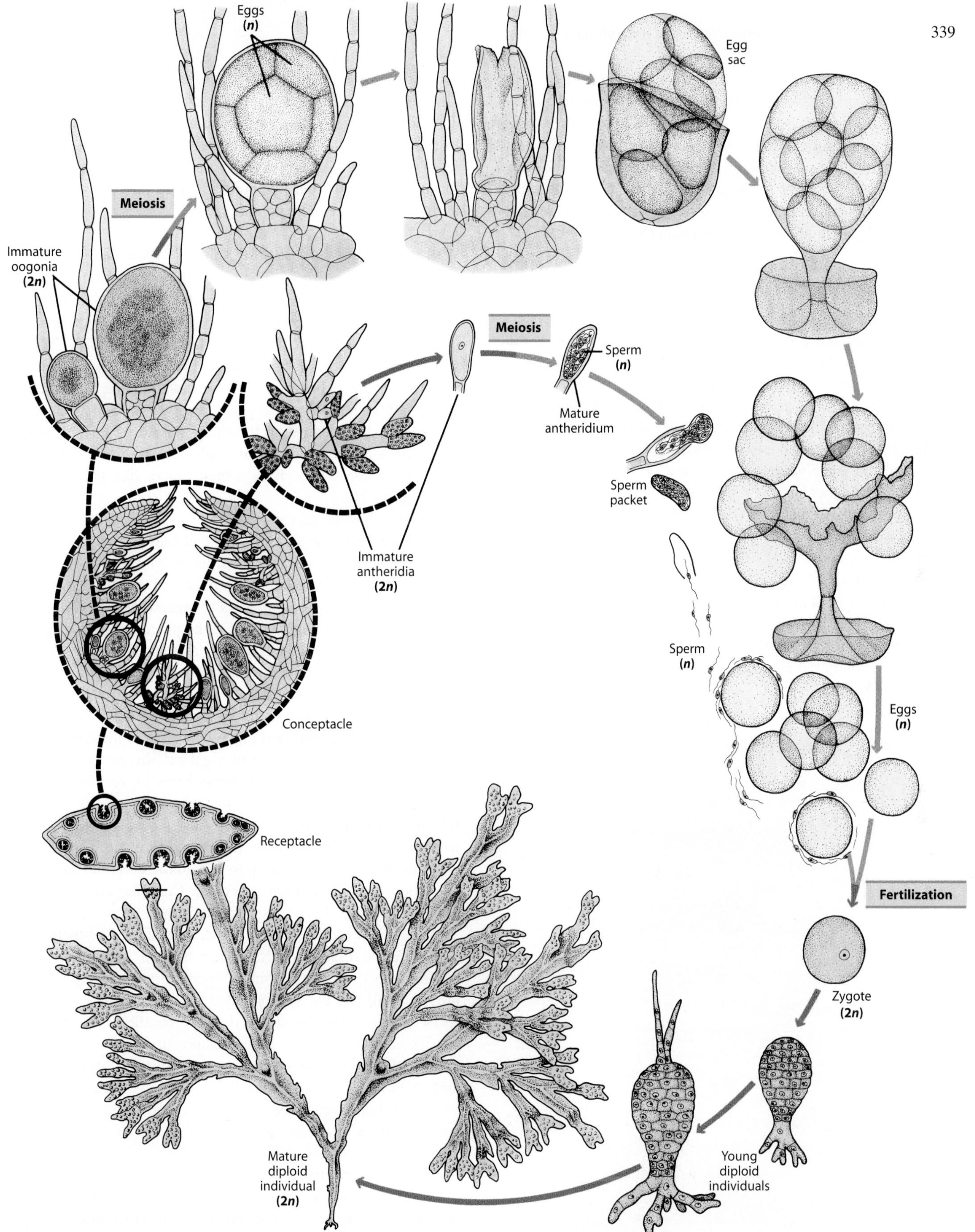

15–23 ***Fucus*** **life cycle** In *Fucus,* gametangia are formed in specialized hollow chambers known as conceptacles, which are found in fertile areas called receptacles at the tips of the branches of diploid individuals (lower left). There are two types of gametangia: oogonia and antheridia. Meiosis is followed immediately by mitosis to give rise to 8 eggs per oogonium and 64 sperm per antheridium. Eventually, the eggs and sperm are set free in the water, where fertilization takes place. Meiosis is gametic, and the zygote grows directly into the new diploid individual.

the early appearance of the gametic life cycle in our own metazoan (Gk. *meta,* "between," and *zōion,* "animal") lineage.

Some brown seaweeds, including *Fucus,* may contain large amounts of phenolic compounds that discourage herbivores. Other brown seaweeds tend to produce terpenes for the same purpose (pages 31 to 33). In some cases, these compounds also have anti-microbial or anti-tumor activities, prompting investigations of their potential use in human medicine.

Red Algae: Phylum Rhodophyta

Red algae are particularly abundant in tropical and warm waters, although many are found in the cooler regions of the world. There are about 6000 known species in 500 to 600 genera, only a few of which are unicellular—such as *Cyanidium,* one of the only organisms capable of growth in acidic hot springs—or microscopic filaments. The vast majority of red algae are more structurally complex, macroscopic seaweeds. Fewer than 100 different species of red algae occur in fresh water (Figure 15–24), but in the sea, the number of species is greater than that of all other types of seaweeds combined (Figure 15–25). Red algae usually grow attached to rocks or to other algae, but there are a few floating forms.

The chloroplasts of red algae contain phycobilins, which mask the color of chlorophyll *a* and give red algae their distinctive color. These pigments are particularly well suited to the absorption of the green and blue-green light that penetrates into deep water, where red algae are well represented. Biochemically and structurally, the chloroplasts of red algae closely resemble the cyanobacteria from which they were almost certainly derived, directly following endosymbiosis. Some red algae have lost most or all of their pigments and grow as parasites on other red algae. Chloroplasts of a few primitive red algae have starch-forming pyrenoids, but pyrenoids seem to have been lost prior to the origin of the more complex forms.

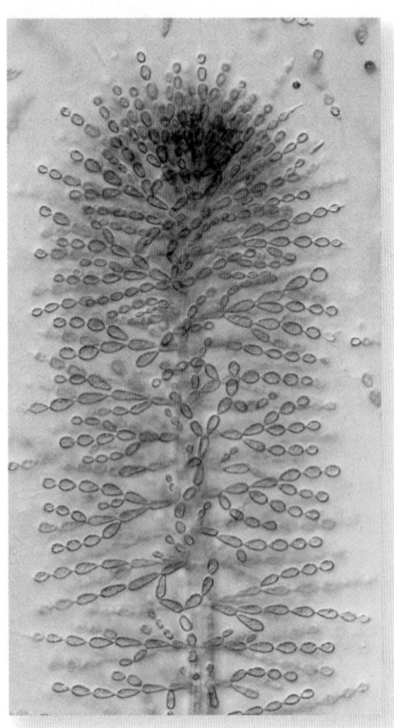

15–24 Freshwater red alga Seen here is the simple, filamentous, freshwater red alga *Batrachospermum* sp. The soft, gelatinous, branched axes of this red alga are most frequently found in cold streams, ponds, and lakes, where they occur throughout the world.

Cells of Red Algae Have Some Unique Features

Red algae are unusual among the algae, and unique among algal phyla, in having neither centrioles nor flagellated cells. It is likely that their ancestors had centrioles and that their loss resulted from a loss of function of the genes necessary for their synthesis. In place of centrioles, which occur in many other eukaryotes, the red algae have microtubule organizing centers called **polar rings.** The main food reserves of the red algae are granules of **floridean starch,** which are stored in the cytosol. Floridean starch, a unique molecule that resembles the amylopectin portion of starch, is actually more like glycogen than starch (page 21).

The cell walls of most red algae are composed of a loose network of cellulose microfibrils embedded in an amorphous gel-like mixture of sulfated galactan polymers and mucilages. The sulfated polygalactans are the major constituents of the commercially valuable **agar** and **carrageenans** (see "Algae and Human Affairs" on page 321). It is the mucilaginous component that gives such red algae their characteristic slippery texture. Continuous production and sloughing of the mucilage helps the algae rid themselves of other organisms that might colonize their surfaces and reduce their exposure to light.

In addition, certain red algae deposit calcium carbonate in their cell walls. The function of algal calcification is uncertain. One hypothesis is that calcification helps algae obtain carbon dioxide from the water for photosynthesis. Many of the calcified red algae are especially tough and stony, and they constitute the family Corallinaceae, the **coralline algae.** Calcification explains the occurrence of possible fossil coralline algae that are more than 700 million years old. Coralline algae are common throughout the oceans of the world, growing on stable surfaces that receive enough light, including 268-meter-deep seabed rocks (Figure 15–26). Other habitats include the rocks of tidal pools, on which jointed coralline red algae grow (Figure 15–25b), and the surf-pounded shoreward surfaces of coral reefs, where crust-forming (crustose) coralline algae (Figure 15–25c) help to stabilize reef structure. Large areas of diverse coral reefs around the world owe their survival, in part, to the architectural strength conferred by coralline algae. In recent years, a bright orange bacterium that causes a lethal disease of coralline algae has been spreading throughout the South Pacific, endangering thousands of kilometers of reef.

Recently, relatively small amounts of lignin have been detected in what have been described as secondary cell walls in the coralline alga *Calliarthron cheilosporioides.* The lignin is most abundant in the noncalcified joints, which are the most mechanically stressed parts of the alga. True lignin, which adds rigidity to cell walls, was previously thought to occur only in vascular plants. According to one hypothesis, the molecular pathway producing lignin emerged long before land plants

(a) (b)

(c) (d)

15–25 Marine red algae ***(a)*** In *Bonnemaisonia hamifera,* the basically filamentous structure of the red algae is clearly evident. The branched filaments are hooked, enabling the alga to cling to other seaweeds. ***(b)*** Jointed coralline algae in a tidal pool in California. ***(c)*** The reef-stabilizing, crustose coralline red alga *Porolithon craspedium.* ***(d)*** Irish moss, *Chondrus crispus,* an important source of carrageenan.

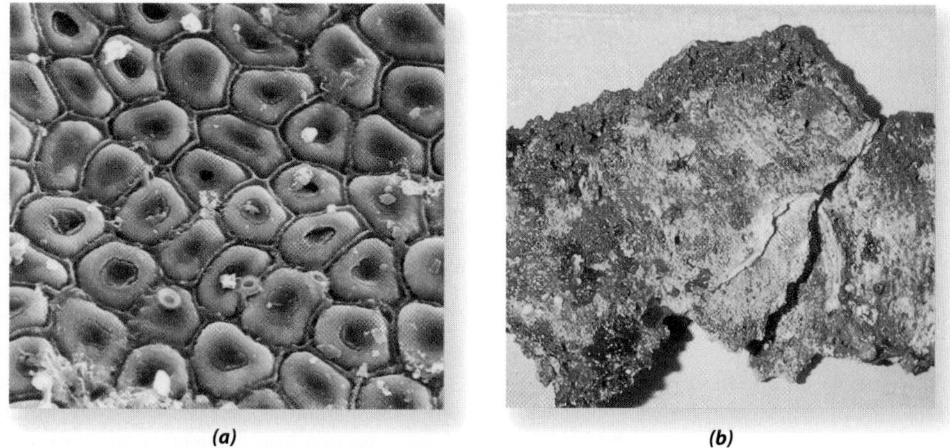

(a) (b)

15–26 Coralline algae ***(a)*** Scanning electron micrograph of an unidentified crustose red algal species from a depth of 268 meters on a seamount in the Bahamas, nearly 100 meters below the lowest limits established for any other photosynthesizing organism. At this depth, the light intensity was estimated as 0.0005 percent of its value at the ocean surface. The alga formed patches about a meter across, covering about 10 percent of the rock surface. When tested in the laboratory, this alga was found to be about 100 times more efficient than its shallow-water relatives in capturing and using light energy. ***(b)*** Purple crustose coralline algae from the same seamount.

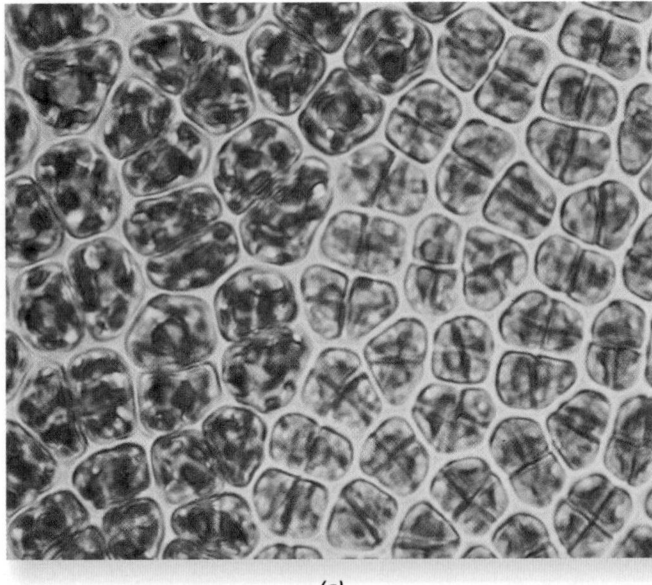

(a)

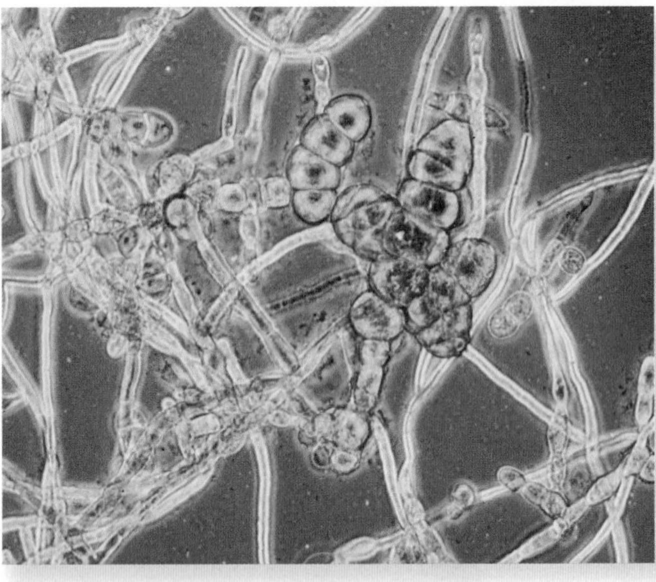

(b)

15–27 ***Porphyra nereocystis*** The life history of this red alga includes both bladelike and filamentous phases. ***(a)*** The bladelike phase, which is a gametophyte, produces both female gametangia, called carpogonia (reddish cell aggregates on left), and male gametangia, called spermatangia (cell aggregates on right). ***(b)*** After fertilization, the diploid carpospores give rise to a system of branched filaments. Meiosis occurs during germination of the spores (conchospores) produced by the filamentous phase. These haploid cells then grow into the bladelike gametophyte.

evolved from green algae, and the pathway may have evolved independently more than once—a case of parallel, or convergent, evolution.

Many red algae produce unusual toxic terpenoids (pages 31 to 33) that may assist in deterring herbivores. Some of these terpenoids have anti-tumor activity and are currently being tested for possible use as anti-cancer drugs.

A few genera, such as *Porphyra,* have cells densely packed together into one- or two-layered sheets (Figure 15–27). However, most red algae are composed of filaments that are often densely interwoven and held together by the mucilaginous layer, which has a rather firm consistency. Growth in filamentous red algae is initiated by a single, dome-shaped apical cell that cuts off segments sequentially to form an axis. This axis, in turn, forms whorls of lateral branches. In most red algae, the cells are interconnected by **primary pit plugs,** or primary pit connections (Figure 15–28), that develop at cytokinesis. Many red algae are multiaxial—that is, made up of many coherent filaments forming a three-dimensional body. In such forms, the filaments are interconnected by the formation of **secondary pit plugs.** These pits form between cells of different filaments when the filaments come into contact with one another.

Red Algae Have Complicated Life Histories

Many red algae reproduce asexually by discharging spores, called **monospores,** into the water. If conditions are suitable, monospores may attach to a substrate. By repeated mitoses, a new seaweed similar to the monospore-producing parent is produced. Sexual reproduction also occurs widely among the multicellular red algae and can involve very complex life histories.

The simplest type of sexual life history in red algae involves an alternation of generations between two separate, multicellular forms of the same species—a haploid, gamete-producing **gametophyte** and a diploid, spore-producing **sporophyte.** The gametophyte produces **spermatangia** (singular: spermatangium), structures that generate and release nonmotile **spermatia** (singular: spermatium), or male gametes, which are carried to female gametes by water currents. The female gamete, or egg, is the lower, nucleus-containing portion of a structure known as the **carpogonium** (plural: carpogonia), which is borne on the same gametophyte as the spermatangia and remains attached to it. The carpogonium develops a protuberance, the **trichogyne,** for reception of spermatia. When a spermatium comes in contact with a trichogyne, the two cells fuse. The male nucleus then migrates down the trichogyne to the female nucleus and fuses with it. The resultant diploid zygote produces a few diploid **carpospores,** which are released from the parent gametophyte into the water. If they survive, the carpospores attach to a surface and grow into the sporophyte, which produces haploid spores by sporic meiosis. If these haploid spores survive, they in turn attach to a surface and grow into gametophytes, thus completing the life cycle.

Experts believe that the red algae acquired an alternation of two multicellular generations early in their evolutionary history, as an adaptive response to their lack of flagellated male gametes. Nonflagellated gametes cannot swim toward female gametes, as do the flagellated male gametes of some other protists,

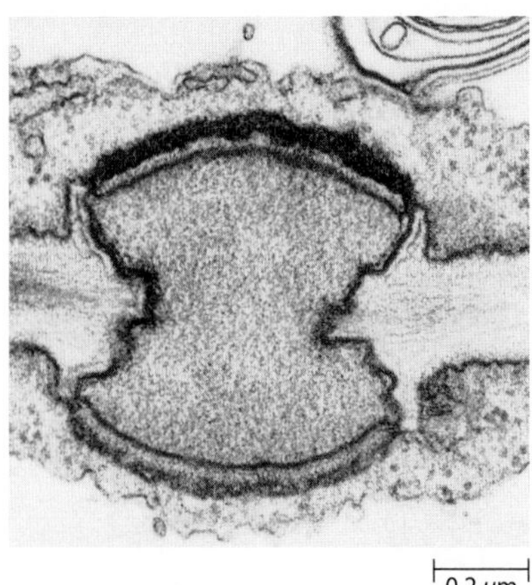

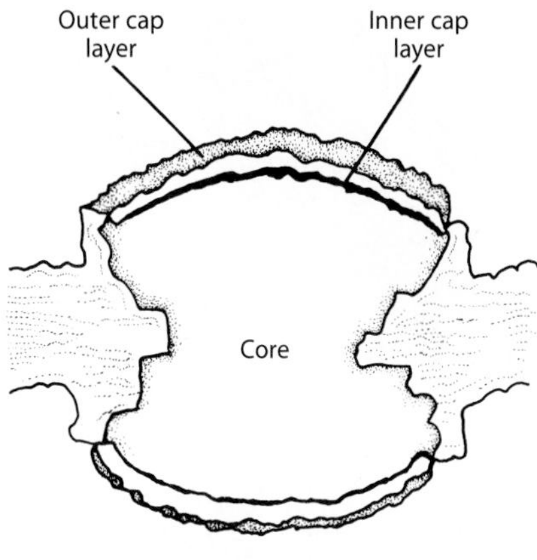

15–28 A primary pit plug in the red alga *Palmaria* Pit plugs are distinct, lens-shaped plugs that form between the cells of red algae as the cells divide. Similar plugs, called secondary pit plugs, also frequently form between the cells of adjacent filaments that come into contact with each other, linking together the bodies of individual red algae. Pit-plug cores are protein, and their outer cap layers are, at least in part, polysaccharide.

animals, and some plants. Hence, fertilization may be more a matter of chance, with the consequence that zygote formation may be relatively rare. Alternation of generations is regarded as an adaptation that increases the number and genetic diversity of the progeny resulting from each individual fertilization event, or zygote. This increase occurs because a multicellular sporophyte can produce many more spores—and more diverse haploid spores—than could a single, meiotic zygote nucleus. Alternation of two multicellular generations also occurs in several other protist groups, such as green and brown seaweeds, and in bryophytes and vascular plants (discussed in Chapters 16 to 19), where it may have similar ecological and genetic benefits.

A further evolutionary advance has occurred in most red algae. Rather than immediately producing spores, the zygote nucleus divides repeatedly by mitosis, generating a third multicellular life-cycle phase, the diploid **carposporophyte** generation. The carposporophyte generation remains attached to its parental gametophyte and probably receives organic nutrients from it. These nutrients help support rapid proliferation of cells by mitosis. When the carposporophyte reaches its mature size, mitosis occurs in apical cells, giving rise to carpospores. The carpospores are released into the water, settle onto a substrate, and grow into separate diploid sporophytes.

In many red algae, a mitotically produced copy of the diploid zygote nucleus is transferred to another cell of the gametophyte. This cell, known as an **auxiliary cell,** serves as a host and nutritional source for repeated mitoses by the adopted nucleus. Proliferation of diploid filaments from the auxiliary cell generates a carposporophyte and carpospores. In many forms, multiple copies of the diploid zygote nucleus are carried by the growth of long tubular cells throughout the algal body and are deposited into many additional auxiliary cells. Each diploid nucleus then produces many carposporophytes, which release very large numbers of carpospores into the water. In one case, each zygote nucleus is known to result in the release of some 4500 carpospores. Each carpospore is capable of growing into a usually free-living, multicellular diploid generation, called the **tetrasporophyte.** Meiosis occurs in specialized cells of the tetrasporophyte, called **tetrasporangia.** Each of the tetraspores produced can germinate into a new gametophyte, if conditions are favorable. *Polysiphonia,* a widely distributed marine red alga, provides an example of this kind of life cycle (Figure 15–29).

The life history of most red algae thus consists of three phases: (1) a haploid gametophyte; (2) a diploid phase, the carposporophyte; and (3) another diploid phase, the tetrasporophyte. The carposporophyte generation is regarded as an additional way to increase the genetic products of sexual reproduction when fertilization rates are low. Alternation of generations involving three multicellular generations is unique to the red algae. The ability to produce many carposporophytes with the resultant larger numbers of carpospores and potentially huge numbers of tetraspores, all from a single zygote, has helped the red algae conquer the sexual disability imposed by their lack of flagella.

In most red algae, the gametophyte and tetrasporophyte generations closely resemble one another and are therefore said to be isomorphic, as in *Polysiphonia* (Figure 15–29). Coralline algae also have isomorphic life cycles. However, an increasing number of heteromorphic life cycles are also being discovered. In these species, the tetrasporophytes are either microscopic and filamentous or consist of a thin crust that is tightly attached to a rock substrate. Phycologists speculate that differences in appearance have selective advantages in responding to seasonal changes or other environmental variations. The development of techniques for cultivating algae in the laboratory has led to the discovery that what appear to be distinct species are in some cases alternating generations of the same species.

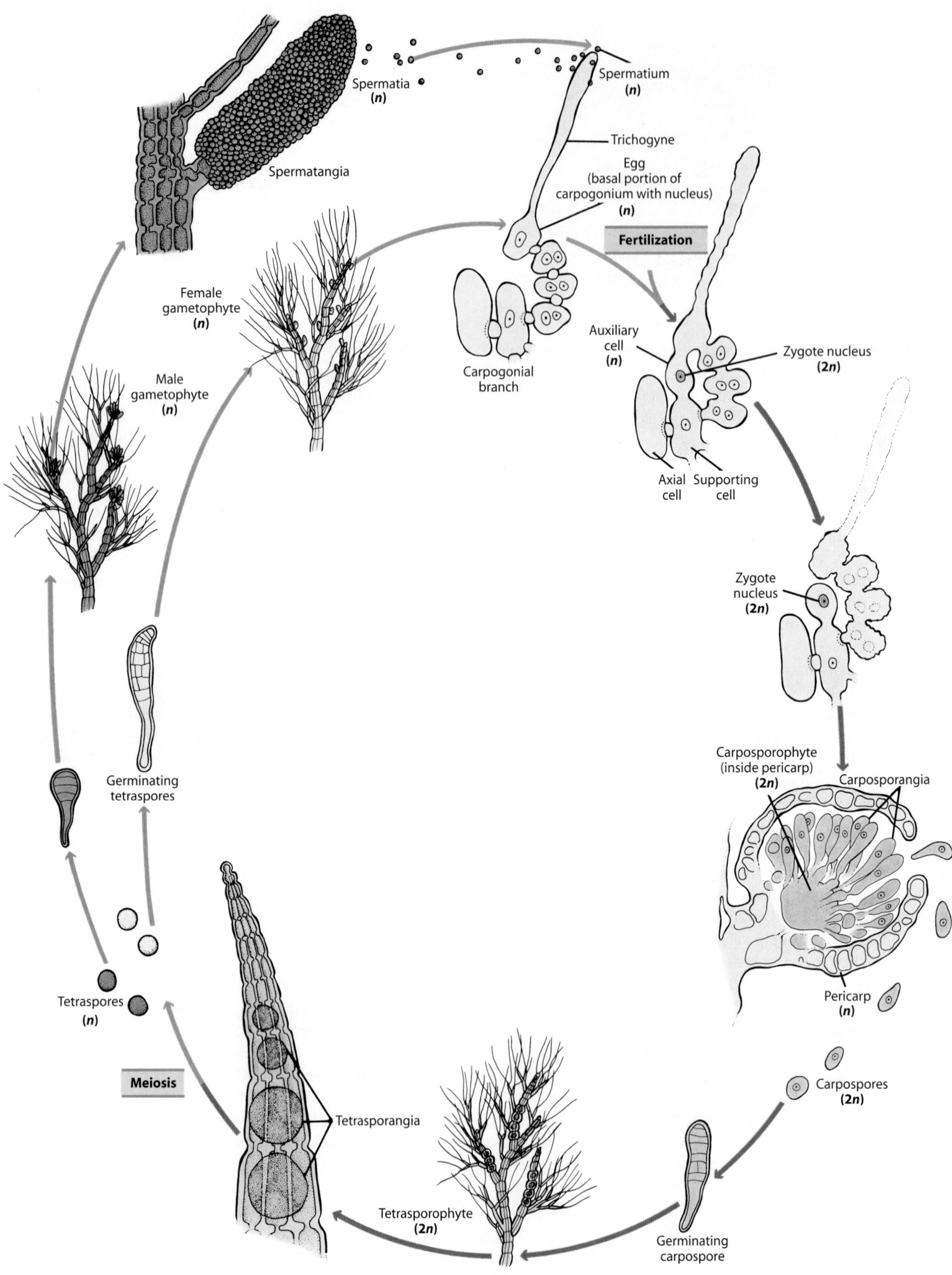
Spermatia
(n)
Spermatium
(n)
Trichogyne
Spermatangia
Egg
(basal portion of
carpogonium with nucleus)
(n)
Fertilization
Female
gametophyte
(n)
Carpogonial
branch
Auxiliary
cell
(n)
Zygote nucleus
(2n)
Male
gametophyte
(n)
Axial
cell
Supporting
cell
Zygote
nucleus
(2n)
Germinating
tetraspores
Carposporophyte
(inside pericarp)
(2n)
Carposporangia
Tetraspores
(n)
Pericarp
(n)
Meiosis
Carpospores
(2n)
Tetrasporangia
Tetrasporophyte
(2n)
Germinating
carpospore

15–29 Life cycle of *Polysiphonia,* a marine red alga The sex organs arise near the tips of the branches of the haploid gametophytes, which arise from the haploid tetraspores (lower left). The spermatangia, which occur in dense clusters, release cells that function directly as spermatia. The enlarged basal portion of the carpogonium contains the nucleus and functions directly as an egg. Following fertilization, diploid carpospores are formed by mitosis within carposporangia. The carpospores are ultimately liberated through an opening in the pericarp, which is an outer structure that develops around them. The pericarp is derived from cells flanking the carpogonial branch. The carpospores germinate, giving rise to the tetrasporophytes, which are similar in size and appearance to the gametophytes. The tetrasporophytes produce tetrasporangia. In each tetrasporangium, one meiotic event gives rise to four haploid tetraspores, and the cycle begins anew.

Green Algae

The green algae, including at least 17,000 species, are diverse in structure and life history. Although most green algae are aquatic, they are found in a wide variety of habitats, including on the surface of snow (Figure 15–30), on tree trunks, in the soil, and in symbiotic associations with fungi (lichens), freshwater protozoa, sponges, and coelenterates. A surprisingly diverse population of microscopic, unicellular, free-living green algae occurs in desert microbiotic crusts worldwide. In addition to the green algae, such crusts comprise communities of cyanobacteria, diatoms, lichens, bryophytes, and other taxa, bound to the upper layer of soil. Molecular systematic and physiological data indicate that these desert algae evolved from freshwater green algae multiple times. Some green algae—such as species of the unicellular genera *Chlamydomonas* and *Chloromonas,* found growing on the surface of snow (Figure 15–30), and the filamentous alga *Trentepohlia,* which grows on rocks and tree trunks or

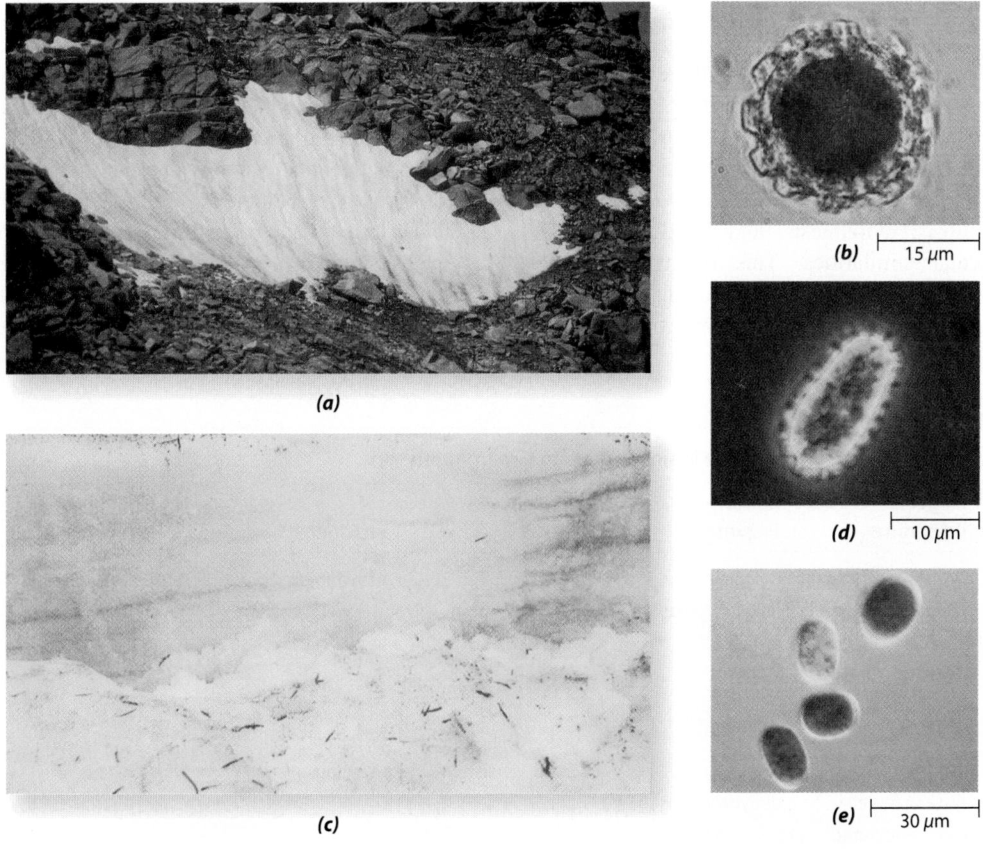

15–30 Snow algae Snow algae are unique because of their tolerance to temperature extremes, acidity, high levels of irradiation, and minimal nutrients for growth. ***(a)*** In many parts of the world, as in alpine areas from northern Mexico to Alaska, the presence of large numbers of snow algae produces "red snow" during the summer. This photograph was taken near Beartooth Pass in Montana. ***(b)*** Dormant zygote of the snow alga *Chlamydomonas nivalis.* The red color results from carotenoids that serve to protect the chlorophyll in the zygote. ***(c)*** Green snow occurs just below the surface, usually near tree canopies in alpine forests. It is widespread, occurring as far south as Arizona and as far north as Alaska and Quebec. This photograph was taken at Cayuse Pass, Mt. Rainier National Park, Washington. ***(d)*** Resting zygote of the alga *Chloromonas brevispina,* found in green snow. ***(e)*** Three brilliant orange resting zygotes of *Chloromonas granulosa,* the alga responsible for orange snow. The single yellow zygote will change to orange as it matures.

branches—produce large amounts of carotenoids that function as a shield against intense light. These accessory pigments give the algae an orange, red, or rust color. Most aquatic green algae are found in fresh water, but a number of groups are marine. Many green algae are microscopic, although some of the marine species are large. *Codium magnum* of Mexico, for example, sometimes attains a breadth of 25 centimeters and a length of more than 8 meters.

The resemblance of green algae to the bryophytes and vascular plants, which are adapted for life on land, has long been recognized. The green algae, bryophytes, and vascular plants are the only groups of organisms that contain chlorophylls *a* and *b* and store starch, their food reserve, inside plastids. Some, but not all, green algae are like bryophytes and vascular plants in having firm cell walls composed of cellulose, hemicelluloses, and pectic substances. In addition, the microscopic structure of the flagellated reproductive cells in some green algae resembles that of plant sperm cells. These features, in conjunction with molecular data, strongly indicate that green algae and land plants (bryophytes and vascular plants) form a monophyletic group known as the Viridiplantae (viridophytes, or green plants).

Traditional classification systems group green algae according to their outer structure: unicellular flagellates are grouped together, filamentous types are grouped together, and so forth. As we observed earlier for the brown algae, however, related green algae cannot always be recognized by their outer structure. Evidence of a relationship is now revealed by ultrastructural studies of mitosis, cytokinesis, and reproductive cells, as well as molecular similarities. This recent information has resulted in a new systematic alignment of the green algae into several classes, three of which—the Chlorophyceae, the Ulvophyceae, and the Charophyceae—are discussed in this book (Table 15–2).

Differences in Cell Division and Motile Cells Exist among Classes of Green Algae

The members of the largest class of green algae, the freshwater Chlorophyceae, have a unique mode of cytokinesis involving a **phycoplast** (Figure 15–31a, b). In these algae, the daughter nuclei move toward one another as the nonpersistent mitotic spindle collapses, and a new system of microtubules, the phycoplast, develops parallel to the plane of cell division. Presumably, the role of the phycoplast is to ensure that the **cleavage furrow,** resulting from infolding of the plasma membrane, will pass between the two daughter nuclei. The nuclear envelope persists throughout mitosis. In motile cells of the Chlorophyceae, there is a cross-shaped pattern of four narrow bands of microtubules known as **flagellar roots,** which are associated with the basal bodies (centrioles) of the flagella (Figure 15–32).

In other classes of green algae, spindles may remain present throughout cytokinesis, until they are disrupted either by furrowing (Figure 15–31c) or by growth of a cell plate. A cell plate originates in the central region of the cell and grows outward to its margins. Some members of the class Charophyceae produce a new cytokinetic microtubular system, the **phragmoplast,** which is nearly identical to that present in bryophytes and vascular plants. The microtubules in a phragmoplast are oriented perpendicular to the plane of cell division (Figure 15–31d). The phragmoplast serves as a "scaffolding" for formation of the cell plate.

The flagellated cells of the Charophyceae are distinct from those of the other classes in that they have an asymmetrical

Table 15-2 Characteristics of Three of the Major Groups of Green Algae

Group	Flagellar Apparatus	Photorespiratory Enzymes	Mitosis	Cytokinesis	Habitat (primary)	Life History
Chlorophyceae	Symmetrical root system; flagellar roots associated with basal bodies (centrioles)	Glycolate dehydrogenase	Closed, nonpersistent spindle	Furrowing phycoplast, some with cell plate and plasmodesmata	Freshwater or terrestrial	Zygotic meiosis
Ulvophyceae	Symmetrical root system; flagellar roots associated with basal bodies (centrioles)	Glycolate dehydrogenase	Closed, persistent spindle	Furrowing	Marine or terrestrial	Zygotic meiosis or alternation of generations with sporic meiosis or gametic meiosis
Charophyceae	Asymmetrical flagellar root system; often associated with multilayered structure	Glycolate oxidase and catalase in peroxisome	Open, persistent spindle	Furrowing, some with cell plate, phragmoplasts and plasmodesmata	Freshwater or terrestrial	Zygotic meiosis

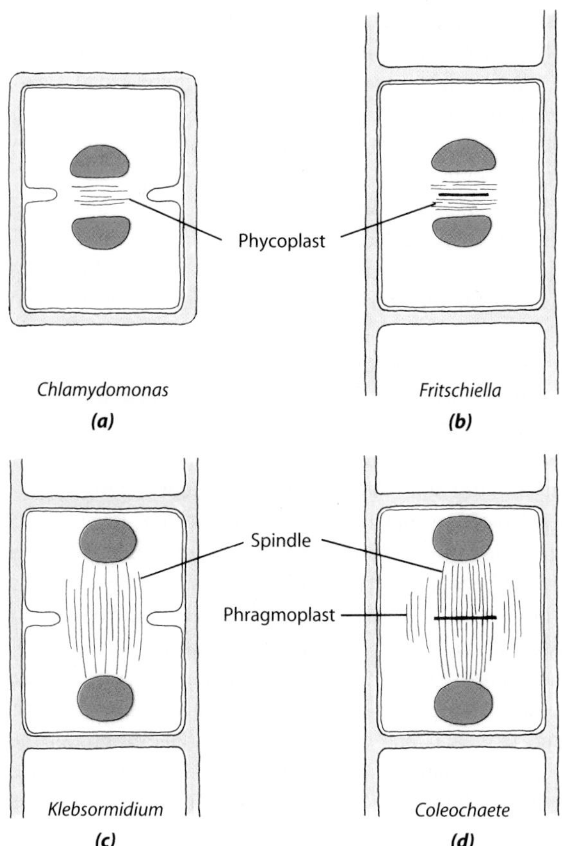

15–31 Cytokinesis in two classes of green algae ***(a), (b)*** In the class Chlorophyceae, the mitotic spindle is nonpersistent and the daughter nuclei, which are relatively close together, are separated by a phycoplast. Cytokinesis in *Chlamydomonas* is by furrowing, whereas cytokinesis in *Fritschiella* occurs by cell plate formation. ***(c)*** In simpler members of the class Charophyceae, such as *Klebsormidium,* the mitotic spindle is persistent and the daughter nuclei are relatively far apart. Cytokinesis occurs by furrowing. ***(d)*** Advanced charophytes such as *Coleochaete* and *Chara* have a plantlike phragmoplast, and cytokinesis occurs by cell plate formation as in plants. Ulvophyceae, like Charophyceae, also have a persistent spindle, but do not have a phragmoplast or a cell plate.

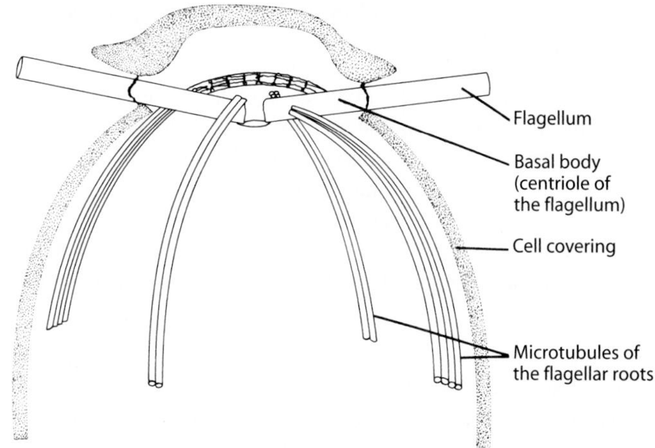

15–32 Flagellar roots Diagram of the cross-shaped arrangement of four narrow bands of microtubules known as flagellar roots. These flagellar roots are associated with the flagellar basal bodies (centrioles) and are characteristic of green algae of the class Chlorophyceae.

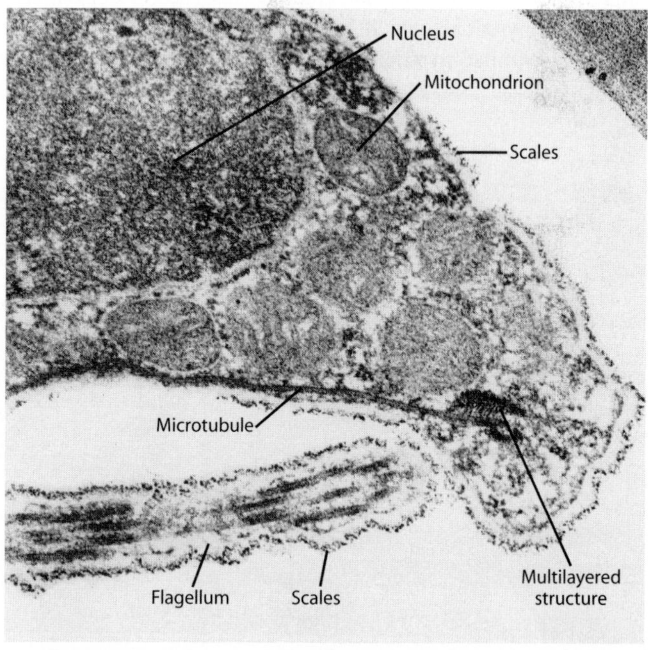

15–33 Charophyte asymmetrical flagellar root system Electron micrograph of the anterior portion of a motile sperm of the green alga *Coleochaete* (class Charophyceae). Shown here is the multilayered structure, which is associated with the flagellar root system at the base of the flagellum. The multilayered structure is also characteristic of the sperm of bryophytes and some vascular plants, and it is one of the features linking these plants with the Charophyceae, their ancestors. As seen here, a layer of microtubules extends from the multilayered structure down into the posterior end of the cell and serves as a cytoskeleton for these wall-less cells. The flagellar and plasma membranes are covered by a layer of small scales.

flagellar root system of microtubules (Figure 15–32). One role of the flagellar root system is to provide the means by which a flagellum is anchored in place. Quite often, an entity known as a **multilayered structure** is associated with one of the flagellar roots (Figure 15–33). The type of flagellar root is often an important taxonomic character. The flagellar root system of the Charophyceae, with its multilayered structure, is very similar to that found in the sperm of bryophytes and some vascular plants. For these and other reasons, including biochemical and molecular similarities, Charophyceae is the group of living green algae considered to be closest to the ancestry of bryophytes and vascular plants.

Class Chlorophyceae, the Chlorophytes, Consists of Mainly Freshwater Species

The class Chlorophyceae includes flagellated and nonflagellated unicellular algae, motile and nonmotile colonial algae, filamentous algae, and algae consisting of flat sheets of cells. The members of this class live mainly in fresh water, although a few unicellular, planktonic species occur in coastal marine waters. A few Chlorophyceae are essentially terrestrial, living in habitats such as on snow (Figure 15–30), in soil, on rocks, and on tree trunks or branches.

***Chlamydomonas* Is an Example of a Motile Unicellular Chlorophyte** The common freshwater green alga *Chlamydomonas* (Figure 15–34), a unicellular form with two equal flagella, has been widely used as a model system for molecular studies of the genes regulating photosynthesis and other cell processes. Molecular analyses have revealed that *Chlamydomonas* is a polyphyletic group; that is, it consists of several distinct lineages, all of which are, coincidentally, unicellular with two equal flagella.

The chloroplast of *Chlamydomonas,* in having a red photosensitive eyespot, or stigma, that aids in the detection of light, is similar to that of many other green flagellates and of the zoospores of multicellular green algae. The chloroplast also contains a pyrenoid, which is typically surrounded by a shell of starch. Similar pyrenoids occur in many other green algal species. The uninucleate protoplast is surrounded by a thin glycoproteinaceous (carbohydrate-protein) cell wall, inside which is the plasma membrane. There is no cellulose in the cell wall of *Chlamydomonas.* At the anterior end of the cell are two contractile vacuoles, which collect excess water and ultimately discharge it from the cell.

Chlamydomonas reproduces both sexually and asexually. During asexual reproduction, the haploid nucleus usually divides by mitosis to produce up to 16 daughter cells within the parent cell wall. Each cell then secretes a wall around itself and develops flagella. The cells secrete an enzyme that breaks down the parental wall, and the daughter cells can then escape, although fully formed daughter cells are often retained for some time within the parent cell wall.

Sexual reproduction in *Chlamydomonas* involves the fusion of individuals belonging to different mating types (Figure 15–35). The vegetative cells are induced to form gametes by nitrogen starvation. The gametes, which resemble the vegetative cells, first become aggregated in clumps. Within these clumps, pairs are formed that stick together, first by their flagellar membranes and later by a slender protoplasmic thread—the conjugation tube—that connects them at the base of their flagella. As soon as this protoplasmic connection is formed, the flagella become free, and one or both pairs of flagella propel the partially fused gametes through the water. The protoplasts of the two gametes fuse completely (plasmogamy), followed by fusion

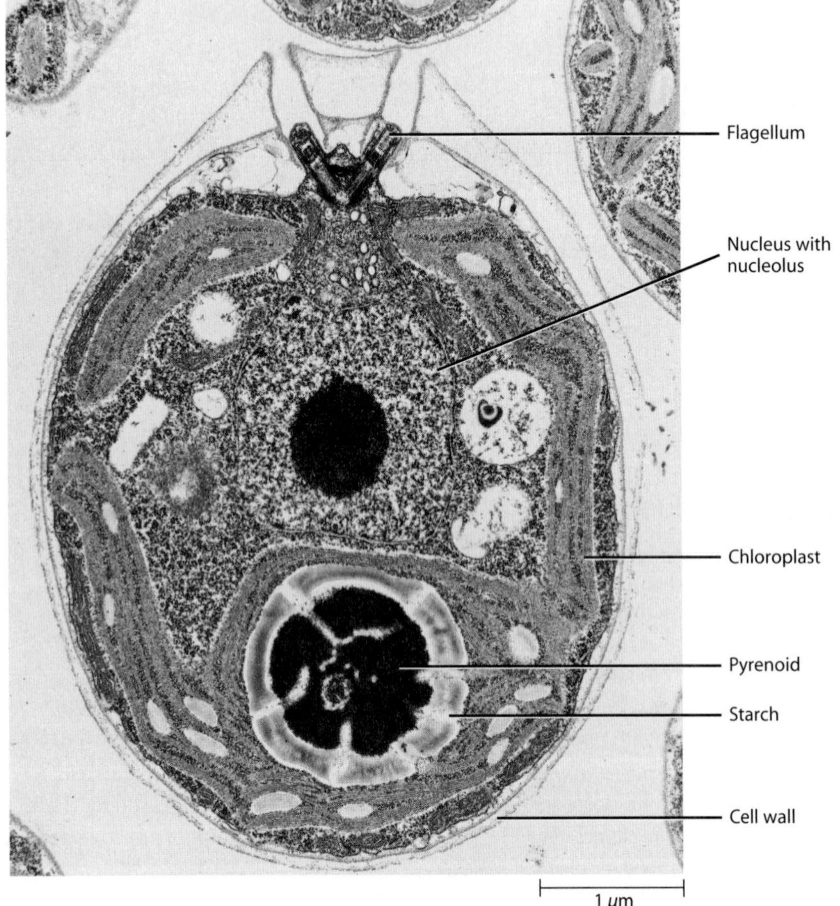

15–34 *Chlamydomonas,* a motile unicellular green alga The stigma was not present in the section in this electron micrograph. Only the bases of the flagella are visible. The nucleus of this uninucleate flagellate contains a prominent nucleolus. A shell of starch surrounds the pyrenoid, which is located within the chloroplast. *Chlamydomonas* is a member of the Chlorophyceae.

of their nuclei (karyogamy), forming the zygote. The four flagella soon shorten and eventually disappear, and a thick cell wall forms around the diploid zygote. This thick-walled, resistant zygote, or zygospore, then undergoes a period of dormancy. Meiosis occurs at the end of the dormant period, producing four haploid cells, each of which develops two flagella and a cell wall. These cells can either divide asexually or mate with a cell of another mating strain to produce a new zygote. Thus, *Chlamydomonas* exhibits zygotic meiosis (see Figure 12–17a), and the haploid phase is the dominant phase in its life cycle.

The Class Chlorophyceae Also Includes Motile Colonies Some Chlorophyceae form motile colonies, the most spectacular of which is *Volvox,* a hollow sphere, known as a **spheroid.** The spheroid is made up of a single layer of 500 to 60,000 vegetative, biflagellated cells that serve primarily a photosynthetic function, as well as a small number of larger, nonflagellated reproductive cells (Figure 15–36). The specialized reproductive cells undergo repeated mitoses to form many-celled, juvenile spheroids, which "hatch" from the parental spheroid by releasing an enzyme that dissolves the transparent parental matrix. As the spheroids first develop, all of the flagella face the hollow center, so the colony must turn inside out before it can become motile.

Sexual reproduction in *Volvox* is always oogamous. In all species that have been studied, sexual reproduction is synchronized within the population of colonies by a sexual inducer molecule, a glycoprotein with a molecular weight of about 30,000. This inducer molecule is produced by a spheroid that has itself

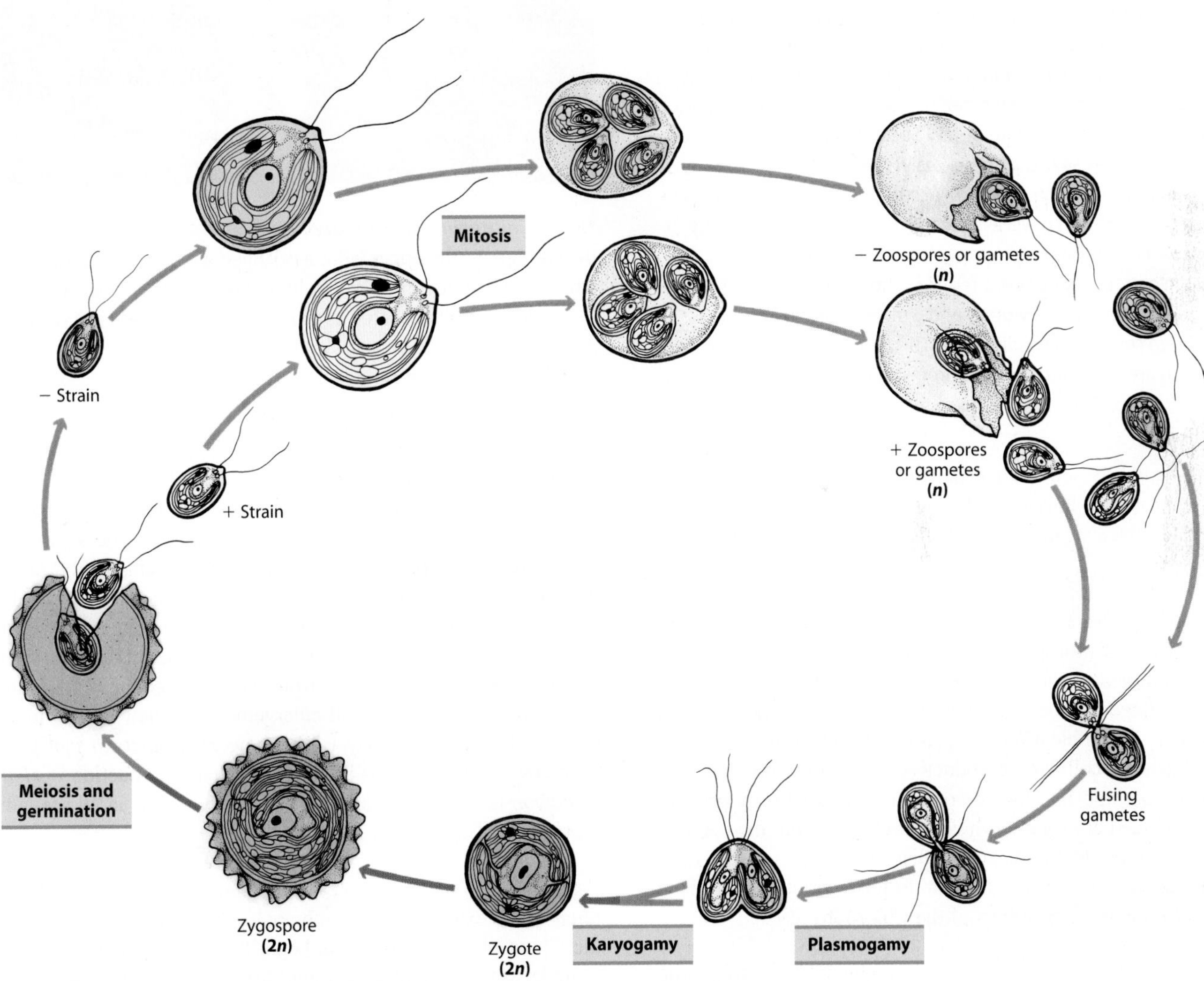

15–35 Life cycle of *Chlamydomonas* Sexual reproduction occurs when gametes of different mating types come together, cohering at first by their flagellar membranes and then by a slender protoplasmic thread—the conjugation tube (lower right). The protoplasts of the two cells fuse completely (plasmogamy), followed by the union of their nuclei (karyogamy). A thick wall is then formed around the diploid zygote, known at this point as a zygospore. After a period of dormancy, meiosis occurs, followed by germination, and four haploid cells emerge. Asexual reproduction of the haploid individuals by cell division is the most frequent mode of reproduction.

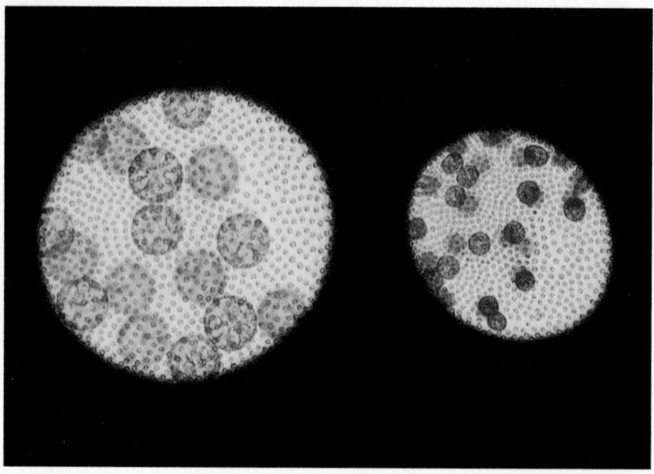

15–36 Two spheroids of *Volvox carteri* The approximately 2000 small cells that dot the periphery of the two spheroids (two individuals) are the *Chlamydomonas*-like somatic cells. In *V. carteri*, somatic cells are not interconnected in mature colonies as they are in some members of the genus. A few cells are capable of becoming reproductive, either sexually or asexually. The spheroid on the left is asexual. Mitotic division has already occurred, producing 16 juvenile spheroids, each of which will eventually digest a passageway out of the parent colony and swim away. The spheroid on the right is sexual. When heat-shocked, a female spheroid produces egg-bearing sexual female individuals and a male spheroid produces sperm-laden sexual male individuals. The zygotes produced by fertilization are heat- and desiccation-resistant.

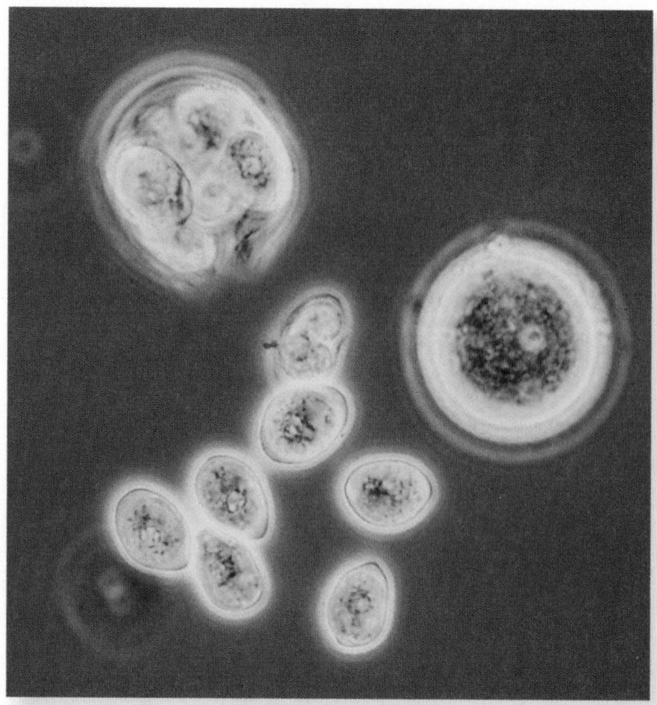

15–37 *Chlorococcum echinozygotum*, a nonmotile unicellular green alga At the upper left is a cell filled with asexual zoospores, which have formed mitotically within the cell. The smaller cells in the middle are biflagellated zoospores (the flagella are not visible). A nonmotile vegetative cell is at the upper right. *Chlorococcum* is a member of the Chlorophyceae.

become sexual by some other, as yet poorly understood mechanism. One male colony of *Volvox carteri* may produce enough inducer to induce sexual reproduction in over half a billion other colonies.

Some Unicellular Chlorophyceae Are Nonmotile One unicellular, nonmotile form of Chlorophyceae is *Chlorococcum,* which is very commonly found in the microbial flora of soils (Figure 15–37). There is a huge number of genera of unicellular soil algae that superficially resemble *Chlorococcum* but can be distinguished on the basis of cellular, reproductive, and molecular features. *Chlorococcum* and its relatives reproduce asexually by producing biflagellated zoospores, which are released from the parental cell. Sexual reproduction is accomplished by the release of flagellated gametes, which fuse in pairs to form zygotes. Meiosis is zygotic, as it is in all members of the Chlorophyceae.

The Class Chlorophyceae Also Includes Nonmotile Colonies Nonmotile colonial members of the Chlorophyceae include *Hydrodictyon,* the "water net" (Figure 15–38). Under favorable conditions, it forms massive surface blooms in ponds, lakes, and gentle streams. Each colony consists of many large, cylindrical cells arranged in the form of a lacy, hollow cylinder. Initially uninucleate, each cell eventually becomes multinucleate. At maturity, each cell contains a large central vacuole and peripheral cytoplasm containing the nuclei and a large reticulate (resembling a net) chloroplast with numerous pyrenoids. *Hydrodictyon* reproduces asexually through the formation of many uninucleate, biflagellated zoospores in each cell of the net. The zoospores are not released from parental cells but, rather amazingly, group themselves into geometric arrays of four to nine (most typically six) within the cylindrical parent cell. Zoospores then lose their flagella and form the component cells of daughter mini-nets. These are eventually released from the parent cell and grow into mature nets by dramatic cell enlargement. In view of this mode of reproduction, it is easy to see how *Hydrodictyon* can form such conspicuous blooms in nature. Sexual reproduction in *Hydrodictyon* is isogamous, and meiosis is zygotic, as in all sexually reproducing Chlorophyceae.

There Are Also Filamentous and Parenchymatous Chlorophyceae *Oedogonium* is an example of an unbranched filamentous member of the Chlorophyceae. Filaments begin their development attached to underwater substrates by a holdfast, but massive growths may later break away to form noticeable floating blooms in lakes. The mode of cell division in *Oedogonium* results in the formation of characteristic "caps" or annular scars with each cell division (Figure 15–39). Thus, these scars reflect the number of divisions that have occurred in a given cell.

The branched filamentous and parenchymatous, or tissue-like, Chlorophyceae include algae that have the most complex structures found in the class. Their cells can be specialized with

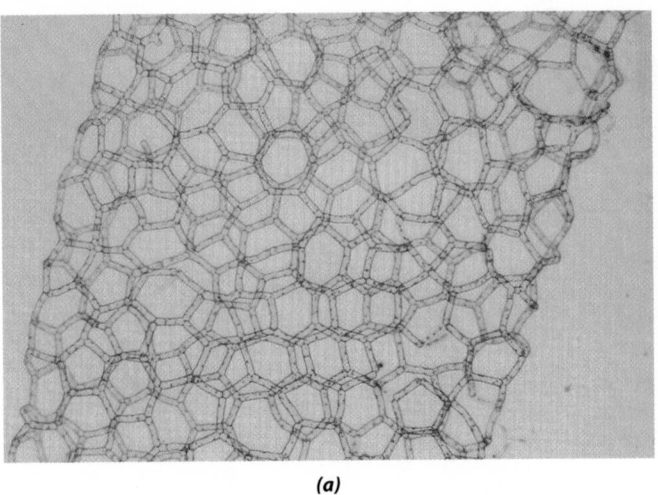

(a)

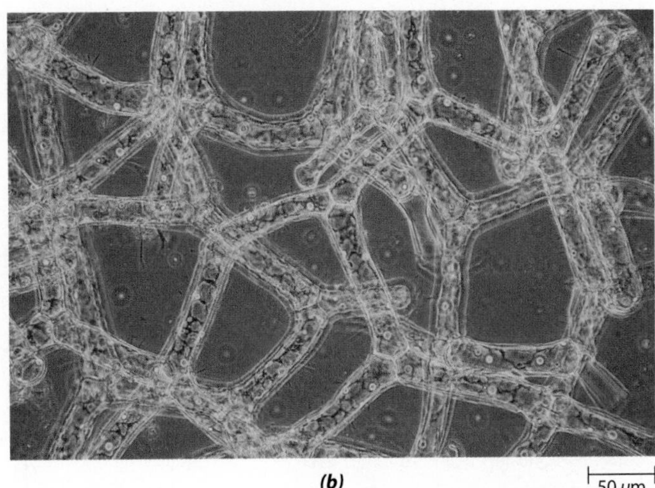

(b)

15–38 ***Hydrodictyon*** ***(a)*** A stained portion of the "water net," *Hydrodictyon*, a colonial member of the Chlorophyceae. ***(b)*** Higher magnification of a portion of *Hydrodictyon reticulatum*.

respect to particular functions or positions in the algal body and, like plant cells, are sometimes connected by plasmodesmata. *Fritschiella,* for example, is composed of subterranean rhizoids, a prostrate system at or just beneath the soil surface, and two kinds of erect branches (Figure 15–40).

Class Ulvophyceae, the Ulvophytes, Consists of Mainly Marine Species

The Ulvophyceae are primarily marine, but a few important representatives are found in fresh water, probably having moved there from the marine habitat at some time in the past. Most ulvophytes are filamentous or composed of flat sheets of cells, or they may be macroscopic and multinucleate. Ulvophytes have a closed mitosis in which the nuclear envelope persists; the spindle is persistent through cytokinesis.

The flagellated cells of ulvophytes are scaly or naked, like those of Charophyceae, but in contrast to motile charophytes, ulvophytes are nearly radially symmetrical and have apical, forward-directed flagella resembling those of the chlorophytes. The flagellated cells of Ulvophyceae may have two, four, or many flagella, as do those of Chlorophyceae. The flagellated cells of Charophyceae are always biflagellate. Ulvophytes are the only green algae that have an alternation of generations

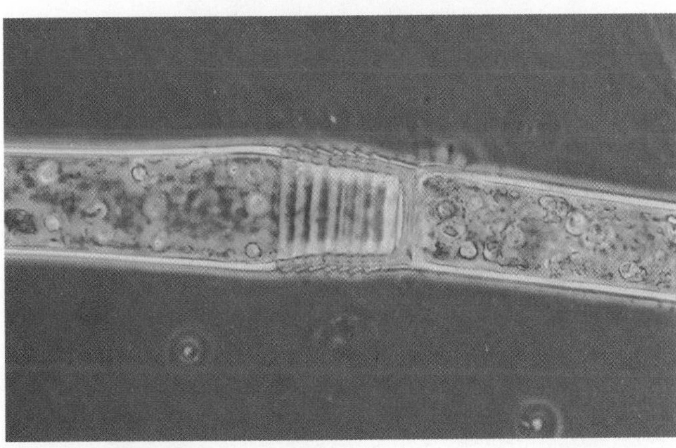

15–39 ***Oedogonium*** *Oedogonium* is an unbranched, filamentous member of the Chlorophyceae. Seen here is a portion of a vegetative filament showing annular scars.

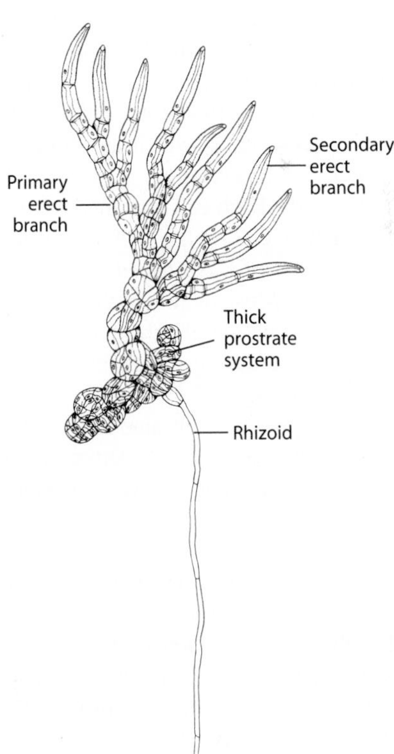

15–40 ***Fritschiella*** Composed of subterranean rhizoids and two types of erect branches, *Fritschiella* is a terrestrial member of the Chlorophyceae. In its adaptation to a terrestrial habitat, *Fritschiella* has independently developed some of the features that are characteristic of plants.

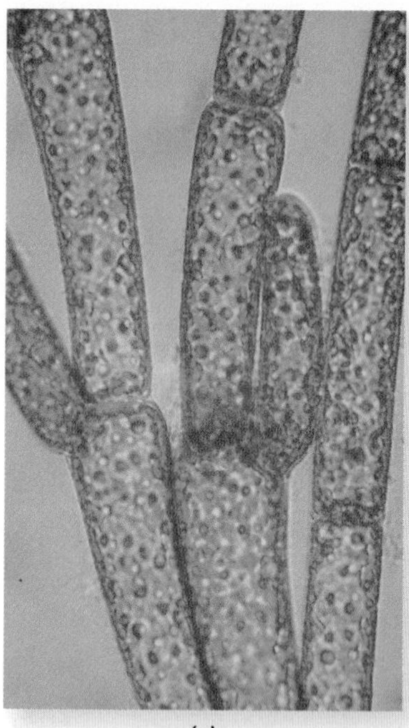

(a)

(b)

(c)

15–41 ***Cladophora*** A member of the class Ulvophyceae, *Cladophora* is widespread in marine and freshwater habitats. The marine species, like most Ulvophyceae, have an alternation of generations, but the freshwater species do not. ***(a)*** Branched filaments of *Cladophora*. ***(b)*** Part of an individual cell, showing the netlike chloroplast. ***(c)*** An individual of *Cladophora* growing in a sluggish stream in California.

with sporic meiosis or a diploid, dominant life history involving gametic meiosis.

One evolutionary line of Ulvophyceae consists of filamentous algae with large, multinucleate, septate cells. One genera of this group, *Cladophora* (Figure 15–41), is widespread in both salt water and fresh water, sometimes forming nuisance blooms, slimy and stinky, in freshwater habitats. Its filaments commonly grow in dense mats, which are either free-floating or attached to rocks and vegetation. The filaments elongate and branch near the ends. Each cell contains many nuclei and a single, peripheral, netlike chloroplast with many starch-forming pyrenoids. Marine species of *Cladophora* have an alternation of isomorphic generations. Most of the freshwater species, however, do not have an alternation of generations, apparently having lost this characteristic during the course of their transition from marine to fresh waters.

A second kind of growth habit among the Ulvophyceae is that of *Ulva*, commonly known as sea lettuce (Figure 15–42). This familiar alga is common along temperate seashores throughout the world. Individuals of *Ulva* consist of a glistening, flat thallus that is two cells thick, and a meter or more long in exceptionally large individuals. The thallus is anchored to the substrate by a holdfast produced by extensions of the cells at its base. Each cell of the thallus contains a single nucleus and chloroplast. *Ulva* is anisogamous, meaning one flagellated gamete is larger than the other; it has an alternation of isomorphic generations like that of many other Ulvophyceae (Figure 15–43).

The siphonous marine algae, so-called because of their tubular cells, constitute additional evolutionary lines in the Ulvophyceae (Figure 15–44). They are characterized by very large, branched, coenocytic cells that are rarely septate. The multinucleate cytoplasm is arranged around a large central vacuole. These algae, which are very diverse, develop through repeated nuclear division without the formation of cell walls, and consist of enormous single cells, some as long as a meter. Cell walls are produced only in the reproductive phases of the siphonous green algae.

Codium, mentioned on page 346, is a member of this group. It is a spongy mass of densely intertwined coenocytic filaments

15–42 ***Ulva*** Sea lettuce, *Ulva*, a common member of the class Ulvophyceae that grows on rocks, pilings, and similar places in shallow seas worldwide.

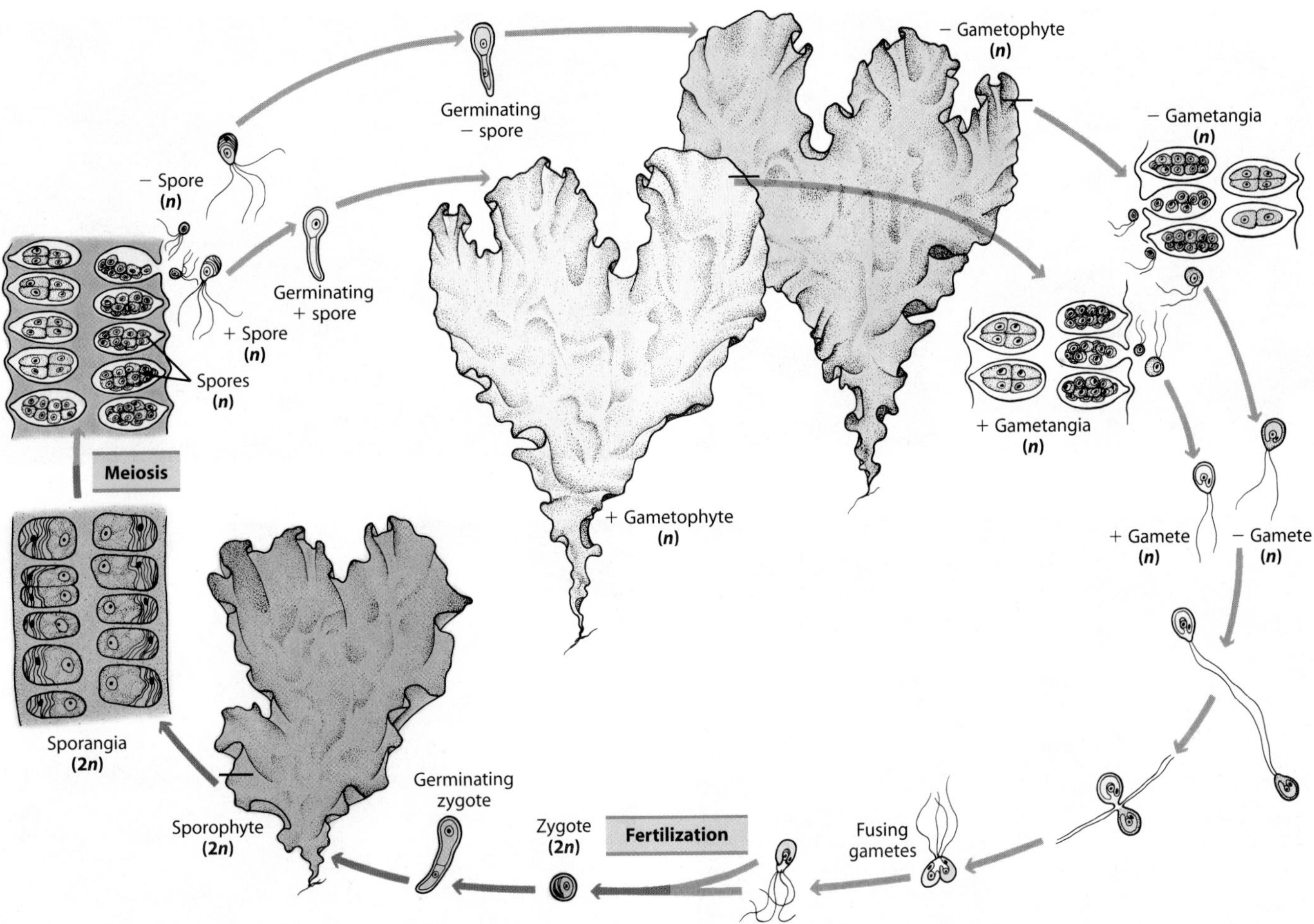

15–43 Life cycle of *Ulva* The sea lettuce, *Ulva,* has an alternation of isomorphic generations, in which the gametophyte and sporophyte are indistinguishable except for their reproductive structures. The haploid gametophyte produces haploid gametes, and the gametes fuse to form a diploid zygote. A sporophyte, in which all cells are diploid, develops from the zygote. The sporophyte produces haploid spores by meiosis. The haploid spores develop into haploid gametophytes, and the cycle begins again.

(Figure 15–44a). Strains of *Codium fragile* may be quite weedy, and nuisance growths of this seaweed are spreading in quiet waters of the temperate zone. *Caulerpa taxifolia,* which was released accidentally from the Oceanographic Museum of Monaco in 1984, has become a serious invasive species, expanding in the Mediterranean by an average of 50 kilometers each year. *Ventricaria* (also known as *Valonia*), which is common in tropical waters, has been widely used in studies of cell walls and in physiological experiments. *Ventricaria* appears to be unicellular but is actually a large, multinucleate vesicle attached to the substrate by several rhizoids (Figure 15–44b). It grows to the size of a hen's egg. Another well-known siphonous green alga is *Acetabularia* (Figure 15–44c), which has been widely used in experiments on the genetic basis of differentiation. The siphonous green algae are primarily diploid. The gametes are the only haploid cells in the life cycle.

Halimeda and related genera of siphonous green algae are notable for their calcified cell walls. When these algae die and disintegrate, they play a major role in the generation of the white carbonate sand that is so characteristic of tropical waters. A number of genera of algae, including *Halimeda* (Figure 15–44d), contain secondary metabolites that significantly reduce feeding by herbivorous fish. The fronds of *Halimeda* expand and grow rapidly at night, building up toxic compounds that deter herbivores, which are active mainly during the day. Within an hour after sunrise, the fronds turn from white to green as chloroplasts move outward from the interior, or lower portions, of the thallus and begin to photosynthesize rapidly. At night, the chloroplasts move to the interior of the thallus, rendering the algae less appetizing to herbivores.

Class Charophyceae Includes Members That Most Closely Resemble the Bryophytes and Vascular Plants

Charophyceae (also known as charophyte algae or streptophyte algae) consist of unicellular, colonial, filamentous, and

15–44 Siphonous marine green algae Four genera of siphonous green algae of the class Ulvophyceae are shown here. ***(a)*** A species of *Codium,* abundant along the Atlantic coast. ***(b)*** *Ventricaria,* common in tropical waters; individuals are often about the size of a hen's egg. ***(c)*** *Acetabularia,* the "mermaid's wine glass," a mushroom-shaped siphonous green alga. The siphonous green alga in the background is *Dasycladus.* This photograph was taken in the Bahamas. ***(d)*** *Halimeda,* a siphonous green alga that is often dominant in reefs in warmer waters throughout the world. This alga produces distasteful compounds that retard grazing by fish and other marine herbivores.

parenchymatous genera. Their relationship with one another, and to the bryophytes and vascular plants, is revealed by many fundamental structural, biochemical, and genetic similarities. These include the presence of asymmetrical flagellated cells, some of which have distinctive multilayered structures (Figure 15–33). Other similarities are breakdown of the nuclear envelope during mitosis, persistent spindles or phragmoplasts at cytokinesis, the presence of phytochrome, flavonoids, and the chemical precursors of cuticle, and other molecular features.

The early divergent members of the charophycean lineage seem to include *Mesostigma, Chlorokybus,* and *Klebsormidium* (Figure 15–45). *Mesostigma* is a unicellular freshwater, scaly flagellate; *Chlorokybus,* a rarely encountered terrestrial or freshwater green alga, consists of packets of cells held together by mucilage; and *Klebsormidium* is a freshwater, unbranched filament. None of these charophyceans form conspicuous growths or blooms in nature, nor are they of any economic importance.

Spirogyra (Figure 15–46) is a well-known genus of unbranched, filamentous Charophyceae that often forms frothy or slimy floating masses in bodies of fresh water. Each filament is surrounded by a watery sheath. The name *Spirogyra* refers to the helical arrangement of the one or more ribbonlike chloroplasts, with numerous pyrenoids, found within each uninucleate cell. Asexual reproduction occurs in *Spirogyra* by cell division and fragmentation. There are no flagellated cells at any stage of the life cycle, but, as noted earlier, flagellated reproductive cells do occur in other genera of Charophyceae. During sexual reproduction in *Spirogyra,* a conjugation tube forms between two filaments. The contents of the two cells that are joined by the tube serve as isogametes. Fertilization may occur in the tube, or

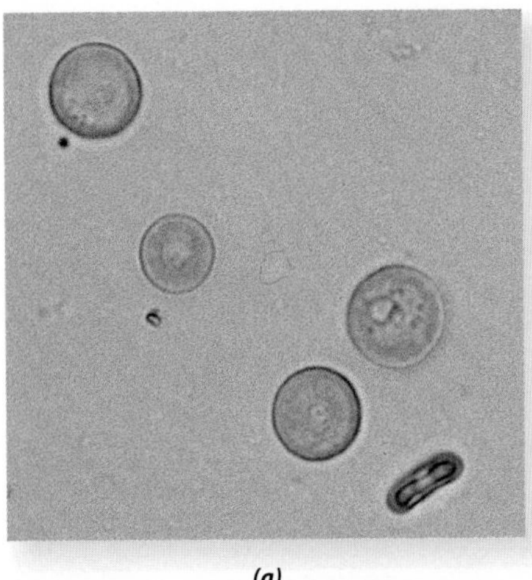
(a)

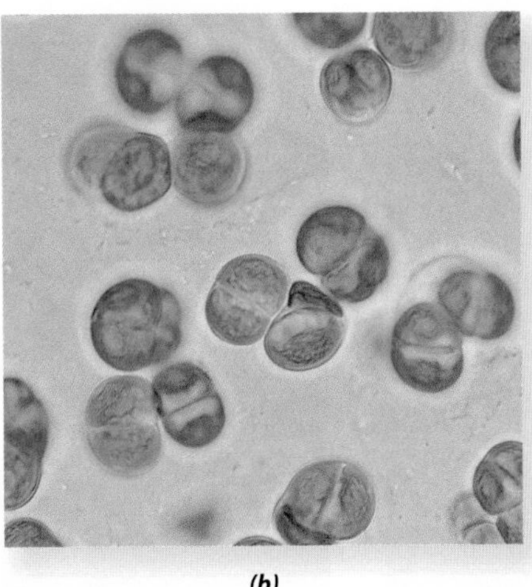
(b)

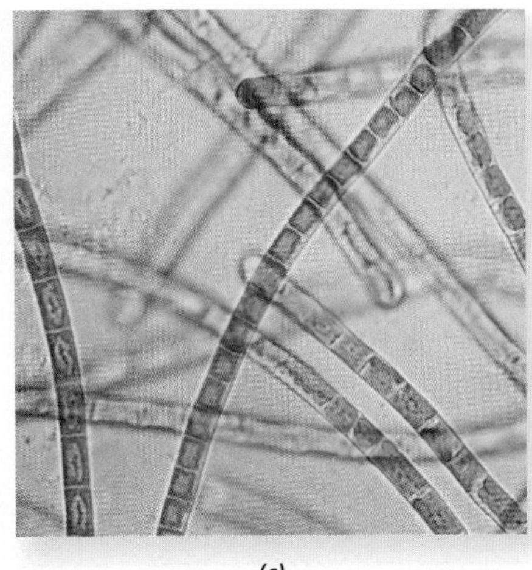
(c)

15–45 Three charophycean algae ***(a)*** *Mesostigma,* ***(b)*** *Chlorokybus,* and ***(c)*** *Klebsormidium* are early divergent members of the charophycean lineage.

one of the gametes may migrate into the other filament, where fertilization then takes place. The zygotes become surrounded by thick walls that contain **sporopollenin.** This protective substance is the most resistant biopolymer known. Sporopollenin enables the zygotes to survive harsh conditions for long periods of time, before germinating when conditions improve. Meiosis is zygotic, as in all members of the Charophyceae.

The **desmids** are a large group of freshwater green algae related to *Spirogyra.* Like *Spirogyra,* they lack flagellated cells. Some desmids are filamentous, but most are unicellular. Most desmid cells consist of two sections, or semi-cells, joined by a narrow constriction—the isthmus (Figure 15–47). Desmid cell division and sexual reproduction are very similar to those of *Spirogyra.* There are thousands of desmid species. They are most abundant and diverse in peat bogs and ponds that are poor in mineral nutrients. Some are associated with distinctive and possibly symbiotic bacteria, which live in the mucilaginous sheaths.

Two orders of charophycean green algae, the Coleochaetales and the Charales, resemble bryophytes and vascular plants more closely than do other charophytes in the details of their cell division and sexual reproduction. These orders have a plantlike microtubular phragmoplast operating during cytokinesis. Like the bryophytes and vascular plants, they are oogamous, and their sperm are ultrastructurally similar to those of bryophytes. The bryophytes and vascular plants were probably derived from an extinct member of the Charophyceae that in many respects resembled living members of the Coleochaetales and Charales.

Morphological and molecular studies indicate that an early basal split in the green algae gave rise to a chlorophyte clade

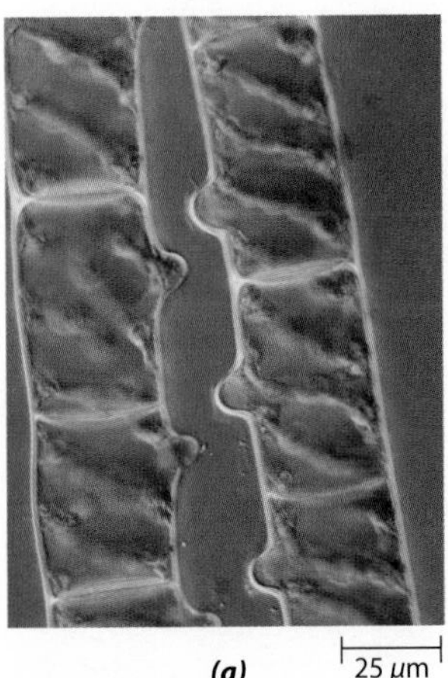

(a)

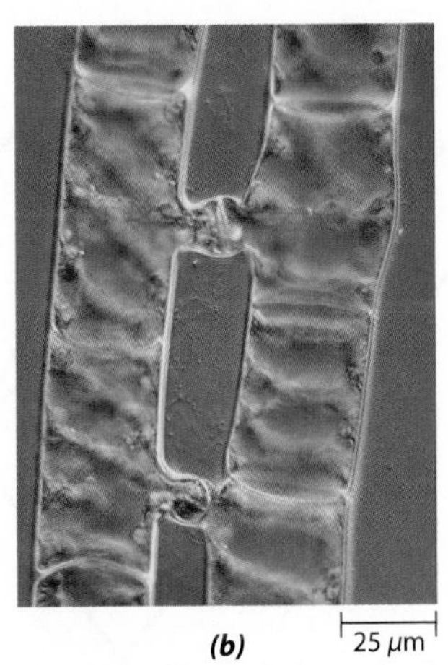

(b)

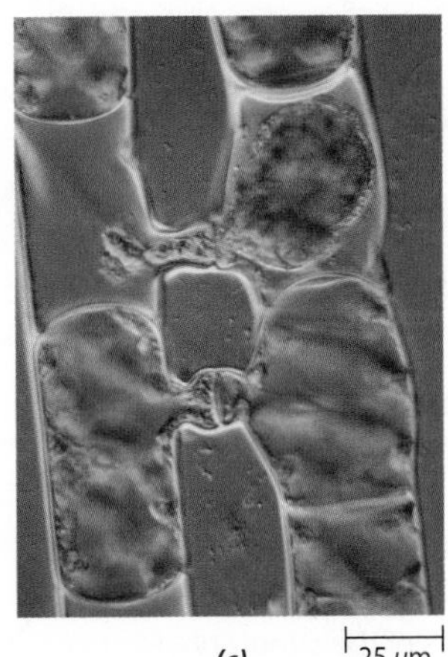

(c)

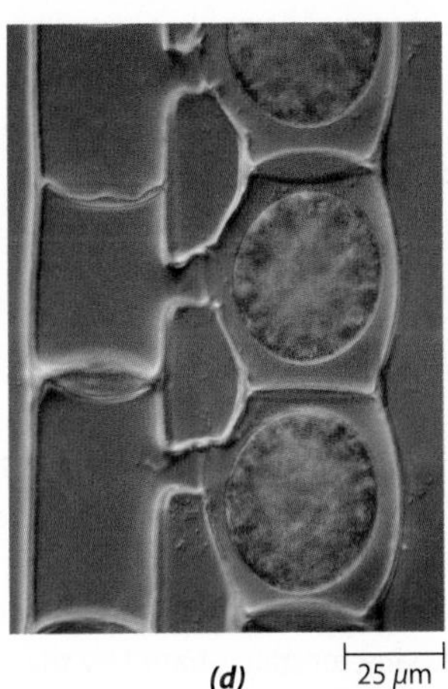

(d)

15–46 Sexual reproduction in *Spirogyra* ***(a), (b)*** The formation of conjugation tubes between the cells of adjacent filaments. ***(c)*** The contents of the cells of the – strain (on the left) pass through these tubes into the cells of the + strain. ***(d)*** Fertilization occurs within these cells. The resulting zygote develops a thick, resistant cell wall and is termed a zygospore. The vegetative filaments of *Spirogyra* are haploid, and meiosis occurs during germination of the zygospores, as it does in all Charophyceae.

(a) *(b)*

15–47 Desmids The desmids are a group of thousands of species of unicellular, freshwater Charophyceae. Most desmids are constricted into two parts, which gives them a very characteristic appearance. ***(a)*** *Xanthidium armatum.* ***(b)*** Cell division in *Micrasterias thomasiana.* The smaller half of each daughter individual will grow to the size of the larger half, forming a full-sized desmid.

and a streptophyte clade (Figure 15–48). The **chlorophyte clade** contains most of the green algae, whereas the **streptophyte clade** consists of the charophycean orders Coleochaetales and Charales, the zygnemataleans (not discussed in this book), some early divergent members of the charophycean lineage, and the bryophytes and vascular plants ("land plants"). *Mesostigma* (Figure 15–45a) appears to belong at the base of the streptophyte clade.

The order Coleochaetales, with about 20 species, includes both branched filamentous genera and discoid (disk-shaped) genera that grow by division of apical or peripheral cells (Figure 15–49). *Coleochaete,* which grows on the surface of submerged

Chlorophytes
Streptophytes
Green algae
Chlorophyceae
Trebouxiophyceae
Ulvophyceae
Mesostigma
Chlorokybus
Klebsormidiales
Zygnematales
Desmidiales
Coleochaetales
Charales
Embryophytes (Land plants)
Multicellular sporophyte, embryo, gametangia, sporangium, cuticle
Phragmoplast, branched, apical growth, oogamy, plasmodesmata, retention of egg
Filamentous growth
Multilayered structure of flagella and cytoskeleton
Prasinophytes
>1000 million years ago (MYA)
Chlorophyll *b*, "stellate" flagellar structure, gene transfers

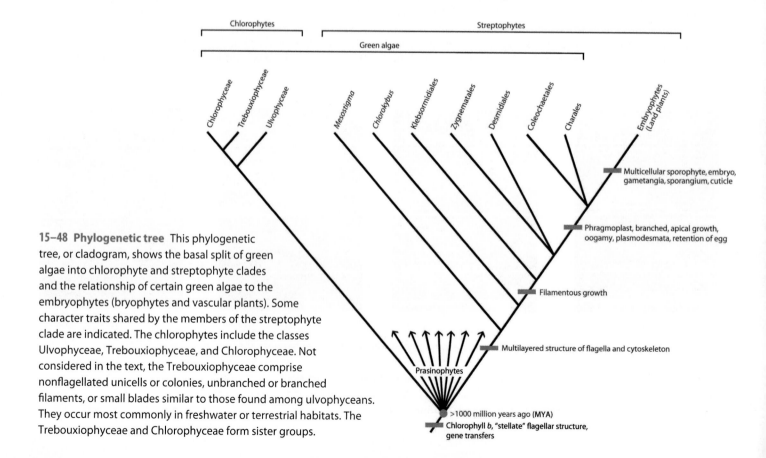

15–48 Phylogenetic tree This phylogenetic tree, or cladogram, shows the basal split of green algae into chlorophyte and streptophyte clades and the relationship of certain green algae to the embryophytes (bryophytes and vascular plants). Some character traits shared by the members of the streptophyte clade are indicated. The chlorophytes include the classes Ulvophyceae, Trebouxiophyceae, and Chlorophyceae. Not considered in the text, the Trebouxiophyceae comprise nonflagellated unicells or colonies, unbranched or branched filaments, or small blades similar to those found among ulvophyceans. They occur most commonly in freshwater or terrestrial habitats. The Trebouxiophyceae and Chlorophyceae form sister groups.

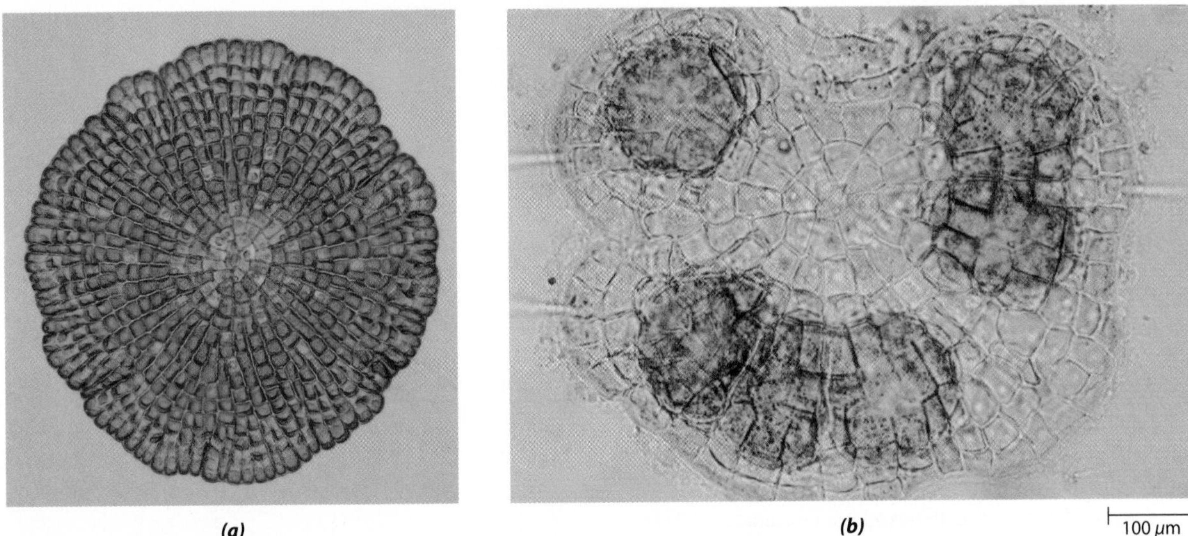

15–49 *Coleochaete* ***(a)*** The charophycean green alga *Coleochaete* grows on rocks and on stems of aquatic flowering plants in shallow lake waters. ***(b)*** Individuals of this species of *Coleochaete* consist of a parenchymatous disk, which is generally one cell thick. The large cells are zygotes, which are protected by a cellular covering. The hair cells extending from the disk are ensheathed at the base; *Coleochaete* means "sheathed hair." These hairs are thought to discourage aquatic animals from feeding on the alga.

rocks or freshwater plants, has uninucleate vegetative cells, each containing one large chloroplast with an embedded pyrenoid. Very similar chloroplasts and pyrenoids occur in some of the hornworts, a group of bryophytes discussed in Chapter 16. *Coleochaete,* like a number of charophytes, reproduces asexually by zoospores that are formed singly within cells. Sexual reproduction is oogamous. The zygotes, which remain attached to the parental thallus, stimulate the growth of a layer of cells that covers the zygotes. In at least one species, these parental cells have wall ingrowths similar to those occurring at the gametophyte-sporophyte junction of bryophytes (see Chapter 16) and many vascular plants. These specialized cells, called transfer cells, are believed to function in nutrient transport between the gametophyte and sporophyte. The wall ingrowths suggest how the distinctive life cycle of plants and the sporophyte generation might have evolved in the long-extinct ancestors of modern plants.

The order Charales includes 81 to 400 living species of distinctive green algae (the number depending on the expert) found primarily in fresh water or sometimes in brackish water. Modern forms such as *Chara* (Figure 15–50) are commonly known as the stoneworts, because some of them have heavily calcified cell walls. Calcification of the distinctive reproductive structures of

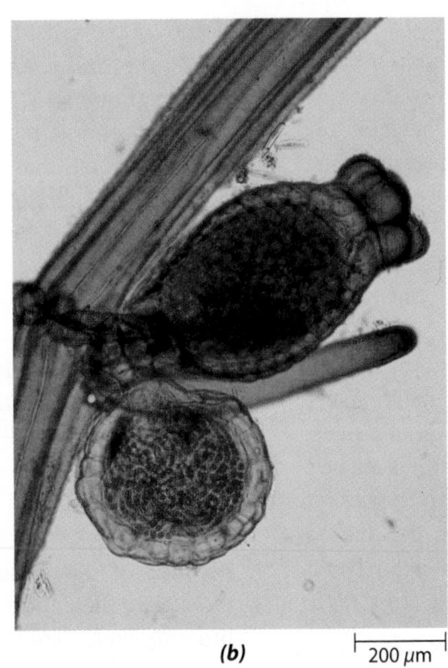

15–50 *Chara* ***(a)*** The stonewort *Chara* (class Charophyceae) grows in shallow waters of temperate lakes. ***(b)*** *Chara* showing gametangia. The top structure is an oogonium, and the round structure below is an antheridium.

ancient relatives has resulted in a good fossil record extending back to Late Silurian times (about 410 million years ago).

The Charales, like *Coleochaete* and the bryophytes and vascular plants, exhibit apical growth. In addition, the thallus is differentiated into nodal and internodal regions. The tissue organization in the nodal regions resembles the parenchyma of plants, as does the pattern of plasmodesmatal connections. From the nodal regions arise whorls of branches. In some species, files of cells grow over the central, filamentous axis, creating a thicker, stronger thallus. The sperm of Charales are produced in multicellular antheridia that are more complex than those of any other group of protists. Their eggs are borne in oogonia enclosed by several long, tubular, twisted cells. These cells are in a position analogous to the distinctive female gametangia of seedless plants, and they may serve similar functions. Sperm are the only flagellated cells in the Charales life cycle, and they very closely resemble bryophyte sperm. Zygotes are believed to germinate by meiosis, though this has been difficult to study because zygotes are enclosed in very tough walls that include sporopollenin. Sporopollenin is also a component of the walls of plant spores and pollen and is responsible for the widespread occurrence of these cells in the fossil record.

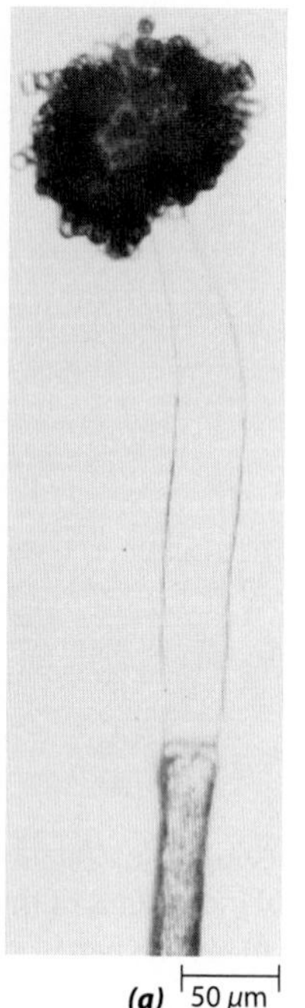

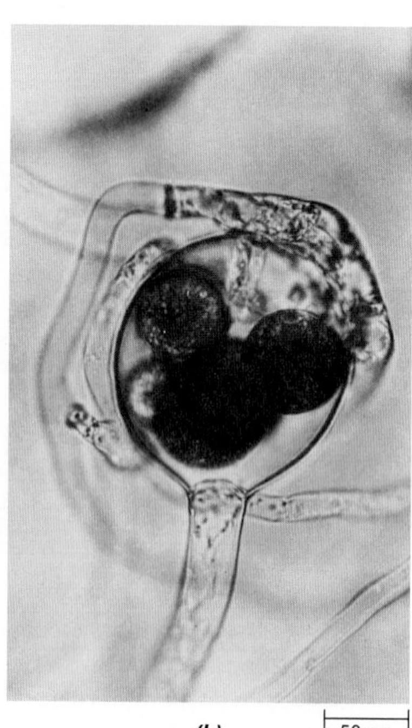

(a) ***(b)***

15–51 Zoospores and oospores *Achlya ambisexualis* is an oomycete that reproduces both asexually and sexually. ***(a)*** Empty sporangium with zoospores encysted about its opening, a distinctive feature of asexual reproduction in *Achlya.* ***(b)*** Sex organs, showing fertilization tubes extending from the antheridium through the wall of the oogonium to the eggs. Fertilization results in the formation of thick-walled zygotes known as oospores.

Heterotrophic Protists

The oomycetes and slime molds are heterotrophic organisms once regarded as fungi. Unlike the slime molds, the oomycetes are fungus-like stramenopiles (page 330). Plasmodial slime molds and cellular slime molds, although unrelated, are treated together here in order to compare their similarities.

Oomycetes: Phylum Oomycota

The phylum Oomycota, with about 700 species, is a distinctive heterotrophic group. Like the dinoflagellates and many green algae, the cell walls of oomycetes are composed largely of cellulose or cellulose-like polymers. The oomycetes range from unicellular to highly branched, coenocytic, and filamentous forms. The latter somewhat resemble the hyphae that are characteristic of fungi, which is why oomycetes were once grouped with the fungi.

Most species of oomycetes can reproduce both sexually and asexually. Asexual reproduction among the oomycetes is by means of motile zoospores (Figure 15–51a), which have the two flagella that characterize the stramenopiles—one ornamented with hairs and one smooth. Sexual reproduction is oogamous, meaning that the female gamete is a relatively large, nonflagellated egg, and the male gamete is noticeably smaller and flagellate.

In the oomycetes, one to many eggs are produced in a structure called an **oogonium** (plural: oogonia), and an **antheridium** (plural: antheridia) contains numerous male nuclei (Figure 15–51b). Fertilization results in the formation of a thick-walled zygote, the **oospore,** for which this phylum is named. The oospore serves as a resting stage that can tolerate stressful conditions. When conditions improve, the oospore germinates. Other stramenopiles, as well as many green algae that live in similar habitats, also produce resistant resting stages that result directly from sexual reproduction.

The Water Molds Are Aquatic Oomycetes One large group of the phylum Oomycota is aquatic. The members of this group, the so-called water molds, are abundant in fresh water and are easy to isolate from it. Most of them are saprotrophic, living on the remains of dead plants and animals, but a few are parasitic, including species that cause diseases of fish and fish eggs.

Some water molds, such as *Saprolegnia* (Figure 15–52), can undergo sexual reproduction, with male and female sex organs borne on the same individual; in other words, these water molds are **homothallic.** Other water molds, such as some species of *Achlya* (Figure 15–51), are **heterothallic**—male and female sex organs are borne by different individuals, or if they are borne on one individual, that individual is genetically incapable of fertilizing itself. Both *Saprolegnia* and *Achlya* can reproduce sexually and asexually.

Some Terrestrial Oomycetes Are Important Plant Pathogens Another group of oomycetes is primarily terrestrial, although the organisms still form motile zoospores when liquid water is available. Among this group, the order Peronosporales, are several forms that are economically important. One of these is *Plasmopara viticola,* which causes downy mildew in grapes. Downy mildew was accidentally introduced into France in the late 1870s on grape stock from the United States, which had been imported because of its resistance to other diseases. The mildew soon threatened the entire French wine industry. It was eventually brought under control by a combination of good fortune and

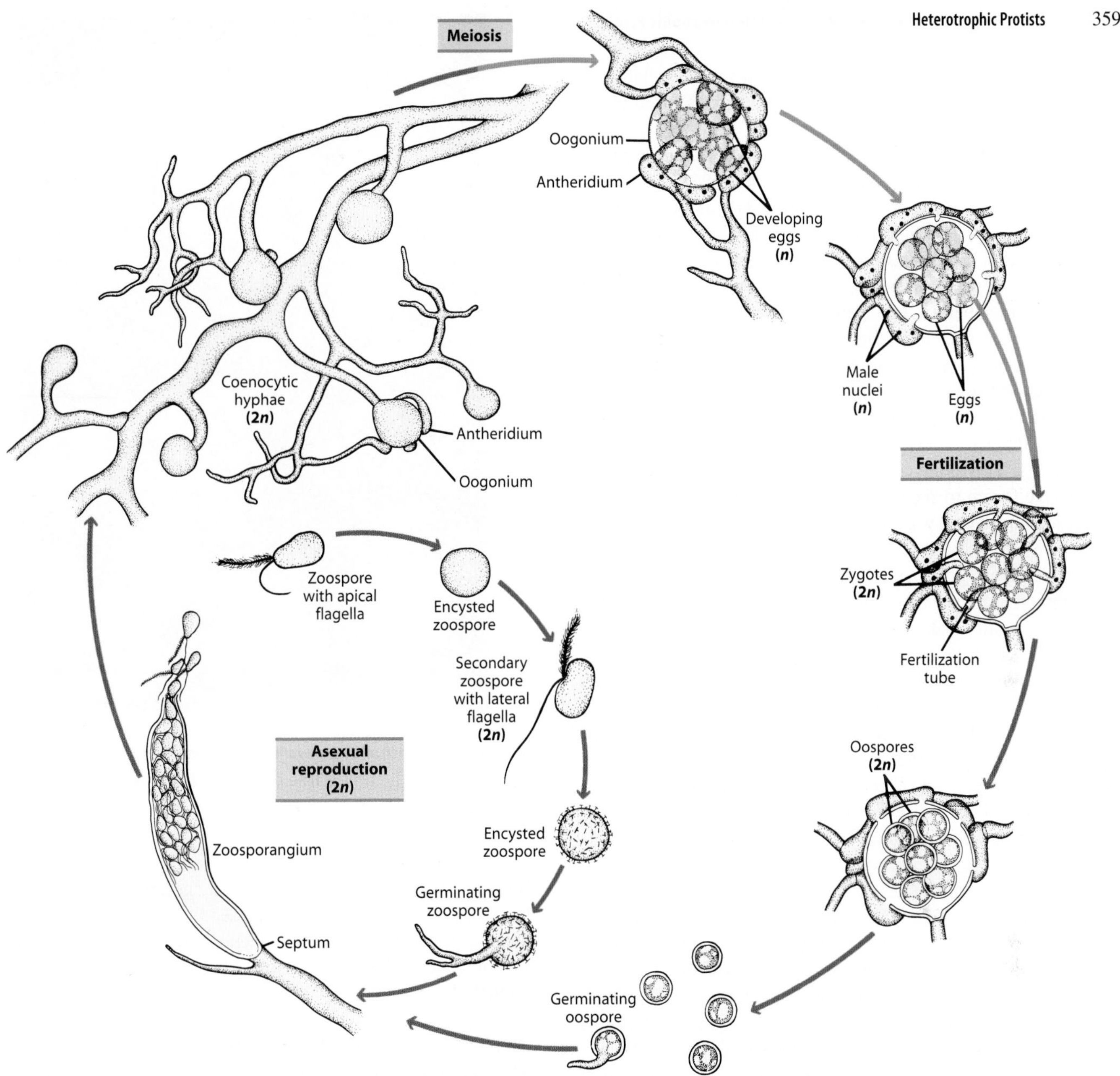

15–52 Life cycle of *Saprolegnia,* an oomycete The mycelium of this water mold is diploid. Reproduction is mainly asexual (lower left). Biflagellated zoospores released from a sporangium, known as a zoosporangium because it produces zoospores, swim for a while and then encyst. Each eventually gives rise to a secondary zoospore, which also encysts and then germinates to produce a new mycelium.

During sexual reproduction in *Saprolegnia,* oogonia and antheridia are formed on the same hypha (upper left). Meiosis occurs within these structures. The oogonia are enlarged cells in which a number of spherical eggs are produced. The antheridia develop from the tips of other filaments of the same individual and produce numerous male nuclei. In mating, the antheridia grow toward the oogonia and develop fertilization tubes, which penetrate the oogonia, as seen in *Achlya* in Figure 15–51b.

Male nuclei travel down the fertilization tubes to the female nuclei and fuse with them. Following each nuclear fusion, a thick-walled zygote—the oospore—is produced. On germination, the oospore develops into a hypha, which eventually produces a zoosporangium, beginning the cycle anew.

skillful observation. French vineyard owners in the vicinity of Medoc customarily put a distasteful mixture of copper sulfate and lime on vines growing along the roadside to discourage passersby from picking the grapes. A professor from the University of Bordeaux who was studying the problem of downy mildew noticed that these plants were free from symptoms of the disease. After conferring with the vineyard owners, the professor prepared his own mixture of chemicals—the

Bordeaux mixture—which was made generally available in 1882. The Bordeaux mixture was the first chemical used in the control of a plant disease.

Another economically important member of this group is the genus *Phytophthora* (meaning "plant destroyer"). *Phytophthora,* with about 35 species, is a particularly important plant pathogen that causes widespread destruction of many crops, including cacao, pineapples, tomatoes, rubber, papayas, onions, strawberries, apples, soybeans, tobacco, and citrus trees. A widespread member of this genus, *Phytophthora cinnamomi,* which occurs in soil, has killed or rendered unproductive millions of avocado trees in southern California and elsewhere. It has also destroyed tens of thousands of hectares of valuable eucalyptus timberland in Australia. The zoospores of *P. cinnamomi* are attracted to the plants they infect by chemicals exuded by the roots. This oomycete also produces resistant spores that can survive for up to six years in moist soil. Extensive breeding efforts are under way with avocados and other susceptible crops to produce strains resistant to this oomycete.

In 2000, a previously unknown *Phytophthora* species, *Phytophthora ramorum,* was identified as the cause of the disease called sudden oak syndrome. The disease is devastating to coastal live oaks, tanoaks, black oaks, and Shreve oaks, from Big Sur in California to southern Oregon. The organism invades the bark and eventually girdles the tree, causing its death. An early symptom of the disease is the oozing of a reddish-black sap from the bark (Figure 15–53). *Phytophthora ramorum* has appeared in at least 26 other species, including rhododendron, Douglas fir, and the coast redwood. Most of these species suffer only twig or leaf infection. Where the pathogen originated is unknown.

15–53 *Phytophthora ramorum,* the cause of sudden oak syndrome By the time a tree oozes sap, the first outward sign of infection, it is already doomed.

The best-known species of *Phytophthora,* however, is *Phytophthora infestans* (Figure 15–54), the cause of the late blight of potatoes, which produced the great potato famine of 1846 to 1847 in Ireland. As a result of this famine, about 800,000 people starved to death, and great numbers of people emigrated, many of them to the United States. Even today, *P. infestans* is still a serious pest of the potato crop, causing upward of $5 billion of damage each year worldwide. A gene has been found in a species of wild potato *(Solanum bulbocastanum)* that protects against late blight. This blight-resistant gene has now been inserted into cells of commercial potatoes *(Solanum tuberosum),* and the engineered plants appear to be resistant to a range of late-blight pathogens. The gene that protects the potatoes from *P. infestans* comes from a plant believed to have coevolved in Mexico alongside the pathogen. Genome sequences of *P. ramorum* and *Phytophthora sojae,* which causes root rot in soybeans, support a photosynthetic ancestry for the stramenopiles.

Also deserving of mention is the genus *Pythium,* the members of which are soil-inhabiting organisms found all over the world. *Pythium* species are the most important causes of **damping-off** diseases, which kill young seedlings. These oomycetes attack a wide variety of economically important crops and can even be a very serious problem in turf grasses used for golf courses and football fields. Some *Pythium* species attack and rot seeds in the field and may destroy seedlings either before they emerge from the soil (preemergence damping-off) or afterward (postemergence damping-off). Postemergence damping-off, in which the young seedling rots near the soil line and then falls over, is a special problem in greenhouses, where large numbers of seedlings are grown in dense stands. Many home gardeners encounter this problem in the spring when renewing their flower beds with recently purchased annuals.

Plasmodial Slime Molds: Phylum Myxomycota

The plasmodial slime molds, or myxomycetes, are a group of about 700 species that seem to have no direct relationship to the cellular slime molds, the fungi, or any other group. Although referred to as molds, as are certain fungi, molecular evidence demonstrates that neither the plasmodial slime molds nor the cellular slime molds are closely related to the fungi. When conditions are appropriate, the plasmodial slime molds exist as thin, streaming, multinucleate masses of protoplasm that creep along in amoeboid fashion. Lacking cell walls, this "naked" mass of protoplasm is called a **plasmodium** (Figure 15–55). As the plasmodia travel, they engulf and digest bacteria, yeast cells, fungal spores, and small particles of decaying plant and animal matter. Plasmodia can also be cultured on media that do not contain particulate matter, suggesting that the plasmodia can also obtain food by absorption of dissolved organic compounds. As the plasmodium grows, the nuclei divide repeatedly and synchronously; that is, all the nuclei in a plasmodium divide at the same time. Centrioles are present, and mitosis is similar to that of plants, although the chromosomes are very small.

Typically, the moving plasmodium is fan-shaped (see Figure 12–15a), with flowing protoplasmic tubules that are thicker

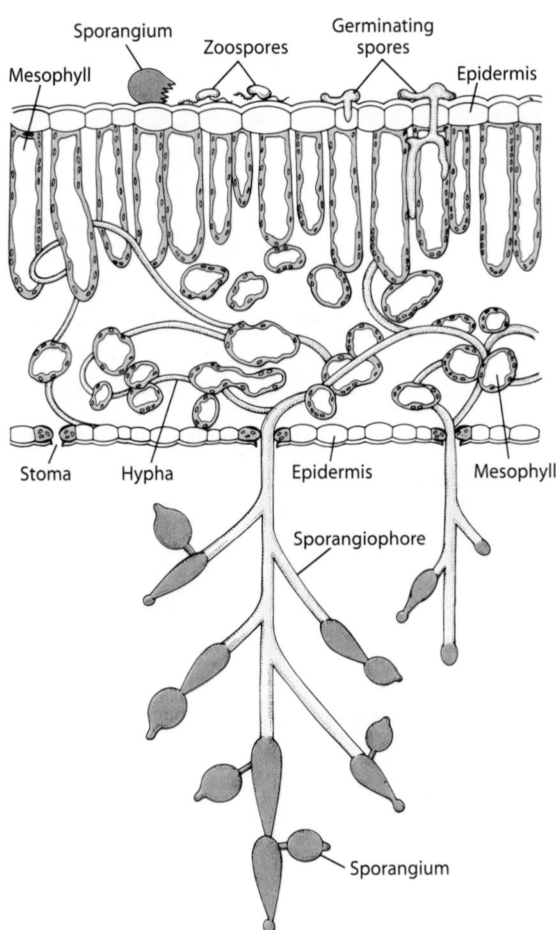

15–54 *Phytophthora infestans,* the cause of late blight of potatoes The cells of the potato leaf are shown in green. In the presence of water and low temperatures, either of two events can occur. Zoospores can be released from the sporangia and swim to the germination site (as shown here), or the sporangia can germinate directly through a germ tube.

15–55 Plasmodium A plasmodium of a plasmodial slime mold, *Physarum,* growing on a tree trunk.

at the base of the fan and spread out, branch, and become thinner toward their outer ends. The tubules are composed of slightly solidified protoplasm through which more liquefied protoplasm flows rapidly. The foremost edge of the plasmodium consists of a thin film of gel separated from the substrate by only a plasma membrane and a slime sheath.

Plasmodial slime molds have a life history that involves sexual reproduction and therefore may be among the earliest divergent protists to have acquired sex. Plasmodial growth continues as long as an adequate food supply and moisture are available. Generally, when either of these is in short supply, the plasmodium migrates away from the feeding area. At such times, plasmodia may be found crossing roads or lawns, climbing trees, or in other unlikely places. In many species, when the plasmodium stops moving, it divides into a large number of small mounds. The mounds are similar in size and volume, so their formation is probably controlled by chemical effects within the plasmodium. The life cycle of a typical plasmodial slime mold is summarized in Figure 15–56. Each mound produces a sporangium, usually borne at the tip of a stalk (Figure 15–57b). The mature sporangium is often extremely ornate (Figure 15–57a). The protoplasm of the young sporangium contains many nuclei, which increase in number by mitosis. Progressively, the protoplasm is cleaved into a large number of spores, each containing a single diploid nucleus. Meiosis then occurs, giving rise to four haploid nuclei per spore. Three of the four nuclei disintegrate, however, leaving each spore with a single haploid nucleus. In some members of this group, discrete sporangia are not produced, and the entire plasmodium may develop either into a **plasmodiocarp** (Figure 15–57c), which retains the former shape of the plasmodium, or into an **aethalium** (Figure 15–57d), in which the plasmodium forms a large mound that is essentially a single large sporangium.

When habitats dry out, plasmodia can quickly form an encysted stage, the **sclerotium.** Sclerotia are easily seen on firewood piles, because these organisms are often brightly colored in shades of yellow and orange. Sclerotia are of great importance for the survival of plasmodial slime molds, especially in fast-drying habitats such as dead cacti or soil in deserts, where these organisms are abundant.

The spores of myxomycetes are also resistant to environmental extremes and can be very long-lived. Some have germinated after being kept in the laboratory for more than 60 years. Thus, spore formation in this group seems to make possible not only genetic recombination but also survival under adverse conditions.

Under favorable conditions, the spores split open and the protoplast slips out (Figure 15–56). The protoplast may remain amoeboid, or it may develop one to four smooth flagella. The amoeboid and flagellated stages are readily interconvertible. The amoebas feed by the ingestion of bacteria and organic material and multiply by mitosis and cell cleavage. If the food supply is used up or conditions are otherwise unfavorable, an amoeba may cease moving about, become round, and secrete a thin wall to form a **microcyst.** These microcysts can remain viable for a year or more, resuming activity when favorable conditions return.

After a period of growth, plasmodia appear in the amoeba population. Their formation is governed by a number of factors, including cell age, environment, density of amoebas, and cyclic

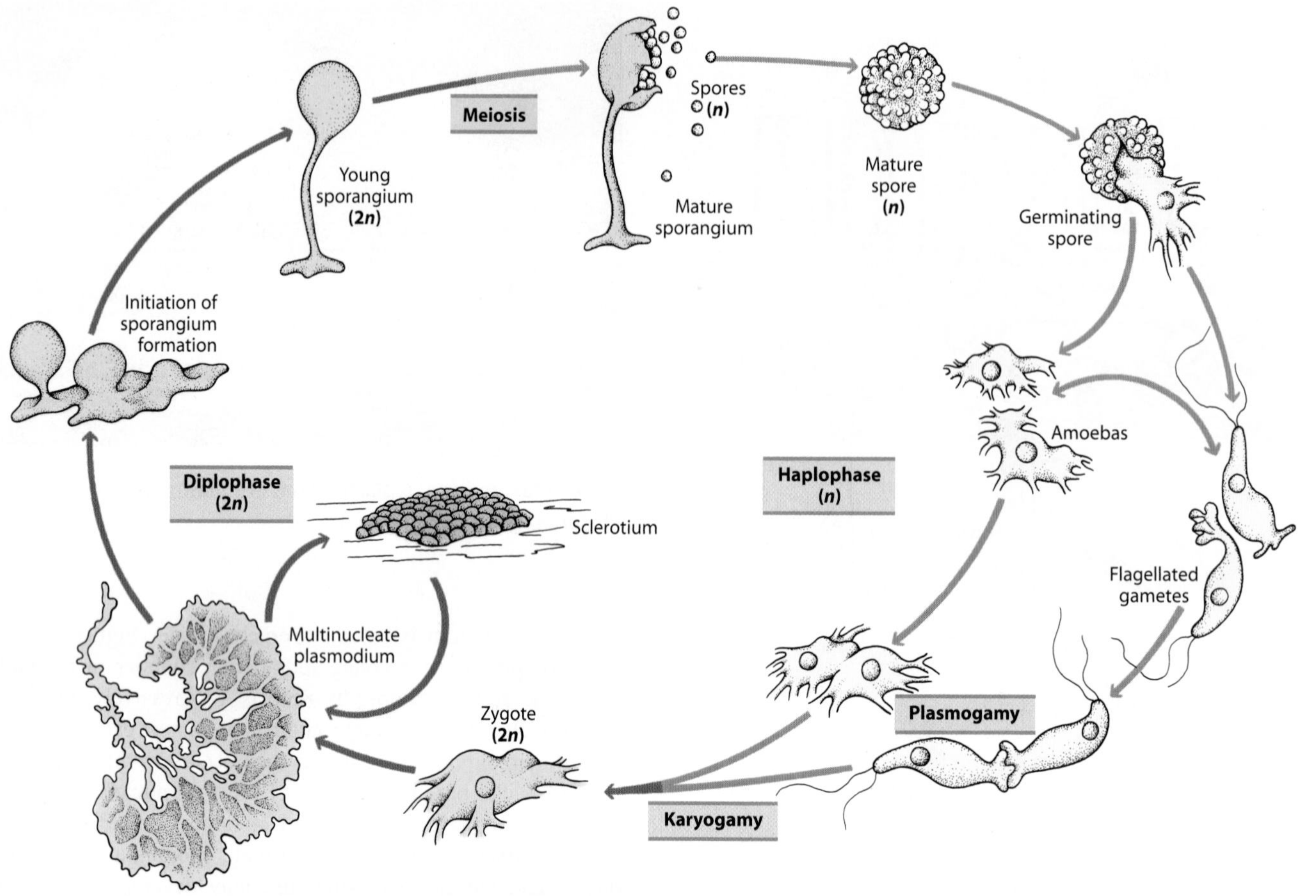

15–56 Life cycle of a typical myxomycete Sexual reproduction in the plasmodial slime molds consists of three distinct phases: plasmogamy, karyogamy, and meiosis. Plasmogamy is the union of two protoplasts, which may be amoebas or flagellated gametes derived from germinated spores; it brings two haploid nuclei together in the same cell. Karyogamy is the fusion of these two nuclei, resulting in the formation of a diploid zygote and initiation of the so-called diplophase of the life cycle. The plasmodium is a multinucleate, free-flowing mass of protoplasm that can pass through a silk cloth or a piece of filter paper and remain unchanged. In nature, plasmodia often form a hardened stage, the sclerotium, and are able to survive dry periods well in this condition. The active plasmodium may ultimately form sporangia. Within the developing sporangia, meiosis restores the haploid condition and thus initiates the haplophase of the life cycle.

adenosine monophosphate (cyclic AMP, or cAMP). These factors play roles similar to those in the cellular slime mold *Dictyostelium discoideum,* discussed next. One method of plasmodium formation is by the fusion of gametes. The gametes are usually genetically different from one another, in which case they are ultimately derived from different haploid spores. These gametes are simply some of the amoebas or flagellated cells that now have a new role. In some species and strains, however, the plasmodium is known to form directly from a single amoeba. Such plasmodia are usually haploid, like the amoebas from which they arose.

Cellular Slime Molds: Phylum Dictyosteliomycota

The cellular slime molds, or dictyostelids—a group of about 50 species in four genera—are probably more closely related to the amoebas (phylum Rhizopoda, not considered in this book) than to any other group. Also known as "social amoebas," they are common inhabitants of most litter-rich soils, where they usually exist as free-living amoeba-like cells, or **myxamoebas.** The myxamoebas feed on bacteria by phagocytosis (Figure 15–58a). Unlike fungi, with which they were once grouped, the cellular slime molds have cellulose-rich cell walls during part of their life cycle (Figure 15–58) and undergo the type of mitosis found in plants and animals, in which the nuclear envelope breaks down. In addition, again unlike the fungi, cellular slime molds have centrioles.

Dictyostelium reproduces by cell division and shows little morphological differentiation until it exhausts the available supply of bacteria. In response to starvation, individual cells aggregate to form a motile, sluglike mass. The uninucleate, haploid myxamoebas retain their individuality in the mass (Figure 15–58b), which usually contains 10,000 to 125,000 individuals. This mass, called a **pseudoplasmodium** (Figure 15–58d), or **slug,** migrates to a new place before differentiating and releasing

15–57 Spore-producing structures in plasmodial slime molds ***(a)*** Sporangium of *Arcyria nutans.* ***(b)*** Sporangia of *Stemonitis splendens.* ***(c)*** Plasmodiocarp of *Hemitrichia serpula.* ***(d)*** Aethalia of *Lycogala* growing on bark.

asexual spores. Thus, it avoids releasing new spores into habitats depleted of bacteria.

The myxamoebas aggregate by **chemotaxis,** migrating toward a source of cAMP, which is secreted by the starved myxamoebas. The cAMP diffuses away from the cells, establishing a concentration gradient along which the surrounding cells move toward the cAMP-secreting cells. The secreting cells, in turn, are stimulated to emit a new "pulse" of cAMP after a period of about five minutes. At least three waves of cells are recruited in this fashion. Binding of the cAMP signal to receptors in the plasma membrane triggers massive rearrangement of actin filaments, enabling the myxamoeba to crawl toward the source of cAMP. As the cells accumulate at the aggregation center, their plasma membranes become sticky; this causes them to adhere to each other and results in formation of a pseudoplasmodium.

The eventual developmental fate of an individual cell is determined at an early stage by its position in the aggregation, which seems to be controlled by the stage of the cell cycle the individual myxamoeba was in when aggregation began. Cells that divide between about 1.5 hours before and 40 minutes after the onset of starvation enter the aggregation last. When migration ceases, cells in the forward, or anterior, end of the slug become the stalk cells of the developing spore-bearing, or fruiting, body. The stalk cells become coated with cellulose, providing rigidity to the stalk, and then die by **apoptosis** (a form of programmed cell death). Meanwhile, the posterior cells of the pseudoplasmodium move to the top of the stalk and become dormant spores. Eventually, the spores are dispersed. If spores fall on a warm, damp surface, they germinate. Each spore releases a single myxamoeba, and the cycle is repeated.

The amoebas that form the stalks have been characterized as *altruistic,* because they give up their chance at reproduction and instead help other amoebas produce spores. The *csaA* gene, which codes for a cell-adhesion protein, enables these amoebas to recognize and cooperate with one another. Laboratory tests have shown that some mutant cells, called cheating cells, are able to avoid becoming part of the stalk and instead move directly to the top, the spore-bearing part of the fruiting body. Studies have revealed that these cheater amoebas lack the *csaA* gene.

Reproduction involving asexual spores is common in the cellular slime molds. Sexual reproduction also occurs frequently and results in the formation of walled zygotes called **macrocysts.** The macrocysts are formed by aggregations of myxamoebas that are smaller than those involved in the formation of slugs. In addition, these aggregations are rounded in outline, rather than elongate. During formation of a macrocyst, two haploid myxamoebas first fuse, forming a single large myxamoeba, the zygote, which becomes actively phagocytic. The zygote continues to feed voraciously until all of the surrounding myxamoebas have been engulfed, becoming a giant cell in the process. At this stage, a thick cell wall, rich in cellulose, is laid down around the giant cell, and a mature macrocyst is formed. Within the macrocyst, the zygote—the only diploid cell of the life cycle—undergoes meiosis and several mitotic divisions before germination and the release of numerous, haploid myxamoebas.

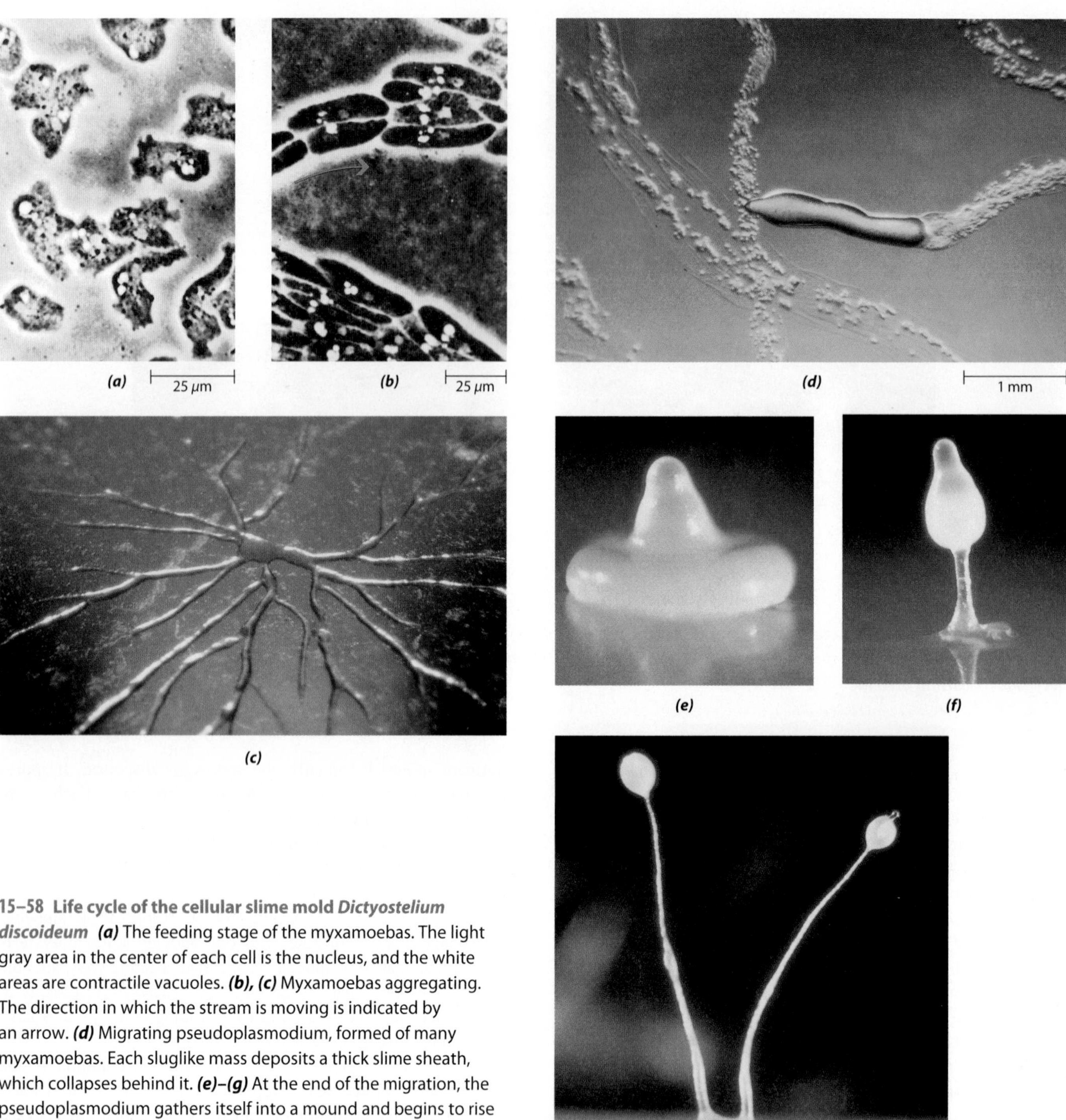

15–58 Life cycle of the cellular slime mold *Dictyostelium discoideum* ***(a)*** The feeding stage of the myxamoebas. The light gray area in the center of each cell is the nucleus, and the white areas are contractile vacuoles. ***(b), (c)*** Myxamoebas aggregating. The direction in which the stream is moving is indicated by an arrow. ***(d)*** Migrating pseudoplasmodium, formed of many myxamoebas. Each sluglike mass deposits a thick slime sheath, which collapses behind it. ***(e)–(g)*** At the end of the migration, the pseudoplasmodium gathers itself into a mound and begins to rise vertically, differentiating into a stalk and a mass of spores.

SUMMARY

See Table 15–1 on pages 318 and 319 for a comparative summary of protist characteristics

The Protists Include a Variety of Autotrophic and Heterotrophic Organisms

Protists are eukaryotic organisms that are not included in the plant, fungal, or animal kingdoms. Those considered in this book include both photosynthetic (autotrophic) organisms, the algae, and heterotrophic organisms once treated as fungi. The latter group includes the Oomycota, Myxomycota, and Dictyosteliomycota.

The algae include photosynthetic protists and their colorless relatives. Algae are important components of aquatic food webs and, together with the cyanobacteria, constitute the phytoplankton. The algae also play significant roles in global carbon and sulfur cycling.

Algae Obtain Nutrients in a Variety of Ways

Although many algae are capable of photosynthesis, uptake and utilization of dissolved organic compounds is also common. Particle feeding occurs in at least some members of the dinoflagellates, euglenoids, cryptomonads, haptophytes, and green algae. Photosynthetic algae that utilize both organic and inorganic carbon sources are called mixotrophs. Phagocytosis is probably the way in which the ancestors of the pigmented members of these algal groups acquired their chloroplasts. Marine haptophytes are significant components of food webs and are important mediators of global climate.

Protists Known as Stramenopiles Have Two Flagella of Different Length and Ornamentation

Five groups of stramenopiles are considered in this book: the diatoms, chrysophytes, xanthophytes, brown algae, and oomycetes. With the exception of the diatoms, their flagella occur in pairs, one long flagellum ornamented with hairs, and one short, smooth flagellum.

The Brown Algae Include the Largest and Structurally Most Complex of the Marine Algae

The brown algae are the most conspicuous seaweeds of temperate, boreal, and polar waters. In many kelps, the vegetative body is well differentiated into holdfast, stipe, and blade. Some kelps have food-conducting tissues that approach, in their complexity, those of vascular plants.

The Red Algae Have Complex Life Histories

The red algae, which are particularly abundant in warm marine waters, typically have life histories consisting of three distinct phases—gametophyte, carposporophyte, and tetrasporophyte. The chloroplasts of red algae are biochemically and structurally very similar to the cyanobacteria from which they very likely were derived.

Several Classes of Green Algae Are Recognized on the Basis of Cell Division, Reproductive Cell Structure, and Molecular Similarities

The Ulvophyceae are primarily marine, and some exhibit an alternation of generations. The Chlorophyceae have a unique mode of cytokinesis: the mitotic spindle collapses at telophase, and a phycoplast develops parallel to the plane of cell division. Some members of the Charophyceae produce a cytokinetic microtubular system, the phragmoplast, which is nearly identical to that of present-day bryophytes and vascular plants. In all probability, the bryophytes and vascular plants evolved from an extinct member of the Charophyceae that, in many respects, resembled extant members of the Coleochaetales and Charales.

The Oomycetes and Slime Molds Are Heterotrophic Protists

The oomycetes include both aquatic and terrestrial members. Some terrestrial oomycetes are important plant pathogens, including *Plasmopara viticola,* which causes downy mildew of grapes; several species of *Phytophthora,* which cause some very important plant diseases such as late blight of potatoes and sudden oak syndrome; and several species of *Pythium,* which cause damping-off diseases of seedlings.

The two unrelated groups of slime molds resemble fungi in their production of spores and resemble protozoa in their ability to move about. Both plasmodial and cellular slime molds ingest particulate food. The plasmodial slime molds are so-named because, under appropriate conditions, they exist as a thin, streaming, multinucleate (diploid) mass of "naked" protoplasm called a plasmodium. The cellular slime molds usually exist as free-living amoeba-like cells called myxamoebas. Under unfavorable conditions, the uninucleate, haploid myxamoebas form a mass called a pseudoplasmodium, or slug, in which the myxamoebas retain their individuality.

QUESTIONS

1. Expound upon the phytoplankton as the "great meadow of the sea."
2. "When the going gets tough, the tough get going." Describe how each of the following groups of organisms adapt to tough times, such as periods of inadequate nutrient or moisture levels: dinoflagellates, plasmodial slime molds, cellular slime molds.
3. Pellicle, stigma, contractile vacuole, paramylon, pyrenoid. These are entities found in *Euglena.* What is the function of each?
4. What do the organisms *Karenia brevis* and *Gonyaulax tamarensis* have in common?
5. What pigments do the diatoms, chrysophytes, xanthophytes, and brown algae have in common? Which of these pigments is responsible for the color of these algae?
6. Identify the plant diseases caused by each of the following oomycetes: *Plasmopara viticola, Phytophthora infestans, Phytophthora ramorum,* and *Pythium spp.*
7. The diatoms may be characterized as "the algae that live in glass houses." Explain.
8. Some kelps have the most highly differentiated bodies among the algae. In what ways?
9. *Fucus* has a life cycle that is, in some ways, similar to our own. Explain.
10. Of what advantage is the diploid carposporophyte generation to the red algae?
11. Distinguish between each of the following: oogonium and antheridium; homothallic and heterothallic; pinnate and centric; phycoplast and phragmoplast.
12. Distinguish among the three classes of green algae: Chlorophyceae, Ulvophyceae, and Charophyceae.
13. What traits do *Coleochaete* and the Charales have in common with the bryophytes and vascular plants?

CHAPTER 16

Bryophytes

◀ **A mossy waterfall in Scotland** Lacking a vascular system, mosses typically live in damp environments from which they absorb water and nutrients through their leaves and stems. Mosses are sometimes used in green roofs, where their minimal nutrient requirements and overall light weight make them good choices for rooftop plantings in moist, shady locations.

CHAPTER OUTLINE

Bryophytes—liverworts, mosses, and hornworts—are small "leafy" or flat plants that most often grow in moist locations in temperate and tropical forests or along the edges of wetlands and streams. But bryophytes are not confined to such habitats. Many species of mosses are found in relatively dry deserts, and several form extensive masses on dry, exposed rocks that can become very hot (Figure 16–1). Mosses sometimes dominate the terrain to the exclusion of other plants over large areas north of the Arctic Circle. Mosses are also the dominant plants on rocky slopes above timberline in the mountains, and a significant number of mosses are able to withstand the long periods of severe cold on the Antarctic continent (Figure 16–2). A few bryophytes are aquatic, and some are even found on rocks splashed by ocean waves. None are truly marine, however, with the exception of the aquatic moss *Fontinalis dalecarlica,* which can grow in the northern Baltic Sea, owing to its low salinity.

Bryophytes contribute significantly to plant biodiversity and are also important in some parts of the world for the large amounts of carbon they store, thereby playing a significant role in the global carbon cycle. Increasing evidence indicates that the first plants were much like extant, or living, bryophytes, and even today, together with lichens, bryophytes are important initial colonizers of bare rock and soil surfaces. Like the lichens, some bryophytes are remarkably sensitive to air pollution, and they are often absent or represented by only a few species in highly polluted areas. The bryophytes are important as models of the earliest land plants, helping us to understand how plants arose and started to affect their environments.

CHECKPOINTS

After reading this chapter, you should be able to answer the following:

1. What are the general characteristics of the bryophytes? In other words, what does it take to be a bryophyte?
2. How are the three phyla of bryophytes similar to and different from one another?
3. How does sexual reproduction occur in bryophytes? What are the principal parts of the resulting sporophyte of most bryophytes?
4. What are the distinguishing features of the two clades of liverworts?
5. What are the distinguishing features of the peat mosses (class Sphagnidae) and of the "true mosses" (class Bryidae)?
6. What are the distinguishing features of the hornworts?

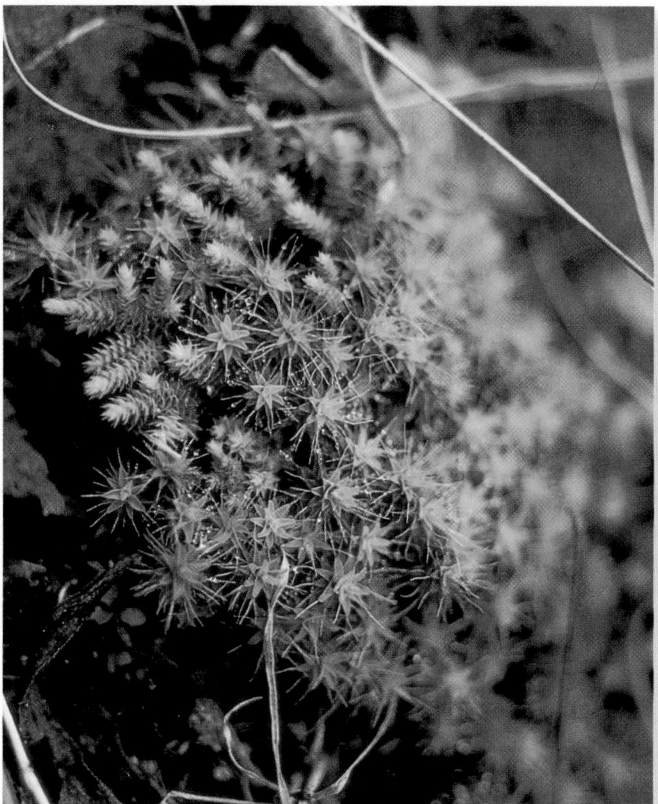

16–1 A dry-land moss *Tortula obtusissima* lives on and around limestone in the central plateau of Mexico. Having no roots, the plants obtain their moisture directly from the external environment in the form of dew or rain. They can recover physiologically from complete dryness in less than five minutes.

The Relationships of Bryophytes to Other Groups

In many respects, bryophytes are transitional between the charophycean green algae (page 353) and the vascular plants (discussed in Chapters 17 through 20). Both "bryophytes" and "charophycean green algae" are paraphyletic groups (groups that do not include all the descendants of a single common ancestor)—hence the use of informal names for these groups. Informal names are useful for discussing organisms having similar habitats or adaptations. In the last chapter, we considered some of the features shared by charophytes and plants (bryophytes and vascular plants, which are adapted for life on the land). Both contain chloroplasts with well-developed grana, and both have motile cells that are asymmetrical, with flagella that extend from the side rather than the end of the cell. During the cell cycle, both charophycean green algae and plants exhibit breakdown of the nuclear envelope at mitosis and persistent spindles or phragmoplasts during division of the cytoplasm (cytokinesis). In addition, you may recall that, among the charophycean green algae, the Coleochaetales and Charales appear to be more closely related to the plants than are any others. For example, members of these groups, such as *Coleochaete* and *Chara,* are like plants in having oogamous sexual reproduction, that is, a non-flagellated egg cell that is fertilized by a flagellated sperm. In *Coleochaete,* the zygotes are retained within the parental thallus and, in at least one species of *Coleochaete,* the cells covering the zygotes develop wall ingrowths. These covering cells apparently function as transfer cells involved with the transport of sugars to the zygotes.

Bryophytes and vascular plants share a number of characters that distinguish them from the charophytes. These shared

(a)

(b)

16–2 Moss in Antarctica ***(a)*** At about 3000 meters elevation on Mount Melbourne, Antarctica, the daily temperatures in summer mainly range from –10° to –30°C. In this incredibly harsh environment, botanists from New Zealand discovered patches of a moss of the genus *Campylopus* ***(b)***, growing in the bare areas visible in the photograph, where volcanic activity produces temperatures that may reach 30°C. The growth of *Campylopus* in this locality demonstrates the remarkable dispersal powers of mosses, as well as their ability to survive in harsh habitats.

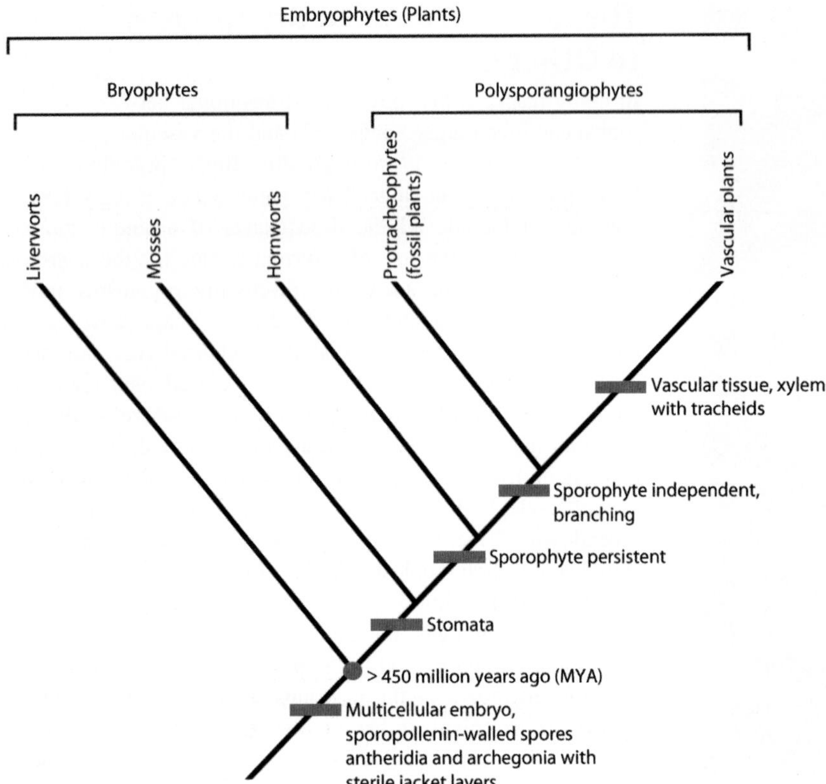

16–3 Cladogram of embryophytes This cladogram reflects one point of view of the phylogenetic relationships among the bryophyte lineages and between the bryophytes and the polysporangiophytes (plants with branching sporophytes and multiple sporangia). The term "embryophytes," a synonym for plants, refers to the fact that a multicellular embryo is retained within the female gametophyte (see page 372). This cladogram indicates that the hornworts share a more recent common ancestor with the polysporangiophytes than do the liverworts or mosses and that the liverworts are a sister group to all other embryophytes.

characteristics include: (1) the presence of male and female gametangia, called **antheridia** and **archegonia,** respectively, with a protective layer called a sterile jacket layer; (2) retention of both the zygote and the developing multicellular embryo, or young sporophyte, within the archegonium or the female gametophyte; (3) the presence of a multicellular diploid sporophyte, which results in an increased number of meioses and an amplification of the number of spores that can be produced following each fertilization event; (4) multicellular sporangia consisting of a sterile jacket layer and internal spore-producing **(sporogenous)** tissue; (5) meiospores with walls containing sporopollenin, which resists decay and drying; and (6) tissues produced by an apical meristem. Charophytes lack all of these shared bryophyte and vascular plant characters, which are correlated with the existence of plants on land. Therefore, only the bryophytes and vascular plants are placed in the kingdom Plantae in this book.

Living bryophytes lack the water- and food-conducting (vascular) tissues called xylem and phloem, respectively, that are present in vascular plants. Although some bryophytes have specialized conducting tissues, the cell walls of the bryophyte water-conducting cells are not lignified, as are those of the vascular plants. Also, there are differences in the life cycles of bryophytes and vascular plants, both of which exhibit alternating heteromorphic gametophytic and sporophytic generations. In the bryophytes, the gametophyte usually is larger as well as free-living, and the sporophyte is smaller and permanently attached to, and nutritionally dependent on, its parental gametophyte. By contrast, the sporophyte of vascular plants is larger than the gametophyte and ultimately free-living. In addition, the bryophyte sporophyte is unbranched and bears only a single sporangium, whereas the sporophytes of extant vascular plants are branched and bear many more sporangia (polysporangiophytes). Vascular plant sporophytes therefore produce a great many more spores than do the sporophytes of bryophytes.

It is quite clear that the bryophytes include the earliest of the extant plant groups. Modern bryophytes can therefore provide important insights into the nature of the earliest land-adapted plants and the process by which plants evolved. A comparison of the structure and reproduction of extant bryophytes with those of ancient fossils and living vascular plants can show how various features of vascular plants may have evolved. Research on the moss *Physcomitrella patens,* whose genome has been sequenced, promises to add substantially to our understanding of plant evolution and diversity (see Figure 16–20). Its role as a model plant system that allows targeting of specific genes is proving to be valuable.

The bryophytes are grouped into three phyla: Marchantiophyta (the liverworts), Bryophyta (the mosses), and Anthocerotophyta (the hornworts). A recent study, involving analyses of three datasets (chloroplast, mitochondrial, and nuclear genes), strongly supports liverworts as sister to all other land plants and also that the hornworts share a more recent ancestor with the vascular plants (Figure 16–3).

Comparative Structure and Reproduction of Bryophytes

Some bryophytes, namely the hornworts and certain liverworts, are described as "thalloid," because their gametophytes, which are generally flat and dichotomously branched (forking repeatedly into two equal branches), are **thalli** (singular: thallus). Thalli are undifferentiated bodies, or bodies not differentiated into roots, leaves, and stems. Such thalli are often relatively thin, which may facilitate the uptake of both water and CO_2. Some bryophyte gametophytes have specialized adaptations on their upper surface for increasing CO_2 permeability while at the same time reducing water loss. The surface pores of the thalloid liverwort *Marchantia* are one such example (Figure 16–4). On the other hand, the gametophytes of some liverworts (the leafy liverworts) and the mosses are said to be differentiated into "leaves" and "stems," but it could be argued that these are not true leaves and stems because they occur in the gametophytic generation and do not contain xylem and phloem. However, the thalli of certain liverworts and mosses do contain centrally located strands of cells that appear to have conducting functions. Such cells may be similar to ancient evolutionary precursors of phloem and lignified vascular tissues. Inasmuch as the terms "leaf" and "stem" are commonly used when referring to the leaflike and stemlike structures of the gametophytes of leafy liverworts and mosses, this practice will be followed in this book. The true leaves and stems of vascular plants are produced by the sporophytes.

Surface layers reminiscent of the waxy cuticles commonly found on surfaces of the true leaves and stems of vascular plants also occur on the surfaces of some bryophytes. The cuticle of sporophytes is closely correlated with the presence of stomata, which function primarily in the regulation of gas exchange. The pores seen in some bryophyte gametophytes, such as those of *Marchantia,* are considered to be analogous to stomata (Figure 16–4). The biochemistry and evolution of the bryophyte cuticle are poorly understood, however, primarily because bryophyte cuticles are more difficult to remove for chemical analysis than are cuticles of vascular plants.

The gametophytes of both thalloid and leafy bryophytes are generally attached to the substrate, such as soil, by **rhizoids** (Figure 16–4). The rhizoids of mosses are multicellular, each consisting of a linear row of cells, whereas those of liverworts and hornworts are unicellular. The rhizoids of bryophytes generally serve only to anchor the plants, because absorption of water and inorganic ions commonly occurs directly and rapidly throughout the gametophyte. Mosses, in particular, often have special hairs and other structural adaptations that aid in external water transport and absorption by leaves and stems. In addition, bryophytes often harbor fungal or cyanobacterial symbionts that may aid in acquisition of mineral nutrients. Rootlike organs are lacking in the bryophytes.

The cells of bryophyte tissues are interconnected by plasmodesmata. Bryophyte plasmodesmata are similar to those of vascular plants in possessing an internal component known as the desmotubule (Figure 16–5). The desmotubule is derived from a segment of tubular endoplasmic reticulum that becomes entrapped in developing cell plates during cytokinesis (see Figure 3–46). Certain charophycean green algae also possess plasmodesmata.

The cells of most bryophytes resemble those of vascular plants in having many small, disk-shaped plastids. All of the cells of some hornwort species, and the apical and/or

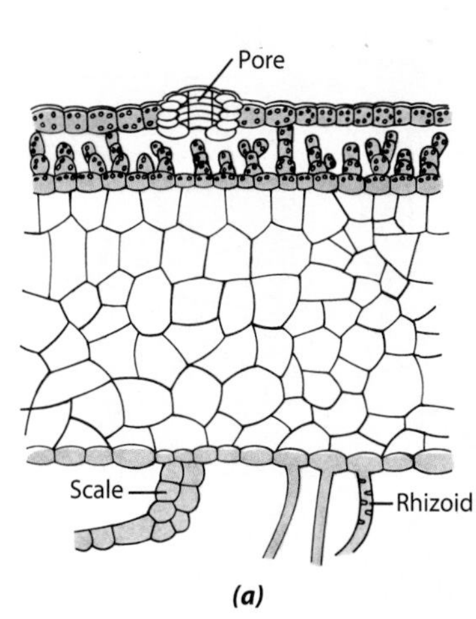

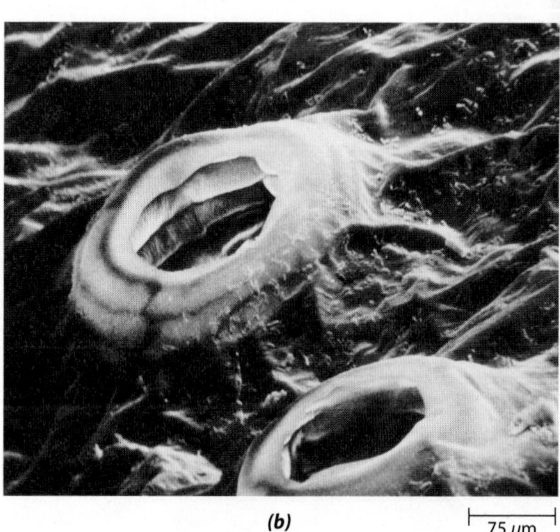

16–4 Surface pores of *Marchantia* ***(a)*** Transverse section of the gametophyte of *Marchantia,* a thalloid liverwort. Numerous chloroplast-bearing cells are evident in the upper layers, and there are several layers of colorless cells below them, as well as rhizoids that anchor the plant body to the substrate. Pores permit the exchange of gases in the air-filled chambers that honeycomb the upper photosynthetic layer. The specialized cells that surround each pore are usually arranged in four or five superimposed circular tiers of four cells each, and the whole structure is barrel-shaped. Under dry conditions, the cells of the bottommost tier, which usually protrude into the chamber, become juxtaposed and retard water loss, whereas under moist conditions they separate. Thus the pores serve a function similar to that of the stomata of vascular plants. ***(b)*** A scanning electron micrograph of two pores on the dorsal surface of a gametophyte of *Marchantia.*

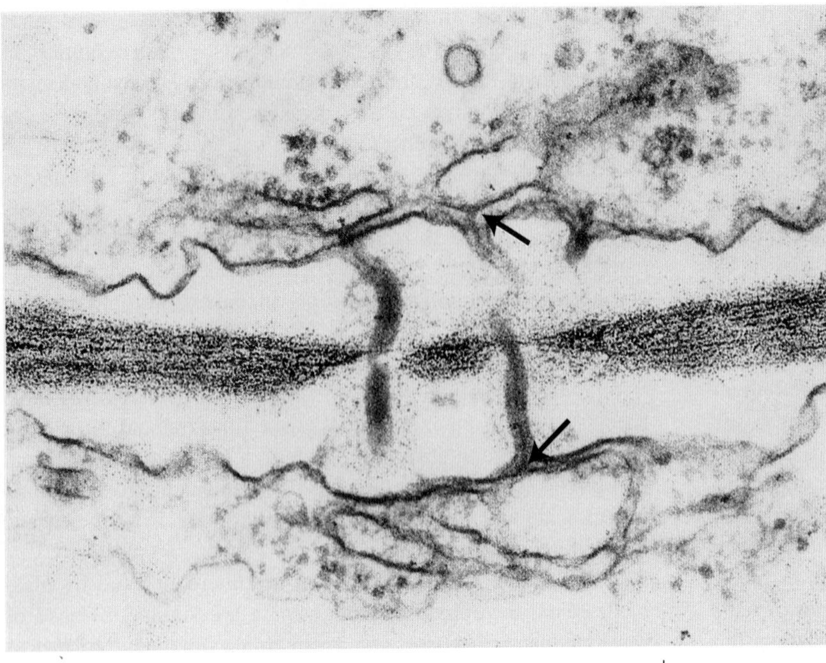

16–5 Bryophyte plasmodesmata Longitudinal view of plasmodesmata in the liverwort *Monoclea gottschei.* Note that the desmotubule in the plasmodesma on the right (arrows) is continuous with the endoplasmic reticulum in the cytosol.

reproductive cells of many bryophytes, by contrast, have only a single large plastid per cell. This characteristic is believed to be an evolutionary holdover from ancestral green algae, which, like modern *Coleochaete,* probably contained only a single large plastid per cell. During cell division, the cells of bryophytes and vascular plants produce preprophase bands consisting of microtubules that specify the position of the future cell wall. Such bands are lacking in the charophycean green algae.

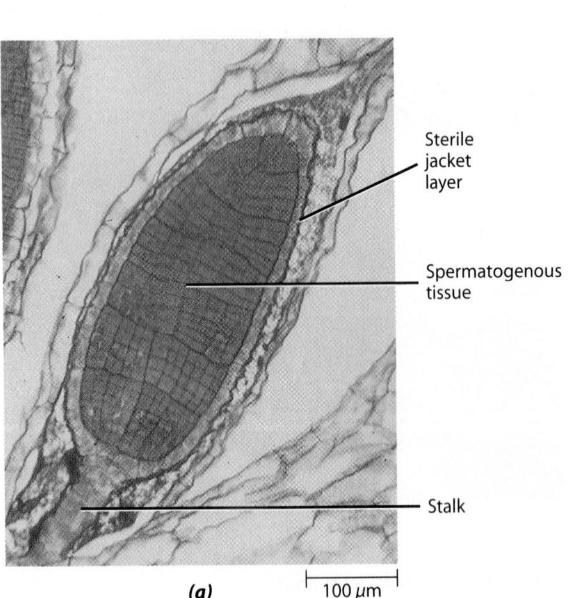

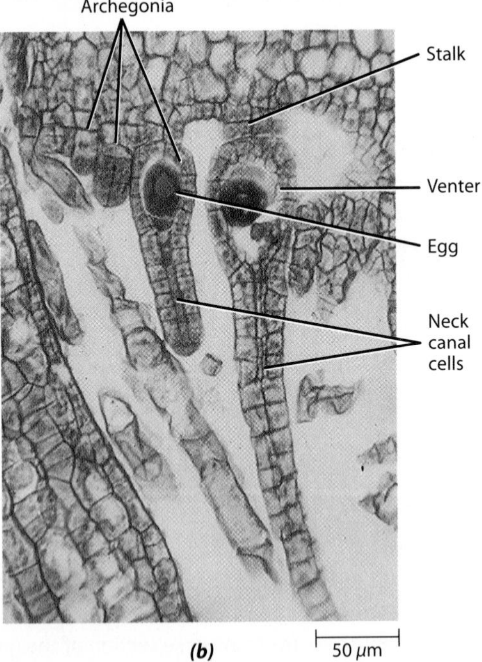

16–6 Gametangia of *Marchantia,* a liverwort ***(a)*** A developing antheridium, consisting of a stalk and a sterile—that is, non-sperm-forming—jacket layer enclosing spermatogenous tissue. The spermatogenous tissue develops into spermatogenous cells, each of which forms a single sperm propelled by two flagella. ***(b)*** Several archegonia at different stages of development. An egg is contained in the venter, a swollen portion at the base of each flask-shaped archegonium. When the egg is mature, the neck canal cells disintegrate, creating a fluid-filled tube through which the biflagellated sperm swim to the egg in response to chemical attractants. In *Marchantia,* the archegonia and antheridia are borne on different gametophytes.

Sperm Are the Only Flagellated Cells Produced by Bryophytes, and They Require Water to Swim to the Egg

Many bryophytes can reproduce asexually by fragmentation (vegetative propagation), whereby small pieces of tissue produce an entire gametophyte. Another widespread means of asexual reproduction in both liverworts and mosses is the production of **gemmae** (singular: gemma)—multicellular bodies that give rise to new gametophytes (see Figure 16–13). Unlike some charophycean green algae, which can generate flagellated zoospores for asexual reproduction, sperm are the only flagellated cells produced by bryophytes. Loss of the ability to produce zoospores, which are likely to be less useful on land than in the water, is probably correlated with the absence of centrioles from the spindles of bryophytes and other plants (page 64). Mitosis in certain liverworts and hornworts shows features that are intermediate between those of charophycean green algae and vascular plants, suggesting evolutionary stages leading to the absence of centrioles in plant mitosis.

Sexual reproduction in bryophytes involves production of antheridia and archegonia, often on separate male and female gametophytes. In some species, sex is known to be controlled by the distribution at meiosis of distinctive sex chromosomes. In fact, sex chromosomes in plants were first discovered in bryophytes. The spherical or elongated antheridium is commonly stalked and consists of a sterile jacket layer, one cell thick, that surrounds numerous **spermatogenous cells,** cells that develop into sperm cells (Figure 16–6a). The "jacket" layer of cells is said to be "sterile" because it cannot produce sperm. Each spermatogenous cell forms a single biflagellated sperm that must swim through water to reach the egg, located inside an archegonium. Liquid water is therefore required for fertilization in bryophytes.

The archegonia of bryophytes are flask-shaped, with a long neck and a swollen basal portion, the **venter,** which encloses a single egg (Figure 16–6b). The outer layer of cells of the neck and venter forms the sterile protective layer of the archegonium. The central cells of the neck, the **neck canal cells,** disintegrate when the egg is mature, resulting in a fluid-filled tube through which the sperm swim to the egg. During this period, chemicals are released that attract sperm. After fertilization, the zygote remains within the archegonium, where it is nourished by sugars, amino acids, and probably other substances provided by the maternal gametophyte. This form of nutrition is known as **matrotrophy** ("food derived from the mother"). Thus supplied, the zygote undergoes repeated mitotic divisions, generating the multicellular embryo (Figure 16–7), which eventually develops into the mature sporophyte (Figure 16–8).

There are no plasmodesmatal connections between cells of the two adjacent generations. Nutrient transport is thus apoplastic—that is, nutrients move along the cell walls. This transport is facilitated by a **placenta** located at the interface between the two generations, sporophyte and parental

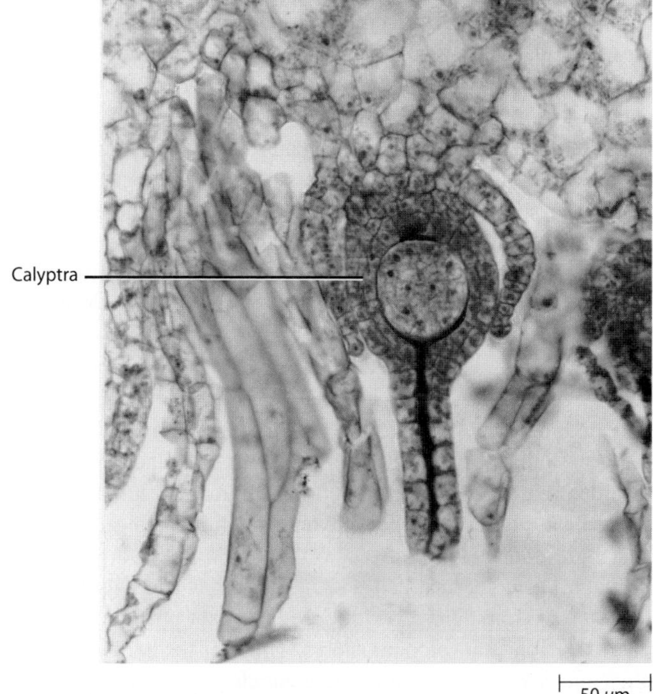

16–7 *Marchantia* embryo An early stage in development of the embryo, or young sporophyte, of *Marchantia.* Here the young sporophyte is nothing more than an undifferentiated spherical mass of cells within the enlarged venter, or calyptra.

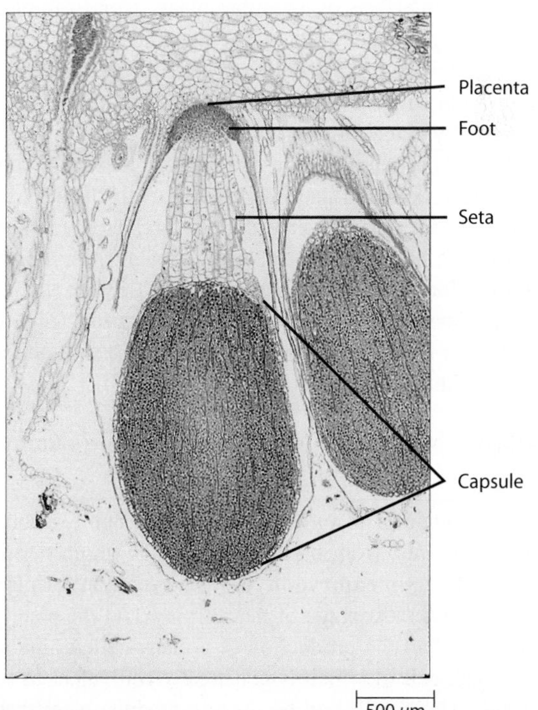

16–8 *Marchantia* sporophyte A nearly mature sporophyte of *Marchantia,* with a distinct foot, seta, and capsule, or sporangium. The placenta is at the interface between the foot and gametophyte and consists of transfer cells of both sporophyte and gametophyte.

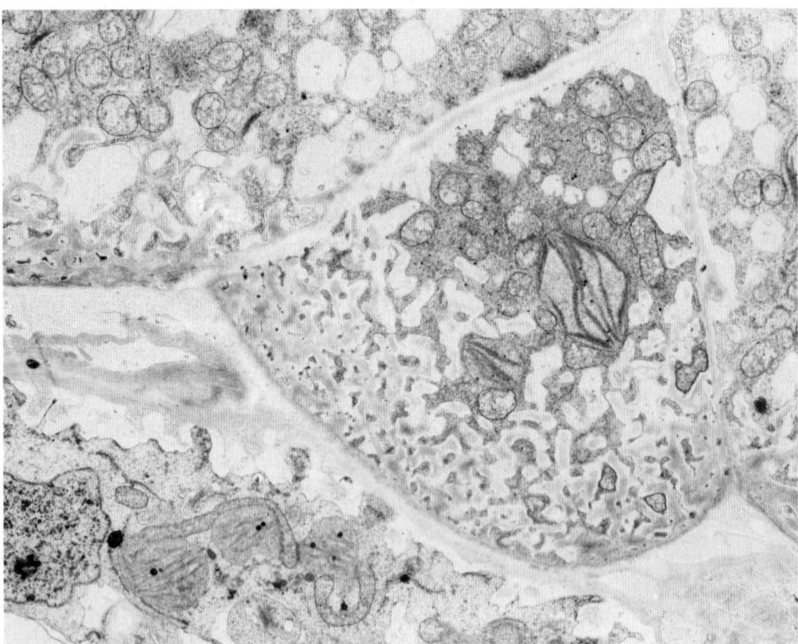

16–9 Bryophyte placenta The gametophyte-sporophyte junction—the placenta—in the liverwort *Carrpos monocarpos*. Extensive wall ingrowths develop in the single cell layer of transfer cells in the sporophyte (upper three cells). There are several layers of transfer cells in the gametophyte (lower left corner), but their wall ingrowths are not as highly branched as those of the sporophyte layer. Numerous chloroplasts and mitochondria are present in the placental cells of both generations.

gametophyte (Figure 16–9), and therefore analogous to the placenta of mammals. The bryophyte placenta is composed of transfer cells with an extensive labyrinth of highly branched cell wall ingrowths that vastly increase the surface area of the plasma membrane across which active nutrient transport takes place. Similar transfer cells occur at the gametophyte-sporophyte interface of vascular plants (for example, *Arabidopsis* and soybean) and at the haploid-diploid junction of *Coleochaete* (page 357). The occurrence of placental cells in *Coleochaete* suggests that matrotrophy had already evolved in the charophyte ancestors of plants.

As the bryophyte embryo develops, the venter undergoes cell division, keeping pace with the growth of the young sporophyte. The enlarged venter of the archegonium is called a **calyptra.** At maturity, the sporophyte of most bryophytes consists of a **foot,** which remains embedded in the archegonium, a **seta,** or stalk, and a **capsule,** or **sporangium** (Figure 16–8). The transfer cells at the junction between the foot and archegonium constitute the placenta.

The Term "Embryophytes" Is an Appropriate Synonym for Plants

The occurrence of a multicellular, matrotrophic embryo in all groups of plants, from bryophytes through angiosperms, is the basis for the term **embryophytes** as a synonym for plants (Figure 16–3). The advantage of matrotrophy and the plant placenta is that they fuel the production of a many-celled diploid sporophyte, each cell of which is genetically equivalent to the fertilized egg. These cells can be used to produce many genetically diverse haploid spores upon meiosis in the sporangium. This condition may have provided a significant advantage to early plants as they began to occupy the land. Production of greater numbers of spores per fertilization event may also have helped compensate for low fertilization rates when water became scarce. The sporophytic generation of plants is thought to have evolved from a zygote, such as those produced by charophytes, in which meiosis was delayed until after at least a few mitotic divisions had occurred. The more mitotic divisions that occur between fertilization and meiosis, the larger the sporophyte that can be formed and the greater the number of spores that can be produced. Throughout the evolutionary history of plants, there has been a tendency for sporophytes to become increasingly larger in relation to the gametophyte generation.

The sporophyte epidermis of mosses and many hornworts contains stomata—each bordered by two guard cells—that resemble the stomata of vascular plants. The moss stomata, however, are able to open and close and thus regulate gas exchange for only a short period after their development. Thereafter they remain open and their function is uncertain. Perhaps they then function to generate a flow of water and nutrients between the sporophyte and gametophyte, induced by the loss of water vapor through the stomata. The hornwort stomata apparently lack the ability to open and close. Once open they remain open. It has been suggested that these stomata are essential for the dehydration and dehiscence (splitting) of the sporangium. The presence of stomata on the sporophytes of mosses and hornworts is regarded as evidence of an important evolutionary link to the vascular plants. Liverwort sporophytes, which are typically smaller and more ephemeral than those of mosses and hornworts, lack stomata. The epidermal cell walls of the moss and liverwort sporophytes are impregnated with decay-resistant phenolic materials that may protect developing spores. Those of the hornwort sporophyte are covered with a protective cuticle.

The Sporopollenin Walls of Bryophyte Spores Have Survival Value

Bryophyte spores, like those of all other plants, are encased in a substantial wall impregnated with the most decay- and chemical-resistant biopolymer known, **sporopollenin.** The sporopollenin walls enable bryophyte spores to survive dispersal through the air from one moist site to another. The spores of charophycean green algae, which are typically dispersed in water, are not enclosed by a sporopollenin wall. Charophyte zygotes, however, are lined with sporopollenin and can therefore tolerate exposure and microbial attack, remaining viable for long periods. The sporopollenin-walled spores of plants are thought to have originated from the zygotes of charophycean green algae by a change in the timing of sporopollenin deposition.

Bryophyte spores germinate to form juvenile developmental stages, which in mosses are called **protonemata** (singular: protonema; from the Greek, *prōtos,* "first," and *nēma,* "thread"). From the protonemata (see Figure 16–20 on page 381), gametophytes and gametangia develop. Protonemata are characteristic of all mosses and are also found in some liverworts, but not in hornworts.

Liverworts: Phylum Marchantiophyta

Liverworts are a group of about 5200 species of plants that are generally small and inconspicuous, although they may form relatively large masses in favorable habitats, such as moist, shaded soil or rocks, tree trunks, or branches. A few kinds of liverworts grow in water. The name "liverwort" dates from the ninth century, when it was thought, because of the liver-shaped outline of the gametophyte in some genera, that these plants might be useful in treating diseases of the liver. According to the medieval "Doctrine of Signatures," the outward appearance of a body signaled the possession of special properties. The Anglo-Saxon ending *-wort* (originally *wyrt*) means "herb"; it appears as a part of many plant names in the English language.

Most liverwort gametophytes develop directly from spores, but some genera first form a protonema-like filament of cells, from which the mature gametophyte develops. Gametophytes continue to grow from an apical meristem. There are three major types of liverworts, differentiated on the basis of structure and grouped into two clades. One clade consists of the complex thalloid liverworts, which have internal tissue differentiation. The other clade contains the leafy liverworts and the simple thalloid liverworts, which consist of ribbons of relatively undifferentiated tissue. Some simple thalloid types contain elongated water-conducting cells with tapering ends and thick walls perforated with numerous pits.

Most liverworts develop close symbiotic associations with the glomeromycetes (see Chapter 14), which enter the thallus through the rhizoids. Liverworts produce abundant mucilage, which is believed to aid in water retention.

Complex Thalloid Liverworts Include *Riccia, Ricciocarpus,* and *Marchantia*

Thalloid liverworts can be found on moist, shaded banks and in other suitable habitats, such as flowerpots in a cool greenhouse. The thallus, which is about 30 cells thick at the midrib and approximately 10 cells thick in the thinner portions, is sharply differentiated into a thin, chlorophyll-rich upper (dorsal) portion and a thicker, colorless lower (ventral) portion (Figure 16–4a). The lower surface bears rhizoids, as well as rows of scales. The upper surface is often divided into raised regions, each with a large pore that leads to an underlying air chamber (Figure 16–4b).

The sporophyte structure of *Riccia* and *Ricciocarpus* is among the simplest seen in the liverworts (Figure 16–10),

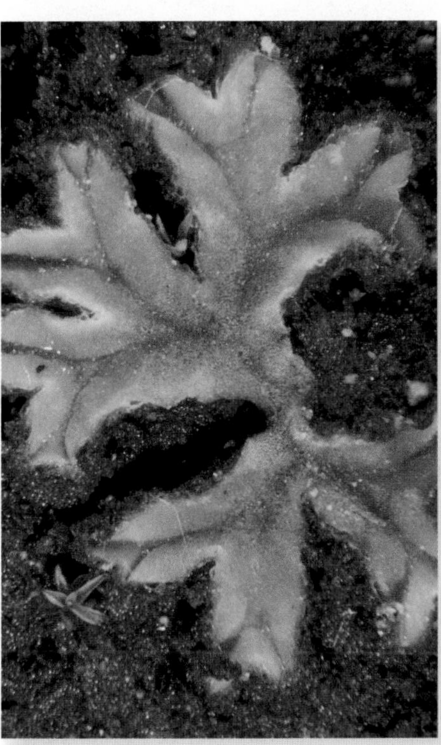
(a)

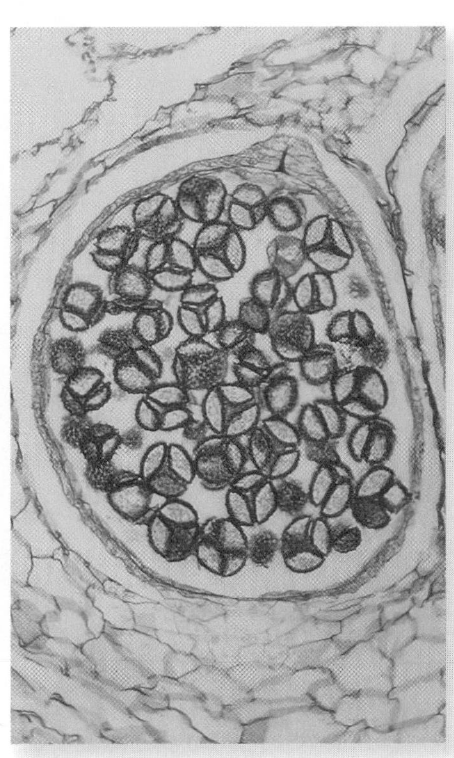
(b)

16–10 ***Riccia,*** **one of the simplest liverworts** ***(a)*** The system of branching of *Riccia* gametophytes is dichotomous, that is, the main and subsequent axes fork into two branches. ***(b)*** The sporophyte, which is embedded within the gametophyte, consists solely of a spherical capsule.

although it is a derived condition, not an ancestral one. *Ricciocarpus,* which grows in water or on damp soil, is bisexual—that is, both sex organs arise on the same plant. Some species of *Riccia* are aquatic, although most are terrestrial. *Riccia* gametophytes may be either unisexual or bisexual. In both *Riccia* and *Ricciocarpus,* the sporophytes are deeply embedded within the dichotomously branched gametophytes and consist of little more than a sporangium. No special mechanism for spore dispersal occurs in these sporophytes. When the portion of the gametophyte containing mature sporophytes dies and decays, the spores are liberated.

One of the most familiar of liverworts is *Marchantia,* a widespread genus that grows on moist soil and rocks (see Figure 12–16b). Its dichotomously branched gametophytes are larger than those of *Riccia* and *Ricciocarpus.* Unlike the latter two genera, in which the sex organs are distributed along the dorsal surface of the thallus, in *Marchantia* the gametangia are borne on specialized structures called **gametophores,** or **gametangiophores.** The gametophore stalks are just rolled-up regular thalli that grow perpendicular to the ground rather than flat against it.

The gametophytes of *Marchantia* are unisexual, and the male and female gametophytes can be readily distinguished by their distinctive gametophores. The antheridia are borne on disk-headed gametophores called **antheridiophores,** whereas the archegonia are borne on umbrella-headed gametophores called **archegoniophores** (Figure 16–11). In *Marchantia,* the sporophyte generation consists of a foot, a short seta, and a capsule (Figure 16–8). In addition to spores, the mature sporangium contains elongated cells called **elaters,** which have helically arranged hygroscopic (moisture-absorbing) wall thickenings (Figure 16–12). The walls of elaters are sensitive to slight changes in humidity, and, after the capsule dehisces (dries out and opens) into a number of petal-like segments, the elaters undergo a twisting action that helps disperse the spores.

Fragmentation is the principal means of asexual reproduction in liverworts, but another widespread mechanism is the production of gemmae. In *Marchantia,* the gemmae are produced in special cuplike structures—called **gemma cups**—located on the dorsal surface of the gametophyte (Figure 16–13). The gemmae are dispersed primarily by splashes of rain.

The life cycle of *Marchantia* is illustrated in Figure 16–14 (pages 376 and 377).

Leafy Liverworts Have a Distinctive Leaf Structure and/or Arrangement

The leafy liverworts are a very diverse group that includes more than 4000 of the 5200 species of the phylum Marchantiophyta (Figure 16–15 on page 378). The leafy liverworts are especially abundant in the tropics and subtropics, in regions of heavy rainfall or high humidity, where they grow on the leaves and the bark of trees, as well as on other plant surfaces (Figure 16–16 on page 378). There are probably many tropical species

(a)

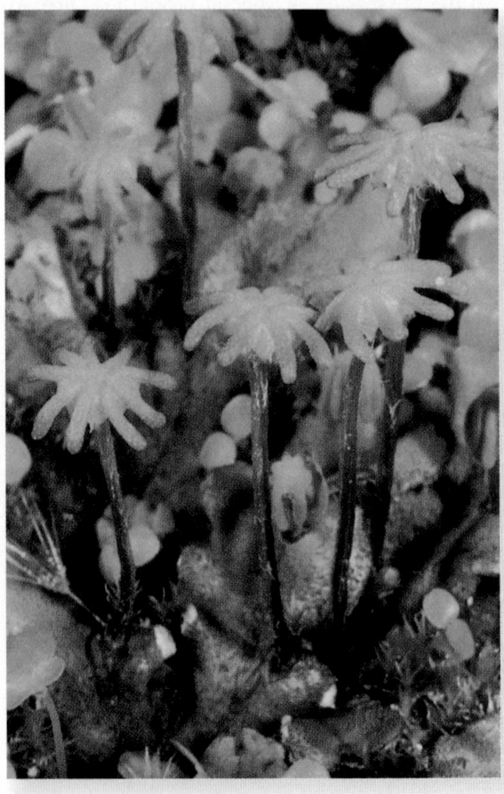

(b)

16–11 Gametophytes of *Marchantia* The antheridia ***(a)*** and archegonia ***(b)*** are elevated on stalks—the antheridiophores and archegoniophores, respectively—above the thallus.

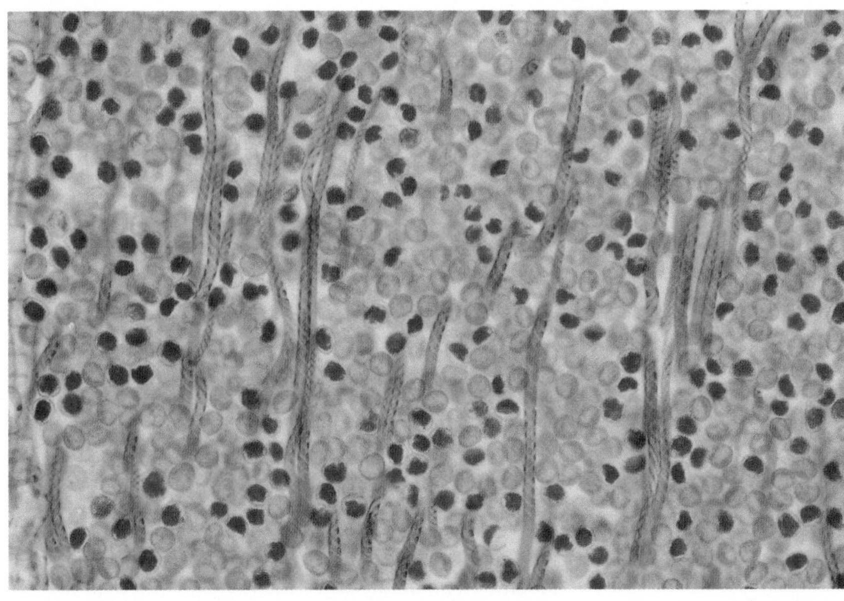

16–12 Spores and elaters Mature spores (red spheres) and elaters (green strands) from a capsule of *Marchantia.*

that have not yet been described. Leafy liverworts are also well represented in temperate regions. The plants are usually well branched and form small mats.

Liverwort leaves, like moss leaves, generally consist of only a single layer of undifferentiated cells. One way to distinguish liverworts from mosses is that moss leaves are usually equal in size and spirally arranged around the stem, whereas many liverworts have two rows of equal-sized leaves and a third row of smaller leaves along the lower surface of the gametophyte. Most moss leaves splay outward from the stem in three dimensions, but a few have leaves flattened into one plane, as do many liverworts. In addition, moss leaves sometimes have a thickened "midrib," but liverwort leaves lack this structure. Moss leaves are most often entire, in contrast to liverwort leaves, which can be highly lobed or dissected.

In the leafy liverworts, the antheridia are generally borne on a short side branch with modified leaves, known as the **androecium.** The developing sporophyte, as well as the archegonium from which it develops, is characteristically surrounded by a tubular sheath known as the **perianth** (Figure 16–15c).

(a)

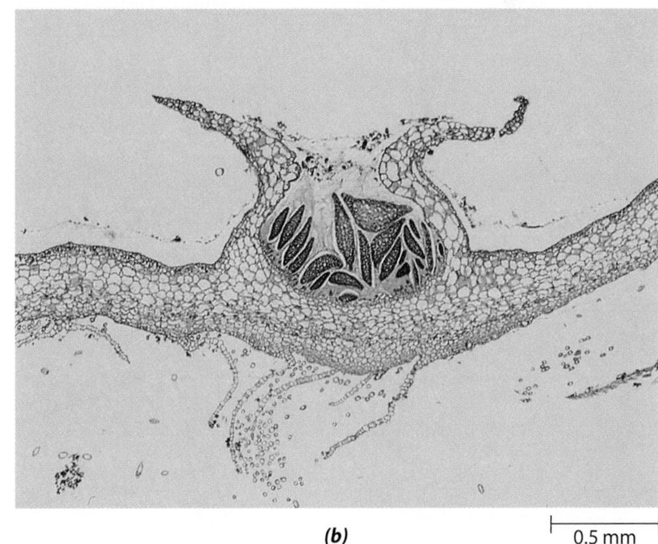

(b)

16–13 Gemma cups ***(a)*** Gametophyte of *Marchantia,* with gemma cups containing gemmae. The gemmae appear as more or less disk-shaped pieces of tissue. The gemmae are splashed out by the rain and may then grow into new gametophytes, each genetically identical to the parent plant from which it was derived by mitosis. ***(b)*** Longitudinal section of a gemma cup. The gemmae are the dark structures, which in sectional view appear more or less lens-shaped.

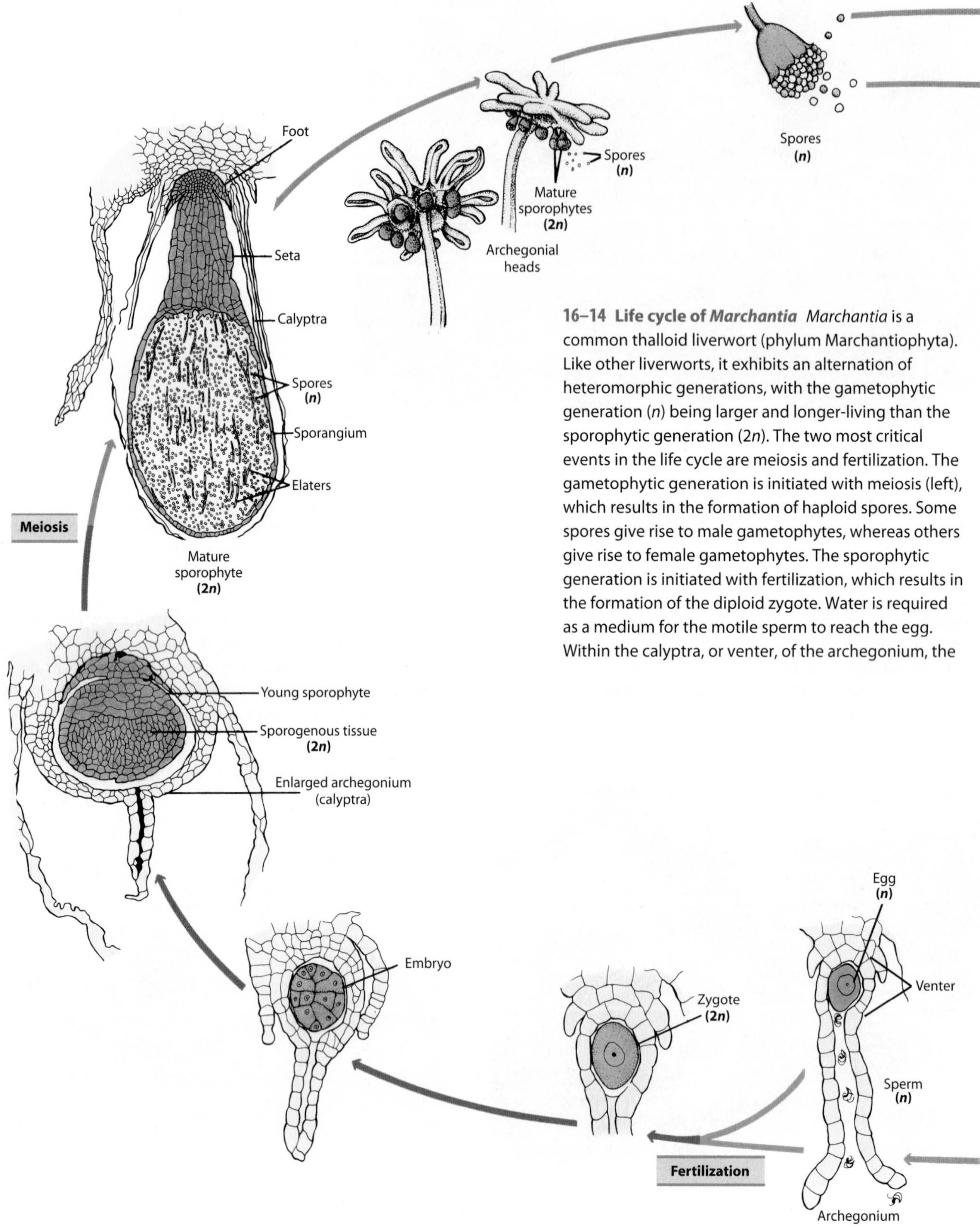

16–14 Life cycle of *Marchantia* *Marchantia* is a common thalloid liverwort (phylum Marchantiophyta). Like other liverworts, it exhibits an alternation of heteromorphic generations, with the gametophytic generation (n) being larger and longer-living than the sporophytic generation ($2n$). The two most critical events in the life cycle are meiosis and fertilization. The gametophytic generation is initiated with meiosis (left), which results in the formation of haploid spores. Some spores give rise to male gametophytes, whereas others give rise to female gametophytes. The sporophytic generation is initiated with fertilization, which results in the formation of the diploid zygote. Water is required as a medium for the motile sperm to reach the egg. Within the calyptra, or venter, of the archegonium, the

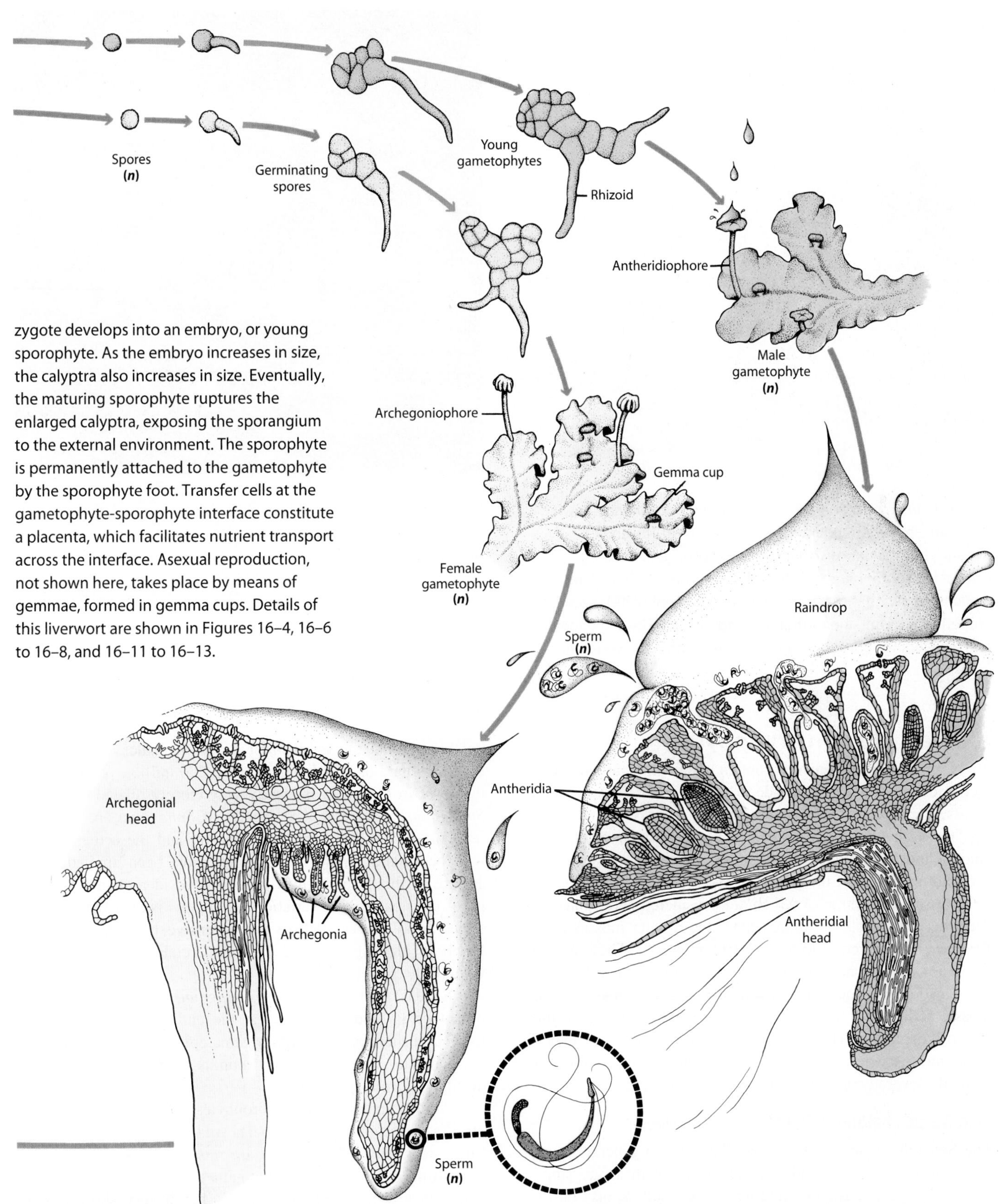

zygote develops into an embryo, or young sporophyte. As the embryo increases in size, the calyptra also increases in size. Eventually, the maturing sporophyte ruptures the enlarged calyptra, exposing the sporangium to the external environment. The sporophyte is permanently attached to the gametophyte by the sporophyte foot. Transfer cells at the gametophyte-sporophyte interface constitute a placenta, which facilitates nutrient transport across the interface. Asexual reproduction, not shown here, takes place by means of gemmae, formed in gemma cups. Details of this liverwort are shown in Figures 16–4, 16–6 to 16–8, and 16–11 to 16–13.

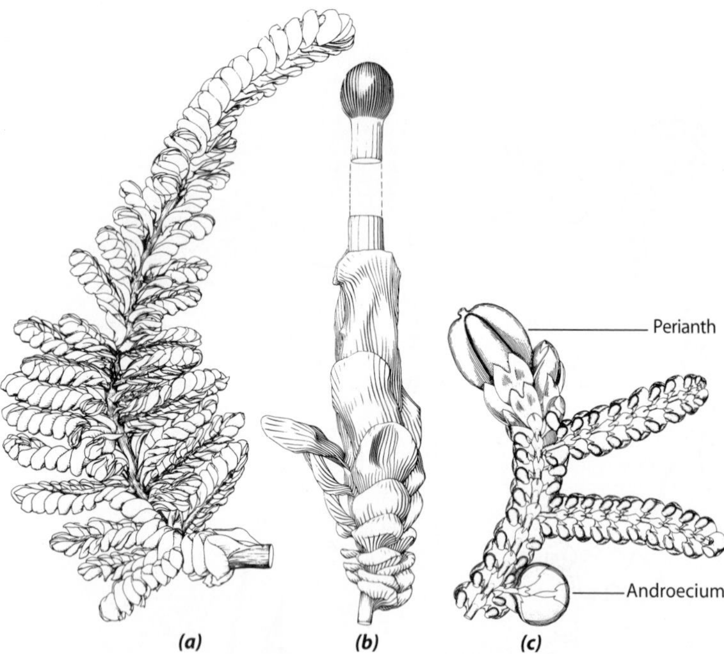

16–15 Leafy liverworts ***(a)*** *Clasmatocolea puccionana,* showing the characteristic arrangement of the leaves. ***(b)*** The end of a branch of *Clasmatocolea humilis.* The capsule and the long stalk of the sporophyte are visible. ***(c)*** A portion of a branch of *Frullania,* showing the characteristic arrangement of its leaves. The antheridia are contained within the androecium. The archegonium and developing sporophyte are contained within the perianth.

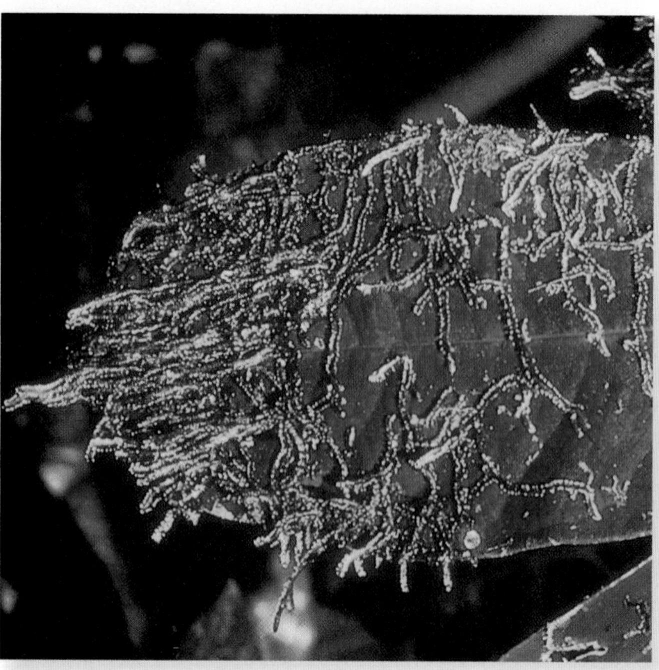

16–16 Leafy liverwort A leafy liverwort is seen here growing on the leaf of an evergreen tree in the rainforest of the Amazon Basin, near Manaus, Brazil.

Mosses: Phylum Bryophyta

Many groups of organisms contain members that are commonly called "mosses"—reindeer "mosses" are lichens, scale "mosses" are leafy liverworts, while club "mosses" and Spanish "moss" belong to different groups of vascular plants. Sea "moss" and Irish "moss" are algae. The genuine mosses, however, are members of the phylum Bryophyta, which consists of five classes, only three of which are considered here: Sphagnidae (the peat mosses), Andreaeidae (the granite mosses), and Bryidae (often referred to as the "true mosses"). These groups are distinct from one another, differing in many important features. Molecular and other information suggests that the peat mosses and granite mosses diverged earlier from the main line of moss evolution. The class Bryidae contains the vast majority of moss species, with some 10,000 species; new forms are being discovered constantly, especially in the tropics.

Peat Mosses Belong to the Class Sphagnidae

The Sphagnidae are currently classified in two genera, *Sphagnum,* the peat mosses, and *Ambuchanania,* with a rare single species found in Tasmania that grows as small "spots" on the surface of wet sand. Distinctive features of *Sphagnum*'s gametophytes and sporophytes (Figure 16–17), as well as comparative DNA sequences, indicate that it diverged early from the main line of moss evolution. The time of its first appearance is not known, but the fossil order Protosphagnales, consisting of several genera of Permian age (about 290 million years ago; see inside of front cover), is clearly very closely related to modern *Sphagnum.* More than 400 species of *Sphagnum* are generally recognized, but the plants are variable, and the actual number may be smaller. Some 4135 names have been applied to the species of this genus, which gives a good idea of the complexity of their patterns of variation. *Sphagnum* is distributed worldwide, in wet areas such as the extensive bog regions of the Northern Hemisphere, and is commercially and ecologically valuable.

Sexual reproduction in *Sphagnum* involves formation of antheridia and archegonia at the ends of special branches located at the tips of the moss gametophyte. Fertilization takes place in late winter, and four months later mature spores are discharged from sporangia.

Among mosses, the sporophytes of *Sphagnum* (Figure 16–17a) are quite distinctive. The red to blackish-brown capsules are nearly spherical and are raised on a stalk, the **pseudopodium,** which is part of the gametophyte and may be up to 3 millimeters long. The sporophyte has a very short seta, or stalk. Spore discharge in *Sphagnum* is spectacular (Figure 16–17c). At the top of the capsule is a lidlike **operculum,** separated from the rest of the capsule by a circular groove. As the capsule matures and dries, its epidermal cells collapse laterally,

(a)

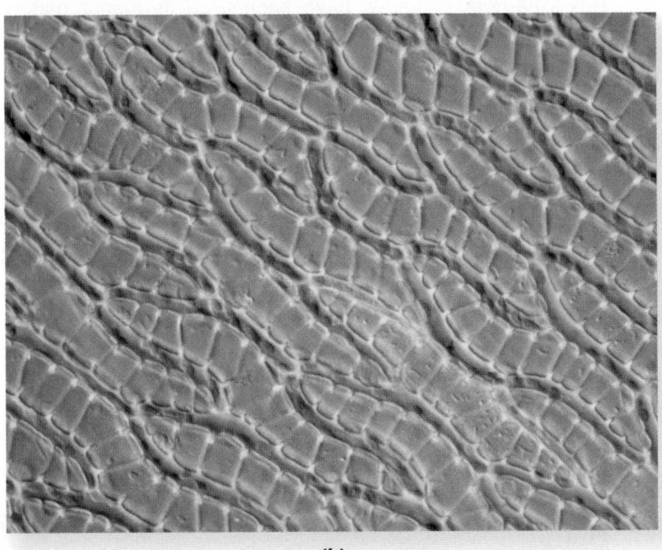

(b)

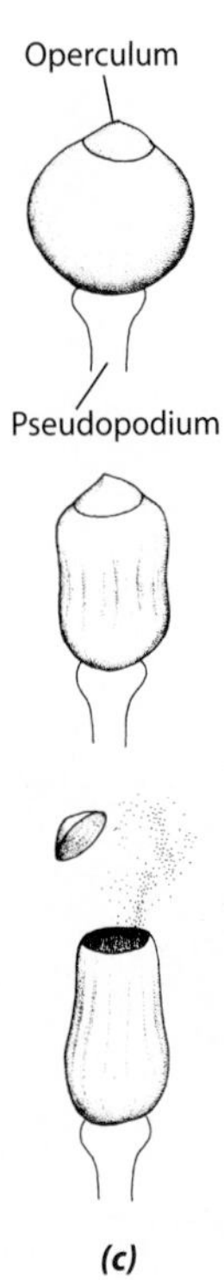

(c)

16–17 A peat moss, *Sphagnum* ***(a)*** A gametophyte, with many attached sporophytes. Some of the capsules, such as the two at the front, have already discharged their spores. ***(b)*** Structure of a leaf. Large, dead hyaline cells (pale blue), with ringlike or spiral wall thickenings, are surrounded by highly elongated, living cells (green), rich in chloroplasts. ***(c)*** Dehiscence of a capsule. As the capsule dries, it contracts, changing from a spherical to a cylindrical shape. This change in shape causes compression of trapped gas within the capsule. When the compressed gas reaches a pressure of about 5 bars, the pressure inside the capsule blows off the operculum, with the explosive release of a cloud of spores.

causing the capsule to change shape from spherical to cylindrical. In the process, the internal air pressure of the capsule builds to about 5 bars, similar to pressures in the tires of trailer trucks. The operculum is eventually blown off with an audible click, and the escaping gas carries the spore mass upward in an explosive burst. In one study, the average launch velocity of gas and spores was 16 ± 7 meters per second, reaching a mean height of 114 ± 9 millimeters.

Asexual reproduction by fragmentation is very common. Young branches and stem pieces that break off from the gametophyte and injured leaves can regenerate new gametophytes. As a result, *Sphagnum* forms large, densely packed clumps.

Three Features Distinguish the Sphagnidae from Other Mosses The most distinctive differences between the class Sphagnidae and other mosses are the unusual protonema, the peculiar morphology of the gametophyte, and the explosive operculum mechanism. The protonema—the first stage of development of the gametophyte—of the Sphagnidae does not consist of an extensive array of multicellular, branched filaments as in most other mosses. Instead, each protonema consists of a plate of cells, one cell-layer thick, that grows by a marginal meristem, in which most of the cells can divide in one of only two possible directions. In these respects, the protonema of *Sphagnum* is remarkably similar to the disk-shaped thalli of *Coleochaete* (Figure 15–49). The erect gametophyte arises from a budlike structure that grows from one of the marginal cells (Figure 16–18). This structure contains an apical meristem that divides in three directions, forming leaf and stem tissues.

The stems of *Sphagnum* gametophytes bear clusters of branches, often five at a node, which are more densely tufted near the tips of the stem, resulting in a moplike head. Both branches and stems bear leaves, but stem leaves often have little or no chlorophyll, whereas most branch leaves are green. The leaves are one cell-layer thick and are composed of two distinctly different types of cells: (1) large, dead cells called

16–18 *Sphagnum* gametophyte A young leafy gametophyte of *Sphagnum* is seen here arising from a platelike protonema.

hyaline cells, or hyalocysts, with ringlike and spiral wall thickenings, and (2) narrow green or occasionally red-pigmented living cells, each containing several discoid chloroplasts (Figure 16–17b). Every dead cell is surrounded by a single layer of photosynthetic cells, which are interconnected throughout the leaf, forming a network. The walls of the dead cells contain pores and readily become filled with water. As a result, the water-holding capacity of the peat mosses is up to 20 times their dry weight. By comparison, cotton absorbs only four to six times its dry weight. Both living and dead cell walls of *Sphagnum* are impregnated with decay-resistant phenolic compounds and have antiseptic properties. In addition, the peat mosses contribute to the acidity of their own environment by releasing hydrogen ions; in the center of peat bogs the pH is often less than 4—very acidic and unusual for a natural environment.

Because of their superior absorptive and antiseptic qualities, *Sphagnum* mosses have been used as diaper material by native peoples and, in Europe from the 1880s to World War I, as dressings for wounds and boils. *Sphagnum* is still very widely used in horticulture as a packing material for plant roots, as a planting medium, and as a soil additive. Gardeners mix peat moss with soil to increase the water-holding capacity of the soil and make it more acidic for acid-loving plants, such as conifers and rhododendrons. The harvesting and processing of *Sphagnum* from peat bogs for these purposes is a multimillion-dollar industry and is of ecological concern because it can result in severe degradation of some wetlands. Efforts are under way to develop techniques for regenerating peatlands because of their ecological importance.

(a)

(b)

16–19 Granite moss ***(a)*** *Andreaea* growing on an alpine rock, where the plants form dense, reddish-brown cushions. ***(b)*** Open sporangia (or capsules) of *Andreaea rupestris*. As the capsule dries out, it contracts and opens by four lateral slits, allowing the spores to fall out.

The Ecology of *Sphagnum* Is of Worldwide Importance *Sphagnum*-dominated peatlands occupy 1 to 3 percent of the Earth's surface, an enormous area equal to about one-half that of the United States. *Sphagnum* is thus one of the most abundant plants in the world. Peatlands are of particular importance in the global carbon cycle because peat stores very large amounts (about 400 gigatons, or 400 billion metric tons, on a global basis) of organic carbon that is not readily decayed to CO_2 by microorganisms. Peat is formed from the accumulation and compression of the mosses themselves, as well as the sedges, reeds, grasses, and other plants that grow among them. In Ireland and some other northern regions, dried peat is burned and used widely as industrial fuel, as well as for domestic heating. Ecologists are concerned that global warming brought about by increasing amounts of CO_2 and other gases in the atmosphere—due in large part to human activities—might result in oxidation of peatland carbon. This could further increase CO_2 levels and global temperatures (see "Global Warming: The Future is Now," on pages 140 and 141).

Granite Mosses Belong to the Class Andreaeidae

The class Andreaeidae comprises two genera, *Andreaea* and *Andreaeobryum.* The genus *Andreaea* consists of about 100 species of small, blackish-green or dark reddish-brown tufted mosses (Figure 16–19a), which in their own way are as peculiar as *Sphagnum. Andreaea* occurs in mountainous or Arctic regions, often on granite rocks—hence its common name "granite moss." *Andreaeobryum* (with a single species) is restricted to northwestern Canada and adjacent Alaska and grows primarily on calcareous (calcium-containing) rocks. In *Andreaea,* the protonema is unusual in having two or more rows of cells, rather than one row as in most mosses. The rhizoids are unusual in consisting of two rows of cells. The minute capsules are marked by four vertical lines of weaker cells along which the capsule splits, and the capsule remains intact above and below these dehiscence lines. The resulting four valves are very sensitive to the humidity of the surrounding air, opening when it is dry—the spores can be carried far by the wind under such circumstances—and closing when it is moist. This mechanism of spore discharge, by means of slits in the capsule, is different from that of any other moss (Figure 16–19b).

"True Mosses" Belong to the Class Bryidae

The class Bryidae contains most of the species of mosses. In this group of mosses—the "true mosses"—the branching filaments of the protonemata are composed of a single row of cells and resemble filamentous green algae (Figure 16–20). They can usually be distinguished from the green algae, however, by their slanted cross walls. Leafy gametophytes develop from minute budlike structures on the protonemata. In a few genera of mosses the protonema is persistent and assumes the major photosynthetic role, whereas the leafy shoots of the gametophyte are minute.

Many Mosses Have Tissues Specialized for Water and Food Conduction Moss gametophytes, which exhibit varying degrees of complexity, can be as small as 0.5 millimeter or up to 50 centimeters or longer (Figure 16–21). All have multicellular rhizoids, and the leaves are normally only one cell-layer thick except at the midrib (which is lacking in some genera). In many mosses, the stems of the gametophytes and sporophytes have a central

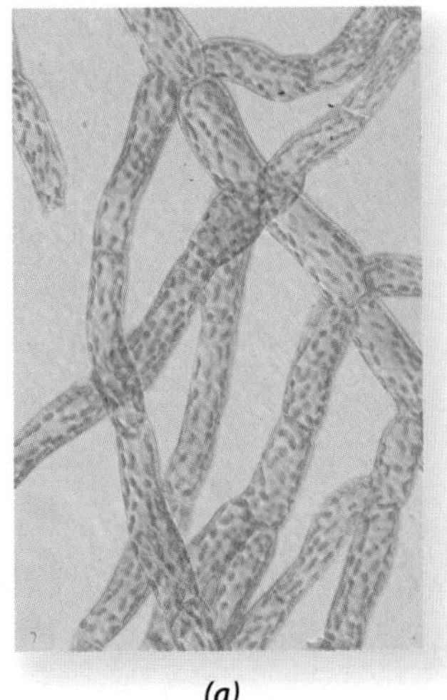

(a)

(b)

16–20 A "true moss" ***(a)*** Protonemata of the model plant *Physcomitrella patens.* The protonemata resemble filamentous green algae. ***(b)*** A leafy gametophyte of *P. patens* with rhizoids (at base), which anchor the gametophyte to the substrate. Protonemata are the first stage of the gametophyte generation of mosses and some liverworts.

16–21 ***Dawsonia superba*** The tallest of the true mosses, *Dawsonia superba* gametophytes can reach heights of 50 centimeters. Those shown here are 20 centimeters tall.

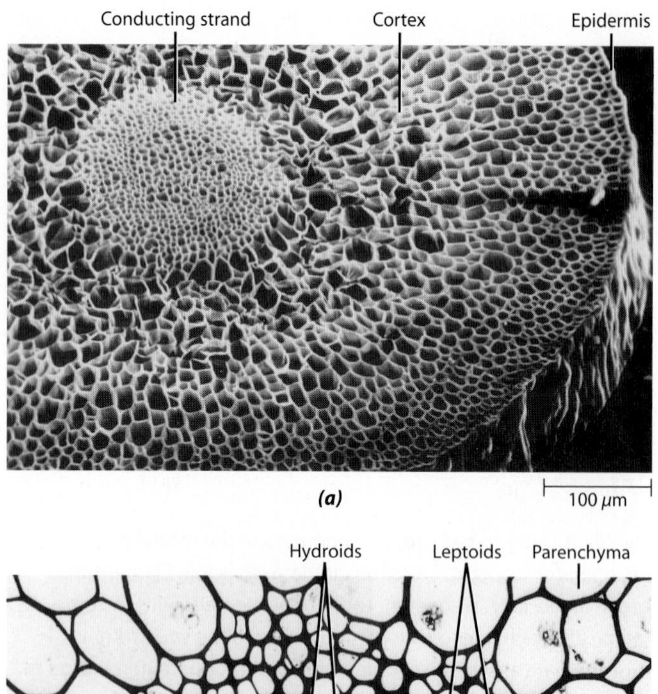

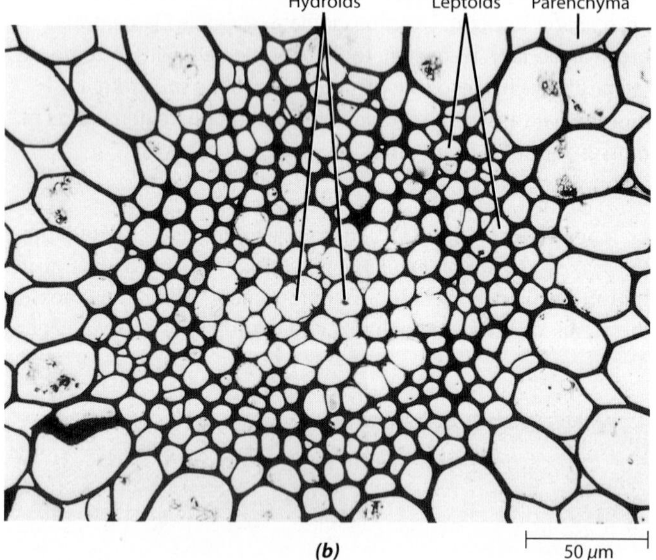

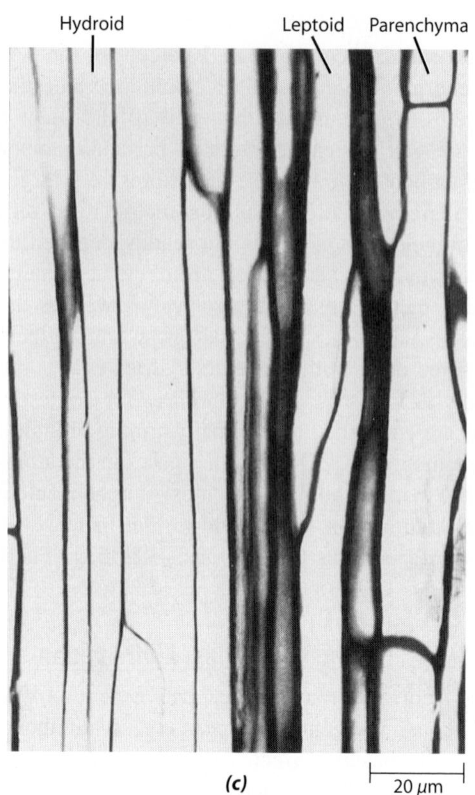

16–22 Hydroids and leptoids Conducting strands in the seta, or stalk, of a sporophyte of the moss *Dawsonia superba*. ***(a)*** General organization of the seta as seen in transverse section with the scanning electron microscope. ***(b)*** Transverse section showing the central column of water-conducting hydroids surrounded by a sheath of food-conducting leptoids and the parenchyma of the cortex. ***(c)*** Longitudinal section of a portion of the central strand, showing (from left to right) hydroids, leptoids, and parenchyma.

strand of water-conducting tissue called **hadrom.** The water-conducting cells are known as **hydroids** (Figure 16–22). Hydroids are elongated cells with inclined end walls that are thin and highly permeable to water, making them the preferred pathways for water and solutes. Hydroids resemble the water-conducting tracheary elements of vascular plants, because both lack a living protoplast at maturity (see Chapter 23). Unlike tracheary elements, however, hydroids lack specialized, lignin-containing wall thickenings. In some moss genera, food-conducting cells, also known as **leptoids,** surround the strand of hydroids (Figure 16–22). The food-conducting tissue is called **leptom.** The leptoids are elongated cells that have some structural and developmental similarities to the food-conducting sieve elements of seedless vascular plants (see page 394). At maturity, both cell types have inclined end walls with small pores and living protoplasts with degenerate nuclei. The conducting cells of mosses—the hydroids and leptoids—apparently are similar to those of certain fossil plants known as **protracheophytes,** which may represent an intermediate stage in the evolution of vascular plants, or **tracheophytes** (Figure 16–3; see also page 402).

Sexual Reproduction in Mosses Is Similar to That of Other Bryophytes The sexual cycle of mosses (see the moss life cycle in Figure 16–28, pages 386 and 387) is similar to that of liverworts and hornworts in that it involves production of male and female gametangia, an unbranched matrotrophic (maternally nourished) sporophyte, and specialized spore dispersal processes.

Gametangia may be produced by mature leafy gametophytes, either at the tip of the main axis or on a lateral branch. In some genera, the gametophytes are unisexual (Figure 16–23), but in other genera both archegonia and antheridia are produced by the same plant. Antheridia are often clustered within leafy structures called splash cups (Figure 16–24). The sperm from several to many antheridia are discharged into a drop of water within each cup and are then dispersed as raindrops fall into the cup. Insects may also carry drops of water, rich in sperm, from plant to plant.

Moss sporophytes, like those of hornworts and liverworts, are borne on the gametophytes, which supply the sporophytes with nutrients. A short foot at the base of the stalklike seta is embedded in the tissue of the gametophyte, and the cells of both foot and adjacent gametophyte function as transfer cells in the placenta. In the moss *Polytrichum,* simple sugars have been shown to move across the junction between generations. The capsules, or sporangia, usually take 6 to 18 months to reach maturity in temperate species and are generally elevated on a seta into the air, thus facilitating spore dispersal. Some mosses produce brightly colored sporangia that attract insects. Setae may reach 15 to 20 centimeters in length in a few species but may be very short or entirely absent in other species. The setae of many moss sporophytes contain a central strand of hydroids, which in some genera is surrounded by leptoids (Figure 16–22). Stomata are normally present on the epidermis of moss sporophytes. Some moss stomata, however, are bordered by only a single, doughnut-shaped guard cell (Figure 16–25).

Generally, the cells of the young and maturing sporophyte contain chloroplasts and carry out photosynthesis. When a moss sporophyte is mature, however, it gradually loses its ability to photosynthesize and turns yellow, then orange, and finally brown. The calyptra, derived from the archegonium, is commonly lifted upward with the capsule as the seta elongates. Prior to spore dispersal, the protective calyptra falls off, and the operculum of the capsule bursts off, revealing a ring of teeth—the **peristome**—surrounding the opening (Figure 16–26). The teeth of the peristome are formed by the splitting, along a zone of weakness, of a cellular layer near the end of the capsule. In most mosses, the teeth uncurl slowly when the air is relatively dry and curl up again when it is moist. The movements of the teeth expose the spores, which are gradually released. A capsule sheds up to 50 million haploid spores, each of which is capable of giving rise to a new gametophyte. The peristome is a characteristic of the class Bryidae and is lacking in the other two classes of mosses. Distinctive features of the peristomes of different moss groups are used in classifying and identifying mosses.

Asexual reproduction usually occurs by fragmentation, because virtually any portion of the moss gametophyte is capable of regeneration. However, some mosses produce specialized asexual reproductive structures.

Mosses Exhibit "Cushiony" or "Feathery" Growth Patterns Two patterns of growth are common among Bryidae (Figure 16–27). In the "cushiony" mosses, the gametophytes are erect and little branched, usually bearing terminal sporophytes. In the "feathery" form, the gametophytes are highly branched, the plants are creeping, and the sporophytes are borne laterally. This second type of growth pattern is commonly found in mosses that hang in masses from the branches of trees in rainforests and tropical cloud forests. Organisms such as these, which grow on other organisms but are not parasitic on them, are called **epiphytes.** Trees provide a diverse array of microhabitats that are occupied by moss and other bryophyte species. Among these microhabitats are tree bases and buttress roots, fissures and ridges of bark, irregular surfaces of twigs, depressions at the bases of branches, and the surfaces of leaves.

A number of moss genera and species are highly endemic, that is, restricted to very limited geographic areas. Many of the endemic mosses grow as epiphytes in high-altitude temperate

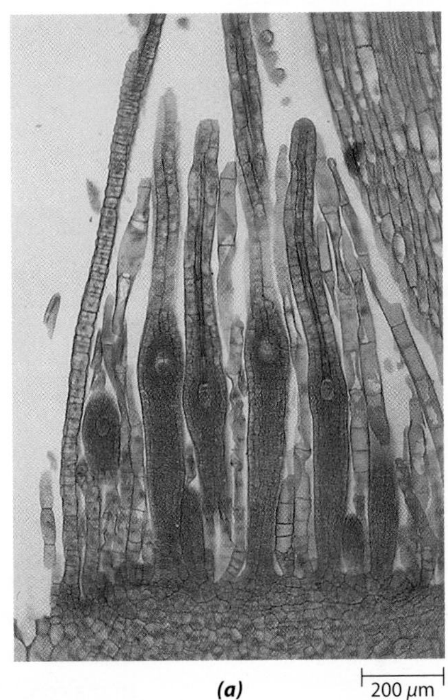

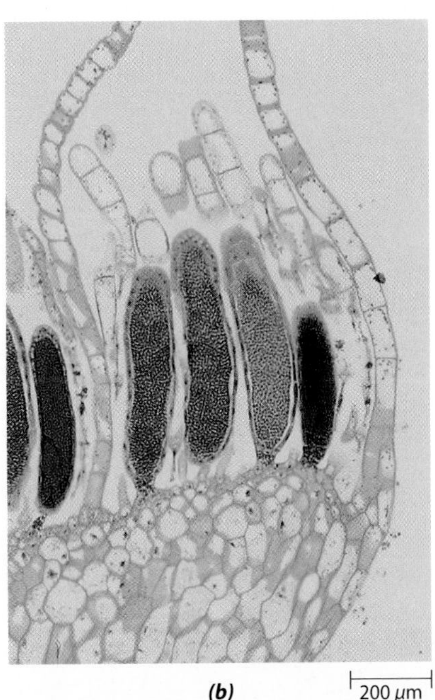

16–23 Gametangia of *Mnium,* a unisexual moss ***(a)*** Longitudinal section through an archegonial head, showing the pink-stained archegonia surrounded by sterile structures called paraphyses. ***(b)*** Longitudinal section through an antheridial head, showing antheridia surrounded by paraphyses.

16–24 Splash cups Leafy male gametophytes of the moss *Polytrichum piliferum,* showing the mature antheridia clustered together in cup-shaped heads, known as splash cups. The sperm are discharged into drops of water held within these leafy heads and are then splashed out of them by raindrops, sometimes reaching the vicinity of archegonia on other gametophytes (Figure 16–28).

and tropical cloud forests, where the bryophyte biodiversity is poorly known. Bryophytes also have important but poorly catalogued interactions with a variety of invertebrates, some of which live, breed, and feed preferentially in mosses. Some experts are concerned that the growth of human populations may drastically alter natural environments, leading to extensive loss of bryophyte species and associated animals before many of the organisms are even described.

The leafy shoots and, in some cases, the sporophytes of mosses are associated with a broad array of fungi, including glomeromycetes, ascomycetes, and Agaricomycotina. Whether such associations are mutualistic remains to be determined. Mosses commonly are associated with cyanobacteria, which grow as epiphytes on leaves and stems. In *Sphagnum,* cyanobacteria even occur inside the hyaline cells, the large dead cells that can hold water.

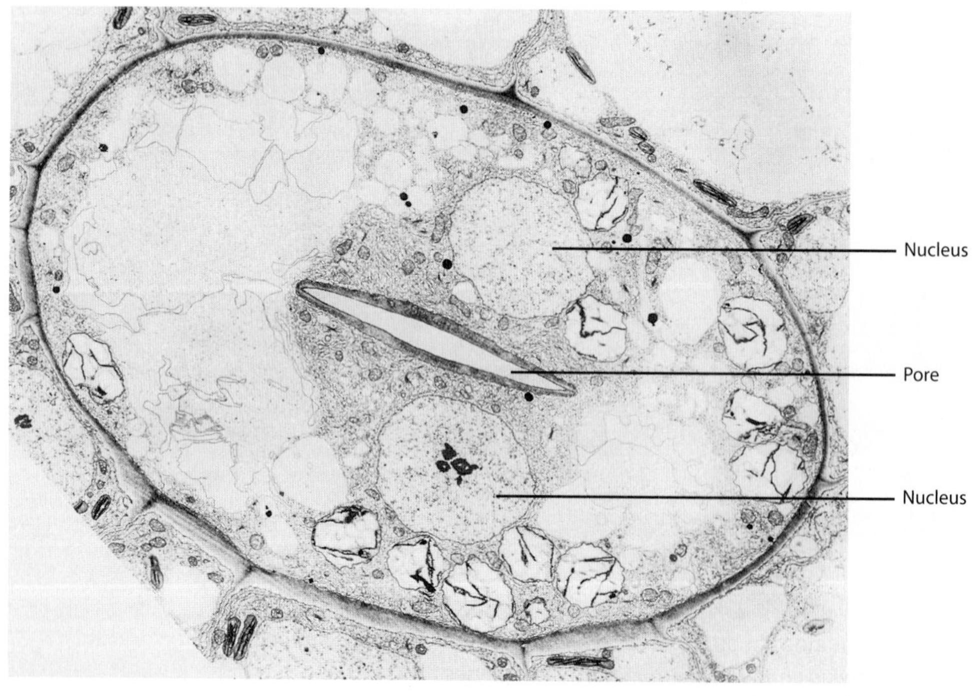

16–25 Moss stoma with single guard cell Mature one-celled stoma of the moss *Funaria hygrometrica.* The stoma consists of a single, binucleate guard cell, as the wall bordering the pore in the middle of the cell does not extend to the ends of the cell.

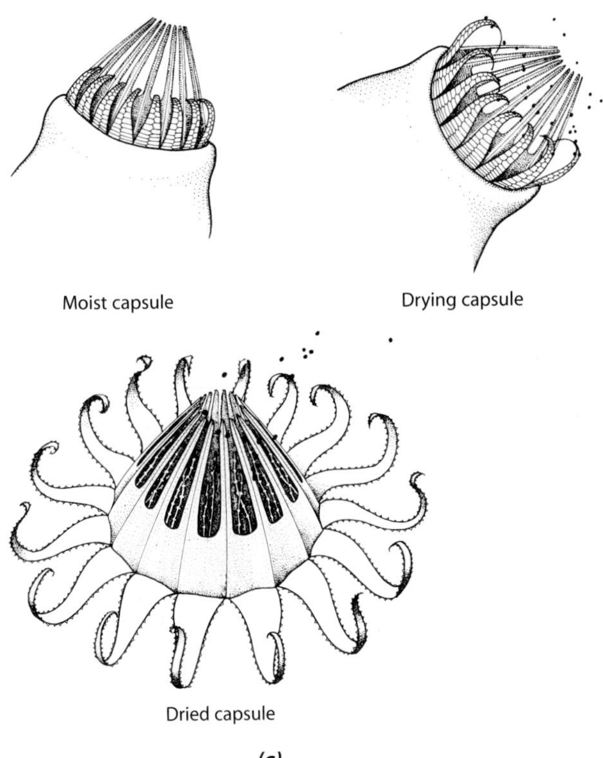

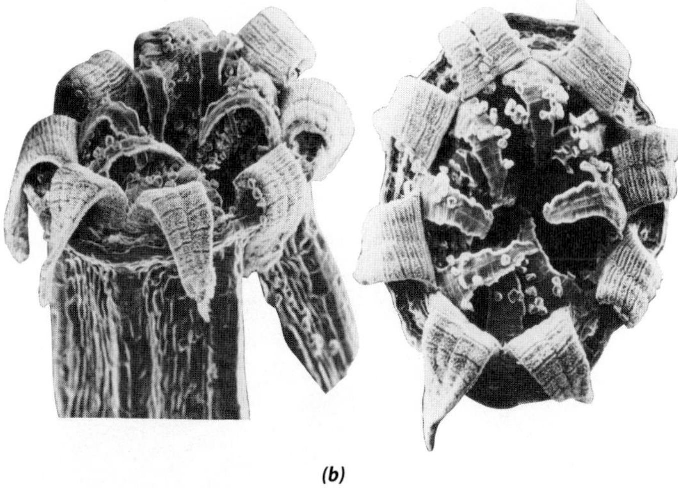

16–26 Peristome teeth in mosses of the class Bryidae
(a) *Brachythecium* has a peristome consisting of two rings of teeth, which open to release the spores in response to changes in moisture. The outer set of peristome teeth interlocks with the inner set under damp conditions. As the capsule dries out, the outer teeth pull away, allowing the dispersal of spores by the wind. ***(b)*** Scanning electron micrographs of the peristome teeth of two capsules of *Orthotrichum*, showing the inner teeth curved inward and the outer teeth curved outward in dry conditions.

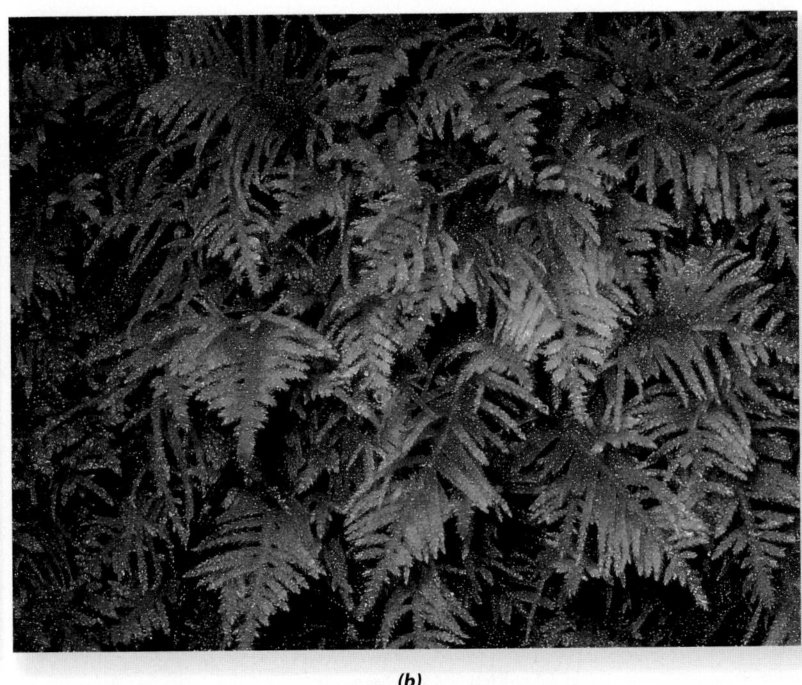

16–27 "Cushiony" and "feathery" growth patterns Seen here are the two common growth forms found in the gametophytes of different genera of mosses in the class Bryidae. ***(a)*** The cushiony form, in which the gametophytes are erect and have few branches, shown here in *Polytrichum juniperinum*. Sporophytes, each consisting of a spore capsule atop a long, slender seta, can be seen rising above the gametophytes. ***(b)*** The feathery form, with matted, creeping gametophytes, is shown here in *Thuidium abietinum*.

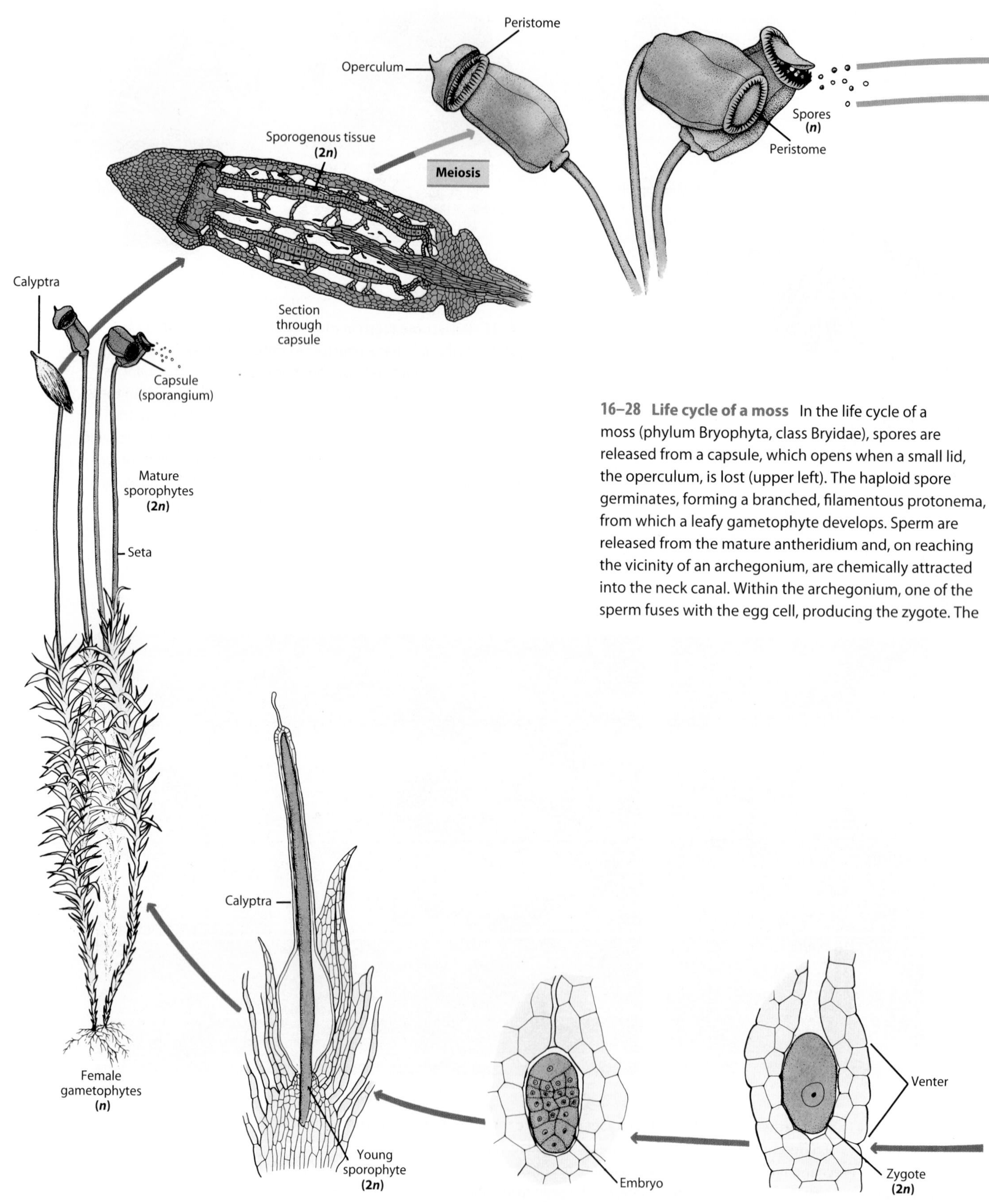

16–28 Life cycle of a moss In the life cycle of a moss (phylum Bryophyta, class Bryidae), spores are released from a capsule, which opens when a small lid, the operculum, is lost (upper left). The haploid spore germinates, forming a branched, filamentous protonema, from which a leafy gametophyte develops. Sperm are released from the mature antheridium and, on reaching the vicinity of an archegonium, are chemically attracted into the neck canal. Within the archegonium, one of the sperm fuses with the egg cell, producing the zygote. The

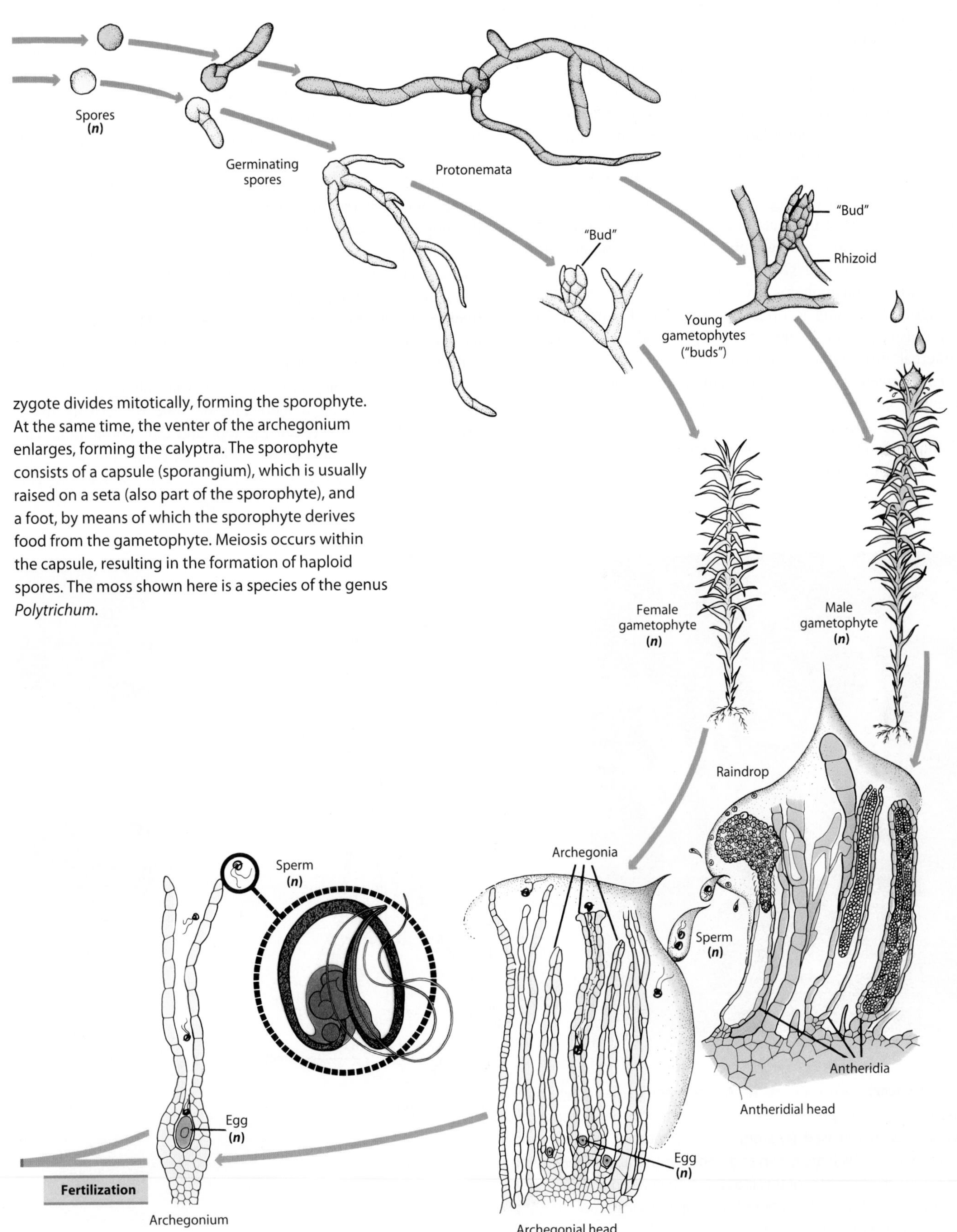

zygote divides mitotically, forming the sporophyte. At the same time, the venter of the archegonium enlarges, forming the calyptra. The sporophyte consists of a capsule (sporangium), which is usually raised on a seta (also part of the sporophyte), and a foot, by means of which the sporophyte derives food from the gametophyte. Meiosis occurs within the capsule, resulting in the formation of haploid spores. The moss shown here is a species of the genus *Polytrichum*.

Hornworts: Phylum Anthocerotophyta

The hornworts, with more than 300 species, constitute the least diverse lineage of bryophytes. The name "hornwort" is derived from their hornlike sporophytes. The members of the genus *Anthoceros* are the most familiar of the 11 genera. The gametophytes of hornworts (Figure 16–29a) superficially resemble those of the thalloid liverworts, but there are many features that indicate a relatively distant relationship. For example, the cells of most species usually have a single large chloroplast with a pyrenoid, as in the green alga *Coleochaete*. Some hornwort species have cells containing many small chloroplasts lacking pyrenoids, as do most plant cells, but even in these hornworts, the apical cell contains a single plastid, reflecting the ancestral condition (pages 369 and 370).

Hornwort gametophytes are often rosettelike, and their dichotomous branching is often not apparent (Figure 16–29). They are usually 1 to 2 centimeters across and lack conspicuous internal differentiation, except for the presence of cavities—large intercellular spaces—containing colonies of the filamentous cyanobacterium *Nostoc* embedded in mucilage. *Nostoc* fixes nitrogen and supplies it to its host. Many cells of the thallus, including those of the epidermis, secrete mucilage, which is essential for water retention. The lower epidermis of the thallus contains numerous small pores bordered by two kidney-shaped cells that resemble the stomatal guard cells of the sporophyte. The pores of the thallus are filled with mucilage; they serve not for gas exchange but rather as entry sites for the filamentous *Nostoc*. Some hornworts have mycorrhiza-like associations with glomeromycetes.

Most hornwort gametophytes are unisexual. Some *Anthoceros* gametophytes are unisexual, others are bisexual. In the bisexual gametophytes, development of the antheridia usually precedes that of the archegonia. The antheridia and archegonia are sunken on the dorsal surface of the gametophyte, with the antheridia clustered in chambers. Numerous sporophytes may develop on the same gametophyte.

The sporophyte of *Anthoceros,* which is an upright, elongated structure, consists of a foot and a long, cylindrical capsule, or sporangium (Figures 16–29 and 16–30). Unlike the liverworts and true mosses, it lacks a seta. The foot penetrates the gametophyte tissue and forms a placenta across which the sporophyte obtains nourishment from the gametophyte (Figure 16–30a). A unique aspect of hornwort sporophytes is that early in their development, a meristem, or zone of actively dividing cells, develops between the foot and the sporangium. This basal meristem remains active as long as conditions are favorable for growth, so that the sporangium continues to elongate

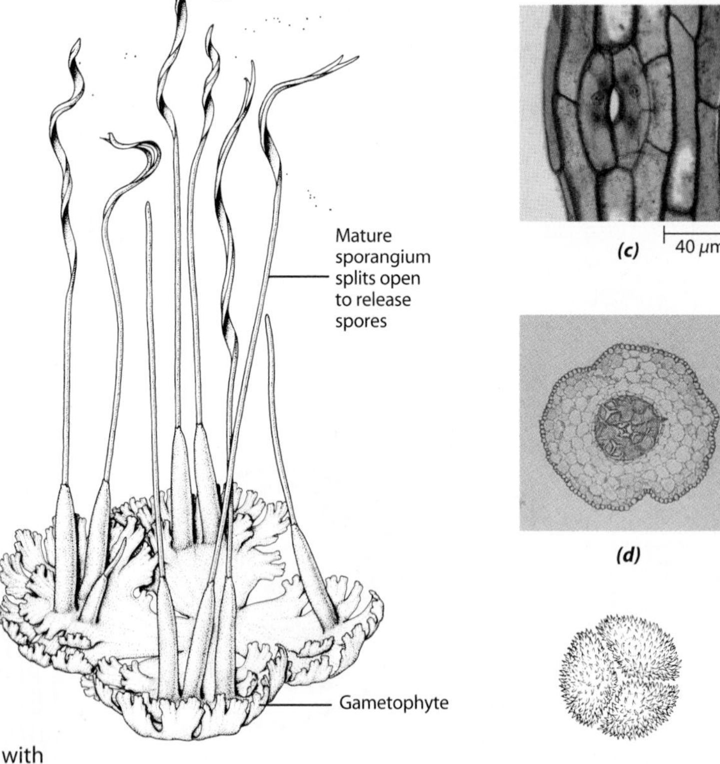

16–29 ***Anthoceros,* a hornwort** ***(a)*** A dark green gametophyte with attached (elongated) sporophytes. ***(b)*** When mature, the sporangium splits, and the spores are released. ***(c)*** Stomata are abundant on the sporophytes of the hornworts, which are green and photosynthetic. ***(d)*** Developing spores, visible in the center of this cross section of a sporangium, and ***(e)*** mature spores still held in a tetrad, a group of four spores—three of which are visible here—formed from a spore mother cell by meiosis.

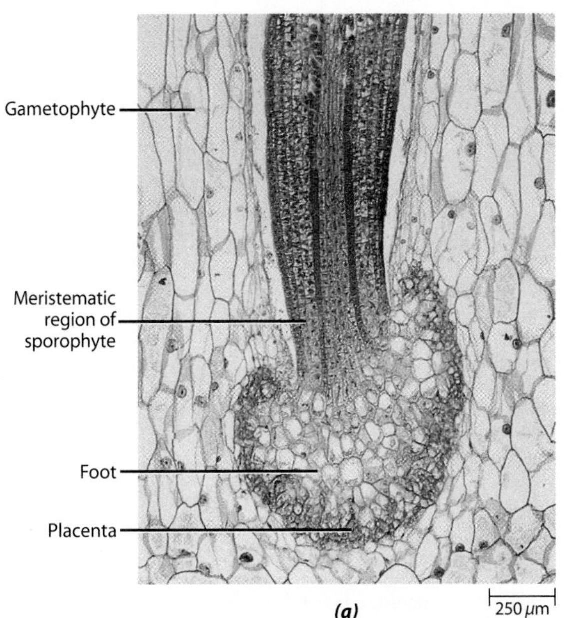

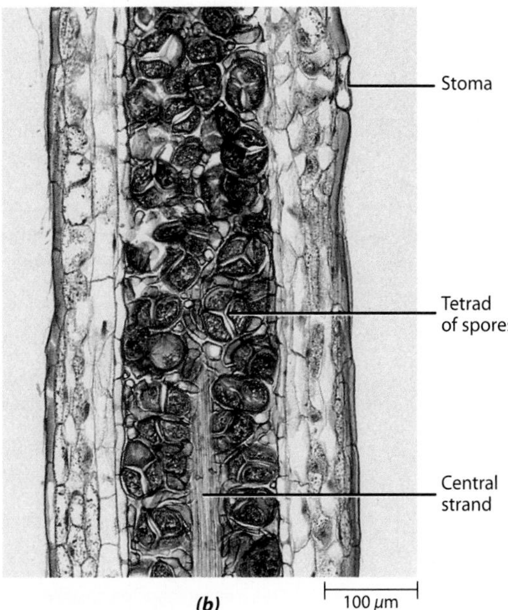

16–30 ***Anthoceros*** ***(a)*** Longitudinal section of the lower portion of a sporophyte, showing its foot embedded in the tissue of the gametophyte. ***(b)*** Longitudinal section of a portion of a sporangium, showing tetrads of spores with elater-like structures among them. The central strand of tissue in the lower part of the sporangium consists of tissue that may function to conduct water and nutrients.

for a prolonged period of time. Consequently, all stages of spore development, from meiosis near its base to mature spores above, can be observed in a single sporangium. Having several layers of photosynthetic cells, the sporophyte is green. It is covered with a cuticle and has stomata (Figures 16–29c and 16–30b) that remain open permanently. Maturation of the spores and, ultimately, dehiscence of the sporangium begins near its tip and extends toward the base as the spores mature (Figure 16–29b and 16–30b). Among the spores are sterile, elongated, often multicellular structures called pseudoelaters that resemble the elaters of liverworts. The dehiscing sporangium splits longitudinally into ribbonlike leaves. Spore dispersal is aided by twisting of the sporangium wall and the pseudoelaters, which disrupt the spore mass on drying.

SUMMARY

Plants Most Likely Evolved from a Charophycean Green Alga

Plants, collectively known as embryophytes, seem to have been derived from a charophycean green alga. The two groups share many unique features, including a phragmoplast and cell plate at cytokinesis. Molecular and other evidence strongly suggests that plants are descended from a single common ancestor and that bryophytes include the earliest living plants to have diverged from the main line of plant evolution. The earliest plants were probably similar in several respects to modern liverworts. The characteristics shared by all plants are tissues produced by an apical meristem, a life history involving alternation of heteromorphic generations, protectively walled gametangia, matrotrophic embryos, and sporopollenin-walled spores.

The Bryophytes Are the Liverworts, Mosses, and Hornworts

The bryophytes consist of three phyla of structurally rather simple, small plants. Their gametophytes are always nutritionally independent of the sporophytes, whereas the sporophytes are permanently attached to the gametophytes and nutritionally dependent on them for at least some time early in embryo development. The male sex organs, antheridia, and female sex organs, archegonia, both have protective jacket layers. Each archegonium contains a single egg, whereas each antheridium produces numerous sperm. The biflagellated sperm are free-swimming and require water to reach the egg. Except in the hornworts, the sporophyte is typically differentiated into a foot, a seta, and a capsule, or sporangium. Hornwort sporophytes lack a seta. Maturing moss and hornwort sporophytes are green and become less nutritionally dependent on their gametophytes than those of liverworts, which usually remain completely dependent on their gametophytes.

The Sporophytes of the Bryophytes Differ from One Another

Liverworts (phylum Marchantiophyta) lack stomata, whereas stomata are present in mosses and hornworts. The liverwort phylum consists of the complex thalloid liverworts, which have internal tissue differentiation, the leafy liverworts, and the simple thalloid liverworts, which consist of ribbons of relatively undifferentiated tissue. Mosses (phylum Bryophyta), at least those in some

SUMMARY TABLE Comparative Summary of Characteristics of Bryophyte Phyla

PHYLUM	NUMBER OF SPECIES	GENERAL CHARACTERISTICS OF GAMETOPHYTE	GENERAL CHARACTERISTICS OF SPOROPHYTE	HABITATS
Marchantiophyta (liverworts)	5200	Free-living generation; both thalloid and leafy genera; pores in some thalloid types; unicellular rhizoids; most cells have numerous chloroplasts; many produce gemmae; protonema stage in some; growth from apical meristem	Small and nutritionally dependent on gametophyte; unbranched; consists of little more than sporangium in some genera, and of foot, short seta, and sporangium in others; phenolic materials in epidermal cell walls; lacks stomata	Mostly moist temperate and tropical; a few aquatic; often as epiphytes
Bryophyta (mosses)	12,800	Free-living generation; leafy; multicellular rhizoids; most cells have numerous chloroplasts; many produce gemmae; protonema stage that grows by marginal meristem followed by further growth from an apical meristem in *Sphagnum;* growth by apical meristem only in Bryidae; some species have leptoids and nonlignified hydroids	Small and nutritionally dependent on gametophyte; unbranched; consists of foot, long seta, and sporangium in Bryidae; phenolic materials in epidermal cell walls; stomata; some species have leptoids and nonlignified hydroids	Mostly moist temperate and tropical; some Arctic and Antarctic; many in dry habitats; a few aquatic
Anthocerotophyta (hornworts)	300	Free-living generation; thalloid; unicellular rhizoids; most have single chloroplast per cell	Small and nutritionally dependent on gametophyte; unbranched; consists of foot and long, cylindrical sporangium, with a meristem between foot and sporangium; cuticle; stomata; no specialized conducting tissues	Moist temperate and tropical

groups, have both specialized conducting tissue and stomata that resemble those of vascular plants. The conducting tissue of mosses, when present, consists of hydroids (water-conducting cells) and leptoids (food-conducting cells). Phylum Bryophyta consists of three classes: the peat mosses, the granite mosses, and the "true mosses." Hornworts (phylum Anthocerotophyta) have a unique basal meristem and lack specialized conducting tissue.

Bryophytes Are Important Ecologically

Bryophytes are particularly abundant and diverse in temperate rainforests and tropical cloud forests. The moss *Sphagnum* occupies 1 to 3 percent of the Earth's surface, is economically valuable, and plays an essential role in the global carbon cycle.

QUESTIONS

1. By means of a simple, labeled diagram, outline a generalized life cycle of a bryophyte. Explain why it is referred to as an alternation of heteromorphic generations.
2. What evidence is there in support of a charophycean green alga ancestry for plants?
3. Bryophytes and vascular plants share a number of characters that distinguish them from charophycean green algae and that adapt them for existence on land. What are those characters?
4. In your opinion, which of the bryophytes has the most highly developed sporophyte? Which has the most highly developed gametophyte? In each case, give the reasons for your answer.
5. What characters shared by vascular plants are lacking in bryophytes?
6. Describe the structural modifications related to water absorption in *Sphagnum.* Why is *Sphagnum* of such great ecological importance?

CHAPTER 17

Seedless Vascular Plants

Arsenic removal The Chinese brake fern *(Pteris vittata)* transports arsenic from contaminated soils through its roots to its fronds, where the toxic metalloid accumulates at high levels. The fronds can then be harvested, removing arsenic from the environment. A gene has been identified that codes for the membrane protein involved with the pumping of arsenic into the plant cell's vacuole, protecting the cell from damage.

CHAPTER OUTLINE

Plants, like all living organisms, had aquatic ancestors. The story of plant evolution is therefore inseparably linked with their progressive occupation of the land and their increasing independence from water for reproduction. In this chapter, we first discuss the general features of vascular plant evolution—features linked with life on land—and the organization of the vascular plant body. We then describe the seedless vascular plants and tell the story of the club mosses, the ferns, and the horsetails.

Evolution of Vascular Plants

In the previous chapter, we noted that the bryophytes and vascular plants share a number of important characters, and that together these two groups of plants—both of which have multicellular embryos—form a monophyletic lineage, the embryophytes. As you will recall, it has been hypothesized that this lineage has ancestors in common with charophycean-algal-like organisms (pages 353 and 354). Both bryophytes and vascular plants have a basically similar life cycle—an alternation of heteromorphic generations—in which the gametophyte differs from the sporophyte. Two important characteristics of bryophytes, however, are the presence of free-living gametophytes, which are usually the more prominent generation, and

CHECKPOINTS

After reading this chapter, you should be able to answer the following:

1. What "pivotal steps" in the early history of plant evolution contributed to the success of vascular plants in their occupation of the land?
2. Explain the evolutionary origin of microphylls and megaphylls. Which groups of seedless vascular plants have microphylls? Which have megaphylls?
3. What is meant by homospory and heterospory? What are the contrasting features of the gametophytes produced by homosporous and heterosporous plants?
4. Describe the characteristics of each of the following phyla of seedless vascular plants: Rhyniophyta, Zosterophyllophyta, Trimerophytophyta, Lycopodiophyta, and Monilophyta. Which of these are exclusively fossil phyla?
5. In terms of their structure and method of development, how do eusporangia differ from leptosporangia?
6. Which ferns are eusporangiate? Which are leptosporangiate?

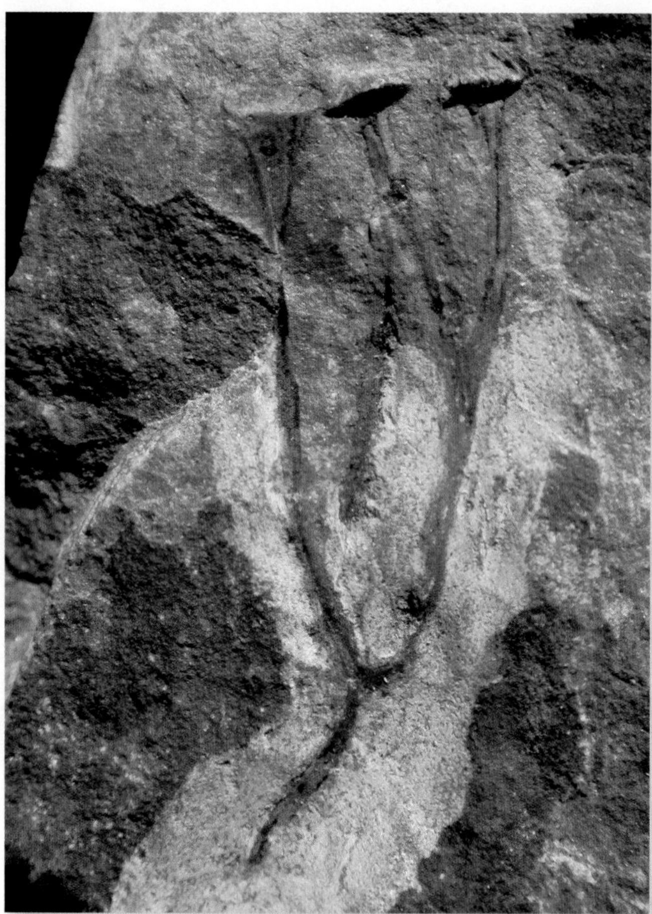

17–1 ***Cooksonia*** The oldest known vascular plant, *Cooksonia*, consisted of dichotomously branching axes. This fossil, found in Shropshire, England, is from the Late Silurian (414 to 408 million years ago). Its leafless aerial stems are 2.5 centimeters high, and they terminate in sporangia, or spore-producing structures. These small plants, which grew no higher than 6.5 centimeters, probably lived in moist environments such as mud flats.

a sporophyte that is permanently attached to, and nutritionally dependent on, its parental gametophyte. By contrast, vascular plants have sporophytes that are more prominent than the gametophytes and are ultimately free-living (Figure 17–1). Thus, the occupation of the land by the bryophytes was undertaken with emphasis on the gamete-producing generation and the requirement for water to enable its motile sperm to swim to the eggs. This requirement for water undoubtedly accounts in part for the small size and ground-hugging form of most bryophyte gametophytes.

Relatively early in the history of plants, the evolution of efficient fluid-conducting systems, consisting of xylem and phloem, solved the problem of water and food transport throughout the plant body—a serious concern for any large organism growing on land. The ability to synthesize lignin, which is incorporated into the cell walls of supporting and water-conducting cells, was also a pivotal step in the evolution of plants. It has been suggested that the earliest plants could stand erect only by means of turgor pressure, which limited not only the environments in which they could live but also the stature of such plants. Lignin adds rigidity to the walls, making it possible for the vascularized sporophyte to reach great heights. Vascular plants are also characterized by the capacity to branch profusely through the activity of apical meristems located at the tips of stems and branches. In the bryophytes, on the other hand, increase in the length of the sporophyte is subapical; that is, it occurs below the tip of the stem. In addition, each bryophyte sporophyte is unbranched and produces a single sporangium. By contrast, the branched sporophytes of vascular plants produce multiple sporangia; they are polysporangiate. Picture a pine tree—a single individual—with its numerous branches and many cones, each of which contains multiple sporangia, and below it a carpet of moss gametophytes—many individuals—each bearing a single unbranched sporophyte tipped by a single sporangium.

The belowground and aboveground parts of the sporophytes of early vascular plants differed little from one another structurally, but ultimately, the ancient vascular plants gave rise to more specialized plants with a more highly differentiated plant body. These plants consisted of roots, which function in anchorage and absorption of water and minerals, and stems and leaves, which provide a system well suited to the demands of life on land—namely, the acquisition of energy from sunlight, of carbon dioxide from the atmosphere, and of water. Meanwhile, the gametophytic generation underwent a progressive reduction in size and gradually became more protected and nutritionally dependent on the sporophyte. Finally, seeds evolved in one evolutionary line. **Seeds** are structures that provide the embryonic sporophyte with nutrients and also help protect it from the rigors of life on land—thus providing a means of withstanding unfavorable environmental conditions. Obviously, the seedless vascular plants lack seeds. Moreover, the gametophytes of most seedless vascular plants, like those of the bryophytes, are free-living, and water is required in the environment for their motile sperm to swim to the eggs.

Because of their adaptations for existence on land, the vascular plants have been ecologically successful and are the dominant plants in terrestrial habitats. They were already numerous and diverse by the Devonian period (408 to 362 million years ago; see inside of front cover) (Figure 17–2). There are seven phyla with living representatives. In addition, there are several phyla that consist entirely of extinct vascular plants. In this chapter we describe some of the characteristic features of vascular plants and discuss five phyla of seedless vascular plants, three of which are extinct. In Chapters 18 through 20 we will discuss the seed plants, which include five phyla with living representatives.

Organization of the Vascular Plant Body

The sporophytes of early vascular plants were dichotomously branched (evenly forked) axes that lacked roots and leaves. With evolutionary specialization, morphological and physiological differences arose between various parts of the plant body,

17–2 Early Devonian landscape By the Early Devonian, some 408 to 387 million years ago, small leafless plants with simple vascular systems were growing upright on land. It is thought that their pioneering ancestors were bryophyte-like plants, seen here near the water at the center, that invaded land sometime in the Ordovician (510 to 439 million years ago). The vascular colonizers shown are, from left to center, very tiny *Cooksonia* with rounded sporangia, *Zosterophyllum* with clustered sporangia, and *Aglaophyton* with solitary, elongated sporangia. During the Middle Devonian (387 to 374 million years ago), larger plants with more complex features became established. Seen here on the right are, from back to front, *Psilophyton,* a robust trimerophyte with plentiful sterile and fertile branchlets, and two lycophytes with simple microphyllous leaves, *Drepanophycus* and *Protolepidodendron.*

bringing about the differentiation of roots, stems, and leaves—the organs of the plant (Figure 17–3). Collectively, the roots make up the **root system,** which anchors the plant and absorbs water and minerals from the soil. The stems and leaves together make up the **shoot system,** with the stems raising the specialized photosynthetic organs—the leaves—toward the sun. The vascular system conducts water and minerals to the leaves and the products of photosynthesis away from the leaves to other parts of the plant.

The different kinds of cells of the plant body are organized into tissues, and the tissues are organized into still larger units called tissue systems. Three tissue systems—dermal, vascular, and ground—which occur in all organs of the plant, are continuous from organ to organ and reveal the basic unity of the plant body. The **dermal tissue system** makes up the outer, protective covering of the plant. The **vascular tissue system** comprises the conductive tissues—**xylem** and **phloem**—and is embedded in the **ground tissue system** (Figure 17–3). The principal differences in the structures of root, stem, and leaf lie in the relative distribution of the vascular and ground tissue systems, as will be discussed in Section 5.

Primary Growth Involves the Extension of Roots and Stems, and Secondary Growth Increases Their Thickness

Primary growth may be defined as the growth that occurs relatively close to the tips of roots and stems. It is initiated by the apical meristems and is primarily involved with extension of the plant body—often the vertical growth of a plant. The tissues arising during primary growth are known as **primary tissues,** and the part of the plant body composed of these tissues is called the **primary plant body.** Ancient vascular plants, and many contemporary ones as well, consist entirely of primary tissues.

In addition to primary growth, many plants undergo additional growth that thickens the stem and root; such growth is

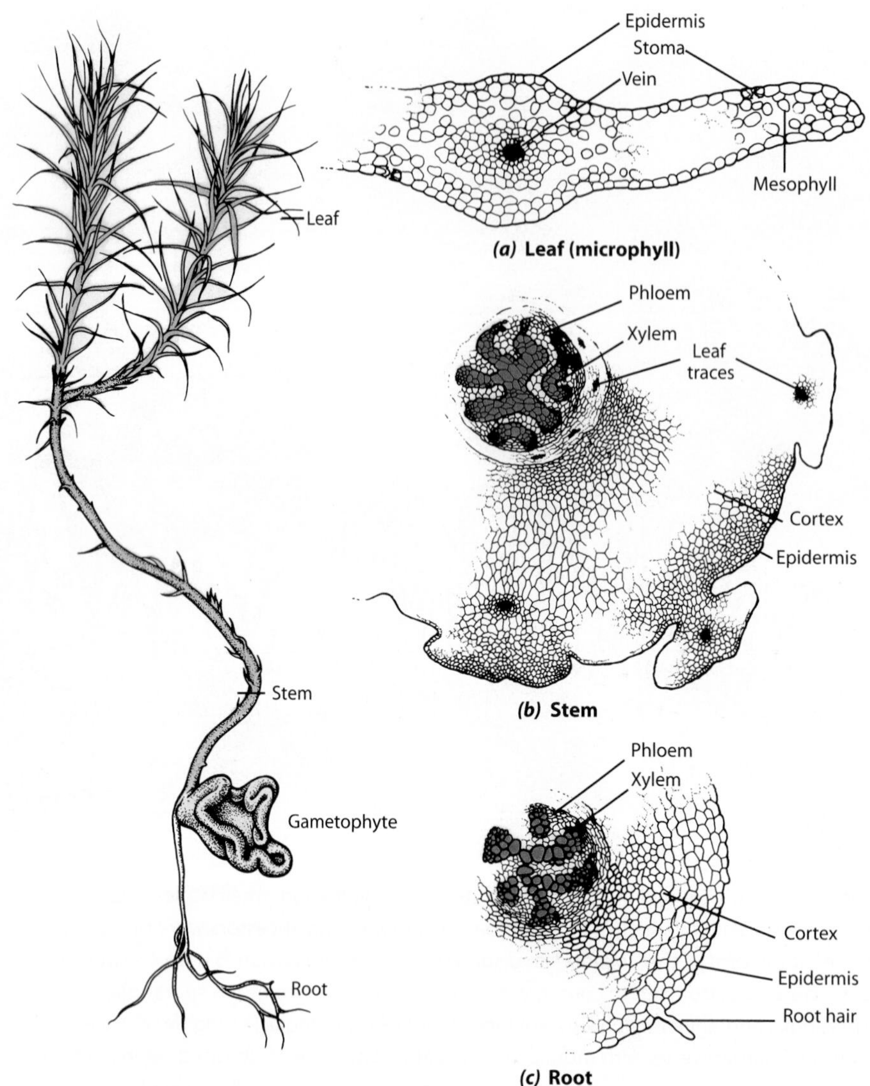

17–3 Club moss sporophyte A diagram of a young sporophyte of the club moss *Lycopodium lagopus,* which is still attached to its subterranean gametophyte. The dermal, vascular, and ground tissues are shown in transverse sections of ***(a)*** leaf, ***(b)*** stem, and ***(c)*** root. In all three organs, the dermal tissue system is represented by the epidermis, and the vascular tissue system, consisting of xylem and phloem, is embedded in the ground tissue system. The ground tissue in the leaf (a microphyll) is represented by the mesophyll, and in the stem and root by the cortex, which surrounds a solid strand of vascular tissue, or protostele. The leaf is specialized for photosynthesis, the stem for support of the leaves and for conduction, and the root for absorption and anchorage.

termed **secondary growth.** It results from the activity of lateral meristems, one of which, the **vascular cambium,** produces **secondary vascular tissues** known as secondary xylem and secondary phloem (see Figure 26–6). The production of secondary vascular tissues is commonly supplemented by the activity of a second lateral meristem, the **cork cambium,** which forms a **periderm,** composed mostly of cork tissue. The periderm replaces the epidermis as the dermal tissue system of the plant. The secondary vascular tissues and periderm make up the **secondary plant body.** Secondary growth appeared in the Middle Devonian period, about 380 million years ago, in several unrelated groups of vascular plants.

Tracheary Elements—Tracheids and Vessel Elements—Are the Conducting Cells of the Xylem

Tracheary elements, the conducting cells of the xylem, have distinctive, lignified wall thickenings (Figure 17–4) and are frequently well preserved in the fossil record. In contrast, sieve elements, the conducting cells of the phloem, have soft walls and often collapse after they die, so they are rarely well preserved as fossils. Because of their various wall patterns, the tracheary elements provide valuable clues to the interrelationships of the different groups of vascular plants.

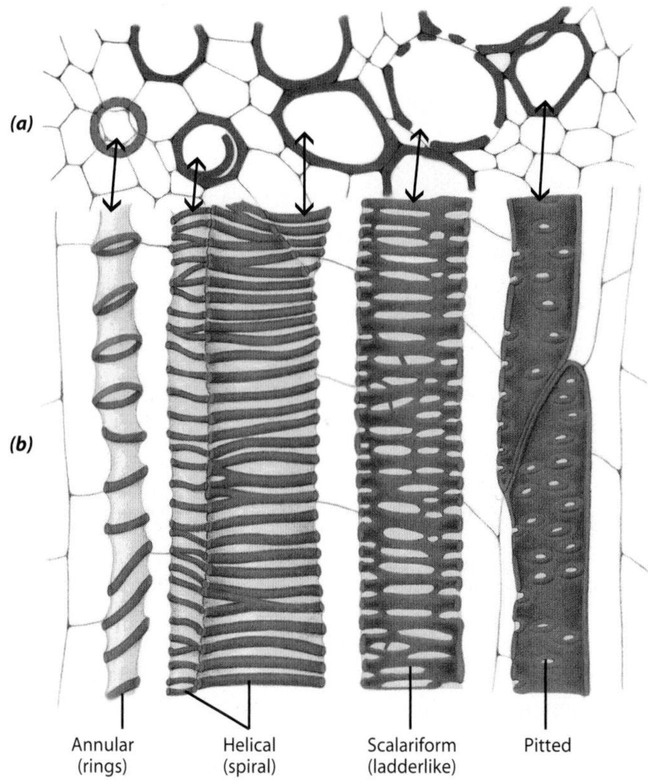

17–4 Tracheary elements The conducting cells of the xylem are the tracheary elements. A portion of the stem of Dutchman's pipe, *Aristolochia* (see Figure 20–6) in ***(a)*** transverse and ***(b)*** longitudinal views, showing some of the distinctive types of wall thickenings exhibited by tracheary elements. Here, the wall thickenings vary, left to right, from those elements formed earliest in the development of the plant to those formed more recently.

In fossil vascular plants from the Silurian and Devonian periods, the tracheary elements are elongated cells with long, tapering ends. Such tracheary elements, called **tracheids,** are the only type of water-conducting cell in most vascular plants, other than angiosperms and a peculiar group of gymnosperms known as the gnetophytes (phylum Gnetophyta; see page 453). Tracheids not only provide channels for the passage of water and minerals, but in many modern plants they also provide support for stems. Water-conducting cells are rigid, mostly because of the lignin in their walls. This rigidity made it possible for an upright habit to evolve in these plants, and eventually for some of them to become trees.

Tracheids are more primitive (that is, less specialized) than **vessel elements,** which are the principal water-conducting cells in angiosperms. Vessel elements apparently evolved independently in several groups of vascular plants. This is an excellent example of convergent evolution—the independent development of similar structures by unrelated or only distantly related organisms (see the essay "Convergent Evolution," on page 239).

Vascular Tissues Are Located in the Vascular Cylinders, or Steles, of Roots and Stems

The primary vascular tissues—primary xylem and primary phloem—and, in some vascular plants, a central column of ground tissue known as the **pith,** make up the central cylinder, or **stele,** of the stem and root in the primary plant body. Several types of steles are recognized, among them the protostele, the siphonostele, and the eustele (Figure 17–5).

The **protostele**—the simplest and most ancient type of stele—consists of a solid cylinder of vascular tissue in which the phloem either surrounds the xylem or is interspersed within it (Figures 17–3 and 17–5a). It is found in the extinct groups

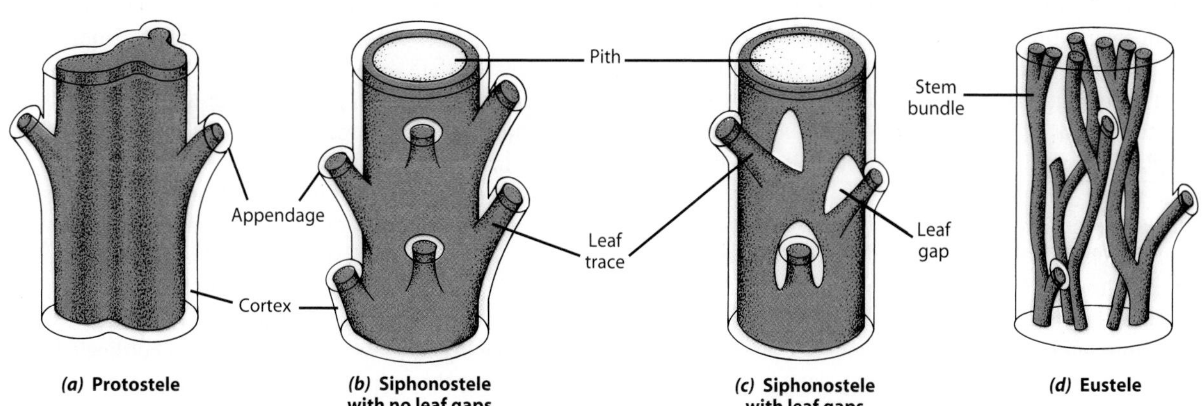

17–5 Steles ***(a)*** A protostele, with diverging traces of appendages, the evolutionary precursors of leaves. ***(b)*** A siphonostele with no leaf gaps; the vascular traces leading to the leaves simply diverge from the solid cylinder of vascular tissue. This sort of siphonostele is found in *Selaginella,* among other plants. ***(c)*** A siphonostele with leaf gaps, commonly found in ferns. ***(d)*** A eustele, found in almost all seed plants. Siphonosteles and eusteles appear to have evolved independently from protosteles.

of seedless vascular plants discussed below, as well as in lycophytes (composed primarily of club mosses) and in the juvenile stems of some other living groups. In addition, it is the type of stele found in most roots.

The **siphonostele**—the type of stele found in the stems of most species of seedless vascular plants—is characterized by a central pith surrounded by the vascular tissue (Figure 17–5b). The phloem may form only outside the cylinder of xylem or on both sides of it. In the siphonosteles of ferns, the departure from the stem of the vascular strands leading to the leaves—the **leaf traces**—generally is marked by gaps known as **leaf gaps,** in the siphonostele (as in Figure 17–5c). These leaf gaps are filled with parenchyma cells just like those that occur within and outside the vascular tissue of the siphonostele. Although the leaf traces in seed plants are associated with parenchymatous areas reminiscent of leaf gaps, these areas are generally not considered to be homologous to leaf gaps. Therefore, we will refer to these areas in seed plants as **leaf trace gaps.**

If the primary vascular cylinder consists of a system of discrete strands around a pith, as it does in almost all seed plants, the stele is called a **eustele** (Figure 17–5d). Comparative studies of living and fossil vascular plants have suggested that the eustele of seed plants evolved directly from a protostele. Eusteles appeared first among the progymnosperms, a group of spore-bearing plants that are discussed in Chapter 18 (see pages 432 and 433). Siphonosteles evidently evolved independently from protosteles. This evidence indicates that none of the groups of seedless vascular plants with living representatives gave rise to any living seed plants.

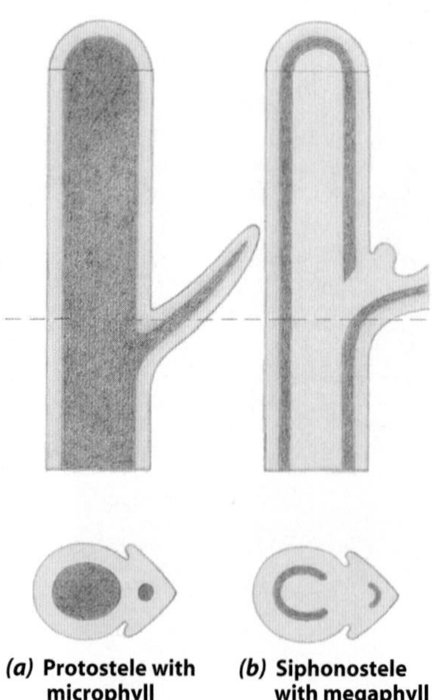

17–6 Microphylls and megaphylls Longitudinal and transverse sections through ***(a)*** a stem with a protostele and a microphyll and ***(b)*** a stem with a siphonostele and a megaphyll, emphasizing the nodes, or regions where the leaves are attached. Note the presence of pith and a leaf gap in the stem with a siphonostele and their absence in the stem with a protostele. Microphylls are characteristic of lycophytes, while megaphylls are found in all other vascular plants.

Roots and Leaves Evolved in Different Ways

Although the fossil record reveals little information on the origins of roots as we know them today, they must have evolved from the lower, often subterranean, portions of the axis of ancient vascular plants. For the most part, roots are relatively simple structures that seem to have retained many of the ancient structural characteristics no longer present in the stems of modern plants.

Leaves are the principal lateral appendages of the stem. Regardless of their ultimate size or structure, they arise as protuberances (leaf primordia) from the apical meristem of the shoot. From an evolutionary perspective, there are two fundamentally distinct kinds of leaves—microphylls and megaphylls.

Microphylls are usually relatively small leaves that contain only a single strand of vascular tissue (Figure 17–6a). Microphylls are typically associated with stems possessing protosteles and are characteristic of the lycophytes (Figure 17–3). The leaf traces leading to microphylls are not associated with gaps, and there is usually only a single vein in each leaf. Even though the name *microphyll* means "small leaf," some species of the lycophyte *Isoetes* have fairly long leaves (see Figure 17–21). In fact, certain Carboniferous and Permian lycophytes had microphylls a meter or more in length.

According to different theories, microphylls may have evolved as superficial lateral outgrowths of the stem (Figure 17–7a) or from the sterilization of sporangia in lycophyte ancestors. According to one theory, microphylls began as small, scalelike or spinelike outgrowths, called enations, devoid of vascular tissue. Gradually, rudimentary leaf traces developed, which initially extended only to the base of the enation. Finally, the leaf traces extended into the enation, resulting in formation of the primitive microphyll.

Most **megaphylls,** as the name implies, are larger than most microphylls. With few exceptions, they are associated with stems that have either siphonosteles or eusteles. The leaf traces leading to the megaphylls from siphonosteles and eusteles are associated with leaf gaps and leaf trace gaps, respectively (Figure 17–6b). Unlike the microphylls, the blade, or lamina, of most megaphylls has a complex system of branching veins.

It seems likely that megaphylls evolved from entire branch systems by a series of steps similar to that shown in Figure 17–7b. The earliest plants had a leafless, dichotomously branching axis, without distinction between axis and megaphylls. Unequal branching resulted in more aggressive branches "overtopping" the weaker ones. The subordinated, overtopped lateral branches represented the beginning of leaves, and the more aggressive portions became stemlike axes. This was followed by flattening out, or "planation," of the lateral branches. The final step was fusion, or "webbing," of the separate lateral branches to form a primitive lamina, or blade. Megaphylls originated independently at least three times (in ferns, horsetails, and seed plants).

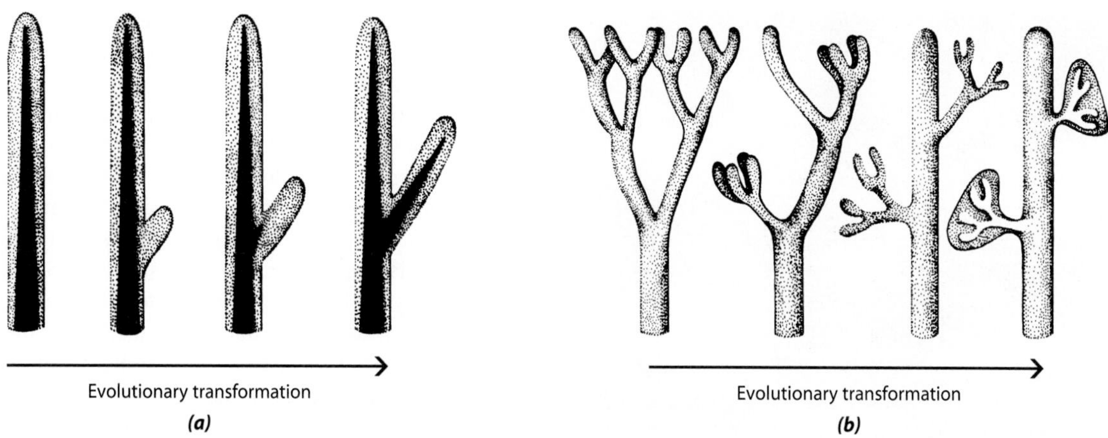

17–7 Evolution of microphylls and megaphylls ***(a)*** According to one widely accepted theory, microphylls evolved as outgrowths, called enations, of the main axis of the plant. ***(b)*** Megaphylls evolved by fusion of branch systems.

Reproductive Systems

As mentioned previously, all vascular plants are oogamous—that is, they have large nonmotile eggs and small sperm that swim to, or are conveyed to, the egg. In addition, all vascular plants have an alternation of heteromorphic generations, in which the sporophyte is larger and structurally much more complex than the gametophyte (Figure 17–8). Oogamy is clearly favored in plants, since only one of the kinds of gamete must navigate across a hostile environment outside the plant.

Homosporous Plants Produce Only One Kind of Spore, Whereas Heterosporous Plants Produce Two Types

Early vascular plants produced only one kind of spore as a result of meiosis; such vascular plants are said to be **homosporous.** Among living vascular plants, homospory is found in almost all ferns, the horsetails (equisetophytes), and some of the lycophytes. It is clearly the basic condition from which heterospory evolved. On germination, the spores of homosporous plants have the potential to produce bisexual gametophytes—that is, gametophytes that bear both sperm-producing antheridia and egg-producing archegonia. Yet the sporophytes of most ferns are heterozygous. If the sperm from one bisexual gametophyte were to fertilize an egg from the same gametophyte, the resulting sporophyte would be homozygous for all gene loci.

Ferns have various mechanisms by which they promote cross-fertilization, or outcrossing. One such mechanism involves maturation that occurs at different times for the antheridia and archegonia. Thus, rather than fertilizing their own eggs, the sperm produced by one gametophyte fertilize the eggs of neighboring, genetically different gametophytes. In many homosporous ferns, sex expression is determined by the age of the plant or by water-soluble substances called **antheridiogens,** or both. Antheridiogens secreted by developing gametophytes induce premature formation of antheridia in smaller, less mature gametophytes. Some of the larger gametophytes become female, while the small gametophytes remain wholly antheridial. Self-fertilization does occur in a few homosporous fern species, and it may be advantageous for pioneer species in the early stages of colonization or if only one spore reaches a new site (Figure 17–9).

Heterospory—the production of two types of spores in two different kinds of sporangia—is found in some of the lycophytes, as well as in a few ferns and all seed plants. Heterospory arose many times in unrelated groups during the evolution of vascular plants. It was common as early as the Devonian period, with the earliest record from about 370 million years ago. The two types of spores are called **microspores**

Gametophyte (n)
Antheridium with sperm (n)
Archegonium with egg (n)
Fertilization
Zygote ($2n$)
Embryo
Sporophyte ($2n$)
Sporangia
Spore mother cells ($2n$)
Meiosis
Tetrads of spores (n)
Spore

17–8 Generalized life cycle of a vascular plant The sporophyte is larger and structurally more complex than the gametophyte and ultimately free-living. This life cycle shows an alternation of heteromorphic generations.

17–9 *Dryopteris expansa* This fern is one of the few homosporous ferns that self-fertilizes.

and **megaspores,** and they are produced in **microsporangia** and **megasporangia,** respectively. The two types of spores are differentiated on the basis of function and not necessarily relative size. Microspores give rise to male gametophytes (microgametophytes), and megaspores give rise to female gametophytes (megagametophytes). Both types of unisexual gametophytes are much reduced in size compared with the gametophytes of homosporous vascular plants. Another difference is that in heterosporous plants, the gametophytes develop within the spore wall (endosporic development), whereas in homosporous plants the gametophytes develop outside the spore wall (exosporic development).

Over Evolutionary Time, the Gametophytes of Vascular Plants Have Become Smaller and Simpler

The relatively large gametophytes of homosporous plants are independent of the sporophyte for their nutrition, although the subterranean gametophytes of some species—such as those of *Botrychium, Psilotum* (see Figure 17–29), and several genera of club mosses (Lycopodiaceae)—are heterotrophic, depending on endomycorrhizal fungi for their nutrients. Other genera of club mosses, as well as the horsetails and most ferns, have free-living, photosynthetic gametophytes. In contrast, the gametophytes of many heterosporous vascular plants, and especially those of the seed plants, are dependent on the sporophyte for their nutrition.

The initial stages of plant evolution from charophycean-algal-like ancestors involved elaboration and modification of both the gametophyte and the sporophyte. Within vascular plants, however, the evolution of the gametophyte is characterized by an overall trend toward reduction in size and complexity, and the gametophytes of flowering plants, the angiosperms, are the most reduced of all (see page 465). The mature megagametophyte of angiosperms commonly consists of only seven cells, one of them an egg cell. When mature, the angiosperm microgametophyte contains only three cells, and two of them are sperm. Archegonia and antheridia, which are found in all seedless vascular plants, have apparently been lost in the lineage leading to angiosperms. All but a few gymnosperms—conifers are a familiar example—produce archegonia but lack antheridia (see Chapter 18). In the seedless vascular plants, the motile sperm swim through water to the archegonium. These plants must therefore grow in habitats where water is at least occasionally plentiful. In angiosperms and in most gymnosperms, entire microgametophytes **(pollen grains)** are carried to the vicinity of the megagametophyte. This transfer of the pollen grains is called **pollination.** Germination of the pollen grains produces special structures called **pollen tubes,** through which motile sperm (in cycads and *Ginkgo*) swim to—or nonmotile sperm (in conifers, gnetophytes, and angiosperms) are transferred to—the egg to achieve fertilization.

The Phyla of Seedless Vascular Plants

Several groups of seedless vascular plants flourished during the Devonian period, of which the three most important are recognized as the Rhyniophyta, Zosterophyllophyta, and Trimerophytophyta. All three groups were extinct by about the end of the Devonian, 360 million years ago. All three phyla consisted of seedless plants that were relatively simple in structure. A fourth phylum of seedless vascular plants, Progymnospermophyta, or progymnosperms, will be discussed in Chapter 18, because the members of that group may have been ancestral to the seed plants, both gymnosperms and angiosperms. In addition to these extinct phyla, we discuss in this chapter the Lycopodiophyta and Monilophyta, the two phyla of seedless vascular plants that have living representatives.

The overall pattern of diversification of plants may be interpreted in terms of the successive rise to dominance of four major plant groups that largely replaced the groups that were dominant earlier. In each instance, numerous species evolved in the groups that were rising to dominance. The major groups are as follows:

1. Early vascular plants, characterized by relatively small stature and a simple and presumably primitive morphology. These plants included the rhyniophytes, zosterophyllophytes, and trimerophytes (Figure 17–10), which were dominant from the Middle Silurian to the Middle Devonian, from about 425 to 370 million years ago.
2. Monilophytes, lycophytes, and progymnosperms. These more complex groups were dominant from the Late Devonian period through the Carboniferous period (Figure 17–11), from about 375 to about 290 million years ago (see the essay "Coal Age Plants" on pages 400 and 401).
3. Seed plants, which arose starting in the Late Devonian period, at least 380 million years ago, and evolved many new lines by the Permian period. Gymnosperms dominated land floras throughout most of the Mesozoic era, until about 100 million years ago.
4. Flowering plants, which appeared in the fossil record at least 135 million years ago. This phylum became abundant in most parts of the world within 30 to 40 million years and has remained dominant ever since.

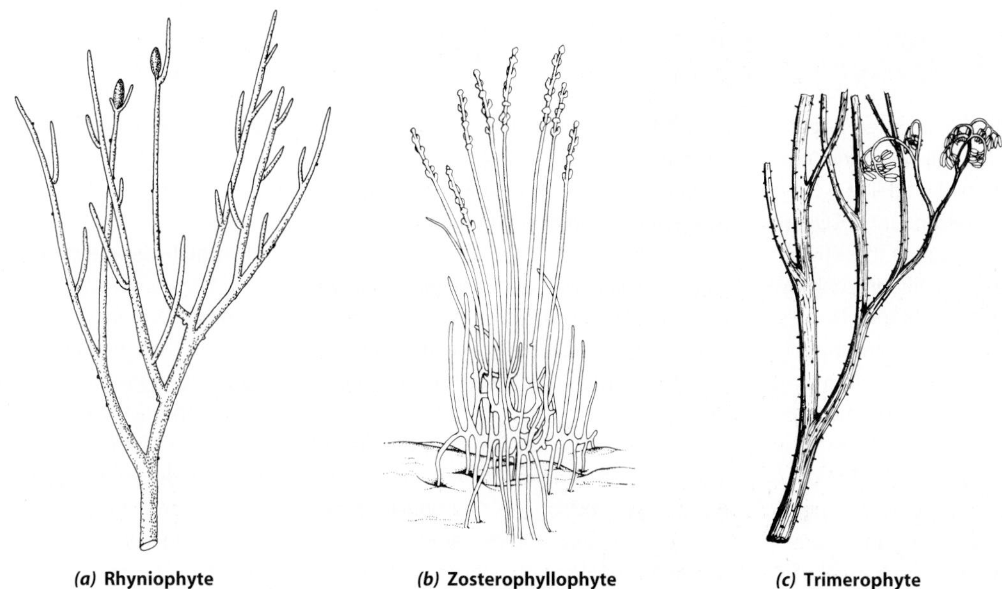

17–10 Early vascular plants ***(a)*** *Rhynia gwynne-vaughanii* is a rhyniophyte, one of the simplest of the known vascular plants. The axis was leafless and dichotomously branched, with numerous lateral branches. The sporangia were terminal on some of the upright main branches and usually overtopped by development of lateral branches. ***(b)*** In *Zosterophyllum* and the other zosterophyllophytes, the generally kidney-shaped sporangia were borne laterally in a helix or in two rows on the stems. Sporangia split along definite slits that formed around the outer margin. The zosterophyllophytes were larger than the rhyniophytes, but like the latter, they were mostly dichotomously branched plants that were naked, spiny, or toothed. ***(c)*** The trimerophytes were larger plants with more complex branching that was generally differentiated into a strong central axis with smaller side branches. The side branches were dichotomously branched and often had terminal masses of paired sporangia that tapered at both ends. The best-known genera included in this group are *Psilophyton* and *Trimerophyton.* A reconstruction of *Psilophyton princeps* is shown here. Individuals of *R. gwynne-vaughanii* ranged up to about 18 centimeters tall, while some of the trimerophytes were up to 1 meter or more in height.

17–11 Carboniferous swamp Seen here is a reconstruction of a swamp forest of the Carboniferous period. Most of the trees shown are lycophytes, but two giant horsetails (equisetophytes) are represented by the two dark trees in the right foreground. Ferns can also be seen on the left.

COAL AGE PLANTS

The amount of carbon dioxide used in photosynthesis is about 100 billion metric tons annually, about a tenth of the total CO_2 present in the atmosphere. The amount of CO_2 returned as a result of oxidation of these living materials is about the same, differing only by 1 part in 10,000. This very slight imbalance is caused by the burying of organisms in sediment or mud under conditions in which oxygen is excluded and decay is only partial. This accumulation of partially decayed plant material is known as peat (page 381). The peat may eventually become covered with sedimentary rock and thus placed under pressure. Depending on time, temperature, and other factors, peat may become compressed into soft or hard coal, one of the so-called fossil fuels.

During certain periods in the Earth's history, the rate of fossil-fuel formation was greater than at other times. One such time was the Carboniferous period, which extended from about 362 to 290 million years ago (see Figures 17–11 and 18–8). The lands were low, covered by shallow seas or swamps, and, in what are now temperate regions of Europe and North America, conditions were favorable for year-round growth. These regions were tropical to subtropical, with the equator then arcing across the Appalachians, over northern Europe, and through Ukraine. Five groups of plants dominated the swamplands, and three of them were seedless vascular plants—lycophytes, equisetophytes (calamites), and ferns. The other two were gymnospermous types of seed plants—seed ferns (Pteridospermales) and cordaites (Cordaitales).

Lycophyte Trees

For most of the "Age of Coal" in the Late Carboniferous period (Pennsylvanian), lycophyte trees, such as *Lepidodendron,* dominated the coal-forming swamps. Most of these plants grew to heights of 10 to 35 meters and were sparsely branched ***(a).*** After the plant attained most of its total height, the trunk branched dichotomously. Successive branching produced progressively smaller branches until, finally, the tissues of the branch tips lost their ability to grow further. The branches bore long microphylls. The tree lycophytes were largely supported by a massive periderm surrounding a relatively small amount of xylem.

Like *Selaginella* and *Isoetes,* lycophyte trees were heterosporous, and their sporophylls were aggregated into cones. Some of these trees produced structures analogous to seeds.

As the swamplands began to dry up and the climate in Euramerica began to change toward the end of the Carboniferous period, the lycophyte trees vanished almost overnight, geologically speaking. The only remaining living relative of the group is the genus *Isoetes.* Herbaceous lycophytes essentially similar to *Lycopodium* and *Selaginella* existed in the Carboniferous period, and representatives of some of them have survived to the present; there are 10 to 15 living genera.

Calamites

Calamites, or giant horsetails, were plants of treelike proportions, reaching heights of 18 meters or more. Like the plant body of *Equisetum,* that of the calamites consisted of a branched aerial portion and an underground rhizome system. In addition, the leaves and branches were whorled at the nodes. Even the stems were remarkably similar to those of *Equisetum,* except for the presence of secondary xylem in the calamites, which accounted for much of the great diameter of the stems (trunks up to one-third of a meter in diameter). The similarities between calamites and living *Equisetum* are so strong that they are now regarded as belonging to the same order.

The fertile appendages, or sporangiophores, of the calamites were aggregated into cones. Although most were homosporous, a few giant horsetails were heterosporous. Like most of the lycophyte trees, the giant horsetails declined in importance toward the end of the Paleozoic but persisted in much reduced form through the Mesozoic and Tertiary. Recent molecular evidence has led botanists to regard the horsetails as an ancient line of monilophytes, but not as ferns (Figure 17–14).

Ferns

Many of the ferns represented in the fossil record are recognizable as members of today's primitive fern families. The "Age of Ferns" in the Late Carboniferous period was dominated by tree ferns such as *Psaronius,* one of the Marattiopsida—a eusporangiate group. Up to 8 meters tall, *Psaronius* had a stele that expanded toward the apex; the stele was covered below with adventitious roots, which played the key role in supporting the plant. The stem of *Psaronius* ended in an aggregate of large, pinnately compound fronds.

Seed Plants

The two remaining plant groups that dominated the tropical lowlands of Euramerica were the seed ferns and the cordaites. The fossil plants that are usually grouped as seed ferns are probably of diverse evolutionary relationships. The remains of seed ferns are common fossils in rocks of Carboniferous age ***(b).*** Their large, pinnately compound fronds were so fernlike that these plants were long regarded as ferns. Then in 1905, F. W. Oliver and D. H. Scott demonstrated that the plants bore seeds and so were gymnosperms. Many species were small, shrubby, or scrambling plants.

Other probable seed ferns were tall, woody trees. The fronds of seed ferns were borne at the top of their stem, or trunk, and the fronds bore microsporangia and seeds. The seed ferns survived into the Mesozoic era.

The cordaites were widely distributed during the Carboniferous period both in swamps and in drier environments. Although some members of the order were shrubs, many were tall (15 to 30 meters), highly branched trees that perhaps formed extensive forests. Their long (up to 1 meter), straplike leaves were spirally arranged at the tips of the youngest branches ***(c).*** The center of the stem was occupied by a large pith, and a vascular cambium gave rise to a complete cylinder of secondary xylem. The root system, located at the base of the plant, also contained secondary xylem. The plants bore pollen-bearing cones and seed-bearing, conelike structures on separate branches. The cordaites persisted into the Permian period (286 to 248 million years ago), the drier and cooler period that followed the Carboniferous, but were apparently extinct by the beginning of the Mesozoic.

In Conclusion

The dominant tropical coal-swamp plants of the Carboniferous period in Euramerica—the lycophyte trees—became extinct during the Late Paleozoic, a time of increasing tropical drought. Only the herbaceous relatives of the tree lycophytes and horsetails of the Carboniferous period continued to flourish and exist today, as do several groups of ferns that appeared in the Carboniferous period. Both the seed ferns and the cordaites eventually disappeared. Only one group of Carboniferous gymnosperms, the conifers (not a dominant group at the time), survived and went on to produce new types during the Permian period. The living conifers are discussed in detail in Chapter 18.

(a) Scientists believe that once a lycophyte tree was stabilized by its shallow, forking, rootlike axes, it pushed rapidly skyward. These underground axes produced spirally arranged rootlets, seen here as slender projections emerging from the forest floor. From left to right, a young leafy form, a pole-like juvenile, and a giant adult of 35 meters. ***(b)*** One of the most interesting gymnosperm groups is the seed ferns, a large, artificial group of primitive seed-bearing plants that appeared in the Late Devonian and flourished for about 125 million years. Fossils of these bizarre plants are common in rocks of Carboniferous age and have been well known to paleobotanists for a century or more. Their vegetative parts are so fernlike that for many years they were grouped with the ferns. This drawing is a reconstruction of the Carboniferous seed fern *Medullosa noei.* The plant was about 5 meters tall. ***(c)*** Tip of a young branch of the primitive conifer-like *Cordaites,* with long, straplike leaves.

Phylum Rhyniophyta

The earliest known vascular plants that we understand in detail belong to the phylum Rhyniophyta, a group that dates back to the Middle Silurian, at least 425 million years ago. The group became extinct in the Middle Devonian (about 380 million years ago). Earlier vascular plants were probably similar; their remains go back at least another 15 million years. Rhyniophytes were seedless plants, consisting of simple, dichotomously branching (evenly forked) axes, or stems, with terminal sporangia. Their plant bodies were not differentiated into roots, stems, and leaves, and they were homosporous. The name of the phylum comes from the good representation of these primitive plants as fossils preserved in chert near the village of Rhynie, in Scotland.

Among the first rhyniophytes to be described is *Rhynia gwynne-vaughanii.* Probably a marsh plant, it consisted of an upright, dichotomously branched aerial system attached to a dichotomously branched rhizome (underground stem) system with rhizoids. Among the distinctive features of *R. gwynne-vaughanii* are the numerous lateral branches that arose from the dichotomized axes (Figures 17–10a and 17–12) and the short branches on which the sporangia were often borne. The aerial branch system, which was about 18 centimeters in height, was covered with a cuticle and contained stomata. Lacking leaves, the aerial axes served as the photosynthetic organs.

The internal structure of *R. gwynne-vaughanii* was similar to that of many of today's vascular plants. A single layer of superficial cells—the epidermis—surrounded the photosynthetic tissue of the cortex, and the center of the axis consisted of a solid strand of xylem surrounded by one or two layers of phloemlike cells. The tracheids were different from those of most vascular plants, and although they had internal thickenings, they share some features with the water-conducting cells of mosses.

Probably the best-known plant found in the Rhynie chert is *Aglaophyton major,* which was originally called *Rhynia major* (Figure 17–13). A more robust plant than *R. gwynne-vaughanii,*

17–12 *Rhynia gwynne-vaughanii* Fossil remains of *Rhynia gwynne-vaughanii* from the Rhynie chert in Scotland, showing well-preserved axes (stems) in the upright growth position.

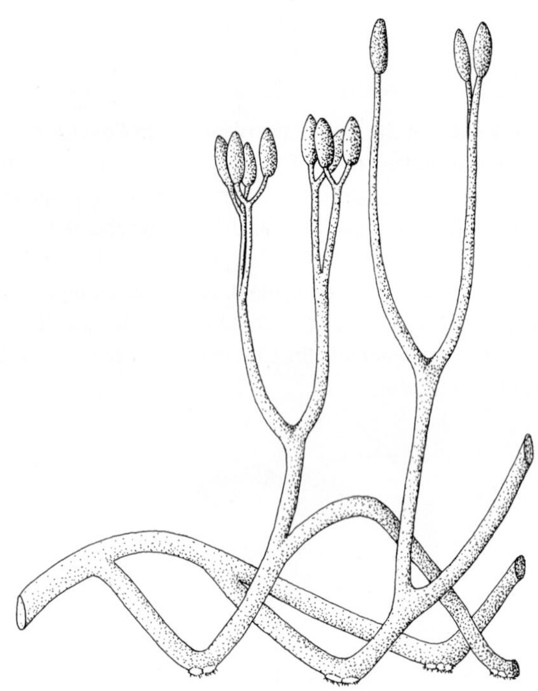

17–13 *Aglaophyton major* The central strand of conducting tissue in *Aglaophyton major* lacks tracheids but contains cells similar to the hydroids of mosses. *Aglaophyton major* may be an intermediate stage, known as a protracheophyte, in the evolution of vascular plants; it was previously known as *Rhynia major* when it was considered to be a vascular plant.

reaching heights of about 50 centimeters, it consisted of an extensive dichotomously branched rhizome system with a limited number of upright stems that branched dichotomously. All axes terminated in sporangia. For over 60 years, *R. major* was considered to be a vascular plant. Then it was shown that the cells forming the central strand of conducting tissue lack the wall thickenings typical of tracheids. Rather than being tracheids, these cells are more like the hydroids of modern mosses. For this reason, this fossil plant has been removed from the genus *Rhynia* and assigned to the new genus *Aglaophyton. Aglaophyton major,* with its branched axes and multiple sporangia, may represent an intermediate stage—known as a **protracheophyte**—in the evolution of vascular plants and should probably not be retained in the phylum Rhyniophyta.

Cooksonia, a rhyniophyte that is believed to have inhabited mud flats, has the distinction of being the oldest known vascular plant (Figure 17–1). Specimens of *Cooksonia* have been found in Wales, Scotland, England, the Czech Republic, Canada, and the United States. *Cooksonia* is the smallest and simplest vascular plant known from the fossil record. Its slender, leafless aerial stems ranged up to about 6.5 centimeters long and terminated in globose sporangia. Tracheids have been identified in the central region of the axes of *Cooksonia pertoni* from the Lower Devonian. Plants similar in form to *C. pertoni* occurred in the Silurian, but it is questionable whether they contained vascular tissues. The genus *Cooksonia* may contain early simple fossil plants of diverse relationships. Some of these plants may be associated with the protracheophyte *Aglaophyton,* but others are almost certainly true vascular plants. *Cooksonia* had become extinct by the Lower Devonian period, about 390 million years ago.

Evidence from the Rhynie chert and the Lower Devonian of Germany indicates that the gametophytes of plants such as *Aglaophyton*—and, by implication, *Rhynia* and *Cooksonia* (among others)—were relatively large, branched structures. Some of these gametophytes apparently had water-conducting cells, cuticles, and stomata. Hence, some of these plants had an alternation of isomorphic generations, in which sporophyte and gametophyte were basically similar except for their sporangia and gametangia, respectively.

Phylum Zosterophyllophyta

The fossils of a second phylum of extinct seedless vascular plants—the Zosterophyllophyta—have been found in strata from the Early to Late Devonian period, from about 408 to 370 million years ago. Like the rhyniophytes, the zosterophyllophytes were leafless and dichotomously branched. The aerial stems were covered with a cuticle, but only the upper ones contained stomata, indicating that the lower branches may have been embedded in mud. In *Zosterophyllum,* it has been suggested that the lower branches frequently produced lateral branches that forked into two axes, one that grew upward, the other downward (Figure 17–10b). The downward-growing branches may have functioned like a root, permitting the plant to spread outward from the center by providing support. The zosterophyllophytes are so named because of their general resemblance to the modern seagrasses of the genus *Zostera,* marine angiosperms that superficially resemble grasses.

Unlike those of the rhyniophytes, the globose or kidney-shaped sporangia of the zosterophyllophytes were borne laterally on short stalks. These plants were homosporous. The internal structure of the zosterophyllophytes was essentially similar to that of the rhyniophytes, except that in the zosterophyllophytes, the first xylem cells to mature were located around the periphery of the xylem strand and the last to mature were located in the center. This process, known as centripetal differentiation, is the opposite of the centrifugal differentiation found in the rhyniophytes.

Early zosterophyllophytes were almost certainly the ancestors of the lycophytes. The sporangia of zosterophyllophytes and early lycophytes are very similar and in both groups were also borne laterally. The xylem in both phyla also differentiated centripetally.

Phylum Trimerophytophyta

The phylum Trimerophytophyta, which probably evolved directly from the rhyniophytes, most likely contains plants of diverse evolutionary relationships and seems to represent the ancestral stock of both the ferns and the progymnosperms; they are a diverse group. Larger and more complex plants than the rhyniophytes or zosterophyllophytes (Figure 17–10c), the trimerophytes first appeared in the Early Devonian period about 395 million years ago and had become extinct by the end of the Middle Devonian, about 20 million years later—a relatively short period of existence.

Although generally larger and evolutionarily more specialized than the rhyniophytes, trimerophytes still lacked leaves. Branching, however, was more complex, with the main axis forming lateral branch systems that dichotomized several times. The trimerophytes, like the rhyniophytes and zosterophyllophytes, were homosporous. Some of their smaller branches terminated in elongated sporangia, while others were entirely vegetative. Besides their more complex branching pattern, the trimerophytes had a more massive vascular strand than the rhyniophytes. Together with a broad band of thick-walled cells in the cortex, the large vascular strand probably was capable of supporting fairly large plants, over a meter in height. As in the rhyniophytes, the xylem of the trimerophytes differentiated centrifugally. The name of the phylum comes from the Greek words *tri-, meros,* and *phyton,* meaning "three-parted plant," because of the organization of the secondary branches into three rows in the genus *Trimerophyton.*

Phylum Lycopodiophyta

The 10 to 15 living genera and approximately 1200 living species of Lycopodiophyta are the representatives of an evolutionary line that extends back to the Devonian period. Morphological and molecular evidence indicate that in the Early to Middle Devonian (before 400 million years ago), a basal split occurred separating a **lycophyte clade** that included the modern lycophyte lineage from a clade known as the euphyllophytes (Figure 17–14). The **euphyllophyte clade** includes all other living vascular plant lineages—the monilophytes (ferns and horsetails) and the seed plants.

There are a number of orders of lycophytes, and at least three of the extinct ones included small to large trees. The three orders of living lycophytes, however, consist entirely of nonwoody herbaceous plants; each order includes a single family. All lycophytes, living and fossil, possess microphylls, and this type of leaf, which shows relatively little diversity in form, is highly characteristic of the phylum; all lycophytes are eusporangiate. Tree lycophytes, such as *Lepidodendron,* were among the dominant plants of the coal-forming forests of the Carboniferous period (see the essay on pages 400 and 401). Most lines of woody lycophytes—those lycophytes that exhibited secondary growth—became extinct before the end of the Paleozoic era, 248 million years ago.

The Club Mosses Belong to the Family Lycopodiaceae

Perhaps the most familiar living lycophytes are the club mosses, family Lycopodiaceae (see Figure 12–16c). All but two genera of living lycophytes belong to this family, most members of which were formerly grouped in the collective genus *Lycopodium.* Seven of these genera are represented in the United States and Canada, but most of the estimated 350 to 400 species in the family are tropical. The taxonomic boundaries of the genera of this family are poorly understood, and as many as 15 genera may ultimately be recognized. Lycopodiaceae extend from Arctic regions to the tropics, but they rarely form conspicuous

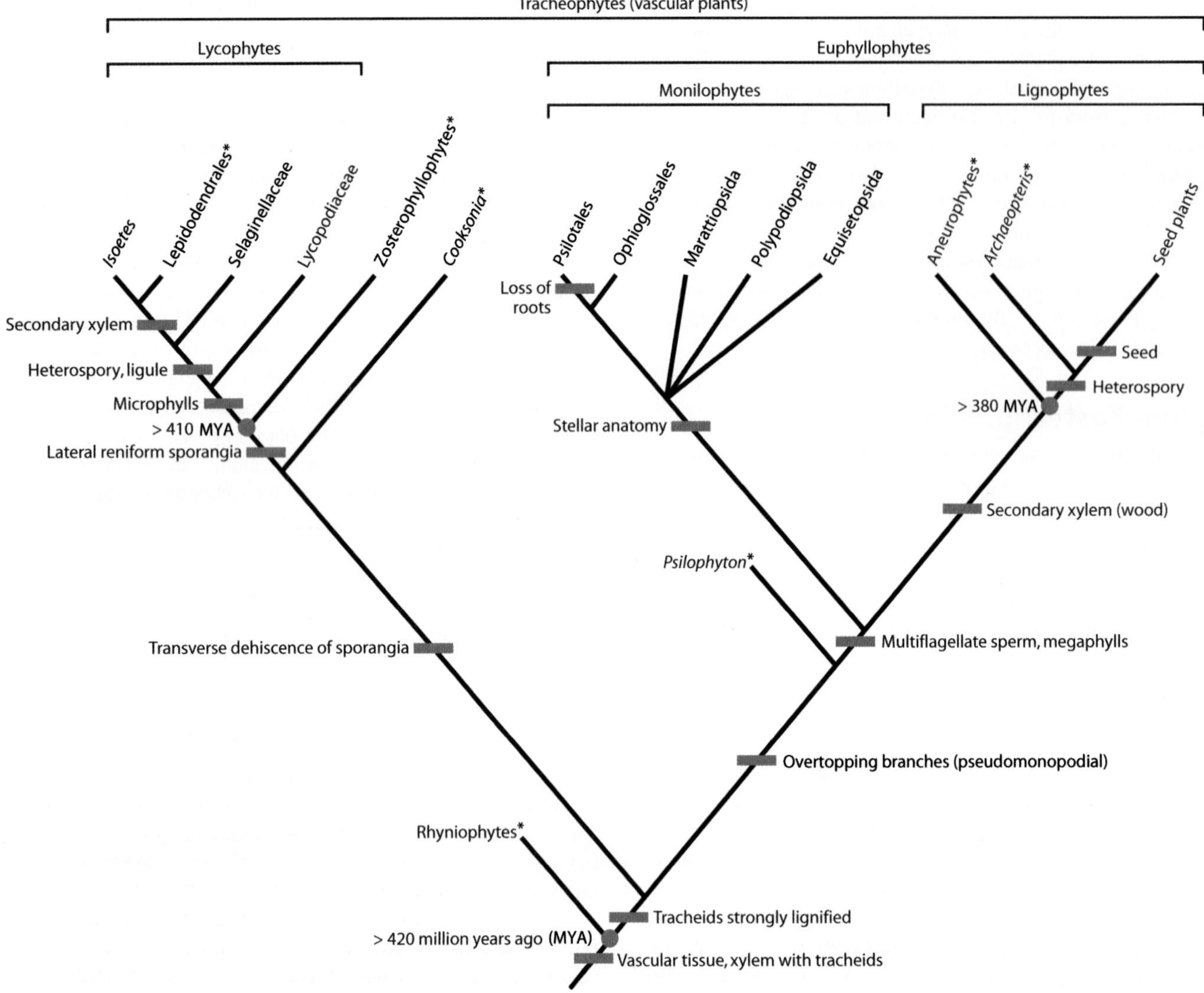

17–14 Phylogenetic relationships of the vascular plants (tracheophytes) This phylogenetic tree shows the basal split between the lycophyte clade and the euphyllophyte clade. Also shown are the relationships of some seedless vascular plants, namely the monilophytes (the Psilotopsida, comprising Psilotales and Ophioglossales, and the Marattiopsida, Polypodiopsida, and Equisetopsida) and the progymnosperms (aneurophytes and *Archaeopteris*), to seed plants. The clade containing the progymnosperms and seed plants has been called the lignophytes, or producers of wood. Characters marking major clades are indicated. Asterisks indicate extinct taxa.

elements in any plant community. Most tropical species, many of which belong to the genus *Phlegmariurus,* are epiphytes and thus rarely seen, but several of the temperate species form mats that may be evident on forest floors.

The sporophytes of most genera of Lycopodiaceae consist of a branching rhizome from which aerial branches and roots arise. Both stems and roots are protostelic (Figure 17–15). The microphylls of Lycopodiaceae are usually spirally arranged, but they appear opposite or whorled in some members of the group. Lycopodiaceae are homosporous; the sporangia occur singly on the upper surface of fertile microphylls called **sporophylls,** which are modified leaves or leaflike organs that bear the spore-producing sporangia (Figure 17–16). In *Huperzia* (Figure 17–17a, page 408) and *Phlegmariurus,* the sporophylls are similar to ordinary microphylls and are interspersed among the sterile microphylls. In the other genera of Lycopodiaceae found in the United States and Canada, including *Diphasiastrum* (see Figure 12–16c) and *Lycopodium,* nonphotosynthetic sporophylls are grouped into **strobili,** or cones, at the ends of the aerial branches (Figure 17–17b).

On germination, the spores of Lycopodiaceae give rise to bisexual gametophytes that, depending on the genus, are either green, irregularly lobed masses (*Lycopodiella, Palhinhaea,* and *Pseudolycopodiella,* among the genera that occur in the United States and Canada) or subterranean, nonphotosynthetic, mycorrhizal structures (*Diphasiastrum, Huperzia, Lycopodium,* and *Phlegmariurus,* among the genera represented in the United States and Canada). The development and maturation of archegonia and antheridia in a gametophyte of Lycopodiaceae may require from 6 to 15 years, and their gametophytes may even

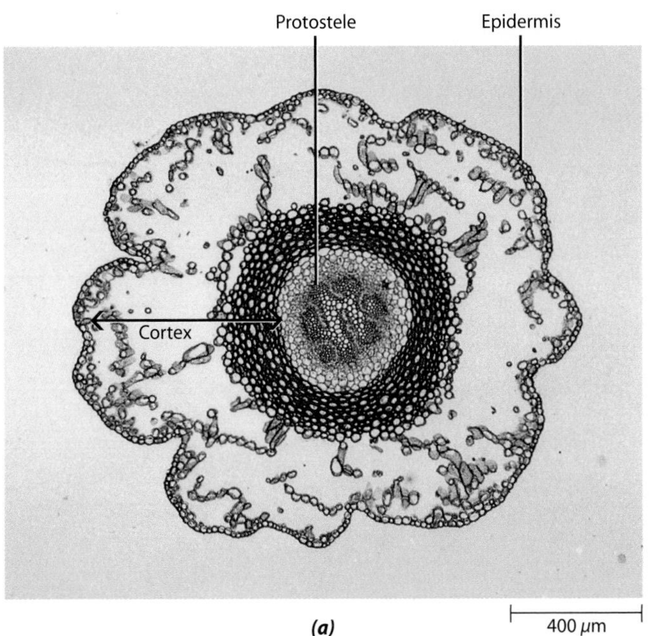

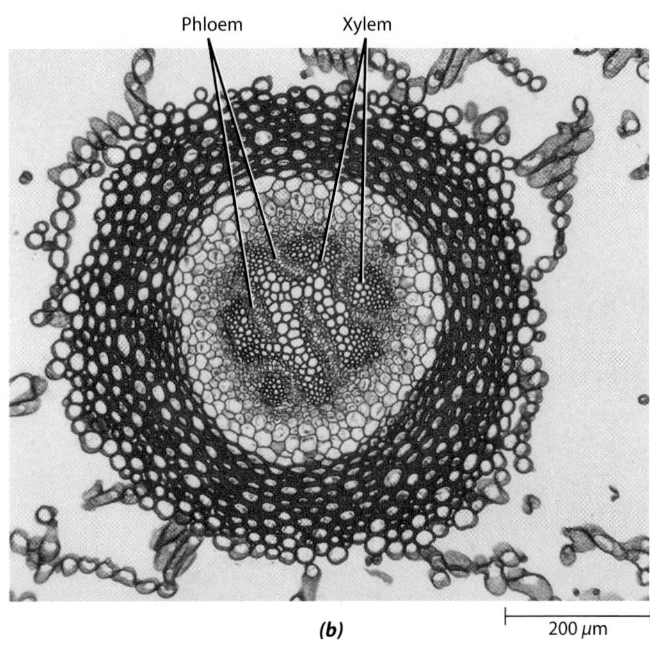

17–15 Protosteles Both the stem and root of members of the family Lycopodiaceae are protostelic. ***(a)*** Transverse section of *Diphasiastrum complanatum* stem, showing mature tissues. Note the large air spaces in the cortex, which surrounds the central protostele. ***(b)*** Detail of protostele of *D. complanatum,* showing xylem and phloem. See also Figure 17–3.

produce a series of sporophytes in successive archegonia as they continue to grow. Even though their gametophytes are bisexual, self-fertilization rates are very low, and the gametophytes of these species predominantly cross-fertilize.

Water is required for fertilization in Lycopodiaceae. The biflagellated sperm swim through water to the archegonium. Following fertilization, the zygote develops into an embryo, which grows within the venter of the archegonium. The young sporophyte may remain attached to the gametophyte for a long time, but eventually it becomes independent. The life cycle of *Lycopodium lagopus,* a representative of those Lycopodiaceae that have a subterranean, mycorrhizal gametophyte and form a strobilus, is illustrated in Figure 17–16.

Among the genera of Lycopodiaceae found in the United States and Canada, *Huperzia* (Figure 17–17a), the fir mosses, consists of 7 species; *Lycopodium* (Figure 17–17b), the tree club mosses, 5 species; *Diphasiastrum* (see Figure 12–16c), the club mosses and running pines, 11 species; and *Lycopodiella,* 6 species. These genera, and the others now recognized in Lycopodiaceae, differ from one another in various technical characteristics, including the arrangement of the sporophylls, the presence of rhizomes, the organization of the vegetative body, the nature of the gametophyte, and basic chromosome numbers.

The Spike Mosses Belong to the Family Selaginellaceae

Among the living genera of lycophytes, *Selaginella,* the only genus of the family Selaginellaceae, has the most species, about 750. Most of this family of spike mosses are tropical in distribution. Many grow in moist places, although a few inhabit deserts, becoming dormant during the driest part of the year. Among the latter is the so-called resurrection plant, *Selaginella lepidophylla,* found in Texas, New Mexico, and Mexico (Figure 17–18a, page 409).

The herbaceous sporophyte of *Selaginella* is basically similar to that of some Lycopodiaceae in that it bears microphylls and its sporophylls are arranged in strobili (Figure 17–18b). Unlike Lycopodiaceae, however, *Selaginella* has a small, scalelike outgrowth, called a **ligule,** near the base of the upper surface of each microphyll and sporophyll (Figure 17–19, pages 410 and 411). The stem and root are protostelic (Figure 17–20, page 412).

Whereas Lycopodiaceae are homosporous, *Selaginella* is heterosporous, with unisexual—male and female—gametophytes. Each sporophyll bears a single sporangium on its upper surface. Megasporangia are borne by **megasporophylls,** and microsporangia are borne by **microsporophylls.** Both kinds of sporangia occur in the same strobilus.

The male gametophytes (microgametophytes) in *Selaginella* develop within the microspore, and they lack chlorophyll. At maturity, the male gametophyte consists of a single prothallial, or vegetative, cell and an antheridium, which gives rise to many biflagellated sperm. The microspore wall must rupture for the sperm to be liberated.

During development of the female gametophyte (megagametophyte), the megaspore wall ruptures and the gametophyte protrudes through the rupture to the outside. This is the portion of the female gametophyte in which the archegonia develop. It has been reported that the female gametophytes sometimes develop chloroplasts, although most *Selaginella* gametophytes derive their nutrition from food stored within the megaspores.

Water is required for the sperm to swim to the archegonia and fertilize the eggs. Commonly, fertilization occurs after the gametophytes have been shed from the strobilus. During

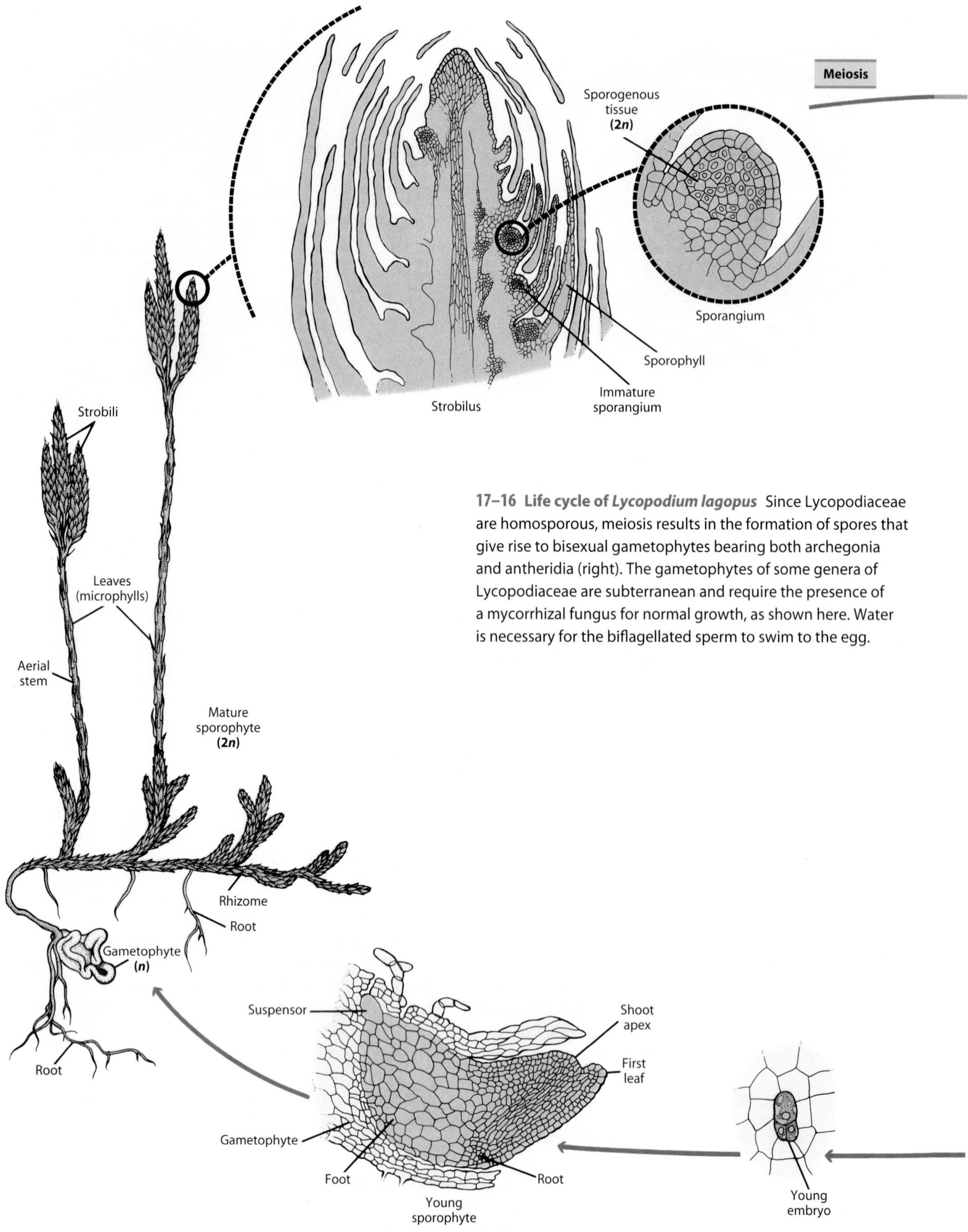

17–16 Life cycle of *Lycopodium lagopus* Since Lycopodiaceae are homosporous, meiosis results in the formation of spores that give rise to bisexual gametophytes bearing both archegonia and antheridia (right). The gametophytes of some genera of Lycopodiaceae are subterranean and require the presence of a mycorrhizal fungus for normal growth, as shown here. Water is necessary for the biflagellated sperm to swim to the egg.

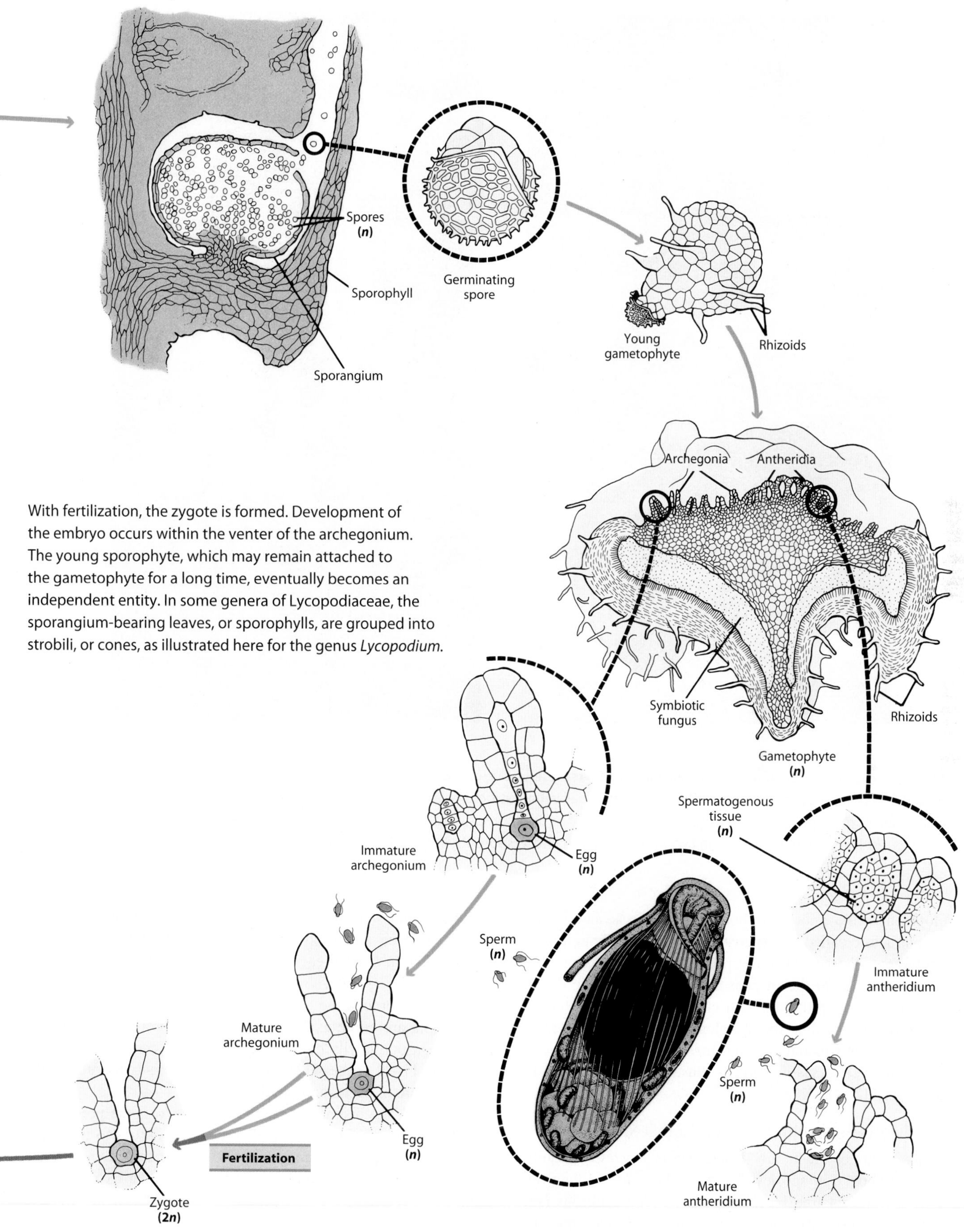

With fertilization, the zygote is formed. Development of the embryo occurs within the venter of the archegonium. The young sporophyte, which may remain attached to the gametophyte for a long time, eventually becomes an independent entity. In some genera of Lycopodiaceae, the sporangium-bearing leaves, or sporophylls, are grouped into strobili, or cones, as illustrated here for the genus *Lycopodium*.

(a)

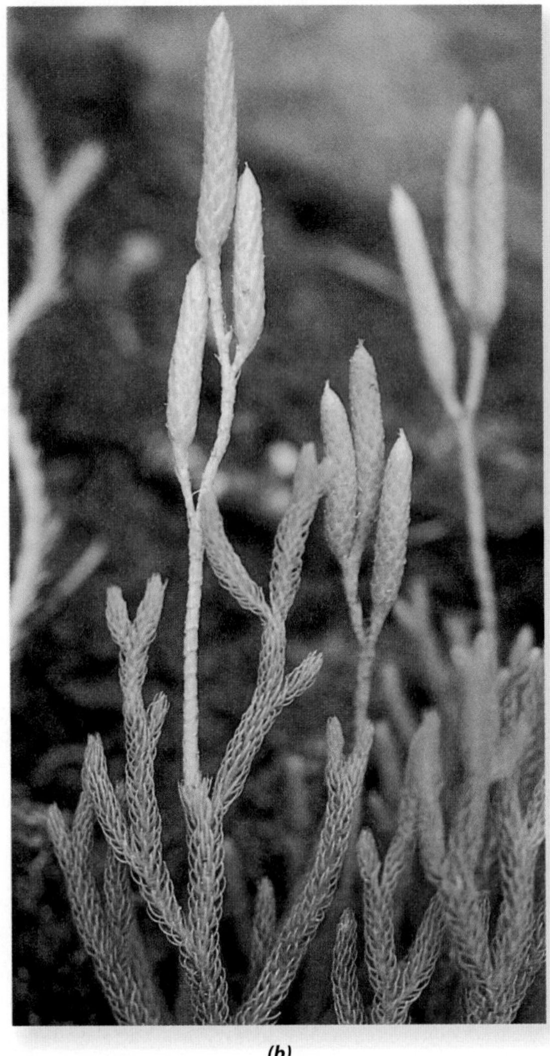

(b)

17–17 Sporophylls and strobili ***(a)*** *Huperzia lucidula* is a representative of those genera of Lycopodiaceae that lack differentiated cones, or strobili. The sporangia (small yellow structures along the stem) are borne in the axils of fertile microphylls known as sporophylls. Areas of fertile sporophylls alternate with regions of sterile microphylls. ***(b)*** The terminal branches in *Lycopodium lagopus* end in sporophylls grouped into strobili.

development of the embryos in both Lycopodiaceae and *Selaginella,* a structure called a **suspensor** is formed. Although inactive in Lycopodiaceae and some species of *Selaginella,* in other species of *Selaginella* the suspensor serves to thrust the developing embryo deep within the nutrient-rich tissue of the female gametophyte. Gradually, the developing sporophyte emerges from the gametophyte and becomes independent.

The life cycle of *Selaginella* is illustrated in Figure 17–19 (pages 410 and 411).

The Quillworts Belong to the Family Isoetaceae

The only genus of the family Isoetaceae is *Isoetes,* the quillworts, with about 150 species. *Isoetes* is the nearest living relative of the ancient tree lycophytes of the Carboniferous (see the essay on pages 400 and 401). Plants of *Isoetes* may be aquatic, or they may grow in pools that become dry at certain seasons. The sporophyte of *Isoetes* consists of a short, fleshy underground stem (corm) bearing quill-like microphylls on its upper surface and roots on its lower surface (Figure 17–21, page 412). In *Isoetes,* each leaf is a potential sporophyll.

Like *Selaginella, Isoetes* is heterosporous. The megasporangia are borne at the base of megasporophylls, and the microsporangia are borne at the base of microsporophylls, similar to the megasporophylls but located nearer the center of the plant (Figure 17–22, page 412). A ligule is present just above the sporangium of each sporophyll.

One of the distinctive features of *Isoetes* is the presence of a specialized cambium that adds secondary tissues to the corm. Externally the cambium produces only parenchyma tissue, whereas internally it produces a peculiar vascular tissue consisting of sieve elements, parenchyma cells, and tracheids in varying proportions.

Some species of *Isoetes* (sometimes assigned to another genus, *Stylites*) growing at high elevations in the tropics have

17–18 Representative *Selaginella* ***(a)*** *Selaginella lepidophylla,* the resurrection plant, a plant that becomes completely dried out when water is not available but quickly revives following a rain. This plant was growing in Big Bend National Park, in Texas. ***(b)*** *Selaginella rupestris,* the rock spike moss, with strobili. ***(c)*** *Selaginella kraussiana,* a prostrate, creeping plant. ***(d)*** *Selaginella willdenowii,* from the Old World tropics. Shade-loving, it climbs to 7 meters and has peacock-blue leaves with a metallic sheen. Note the clearly evident rhizomes.

the unique characteristic of obtaining their carbon for photosynthesis from the sediment in which they grow, rather than from the atmosphere. The leaves of these plants lack stomata, have a thick cuticle, and carry on essentially no gas exchange with the atmosphere. Like at least some other species of *Isoetes* in which the plants dry out for part of the year, these species have CAM photosynthesis (pages 145 to 147).

Phylum Monilophyta

The Great Majority of Monilophytes Are Ferns

The Monilophyta comprise the ferns and horsetails. These groups were once regarded as belonging to separate phyla, but recent analyses of morphological characters and of both chloroplast and nuclear genes indicate that the ferns and horsetails compose a clade (Monilophyta), with four major lineages (Figure 17–14): (1) Psilotopsida, (2) Marattiopsida, (3) Polypodiopsida, and (4) Equisetopsida. The relationships among these lineages are still under active investigation. The common name "fern" is applied to members of the lineages Psilotopsida, Marattiopsida, and Polypodiopsida.

Ferns have been relatively abundant in the fossil record from the Carboniferous period to the present (see the essay on pages 400 and 401). There are more than 12,000 living species of ferns, the largest and most diverse group of plants other than the angiosperms (Figures 17–23 and 17–24, pages 413 and 414). It appears likely that the differentiation of modern ferns took place in the Upper Cretaceous period, after the formation of diverse forests of angiosperms increased the range of habitats into which ferns could radiate.

The diversity of ferns is greatest in the tropics, where about three-fourths of the species are found. Here, not only are there many species of ferns, but ferns are abundant in many

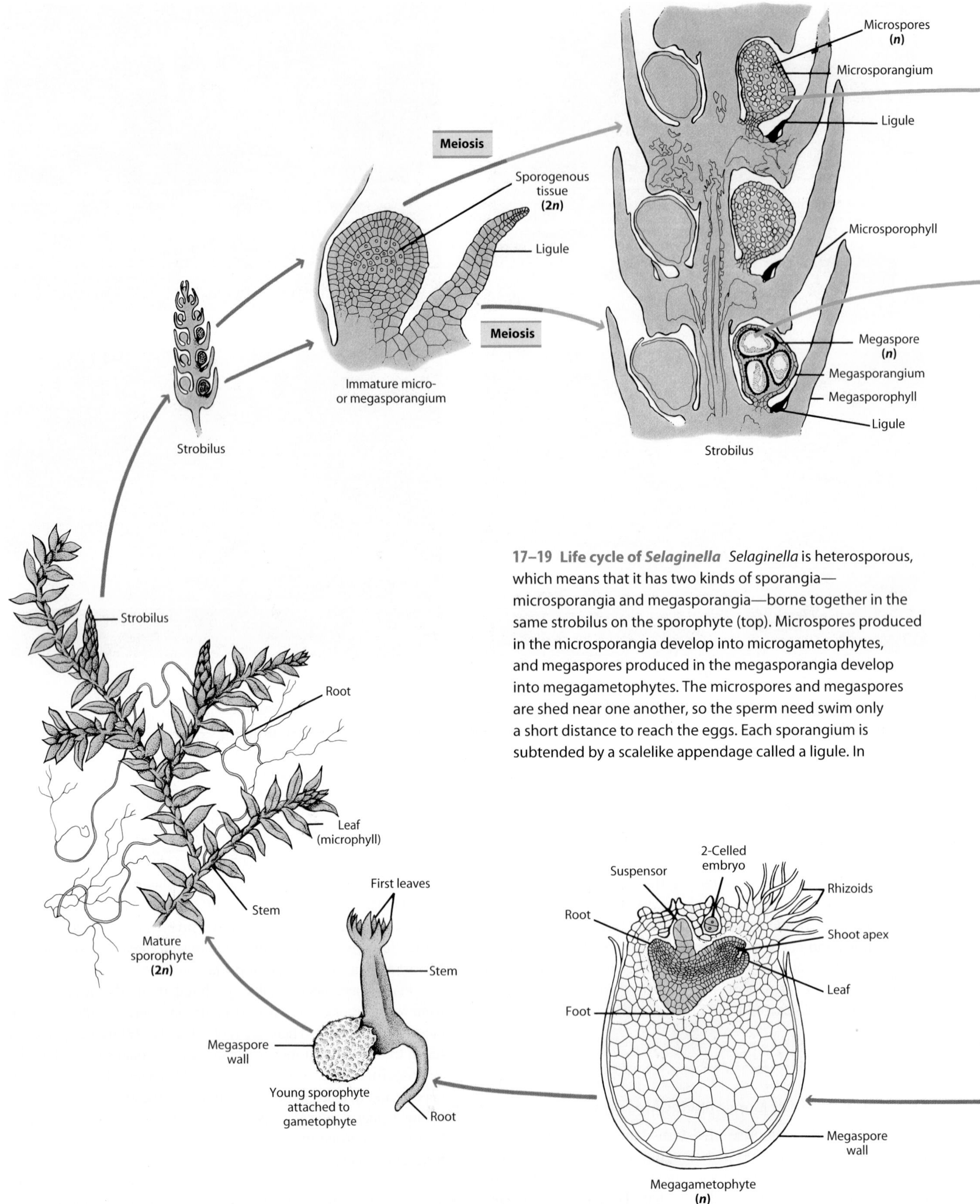

17–19 Life cycle of *Selaginella* *Selaginella* is heterosporous, which means that it has two kinds of sporangia—microsporangia and megasporangia—borne together in the same strobilus on the sporophyte (top). Microspores produced in the microsporangia develop into microgametophytes, and megaspores produced in the megasporangia develop into megagametophytes. The microspores and megaspores are shed near one another, so the sperm need swim only a short distance to reach the eggs. Each sporangium is subtended by a scalelike appendage called a ligule. In

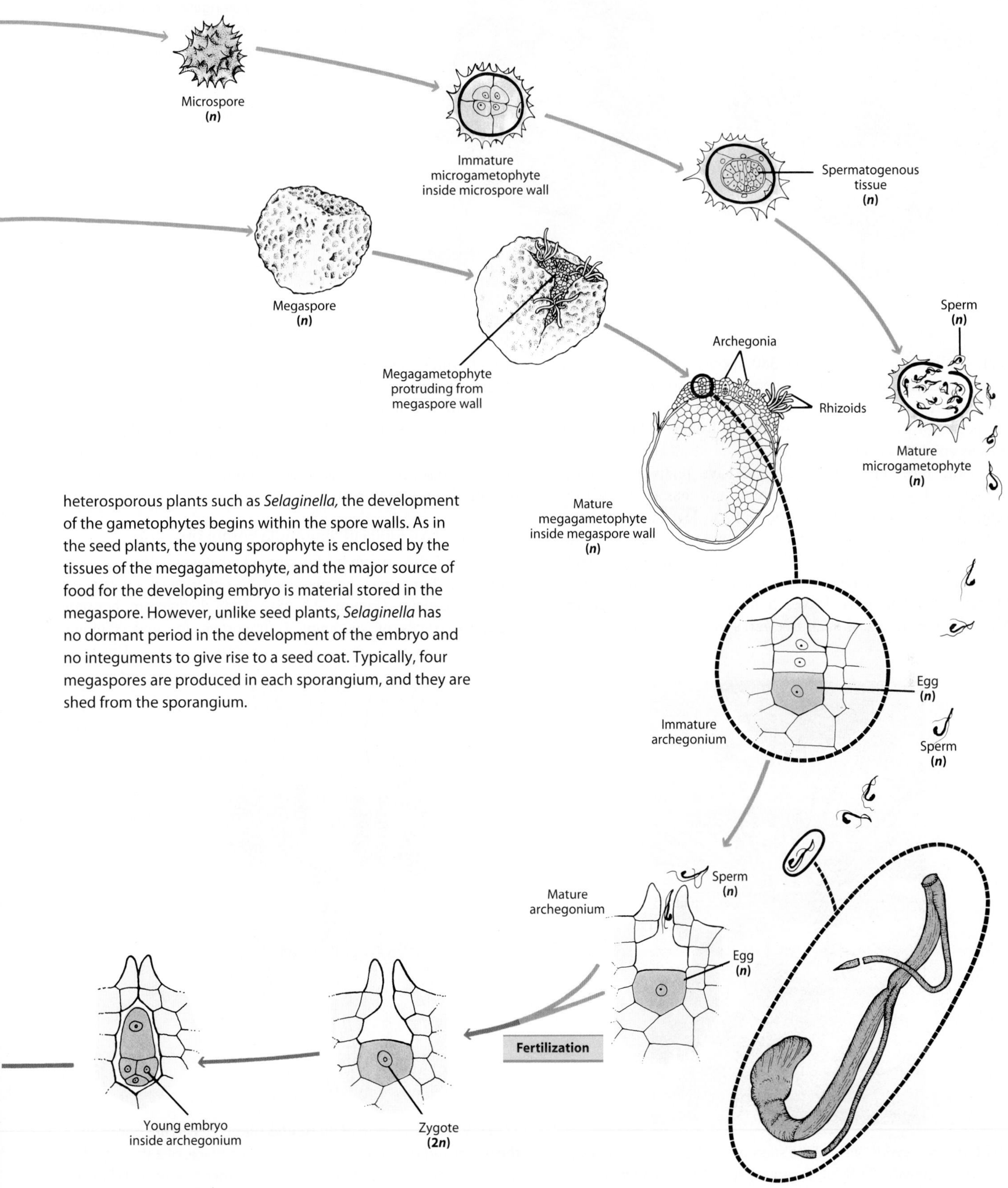

heterosporous plants such as *Selaginella,* the development of the gametophytes begins within the spore walls. As in the seed plants, the young sporophyte is enclosed by the tissues of the megagametophyte, and the major source of food for the developing embryo is material stored in the megaspore. However, unlike seed plants, *Selaginella* has no dormant period in the development of the embryo and no integuments to give rise to a seed coat. Typically, four megaspores are produced in each sporangium, and they are shed from the sporangium.

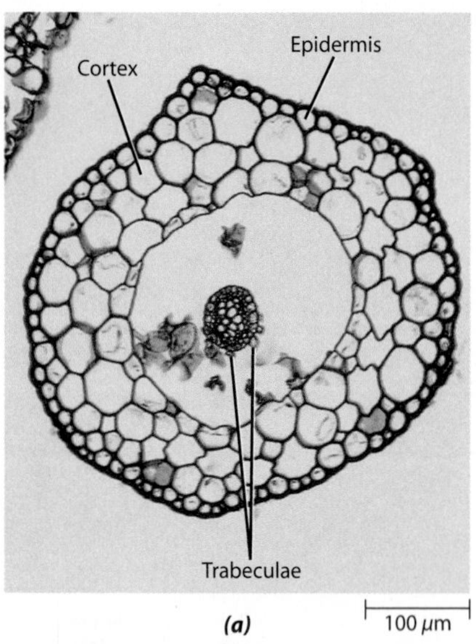

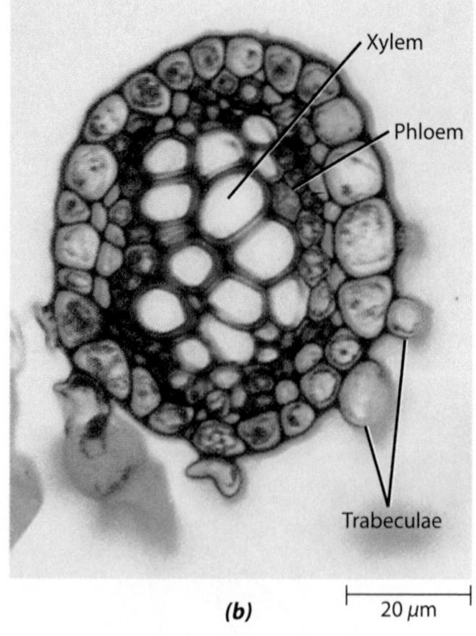

17–20 ***Selaginella* protostele** ***(a)*** Transverse section of stem, showing mature tissues. The protostele is suspended in the middle of the hollow stem by elongated cortical cells (endodermal cells) called trabeculae. Only a portion of each trabecula can be seen here. ***(b)*** Detail of protostele.

plant communities. Only about 380 species of ferns occur in the United States and Canada, whereas about 1000 occur in the small tropical country of Costa Rica. Approximately a third of all species of tropical ferns grow on the trunks or branches of trees as epiphytes (Figure 17–23).

Some ferns are very small and have undivided leaves. *Lygodium* (Figure 17–24), a climbing fern, has leaves with a long, twining rachis (an extension of the leaf stalk, or petiole) that may be up to 30 meters or more in length. Some tree ferns, such as those of the genus *Cyathea* (Figure 17–23b), have been recorded to reach heights of more than 24 meters and to have leaves 5 meters or more in length. Although the trunks of such tree ferns may be 30 centimeters or more thick, their tissues are entirely primary in origin. Most of this thickness is the fibrous root mantle; the true stem is only 4 to 6 centimeters in diameter. The herbaceous genus *Botrychium* (see Figure 17–26a, page 416) has long been cited as the only living fern known to form a vascular cambium, but the presence of a vascular cambium in *Botrychium* has recently been questioned.

There Are Two Kinds of Sporangia within the Ferns

In terms of the structure and method of development of their sporangia, ferns may be classified as either eusporangiate or leptosporangiate (Figure 17–25). The distinction between these two types of sporangia is important for understanding

17–21 ***Isoetes storkii*** View of the sporophyte showing its quill-like leaves (microphylls), fleshy underground stem (corm), and roots. *Isoetes* is the last living representative of the group that included the extinct tree lycophytes of the Carboniferous coal swamps.

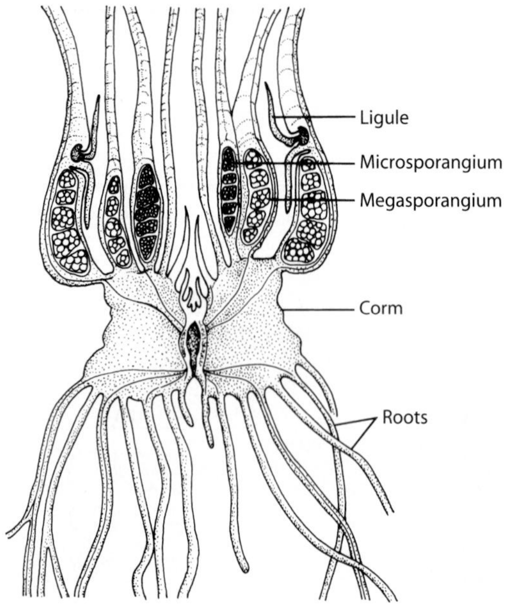

17–22 Vertical section of an *Isoetes* plant Leaves are borne on the upper surface, and roots on the lower surface, of a short, fleshy underground stem, or corm. Some leaves (megasporophylls) bear megasporangia, and others (the microsporophylls), which are located nearer the center of the plant, bear microsporangia.

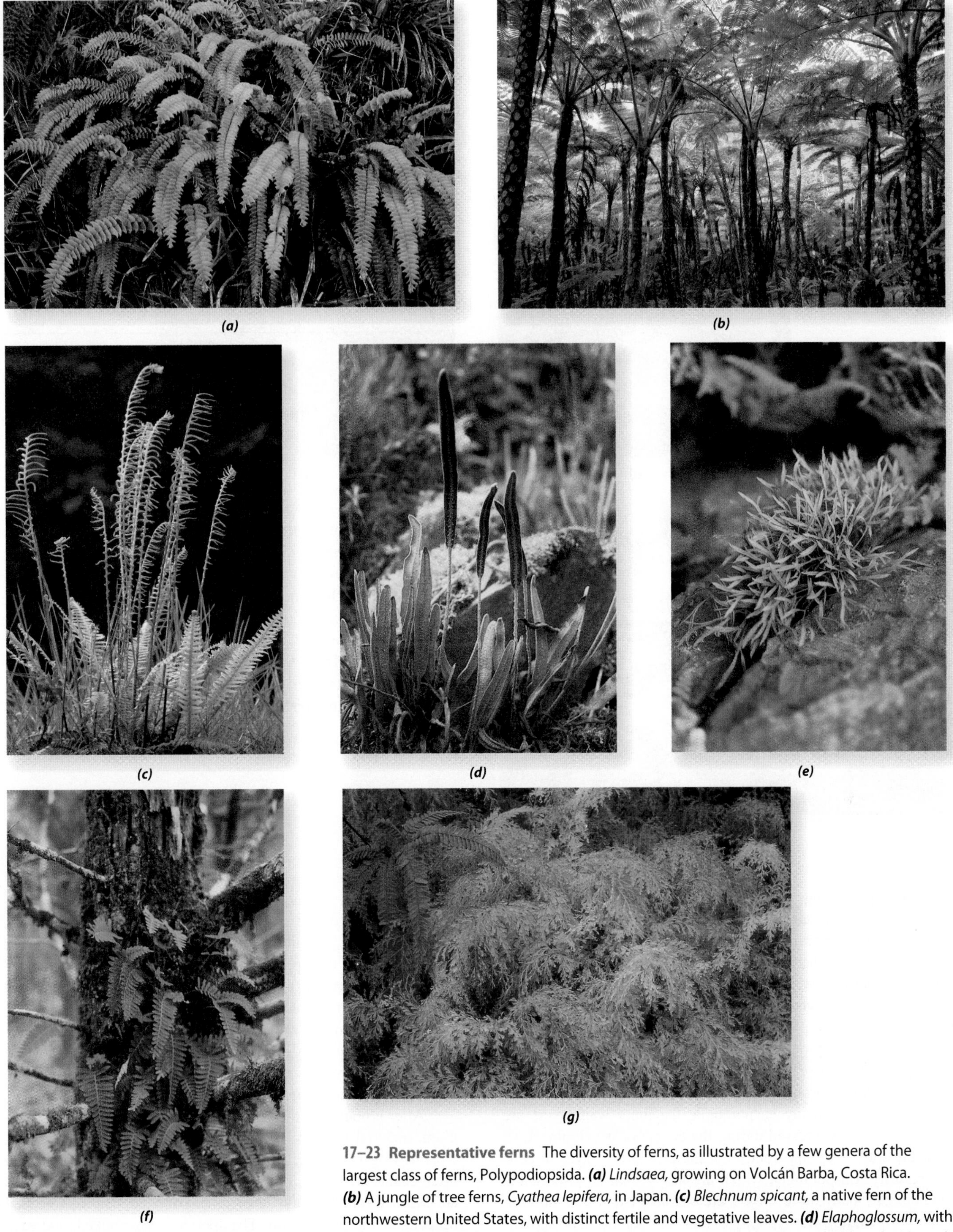

(a) (b) (c) (d) (e) (f) (g)

17–23 Representative ferns The diversity of ferns, as illustrated by a few genera of the largest class of ferns, Polypodiopsida. ***(a)*** *Lindsaea,* growing on Volcán Barba, Costa Rica. ***(b)*** A jungle of tree ferns, *Cyathea lepifera,* in Japan. ***(c)*** *Blechnum spicant,* a native fern of the northwestern United States, with distinct fertile and vegetative leaves. ***(d)*** *Elaphoglossum,* with thick, undivided leaves, near Cuzco, Peru. ***(e)*** *Asplenium septentrionale,* a small fern distributed widely in the Northern Hemisphere, growing on metal-rich soil near a lead-silver mine in Wales. ***(f)*** *Pleopeltis polypodioides,* growing as an epiphyte on a juniper trunk in Arkansas. ***(g)*** A species of *Hymenophyllum,* one of the filmy ferns, so-called because of their delicate leaves. Filmy ferns occur as epiphytes primarily in tropical rainforests or wet temperate regions.

(a)

(b)

(c)

17–24 ***Lygodium microphyllum*** ***(a)*** Growing on cypress trees in southern Florida, the climbing fern seen here is an invasive plant from Australia. The extensive fern growth at the base of the trees allows ground fires, which are normally beneficial, to develop into destructive crown fires that kill trees. ***(b)*** Each fertile leaf can produce 20,000 spores, which are dispersed by wind from the tree canopy, an efficient reproductive strategy that thwarts efforts to control the spread of these ferns. ***(c)*** Nonfertile vegetative leaves.

relationships among vascular plants. In a **eusporangium,** the parent cells, or initials, are located at the surface of the tissue from which the sporangium is produced (Figure 17–25a). These initials divide by the formation of walls parallel to the surface, producing an inner and an outer series of cells. The outer cell layer, by further divisions in both planes, builds up the several-layered wall of the sporangium. The inner layer gives rise to a mass of irregularly oriented cells from which the spore mother cells ultimately arise. The innermost wall layer comprises the **tapetum,** which probably provides nourishment to the developing spores. In many eusporangia, the inner wall layers are stretched and compressed during the course of development, so that the walls may apparently consist of a single layer of cells at maturity. Eusporangia, which are larger than leptosporangia and contain many more spores, are characteristic of all vascular plants—including the lycophytes—except for the leptosporangiate ferns.

In contrast to the multicellular origin of eusporangia, **leptosporangia** arise from a single superficial initial cell, which divides transversely or obliquely (Figure 17–25b). The inner of the two cells produced by this division may either contribute cells that produce a large part of the sporangial stalk or, more commonly, remain inactive and play no role in the further development of the sporangium. By a precise pattern of divisions, the outer cell ultimately gives rise to an elaborate, stalked sporangium, with a spherical capsule having a wall that is one cell thick. Within this wall is the two-layered tapetum characteristic of leptosporangia. The inner mass of the leptosporangium eventually differentiates into spore mother cells, which undergo meiosis to produce four spores each.

After it nourishes the young dividing cells within the sporangium, material from the tapetum is deposited around the spores, creating ridges, spines, and other types of surface features that are often characteristic for individual families and genera. The spores are exposed following the development of a crack in the so-called lip cells of the sporangium. The sporangia are stalked, and each contains a special layer of unevenly thick-walled cells called an **annulus.** As the sporangium dries out, contraction of the annulus causes tearing in the middle of the capsule. The sudden explosion and snapping back of the annulus to its original position then

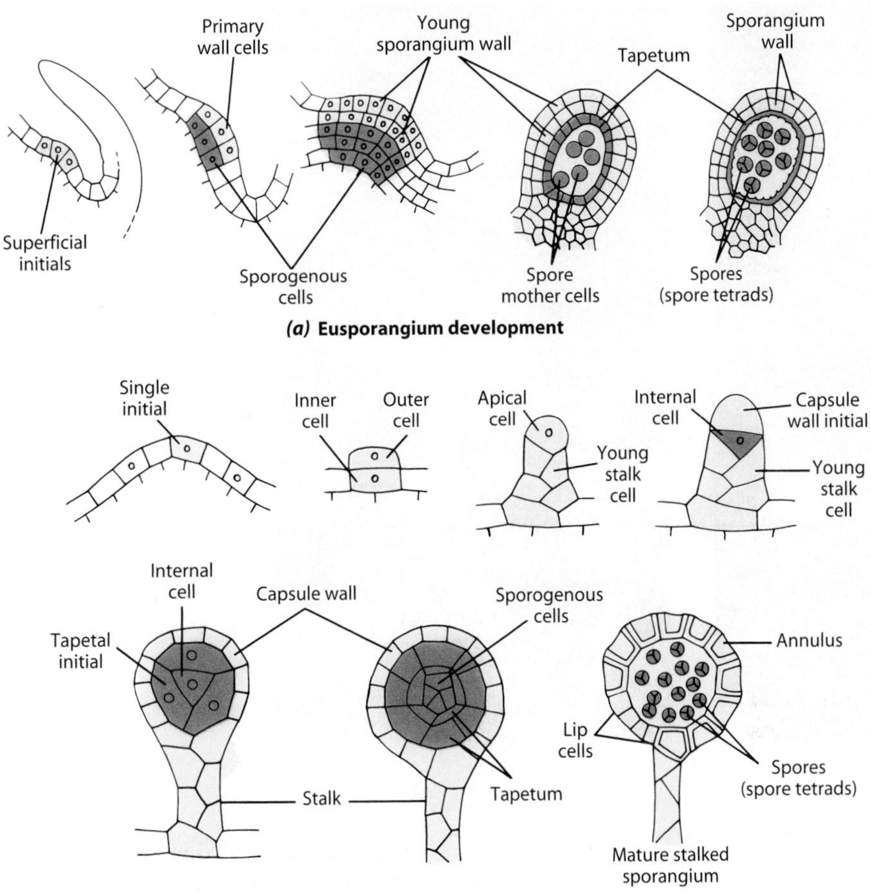

17–25 Eusporangia and leptosporangia Development and structure of the two principal types of fern sporangia. ***(a)*** The eusporangium originates from a series of superficial parent cells, or initials. Each eusporangium develops a wall two or more layers thick (although at maturity the inner wall layers may be crushed) and a high number of spores. ***(b)*** The leptosporangium originates from a single initial cell, which first produces a stalk and then a capsule. Each leptosporangium gives rise to a relatively small number of spores.

results in a catapult-like discharge of the spores. In eusporangia, the stalks are more massive and, though there may be preformed lines of dehiscence, there is no annulus and no catapult-like discharge of spores.

Most living ferns are homosporous, producing only one kind of spore. Heterospory, with both microspores and megaspores produced, is restricted to the water ferns (see Figure 17–36), which are further discussed below. A few extinct ferns also were heterosporous. We consider here some examples of each of the four major lineages of monilophytes.

The Classes Psilotopsida and Marattiopsida Are Eusporangiate Ferns

The class Psilotopsida consists of two orders of homosporous ferns, Ophioglossales and Psilotales. Of the four genera of Ophioglossales, *Botrychium,* the grape ferns (Figure 17–26a), and *Ophioglossum,* the adder's tongues (Figure 17–26b), are widespread in the north temperate region. In both of these genera, a single leaf typically is produced each year from the stem. Each leaf consists of two parts: (1) a vegetative portion, or blade, which is deeply dissected in *Botrychium* and undivided in most species of *Ophioglossum,* and (2) a fertile segment. In *Botrychium,* the fertile segment is dissected in the same way as the vegetative portion and bears two rows of eusporangia on the outermost segments. In *Ophioglossum,* the fertile portion is undivided and bears two rows of sunken eusporangia.

The Psilotales include two living genera, *Psilotum* and *Tmesipteris. Psilotum,* the whisk fern, is tropical and subtropical in distribution. In the United States, it occurs in Alabama, Arizona, Florida, Hawaii, Louisiana, North Carolina, and Texas, as well as in Puerto Rico, and it is a common greenhouse weed. *Tmesipteris* is restricted in distribution to Australia, New Caledonia, New Zealand, and other regions of the South Pacific. Both genera are very simple plants. Their simple structure—tiny leaves and absence of roots—appears to be a derived condition.

The sporophyte of *Psilotum* consists of a dichotomously branching aerial portion with small scalelike "foliar" structures and a branching underground portion, or a system of rhizomes with many rhizoids. A symbiotic fungus—an endomycorrhizal glomeromycete (pages 290 and 291)—is present in the outer cortical cells of the rhizomes. The sporangia of *Psilotum* are

(a)

(b)

17–26 Ophioglossales Representatives of the two genera of Ophioglossales that occur in North America. ***(a)*** *Botrychium dissectum.* In the genus *Botrychium,* the lower, vegetative portion of the leaf is divided. ***(b)*** In *Ophioglossum vulgatum,* the lower portion of the leaf is undivided. In both genera, the erect, fertile, upper part of the leaf is sharply distinct from the vegetative portion.

generally aggregated in groups of three on the ends of short, lateral branches (Figure 17–27).

Tmesipteris grows as an epiphyte on tree ferns and other plants (Figure 17–28) and in rock crevices. The leaves of *Tmesipteris,* which are larger than the scalelike structures of *Psilotum,* are supplied with a single, unbranched vascular bundle. In other respects *Tmesipteris* is essentially similar to *Psilotum.*

The gametophytes of *Botrychium, Ophioglossum,* and *Psilotum* (Figure 17–29) are subterranean, tuberous, elongated structures with numerous rhizoids; they contain symbiotic fungi. Some *Psilotum* gametophytes contain vascular tissue. The gametophytes are bisexual, bearing both antheridia and

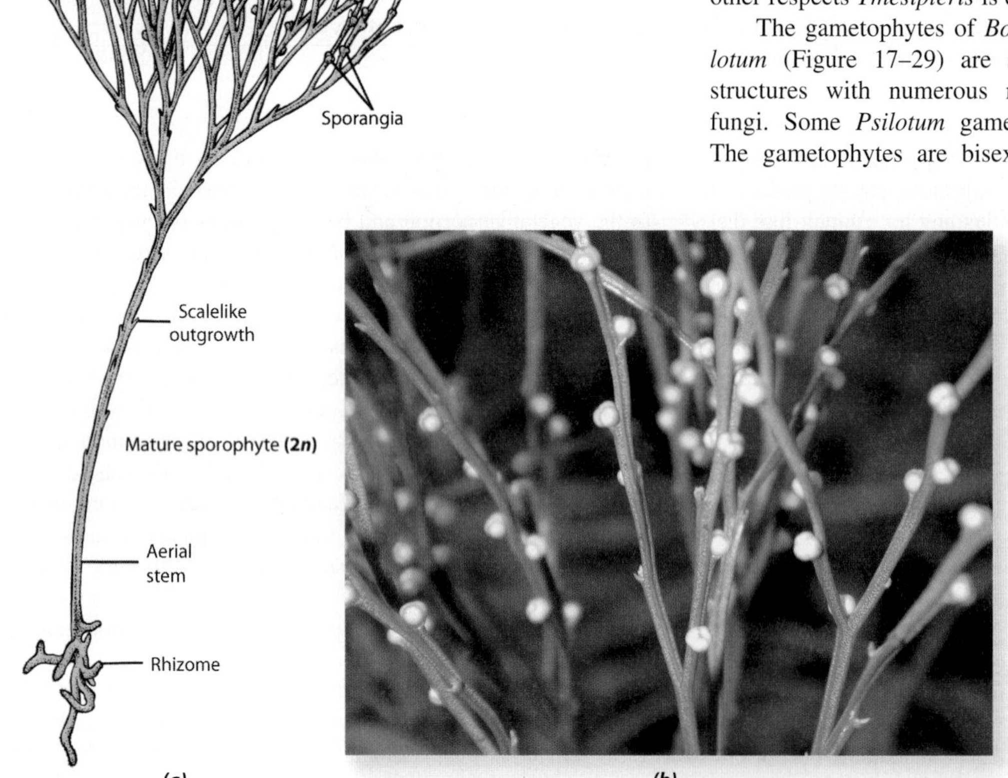

17–27 *Psilotum nudum* sporophyte ***(a)*** In *Psilotum,* the sporophyte consists of a dichotomously branching aerial portion, with small scalelike outgrowths, and a system of rhizomes. Sporangia are borne in united groups of three in the axils of some scalelike outgrowths. ***(b)*** The dichotomously branching aerial portion of the sporophyte with numerous yellow sporangia.

(a) (b)

17–28 *Tmesipteris* ***(a)*** *Tmesipteris parva* is seen here growing on the trunk of the tree fern *Cyathea australis* in New South Wales, Australia. ***(b)*** *Tmesipteris lanceolata,* in New Caledonia, an island of the southwest Pacific. The leaves contain a single unbranched vascular bundle.

archegonia. The sperm are multiflagellated and require water to swim to the eggs.

In the nature of their gametophytes, the structure of their leaves, and several other anatomical details, the Ophioglossales are sharply distinct from other living ferns and are clearly an early diverging and distinct group. Unfortunately, the group has no well-established fossil record before about 50 million years ago. Fossils of *Psilotum* and *Tmesipteris* are lacking entirely. One member of the Ophioglossales, *Ophioglossum reticulatum,* has the highest chromosome number known in any living organism, with a diploid complement of about 1260 chromosomes.

The only other group of eusporangiate ferns, the tropical Marattiopsida, is an ancient group with a fossil record that extends back to the Carboniferous period. The members of this order resemble more familiar groups of ferns more closely than they resemble Ophioglossales. *Psaronius,* an extinct tree fern, was a member of this order (see the essay on pages 400 and 401). The six living genera of Marattiales include about 200 species.

Most Polypodiopsida Are Homosporous Leptosporangiate Ferns

Nearly all familiar ferns are members of the class Polypodiopsida, with at least 10,500 species. About 35 families and 320 genera are recognized in the class. Polypodiopsida differ from Psilotopsida and Marattiopsida in being leptosporangiate; they differ from the water ferns, which we discuss next, in being homosporous. All ferns other than Psilotopsida and Marattiopsida, in fact, are leptosporangiate, and very few have the subterranean gametophytes with symbiotic fungi that are characteristic of the Psilotopsida and Marattiopsida. Clearly, leptosporangia and the other distinctive features of most ferns are specialized characters, because they occur nowhere else among the vascular plants, including Psilotopsida and Marattiopsida, which share more features in common with other groups of ancient plants.

Most garden and woodland ferns of temperate regions have siphonostelic rhizomes (Figure 17–30) that produce new sets of leaves each year. The fern embryo produces a true root, but this soon withers, and the rest of the roots arise from the rhizomes near the bases of the leaves. The leaves, or **fronds,** are megaphylls, and they represent the most conspicuous part of the sporophyte. Their high surface-to-volume ratio allows them to capture sunlight much more effectively than the microphylls of the lycophytes. The ferns are the only seedless vascular plants to possess well-developed megaphylls. Commonly, the fronds are compound; that is, the lamina is divided into leaflets, or **pinnae,** which are attached to the **rachis,** an extension of the leaf stalk, or petiole. In nearly all ferns, the young leaves are coiled (circinate); they are commonly referred to as "fiddleheads" (Figure 17–31). This type of leaf development is known as **circinate vernation.** Uncoiling of the fiddlehead results from more rapid growth on the inner than on the outer surface of the leaf early in development and is mediated by the hormone auxin (see Chapter 27), produced by the young pinnae on the inner side of the fiddlehead. This type of vernation protects the delicate embryonic leaf tip during development. Both fiddleheads and rhizomes are usually clothed with either hairs or scales, both of which are epidermal outgrowths; the characteristics of these structures are important in fern classification.

The sporangia of Polypodiopsida occur on the margins or lower surfaces of the leaves, on specially modified leaves, or on separate stalks. The sporangia commonly occur in clusters called **sori** (singular: sorus) (Figures 17–32 and 17–33, pages 420 and 421), which may appear as yellow, orange, brownish, or blackish lines, dots, or broad patches on the lower surface of a frond. In many genera, the young sori are covered by specialized outgrowths of the leaf, the **indusia** (singular: indusium; Figures

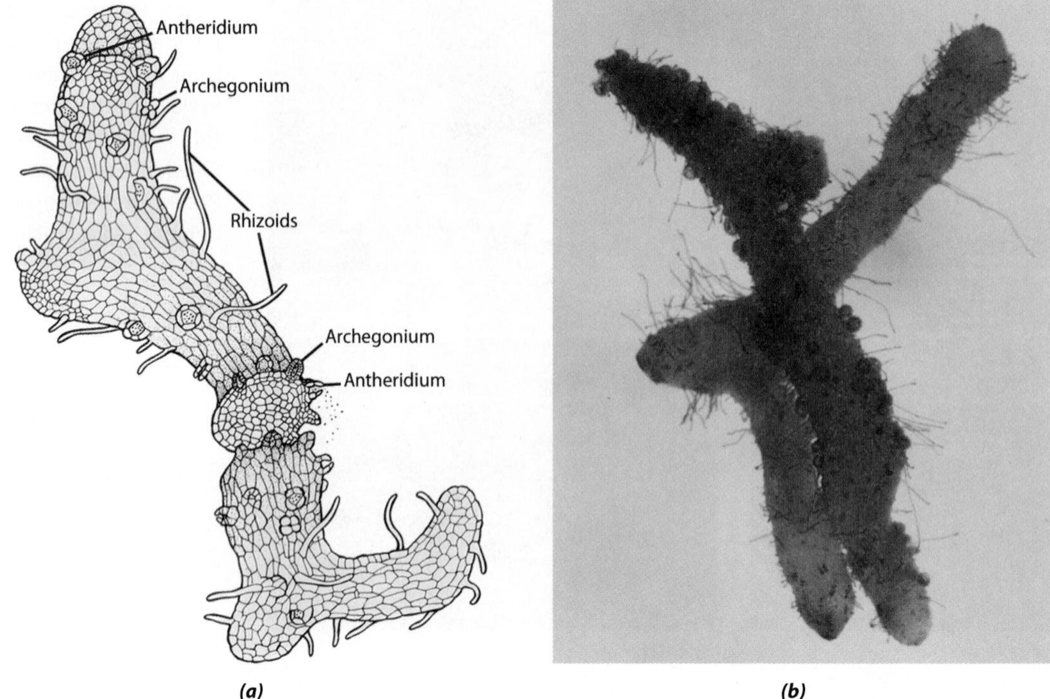

17–29 *Psilotum nudum* gametophyte ***(a)*** The *Psilotum* gametophyte is bisexual, bearing both antheridia and archegonia. ***(b)*** The gametophytes, which are subterranean, resemble portions of the rhizome.

17–32c and 17–33), which may shrivel when the sporangia are ripe and ready to shed their spores. The shape of the sorus, its position, and the presence or absence of an indusium are important characteristics in the taxonomy of the Polypodiopsida.

The spores of ferns in the class Polypodiopsida give rise to free-living, potentially bisexual gametophytes, which are often found in moist places, such as the sides of pots in greenhouses. The gametophyte typically develops rapidly into a flat, heart-shaped, usually membranous structure, the **prothallus,** that has numerous rhizoids on its central lower surface. Both antheridia and archegonia develop on the ventral (lower) surface of the prothallus. The antheridia more typically occur

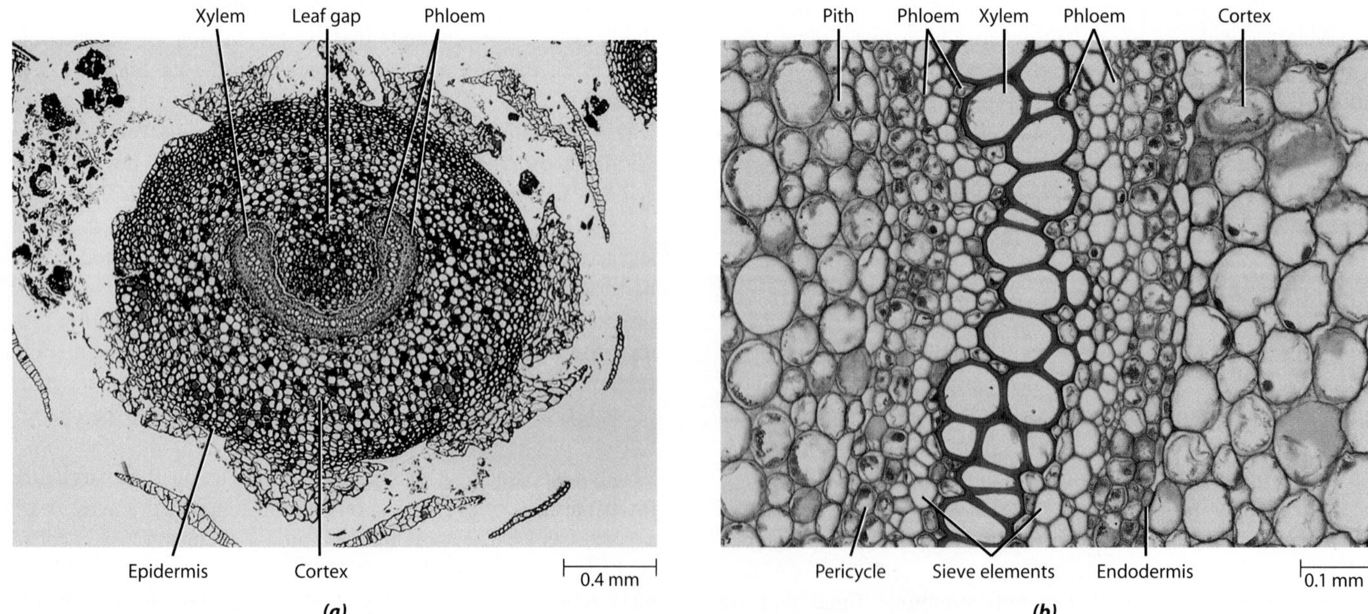

17–30 Anatomy of fern rhizomes ***(a)*** Transverse section of a rhizome of *Adiantum,* or maidenhair fern, showing the siphonostele. Note the wide leaf gap. ***(b)*** Transverse section of part of the vascular region of a rhizome of the tree fern *Dicksonia.* The phloem is composed mainly of sieve elements; the xylem is composed entirely of tracheids.

17–31 Fiddleheads Ostrich fern *(Matteuccia struthiopteris)* fiddleheads are gathered commercially in upper New England and New Brunswick; they are marketed fresh, canned, and frozen. These fiddleheads, which taste somewhat like crisp asparagus, should be picked when they are less than 15 centimeters long. Although ostrich fern fiddleheads are safe to eat, the fiddleheads of many ferns are considered toxic.

among the rhizoids, while the archegonia are usually formed near the notch, an indentation at the anterior end of the gametophyte. The order of appearance of these gametangia is controlled genetically and can be mediated by special chemicals produced by the gametophytes. The timing of the maturation of the antheridia and archegonia can determine whether self-fertilization or cross-fertilization takes place (page 397). Water is required for the multiflagellated sperm to swim to the eggs.

Early in its development, the embryo, or young sporophyte, receives nutrients from the gametophyte through a foot. Development is rapid, and the sporophyte soon becomes an independent plant, at which time the gametophyte commonly disintegrates.

Typically, the sporophyte is the perennial stage in ferns, and the small, thalloid gametophyte is short-lived. Remarkably, the strap-shaped or filamentous gametophytes of some species of ferns, including three genera with six tropical species found in the southern Appalachians, persist indefinitely without ever producing sporophytes (Figure 17–34). These species have never yet been induced to produce sporophytes in the laboratory. They reproduce by vegetative outgrowths called gemmae that fall off and are blown away to found new colonies. These ferns appear to be distinct from the other, sporophyte-producing species of their respective genera, as judged by differences in their enzymes, and probably should be treated as distinct species. Such situations are common in mosses and are being discovered in ferns much more widely than was previously expected. Populations of perennial, free-living gametophytes of *Trichomanes speciosum,* discovered in the Elbsandsteingebirge (a mountain range shared by Germany and the Czech Republic), are estimated to be over 1000 years old. The possibility exists that they are relics of former populations that included both sporophytes and gametophytes. The extinction of the sporophytes possibly occurred as a result of climatic changes during the glacial intervals of the last 2 million years.

The life cycle of one of the Polypodiopsida is shown in Figure 17–35 (pages 422 and 423).

The Water Ferns, of the Class Polypodiopsida, Are Heterosporous Leptosporangiate Ferns

The water ferns consist of one order, the Salviniales, and two families, Marsileaceae and Salviniaceae. Although they are structurally very different from each other, recent evidence from molecular analyses indicates that the two families were derived from a common terrestrial ancestor. All water ferns are heterosporous, and they are the only living heterosporous ferns. There are five genera of water ferns. The slender rhizomes of the three genera of Marsileaceae, including *Marsilea* (which has about 50 to 70 species), grow in mud, on damp soil, or often with the leaves floating on the surface of water (Figure 17–36a, page 424). The leaves of *Marsilea* resemble those of a four-leaf clover. Drought-resistant, bean-shaped reproductive structures called **sporocarps,** which may remain viable even after 100 years of dry storage, germinate when placed in water to produce chains of sori, each bearing series of megasporangia and microsporangia (Figure 17–36b).

The two genera of Salviniaceae, *Azolla* (see Figure 29–12) and *Salvinia* (Figure 17–36c), are small plants that float on the surface of water. Both genera produce their sporangia in sporocarps that are quite different in structure from those of Marsileaceae. In *Azolla,* the tiny, crowded, bilobed leaves are borne on slender stems. A pouch that forms on the upper, photosynthetic lobe of each leaf is inhabited by colonies of the cyanobacterium *Anabaena azollae.* The lower, smaller lobe of each leaf is often nearly colorless. Because of the nitrogen-fixing abilities of *Anabaena, Azolla* has been used to maintain the fertility of rice paddies and of certain natural ecosystems. The undivided leaves of *Salvinia,* which are up to 2 centimeters long, are borne in whorls of three on the floating rhizome. One of the three leaves hangs down below the surface of the water and is highly dissected, resembling a mass of whitish roots. These "roots," however, bear sporangia, which reveals that they are actually leaves. The two upper leaves, which float on the water, are covered by hairs that protect their surface from getting wet, and the leaves float back to the surface if they are temporarily submerged.

The Equisetopsida Have Jointed Stems and Are Eusporangiate

The equisetophytes, like the lycophytes, extend back to the Devonian period. They reached their maximum abundance and diversity later in the Paleozoic era, about 300 million years ago. During the Late Devonian and Carboniferous periods, they were represented by the calamites (see the essay on pages 400 and 401), a group of trees that reached 18 meters or more in height,

17–32 Sori Clusters of sporangia, or sori, are found on the undersides or margins of the leaves of ferns. ***(a)*** In *Polypodium virginianum* and other ferns of this genus, the sori are bare. ***(b)*** In the bracken fern, *Pteridium aquilinum,* shown here, as well as in the maidenhair ferns *(Adiantum),* the sori are located along the margins of the leaf blades, which are rolled back over them. ***(c)*** In this *Polystichum* species, the indusia, which completely cover the sori, eventually contract, exposing the sporangia as they near maturity. ***(d)*** In the cutleaf grapefern, *Botrychium dissectum,* the sori are enfolded by globular lobes of the pinna (leaflet) and therefore not visible. After overwintering, the lobes separate slightly, and the spores are released early in the spring, often over the snow.

with a trunk that could be more than 45 centimeters thick. Today the equisetophytes are represented by a single herbaceous genus, *Equisetum* (Figure 17–37, page 424), which consists of 15 species. Because *Equisetum* is essentially identical to *Equisetites,* a plant that appeared about 300 million years ago, in the Carboniferous period, *Equisetum* may be the oldest surviving genus of plants on Earth. The position of the equisetophytes within the Monilophyta remains uncertain, although they appear to have a connection with the Marattiopsida and Polypodiopsida (leptosporangiate ferns) as a clade (Figure 17–14).

The species of *Equisetum,* known as the horsetails, are widespread in moist or damp places, by streams, and along the edge of woods. Horsetails are easily recognized because of their conspicuously jointed stems and rough texture. The small, scalelike leaves are whorled at the nodes. When present, branches arise laterally at the nodes and alternate with the leaves. The internodes (the portions of the stems between successive nodes) are ribbed, and the ribs are tough and strengthened with siliceous deposits in the epidermal cells. Horsetails have been used to scour pots and pans, particularly in colonial

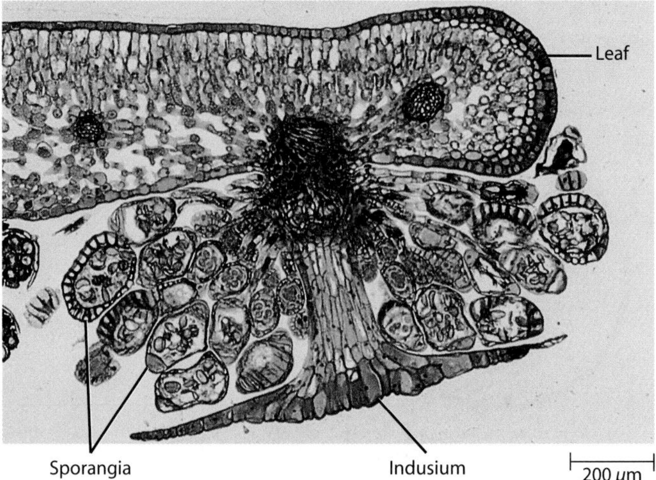

17–33 Sorus with an indusium Transverse section of a leaf of *Cyrtomium falcatum,* a homosporous fern, showing a sorus on the lower surface. The sporangia are in different stages of development and are protected by an umbrella-like indusium.

and frontier times, and have thus earned the name "scouring rushes." The roots originate at the nodes of the rhizomes, which are important in vegetative propagation.

The aerial stems of *Equisetum* arise from branching underground rhizomes, and, although the plants may die back during unfavorable seasons, the rhizomes are perennial. The aerial stem is complex anatomically (Figure 17–38, page 425). At maturity, its internodes contain a hollow pith surrounded by a ring of smaller canals called **carinal canals.** Each of these smaller canals is associated with a strand of xylem and phloem.

Equisetum is homosporous. Sporangia are borne in groups of 5 to 10 along the margins of small, umbrella-like structures known as **sporangiophores** (sporangia-bearing branches), which are clustered into strobili (cones) at the apex of the stem (Figure 17–37a; see also 17–41). The fertile stems of some species do not contain much chlorophyll. In these species, the fertile stems are sharply distinct from the vegetative stems, often appearing before the latter early in the spring (Figure 17–37). In other species of *Equisetum,* the strobili are borne at the tips of otherwise vegetative stems (see Figure 12–16d). When the spores are mature, the sporangia contract and split along their inner surface, releasing numerous spores. **Elaters**—thickened bands that arise from the outer layer of the spore wall—coil when moist and uncoil when dry, thus playing a role in spore

(a)

(b)

(c)

17–34 Asexually reproducing ferns In some ferns growing in widely scattered parts of the world, the gametophytes reproduce asexually and persist; sporophytes are not formed, either in the field or in the laboratory. These photographs show two of the three fern genera known to exhibit this habit in the eastern United States. ***(a)*** Typical habitat of persistent gametophytes of *Vittaria* and *Trichomanes,* Ash Cave, Hocking County, Ohio. ***(b)*** *Trichomanes* gametophytes, Lancaster County, Pennsylvania. ***(c)*** *Vittaria* gametophytes, Franklin County, Alabama.

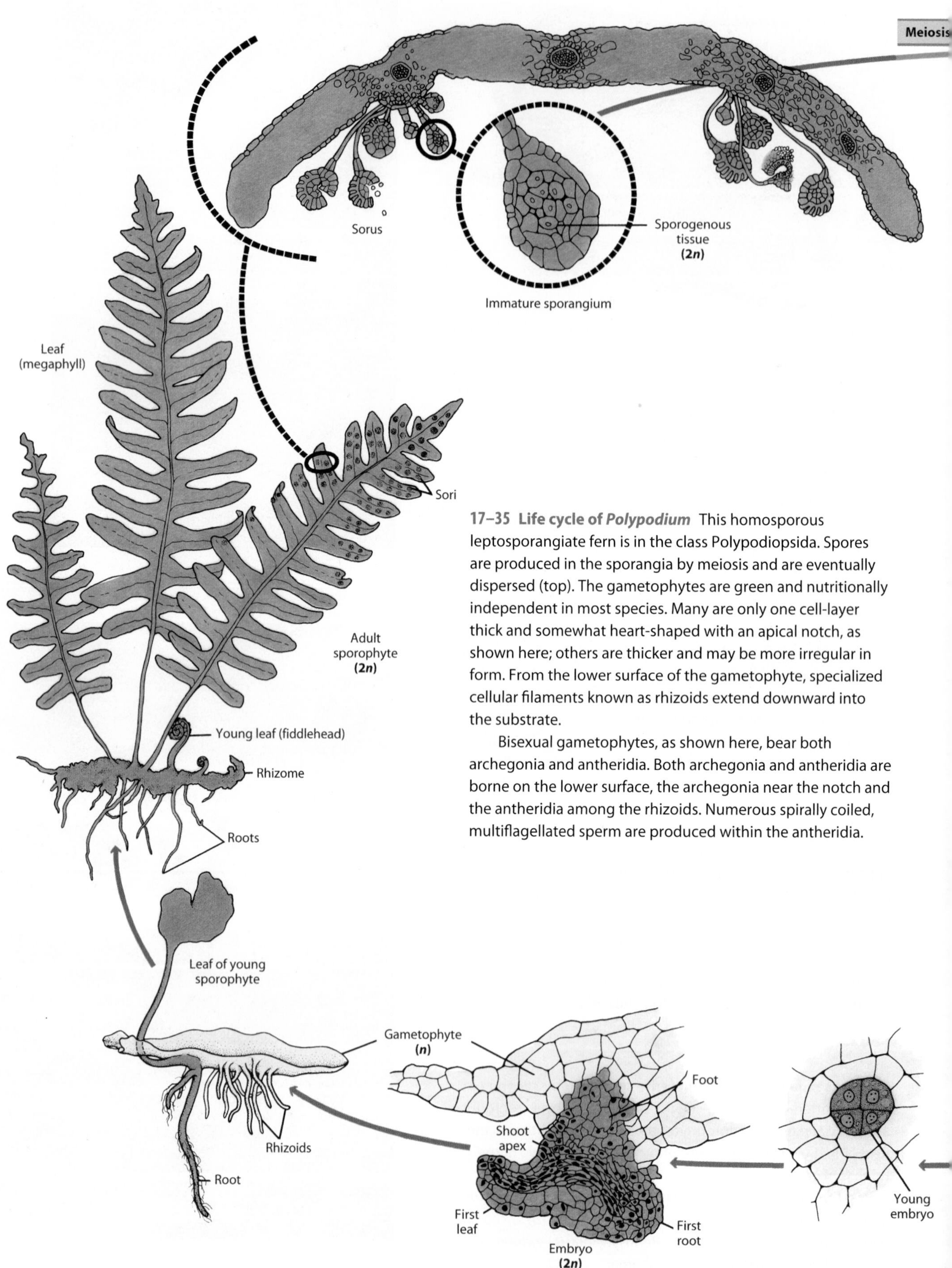

17–35 Life cycle of *Polypodium* This homosporous leptosporangiate fern is in the class Polypodiopsida. Spores are produced in the sporangia by meiosis and are eventually dispersed (top). The gametophytes are green and nutritionally independent in most species. Many are only one cell-layer thick and somewhat heart-shaped with an apical notch, as shown here; others are thicker and may be more irregular in form. From the lower surface of the gametophyte, specialized cellular filaments known as rhizoids extend downward into the substrate.

Bisexual gametophytes, as shown here, bear both archegonia and antheridia. Both archegonia and antheridia are borne on the lower surface, the archegonia near the notch and the antheridia among the rhizoids. Numerous spirally coiled, multiflagellated sperm are produced within the antheridia.

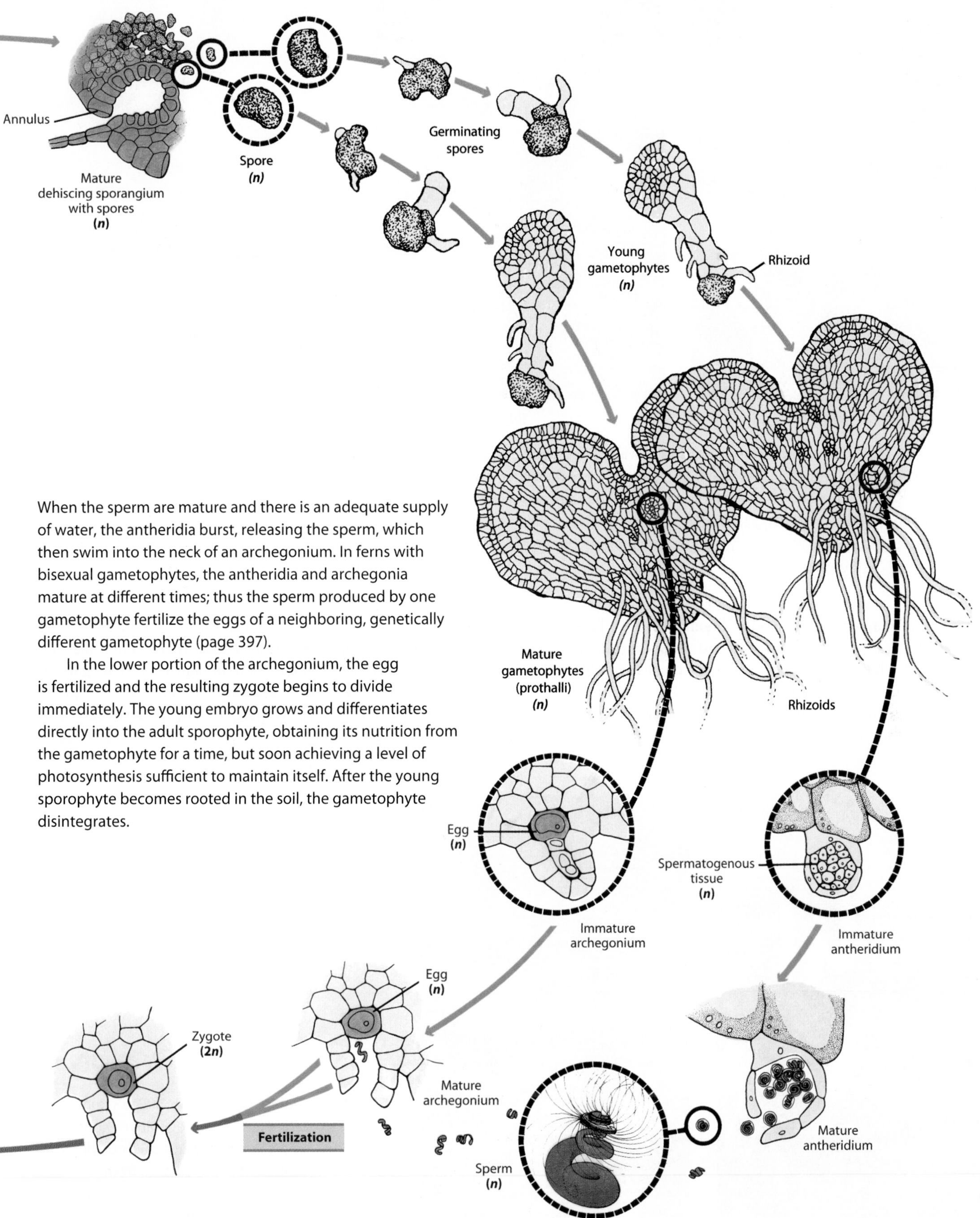

When the sperm are mature and there is an adequate supply of water, the antheridia burst, releasing the sperm, which then swim into the neck of an archegonium. In ferns with bisexual gametophytes, the antheridia and archegonia mature at different times; thus the sperm produced by one gametophyte fertilize the eggs of a neighboring, genetically different gametophyte (page 397).

In the lower portion of the archegonium, the egg is fertilized and the resulting zygote begins to divide immediately. The young embryo grows and differentiates directly into the adult sporophyte, obtaining its nutrition from the gametophyte for a time, but soon achieving a level of photosynthesis sufficient to maintain itself. After the young sporophyte becomes rooted in the soil, the gametophyte disintegrates.

(a)

(b)

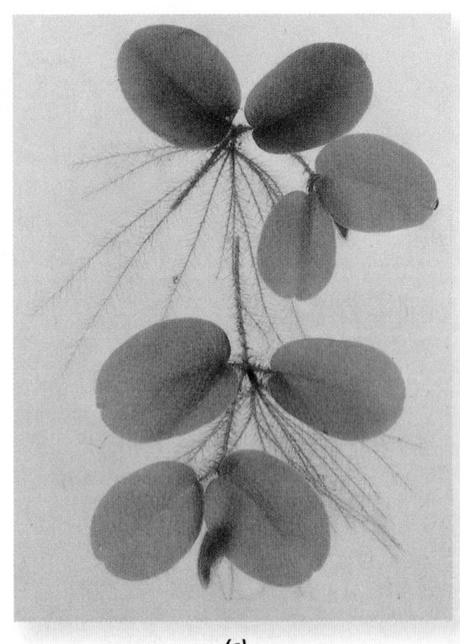

(c)

17–36 Water ferns The two very distinct orders of water ferns are the only living heterosporous ferns. ***(a)*** *Marsilea polycarpa,* with its leaves floating on the surface of the water, photographed in Venezuela. ***(b)*** *Marsilea,* showing the germination of a sporocarp, with chains of sori. Each sorus contains a series of megasporangia and microsporangia. ***(c)*** *Salvinia,* with two floating leaves and one feathery dissected submerged leaf at each node. These two genera are representatives of the order Salviniales.

dispersal (Figure 17–39; see also Figure 17–41). These are quite distinct from the elaters that aid in spore dispersal in *Marchantia.* There, the elaters are elongated cells with helically arranged wall thickenings (see Figure 16–12).

The gametophytes of *Equisetum* are green and free-living, and they range in diameter from a few millimeters to 1 centimeter or even 3 to 3.5 centimeters in some species. Gametophytes become established mainly on mud that has recently been flooded and is rich in nutrients. The gametophytes, which reach sexual maturity in three to five weeks, are either bisexual or male (Figure 17–40). In bisexual gametophytes, the archegonia develop before the antheridia, a

(a)

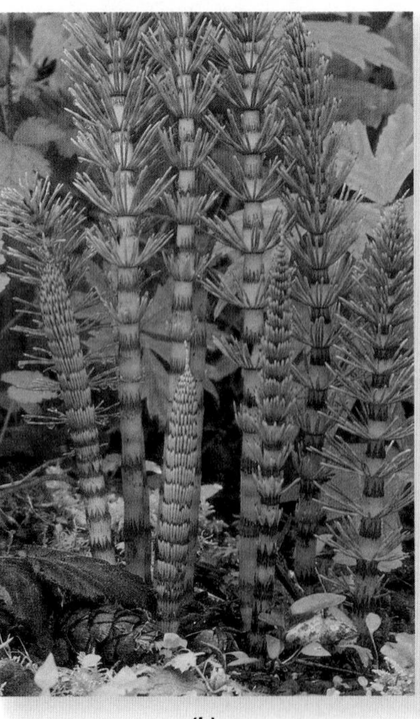

(b)

17–37 *Equisetum* ***(a)*** A species of *Equisetum* that has separate fertile and vegetative shoots. The fertile shoots essentially lack chlorophyll and are very different in appearance from the vegetative shoots. Each fertile shoot has a terminal strobilus. Notice the whorls of scale-like leaves at each node. ***(b)*** Branching vegetative shoots of *Equisetum arvense.*

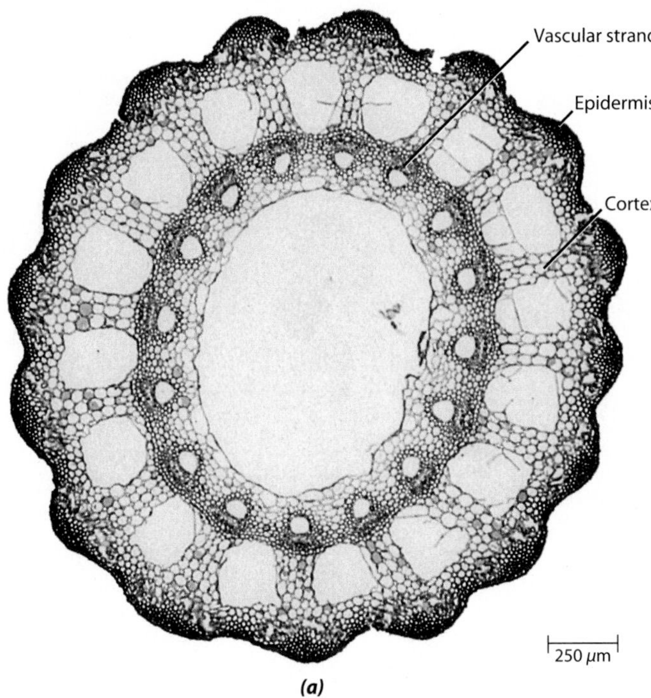

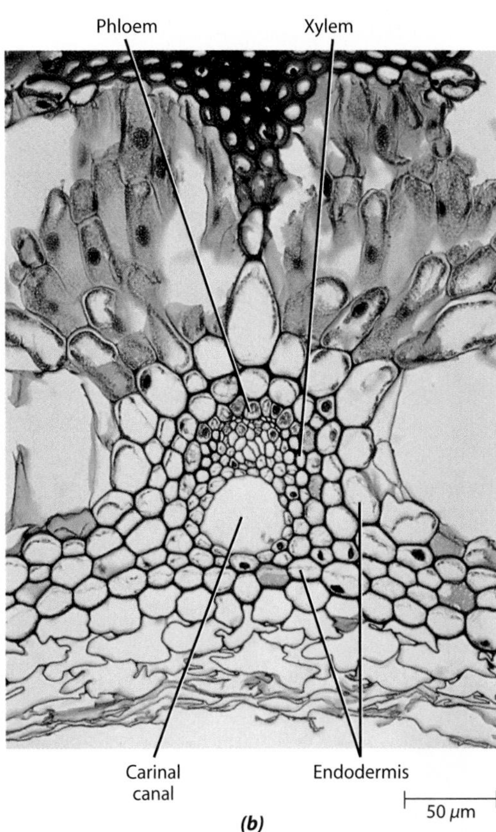

17–38 Stem anatomy of *Equisetum* ***(a)*** Transverse section of an *Equisetum* stem, showing mature tissues. ***(b)*** Detail of a vascular strand, showing xylem and phloem.

17–39 *Equisetum* spores Spores of the horsetail *Equisetum arvense,* as seen in a scanning electron micrograph. Shown here are moist spores wrapped tightly by thickened bands, known as elaters, which are attached to the spore walls. As the spores dry, the elaters uncoil, helping to disperse the spores from the sporangium.

developmental pattern that increases the probability of cross-fertilization. The sperm are multiflagellated and require water to swim to the eggs. The eggs of several archegonia on a single gametophyte may be fertilized and develop into embryos, or young sporophytes.

The life cycle of Equisetum is illustrated in Figure 17–41 (pages 426 and 427).

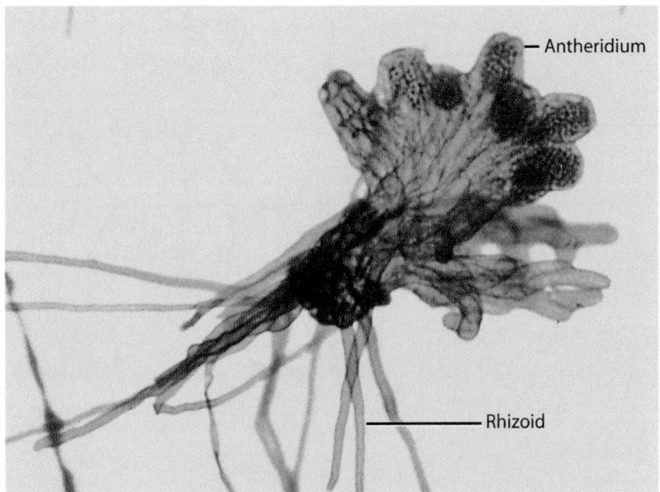

17–40 *Equisetum* gametophyte Bisexual gametophyte of *Equisetum,* showing antheridia. The archegonia are deeply embedded within the gametophyte and are not discernible here. Rhizoids can be seen extending from the lower surface of the gametophyte.

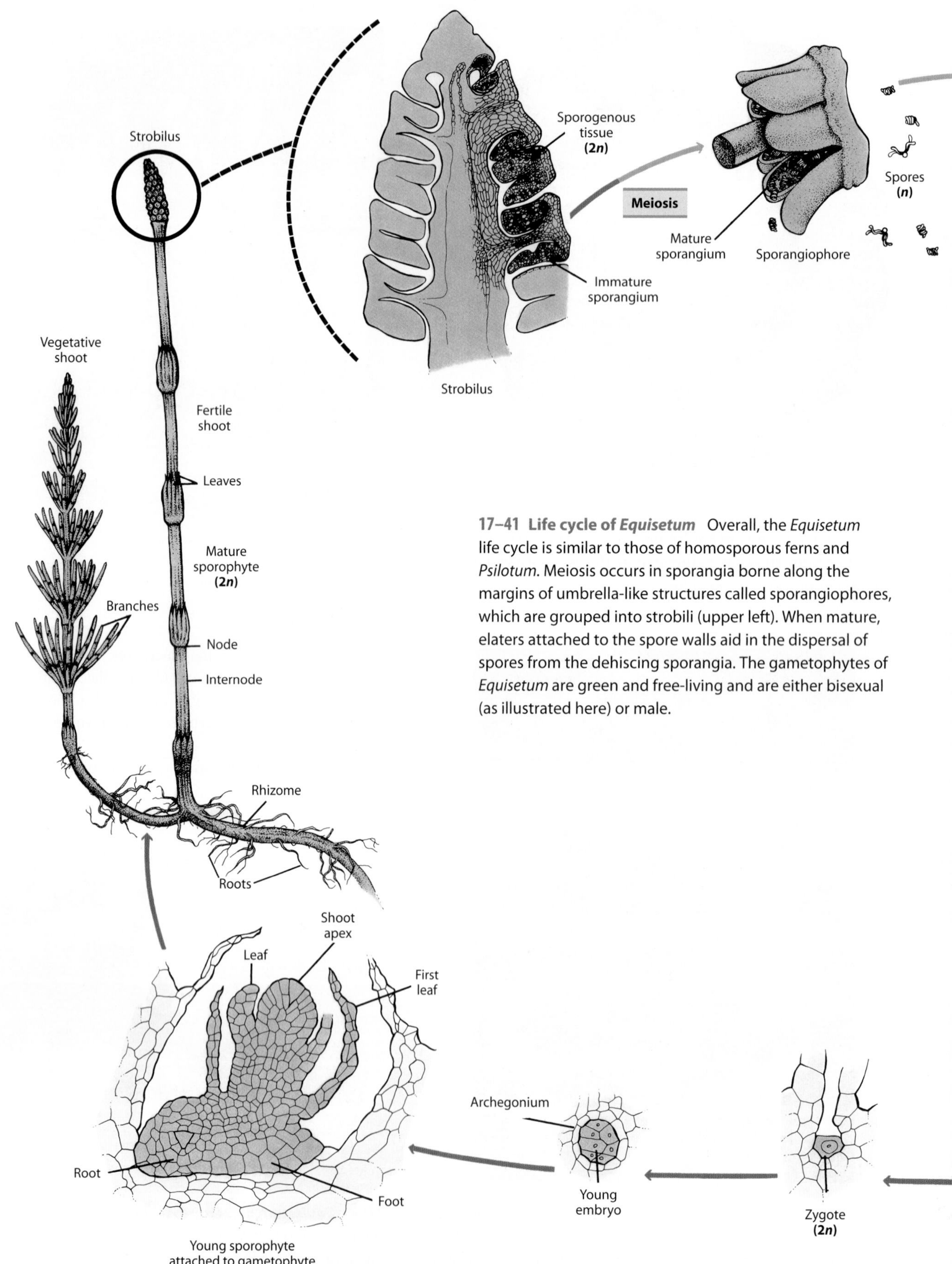

17–41 Life cycle of *Equisetum* Overall, the *Equisetum* life cycle is similar to those of homosporous ferns and *Psilotum*. Meiosis occurs in sporangia borne along the margins of umbrella-like structures called sporangiophores, which are grouped into strobili (upper left). When mature, elaters attached to the spore walls aid in the dispersal of spores from the dehiscing sporangia. The gametophytes of *Equisetum* are green and free-living and are either bisexual (as illustrated here) or male.

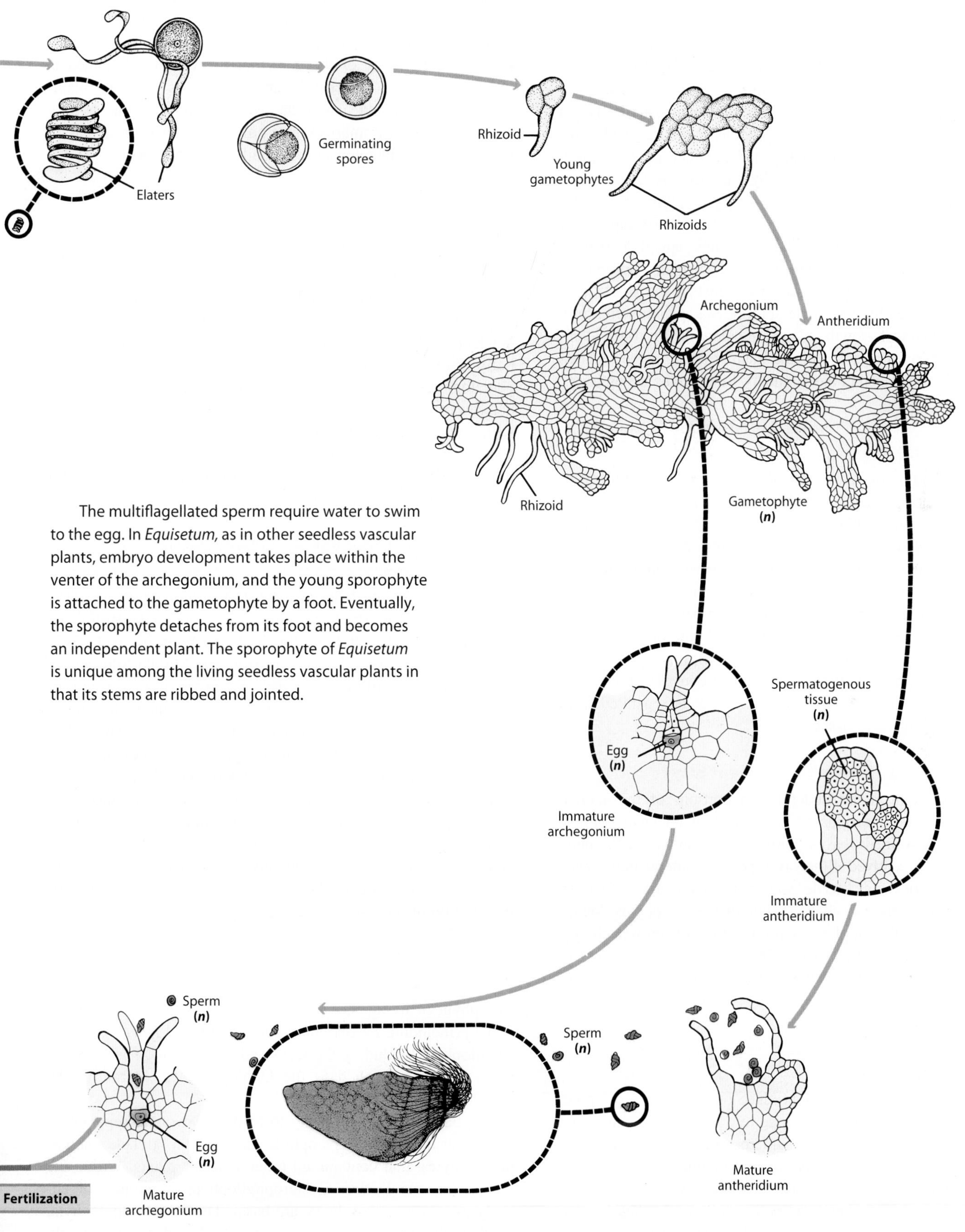

The multiflagellated sperm require water to swim to the egg. In *Equisetum,* as in other seedless vascular plants, embryo development takes place within the venter of the archegonium, and the young sporophyte is attached to the gametophyte by a foot. Eventually, the sporophyte detaches from its foot and becomes an independent plant. The sporophyte of *Equisetum* is unique among the living seedless vascular plants in that its stems are ribbed and jointed.

SUMMARY

Vascular plants are characterized by the possession of the vascular tissues xylem and phloem, which contributed to the success of these plants in their occupation of the land. All vascular plants exhibit an alternation of heteromorphic (dissimilar in form) generations in which the sporophyte is large and complex and ultimately nutritionally independent of the gametophyte. The ability of the sporophyte to become large and to reach great heights was made possible by its ability to synthesize lignin, which adds rigidity to the cell walls of supporting cells and water-conducting cells. With their apical meristems and branched pattern of growth, the vascular plant sporophytes can produce multiple sporangia, in contrast to the single sporangium produced by bryophyte sporophytes, which increase in length by subapical growth, occurring below the tip of the stem.

A comparison of some of the main features of the seedless vascular plants is given in the Summary Table on the facing page.

The Primary Vascular Tissues Are Arranged in Steles of Three Basic Types

The plant bodies of many vascular plants consist entirely of primary tissues. Today, secondary growth is confined largely to the seed plants, although it occurred in several unrelated extinct groups of seedless vascular plants. Among the living seedless vascular plants, secondary growth is found with certainty in only one genus, *Isoetes*. The primary vascular tissues and associated ground tissues exhibit three basic arrangements: (1) the protostele, which consists of a solid core of vascular tissue; (2) the siphonostele, which contains a pith surrounded by vascular tissue; and (3) the eustele, which consists of a system of strands surrounding a pith, with the strands separated from one another by ground tissue.

Roots and Leaves Evolved in Different Ways

Roots evolved from the underground portions of the ancient plant body. Leaves originated in more than one way. Microphylls, single-veined leaves with leaf traces not associated with leaf gaps, evolved either as superficial lateral outgrowths of the stem or from sterile sporangia. They are associated with protosteles and are characteristic of the lycophytes. Megaphylls, leaves with complex venation, evolved from branch systems. They are associated with siphonosteles and eusteles. In the siphonosteles of ferns, the leaf traces are associated with leaf gaps.

Vascular Plants May Be Either Homosporous or Heterosporous

Homosporous vascular plants produce only one type of spore, which has the potential to give rise to a bisexual gametophyte. Heterosporous plants produce microspores and megaspores, which germinate and give rise to male gametophytes and female gametophytes, respectively. Most seedless vascular plants are homosporous, but heterospory is exhibited by *Selaginella, Isoetes,* and the water ferns (Salviniales). The gametophytes of heterosporous plants are much reduced in size compared with those of homosporous plants. In the history of the vascular plants, heterospory has evolved several times. There has been a long, continuous evolutionary trend toward a reduction in the size and complexity of the gametophyte, which culminated in the angiosperms (flowering plants). Seedless vascular plants have archegonia and antheridia, whereas most of the gymnosperms (conifers, primarily) have only archegonia. In the angiosperms, both archegonia and antheridia have been lost.

Seedless Vascular Plants Exhibit an Alternation of Heteromorphic Generations

The life cycles of the seedless vascular plants all represent modifications of an essentially similar alternation of heteromorphic generations in which the adult sporophyte is large and free-living. The gametophytes of the homosporous species are independent of the sporophyte for their nutrition. Although potentially bisexual, producing both antheridia and archegonia, these gametophytes are functionally unisexual. The gametophytes of heterosporous species are unisexual, much reduced in size, and, with few exceptions, dependent on stored food derived from the sporophyte for their nutrition. All of the seedless vascular plants have motile sperm, and the presence of water is necessary for the sperm to swim to the eggs.

The Oldest Fossils of Vascular Plants Belong to the Phylum Rhyniophyta

Vascular plants date from at least 440 million years ago. The earliest plants for which we know a number of structural details belong to the phylum Rhyniophyta, made up exclusively of fossils, the oldest of which are from the Middle Silurian period, about 425 million years ago. Some fossils, once considered rhyniophytes, have conducting cells similar to bryophyte hydroids rather than to tracheids. These plants, which had branched axes and multiple sporangia, may represent an intermediate stage in the evolution of vascular plants. They are called protracheophytes. The plant bodies of the rhyniophytes and other contemporary plants were simple, dichotomously branching axes lacking roots and leaves. With evolutionary specialization, morphological and physiological differences arose between various parts of the plant body, bringing about the differentiation of root, stem, and leaf.

The Living Seedless Vascular Plants Are Classified in Two Phyla

The phyla of living seedless vascular plants are the Lycopodiophyta (including *Lycopodium, Selaginella,* and *Isoetes*) and the Monilophyta (including the ferns and *Equisetum*). The common name "fern" is applied to the Ophioglossales, Marattiopsida, and Polypodiopsida.

A basal split in the Early to Middle Devonian separated a clade that included the lycophyte lineage from a clade, the euphyllophytes, that contains all other vascular plants. Within the lycophytes, the leafless Zosterophyllophyta, a phylum of entirely extinct vascular plants, are basal. The Trimerophytophyta, another phylum of entirely extinct vascular plants, apparently are ancestral to the monilophytes and progymnosperms.

SUMMARY TABLE A Comparison of Some of the Main Features of the Seedless Vascular Plants

PHYLUM	HOMOSPOROUS OR HETEROSPOROUS	TYPE OF LEAVES	TYPE OF STELE	SPORANGIA	MISCELLANEOUS CHARACTERISTICS
Rhyniophyta (rhyniophytes)	Homosporous	None	Protostele	Terminal	Exclusively fossils; likely ancestors of trimerophytes
Zosterophyllophyta (zosterophyllophytes)	Many homosporous; some heterosporous	None	Protostele	Lateral	Exclusively fossils; closely related to lycophytes
Trimerophytophyta (trimerophytes)	Homosporous	None	Protostele	Terminal on ultimate dichotomies	Exclusively fossils; likely ancestors of ferns, progymnosperms, and perhaps horsetails
Lycopodiophyta (lycophytes)	Lycopodiaceae homosporous; Selaginellaceae and Isoetaceae heterosporous	Microphyll	Most with protostele or modified protostele	On or in the axils of sporophylls	Members of the Selaginellaceae and Isoetaceae have ligules; many extinct representatives
Monilophyta					
Psilotopsida	Homosporous	Megaphyll or absent through reduction	Protostele or more complex	Lateral; eusporangiate	Diverse in structure and anatomy; gametophytes subterranean, mycorrhizal
Marattiopsida	Homosporous	Megaphyll	Siphonostele or more complex	On sporophylls; eusporangiate	Large plants with complex leaves; gametophytes on soil surface, photosynthetic
Polypodiopsida (leptosporangiate ferns)	All homosporous except for Marsileales and Salviniales, which are heterosporous	Megaphyll	Protostele in some; siphonostele or more complex types in others	On sporophylls; clustered in sori; leptosporangiate	Diverse in habit and habitat; gametophytes on soil surface, photosynthetic
Equisetopsida (equisetophytes)	Homosporous; some fossils heterosporous	Microphyll-like through reduction	Eustele-like siphonostele	On sporangiophores in strobili	Represented today by single genus, Equisetum, the horsetails

Lycophytes Are Characterized by Microphylls; the Members of the Other Phyla of Vascular Plants Have Megaphylls

Two classes of ferns (Psilotopsida and Marattiopsida) possess eusporangia. In eusporangia, the walls are several cell layers thick, and a number of cells participate in the initial stages of sporangial development. Other ferns—Polypodiopsida, including the water ferns (order Salviniales)—form leptosporangia, specialized structures in which the wall consists of a single layer of cells that develop from a single initial cell. Leptosporangia have evolved only in the Polypodiopsida; all other vascular plants possess eusporangia.

Five groups of vascular plants dominated the swamplands of the Carboniferous period ("Age of Coal"), and three of them were seedless vascular plants—lycophytes, equisetophytes (calamites, or giant horsetails), and ferns. The other two were gymnosperms—the seed ferns and the cordaites.

QUESTIONS

1. What basic structural features do the Rhyniophyta, Zosterophyllophyta, and Trimerophytophyta have in common?
2. The vessel elements and heterospory present in several unrelated groups of vascular plants represent excellent examples of convergent evolution. Explain.
3. With the use of simple, labeled diagrams, describe the structure of the three basic types of steles.
4. Compare the life cycle of a moss with that of a homosporous leptosporangiate fern.
5. What is coal? How was it formed? What plants were involved in its formation?
6. The bryophytes often are referred to as the "amphibians of the plant kingdom," but that characterization might also be applied to the seedless vascular plants. Suggest why.
7. Distinguish between the lycophyte clade and the euphyllophyte clade?

CHAPTER 18

Gymnosperms

◀ **A threatened pine** Growing at high elevations on wind-swept slopes, the whitebark pine *(Pinus albicaulis)* plays an important ecological role by providing essential food for a wide variety of species, including grizzlies, squirrels, and birds. Although certain whitebark pines are resistant, blister rust has killed many of the trees, and climate change and years of fire suppression have contributed to a new threat—the mountain pine beetle.

CHAPTER OUTLINE

One of the most dramatic innovations to arise during the evolution of the vascular plants was the seed. Seeds are one of the principal factors responsible for the dominance of seed plants in today's flora—a dominance that has become progressively greater over a period of several hundred million years. The reason is simple: the seed has great survival value. The protection that a seed affords the enclosed embryo, as well as the stored food that is available to that embryo at the critical stages of germination and establishment, give seed plants a great selective advantage over their free-sporing relatives and ancestors, that is, over plants that shed their spores.

Evolution of the Seed

All seed plants are heterosporous, producing **megaspores** and **microspores** that give rise, respectively, to **megagametophytes** (female gametophytes) and **microgametophytes** (male gametophytes). Heterospory, however, is not unique to seed plants. As we discussed in Chapter 17, some seedless vascular plants are also heterosporous. The production of seeds is, however, a particularly extreme form of heterospory that has been modified to form an **ovule,** the structure that develops into the seed. Indeed, a **seed** is simply a matured ovule containing an embryo. The immature ovule consists of a **megasporangium,** the structure in which the megaspores are produced, surrounded by one or two additional layers of tissue, the **integuments** (Figure 18–1).

Several events led to the evolution of an ovule, including:

1. Retention of the megaspores within the megasporangium, which is fleshy and called the **nucellus** in seed plants—in other words, the megasporangium no longer releases the spores.

CHECKPOINTS

After reading this chapter, you should be able to answer the following:

1. What is a seed, and why was the evolution of the seed such an important innovation for plants?
2. According to the current hypothesis, from which group of plants did seed plants evolve? What is the evidence for this hypothesis?
3. How do the mechanisms by which sperm reach the eggs in gymnosperms and in seedless vascular plants differ?
4. Give the distinguishing features of the four phyla of living gymnosperms.
5. In what ways do gnetophytes resemble angiosperms?

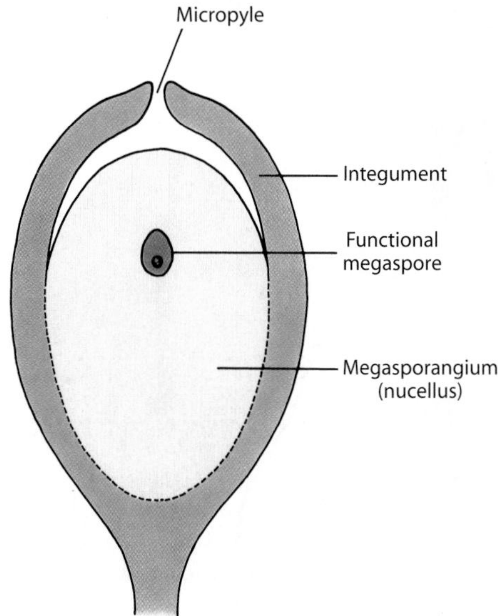

18–1 Longitudinal section of an ovule The ovule consists of a megasporangium (nucellus) enveloped by an integument with an opening, the micropyle, at its apical end. A single functional megaspore is retained within the megasporangium and will give rise to a megagametophyte that is retained within the megasporangium. Following fertilization, the ovule matures into a seed, which becomes the unit of dispersal. Gymnosperm ovules have a single integument, whereas angiosperm ovules typically have two.

2. Reduction in the number of megaspore mother cells in each megasporangium to one.
3. Survival of only one of the four megaspores produced by the spore mother cell, leaving a single functional megaspore in the megasporangium.
4. Formation of a female gametophyte inside the single functional megaspore—that is, formation of an endosporic (within the wall) female gametophyte that is no longer free living and is retained within the megasporangium.
5. Development of the embryo, or young sporophyte, within the female gametophyte retained within the megasporangium.
6. Formation of an integument that completely envelops the megasporangium, except for an opening at its apex called the **micropyle.**
7. Modification of the apex of the megasporangium to receive microspores, or pollen grains.

Related to these events is a basic shift in the unit of dispersal from the megaspore to the seed, the embryo-containing megasporangium with integuments.

The Fossil Record Provides Clues to Ovule Evolution

The exact order in which the events occurred is unknown because of the incompleteness of the fossil record. We know that they occurred fairly early in the history of vascular plants, however, because the oldest ovules or seeds are from the Late Devonian (about 365 million years ago; see inside of front cover). One of these early seed plants is *Elkinsia polymorpha* (Figure 18–2). The ovule of *Elkinsia* consisted of a nucellus and four or five integumentary lobes with little or no fusion between the lobes. The integumentary lobes curved inward at their tips, forming a ring around the apex of the nucellus. The ovules were surrounded by dichotomously branched, sterile structures called cupules. The integuments of ovules apparently evolved through the gradual fusion of integumentary lobes until the only opening left was the micropyle (Figure 18–3).

A Seed Consists of an Embryo, Stored Food, and a Seed Coat

In modern seed plants, the ovule consists of a nucellus enveloped by one or two integuments with a micropyle. When the ovules of most gymnosperms are ready for fertilization, the nucellus contains a megagametophyte composed of nutritive tissue and archegonia. After fertilization, the integuments develop into a **seed coat,** and a seed is formed. In most modern seed plants, an embryo develops within the seed before dispersal—exceptions include *Ginkgo* (see page 450) and many cycads. In addition, all seeds contain stored food.

There Are Five Phyla of Seed Plants with Living Representatives

Seed plants arose starting in the Late Devonian period, at least 365 million years ago. During the next 50 million years, a wide array of seed-bearing plants evolved, many of which are grouped together as the so-called seed ferns, while others are recognized as cordaites and conifers (see the essay "Coal Age Plants" on pages 400 and 401).

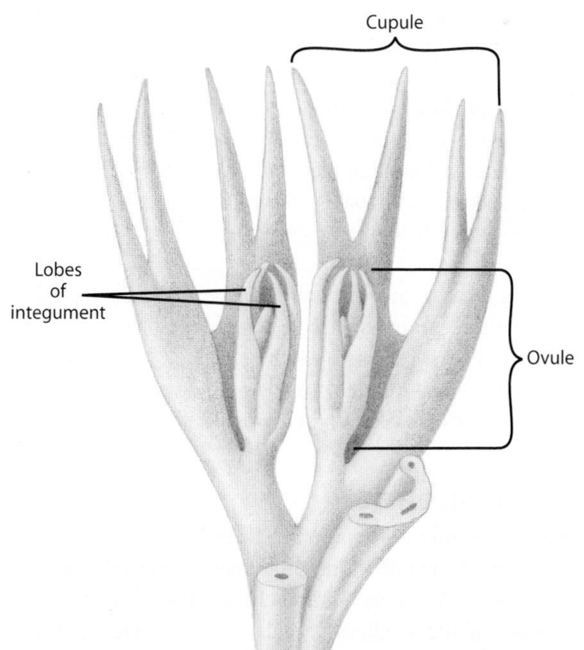

18–2 *Elkinsia polymorpha* Reconstruction of a fertile branch of the Late Devonian plant *Elkinsia polymorpha,* showing its ovules. Each ovule was overtopped by a dichotomously branched, sterile structure called a cupule. Note the more or less free lobes of the integument.

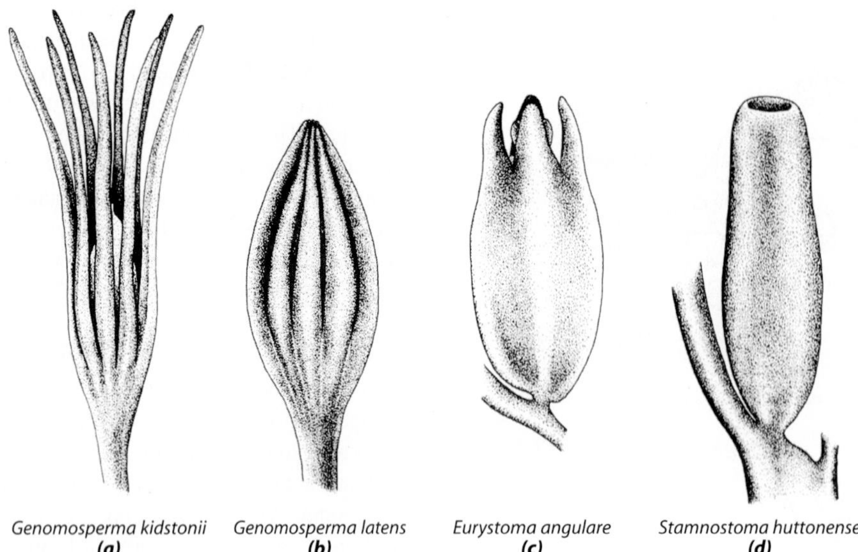

18–3 Evolution of integuments Seedlike structures in several Paleozoic plants, showing some potential stages in the evolution of the integument. ***(a)*** In *Genomosperma kidstonii* (Gk. *genomein,* "to become," and *sperma,* "seed"), eight fingerlike projections arise at the base of the megasporangium and are separate for their entire length. ***(b)*** In *Genomosperma latens,* the integumentary lobes are fused from the base of the megasporangium for about a third of their length. ***(c)*** In *Eurystoma angulare,* fusion is almost complete, and ***(d)*** in *Stamnostoma huttonense,* it is complete, with only the micropyle remaining open at the top.

All of the seed plants typically possess megaphylls, which are generally large leaves with several to many veins, but which are modified to needles or scales in some groups. There are five phyla with living representatives: Coniferophyta, Cycadophyta, Ginkgophyta, Gnetophyta, and Anthophyta. The phylum Anthophyta comprises the angiosperms, or flowering plants; the remaining four phyla are commonly referred to as gymnosperms. The gymnosperms represent a series of evolutionary lineages of seed-bearing plants. Although there are only about 840 species of living gymnosperms—compared with at least 300,000 species of angiosperms—individual gymnosperm species are often dominant over wide areas.

Before beginning our discussion of seed plants, we briefly examine one more group of seedless vascular plants—the progymnosperms. They are discussed here, rather than in Chapter 17, because they may be the progenitors of seed plants, or, at least, they are closely related to the first seed plants.

Progymnosperms

In the Late Paleozoic era, around 290 million years ago, there was a group of plants called the progymnosperms (phylum Progymnospermophyta), which had characteristics intermediate between those of the seedless vascular trimerophytes and those of the seed plants. Although the progymnosperms reproduced by means of freely dispersed spores, they produced secondary xylem (wood) remarkably similar to that of living conifers (Figure 18–4). The progymnosperms were unique among woody Devonian plants in that they also produced secondary phloem. Both the progymnosperms and Paleozoic ferns probably evolved from the more ancient trimerophytes (see Figure 17–10c), from which they differed primarily by having more elaborate and more highly differentiated branch systems and correspondingly more complex vascular systems.

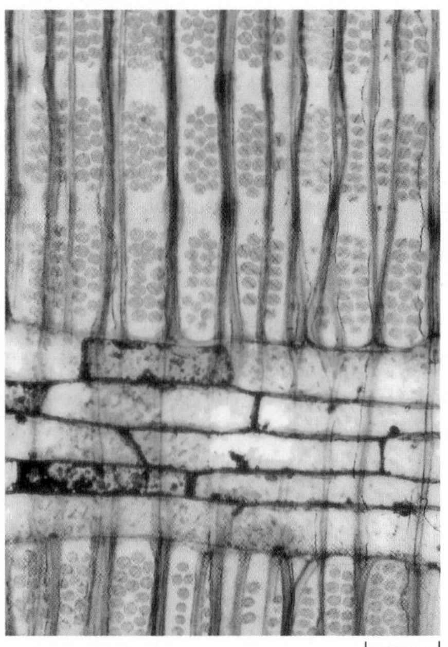

18–4 Progymnosperm wood Radial view of the secondary xylem, or wood, of the progymnosperm *Callixylon newberryi.* This fossil wood, with its regular series of pitted tracheids, is remarkably similar to the wood of certain conifers.

18–5 ***Archaeopteris*** Reconstruction of the progymnosperm *Archaeopteris,* which is common in the fossil record of eastern North America. Specimens of *Archaeopteris* attained heights of 17 meters or more, and some of them seem to have formed forests.

18–6 Lateral branch and leaves of *Archaeopteris* Reconstruction of a frondlike lateral branch system of the progymnosperm *Archaeopteris macilenta.* Fertile leaves can be seen bearing maturing sporangia (shown here as brown) on centrally located primary branches.

In the progymnosperms, the most important evolutionary advance over both the trimerophytes and the ferns is the presence of a **bifacial vascular cambium**—that is, one that produces both secondary xylem (internally) and secondary phloem (externally). This type of vascular cambium is characteristic of seed plants and is believed to have evolved first in the progymnosperms.

A major kind of progymnosperm, the *Archaeopteris* type, appeared in the Devonian period, about 370 million years ago, and extended into the Mississippian period, about 340 million years ago (Figure 18–5). It was the main component of the earliest forests until its extinction. In this group, the lateral branch systems were flattened in one plane and bore structures considered to be leaves (Figure 18–6). Apparently, a **eustele**—that is, a system of vascular tissues in discrete strands arranged in a ring around a pith (see Figure 17–5d)—evolved in this group of progymnosperms. The eustele is the strong similarity linking this group with the living seed plants. The larger branches of *Archaeopteris*-type progymnosperms had a pith. Although most progymnosperms were homosporous, some species of *Archaeopteris* were heterosporous. Thus, both the production of wood and heterospory predate the evolution of the seed.

Fossil logs of *Archaeopteris,* called *Callixylon,* may be up to a meter or more in diameter and 10 meters long, indicating that at least some species of this group were large trees. They appear to have formed extensive forests in some regions. As the reconstruction in Figure 18–5 suggests, individuals of *Archaeopteris* may have resembled conifers in their branching patterns.

Evidence accumulated during the past several decades strongly indicates that the seed plants evolved from plants similar to the progymnosperms, following the appearance of the seed in what now seems to have been the common ancestor of all seed plants (Figure 18–7). However, many problems still remain to be solved in developing a more detailed understanding of the early evolution of the seed plants.

Extinct Gymnosperms

Two groups of extinct gymnosperms—the seed ferns (Pteridospermales) and the cordaites (Cordaitales, primitive conifer-like plants)—were discussed and illustrated in Chapter 17. The seed ferns, or pteridosperms, are a very diverse and highly artificial group, ranging in age from the Devonian to the Jurassic. They varied in form from slender, branched plants to plants with the appearance of tree ferns (Figure 18–8; see also the essay on pages 400 and 401). Several groups of extinct Mesozoic plants are also sometimes included with the seed ferns. It appears that a series of Devonian-Carboniferous seed ferns, including medullosans, are situated at the base of seed plant phylogeny. Exactly how these different groups of seed ferns relate to the living gymnosperms remains uncertain.

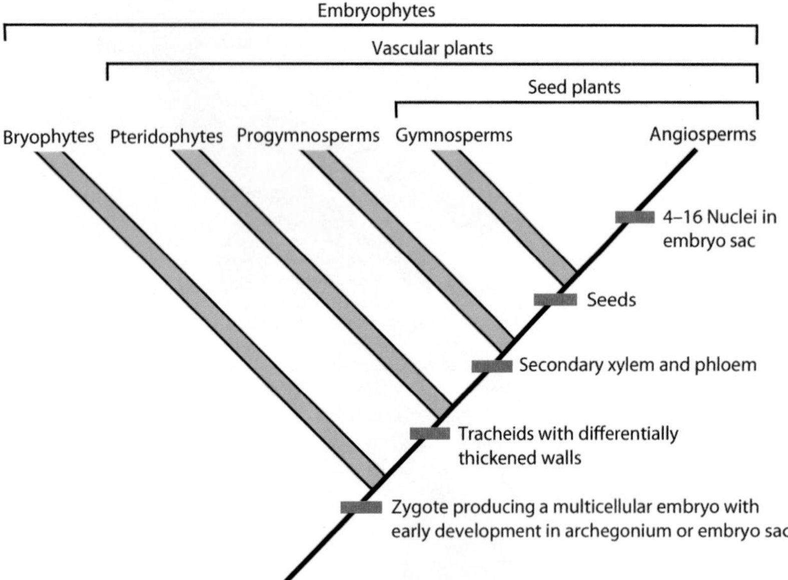

18–7 Phylogenetic relationships among the major groups of embryophytes A simplified summary showing the phylogenetic relationships among the major groups of embryophytes (organisms with multicellular embryos). The embryophytes, vascular plants, seed plants, and angiosperms are monophyletic groups, whereas the bryophytes, pteridophytes (seedless vascular plants), progymnosperms, and gymnosperms each contain several lineages, indicated here by a broad band, and are paraphyletic (page 238). Note, however, that the extant groups of gymnosperms are monophyletic. A single example is given of the characters that define each of the monophyletic groups.

Another group of extinct gymnosperms—the cycadeoids, or Bennettitales—consisted of plants with palmlike leaves, somewhat resembling the living cycads (see page 448). The Bennettitales are an enigmatic group of Mesozoic gymnosperms that disappeared from the fossil record during the Cretaceous. Some paleobotanists believe that the Bennettitales may have been members of the same evolutionary line as angiosperms, but it is still uncertain exactly where the Bennettitales fit in phylogenetically. Reproductively, Bennettitales were distinct from the cycads in several respects, including the presence of flowerlike reproductive structures that were bisexual in some species (Figure 18–9).

Living Gymnosperms

There are four phyla of gymnosperms with living representatives: Coniferophyta (conifers), Cycadophyta (cycads), Ginkgophyta (maidenhair tree, or ginkgo), and Gnetophyta (gnetophytes). The name *gymnosperm,* which literally means "naked seed," points to one of the principal characteristics of plants belonging to these four phyla—namely, that their ovules and seeds are exposed on the surface of modified leaves known as sporophylls, or analogous structures.

The phylogenetic relationships among the four groups of living, or extant, gymnosperms remain quite uncertain. Recent molecular analyses indicate that these extant groups are monophyletic—that is, the groups consist of an ancestor and all of its descendants. (Note, however, that when fossil and extant groups are considered, the gymnosperms are paraphyletic, containing a common ancestor and some, but not all, of its descendants.)

Some molecular analyses supporting the monophyletic relationship of extant gymnosperms indicate a close relationship between the gnetophytes and conifers, the two groups forming a clade in which monophyletic conifers are sister (most closely related) to monophyletic gnetophytes (the gnetifer hypothesis; Figure 18–10a). Other molecular analyses unite the gnetophytes more specifically with the Pinaceae (the pine family) and place the gnetophytes as sister to a clade of other conifer families (the gnepine hypothesis; Figure 18–10b). Another, earlier possibility, based on the analyses of morphological characters of seed plants, indicated that the gnetophytes and Bennettitales together with the angiosperms form a clade referred to as the "anthophytes" (not to be confused with the term Anthophyta, the phylum of angiosperms), to emphasize their flowerlike reproductive structures (the anthophyte hypothesis; Figure 18–10c). Subsequent molecular studies, however, do not support the existence of an anthophyte clade. The phylogenetic relationships among seed plant lineages remain uncertain.

In the Gymnosperms, the Microgametophytes (Male Gametophytes) Develop as Pollen Grains

In the ferns and other seedless vascular plants, water is required for the motile, flagellated sperm to reach and fertilize the eggs.

18–8 Late Carboniferous landscape Tropical swamps of the Late Carboniferous were dominated by several genera of giant lycophyte trees, which, when mature, formed a forest canopy of airy, diffusely branched crowns, seen here in the background. These trees had massive trunks (left foreground and elsewhere) that were stabilized in the swampy mire by long stigmarian axes, from which numerous rootlets, possibly photosynthetic, extended. Notice also some low-growing lycophytes with cones (center foreground). Flooded and boggy areas of the forest floor favored other lycophyte types, such as branchless *Chaloneria* (center and right foreground), and horsetail types, such as the shrub *Sphenophyllum* (lower right corner) and the Christmas-tree-shaped *Diplocalamites* (far right midground).

Less wet, or slightly elevated ground, as seen at the left, fostered mixed vegetation, including early conifers, ground-cover ferns, tall tree ferns, and seed ferns. Among the seed plants were *Cordaixylon,* a shrubby conifer relative with strap-shaped leaves (left foreground), and *Callistophyton,* a scrambling seed fern, seen growing at the base of the largest lycophyte tree. In the left background is *Psaronius,* an early tree fern. Elsewhere, robust plants such as the upright umbrella-shaped seed fern *Medullosa* are seen occupying sunnier, more disturbed sites opened up by previous channel flooding (right midground and background).

(a)

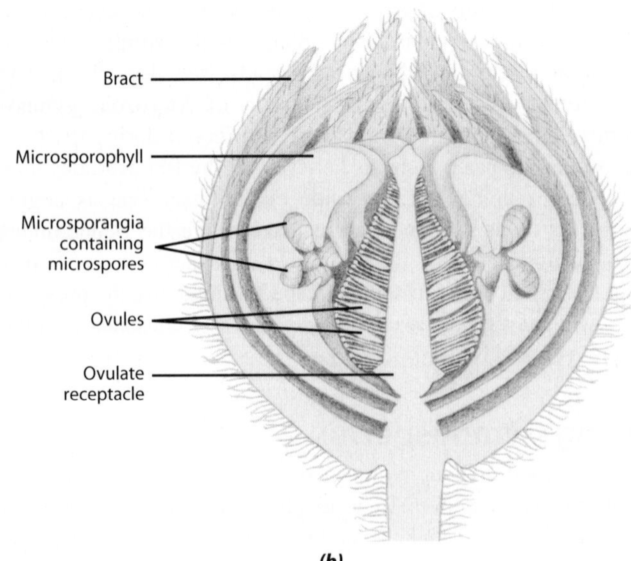

(b)

18–9 Bennettitales ***(a)*** Reconstruction of *Wielandiella*, an extinct gymnosperm from the Triassic. *Wielandiella* has a forked branching pattern. A single strobilus, or cone, is borne at each fork. ***(b)*** Diagrammatic reconstruction of the bisporangiate, or bisexual, strobilus of *Williamsoniella coronata* from the Jurassic. The strobilus consists of a central ovulate receptacle surrounded by a whorl of microsporophylls bearing microsporangia containing microspores, which develop into microgametophytes (pollen grains). Hairy bracts enclose the reproductive parts.

In the gymnosperms, however, water is not required as a medium for transporting the sperm to the eggs. Instead, the partly developed microgametophyte, the **pollen grain,** is transferred bodily (usually passively, by the wind) to the vicinity of a megagametophyte within an ovule. This process is called **pollination.** After pollination, the endosporic microgametophyte produces a tubular outgrowth, the **pollen tube.** The microgametophytes of gymnosperms and other seed plants do not form antheridia.

In the cycads and *Ginkgo,* fertilization is transitional between the condition found in ferns and other seedless plants, in which free-swimming sperm occur, and the condition found in other seed plants, which have nonmotile sperm. The microgametophytes of cycads and *Ginkgo* produce a pollen tube, but this does not penetrate the archegonium (Figure 18–11). Instead, it is haustorial and may grow for several months in the tissue of the nucellus, where it absorbs nutrients. Eventually, the pollen grain bursts in the vicinity of the archegonium, releasing multiflagellated, swimming sperm cells (see Figure 18–37). The sperm then swim to an archegonium, and one of them fertilizes the egg.

In conifers, gnetophytes, and angiosperms, the sperm are nonmotile and the pollen tubes convey the sperm directly to the egg cell. With this innovation, seed plants were no longer

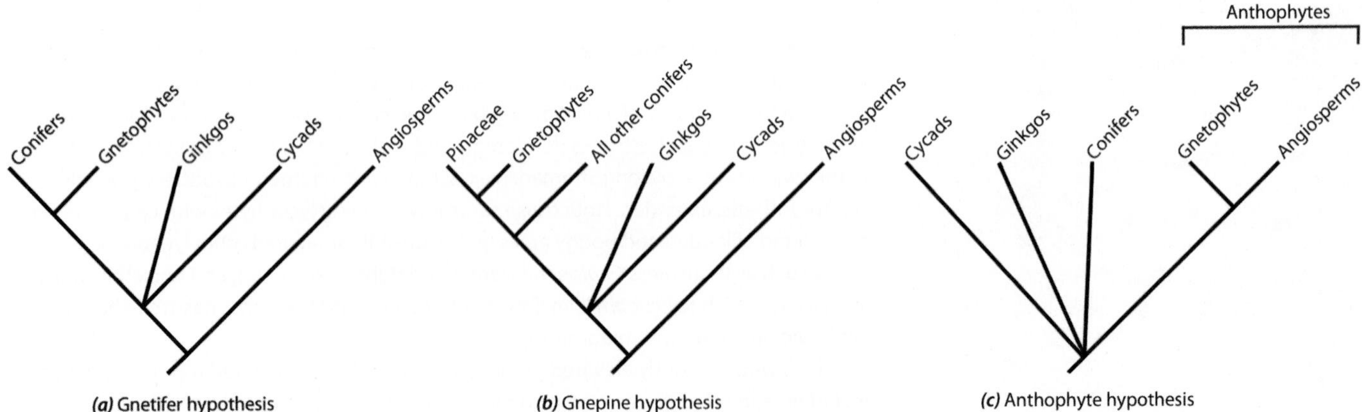

18–10 Alternative hypotheses of relationships among the five major extant lineages of seed plants ***(a)*** The gnetifer hypothesis proposes that the gnetophytes are most closely related to conifers. ***(b)*** The gnepine hypothesis proposes that the gnetophytes are nested within the conifers as the sister group of the Pinaceae. ***(c)*** According to the anthophyte hypothesis, the gnetophytes are most closely related to angiosperms.

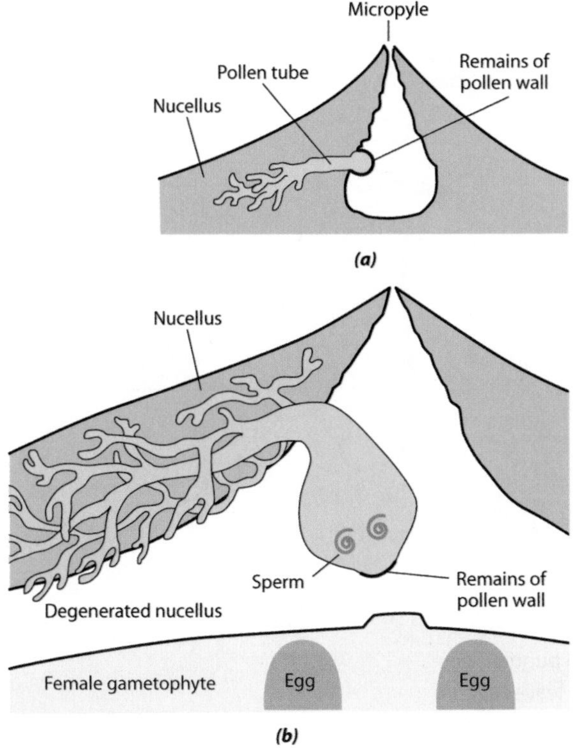

18–11 Development of the microgametophyte of *Ginkgo biloba* ***(a)*** Early in its development, the pollen tube grows by tip growth and begins to form what will become a highly branched haustorial structure. The pollen tube in *Ginkgo* grows intercellularly in the nucellus. ***(b)*** Late in development, the basal end of the pollen tube enlarges into a saclike structure that contains the two multiflagellated sperm. Subsequently, the basal end of the pollen tube ruptures, releasing the two sperm, which then swim to the eggs contained in the archegonia of the megagametophyte.

dependent on the presence of free water to ensure fertilization—a necessity for all seedless plants. The presence of haustorial pollen tubes in *Ginkgo* and cycads suggests that, originally, the pollen tube evolved to absorb nutrients for the production of sperm by the microgametophyte during its growth within the ovule. From this perspective, the conveyance of nonmotile sperm by a pollen tube that grows directly to an egg can be seen as a later evolutionary modification of a structure initially developed for another purpose.

With few exceptions, the megagametophyte (female gametophyte) of gymnosperms produces several archegonia. As a result, more than one egg may be fertilized, and several embryos may begin to develop within a single ovule—a phenomenon known as **polyembryony.** In most cases, however, only one embryo survives and therefore relatively few fully developed seeds contain more than one embryo.

Phylum Coniferophyta

By far the most numerous, most widespread, and most ecologically important of the gymnosperm phyla living today are the Coniferophyta, which comprise some 70 genera with about 630 species. The tallest vascular plant, the redwood *(Sequoia sempervirens)* of coastal California and southwestern Oregon, is a conifer. Redwood trees attain heights of up to 115.6 meters and trunk diameters in excess of 11 meters. The conifers, which also include pines, firs, and spruces, are of great commercial value. Their stately forests are one of the most important natural resources in vast regions of the north temperate zone. During the Early Tertiary period, some genera were more widespread than they are now, and a diverse conifer flora was present across huge expanses on all of the northern continents.

The history of the conifers extends back at least to the Late Carboniferous period, some 300 million years ago. The needle-like leaves of modern conifers have many drought-resistant features, which may bring ecological advantages in certain habitats and may also be related to the diversification of the phylum during the relatively dry and cold Permian period (290 to 245 million years ago). At that time, increasing worldwide aridity may have favored structural adaptations such as those of conifer leaves.

We begin our discussion of the conifers with *Pinus,* from which the family Pinaceae receives its name. Molecular data indicate a basal split between the Pinaceae and a clade that includes all of the other conifers.

Pines Are Conifers with a Unique Leaf Arrangement

The pines (genus *Pinus*), which include perhaps the most familiar of all gymnosperms (Figure 18–12), dominate broad

18–12 Longleaf pines These longleaf pines, *Pinus palustris,* were photographed in North Carolina.

(a)

(b)

18–13 Needlelike leaves of the pines ***(a)*** The adult leaves of the pines occur in bundles, or fascicles. The fascicles of Bhutan pine, *Pinus wallichiana,* shown here, contain five leaves each. ***(b)*** A seedling of pinyon pine, *Pinus edulis,* showing spirally arranged juvenile leaves and a young taproot system. The mature leaves of this species are borne in fascicles of two needles each.

stretches of North America and Eurasia and are widely cultivated even in the Southern Hemisphere. There are about 100 species of pines, all of which are characterized by an arrangement of the leaves that is unique among living conifers. In pine seedlings, the needlelike leaves are spirally arranged and borne singly on the stems (Figure 18–13b). After a year or two of growth, a pine begins to produce its leaves in bundles, or fascicles, each of which contains a specific number of leaves (needles)—from one to eight, depending on the species (Figure 18–13a). These fascicles, wrapped at the base by a series of small, scalelike leaves, are actually short shoots in which the activity of the apical meristem is restricted. Thus, a fascicle of needles in a pine is morphologically a **determinate** (restricted in growth) branch. Under unusual circumstances, the apical meristem within this fascicle of needles may be reactivated and grow into a new shoot with **indeterminate** growth, or sometimes may even produce roots and grow into an entire pine tree.

The leaves of pines, like those of many other conifers, are impressively suited for growth under conditions where water may be scarce or difficult to obtain (Figure 18–14). A thick cuticle, which reduces evaporation from the interior of the leaf, covers the epidermis, beneath which are one or more layers of compactly arranged, thick-walled cells—the hypodermis. The stomata are sunken below the surface of the leaf. The mesophyll, or ground tissue of the leaf, consists of parenchyma cells with conspicuous wall ridges that project into the cells, increasing their surface area. Commonly, the mesophyll is penetrated by two or more resin ducts. One vascular bundle, or two bundles side by side, are found in the center of the leaf. The vascular bundles, made up of xylem and phloem, are surrounded by transfusion tissue, composed of living parenchyma cells and short, nonliving tracheids. The transfusion tissue is believed to conduct materials between the mesophyll and the vascular bundles. A single layer of cells known as the endodermis surrounds the transfusion tissue, separating the transfusion tissue from the mesophyll.

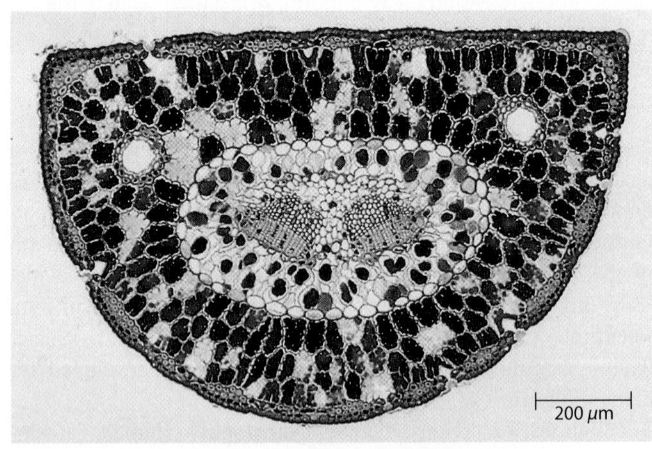

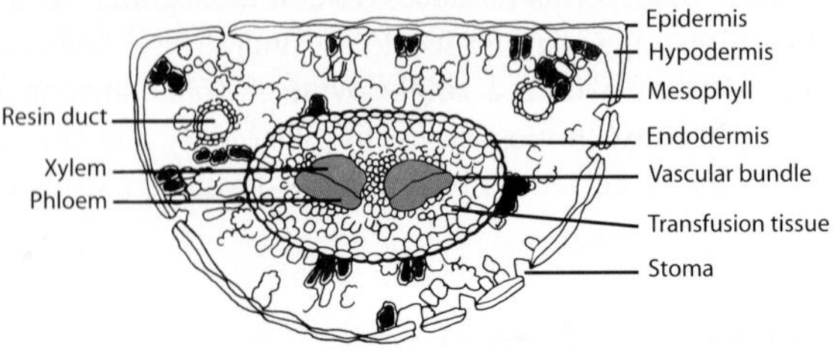

18–14 Pine needle Transverse section of a *Pinus* needle, showing mature tissues.

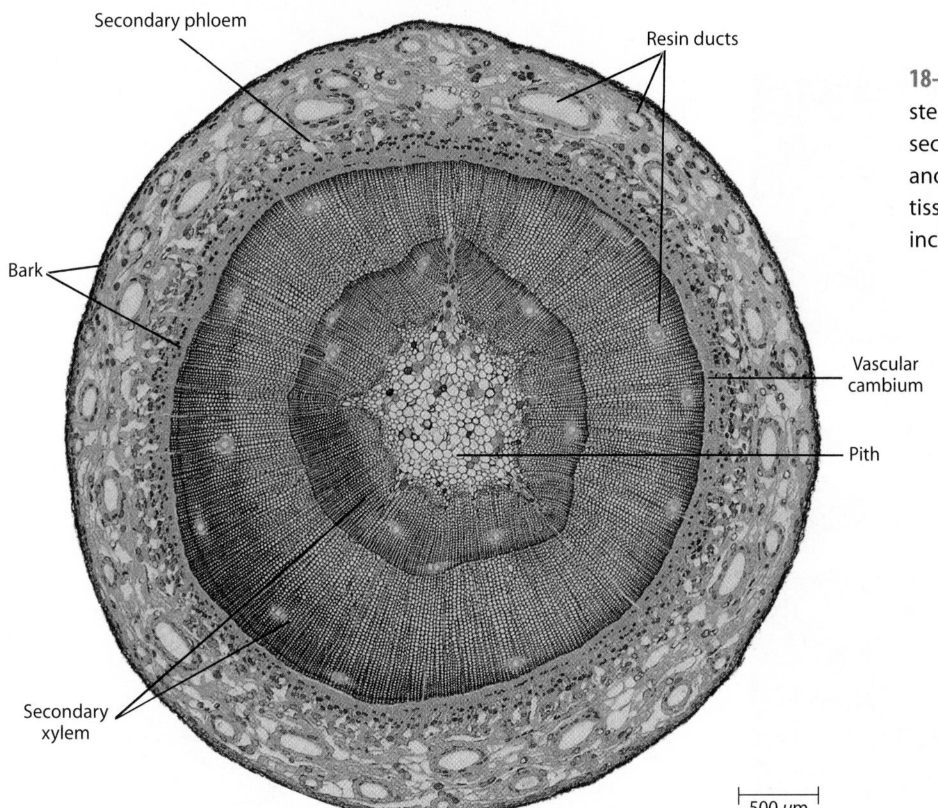

18–15 Pine stem Cross section of a *Pinus* stem, showing secondary xylem and secondary phloem separated from one another by vascular cambium. All of the tissues outside the vascular cambium, including the phloem, make up the bark.

Most pine species retain their needles for two to four years, and the overall photosynthetic balance of a given plant depends on the health of several years' crops of needles. In bristlecone pine *(Pinus longaeva),* the longest-lived tree (see Figure 26–27), the needles are retained for up to 45 years and remain photosynthetically active the entire time. Because the leaves of pines and other evergreens function for more than one season, they are exposed to possible damage by drought, freezing, or air pollution for much longer than the leaves of deciduous plants, which are replaced each year.

In the stems of pines and other conifers, secondary growth begins early and leads to the internal formation of substantial amounts of secondary xylem, or wood (Figure 18–15). Secondary xylem is produced toward the inside of the vascular cambium, and secondary phloem is produced toward the outside. The xylem of conifers consists primarily of tracheids, while the phloem consists of sieve cells, which are the only type of food-conducting cell in gymnosperms (see pages 547 and 548). Both kinds of tissue are traversed radially by narrow rays (see page 617). With the initiation of secondary growth, the epidermis is eventually replaced with a periderm, which is a protective tissue that has its origin in the outer layer of cortical cells. As secondary growth continues, subsequent periderms are produced by active cell division deeper in the bark.

The Pine Life Cycle Extends over a Period of Two Years

As you read this account of reproduction in pines, you may find it useful to refer, from time to time, to the pine life cycle (see Figure 18–19 on pages 442 and 443).

The microsporangia and megasporangia of pines and most other conifers are borne in separate cones, or strobili, on the same tree. Ordinarily, the microsporangiate (pollen-producing) cones are borne on the lower branches of the tree and the megasporangiate, or ovulate, cones are borne on the upper branches. In some pines, they are borne on the same branch, with the ovulate cones closer to the tip. Because the pollen is not normally blown straight upward, the ovulate cones are usually pollinated by pollen from another tree, thus enhancing outcrossing.

Microsporangiate cones of the pines are relatively small, usually 1 to 2 centimeters long (Figure 18–16). The microsporophylls (Figure 18–17) are spirally arranged on the cones and are more or less membranous. Each bears two microsporangia on its lower surface. A young microsporangium contains many **microsporocytes,** or **microspore mother cells.** In early spring, the microspore mother cells undergo meiosis and each produces four haploid microspores. Each microspore develops into a winged pollen grain, consisting of two **prothallial cells,** a **generative cell,** and a **tube cell** (Figures 18–18 and 18–19). This four-celled pollen grain is the immature microgametophyte. It is at this stage that the pollen grains are shed in enormous quantities; some are carried by the wind to the ovulate cones.

The ovulate cones of pines are much larger and more complex in structure than the pollen-bearing cones (Figure 18–20, page 444). The **ovuliferous scales** (cone scales), which bear the ovules, are not simply megasporophylls. Instead, they are entire modified determinate branch systems properly known as **seed-scale complexes.** Each seed-scale complex consists of the ovuliferous scale—which bears two ovules on its upper surface—and a subtending sterile bract (Figure 18–21, page 444). The scales are arranged spirally around the axis of the cone. (The ovulate cone is, therefore, a compound structure, whereas the

18–16 Microsporangiate cones of *Pinus* Microsporangiate cones of Monterey pine, *Pinus radiata,* are seen here shedding pollen, which is blown about by the wind. Some of the pollen reaches the vicinity of the ovules in the ovulate cones and then germinates, producing pollen tubes and eventually bringing about fertilization.

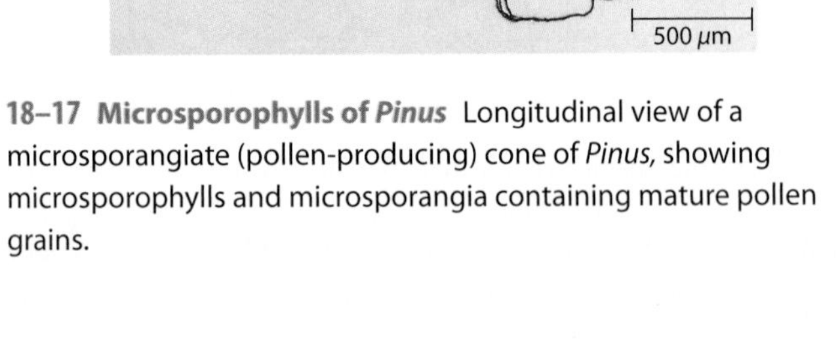

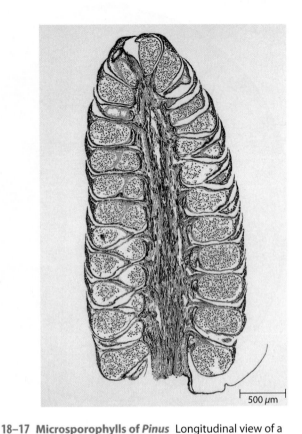

18–17 Microsporophylls of *Pinus* Longitudinal view of a microsporangiate (pollen-producing) cone of *Pinus,* showing microsporophylls and microsporangia containing mature pollen grains.

microsporangiate cone is a simple one, with the microsporangia directly attached to the microsporophylls.) Each ovule consists of a multicellular nucellus (the megasporangium) surrounded by a massive integument with an opening, the micropyle, facing the cone axis (Figure 18–21). Each megasporangium contains a single **megasporocyte,** or **megaspore mother cell,** which ultimately undergoes meiosis, giving rise to a linear series of four megaspores. However, only one of these megaspores is functional; the three nearest the micropyle soon degenerate.

Pollination in pines occurs in the spring (Figure 18–22, page 445). At this stage, the scales of the ovulate cone are widely separated. As the pollen grains settle on the scales, many adhere to pollination drops, which exude from the micropylar canals at the open ends of the ovules. In addition to simple water-soluble compounds, such as sugars, amino acids, and organic acids, pollination drops contain several proteins that are believed to function in both pathogen defense and pollen development. As the pollination drops contract, they carry the pollen grains through the micropylar canal and into contact with the nucellus. At its micropylar end, the nucellus has a slight depression. The pollen grains come to lie in this shallow cavity. After pollination, the scales grow together and help protect the developing ovules. Shortly after coming into contact with the nucellus, the pollen grain germinates to form a pollen tube. At this time, meiosis has not yet occurred in the nucellus, or megasporangium. Failure of pollination results in abortion of the ovules, which occurs about 95 percent of the time in gymnosperms.

About a month after pollination, the four megaspores are produced, only one of which develops into a megagametophyte. The development of the megagametophyte is sluggish. It often does not begin until some six months after pollination, and even then may require another six months for completion. In the early stages of megagametophyte development, mitosis proceeds without immediate cell wall formation. About 13 months after pollination, when the megagametophyte contains some 2000 free nuclei, cell wall formation begins. Then, approximately 15 months after pollination, archegonia, usually two or three in number, differentiate at the micropylar end of the megagametophyte, and the stage is set for fertilization.

About 12 months earlier, the pollen grain had germinated, producing a pollen tube that slowly digested its way through the tissues of the nucellus toward the developing megagametophyte. About a year after pollination, the generative cell of the four-celled microgametophyte undergoes division, giving rise to two kinds of cells—a **sterile cell** (stalk cell) and a **spermatogenous cell** (body cell). Subsequently, before the pollen tube reaches the megagametophyte, the spermatogenous cell divides to produce two sperm. The microgametophyte, or germinating pollen grain, is now mature. Recall that seed plants do not form antheridia.

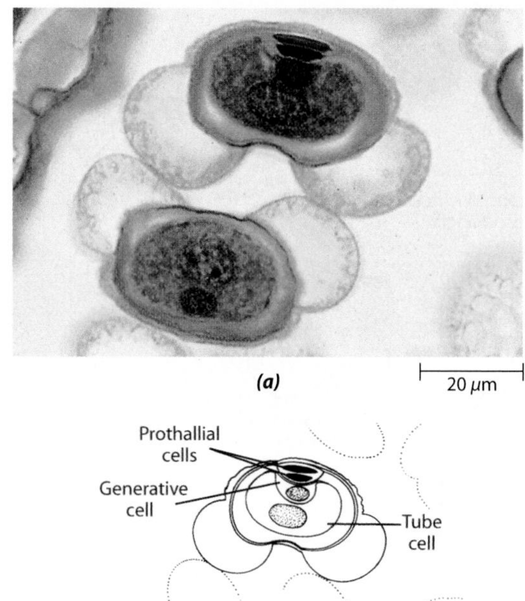

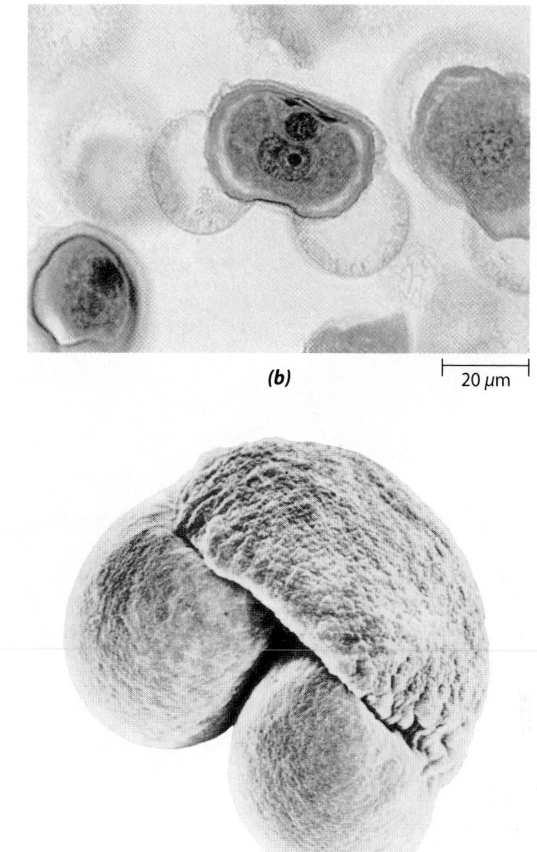

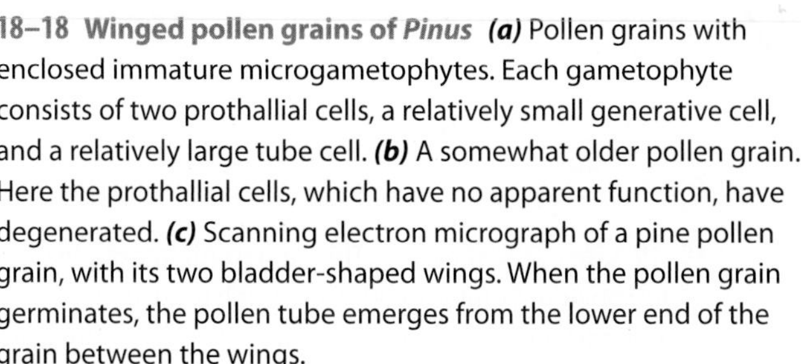

18–18 Winged pollen grains of *Pinus* ***(a)*** Pollen grains with enclosed immature microgametophytes. Each gametophyte consists of two prothallial cells, a relatively small generative cell, and a relatively large tube cell. ***(b)*** A somewhat older pollen grain. Here the prothallial cells, which have no apparent function, have degenerated. ***(c)*** Scanning electron micrograph of a pine pollen grain, with its two bladder-shaped wings. When the pollen grain germinates, the pollen tube emerges from the lower end of the grain between the wings.

Some 15 months after pollination, the pollen tube reaches the egg cell of an archegonium, where it discharges much of its cytoplasm and both of its sperm into the egg cytoplasm (Figure 18–23, page 445). One sperm nucleus unites with the egg nucleus, and the other degenerates. Commonly, the eggs of all archegonia are fertilized and begin to develop into embryos (the phenomenon of **polyembryony**). Only one embryo usually develops fully, but about 3 to 4 percent of pine seeds contain more than one embryo and produce two or three seedlings on germination.

During early embryo development, four tiers of cells are produced near the lower end of the archegonium. Each of the four cells of the uppermost tier (that is, the tier farthest from the micropylar end of the ovule) begins to form an embryo. Simultaneously, the four cells of the tier below the embryos, the suspensor cells, elongate greatly and force the four developing embryos through the wall of the archegonium and into the megagametophyte. Thus, a second type of polyembryony is found in the pine life cycle. As many as 16 embryos may be initiated in a given seed, but usually only one of them develops fully. During embryogeny, the integument develops into a seed coat.

The conifer seed is a remarkable structure, because it consists of a combination of two different diploid sporophyte generations—the seed coat (and remnants of the nucellus) and the embryo—and one haploid gametophyte generation (Figure 18–24, page 446). The gametophyte serves as a food reserve or nutritive tissue. The embryo consists of a hypocotyl-root axis, with a rootcap and apical meristem at one end and an apical meristem and several (generally eight) cotyledons, or seed leaves, at the other. The integument consists of three layers, of which the middle layer becomes hard and serves as the seed coat.

The seeds of pines are often shed from the cones during the autumn of the second year following the initial appearance of the cones and pollination. At maturity, the cone scales separate, and the winged seeds of most species flutter through the air, sometimes blown considerable distances by the wind. In some pine species, such as lodgepole pines *(Pinus contorta),* the scales do not separate until the cones are subjected to extreme heat. When a forest fire sweeps rapidly through the pine grove and burns the parent trees, most of the fire-resistant cones are only scorched. These cones open, releasing the seed crop accumulated over many years, and reestablish the species. In other species of pines, including the limber pine *(Pinus flexilis),* whitebark pine (*Pinus albicaulis;* page 430), and pinyon pines of western North America, as well as a few similar species of Eurasia, the wingless, large seeds are harvested, transported, and stored for later eating by large, crowlike birds called nutcrackers. The birds miss many of the seeds they store, aiding in the dispersal of the pines.

The pine life cycle is summarized in Figure 18–19 (pages 442 and 443).

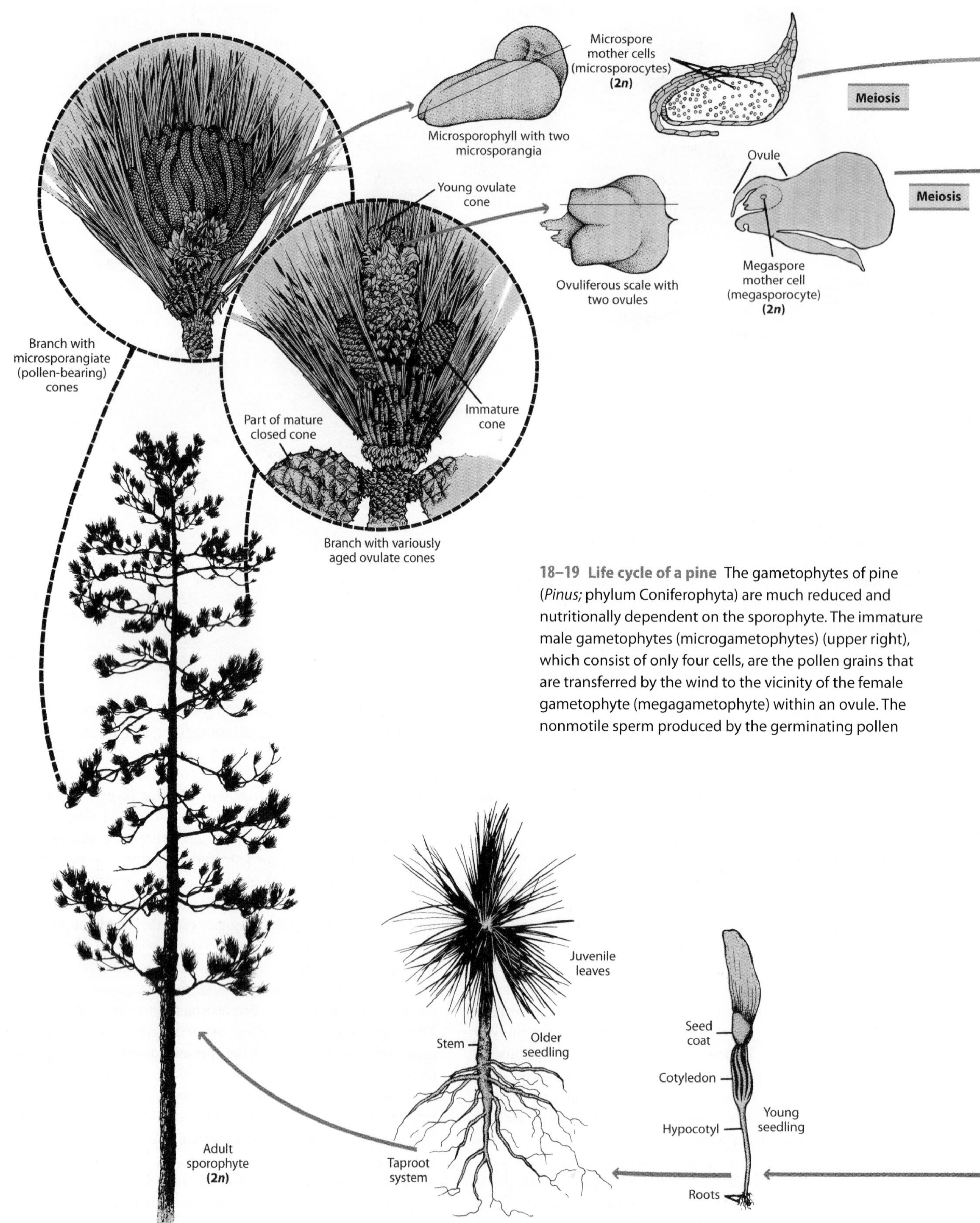

18–19 Life cycle of a pine The gametophytes of pine (*Pinus;* phylum Coniferophyta) are much reduced and nutritionally dependent on the sporophyte. The immature male gametophytes (microgametophytes) (upper right), which consist of only four cells, are the pollen grains that are transferred by the wind to the vicinity of the female gametophyte (megagametophyte) within an ovule. The nonmotile sperm produced by the germinating pollen

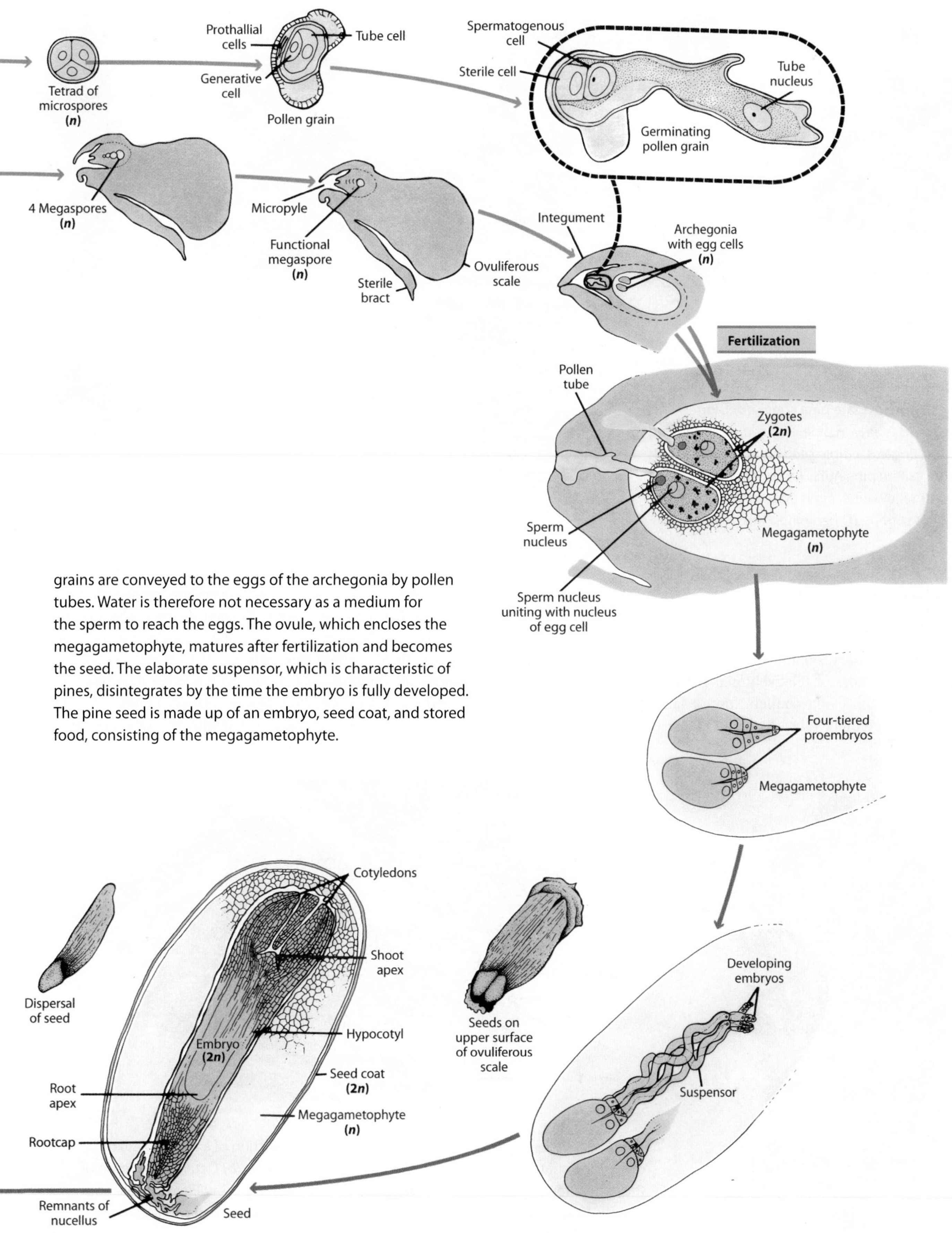

grains are conveyed to the eggs of the archegonia by pollen tubes. Water is therefore not necessary as a medium for the sperm to reach the eggs. The ovule, which encloses the megagametophyte, matures after fertilization and becomes the seed. The elaborate suspensor, which is characteristic of pines, disintegrates by the time the embryo is fully developed. The pine seed is made up of an embryo, seed coat, and stored food, consisting of the megagametophyte.

(a) *(b)* *(c)* *(d)* *(e)* *(f)*

18–20 Ovulate pine cones Shown here are the relative sizes of some mature ovulate pine cones. ***(a)*** Digger pine, *Pinus sabiniana.* ***(b)*** A pinyon pine, *Pinus edulis,* in top and side views. The wingless, edible seeds of this and certain other pines are called "pine nuts." ***(c)*** Sugar pine, *Pinus lambertiana.* ***(d)*** Yellow pine, *Pinus ponderosa.* ***(e)*** Eastern white pine, *Pinus strobus.* ***(f)*** Red pine, *Pinus resinosa.*

Other Important Conifers Occur throughout the World

Although other conifers lack the needle clusters of pines and may also differ in a number of relatively minor details of their reproductive systems, the living conifers form a fairly homogeneous group. In most conifers other than the pines, the reproductive cycle takes only a year; that is, the seeds are produced in the same season that the ovules are pollinated. In such conifers, the time between pollination and fertilization usually ranges from three days to three or four weeks, instead of 15 months or so.

Among the important genera of conifers other than pines are the firs (*Abies;* Figure 18–25a), larches (*Larix;* Figure 18–25b), spruces *(Picea),* hemlocks *(Tsuga),* Douglas firs *(Pseudotsuga),*

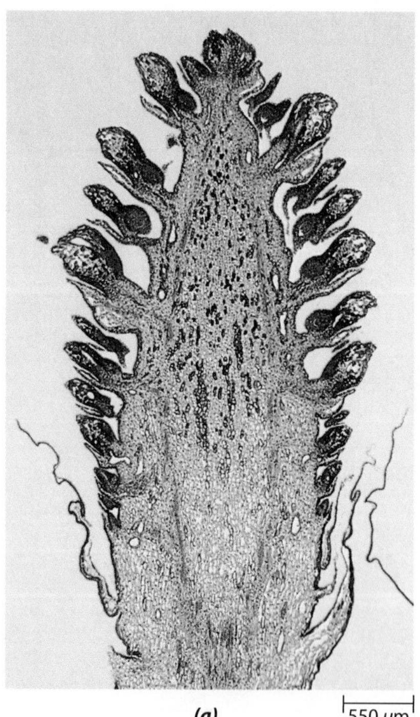

(a)

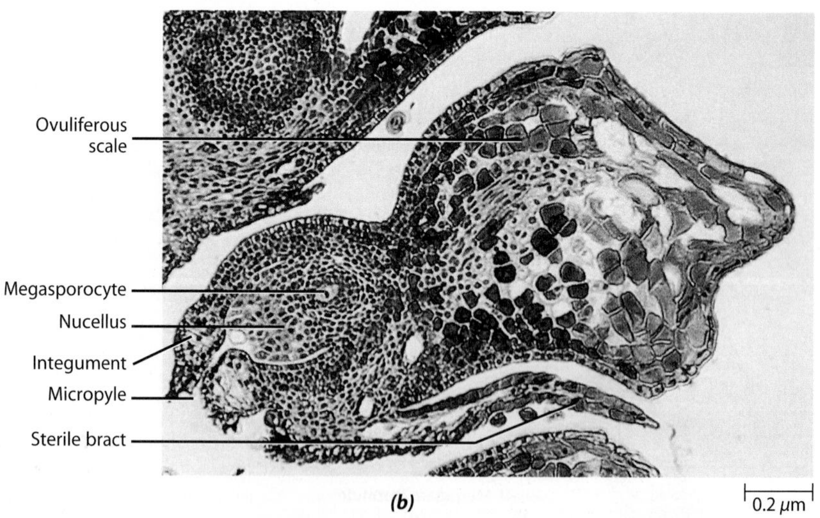

(b)

18–21 Seed-scale complexes *Pinus.* ***(a)*** Longitudinal section of a young ovulate cone, showing the seed-scale complexes along the margins. ***(b)*** Detail of a portion of the cone, showing a seed-scale complex, which consists of the ovuliferous scale and a sterile bract. Note the megasporocyte (megaspore mother cell) surrounded by the nucellus. (See also the pine life cycle, Figure 18–19.)

18–22 Ovulate and pollen-bearing cones Ovulate cones of lodgepole pine, *Pinus contorta,* are visible as tiny red structures at the top of a tall central branch in late spring of their first year, the time at which pollination occurs. One-year-old ovulate cones are visible at the base of this branch. Orange, pollen-bearing microsporangiate cones cluster around the shorter branches.

cypresses (*Cupressus;* Figure 18–26), and junipers (*Juniperus;* Figure 18–27), often misleadingly called "cedars" in North America. *Abies, Larix, Picea, Tsuga,* and *Pseudotsuga* are all Pinaceae; *Cupressus* and *Juniperus* belong to the Cupressaceae. Recently, a new species of Cupressaceae was reported from the northern Vietnamese border province of Ha Giang. Named *Callitropsis vietnamensis,* it is rare among conifers in that both juvenile and mature leaves occur simultaneously on normal plagiotropic (horizontally growing) branching systems of adult trees. In the yews (family Taxaceae), a solitary ovule is borne in a highly reduced cone and surrounded by a fleshy, cuplike structure—the **aril** (Figure 18–28a).

One of the most interesting groups of conifers is the family Araucariaceae, the living members of which occur naturally only in the Southern Hemisphere. The family achieved its greatest diversity in the Jurassic and Cretaceous periods, between 200 and 65 million years ago, but became extinct in the Northern Hemisphere in the Late Cretaceous. Only three surviving genera—*Agathis, Araucaria,* and *Wollemia*—are known to exist. *Wollemia* was discovered in 1994 in a canyon about 150 kilometers northwest of Sydney, Australia (Figure 18–29, page 448). Fewer than 40 trees were found in two small groves, making *Wollemia nobilis* the world's rarest plant species. A species of *Araucaria* called Panama pine is one of the most valuable timber trees in South America. Some species of *Araucaria,* such as the monkey-puzzle tree *(Araucaria araucana)* of Chile and the Norfolk Island pine *(Araucaria heterophylla),* are frequently cultivated in areas where the climate is mild (Figure 18–30, page 449). Seedlings of Norfolk Island pine are also grown as houseplants.

Another interesting group of conifers is the redwoods and their relatives (previously placed in the family Taxodiaceae, which is now included in the Cupressaceae), the wood of which

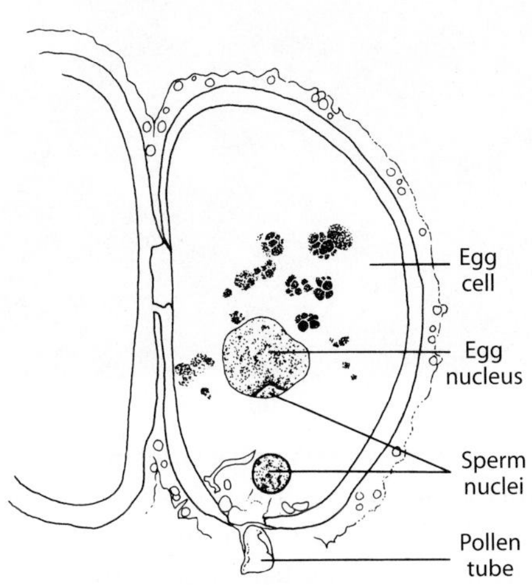

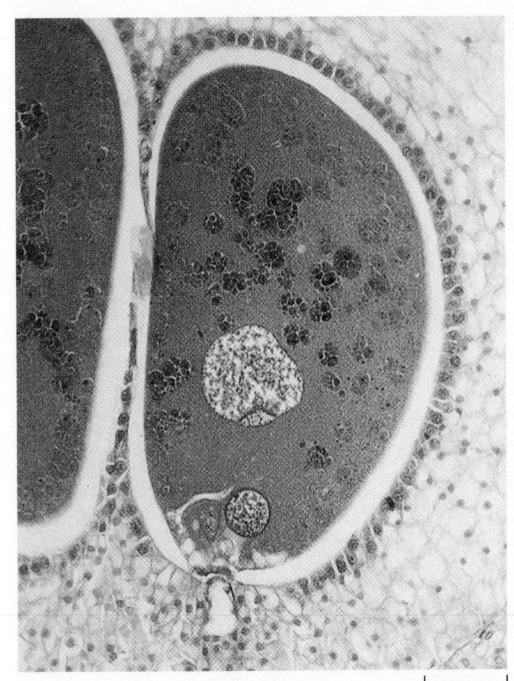

18–23 Fertilization in *Pinus* Fertilization occurs as a sperm nucleus unites with the egg nucleus. The second sperm nucleus (below) is nonfunctional and will eventually disintegrate.

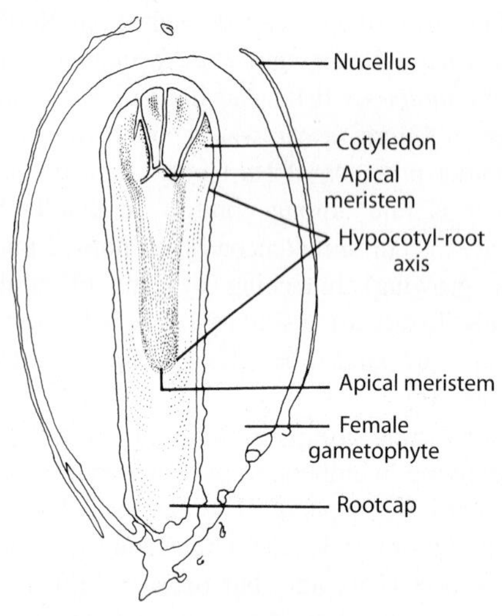

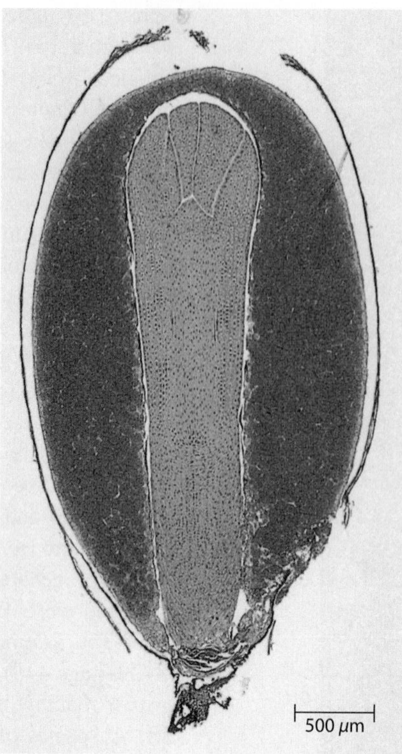

18–24 *Pinus* seed, longitudinal section The hard protective seed coat (here removed) and embryo represent successive sporophyte (2*n*) generations, with a gametophyte generation intervening. A remnant of the nucellus (megasporangium) forms a papery shell around the gametophyte.

is found in the Triassic; a number of fossils (leaves and cones) date from the Middle Jurassic (185 to 165 million years ago). These conifers are represented today by widely scattered species that are the remnants of populations that were much more widespread during the Tertiary period (Figure 18–31). One of the most remarkable of these is the coast redwood, *Sequoia sempervirens,* the tallest living plant. The famous "big tree," *Sequoiadendron giganteum* (Figure 18–32, page 450), which forms

(a)

(b)

18–25 Fir and larch These are two genera of the pine family (Pinaceae). ***(a)*** Ovulate cones of balsam fir, *Abies balsamea.* The upright cones, which are 5 to 10 centimeters long, do not fall to the ground whole, as they do in the pines. Instead, these cones shatter and fall apart while still attached, scattering the winged seeds. ***(b)*** European larch, *Larix decidua.* The needlelike leaves of larch, which are borne singly on both long shoots and short branch shoots, are spirally arranged. Unlike most conifers, the larches are deciduous; that is, they shed their leaves at the end of each growing season.

18–26 Cypress In the subglobose cones of the cypresses, the scales are crowded together, as in this Gowen cypress, *Cupressus goveniana.* The small trees of this species, only about 6 meters tall at maturity, are extremely local—found only near Monterey, California.

18–27 Juniper The common juniper, *Juniperus communis,* has spherical ovulate cones like those of the cypresses, but in juniper the scales are fleshy and fused together. Juniper "berries" give gin its distinctive taste and aroma.

spectacular, widely scattered groves along the western slope of the Sierra Nevada of California, and the bald cypresses *(Taxodium)* of the southeastern United States and Mexico (Figure 18–33, page 450) also belong to this family.

Like most of the living genera of this group, *Metasequoia,* the dawn redwood, was much more widespread in the Tertiary period than it is now (Figure 18–34, page 451). *Metasequoia* occurred widely across Eurasia and was the most abundant conifer in western and Arctic North America from the Late Cretaceous period to the Miocene epoch (from about 90 to about 15 million years ago). It survived both in Japan and in eastern Siberia until a few million years ago. The genus *Metasequoia* was first described from fossil material by the Japanese paleobotanist Shigeru Miki in 1941 (Figure 18–31). Three years

(a)

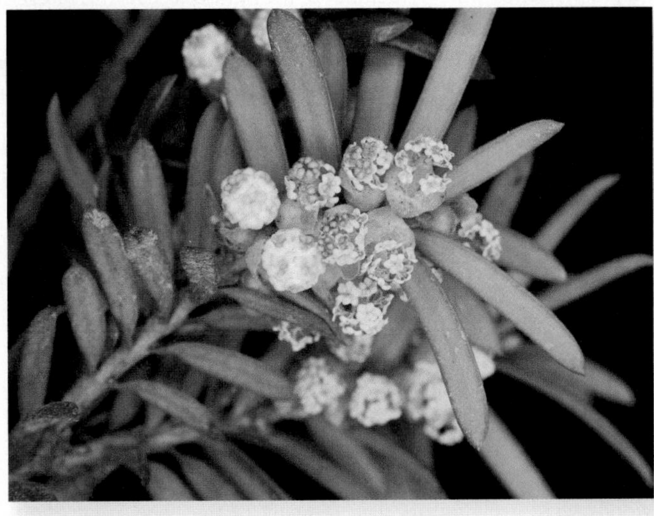

(b)

18–28 Yew The conifers of the yew family (Taxaceae) have seeds that are surrounded by a fleshy cup—the aril. The arils attract birds and other animals, which eat them and thus spread the seeds. ***(a)*** Members of the genus *Taxus,* the yews—which occur throughout the Northern Hemisphere—produce red, fleshy ovulate structures. ***(b)*** Sporophylls and microsporangia of the pollen-bearing cones of a yew. Ovulate cones and pollen-bearing cones are found on separate individuals. The seeds and leaves of yews contain a toxic substance and in the United States are an important cause of poisoning by plants in children, although fatalities are extremely rare.

18–29 ***Wollemia* pine** A member of the Araucariaceae, this rarest of plant species—*Wollemia nobilis*—grows in a rainforest of broad-leaved evergreen trees, over which it rises as an emergent tall (up to 40 meters) tree. ***(a)*** King Billy, the tallest *W. nobilis,* growing in a deep sandstone canyon in the Blue Mountains, northwest of Sydney, Australia. ***(b)*** Spherical ovulate cones appear above the pollen-bearing cones, which hang downward. ***(c)*** Two branches with leaves arranged in four rows.

later, the Chinese forester Tsang Wang, while visiting remote Sichuan Province in south-central China, discovered a huge tree of a kind he had never seen before. The natives of the area had built a temple around the base of the tree. Tsang collected specimens of the tree's needles and cones, and studies of these samples revealed that the fossil *Metasequoia* had "come to life." In 1948, paleobotanist Ralph Chaney of the University of California, Berkeley, led an expedition down the Yangtze River and across three mountain ranges to valleys where dawn redwoods were growing, the last remnant of the once great *Metasequoia* forest. In 1980, about 8000 to 10,000 trees still existed in the *Metasequoia* valley, about 5000 of which had diameters over 20 centimeters. Unfortunately, the trees were not reproducing there, because of harvesting of the seeds for cultivation and the lack of a suitable habitat in which seedlings could become established. Thousands of the seeds have been distributed widely, however, and this "living fossil" can now be seen growing in parks and gardens all over the world.

Other Living Gymnosperm Phyla: Cycadophyta, Ginkgophyta, and Gnetophyta

Cycads Belong to the Phylum Cycadophyta

The other groups of living gymnosperms are remarkably diverse and scarcely resemble one another at all. Among them are the cycads, phylum Cycadophyta, which are palmlike plants found mainly in tropical and subtropical regions. These unique plants, which appeared at least 250 million years ago during the Permian period, were so numerous in the Mesozoic era, along with the superficially similar Bennettitales, that this period is often called the "Age of Cycads and Dinosaurs." Living cycads comprise 11 genera, with about 300 species. *Zamia integrifolia,* found commonly in the sandy woods of Florida, is the only cycad native to the United States (Figure 18–35).

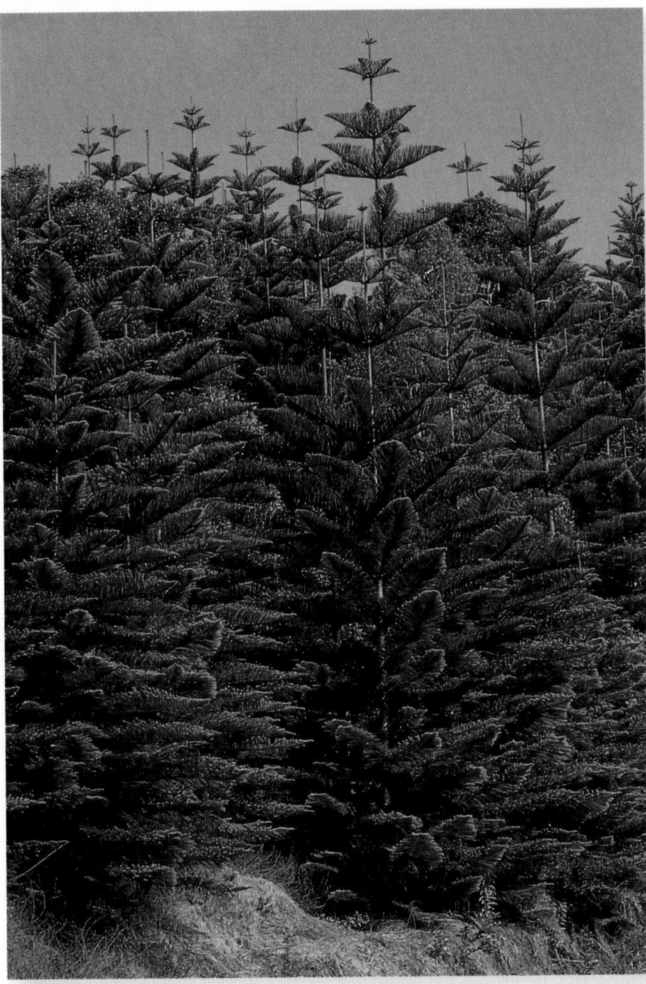

18–30 Norfolk Island pine A member of the family Araucariaceae, Norfolk Island pine, *Araucaria heterophylla,* is native to Norfolk Island in the southern Pacific Ocean, where it grows to a height of 60 meters or more.

Most cycads are fairly large plants; some reach 18 meters or more in height. Many have a distinct trunk that is densely covered with the bases of shed leaves. The functional leaves characteristically occur in a cluster at the top of the stem; thus the cycads resemble palms. (Indeed, a common name for some cycads is "sago palms.") Unlike palms, however, cycads exhibit true, if sluggish, secondary growth from a vascular cambium; the central portion of the cycad trunk consists of a great mass of pith. Cycads are often highly toxic, containing abundant amounts of both neurotoxins and carcinogenic compounds. All cycads form roots that grow upward and branch dichotomously near the soil surface. Because of their resemblance to marine coral, these roots are called **coralloid roots.** Cortical cells of the coralloid roots host the cyanobacterium *Anabaena cycadeae,* which fixes atmospheric nitrogen and presumably contributes nitrogenous substances to the host plant.

The reproductive units of cycads are more or less reduced leaves with attached sporangia that are loosely or tightly clustered into conelike structures near the apex of the plant. The pollen and ovulate cones of cycads are borne on different plants (Figure 18–36). The pollen tubes formed by the microgametophytes are typically unbranched or only slightly branched. In most cycads, growth of the pollen tube results in significant destruction of the nucellar tissue. Before fertilization, the basal end of the microgametophyte swells and elongates, bringing the sperm close to the eggs. The basal end then ruptures, and the released multiflagellated sperm swim to the eggs (Figure 18–37, page 452). Each microgametophyte produces two sperm.

The role of insects in the pollination of cycads is of special interest. Beetles of several groups have frequently been found to be associated with the male cones, and less frequently with the female cones, of the members of several genera of cycads. For example, weevils (family Curculionidae) of the genus *Rhopalotria* carry out their entire life cycle on and within male cones of *Zamia* and also visit the female cones. Pollen-eating beetles, although not weevils, have certainly been present throughout the

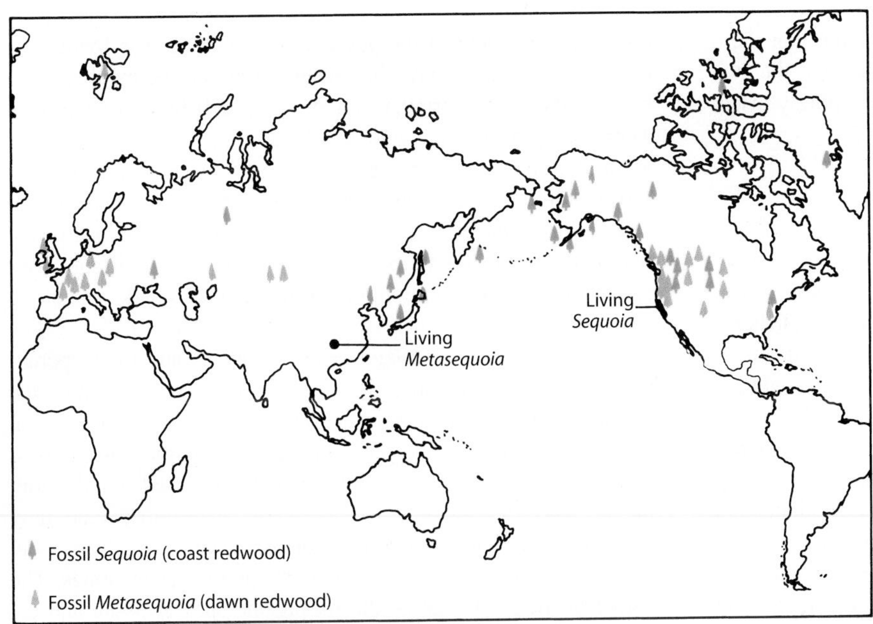

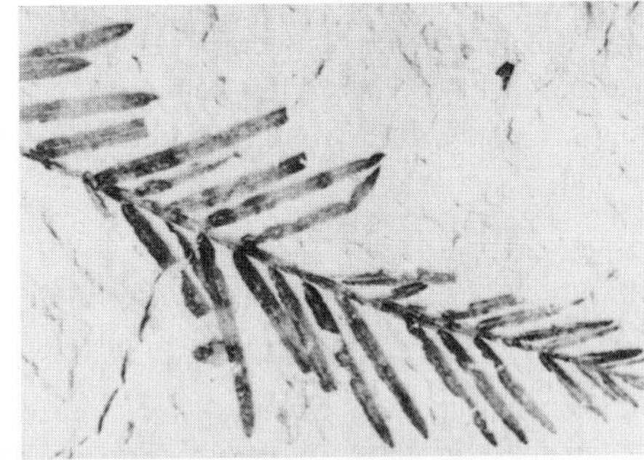

18–31 Fossil branchlet of *Metasequoia* The fossil shown here of *Metasequoia* is about 50 million years old. The accompanying map shows the geographic distribution of some living and fossil members of the redwoods (Cupressaceae).

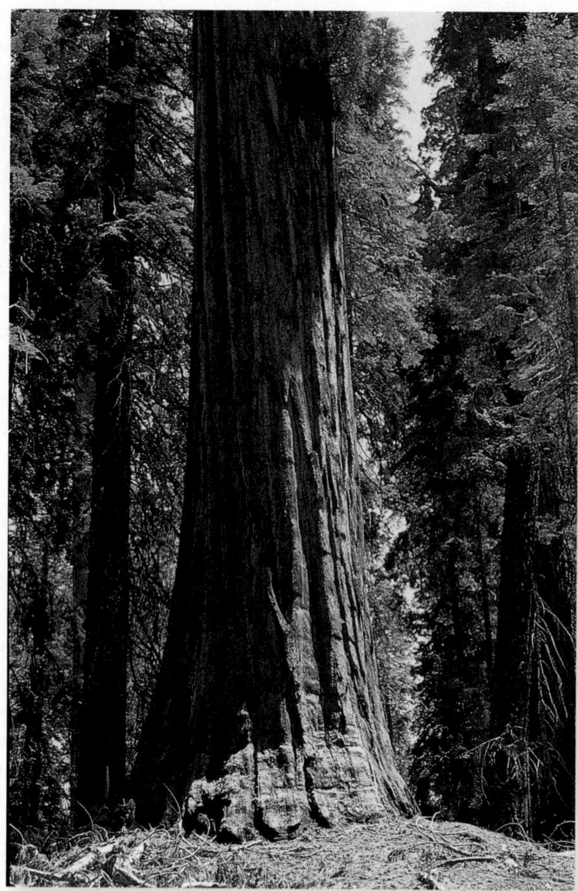

18–32 Sequoias The "big trees," *Sequoiadendron giganteum,* of the western slope of California's Sierra Nevada are the largest of gymnosperms. The largest specimen, the General Sherman sequoia, is more than 80 meters high and is estimated to weigh at least 2500 metric tons. The largest of living animals, the blue whale, pales by comparison. Blue whales rarely exceed 35 meters in length and 180 metric tons in weight.

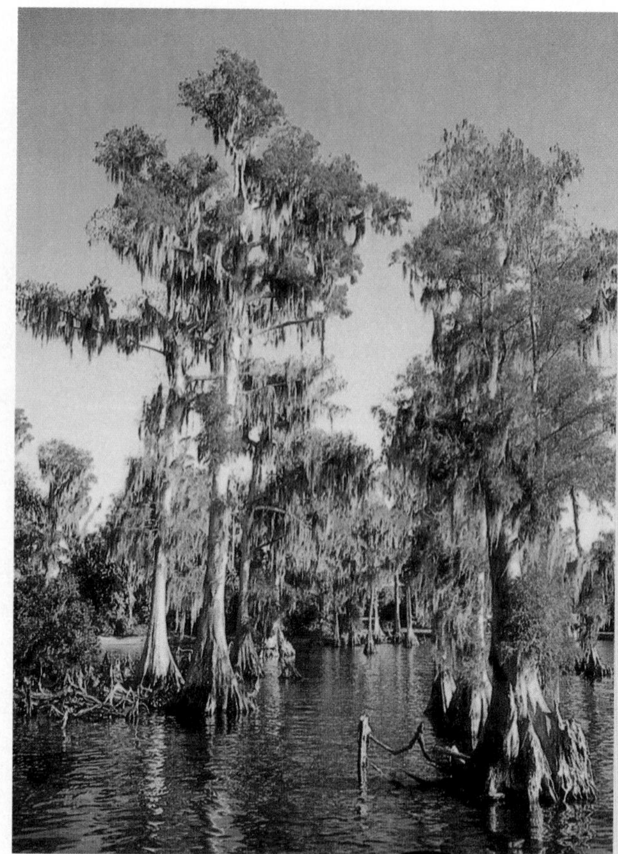

18–33 Bald cypresses The bald cypress, *Taxodium distichum,* is a deciduous member of the Cupressaceae that grows in the swamps of the southeastern United States. Like the larch, it is one of the few conifers that shed their leaves (actually leafy shoots) at the end of each growing season. In this autumn photograph, the leaves have begun to change color. Spanish "moss" *(Tillandsia usneoides),* actually a flowering plant related to the pineapple, is seen hanging in masses from the branches of these trees.

history of the cycads. It seems reasonable to assume that there has been a long relationship between the members of the two groups. Cycads are now considered to be overwhelmingly insect pollinated.

Ginkgo biloba Is the Only Living Member of the Phylum Ginkgophyta

The maidenhair tree, *Ginkgo biloba,* is easily recognized by its fan-shaped leaves with their openly branched, dichotomous (forking) pattern of veins (Figure 18–38). It is an attractive, stately, slow-growing tree that may reach a height of 30 meters or more. The leaves on the numerous slowly growing spur shoots, or short shoots, of *Ginkgo* are more or less entire, whereas those on the long shoots and seedlings are often deeply lobed. Unlike most other gymnosperms, *Ginkgo* is deciduous; its leaves turn a beautiful golden color before falling in autumn.

Ginkgo biloba is the sole living survivor of a genus that has changed little for more than 150 million years, and it is the only living member of the phylum Ginkgophyta. The living species shares features with other genera of gymnosperms that range back to the Early Permian period, some 270 million years ago. There are probably no wild stands of *Ginkgo* anywhere in the world, but the tree was preserved in temple grounds in China and Japan. Introduced into other countries, it has been an important feature of the parks and gardens of the temperate regions of the world for about 200 years. *Ginkgo* is especially resistant to air pollution and so is commonly cultivated in urban parks and along city streets.

Like the cycads, *Ginkgo* bears its ovules and microsporangia on different individuals. The ovules are borne in pairs on the end of short stalks and ripen to produce fleshy-coated seeds in autumn (Figure 18–38b). The rotting flesh of the *Ginkgo* seed coat is infamous for its vile odor, which is derived mainly from the presence of butanoic and hexanoic acids—the same fatty acids found in rancid butter and Romano cheese. For this reason, the male *Ginkgo* is preferred for parks and street plantings. The kernel of the seed (that is, the megagametophytic tissue and the

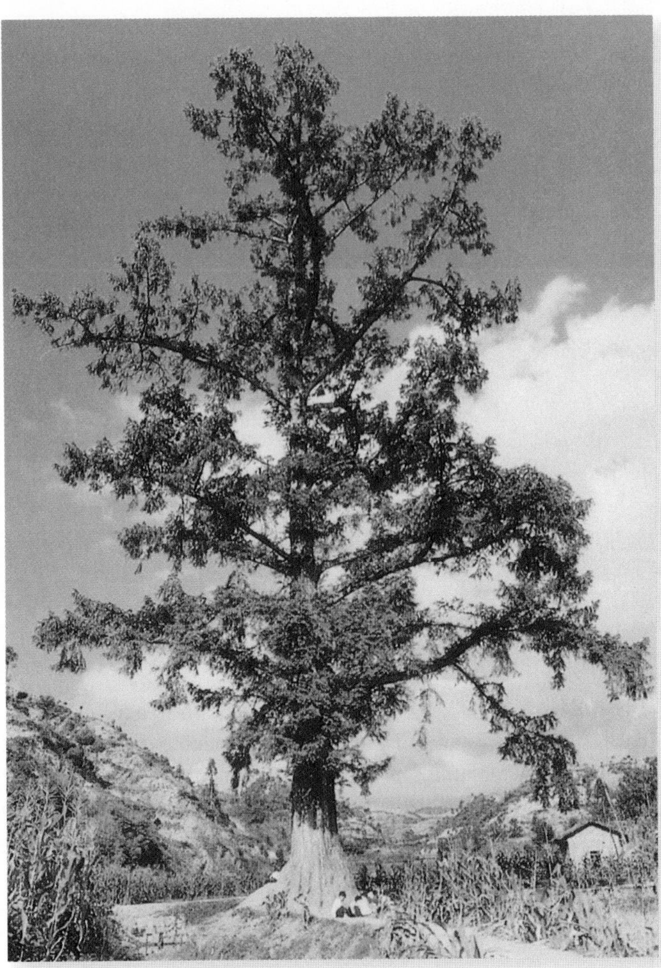

18–34 Dawn redwood This particular dawn redwood *(Metasequoia glyptostroboides)*, seen here growing in Hubei Province in central China, is more than 400 years old.

18–35 *Zamia integrifolia* Male and female plants of *Zamia integrifolia*, the only species of cycad native to the United States. The stems are mostly or entirely underground and, along with the storage roots, were used by Native Americans as food and as a source of starch. The two large gray cones in the foreground are ovulate cones; the smaller brown cones are microsporangiate cones.

embryo), however, has a fishy taste and is a prized delicacy in China and Japan.

In *Ginkgo*, fertilization within the ovules may not occur until they have been shed from their parent tree. As in cycads, the microgametophyte forms an extensively branched, haustorium-like system that develops from an initially unbranched pollen tube (Figure 18–11). Growth of the pollen tube within the nucellus is strictly intercellular, without any apparent

(a)

(b)

18–36 Cycads ***(a)*** *Encephalartos ferox*, a cycad native to Africa. Shown here is a female plant with ovulate cones. ***(b)*** A female plant of *Cycas siamensis*. In this genus, the seeds are borne along the outer edges of the leathery megasporophylls.

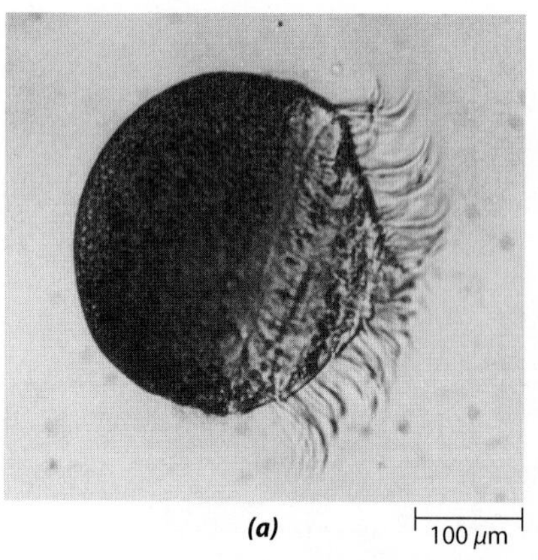

(a)

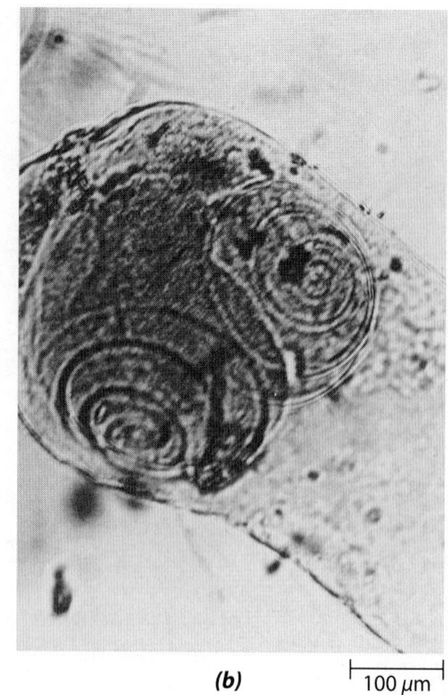

(b)

18–37 Cycad sperm Sexual reproduction in cycads and *Ginkgo* is unusual in combining motile sperm with pollen tubes. ***(a)*** The sperm of the cycad *Zamia pumila,* shown here, swim by virtue of an estimated 40,000 flagella. ***(b)*** Sperm are transported to the vicinity of the egg cells in the ovule by means of a pollen tube (as shown for *Ginkgo* in Figure 18–11).

damage to the adjacent nucellar cells. Eventually, the basal end of this system develops into a saclike structure that, at maturity, contains two large, multiflagellated sperm. Rupture of the saclike portion of the pollen tube releases these sperm to swim to the eggs within the megagametophyte of the ovule.

Ginkgo has recently been found to harbor a *Coccomyxa*-like green alga. Within viable *Ginkgo* host cells, the alga is said to exist in an immature "precursor" state: neither the nucleus nor the mitochondria are discernible and the chloroplast appears to be nonfunctional, with diffuse electron-dense regions marking

(a)

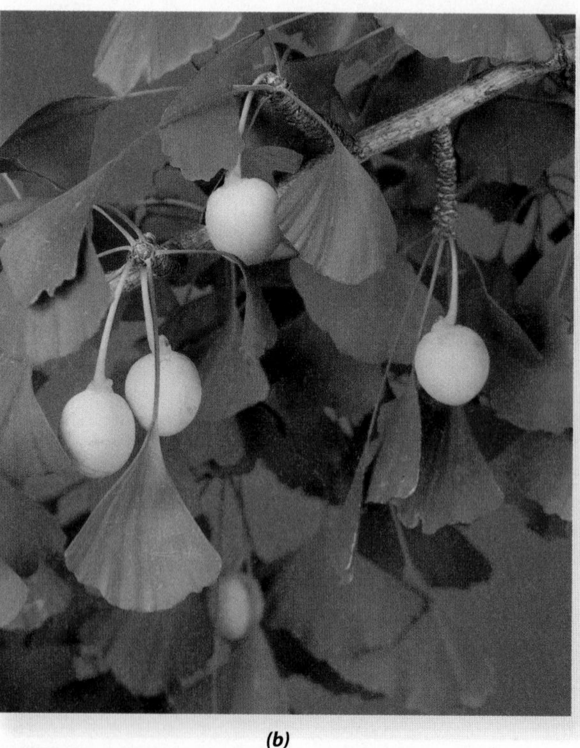
(b)

18–38 *Ginkgo biloba* ***(a)*** The *Ginkgo* tree was given its common name "the maidenhair tree" because of the resemblance between its leaves and the leaflets of maidenhair fern *(Adiantum).* ***(b)*** *Ginkgo* leaves and fleshy seeds attached to spur shoots.

(a)

(b)

(c)

18–39 *Gnetum* The large, leathery leaves of the tropical gnetophyte *Gnetum* resemble those of certain eudicots. The species of *Gnetum* grow as shrubs or woody vines in tropical or subtropical forests. ***(a)*** Megasporangiate inflorescences and leaves of *Gnetum gnemon*. ***(b)*** Microsporangiate inflorescence and leaves and ***(c)*** mature, red fleshy seeds with leaves of *Gnetum urens*, photographed in French Guiana.

thylakoid-like membranous structures. Mature algae with eukaryotic traits and a normal functional chloroplast are almost exclusively found only in host cells that are dying. This association between *Ginkgo* and algae has been found in tissues of *Ginkgo* trees from Asia, Europe, and North America.

The Phylum Gnetophyta Contains Members with Angiosperm-like Features

The gnetophytes comprise three living genera and about 75 species of very unusual gymnosperms: *Gnetum, Ephedra,* and *Welwitschia. Gnetum,* a genus of about 35 species, consists of trees and climbing vines with large, leathery leaves that closely resemble those of eudicots (Figure 18–39). It is found throughout the moist tropics.

Most of the approximately 40 species of *Ephedra* are profusely branched shrubs with inconspicuous, small, scalelike leaves (Figure 18–40). With its small leaves and apparently jointed stems, *Ephedra* superficially resembles *Equisetum.* Most species of *Ephedra* inhabit arid or desert regions of the world.

Welwitschia (with only one species, *Welwitschia mirabilis*) is probably the most bizarre vascular plant (Figure 18–41). Most of the plant is buried in sandy soil. The exposed part consists of a massive, woody, concave disk that typically produces only two strap-shaped leaves that split lengthwise with age. Some plants grow one or two additional leaves. The cone-bearing branches arise from meristematic tissue on the margin of the disk. *Welwitschia* grows in the coastal desert of southwestern Africa, in Angola, Namibia, and South Africa.

Although the genera of Gnetophyta are clearly related to one another and are appropriately placed together (molecular studies strongly support that gnetophytes are monophyletic, with *Ephedra* basal to the derived *Gnetum* and *Welwitschia*), they differ greatly in their characteristics. These genera do, however, have many angiosperm-like features, such as the similarity of their strobili to some angiosperm inflorescences (flower clusters), the presence of vessels in their xylem, and the lack of archegonia in *Gnetum* and *Welwitschia.* (Current analysis favors the idea that the last two features, while similar to those in the angiosperms, were independently derived in gnetophytes and angiosperms.) The mature megagametophyte of *Ephedra,* like those of pine, typically contains two or three archegonia.

You will recall that in pine, as in most other gymnosperms, only one of the two sperm, or sperm nuclei, produced by the germinating pollen grain is functional; one sperm nucleus fertilizes the egg nucleus, and the other degenerates. In the 1990s, it

(a) (b) (c) (d)

18–40 ***Ephedra*** Of the three living genera of Gnetophyta, *Ephedra* is the only one found in the United States. ***(a)*** A male shrub of *Ephedra viridis*, in California. It is a densely branched shrub with scalelike leaves, like other members of the genus. ***(b)*** Microsporangiate strobili of *E. viridis*. Note the scalelike leaves on the stem. ***(c)*** Female plant of *E. viridis* with seeds. ***(d)*** *Ephedra trifurca*, in Arizona, with microsporangiate strobili.

was reported that double fertilization—defined as two fertilization events in a single megagametophyte by two sperm from a single pollen tube—occurs in *Ephedra* and *Gnetum*. In *Ephedra*, the egg cell of each archegonium contains two female nuclei, the egg nucleus and its sister nucleus, the ventral canal nucleus. Each microgametophyte of *Ephedra* produces a single binucleate sperm cell. When the pollen tube reaches an archegonium, one sperm nucleus fertilizes the egg nucleus and the other may fuse with the ventral canal nucleus. (A similar phenomenon has been reported in Douglas fir.) In *Gnetum*, each pollen tube also contains a single binucleate sperm cell. Double fertilization occurs when each of the two sperm nuclei released from the pollen tube fuses with a separate, undifferentiated female nucleus within the free-nuclear megagametophyte. Thus double fertilization, hitherto considered unique to angiosperms, also occurs in *Ephedra* and *Gnetum*, although it is not the norm in *Ephedra*. Unlike in flowering plants, however, where double fertilization produces a distinctive embryo-nourishing tissue called endosperm (in addition to an embryo), the second fertilization event in *Ephedra* and *Gnetum* produces an extra embryo, which ultimately aborts. As is the case with all gymnosperms, in *Ephedra* and *Gnetum* a large megagametophyte serves to nourish the surviving embryo as it develops within the seed.

None of the living gnetophytes could possibly be an ancestor of any angiosperm—each of the three living genera of gnetophytes has its own unique specializations. Interestingly, however, the reproductive structures of at least some species of all three genera of gnetophytes produce nectar and are visited by insects. Wind pollination is clearly important—at least in *Ephedra*—but insects also play a role in the pollination of these plants.

(a)

(b)

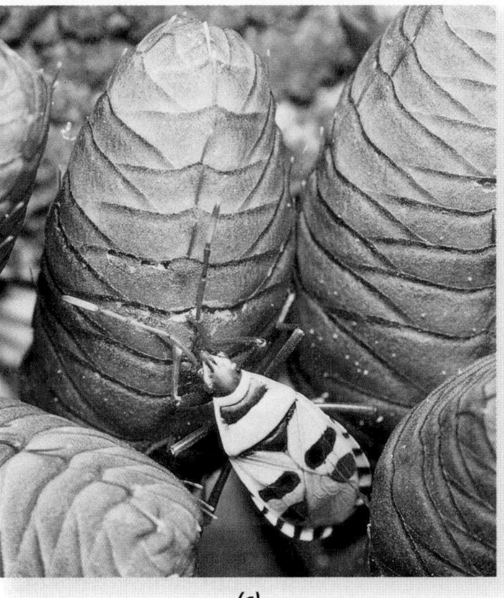

(c)

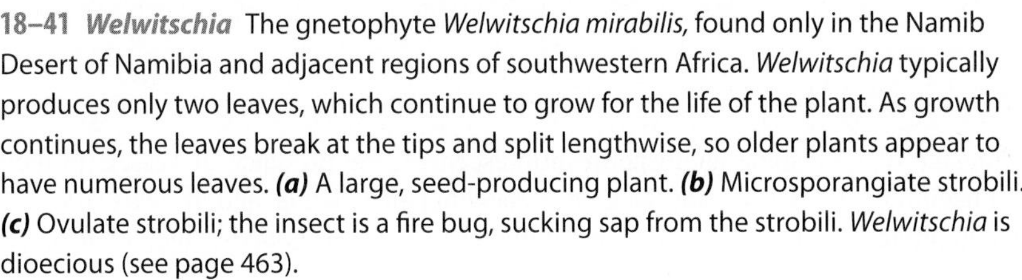

18–41 ***Welwitschia*** The gnetophyte *Welwitschia mirabilis,* found only in the Namib Desert of Namibia and adjacent regions of southwestern Africa. *Welwitschia* typically produces only two leaves, which continue to grow for the life of the plant. As growth continues, the leaves break at the tips and split lengthwise, so older plants appear to have numerous leaves. ***(a)*** A large, seed-producing plant. ***(b)*** Microsporangiate strobili. ***(c)*** Ovulate strobili; the insect is a fire bug, sucking sap from the strobili. *Welwitschia* is dioecious (see page 463).

SUMMARY

Seed plants consist of five phyla with living representatives. One of these is the overwhelmingly successful angiosperms (phylum Anthophyta). The remaining four are commonly grouped together as gymnosperms.

A Seed Develops from an Ovule

Seeds, with their great survival characteristics, give the plants that produce them a major selective advantage. The prerequisites of the seed habit include heterospory (producing both male and female spores); retention of a single megaspore; development of the embryo, or young sporophyte, within the megagametophyte; and integuments. All seeds consist of an embryo, stored food, and a seed coat derived from the integument(s). In gymnosperms, the stored food of the seed is provided by the haploid megagametophyte. In addition to producing seeds, all seed plants bear megaphylls.

Seed Plants Most Likely Evolved from the Progymnosperms

The oldest known seedlike structures occur in strata of the Late Devonian period, about 365 million years old. A possible progenitor of the gymnosperms and angiosperms is the progymnosperms, an extinct group of seedless Paleozoic vascular plants. Among the major extinct groups of gymnosperms are the seed ferns (Pteridospermales), a diverse and unnatural group, and the cycadeoids, or Bennettitales, which had leaves resembling those of cycads but very different reproductive structures.

All Gymnosperms Have the Same Basic Life Cycle

Living gymnosperms comprise four phyla: Coniferophyta, Cycadophyta, Ginkgophyta, and Gnetophyta. Their life cycles are essentially similar: an alternation of heteromorphic generations with large, independent sporophytes and greatly reduced gametophytes. The ovules (megasporangia plus integuments) are exposed on the surfaces of the megasporophylls or analogous structures. At maturity, the megagametophyte of most gymnosperms is a multicellular structure with several archegonia. The microgametophytes develop inside pollen grains. Antheridia are lacking in all seed plants. In gymnosperms, the male gametes, or sperm, arise directly from the spermatogenous cell. Except for the cycads and Ginkgo, which have flagellated sperm, the sperm of seed plants are nonmotile.

Pollination and Pollen Tube Formation Eliminate the Need for Water for the Sperm to Reach the Egg

In seed plants, water is not necessary for the sperm to reach the eggs, as it is in seedless vascular plants. Instead, the sperm are conveyed to the eggs by a combination of pollination and pollen

SUMMARY TABLE Gymnosperm Phyla with Living Representatives

PHYLUM	REPRESENTATIVE GENUS OR GENERA	TYPE OF TRACHEARY ELEMENT(S)	PRODUCE MOTILE SPERM?	POLLEN TUBE A TRUE SPERM CONVEYOR?	TYPE OF LEAVES PRODUCED	MISCELLANEOUS FEATURES
Coniferophyta (conifers)	*Abies, Picea, Pinus,* and *Tsuga*	Tracheids	No	Yes	Most needlelike or scalelike	Ovulate and microsporangiate cones on same plant; ovulate cones compound; pine needles in fascicles
Cycadophyta (cycads)	*Cycas* and *Zamia*	Tracheids	Yes	No	Palmlike	Ovulate and microsporangiate cones simple and on separate plants
Ginkgophyta (maidenhair tree)	*Ginkgo*	Tracheids	Yes	No	Fan-shaped	Ovules and microsporangia on separate plants; fleshy-coated seeds
Gnetophyta (gnetophytes)	*Ephedra, Gnetum,* and *Welwitschia*	Tracheids and vessel elements	No	Yes	*Ephedra:* small scalelike leaves; *Gnetum:* relatively broad, leathery leaves arranged in pairs; *Welwitschia:* two enormous, strap-shaped leaves	Ovulate and microsporangiate cones compound, borne on separate plants, except for some species of *Ephedra;* plants have conifer-like and angiosperm-like features; leaves borne in opposite pairs

tube formation. Pollination in gymnosperms is the transfer of pollen from microsporangium to megasporangium (nucellus). Fertilization occurs when one sperm of the microgametophyte (germinated pollen grain) unites with the egg, which in most gymnosperms is located in an archegonium. The second sperm has no apparent function (except in *Ephedra* and *Gnetum*), and it disintegrates. After fertilization, each ovule develops into a seed. Overall, the seed is a matured ovule containing an embryo.

There Are Four Phyla of Gymnosperms with Living Representatives

The conifers, phylum Coniferophyta, are the largest and most widespread phylum of living gymnosperms, with about 70 genera and some 630 species. They dominate many plant communities throughout the world, with pines, firs, spruces, and other familiar trees over wide stretches of the north. Living cycads, phylum Cycadophyta, consist of 11 genera and about 300 species, mainly tropical but extending away from the equator in warmer regions. Cycads are palmlike plants with trunks and sluggish secondary growth. There is only one living species of the phylum Ginkgophyta, the maidenhair tree *(Ginkgo biloba),* which is found only in cultivation. The three genera of the phylum Gnetophyta show features of conifers and of angiosperms (phylum Anthophyta), such as the similarity of their strobili to some angiosperm inflorescences, the presence of vessels in their xylem, the lack of archegonia in *Gnetum* and *Welwitschia,* and the occurrence of double fertilization in *Ephedra* and *Gnetum.*

QUESTIONS

1. One of the most important evolutionary advances in the progymnosperms was the presence of a bifacial vascular cambium. What is a bifacial vascular cambium, and where is it found besides in the progymnosperms?
2. In what way do the Bennettitales, or cycadeoids, resemble cycads? How do they differ from the cycads?
3. The potential for polyembryony occurs twice in the pine life cycle. Explain.
4. Diagram and label the components of each of the following: a pine ovule with a mature megagametophyte; a mature pine microgametophyte (germinated pollen grain with sperm); and a mature pine seed.
5. Evidence exists in the cycads and *Ginkgo* that the first pollen tubes were haustorial structures, not true sperm conveyors. Explain.
6. Explain how the fertilization events in *Ephedra* differ from those in other gymnosperms.

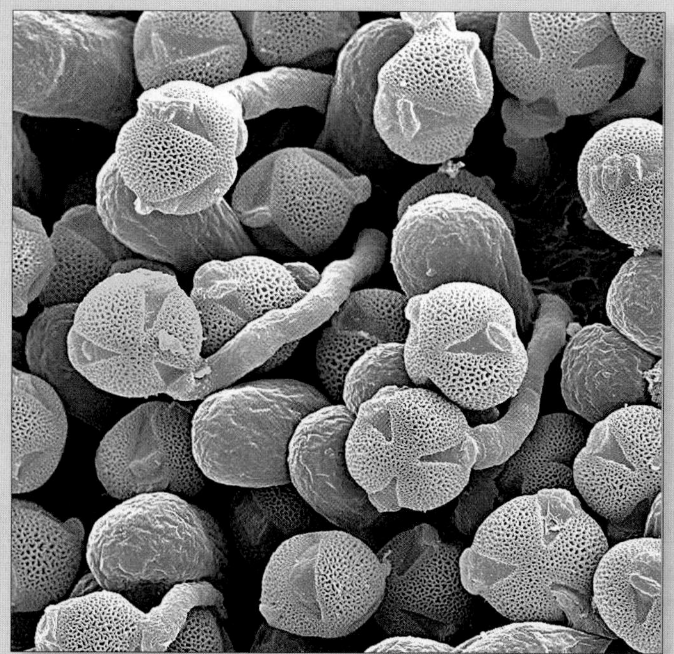

CHAPTER 19

Introduction to the Angiosperms

◀ **Pollination** Bumblebees have deposited pollen grains from several monkeyflower *(Mimulus ringens)* plants onto the stigma (gray) of a single flower. Some of the pollen grains (yellow) have germinated to form pollen tubes that will transport sperm cells to the ovules within the flower. Pollen grains from different donor plants often fertilize adjacent ovules, resulting in genetically diverse offspring within a fruit.

CHAPTER OUTLINE

Angiosperms—the flowering plants—make up much of the visible world of modern plants. Trees, shrubs, lawns, gardens, fields of wheat and corn, wildflowers, fruits and vegetables on the grocery shelves, the bright splashes of color in a florist's window, the geranium on a fire escape, duckweed and water lilies, eel grass and turtle grass, saguaro cacti and prickly pears—wherever you are, flowering plants are there also.

Diversity in the Phylum Anthophyta

Angiosperms make up the phylum Anthophyta, which includes at least 300,000 and possibly as many as 450,000 species and is thus, by far, the largest phylum of photosynthetic organisms. In their vegetative and floral features, angiosperms are enormously diverse. In size, they range from species of *Eucalyptus* trees over 100 meters tall with trunks nearly 20 meters in girth (Figure 19–1) to some species of duckweeds, which are simple, floating plants often scarcely 1 millimeter long (Figure 19–2). Some angiosperms are vines that climb high into the canopy of the tropical rainforest, while others are epiphytes that grow in that canopy. Many angiosperms, such as cacti, are adapted for growth in extremely arid regions. For over 100 million years, the flowering plants have dominated the land.

In terms of their evolutionary history, the angiosperms are a group of seed plants with special characteristics: flowers, fruits, and distinctive life-cycle features that differ from those of all other plants. In this chapter, we outline these characteristics and place them in perspective, and in the next chapter, we discuss the evolution of the angiosperms. We consider the structure and development of the angiosperm plant body (sporophyte) in some detail in Section 5.

Angiosperms share so many unique features that it is clear they are monophyletic (derived from a single common ancestor). They comprise a number of evolutionary lines, some

CHECKPOINTS

After reading this chapter, you should be able to answer the following:

1. What is a flower, and what are its principal parts?
2. Describe some of the variations in flower structure.
3. By what processes do angiosperms form microgametophytes (male gametophytes)? How are these processes both similar to and different from those that give rise to megagametophytes (female gametophytes)?
4. What is the structure, or composition, of the mature microgametophyte in angiosperms? Of the mature megagametophyte?
5. Describe double fertilization in angiosperms. What are the products of this process?

19–1 Giant eucalyptus trees Mountain ash *(Eucalyptus regnans)* trees are shown here growing in the Dandenong Ranges of southeastern Australia. At the end of the nineteenth century, heights over 150 meters were recorded for *E. regnans* trees. During a single season, each massive tree can produce more than a million white flowers. Many of the remaining groves of *E. regnans* are being destroyed for wood chips.

with only a few members, and two very large ones, the classes Monocotyledonae (monocots), with at least 90,000 species (Figure 19–3), and Eudicotyledonae (eudicots), with at least 200,000 species (Figure 19–4). Monocots include such familiar plants as the grasses, lilies, irises, orchids, cattails, and palms, as well as rice and bananas. Eudicots are more diverse, including almost all of the familiar trees and shrubs (other than the conifers) and many of the herbs (nonwoody plants). Other groups of archaic flowering plants that are neither monocots nor eudicots are discussed in Chapter 20. These plants were formerly grouped with the eudicots as "dicots," but we now know that this is an artificial system of classification that simply overemphasizes the distinctiveness of monocots from other angiosperms. The major features of monocots and eudicots are summarized in Table 19–1.

In terms of their mode of nutrition, almost all angiosperms are free-living, but a few parasitic and myco-heterotrophic forms exist. There are about 200 species of parasitic monocots and about 2800 species of parasitic eudicots, including the mistletoes (see Figure 20–27), *Cuscuta* (Figure 19–5a), and *Rafflesia* (Figure 19–5b). Parasitic plants form specialized absorptive organs called haustoria that penetrate the tissues of their hosts. **Myco-heterotrophic** plants, which lack chlorophyll and hence are nonphotosynthetic, have obligate relationships with

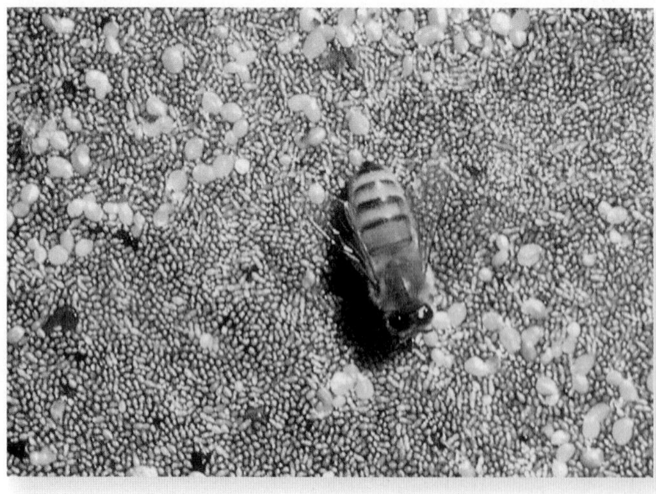

(a)

(b)

19–2 Duckweed The duckweeds (family Lemnaceae) are the smallest flowering plants. ***(a)*** A honeybee is seen here resting on a dense floating mat of three species of duckweed. The larger plants are *Lemna gibba,* about 2 to 3 millimeters long; the smaller ones are two species of *Wolffia,* up to 1 millimeter long. ***(b)*** Flowering plant of *L. gibba;* two stamens and a style protrude from a pocket on the upper surface of the plant.

(a) (b) (c)

19–3 Monocots ***(a)*** The sepals and petals of the iris flower are similar in color. Extensively used as an ornamental in gardens and as cut flowers, the iris belongs to the family Iridaceae. ***(b)*** Flowers and fruits of the banana plant *(Musa × paradisiaca)*. The banana flower has an inferior ovary, and the tip of the fruit bears a large scar left by the fallen flower parts. ***(c)*** In *Trillium erectum*, a member of the death camas family, Melanthiaceae, the sepals are green and the petals red. As is typical of monocot flowers, the sepals and petals are in threes.

(a) (b) (c)

19–4 Eudicots ***(a)*** Saguaro cactus *(Carnegiea gigantea)*. The cacti, of which there are about 2000 species, are almost exclusively a New World family. The thick, fleshy stems, which store water, contain chloroplasts and have taken over the photosynthetic function of the leaves. ***(b)*** Round-lobed hepatica *(Anemone americana)*, which flowers in deciduous woodlands in the early spring. The flowers have no petals but have 6 to 10 sepals and numerous spirally arranged stamens and carpels. ***(c)*** California poppy *(Eschscholzia californica)*, with its brilliant orange petals, is the state flower of California and is protected by law.

Table 19-1 Main Differences between Monocots and Eudicots

Characteristic	Monocots	Eudicots
Flower parts	In threes (usually)	In fours or fives (usually)
Pollen	Monoaperturate (having one pore or furrow)	Triaperturate (having three pores or furrows)
Cotyledons	One	Two
Leaf venation	Usually parallel	Usually netlike
Primary vascular bundles in stem	Scattered arrangement	In a ring
True secondary growth, with vascular cambium	Rare	Commonly present

mycorrhizal fungi that are also associated with a second plant, in this case, a green, actively growing photosynthetic angiosperm. The fungus forms a connecting bridge that transfers carbohydrates from the photosynthetic (autotrophic) plant to the myco-heterotroph—for example, Indian pipe (Figure 19–5c).

The Flower

The flower is a determinate shoot—that is, a shoot with growth of a limited duration—that bears sporophylls, which are sporangium-bearing leaves (Figure 19–6). The name "angiosperm" is derived from the Greek words *angeion,* meaning "vessel," and *sperma,* meaning "seed." The definitive structure of the flower is the carpel—the "vessel." The **carpel** contains the ovules, which develop into seeds after fertilization, while the carpel itself develops into the fruit wall.

Flowers may be clustered in various ways into aggregations called **inflorescences** (Figures 19–7 and 19–8). The stalk of an inflorescence or of a solitary flower is known as a **peduncle,** and the stalk of an individual flower in an inflorescence is called a **pedicel.** The part of the flower stalk to which the flower parts are attached is termed the **receptacle.**

The Flower Consists of Sterile and Fertile, or Reproductive, Parts Borne on the Receptacle

Many flowers include two sets of sterile appendages, the **sepals** and **petals,** which are attached to the receptacle below the fertile parts of the flower, the **stamens** and **carpels.** The sepals arise below the petals, and the stamens arise below the carpels. Collectively, the sepals form the **calyx,** and the petals form the **corolla.** The sepals and petals are essentially leaflike in structure. Commonly the sepals are green and relatively thick, and the petals are brightly colored and thinner, although

(a)

(b)

(c)

19–5 Parasitic and myco-heterotrophic angiosperms These plants have little or no chlorophyll; they obtain their food as a result of the photosynthesis of other plants. ***(a)*** Dodder *(Cuscuta salina),* a parasite that is bright orange or yellow. Dodder is a member of the morning glory family (Convolvulaceae). ***(b)*** The world's largest flower, *Rafflesia arnoldii,* growing in Sumatra. Plants of this genus are parasitic on the roots of a member of the grape family (Vitaceae). There are more than 3000 species of parasitic angiosperms, representing 17 families. ***(c)*** Indian pipe *(Monotropa uniflora),* a myco-heterotroph, lacks chloroplasts and obtains its food from the roots of other, photosynthetic plants via the fungal hyphae associated with its roots.

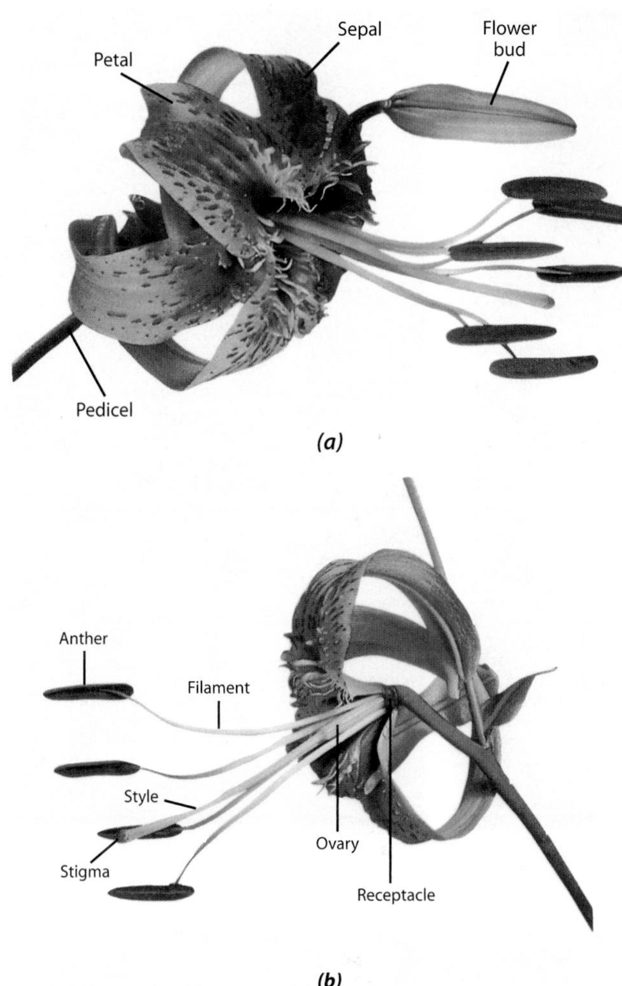

19–6 Parts of a lily flower ***(a)*** An intact flower of a lily *(Lilium henryi)*. In some flowers, such as lilies, the sepals and petals are similar to one another, and the perianth parts—the sepals and petals together—may then be referred to as tepals. Note that the sepals are attached to the receptacle below the petals. ***(b)*** Two tepals and two stamens have been removed to reveal the ovary. The gynoecium consists of the ovary, style, and stigma. Each stamen consists of a filament and an anther. The sepals, petals, and stamens are attached to the receptacle below the ovary, which is composed, in the lily flower, of three fused carpels. Such a flower is said to be hypogynous.

in many flowers the members of both whorls (a whorl is a circle of flower parts of one kind) are similar in color and texture. Together, the calyx (the sepals) and corolla (the petals) form the **perianth.**

The stamens—the pollen-bearing parts of the flower, collectively called the **androecium** ("house of man")—are microsporophylls. In all but a few living angiosperms, the stamen consists of a slender stalk, or **filament,** which bears a two-lobed **anther** containing four microsporangia, or **pollen sacs,** in two pairs—a defining characteristic of angiosperms.

The carpels—the ovule-bearing parts of the flower, collectively known as the **gynoecium** ("house of woman")—are megasporophylls that are folded lengthwise, enclosing one or

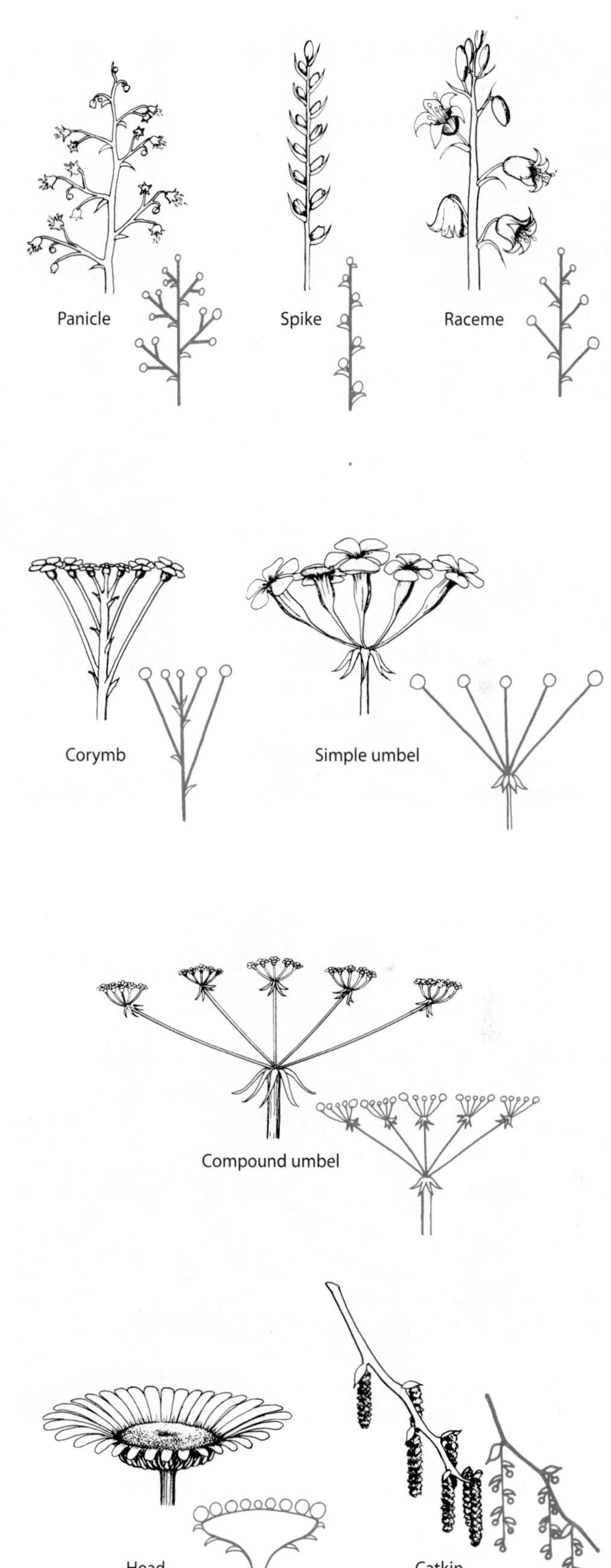

19–7 Types of inflorescences Illustrations of some of the common types of inflorescences found in the angiosperms, accompanied by simplified diagrams (in color).

(a)

(b)

(c)

(d)

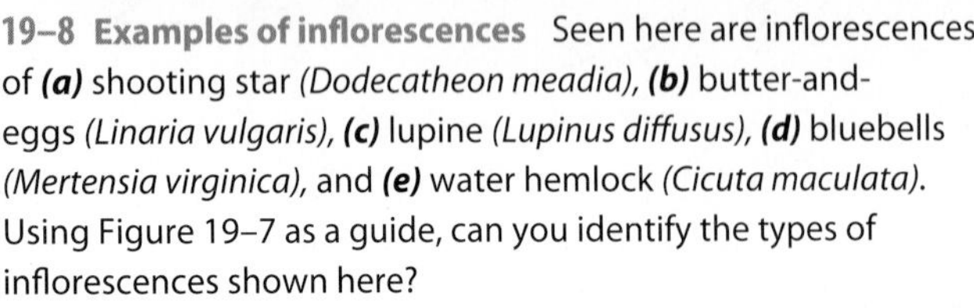

19–8 Examples of inflorescences Seen here are inflorescences of ***(a)*** shooting star *(Dodecatheon meadia)*, ***(b)*** butter-and-eggs *(Linaria vulgaris)*, ***(c)*** lupine *(Lupinus diffusus)*, ***(d)*** bluebells *(Mertensia virginica)*, and ***(e)*** water hemlock *(Cicuta maculata)*. Using Figure 19–7 as a guide, can you identify the types of inflorescences shown here?

(e)

more ovules. A given flower may contain one or more carpels, which may be separate or partly or entirely fused together. Sometimes the individual carpel or the group of fused carpels is called a **pistil.** The word "pistil" comes from the same root as "pestle," the instrument with a similar shape that pharmacists use for grinding substances into a powder in a mortar.

To sum up, individual flowers may have up to four whorls of appendages. From the outside in, the whorls are the sepals (collectively, the calyx); the petals (collectively, the corolla); the stamens (collectively, the androecium); and the carpels (collectively, the gynoecium).

In most flowers, the individual carpels or groups of fused carpels are differentiated into three parts: a lower part, the **ovary,** which encloses the ovules; a middle part, the **style,** through which the pollen tubes grow; and an upper part, the **stigma,** which receives the pollen. In some flowers, a distinct style is absent. If the carpels are fused, there may be a common style, or each carpel may retain a separate one. The common ovary of such fused carpels is generally (but not always) partitioned into two or more **locules**—chambers of the ovary that contain the ovules. The number of locules is usually related to the number of carpels in the gynoecium.

The Ovules Are Attached to the Ovary at the Placenta

The portion of the ovary where the ovules originate and to which they remain attached until maturity is called the **placenta.** The arrangement of the placentae—known as the **placentation**—and consequently of the ovules varies among the different groups of flowering plants (Figure 19–9). In some, the placentation is *parietal;* that is, the ovules are borne on the ovary wall or on extensions of it. In other flowers, the ovules are borne on a central column of tissue in a partitioned ovary with as many locules as there are carpels. This is *axile* placentation. In still others, the placentation is *free central,* the ovules being borne on a central column of tissue not connected by partitions with the ovary wall. And finally, in some flowers, a single ovule occurs at the base (*basal* placentation) or apex (*apical* placentation) of a unilocular ovary. These differences are important in the classification of the flowering plants.

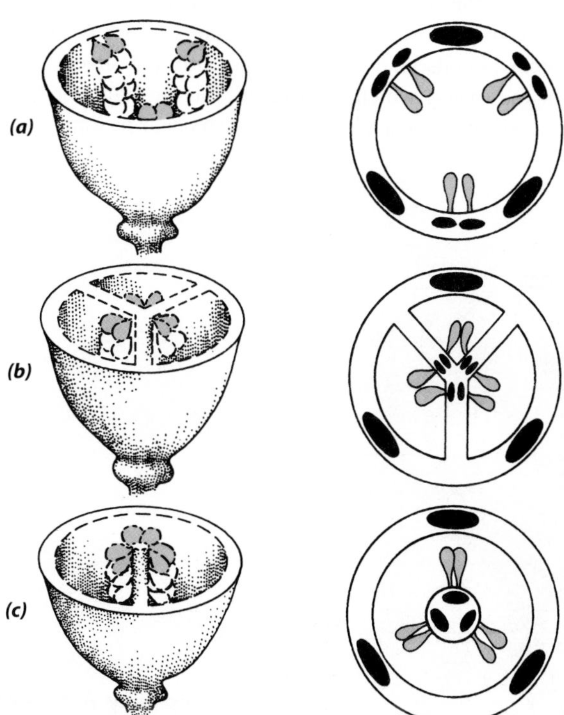

19–9 Placentation Three types of placentation are shown here, with the ovules indicated in color: ***(a)*** parietal, ***(b)*** axile, and ***(c)*** free-central. The vascular bundles are shown as solid structures in the ovary walls. Not shown are basal and apical placentation, with a single ovule at the base or apex, respectively, of a unilocular ovary.

There Are Many Variations in Flower Structure

The majority of flowers include both stamens and carpels, and such flowers are said to be **perfect** (bisexual). If either stamens or carpels are missing, the flower is **imperfect** (unisexual) and, depending on the part that is present, is either **staminate** or **carpellate** (or pistillate) (Figure 19–10). If both staminate and carpellate flowers occur on the same plant, as in maize (see Figure 20–18b, c) and the oaks, the species is said to be **monoecious** (from the Greek words *monos,* "single," and *oikos,* "house"). If staminate and carpellate flowers are found on separate plants, the species is said to be **dioecious** ("two houses"), as in the willows and hemp *(Cannabis sativa).*

Any one of the four floral whorls—sepals, petals, stamens, or carpels—may be lacking. Flowers that have all four floral whorls are called **complete** flowers. If any whorl is lacking, the flower is said to be **incomplete.** Thus an imperfect flower, which is lacking either stamens or carpels, is also incomplete, but not all incomplete flowers are imperfect, because they may well have both stamens and carpels.

The arrangement of the floral parts may be spiral on a more or less elongated receptacle, or similar parts, such as

19–10 Oak flowers Staminate flowers borne on elongated yellow catkins are seen here on a branch of tanbark oak *(Lithocarpus densiflorus).* The two acorns were derived from carpellate flowers. Most members of the oak, or beech, family (Fagaceae), including the true oaks *(Quercus),* are monoecious, meaning that the staminate and carpellate flowers are separate but are borne on the same tree.

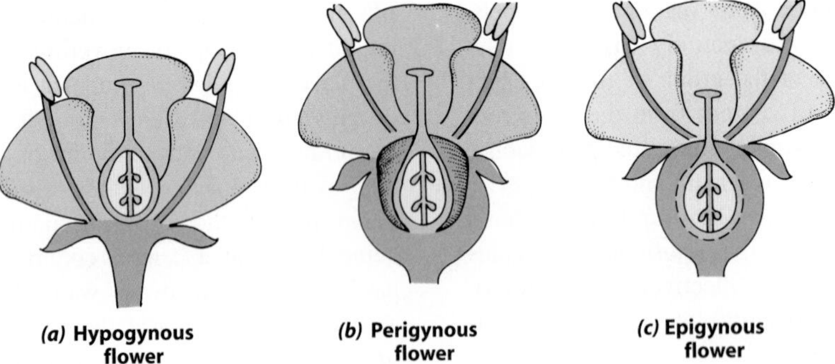

19–11 Position of ovary Types of flowers in common families of eudicots, showing differences in the position of the ovary. ***(a)*** In Ranunculaceae, the buttercup family, the sepals, petals, and stamens are attached below the ovary and there is no fusion; such flowers are said to be hypogynous. ***(b)*** In contrast, many Rosaceae, such as cherries, have superior ovaries, with the sepals, petals, and stamens fused together to form a cup-shaped extension of the receptacle called the hypanthium. Such flowers are said to be perigynous. ***(c)*** The flowers of other plants, such as Apiaceae, the parsley family, have inferior ovaries; that is, the sepals, petals, and stamens appear to be attached above the ovaries. Such flowers are said to be epigynous.

petals, may be attached in a whorl. The floral parts may be united with other members of the same whorl **(connation)** or with members of other whorls **(adnation).** An example of adnation is the union of stamens with the corolla (stamens adnate to the corolla), which is fairly common and occurs, for example, in members of the primrose (Primulaceae), morning glory (Convolvulaceae), gentian (Gentianaceae), milkweed (Apocynaceae), snapdragon (Plantaginaceae), mint (Lamiaceae), honeysuckle (Caprifoliaceae), and aster (Asteraceae), or composite, families. When the floral parts of the same whorl are not joined, the prefixes *apo-* (meaning "separate") or *poly-* may be used to describe the condition. When the parts are joined, or connate, either *syn-* or *sym-* ("together") is used. For example, in an aposepalous or polysepalous calyx, the sepals are not joined; in a synsepalous one, they are.

In addition to this variation in arrangement of flower parts (spiral or whorled), the level of insertion of the sepals, petals, and stamens on the floral axis varies in relation to the ovary or ovaries (Figure 19–11). If the sepals, petals, and stamens are attached to the receptacle below the ovary, as they are in lilies, the ovary is said to be **superior** (Figure 19–6). In other flowers the sepals, petals, and stamens apparently are attached near the top of the ovary, which is **inferior.** Intermediate conditions, in which part of the ovary is inferior, also occur in some plants.

(a)

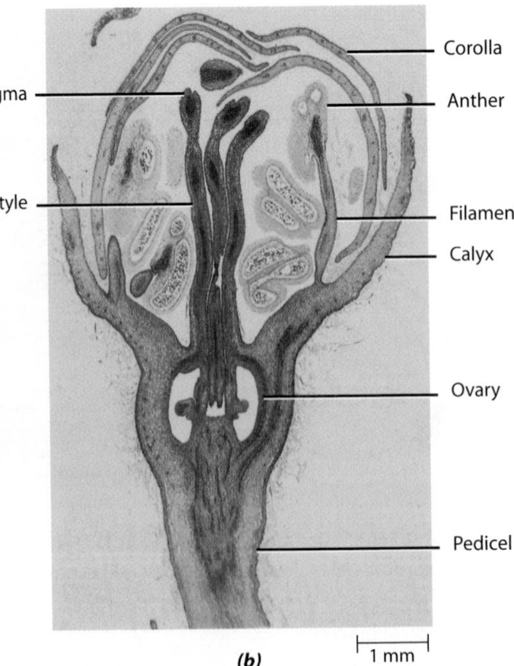

(b)

19–12 Epigyny Apple *(Malus domestica)* flowers, ***(a)*** and ***(b)***, exhibit epigyny—their sepals, petals, and stamens apparently arise from the top of the ovary. In *(b)* the flower is nearly open, but the stamens are not yet erect.

(a)

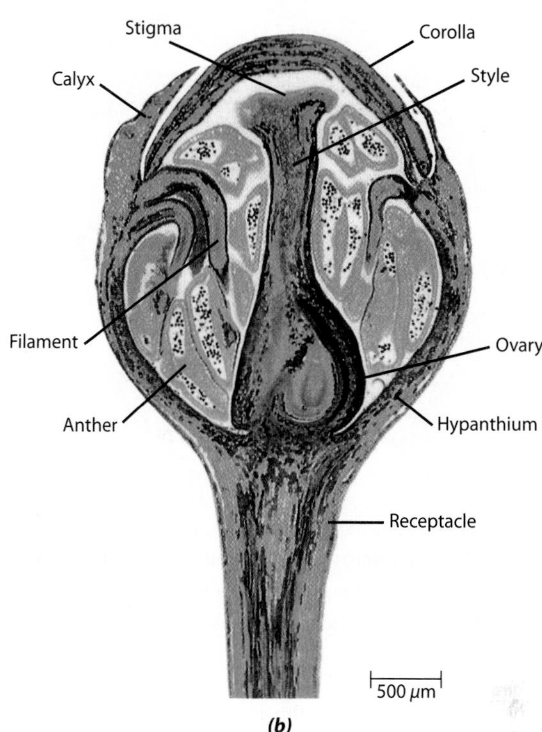

(b)

19–13 Perigyny Cherry *(Prunus)* flowers, ***(a)*** and ***(b)***, exhibit perigyny—their sepals (calyx), petals (corolla), and stamens are attached to a hypanthium. In *(b)* the filaments of the stamens are bent and crowded in the hypanthium, because the flower has not yet opened.

In terms of the points of insertion of the perianth and stamens, there are three categories. In **hypogynous** flowers, perianth and stamens are situated on the receptacle beneath the ovary and free from it and from the calyx, as in lilies (Figure 19–6); in **epigynous** flowers, perianth and stamens arise from the top of the ovary, as in apple blossoms (Figure 19–12); and in **perigynous** flowers, the stamens and petals are adnate to the calyx, thus forming a short tube *(hypanthium)* arising from the base of the ovary, as in cherry flowers (Figure 19–13).

Finally, symmetry in flower structure has two main forms. In **radially symmetrical** flowers, the different whorls are made up of members of similar shape that radiate from the center of the flower and are equidistant from each other. Such flowers—roses and tulips are examples—are said to be **regular,** or actinomorphic (from the Greek root *aktin-,* "ray"). In **bilaterally symmetrical** flowers, one or more members of at least one whorl are different from other members of the same whorl. Bilaterally symmetrical flowers—for example, snapdragons and garden peas—are said to be **irregular,** or zygomorphic (Gk. *zygon,* "yoke" or "pair"). Some regular flowers have irregular color patterns, which give their pollinators an image similar to that of a structurally irregular flower.

The Angiosperm Life Cycle

Angiosperm gametophytes are much reduced in size—more so than those of any other heterosporous plants, including the other seed plants (gymnosperms). The mature microgametophyte consists of only three cells. The mature megagametophyte (embryo sac), which is retained for its entire existence within the tissues of the sporophyte, or specifically of the ovule, typically consists of only seven cells (see Figures 19–19 and 19–23, pages 470 and 474). Both antheridia and archegonia are lacking. Pollination is indirect; that is, pollen is deposited on the stigma, after which the pollen tube grows through or on the surface of tissues of the carpel to convey two nonmotile sperm to the female gametophyte. After fertilization, the ovule develops into a seed, which is enclosed in the ovary. At the same time, the ovary (and sometimes additional structures associated with it) develops into a fruit. (See Figure 19–22 on pages 472 and 473.)

Microsporogenesis and Microgametogenesis Culminate in the Formation of Sperm

Two distinct processes—microsporogenesis and microgametogenesis—lead to the formation of the microgametophyte. **Microsporogenesis** is the formation of microspores (single-celled precursors of pollen grains) within the microsporangia, or pollen sacs, of the anther. **Microgametogenesis** is the later development of the microgametophyte to the three-celled stage.

When it is first formed, the anther consists of a uniform mass of cells, except for the partly differentiated epidermis. Eventually, four columns of fertile, or **sporogenous,** cells become discernible within the anther. Each column is surrounded by several layers of sterile cells, which develop into the wall of the pollen sac. The outermost layers will later trigger opening of

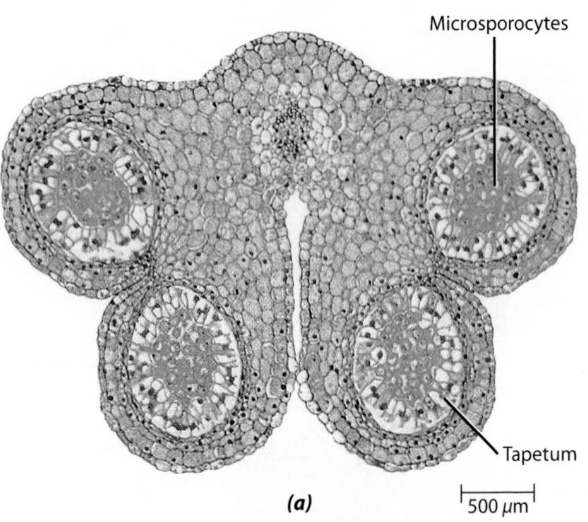

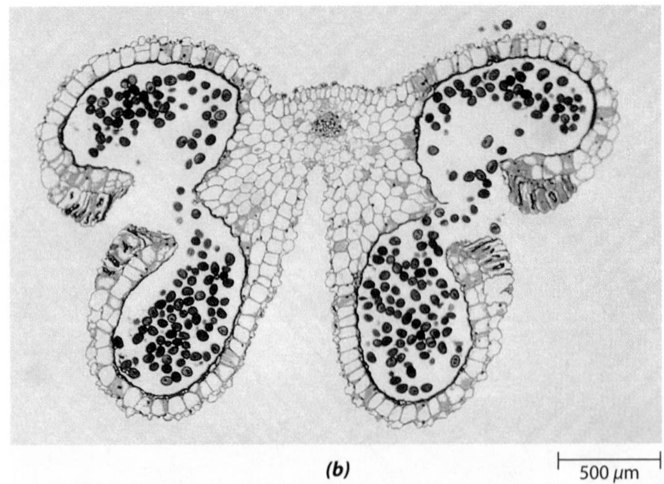

19–14 Transverse sections of lily *(Lilium)* anthers ***(a)*** Immature anther, showing the four pollen sacs containing microsporocytes surrounded by the nutritive tapetum. ***(b)*** Mature anther containing pollen grains. The partitions between the adjacent pollen sacs break down during dehiscence, or shedding of the pollen, as shown here.

the anther (anthesis), whereas the innermost layer of the pollen sac wall forms the nutritive **tapetum** (Figure 19–14a). The tapetum adds a lipid-rich coat to the surface layer of the pollen grain and the spaces within it. The sporogenous cells become microsporocytes (pollen mother cells), which divide meiotically. Each diploid microsporocyte gives rise to a tetrad of haploid microspores. Microsporogenesis is completed with formation of the single-celled microspores, or pollen grains.

During meiosis, each nuclear division may be followed immediately by cell wall formation (common in monocots), or the four microspore protoplasts may be walled off after the second meiotic division (common in eudicots). Subsequently, the major features of the pollen grains are established (Figure 19–15).

The pollen grains develop a resistant outer wall, the **exine,** and an inner wall, the **intine.** The surface of the exine may be smooth or variously sculptured. The exine often is interrupted by pores or linear apertures, which are preferential sites for pollen tube initiation. Pollen grains that lack apertures generally form pollen tubes in portions of the exine that are thinner than others. Apertures are also sites of uptake of various substances, and they have the ability to contract or expand in response to variations in osmotic pressure. Contraction of the apertures, which protects the pollen grain from drying out, may be accompanied by folding or buckling of the pollen grain, further minimizing water loss. The exine is composed of the resistant substance **sporopollenin** (pages 372 and 373), which apparently is synthesized primarily by the tapetum. This polymer, composed chiefly of carotenoids, is present in the spore walls of all plants. It provides the microgametophyte with a strong protective barrier against UV irradiation, dehydration, and pathogen attack. The intine, which is composed of cellulose and pectin, is laid down by the microspore protoplasts. An angiosperm pollen coat, which is often scented, pigmented, and enzyme-rich, is secreted onto the textured exine by the tapetum and is unique to all but a few angiosperms.

Microgametogenesis in angiosperms is uniform and begins when the microspore divides mitotically, forming two asymmetrical cells within the original microspore wall. The division forms a large **vegetative cell,** or **tube cell,** and a small **generative cell,** which moves to the interior of the pollen grain. This two-celled pollen grain is an immature microgametophyte. In about two-thirds of angiosperm species, the microgametophyte is in this *two-celled* stage at the time the pollen grains are liberated from the anther (Figure 19–16). In the remaining angiosperm species, the generative nucleus divides prior to release of the pollen grains, giving rise to two male gametes, or sperm, and resulting in a *three-celled* microgametophyte (Figure 19–17). Mature pollen grains may be packed with starch or oils, depending on the taxon, and these substances are a nutritious source of food for animals.

Pollen grains, like the spores of seedless plants, vary considerably in size and shape. The smallest pollen grains are about 10 micrometers in diameter, and the largest (in the pawpaw family, Annonaceae) are 350 micrometers in diameter. They range from spherical to rod-shaped. They also differ in the number and arrangement of apertures. The apertures may be long and grooved (furrowed) or round and porelike, or a combination of the two. Nearly all families, many genera, and a fair number of species of flowering plants can be identified solely by their pollen grains, on the basis of such characteristics as size, number, arrangement, and type of apertures, and exine sculpturing. In contrast to larger parts of plants—such as leaves, flowers, and fruits—pollen grains, because of their tough, highly resistant exine, are very widely represented in the fossil record. Studies of fossil pollen can provide valuable insights into the kinds of plants and plant communities, and thus the nature of the climates, that existed in the past.

In contrast to the spores of most seedless plants, which are also products of meiosis, pollen grains undergo mitosis before dispersal. Pollen grains, therefore, have two or three nuclei

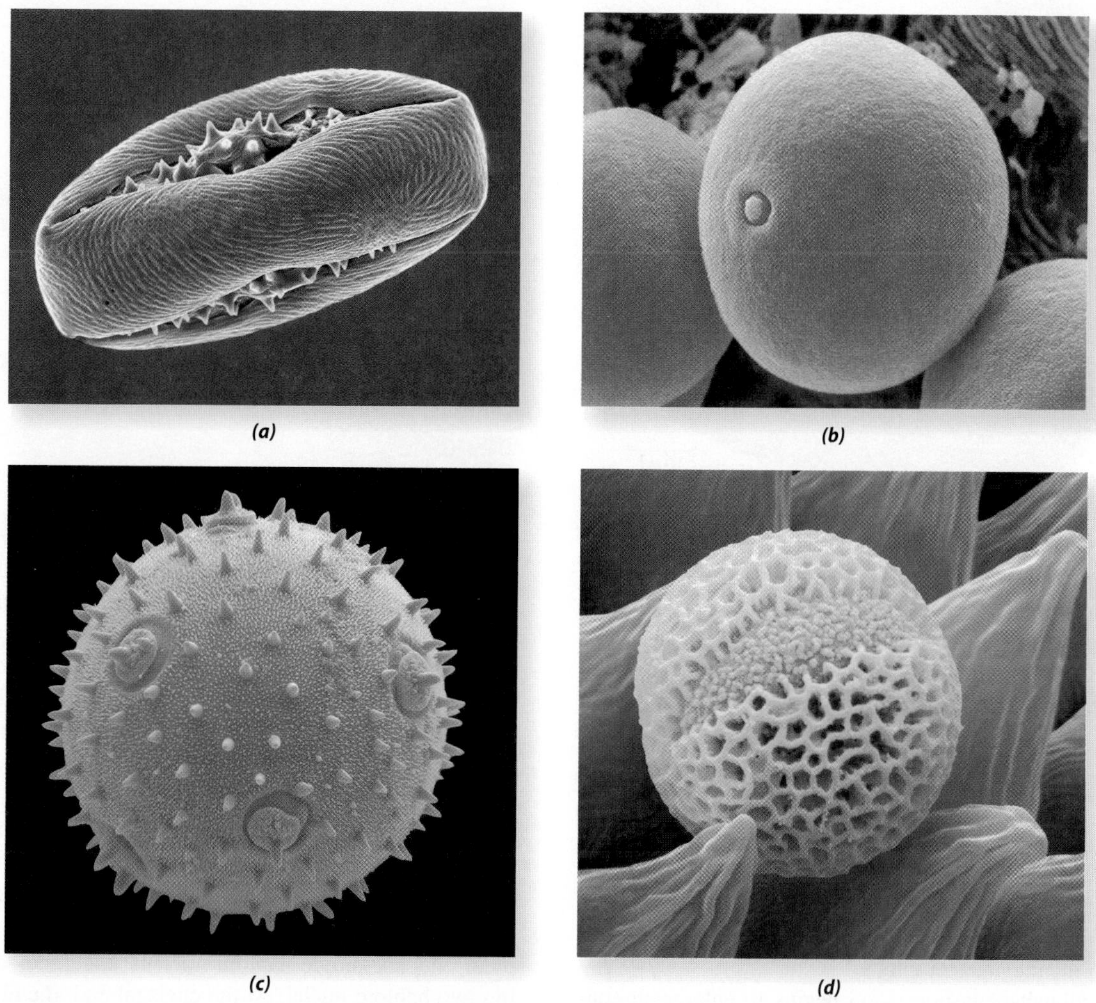

19–15 **Pollen grains** The wall of the pollen grain protects the male gametophyte on its often hazardous journey from the anther to the stigma. The sculpturing of the wall is distinctly different from one species to another, as revealed in these colorized scanning electron micrographs. ***(a)*** Pollen grain of the horse chestnut *(Aesculus hippocastanum)*. Each grain has three lobes, separated by deep furrows. When the pollen grain germinates, the pollen tube will emerge from a pore in one of the furrows. ***(b)*** The smooth pollen grain of timothy grass *(Phleum pratense)* has a single porelike aperture. ***(c)*** The spiny pollen grain of pumpkin *(Cucurbita pepo)* has multiple pores. ***(d)*** A pollen grain of French lavender *(Lavandula dentata)*, a member of the mint family, Lamiaceae, nestled among petal cells of the flower. A furrow interrupts the reticulate sculpturing of the exine.

when shed, whereas most spores have only one. In addition, whereas spores germinate through a characteristic Y-shaped suture on their surface, pollen grains germinate through their apertures.

Megasporogenesis and Megagametogenesis Culminate in Formation of an Egg and Polar Nuclei

Two distinct processes—megasporogenesis and megagametogenesis—lead to the formation of the megagametophyte, or embryo sac. **Megasporogenesis** involves meiosis and results in the formation of megaspores within the ovule in the nucellus (megasporangium). **Megagametogenesis** is the development of the megaspore into the embryo sac.

The ovule is a relatively complex structure, consisting of a stalk, or **funiculus,** bearing a nucellus enclosed by one or two integuments. Depending on the species, one to many ovules may arise from the placentae, or ovule-bearing regions of the ovary wall (Figure 19–9). Initially, the developing ovule is entirely nucellus (Figure 19–18a), but it soon develops one or two enveloping layers, the integuments, which envelop the nucellus but leave a small opening, known as the **micropyle,** at one end of the ovule (Figures 19–18b and 19–19, page 470).

About 70 percent of extant angiosperms undergo a pattern of megasporogenesis and megagametogenesis referred to as the *Polygonum* type (Figure 19–20a, page 470); it proceeds as follows. Early in development of the ovule, a single megasporocyte arises in the nucellus. The diploid megasporocyte divides

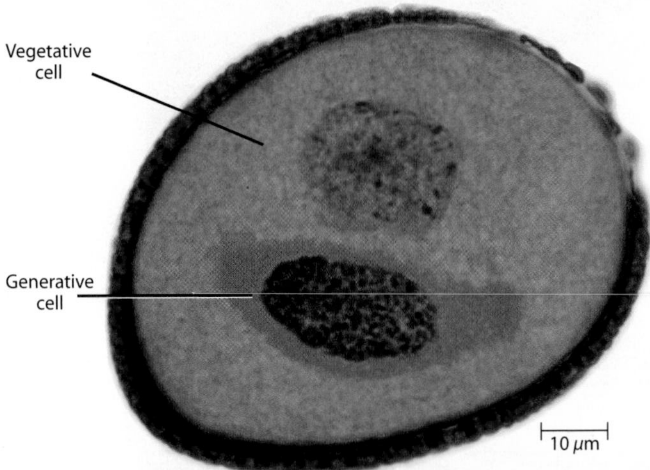

19–16 Two-celled microgametophyte Mature pollen grain of *Lilium,* containing a two-celled male gametophyte. The spindle-shaped generative cell will divide mitotically after germination of the pollen grain. The larger vegetative cell, which contains the generative cell, will form the pollen tube. The round structure above the generative cell is the vegetative cell nucleus.

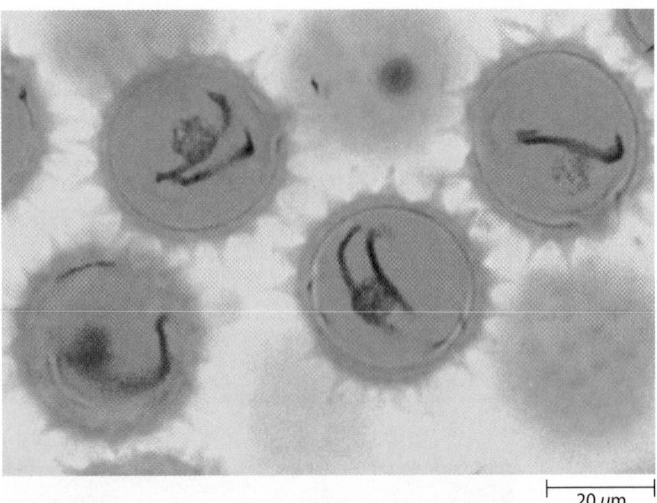

19–17 Three-celled microgametophyte Mature pollen grains—three-celled male gametophytes—of the telegraph plant *(Silphium terebinthinaceum,* family Asteraceae). Prior to pollination, each pollen grain contains two filamentous sperm cells, which are suspended in the cytoplasm of the larger vegetative cell. The pollen of *Silphium* is shed at the three-celled stage, whereas that of *Lilium,* shown in Figure 19–16, is shed at the two-celled stage.

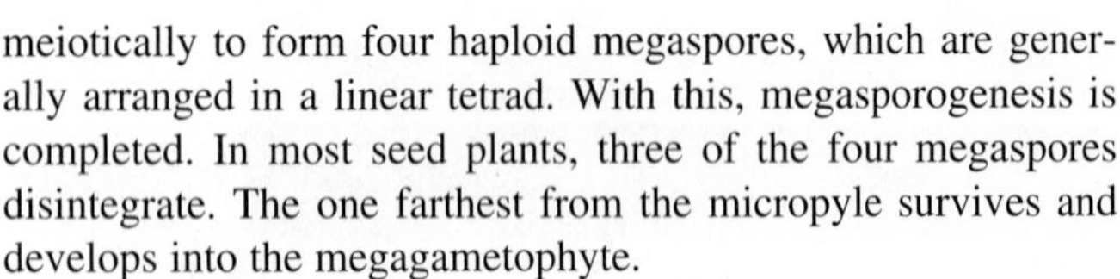

meiotically to form four haploid megaspores, which are generally arranged in a linear tetrad. With this, megasporogenesis is completed. In most seed plants, three of the four megaspores disintegrate. The one farthest from the micropyle survives and develops into the megagametophyte.

The functional megaspore soon begins to enlarge in concert with the expansion of the nucellus, and the nucleus of the megaspore divides mitotically. Each of the resulting nuclei divides mitotically, followed by yet another mitotic division of the four resultant nuclei. At the end of the third mitotic division, the eight nuclei are arranged in two groups of four, one group near the micropylar end of the megagametophyte and the other at the opposite, or **chalazal,** end (Figure 19–18c). One nucleus from each group migrates into the center of the eight-nucleate cell; these two nuclei are then called **polar nuclei.** The remaining three nuclei at the micropylar end become organized as the **egg apparatus,** consisting of an **egg cell** and two short-lived cellular **synergids,** each with a thickened and convoluted cell wall, called the filiform apparatus, at its extreme micropylar end. Cell wall formation also occurs around the three nuclei left at the chalazal end, forming the **antipodals.** The **central cell** contains the two polar nuclei. This eight-nucleate, seven-celled structure is the mature megagametophyte, or **embryo sac** (Figure 19–19).

Other patterns of megasporogenesis and megagametogenesis occur in about a third of the flowering plants. One unusual pattern, called the *Fritillaria* type, occurs in *Lilium,* which is illustrated in Figures 19–18 and 19–20b. In *Lilium,* no wall formation occurs during megasporogenesis; all four megaspore nuclei participate in formation of the embryo sac. Three of the nuclei move to the chalazal end of the embryo sac, while the remaining nucleus becomes situated at the micropylar end. This arrangement of the nuclei represents the *first four-nucleate stage* in development of the embryo sac. What happens next is quite different at the two ends of the embryo sac. At the micropylar end, the single haploid nucleus undergoes mitosis, yielding two haploid nuclei. At the chalazal end, the mitotic spindles of the three sets of chromosomes unite, and mitosis results in two nuclei that are $3n$ (triploid) in chromosome number. As a result of these events, a *second four-nucleate stage* is produced, with two haploid nuclei at the micropylar end of the embryo sac and two triploid nuclei at the chalazal end. Embryo sac development then proceeds in the manner described above for the more frequent kind of embryo sac formation, which has a single four-nucleate stage.

19–18 Ovule and embryo sac development (on facing page) ▶ Some stages in the development of an ovule and embryo sac in *Lilium.* ***(a)*** Two young ovules, each with a single, large megasporocyte surrounded by the nucellus. Integuments have not begun to develop. ***(b)*** The ovule has now developed integuments with a micropyle. The megasporocyte is in the first prophase of meiosis. ***(c)*** An ovule with an eight-nucleate embryo sac (only six of the nuclei can be seen here, four at the micropylar end and two at the opposite, chalazal end). The polar nuclei have not yet migrated to the center of the sac. The funiculus is the stalk of the ovule.

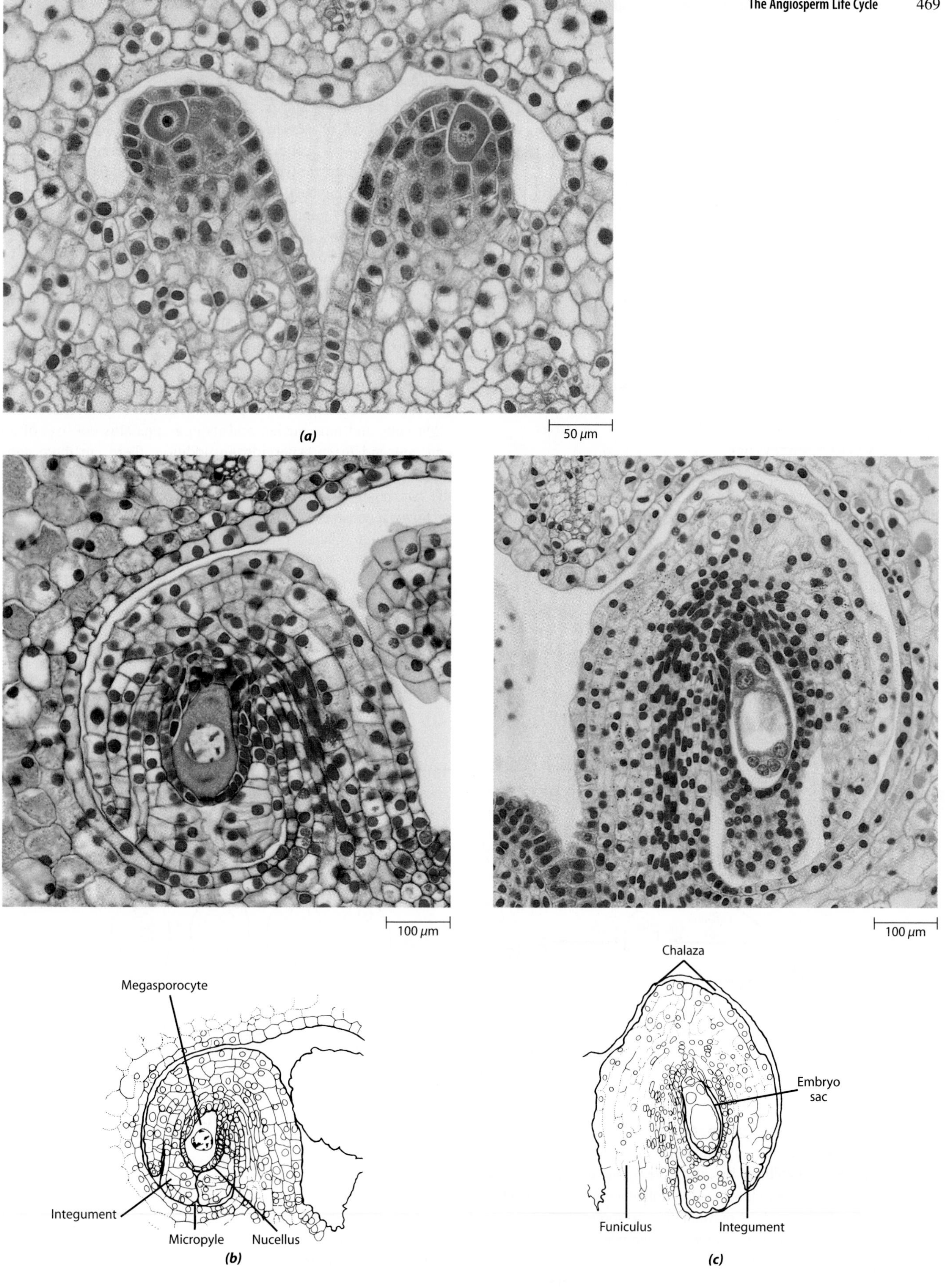
50 μm
(a)
100 μm
100 μm
Megasporocyte
Integument
Micropyle
Nucellus
(b)
Chalaza
Embryo sac
Funiculus
Integument
(c)

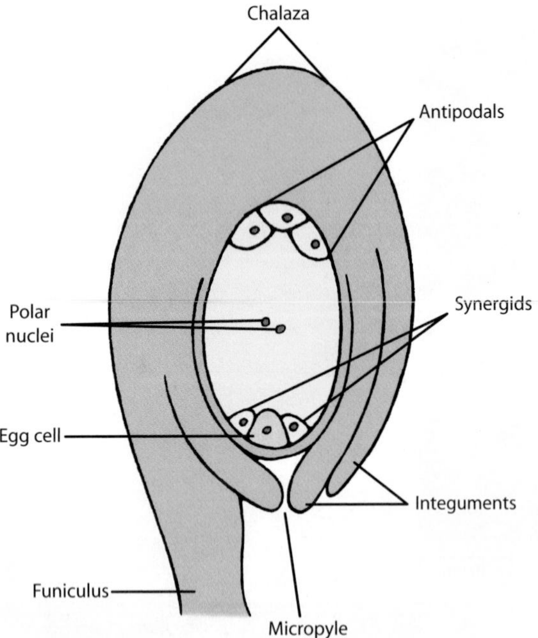

19–19 Mature embryo sac Longitudinal section of an ovule with a mature embryo sac (megagametophyte, or female gametophyte). The mature embryo sac is an eight-nucleate, seven-celled structure, consisting of three antipodals at the chalazal end of the ovule, an egg apparatus (egg cell and two synergids) at the micropylar end, and a large central cell with two polar nuclei.

Most Ancient Angiosperm Lineages Do Not Have *Polygonum*-type Embryo Sacs

Because the great majority of angiosperms have mature embryo sacs consisting of seven cells and eight genetically identical nuclei, it was long believed that the first flowering plants must have had *Polygonum*-type embryo sacs. A series of molecular studies, beginning in 1999, changed that perception. The studies identified a "basal grade" of three angiosperm lineages: the monotypic Amborellaceae, the Nymphaeales, and the Austrobaileyales, with *Amborella* (or *Amborella* plus Nymphaeales) sister to all other angiosperms. Subsequent studies revealed that the mature embryo sacs of members of the Nymphaeales and Austrobaileyales, the *Oenothera*-type embryo sac, contain four cells and four nuclei at maturity: an egg apparatus, consisting of an egg cell and two cellular synergids, as well as a uninucleate central cell (Figure 19–20c). The mature embryo sac of *Amborella* resembles a mature *Polygonum*-type embryo sac, but it consists of eight cells and nine nuclei, and its egg apparatus consists of an egg cell and *three* synergids. In addition, there are three antipodals and a binucleate central cell. (Just before fertilization, the two polar nuclei of the central cell fuse, as is characteristic of many angiosperms.) Thus, none of the most ancient lineages of flowering plants produce a seven-celled, eight-nucleate embryo sac.

Pollination and Double Fertilization in Angiosperms Are Unique

With **dehiscence** of the anther—that is, opening of the pollen sacs—the pollen grains are transferred to the stigmas in a

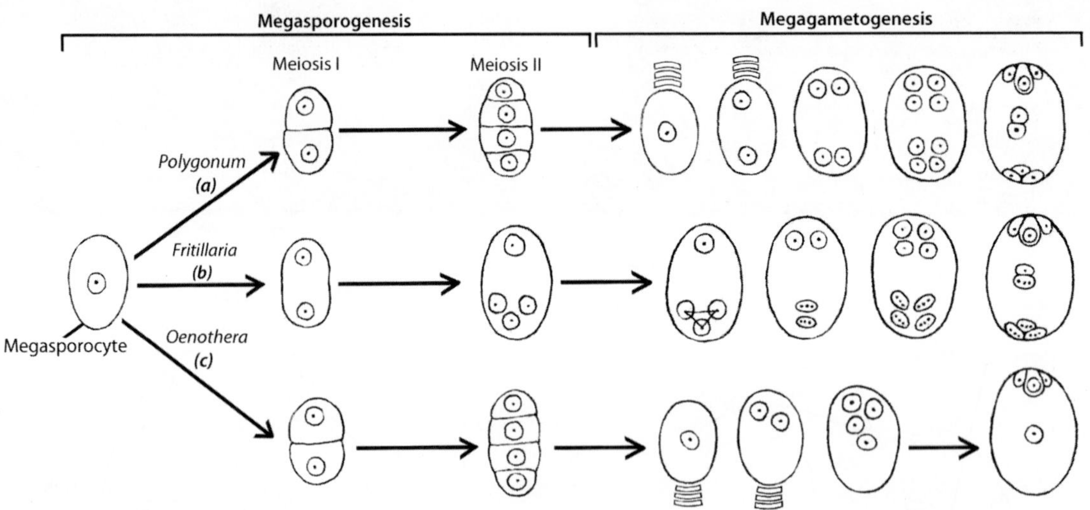

19–20 Comparison of megasporogenesis and megagametogenesis in selected angiosperms ***(a)*** The most common type of embryo sac is the *Polygonum* type. ***(b)*** Much less common is the type exhibited by *Lilium* (the *Fritillaria* type). ***(c)*** The *Oenothera*-type embryo sac is exhibited by two ancient lineages (Nymphaeales and Austrobaileyales) and by the eudicot *Oenothera*. On the basis of the number of megaspores that participate in formation of the embryo sac, both *(a)* and *(c) exhibit monosporic development* (from a single megaspore) and *(b)* exhibits *tetrasporic development* (from four megaspore nuclei). Not shown here is an example of the third category, *bisporic development* (from two megaspore nuclei).

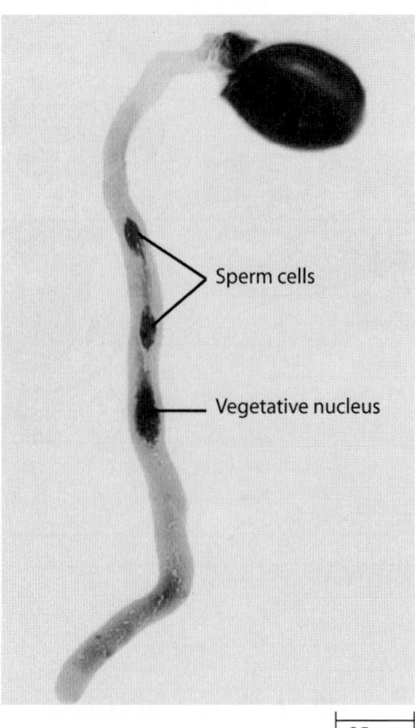

19–21 Mature microgametophyte A microgametophyte, or mature male gametophyte, of Solomon's seal *(Polygonatum)*. The vegetative, or tube, nucleus leads the two sperm cells during their journey in the pollen tube. The vegetative nucleus and two sperm cells are linked, forming the male germ unit.

variety of ways (see Chapter 20). The process whereby this transfer occurs is called **pollination.** Once in contact with the stigma, viable, compatible pollen grains take up water from the cells of the stigma surface. Following this hydration, the pollen grain germinates, forming a pollen tube. If the generative cell has not already divided, it does so within the elongating pollen tube, forming the two sperm. The germinated pollen grain, with its vegetative nucleus and two sperm cells, is the mature microgametophyte (Figure 19–21).

The stigma and style are modified both structurally and physiologically to facilitate germination of the pollen grain and growth of the pollen tube. The surface of many stigmas—so-called **wet stigmas**—consists of glandular tissue that secretes copious amounts of proteins, amino acids, and lipids. The cuticle on the surface of **dry stigmas** contains a hydrated layer consisting of proteins, carbohydrates, and a small amount of lipid. The pollen tubes produced by the germinated pollen grains grow downward between cells of the stigma and enter the style. There, they grow between the cells of a specialized tissue termed **transmitting tissue.** Many monocots and certain groups of eudicots have **open** (or hollow) **stigmas,** which are lined by a glandular epidermis on which the pollen tube grows. After entering the ovary and reaching an ovule, the pollen tube grows out of the transmitting tissue, follows the surface of the funiculus, and enters the micropyle, conveying the two sperm cells and the vegetative nucleus in the process. Throughout their journey within the pollen tube, the two sperm cells are physically associated with the vegetative nucleus, forming a linked unit termed the **male germ unit** (Figure 19–21).

Compared with those of gymnosperms, the pollen tubes of most angiosperms have considerably greater distances to grow from the site of reception to that of fertilization. To compensate for the greater distances, evolution has favored greater growth rates in angiosperms—some 1000-fold greater than those of most gymnosperms. The walls of angiosperm pollen tubes developed a unique structure: a plastic and rapidly extending tip and a reinforced lateral wall composed of callose (a polysaccharide consisting of spirally wound chains of glucose residues). The callose strengthens the tube and provides greater resistance to tensile (stretching) stress. In addition, as they grow, many angiosperm pollen tubes deposit callose plugs that seal off older portions of the tube from the sperm-containing apical portion. The plugs may help maintain positive turgor in the apical, growing portion of the pollen tubes, allowing them to reach greater distances.

Guidance of the pollen tube through the style appears to be governed by the cells of the transmitting tissue. After the pollen tube enters the ovary, it is guided by diffusible chemoattractants produced at the micropylar end of the ovule. Studies on *Torenia* spp. embryo sacs indicate that the synergids are the source of the attractant.

When the pollen tube reaches the embryo sac, it enters one of the synergids near the filiform apparatus and discharges its contents into the degenerated synergid. During degeneration of the synergid, aggregations of actin, called actin "coronas," form near the sperm cells and extend to the fertilization targets, the egg cell and the central cell. The coronas mark the pathways of sperm cell and sperm nucleus migration, indicating that the migration involves actin and myosin interactions, as occurs in cytoplasmic streaming (see page 55). Ultimately, one sperm nucleus unites with the egg nucleus and the other with the polar cell nuclei.

Recall that in most gymnosperms, only one of the two sperm cells is functional; one unites with the egg and the other degenerates. The involvement of both sperm cells—the union of one with the egg and the other with the polar nuclei—is called **double fertilization** (Figure 19–22). Double fertilization in angiosperms, which leads to formation of an embryo and **endosperm,** is a defining characteristic of angiosperms. Although, by definition, double fertilization also occurs in *Ephedra* and *Gnetum* (phylum Gnetophyta; see Chapter 18), the second fertilization event in these gnetophytes does not yield endosperm but rather an extra embryo that eventually aborts.

In angiosperms with the most common type of embryo sac formation (the *Polygonum* type), the fusion of one of the sperm nuclei with the two polar nuclei, called **triple fusion,** results in a triploid ($3n$) **primary endosperm nucleus.** For *Lilium,* illustrated in Figures 19–18 and 19–23, in which one of the polar nuclei is triploid and the other haploid, triple fusion results in a pentaploid ($5n$) primary endosperm nucleus. And, in the Nymphaeales and Austrobaileyales, each of which has a single haploid central cell nucleus, fusion of that nucleus with a sperm nucleus results in a diploid primary endosperm nucleus. Other situations occur in various groups of angiosperms. In any case, the vegetative nucleus degenerates during the process of double fertilization, and the remaining synergid and antipodals also degenerate near the time of fertilization or early in the course of differentiation of the embryo.

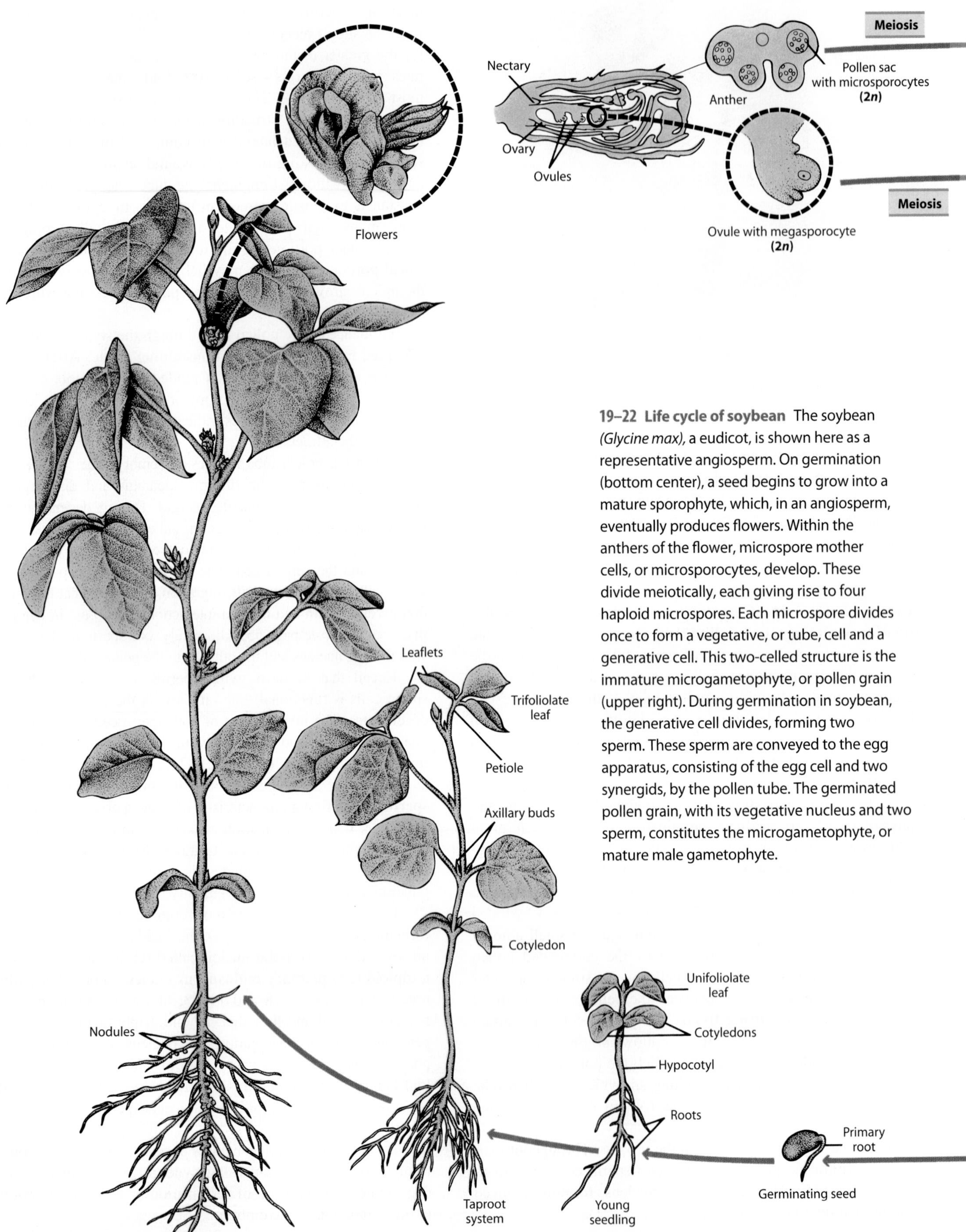

19–22 Life cycle of soybean The soybean *(Glycine max),* a eudicot, is shown here as a representative angiosperm. On germination (bottom center), a seed begins to grow into a mature sporophyte, which, in an angiosperm, eventually produces flowers. Within the anthers of the flower, microspore mother cells, or microsporocytes, develop. These divide meiotically, each giving rise to four haploid microspores. Each microspore divides once to form a vegetative, or tube, cell and a generative cell. This two-celled structure is the immature microgametophyte, or pollen grain (upper right). During germination in soybean, the generative cell divides, forming two sperm. These sperm are conveyed to the egg apparatus, consisting of the egg cell and two synergids, by the pollen tube. The germinated pollen grain, with its vegetative nucleus and two sperm, constitutes the microgametophyte, or mature male gametophyte.

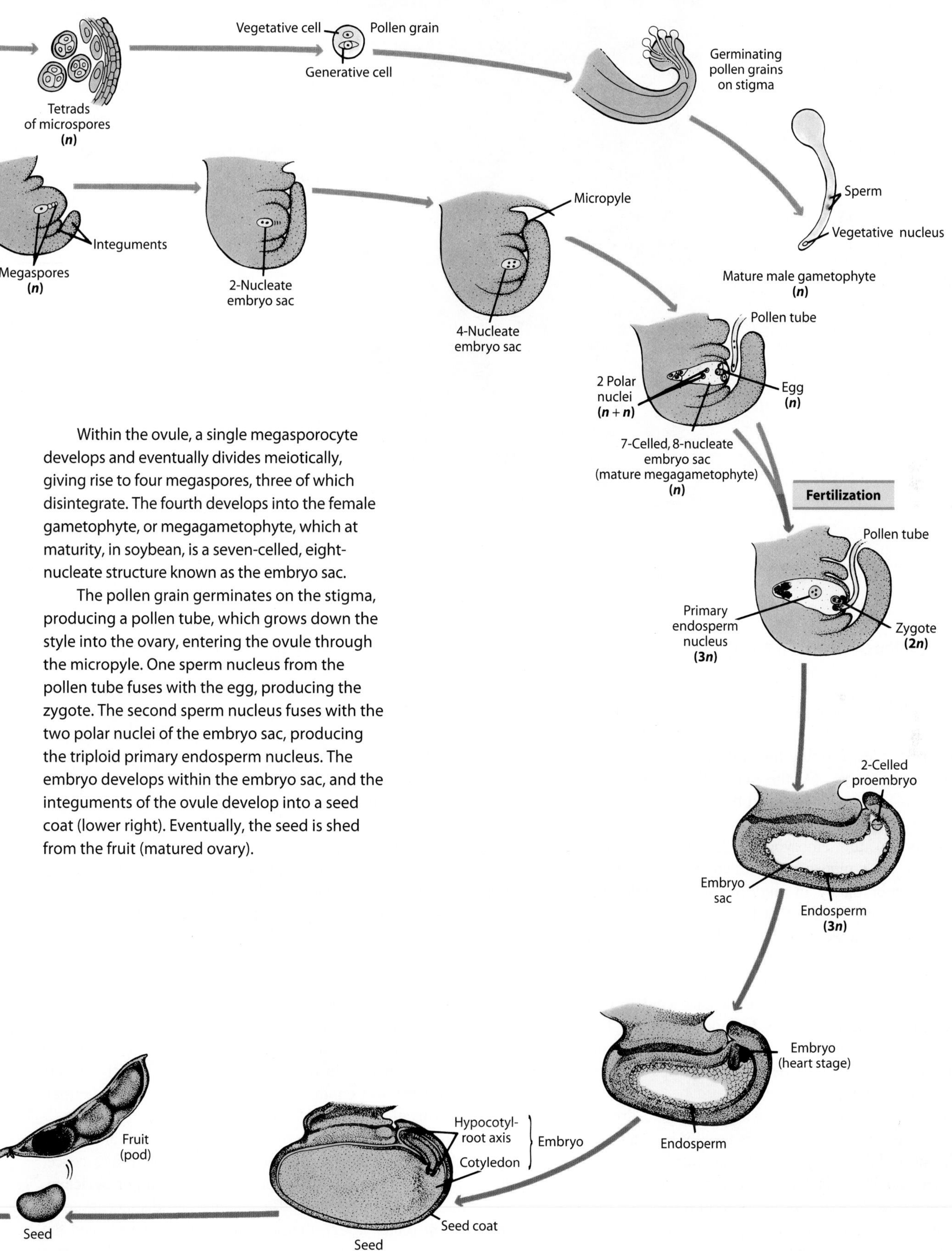

Within the ovule, a single megasporocyte develops and eventually divides meiotically, giving rise to four megaspores, three of which disintegrate. The fourth develops into the female gametophyte, or megagametophyte, which at maturity, in soybean, is a seven-celled, eight-nucleate structure known as the embryo sac.

The pollen grain germinates on the stigma, producing a pollen tube, which grows down the style into the ovary, entering the ovule through the micropyle. One sperm nucleus from the pollen tube fuses with the egg, producing the zygote. The second sperm nucleus fuses with the two polar nuclei of the embryo sac, producing the triploid primary endosperm nucleus. The embryo develops within the embryo sac, and the integuments of the ovule develop into a seed coat (lower right). Eventually, the seed is shed from the fruit (matured ovary).

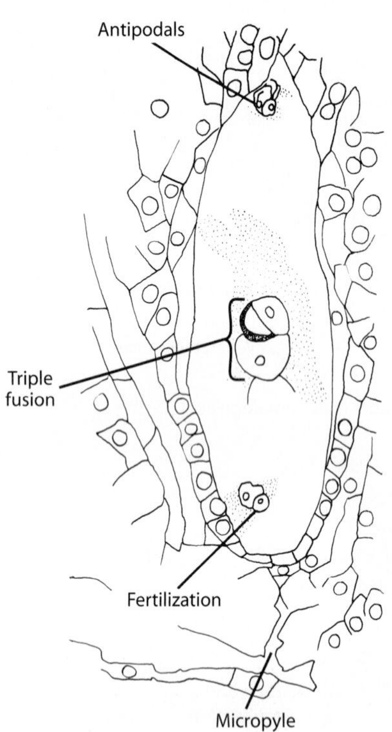

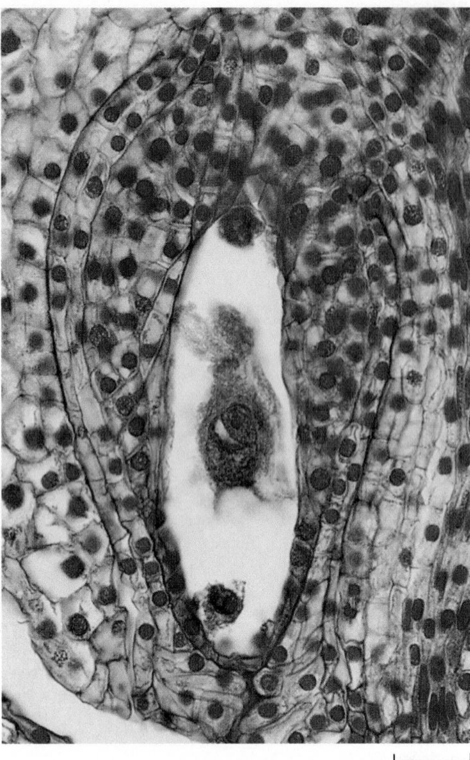

19–23 Double fertilization Union of sperm and egg nuclei can be seen in the lower half of this micrograph of *Lilium*. Triple fusion of the other sperm nucleus and the two polar nuclei has taken place above. The three cells known as the antipodals can be seen at the chalazal end of the embryo sac, opposite the micropyle.

The Ovule Develops into a Seed and the Ovary Develops into a Fruit

In double fertilization, several processes leading to development of the seed and fruit are initiated: (1) the primary endosperm nucleus divides, forming the **endosperm;** (2) the zygote develops into an embryo; (3) the integuments develop into a seed coat; and (4) the ovary wall and related structures develop into a fruit.

In contrast to the embryogeny (embryo development) in the majority of gymnosperms, which begins with a free-nuclear stage, embryogeny in angiosperms resembles that of the seedless vascular plants in that the first nuclear division of the zygote is accompanied by cell wall formation. In the early stages of development, the embryos of monocots undergo sequences of cell division somewhat similar to those of other angiosperms, and the embryo becomes a spherical ball of cells. With formation of the cotyledons, monocot embryos become distinctive, forming only one cotyledon, whereas other angiosperm embryos form two. The details of angiosperm embryogeny are presented in Chapter 22.

Endosperm formation begins with the mitotic division of the primary endosperm nucleus and is usually initiated prior to the first division of the zygote. In some angiosperms, a variable number of free-nuclear divisions precede cell wall formation in a process known as *nuclear-type* endosperm formation. In some other species, the initial and subsequent mitoses are always followed by cytokinesis, which is known as *cellular-type* endosperm formation. Although endosperm development may occur in a variety of ways, the function of the resulting tissue remains the same: to provide essential food materials for the developing embryo and, in many cases, for the young seedling as well. In the seeds of some groups of angiosperms, the nucellus proliferates into a food-storage tissue known as **perisperm.** Some seeds may contain both endosperm and perisperm, as in the beet *(Beta)*. In many eudicots and some monocots, however, most or all of these storage tissues are absorbed by the developing embryo before the seed becomes dormant, as in peas or beans. The embryos of such seeds typically develop fleshy, food-storing cotyledons. The principal food materials stored in seeds are carbohydrates, proteins, and lipids.

Angiosperm seeds differ from those of gymnosperms in the origin of their stored food. In gymnosperms, the stored food is provided by the female gametophyte. In angiosperms, it is provided, at least initially, by endosperm, which is neither gametophytic nor sporophytic tissue. Interestingly, the nutritive tissue is built up *after* fertilization occurs in *Gnetum* and angiosperms, whereas in other seed plants, the nutritive tissue is formed, partly (in conifers) or entirely (other gymnosperms), *before* fertilization occurs.

With development of the ovule into a seed, the ovary, sometimes along with other portions of the flower or inflorescence, develops into a fruit. As this occurs, the ovary wall, or **pericarp,** often thickens and becomes differentiated into distinct layers—the *exocarp* (outer layer), the *mesocarp* (middle layer), and the *endocarp* (inner layer), or exocarp and endocarp only. These

HAY FEVER

In the temperate areas of the Northern Hemisphere, it is estimated that between 10 and 18 percent of all people suffer, at some point in their lives, from hay fever, which can be highly debilitating. Some of the proteins that occur in hollow spaces within pollen grain walls, and which can be released immediately following contact with a moist surface, are generally the culprits. Among these proteins are some that can act as very powerful allergens and antigens, provoking strong reactions by the human immune system. These are often also involved in genetic self-incompatibility. Proteins may also be released in tiny particles of the tapetum, smaller than pollen grains, which may become airborne as the anther splits open.

Wind-borne pollen, such as that of grasses, birch trees, and ragweed, is particularly important as an agent of hay fever, because it is shed in large amounts directly into the air and is thus more likely to reach susceptible victims than are the larger pollen grains of insect-pollinated plants. The quantity of pollen inhaled seems to be the most important factor in determining whether there will be an allergic response, but surprisingly, some wind-borne pollen that is shed in huge amounts, such as that of maize and pines, rarely causes any difficulty. The scent of certain flowers can also cause reactions that resemble hay fever, perhaps in part by increasing the sensitivity of the nasal membranes.

In temperate North America, the hay fever season can be divided into three parts. In spring, most hay fever is associated with tree pollen from such sources as oaks, elms, maples, poplars, pecans, and birches. In summer, grass pollen predominates, with Bermuda grass, timothy grass, and orchard grass being important in different regions. By fall, ragweed and grasses different from those that predominate in summer become major irritants. The susceptibility of individuals to different kinds of plants varies greatly.

When newly introduced plants, such as rapeseed (the source of canola oil), become widely cultivated, they may become important new causes of hay fever. In the arid southwestern United States, the irrigation of large areas for lawns and golf courses, and the introduction of many kinds of weeds to the area, have made hay fever common where it was once virtually unknown.

The incidence of hay fever in the United States has been rising rapidly for more than 60 years, even though the pollen count is actually falling in many areas. Part of the increase is related to better detection of the problem, but there is clearly a genuine increase in hay fever. Understanding why this has occurred will require better knowledge of the human immune system.

layers are generally more conspicuous in fleshy fruits than in dry ones. Fruits are discussed in greater detail in Chapter 20.

An angiosperm life cycle is summarized in Figure 19–22 (pages 472 and 473).

SUMMARY

Angiosperms, the Flowering Plants, Constitute the Phylum Anthophyta

The two largest classes of the phylum Anthophyta are the Monocotyledonae (at least 90,000 species) and the Eudicotyledonae (at least 200,000 species). Flowering plants differ from other seed plants in various distinctive characteristics, such as the enclosure of their ovules within megasporophylls, called the carpels, which defines angiosperms; the production of nutritive endosperm in their seeds; and their distinctive reproductive structure, the flower.

The Flower Is a Determinate Shoot That Bears Sporophylls

Flowers have up to four whorls of appendages. From the outside in, these whorls are sepals (collectively, the calyx); petals (collectively, the corolla); stamens (collectively, the androecium); and carpels (collectively, the gynoecium). Sepals and petals are sterile, with the sepals often green and protective, covering the flower in bud. Petals are often colored and function in attracting pollinators. Individual stamens are generally divided into a filament and an anther, containing four pollen sacs (two pairs). Carpels are usually differentiated into a swollen basal part, the ovary, and a slender upper part, the style, terminating in the receptive stigma.

There are many variations in flower structure. One or more of the four whorls are missing in some flowers, which are called incomplete; complete flowers possess all four whorls. Perfect flowers bear both stamens and carpels in the same flower; imperfect flowers are unisexual and bear either stamens or carpels. Flowers can be regular (radially symmetrical) or irregular (bilaterally symmetrical).

In Angiosperms, Pollination Is Followed by Double Fertilization

Pollination in angiosperms takes place by the transfer of pollen from anther to stigma. A pollen grain is an immature male gametophyte (microgametophyte). At the time of dispersal, the male gametophyte may contain either two or three cells. Initially, there is a vegetative cell and a generative cell, the latter dividing before or after dispersal to give rise to two sperm. The germinated pollen grain with its vegetative nucleus and two sperm is the mature microgametophyte.

The mature female gametophyte (megagametophyte) of an angiosperm is called an embryo sac. In most angiosperms, the embryo sac has seven cells and eight nuclei: an egg apparatus consisting of an egg cell and two synergids at the micropylar end, three antipodals at the opposite, chalazal end, and a central cell with two nuclei. The numbers of cells and nuclei vary in different groups. Both sperm cells function during angiosperm fertilization (double fertilization). One sperm unites with the egg, producing a diploid zygote. The other unites with the two polar nuclei, giving rise to the primary endosperm nucleus, which is often triploid ($3n$). The primary endosperm nucleus divides to produce a unique kind of nutritive tissue, the endosperm, which may be absorbed by the embryo before the seed is mature or may persist in the mature seed. None of the ancient basal-grade lineages of angiosperms have seven-celled, eight-nucleate embryo sacs. The embryo sac of *Amborella* has three synergids and consists of eight cells and nine nuclei; the embryo sacs of the Nymphaeales and Austrobaileyales have only four cells and four nuclei. The angiosperms share double fertilization with the gnetophytes *Ephedra* and *Gnetum,* but in these gymnosperms, the process results in the formation of two embryos, only one of which survives

An Ovule Develops into a Seed, and an Ovary Develops into a Fruit

The ovaries (sometimes with some associated floral parts) develop into fruits, which enclose the seeds. Along with the flower from which it is derived, the fruit is a defining characteristic of the angiosperms.

QUESTIONS

1. Distinguish among or between the following: calyx, corolla, and perianth; stigma, style, and ovary; complete and incomplete; perfect and imperfect; androecium and gynoecium.
2. Diagram and label as completely as possible a complete hypogynous flower, in which none of the floral parts are joined.
3. An imperfect flower is automatically incomplete, but not all incomplete flowers are imperfect. Explain.
4. Diagram and label completely a mature male gametophyte (germinated pollen grain) and a mature seven-celled, eight-nucleate female gametophyte (embryo sac) of an angiosperm. Compare these gametophytes with their counterparts in pine.
5. Double fertilization followed by the formation of endosperm is unique to angiosperms. How does double fertilization in the gnetophytes *Ephedra* and *Gnetum* differ from that in angiosperms?

CHAPTER 20

Evolution of the Angiosperms

◀ **A noxious weed** Originally planted as an attractive garden ornamental requiring little water—and therefore protective of the environment—myrtle spurge *(Euphorbia myrsinites)* is now listed as a noxious weed in parts of the western United States. Not only does it crowd out native plants, but, as an example of biochemical coevolution, it also produces a highly toxic sap that causes blistering of the skin and severe burns of the eye.

In a letter to a friend, Charles Darwin once referred to the apparently sudden appearance of the angiosperms in the fossil record as "an abominable mystery." In the early fossil-bearing strata, about 400 million years old, one finds simple vascular plants, such as rhyniophytes and trimerophytes. Then there is a Devonian and Carboniferous proliferation of ferns, lycophytes, sphenophytes, and progymnosperms, which were dominant until about 300 million years ago (see inside of front cover). The early seed plants first appeared in the Late Devonian period and led to the gymnosperm-dominated Mesozoic floras. Finally, early in the Cretaceous period, about 135 million years ago, angiosperms appear in the fossil record, gradually achieving worldwide dominance in the vegetation by about 90 million years ago. Approximately 75 million years ago, many modern families and some modern genera of this phylum already existed.

Given their relatively late appearance in the fossil record, why did the angiosperms rise to world dominance and then continue to diversify to such a spectacular extent? In this chapter, we attempt to answer this question, centering on possible angiosperm ancestors, the time of origin and diversification of the angiosperms, phylogenetic relationships within the angiosperms, the evolution of the flower and its pollinators (Figure 20–1), the evolution of fruits, and the role of certain chemical substances in angiosperm evolution. This chapter illustrates some of the reasons for the evolutionary success of the flowering plants.

Angiosperm Ancestors

Since the time of Darwin, scientists have attempted to understand the ancestry of the angiosperms. One approach has been

CHECKPOINTS

After reading this chapter, you should be able to answer the following:

1. What are the major lineages of angiosperms, and what are their relationships to one another?
2. Describe the four principal evolutionary trends among flowers.
3. What feature has evolved in angiosperms that allows them directed mobility in seeking a mate?
4. How do beetle-, bee-, moth-, and bat-pollinated flowers differ from one another?
5. What are some of the adaptations of fruits, in relation to their dispersal agents?
6. How might secondary metabolites have influenced angiosperm evolution?

20–1 Angiosperms and pollinators The evolution of the flowering plants is, to a large extent, the story of increasingly specialized relationships between flowers and their insect pollinators, in which beetles played an important early role. The black locust borer beetle *(Megacylene robiniae)* attacks only the black locust *(Robinia pseudoacacia)*. The pollen-laden beetle shown here is visiting the flowers of a goldenrod (*Solidago* sp.).

to search for their possible ancestors in the fossil record. In this effort, particular emphasis has been placed on assessing the ease with which the ovule-bearing structures of various gymnosperms could be transformed into a carpel. Recently, phylogenetic analyses (cladistics) based on fossil, morphological, and molecular data have revitalized attempts to define the major natural groups of seed plants and to understand their interrelationships.

It has long been hypothesized that a link exists between the seed-bearing organs (cupules) of the Caytoniales—a group of Mesozoic seed ferns—and the carpels of angiosperms. (One early investigator initially regarded the Caytoniales as a new group of angiosperms.) The Bennettitales (Triassic to Cretaceous) also have been considered as possible angiosperm ancestors. Some (for example, *Wielandiella;* page 436) had flowerlike bisexual strobili, with separate ovulate and pollen-bearing sporophylls (see Figure 18–9).

For much of the 1980s and 1990s, the anthophyte hypothesis (pages 435 and 436) dominated other concepts of angiosperm ancestry. According to this hypothesis, the gnetophytes are the closest relatives of the angiosperms. This idea was based originally on the analysis of morphological characters, but subsequent molecular analyses cast serious doubt on the existence of an anthophyte clade, indicating instead that the gnetophytes are nested within the conifers, and that the angiosperms and extant gymnosperms are each monophyletic. The latter would preclude the angiosperms having any close relatives among the living gymnosperms. Although the origin of angiosperms remains a mystery, recent contributions from paleobotany, phylogenetics, classical developmental biology, and modern developmental genetics (evo-devo) have contributed greatly to our understanding of the time of origin of the angiosperms and their diversification.

Time of Origin and Diversification of the Angiosperms

The unique characteristics of the angiosperms include flowers, seeds enclosed by a carpel, double fertilization leading to endosperm formation, a reduced three-nucleate microgametophyte, a reduced megagametophyte (consisting of seven cells and eight nuclei in most angiosperms), stamens with two pairs of pollen sacs, and the presence of sieve-tube elements and companion cells in the phloem (see Chapter 23). These similarities clearly suggest that the members of the Anthophyta were derived from a single common ancestor, which would have been derived from a seed plant that lacked flowers, carpels, and fruits.

The earliest unequivocally angiosperm fossils are pollen grains from about 135 million years ago. The earliest angiosperm fossils from which we can gain a good impression of the entire plant are those of *Archaefructus* (see Figure 20–8), which has been dated as from the Early Cretaceous, about 125 million years ago. Evidence of the first flowering plants appears as pollen in the fossil record, around 125 to 130 million years ago. Most recent molecular estimates for the age of angiosperms, however, are in the range of 140 to 180 million years old. Clearly, all of the characteristic features of angiosperms did not appear together in one ancestral plant—evolution proceeds at different rates in different plant organs—so the time and nature of the origin of the group will certainly be a matter of definition when we have accumulated enough evidence. By the Middle Cretaceous, many major lineages of angiosperms appear in the fossil record, and by the end of the Cretaceous, further extensive diversification had occurred. The angiosperms had become the dominant plants in many terrestrial environments.

Like the gymnosperms, the earliest angiosperms produced pollen with a single aperture (monoaperturate), as is still found today among the basal angiosperms and the monocots. This feature can therefore be considered an ancestral one that has been retained in the course of evolution.

Phylogenetic Relationships of the Angiosperms

In Chapter 19, we discussed the two largest classes of angiosperms, the monocots and the eudicots, which between them comprise 97 percent of the species of the Anthophyta. The monocots clearly had a common ancestor, as indicated by their single cotyledon and a number of other unique features. The same is true of the eudicots, which have a characteristic derived feature, their triaperturate pollen (pollen with three furrows or pores, as well as pollen types derived from the triaperturate group).

The remaining 3 percent of living angiosperm species include those with some of the most archaic features. They consist of several evolutionary lines that are quite distinct from one another. Their relationships with the other groups of angiosperms have been determined more precisely in recent years, thanks to macromolecular comparisons and strict analysis of the relationships between evolutionary lines based on ancestral and derived characteristics.

Several evolutionary lines of angiosperms arose before the split between the monocots and the eudicots. Until recently, all of these archaic plants were regarded as "dicots," but they are no more dicots than monocots. All of these plant groups, like the monocots, have monoaperturate pollen or some modification of this type of pollen, indicating that the triaperturate pollen of the eudicots is a derived characteristic that marks that group.

The most ancient angiosperm lineages—the so-called *basal grade* angiosperms—are *Amborella trichopoda* (the sole species of the family Amborellaceae), the order Nymphaeales (water lilies), and the order Austrobaileyales. Molecular phylogenetic studies clearly indicate that *Amborella,* then Nymphaeales, then Austrobaileyales are sister groups to all other flowering plants, termed the Mesangiospermae.

Amborella is a shrubby plant from the island of New Caledonia. Its small flowers, which lack distinct sepals and petals, are imperfect (unisexual), with staminate and carpellate flowers borne on separate plants (Figure 20–2). The carpellate flowers, however, contain sterile stamens (staminodes), an indication that *Amborella* may have evolved from ancestors with perfect

20–2 *Amborella trichopoda* ***(a)*** The woody, evergreen plant *Amborella trichopoda* is the only species in the family Amborellaceae. It is a sprawling shrub with small flowers in which the segments of the perianth (petals and sepals, in many groups of plants) are undifferentiated. *Amborella* is dioecious, meaning that the ***(b)*** staminate and ***(c)*** carpellate flowers are found on different plants. The stamens are rather undifferentiated, with no stalks, and the few carpels in the carpellate flowers develop into ***(d)*** small resinous fruits (drupes), each containing a pockmarked stony seed. New Caledonia, which split from Australia-Antarctica some 80 million years ago, is home to an unusually high number of angiosperm groups with archaic features.

20–3 Water lilies The fragrant water lily, *Nymphaea odorata,* is native to the eastern half of the United States and ranges south through the Caribbean into South America. The water lily family (Nymphaeaceae) is a small group of distinctive and beautiful land plants that became adapted to an aquatic habitat during Cretaceous times and has retained those characters ever since.

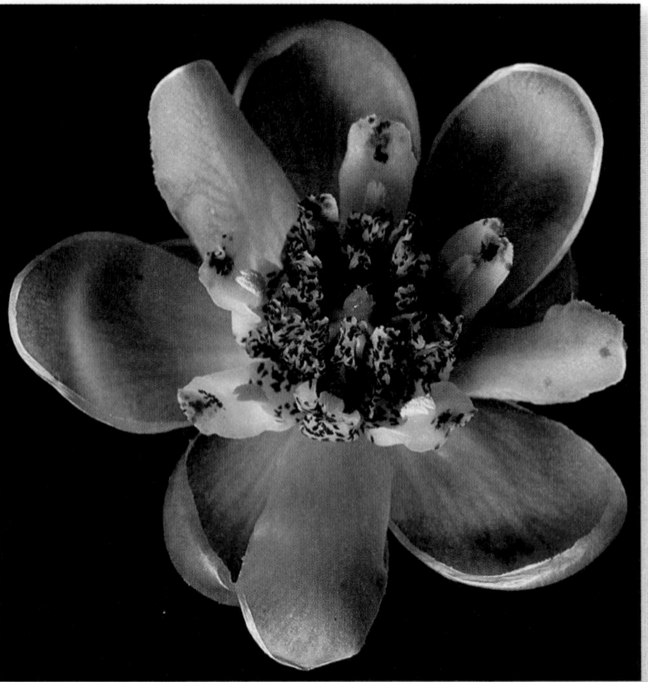

20–4 *Austrobaileya scandens* The only species in its genus and family, *Austrobaileya scandens* also lends its name to the order containing it, the Austrobaileyales. Its large flowers exhibit many features that point to what ancient bisexual flowers might have looked like. The outer floral whorl consists of spirally arranged tepals (perianth parts not differentiated into sepals and petals). The next whorl consists of spirally arranged stamens consisting of anthers (yellow) supported by flattened structures rather than by filaments. As the stamens progress spirally toward the center of the flower, they become progressively more sterile, losing their anthers to form a novel whorl between stamens and carpels. These purple-spotted staminodes apparently isolate the pollen-bearing stamens from the stigmatic surfaces at the carpel tips in the center of the flower. Receptivity of the stigma to pollen ends, however, before the anthers begin shedding pollen. A single flower contains about 15 carpels, each with about 8 to 12 ovules. This flower was photographed by Dr. Joseph Williams in the heart of the wet tropics of North East Queensland, Australia.

(bisexual) flowers. Unlike in the vast majority of angiosperms, the xylem of *Amborella* lacks vessels, tracheids being its only water conduits (pages 394 and 395). As discussed earlier (page 470), the embryo sac (mature female gametophyte) of *Amborella* is unique in that it is eight-celled and nine-nucleate.

The Nymphaeales are herbaceous, aquatic plants adapted to high light intensity (Figure 20–3). The Austrobaileyales, by contrast, are mostly shrubs or small trees adapted to low light intensity; they live in the moist tropical forest understory (Figure 20–4). Recall that both Nymphaeales and Austrobaileyales produce four-celled, four-nucleate embryo sacs, and they form diploid endosperm. Most Nymphaeales either lack vessels or have vessels that resemble tracheids.

The first lineage to diverge within the Mesangiospermae was the magnoliids, or magnoliid clade, including the magnolia family (Magnoliaceae; order Magnoliales; Figure 20–5), in which the flowers have numerous, spirally arranged flower parts. Also included in the magnoliid clade are the Laurales, Canellales, and Piperales. The Laurales include the laurel (Lauraceae) and spicebush (Calycanthaceae) families; the Piperales, the pepper (Piperaceae) and pipevine (Aristolochiaceae; Figure 20–6) families; and the Canellales, the winter's bark (Winteraceae) family. A number of other families of plants with similar features are also included, many of them concentrated in or restricted to the Australasian region and the Southern Hemisphere generally. One of their characteristic features is that the leaves of most magnoliids contain oil cells with ethereal (ether-containing) oils, the basis of the characteristic scents of nutmeg, pepper, and bay leaves. The Winteraceae lack vessels, apparently as a result of evolutionary loss. There are about 20 families of plants in the magnoliids with living representatives.

The monocots constitute a second major lineage of mesangiosperms that retain some of the basal angiosperm features, such as monoaperturate pollen and 3-merous (floral parts in threes) flowers. A third and final major clade of mesangiosperms is represented by the eudicots. Figure 20–7 depicts the relationships among the major groups of angiosperms.

As mentioned above, the earliest representation of an angiosperm that is well documented in the fossil record is *Archaefructus* (Figure 20–8), which was discovered in China in the late 1980s and dated at 125 million years old. *Archaefructus* was a small herbaceous aquatic plant with nonshowy flowers, lacking a perianth (sepals and petals). Branches bearing stamens and carpels extended above the water surface. The numerous stamens may have attracted pollinators. The aquatic nature of this early angiosperm may indicate that the early evolution of angiosperms took place in open wet or aquatic environments subject

(a)

(b)

(c)

20–5 *Magnolia* Flowers and fruits of the southern magnolia, *Magnolia grandiflora,* a woody magnoliid. ***(a)*** The cone-shaped receptacle bears numerous spirally arranged carpels from which curved styles emerge. Below the styles are the cream-white stamens. The anthers have not yet shed their pollen, whereas the stigmas are receptive. Such flowers are said to be protogynous. ***(b)*** The floral axis of a second-day flower, showing stigmas that are no longer receptive and stamens that are shedding pollen. ***(c)*** Fruit, showing carpels and bright red seeds, each protruding on a slender stalk.

20–6 Dutchman's pipe *Aristolochia grandiflora,* the Dutchman's pipe, belongs to the pipevine, or birthwort, family (Aristolochiaceae), which are magnoliids. The flowers of *Aristolochia* emit odors ranging from lemony to the foul smell of rotting meat, all of which attract their insect pollinators.

to frequent disturbances. Such conditions would have favored small, fast-growing plants with a short generation time, a set of features that is still characteristic of many angiosperms today.

Recently, the first intact fossil (aboveground portion) of a mature eudicot was found, also in China. Named *Leefructus mirus,* this 125-million-year-old find has been placed among the Ranunculaceae (the buttercup family).

Throughout most of the last half of the twentieth century, most botanists thought the earliest angiosperms had large flowers with numerous, spirally arranged (rather than whorled) flower parts, thus resembling the flowers of a magnolia (Figure 20–5). With the discovery of *Archaefructus,* however, and the most recent molecular-based phylogenies that place *Amborella* and the Nymphaeales as the earliest-diverging of the living angiosperms, it is now clear that plants with characteristics similar to those of *Amborella* or diverse aquatic angiosperms predate the appearance of plants with magnolia-like flowers by at least 10 to 20 million years. Thus it now seems much more likely that the flowers of the original angiosperms were smaller, simpler, and nonshowy, with simple pollination systems. Also, magnolia flowers and all of the floral diversity that marks modern angiosperms arose long after the first appearance of the angiosperm group. Angiosperms spread rapidly around the world and differentiated as the boundaries between the major climatic zones became more pronounced, especially during the last 70 million years.

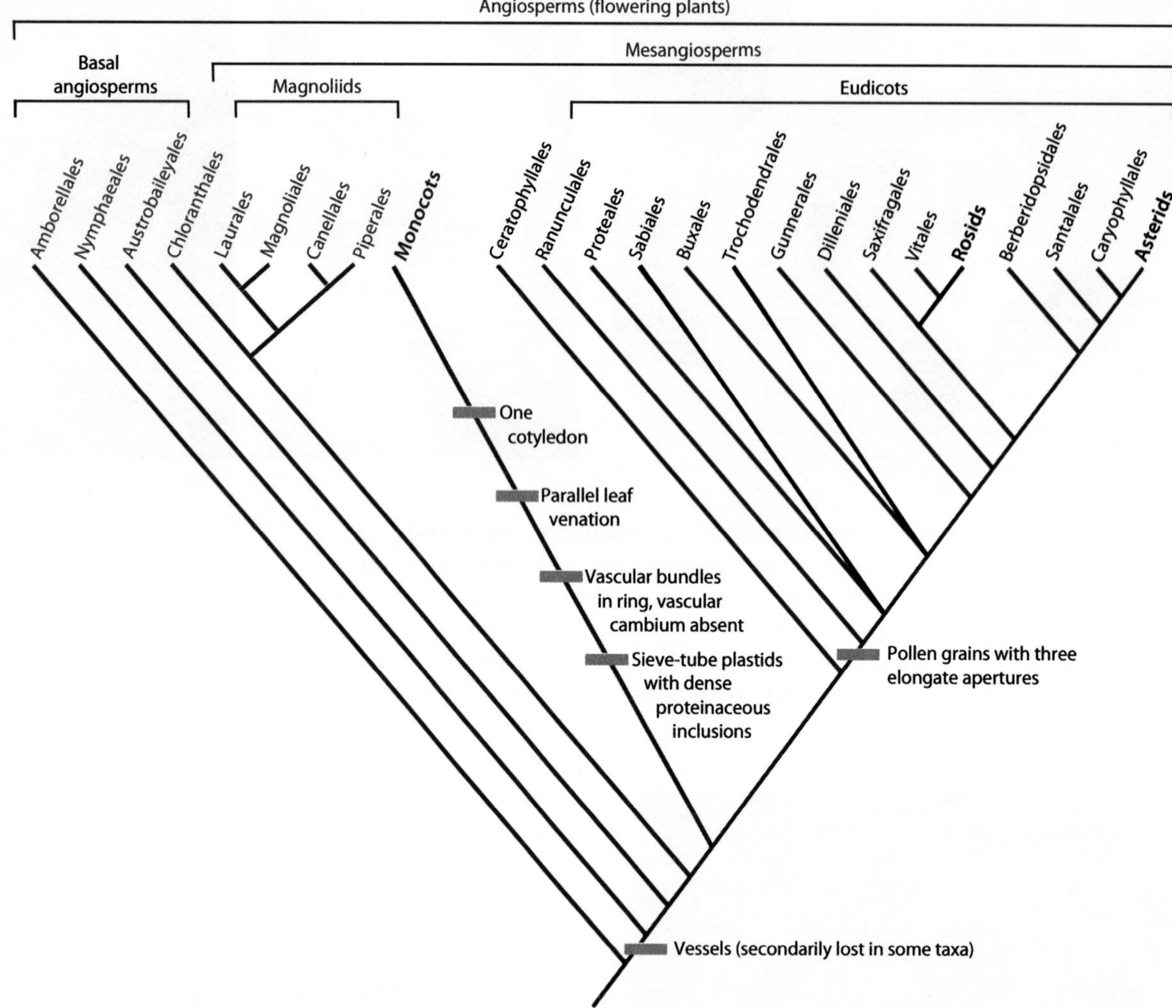

20–7 Cladogram showing phylogenetic relationships of the angiosperms The most ancient angiosperm lineages—the basal grade angiosperms—are placed below the rest of the flowering plants, the Mesangiosperms. The Mesangiosperms include the Chloranthales, magnoliids, monocots, Ceratophyllales, and eudicots. Within the eudicots are two major monophyletic groups: the rosids, with 16 orders, and the asterids, with 14 orders. Rosids typically have ovules with two integuments and a nucellus consisting of two or more layers of cells. The asterids typically have ovules with a single integument and a nucellus composed of a single layer of cells. Some common rosids are violets, begonias, legumes, crucifers (including *Arabidopsis thaliana*), and cucumbers, as well as linden, cottonwood, and elm trees. Asterids include such well-known plants as blueberries, snapdragons, dogwoods, tomatoes, potatoes, carrots, bluebells, mints, and daisies.

Evolution of the Flower

What were the flowers of the earliest angiosperms like? Of course, we do not know this from direct observation, but we can deduce their nature from what we know of certain living plants and from the fossil record. In general, the flowers of these plants were diverse both in the numbers of floral parts and in the arrangement of these parts. Most modern families of angiosperms tend to have more fixed floral patterns that do not vary much in their basic structural features within a particular family. We discuss here the derivation of these patterns over the course of evolution, considering the different whorls of the flower from the outside in, moving from the perianth inward to the androecium and the carpel.

The Parts of the Flower Provide Clues to Angiosperm Evolution

The Perianth of Early Angiosperms Did Not Have Distinct Sepals and Petals

In the earliest angiosperms, the perianth, if present, was never sharply divided into calyx and corolla. Either the sepals and petals were identical, or there was a gradual transition in appearance between these two whorls, as in modern magnolias and water lilies. In some angiosperms, including the water lilies, petals seem to have been derived from sepals. In other words, the petals can be viewed as modified leaves that have become specialized for attracting pollinators. In most angiosperms, however, the petals were probably derived originally from stamens that lost their sporangia—becoming "sterilized"—and then were specially

(a)

(b)

(c)

20–8 ***Archaefructus sinensis*** The fossils of *Archaefructus sinensis*, the earliest well-documented flowering plant, are approximately 125 million years old. They were recovered from semi-aquatic fossil beds preserved in Northern China. ***(a)*** Reconstruction of a whole plant, showing the slender roots, dissected leaves, and floral axes with closed carpels above, closed stamens below. ***(b)*** Whole specimen minus the roots. ***(c)*** Full view of a fertile fruiting axis, showing the closed carpels and closed stamens and some leaf material.

modified for their new role. Most petals, like stamens, are supplied by just one vascular strand. In contrast, sepals are normally supplied by the same number of vascular strands as are the leaves of the same plant (often three or more). Within sepals and petals alike, the vascular strands usually branch, so the number of strands that enter them cannot be determined from the number of veins in the main body of these structures.

Petal fusion has occurred a number of times during the evolution of the angiosperms, resulting in the familiar tubular corolla that is characteristic of many families (Figure 20–9c). In these flowers, the stamens often fuse with the tubular corolla and appear to arise from it. In some families, the sepals are also fused into a tube.

The Stamens of Early Angiosperms Were Diverse in Structure and Function The stamens of some families of woody magnoliids are broad, colored, and often scented, playing an obvious role in attracting floral visitors. In other archaic angiosperms, the stamens, although relatively small and often greenish, may also be fleshy. Most monocots and eudicots, in contrast, have stamens with generally thin filaments and thick, terminal anthers (for example, see Figure 19–6).

In some specialized flowers the stamens are fused together. Their fused filaments may then form columnar structures, as in the members of the pea, melon, mallow (Figure 20–9d), and sunflower (Figure 20–10d) families, or they may be fused with the corolla, as in the phlox, snapdragon, and mint families. In certain plant families, some of the stamens have become secondarily sterile: they have lost their sporangia and become transformed into specialized structures, such as nectaries. Nectaries are glands that secrete **nectar,** a sugary fluid that attracts pollinators and provides food for them. Most nectaries are not modified stamens, however, and arose in other ways. During the course of evolution of the angiosperm flower, the sterilization of stamens, as noted above, also played an important role in the evolution of petals.

The Carpels of Many Early Angiosperms Were Unspecialized Some archaic angiosperms have somewhat leaflike carpels, with no specialized areas for the entrapment of pollen grains comparable to the specialized stigmas of most living angiosperms. The carpels of many woody magnoliids and other plants that retain archaic features are free from one another, instead of fused together as in most other angiosperms. In a few living angiosperms, the carpels are incompletely closed, although pollination is always indirect—that is, the pollen does not come into direct contact with the ovules. In the vast majority of living angiosperms, the carpels are closed and sharply differentiated into stigmas, styles, and ovaries. There is also more variation in the arrangement of the ovules among eudicots: the ovules can be positioned along the outer wall of the ovary, or along the central axis, or sometimes reduced to a single ovule attached to the base or even the apex of the ovary (see Figure 19–9).

Four Evolutionary Trends among Flowers Are Evident

Insect pollination quite probably accelerated the early evolution of angiosperms, both through the possibilities it provided for isolating small populations and, with indirect pollination,

20–9 Examples of specialized flowers ***(a)*** Wintergreen, *Chimaphila umbellata*. The sepals (not visible) and petals are reduced to five each, the stamens to 10, and the five carpels are fused into a compound gynoecium with a single stigma. ***(b)*** Lotus, *Nelumbo lutea*. The numerous tepals (perianth not differentiated into sepals and petals) and stamens are spirally arranged; the carpels are embedded in a flat-topped receptacle. ***(c)*** Chaparral honeysuckle, *Lonicera hispidula*. The ovary is inferior and has two or three locules; the sepals are reduced to small teeth at its apex. The petals are fused into a corolla tube in the zygomorphic (bilaterally symmetrical) flower, and the five stamens, which protrude from the tube, are attached to its inner wall. The style is longer than the stamens, and the stigma is elevated above them. A pollinator visiting this flower would contact the stigma first, so that if it were carrying pollen from another flower, it would deposit that pollen on the stigma before reaching the anthers. Fruits of this species are shown in Figure 20–28b. ***(d)*** Diagram of a longitudinal section of a cotton *(Gossypium)* flower, which is in the mallow family, with the sepals and petals removed and showing the column of stamens fused around the style.

through the competition fostered among many pollen grains as they grew through the stigmatic tissue. Both bisexual and unisexual flowers appeared early in angiosperm evolution, so we cannot assume that one or the other is ancestral. The undifferentiated perianth of early angiosperms soon gave rise to distinct petals and sepals. As the angiosperms continued to diversify, and relationships with specialized pollinators became more tightly linked, the number and arrangement of floral patterns became more stereotyped. The following four generalized trends are evident (as seen in the examples in Figure 20–9):

20–10 Composites (family Asteraceae) ***(a)*** Diagram showing the organization of the head of a member of this family. The disk and ray flowers are subordinated to the overall display of the head, which functions as a single large flower in attracting pollinators. ***(b)*** Thistle, *Cirsium pastoris.* Members of the thistle tribe have only disk flowers. This particular species of thistle has bright red flowers and is regularly visited by hummingbirds, which are its primary agents of pollination. ***(c)*** *Agoseris,* a wild relative of the dandelion, *Taraxacum.* In the inflorescences of the chicory tribe (the group of composites to which dandelions and their relatives belong), there are no disk flowers. The marginal ray flowers, however, are often enlarged. ***(d)*** Sunflower, *Helianthus annuus.*

1. From flowers with few to many parts that are indefinite in number, flowers have evolved toward having few parts that are definite in number.
2. The floral axis has become shortened so that the original spiral arrangement of parts is no longer evident, and the floral parts often have become fused.
3. The ovary has become inferior rather than superior in position, and the perianth has become differentiated into a distinct calyx and corolla.
4. The radial symmetry (regularity), or actinomorphy, of early flowers has given way to bilateral symmetry (irregularity), or zygomorphy, in more advanced ones.

The Asteraceae and Orchidaceae Are Examples of Specialized Families

Among the most specialized of flowers are those of the family Asteraceae (Compositae), which are eudicots, and those of the family Orchidaceae, which are monocots. In number of species, these are the two largest families of angiosperms.

The Flowers of the Asteraceae Are Closely Bunched together into a Head In the Asteraceae (the composites), the epigynous flowers are relatively small and closely bunched together to form a head. Each of the tiny flowers has an inferior ovary composed of two fused carpels with a single ovule in one locule (Figure 20–10).

In composite flowers, the stamens are reduced to five in number and are usually fused to one another (connate) and to the corolla (adnate). The petals, also five in number, are fused to one another and to the ovary, and the sepals are absent or reduced to a series of bristles or scales known as the **pappus.** The pappus often serves as an aid to dispersal by wind, as it does in the familiar dandelion, a member of the Asteraceae (see Figures 20–25 and 20–26b). In other members of this family, such as beggar-ticks *(Bidens),* the pappus may be barbed, serving to attach the fruit to a passing animal and thus to enhance its chances of being dispersed from place to place. In many members of the family Asteraceae, each head includes two types of flower: (1) disk flowers, which make up the central portion of the aggregate, and (2) ray flowers, which are arranged on the outer periphery. The ray flowers are often carpellate, but sometimes they are completely sterile. In some members of the Asteraceae, such as sunflowers, daisies, and black-eyed Susans, the fused bilaterally symmetrical (zygomorphic) corolla of each ray flower forms a long, strap-shaped "petal."

In general, the composite head has the appearance of a single large flower. Unlike many single flowers, however, the head matures over a period of days, with the individual flowers opening serially in an inward-moving spiral pattern. As a consequence, the ovules in a given head may be fertilized by several different pollen donors. The success of this plan as an evolutionary strategy is attested to by the great abundance of the members of the Asteraceae and also their great diversity, which, with about 22,000 species, makes them the second largest family of flowering plants.

Orchidaceae Is the Largest Angiosperm Family Another successful flower plan is that of the orchids (Orchidaceae), which, unlike the composites, are monocots. There are probably at least 24,000 species of orchids, making them the largest family of flowering plants. In contrast to the composites, however, individual species of orchids are rarely very abundant. Most species of orchids are tropical, and only about 140 are native to the United States and Canada. In the orchids, the three carpels are fused and, as in the composites, the ovary is inferior (Figure 20–11). Unlike the composites, however, each orchid ovary contains many thousands of minute ovules. Consequently, each pollination event may result in the production of a huge number of seeds. Usually, only one stamen is present (in one subfamily, the lady-slipper orchids, there are two) and is characteristically fused with the style and stigma into a single complex structure—the **column.** The entire contents of an anther are held together and dispersed as a unit—the **pollinium.** The three petals of orchids are modified so that the two lateral ones form wings and the third forms a cuplike lip that is often very large and showy. The sepals, also three in number, are often colored and look much like petals. The flower is always bilaterally symmetrical and often bizarre in appearance.

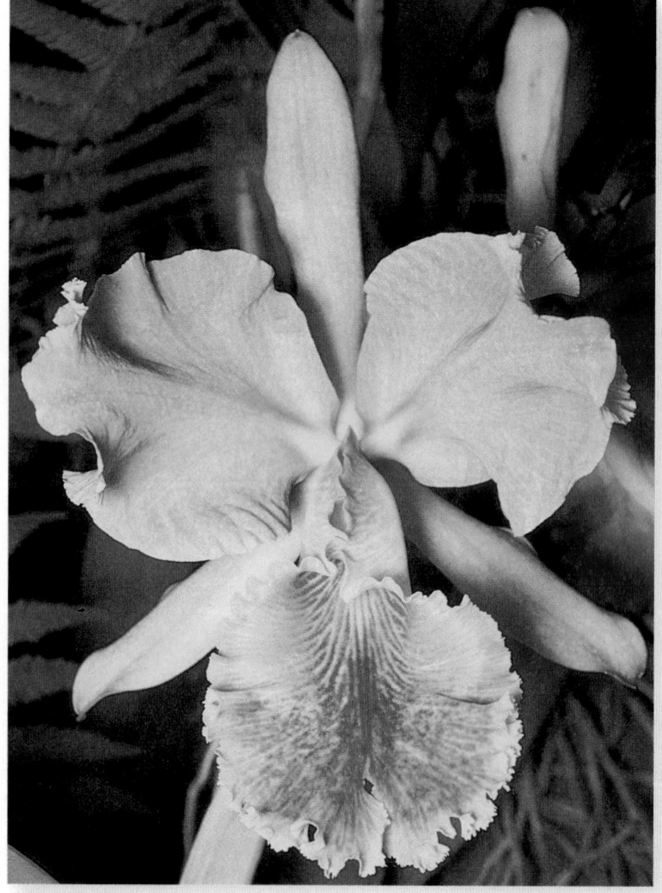

(a)

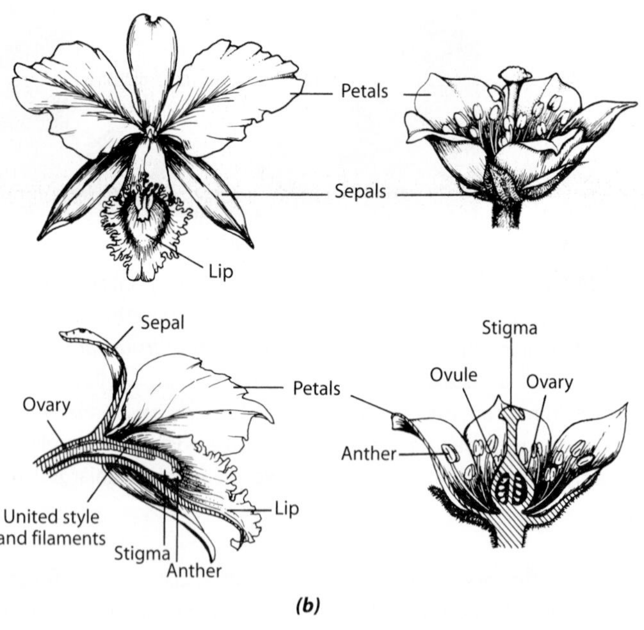

20–11 Orchids (family Orchidaceae) ***(a)*** An orchid of the genus *Cattleya.* Orchids have extremely specialized flowers. ***(b)*** A comparison of the parts of an orchid flower, shown on the left, with those of a radially symmetrical flower, shown on the right. The orchid "lip" is a modified petal that serves as a landing platform for insects.

Among the orchids are some species with flowers the size of a pinhead and others with flowers more than 20 centimeters in diameter. Several genera lack chlorophyll and survive as myco-heterotrophs (they are symbiotic with fungi in their roots). Two Australian species grow entirely underground, their flowers appearing in cracks in the ground, where they are pollinated by flies. In the commercial production of orchids, the plants are cloned by making divisions of meristematic tissue, and thousands of identical plants can be produced rapidly and efficiently (see Chapter 10). There are more than 60,000 registered hybrids of orchids, many of them involving two or more genera. The seed pods of orchids of the genus *Vanilla* are the natural source of the popular flavoring of the same name.

Animals Serve as the Primary Agents of Floral Evolution

Plants, unlike most animals, cannot move from place to place to find food or shelter or to seek a mate. In general, plants must satisfy those needs by growth responses and by the structures that they produce. In many angiosperms, however, a set of features has evolved that, in effect, allows them directed mobility in seeking a mate. This set of features is embodied in the flower. By attracting insects and other animals with their flowers, and by directing the behavior of these animals so that cross-pollination (and therefore cross-fertilization) will occur at a high frequency, the angiosperms have transcended their rooted condition. In this one respect, they have become just as mobile as animals. How was this achieved?

Flowers and Insects Have Coevolved The earliest seed-bearing plants were pollinated passively. Large amounts of pollen were blown about by the wind, reaching the vicinity of the ovules only by chance. The ovules, which were borne on the leaves or within cones, exuded sticky drops of sap from their micropyles. These drops served to catch the pollen grains and to draw them toward the micropyle. As in most modern cycads (page 449) and gnetophytes, insects feeding on the pollen and other flower parts began returning to these new-found sources of food and thus transferred pollen from plant to plant. Such a system is more efficient than passive pollination by the wind. It allows much more accurate pollination with many fewer pollen grains.

Attraction of insects to the naked ovules of these plants sometimes resulted in the loss of ovules to the insects. The evolution of a closed carpel, therefore, gave certain seed plants—the ancestors of the angiosperms—a reproductive, and thus a selective, advantage. Further changes in the shape of the flower, such as the evolution of the inferior ovary, may have been additional means of protecting the ovules from being eaten by insects and other animals, thus providing a further reproductive advantage.

Another important evolutionary development was the bisexual flower. The presence of both carpels and stamens in a single flower (in contrast, for instance, to the separate microsporangiate and megasporangiate cones of living conifers) offers a selective advantage by making each visit by a pollinator more effective. The pollinator can both pick up and deliver pollen at each stop.

If a given plant species is pollinated by only one or a few kinds of visitors, selection favors specializations related to the characteristics of these visitors. Many of the modifications that have evolved in flowers promoted consistency in the specific type of visitor to that particular kind of flower. For example, many modern angiosperms are pollinated solely or chiefly by beetles, others by flies (Figure 20–12). Both depend on floral odors that are fruity or resemble dung or carrion, and in

(a)

(b)

20–12 Beetle- and fly-pollinated flowers ***(a)*** A pollen-eating beetle *(Asclera ruficornis)* at the open, bowl-shaped flower of round-leaved hepatica, *Anemone americana.* The species of the family (Oedemeridae) to which this beetle belongs feed only on pollen as adults. ***(b)*** The foul-scented and often dark-colored flowers of many species of milkweeds (Apocynaceae), such as those of this African succulent plant, *Stapelia schinzii,* are pollinated by carrion flies.

20–13 Bee pollination Bees have become as highly specialized as the flowers they have been associated with during the course of their evolution. Their mouthparts have become fused into a sucking tube containing a tongue. The first segment of each of the three pairs of legs has a patch of bristles on its inner surface. Those of the first and second pairs are pollen brushes that gather the pollen that sticks to the bee's hairy body. On the third pair of legs, the bristles form a pollen comb that collects pollen from these brushes and from the abdomen. From the comb, the pollen is forced up into pollen baskets, concave surfaces fringed with hairs on the upper segment of the third pair of legs. Shown here is a honeybee *(Apis mellifera)* foraging in a flower of rosemary, *Rosmarinus officinalis.* In rosemary, the stamens and stigma arch upward out of the flower, and both come into contact with the hairy back of any visiting bee of the proper size. In this photograph, the anthers can be seen depositing white pollen grains on the bee.

beetle-pollinated flowers, essential floral parts are often covered and thus protected from their gnawing visitors.

Bees, however, are the most important group of flower-visiting insects, responsible for the pollination of more types of plants than any other group. In fact, bees and flowering plants have become diverse together over the course of the past 80 million years.

Both male and female bees live on nectar, and the females also collect pollen to feed the larvae. Bees have mouthparts, body hairs, and other appendages with special adaptations that make them suitable for collecting and carrying nectar and pollen (Figure 20–13). As Karl von Frisch and other investigators of insect behavior have shown, bees can learn quickly to recognize colors, odors, and outlines. The portion of the light spectrum that is visible to most insects, including bees, is somewhat different from the portion visible to humans. Unlike human beings, bees perceive ultraviolet as a distinct color; however, they don't perceive red, which therefore tends to merge with the background.

Bee flowers have showy, brightly colored petals that are usually blue or yellow. They often have distinctive patterns by which bees can efficiently recognize them. Such patterns include "honey guides," which are special markings that indicate the position of the nectar (Figure 20–14).

Flowers that are regularly pollinated by butterflies and diurnal moths (those that are active in the day rather than at night) are similar, in general, to bee-pollinated flowers, but they often have "landing platforms" built into the structure of the flower (Figures 20–11b and 20–15). Those that are pollinated by nocturnal (night-flying) moths are generally white or pale in color, and often have a sweet, penetrating scent that is emitted only after sunset. The nectary in a moth or butterfly flower is often located at the base of a long, slender corolla tube or a spur and is usually accessible only to the long, sucking mouthparts of these insects.

Some flowering plants, including about one-third of orchid species, attract insect pollinators by "food-deception." Such plants signal the presence of a food reward, such as nectar or pollen, without providing one. More often, food-deceivers lure pollinators by mimicking the appearance of food-rewarding flowers. A less common form of deception, known exclusively in orchids, is "sexual-deception." Here, pollination is carried out by deceived male insects that attempt to copulate with the flower. Such flowers mimic the appearance of female insects and give off a fragrance that closely resembles the pheromones of the female insect.

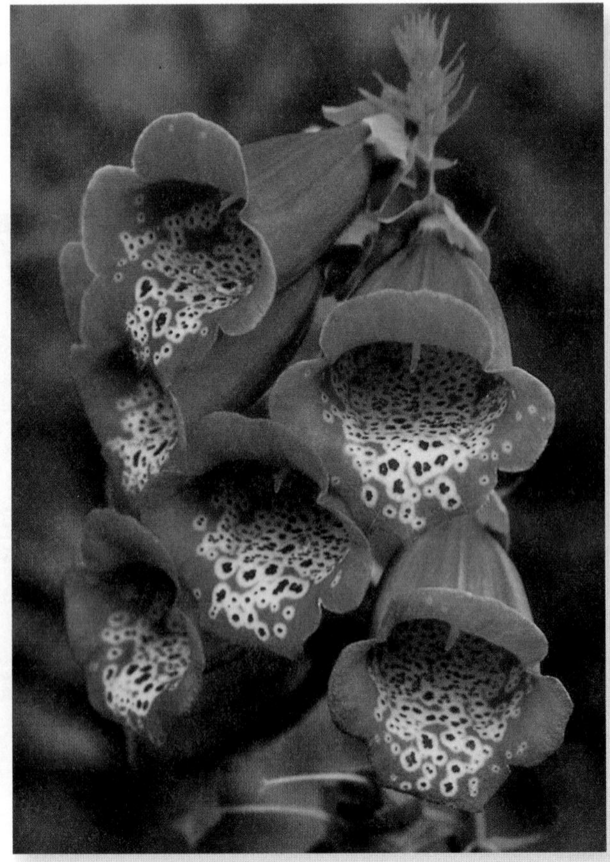

20–14 Honey guides So-called honey guides on the flowers of the foxglove, *Digitalis purpurea,* serve as distinctive signals to insect visitors. The lower lip of the fused corolla serves as a landing platform of the kind that is commonly found in bee flowers.

20–15 Butterfly pollination A copper butterfly *(Lycaena gorgon)* sucking nectar from the flowers of a daisy. The long, sucking mouthparts of moths and butterflies are coiled up at rest and extended when feeding. They vary in length from species to species. Only a few millimeters long in some of the smaller moths, they are 1 to 2 centimeters long in many butterflies, 2 to 8 centimeters long in some hawkmoths of the North Temperate zone, and as long as 25 centimeters in a few kinds of tropical hawkmoths.

Bird- and Bat-Pollinated Flowers Produce Copious Nectar Bird-pollinated flowers generally have a copious, thin nectar but usually have little odor, because the sense of smell is poorly developed in birds. Birds, however, have a keen sense of color, and the flowers they visit regularly are colorful, with red and yellow ones being the most common (Figure 20–16). Such flowers include red columbine, scarlet passion flower, and hibiscus.

Some plant groups, especially in the tropics, have flowers that are regularly pollinated by bats. Bats that obtain most of their food from flowers have slender, elongated muzzles and long, extensible tongues, sometimes with a brushlike tip (Figure 20–17). Most of the flowers that are pollinated by bats produce

20–16 Bird pollination A male Anna's hummingbird *(Calypte anna)* at a flower of the scarlet monkey-flower, *Mimulus cardinalis,* in southern California. Note the pollen on the bird's forehead, which is in contact with the stigma of the flower.

20–17 Bat pollination By thrusting its face into the tubular corolla of a flower of an organ-pipe cactus, *Stenocereus thurberi,* this bat *(Leptonycteris curasoae)* is able to lap up nectar with its long, bristly tongue. Some of the pollen clinging to the bat's face and neck is transferred to the next flower it visits. This species of bat, which is one of the more specialized nectar-feeding bats, migrates during late spring and early summer from central and southern Mexico to the deserts of the southwestern United States, where it feeds on the nectar and pollen of organ-pipe and saguaro cacti and on the flowers of agaves.

copious nectar and are dull colored, and many of them open only at night. Such flowers often hang down on long stalks below the foliage, or are borne on the trunks of trees, where the bats can get to them easily. Bat flowers characteristically have either very strong fermenting or fruitlike odors or musty scents resembling those produced by bats to attract one another.

Wind-Pollinated Flowers Produce No Nectar Wind-pollinated flowers, which do not rely on animal pollinators, produce no nectar, have dull colors, and are relatively odorless. The petals of these flowers are either small or absent, and the sexes are often separated on the same plant. They are best represented in temperate regions, with many plants of the same species often growing together and pollen dispersal often occurring in early spring, before the plants have formed their leaves. Oaks, birches (Figure 20–18a), and grasses (Figure 20–18b, c) are familiar examples of wind-pollinated plants. Wind-pollinated flowers usually have well-exposed anthers that readily lose their pollen to the wind. The large stigmas are characteristically exposed, and they often have branches or feathery outgrowths adapted for intercepting wind-borne pollen grains.

The Most Important Pigments in Floral Coloration Are the Flavonoids

Color is one of the most conspicuous features of angiosperm flowers—a characteristic by which members of the phylum are easily recognized. The variety of flower colors evolved in relation to the plants' pollination systems and, in general, are advertisements for particular kinds of animals, as we have just seen.

The pigments responsible for the colors of angiosperm flowers are generally common in all vascular plants. However, it is the way in which they are concentrated in flowers, and particularly in their corollas, that is a special characteristic of the flowering plants. Surprisingly, all flower colors are produced by a small number of pigments. Many red, orange, or yellow flowers owe their color to the presence of carotenoid pigments similar to those that occur in leaves (pages 33 and 34; they are also found in all plants, in green algae, and in some other organisms). The most important pigments in floral coloration, however, are **flavonoids,** which are compounds with two six-carbon rings linked by a three-carbon unit. Flavonoids probably occur in all angiosperms, and they are sporadically distributed among the members of other groups of plants. In leaves, flavonoids block UV radiation, which is destructive to nucleic acids and proteins. They usually selectively admit light of blue-green and red wavelengths, which are important for photosynthesis.

Pigments belonging to one major class of flavonoids, the **anthocyanins,** are major determinants of flower color (Figure 20–19). Most red and blue plant pigments are anthocyanins, which are water-soluble and are found in vacuoles. By contrast, the carotenoids are oil-soluble and are found in plastids. The color of an anthocyanin pigment depends on the acidity of the cell sap of the vacuole. Cyanidin, for example, is red in acidic

(a)

(b)

(c)

20–18 Wind pollination Shown here are flowers of two kinds of wind-pollinated plants. ***(a)*** The staminate flowers of the paper birch, *Betula papyrifera,* hang down in flexible, thin tassels several centimeters long. These tassels are whipped by passing breezes, and the pollen, when mature, is scattered about by the wind. ***(b)*** Grasses, such as maize *(Zea mays),* characteristically have enlarged, often feathery stigmas that efficiently catch the wind-blown pollen shed by the hanging anthers. The "silk" on the ears of corn, or maize, consists of many stigmas, each leading to a grain on the young cob below. ***(c)*** Stamen-bearing (staminate) inflorescences, the "tassels," at the top of the maize stem.

Pelargonidin

Cyanidin

Delphinidin

20–19 Anthocyanins These three anthocyanin pigments are the basic pigments on which flower colors in many angiosperms depend: pelargonidin (red), cyanidin (violet), and delphinidin (blue). Related compounds known as flavonols are yellow or ivory, and the carotenoids are red, orange, or yellow. Betacyanins (betalains) are red pigments that occur in one group of eudicots. Mixtures of these different pigments, together with changes in cellular pH, produce the entire range of flower color in the angiosperms. Changes in flower color provide "signals" to pollinators, telling them which flowers have opened recently and are more likely to provide food.

solution, violet in neutral solution, and blue in alkaline solution. In some plants, the flowers change color after pollination, usually because of the production of large amounts of anthocyanins, and then become less conspicuous to insects.

The **flavonols,** another group of flavonoids, are very commonly found in leaves and also in many flowers. A number of these compounds are colorless or nearly so, but they may contribute to the ivory or white hues of certain flowers.

For all flowering plants, different mixtures of flavonoids and carotenoids (as well as changes in cellular pH) and differences in the structural, and thus the reflective, properties of the flower parts produce the characteristic colors. The bright fall colors of some leaves come about when large quantities of colorless flavonols are converted into anthocyanins as the chlorophyll breaks down (see also page 51). In the all-yellow flowers of the marsh marigold *(Caltha palustris),* the UV-reflective outer portion is colored by carotenoids, whereas the UV-absorbing inner portion is yellow to our eyes because of the presence of a yellow chalcone, one of the flavonoids. To a bee or other insect, the outer portion of the flower appears to be a mixture of yellow and ultraviolet, a color called "bee's purple," whereas the inner portion appears pure yellow (Figure 20–20). Most, but not all, UV reflectivity in flowers is related to the presence of carotenoids, and thus ultraviolet patterns are more common in yellow flowers than in others.

In the goosefoot, cactus, and portulaca families and in other members of the order Caryophyllales, the reddish pigments are not anthocyanins or even flavonoids but a group of more complex aromatic compounds known as **betacyanins** (or betalains). The red flowers of *Bougainvillea* and the red color of beets are

(a)

(b)

20–20 Bee's purple The color perception of most insects is somewhat different from that of human beings. To a bee, for example, ultraviolet light (which is invisible to humans) is seen as a distinct color. These photographs show a flower of marsh marigold, *Caltha palustris,* ***(a)*** in natural light, showing the solid yellow color as the flower appears to humans, and ***(b)*** in ultraviolet light. The portions of the flower that appear light in *(b)* reflect both yellow and ultraviolet light, which combine to form a color known as "bee's purple," whereas the dark portions of the flower absorb ultraviolet and therefore appear pure yellow when viewed by a bee.

due to the presence of betacyanins. No anthocyanins occur in these plants, and the families characterized by betacyanins are closely related to one another.

Evolution of Fruits

Just as flowers have evolved in relation to pollination by many different kinds of animals and other agents, so have fruits evolved for dispersal in many different ways. Fruit dispersal, like pollination, is a fundamental aspect of the evolutionary radiation of the angiosperms. Before we consider this subject in more detail, however, we must present some basic information about fruit structure.

Strictly Defined, a Fruit Is a Matured Ovary

A fruit is a matured ovary, but more broadly defined it is a matured ovary along with whatever noncarpellary tissue, or **accessory tissue,** becomes united with the ovary during its maturation. In some taxa, the accessory tissue dominates over the carpellary tissue in the matured fruit, as in the strawberry, which consists largely of the expanded receptacle. Fruits may develop without fertilization and seed development. This phenomenon is known as parthenocarpy, and such fruits are known as **parthenocarpic fruits.** Parthenocarpy is widespread, especially in species with large numbers of ovules, such as banana, citrus, pumpkin, fig, and pineapple.

Fruits are generally classified as simple, multiple, or aggregate, depending on the arrangement of the carpels from which the fruit develops. **Simple fruits** develop from a single carpel or from two or more united carpels (bean pod, cherry, tomato). **Aggregate fruits** are formed from a gynoecium—an apocarpous gynoecium—in which each carpel retains its identity in the mature state (magnolias, raspberries, strawberries). The individual matured carpels, or ovaries, are called **fruitlets** (Figure 20–5c). **Multiple fruits** are derived from an inflorescence, that is, from the combined gynoecia of many flowers (fig, mulberry, pineapple). Any fruit that contains accessory tissue is also called an **accessory fruit,** whether simple, aggregate, or multiple. Thus, the fruits of apples and pears, in which the floral tube becomes the major fleshy part, is a simple accessory fruit; the strawberry is an aggregate accessory fruit; and the pineapple is a multiple accessory fruit.

Simple fruits are by far the most diverse of the three groups. When ripe they may be fleshy or more or less dry. The three main types of fleshy fruits are berries, drupes, and pomes. **Berries** are fleshy fruits with one to many seeds; all parts are fleshy or pulpy except the exocarp, which may be skinlike or rindlike (tomatoes, grapes, dates, citrus, cucurbits). **Drupes** (stone fruits) are generally one-seeded, with usually a thin and skinlike exocarp, fleshy mesocarp, and stony endocarp, which encloses the seed (peaches, cherries, olives, plums) (Figure 20–21). Coconuts are also drupes but with an outer fibrous layer rather than a fleshy one; what is usually seen in the grocery stores of temperate regions is the stony inner layer (endocarp) of the fruit. **Pomes** develop from a compound inferior ovary. The bulk of the flesh is derived from noncarpellary tissue (the floral tube), and the endocarp enclosing the seed is cartilaginous (apple, pear, quince, and related rosaceous genera).

Dry, simple fruits are classified as either dehiscent (Figures 20–22 and 20–23) or indehiscent (Figure 20–24). **Dehiscent fruits** split open at maturity and commonly contain several seeds. **Indehiscent fruits** do not split open at maturity and usually originate from an ovary in which only one seed develops, though more than one ovule may be present.

There are several kinds of dehiscent simple dry fruits. The **follicle** is derived from a single carpel that splits along one side at maturity, as in the columbines, milkweeds (Figure 20–22a), and magnolias (Figure 20–5c). The **legume,** which is characteristic of the pea family (Fabaceae), resembles a follicle, but it splits along both sides (Figure 22–23). The fruit of the mustard family (Brassicaceae) is called a **silique.** It is formed of two carpels. At maturity the two halves split away from a persistent central partition to which the seeds are attached (Figure 20–22c). The most common of the dehiscent simple dry fruits is the **capsule,** which is derived from a compound ovary (of more than one carpel). Capsules release their seeds in a variety of ways. In the poppy family (Papaveraceae), for example, the seeds are often shed when the capsule splits longitudinally, but in some

(a)

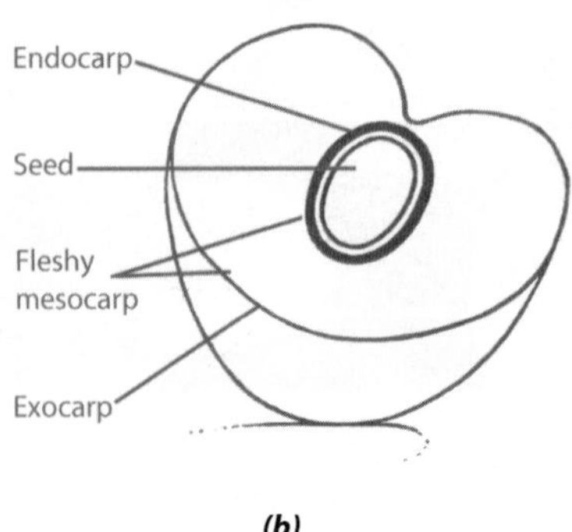

(b)

20–21 Drupes (stone fruits) Drupes are generally one-seeded, with the mature fruit composed of three distinct parts: a relatively thin exocarp or skin, a thick fleshy mesocarp, and a stony endocarp. The endocarp is composed of tightly packed sclereids (stone cells) and forms the pit, or stone, of the fruit. The seed is enclosed within the endocarp. ***(a)*** A section through a nectarine *(Prunus persica nectarina)* fruit, revealing the pit, which is surrounded by the fleshy mesocarp. The exocarp is leathery. ***(b)*** Diagram of a drupe.

(a)

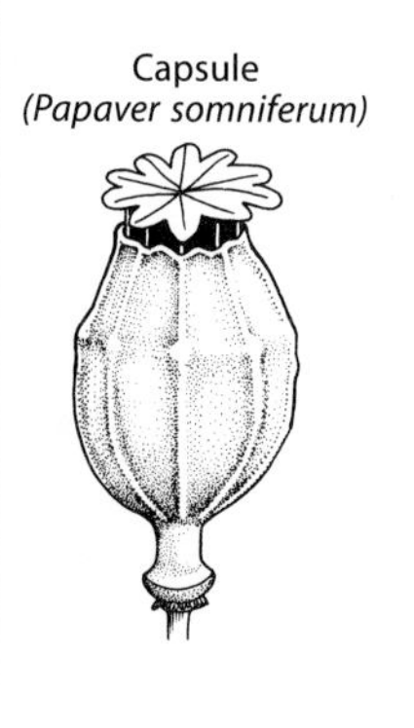

(b)

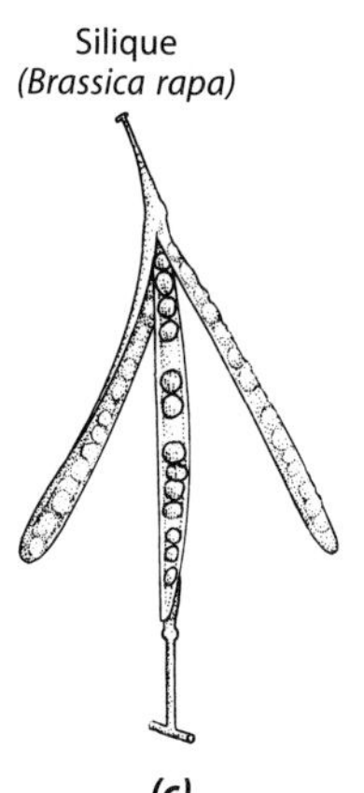

(c)

20–22 Dehiscent fruits
(a) Bursting follicles of a milkweed (*Asclepias*). ***(b)*** In some members of the poppy family (Papaveraceae), such as the red poppy (genus *Papaver*), the capsule sheds its seeds through pores near the top of the fruit. ***(c)*** Plants of the mustard family (Brassicaceae) have a characteristic fruit known as a silique, in which the seeds arise from a central partition, and the two enclosing valves fall away at maturity.

members of the family they are shed through openings near the top of the capsule (Figure 20–22b).

Indehiscent simple dry fruits are found in many plant families. Most common is the **achene,** a small, one-seeded fruit, with the seed attached to the pericarp at one point only (by the funiculus). Consequently, the pericarp is readily separated from the seed coat. Achenes are characteristic of the buttercup family (Ranunculaceae) and the buckwheat family (Polygonaceae). Winged achenes, such as those found in elms and ashes, are commonly known as **samaras** (Figure 20–24). In the Asteraceae, achene-like fruit is derived from an inferior ovary; technically, it is called a **cypsela** (Figure 20–25). The achenelike fruit that occurs

(a)

(b)

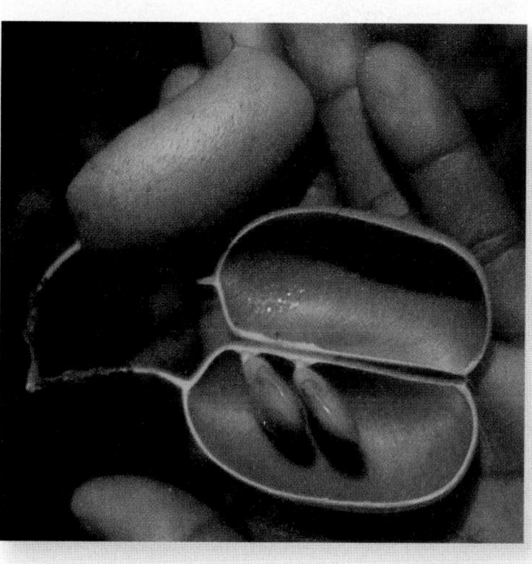

(c)

20–23 Legumes, dehiscent fruits The legume, a kind of fruit that is usually dehiscent, is the characteristic fruit of the pea family, Fabaceae (also called Leguminosae). With about 18,000 species, Fabaceae is one of the largest families of flowering plants. Many members of the family are capable of nitrogen fixation because of the presence of nodule-forming bacteria of the genera *Bradyrhizobium* and *Rhizobium* on their roots (see Chapter 29). For this reason, these plants are often the first colonists on relatively infertile soils, as in the tropics, and they may grow rapidly there. The seeds of a number of plants of this family, such as peas, beans, and lentils, are important foods. ***(a)*** Legumes of the garden pea, *Pisum sativum.* ***(b)*** Legumes of *Albizzia polyphylla,* growing in Madagascar. Each seed is in a separate compartment of the fruit. ***(c)*** Legume of *Griffonia simplicifolia,* a West African tree. The two valves of the legume are split apart, revealing the two seeds inside.

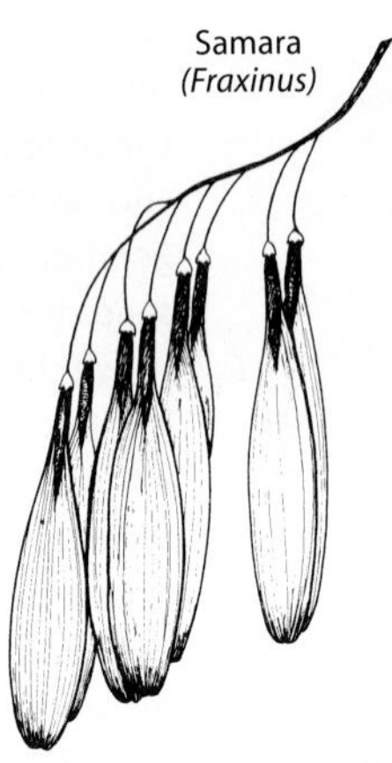

20–24 Indehiscent fruit The samara, a winged indehiscent fruit characteristic of ashes *(Fraxinus)* and elms *(Ulmus)*, retains its single seed at maturity. Samaras are dispersed by wind.

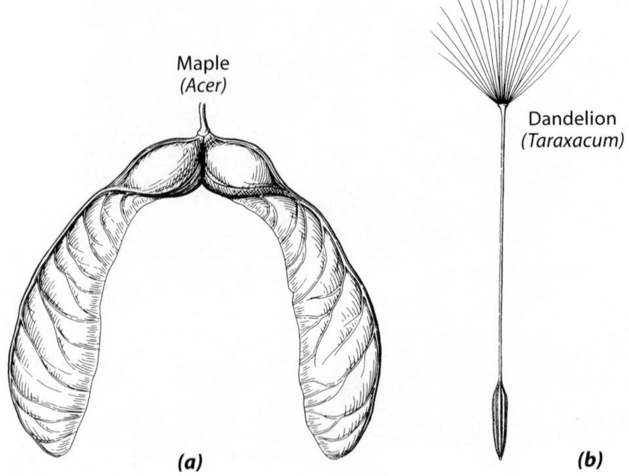

20–26 Wind-dispersed fruits ***(a)*** In maples *(Acer)*, each half of the schizocarp has a long wing. ***(b)*** The fruits of the dandelion *(Taraxacum)* and many other composites have a modified calyx, called the pappus, which adheres to the mature cypsela and may form a plumelike structure that aids in wind dispersal.

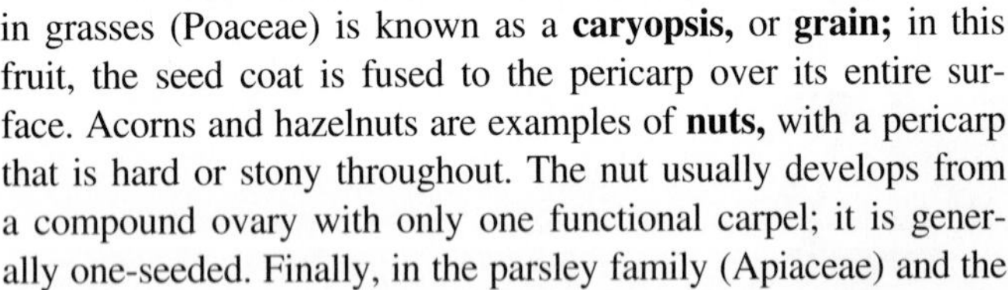

in grasses (Poaceae) is known as a **caryopsis,** or **grain;** in this fruit, the seed coat is fused to the pericarp over its entire surface. Acorns and hazelnuts are examples of **nuts,** with a pericarp that is hard or stony throughout. The nut usually develops from a compound ovary with only one functional carpel; it is generally one-seeded. Finally, in the parsley family (Apiaceae) and the maples (Sapindaceae), as well as in a number of other, unrelated groups, the fruit is a **schizocarp,** which splits at maturity into two or more one-seeded portions (Figure 20–26a).

20–25 Cypselas The familiar small, indehiscent fruits of dandelions, which are technically known as cypselas (but often loosely called achenes), have a plumelike, modified calyx (the pappus) and are spread by the wind. This photograph shows the fruiting heads of a plant of the genus *Agoseris*, which is closely related to the dandelions.

Fruits and Seeds Have Evolved in Relation to Their Dispersal Agents

Just as flowers have evolved according to the characteristics of the pollinators that visit them regularly, so have fruits evolved in relation to their dispersal agents. In both coevolutionary systems, there have, in general, been many changes within individual families in relation to the different dispersal agents. There is also a great deal of convergent evolution toward similar-appearing structures with similar functions. We review here some of the adaptations of fruits in relation to their modes of dispersal.

Many Plants Have Wind-Borne Fruits and Seeds Some plants have light fruits or seeds that are dispersed by the wind (Figures 20–22a and 20–24 through 20–26). The dustlike seeds of all members of the orchid family, for example, are wind-borne. Other fruits have wings, sometimes formed from perianth parts, that allow them to be blown from place to place. In the schizocarps of maples, for example, each carpel develops a long wing (Figure 20–26a). The two carpels separate and fall when mature. Many members of the Asteraceae—dandelions, for example—develop a plumelike pappus, which aids in keeping the light fruits aloft (Figures 20–25 and 20–26b). In some plants, the seed itself, rather than the fruit, bears the wing or plume. The familiar butter-and-eggs *(Linaria vulgaris)* has a winged seed, and both fireweed *(Chamaenerion)* and milkweed (*Asclepias;* Figure 20–22a) have plumed seeds. In willows and poplars (family Salicaceae), the seed coat is covered with woolly hairs. In tumbleweeds *(Salsola)*, the whole plant (or a portion of it) is blown along by the wind, scattering seeds as it moves.

Other plants shoot their seeds aloft. In touch-me-not *(Impatiens)*, the valves of the capsules separate suddenly, throwing seeds for some distance. In the witch hazel *(Hamamelis),* the endocarp contracts as the fruit dries, discharging the seeds so forcefully that they sometimes travel as far as 15 meters from the plant. Another example of self-dispersal is shown in Figure 20–27. In contrast to these active methods of dispersal, the seeds or fruits of many plants simply drop to the ground and are dispersed more or less passively (or sporadically, such as by rainwater or floods).

Fruits and Seeds Adapted for Floating Are Dispersed by Water The fruits and seeds of many plants, especially those growing in or near water sources, are adapted for floating; they can float either because air is trapped in some part of the fruit or because the fruit contains tissue that includes large air spaces. Some fruits are especially adapted for dispersal by ocean currents. Notable among these is the coconut, which is why almost every newly formed Pacific atoll quickly acquires its own coconut tree. Rain, also a common means of fruit and seed dispersal, is particularly important for plants that live on hillsides or mountain slopes.

Fruits and Seeds That Are Fleshy or Have Adaptations for Attachment Are Dispersed by Animals The evolution of sweet and often highly colored, fleshy fruits was clearly involved in the coevolution of animals and flowering plants. The majority of fruits in which much of the pericarp is fleshy—bananas, cherries, raspberries, dogwoods, grapes—are eaten by vertebrates. When such fruits are eaten by birds or mammals, the seeds contained by the fruits are spread by being passed unharmed through the digestive tract or, in birds, by being regurgitated at a distance from the place where they were ingested (Figure 20–28). Sometimes, partial digestion aids the germination of seeds by weakening their seed coats.

When fleshy fruits ripen, they undergo a series of characteristic changes, mediated by the hormone ethylene, which we discuss in Chapter 27. Among these changes are a rise in sugar content, a softening of the fruit caused by the breakdown of pectic substances, and, often, a change in color from inconspicuous, leaflike green to bright red (Figure 20–28a), yellow, blue, or black. The seeds of some plants, especially tropical ones, often have fleshy appendages, or arils, with the bright colors characteristic of fleshy fruits and, like these fruits, are aided in their dispersal by vertebrates.

A number of angiosperms have fruits or seeds that are dispersed by adhering to fur or feathers (Figure 20–29). These fruits and seeds have hooks, barbs, spines, hairs, or sticky coverings that allow them to be transported, often for great distances, attached to the bodies of animals.

Ants are another important agent of seed dispersal in some plants. These plants have a special adaptation on the exterior of their seeds, called an elaiosome, a fleshy pigmented appendage that contains lipids, protein, starch, sugars, and vitamins. The ants usually carry such seeds back to their nests, where the elaiosomes are consumed by other workers or larvae, and the seeds are left intact. The seeds readily germinate in this location, and seedlings often become established, protected from their predators and perhaps benefiting from nutrient enrichment as well. Up to a third of the species in some plant communities, such as the herb understory communities in the deciduous forests of the central and eastern United States, are dispersed by ants in this way. These plants include such familiar species as spring beauty *(Claytonia virginica),* Dutchman's-breeches *(Dicentra cucullaria),* bloodroot *(Sanguinaria canadensis),* many species of violets *(Viola;* see Figure 12–2), and *Trillium* (Figures 1–5 and 19–3c).

(a)

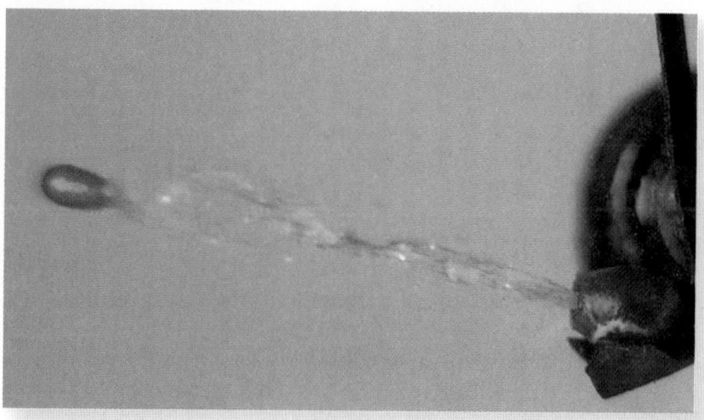

(b)

20–27 Self-dispersal of seeds Dwarf mistletoe *(Arceuthobium),* a parasitic eudicot that causes serious loss of forest productivity in the western United States. ***(a)*** A plant growing on a pine branch in California. ***(b)*** Seed discharge. Very high hydrostatic pressure builds up in the fruit and shoots the seeds as far as 15 meters laterally. The seeds have an initial velocity of about 100 kilometers per hour. This is one of the ways in which the seeds are spread from tree to tree, although they are also sticky and can be carried from one tree to another over much longer distances by adhering to the feet or feathers of birds.

(a)

(b)

20–28 Fleshy fruits The seeds of fleshy fruits are usually dispersed by vertebrates that eat the fruits and either regurgitate the seeds or pass them as part of their feces. Examples of vertebrate-dispersed fruits are shown here. ***(a)*** Strawberries *(Fragaria),* an example of both an aggregate fruit and an accessory fruit. The achenes are borne on the surface of a fleshy receptacle, which constitutes most of the fruit. Immature strawberries, like the immature stages of many bird- or mammal-dispersed fruits, are green, but they become red when the seeds are mature and thus ready for dispersal. ***(b)*** Berries of chaparral honeysuckle, *Lonicera hispidula.* These berries develop from inferior ovaries, and they therefore incorporate fused portions of the outer floral whorls—accessory tissues. A flower of this species is shown in Figure 20–9c.

(a)

(b)

20–29 Hooks and spines ***(a)*** The fruits of the African plant *Harpagophytum,* a member of the sesame family (Pedaliaceae), are equipped with "grappling hooks," which catch in the fur on the legs of large mammals. In this way, the fruits are spread from place to place. ***(b)*** Mature inflorescences of cocklebur *(Xanthium),* which attach themselves to passing animals and are dispersed. In this case, the entire inflorescence is the dispersal unit, rather than the fruits alone, as in *Harpagophytum.* Cocklebur is a member of the family Asteraceae.

Biochemical Coevolution

Also important in the evolution of angiosperms are the so-called **secondary metabolites,** or **secondary plant products** (pages 30 to 35). Once thought of as waste products, these include an array of chemically unrelated compounds, such as **alkaloids** (including morphine, cocaine, caffeine, and nicotine; see Figure 2–22); **terpenoids** (including essential oils, taxol, rubber, cardiac glycosides, and isoprene; see Figure 2-23); **phenolics** (including flavonoids, tannins, lignins, catechols, and salicylic acid; see Figure 2–28); **quinones** (including coenzyme Q; see Figure 6–9); and even raphides (needlelike crystals of calcium oxalate; see Figure 3–20). The presence of certain of these compounds can characterize whole families, or groups of families, of flowering plants.

In nature, these chemicals appear to play a major role either in restricting the palatability of the plants that produce them or in causing animals to avoid the plants altogether (Figure 20–30). When a given family of plants is characterized by a distinctive group of secondary plant products, those plants are apt to be eaten only by insects belonging to certain families. The mustard family (Brassicaceae), for example, is characterized by the presence of mustard-oil glycosides, as well as the associated enzymes that break down these glycosides to release the pungent odors associated with cabbage, horseradish, and mustard. Plant-eating insects of most groups ignore plants of the mustard family and will not feed on them even if they are starving. However, certain groups of true bugs and beetles, and the larvae of some groups of moths, feed only on the leaves of plants of the mustard family. The larvae of most members of the butterfly subfamily Pierinae (which includes the cabbage butterflies and orange-tips) also feed only on these plants. The same chemicals that act as deterrents to most groups of insect herbivores often act as feeding stimuli for these narrowly restricted feeders.

Clearly, the ability to manufacture the mustard-oil glycosides and to retain them in their tissues is an important evolutionary step that protects the plants of the mustard family from most herbivores. From the standpoint of herbivores in general, such protected plants represent an unexploited food source for any group of insects that can tolerate or break down the poisons manufactured by the plants.

Herbivorous insects that are narrowly restricted in their feeding habits to groups of plants with certain secondary plant products are often brightly colored. This coloration serves as a signal to their predators that they carry the noxious chemicals such as alkaloids and cardiac glycosides in their bodies and hence are unpalatable (page 33). Various drugs and psychedelic chemicals, such as the active ingredients in marijuana *(Cannabis sativa)* and the opium poppy *(Papaver somniferum),* among others, are also secondary plant products that, in nature, presumably play a role in discouraging the attacks of herbivores (Figure 20–31).

Still more complex systems are known. When the leaves of potato or tomato plants are wounded, as by the Colorado potato beetle, the concentration of proteinase inhibitors, which interfere with the digestive enzymes in the beetle's gut, rapidly increases in the wounded tissues. Other plants manufacture molecules that resemble the hormones of insects or other predators and thus interfere with the predators' normal growth and development.

As mentioned earlier in the chapter, pollination and fruit-dispersal systems have developed particular coevolutionary patterns in which many of the possible variants have evolved not once but several times within a particular plant family or even genus. The resulting array of forms gives many groups of angiosperms a wide variety of pollination and fruit-dispersal mechanisms. In the case of biochemical relationships, however, the evolutionary steps appear to have been large and definitive, and whole families of plants can be characterized biochemically and associated with major groups of plant-eating insects. These biochemical relationships seem to have played a key role in the success of the angiosperms, which have a vastly more diverse array of secondary plant products than any other group of plants.

20–30 Poison ivy A secondary plant product, 3-pentadecanedienyl catechol, produced by poison ivy, *Toxicodendron radicans,* causes an irritating rash on the skin of many people. The ability to produce this phenolic presumably evolved under the selective pressure exerted by herbivores. Fortunately, the plant is easily identifiable by its characteristic compound leaves with their three leaflets.

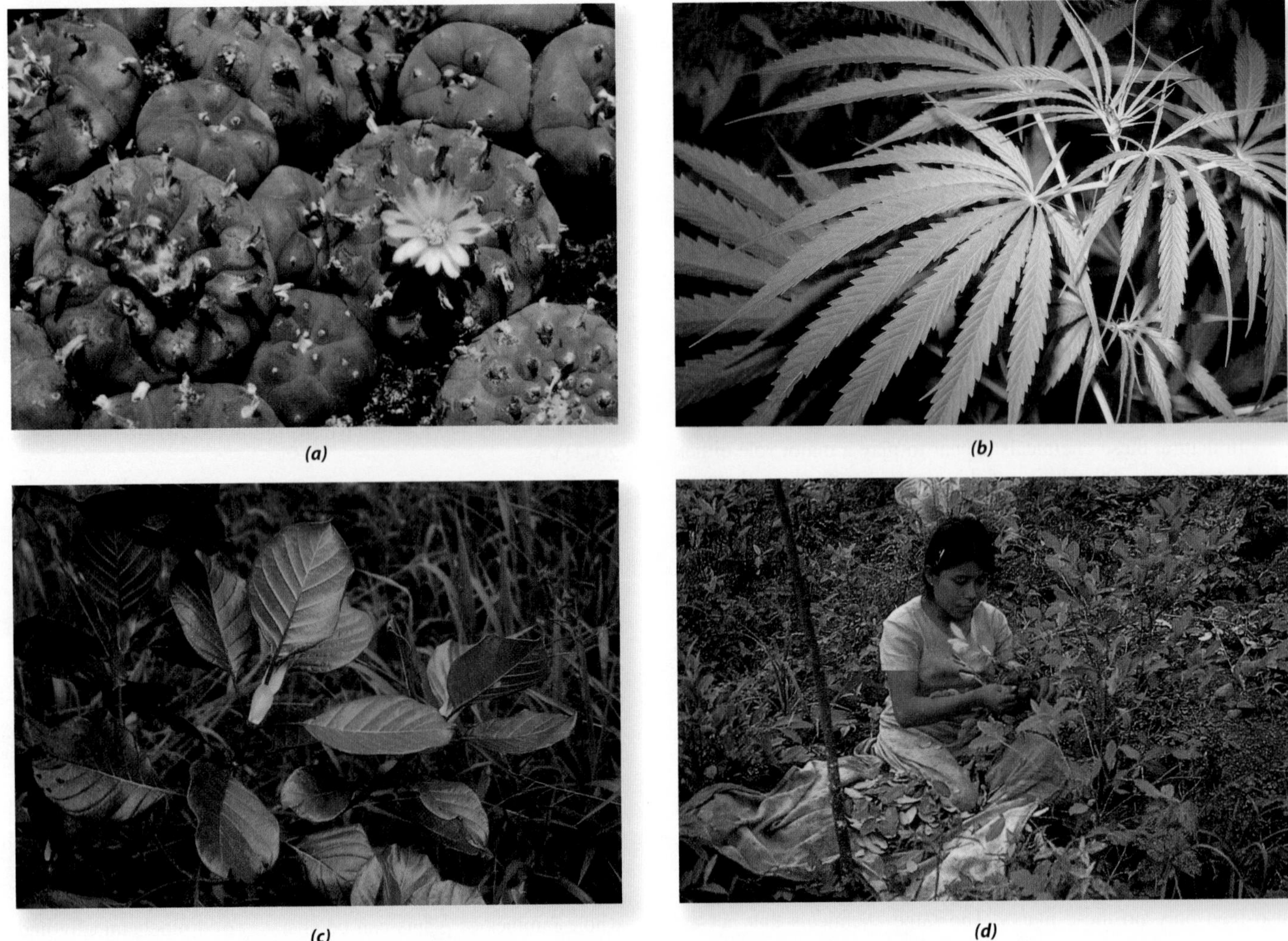

20–31 Hallucinogenic and medicinal compounds ***(a)*** Mescaline, from the peyote cactus *(Lophophora williamsii)*, is used ceremonially by many Native American groups of northern Mexico and the southwestern United States. ***(b)*** Tetrahydrocannabinol (THC) is the most important active molecule in marijuana *(Cannabis sativa)*. ***(c)*** Quinine, a valuable drug used in the treatment and prevention of malaria, is derived from tropical trees and shrubs of the genus *Cinchona*. ***(d)*** Cocaine (page 31), a drug that has been extensively abused, is derived from coca *(Erythroxylum coca)*, a cultivated plant of northwestern South America. A Peruvian woman is shown here harvesting the leaves of cultivated coca. The secondary plant products identified in these plants presumably protect them from the depredations of insects, but they are also physiologically active in their vertebrate predators, including humans.

SUMMARY

The Ancestry of the Angiosperms Remains to Be Determined

Several groups of seed plants—Caytoniales, Bennettitales, and gnetophytes—have at one time or another been hypothesized as angiosperm ancestors.

Several Factors Help Explain the Worldwide Success of Angiosperms

The earliest definitely angiosperm remains are from the Early Cretaceous period, around 125 to 130 million years ago; they include both flowers and pollen. The flowering plants became dominant worldwide between 90 and 80 million years ago. Possible reasons for their success include various adaptations for drought resistance, such as the evolution of the deciduous habit, as well as the evolution of efficient and often specialized mechanisms for pollination and seed dispersal.

A Few, Relatively Small Groups of Angiosperms Retain Archaic Features

Some angiosperms have features retained from the early history of the group. These include evolutionary lines such as the New Caledonian shrub *Amborella*, the water lilies, and *Austrobaileya*. Another such evolutionary line consists of the magnoliids, with about 20 families, including the magnolia and laurel families. All of these plant groups have pollen with a single pore (or furrow), as do the monocots, which constitute about 22 percent of living angiosperms. The eudicots, with pollen that has three apertures (pores or furrows), comprise about three-quarters of the species of angiosperms.

The Four Whorls of Flower Parts Have Evolved in Different Ways

Most angiosperm flowers consist of four whorls. The outermost whorl consists of sepals, which are specialized leaves that protect the flower in bud. In contrast, the petals of most angiosperms have evolved from stamens that have lost their sporangia during the course of evolution. Stamens with anthers that comprise two pairs of pollen sacs are one of the diagnostic features of angiosperms. In the course of evolution, differentiation between the anther and the slender filament seems to have increased. Carpels are somewhat leaflike structures that have been transformed during the course of evolution to enclose the ovules. In most plants, the carpels have become specialized and differentiated into a swollen, basal ovary, a slender style, and a receptive terminal stigma. The loss of individual floral whorls or fusion within and between adjacent whorls have led to the evolution of many specialized floral types, which are often characteristic of particular families.

Angiosperms Are Pollinated by a Variety of Agents

Pollination by insects is basic in the angiosperms, and the first pollinating agents were probably beetles. The closing of the carpel, in an evolutionary sense, may have contributed to protection of the ovules from visiting insects. Pollination interactions with more specialized groups of insects seem to have evolved later in the history of the angiosperms, and flies, butterflies, and moths have each left their mark on the morphology of certain angiosperm flowers. The bees, however, are the most specialized and constant of flower-visiting insects and have probably had the greatest effect on the evolution of flowers. Each group of flower-visiting animals is associated with a particular group of floral characteristics related to the animals' visual and olfactory senses. Some angiosperms have become wind-pollinated, shedding copious quantities of small, nonsticky pollen and having well-developed, often feathery stigmas that are efficient in collecting pollen from the air. Water-pollinated plants have either filamentous pollen grains that float to submerged flowers or various ways of transmitting pollen through or across the surface of the water.

Various Factors Affect the Relationship between Plant and Pollinator

Flowers that are regularly visited and pollinated by animals with high energy requirements, such as hummingbirds, hawkmoths, and bats, must produce large amounts of nectar. These sources of nectar must then be protected and concealed from other potential visitors with lower energy requirements. Such visitors might satiate themselves with nectar from a single flower (or from the flowers of a single plant) and therefore fail to move on to another plant of the same species to effect cross-pollination. Wind pollination is most effective when individual plants grow together in large groups, whereas insects, birds, or bats can carry pollen great distances from plant to plant.

Flower Colors Are Determined Mainly by Carotenoids and Flavonoids

Carotenoids are yellow, oil-soluble pigments that occur in plastids (chloroplasts and chromoplasts) and act as accessory pigments in photosynthesis. Flavonoids are water-soluble ring compounds present in the vacuole. Anthocyanins, which are blue or red pigments that constitute one major class of flavonoids, are especially important in determining the colors of flowers and other plant parts.

Fruits Are Basically Matured Ovaries

Fruits are just as diverse as the flowers from which they are derived, and they can be classified either morphologically, in terms of their structure and development, or functionally, in terms of their methods of dispersal. Fruits are matured ovaries; if additional flower parts are retained in their mature structure, they are said to be accessory fruits. Simple fruits are derived from a single carpel or from two or more united carpels, aggregate fruits from the free carpels of one flower, and multiple fruits from an inflorescence. Dehiscent fruits split open to release the seeds, and indehiscent fruits do not split.

Fruits and Seeds Are Dispersed by Wind, Water, or Animals

Wind-borne fruits or seeds are light and often have wings or tufts of hairs that aid in dispersal. The fruits of some plants expel their seeds explosively. Some seeds or fruits are borne away by water, in which case they must be buoyant and have water-resistant coats. Others are disseminated by birds or mammals and frequently have tasty, fleshy coverings or hooks, spines, or other devices that adhere to fur or feathers. Ants disperse the seeds and fruits of many plants; the dispersal units typically have an oily appendage called an elaiosome, which the ants consume.

Secondary Plant Products Are Important in the Evolution of Angiosperms

Biochemical coevolution has been an important aspect of the evolutionary success and diversification of the angiosperms. Certain groups of angiosperms produce various secondary products, or secondary metabolites, such as alkaloids, which protect them from most foraging herbivores. However, certain herbivores (usually those with narrow feeding habits) are able to feed on these plants and are regularly found associated with them. Potential competitors are excluded from the same plants because of their inability to handle the toxic substances. This pattern indicates that a stepwise pattern of coevolutionary interaction has occurred, and the early angiosperms most likely were similarly protected by their ability to produce some chemicals that functioned as poisons for herbivores.

QUESTIONS

1. What unique characteristics of Anthophyta (angiosperms) indicate that the members of this phylum were derived from a single common ancestor?
2. With the discovery of *Archaefructus* and the results of molecular phylogenetic studies indicating that *Amborella* is sister to all other flowering plants, our concept of flower structure in the earliest angiosperms has changed. Explain.
3. Evolutionarily, petals apparently have been derived from two different sources. What are they?
4. Explain what is meant by coevolution, and provide two examples involving different insects and plants.
5. Why are wind-pollinated angiosperms best represented in temperate regions and relatively rare in the tropics?
6. Distinguish among simple, aggregate, and multiple fruits, and give an example of each.

CHAPTER 21

Plants and People

Black pepper As an accepted form of currency throughout the Middle Ages, pepper was once traded ounce for ounce with precious metals. This fifteenth-century French illustration shows harvesting of the peppercorns and presentation of the precious commodity to the king. Sumptuous banquets were known for their extravagant use of pepper, a spice used in cooking but also valued for its medicinal properties.

CHAPTER OUTLINE

Our own species, *Homo sapiens,* has been in existence for at least 500,000 years and has been abundant for approximately 150,000 years. Like all other living organisms, we represent the product of at least 3.5 billion years of evolution. Our immediate antecedents, members of the genus *Australopithecus,* first appeared no less than 5 million years ago. At about that time, they apparently diverged within Africa from the evolutionary line that gave rise to the chimpanzees and gorillas, our closest living relatives. The members of the genus *Australopithecus* were relatively small apes that often walked on the ground on two legs.

Larger, tool-using human beings—members of the genus *Homo*—first appeared about 2 million years ago. They almost certainly evolved from members of the genus *Australopithecus,* but they had much larger brains, apparently associated with, and reinforced by, their ability to use tools. Early members of the genus *Homo* probably subsisted mainly by gathering food (picking fruits, seeds, and nuts; harvesting edible shoots and leaves; digging roots), scavenging dead animals, and occasionally hunting. They learned how to use fire no less than 1.4 million years ago. Their hunter-gatherer methods for obtaining food and shelter seem to have resembled those of some modern-day groups of humans (Figure 21–1). Our species, *Homo sapiens,* appeared in Africa by about 500,000 years ago and in Eurasia by about 250,000 years ago.

About 34,000 years ago, the powerfully built, short, stocky Neanderthal people, who had been abundant in Europe and western Asia, disappeared completely. They were replaced by human beings who were essentially just like us. From that time onward, our ancestors made increasingly complex tools, as well as personal ornaments, of stone and also of bone, ivory, and antler—materials that had not been used earlier. These people were excellent hunters, preying on the herds of large animals with which they shared their environment. They also began to create magnificent ritual paintings on the walls of caves. The foundations of modern society had been laid.

CHECKPOINTS

After reading this chapter, you should be able to answer the following:

1. When and where did agriculture in the Fertile Crescent begin? What plants were particularly important as early crops?
2. What plants were important in New World agriculture? How do these plants differ from those first cultivated in the Old World?
3. How does a spice differ from an herb? Where did spices and herbs originate?
4. Name the world's principal crop plants today.
5. How has the growth of the human population changed since 1600? What problems have arisen because of this growth?

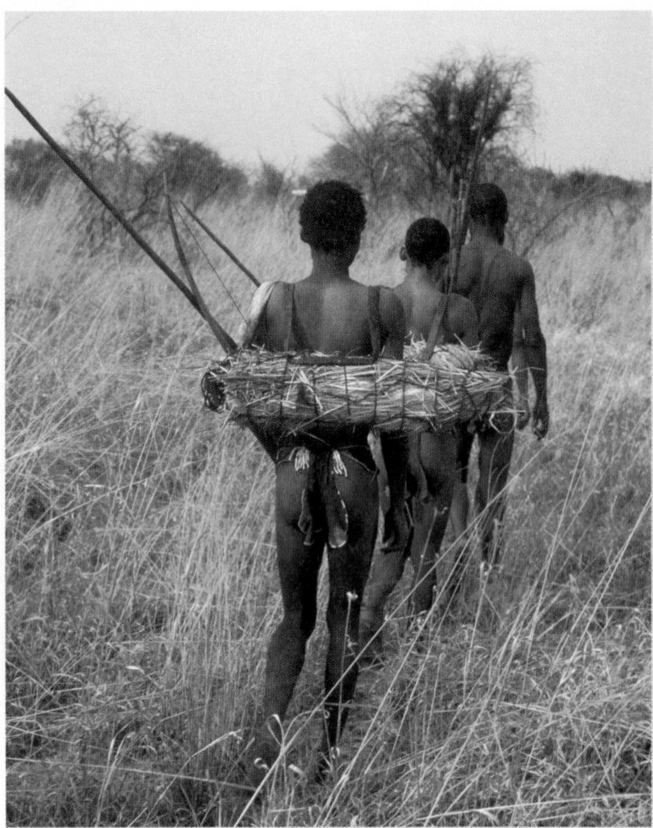

21–1 Hunter-gatherers Kalahari San (formerly known as Bushmen) looking for ostrich eggs, each of which is equivalent in size to 22 chicken eggs. These hunter-gatherers live in the northwest Kalahari Desert of Namibia.

The Rise of Agriculture

The Beginnings of Agriculture Involved the Deliberate Planting of Wild Seeds

The modern human beings who replaced the Neanderthals soon migrated over the entire surface of the globe. They colonized Siberia soon after they appeared in Europe and western Asia, and they reached North America by at least 14,000 years ago. Their migration eastward took place during one of the cold periods of the Pleistocene epoch, when savannas, with their large herds of grazing mammals, were widespread. As these people migrated, they seem to have been responsible for the extinction of many species of these animals. In any event, extensive hunting by humans, together with major changes in climate, occurred at the same time that these animals were disappearing from many parts of the world.

At the end of the last glacial period, about 18,000 years ago, the glaciers began to retreat, just as they had 18 or 20 times during the preceding 2 million years. Forests migrated northward across Eurasia and North America, while grasslands became less extensive and the large animals associated with them dwindled in number. Probably no more than 5 million humans existed throughout the world, and they gradually began to utilize new sources of food. Some of them lived along the seacoasts, where animals that could be used as sources of food were locally abundant; others, including some coastal peoples, began to cultivate plants, thus gaining a new, relatively secure source of food.

The first deliberate planting of seeds was probably the logical consequence of a simple series of events. For example, the wild cereals (grain-producing members of the grass family, Poaceae) grow readily on open or disturbed areas, patches of bare land where there are few other plants to compete with them. People who gathered these grains regularly might have accidentally spilled some of them near their campsites, or deliberately planted them, thus creating a more dependable source of food. In places where wild grains and legumes were abundant and readily gathered, humans would have remained for long periods of time, eventually learning how to increase their yields by saving and planting seeds, by watering and fertilizing their crops, and by protecting them from mice, birds, and other pests. Such continuous management of "wild" resources may have gradually intensified toward what is now considered **cultivation** (Figure 21–2).

Over time, as humans began to select particular genetic variants, the characteristics of these plants would change gradually as more seeds were selected from plants that were easier to gather and store. This is the process of **domestication,** by which genetic changes in plant populations evolved because of cultivation and selection by humans. With the cereals—barley, wheat, rice, and maize—domestication led to more and larger grains, thicker stalks, seeds that separate freely from the chaff, and improved flavor. An additional feature shared by cereals, and most other crops—and key to their domestication—was the loss of natural seed dispersal, which is required by wild species to produce the next generation. Domesticated plants hold on to ripe seeds, allowing them to be harvested by humans for food and replanting. Thus, as the domestication process continued, a crop plant steadily became more and more dependent on the humans who cultivated it, just as humans became more and more dependent on the plant.

Until relatively recently, most archaeobotanists (botanists who study plant remains at archaeological sites) regarded the advent of agriculture as an abrupt break—200 or fewer years—from the hunter-gatherer lifestyle practiced by humans for millions of years: an "agricultural revolution." It was thought that domesticated crops appeared very soon after people began to cultivate fields. Today, however, many archaeobotanists believe that the full domestication of wheat and other crops might actually have taken thousands of years. New data suggest that the road from gathering wild plants to domesticating them was a "long and winding one." Under the circumstances, it is questionable whether this very long process should be characterized as a "revolution."

Implements associated with the harvesting and processing of grains, including flint sickle blades, grinding stones, and stone mortars and pestles, were in use long before humans began cultivating plants. Sickle blades have been found in deposits dated at 12,000 years ago, and a grindstone found in what today is Israel has been dated at 23,000 years ago. The grindstone contained starch grains characteristic of wild barley. Starch grains retrieved from the surfaces of Middle Stone Age tools from Mozambique indicate that humans there relied on grass seeds at least 105,000 years ago; 89 percent of the starch grains were of *Sorghum* species.

(a) (b)

21–2 Wheat cultivation ***(a)*** Harvesting and ***(b)*** winnowing of wheat *(Triticum)* in Tunisia, North Africa. Similar small-scale cultivation of wheat has been taking place around the Mediterranean basin for more than 10,000 years.

The Fertile Crescent Is the Oldest Known Center of Plant Domestication

Centers of plant domestication, or of agriculture, arose independently in at least 11 different centers throughout the world (Figure 21–3). In the Old World, the domestication of plants began 13,000 to 11,000 years ago in an area of the Near East known as the **Fertile Crescent,** in lands that extend through parts of what are now Iran, Iraq, Turkey, Syria, Lebanon, Jordan, and Israel. In this region, wild barley (*Hordeum vulgare*

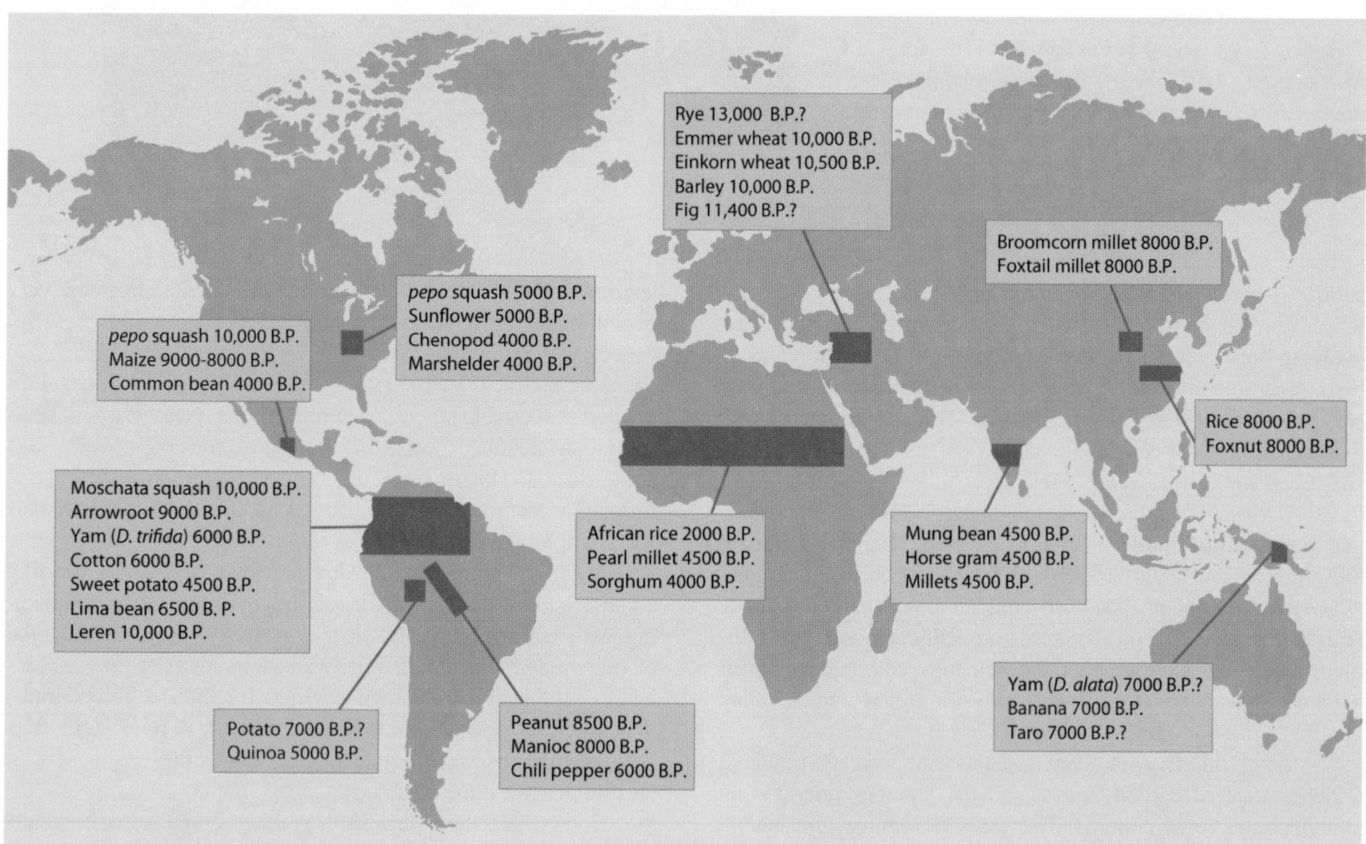

21–3 Independent centers of plant domestication People in at least 11 parts of the world independently brought plants under domestication. The principal crop plants and estimates of the periods when they were first domesticated are indicated for each location. B.P., before the present.

(a)

(b)

(c)

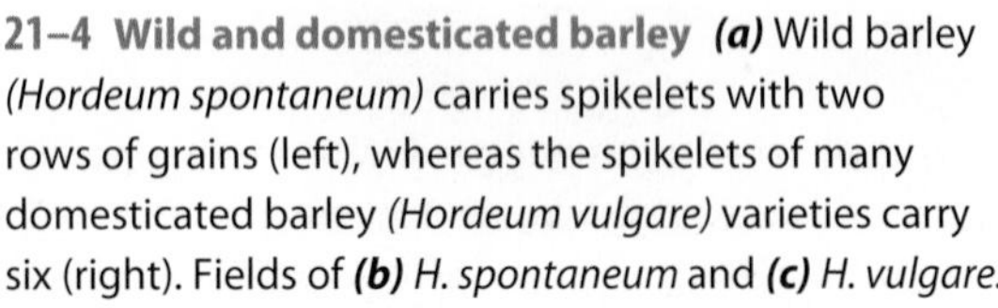

21–4 Wild and domesticated barley ***(a)*** Wild barley *(Hordeum spontaneum)* carries spikelets with two rows of grains (left), whereas the spikelets of many domesticated barley *(Hordeum vulgare)* varieties carry six (right). Fields of ***(b)*** *H. spontaneum* and ***(c)*** *H. vulgare.*

subsp. *spontaneum,* the progenitor of *Hordeum vulgare*) and wheat (emmer and einkorn wheat, earlier forms than the later-evolving bread wheat, *Triticum aestivum*) were the first plants to be brought into cultivation, with lentils *(Lens culinaris)* and peas *(Pisum sativum)* soon following (Figure 21–4). Additional plants that were cultivated early in this area were chickpeas, or garbanzos *(Cicer arietinum),* vetch (*Vicia* spp.), olives *(Olea europaea),* dates *(Phoenix dactylifera),* pomegranates *(Punica granatum),* and grapes *(Vitis vinifera).* Wine made from the grapes and beer brewed from the grains were used from very early times. Flax *(Linum usitatissimum)* was also cultivated very early, probably both as a source of food (the seeds are still eaten today in Ethiopia) and as a source of fiber for weaving cloth. Archeological evidence indicates that figs were cultivated in the Jordan Valley 11,400 to 11,200 years ago.

Among the first cultivated plants, the cereals provided a rich source of carbohydrates, and the legumes provided an abundant source of proteins. The seeds of legumes are among the most protein-rich of all plant parts, and their proteins, in turn, often are rich in the particular amino acids that are poorly represented in the cereals. It is not surprising, therefore, that legumes have been cultivated, along with cereals, from the very beginnings of agriculture in various parts of the world.

The Domestication of Animals Followed the Domestication of Plants Following the establishment of crops in the Near East, various animals—including goats, sheep, cattle, and pigs—were also domesticated. Horses were domesticated later, in southwestern Europe, and chickens in Southeast Asia, but all of these animals spread rapidly throughout the world.

The first animals to be domesticated were dogs, starting about 15,000 years ago. Their role in guarding people and helping with hunting makes it clear why they were so useful. All of our modern breeds of dogs were derived from gray wolves in the Middle East and then selected for their different traits. When people first came to the New World, they brought dogs with them. Cats were domesticated about 5000 years ago, in North Africa and the near East.

Everywhere they were kept, grazing animals ate the plants that were available to them, whether cultivated or not. In addition, the domesticated animals produced wool, hides, milk, cheese, and eggs, and they could themselves be eaten by their

21–5 Grazing and desertification Herds of domesticated animals, like these karakul sheep in Afghanistan, devastated large areas of the eastern Mediterranean region as their numbers increased. In many areas, only spiny or poisonous plants survived, while previously fertile fields were converted to desert.

21–6 Soybeans Although a major crop in the United States for only about 60 years, soybeans *(Glycine max)* are one of the richest sources of nutrients among food plants. The seeds consist of 40 to 45 percent protein and 18 percent fats and oils. Like many legumes, soybeans harbor nitrogen-fixing bacteria in nodules on their roots, by means of which they are able to obtain nitrogen for their own growth and to enrich the soil (see Chapter 29).

owners. As human beings increased in number, they built up herds of grazing animals so large that they began to destroy their own pastures, causing widespread ecological destruction (Figure 21–5). Much of the Near East and other arid areas around the Mediterranean Sea are still badly overgrazed, and deserts continue to spread, as they have from the initial formation of large herds of domestic animals.

Evidence Exists for Early Domestication of Food Plants in China, Tropical Asia, and Africa The agriculture that had originated in the Near East spread northwestward, extending over much of Europe and reaching Britain by about 4000 B.C.E. At the same time, agriculture was developed independently in other parts of the world. There is evidence that agriculture may have been practiced in the subtropical Yellow River area of China nearly as early as in the Near East. Several genera of cereal grasses known as millets were cultivated for their grain, and eventually rice (*Oryza sativa,* domesticated from the wild species *Oryza rufipogon*), now one of the most valuable cereals in the world, was added. Later, rice displaced the millets as a crop throughout much of their earlier range. Soybeans *(Glycine max)* have been cultivated in China for at least 3100 years and used to make bean curd and soy sauce (Figure 21–6).

In other parts of subtropical Asia, an agriculture based on rice and different legumes and root crops was developed. Such animals as water buffalo, camels, and chickens were domesticated in Asia long ago, and they became important elements in the systems of cultivation practiced there (Figure 21–7).

Eventually, plants such as the mango *(Mangifera indica)* and the various kinds of citrus (*Citrus* spp.) were brought into cultivation in tropical Asia. Taro *(Colocasia esculenta)* is a very important food plant in tropical Asia, where it is grown for its starchy corms; *Xanthosoma* is a related food plant of the New World tropics. *Colocasia* and other similar genera, including *Xanthosoma,* are the source of poi, a staple starch food of the Pacific Islands, including Hawaii, where these plants were brought by Polynesian settlers about 1500 years ago.

Bananas *(Musa × paradisiaca)* are among the most important domesticated plants to come from tropical Asia; their fruits are a staple food throughout the tropical parts of the world. Starchy varieties of bananas, known as plantains, are much more important as a source of food in tropical countries (where two-thirds of the total banana crop is consumed) than are the sweet varieties, which are a more familiar food in temperate regions. Wild bananas have large, hard seeds, but the domesticated varieties, like many kinds of cultivated citrus fruits, are seedless. The banana reached Africa over 2500 years ago, and it was brought to the New World soon after the voyages of Columbus. New Guinea has been identified as the center of origin for taro, bananas, sugarcane *(Saccharum officinale),* and the yam *(Dioscorea × alata).*

Domestication of food plants also took place in Africa, but we have little direct evidence of the timing of its origins. In any event, there is a gap of at least 5000 years between the origins of agriculture in the Fertile Crescent and its appearance at the southern tip of the African continent. Such grains as sorghum (*Sorghum* spp.) and various kinds of millet (*Pennisetum* spp. and *Panicum* spp.), various kinds of vegetables, including

21–7 Rice Half of the food consumed by some 1.6 billion people and more than a quarter of the food consumed by another 400 million is provided by rice *(Oryza sativa)*. It has been cultivated for at least 6000 years, and it is currently grown on about 11 percent of the world's arable land. Water buffalo are used to plow rice terraces, as seen here in Bali, Indonesia.

cowpeas *(Vigna unguiculata)* and okra *(Hibiscus esculentus)*, and several kinds of root crops, including yams (*Dioscorea* spp.), were all initially brought into cultivation in Africa. New archaeological evidence indicates the presence of domesticated pearl millet in Mali 4500 years ago.

A species of cotton *(Gossypium)* was also initially brought into cultivation in Africa. The various species of cotton are widespread as wild plants, mainly in seasonally arid but mild climates, and they are obviously useful; the long hairs on their seeds are easily woven into cloth (Figure 21–8). Cotton seeds are also used as a source of oil, and the seed meal from which the oil has been expressed is used for feeding animals. Coffee *(Coffea arabica)* is another crop of African origin. It was brought into cultivation much later than the other plants mentioned here, but it is now a very important commercial crop in the tropics.

Agriculture in the New World Utilized Many New Species

A parallel development of agriculture took place in three New World centers of origin: Mesoamerica (Mexico and Central America), South America, and North America. No domesticated plants seems to have been brought by human beings from the Old World to the New World prior to 1492, although ancient-DNA sequence analysis and archaeological evidence

21–8 Cotton One of the most useful plants cultivated for fiber, cotton *(Gossypium)* appears to have been domesticated independently in India and China (the same species of cotton is cultivated in both areas), as well as in Africa, Mexico, and western South America, all of which grow different species of cotton. Cotton has been cultivated for thousands of years for cloth and has also become an important source of edible oil during the past century. The cotton that is grown widely throughout the world today is a polyploid from the New World; the diploid species of the Old World are more locally cultivated.

indicate that the bottle gourd *(Lagenaria siceraria),* which is indigenous to Africa, was widely distributed in the New World 8000 years ago. It has been suggested that the bottle gourd was brought to the Americas from Asia by Paleoindians (the oldest known humans of the Americas) as they colonized the New World. Dogs were certainly brought by people migrating to North America across the Bering Straits but were the only kind of domestic animal that these humans brought with them. This attests to the great value of dogs for protection, hunting, herding, and, commonly, as a source of meat. As we have seen, dogs were domesticated in Asia starting at least 15,000 years ago, well before plants were first cultivated as crops.

The plants brought into cultivation in the New World were different from those first cultivated in the Old World. Pumpkins and squashes (*Cucurbita* spp.) were the first to be cultivated in the Americas. In place of wheat, barley, and rice, there was maize (*Zea mays;* Figure 21–9); and in place of lentils, peas, and chickpeas, the inhabitants of the New World cultivated common beans *(Phaseolus vulgaris),* lima beans *(Phaseolus lunatus),* and peanuts *(Arachis hypogaea),* among other legumes. The other important crop plants of Mexico included cotton (*Gossypium* spp.), chili peppers (*Capsicum* spp.), tomatoes *(Solanum lycopersicum),* tobacco *(Nicotiana tabacum),* pigweed *(Amaranthus),* cacao (*Theobroma cacao,* which yields cocoa, the major ingredient in chocolate), pineapple *(Ananas comosus),* and avocados *(Persea americana).*

Cotton was domesticated independently in the New and Old Worlds, with different species cultivated in the different centers of domestication. Its cultivation in Peru dates back at least 6000 years, and in Mexico at least 4000 years. The New World cottons are polyploid, and they are the source of nearly all the cotton cultivated throughout the world today; in contrast, the Old World cottons are diploid.

After the diverse cultivated plants of the New World were discovered by Europeans, following the voyages of Columbus, many of them were brought into cultivation in Europe, spreading from there to the rest of the world. Not only were some of these plants totally new to Europeans, but others, such as the New World species of cotton, were better than the forms already in cultivation in Eurasia and soon replaced them.

The Early Domestication of Plants Occurred in Both Mesoamerica and South America Squashes were the first plants to be domesticated in the New World, about 10,000 years ago in Mexico and South America, with maize domestication taking place somewhat later, nearly 9000 years ago in Mexico (see the essay on page 510). The New World crops eventually became widespread in both North and South America. Similar crops were also widely grown throughout the lowlands and middle elevations of South America. In fact, it is likely that agriculture was developed independently in Mexico and in Peru, although the question cannot be settled with the evidence now available. The earliest evidence of agriculture in Peru is almost as ancient as that from Mexico, and some domestic plants, such as peanuts, were clearly brought from South America to Mexico by humans.

In the south central Andes of South America, a distinctive kind of agriculture was developed (Figure 21–10), based on tuberous crops such as potatoes (*Solanum tuberosum* and related species) and seed crops such as quinoa *(Chenopodium quinoa)* and lupines (*Lupinus* spp., of the family Fabaceae). Potatoes were cultivated throughout the highlands of South America at the time of Columbus, but they did not reach Central America or Mexico until they were taken there by the Spaniards. They have been one of the most important foods in Europe for two centuries, yielding more than twice as many calories per hectare as wheat.

Although the familiar white, or Irish, potato is a member of the nightshade family (Solanaceae), the sweet potato *(Ipomoea batatas),* as you might know if you have seen a field of it in flower, is a member of the morning glory family

21–9 Maize The cereal known as maize *(Zea mays)* is the most important crop plant in the United States, where some 80 percent of the crop is consumed by livestock. At the time of Columbus, maize was cultivated from southern Canada to southern South America. Five main types are recognized: popcorn, flint corns, flour corns, dent corns, and sweet corn. Dent corns, which have a dent in each kernel, are primarily responsible for the productivity of the U.S. corn belt and are mainly used as animal food. Dent corn is increasingly important as a source of high-fructose corn syrup, which is used in canned soft drinks, and of ethanol.

(a)

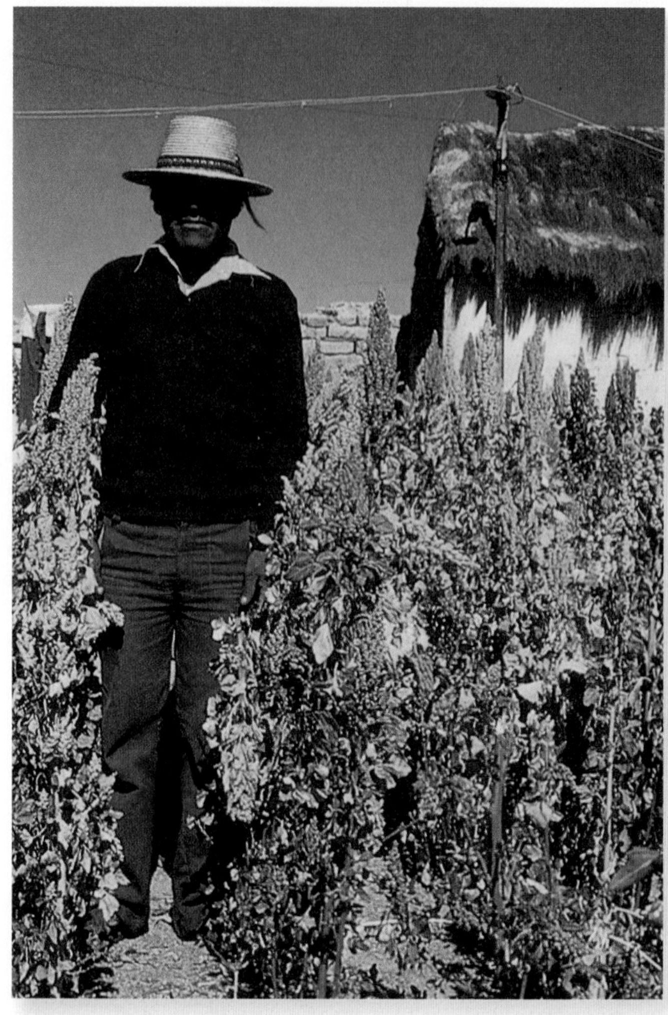

(b)

21–10 Andean agriculture A distinctive form of agriculture based on tuberous crops was developed at high elevations in the Andes of South America. ***(a)*** Three of the four major root crops that are cultivated in the Andes, shown here for sale in the market at Tarma, Peru: potatoes *(Solanum tuberosum)*, añu *(Tropaeolum tuberosum)*, and ullucu *(Ullucus tuberosus)*. Ullucu, which can be grown at higher elevations than potatoes, forms large, nutritious tubers; it might well become a useful crop elsewhere in the world. The fourth root crop that is common in the Andes is oca *(Oxalis tuberosa)*, which is also cultivated to a limited extent in New Zealand and elsewhere. The tubers of all four of these plants are naturally frozen and dried by Andean farmers, after which the tubers can be easily stored for later consumption. ***(b)*** Quinoa *(Chenopodium quinoa)*, a high-protein grain of the pigweed family (Chenopodiaceae), is an important crop in the Andes—here shown in cultivation in northern Chile—and is being tested for more widespread cultivation.

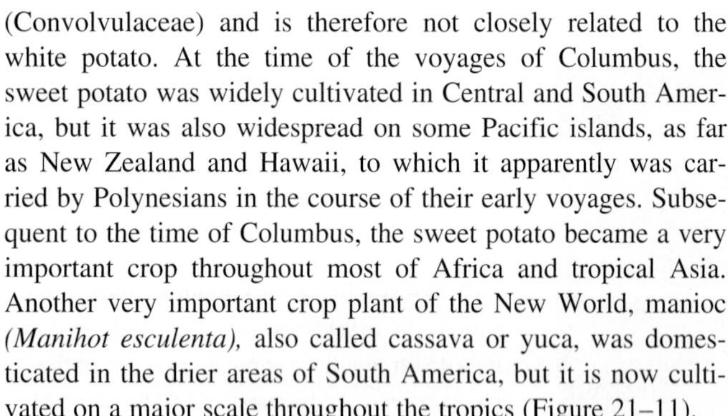

(Convolvulaceae) and is therefore not closely related to the white potato. At the time of the voyages of Columbus, the sweet potato was widely cultivated in Central and South America, but it was also widespread on some Pacific islands, as far as New Zealand and Hawaii, to which it apparently was carried by Polynesians in the course of their early voyages. Subsequent to the time of Columbus, the sweet potato became a very important crop throughout most of Africa and tropical Asia. Another very important crop plant of the New World, manioc *(Manihot esculenta)*, also called cassava or yuca, was domesticated in the drier areas of South America, but it is now cultivated on a major scale throughout the tropics (Figure 21–11).

Plants Were Domesticated Later in North America The domestication of plants by Native Americans in what is now the United States began later than in Mesoamerica and South America. Several indigenous plants, including sunflower (*Helianthus annuus* var. *macrocarpus;* Figure 21–12), squash *(Cucurbita pepo* subsp. *ovifera)*, marsh elder *(Iva annua* var. *macrocarpa)*, and chenopod *(Chenopodium berlandieri* subsp. *jonesianum)*, were domesticated between 5000 and 4000 years ago. Several others were cultivated but never became domesticated. With the introduction of maize and later the common bean, only sunflower and squash, among the indigenous plants, continued to be grown.

Very Few Animals Were Domesticated in the New World The so-called Muscovy duck (which, despite its name, does not come from Moscow), turkeys, guinea pigs, llamas, and alpacas are among the few domesticated animals to have originated in the New World. Towns and, eventually, large cities were organized, just as they had been throughout Eurasia, wherever the agricultural systems were well enough developed to support them. Crops were extensively cultivated around such centers, but there were no large herds of domestic animals comparable to those that had become so prominent throughout Europe and Asia.

When Europeans colonized the Western Hemisphere, however, they took their herds with them. In time, these herds brought about the same sort of widespread ecological devastation in parts of the New World that had taken place millennia earlier in the Near East and other parts of Eurasia. Many natural plant communities cannot readily be converted to pastures. For example, the widespread clearing of moist tropical forests to make pastures has been enormously destructive wherever it has been attempted. For the most part, pastures in such areas are

(a)

(b)

21–11 Manioc Throughout the tropics, manioc *(Manihot esculenta)* is one of the most important root crops. Tapioca is produced from the starch extracted from the roots of this plant. Some of the cultivated strains, the bitter maniocs, contain poisonous cyanide compounds, which must be removed before the tuberous root is eaten. ***(a)*** Manioc cultivated in a forest clearing in southern Venezuela. ***(b)*** A Tirió woman in southern Surinam peeling a manioc root to prepare it for cooking.

productive only in the short term, until the nutrients in the soil have been exhausted, and they must often be abandoned after no more than 10 or 15 years.

Spices and Herbs Are Plants Prized for Their Flavors

The substances that are produced by plants primarily to defend themselves from insects and other herbivores were discussed in Chapter 2, pages 30 to 35, and Chapter 20, pages 497 and 498. These substances contribute to the flavors, odors, and tastes of many plants and thus add features that have been utilized by human beings since prehistoric times. Some of these substances make plants poisonous to humans and other animals, but others add characteristics that humans find desirable.

Spices—strongly flavored parts of plants, usually rich in essential oils—may be derived from the roots, bark, seeds, fruits, or buds. **Herbs,** on the other hand, are usually the leaves of non-woody plants, although laurel, or bay, leaf and a few other condiments derived from trees or shrubs are considered herbs also. In practice, herbs and spices intergrade completely. Although it is widely believed that herbs and spices were used to cover the taste of spoiled meat or to preserve meat during the Middle Ages, there is no evidence for this. Spices were much more expensive than meat, which was readily available at the time. Quite clearly, herbs and spices were added to fresh foods as flavors to add zest and variety to an otherwise bland diet; in addition, they were used both as therapeutic foods and as medicines.

Historically, the Most Important Spices Were Cultivated in Tropical Asia

Spices and herbs have been used widely in cooking for as long as we have records. The search for spices played a major role in the great Portuguese, Dutch, and English voyages that began in the thirteenth century and eventually led to Europeans encountering lands in the Western Hemisphere. The most important

21–12 Sunflower Considered an important crop because of the oil obtained from their seeds, sunflowers *(Helianthus annuus)* were first domesticated about 4840 years ago in what is now the United States.

THE ORIGIN OF MAIZE

Ears of maize, or corn, differ so greatly from those of its ancestor that for many years the wild progenitor species was not recognized. We now know that maize *(Zea mays* subsp. *mays)* is simply the domesticated form of a large wild grass, a southern Mexican annual teosinte, now named *Zea mays* subsp. *parviglumis.* Previously, this ancestor was assigned to a different genus and named *Euchlaena mexicana.* Teosinte plants have hundreds of small, narrow, two-rowed ears, each made up of 5 to 12 fruit cases (each permanently enclosing one grain) that disarticulate and fall free at maturity. Because the fruit cases are woody and not separable from the grains, the flour (meal) that results from grinding them is unpalatable. The several species of teosinte grow from southern Chihuahua, Mexico, to Nicaragua. They are capable of forming hybrids with maize (the maíz de coyote, or wild maize, as the hybrids have been called) and may do so spontaneously when maize and teosinte grow together.

Maize is known only as a cultivated plant and could not persist on its own in the wild. Not only are its large naked grains vulnerable, but they are permanently attached to a central axis (the "corn cob") and permanently protected by many overlapping leaf sheaths (the "husks") that would not permit dissemination even if the grains were free. In short, the maize ear is a well-packaged, high-yielding, and easily harvested agricultural artifact, with all its principal characteristics selected by the early farmers.

Maize was domesticated about 9000 years ago, probably only once, in the Balas River Valley of southern Mexico. Its domestication obviously involved making the teosinte grains accessible as food, as well as increasing grain and ear size and harvestability. As a result of human selection for these characteristics, maize came to differ from teosinte in much the same way that the unbranched and monocephalic (single-headed) cultivated sunflower differs from its highly branched and many-headed wild ancestor. Both maize and the cultivated sunflower have all the reproductive resources of a single branch concentrated in one gigantic, many-seeded, terminal structure, the maize ear or the sunflower, which results in easier harvesting and higher yield.

John Doebley, of the University of Wisconsin–Madison, and others have identified the major genetic changes involved in the transformation of teosinte to maize. Chief among them is the *teosinte branched 1 (tb1)* gene, which largely controls the architecture of the plant, changing it from branched and grasslike to the single-stalk form of domesticated maize. Two others are the *teosinte glume architecture 1 (tga1)* gene, which acts to eliminate the hard casing on teosinte kernels, and *zea floricaula/Leafy 2 (zfl2),* which influences flowering time, leaf number, and inflorescence structure. It has been proposed that *zfl2* patterns inflorescence meristems so that more than two ranks of reproductive organs are formed.

In 1977, a new species of wild perennial teosinte, *Zea diploperennis,* was found in southwestern Mexico by Rafael Guzmán, then an undergraduate student at the University of Guadalajara. In 1978, it was recognized as a new species by Hugh Iltis, John Doebley, and their associates in Madison. Interfertile with annual maize, *Z. diploperennis* carries the genes for resistance to seven of the nine major viruses that infect maize in the United States; for five of these, no other source of resistance is known. The economic implications are obvious when one considers the worldwide gross value of maize—nearly $60 billion in 1991 alone.

Zea mays subsp. *parviglumis* (left) and *Zea mays* subsp. *mays* (right).

spices came from the tropics of Asia and nearby islands; they were responsible for the voyages and for a great deal of warfare as well. By the third century B.C.E., caravans of camels—often requiring two years for the trip—were carrying spices from tropical Asia to the civilizations of the Mediterranean region. Included among these spices were cinnamon (the bark of *Cinnamomum zeylanicum*), black pepper (the dried, ground fruits of *Piper nigrum;* Figure 21–13), cloves (the dried flower buds of *Eugenia aromatica*), cardamom (the seeds of *Elettaria cardamomum*), ginger (the rhizomes of *Zingiber officinale*), and nutmeg and mace (the seeds and the dried outer seed coverings, respectively, of *Myristica fragrans;* Figure 21–14). When the Romans learned that by taking advantage of the seasonal shifts of the monsoon winds they could reach India by sea from Aden, they shortened the trip to about a year—but it was still a highly dangerous and uncertain enterprise. A smaller number of additional spices, including vanilla (the dried, fermented seedpods of the orchid *Vanilla planifolia*), both hot and sweet peppers (*Capsicum* spp.), and allspice (the dried, unripe berries of *Pimenta officinalis*), so-called because it was considered to combine the flavors of cinnamon, clove, and nutmeg, came from the New World tropics after the voyages of Columbus.

Herbs Originated in Many Parts of the World In Europe and the Mediterranean region generally, there were many different kinds of native herbs, ones that were more familiar locally and, perhaps for that reason, not as highly prized as some of the spices available only from distant lands. Especially prominent among these

21–13 Black pepper For thousands of years, black pepper *(Piper nigrum)* has been known as an important spice. On a worldwide basis, the use and consumption of black pepper is approximately equal to that of all other spices combined.

herbs are members of the mint family (Lamiaceae). These include thyme (*Thymus* spp.), mint (*Mentha* spp.), basil *(Ocimum vulgare),* oregano *(Origanum vulgare),* and sage (*Salvia* spp.). Also important were members of the parsley family (Apiaceae), including parsley *(Petroselinum crispum),* dill *(Anethum graveolens),* caraway *(Carum carvi),* fennel *(Foeniculum vulgare),* coriander *(Coriandrum sativum),* and anise *(Pimpinella anisum).* Some members of this family (parsley, for example) are grown primarily for their leaves, and some (such as caraway) for their seeds, but many (such as dill and coriander) are valued for both.

21–14 Nutmeg The spice nutmeg is derived from the ground seeds of the nutmeg *(Myristica fragrans)* tree, whereas the spice mace comes from the fleshy aril, seen here as bands of red tissue. Birds act as natural seed distributors by removing the seed and aril to consume the aril for its high protein content. The seed is soon dropped and thus planted on the forest floor away from the mother tree of this dioecious species.

Tarragon *(Artemisia dracunculus)* is an herb that consists of the leaves of a plant belonging to the same genus as wormwood and the sagebrush of the western United States and Canada. Mustard *(Brassica nigra)* seed, which can be ground into the spice we call mustard, likewise comes from a plant native to Eurasia. Bay leaves, an herb derived from temperate members of the largely tropical laurel family (Lauraceae), are traditionally taken from the tree *Laurus nobilis* of the Mediterranean region but now also often taken from *Umbellularia californica* of California and Oregon. Saffron, popular in the Near East and adjacent regions, consists of the dried stigmas of *Crocus sativus,* a small, bulbous plant of the iris family (Iridaceae). The stigmas are gathered laboriously by hand, which accounts for the extremely high price of saffron and the fact that it is so prized—as much for its color as for its taste.

Although they are neither herbs nor spices, mention should be made of coffee and tea. Coffee (Figure 21–15) and tea *(Camellia sinensis)* provide the two most important beverages in the world; both are consumed primarily because of the stimulating alkaloid, caffeine, that they contain. Coffee is made from seeds of the coffee plant that have been dried, roasted, and ground, whereas tea is prepared from the dried leafy shoots of the tea plant. Coffee, as mentioned earlier, was domesticated in the mountains of northeastern Africa, whereas tea was first

21–15 Coffee Domesticated in Africa, coffee *(Coffea arabica)* is an important cash crop throughout the tropics. It is a member of the madder family (Rubiaceae), as is cinchona *(Cinchona),* which produces the medically useful alkaloid quinine. Rubiaceae is one of the larger families of flowering plants, with about 6000 species, mainly tropical.

cultivated in the mountains of subtropical Asia. Both are now widespread crops throughout the warm regions of the world. Today, coffee provides the livelihood for some 25 million people, and it provides a major source of income for the 50 tropical nations that export it. A third of the world's supply of coffee comes from Brazil.

Agriculture Is a Global Phenomenon

For the past 500 years, the important crops have been carried throughout the world and cultivated wherever they grow best. The major grains—wheat, rice, and maize—are grown everywhere that the climate will permit. Plants unknown in Europe before the voyages of Columbus, including maize, tomatoes, and *Capsicum* peppers, are now cultivated throughout the world. Sunflowers, first domesticated in the area that is now the United States, today produce more than half of their total worldwide yield in Russia. Sunflower seeds are widely consumed by humans, and sunflower meal is important for animal feed. Sunflowers are also displacing the traditional olives as a source of oil in many parts of Spain and elsewhere in the Mediterranean region. Throughout the world, the sunflower is being extensively grown for its oil, and it is second only to the soybean among plants grown for this purpose.

Some tropical crops have also become widespread. For example, rubber (derived from several species of trees of the genus *Hevea,* of the family Euphorbiaceae) was brought into cultivation on a commercial scale about 160 years ago. The main area of rubber production is in tropical Asia (see Figure 2–24). For rubber, as for many other crops, cultivation in areas other than those where the plants are native seems to be favored. Here, the plants are often free of the pests and diseases that attack them in their native lands—unless, of course, the pest has been transported too, such as in stored seed. The need for strict quarantines to avoid transferring pests and diseases of plants between countries or continents is often overlooked, however, in the eagerness to establish trade links between nations.

Oil palms *(Elaeis guineensis)* are native to West Africa but are now grown in all tropical regions. Although they have been grown on a commercial scale for only about 80 years, oil palms are among the most important cash crops of the tropics today. Among the others are coffee and bananas, both widespread. Cacao, which was first semidomesticated in tropical Mexico and Central America, is now most important as a crop in West Africa (Figure 21–16). Sugarcane was domesticated in New Guinea and adjacent regions, whereas sugar beets were developed from other cultivated members of their species in Europe. Yams (*Dioscorea* spp.) are an important tropical root crop. A number of yam species are cultivated throughout the tropics, some of them from West Africa, others from Southeast Asia, and a few from Latin America. The higher quality yams have now spread throughout the tropics, and they provide a staple food over wide areas. Manioc is important as an industrial crop in addition to being one of the most important human food crops (Figure 21–11). As a result of large-scale plantings, manioc has become a significant source of industrial starch and animal feed. Tons of processed, dried, and pelletized manioc are exported from Southeast Asia, particularly Thailand, to Europe to be used as a major food supplement in the swine and dairy cattle industries. From small plantings to large machine-operated farms in Central and South America, as well as in Africa and Asia, manioc provides one of the most important food sources for the expanding populations of the tropics.

21–16 Cacao The cacao *(Theobroma cacao)* tree is the source of chocolate and cocoa. Each of the large fruits, called "pods," contains about 40 large seeds, or "beans," which are dried, roasted, and ground to form an oily mass known as cocoa liquor. Further grinding and blending produces cocoa butter, which is further processed to produce chocolate. Cacao was first domesticated in Mexico, where chocolate was a prized drink among the Aztecs—at times, the beans were used as currency.

Another of the most important cultivated plants of the tropics is the coconut palm *(Cocos nucifera),* which seems to have originated in the region of the western Pacific and tropical Asia but was widespread in the western and central Pacific Ocean area before the European voyages of exploration. Natural stands of coconuts are rare in the eastern Pacific, with a few known from Central America. The vast range of the coconut may be the result of natural dispersal of the fruits floating in the sea, rather than human intervention. Each tree produces about 50 to 100 fruits (drupes) each year, and they are a rich source of protein, oils, and carbohydrates. Coconut shells, leaves, husk fibers, and trunks are used to make many useful items, including clothing, building materials, and utensils. The parts consumed by people are the solid and liquid endosperm of the coconut.

21–17 Sugar beet The sugar beet *(Beta vulgaris)* is simply a variety of the ordinary beet—selected from strains grown earlier for fodder, not those cultivated as a root crop—in which the sucrose content has been increased by selection, from about 2 percent to more than 20 percent. Beets were domesticated in Europe, where their leaves have long been used for food, and the swollen red root more recently. For about 300 years, sugar beets have been used as a source of sugar that has been competitive with sugarcane, which must be grown in the tropics and imported to the countries of the developed world. In the United States, production of raw sugar from sugar beets amounts to about a third of the total sugar consumed domestically.

The World's Food Supply Is Based Mainly on Fourteen Kinds of Crop Plants

Modern agriculture has become highly mechanized in temperate regions and in some parts of the tropics. It has also become highly specialized, with just six kinds of plants—wheat, rice, maize, potatoes, sweet potatoes, and manioc—directly or indirectly (that is, after feeding to animals) providing more than 80 percent of the total calories consumed by human beings. These plants are rich in carbohydrates, but they do not provide a balanced diet for humans. They are usually eaten with legumes, such as common beans, peas, lentils, peanuts, or soybeans, which are rich in protein, and with leafy vegetables, such as lettuce, cabbage, and spinach *(Spinacia oleracea)*, which are abundant sources of vitamins and minerals. Such plants as sunflowers, soybeans, canola, and olives provide oils, which are also necessary in the human diet.

In addition to the six major food crops, there are eight others of considerable importance to human beings: sugarcane, sugar beet (Figure 21–17), common beans, soybeans, barley, sorghum, coconuts, and bananas. Taken together, these 14 kinds of plants constitute the great majority of crops widely cultivated as sources of food.

There are great regional differences in the human diet. For example, rice (Figure 21–7) provides more than three-quarters of the diet in many parts of Asia, and wheat (Figure 21–18) is equally dominant in parts of North America and Europe. In regions with scarce rainfall, maize can be grown successfully

21–18 Wheat First domesticated in the Near East, wheat has become the most widely grown crop in the world today. Along with barley, which is now largely used for animal food and as the source of malt for making beer, wheat was probably one of the first two plants to be cultivated. Because of the leavening properties of some of its gluten proteins, the so-called hard red winter and hard red spring types of bread wheat *(Triticum aestivum)* are used for making bread.

only with supplementary irrigation, which is usually not necessary for wheat. Extending the productive ranges of these major grains and finding additional crop species are tasks of the greatest importance to the human species, as we shall see.

The Growth of Human Populations

The roughly 5 million humans who lived 10,500 years ago were already the most widely distributed large land mammal in the world. Subsequently, however, with the development of agriculture, this number has grown at an accelerating pace. Today, humans and their domesticated animals make up 97 percent of the biomass of vertebrates on land; 10,500 years ago, it was less than 1 percent.

Various mechanisms come into play to limit the numbers of individuals in groups of humans that make their living by hunting. A woman on the move cannot carry more than one infant along with her household baggage, minimal though that baggage may be. Consequently, she may resort to abortion or, more commonly, infanticide. In addition, there is a high natural mortality in these populations, particularly among the very young, the old, the ill, the disabled, and women in childbirth. As a result of these factors, populations that are dependent on hunting tend to remain small. In addition, there is not much incentive for specialization of knowledge or skills, because the basic skills on which individual survival depends are of the greatest importance.

The Development of Agriculture Dramatically Affected Population Growth

Once most groups of humans became sedentary, there was no longer the same urgent need to limit the number of births, and children may have become more of an asset to their families than previously, because they were able to help with agricultural and other chores. In the cities and towns that agricultural productivity made possible, human knowledge became increasingly specialized. Because the efforts of a few people could produce enough food for everyone, patterns of life became more and more diversified. People became merchants, artisans, bankers, scholars, poets—all the rich mixture of which a modern community is composed. The development of agriculture set human society on its modern course.

As a consequence of the development of agriculture, the numbers of human beings grew to between 100 and 300 million, distributed all over the Earth, by 2000 years ago. Over a period of about 8000 years, the human population had increased about 25-fold. By 1650, the world population had reached 500 million, with many people living in urban centers. In the late 1700s, it had reached 900 million. The human birth rate has remained essentially constant throughout the world from the seventeenth century onward, but the death rate has decreased dramatically in some regions, with the result that the population overall has grown at an unprecedented rate. At the end of the twentieth century, the birth rate itself dropped in developed countries.

How Will the World's Rapidly Growing Population Be Fed?

As the twenty-first century began, there were more than 6 billion humans on our planet; this compares with 2.5 billion in 1950 (see Figure 1–12). The global population doubled in less than 40 years. For the planet as a whole, the population is growing at about 1.3 percent per year. This means that about 158 humans are being added to the world population every minute—more than 220,000 each day, or about 83 million every year—about equal to the total population of Germany or the Philippines. A high proportion—usually about 35 to 45 percent—of the people living in the developing countries are under 15 years of age. The comparable percentage for developed countries is about 12 to 18 percent in Europe and Japan, 18 percent in Canada, and 21 percent in the United States. These young people have not yet reached the age at which humans usually bear children. Consequently, population growth in developing countries cannot soon be brought under control, even though government policy and individual choice often favor such a trend.

21–19 Poverty A family crosses a garbage-filled puddle in a slum district of Manila, the capital of the Philippines. The district, called Baseco, is well known as a community of living kidney donors, forced to sell their kidneys in order to survive. Impoverished people living under such difficult conditions represent more than a billion of the world's population. Their prospects for the future depend directly on limiting population growth, incorporating the poor into the global economy, and finding new and improved methods for productive agriculture in the tropics and subtropics.

BIOFUELS: PART OF THE SOLUTION, OR ANOTHER PROBLEM?

When we burn fossil fuels, we are, in essence, burning ancient plants, releasing back into the atmosphere carbon that was sequestered by those plants millions of years ago. That extra carbon is the principal cause of global climate change. But what if, instead of burning ancient carbon, we could use today's plants for our energy needs? Their carbon has been freshly removed from the atmosphere and will just as quickly cycle back to today's plants after we burn that carbon. According to this scenario, there would be no net increase in atmospheric carbon from using such a fuel source. As an added bonus, we would be less dependent on imported oil from politically unstable regions of the world.

This scheme is the lure of biofuels, but unfortunately, it is not that simple. To begin with, the most versatile fossil fuels—oil, gasoline, or diesel—are in liquid form, while plant biomass is solid. Fermentation of plant sugars produces liquid ethanol, which itself can be a versatile fuel, although it carries only two-thirds as much energy as an equivalent volume of gasoline. The real problem with ethanol, however, is the source of plants used to make it. Currently in the United States, ethanol is made from feed corn, the primary food for many livestock. Diverting corn in this way from the human food chain for even the relatively tiny amount of ethanol currently produced has already driven up food prices worldwide. Agriculture uses more than one-third of all land in the temperate regions, and more than two-thirds of the fresh water. Significantly increasing those amounts to devote to biofuels would have major impacts on the environment, including increased release of carbon from newly cleared land. Ethanol may not even reduce greenhouse gases, once the impacts of fertilizer, tractor fuel, purification, and transport have been factored in.

If using kernels of corn does not make sense, how about the rest of the plant? Like the kernel, cornstalks are primarily sugar. But they are in the form of cellulose, a complex carbohydrate that only a few microorganisms can break down. While research is ongoing, there is currently no economically practical way to release the sugars from cellulose for fermentation to ethanol. Should this become feasible, we'll face another problem. Diversion of all that organic matter away from soil replenishment could have dire consequences on soil health and productivity.

A more environmentally friendly source of biomass for fuels may be algae. Algae can be grown in shallow, briny pools on land useless for agriculture, and they can be used to clean up polluted water, producing a biomass that can be fermented to produce ethanol, methanol, and butanol (see Figure 15–4). The ability of algae to convert sunlight to biomass is unmatched by higher plants. Many algae, such as diatoms, also produce oils that can be used directly for fuel, without the need for fermentation. Nonetheless, as of 2011, there is no way to grow the algae and harvest the oil economically, and the hurdles to doing so are formidable. Research is under way, but for the foreseeable future, biofuels seem unlikely to become a solution to our energy needs.

The global population is expected to reach 9 billion by 2050. Feeding that number of people will be an unprecedented challenge. In 2008, the Food and Agriculture Organization of the United Nations reported that more than one billion people—more than one in seven of us—suffer malnutrition. Living in absolute poverty, these people are unable to obtain food, shelter, or clothing dependably (Figure 21–19).

Humans are presently estimated to be consuming, wasting, or diverting more than 40 percent of the total net photosynthetic productivity on land, and that productivity has been greatly reduced by our burning and clearing of natural vegetation in the past. Although we have achieved a 2.6-fold increase in world grain production since 1950, this increase has been accomplished at the expense of no less than 25 percent of the topsoil and more than 15 percent of the land that is under cultivation. We lack the agricultural technology to convert most of the lands in the tropics to sustainable productivity, and most of the world's lands that can be cultivated using available techniques are already in cultivation. The solution to the problem of feeding the world's population must be found in the regions where most people live—the tropics and subtropics.

Substantial food increases are needed if the world's population is to be fed adequately. Recent studies have suggested that, by 2050, the world will need 70 to 100 percent more food than is now available. Realistically, there appears to be little hope of attaining this. Between 1990 and 2007, global yields of maize, rice, wheat, and soybeans (in metric tons per harvested hectare) actually declined in half of the countries (both developed and developing) that grow these four crops. For people living in the United States, who spend, on average, less than a fifth of their personal income on food, the increasing cost of food already occasions serious concern. For people in developing nations, who may spend 80 to 90 percent of their income on food, it can be a death sentence.

Agriculture in the Future

Advances in Agriculture Have Brought Problems as Well as Benefits

The first significant advance that led to a massive increase in agricultural productivity was the development of irrigation (Figure 21–20). The necessity of providing water to crops has always been so evident that irrigation was practiced in the Near East as early as 7000 years ago, and it was developed independently in Mexico apparently about 5000 years ago.

During the twentieth century, two factors played major roles in bringing about an enormous increase in crop production. The first was the discovery during World War I by two German chemists of a way to make ammonia by combining nitrogen from the air with hydrogen from fossil fuels. The ammonia was then converted to nitrate, which was used in the

21–20 Irrigation This irrigated cotton field in Texas is representative of modern, intensive agriculture. Irrigation may pose severe environmental problems over the long run, however, especially if it is coupled with the intensive use of pesticides and herbicides. More pesticides are used on cotton than on any other crop in the world.

manufacture of explosives. This process, known as the Haber-Bosch process, is still used in the manufacture of most fertilizers, a high-energy-requiring process (see Chapter 29).

The second factor was the introduction of hybrid maize seed. Inbred lines of maize (themselves originally of hybrid origin) are used as parents. When they are crossed, the result is seed that produces very vigorous hybrid plants. The strains to be crossed are grown in alternating rows, with the tassels (staminate inflorescences) being removed from one row of plants by machine or by hand, so that all of the seeds set on those plants will be of hybrid origin. Through a careful selection of the best inbred lines, vigorous strains of hybrid maize that are suitable for cultivation at any given locality can be produced. Because of the uniform characteristics of the hybrid plants, they are easier to harvest, and they uniformly produce much higher yields than nonhybrid individuals. Less than 1 percent of the maize grown in the United States in 1935 was hybrid maize, but now virtually all of it is. At the same time, increasingly specialized and efficient machinery was developed for agriculture.

One of the great problems worldwide is how to utilize the gains in productivity that are made possible by increased mechanization, irrigation, and fertilization without simultaneously displacing millions of workers. Throughout much of the developing world, well over three-quarters of the people are directly engaged in food production, compared with less than 3 percent in the United States in the early twenty-first century.

Research has already done a great deal to improve agriculture. In the United States, the land-grant college system and the associated state agricultural experiment stations have made major contributions in this area. In recent years, public funding has been redirected away from agricultural productivity toward other concerns, such as environmental effects of agriculture, food safety, and other aspects of food quality. Similar shifts in public funding from research on agricultural productivity have also taken place in other developed countries.

The energy cost of producing crops in the United States and other developed countries is very high. Modern agriculture likewise depends on an elaborate distribution system that is very expensive in terms of energy and is easily disrupted. A significant proportion of each crop, the exact amount depending on the region and the year, is lost to insects and other pests. In many regions, an additional significant proportion is lost after harvest, either to spoilage or to insects, rats, mice, and other pests. Water is increasingly expensive in many areas, and the quality of local water supplies is often degraded by runoff from fields to which fertilizers and pesticides have been applied. Soil erosion is a problem everywhere and increases with the intensity of the agriculture (one solution is shown in Figure 21–21). Many efforts are under way to achieve improvements in crop productivity, to protect crops from pests, and to increase the efficiency with which crops use water. Each of these are discussed below.

Improving the Quality of Existing Crops Is an Important Goal

The most promising approach to alleviating the world's food problem seems to lie in the further development of existing crops grown on land that is already in cultivation. Most land suitable for plant agriculture is already being cultivated, and increasing the supplies of water, fertilizers, and other chemicals for crops is not economically feasible in many parts of the world. Thus the genetic improvement of existing crops is of exceptional importance. Such development entails improving not only the yield of these crops but also the quantity of proteins and other nutrients that they contain. The *quality* of the protein in food plants is also of the greatest importance to human nutrition: animals, including human beings, must be able to obtain from food the right amounts of all the essential amino acids—the ones they cannot manufacture themselves. Nine of the 20 amino acids required by human adults must be obtained from food (see the essay on page 25); the other 11 can be manufactured in the human body. Plants that have been selected for an improved protein content, however, inevitably have higher requirements for nitrogen and other nutrients than do their less modified ancestors. For this reason, improved crops cannot always be grown on the marginal lands where they would be especially useful.

The quality of crops can also be improved in many respects other than yield and protein composition and quantity. Newly developed crop varieties may be more resistant to disease, as when they contain secondary plant products that their herbivores find distasteful; more interesting in form, shape, or color (such as apples of a brighter red); more amenable to storage and shipping (such as tomatoes that stack better in boxes); or improved in other features that are important for the crop in question.

Research at International Agricultural Research Centers Made Possible the Green Revolution Despite all the progress made through the use of fertilizers and more efficient farm machinery, by 1950 food production had fallen behind a rapidly growing population. In response to this challenge, extensive efforts were begun to increase the yields of wheat and other grains, and to develop wheat and rice genotypes that would not lodge, or fall over, at

21–21 Control of soil erosion Use of the no-tillage cropping system, which combines ancient and modern agricultural practices, has been increasing rapidly. Soil erosion is virtually eliminated in this system. Here, no-tillage soybeans planted between rows of corn (maize) stubble hold the soil in place and minimize the runoff of water and nutrients during rainfalls. When no-tillage methods are used, the energy input into maize and soybean production is reduced by 7 and 18 percent, respectively, and the crop yields are as high as or higher than those obtained by conventional plowing and disking methods.

high fertility. Through the use of conventional breeding methods, Norman Borlaug and others succeeded in developing semi-dwarf, disease-resistant varieties of wheat and rice that were capable of responding to fertilizers without lodging (Figure 21–22). These efforts, carried out largely at international crop-improvement centers located in subtropical regions, led to dramatic improvements in crop yields. When the improved varieties of wheat, maize, and rice produced at these centers were grown in such countries as Mexico, India, and Pakistan, they made possible the pattern of improved agricultural productivity that has been called the Green Revolution. The techniques of breeding, fertilization, and irrigation developed as part of the Green Revolution have been applied in many countries of the developing world.

Although the Green Revolution greatly increased grain production and averted disaster, it did not reach many of the neediest people, most notably those of sub-Saharan Africa. Fertilization, mechanization, and irrigation are all necessary components of the success of the Green Revolution. Consequently, only relatively wealthy landowners can cultivate the new crop varieties, and in many areas the net effect has been to accelerate the consolidation of farmlands into a few large holdings by the very wealthy. Such consolidation has not necessarily provided either jobs or food for the majority of local peoples. Money necessary to purchase fertilizer and farm machinery was and still is not available in sub-Saharan Africa, nor is an adequate supply of water for irrigation. Moreover, the principal plant foodstuffs of sub-Saharan Africa are the indigenous sorghum, cowpea, millet, and African rice *(Oryza glaberrima),* in addition to the more recently adopted cassava and maize, not wheat and rice, the miracle varieties of which were largely responsible for igniting the Green Revolution.

In 2006, the Bill and Melinda Gates and Rockefeller Foundations jointly launched the Alliance for a Green Revolution in Africa (AGRA), with three objectives: (1) to increase farm yield through the use of agronomic improvements, (2) to reduce losses and enhance the quality of crops through genetic improvements, and (3) to make sure farmers benefit from increases in production. Prior to the formation of AGRA, plant geneticists at the Rockefeller Foundation concluded that it would be a mistake to rely on just a few crop varieties in Africa. Thus, the African Green Revolution will involve ways of increasing yields for a wide range of crops, grown under a wide range of farming conditions, with rainfall as the main source of water. An agroecology-based approach has been adopted, in which farmers participate in each step of the technology development process. As defined by

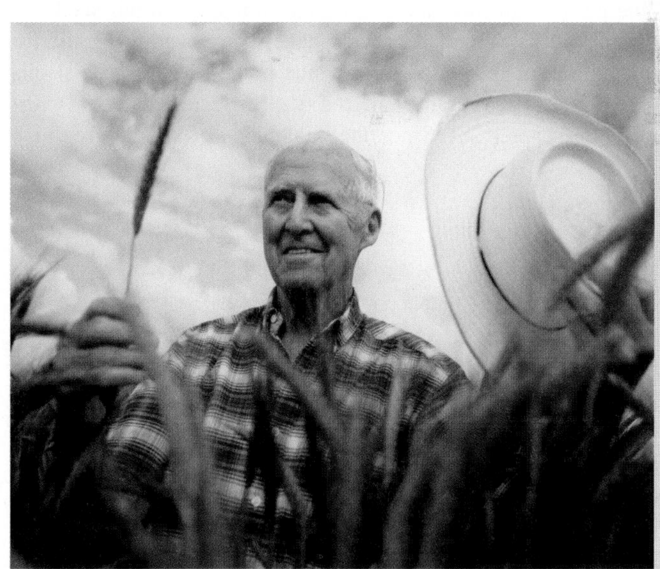

21–22 Norman Borlaug Awarded the Nobel Peace Prize in 1970, Norman Borlaug was the leader of a research project sponsored by the Rockefeller Foundation under which new strains of wheat were developed at the International Center for the Improvement of Maize and Wheat in Mexico. Widely planted, these new strains changed the status of Mexico from that of a wheat importer when the program began in 1944 to that of an exporter by 1964. Borlaug also played a vital role in introducing higher-yielding wheat varieties to India and Pakistan.

Joe DeVries, agroecologies are farming areas "with a more or less common set of constraints and advantages, that is, . . . rainfall amounts and distributions, temperatures, soils, and other factors important to crop growth."

To Provide Protection against Pathogens, the Genetic Diversity of Crop Plants Must Be Preserved and Utilized Intensive programs of breeding and selection have tended to narrow the genetic variability of crop plants. Much artificial selection, for obvious reasons, has been concerned with yield, and sometimes disease resistance has been lost among the highly uniform members of a progeny that has been selected strongly for increased yield. Overall, individual crop plants have tended to become increasingly uniform genetically, as particular traits have been stressed more strongly than others. The more genetically uniform crop plants become, the more susceptible they become to diseases and pests. An excellent example is the southern corn leaf blight epidemic, caused by the fungus *Cochliobolus heterostrophus,* that swept the United States in 1970. Approximately 15 percent of the country's maize crop was destroyed—a loss of approximately $1 billion (Figure 21–23). These losses were apparently related to the appearance of a new race of the fungus and to the high degree of genetic uniformity of the crop. Had the crop been genetically *heterogeneous,* it is likely that the losses would have been less severe.

To help guard against such losses, it is necessary to locate and to preserve distinct strains of our important crops, both presently cultivated varieties and earlier cultivars, because these strains—even though their overall characteristics may not be attractive economically—may contain genes useful in the continuing fight against pests and diseases (Figure 21–24). The germplasm (hereditary material) in reserve in our **seed** and **clonal banks** may also provide genes for elevating yields, for adaptation, or for certain high-value traits, such as special oils (see the essay "Doomsday Seed Vault: Securing Crop Diversity" in Chapter 28). Since the beginnings of agriculture, huge reserves of variability have accumulated in all crop plants by the processes of mutation, hybridization, artificial selection, and adaptation to a wide range of conditions. For such crops as wheat, potatoes, and maize, there are literally thousands of known strains. In addition, even more genetic variability exists among the wild relatives of the cultivated crops. The problem lies in finding, preserving, and using the genetic variability of cultivated plants and their wild relatives before they are lost.

21–23 Loss of disease resistance Southern leaf blight of maize, caused by the fungus *Cochliobolus heterostrophus*. This fungus devastated crops that had a reduced genetic variability as a result of artificial selection for increased yield.

As an example of the role of genetic diversity in the history of, and the prospects for, a single crop plant, we consider the potato (Figure 21–10). There are more than 60 species of potatoes, most of them never cultivated, and thousands of different strains of cultivated potatoes are known. Despite this, most of the cultivated potatoes in the United States and Europe are descended from a very few strains that were brought to Europe in the late sixteenth century. This genetic uniformity led directly to the Irish potato famine of 1846 and 1847, in which the potato crop was nearly wiped out by a blight caused by the water mold (oomycete) *Phytophthora infestans* (see Chapter 15). The subsequent breeding of strains of potatoes resistant to the blight restored the status of the plant as a crop in Ireland and elsewhere. Looking to the future, the potential for further developing potatoes as a crop through the use of additional cultivated and wild strains is enormous.

Striking examples of the potential of obtaining additional genetic material from the wild have already been achieved by tomato breeders. The collection of strains of tomatoes has led to effective control of many of the important diseases of tomatoes, such as the rots caused by the asexual fungi *Fusarium* and *Verticillium* and by several viral diseases. The nutritive value and flavor (sweetness) of tomatoes have been greatly enhanced, and their ability to tolerate saline or other unfavorable conditions has been increased through the systematic collection, analysis, and use of strains of wild tomatoes in breeding programs.

Wheat is presently under assault by a new and virulent race of *Puccinia graminis,* the cause of black stem rust. This new race is capable of overcoming the gene *Sr31,* which until now has provided wheat's resistance to the fungus (page 302).

Various Wild Plants Have Great Potential for Becoming Important Crops

In addition to the plant species already widely cultivated, there are many wild plants, or cultivated plants that are locally important, that could, if brought into more general cultivation, make important contributions to the world economy. For example, as mentioned earlier, we still derive more than 80 percent of our calories from only six kinds of angiosperms of the approximately 300,000 known species. Only about 3000 angiosperm species have ever been cultivated for food, and the great majority of those either are no longer used or are used very locally. Only about 150 plant species have ever been cultivated widely. Many plants other than this limited number, however—especially those that have been used before but have been completely abandoned or seen as having less importance—may prove to be very useful. Some of these species are still in cultivation in different areas of the world.

Although we are accustomed to thinking of plants primarily as important sources of food, they also produce oils, drugs,

(a) (b)

21–24 Preservation of crop strains ***(a)*** U.S. Department of Agriculture seed bank at Fort Collins, Colorado, showing seeds being sorted out and sealed for long-term storage. Approximately 200,000 germplasm lines are stored at this facility. ***(b)*** The seed potato farm, a clonal bank, at Three Lakes, Wisconsin, is the national repository for strains of potatoes.

pesticides, perfumes, and many other products that are important to our modern industrial society. We have come to regard the production of these nonfood substances from plants, however, as a rather archaic method that has now been fully supplanted by chemical synthesis in commercial laboratories. Yet their production by plants requires no energy other than that of the sun: it occurs naturally. As our sources of nonrenewable energy approach exhaustion and prices increase, it becomes increasingly important to find ways to produce complex chemical molecules less expensively. Further, the vast majority of plants have never been examined or tested to determine their usefulness.

Efforts are in progress to develop perennial versions of our important grain crops, all of which are annuals. Perennial grain crops would have significant advantages over their annual counterparts. For example, perennial plants tend to have longer growing seasons and deeper root systems and thus are more efficient users of nutrients and water. With their greater root mass, perennials reduce erosion risks and therefore more effectively maintain topsoil. The longer growing seasons and leaf duration of perennials mean they have longer photosynthetic seasons and greater productivity. In addition, perennial crops store more carbon in the soil than do annual crops (Figure 21–25).

Programs are now under way in Argentina, Australia, China, India, Sweden, and the United States to identify and improve perennials and hybrid plant populations (derived from annual and perennial parents) for use as grain crops. These include wheat, rice, maize, sorghum, and pigeon pea *(Cajanus cajan).* Studies are also in progress on three oilseed crops from the flax, sunflower, and mustard families.

One important area of investigation in the search for valuable new crops focuses on drought- and salt-tolerant species. Increasingly, intensive agriculture is being practiced in the arid and semiarid areas of the world, primarily in response to the demands of the ever-increasing human population. Such practices have placed enormous demands on very limited local water supplies, which have become increasingly brackish as the water is used, reused, and polluted with fertilizer from adjacent cultivated areas. In addition, there are many parts of the world, especially near seashores, where the soil and the local water supplies are naturally saline. Unproductive when farmed using

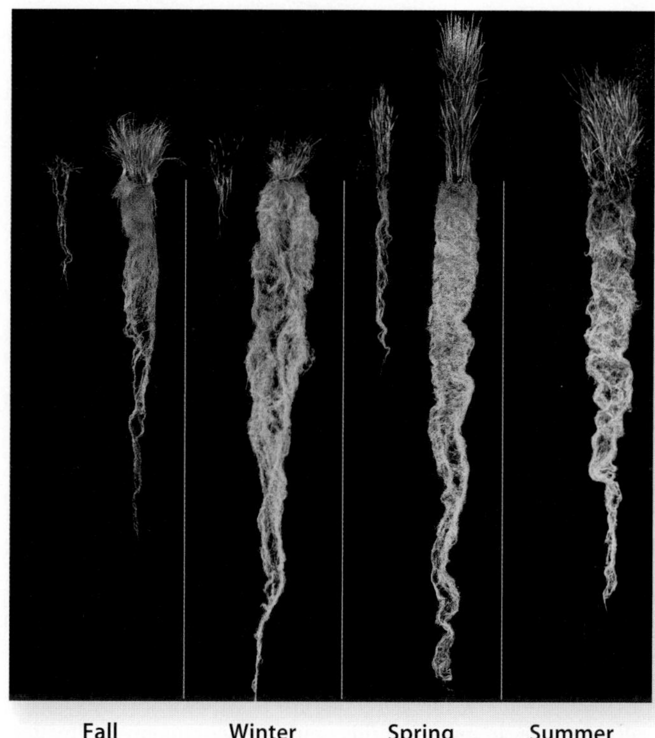

Fall Winter Spring Summer

21–25 Wheat comparisons Throughout the year, annual winter wheat (on the left in each panel) is less robust than its wild perennial relative, intermediate wheatgrass (on the right in each panel). Moreover, by summer, the annual winter wheat has disappeared.

(a)

(b)

21–26 Salt tolerance The development of new crops, or new strains of presently cultivated crops, that can succeed at relatively high salt concentrations is important in many areas of the world, particularly those that are arid or semiarid. ***(a)*** Eggplants *(Solanum melongena)* cultivated in the Arava Valley of Israel using drip irrigation (water reaches the plants through plastic tubes and is greatly conserved in this way) with highly saline water (1800 parts per million). Twenty years ago, this amount of salinity was thought to be incompatible with commercial crop production. ***(b)*** At an experimental site on the Mediterranean coast of Israel, a highly nutritious fodder shrub, *Atriplex nummularia,* is grown with 100 percent seawater. Yields are similar to those achieved with alfalfa under normal conditions, but the leaves and stems of *Atriplex* are highly salty and thus not as valuable for fodder as alfalfa. Current research is aimed at overcoming the problems of seawater irrigation, and this type of irrigation may one day make possible the cultivation of huge areas that are now coastal deserts.

traditional agricultural methods, such areas might be brought under cultivation if the right kinds of plants could be found (Figure 21–26).

A plant that is both drought- and salt-tolerant is *Simmondsia chinensis,* the jojoba (pronounced "ho-*ho*-ba"), which, despite its Latin name, is a shrub native to the deserts of northwestern Mexico and the adjacent United States (Figure 21–27). The large seeds of jojoba contain about 50 percent liquid wax, a substance that has impressive industrial potential. Wax of this sort is indispensable as a lubricant where extreme pressures are developed, as in the gears of heavy machinery and in automobile transmissions. It is difficult to produce synthetic liquid wax commercially, and the endangered sperm whale is the only alternative natural source. Crops of jojoba are being cultivated in arid areas throughout the world.

Finding suitable plants for saline areas entails not only seeking out entirely new crop plants but also breeding salt-tolerance into traditional crops (see the essay on page 703). In tomatoes, for example, a wild species that grows on the sea cliffs of the Galápagos Islands and is therefore very salt-resistant, *Solanum cheesmaniae,* has been used as a source of salt-tolerance in new hybrids with the commonly cultivated tomato, *Solanum lycopersicum.* The selection of these hybrids was carried out in a medium having half the salinity of seawater. Researchers have succeeded in selecting hybrid individuals that will complete their development in water of that salinity. Salt-tolerant strains of barley have been developed using similar methods. Genetic engineering is also being used to develop salt-tolerance in plants.

Plants Continue to Be Important Sources of Drugs

In addition to their other uses, plants are important sources of medicinal drugs. In fact, about a quarter of the prescriptions

21–27 Jojoba Crops of jojoba *(Simmondsia chinensis)* are increasingly being planted throughout the arid areas of the world. Jojoba is a source of waxes with special lubricating characteristics and other useful properties.

21–28 Rosy periwinkle The rosy periwinkle *(Catharanthus roseus)* is the natural source of the drugs vinblastine and vincristine. These drugs, which were developed in the 1960s, are highly effective against certain cancers. Vinblastine is typically used to treat Hodgkin's disease, a form of lymphoma, and vincristine is used in cases of acute leukemia. Before the development of vinblastine, a person with Hodgkin's disease had a 1 in 5 chance of survival; now, the odds have been increased to 9 in 10. This periwinkle is widespread throughout the warmer regions of the Earth, but it is native only to Madagascar, an island where a small proportion of the natural vegetation remains undisturbed.

written in the United States contain at least one product that is derived from a plant. For millennia, humans have been using plants for medicinal purposes. Botany, in fact, was traditionally regarded as a branch of medicine, and only in the past 160 years or so have there been professional botanists, as distinct from physicians. There has not, however, been any comprehensive effort to identify and bring into use previously unexploited secondary metabolites of the sort discussed in Chapters 2 and 20.

One of the reasons that plants will continue to be of importance as sources of drugs, despite the ease with which many drugs can be synthesized in the laboratory, is that plants manufacture the drugs inexpensively, without requiring additional energy. Furthermore, the structure of some molecules—such as the steroids, including cortisone and the hormones used in birth-control pills—is so complex that, although they can be synthesized chemically, the production methods make them prohibitively expensive. For this reason, anti-fertility pills and cortisone were manufactured in the past principally from substances extracted from the roots of wild yams *(Dioscorea)*, obtained largely from Mexico. When these sources were virtually exhausted, other plants, including *Solanum aviculare*, a member of the same genus as the nightshades and potatoes, were developed into crops for the same purpose.

In addition to considerations of cost, the marvelous variety of substances that plants produce is important: a source of new products that is seemingly inexhaustible (Figure 21–28). One way in which scientists have learned about potential new drugs is by studying the medical uses of various plants by rural or indigenous peoples (Figure 21–29). For example, the contraceptive properties of Mexican yams were "discovered" in this way.

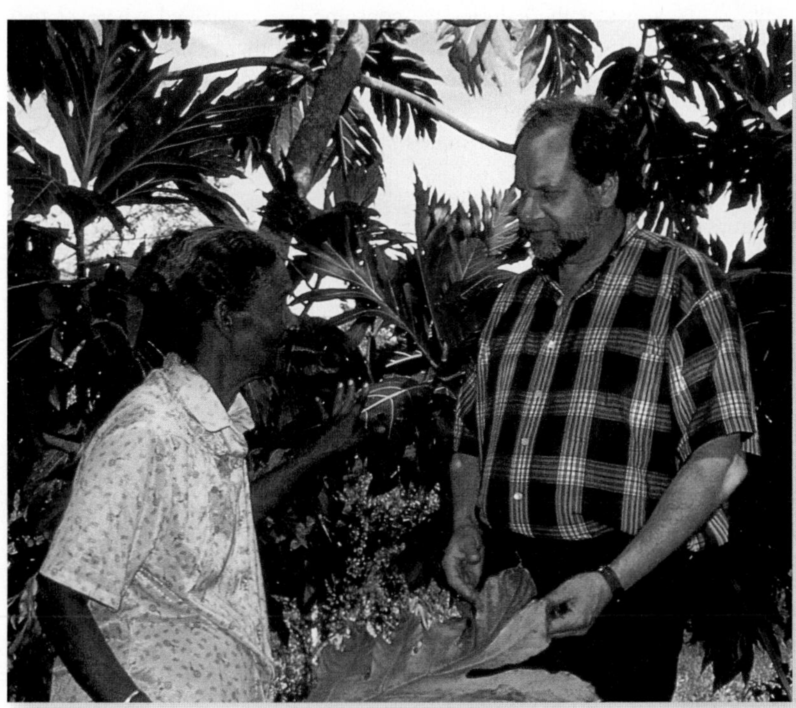

21–29 Medicinal use of forest plants Hortense Robinson, a traditional healer from Belize, explains the medicinal uses of breadfruit leaves to Michael Balick, Director of the Institute of Economic Botany at the New York Botanical Garden. Boiling the leaves in water results in an extract that is used to treat diabetes and high blood pressure. Although the study of indigenous uses of forest plants has led to the discovery of important drugs, such as D-tubocurarine chloride (used as a muscle relaxant in open-heart surgery) and ipecac (employed in the treatment of amoebic dysentery), the opportunities for gaining such knowledge are rapidly being lost as indigenous cultures disappear. Whole groups of people lose their traditional life styles, and forests as well as other wilderness areas that contain medicinal plants are destroyed. Valuable knowledge that has accumulated through thousands of years of trial-and-error experimentation is being forgotten in a very short period of time. Much of this information has been transmitted orally, because written records do not exist.

As we expand our search for useful plants, we should keep in mind the rapid loss of plant species that is related to (1) the rapid growth of the human population; (2) poverty, especially in the tropics, where about two-thirds of the world's species of plants occur; and (3) our incomplete understanding of how to construct productive agricultural systems in the tropics. With the complete destruction of undisturbed tropical forests, which seems almost certain to occur within a century, many kinds of plants, animals, and microorganisms will become extinct. Because our knowledge of plants, especially those of the tropics, is so rudimentary, we are faced with the prospect of losing many plant species before we even learn of their existence, much less have the opportunity to find out whether they might be of some use to us. The examination of wild plants for potential human use must be accelerated, and promising species must be preserved in seed banks, in cultivation, or, preferably, in natural reserves.

SUMMARY

Humans Evolved Relatively Late in the Earth's History

The human race originated in Africa. The genus *Australopithecus* existed there at least 5 million years ago. Our genus, *Homo,* apparently evolved from *Australopithecus* about 2 million years ago, and our species, Homo sapiens, has existed for at least 500,000 years.

Plant Domestication First Appeared In the Fertile Crescent

Starting about 13,000 to 11,000 years ago, in the Fertile Crescent—an area that extends from Jordan and Israel through Iraq to Iran—humans began to cultivate such plants as wheat, barley, lentils, and peas. In cultivating and caring for their crops, early farmers changed the characteristics of these plants—the plants became domesticated—so that they became more nutritious and gradually differed in many ways from their wild relatives. Agriculture spread from this center across Europe, reaching Britain by about 6000 years ago. Evidence indicates that agriculture originated independently in several centers, including Africa, Mesoamerica, South America, North America, China, India, and New Guinea. Many crops were first brought into domestication in Africa, including yams, okra, coffee, and cotton, which was also domesticated independently in the New World and perhaps in Asia. In China, agriculture based on staples such as rice and soybeans—and, farther south, citrus, mangoes, taro, bananas, and other crops—was developed.

The Domestication of Animals Followed the Domestication of Plants

Domestic animals were an important feature of agriculture in the Old World from the earliest times. Herds of grazing animals were ecologically destructive to many of the semiarid areas of the Old World, especially as the animals increased in number, but they were also important sources of food. When these grazing animals were introduced into Latin America following the voyages of Columbus, they proved to be enormously destructive in many habitats, including tropical forests.

Agriculture in the New World Utilized Many New Species

Agriculture in the New World began about 10,000 years ago, with the cultivation of squash at least 9000 years ago in southern Mexico. Evidence indicates that agriculture in Peru is nearly as ancient as that of Mexico. Dogs were brought to the New World by people migrating from Asia, but no other domestic animals or plants, with the exception of the bottle gourd, seem to have been introduced in this way. Columbus and those who followed him found a virtual cornucopia of new crops to take back to the Old World. These crops included maize, common beans, lima beans, tomatoes, tobacco, chili peppers, potatoes, sweet potatoes, manioc, pumpkins and squashes, avocados, cacao, and the major cultivated species of cotton.

Spices and Herbs Are Plants Prized for Their Flavors and Odors

Spices, which are strongly flavored plant parts usually rich in essential oils, may be derived from the roots, bark, seeds, fruits, or buds, whereas herbs are usually the leaves of nonwoody plants. Cinnamon, black pepper, and cloves are examples of spices; mint, dill, and tarragon are herbs.

The World's Food Supply Is Based Mainly on a Relatively Small Number of Crop Plants

For the past 500 years, the important crops have been cultivated throughout the world. Wheat, rice, and maize, which provide most of the calories we consume, are cultivated wherever they will grow, and a limited number of other plants have achieved worldwide commercial status. Six crops—wheat, rice, maize, potatoes, sweet potatoes, and manioc—provide over 80 percent of the calories consumed by the world's human population.

Rapid Population Growth Has Caused Numerous Problems

The human population has grown from an estimated 500 million people in 1650 to more than 6 billion by the beginning of the twenty-first century. It is expected to reach 9 billion by 2050. About a seventh of the world population lives in absolute poverty. As a result of the growth in population and of widespread poverty, and because relatively little has been done to develop agricultural practices suitable for tropical regions, the tropics are being devastated ecologically, with up to 20 percent of the world's species likely to be lost over the next 30 years.

QUESTIONS

1. Among the first cultivated plants were cereals and legumes. What food value does each of these plant groups contribute to the human diet?
2. Explain how the development of agriculture affected population growth.
3. What is meant by the Green Revolution?
4. Explain the importance of preserving the genetic diversity of crop plants as a safeguard against pathogens.
5. Comment on the importance of breeding drought- and/or salt-tolerance into traditional crop plants.
6. Although many medicinal drugs can be synthesized in the laboratory, plants will continue to be an important source of such drugs. Why?
7. What developments in agriculture have the potential for helping solve the problem of world hunger?

SECTION 5

THE ANGIOSPERM PLANT BODY: Structure and Development

◀ Soon after blooming, the endangered pima pineapple cactus (*Coryphantha scheeri* var. *robustispina*) of the Sonoran Desert produces sweet, green fruits that provide food and water for desert animals, such as rodents and rabbits. As the fruits are consumed, the large seeds are dispersed. The sturdy spines are modified leaves that provide protection from predation.

CHAPTER 22

Early Development of the Plant Body

◀ **Curing olives** The olive—a drupe containing a seed within a pit—has been prized for thousands of years, and ancient trees dot the landscape around the Mediterranean. When first harvested, the olives have a strongly bitter taste due to tannins, which are removed by a variety of curing methods. The olives can then be pressed into fragrant olive oil, known for its distinctive flavor and cholesterol-lowering properties.

CHAPTER OUTLINE

In Chapter 20, we traced the long evolutionary development of the angiosperms, beginning with their presumed ancestor, a relatively complex, multicellular green alga. As we pointed out, the forked axes of early vascular plants were the forerunners of the shoots and roots of most modern vascular plants.

In this section, we are concerned with the structure and development of the angiosperm plant body, or sporophyte, which is a result of that long period of evolutionary specialization. This chapter begins with the formation of the embryo, a process known as **embryogenesis,** the first of two phases of seed development. Embryogenesis establishes the body plan of the plant, consisting of two superimposed patterns: an **apical-basal pattern** along the main axis and a **radial pattern** of concentrically arranged tissue systems (Figure 22–1). Embryogenesis is accompanied by seed development. The seed, with its mature embryo, stored food, and protective seed coat, confers significant selective advantages over plants lacking seeds: the seed enhances the plant's ability to survive adverse environmental conditions and facilitates dispersal of the species.

As we follow the development of the angiosperm plant body, keep in mind the evolution of the vascular plants that we explored in the previous section. Developmental and evolutionary biologists have been fascinated by the evolution of developmental patterns (evolutionary development, also known as evo-devo). Great strides have been made by studying highly conserved genes—genes with similar DNA sequences in distantly related organisms—that regulate key developmental pathways. Much of what we know about this regulation comes from studying mutations that disrupt normal embryo development.

Formation of the Embryo

The early stages of embryogenesis are essentially the same in all angiosperms (Figures 22–2 and 22–3). Formation of the embryo begins with division of the zygote within the embryo sac of the ovule. In most flowering plants, the first division of the zygote is asymmetrical and transverse with regard to the long axis of the

CHECKPOINTS

After reading this chapter, you should be able to answer the following:

1. How is polarity important in embryonic development of plants?
2. What are the three primary meristems of plants, and which tissues do they form?
3. Through what stages do embryos of eudicots develop? How does embryo development in monocots differ from that in eudicots?
4. What are the main parts of a mature eudicot or monocot embryo?
5. What phenomena, or processes, characterize seed maturation, the second phase of seed development?
6. Of what significance is seed dormancy to a plant?

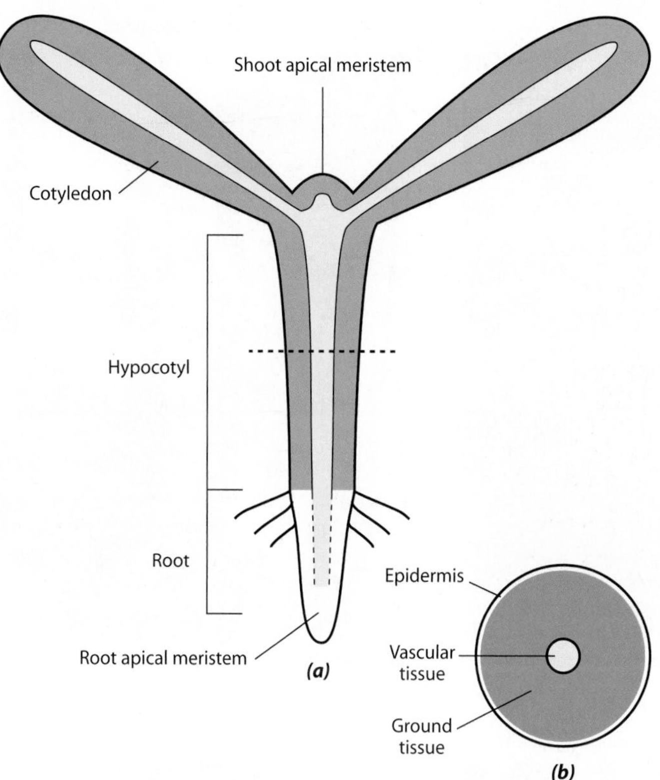

22–1 Body plan of an *Arabidopsis* seedling ***(a)*** The apical-basal pattern consists of an axis with a shoot tip at one end and a root tip at the other. ***(b)*** A transverse section through the hypocotyl reveals the radial pattern, consisting of the three tissue systems, as represented by the epidermis, ground tissue, and vascular tissue.

zygote (Figures 22–2a and 22–3a). With this division, the apical-basal **polarity** of the embryo is established. The upper (chalazal) pole, consisting of a small *apical cell,* gives rise to most of the mature embryo. The lower (micropylar) pole, consisting of a large *basal cell,* produces a stalklike **suspensor** that anchors the embryo at the micropyle, the opening in the ovule through which the pollen tube enters.

Polarity is a key component of biological pattern formation. The term arises by analogy with a magnet, which has plus and minus poles. "Polarity" means simply that whatever is being discussed—a plant, an animal, an organ, a cell, or a molecule—has one end that is different from the other. Polarity in plant stems is a familiar phenomenon. In plants that are propagated by stem cuttings, for example, roots will form at the lower end of the stem, with leaves and buds at the upper end.

The establishment of polarity is an essential first step in the development of all higher organisms, because it fixes the

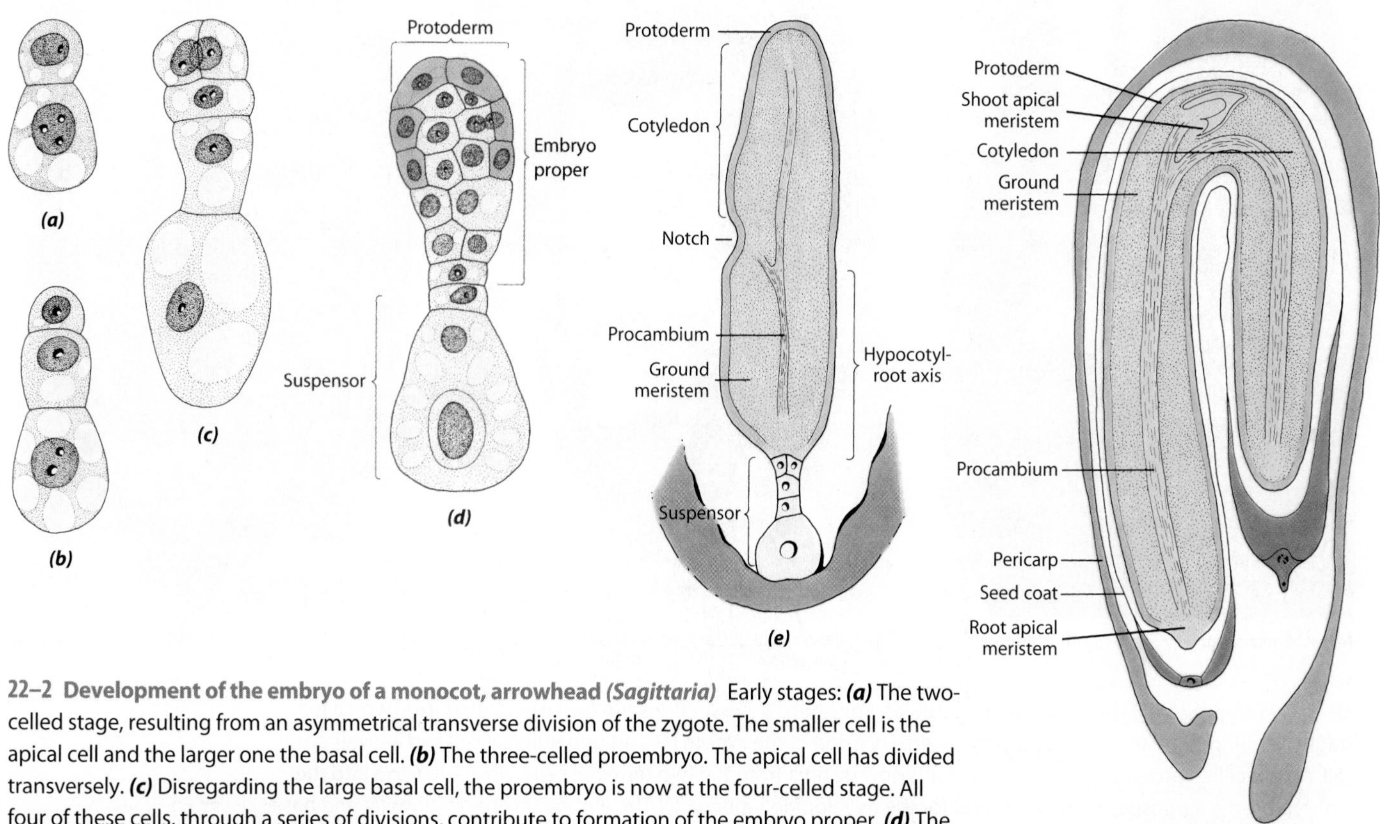

22–2 Development of the embryo of a monocot, arrowhead *(Sagittaria)* Early stages: ***(a)*** The two-celled stage, resulting from an asymmetrical transverse division of the zygote. The smaller cell is the apical cell and the larger one the basal cell. ***(b)*** The three-celled proembryo. The apical cell has divided transversely. ***(c)*** Disregarding the large basal cell, the proembryo is now at the four-celled stage. All four of these cells, through a series of divisions, contribute to formation of the embryo proper. ***(d)*** The protoderm has been initiated at the terminal end of the embryo proper. At this stage, the suspensor consists of only two cells, one of which is the large basal cell. Late stages: ***(e)*** A depression, or notch (the site of the future apical meristem of the shoot), has formed at the base of the emerging cotyledon. ***(f)*** The cotyledon curves, and the embryo is approaching maturity. The suspensor is no longer present.

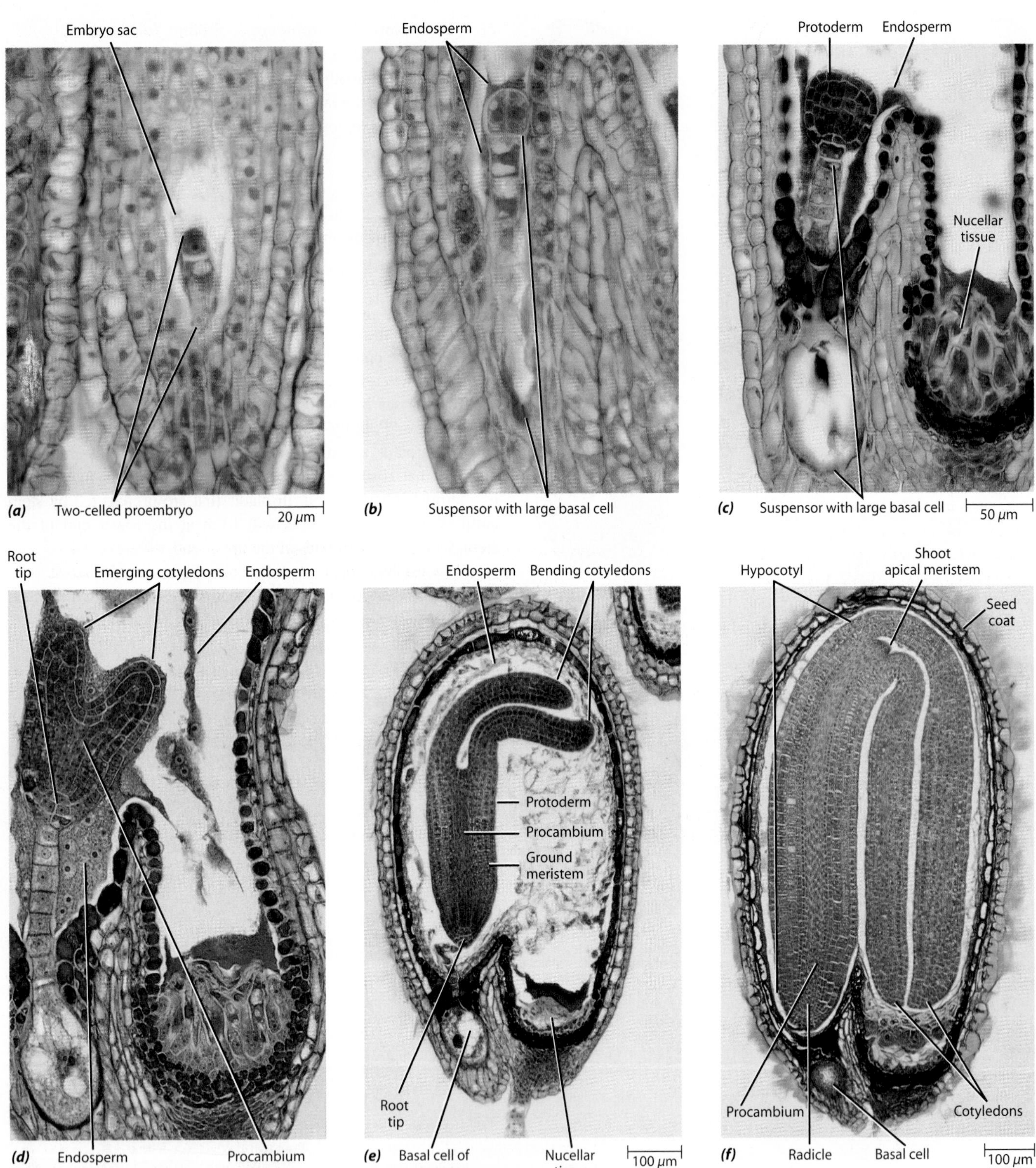

22–3 Development of the embryo of a eudicot, shepherd's purse *(Capsella bursa-pastoris)* ***(a)*** Two-celled stage, resulting from an asymmetrical transverse division of the zygote into an upper apical cell and a lower basal cell. ***(b)*** Six-celled proembryo. The suspensor is now distinct from the two terminal cells, which develop into the embryo proper. Endosperm provides food for the developing embryo. ***(c)*** The embryo proper is globular and has a protoderm, which will develop into the epidermis. The large cell near the bottom is the basal cell of the suspensor. ***(d)*** Embryo at the heart stage, when the cotyledons, the first leaves of the plant, begin to emerge. ***(e)*** Embryo at the torpedo stage. In *Capsella,* the embryos curve. Ground meristem, the precursor of the ground tissue, surrounds the procambium, which will develop into the vascular tissues, xylem and phloem. ***(f)*** Mature embryo. The part of the embryo below the cotyledons is the hypocotyl. At the lower end of the hypocotyl is the root, or radicle.

structural **axis** of the body, the "backbone" on which the lateral appendages will be arranged. In some angiosperms, polarity is already established in the egg cell and zygote, where the nucleus and most of the cytoplasmic organelles are located in the upper portion of the cell, and the lower portion is dominated by a large vacuole.

Through an orderly progression of divisions, the embryo eventually forms a nearly spherical structure—the **embryo proper**—and the suspensor (Figures 22–2d and 22–3b, c). Before this stage is reached, the developing embryo is referred to as the **proembryo.**

The Protoderm, Procambium, and Ground Meristem Are the Primary Meristems

When first formed, the embryo proper consists of a mass of relatively undifferentiated cells. Soon, however, changes in the internal structure of the embryo proper result in the initial development of the concentrically arranged tissue systems, the first expression of radial polarity during embryogenesis. The future epidermis, the **protoderm,** is formed by periclinal divisions—divisions parallel to the surface—in the outermost cells of the embryo proper (Figures 22–2d and 22–3c). Subsequently, vertical divisions within the embryo proper result in the initial distinction between **ground meristem** and **procambium** (Figure 22–3d, e). The ground meristem, which is the precursor of the ground tissue, surrounds the procambium, which is the precursor of the vascular tissues known as xylem and phloem. The protoderm, ground meristem, and procambium—the so-called **primary meristems,** or primary meristematic tissues—extend into the other regions of the embryo as embryogenesis continues (Figures 22–2e, f and 22–3e, f).

Embryos Progress through a Sequence of Developmental Stages

The stage of embryo development preceding cotyledon development—that is, when the embryo proper is spherical—is often referred to as the *globular stage.* Development of the cotyledons, the first leaves of the plant, may begin either during or after the time when the procambium becomes discernible. As the cotyledons develop in eudicots, the globular embryo gradually assumes a two-lobed form, or heart shape. This stage is known as the *heart stage* (Figure 22–3d). The globular embryos of monocots form only one cotyledon and become cylindrical in shape (Figure 22–2e). In both monocots and eudicots, the apical-basal pattern of the embryo proper first becomes discernible just prior to the emergence of the cotyledon(s). The axis is partitioned into shoot apical meristem, cotyledon(s), hypocotyl (the stemlike axis below the cotyledon or cotyledons), embryonic root, and root apical meristem.

The so-called *torpedo stage* of embryo development occurs as the cotyledon(s) and axis elongate, and the primary meristems extend along with them (Figures 22–2f and 22–3e). During elongation, the embryo may remain straight or become curved. The single cotyledon of the monocot often becomes so large in comparison with the rest of the embryo that it is the dominating structure (see Figure 22–6c).

During early embryogenesis, cell division takes place throughout the young sporophyte. However, as the embryo develops, the addition of new cells gradually becomes restricted to the apical meristems of the shoot and root. **Apical meristems** are found at the tips of all shoots and roots and consist of cells that are capable of repeated division. Meristems are embryonic tissue regions; apical meristems are involved with extension of the plant body. In angiosperm (except for monocot) embryos, the apical meristem of the shoot arises between the two cotyledons (Figure 22–3f). In monocot embryos, on the other hand, the shoot apical meristem arises on one side of the cotyledon and is completely surrounded by a sheathlike extension from the base of the cotyledon (Figure 22–2f). The apical meristems of both shoot and root are of great importance, because these tissues are the source of virtually all of the new cells responsible for development of the seedling and the adult plant.

During Its Short Life, the Suspensor Supports Early Development of the Embryo Proper

Unlike the suspensors of *Selaginella,* a seedless vascular plant (page 408), and pine (page 441), which function merely to push the developing embryos into nutritive tissues, angiosperm suspensors are metabolically active. They support early development of the embryo proper by providing it with nutrients and growth regulators, particularly gibberellins. Numerous plasmodesmata connect the cells of the suspensor with those of the developing embryo proper. Suspensors vary widely in structure and size, ranging from a single cell in orchids to a massive collection of cells, as in scarlet runner bean *(Phaseolus coccineus).* The suspensor is short-lived, undergoing *programmed cell death* (pages 48 and 49) at the same time as the torpedo stage of development. It is therefore not present in the mature seed, although remnants of the basal cell of the suspensor often are still discernible (Figure 22–3f).

Several lines of evidence indicate that the normally developing embryo proper limits the growth and differentiation of the suspensor, the cells of which have the potential to generate embryos. Such evidence is provided by several *Arabidopsis* embryo loss-of-function mutants, such as *raspberry1, sus,* and *twn,* in which disruption in the development of the embryo proper leads to proliferation of suspensor cells. Some of these suspensor cells acquire characteristics normally restricted to cells of the embryo proper. The *twn* mutants are the most striking of the *Arabidopsis* embryo loss-of-function mutants. In these mutants, the cells of the suspensor undergo embryogenic transformation, forming viable twin and occasionally triplet embryos within the seed (Figure 22–4). These studies of mutant loss-of-function phenotypes reveal that, in wild-type plants, interactions occur between the suspensor and embryo proper. The embryo proper is presumed to transmit specific inhibitory signals to the suspensor that suppress its development into an embryo.

Genes Have Been Identified That Determine Major Events in Embryogenesis

The description of embryo formation tells us how the primary body plan of the plant develops, but it reveals little about the underlying mechanisms. Large populations of mutagen-treated *Arabidopsis* plants have been systematically screened for mutations that have an effect on plant development. When phenotypes

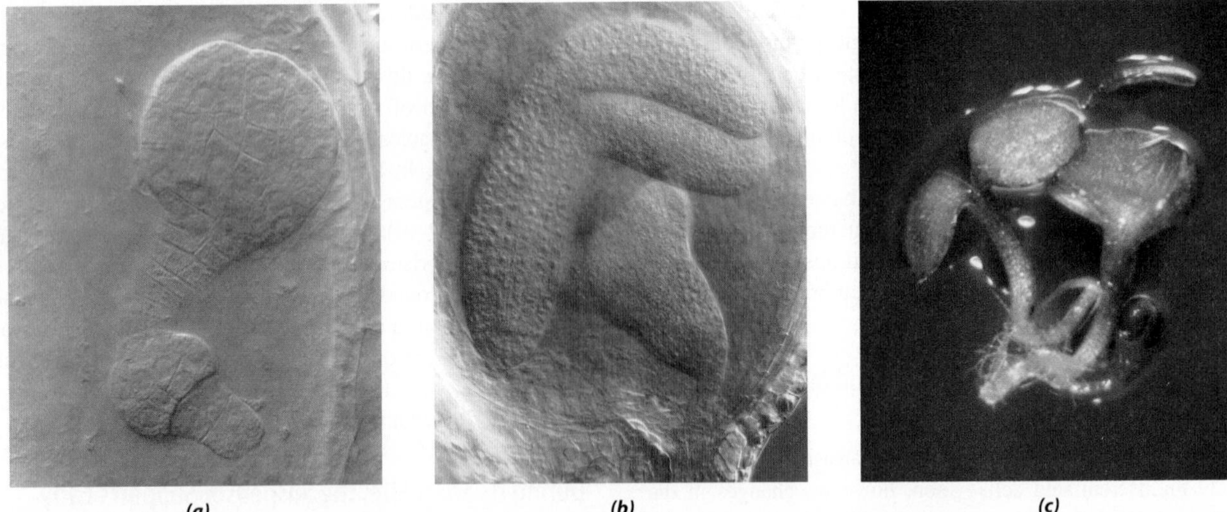

(a) *(b)* *(c)*

22–4 Twin embryo development in the *twn* mutant of *Arabidopsis thaliana,* a eudicot ***(a)*** A secondary embryo can be seen developing from the suspensor of the larger, primary embryo. Both embryos are at the globular stage. ***(b)*** The cotyledons of the primary embryo are partly developed. Cotyledon development in the secondary embryo is at an early stage. ***(c)*** Twin seedlings from a germinated seed. The seedling on the left resembles the wild-type (nonmutant) seedling. Its sibling has one large cotyledon.

are successfully altered, it is possible to identify corresponding genes that govern plant development, a first step in determining how the genes function.

Very promising results have been obtained in identifying genes responsible for major events in *Arabidopsis* embryogenesis. A minimal set of around 750 distinct genes is thought to coordinate embryo development in *Arabidopsis.* Some of these regulatory genes affect the apical-basal pattern of the embryo and seedling. Mutations in these genes delete different regions of the apical-basal pattern (Figure 22–5). Another group of genes in *Arabidopsis* is involved in determining the radial pattern of tissue differentiation. Mutations in one of these genes, for example, prevents formation of the protoderm. Still another group of genes regulates the changes in cell shape that give the embryo and seedling their characteristic elongated shape.

The Mature Embryo

The mature embryo of flowering plants consists of an axis bearing two cotyledons (Figures 22–3f and 22–6a, b), or one cotyledon if it is a monocot (Figures 22–2f and 22–6c, d). In the following pages we compare eudicots and monocots, the two largest groups of angiosperms.

At opposite ends of the embryo axis are the apical meristems of the shoot and root. In some embryos, only an apical meristem occurs above attachment of the cotyledon(s) (Figures

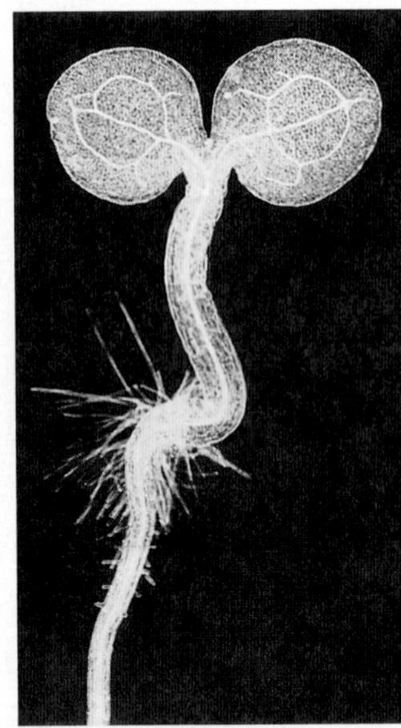

(a) **Wild type**

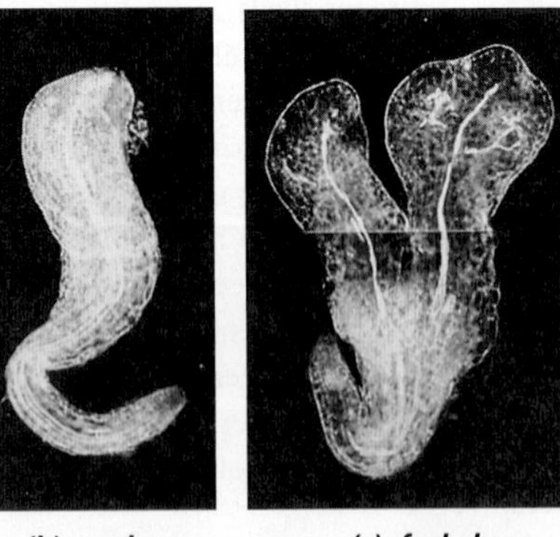

(b) gurke ***(c) fackel***

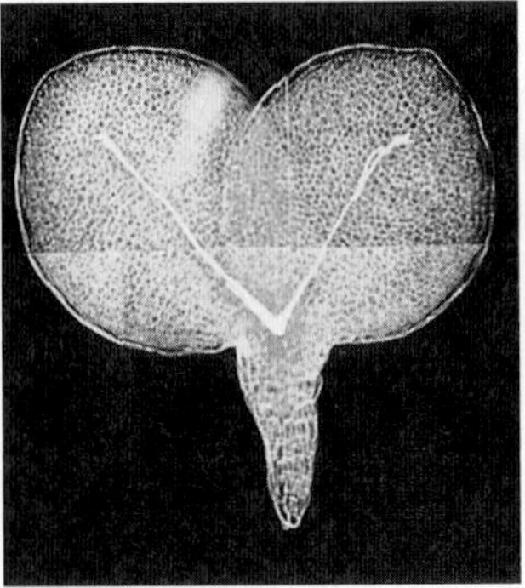

(d) monopteros

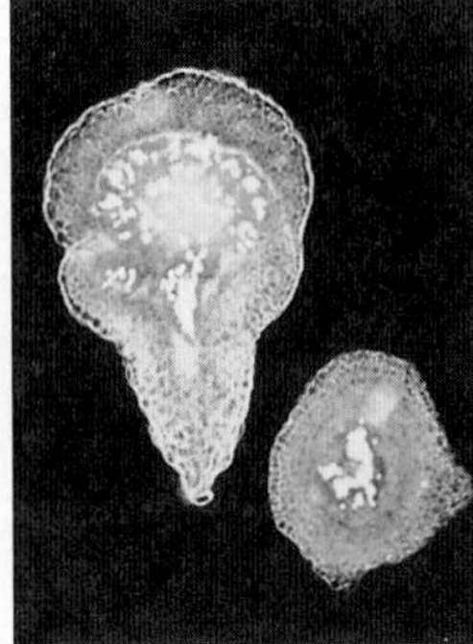

(e) gnom

22–5 Mutant seedlings of *Arabidopsis* ***(a)*** A normal (wild-type) seedling for comparison with four types of mutants that lack major parts of the seedling structure. In the mutant seedlings shown here, ***(b)*** *gurke* lacks a shoot apical meristem and cotyledons; ***(c)*** *fackel* lacks a hypocotyl, so the shoot meristem and cotyledons are attached directly to the root; ***(d)*** *monopteros* lacks a root; and ***(e)*** *gnom* lacks both apical and basal portions—the remaining stem contains epidermis, ground tissue, and vascular tissue. The seedlings have been "cleared" to reveal the vascular tissue within them.

22–2f, 22–3f, and 22–6b, c). In others, the embryonic shoot, consisting of a stemlike axis called the **epicotyl,** with one or more young leaves and an apical meristem, occurs above *(epi-)* the cotyledon(s). This embryonic shoot, the first bud, is called a **plumule** (Figures 22–6a, d and 22–7).

The stemlike axis below *(hypo-)* the cotyledon(s) is known as the **hypocotyl.** At the lower end of the hypocotyl there may be an embryonic root, or **radicle,** with distinct root characteristics (Figures 22-3f and 22–7b). In many plants, however, the lower end of the axis consists of little more than an apical meristem covered by a rootcap. If a radicle cannot be distinguished in the embryo, the embryonic axis below the cotyledon(s) is called the **hypocotyl-root axis** (Figures 22–2e and 22–6a, b, c).

In discussing the development of the angiosperm seed in Chapter 19, we noted that in many eudicots, most or all of the food-storing endosperm and the perisperm (which develops from nucellar tissue), if present, is absorbed by the developing embryo (page 472). The embryos of such seeds develop large, fleshy, food-storing cotyledons that nourish the embryo as it resumes growth. Familiar examples of seeds with sizable cotyledons and no remaining endosperm are sunflower, walnut, pea, and garden bean (Figure 22–6a). In eudicots with large amounts of endosperm, such as the castor bean, the cotyledons are thin and membranous (Figure 22–6b). These cotyledons serve to absorb stored food from the endosperm itself during resumption of growth of the embryo.

In the monocots, the single cotyledon, in addition to functioning as a food-storing or photosynthetic organ, also performs an absorptive function (Figure 22–6c, d). Embedded in endosperm, the cotyledon absorbs food digested from the endosperm by enzymatic activity. The digested food is then moved by way of the cotyledon to the growing regions of the embryo. Among the most highly differentiated of monocot embryos are those of grasses (Figures 22–6d and 22–7). When fully formed, the grass embryo possesses a massive cotyledon, the **scutellum.** The scutellum is attached to one side of the axis of the embryo, which has a radicle at its lower end and a plumule at its upper end. In the grasses, both the radicle and the plumule are additionally enclosed by sheathlike protective structures called the **coleorhiza** and the **coleoptile,** respectively (Figures 22–6d and 22–7).

All seeds have an outer coat, the **seed coat,** which develops from the integument(s) of the ovule and provides protection for the embryo. The seed coat is usually much thinner than the integument(s) from which it is formed. The thin, dry seed coat may have a papery texture, but in many seeds it is very hard and highly impermeable to water. In the grasses, the outer covering

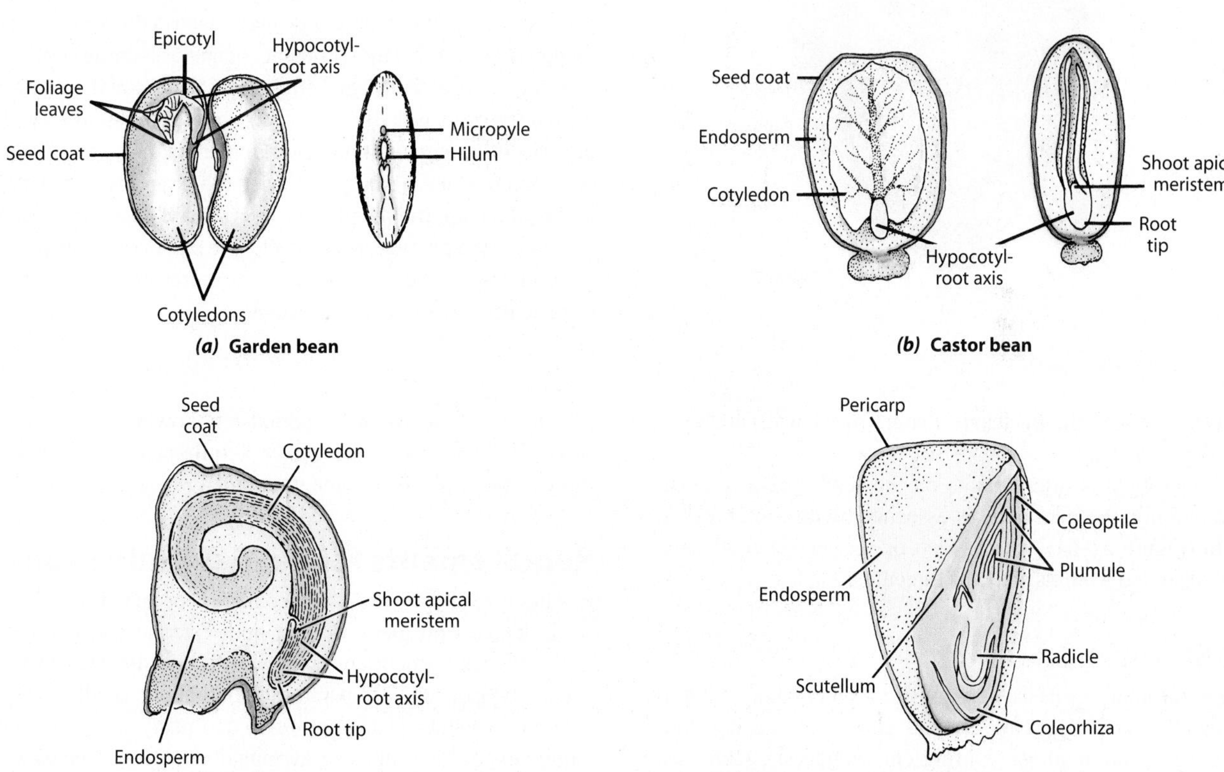

22–6 Seeds of some common eudicots and monocots ***(a)*** Seeds of the garden bean *(Phaseolus vulgaris),* a eudicot, shown open and from an external edge view. The garden bean embryo has a plumule above the cotyledons, consisting of a short stem (epicotyl), a pair of foliage leaves, and an apical meristem. The apical meristem is found between the foliage leaves and cannot be seen here. The fleshy cotyledons of the garden bean embryo contain the stored food. ***(b)*** A seed of the castor bean *(Ricinus communis),* another eudicot, opened to show both flat and edge views of the embryo. The castor bean embryo has only an apical meristem above the attachment of the cotyledons. The stored food is in the endosperm. The embryos of the monocots ***(c)*** onion *(Allium cepa)* and ***(d)*** maize *(Zea mays)* are shown in longitudinal view. The shoot apical meristem of the onion embryo lies on one side and at the base of the cotyledon, which is much larger than the rest of the embryo. The maize embryo has a well-developed scutellum (cotyledon) and radicle. The stored food in both seeds is in the endosperm.

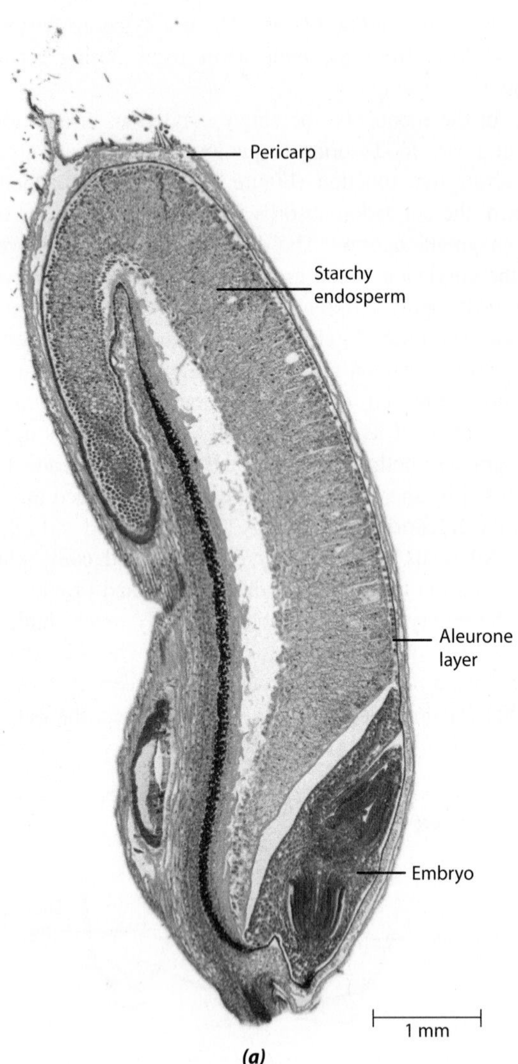

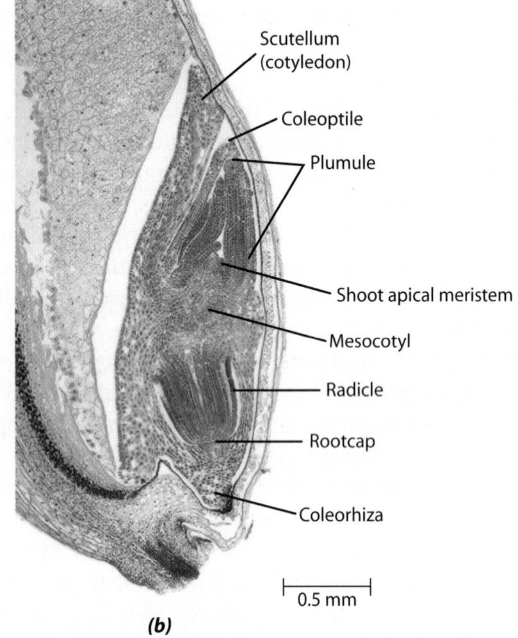

22–7 Mature grain, or kernel, of wheat *(Triticum aestivum)*, a monocot ***(a)*** The starchy endosperm seen in this longitudinal section is surrounded by a protein-containing aleurone layer. The covering layers of the wheat kernel consist largely of the pericarp, the mature ovary wall. The seed coat, which becomes fused with the pericarp, disintegrates during development of the kernel. ***(b)*** Detail of the mature wheat embryo, showing the large cotyledon known as the scutellum. The sheathlike coleorhiza and coleoptile enclose the radicle and plumule, respectively. The part of the embryo axis between the point of attachment of the scutellum and that of the coleoptile is known as the mesocotyl.

layers of the seed are the **pericarp** (mature ovary wall) and remnants of the seed coat.

The micropyle is often visible on the seed coat as a small pore. Commonly, the micropyle is associated with a scar, called the **hilum** (Figure 22–6a), which is left on the seed coat after the seed has separated from its stalk, or funiculus.

Seed Maturation

At the end of embryogenesis, cell division in the mature embryo ceases and the seed enters the second phase of seed development, the **maturation phase.** During embryogenesis, there is a continuous flow of nutrients from the parent plant tissues to the tissues of the ovule. It is during the seed maturation phase, however, that a massive buildup of food reserves (starch, storage proteins, oils) occurs in the endosperm, perisperm, or cotyledons of the developing seed. In addition, during the maturation phase, the seed undergoes **desiccation** as it loses water (90 percent or more) to the surrounding environment. Finally, the seed coat, derived from the integuments, hardens, encasing the embryo and stored food in "protective armor."

As a result of desiccation, metabolism within the seed decreases to an almost imperceptible level, allowing the embryo to remain viable for long periods. Following desiccation, the seeds of some plants then enter a **quiescent** ("resting") state; those of other plants become **dormant.**

Requirements for Seed Germination

Resumption of growth of the embryo, or **germination** of the seed, depends on many factors, both external and internal. The three especially important external, or environmental, factors are water, oxygen, and temperature. In addition, small seeds, such as those of lettuce *(Lactuca sativa)* and many weeds, commonly require exposure to light for germination to occur (see page 670).

Because most mature seeds are extremely dry, germination is not possible until the seed imbibes the water required for metabolic activities. Enzymes already present in the seed are then activated, and new ones are synthesized for the digestion and utilization of the stored foods accumulated in the cells during seed maturation. The same cells that had earlier synthesized enormous amounts of reserve materials now completely reverse their metabolic processes and digest the stored food. Cell enlargement and cell division are initiated in the embryo and follow patterns characteristic of the species. Further growth requires a continuous supply of water and nutrients. As the seed imbibes water, it

WHEAT: BREAD AND BRAN

Like all grasses, bread wheat *(Triticum aestivum)* is a monocot, and its fruit—the grain, or kernel—is one-seeded. As shown in Figure 22–7, the endosperm and the embryo are surrounded by covering layers composed of the pericarp and the remains of the seed coat. More than 80 percent of the volume of the wheat kernel is made up of starchy endosperm. The outermost layer of the endosperm, called the aleurone layer, contains lipid and protein reserves. The aleurone layer surrounds both the starchy endosperm and the embryo.

White flour is made from the starchy endosperm. In the milling of wheat, the bran—consisting of the covering layers and the aleurone layer—is removed. The bran constitutes about 14 percent of the kernel. Actually, the bran somewhat decreases the nutritional value of the wheat kernel. Because bran is mostly cellulosic, it cannot be digested by humans and tends to speed the passage of food through the intestinal tract, resulting in lower absorption. The embryo ("wheat germ"), which represents about 3 percent of the kernel, is also removed during milling, because its high oil content would reduce the length of time that the flour can be stored. The bran and wheat germ, which contain most of the vitamins found in wheat, are increasingly being used for human consumption, as well as for livestock feed.

Oat *(Avena sativa)* bran has become a popular food item among health-conscious people in many parts of the world. Research results indicate that oat bran, as part of a low-fat diet, can help reduce the amount of cholesterol in the blood. Apparently, the water-soluble fibers in the oat bran form a gel in the small intestine that traps cholesterol and prevents it from being reabsorbed into the bloodstream. Instead, the trapped cholesterol is excreted with other bodily wastes. Cholesterol is also reduced by the water-soluble fibers found in rice *(Oryza sativa)* and barley *(Hordeum vulgare)* bran.

Flaxseed meal (whole ground seed of *Linum usitatissimum*) is rapidly surpassing bran as a dietary supplement among people concerned with cholesterol. In addition to its high fiber content, flaxseed meal is a rich source of omega-3 and omega-6 fats, all of which are believed to reduce cholesterol levels in the blood.

swells, and considerable pressure may develop within it (see the essay "Imbibition" on page 80).

During the early stages of germination, glucose breakdown may be entirely anaerobic, but as soon as the seed coat is ruptured, the seed switches to an aerobic pathway, which requires oxygen (see Chapter 6). If the soil is waterlogged, the amount of oxygen available to the seed may be inadequate for respiration (the complete oxidation of glucose), and the seed will fail to grow into a seedling.

Many seeds germinate over a fairly wide range of temperatures, but they usually will not germinate below or above a certain temperature range specific for the species. The minimum temperature for many species is 0° to 5°C, the maximum is 45° to 48°C, and the optimum range is 25° to 30°C.

Dormant Seeds Will Not Germinate Even under Favorable External Conditions

In contrast to quiescent seeds, which, given favorable external conditions, will germinate on rehydration, dormant seeds fail to germinate even when external conditions are favorable (see Chapter 28). Both seed coat and embryo can cause dormancy. Several factors, or mechanisms, have been identified as causing **coat-imposed dormancy:** impermeability of the seed coat to water or oxygen; rigidity of the seed coat (mechanical constraint), which prevents the embryonic root from penetrating the coat; prevention by the seed coat of the escape of growth inhibitors from the seed; and the presence in the seed coat of inhibitors that can suppress growth of the embryo. Coat-imposed dormancy is exhibited by conifers, most cereals, and several eudicots.

The ratio of the plant hormones abscisic acid and gibberellic acid has been shown to play a major role in **embryo dormancy.** Whereas abscisic acid promotes dormancy, gibberellic acid promotes germination (see pages 650 and 651). Embryo dormancy also has been attributed to physiological immaturity of the embryo. Some physiologically immature seeds must undergo a complex series of enzymatic and biochemical changes, collectively called **after-ripening,** before they will germinate. In temperate regions, after-ripening is triggered by the low temperatures of winter. Thus germination of the seed is inhibited during the coldest part of winter, when the seedling would be unlikely to survive. During after-ripening, the seed maintains a low level of metabolic activity and thus preserves its viability. Embryo dormancy is common in the Rosaceae (roses, apples, and cherries are a few examples of this large family), as well as other woody species, and in some grasses. Some degree of both kinds of dormancy can exist simultaneously or successively in many species.

The dormancy acquired during seed maturation is called *primary dormancy.* Seed released from the plant may be in the primary state or nondormant. Seeds that are no longer dormant but encounter unfavorable conditions for germination (for example,

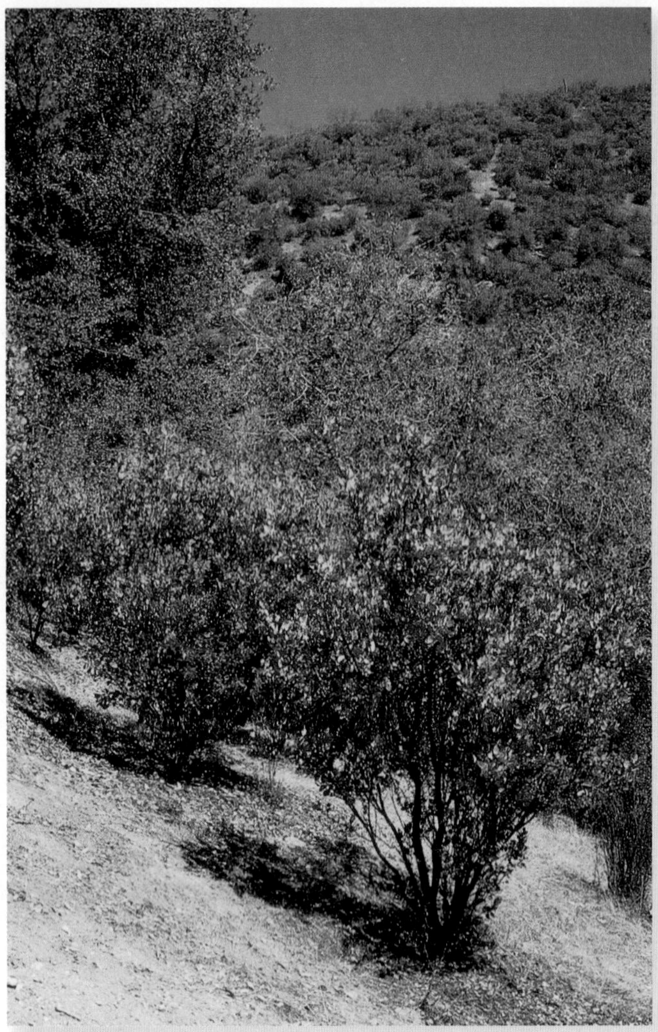

22–8 Manzanita The long-lived seeds of manzanita *(Arctostaphylos viscida)* of the California chaparral community remain viable in the soil for years. Scarification, or breakdown of the seed coat, by fire or other means is necessary to break dormancy and induce germination.

unsuitable temperature or light) may be induced to enter a state of dormancy again, called *secondary dormancy.*

Dormancy is of great survival value to the plant. As in the example of after-ripening, it is a method of ensuring that conditions will be favorable for growth of the seedling when germination does occur. Some seeds must pass through the digestive tracts of birds or mammals before they will germinate, resulting in wider dispersal of the plant species. Some seeds of desert species germinate only when inhibitors in their coats are leached away by rainfall; this adaptation ensures that the seed will germinate only during those rare intervals when desert rainfall provides sufficient water for the seedling to mature. Other seeds must be cracked mechanically, as by tumbling along in the rushing water of a gravelly streambed. Still other seeds lie dormant in cones or fruits until the heat of a fire releases them.

The Mediterranean-type-climate vegetation of the California chaparral community is dominated by shrubs and small trees that produce woody outgrowths called **burls,** or **lignotubers,** at the base of their stems. The burls contain dormant buds that sprout after the shoots have been mechanically damaged or consumed by fire. Among the manzanitas *(Arctostaphylos),* however, there are a number of species that do not regenerate in this way, but instead depend on dormant seeds that germinate only after a fire (Figure 22–8), as do many herbaceous plants.

Finally, the seeds of species that live in woodland clearings require either the death of a canopy tree or some other disturbance to form an opening in the canopy before they will germinate. These typically include the seeds of shade-intolerant species—for example, members of the pokeweed family (Phytolaccaceae). In summary, the germination strategies of plants are closely linked to the ecological conditions existing in their particular habitats. (See Chapter 27, pages 649 to 651, and Chapter 28, pages 674 to 676, for further discussion of dormancy in seeds.)

From Embryo to Adult Plant

When germination occurs, the first structure to emerge from most seeds is the root, which enables the developing seedling to become anchored in the soil and to absorb water (Figure 22–9). As this first root, called the **primary root,** or **taproot,** continues to grow, it develops branch roots, or **lateral roots.** These roots in turn may give rise to additional lateral roots. In this manner, a much-branched root system develops. Commonly, the primary root in monocots is short-lived, and the root system of the adult plant develops from **stem-borne roots,** which arise at the nodes (the parts of the stem where the leaves are attached) and then produce lateral roots.

Seed Germination May Be Epigeous or Hypogeous

The way in which the shoot emerges from the seed during germination varies among different plant groups. For example, after

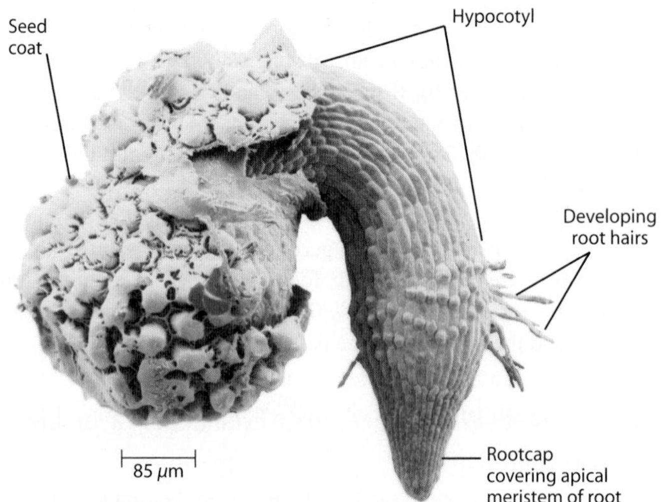

22–9 A germinating *Arabidopsis* seed Hypocotyl and embryonic root (radicle) are seen here protruding from the seed coat. The rest of the embryo, including the cotyledons and shoot apical meristem, is enclosed within the seed coat. A rootcap covers the root apical meristem. Root hairs in various stages of development can be seen just above the rootcap. Note that the hypocotyl has begun to bend, that is, to form a hook.

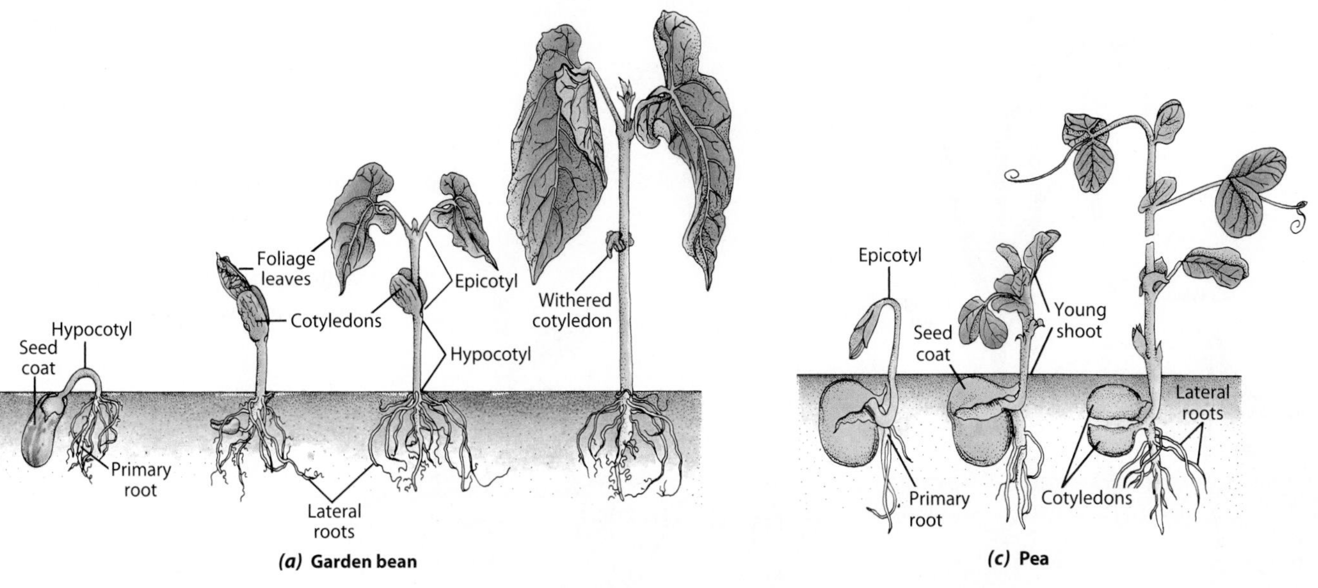

(a) Garden bean

(c) Pea

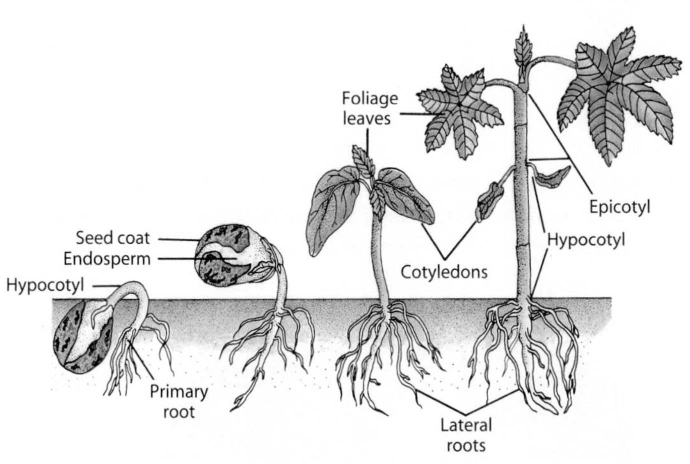

(b) Castor bean

22–10 Stages in the germination of some common eudicots Seed germination in both ***(a)*** the garden bean *(Phaseolus vulgaris)* and ***(b)*** the castor bean *(Ricinus communis)* is epigeous. During germination, the cotyledons are carried above ground by the elongating hypocotyl. In both of these seedlings, the elongating hypocotyl forms a hook, which then straightens out, pulling the cotyledons and plumule or shoot apex above ground. ***(c)*** By contrast, seed germination in the pea *(Pisum sativum)* is hypogeous. The cotyledons remain underground and the hypocotyl does not elongate. In hypogeous germination, typified by the pea seedling, it is the epicotyl that elongates and forms a hook, which pulls the plumule above ground as it straightens out.

the root emerges from the seed of the garden bean *(Phaseolus vulgaris),* the hypocotyl elongates and becomes bent in the process (Figure 22–10a). Thus, the delicate shoot tip is protected from injury by being pulled rather than pushed through the soil. When the bend, or **hook,** as it is called, reaches the soil surface, it straightens out and pulls the cotyledons and plumule up into the air. This type of seed germination, in which the cotyledons are carried above ground level, is called **epigeous.** During germination and subsequent development of the seedling, the food stored in the cotyledons is digested and the products are transported to the growing parts of the young plant. The cotyledons gradually decrease in size, wither, and eventually drop off. By this time, the seedling has become *established;* that is, it is no longer dependent on the stored food of the seed for its nourishment. The plant is now a photosynthesizing, autotrophic organism; the seedling phase has come to an end.

Germination of the castor bean *(Ricinus communis)* is similar to that of the garden bean, except that in the castor bean the stored food is found in the endosperm. As the hook straightens out, the cotyledons and plumule are carried upward, along with the endosperm and often the seed coat (Figure 22–10b). During this period, the digested foods of the endosperm are absorbed by the cotyledons and are transported to the growing parts of the seedling. In both the garden bean and castor bean, the cotyledons become green on exposure to light and photosynthesize, but they have a limited lifespan. In some plants, such as squash *(Cucurbita maxima),* the cotyledons become important photosynthetic organs.

In the pea *(Pisum sativum),* the epicotyl is the structure that elongates and forms the hook. This protects the shoot tip and young leaves, as does the hooked hypocotyl in bean. As the epicotyl straightens out, the plumule is raised above the soil surface. Because elongation occurs above the cotyledons, they remain in place in the soil (Figure 22–10c), where they eventually decompose after the stored food has been mobilized for growth of the seedling. This type of seed germination, in which the cotyledons remain underground, is called **hypogeous.**

In the majority of monocot seeds, the stored food is found in the endosperm (Figure 22–6c, d). In relatively simple monocot seeds, such as those of the onion *(Allium cepa),* it is elongation of the single tubular cotyledon that results in formation of a hooked cotyledon (Figure 22–11a). When the cotyledon straightens, it carries the seed coat and enclosed endosperm

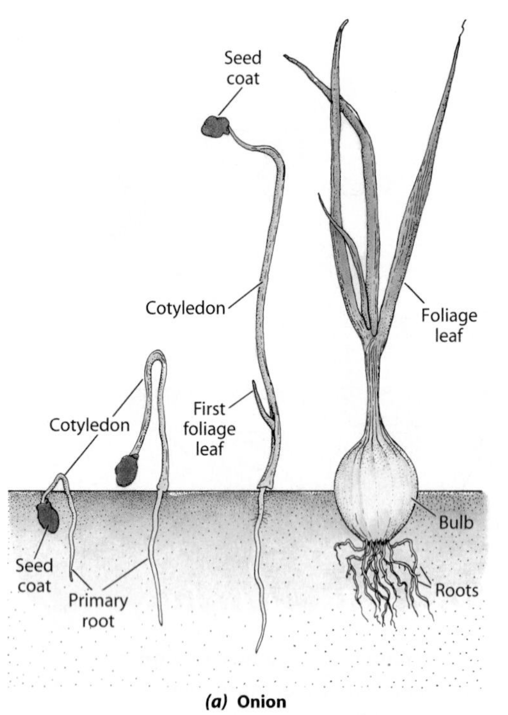

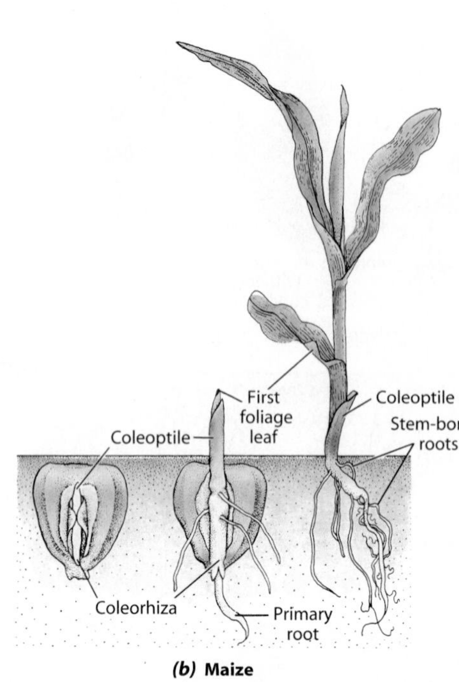

22–11 Stages in the germination of two common monocots ***(a)*** Seed germination in onion *(Allium cepa)* is epigeous; ***(b)*** in maize *(Zea mays)* it is hypogeous. The three seminal roots shown in the middle image are short-lived, while the stem-borne roots shown at the right contribute to the mature root system.

upward. Throughout this period, and for some time afterward, the developing seedling obtains nutrients from the endosperm by way of the cotyledon. Furthermore, the green cotyledon in onion functions as a photosynthetic leaf, contributing significantly to the food supply of the developing seedling. Soon the plumule emerges from the protective, sheathlike base of the cotyledon, elongates, and forms the foliage leaves of the seedling.

Our last example of seedling development is provided by maize *(Zea mays),* a monocot that has a highly differentiated embryo (Figure 22–12). The coleorhiza, which encloses the radicle, is the first structure to grow through the pericarp (the mature ovary wall of the grain). (In maize, the seed and fruit wall are fused, and hence the pericarp functions as a "seed coat.") The coleorhiza is followed by the radicle, which elongates very rapidly and quickly penetrates the coleorhiza (Figure 22–11b). After the primary root emerges, the coleoptile, which surrounds the plumule, is pushed upward by elongation of the first stem internode, called the *mesocotyl* (as seen in the wheat kernel in Figure 22–7b). (An internode is the part of the stem between two successive nodes; see Figure 1–9.) Shortly after the coleoptile reaches the soil surface, it ceases to elongate, and the first leaves of the plumule emerge through an opening at its tip. In addition to the primary root, two or more seminal roots, which arise from the cotyledonary node (the part of the axis where the cotyledons are attached), grow through the pericarp and then bend downward (Figure 22–12).

Regardless of how the shoot emerges from the seed, the activity of the apical meristem of the shoot results in the formation of an orderly sequence of leaves, nodes, and internodes. Additional apical meristems (buds) develop in the leaf axils (the upper angles between leaves and stems) and produce axillary shoots. These, in turn, may form additional axillary shoots.

The period from germination to the time the seedling becomes established as an independent organism constitutes the most crucial phase in the life history of the plant. During this period, the plant is most vulnerable to injury by a wide range of insect pests and parasitic fungi, and water stress can very rapidly prove fatal.

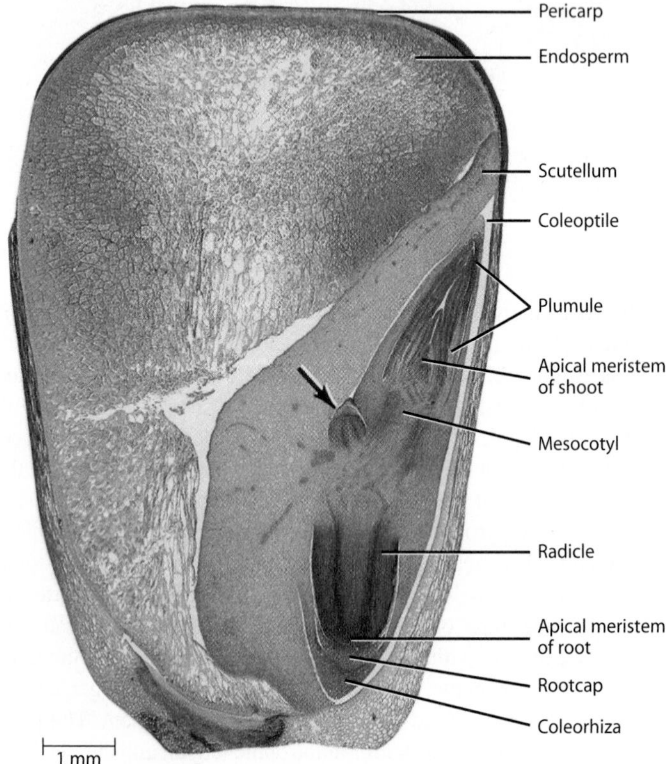

22–12 Mature grain, or kernel, of maize *(Zea mays)* Grain embryos commonly contain two or more seminal (seed) roots. One seminal root can be seen in this longitudinal section (arrow). Although at first pointed upward, these roots become inclined downward with further growth.

SUMMARY

During Embryogenesis, the Body Plan of the Plant, Consisting of an Apical-Basal Pattern and a Radial Pattern, Is Established

Beginning with the zygote, the shoot and root of the young plant are initiated as one continuous structure. With the asymmetrical division of the zygote, the apical-basal polarity of the embryo is established. Through an orderly progression of divisions, the embryo differentiates into a suspensor and an embryo proper. With the appearance of the primary meristems—the protoderm, ground meristem, and procambium—within the embryo proper, the radial pattern emerges. The outer layer, the protoderm, is the precursor of the epidermis; the ground meristem, the precursor of the ground tissue, surrounds the procambium, the precursor of the vascular tissues (xylem and phloem). During the transition between the globular and heart stages of eudicots, the apical-basal pattern of the embryo becomes discernible. Development of the cotyledons may begin either during or after the time when the procambium becomes discernible. In monocots, the globular embryo forms only one cotyledon and becomes cylindrical in shape. As the embryo develops, apical meristems eventually are established at the shoot and root tips.

Mutations Disrupt Normal Embryo Development

The suspensors of angiosperm embryos are metabolically active and play a role in supporting early development of the embryo proper. In some embryo loss-of-function mutants of *Arabidopsis*, the cells of the suspensor may form secondary embryos. Other mutations affect the apical-basal pattern of the embryo and seedling, resulting in the deletion of major portions of the plant.

At the End of Embryogenesis the Seed Enters the Maturation Phase

During the maturation phase, food reserves accumulate, the seed undergoes desiccation, and the seed coat hardens. Desiccation allows the dormant seed to remain viable for long periods.

The Mature Embryo Consists of a Hypocotyl-Root Axis and One or Two Cotyledons

The seeds of flowering plants consist of an embryo, a seed coat, and stored food. When fully formed, the embryo consists basically of a hypocotyl-root axis bearing either one or two cotyledons and an apical meristem at the shoot tip and at the root tip. The cotyledons of most eudicots are fleshy and contain the stored food of the seed. In other eudicots and in most monocots, food is stored in the endosperm, and the cotyledons function to absorb the simpler compounds resulting from the digestion of that food. These compounds are then transported to growing regions of the embryo.

A Dormant Seed Will Not Germinate Even When External Conditions Are Favorable

Germination of the seed—a resumption of growth of the embryo—depends on environmental factors, including water, oxygen, and temperature. Many seeds must pass through a period of dormancy before they are able to germinate. Both the seed coat and the embryo can cause dormancy. Dormancy is of great survival value to the plant because it is a method of ensuring conditions will be favorable for growth of the seedling when germination occurs.

Following Emergence of the Root and Shoot, the Seedling Becomes Established

The root is the first structure to emerge from most germinating seeds, enabling the seedling to become anchored in the soil and to absorb water. The way in which the shoot emerges from the seed varies from species to species. The embryos of many angiosperms other than monocots form a bend, or hook, in their hypocotyls or epicotyls. As the hook straightens out, the delicate shoot tip is pulled upward from the soil, thus preventing injury that might occur if the shoot tip were pushed through the soil.

QUESTIONS

1. Distinguish among or between the following: proembryo, embryo proper, and suspensor; globular stage, heart stage, and torpedo stage; epicotyl and plumule; hypocotyl-root axis and radicle; coleorhiza and coleoptile.
2. Explain what is meant by the apical-basal pattern and radial pattern of the plant.
3. What role is played by the suspensor in angiosperms, and what evidence indicates that the embryo proper suppresses the embryonic pathway in the suspensor?
4. How have mutations helped us to understand embryo development?
5. What factors, or mechanisms, have been identified as causing coat-imposed dormancy of seeds?
6. Suggest why the root is the first structure to emerge from a germinating seed.
7. What environmental factors are especially important for seed germination?
8. What are some of the ways in which the shoot emerges from the seed during germination? What is meant by epigeous and hypogeous germination?

CHAPTER 23

Cells and Tissues of the Plant Body

Linen from flax Following harvesting and drying, common flax *(Linum usitatissimum)* is treated to remove unwanted plant material, and the remaining fibers from the stems are combed and spun into linen yarn for weaving into cloth. Cultivated widely in ancient Egypt, linen was used to wrap mummies, and in a prehistoric cave in the Republic of Georgia, dyed flax fibers were reportedly found that date to 30,000 B.C.E.

CHAPTER OUTLINE

As we discussed in the previous chapter, the process of embryogenesis establishes the apical-basal axis of the plant, with a shoot apical meristem at one end and a root apical meristem at the other. During embryogenesis, the radial pattern of the tissue systems within the axis is also determined. Embryogenesis, however, is only the beginning of the development of the plant body. Most plant development occurs after embryogenesis, through the activity of the **meristems.** These embryonic regions or populations of cells retain the potential to divide long after embryogenesis is over. With germination of the seed, the **root** and **shoot apical meristems** of the embryo generate cells that give rise to the roots, stems, leaves, and flowers of the adult plant.

Apical Meristems and Their Derivatives

Apical meristems are found at the tips of all roots and stems and are involved primarily with extension of the plant body (Figure 23–1). The term "meristem" (Gk. *merismos,* "division") emphasizes cell division activity as a characteristic of a meristematic tissue. The cells that maintain the meristem as a continuing source of new cells are called **initials.** Initials divide in such a way that one of the sister cells remains in the meristem as an initial while the other becomes a new body cell or **derivative.** Derivatives, in turn, may divide several times near the root or shoot tip before undergoing differentiation. Divisions are not limited, however, to the initials and their immediate derivatives. The primary meristems—protoderm, procambium, and ground meristem—which are initiated during embryogenesis, are

CHECKPOINTS

After reading this chapter, you should be able to answer the following:

1. What is a meristem, and what is its composition?
2. Describe the three overlapping processes of plant development and the way they overlap.
3. What are the three tissue systems of the plant body? Of what tissues are they composed?
4. How do parenchyma, collenchyma, and sclerenchyma cells differ from one another? What are their respective functions?
5. What are the principal conducting cells in the xylem? In the phloem? Describe the characteristics of each cell type.
6. Describe the cell types that occur in the epidermis and the roles they play.

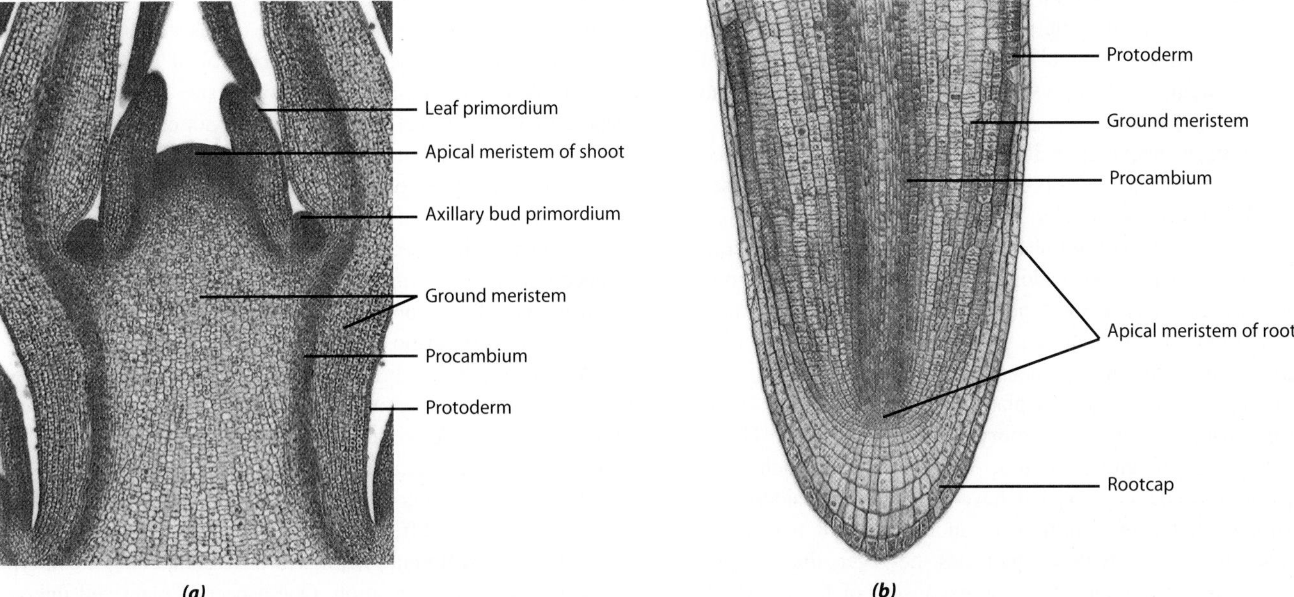

23–1 Shoot and root apical meristems ***(a)*** Longitudinal section of a shoot tip of lilac *(Syringa vulgaris)*, showing the shoot apical meristem and primordia of leaves and axillary buds. ***(b)*** Root tip of radish *(Raphanus sativus)*, in longitudinal section, showing the root apical meristem covered by a rootcap. Note the files, or lineages, of cells behind the root apical meristem. Protoderm, ground meristem, and procambium are partly differentiated tissues known as primary meristems.

extended throughout the plant body by the activity of the apical meristems. These primary meristems are partly differentiated tissues that remain meristematic for some time before they begin to differentiate into specific cell types in the primary tissues (Figure 23–2). Growth of this type, which involves extension of the plant body and formation of the primary tissues, is called **primary growth,** and the part of the plant body composed of these tissues is called the **primary plant body** (page 393). The primary growth of root and shoot are discussed in greater detail in Chapters 24 and 25, respectively, and secondary growth, which involves thickening of the stem and root, is discussed in Chapter 26.

The existence of meristems, which add to the plant body throughout its life, underlies one of the principal differences between plants and animals. Birds and mammals, for example, form all of their organs early in life and stop growing when they reach maturity, although the cells of certain "turnover" tissues, such as skin and the lining of the intestine, continue to divide. Plants, however, form new organs and continue to grow during their entire life span. Such unlimited or prolonged growth of the apical meristems is described as **indeterminate.**

Growth in plants is the counterpart, to some extent, of motility in animals. Plants "move" by extending their roots and shoots, both of which involve changes in size and form. As a result of these changes, a plant is able to modify its relationship with the environment, for example, by curving toward the light and extending its roots toward water. This *developmental plasticity* of plants therefore roughly corresponds to a whole series of motor acts in animals, especially those associated with obtaining food and water. In fact, growth in plants serves many of the functions that we group under the term "behavior" in animals.

Growth, Morphogenesis, and Differentiation

Development, the sum total of events that progressively form an organism's body, involves three overlapping processes: growth, morphogenesis, and differentiation. Development occurs in response to instructions contained in the genetic information that an organism inherits from its parents. In plants, the specific developmental pathway followed is determined in large

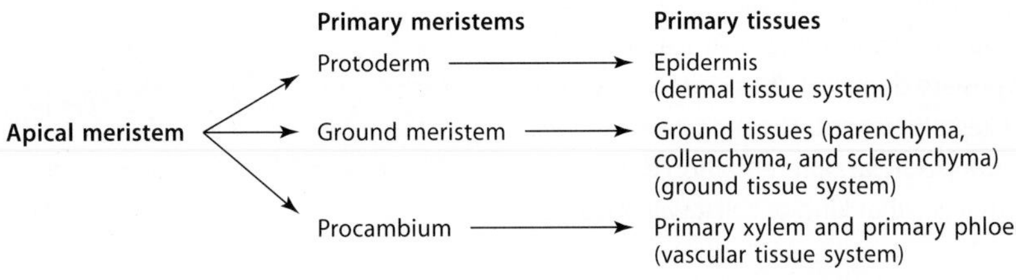

23–2 From apical meristem to primary tissues The apical meristem gives rise to the primary meristems, which give rise to the tissues and tissue systems of the primary plant body.

part by the position (location) of cells and tissues within the apical-basal and radial patterns. In addition, development is determined by environmental factors such as, in the case of plants, daylength, light quality and quantity, temperature, and gravity (see Chapter 28).

Growth, an irreversible increase in size, is accomplished by a combination of cell division and cell enlargement (see Chapter 3). Cell division does not of itself constitute growth. It may simply increase cell number without increasing the overall volume of a structure. The addition of cells to the plant body by cell division increases the potential for growth by increasing the number of cells that can grow, but most plant growth is brought about by cell enlargement.

During its development, a plant assumes a particular shape or form—that is, it undergoes **morphogenesis** (from the Greek words for "shape" and "origin"). The planes in which cells divide and subsequently expand have long been considered as the primary factors determining the morphology of a plant or plant part. Increasing evidence indicates, however, that the primary event in morphogenesis is the expansion of tissue, which then is subdivided into smaller units by cell division—that is, cell and tissue differentiation follows morphogenesis.

Differentiation is the process by which cells with identical genetic constitution become different from one another and from the meristematic cells from which they originated (Figure 23–3), and this process often begins while the cell is still enlarging. Cellular differentiation depends on the control of gene expression. Different types of cells and tissues synthesize different proteins, because they express certain sets of genes not expressed by other types of cells and tissues. For example, both fibers and collenchyma cells are types of supportive cells, but the cell walls of fibers typically are rigid whereas those of collenchyma cells are flexible. During its development, the fiber makes enzymes that produce lignin, which imparts the rigid property to its walls. The collenchyma cell, in contrast, makes enzymes that produce pectins, which impart plastic properties to its walls.

Although cellular differentiation depends on the control of gene expression, the fate of a plant cell—that is, what kind of cell it will become—is determined by its *final position* in the developing organ. Cell lineages, such as those in the root (Figure 23–1b), may be established, but cell differentiation is not dependent on cell lineage. If an undifferentiated cell is displaced from its original position to a different one, it will differentiate into a cell type appropriate to its new position. One aspect of plant cell interaction is the communication of *positional information* from one cell to another.

Discussions of differentiation often include a reference to determination and competence. **Determination** means progressive

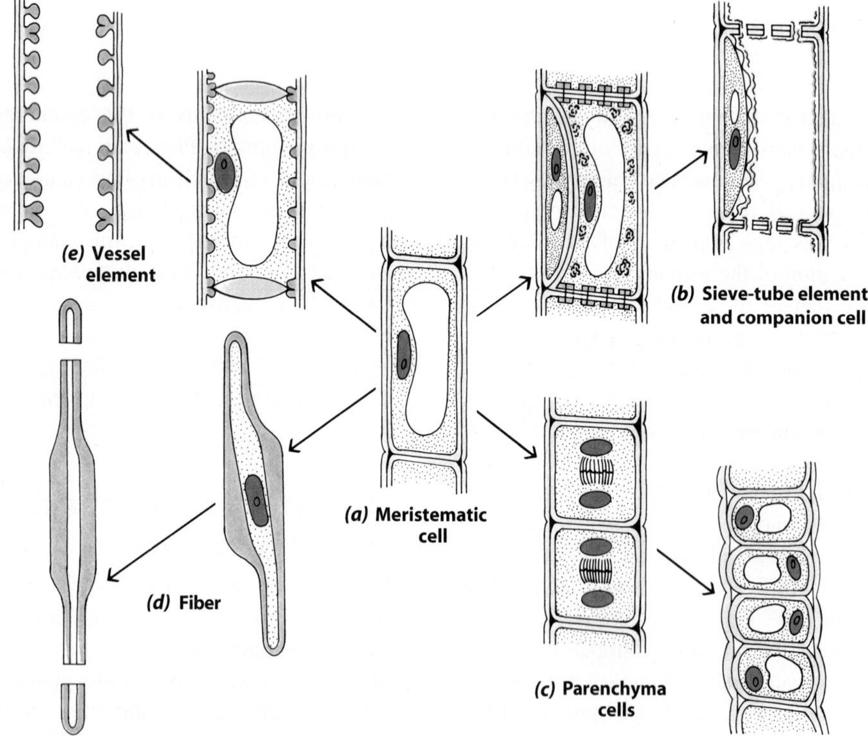

23–3 Differentiation of cells Various cell types may originate from a meristematic cell of the procambium or the vascular cambium. ***(a)*** The meristematic cell depicted here (at the center), with a single large vacuole, is typical of the meristematic cells of the vascular cambium. Procambial cells typically contain several small vacuoles. Five different cell types are depicted: ***(b)*** sieve-tube element with a companion cell, ***(c)*** parenchyma cell, ***(d)*** fiber, and ***(e)*** vessel element. The meristematic cells or precursors of these cells had an identical genetic constitution. Cellular differentiation depends on the control of gene expression, but the fate of a plant cell—that is, what kind of cell it will become—is determined by its final position in the developing organ.

commitment to a specific course of development that brings about a weakening or loss of capacity to resume growth. Some cells are determined earlier and more completely than others, whereas some maintain the capacity to dedifferentiate and divide, producing progeny that can differentiate into virtually any cell type. **Competency** refers to the ability of a cell to develop in response to a specific signal, such as light.

Internal Organization of the Plant Body

Cells are associated with each other in various ways, forming structural and functional units called **tissues.** Moreover, the principal tissues of vascular plants are grouped together into larger units on the basis of their continuity throughout the plant body. These larger units, known as **tissue systems,** are readily recognized, often with the unaided eye. There are three tissue systems, and their presence in root, stem, and leaf reveals both the basic similarity of the plant organs and the continuity of the plant body. The three tissue systems are (1) the **ground** (or **fundamental**) **tissue system,** (2) the **vascular tissue system,** and (3) the **dermal tissue system.** As mentioned in Chapter 22, the tissue systems are initiated during development of the embryo, where their precursors are represented by the primary meristems—ground meristem, procambium, and protoderm, respectively (Figure 23–2). Each tissue system consists of one or more distinct tissues.

The ground tissue system consists of the three ground tissues—parenchyma, collenchyma, and sclerenchyma. Parenchyma is by far the most common of the ground tissues. The vascular tissue system consists of the two conducting tissues—xylem and phloem. The dermal tissue system is represented by the epidermis, a single tissue, which is the outer protective covering of the primary plant body, and later by the periderm in plant parts that undergo a secondary increase in thickness (pages 393 and 394).

Within the plant body, the various tissues are distributed in a radial pattern, with specific arrangements, depending on the plant part or plant taxon or both. The patterns are essentially alike from one plant part to another in that the vascular tissues are embedded within the ground tissue, with the dermal tissue forming the outer covering. The principal differences in patterns depend largely on the relative distribution of the vascular and ground tissues (Figure 23–4). In the stem of a eudicot, for example, the vascular system may form a system of interconnected strands embedded within the ground tissue. The region formed internal to the strands is called the pith, and the region external to them is called the cortex. In the root of the same plant, the vascular tissues may form a solid cylinder (vascular cylinder, or stele) surrounded by a cortex. In the leaf, the vascular system typically forms a system of vascular bundles (veins), embedded in photosynthetic ground tissue (mesophyll).

Tissues may be defined as groups of cells that are structurally and/or functionally distinct. Tissues composed of only one type of cell are called **simple tissues,** whereas those composed of two or more types of cells are called **complex tissues.** The ground tissues—parenchyma, collenchyma, and sclerenchyma—are simple tissues; xylem, phloem, epidermis, and periderm are complex tissues.

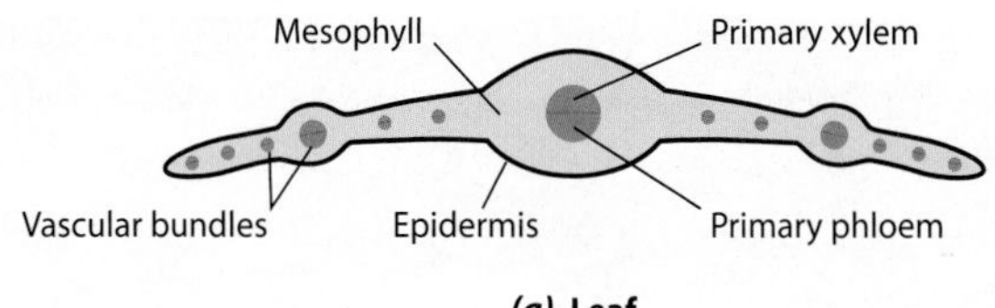

***(a)* Leaf**

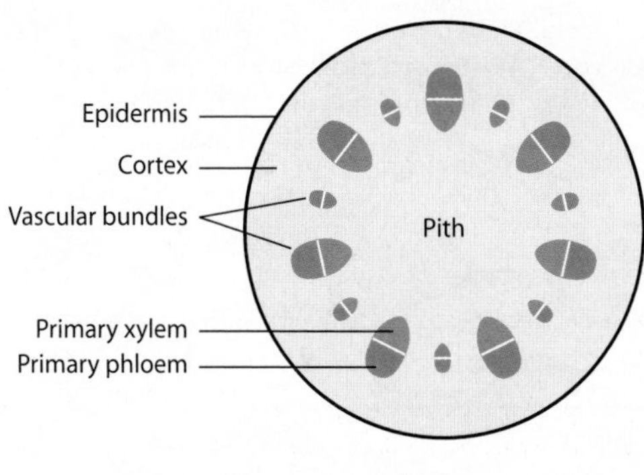

***(b)* Stem**

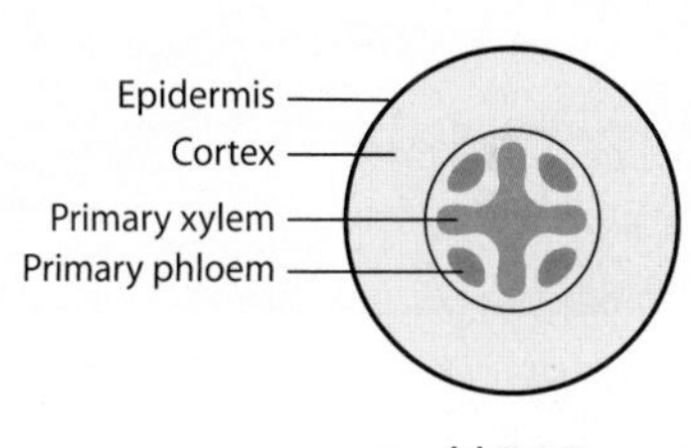

***(c)* Root**

23–4 Distribution of primary tissues in leaves, stems, and roots of eudicots ***(a)*** In the leaf, the ground tissue is represented by the mesophyll, which is specialized for photosynthesis. Embedded in the mesophyll are the vascular bundles, made up of primary xylem for water transport and primary phloem for sugar transport. ***(b)*** In the stem, the ground tissue is represented by the pith and cortex, and the vascular bundles are arranged in a circular pattern between the cortex and pith. ***(c)*** In the root, the ground tissue is represented by cortex only, which surrounds the primary xylem and phloem. The epidermis forms the outer covering of all three plant parts.

Ground Tissues

Parenchyma Tissue Is Involved in Photosynthesis, Storage, and Secretion

Of variable shape and size, **parenchyma cells** are the most numerous cells in the plant body. In the primary plant body, parenchyma cells commonly occur as continuous masses—as **parenchyma tissue**—for example, in the cortex (Figure 23–5) and pith of stems and roots, in leaf mesophyll (see Figure 23–27), and in the flesh of fruits. In addition, parenchyma cells occur as vertical strands of cells in the primary and secondary vascular tissues and as horizontal strands called **rays** in the secondary vascular tissues (see Chapter 26).

Characteristically living at maturity, parenchyma cells are capable of cell division, and although their walls are commonly primary, some parenchyma cells also have secondary walls. Because they retain their meristematic ability, parenchyma cells

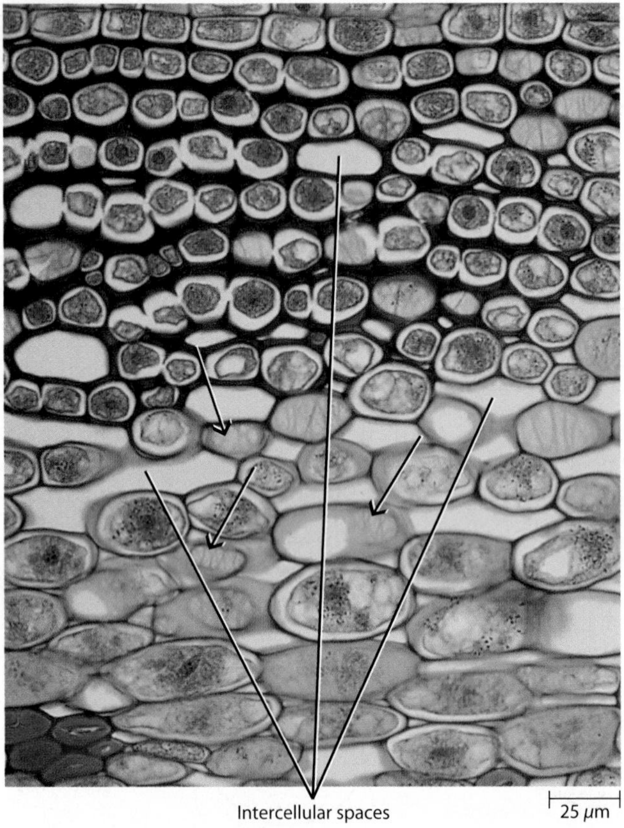

23–5 Parenchyma and collenchyma Parenchyma cells (below) and collenchyma cells, with unevenly thickened walls (above), are seen here in a transverse section of the cortex of an elderberry *(Sambucus canadensis)* stem. In some of the parenchyma cells, a meshwork of lines is visible in the facing walls. The light areas within the meshwork are primary pit-fields (arrows), which are thin areas in the walls. The protoplasts of the collenchyma cells are plasmolyzed and therefore appear withdrawn from the walls. The clear areas between the cells are intercellular spaces.

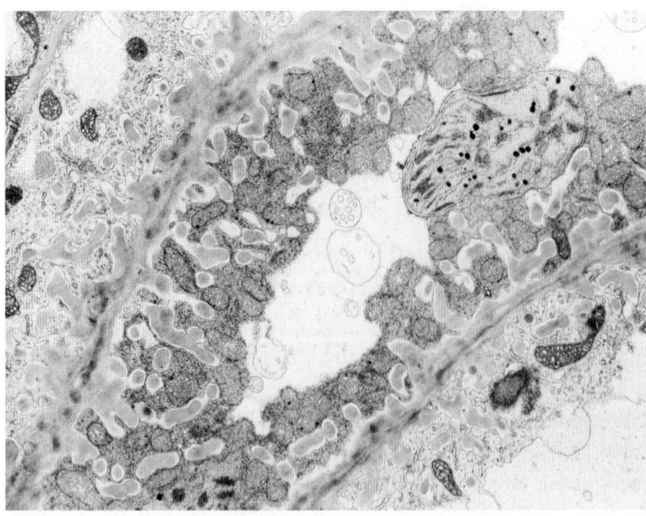

23–6 Transfer cells Transverse section of a portion of the phloem from a small vein of a sow thistle *(Sonchus)* leaf, showing transfer cells with their numerous wall ingrowths, which facilitate solute movement by increasing the surface of the plasma membrane.

having only primary walls play an important role in regeneration and wound healing. It is these cells that initiate adventitious structures, such as adventitious roots on stem cuttings. In addition, given proper conditions for growth and development, these cells have the ability to become embryonic cells and to develop into an entire plant. Such cells are said to be **totipotent** (see page 202). Parenchyma cells are involved in such activities as photosynthesis, storage, and secretion—activities dependent on living protoplasts. Parenchyma cells may also play a role in the movement of water and the transport of food substances in plants. In many succulents, such as the Cactaceae, *Sansevieria,* and *Peperomia,* parenchyma is specialized as a water-storage tissue.

Transfer Cells Are Parenchyma Cells with Wall Ingrowths

The cell wall ingrowths of **transfer cells** often greatly increase the surface area of the plasma membrane (Figure 23–6), and transfer cells are believed to facilitate the movement of solutes over short distances. The presence of transfer cells is generally correlated with the existence of intensive movement of solutes—either inward (uptake) or outward (secretion)—across the plasma membrane.

Transfer cells are exceedingly common and probably serve a similar function throughout the plant body. They occur in association with the xylem and phloem of small, or minor, veins in cotyledons and in the leaves of many herbaceous eudicots. Transfer cells are also associated with the xylem and phloem of leaf traces at the nodes in both eudicots and monocots. In addition, they are found in various tissues of reproductive structures (placentae, embryo sacs, endosperm) and in various glandular structures (nectaries, salt glands, the glands of carnivorous plants) where intensive short-distance transfer of solutes takes place.

Collenchyma Tissue Supports Young, Growing Organs

Collenchyma cells, like parenchyma cells, are living at maturity (Figures 23–5 and 23–7). **Collenchyma tissue** commonly occurs in discrete strands or as continuous cylinders beneath the epidermis in stems and petioles (leaf stalks). It is also found bordering the veins in eudicot leaves. (The "strings" on the outer surface of celery stalks, or petioles, consist almost entirely of collenchyma.) Collenchyma cells typically are elongated. Characteristically, their most distinctive feature is their unevenly thickened, nonlignified primary walls, which are soft and pliable and have a glistening appearance in fresh tissue (Figure 23–7). Because collenchyma cells are living at maturity, they can continue to develop thick, flexible walls while the organ is still elongating, making these cells especially well adapted for the support of young, growing organs.

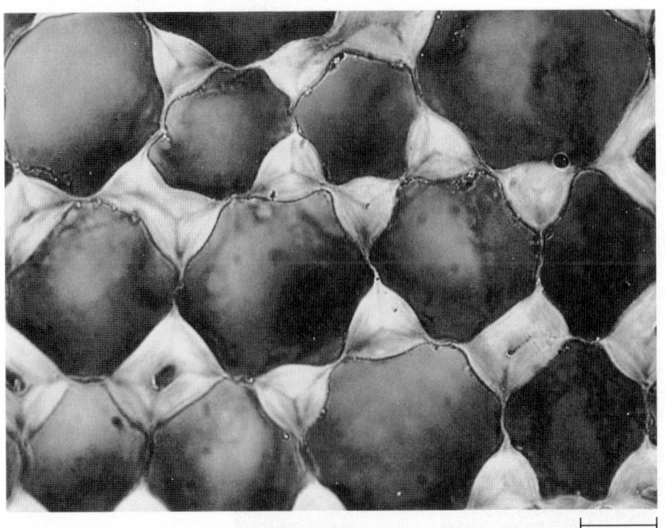

23–7 Collenchyma, fresh tissue Transverse section of collenchyma tissue from a petiole of rhubarb *(Rheum rhabarbarum)*. In fresh tissue like this, the unevenly thickened collenchyma cell walls have a glistening appearance.

Collenchyma is the typical supporting tissue of growing stems, leaves, and floral parts and of most herbaceous (nonwoody) organs that undergo little or no secondary growth. Roots rarely have collenchyma. Collenchyma is absent in stems and leaves of many monocots that form sclerenchyma early in their development.

Sclerenchyma Tissues Strengthen and Support Plant Parts No Longer Elongating

Sclerenchyma cells may form continuous masses—as **sclerenchyma tissue**—or they may occur in small groups, or individually, or among other cells. They may develop in any or all parts of the primary and secondary plant bodies and often lack protoplasts at maturity. The principal characteristic of sclerenchyma cells is their thick, often lignified, secondary walls. Because of these walls, sclerenchyma cells are important strengthening and supporting elements in plant parts that have ceased elongating (see the description of the secondary wall on pages 59 and 60).

Two types of sclerenchyma cells are recognized: fibers and sclereids. **Fibers** are generally long, slender cells that occur in strands or bundles (Figure 23–8). The so-called bast

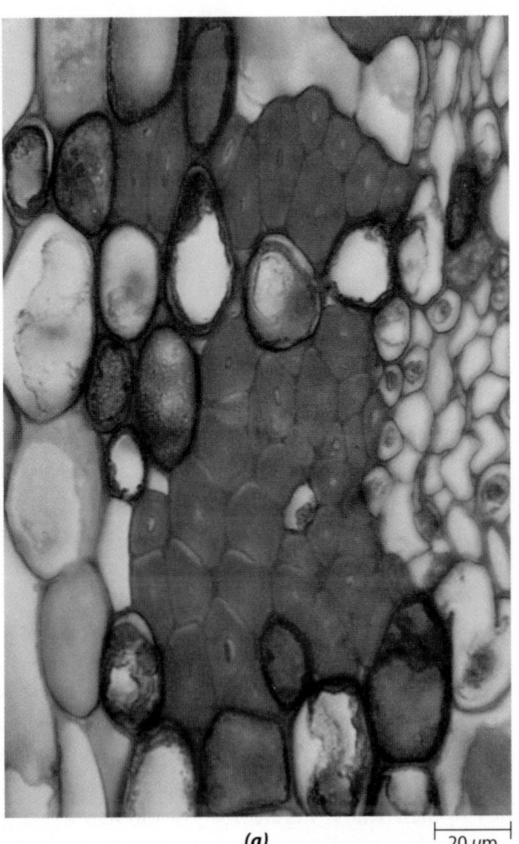

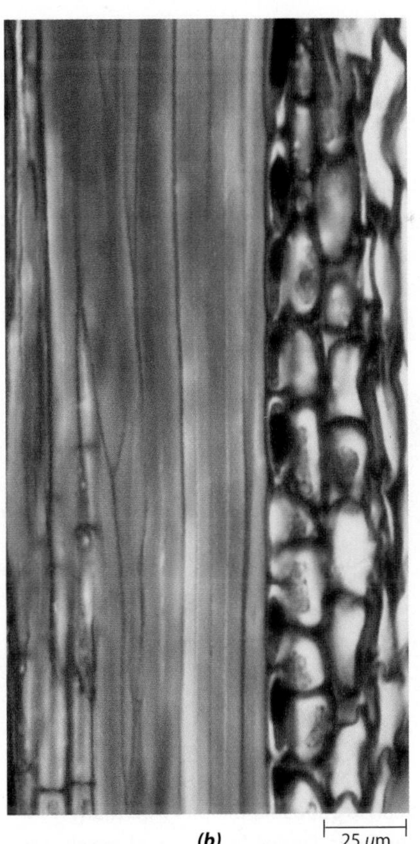

23–8 Fibers from the stem of basswood *(Tilia americana)* ***(a)*** Cross-sectional and ***(b)*** longitudinal views of primary phloem fibers. The secondary walls of these long, thick-walled fibers contain relatively inconspicuous pits. Only a portion of the length of these fibers can be seen in *(b)*. The fibers are bordered on the right by parenchyma cells.

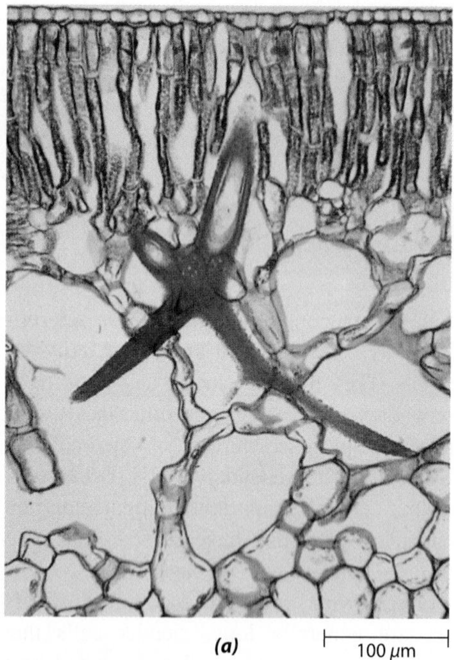

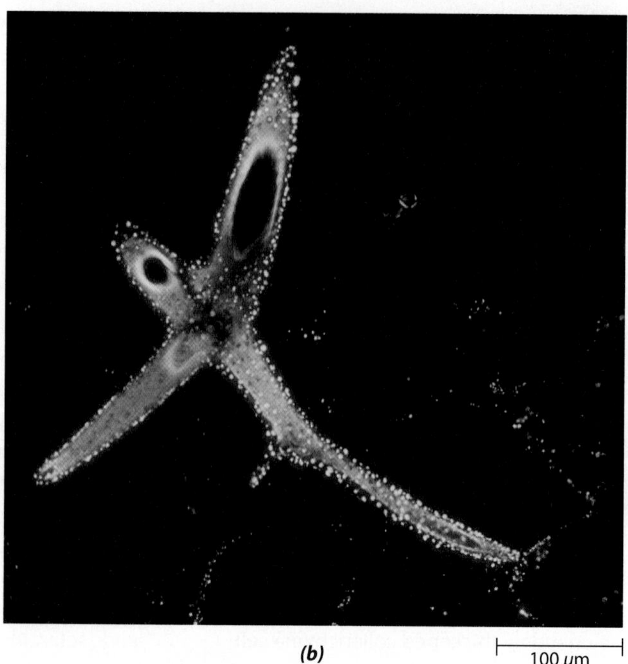

23–9 Branched sclereid Seen here in ***(a)*** ordinary light and ***(b)*** polarized light is a branched sclereid from a leaf of the water lily *(Nymphaea odorata)*. Numerous small, angular crystals of calcium oxalate are embedded in the wall of this sclereid.

fibers—such as hemp, jute, and flax—are derived from the stems of eudicots. Other economically important fibers, such as Manila hemp, are extracted from the leaves of monocots. Fibers vary in length, ranging from 0.8 to 6 millimeters in jute, 5 to 55 millimeters in hemp, and 9 to 70 millimeters in flax. **Sclereids** are variable in shape and are often branched (Figure 23–9), but compared with most fibers, they are relatively short cells. Sclereids may occur singly or in aggregates throughout the ground tissue. They make up the seed coats of many seeds, the shells of nuts, and the stone (endocarp) of stone fruits, such as olives, peaches, and cherries, and they give pears their characteristic gritty texture (Figure 23–10).

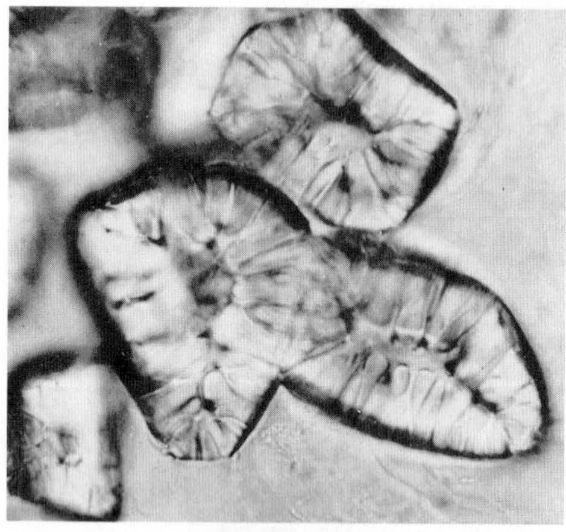

23–10 Stone cells (sclereids) of the pear The secondary walls of these sclereids contain conspicuous simple pits with many branches, known as ramiform pits. During formation of the clusters of stone cells in the flesh of the pear *(Pyrus communis)* fruit, cell divisions occur concentrically around some of the sclereids formed earlier. The newly formed cells differentiate as stone cells, adding to the cluster. See Figure 3–30.

Vascular Tissues

Xylem Is the Principal Water-Conducting Tissue in Vascular Plants

Besides its role as the principal water-conducting tissue, **xylem** is also involved in the conduction of minerals, in support, and in food storage. Together with the phloem, the xylem forms a continuous system of vascular tissue extending throughout the plant body (Figures 23–4 and 23–11). In the primary plant body, the xylem is derived from the procambium. During secondary growth, the xylem is derived from the vascular cambium (see Chapter 26).

The principal conducting cells of the xylem are the **tracheary elements,** of which there are two types: the **tracheids** and the **vessel elements.** Both are elongated cells that have secondary walls and lack protoplasts at maturity, and both types may have *pits* in their walls (Figure 23–12a through d; see page 60 for an introduction to pit structure). Unlike tracheids, vessel elements contain *perforations,* which are areas lacking both primary and secondary walls. The part of the wall bearing the perforation or perforations is called the **perforation plate** (Figure 23–13). Perforations generally occur on the end walls, with the vessel ele-

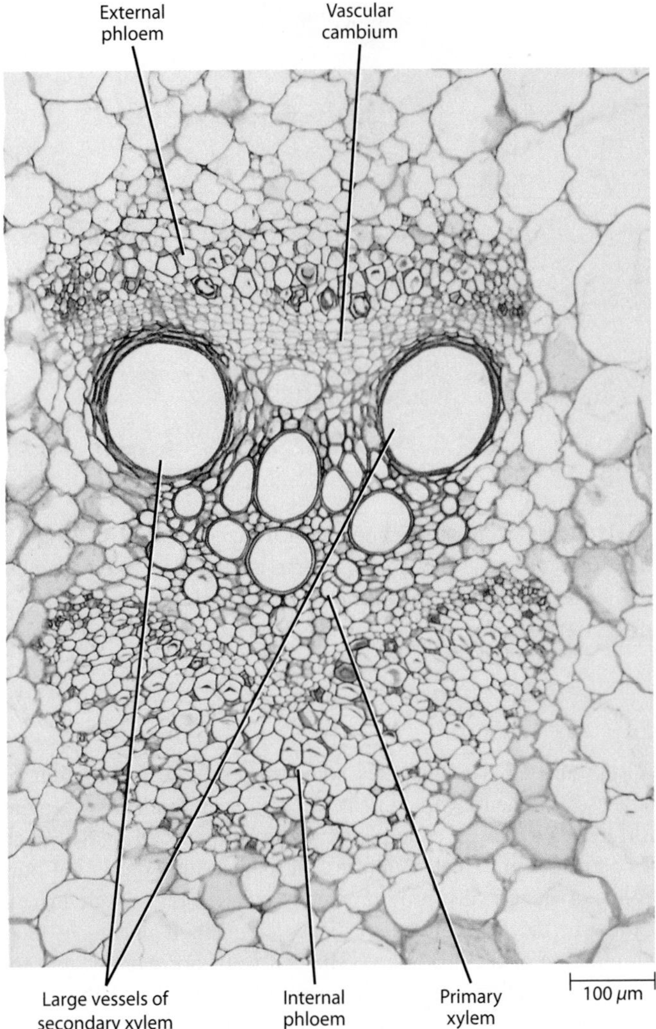

23–11 Vascular bundle Transverse section of a vascular bundle from the stem of a squash *(Cucurbita maxima)*, a favorite species for the study of phloem. Phloem occurs both exterior to and interior to the xylem in squash vascular bundles. Typically, a vascular cambium develops between the external phloem and the xylem but not between the internal phloem and the xylem. Here the vascular cambium has produced some secondary phloem (two to three layers of cells) to the outside and some secondary xylem to the inside. The secondary xylem is largely represented by the two large vessels. All of the internal phloem is primary.

ments joined end-to-end, forming long, continuous columns, or tubes, called **vessels** (Figure 23–14; see also Figure 26–24b).

The tracheid, lacking perforations, is a less specialized type of cell than the vessel element, which is the principal water-conducting cell in angiosperms. Vessel elements have evolved independently in several groups of vascular plants. The tracheid is the only type of water-conducting cell found in most seedless vascular plants and gymnosperms. The xylem of many angiosperms, however, contains both vessel elements and tracheids.

Vessel elements are generally thought to be more efficient conductors of water than the tracheids, because water can flow relatively unimpeded from vessel element to vessel element through the perforations. Vessel elements, however, with their open systems, are less safe for the plant than tracheids. Water flowing from tracheid to tracheid must pass through the pit membranes—the thin, modified primary cell walls—of the pit-pairs (page 60). Although the porous pit membranes offer relatively little resistance to the flow of water, they block even the smallest of air bubbles (see Chapter 30). Thus, air bubbles that form in a tracheid—for example, during the alternate freezing and thawing of xylem water in the spring—are restricted to that tracheid, and any resulting obstruction to water flow is also limited. In contrast, air bubbles formed in a vessel element can potentially obstruct the flow of water for the entire length of the vessel. Wide vessels are more efficient at water conduction than narrow vessels, but wide vessels also tend to be longer and may not be as casualty-free as narrow vessels.

The tracheary elements of the primary xylem have a variety of secondary wall thickenings. During the period of elongation or expansion of the roots, stems, and leaves, the secondary walls of the first-formed tracheary elements of the early-formed primary xylem (protoxylem; *proto-* meaning "first") are

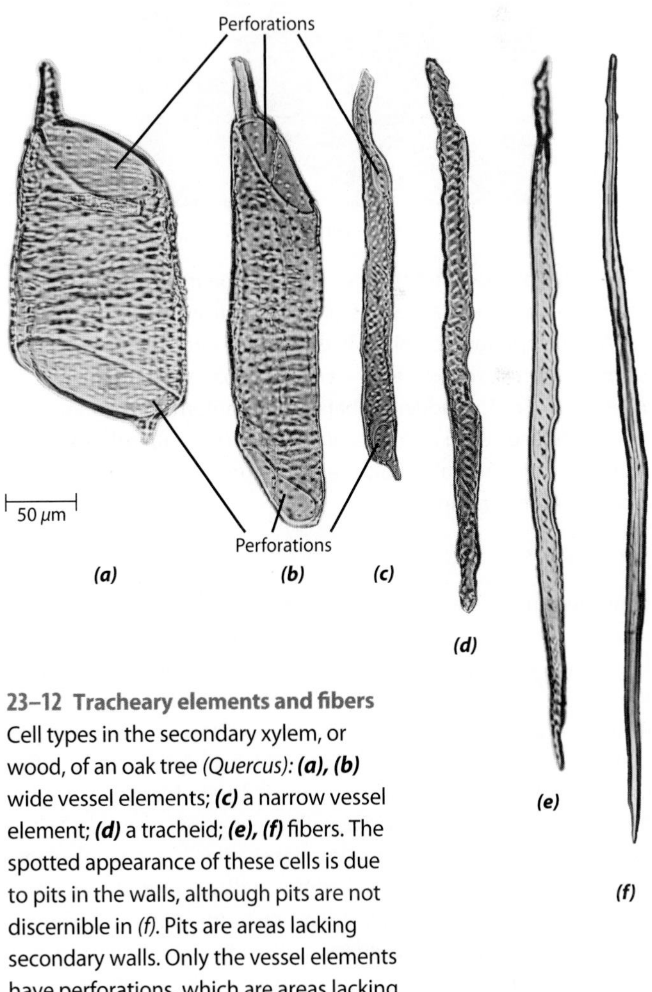

23–12 Tracheary elements and fibers Cell types in the secondary xylem, or wood, of an oak tree *(Quercus)*: ***(a), (b)*** wide vessel elements; ***(c)*** a narrow vessel element; ***(d)*** a tracheid; ***(e), (f)*** fibers. The spotted appearance of these cells is due to pits in the walls, although pits are not discernible in *(f)*. Pits are areas lacking secondary walls. Only the vessel elements have perforations, which are areas lacking both primary and secondary cell walls (see Figure 23–13).

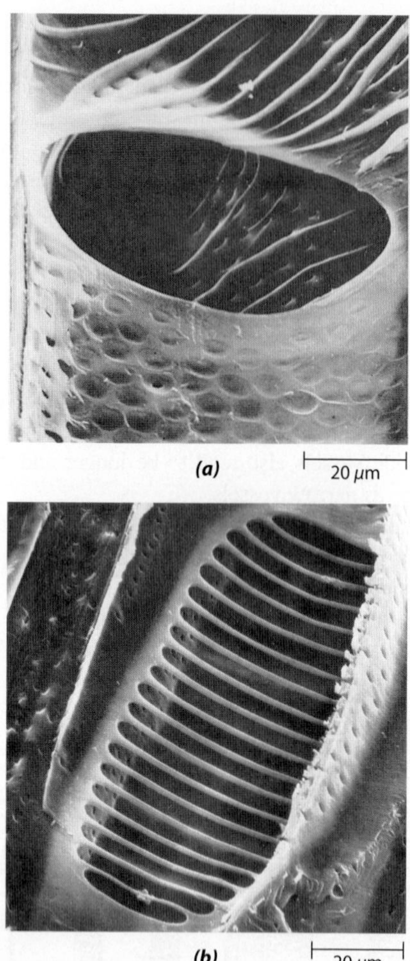

23–13 Perforation plates Scanning electron micrographs of the perforated end walls of vessel elements from secondary xylem. ***(a)*** A simple perforation plate, with its single large opening, seen here between two vessel elements in basswood *(Tilia americana)*. ***(b)*** The ladderlike bars of a scalariform perforation plate between vessel elements of red alder *(Alnus rubra)*. Pits can be seen in the wall below the perforation plate in *(a)* and in portions of the wall in *(b)*.

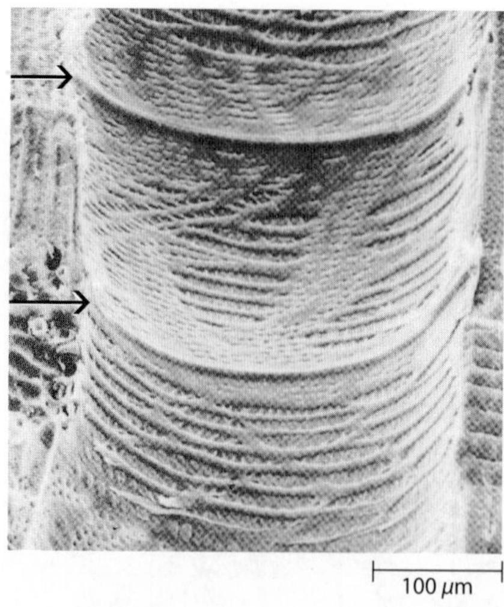

23–14 Vessel elements Scanning electron micrograph showing parts of three vessel elements of a vessel in secondary xylem of red oak *(Quercus rubra)*. Notice the rims (arrows) of simple perforation plates between the vessel elements, which are arranged end-on-end.

deposited in the form of rings or spirals (Figure 23–15). These annular (ringlike) or helical (spiral) thickenings make it possible for the tracheary elements to be stretched or extended after the cells have differentiated, although the cells are frequently destroyed during the overall elongation of the organ. In the primary xylem, the nature of the wall thickening is greatly influenced by the amount of elongation. If little elongation occurs, nonextensible rather than extensible elements appear. On the other hand, if much elongation takes place, many elements with annular and helical thickenings will develop. In the late-formed primary xylem (metaxylem; *meta-* meaning "after") and secondary xylem, the secondary cell walls of the tracheids and vessel elements cover the entire primary walls, except at the pit membranes and at the perforations of the vessel elements (Figure 23–12a through d). These cells, called pitted elements, are rigid and cannot be stretched.

Figure 23–16 illustrates some stages of differentiation of a vessel element with helical thickenings. Tracheary element differentiation is an example of **programmed cell death** (page 48). Programmed cell death, as the name suggests, is the outcome of genetically programmed processes that result in cell death. In the case of the tracheary element, it results in total elimination of the protoplast. The cell walls are retained, except at the perforation sites of the vessel elements. There the entire primary wall disappears, providing uninterrupted conduits for the transport of water and dissolved substances through the vessels (see Chapter 30).

In addition to tracheids and vessel elements, the xylem tissue contains parenchyma cells that store various substances. Xylem parenchyma cells commonly occur in vertical strands, but in the secondary xylem, they are also found in rays. Xylem can also contain fibers (Figure 23–12e, f), some of which are living at maturity and serve a dual function of storage and support. Sclereids are sometimes also present in xylem. The cell types of the xylem and their principal functions are listed in Table 23–1, page 548.

Phloem Is the Principal Food-Conducting Tissue in Vascular Plants

Although correctly characterized as the principal food-conducting tissue of vascular plants, the **phloem** plays a much greater role in the life of a plant. In addition to sugars, a great many other substances are transported in the phloem, including amino acids, lipids, micronutrients, hormones, the floral stimulus (florigen; see Chapter 28), and numerous proteins and RNAs, some of which serve as signaling molecules. Indeed, long-distance signaling in plants occurs predominantly through the phloem, which has been dubbed the "superinformation highway." The phloem is also the route for movement of a range of plant viruses.

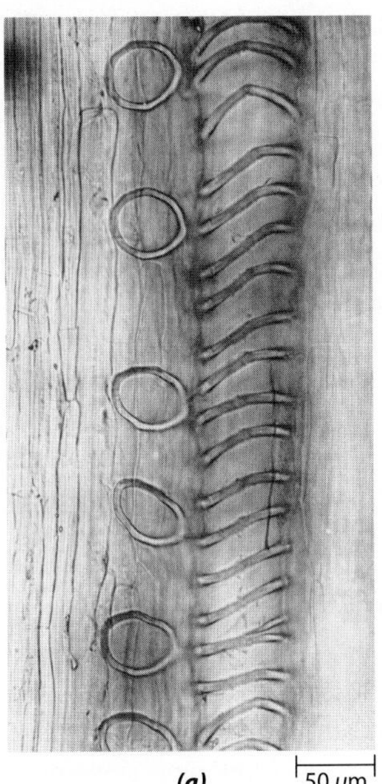

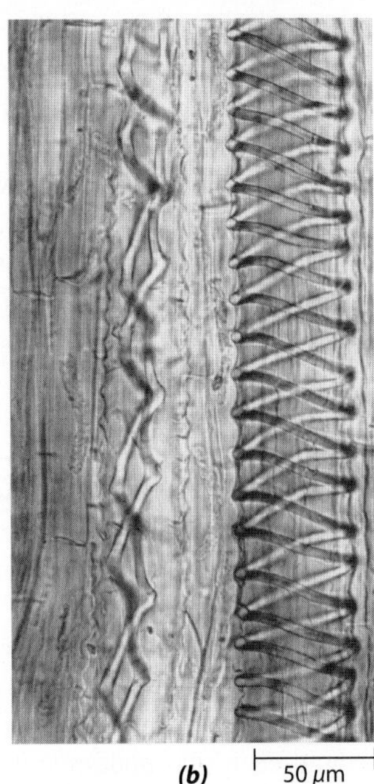

23–15 Tracheary elements Parts of tracheary elements from the first-formed primary xylem (protoxylem) of the castor bean *(Ricinus communis)*. ***(a)*** Annular (the ringlike shapes at left) and helical wall thickenings in partly extended elements. ***(b)*** Double helical thickenings in elements that have been extended. The element on the left has been greatly extended, and the coils of the helices have been pulled far apart.

The phloem may be primary or secondary in origin (Figure 23–11). As with primary xylem, the first-formed primary phloem (protophloem) is frequently stretched and destroyed during elongation of the organ. The metaphloem differentiates later and, in plants without secondary growth, constitutes the only conducting phloem in adult plant parts.

The principal conducting cells of the phloem are the **sieve elements.** The term "sieve" refers to the clusters of pores, known as **sieve areas,** through which the protoplasts of adjacent sieve elements are interconnected. Among seed plants, two types of sieve element are recognized: **sieve cells** (Figures 23–17 and 23–18) and **sieve-tube elements** (Figures 23–19 through 23–24). Sieve cells are the only type of food-conducting cell in gymnosperms, whereas only sieve-tube elements occur in angiosperms. The sieve elements of seedless vascular plants are variable in structure and hence are referred to simply as "sieve elements."

In sieve cells, the sieve area pores are narrow and the sieve areas are rather uniform in structure on all walls. Most of the sieve areas are concentrated on the overlapping ends of the long, slender sieve cells (Figure 23–17a). In sieve-tube elements, however, the sieve areas on some walls have larger pores than those on other walls of the same cell. The part of the wall bearing the sieve areas with larger pores is called a **sieve**

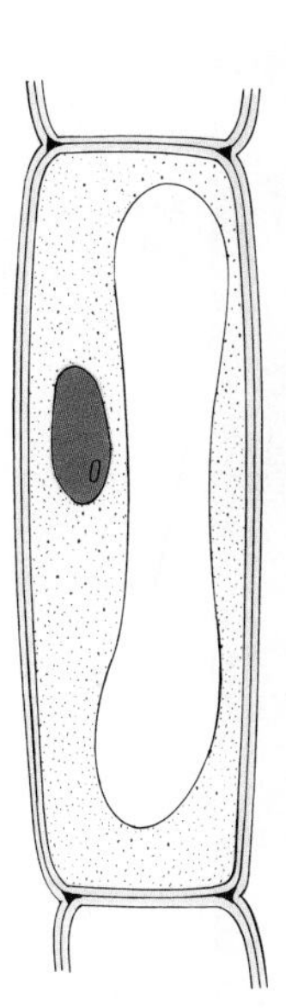

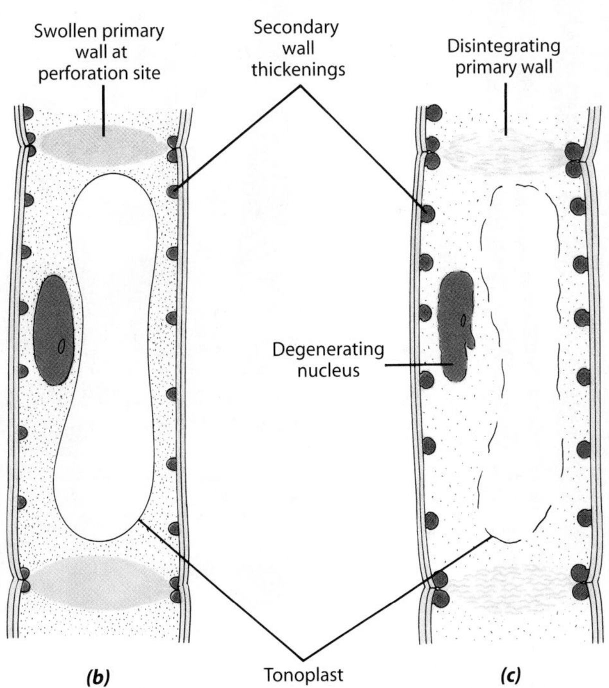

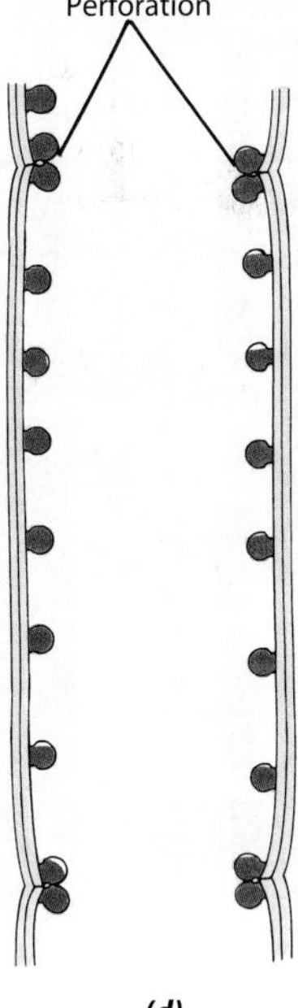

23–16 Differentiation of a vessel element ***(a)*** Young, highly vacuolated vessel element without a secondary wall. ***(b)*** The cell has expanded laterally, secondary wall deposition—in the form of a helix when seen in three dimensions—has begun, and the primary wall at the perforation site has increased in thickness. ***(c)*** Secondary wall deposition has been completed, and the cell is at the lysis stage of programmed cell death. The nucleus is degenerating, the tonoplast is ruptured, and the wall at the perforation site has partly disintegrated. ***(d)*** The cell is now mature; it lacks a protoplast and perforation is complete at both ends.

Table 23-1 Cell Types of the Xylem and Phloem

Cell Types	Principal Function
Xylem	
Tracheary elements	Conduction of water and minerals
Tracheids	
Vessel elements	
Fibers	Support; sometimes storage
Parenchyma	Storage
Phloem	
Sieve elements	Long-distance transport of food materials and signaling molecules
Sieve cells (with albuminous cells*)	
Sieve-tube elements (with companion cells*)	
Sclerenchyma	Support; sometimes storage
Fibers	
Sclereids	
Parenchyma	Storage

*Albuminous cells and companion cells are specialized parenchyma cells.

plate (Figures 23–19 and 23–20). Although sieve plates may occur on any wall, they generally are located on the end walls. Sieve-tube elements area arranged end-on-end in longitudinal series called **sieve tubes.** Thus, one of the principal distinctions between the two types of sieve element is the presence of sieve plates in sieve-tube elements and their absence in sieve cells.

The walls of sieve elements are described as primary. In cut sections of phloem tissue, the pores of the sieve areas and sieve plates of mature sieve elements are generally blocked by or lined with a wall substance called **callose,** which is a polysaccharide composed of spirally wound chains of glucose residues (page 58; Figures 23–17 and 23–19). Most, if not all, of the callose seen in the pores of conducting sieve elements is deposited there in response to injury during preparation of the tissue for microscopy. This callose, and any callose resulting from wounding, is referred to as "wound callose." Callose typically is also deposited at the sieve areas and sieve plates of senescing sieve elements; this is referred to as "definitive callose." In addition, callose in the form of platelets appears beneath the plasma membrane around each plasmodesma at the sites of developing sieve-plate pores (Figure 23–23).

Unlike tracheary elements, sieve elements have living protoplasts at maturity (Figures 23–18 and 23–20). As the sieve

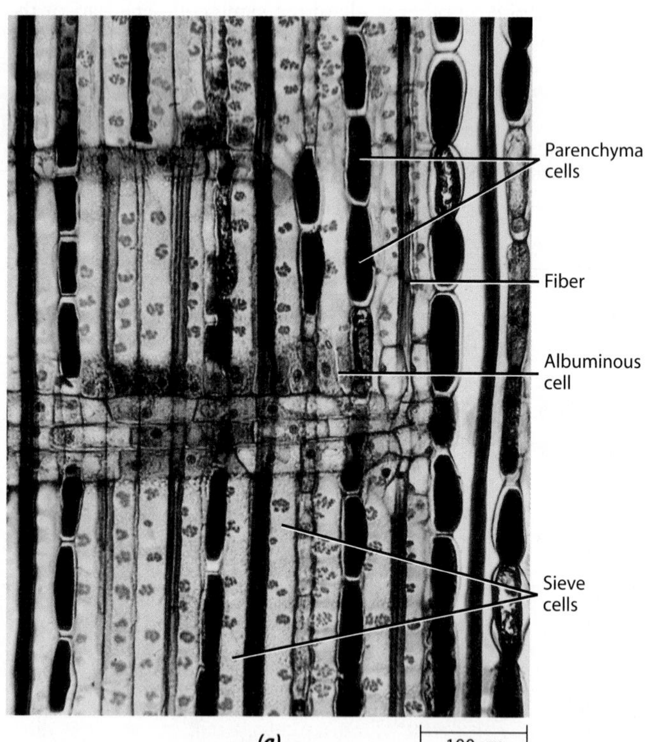

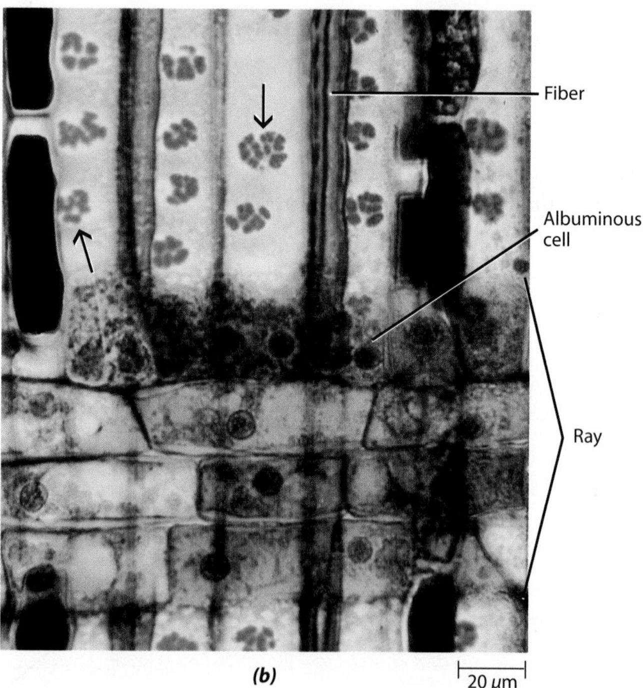

23–17 Sieve cells ***(a)*** Longitudinal (radial) view of secondary phloem of yew *(Taxus canadensis)*, a conifer, showing vertically oriented sieve cells, strands of parenchyma cells, and fibers. Parts of two horizontally oriented rays can be seen traversing the vertical cells. Specialized parenchyma cells known as albuminous cells, or Strasburger cells (see page 552), are characteristically associated with the sieve cells of gymnosperms. ***(b)*** Detail of a portion of the secondary phloem of yew, showing sieve areas (arrows), with callose (stained blue) on the walls of the sieve cells, and albuminous cells, which here constitute the top row of cells in the ray.

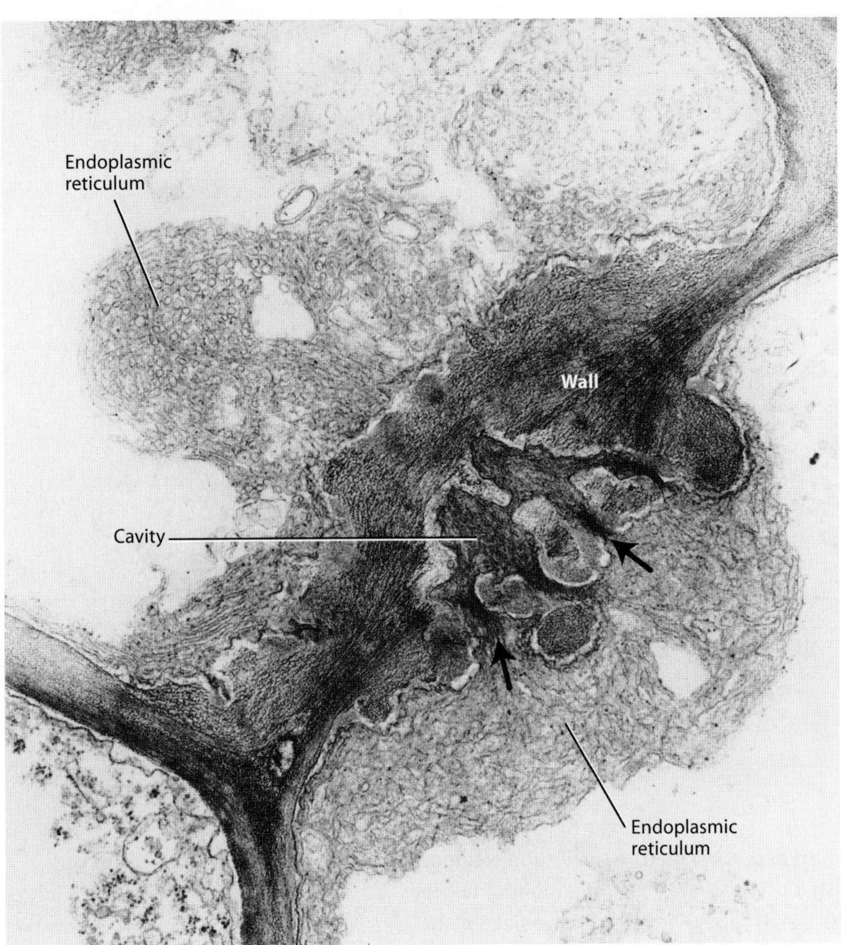

23–18 Sieve area in wall between two mature sieve cells Electron micrograph of a glancing section of a sieve area between two sieve cells in the hypocotyl of red pine *(Pinus resinosa).* Large amounts of tubular endoplasmic reticulum are seen on both sides of the wall. Endoplasmic reticulum can be seen traversing the pores (arrows) and entering a large cavity in the middle of the wall that contains much endoplasmic reticulum. Unlike the angiosperm sieve pores, which are continuous across the common wall, the sieve pores of gymnosperms extend only half way to the "median cavity."

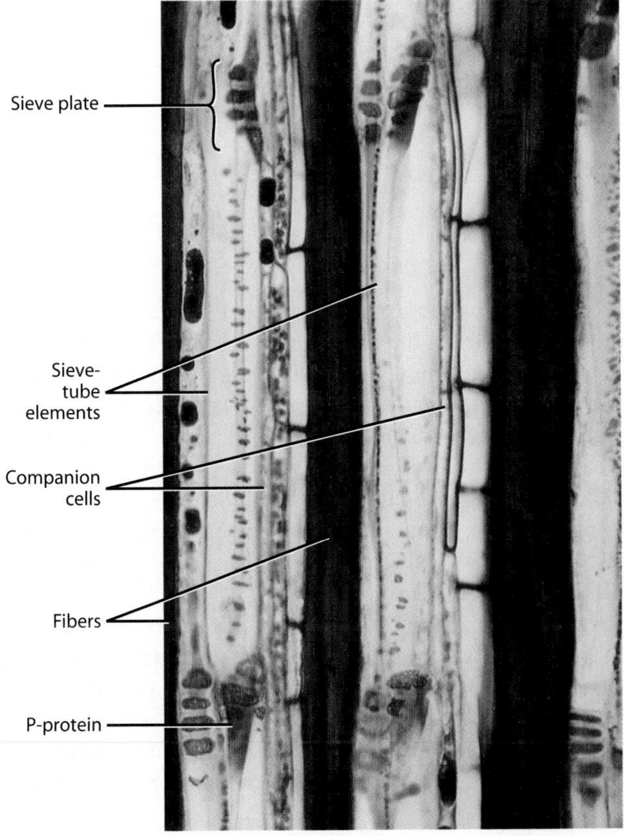

element differentiates, however, it undergoes profound changes, the major ones being breakdown of the nucleus and the tonoplast and formation of the sieve areas. Sieve element differentiation also results in loss of the ribosomes, the Golgi apparatus, and the cytoskeleton. At maturity, all of the remaining components of the sieve-element protoplast are distributed along the wall. The components are the plasma membrane, a network of smooth, endoplasmic reticulum—which is quite abundant in sieve cells,

23–19 Sieve-tube elements Longitudinal (radial) view of secondary phloem of basswood *(Tilia americana),* showing sieve-tube elements, with sieve plates, and conspicuous groups of thick-walled fibers. Specialized parenchyma cells known as companion cells (see page 552) are characteristically associated with sieve-tube elements. P-protein, a characteristic component of all angiosperm sieve-tube elements except those of some monocots, has accumulated at the sieve plates in these elements as slime plugs. Callose at the sieve plates and lateral sieve areas is stained blue.

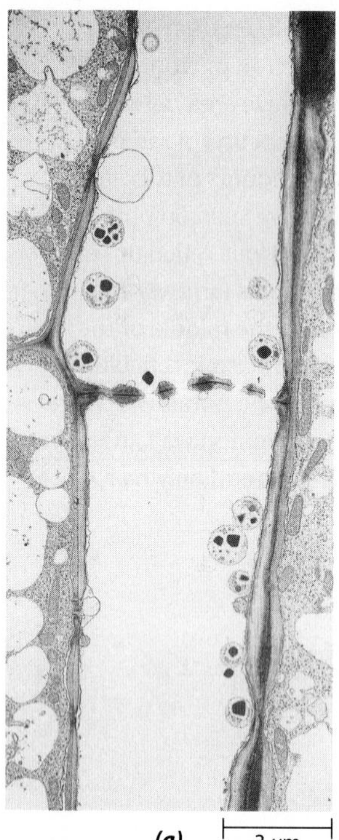

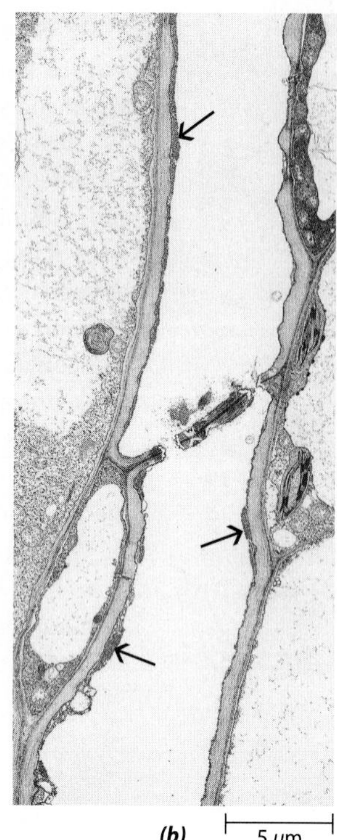

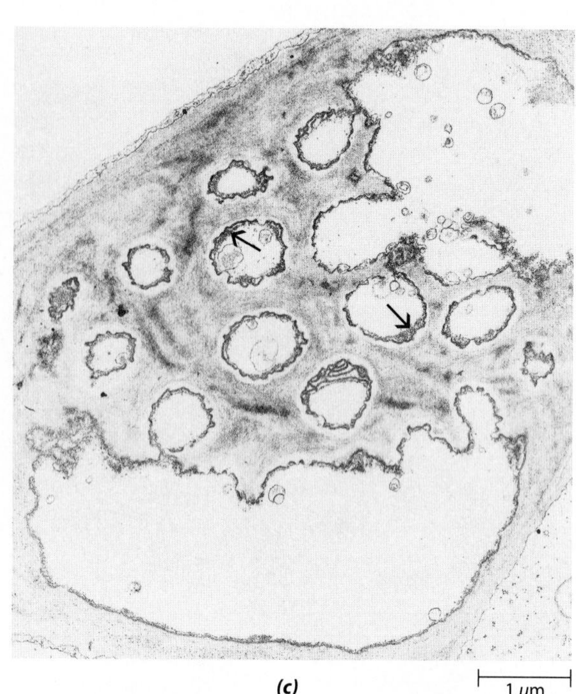

23–20 Mature sieve-tube elements Electron micrographs of parts of mature sieve-tube elements from phloem in the stems of maize *(Zea mays)* and squash *(Cucurbita maxima)*. ***(a)*** Longitudinal view of parts of two mature sieve-tube elements and a sieve plate of maize. The pores of the sieve plate are open. The numerous round organelles with dense inclusions composed of protein are plastids. This type of plastid is typical of monocot sieve-tube elements. The sieve-tube elements of maize, like those of some other monocots, lack P-protein. ***(b)*** Similar, longitudinal view of parts of two sieve-tube elements of squash. The sieve-tube elements of squash, like those of most angiosperms, contain P-protein. In these sieve-tube elements, the P-protein is distributed along the wall (arrows) and the sieve-plate pores are open. At the lower left and upper right (partial view) are companion cells. ***(c)*** Face view of part of a sieve plate between two mature sieve-tube elements of squash. As in *(b)*, the sieve-plate pores are open, lined in part with P-protein (arrows).

especially at the sieve areas (Figure 23–18)—and some plastids and mitochondria. Thus, unlike the tracheary element protoplast, which undergoes programmed cell death that results in a total breakdown during differentiation, the sieve element protoplast undergoes a selective breakdown. As we will see, for the sieve element to perform its role as a food-conducting conduit, it must remain alive (see Chapter 30).

The protoplasts of the sieve-tube elements of angiosperms, with the exception of some monocots, are characterized by the presence of a proteinaceous substance once called "slime" and now known as **P-protein** (the "P" stands for phloem). P-protein has its origin in the young sieve-tube element in the form of discrete bodies called P-protein bodies (Figure 23–22a, c). During the late stages of differentiation, the P-protein bodies in most species elongate and disperse, and the P-protein, as well as other surviving components of the mature cell, becomes distributed along the walls. In cut sections of phloem tissues, the P-protein usually accumulates at the sieve plates as "slime plugs" (Figures 23–19 and 23–22b). Slime plugs, which are not found in undisturbed cells, result from the surging of the contents of sieve tubes that are severed as the tissue is cut. In undisturbed, mature sieve-tube elements, the sieve-plate pores are lined with P-protein but are not plugged by it (Figure 23–20b, c). Some botanists believe that, together with wound callose, P-protein serves to seal the sieve-plate pores at the time of wounding, thus preventing the loss of contents from the sieve tubes.

Thus, with open sieve plate pores, the mature sieve tube provides an obstruction-free pathway for the movement of water and dissolved substances in angiosperm phloem (see Chapter 30). A large number of viruses also find the sieve tube a suitable conduit for rapid movement throughout the plant (see Chapter 13). The role of the large amounts of tubular endoplasmic reticulum

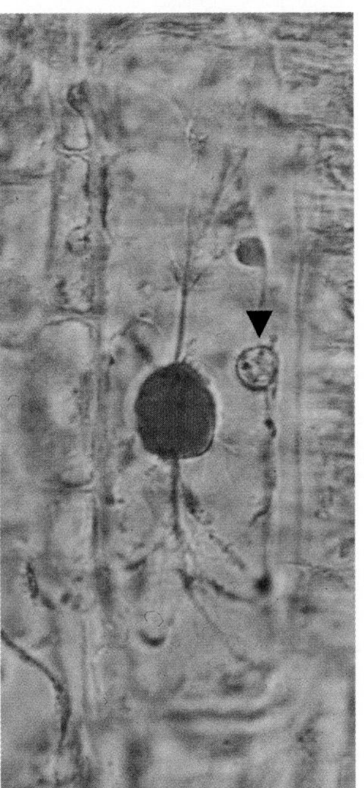

23–21 Forisome This longitudinal section shows a single nondispersive P-protein body, or forisome, in an immature sieve-tube element of black locust *(Robinia pseudoacacia)*. This section has been treated with a dye that stains protein red. The arrowhead points to the nucleus.

that apparently occlude the sieve area pores of gymnosperm sieve cells remains to be explained (Figure 23–18).

The sieve-tube elements in some legumes produce a single, relatively large P-protein body that does not disperse during the later stages of maturation (Figure 23–21). Originally called "nondispersive P-protein bodies," these bodies have been shown to undergo rapid and reversible calcium-controlled changes from the condensed "resting stage" to a dispersed stage, in which they occlude the sieve-plate pores. Now referred to as **forisomes** ("gate-keepers"), their behavior lends support to the proposal that dispersive P-protein bodes function to seal the sieve-plate pores in disturbed sieve tubes. Figure 23–23 illustrates some stages of

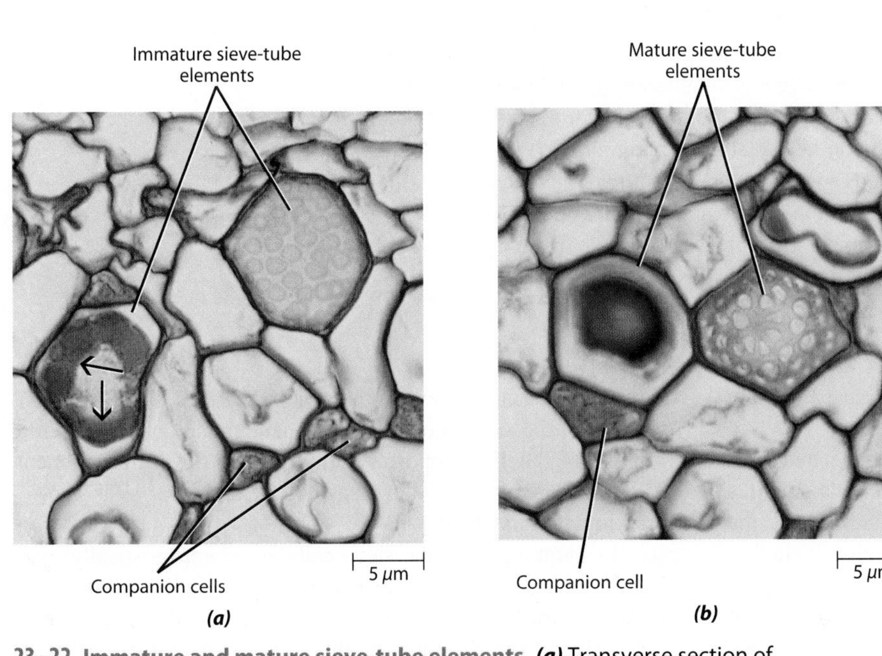

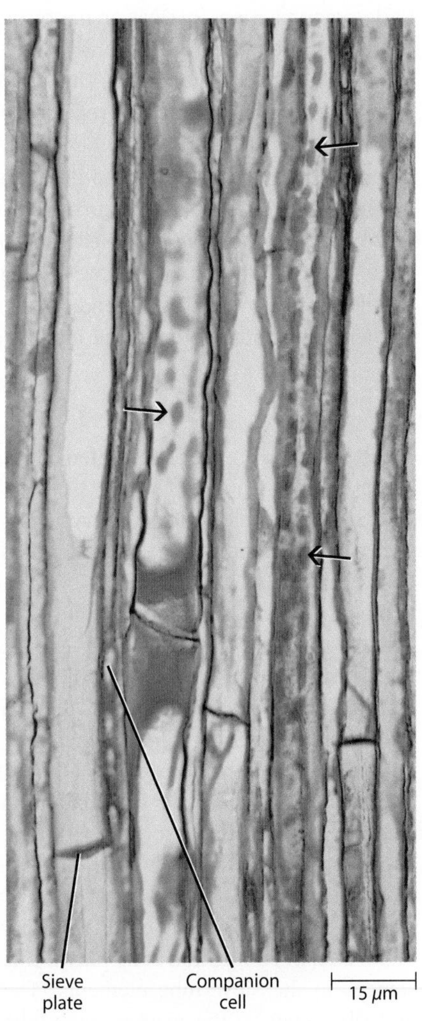

23–22 Immature and mature sieve-tube elements ***(a)*** Transverse section of phloem from the stem of squash *(Cucurbita maxima)* showing two immature sieve-tube elements. P-protein bodies (arrows) can be seen in the sieve-tube element on the left; an immature sieve plate with callose platelets (stained blue) is visible in the one to the right, above. The sieve plates of squash are simple sieve plates (one sieve area per plate). The small, dense cells are companion cells. ***(b)*** Transverse section showing two mature sieve-tube elements. A slime plug can be seen in the sieve-tube element on the left; a mature sieve plate can be seen in the one on the right. The small, dense cells are companion cells. ***(c)*** Longitudinal section showing mature and immature sieve-tube elements (arrows point to P-protein bodies in the immature cells).

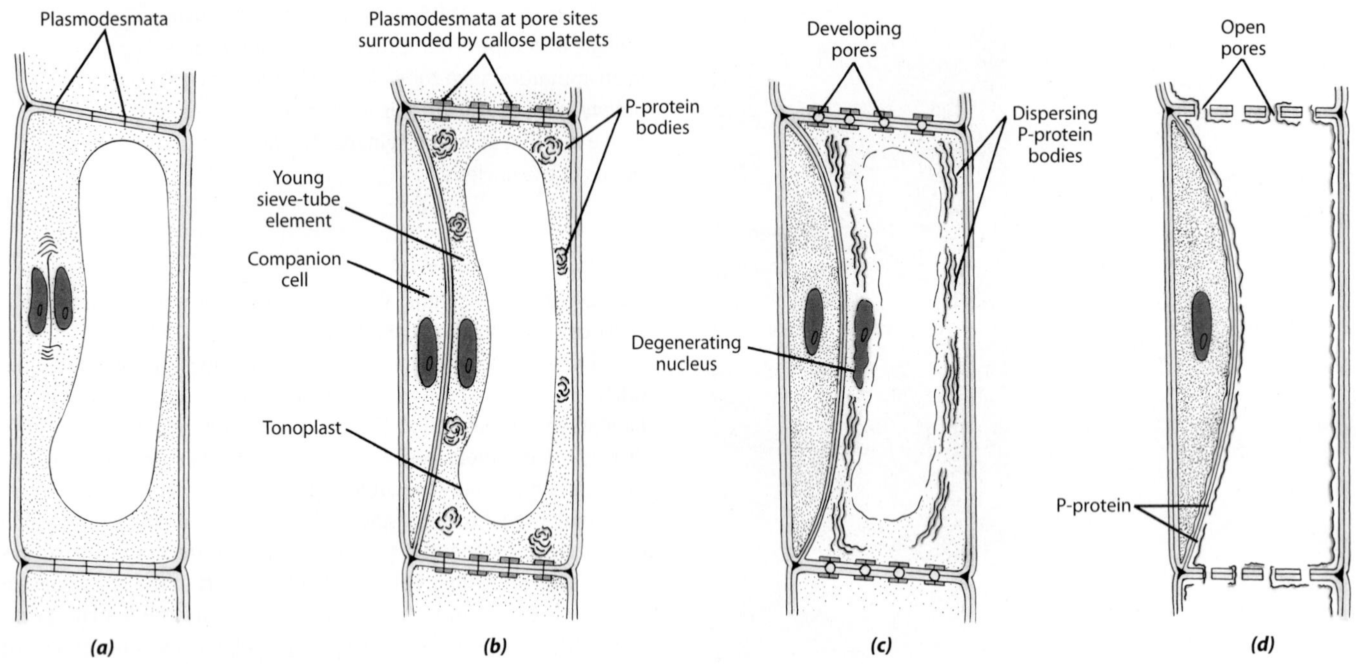

23–23 Differentiation of a sieve-tube element ***(a)*** The mother cell of the sieve-tube element undergoing division. ***(b)*** Division has resulted in formation of a young sieve-tube element and a companion cell. After division, one or more P-protein bodies arise in the cytoplasm, which is separated from the vacuole by a tonoplast. The wall of the young sieve-tube element has thickened, and the sites of the future sieve-plate pores are represented by plasmodesmata. Each plasmodesma is now surrounded by a platelet of callose on either side of the wall. ***(c)*** The nucleus is degenerating, the tonoplast is breaking down, and the P-protein bodies are dispersing in the cytoplasm lining the wall of the sieve-tube element. At the same time, the plasmodesmata of the developing sieve plates are beginning to widen into pores. ***(d)*** At maturity, the sieve-tube element lacks a nucleus and a vacuole. All of the remaining protoplasmic components, including the P-protein, line the walls, and the sieve-plate pores are open. The callose platelets were removed as the pores widened. Not shown here but also present in the mature sieve-tube element are smooth endoplasmic reticulum, mitochondria, and plastids.

differentiation of a sieve-tube element with dispersive P-protein bodies.

Sieve-tube elements are characteristically associated with specialized parenchyma cells called **companion cells** (Figures 23–19 and 23–22 through 23–24), which contain all of the components commonly found in living plant cells, including a nucleus. Sieve-tube elements and their associated companion cells are closely related developmentally (they are derived from the same mother cell), and they have numerous cytoplasmic connections with one another. The connections typically consist of a small pore on the side of the sieve-tube element and much-branched plasmodesmata on the companion cell side (Figure 23–24). Because of their numerous plasmodesmatal connections with the sieve-tube elements, as well as their general ultrastructural resemblance to secretory cells (high ribosome population and numerous mitochondria), it has long been believed that companion cells play a role in the delivery of substances to the sieve-tube elements. The absence of a nucleus and ribosomes in the mature sieve-tube element suggests that the substances delivered by the companion cells include the informational molecules, proteins, and ATP necessary for maintenance of the sieve-tube element. The companion cell therefore represents the life-support system for the sieve-tube element. The mechanism of phloem transport in angiosperms is considered in detail in Chapter 30.

In gymnosperms, the sieve cells are characteristically associated with specialized parenchyma cells called **albuminous cells,** or Strasburger cells (Figure 23–17b). Although generally not derived from the same mother cell as their associated sieve cells, albuminous cells are believed to perform the same roles as companion cells. Like the companion cell, the albuminous cell contains a nucleus, in addition to other cytoplasmic components characteristic of living cells. Both albuminous cells and companion cells die when their associated sieve elements die—one more indication of the interdependence between sieve elements and their albuminous cells or companion cells.

Other parenchyma cells occur in the primary and secondary phloem (Figures 23–17 and 23–19). They are largely concerned with the storage of various substances. Fibers (Figures 23–17 and 23–19) and sclereids may also be present. The cell types of the

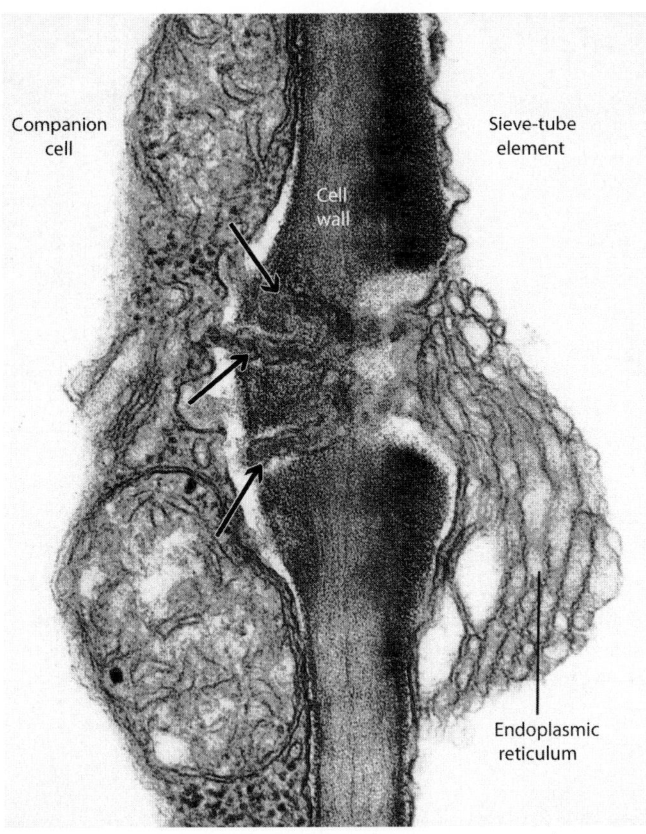

23–24 Sieve-tube element–companion cell connections Electron micrograph showing pore-plasmodesmata connections between a companion cell and a sieve-tube element in a leaf vein of barley *(Hordeum vulgare)*. An aggregate of endoplasmic reticulum is associated with the pore on the sieve-tube element side. The arrows point to branched plasmodesmata on the companion cell side.

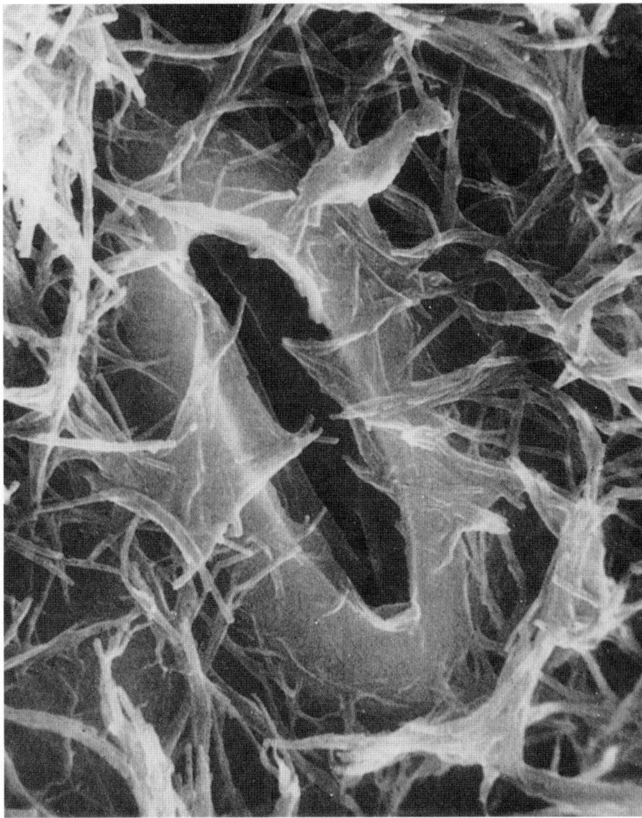

23–25 Stoma, surface view Scanning electron micrograph of the lower epidermis of a *Eucalyptus globulus* leaf, showing a single stoma with its two guard cells. Numerous filaments of epicuticular wax deposits can be seen.

phloem and their principal functions are listed, along with those of the xylem, in Table 23–1.

Dermal Tissues

The Epidermis Is the Outermost Cell Layer of the Primary Plant Body

The **epidermis** constitutes the dermal tissue system of leaves, floral parts, fruits, and seeds—and also of stems and roots until they undergo considerable secondary growth. Epidermal cells are quite diverse both functionally and structurally. In addition to the relatively unspecialized cells that form the bulk of the epidermis, the epidermis may contain **guard cells** (Figures 23–25 and 23–26), many types of appendages, or **trichomes** (Figures 23–27 and 23–28), and other kinds of specialized cells.

Most of the epidermal cells are compactly arranged, providing considerable mechanical protection to the plant part. The walls of the epidermal cells of the aerial parts are covered with a cuticle, which minimizes water loss. The cuticle consists mainly of cutin and wax (pages 23 and 24). In many plants the wax is exuded over the surface of the cuticle, either in smooth sheets or as rods or filaments extending upward from the surface, the so-called epicuticular wax (Figure 23–25; see also Figure 2–10). It is this wax that is responsible for the whitish or bluish "bloom" on the surface of some leaves and fruits.

In stems and coleoptiles, the epidermis, which is under tension, has been regarded as the tissue that controls elongation of the entire organ. The epidermis is the site of the light perception involved in circadian leaf movement and photoperiodic induction (see Chapter 28).

Interspersed among the flat, tightly packed epidermal cells, which typically lack chloroplasts, are the chloroplast-containing guard cells (Figures 23–25 and 23–26). The guard cells regulate the small pores, or **stomata** (singular: stoma), in the aerial parts of the plant and hence control the movement of gases, including water vapor, into and out of those parts. (The term "stoma" commonly is applied to the pore and the two guard cells. The mechanism of stomatal opening and closing is discussed in Chapters 27 and 30.) Although stomata occur on all aerial parts, they are most abundant on leaves. Guard cells are often associated with

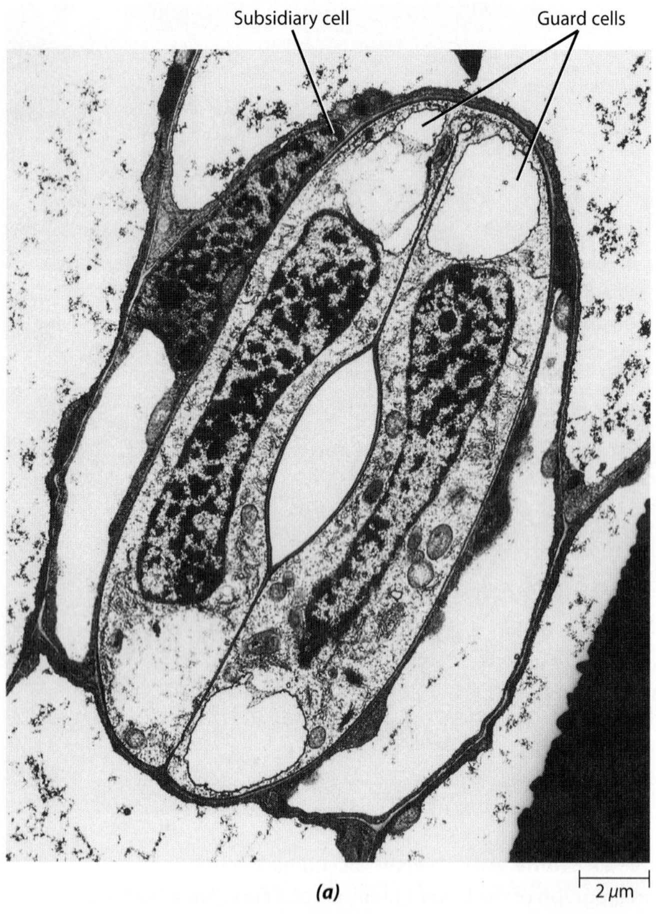

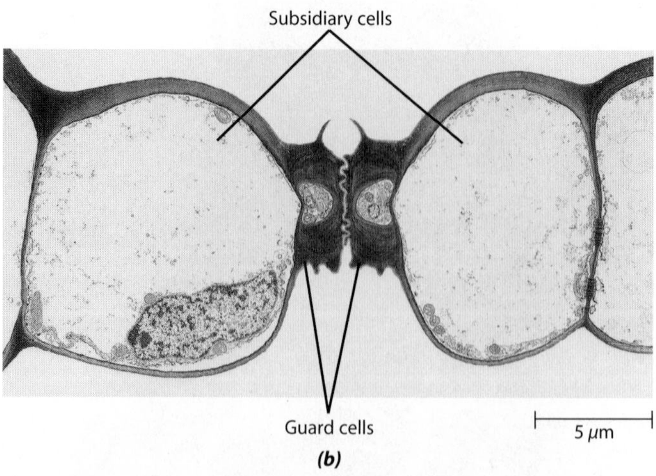

23–26 Stomata of a maize leaf ***(a)*** Section taken parallel to the surface of a leaf of maize *(Zea mays)* showing the open pore of a stoma with its immature guard cells, whose walls have not yet thickened, and two associated subsidiary cells. ***(b)*** Transverse section through a closed stoma. Each thick-walled guard cell is attached to a subsidiary cell. The interior of the leaf is below.

epidermal cells, called **subsidiary cells,** that differ in shape from other epidermal cells (Figure 23–26).

Trichomes have a variety of functions. Root hairs are trichomes that facilitate the absorption of water and minerals from the soil. Studies of plants from arid locations indicate that an increase in leaf hairiness (pubescence), which is due to leaf trichomes, results in increased reflection of solar radiation, lower leaf temperatures, and lower rates of water loss. Many "air plants," such as epiphytic bromeliads, utilize leaf trichomes for the absorption of water and minerals. In contrast, in the saltbush *(Atriplex),* which grows in soil containing high levels of salt, the trichomes secrete salty solutions from the leaf tissue, preventing an accumulation of these toxic substances in the plant. Trichomes may also provide a defense against insects. In many species, increased hairiness correlates with increased resistance to attack by insects. The hooked hairs of some plant

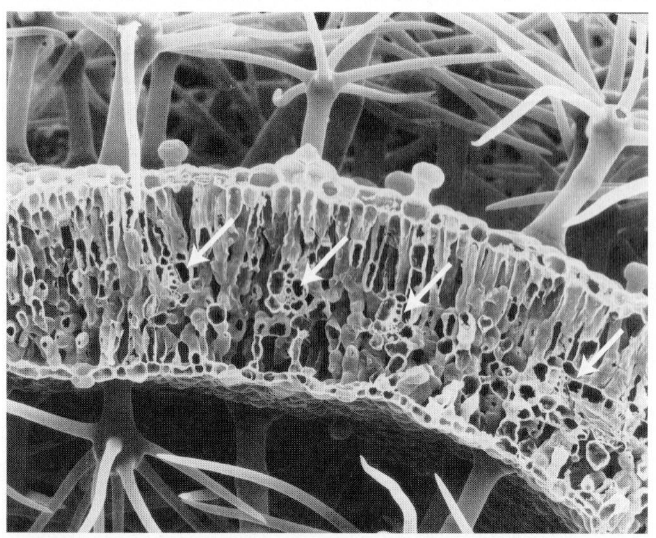

23–27 Branched and unbranched trichomes Transverse section of a common mullein *(Verbascum thapsus)* leaf, observed with a scanning electron microscope. The leaves and stems of the common mullein appear densely woolly because of the presence of great numbers of highly branched trichomes, such as those seen here on both the upper and lower surfaces of the leaf. Short, unbranched glandular trichomes are also present. The ground tissue of the leaf is represented by the mesophyll, which is permeated by numerous vascular bundles, or veins (arrows).

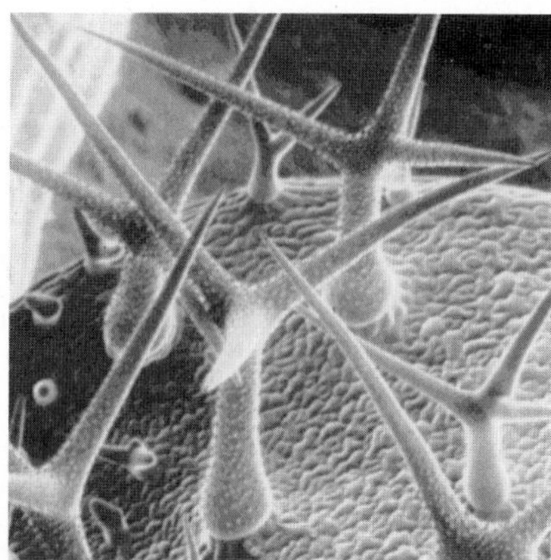

23–28 Branched trichomes of *Arabidopsis* Scanning electron micrograph of branched trichomes on the leaf of *Arabidopsis thaliana*. Genetic analysis has identified genes that are essential for the initiation of trichome development in *Arabidopsis*.

Dead epidermis
Cork
Cork cambium
Phelloderm
Cortex

23–29 Periderm Transverse section of the periderm, which consists of cork cells, cork cambium, and phelloderm. This periderm from the stem of apple *(Malus domestica)* consists largely of cork cells, which are laid down to the outside (above) in radial rows by the cells of the cork cambium. One or two layers of phelloderm cells lie below the cork cambium.

species impale insects and their larvae, and the trichomes of carnivorous plants play an important role in trapping insect prey (see Chapter 29). Glandular (secretory) hairs may provide a chemical defense.

Because of their simplicity and visibility, the leaf trichomes of *Arabidopsis* have provided an ideal genetic model system for the study of cell fate and morphogenesis in plants (Figure 23–28). Several genes involved in trichome development on *Arabidopsis* leaves have been identified, including *GL1 (GLABROUS1)* and *TTG (TRANSPARENT TESTA GLABRA)*. An early event in commitment of a protodermal cell to trichome development involves an increase in the expression of *GL1*. By contrast, evidence indicates that *TTG* may play a role in inhibiting protodermal cells bordering the trichomes from becoming trichomes. The *TTG* gene has also been implicated in cell fate in the root epidermis (see Chapter 24); *ttg* mutants lack trichomes on aerial plant parts but have extra root hairs. (By custom, the mutant gene is indicated by lowercase letters.)

Periderm Is Secondary Protective Tissue

A **periderm** commonly replaces the epidermis in stems and roots that undergo secondary growth. Although the cells of the periderm are generally arranged compactly, portions—the lenticels—are loosely arranged and provide for aeration of the internal tissues of roots and stems. The periderm consists largely of protective **cork,** or *phellem,* which is nonliving and has walls that are heavily suberized at maturity. The periderm also includes **cork cambium,** or *phellogen,* and **phelloderm,** a living parenchyma-like tissue (Figure 23–29). The cork cambium forms cork tissue on its outer surface and phelloderm on its inner surface. The origin of the cork cambium is variable, depending on the species and plant part. The periderm is considered in detail in Chapter 26.

SUMMARY

Primary Growth Results from the Activity of Apical Meristems

After embryogenesis, most plant development occurs through the activity of meristems, which consist of initials and their immediate derivatives. The apical meristems are involved primarily with extension of roots and stems. Also called primary growth, this growth results in the formation of primary tissues, which constitute the primary body of the plant. Some tissues—parenchyma, collenchyma, and sclerenchyma—are composed of a single type of cell and are called simple tissues. Others—xylem, phloem, epidermis, and periderm—are composed of two or more cell types and are called complex tissues.

Development Involves Three Overlapping Processes: Growth, Morphogenesis, and Differentiation

Growth, an irreversible increase in size, is accomplished primarily by cell enlargement. Morphogenesis is the acquisition of a particular shape or form, and differentiation is the process by which cells with identical genetic constitution become different from one another through differential gene expression. Although cellular differentiation depends on the control of gene expression, the fate of a plant cell is determined by its final position in the developing organ.

Vascular Plants Are Composed of Three Tissue Systems

The tissue systems—dermal, vascular, and ground—which are present in root, stem, and leaf, reveal the basic similarity of the plant organs and the continuity of the plant body. Initiated

SUMMARY TABLE Tissue Systems, Tissues, and Cell Types

TISSUE SYSTEM	TISSUE	CELL TYPE	COMPOSITION OR CHARACTERISTICS	LOCATION	FUNCTIONS
Dermal	Epidermis		Unspecialized cells; guard cells; trichomes; other kinds of specialized cells	Outermost layer of cells of the primary plant body	Mechanical protection; minimizes water loss (cuticle); aeration of internal tissue via stomata
	Periderm		Cork cells; cork cambium cells; parenchyma-like cells of the phelloderm; sclereids	Initial periderm beneath epidermis; subsequently formed periderms occur deeper in bark	Replaces epidermis as protective tissue in roots and stems; aeration of internal tissue via lenticels
Ground	Parenchyma tissue	Parenchyma	Shape: commonly polyhedral (many-sided); variable Cell wall: primary, or primary and secondary; may contain lignin, suberin, or cutin Living when functional	Throughout the plant body, as parenchyma tissue in cortex; as pith and pith rays; in xylem and phloem	Metabolic processes such as respiration, secretion, and photosynthesis; storage and conduction; wound healing and regeneration
	Collenchyma tissue	Collenchyma	Shape: elongated Cell wall: unevenly thickened primary only; lacks lignin Living when functional	On the periphery (beneath the epidermis) in young elongating stems; often as a cylinder of tissue or only in patches; in ribs along veins in some leaves	Support in primary plant body
	Sclerenchyma tissue	Fiber	Shape: generally very long Cell wall: primary and thick secondary; often lignified Often (not always) functional cells are dead	Sometimes in cortex of stems, most often associated with xylem and phloem; in leaves of monocots	Support; storage
		Sclereid	Shape: variable; generally shorter than fibers Cell wall: primary and thick secondary; typically lignified May be living or dead when functional	Throughout the plant	Mechanical; protective

during development of the embryo, the tissue systems are derived from the primary meristems: protoderm, procambium, and ground meristem, respectively. A summary of plant tissues and their cell types is given in the Summary Table.

QUESTIONS

1. Distinguish between the following: collenchyma cell and sclerenchyma cell; tracheid and vessel element; perforation plate and pit; sieve cell and sieve-tube element; callose and P-protein.
2. What is meant by "growth"?
3. Where might one find transfer cells in a plant? What role do they play?
4. How do simple tissues differ from complex tissues? Give examples of each.
5. How do sclereids differ from fibers?
6. What is the developmental and/or functional relationship between a sieve-tube element and its companion cell(s)?
7. Explain the following: Tracheary elements undergo programmed cell death, but sieve elements undergo a selective breakdown.
8. What is the probable role of P-protein in mature sieve-tube elements?

TISSUE SYSTEM	TISSUE	CELL TYPE	COMPOSITION OR CHARACTERISTICS	LOCATION	FUNCTIONS
Vascular	Xylem	Tracheid	Shape: elongated and tapering Cell wall: primary and secondary; lignified; contains pits but not perforations Dead when functional	Xylem	Chief water-conducting element in gymnosperms and seedless vascular plants; also found in some angiosperms
		Vessel element	Shape: elongated, generally not as long as tracheids; vessel elements end-on-end constitute a vessel Cell wall: primary and secondary; lignified; contains pits and perforations Dead when functional	Xylem	Chief water-conducting element in angiosperms
	Phloem	Sieve cell	Shape: elongated and tapering Cell wall: primary in most species; with sieve areas; callose often associated with wall and pores Living at maturity; either lacks or contains remnants of a nucleus at maturity; lacks distinction between vacuole and cytoplasm; contains large amounts of tubular endoplasmic reticulum; lacks P-protein	Phloem	Food-conducting element in gymnosperms
		Albuminous cell	Shape: generally elongated Cell wall: primary Living at maturity; associated with sieve cell, but generally not derived from same mother cell as sieve cell; has numerous plasmodesmatal connections with sieve cell	Phloem	Believed to play a role in the delivery of substances to the sieve cell, including informational molecules and ATP
		Sieve-tube element	Shape: elongated Cell wall: primary, with sieve areas; sieve areas on end walls (sieve plates) with larger pores than those on side walls; callose often associated with walls and pores Living at maturity; either lacks a nucleus at maturity or contains only remnants of a nucleus; in angiosperms, except for some monocots, contains a proteinaceous substance known as P-protein; several sieve-tube elements in a vertical series constitute a sieve tube	Phloem	Food-conducting element in angiosperms
		Companion cell	Shape: variable, generally elongated Cell wall: primary Living at maturity; closely associated with sieve-tube element; derived from same mother cell as sieve-tube element; has numerous pore-plasmodesmata connections with its associated sieve-tube element	Phloem	Believed to play a role in the delivery of substances to the sieve-tube element, including informational molecules and ATP

CHAPTER 24

The Root: Structure and Development

◀ **Strangler fig** Beginning life as a seed deposited high in a tree by a bird or monkey, the strangler fig *(Ficus)* grows first as an epiphyte and then extends its long roots to the ground as it surrounds its host. Here, the host is Ta Prohm, a temple at Angkor Wat in Cambodia. The growth habit of the strangler fig is an adaptation for thriving in dense forests, where there is intense competition for sunlight and nutrients.

CHAPTER OUTLINE

The first structure to emerge from the germinating seed is the root, enabling the developing seedling to become anchored in the soil and to absorb water. This reflects the two primary functions of roots: **anchorage** and **absorption** (Figure 24–1). Two other functions associated with roots are **storage** and **conduction.** Most roots are important storage organs, and some, such as those of the carrot, sugar beet, and sweet potato, are specifically adapted for the storage of food. Foods manufactured above ground, in photosynthesizing portions of the plant, move down through the phloem to the storage tissues of the root. This food may eventually be used by the root itself, but more often the stored food is digested and the products transported back through the phloem to the aboveground parts. In biennial plants, such as the sugar beet and carrot, which complete their life cycle over a two-year period, large food reserves accumulate in the storage regions of the root during the first year. These food reserves are then used during the second year to produce flowers, fruits, and seeds. Water and minerals, or inorganic ions, absorbed by the roots move through the xylem to the aerial parts of the plant.

Hormones (particularly cytokinins and gibberellins) synthesized in meristematic regions of the roots are transported upward in the xylem to the aerial parts, where they stimulate growth and development (see Chapter 27). Roots also synthesize a wide variety of secondary metabolites, such as nicotine, which in tobacco is transported to the leaves (see Chapter 2). In addition, roots function in **clonal regeneration** (the roots of certain eudicots produce buds that can develop into new shoots), redistribution of water within the soil (see Chapter 30), and secretion of a vast array of substances (root exudates) into

CHECKPOINTS

After reading this chapter, you should be able to answer the following:

1. Name the two principal types of root systems, and describe how they differ in both origin and structure.
2. What changes occur to the rootcap during elongation of the root, and what are some functions of the rootcap?
3. What tissues are found in a root at the end of primary growth, and how are they arranged?
4. Describe the effect of secondary growth on the primary body of the root.
5. Where do lateral roots originate, and why are they said to be endogenous?

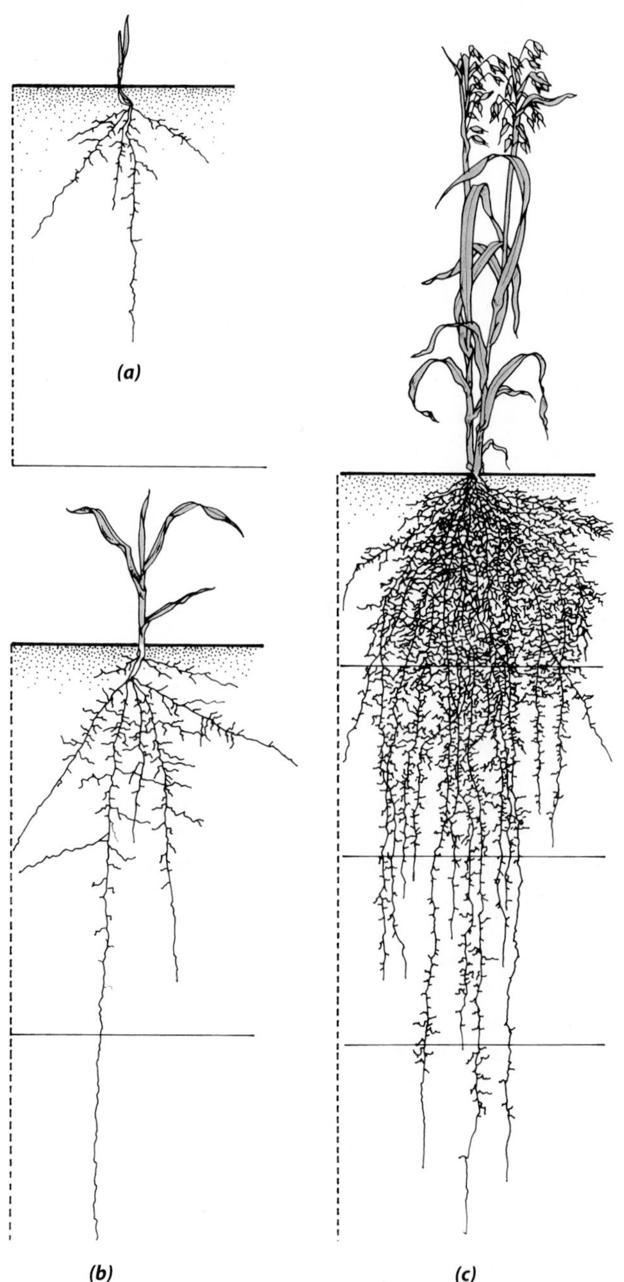

24–1 Development of root and shoot in a monocot Diagrams of oat *(Avena sativa)* plants, showing relative sizes of the root and shoot systems at ***(a)*** 31 days, ***(b)*** 45 days, and ***(c)*** 80 days after planting. The oat plant, a monocot, has a fibrous root system. The roots are involved primarily with anchorage and absorption. Each of the vertical units depicted here represents 1 foot (approximately 30 centimeters).

the **rhizosphere,** the volume of soil around living plant roots that is influenced by root activity.

Root Systems

The first root of the plant originates in the embryo and is usually called the **primary root.** In all seed plants except the monocots, the primary root, termed the **taproot,** grows directly downward, giving rise to branch roots, or **lateral roots.** The older lateral roots are found nearer the base of the root (where the root and stem meet), and the younger ones nearer the root tip. This type of root system—that is, one formed from a strongly developed primary root and its branches—is called a **taproot system** (Figure 24–2a).

In monocots, the primary root is usually short-lived, and, instead, the main root system develops from roots that arise from the stem. These stem-borne roots, commonly called **adventitious roots,** and their lateral roots give rise to a **fibrous root system,** in which no one root is more prominent than the others (Figures 24–1 and 24–2b).

The spatial configuration, or architecture, of a root system can show considerable variation even within different parts of a single root system. Growing roots are extremely sensitive to a wide range of environmental parameters, including gravity, light, gradients of moisture, temperature, and nutrients in the soil. A striking example of the developmental plasticity, or adaptability, of the root system of many species is its response to the uneven distribution of nitrogen and inorganic phosphate by the preferential and rapid development of lateral roots into nutrient-rich zones (see Chapter 29).

The extent of a root system—that is, the depth to which it penetrates the soil and the distance it spreads laterally—is dependent on several factors, including the environmental parameters just mentioned. Taproot systems generally penetrate deeper into the soil than fibrous root systems. The shallowness of fibrous root systems and the tenacity with which they cling to soil particles make such plants especially well suited as ground cover for the prevention of soil erosion. Most trees have surprisingly shallow root systems, with 90 percent or more of all roots located in the upper 60 centimeters (about 2 feet) of soil. The bulk of most **fine roots,** or so-called **feeder roots,** the roots actively engaged in the uptake of water and minerals, occurs in the upper 15 centimeters of soil, the part of the soil normally richest in nutrients. Many fine roots are heavily infected with mycorrhizal fungi (see Chapter 14). Some trees, such as spruces, beeches, and poplars, rarely produce deep taproots, whereas others, such as oaks and many pines, commonly produce relatively deep taproots, making these trees difficult to transplant.

The record for depth of penetration by roots probably belongs to *Boscia albitrunca* (family Capparidaceae), at 68 meters (over 220 feet), in the central Kalahari Desert. Roots of the desert shrub mesquite *(Prosopis juliflora)* were found growing at a depth of 53 meters (nearly 175 feet) at an open-pit mine near Tucson, Arizona. During the digging of the Suez Canal in Egypt, roots of *Tamarix* and *Acacia* trees were found at a depth of 30 meters. The roots of alfalfa *(Medicago sativa)* may extend to depths of 6 meters or more. The lateral spread of tree roots is usually greater—frequently four to seven times greater—than the spread of the crown of the tree. The root systems of maize plants *(Zea mays)* often reach a depth of about 1.5 meters, with a lateral spread of about a meter on all sides of the plant.

One of the most detailed studies on the extent of the root and shoot systems of any one plant was conducted on a four-month-old rye plant *(Secale cereale).* The total surface area of the root system, including root hairs, was 639 square meters,

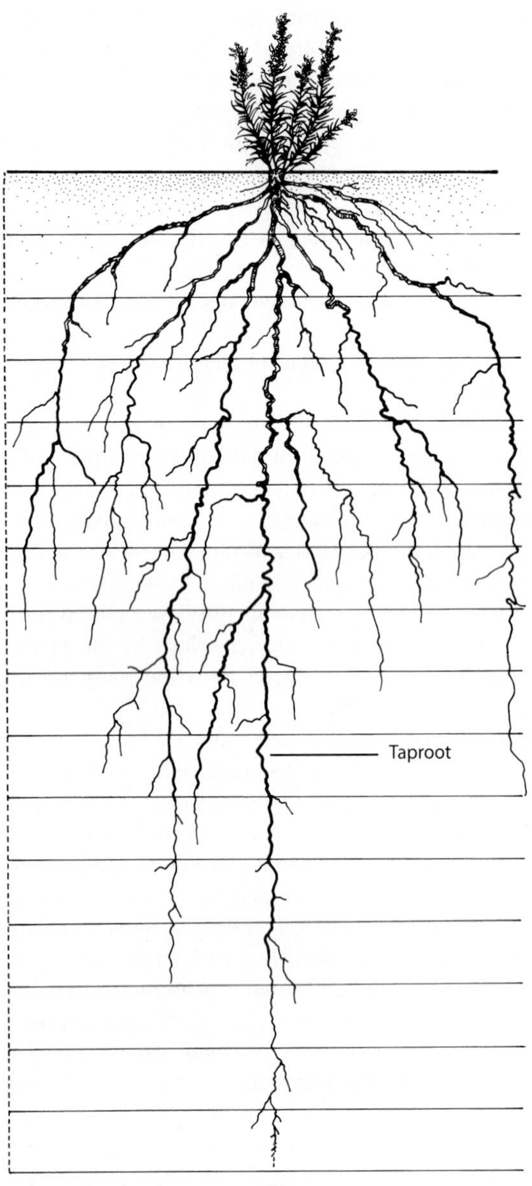

(a)

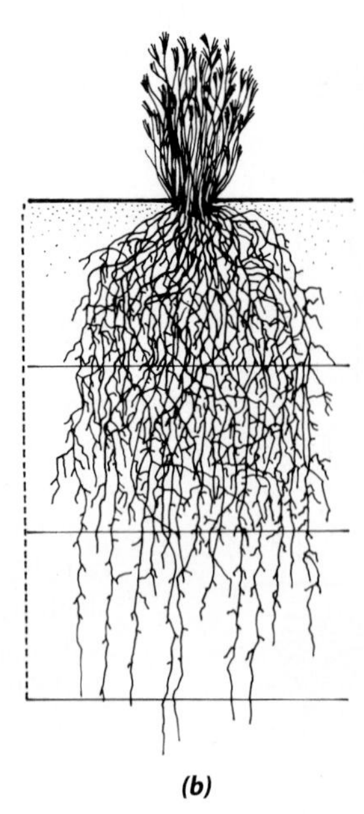

(b)

24–2 Taproot and fibrous root systems Two types of root systems are represented here by two prairie plants. ***(a)*** Taproot system of the blazing star *(Liatris punctata)*, a eudicot. ***(b)*** Fibrous root system of wire grass *(Aristida purpurea)*, a monocot. Each of the vertical units depicted here represents 30 centimeters. Taproot systems generally penetrate deeper into the soil than fibrous root systems.

or 130 times the surface area of the shoot system. Even more amazing is that the roots occupied only about 6 liters of soil.

The Plant Maintains a Balance between Its Shoot and Root Systems

In a growing plant, a balance is maintained between the total surface area available for the manufacture of food (the photosynthesizing surface) and the surface area available for the absorption of water and minerals. This functional balance, which is between the fine roots and leaf area, can be expressed as a ratio, the root:shoot ratio. In seedlings, the total water- and mineral-absorbing surface usually far exceeds the photosynthesizing surface. However, the root:shoot ratio decreases gradually as a plant ages.

If damage to the root system seriously reduces its absorbing surfaces, shoot growth is reduced by lack of water, minerals, and root-produced hormones. In turn, reduction in the size of the shoot system limits root growth by decreasing the availability of carbohydrates and shoot-produced hormones to the roots. Fine roots are generally short-lived, persisting on average for just a few months, although some may live much longer. The short-lived fine roots of trees are in a state of constant flux, with death and replacement occurring simultaneously; therefore, the fluctuation in the population and concentration of roots in the soil is as dynamic as that of the leaves and twigs in the air above. It has been estimated that as much as 33 percent of global annual net productivity in terrestrial ecosystems is devoted to the production of fine roots. Even when plants are carefully transplanted, the balance between shoot and root is invariably disturbed. Most of the fine roots are left behind when the plant is removed from the soil. Cutting back the shoot helps to reestablish a balance between the root system and the shoot system, as does repotting a plant that is root bound into a larger container.

Origin and Growth of Primary Tissues

The growth of many roots is apparently a continuous process that stops only under such adverse conditions as drought and low temperatures. During their growth through the soil, roots follow the path of least resistance and frequently follow spaces left by earlier roots that have died and rotted.

The Tip of the Root Is Covered by a Rootcap, Which Produces Mucilage

The **rootcap** is a thimblelike mass of living parenchyma cells that protects the apical meristem behind it and aids the root in its penetration of the soil (Figures 24–3 through 24–6). As the root grows longer and the rootcap is pushed forward, cells on the periphery of the rootcap secrete large amounts of **mucilage**

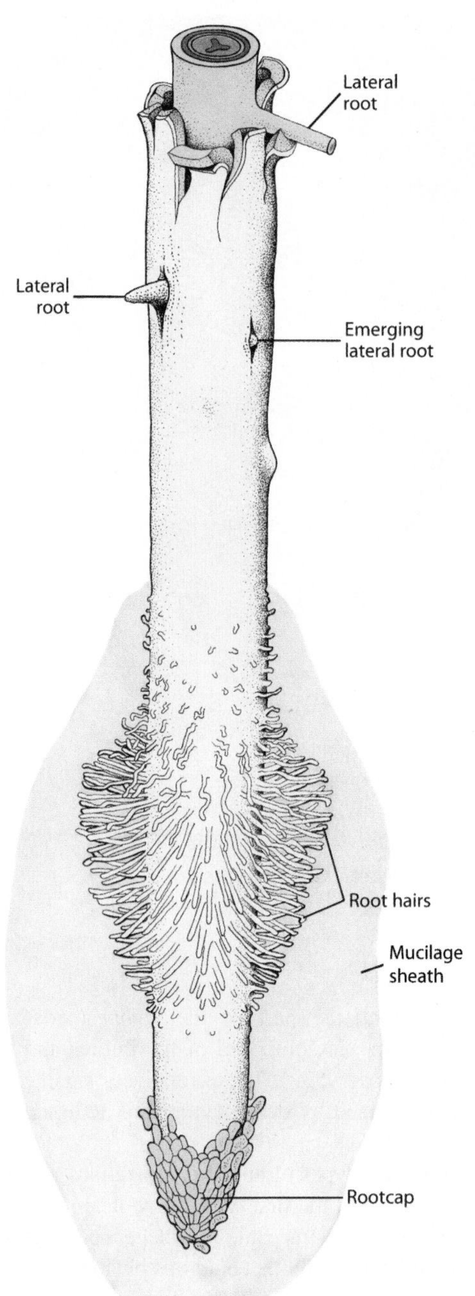

24–3 Eudicot root A portion of a eudicot root, showing the spatial relationship between the rootcap and the region of root hairs, and (near the top) the sites of emergence of the lateral roots, which arise from deep within the parent root. New root hairs arise just behind the region of elongation at about the same rate as the older hairs die off. The root tip is covered by a mucilage sheath, which lubricates the root during its passage through the soil.

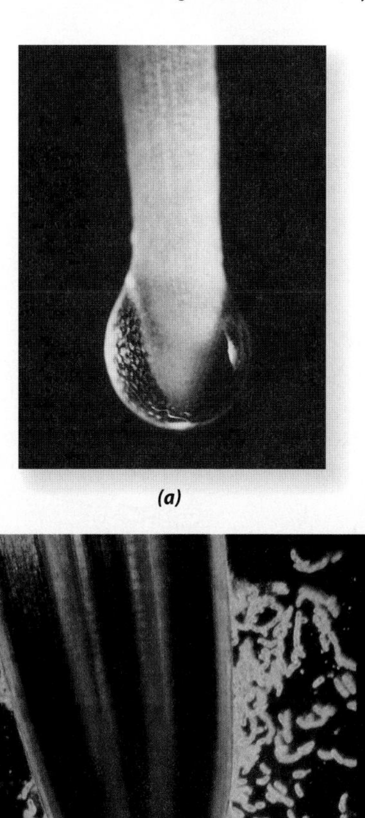

24–4 Mucilage sheath on root tip ***(a)*** Mucilage on the rootcap of a maize *(Zea mays)* root, containing border cells. ***(b)*** Dark-field image of a living root showing a "cloud" of border cells suspended in the mucilage sheath, which is not apparent in this view.

(a highly hydrated polysaccharide), which lubricates the root during its passage through the soil (Figure 24–4). Eventually, these peripheral cells are released from the rootcap. Called **border cells,** these rootcap cells are programmed to separate from the rootcap and from each other as they reach the rootcap periphery. On their release, the border cells—which may remain alive in the rhizosphere for several weeks—undergo changes in gene expression that enable them to produce and exude specific proteins that are completely different from those of the rootcap. As the border cells are released, new cells are added to the rootcap. The number of border cells released each day varies, in part, with the plant family. For example, as few as 10 border cells are shed daily for tobacco (family Solanaceae), compared with as many as 10,000 for cotton (family Malvaceae). Border cells and their products may contribute up to 98 percent of the weight of carbon-rich material released into the soil as root exudates. With rising levels of carbon dioxide in the atmosphere and associated climate change, the role of soil in sequestering carbon is of great interest.

Several functions have been attributed to border cells and their exudates. Among these functions are protection of the apical

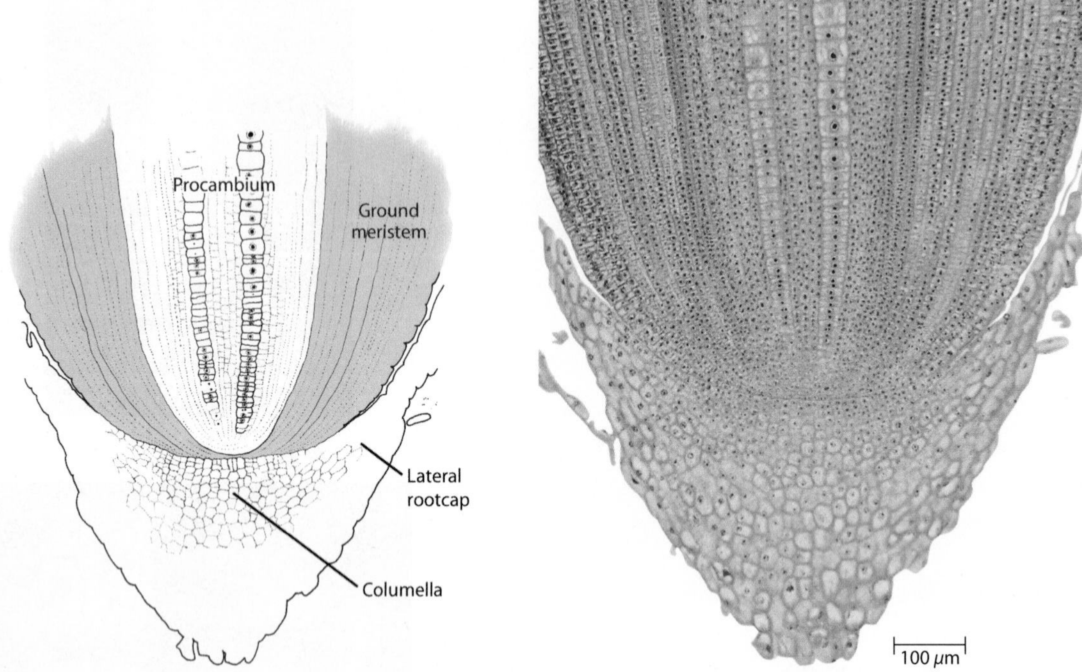

24–5 Closed type of root apical organization Three distinct layers of initials can be seen in this longitudinal section of an apical meristem of a maize *(Zea mays)* root tip. The lower layer gives rise to the rootcap, consisting of the columella and lateral rootcap; the middle layer to the protoderm and ground meristem, which develops into the cortex; and the upper layer to the procambium, which develops into the vascular cylinder. The protoderm differentiates from the outer layer of the ground meristem. Compare the organization of this apical meristem with that of the onion root tip shown in Figure 24–6b.

meristem from infection, maintenance of intimate root-soil contact, mobilization of essential elements for uptake by the roots, short-term protection from drying out (desiccation), and specific attraction or repulsion of bacteria. It also has been proposed that border cells, functioning like ball bearings, decrease the frictional resistance for the growing root.

The rootcap itself does far more than provide protection for the apical meristem and aid the root in its penetration of the soil. It has been characterized as a "multifunctional molecular relay station," because it is the rootcap that senses, processes, and transmits signals to the meristem and the elongation region of the root and therefore controls the direction of root movement through the soil. The rootcap typically consists of a central column of cells, the **columella,** and a lateral portion, the **lateral rootcap,** which surrounds the columella (Figures 24–5 and 24–6). The columella is the site that perceives gravity (gravitropism) and water potential gradients (hydrotropism) (see Chapter 28).

Apical Organization in Roots May Be Either Open or Closed

The root apical meristem—the region of actively dividing cells—extends for a considerable distance from the apex toward the older part of the root. Apart from the rootcap, the most striking structural feature of the root apex is the arrangement of longitudinal files, or lineages, of cells that emanate from the apical meristem. The most distal (closest to the root tip) and least differentiated part of the apical meristem—termed the **promeristem**—is composed of the initials and their immediate derivatives (page 538). These relatively small, many-sided cells are characterized by dense cytoplasm and large nuclei (Figures 24–5 and 24–6).

Two main types of apical organization are found in the roots of seed plants. In the first type, the rootcap, the vascular cylinder, and the cortex are traceable to independent layers of cells in the apical meristem, with the epidermis having a common origin with either the rootcap or the cortex (Figure 24–5). This type of root apical organization is referred to as the "closed type," and each of the three regions—rootcap, vascular cylinder, and cortex—is interpreted as having its own initials. In the second type of root apical organization, all regions, or at least the cortex and the rootcap, arise from one group of initials (Figure 24–6). This is called the "open type."

Although the region of initials in the root apical meristems is mitotically active early in root development, divisions become infrequent in this region later in the growth of the root. Most cell division then occurs a short distance beyond the quiescent initials. The relatively inactive region of the apical meristem, which corresponds to the promeristem, is known as the **quiescent center** (Figure 24–7). The quiescent center does not include the initials of the rootcap.

As indicated by the word "relatively," the quiescent center is not totally devoid of divisions under normal conditions.

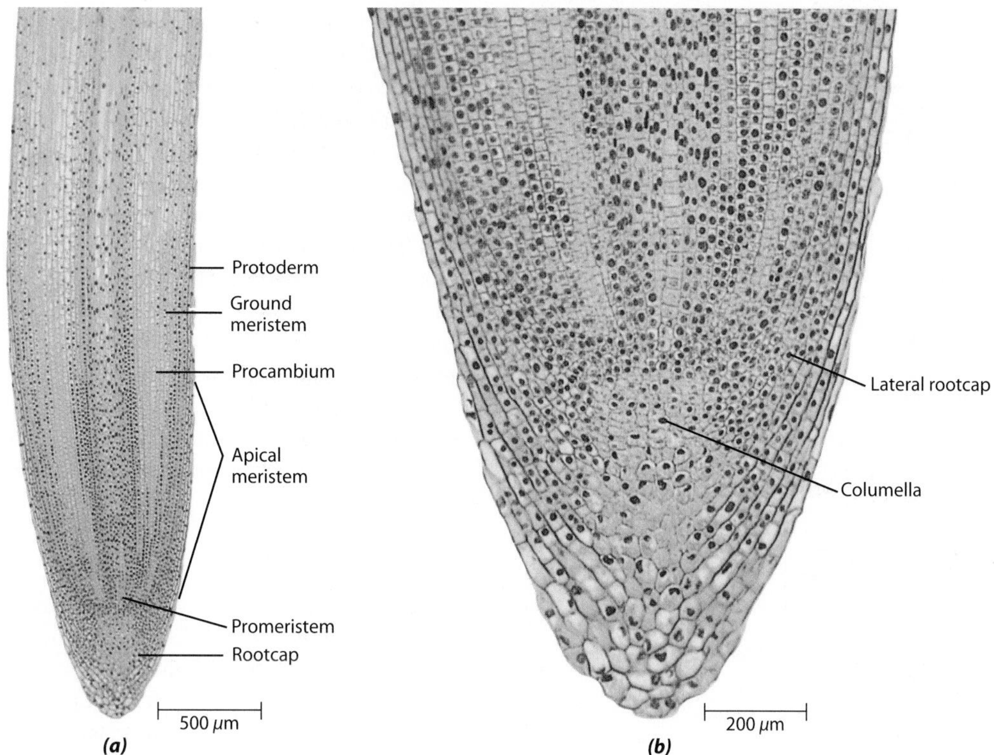

24–6 Open type of root apical organization ***(a)*** In this longitudinal section of an onion *(Allium cepa)* root tip, the primary meristems—protoderm, ground meristem, and procambium—can be distinguished close to the apical meristem. The procambium develops into the vascular cylinder. ***(b)*** Detail of the apical meristem. Compare the organization of this apical meristem with that of the maize root tip shown in Figure 24–5.

24–7 Quiescent center of root apical meristem Autoradiograph of a quiescent center (dashed oval), as seen in longitudinal section through the apical meristem of a maize *(Zea mays)* root tip. To prepare this autoradiograph, the root tip was supplied for one day with thymidine (a DNA precursor) labeled with the radioactive hydrogen isotope tritium (^{3}H). In the rapidly dividing cells around the quiescent center, the radioactive material was quickly incorporated into the nuclear DNA, while less radioactive material was taken up by the relatively inactive cells of the quiescent center.

Moreover, the quiescent center is able to repopulate the bordering meristematic regions when they are injured. In one study, for example, it was found that isolated quiescent centers of maize *(Zea mays)* grown in sterile culture could form whole roots without first forming callus, or wound, tissue. In another study of maize roots, a striking correlation was found between the size of the quiescent center and the complexity of the primary vascular pattern of the root. These and other studies suggest that the quiescent center plays an essential role in organization and development of the root.

Growth in the Length of Roots Occurs Near the Root Tip

The distance behind the promeristem at which most cell division takes place differs from species to species and also within the same species, depending on the age of the root. The region of actively dividing cells—the apical meristem—commonly is called the **region of cell division** (Figure 24–8).

Behind the region of cell division, but not sharply delimited from it, is the **region of elongation,** which usually is only a few millimeters in length (Figure 24–8). The elongation of cells in this region results in most of the increase in length of the root. Above this region the root does not increase in length. Thus, the growth in length of the root occurs near the root tip and results in a very limited portion of the root constantly being pushed through the soil.

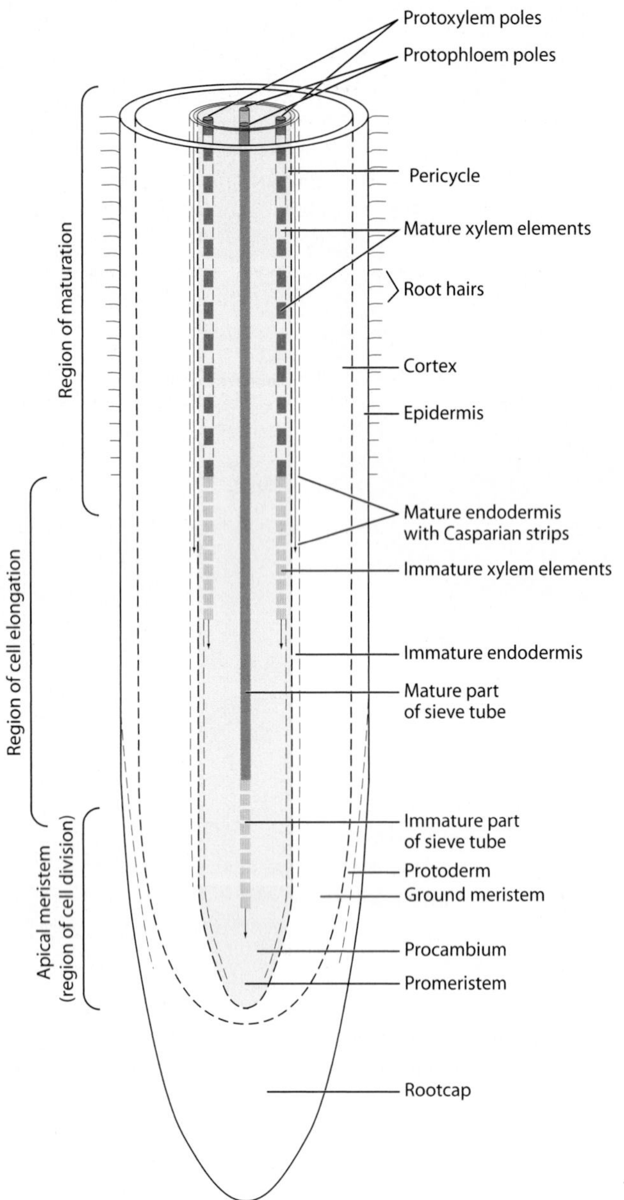

24–8 Early stages in the primary development of a root tip The apical meristem, or region of cell division, extends for a considerable distance behind the promeristem. These cell divisions overlap with the region of cell elongation and with the region of maturation. At various distances from the apical meristem, cells enlarge and develop as specific cell types according to their position in the root. The three primary meristems—protoderm, ground meristem, and procambium—occur close to the apical meristem. The mature portion of the sieve tube is closer to the apical meristem than are mature portions of xylem elements, indicating that the protophloem sieve tubes reach maturity earlier. Maturation of the endodermis (with Casparian strips) occurs prior to maturation of the xylem elements and development of root hairs. The sites of the first primary xylem elements and primary phloem elements are referred to as the protoxylem and protophloem poles, respectively.

Beyond the region of elongation is the **region of maturation,** or differentiation, in which most of the cells of the primary tissues mature (Figure 24–8). Root hairs are also produced in this region, and sometimes this part of the root is called the root-hair zone (Figure 24–3). Obviously, if root hairs were to arise in the region of elongation, they would soon be sheared off by abrasion as the root is pushed through the soil.

It is important to note the gradual transition from one region of the root to another. The regions are not sharply delimited. At the same level of the root, these processes overlap not only in different tissue regions but also in different cell files, or rows, of the same tissue region. Cell division may continue well into the region where cell elongation occurs rapidly. Some cells begin to elongate and differentiate in the region of cell division, whereas others reach maturity in the region of elongation. For example, the first-formed elements of the phloem and xylem mature in the region of cell elongation and are often stretched and destroyed during elongation of the root. As can be seen in Figure 24–8, the first-formed primary phloem elements (known as **protophloem sieve tubes**) reach maturity nearer the root tip than do the first-formed primary xylem elements (known as **protoxylem elements**), an indication of the need for food substances to be transported in the sieve tubes for root growth.

The protoderm, ground meristem, and procambium can be distinguished in very close proximity to the apical meristem (Figures 24–6 and 24–8). These are the primary meristems that differentiate into the epidermis, the cortex, and the primary vascular tissues, respectively (see Chapter 23).

Primary Structure

Compared with the stem, the internal structure of the root is usually relatively simple. This is due in large part to the absence of leaves and the corresponding absence of nodes and internodes (see Figure 1–9). Thus, the arrangement of the primary tissues shows very little difference from one level of the root to another.

The three tissue systems of the root in the primary stage of growth—the epidermis (dermal tissue system), the cortex (ground tissue system), and the vascular tissues (vascular tissue system)—can be readily distinguished from one another. In most roots, the vascular tissues form a solid cylinder (see Figure 24–10), but many roots have a pith or pithlike region in the center (see Figure 24–11).

The Epidermis in Young Roots Absorbs Water and Minerals

The root epidermis consists of closely packed elongated cells with thin walls that lack a cuticle and offer little resistance to the passage of water and minerals into the root. In young roots the epidermis is specialized as an absorbing tissue. The uptake of water and minerals by the root is facilitated by **root hairs**—tubular extensions of epidermal cells—which greatly increase the absorptive surface of the root (Figure 24–9). In the study of a four-month-old rye plant, it was estimated that the plant contained approximately 14 billion root hairs, with an absorbing surface of 401 square meters. Placed end-to-end, these root hairs would extend well over 10,000 kilometers.

Root hairs are relatively short-lived and are confined largely to the region of maturation. The production of new root hairs occurs just behind the region of cell elongation (Figure 24–8) and

(a)

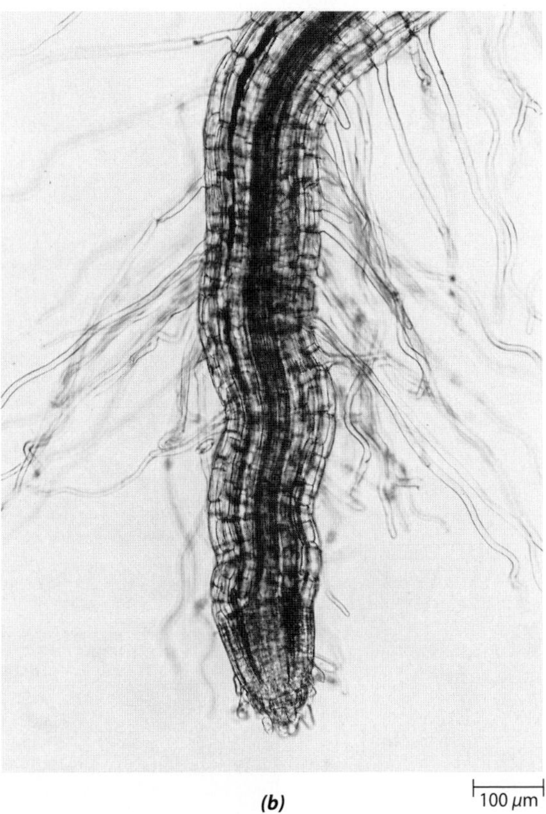

(b)

24–9 Root hairs ***(a)*** A radish *(Raphanus sativus)* seedling. Note the shed seed coat, the cotyledons, curved hypocotyl, and primary root, with numerous root hairs. Most of the uptake of water and minerals occurs through the root hairs, which form just behind the growing tip of the root. ***(b)*** Root of a bentgrass *(Agrostis tenuis)* seedling. The root hairs may be as much as 1.3 centimeters long and may attain their full size within hours. Each hair is comparatively short-lived, but the formation of new root hairs and the death of old ones continue as long as the root is growing.

at a rate that nearly matches the rate at which older root hairs are dying off at the upper end of the root-hair zone. As the tip of the root penetrates the soil, new root hairs develop behind it, providing the root with surfaces capable of absorbing new supplies of water and minerals, or inorganic ions. (See Chapter 30 for a discussion of the absorption of water and inorganic ions by roots.) It is, of course, the new and growing roots—the fine roots—that are primarily involved in the absorption of water and minerals. For this reason, great care must be taken by gardeners to move as much soil as possible along with the root system during transplantation. If the plant is simply "torn" from the soil, most of the fine roots are left behind and the plant probably will not survive.

Mutually beneficial symbiotic associations—mycorrhizas—occur between fungi and the roots of most vascular plants (see Chapter 14). The hyphal network of the fungus may extend far beyond the mycorrhizas, making it possible for the plant to obtain water and nutrients from a much larger volume of soil than is made possible by root hairs. Typically, root hairs do not develop on ectomycorrhizas (pages 312 to 314).

The Cortex Represents the Ground Tissue System in Most Roots

As seen in transverse section, the cortex occupies by far the greatest area of the primary body of most roots (Figure 24–10a). The plastids of the cortical cells commonly store starch but are usually devoid of chlorophyll. Gymnosperm and angiosperm roots that undergo considerable amounts of secondary growth shed their cortex early. In such roots, the cortical cells remain parenchymatous. By contrast, in many monocots and in extremely herbaceous eudicots, which consist entirely of primary tissues, the cortex is retained for the life of the root, and many of the cortical cells develop secondary walls that become lignified (page 34).

Regardless of the degree of differentiation, cortical tissue contains numerous intercellular spaces—air spaces essential for aeration of the root cells (Figures 24–10b and 24–11b). In many aquatic and wetland plants the intercellular spaces become large, resulting in the formation of **aerenchyma,** the term applied to parenchyma tissue with large and abundant intercellular spaces. The cortical cells have numerous contacts with one another, and their protoplasts are connected by plasmodesmata. Substances moving across the cortex may thus follow a *symplastic pathway,* moving from one protoplast to another by way of plasmodesmata, or an *apoplastic pathway* via the cell walls, or both. (The concept of symplast and apoplast is discussed on page 88, and the uptake of water and minerals by roots in Chapter 30.)

Unlike the rest of the cortex, the innermost layer is compactly arranged and lacks air spaces. This layer, the **endodermis** (Figures 24–10 and 24–11), is characterized by the presence of **Casparian strips** in its anticlinal walls (the radial and transverse cell walls, which are perpendicular to the surface of the root). The Casparian strip is not merely a wall thickening but an integral bandlike portion of the primary wall and middle lamella (the layer of intercellular material joining adjacent cells) that is impregnated with suberin and sometimes lignin. The suberin and lignin infiltrate the spaces in the wall usually occupied by water, thus imparting a hydrophobic property to these specific regions

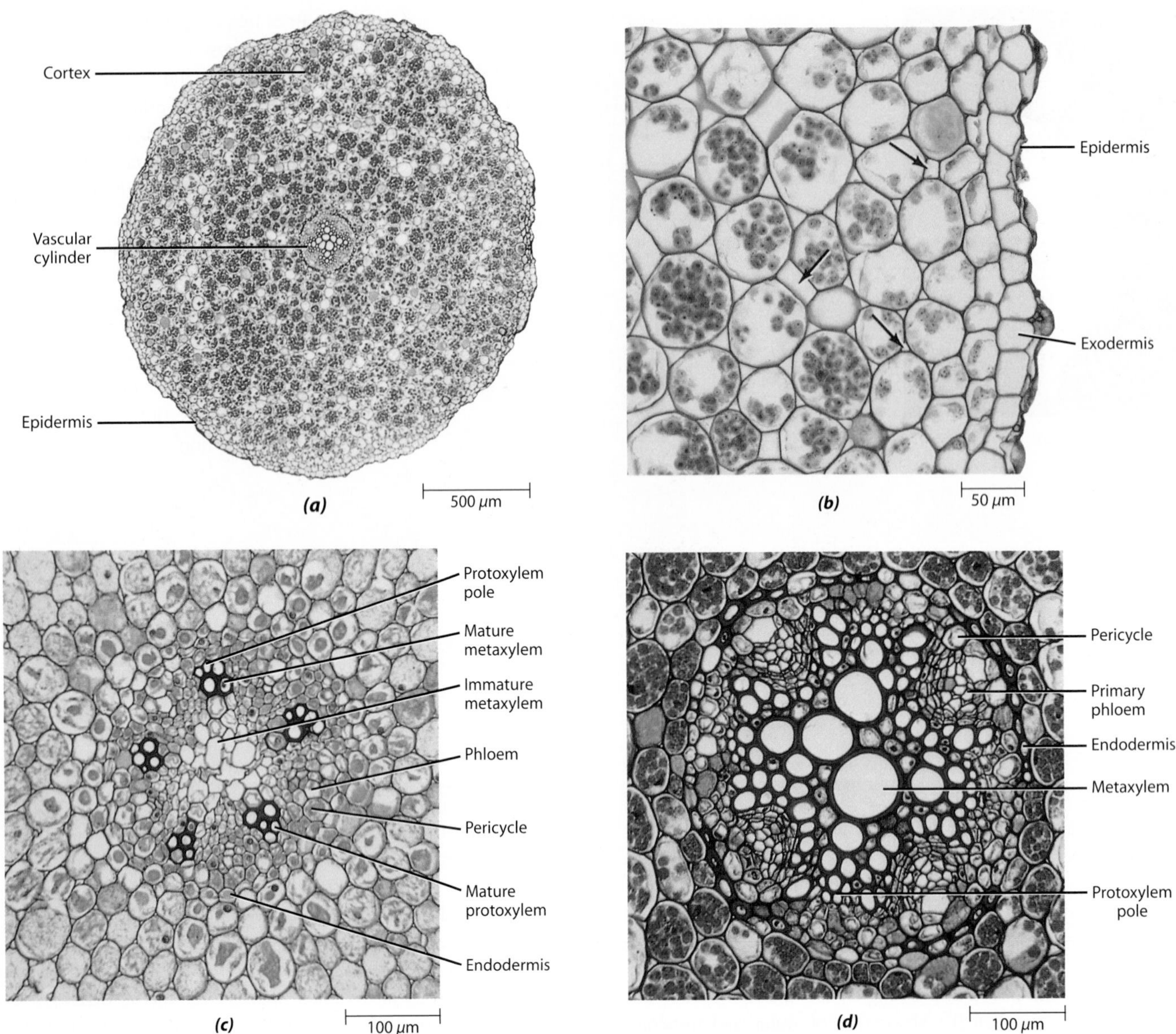

24–10 Structure of a eudicot root Transverse sections of the root of a buttercup *(Ranunculus)*. ***(a)*** Overall view of a mature root. ***(b)*** Detail of the outer portion of a mature root. In this root, the epidermis has died and the exodermis, the outer layer of cells of the cortex, has replaced the epidermis as the functional surface layer. Note the intercellular spaces (arrows) among the cortical cells interior to the compactly arranged exodermis. The intercellular spaces are essential for aeration of the root cells. ***(c)*** Detail of an immature vascular cylinder. Notice the intercellular spaces among the cortical cells. ***(d)*** Detail of a mature vascular cylinder. Numerous starch grains are evident in the cortical cells.

of the cell wall. The plasma membranes of endodermal cells are quite firmly attached to the Casparian strips (Figure 24–12). Because the endodermis is compact and the Casparian strips are impermeable to water and ions, apoplastic movement of water and solutes across the endodermis is blocked by the strips. Hence, all substances entering and leaving the vascular cylinder must pass through the protoplasts of the endodermal cells. This is accomplished either by crossing the plasma membranes of these cells or by moving symplastically via the numerous plasmodesmata connecting the endodermal cells with the protoplasts of neighboring cells of the cortex and vascular cylinder.

As mentioned previously, in roots that undergo secondary growth, the cortex and its innermost layer, the endodermis, are shed early. Although these older roots may still absorb water and minerals from the soil, they largely transport the water and minerals absorbed by younger roots, with which they are connected. Roots in which growth has ceased senesce, die, and decay. During vigorous growth, the roots are replaced at least as rapidly

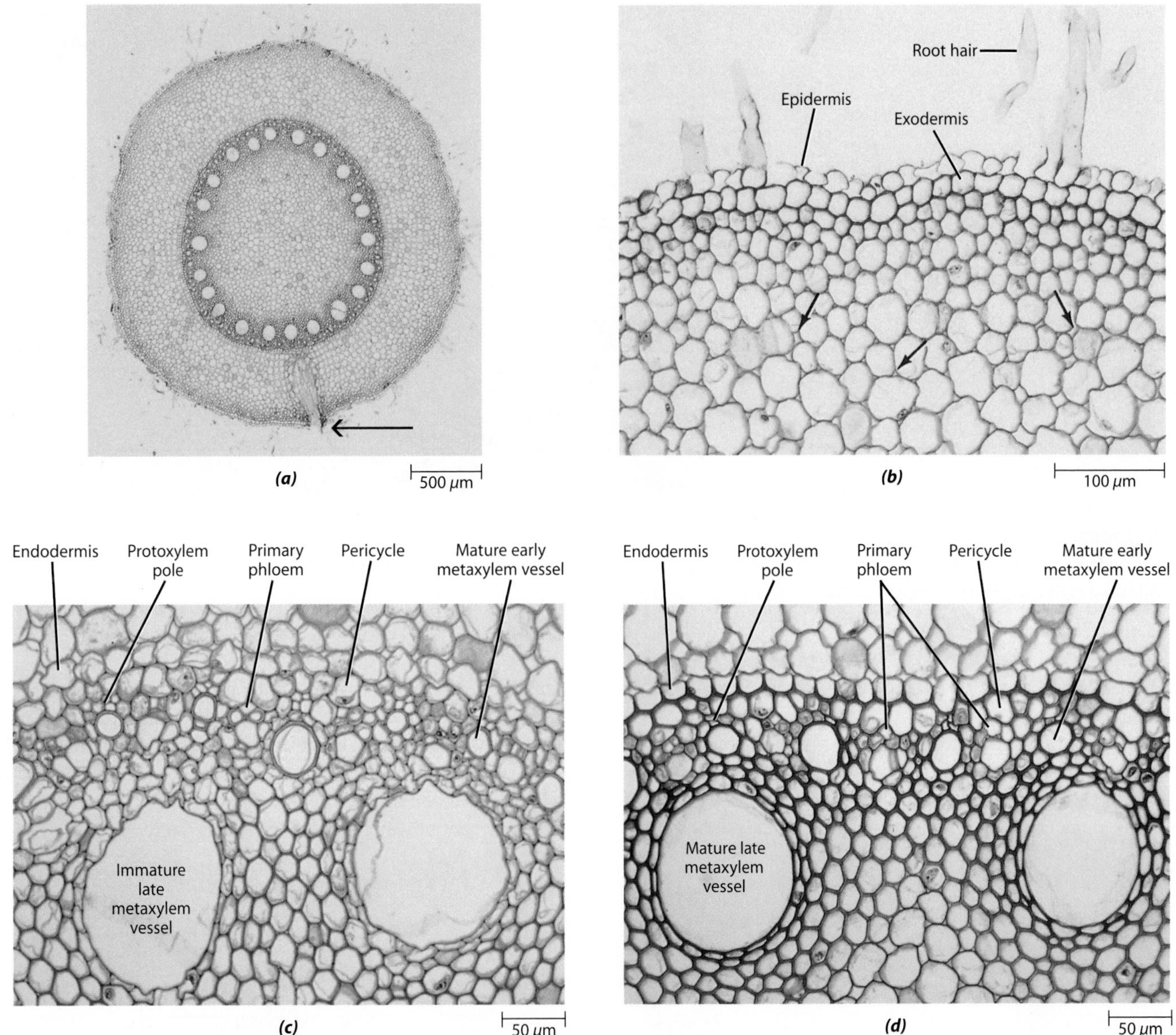

24–11 Structure of a monocot root Transverse sections of a maize *(Zea mays)* root. ***(a)*** Overall view of a mature root. Part of a lateral root is indicated by the arrow. The vascular cylinder, with its pith, is quite distinct. ***(b)*** Detail of the outer portion of a mature root, showing the epidermis with root hairs and part of the cortex. The outer layer of cortical cells is differentiated as a compactly arranged exodermis. Note the intercellular spaces (arrows) among the other cortical cells. ***(c)*** Detail of an immature vascular cylinder. ***(d)*** Detail of a mature vascular cylinder.

as they die back. In older roots in which the cortex is retained, a suberin lamella, consisting of alternating layers of suberin and wax, is eventually deposited internally over all wall surfaces of the endodermis. This is followed by the deposition of cellulose, which may become lignified (Figures 24–13 and 24–14). These changes in the endodermis begin opposite the phloem strands and spread toward the protoxylem (Figure 24–10d).

Opposite the protoxylem, some of the endodermal cells may remain thin-walled and retain their Casparian strips for a prolonged period. Such cells are called **passage cells.** In some species these endodermal cells remain as passage cells, whereas in others they eventually become suberized and deposit additional cellulose. A misconception concerning the endodermis is that development of the suberin lamellae prevents the movement of substances across this innermost layer of cortical cells. This is not the case. As long as the endodermal cells remain alive, their plasmodesmata remain intact, providing a symplastic pathway for the movement of water and minerals. The uptake of water

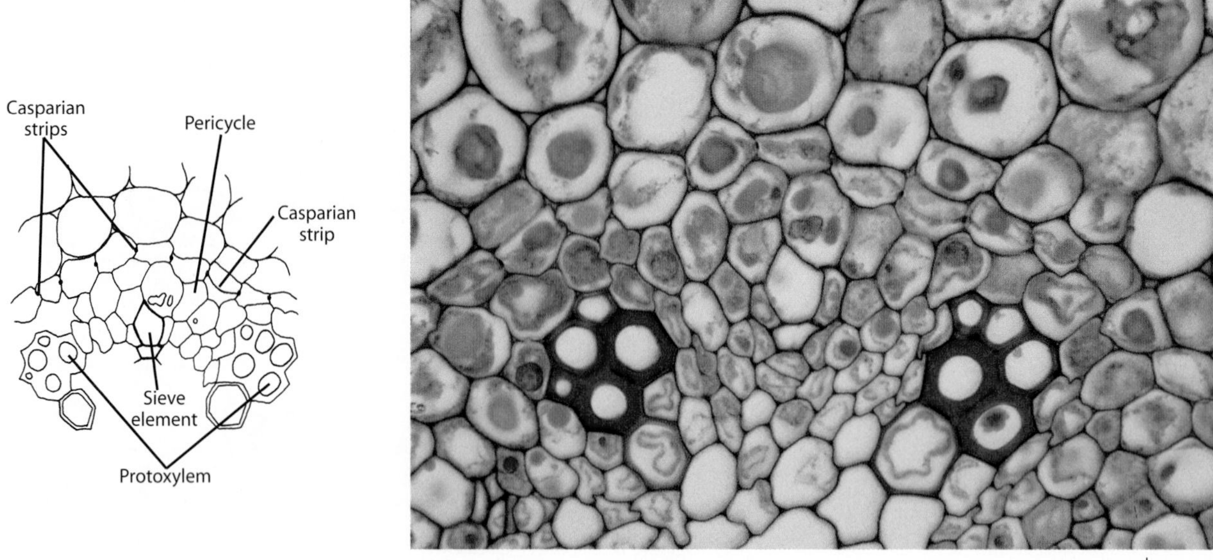

24–12 Casparian strips High-magnification view of a portion of an immature buttercup *(Ranunculus)* root, showing Casparian strips in the endodermal cells. Notice that the plasmolyzed protoplasts of the endodermal cells cling to the strips, a phenomenon called band plasmolysis.

and minerals by roots may take place well behind the region of root hairs.

The roots of most angiosperms have a second compact layer of cells with Casparian strips. This layer, called the **exodermis,** develops from the outermost layer or layers of cells of the cortex. Development of the Casparian strips is quickly followed by deposition of a suberin lamella and, at least in some species, of a cellulosic layer as well. The suberized cell walls of the exodermis apparently reduce water loss from the root to the soil and provide a defense against attack by microorganisms. (See Chapter 30 for further discussion of the role of the exodermis and endodermis in the movement of water and solutes across the root.)

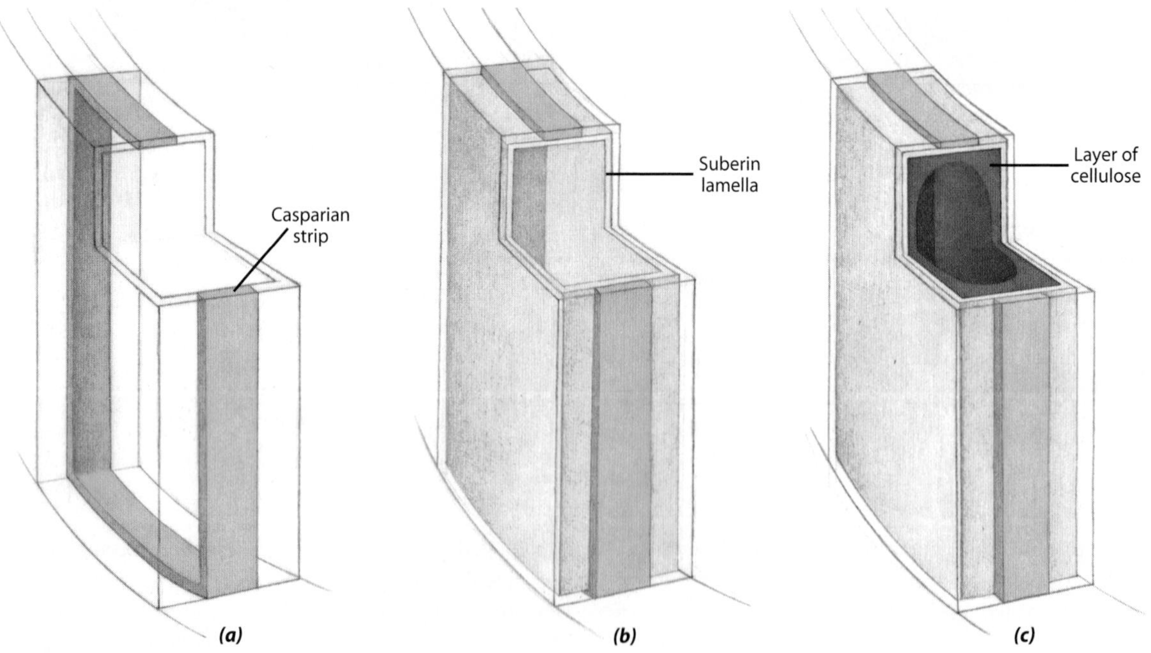

24–13 Root endodermal cell Three-dimensional diagrams showing three developmental stages of an endodermal cell in a root that remains in a primary state. ***(a)*** Initially, the endodermal cell is characterized by the presence of a Casparian strip in its anticlinal walls. ***(b)*** A suberin lamella is then deposited internally over all wall surfaces. ***(c)*** Finally, the suberin lamella is covered internally by a thick, often lignified, layer of cellulose. The outside of the root is to the left in all three diagrams.

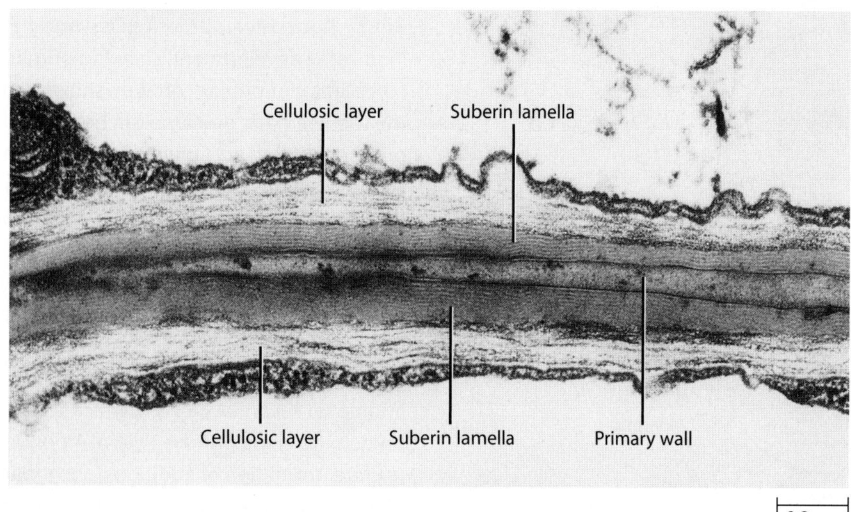

24–14 Structure of endodermal cell wall in an older root Electron micrograph showing a section through the cell wall between two adjacent endodermal cells of a squash *(Cucurbita pepo)* root. In this late stage of differentiation of the endodermis, suberin lamellae are covered by cellulosic wall layers on both sides of the primary wall. Note the alternating light and dense bands in the suberin lamellae, which are interpreted as consisting of wax and suberin, respectively.

The Vascular Cylinder Includes the Primary Vascular Tissues and the Pericycle

The **vascular cylinder** of the root consists of primary vascular tissues and one or more layers of nonvascular cells that constitute the **pericycle,** which completely surrounds the vascular tissues (Figures 24–10 and 24–11). The pericycle is considered part of the vascular cylinder because, like the vascular tissues, it originates from the procambium. In the young root, the pericycle is composed of parenchyma cells with primary walls, but as the root ages, the cells of the pericycle may develop secondary walls (Figure 24–11d).

The pericycle plays several important roles. In most seed plants, lateral roots arise in the pericycle. In plants undergoing secondary growth, the pericycle contributes to the vascular cambium opposite the protoxylem and generally gives rise to the first cork cambium. Pericycle often proliferates—that is, gives rise to more pericycle.

The center of the vascular cylinder of most roots is occupied by a solid core of primary xylem from which ridgelike projections extend toward the pericycle (Figure 24–10). Nestled between the ridges of xylem are strands of primary phloem. (Lacking a pith, the vascular cylinder in such roots is a protostele, a solid cylinder of vascular tissue; pages 395 and 396.)

The number of ridges of primary xylem differs from species to species, sometimes even varying along the axis of a given root. If two ridges are present, the root is said to be diarch; if three are present, triarch; if four, tetrarch (Figure 24–10d); if five, pentarch (Figure 24–10c); and if many are present, polyarch (Figure 24–11). The first *(proto-)* primary xylem elements to mature in roots are located next to the pericycle, and the tips of the ridges are commonly referred to as **protoxylem poles** (Figures 24–10 and 24–11). The **metaxylem** (*meta-* meaning "after")—the part of the primary xylem that differentiates after the protoxylem—occupies the inner portions of the ridges and the center of the vascular cylinder. The roots of some angiosperms, both eudicots and monocots (such as maize), have a pith or pithlike region (Figure 24–11), which some botanists regard as part of the vascular cylinder because they consider it to be of procambial origin.

Effect of Secondary Growth on the Primary Body of the Root

As we have mentioned, secondary growth in roots and stems consists of the formation of (1) secondary vascular tissues, that is, secondary xylem and secondary phloem, from a vascular cambium and (2) periderm, composed mostly of cork tissue, from a cork cambium. Commonly, the roots of monocots lack secondary growth and so consist entirely of primary tissues. In addition, the roots of many herbaceous eudicots undergo little or no secondary growth and remain largely primary in composition. (See Chapter 26 for a discussion of the vascular cambium.)

In roots that exhibit secondary growth, formation of the vascular cambium is initiated by divisions of procambial cells that remain meristematic and are located between the primary xylem and primary phloem in portions of the root that are no longer elongating. Thus, depending on the number of phloem strands present in the root, two or more independent regions of cambial activity are initiated more or less simultaneously (Figure 24–15). Soon afterward, the pericycle cells opposite the protoxylem poles also divide, and the inner sister cells resulting from these divisions contribute to the vascular cambium. Now the cambium completely surrounds the core of xylem.

As soon as it is formed, the vascular cambium opposite the phloem strands begins to produce secondary xylem toward the

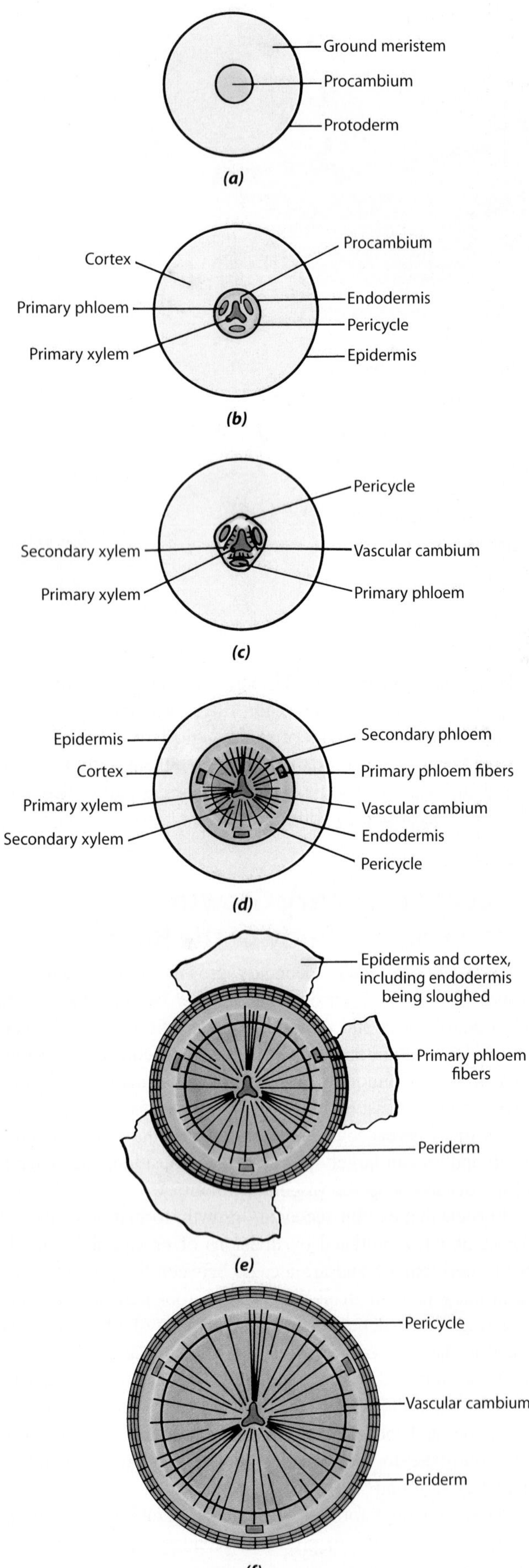

24–15 Root development in a woody eudicot ***(a)*** Early stage in primary development, showing primary meristems. ***(b)*** At the completion of primary growth, showing the primary tissues and the meristematic procambium between the primary xylem and primary phloem. ***(c)*** Origin of vascular cambium. In the triarch root represented here, cambial activity has been initiated in three independent regions from procambium between the three primary phloem strands and the primary xylem. The pericycle cells opposite the three ridges of primary xylem will also contribute to the vascular cambium. Some secondary xylem has already been produced by the newly formed vascular cambium of procambial origin. ***(d)*** After formation of some secondary phloem and additional secondary xylem, which further separate the primary phloem from the primary xylem. A periderm has not yet formed. ***(e)*** After formation of additional secondary xylem and secondary phloem and of periderm. ***(f)*** At the end of the first year's growth, showing the effect of secondary growth—including periderm formation—on the primary plant body. In *(d)* through *(f)*, the radiating lines represent rays, which consist of radial rows of parenchyma cells.

inside, and in the process the strands of primary phloem are displaced outward from their positions between the ridges of primary xylem. By the time the cambium opposite the protoxylem poles is actively dividing, the cambium is circular in outline, and the primary phloem has been separated from the primary xylem (Figure 24–15).

By repeated divisions toward the inside and outside, secondary xylem and secondary phloem are added to the root (Figures 24–15 and 24–16). Files of parenchyma cells that extend radially in the secondary xylem and secondary phloem form rays. In some roots, the vascular cambium derived from the pericycle forms wide rays, whereas narrower rays are produced in other parts of the secondary vascular tissues.

With increases in width of the secondary xylem and phloem, most of the primary phloem is crushed or obliterated. Primary phloem fibers, if present, may be the only remaining distinguishable components of the primary phloem in roots that have undergone secondary growth.

In most woody roots, a protective layer of secondary origin, called the **periderm,** replaces the epidermis as the protective covering on this portion of the root. Periderm formation usually follows initiation of secondary xylem and secondary phloem production. Divisions of pericycle cells cause an increase in the number of layers of pericycle cells in radial extent. A complete cylinder of **cork cambium,** which arises in the outer part of the proliferated pericycle, produces **cork** toward its outer surface and **phelloderm** toward its inner surface. Collectively, these three tissues—cork, cork cambium, and phelloderm—make up the periderm. The remaining cells of the proliferated pericycle may form tissue that resembles a cortex. Some regions of the periderm allow gas exchange between the root and the soil atmosphere. These are the **lenticels,** spongy areas in the periderm with numerous intercellular spaces that permit the passage of air.

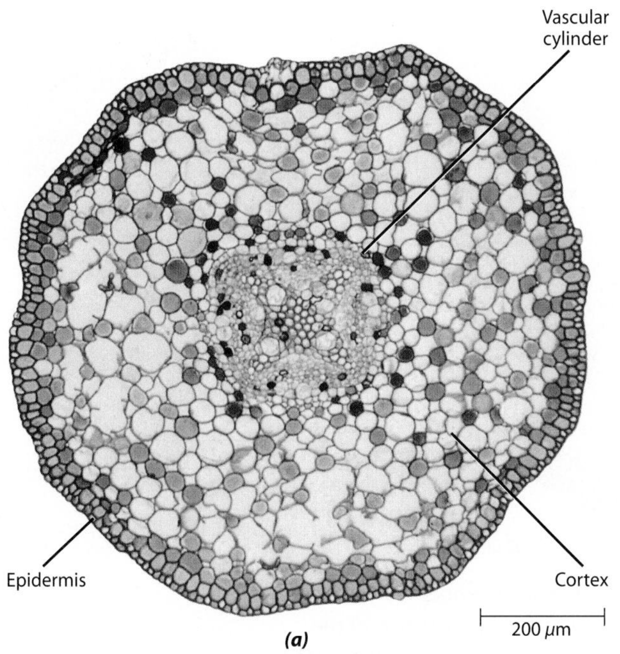

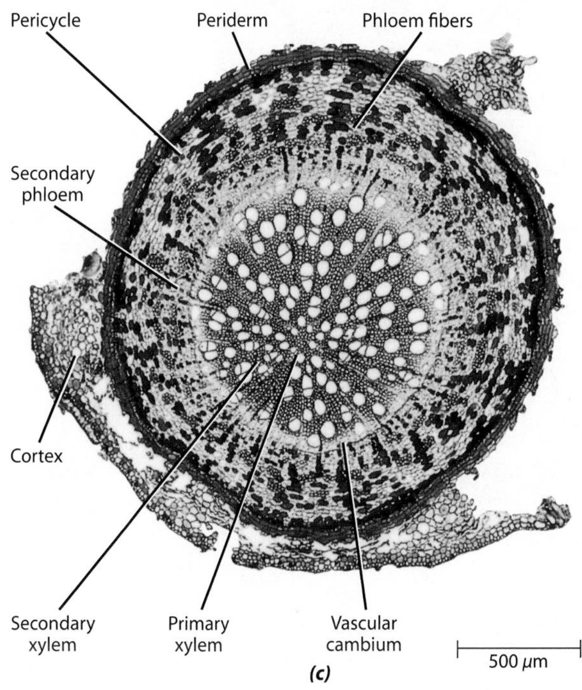

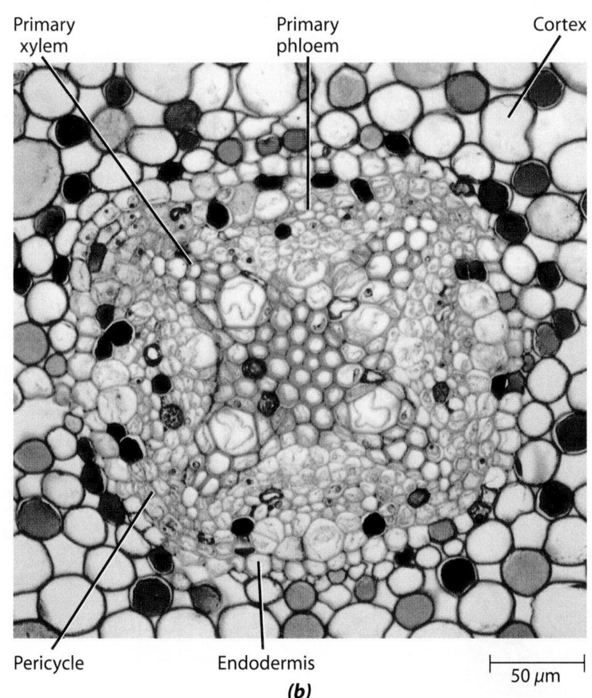

24–16 Primary and secondary growth in a woody eudicot root Transverse sections of a willow *(Salix)* root, which becomes woody. ***(a)*** Overall view of the root near completion of primary growth. ***(b)*** Detail of the primary vascular cylinder. ***(c)*** Overall view of the root at the end of the first year's growth, showing the effect of secondary growth on the primary plant body.

With formation of the first periderm in the root, the cortex (including the endodermis) and the epidermis are isolated from the rest of the root; they eventually die and are sloughed. At the end of the first year's growth, the following tissues are present in a woody root (from outside to inside): possible remnants of the epidermis and cortex; periderm; pericycle; primary phloem (fibers, if present, and crushed soft-walled cells); secondary phloem; vascular cambium; secondary xylem; and primary xylem (Figures 24–15f and 24–16c).

Origin of Lateral Roots

In most seed plants, lateral roots (branch roots) originate in the pericycle opposite the protoxylem poles. Because lateral roots originate from deep within the parent root, they are said to be **endogenous,** meaning "originating within" (Figures 24–3 and 24–17).

Divisions in the pericycle that initiate lateral roots occur some distance behind the region of elongation in partially or fully differentiated root tissues. In angiosperm roots, derivatives of both the pericycle and the endodermis commonly contribute to the new lateral root primordium, although in many cases, divisions in the derivatives of the endodermis are short-lived. As the young lateral root, or **root primordium,** increases in size, it pushes its way through the cortex (Figure 24–17), possibly secreting enzymes that digest some of the cortical cells lying in its path. While still very young, the root primordium develops a rootcap and apical meristem, and the primary meristems appear. Initially, the vascular cylinders of lateral root and parent root are not connected to one another. The two vascular cylinders are joined later, when derivatives of intervening pericycle and vascular parenchyma cells differentiate into xylem and phloem. (See Chapter 27 for further discussion of lateral root development.)

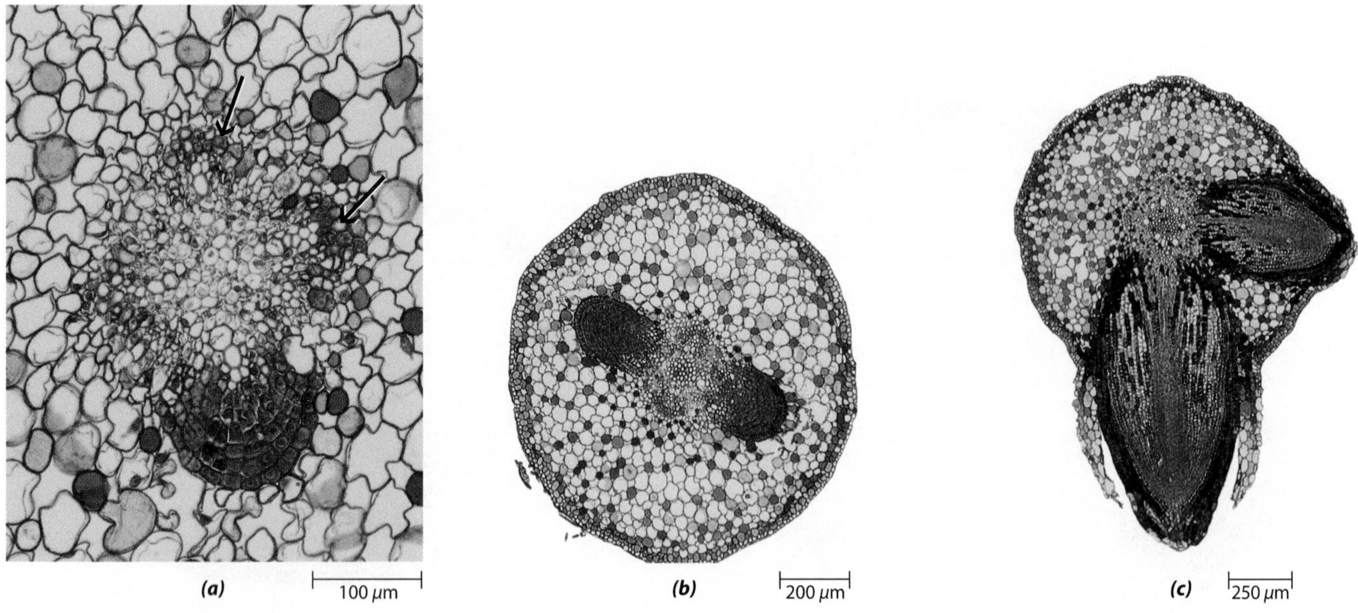

24–17 Development of lateral roots Seen here are three stages in the origin of lateral roots in a willow *(Salix)*. ***(a)*** One root primordium is present (below), and two others are being initiated in the region of the pericycle (arrows). The vascular cylinder is still very young. ***(b)*** Two root primordia are penetrating the cortex. ***(c)*** One lateral root has reached the outside, and the other is about to break through.

Aerial Roots and Air Roots

Aerial roots are roots produced from aboveground structures. The aerial roots of some plants serve as **prop roots** for support, as in maize (Figure 24–18). When these stem-borne roots come in contact with the soil, they branch and begin to function in the absorption of water and minerals. **Stilt roots** are produced from the stems and branches of many tropical trees, such as the red mangrove *(Rhizophora mangle)*, the banyan tree *(Ficus benghalensis)*, and some palms (Figure 24–19). Other aerial roots, as in the ivy *(Hedera helix)*, cling to the surface of objects such as walls and provide support for the climbing stem.

Roots require oxygen for respiration, which is why most plants cannot live in soil that is inadequately drained and consequently lacks air spaces. In some trees that grow in swampy habitats, portions of the roots grow out of the water. Hence, the roots of such trees serve not only to anchor but also to aerate the root system. For example, the root systems of the black mangrove *(Avicennia germinans)* and white mangrove *(Laguncularia racemosa)* develop negatively gravitropic extensions called **air roots,**

24–18 Prop roots These prop roots of maize *(Zea mays)* are a type of aerial root.

(a)

(b)

24–19 Stilt roots Seen here are stilt roots of ***(a)*** the neotropical palm (*Iriartea deltoidea,* family Arecaceae), and ***(b)*** the banyan tree (*Ficus benghalensis,* family Moraceae).

24–20 Pneumatophores The white mangrove *(Laguncularia racemosa)* produces air roots, or pneumatophores, which protrude from the mud near the base of the tree.

or **pneumatophores,** which grow upward (against gravity) out of the mud and so provide adequate aeration (Figure 24–20) via numerous lenticels and a broad cortex composed of aerenchyma tissue.

Many special adaptations of roots are found among epiphytes—plants that grow on other plants but are not parasitic on them. The root epidermis of epiphytic orchids, for example, is several layers thick (Figure 24–21, page 576) and, in some species, is the only photosynthetic tissue of the plant. This multiple epidermis, called velamen, provides mechanical protection for the cortex and reduces water loss. The velamen may also function in the absorption of water and nutrients and in the storage of water.

Adaptations for Food Storage: Fleshy Roots

Most roots are storage organs, and in some plants the roots are specialized for this function. Such roots are fleshy, because of an abundance of storage parenchyma, which is permeated by

GETTING TO THE ROOT OF ORGAN DEVELOPMENT

Much research on organ development in plants has focused on the root of *Arabidopsis thaliana.* The three foremost reasons for this are: (1) the simple structure of the *Arabidopsis* root; (2) the ease with which *Arabidopsis* seedlings can be grown on nutrient agar in Petri plates; and (3) the ease with which *Arabidopsis* may be transformed, mutated, and genetically characterized, as a molecular model system for which the entire genome has been sequenced and is available. When the plates containing seedlings are oriented vertically, the roots grow along the surface of the solidified agar media, where any abnormalities, which represent mutations, can be easily observed.

The organization of the mature *Arabidopsis* root is essentially similar to that of the embryonic root, which becomes distinct during the heart stage of embryogenesis (see Figure 22–3d). The radial pattern is as follows: an outer region of epidermal cells, a middle region consisting of one layer each of parenchymatous cortical cells and endodermal cortical cells, and an innermost region (the vascular cylinder, or stele) consisting of pericycle and vascular tissue. The two cortical layers invariably consist of eight cells each. Two types of cells occur in the epidermis: those that produce root hairs and those that remain hairless. The hair cells are always located over the junction of the radial walls between two cortical cells (H position), and the hairless (non–root hair) cells are always directly over cortical cells (N position). This pattern is established in the embryo and is retained during the seedling stage.

The apical meristem of the *Arabidopsis* root has a closed type of organization, with three layers of initials. The lower layer consists of columella rootcap initials, as well as the initials of the lateral rootcap cells and epidermis; the middle layer consists of the initials of the cortex (parenchymatous and endodermal cortical cells); and the upper layer consists of initials of the vascular cylinder (pericycle and vascular tissue). At the center of the middle layer is a set of four cells that rarely divide. This set of cells constitutes the quiescent center of the root.

A great many root mutants of *Arabidopsis* have been identified. In particular, the control of cell fate in the epidermis has been examined in some detail. In one study, it was found that the *TTG (TRANSPARENT TESTA GLABRA)* gene is required to specify epidermal cell fate and cell patterns. The normal positional relationship—with hair cells in the H position and hairless cells in the N position—was absent in *ttg* mutants. In such mutants, epidermal cells in all positions differentiate into hair cells, which develop root hairs that are indistinguishable from normal, wild-type hairs. In transverse sections from the mature portion of the root, the number of epidermal and cortical cell files (lineages) is similar to that of the wild-type root. Overall, the results of this study indicate that the *ttg* mutation alters the positional control of root-hair-cell differentiation but affects neither root-hair formation nor the structure of the mature root. Apparently, in wild-type roots, the *TTG* gene either provides or responds to positional signals that cause differentiating epidermal cells lying directly over cortical cells to remain hairless.

Positional control through cell-cell interaction, rather than cell lineage relationships (as determined by organization of the apical meristem), plays the most important role in determining cell fate in the *Arabidopsis* root. This has been clearly demonstrated by laser ablation experiments. In these experiments, specific cells were ablated, or removed, with a laser, and the effect on neighboring cells was observed. For example, when the cells of the quiescent center were ablated, they were replaced and displaced toward the rootcap by cells derived from the vascular cylinder (procambium). These new cells subsequently acquired rootcap characteristics. Ablated cortical initials were replaced by pericycle cells, which then switched fate and behaved as cortical initials. Ablation of a single daughter cell of a cortical initial had no effect on subsequent divisions of that initial, which was in contact with other cortical daughter cells of neighboring cortical initials. When all cortical daughter cells bordering a cortical initial were ablated, however, that initial was unable to generate files of parenchymatous and endodermal cortical cells. Apparently, the cortical initials, and perhaps all initials, depend on positional information from more mature daughter cells within the same cell layer. In other words, the initials of the root apical meristem seem to lack intrinsic pattern-generating information. This is contrary to the traditional view of meristems as autonomous pattern-generating machines.

vascular tissue. The development of some storage roots, such as that of the carrot *(Daucus carota),* is essentially similar to that of nonfleshy roots, except for a predominance of parenchyma cells in the secondary xylem and phloem of the storage roots. The root of the sweet potato *(Ipomoea batatas)* develops much like that of the carrot; however, in the sweet potato, additional vascular cambium cells develop within the secondary xylem around individual vessels or groups of vessels (Figure 24–22). These additional cambia (plural of cambium), while producing a few tracheary elements toward and a few sieve tubes away

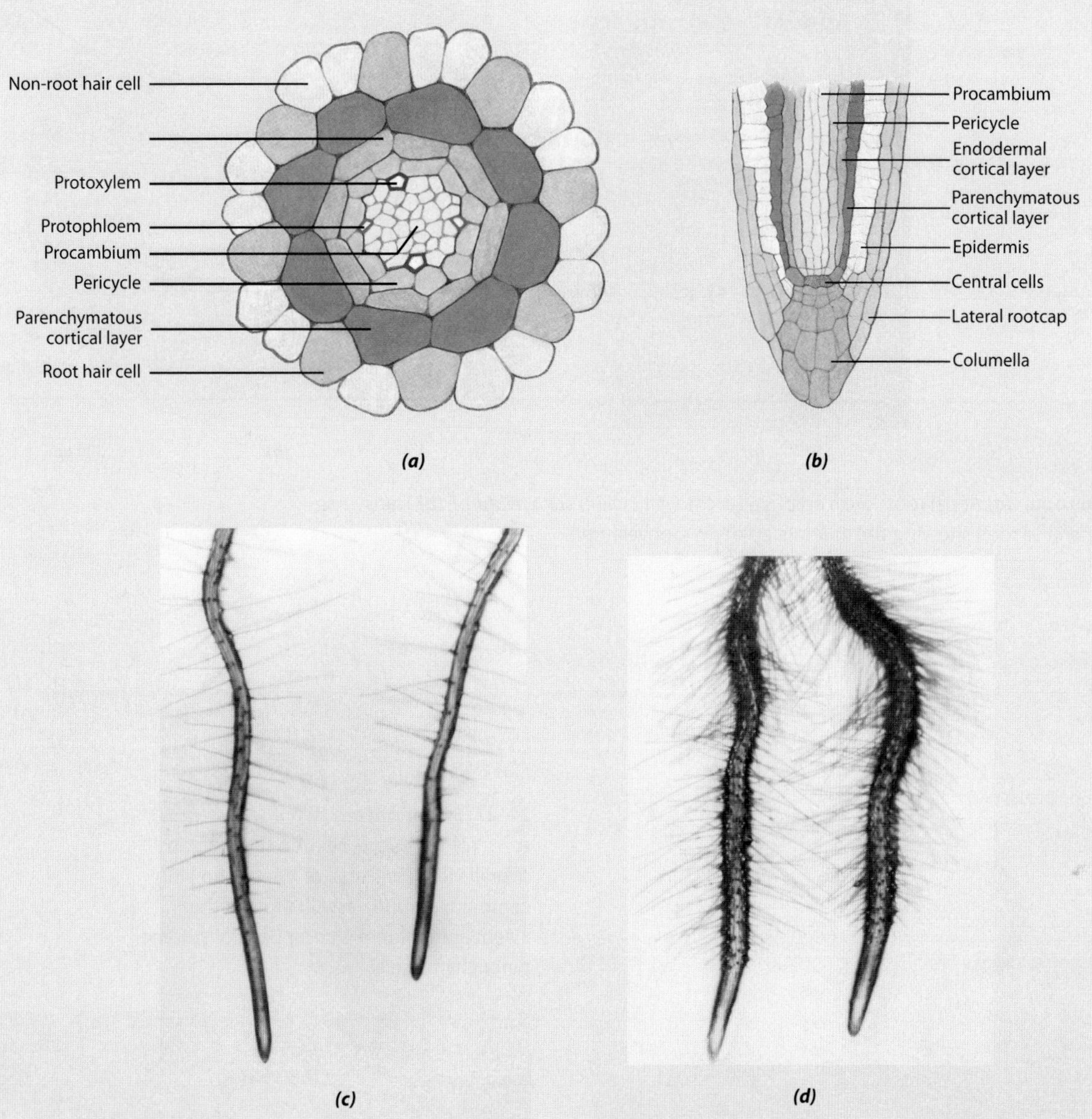

The root of *Arabidopsis thaliana* ***(a)*** Transverse section of a root prior to the development of root hairs, showing the outer epidermal cells, the middle cortical cells, and the innermost vascular cylinder. Note that the epidermal cells destined to form root hairs are located over radial walls of cortical cells, whereas those that will remain hairless are located directly over cortical cells. ***(b)*** Longitudinal section of a root tip showing the relationship of the different layers or regions of the root to the layers of initials in the apical meristem. ***(c)*** Seedling roots of wild-type *Arabidopsis thaliana,* showing normal frequency and arrangement of root hairs. ***(d)*** Seedling roots of a *ttg* mutant with an excessive number of root hairs.

from the vessels, mainly produce storage parenchyma cells in both directions. In the sugar beet *(Beta vulgaris),* most of the increase in thickness of the root results from the development of extra cambia (supernumerary, or successive, cambia) around the original vascular cambium (Figure 24–23). These concentric layers of cambia, which superficially resemble growth rings in woody roots and stems, produce parenchyma-dominated xylem toward the inside and phloem toward the outside. The upper portion of most fleshy roots actually develops from the hypocotyl.

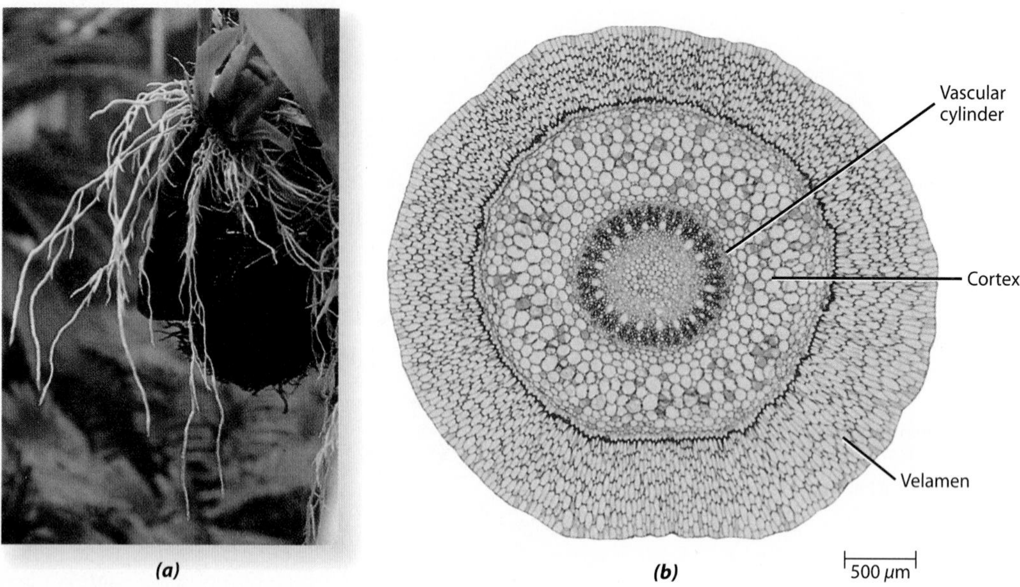

24–21 Aerial roots ***(a)*** Aerial roots of an epiphytic orchid *(Oncidium sphacelatum)*. ***(b)*** Transverse section of an orchid root, showing the multiple epidermis, or velamen.

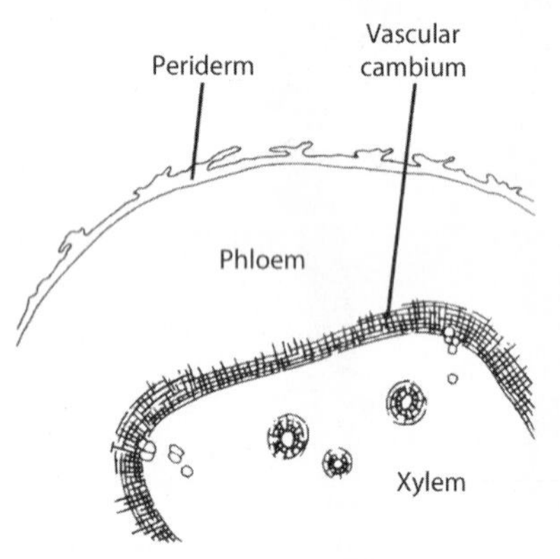

24–22 Sweet potato *(Ipomoea batatas)* root ***(a)*** Transverse section showing an overall view. ***(b)*** Detail of xylem, showing cambium around vessels. Most of the xylem and phloem is composed of storage parenchyma cells.

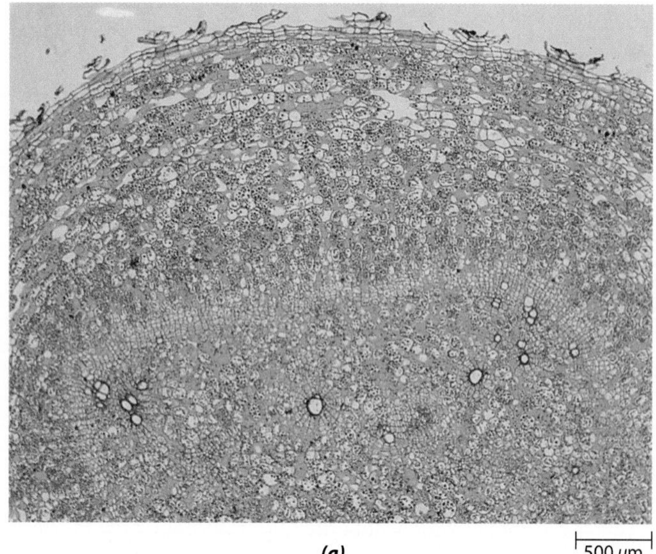

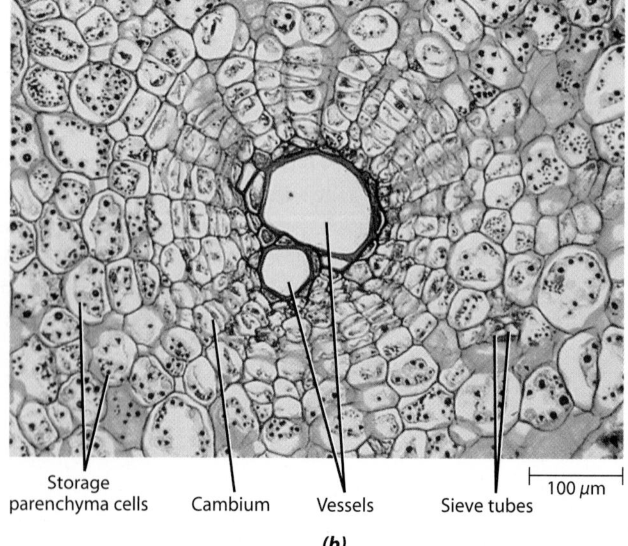

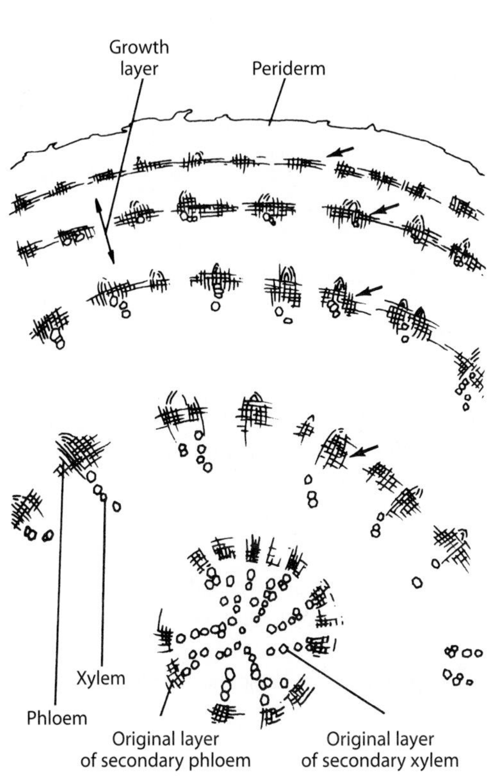

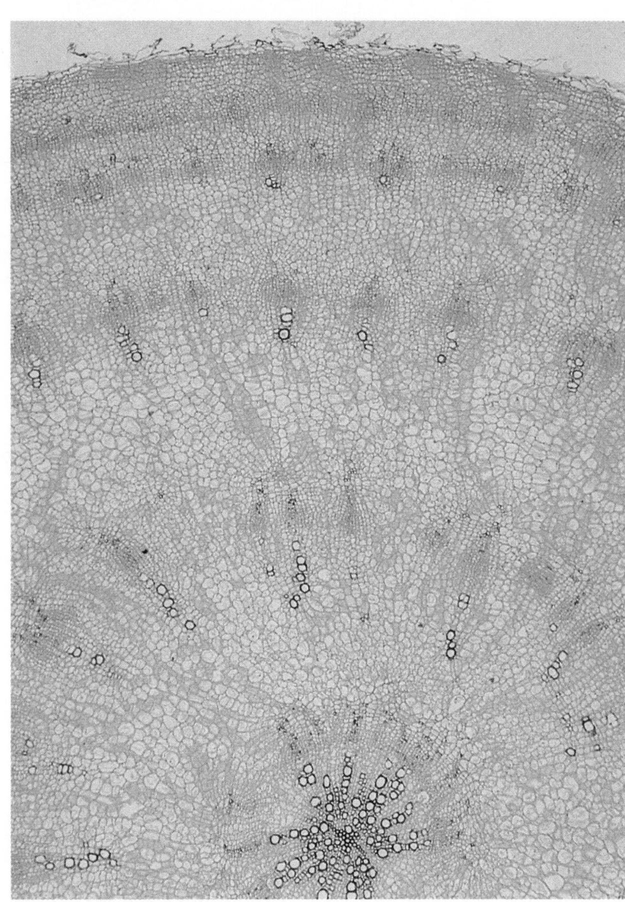

24–23 Sugar beet *(Beta vulgaris)* root Increase in thickness of the sugar beet root results largely from the activity of supernumerary cambia (arrows). The original vascular cambium produces relatively little secondary xylem and secondary phloem (in the center of the root).

SUMMARY

Roots Are Organs Specialized for Anchorage, Absorption, Storage, and Conduction

Seed plants other than monocots commonly produce taproot systems, which form from a strongly developed primary root and its branches. Monocots usually produce fibrous root systems, in which no one root is more prominent than the others. The extent of the root system depends on several factors, but the bulk of most fine roots—the roots actively engaged in the uptake of water and minerals—is found in the upper 15 centimeters of the soil.

The Root Tip Can Be Roughly Divided into Regions of Cell Division, Elongation, and Maturation

The apical meristems of most actively growing roots contain a quiescent center, which corresponds to the promeristem (the initials and their immediate derivatives); most cell division occurs a short distance from the quiescent initials. In addition to the apical meristem, or region of cell division, two other growth regions can be recognized in growing roots: the region of elongation and the region of maturation. During primary growth, the apical meristem gives rise to the three primary meristems—protoderm, ground meristem, and procambium—which differentiate into epidermis, cortex, and vascular cylinder, respectively. In addition, the apical meristem produces the rootcap, which controls the direction of root movement and serves to protect the meristem and aid the root in its penetration of the soil. Mucilage produced by the outer rootcap cells lubricates the root during its passage through the soil. Cells on the periphery of the rootcap are released into the rhizosphere as the rootcap is pushed through the soil. These border cells, which may remain alive for several weeks, exude specific proteins into the rhizosphere. Several functions have been attributed to the border cells and their exudates. The columella of the rootcap plays an important role in the response of the root to gravity (gravitropism) and water potential gradients (hydrotropism).

The Root Epidermis and Cortex May Be Modified with Age

Many epidermal cells of the root develop root hairs, which greatly increase the absorbing surface of the root. With the exception of the endodermal layer, the cortex contains numerous intercellular spaces. The compactly arranged cells of the endodermis, which forms the inner boundary of the cortex, contain

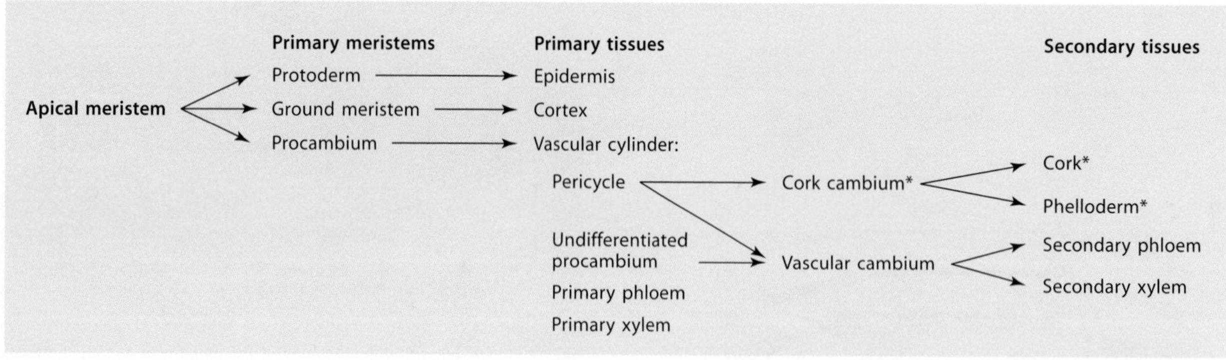

24–24 Summary of root development in a woody eudicot during the first year of growth

Casparian strips on their anticlinal walls. Consequently, all substances moving between the cortex and the vascular cylinder must pass through the protoplasts of the endodermal cells. The roots of most angiosperms also have an exodermis, which forms the outer boundary of the cortex and also consists of a compact layer of cells with Casparian strips.

The Vascular Cylinder Consists of the Primary Vascular Tissues and the Encircling Pericycle

The primary xylem usually occupies the center of the vascular cylinder and has radiating ridges that alternate with strands of primary phloem. In addition to the vascular cylinder, the root at the end of primary growth consists of the cortex and epidermis. Lateral roots, or branch roots, originate in the pericycle—they are said to be endogenous because they originate from within—and push their way to the outside through the cortex and epidermis.

Secondary Growth in Roots Involves Both the Vascular Cambium and the Cork Cambium

Secondary growth results in disruption of the primary body of the root as the strands of primary phloem are separated from the primary xylem. This occurs through the formation of secondary vascular tissues by the vascular cambium. In roots, the vascular cambium arises partly from procambium that remains undifferentiated between the primary xylem and the primary phloem strands, and partly from pericycle opposite the ridges of primary xylem. In most woody roots, the cork cambium of the first periderm originates from the pericycle. Consequently, formation of the periderm results in isolation and eventual separation of the cortex and epidermis from the rest of the root. Figure 24–24 presents a summary of root development in a woody eudicot, beginning with the apical meristem and ending with the secondary tissues produced during the first year's growth.

Root Modifications Include Aerial Roots, Air Roots, and Fleshy Roots

Most roots are storage organs, and in some plants, such as the carrot, sweet potato, and sugar beet, the roots are specialized for this function. The fleshy roots have an abundance of storage parenchyma permeated by vascular tissue.

QUESTIONS

1. Distinguish between the following: endodermal cells and passage cells; endodermis and exodermis; protoxylem and metaxylem; aerial roots and air roots.
2. What are the principal functions of roots?
3. Discuss the need for a plant to maintain a balance between its shoot and root systems.
4. What are border cells, and what are some functions attributed to them?
5. During growth in the length of a root, a very limited portion of the root is constantly being pushed through the soil. Explain.
6. How do the Casparian strips of endodermal cells affect the movement of water and solutes across the endodermis?
7. Distinguish between the promeristem and the apical meristem of a root. Which part corresponds to the quiescent center?
8. What structural features do all fleshy storage roots have in common?

CHAPTER 25

The Shoot: Primary Structure and Development

◀ **Expanding shoot** During the winter, this shoot of the shagbark hickory *(Carya ovata)* was greatly telescoped and existed as a terminal bud. Expansion of the bud has caused the protective brown bud scales to separate and fold back. After the shoot is fully expanded, a new terminal bud will form and pass through a dormant period before it, too, is capable of expanding and repeating the cycle.

CHAPTER OUTLINE

The **shoot,** which consists of the stem and its leaves, is the aboveground portion of the plant that is familiar to us. It is initiated during development of the embryo, where it may be represented by a plumule, consisting of an epicotyl (the stem above the level of cotyledon attachment), one or more young leaves, and an apical meristem, or by an apical meristem only. As we shall see, the shoot is structurally more complex than the root. In contrast to the root, the shoot has nodes and internodes, with one or more leaves attached at each node (Figures 25–1 and 25–2). Whereas the shoot apex produces leaves and axillary buds, which develop into lateral shoots, the root apex produces no lateral organs. (You will recall from Chapter 24 that lateral roots arise in the region of maturation, which occurs behind the root apex.) At each node, one or more strands of the vascular cylinder of the stem turn outward and extend into the leaf, leaving one or more gaps in the vascular cylinder opposite the leaf. There are no such gaps in the vascular cylinders (protosteles) of roots.

CHECKPOINTS

After reading this chapter, you should be able to answer the following:

1. Describe the structure of the shoot apical meristem of angiosperms. What is the relationship between the zones of the shoot apical meristem and the primary meristems of the shoot?
2. Name the three basic types of organization found in the primary structure of the stems of seed plants.
3. What are leaf traces, and how are they indicative of the intimate relationship that exists between the stem and the leaf?
4. What hypotheses have been proposed to explain the pattern of leaf arrangement on stems?
5. Describe the structural differences between the leaves of monocots and those of other angiosperms.
6. How have homeotic mutations contributed to our understanding of the genetic control of flower development?

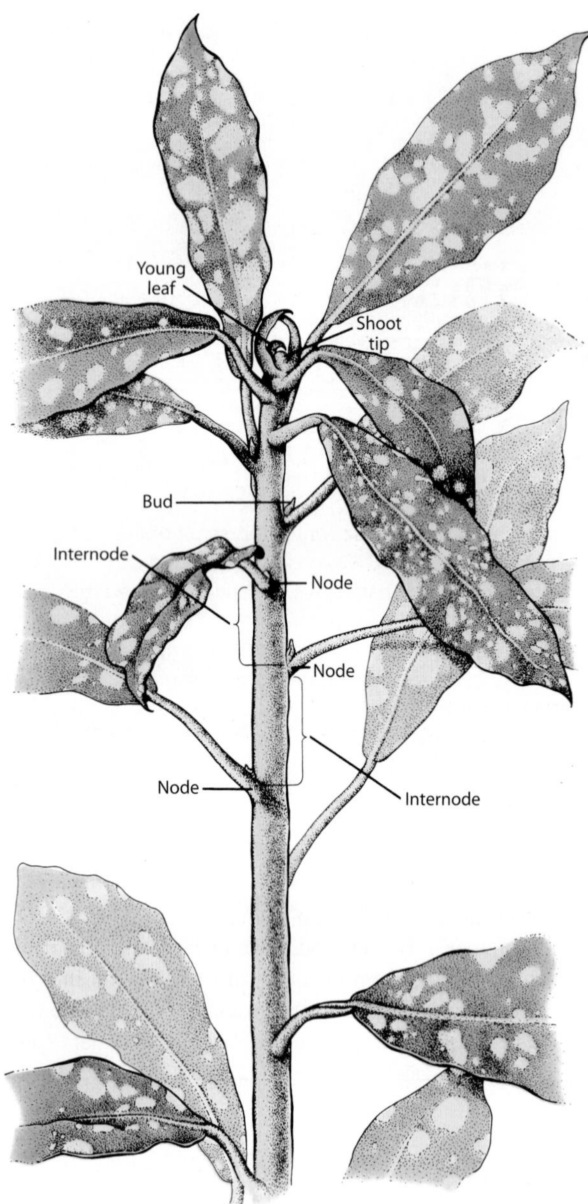

25–1 Portion of a *Croton* shoot The leaves of *Croton,* a eudicot, have a mottled appearance due to clonal variations in the ability of leaf cells to produce chlorophyll, and they are spirally arranged along the stem. At the shoot tip, the leaves are so close together that nodes and internodes are not distinguishable as separate regions of the stem. Growth in length of the stem between successive leaves, which are attached to the stem at the nodes, results in formation of the internodes.

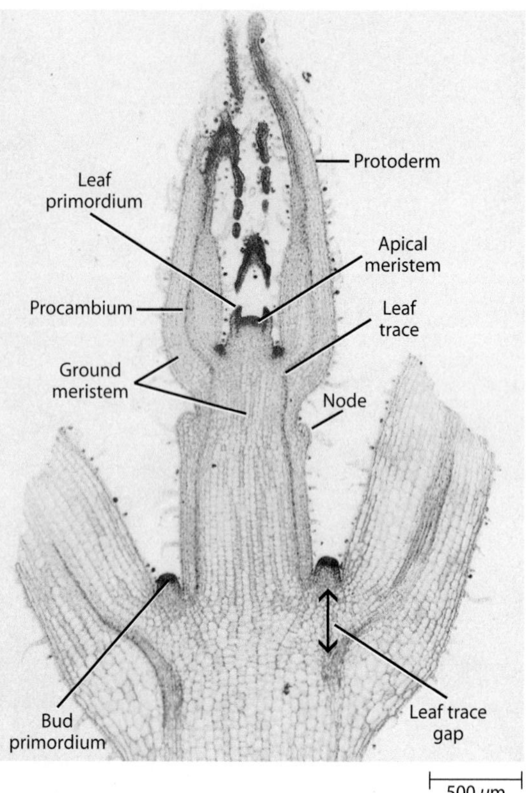

25–2 *Coleus* shoot tip The leaves of the common houseplant *Coleus blumei,* a eudicot, are arranged opposite one another at the nodes. Each successive pair is at right angles to the previous pair (decussate phyllotaxis), and therefore the leaves of the labeled node are at right angles to the plane of section. (See pages 588 and 589 for a discussion of leaf traces, leaf trace gaps, and phyllotaxis.)

The two principal functions associated with stems are **support** and **conduction.** The leaves—the principal photosynthetic organs of the plant—are supported by the stems, which place the leaves in favorable positions for exposure to light. Substances manufactured in the leaves are transported downward through the stems by way of the phloem to sites where they are needed, such as developing plant parts and storage tissues of both stems and roots. At the same time, water and minerals are transported upward in the xylem from the roots and into the leaves via the stem.

Origin and Growth of the Primary Tissues of the Stem

The apical meristem of the shoot is a dynamic structure that, in addition to adding cells to the primary plant body, repetitively produces leaf primordia and bud primordia, resulting in a succession of repeated units called **phytomeres** (Figure 25–3). The **leaf primordia** develop into leaves and the **bud primordia** into lateral shoots. Unlike the root apical meristem, the vegetative shoot apical meristem lacks a specialized protective covering comparable to the rootcap. On the other hand, it commonly is surrounded by young leaves that fold over it and provide it with protection. Although the term "shoot apex" often is used as a synonym for shoot apical meristem, more correctly, the term **apical meristem** denotes only the part of the shoot lying distal to, or above, the youngest leaf primordium. The **shoot apex** includes the apical meristem together with the subapical region bearing young leaf primordia.

The vegetative shoot apical meristem of most flowering plants has what is called a **tunica-corpus** type of organization (Figure 25–4). The two regions—tunica and corpus—are distinguished by the planes of cell division that occur in them. The tunica consists of the outermost layer (or layers) of cells that divide **anticlinally,** that is, in planes perpendicular to the surface of the meristem (Figure 25–4c). These divisions contribute

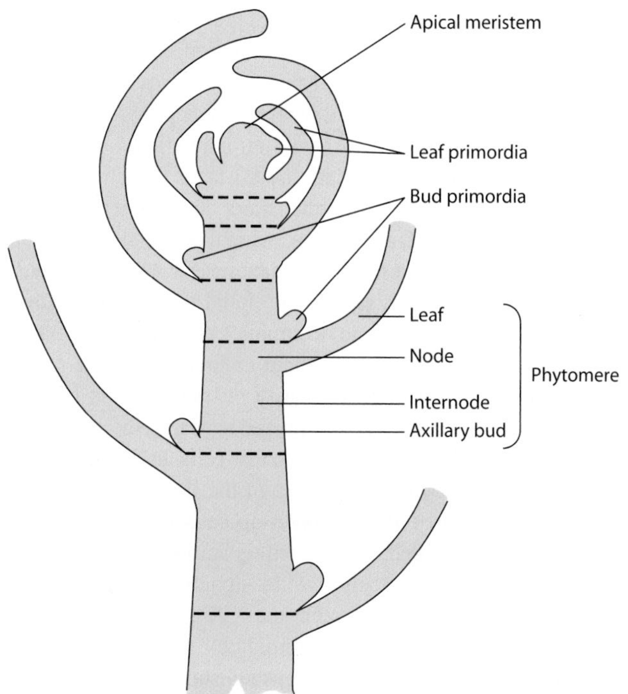

25–3 Phytomeres The apical meristem at the tip of the shoot is protected by young leaves that fold over it, as seen in this longitudinal section of a eudicot shoot. Activity of the apical meristem, which repetitively produces leaf and bud primordia, results in a succession of repeated units called phytomeres. Each phytomere consists of a node with its attached leaf, the internode below that leaf, and the bud at the base of the internode. The boundaries of the phytomeres are indicated by the dashed lines. Note that the internodes are of increasing length the farther they are from the apical meristem. Internodal elongation accounts for most of the increase in length of the stem.

to surface growth, without increasing the number of cell layers in the meristem. The corpus consists of a body of cells that lie beneath the tunica layers. In the corpus, the cells divide in various planes and add bulk to the developing shoot. The corpus and each layer of the tunica have their own initials. The initials of the corpus lie beneath those of the tunica and add cells to the corpus by dividing **periclinally,** that is, parallel with the apical surface (Figure 25–4c). Hence, the number of layers of initials in any given meristem is equal to the number of layers of tunica plus one.

The number of tunica layers differs from species to species. Most eudicots have apices consisting of three superimposed layers of cells: two tunica layers and the initial layer of the corpus. These three cell layers commonly are designated L1 (outermost), L2, and L3 (innermost) (Figure 25–4a). Although the L1 layer divides almost exclusively anticlinally, L1 cells occasionally divide periclinally. When this happens, the inner daughter cell

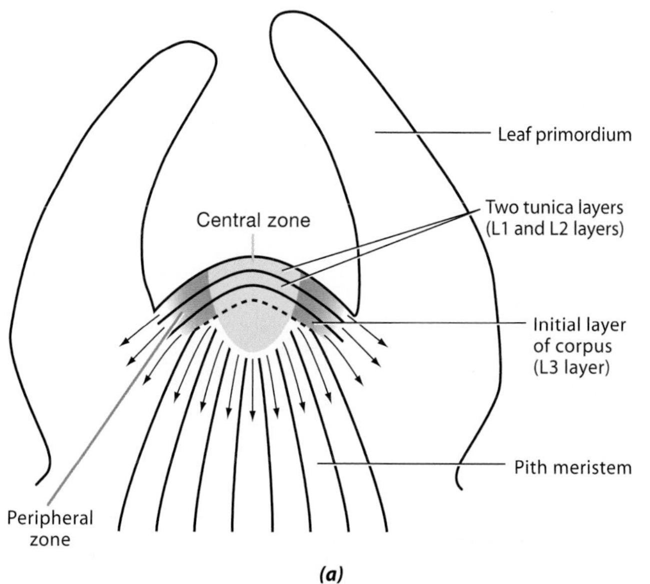

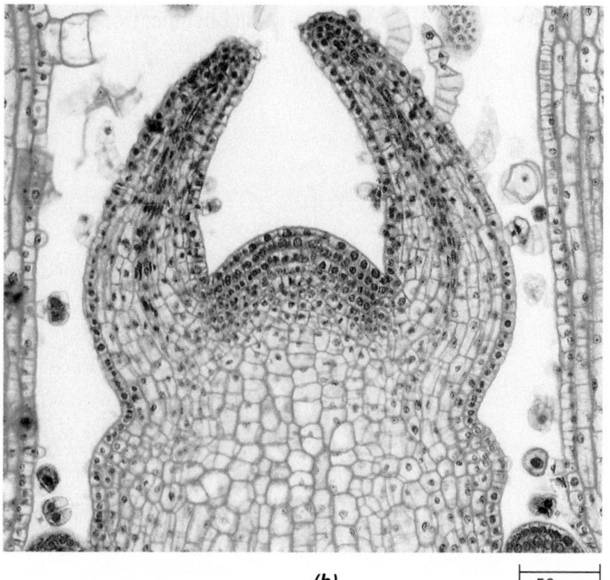

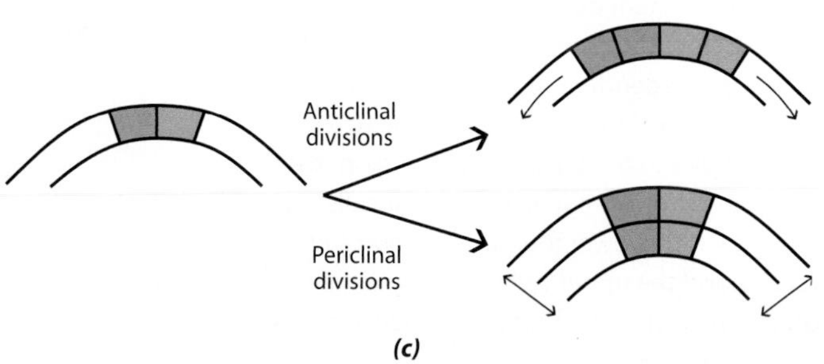

25–4 Tunica-corpus organization ***(a), (b)*** Detail of a *Coleus blumei* shoot apex. *Coleus* has a two-layered tunica, represented by the L1 and L2 layers of the apical meristem. The initial layer of the corpus is represented by the L3 layer. The corpus and portions of overlying tunica layers correspond to the central zone. The most mitotically active part of the apical meristem is the peripheral zone. ***(c)*** Diagram illustrating anticlinal and periclinal divisions. Cell divisions in the tunica layers are almost exclusively anticlinal, whereas those in the initial layer of the corpus are both anticlinal and periclinal. By dividing periclinally, the cells of the initial layer of the corpus add cells to the corpus.

is displaced into the L2 layer, where it differentiates as though it were derived from the L2 layer. Similar displacements may occur between derivatives of the L2 and L3 layers, with comparable results. Here again is evidence that cell differentiation is not dependent on cell lineage but rather on the final position of a cell in the developing organ (see the essay on pages 574 and 575).

In the shoot apices of angiosperms, the corpus and the portions of the tunica layers overlying the corpus constitute the so-called **central zone,** which corresponds to the promeristem of the shoot apical meristem. The central zone is surrounded by the **peripheral zone,** or peripheral meristem, which originates partly from the tunica (L1 and L2 layers) and partly from the corpus, whose origin can be traced to the L3 layer (Figure 25–4). Three-dimensionally, the peripheral zone forms a ring around the central zone. To the inside of this ring and just beneath the central zone is the **pith meristem.** Cell divisions are relatively infrequent in the central zone, and in this respect, the central zone is analogous to the quiescent center of the root apical meristem. However, the peripheral zone is mitotically very active.

Genetic and molecular studies have identified genes required for the establishment of and regulation of the size of the vegetative shoot apical meristem in *Arabidopsis* (Figure 25–5). For example, establishment of the shoot apical meristem requires activity of the *SHOOTMERISTEMLESS (STM)* gene, which is first expressed in one or two cells of the late globular stage embryo. Severe loss-of-function *stm* mutations result in seedlings with normal roots, hypocotyls, and cotyledons, but lacking apical meristems. A second gene, *WUSCHEL (WUS),* first expressed at the 16-cell stage of embryo development—long before the meristem is evident—is necessary both for establishment of the shoot apical meristem and for maintenance of initial-cell function. In *wus* mutants, the initials undergo differentiation. STM messenger RNA is present in the central and peripheral zones of the shoot apical meristem but is absent in developing leaf primordia. In the fully developed embryo, *WUS* expression is restricted to a small group of central-zone cells beneath the L3 layer (the initial layer of the corpus) and persists throughout shoot development.

In addition to *STM* and *WUS,* which promote initial-cell function, other genes—the *CLAVATA* genes *(CLV1, CLV2, CLV3)* —regulate meristem size by repressing initial-cell activity. Mutations in the *CLV* genes cause an accumulation of undifferentiated cells in the central zone, bringing about an increase in size of the meristem. *CLV3* expression is primarily restricted to the L1 and L2 layers and a few L3 cells, and probably marks the initials in these layers. The *CLV1-CLV2*-expressing cells underlie the L1 and L2 layers, and *WUS* is expressed in the deepest part of the meristem. The region of *WUS*-expressing cells has been termed the *organizing center,* because it confers initial-cell identity to the overlying neighboring cells, while signals from the region of *CLV3*-expressing cells, through the *CLV1-CLV2*-expressing region, act negatively to dampen initial-cell activity. Thus, the interaction between *WUS* and *CLV3* establishes a feedback loop with the potential to adjust the size of the initial-cell population.

What is the relationship between these zones and the primary meristems of the shoot? The protoderm always originates from the outermost tunica (L1) layer, whereas the procambium and part of the ground meristem (the cortex and sometimes part of the pith) are derived from the peripheral meristem. The rest of the ground meristem (all or most of the pith) is formed by the pith meristem.

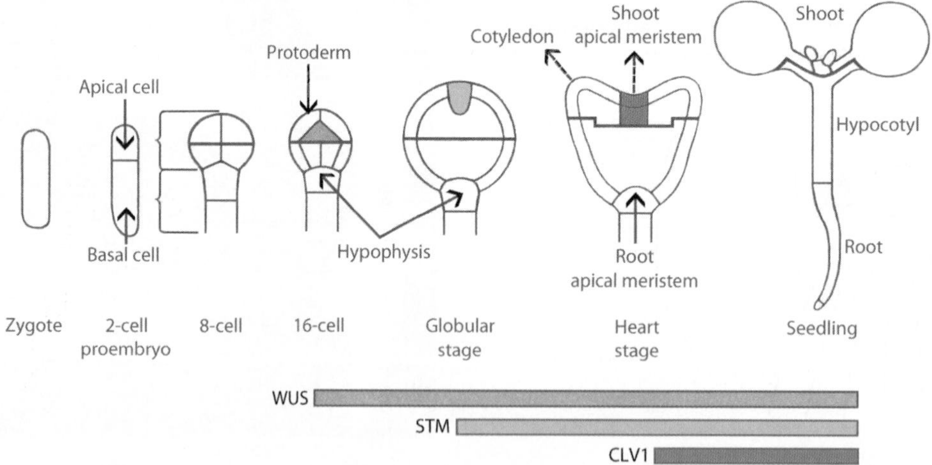

25–5 Formation of the shoot apical meristem The horizontal bars denote the stages at which messenger RNA for each of the genes is detected. The first indication of formation of the shoot apical meristem during embryo development in *Arabidopsis* is the initiation of *WUS* expression at the 16-cell stage. This is long before the meristem is discernible. Subsequently, *STM* and *CLV1* expression is initiated. The onset of *STM* expression is independent of *WUS* activity, and the initiation of *CLV1* expression is independent of *STM*. Note that an asymmetrical division of the zygote gives rise to a small apical cell and a large basal cell. The apical cell is the precursor of the embryo proper and the basal cell of the suspensor. Vertical and transverse divisions of the apical cell result in an 8-cell proembryo, consisting of two layers of four cells each. The upper four cells are the source of the shoot apical meristem and cotyledons, and the lower four are the source of the hypocotyl. The uppermost cell of the filamentous suspensor divides transversely, and the upper cell (hypophysis) gives rise to the central cells of the root apical meristem and the columella of the rootcap. The rest of the root apical meristem and lateral rootcap are derived from the embryo proper.

(a)

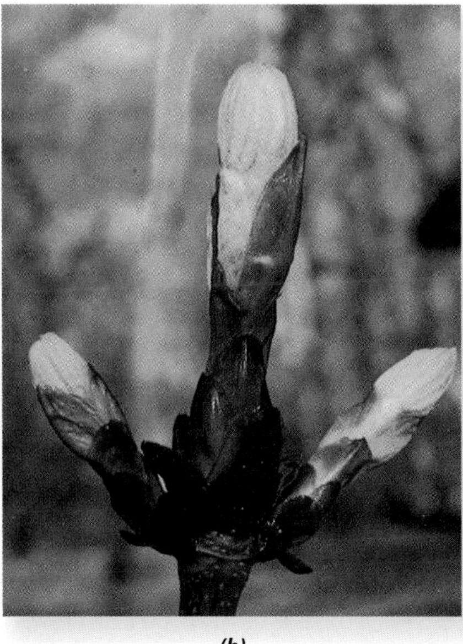

(b)

(c)

25–6 Growth of horse chestnut buds Stages in growth of the terminal bud and two lateral buds of the horse chestnut *(Aesculus hippocastanum)*. ***(a)*** The young shoots are tightly packed in the buds and are protected by bud scales, which are highly modified leaves initiated late in the previous growing season. ***(b)*** The buds open to reveal the oldest rudimentary leaves. ***(c)*** Internodal elongation has separated the nodes from one another. The terminal bud of the horse chestnut is a mixed bud, containing both leaves and flowers, although the flowers are not visible here. The lateral buds produce only leaves.

Although the primary tissues of the stem pass through periods of growth similar to those of the root, the stem cannot be divided along its axis into regions of cell division, elongation, and maturation as in the case of roots. When actively growing, the apical meristem of the shoot gives rise to leaf primordia in such rapid succession that nodes and internodes cannot at first be distinguished. As growth begins between the levels of leaf attachment, the elongated parts of the stem take on the appearance of internodes, and the portions of the stem where the leaves are attached become recognizable as nodes (Figures 25–1 and 25–6). Thus, increase in length of the stem occurs largely by internodal elongation, which may take place simultaneously over several internodes.

The meristematic activity causing elongation of the internode may be fairly uniform throughout the internode. In some species it occurs as a wave progressing from the base of the internode upward, while in others, such as grasses, it is restricted largely to the base of the internode. A localized meristematic region in the elongating internode is called an **intercalary meristem** (a meristematic region between two more highly differentiated regions). Certain elements of the primary xylem and primary phloem—specifically the protoxylem and protophloem—differentiate within the intercalary meristem and connect the more highly differentiated regions of the stem above and below the meristem.

Increase in stem thickness during primary growth involves both longitudinal (periclinal) divisions and cell enlargement. In plants with secondary growth, this primary thickening is moderate. Monocots usually lack secondary growth, but a few, such as palms, have massive primary growth. This growth occurs so close to the apical meristem that the shoot apex appears inserted on a shallow cone or even in a depression (Figure 25–7). Activity in the apex proper is not great, but immediately behind it cell division is intensive. The meristem responsible for the abrupt expansion of the apical region into a wide crown lies below the young leaf bases. Within this meristematic region, localized cell divisions result in the formation of procambial strands. This zone of procambium formation is known as the *meristematic cap*. The bulk of the meristem responsible for stem thickening is located below the cap, although the ground tissue between the procambial strands of the cap also contributes to the thickening.

The apical meristem of the shoot gives rise to the same primary meristems found in the root: protoderm, procambium, and ground meristem (Figure 25–2). These primary meristems in turn develop into the mature tissues of the primary plant body: protoderm becomes epidermis, procambium becomes primary vascular tissues, and ground meristem becomes ground tissue.

Primary Structure of the Stem

Considerable variation exists in the primary structure of stems of seed plants, but three basic types of organization can be recognized. (1) In some seed plants other than monocots, the vascular system of the internode appears as a more or less continuous cylinder within the ground tissue (Figure 25–8a). (2) In others, the primary vascular tissues develop as a cylinder of discrete strands, or bundles, separated by ground tissue (Figure 25–8b). (3) In the stems of most monocots and of some herbaceous (nonwoody) eudicots, the arrangement of the procambial strands and vascular bundles is more complex. As seen in transverse sections, the

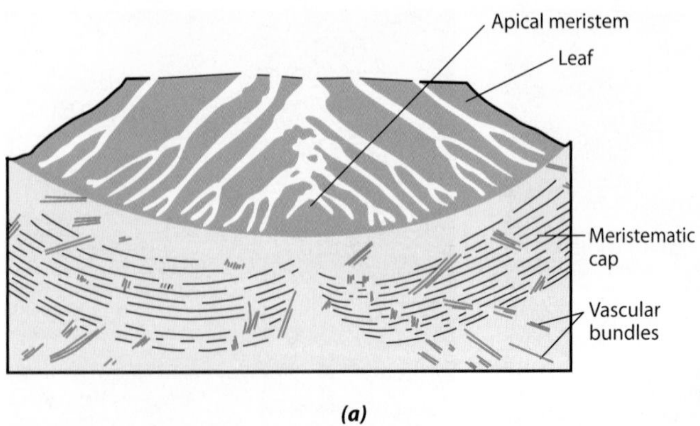

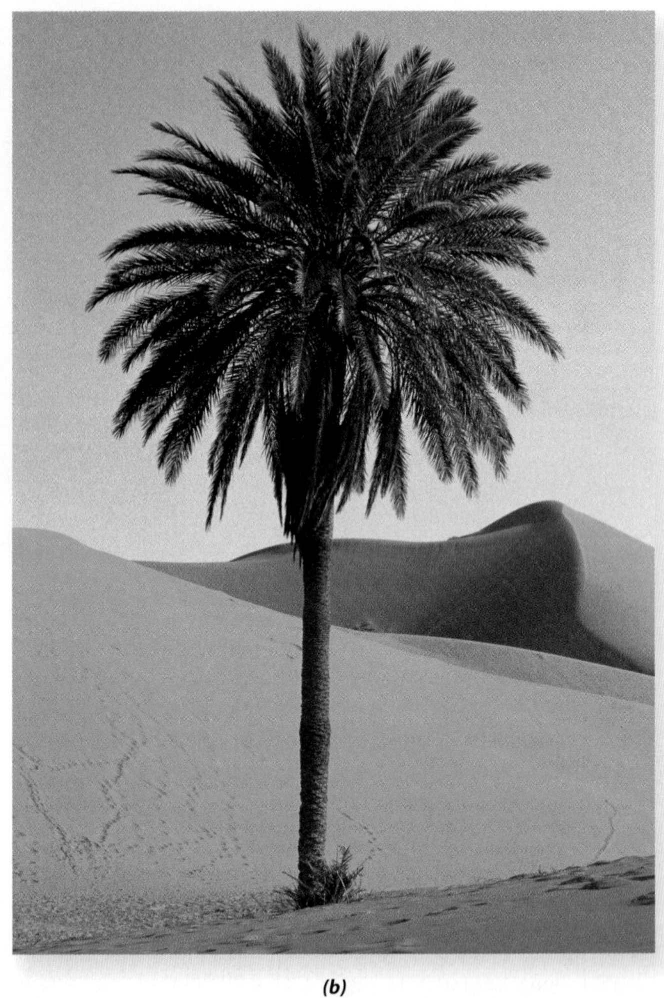

25–7 Increase in stem thickness in monocots ***(a)*** Diagrammatic representation of the anatomy of the top, or crown, of a thick-stemmed monocot without secondary growth, such as a palm tree ***(b)***. Increase in thickness is due to meristematic activity below the young leaf bases. The apical meristem and youngest leaf primordia are conventional in size, although they appear sunken below broad stem tissues. The zone of procambium formation is called the meristematic cap.

vascular bundles occur in more than one ring of bundles or are scattered throughout the ground tissue. In the latter instance, the ground tissue cannot be distinguished as cortex and pith (Figure 25–8c).

In the discussion that follows, the stem of basswood *(Tilia americana)* is used to exemplify the first type of organization. The second type is exemplified by the elderberry *(Sambucus canadensis),* alfalfa *(Medicago sativa),* and buttercup *(Ranunculus)* stems, and the third type by maize *(Zea mays).* The basswood and elderberry stems are also examples of stems that undergo much secondary growth. (They will be revisited during our discussion of secondary growth in Chapter 26.) By contrast, the alfalfa stem undergoes relatively little secondary growth, and the stems of buttercup (a eudicot) and maize (a monocot) undergo none at all.

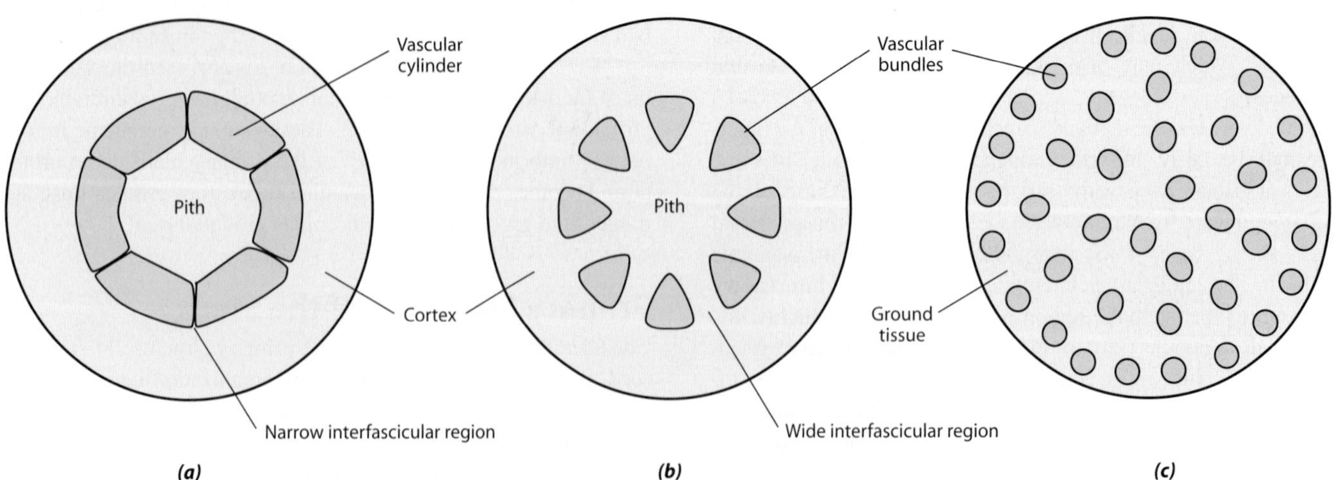

25–8 Three basic types of organization in the primary structure of stems, as seen in transverse section ***(a)*** The vascular system appears as a continuous hollow cylinder around the pith. ***(b)*** Discrete vascular bundles form a single ring around the pith. ***(c)*** The vascular bundles are scattered throughout the ground tissue.

The Primary Vascular Tissues of the Basswood Stem Form an Almost Continuous Vascular Cylinder

Figure 25–9 shows the basswood *(Tilia americana)* stem, with what appears to be a continuous cylinder of primary vascular tissues. In fact, the vascular cylinder is composed of vascular bundles that are separated from one another by very narrow, inconspicuous regions of ground parenchyma. These parenchymatous regions, called **interfascicular regions,** interconnect the cortex and pith. (Interfascicular means "between the bundles, or fascicles.")

As in most stems, the epidermis is a single layer of cells covered by a cuticle. The stem epidermis generally contains far fewer stomata than the leaf epidermis.

The cortex consists of collenchyma and parenchyma cells. The several layers of collenchyma cells, which provide support to the young stem, form a continuous cylinder beneath the epidermis. The rest of the cortex consists of parenchyma cells that will contain chloroplasts when mature. The innermost layer of cortical cells, which have deeply colored contents, sharply delimits the cortex from the cylinder of primary vascular tissues.

In the great majority of stems, including those of *Tilia,* the primary phloem develops from the outer cells of the procambium, and the primary xylem develops from the inner cells. However, not all procambial cells mature as primary tissues. A single layer of cells between the primary xylem and the primary phloem remains meristematic and becomes the vascular cambium. *Tilia* is also an example of a woody stem—a stem that produces much secondary xylem later in growth. After internodal elongation is completed in the *Tilia* stem, fibers develop in the primary phloem. These fibers are called *primary phloem fibers* (see Figure 26–9).

The inner boundary of the primary xylem in *Tilia* is sharply delimited by one or two layers of pith cells with deeply colored contents. The pith is composed primarily of parenchyma cells and contains numerous large ducts, or canals, containing mucilage (a slimy carbohydrate). Similar ducts are formed in the cortex (Figure 25–9). As the cortical and pith cells increase in size, numerous intercellular spaces develop among them; these air spaces are essential for exchange of gases with the atmosphere. The cortical and pith parenchyma cells store various substances.

The Primary Vascular Tissues of the Elderberry Stem Form a System of Discrete Strands

In the stem of the elderberry *(Sambucus canadensis),* the interfascicular regions—also called **pith rays**—are relatively wide, and hence the procambial strands and primary vascular bundles form a system of discrete strands around the pith. The epidermis, cortex, and pith are similar in organization to those of *Tilia.* For this reason, we use the following discussion of the *Sambucus* stem to explain in more detail the development of the primary vascular tissues of stems.

Figure 25–10a shows three procambial strands in which the primary vascular tissues have just begun to differentiate. The strand on the left is somewhat older than the two on the right and contains at least one mature sieve element and one mature tracheary element. Notice that the first mature sieve element appears in the outer part of the procambial strand (next to the cortex) and that the first mature tracheary element appears in the innermost

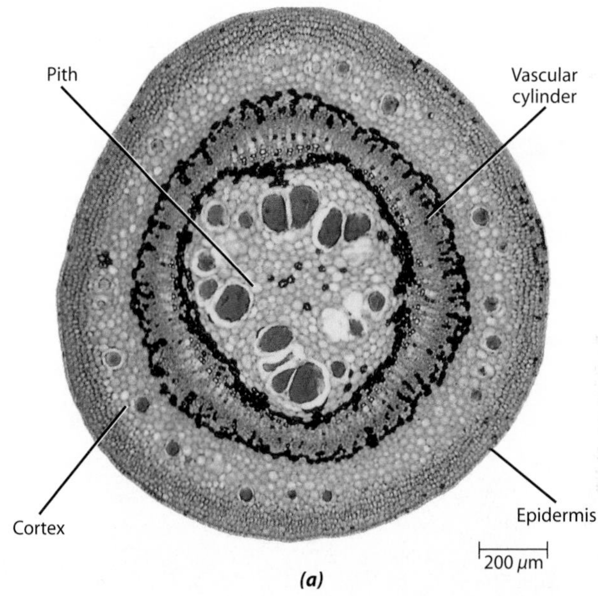

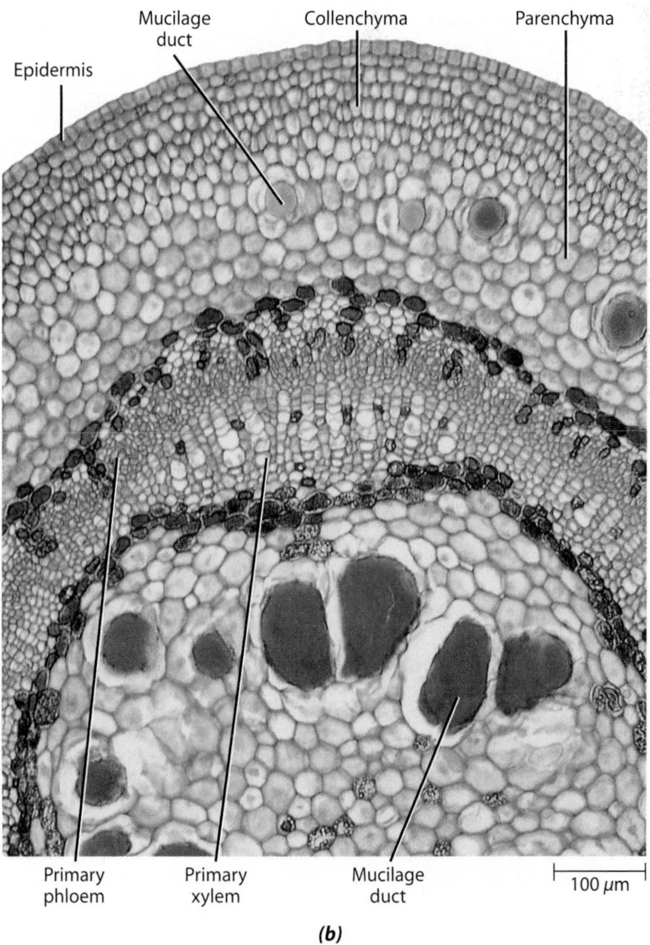

25–9 Primary growth in a basswood stem ***(a)*** Transverse section of the basswood *(Tilia americana)* stem in a primary stage of growth. The vascular tissues appear as a continuous hollow cylinder that divides the ground tissue into pith and cortex. ***(b)*** Detail of a portion of the same basswood stem.

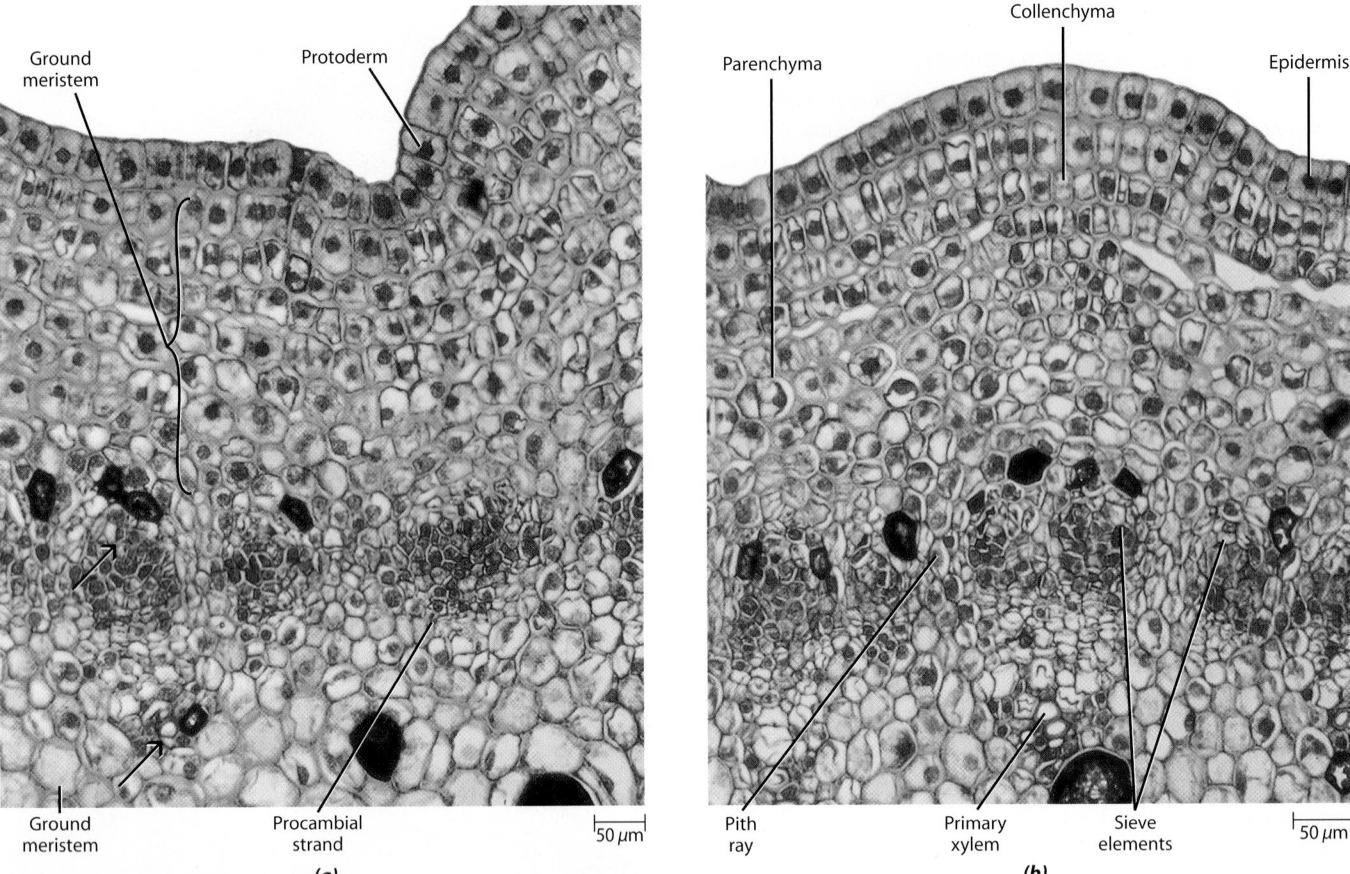

25–10 Primary growth in an elderberry stem Transverse sections of the stem of the elderberry *(Sambucus canadensis)* in a primary stage of growth. ***(a)*** A very young stem, showing protoderm, ground meristem, and three discrete procambial strands. The procambial strand on the left contains one mature sieve element (upper arrow) and one mature tracheary element (lower arrow). ***(b)*** Primary tissues further along in development. ***(c)*** Stem near completion of primary growth. Fascicular and interfascicular cambia are not yet formed. (For further stages in the growth of the elderberry stem, see Figures 26–7, 26–8, and 26–10.)

part (next to the pith). Comparing Figures 25–10a and 25–10c, we see that the more recently formed sieve elements appear closer to the center of the stem and that the xylem differentiates in the opposite direction.

The first-formed primary xylem and primary phloem elements (protoxylem and protophloem, respectively) are stretched during elongation of the internode and are frequently destroyed. As in the *Tilia* stem, fibers develop in the primary phloem after internodal elongation is completed (see Figure 26–8).

Like the stems of *Tilia,* those of *Sambucus* become woody. In *Tilia,* because the interfascicular regions are very narrow, almost all of the vascular cambium originates from procambial cells between the primary xylem and primary phloem. In *Sambucus,* with its relatively wide interfascicular regions, a substantial portion of the vascular cambium develops from the interfascicular parenchyma.

The Stems of Alfalfa and Buttercup Are Herbaceous

The stems of many eudicots undergo little or no secondary growth and therefore are **herbaceous,** or nonwoody (see Chapter 26). Examples of herbaceous eudicot stems are found in alfalfa *(Medicago sativa)* and in the buttercups *(Ranunculus).*

Medicago is an example of an herbaceous eudicot that exhibits some secondary growth (Figure 25–11). The structure and development of the primary tissues of the *Medicago* stem are similar to those of *Sambucus* and other woody angiosperms. The vascular bundles are separated by wide interfascicular regions and surround a large pith. The vascular cambium is partly fascicular (procambial) and partly interfascicular (interfascicular parenchyma) in origin. During secondary growth, the secondary vascular tissues are formed mainly from the cambial cells derived from procambium. The interfascicular cambium generally produces only sclerified parenchyma (parenchyma with secondary walls) on the xylem side.

The stem of *Ranunculus* is an example of extreme herbaceousness, and its vascular bundles resemble those of many monocots. The vascular bundles retain no procambium after the primary vascular tissues mature; hence, the bundles never develop a vascular cambium and they lose their potential for further growth. Vascular bundles such as those of *Ranunculus* (Figure 25–12) and the monocots, in which all the procambial

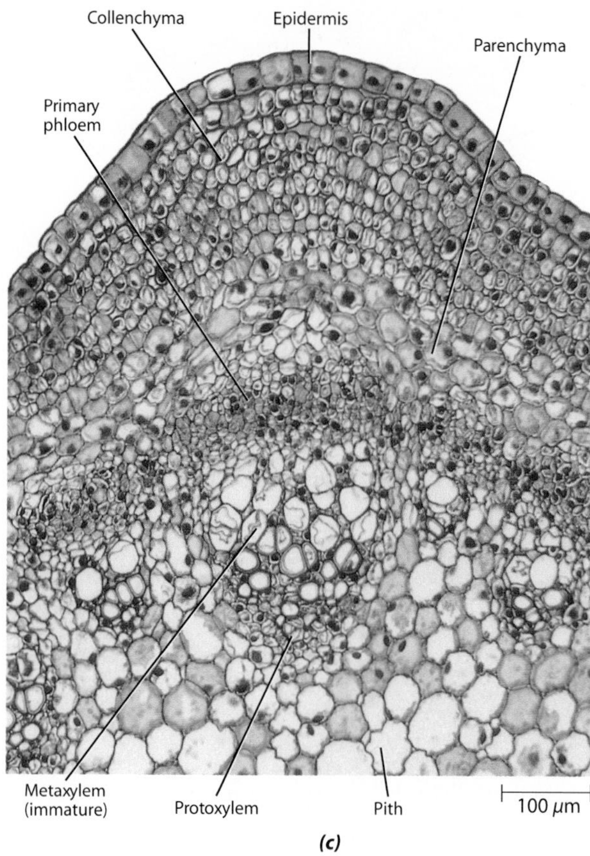

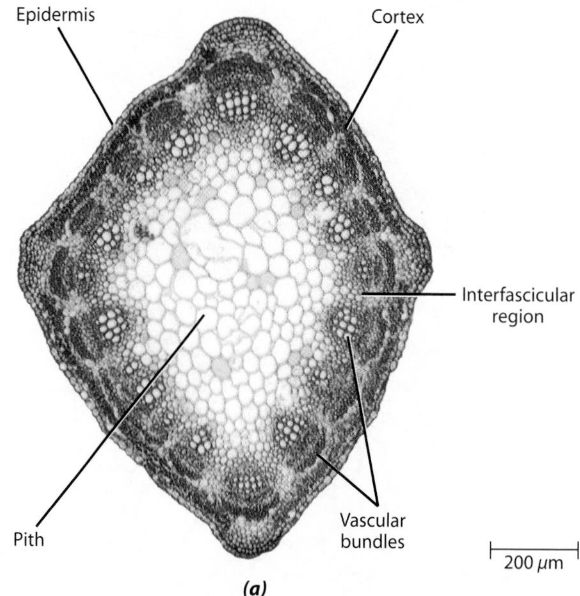

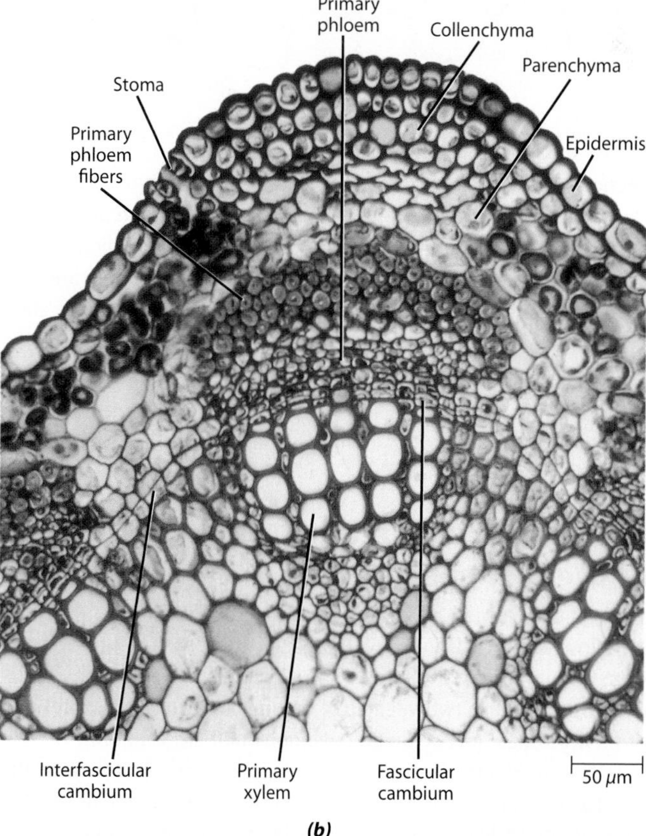

25–11 Stem of alfalfa ***(a)*** Transverse section of the stem of alfalfa *(Medicago sativa)*, a eudicot with discrete vascular bundles. ***(b)*** Detail of a portion of the same alfalfa stem.

cells mature and the potential for further growth within the bundle is lost, are said to be **closed.** Closed vascular bundles are usually entirely surrounded by a sheath composed of sclerenchyma cells. Vascular bundles that do give rise to a cambium are said to be **open.** In most eudicots, the vascular bundles are of the open type; they produce some secondary vascular tissues.

The Vascular Bundles of Maize Stems Appear Scattered in Transverse Section

The herbaceous stem of maize *(Zea mays)* exemplifies the stems of monocots in which the vascular bundles appear scattered throughout the ground tissue in transverse section (Figure 25–13). As in other monocots, the vascular bundles of maize are closed.

Figure 25–14 shows three stages in the development of a maize vascular bundle. As in the bundles of eudicot stems, the phloem develops from the outer cells of the procambial strand, and the xylem develops from the inner cells. Also, as described previously, the phloem and the xylem differentiate in opposite directions. The first-formed phloem and xylem elements (the protophloem and protoxylem) are stretched and destroyed during elongation of the internode. This results in the formation of a large space, called a protoxylem lacuna, on the xylem side of the bundle (Figure 25–14c). The mature vascular bundle contains two large metaxylem vessels, and the phloem (metaphloem) is composed of a large, clearly defined group of sieve elements and companion cells. The entire bundle is enclosed in a sheath of sclerenchyma cells.

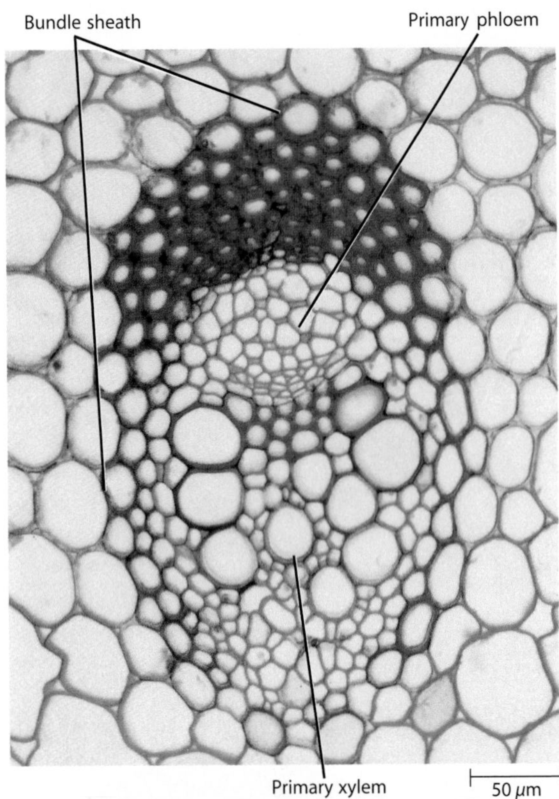

25–12 A vascular bundle in the buttercup stem Transverse section of a vascular bundle of the buttercup *(Ranunculus)*, an herbaceous eudicot. The vascular bundles of the buttercup are closed; that is, all of the procambial cells mature, precluding secondary growth. The primary phloem and primary xylem are surrounded by a bundle sheath of thick-walled sclerenchyma cells. Compare the vascular bundle shown here with the mature vascular bundle of maize shown in Figure 25–14c.

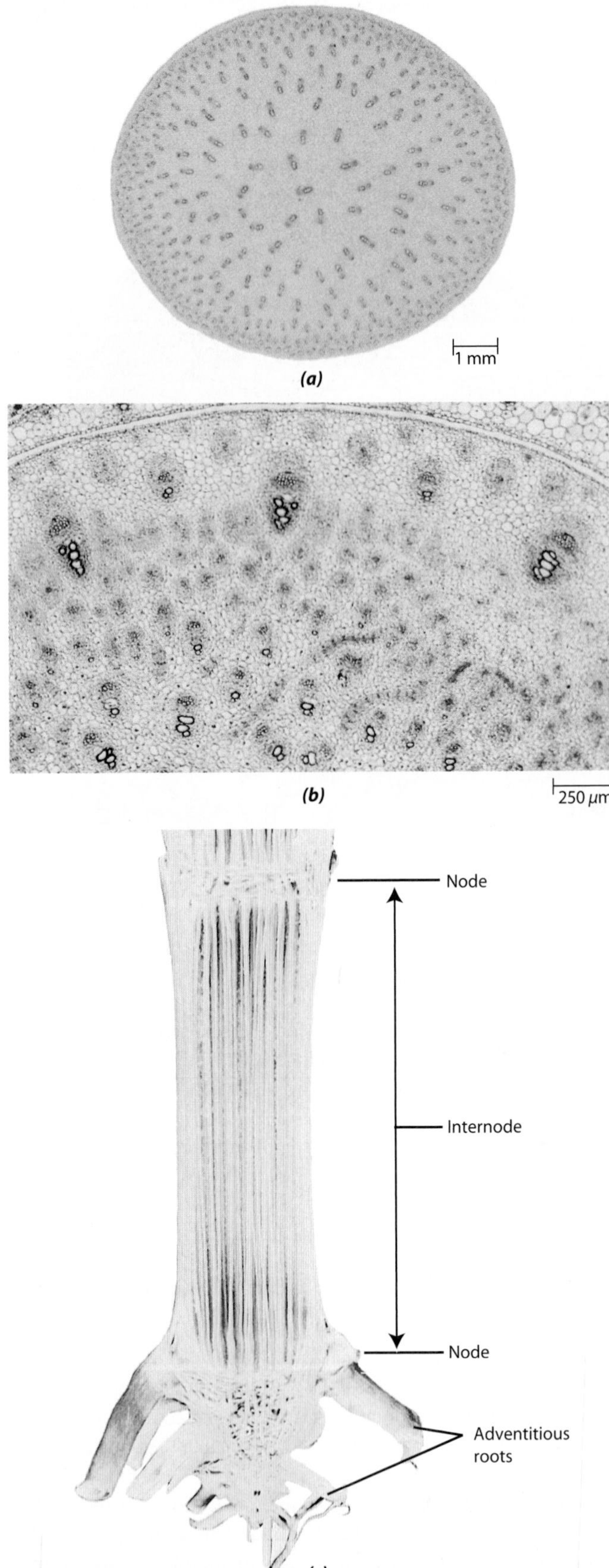

25–13 Stem of maize ***(a)*** Transverse section of the internodal region of a maize *(Zea mays)* stem, showing numerous vascular bundles scattered throughout the ground tissue. ***(b)*** Transverse section of the nodal region of a young maize stem, showing horizontal procambial strands that interconnect the vertical bundles. ***(c)*** A mature stem split longitudinally; the ground tissue has been removed to expose the vascular system.

Relation between the Vascular Tissues of the Stem and Leaf

The pattern formed by the vascular bundles in the stem reflects the close structural and developmental relationship between the stem and its leaves. The term "shoot" serves not only as a collective term for these two vegetative organs but also as an expression of their intimate physical and developmental association.

The procambial strands of the stem arise just below the developing leaf primordia and sometimes are present below the sites of future leaf primordia even before the primordia are discernible. As the leaf primordia increase in length, the procambial strands differentiate upward within them. From its inception, the procambial system of the leaf is continuous with that of the stem.

At each node, one or more vascular bundles diverge from the cylinder of strands in the stem, cross the cortex, and enter the leaf or leaves attached at that node (Figure 25–15). The extensions from the vascular system in the stem toward the leaves are called **leaf traces,** and the wide interfascicular regions, or gaps of ground tissue, in the vascular cylinder located above the level where leaf traces diverge toward the leaves are called **leaf trace**

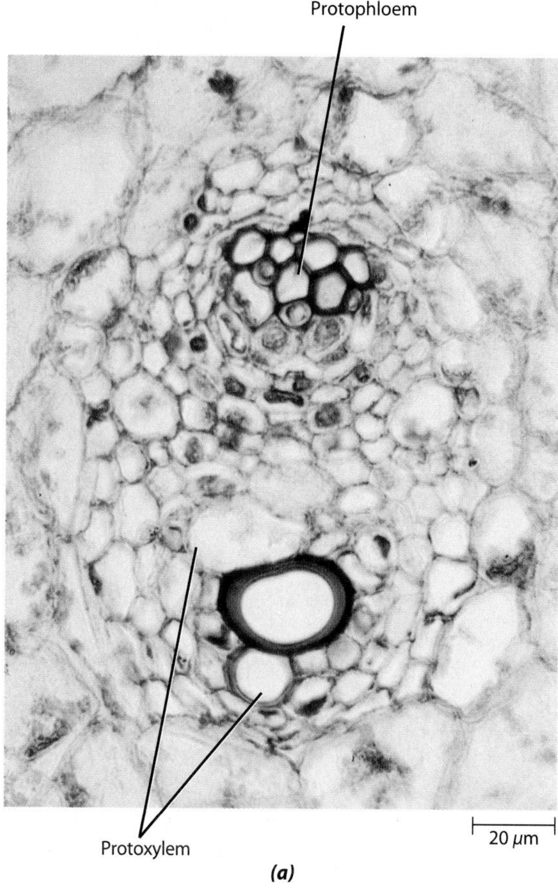

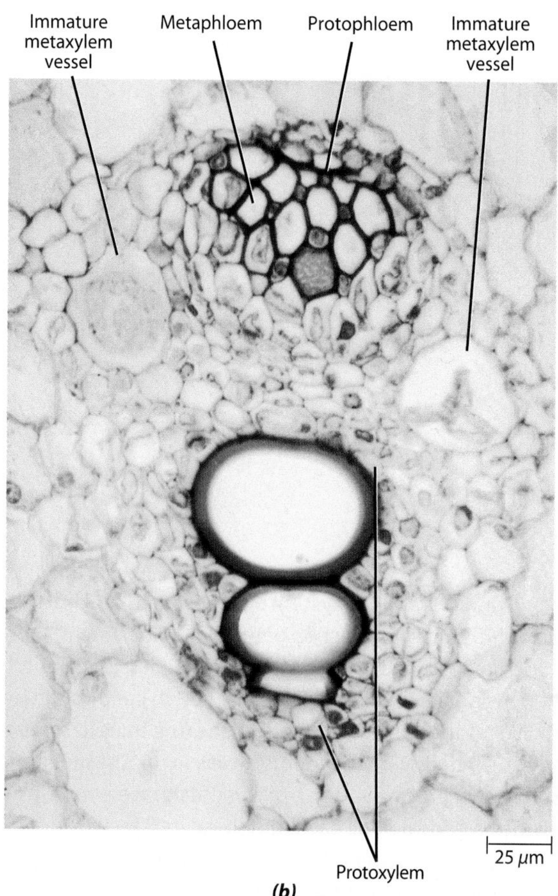

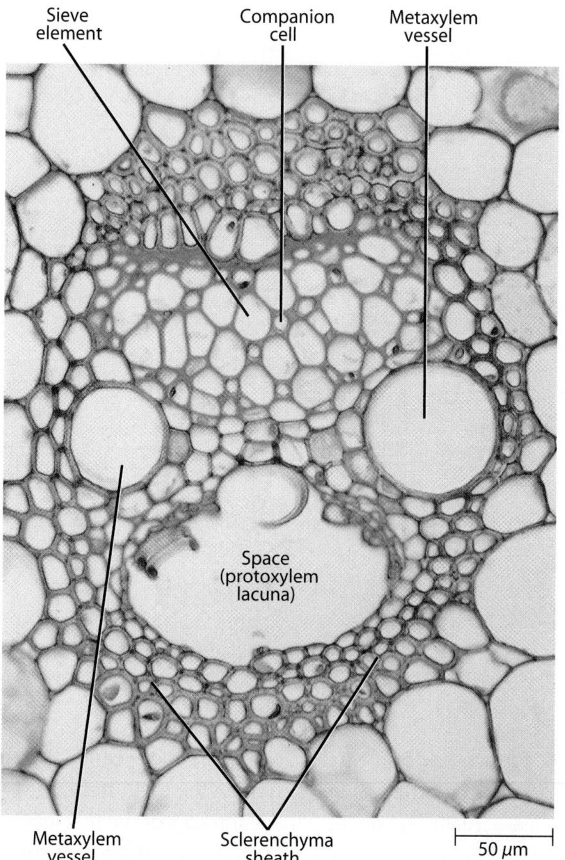

25–14 Differentiation of the vascular bundles of maize Three stages in the development of maize *(Zea mays)* vascular bundles are seen here in transverse sections of the stem. ***(a)*** The protophloem elements and two protoxylem elements are mature. ***(b)*** The protophloem sieve elements are now crushed, and much of the metaphloem is mature. Three protoxylem elements are now mature, and the two metaxylem vessel elements are almost fully expanded. ***(c)*** Mature vascular bundle surrounded by a sheath of thick-walled sclerenchyma cells. The metaphloem is composed entirely of sieve-tube elements and companion cells. The portion of the vascular bundle once occupied by the protoxylem elements is now a large space known as the protoxylem lacuna. Note the wall thickenings of destroyed protoxylem elements bordering the air space.

gaps. A leaf trace extends from its connection with a bundle in the stem—a **stem bundle**—to the level at which it enters the leaf. A single leaf may have one or more leaf traces connecting its vascular system with that in the stem. The number of internodes that leaf traces traverse before they enter a leaf differs, so the traces vary in length.

If the stem bundles are followed either upward or downward in the stem, they are found to be associated with several leaf traces. A stem bundle and its associated leaf traces are called a **sympodium** (plural: sympodia) (Figure 25–15). In some stems, some or all of the sympodia are interconnected, whereas in others, all the sympodia are independent units of the vascular system. Regardless, the pattern of the vascular system in the stem is

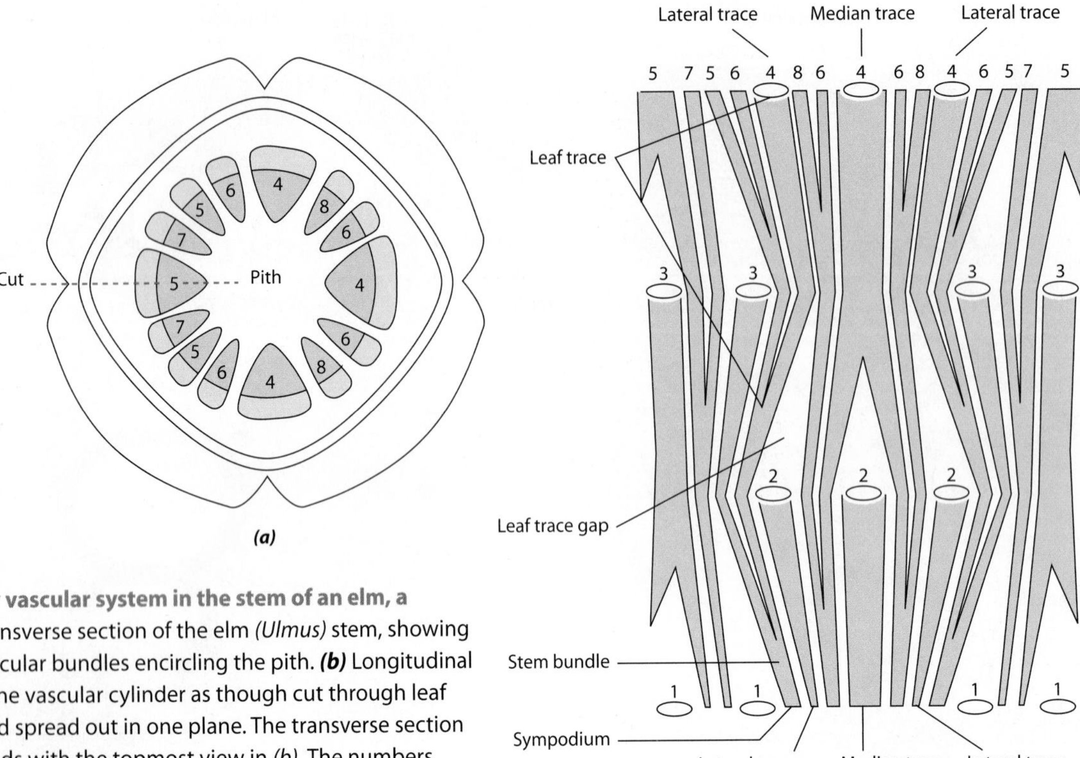

25–15 Primary vascular system in the stem of an elm, a eudicot ***(a)*** Transverse section of the elm *(Ulmus)* stem, showing the discrete vascular bundles encircling the pith. ***(b)*** Longitudinal view showing the vascular cylinder as though cut through leaf trace 5 in *(a)* and spread out in one plane. The transverse section in *(a)* corresponds with the topmost view in *(b)*. The numbers in both views indicate leaf traces. Three leaf traces—one in a median position and one each (the so-called lateral traces) on either side of it—connect the vascular system of the stem with that of the leaf. A stem bundle and its associated leaf traces are called a sympodium.

a reflection of the arrangement of the leaves on the stem. Buds commonly develop in the axils of leaves, and their vascular system is connected with that of the main stem by **branch traces.** Hence, at each node, both leaf traces and branch traces (commonly two per bud) diverge outward from the main stem (Figure 25–16).

Leaves Are Arranged in Orderly Patterns on the Stem

The arrangement of leaves on a stem is called **phyllotaxis,** or phyllotaxy. The most common type of phyllotaxis is **spiral,** or helical, with one leaf at each node and the leaves forming a spiral pattern around the stem. For example, the oaks *(Quercus), Croton* (Figure 25–1), and mulberry *(Morus alba)* (Figure 25–17a) have spirally arranged leaves. In other plants with a single leaf at each node, the leaves are disposed in two opposite ranks, as in the grasses. This type of phyllotaxis is called **distichous.** In some plants the leaves are formed in pairs at each node, and the phyllotaxis is said to be **opposite,** as in the maples *(Acer)* and honeysuckle *(Lonicera).* If each successive pair of leaves is at a right angle to the previous pair, the arrangement is termed **decussate.** Decussate phyllotaxis is exemplified by the members of the mint family (Lamiaceae), including *Coleus* (Figure 25–2). Plants with **whorled** phyllotaxis, such as Culver's-root *(Veronicastrum virginicum),* have three or more leaves at each node (Figure 25–17b).

The mechanism underlying the orderly initiation of leaves around the circumference of the shoot apical meristem has been

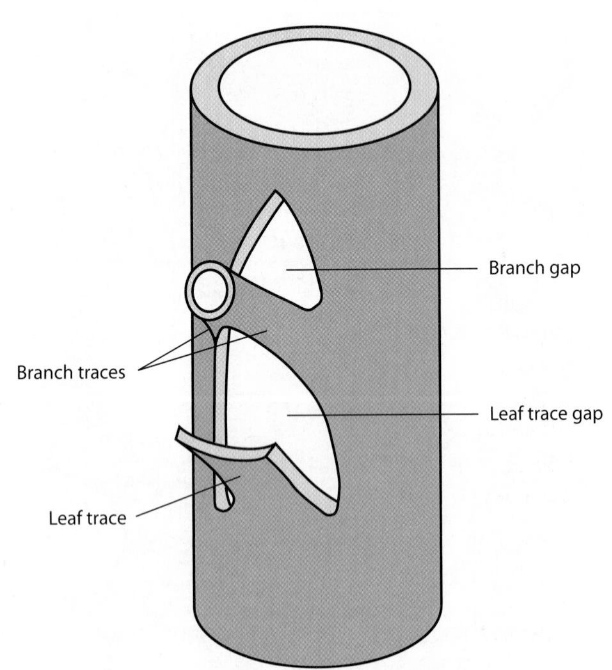

25–16 Branch and leaf traces The relation between branch traces and a leaf trace to the vascular system in the main stem. The branch traces are actually leaf traces—the leaf traces of the first leaves of the bud or lateral branch. Except in monocots, there are usually two branch traces per bud.

25–17 Examples of simple leaves ***(a)*** Mulberry *(Morus alba).* ***(b)*** Culver's-root *(Veronicastrum virginicum).* ***(c)*** Sugar maple *(Acer saccharum).* ***(d)*** Silver maple *(Acer saccharinum).* ***(e)*** Red oak *(Quercus rubra).* Note the spiral arrangement of the leaves in mulberry, and the whorled arrangement of those in Culver's-root. Leaf arrangement in the maples is opposite, and in oak it is spiral, although only single leaves of these trees are shown here.

of interest to botanists for a long time. An early view—based on results of surgical manipulations—was that a new leaf primordium arises in the *first available space,* that is, when sufficient width and distance from the apex is attained. Another explanation for the arrangement of leaves on a stem is the *inhibitory field hypothesis.* According to this hypothesis, preexisting leaf primordia inhibit the formation of new ones in their immediate vicinity. It has also been suggested that *biophysical forces* in the growing apex determine the sites of leaf initiation. In this hypothesis, a leaf primordium is initiated when a portion of the tunica surface bulges or buckles, a condition brought about by a localized reduction in the surface layer's ability to resist pressure from the tissues below.

It has become increasingly clear that the plant hormone auxin, which is directed to the primordium initiation site by the PIN1 auxin efflux carrier (see Chapter 27), is the primary signal involved in the initiation of leaf primordia. According to this *auxin-based model* of phyllotaxis, high concentrations of auxin are required for the initiation of a new primordium, and PIN1 carriers expressed in the L1-layer cells are required to generate such high levels of auxin. The young primordia initially act as sinks and deplete the auxin in neighboring cells, thus inhibiting the initiation of additional primordia in their vicinity. Eventually, a new primordium grows out at the site of the high auxin level and then begins to produce its own auxin. Note that this hypothesis closely resembles the inhibitory field hypothesis mentioned above.

Although some researchers argue that cell division is the primary target for auxin, others contend that the cell wall is the principal target. Support for the latter view comes partly from studies in which the localized application of the cell wall protein expansin to the shoot apical meristem of tomato induced the formation of leaflike outgrowths. Apparently, the expansin promoted cell wall extensibility (see Chapter 3) in the outer cell layer of the tunica, resulting in outward bulging of the tissue. Further studies have shown that expansin genes are specifically expressed at the site of primordium initiation in both tomato and rice. Moreover, expansin expressed in transformed plants induced primordia

25–18 Pinnately compound leaf of the pea *(Pisum sativum)* Notice the stipules at the base of the leaf and the slender tendrils at the tip of the leaf. In the pea leaf, the stipules are often larger than the leaflets.

capable of developing into normal leaves. In addition, it has been shown that modifications in pectin composition and in the cross-linking of cellulose microfibrils (see Chapter 3) are associated with the rapid outgrowth of young primordia. These studies support the view, presented in Chapter 22, that the primary event in morphogenesis is the expansion of tissue, which is then subdivided into smaller units by cell division, much as a house is subdivided into smaller units, the rooms.

Morphology and Structure of the Leaf

Leaves vary greatly in form and in internal structure. In magnoliids and eudicots, the leaf commonly consists of an expanded portion, the **blade,** or lamina, and a stalklike portion, the **petiole** (Figure 25–17). Scalelike or leaflike appendages called **stipules** develop at the base of some leaves (Figure 25–18). Many leaves lack petioles and are said to be **sessile** (Figure 25–19). In most monocots and certain eudicots, the base of the leaf is expanded

(a)

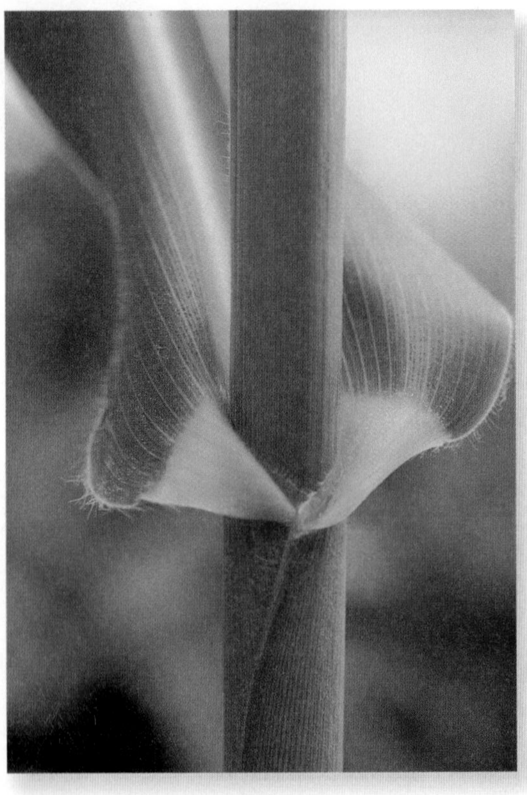

(b)

25–19 Sessile leaves ***(a)*** Leaves without a petiole, known as sessile leaves, are often found in eudicots, such as *Moricandia*, a member of the mustard family. ***(b)*** Sessile leaves are particularly characteristic of grasses and other monocots. In maize *(Zea mays)*, a monocot, the base of the leaf forms a sheath around the stem. The ligule, a small flap of tissue extending upward from the sheath, is visible. The parallel arrangement of the larger longitudinal veins is conspicuous in the portion of the blade shown here.

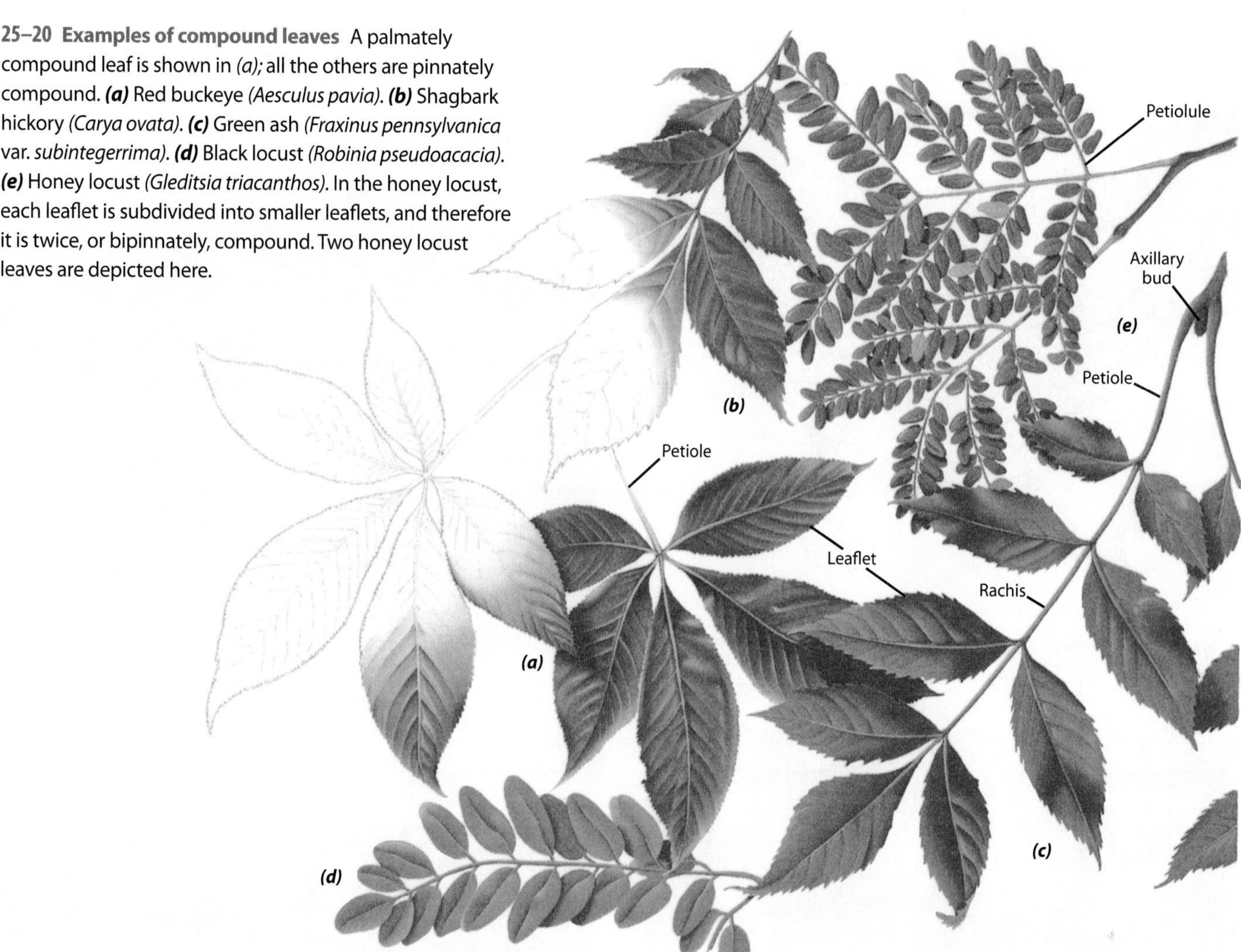

25–20 Examples of compound leaves A palmately compound leaf is shown in *(a)*; all the others are pinnately compound. ***(a)*** Red buckeye *(Aesculus pavia)*. ***(b)*** Shagbark hickory *(Carya ovata)*. ***(c)*** Green ash *(Fraxinus pennsylvanica* var. *subintegerrima)*. ***(d)*** Black locust *(Robinia pseudoacacia)*. ***(e)*** Honey locust *(Gleditsia triacanthos)*. In the honey locust, each leaflet is subdivided into smaller leaflets, and therefore it is twice, or bipinnately, compound. Two honey locust leaves are depicted here.

into a **sheath,** which encircles the stem (Figure 25–19b). In some grasses, the sheath extends the length of an internode.

The leaves of magnoliids and eudicots are either simple or compound. In **simple leaves,** the blades are not divided into distinct parts, although they may be deeply lobed (Figure 25–17). The blades of **compound leaves** are divided into leaflets, each usually with its own small petiole (called a petiolule). Two types of compound leaves can be distinguished: pinnately compound and palmately compound leaves (Figure 25–20). In pinnately compound leaves, the leaflets arise from either side of an axis, the **rachis,** like the pinnae of a feather. (The rachis is an extension of the petiole.) The leaflets of a palmately compound leaf diverge from the tip of the petiole, and a rachis is lacking.

Because leaflets are similar in appearance to simple leaves, it is sometimes difficult to determine whether the structure is a leaflet or a leaf. Two criteria may be used to distinguish leaflets from leaves: (1) buds are found in the axils of leaves—both simple and compound—but not in the axils of leaflets, and (2) leaves extend from the stem in various planes, whereas the leaflets of a given leaf all lie in the same plane.

Variations in the structure of angiosperm leaves are to a great extent related to the habitat, and the availability of water is an especially important factor affecting leaf form and structure. On the basis of their water requirements or adaptations, plants are commonly characterized as **mesophytes** (plants that require an environment that is neither too wet nor too dry), **hydrophytes** (plants that require a large supply of water or grow wholly or partly submerged in water), and **xerophytes** (plants that are adapted to arid habitats). Such distinctions are not sharp, however, and leaves often exhibit a combination of features that are characteristic of different ecological types. Regardless of their differing forms, the foliage leaves of angiosperms are specialized as photosynthetic organs and, like roots and stems, consist of dermal, ground, and vascular tissue systems (see Figure 23–4).

The Epidermis, with Its Compact Structure, Provides Strength to the Leaf

The mass of epidermal cells of the leaf, like those of the stem, are compactly arranged and covered with a cuticle that reduces water loss (page 23). Stomata may occur on both sides of the leaf or on only one side, either the upper or, more commonly, the lower side (Figure 25–21). In leaves of hydrophytes that float on the surface of the water, stomata may occur in the upper

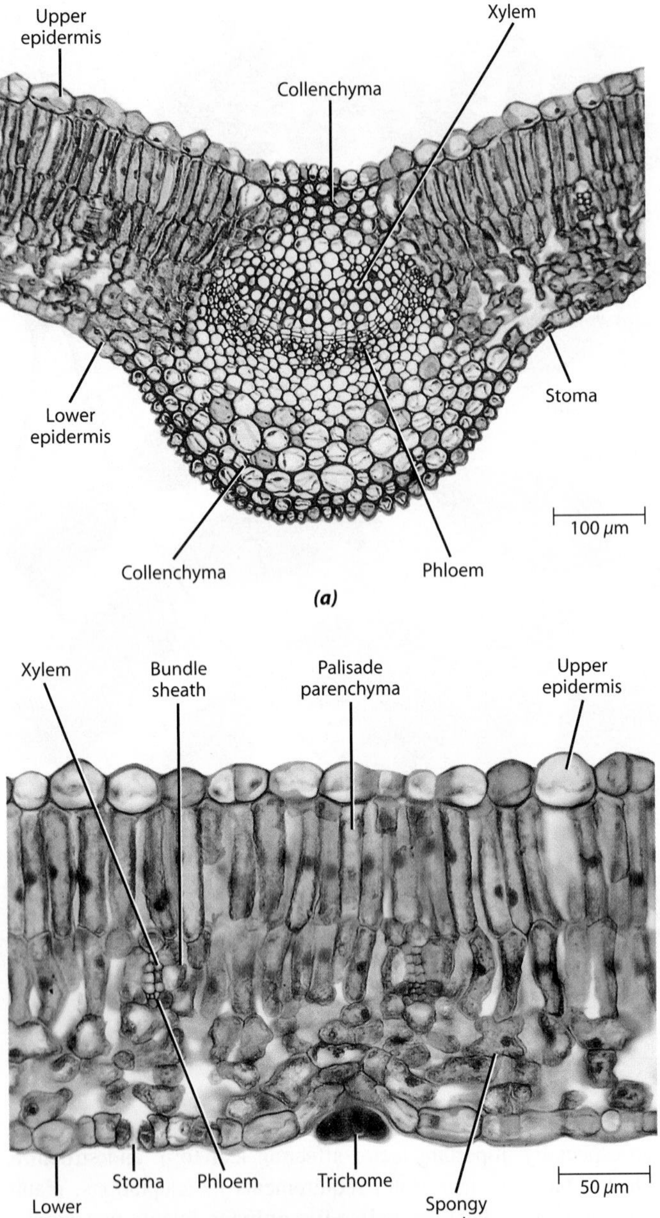

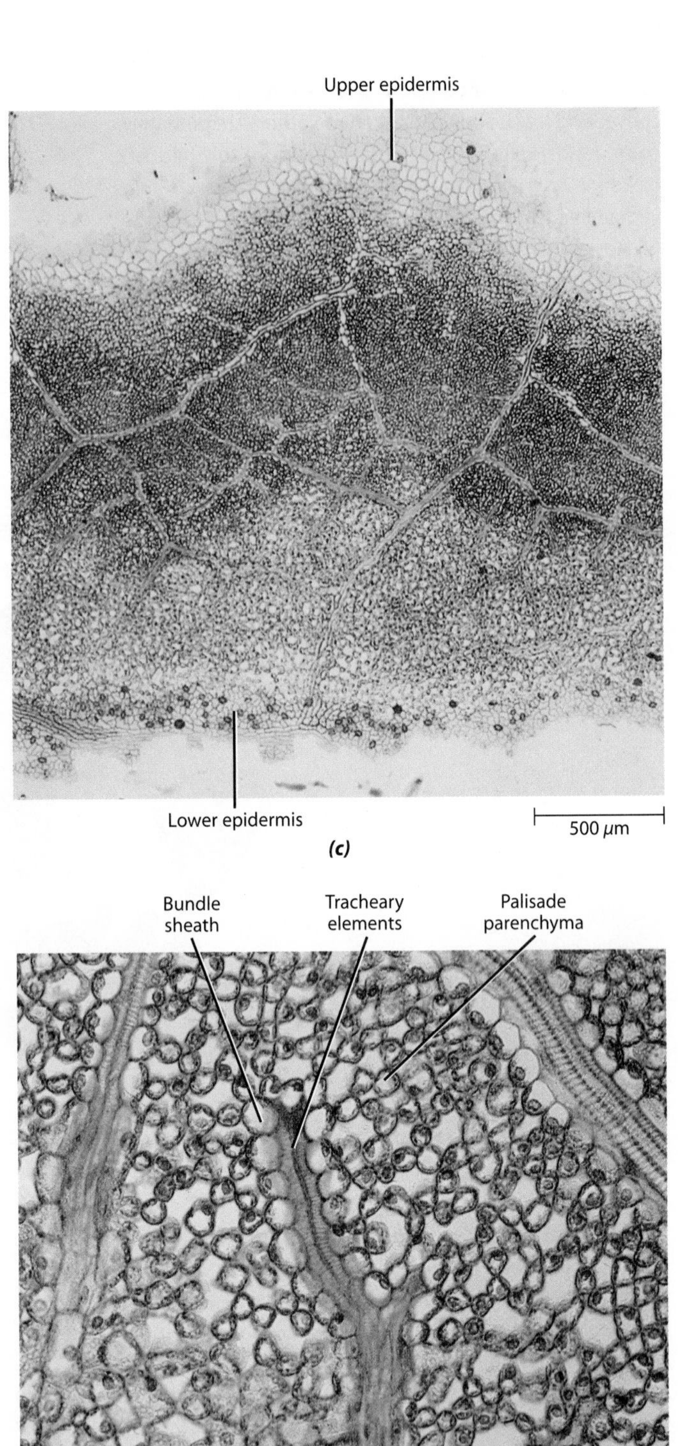

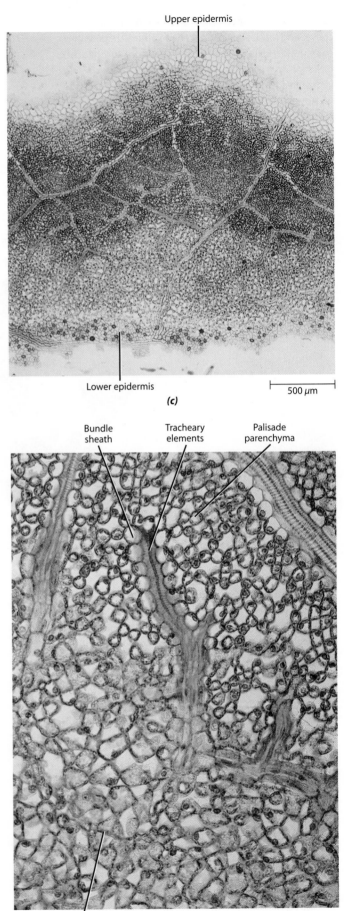

25–21 Sections of a lilac *(Syringa vulgaris)* leaf ***(a)*** A transverse section through a midrib showing the midvein. ***(b)*** A transverse section through a portion of the blade. Two small veins (minor veins) can be seen in this view. ***(c)*** A "paradermal" section of the leaf. Strictly speaking, a paradermal section is one cut parallel to the epidermis. In practice, such sections are more or less oblique and extend from the upper to the lower epidermis. Thus, part of the upper epidermis can be seen in the light area at the top of this micrograph, and part of the lower epidermis at the bottom. Notice the greater number of stomata in the lower epidermis, as evidenced by the number of red-stained guard cells. (The few darkly stained spots are trichomes, or epidermal hairs.) The venation in lilac is netted. ***(d)*** Enlargement of a portion of the section in *(c)*, showing palisade parenchyma and spongy parenchyma with a vein ending, sectioned through some tracheary elements and surrounded by a bundle sheath. ***(e)*** This enlargement (at the top of the facing page) shows a portion of the lower epidermis with two trichomes and several stomata.

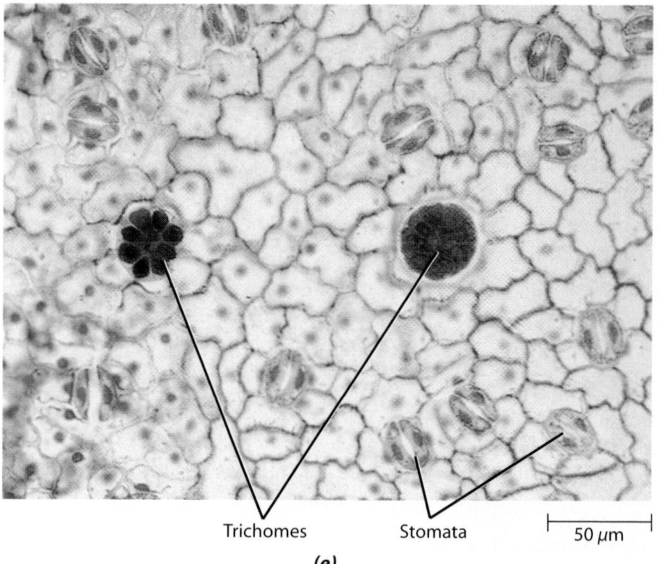

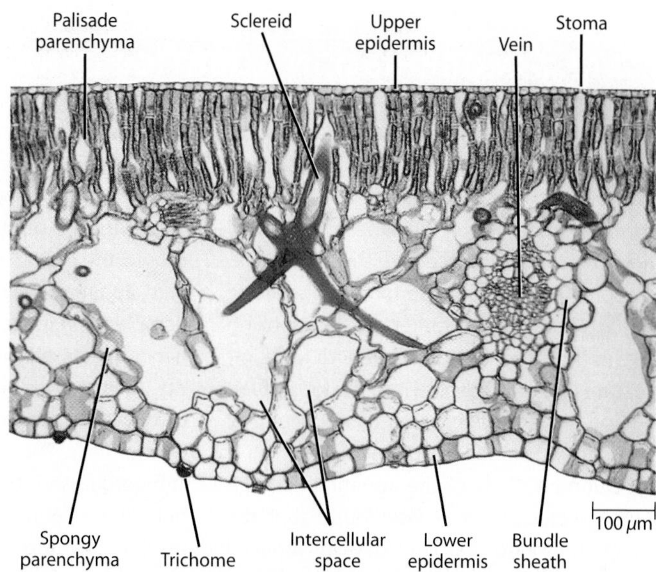

25–22 Leaf of the water lily *(Nymphaea odorata)* The *Nymphaea* leaf, seen here in transverse section, floats on the surface of the water and has stomata in the upper epidermis only. As is typical of hydrophytes, the vascular tissue of the leaf is much reduced, especially the xylem. The palisade parenchyma consists of several layers of cells above the spongy parenchyma. Note the large intercellular (air) spaces, which add buoyancy to this floating leaf.

epidermis only (Figure 25–22); the submerged leaves of hydrophytes usually lack stomata entirely. The leaves of xerophytes generally contain greater numbers of stomata than those of other plants. Presumably, these numerous stomata permit a higher rate of gas exchange during the relatively rare periods of favorable water supply. In many xerophytes, the stomata are sunken in depressions on the lower surface of leaves (Figure 25–23). The depressions may also contain many epidermal hairs, or trichomes. Together these two features may serve to reduce water loss from the leaf. Trichomes may be found on either or both surfaces of a leaf. Thick coats of epidermal hairs and the secreted resins of some hairs may also retard water loss from leaves.

In the leaves of most angiosperms other than monocots, the stomata are usually scattered over the surface (Figure 25–24a); their development is mixed—that is, mature and immature stomata occur side by side in a partially developed leaf. In most monocots, the stomata are arranged in rows parallel with the long axis of the leaf (Figure 25–24b). Their formation begins at the tips of the leaves and progresses downward toward the base.

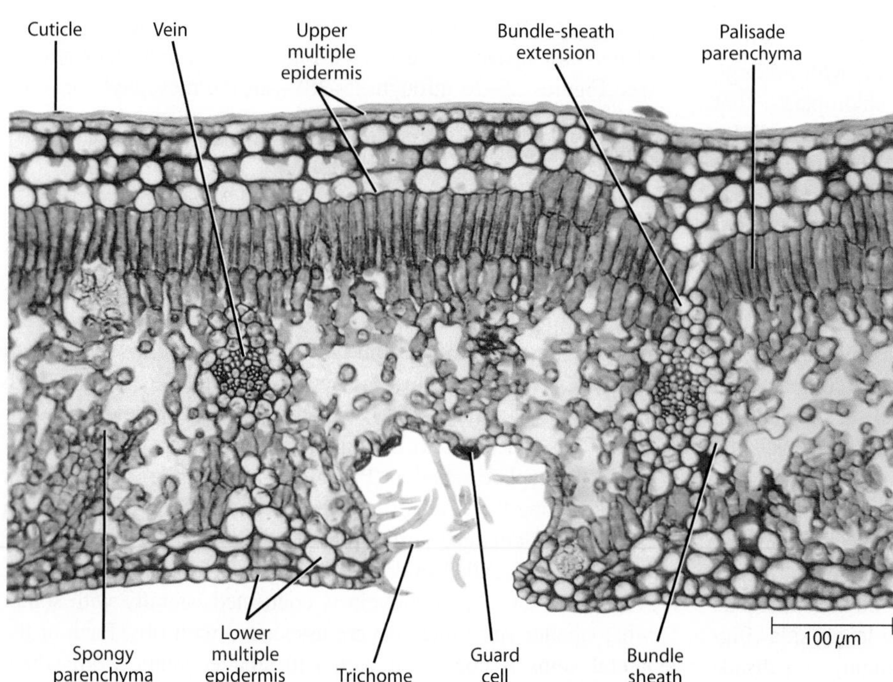

25–23 Oleander *(Nerium oleander)* leaf *Oleander* is a xerophyte, and this is reflected in the structure of the leaf, seen here in transverse section. Note the very thick cuticle covering the multiple (several-layered) epidermis on the upper and lower surfaces of the leaf. The stomata and trichomes are restricted to invaginated portions of the lower epidermis called stomatal crypts.

LEAF DIMORPHISM IN AQUATIC PLANTS

In the natural environment, the leaves of aquatic flowering plants may develop into two distinct forms (see Figure 8–22). Leaves forming under water develop into narrow and often highly dissected structures (water forms), whereas those formed above the surface develop into rather ordinary-looking leaves (land forms). It is possible to force immature leaves to develop into the form atypical for a given environment by applying any one of a wide range of treatments.

During a study of the aquatic plant *Callitriche heterophylla,* it was found that the plant hormone gibberellic acid induced shoots growing in the air (emergent shoots) to produce water-form leaves. Abscisic acid, another plant hormone (see Chapter 27), led to the formation of land-form leaves on submerged shoots. Higher temperatures or the addition of mannitol, a sugar alcohol, to the water also caused submerged shoots to produce land-form leaves.

In nature, the cellular turgor pressure of the submerged (water-form) leaves is relatively high, while that of the emergent (land-form) leaves is relatively low. The lower values in the emergent leaves may be due in part to transpirational loss of water vapor through the numerous stomata on the leaf surfaces. Associated with the higher turgor pressures in developing water-form leaves are long epidermal cells in the mature leaves.

In the experiments, gibberellic acid caused the cells of emergent leaves to elongate, apparently by increasing cell wall plasticity, leading to increased water uptake and hence increased turgor pressure. At maturity, these leaves had all the characteristics of typical water forms, including long epidermal cells. The limited cell expansion in submerged shoots exposed to abscisic acid or high temperatures did not seem to be the result of low turgor; instead, the walls of the treated cells became less yielding so that high turgor failed to promote cell expansion. Growing submerged shoots in a mannitol solution resulted in turgor pressures similar to those of the emergent controls and the production of leaves with short epidermal cells.

The results of these experiments suggest that the relative magnitude of cellular turgor pressure determines the final leaf size and form in *C. heterophylla.* Thus, merely the presence or absence of water surrounding the developing leaf ensures that the leaf will be suitably adapted to life above or below the water.

Land-form leaves

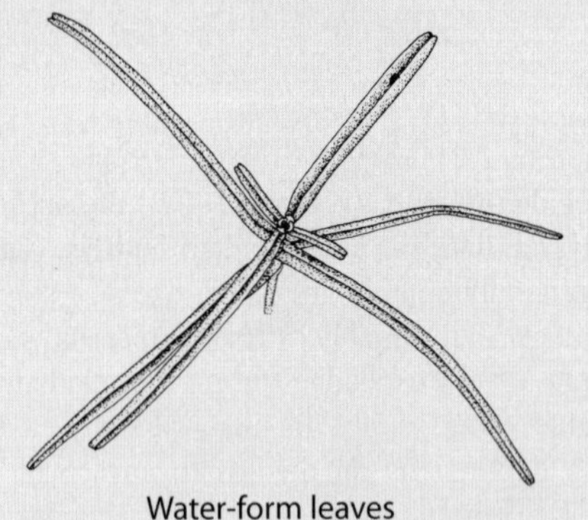

Water-form leaves

The Mesophyll Is Specialized for Photosynthesis

It is the **mesophyll,** the ground tissue of the leaf, with its large volume of intercellular spaces and numerous chloroplasts, that is particularly specialized for photosynthesis. The intercellular spaces are connected with the outer atmosphere through the stomata, which facilitate rapid gas exchange, an important factor in photosynthetic efficiency. In mesophytes, the mesophyll commonly is differentiated into **palisade parenchyma** and **spongy parenchyma.** The cells of the palisade tissue are columnar, with their long axes oriented at a right angle to the epidermis, and the spongy parenchyma cells are irregular in shape (Figure 25–21b, d). Although the palisade parenchyma appears more compact than the spongy parenchyma, most of the vertical walls of the palisade cells are exposed to intercellular spaces, and in some leaves, the palisade surface may be two to four times greater than the spongy surface. Chloroplasts are also more numerous in palisade cells than in spongy cells, and thus most of the photosynthesis in the leaf seems to take place within the palisade parenchyma.

The palisade parenchyma is usually located on the upper side of the leaf, and the spongy parenchyma on the lower side (Figure 25–21). In certain plants, including many xerophytes, palisade parenchyma often occurs on both sides of the leaf. In some plants—for instance, maize (see Figure 7–23) and other grasses (see Figures 25–26 through 25–28)—all the mesophyll cells are of more or less similar shape, and there is no distinction between spongy and palisade parenchyma.

Vascular Bundles Are Distributed throughout the Mesophyll

The mesophyll of the leaf is thoroughly permeated by numerous vascular bundles, or **veins,** which are continuous with the vascular system of the stem. In most angiosperms other than monocots, the veins are arranged in a branching pattern, with successively smaller veins branching from somewhat larger ones. This type of vein arrangement is known as **netted venation,** or reticulate venation (Figure 25–25). The largest vein often extends along the long axis of the leaf as a midvein. The midvein occurs in an enlarged portion of the blade that appears as a ridge or rib—the so-called midrib—on the lower surface of the blade (Figure 25–21a). The midvein is connected laterally with somewhat smaller veins that also are associated with ribs. Each of the lateral veins is connected with still smaller veins, from which

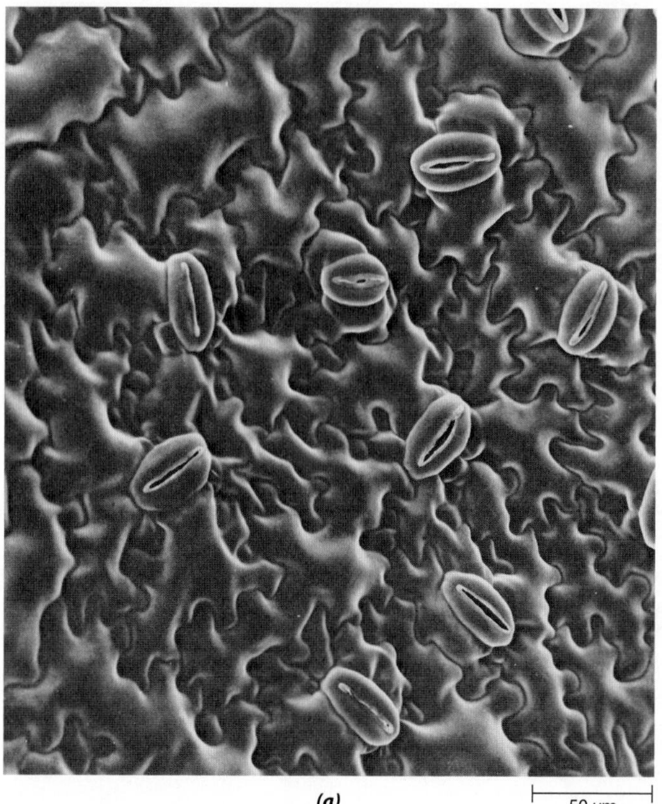

25–24 Stomata in a eudicot and a monocot ***(a)*** Potato *(Solanum tuberosum)* leaf, showing the random arrangement of stomata typical of the leaves of eudicots. As seen in this scanning electron micrograph, the guard cells in potato are crescent-shaped and are not associated with subsidiary cells. ***(b)*** Scanning electron micrograph of a maize *(Zea mays)* leaf, showing the parallel arrangement of stomata typical of the leaves of monocots. In maize, each pair of narrow guard cells is associated with two subsidiary cells, one on each side of the stoma (see Figure 23–26).

other small veins diverge. By contrast, most monocots have many veins that extend along the long axis of the leaf. These veins may be almost equal in size or they may vary in size, with larger veins alternating with smaller ones. This vein arrangement is called **parallel venation,** or striate venation (Figure 25–19b), although the longitudinally oriented veins converge and join at the apex of the leaf. The longitudinal veins are interconnected by considerably smaller transverse veins, forming a complex network.

The veins contain xylem and phloem, which generally are entirely primary in origin. The midvein and sometimes the coarser veins, however, may undergo limited secondary growth in the leaves of some angiosperms, but not in those of monocots. In the vein endings of some leaves, the xylem elements often extend farther than the phloem elements, but in some plants both xylem and phloem elements extend to the ends of the vein. Commonly, the xylem is on the upper side of the vein and the phloem on the lower side (Figure 25–21a, b).

The small veins of the leaf that are more or less completely embedded in mesophyll tissue are called **minor veins,** while the large veins associated with ribs are called **major veins.** It is the minor veins that play the principal role in the collection of photosynthates (organic compounds produced by photosynthesis) from the mesophyll cells. With increasing vein size, the veins become less closely associated spatially with the mesophyll and increasingly embedded in nonphotosynthetic rib tissues. Hence, as the veins increase in size, their primary function changes from collection of photosynthates to transport of photosynthates out of the leaf.

The vascular tissues of the veins are rarely exposed to the intercellular spaces of the mesophyll. The large veins are surrounded by parenchyma cells that contain few chloroplasts, whereas the small veins usually are enclosed by one or more layers of compactly arranged cells that form a **bundle sheath** (Figures 25–21b, d and 25–22). In some plants, the cells of the bundle sheath resemble the mesophyll cells in which the small veins are located. The bundle sheaths extend to the ends of the veins, ensuring that no part of the vascular tissue is exposed to air in the intercellular spaces and that all substances entering and leaving the vascular tissues must pass through the sheath (Figure 25–21d). The bundle sheath of angiosperm leaves is positioned similarly to the endodermis of the root. Like the root endodermis, the bundle sheath has Casparian strips in its anticlinal walls. Hence, the bundle sheath may be regarded as an endodermis, which may similarly control movement of substances to and from the vascular tissues. In most leaf preparations the Casparian strips are not distinguishable.

In many leaves, the bundle sheaths are connected with either upper or lower epidermis, or with both, by panels of cells

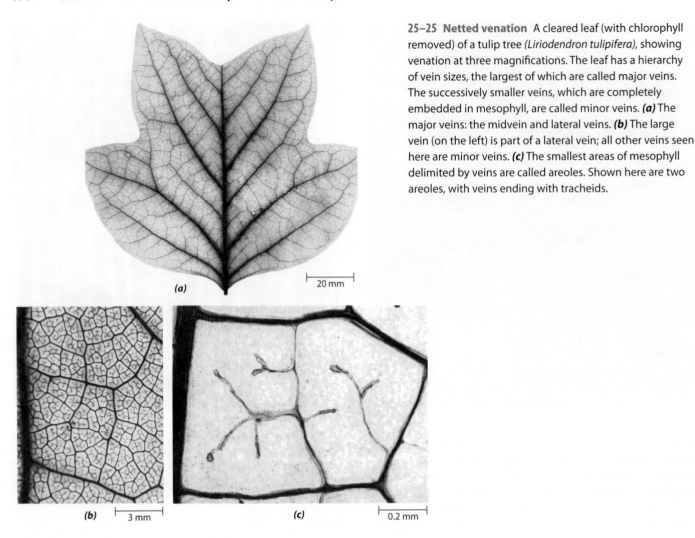

25–25 Netted venation A cleared leaf (with chlorophyll removed) of a tulip tree *(Liriodendron tulipifera)*, showing venation at three magnifications. The leaf has a hierarchy of vein sizes, the largest of which are called major veins. The successively smaller veins, which are completely embedded in mesophyll, are called minor veins. ***(a)*** The major veins: the midvein and lateral veins. ***(b)*** The large vein (on the left) is part of a lateral vein; all other veins seen here are minor veins. ***(c)*** The smallest areas of mesophyll delimited by veins are called areoles. Shown here are two areoles, with veins ending with tracheids.

resembling the bundle-sheath cells (Figure 25–23). These panels are called **bundle-sheath extensions.** Besides offering mechanical support to the leaf, in some leaves they apparently conduct water from the xylem to the epidermis.

The epidermis itself provides considerable strength to the leaf, because of its compact structure and its cuticle. In addition, collenchyma or sclerenchyma may be present beneath the epidermis of the vein ribs of the larger veins in many angiosperm leaves, providing further support to the leaf. In monocot leaves, the veins and leaf margins may be bordered by fibers. In other angiosperm leaves, collenchyma cells may be found along the leaf margins.

Grass Leaves

After the discovery of the C_4 pathway of photosynthesis in sugarcane (pages 140 to 142), a great many studies were devoted to the comparative anatomy of grass leaves in relation to photosynthetic pathways. It was discovered that the leaves of C_3 and C_4 grasses have rather consistent anatomical differences. For example, in the leaves of C_4 grasses, the mesophyll cells and bundle-sheath cells typically form two concentric layers around the vascular bundles, as seen in transverse sections (Figure 25–26). The compactly arranged bundle-sheath cells of the C_4 grasses are very large parenchyma cells that contain many large, conspicuous chloroplasts. This concentric arrangement of the mesophyll and bundle-sheath layers in C_4 plants is referred to as **Kranz** (German for "wreath") **anatomy.** The significance of the Kranz anatomy in relation to C_4 photosynthesis is discussed on pages 143 to 145.

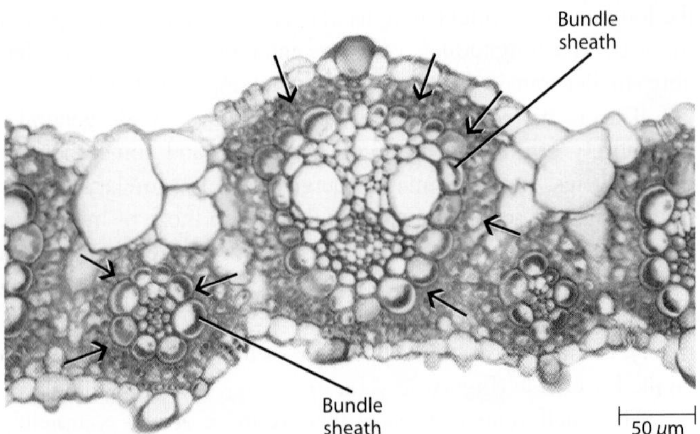

25–26 Leaf of sugarcane, a C_4 grass Transverse section of a sugarcane *(Saccharum officinarum)* leaf. As is typical of C_4 grasses, the mesophyll cells (arrows) are radially arranged around the bundle sheaths, which consist of large cells containing many large chloroplasts.

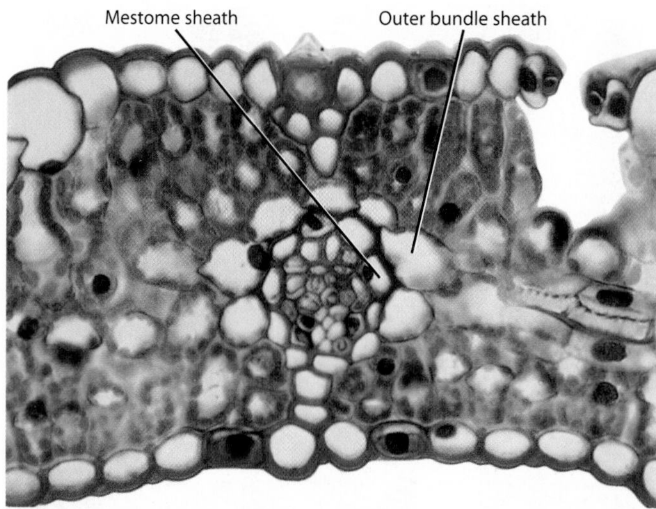

25–27 Leaf of wheat, a C_3 grass Transverse section of a wheat *(Triticum aestivum)* leaf. As is typical of C_3 grasses, the mesophyll cells are not radially arranged around the bundle sheaths. In wheat, the veins are surrounded by two bundle sheaths: an outer sheath of relatively thin-walled parenchyma cells and an inner sheath, the mestome sheath, consisting of thick-walled cells.

In the leaves of C_3 grasses, by contrast, the mesophyll cells and bundle-sheath cells are not concentrically arranged. Moreover, the relatively small cells of the parenchymatous bundle sheaths in these plants have rather small chloroplasts, and at low magnifications the cells appear empty and clear. Commonly, an inner, more or less thick-walled sheath (the so-called mestome sheath) is also present in the C_3 grasses (Figure 25–27).

Another consistent structural difference between the leaves of C_3 and C_4 grasses is their interveinal distances, that is, the distances between laterally adjacent bundle sheaths. In C_4 grasses, only two to four mesophyll cells intervene between laterally adjacent bundle sheaths, whereas in C_3 grasses, more than four (an average of 12 for the C_3 species in one study) mesophyll cells intervene.

The leaves of C_4 plants generally export photosynthates both more rapidly and more completely than the leaves of C_3 plants. The reasons for these differences are unknown, but it has been suggested that the differences in physical distance between the mesophyll cells and the phloem of the vascular bundles may influence the rate and efficiency of loading of photosynthates into the sieve tubes.

The epidermis of grasses is made up of a variety of cell types. Most of the epidermal cells are narrow, elongated cells. Some especially large ones, termed **bulliform cells,** or motor cells, occur in longitudinal rows and are believed by some botanists to participate in the mechanism of folding and unfolding, or rolling and unrolling, of the leaves, responses resulting from changes in water potential (Figure 25–28). During excessive loss of water, the bulliform cells in some leaves become flaccid and the leaf folds or rolls. The epidermis also contains narrow, thick-walled guard cells, which are associated with subsidiary cells (Figure 25–24b; see also Figure 23–26).

Development of the Leaf

Clonal analysis—the analysis of lineages, or clones, of genetically different cells—has revealed that leaf primordia are initiated by groups of cells in the peripheral zone of the apical meristem.

100 μm
(a)
Upper epidermis
Bulliform cell
Mesophyll cell
Stoma
Sclerenchyma strands
Midvein
(b)
200 μm

25–28 Leaf of annual bluegrass, a C_3 grass Portions of ***(a)*** folded and ***(b)*** unfolded leaves, including the midvein, of annual bluegrass *(Poa annua)*. In the grass leaf, the mesophyll is not differentiated as palisade and spongy parenchyma. Strands of sclerenchyma cells commonly occur above and below the veins. The epidermis of the grass leaf contains bulliform cells—large epidermal cells thought to play a part in the folding and unfolding (rolling and unrolling) of grass leaves. In the *Poa* leaf shown in *(a)*, the bulliform cells in the upper epidermis are partly collapsed and the leaf is folded. An increase in turgor in the bulliform cells would presumably cause the leaf to unfold, as in *(b)*.

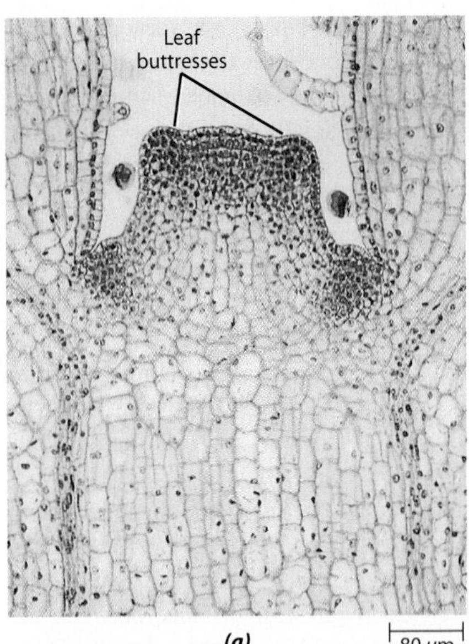

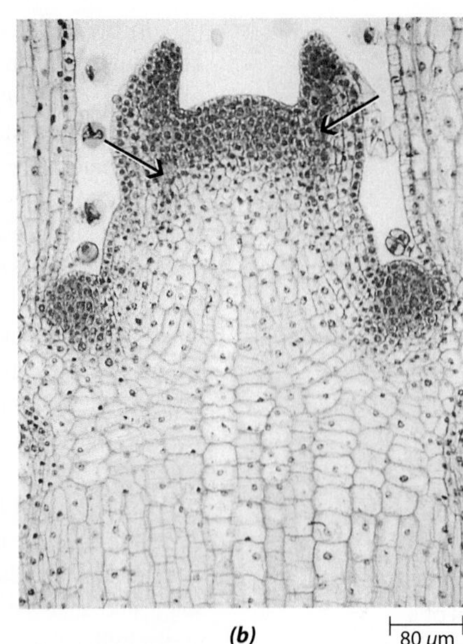

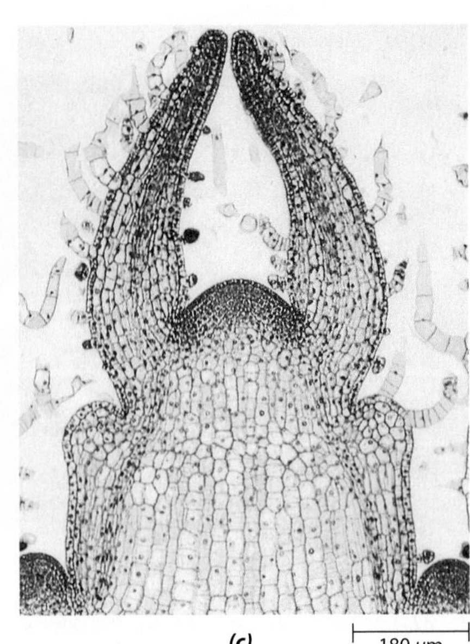

25–29 Early stages of leaf development in *Coleus* As seen in these longitudinal sections of the shoot tip, the leaves of *Coleus blumei* occur in pairs, opposite one another at the nodes (Figure 25–2). ***(a)*** Two small bulges, or leaf buttresses, can be seen opposite each other on the flanks of the apical meristem. Also visible is a bud primordium arising in the axil of each of the two young leaves, below. ***(b)*** Two erect, peglike leaf primordia have developed from the leaf buttresses. Notice the procambial strands (arrows) extending upward into the leaf primordia. The bud primordia, below, are more developed than those in *(a)*. ***(c)*** As the leaf primordia elongate, the procambial strands, which are continuous with the leaf trace procambium in the stem, continue to develop into the leaves. Trichomes, or epidermal hairs, seen here on the leaf primordia, develop from certain protodermal cells very early, long before the protoderm matures to become the epidermis.

The group of cells spans all three layers of the meristem—L1, L2, and L3 (Figure 25–4a)—and ranges from about 5 to 10 cells per layer in *Arabidopsis* to somewhere between 50 and 100 cells per layer in tobacco, cotton, and maize. These cells are referred to as the **founder cells** of the leaf.

The earliest structural evidence of leaf initiation is a change in the orientation of cell division and expansion within the founder cells. This results in the formation of a bulge, often called a "leaf buttress" (Figure 25–29a). With continued growth, each buttress develops into a leaf primordium, which is generally flatter on the surface facing the apical meristem (the future upper surface of the leaf) than on the opposite side (the future lower surface) (Figure 25–29b).

Shortly after the leaf primordium emerges from the buttress, a distinctive, dense band of cells forms on opposite sides (along the margins) of the primordium. Formation of the blade is initiated in these narrow bands—termed "marginal meristems" or "marginal blastozones"—whereas the central region of the primordium differentiates into the midrib or rachis (Figure 25–30). In simple leaves with entire margins (margins not in any way lobed or indented), activity of the marginal meristem is of very short duration. By contrast, in simple leaves with more complex shapes, subdivision of the marginal meristem into regions of growth enhancement and growth suppression results in the formation of lobes and marginal serrations. Regions of prolonged growth enhancement are also involved in formation of the leaflets of compound leaves. Differences in the duration of meristem activity and in the amount of expansion within the plane of the lamina account for much of the diversity of leaf shape.

In leaves derived from apical meristems with two tunica layers, the L1 layer gives rise to the epidermis, and the L2 and L3 layers contribute to the internal tissues. Expansion and elongation of the leaf occur largely by *intercalary growth,* that is, by cell division and cell enlargement throughout the blade, with cell enlargement contributing the most. Differences in rates of cell division and cell enlargement in the various layers of the blade result in the formation of numerous intercellular spaces and produce the mesophyll form characteristic of the leaf. Typically, the leaf stops growing first at the tip and last at the base. Compared with growth of the stem, the growth of most leaves is of short duration. The restricted type of growth exhibited by the leaf and by floral apices is said to be *determinate,* in contrast to the unlimited, or *indeterminate,* type of growth of the vegetative apical meristems.

Vascular development in the leaves of angiosperms other than monocots begins with differentiation of the procambium of the future midvein. This procambium differentiates upward into the primordium as an extension of the leaf trace procambium (Figure 25–29c). All of the coarse, or major, veins develop upward and/or outward toward the margins of the

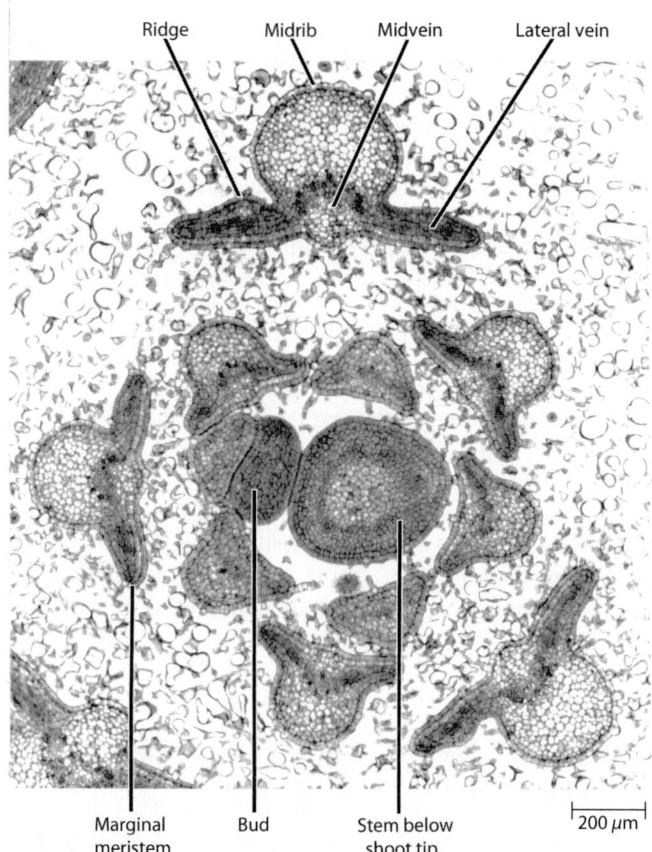

25–30 Developing leaves of tobacco Transverse section of developing leaves of tobacco *(Nicotiana tabacum)* grouped around the shoot tip, sectioned below the apical meristem. The younger leaves are nearer the axis. A leaf primordium at first lacks differentiation into midrib and blade. Some early stages in development of blade and midrib can be seen here. Portions of numerous trichomes surround the developing leaves.

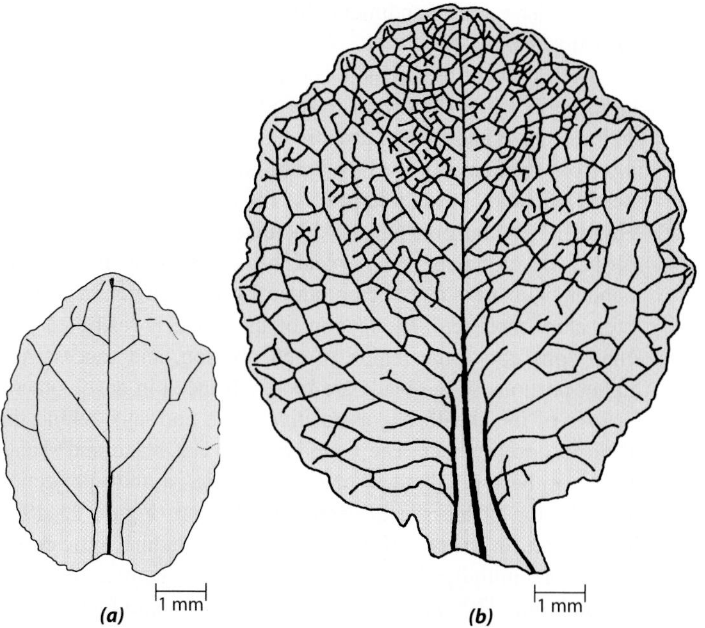

25–31 Two stages in the development of the vascular system in the leaf of lettuce In lettuce *(Lactuca sativa)*, ***(a)*** the major veins develop upward in the blade, while ***(b)*** the minor, or smaller, veins develop from the tip to the base of the blade. Thus, the tip of the leaf is the first region to have a complete system of veins.

leaf, in continuity with the procambium of the midvein (Figure 25–31a). The smaller, or minor, veins are initiated at the tip of the leaf (Figure 25–31b) and develop from the tip to the base, in continuity with the coarser veins. Thus, the tip of the leaf is the first part to have a complete system of veins. This course of development reflects the overall maturation of the leaf, which is from the tip to the base.

The pattern of leaf vein development described above has been corroborated in *Arabidopsis* with use of the molecular marker for procambium identity, *AtHB8::GUS.* Note in Figure 25–32 that the veins appear to be continuous in their development and that *AtHB8* expression follows a progressive pattern, with

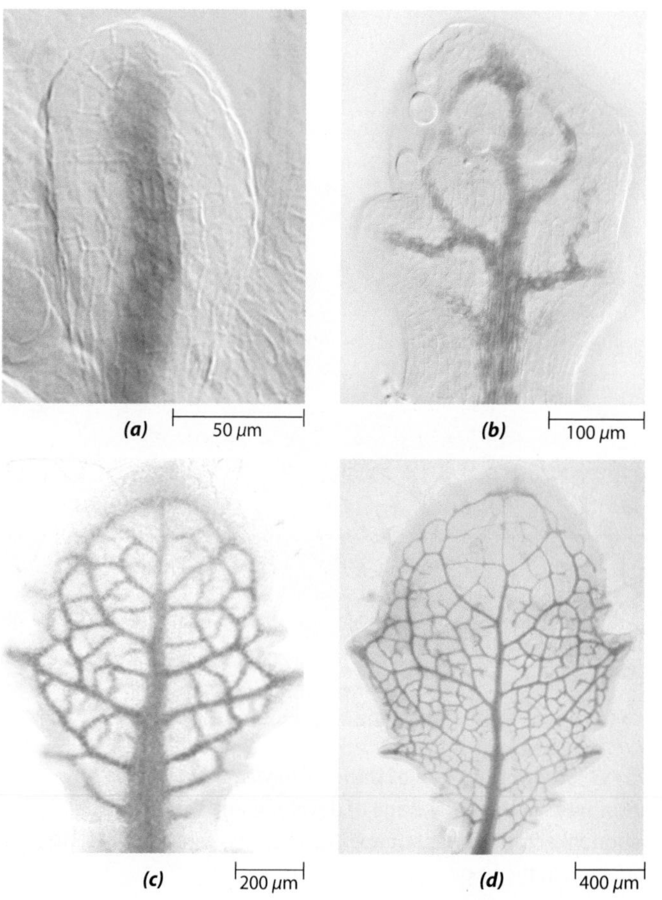

25–32 Leaf vein development in *Arabidopsis* Expression of the *AtHB8::GUS* marker for procambium identity in leaves of various ages indicates that the veins are continuous in their development and that larger veins appear before smaller veins. ***(a)*** and ***(b)*** show the pattern of vein development in cleared whole leaves after 6 and 24 hours, respectively; ***(c)*** and ***(d)*** on days 2 and 3, respectively.

new, smaller veins extending unidirectionally from larger, earlier-formed veins.

Development of the monocot leaf differs from that of other angiosperm leaves in several respects. For example, in grasses such as maize and barley, growth activity quickly spreads laterally from the flanks of the developing leaf primordium and completely encircles the shoot apex. As the primordium increases in length, it gradually acquires a hoodlike shape (Figure 25–33). Further development of the leaf blade proceeds in a linear manner, with new cells being added by the activity of a basal intercalary meristem. Elongation of the blade is restricted to a small zone above the region of cell division, and successively higher portions of the blade are more advanced in development. Growth of the sheath begins relatively late and lags behind the blade in development. The boundary between blade and sheath does not become distinct until the ligule, a thin projection from the top of the sheath, begins to develop (Figure 25–19b). In addition, in grass leaves, all of the longitudinal procambial strands are initially discontinuous with the leaf trace, or stem, procambium. The midvein and other large longitudinal strands arise in the stem segment at the base of the primordium and develop upward into the primordium and downward into the stem below. The small longitudinal strands, which arise in the primordium and form equidistantly between previously formed adjacent longitudinal strands, do not extend downward into the sheath. Instead, they merge with other strands above the sheath. As mentioned previously, all longitudinal bundles are interconnected by transverse veins.

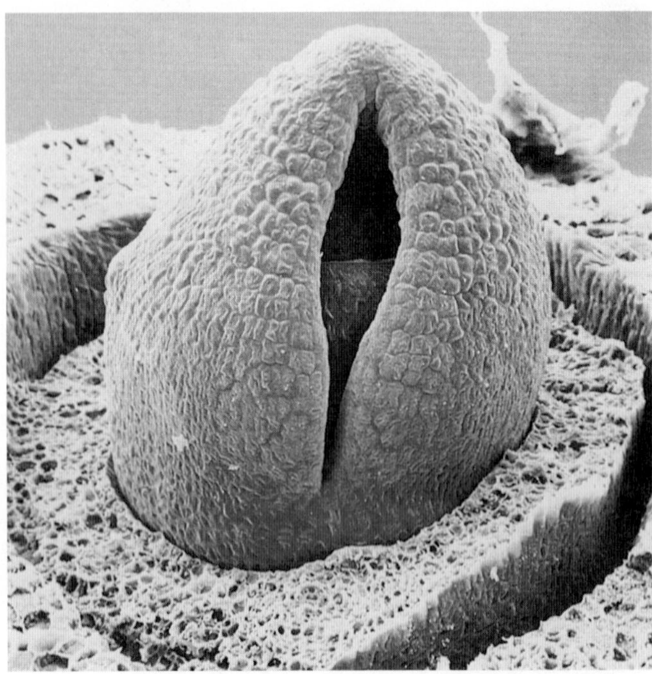

25–33 Developing leaf of barley *(Hordeum vulgare)*, a monocot At this early stage of development, the leaf blade has a hoodlike form. The shoot apex can be seen through the slitlike opening in the hood.

The vascular pattern is established early in the development of angiosperm leaves, often when the developing leaf is only a few millimeters long—for example, in *Arabidopsis* when the leaf is about 2 millimeters long, and in barley when the primordium is 3 to 4 millimeters long. (See Chapter 27 for discussion of the role of auxin in the development of leaf vasculature.)

Sun and Shade Leaves Develop under Different Light Intensities

Environmental factors, especially light, can have substantial developmental effects on the size and thickness of leaves. In many species, leaves grown under high light intensities, the so-called **sun leaves,** are smaller and thicker than the **shade leaves** that develop under low light intensities (Figure 25–34). The increased thickness of sun leaves is due mainly to a greater development of the palisade parenchyma. The vascular system of sun leaves is more extensive and the walls of epidermal cells are thicker than those of shade leaves. In addition, the ratio of the internal surface area of the mesophyll to the area of the leaf blade

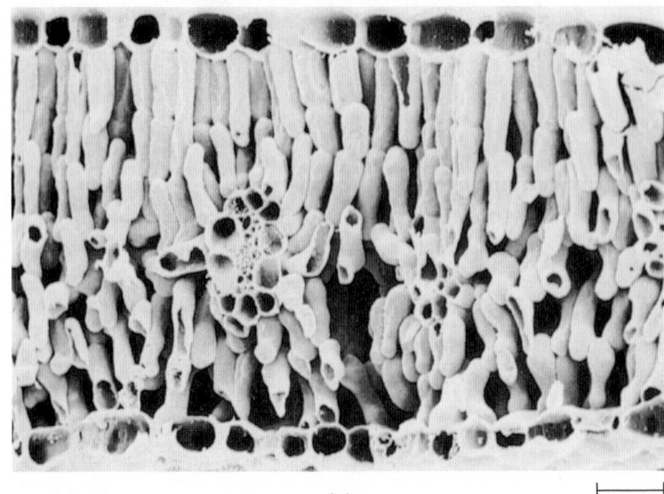

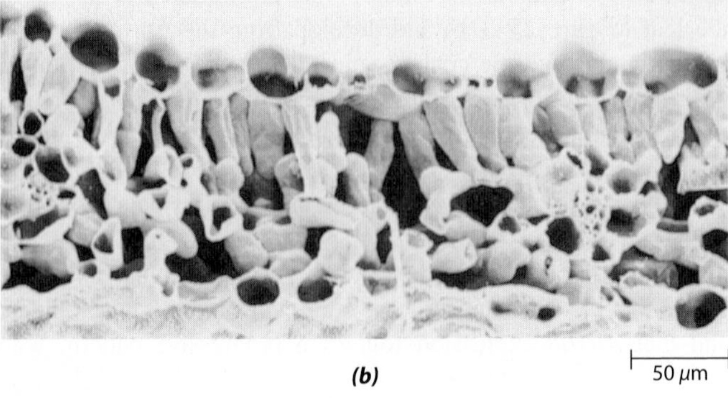

25–34 Sun and shade leaves Scanning electron micrographs showing transverse sections of veins (arrows) of ***(a)*** sun leaves and ***(b)*** shade leaves of *Thermopsis montana,* a member of the pea family (Fabaceae). Note that the sun leaf is considerably thicker than the shade leaf, a condition due mainly to a greater development of palisade parenchyma in the sun leaf.

STRONG, VERSATILE, SUSTAINABLE BAMBOO

There aren't many grasses strong enough to support people as they climb over a hundred meters (roughly 300 feet) into the air. In fact, there may only be one—bamboo. This versatile grass can be carved into utensils, split and woven into mats, bent and shaped into all manner of objects both graceful and utilitarian, and laminated to make flooring harder than oak. Bamboo can be bound together to form the posts and beams to frame a house or the cables of a suspension bridge. And lashed at right angles into a three-dimensional framework, bamboo forms a strong and lightweight scaffolding that ascends skyscrapers. In some parts of Asia, bamboo, not steel, is the major material used to build construction-site scaffolding.

Bamboo is in fact not just one species—there are over 1000, all in the grass (Poaceae) family—and each species has slightly different properties. Of these, the most economically important is *Phyllostachys edulis,* or Mao Zhu bamboo, which can grow to 20 meters and is the principal source of both edible bamboo shoots and structural poles. Forests of Mao Zhu bamboo cover more than 3 million hectares (7 million acres) in China.

Like many other grasses, bamboo spreads by sending out underground stems called rhizomes. Bamboos can be loosely classified as "clumpers" or "runners," depending on how far and how fast the rhizomes spread. *Phyllostachys edulis* is a runner, quickly forming large groves. Runners tend to be invasive and hard to control, but their dense rhizome mats make them excellent for stabilizing soil and preventing erosion. Clumpers are not invasive and are favored for decorative landscaping.

In all species, the underground rhizome forms a bud, which develops into a soft, fleshy shoot that emerges from the ground. It is harvested at this stage and cut into edible bamboo shoots, a familiar ingredient in Asian cuisine. As the shoot grows, it begins to harden and sprout leaves along its length; at this stage, the shoot is called a culm. Even the tallest bamboo culms grow to their final height in only a few months, and at the peak of their growth spurt, some species grow by a meter per day. The culm remains pliant and flexible early on, suitable for splitting and weaving. Over the next several years, the culm fully hardens, and at this stage it can be harvested and used for structural applications.

With their strong outer shell and hollow core, fully hardened bamboo poles have a strength-to-weight ratio similar to steel (one architect in fact calls bamboo "vegetal steel"). Its strength and light weight give it advantages as a building material, complemented by its environmental benefits—it sequesters atmospheric carbon, requires little fossil fuel to produce, and can be composted at the end of its useful life.

Bamboo also offers poor rural communities an indigenous, low-cost, and versatile building material with some surprising applications. The Earth Institute at Columbia University is working to create a small-scale bamboo bicycle frame industry in Africa, with the goal of developing a local business—based on sustainable resources—that can provide affordable and adaptable transportation to a large population.

Like all grasses, bamboo eventually flowers, and in many species, the plant subsequently dies. *Phyllostachys edulis* flowers sporadically, so that at any given time, some portions of a forest are in bloom while others are not. But many bamboo species engage in "gregarious" flowering, where all plants of the species, even when separated by hundreds of kilometers, flower at once. The intervals between flowering events differ by species, and range from several years to over a hundred.

The mass dying-off of the bamboo after gregarious flowering can be catastrophic for local human populations, and not only because of the loss of food and livelihood. The sudden abundance of bamboo seed leads to population explosions of creatures that feed on it, especially rats. As their numbers swell and the seed is consumed or rots, the rats may overrun local grain fields and storage supplies, inducing famine. While it is currently impossible to alter the bamboo's flowering cycle to prevent gregarious flowering, coordinated governmental responses can mitigate its effects by providing food, protecting grain supplies, and preventing or treating rodent-borne diseases.

Workers assemble bamboo scaffolding on a skyscraper at a construction site in Hong Kong.

is much higher in sun leaves. One effect of these differences is that, although the two leaf types have similar photosynthetic rates at low light intensities, shade leaves are not adapted to high light intensities and consequently have considerably lower maximum photosynthetic rates under these conditions.

Because light intensities vary greatly in different parts of the crowns of trees, extreme forms of sun and shade leaves can be found there. Sun and shade leaves also occur in shrubs and in herbaceous plants. They can be induced to form by growing the plants under high or low light intensities.

Leaf Abscission

In many plants, the normal separation of the leaf from the stem—the process of **abscission**—is preceded by certain structural and chemical changes near the base of the petiole. These changes result in the formation of an **abscission zone** (Figure 25–35). In woody angiosperms, two layers can be distinguished in the abscission zone: a separation (abscission) layer and a protective layer. The *separation layer* consists of relatively short cells with poorly developed wall thickenings that make it structurally weak. Prior to abscission, certain reusable ions and molecules are returned to the stem, among them magnesium ions, amino acids (derived from proteins), and sugars (some derived from starch). Enzymes then break down the cell walls in the separation layer. The cell wall changes may include weakening of the middle lamella and hydrolysis of the cellulosic walls themselves. Cell division may precede the actual separation. If cell division takes place, the newly formed cell walls are the ones largely affected by the degradation processes. Beneath the separation layer, a *protective layer* composed of heavily suberized cells is formed, further isolating the leaf from the main body of the plant before the leaf drops. Tyloses may form in the tracheary elements before abscission (see pages 632 and 633).

Eventually, the leaf is held to the plant by only a few strands of vascular tissue, which may be broken by the enlargement of parenchyma cells in the separation layer. After the leaf falls, the protective layer is evident as a **leaf scar** on the stem (see Figure 26–16). Hormonal factors associated with abscission are discussed in Chapter 27.

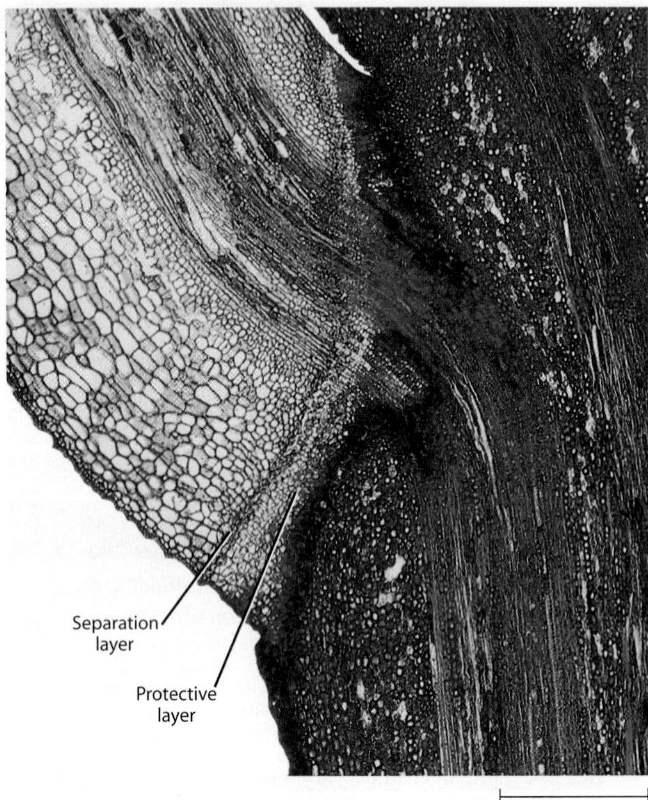

25–35 Abscission zone in maple *(Acer)* leaf Longitudinal section through the base of the petiole showing the separation layer and protective layer of cells that make up the abscission zone.

Transition between the Vascular Systems of the Root and Shoot

As discussed in previous chapters, the distinction between plant organs is based primarily on the relative arrangement of the vascular tissues within the ground tissues. For example, in eudicot roots the vascular tissues generally form a solid cylinder surrounded by the cortex, and the strands of primary phloem alternate with the radiating ridges of primary xylem. By contrast, in the stem the vascular tissues often form a cylinder of discrete strands around a pith, with the phloem on the outside of the vascular bundles and the xylem on the inside. Obviously, somewhere in the primary plant body, a change must take place from the type of structure found in the root to that found in the shoot. This change is a gradual one, and the region of the plant axis through which it occurs is called the **transition region.**

As described in Chapter 22, the shoot and root are initiated as a single continuous structure during development of the embryo. Consequently, vascular transition occurs in the axis of the embryo or young seedling. The transition is initiated during the appearance of the procambial system in the embryo and is completed with the differentiation of the variously distributed procambial tissues in the seedling. Vascular continuity between the root and shoot systems is maintained throughout the life of the plant.

The structure of the transition region can be very complex, and much variation exists in the transition regions of different kinds of plants. In most seed plants other than monocots, the transition occurs within the vascular system connecting the root and the cotyledons. Figure 25–36 depicts a type of transition region commonly found in eudicots. Notice the diarch (having two protoxylem poles) structure of the root; the branching and reorientation of the primary xylem and the primary phloem, which in the upper part of the hypocotyl-root axis result in the formation of a pith; and the traces of the first leaves of the epicotyl.

Development of the Flower

The development of a flower or inflorescence ends the meristematic activity of the vegetative shoot apex. During the transition to flowering, the vegetative shoot apex undergoes a sequence of physiological and structural changes and is transformed into a reproductive apex. Consequently, flowering may be considered a stage in the development of the shoot apex and of the plant as a whole. Inasmuch as the reproductive apex exhibits a determinate growth pattern, flowering in annuals indicates that the plant is approaching completion of its life cycle. By contrast, flowering in perennials can be repeated annually or more frequently. Various environmental factors, including the length of day and the temperature, are known to be involved in the induction to flowering (see Chapter 28).

The transition from a vegetative shoot apex to a floral apex is often preceded by elongation of the internodes and early

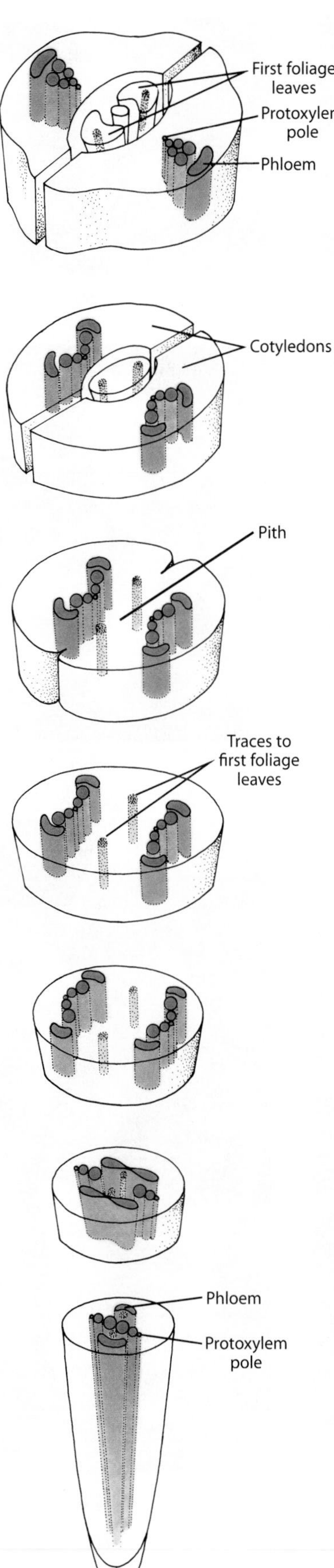

25–36 Transition region Shown here is the transition region—the connection between root and cotyledons—in the seedling of a eudicot with a diarch root, that is, a root with two protoxylem poles. In the root, the primary vascular system, consisting of xylem (red) and phloem (blue), is represented by a single cylinder of vascular tissue. In the hypocotyl-root axis (the part of the seedling below the cotyledons), the xylem and phloem become reoriented, and the vascular system branches and diverges into the cotyledons. Leaf traces to the first foliage leaves are at a procambial stage of development.

development of lateral buds below the shoot apex. The apex itself undergoes a marked increase in mitotic activity, accompanied by changes in dimensions and organization: the relatively small apex with a tunica-corpus type of organization becomes broad and domelike.

The initiation and early stages of development of the sepals, petals, stamens, and carpels are quite similar to those of leaves, their evolutionary precursors. Commonly, initiation of the floral parts begins with the sepals, followed by the petals, then the stamens, and finally the carpels (Figure 25–37). This usual order of appearance may be modified in certain flowers, but the floral parts always have the same relative spatial relation to one another. The floral parts may remain separate during their development, or they may become united within whorls (connation) and between whorls (adnation). We discussed some variations in flower structure in Chapter 20.

A Small Set of Regulatory Genes Determines Organ Identity in Flowers

Our understanding of the genetic control of flower development has been greatly advanced through the study of mutations that change floral part, or floral organ, identity. Such mutations, which result in the formation of the wrong organ in the wrong place, are called **homeotic mutations.** Horticulturists long have selected and propagated mutations that cause "double flowers," including the rose varieties grown in most gardens today. Whereas the wild rose contains only 5 petals, the rose doubles have 20 or more, the result of homeotic mutations that change the stamens of the wild-type rose into petals. Mutations of floral organ **homeotic genes**—genes affecting floral organ identity—have been studied most intensively in snapdragon *(Antirrhinum majus)* and *Arabidopsis thaliana.* Most plant homeotic genes belong to a class of related sequences known as **MADS box genes,** many of which control aspects of development.

In the wild-type, or normal, *Arabidopsis* flower, the floral parts are arranged in four whorls (Figure 25–38a, b, page 608). The first (outermost) whorl consists of four green sepals; the second consists of four white petals; the third contains six stamens, two of which are shorter than the other four; and the fourth is represented by a single pistil (apocarpous gynoecium) composed of two fused carpels forming an elongated, two-loculed ovary, with numerous ovules, a short style, and a stigma.

The study of homeotic mutations in *Arabidopsis* first identified three classes of genes—designated A, B, and C—that are essential to the normal development and order of appearance of

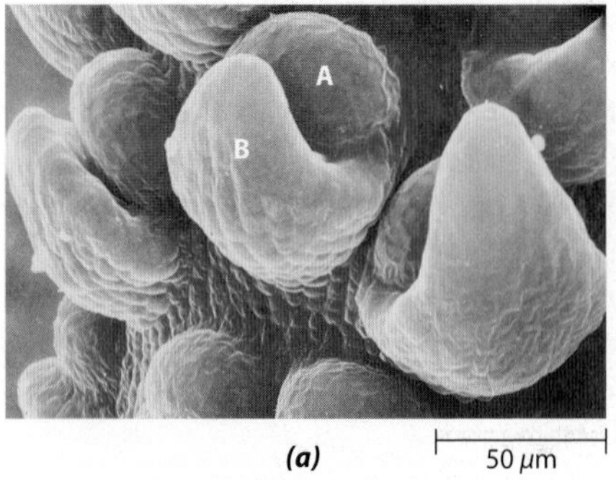

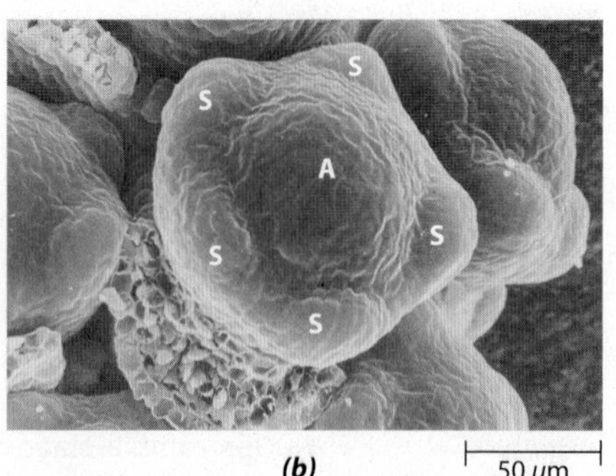

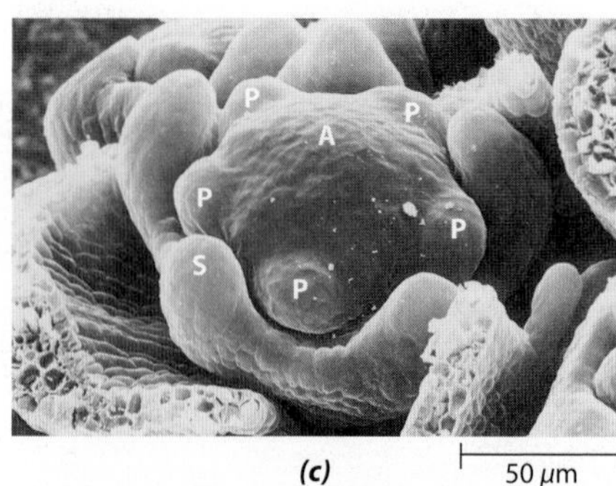

25–37 Flower development in a legume Scanning electron micrographs showing some stages in the development of the perfect flower of *Neptunia pubescens,* a legume with radial symmetry and whorled floral parts. ***(a)*** Floral apex (A) in the axil of a bract (B). ***(b)*** Five sepal primordia (S) have been initiated around the floral apex. ***(c)*** Five petal primordia (P) have been initiated around the floral apex, alternating with the sepals (S). During their development, the sepals will form a calyx tube. ***(d)*** Five stamen primordia (two are indicated by arrows) have been initiated around the floral apex, alternating with the petals (P). ***(e)*** A second whorl of stamens (arrow) has been initiated, its members alternating with members of the first stamen whorl (ST_1). The carpel (C) has been initiated at the center of the floral apex. All floral parts are now present. ***(f)*** The carpel has developed a cleft, which will form the locule of the ovary. The stamens (ST_1) of the outer, or first, whorl are beginning to differentiate anthers and filaments. (ST_2 designates the inner, or second, whorl of stamens.) ***(g)*** The carpel is beginning to differentiate into a style and ovary. ***(h)*** Older flower with both whorls of stamens in view. ***(i)*** Older flower with some stamens removed to reveal the carpel, with differentiated ovary (O), style, and stigma (arrow). Tips of the subtending bracts have been removed in *(b)* through *(i)*. In *(f)* through *(i),* most sepals and petals have also been removed.

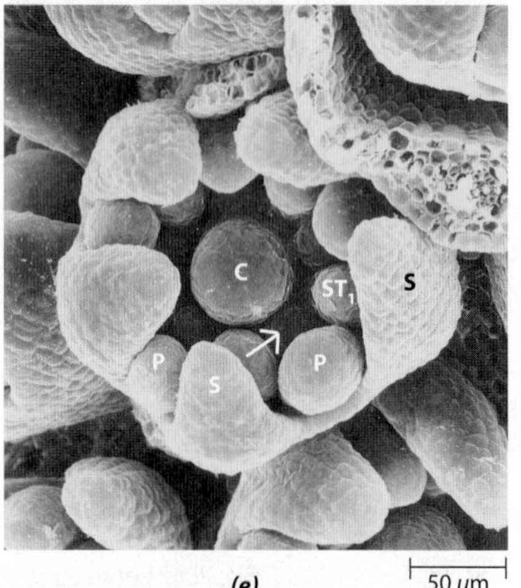

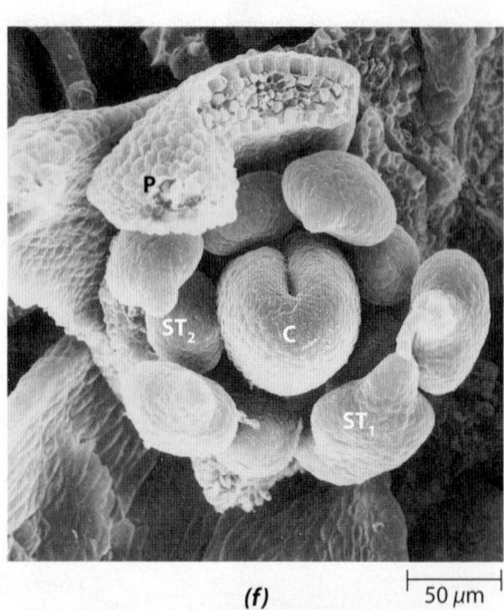

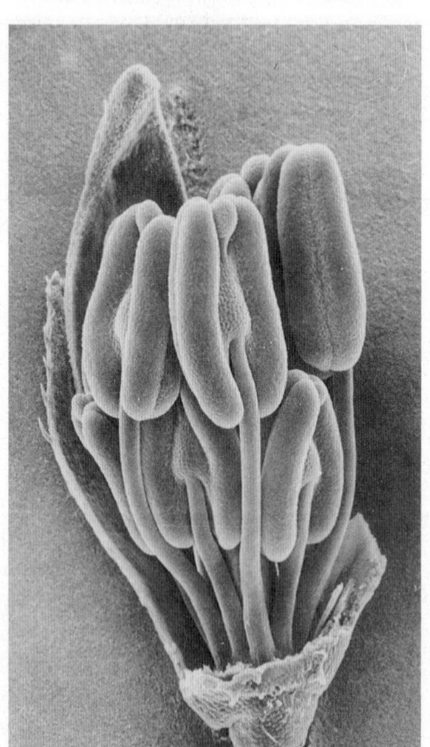

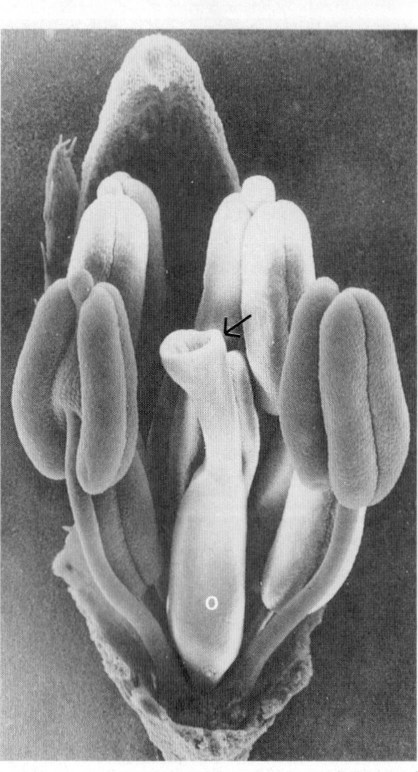

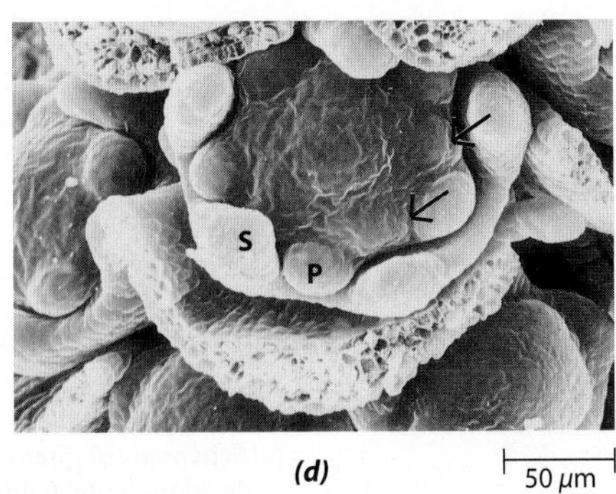

(d)

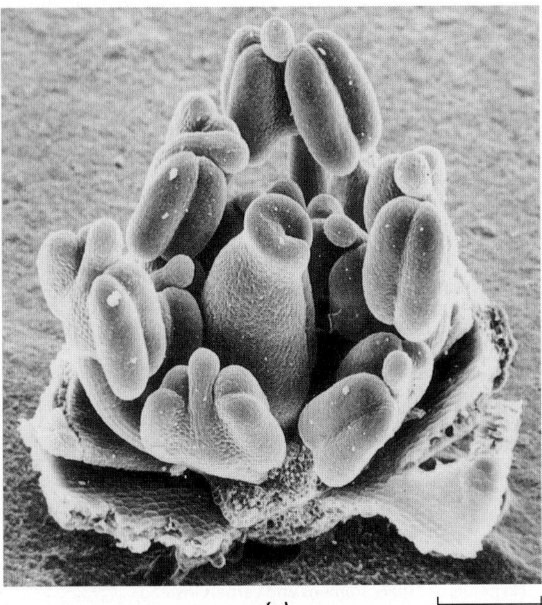

(g)

the floral organs. The three classes are expressed in overlapping fields within the floral meristem; that is, each class acts in two adjacent whorls and is partly or wholly responsible for the identity of the organs in those two whorls (Figure 25–38c). Class A genes function in field 1, the first and second whorls (which normally form sepals and petals, respectively); class B genes in field 2, the second and third whorls (which normally form petals and stamens, respectively); and class C genes in field 3, the third and fourth whorls (which normally form stamens and carpels, respectively). On that basis, a genetic model—the **ABC model**—was proposed to account for the identity of the four whorls of the typical eudicot flower (Figure 25–39). The expression of class A genes alone determines the development of sepals; class A + B specify petals; class B + C, stamens; and class C alone, carpels.

In the ABC model, five different genes specify floral organ identity in *Arabidopsis: APETALA1 (AP1), APETALA2 (AP2), APETALA3 (AP3), PISTILLATA (PI),* and *AGAMOUS (AG).* Class A activity is encoded by *AP1* and *AP2.* (*AP2* is the only floral organ identity gene that is not a MADS-box gene.) Loss of class A function results in the formation of carpels instead of sepals in the first whorl and of stamens instead of petals in the second whorl (Figure 25–39). Class B activity is encoded by *AP3* and *PI.* Loss of class B function results in the formation of sepals instead of petals in the second whorl and of carpels instead of stamens in the third whorl. Class C activity is encoded by *AG.* Loss of its function results in the formation of petals instead of stamens in the third whorl and loss of the carpels (the fourth whorl), which are replaced by a new floral meristem that generates additional sepals and petals. An important facet of the ABC model is that the A and C functions are mutually antagonistic, such that class A activity expands in class C mutant flowers and vice versa.

Development of the ABC model contributed greatly to our understanding of floral development. The model has recently been expanded to include five gene classes (A, B, C, D, E) and is therefore known as the **ABCDE model** of flower development (Figure 25–40). In *Arabidopsis,* the D-function gene *SEEDSTICK (STK)* is involved in ovule development and is required for the dispersal of seed in mature flowers. The D function was originally discovered in petunia. With discovery of the E function (originally found in petunia and tomato), it became clear that the A-, B-, and C-function genes need E-function genes to produce floral organs. In *Arabidopsis,* the E function is encoded by a set of MADS-box genes called *SEPALLATA (SEP),* of which there are four: *SEP1–4; SEP1–3* are expressed in the second, third, and fourth whorls, and *SEP4* is expressed in all whorls. The *SEP* genes mediate interactions between the organ identity proteins. The *sep1 sep2 sep3* triple mutant produces sepals in all floral whorls. Addition of the *sep4* mutation results in conversion of all floral organs into leaves, and hence a complete loss of floral organ identity. According to the ABCDE model, class A + E genes specify sepals; class A + B + E, petals; class B + C + E, stamens; class C + E, carpels; and class C + D + E, ovules (Figure 25–40).

In *Arabidopsis,* a gene called *LEAFY (LFY)* encodes the transcription factor, LFY, that assigns the floral fate of meristems. This transcription factor is expressed at a high level throughout young floral meristems and activates the genes that give the floral organ primordia their identities. Plants with the *leafy* mutation do not make flowers (Figure 25–41).

Stem and Leaf Modifications

Stems and leaves may undergo modifications and perform functions quite different from those commonly associated with these two components of the shoot. One of the most common modifications is the formation of **tendrils,** which aid in support. Some tendrils are modified stems. The tendrils of the grape *(Vitis)* (Figure 25–42), for example, are modified stems that coil around a supporting structure. In the grape, the tendrils sometimes produce small leaves or flowers. In Boston ivy *(Parthenocissus tricuspidata)* and Virginia creeper *(Parthenocissus quinquefolia),* the tendrils are also modified stems, which in these plants form adhesive disks at their tips. Most tendrils, however, are leaf modifications. In legumes, such as the garden pea *(Pisum sativum),* the tendrils constitute the terminal part of the pinnately compound leaf (Figure 25–18).

Modified stems that assume the form of and closely resemble foliage leaves are called **cladophylls.** The feathery, leaflike branches of asparagus *(Asparagus officinalis)* are familiar

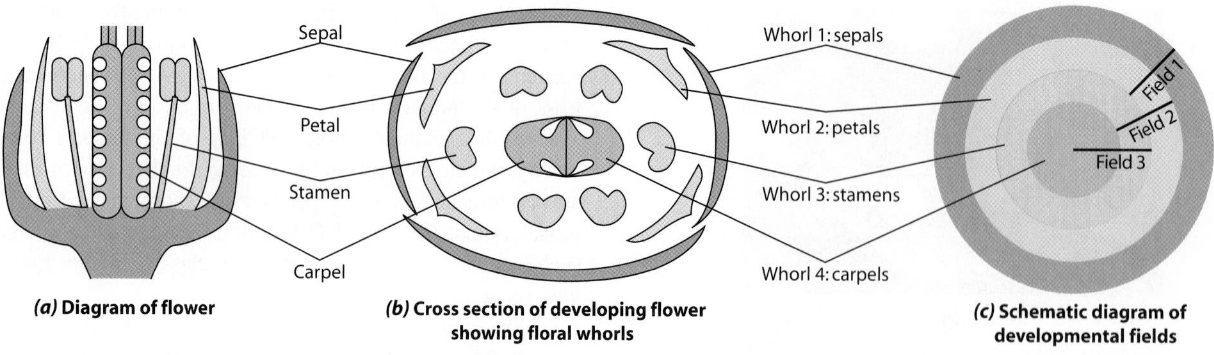

***(a)* Diagram of flower**

***(b)* Cross section of developing flower showing floral whorls**

***(c)* Schematic diagram of developmental fields**

25–38 Wild-type, or normal, *Arabidopsis* flower ***(a)*** Diagram of a median longitudinal section through a flower. ***(b)*** Floral diagram showing the basic plan of the flower, which consists of four whorls. ***(c)*** Schematic diagram showing overlapping developmental fields, which correspond to the expression patterns of specific floral organ identity genes. Class A genes function in field 1, class B genes in field 2, and class C genes in field 3.

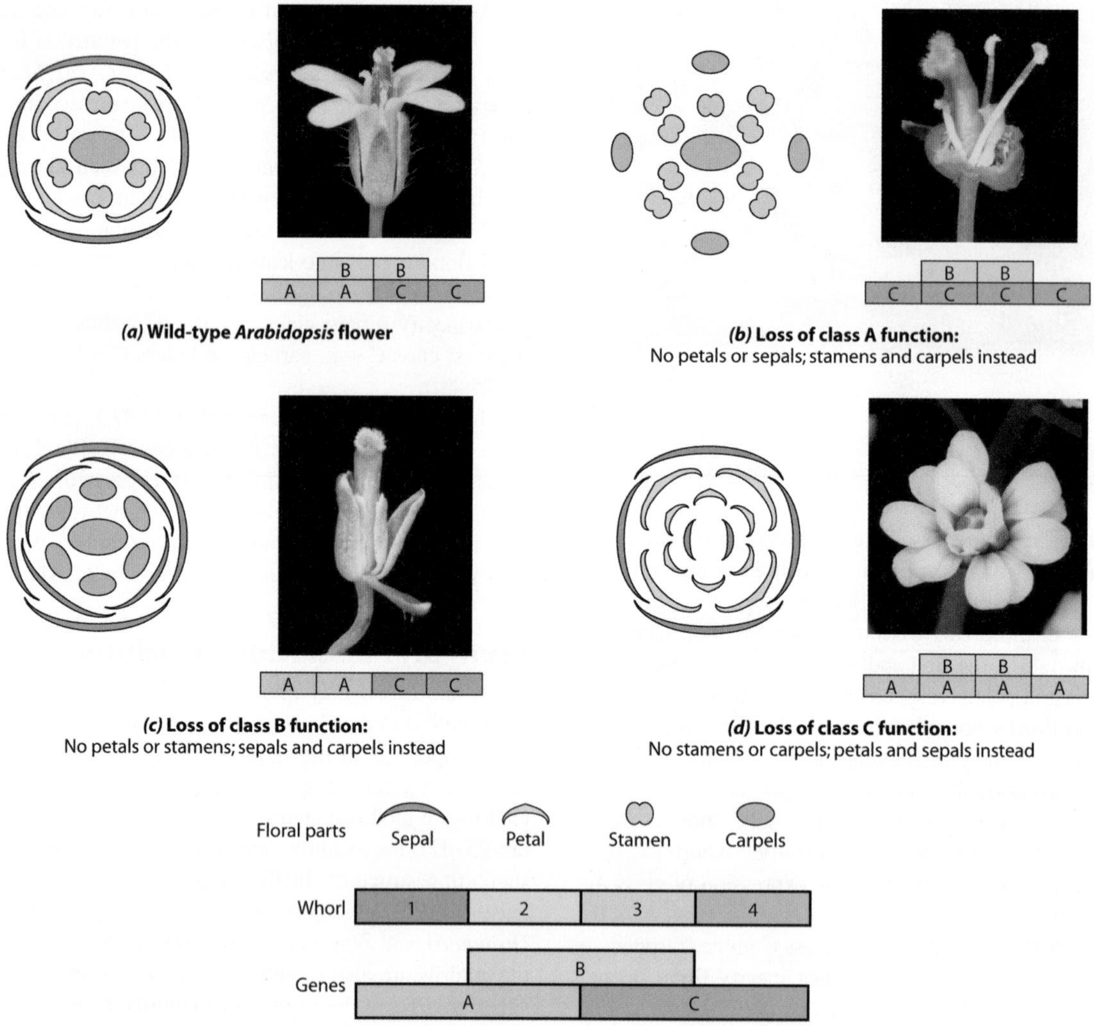

***(a)* Wild-type *Arabidopsis* flower**

***(b)* Loss of class A function:**
No petals or sepals; stamens and carpels instead

***(c)* Loss of class B function:**
No petals or stamens; sepals and carpels instead

***(d)* Loss of class C function:**
No stamens or carpels; petals and sepals instead

25–39 The ABC model of floral organ determination in *Arabidopsis* ***(a)*** In the wild-type flower, expression of class A genes alone determines the development of sepals; class A + B specify petals; class B + C, stamens; and class C alone, carpels. The effects on floral organ identity resulting from mutation and loss of function of class A, B, and C homeotic genes are shown in *(b)–(d)*. ***(b)*** Loss of class A function results in expansion of class C function throughout the floral meristem. ***(c)*** Loss of class B function results in expression of only A and C. ***(d)*** Loss of class C function results in spread of A function throughout the meristem.

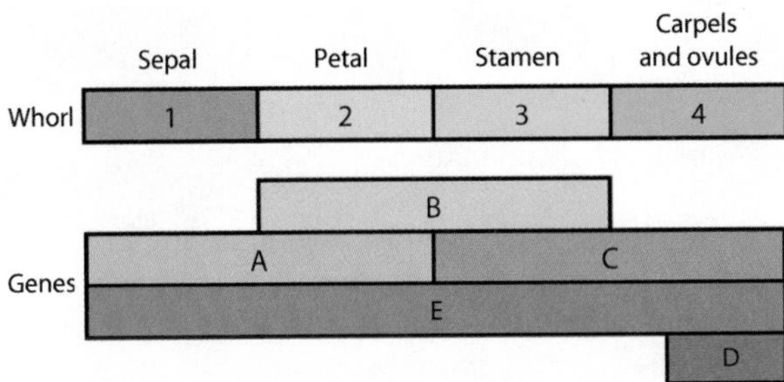

25–40 The ABCDE model of floral organ determination in *Arabidopsis* In addition to the A-, B-, and C-function genes of the ABC model, this model includes two additional gene classes, D and E. In the ABCDE model, class A + E genes specify sepals; class A + B + E, petals; class B + C + E, stamens; class C + E, carpels; and class C + D + E, ovules.

examples of cladophylls (Figure 25–43). The thick and fleshy aerial shoots ("spears") of asparagus are the edible portion of the plant. The scales on the spears are true leaves. If asparagus plants are allowed to continue growing, cladophylls develop in the axils of the minute, inconspicuous scales and then act as photosynthetic organs. In some cacti, the branches resemble leaves but are, in fact, cladophylls (Figure 25–44). As mentioned previously, true leaves typically have buds in their axils, whereas cladophylls do not. This characteristic may be used to distinguish between the two.

In some plants, the leaves are modified as spines, which are hard, dry, and nonphotosynthetic. The terms "spine" and "thorn" are frequently used interchangeably, but technically, thorns are modified branches that arise in the axils of leaves (Figure 25–45). Another term commonly used interchangeably with thorn and

(a)

(b)

25–41 The *leafy* mutant of *Arabidopsis* Mutations in the *LEAFY* gene of *Arabidopsis* prevent transcription of the ABC genes and hence flower production. ***(a)*** Wild-type plant. ***(b)*** The *leafy* mutant.

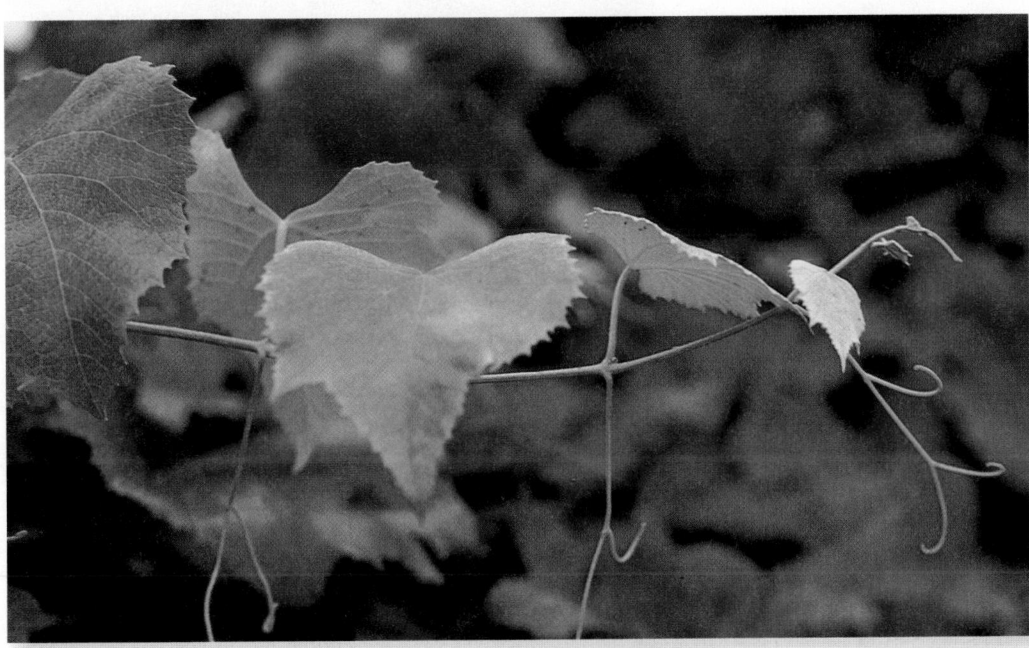

25–42 Grape tendrils The tendrils of grape *(Vitis)* are modified stems.

25–43 Cladophylls of asparagus The filmy branches of the common edible asparagus *(Asparagus officinalis)* resemble leaves. Such modified stems are called cladophylls.

25–44 Cladophylls of a cactus The branches of the spineless cactus *Epiphyllum* resemble leaves but are actually modified stems.

spine is "prickle." A prickle, however, is neither a stem nor a leaf but a small, more or less slender, sharp outgrowth from the cortex and epidermis. The so-called thorns on rose stems are prickles. All three structures—spines, thorns, and prickles—may serve in defense, reducing predation by herbivores. In one special plant-herbivore interaction, the spines of the bull's-horn acacia provide shelter for ants that kill other insects attempting to feed on the acacia plant.

(a)

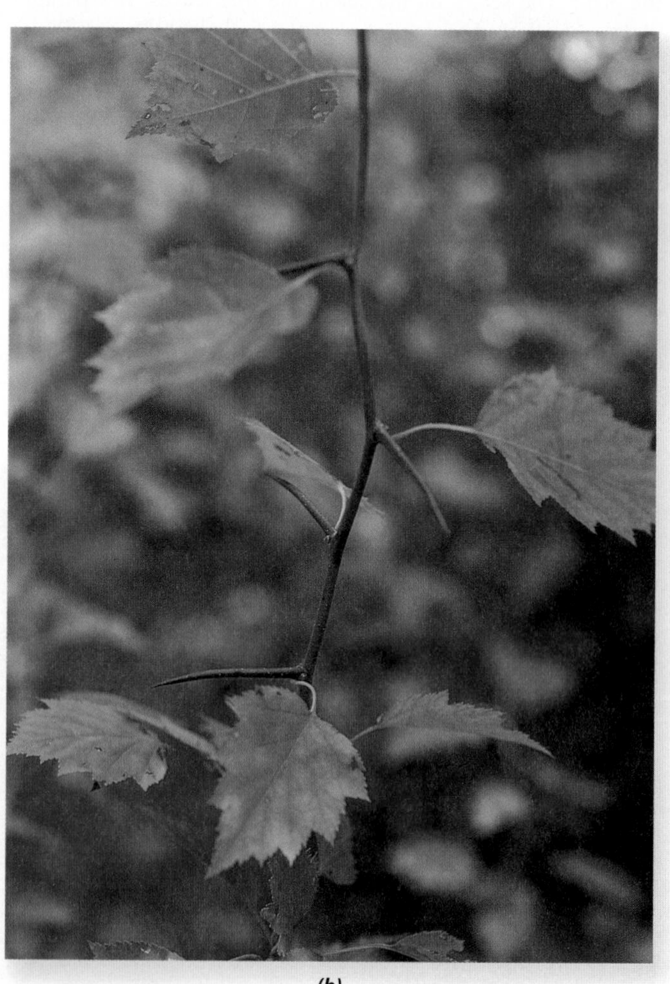

(b)

25–45 Spines and thorns ***(a)*** Spines, as in *Ferocactus melocactiformis,* are modified leaves. The spines originate in the position of the bud scales. ***(b)*** Thorns are modified branches, which, as this photograph of a hawthorn *(Crataegus)* shows, arise in the axils of the leaves.

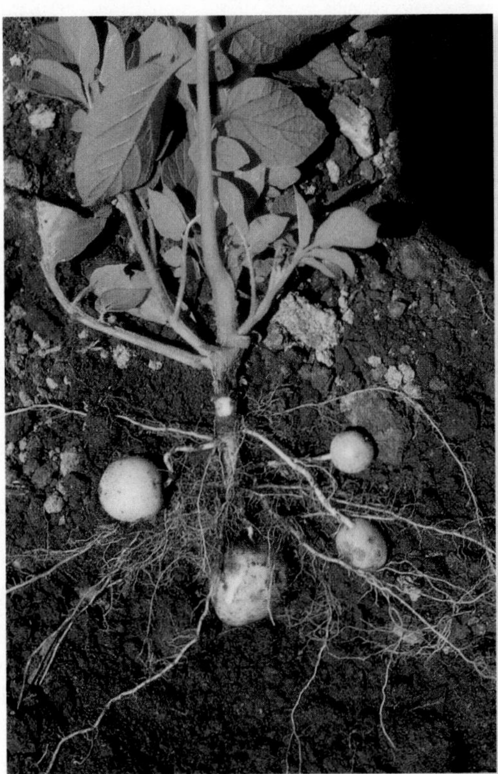

25–46 Rhizome White potato *(Solanum tuberosum),* with tubers attached to a rhizome, or underground stem.

Among the most spectacular of modified or specialized leaves are those of the carnivorous plants, such as the pitcher plant, the sundew, and the Venus flytrap, which capture insects and digest them with secreted enzymes. The nutrients thus made available are then absorbed by the plant (see the essay on page 694).

Some Stems and Leaves Are Specialized for Food Storage

Stems, like roots, serve food-storage functions. Probably the most familiar type of specialized storage stem is the **tuber,** as exemplified by the Irish, or white, potato *(Solanum tuberosum).* In the white potato, tubers arise at the tips of **stolons** (slender stems growing along the surface of the ground) of plants grown from seed. However, when cuttings of a tuber—the "seed pieces"—are used for propagation, the tubers arise at the ends of long, thin **rhizomes,** or underground stems (Figure 25–46). Except for vascular tissue, almost the entire mass of the tuber inside the periderm (the "skin") is storage parenchyma. The "eyes" of the white potato are nodal depressions containing groups of buds, and each seed piece cut from a potato tuber must include at least one eye. The depression is the axil of a scalelike leaf. The scalelike leaves are helically arranged on the tuber, as are the leaves of the aboveground stems.

A **bulb** is a large bud consisting of a small, conical stem with numerous modified leaves attached to it. The leaves are scalelike and have thickened bases where food is stored. Roots arise from the bottom of the stem. Familiar examples of plants with bulbs are the onion (Figure 25–47a) and the lily.

Although superficially similar to bulbs, **corms** consist primarily of thickened, fleshy stem tissue. Their leaves commonly are thin and much smaller than those of bulbs; consequently, the

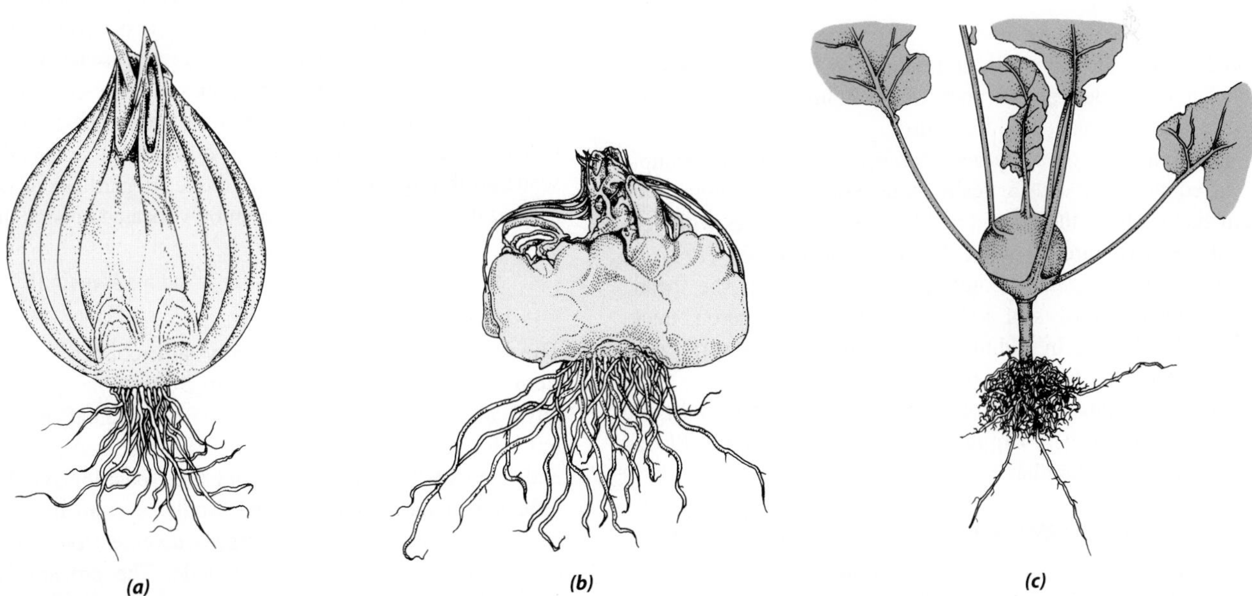

25–47 Examples of modified leaves or stems ***(a)*** An onion *(Allium cepa)* bulb, which consists of a conical stem with scalelike, food-containing leaves. The leaves are the edible part of the onion. ***(b)*** A gladiolus *(Gladiolus grandiflorus)* corm, which is a fleshy stem with small, thin leaves. ***(c)*** The fleshy storage stem of kohlrabi (*Brassica oleracea* var. *caulorapa*).

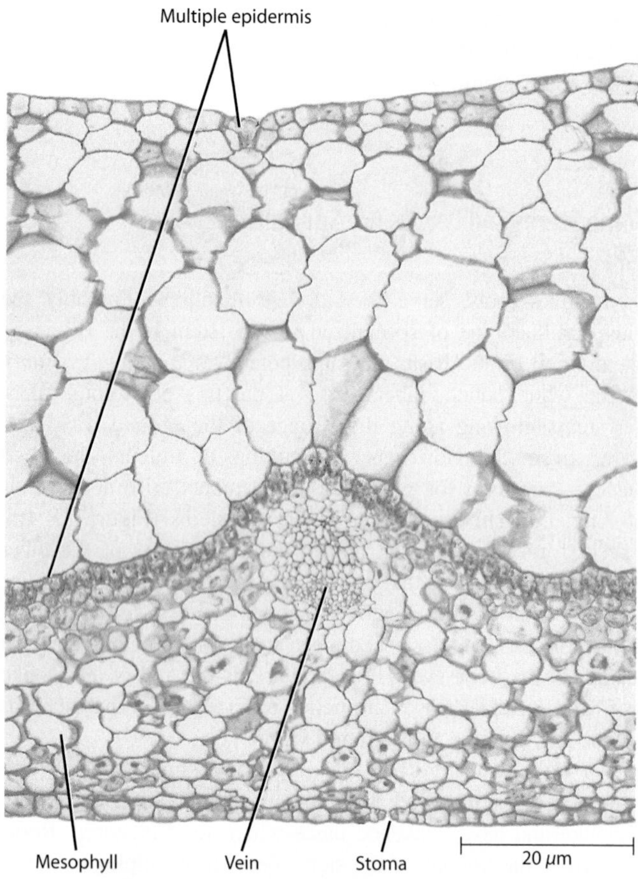

25–48 Multiple epidermis Transverse section of a leaf blade of *Peperomia.* The very thick multiple epidermis, visible on its upper surface, presumably functions as water-storage tissue.

stored food of the corm is found within the fleshy stem. Several well-known garden plants, such as gladiolus (Figure 25–47b), crocus, and cyclamen, produce corms.

Kohlrabi *(Brassica oleracea* var. *caulorapa)* is one example of an edible plant with a fleshy storage stem. The short, thick stem stands above the ground and bears several leaves with very broad bases (Figure 25–47c). The common cabbage *(Brassica oleracea* var. *capitata)* is closely related to kohlrabi. The "head" of the cabbage consists of a short stem bearing numerous thick, overlapping leaves. In addition to a terminal bud, several well-developed axillary buds may be found within the head.

The leaf stalks, or petioles, of some plants become quite thick and fleshy. Celery *(Apium graveolens)* and rhubarb *(Rheum rhabarbarum)* are familiar examples.

Some Stems and Leaves Are Specialized for Water Storage

Succulent plants are plants that have tissues specialized for the storage of water. Most of these plants—such as the cacti of the American deserts, the cactuslike euphorbias of the African deserts (see "Convergent Evolution" on page 239), and the century plant *(Agave)*—normally grow in arid regions, where the ability to store water is necessary for their survival. The green, fleshy stems of the cacti serve as both photosynthetic and storage organs. The water-storing tissue consists of large, thin-walled parenchyma cells that lack chloroplasts.

In the century plant, the leaves are succulent. As in succulent stems, nonphotosynthetic parenchyma cells of the ground tissue constitute the water-storing tissue. Other examples of plants with succulent leaves are the ice plant *(Mesembryanthemum crystallinum),* the stonecrops *(Sedum),* and certain species of *Peperomia.* In the ice plant, large epidermal cells with appendages (trichomes) called water vesicles, which superficially resemble beads of ice, serve to store water. The water-storing cells of the *Peperomia* leaf are part of a multiple (several-layered) epidermis derived by periclinal divisions of the protoderm (Figure 25–48).

SUMMARY

The Shoot Apical Meristem Produces Leaf Primordia, Bud Primordia, and Stem Primary Tissues

The vegetative shoot apical meristems of most flowering plants have a tunica-corpus type of organization consisting of one or more peripheral layers of cells (the tunica) and an interior body of cells (the corpus). The apices consist of three superimposed layers of cells—two tunica layers and the initial layer of the corpus—which are designated L1, L2, and L3. The primary tissues of the stem pass through periods of growth similar to those of the root, but unlike the root, the stem cannot be divided into regions of cell division, elongation, and maturation. The stem increases in length largely by internodal elongation.

Three Basic Types of Organization Exist in the Primary Structure of Stems

As in the root, the apical meristem of the shoot gives rise to protoderm, procambium, and ground meristem, which develop into the primary tissues. There are three basic patterns of the relative distribution of ground and primary vascular tissues in the primary structure of stems: the primary tissues may develop (1) as a more or less continuous cylinder within the ground tissue, (2) as a cylinder of discrete strands, or (3) as a system of strands that appear scattered throughout the ground tissue. Regardless of the type of organization, the phloem is commonly located outside the xylem.

Leaves and Stems Are Closely Related Physically and Developmentally

The term "shoot" serves not only as a collective term for the stem and its leaves but also as an expression of their intimate physical and developmental association. Leaf primordia are initiated by groups of cells, called founder cells, in the peripheral zone of the shoot apex, and their position on the stem is reflected in the pattern of the vascular system in the stem. At each node, one or more leaf traces diverge from the stem and enter the leaf or leaves at that node. The procambial strands from which the leaf traces are formed develop just below the developing leaf primordia, and they are sometimes present below the sites of the future leaf primordia even before the primordia are discernible. Several hypotheses have been advanced to explain the mechanism underlying the pattern of leaf arrangement, or phyllotaxis; among these are the first available space hypothesis, the inhibitory field hypothesis, and the biophysical forces hypothesis. These hypotheses have been superseded by an auxin-based model (see Chapter 27).

Variations in Leaf Structure Are Largely Related to Habitat

Except for grasses and related monocots, most angiosperm leaves consist of a blade and a petiole. The blades of some leaves (compound leaves) are divided into leaflets, whereas those of others (simple leaves) are not. Stomata are commonly more numerous on the lower than on the upper surface of the leaf. The ground tissue, or mesophyll, of the leaf is specialized as a photosynthetic tissue and, in mesophytes, is differentiated into palisade parenchyma and spongy parenchyma. The mesophyll is thoroughly permeated by air spaces and by veins, which are composed of xylem and phloem surrounded by a parenchymatous bundle sheath. The xylem commonly occurs on the upper side of the vein, and the phloem on the lower side.

Most C_3 and C_4 Grasses Can Be Distinguished on the Basis of Leaf Anatomy

In most monocots, including the grasses, the leaf consists of a blade and a sheath, which encircles the stem. The leaves of C_3 and C_4 grasses have some distinct anatomical differences, the most notable being the presence in C_4 grasses and absence in C_3 grasses of a Kranz anatomy—an arrangement in which the mesophyll cells and the bundle-sheath cells form concentric layers around the vascular bundles.

Leaves Exhibit Determinate Growth, and Stems Exhibit Indeterminate Growth

Leaves are determinate in growth; that is, their development is of relatively short duration. By contrast, the vegetative shoot apices that initiate leaves may exhibit unlimited, or indeterminate, growth. In many species, leaves grown under high light intensities (sun leaves) are smaller and thicker than those grown under low light intensities (shade leaves).

The Separation of a Leaf from a Branch by Abscission Is a Complex Process

In many plants, leaf abscission is preceded by formation of an abscission zone, consisting of a separation layer and a protective layer, at the base of the petiole. After the leaf falls, the protective layer is evident as a leaf scar on the stem.

The Transition Region Is Where the Root and Shoot Unite

The change in distribution of the vascular and ground tissues between that found in the root and that in the shoot occurs in a region of the plant axis of the embryo and young seedling called the transition region. In most seed plants other than monocots, vascular transition is located within the vascular system connecting the root and the cotyledons.

Flower Development May Be Explained by Overlapping Gene Expression

In *Arabidopsis,* the overlapping expression of five classes of homeotic genes (A, B, C, D, E) determines the identity of the floral organs. Class A + E genes specify sepals; class A + B + E, petals; class B + C + E, stamens; class C + E, carpels; and class D + E, ovules.

Stems May Serve Food-Storage or Water-Storage Functions

Like roots, stems may be specialized for food storage. Examples are tubers and corms. Water-storing plants are known as succulents. The water-storage tissue of succulent plants is made up of large, vacuolated parenchyma cells. Stems or leaves or both may be succulent.

QUESTIONS

1. Distinguish between the following: leaf primordium and bud primordium; leaf trace and leaf trace gap; simple leaf and compound leaf; separation layer and protective layer; closed vascular bundle and open vascular bundle.
2. By means of simple, labeled diagrams, compare the structure of a eudicot root with that of a eudicot stem at the end of primary growth. Assume that the root is triarch and the vascular cylinder of the stem consists of a system of discrete vascular bundles.
3. The term "shoot" serves as more than just a collective term for the stem and its leaves. Explain.
4. How might the distribution of stomata differ among the leaves of mesophytes, hydrophytes, and xerophytes?
5. Explain why the mesophyll tissue is particularly suited for photosynthesis.
6. What are the principal roles of the major and minor veins of leaves?
7. In what ways does the leaf anatomy of C_3 grasses differ from that of C_4 grasses?
8. What are the main events in the initiation and development of a leaf?
9. Structurally, how do sun leaves differ from shade leaves?
10. What is the role of the *LEAFY* gene in floral development?

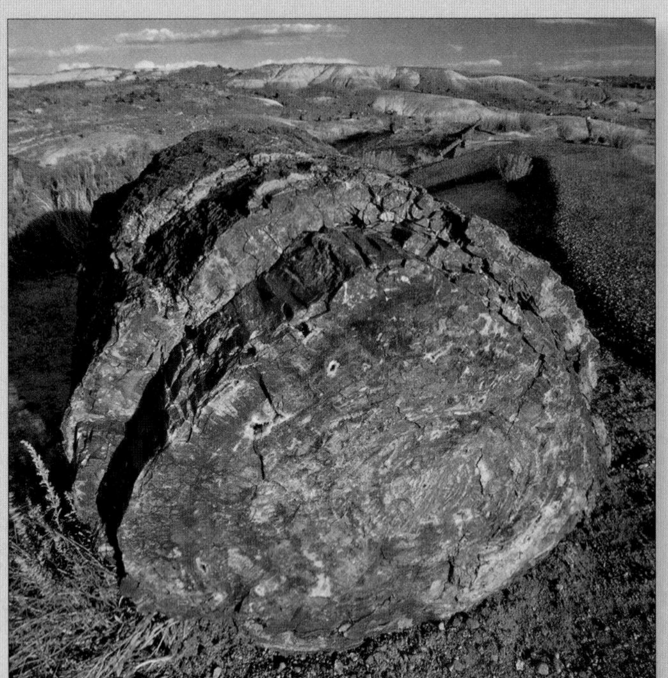

CHAPTER 26

Secondary Growth in Stems

◀ **Petrified wood** Some 225 million years ago, in the late Triassic when dinosaurs began to appear, a volcanic eruption is thought to have toppled now extinct conifers in the lush tropical area that became northeastern Arizona. Washed into low-lying areas, the trees became water-logged and sank. Volcanic ash dissolved in the water supplied silica, which replaced the plant cells of this section of tree trunk—primarily secondary xylem—with colorful quartz.

CHAPTER OUTLINE

In many plants—that is, most monocots and certain herbaceous eudicots, such as buttercup *(Ranunculus)*—growth in a given part of the plant body ceases with maturation of the primary tissues. At the other extreme are the gymnosperms and woody magnoliids and woody eudicots, in which the roots and stems continue to increase in diameter in regions that are no longer elongating (Figure 26–1). This increase in thickness or girth of the plant body, termed **secondary growth,** results from activity of the two **lateral meristems:** the **vascular cambium** and the **cork cambium.**

Herbs, or herbaceous plants, are plants with shoots that undergo little or no secondary growth. In temperate regions, either the herbaceous shoot or the entire plant, depending on the species, lives for only one season. Woody plants—trees and shrubs—can live for many years. At the start of each growing season, primary growth is resumed, and additional secondary tissues are added to the older plant parts through reactivation of the lateral meristems. Although most monocots lack secondary growth, some, such as the palms, may develop thick stems by primary growth alone (pages 583 and 584). Some palms undergo a type of secondary growth called diffuse secondary growth, which takes place in older parts of the stem a considerable distance from the shoot apex. There the parenchyma cells of the ground tissue continue to divide and expand for a long period, accompanied by a proportional increase in the size of the intercellular spaces.

Annuals, Biennials, and Perennials

Plants are often classified according to their seasonal growth cycles as annuals, biennials, or perennials. In the **annuals**—which include many weeds, wildflowers, garden flowers, and vegetables—the entire cycle from seed to vegetative plant to flowering plant to seed occurs within a single growing season, which may be only a few weeks in length. Only the dormant seed bridges the gap between one season of growth and the next.

CHECKPOINTS

After reading this chapter, you should be able to answer the following:

1. How do annuals, biennials, and perennials differ from one another?
2. What types of cells make up the vascular cambium, and how do these cells function?
3. How does secondary growth affect the primary body of the stem?
4. What tissues are produced by the cork cambium, and what is the function of the periderm?
5. What is bark, and how does its composition change during the life of a woody plant?
6. What is wood, and how does conifer wood differ from angiosperm wood?

26–1 Shagbark hickory *(Carya ovata)* in the summer condition Plants have been able to achieve such great stature because of the ability of their roots and stems to increase in girth, that is, to undergo secondary growth. Most of the tissue produced in this manner is secondary xylem, or wood, which not only conducts water and minerals to the far reaches of the shoot but also provides great strength to the roots and stems.

Biennials are annuals with life cycles—from seed germination to seed formation—that cover two growing seasons. The first season of growth of some biennials results in formation of a root, a short stem, and a rosette of leaves near the soil surface. In the second growing season, the plant flowers, fruits, forms seed, and dies, completing the life cycle. In the northern hemisphere, the life cycle crosses a calendar year. In the southern hemisphere, however, biennials are regular annuals whose life cycle is interrupted by winter. Had botanists from the southern hemisphere been the first to describe the life cycles, it is likely that the term "winter annual" would have prevailed rather than biennial. In temperate regions, annuals and biennials seldom become woody, although both their stems and their roots may undergo a limited amount of secondary growth.

Perennials are plants in which the vegetative structures live year after year. The herbaceous perennials pass unfavorable seasons as dormant underground roots, rhizomes, bulbs, or tubers. The woody perennials, which include vines, shrubs, and trees, survive above ground but usually stop growing during the unfavorable seasons. Woody perennials flower only when they become adult plants, which may take many years. For example, the horse chestnut, *Aesculus hippocastanum,* does not flower until it is about 25 years old. *Puya raimondii,* a large (up to 10 meters high) relative of the pineapple that is found in the Andes, takes about 150 years to flower. Many woody plants in temperate regions are deciduous, losing all their leaves at the same time and developing new leaves from buds when the season again becomes favorable for growth. In evergreen trees and shrubs, leaves are also lost and replaced, but not all leaves simultaneously.

The Vascular Cambium

Unlike the many-sided initials of the apical meristems, which contain dense cytoplasm and large nuclei, the meristematic cells of the vascular cambium are highly vacuolated. They exist in two forms: as vertically oriented **fusiform initials,** which are several to many times longer than they are wide, and as horizontally oriented **ray initials,** which are slightly elongated or squarish (Figures 26–2 and 26–3). The fusiform initials appear flattened or brick-shaped in transverse section.

Secondary xylem and secondary phloem are produced through periclinal divisions of the fusiform and ray initials and their immediate derivatives. In other words, the cell plate that forms between these dividing cambial initials is parallel to the surface of the root or stem (Figure 26–4a). If the derivative of a cambial initial is produced toward the outside of the root or stem, it eventually becomes a phloem cell; if it is produced toward the inside, it becomes a xylem cell. In this manner, a long, continuous radial file, or row, of cells is formed, extending from the cambial initial outward into the phloem and inward into the xylem (Figure 26–5).

The xylem and phloem cells produced by the fusiform initials have their long axes oriented vertically and make up what is known as the **axial system** of the secondary vascular tissues. The ray initials produce horizontally oriented **ray cells,** which form the **vascular rays** or **radial system** (Figure 26–5). Composed largely of parenchyma cells, the vascular rays are variable in length. They serve as pathways for the movement of food substances from the secondary phloem to the secondary xylem, and the movement of water from the secondary xylem to the secondary phloem. Vascular rays also serve as storage centers for such substances as starch, proteins, and lipids and may also synthesize some secondary metabolites.

In a restricted sense, the term "vascular cambium" is used to refer only to the cambial initials, of which there is one per radial file. However, it is often difficult, if not impossible, to distinguish between the initials and their immediate derivatives, which may remain meristematic for a considerable time before differentiating. Even in the winter condition, when the cambium is dormant, or inactive, several layers of similar-appearing undifferentiated cells can be seen between the xylem and phloem. Consequently, some botanists use the term "vascular cambium" in a broader sense to refer to the initials and their immediate derivatives, which are indistinguishable from the initials. Other botanists refer to this region of initials and derivatives as the **cambial zone.**

As the vascular cambium adds cells to the secondary xylem and the core of xylem increases in diameter, the cambium is displaced outward, thus increasing in circumference. To accommodate the increased circumference, new cells are added to the vascular cambium by anticlinal divisions of the initials (Figure 26–4b). Along with an increase in the number of fusiform initials, new ray initials and rays are added, so that a fairly constant ratio of rays to fusiform cells is maintained in the secondary vascular tissues. New ray initials that give rise to new rays have their origin in subdivision of fusiform initials. Clearly,

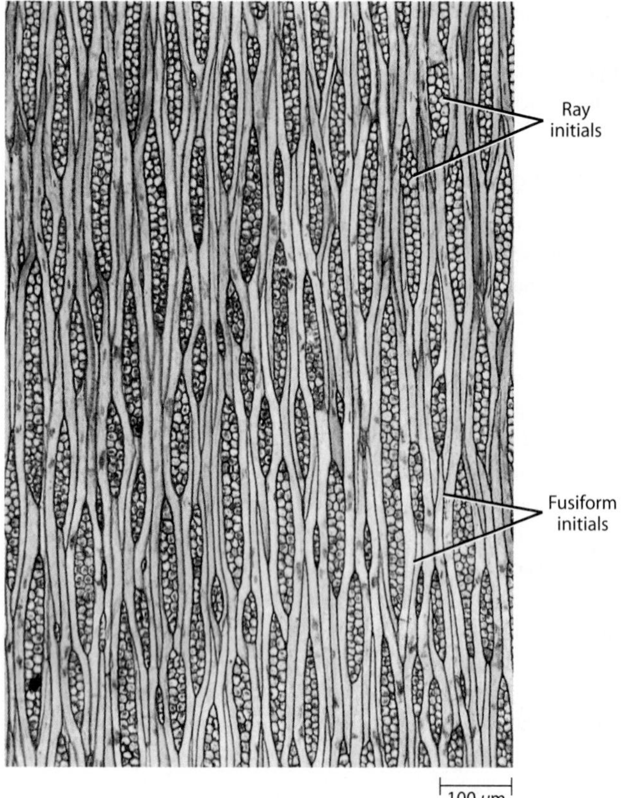

26–2 Vascular cambium of the apple *(Malus domestica)* tree Tangential longitudinal sections are cut at right angles to the rays, so we see the rays here in transverse section (see Figure 26–12). A cambium such as this, in which the fusiform initials are not arranged in horizontal tiers on tangential surfaces, is said to be nonstoried. Fusiform initials average 0.53 millimeter in length in the apple.

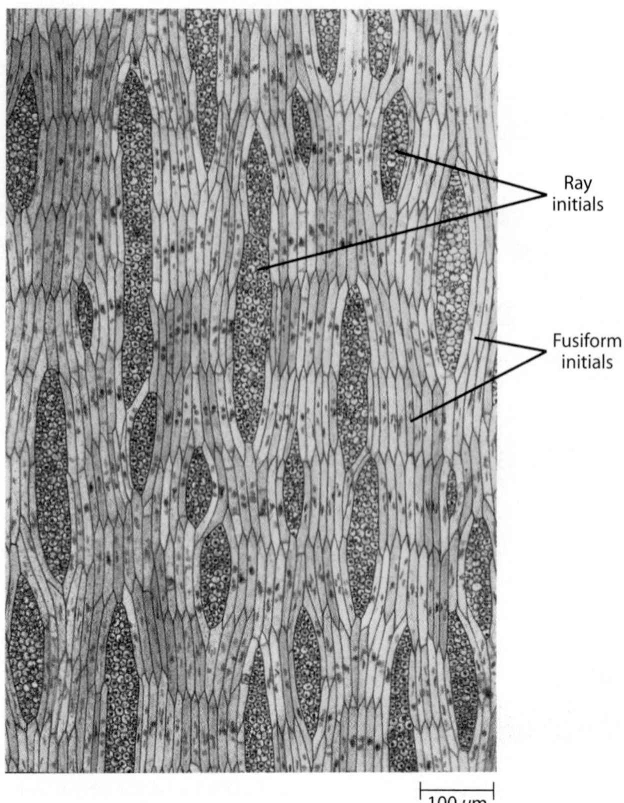

26–3 Vascular cambium of the black locust *(Robinia pseudoacacia)* tree A cambium such as this, in which the fusiform initials are arranged in horizontal tiers on tangential surfaces (see Figure 26–12), is said to be storied. Fusiform initials average 0.17 millimeter in length in black locust.

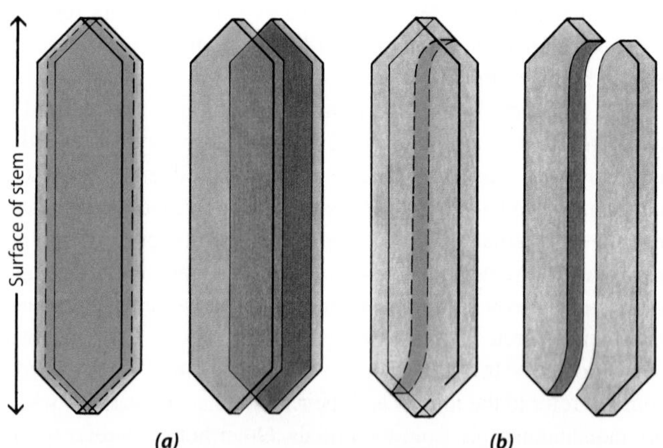

26–4 Periclinal and anticlinal divisions of fusiform initials ***(a)*** Periclinal divisions, which occur parallel to the surface, are involved in the formation of secondary xylem and secondary phloem cells, and result in the formation of radial rows of cells (see Figure 26–5). When an initial divides periclinally, the two daughter cells appear one behind (or in front of) the other. ***(b)*** Anticlinal divisions, which occur perpendicular to the surface, are involved in the multiplication of fusiform initials. When an initial divides anticlinally, the two daughter cells appear side by side.

the developmental changes that occur in the cambium are exceedingly complex.

In temperate regions, the vascular cambium is dormant during winter and becomes reactivated in the spring. During reactivation, the cambial cells take up water, expand radially, and begin to divide periclinally. During expansion, the radial walls of the cambial cells and their derivatives become thinner, and as a result, the bark (all tissues to the outside of the vascular cambium) may be easily peeled or "slipped" off the stem. New growth layers, or increments, of secondary xylem and secondary phloem are laid down during the growing season. Reactivation of the vascular cambium is triggered by expansion of the buds and resumption of their growth. The hormone auxin, produced by the developing shoots, moves downward in the stems and stimulates the resumption of cambial activity. Other factors are also involved in cambial reactivation and in continued normal growth of the cambium (see Chapter 27).

In some plants, the cambial cells divide more or less continuously, and the xylem and phloem elements undergo gradual differentiation. This type of cambial activity is found in plants growing in tropical regions. Not all tropical plants exhibit continuous cambial activity, however. About 75 percent of the trees growing in the rainforest of India exhibit continuous cambial activity. The percentage of such trees drops to 43 percent in the rainforest of the Amazon basin and to only 15 percent in that of Malaysia.

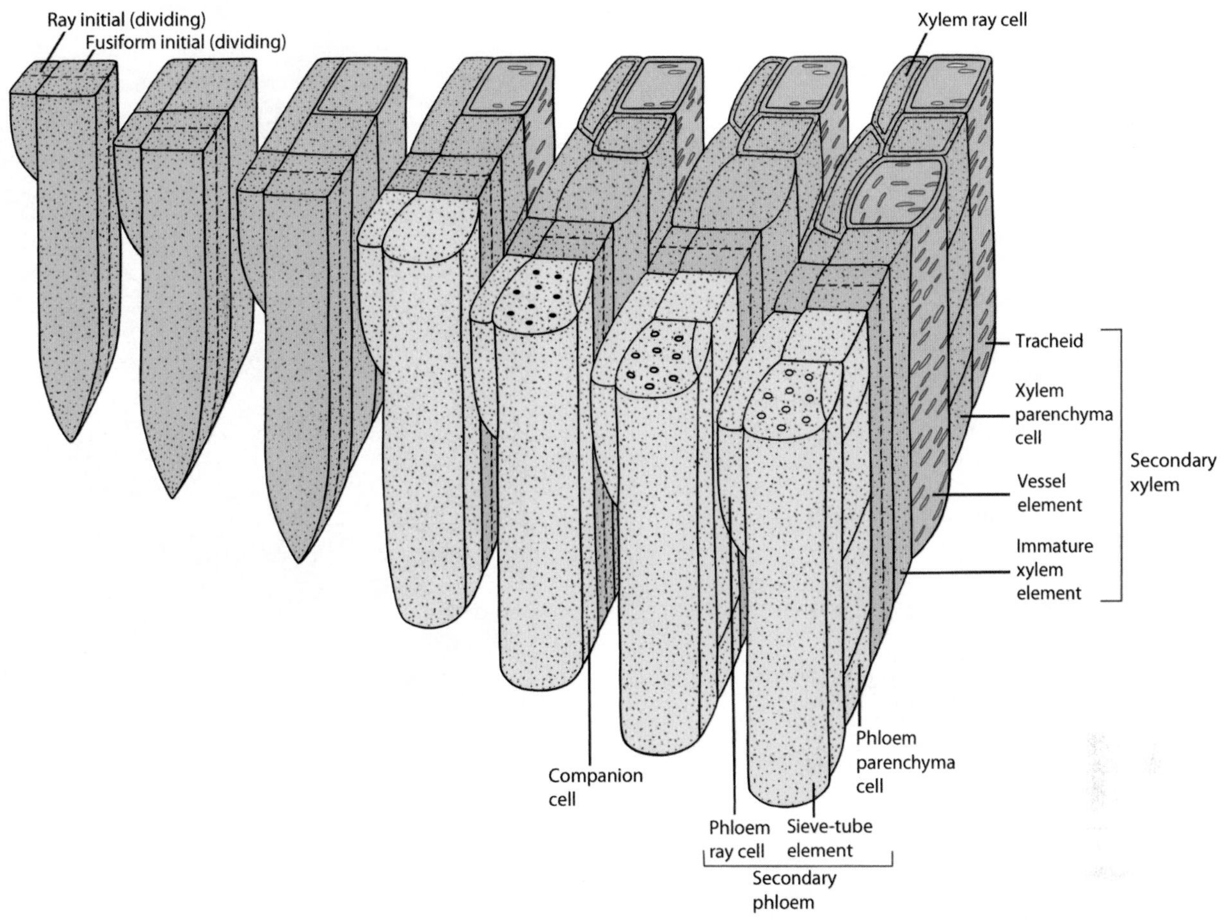

26–5 Relationship of the vascular cambium to secondary xylem and secondary phloem The vascular cambium is made up of two types of cells: fusiform initials, which form the axial system, and ray initials, which form the radial system. When the cambial initials produce secondary xylem and secondary phloem, they divide periclinally, that is, parallel to the surface of the stem or root. Following division of an initial, one daughter cell (the initial) remains meristematic, and the other (the derivative of the initial) eventually develops into one or more cells of the vascular tissue. Cells produced toward the inner surface of the vascular cambium become xylem elements, and those produced toward the outer surface become phloem elements. With the production of additional secondary xylem, the vascular cambium and secondary phloem are displaced outward. The ray initials divide to form vascular rays, which lie at a right angle to the derivatives of the fusiform initials.

Genomic studies on *Arabidopsis thaliana* and on black cottonwood *(Populus trichocarpa),* which has become the consensus model taxon for woody plant genomics and biotechnology, indicate that many of the key genes and mechanisms that regulate secondary growth are also required for primary growth and function of the shoot apical meristem. This may explain how new species with woody habit can be derived from herbaceous species within a relatively short period of time. Several plant hormones, including auxin, cytokinin, gibberellin, and ethylene, have been implicated in regulating the maintenance and activity of the cambial initials (see Chapter 27).

Effect of Secondary Growth on the Primary Body of the Stem

As mentioned in Chapter 25, the vascular cambium of the stem arises from the procambium that remains undifferentiated between the primary xylem and primary phloem, as well as from parenchyma of the interfascicular regions (regions between the vascular bundles, or fascicles). That portion of the cambium arising within the vascular bundles is known as **fascicular cambium,** and that arising in the interfascicular regions, or pith rays, is called **interfascicular cambium.** The vascular cambium of the stem, unlike that of the root, is essentially circular in outline from its inception (Figure 26–6).

In woody stems, the production of secondary xylem and secondary phloem results in the formation of a cylinder of secondary vascular tissues, with rays extending radially through the cylinder (Figure 26–6). Commonly, much more secondary xylem than secondary phloem is produced in the stem in any given year; this is also true for the root. With secondary growth, the primary phloem is pushed outward and its thin-walled cells are destroyed. Only the thick-walled primary phloem fibers, if present, remain intact (see Figure 26–8). For a discussion of secondary growth in roots, see pages 569 to 571.

An elderberry *(Sambucus canadensis)* stem in two stages of secondary growth is shown in Figures 26–7 and 26–8. (Primary growth in the *Sambucus* stem is described on pages 585 and 586.)

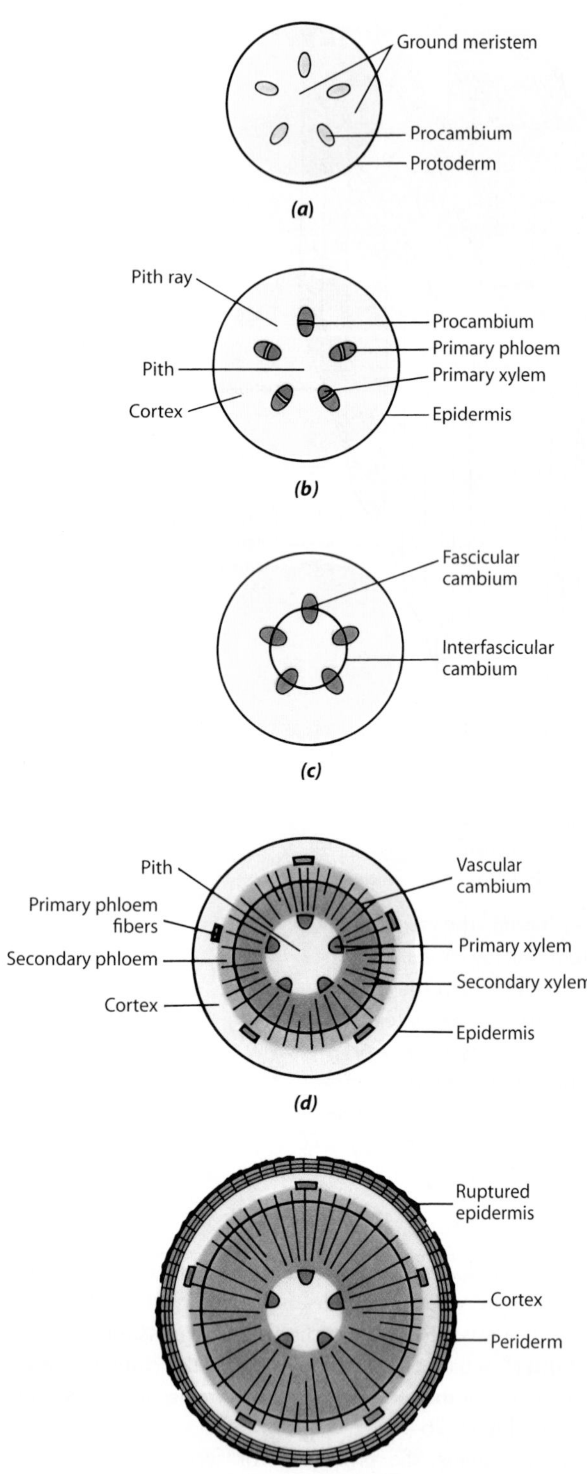

26–6 Stem development in a woody angiosperm ***(a)*** Early stage in primary development, showing the three primary meristems. ***(b)*** At completion of primary growth. ***(c)*** Origin of vascular cambium. ***(d)*** After formation of some secondary xylem and secondary phloem. ***(e)*** At the end of the first year's growth, showing the effect of secondary growth—including periderm formation—on the primary plant body. In *(d)* and *(e)*, the radiating lines represent rays. (Compare this with root development as depicted in Figure 24–15.)

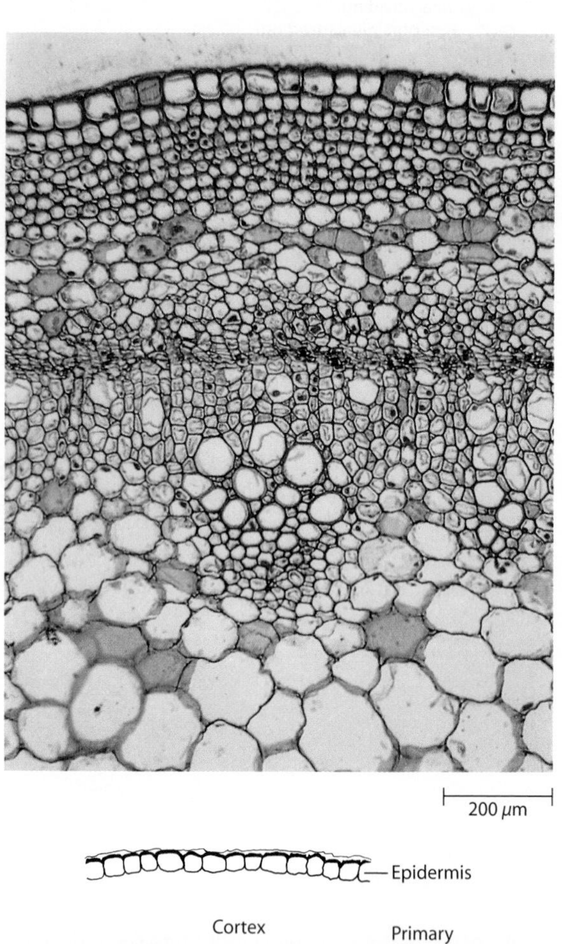

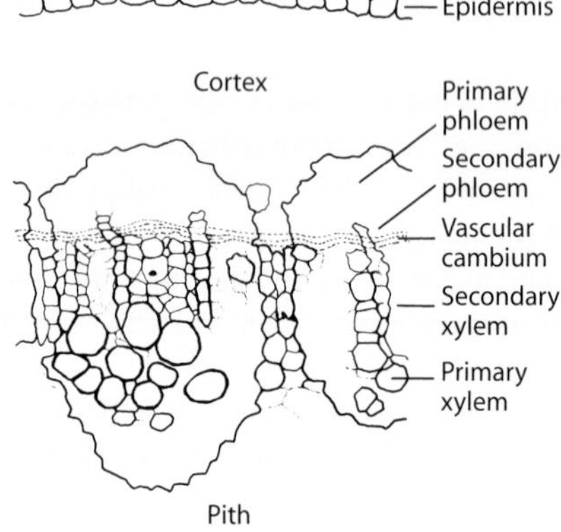

26–7 Elderberry *(Sambucus canadensis)* stem with a small amount of secondary growth As seen in this transverse section, only a small amount of secondary xylem and secondary phloem have been produced, and a cork cambium has not yet formed.

Only small amounts of secondary xylem and secondary phloem have been produced in the stem shown in Figure 26–7. By the end of the first year's growth, considerably more secondary xylem than secondary phloem has formed (Figure 26–8). Note that secondary growth begins in year one. The thick-walled cells outside the secondary phloem are primary phloem fibers. The soft-walled cells—sieve-tube elements and companion cells—are no longer discernible. They were obliterated during development of the primary phloem fibers.

Figure 26–9 shows one-year-old, two-year-old, and three-year-old stems of basswood *(Tilia americana)*. In Chapter 25,

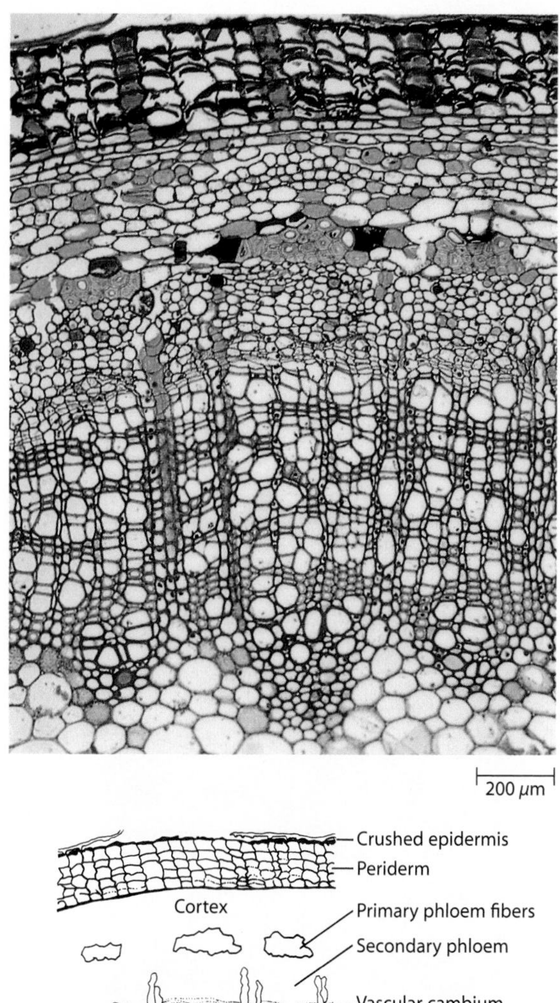

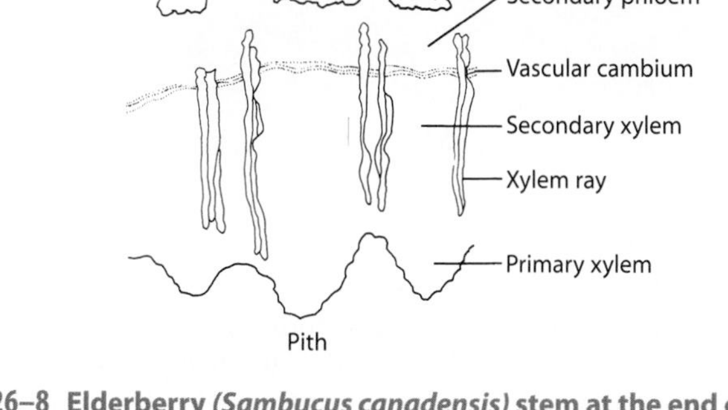

26–8 Elderberry *(Sambucus canadensis)* stem at the end of the first year's growth As this transverse section shows, considerably more secondary xylem than secondary phloem has been produced. Secondary growth begins in the first year.

the stem of *Tilia* was given as an example of one in which the primary vascular system appears as a more or less continuous cylinder (see Figures 25–8 and 25–9), the vascular bundles being separated from one another by very narrow interfascicular regions, or pith rays. Thus most of the vascular cambium in the *Tilia* stem is fascicular in origin. Some of the rays in the secondary phloem of *Tilia* become dilated ("flaring") toward the periphery as the stem increases in girth (Figure 26–9a). This is one way in which the tissues outside the vascular cambium keep up with the increased girth of the core of xylem.

The vascular cambia and secondary tissues of root and stem are continuous with one another. There is no transition region in the secondary plant body as there is in the primary plant body (pages 604 and 605).

The Periderm Is the Dermal Tissue System of the Secondary Plant Body

In most woody stems, as in most woody roots, periderm formation usually follows the initiation of secondary xylem and secondary phloem production. The **periderm** replaces the epidermis as the protective covering on those portions of the plant. Structurally, the periderm consists of three parts: the **cork cambium,** or phellogen, the meristem that produces the periderm; **cork,** or phellem, the protective tissue formed to the outside by the cork cambium; and the **phelloderm,** a tissue that resembles cortical parenchyma, formed to the inside by the meristem (Figure 26–10).

In the stems of most woody plants, the first periderm usually appears during the first year of growth, most commonly originating in a layer of cortical cells immediately below the epidermis, occasionally in the epidermis (Figure 26–8). In some species, the first periderm arises deeper in the stem, usually in the primary phloem.

Repeated divisions of the cork cambium result in the formation of radial rows of compactly arranged cells, most of which are cork cells (Figures 26–10 and 26–11, page 622; see also Figure 23–29). During differentiation of the cork cells, their inner wall surfaces are lined by suberin lamellae, consisting of alternating layers of suberin and wax, which make the tissue highly impermeable to water and gases. The walls of the cork cells may also become lignified. The cork cells die once they complete their differentiation.

The cells of the phelloderm are living at maturity, lack suberin lamellae, and, as mentioned above, resemble cortical parenchyma cells. The phelloderm cells may be distinguished from cortical cells by their inner position in the radial rows of other periderm cells (Figure 26–11). Because the first periderm of the stem usually arises in the outer layer of cortical cells, the cortex of the stem is not sloughed during the first year as is the cortex of woody roots (Figures 26–6 and 26–8; compare with Figures 24–15 and 24–16c), although the epidermis dries up and peels off.

At the end of the first year's growth, the following tissues are present in the stem (outside to inside): remnants of the epidermis; periderm; cortex; primary phloem (fibers and crushed soft-walled cells); secondary phloem; vascular cambium; secondary xylem; primary xylem; and pith (Figure 26–6). (Compare this list of tissues with that for the woody root at the end of the first year's growth on page 571).

The Lenticels Allow Gas Exchange through the Periderm

In the preceding discussion, we noted that the suberized cork cells are compactly arranged and, as a tissue, present an impermeable barrier to water and gases. However, the inner tissues of the stem, like all metabolically active tissues, need to exchange gases with the surrounding air, just as the inner tissues of the root need to exchange gases with the surrounding air spaces between soil particles. In stems and roots containing periderms, this necessary gas exchange is accomplished by means of **lenticels**

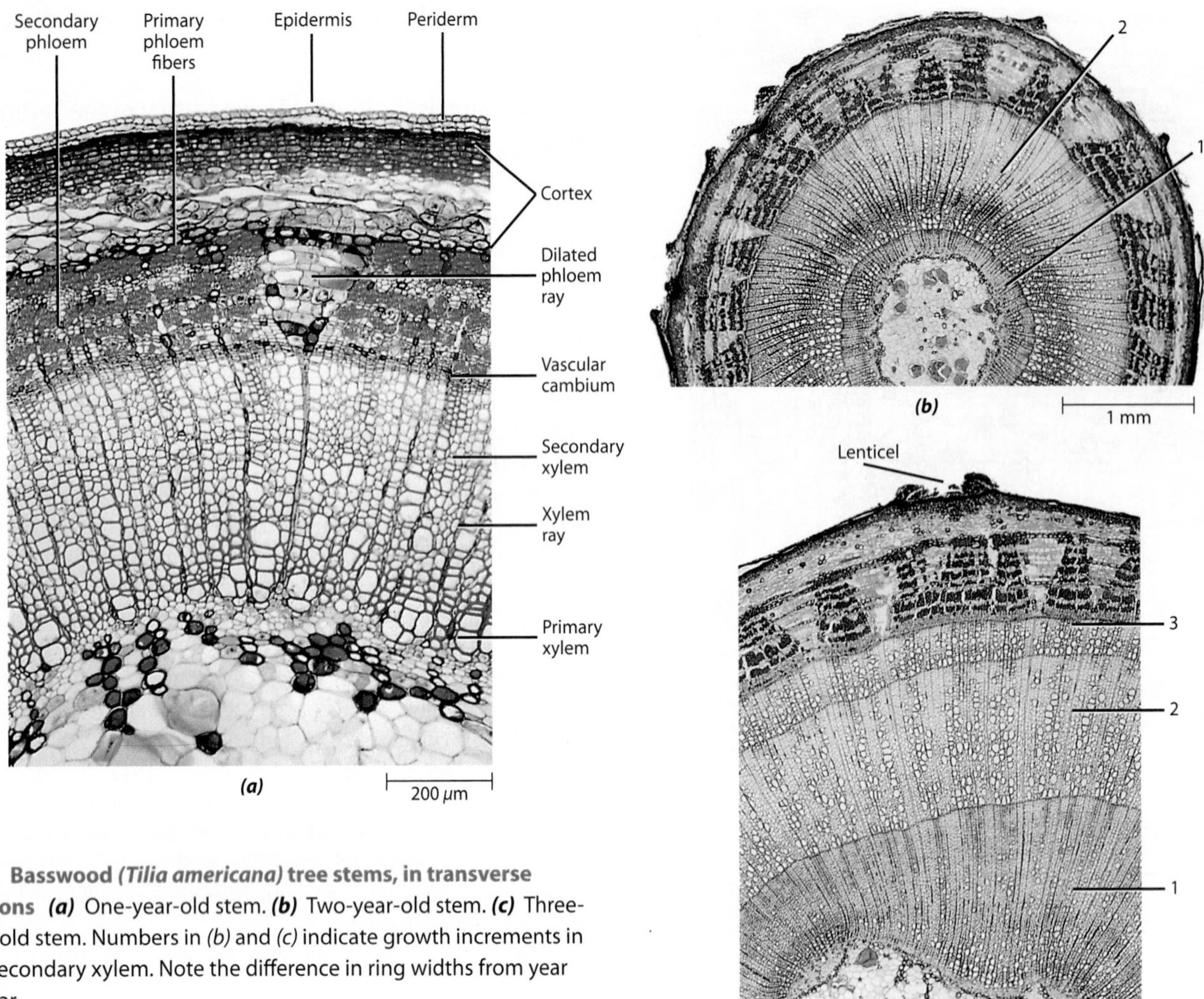

26–9 Basswood *(Tilia americana)* tree stems, in transverse sections ***(a)*** One-year-old stem. ***(b)*** Two-year-old stem. ***(c)*** Three-year-old stem. Numbers in *(b)* and *(c)* indicate growth increments in the secondary xylem. Note the difference in ring widths from year to year.

(Figures 26–10c, d and 26–11)—portions of the periderm with numerous intercellular spaces.

Lenticels begin to form during development of the first periderm (Figure 26–10), and in the stem they generally appear below a stoma or group of stomata. On the surface of the stem or root, the lenticels appear as raised circular, oval, or elongated areas (see Figure 26–16). Lenticels are also formed on some fruits—for instance, the small dots on the surface of apples and pears are lenticels. As the roots and stems age, lenticels continue to develop at the bottom of cracks in the bark in newly formed periderms.

The Bark Includes All Tissues outside the Vascular Cambium

The terms "periderm," "cork," and "bark" are often unnecessarily confused with one another. As we have discussed, cork is one of three parts of the periderm; it is a secondary tissue that replaces the epidermis in most woody roots and stems. The term **bark** refers to all the tissues outside the vascular cambium, including the periderm or periderms when present (Figures 26–12 and 26–13). When the vascular cambium first appears and secondary phloem has not yet been formed, the bark consists entirely of primary tissues. At the end of the first year's growth, the bark includes any primary tissues still present, the secondary phloem, the periderm, and any dead tissues remaining outside the periderm.

Each growing season, the vascular cambium adds secondary phloem to the bark, as well as secondary xylem, or wood, to the core of the stem or root. Usually the vascular cambium produces less secondary phloem than secondary xylem. In addition, the soft-walled cells (sieve elements and various kinds of parenchymatous elements) of the old secondary phloem are usually crushed (Figures 26–14 and 26–15). Eventually, the old secondary phloem is separated from the rest of the phloem by newly formed periderms and is ultimately sloughed. As a result, considerably less secondary phloem accumulates in the stem or root than secondary xylem, which continues to accumulate year after year.

As the stem or root increases in girth, considerable stress is placed on the older tissues of the bark. In some plants, tearing of these tissues results in the formation of large air spaces. In many plants, the parenchyma cells of the axial system and rays divide

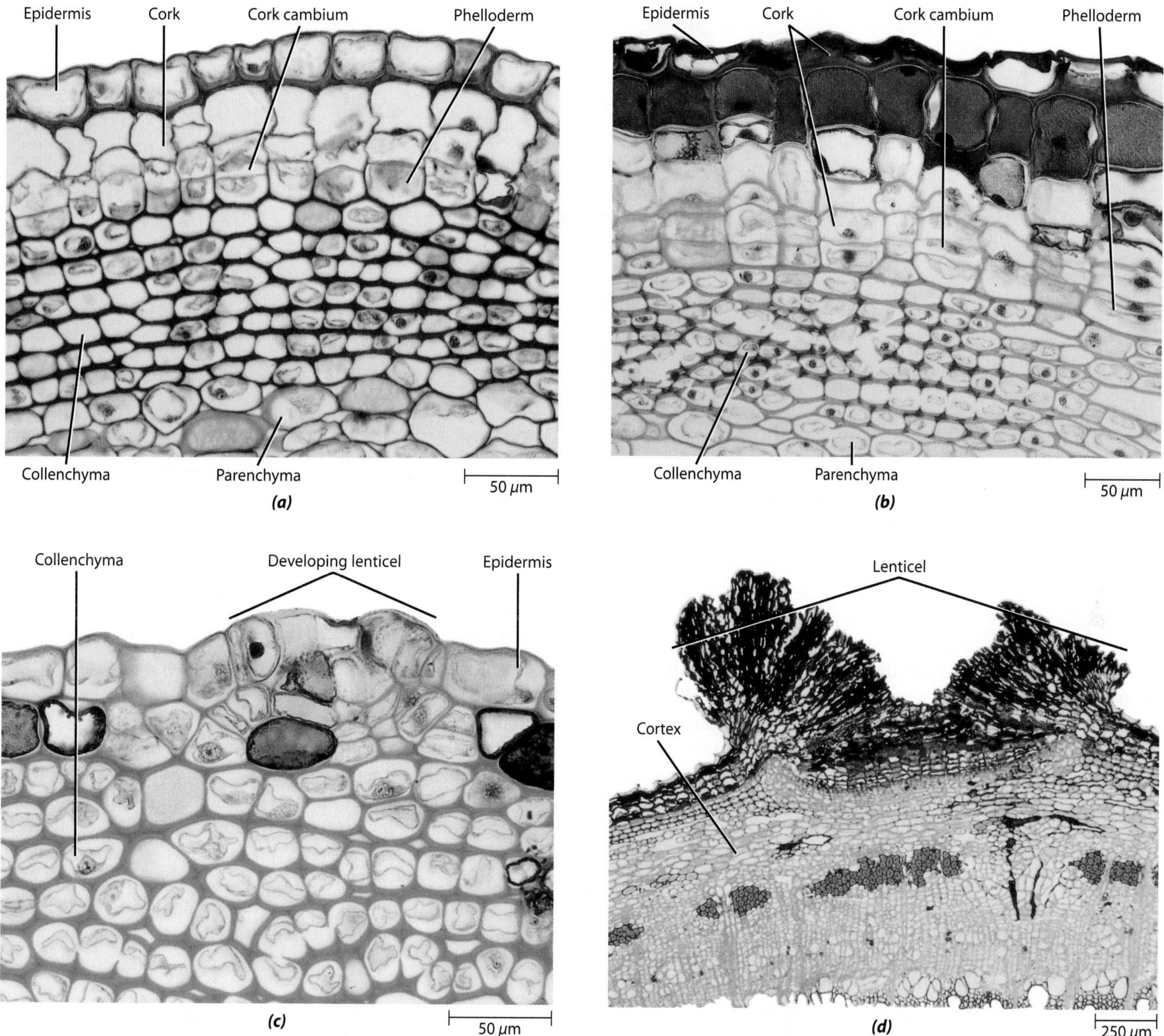

26–10 Some stages of periderm and lenticel development in elderberry *(Sambucus canadensis)*
(a) Newly formed periderm, which consists of cork cambium, cork, and phelloderm, is seen beneath the epidermis in this transverse section. The periderm has separated the epidermis from the cortex, which consists of collenchyma and parenchyma. ***(b)*** Periderm in a more advanced stage of development, with an increase in the size of the cork layer. The phelloderm in *Sambucus* generally consists of a single layer of cells. Note that the epidermis is degenerating. ***(c)*** Initiation of a lenticel. Collenchyma cells of the cortex can be seen beneath the developing lenticel. ***(d)*** Well-developed lenticel.

and enlarge. In this manner, the old secondary phloem keeps up for a while with the increase in circumference of the stem or root. We noted earlier that, in the *Tilia* stem, certain rays, known as dilated rays, become very wide as the stem increases in girth (Figure 26–9a).

The first-formed periderm may keep up with the increasing girth of the root or stem for several years, with the cork cambium exhibiting periods of activity and inactivity that may or may not correspond to the periods of activity of the vascular cambium. In the stems of apple *(Malus domestica)* and pear *(Pyrus communis)* trees, the first cork cambium may remain active for up to 20 years. In most woody roots and stems, additional periderms are formed as the axis increases in circumference. After the first periderm, subsequently formed periderms originate deeper and deeper in the bark (Figures 26–12 and 26–13) from parenchyma cells of the phloem no longer actively engaged in the transport of

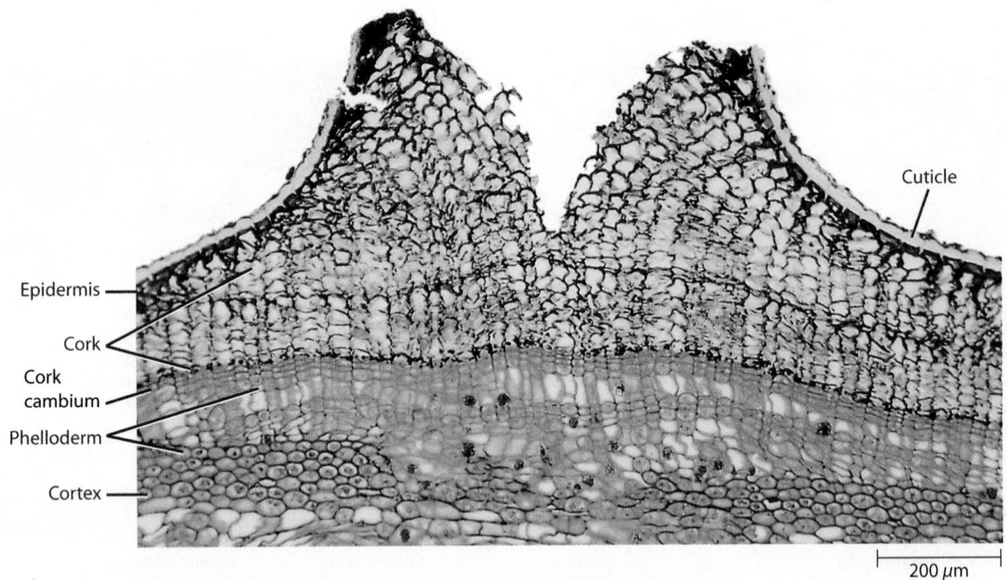

26–11 Lenticel of the stem of Dutchman's pipe *(Aristolochia)* Unlike that of *Sambucus*, the phelloderm of *Aristolochia* consists of several layers of cells, as seen in this transverse section.

food substances. These parenchyma cells become meristematic and form new cork cambia.

All of the tissues outside the innermost cork cambium—all of the periderms, together with any cortical and phloem tissues included among them—make up the *outer bark* (Figures 26–12 and 26–13). With maturation of the suberized cork cells, the tissues outside them are separated from the supply of water and nutrients. Hence the outer bark consists entirely of dead tissues. The living part of the bark, which is inside the innermost cork cambium and extends inward to the vascular cambium, is called the *inner bark* (Figures 26–12 and 26–13).

The manner in which new periderms are formed and the kinds of tissues isolated by them have a marked influence on the appearance of the outer surface of the bark (Figure 26–17). In some barks the newly formed periderms develop as discontinuous overlapping layers, resulting in formation of a scaly type of bark called scale bark (Figures 26–12 and 26–13). Scale barks are found, for example, on relatively young stems of pine *(Pinus)* and pear *(Pyrus communis)* trees. Less commonly, the newly formed periderms arise as more or less continuous, concentric rings around the axis, resulting in formation of a ring bark. Grape *(Vitis)* and honeysuckle *(Lonicera)* are examples of plants with ring barks. The barks of many plants are intermediate between ring and scale barks.

Commercial cork is obtained from the bark of the cork oak, *Quercus suber,* which is native to the Mediterranean region

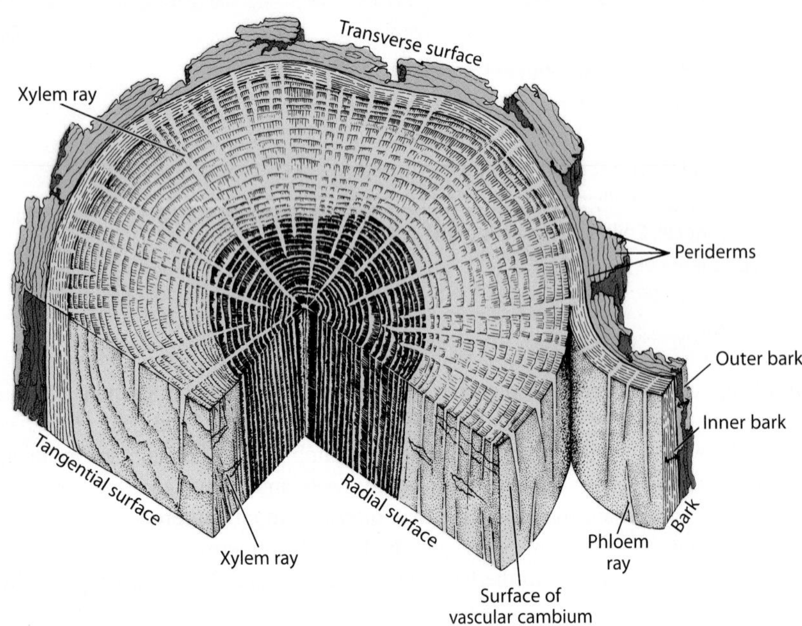

26–12 Red oak *(Quercus rubra)* stem, showing the transverse, tangential, and radial surfaces The dark area in the center is heartwood, and the lighter part of the wood is sapwood.

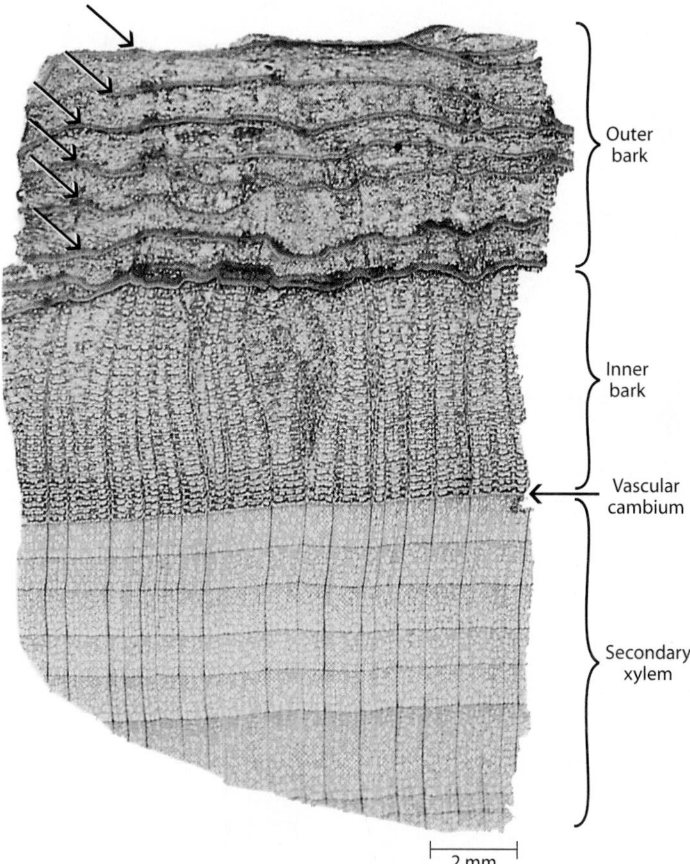

26–13 Bark and some secondary xylem from an old stem of basswood *(Tilia americana)* Several periderms (arrows) can be seen traversing the mostly brownish outer bark in the upper third of this transverse section. Below the outer bark is the inner bark, which appears quite distinct from the more lightly stained secondary xylem in the lower third of the section.

(Figure 26–18, page 626). The first cork cambium of this tree has its origin in the epidermis, and the cork produced by it is of little commercial value. When the tree is about 20 years old, the original periderm is removed, and a new cork cambium is formed in the cortex, just a few millimeters below the site of the first one. The cork produced by the new cork cambium accumulates very rapidly and after about 10 years is thick enough to be stripped off the tree. Once again a new cork cambium arises beneath the previous one, and after about another 10 years the cork can be stripped again. After several strippings the new cork cambia are formed in the secondary phloem. This procedure may be repeated at about 10-year intervals until the tree is 150 or more years old. The spots and long dark streaks seen on the surfaces of commercial cork are lenticels.

In most woody roots and stems, very little secondary phloem is actually involved in the conduction of food. In many species, only the current year's growth increment, or growth ring, of secondary phloem is active in the long-distance transport of food

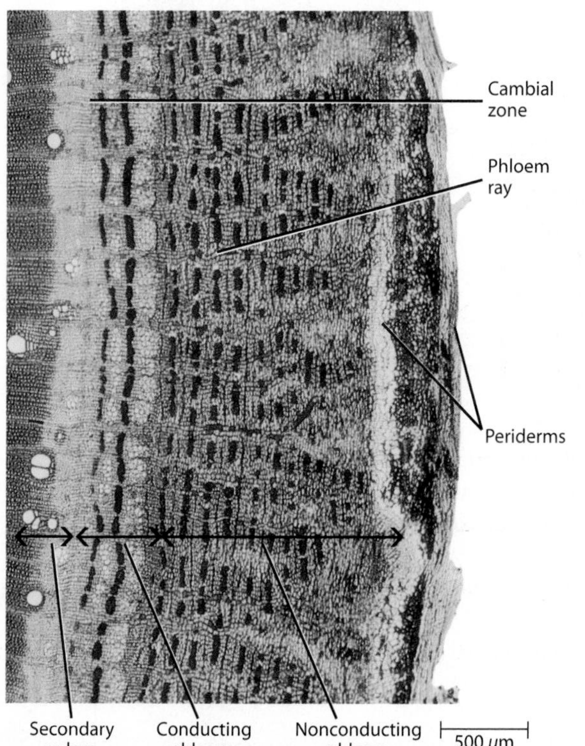

26–14 Bark of a black locust *(Robinia pseudoacacia)* stem The bark consists mostly of nonconducting phloem, as is visible in this transverse section.

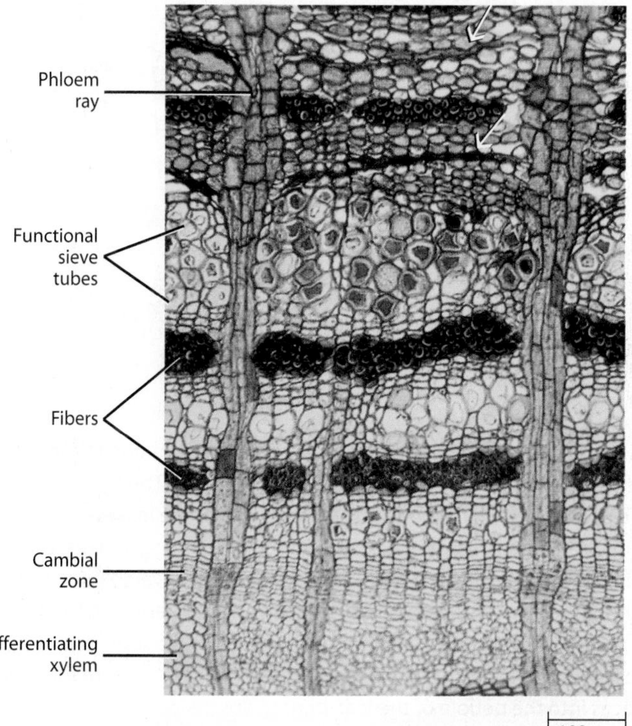

26–15 Secondary phloem of the black locust *(Robinia pseudoacacia)* The transverse section shows mostly conducting phloem. Sieve tubes (indicated by arrows) of the nonconducting phloem have collapsed.

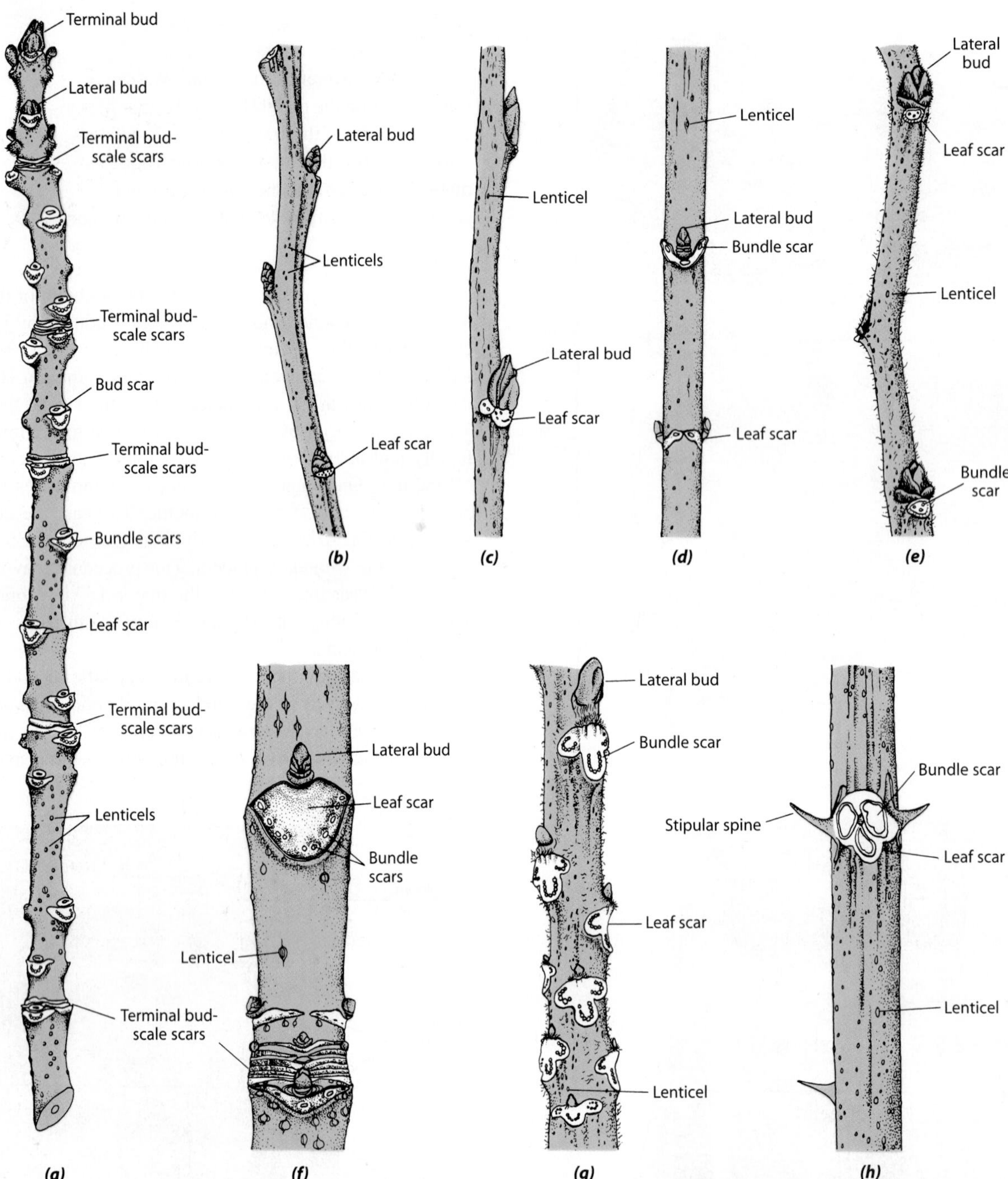

26–16 External features of woody stems Examination of the twigs of deciduous woody plants reveals many important developmental and structural features of the stem. The most conspicuous structures on the twigs are the buds. Buds occur at the tips—the terminal buds—and in the axils of the leaves—the lateral, or axillary, buds of the twigs. In addition, accessory buds are found in some species. Commonly occurring in pairs, the accessory buds are located one on each side of a lateral bud. In some species, the accessory buds do not develop if their associated lateral bud undergoes normal development. In others, the accessory buds form flowers and the lateral bud forms a leafy shoot.

After the leaves fall, leaf scars, with their bundle scars, can be seen beneath the lateral buds. The protective layer of the abscission zone produces the leaf scar. The bundle scars are the severed ends of vascular bundles that extended from the leaf traces into the petiole of the leaf, prior to abscission.

Groups of terminal bud-scale scars reveal the locations of previous terminal buds, and until they are obscured by secondary growth, these groups of scars may be used to determine the age of portions of the stem. The portion of stem between two groups of such scars represents one year's growth. The lenticels appear as slightly raised areas on the stem. ***(a)*** Green ash *(Fraxinus pennsylvanica* var. *subintegerrima).* ***(b)*** White oak *(Quercus alba).* ***(c)*** Basswood *(Tilia americana).* ***(d)*** Box elder *(Acer negundo).* ***(e)*** American elm *(Ulmus americana).* ***(f)*** Horse chestnut *(Aesculus hippocastanum).* ***(g)*** Butternut *(Juglans cinerea).* ***(h)*** Black locust *(Robinia pseudoacacia).*

26–17 Bark of four species of trees ***(a)*** Thin, peeling bark of the paper birch *(Betula papyrifera)*. Horizontal lines on the surface of the bark are lenticels. ***(b)*** Shaggy bark of the shagbark hickory *(Carya ovata)*. ***(c)*** Scaly bark of a sycamore, or buttonwood *(Platanus occidentalis)*. ***(d)*** Deeply furrowed bark of the black oak *(Quercus velutina)*.

through the stem. This is because the sieve elements are short-lived (see Chapter 23), and most of them die by the end of the same year in which they are derived from the vascular cambium. In some plants, such as black locust *(Robinia pseudoacacia)*, the sieve elements collapse and are crushed relatively soon after they die (Figures 26–14 and 26–15).

The part of the inner bark that contains living, functional sieve elements and is actively engaged in the transport of food substances is called **conducting phloem.** Although the sieve elements outside the conducting phloem are dead, the phloem parenchyma cells (axial parenchyma) and the parenchyma cells of the rays may remain alive and continue to function as storage

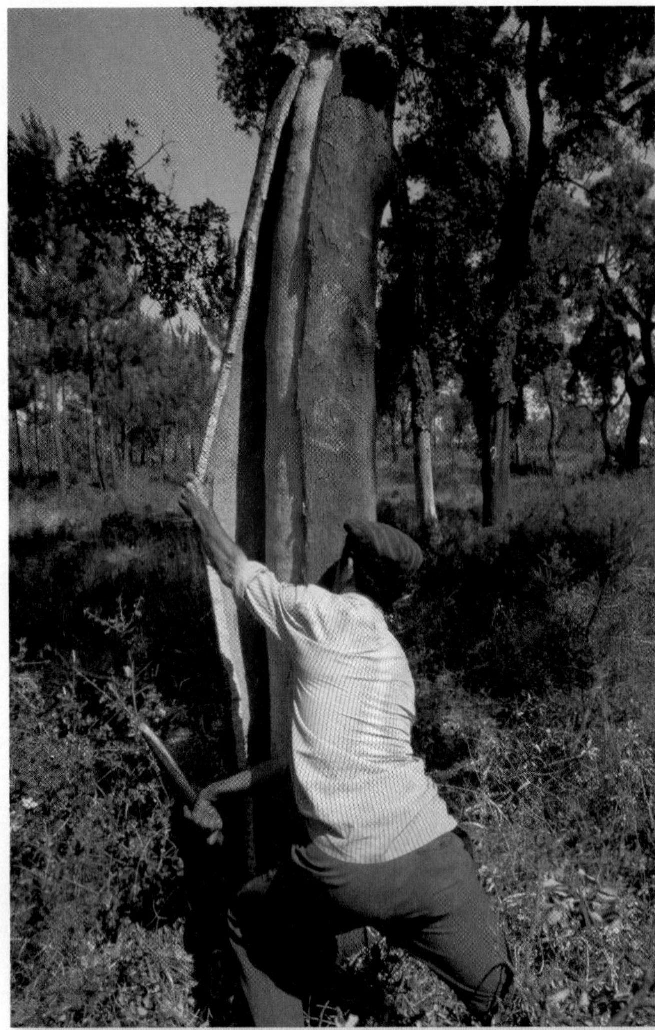

26–18 Harvesting cork Wielding an ax, a skilled worker strips a thick layer of cork from a cork oak *(Quercus suber)* in a forest in southern Portugal. This sustainable, low-impacting harvesting, which is threatened by the increasing use of plastic wine corks, helps support one of the most important wildlife habitats in Europe. Cork is valued commerically for its imperviousness to liquids and gases, as well as its strength, elasticity, and lightness.

cells for many years. This part of the inner bark is known as **non-conducting phloem** (Figures 26–14 and 26–15). Only the outer bark is composed entirely of dead tissue (Figure 26–13).

Wood: Secondary Xylem

Apart from the use of various plant tissues as food, no single plant tissue has played a more indispensable role in human survival throughout recorded history than **wood,** or secondary xylem. Among other things, wood has been used for shelter, for fire to provide warmth and to cook, for weapons, furniture, tools, and toys, for paper pulp, and as a means of transportation in the form of rafts, boats, and wheels. Commonly, woods are classified as either **hardwoods** or **softwoods.** The so-called hardwoods are angiosperm (magnoliid and eudicot) woods, and the softwoods are conifer woods. The two kinds of wood have basic structural differences, but the terms "hardwood" and "softwood" do not accurately express the relative density (weight per unit volume) or hardness of the wood. For example, one of the lightest and softest of woods is balsa *(Ochroma lagopus),* a tropical eudicot. By contrast, the woods of some conifers, such as slash pine *(Pinus elliottii),* are harder than some hardwoods. Although this chapter deals largely with secondary growth in angiosperms, for practical reasons we consider conifer and angiosperm woods together.

Conifer Wood Lacks Vessels

The structure of conifer wood is simpler than that of most angiosperms. The principal features of conifer wood are its lack of vessels (see Chapter 23 for a discussion of tracheary elements) and its relatively small amount of axial, or wood, parenchyma. Long, tapering tracheids constitute the dominant cell type in the axial system. In certain genera, such as *Pinus,* the only parenchyma cells of the axial system are those associated with **resin ducts.** Resin ducts are relatively large intercellular spaces lined with thin-walled parenchyma cells that secrete resin into the duct. In *Pinus,* resin ducts occur in both the axial system and the rays (Figures 26–19 and 26–20). Wounding, pressure, and injuries by frost and wind can stimulate the formation of resin ducts in

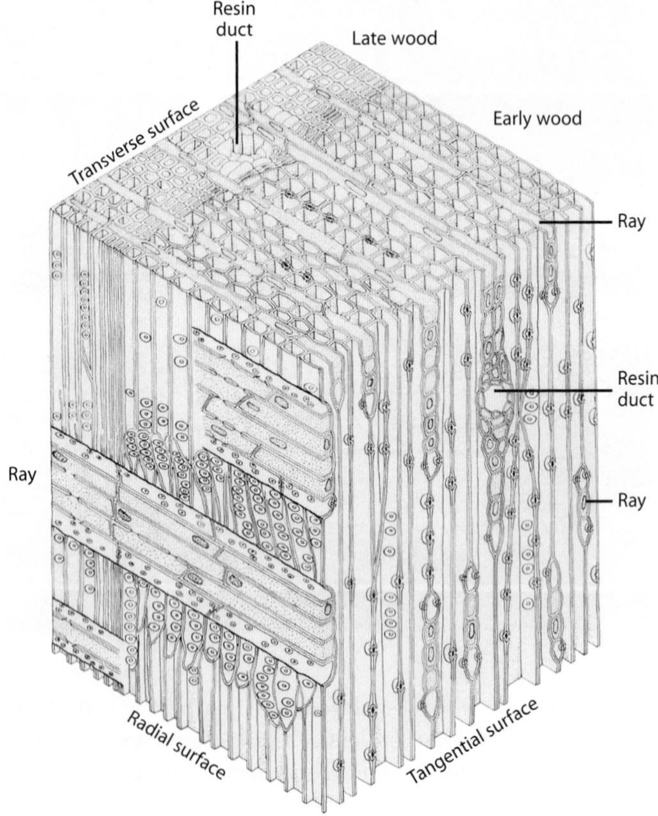

26–19 Secondary xylem of white pine *(Pinus strobus)* With the exception of the parenchyma cells associated with the resin ducts, the axial system of this conifer consists entirely of tracheids, as seen in this block diagram. The rays are only one cell wide, except for those containing resin ducts. (Early and late wood are described on page 631.)

THE TRUTH ABOUT KNOTS

Young trees typically have numerous branches growing from their trunks, and yet branches often are entirely lacking on the lower part of the trunks of older trees. What becomes of the branches found on the trunks of younger trees? Providing they remain on the tree, they are buried within the growing trunk, forming knots.

Branches have their origin from buds, and as long as they are living, they undergo a periodic increase in length and thickness just like the trunk in which they are embedded. The vascular cambium of the living branch is continuous with that of the trunk. Thus, during periods of cambial activity, new wood is added as a continuous layer over branch and trunk, tightly fixing the knot in the wood of the trunk ***(a).*** Such knots are called tight knots ***(c).*** They remain in place when the trunk is sawed for lumber.

When a branch dies, it stops growing and gradually is engulfed, bark and all, by the wood of the still growing trunk. Inasmuch as its cambium is no longer active, from that point onward the branch lacks physical continuity with the trunk ***(b).*** The dead branch may lose its bark, but if it remains on the trunk it will be engulfed by wood and continue as a knot. Such knots, called loose knots, may drop out of sawed lumber ***(d), (e).*** Sometimes changes in the wood of dead branches lead to the formation of substances that make the knots extremely hard and the wood very difficult to work with hand tools. On the other hand, woods with prominent knots, such as knotty pine, are valued for their decorative quality.

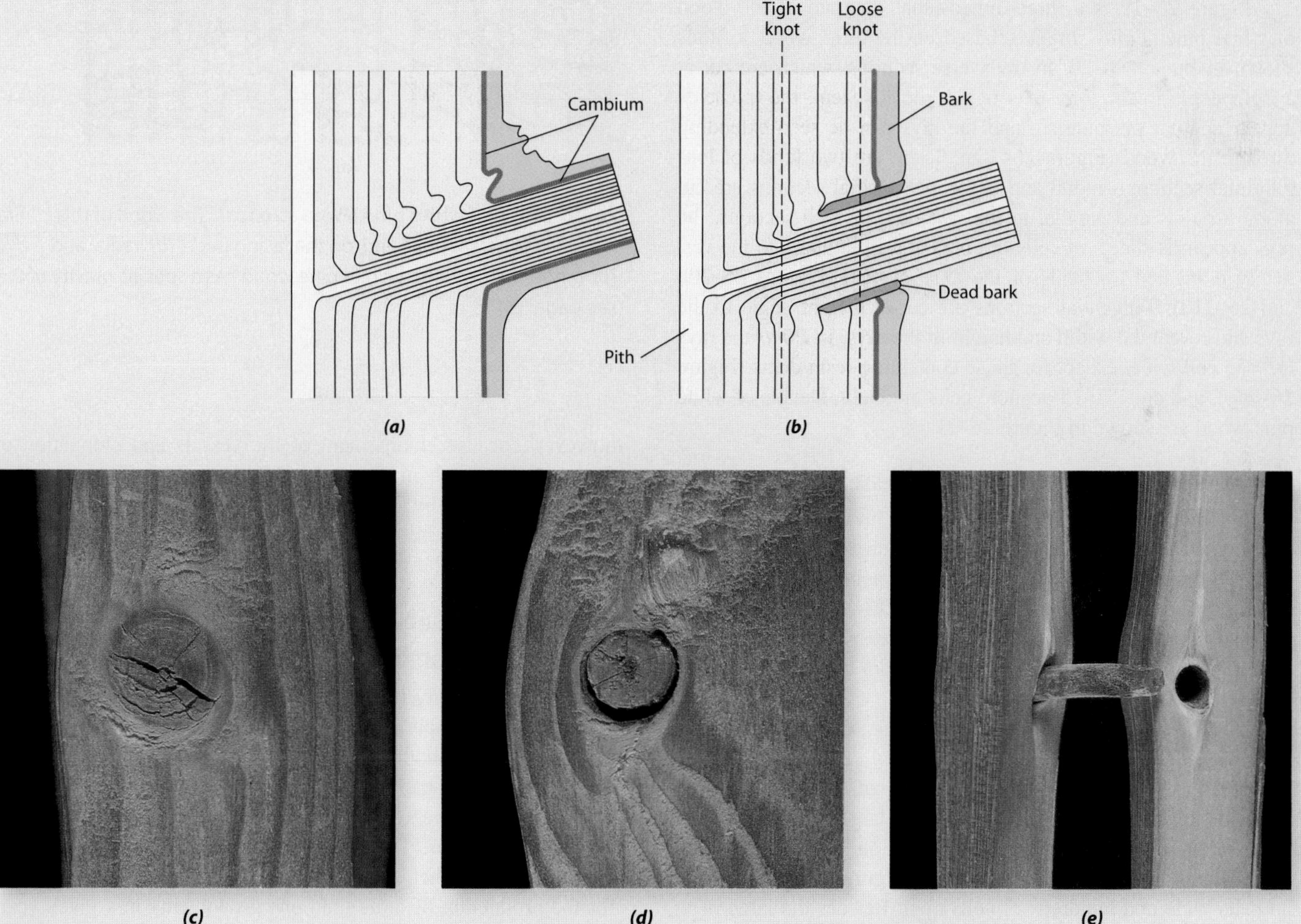

Radial sections of wood containing knots ***(a)*** Trunk with tight knot. The cambium and growth rings are continuous between wood and knot. ***(b)*** View of trunk after branch has died. The dashed line on the left extends through the region of the tight knot. The dashed line on the right extends through the region of the loose knot.

Different portions of the same knot from white pine *(Pinus strobus)* wood: ***(c)*** a tight knot; ***(d)*** a loose knot; ***(e)*** the loose knot removed from the branch wood.

conifer wood, leading some investigators to suggest that all resin ducts result from trauma. Resin apparently protects the plant from attack by decay-producing fungi and bark beetles.

The tracheids of conifers are characterized by large, circular, bordered pits on their radial walls. Pits are most abundant on the ends of the cells, where the cells overlap with other tracheids (Figures 26–19 through 26–21). The pairs of pits, known as pit-pairs (page 60), between conifer tracheids are each characterized by the presence of a **torus** (plural: tori), a thickened central portion of the pit membrane (Figure 26–22) that consists mainly of middle lamella and two primary walls. The torus is slightly larger than the openings, or apertures, in the pit-borders (Figure 26–21). The pit membrane is flexible, and under certain conditions, the torus may block one of the apertures and prevent the movement of water or gases through the pit-pair. Although long thought to occur only in certain genera of gymnosperms, tori have been reported in the bordered pit-pairs of tracheids and vessel elements in several genera of eudicots.

Figure 26–19 is a three-dimensional diagram of the wood of white pine *(Pinus strobus),* based on the three wood sections shown in Figure 26–20. In transverse sections, which are cut at a right angle to the long axis of the root or stem, the tracheids appear angular or squarish, and the rays can be seen extending through the wood (Figure 26–20a). There are two kinds of longitudinal sections—radial and tangential. Radial sections are cut along a radius and parallel to the rays, and in such sections, the rays appear as sheets of cells oriented at a right angle to the vertically elongated tracheids of the axial system (Figures 26–20b and 26–21d). Tangential sections are cut at a right angle to the rays and reveal the width and height of the rays. In *Pinus* the rays are one cell wide, except for those containing resin ducts (Figure 26–20c), and are 1 to 15 or more cells in height. Details of white pine wood are shown in Figure 26–21.

Angiosperm Woods Typically Contain Vessels

Wood structure in angiosperms is much more varied than in conifers, owing in part to a greater number of cell types in the axial system, including vessel elements, tracheids, several types of fibers, and parenchyma cells (Figures 26–23 through 26–25; see also Figure 23–12). The presence of vessel elements, in particular, distinguishes angiosperm woods from conifer wood, with only a few exceptions.

The rays of angiosperm woods range from one to many cells wide and from one to several hundred cells high, and thus are often considerably larger than those of conifer wood. In some angiosperm woods, such as oak, the large rays can be seen with the unaided eye (Figure 26–12). The large rays of the red oak *(Quercus rubra)* wood illustrated in Figure 26–24c are 12 to 30 cells wide and hundreds of cells high. Besides the large rays, oak wood has numerous rays that are only one cell wide. In red oak wood, the rays make up, on average, about 21 percent of the volume of the wood. Overall, the rays of hardwoods average about 17 percent of the volume of the wood, while the average for conifer wood is about 8 percent.

As in conifer wood, transverse sections of angiosperm woods reveal radial files of cells of both the axial and radial systems derived from the cambial initials (Figures 26–23 and 26–24a). The files may not be as orderly as in conifer wood, however, because enlargement of the vessels and elongation of fibers tend to push many of the cells out of position. The displacement of rays by vessel elements is particularly conspicuous in the transverse section of red oak wood, as shown by the wavy rays to the left of the vessels in Figure 26–24a.

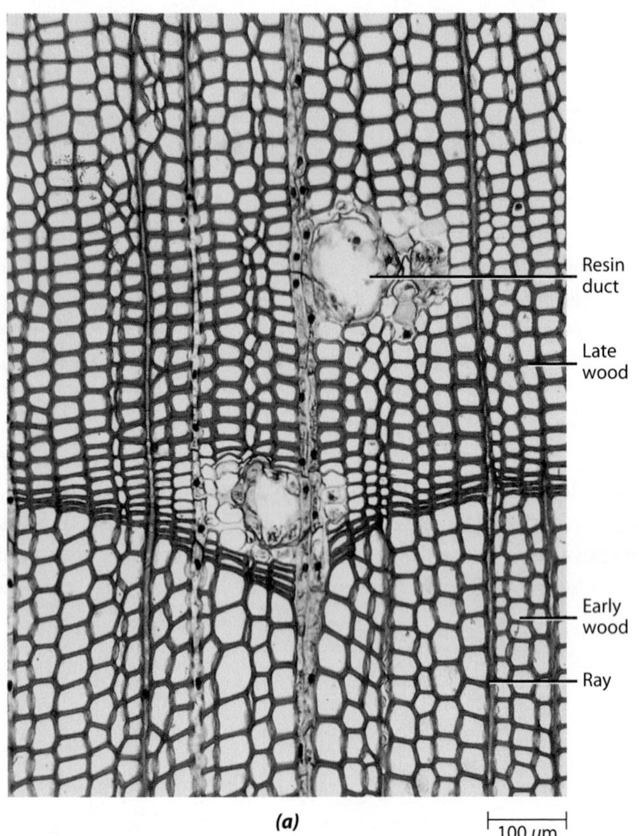

26–20 Wood of white pine *(Pinus strobus)* The wood of this conifer is seen in ***(a)*** transverse and, on the facing page, ***(b)*** radial and ***(c)*** tangential sections. (White pine wood has a specific gravity of 0.34; see page 634.)

Growth Rings Result from the Periodic Activity of the Vascular Cambium

The periodic activity of the vascular cambium, which is seasonally related in temperate zones, produces growth increments, or **growth rings,** in both secondary xylem and secondary phloem (in the phloem the increments are not always readily discernible). If a growth layer represents one season's growth, it is called an **annual ring** (Figure 26–26, page 632). Abrupt changes in available water and other environmental factors may be responsible for the production of more than one growth ring in a given year; such rings are called *false annual rings.* Thus the age of a given portion of an old woody stem can be estimated by counting the growth rings, but the estimates may be inaccurate if false annual rings are included. Trees that exhibit continuous cambial activity, such as many tropical rainforest trees, may lack growth rings entirely. It is therefore difficult to judge their age based on analysis of their wood.

The width of individual growth rings may differ greatly from year to year as a function of such environmental factors as

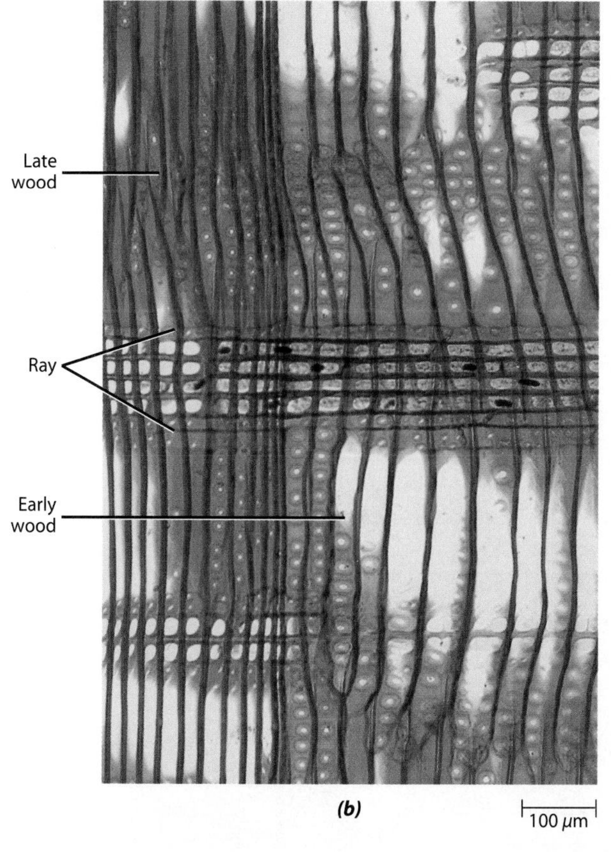

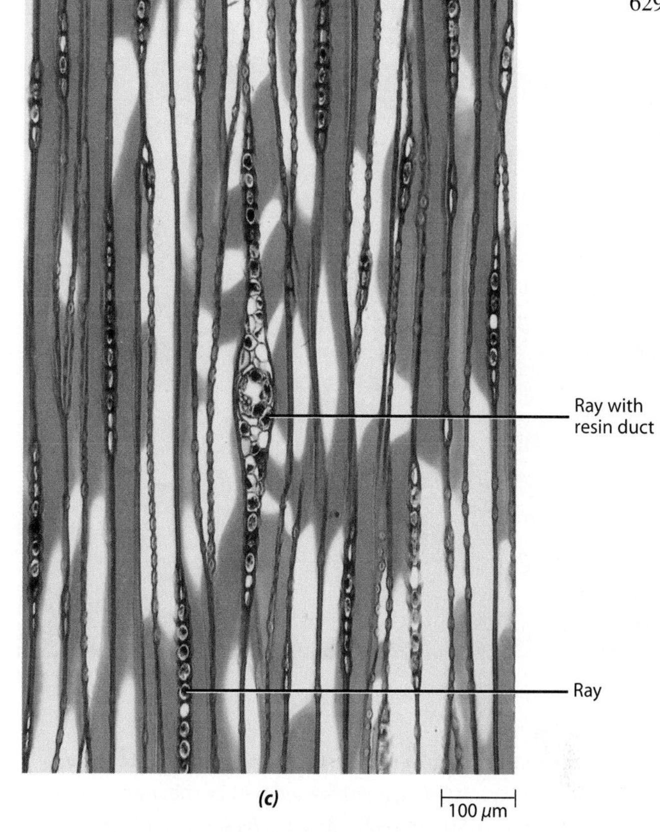

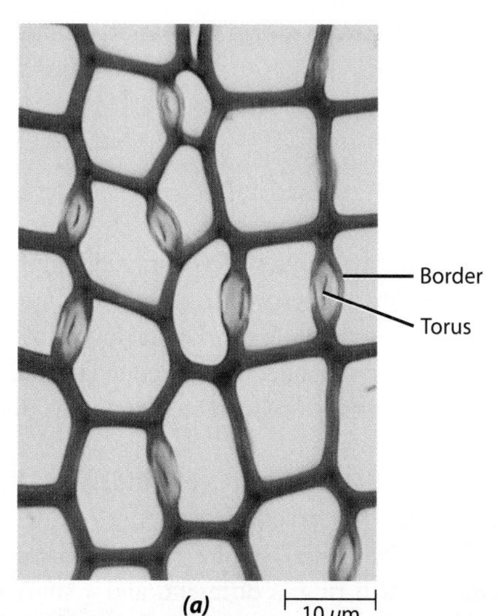

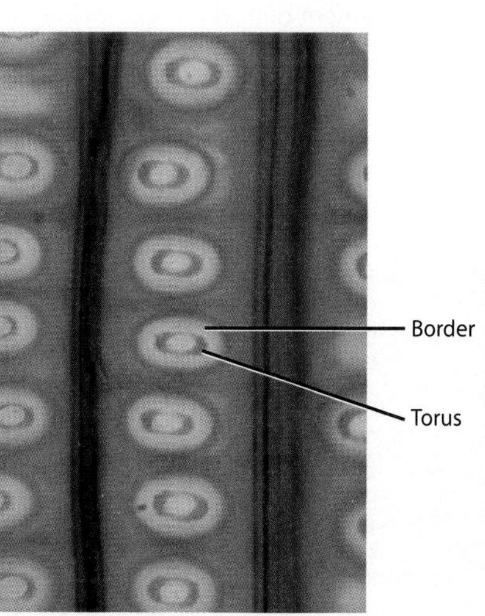

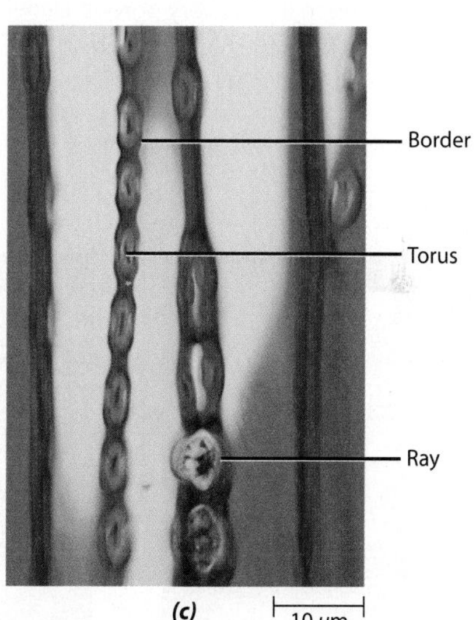

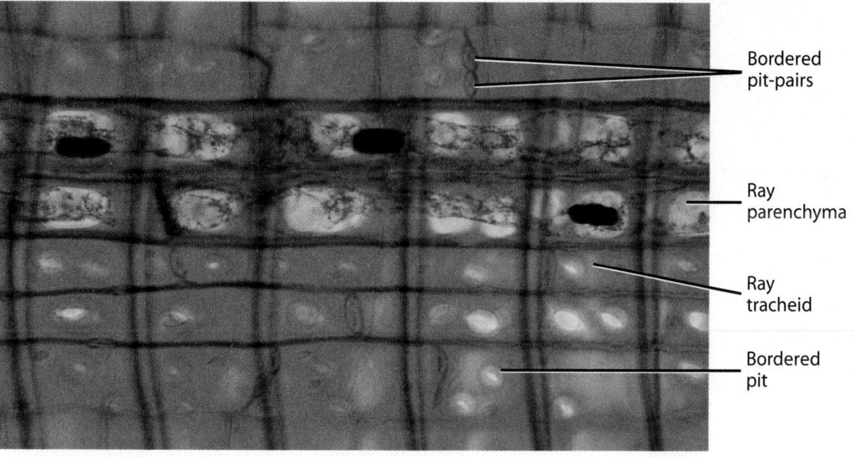

26–21 Details of white pine *(Pinus strobus)* wood ***(a)*** Transverse section, showing bordered pit-pairs on radial walls of tracheids. ***(b)*** Radial section, showing the face view of bordered pit-pairs in walls of tracheids. ***(c)*** Tangential section, showing bordered pit-pairs of tracheids. ***(d)*** Radial section, showing ray. The rays of pine and other conifers are composed of ray tracheids and ray parenchyma cells. In *(d)*, ray tracheids occur at the top and bottom of the ray and ray parenchyma cells in the middle. Notice the bordered pits of ray tracheids. Above the ray parenchyma are two adjacent bordered pit-pairs.

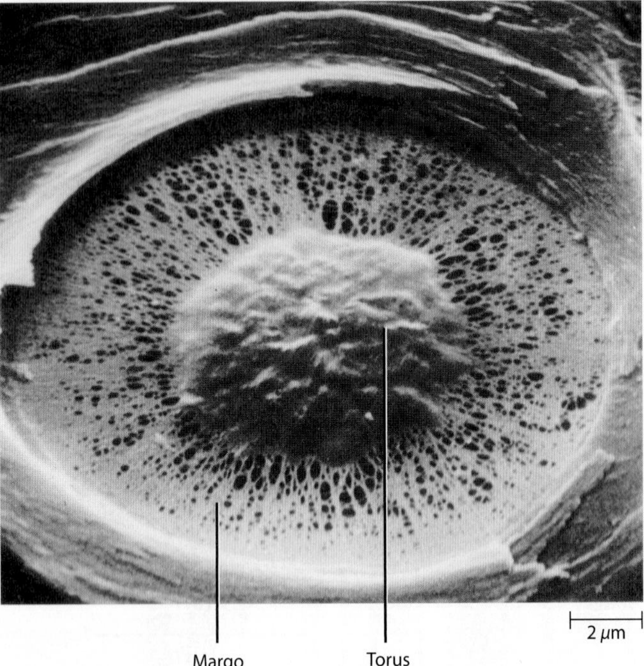

26–22 Pit membrane of a bordered pit-pair in a white pine *(Pinus strobus)* tracheid The thickened part of the membrane is the torus, which is impermeable to water. As seen in this scanning electron micrograph, the part of the membrane surrounding the torus, called the margo, is very porous, allowing the movement of water and ions from tracheid to tracheid.

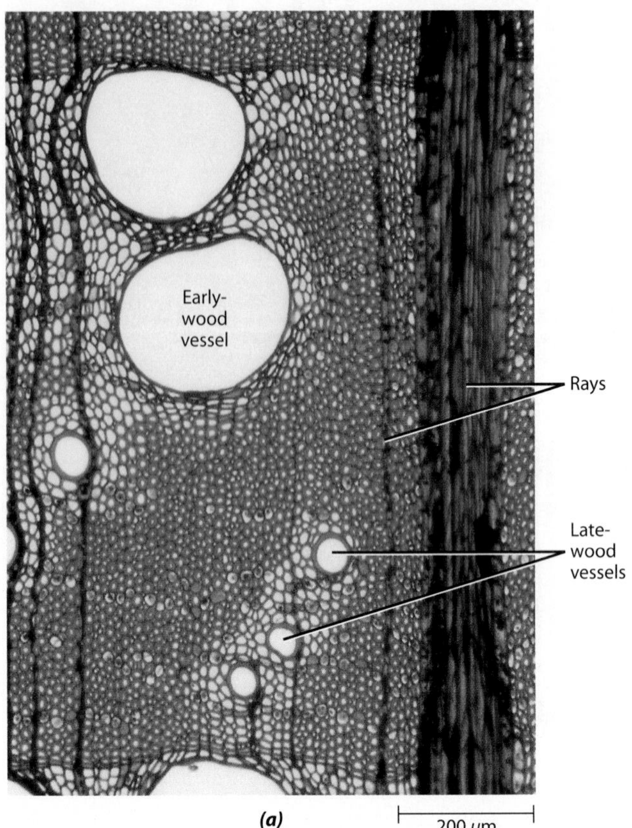

26–24 Wood of red oak *(Quercus rubra)* The wood is seen here in ***(a)*** transverse and, on the facing page, ***(b)*** radial and ***(c)*** tangential sections. (Red oak wood has a specific gravity of 0.57.)

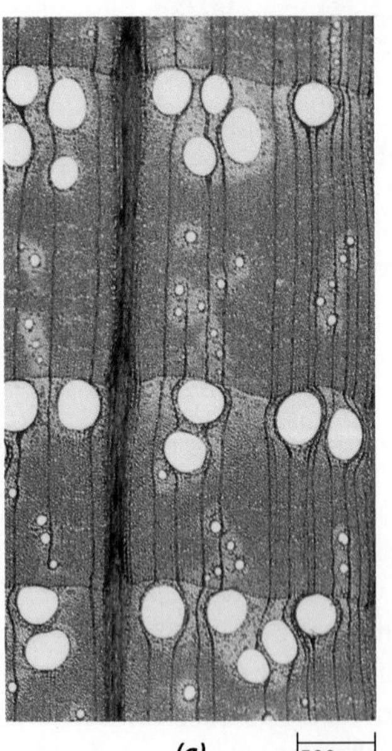

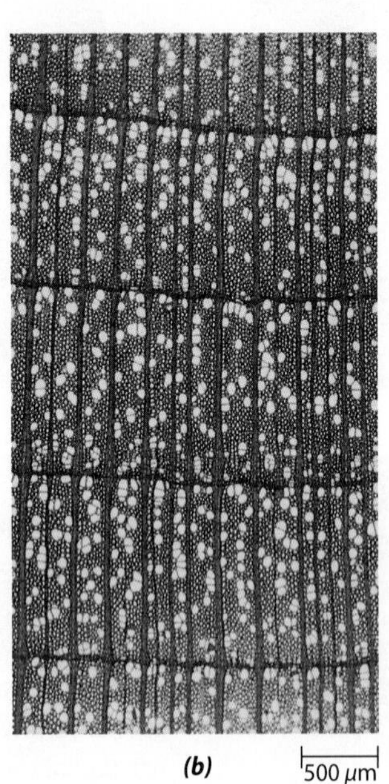

26–23 Growth layers of wood, in transverse sections *(a)* Red oak *(Quercus rubra)*. The large vessels of ring-porous wood such as red oak are found in the early wood. The dark vertical lines are rays. ***(b)*** Tulip tree *(Liriodendron tulipifera)*, a diffuse-porous wood.

light, temperature, rainfall, available soil water, and length of the growing season. The width of a growth ring is a fairly accurate index of the rainfall of a particular year. Under favorable conditions—that is, during periods of adequate or abundant rainfall—the growth rings are wide; under unfavorable conditions, they are narrow.

In semiarid regions, where there is very little rain, the tree is a sensitive rain gauge. An excellent example of this is the bristlecone pine *(Pinus longaeva)* of the western Great Basin (Figure 26–27). Each growth ring is different, and a study of the rings tells a story that dates back thousands of years. The oldest-known living specimen of bristlecone pine is 4845 years old. Dendrochronologists—scientists who conduct historical research by studying the growth rings of trees—have been able to match samples of wood from living and dead trees, and in this way they have built up a continuous series of rings dating back more than 8200 years. The widths of the growth rings of bristlecone pines at higher elevations (the upper tree line) have also been found to be closely related to temperature changes, which dramatically affect the length of the growing season in alpine climates. A record of average ring width in these trees provides a valuable guide to past temperatures and climatic conditions. For example, in the White Mountains of California, the summers were relatively warm from 3500 B.C.E. to 1300 B.C.E., and the tree line was about 150 meters

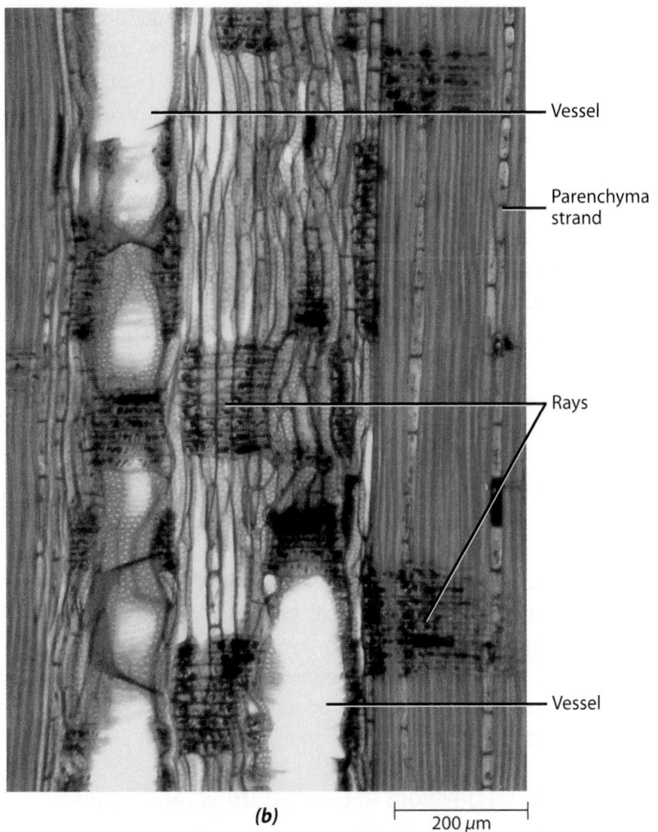

(b)

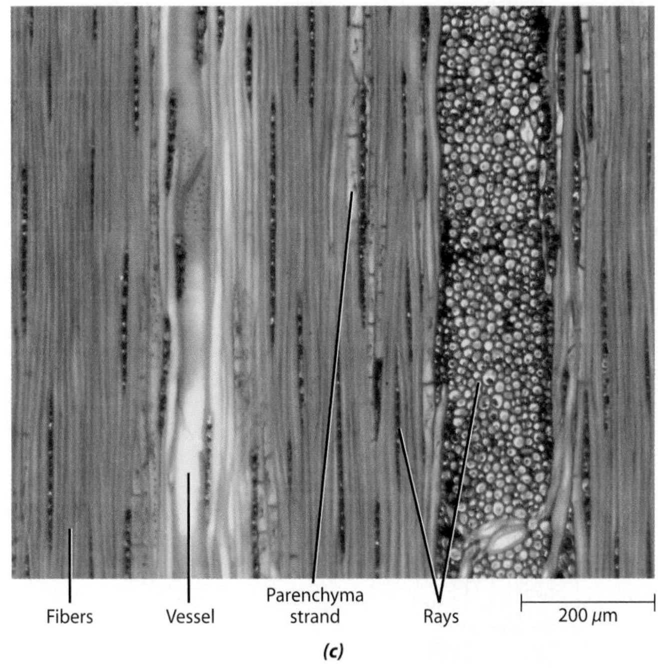

(c)

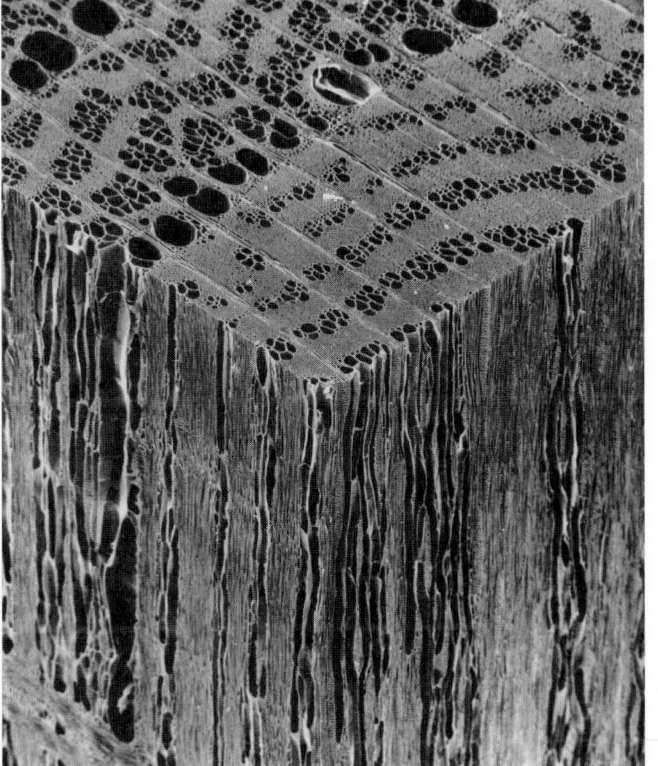

above its present level. Summers were cool from 1300 B.C.E. to 200 B.C.E.

The structural basis for the visibility of growth layers in wood is the difference in the density of the wood produced early in the growing season and that produced later (Figures 26–20, 26–23, and 26–24). The **early wood** is less dense (with wider cells and proportionally thinner walls) than the **late wood** (with narrower cells and proportionally thicker walls). In a given growth layer, the change from early to late wood may be very gradual and almost imperceptible. However, where the late wood of one growth layer abuts the early wood of the following growth layer, the change is abrupt and thus clearly discernible.

26–25 Block of American elm *(Ulmus americana)* wood, showing its three surfaces By comparing with Figures 26–19, 26–23, and 26–24, you should be able to identify each face, or surface, in this scanning electron micrograph. This is a semi-ring-porous wood, with the late-wood vessels arranged in wavy lines, a characteristic feature of the elms. Identify the early-wood and late-wood vessels and the rays in all three faces. The dense portion of the wood is composed largely of fibers. Axial parenchyma cells are also present but are not distinguishable at this magnification. (American elm wood has a specific gravity of 0.46.)

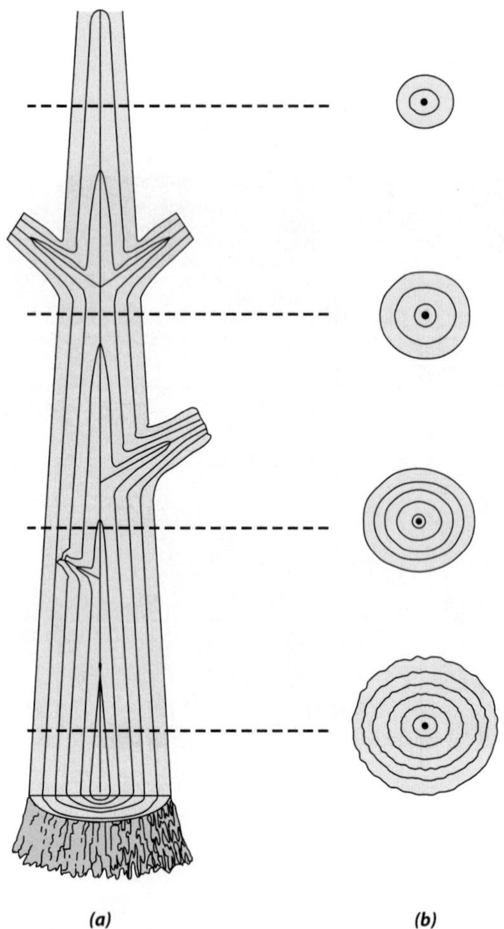

26–26 Annual rings Each annual ring in wood generally represents one year's increment of growth. The number of rings varies with the distance above ground, the oldest part of the trunk occurring at ground level. ***(a)*** Diagram of a median longitudinal section of a tree trunk, and ***(b)*** transverse sections taken at four different levels. Once secondary growth has begun in a portion of stem (or root), that portion no longer increases in length.

In some angiosperm woods, size differences of the vessels, or pores, in early and late woods are quite marked. In these trees, the pores (vessels) of the early wood are distinctly larger than those of the late wood. Such woods are called **ring-porous** woods (Figures 26–23a and 26–24a). In other angiosperm woods, the pores are fairly uniform in distribution and size throughout the growth layer. These are called **diffuse-porous** woods (Figure 26–23b). In ring-porous woods, almost all the water is conducted in the outermost growth layer, at speeds about 10 times greater than in diffuse-porous woods.

Sapwood Conducts and Heartwood Does Not

As wood ages, it gradually becomes nonfunctional in conduction and storage. Before this happens, however, the wood often undergoes visible changes, which involve the loss of reserve foods and infiltration of the wood by various substances (such as oils, gums, resins, and tannins) that color it and sometimes make it aromatic. This often darker, nonconducting wood is called **heartwood,** while the generally lighter, conducting wood is called sapwood (Figure 26–12). **Sapwood,** by definition, is the part of the wood in a living tree that contains living cells and reserve materials. It may or may not be entirely functional in the conduction of water. Heartwood formation is believed to be a process that enables the plant to remove from regions of growth any secondary metabolites that may be inhibitory or even toxic to living cells. The accumulation of these substances in the heartwood results in death of the living cells of the wood.

The proportion of sapwood to heartwood and the degree of visible difference between them vary greatly from species to species. Some trees, such as maple *(Acer),* birch *(Betula),* and ash *(Fraxinus),* have thick sapwoods; others, such as locust *(Robinia),* catalpa *(Catalpa),* and yew *(Taxus),* have thin sapwoods. Still other trees, such as the poplars *(Populus),* willows *(Salix),* and firs *(Abies),* have no clear distinction between sapwood and heartwood.

In many woods, tyloses are formed in the vessels when they become nonfunctional (Figure 26–28). **Tyloses** are balloonlike outgrowths from ray or axial parenchyma cells through pit cavities in the vessel wall. They may completely block the vessel. Tyloses are often induced in response to plant pathogens and may serve as a defensive mechanism by inhibiting spread of the pathogen throughout the plant via the xylem.

Reaction Wood Develops in Leaning Trunks and Limbs

The formation of **reaction wood** is a developmental response by a leaning branch or stem to counteract the force of gravity. In conifers, the reaction wood develops on the underside of the leaning part and is called *compression wood.* In angiosperms, it develops on the upper side and is called *tension wood.*

Compression wood is produced by the increased activity of the vascular cambium on the lower side of the bent stem, which results in the formation of eccentric growth rings. Portions of growth rings located on the lower side are generally much wider than those on the upper side (Figure 26–29). Hence, production of compression wood causes straightening by expanding or pushing the trunk or limb upright. Compression wood has more lignin and less cellulose than normal wood, and its lengthwise shrinkage on drying is often 10 or more times greater than that of normal wood. (Normal wood usually shrinks lengthwise not more than 0.1 to 0.3 percent.) The difference in relative lengthwise shrinkage of normal and compression wood in a drying board often causes the board to twist and cup. Such wood is virtually useless except as fuel.

Tension wood is produced by increased activity of the vascular cambium on the upper side of the stem and, as in the case of compression wood, is recognized by the presence of eccentric growth rings. To straighten the stem, the tension wood must exert a pull; hence the name "tension wood." Positive identification of tension wood requires microscopic examination of wood sections. Anatomically, the principal distinguishing feature is the presence of gelatinous fibers, which are identified by the presence of a so-called gelatinous layer. This innermost secondary wall layer can be distinguished from the outer secondary wall layer(s) by its high cellulose content and lack of lignin. Lengthwise shrinkage of tension wood rarely exceeds 1 percent, but boards containing it twist out of shape on drying. When logs with tension wood are

26–27 Bristlecone pine *(Pinus longaeva)* ***(a)*** Seen here growing in the White Mountains of eastern California, these bristlecone pines, which grow near the timberline, are the world's oldest continuously standing living trees; one tree has reached an age of 4845 years. ***(b)*** Transverse section of wood from a bristlecone pine, showing the variation in width of annual rings. This section of wood, only a portion of which is shown here, begins approximately 6260 years ago; the shaded band of rings represents the 30 years from 4240 B.C.E. to 4210 B.C.E. The overlapping patterns of rings in dead trees have made it possible to determine relative precipitation extending back some 8200 years.

Recently, spruce clones *(Picea abies)* have been found in Sweden growing from root stock estimated to be 9550 years old. Several years earlier, a ring of creosote bushes *(Larrea divaricata)*, all apparently derived from one seed, was estimated to be about 12,000 years old. Growing in the Mojave Desert 150 miles northeast of Los Angeles, the ring of bushes has been dubbed "King Clone." The record for longevity may belong, however, to a shrub known as king's holly (*Lomatia tasmanica,* family Proteaceae), which was discovered by a group of Tasmanian botanists and estimated by them to be more than 43,000 years old.

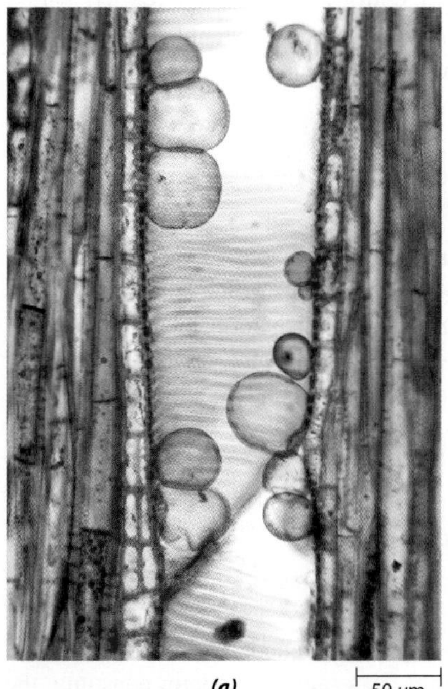

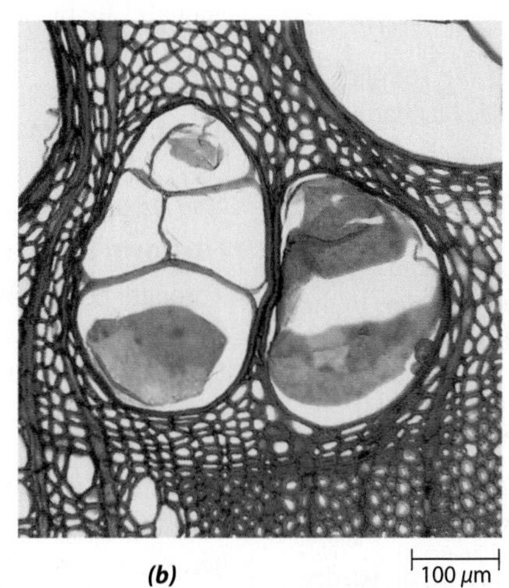

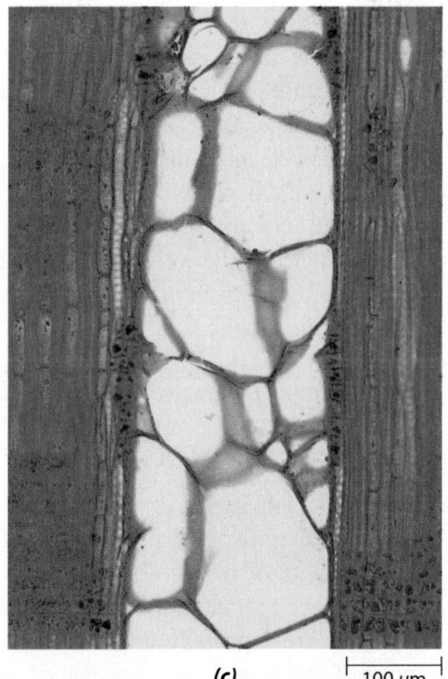

26–28 Tyloses Tyloses are balloonlike outgrowths of parenchyma cells that partially or completely block the lumen of a vessel. ***(a)*** Longitudinal section showing tyloses protruding into a vessel of a wounded grape *(Vitis vinifera)* stem. ***(b)*** Transverse and ***(c)*** longitudinal sections showing vessels of white oak *(Quercus alba)* that are occluded by tyloses.

26–29 Reaction wood in a conifer Transverse section of the stem of a pine (*Pinus* sp.), showing compression wood with wider growth rings on the lower side.

sawed green, the tension wood tears loose in bundles of fibers, imparting a wooly appearance to the boards.

Density and Specific Gravity Are Good Indicators of the Strength of Wood

Density is the single most important indicator of the strength of wood and can be used to predict such characteristics as hardness, nailing resistance, and ease of machining. Dense woods generally shrink and swell more than light woods. In addition, the densest woods make the best fuel.

The *specific gravity* of a substance is the ratio of the weight of the substance to the weight of an equal volume of water. In the case of wood, the oven-dry weight of the wood is used in figuring its specific gravity:

$$\text{Specific gravity} = \frac{\text{Oven-dry weight of wood}}{\text{Weight of the displaced volume of water}}$$

The specific gravity of dry solid wood substance (that is, dry cell wall material) of all plants is about 1.5. The differences in specific gravity of woods depend, therefore, on the proportion of wall substance to lumen (the space bounded by the cell wall).

Fibers are especially important in determination of specific gravity. If the fibers are thick-walled and narrow-lumened, the specific gravity tends to be high. Conversely, if the fibers are thin-walled and wide-lumened, it tends to be low. The presence of numerous thin-walled vessels also tends to lower the specific gravity.

Density is expressed as weight per unit volume, either as pounds per cubic foot (English) or as grams per cubic centimeter (metric). Water has a density of 62.4 lb/ft^3, or 1 g/cm^3. A wood weighing 31.2 lb/ft^3, or 0.5 g/cm^3, is therefore one-half as dense as water and has a specific gravity of 0.5. The *Guinness Book of World Records* lists black ironwood *(Olea capensis)* of South Africa as the heaviest wood (93 lb/ft^3, or 1.49 g/cm^3) and *Aeschynomene hispida* of Cuba as the lightest (2.75 lb/ft^3, or 0.044 g/cm^3). Their respective specific gravities are 1.49 and 0.044. The specific gravities of most commercially useful woods are between 0.35 and 0.65.

SUMMARY

Secondary Growth Causes an Increase in the Girth of Stems and Roots

Secondary growth, the increase in girth in regions that are no longer elongating, occurs in all gymnosperms and in most angiosperms other than monocots, and it involves the activity of the two lateral meristems—the vascular cambium and the cork cambium, or phellogen. Herbaceous plants undergo little or no secondary growth, whereas woody plants—trees and shrubs—may continue to increase in thickness for many years. Figure 26–30 presents a summary of stem development in a woody plant, beginning with the apical meristem and ending with the secondary tissues produced during the first year's growth.

Plants Are Often Classified According to Their Seasonal Growth Cycles

Plants that undergo the entire cycle from seed to vegetative plant to flowering plant to seed in a single growing season are called annuals. In biennials, two seasons are required for the period from seed germination to seed formation. Perennials are plants in which the vegetative structures live year after year. Some perennials are herbaceous, others are woody.

The Vascular Cambium Contains Two Types of Initials: Fusiform Initials and Ray Initials

Through periclinal (parallel to the surface) divisions, the fusiform initials give rise to the components of the axial system, and the ray initials produce ray cells, which form the vascular rays, or radial system. Increase in circumference of the cambium is accomplished by anticlinal (perpendicular to the surface) divisions of the initials.

The Cork Cambium Produces a Protective Covering on the Secondary Plant Body

The first cork cambium in most stems originates in a layer of cells immediately below the epidermis. The cork cambium produces cork toward the outside and phelloderm toward the inside. Together, the cork cambium, cork, and phelloderm constitute the periderm. Although most of the periderm consists of compactly arranged cells, isolated areas called lenticels have numerous intercellular spaces and play an important role in the exchange of gases through the periderm.

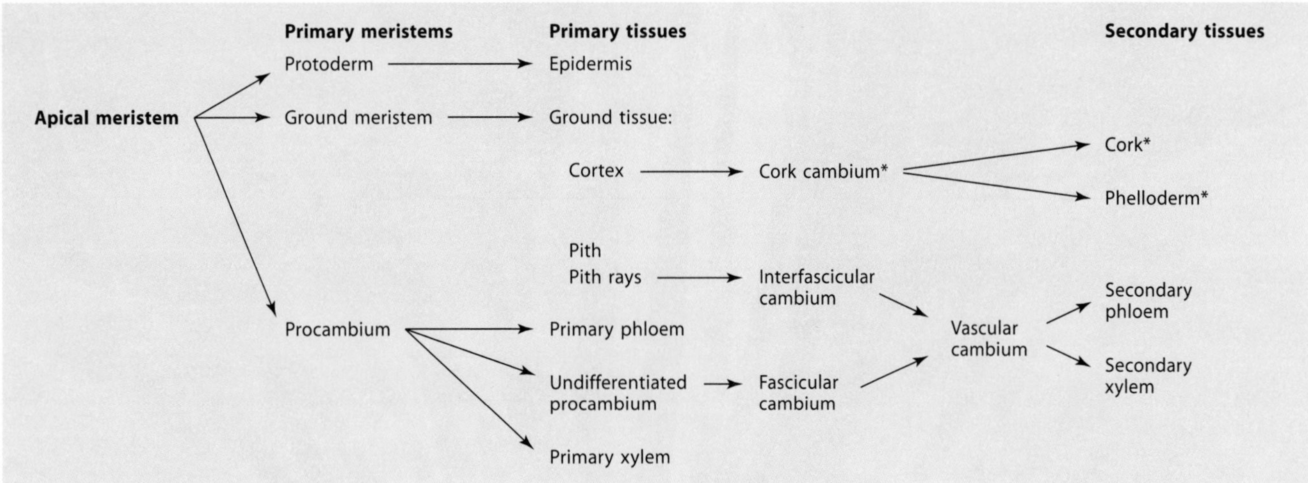

26–30 Summary of stem development in a woody angiosperm during the first year of growth (Compare this with root development, as summarized in Figure 24–24.)

Bark Consists of All Tissues outside the Vascular Cambium

In old roots and stems, most of the phloem of the bark is nonconducting. Sieve elements are short-lived, and in many species only the present year's growth increment contains functional sieve elements. After the first periderm, subsequently formed periderms originate deeper and deeper in the bark from parenchyma cells of nonconducting phloem.

Wood Is Secondary Xylem

Woods are classified as either softwoods or hardwoods. All so-called softwoods are conifers and all so-called hardwoods are angiosperms (woody magnoliids and woody eudicots). Conifer woods, which are structurally simpler than angiosperm woods, consist of tracheids and parenchyma cells. Some contain resin ducts. Angiosperm woods may contain a combination of all the following cell types: vessel elements, tracheids, several types of fibers, and parenchyma cells.

Growth Rings Result from the Periodic Activity of the Vascular Cambium

Growth layers that correspond to yearly increments of growth are called annual rings. The difference in density between the late wood of one growth increment and the early wood of the following increment makes it possible to distinguish the growth layers. Density and specific gravity are good indicators of the strength of wood. In many plants, the nonconducting heartwood is visibly distinct from the actively conducting sapwood.

Reaction Wood Develops in Response to the Force of Gravity in a Leaning Branch or Stem

Commonly, reaction wood develops on the underside of leaning trunks and limbs of conifers and on the upper side of similar parts in angiosperms; its formation causes straightening of the trunk or limb. Reaction wood is called compression wood in conifers and tension wood in angiosperms.

QUESTIONS

1. Distinguish between the following: axial system and radial system; fascicular cambium and interfascicular cambium; inner bark and outer bark; conducting phloem and nonconducting phloem.
2. By means of simple, labeled diagrams, compare the structure of a woody eudicot root with that of a woody eudicot stem at the end of the first year's growth. Assume that the root is triarch and the primary vascular system of the stem consists of discrete vascular bundles.
3. If a nail were driven into a tree at a height of five feet from ground level, and the tree increased in height by, on average, two feet per year, approximately how high above ground level would the nail be 10 years later? Explain your answer.
4. What structural feature of wood is responsible for the visibility of growth rings?
5. Of what importance are lenticels to the plant?
6. The terms "hardwood" and "softwood" do not accurately express the degree of density or hardness of wood. Explain.
7. The age of a woody stem cannot always be accurately estimated by counting the growth rings. Why?
8. Of what "value" is heartwood to a plant?
9. Why do some woods sink in water and others do not?
10. What are the knots in wood?

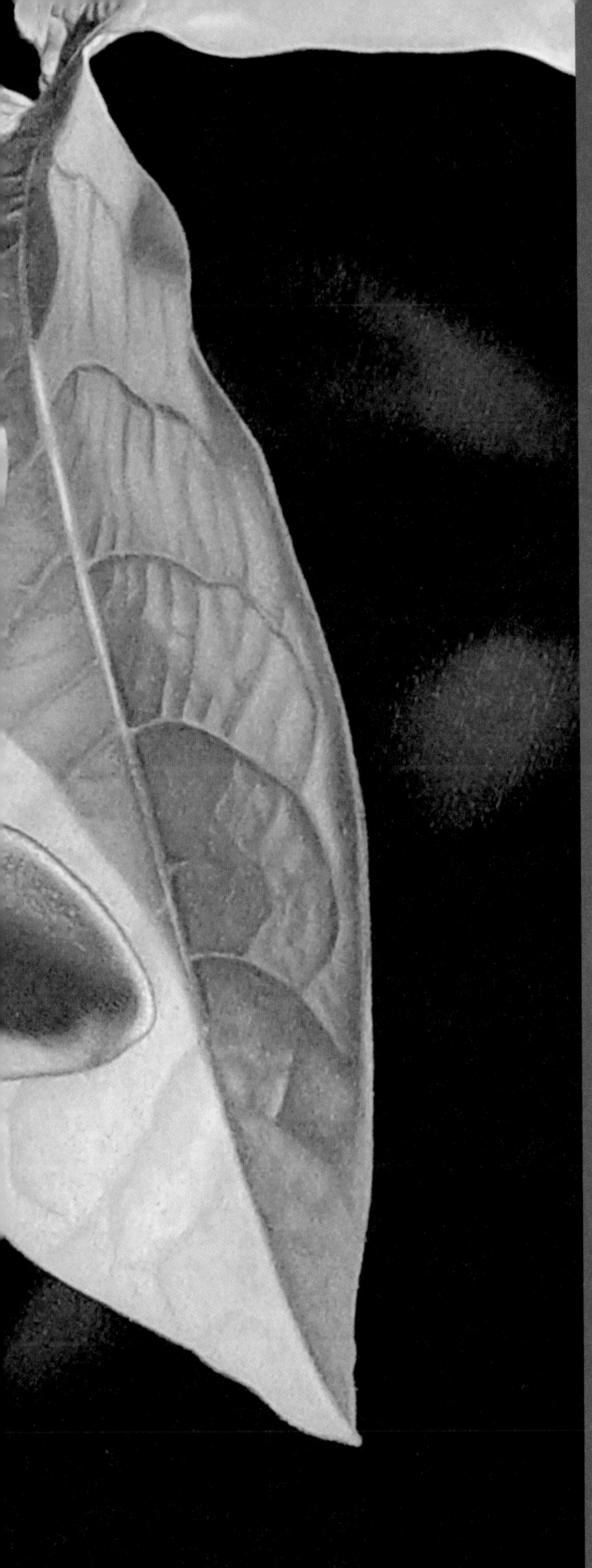

SECTION 6

PHYSIOLOGY OF SEED PLANTS

◀ The green, photosynthesizing leaves of the passionflower (*Passiflora* sp.) are its principal source of sugars, and its nectar-filled flowers are an important destination for those sugars. The nectar attracts hummingbirds, bees, and other insects, which forage for the sugary liquid and, in the process, pollinate the flowers. The passionflower was used by Native Americans as an herb for calming anxiety and insomnia.

CHAPTER 27

Regulating Growth and Development: The Plant Hormones

◀ **Growing cotton** Researchers have found that cotton yields can be increased 5 to 10 percent by adding the naturally occurring class of hormones called cytokinins to cotton seeds or young cotton plants growing in water-limited environments. The hormones stimulate more extensive root development, allowing access to deeper soil moisture, and they also cause the growth of a waxy surface on the leaves that reduces water loss.

CHAPTER OUTLINE

A plant needs light from the sun, carbon dioxide from the air, and water and minerals from the soil in order to grow. As discussed in Section 5, the plant does far more than simply increase its mass and volume as it grows. It differentiates, develops, and takes shape, forming a variety of cells, tissues, and organs. Many of the details of how these processes are regulated are not known. It has become clear, however, that normal development depends on the interplay of a number of internal and external factors. The principal internal factors that regulate plant growth and development are chemical, and they are the subject of this chapter. Some of the external factors affecting plant growth—such as light, temperature, daylength, and gravity—are discussed in Chapter 28.

The growth and development of a multicellular organism would not be possible without effective communication among its cells, tissues, and organs. In both plants and animals, the regulation and coordination of metabolism, growth, and morphogenesis depend on *chemical signals* called **hormones.** The term "hormone" comes from the Greek *horman,* meaning "to stimulate."

The concept of plant hormones, or phytohormones, has long been colored by the concept of hormones that arose from physiologists' studies of mammals. In its original sense, the hormone concept had three basic elements: (1) synthesis of the hormone in one part of the body, (2) its transport (via the bloodstream in mammals) to another part of the body, where (3) it induces a chemical response to control a specific physiological event.

Although some plant hormones are produced in one tissue and transported to another tissue, where they produce specific

CHECKPOINTS

After reading this chapter, you should be able to answer the following:

1. Name the six major groups of plant hormones. Why should hormones be considered regulators rather than stimulators?
2. What are the sites of biosynthesis of each of the major groups of plant hormones?
3. Describe some of the effects caused by each of the major groups of plant hormones.
4. How was the discovery of cytokinins linked with the development of tissue culture?
5. What are some of the mechanisms by which plant hormones exert their effects at the molecular level?

physiological responses, others act within the tissues where they are produced. In either case, these chemical signals communicate information about the developmental or physiological state of cells, tissues, and, in some cases, widely separated organ systems. Plant hormones are synthesized at several different locations in the plant body, not in any one specific gland or tissue. They are active in very small quantities. The shoot of a pineapple *(Ananas comosus),* for example, contains only 6 micrograms of indole-3-acetic acid, a common plant hormone, per kilogram of plant material. One enterprising plant physiologist calculated that the weight of the hormone in relation to that of the shoot is comparable to the weight of a needle in 20 metric tons of hay.

It is now clear that some plant hormones, rather than acting as stimulators, have inhibitory influences. Moreover, the response to a particular hormone depends not only on its chemical structure but also on how it is "read" by the target tissue. The same hormone can elicit different responses in different tissues or at different times of development in the same tissue. Some phytohormones are able to influence the biosynthesis of another hormone, or even to interfere in the signal transduction of another hormone. Tissues may require different amounts of hormones. Such differences are referred to as differences in **sensitivity.** Thus, plant systems may vary the intensity of hormone signals by altering hormone concentrations or by changing the sensitivity to hormones that are already present.

At any given time, much, if not most, of a given plant hormone may occur in an inactive, conjugated form—that is, bound to such substances as sugars or sugar alcohols or to amino acids, peptides, or even proteins. Hydrolysis of the inactive conjugates by the plant tissue releases the "free," or active, form of the hormone.

It is important to note that plant hormones rarely, if ever, act alone. Rather, the large array of responses mediated by these compounds is achieved by a combination of interactions—or "cross-talk"—between or among the hormones and other signals.

Traditionally, five classes of plant hormones—the "classic five"—have received the most attention: auxins, cytokinins, ethylene, abscisic acid, and gibberellins (Table 27–1). More recently, it has become clear that additional chemical signals are used by plants. For example, the **brassinosteroids**—a group of naturally occurring polyhydroxylated steroids—appear to be required for the normal growth of most plant tissues and now rank with the "classic five" as a major class of plant hormones (Table 27–1). Several other signaling molecules have been identified as playing roles in plants' resistance to pathogens and defense against herbivores; these include salicylic acid, jasmonic acid, and systemin.

Salicylic acid, a phenolic compound with a structure similar to aspirin (see Figure 2–28), has been implicated in the activation of disease resistance following pathogen invasion. It is also a key signal in regulating thermogenesis (heat production) in certain members of the arum family (Araceae), such as the skunk cabbage *(Symplocarpus foetidus)* (page 94) and the voodoo lily *(Sauromatum guttatum).* During the rapid growth associated with flowering in these species, the temperature in parts of the inflorescence may increase by as much as 25°C above air temperature.

Jasmonic acid, or **jasmonate,** a derivative of linolenic acid, activates plant defenses against insect herbivores and many microbial pathogens. In addition, it may mediate the response to drought, ozone, UV radiation, and other abiotic stresses.

Systemin, a short polypeptide (18 amino acids), mediates the response to wounding by insects, a response that involves production of defense proteins—protease inhibitors—in leaves and stems of solanaceous plants, such as potato and tomato. The protease inhibitors interfere with protein digestion in the attacking insects, retarding their growth and development. These protease inhibitors accumulate not only in wounded leaves but also in undamaged leaves far from the damaged sites, indicating that long-distance signal transmission from the damaged site induces a systemic defense. Evidence indicates that the signal is jasmonic acid, the biosynthesis of which is activated by systemin. The jasmonic acid is transported through the phloem to other parts of the plant, where it activates the expression of genes that encode protease inhibitors; in this way, undamaged plant tissues are able to defend against subsequent insect attacks. Another recently discovered polypeptide hormone, **florigen,** which is well over 100 amino acids long, is produced in leaves and causes flowering in the shoot apical meristem (see Chapter 28).

We begin our discussion of the six major groups of hormones with auxin, the first substance to be identified as a plant hormone.

Auxins

Some of the first recorded experiments on growth-regulating substances were performed by Charles Darwin and his son Francis and reported in *The Power of Movement in Plants,* published in 1881. The Darwins first made systematic observations of the bending of plant shoots toward light (phototropism; see Chapter 28), using seedlings of canary grass *(Phalaris canariensis)* and oat *(Avena sativa).* They found that bending of the coleoptile (the sheathlike, protective structure covering the shoot of grass seedlings) did not occur if they covered its upper portion with a cylinder of metal foil or a hollow tube of blackened glass (or a blackened quill) and illuminated the plant from one side (Figure 27–1). If, however, the tip was enclosed in a transparent glass tube (or quill), it did bend toward the light. From these experiments, the Darwins concluded that "when seedlings are freely exposed to a lateral light some influence is transmitted from the upper to the lower part, causing the latter to bend." We now know that this bending is the result of differential cell elongation (see Chapter 28).

In 1926, the plant physiologist Frits W. Went succeeded in isolating the "influence" from the coleoptile tips of oat *(Avena)* seedlings. Went named this chemical substance **auxin,** from the Greek *auxein,* meaning "to increase."

As can be seen in Figure 27–2, the principal naturally occurring auxin, **indole-3-acetic acid** (abbreviated **IAA**), closely resembles the amino acid tryptophan (see Figure 2–13c). Both tryptophan and IAA are synthesized from indole, and several biosynthetic pathways using tryptophan as a precursor have been shown to produce IAA in plants. However, maize and *Arabidopsis* mutants that are unable to synthesize tryptophan can

Table 27-1 Major Plant Hormones: Their Nature, Occurrence, and Effects

Hormone(s)	Chemical Nature	Sites of Biosynthesis	Transport	Effects
Auxins	Indole-3-acetic acid (IAA) is the principal naturally occurring auxin. Possibly synthesized via tryptophan-dependent and tryptophan-independent pathways.	Primarily in leaf primordia and young leaves and in developing seeds.	Auxin is transported both polarly (unidirectionally) and nonpolarly.	Apical dominance; tropic responses; vascular tissue differentiation; promotion of cambial activity; induction of adventitious roots on cuttings; inhibition of leaf and fruit abscission; stimulation of ethylene synthesis; inhibition or promotion (in pineapples) of flowering; stimulation of fruit development.
Cytokinins	Cytokinins are N^6-adenine derivatives, phenyl urea compounds. Zeatin is the most common cytokinin in plants.	Primarily in root tips.	Cytokinins are transported in the xylem from roots to shoots.	Promotion of cell division; promotion of shoot formation in tissue culture; delay of leaf senescence; application of cytokinin can cause release of lateral buds from apical dominance and can increase root development in arid conditions.
Ethylene	The gas ethylene (C_2H_4) is synthesized from methionine. It is the only hydrocarbon with a pronounced effect on plants.	In most tissues in response to stress, especially in tissues undergoing senescence or ripening.	Ethylene, a gas, moves by diffusion from its site of synthesis.	Fruit ripening (especially in climacteric fruits, such as apples, bananas, and avocados); leaf and flower senescence; leaf and fruit abscission.
Abscisic acid	Abscisic acid is synthesized from a carotenoid intermediate. The name is a misnomer because the hormone has little to do with abscission.	In mature leaves and roots, especially in response to water stress. May be synthesized in seeds.	Abscisic acid is exported from leaves in the phloem; from roots in the xylem.	Stomatal closure; induction of photosynthate transport from leaves to developing seeds; induction of storage-protein synthesis in seeds; embryogenesis; may affect induction and maintenance of dormancy in seeds and buds of certain species.
Gibberellins	Gibberellic acid (GA_3), a fungal product, is the most widely studied. Gibberellins are synthesized via the terpenoid pathway.	In young tissues of the shoot and developing seeds. It is uncertain whether synthesis also occurs in roots.	Gibberellins are probably transported in the xylem and phloem.	Hyperelongation of shoots by stimulating both cell division and cell elongation, producing tall, as opposed to dwarf, plants; induction of seed germination; stimulation of flowering in long-day plants and biennials; regulation of production of seed enzymes in cereals.
Brassinosteroids	Brassinosteroids are polyhydroxylated steroid compounds, synthesized as a branch of the terpenoid pathway.	Throughout the plant, especially in young growing tissues.	Endogenous brassinosteroids act locally, at or near their sites of synthesis.	A wide range of developmental and physiological processes, including cell division and cell expansion; branching; vascular tissue differentiation; development of lateral roots; seed germination; leaf senescence.

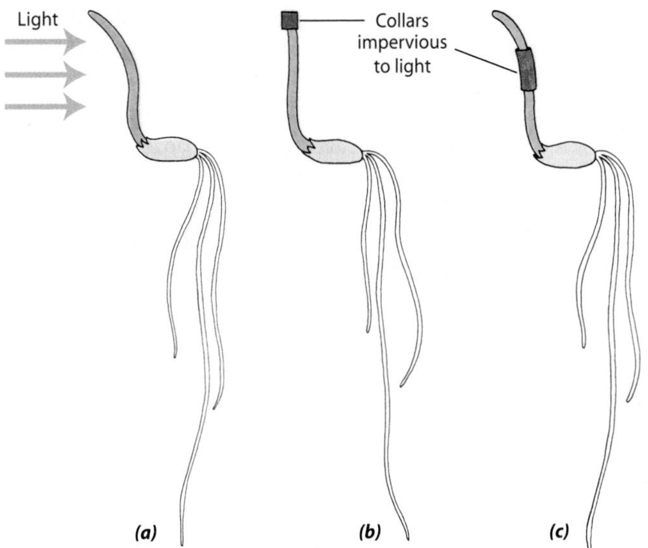

27–1 The Darwins' experiment ***(a)*** Seedlings normally bend toward the light. ***(b)*** When the tip of a seedling was covered by a lightproof collar, this bending did not occur. (Bending did occur when the tip of a seedling was covered with a transparent collar.) ***(c)*** When a lightproof collar was placed below the tip, the characteristic light response took place. From these experiments, the Darwins concluded that, in response to light, an "influence" that causes bending is transmitted from the tip of the seedling to the area below the tip where bending normally occurs.

still make IAA. Thus, plants are apparently capable of producing this essential plant growth regulator by a variety of pathways. Mutants lacking either auxin or cytokinin have yet to be found, suggesting that these hormones are required for viability; mutations eliminating them are lethal.

Shoot apical meristems and young leaves are the primary sites of auxin synthesis. The apical meristems of roots are also important sites of auxin synthesis, although the root is dependent on the shoot for much of its auxin. Developing fruits and seeds contain high levels of auxin.

Auxin Is the Only Plant Hormone Known to be Transported Polarly

Auxin has been implicated in many aspects of plant development, including the overall polarity of the plant root-shoot axis, which is established during embryogenesis (see Chapter 22). This structural polarity is traceable to the polar, or unidirectional, transport of auxin in the plant.

In shoots, the **polar transport** is always **basipetal,** that is, from the shoot tip and leaves downward in the stem (Figure 27–3). However, once this downward-directed stream moves beyond the base of the root (the root-shoot junction) and toward the root tip, its direction is described as **acropetal** (toward the apex). The velocity of polar auxin transport—2 to 20 centimeters per hour—is faster than the rate of passive diffusion. Although transport of auxin is characteristic of shoot apical meristems and young leaves, most of the auxin synthesized in mature leaves seems to be transported throughout the plant *nonpolarly* via the phloem, at rates considerably higher than those of polar transport.

The principal route of polar auxin transport in stems and leaves is via vascular parenchyma cells, most likely those associated with the xylem. Most of the auxin reaching the root tip is transported nonpolarly in the sieve tubes of the phloem. In the root tip, the auxin is redirected basipetally (polarly) in the epidermal and cortical parenchyma cells (see Figure 27–7).

Polar Transport of Auxin Is Mediated by Efflux Carriers Precisely Aligned with the Direction of Auxin Transport

According to the prevailing model for polar auxin transport (Figure 27–4), IAA enters (influx) the conducting (parenchyma) cells in a protonated (IAAH) form by passive diffusion and/or as an anion (IAA^-) by secondary active cotransport (see Chapter 4) via influx carriers (identified as AUX1 protein in *Arabidopsis* roots); such carriers appear to be uniformly distributed around the conducting cells. Inside the cell, IAAH dissociates at the higher pH of the cytoplasm into IAA^- and H^+ and can exit (efflux) the cell only through the activity of the auxin efflux carriers. These carriers, which are precisely aligned with the direction of auxin transport, are known as **PIN proteins.** They are named after the pin-shaped inflorescences formed by the *pin1* mutant of *Arabidopsis.* PIN proteins are constantly being cycled

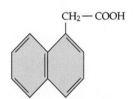

Indole ring
Acetic acid side chain
$CH_2—COOH$
N
H
(a) Indole-3-acetic acid (IAA)

$O—CH_2—COOH$
Cl
Cl
(b) 2,4-Dichlorophenoxyacetic acid (2,4-D)

$CH_2—COOH$
(c) 1-Naphthaleneacetic acid (NAA)

27–2 Auxins ***(a)*** Indole-3-acetic acid (IAA) is the principal naturally occurring auxin. ***(b)*** 2,4-Dichlorophenoxyacetic acid (2,4-D), a synthetic auxin, is widely used as an herbicide. ***(c)*** 1-Naphthaleneacetic acid (NAA), another synthetic auxin, is commonly employed to induce the formation of adventitious roots in cuttings and to reduce fruit drop in commercial crops. The synthetic auxins, unlike IAA, are not readily broken down by natural plant enzymes and microbes and so are better suited than IAA for commercial purposes.

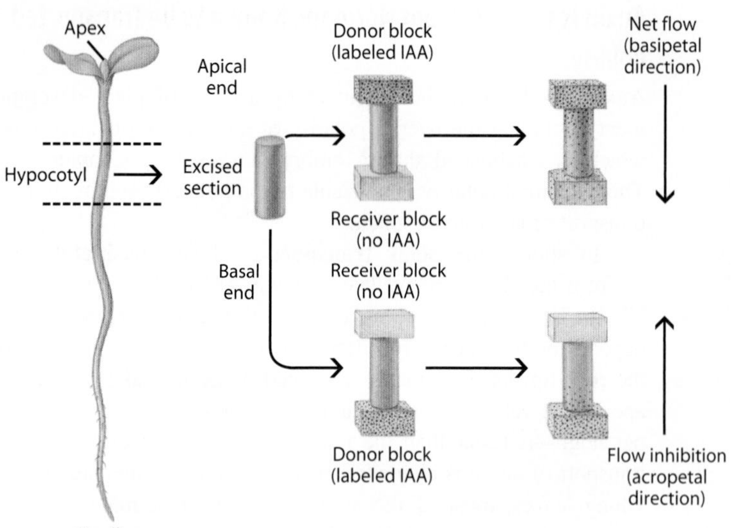

27–3 Auxin transport Experimental demonstration of polar auxin transport in stems, here represented by a segment of hypocotyl from a seedling. Hypocotyl segments are placed between agar blocks. The donor block contains radioactively labeled auxin (IAA). The rate of auxin transport is measured as radioactivity accumulating in the receiver block after a set time. The rate is much faster in the basipetal (polar) than in the acropetal (nonpolar) direction.

between the plasma membrane and internal secretory compartments of the cell, and this cycling enables the direction of auxin transport to be changed in response to environmental or developmental signals.

In addition to PIN proteins, members of the so-called ABCB family of protein carriers have been found to function in auxin efflux. Although the ABCBs are generally uniformly, rather than polarly, distributed in the plasma membranes of cells in shoot and root apices, when specific ABCB and PIN proteins co-occur in the same location of the cell, they function synergistically to stimulate directional auxin transport.

Auxin Plays a Role in the Differentiation of Vascular Tissue

The gradient of auxin caused by its polar transport influences differentiation of the vascular tissue in developing leaves and the elongating shoot. In very young leaf primordia of *Arabidopsis,* auxin synthesized at the tip of the leaf induces formation of the midvein. As the leaf develops, the site of auxin production gradually shifts toward the base along the margins, and later moves toward the central region of the blade. Sites of maximal auxin activity along the margins induce formation of the lateral veins. As noted in Chapter 25, the smaller, or minor, veins are initiated at the tip of the leaf and develop basipetally in continuity with previously formed, coarser veins. Thus, the shift in auxin production from the tip of the leaf to its base is a reflection of the overall maturation of the vascular system and of the leaf.

Using a *GUS* reporter gene (pages 195 and 196) to detect auxin synthesis in a developing *Arabidopsis* leaf, it can be shown that auxin is produced at sites along the leaf margins by groups of cells that will differentiate into hydathodes (Figure 27–5). Hydathodes are glandlike structures through which the plant releases water (see Chapter 30). The area with the dark blue *GUS* stain, as in Figure 27–5, indicates intense auxin activity at the site of a future hydathode. The more diffuse stain leading down to a developing vascular strand shows that auxin is playing a role in the differentiation of vascular tissues.

When a stem of cucumber *(Cucumis sativus)* or some other herbaceous eudicot is wounded in such a way as to sever and

27–4 Schematic model for polar auxin transport The auxin indole-3-acetic acid (IAA) enters the cell by passive diffusion in a protonated (IAAH) form and/or by secondary active cotransport in its anionic form (IAA^-), via an influx carrier (AUX1 protein). The anionic form (IAA^-) predominates in the cytosol, which has a neutral pH. The anions exit the cells only at the basal end, via efflux carriers (PIN proteins).

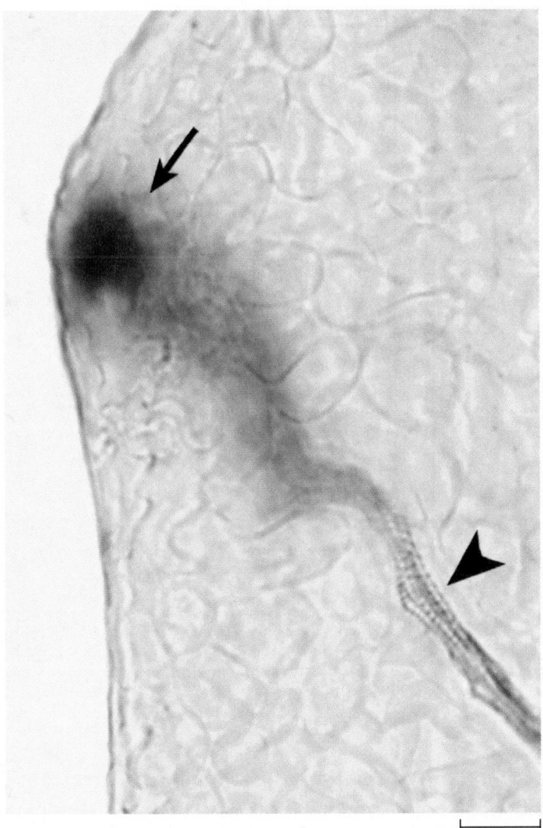

27–5 Auxin synthesis The dark blue *GUS* reporter gene stain shown here (arrow) marks the site of auxin synthesis along the margin of a young *Arabidopsis* leaf. The site indicated by the arrow corresponds to the location of a young hydathode, a glandlike structure through which water is released once the hydathode is mature (see Chapter 30). A gradient of *GUS* activity extends from the hydathode to a differentiating vascular strand (arrowhead). The *GUS* reporter gene was linked to an auxin-sensitive promoter.

remove portions of the vascular bundles, new vascular tissues form from cells in the pith and connect the severed bundles. However, if the leaves and buds above the wound are removed, the formation of new cells is delayed. With the addition of IAA to the stem just above the wound, new vascular tissue begins to form (Figure 27–6). Auxin similarly has an important role in the joining of vascular traces from developing leaves to the bundles in the stem.

Auxin Plays a Role in the Induction and Arrangement of Leaves

In Chapter 25, we discussed some aspects of the mechanism underlying the orderly initiation and arrangement of leaves. Studies have shown that, in tomato, inhibition of polar auxin transport blocks leaf formation at the vegetative apical meristem, resulting in pinlike naked stems with an intact meristem at the tip. Microapplication of IAA to such tips restores leaf formation. Externally applied (exogenous) IAA also induced flower formation on inflorescence apices of the *Arabidopsis* mutant *pin1-1*. In this mutant, flower formation is blocked because of a mutation in the auxin efflux carrier that moves auxin out of the cell.

Auxin Provides Chemical Signals That Communicate Information over Long Distances

In many plant species, the basipetal flow of auxin from the growing apical bud inhibits the growth of axillary (lateral) buds. If growth at the shoot tip is interrupted, the flow of auxin is diminished and the lateral shoots begin to develop. The inhibitory influence of an apical bud on the lateral buds is referred to as **apical dominance.** The role of auxin in apical dominance can be demonstrated experimentally. For instance, when the shoot tip of a bean plant *(Phaseolus vulgaris)* is removed, the lateral buds begin to grow. However, when auxin is applied to the cut surface, the growth of the lateral buds is inhibited.

Originally, auxin synthesized in the shoot tip was thought to be transported basipetally to the axillary buds, where it inhibited bud outgrowth. It soon became apparent, however, that auxin does not act directly at the bud; rather, it affects the bud from

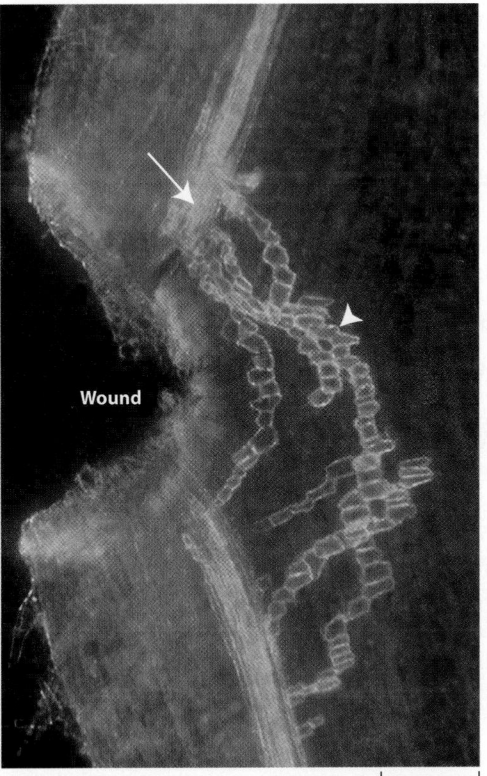

27–6 IAA-induced xylem regeneration around a wound Longitudinal view of vascular tissue (xylem) regeneration (arrowhead) around a seven-day-old wound in a young internode of cucumber *(Cucumis sativus)*. The leaves and buds had been removed from the plant in order to reduce the amount of auxin produced. IAA (0.1 percent in lanolin) was applied to the upper end of the internode immediately after wounding. The micrograph shows the typical pattern of xylem differentiation induced by the basipetal polar movement of auxin arriving from above (arrow) and flowing around the wound.

the xylem and sclerenchyma cells between the vascular bundles (interfascicular sclerenchyma) in the stem. It was then proposed that a long-distance second messenger (pages 87 and 88) of auxin action (rather than auxin itself), moving upward into the branches via the transpiration stream, inhibits bud outgrowth.

A strong candidate for this second messenger is **strigolactone,** a newly recognized hormone that was recently discovered to interact with auxin in regulating apical dominance. Strigolactones (a group of terpenoid lactones derived from carotenoids) were originally identified as compounds in root exudates that stimulate germination of the seed of harmful root-parasitic plants, such as witchweeds (*Striga* spp.) and broomrapes (*Orobanche* and *Phelipanche* spp.). The seedlings of these parasites attach to the roots of host plants and use nutrients from the hosts for their own nutrition and reproduction, often killing the host. Mutants incapable of producing strigolactones showed lower germination of root parasites. More recently, strigolactones were identified as signals exuded from roots that promote the formation of arbuscular mycorrhizas, which increases a plant's ability to obtain water and minerals (see Chapter 14). Mutants deficient in strigolactone were less capable of forming these beneficial associations with fungi. Strigolactones also play a role in the control of branching. Mutant peas incapable of producing strigolactones branch without restraint. When strigolactones are administered to these mutant plants, the unrestrained branching stops. Strigolactone is produced in both shoots and roots.

In woody plants, auxin promotes activity of the vascular cambium. With expansion of buds and their resumption of growth in the spring, auxin moves downward in the stems and stimulates the cambial cells to divide, forming secondary vascular tissue.

Auxin Promotes the Formation of Lateral and Adventitious Roots

As discussed in Chapter 24, lateral roots are initiated in the pericycle. Auxin is the key signal controlling development of these lateral roots. In *Arabidopsis* seedlings, the initiation of lateral roots depends on basipetal transport of auxin from the tip of the parent root, whereas auxin transported from the shoot (acropetal transport in the root) is required for the emergence, or outgrowth, of the lateral roots (Figure 27–7). The pericycle cells involved in lateral root formation are called **lateral root founder cells.** It is believed that pericycle cells destined to become lateral root founder cells are specified, or primed for future lateral root formation, in the basal half of the parent-root apical meristem.

The first practical application of auxin involved its use in initiating adventitious roots in cuttings. The practice of treating cuttings with auxin is commercially important, especially for the vegetative propagation of woody plants. Application of a high concentration of auxin to already-growing roots, however, usually inhibits their growth.

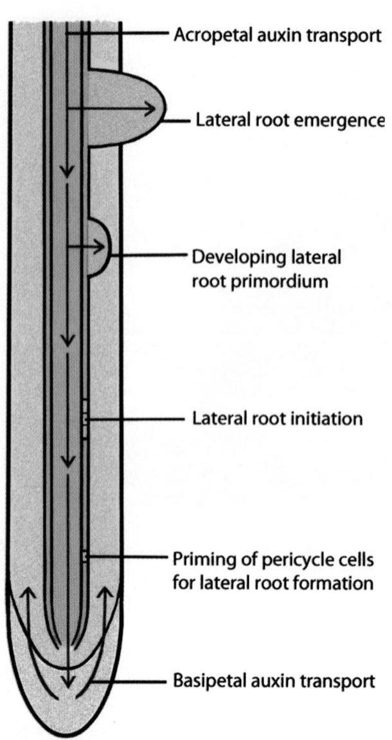

27–7 Lateral root formation The formation of lateral roots is dependent on both acropetal (from base to tip) and basipetal (from tip to base) transport of auxin in the parent root. Most of the auxin reaching the root tip is transported acropetally and nonpolarly in the sieve tubes of the phloem. At the root tip, the auxin is redirected to the basal part of the root (basipetal polar transport). The priming of pericycle cells as lateral root founder cells is believed to occur in the basal half of the apical meristem. Lateral root initiation is dependent on auxin from the root tip, and emergence of the lateral root requires auxin from the shoot.

Auxin Promotes Fruit Development

Auxin is involved in the formation of fruit. Ordinarily, if a flower is not pollinated and fertilized, the fruit will not develop. By treating the female flower parts (carpels) of certain species with auxin, it is possible to produce **parthenocarpic fruit** (Gk. *parthenos,* meaning "maiden, virgin"), which is fruit produced without fertilization—for example, seedless tomatoes, cucumbers, and eggplants. Many or most of these seedless fruits, however, still contain immature ovules.

Developing seeds are a source of auxin. If, during development of the aggregate fruit of the strawberry *(Fragaria × ananassa),* all of the achenes (one-seeded fruits) are removed, the receptacle stops growing altogether. If a narrow ring of achenes is left intact, the fruit forms a bulging girdle of growth in the area of the achenes (Figure 27–8). On application of auxin to the deseeded receptacle, growth proceeds normally.

Synthetic Auxins Are Used to Kill Weeds

Synthetic auxins such as 2,4-dichlorophenoxyacetic acid (2,4-D) (Figure 27–2) have been used extensively for the control of weeds on agricultural lands. In economic terms, this is the major practical use for plant growth regulators. How these herbicides kill weeds is not completely clear, although some herbicides, such as dichlorophenyldimethylurea and paraquat, are known to block photosynthetic electron flow. 2,4-D is not broken down in plants as readily as natural auxins, and the resulting artificially high levels of auxinlike compounds certainly contribute to the lethal

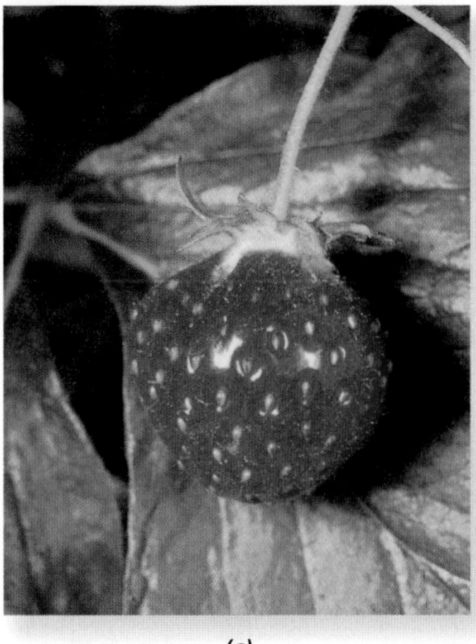

(a)

(b)

(c)

27–8 Auxin and fruit development Auxin, produced by developing embryos, promotes maturation of the ovary wall and the development of fleshy fruits. ***(a)*** Normal strawberry *(Fragaria × ananassa)*, ***(b)*** strawberry from which all seeds have been removed, and ***(c)*** strawberry from which a horizontal band of seeds has been removed. If a paste containing auxin were applied to *(b)*, the strawberry would grow normally.

effects. The mechanism by which herbicides kill only *certain* weeds is also largely unknown. The selectivity of these compounds for killing so-called broad-leaf weeds is due in part to the greater absorption and rate of transport of the herbicides by broad-leaf weeds than by grasses, which can rapidly inactivate synthetic auxins by conjugation (binding them to another substance).

Cytokinins

In 1941, Johannes van Overbeek found that coconut *(Cocos nucifera)* milk, which is liquid endosperm, contained a potent growth factor different from anything known at that time. This factor (or factors) greatly accelerated the development of plant embryos and promoted the growth of isolated tissues and cells in the test tube. Van Overbeek's discovery had two effects: it gave impetus to studies of isolated plant tissues, and it launched the search for another major group of plant growth regulators.

The basic medium used for tissue culture of plant cells contains sugar, vitamins, and various salts. In the early 1950s, Folke Skoog and his co-workers showed that a stem segment of tobacco *(Nicotiana tabacum)* initially grew in such a culture medium, but its growth soon slowed or stopped. Apparently, some growth stimulus originally present in the tobacco stem became exhausted. The addition of auxin had no effect. When coconut milk was added to the medium, however, the cells began to divide and growth of the tobacco stem resumed.

Skoog and his co-workers set out to identify the growth factor in the coconut milk. After many years of effort, they succeeded in producing a thousandfold purification of a growth factor, but they could not isolate it. So, changing course, they tested a variety of purine-containing substances—largely nucleic acids—in the hope of finding a new source of the growth factor. This led to the discovery by Carlos O. Miller that a breakdown product of DNA contained material that was highly active in promoting cell division.

Subsequently, Miller, Skoog, and their co-workers succeeded in isolating the growth factor from a DNA preparation and identifying its chemical nature. They called this substance **kinetin** and named the group of growth regulators to which it belongs the **cytokinins,** because of their involvement in cytokinesis, or cell division. As shown in Figure 27–9, kinetin resembles the purine adenine, which was the clue that led to its discovery. Kinetin, which probably does not naturally occur in plants, has a relatively simple structure, and biochemists were soon able to synthesize some related compounds that behaved like cytokinins. Eventually, Miller isolated a natural cytokinin from kernels of maize *(Zea mays)*, which he called **zeatin;** it is the most active of the naturally occurring cytokinins.

Free cytokinins have now been isolated from many different species of angiosperms, where they are found primarily in actively dividing tissues, including seeds, fruits, and leaves, and in root tips. They have also been found in bleeding sap—the sap that drips out of pruning cuts, cracks, and other wounds in many types of plants. Cytokinins also have been identified in algae, mosses, horsetails, ferns, and conifers.

Although practical applications for cytokinins are not as extensive as those for auxin, they have been important in plant development research. Cytokinins are central to tissue culture methods and are extremely important for plant biotechnology (see Figure 10–9). Treatment of lateral buds with cytokinin often causes the buds to grow, even in the presence of auxin, thus modifying apical dominance.

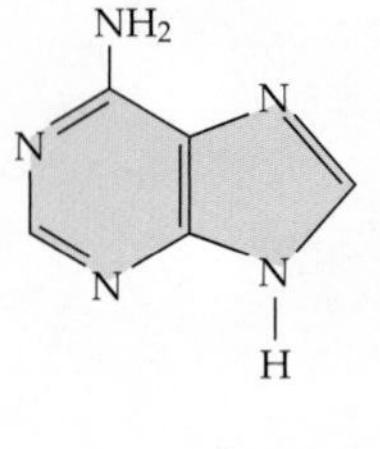

27–9 Cytokinins Note the similarities between the purine adenine ***(a)*** and these four cytokinins. ***(b)*** Kinetin and 6-benzylamino purine (BAP) are commonly used synthetic cytokinins. ***(c)*** Zeatin and isopentenyl adenine (i^6Ade) have been isolated from plants.

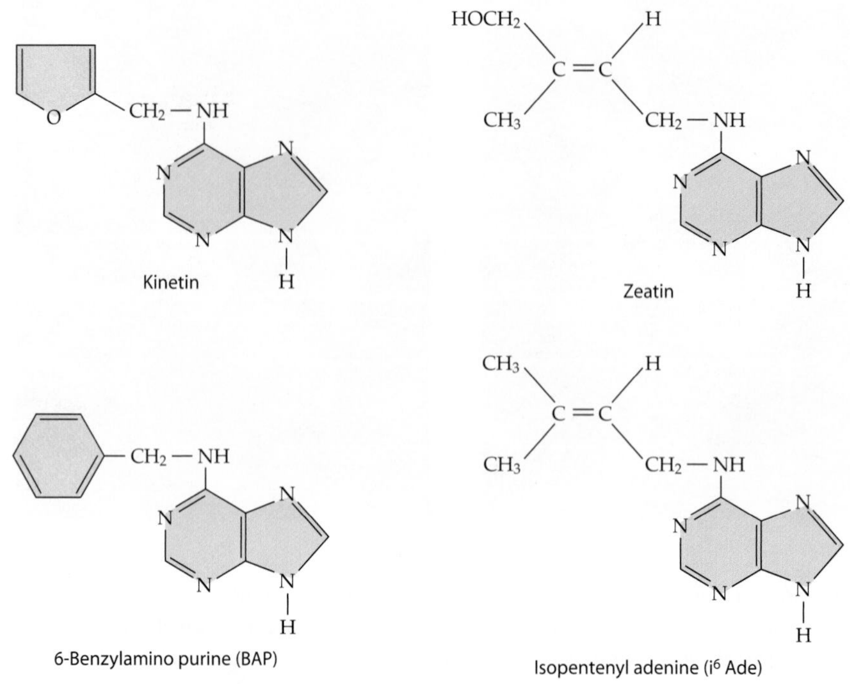

The Cytokinin:Auxin Ratio Regulates the Production of Roots and Shoots in Tissue Cultures

The undifferentiated plant cell has two courses open to it: either it can enlarge, divide, enlarge, and divide again, or, without undergoing cell division, it can elongate. The cell that divides repeatedly remains essentially undifferentiated, or meristematic, whereas the elongating cell ultimately differentiates. In studies of tobacco stem tissues, the addition of auxin to the tissue culture produces rapid cell expansion, so that giant cells are formed. Kinetin alone has little or no effect, but auxin plus kinetin results in rapid cell division, and large numbers of relatively small, undifferentiated cells are formed. In other words, cells remain meristematic in the presence of certain concentrations of both cytokinin and auxin.

In the presence of a high concentration of auxin, **callus tissue**—a growth of undifferentiated plant cells in tissue culture—frequently gives rise to organized roots. In tobacco pith callus, the relative concentrations of auxin and kinetin determine whether roots or buds form (Figure 27–10). With higher concentrations of auxin, roots are formed, and with higher concentrations of kinetin, buds are formed. When both auxin and kinetin are present in roughly equal concentrations, the callus continues to produce undifferentiated cells.

Cytokinin and auxin also act antagonistically in maintenance of the root apical meristem. For meristem maintenance, the rate of cell division must equal the rate of cell differentiation. In *Arabidopsis,* cytokinins have been shown to determine root meristem size by controlling the rate of cell differentiation at the boundary (referred to as the transition zone) between dividing and expanding cells in the different files of vascular tissue, thereby antagonizing, or counteracting, the effects of auxin, which mediates cell division. In addition, cytokinins antagonize auxin distribution during lateral root initiation. Acting in the pericycle cells at the protoxylem pole, cytokinin is a negative regulator of lateral root formation, preventing establishment of the auxin gradient required for normal initiation of lateral roots.

Cytokinins Delay Leaf Senescence

In most species of plants, leaves begin to turn yellow as soon as they are removed from the plant. This yellowing, which is due to a loss of chlorophyll, can be delayed by cytokinins. For example, when excised and floated on plain water, leaves of the cocklebur *(Xanthium strumarium)* turn yellow in about 10 days. With kinetin (10 milligrams per liter) present in the water, much of the green color and fresh appearance of the leaf are maintained. If an excised leaf is spotted with kinetin-containing solutions, the spots remain green while the rest of the leaf yellows. Furthermore, if the cytokinin-spotted leaf contains radioactive amino acids, it can be shown that the amino acids migrate from other parts of the leaf to the cytokinin-treated areas. Another dramatic demonstration of the effect of cytokinins on leaf longevity is provided in Figure 10–14e.

One interpretation of the senescence of detached leaves is that cytokinins are limited in the detached leaf. This interpretation leads to an important and still unanswered question as to the site(s) of cytokinin production within plants. As mentioned previously, cytokinins are most abundant in actively dividing seeds, fruits, and leaves, and in root tips. This is not evidence, however, that cytokinins are synthesized in these organs, because the cytokinins might be transported there from some other site(s). Based on several lines of evidence, root tips most certainly are involved

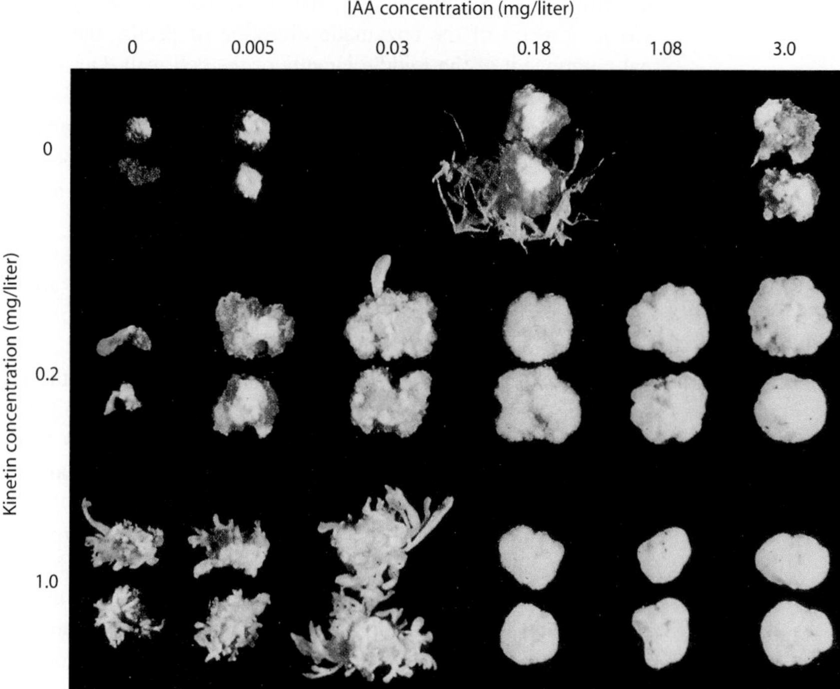

27–10 Callus development Effect of increasing IAA concentration at different kinetin levels on growth and organ formation of tobacco *(Nicotiana tabacum)* callus cultured on nutrient agar. Note that very little growth occurred without the addition of either IAA or kinetin (top left). Higher levels of IAA alone (top rows) promoted root formation, whereas higher IAA repressed bud formation when used alone or in combination with kinetin (right-hand columns). The higher level of kinetin (lower rows) was more effective than the lower level (middle rows) in promoting bud development, but both kinetin levels were too high for promotion of root growth.

in cytokinin synthesis (see Figure 28–7a). It is widely accepted that cytokinins synthesized in root tips are then transported in the xylem to all other parts of the plant. However, the results of experiments with transgenic plants in which systemic and local cytokinin synthesis are controllable indicate that locally synthesized cytokinin, not root-derived cytokinin, is needed to release buds from dormancy.

Ethylene

Ethylene was known to have effects on plants long before the discovery of auxin. The "botanical" history of **ethylene,** a simple hydrocarbon ($H_2C{=}CH_2$), goes back to the 1800s, when city streets were lighted by lamps that burned illuminating gas. In Germany, leaks from illuminating gas mains was found to cause defoliation of shade trees along the streets.

In 1901, Dimitry Neljubov demonstrated that ethylene was the active component of illuminating gas. He noticed that exposure of pea seedlings to illuminating gas caused stems to grow horizontally. When the components of illuminating gas were individually tested for effects, all were inactive except ethylene, which was active at concentrations as low as 0.06 part per million (ppm) in air. Neljubov's findings have led to the realization that ethylene exerts a major influence on many, if not all, aspects of growth and development in plants, including growth of most tissues, fruit maturation, fruit and leaf abscission, and senescence.

The biosynthesis of ethylene begins with the amino acid methionine, which reacts with ATP to form a compound known as *S*-adenosylmethionine (Figure 27–11). This compound is split into two different products, one of which is called 1-aminocyclopropane-1-carboxylic acid (ACC). Enzymes on the tonoplast then convert ACC into ethylene, CO_2, and ammonium ion. Apparently, the formation of ACC is the step that is affected by those treatments—for example, high auxin concentrations, air pollution damage, wounding—that stimulate ethylene production.

Ethylene May Inhibit or Promote Cell Expansion

In most plant species, ethylene has an inhibitory effect on cell expansion. The **triple response** in pea seedlings is a classic example. Ethylene treatment of dark-grown (etiolated) pea seedlings results in (1) a decrease in longitudinal growth; (2) an increase in radial expansion of the epicotyls and roots; and (3) the horizontal orientation of epicotyls, as observed by Neljubov a century ago (Figure 27–12). The triple response is an adaptation that enables seedlings to overcome obstacles, such as debris, and successfully emerge into the light during germination. Ethylene also elicits rapid stem growth in some semiaquatic species. In deep-water or floating varieties of rice, for example, submergence of young rice plants during the monsoon season triggers an increase in ethylene biosynthesis. The resulting ethylene-induced internodal elongation provides a mechanism for the rice plants to keep pace with rising flood waters. Another ethylene-mediated response to flooding, occurring in mesophytic plants (plants requiring an environment that is neither too wet nor too dry), is an increase in the development of air spaces in submerged tissues. Formation of these air spaces results from the ethylene-mediated degeneration of cortical parenchyma tissues.

Ethylene Plays a Role in Fruit Ripening

Ripening of fruit involves many changes. In fleshy fruits, the chlorophyll is degraded and other pigments may form, changing

$$CH_3-S-CH_2-CH_2-CH(NH_3^+)-COO^-$$

Methionine

↓ ATP → $PP_i + P_i$

$$CH_3-\overset{+}{S}(\text{Adenine-ribose})-CH_2-CH_2-CH(NH_3^+)-COO^-$$

S-Adenosylmethionine (SAM)

↓ Stimulated by high auxin concentrations, air pollution, wounding

H_2C, H_2C (ring) — C — NH_3^+, COO^-

1-Aminocyclopropane-1-carboxylic acid (ACC)

↓ → $CO_2 + NH_4^+$

$$CH_2{=}CH_2$$

Ethylene

27–11 Biosynthesis of ethylene Methionine serves as the precursor of ethylene in all higher plant tissues. *S*-Adenosylmethionine (SAM) is converted to 1-aminocyclopropane-1-carboxylic acid (ACC), the immediate precursor of ethylene, by ACC synthase.

the fruit color. Simultaneously, the fleshy part of the fruit softens as a result of the enzymatic digestion of pectin, the principal component of the middle lamella of the cell wall. During this same period, starches and organic acids or, as in the case of the avocado *(Persea americana),* oils are metabolized into sugars. As a consequence of these changes, fruits become conspicuous and palatable and thus attractive to animals that eat the fruit and so scatter the seed.

During the ripening of many fruits—including tomatoes, avocados, bananas, peaches, and pome fruits such as apples and pears—there is a large increase in cellular respiration, as evidenced by an increased uptake of oxygen and production of carbon dioxide (Figure 27–13). This phase is known as the **climacteric,** and such fruits are called climacteric fruits. Fruits that show a steady decline, or gradual ripening, such as citrus fruits, grapes, cherries, and strawberries, are called **nonclimacteric fruits.**

In climacteric fruits, increased ethylene synthesis precedes and is responsible for accelerating the rate of many of the ripening processes described above. The effect of ethylene on fruit ripening has agricultural importance. A major use is in promoting the ripening of tomatoes that are picked green and stored in the absence of ethylene until just before marketing. Ethylene is also used to hasten the ripening of walnuts and grapes. Using gene transfer techniques, biotechnologists are now able to genetically alter both ethylene synthesis and sensitivity (see Figure 10-14c, d).

Whereas Ethylene Promotes Abscission, Auxin Prevents Abscission

Ethylene promotes abscission, or shedding, of leaves, flowers, and fruits in a variety of plant species. In leaves, ethylene

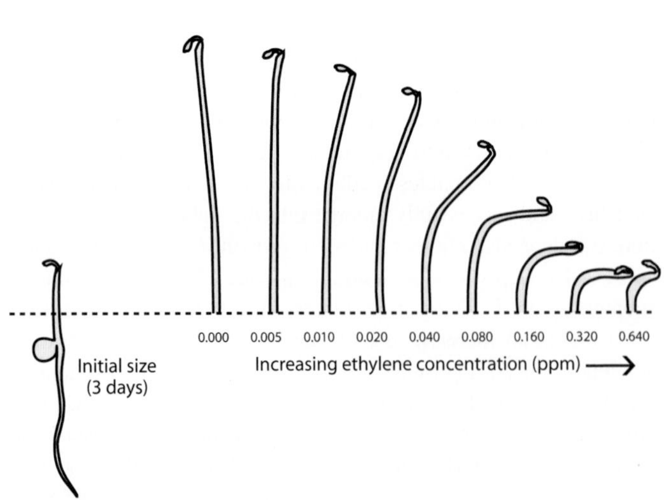

27–12 Triple response of pea seedlings to ethylene The effects of increasing ethylene concentrations on the growth of dark-grown pea *(Pisum sativum)* seedlings are shown here for seven-day-old seedlings. The so-called triple response includes a decrease in epicotyl elongation, a thickening of the shoot, and a change in orientation of growth from vertical to horizontal.

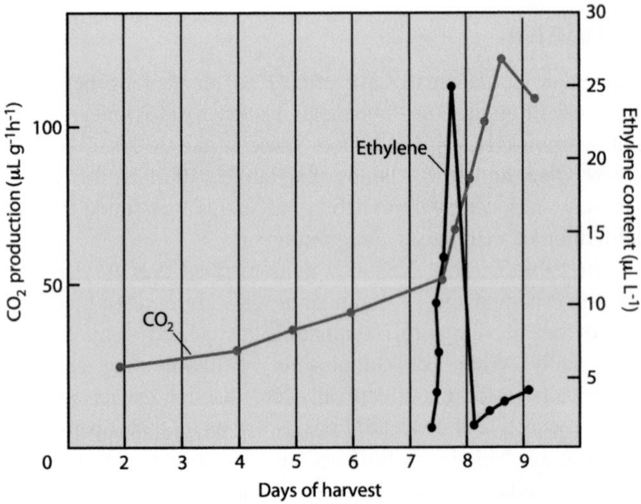

27–13 Ethylene production and respiration in banana Fruit ripening in banana, a climacteric fruit, is characterized by a rise in the respiration rate, as evidenced by increased production of CO_2. Note that the production of ethylene precedes the increase in CO_2 production. ($\mu L\ g^{-1}h^{-1}$ means microliter/gram/hour; $\mu L\ L^{-1}$ means microliter/liter)

presumably triggers the enzymes that cause the cell wall dissolution associated with abscission (page 604). Ethylene is used commercially to promote fruit loosening in cherries, blackberries, grapes, and blueberries, thus making mechanical harvesting possible. It is also used as a fruit-thinning agent in commercial orchards of prunes and peaches.

In many systems, abscission is controlled by an interaction of ethylene and auxin. Whereas ethylene triggers abscission, auxin seems to reduce the sensitivity of abscission zone cells to ethylene, thus preventing abscission. This auxin effect has also been used commercially. For instance, auxin treatment prevents preharvest drop of citrus fruits. In some cases, however, high concentrations of auxin stimulate abscission; this effect is thought to be due to an auxin stimulation of ethylene production.

Ethylene Seems to Play a Role in Sex Expression in Cucurbits

Ethylene seems to play a major role in determining the sex of flowers in some monoecious plants (plants having male and female flowers borne on the same individual). In cucurbits (family Cucurbitaceae; cucumber, squash), male and female flowers develop from the same immature flowers with primordia for both stamens and carpels. High levels of gibberellins (see below) are associated with maleness, and treatment with ethylene changes the expression of sex to femaleness. Under the influence of ethylene, the stamens of immature bisexual cucumber *(Cucumis sativus)* flowers undergo programmed cell death, resulting in the formation of female flowers with a functional pistil. In addition, cucumbers grown under short light periods, or short-day conditions, which promote femaleness, were found to produce more ethylene than those grown under long-day conditions (see Chapter 28). Hence, in cucurbits, ethylene apparently participates in the regulation of sex expression and is associated with the promotion of femaleness.

Abscisic Acid

At certain times, the survival of the plant depends on its ability to restrain its growth or its reproductive activities. In 1949, Paul F. Wareing discovered that the dormant buds of ash and potatoes contain large amounts of a growth inhibitor, which he called *dormin.* During the 1960s, Frederick T. Addicott reported the discovery in leaves and fruits of a substance capable of accelerating abscission, which he called *abscisin.* Abscisin and dormin were soon found to be chemically identical. The compound is now known as **abscisic acid,** or **ABA** (Figure 27–14). This is an unfortunate name choice because it now appears that this substance has no direct role in abscission. Abscisic acid synthesis begins in chloroplasts and other plastids via the terpenoid pathway.

27–14 Abscisic acid Exogenous (external) applications of abscisic acid (ABA) may inhibit plant growth, but the hormone also seems to act as a promoter (for instance, of storage-protein synthesis in seeds).

Abscisic Acid Prevents Seed Germination

Abscisic acid levels increase during early seed development in many plant species. This increase in abscisic acid stimulates the production of seed storage proteins and is also responsible for preventing premature germination. The breaking of dormancy in many seeds is correlated with declining abscisic acid levels in the seed. In maize, there are single-gene mutants that either lack the ability to make the hormone or exhibit a reduced sensitivity to it. As a result, mutant embryos cannot become dormant, and they germinate, in the case of maize, directly on the cob. Such mutants are called viviparous mutants (Figure 27–15).

Abscisic Acid Plays a Role as a Root-to-Shoot Signal

Plants frequently are exposed to abiotic conditions—drought, salt, freezing—that generate water stress or water deficiency. Under such conditions, the roots respond by increasing the biosynthesis of abscisic acid and releasing it into the xylem, through which it moves rapidly to the leaves. In the leaves, the stomata respond to the increased abscisic acid concentration by closing and hence reducing the loss of water by transpiration (see Chapter 30). Mutant plants *(wilty)* incapable of synthesizing abscisic

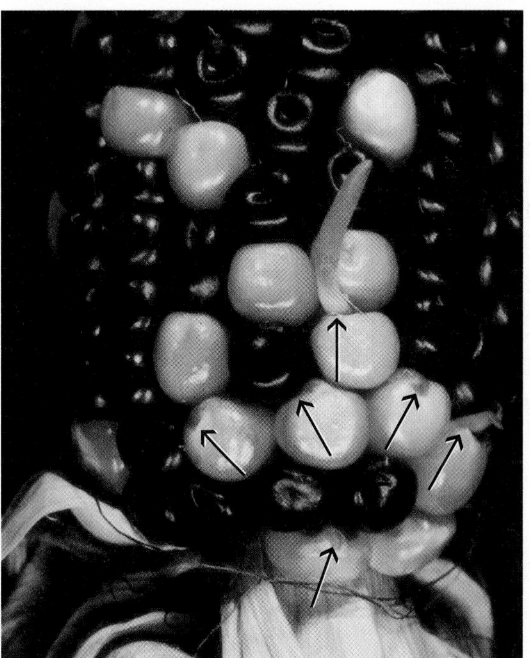

27–15 *Viviparous* mutants of maize *(Zea mays)* The *viviparous-1 (vp1)* gene reduces the sensitivity of the mutant embryos to abscisic acid, which in normal plants prevents premature germination. The mutant embryos undergo precocious germination of the seed (arrows) on the parent plant, because they have failed to enter the developmental arrest caused by abscisic acid in normal plants.

acid show a wilting phenotype; that is, they are capable of growing normally only in very humid conditions. As noted, the hormone induces stomatal closing in most plant species and can thus promote resistance to pathogens by inhibiting their entry via stomata (see pages 657 and 658).

Abscisic acid is transported in the phloem as well as in the xylem and is normally more abundant in the phloem sap than in the xylem sap.

Gibberellins

The early history of gibberellin research was an exclusive product of Japanese scientists. In 1926, the same year that Went isolated auxin from oat *(Avena)* coleoptile tips, E. Kurosawa of Japan was studying a disease of rice *(Oryza sativa)* called "foolish seedling disease." The condition causes plants to grow rapidly, fall over, and appear spindly, pale-colored, and sickly. The cause of these symptoms, Kurosawa discovered, was a substance produced by a fungus, *Gibberella fujikuroi,* which was parasitic on the seedlings.

Gibberellin (GA) was named and isolated in 1934 by the chemists T. Yabuta and Y. Sumiki. The discovery attracted little interest in the western world until after World War II. In 1956, J. MacMillan, in England, first successfully isolated gibberellin from a plant (the seed of the bean *Phaseolus vulgaris).* Since then, gibberellins have been identified in many species of plants, and it is now believed that they occur in all plants. They are present in various amounts in all parts of the plant, with the highest concentrations in immature seeds. More than 136 naturally occurring gibberellins have been isolated and identified chemically in a range of organisms, and a given species of plant typically contains several gibberellins that vary slightly in structure (Figure 27–16), as well as in biological activity. The best studied of the group is GA3 (known as gibberellic acid), which is also produced by the fungus *G. fujikuroi.* The definition of a gibberellin is a compound with a structure like that shown in Figure 27–16 that elicits a biological response. Part of the reason there are so many gibberellins is that the precursors to the most active form of the hormone, as well as compounds in the pathway of inactivation, also elicit a biological response.

The gibberellins have dramatic effects on stem and leaf elongation in intact plants by stimulating both cell division and cell elongation.

Application of Gibberellin Can Cause Dwarf Mutants to Grow Tall

The role of gibberellins in stem growth is most clearly demonstrated when these hormones are applied to many types of dwarf mutants (Figure 27–17). Under gibberellin treatment, such plants become indistinguishable from normal tall, nonmutant plants, indicating that these mutants are unable to synthesize gibberellin and that tissue growth requires this hormone.

Gibberellin-related mutants can be subdivided into two groups: (1) those that are gibberellin-deficient, in which genes that regulate biosynthesis of the hormone are affected, and (2) those in which the gibberellin response is affected. One of the best-studied gibberellin-deficient mutants is the *Arabidopsis ga1-3* mutant. This mutation is due to a large deletion that abolishes the function of a gene *(GA1)* that encodes an enzyme involved in an early step of gibberellin biosynthesis. The addition of gibberellin to this mutant restores the tall stature of the wild type. The *Arabidopsis gai* mutant is an example of a gibberellin response mutant. The mutant plants are dwarfed, but they are not deficient in the hormone. *GAI* encodes a protein that is part of the gibberellin response system, and thus the *gai* mutant phenotype is not reversed by the addition of gibberellin.

The increase in yields of rice and wheat during the 1960s and 1970s, characterized as the "Green Revolution," was largely due to the incorporation of dwarfing genes for both crops carried out by plant breeders (see Chapter 21). These genes allowed the development of shorter, stiff-stemmed varieties that devoted more of their energy to the production of grain and less to the production of straw and leaf material. In addition, the shorter plants were more resistant to damage by wind and rain than their taller, wild-type versions.

Gibberellins Play Multiple Roles in Breaking Seed Dormancy and in Germination

The seeds of many plants require a period of dormancy before they will germinate. In certain plants, dormancy usually cannot be broken except by exposure to cold or to light (see Chapters 22 and 28). In many species, including lettuce, tobacco, and wild oats, gibberellins will substitute for the dormancy-breaking cold or light requirement and promote growth of the embryo and emergence of the seedling. Specifically, gibberellins enhance cell elongation, making it possible for the root to penetrate the

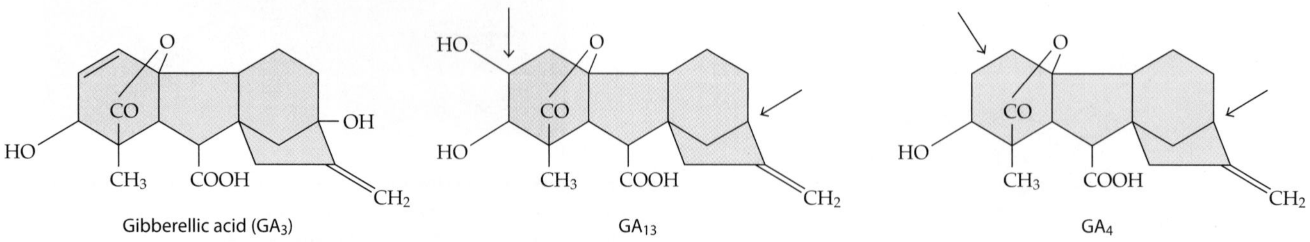

27–16 Gibberellins Shown here are 3 of the more than 136 gibberellins that have been isolated from natural sources. GA_3 (gibberellic acid) is the most abundant in fungi and most biologically active in many tests. The arrows indicate where minor structural differences occur that distinguish the other two examples of gibberellins, GA_{13} and GA_4.

27–17 **Gibberellin effects on dwarf mutants** Contender beans, a dwarf cultivar of the common bean *(Phaseolus vulgaris),* were treated with gibberellin (right) to produce normal tall growth. The plant on the left was not treated and served as a control.

growth-restricting seed coat or fruit wall. This effect of gibberellin has at least one practical application. Gibberellic acid hastens seed germination and thus ensures uniform germination for the production of the barley malt used in brewing beer and ale.

In barley *(Hordeum vulgare)* and other grass seeds, a specialized layer of endosperm cells, called the **aleurone** (see Figure 22–7), lies just inside the seed coat. The cells in the aleurone layer are rich in protein. When the seeds begin to germinate (triggered by the uptake of water), the embryo releases gibberellins, which diffuse to the aleurone cells and stimulate them to synthesize hydrolytic enzymes. One of these enzymes is alpha-amylase (α-amylase), which hydrolyzes starch. The enzymes produced by the aleurone cells digest the stored food reserves of the starchy endosperm. The resulting sugars, amino acids, and nucleic acids are absorbed by the scutellum and then transported to the growing regions of the embryo (Figures 27–18 and 27–19).

Gibberellin Can Cause Bolting and Can Affect Fruit Development

Some plants, such as cabbages *(Brassica oleracea* var. *capitata)* and carrots *(Daucus carota),* form rosettes before flowering. (In a rosette, leaves develop but the internodes between them do not elongate.) In these plants, flowering can be induced by exposure to long days, to cold (as in the biennials), or to both. Following the appropriate exposure, the stems elongate—a phenomenon known as *bolting*—and the plants flower (Figure 27–20). Application of gibberellin to such plants causes bolting and flowering without the appropriate cold or long-day exposure. Bolting is brought about by an increase both in cell number and in cell

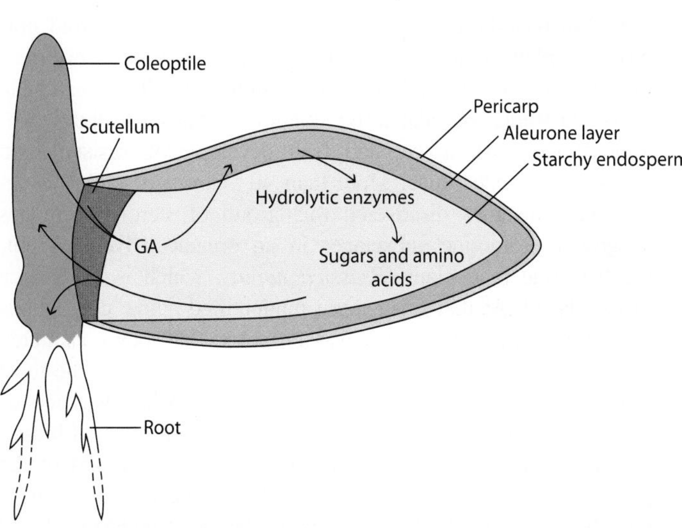

(a)

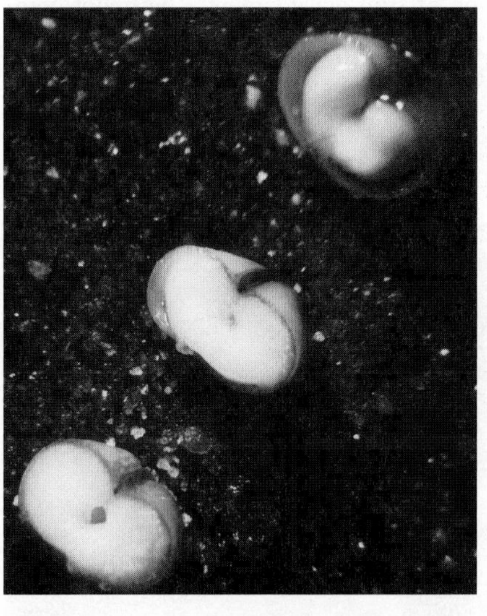

(b)

27–18 **Action of gibberellin in barley seeds** ***(a)*** Gibberellin (GA) produced by the embryo migrates into the aleurone layer, stimulating the synthesis of hydrolytic enzymes. These enzymes are released into the starchy endosperm, where they break down the endosperm reserves into sugars and amino acids, which are soluble and diffusible. The sugars and amino acids are then absorbed by the scutellum (cotyledon) and transported to the shoot and root for growth. ***(b)*** Each of these three seeds has been cut in half and the embryo removed. Forty-eight hours before the picture was taken, the seed at the lower left was treated with plain water, the seed in the center was treated with a solution of 1 part per billion of gibberellin, and the seed at the upper right was treated with 100 parts per billion of gibberellin. Digestion of the starchy storage tissue has begun to take place in the treated seeds.

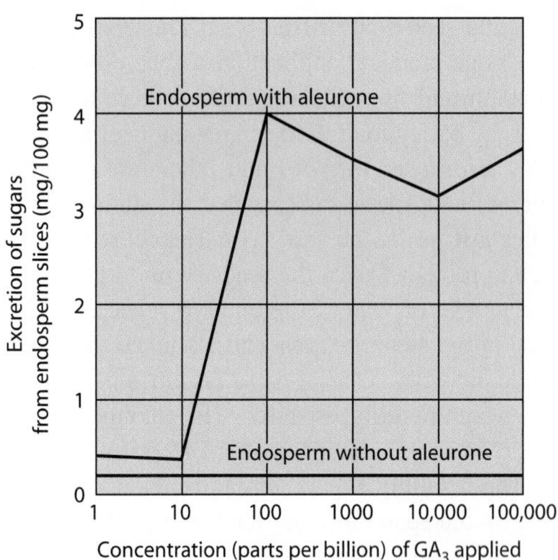

27–19 Gibberellin acts on the aleurone layer The release of sugar from endosperm can be induced by gibberellic acid (GA_3) treatment. These data show that sugars are produced only when the aleurone layer is present. Gibberellin stimulates the aleurone layer to produce the enzyme α-amylase, which digests the starches stored in the endosperm.

27–21 Gibberellin and grapes The effect of gibberellic acid (GA_3) on the growth of Thompson Seedless grapes, a cultivar of *Vitis vinifera.* The bunch of grapes on the left was untreated, whereas that on the right was treated with the hormone. The result is looser clusters of larger grapes.

elongation. Gibberellin can thus be used for early seed production of biennial plants.

Gibberellins, like auxin, can cause the development of parthenocarpic fruits, including apples, currants, cucumbers, and eggplants. In some fruits, such as mandarin oranges, almonds, and peaches, the gibberellins have been effective, where auxin has not, in the promotion of fruit development. The major commercial application of gibberellins, however, is in the production of table grapes. In the United States, large amounts of gibberellic acid (GA3) are applied annually to the Thompson Seedless grapes, a cultivar of *Vitis vinifera.* Treatment causes larger fruit and much looser clusters, making them more attractive to the consumer (Figure 27–21).

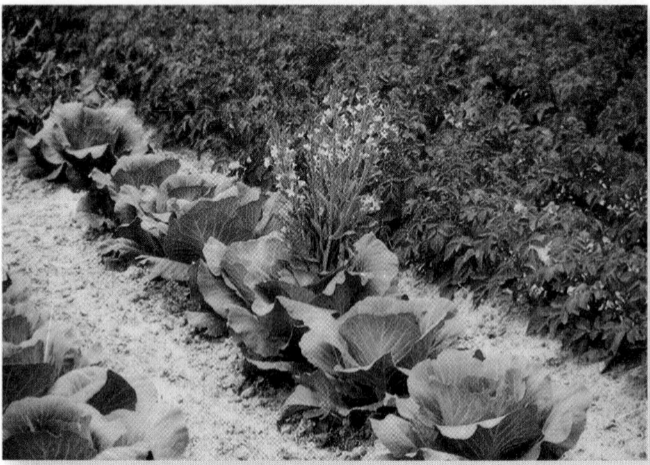

27–20 Gibberellin and bolting In this row of cabbages *(Brassica oleracea* var. *capitata),* the plant in the middle has bolted and flowered naturally. Bolting and flowering can also be induced artificially by treatment with a gibberellin.

Brassinosteroids

Brassinosteroids are a group of growth-promoting steroid hormones in plants that play essential roles in a wide range of developmental processes, such as cell division and cell elongation in roots and stems, vascular differentiation, responses to light (photomorphogenesis), flower and fruit development, resistance to stresses, and senescence. More than 40 years ago, J. W. Mitchell and colleagues discovered this group of steroid hormones as growth-promoting substances in an organic solvent of pollen from the rape plant *(Brassica napus),* which is the source of canola oil. At that time, these unidentified active compounds were named *brassins,* and it was proposed that they constituted a new group of plant hormones. Some nine years later, the most bioactive brassin compound in purified rape pollen was identified and named *brassinolide.* Since then, the occurrence of brassinosteroids has been demonstrated in virtually every part of the plant. About 70 related phytosteroids—brassinosteroids—have now been identified. **Brassinolide** is the most widespread and active brassinosteroid in plants (Figure 27–22a). Its immediate biosynthetic precursor, **castasterone,** has weak brassinosteroid activity (Figure 27–22b).

Exogenously applied 24-epibrassinolide has been shown to undergo long-distance transport from root to shoot via the transpiration stream in the xylem, but experimental evidence reveals that endogenous brassinosteroids do not undergo such transport. Moreover, intermediates in the synthesis of these hormones have

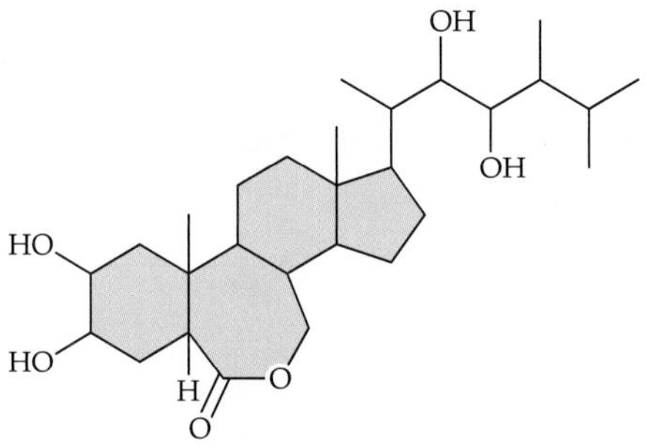

Brassinolide

(a)

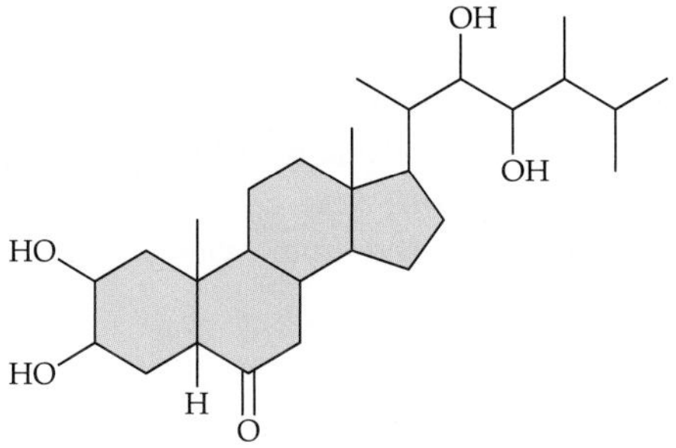

Castasterone

(b)

27–22 Brassinosteroids ***(a)*** Brassinolide is the most widespread and active brassinosteroid in plants. ***(b)*** Castasterone, the immediate biosynthetic precursor of brassinolide, has weak brassinosteroid activity.

been found in all plant organs (although different intermediates predominate in different organs), indicating that endogenous brassinosteroids act locally at or near their sites of synthesis, with each organ synthesizing and responding to its own such hormones.

Brassinosteroids Are Required for Normal Plant Growth

The requirement of brassinosteroids for normal plant growth is clearly demonstrated by the dwarf phenotype of *Arabidopsis* mutants that lack the hormone. The leaves of these mutants have both smaller and fewer cells than those of their wild-type counterparts. Overexpression of the gene *DWF4* for the biosynthesis of a brassinosteroid, on the other hand, results in elevated levels of the hormone and an increase in plant size.

Brassinosteroids Are Essential for Tracheary Element Differentiation

Particularly useful for the study of tracheary element differentiation is the *Zinnia elegans* experimental system. In this system, isolated mesophyll cells nurtured in cultures containing auxin and cytokinin can be made to dedifferentiate and redifferentiate into tracheary elements (Figure 27–23). The tracheary element maturation phase, which involves secondary cell wall formation and programmed cell death, is preceded by a rapid increase in concentrations of brassinosteroids, which are necessary for this final stage of tracheary element differentiation.

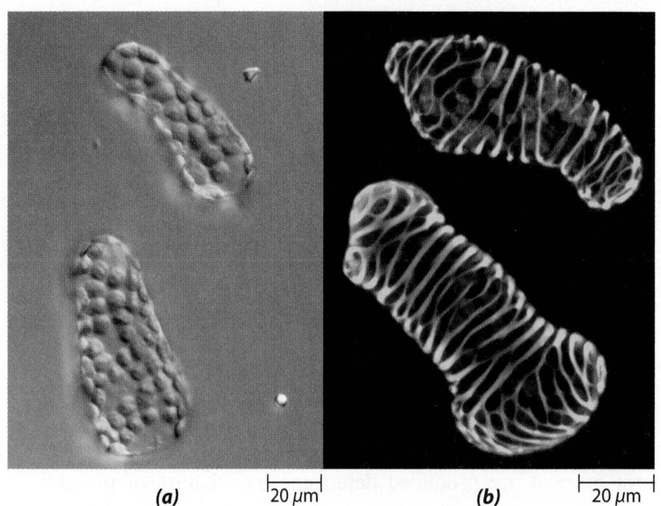

27–23 In vitro culture of zinnia leaf mesophyll cells Mesophyll cells ***(a)*** before and ***(b)*** after dedifferentiation and redifferentiation into tracheary elements. Brassinosteroids are essential for the final stage (maturation) of the tracheary elements to proceed.

The Molecular Basis of Hormone Action

Up to this point, we have largely discussed plant hormones in terms of their effects on plant development. We now turn to some of the molecular mechanisms by which these chemical regulators influence growth, development, and rapid responses at the cellular level.

The development of organs (organogenesis) can be described in terms of a coordinated series of cell divisions and subsequent cell enlargements. Specialization of cell types within an organ (differentiation) is the result of selective expression of a particular set of genes within the genome of each individual cell. Clearly, to coordinate these cellular processes during development, individual cells must communicate with each other. Such communication is achieved by plant hormones, which help to coordinate growth and development by acting as chemical messengers between cells. This concept is supported in part by numerous observable influences of plant hormones on the rate of cell division and on the rate and direction of cell expansion (Table 27–2; Figure 27–24). In addition, it has become clear that the "traditional" plant hormones, as well as those newly discovered, can act either to stimulate or to repress specific genes within the nucleus. In fact, many observable hormone responses seem to be the result of such differential gene expression.

Hormones Control the Expression of Specific Genes

The **totipotency** of plant cells—that is, the potential of plant cells to give rise to entire plants—is clear evidence that all of the genes present in the zygote are also present in each living cell of the adult plant (see the essay on page 202). In any one cell, however, only selected genes are expressed—that is, transcribed into mRNA and subsequently translated into proteins. The specific

Table 27-2 Hormonal Influences on Basic Cellular Processes

Hormone(s)	Rate of Cell Division	Rate of Cell Expansion	Direction of Cell Expansion	Differentiation (Gene Expression)
Auxins	+	+	Longitudinal	+
Cytokinins	+	Little or no effect	None	+
Ethylene	+ or –	+ or –	Lateral	+
Abscisic acid	–	–	None	+
Gibberellins	+	+	Longitudinal	+

Key: + positive effect; – negative effect

proteins that are produced determine the identity of the cell. It is proteins, specifically enzymes, that catalyze most of the cell's chemical reactions, and it is also proteins that form or produce most of the structural elements within and around the cell. Thus, for example, a cortical cell in a root and a mesophyll cell in a leaf differ structurally and functionally because of differences in gene expression during the course of their development.

The molecular mechanisms by which individual genes are switched on and off can be very complex. Some general principles have been established from studies in both plants and

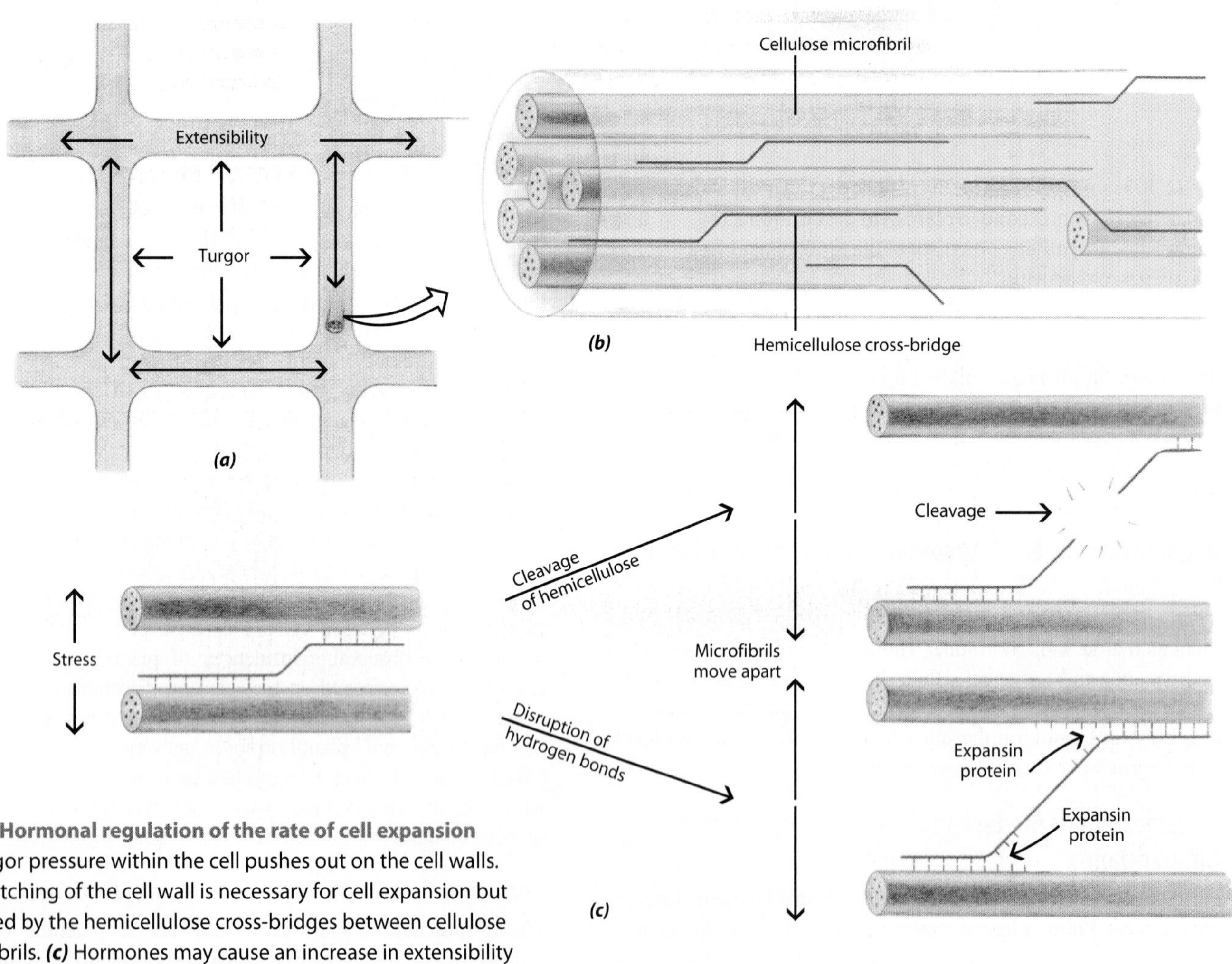

27–24 Hormonal regulation of the rate of cell expansion ***(a)*** Turgor pressure within the cell pushes out on the cell walls. ***(b)*** Stretching of the cell wall is necessary for cell expansion but is limited by the hemicellulose cross-bridges between cellulose microfibrils. ***(c)*** Hormones may cause an increase in extensibility by stimulating a reversible cleavage of the hemicellulose cross-bridges or by disrupting hydrogen bonds between the cellulose microfibrils and the hemicellulose cross-bridges. Disruption of the hydrogen bonds is mediated by a cell wall protein called expansin. These modifications allow cellulose microfibrils to move apart, resulting in an irreversible expansion of the wall.

animals. A eukaryotic gene is composed of a *coding sequence,* which specifies the amino acid sequence of the gene's protein product, as well as *regulatory sequences,* which are regions of DNA bordering the coding sequence that play a regulatory role in gene transcription. Proteins called *regulatory transcription factors* can bind directly to specific DNA sequences within a regulatory sequence, activating (switching on) or repressing (switching off) that particular gene.

Plant molecular biologists are currently studying a number of plant genes that are either activated or repressed by such factors as light, environmental stress (see Chapter 28), and hormones. A classic example of how hormones control gene expression is provided by studies on the antagonistic effects of gibberellin and abscisic acid on the synthesis of the starch-degrading enzyme α-amylase in the aleurone layer of barley seeds (Figure 27–18). If aleurone tissues are incubated in a medium containing various hormones and radioactive amino acids, the amino acids are incorporated into any new proteins synthesized during the treatment. These proteins can then be visualized and identified by separating them by size using gel electrophoresis (see Figure 10–7) and detecting radioactive proteins with X-ray film. Treatment of aleurone tissues with gibberellin causes an increased abundance of the α-amylase protein. This effect is counteracted by simultaneous application of abscisic acid, which represses α-amylase gene expression while gibberellin activates the gene. More recently, studies using microarray technology (see Chapter 10) indicate that hormone-treated plant tissues typically show alterations in expression of hundreds of genes. These results explain how hormones can have such fundamental effects on so many aspects of growth and development.

Hormones Can Regulate Cell Expansion and Cell Division

The specific forms that plant organs assume as they emerge from meristems involve a coordination of cell division and cell expansion of each cell in the developing organ. Not surprisingly, plant hormones exert influences on these processes throughout development. Hormones seem to regulate the timing of cell division through interaction with the cellular machinery that establishes checkpoints in the cell cycle (pages 62 to 64). Hormones can also affect growth by influencing the rate of cell expansion. The rate at which an individual cell expands is controlled by (1) the amount of turgor pressure inside the cell, pushing against the cell wall (page 80), and (2) the extensibility of the cell wall (Figure 27–24; see also pages 57 and 58). Extensibility, a physical property of the wall, is a measure of how much the wall will stretch permanently when a force is applied to it. As indicated in Table 27–2, the five classical groups of hormones are capable of influencing the rate of cell expansion. In most cases examined, hormones affect the extensibility of the cell wall but have little direct influence on the turgor pressure. Auxin and gibberellin stimulate plant growth by increasing the extensibility of cell walls, whereas abscisic acid and ethylene inhibit plant growth by causing a decrease in extensibility.

The mechanisms by which hormones alter the extensibility of cell walls are not well understood. Two hypotheses are currently in favor. In the *acid growth hypothesis,* hormones—particularly auxin—activate a proton-pumping enzyme in the plasma membrane that pumps protons from the cytosol into the cell wall. The resulting drop in pH is thought to cause a loosening of the cell wall structure. This occurs either through the breakage and re-formation of noncellulosic polysaccharides, which normally cross-link the cellulose microfibrils, or through the action of a class of proteins called **expansins,** which disrupt hydrogen bonds between polysaccharides in the wall (Figure 27–24). An alternative hypothesis is based on discoveries that auxin activates the expression of specific genes within a few minutes of its application. The products of these genes are thought to influence the delivery of new wall materials in such a way as to affect cell wall extensibility. These two hypotheses are not mutually exclusive, and both may be required to explain the influence of hormones on cell expansion.

In addition to affecting the *rate* of cell expansion, plant hormones can also influence the *direction* of expansion. Once a cell has divided, the shape assumed by the daughter cells as they enlarge will determine the ultimate form of the developing tissue or organ. For example, many of the cells in a developing leaf tend to expand primarily in a lateral direction. This lateral expansion, in addition to the pattern of cell division, results in formation of a platelike organ. In contrast, the cells in growing stem tissues tend to expand longitudinally, resulting in the "unidirectional" growth characteristic of an elongating stem. These differences in the direction of cell expansion are apparently determined by the orientation of the cellulose microfibrils as they are deposited in the developing cell wall. If cellulose microfibrils are deposited in a random orientation, the cells tend to expand in all directions. If the fibrils are laid down in a primarily transverse orientation, the cells tend to expand longitudinally (just as a coiled spring is much easier to stretch in the direction perpendicular to the orientation of the coils).

The orientation of cellulose microfibrils appears to be governed by the orientation of microtubules lying just inside the plasma membrane (Figure 27–25; see also pages 60 and 61), and this arrangement of microtubules is influenced by hormones. Gibberellins, for example, promote a transverse arrangement of microtubules, resulting in greater longitudinal growth,

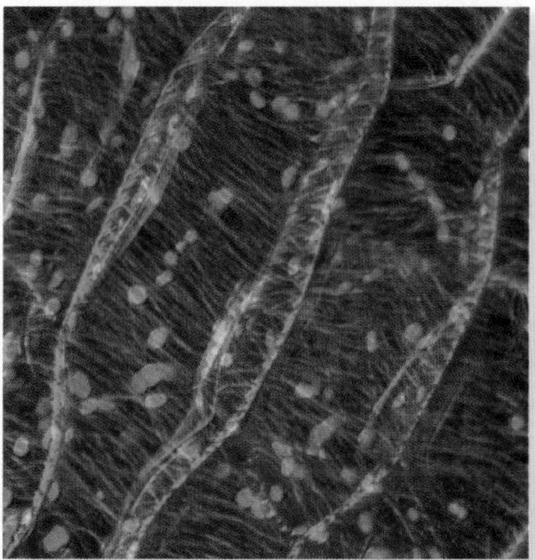

27–25 Cortical microtubules These transverse arrays of cortical microtubules in living hypocotyl cells of *Arabidopsis* have been made visible by the use of green fluorescent protein (GFP); the numerous red-colored structures are chloroplasts. The transverse arrangement of microtubules is promoted by gibberellins.

or elongation. On the other hand, in stems, ethylene treatment causes some degree of reorientation of microtubules to the longitudinal direction, which promotes a more lateral (radial) expansion of the cells. This response to ethylene results in a stem that is shorter and thicker.

Hormones Alter Cell Growth and Gene Expression through Signal-Transduction Pathways

As discussed in Chapter 4 (pages 87 and 88), the signal-transduction pathways begin with protein receptors that bind to specific hormones. Binding of hormones to receptors initiates a series of biochemical events (signal transduction) that transmits information from hormone-occupied receptors to the cellular machinery that carries out specific responses.

Signal-transduction pathways can be very difficult to study, because the protein components are present at very low concentrations within the cell. Important breakthroughs in our understanding of these processes have come from genetic studies of the model organism *Arabidopsis* (page 199). Mutants of *Arabidopsis* that fail to respond to hormone application may be defective in a signal-transduction component. Cloning of the genes associated with such mutations has led to the discovery of hormone receptors and signaling components.

Using the triple response as an assay (Figure 27–12), studies showed that mutations that either render *Arabidopsis* insensitive to ethylene or cause the responses to occur even in the absence of ethylene allowed the identification of both positive and negative regulators of this signaling pathway (Figure 27–26). The receptors identified in these studies are integral membrane proteins of the endoplasmic reticulum that are evolutionarily related to receptors in bacteria. Plants have multiple ethylene receptors that directly interact with the CTR1 (CONSTITUTIVE TRIPLE RESPONSE1) protein, which is a negative regulator (Figure 27–27). *Arabidopsis* has five ethylene receptors, but ETR1 (ETHYLENE-RESPONSE1) may play the predominant role. In the absence of ethylene, the ETR1 receptor activates the CTR1 protein, which in turn represses ethylene signaling. When ethylene is present, it interacts with copper at the binding site of the receptor and changes the conformation (that is, the structure) of the receptor. The ethylene-bound receptor, with its altered conformation, no longer interacts with CTR1. This causes CTR1 to become inactive, allowing activation of the transmembrane protein EIN2 (ETHYLENE-INSENSITIVE2). The EIN2 protein then transmits the ethylene signal into the nucleus. There the signal turns on EIN3 transcription factors, which induce the expression of ERF1 (ETHYLENE RESPONSE FACTOR1) transcription factors. (EIN3 and ERF1 are positive regulators in the ethylene signaling pathway.) Activation of this transcriptional cascade leads to extensive changes in gene expression. Our understanding of the ethylene signaling pathway is far from complete. For example, it remains to be determined how the CTR1 protein affects the membrane-associated EIN2 protein and how EIN2 transmits the ethylene signal into the nucleus, although it appears that Ca^{2+} signaling may be involved. With further study, additional components will most likely be discovered.

Similarities exist in the hormone-receptor interactions and signaling pathways for auxin and gibberellin. For both hormones, the receptors function in the nucleus, and binding to the receptors directly affects the activity of repressor proteins. Degradation of the repressor proteins leads to hormone-regulated gene expression. Despite these similarities, the binding of auxin and gibberellin to their respective receptors differs. Whereas auxin acts as "molecular glue" for the interaction with its receptor, the interaction of gibberellin with its receptor results in a conformational change in the receptor, which then folds over the hormone like a lid, completely burying it inside the receptor.

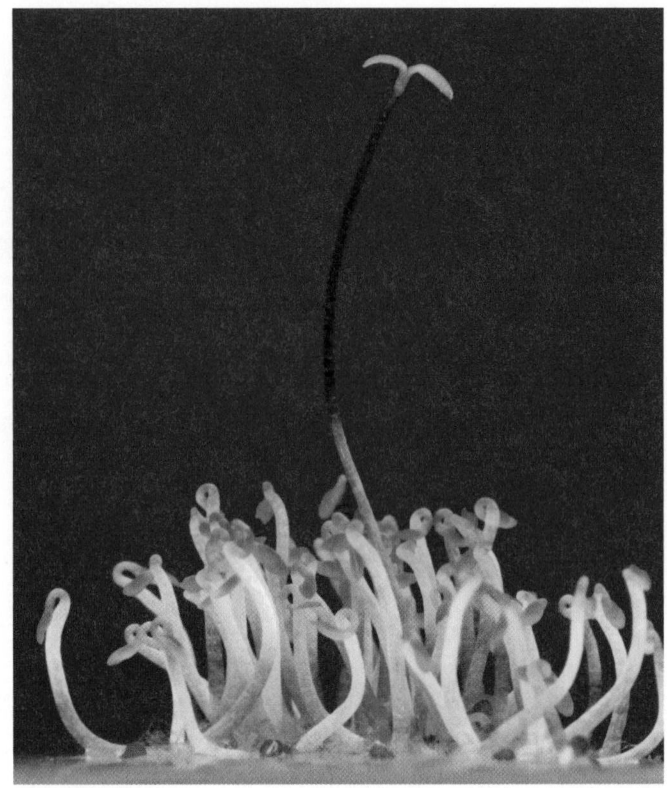

27–26 The *etr* mutant of *Arabidopsis* These *Arabidopsis* seedlings were grown in the dark in the presence of ethylene. All but one of the seedlings—the *etr* mutant, which is completely insensitive to ethylene—show the triple response.

Three cytokinin receptors have been identified in *Arabidopsis.* They are transmembrane protein kinases (enzymes that transfer phosphate from ATP to other molecules) associated with the plasma membrane. On cytokinin binding, the receptors autophosphorylate (that is, phosphorylate themselves), and the phosphate is subsequently transferred to a series of target proteins. The phosphate ultimately ends up on a transcription factor in the nucleus, where target genes are activated.

The brassinosteroid receptor BRI1 (BR INSENSITIVE1) differs from those already considered in that it contains an extracellular region involved in signal recognition, as well as a single transmembrane region and a cytoplasmic region that initiates intracellular signal transduction (Figure 27–28). BRI1 is associated with a coreceptor kinase named BAK1 (BRI1-associated kinase). Binding of the brassinosteroid to BRI1 activates the receptor and induces association with its coreceptor, BAK1. By sequential transphosphorylation between BRI1 and BAK1, BRI1 becomes fully activated and initiates a signaling cascade. The cascade involves phosphorylation and dephosphorylation events that lead to brassinosteroid-regulated gene transcription.

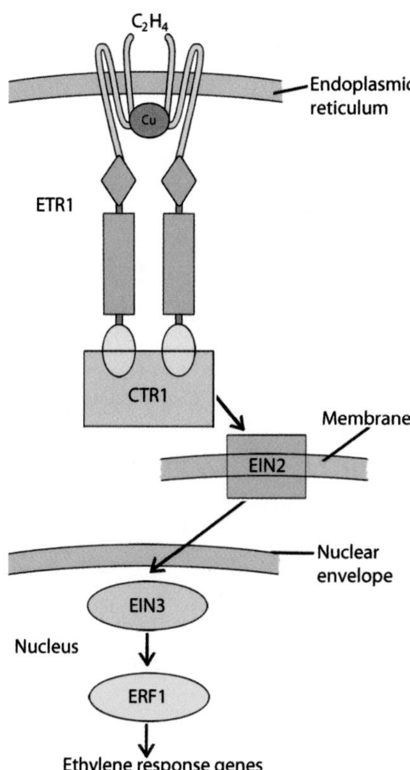

27–27 Simplified diagram of the ethylene signaling network Binding of ethylene (C_2H_4) to the ethylene receptor ETR1, which is located on the endoplasmic reticulum, results in inactivation of the receptor and of the negative regulator CTR1. Inactivation of CTR1 allows activation of the transmembrane protein EIN2, which then transmits the ethylene signal into the nucleus. There the signal turns on EIN3 transcription factors, which induce expression of ERF1 transcription factors. Activation of the transcription factors results in the production of new classes of mRNA, and translation of the new mRNAs results in novel proteins that mediate hormonal responses.

Second Messengers Mediate Hormonal Responses

The calcium ion is of particular interest in hormone action. Generally, Ca^{2+} levels in the cytoplasm are very low. Hormonal stimulation of calcium ion channels results in a temporary elevation of Ca^{2+} levels. Binding of calcium ions to the calcium-binding sites of certain proteins alters the activity of these proteins, much as hormones activate receptor proteins. Protein kinases may be activated by Ca^{2+} or other "second messengers." The activated protein kinases may then modify "target" proteins, transferring phosphate groups onto certain amino acids of the target protein and thus altering its activity.

Substances, such as calcium ions, that mediate hormonal responses are often referred to as **second messengers.** Second messengers perform two important functions: (1) they are involved in the transfer of information from the hormone-receptor complex to target proteins, and (2) they amplify the signal produced by the hormone. Receptor activation of a single calcium ion channel may result in the release of hundreds of calcium ions into the cytosol. Each calcium ion can, in turn, activate a protein kinase molecule, and each protein kinase molecule can phosphorylate many molecules of target protein. Complex response pathways involving second messengers also contribute to the diversity of possible responses to a given hormone. Different cell types may have the same plasma membrane receptor but may respond quite differently to the same hormone if they carry a different complement of protein kinases and target proteins. (For a more detailed discussion of signal-transduction pathways and the role of second messengers, see pages 87 and 88.)

Stomatal Movement Involves a Specific Hormone Response Pathway

In addition to regulating gene expression, hormones can have a direct effect on cellular physiology. An example is provided by abscisic acid. Stomata are small openings in the epidermis, each surrounded by two guard cells that change their shape to bring about the opening and closing of the pores. The term *stoma* (Greek for "mouth") is conventionally used to designate both the pore and the two guard cells. The degree of stomatal opening determines to a great extent the rate of gas exchange across the epidermis. A number of endogenous and environmental signals influence stomatal pore size (see Chapter 28). All of these signals work by regulating the water content, or turgor pressure, of the guard cells (Figure 27–29). Abscisic acid is one endogenous signal that is particularly important in the control of stomatal movement.

Several possible classes of abscisic acid receptors have been proposed. Although most proposals are controversial, the details of the cellular events that occur within minutes of the addition of abscisic acid to isolated guard cell protoplasts indicate that rapid changes in the osmotic potential of guard cells are mediated by

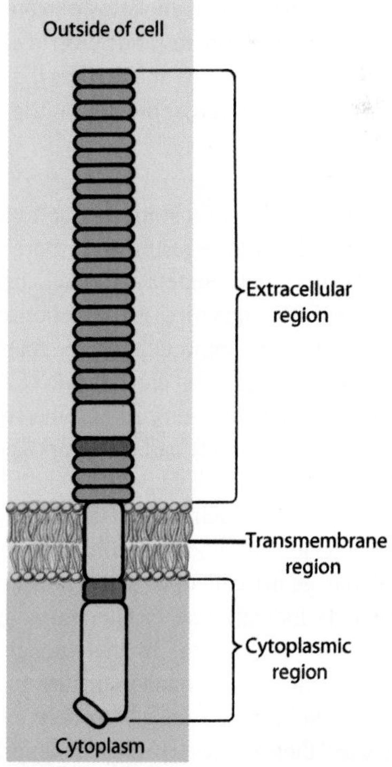

27–28 Brassinosteroid receptor BRI1 The BRI1 receptor is localized on the plasma membrane and consists of an extracellular region that is involved in signal recognition, a single transmembrane region, and a cytoplasmic region that initiates signal transduction within the cell.

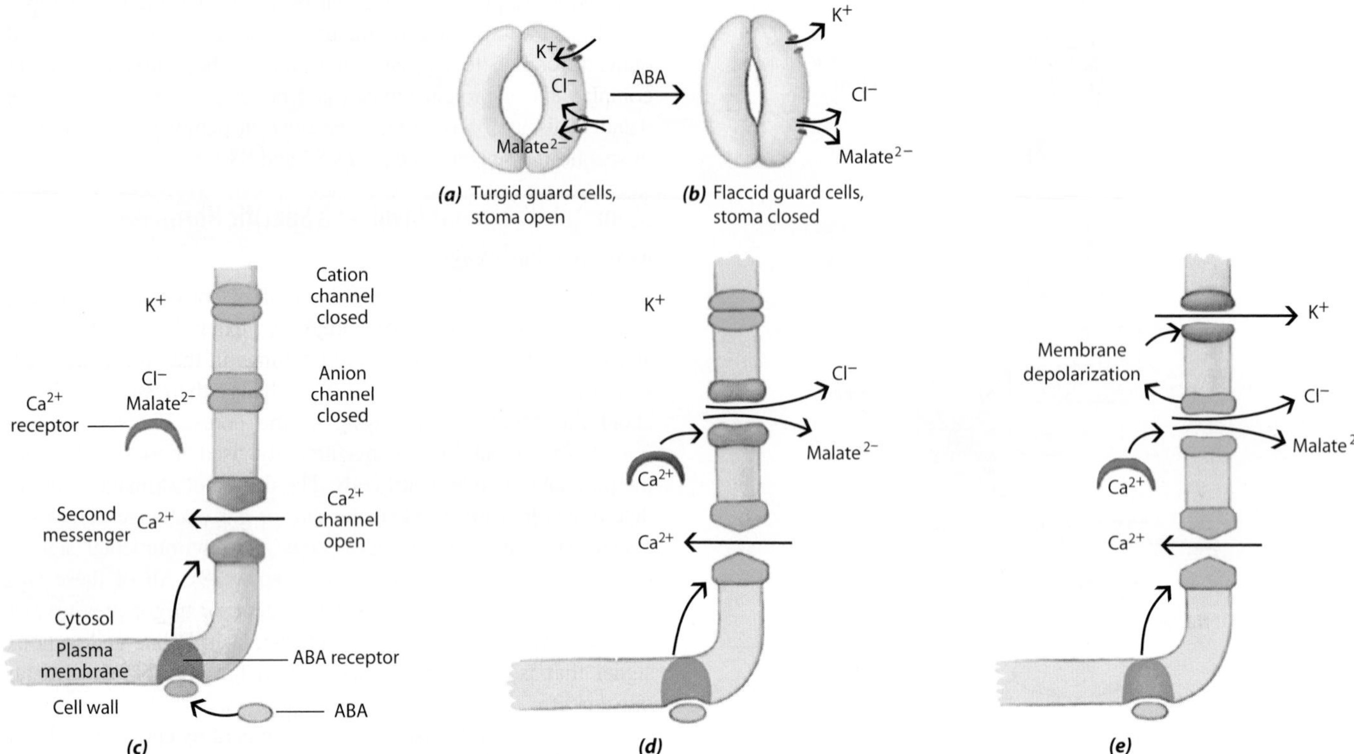

27–29 Regulation of stomatal movement ***(a)*** Turgor pressure caused by a high solute concentration in guard cells bordering the stomatal pore keeps the pore open. ***(b)*** Release of solutes in response to abscisic acid (ABA) reduces turgor pressure in the guard cells, resulting in closure of the stomatal pore. The sequence of events that leads to ABA-mediated closure of stomata involves ion channels through the guard cell plasma membrane. In this model, ***(c)*** binding of ABA to its receptor in the plasma membrane causes Ca^{2+} channels to open. ***(d)*** The Ca^{2+} released into the cytosol acts as a second messenger to open anion channels through which Cl^- and malate^{2-} ions flow from the cytosol into the cell wall. ***(e)*** The resulting drop in electrical potential (membrane depolarization) across the plasma membrane opens the K^+ channels and permits the release of K^+ into the cell wall. The flow of solutes from cytosol to cell wall results in decreased turgor pressure in the guard cells, causing closure of the stoma.

the hormone. The opening of a stoma is due to the uptake of solutes by its guard cells, which results in a more negative osmotic potential of the guard cell contents. (Recall that osmotic potential, which is a negative quantity, is a function of solute concentration; see page 79.) The important solutes that contribute to the osmotic potential of guard cells are chloride (Cl^-) and potassium (K^+) ions, which are actively pumped into the cells, and malate^{2-}, a negatively charged (anionic) carbon compound synthesized by the guard cells.

Studies using the patch-clamp technique (see the essay on page 84) indicate that anion-specific channels in the plasma membrane of the guard cell open in response to abscisic acid. Some experiments indicate that calcium ions may act as a second messenger in this system. In this model (Figure 27–29), abscisic acid activates Ca^{2+} channels in the plasma membrane, which results in an influx of Ca^{2+} from the cell wall to the cytosol. The Ca^{2+} then causes opening of anion channels in the plasma membrane by activating protein kinases. The opening of these channels results in the rapid movement of anions, primarily Cl^- and malate^{2-}, from the cytosol to the cell wall. The subsequent depolarization of the plasma membrane—that is, the loss in electrical charge difference across the membrane—triggers the opening of K^+ channels. The result is a movement of K^+ from the cytosol to the cell wall. This rapid movement of Cl^-, malate^{2-}, and K^+ results in a less negative osmotic potential (higher water potential) of the guard cell cytosol and a more negative osmotic potential (lower water potential) of the wall. Water then moves down its water potential gradient from the cytosol to the cell wall, reducing the turgor of the guard cells and causing closure of the stomatal pore. When the abscisic acid signal is removed, the guard cells slowly transport the K^+ and Cl^- back into the cell, using an electrochemical proton gradient generated by a proton pump (H^+-ATPase) in the plasma membrane. A more negative osmotic potential is reestablished within the guard cells, water flows into the cells by osmosis, and the resulting increase in turgor causes the stomatal pore to reopen.

SUMMARY

Plant Hormones Play a Major Role in Regulating Growth

Plant hormones are chemical signals that regulate and coordinate metabolism, growth, and morphogenesis and are active in extremely small amounts. The same hormone can elicit different responses in different tissues or at different times in development of the same tissue. Six groups of phytohormones (plant hormones), the "classic five"—auxins, cytokinins, ethylene, abscisic acid, gibberellins—and the brassinosteroids, are considered major classes of phytohormones. It has become increasingly clear, however, that plants also use additional chemical signals and growth regulators.

Auxins Are the Only Phytohormones Known to Be Transported in a Polar Manner

Shoot apical meristems and young leaves are the primary sites of auxin synthesis, but root apical meristems are also important sites of auxin synthesis. In shoots, the polar transport is always basipetal—downward in the stem and toward the base of the root. Auxin controls the lengthening of the shoot, chiefly by promoting cell elongation. Polar transport of auxin is mediated by efflux carriers (PIN proteins) precisely aligned with the direction of transport. Auxin also plays a role in differentiation of vascular tissues and initiates cell division in the vascular cambium. It often inhibits growth in lateral buds, thus maintaining apical dominance. The same quantity of auxin that promotes growth in the stem inhibits growth in the main root system. Auxin promotes the formation of adventitious roots in cuttings and retards abscission in leaves, flowers, and fruits. In fruits, auxin produced by seeds stimulates growth of the ovary wall.

Cytokinins Are Involved in Cytokinesis

The cytokinins are chemically related to certain components of nucleic acids. They are most abundant in actively dividing tissues such as seeds, fruits, leaves, and root tips. Cytokinins act in concert with auxin to cause cell division in plant tissue cultures. In tobacco pith cultures, a high concentration of auxin promotes root formation, whereas a high concentration of cytokinins promotes bud formation. In intact plants, cytokinins promote the growth of lateral buds, opposing the effects of auxin. Cytokinins prevent senescence in leaves by stimulating protein synthesis.

Ethylene Is a Gaseous Plant Hormone

Ethylene, a simple hydrocarbon, exerts a major influence on many, if not all, aspects of plant growth and development, including growth of most tissues, fruit and leaf abscission, and fruit ripening. In most plant species, ethylene has an inhibitory effect on cell expansion. In some semiaquatic species, however, it elicits rapid stem growth. Ethylene apparently plays a major role in determining the sex of flowers in some monoecious plants.

Abscisic Acid Induces Dormancy and Stomatal Closure

Abscisic acid is a growth-inhibiting hormone found in dormant buds and in fruits, stimulating the production of seed proteins and the closing of stomata. Studies using the patch-clamp technique indicate that calcium ion channels in the guard cell plasma membrane open in response to abscisic acid, triggering a sequence of events that lowers the turgor of the guard cells and causes stomatal closure.

Gibberellins Stimulate the Growth of Stems and the Germination of Seeds

The gibberellins control shoot elongation, as is especially evident in dwarf plants. When such plants are treated with gibberellin, normal growth is restored. In many species, gibberellins can substitute for the dormancy-breaking cold or light required for their seeds to germinate. In barley and other grass seeds, the embryo releases gibberellins that cause the aleurone layer of the endosperm to produce α-amylase. This enzyme breaks down the starch stored in the endosperm, releasing sugar that nourishes the embryo and promotes germination. Whereas gibberellins activate α-amylase gene expression, abscisic acid represses the gene.

Brassinosteroids Play a Role in a Wide Range of Developmental Processes

Brassinosteroids play essential roles in cell division and cell elongation, vascular tissue differentiation, flower and fruit development, resistance to stress, and senescence. The requirement for brassinosteroids in normal plant growth is exemplified by the dwarf phenotype of *Arabidopsis.*

Hormones Alter Cell Growth and Gene Expression through Signal-Transduction Pathways

At the molecular level, hormones influence developmental processes through their interaction with receptors in the plant cell. The plant hormones mediate many processes by activating or repressing sets of genes in the cell's nucleus. At the cellular level, hormones influence the rate and direction of cell expansion and the rate of cell division. These effects are mediated through complex biochemical response pathways in the cell, often involving second messengers.

QUESTIONS

1. Home gardeners commonly pinch off the shoot tips of certain plants in order to promote fuller, bushier growth. Explain why removal of the shoot tip should promote such growth.
2. What is meant by parthenocarpic fruit? Which two plant hormones are used to produce such fruit?
3. Compare and/or contrast auxins and cytokinins in each of the following: principal sites of biosynthesis; polarity of transport; cell types or tissues involved in transport; effect on cell division; effect on the production of roots and shoots in tissue cultures.
4. In what ways is ethylene a unique plant hormone?
5. In what way are gibberellins of value to the brewing industry?
6. Explain how abscisic acid regulates the opening and closing of stomata.
7. Explain how the orientation of cellulose microfibrils determines the direction of cell expansion and the role of microtubules in this process.
8. How does the brassinosteroid receptor differ from those of the other major plant hormones?

CHAPTER 28

External Factors and Plant Growth

Vigorous and weedy Resembling the ornamental morning glory, field bindweed *(Convolvulus arvensis)* is instead a rampant, invasive intruder with slender twining or trailing stems that frequently form thick mats. The plant grows from seeds, rhizomes, roots, and even root fragments in the soil, climbing aggressively, often using other plants for scaffolding as it spreads out and searches for light.

CHAPTER OUTLINE

Living things must regulate their activities in accordance with the world around them. Many animals, being mobile, can change their circumstances to some extent—foraging for food, courting a mate, and seeking or even making shelters in bad weather. A plant, by contrast, is immobilized once it sends down its first root. Yet plants do have the ability to respond to and make adjustments to a wide range of changes in their external environment. This ability is expressed primarily in changing patterns of growth. The large variation in form that is often observed among genetically identical individuals is due to the impact of the environment on the plant's development.

The Tropisms

A growth response involving bending, or curving, of a plant part toward or away from an external stimulus that determines the direction of movement is called a **tropism.** A response toward the stimulus is said to be *positive;* a response away from the stimulus is *negative.*

Phototropism Is Growth in Response to Directional Light

Perhaps the most familiar interaction between plants and the external world is the curving of growing shoot tips toward light (see Figure 27–1). This growth response, **phototropism,** is due to the hormone auxin and results in elongation of the cells on the shaded side of the tip. What role does the light play in the phototropic response? Three possibilities are: (1) light decreases the auxin sensitivity of the cells on the lighted side, (2) light destroys auxin, or (3) light drives auxin to the shaded side of the growing tip.

To test these hypotheses, Winslow Briggs and his co-workers carried out a series of experiments based on earlier work by Frits Went (Figure 28–1). These investigators

CHECKPOINTS

After reading this chapter, you should be able to answer the following:

1. What is a tropism? By what mechanisms do plants respond to light? To gravity? To a moisture gradient?
2. Why is it important that plants be able to "tell time"? What are some characteristics of the circadian clock in plants?
3. Describe the effect of daylength on flowering.
4. What is phytochrome, and how is it involved in flowering, seed germination, and stem growth?
5. What is dormancy, and what environmental cues may be necessary to break the dormancy of seeds and buds?

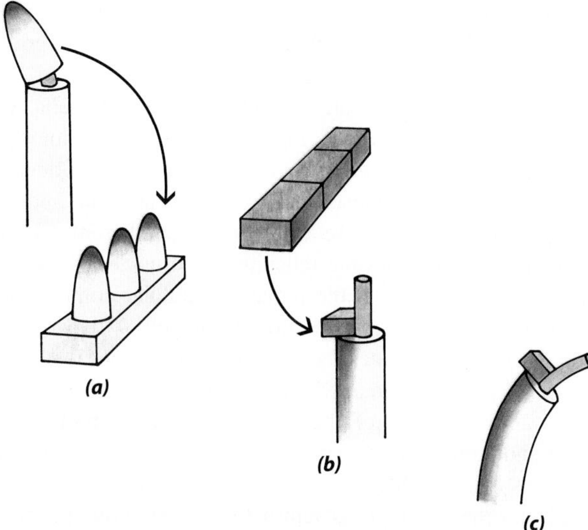

28–1 **Went's experiment** ***(a)*** Went cut the coleoptile tips from oat *(Avena sativa)* seedlings and placed the tips on a slice of agar for about an hour. ***(b)*** He then cut the agar into small blocks and placed a block on one side of each decapitated shoot. ***(c)*** The seedlings, which were kept in the dark during the entire experiment, subsequently were observed to bend away from the side on which the agar block was placed. From this result, Went concluded that the "influence" that caused the seedling to bend was chemical and that it accumulated on the side away from the light. In Went's experiment, auxin molecules (blue) produced in the coleoptile tip were first transferred to the agar and then to one side of the seedling by means of an agar block.

first showed that the same total amount of auxin is obtained from the tips of coleoptiles whether they remain in the dark or in the light. However, following exposure to light on one side, more auxin is obtained from the shaded side than from the lighted side. If the tip is split and a barrier, such as a thin piece of glass, is placed between the two halves (the light and dark sides), this differential distribution of auxin is no longer observed (Figure 28–2). In other words, Briggs demonstrated that auxin migrates laterally from the light side to the dark side. Experiments using ^{14}C-labeled auxin (indole-3-acetic acid, or IAA) have shown that it is migration of auxin, not destruction of auxin, that accounts for the different amounts obtained from the light and dark sides of the coleoptile. These experiments are consistent with the hypothesis that auxin redistribution is responsible for phototropic curvature.

Lateral movement of the auxin toward the shaded side of the coleoptile occurs at the tip. The auxin then moves basipetally from the tip to the elongation zone below, where it stimulates cell elongation. With the acceleration of growth on the shaded side and the slowing of growth on the illuminated side, the differential growth results in curvature toward the light.

The redistribution of auxin in response to light is mediated by a photoreceptor, a pigment-containing protein that absorbs light and converts the signal into a biochemical response (for a discussion of receptors and signal processing, see pages 656 and 657).

The phototropic response is triggered by blue wavelengths of light (400 to 500 nanometers). Briggs and his co-workers used mutants of *Arabidopsis* that are insensitive to blue light to clone the genes that code for photoreceptors. The photoreceptor proteins are related to other well-studied proteins that bind pigments called flavins. Flavins absorb light primarily in the blue wavelengths of the spectrum, providing an explanation for why blue light is most effective in phototropic responses. Two flavoproteins, **phototropins 1** and **2,** have been identified

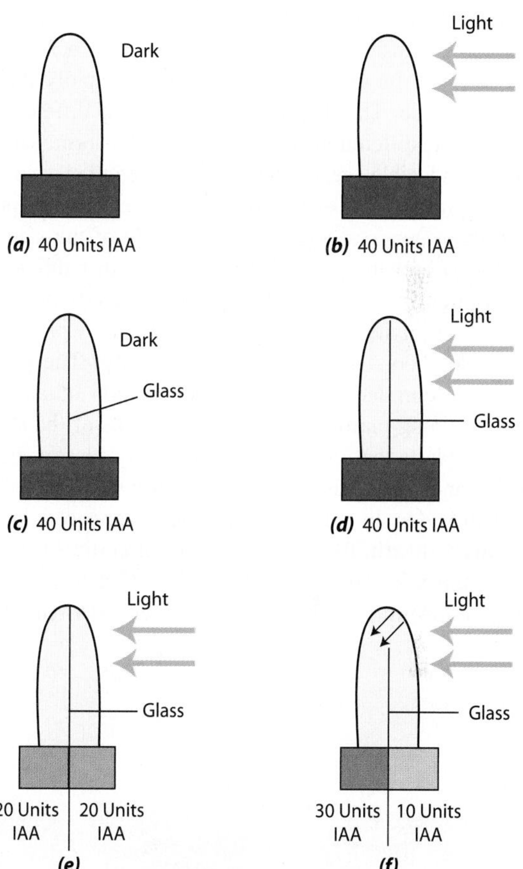

28–2 **Briggs's experiments** Experiments by Briggs and co-workers showing lateral displacement of IAA (auxin) in unilaterally illuminated maize *(Zea mays)* coleoptile tips. The diffusible IAA was collected in agar blocks. The units indicate the relative amounts of IAA present in each block. ***(a), (b)*** The same amount of IAA is collected from intact coleoptiles whether they remain in the dark or are illuminated. ***(c), (d)*** If the coleoptiles are split in half by a thin piece of glass but remain intact, the same amount of IAA is collected as in *(a)* and *(b)*. ***(e)*** If the two halves and the agar block below are completely separated, the same amounts of IAA are collected from the dark and illuminated sides. ***(f)*** When the coleoptile is not split entirely to the top, more IAA diffuses from the darkened half than from the illuminated half. These results support the idea that unilateral illumination induces a lateral movement of IAA from the illuminated to the darkened half of the coleoptile.

as the photoreceptors for the blue-light signaling pathway in *Arabidopsis* hypocotyls and oat coleoptiles.

Gravitropism Is Growth in Response to Gravity

Another familiar tropism is **gravitropism,** a response to gravity (Figure 28–3). If a seedling is placed on its side, its root will grow downward (positive gravitropism) and its shoot will grow upward (negative gravitropism). The original explanation for this mechanism involved the redistribution of auxin from the upper to the lower side of the shoot or root. In shoots, the higher auxin concentration on the lower side stimulates more rapid cell expansion on that side of the stem, resulting in an upward curving of the stem. For the root, however, which is more sensitive to auxin, the increased auxin concentration on the lower side actually inhibits cell expansion, resulting in a downward bending of the root as cells on the upper side expand more rapidly than those on the lower side. This is a classic example of different tissues exhibiting quite different responses to the same hormonal signal.

Auxin-inducible genes have been identified and cloned from a number of species. Auxin-activated transcription of some of these genes occurs only on the side of the stem showing increased growth. It is still not clear whether this activation of transcription is due to an increase in auxin concentration or an increase in sensitivity of the tissue to auxin already present.

How do shoots and roots perceive gravity? The perception of gravity is correlated with the sedimentation of amyloplasts (starch-containing plastids) within specific cells of the shoot and root. Amyloplasts that play the role of gravity sensors are called **statoliths,** and the gravity-sensing cells in which they occur are termed **statocytes.** In coleoptiles and stems, gravity is perceived in the **starch sheath,** the innermost layer of cortical cells, which surrounds the vascular tissues. The starch sheath is continuous with the endodermis of the root, but the endodermal cells lack amyloplasts. *Arabidopsis* mutants lacking amyloplasts in the starch sheath display shoot growth that does not respond to gravity, but their roots exhibit normal gravitropic growth.

In roots, the statocytes are localized in the columella, or central column, of the rootcap (Figure 28–4a). The columella cells are highly polarized (Figure 28–4b). In vertically oriented columella cells, the amyloplasts are sedimented near the bottom of each cell, near the transverse wall but separated from it by a network of tubular endoplasmic reticulum. The nucleus is at the opposite end of the cell, and the remaining cellular components are largely found in the central portion of the cell. A network of fine actin filaments permeates the cytosol. When a root is placed in a horizontal position, the amyloplasts, which were sedimented near the transverse walls of the vertically growing root, slide downward and come to rest near what were previously vertically oriented walls (Figure 28–5). After several hours, the root curves downward and the amyloplasts return to their previous position along the transverse walls.

Evidence in support of the **starch-statolith hypothesis**—that the perception of gravity is mediated by the sedimentation of amyloplasts within statocytes—comes in part from studies in which the removal (by laser ablation) of columella cells in the *Arabidopsis* rootcap had an inhibitory effect on root curvature in response to gravity. Other support comes from studies on starchless or starch-deficient *Arabidopsis* mutants. The starch-deficient mutants were less sensitive to gravity than the wild-type plants, and the starchless mutants much less sensitive. How the movement of the starch-containing amyloplasts is translated into differential growth rates has yet to be explained.

Although the downward curvature of the root in response to gravity (gravistimulation) was originally thought to be driven by suppression of elongation along the lower side of the root, we now know that the curvature is initiated in a region distal (toward

(a)

(b)

28–3 Gravitropism Gravitropic responses in the shoot of a young tomato *(Solanum lycopersicum)* plant. ***(a)*** The plant was placed on its side and kept in a stationary position. ***(b)*** The plant was placed upside down and held in position in a ring stand. Although originally straight, the stem bent and grew upward in both cases. If a plant is held horizontally and slowly rotated around its horizontal axis, no curvature (gravitropic response) will occur; that is, the plant will continue to grow horizontally.

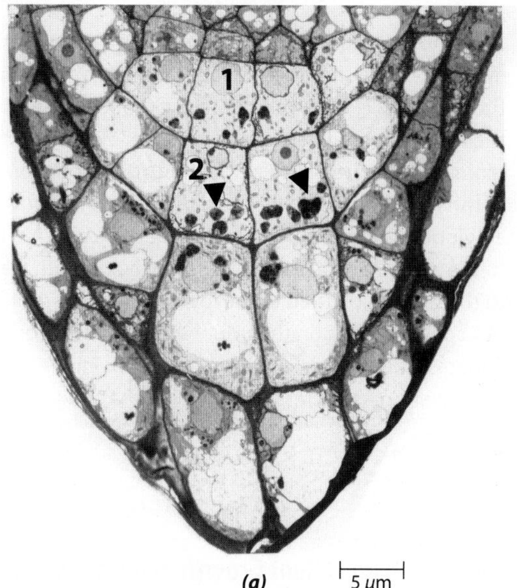

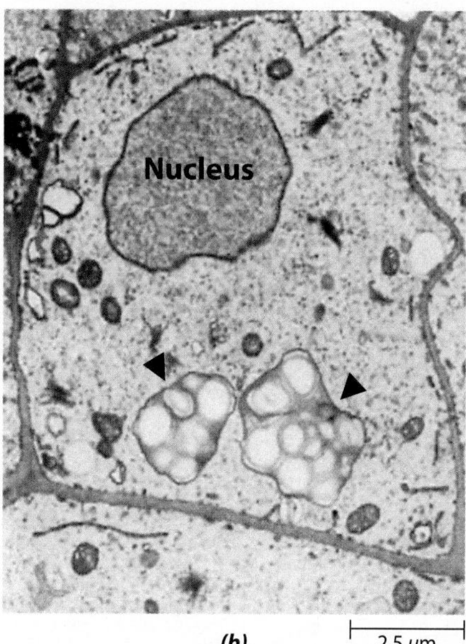

28–4 Statocytes and statoliths Electron micrographs of ***(a)*** a median longitudinal section of the rootcap of the primary root of *Arabidopsis,* showing statocytes, or gravity-sensing cells (1, 2), in the columella (central column of cells of the rootcap), and ***(b)*** an enlarged view of a statocyte. The gravity "pull" is toward the bottom in both micrographs. In *(a)* and *(b)* the statoliths, which are amyloplasts that act as gravity sensors (arrowheads), are sedimented along the transverse walls.

the root tip) to the main zone of elongation. In this so-called *distal elongation zone,* elongation is *stimulated* along its upper side, while elongation is *suppressed* along the lower side of both the distal and main elongation zones. Thus, the development of curvature is initiated very near the tip of the root.

Polar auxin transport through cell files is mediated by a combination of auxin influx and auxin efflux carriers—the AUX1 proteins and PIN proteins, respectively. In *Arabidopsis* roots, PIN1 in the distal plasma membrane mediates the acropetal transport of auxin through the vascular cylinder toward the root tip. In vertically oriented *Arabidopsis* roots, PIN3 protein typically is distributed symmetrically in the plasma membrane of the statocytes. When the root is placed on its side and the amyloplasts settle to the lower side, PIN3 quickly moves to the plasma membrane on the lower side of the columella cells, driving the asymmetrical redistribution of auxin toward the lower lateral rootcap cells. The auxin is then transported basipetally on the lower side of the root from the tip to the distal elongation zone, the first region of the root to respond to gravity (Figure 28–6). Several other growth regulators in addition to auxin have been implicated in root gravitropic responses, including abscisic acid, brassinosteroids, ethylene, nitric oxide, and cytokinins (Figure 28–7).

Although the starch-statolith hypothesis is widely accepted, it has not gone unchallenged. Most notable among other models is the hydrostatic pressure model, which proposes that the

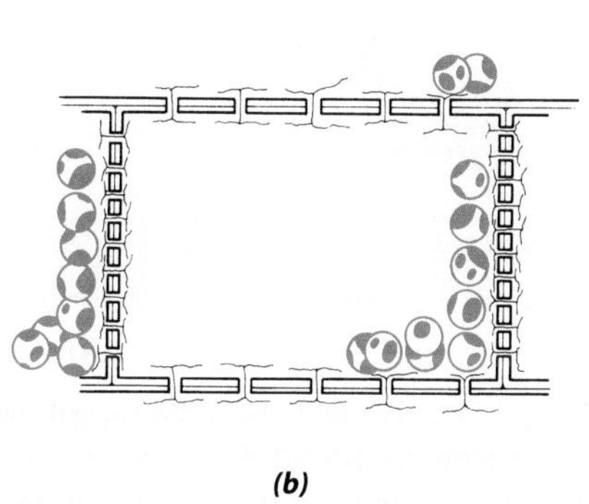

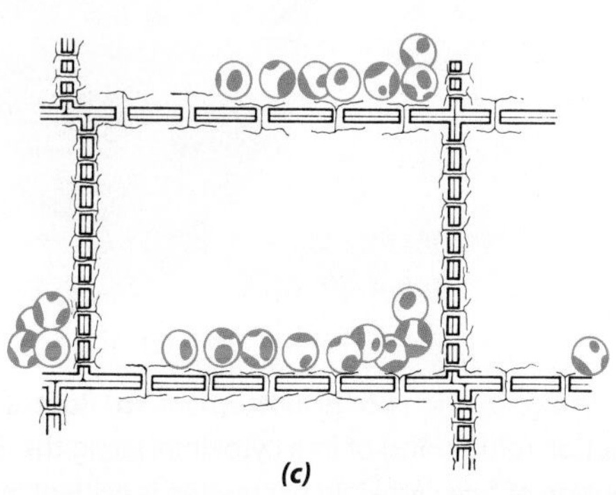

28–5 Amyloplasts and gravity The response to gravity of amyloplasts (statoliths) in the columella cells (statocytes) of a rootcap. ***(a)*** The amyloplasts are normally sedimented near the transverse walls in the rootcap of a root growing vertically downward. ***(b), (c)*** When the root is placed on its side, the amyloplasts slide toward the normally vertical walls that are now parallel with the soil surface. This movement of amyloplasts may play an important role in the perception of gravity by roots.

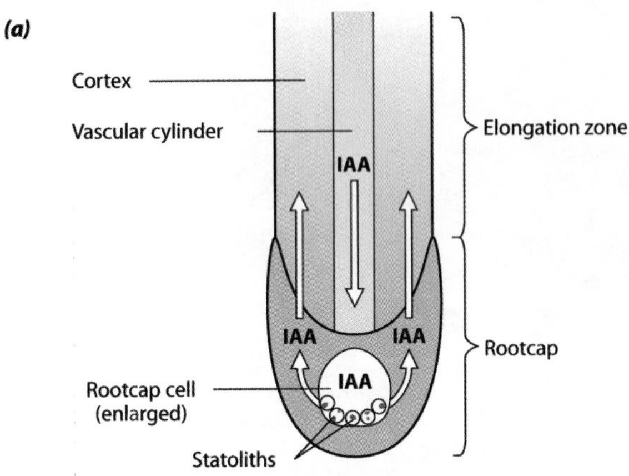

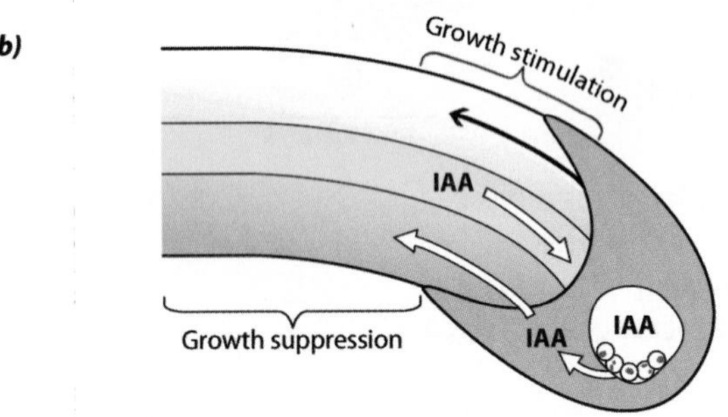

28–6 Auxin and root gravitropism ***(a)*** Auxin (IAA) is transported down from the shoot to the root tip in the vascular cylinder. From the root tip it is redistributed to the root cortex and epidermis, and then transported back up the root to the elongation zone, where it regulates the rate of cell elongation. ***(b)*** A horizontal reorientation of the root is detected by the sedimentation of starch-containing amyloplasts (statoliths) in cells of the rootcap. This, in turn, leads to the asymmetrical redistribution of auxin, with less auxin going to the upper side than to the lower side of the root. The decreased IAA concentration on the upper side stimulates growth there and initiates a downward curvature, while the higher concentration on the lower side suppresses growth on that side.

hydrostatic pressure exerted by the total weight of the protoplast, rather than by the amyloplasts alone, mediates gravity perception. On stimulation by gravity, calcium ion channels would be activated, triggering a signaling cascade within the statocytes.

Hydrotropism Is Root Growth in Response to a Moisture Gradient

Although **hydrotropism,** the directed growth of plant roots in response to a moisture gradient, has long been recognized, study of this tropism was difficult because the competing gravitropic response normally is much greater. The discovery of a mutant pea *(Pisum sativum),* known as *ageotropum,* that does not respond to gravity provided a breakthrough. This so-called agravitropic mutant displays a distinct hydrotropic response. Furthermore, the failure of decapped roots to curve hydrotropically provided definite proof of hydrotropism and of the role of the rootcap in this phenomenon. Additional evidence that the hydrotropic response in roots is suppressed by gravitropism under normal gravity on

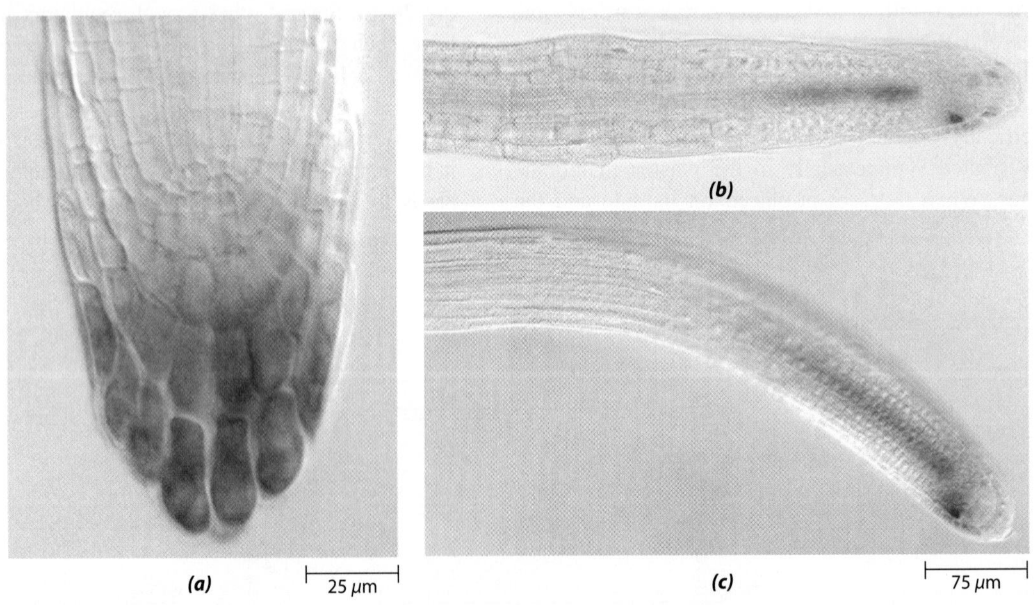

28–7 Cytokinin and root gravitropism ***(a)*** Rootcap of a vertical root of *Arabidopsis* showing symmetrical distribution (blue color) of free cytokinin, using the *GUS* reporter gene (see page 195). Asymmetrical distribution of free cytokinin expression is evident in a horizontally oriented root ***(b)*** after 30 minutes and ***(c)*** after 60 minutes of gravistimulation. Both *(b)* and *(c)* show the fast asymmetrical cytokinin activation pattern, detected as a dark blue concentrated spot at the lower side of the rootcap. The concentrated cytokinin on the lower side of the root inhibits growth on that side, whereas the upper side, with decreased cytokinin, continues to elongate, resulting in downward curvature.

Earth comes from studies on the roots of wild-type plants grown in microgravity (a condition in space in which only minuscule forces are experienced). Under such conditions, the gravitropic response was nullified and the root exhibited positive hydrotropism.

The underlying mechanisms that regulate hydrotropism remain unknown. Auxin apparently plays an indispensable role in both hydrotropic and gravitropic responses. Inhibitors of auxin influx and efflux had no effect, however, on hydrotropism of *Arabidopsis* roots, but they severely inhibited gravitropism. These results indicate that although polar transport of auxin is required for gravitropism, it is not required for hydrotropism, or perhaps the role of auxin differs in the two tropisms.

Thigmotropism Is Growth in Response to Touch

Another common tropism is **thigmotropism** (from the Greek *thigma,* meaning "touch"), a response to contact with a solid object. It enables roots to navigate around rocks and shoots of climbing plants to wrap around other structures for support. One of the most common examples of thigmotropism is seen in tendrils, which are modified leaves or stems (page 607). The tendrils wrap around any object with which they come in contact (Figure 28–8) and so enable the plant to cling and climb. The response can be rapid; a tendril may wrap around a support one or more times in less than an hour. Cells touching the support shorten slightly, and those on the other side elongate. Whether the mechanism of thigmotropism involves auxin gradients is uncertain.

Studies by M. J. Jaffe have shown that tendrils can store the "memory" of tactile stimulation. For example, if pea *(Pisum sativum)* tendrils are kept in the dark for three days and then rubbed, they will not coil, perhaps because of the requirement for ATP. If, however, they are illuminated within two hours after the rubbing, they will show the coiling response.

28–8 Thigmotropism Tendrils of bur cucumber *(Cucumis anguria).* Twisting is caused by different growth rates on the inside and outside of the tendril.

Circadian Rhythms

It is a common observation that the flowers of some plants open in the morning and close at dusk. Also, many leaves unfold in the sunlight and refold at night (Figure 28–9), a phenomenon first recorded by Androsthenes, a Greek soldier, in the fourth century B.C.E. In 1729, the French scientist Jean-Jacques de Mairan noticed that these diurnal (daily) movements continue even when the plants are kept in dim light (Figure 28–10). Since then, a great many more plant phenomena have been found to exhibit regular daily rhythms, among them photosynthetic activity, seed germination, enzyme and auxin production, stomatal movement, flower opening, cell division rates, stem elongation, and fragrance emission. Also, such rhythms continue even when all environmental conditions are kept constant. These regular, approximately 24-hour cycles are called **circadian rhythms** (from the Latin *circa,* meaning "approximately," and *dies,* meaning "a day"). Circadian rhythms are universal among eukaryotes and are also present in photosynthetic bacteria such as cyanobacteria, coordinating their metabolism with the times of day when photosynthesis occurs.

Circadian Rhythms Are Controlled by Circadian Clocks

Because circadian rhythms persist when all environmental conditions are kept constant and in the absence of environmental cues, the timing mechanism that anticipates diurnal cycles is considered to be **endogenous.** This internal timing mechanism is called a **circadian clock.** In both unicellular and multicellular organisms, circadian rhythms occur at the level of the cell; hence, multicellular organisms have multiple clocks.

Conceptually, circadian clocks have been thought of as consisting of three parts: a central **oscillator** that generates rhythmic behavior, input pathways that carry environmental information to synchronize the central oscillator, and output pathways that regulate physiological and biochemical processes. (It is estimated that circadian clocks cause the regular oscillation of between 30 and 40 percent of the genes in *Arabidopsis,* even when grown under constant light and temperature.) This model, although useful, is rather simplistic, because interactions occur between various parts of the clock, particularly between the output and input pathways.

Circadian Clocks Are Synchronized by the Environment

Under constant environmental conditions, such as can be maintained only in a laboratory, the period of the circadian rhythm is said to be **free-running**—that is, its intrinsic, or natural, period (between 22 and 29 hours in *Arabidopsis*) does not have to be reset at each cycle. Indeed, by definition, a true circadian-regulated process is free-running. In the natural world, the environment acts as a synchronizing agent (or *Zeitgeber,* German for "time giver"). In fact, the environment is responsible for

(a) (b)

28–9 Diurnal movements Leaves of the wood sorrel *(Oxalis)*, ***(a)*** during the day and ***(b)*** at night. One hypothesis for the function of such sleep movements is that they prevent the leaves from absorbing moonlight on bright nights, thus protecting the photoperiodic phenomena discussed later in this chapter. Another hypothesis, proposed by Charles Darwin more than a century ago, is that the folding reduces heat loss from the leaves at night.

keeping a circadian rhythm in step with the daily 24-hour light-dark cycle. If a circadian rhythm of a plant were greater or less than 24 hours, the rhythm would soon get out of step with the 24-hour light-dark cycle. Environmental synchronizing also permits adjustment of processes to coincide with seasonally changing daylengths in regions other than at the equator. For example, flowers of many plant species open at dusk or dawn to coincide with the circadian feeding patterns of their pollinators (for example, moth-pollinated flowers open in the evening, and the time when dusk or dawn occurs can change throughout a growing season). Thus, many plants not only must become resynchronized—that is, *entrained*—to the 24-hour day but must adapt to changing photoperiods.

Entrainment is the process by which a periodic repetition of light and dark (or some other external cycle) causes a circadian rhythm to synchronize with the same cycle as the entraining factor. Light-dark cycles and temperature cycles are the principal factors in entrainment (Figure 28–11). The light is perceived by two families of photoreceptors: the red-light-sensing **phytochromes** and the blue-light-sensing **cryptochromes** CRY1 and CRY2.

Two other characteristics of the circadian clock are temperature compensation and gating. **Temperature compensation** allows the clock to oscillate at approximately the same frequency over a broad range of physiological temperatures (12° to 27°C). It serves as a compensatory mechanism that buffers the clock

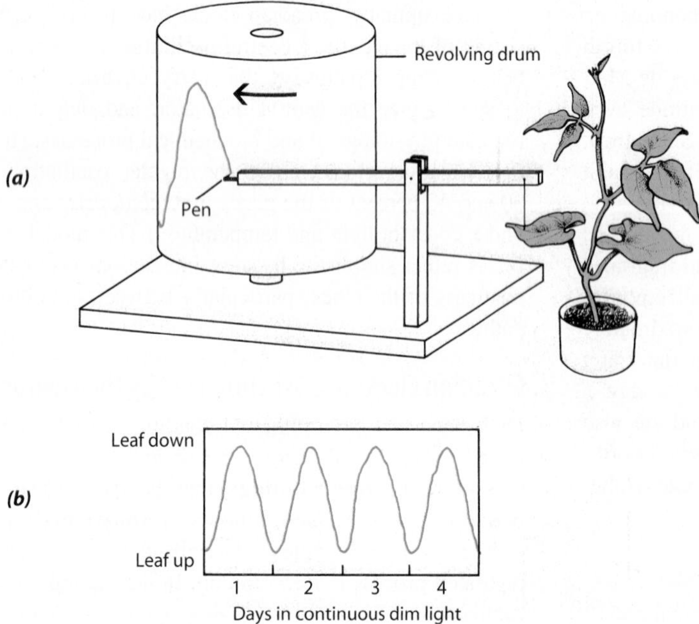

28–10 Circadian rhythms In many plants, the leaves move outward during the day, to a position perpendicular to the stem and the sun's rays; at night, the leaves move upward, toward the stem. ***(a)*** These sleep movements can be recorded on a revolving drum, using a delicately balanced pen-and-lever system attached to a leaf by a fine thread. Many plants, such as the common bean *(Phaseolus vulgaris)* shown here, will continue to exhibit these movements for several days, even when kept in continuous dim light. ***(b)*** A recording of this circadian rhythm, showing its persistence under constant dim light.

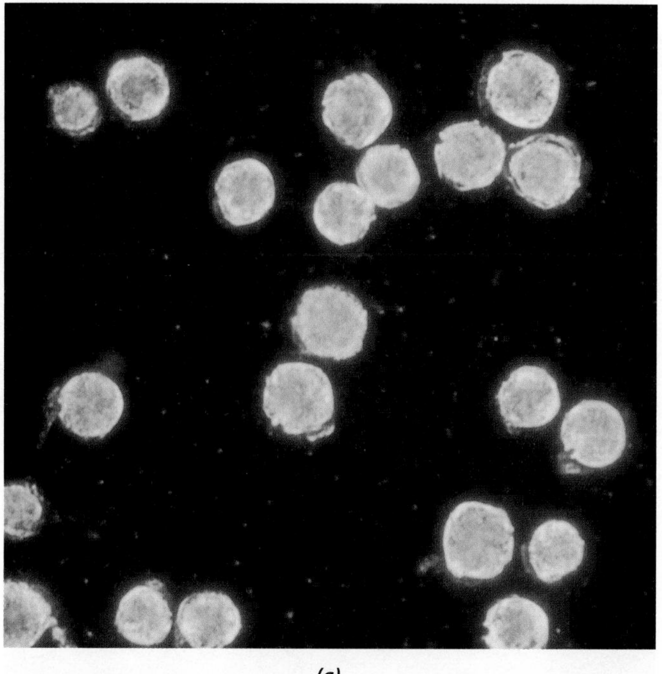

(a)

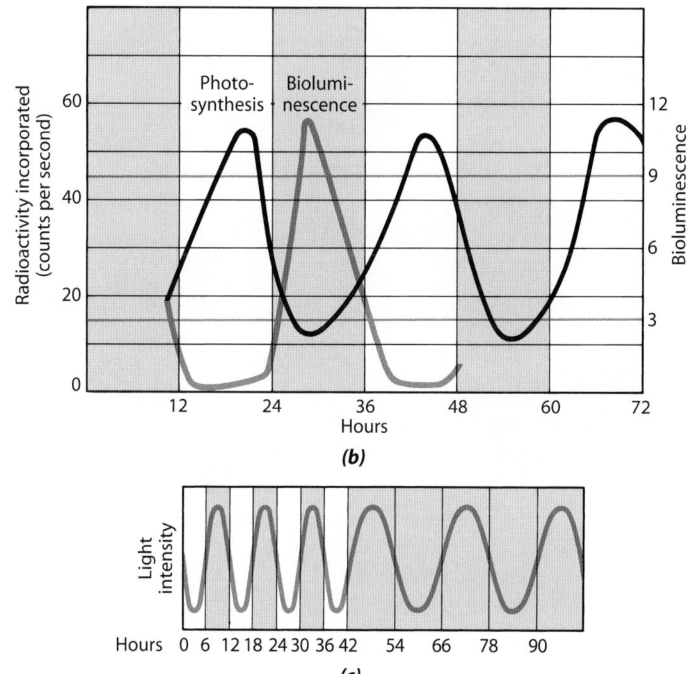

28–11 Entrainment ***(a)*** Photomicrograph of the dinoflagellate *Gonyaulax polyedra,* a single-celled marine alga. ***(b)*** In *G. polyedra,* three different functions follow separate circadian rhythms: bioluminescence, which reaches a peak in the middle of the night (colored curve); photosynthesis, which reaches a peak in the middle of the day (black curve); and cell division (not shown), which is restricted to the hours just before dawn. If *Gonyaulax* is kept in continuous dim light, these three functions continue to occur with the same rhythm for days and even weeks, long after a number of cell divisions have taken place. ***(c)*** The rhythm of bioluminescence in *Gonyaulax,* like most circadian rhythms, can be altered by modifying the cycles of illumination. For example, if cultures of the alga are exposed to alternating light and dark periods, each 6 hours in duration, the rhythmic function becomes entrained to this imposed cycle (left). If the cultures are then placed in continuous dim light, the organisms return to their original rhythm of about 24 hours.

against temperature changes. Another effect of circadian clock control is that stimuli of equal strength employed at different times of the day can result in a different intensity of response. This phenomenon is called **gating.**

In *Arabidopsis,* the circadian oscillator is thought to include at least three interlocking feedback loops. The central loop comprises three major clock genes, *TOC1, LHY,* and *CCA1,* and their protein products, all of which regulate transcription. Whereas the TOC1 protein is a positive regulator of the *LHY* and *CCA1* genes, the LHY and CCA1 proteins are negative regulators of the *TOC1* gene. In this feedback loop, the light of dawn activates expression of the *LHY* and *CCA1* genes, resulting in increased levels of the proteins LHY and CCA1, which repress the expression of the *TOC1* gene. Given that TOC1 is a positive regulator of *LHY* and *CCA1* expression, repression of the *TOC1* gene causes a progressive reduction in levels of LHY and CCA1 proteins, and a concomitant increase in *TOC1* expression. By the end of the day, at dusk, with LHY and CCA1 levels at their minimum, TOC1 reaches its maximum and indirectly stimulates the expression of *LHY* and *CCA1.* The cycle then begins again (Figure 28–12).

An excellent example of a clock's control of the expression of specific plant genes is found in the genes that encode chlorophyll *a/b*-binding (CAB) proteins of the light-harvesting complexes. This was elegantly demonstrated by fusing a fragment of the *Arabidopsis* chlorophyll *a/b*-binding protein 2 (CAB2) promoter and the coding region of the firefly luciferase gene, providing a bioluminescent reporter of the transcriptional activity of the *CAB* genes (see page 195 for the use of reporter genes). When this reporter gene was transferred into *Arabidopsis* plants, the influence of the clock on the expression of the *CAB* genes could be monitored by detecting the luminescence resulting from the activity of luciferase (Figure 28–13). This reporter system has been used to screen for mutants that affect the clock mechanism in *Arabidopsis.* In fact, one of the genes of the plant clock, *TOC1,* was identified from such a mutant screen.

In addition to coordinating daily events, the primary usefulness of the circadian clock is that it enables the plant or animal to respond to the changing seasons by accurately measuring changing daylength. Even at tropical latitudes, many plants are responsive to daylength and may use this signal to synchronize flowering or other activities with such seasonal events as wet or dry periods. In this way, changes in the environment trigger responses that result in adjustments of growth, reproduction, and other activities of the organism. When correctly tuned,

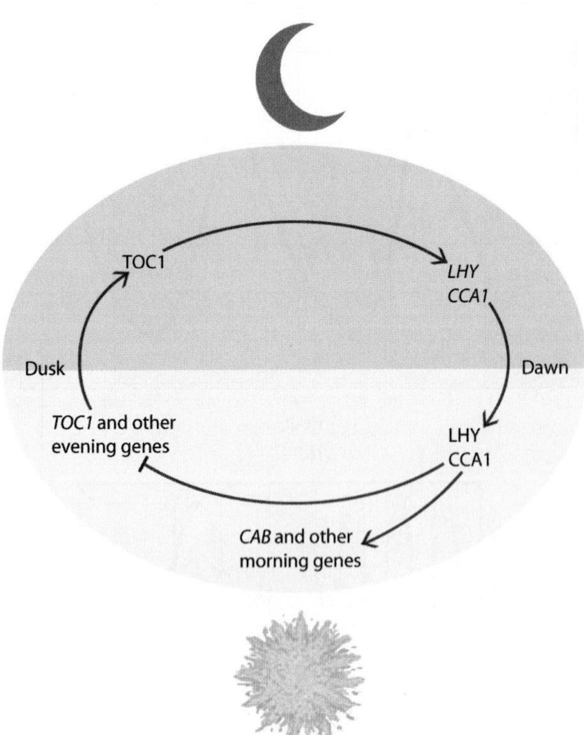

28–12 Circadian oscillator In this hypothetical model of the central loop of the circadian oscillator in *Arabidopsis,* the TOC1 protein (upper left) is a positive regulator of the *LHY* and *CCA1* genes, which reach their maximum expression levels at dawn. The proteins they produce, LHY and CCA1, activate the expression of *CAB* and other morning genes, while suppressing the expression of *TOC1* and other evening genes. A steady reduction in the levels of LHY and CCA1 proteins during the day allows *TOC1* transcription to increase and reach maximum levels by the end of the day.

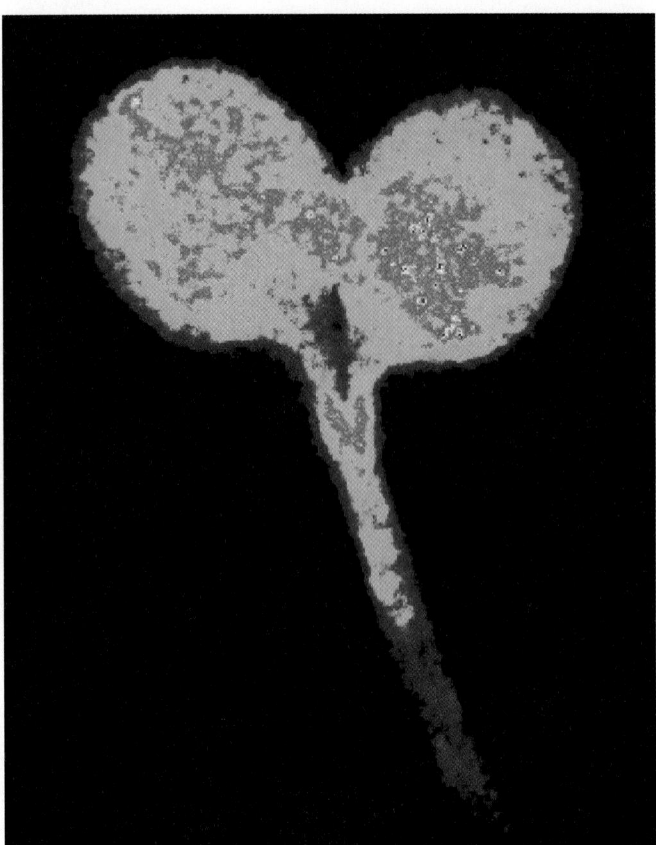

28–13 Control of gene expression by the circadian clock Shown here is an *Arabidopsis* seedling in which the firefly luciferase gene was fused to the promoter region of the *CAB2* gene. Expression of the *CAB2* gene, which is regulated by light and the circadian clock, is greatest (red) in the cotyledons and absent (blue) in the root.

the circadian clock leads to improved survival and competitive advantage.

Photoperiodism

The effect of daylength on flowering was discovered in the 1920s by two investigators at the U.S. Department of Agriculture, W. W. Garner and H. A. Allard, who found that neither a spontaneously arising mutant line of tobacco *(Nicotiana tabacum)* called Maryland Mammoth nor the Biloxi variety of soybean *(Glycine max)* would flower unless the daylength was shorter than a critical number of hours. Garner and Allard called this daylength-dependent phenomenon **photoperiodism.** Plants that flower only under certain daylength conditions are said to be *photoperiodic.* Photoperiodism is a biological response to a change in the proportions of light and dark in a 24-hour daily cycle. It allows organisms to detect the time of year and to undergo seasonal developmental changes. The circadian clock provides a key feature of the timing component of photoperiodism. Although the concept of photoperiodism began with studies of plants, it has now been demonstrated in various fields of biology, including the mating, migrating, and hibernating behaviors of such diverse animals as codling moths, spruce budworms, aphids, potato worms, fish, birds, and mammals.

Daylength Is a Major Determinant of Flowering Time

Garner and Allard went on to test and confirm their discovery with many other species of plants. They found that plants are of three general types, which they called short-day, long-day, and day-neutral. **Short-day plants** flower in early spring or fall; they must have a light period *shorter* than a critical length. For instance, the common cocklebur *(Xanthium strumarium)* is induced to flower by 16 hours or less of light (Figures 28–14 and 28–15). Other short-day plants are some chrysanthemums (Figure 28–16a), poinsettias, strawberries, and primroses.

Long-day plants, which flower chiefly in the summer, flower only if the light periods are *longer* than a critical length. Spinach, some potatoes, some wheat varieties, lettuce, and henbane *(Hyoscyamus niger)* are examples of long-day plants (Figures 28–15 and 28–16b).

Both cocklebur and spinach will bloom if exposed to 14 hours of daylight, yet only one is designated as a long-day plant. The important factor is not the absolute length of the photoperiod but whether it is longer or shorter than some critical interval. The short-day cocklebur also flowers readily when exposed to 8 hours of daylight, but the long-day spinach does not; conversely, spinach blooms if exposed to 16 hours of daylight, but cocklebur does not.

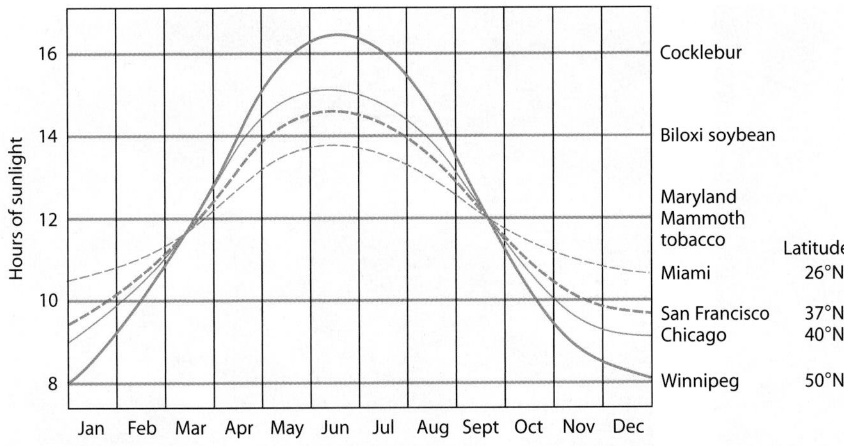

28–14 Daylength and flowering The relative length of day and night determines when plants flower. The four curves depict the annual change of daylength in four North American cities at four different latitudes. The horizontal green lines indicate the effective photoperiod of three different short-day plants. The common cocklebur *(Xanthium strumarium)*, for instance, requires 16 hours or less of light. In Miami, San Francisco, and Chicago, it can flower as soon as it matures, but in Winnipeg, the buds do not appear until early August, so late that the frost will probably kill the plants before the seed is set.

Day-neutral plants flower without respect to daylength. Some examples of day-neutral plants are cucumber, sunflower, rice, maize, and garden pea.

Within individual species of plants that cover a large north-south range, different photoperiodic ecotypes (locally adapted variants of an organism) are often observed; different populations are precisely adjusted to the demands of the local day-night regimen. Thus, in many species of prairie grasses, which range from southern Canada to Texas, northern ecotypes flower before southern ones when they are grown together in a common environment.

The photoperiodic response can be remarkably precise. At 22.5°C, the long-day plant henbane will flower when exposed to photoperiods of 10 hours and 20 minutes (Figure 28–15), but not with a photoperiod of 10 hours. Environmental conditions also affect photoperiodic behavior. For instance, at 28.5°C henbane requires 11.5 hours of light, whereas at 15.5°C it requires only 8.5 hours.

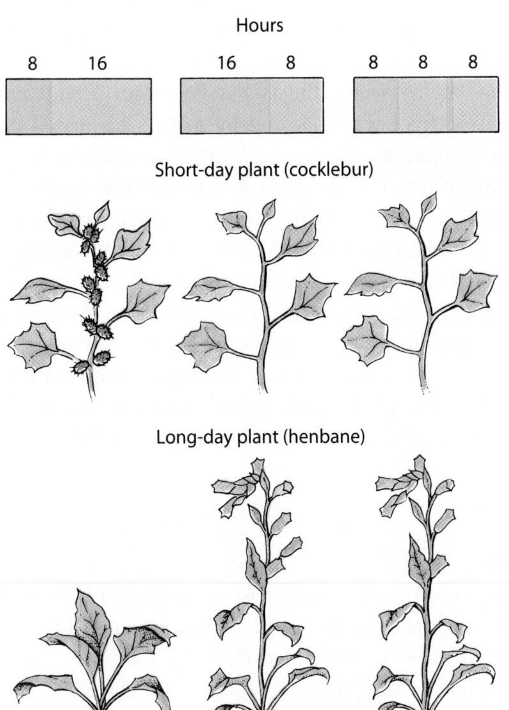

The response varies among species. Some plants require only a single exposure to the critical day-night cycle, whereas others, such as spinach, require several weeks of exposure. In many plants, there is a correlation between the number of induction cycles and the rapidity of flowering or the number of flowers formed. Some plants have to reach a certain degree of maturity before they flower, whereas others respond to the appropriate photoperiod when they are seedlings. Some plants, when they get older, will eventually flower even if not exposed to the appropriate photoperiod, although they flower much earlier with the proper exposure.

Plants Keep Track of Daylength by Measuring the Length of Darkness

In 1938, Karl C. Hamner and James Bonner began a study of photoperiodism, using the short-day plant cocklebur, which requires 16 hours or less of light per 24-hour cycle to flower. This plant is particularly useful for experimental purposes because, under laboratory conditions, a single exposure to a short-day cycle will induce flowering two weeks later, even if the plant is immediately returned to long-day conditions. Hamner and Bonner showed that it is the leaf blade of the cocklebur that perceives the photoperiod. A completely defoliated plant cannot be induced to flower. But if as little as one-eighth of a young, fully expanded leaf is left on the stem, the single short-day exposure induces flowering.

28–15 Photoperiods and flowering Short-day plants flower only when the photoperiod is less than some critical value. The common cocklebur *(Xanthium strumarium)*, a short-day plant, requires less than 16 hours of light to flower. Long-day plants flower only when the photoperiod is longer than some critical value. The long-day plant henbane *(Hyoscyamus niger)* requires about 10 hours or more (depending on temperature) to flower. If the dark period is interrupted by a flash of light, *Hyoscyamus* will flower even when the daily light period is shorter than 10 hours. However, a pulse of light during the dark period has the opposite effect in a short-day plant—it prevents flowering. The bars at the top indicate the duration of light and dark periods in a 24-hour day.

(a)

(b)

28–16 Short-day and long-day plants Seen here are ***(a)*** chrysanthemum (*Chrysanthemum* sp.) plants, which are short-day plants, and ***(b)*** spinach *(Spinacia oleracea)* plants, which are long-day plants. The plants on the left in each photograph were grown under short-day conditions, while those on the right in each set were grown under long-day conditions. Note that the plants exposed to long days have longer stems than those exposed to short days, regardless of whether or not they flower.

In the course of these studies, in which they tested a variety of experimental conditions, Hamner and Bonner made a crucial and totally unexpected discovery. If the period of darkness is interrupted by as little as a one-minute exposure to light from a 25-watt bulb, flowering does not occur. Interruption of the light period by darkness has absolutely no effect on flowering. Subsequent experiments with other short-day plants showed that they, too, required periods of uninterrupted darkness rather than uninterrupted light.

The most sensitive part of the dark period with respect to light interruption is in the middle of the period. If a short-day plant, such as cocklebur, is given an 8-hour light period and then an extended period of darkness, the plant passes through a stage of increasing sensitivity to light interruption, lasting about 8 hours, followed by a period when light interruption has a diminishing effect. In fact, a one-minute light exposure after 16 hours of darkness stimulates flowering.

On the basis of the findings of Garner and Allard, described earlier, commercial growers of chrysanthemums had found that they could hold back blooming in short-day plants by extending the daylight with artificial light. On the basis of the new experiments by Hamner and Bonner, growers were able to delay flowering simply by switching on the light for a short period in the middle of the night.

What about long-day plants? They also measure darkness. A long-day plant that will flower if kept under light for 16 hours and dark for 8 hours will also flower on 8 hours of light and 16 hours of dark if the dark is interrupted by a one-hour exposure to light.

An important clue to the mechanism of a plant's response to the relative proportions of light and darkness came from a team of research workers at the U.S. Department of Agriculture Research Station in Beltsville, Maryland. The clue came from an earlier study performed with lettuce *(Lactuca sativa)* seeds, which germinated only if exposed to light. This is a requirement for many small seeds, which must germinate in loose soil and near the surface for the seedling to emerge. In studying the light requirement of germinating lettuce seeds, earlier workers had shown that red light stimulates germination and that light of a slightly longer wavelength (far-red) inhibits germination even more effectively than does no illumination at all.

Hamner and Bonner had shown that when the dark period was interrupted by a single flash of light from an ordinary light bulb, the cocklebur did not flower. The Beltsville group, following this lead, began to experiment with light of different wavelengths, varying the intensity and duration of the flash. They found that red light with a wavelength of about 660 nanometers was most effective in preventing flowering in the cocklebur and other short-day plants and in promoting flowering in long-day plants.

Using lettuce seeds, the Beltsville group found that when a flash of red light was followed by a flash of far-red light, the seeds did not germinate. The red light most effective in inducing germination in the lettuce seeds was of the same wavelength as that involved in the flowering response—about 660 nanometers. Furthermore, the light most effective in inhibiting the effect produced by red light had a wavelength of 730 nanometers, in the far-red region. The sequence of red and far-red flashes could be repeated over and over; the number of flashes did not matter, but the nature of the final one did. If the sequence ended with a red flash, the majority of the seeds germinated. If it ended with a far-red flash, the majority did not germinate (Figure 28–17).

Phytochrome Is the Primary Photoreceptor Involved in Photoperiodism

The photoreceptors involved both in flowering and in lettuce-seed germination exist in two different, interconvertible forms: $\mathbf{P_r}$, which absorbs red light, and $\mathbf{P_{fr}}$, which absorbs far-red light. When a molecule of P_r absorbs a photon of 660-nanometer light,

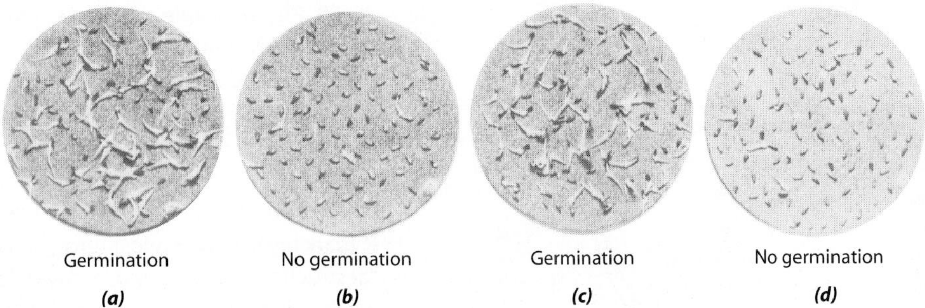

28–17 Light and the germination of lettuce seeds ***(a)*** Seeds exposed briefly to red light; ***(b)*** seeds exposed to red light followed by far-red light; ***(c)*** seeds exposed to a sequence of red, far-red, red; ***(d)*** seeds exposed to a sequence of red, far-red, red, far-red. Whether or not the seeds germinated, as seen in *(a)* and *(c)*, depended on the final wavelength in the series of exposures, with red light promoting and far-red light inhibiting germination.

the molecule is converted to P_{fr} in a matter of milliseconds; when a molecule of P_{fr} absorbs a photon of 730-nanometer light, it is quickly reconverted to the P_r form. These are called *photoconversion reactions.* Phytochromes are synthesized in the dark in the P_r form; following conversion to the P_{fr} form, the phytochromes move to the nucleus. The P_{fr} form is biologically active (that is, it will trigger a response such as seed germination), whereas P_r is inactive. The photoreceptors can thus function as a biological switch, turning responses on or off.

The lettuce-seed germination experiments are easily understood in these terms. Since P_r absorbs red light most efficiently (Figure 28–18), this light will convert a high proportion of the molecules to the P_{fr} form, thereby inducing germination. Subsequent far-red light absorbed by P_{fr} will convert essentially all of the molecules back to P_r, canceling the effect of the prior red light.

What about flowering under natural day-night cycles? Since white light contains both red and far-red wavelengths, both forms of the photoreceptor are exposed simultaneously to photons that are efficient in promoting their photoconversion. After a few minutes in the light, then, a photoequilibrium is established in which the rates of conversion of P_r to P_{fr} and P_{fr} to P_r are equal, and the proportion of each type of photoreceptor molecule is constant (about 60 percent P_{fr} in noontime sunlight).

When plants are switched to darkness, the level of P_{fr} steadily declines over a period of several hours. If a high level of P_{fr} is regenerated by pulse irradiation with red light in the middle of the dark period (Figure 28–15), it will inhibit flowering in short-day (that is, "long-night") plants that otherwise would have flowered and promote flowering in long-day (that is, "short-night") plants that otherwise would not have flowered. In either case, the effect of the red pulse can be canceled by an immediate far-red pulse, which reconverts the P_{fr} to P_r.

In 1959, Harry A. Borthwick and his co-workers at Beltsville named this photoreceptor **phytochrome,** meaning "plant color," and presented conclusive physical evidence for its existence. The principal characteristics of the receptor, as they are presently understood, are summarized schematically in Figure 28–19.

Phytochrome in Tissues Can Be Detected with a Spectrophotometer Phytochrome is present in plants in much smaller amounts than pigments such as chlorophyll. To detect phytochrome, one needs to use a spectrophotometer that is sensitive to extremely small changes in light absorbance. Such an instrument was not available until some seven years after the existence of phytochrome was proposed; in fact, the first use of this spectrophotometer was to detect and isolate phytochrome.

To avoid interference from chlorophyll, which also absorbs light of about 660 nanometers, dark-grown seedlings (in which chlorophyll has not yet developed) were chosen as the source of phytochrome. The pigment associated with the phytochrome receptor proved to be blue (why might you expect this color?), and it showed the characteristic P_r to P_{fr} conversion in the test tube by reversibly changing color slightly in response to red or far-red light.

The phytochrome molecule contains two distinct parts: a light-absorbing portion (the **chromophore**) and a large protein portion (Figure 28–20). The chromophore is much like the phycobilins that serve as accessory pigments in cyanobacteria and red algae. Genes for the protein portion have been isolated from several species, and the amino acid sequence has been deduced from the nucleotide sequence. Most plants have several different phytochromes encoded by a family of divergent genes. In *Arabidopsis,* five such genes are known (*PHYA* through *PHYE*).

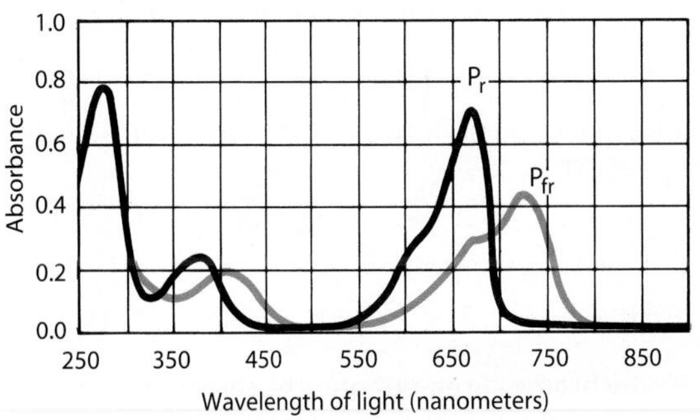

28–18 Absorption spectra of P_r and P_{fr} The difference in the absorption spectra of the two forms of phytochrome, P_r and P_{fr}, made it possible to isolate the photoreceptors.

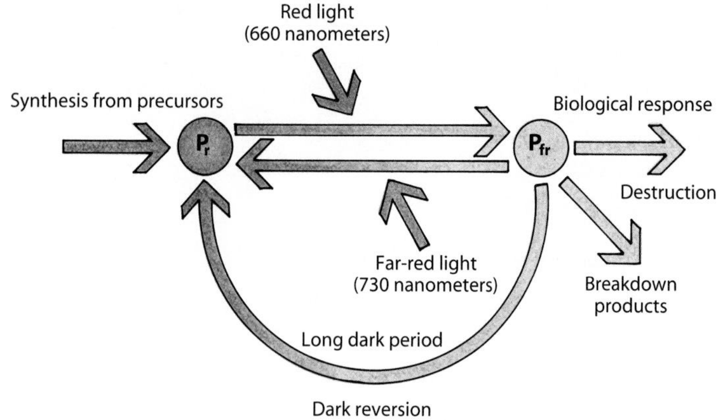

28–19 Phytochrome synthesis Phytochrome is continuously synthesized in the P_r form from its amino acid precursors and accumulates in this form in dark-grown plants. P_r changes to P_{fr} when exposed to red light, which is present in sunlight. P_{fr} is the active form that induces a biological response. P_{fr} is converted back to P_r by photoconversion when exposed to far-red light. In darkness, P_{fr} reverts to P_r (dark reversion) or is lost through a process termed "destruction" that occurs over several hours and probably involves hydrolysis by a protease. All three alternative pathways for P_{fr} removal provide the potential for reversing induced responses. It should be noted, however, that dark reversion has been detected only in eudicots.

A primary mechanism of phytochrome signaling involves physical interaction of the phytochrome photoreceptor with proteins called **phytochrome-interacting factors (PIFs).** The PIFs, which are found in the nucleus and can bind to DNA, display a diverse array of regulatory functions controlling **photomorphogenesis** (the regulation of growth by light signals). Apparently, the PIFs act primarily as negative regulators of the phytochrome response. The physical interaction of phytochromes with the PIF proteins leads to rapid degradation of the latter, enabling the plant to alter gene expression rapidly in response to fluctuations in the light environment. As yet there is no evidence that phytochrome interacts directly with these genes.

Phytochrome Is Involved in a Wide Variety of Plant Responses The germination of many seeds occurs in the dark. In the seedlings, the stem elongates rapidly, pushing the shoot (or, in grasses, the coleoptile) up through the dark soil. During this early stage of growth, there is essentially no enlargement of the leaves, which would interfere with passage of the shoot through the soil. Soil is not necessary for this growth pattern; any seedling grown in the dark will be elongated and spindly and will have a closed hook and small leaves. It will also be yellow to colorless, because plastids do not turn green until exposed to light. Such a seedling is said to be **etiolated** (Figure 28–21a).

28–20 Phytochrome chromophore The chromophore, the light-absorbing portion of the phytochrome molecule, is shown here in its P_r form; its attachment to the protein part of the molecule is indicated.

When the seedling tip emerges into the light, the etiolated growth pattern gives way to normal plant growth and response, such as phototropism (Figure 28–21b). In eudicots, the hook unbends, the growth rate of the stem may slow down somewhat, and leaf growth begins (see Figure 22–10). In grasses, growth of the mesocotyl (the part of the embryo axis between scutellum and coleoptile; Figure 22–12) stops, the stem elongates, and the leaves open (see Figure 22–11). Such light-triggered growth and developmental responses are called *photomorphogenic responses.*

A dark-grown bean seedling that receives five minutes of red light a day, for instance, will show these light effects beginning on the fourth day. If the exposure to red light is followed by a five-minute exposure to far-red, none of the changes usually produced by the red light appear. Similarly, in the seedlings of grains, termination of mesocotyl growth is triggered by exposure to red light, and the effect of red light is canceled by far-red.

Studies of the *long-hypocotyl* mutants of *Arabidopsis* have provided information on the roles of specific photoreceptor genes in photomorphogenesis. In these mutants, light does not arrest hypocotyl elongation, indicating that the wild-type *HY* genes are involved in light perception or response. In the *hy1* and *hy2* mutants, biosynthesis of the phytochrome chromophore is blocked.

In another mutant, the *hy4* mutant, the ability to respond to blue light is blocked. Cloning of the gene responsible for the *hy4* mutant revealed that it is a blue-light receptor gene. This finding illustrates the complexity of responses to light that involve the contributions of several specific photoreceptors, each responsible for a different aspect of photomorphogenesis, and it also illustrates the value of genetic analyses in sorting out such complexity.

An important feature of phytochrome in plants growing in a natural environment is the detection of shading by other plants. Light reflected or transmitted through living vegetation is depleted in the red and blue wavelengths, which are absorbed by chlorophyll and carotenoid pigments and used for photosynthesis. This results in a reduction of the red/far-red ratio; thus, plants growing in the shade receive more far-red light than red light. In shade-avoiding species, the perception of a low red/far-red light ratio initiates a suite of developmental responses termed the *shade-avoidance syndrome.* These responses include elongation

28–21 Etiolation ***(a)*** Dark-grown seedlings, such as the bean plant on the left, are thin and pale; they have longer internodes and smaller leaves than light-grown seedlings, such as that on the right. The group of physical characteristics exhibited by dark-grown seedlings, known as etiolation, has survival value for the seedlings because it increases their chance of reaching light before their stored energy supplies are exhausted. ***(b)*** Phototropism in etiolated soybean seedlings newly exposed to light.

of hypocotyls, internodes, and petioles, upward leaf movement, and increased apical dominance (Figure 28–22). Such responses serve to elevate the leaves within the canopy, enhancing a plant's ability to forage for light and to overtop competing vegetation. If the reduced red/far-red light ratio persists and the plant fails to overtop its neighbors, flowering is accelerated, thereby promoting seed production and increasing chances for reproductive success. Some studies indicate that a reduction of the blue wavelengths can induce pronounced shade-avoidance responses.

28–22 Shade-avoidance responses In each of the species shown here, the plant on the left is a control and the plant on the right was grown under low red/far-red light conditions. Note that the plants exposed to low red/far-red light exhibit more vertical leaf angles, elongated stems and/or petioles, and early flowering.

The Floral Stimulus

With the discovery that it is the leaf that perceives daylength, it became clear that a flowering-promoting signal, or floral stimulus, must be transmitted from the leaf to the shoot apical meristem, where flowering occurs.

The early experiments on the floral stimulus were carried out independently in several laboratories in the 1930s. The plant physiologist M. Kh. Chailakhyan conducted some of the earliest experiments. Using the short-day *Chrysanthemum indicum,* he showed that if the upper portion of the plant was defoliated and the leaves on the lower part were exposed to a short-day induction period, the plant would flower. If, however, the upper, defoliated part was kept on short days and the lower, leafy part on long days, no flowering occurred. He interpreted these results as indicating that the leaves form a substance (he thought it was a hormone) that moves to the shoot apex and initiates flowering. Chailakhyan named this hypothetical substance **florigen,** the "flower maker."

Further experiments showed that the flowering response will not take place if the leaf is removed immediately after photoinduction. But if left on the plant for a few hours after the induction cycle is complete, the leaf can then be removed without affecting flowering. The floral stimulus can pass through a graft from a photoinduced plant to a noninduced plant. Unlike auxin, however, which can pass through agar or nonliving tissue, florigen can move from one plant tissue to another only if they are connected by living tissue. If a branch is girdled, that is, if a circular strip of bark is removed, florigen movement ceases. On the basis of these data, it was concluded that florigen moves via the phloem, the pathway followed by most organic substances and mobile macromolecules in plants.

After seven more decades of research, the identity of florigen—at times referred to as the "Holy Grail" of flowering research—became known, especially from research on *Arabidopsis. Arabidopsis* is a facultative long-day plant—that is, it flowers more rapidly in long days but will eventually flower in short days. Based on the flowering behavior of *Arabidopsis* mutants (Figure 28–23), several genes involved in the production of a flowering promoter in long days were identified. These genes include *GI (GIGANTEA), CO (CONSTANS),* and *FT (FLOWERING LOCUS1),* which act in that order.

Briefly, the GI protein is involved in functioning of the circadian clock, and *CO* gene expression is regulated by the clock. Thus, mutations in clock components, such as in the gene *GI,* affect *CO* expression. The CO protein is required for induction of *FT* expression in long days, and the *FT* gene in turn encodes the small protein FT—which is now known to be the florigen.

In young *Arabidopsis* plants, the genes *CO* and *FT* are expressed in companion cells of minor-vein phloem of source (exporting) leaves. The FT protein produced in the companion cells is transferred to the sieve tubes, moves out of the leaf in the assimilate stream (see Chapter 30), and is transported via the phloem to the shoot apical meristem. In the apical meristem, the FT protein forms a complex with the transcription factor FD. The FT-FD complex initiates flowering by activating such floral meristem identity genes as *APETALA1* (see Chapter 25) and other floral promoters, such as *SOC1 (SUPPRESSOR OF EXPRESSION OF CONSTANS1).* Once flowering commences in the *Arabidopsis* shoot apical meristem, it is irreversible.

Vernalization: Cold and the Flowering Response

Cold may affect the flowering response. For example, if winter rye *(Secale cereale)* is planted in the autumn, it germinates during the winter and flowers in the following summer, 7 weeks after growth resumes. If it is planted in the spring, it does not flower for 14 weeks. In 1915, the plant physiologist Gustav Gassner discovered that he could influence the flowering of winter rye and other cereal plants by controlling the temperature of the germinating seeds. If imbibed seeds (see essay on page 80) of the winter strain were kept at near-freezing (1°C) temperatures during germination, the winter rye, even when planted in late spring, would flower in the same summer that it was planted. This process, by which exposure to cold renders plants competent to flower, is called **vernalization** (from the Latin *vernus,* meaning "of the spring").

Even after vernalization, the plant must be subjected to a suitable photoperiod, usually long days. The vernalized winter rye behaves like a typical long-day plant, flowering in response to the long days of summer. A vernalization requirement is common in winter annuals (which are sown in the fall and flower the following summer) and in biennials, such as radish *(Raphanus sativus),* celery *(Apium graveolens),* and carrot *(Daucus carota),* that produce a short stem and rosette of leaves during the first season and flower the following summer. This requirement ensures that such plants will not flower prematurely in response to short fluctuations in temperature in the fall.

Unlike photoperiod, which is perceived in leaves, vernalization takes place directly in the cells of the shoot apical meristem. In winter-annual types of *Arabidopsis,* two genes, *FLOWERING LOCUS C (FLC)* and *FRIGIDA (FRI),* are associated with the vernalization requirement. The *FLC* gene inhibits flowering directly by repressing the floral activator genes *FT, FD,* and *SOC1.* The *FRI* gene is required for high levels of *FLC* expression. Vernalization turns off *FLC* expression, resulting in a shoot apical meristem that is competent to flower in response to long days. It was recently discovered that a long, noncoding RNA molecule, named COLDAIR, is responsible for silencing the *FLC* gene. After exposure to 20 days of near-freezing temperatures, COLDAIR becomes active. The process of silencing *FLC* is completed after about 30 to 40 days of cold.

In most biennial long-day plants that form rosettes, gibberellin treatment can substitute for the cold requirement. If gibberellin is applied, such plants will elongate rapidly and then flower. Application of gibberellin to short-day plants or to nonrosette long-day plants has little effect on (or inhibits) the flowering response. However, if gibberellin synthesis is inhibited when the plant is being exposed to the appropriate inductive cycle, the plant will not flower unless gibberellin is added to it.

Dormancy

Plants do not grow at the same rate at all times. During unfavorable seasons, they limit their growth or cease to grow altogether.

(a)

(b)

28–23 *Arabidopsis thaliana* wild-type plants and flowering-time mutants
(a) Plants grown under long-day, inductive conditions. On the right is a wild-type plant and on the left is a plant containing a mutation that delays flowering. The mutant plant has greatly increased in size vegetatively because of the delay in flowering. The mutant is several weeks older than the wild type. ***(b)*** Plants grown under short-day, noninductive conditions. On the right is a wild-type plant and on the left is a plant containing a mutation that causes rapid flowering independent of photoperiod. In this case, the wild type is several weeks older than the mutant.

This ability enables plants to survive periods of water scarcity or low temperature.

Dormancy is a special condition of arrested growth. After periods of ordinary rest, growth resumes when the temperature becomes milder or when water or any other limiting factor becomes available again. A dormant bud or embryo, however, can be "activated" only by certain, often quite precise, environmental cues. This adaptation is of great survival importance to the plant. For example, the buds of plants expand, flowers form, and seeds germinate in the spring—but how do they recognize spring? If warm weather alone were enough, in many years all the plants would flower and all the seedlings would start to grow during warm autumn weather, only to be destroyed by the winter frost. The same could be said for any one of the warm spells that often punctuate the winter season. The reason that the dormant seed or bud does not respond to these apparently favorable conditions is that it contains endogenous inhibitors, which must be removed or neutralized before the period of dormancy can be terminated. In contrast to this "reluctance" to grow too rapidly, commercial seeds are artificially selected for their readiness to germinate promptly when they are exposed to favorable conditions, a trait that would be a great hazard for wild seeds.

Seeds Require Specific Environmental Cues to Break Dormancy

The seeds of almost all plants growing in areas with marked seasonal temperature variations require a period of cold prior to germination. This requirement is normally satisfied by winter temperatures. The seeds of many ornamental plants have a similar cold requirement. If moist seed is exposed to a low temperature for many days (average optimum temperature and time, 5°C for 100 days), dormancy may be broken and the seed germinated. This horticultural procedure is termed **stratification.** Many seeds require drying before they germinate; this prevents germination within the moist fruit of the parent plant. As discussed earlier, some seeds, such as lettuce seeds, require exposure to light, but others are inhibited by light.

Some seeds will not germinate until they have become abraded, as by soil action. Such abrasion wears away the seed coat, permitting water or oxygen to enter the seed and, in some cases, removing the source of inhibitors. Hard seed coats that interfere with water absorption and embryo enlargement are common among legumes. Microbial digestion sometimes contributes to the softening of a hard seed coat.

The seeds of some desert species germinate only when sufficient rain has fallen to leach away inhibitory chemicals in the seed coat. The amount of rainfall necessary to wash off these germination inhibitors is directly related to the supply of water that the desert plant needs to become established as a seedling. Mechanical abrasion or breaking of the seed coat **(scarification)** with a knife, file, or sandpaper may allow removal of the "hard seed" condition or inhibitor or initiation of the metabolic activity required for germination. Germination may also be induced by soaking the seeds in alcohol or some other fat solvent (to dissolve waxy substances that impede entry of water) or in concentrated acids. These procedures are widely used by horticulturalists. (See also Chapter 22, page 534.)

Some seeds may remain viable for a long time in the dormant condition, enabling them to exist for many years, decades, and even centuries under favorable conditions. The oldest, directly (radiocarbon) dated seed known to have germinated is that of an approximately 2000-year-old date palm *(Phoenix dactylifera)* excavated at Masada near the Dead Sea. The previous record holder was a sacred lotus *(Nelumbo nucifera)* seed found in an ancient lake bed at Pulantien in Liaoning Province, China. Radiocarbon dating indicated that seed to be about 1300 years old. A report that Arctic lupine *(Lupinus arcticus)* plants were grown from seeds considered to be at least 10,000 years old has been discredited.

Some seeds buried in soil can slip into and out of dormancy (cycling seed dormancy), keeping rhythm with the changing temperatures of the seasons. For example, some summer annual seeds dug up from the soil in the fall will fail to germinate, no matter how favorable the environmental conditions (light, temperature, water). In the spring, however, the same types of viable seeds, under similar favorable conditions, will germinate readily.

Plant scientists have become increasingly interested in the factors involved in maintaining seed viability. Various enzyme systems may progressively fail in the stored seed, eventually leading to a complete loss of viability. Under what circumstances might viability be prolonged? Such questions are relevant to the worldwide interest in developing **seed banks,** with the goal of preserving the genetic characteristics of the wild and early cultivated varieties of various crop plants for use in future breeding programs (see the essay on page 677). The need for such seed banks arises from progressive replacement of older varieties with newer ones and elimination of the replaced varieties, chiefly through destruction of their habitats. In addition, many wild species of plants are in danger of becoming extinct, and the seeds of these plants should be preserved if possible.

The Dormant Condition in Buds Is Preceded by Acclimation

Dormancy in buds is essential to the survival of temperate herbaceous and woody perennials that are exposed to low temperatures during winter. Although dormant buds do not elongate—that is, do not exhibit observable growth—they may undergo meristematic activity during various phases of dormancy.

Bud dormancy in many trees is initiated in midsummer, long before leaf fall in autumn. The dormant bud is an embryonic shoot consisting of an apical meristem, nodes and internodes (not yet extended), and small rudimentary leaves or leaf primordia with buds or bud primordia in their axils, all enclosed by *bud scales* (Figure 28–24). The bud scales are very important because they help prevent desiccation, restrict the movement of oxygen into the bud, and insulate the bud from heat loss. Growth inhibitors are known to accumulate in bud scales as well as in bud axes and leaves within buds. In many respects, therefore, the roles of the bud scales parallel those of the seed coat.

During and following the cessation of growth leading to the dormant condition, the plant tissues begin to undergo numerous physical and physiological changes to prepare the plant for winter, a process known as **acclimation.** Decreasing daylength is the primary factor involved in the induction of dormancy in buds (Figure 28–25). Generally, more inhibitors are found in leaves and buds under short-day conditions than under long-day conditions. Acclimation to cold leads to **cold hardiness**—the ability of the plant to survive the extreme cold and drying effects of winter weather.

As with seeds, the buds of many plant species require cold to break dormancy. If branches of flowering trees and shrubs are cut and brought indoors in autumn, they do not flower; but if the same branches are left outside until late winter or early spring, they will bloom in the warmer indoor temperatures. Deciduous

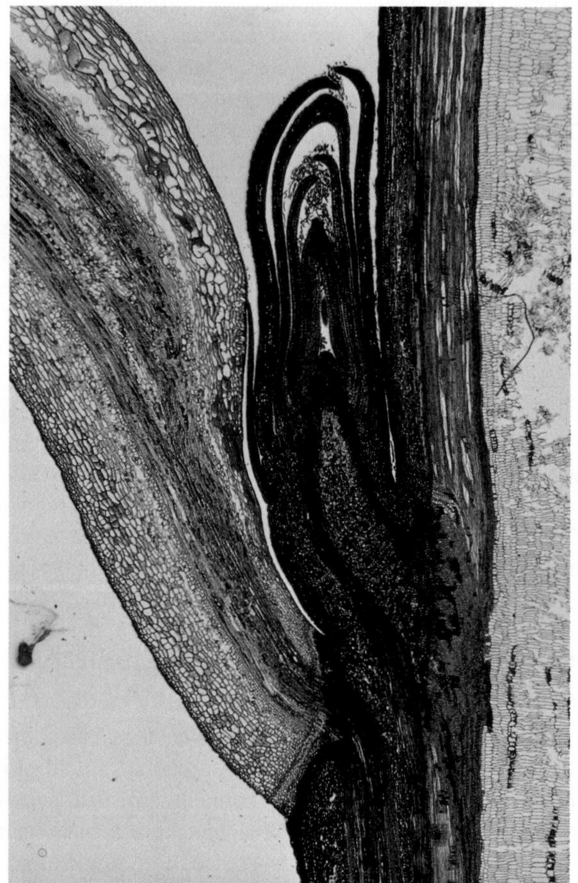

28–24 Dormant bud Longitudinal section of a dormant axillary bud of a maple *(Acer).* The bud consists of an embryonic shoot enclosed by bud scales.

DOOMSDAY SEED VAULT: SECURING CROP DIVERSITY

Four hundred miles off the northern coast of Norway, on the side of an icy mountain overlooking the Greenland Sea, a stainless steel portal rises like a mysterious monument out of the snow. Beyond its doors, a concrete passageway pierces the permafrost and burrows 100 meters deep into the mountain. At the end are three identical chambers, each filled with hundreds of identical crates. Each crate contains hundreds of small, neatly labeled, airtight aluminum foil pouches. Within each pouch are several hundred seeds, each set of seeds from a unique crop variety, collected from all over the world and sent to this lonely island for safekeeping. Collectively, the more than half-million samples represent—and safeguard—a large fraction of the agricultural genetic diversity on the planet.

The Svalbard Global Seed Vault, a project of the Norwegian government, opened in 2008. Its mission is to provide a back-up storage facility for the hundreds of crop-plant gene banks around the world. These banks are on the front lines of preservation for the enormous, but dwindling, diversity of plant varieties used in agriculture. There are over 40,000 varieties of rice, for example, but only a handful are grown commercially. Many of the rest are in danger of being lost, as more and more farmers grow fewer and fewer varieties. And yet the rarer varieties may contain unique characteristics of disease resistance, drought tolerance, or flavor that may one day be used to improve commercially grown rice or to aid local farmers in growing a crop suitable to their needs.

The Svalbard Vault stores duplicates of seed samples, in case the sample at a primary gene bank is damaged or lost. Seeds last longest when they are kept dry and cold; the chambers at Svalbard are cooled to −18°C, and the storage pouches are moisture-proof. Global warming may eventually melt the permafrost, but, sunk deep within the mountain, the chambers are naturally buffered against temperature extremes and are out of reach of rising sea levels. When the vault is full, it will contain 4.5 million samples, and it will be the largest collection of seeds in the world.

Under ideal storage conditions, seeds of some crops may remain viable for hundreds or even thousands of years. But even the Doomsday Vault, as it is sometimes called, cannot completely halt the slow loss of a seed's vitality. Thus, primary gene banks must continually exhume their samples, germinate them to grow mature plants, and create new seed samples, which can be stored away again.

Cold storage The Svalbard Global Seed Vault extends deep inside solid rock on the Arctic coast of Spitzbergen, the largest island of the Svalbard archipelago, which constitutes the northernmost part of Norway.

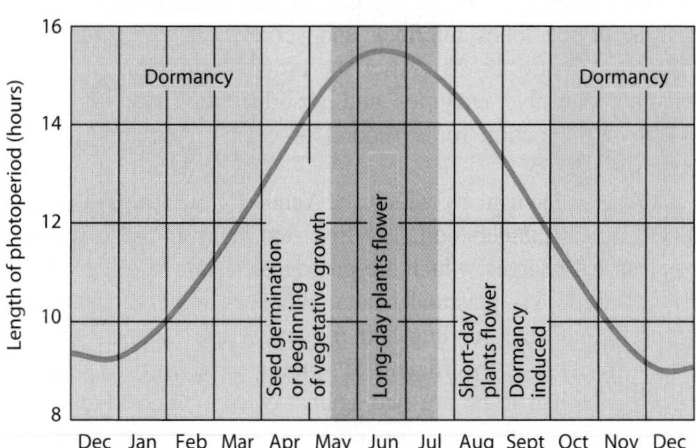

28–25 Daylength and dormancy Relationship between daylength and the developmental cycle of plants in the North Temperate Zone.

fruit trees, such as the apple, chestnut, and peach, cannot be grown in climates where the winters are not cold. Similarly, bulbs such as those of tulips, hyacinths, and narcissus can be "forced," that is, made to bloom indoors in the winter, but only if they have previously been outside or in a cold place. Such bulbs are actually large buds in which the leaves are modified for storage (page 611).

Cold is not required to break dormancy in all cases. In the potato, for instance, in which the "eyes" are dormant buds, at least two months of dry storage are the chief requirement; temperature is not a factor. In many plants, particularly trees, the photoperiodic response breaks winter dormancy, with the dormant buds being the receptor organs.

Ethylene is sometimes used to promote bud sprouting in potato and other tubers. Application of gibberellins sometimes breaks bud dormancy. For instance, gibberellin treatment of a peach bud may induce development after the bud has been kept for 164 hours below 8°C. Does this mean that under normal conditions, an increase in gibberellin terminates the dormancy? Not necessarily. Dormancy may be a state of balance between growth inhibitors and growth stimulators. Addition of any growth stimulator (or removal of inhibitors, such as abscisic acid) may alter the balance so that growth begins.

Nastic Movements and Solar Tracking

Nastic movements are plant movements that occur in response to a stimulus but with a direction of movement that is independent of the position of the origin of the stimulus. Probably the most widely occurring nastic movements are the *sleep movements.* Known technically as **nyctinastic movements** (meaning "night closure," from the Greek *nyx,* "night," and *nastos,* "close-pressed"), they constitute the up and down movements of leaves in response to the daily rhythms of light and darkness. The leaves are oriented horizontally during the day and vertically at night (Figure 28–9). At night, the leaflets of compound leaves fold and the edges of opposite leaves come together (Figure 28–26). These movements are especially common in leguminous plants.

Most nyctinastic leaf movements result from changes in the size of parenchyma cells in the jointlike thickenings at the base of each leaf (and, if the leaf is compound, at the base of each leaflet). This thickening, known as a **pulvinus,** is a flexible cylinder with the vascular system concentrated in the center. Anatomically, all pulvini consist of a core of vascular tissue surrounded by a bulky cortex of thin-walled parenchyma cells, the inner layer of which is a true endodermis with Casparian strips (Figure 28–26). Pulvinar movement is associated with reversible changes in turgor and with concomitant contractions and expansions of the ground parenchyma on opposite sides of the pulvinus. Turgor changes in the contracting and expanding cells, termed *motor cells,* are brought about by fluxes of potassium and chloride ions across the plasma membrane of the motor cells, followed by water fluxes in the same direction. These changes are under the control of the circadian clock and phytochrome.

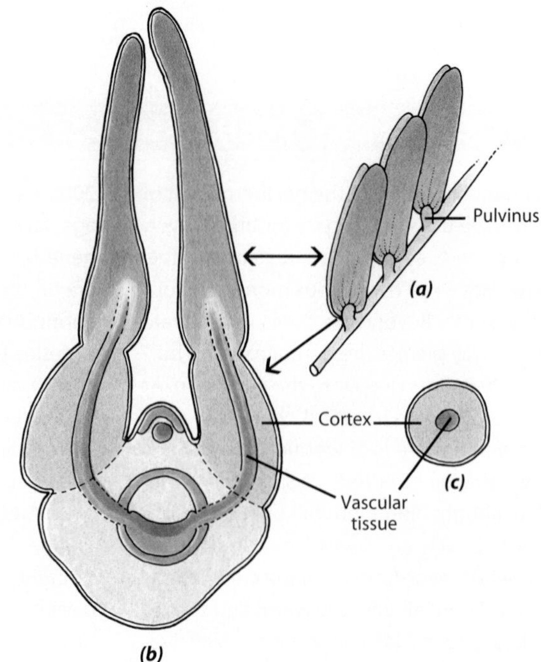

28–26 Pulvini Diagrammatic representation of pulvini in *Mimosa pudica.* ***(a)*** Portion of rachis showing three leaflets, each with a pulvinus at its base. ***(b)*** Transverse section through a rachis with two leaflets in the closed condition, showing a longitudinal view of the pulvini. ***(c)*** Transverse section through a pulvinus showing the core of vascular tissue surrounded by a cortex, which consists largely of thin-walled parenchyma cells.

Thigmonastic Movements Are Nastic Movements Resulting from Mechanical Stimulation

Thigmonastic movements are exemplified by the familiar sensitive plant *(Mimosa pudica),* whose leaflets and sometimes entire leaves droop suddenly in response to touch, shaking, or electrical or thermal stimulation (Figure 28–27). As with sleep movements (also exhibited by *M. pudica),* this response is a result of a sudden change in turgor pressure in motor cells of the pulvinus at the base of each leaflet and leaf. The loss of water via aquaporins in the plasma membrane of these cells (page 84) follows the efflux of chloride, potassium, and calcium ions from the cells into the apoplast. The accumulation of ions in the apoplast seems to be initiated by a decrease of water potential triggered by an apoplastic accumulation of sucrose unloaded from the phloem. ATPase activity seems to be strongly involved in the thigmonastic movement in *M. pudica.* A high level of H^+-ATPase has been found in both the pulvini and the phloem. Only a single leaflet needs to be stimulated; the stimulus then moves to other parts of the leaf and throughout the plant. Two distinct mechanisms, one electrical and the other chemical, appear to be involved in the spread of the stimulus in the sensitive plant. The rapid movements of the leaflets are believed to startle potential herbivores.

The carnivorous Venus flytrap *(Dionaea muscipula),* found only in wet pine savannas of southern North Carolina and northern South Carolina, is one of only two plants worldwide that actively capture animal prey by traps that spring shut; the other is the European waterwheel *(Aldrovanda vesiculosa).* The leaves of the flytrap consist of two parts, a relatively broad petiole (the "footstalk") and the trap, a lobed blade held together by the midrib, which serves as the hinge of the trap. The upper, or inner, surface of each lobe contains three "trigger hairs," which act as mechanosensors. When the trap closes, teeth along the outer margin of the lobes interlock and serve to prevent prey from escaping (Figure 28–28). After the prey is captured, the flytrap excretes digestive enzymes and absorbs the liquefying meal. Traps may function for a second or even a third time before becoming defunct.

The mechanism by which the Venus flytrap snaps is gradually becoming understood. The first step involves the generation of electrical charges, which are propagated from the trigger hairs to the midrib via plasmodesmata. When an insect first brushes against a hair, the bending hair triggers a tiny electrical charge, which by itself is insufficient to stimulate the trap. It is necessary for the hair to be brushed against twice or for a second hair to be touched within an interval of 0.75 to 40 seconds. The first charge is "memorized" by the flytrap, and as soon as the second charge propagates to the midrib, electrical signals activate ATP hydrolysis, starting proton transport, which initiates the opening

(a)

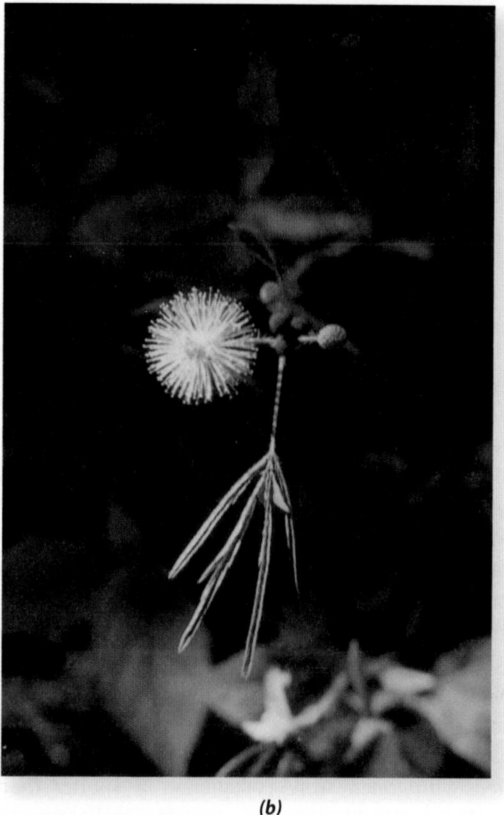

(b)

28–27 Thigmonastic movements Thigmonasty in the sensitive plant *(Mimosa pudica)*. ***(a)*** Normal position of leaves and leaflets. ***(b)*** Responses to touch result from changes in turgor pressure in certain cells of the jointlike thickenings (pulvini) at the base of the leaflets. Only a single leaflet need be stimulated for the response in *(b)* to occur.

of aquaporins and a surge of water from motor cells beneath the upper epidermis to those beneath the lower epidermis. The loss of turgor by cells beneath the upper epidermis, accompanied by expansion of cells under the lower epidermis, closes the trap. Once the second charge is propagated, the trap snaps closed in 0.3 second. Charles Darwin was quite taken by the remarkable quickness and power of the Venus flytrap, a plant he called "one of the most wonderful in the world."

(a)

(b)

28–28 Touch response in the Venus flytrap *(Dionaea muscipula)* Here, an unwary fly, attracted by nectar secreted on the leaf surface, can be seen on a leaf ***(a)*** before and ***(b)*** after its closure. Each leaf half is equipped with three sensitive hairs. When an insect walks on one of the leaves, it brushes against the hairs, triggering the traplike closing of the leaf. The toothed edges mesh, the leaf halves gradually squeeze shut, and the insect is pressed against digestive glands on the inner surface of the trap.

The trapping mechanism is so specialized that it can distinguish between living prey and inanimate objects, such as pebbles and small sticks, that fall on the leaf: the leaf will not close unless two of its hairs are touched in succession or one hair is touched twice.

The Generalized Effects of Mechanical Stimuli on Plant Growth and Development Occur by Thigmomorphogenesis

In addition to the specialized responses of some plants—such as the sensitive plant and Venus flytrap—to touch and other mechanical stimuli, plants also respond to mechanical stimuli by altering their growth patterns, a phenomenon known as **thigmomorphogenesis.** Although botanists have long known that plants grown in a greenhouse tend to be taller and more spindly than plants grown outside, it was not until the 1970s that systematic studies undertaken by M. J. Jaffe revealed that regular rubbing or bending of stems inhibits their elongation and stimulates their radial expansion, resulting in shorter, stockier plants. Plants in a natural environment are, of course, subjected to similar stimuli, in the form of wind, raindrops, and rubbing by passing animals and machines. The swaying of plants by wind is a powerful causal factor in thigmomorphogenesis.

As we saw in Chapter 27, such growth responses are caused by changes in gene expression. Studies on *Arabidopsis* show that some of the genes whose expression is induced by touch encode proteins related to the calcium-binding protein calmodulin, suggesting a role for Ca^{2+} in mediating the growth responses. (Ca^{2+} has been implicated in the regulation of a number of plant processes, including mitosis, polarized cell growth, cytoplasmic streaming, and both tropic and nastic movements.) *Arabidopsis* plants stimulated with touch were conspicuously shorter than untreated plants (Figure 28–29).

28–29 Thigmomorphogenesis *Arabidopsis thaliana* plants at six weeks of age. The plant on the left was touched twice daily; the plant on the right is the untreated control. The inhibition of growth by touch or other noninjurious mechanical stimulation that results in reduced length and increased width is referred to as thigmomorphogenesis.

Some Plants Orient Their Leaves to the Sun by Solar Tracking

The leaves and flowers of many plants have the ability to move diurnally, orienting themselves either perpendicular or parallel to the sun's direct rays. Commonly known as **solar tracking** (Figure 28–30), this phenomenon is technically called **heliotropism** (from

(a)

(b)

28–30 Solar tracking ***(a)*** The leaves of a lupine *(Lupinus arizonicus)* orient to track the course of the sun throughout the day. This phenomenon is commonly known as solar tracking, or heliotropism. ***(b)*** Solar tracking by a field of sunflowers *(Helianthus annuus).*

the Greek *helios,* meaning "sun"). Unlike stem phototropism, the leaf movement of heliotropic plants is not the result of asymmetrical growth. In most cases the movements involve pulvini at the bases of leaves and/or leaflets, and apparently entail mechanisms similar to those associated with nyctinastic movements. Some petioles appear to have pulvinal characteristics along most or all of their length. Some common plants exhibiting heliotropic leaf movements are cotton, soybeans, cowpeas, lupines, and sunflowers. It has been suggested that floral heliotropism is closely related to phototropism. However, the stem of the flower or inflorescence is essentially an upright beam, one end of which is fixed and the other of which, the flower or inflorescence, tracks the sun over an angle of close to 180°. Motor cells such as those found in pulvini have not been found in flower stems.

SUMMARY

Plants Have a Variety of Adaptations That Enable Them to Detect and Respond to Alterations in Their Environment

Phototropism is the curving of a growing shoot toward light. The differential growth of the seedling is caused by lateral migration of the growth hormone auxin under the influence of light. The photoreceptors for this response are pigment-containing proteins that absorb blue light and convert the signal into a biochemical response. Gravitropism is the response of a shoot or a root to gravity. The movement of auxin to the lower surface of a horizontally oriented shoot or root plays a role in the upward curvature of the shoot and the downward curvature of the root, although in the root the downward curvature is initiated by the stimulation of elongation along its upper side. The site of perception of gravity in shoots is in the starch sheath and in roots is in columella cells of the rootcap, where starch-containing amyloplasts act as gravity sensors. Hydrotropism is directed growth of roots in response to a moisture gradient. Thigmotropism is a response to contact with a solid object, as in the winding of tendrils.

Living Organisms Exhibit Circadian Rhythms

Circadian rhythms are cycles of activity that recur at intervals of about 24 hours in an organism under constant environmental conditions. These rhythms are endogenous, caused not by environmental factors but by an internal timing mechanism called the circadian clock, which controls the expression of specific genes. The clock is synchronized, or entrained, by the environment, principally by light-dark cycles and temperature cycles.

Photoperiodism Is the Response of Organisms to Changing 24-Hour Cycles of Light and Darkness

Photoperiodic responses control the onset of flowering in many plants. Some plants, known as long-day plants, flower only when the periods of light exceed a critical length. Other plants, short-day plants, flower only when the periods of light are less than some critical length. Day-neutral plants flower regardless of photoperiods. Experiments have shown that the dark period rather than the light period is the critical factor.

Phytochrome Is Involved in Photoperiodism

Phytochrome, a photoreceptor commonly present in the tissues of plants, is the molecule that detects transitions between light and darkness. This photoreceptor exists in two forms, P_r and P_{fr}. P_r absorbs red light and is thereby converted to P_{fr}. The P_{fr} absorbs far-red light and is converted to P_r. P_{fr} is the active form of the photoreceptor; it promotes flowering in long-day plants, inhibits flowering in short-day plants, promotes germination in lettuce seeds, and promotes normal growth in seedlings. The phytochrome molecule contains two distinct parts: a light-absorbing portion (the chromophore) and a large protein portion.

The Photoperiod Is Perceived in the Leaves, but the Response Takes Place in the Shoot Apical Meristem

Analysis of flowering in *Arabidopsis* indicates that the FT (FLOWERING LOCUS1) protein, which is synthesized in companion cells and transferred to sieve tubes in the phloem of minor veins in source leaves, is the signal that induces floral development in the shoot apical meristem. FT is transported in the sieve tubes to the apical meristem, where it forms a complex with the transcription factor FD. The FT-FD complex initiates flowering by activating floral meristem genes.

Flowering in Some Plants Is Promoted by Prolonged Cold Temperatures

The process by which exposure to prolonged cold renders a plant competent to flower is called vernalization. A requirement for vernalization is common in many winter annuals and biennials in temperate climates and prevents them from flowering during short cold periods in the fall. Unlike photoperiod, which is perceived in the leaves, vernalization takes place directly in cells of the shoot apical meristem.

Dormancy Permits the Plant to Survive Water Shortages and Extremes of Hot or Cold

Dormancy is a special condition of arrested growth in which entire plants, or such structures as seeds or buds, do not renew growth without special environmental cues. The requirement for such cues, which include cold exposure, dryness, and a suitable photoperiod, prevents the tissues from breaking dormancy during superficially favorable conditions. Decreasing daylength is the primary factor involved in the induction of dormancy in buds. Acclimation to cold leads to cold hardiness, the ability of the plant to survive the extreme cold of winter weather.

Many Plants Exhibit Nastic Movements

Plant movements that occur in response to a stimulus, but in a direction independent of the direction of the stimulus, are called nastic movements. Among these are the widely occurring sleep, or nyctinastic, movements—the up and down movements of leaves in response to the daily rhythm of light and darkness. Nastic movements resulting from touch (thigmonastic movements) include the triggered closing of the carnivorous Venus flytrap.

SUMMARY TABLE The Major Types of Plant Movements or Growth Responses to External Stimuli

TYPE OF MOVEMENT	DESCRIPTIONS AND EXAMPLES	MECHANISMS AND/OR OTHER FEATURES
Tropism: directional growth in response to an external stimulus	Phototropism: growth of shoot, coleoptile, or petiole of leaf toward the light	Caused by a light-induced lateral redistribution of auxin to the shaded side of the structure. Auxin then stimulates cell elongation on the shaded side. Redistribution of auxin is mediated by two flavoprotein photoreceptors.
	Gravitropism: downward growth of roots and upward growth of shoots	Apparently caused by the gravity-induced redistribution of auxin to the lower side of the stem and root. Higher auxin concentration in stems stimulates cell elongation; in roots, it inhibits cell elongation. In coleoptiles and stems, gravity is perceived in the starch sheath; in roots, in columella cells of the rootcap.
	Hydrotropism: directed growth of roots in response to a moisture gradient	Perceived in the rootcap, and auxin plays a role.
	Thigmotropism: response to contact with a solid object	Responsible for the coiling of tendrils around a support. Uncertain whether auxin gradients are involved.
Nastic movement: movement in response to an external stimulus, with direction of movement unrelated to the direction of the stimulus	Nyctinasty: sleep movements of leaves	Results from turgor changes in motor cells of the pulvinus. Under the control of the circadian clock and phytochrome.
	Thigmonasty: movement, such as closing of leaves of sensitive plant and Venus flytrap, resulting from mechanical stimulation	Results from turgor changes in motor cells of the pulvinus. Involves both electrical and mechanical mechanisms. ATPase activity strongly involved.
Thigmomorphogenesis: growth response unrelated to the direction of the external stimulus	Mechanical stimulus such as rubbing or bending of stems inhibits elongation and stimulates radial expansion	Involves changes in expression of genes encoding proteins related to the calcium-binding protein calmodulin. Thus calcium ions are implicated in the mechanism.
Solar tracking: also known as heliotropism	Leaves and flowers orient themselves with respect to the sun's rays during the day	For leaves, apparently results from turgor changes in cells of the pulvinus, as with nyctinastic movements. The mechanism of floral heliotropism uncertain.

Thigmomorphogenesis and Solar Tracking Are Familiar Phenomena

A more generalized effect of mechanical stimuli on plant growth and development is thigmomorphogenesis, whereby plants may respond to mechanical stimuli by altering their growth patterns. Such plants typically are shorter and stockier than nonstimulated plants. In addition, the leaves and flowers of many plants follow or track the sun—a phenomenon known as solar tracking, or heliotropism—during the course of the day, either maximizing or minimizing absorption of solar radiation.

QUESTIONS

1. Explain the starch-statolith hypothesis, as it applies to roots. What evidence is there in support of the hypothesis?
2. Distinguish between the following: circadian rhythm and circadian clock; phototropism and photoperiodism; thigmotropism and thigmonastic movement.
3. Plants keep track of daylength by measuring the length of darkness. Explain.
4. Explain how it is possible for a long-day plant and a short-day plant growing at the same location to flower on the same day of the year.
5. Suppose you are given a chrysanthemum plant, in bloom, one autumn and you decide to keep it indoors as a houseplant. What precautions will you need to take to ensure that it blooms again?
6. Explain how the floral stimulus, which originates in the minor veins of source leaves, has its effect in the shoot apical meristem.
7. The flowers of some plants open in the morning and close at dusk. Design and describe an experiment that could be used to determine whether these daily rhythms are controlled by the plant's circadian clock or by the presence or absence of light.
8. What mechanisms are responsible for the sleep movements of leaves and for the closure of the Venus flytrap?

CHAPTER 29

Plant Nutrition and Soils

◀ **Subterranean orchid** Living completely underground, this rare Australian orchid *(Rhizanthella gardneri)* is seen here with soil partially removed. Leafless and unable to photosynthesize, these orchids have developed a parasitic relationship with a mycorrhizal fungus associated with the roots of a shrub, the broom honeymyrtle. The orchids receive their nutrients from the photosynthesizing shrub via the fungal hyphae.

CHAPTER OUTLINE

Essential Elements
Functions of Essential Elements
The Soil
Nutrient Cycles
Nitrogen and the Nitrogen Cycle
Phosphorus and the Phosphorus Cycle
Human Impact on Nutrient Cycles and Effects of Pollution
Soils and Agriculture
Plant Nutrition Research

Plants must obtain from the environment the specific raw materials required in the complex biochemical reactions necessary for the maintenance of their cells and for growth. In addition to light, plants require water and certain chemical elements for metabolism and growth. Much of the evolutionary development of plants has involved structural and functional specialization necessary for the efficient uptake of these raw materials and their distribution to living cells throughout the plant.

In contrast to those of animals, the nutritional demands of plants are relatively simple. Under favorable environmental conditions, most green plants can use light energy in the process of photosynthesis to transform CO_2 and H_2O into organic compounds for their energy source. They can also synthesize all of their required amino acids and vitamins, using the products of photosynthesis and inorganic nutrients, such as nitrogen, drawn from the environment.

Plant nutrition involves the uptake from the environment of all the raw materials required for essential biochemical processes, the distribution of these materials within the plant, and their utilization in metabolism and growth.

More than 60 chemical elements have been identified in plants, including gold, silver, lead, mercury, arsenic, and uranium. Obviously, not all of the elements present in plants are essential or even useful. The presence of nonuseful and potentially toxic elements such as cadmium is, to a certain extent, a reflection of the composition of the soil in which the plants are

CHECKPOINTS

After reading this chapter, you should be able to answer the following:

1. What elements are essential for plant growth, and what are their functions?
2. Describe some of the common symptoms associated with nutrient deficiencies. How does the mobility of a nutrient affect the symptoms associated with its deficiency?
3. What are the sources of the inorganic nutrients utilized by plants?
4. Why are nutrient cycles so important for plants? What are the main components of the nitrogen and phosphorus cycles?
5. In what ways have humans disrupted nutrient cycles? How is plant nutrition research contributing to solutions for problems associated with agriculture and horticulture?

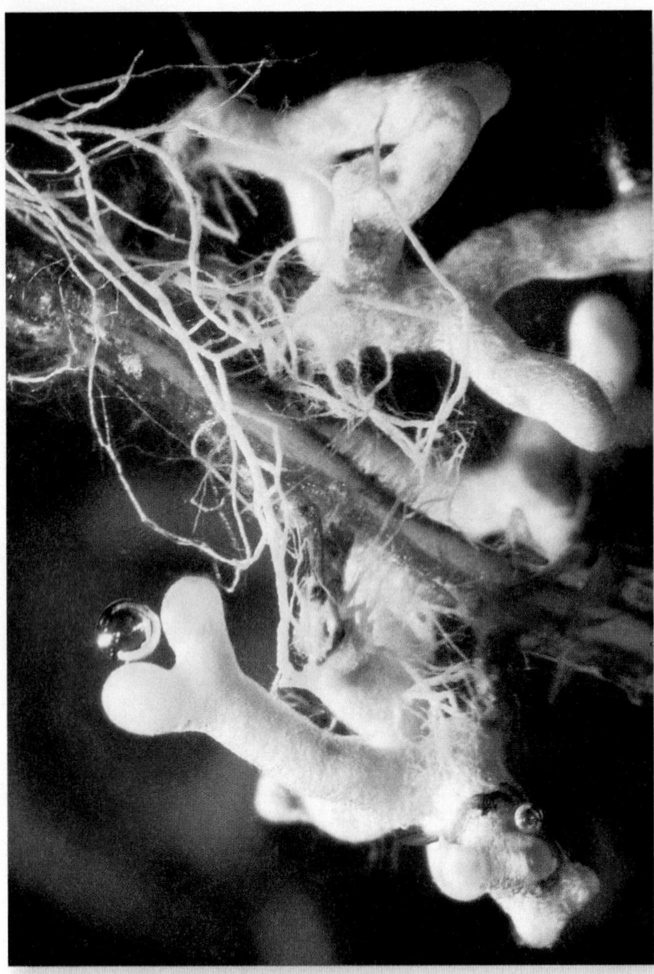

29–1 Mycorrhizas The fungus *Boletus parasiticus* forms ectomycorrhizas, seen here as sheaths surrounding the roots of a red pine *(Pinus resinosa)*. The narrow strands of mycelium serve as extensions of the roots. Mycorrhizas benefit their host plants by increasing uptake of water and essential elements, especially phosphorus. Mycorrhizas also provide protection against attack by pathogenic fungi and nematodes. This striking image was achieved by photographing the roots under water—hence the air bubbles.

growing (see Figure 1–13). Most of the chemical elements found in plants are absorbed as inorganic ions from the soil solution.

The roles played by root hairs and by the fungal hyphae of mycorrhizas (Figure 29–1) in the absorption of inorganic ions are discussed in Chapters 30 and 14, respectively. In Chapter 30, we consider the mechanism of ion absorption by roots and the pathway followed by the ions from the soil solution to the tracheary elements in the vascular cylinder of the root.

Essential Elements

By as early as 1800, chemists and plant biologists had analyzed plants and demonstrated that certain chemical elements were absorbed from the environment. Opinions differed, however, on whether the absorbed elements were impurities or constituents required for essential functions. By the mid-1880s, through the use of **hydroponics**—the technique of growing plants with their roots immersed in a nutrient solution without soil—it had been established that at least 10 of the chemical elements present in plants were necessary for normal growth. In the absence of any one of these elements, plants displayed characteristic abnormalities of growth, or deficiency symptoms, and often did not reproduce normally. These 10 elements—carbon, hydrogen, oxygen, potassium, calcium, magnesium, nitrogen, phosphorus, sulfur, and iron—were designated as **essential elements** for plant growth. They are also referred to as essential inorganic nutrients.

In the early 1900s, manganese was also found to be an essential element. During the next 50 years, with the aid of improved techniques for removing impurities from nutrient cultures, five additional elements—zinc, copper, chlorine, boron, and molybdenum—were determined to be essential, bringing the total to 16 elements. A seventeenth element, nickel, was added to the list in the 1980s.

There are two primary criteria by which an element is judged to be essential to a plant: (1) whether it is needed for the plant to complete its life cycle (that is, to produce viable seed) and/or (2) whether it is part of any molecule or constituent of the plant that is itself essential to the plant, such as the magnesium in the chlorophyll molecule or the nitrogen in proteins. A third criterion also used by many plant nutritionists is whether deficiency symptoms appear in the absence of the element, even though the plant may be capable of producing viable seed.

The Essential Elements Can Be Divided into Macronutrients and Micronutrients

Chemical analyses for essential elements are a useful means of determining the relative amounts of various elements required for the normal growth of different plant species. One way to determine which elements are essential and in what concentrations is to chemically analyze healthy plants. The freshly harvested plants or plant parts are heated in an oven to drive off the water, then the remaining material, or **dry matter,** is analyzed.

Table 29–1 lists the 17 elements currently believed to be essential to all vascular plants and the approximate internal concentrations considered adequate for normal plant growth and development. Note that the concentrations vary over a wide range. The first nine elements are called **macronutrients,** because they are required in large amounts (more than 1000 milligrams per kilogram of dry weight). The last eight elements are called **micronutrients,** or trace elements, because they are required in very small, or trace, amounts (less than 100 milligrams per kilogram of dry weight). Many ions are accumulated inside cells at concentrations much higher than those in the soil solution, and often at concentrations much higher than those minimally required. As discussed in Chapter 4, plant cells must use energy to accumulate solutes, that is, to move solutes against an electrochemical gradient.

Certain plant species and taxonomic groups are characterized by unusually high or low amounts of specific elements (Figure 29–2). As a result, plants growing in the same nutrient medium may differ markedly in nutrient content. Eudicots generally require greater amounts of calcium and boron than do monocots. Analyses of this sort are particularly valuable in agriculture

Table 29-1 **Essential Elements for Most Vascular Plants and Internal Concentrations Considered Adequate**

Element	Chemical Symbol	Form Available to Plants	Usual Concentration in Healthy Plants (% or ppm of dry weight)	Number of Atoms Relative to Molybdenum
Macronutrients				
Obtained from water or carbon dioxide				
Hydrogen	H	H_2O	6%	60,000,000
Carbon	C	CO_2	45%	40,000,000
Oxygen	O	O_2, H_2O, CO_2	45%	30,000,000
Obtained from the soil				
Nitrogen	N	NO_3^-, NH_4^+	1.5%	1,000,000
Potassium	K	K^+	1.0%	250,000
Calcium	Ca	Ca^{2+}	0.5%	125,000
Magnesium	Mg	Mg^{2+}	0.2%	80,000
Phosphorus	P	$H_2PO_4^-$, HPO_4^{2-}	0.2%	60,000
Sulfur	S	SO_4^{2-}	0.1%	30,000
Micronutrients				
Chlorine	Cl	Cl^-	100 ppm	3000
Iron	Fe	Fe^{3+}, Fe^{2+}	100 ppm	2000
Boron	B	H_3BO_3, $H_2BO_3^-$	20 ppm	2000
Manganese	Mn	Mn^{2+}	50 ppm	1000
Zinc	Zn	Zn^{2+}	20 ppm	400
Copper	Cu	Cu^+, Cu^{2+}	6 ppm	100
Nickel	Ni	Ni^{2+}	0.1 ppm	2
Molybdenum	Mo	MoO_4^{2-}	0.1 ppm	1

After E. Epstein, *Mineral Nutrition of Plants: Principles and Perspectives* (New York: John Wiley & Sons, Inc., 1972).

as a guide to the nutritional well-being of plants and the need for applications of fertilizer. Potential nutritional deficiencies in livestock that consume specific plants can also be predicted by inorganic analyses.

Nutritional studies have established that some elements are essential for only limited groups of plants or for plants grown under specific environmental conditions. Such elements are referred to as **beneficial elements.** The five most investigated beneficial elements are aluminum (Al), cobalt (Co), sodium (Na), selenium (Se), and silicon (Si). Aluminum, the third most abundant element in the Earth's crust, is toxic to both plants and animals at elevated levels. It is beneficial at low concentrations, however, to some plants, such as the tea plant *(Camellia sinensis),* in which it induces increased antioxidant enzyme activity. Plants that accumulate high levels of aluminum may use the element in their tissues to deter herbivores. Cobalt, which is not very abundant (between 15 and 25 parts per million, or ppm, in soils), benefits legumes such as alfalfa *(Medicago sativa).* It is not the alfalfa, however, that requires the cobalt but the symbiotic nitrogen-fixing bacteria growing in association with its roots. Sodium is known to be an essential element for certain halophytes (see the essay on page 703) and for C_4 or CAM photosynthetic pathways (see Chapter 7), and selenium (typical soil levels below 1 ppm) is beneficial, if not essential, for plants that accumulate high levels of this element. Silicon, which makes up more than 25 percent of the Earth's crust, is a major constituent of plants. So far it has proved to be essential only in the horsetails (Figure 29–2b), sedges, and grasses. The silicon accumulates at high levels in epidermal cell walls and in intercellular spaces between subepidermal cells, providing support for the stems and leaves as well as resistance to insects, pathogenic fungi, and bacteria.

Functions of Essential Elements

Essential elements have many roles in plants, including structural, enzymatic, regulatory, and ionic. For example, nitrogen and sulfur are major components of both proteins and coenzymes, and magnesium, in addition to being part of the chlorophyll molecule, is required for the activity of many enzymes. Calcium is an

(a)

(b)

29–2 Some plants contain large amounts of specific elements *(a)* Plants of the mustard family, such as wintercress *(Barbarea vulgaris)*, use sulfur in the synthesis of the mustard oils that give the plants their sharp taste. These mustard plants were growing in Ithaca, New York. *(b)* Horsetails incorporate silicon into their cell walls, making the plants indigestible to most herbivores but useful, at least in colonial America, for scouring pots and pans. These bushy vegetative shoots, as well as several stalklike fertile shoots, of the horsetail *Equisetum telemateia* were photographed in California.

important second messenger that controls stomatal opening and closing by regulating ionic gradients established by potassium ions and several anions (page 658). All nutrients have well-known specific functions (Table 29–2) that are impaired when the supply of the nutrient is inadequate. Because the nutrients are involved in such fundamental processes, the deficiencies affect a wide variety of structures and functions in the plant body.

Nutrient Deficiency Symptoms Depend on the Function(s) and Mobility of the Essential Element

Because they are the most easily observed, most well-described nutrient deficiency symptoms are associated with the shoot. They include such symptoms as stunted growth of stems and leaves, localized death of tissues **(necrosis),** and yellowing of the leaves due to loss or reduced development of chlorophyll **(chlorosis)** (Figure 29–3). The deficiency symptoms for any essential element depend not only on the element's role in the plant but also on its mobility within the plant—that is, the relative ease with which the element is transported in the phloem from older to younger plant parts, particularly leaves.

Take, for example, magnesium, which is an essential part of the chlorophyll molecule. Without magnesium, chlorophyll cannot be formed, resulting in chlorosis. Chlorosis of the older leaves of magnesium-deficient plants typically becomes more severe than that of the younger leaves. The reason for this is the ability of plants to mobilize limiting elements to where they are most needed, for example, from older to younger leaves. The withdrawal of magnesium from older leaves to younger leaves also depends on magnesium's mobility in the phloem. Elements that readily move through the phloem are said to be **phloem-mobile.** In addition to magnesium, phosphorus, potassium, and nitrogen are phloem-mobile. Other elements, such as boron, iron, and calcium, are relatively immobile, whereas copper, manganese, molybdenum, sulfur, and zinc are usually intermediate in mobility. Deficiency symptoms of phloem-mobile elements appear earliest and are most pronounced in older leaves, whereas those of phloem-immobile elements appear first in younger leaves.

Some of the more common nutrient deficiency symptoms are given in Table 29–2. The table does not include the macronutrients carbon, oxygen, and hydrogen, which are acquired primarily from CO_2 and H_2O during photosynthesis and are the major components of the plant's organic compounds.

The Soil

Soil is the primary nutrient medium for plants. Soils provide plants not only with physical support but also with inorganic nutrients, as well as with water and a suitable gaseous environment for root systems. Understanding the origins of soils, as well as their chemical and physical properties in relation to plant growth requirements, is critical in planning for the nutrition of field crops.

Table 29-2 Essential Mineral Elements: Functions and Deficiency Symptoms

Element	Functions	Deficiency Symptoms
Macronutrients		
Nitrogen	Component of such molecules as amino acids, proteins, nucleotides, nucleic acids, chlorophylls, coenzymes	General chlorosis, especially in older leaves; in severe cases, leaves become completely yellow and then become tan as they die; some plants exhibit purple coloration due to accumulation of anthocyanins
Potassium	Involved in osmosis and ionic balance and in opening and closing of stomata; cofactor for many enzymes	Mottled or chlorotic leaves with small spots of necrotic tissue at tips and margins; weak, narrow stems; mostly older leaves affected
Calcium	Component of middle lamella of cell walls; enzyme cofactor; involved in cellular membrane permeability; second messenger in signal transduction	Shoot and root tips die; young leaves at first hooked, then die back at tips and margins, developing cut-out appearance at these sites
Magnesium	Component of the chlorophyll molecule; activator of many enzymes	Mottled or chlorotic leaves; may redden; sometimes with necrotic spots; leaf tips and margins turned upward; mostly older leaves affected; stems slender
Phosphorus	Component of energy-carrying phosphate compounds (ATP and ADP), nucleic acids, several coenzymes, phospholipids	Plants dark green, often accumulating anthocyanins and becoming red or purple; in later stages of growth, stems stunted; oldest leaves become dark brown and die
Sulfur	Component of some amino acids and proteins and of coenzyme A	Young leaves with light green veins and interveinal (between the veins) areas; chlorosis initially in mature and young leaves, rather than in older leaves as with nitrogen deficiency
Micronutrients		
Chlorine	Involved in osmosis and ionic balance; required for photosynthetic reactions that produce oxygen	Wilted leaves with chlorotic and necrotic spots; leaves often become bronze color; roots stunted in length and thickened near tips
Iron	Required for chlorophyll synthesis; component of cytochromes and nitrogenase	Interveinal chlorosis of young leaves; stems short and slender
Boron	Influences Ca^{2+} utilization, nucleic acid synthesis, and membrane integrity; related to cell wall stability	Earliest symptom is failure of root tips to elongate; young leaves light green at bases; leaves become twisted and shoot dies back at terminal bud
Manganese	Activator of some enzymes; required for integrity of chloroplast membrane and for oxygen release in photosynthesis	Initially, interveinal chlorosis on younger or older leaves, depending on species, followed by or associated with interveinal necrotic spots; disorganization of thylakoid membranes of chloroplasts
Zinc	Activator or component of many enzymes	Reduction in leaf size and length of internodes; leaf margins often distorted; interveinal chlorosis; mostly older leaves affected
Copper	Activator or component of some enzymes involved in oxidation and reduction	Young leaves dark green, twisted, misshapen, and often with necrotic spots
Nickel	Essential part of enzyme functioning in nitrogen metabolism	Necrotic spots in leaf tips
Molybdenum	Required for nitrogen fixation and nitrate reduction	Interveinal chlorosis appearing first on older leaves and then progressing to youngest leaves; chlorosis followed by gradual necrosis of interveinal areas and then of remaining tissues

After E. Epstein, *Mineral Nutrition of Plants: Principles and Perspectives* (New York: John Wiley & Sons, Inc., 1972).

(a)

(b)

29–3 Chlorosis The loss or reduced development of chlorophyll resulting from mineral deficiency is known as chlorosis. ***(a)*** Deficiency of magnesium, a phloem-mobile element, in maize *(Zea mays)*. The older leaves are more affected than the younger leaves, which are able to translocate magnesium from the older leaves. ***(b)*** Deficiency of iron, a so-called phloem-immobile element, results in symptoms of chlorosis in younger leaves, as seen here in sorghum *(Sorghum bicolor)*.

The Weathering of Rocks Produces the Inorganic Nutrients Utilized by Plants

The Earth is composed of 92 naturally occurring elements, which are often found in the form of minerals. **Minerals** are naturally occurring inorganic compounds that are usually composed of two or more elements in definite proportions by weight. Quartz (SiO_2), calcite ($CaCO_3$), and kaolinite ($Al_4Si_4O_{10}(OH)_8$) are examples of minerals.

Weathering processes, involving the physical disintegration and chemical decomposition of minerals and rocks at or near the Earth's surface, produce the inorganic materials from which soils are formed. Weathering can be initiated by heating and cooling, which cause substances in rocks to expand and contract, splitting the rocks apart. Water, wind, and glaciers often carry the rock fragments great distances, exerting a scouring action that breaks and wears the fragmented rock into even smaller particles.

Soils also contain organic materials. If light and temperature conditions permit, bacteria, fungi, algae, lichens, and bryophytes, as well as small vascular plants, gain a foothold on or among the weathered rocks and minerals. Growing roots also split the rocks, and the disintegrating bodies of plants and of the animals associated with them add to the accumulating organic material. Finally, larger plants move in, anchoring the soil in place with their root systems (Figure 29–4), and a new community begins.

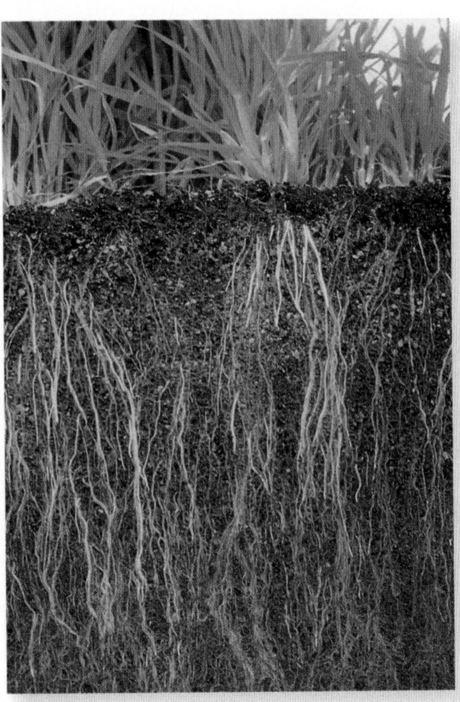

29–4 Fibrous roots The fibrous root systems of grasses bind and anchor prairie soil in place.

Soils Consist of Layers Called Horizons

Examining a vertical section of soil, one can see variations in the color, the amount of living and dead organic matter, the porosity, the structure, and the extent of weathering. These variations generally result in a succession of rather distinct layers that soil scientists refer to as **horizons.** At least three horizons—designated A, B, and C—are recognized (Figure 29–5).

The A horizon (sometimes called the "topsoil") is the upper region, that of the greatest physical, chemical, and biological activity. The A horizon contains the greatest portion of the soil's organic material, both living and dead. It is the horizon where **humus**—the dark-colored mixture of colloidal organic decay products—accumulates. It is alive with populations of roots, insects and other small arthropods, earthworms, protists, nematodes, and decomposer organisms (Figure 29–6).

The B horizon (sometimes called the "subsoil") is a region of deposition. Iron oxide, clay particles, and small amounts of organic matter are among the materials leached from the A horizon into the B horizon by water percolating, or moving down, through the soil. The B horizon contains much less organic material and is less weathered than the A horizon above it. Human activity has often mixed together the A and B horizons through tillage operations, forming an Ap ("p" for "plow") horizon, which blends into the B horizon.

The C horizon, or soil base, is composed of the broken-down and weathered rocks and minerals from which the true soil in the upper horizons is formed.

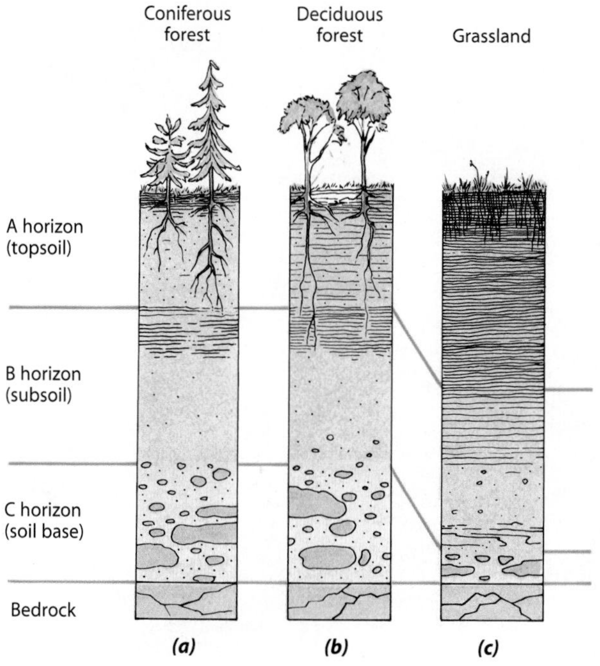

29–5 Three major soil types ***(a)*** The litter of the northern coniferous forest is acidic and slow to decay, and the soil has little accumulation of humus, is very acidic, and is leached of minerals. ***(b)*** In the cool, temperate deciduous forest, decay is somewhat more rapid, leaching less extensive, and the soil more fertile. Such soils have been widely used for agriculture, but they need to be prepared by adding lime (to reduce acidity) and fertilizer. ***(c)*** In the grasslands, almost all of the plant material above the ground dies each year, as do many of the roots, and thus large amounts of organic matter are constantly returned to the soil. In addition, the finely divided roots penetrate the soil extensively. The result is highly fertile soil, often black in color, with a topsoil sometimes more than a meter in depth. Natural soils, particularly forest soils, often contain a layer of decomposing litter on top of the A horizon. This layer is called the O horizon.

Soils Are Composed of Solid Matter and Pore Space

The pore space is the space around the soil particles. Different proportions of air and water occupy the pore space, depending on prevailing moisture conditions. The soil water is present primarily as a film on the surfaces of soil particles. The fragments of rocks and minerals in the soil vary in size from sand grains, which can be seen easily with the naked eye, to clay particles too small to be seen even under the low power of a light microscope. The following classification is one scheme for categorizing soil particles according to size:

Particles	Diameter (in micrometers)
Coarse sand	200 to 2000
Fine sand	20 to 200
Silt	2 to 20
Clay	Less than 2

Soils contain a mixture of particles of different sizes and are divided into textural classes according to the proportions of different particles present in the mixture. For example, soils that contain 35 percent or less clay and 45 percent or more sand are sandy; those containing 40 percent or less clay and 40 percent or more silt are silty. **Loam soils** contain sand, silt, and clay in proportions that result in ideal agricultural soils. The coarser soil particles of loam aid drainage, while the finer soil particles have high nutrient-retention capabilities.

The solid matter of soils consists of both inorganic and organic materials, with the proportions differing greatly in different soils. The organic component includes the remains of organisms in various stages of decomposition, a highly decomposed fraction (humus), and a wide range of living plants and animals. Structures as large as tree roots may be included, but the living phase is dominated by fungi, bacteria, and other microorganisms (Figure 29–6).

The Pore Space of Soils Is Occupied by Air and Water

Approximately 50 percent of the total soil volume is represented by pore space, which is occupied by varying proportions of air and water, depending on moisture conditions. When no more than half the pore space is occupied by water, adequate oxygen is available for root growth and other biological activity.

Following a heavy rain or irrigation, soils retain a certain amount of water and remain moist even after gravity has removed loosely bound water. If the soil is made up of large fragments,

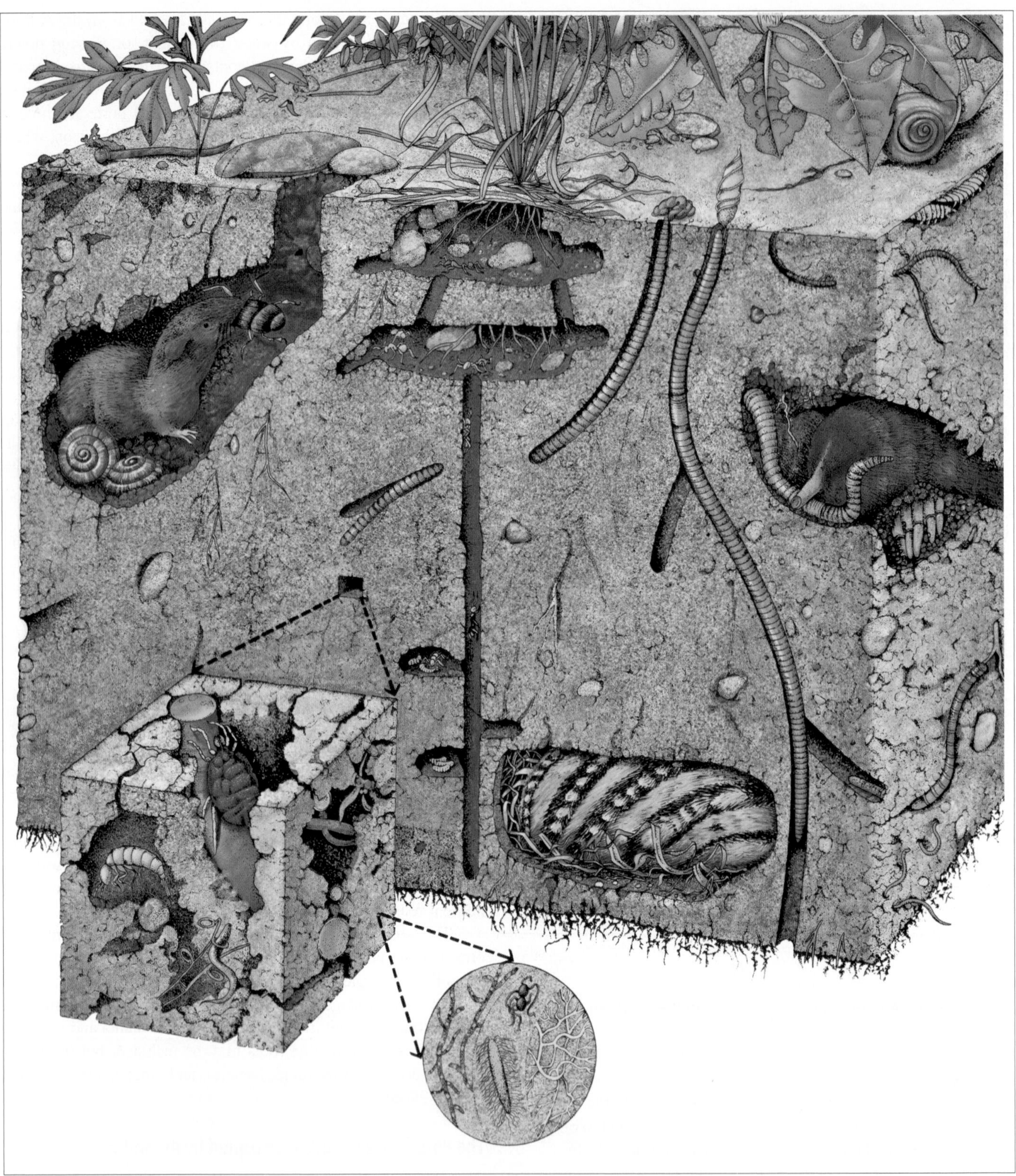

29–6 Living organisms of the A horizon Plants share the soil with a vast number of living organisms, ranging from microbes to small mammals such as moles, shrews, and ground squirrels. Multitudes of burrowing creatures—most notably ants and earthworms—aerate the soil and improve its ability to absorb water. Called by Aristotle "the intestines of the Earth," earthworms refine the soil by processing it through their gut. The refined soil is then deposited on the soil surface in the form of castings. In a single year, the combined activities of earthworms may produce as much as 500 metric tons of castings per hectare. The castings are very fertile, containing 5 times the nitrogen content of the surrounding soil, 7 times the phosphorus, 11 times the potassium, 3 times the magnesium, and twice the calcium. Bacteria and fungi are the principal decomposers of the organic matter in soil.

the pores and spaces between them are large. Water will drain through such soil rapidly, and relatively little will be available for plant growth in the A and B horizons. Because of their finer pores and the attractive forces that exist between water molecules and small-size clay particles, clay soils are able to hold a much greater amount of water against the action of gravity. Clay soils may retain three to six times more water than a comparable volume of sand; hence, soils with more clay can hold more water that is available to plants. The percentage of water that a soil can hold against the action of gravity is called its **field capacity.**

If a plant is allowed to grow indefinitely in a sample of soil and no water is added, the plant will eventually be unable to absorb water rapidly enough to meet its needs, and it will droop and wilt. When wilting is severe, plants fail to recover even when placed in a humid chamber. The percentage of water remaining in a soil when such irreversible wilting occurs is called the **permanent wilting percentage** of that soil.

Figure 29–7 shows the relationship between the soil water content and the potential at which water is held by sandy, loam, and clay soils. The forces that retain water in the soil can be expressed in the same terms (in this case, water potential) as the forces for water uptake in cells and tissues (page 76). The soil water potential decreases gradually with a decrease in the soil moisture below field capacity. Soil scientists have agreed to consider soil with a water potential of –1.5 megapascals to be at its permanent wilting percentage.

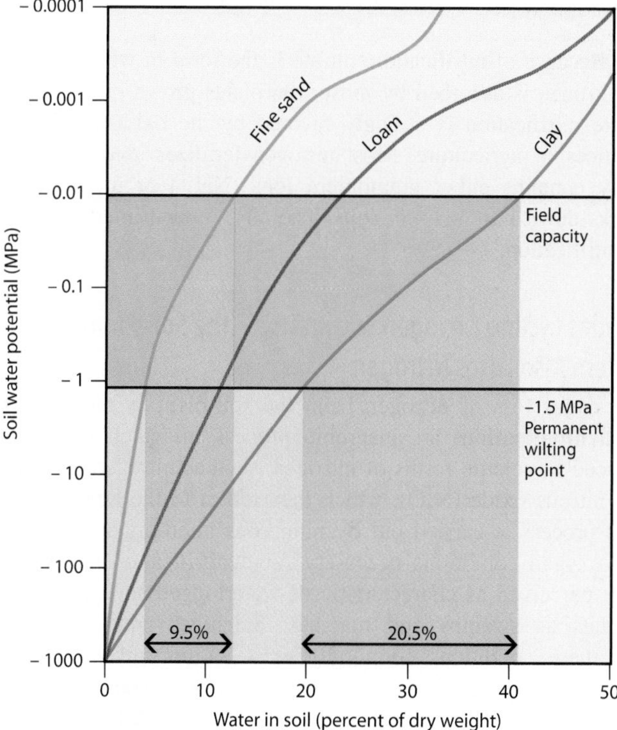

29–7 Soil water content and potential Relationship between soil water content and soil water potential in fine sand, loam, and clay soils. (The curves are plotted on a logarithmic scale.) Note that whereas water available to plants in a fine sand soil is only about 9.5 percent of the dry weight of the sand, the water available to plants in a clay soil is considerably greater, about 20.5 percent of the dry weight of clay. Soil with a water potential of –1.5 megapascals (MPa) is considered to be at its permanent wilting percentage.

Soils Retain Cations but Lose Anions to Leaching

The inorganic nutrients taken in through the roots of plants are present in the soil solution as ions. Most metals form positively charged ions, that is, cations, such as Ca^{2+}, K^+, Na^+, and Mg^{2+}. Both clay particles and humus may have an excess of negative charges on their colloidal surfaces where cations can be bound and thus held against the leaching action of percolating soil water.

These weakly bound cations can be replaced by other cations and thus released into the soil solution, where they become available for plant growth. This process is called **cation exchange.** For example, when CO_2 is released by respiring roots, it dissolves in the soil solution to become carbonic acid (H_2CO_3). The carbonic acid then ionizes to produce bicarbonate (HCO_3^-) and hydrogen (H^+) ions. The H^+ produced in this way may *exchange* for the nutrient cations on the clay and humus.

The principal negatively charged ions, or anions, found in soil are NO_3^-, SO_4^{2-}, HCO_3^-, and OH^-. Anions are leached out of the soil more rapidly than cations, because anions do not attach to clay particles. Leached nitrate ions, in particular, have polluted ground and surface water supplies. An exception is phosphate, which is retained against leaching because it forms insoluble precipitates. Phosphate is specifically adsorbed by, or held on the surface of, compounds containing iron, aluminum, and calcium.

Even though iron ranks fourth in abundance among all elements on the surface of the Earth, it normally is oxidized in the ferric (Fe^{3+}) form, which is insoluble and thus not available to plants. Two distinct mechanisms, or strategies, have evolved in plants to maximize iron mobilization and uptake from the soil. All plants except grasses use what is referred to as **Strategy I.** This strategy includes the induction of three activities localized at the plasma membrane: (1) a proton pump acidifies the rhizosphere, driving more iron into solution; (2) after acidification, Fe^{3+} is reduced to Fe^{2+}; and (3) the Fe^{2+} is then moved across the plasma membrane by a Fe^{2+} transporter. In **Strategy II,** the grasses produce and release into the soil special chelating compounds called **phytosiderophores** that have a high affinity for Fe^{3+}. The Fe^{3+}-phytosiderophores complexes are taken up into the root via transporters in the plasma membrane.

The acidity or alkalinity of soil is related to the availability of inorganic nutrients for plant growth. Soils vary widely in pH, and many plants have a narrow range of tolerance on this scale. In alkaline soils, some cations are precipitated, and such elements as iron, manganese, copper, and zinc may thereby become unavailable to plants. Mycorrhizas (Figure 29–1; see also pages 312 and 700) are especially important in the absorption and transfer of phosphorus in most plants, but these structures have also been implicated in the increased absorption of nitrogen, manganese, copper, and zinc.

Nutrient Cycles

As we now know, virtually all vascular plants require 17 essential elements for normal growth and development. Inasmuch as the Earth is essentially a closed system, the elements are available only in limited supply. Life on Earth depends, therefore, on the recycling of these elements. Both macronutrients and micronutrients are recycled through plant and animal bodies, returned to the soil, broken down, and taken up into plants again. Each element has a different cycle, involving many different organisms and different enzyme systems. Some cycles, such as those of carbon, oxygen, sulfur, and nitrogen, which exist in gaseous forms (as elements or compounds) in the atmosphere, are essentially global. Other cycles, such as those of phosphorus, calcium, potassium, and the micronutrients, which are not found in a gaseous state, are generally more localized. Because nutrient cycles involve both living organisms and the physical environment, they are also called **biogeochemical cycles.**

Nutrient cycles are said to be "leaky," because not all of the nutrients returned to the soil become available for plant use. Some are lost from the system. Soil erosion, for instance, removes the nutrient-rich (especially phosphorus and nitrogen) topsoil, which is carried away in streams and rivers and eventually ends up in the ocean. The removal of nutrients (especially nitrogen and potassium) with harvesting, along with the losses of nitrogen- and sulfur-containing gases to the atmosphere when plants are burned, also contributes to nutrient loss. In addition, leaching accounts for losses of all types of soluble nutrients, most notably potassium, nitrate, and sulfate.

Nitrogen and the Nitrogen Cycle

The chief reservoir of nitrogen is the atmosphere; in fact, nitrogen gas (N_2) makes up about 78 percent of the atmosphere. Most living things, however, cannot use elemental atmospheric nitrogen to make amino acids and other nitrogen-containing compounds. They are therefore dependent on the more reactive nitrogenous compounds, such as ammonium and nitrate, present in soil. Unfortunately, these compounds are not nearly as abundant as nitrogen gas. So, despite the abundance of nitrogen in the atmosphere, a shortage of nitrogen in the soil is often the major limiting factor in plant growth.

The process by which this limited amount of nitrogen is circulated and recirculated through the world of living organisms is known as the nitrogen cycle (Figure 29–8). The three principal stages of this cycle are (1) ammonification, (2) nitrification, and (3) assimilation.

Ammonium Is Released as Organic Material Decays

Much of the soil nitrogen is derived from dead organic materials in the form of complex organic compounds such as proteins, amino acids, nucleic acids, and nucleotides. The nitrogenous compounds usually are rapidly decomposed into simpler compounds by soil-dwelling saprotrophic bacteria and various fungi. These organisms incorporate the nitrogen into amino acids and proteins and release excess nitrogen in the form of ammonium ions (NH_4^+). This process is known as **ammonification** or nitrogen mineralization. In alkaline media, the nitrogen may be converted to ammonia gas (NH_3). This conversion usually occurs, however, only during the decomposition of large amounts of nitrogen-rich material, as in a manure pile or a compost heap that has contact with the atmosphere. Within the soil, the ammonia produced by ammonification dissolves in the soil water, where it combines with protons to form the ammonium ion. In some ecosystems, the ammonium ion is not rapidly oxidized but remains in the soil. Plants growing in these soils are able to take up NH_4^+ and use it in the synthesis of plant protein.

In Some Soils, Nitrifying Bacteria Convert Ammonium into Nitrite and Then into Nitrate

Several species of bacteria common in soils are able to oxidize ammonia or ammonium ions. The oxidation of ammonium, or **nitrification,** is an energy-yielding process, and the energy released is used by these bacteria to reduce carbon dioxide (in much the same way that photosynthetic autotrophs use light energy in the reduction of carbon dioxide). Such organisms are known as chemosynthetic autotrophs (as distinct from photosynthetic autotrophs). The chemosynthetic nitrifying bacterium *Nitrosomonas* is primarily responsible for oxidation of ammonium to nitrite ions (NO_2^-):

$$2NH_4^+ + 3O_2 \longrightarrow 2NO_2^- + 4H^+ + 2H_2O$$

Nitrite is toxic to plants, but it rarely accumulates in the soil. *Nitrobacter,* another genus of bacteria, oxidizes the nitrite to form nitrate ions (NO_3^-), again with a release of energy:

$$2NO_2^- + O_2 \longrightarrow 2NO_3^-$$

Because of nitrification, nitrate is the form in which almost all nitrogen is absorbed by most crop plants grown on dry land, where nitrification is strongly favored by the oxidizing tillage practices of agriculture. Most nitrogen fertilizer used commercially contains either ammonium ions (NH_4^+) or urea, which breaks down into NH_4^+ in soils. The NH_4^+ is converted to NO_3^- by nitrification.

Besides Cycling Nitrogen within Itself, the Soil-Plant System Also Loses Nitrogen

The major loss of nitrogen from the soil-plant system occurs by **denitrification,** an anaerobic process in which nitrate is reduced to volatile forms of nitrogen, such as nitrogen gas (N_2) and nitrous oxide (N_2O), which then return to the atmosphere. This process is carried out by numerous microorganisms. The low oxygen conditions necessary for denitrification have long been perceived as characteristic of waterlogged soils and such habitats as swamps and marshes. Scientists now recognize that these conditions commonly exist within soil aggregates even in the absence of excessive water. Consequently, denitrification is virtually a universal process in soils. A fresh supply of readily decomposable organic matter provides the energy source required by denitrifying bacteria and, if other conditions are appropriate, promotes denitrification. Lack of an energy source allows nitrate concentrations to build up to high levels in groundwater.

Nitrogen is also lost from an ecosystem by removal (harvesting) of plants from the soil, by soil erosion, by burning of plants, and by leaching. Nitrates and nitrites, both of which are anions, are particularly susceptible to being washed from the root zone by water percolating through the soil.

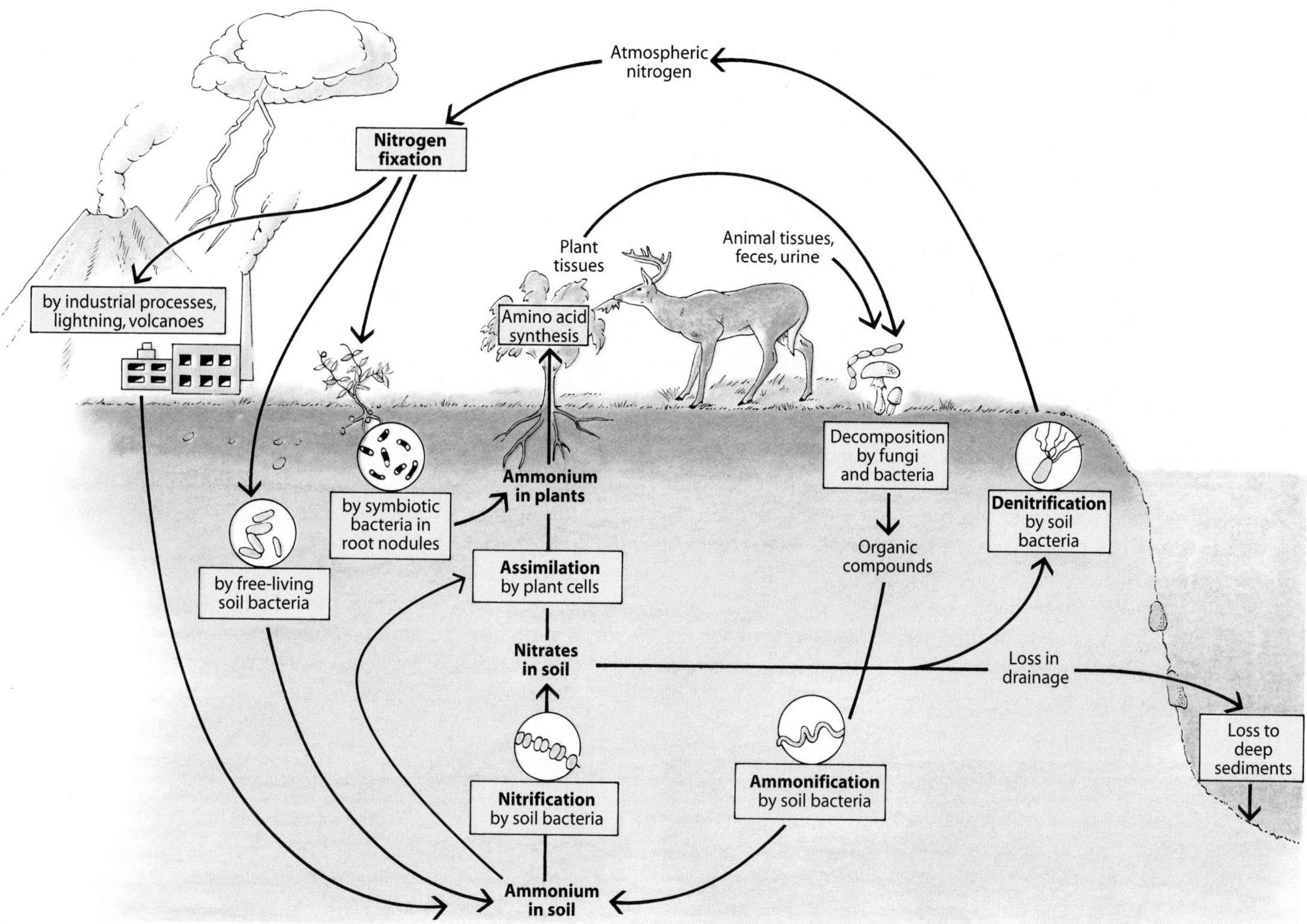

29–8 The nitrogen cycle in a terrestrial ecosystem The primary reservoir of nitrogen is the atmosphere, where nitrogen makes up 78 percent of dry air. Only a few microorganisms, some symbiotic and others free-living, are capable of fixing nitrogen gas into inorganic compounds that can be used by plants in the synthesis of amino acids and other organic nitrogenous compounds.

The nitrogen cycle in aquatic ecosystems is similar to the terrestrial cycle. Although the specific organisms differ, the processes are essentially the same, with specific groups of bacteria carrying out the various chemical reactions on which the cycle depends. Loss of nitrogen in drainage from soil can lead to eutrophication, or the buildup of high levels of nitrogen in the water, which can result in massive growth of algae and flowering plants.

Replenishment of Nitrogen Occurs Primarily by Nitrogen Fixation

If the nitrogen that is removed from the soil were not steadily replaced, virtually all life on this planet would slowly disappear. Nitrogen is replenished in the soil primarily by nitrogen fixation. Much lesser amounts are added by precipitation and the weathering of rocks.

Nitrogen fixation is the process by which atmospheric N_2 is reduced to NH_4^+ and made available for transfer to carbon-containing compounds to produce amino acids and other nitrogen-containing organic compounds. Nitrogen fixation, which can be carried out only by certain bacteria, is a process on which all living organisms depend, just as most organisms ultimately depend on photosynthesis as the source of their energy.

The enzyme that catalyzes the fixation of nitrogen is called **nitrogenase.** It is similar in all organisms from which it has been isolated. Nitrogenase contains molybdenum, iron, and sulfide prosthetic groups, and therefore these elements are essential for nitrogen fixation. Nitrogenase also uses large amounts of ATP as an energy source, making nitrogen fixation an expensive metabolic process.

Nitrogen-fixing bacteria may be classified according to their mode of nutrition: those that are free-living (nonsymbiotic) and those that live in symbiotic association with certain vascular plants.

The Most Effective Nitrogen-Fixing Bacteria Form Symbiotic Relationships with Plants

Of the two classes of nitrogen-fixing organisms, the symbiotic bacteria are by far the most important in terms of total amounts of nitrogen fixed. The most common of the nitrogen-fixing bacteria are *Rhizobium* and *Bradyrhizobium,* both of which invade

CARNIVOROUS PLANTS

A few species of plants are able to use animal proteins directly as a nitrogen source. These carnivorous plants have special adaptations for luring and trapping insects and other very small animals. The plants digest the trapped organisms, absorbing the nitrogenous compounds they contain, as well as other organic compounds and minerals such as potassium and phosphate. Most of the carnivores of the plant world are found in bogs, a habitat that is usually quite acidic and thus not favorable for the growth of nitrifying bacteria. Carnivorous plants entrap their prey in a variety of ways. The common bladderwort *(Utricularia vulgaris)* ***(a)*** is a free-floating aquatic plant with traps that are tiny, flattened, pear-shaped bladders. Each bladder has a mouth guarded by a hanging door. The tripping mechanism consists of four stiff bristles near the lower free edge of the door. When a small animal brushes against these bristles, the hairs distort the lower edge of the door, causing it to spring open. Water then rushes into the bladder, sucking the animal inside, and the door snaps shut behind it. A range of enzymes secreted by the internal wall of the bladder and by the resident bacterial population digest the animal. Released minerals and organic compounds are taken up through the cell walls of the trap, while the undigested exoskeleton remains within the bladder.

The butterwort *(Pinguicula)* ***(b)*** captures small insects with its numerous stalked glands, scattered over the leaf. Each of these glands bears a globular droplet of mucilaginous secretion, so the leaf is sticky, like flypaper ***(c)***. Insects coming in contact with the secretion pull it out into strands that set to form strong cables. The more the insect struggles, the more glands it touches, and the more tightly it is held to the leaf. Scattered among the stalked glands are sessile glands (not visible in the figure) that bear no surface material until they are stimulated by the struggling of the captured prey. The sessile glands then pour out an enzyme-containing secretion that quickly forms a pool around the insect. These enzymes digest the prey, and the products accumulate in the secretion pool. After digestion is complete, the pool is absorbed into the leaf, and the digestion products are distributed to the growing parts of the plant.

Similar secretion mechanisms are found in other carnivorous plants, including the sundew *(Drosera intermedia)* (page 209) and the Venus flytrap *(Dionaea muscipula)* (see Figure 28–28).

(a)

(c)

(b)

(a) The common bladderwort *(Utricularia vulgaris)*. ***(b)*** A butterwort *(Pinguicula vulgaris)*. ***(c)*** View of a butterwort leaf, showing glands with mucilaginous droplets.

the roots of legumes, such as alfalfa *(Medicago sativa),* clovers *(Trifolium),* peas *(Pisum sativum),* soybean *(Glycine max),* and beans *(Phaseolus).* In the symbiotic association between bacteria and legumes, the bacteria provide the plant with a form of nitrogen that it can use to make protein. The plant, in turn, provides the bacteria with carbon-containing molecules that are used both as an energy source for their nitrogen-fixing activity and for the production of nitrogenous compounds.

The beneficial effects on the soil of growing leguminous plants have been recognized for centuries. Theophrastus, who lived in the third century B.C.E., wrote that the Greeks used crops of fava, or broad, beans *(Vicia faba)* to enrich the soils. In modern agriculture, it is common practice to rotate a nonleguminous crop, such as maize *(Zea mays),* with a leguminous crop, such as alfalfa, or maize with soybeans and then wheat. The leguminous plants are either harvested for hay, leaving behind the nitrogen-rich roots, or, better still, simply plowed under. These plants are often referred to as "green manure." A crop of alfalfa that is plowed back into the soil may add as much as 300 to 350 kilograms of nitrogen per hectare of soil. As a conservative estimate, 150 to 200 million metric tons of fixed nitrogen are added to the Earth's surface each year by such biological systems.

Nodules Are Produced by the Host-Plant Root on Infection by Bacteria The **root nodules** are unique nitrogen-fixing organs that result from a symbiotic interaction between the plant and nitrogen-fixing bacteria. The nodules provide a favorable environment for nitrogenase.

In most legumes, the infection process begins with the attachment of **rhizobia**—the generic term used to designate any bacterium that can nodulate legumes—to the tip of emerging root hairs in response to chemical attractants, **flavonoids,** released by the root hairs (Figures 29–9a and 29–10a). This leads to the induction of nodulating *(nod)* genes and the synthesis of bacterial signaling molecules called **Nod factors.** The Nod factors then activate a cascade of expression of plant genes required for root nodule formation. The symbiosis between a species of *Rhizobium* or *Bradyrhizobium* and a legume is quite specific; for example, bacteria that invade and induce nodule formation in clover *(Trifolium)* roots will not induce nodules on the roots of soybeans *(Glycine).*

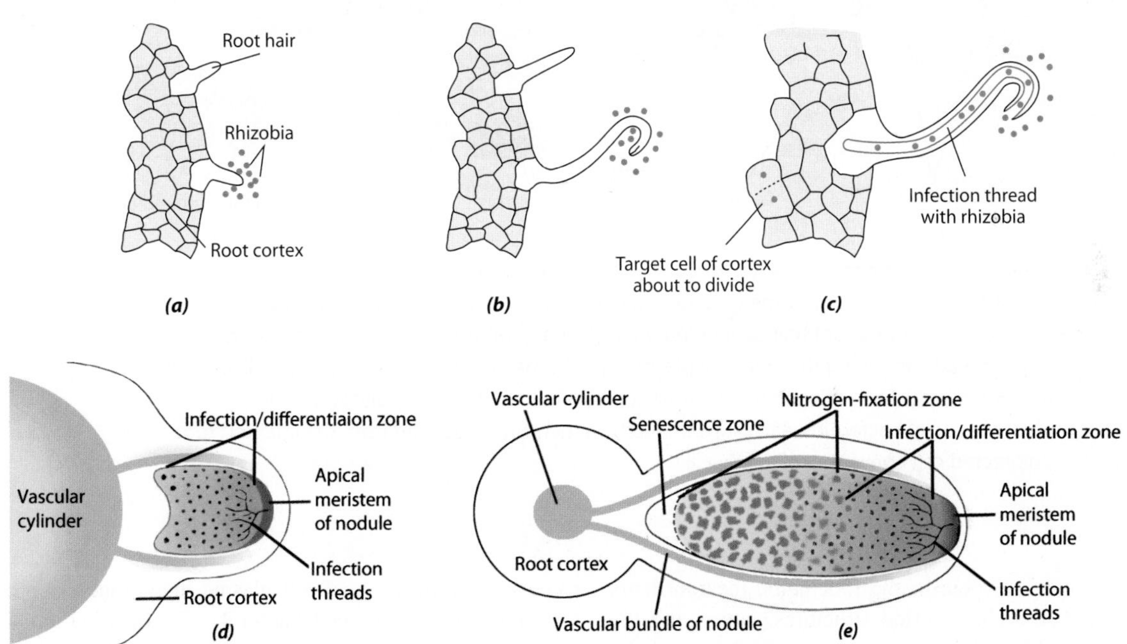

29–9 Stages in development of an indeterminate root nodule ***(a)*** Rhizobia bind to an emerging root hair, to which they are attracted by chemicals released by the hair. ***(b)*** The root hair curls in response to factors produced by the rhizobia. ***(c)*** An infection thread is formed. It will carry the dividing rhizobia down the root hair and into the cortical cells that have responded by dividing. ***(d)*** A nodule primordium forms from the rapidly dividing infected cortical cells. Derivatives of the cortical cells form an apical meristem, which gives rise to the tissues of the nodule as it grows out through the cortex. ***(e)*** The mature nodule consists of a persistent apical meristem that continuously produces new cells, which become infected with rhizobia, and gives rise to a gradient of developmental zones (nitrogen-fixation and infection/differentiation zones). The entire nodule is surrounded by a narrow cortex. Vascular bundles of the nodule are connected with the vascular tissues of the root vascular cylinder.

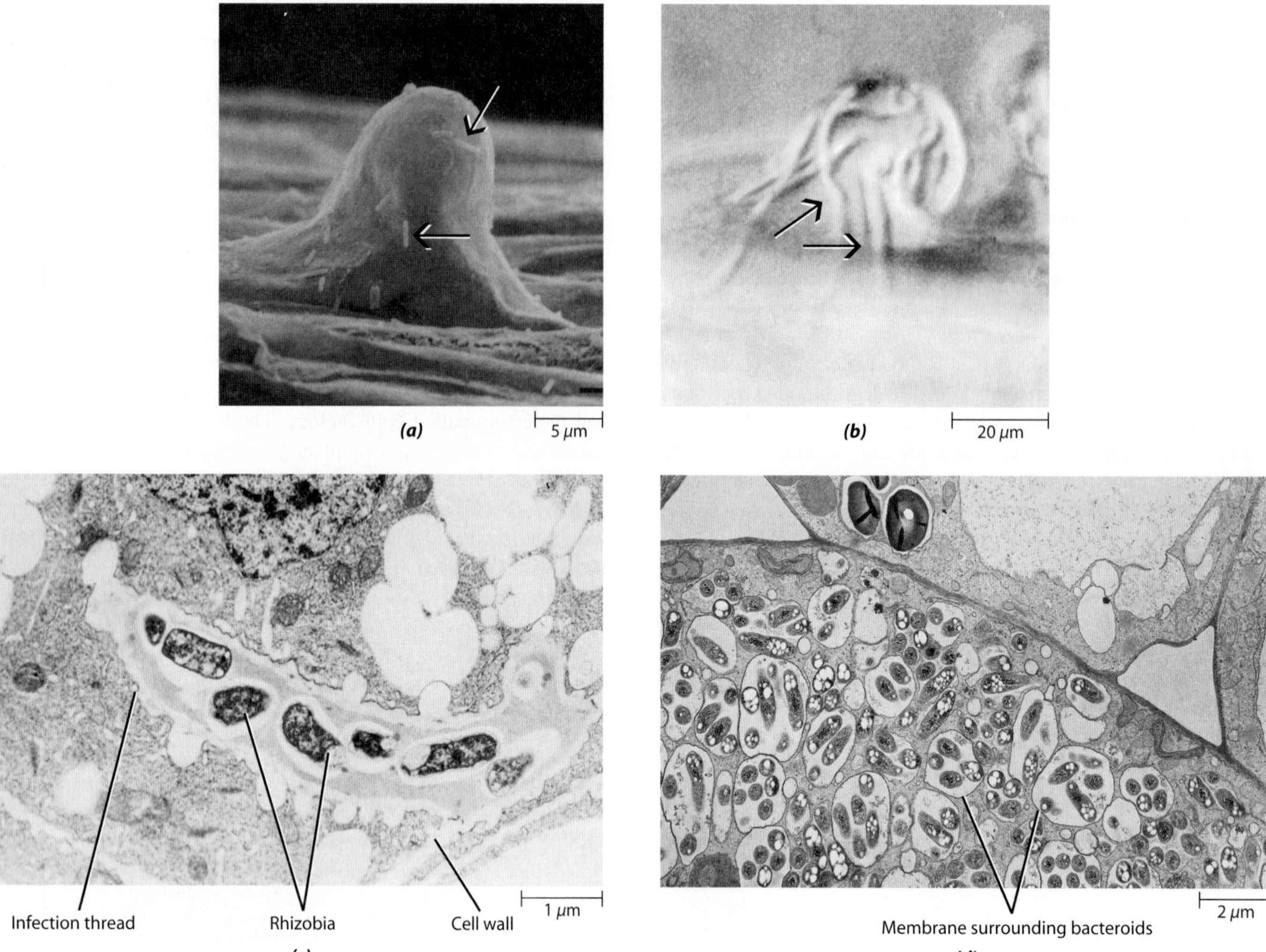

29–10 Events in the infection of soybean by *Bradyrhizobium japonicum* ***(a)*** Scanning electron micrograph showing rhizobia (arrows) attached to a recently emerged root hair. ***(b)*** Differential interference contrast photomicrograph showing a short, curled root hair containing multiple infection threads (arrows). ***(c)*** Electron micrograph of an infection thread containing rhizobia. The plasma membrane and cell wall of the root cell are continuous with the infection thread. Each rhizobium is surrounded by a halo of capsular polysaccharides. ***(d)*** Electron micrograph of groups of bacteroids, each surrounded by a membrane derived from the infected root-nodule cell. Note the uninfected cell above the infected cell.

On attachment of rhizobia to the root hairs, the root hairs typically develop into tightly curled structures, entrapping the bacteria (Figures 29–9b and 29–10b). Invasion of the root hairs and underlying cortical cells by the rhizobia occurs via **infection threads,** tubular structures formed by the progressive inward growth of the root-hair cell walls from the sites of penetration (Figures 29–9c and 29–10c). The rhizobia divide at the tip of the growing infection thread, forming a column of rhizobia within the thread (Figures 29–9c and 29–10c). A single root hair may be penetrated by several rhizobia and thus may contain several infection threads (Figure 29–10b). Infection occurs only in root hairs located opposite cortical cells that have simultaneously been stimulated to undergo cell division to form a **nodule primordium** (Figure 29–9d). When the infection thread reaches the developing primordium, the rhizobia are released into envelopes derived from the host-cell plasma membrane, a process resembling endocytosis (see Chapter 4). The rhizobia remain enclosed in this membrane, where they continue to divide and develop into **bacteroids,** the name now given to the enlarged nitrogen-fixing rhizobia (Figure 29–10d). The membrane, together with the bacteroids, forms a functional nitrogen-fixing unit known as a **symbiosome.** Proliferation of the membrane-enclosed bacteroids and cells of the nodule primordium results in formation of the nodules.

Two Types of Nodules Can Be Distinguished: Determinate and Indeterminate

There are two types of root nodules. **Indeterminate nodules** are cylindrical and elongated due to the presence of a persistent meristem. At maturity, indeterminate nodules consist of several zones, including the meristem, followed by zones of infection/

differentiation, nitrogen fixation, and senescence (Figure 29–9e). Typical indeterminate nodules are found in alfalfa *(Medicago sativa),* fava bean *(Vicia faba),* pea *(Pisum sativum),* and clover *(Trifolium).* **Determinate nodules** lack a persistent meristem, are usually spherical in form, and do not exhibit an obvious developmental gradient. Such nodules are found, for instance, in the common bean *(Phaseolus vulgaris),* soybean *(Glycine max)* (Figure 29–11), cowpea *(Vigna unguiculata),* and *Lotus japonicus.* Organogenesis of determinate nodules begins with cell division in the outer cortex, whereas in indeterminate nodules, cell division begins in the inner cortex. The type of nodule development is characteristic of the plant host, not of the nodule bacteria. The hormones auxin, cytokinins, gibberellins, and brassinosteroids positively regulate nodule formation, while abscisic acid, ethylene, jasmonic acid, and salicylic acid act as negative regulators to repress nodulation. The number of root nodules in leguminous plants is tightly regulated so as to prevent excessive nodule production and a resultant overconsumption of sugars formed by photosynthesis, a condition that would be detrimental to the host plant.

The fully developed legume nodule consists of a relatively narrow cortex surrounding a large central zone, which contains bacteroid-infected and uninfected cells (Figure 29–11b). Vascular bundles, which radiate from the point of attachment of the nodule to the root, occur in the inner cortex. Determinate nodules also form lenticels, which serve to enhance gas exchange (see Chapter 26). In the nodules, atmospheric nitrogen (N_2) is converted into ammonia by the bacteroids and then rapidly converted into organic forms (amides or ureides). Bacterial *nif* genes are involved in nitrogen fixation. In temperate-region legumes, such as white clover *(Trifolium repens),* fava bean *(Vicia faba),* and pea *(Pisum sativum),* amides (principally asparagine or glutamine, both amino acids) are exported into the roots and transported to the shoots in the xylem. In legumes of tropical origin, such as common bean *(Phaseolus vulgaris),* cowpea *(Vigna unguiculata),* and soybean *(Glycine max),* ureides (derivatives of urea) are exported from the nodules.

The oxygen concentration in the bacteria-infected cells must be carefully regulated, because oxygen is a potent, irreversible inhibitor of nitrogenase. Oxygen is required, however, for aerobic respiration to meet the large ATP demands of nitrogenase. Oxygen is also needed for other metabolic activities in both the bacteria and the plant cells. The regulation of oxygen levels is accomplished in large part by an oxygen-binding heme protein,

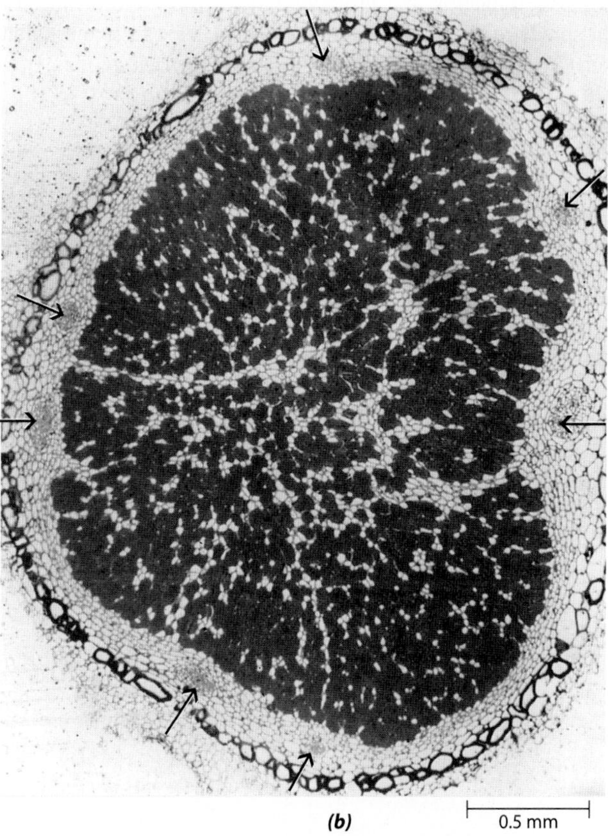

29–11 Determinate root nodules of soybean ***(a)*** Nitrogen-fixing nodules on the roots of a soybean *(Glycine max)* plant, a legume. These nodules are the result of a symbiotic relationship between the soil bacterium *Bradyrhizobium japonicum* and the cortical cells of the root. ***(b)*** Section through a mature soybean root nodule. Vacuolated uninfected cells can be seen among the darkly stained infected cells in the central zone of the nodule. The nodule cortex, containing vascular bundles (arrows) and a layer of darkly stained sclerenchyma cells, surrounds the central zone.

leghemoglobin, found in the cytosol of infected cells at rather high concentrations. This protein, which imparts a pink color to the central region of the nodule, just as hemoglobin imparts a red color to blood, is produced partly by the bacteroid (the heme portion) and partly by the plant (the globin portion). The leghemoglobin is thought to buffer the oxygen concentration within the nodule, allowing respiration without inhibiting the nitrogenase. Leghemoglobin also acts as an oxygen carrier, facilitating O_2 diffusion to the bacteroids.

Legumes Are Not the Only Plants That Form Nodules In addition to plants of the family Fabaceae, the genus *Parasponia* (Cannabaceae) forms nodules with rhizobia. Members of eight other families of flowering plants form nodules with actinomycetes of the genus *Frankia.* These include such familiar plants as alder trees (*Alnus;* family Betulaceae), sweet gale (*Myrica gale;* Myricaceae), sweet fern (*Comptonia;* Myricaceae), and mountain lilacs (*Ceanothus;* Rhamnaceae).

Beneficial Interactions Also Occur between Nitrogen-Fixing Bacteria and Plants That Do Not Form Nodules *Rhizobium,* one of the principal root-nodule endosymbionts of legumes, also colonizes the roots of several important cereal crops that do not form nodules, among them barley, maize, rice, wheat, and sorghum. The rhizobia gain entrance to the root at sites of lateral root emergence and in the regions of elongation and differentiation. From the root, the rhizobia spread systemically to the stems and leaves via the xylem vessels. Another nitrogen-fixing bacterium, *Azoarcus,* has been isolated from rice and kallar grass *(Leptochloa fusca).* In addition, *Acetobacter diazotrophicus* and *Herbaspirillum seropedicae* have been isolated from sugarcane, and *Burkholderia* from the grape *(Vitis vinifera)* vine. The list continues to grow. In all cases, the bacterial endophytes inhabit the intercellular spaces and xylem vessels of the roots, stems, and leaves, not living cells. Several studies indicate that the host plants can obtain a substantial amount—in some cases 60 percent or more—of their nitrogen needs from the nitrogen-fixing endophytes.

Most endophyte taxa are also found in the rhizosphere, and it has been estimated that these free-living nitrogen-fixing bacteria probably add 7 kilograms of nitrogen to a hectare of soil each year. Other endophytes, such as *Acetobacter, Azoarcus,* and *Herbaspirillum,* have been isolated only from plants.

Many Photosynthetic Bacteria, Such as Cyanobacteria, Are Important Nitrogen-Fixing Organisms Another symbiotic relationship is of considerable practical interest in certain parts of the world. *Azolla* is a small floating water fern, and *Anabaena* is a nitrogen-fixing cyanobacterium that lives in the cavities of the *Azolla* fronds (Figure 29–12). The *Azolla-Anabaena* symbiosis is unique among nitrogen-fixing symbioses in that the relationship is sustained throughout the life cycle of the host. *Azolla* infected with *Anabaena* may contribute as much as 50 kilograms of nitrogen per hectare. In the Far East, for example, heavy growths of *Azolla-Anabaena* are permitted to develop on rice paddies. The rice plants eventually shade out the *Azolla,* and as the fern dies, nitrogen is released for use by the rice plants.

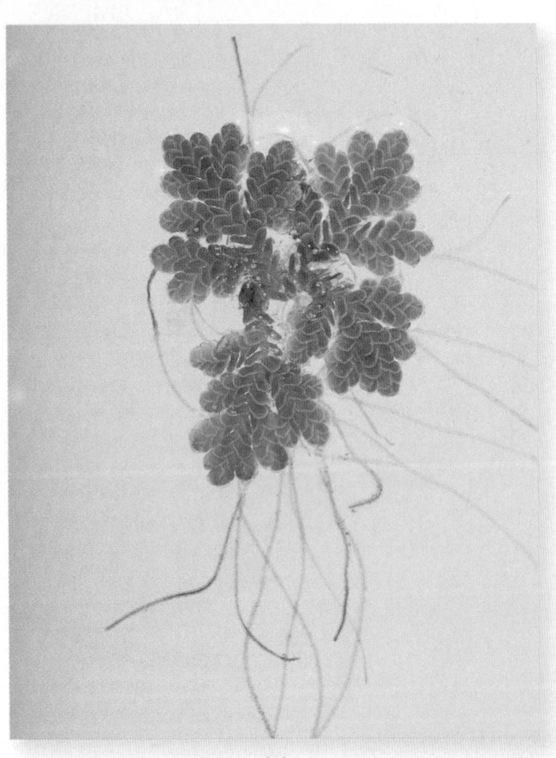

(a)

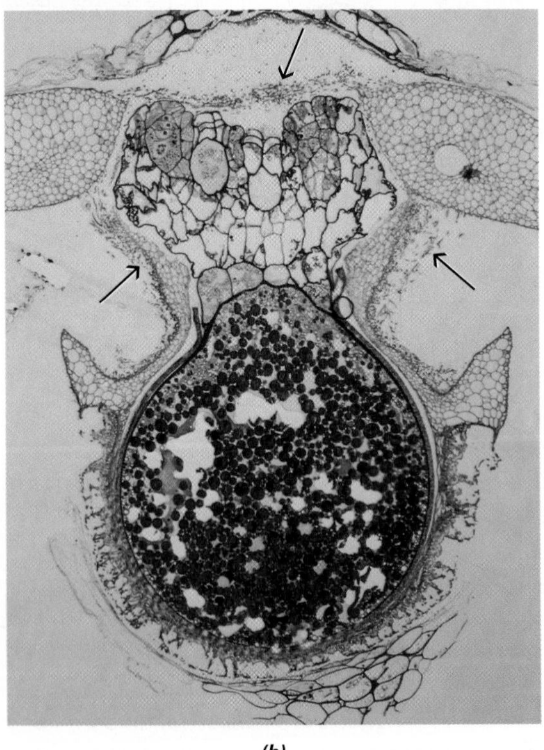

(b)

29–12 *Azolla-Anabaena* symbiosis ***(a)*** *Azolla filiculoides,* a water fern that grows in symbiotic association with the cyanobacterium *Anabaena.* ***(b)*** Filaments of *Anabaena* can be seen (arrows) associated with a female gametophyte (megagametophyte) that has developed from a germinated megaspore of *Azolla.*

Nitrogen-fixing cyanobacteria (primarily *Nostoc* sp.) that live epiphytically on the leaves of the feather moss *(Pleurozium schreberi)* have been found to be the main source of nitrogen in northern boreal forest ecosystems. In the boreal biome, which accounts for 17 percent of Earth's land surface, nitrogen is the primary growth-limiting nutrient.

Industrial Nitrogen Fixation Has High Energy Costs

The industrial, or commercial, production of fixed nitrogen (called the "Haber-Bosch process") was first developed in 1914, and its use has increased steadily to the current level of approximately 50 million metric tons per year, which is equivalent to almost half of the planet's biological fixation. Most of this nitrogen is used in agricultural fertilizers. Industrial fixation, unfortunately, is accomplished at a high energy cost in terms of fossil fuels. In this process, N_2 is reacted with H_2 at high temperature and pressure, in the presence of metal catalysts, to form ammonia. The energy-expensive components are hydrogen, which is derived from natural gas, petroleum, or coal, and the high temperature and pressure required to promote the reaction, which also require great amounts of fossil fuels. In such developed countries as the United States, this process, despite its high cost, can account for as much as one-third of the annual newly fixed nitrogen.

Additional Strategies for Nitrogen Uptake Have Been Adopted by Plants

Not all plants rely on nitrogen fixation for their nitrogen needs. Both ectomycorrhizas and ericoid (family Ericaceae) mycorrhizas (see Chapter 14), which are common components of plants growing in infertile soils, directly break down proteins in organic matter found in the soil. They then absorb and transfer amino acids to the host plant directly, without mineralization to nitrate and ammonia. Another strategy has been adopted by carnivorous plants, such as the Venus flytrap (see Figure 28–28), which use animal proteins directly as a nitrogen source (see the essay on page 694). Parasitic angiosperms, such as the mistletoes (families Loranthaceae and Viscaceae), exhibit yet another strategy. They attach themselves to the host by highly modified roots called haustoria, which permeate the host phloem and xylem (Figure 29–13). Many species adopt more than one strategy to obtain nitrogen; it is common, for example, to find arbuscular mycorrhizas (page 312) and nitrogen-fixing symbioses occurring in the same plant.

Assimilation of Nitrogen Is the Conversion of Inorganic Nitrogen into Organic Compounds

The assimilation of inorganic nitrogen (nitrates and ammonia) into organic compounds is one of the most important processes in the biosphere, on an almost equal footing with photosynthesis and respiration. The principal source of nitrogen that is available to crop plants grown under field conditions is nitrate. Once nitrate enters a cell, it is reduced to ammonia, which is then rapidly incorporated into organic compounds via the glutamine synthetase–glutamate synthase pathway shown in Figure 29–14. In most herbaceous plants, this process takes place in the cytosol and in leaf chloroplasts, in close association with photosynthesis. (Many plant physiologists consider it to be an extension of photosynthesis.) When the amount of nitrate supplied to the roots is low, nitrate reduction in many plants takes place primarily in the plastids of the roots.

29–13 Mistletoe, a parasitic angiosperm ***(a)*** *Phoradendron leucarpum* (family Viscaceae), sold in North America as a Christmas decoration, growing on a branch of a host tree. ***(b)*** Transverse section of a juniper *(Juniperus occidentalis)* stem, which has been penetrated by the haustorium (modified root) of the mistletoe *P. leucarpum.*

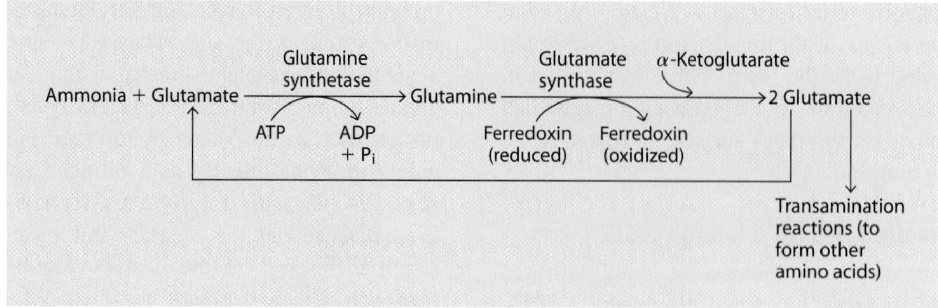

29–14 The glutamine synthetase–glutamate synthase pathway (abbreviated) in leaves The assimilation of nitrogen into organic compounds requires ATP and reduced ferredoxin, both of which are readily available in photosynthesizing cells. Of the two glutamate molecules produced, one recycles to bind with ammonia, perpetuating the pathway, whereas the other is transaminated to form other amino acids. In roots, either NADH or NADPH replaces the ferredoxin in this pathway.

The organic nitrogen made available by root-metabolized nitrate is transported in the xylem, primarily as amino acids.

Phosphorus and the Phosphorus Cycle

Compared with nitrogen, the amount of phosphorus required by plants is relatively small (Table 29–1). Nevertheless, of all the elements for which the Earth's crust is the primary reservoir, phosphorus is the most likely to limit plant growth. Phosphorus is present at very low concentrations, unevenly distributed, and almost immobile in soils. In Australia, where the soils are extremely weathered and deficient in phosphorus, the distribution and limits of native plant communities are often determined by the available soil phosphate.

The Phosphorus Cycle Seems Simpler Than the Nitrogen Cycle

The phosphorus cycle (Figure 29–15) seems simple when compared with the nitrogen cycle, because there are fewer steps. The phosphorus cycle also differs from the nitrogen cycle in that the Earth's crust, rather than the atmosphere, is the primary reservoir for replenishing phosphorus. There are no significant phosphorus-containing gases. Because humus and soil particles bind phosphate ions (PO_4^{3-})—the only important inorganic form of phosphorus—recycling of phosphorus tends to be quite localized. As discussed earlier in the chapter, the weathering of rocks and minerals over long periods is the source of most of the phosphorus in the soil solution.

Phosphorus circulates from plants to animals and is returned to the soil in organic forms in residues and wastes. These organic forms are converted to inorganic phosphate through the activities of microorganisms, and the phosphorus again becomes available to plants (Figure 29–15).

Some phosphorus is lost from the terrestrial ecosystem by leaching and erosion, but in most natural ecosystems, the weathering of rock can keep pace with this loss. The lost phosphorus eventually reaches the oceans, where it is deposited in sediments as precipitates and in the remains of organisms. In the past, the use of guano (deposits of seabird feces) as agricultural fertilizer returned some of the ocean phosphorus to terrestrial ecosystems. However, most of the phosphorus in deep-sea sediments becomes available only as a result of major geologic uplifts. Phosphate fertilizers are used for agricultural soils that have inadequate levels of phosphorus available to plants. Deposits of phosphate rock are being mined on a large scale for this use, and the depletion of readily available phosphate deposits for use as fertilizer is likely to be a major issue in agriculture in the next century.

Plants Have Several Strategies for Acquiring Phosphate

Because of phosphate's low solubility and mobility, effective uptake of phosphate usually requires scavenging from large volumes of soil. This is achieved either through rapid root growth, with the proliferation of lateral roots and root hairs, or by symbioses with mycorrhizal fungi, particularly arbuscular mycorrhizal fungi (see Chapter 14). Development by the fungus of an extensive network of hyphae, called extraradical hyphae, in the soil is especially important in the acquisition of phosphorus. The extraradical hyphae extend far beyond the phosphorus-depletion zone of the host root, greatly expanding the volume of soil from which the plants can access phosphate.

Plants that never, or rarely, form mycorrhizas have adopted an alternative strategy to survive on phosphate-impoverished soils, such as those of Australia. There, members of the Proteaceae, which seldom form mycorrhizas, develop specialized structures called **cluster roots** (Figure 29–16). A cluster root consists of a portion of a lateral root bearing many closely spaced, hairy rootlets (50 to 1000 per centimeter of root axis). Cluster roots are commonly characterized as bottlebrush-like. These structures were once believed to occur only in the Proteaceae and were originally called "proteoid roots," but cluster-rooted plants have now been found in members of nine additional families, among them the Betulaceae, Casuarinaceae, Cyperaceae, and Fabaceae. Because of its ease of cultivation, white lupine (*Lupinus albus;* Fabaceae) has been used most often to study the function of cluster roots. Cluster roots are ephemeral and must continually be replaced by extension of the parent root axis. When fully formed,

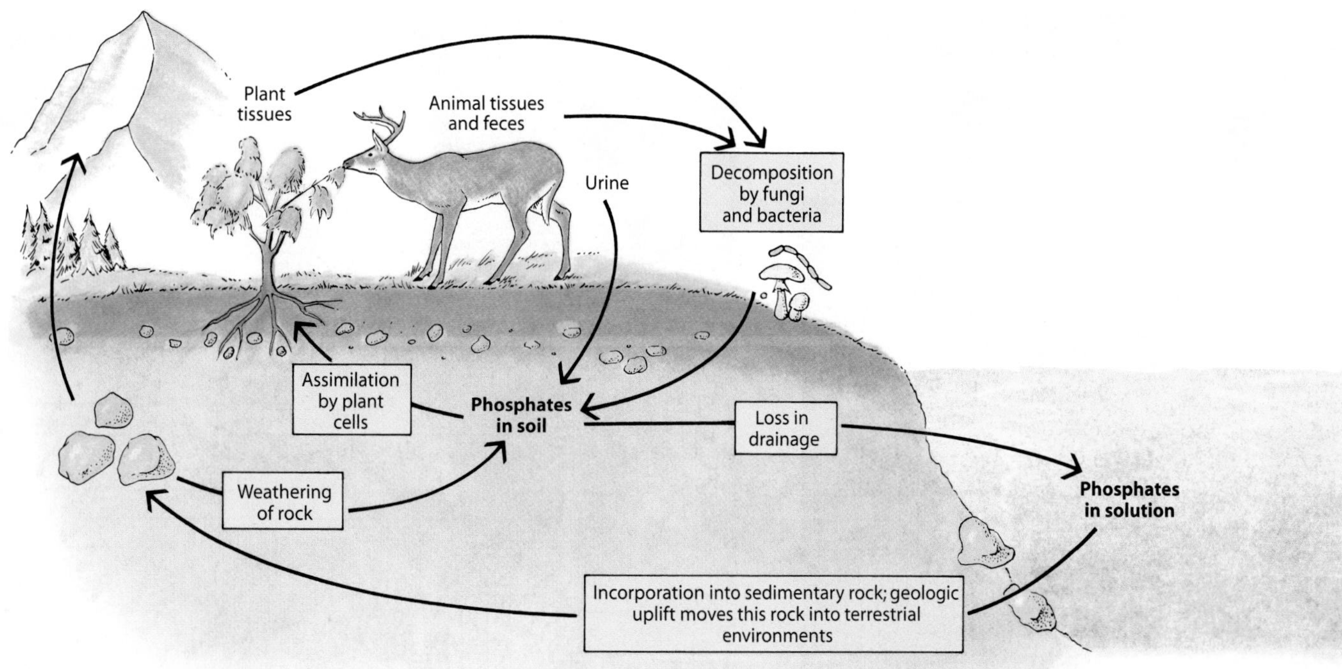

29–15 The phosphorus cycle in a terrestrial ecosystem Phosphorus is essential to all living organisms as a component of energy-carrier molecules, such as ATP, and of the nucleotides of DNA and RNA. Like other minerals, phosphorus is released from dead tissues by the activities of decomposers, taken up from the soil by plants, and cycled through the ecosystem.

The phosphorus cycle in an aquatic ecosystem involves different organisms but is, in most respects, similar to the terrestrial cycle shown here. In aquatic ecosystems, however, a considerable amount of phosphorus is incorporated into the shells and skeletons of aquatic organisms. This phosphorus, along with phosphates that precipitate out of the water, is subsequently incorporated into sedimentary rock. Such rock, returned to the land surface by geologic uplift, is the primary terrestrial reservoir of phosphorus.

they release large amounts of malate and citrate over a period of 2 to 3 days. This remarkable phenomenon, during which phosphatases and protons are also released, is referred to as "exudative burst." It is the anions, particularly citrate, that mobilize phosphorus by bonding with soil minerals such as aluminum, iron, and calcium.

Human Impact on Nutrient Cycles and Effects of Pollution

Normal functioning of the phosphorus, nitrogen, and other nutrient cycles requires the orderly transfer of elements between steps in the cycle in order to prevent buildup or depletion of nutrients at any one stage. Over millions of years, organisms have been provided with needed quantities of essential inorganic nutrients through the normal functioning of such cycles. However, the need to adequately feed an exponentially growing human population has drastically affected some cycles, and in time this may lead to harmful accumulations and depletions of nutrients. For example, crop removal and increased soil erosion have accelerated phosphorus loss from soils. Phosphorus added to aquatic habitats in the form of sewage, as well as drainage from fertilized agricultural fields, has caused massive growths of algae and flowering plants, thus seriously reducing the recreational value of the affected areas. Many states have banned phosphorus-containing detergents for this reason.

Normal functioning of the nitrogen cycle, which has a relatively high degree of efficiency, involves a balance between fixation processes, which remove nitrogen from the atmosphere, and denitrification processes, which return nitrogen to the atmosphere. Unfortunately, with the increasing introduction of fixed nitrogen, in the form of nitrates, into the environment through the extensive use of commercial fertilizers, the efficiency of the cycle declines. In addition, marshes and wetlands, the primary sites of denitrification, are being destroyed at an alarming rate through their conversion to building sites, agricultural land, and dump sites.

In some areas where nitrate fertilizers are heavily used, nitrate pollution has contaminated drinking water supplies. In California, more water wells are contaminated with nitrate than any other pollutant, exceeding the maximum amount (10 ppm) allowed by the Environmental Protection Agency.

Until recently, it was assumed that nitrogen loss from the soil by leaching in desert ecosystems was negligible. It is now

(a)

(b)

29–16 Cluster roots Cluster roots on the root systems of two *Hakea* species (Proteaceae), ***(a)*** *Hakea prostrata* and ***(b)*** *Hakea sericea,* both from Australia. Also called proteoid roots, they greatly increase the absorbing surface of the root system.

known that long-term leaching of nitrate from desert soils has resulted in huge nitrate reservoirs (up to about 10,000 kilograms of nitrogen as nitrate per hectare) in subsoil zones (Figure 29–5). There is concern that if, in the future, such deserts are irrigated or the climate becomes wetter, these huge nitrate reservoirs could percolate into aquifers and contaminate groundwater.

Soils and Agriculture

In natural situations, the elements present in soil recirculate and so become available again for plant growth. As discussed above, negatively charged clay particles and organic matter can bind such positively charged ions as Ca^{2+}, Na^+, K^+, and Mg^{2+}. These ions are then displaced from the clay particles by other cations (cation exchange) and absorbed by roots. In general, the cations required by plants are present in large amounts in fertile soils, and the amounts removed by a single crop are small. However, when a series of crops is grown on a particular field and the nutrients are continuously removed from their cycles as the crops are harvested, some of these cations may no longer be present in sufficient plant-available forms. For example, in almost all soils, most of the potassium is present in forms that are not exchangeable and not plant-available. The same is true for phosphorus and nitrogen. Soils that are transferred from natural ecosystems to agricultural systems often do not have enough plant-available nutrients to support crops for commercial harvest, although the quantities are sufficient for native plant communities.

Programs for supplementing nutrient supplies for agricultural and horticultural crops should be based on soil testing, which is used to diagnose nutrient deficiencies and to predict a likely response to the addition of fertilizer in a recommended amount. Nitrogen, phosphorus, and potassium are the three elements that are commonly included in commercial fertilizers. Fertilizers are usually labeled with a formula that indicates the percentage of each of these elements. A 10-5-5 fertilizer, for example, is one that contains 10 percent nitrogen (N), 5 percent phosphorus (as phosphorus pentoxide, P_2O_5), and 5 percent potassium (as potassium oxide, K_2O). This method of stating phosphorus and potassium contents of fertilizer is a historical relic dating from the days when analytical chemists reported all of their element analyses as oxides.

Other essential inorganic nutrients, although required in very small amounts, can sometimes become limiting factors in soils on which crops are grown. Experience has shown that the most common deficiencies are those for iron, sulfur, magnesium, zinc, and boron.

HALOPHYTES: A FUTURE RESOURCE?

Unlike most animals, most plants generally do not require sodium and, moreover, cannot survive in brackish waters or saline soils. In such environments, the solution surrounding the roots often has a higher solute concentration than the plant cells, causing water to move out of the roots by osmosis. Even if the plant is able to absorb water, it faces additional problems from the high level of sodium ions (Na^+). If the plant takes up water and excludes sodium ions, the solution surrounding the roots becomes even saltier, increasing the likelihood of water loss through the roots. The salt may eventually become so concentrated that it forms a crust around the roots, effectively blocking the supply of water to the plant. Another problem is that sodium ions may enter the plant in preference to potassium ions (K^+), depriving the plant of an essential nutrient, as well as inhibiting some enzyme systems.

Some plants—known as halophytes—can grow in saline environments such as deserts, salt marshes, and coastal areas. In all of these plants, mechanisms have evolved for dealing with high sodium concentrations, and for some of the plants, sodium appears to be a required nutrient. The adaptations of halophytes vary. In many halophytes, a sodium-potassium pump seems to play a major role in maintaining a low sodium concentration within the cells, while simultaneously ensuring that a sufficient supply of potassium ions enters the plant. In some species, the pump operates primarily in the root cells, pumping sodium back to the environment and potassium into the root. The presence of calcium ions (Ca^{2+}) in the soil solution is thought to be essential for the effective functioning of this mechanism.

Other halophytes take in sodium through the roots, but then either secrete it or isolate it from the living cytoplasm of the plant body. In *Salicornia* (pickleweed), a sodium-potassium pump (or a variant of it) operates in the tonoplasts (vacuolar membranes) of leaf cells. Sodium ions enter the cells but are immediately pumped into vacuoles and isolated from the cytoplasm. In such plants, the solute concentration of the vacuoles is higher than that of the environment, establishing the necessary osmotic potential for the movement of water into the roots. In other genera, the salt is pumped into the intercellular spaces of the leaves and then secreted from the plant. In *Distichlis palmeri* (Palmer's grass), the salt exudes through specialized cells (not the stomata) onto the surface of the leaf. In *Atriplex* (saltbush), it is concentrated by special salt glands and pumped into bladders. The bladders expand as the salt accumulates, finally bursting. Rain or the passing tide washes the salt away.

Halophytes are of current interest not only because of the light they may shed on the osmoregulatory mechanisms of plants but also because of their potential as crop plants. In a world with an ever-increasing need for food, vast areas are unsuitable for agricultural purposes because of the salinity of the soil. For example, there are over 30,000 kilometers of desert coastline and about 400 million hectares of desert with potential water supplies that are too salty for crop plants. Moreover, each year, about 200,000 hectares of irrigated crop land become so salty that further agriculture is impossible. When arid land is heavily irrigated, as in large areas of the western United States, salts from the irrigation waters accumulate in the soil. This accumulation occurs because, in both evaporation from the soil and transpiration from plants, essentially pure water is given off, leaving all solutes behind. Over many years, the salt concentration of the soil increases, eventually reaching levels that cannot be tolerated by most plants. It has been suggested that the ancient civilizations of the Near East ultimately fell because their heavily irrigated land became so salty that food could no longer be grown there.

One way to extend the life of irrigated crop lands and to bring barren areas into agricultural use would be to breed salt tolerance into conventional crop plants. Thus far, however, such efforts have met with little success. Other approaches, including genetic engineering, have met with limited success. This is because, as yet, only a single gene has been transformed, but salt tolerance is a multigenic property that involves an array of physiological, biochemical, and molecular processes. Nevertheless, the quest continues for halophyte improvement.

(a)

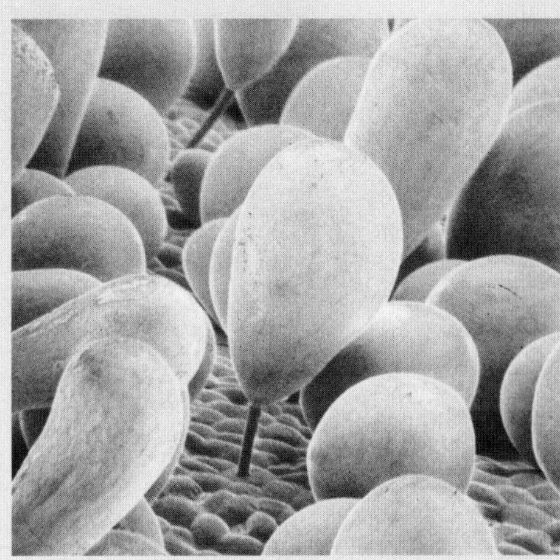

(b)

(a) *Atriplex* (saltbush) is one of several halophytes being evaluated as potential crop plants. ***(b)*** The surface of an *Atriplex* leaf. Salt is pumped from the leaf tissues through narrow stalk cells into the large, expandable bladder cells.

COMPOST

Composting, a practice as old as agriculture itself, has attracted increased interest as a means of utilizing organic wastes by converting them to fertilizer. The starting material is any collection of organic matter—leaves, kitchen garbage, animal manure, straw, lawn clippings, sewage sludge, sawdust—and the population of bacteria and other microorganisms normally present. The only other requirements are oxygen and moisture. Grinding of the organic matter is not essential, but it provides a greater surface area for microbial attack and thus speeds the process.

In a compost heap, microbial growth accelerates quite rapidly, generating heat, much of which is retained, because the outer layers of organic matter act as insulation. In a large heap (2 meters × 2 meters × 1.5 meters, for instance), the interior temperature rises to about 70°C; in small heaps, it usually reaches 40°C. As the temperature rises, the population of decomposers changes, with thermophilic and thermotolerant forms replacing the earlier organisms. As the original forms die, their organic matter also becomes part of the product. A useful side effect of the temperature increase is that it destroys most of the common pathogenic bacteria that may have been present, for example, in sewage sludge, as well as cysts, eggs, and other immature forms of plant and animal parasites.

With the passage of time, changes in pH also occur in a compost heap. The initial pH is usually slightly acidic, about 6.0, which is comparable to the pH of most plant fluids. During the early stages of decomposition, the production of organic acids causes a further acidification, with the pH decreasing to about 4.5 to 5.0. However, as the temperature rises, the pH also increases; the composted material eventually levels off at slightly alkaline values (7.5 to 8.5).

An important factor in composting (as in any biological growth process) is the ratio of carbon to nitrogen, which is ideally about 30:1 (by weight). If the C/N ratio is too high, microbial growth slows; if too low, some nitrogen escapes as ammonia. If the compost materials are quite acidic, limestone (calcium carbonate) may be added to balance the pH; however, if too much is added, it will increase the nitrogen loss.

Studies of municipal compost piles in Berkeley, California, demonstrated that if large piles were kept moist and aerated, composting could be completed in as little as two weeks. Generally, though, three months or more during the winter are needed to complete the process. If compost is added to the soil before the composting process is complete, it may temporarily deplete the soil of soluble nitrogen.

Because it greatly reduces the bulk of plant wastes, composting can be a very useful means of waste disposal. In Scarsdale, New York, for example, leaves composted in a municipal site were reduced to one-fifth their original volume. At the same time, they formed a useful soil conditioner, improving both aeration and water-holding capacity. Chemical analyses indicate, however, that a rich compost commonly contains, in dry weight, only about 1.5 to 3.5 percent nitrogen, 0.5 to 1.0 percent phosphorus, and 1.0 to 2.0 percent potassium, far less than commercial fertilizers. Yet, unlike commercial fertilizers, compost can be a source of nearly all the elements known to be needed by plants. Compost provides a continuous balance of nutrients, releasing them gradually as it continues to decompose in the soil.

Today, the main driving force behind composting is the increasing cost of waste disposal and the increasing difficulty of finding suitable disposal sites—sites that will not adversely affect the environment or pollute our waters. Unfortunately, composting is a long way from becoming economically viable as a substitute for manufacturing fertilizers for commercial agriculture.

Plant Nutrition Research

Research on the inorganic nutrients essential for crop plants—particularly the quantities of nutrients required for optimal crop yields and the capacities of various soils to provide the nutrients—has been of great practical value in agriculture and horticulture. Because of the steady increase in worldwide food needs, this type of research undoubtedly will continue to be essential.

Ways Are Being Sought to Overcome Soil Deficiencies and Toxicities

Modification and manipulation of soils by adding nutrients as fertilizers, by raising the pH with lime, or by removing excess salts by leaching with water may not be the only means of improving and maintaining crop production in below-optimum soils. By using the knowledge and techniques of plant breeding,

THE WATER CYCLE

The Earth's supply of water is stable and is used over and over again. Most of the water (98 percent) is present in oceans, lakes, and streams. Of the remaining 2 percent, some is frozen in polar ice and glaciers, some is found in the soil, some is in the atmosphere as water vapor, and some is in the bodies of living organisms.

Sunshine evaporates water from the oceans, lakes, and streams, from moist soil surfaces, and from the bodies of living organisms, drawing the water back up into the atmosphere, from which it falls again as rain or snow. Evaporation exceeds precipitation over the oceans, resulting in a net movement of water vapor, carried by wind, from the ocean to the land. Over 90 percent of the water lost on the land is lost by plant transpiration (evaporation of water from the soil plus transpiration from plants is called evapotranspiration). This constant movement of water from the Earth into the atmosphere and back again is known as the water cycle. The water cycle is driven by solar energy.

Some of the water that falls on the land percolates down through the soil until it reaches a zone of saturation. In the zone of saturation, all hollows and cracks in the rock are filled with water. Below the zone of saturation is solid rock, through which water cannot penetrate. The upper surface of the zone of saturation is known as the water table.

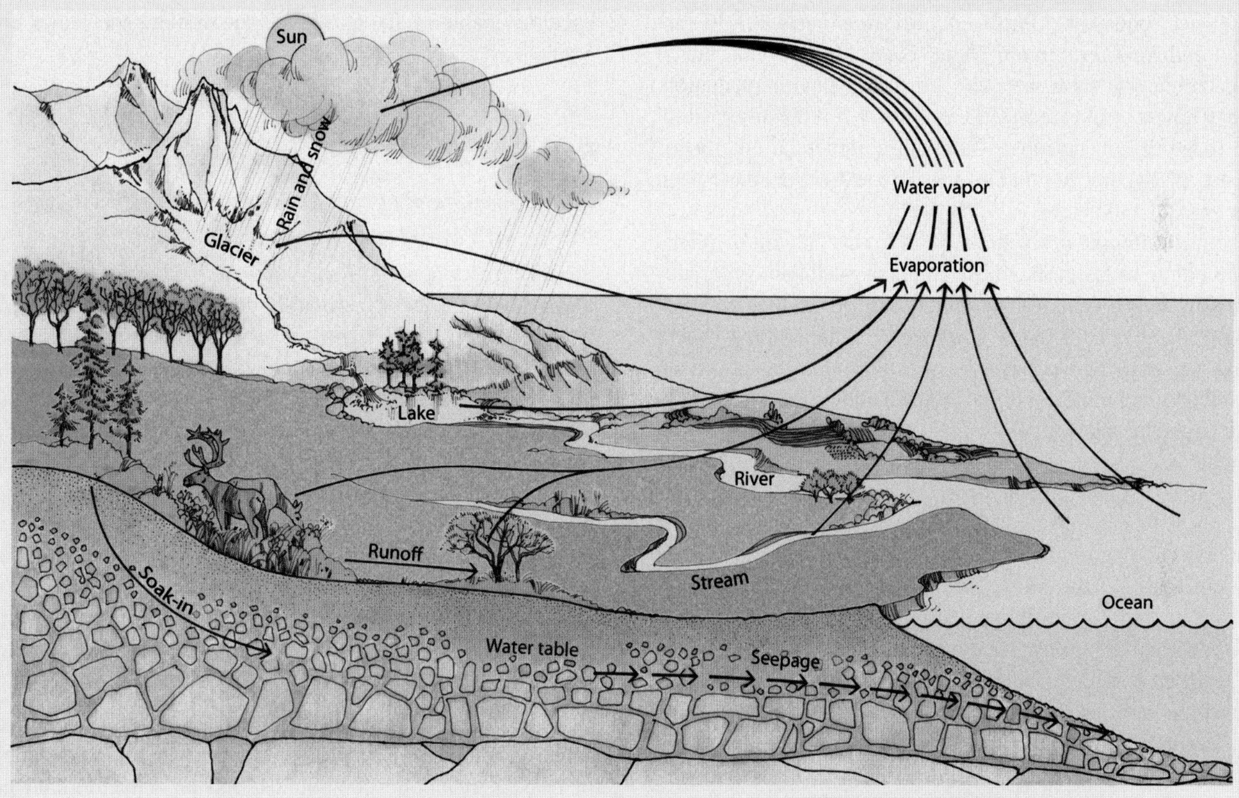

biotechnology, and plant nutrition, it may be possible to select and develop cultivars of crop species that are better adapted for growth in nutrient-deficient environments. The validity of this research is confirmed by the natural occurrence of wild plants in nutritional environments that are very different from the average soil environments in which crop plants are grown. Examples of such less optimal environments are acidic *Sphagnum* bogs, in which the pH may be less than 4.0, and mine tailings, which often contain high concentrations of potentially toxic metals.

Plants that tolerate high concentrations of potentially toxic metals, such as zinc, nickel, chromium, cadmium, and lead, are being investigated as part of phytoremediation strategies to restore the soils around old smelters, metal finishers, and nuclear weapons plants and other polluted sites (see Figure 1–13). Of special interest are plants known as **hyperaccumulators,** which can accumulate extraordinarily high amounts of heavy metals (at concentrations 100- to 1000-fold higher than non-hyperaccumulators) in their shoots, principally the leaves, without suffering phytotoxic damage. Hyperaccumulators are especially efficient in detoxification and sequestration of heavy metals. Preferentially, these processes occur in the epidermis, trichomes, and the cuticle of leaves, sites where the heavy metals will do the least damage to the photosynthetic apparatus in the mesophyll cells.

Studies of hyperaccumulators and related non-hyperaccumulators indicate that most of the key steps in hyperaccumulation rely not on novel genes but on genes that are held in common but are expressed and regulated differently in the two kinds of plants. In hyperaccumulators, the enhanced uptake of heavy metals, their rapid and efficient transport via the xylem to the shoot, and their sequestration in the leaves apparently rely, in part, on an overexpression of the same genes found in non-hyperaccumulators. It is presumed that the high concentrations of heavy metals in the shoot function in defense against herbivores and pathogens.

About 450 species of angiosperms have been identified as heavy-metal hyperaccumulators, which is less than 0.2 percent of all known species. Of these, about 25 percent belong to the mustard family (Brassicaceae), and the greatest number of studies have been carried out on two of its members, alpine penny cress *(Thlaspi caerulescens)* and *Arabidopsis halleri.* It has been estimated that a crop of *T. caerulescens* or *A. halleri,* after harvesting, could remove an amount equal to decades of cadmium accumulation from pastures that have been treated with cadmium-rich phosphate fertilizers. *Alyssum bertolonii* (Brassicaceae) and *Berkheya coddii* (Asteraceae), fast-growing nickel hyperaccumulators, show promise for nickel phytoremediation, and the Chinese brake fern *(Pteris vittata)* for the removal of arsenic (page 391). The most successful commercial enterprises in the use of hyperaccumulators for phytoextraction have been those for nickel and arsenic.

A quite different problem is to find ways to make plants tolerate certain metals. The most common metal in soils, aluminum, causes problems for agriculture on 30 to 40 percent of the world's arable land, most commonly in the tropics, where the soils are acidic. In nonacidic soils, aluminum is locked up in insoluble compounds, but in acidic soils it becomes soluble (Al^{3+}) and is taken up by roots, rapidly inhibiting growth, or elongation, of the root apex. Aluminum causes extensive injury to the root, leading to poor ion and water uptake. Aluminum resistance in most plant species relies on the efflux of organic anions that bind Al^{3+}, forming complexes that are not readily absorbed by the roots. Efflux of the anions, most commonly malate and citrate, is generally restricted to the root apices and requires Al^{3+} to trigger the response. Several genes have been identified that contribute to Al^{3+} resistance in both cereal and noncereal species. The genes encode proteins bound to the plasma membrane that function as anion channels activated by Al^{3+}, releasing either malate or citrate. The goal of most molecular studies on aluminum toxicity is to develop plants that are more resistant to aluminum—especially cultivars of rice, wheat, and maize—and to discover more aluminum-resistant genes by using a modern genomics approach (see the essay "Model Plants: *Arabidopsis thaliana* and *Oryza sativa,*" page 199).

Biological Nitrogen Fixation Is Being Manipulated to Improve Efficiency

Manipulation of biological nitrogen fixation also offers tremendous potential for improved efficiency in nitrogen utilization. One aspect of research in this area is concerned with improving the efficiency of the association of legumes with either *Rhizobium* or *Bradyrhizobium,* for example, through the genetic screening of both legumes and bacteria. Such screening could identify combinations that would result in increased nitrogen fixation in specific environments. This could result from a greater photosynthetic efficiency in legumes, so that more carbohydrate is available for bacterial nitrogen fixation and growth. However, nitrogen fixation requires considerable energy, and any increase in fixation might be at the expense of shoot productivity.

A second research approach is to develop additional and more effective associations of free-living and endophytic nitrogen-fixing bacteria with grain food crops. The benefits of inoculating cereal crops with selected strains of endophytic rhizobia have been examined in the Nile delta of Egypt. Through the use of such "biofertilizers," the crop yield for rice and wheat increased up to 30 percent above that reached by using inorganic fertilizers alone. This was also achieved at considerably less cost than with the use of inorganic fertilizers, and with decreased environmental pollution.

Probably the most exciting research approach is to be found in genetic engineering. One example is genetic modifications of the *nif* genes necessary for nitrogen fixation and their transfer to bacteria unable to fix nitrogen, transforming them into nitrogen fixers.

SUMMARY

Plants Require Macronutrients and Micronutrients for Growth and Development

Seventeen inorganic nutrients are required by most plants for normal growth. Of these, carbon, hydrogen, and oxygen are derived from air and water. The rest are absorbed by roots in the form of ions. These 17 elements are categorized as either macronutrients or micronutrients, depending on the amounts in which they are required. The macronutrients are sulfur, phosphorus, magnesium, calcium, potassium, nitrogen, hydrogen, carbon, and oxygen. The micronutrients are molybdenum, nickel, copper, zinc, manganese, boron, iron, and chlorine. Some inorganic nutrients, such as aluminum, cobalt, sodium, selenium, and silicon, are essential only for specific organisms and are referred to as beneficial elements.

Inorganic Nutrients Perform a Number of Important Roles in Cells

Inorganic nutrients regulate osmosis and affect cell permeability. Some also serve as structural components of cells, as components of critical metabolic compounds, and as activators and components of enzymes. These functions are impaired when the supply of the essential nutrient is inadequate, and this results in nutrient deficiency symptoms such as stunted growth of stems and leaves, localized death of tissues (necrosis), and yellowing of leaves (chlorosis).

Soils Provide Both a Chemical and a Physical Environment for Plant Growth

The chemical and physical properties of soils are critical in determining their ability to provide the inorganic nutrients, water, and other conditions necessary for maximum crop plant production. The weathering of rocks and minerals supplies the inorganic component of soils. In addition to inorganic nutrients, soils contain organic matter and pore space occupied by varying proportions of water and gases. At least three layers, or horizons—designated A, B, and C—exist in all soils. The A horizon contains the greatest portion of the soil's organic matter, including humus and large numbers of living organisms. Loam soils contain sand, silt, and clay in proportions that result in ideal agricultural soils. Under agricultural conditions, nitrogen, phosphorus, and potassium are the nutrients most limiting to plant growth and most frequently added to soils in fertilizers.

The Elements Required by Plants Are Recycled Locally and Globally

Each essential inorganic nutrient is circulated, in a complex cycle, among organisms and between organisms and the environment. Because nutrient cycles involve both living organisms and the physical environment, they are also called biogeochemical cycles. Nutrient cycles are leaky; that is, not all of the nutrients returned to the soil become available for plant use.

Ammonification, Nitrification, and Denitrification Are Reactions in the Nitrogen Cycle Carried out by Soil Bacteria

The circulation of nitrogen through the soil, through the bodies of plants and animals, and back to the soil again is known as the nitrogen cycle. Most soil nitrogen is derived from dead organic materials of plant and animal origin. These substances are decomposed by soil organisms. Ammonification—the release of ammonium ions (NH_4^+) from nitrogen-containing compounds—is carried out by soil bacteria and fungi. Nitrification is the oxidation of ammonia or ammonium ions to form nitrites and nitrates. One type of bacterium is responsible for the oxidation of ammonia to nitrite, and another for the oxidation of nitrite to nitrate. Nitrogen enters crop plants almost entirely in the form of nitrates. The major loss of nitrogen from the soil is by denitrification. Nitrogen is also lost by crop removal, soil erosion, fire, and leaching.

Nitrogen Fixation Is a Crucial Aspect of the Nitrogen Cycle

Nitrogen is replenished in the soil primarily by nitrogen fixation, the process by which N_2 is reduced to ammonium and made available for assimilation into amino acids and other organic nitrogenous compounds. Biological nitrogen fixation is carried out only by bacteria, including those (*Rhizobium* and *Bradyrhizobium*) that are symbionts of leguminous plants, as well as free-living bacteria and actinomycetes (*Frankia*) in symbiotic relationships with a few genera of nonleguminous plants. The most efficient nitrogen-fixing bacteria are those that form symbiotic relationships with plants, which produce root nodules on infection by the bacteria. Some nitrogen-fixing bacteria inhabit the intercellular spaces and xylem vessels of their host plants as endophytes. In agriculture, plants are removed from the soil and thus nitrogen and other elements are not recycled, as they are in natural ecosystems; these elements must be replenished in either organic or inorganic form.

Phosphorus Recycling Is Quite Localized

The phosphorus cycle differs from the nitrogen cycle in part because the Earth's crust, rather than the atmosphere, is the primary replenishment reservoir of phosphorus. The weathering of rocks and minerals over long periods of time is the source of most phosphorus in the soil solution. Phosphorus circulates from plants to animals and is returned to the soil in organic forms, which are then converted to inorganic forms by microorganisms and thus made available to plants. Although some phosphorus is lost from the terrestrial ecosystem by leaching and erosion, weathering of rock usually keeps pace with this loss. Plants have several strategies for the uptake of phosphate, the only important inorganic form of phosphorus: rapid root growth, symbioses with mycorrhizal fungi, and cluster roots.

Human Activities Have Had Drastic Effects on Some Nutrient Cycles

Much of the damage to nutrient cycles is related to the need to adequately feed an exponentially growing human population. Crop removal and increased soil erosion have accelerated phosphorus loss from soils. Marshes and wetlands, the primary sites of denitrification, are being destroyed. Soils transferred from natural ecosystems to agricultural systems often do not have enough plant-available nutrients to support crops for commercial harvest, although nutrient quantities are sufficient for native plant communities.

Plant Nutrition Research Is of Great Practical Value to Agriculture

The knowledge and techniques of plant breeding and plant nutrition are being used to select and develop cultivars of crop species that are better adapted for growth in nutrient-deficient environments. Hyperaccumulators, which concentrate heavy metals at levels much higher than normal, are being identified and used to restore the soils around old smelters, metal finishers, and nuclear weapons plants. Both legumes and bacteria are being genetically screened for combinations that could result in increased nitrogen fixation in specific environments. Genetic engineering offers the possibility of transferring the genes necessary for nitrogen fixation from one organism to another.

QUESTIONS

1. Distinguish among or between the following: macronutrients, micronutrients, and beneficial elements; A horizon, B horizon, and C horizon; necrosis and chlorosis; field capacity and permanent wilting percentage; ammonification, nitrification, and denitrification; symbiotic nitrogen-fixing bacteria and free-living nitrogen-fixing bacteria; determinate nodules and indeterminate nodules.
2. What attributes of loam soils help make them ideal agricultural soils?
3. How does the size of the spaces around soil particles influence the amount of water that is available to plants?
4. Why is cation exchange important to plants?
5. Describe the sequence of events leading to nodule formation on legume roots.
6. Some plants are aluminum hyperaccumulators, whereas others are aluminum-resistant. Explain how one differs from the other.

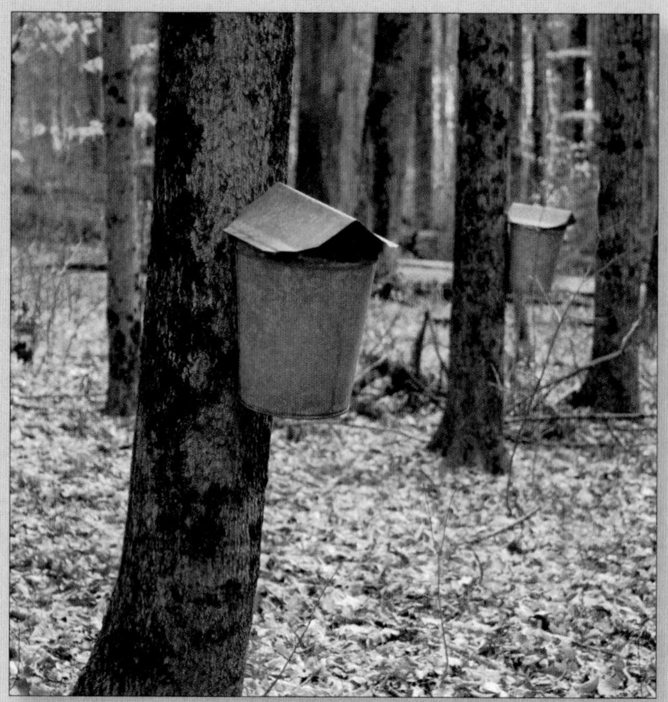

CHAPTER 30

The Movement of Water and Solutes in Plants

◀ **Maple sugaring** As the spring thaw arrives, starch stored in the roots and trunks of sugar maples *(Acer saccharum)* is converted to a watery, sucrose-containing sap that rises in the xylem of the tree to nourish the leaf buds. Sap is collected through holes bored into the trunk, and boiling reduces the sap to syrup—40 liters of sap are needed to produce 1 liter of maple syrup. The sap stops flowing when leaves appear.

CHAPTER OUTLINE

Plants transport both organic and inorganic nutrients and water throughout the plant body. The capacity to do so is critical in determining the ultimate structure and function of the plant's component parts, as well as its development and overall form. The tissues involved in the long-distance transport of substances in the plant are the xylem and the phloem. As discussed in Chapters 23 through 25, these two tissues form a continuous vascular system that penetrates practically every part of the plant. Xylem and phloem are closely associated both spatially and functionally. Although we often think of the xylem as *the* water-conducting tissue and the phloem as *the* food-conducting tissue, their functions overlap. The phloem, for example, moves large volumes of water throughout the plant. The phloem, in fact, is the principal source of water for many developing plant parts, such as the developing fruits of a wide variety of crops. In addition, substances are transferred from the phloem to the xylem and recirculated throughout the plant.

The first investigators of "circulation" in plants were seventeenth-century physicians who searched for pathways and pumping mechanisms analogous to the circulation of blood in animals. Great strides were made during the eighteenth century toward understanding the movement of water and dissolved minerals in xylem. By the end of the nineteenth century, a plausible mechanism was proposed for the rise of water in tall plants. It was not until the 1920s and 1930s, however, that the role of phloem as a food-conducting tissue was widely accepted. We have come a long way since then in our understanding of phloem transport.

CHECKPOINTS

After reading this chapter, you should be able to answer the following:

1. What is transpiration, and why is it dubbed an "unavoidable evil"?
2. Describe the role of turgor and the orientation of cellulose microfibrils in the guard cell walls in the opening and closing of stomata.
3. How does the cohesion-tension theory account for the movement of water to the top of tall trees?
4. How does the osmotically generated pressure-flow mechanism account for the movement of sugars from source to sink?
5. By what mechanisms do apoplastic and symplastic phloem-loading plants move sugars into the sieve tube–companion cell complexes?

Movement of Water and Inorganic Nutrients through the Plant Body

Plants Lose Large Quantities of Water to Transpiration

In the early eighteenth century, Stephen Hales, an English physician, noted that plants "imbibe" a much greater amount of water than animals do. He calculated that one sunflower plant, bulk for bulk, "imbibes" and "perspires" 17 times more water than a human every 24 hours (Figure 30–1). Indeed, the total quantity of water absorbed by any plant is enormous—far greater than that used by any animal of comparable weight. An animal uses less water because much of its water is recirculated through its body over and over again, in the form (in vertebrates) of blood plasma and other fluids. In plants, nearly 99 percent of the water taken in by the roots is released into the air as water vapor. Table 30–1 shows the amounts of water released by several crop plants in a single growing season. These amounts pale, however, when compared with the water lost *in a single day*—200 to 400 liters (50 to 100 gallons)—by a single tree growing in a deciduous forest of southwestern North Carolina. This loss of water vapor from plants, known as **transpiration,** may involve any aboveground part of the plant body, but leaves are by far the most important organs of transpiration.

30–1 Hales's sunflower Diagram of the sunflower plant from Stephen Hales's description of his experiments in the early 1700s on the movement of water through the plant body. Hales found that most of the water "imbibed" by the plant was lost by "perspiration."

Table 30-1 Water Loss by Transpiration in One Plant in a Single Growing Season

Plant	Water Loss (liters)
Cowpea *(Vigna sinensis)*	49
Potato *(Solanum tuberosum)*	95
Wheat *(Triticum aestivum)*	95
Tomato *(Solanum lycopersicum)*	125
Maize *(Zea mays)*	206

After J. F. Ferry, *Fundamentals of Plant Physiology* (New York: Macmillan Publishing Company, 1959).

Why do plants lose such large quantities of water to transpiration? This question can be answered by considering the requirements for the chief function of the leaf: photosynthesis, the source of all the food for the entire plant body. The energy necessary for photosynthesis comes from sunlight. Therefore, for maximum photosynthesis, a plant must spread a maximum surface area to the sunlight—at the same time creating a large transpiring surface. But sunlight is only one of the requirements for photosynthesis; the chloroplasts also need carbon dioxide. Under most circumstances, carbon dioxide is readily available in the air surrounding the plant. However, for the carbon dioxide to enter a plant cell, which it does by diffusion, it must go into solution, because the plasma membrane is nearly impervious to the gaseous form of carbon dioxide. The gas must therefore come into contact with a moist cell surface—but wherever water is exposed to unsaturated air, evaporation occurs. In other words, the uptake of carbon dioxide for photosynthesis and the loss of water by transpiration are inextricably bound together in the life of the green plant.

Water Vapor Diffuses from the Leaf to the Atmosphere through Stomata

Transpiration—sometimes called an "unavoidable evil"—can be extremely injurious to a plant. Excessive transpiration (water loss exceeding water uptake) retards the growth of many plants and kills many others by dehydration. Despite their long evolutionary history, plants have not developed a structure that is both favorable to the entrance of the carbon dioxide essential in photosynthesis and unfavorable to the loss of water vapor by transpiration. However, a number of special adaptations minimize water loss while promoting carbon dioxide gain.

The Cuticle Serves as an Effective Barrier to Water Loss Leaves are covered by a waxy **cuticle** that makes their surfaces largely impervious to both water and carbon dioxide (page 23). Only a small fraction of the water transpired by plants is lost through this protective outer coating, and another small fraction is lost through the lenticels in the bark (see Figures 26–10d and 26–11). By far the largest amount of water transpired by a vascular plant is lost through

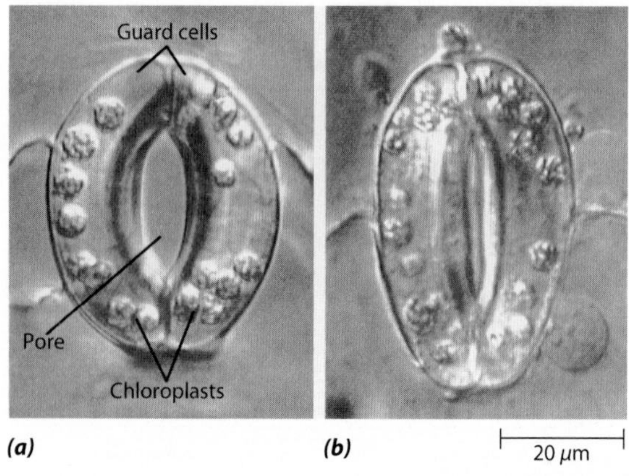

30–2 Stomata ***(a)*** Open and ***(b)*** closed stomata in the epidermis of a broad bean *(Vicia faba)* leaf. Note the different shape of the guard cells in the open and closed conditions. The change in shape is largely due to the structure of the guard cell wall.

the stomata (Figure 30–2; see also Figures 23–25, 23–26, and 25–24). Stomatal transpiration involves two steps: (1) evaporation of water from cell wall surfaces bordering the intercellular (air) spaces of the leaf, and (2) diffusion of the resultant water vapor from the intercellular spaces into the atmosphere by way of the stomata (see Figure 30–23).

The Opening and Closing of Stomata Controls the Exchange of Gases across the Leaf Surfaces As discussed previously (pages 657 and 658), changes in the shape of the guard cells bring about the opening and closing of the stomatal pores (Figure 30–2). Although stomata occur on all aboveground parts of the primary plant body, they are, by far, most abundant in leaves. The stomatal density—number of stomata per square millimeter—may be quite large, and differs from one part of the leaf to another. Some examples of stomatal densities, comparing lower epidermis to upper epidermis, are: oat *(Avena sativa)*, 45/50; maize *(Zea mays)*, 108/98; tobacco *(Nicotiana tabacum)*, 190/50; black oak *(Quercus velutina)*, 405/0; and basswood *(Tilia americana)*, 891/0. The stomata lead into a honeycomb of air spaces surrounding the thin-walled mesophyll cells within the leaf. These spaces, which make up 15 to 40 percent of the total volume of the leaf, contain air saturated with water vapor that has evaporated from the damp surfaces of the mesophyll cells. Although the stomatal pores account for only about 1 percent of the total leaf surface, more than 90 percent of the water transpired by the plant is lost through the stomata. The rest is lost through the cuticle.

Closing of stomata not only prevents loss of water vapor from the leaf but also prevents entry of carbon dioxide into the leaf. A certain amount of carbon dioxide, however, is produced by the plant during respiration, and as long as light is available, this carbon dioxide can be used to sustain a very low level of photosynthesis even when the stomata are closed.

Stomatal Movements Result from Changes in Turgor Pressure within the Guard Cells In Chapter 27, we discussed the role of abscisic acid in stomatal movement during periods of water stress. In the leaves of most well-watered plants, light is the dominant signal controlling stomatal movements. Stomata open in the morning as light levels reaching the leaf surfaces increase, and close as light levels decrease. Stomatal opening occurs when solutes are actively accumulated in the guard cells. This accumulation of solutes (and resultant decrease in guard cell water potential) causes osmotic movement of water into the guard cells and a buildup of turgor pressure in excess of that in the surrounding epidermal cells. Stomatal closing is brought about by the reverse process: with a decline in guard cell solutes (and resultant increase in guard cell water potential), water moves out of the guard cells and the turgor pressure decreases.

Turgor is thus maintained or lost due to the passive osmotic movement of water into or out of the guard cells along a gradient of water potential that itself is created by active solute transport. As noted in Chapter 27, the potassium ion (K^+) is the principal solute involved in this mechanism. Uptake of K^+ by the guard cells is driven by a proton (H^+) gradient mediated by a blue-light-activated plasma membrane H^+-ATPase. This K^+ uptake is accompanied by the uptake of chloride ions (Cl^-) and the accumulation of malate^{2-}, which is synthesized from starch in the guard cell chloroplasts.

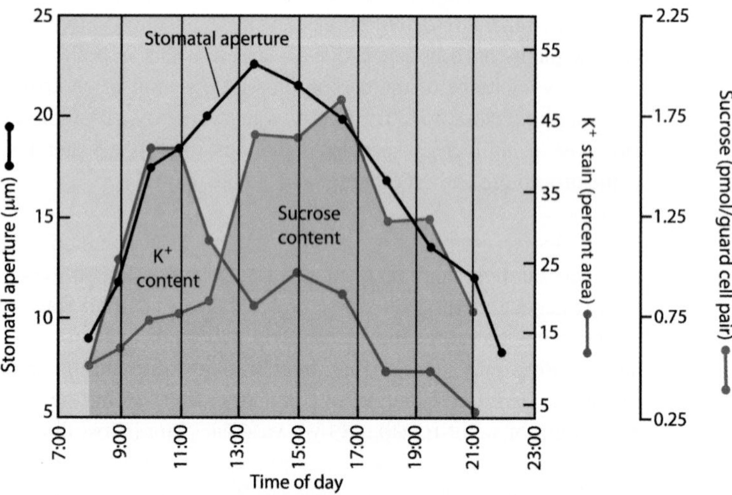

30–3 Stomatal opening and closing This graph shows the daily course of changes in the size of the stomatal pore in intact leaves of broad bean *(Vicia faba)*, in relation to the potassium (K^+) and sucrose content of the guard cells. Whereas potassium is the dominant osmoticum, or osmotically active solute, involved in stomatal opening in the morning, sucrose is the dominant osmoticum involved in changes in the afternoon and evening. (1 picomol, or pmol, = 10^{-12} mols.)

Phototropins localized in the guard cell plasma membrane are the photoreceptors of the blue light. The chloroplast carotenoid pigment zeaxanthin has also been implicated in blue-light photoreception in guard cells. Some studies indicate that sucrose, in addition to K^+, is a primary **osmoticum,** or osmotically active solute, in guard cells, with K^+ being the dominant osmoticum in the early opening stages in the morning, and sucrose becoming the dominant osmoticum in the early afternoon (Figure 30–3). The sucrose has its origin from glucose and fructose derived from starch hydrolysis in the guard cell chloroplasts. Stomatal closing at day's end parallels a loss in the sucrose content. Hence, stomatal opening apparently is associated with K^+ uptake, and stomatal closing with a decrease in sucrose content.

Radial Orientation of Cellulose Microfibrils in the Guard Cell Walls Is Required for Pore Opening The structure of the guard cell walls plays a crucial role in stomatal movements. During expansion of the paired guard cells, two physical constraints cause the cells to bend and thus to open the pore. One of these constraints is the radial orientation of cellulose microfibrils in the guard cell walls (Figure 30–4a). This **radial micellation** allows the guard cells to lengthen while preventing them from expanding laterally. The second constraint is found at the ends of the guard cells, where they are attached to one another. This common wall remains almost constant in length during opening and closing of the stoma. Consequently, increase in turgor pressure causes the outer (dorsal) walls of the guard cells to move outward relative to their common walls. As this happens, the radial micellation transmits this movement to the wall bordering the pore (the ventral wall), and the pore opens. Figure 30–4b through d depicts the results of some experiments with balloons that have been used in support of the role of radial micellation in stomatal movement.

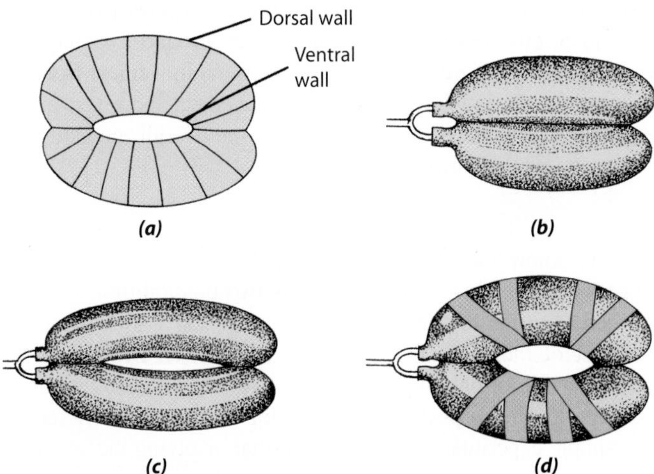

30–4 Radial micellation in guard cells ***(a)*** A pair of guard cells, with lines indicating the radial arrangement of cellulose microfibrils in the guard cell walls. ***(b)*** Two partially inflated balloons were glued together near their ends to model the effect of radial micellation on the opening of stomata. ***(c)*** The same balloons at higher pressure, that is, fully inflated. A narrow slit is visible. ***(d)*** A pair of fully inflated balloons after bands of tape were added to simulate radial micellation. As a result, the opening is greatly increased.

Carbon Dioxide Concentration and Temperature Also Affect Stomatal Movement In most species, an increase in carbon dioxide concentration causes stomata to close. The magnitude of this response varies greatly from species to species and with the degree of water stress a plant has undergone or is undergoing. In maize *(Zea mays),* the stomata may respond to changes in carbon dioxide in a matter of seconds. The site for sensing the level of carbon dioxide is located in the guard cells.

Within normal ranges (10° to 25°C), changes in temperature have little effect on stomatal behavior, but temperatures higher than 30° to 35°C can lead to closure. The closing can be prevented, however, if the plant is kept in air without carbon dioxide, which suggests that temperature changes work primarily by affecting carbon dioxide concentration in the leaf. An increase in temperature results in an increase in respiration and a concomitant increase in concentration of intercellular carbon dioxide, which may be the actual cause of stomatal closure in response to heat. Many plants in hot climates close their stomata regularly at midday, apparently because carbon dioxide has accumulated in the leaf and the leaves are dehydrated as water loss by transpiration exceeds water uptake by absorption.

Stomata not only respond to environmental factors but also exhibit daily rhythms of opening and closing that appear to be controlled from within the plant—that is, they also exhibit circadian rhythms (page 665).

A wide variety of succulents—including cacti, the pineapple *(Ananas comosus),* and members of the stonecrop family (Crassulaceae)—open their stomata at night, when conditions are least favorable to transpiration (Figure 30–5). The crassulacean acid metabolism (CAM) characteristic of such plants has a pathway for carbon flow not substantially different from that of C_4 plants (page 145). At night, when their stomata are open, CAM plants take in carbon dioxide and convert it to organic acids. During the day, when their stomata are closed, the carbon dioxide is released from these organic acids for use in photosynthesis. This is obviously advantageous in the conditions of high light intensity and water stress under which CAM plants live.

It has long been presumed that the stomata of C_3 plants are closed in the dark, causing transpiration to decrease to zero, but growing evidence suggests that many species of trees and shrubs do not close their stomata completely in the dark. For some species, the nighttime loss of water from the leaves, called **nocturnal transpiration,** constitutes a significant fraction of the daily water use. Nocturnal water loss by trees in the nutrient-poor Brazilian cerrado (savanna), for example, may account for 28 percent or more of total daily transpiration. It has been suggested that nutrient uptake is enhanced by nocturnal transpiration—occurring at a time when evaporation is lower than during the day, with its increased stomatal opening and increased evaporation.

Environmental Factors Affect the Rate of Transpiration

Although stomatal opening and closing are the major plant factors affecting the rate of transpiration, a number of other factors, both in the environment and in the plant itself, influence transpiration. One of the most important of these is *temperature.* The rate of water evaporation doubles for every 10°C rise in temperature. However, because evaporation cools the leaf surface, the

(a)

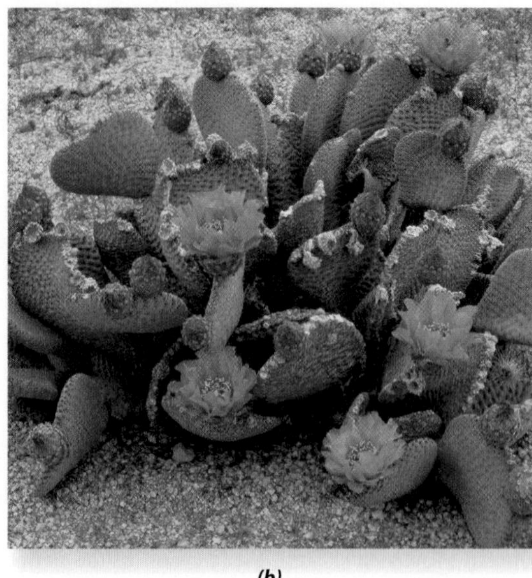

(b)

30–5 CAM plants ***(a)*** Crassulacean acid metabolism (CAM) was first recognized in stonecrops, such as the *Sedum* (family Crassulaceae) shown here. ***(b)*** CAM is characteristic of plants growing in hot, dry environments, in conditions of water and temperature stress. This beavertail cactus *(Opuntia basilaris)* was photographed in Joshua Tree National Park in California.

surface temperature does not rise as rapidly as that of the surrounding air. As noted previously, stomata close when temperatures exceed 30° to 35°C.

Humidity is also important. This is because the rate of transpiration is proportional to the vapor pressure difference, which is the difference in water vapor pressure between the intercellular spaces and the surface of the leaf. Water is lost much more slowly into air already laden with water vapor. Leaves of plants growing in shady forests, where the humidity is generally high, typically spread large, luxuriant leaf surfaces, because for these plants the main problem is getting enough light, not losing water. In contrast, plants of grasslands or other exposed areas often have narrow leaves characterized by relatively little leaf surface, thick cuticles, and sunken stomata. Grassland plants obtain all the light they can use but are constantly in danger of excessive water loss due to the low humidity in the surrounding air.

Air currents also affect the rate of transpiration. A breeze cools your skin on a hot day because it blows away the water vapor that has accumulated near the skin surface and so accelerates the rate of evaporation of water from your body. Similarly, wind blows away the water vapor from leaf surfaces, affecting the vapor pressure difference across the surface. Sometimes, if the air is very humid, wind may decrease transpiration by cooling the leaf, but a dry breeze will greatly increase evaporation and, consequently, transpiration.

Water Is Conducted through the Vessels and Tracheids of the Xylem

Water enters the plant by the roots and is given off, in large quantities, by the leaves. How does the water get from one place to another, often over large vertical distances? This question has intrigued many generations of botanists.

The general pathway that water follows in its ascent has been clearly identified. One can trace this pathway simply by placing a cut stem in water that is colored with any harmless dye (preferably, the stem should be cut under the water to prevent air from entering the conducting elements of the xylem) and then tracing the path of the liquid into the leaves. The stain quite clearly delineates the conducting elements of the xylem. Experiments using radioactive isotopes confirm that the isotope and, presumably, the water do indeed travel by way of vessels (or tracheids) in the xylem. In the experiment shown in Figure 30–6, care had to be taken to separate the xylem from the phloem. Earlier experiments in which this separation was not made produced ambiguous results, because there is a great deal of lateral movement from the xylem into the phloem. This lateral movement, however, as the experiment shows, is not necessary for the overall movement of water and minerals from soil to leaf.

Water Is Pulled to the Top of Tall Trees: The Cohesion-Tension Theory Now that we know the path water takes, the next question is, how does the water move? Logic suggests two possibilities: it can be pushed from the bottom or pulled from the top. As we shall see, the first of these possibilities is not the answer. Briefly, not all plants have root pressure (see page 719), and in those that do, it is not sufficient to push water to the top of a tall tree. Moreover, the simple experiment just described (that involving the cut stem) rules out root pressure as a crucial factor. So we are left with the hypothesis that water is pulled up through the plant body, and this hypothesis is correct according to all present evidence.

When water evaporates from the cell wall surfaces bordering the intercellular spaces in the interior of a leaf during transpiration, it is replaced by water from within the cells. This water diffuses across the plasma membrane, which is freely permeable to water but not to the cell solutes. As a result, the concentration of solutes within the cell increases, and the water potential

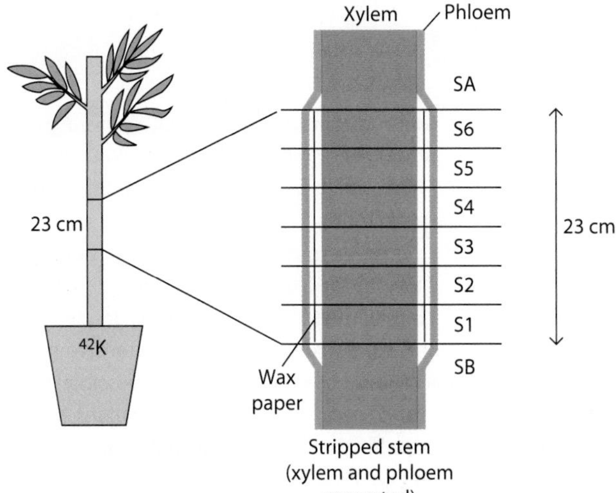

	Stem Segment	^{42}K in Xylem (ppm)	^{42}K in Phloem (ppm)
Above strip	SA	47	53
Stripped section	S6	119	11.6
	S5	122	0.9
	S4	112	0.7
	S3	98	0.3
	S2	108	0.3
	S1	113	20
Below strip	SB	58	84

30–6 Water and ion movement in xylem Distribution of radioactive potassium (^{42}K) added to the soil water shows that xylem is the channel for the upward movement of both water and inorganic ions. Wax paper was inserted between the xylem and the phloem to prevent lateral transport of the isotope. The relative amounts of ^{42}K detected in each segment of the stem are given in the table. Note the reduced amounts of ^{42}K in the phloem of segments from the stripped section of the stem.

of the cell decreases. A gradient of water potential then becomes established between this cell and adjacent, more saturated cells. These cells, in turn, gain water from other cells until, eventually, this chain of events reaches a vein and exerts a "pull," or tension, on the water of the xylem. Because of the extraordinary cohesiveness among water molecules, this tension is transmitted all the way down the stem to the roots. As a result, water is withdrawn from the roots, pulled up the xylem, and distributed to the cells that are losing water vapor to the atmosphere (Figure 30–7). This water loss makes the water potential of the roots more negative and increases their capacity to extract water from the soil. Hence, the lowered water potential in the leaves, brought about by transpiration and/or by the use of water in the leaves, results in a gradient of water potential from the leaves to the soil solution at the surface of the roots. This water potential gradient provides the driving force for the movement of water along the soil-plant-atmosphere continuum.

This theory of water movement is called the **cohesion-tension theory,** because it depends on the cohesiveness of water, the property of water that permits it to withstand tension (Figure 30–8). However, the theory might also be called the

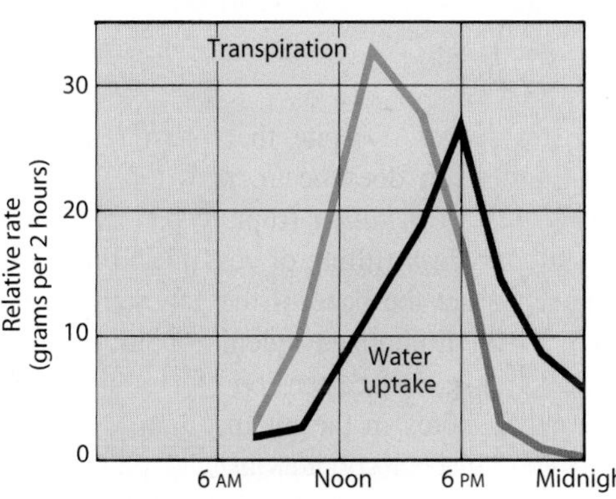

30–7 Transpiration and water uptake Measurements of water movement in ash *(Fraxinus)* trees show that an increase in water uptake follows an increase in transpiration. These data suggest that the loss of water generates the force needed for its uptake.

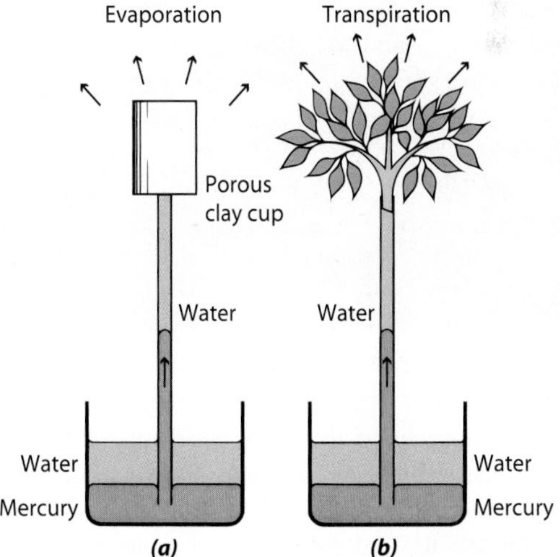

30–8 Demonstration of the cohesion-tension theory ***(a)*** In this simple physical system, a porous clay cup is filled with water and attached to the end of a long, narrow glass tube that is also filled with water. The water-filled tube is placed with its lower end below the surface of a volume of mercury in a beaker. Water evaporates from the pores in the cup and is replaced by water "pulled up" through the tube in a continuous column. As the water rises, mercury is pulled up into the tube to replace it. ***(b)*** Transpiration from leaves results in sufficient water loss to create a similar negative pressure.

GREEN ROOFS: A COOL ALTERNATIVE

When viewed from above, the typical city is a landscape of roofs, mostly flat, black, and lifeless. These roofs, along with the paved streets and sidewalks, are responsible for absorbing much of the heat that turns a city into an "urban heat island," warmer than the surrounding countryside by several degrees or more. A hot roof heats the building below it, increasing fuel use for air conditioning, which in turn contributes to global warming. Rain running off the roof can overload storm-water systems, polluting rivers and other waterways.

A move is under way, however, to convert urban roofs so that they create solutions rather than problems. A "green roof" is covered with growing plants, simultaneously insulating the building below, cooling the air above, and purifying rainwater before slowly discharging it. Green roofs may be simple and almost maintenance free, or as complex and rich as the urban gardener's imagination.

The first step in establishing a green roof is the installation of a waterproof membrane. Drainage structures and a root barrier are laid down next, followed by a layer of soil. These various layers themselves provide some insulation, keeping the building warmer in the winter and cooler in the summer. But the most significant cooling effect comes from the action of the plants. Plants transpire water, which absorbs heat from the air as it evaporates, just as human sweat does. The cooling from this process can keep the roof temperature several degrees below the outside air temperature, compared with many degrees above it for a standard roof. Soil and roots absorb rainwater, filtering out atmospheric pollutants and slowing runoff. They also absorb noise, so a building with a green roof is quieter.

A green roof can be an urban garden, with plots for the building's occupants, or a horticultural meadow full of flowers, grasses, and shrubs. Such a roof requires intensive management, but simple roofs are possible. Species of sedum, a group of succulent alpine plants in the Crassulaceae family, are popular choices for low-maintenance roofs, based on their wide climatic adaptability, tolerance for drought, and slow growth rate. Once such a roof is established, it may require little more than occasional weeding and pruning, even while it works all year round to improve the environment both inside and outside the building.

Urban meadow In 2000, the city of Chicago planted a green roof atop City Hall, replacing a black asphalt roof with an extensive garden consisting of 20,000 plants representing more than 150 species. The plants, chosen for their ability to tolerate wind, drought, and poor soil, are primarily prairie grasses, shrubs, vines, and flowering plants, including various sedums. In addition to improving air quality and reducing storm-water runoff, the green roof has reduced cooling costs for the building. The garden teems with wildlife—especially birds and butterflies, some of them endangered—and colonies of Italian honeybees in the two beehives produce hundreds of pounds of high-quality "rooftop honey" annually.

cohesion-adhesion-tension theory, because adhesion of the water molecules to the walls of the tracheids and vessels of the xylem, as well as to the cell walls of the leaf and root cells, is just as important for the upward movement of water as are cohesion and tension. The cell walls along which the water moves have evolved as a very effective water-attracting surface. This surface is capable of holding water with a force great enough to support a column of water several kilometers high against gravity.

Air Bubbles Can Break the Continuity of Water in the Xylem The cohesion of water in the xylem is increased by the filtering effect of the roots, which remove fine particles that can act as nuclei for bubble formation. Water cohesion is also enhanced by the small diameter of the xylem conduits—vessels and tracheids—through which the water moves. Despite the filtering provided by the roots, bubble formation does occur, and it is a normal occurrence in many trees. *Cavitation* (rupture of the water columns) and subsequent *embolism* (filling of vessels and/or tracheids with air or water vapor) are the bane of the cohesion-tension mechanism (Figure 30–9). Embolized xylem conduits cannot conduct water. Fortunately, the surface tension of the air-water meniscus spanning the small pores in the pit membranes of the bordered pit-pairs between adjacent conduits usually prevents air bubbles from squeezing through the pores, helping to isolate them in a single vessel or tracheid (Figure 30–10). In conifer tracheids, the passage of air is prevented by the lateral displacement of the pit membrane, such that the torus blocks one of the apertures, or openings, of the bordered pit-pair, trapping the air bubble

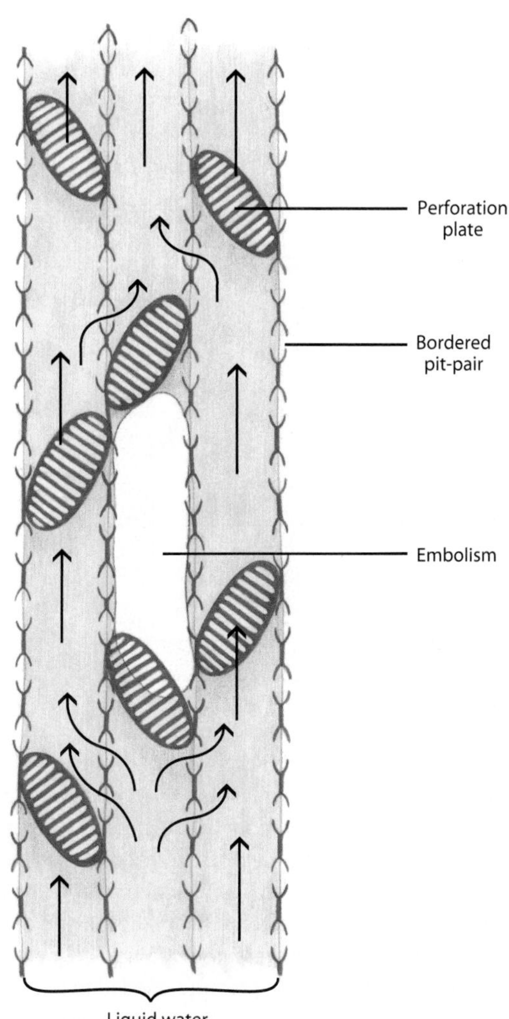

30–9 Embolized vessel element An embolism consisting of water vapor has blocked the movement of water through a single vessel element. However, water is able to detour around the embolized element via the bordered pit-pairs between adjacent vessels. The vessel elements shown here are characterized by ladderlike perforation plates.

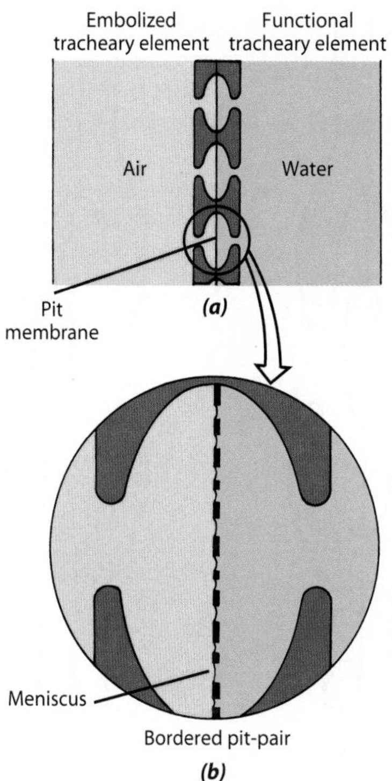

30–10 Embolized tracheary element ***(a)*** Diagram showing bordered pit-pairs between tracheary elements, one of which is embolized (filled with air) and thus nonfunctional. ***(b)*** Detail of a pit membrane. When a tracheary element is embolized, air is prevented from spreading to the adjacent functional tracheary element by the surface tension of the air-water meniscus spanning the pores in the pit membrane. The bordered pit-pairs diagrammed here do not have a torus, the impermeable central thickening typical of the pit membranes of conifer tracheids. Tori are also found in the pit membranes between vessel members of some angiosperms.

(Figure 30–11; see also Figure 26–22). Thus, the pit membranes are very important to the safety of water transport. (See also the discussion on pages 544 and 545.)

Most embolisms are triggered by air sucked into the vessel or tracheid through a pore in the wall or pit membrane adjacent to an already embolized conduit. This process, known as *air seeding,* occurs only when the pressure difference across the wall or pit membrane exceeds the surface tension at the air-water meniscus spanning the pores (Figure 30–10b). The largest pores are the most vulnerable to the penetration of air. A plant is susceptible to this mode of embolism any time even one of its vessels or tracheids becomes air-filled by physical damage (for example, by an insect bite or broken branch). Freezing can also induce embolisms, because air is not soluble in ice, and the xylem sap—that is, the fluid contents of the xylem—contains dissolved air. In addition, it is now becoming clear that xylem dysfunction induced by drought is a serious problem to plants.

The Cohesion-Tension Theory Holds up to the Test How can the cohesion-tension theory be tested? One way to test this theory directly is by measuring the water potential of large pieces of tissues, such as whole shoots, with a *pressure chamber* (also called a pressure bomb). The pressure chamber measures the negative hydrostatic pressure, or tension, in the whole plant part. The assumption is made that the water potential of the xylem is fairly close to that of the whole plant part. For example, when a twig is cut from a transpiring tree, the columns of water, which were under tension in the vessels, abruptly recede below the cut surface. At this time, the cut surface appears dry. To make a measurement, the twig is mounted in a pressure chamber, as shown in

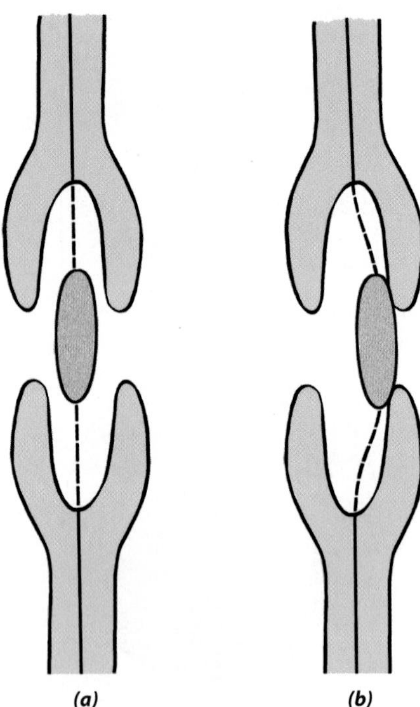

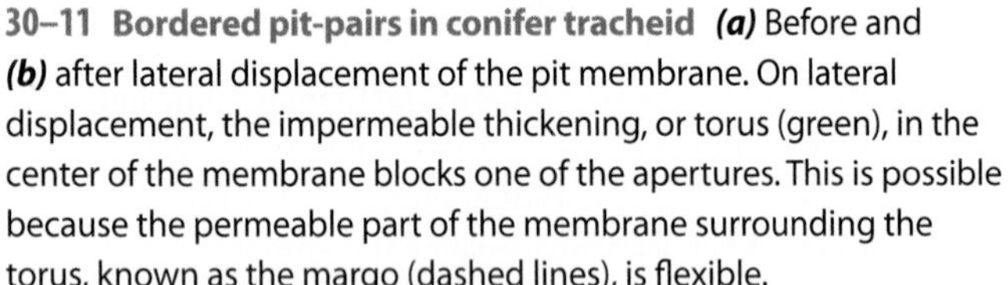

30–11 Bordered pit-pairs in conifer tracheid ***(a)*** Before and ***(b)*** after lateral displacement of the pit membrane. On lateral displacement, the impermeable thickening, or torus (green), in the center of the membrane blocks one of the apertures. This is possible because the permeable part of the membrane surrounding the torus, known as the margo (dashed lines), is flexible.

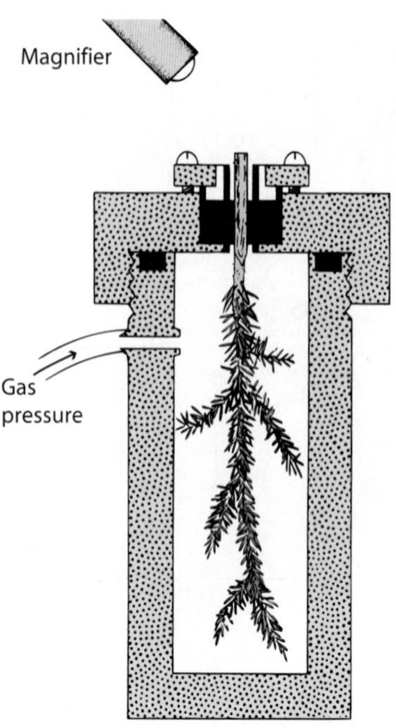

30–12 Measuring plant water potential in a cut branch When the branch is cut, some of the xylem sap—which was under negative pressure, or tension, before the branch was cut—recedes into the xylem below the cut surface. It is assumed that the water potential of the xylem is fairly close to that of the whole branch. The branch is placed in a pressure chamber, where the pressure is increased until sap just emerges from the cut end of the stem. If one assumes that equal pressure is required to force the sap in either direction, the positive pressure needed to force out the sap is, ideally, equal to the tension in the branch before it was severed.

Figure 30–12. The chamber is then pressurized with gas (air or nitrogen) until the curved upper surfaces of the water columns just appear (when magnified) at the cut surface of the twig. The magnitude of the pressure required to return the water to the cut surface is called the "balance pressure" and is equal in magnitude (but opposite in sign) to the negative pressure, or tension, that existed in the xylem before the twig was excised. Results obtained by this method are entirely consistent with the predictions of the cohesion-tension theory. Although the validity of pressure-chamber estimates of xylem tension has been questioned, other studies seem to confirm the existence of large tensions in the xylem.

A second set of data that is in accord with the cohesion-tension theory indicates that the movement of water begins at the top of the tree. The velocity of sap flow in various parts of a tree has been measured by an ingenious method involving a small heating element to warm the xylem contents for a few seconds and a sensitive thermocouple to detect the moment at which the heated xylem sap moves past a specific point (Figure 30–13). As shown in the graph, in the morning the sap begins to flow first in the twigs, as tension arises close to the leaves, and later in the trunk. In the evening, the flow diminishes first in the twigs, as water loss from the leaves diminishes, and later in the trunk. Trees with wide vessels (200 to 400 micrometers in diameter) showed midday peak velocities of 16 to 45 meters per hour (measured at breast height), while those with narrow vessels (50 to 150 micrometers in diameter) had slower midday peak velocities, ranging from 1 to 6 meters per hour.

A third set of supporting data comes from measurements of minute changes in the diameter of the tree trunk (Figure 30–14). The shrinking of the trunk is interpreted by some investigators as resulting from the negative pressures in the water passages of the xylem. Presumably, the water molecules adhering to the sides of the vessels pull the sides inward. When transpiration begins, in the morning, first the upper part of the trunk shrinks, as water is pulled out of the xylem before it can be replenished from the roots; then the lower part shrinks. Later in the day, as the transpiration rate decreases, the upper trunk expands before the lower trunk does. There is some evidence that changes in trunk diameter are due, partly or largely, to shrinkage and expansion of bark tissues as water moves laterally to and from the xylem in response to changing pressures in the xylem.

Note that the energy for evaporation of water molecules—and thus for movement of water and inorganic nutrients through the plant body—is supplied not by the plant but directly by the sun. Note also that movement is possible because of the extraordinary cohesive and adhesive properties of water, to which the plant is so exquisitely adapted.

The cohesion-tension theory is sometimes called the "transpiration-pull theory." This is unfortunate, because "transpiration-pull" implies that transpiration is essential for water to move into leaves. Although transpiration may increase the rate

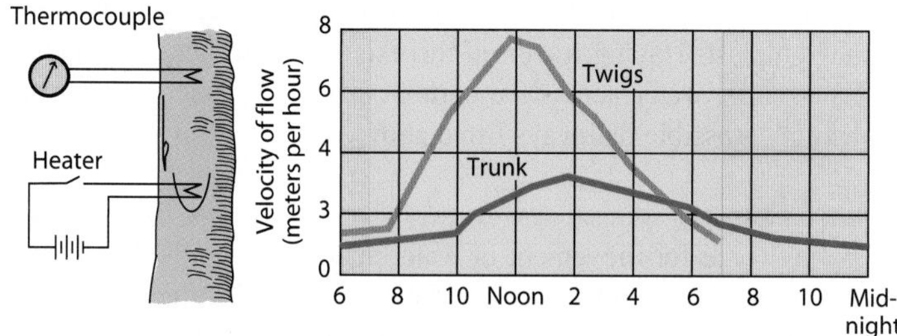

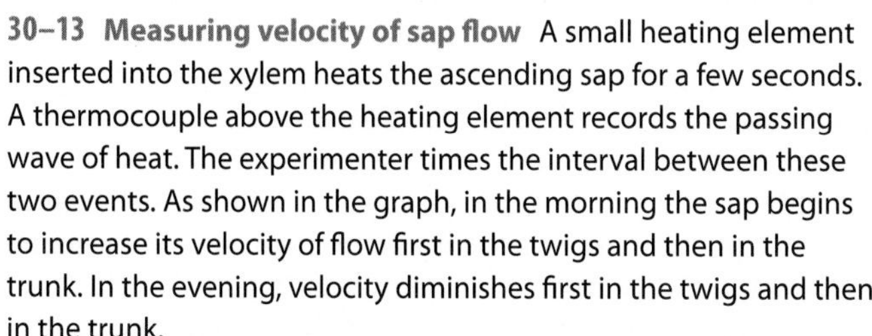

30–13 Measuring velocity of sap flow A small heating element inserted into the xylem heats the ascending sap for a few seconds. A thermocouple above the heating element records the passing wave of heat. The experimenter times the interval between these two events. As shown in the graph, in the morning the sap begins to increase its velocity of flow first in the twigs and then in the trunk. In the evening, velocity diminishes first in the twigs and then in the trunk.

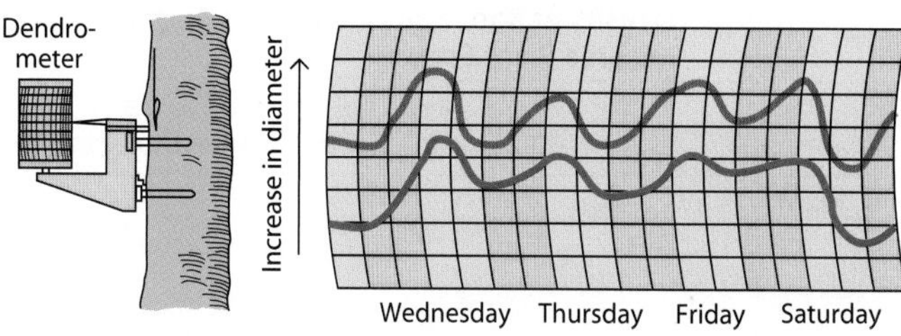

30–14 Fluctuations in tree trunk diameter A dendrometer (left) records small daily fluctuations in the diameter of a tree trunk at two different heights. As shown in the graph, in the morning, shrinkage occurs in the upper trunk slightly before it occurs in the lower trunk. These data suggest that transpiration from the leaves "pulls" water out of the trunk before it can be replenished from the roots. The gray strips signify nighttime, and the yellow strips represent daytime.

at which water moves, any use of water in the leaves produces forces causing the movement of water into them.

Absorption of Water and Ions by Roots

There Appear to Be Limits to Tree Height

As we have seen, the tensile strength of water is great enough to prevent the pulling apart of adjacent water molecules under the tension required to move water up the xylem of tall trees, the tallest of which is a giant redwood *(Sequoia sempervirens)* at 115.6 meters (379.3 feet). A study on the limits to tree height in *S. sempervirens* indicates, however, that the maximum tension exerted on the water columns in eight of the tallest redwoods (including the 115.6-meter-tall tree) is close to the point of embolism. If so, this value would be a major factor in the control of tree height. In the same study, it was noted that as trees grow taller, increasing water stress in the leaves due to gravity and increasing pathlength resistance may ultimately limit leaf expansion and lead to a decline in photosynthesis in the leaves. Such a decline would impose another constraint on tree height. Barring mechanical damage, a maximum possible tree height of 122 to 130 meters was predicted.

Water Absorption by Roots Is Facilitated by Root Hairs

The root system serves to anchor the plant in the soil and, above all, to meet the tremendous water requirements of the leaves resulting from transpiration. Most of the water that a plant takes from the soil enters through the younger parts of the root (see Chapter 24). Absorption takes place directly through the epidermis of younger roots. The root hairs, located several millimeters above the root tip, provide an enormous surface area for absorption (Figure 30–15; Table 30–2). From the root hairs, the water moves through the cortex, the outer layer or layers of which may be differentiated as an exodermis (a subepidermal layer of cells with Casparian strips). From there, the water progresses through the endodermis (the innermost layer of cortical cells) and into the vascular cylinder. Once in the conducting elements of the xylem, the water moves upward through the root and stem and into the leaves, from which most of it is lost to the atmosphere by transpiration. Hence, the soil-plant-atmosphere pathway can be viewed as a continuum of water movement.

Water May Follow One or More of Three Possible Pathways across the Root

The pathway followed by water across the root depends largely on the degree of differentiation of the various tissues that make up the root. In each of the tissues, water may follow one or more of three possible pathways: (1) **apoplastic** (around the protoplasts via the cell walls), (2) **symplastic** (from protoplast to protoplast via plasmodesmata), and/or (3) **transcellular** (from cell to cell, across the plasma membranes and tonoplasts) (Figure 30–16). For example, in a root without an exodermis, water can

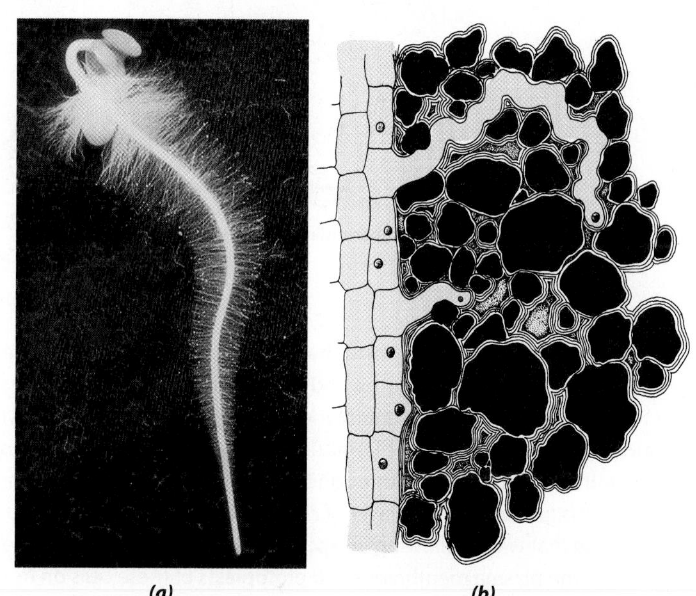

30–15 Root hairs ***(a)*** Primary root of a radish *(Raphanus sativus)* seedling, showing its numerous root hairs. ***(b)*** Root hairs surrounded by soil particles; each particle is covered by a layer of adhered water.

Table 30-2 Density of Root Hairs on the Root Surface in Three Plant Species

Plant	Root Hair Density (per square centimeter)
Loblolly pine *(Pinus taeda)*	217
Black locust *(Robinia pseudoacacia)*	520
Rye *(Secale cereale)*	2500

After J. F. Ferry, *Fundamentals of Plant Physiology* (New York: Macmillan Publishing Company, 1959).

move apoplastically as far as the endodermis. At the endodermis, however, because of the water-impermeable Casparian strips in the radial and transverse walls of the endodermal cells, water is forced to traverse the plasma membranes and protoplasts of these tightly packed cells (pages 565 to 568). In roots having an exodermis, by contrast, the Casparian strips in the radial and transverse walls of the compactly arranged exodermal cells prevent apoplastic movement of water across that cell layer. Water could follow either a symplastic or transcellular pathway across such cells. (The permeability of the transcellular pathway is determined in large part by aquaporins in the plasma membranes.) If, however, the outer tangential walls of the exodermal cells possess suberin lamellae, movement across that surface may be limited to the symplast. Having traversed the exodermis, subsequent movement of the water across the cortex could be by one or more of the three possible pathways listed above.

In the Absence of Transpiration, Roots Can Generate Positive Pressure The driving force for movement of water across the root is the difference in water potential between the soil solution at the surface of the root and the xylem sap. When transpiration is very slow or absent, as it is at night, the gradient of water potential is brought about by the secretion of ions into the xylem. Because the vascular tissue of the root is surrounded by the endodermis, ions do not tend to leak back out of the xylem. Therefore, the water potential of the xylem becomes more negative, and water moves into the xylem by osmosis through the surrounding cells. In this manner, a positive pressure, called **root pressure,** is created, and it forces both water and dissolved ions up the xylem (Figure 30–17).

Dewlike droplets of water at the tips of grass and other leaves in the early morning demonstrate the effects of root pressure (Figure 30–18). These droplets are not dew—which is water that has condensed from the air—but come from within the leaf by a process known as **guttation** (from the Latin *gutta,* meaning "drop"). They exude through openings—commonly, stomata that lack the capacity to close and open—in special structures called **hydathodes,** which occur at the tips and margins of leaves

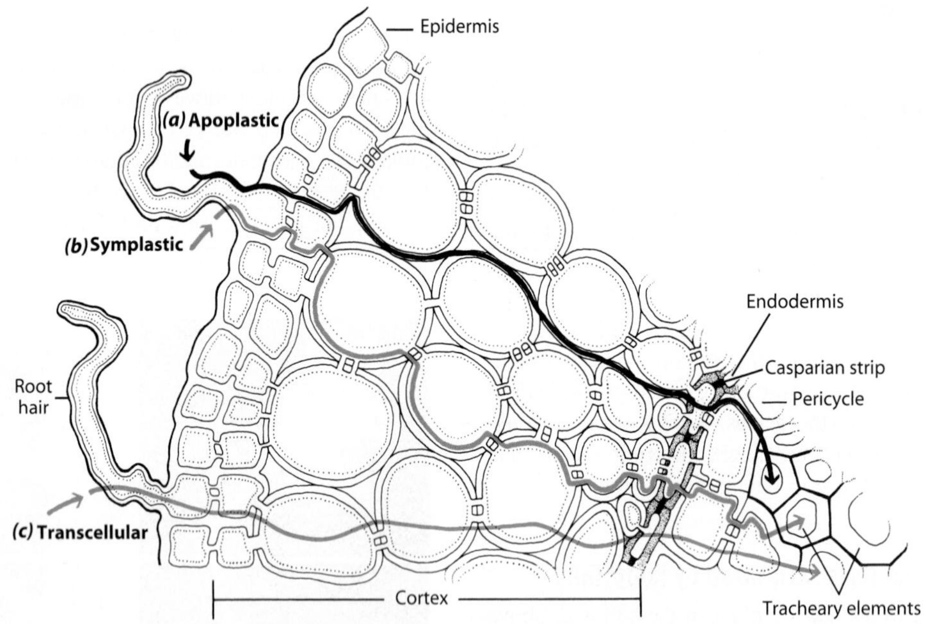

30–16 Movement of water into the root Possible pathways for the movement of water from the soil, across the epidermis and cortex, and into the tracheary elements, or water-conducting elements, of the root. ***(a)*** Apoplastic movement (black line) occurs via cell walls; ***(b)*** symplastic movement (blue line) is from protoplast to protoplast via plasmodesmata; and ***(c)*** transcellular movement (red line) occurs from cell to cell, with the water passing through the plasma membranes and tonoplasts. The root depicted here lacks an exodermis (a subepidermal layer of cells with Casparian strips).

Note that water following an apoplastic pathway is forced by the Casparian strips of the endodermal cells to cross the plasma membranes and protoplasts of these cells on its way to the xylem. Having crossed the plasma membrane on the inner surface of the endodermis, the water may once again enter the apoplastic pathway and make its way into the lumina of the tracheary elements. Inorganic ions are actively absorbed by the epidermal cells and then follow a symplastic pathway across the cortex and into parenchyma cells, from where they may be secreted into the tracheary elements.

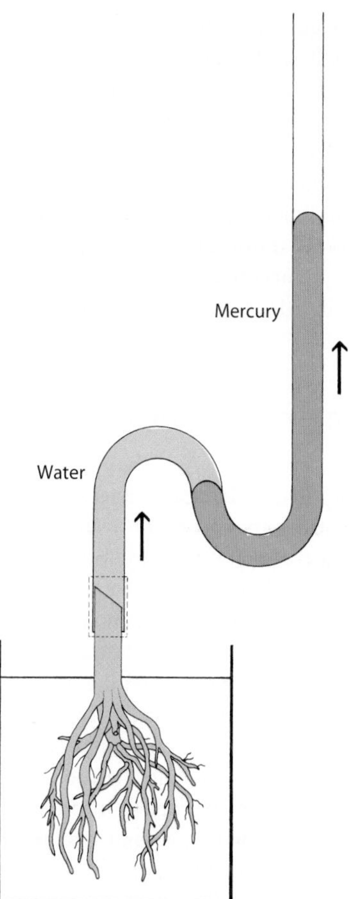

30–17 **Root pressure** Seen here is a demonstration of root pressure in the cut stump of a plant. Uptake of water by the plant's roots causes the mercury to rise in the column. Pressures of 0.3 to 0.5 megapascal have been demonstrated by this method.

30–18 **Guttation** Another demonstration of root pressure is guttation droplets, seen here at the margins of a leaf of lady's mantle *(Alchemilla vulgaris)*. These droplets are not condensation from water vapor in the surrounding air; rather, they are forced out of the leaf through openings in structures called hydathodes at the tips along the leaf margins (see Figure 30–19).

(Figure 30–19). The water of guttation is literally forced out of the leaves by root pressure.

Root pressure is least evident during the day, when the movement of water through the plant is the fastest, and the pressure never becomes high enough to force water to the top of a tall tree. Moreover, many plants, including conifers such as pine, do not develop root pressure. Thus, root pressure can perhaps be regarded as a by-product of the mechanism of pumping ions into the xylem and only indirectly as a mechanism of moving water into the shoot under special conditions.

Many Plants Redistribute Soil Water Hydraulically The passive movement of water from wet to dry soil via roots is called **hydraulic redistribution.** It occurs at night or during periods of low transpiration (during the day in CAM plants) and is driven by water potential gradients in roots and soil. Water can be redistributed hydraulically upward *(hydraulic lift),* downward *(hydraulic descent),* or horizontally. Hydraulic redistribution has been detected in a wide variety of plants, including grasses, succulents, trees, and shrubs.

The benefits of hydraulic distribution are many. For example, water transferred from the subsoil to the surface by deep-rooted plants becomes available to neighboring plants that are

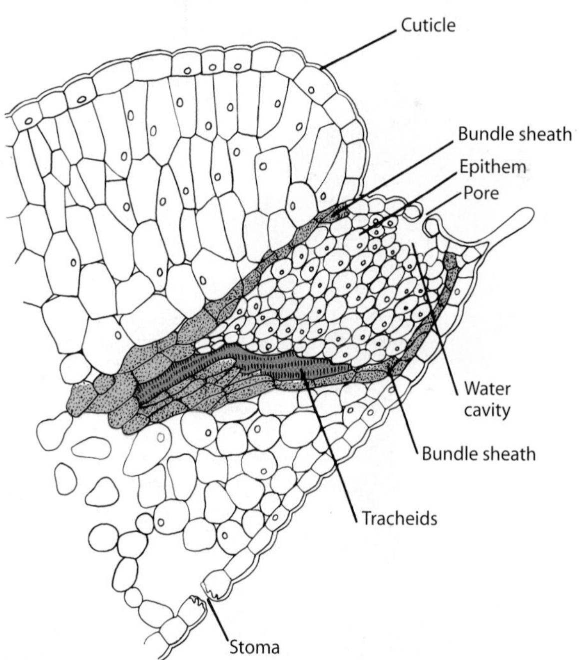

30–19 **Hydathode** Longitudinal view of a hydathode of the leaf of saxifrage *(Saxifraga lingulata)*. The hydathode consists of the terminal tracheids at the end of a vascular bundle, thin-walled parenchyma (the epithem) with numerous intercellular spaces, and an epidermal pore. The tracheids are in direct contact with epithem, which empties into a water cavity behind the pore. The epidermal pores commonly are stomata that lack the capacity to close and open. The bundle sheath surrounds the vascular bundle.

less deeply rooted (Figure 30–20). Conversely, large amounts of water may be transferred downward to dry soil layers by tree roots when the surface layers become wet following rain. The downward movement of water may also benefit the plant by reducing the possibility of water logging in surface soils. Hydraulic redistribution of water has been shown to occur between live oak trees *(Quercus agrifolia)* and their mycorrhizal symbionts, allowing the mycorrhizal hyphae to persist in soils after prolonged drought. Hydraulic redistribution in Amazonian trees apparently has a major impact on climate over the Amazon. During the dry season (July to November), photosynthesis and transpiration increase significantly. Much of the water that sustains the dry-season transpiration comes directly from water in the surface soil layers—water hydraulically lifted from deep storage layers that were recharged during the previous wet season. It has been estimated that the hydraulically lifted water increases dry-season transpiration by about 40 percent over the Amazon Basin.

Water Absorption by Roots of Transpiring Plants May Be Passive During periods of high transpiration rates, ions accumulated in the xylem of the root are swept away in the **transpiration stream,** or flow of water, and the amount of osmotic movement across the endodermis decreases. At such times, the roots become passive absorbing surfaces through which water is pulled by bulk flow generated in the transpiring shoots. Some investigators believe that almost all of the water absorption by the roots of transpiring plants occurs in this passive manner.

During periods of rapid transpiration, water may be removed from around the root hairs so quickly that the local soil becomes depleted of water. Water will then move from some distance away, toward the root hairs, through fine pores in the soil. In general, however, roots come into contact with additional water by growing, although roots will not grow in dry soil. For example, under normal conditions, the roots of apple trees grow an average of about 3 to 9 millimeters a day, the roots of prairie grasses may grow more than 13 millimeters a day, and the main roots of maize plants average 52 to 63 millimeters a day. The results of such rapid growth can be remarkable: a four-month-old rye *(Secale cereale)* plant has over 10,000 kilometers of roots and many billions of root hairs.

The Uptake of Inorganic Nutrients by Roots Is an Energy-Dependent Process

The uptake, or absorption, of inorganic ions takes place through the epidermis, largely the root hairs, of younger roots. Current evidence suggests that the major pathway followed by ions from the epidermis to the endodermis of the root is symplastic—that is, from protoplast to protoplast via plasmodesmata. Ion uptake by the symplastic route begins at the plasma membrane of the epidermal cells. The ions then move from the epidermal cell protoplasts to the first layer of cortical cells (probably an exodermis) through plasmodesmata in the epidermal-cortical cell walls (Figure 30–16). Movement of ions into the root continues in the cortical symplast—again, from protoplast to protoplast via plasmodesmata—through the endodermis, and into the parenchyma cells of the vascular cylinder by diffusion.

Nutrient uptake from soil by most seed plants is greatly enhanced by naturally occurring mycorrhizal fungi associated with the root systems (page 312). Mycorrhizas are especially important in the absorption and transfer of phosphorus, but the increased absorption of zinc, manganese, and copper has also been demonstrated. These nutrients are relatively immobile in soil, and depletion zones for these minerals quickly develop

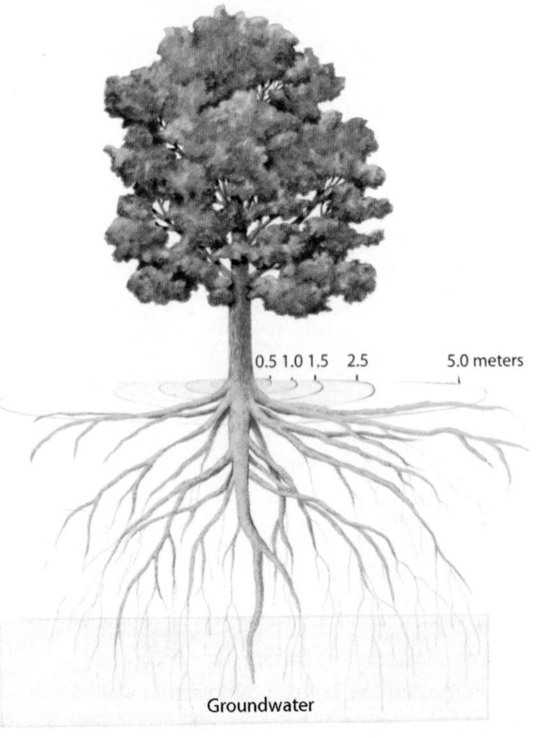

Species	Distance from Base of Tree 0.5 m	1.0 m	1.5 m	2.5 m	5.0 m
May apple *(Podophyllum peltatum)*	61%	53%	50%	9%	0%
False Solomon's seal *(Smilacina racemosa)*	60	58	5	0	0
Wild strawberry *(Fragaria virginiana)*	58	54	50	13	1
Meadow rue *(Thalictrum dioicum)*	55	50	11	2	0
Wild ginger *(Asarum canadense)*	31	21	8	0	0
Large-flowered trillium *(Trillium grandiflorum)*	25	18	0	0	0
Goldenrod *(Solidago flexicaulis)*	20	19	6	0	0
Lowbush blueberry *(Vaccinium vacillans)*	19	10	5	1	0
Velvet grass *(Holcus lanatus)*	21	7	0	0	0
Spicebush *(Lindera benzoin)*	11	6	0	0	0
Linden tree *(Tilia heterophylla)*	0	1	0	1	0
Beech tree *(Fagus grandifolia)*	1	0	0	0	0

30–20 Effects of hydraulic lift on nearby plants The table shows, for each type of plant growing in the vicinity of a sugar maple *(Acer saccharum)* tree, the mean percentage of total xylem water that was hydraulically lifted by this tree. As you can see, the closer the plants were to the sugar maple, the more they benefited from the tree's hydraulic lift.

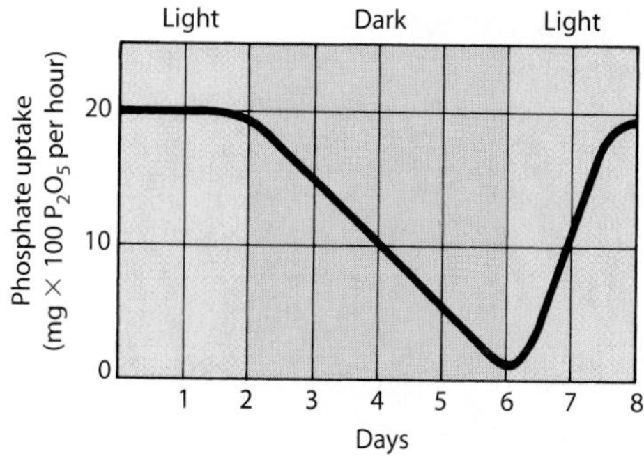

30–21 Energy requirement for nutrient uptake The rate of phosphate absorption (measured as P_2O_5; see page 702) by maize *(Zea mays)* plants fell to near zero after four days of continuous darkness. The rate began to rise again when light was restored. These and other data indicate that mineral ion uptake in plants is a process that requires energy.

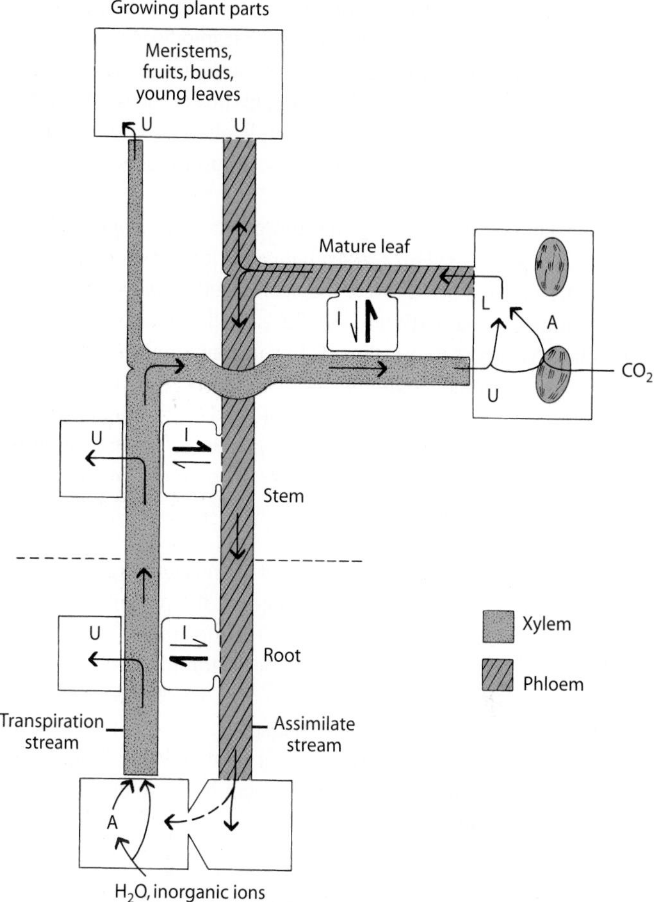

30–22 Exchange between transpiration and assimilate streams Diagram showing the circulation of water, inorganic ions, and assimilates in the plant. Water and inorganic ions taken up by the root move upward in the xylem in the transpiration stream. Some water and ions move laterally into tissues of the root and stem, while some are transported to growing plant parts and mature leaves. In the leaves, substantial amounts of water and inorganic ions are transferred to the phloem and are exported with sucrose in the assimilate stream. The growing plant parts, which are relatively ineffective in acquiring water through the transpiration stream, receive much of their nutrients and water via the phloem. Both water and solutes entering the roots in the phloem may be transferred to the xylem and recirculated in the transpiration stream. The symbol A designates sites specialized for absorption and assimilation of raw materials from the environment; L and U designate sites of loading and unloading, respectively; and I designates principal points of interchange between the xylem and phloem.

around the root and root hairs. The hyphal network of mycorrhizas extends several centimeters out from colonized roots, thus exploiting a large volume of soil more efficiently.

The mineral composition of root cells is far different from that of the medium in which the plant grows. For example, in one study, cells of pea *(Pisum sativum)* roots had a K^+ ion concentration 75 times greater than that of the nutrient solution. Another study showed that the vacuoles of rutabaga *(Brassica napus* var. *napobrassica)* cells contained 10,000 times more K^+ ions than the external solution.

Because substances do not diffuse against a concentration gradient, it is clear that minerals are absorbed by **active transport** (page 85). Indeed, the uptake of minerals is known to be an energy-dependent process. For instance, if roots are deprived of oxygen, or poisoned so that respiration is curtailed, mineral uptake is drastically decreased. Also, if a plant is deprived of light, it will cease to absorb salts after its carbohydrate reserves have been depleted (Figure 30–21) and will finally release minerals back into the soil solution. How the ions enter the mature vessels (or tracheids) of the xylem from the parenchyma cells of the vascular cylinder has been the subject of considerable debate. At one time it was suggested that the ions leak passively from the parenchyma cells into the vessels, but substantial evidence now indicates that the loading, or secretion, of ions into the vessels from the parenchyma cells is a highly regulated, energy-dependent process. Hence, ion transport from the soil to the vessels of the xylem requires two active, or energy-dependent, events: (1) uptake at the plasma membrane of the epidermal cells and (2) secretion into the vessels at the plasma membrane of the parenchyma cells bordering the vessels.

Inorganic Nutrients Are Exchanged between Transpiration and Assimilate Streams

Once secreted into the xylem vessels (or tracheids), inorganic ions are rapidly transported upward and throughout the plant in the transpiration stream. Some ions move laterally from the xylem into surrounding tissues of the roots and stems, while others are transported into the leaves (Figure 30–22).

Much less is known about the pathways followed by ions in the leaves than about those in the roots. Within the leaf, the ions are transported along with the water in the leaf apoplast, that is, in the cell walls. Some ions may remain in the transpiration stream and reach the main regions of water loss—the stomata and other epidermal cells. Most ions eventually enter the protoplasts of the leaf cells, probably by carrier-mediated transport mechanisms similar to those in roots. The ions may then move symplastically (through protoplasts via plasmodesmata) to other parts of the leaf, including the phloem. Inorganic ions may also be absorbed in small amounts through the surfaces of leaves; consequently, fertilization by direct application of micronutrients to the foliage has become standard agricultural practice for some crop plants.

Substantial amounts of the inorganic ions imported into leaves through the xylem are exchanged with the phloem of the leaf veins and exported from the leaf together with sucrose in the assimilate stream (Figure 30–22; see also the discussion of assimilate transport that follows). For instance, in a study of the annual white lupine *(Lupinus albus),* transport in the phloem accounted for more than 80 percent of the fruit's vascular intake of nitrogen and sulfur, and 70 to 80 percent of its phosphorus, potassium, magnesium, and zinc intake. The uptake of such inorganic ions by developing fruits is undoubtedly coupled to the flow of sucrose in the phloem.

Recycling may occur in the plant as nutrients reaching the roots in the descending assimilate stream of the phloem are transferred to the ascending transpiration stream of the xylem (Figure 30–22). Only those ions that can move in the phloem—*phloem-mobile ions*—can be exported from the leaves to any great extent. For example, K^+, Cl^-, and phosphate (HPO_4^{2-}) are readily exported from leaves, whereas Ca^{2+} is relatively immobile. Solutes such as calcium, as well as boron and iron, are characterized as *phloem-immobile.*

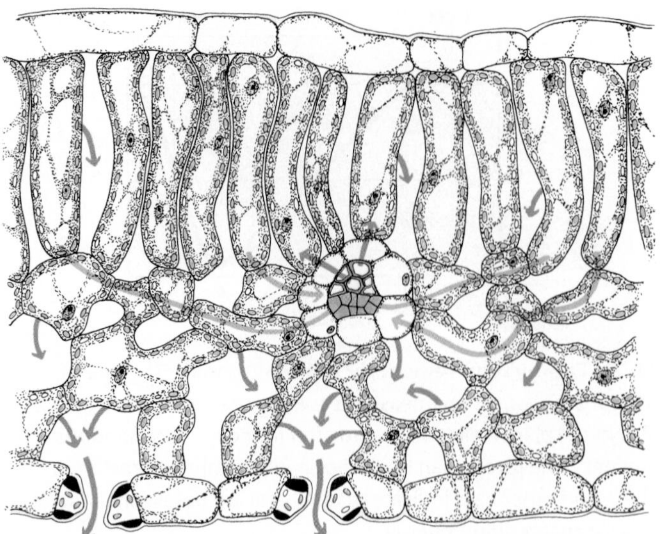

30–23 Water and photosynthate pathways through the leaf Diagram of a leaf showing the pathways followed by water molecules of the transpiration stream as they move from the xylem of a minor vein to the mesophyll cells, then evaporate from the surface of the mesophyll cell walls, and finally diffuse out of the leaf through an open stoma (blue arrows).

Also shown are the pathways followed by sugar molecules manufactured during photosynthesis (photosynthate) as they move from mesophyll cells to the phloem of the same minor vein and enter the assimilate stream (gold arrows). The sugar molecules manufactured in the palisade parenchyma cells are believed to move to the spongy parenchyma cells and then laterally to the phloem via the spongy cells.

Assimilate Transport: Movement of Substances through the Phloem

Whereas water and inorganic solutes ascend the plant in the transpiration stream of the xylem, the sugars manufactured during photosynthesis move out of the leaf in the **assimilate stream** of the phloem (Figure 30–23). The sugars are transported not only to sites where they are used, such as growing shoot and root tips, but also to sites of storage, such as fruits, seeds, and the storage parenchyma of stems and roots (Figure 30–22). The transport of substances in the phloem is referred to as **translocation.**

Assimilate movement is said to follow a source-to-sink pattern. The principal **sources,** or exporters, of assimilate solutes are the photosynthesizing leaves, but storage tissues may also serve as important sources. All plant parts unable to meet their own nutritional needs may act as **sinks,** that is, importers of assimilates. Thus, storage tissues act as sinks when they are importing assimilates and as sources when they are exporting assimilates.

Source-sink relations may be relatively simple and direct, as in some young seedlings, where cotyledons containing reserve food often represent the major source, and growing roots represent the major sink. In older plants, the upper, most recently formed mature leaves commonly export assimilates primarily toward the shoot tip, the lower leaves export assimilates primarily to the roots, and those in between export assimilates in both directions (Figure 30–24a). This pattern of assimilate distribution is markedly altered during the change from vegetative to reproductive growth. Developing fruits are highly competitive sinks that monopolize assimilates from the nearest leaves and frequently from distant leaves as well, often causing a marked decline of vegetative growth (Figure 30–24b).

Experiments with Radioactive Tracers Provide Evidence of Sugar Transport in Sieve Tubes

Early evidence supporting the role of the phloem in assimilate transport came from observations of trees from which a complete ring of bark had been removed. As noted in Chapter 26, the bark in older stems is composed largely of phloem. When a photosynthesizing tree is ringed, or "girdled," in this manner, the bark above the ring becomes swollen, indicating the formation of new wood and bark tissues stimulated by an accumulation of assimilates moving downward in the phloem from the photosynthesizing leaves.

Much convincing evidence of the role of the phloem in assimilate transport has been obtained with radioactive tracers. Experiments with radioactive assimilates (such as ^{14}C-labeled sucrose) not only have confirmed the movement of such substances in the phloem but have also shown conclusively that sugars are transported in the sieve tubes (Figure 30–25).

Aphids Have Proved of Great Value in Phloem Research

Much valuable information on movement of substances in the phloem has come from studies utilizing aphids—small insects that suck the juices of plants. Most species of aphids are phloem-feeders. These aphids insert their modified mouthparts, or stylets, into a stem or leaf, extending them until the tips of the stylets puncture a conducting sieve tube (Figure 30–26). The turgor

30–24 Assimilate transport in young and older plants Diagrams of assimilate transport in a plant at ***(a)*** the vegetative stage and ***(b)*** the fruiting stage. The arrows indicate the direction of assimilate transport in each stage.

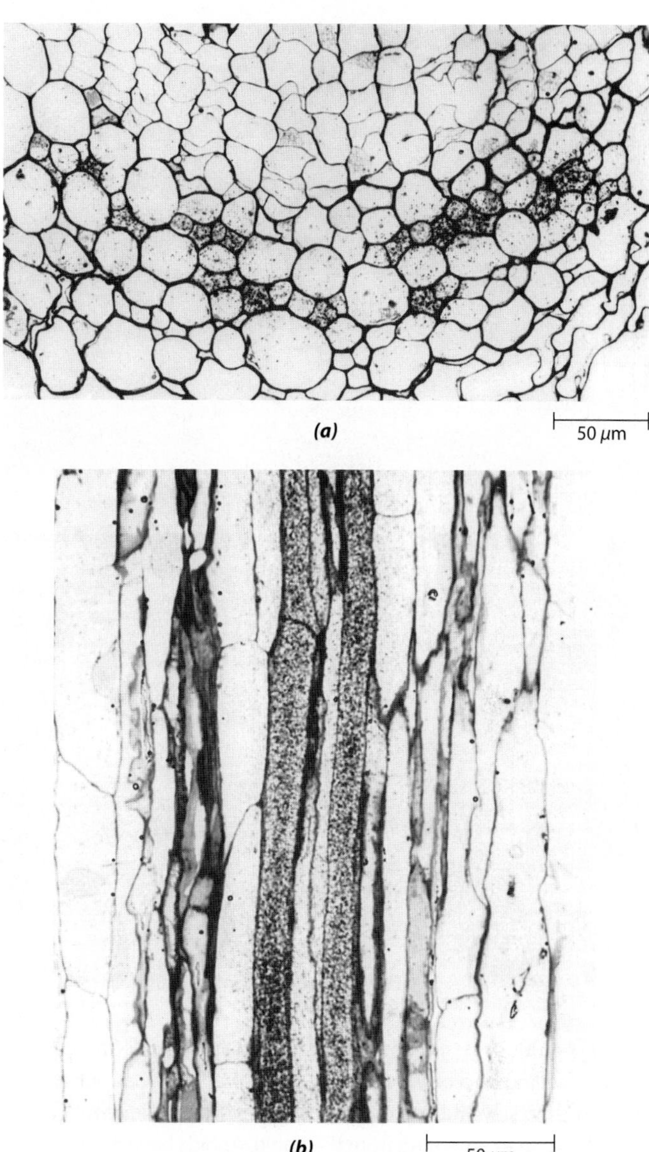

30–25 Sugar transport in sieve tubes Microautoradiographs of ***(a)*** transverse and ***(b)*** longitudinal sections of a vascular bundle from a stem of the broad bean *(Vicia faba)*. A leaf of this stem was exposed to $^{14}CO_2$ for 35 minutes. During that time, $^{14}CO_2$ was incorporated into sugars, which were then transported to other parts of the plant. The leaf was sectioned, and the sections were placed in contact with autoradiographic film for 32 days. When the film was developed and compared with the underlying tissue sections, the radioactivity (visible as dark grains on the film) was seen to be confined almost entirely to the sieve tubes.

pressure of the sieve tube then forces the sieve-tube sap through the aphid's digestive tract and out through its posterior end as droplets of "honeydew." If feeding aphids are anesthetized (to prevent them from withdrawing their stylets from the sieve tube) and severed from their stylets, exudation of sieve-tube sap often continues from the cut stylets for many hours. The exudate can be collected with a micropipette. Analyses of exudates obtained in this manner reveal that sieve-tube sap contains 10 to 25 percent dry matter, 90 percent or more of which is sugar—*mainly sucrose* in most plants. In addition, other solutes are translocated in the phloem, including amino acids, numerous proteins and RNAs, several hormones, and some inorganic solutes, including magnesium, phosphate, potassium, and chloride.

Data obtained from studies utilizing aphids and radioactive tracers indicate that the rates of longitudinal movement of assimilates in the phloem are remarkably fast. For example, sieve-tube sap moved at a rate of about 100 centimeters per hour at the sites of the stylet tips.

Phloem Transport Is Driven by Osmotically Generated Pressure Flow

Several mechanisms have been proposed to explain assimilate transport in the sieve tubes of the phloem. Probably the earliest explanation was that of diffusion, followed by that of cytoplasmic streaming. Normal diffusion and cytoplasmic streaming of the kind found in plant cells were largely abandoned as possible transport mechanisms, however, when it became known that the velocities of assimilate transport (typically 50 to 100 centimeters per hour) are far too great for either of these phenomena to account for long-distance transport via sieve tubes.

Alternative hypotheses have been advanced to explain the mechanism of phloem transport, but only one, the pressure-flow hypothesis, satisfactorily accounts for practically all of the data obtained in experimental and structural studies on phloem.

Originally proposed in 1927 by the German plant physiologist Ernst Münch, and since modified, the pressure-flow

(a)

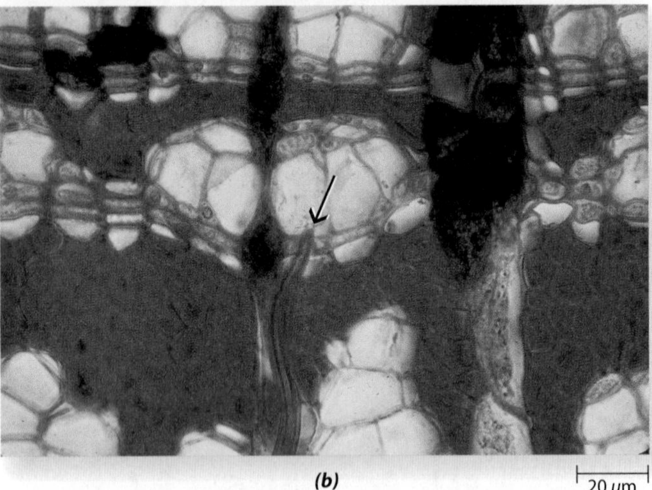

(b)

30–26 Aphids and phloem studies ***(a)*** An aphid *(Longistigma caryae)* feeding on a basswood *(Tilia americana)* stem has punctured a conducting sieve tube. A droplet of "honeydew," or sieve-tube sap, can be seen emerging from the aphid. Aphids have proven to be valuable research "tools" for studies on phloem function. ***(b)*** A photomicrograph showing part of the modified mouthparts (stylets) of an aphid in a sieve tube of the secondary phloem of the basswood stem. An arrow points to the tips of the stylets.

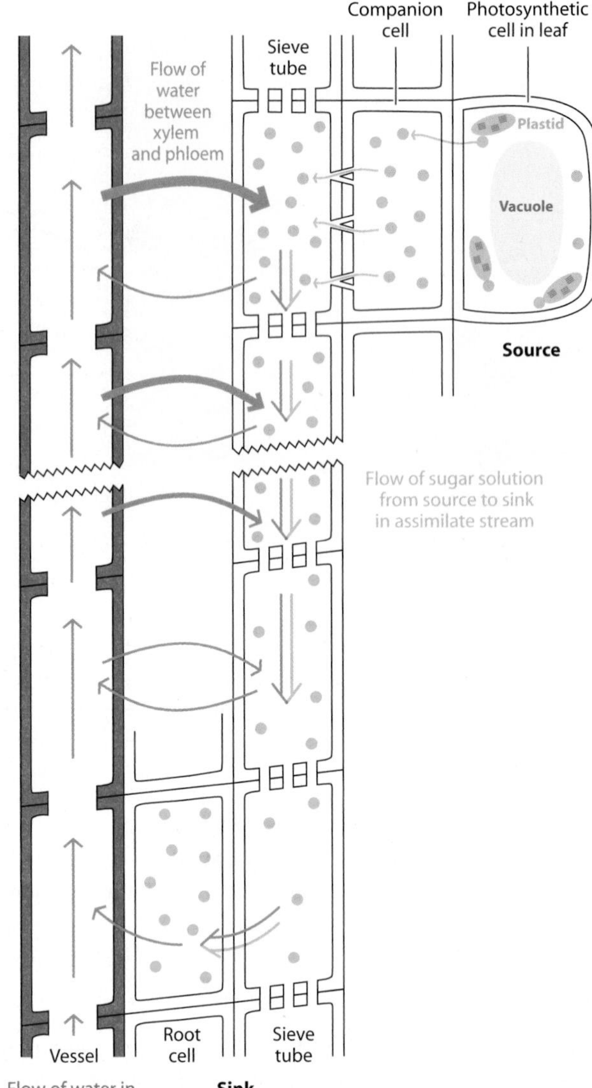

30–27 Pressure-flow hypothesis A diagram of the osmotically generated pressure-flow mechanism for the transport of sugars in the phloem. Sugar molecules (yellow dots) are loaded into the sieve tube–companion cell complex at the source (upper right). With the increased concentration of sugar, the water potential is decreased, and water from the xylem enters the sieve tube by osmosis. Sugar is removed (unloaded) at the sink, and the sugar concentration falls; as a result, the water potential is increased, and water leaves the sieve tube. With the movement of water into the sieve tube at the source and out of it at the sink, the sugar molecules are carried passively by the water along the osmotically generated gradient between source and sink. Note that the sieve tube between source and sink is bounded by a selectively permeable membrane, the plasma membrane. Consequently, water enters and leaves the sieve tube not only at the source and sink but all along the pathway. Evidence indicates that few if any of the original water molecules entering the sieve tube at the source end up in the sink, because they are exchanged with other water molecules that enter the sieve tube from the phloem apoplast along the pathway.

hypothesis is the simplest and the most widely accepted explanation for long-distance assimilate transport in sieve tubes. It is the simplest explanation because it depends only on osmosis as the driving force for assimilate transport.

Briefly stated, the **pressure-flow hypothesis** asserts that assimilates are transported from sources to sinks along a gradient of turgor pressure developed osmotically. For example, sucrose produced in mesophyll cells by photosynthesizing leaves is transported from the mesophyll to the minor veins, where it enters the sieve tubes, resulting in sucrose concentrations in the sieve tubes that are higher than in the mesophyll cells (Figure 30–27). This process, called **phloem loading,** decreases the water potential in a sieve tube and causes water entering the leaf in the transpiration stream to move into the sieve tube by osmosis. With the movement of water into the sieve tube at this source, the sucrose

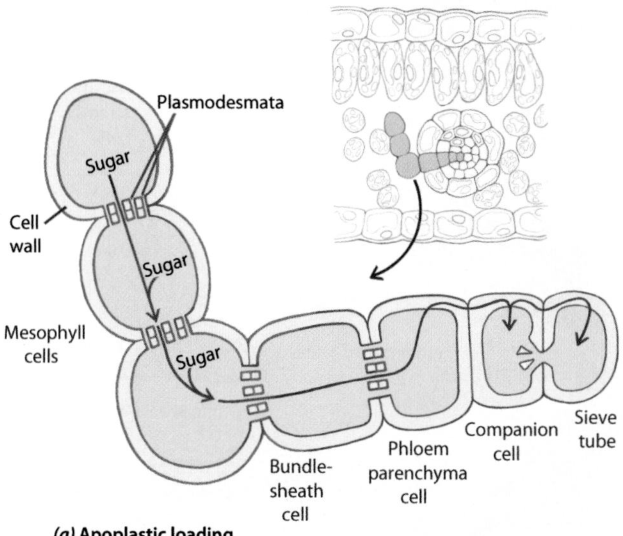

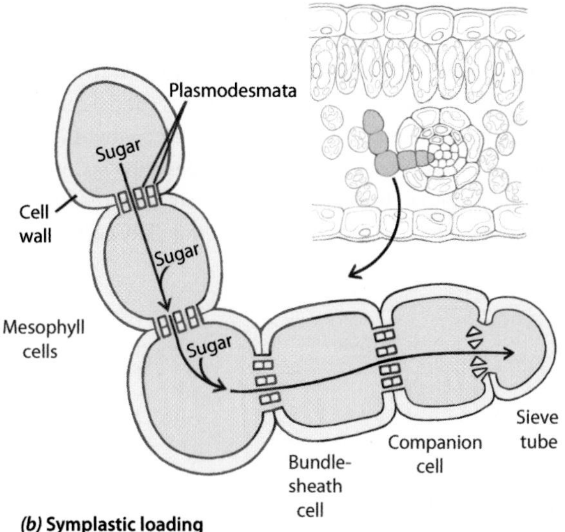

30–28 Pathways of phloem loading in source (photosynthesizing) leaves ***(a)*** In apoplastic phloem-loading species, sugars manufactured in the mesophyll cells of source (photosynthesizing) leaves initially follow a symplastic pathway (via plasmodesmata) but enter the apoplast (cell walls) just prior to being actively loaded into the sieve tube–companion cell complexes. ***(b)*** In symplastic phloem-loading species, the sugars move entirely via plasmodesmata from the mesophyll to the sieve tubes.

is carried passively by the water to a sink, such as growing tissue or a storage root, where the sucrose is unloaded, or removed, from the sieve tube. The removal of sucrose results in an increased water potential in the sieve tube at the sink and subsequent movement of water out of the sieve tube there. The sucrose may be used in growth or respiration or may be stored at the sink, but most of the water returns to the xylem and is recirculated in the transpiration stream.

Note that the pressure-flow hypothesis casts the sieve tubes in a passive role in the movement of sugar solution through them. Active transport may also be involved in the pressure-flow mechanism, but not directly in the long-distance transport through the sieve tubes. Active transport is involved in apoplastic phloem loading and possibly in unloading of sugars at the sinks (see below).

Phloem Loading Can Be Apoplastic or Symplastic The pathway followed by sucrose from the mesophyll cells to the sieve tube–companion cell complexes of the minor veins can be entirely symplastic—that is, from cell to cell via plasmodesmata. Alternatively, the sucrose can enter the apoplast (cell walls) prior to being actively loaded into the sieve tube–companion cell complexes. In apoplastic loaders, the sieve tube–companion cell complexes have virtually no plasmodesmatal connections with other cell types of the leaf (Figure 30–28a), whereas the companion cells of symplastic loaders are connected to the mesophyll symplast via numerous plasmodesmata (Figure 30–28b).

Apoplastic loading is highly correlated with the herbaceous habit and requires energy to pump sucrose into the sieve tube–companion cell complex against a concentration gradient (Figure 30–29). Apoplastic loading of sucrose is driven by an H^+-ATPase that is bound to the plasma membrane. This enzyme utilizes the energy from the hydrolysis of ATP to pump protons (H^+) across the membrane. The proton motive force established by this primary active transport is then used to carry sucrose, and in some cases sugar alcohols (sorbitol and mannitol), into the sieve tube–companion cell complex by specific transporters. This type of secondary active transport is known as **sucrose-proton symport,** or sucrose-proton cotransport (page 85).

One of the symplastic strategies, which uses a so-called **polymer trapping** mechanism for phloem loading, is also an active process, but it does not involve symporters. In this mechanism, sucrose synthesized in the mesophyll diffuses via plasmodesmata to the bundle-sheath cells and from there, via abundant plasmodesmata, into specialized companion cells called **intermediary cells** (Figure 30–30). In the intermediary cells, the sucrose is used to synthesize raffinose and stachyose, a process that increases the concentrations of these sugars to levels close to or equivalent to those of the sugars in plants that load apoplastically. Because of the relatively large size of raffinose and stachyose molecules, these polymers are unable to diffuse back into the bundle sheath, but they can diffuse into the sieve tubes via the abundant pore-plasmodesmata connections in the intermediary cell–sieve tube walls. In this mechanism, energy is used to create a concentration difference between the mesophyll and the phloem. There is no correlation between polymer-trapping species and growth form.

The minor veins of some species (for example, *Amborella trichopoda, Coleus blumei,* and *Cucurbita maxima*) contain both intermediary cells and "ordinary" companion cells. Apparently, both apoplastic and symplastic loading can occur in the same species.

In contrast to these active mechanisms, it has recently been recognized that many tree species exhibit passive symplastic loading. These species do not have a concentrating step in the pathway from the mesophyll to the sieve tube–companion cell complexes, nor do they have intermediary cells or transport raffinose or stachyose. In these plants, sugar concentrations are

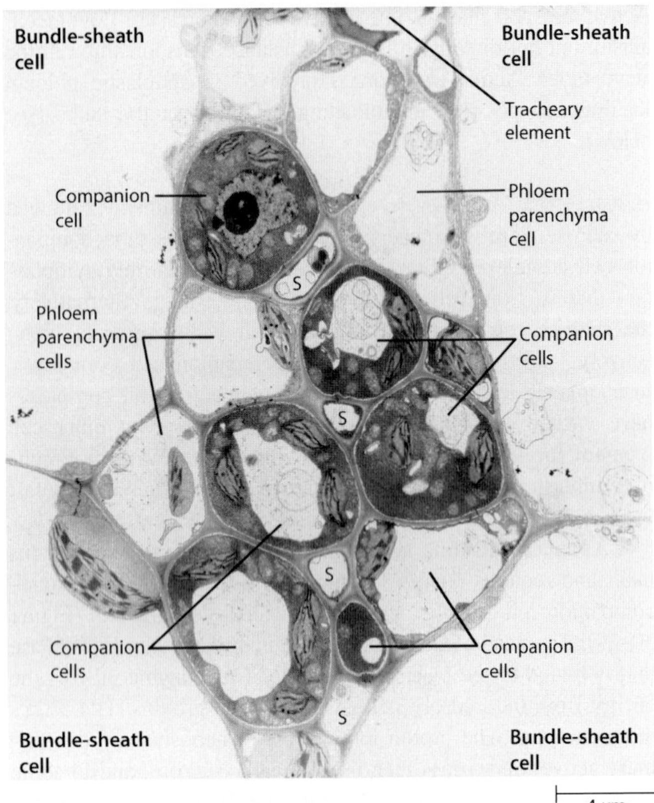

30–29 Minor vein of sugar beet *(Beta vulgaris)* leaf Sugar beet, an herb, is an apoplastic loader. In this transverse section, the vein contains four sieve tubes (S) and seven "ordinary" companion cells. As is typical of minor veins, the sieve tubes are diminutive, much smaller than their associated companion cells. This is a reflection of the role of the companion cells in the active uptake of sucrose from the apoplast for delivery to the sieve tubes. Except for part of a tracheary element (above), the xylem of the vein is not shown.

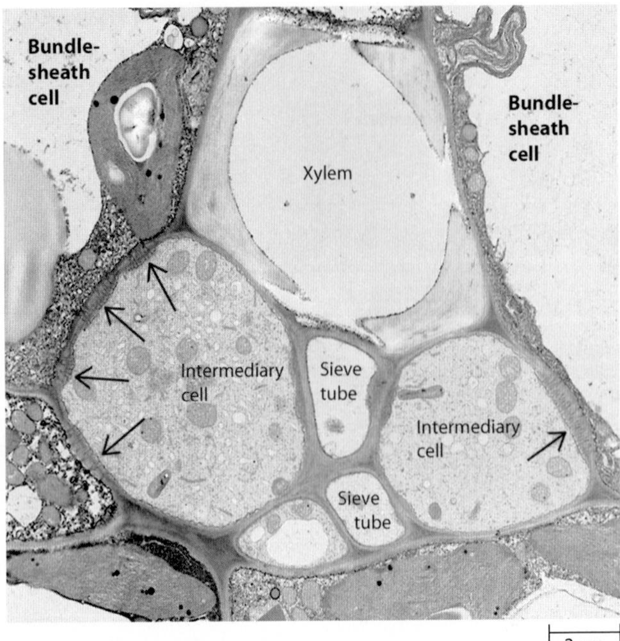

30–30 Minor vein of *Fuchsia triphylla* leaf with intermediary cells *Fuchsia* is a symplastic loader that utilizes specialized companion cells, called intermediary cells, in a polymer-trapping mechanism. This vein contains two sieve tubes and two intermediary cells. The arrows point to fields of plasmodesmata in the thickened cell walls between bundle-sheath and intermediary cells.

higher in the mesophyll cells than in the sieve tube–companion cell complexes, and the concentration gradient between the two groups of cells provides the driving force for sugar molecules to diffuse through the plasmodesmata into the sieve tubes, where the resultant turgor pressures are high enough to generate pressure flow and long-distance transport. Interestingly, this was part of Münch's pressure-flow hypothesis.

Phloem Unloading and Transport into Recipient Sink Cells Can Be Apoplastic or Symplastic The process by which sugars and other assimilates transported in the phloem exit the sieve tubes of sink tissues is called **phloem unloading.** The transport events that immediately follow the unloading of assimilates are referred to as *post-phloem* or *post-sieve-tube transport.*

In growing vegetative sinks, such as young leaves and roots, unloading and transport into sink cells are usually symplastic. In other sinks, unloading is apoplastic. Although the actual unloading process is probably passive, transport into sink tissues depends on metabolic activity. For example, in symplastic unloaders, energy is needed to maintain the concentration gradient between the sieve tubes and sink cells. In storage organs, such as sugar beet roots and sugarcane stems, where apoplastic unloading takes place, energy is needed to accumulate sugars to high concentrations in the sink cells.

SUMMARY

Most of the Water Transpired by a Vascular Plant Is Lost through the Stomata

Most of the water taken up by plant roots is lost to the air as water vapor. This process, called transpiration, is inextricably linked to the uptake by the leaf of CO_2, which is essential in photosynthesis.

A pair of guard cells can change their shape to bring about the opening and closing of the stoma, or pore. Closing of stomata prevents the loss of water vapor from the leaf. Stomatal movements result from changes in turgor pressure within the guard cells and the radial micellation of the guard cell walls. The turgor changes are closely correlated with changes in the solute level in the guard cells. A stoma opens when its guard cells become turgid and closes when they become flaccid.

In Well-Watered Plants, Light Is the Dominant Signal Controlling Stomatal Movements

It is blue light that regulates stomatal movements, through its activation of a proton-pumping ATPase in the guard cell plasma membrane. In the morning, potassium is the principal solute involved in stomatal opening. On potassium efflux, sucrose becomes the dominant osmoticum. Stomatal closing at the end of the day parallels a decrease in sucrose and a resultant loss of guard cell turgor.

Water Moves from the Roots to the Leaves through the Conduits—Vessels and Tracheids—of the Xylem

The current and widely accepted theory for the mechanism of water movement to the top of tall plants through the xylem is the cohesion-tension theory. According to this theory, water is pulled up through the plant body. This pull, or tension, is brought about by transpiration and/or by the use of water in the leaves, which results in a gradient of water potential from the leaves to the soil solution at the surface of the roots. It is the cohesiveness of water that permits it to withstand tension. Embolisms—the presence of air or water vapor—in xylem conduits are the bane of the cohesion-tension mechanism. Fortunately, the pit membranes of the bordered pit-pairs between adjacent tracheary elements usually prevent the passage of air from an embolized conduit into a functional one.

Uptake of Water by Roots Takes Place Largely through the Root Hairs

The root hairs provide an enormous surface area for water uptake. In some plants, the uptake of water from the soil results in the buildup of positive pressure, or root pressure, when transpiration is very slow or absent. This osmotic uptake depends on transport of inorganic ions from the soil into the xylem by the living cells of the root, and it can result in guttation, a process in which liquid water is forced out through special structures (hydathodes) in the tips or margins of leaves. The pathway followed by water across the root may be apoplastic, symplastic, or transcellular; however, apoplastic movement is blocked at the endodermis by the Casparian strips. The water must pass through the plasma membranes and protoplasts of endodermal cells on its way to the xylem.

Many plants redistribute soil water hydraulically. Hydraulic redistribution is a nighttime process by which the roots of plants transfer water from moist soil regions to dry soil regions. Water can be redistributed hydraulically upward, downward, or horizontally.

Inorganic Nutrients Become Available to Plants in Soil Solution in the Form of Ions

Plants employ metabolic energy to concentrate the ions they require. Ion transport from the soil to the vessels of the xylem requires two active, or energy-dependent, events: uptake at the plasma membrane of epidermal cells and secretion into the vessels at the plasma membrane of parenchyma cells bordering the vessels. Inorganic ions follow a mostly symplastic pathway from the epidermis to the xylem. Substantial amounts of the inorganic ions imported into leaves through the xylem are exchanged with the phloem of the leaf veins and exported from the leaf in the assimilate stream. Nutrient uptake from the soil by most seed plants is greatly enhanced by mycorrhizal fungi.

Assimilate Movement in the Phloem Is from Source to Sink

Research on movement of substances in the phloem has been greatly aided by the use of aphids and radioactive tracers. Sieve-tube sap contains sugar (mainly sucrose) and a complex mix of organic and inorganic materials, including amino acids, proteins, RNAs, hormones, and phloem-mobile ions. Rates of longitudinal movement of substances in the phloem greatly exceed the normal rate of diffusion of sucrose in water; the phloem rates typically range from 50 to 100 centimeters per hour.

According to the pressure-flow hypothesis, assimilates move from sources to sinks along turgor pressure gradients developed osmotically. Sugars are loaded into the sieve tube–companion cell complexes at a source. This decreases the water potential in the sieve tube and causes water to move into the sieve tube by osmosis. Meanwhile, removal of sugar at the sink increases the water potential of the sieve tube there. With the movement of water into the sieve tube at the source and out of it at the sink, sugar molecules are carried passively by water along the concentration gradient from source to sink.

Phloem loading can be apoplastic or symplastic. Apoplastic loading is an active process, involving sucrose-proton symport. Species that transport raffinose and stachyose are active symplastic loaders. They have specialized companion cells called intermediary cells and employ a polymer-trapping mechanism. In passive symplastic loaders, sugars produced at high levels in the mesophyll diffuse along concentration gradients to the sieve tubes of minor veins.

QUESTIONS

1. Explain the role of light, potassium ions, and sucrose in stomatal movement.
2. Explain how each of the following factors affects the rate of transpiration: temperature, humidity, air currents.
3. By means of a simple diagram, outline the pathway followed by a water molecule in the transpiration stream, beginning with a root hair and ending in the atmosphere outside a leaf. Label all pertinent tissues and cell layers along the way.
4. Pit membranes are exceedingly important to the safety of water transport. Explain.
5. What evidence supports the cohesion-tension theory?
6. Explain the relationship between root pressure and guttation.
7. Explain the benefit of hydraulic redistribution to plants.
8. The uptake of inorganic nutrients is an energy-requiring process. Explain.
9. What evidence supports the role of sieve tubes as the food-conducting conduits of the phloem?
10. Distinguish between apoplastic and symplastic phloem loading.

Classification of Organisms

There are several different ways to classify organisms. The one presented here follows the overall scheme described in Chapter 12, in which organisms are divided into three domains: Bacteria, Archaea, and Eukarya. Bacteria and Archaea are distinct lineages of prokaryotic organisms. Eukarya, which consists entirely of eukaryotic organisms, includes the protists and the kingdoms Animalia, Fungi, and Plantae. The chief taxonomic categories are domain, kingdom, phylum, class, order, family, genus, species. Recently, it has been hypothesized that the eukaryotes comprise seven supergroups. A supergroup lies between a domain and a kingdom.

The classification that follows includes the protists, except those considered protozoa, as well as the Fungi and Plantae. Certain subphyla and classes given prominence in this book are also included, but the listings are far from complete. The number of species given for each group is the estimated number of living species that have been described and named. Only groups that include living species are described. Viruses are not included in this appendix but are discussed in Chapter 13.

DOMAIN BACTERIA

Bacteria are prokaryotic cells. They lack a nuclear envelope, plastids, mitochondria, and other membrane-bounded organelles, and 9-plus-2 flagella. They are unicellular, but many form aggregates. Their predominant mode of nutrition is absorption, but some groups are photosynthetic or chemosynthetic. Reproduction is predominantly asexual, by fission or budding, but portions of DNA molecules may also be exchanged between cells under certain circumstances. Bacteria are motile by simple flagella or by gliding, or they may be nonmotile.

About 5000 species of bacteria are recognized at present, but this is probably only a small fraction of the actual number. The recognition of species is not comparable with that in eukaryotes and is based largely on metabolic features. One group, the class Rickettsiae—very small bacteria—occurs widely as parasites in arthropods and may consist of tens of thousands of species, depending on the classification criteria used; they are not included in the estimate given here.

The Bacteria can be divided into 17 major groups, or kingdoms. Among these, cyanobacteria are an ancient group that is abundant and important ecologically. Formerly and misleadingly called "blue-green algae," cyanobacteria have a type of photosynthesis that is based on chlorophyll *a*. Cyanobacteria, like the red algae, also have accessory pigments called phycobilins. Many cyanobacteria can fix atmospheric nitrogen, often in specialized cells called heterocysts. Some cyanobacteria form complex filaments or other colonies. Although some 7500 species of cyanobacteria have been described, a more reasonable estimate puts the number of these specialized bacteria at about 200 distinct, nonsymbiotic species.

DOMAIN ARCHAEA

Like Bacteria, Archaea are prokaryotic cells. They lack a nuclear envelope, plastids, mitochondria, and other membrane-bounded organelles, and 9-plus-2 flagella. They are unicellular but sometimes aggregate into filaments or other superficially multicellular bodies. Their predominant mode of nutrition is absorption, but one group of genera obtain their energy by metabolizing sulfur, and another genus, *Halobacterium,* does so through the operation of a proton pump. Many archaea are methanogens, generators of methane. Others are among the most "salt-loving" (extreme halophiles) and "heat-loving" (extreme thermophiles) of all known prokaryotes. Reproduction is asexual, by fission; genetic recombination has not been observed. Archaea are diverse morphologically, ranging from motile flagellated to nonmotile rods, cocci, and spirilla. They differ fundamentally from Bacteria in the base sequences of their ribosomal RNAs and the lipid composition of their plasma membranes. They also differ from Bacteria in lacking peptidoglycans in their cell walls. There are fewer than 100 named species.

DOMAIN EUKARYA

Kingdom Fungi

Fungi are eukaryotic multicellular or, rarely, unicellular organisms in which the nuclei occur in a basically continuous mycelium; this mycelium becomes septate in certain groups and at certain stages of the life cycle. Fungi are heterotrophic; they obtain their nutrition by absorption. Members of all but two groups (Microsporidia and chytrids) form important symbiotic relationships, called mycorrhizas, with the roots of plants. Reproductive cycles typically include both sexual and asexual phases. There are over 100,000 valid species of fungi to which names have been given, and many more will eventually be found. Some have been named two or more times; this is particularly so for fungi that may be classified both as ascomycetes and as asexual fungi. The important characteristics of the major groups of Fungi are provided in Table 14–1.

Phylum Microsporidia: Spore-forming unicellular parasites of animals. Microsporidia are characterized by the presence of a polar tube that penetrates host cells and leads to infection. There are about 1500 species.

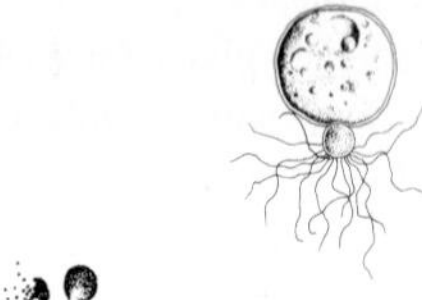

Chytrids: A polyphyletic group of predominantly aquatic heterotrophic organisms with motile cells characteristic of certain stages in their life cycle. The motile cells of most species have a single, posterior, smooth (whiplash) flagellum. Their cell walls are composed of chitin, but other polymers may also be present; they store their food as glycogen. There are about 790 species.

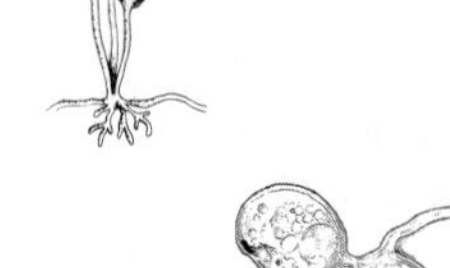

Zygomycetes: A polyphyletic group of terrestrial fungi with hyphae septate only during the formation of reproductive bodies; chitin is predominant in the cell walls. Zygomycetes can usually be recognized by their profuse, rapidly growing hyphae. There are about 1000 described species.

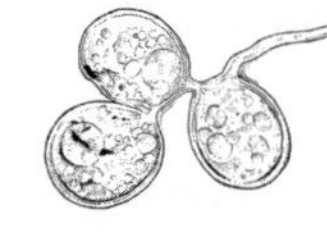

Phylum Glomeromycota: Fungi having mostly coenocytic hyphae and reproducing by means of large, multinucleate asexual spores. Glomeromycetes occur as components of endomycorrhizas that are found in about 80 percent of all vascular plants. They cannot be grown independently of the plant. There are about 200 species.

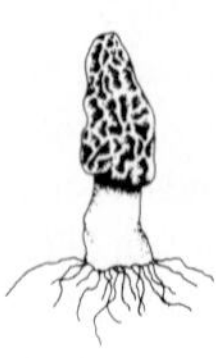

Phylum Ascomycota: Terrestrial and aquatic fungi with the hyphae septate but the septa perforated; complete septa cut off the reproductive bodies, such as spores or gametangia. Chitin is predominant in the cell walls. Sexual reproduction involves the formation of a characteristic cell, the ascus, within which meiosis takes place and ascospores are formed. The hyphae in many ascomycetes are packed together into complex "bodies" known as ascomata. There are about 32,300 species of ascomycetes. The Ascomycota, together with the Basidiomycota, belong to the subkingdom Dikarya.

Phylum Basidiomycota: Terrestrial fungi with the hyphae septate but the septa perforated; complete septa cut off reproductive bodies, such as spores. Chitin is predominant in the cell walls. Sexual reproduction involves formation of basidia, in which meiosis takes place and on which the basidiospores are borne. Basidiomycota are dikaryotic during most of their life cycle, and there is often complex differentiation of "tissues" within their basidiomata. They are the fungal components of most ectomycorrhizas. There are some 22,300 described species.

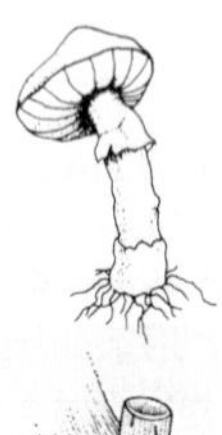

Subphylum Agaricomycotina: Includes the hymenomycetes and gasteromycetes, which are not formal taxonomic groups. The hymenomycetes produce basidiospores in a hymenium exposed on a basidioma; include the mushrooms, coral fungi, and shelf, or bracket, fungi. The gasteromycetes produce basidiospores inside basidiomata, where they are completely enclosed for at least part of their development; include puffballs, earthstars, stinkhorns, and their relatives. The basidia of most Agaricomycotina are aseptate (internally undivided).

Subphylum Pucciniomycotina: Consists of fungi commonly referred to as rusts. Unlike the Agaricomycotina, few rusts form basidiomata; they have septate basidia. There are about 8000 species.

Subphylum Ustilaginomycotina: Commonly referred to as smuts. Like the Pucciniomycotina, they do not form basidiomata, and they form septate basidia. There are about 1070 species.

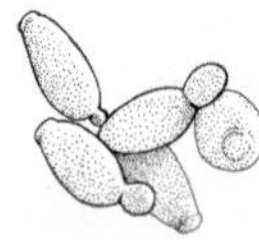

Yeasts: Not a formal taxonomic group; by definition, simply unicellular fungi that reproduce primarily by budding. The yeast growth form is exhibited by a broad range of unrelated fungi encompassing the zygomycetes, Ascomycota, and Basidiomycota. Most yeasts are ascomycetes, but at least a quarter of the genera are Basidiomycota.

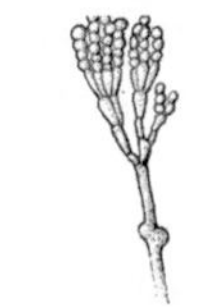

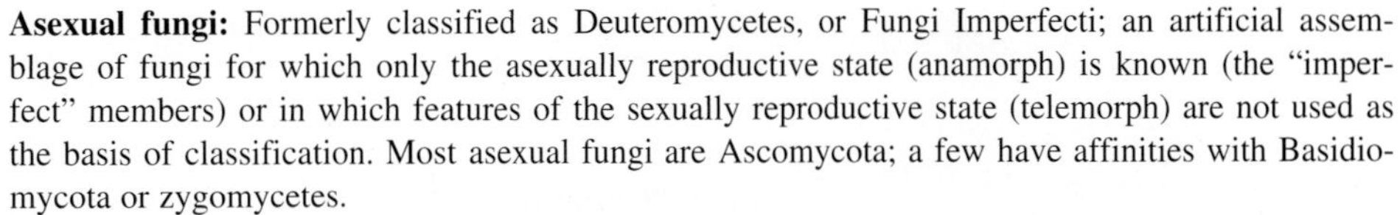

Asexual fungi: Formerly classified as Deuteromycetes, or Fungi Imperfecti; an artificial assemblage of fungi for which only the asexually reproductive state (anamorph) is known (the "imperfect" members) or in which features of the sexually reproductive state (telemorph) are not used as the basis of classification. Most asexual fungi are Ascomycota; a few have affinities with Basidiomycota or zygomycetes.

Lichens: Mutualistic symbiotic associations between a fungal partner and a certain genus of green algae or of cyanobacteria. The fungal component of a lichen is called the mycobiont, and the photosynthetic component is called the photobiont. About 98 percent of the mycobionts belong to the Ascomycota, the remainder to the Basidiomycota. About 13,250 species of lichen-forming fungi have been described.

Protists

The protists are eukaryotic unicellular or multicellular organisms. Their modes of nutrition include ingestion, photosynthesis, and absorption. True sexuality is present in most phyla. They move by means of 9-plus-2 flagella or are nonmotile. Fungi, plants, and animals are specialized multicellular groups derived from protists. The protist groups treated in this book are categorized as algae (photosynthetic protists) and heterotrophic protists (oomycetes and slime molds). The characteristics of the protist groups are outlined in Table 15–1.

Euglenoids: More than 80 genera, of which a third have chloroplasts, with chlorophylls *a* and *b* and carotenoids; the others are heterotrophic. They store food as paramylon, an unusual carbohydrate. Euglenoids usually have two apical flagella and a contractile vacuole. The flexible pellicle is rich in proteins. Sexual reproduction is unknown. There are 800 to 1000 species, most of which occur in fresh water.

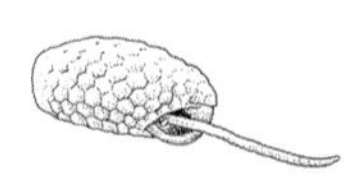

Phylum Cryptophyta: Cryptomonads; photosynthetic organisms that possess chlorophylls *a* and *c* and carotenoids. In addition, some contain a phycobilin, either phycocyanin or phycoerythrin. They are rich in polyunsaturated fatty acids. In addition to a regular nucleus, the cryptomonads contain a reduced nucleus called a nucleomorph. There are 200 known species.

Phylum Haptophyta: Haptophytes; mostly photosynthetic organisms that contain chlorophyll *a* and some variation of chlorophyll *c*. Some have the accessory pigment fucoxanthin. The most distinctive feature of the haptophytes is the haptonema, a threadlike structure that extends from the cell along with two flagella. There are about 300 known species of haptophytes.

Dinoflagellates: Autotrophic organisms, about half of which possess chlorophylls *a* and *c* and carotenoids; the other half lack a photosynthetic apparatus and hence obtain their nutrition either by ingesting solid food particles or by absorbing dissolved organic compounds. Food is stored as starch. A layer of vesicles, often containing cellulose, lies beneath the plasma membrane. This group contains some 4000 known species, mostly biflagellated organisms. These all have lateral flagella, one of which beats in a groove that encircles the organism. Sexual reproduction is generally isogamous, but anisogamy is also present. The mitosis of dinoflagellates is unique. Many dinoflagellates—in a form called zooxanthellae—are symbiotic in marine animals, and they make important contributions to the productivity of coral reefs.

Photosynthetic stramenopiles: The following four classes—Bacillariophyceae, Chrysophyceae, Xanthophyceae, and Phaeophyceae—are photosynthetic stramenopiles, the motile cells of which are characterized by the presence of two flagella, one long with distinctive hairs, the other shorter and smooth.

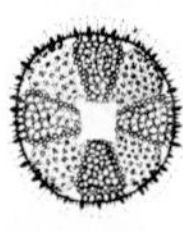

Class Bacillariophyceae: Diatoms; unicellular or colonial organisms with two-part siliceous cell walls, the two halves of which fit together like a Petri dish. Diatoms have chlorophylls *a* and *c*, as well as fucoxanthin. Reserve storage materials include lipids and chrysolaminarin. Diatoms lack flagella, except on some male gametes. There are 10,000 to 12,000 recognized living species.

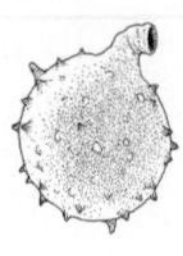

Class Chrysophyceae: Golden algae; primarily unicellular or colonial organisms that possess chlorophylls *a* and *c* and carotenoids, mainly fucoxanthin. Food is stored as the water-soluble carbohydrate chrysolaminarin. The cell walls are absent or consist of cellulose that may be impregnated with minerals; some are covered with silica scales. There are about 1000 known living species.

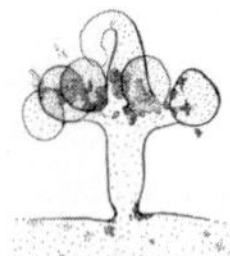

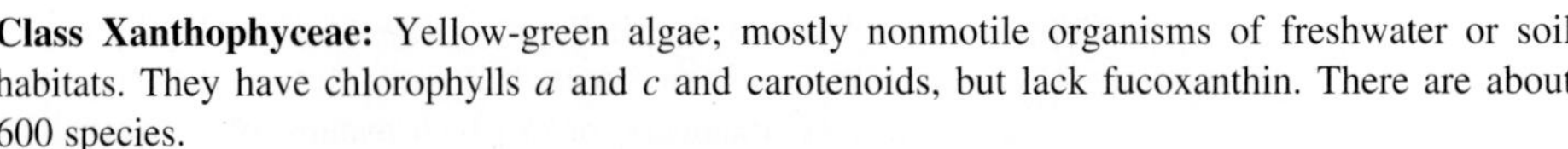

Class Xanthophyceae: Yellow-green algae; mostly nonmotile organisms of freshwater or soil habitats. They have chlorophylls *a* and *c* and carotenoids, but lack fucoxanthin. There are about 600 species.

Class Phaeophyceae: Brown algae; multicellular, nearly entirely marine algae characterized by the presence of chlorophylls *a* and *c* and fucoxanthin. The carbohydrate food reserve is laminarin, and the cell walls have a cellulose matrix containing algin. A considerable amount of differentiation is found in some of the kelps (some of the large brown algae of the order Laminariales), with specialized conducting cells for transporting the products of photosynthesis to the regions of the body that receive little light. There is, however, no differentiation into roots, leaves, and stems, as in the vascular plants. Although there are only about 1500 species, the brown algae dominate rocky shores throughout the cooler regions of the world.

Phylum Rhodophyta: Red algae; primarily marine algae characterized by the presence of chlorophyll *a* and phycobilins. They are particularly abundant in tropical and warm waters. Their carbohydrate food reserve is floridean starch, and the cell walls are composed of cellulose or pectins, with calcium carbonate in many. No motile cells are present at any stage in the complex life cycle. The vegetative body is built up of closely packed filaments in a gelatinous matrix and is not differentiated into roots, leaves, and stems. It lacks specialized conducting cells. There are about 6000 known species.

Green algae: Unicellular or multicellular photosynthetic organisms characterized by the presence of chlorophylls *a* and *b* and various carotenoids. The carbohydrate food reserve is starch; only green algae and plants, which are clearly descended from the green algae, store their reserve food inside plastids. The cell walls of green algae are formed of polysaccharides, sometimes cellulose. Motile cells generally have two apical smooth (whiplash) flagella. True multicellular genera do not exhibit complex patterns of differentiation. Multicellularity has arisen at least twice. There are about 17,000 known species.

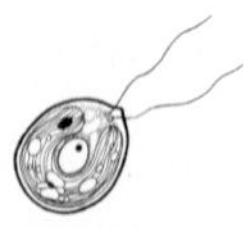

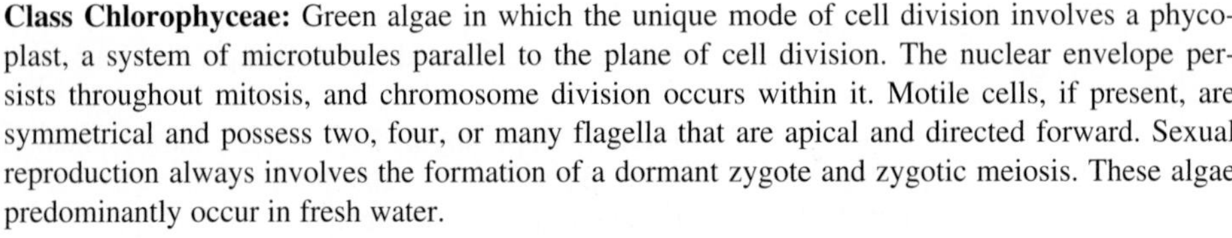

Class Chlorophyceae: Green algae in which the unique mode of cell division involves a phycoplast, a system of microtubules parallel to the plane of cell division. The nuclear envelope persists throughout mitosis, and chromosome division occurs within it. Motile cells, if present, are symmetrical and possess two, four, or many flagella that are apical and directed forward. Sexual reproduction always involves the formation of a dormant zygote and zygotic meiosis. These algae predominantly occur in fresh water.

Class Ulvophyceae: Green algae with a closed mitosis in which the nuclear envelope persists; the spindle is persistent through cytokinesis. Motile cells, if present, are symmetrical and possess two, four, or many flagella that are apical and directed forward. Sexual reproduction often involves alternation of generations and sporic meiosis, and dormant zygotes are rare. These are predominantly marine algae.

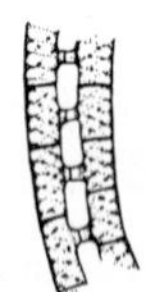

Class Charophyceae: Unicellular, few-celled, filamentous, or parenchymatous green algae in which cell division involves a phragmoplast, a system of microtubules perpendicular to the plane of cell division. The nuclear envelope breaks down during mitosis. Motile cells, if present, are asymmetrical and possess two flagella that are subapical and extend laterally at right angles from the cell. Sexual reproduction always involves the formation of a dormant zygote and zygotic meiosis. Certain members of this class resemble plants more closely than do any other organisms. These algae predominantly occur in fresh water.

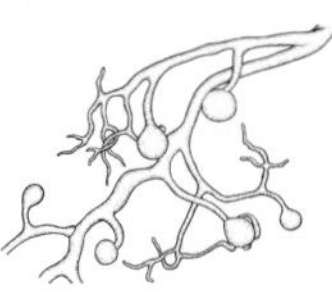

Phylum Oomycota: Oomycetes; aquatic or terrestrial organisms with motile cells characteristic of certain stages of their life cycle. The flagella are two in number—one long with hairs and the other shorter and smooth, as is characteristic of stramenopiles. Cell walls are composed of cellulose or celluloselike polymers, and oomycetes store their food as glycogen. There are about 700 species.

Phylum Myxomycota: Plasmodial slime molds; heterotrophic amoeboid organisms that form a multinucleate plasmodium that creeps along as a mass and eventually differentiates into sporangia, each of which is multinucleate and gives rise to many spores. Sexual reproduction is occasionally observed. The predominant mode of nutrition is by ingestion. There are about 700 species.

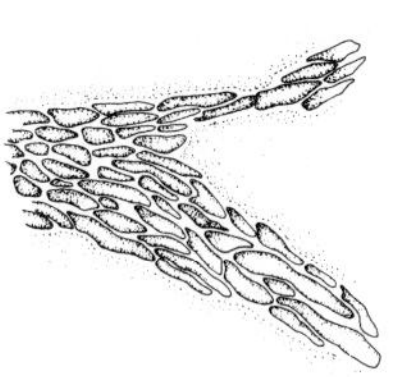

Phylum Dictyosteliomycota: Cellular slime molds, or dictyostelids; heterotrophic organisms that exist as separate amoebas, called myxamoebas, which eventually swarm together to form a pseudoplasmodium, within which they retain their individual identities. Ultimately, the pseudoplasmodium differentiates into a fruiting body. Sexual reproduction involves structures known as macrocysts. Pairs of amoebas first fuse, forming zygotes. Subsequently, these zygotes attract and then engulf nearby amoebas. The principal mode of nutrition is by ingestion. There are about 50 known species.

Kingdom Plantae

Plants are autotrophic (some are derived heterotrophs), multicellular organisms possessing advanced tissue differentiation. All plants have an alternation of generations, in which the diploid phase (sporophyte) includes an embryo and the haploid phase (gametophyte) produces gametes by mitosis. Their photosynthetic pigments and food reserves are similar to those of the green algae. Plants are primarily terrestrial. All but the liverworts, mosses, and hornworts are vascular plants. Summaries of some of the characteristics of the various phyla are found in the Summary Tables of Chapters 16 to 18 and in Table 19–1.

Phylum Marchantiophyta: Liverworts; plants of this and the two following phyla, all of which constitute the bryophytes, have multicellular gametangia with a sterile jacket layer; their sperm are biflagellated. In all three phyla, most photosynthesis is carried out in the gametophyte, on which the sporophyte is dependent. Liverworts lack specialized conducting tissue (with possibly a few exceptions) and stomata; they are the simplest of all living plants. The gametophytes are thallose or leafy, and the rhizoids are single-celled. There are about 5200 species.

Phylum Bryophyta: Mosses; bryophytes with leafy gametophytes; the sporophytes have complex patterns of dehiscence. Specialized conducting tissue is present in both gametophytes and sporophytes of some species. Rhizoids are multicellular. Stomata are present on the sporophytes. There are more than 12,800 species.

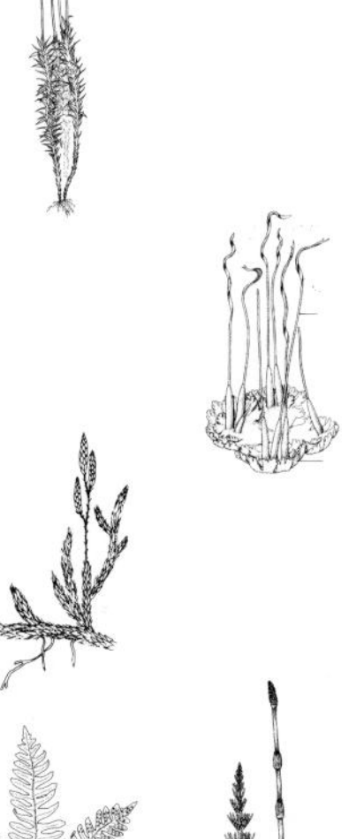

Phylum Anthocerotophyta: Hornworts; bryophytes with thallose gametophytes; the sporophyte grows from a basal intercalary meristem for as long as conditions are favorable. Stomata are present on the sporophyte; there is no specialized conducting tissue. There are more than 300 species.

Phylum Lycopodiophyta: Lycophytes; homosporous and heterosporous vascular plants characterized by the presence of microphylls. The lycophytes are extremely diverse in appearance. All have motile sperm. There are about 1200 species.

Phylum Monilophyta: Ferns and horsetails; all are homosporous, except for the water ferns, which are heterosporous. All ferns possess a megaphyll. The horsetails have small, scalelike leaves (microphyll-like through reduction). The gametophyte is more or less free-living and usually photosynthetic. There are more than 12,000 living species.

Phylum Coniferophyta: Conifers; this and the following three phyla make up the gymnosperms. Conifers are gymnosperms with substantial cambial growth and simple leaves; ovules and seeds are exposed; sperm are nonflagellated. This is the most familiar group of gymnosperms. There are about 630 species.

Phylum Cycadophyta: Cycads; gymnosperms with sluggish cambial growth and pinnately compound, palmlike or fernlike leaves; ovules and seeds are exposed. The sperm are flagellated and motile but are carried to the vicinity of the ovule in a pollen tube. There are about 300 species.

Phylum Ginkgophyta: *Ginkgo;* gymnosperm with considerable cambial growth and fan-shaped leaves with open dichotomous venation; ovules and seeds are exposed; seed coats are fleshy. Sperm are carried to the vicinity of the ovule in a pollen tube but are flagellated and motile. There is only one species.

Phylum Gnetophyta: Gnetophytes; gymnosperms with many angiospermlike features, such as vessels and double fertilization. The gnetophytes are the only gymnosperms that have vessels. They are the group of gymnosperms most closely related to the angiosperms. Motile sperm are absent. There are three very distinctive genera with about 75 species.

Phylum Anthophyta: Flowering plants; angiosperms. These are seed plants in which the ovules are enclosed in a carpel and seeds are borne within fruits. The angiosperms are extremely diverse vegetatively but are characterized by the flower, which is basically insect-pollinated. Other modes of pollination, such as wind pollination, have been derived in a number of different lines. The gametophytes are much reduced, with the female gametophyte often consisting of seven cells at maturity. Double fertilization involving the two sperm of the mature microgametophyte gives rise to the zygote (sperm and egg) and to the primary endosperm nucleus (sperm and polar nuclei); the former becomes the embryo and the latter becomes a special nutritive tissue called endosperm. There are at least 300,000 species.

Class Monocotyledonae: Monocots. The flower parts are usually in threes; leaf venation is usually parallel; primary vascular bundles in the stem are scattered; true secondary growth is not present; one cotyledon. There are at least 90,000 species.

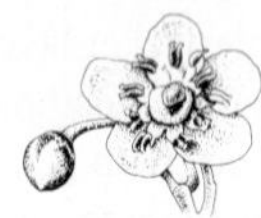

Class Eudicotyledonae: Eudicots. The flower parts are usually in fours or fives; leaf venation is usually netlike; primary vascular bundles in the stem are in a ring; many have a vascular cambium and true secondary growth; two cotyledons. There are at least 200,000 species.

Together, the monocots and eudicots represent about 97 percent of angiosperms. The remaining 3 percent of living angiosperms include the basal grade angiosperms and the magnoliids, the angiosperms with the most primitive features and the ancestors of both monocots and eudicots.

CHAPTER 1

Baskin, Y. 1997. *The Work of Nature: How the Diversity of Life Sustains Us.* Island Press, Washington, DC.

Beerling, D. J. 2007. *The Emerald Planet: How Plants Changed Earth's History.* Oxford University Press, Oxford, New York.

Deamer, D., and **A. L. Weber.** 2010. Bioenergetics and life's origins. *Cold Spring Harbor Perspectives in Biology* 2010;2:a004929.

Fischer, W. W. 2008. Life before the rise of oxygen. *Nature* 455, 1051–1052.

Gaucher, E. A., J. T. Kratzer, and **R. N. Randall.** 2010. Deep phylogeny—how a tree can help characterize early life on Earth. *Cold Spring Harbor Perspectives in Biology* 2010;2:a002238.

Goetz, W. 2010. Phoenix on Mars. *American Scientist* 98, 40–47.

Monastersky, R. 1998. The rise of life on Earth. *National Geographic* 193 (3), 54–81.

Morton, O. 2008. *Eating the Sun: How Plants Power the Planet.* Harper Collins, New York.

Niklas, K. J. 1997. *The Evolutionary Biology of Plants.* The University of Chicago Press, Chicago, London.

Pizzarello, S., and **E. Shock.** 2010. The organic composition of carbonaceous meteorites: the evolutionary story ahead of biochemistry. *Cold Spring Harbor Perspectives in Biology* 2010;2:a002105.

Sleep, N. H. 2010. The Hadean–Archaean environment. *Cold Spring Harbor Perspectives in Biology* 2010;2:a002527.

Trefil, J., H. J. Morowitz, and **E. Smith.** 2009. The origin of life. *American Scientist* 97, 206–213.

Zimmer, C. 2009. On the origin of life on Earth. *Science* 323, 198–199.

SECTION 1: CHAPTERS 2–4

CHAPTER 2

Baldwin, I. T. 2010. Plant volatiles. *Current Biology* 20, R392–R397.

Buchanan, B. B., W. Gruissem, and **R. L. Jones** (eds.). 2000. *Biochemistry and Molecular Biology of Plants.* American Society of Plant Physiologists, Rockville, MD.

Hardin, J., G. Bertoni, L. J. Kleinsmith, and **W. M. Becker.** 2012. *Becker's World of the Cell,* 8th ed. Benjamin Cummings, Boston.

Lodish, H., A. Berk, C. A. Kaiser, M. Krieger, M. P. Scott, A. Bretscher, H. Ploegh, and **P. Matsudaira.** 2008. *Molecular Cell Biology,* 6th ed. W. H. Freeman and Company, New York.

Nelson, D. L., and **M. M. Cox.** 2008. *Lehninger Principles of Biochemistry,* 5th ed. W. H. Freeman and Company, New York.

Sackheim, G. I. 2008. *An Introduction to Chemistry for Biology Students,* 9th ed. Benjamin/Cummings Publishing Company, Menlo Park, CA.

Taiz, L., and **E. Zeiger.** 2010. *Plant Physiology,* 5th ed. Sinauer Associates, Inc., Publishers, Sunderland, MA.

Tanaka, Y., N. Sasaki, and **A. Ohmiya.** 2008. Biosynthesis of plant pigments: anthocyanins, betalains and carotenoids. *The Plant Journal* 54, 733–749.

Treutter, D. 2005. Significance of flavonoids in plant resistance and enhancement of their biosynthesis. *Plant Biology* 7, 581–591.

Unsicker, S. B., G. Kunert, and **J. Gershenzon.** 2009. Protective perfumes: the role of vegetative volatiles in plant defense against herbivores. *Current Opinion in Plant Biology* 12, 479–485.

Wenke, K., M. Kai, and **B. Piechulla.** 2010. Belowground volatiles facilitate interactions between plant roots and soil organisms. *Planta* 231, 499–506.

CHAPTER 3

Alberts, B., A. Johnson, J. Lewis, M. Raff, K. Roberts, and **P. Walter.** 2008. *Molecular Biology of the Cell,* 5th ed. Garland Science, New York, Oxford.

Bloom, K., and **A. Joglekar.** 2010. Towards building a chromosome segregation machine. *Nature* 463, 446–456.

Bock, R., and **J. N. Timmis.** 2008. Reconstructing evolution: gene transfer from plastids to the nucleus. *BioEssays* 30, 556–566.

Delmer, D. P., and **Y. Amor.** 1995. Cellulose biosynthesis. *The Plant Cell* 7, 987–1000.

Emons, A. M. C., H. Höfte, and **B. M. Mulder.** 2007. Microtubules and cellulose microfibrils: how intimate is their relationship? *Trends in Plant Science* 12, 279–281.

Evert, R. F. 2006. *Esau's Plant Anatomy. Meristems, Cells, and Tissues of the Plant Body: Their Structure, Function, and Development,* 3rd ed. John Wiley & Sons, Inc., Hoboken, NJ.

Francis, D. 2007. The plant cell cycle—15 years on. *New Phytologist* 174, 261–278.

Hardin, J., G. Bertoni, L. J. Kleinsmith, and **W. M. Becker.** 2012. *Becker's World of the Cell,* 8th ed. Benjamin Cummings, Boston.

Hepler, P. K., and **J. M. Hush.** 1996. Behavior of microtubules in living plant cells. *Plant Physiology* 112, 455–461.

Inzé, D., and **L. De Veylder.** 2006. Cell cycle regulation in plant development. *Annual Review of Genetics* 40, 77–105.

John, P. C. L., and **R. Qi.** 2008. Cell division and endoreduplication: doubtful engines of vegetative growth. *Trends in Plant Science* 13, 121–127.

Johnson, M. K., and **D. A. Wise.** 2009. The kinetochore moves ahead: contributions of molecular and genetic techniques to our understanding of mitosis. *BioScience* 59, 933–943.

Lodish, H., A. Berk, C. A. Kaiser, M. Krieger, M. P. Scott, A. Bretscher, H. Ploegh, and **P. Matsudaira.** 2008. *Molecular Cell Biology,* 6th ed. W. H. Freeman and Company, New York.

McBride, H. M., M. Neuspiel, and **S. Wasiak.** 2006. Mitochondria: more than just a powerhouse. *Current Biology* 16, R551–R560.

Miyagishima, S.-y. 2005. Origin and evolution of the chloroplast division machinery. *Journal of Plant Research* 118, 295–306.

Miyagishima, S.-y., and **Y. Kabeya.** 2010. Chloroplast division: squeezing the photosynthetic captive. *Current Opinion in Microbiology* 13, 738–746.

Müller, S., A. J. Wright, and **L. G. Smith.** 2009. Division plane control in plants: new players in the band. *Trends in Cell Biology* 19, 180–188.

Pizzo, P., and **T. Pozzan.** 2007. Mitochondria–endoplasmic reticulum choreography: structure and signaling dynamics. *Trends in Cell Biology* 17, 511–517.

Pogson, B. J., N. S. Woo, B. Förster, and **I. D. Small.** 2008. Plastid signalling to the nucleus and beyond. *Trends in Plant Science* 13, 602–609.

Pollard, T. D., and **J. A. Cooper.** 2009. Actin, a central player in cell shape and movement. *Science* 326, 1208–1212.

Rambold, A. S., and **J. Lippincott-Schwartz.** 2011. SevERing mitochondria. *Science* 334, 186–187.

Reape, T. J., E. M. Molony, and **P. F. McCabe.** 2008. Programmed cell death in plants: distinguishing between different modes. *Journal of Experimental Botany* 59, 435–444.

Schrader, M., and **Y. Yoon.** 2007. Mitochondria and peroxisomes: are the "Big Brother" and the "Little Sister" closer than assumed? *BioEssays* 29, 1105–1114.

Staehelin, L. A. 1997. The plant ER: a dynamic organelle composed of a large number of discrete functional domains. *The Plant Journal* 11, 1511–1165.

Taiz, L., and **E. Zeiger.** 2010. *Plant Physiology,* 5th ed. Sinauer Associates, Inc., Publishers, Sunderland, MA.

Terry, L. J., E. B. Shows, and **S. R. Wente.** 2007. Crossing the nuclear envelope: hierarchical regulation of nucleocytoplasmic transport. *Science* 318, 1412–1416.

Verchot-Lubicz, J., and **R. E. Goldstein.** 2010. Cytoplasmic streaming enables the distribution of molecules and vesicles in large plant cells. *Protoplasma* 240, 99–107.

Vianello, A., M. Zancani, C. Peresson, E. Petrussa, V. Casolo, J. Krajňáková, S. Patui, E. Braidot, and **F. Macrì.** 2007. Plant mitochondrial pathway leading to programmed cell death. *Physiologia Plantarum* 129, 242–252.

CHAPTER 4

Bell, K., and **K. Oparka.** 2011. Imaging plasmodesmata. *Protoplasma* 248, 9–25.

Burch-Smith, T. M., S. Stonebloom, M. Xu, and **P. C. Zambryski.** 2011. Plasmodesmata during development: re-examination of the importance of primary, secondary, and branched plasmodesmata structure versus function. *Protoplasma* 248, 61–74.

Engelman, D. M. 2005. Membranes are more mosaic than fluid. *Nature* 438, 578–580.

Evert, R. F. 2006. *Esau's Plant Anatomy. Meristems, Cells, and Tissues of the Plant Body: Their Structure, Function, and Development,* 3rd ed. John Wiley & Sons, Inc., Hoboken, NJ.

Hardin, J., G. Bertoni, L. J. Kleinsmith, and **W. M. Becker.** 2012. *Becker's World of the Cell,* 8th ed. Benjamin Cummings, Boston.

Hyun, T. K., M. N. Uddin, Y. Rim, and **J.-Y. Kim.** 2011. Cell-to-cell trafficking of RNA and RNA silencing through plasmodesmata. *Protoplasma* 248, 101–116.

Irani, N. G., and **E. Russinova.** 2009. Receptor endocytosis and signaling in plants. *Current Opinion in Plant Biology* 12, 653–659.

Lingwood, D., and **K. Simons.** 2010. Lipid rafts as a membrane-organizing principle. *Science* 327, 46–50.

Lodish, H., A. Berk, C. A. Kaiser, M. Krieger, M. P. Scott, A. Bretscher, H. Ploegh, and **P. Matsudaira.** 2008. *Molecular Cell Biology,* 6th ed. W. H. Freeman and Company, New York.

Lucas, W. J., B.-K. Ham, and **J.-Y. Kim.** 2009. Plasmodesmata—bridging the gap between neighboring plant cells. *Trends in Cell Biology* 19, 495–503.

Maurel, C., L. Verdoucq, D.-T. Luu, and **V. Santoni.** 2008. Plant aquaporins: membrane channels with multiple integrated functions. *Annual Review of Plant Biology* 59, 595–624.

Taiz, L., and **E. Zeiger.** 2010. *Plant Physiology,* 5th ed. Sinauer Associates, Inc., Publishers, Sunderland, MA.

Verma, D. P. S. (ed.). 1996. *Signal Transduction in Plant Growth and Development.* Springer, Vienna, New York.

Wudick, M. M., D.-T. Luu, and **C. Maurel.** 2009. A look inside: localization patterns and functions of intracellular plant aquaporins. *New Phytologist* 184, 289–302.

Xu, X. M., and **D. Jackson.** 2010. Lights at the end of the tunnel: new views of plasmodesmal structure and function. *Current Opinion in Plant Biology* 13, 684–692.

Zavaliev, R., S. Ueki, B. L. Epel, and **V. Citovsky.** 2011. Biology of callose (β-1,3-glucan) turnover at plasmodesmata. *Protoplasma* 284, 117–130.

SECTION 2: CHAPTERS 5–7

CHAPTERS 5 AND 6

Alberts, B., A. Johnson, J. Lewis, M. Raff, K. Roberts, and **P. Walter.** 2008. *Molecular Biology of the Cell,* 5th ed. Garland Science, New York, Oxford.

Berg, J. M., J. L. Tymoczko, and **L. Stryer.** 2012. *Biochemistry,* 7th ed. W. H. Freeman and Company, New York.

Garby, L., and **P. S. Larsen.** 1995. *Bioenergetics: Its Thermodynamic Foundations.* Cambridge University Press, Cambridge, New York.

Hardin, J., G. Bertoni, L. J. Kleinsmith, and **W. M. Becker.** 2012. *Becker's World of the Cell,* 8th ed. Benjamin Cummings, Boston.

Lodish, H., A. Berk, C. A. Kaiser, M. Krieger, M. P. Scott, A. Bretscher, H. Ploegh, and **P. Matsudaira.** 2008. *Molecular Cell Biology,* 6th ed. W. H. Freeman and Company, New York.

Millar, A. H., J. Whelan, K. L. Soole, and **D. A. Day.** 2011. Organization and regulation of mitochondrial respiration in plants. *Annual Review of Plant Biology* 62, 79–104.

Nelson, D. L., and **M. M. Cox.** 2008. *Lehninger Principles of Biochemistry,* 5th ed. W. H. Freeman and Company, New York.

Plaxton, W. C., and **F. E. Podestá.** 2006. The functional organization and control of plant respiration. *Critical Reviews in Plant Sciences* 25, 159–198.

Sweetlove, L. J., K. F. M. Beard, A. Nunes-Nesi, A. R. Fernie, and **R. G. Ratcliffe.** 2010. Not just a circle: flux modes in the plant TCA cycle. *Trends in Plant Science* 15, 462–470.

Taiz, L., and **E. Zeiger.** 2010. *Plant Physiology,* 5th ed. Sinauer Associates, Inc., Publishers, Sunderland, MA.

CHAPTER 7

Abelson, J. 2007. The birth of oxygen. *Bulletin of the American Academy of Arts and Sciences* Spring, 28–33.

Alberts, B., A. Johnson, J. Lewis, M. Raff, K. Roberts, and **P. Walter.** 2008. *Molecular Biology of the Cell,* 5th ed. Garland Science, New York, Oxford.

Bauwe, H., M. Hagemann, and **A. R. Fernie.** 2010. Photorespiration: players, partners and origin. *Trends in Plant Science* 15, 330–336.

Bloom, A. J. 2010. *Global Climate Change: Convergence of Disciplines.* Sinauer Associates, Inc., Publishers, Sunderland, MA.

Boyd, C. N., V. R. Franceschi, S. D. X. Chuong, H. Akhani, O. Kiirats, M. Smith, and **G. E. Edwards.** 2007. Flowers of *Bienertia cycloptera* and *Suaeda aralocaspica* (Chenopodiaceae) complete the life cycle performing single-cell C_4 photosynthesis. *Functional Plant Biology* 34, 268–281.

Chapin, F. S., III, J. McFarland, A. D. McGuire, E. S. Euskirchen, R. W. Ruess, and **K. Kielland.** 2009. The changing global carbon cycle: linking plant–soil carbon dynamics to global consequences. *Journal of Ecology* 97, 840–850.

Gerhart, L. M., and **J. K. Ward.** 2010. Plant responses to low [CO_2] of the past. *New Phytologist* 188, 674–695.

Hardin, J., G. Bertoni, L. J. Kleinsmith, and **W. M. Becker.** 2012. *Becker's World of the Cell,* 8th ed. Benjamin Cummings, Boston.

Hohmann-Marriott, M. F., and **R. E. Blankenship.** 2011. Evolution of photosynthesis. *Annual Review of Plant Biology* 62, 515–548.

Leegood, R. C. 2008. Roles of the bundle sheath cells in leaves of C_3 plants. *Journal of Experimental Botany* 59, 1663–1673.

Le Quéré, C. 2010. Trends in the land and ocean carbon uptake. *Current Opinion in Environmental Sustainability* 2, 219–224.

Lev-Yadun, S. 2010. The shared and separate roles of aposematic (warning) coloration and the co-evolution hypothesis in defending autumn leaves. *Plant Signaling & Behavior* 5, 937–939.

Lev-Yadun, S., and **K. S. Gould.** 2007. What do red and yellow autumn leaves signal? *The Botanical Review* 73, 279–289.

Lodish, H., A. Berk, C. A. Kaiser, M. Krieger, M. P. Scott, A. Bretscher, H. Ploegh, and **P. Matsudaira.** 2008. *Molecular Cell Biology,* 6th ed. W. H. Freeman and Company, New York.

Muhaidat, R., R. F. Sage, and **N. G. Dengler.** 2007. Diversity of Kranz anatomy and biochemistry in C_4 eudicots. *American Journal of Botany* 94, 362–381.

Nelson, D. L., and **M. M. Cox.** 2008. *Lehninger Principles of Biochemistry,* 5th ed. W. H. Freeman and Company, New York.

Peterhansel, C., I. Horst, M. Niessen, C. Blume, R. Kebeish, S. Kürkcüoglu, and **F. Kreuzaler.** 2010. Photorespiration. *The Arabidopsis Book* 10.1199/tab.0130.

Reich, P. B. 2010. The carbon dioxide exchange. *Science* 329, 774–775.

Saveyn, A., K. Steppe, N. Ubierna, and **T. E. Dawson.** 2010. Woody tissue photosynthesis and its contribution to trunk growth and bud development in young plants. *Plant, Cell and Environment* 33, 1949–1958.

Silvera, K., K. M. Neubig, W. M. Whitten, N. H. Williams, K. Winter, and **J. C. Cushman.** 2010. Evolution along the crassulacean acid metabolism continuum. *Functional Plant Biology* 37, 995–1010.

Taiz, L., and **E. Zeiger.** 2010. *Plant Physiology,* 5th ed. Sinauer Associates, Inc., Publishers, Sunderland, MA.

Trumbore, S. E., and **C. I. Czimczik.** 2008. An uncertain future for soil carbon. *Science* 321, 1455–1456.

SECTION 3: CHAPTERS 8–11

CHAPTER 8

Chase, C. D. 2007. Cytoplasmic male sterility: a window to the world of plant mitochondrial–nuclear interactions. *Trends in Genetics* 23, 81–90.

Griffiths, A. J. F., S. R. Wessler, S. B. Carroll, and **J. Doebley.** 2012. *Introduction to Genetic Analysis,* 10th ed. W. H. Freeman and Company, New York.

Hardin, J., G. Bertoni, L. J. Kleinsmith, and **W. M. Becker.** 2012. *Becker's World of the Cell,* 8th ed. Benjamin Cummings, Boston.

Hopkin, K. 2009. The evolving definition of a gene. *BioScience* 59, 928–931.

Lodish, H., A. Berk, C. A. Kaiser, M. Krieger, M. P. Scott, A. Bretscher, H. Ploegh, and **P. Matsudaira.** 2008. *Molecular Cell Biology,* 6th ed. W. H. Freeman and Company, New York.

Mogie, M. 1992. *The Evolution of Asexual Reproduction in Plants.* Chapman & Hall, London, New York.

Nelson, D. L., and **M. M. Cox.** 2008. *Lehninger Principles of Biochemistry,* 5th ed. W. H. Freeman and Company, New York.

Pierce, B. A. 2008. *Genetics: A Conceptual Approach,* 3rd ed. W. H. Freeman and Company, New York.

Schurko, A. M., M. Neiman, and **J. M. Logsdon, Jr.** 2008. Signs of sex: what we know and how we know it. *Trends in Genetics* 24, 208–217.

CHAPTER 9

Alberts, B., A. Johnson, J. Lewis, M. Raff, K. Roberts, and **P. Walter.** 2008. *Molecular Biology of the Cell,* 5th ed. Garland Science, New York, Oxford.

Babbitt, G. 2011. Chromatin evolving. *American Scientist* 99, 48, 50–55.

Bennett, M. D., and **I. J. Leitch.** 2011. Nuclear DNA amounts in angiosperms: targets, trends and tomorrow. *Annals of Botany* 107, 467–590.

Chen, M., S. Lv, and **Y. Meng.** 2010. Epigenetic performers in plants. *Development, Growth & Differentiation* 52, 555–566.

Chen, X. 2009. Small RNAs and their roles in plant development. *Annual Review of Cell and Developmental Biology* 35, 21–44.

Chuck, G., H. Candela, and **S. Hake.** 2009. Big impacts by small RNAs in plant development. *Current Opinion in Plant Biology* 12, 81–86.

De Lange, T. 2009. How telomeres solve the end-protection problem. *Science* 326, 948–952.

Griffiths, A. J. F., S. R. Wessler, S. B. Carroll, and **J. Doebley.** 2012. *Introduction to Genetic Analysis,* 10th ed. W. H. Freeman and Company, New York.

Hardin, J., G. Bertoni, L. J. Kleinsmith, and **W. M. Becker.** 2012. *Becker's World of the Cell,* 8th ed. Benjamin Cummings, Boston.

He, G., A. A. Elling, and **X. W. Deng.** 2011. The epigenome and plant development. *Annual Review of Plant Biology* 62, 411–435.

Heslop-Harrison, J. S., and **T. Schwarzacher.** 2011. Organisation of the plant genome in chromosomes. *The Plant Journal* 66, 18–33.

Joyce, G. F. 2007. A glimpse of biology's first enzyme. *Science* 315, 1507–1508.

Lisch, D. 2009. Epigenetic regulation of transposable elements in plants. *Annual Review of Plant Biology* 60, 43–66.

Lodish, H., A. Berk, C. A. Kaiser, M. Krieger, M. P. Scott, A. Bretscher, H. Ploegh, and **P. Matsudaira.** 2008. *Molecular Cell Biology,* 6th ed. W. H. Freeman and Company, New York.

Makeyev, E. V., and **T. Maniatis.** 2008. Multilevel regulation of gene expression by microRNAs. *Science* 319, 1789–1790.

Mélèse, T., and **Z. Xue.** 1995. The nucleolus: an organelle formed by the act of building a ribosome. *Current Opinion in Cell Biology* 7, 319–324.

Pierce, B. A. 2008. *Genetics: A Conceptual Approach,* 3rd ed. W. H. Freeman and Company, New York.

Sugiura, M., and **Y. Takeda.** 2000. Nucleic acids. In B. B. Buchanan, W. Gruissem, and R. L. Jones (eds.), *Biochemistry and Molecular Biology of Plants,* pp. 260–310. American Society of Plant Physiologists, Rockville, MD.

Voinnet, O. 2009. Origin, biogenesis, and activity of plant microRNAs. *Cell* 136, 669–687.

Zhang, Z., C. J. Wippo, M. Wal, E. Ward, P. Korber, and **B. F. Pugh.** 2011. A packing mechanism for nucleosome organization reconstituted across a eukaryotic genome. *Science* 332, 977–980.

CHAPTER 10

Alberts, B., A. Johnson, J. Lewis, M. Raff, K. Roberts, and **P. Walter.** 2008. *Molecular Biology of the Cell,* 5th ed. Garland Science, New York, Oxford.

Bennett, M. D., and **I. J. Leitch.** 2011. Nuclear DNA amounts in angiosperms: targets, trends and tomorrow. *Annals of Botany* 107, 467–590.

Chaves, M., and **B. Davies.** 2010. Drought effects and water use efficiency: improving crop production in dry environments. *Functional Plant Biology* 37, iii–vi.

Collinge, D. B., H. J. L. Jørgensen, O. S. Lund, and **M. F. Lyngkjaer.** 2010. Engineering pathogen resistance in crop plants: current trends and future prospects. *Annual Review of Phytopathology* 48, 269–291.

Damude, H. G., and **A. J. Kinney.** 2008. Enhancing plant seed oils for human nutrition. *Plant Physiology* 147, 962–968.

Flowers, J. M., and **M. D. Purugganan.** 2008. The evolution of plant genomes—scaling up from a population perspective. *Current Opinion in Genetics & Development* 18, 565–570.

Griffiths, A. J. F., S. R. Wessler, S. B. Carroll, and **J. Doebley.** 2012. *Introduction to Genetic Analysis,* 10th ed. W. H. Freeman and Company, New York.

Hardin, J., G. Bertoni, L. J. Kleinsmith, and **W. M. Becker.** 2012. *Becker's World of the Cell,* 8th ed. Benjamin Cummings, Boston.

Heslop-Harrison, J. S., and **T. Schwarzacher.** 2011. Organisation of the plant genome in chromosomes. *The Plant Journal* 66, 18–33.

Hibberd, J. M., J. E. Sheehy, and **J. A. Langdale.** 2008. Using C_4 photosynthesis to increase the yield of rice—rationale and feasibility. *Current Opinion in Plant Biology* 11, 228–231.

Hilbeck, A., M. Meier, J. Römbke, S. Jänsch, H. Teichmann, and **B. Tappeser.** 2011. Environmental risk assessment of genetically modified plants—concepts and controversies. *Environmental Sciences Europe* 23, art13.

Lodish, H., A. Berk, C. A. Kaiser, M. Krieger, M. P. Scott, A. Bretscher, H. Ploegh, and **P. Matsudaira.** 2008. *Molecular Cell Biology,* 6th ed. W. H. Freeman and Company, New York.

Meng, L., S. Zhang, and **P. G. Lemaux.** 2010. Toward molecular understanding of *in vitro* and *in planta* shoot organogenesis. *Critical Reviews in Plant Sciences* 29, 108–122.

Meyer, H. 2011. Systemic risks of genetically modified crops: the need for new approaches to risk assessment. *Environmental Sciences Europe* 23, art7.

Paterson, A. H., M. Freeling, H. Tang, and **X. Wang.** 2010. Insights from the comparison of plant genome sequences. *Annual Review of Plant Biology* 61, 349–372.

Plant Genomes. 2008. *Science* (Special Issue) 320, 465–497.

Powles, S. B., and **Q. Yu.** 2010. Evolution in action: plants resistant to herbicides. *Annual Review of Plant Biology* 61, 317–347.

Raines, C. A. 2011. Increasing photosynthetic carbon assimilation in C_3 plants to improve crop yield: current and future strategies. *Plant Physiology* 155, 36–42.

Rampitsch, C., and **M. Srinivasan.** 2006. The application of proteomics to plant biology: a review. *Canadian Journal of Botany* 84, 883–892.

Takáč, T., T. Pechan, and **J. Šamaj.** 2011. Differential proteomics of plant development. *Journal of Proteomics* 74, 577–588.

Tranel, P. J., and **D. P. Horvath.** 2009. Molecular biology and genomics: new tools for weed science. *BioScience* 59, 207–215.

Van Montagu, M. 2011. It is a long way to GM agriculture. *Annual Review of Plant Biology* 62, 1–23.

Yang, X., and **X. Zhang.** 2010. Regulation of somatic embryogenesis in higher plants. *Critical Reviews in Plant Sciences* 29, 36–57.

Zhu, X.-G., L. Shan, Y. Wang, and **W. P. Quick.** 2010. C_4 rice—an ideal arena for systems biology research. *Journal of Integrative Plant Biology* 52, 762–770.

CHAPTER 11

Barringer, B. C. 2007. Polyploidy and self-fertilization in flowering plants. *American Journal of Botany* 94, 1527–1533.

Bowler, P. J. 2009. Darwin's originality. *Science* 323, 223–226.

Costa, J. T. 2009. The Darwinian revelation: tracing the origin and evolution of an idea. *BioScience* 59, 886–894.

Garfield, D. A., and **G. A. Wray.** 2010. The evolution of gene regulatory interactions. *BioScience* 60, 15–23.

Givnish, T. J. 1998. Adaptive plant evolution on islands: classical patterns, molecular data, new insights. In P. R. Grant (ed.), *Evolution on Islands,* pp. 281–304. Oxford University Press, Oxford, New York.

Godfrey-Smith, P. 2009. *Darwinian Populations and Natural Selection.* Oxford University Press, Oxford.

Griffiths, A. J. F., S. R. Wessler, S. B. Carroll, and **J. Doebley.** 2012. *Introduction to Genetic Analysis,* 10th ed. W. H. Freeman and Company, New York.

Harvey, J. A., T. Bukovinszky, and **W. H. van der Putten.** 2010. Interactions between invasive plants and insect herbivores: a plea for a multitrophic perspective. *Biological Conservation* 143, 2251–2259.

Hayden, T. 2009. What Darwin didn't know. *Smithsonian* 39, 40–48.

Lind, E. M., and **J. D. Parker.** 2010. Novel weapons testing: are invasive plants more chemically defended than native plants? *PLoS One* 5, e10429.

Murrell, C., E. Gerber, C. Krebs, M. Parepa, U. Schaffner, and **O. Bossdorf.** 2011. Invasive knotweed affects native plants through allelopathy. *American Journal of Botany* 98, 38–43.

Pennisi, E. 2007. Natural selection, not chance, paints the desert. *Science* 318, 376.

Pierce, B. A. 2008. *Genetics: A Conceptual Approach,* 3rd ed. W. H. Freeman and Company, New York.

Rieseberg, L. H., and **J. H. Willis.** 2007. Plant speciation. *Science* 317, 910–914.

Ruse, M., and **J. Travis** (eds.). 2009. *Evolution: The First Four Billion Years.* Belknap Press of Harvard University Press, Cambridge, MA.

Thomson, K. 2009. Darwin's enigmatic health. *American Scientist 97,* 198–200.

Van Doorn, G. S., P. Edelaar, and **F. J. Weissing.** 2009. On the origin of species by natural and sexual selection. *Science* 326, 1704–1707.

Willmore, K. E. 2010. Development influences evolution. *American Scientist* 98, 220–227.

Zuppinger-Dingley, D., B. Schmid, Y. Chen, H. Brandl, M. G. A. van der Heijden, and **J. Joshi.** 2011. In their native range, invasive plants are held in check by negative soil-feedbacks. *Ecosphere* 2, art54.

SECTION 4: CHAPTERS 12–21

CHAPTER 12

Angiosperm Phylogeny Group. 2009. An update of the Angiosperm Phylogeny Group classification for the orders and families of flowering plants: APGIII. *Botanical Journal of the Linnaean Society* 161, 105–121.

Green, B. R. 2011. Chloroplast genomes of photosynthetic eukaryotes. *The Plant Journal* 66, 34–44.

Hollingsworth, P. M., S. W. Graham, and **D. P. Little.** 2011. Choosing and using a plant DNA barcode. *PLoS One* 6, e19254.

Hörandl, E., and **T. F. Stuessy.** 2010. Paraphyletic groups as natural units of biological classification. *Taxon* 59, 1641–1653.

Judd, W. S., C. S. Campbell, E. A. Kellogg, P. F. Stevens, and **M. J. Donoghue.** 2008. *Plant Systematics: A Phylogenetic Approach,* 3rd ed. Sinauer Associates, Inc., Publishers, Sunderland, MA.

Keeling, P. J. 2010. The endosymbiotic origin, diversification and fate of plastids. *Philosophical Transactions of the Royal Society B* 365, 729–748.

Lake, J. A., R. G. Skophammer, C. W. Herbold, and **J. A. Servin.** 2009. Genome beginnings: rooting the tree of life. *Philosophical Transactions of the Royal Society B* 364, 2177–2185.

Lane, C. E., and **J. M. Archibald.** 2008. The eukaryotic tree of life: endosymbiosis takes its TOL. *Trends in Ecology and Evolution* 23, 268–275.

Miyagishima, S.-y., and **Y. Kabeya.** 2010. Chloroplast division: squeezing the photosynthetic captive. *Current Opinion in Microbiology* 13, 738–746.

Podani, J. 2010. Monophyly and paraphyly: a discourse without end? *Taxon* 59, 1011–1015.

Rasmussen, B., I. R. Fletcher, J. J. Brocks, and **M. R. Kilburn.** 2008. Reassessing the first appearance of eukaryotes and cyanobacteria. *Nature* 455, 1101–1104.

Sanderson, M. J. 2008. Phylogenetic signal in the eukaryotic tree of life. *Science* 321, 121–123.

Simpson, M. G. 2010. *Plant Systematics,* 2nd ed. Academic Press, San Diego.

Tekle, Y. I., L. W. Parfrey, and **L. A. Katz.** 2009. Molecular data are transforming hypotheses on the origin and diversification of eukaryotes. *BioScience* 59, 471–481.

Valentini, A., F. Pompanon, and **P. Taberlet.** 2009. DNA barcoding for ecologists. *Trends in Ecology and Evolution* 24, 110–117.

Vellai, T., and **G. Vida.** 1999. The origin of eukaryotes: the difference between prokaryotic and eukaryotic cells. *Proceedings of the Royal Society London B* 266, 1571–1577.

Zimmer, C. 2009. On the origin of eukaryotes. *Science* 325, 666–668.

CHAPTER 13

Bloom, K., and **A. Joglekar.** 2010. Towards building a chromosome segregation machine. *Nature* 463, 446–456.

Bonfante, P., and **I.-A. Anca.** 2009. Plants, mycorrhizal fungi, and bacteria: a network of interactions. *Annual Review of Microbiology* 63, 363–383.

DasSarma, S. 2007. Extreme microbes. *American Scientist* 95, 224–231.

Dekas, A. E., R. S. Poretsky, and **V. J. Orphan.** 2009. Deep-sea archaea fix and share nitrogen in methane-consuming microbial consortia. *Science* 326, 422–426.

Emerson, D., L. Agulto, H. Liu, and **L. Liu.** 2008. Identifying and characterizing bacteria in an era of genomics and proteomics. *BioScience* 58, 925–936.

Fonseca, J. M., and **S. Ravishankar.** 2007. Safer salads. *American Scientist 95,* 494–501.

Graham, L. E., J. M. Graham, and **L. W. Wilcox.** 2009. *Algae,* 2nd ed. Benjamin Cummings, San Francisco, New York.

Harrison, J. J., R. J. Turner, L. L. R. Marques, and **H. Ceri.** 2005. Biofilms. *American Scientist* 93, 508–515.

Hull, R. 2002. *Matthews' Plant Virology,* 4th ed. Academic Press, San Diego.

Ingraham, J. L. 2010. *March of the Microbes: Sighting the Unseen.* Belknap Press of Harvard University Press, Cambridge, MA.

Khan, J. A., and **J. Dijkstra.** 2002. *Plant Viruses as Molecular Pathogens.* Food Products Press, New York.

Logue, J. B., H. Bürgmann, and **C. T. Robinson.** 2008. Progress in the ecological genetics and biodiversity of freshwater bacteria. *BioScience* 58, 103–113.

Lugtenberg, B., and **F. Kamilova.** 2009. Plant-growth-promoting rhizobacteria. *Annual Review of Microbiology* 63, 541–556.

Madigan, M. T., J. M. Martinko, D. A. Stahl, and **D. P. Clark.** 2012. *Brock Biology of Microorganisms,* 13th ed. Pearson/Benjamin Cummings, San Francisco.

Microbial Ecology. 2008. *Science* 320 (Special Section), 1027, 1031–1045.

Niehl, A., and **M. Heinlein.** 2011. Cellular pathways for viral transport through plasmodesmata. *Protoplasma* 248, 75–99.

Shapiro, L., H. H. McAdams, and **R. Losick.** 2009. Why and how bacteria localize proteins. *Science* 326, 1225–1228.

Stewart, W. N., and **G. W. Rothwell.** 1993. *Paleobotany and the Evolution of Plants,* 2nd ed. Cambridge University Press, New York.

Strauss, E. 2009. Phytoplasma research begins to bloom. *Science* 325, 388–390.

Taylor, T. N., E. L. Taylor, and **M. Krings.** 2009. *Paleobotany: The Biology and Evolution of Fossil Plants,* 2nd ed. Academic Press, Amsterdam, Boston.

Toro, E., and **L. Shiparo.** 2010. Bacterial chromosome organization and segregation. *Cold Spring Harbor Perspectives in Biology* 2010;2:a000349.

CHAPTER 14

Aanen, D. K., H. H. de Fine Licht, A. J. M. Debets, N. A. G. Kerstes, R. F. Hoekstra, and **J. J. Boomsma.** 2009. High symbiont relatedness stabilizes mutualistic cooperation in fungus-growing termites. *Science 326,* 1103–1106.

Alexopoulos, C. J., C. W. Mims, and **M. Blackwell.** 1996. *Introductory Mycology,* 4th ed. John Wiley & Sons, Inc., New York.

Allen, M. F., W. Swenson, J. I. Querejeta, L. M. Egerton-Warburton, and **K. K. Treseder.** 2003. Ecology of mycorrhizae: a conceptual framework for complex interactions among plants and fungi. *Annual Review of Phytopathology* 41, 271–303.

Benjamin, D. R. 1995. *Mushrooms: Poisons and Panaceas—A Handbook for Naturalists, Mycologists, and Physicians.* W. H. Freeman and Company, New York.

Blackwell, M., D. S. Hibbett, J. W. Taylor, and **J. W. Spatafora.** 2006. Research coordination networks: a phylogeny for kingdom Fungi (Deep Hypha). *Mycologia* 98, 829–837.

Bonfante, P., and **I.-A. Anca.** 2009. Plants, mycorrhizal fungi, and bacteria: a network of interactions. *Annual Review of Microbiology* 63, 363–383.

Davis, R. H. 2000. *Neurospora: Contributions of a Model Organism.* Oxford University Press, Oxford, New York.

Heckman, D. S., D. M. Geiser, B. R. Eidell, R. L. Stauffer, N. L. Kardos, and **S. B. Hedges.** 2001. Molecular evidence for the early colonization of land by fungi and plants. *Science* 293, 1129–1133.

Hibbett, D. S., M. Binder, J. F. Bischoff, M. Blackwell, P. F. Cannon, et al. 2007. A higher-level phylogenetic classification of the Fungi. *Mycological Research* 111, 509–547.

Hudler, G. W. 1998. *Magical Mushrooms, Mischievous Molds.* Princeton University Press, Princeton, NJ.

Hughes, J. K., A. Hodge, A. H. Fitter, and **O. K. Atkin.** 2008. Mycorrhizal respiration: implications for global scaling relationships. *Trends in Plant Science* 13, 583–588.

Jones, M. D. M., I. Forn, C. Gadelha, M. J. Egan, D. Bass, R. Massana, and **T. A. Richards.** 2011. Discovery of novel intermediate forms redefines the fungal tree of life. *Nature* 474, 200–203.

Kirk, P. M., P. F. Cannon, D. W. Minter, and **J. A. Stalpers** (eds.). 2008. *Ainsworth & Bisby's Dictionary of the Fungi,* 10th ed. CSIRO Publishing, Collingwood, Victoria, Australia.

Leake, J. R. 1994. The biology of myco-heterotrophic ("saprophytic") plants. *New Phytologist* 127, 171–216.

Peay, K. G., P. G. Kennedy, and **T. D. Bruns.** 2008. Fungal community ecology: a hybrid beast with a molecular master. *BioScience* 58, 799–810.

Reinhardt, D. 2007. Programming good relations—development of the arbuscular mycorrhizal symbiosis. *Current Opinion in Plant Biology* 10, 98–105.

Richmond, J. Q., A. E. Savage, K. R. Zamudio, and **E. B. Rosenblum.** 2009. Toward immunogenetic studies of amphibian chytridiomycosis: linking innate and acquired immunity. *BioScience* 59, 311–320.

Schardl, C. L., A. Leuchtmann, and **M. J. Spiering.** 2004. Symbioses of grasses with seedborne fungal endophytes. *Annual Review of Plant Biology* 55, 315–340.

Steenkamp, E. T., J. Wright, and **S. L. Baldauf.** 2006. The protistan origins of animals and fungi. *Molecular Biology and Evolution* 23, 93–106.

Stephenson, S. L. 2010. *The Kingdom Fungi: The Biology of Mushrooms, Molds, and Lichens.* Timber Press, Portland, OR.

Taylor, T. N., E. L. Taylor, and **M. Krings.** 2009. *Paleobotany: The Biology and Evolution of Fossil Plants,* 2nd ed. Academic Press, Amsterdam, Boston.

Webster, J., and **R. Weber.** 2007. *Introduction to Fungi,* 3rd ed. Cambridge University Press, Cambridge, New York.

Youngsteadt, E. 2008. All that makes fungus gardens grow. *Science* 320, 1006–1007.

CHAPTER 15

Adey, W. H., P. C. Kangas, and **W. Mulbry.** 2011. Algal turf scrubbing: cleaning surface waters with solar energy while producing a biofuel. *BioScience* 61, 434–441.

Bold, H. C., and **M. J. Wynne.** 1985. *Introduction to the Algae: Structure and Reproduction,* 2nd ed. Prentice-Hall, Englewood Cliffs, NJ.

Cardon, Z. G., D. W. Gray, and **L. A. Lewis.** 2008. The green algal underground: evolutionary secrets of desert cells. *BioScience* 58, 114–122.

Graham, L. E., J. M. Graham, and **L. W. Wilcox.** 2009. *Algae,* 2nd ed. Benjamin Cummings, San Francisco, New York.

Pienkos, P. T., L. Laurens, and **A. Aden.** 2011. Making biofuel from microalgae. *American Scientist* 99, 474–481.

Roberts, J. M., A. J. Wheeler, and **A. Freiwald.** 2006. Reefs of the deep: the biology and geology of cold-water coral ecosystems. *Science* 312, 543–547.

Saade, A., and **C. Bowler.** 2009. Molecular tools for discovering the secrets of diatoms. *BioScience* 59, 757–765.

Silver, M. W. 2006. Protecting ourselves from shellfish poisoning. *American Scientist* 94, 316–325.

Stanley, G. D., Jr. 2006. Photosymbiosis and the evolution of modern coral reefs. *Science* 312, 857–858.

Stewart, W. N., and **G. W. Rothwell.** 1993. *Paleobotany and the Evolution of Plants,* 2nd ed. Cambridge University Press, New York.

Strassmann, J. E., and **D. C. Queller.** 2007. Altruism among amoebas. *Natural History* 116 (7), 24–29.

Taylor, T. N., E. L. Taylor, and **M. Krings.** 2009. *Paleobotany: The Biology and Evolution of Fossil Plants,* 2nd ed. Academic Press, Amsterdam, Boston.

Vroom, P. S., K. N. Page, J. C. Kenyon, and **R. E. Brainard.** 2006. Algae-dominated reefs. *American Scientist* 94, 430–437.

Vroom, P. S., and **C. M. Smith.** 2001. The challenge of siphonous green algae. *American Scientist* 89, 524–531.

CHAPTER 16

Adams, D. G., and **P. S. Duggan.** 2008. Cyanobacteria-bryophyte symbioses. *Journal of Experimental Botany* 59, 1047–1058.

Berbee, M. L., and **J. W. Taylor.** 2007. Rhynie chert: a window into a lost world of complex plant-fungus interactions. *New Phytologist* 174, 475–479.

Davey, M. L., and **R. S. Currah.** 2006. Interactions between mosses (Bryophyta) and fungi. *Canadian Journal of Botany* 84, 1509–1519.

Goffinet, B., W. R. Buck, and **A. J. Shaw.** 2009. Morphology and classification of the Bryophyta. In B. Goffinet and A. J. Shaw (eds.), *Bryophyte Biology,* 2nd ed., pp. 55–138. Cambridge University Press, New York.

Kenrick, P., and **P. R. Crane.** 1997. The origin and early evolution of plants on land. *Nature* 389, 33–39.

Ligrone, R., A. Carafa, J. G. Duckett, K. S. Renzaglia, and **K. Ruel.** 2008. Immunocytochemical detection of lignin-related epitopes in cell walls in bryophytes and the charalean alga *Nitella. Plant Systematics and Evolution* 270, 257–272.

Ligrone, R., J. G. Duckett, and **K. S. Renzaglia.** 2000. Conducting tissues and phyletic relationships of bryophytes. *Philosophical Transactions of the Royal Society B* 355, 795–813.

Malcolm, W. M., and **N. Malcolm.** 2006. *Mosses and Other Bryophytes: An Illustrated Glossary.* Micro-Optics Press, Nelson, New Zealand.

Niklas, K. J. 1997. *The Evolutionary Biology of Plants.* The University of Chicago Press, Chicago, London.

Renzaglia, K. S., S. Schuette, R. J. Duff, R. Ligrone, A. J. Shaw, B. D. Mishler, and **J. G. Duckett.** 2007. Bryophyte phylogeny: advancing the molecular and morphological frontiers. *The Bryologist* 110, 179–213.

Rydin, H., and **J. K. Jeglum.** 2006. *The Biology of Peatlands.* Oxford University Press, Oxford, New York.

Schofield, W. B. 1985 (reprinted 2001). *Introduction to Bryology.* The Blackburn Press, Caldwell, NJ.

Tanurdzic, M., and **J. A. Banks.** 2004. Sex-determining mechanisms in land plants. *The Plant Cell* 16, S61–S71.

Taylor, T. N., H. Kerp, and **H. Hass.** 2005. Life history biology of early land plants: deciphering the gametophyte phase. *Proceedings of the National Academy of Sciences USA* 102, 5892–5897.

Taylor, T. N., E. L. Taylor, and **M. Krings.** 2009. *Paleobotany: The Biology and Evolution of Fossil Plants,* 2nd ed. Academic Press, Amsterdam, Boston.

Vanderpoorten, A., and **B. Goffinet** (eds.). 2009. *Introduction to Bryophytes.* Cambridge University Press, Cambridge, New York.

Whitaker, D. L., and **J. Edwards.** 2010. *Sphagnum* moss disperses spores with vortex rings. *Science* 329, 406.

CHAPTER 17

Banks, J. A. 1999. Gametophyte development in ferns. *Annual Review of Plant Physiology and Plant Molecular Biology* 50, 163–186.

Cantino, P. D., J. A. Doyle, S. W. Graham, W. S. Judd, R. G. Olmstead, D. E. Soltis, P. S. Soltis, and **M. J. Donoghue.** 2007. Towards a phylogenetic nomenclature of *Tracheophyta. Taxon* 56, 822–846.

Chiou, W.-L., and **D. R. Farrar.** 1997. Antheridiogen production and response in Polypodiaceae species. *American Journal of Botany* 84, 633–640.

Cleal, C. J., and **B. A. Thomas.** 2009. *An Introduction to Plant Fossils.* Cambridge University Press, Cambridge, New York.

Eriksson, T. 2004. Ferns reawakened. *Nature* 428, 480–481.

Galtier, J. 2010. The origins and early evolution of the megaphyllous leaf. *International Journal of Plant Sciences* 171, 641–661.

Gifford, E. M., and **A. S. Foster.** 1989. *Morphology and Evolution of Vascular Plants,* 3rd ed. W. H. Freeman and Company, New York.

Hamilton, R. G., and **R. M. Lloyd.** 1991. Antheridiogen in the wild: the development of fern gametophyte communities. *Functional Ecology* 5, 804–809.

Judd, W. S., C. S. Campbell, E. A. Kellogg, P. F. Stevens, and **M. J. Donoghue.** 2008. *Plant Systematics: A Phylogenetic Approach,* 3rd ed. Sinauer Associates, Inc., Publishers, Sunderland, MA.

Kenrick, P., and **P. R. Crane.** 1991. Water-conducting cells in early fossil land plants: implications for the early evolution of tracheophytes. *Botanical Gazette* 152, 335–356.

Kenrick, P., and **P. R. Crane.** 1997. The origin and early evolution of plants on land. *Nature* 389, 33–39.

Ranker, T. A., and **C. H. Haufler** (eds.). 2008. *Biology and Evolution of Ferns and Lycophytes.* Cambridge University Press, Cambridge, New York.

Stewart, W. N., and **G. W. Rothwell.** 1993. *Paleobotany and the Evolution of Plants,* 2nd ed. Cambridge University Press, New York.

Taylor, T. N., H. Kerp, and **H. Hass.** 2005. Life history biology of early land plants: deciphering the gametophyte phase. *Proceedings of the National Academy of Sciences USA* 102, 5892–5897.

Taylor, T. N., E. L. Taylor, and **M. Krings.** 2009. *Paleobotany: The Biology and Evolution of Fossil Plants,* 2nd ed. Academic Press, Amsterdam, Boston.

CHAPTER 18

Cairney, J., and **G. S. Pullman.** 2007. The cellular and molecular biology of conifer embryogenesis. *New Phytologist* 176, 511–536.

Doyle, J. A. 1998. Phylogeny of vascular plants. *Annual Review of Ecology and Systematics* 29, 567–599.

Fernando, D. D., M. D. Lazzaro, and **J. N. Owens.** 2005. Growth and development of conifer pollen tubes. *Sexual Plant Reproduction* 18, 149–162.

Friedman, W. E., and **J. S. Carmichael.** 1996. Double fertilization in Gnetales: implications for understanding reproductive diversification among seed plants. *International Journal of Plant Sciences* 157 (Suppl. 6), S77–S94.

Gifford, E. M., and **A. S. Foster.** 1989. *Morphology and Evolution of Vascular Plants,* 3rd ed. W. H. Freeman and Company, New York.

Graham, S. W., and **W. J. D. Iles.** 2009. Different gymnosperm outgroups have (mostly) congruent signal regarding the root of flowering plant phylogeny. *American Journal of Botany* 96, 216–227.

Judd, W. S., C. S. Campbell, E. A. Kellogg, P. F. Stevens, and **M. J. Donoghue.** 2008. *Plant Systematics: A Phylogenetic Approach,* 3rd ed. Sinauer Associates, Inc., Publishers, Sunderland, MA.

Lake, J. A., R. G. Skophammer, C. W. Herbold, and **J. A. Servin.** 2009. Genome beginnings: rooting the tree of life. *Philosophical Transactions of the Royal Society B* 364, 2177–2185.

Linkies, A., K. Graeber, C. Knight and G. Leubner-Metzger. 2010. The evolution of seeds. *New Phytologist* 186, 817–831.

Nagalingum, N. S., C. R. Marshall, T. B. Quental, H. S. Rai, D. P. Little, and **S. Mathews.** 2011. Recent synchronous radiation of a living fossil. *Science* 334, 796–799.

Norstog, K. J., and **T. J. Nicholls.** 1997. *The Biology of the Cycads.* Comstock Publishing, Ithaca, NY.

Rothwell, G. W., W. L. Crepet, and **R. A. Stockey.** 2009. Is the anthophyte hypothesis alive and well? New evidence from the reproductive structures of Bennettitales. *American Journal of Botany* 96, 296–322.

Sanderson, M. J. 2008. Phylogenetic signal in the eukaryotic tree of life. *Science* 321, 121–123.

Stewart, W. N., and **G. W. Rothwell.** 1993. *Paleobotany and the Evolution of Plants,* 2nd ed. Cambridge University Press, New York.

Taylor, T. N., E. L. Taylor, and **M. Krings.** 2009. *Paleobotany: The Biology and Evolution of Fossil Plants,* 2nd ed. Academic Press, Amsterdam, Boston.

Tekle, Y. I., L. W. Parfrey, and **L. A. Katz.** 2009. Molecular data are transforming hypotheses on the origin and diversification of eukaryotes. *BioScience* 59, 471–481.

CHAPTER 19

Berger, F. 2008. Double-fertilization, from myths to reality. *Sexual Plant Reproduction* 21, 3–5.

Berger, F. 2011. Imaging fertilization in flowering plants, not so abominable after all. *Journal of Experimental Botany* 62, 1651–1658.

Berger, F., Y. Hamamura, M. Ingouff, and **T. Higashiyama.** 2008. Double fertilization—caught in the act. *Trends in Plant Science* 13, 437–443.

Bidartondo, M. J. 2005. The evolutionary ecology of myco-heterotrophy. *New Phytologist* 167, 335–352.

Dresselhaus, T. 2006. Cell–cell communication during double fertilization. *Current Opinion in Plant Biology* 9, 41–47.

Friedman, W. E. 2006. Sex among the flowers. *Natural History* 115 (9), 48–53.

Friedman, W. E. 2007. Embryological evidence for developmental lability during early angiosperm evolution. *Nature* 441, 337–340.

Friedman, W. E., E. N. Madrid, and **J. H. Williams.** 2008. Origin of the fittest and survival of the fittest: relating female gametophyte development to endosperm genetics. *International Journal of Plant Sciences* 169, 79–92.

Friedman, W. E., and **J. H. Williams.** 2004. Developmental evolution of the sexual process in ancient flowering plant lineages. *The Plant Cell* 16, S129–S132.

Gifford, E. M., and **A. S. Foster.** 1989. *Morphology and Evolution of Vascular Plants,* 3rd ed. W. H. Freeman and Company, New York.

Higashiyama, T., and **Y. Hamamura.** 2008. Gametophytic pollen tube guidance. *Sexual Plant Reproduction* 21, 17–26.

Judd, W. S., C. S. Campbell, E. A. Kellogg, P. F. Stevens, and **M. J. Donoghue.** 2008. *Plant Systematics: A Phylogenetic Approach,* 3rd ed. Sinauer Associates, Inc., Publishers, Sunderland, MA.

Kessler, S. A., H. Shimosato-Asano, N. F. Keinath, S. E. Wuest, G. Ingram, R. Panstruga, and **U. Grossniklas.** 2010. Conserved molecular components for pollen tube reception and fungal invasion. *Science* 330, 968–971.

Matsunaga, S., and **S. Kawano.** 2001. Sex determination by sex chromosomes in dioecious plants. *Plant Biology* 3, 481–488.

McCue, A. D., M. Cresti, J. A. Feijó, and **R. K. Slotkin.** 2011. Cytoplasmic connection of sperm cells to the pollen vegetative cell nucleus: potential roles of the male germ unit revisited. *Journal of Experimental Botany* 62, 1621–1631.

Punwani, J. A., and **G. N. Drews.** 2008. Development and function of the synergid cell. *Sexual Plant Reproduction* 21, 7–15.

Williams, J. H., and **W. E. Friedman.** 2004. The four-celled female gametophyte of *Illicium* (Illiciaceae; Austrobaileyales): implications for understanding the origin and early evolution of monocots, eumagnoliids, and eudicots. *American Journal of Botany* 91, 332–351.

Wilsen, K. L., and **P. K. Hepler.** 2007. Sperm delivery in flowering plants: the control of pollen tube growth. *BioScience* 57, 835–844.

Yang, W.-C., D.-Q. Shi, and **Y.-H. Chen.** 2010. Female gametophyte development in flowering plants. *Annual Review of Plant Biology* 61, 89–108.

CHAPTER 20

Angiosperm Phylogeny Group. 2009. An update of the Angiosperm Phylogeny Group classification for the orders and families of flowering plants: APG III. *Botanical Journal of the Linnean Society* 161, 105–121.

Angiosperm Phylogeny Website. www.mobot.org/mobot/research/apweb.

Endress, P. K., and **J. A. Doyle.** 2009. Reconstructing the ancestral angiosperm flower and its initial specializations. *American Journal of Botany* 96, 22–66.

Engelman, R. 2011. Revisiting population growth: the impact of ecological limits. *Environment 360* October 13.

Friedman, W. E. 2009. The meaning of Darwin's "Abominable Mystery." *American Journal of Botany* 96, 5–21.

Friis, E. M., J. A. Doyle, P. K. Endress, and **Q. Leng.** 2003. *Archaefructus*—angiosperm precursor or specialized early angiosperm? *Trends in Plant Science* 8, 369–373.

Frohlich, M. W., and **M. W. Chase.** 2007. After a dozen years of progress the origin of angiosperms is still a great mystery. *Nature* 450, 1184–1189.

Judd, W. S., C. S. Campbell, E. A. Kellogg, P. F. Stevens, and **M. J. Donoghue.** 2008. *Plant Systematics: A Phylogenetic Approach,* 3rd ed. Sinauer Associates, Inc., Publishers, Sunderland, MA.

Leins, P., and **C. Erbar.** 2010. *Flower and Fruit: Morphology, Ontogeny, Phylogeny, Function, and Ecology,* 2nd ed. Schweizerbart Science Publishers, Stuttgart.

Smith, S. A., J. M. Beaulieu, and **M. J. Donoghue.** 2010. An uncorrelated relaxed-clock analysis suggests an earlier origin for flowering plants. *Proceedings of the National Academy of Sciences USA* 107, 5897–5902.

Soltis, D. E., C. D. Bell, S. Kim, and **P. S. Soltis.** 2008. Origin and early evolution of angiosperms. *Annals of the New York Academy of Sciences* 1133, 3–25.

Soltis, P. S., S. F. Brockington, M.-J. Yoo, A. Piedrahita, M. Latvis, M. J. Moore, A. S. Chanderball, and **D. E. Soltis.** 2009. Floral variation and floral genetics in basal angiosperms. *American Journal of Botany* 96, 110–128.

Specht, C. D., and **M. E. Bartlett.** 2009. Flower evolution: the origin and subsequent diversification of the angiosperm flower. *Annual Review of Ecology, Evolution, and Systematics* 40, 217–243.

Stuessy, T. F. 2009. *Plant Taxonomy: The Systematic Evaluation of Comparative Data,* 2nd ed. Columbia University Press, New York.

Taylor, T. N., E. L. Taylor, and **M. Krings.** 2009. *Paleobotany: The Biology and Evolution of Fossil Plants,* 2nd ed. Academic Press, Amsterdam, Boston.

CHAPTER 21

Abbo, S., S. Lev-Yadun, and **A. Gopher.** 2010. Agricultural origins: centers and noncenters—a Near Eastern reappraisal. *Critical Reviews in Plant Sciences* 29, 317–328.

Allaby, R. 2010. Integrating the processes in the evolutionary system of domestication. *Journal of Experimental Botany* 61, 935–944.

Balick, M. J., and **P. A. Cox.** 1996. *Plants, People, and Culture: The Science of Ethnobotany.* Scientific American Library, New York.

Balter, M. 2007. Seeking agriculture's ancient roots. *Science* 316, 1830–1835.

Bartels, D., and **R. Sunkar.** 2005. Drought and salt tolerance in plants. *Critical Reviews in Plant Sciences* 24, 23–58.

Blaustein, R. J. 2008. The green revolution arrives in Africa. *BioScience* 58, 8–14.

Buhner, S. H. 1996. *Sacred Plant Medicine: Explorations in the Practice of Indigenous Herbalism.* Roberts Rinehart Publishers, Boulder, CO.

Doebley, J. 2006. Unfallen grains: how ancient farmers turned weeds into crops. *Science* 312, 1318–1319.

Doebley, J. F., B. S. Gaut, and **B. D. Smith.** 2006. The molecular genetics of crop domestication. *Cell* 127, 1309–1321.

Fuller, D. Q., L. Qin, Y. Zheng, Z. Zhao, X. Chen, L. A. Hosoya, and **G.-P. Sun.** 2009. The domestication process and domestication rate in rice: spikelet bases from the Lower Yangtze. *Science* 323, 1607–1610.

Glover, J. D., J. P. Reganold, L. W. Bell, J. Borevitz, E. C. Brummer, et al. 2010. Increased food and ecosystem security via perennial grains. *Science* 328, 1638–1639.

Gross, B. L., and **K. M. Olsen.** 2010. Genetic perspectives on crop domestication. *Trends in Plant Science* 15, 529–537.

Li, C., A. Zhou, and **T. Sang.** 2006. Rice domestication by reducing shattering. *Science* 311, 1936–1939.

Newell-McGloughlin, M. 2008. Nutritionally improved agricultural crops. *Plant Physiology* 147, 939–953.

Schultes, R. E., and **S. von Reis.** 1995. *Ethnobotany: Evolution of a Discipline.* Dioscorides Press, Portland, OR.

Siebert, C. 2011. *National Geographic* 220 (July), 108–131.

Simpson, B. B., and **M. C. Ogorzaly.** 2001. *Economic Botany: Plants in Our World,* 3rd ed. McGraw-Hill, Boston.

Smith, B. D. 1995. *The Emergence of Agriculture.* Scientific American Library, New York.

Zeder, M. A., E. Emshwiller, B. D. Smith, and **D. G. Bradley.** 2006. Documenting domestication: the intersection of genetics and archaeology. *Trends in Genetics* 22, 139–155.

SECTION 5: CHAPTERS 22–26

CHAPTER 22

Angelovici, R., G. Galili, A. R. Fernie, and **A. Fait.** 2010. Seed desiccation: a bridge between maturation and germination. *Trends in Plant Science* 15, 211–218.

Bradford, K. J., and **H. Nonogaki** (eds.). 2007. *Seed Development, Dormancy, and Germination.* Annual Plant Reviews, vol. 27. Blackwell Publishing, Oxford.

De Smet, I., S. Lau, U. Mayer, and **G. Jürgens.** 2010. Embryogenesis—the humble beginnings of plant life. *The Plant Journal* 61, 959–970.

Esau, K. 1977. *Anatomy of Seed Plants,* 2nd ed. John Wiley & Sons, Inc. New York.

Evert, R. F. 2006. *Esau's Plant Anatomy. Meristems, Cells, and Tissues of the Plant Body: Their Structure, Function, and Development,* 3rd ed. John Wiley & Sons, Inc., Hoboken, NJ.

Finkelstein, R., W. Reeves, T. Ariizumi, and **C. Steber.** 2008. Molecular aspects of seed dormancy. *Annual Review of Plant Biology* 59, 387–415.

Gutierrez, L., O. Van Wuytswinkel, M. Castelain, and **C. Bellini.** 2007. Combined networks regulating seed maturation. *Trends in Plant Science* 12, 294–300.

Hamann, T. 2001. The role of auxin in apical-basal pattern formation during *Arabidopsis* embryogenesis. *Journal of Plant Growth Regulation* 20, 292–299.

Kawashima, T., and **R. B. Goldberg.** 2010. The suspensor: not just suspending the embryo. *Trends in Plant Science* 15, 23–30.

Raghavan, V. 1997. *Molecular Embryology of Flowering Plants.* Cambridge University Press, Cambridge, New York.

Taiz, L., and **E. Zeiger.** 2010. *Plant Physiology,* 5th ed. Sinauer Associates, Inc., Publishers, Sunderland, MA.

CHAPTERS 23–26

Aloni, R., K. Schwalm, M. Langhans, and **C. I. Ullrich.** 2003. Gradual shifts in sites of free-auxin production during leaf-primordium development and their role in vascular differentiation and leaf morphogenesis in *Arabidopsis. Planta* 216, 841–853.

Beck, C. B. 2010. *An Introduction to Plant Structure and Development: Plant Anatomy for the Twenty-first Century,* 2nd ed. Cambridge University Press, Cambridge, New York.

Beeckman, T. (ed.). 2010. *Root Development.* Annual Plant Reviews, vol. 37. Wiley-Blackwell, Chichester, UK.

Core, H. A., W. A. Côté, and **A. C. Day.** 1979. *Wood Structure and Identification,* 2nd ed. Syracuse University Press, Syracuse, NY.

Déjardin, A., F. Laurans, D. Arnaud, C. Breton, G. Pilate, and **J.-C. Leplé.** 2010. Wood formation in Angiosperms. *Comptes Rendus Biologies* 333, 325–334.

De Smet, I., S. Vanneste, D. Inzé, and **T. Beeckman.** 2006. Lateral root initiation or the birth of a new meristem. *Plant Molecular Biology* 60, 871–887.

Dickison, W. C. 2000. *Integrative Plant Anatomy.* Harcourt/Academic Press, San Diego.

Driouich, A., C. Durand, and **M. Vicré-Gibouin.** 2007. Formation and separation of root border cells. *Trends in Plant Science* 12, 14–19.

Du, J., and **A. Groover.** 2010. Transcriptional regulation of secondary growth and wood formation. *Journal of Integrative Plant Biology* 52, 17–27.

Efroni, I., Y. Eshed, and **E. Lifschitz.** 2010. Morphogenesis of simple and compound leaves: a critical review. *The Plant Cell* 22, 1019–1032.

Enstone, D. E., C. A. Peterson, and **F. Ma.** 2003. Root endodermis and exodermis: structure, function, and responses to the environment. *Journal of Plant Growth Regulation* 21, 335–351.

Esau, K. 1977. *Anatomy of Seed Plants,* 2nd ed. John Wiley & Sons, Inc., New York.

Evert, R. F. 2006. *Esau's Plant Anatomy. Meristems, Cells, and Tissues of the Plant Body: Their Structure, Function, and Development,* 3rd ed. John Wiley & Sons, Inc., Hoboken, NJ.

Fahn, A. 1990. *Plant Anatomy,* 4th ed. Pergamon Press, Oxford, New York.

Gartner, B. L. (ed.). 1995. *Plant Stems: Physiology and Functional Morphology.* Academic Press, San Diego.

Gregory, P. J. 2006. *Plant Roots: Growth, Activity, and Interactions with Soils.* Blackwell Publishing, Oxford.

Gunawardena, A. H. L. A. N., and **N. G. Dengler.** 2006. Alternative modes of leaf dissection in monocotyledons. *Botanical Journal of the Linnean Society* 150, 25–44.

Hoadley, R. B. 2000. *Understanding Wood: A Craftsman's Guide to Wood Technology,* 2nd ed. Taunton Press, Newtown, CT.

Jönsson, H., M. G. Heisler, B. E. Shapiro, E. M. Meyerowitz, and **E. Mjolsness.** 2006. An auxin-driven polarized transport model for phyllotaxis. *Proceedings of the National Academy of Sciences USA* 103, 1633–1638.

Kang, J., and **N. Dengler.** 2004. Vein pattern development in adult leaves of *Arabidopsis thaliana. International Journal of Plant Sciences* 165, 231–242.

Knoblauch, M., and **W. S. Peters.** 2004. Forisomes, a novel type of Ca^{2+}-dependent contractile protein motor. *Cell Motility and the Cytoskeleton* 58, 137–142.

Kolek, J., and **V. Kozinka** (eds.). 1992. *Physiology of the Plant Root System.* Kluwer Academic Publishers, Dordrecht, Boston.

Kozlowski, T. T., and **S. G. Pallardy.** 1997. *Growth Control in Woody Plants.* Academic Press, San Diego.

Lake, J. V., P. J. Gregory, and **D. A. Rose** (eds.). 2009. *SEBS 30 Root Development and Function.* Cambridge University Press, Cambridge.

Lenhard, M., and **T. Laux.** 1999. Shoot meristem formation and maintenance. *Current Opinion in Plant Biology* 2, 44–50.

Liu, C., W. Xi, L. Shen, C. Tan, and **H. Yu.** 2009. Regulation of floral patterning by flowering time genes. *Developmental Cell* 16, 711–722.

McCully, M. 1995. How do real roots work? *Plant Physiology* 109, 1–6.

McKown, A. D., and **N. G. Dengler.** 2010. Vein patterning and evolution in C_4 plants. *Botany* 88, 775–786.

Melzer, R., Y.-Q. Wang, and **G. Theissen.** 2010. The naked and the dead: the ABCs of gymnosperm reproduction and the origin of the angiosperm flower. *Seminars in Cell & Developmental Biology* 21, 118–128.

Metcalfe, C. R., and **L. Chalk.** 1979. *Anatomy of the Dicotyledons,* vol. 1, *Systematic Anatomy of Leaf and Stem, with a Brief History of the Subject,* 2nd ed. Clarendon Press, Oxford.

Metcalfe, C. R., and **L. Chalk.** 1983. *Anatomy of the Dicotyledons,* vol. 2, *Wood Structure and Conclusion of the General Introduction,* 2nd ed. Clarendon Press, Oxford.

Mokany, K., R. J. Raison, and **A. S. Prokushkin.** 2006. Critical analysis of root:shoot ratios in terrestrial biomes. *Global Change Biology* 12, 84–96.

Moyroud, E., E. Kusters, M. Monniaux, R. Koes, and **F. Parcy.** 2010. *LEAFY* blossoms. *Trends in Plant Science* 15, 346–352.

Munné-Bosch, S. 2008. Do perennials really senesce? *Trends in Plant Science* 13, 216–220.

Panshin, A. J., and **C. de Zeeuw.** 1980. *Textbook of Wood Technology: Structure, Identification, Properties, and Uses of the Commercial Woods of the United States and Canada,* 4th ed. McGraw-Hill, New York.

Peterson, R. L. 1992. Adaptations of root structure in relation to biotic and abiotic factors. *Canadian Journal of Botany* 70, 661–675.

Raven, J. A., and **D. Edwards.** 2001. Roots: evolutionary origins and biogeochemical significance. *Journal of Experimental Botany* 52 (Suppl. 1), 381–401.

Rijpkema, A. S., M. Vandenbussche, R. Koes, K. Heijmans, and **T. Gerats.** 2010. Variations on a theme: changes in the floral ABCs in angiosperms. *Seminars in Cell & Developmental Biology* 21, 100–107.

Risopatron, J. P. M., Y. Sun, and **B. J. Jones.** 2010. The vascular cambium: molecular control of cellular structure. *Protoplasma* 247, 145–161.

Rolland-Lagan, A.-G. 2008. Vein patterning in growing leaves: axes and polarities. *Current Opinion in Genetics & Development* 18, 348–353.

Srivastava, L. M. 2002. *Plant Growth and Development: Hormones and Environment.* Academic Press, Amsterdam, Boston.

Steeves, T. A., and **I. M. Sussex.** 1989. *Patterns in Plant Development,* 2nd ed. Cambridge University Press, Cambridge, New York.

Taiz, L., and **E. Zeiger.** 2010. *Plant Physiology,* 5th ed. Sinauer Associates, Inc., Publishers, Sunderland, MA.

Vernoux, T., F. Besnard, and **J. Traas.** 2010. Auxin at the shoot apical meristem. *Cold Spring Harbor Perspectives in Biology* 2010;2:a001487.

Waisel, Y., A. Eshel, and **U. Kafkafi** (eds.). 2002. *Plant Roots: The Hidden Half,* 3rd ed. Marcel Dekker, New York.

SECTION 6: CHAPTERS 27–30

CHAPTER 27

Acharya, B. R., and **S. M. Assmann.** 2009. Hormone interactions in stomatal function. *Plant Molecular Biology* 69, 451–462.

Acosta, I. F., and **E. E. Farmer.** 2010. Jasmonates. *The Arabidopsis Book* 10.1199/tab.0129.

Bajguz, A., and **S. Hayat.** 2009. Effects of brassinosteroids on the plant responses to environmental stress. *Plant Physiology and Biochemistry* 47, 1–8.

Benjamins, R., and **B. Scheres.** 2008. Auxin: the looping star in plant development. *Annual Review of Plant Biology* 59, 443–465.

Browse, J. 2009. Jasmonate passes muster: a receptor and targets for the defense hormone. *Annual Review of Plant Biology* 60, 183–205.

Cutler, S. R., P. L. Rodriguez, R. R. Finkelstein, and **S. R. Abrams.** 2010. Abscisic acid: emergence of a core signaling network. *Annual Review of Plant Biology* 61, 651–679.

Donner, T. J., and **E. Scarpella.** 2009. Auxin-transport-dependent leaf vein formation. *Botany* 87, 678–684.

Dugardeyn, J., F. Vandenbussche, and **D. Van Der Straeten.** 2008. To grow or not to grow: what can we learn on ethylene–gibberellin cross-talk by *in silico* gene expression analysis? *Journal of Experimental Botany* 59, 1–16.

Fuerst, E. P., and **M. A. Norman.** 1991. Interactions of herbicides with photosynthetic electron transport. *Weed Science* 39, 458–464.

Fukaki, H., and **M. Tasaka.** 2009. Hormone interactions during lateral root formation. *Plant Molecular Biology* 69, 437–449.

Hartung, W. 2010. The evolution of abscisic acid (ABA) and ABA function in lower plants, fungi and lichen. *Functional Plant Biology* 37, 806–812.

Hartweck, L. M. 2008. Gibberellin signaling. *Planta* 229, 1–13.

Kamiya, Y. 2010. Plant hormones: versatile regulators of plant growth and development. *Annual Review of Plant Biology* 61, Special Online Compilation.

Kim, T.-W., and **Z.-Y. Wang.** 2010. Brassinosteroid signal transduction from receptor kinases to transcription factors. *Annual Review of Plant Biology* 61, 681–704.

Matsubayashi, Y., and **Y. Sakagami.** 2006. Peptide hormones in plants. *Annual Review of Plant Biology* 57, 649–674.

Melotto, M., W. Underwood, and **S. Y. He.** 2008. Role of stomata in plant innate immunity and foliar bacterial diseases. *Annual Review of Phytopathology* 46, 101–122.

Ming, R., J. Wang, P. H. Moore, and **A. H. Paterson.** 2007. Sex chromosomes in flowering plants. *American Journal of Botany* 94, 141–150.

Moubayidin, L., R. Di Mambro, and **S. Sabatini.** 2009. Cytokinin–auxin crosstalk. *Trends in Plant Science* 14, 557–562.

Péret, B., B. De Rybel, I. Casimiro, E. Benková, R. Swarup, L. Laplaze, T. Beeckman, and **M. J. Bennett.** 2009. *Arabidopsis* lateral root development: an emerging story. *Trends in Plant Science* 14, 399–408.

Perilli, S., L. Moubayidin, and **S. Sabatini.** 2010. The molecular basis of cytokinin function. *Current Opinion in Plant Biology* 13, 21–26.

Powles, S. B., and **Q. Yu.** 2010. Evolution in action: plants resistant to herbicides. *Annual Review of Plant Biology* 61, 317–347.

Pruneda-Paz, J. L., and **S. A. Kay.** 2010. An expanding universe of circadian networks in higher plants. *Trends in Plant Science* 15, 259–265.

Ross, J. J., and **J. B. Reid.** 2010. Evolution of growth-promoting plant hormones. *Functional Plant Biology* 37, 795–805.

Santos, F., W. Teale, C. Fleck, M. Volpers, B. Ruperti, and **K. Palme.** 2010. Modelling polar auxin transport in developmental patterning. *Plant Biology* 12 (Suppl. 1), 3–14.

Scarpella, E., D. Marcos, J. Friml, and **T. Berleth.** 2006. Control of leaf vascular patterning by polar auxin transport. *Genes & Development* 20, 1015–1027.

Spartz, A. K., and **W. M. Gray.** 2008. Plant hormone receptors: new perceptions. *Genes & Development* 22, 2139–2148.

Stepanova, A. N., and **J. M. Alonso.** 2009. Ethylene signaling and response: where different regulatory modules meet. *Current Opinion in Plant Biology* 12, 548–555.

Taiz, L., and **E. Zeiger.** 2010. *Plant Physiology,* 5th ed. Sinauer Associates, Inc., Publishers, Sunderland, MA.

Vernoux, T., F. Besnard, and **J. Traas.** 2010. Auxin at the shoot apical meristem. *Cold Spring Harbor Perspectives in Biology* 2010;2:a001487.

Vlot, A. C., D. A. Dempsey, and **D. F. Klessig.** 2009. Salicylic acid, a multifaceted hormone to combat disease. *Annual Review of Phytopathology* 47, 177–206.

Vyskot, B., and **R. Hobza.** 2004. Gender in plants: sex chromosomes are emerging from the fog. *Trends in Genetics* 20, 432–438.

Yoo, S.-D., Y. Cho, and **J. Sheen.** 2009. Emerging connections in the ethylene signaling network. *Trends in Plant Science* 14, 270–279.

CHAPTER 28

Amasino, R. 2010. Seasonal and developmental timing of flowering. *The Plant Journal* 61, 1001–1013.

Amasino, R. M., and **S. D. Michaels.** 2010. The timing of flowering. *Plant Physiology* 154, 516–520.

Angelovici, R., G. Galili, A. R. Fernie, and **A. Fait.** 2010. Seed desiccation: a bridge between maturation and germination. *Trends in Plant Science* 15, 211–218.

Bae, G., and **G. Choi.** 2008. Decoding of light signals by plant phytochromes and their interacting proteins. *Annual Review of Plant Biology* 59, 281–311.

Bentsink, L., and **M. Koornneef.** 2008. Seed dormancy and germination. *The Arabidopsis Book* 10.1199/tab.0119.

Bisgrove, S. R. 2008. The roles of microtubules in tropisms. *Plant Science* 175, 747–755.

Bradford, K. B., and **H. Nonogaki** (eds.). 2007. *Seed Development, Dormancy, and Germination.* Annual Plant Reviews, vol. 27. Blackwell Publishing, Oxford.

Briggs, W. R. 2010. A wandering pathway in plant biology: from wildflowers to phototropins to bacterial virulence. *Annual Review of Plant Biology* 61, 1–20.

Christie, J. M. 2007. Phototropin blue-light receptors. *Annual Review of Plant Biology* 58, 21–45.

Covington, M. F., and **S. L. Harmer.** 2007. The circadian clock regulates auxin signaling and responses in *Arabidopsis. PLoS Biology* 5, e222.

Franklin, K. A. 2008. Shade avoidance. *New Phytologist* 179, 930–944.

Franklin, K. A. 2009. Light and temperature signal crosstalk in plant development. *Current Opinion in Plant Biology* 12, 63–68.

Franklin, K. A., and **P. H. Quail.** 2010. Phytochrome functions in *Arabidopsis* development. *Journal of Experimental Botany* 61, 11–24.

Gilroy, S., and **P. H. Masson** (eds.). 2008. *Plant Tropisms.* Blackwell Publishing, Ames, IA.

Harmer, S. L. 2009. The circadian system in higher plants. *Annual Review of Plant Biology* 60, 357–377.

Hotta, C. T., M. J. Gardner, K. E. Hubbard, S. J. Baek, N. Dalchau, D. Suhita, A. N. Dodd, and **A. A. R. Webb.** 2007. Modulation of environmental responses of plants by circadian clocks. *Plant, Cell and Environment* 30, 333–349.

Imaizumi, T. 2010. *Arabidopsis* circadian clock and photoperiodism: time to think about location. *Current Opinion in Plant Biology* 13, 83–89.

Jones, M. A. 2009. Entrainment of the *Arabidopsis* circadian clock. *Journal of Plant Biology* 52, 202–209.

Matía, I., F. González-Camacho, R. Herranz, J. Z. Kiss, G. Gasset, J. J. W. A. van Loond, R. Marcoe, and **F. J. Medina.** 2010. Plant cell proliferation and growth are altered by microgravity conditions in spaceflight. *Journal of Plant Physiology* 167, 184–193.

McClung, C. R. 2006. Plant circadian rhythms. *The Plant Cell* 18, 792–803.

Miyazawa, Y., Y. Ito, T. Moriwaki, A. Kobayashi, N. Fujii, and **H. Takahashi.** 2009. A molecular mechanism unique to hydrotropism in roots. *Plant Science* 177, 297–301.

Molas, M. L., and **J. Z. Kiss.** 2008. PKS1 plays a role in red-light-based positive phototropism in roots. *Plant, Cell and Environment* 31, 842–849.

Morita, M. T. 2010. Directional gravity sensing in gravitropism. *Annual Review of Plant Biology* 61, 705–720.

Ponce, G., F. Rasgado, and **G. I. Cassab.** 2008. How amyloplasts, water deficit and root tropisms interact? *Plant Signaling & Behavior* 3, 460–462.

Resco, V., J. Hartwell, and **A. Hall.** 2009. Ecological implications of plants' ability to tell the time. *Ecology Letters* 12, 583–592.

Song, Y. H., S. Ito, and **T. Imaizumi.** 2010. Similarities in the circadian clock and photoperiodism in plants. *Current Opinion in Plant Biology* 13, 594–603.

Swarup, R., E. M. Kramer, P. Perry, K. Knox, H. M. O. Leyser, J. Haseloff, G. T. S. Beemster, R. Bhalerao, and **M. J. Bennett.** 2005. Root gravitropism requires lateral root cap and epidermal cells for transport and response to a mobile auxin signal. *Nature Cell Biology* 7, 1057–1065.

Taiz, L., and **E. Zeiger.** 2010. *Plant Physiology,* 5th ed. Sinauer Associates, Inc., Publishers, Sunderland, MA.

Takahashi, H., Y. Miyazawa, and **N. Fujii.** 2009. Hormonal interactions during root tropic growth: hydrotropism versus gravitropism. *Plant Molecular Biology* 69, 489–502.

Turck, F., F. Fornara, and **G. Coupland.** 2008. Regulation and identity of florigen: FLOWERING LOCUS T moves center stage. *Annual Review of Plant Biology* 59, 573–594.

Valladares, F., and **Ü. Niinemets.** 2008. Shade tolerance, a key plant feature of complex nature and consequences. *Annual Review of Ecology, Evolution, and Systematics* 39, 237–257.

Vitha, S., M. Yang, F. D. Sack, and **J. Z. Kiss.** 2007. Gravitropism in the *starch excess* mutant of *Arabidopsis thaliana. American Journal of Botany* 94, 590–598.

Volkov, A. G., H. Carrell, A. Baldwin, and **V. S. Markin.** 2009. Electrical memory in Venus flytrap. *Bioelectrochemistry* 75, 142–147.

Volkov, A. G., J. C. Foster, K. D. Baker, and **V. S. Markin.** 2010. Mechanical and electrical anisotropy in *Mimosa pudica* pulvini. *Plant Signaling & Behavior* 5, 1211–1221.

Yeang, H.-Y. 2009. Circadian and solar clocks interact in seasonal flowering. *BioEssays* 31, 1211–1218.

CHAPTER 29

Amtmann, A., and **P. Armengaud.** 2009. Effects of N, P, K and S on metabolism: new knowledge gained from multi-level analysis. *Current Opinion in Plant Biology* 12, 275–283.

Baxter, I. 2009. Ionomics: studying the social network of mineral nutrients. *Current Opinion in Plant Biology* 12, 381–386.

Bonfante, P., and **I.-A. Anca.** 2009. Plants, mycorrhizal fungi, and bacteria: a network of interactions. *Annual Review of Microbiology* 63, 363–383.

Bowen, G. J. 2011. A faster water cycle. *Science* 332, 430–431.

Canfield, D. E., A. N. Glazer, and **P. G. Falkowski.** 2010. The evolution and future of Earth's nitrogen cycle. *Science* 330, 192–196.

Elser, J., and **E. Bennett.** 2011. Phosphorus cycle: a broken biogeochemical cycle. *Nature* 478, 29–31.

Epstein, E., and **A. J. Bloom.** 2005. *Mineral Nutrition of Plants: Principles and Perspectives,* 2nd ed. Sinauer Associates, Inc., Publishers, Sunderland, MA.

Ferguson, B. J., A. Indrasumunar, S. Hayashi, M.-H. Lin, Y.-H. Lin, D. E. Reid, and **P. M. Gresshoff.** 2010. Molecular analysis of legume nodule development and autoregulation. *Journal of Integrative Plant Biology* 52, 61–76.

Flowers, T. J., H. K. Galal, and **L. Bromham.** 2010. Evolution of halophytes: multiple origins of salt tolerance in land plants. *Functional Plant Biology* 37, 604–612.

Gibson, T. C., and **D. M. Waller.** 2009. Evolving Darwin's "most wonderful" plant: ecological steps to a snap-trap. *New Phytologist* 183, 575–587.

Hänsch, R., and **R. R. Mendel.** 2009. Physiological functions of mineral micronutrients (Cu, Zn, Mn, Fe, Ni, Mo, B, Cl). *Current Opinion in Plant Biology* 12, 259–266.

Kraiser, T., D. E. Gras, A. G. Gutiérrez, B. González, and **R. A. Gutiérrez.** 2011. A holistic view of nitrogen acquisition in plants. *Journal of Experimental Botany* 62, 1455–1466.

Liu, T.-Y., C.-Y. Chang, and **T.-J. Chiou.** 2009. The long-distance signaling of mineral macronutrients. *Current Opinion in Plant Biology* 12, 312–319.

Lugtenberg, B., and **F. Kamilova.** 2009. Plant-growth-promoting rhizobacteria. *Annual Review of Microbiology* 63, 541–556.

Morrissey, J., and **M. L. Guerinot.** 2009. Iron uptake and transport in plants: the good, the bad, and the ionome. *Chemical Reviews* 109, 4553–4567.

Mudgal, V., N. Madaan, and **A. Mudgal.** 2010. Biochemical mechanisms of salt tolerance in plants: a review. *International Journal of Botany* 6, 136–143.

Ohkama-Ohtsu, N., and **J. Wasaki.** 2010. Recent progress in plant nutrition research: cross-talk between nutrients, plant physiology and soil microorganisms. *Plant & Cell Physiology* 51, 1255–1264.

Oldroyd, G. E. D., and **J. A. Downie.** 2008. Coordinating nodule morphogenesis with rhizobial infection in legumes. *Annual Review of Plant Biology* 59, 519–546.

Pilon-Smits, E. A. H., C. F. Quinn, W. Tapken, M. Malagoli, and **M. Schiavon.** 2009. Physiological functions of beneficial elements. *Current Opinion in Plant Biology* 12, 267–274.

Rascio, N., and **F. Navari-Izzo.** 2011. Heavy metal hyperaccumulating plants: how and why do they do it? And what makes them so interesting? *Plant Science* 180, 169–181.

Rosenblueth, M., and **E. Martínez-Romero.** 2006. Bacterial endophytes and their interactions with hosts. *Molecular Plant–Microbe Interactions* 19, 827–837.

Ruan, C.-J., J. A. Teixeira da Silva, S. Mopper, P. Qin, and **S. Lutts.** 2010. Halophyte improvement for a salinized world. *Critical Reviews in Plant Sciences* 29, 329–359.

Rubio, V., R. Bustos, M. L. Irigoyen, X. Cardona-López, M. Rojas-Triana, and **J. Paz-Ares.** 2009. Plant hormones and nutrient signaling. *Plant Molecular Biology* 69, 361–373.

Ryan, P. R., and **E. Delhaize.** 2010. The convergent evolution of aluminium resistance in plants exploits a convenient currency. *Functional Plant Biology* 37, 275–284.

Wesley, L. D. 2010. *Fundamentals of Soil Mechanics for Sedimentary and Residual Soils.* John Wiley & Sons, Inc., Hoboken, NJ.

CHAPTER 30

Atkins, C. A., P. M. C. Smith, and **C. Rodriguez-Medina.** 2011. Macromolecules in phloem exudates—a review. *Protoplasma* 248, 165–172.

Bleby, T. M., A. J. McElrone, and **R. B. Jackson.** 2010. Water uptake and hydraulic redistribution across large woody root systems to 20 m depth. *Plant, Cell and Environment* 33, 2132–2148.

Davidson, A., F. Keller, and **R. Turgeon.** 2011. Phloem loading, plant growth form, and climate. *Protoplasma* 248, 153–163.

Delzon, S., C. Douthe, A. Sala, and **H. Cochard.** 2010. Mechanism of water-stress induced cavitation in conifers: bordered pit structure and function support the hypothesis of seal capillary-seeding. *Plant, Cell and Environment* 33, 2101–2111.

Dinant, S., and **R. Lemoine.** 2010. The phloem pathway: new issues and old debates. *Comptes Rendus Biologies* 333, 307–319.

Domec, J.-C., J. S. King, A. Noormets, E. Treasure, M. J. Gavazzi, G. Sun, and **S. G. McNulty.** 2010. Hydraulic redistribution of soil water by roots affects whole-stand evapotranspiration and net ecosystem carbon exchange. *New Phytologist* 187, 171–183.

Domec, J.-C., B. Lachenbruch, F. C. Meinzer, D. R. Woodruff, J. M. Warren, and **K. A. McCulloh.** 2008. Maximum height in a conifer is associated with conflicting requirements for xylem design. *Proceedings of the National Academy of Sciences USA* 105, 12069–12074.

Evert, R. F. 2006. *Esau's Plant Anatomy. Meristems, Cells, and Tissues of the Plant Body: Their Structure, Function, and Development,* 3rd ed. John Wiley & Sons, Inc., Hoboken, NJ.

Harada, A., and **K.-i. Shimazaki.** 2009. Measurement of changes in cytosolic Ca^{2+} in *Arabidopsis* guard cells and mesophyll cells in response to blue light. *Plant and Cell Physiology* 50, 360–373.

Le Hir, R., J. Beneteau, C. Bellini, F. Vilaine, and **S. Dinant.** 2008. Gene expression profiling: keys for investigating phloem functions. *Trends in Plant Science* 13, 273–280.

Liesche, J., H. J. Martens, and **A. Schulz.** 2011. Symplastic transport and phloem loading in gymnosperm leaves. *Protoplasma* 248, 181–190.

Park, J., Y.-Y. Kim, E. Martinoia, and **Y. Lee.** 2008. Long-distance transporters of inorganic nutrients in plants. *Journal of Plant Biology* 51, 240–247.

Renninger, H. J., N. Phillips, and **D. R. Hodel.** 2009. Comparative hydraulic and anatomic properties in palm trees *(Washingtonia robusta)* of varying heights: implications for hydraulic limitation to increased height growth. *Trees* 23, 911–921.

Rewald, B., J. E. Ephrath, and **S. Rachmilevitch.** 2011. A root is a root is a root? Water uptake rates of *Citrus* root orders. *Plant, Cell and Environment* 34, 33–42.

Steudle, E., and **C. A. Peterson.** 1998. How does water get through roots? *Journal of Experimental Botany* 49, 775–788.

Taiz, L., and **E. Zeiger.** 2010. *Plant Physiology,* 5th ed. Sinauer Associates, Inc., Publishers, Sunderland, MA.

Turgeon, R. 2010. The role of phloem loading reconsidered. *Plant Physiology* 152, 1817–1823.

Turgeon, R., and **R. Medville.** 2011. *Amborella trichopoda,* plasmodesmata, and the evolution of phloem loading. *Protoplasma* 248, 173–180.

Wang, H., Y. Inukai, and **A. Yamauchi.** 2006. Root development and nutrient uptake. *Critical Reviews in Plant Sciences* 25, 279–301.

Westhoff, M., H. Schneider, D. Zimmermann, S. Mimietz, A. Stinzing, et al. 2008. The mechanisms of refilling of xylem conduits and bleeding of tall birch during spring. *Plant Biology* 10, 604–623.

A

Å: *See* ångstrom.

a- [Gk. *a-,* not, without]: Prefix that negates the succeeding part of the word; "an-" before vowels and "h."

abscisic acid [L. *abscissus,* to cut off]: A plant hormone that brings about dormancy in buds, maintains dormancy in seeds, and brings about stomatal closing, among other effects.

abscission (ăb•sizh´ŭn): The dropping off of leaves, flowers, fruits, or other plant parts, usually following the formation of an abscission zone.

abscission zone: The area at the base of a leaf, flower, fruit, or other plant part containing tissues that play a role in the separation of the plant part from the plant body.

absorption spectrum: The spectrum of light waves absorbed by a particular pigment.

accessory bud: A bud generally located above or on either side of the main axillary bud.

accessory cell: *See* subsidiary cell.

accessory fruit: A fruit, or assemblage of fruits, with fleshy parts derived largely or entirely from tissues other than the ovary. An example is the strawberry, with a fleshy receptacle and fruits (achenes) embedded in its surface.

accessory pigment: A pigment that captures light energy and transfers it to chlorophyll *a.*

acclimation: The process by which numerous physical and physiological processes prepare a plant for winter.

achene: A simple, dry, one-seeded, indehiscent fruit in which the seed coat is not adherent to the pericarp.

acid: A substance that dissociates in water, releasing hydrogen ions (H^+) and thus causing a relative increase in the concentration of these ions; a substance with a pH, in solution, of less than 7; a proton donor; the opposite of base.

acid growth hypothesis: The hypothesis that acidification of the cell wall leads to hydrolysis of restraining bonds within the wall and, consequently, to cell elongation driven by the turgor pressure on the wall.

acropetal development (or **differentiation**)**:** Development (or differentiation) that proceeds toward the apex of an organ; the opposite of basipetal development (or differentiation).

actin filament: A helical protein filament, 5 to 7 nanometers thick, composed of globular actin molecules; a major constituent of the cytoskeleton of all eukaryotic cells. Also called a microfilament.

actinomorphic [Gk. *aktis,* ray of light, + *morphē,* form]: Pertaining to a type of flower that can be divided into two equal halves in more than one longitudinal plane. Also called radially symmetrical *or* regular. *See also* zygomorphic.

action spectrum: The spectrum of light waves that elicits a particular reaction.

active site: The region of an enzyme surface that binds the substrate during the reaction catalyzed by the enzyme.

active transport: Energy-requiring transport of a solute across a membrane in the direction of increasing concentration (against the concentration gradient).

ad- [L. *ad-,* toward, to]: Prefix meaning "toward" or "to."

adaptation [L. *adaptare,* to fit]: A peculiarity of structure, physiology, or behavior that aids in fitting an organism to its environment.

adaptive radiation: The evolution from one kind of organism to several divergent forms, each specialized to fit a distinct and diverse way of life.

adenine (ăd´e•nēn): A purine base present in DNA, RNA, and nucleotide derivatives, such as ADP and ATP.

adenosine triphosphate (ATP): A nucleotide consisting of adenine, ribose sugar, and three phosphate groups; the major source of usable chemical energy in metabolism. In hydrolysis, ATP loses one phosphate to become adenosine diphosphate (ADP), releasing usable energy.

adhesion [L. *adhaerere,* to stick to]: The sticking together of unlike objects or materials.

adnate [L. *adnatus,* grown together]: Describing fused unlike parts, such as stamens and petals. *See also* connate.

ADP: *See* adenosine triphosphate.

adsorption [L. *ad-,* to, + *sorbere,* to suck in]: The adhesion of a liquid, gaseous, or dissolved substance to a solid, resulting in a higher concentration of the substance.

adventitious [L. *adventicius,* not properly belonging to]: Referring to a structure arising from an unusual place, such as buds growing at places other than leaf axils, or roots growing from stems or leaves.

aeciospore (ē´sĭ•o•spor) [Gk. *aikia,* injury, + *spora,* seed]: A binucleate spore of rust fungi; produced in an aecium.

aecium, *pl.* **aecia:** In rust fungi, a cuplike structure in which aeciospores are produced.

aerenchyma: Parenchyma tissue containing particularly large intercellular (air) spaces.

aerobic [Gk. *aer,* air, + *bios,* life]: Requiring free oxygen.

aerobic respiration: *See* respiration.

after-ripening: The metabolic changes that must occur in some dormant seeds before germination can take place.

agar: A gelatinous substance derived from certain red algae; used as a solidifying agent in the preparation of nutrient media for the growth of microorganisms.

aggregate fruit: A fruit developing from the several separate carpels of a single flower.

akinete: A vegetative cell that is transformed into a thick-walled resistant spore in cyanobacteria.

albuminous cells: Certain ray and axial parenchyma cells in gymnosperm phloem that are spatially and functionally associated with the sieve cells. Also called Strasburger cells.

aleurone [Gk. *aleuron,* flour]: A proteinaceous material, usually in the form of small granules, occurring in the outermost cell layer of the endosperm of wheat and other grains.

alga, *pl.* **algae** (ăl´ga, ăl´jē): Traditional term for a series of unrelated groups of photosynthetic eukaryotic organisms lacking multicellular sex organs (except for the charophytes); the misnamed "blue-green algae" are cyanobacteria, one of the groups of photosynthetic bacteria.

algin: An important polysaccharide component of brown algal cell walls; used as a stabilizer and emulsifier for some foods and for paint.

alkali [Arabic *algili,* the ashes of the plant saltwort]: A substance with marked basic properties. Also called a base.

alkaline: Pertaining to substances that release hydroxyl ions (OH^-) in water; having a pH greater than 7.

alkaloids: Bitter-tasting nitrogenous compounds that are basic (alkaline) in their chemical properties; include morphine, cocaine, caffeine, nicotine, and atropine.

allele (ă•lēl´) [Gk. *allēlōn,* of one another, + *morphē,* form]: One of the two or more alternative forms of a gene.

allelopathy [Gk. *allēlōn*, of one another, + *pathos*, suffering]: The inhibition of one species of plant by chemicals produced by another species of plant.

allopatric speciation [Gk. *allos*, other, + *patra*, fatherland, country]: Speciation that occurs as the result of the geographic separation of a population of organisms.

allopolyploid: A polyploid formed from the union of two separate chromosome sets and their subsequent doubling.

allosteric interaction [Gk. *allos*, other, + *steros*, shape]: A change in the shape of a protein resulting from binding to the protein of a nonsubstrate molecule; in its new shape, the protein typically has different properties.

alternate phyllotaxy: Leaf arrangement in which there is one bud or one leaf at a node.

alternation of generations: A reproductive cycle in which a haploid (n) phase, the gametophyte, produces gametes, which fuse in pairs to form a zygote, which then germinates to produce a diploid ($2n$) phase, the sporophyte. Spores produced by meiotic division in the sporophyte give rise to new gametophytes, completing the cycle.

amino acids [Gk. *Ammon*, referring to the Egyptian sun god, near whose temple ammonium salts were first prepared from camel dung]: Nitrogen-containing organic acids, the units, or "building blocks," from which protein molecules are built.

ammonification: Decomposition of amino acids and other nitrogen-containing organic compounds, resulting in the production of ammonia (NH_3) and ammonium ions (NH_4^+).

amoeboid [Gk. *amoibē*, change]: Moving or eating by means of pseudopodia (temporary cytoplasmic protrusions from the cell body).

amphi- [Gk. *amphi-*, on both sides]: Prefix meaning "on both sides," "both," or "of both kinds."

amylase (ăm'ĭ·lās): An enzyme that breaks down starch into smaller units.

amyloplast: A leucoplast (colorless plastid) that forms starch grains.

an- [Gk. *an-*, not, without]: Prefix equivalent to "a-," meaning "not" or "without"; used before vowels and "h."

anabolism [Gk. *ana-*, up, + *-bolism* (as in metabolism)]: The constructive part of metabolism; the total chemical reactions involved in biosynthesis.

anaerobic [Gk. *an-*, without, + *aer*, air, + *bios*, life]: Referring to any process that can occur without oxygen, or to the metabolism of an organism that can live without oxygen; strict anaerobes cannot survive in the presence of oxygen.

analogous [Gk. *analogos*, proportionate]: Applied to structures similar in function but different in evolutionary origin, such as the phyllodes of an Australian *Acacia* and the leaves of an oak.

anaphase [Gk. *ana*, away, + *phasis*, form]: A stage in mitosis in which the chromatids of each chromosome separate and move to opposite poles; also, similar stages in meiosis in which chromatids or paired chromosomes move apart.

anatomy: Study of the internal structure of organisms; morphology is the study of their external structure.

andro- [Gk. *andros*, man]: Prefix meaning "male."

androecium [Gk. *andros*, man, + *oikos*, house]: (1) The floral whorl that comprises the stamens; (2) in leafy liverworts, a packetlike swelling containing the antheridia.

aneuploid: A chromosomal aberration in which the chromosome number differs slightly from the normal chromosome number for the species.

angiosperm [Gk. *angion*, vessel, + *sperma*, seed]: Literally, a seed borne in a vessel (carpel); thus, one of a group of plants whose seeds are borne within a mature ovary (fruit).

ångstrom [after A. J. Ångstrom, a Swedish physicist, 1814–1874]: A unit of length equal to 10^{-10} meter; abbreviated Å.

anion [Gk. *anienai*, to go up]: A negatively charged ion.

anisogamy [Gk. *aniso*, unequal, + *gamos*, marriage]: The condition of having dissimilar motile gametes.

annual [L. *annulus*, year]: A plant whose life cycle is completed in a single growing season.

annual ring: In wood, the growth layer formed during a single year. *See also* growth layer.

annulus [L. *anus*, ring]: In ferns, a row of specialized cells in a sporangium; in gill fungi, the remnant of the inner veil forming a ring on the stalk.

antenna complex: The portion of a photosystem that consists of pigment molecules (antenna pigments) that gather light and "funnel" it to the reaction center.

anterior: Situated before or toward the front.

anther [Gk. *anthos*, flower]: The pollen-bearing portion of a stamen.

antheridiophore [Gk. *anthos*, flower, + *phoros*, bearing]: In some liverworts, a stalk that bears antheridia.

antheridium, *pl.* **antheridia:** A sperm-producing structure that may be multicellular or unicellular.

anthocyanin [Gk. *anthos*, flower, + *kyanos*, dark blue]: A water-soluble blue or red pigment found in the cell sap.

Anthophyta: The phylum of angiosperms, or flowering plants.

anthophyte clade: Hypothetical clade consisting of the gnetophytes, Bennettitales, and angiosperms. Molecular studies do not support the existence of an anthophyte clade.

antibiotic [Gk. *anti*, against or opposite, + *biotikos*, pertaining to life]: Natural organic substance that retards or prevents the growth of organisms; generally used to designate substances formed by microorganisms that prevent the growth of other microorganisms.

anticlinal: Perpendicular to the surface.

anticodon: In a tRNA molecule, the three-nucleotide sequence that base-pairs with the mRNA codon for the amino acid carried by that particular tRNA; the anticodon is complementary to the mRNA codon.

antipodals: Three (sometimes more) cells of the mature embryo sac, located at the end opposite the micropyle.

apical dominance: The influence exerted by a terminal bud in suppressing the growth of lateral, or axillary, buds.

apical meristem: The meristem at the tip of the root or shoot in a vascular plant.

apomixis [Gk. *apo*, separate, away from, + *mixis*, mingling]: Reproduction without meiosis or fertilization; vegetative reproduction.

apoplast [Gk. *apo*, away from, + *plastos*, molded]: The cell wall continuum of a plant or organ; the movement of substances via the cell walls is called apoplastic movement or apoplastic transport.

apothecium [Gk. *apothēkē*, storehouse]: A cup-shaped or saucer-shaped open ascoma.

aquaporins: Integral membrane proteins that form water channels across the membrane, facilitating the movement of water and/or small neutral solutes or gases across the membrane.

arbuscular mycorrhizas: Mycorrhizas in which the fungal hyphae penetrate the cortical cells of the plant root, where they form branched structures, the arbuscules. Also called endomycorrhizas.

arch-, archeo- [Gk. *archē*, *archos*, beginning]: Prefix meaning "first," "main," or "earliest."

Archaea: A phylogenetic domain of prokaryotes consisting of the methanogens, most extreme halophiles and hyperthermophiles, and *Thermoplasma*.

archaeon, *pl.* **archaea:** A prokaryotic organism. *See also* Archaea.

archegoniophore [Gk. *archegonos*, the first of a race, + *phoros*, bearing]: In some liverworts, a stalk that bears archegonia.

archegonium, *pl.* **archegonia:** A multicellular structure in which a single egg is produced; found in the bryophytes and some vascular plants.

aril (ăr'ĭl) [L. *arillus*, grape, seed]: An accessory seed covering, often formed by an outgrowth at the base of the ovule; often brightly colored, which may aid in dispersal by attracting animals that eat it and, in the process, carry the seed away from the parent plant.

artifact [L. *ars,* art, + *facere,* to make]: A product that exists because of an extraneous, especially human, agency and does not occur in nature.

artificial selection: The breeding of selected organisms to produce strains with desired characteristics.

ascogenous hyphae [Gk. *askos,* bladder, + *genous,* producing]: Hyphae containing paired haploid male and female nuclei; they develop from an ascogonium and eventually give rise to asci.

ascogonium: The oogonium or female gametangium of the ascomycetes.

ascoma, *pl.* **ascomata:** A multicellular structure in ascomycetes that is lined with specialized cells called asci, in which nuclear fusion and meiosis occur; ascomata may be open or closed. Also called an ascocarp.

ascospore: A spore produced within an ascus; found in ascomycetes.

ascus, *pl.* **asci:** A specialized cell, characteristic of the ascomycetes, in which two haploid nuclei fuse to produce a zygote that immediately divides by meiosis; at maturity, an ascus contains ascospores.

aseptate [Gk. *a-,* not, + L. *septum,* fence]: Nonseptate; lacking cross walls.

asexual reproduction: Any reproductive process, such as fission or budding, that does not involve the union of gametes.

assimilate stream: The flow of assimilates, or food materials, in the phloem; moves from source to sink.

atom [Gk. *atomos,* indivisible]: The smallest unit into which a chemical element can be divided and still retain its characteristic properties.

atomic nucleus: The central core of an atom, containing protons and neutrons, around which electrons orbit.

atomic number: The number of protons in the nucleus of an atom.

atomic weight: The weight of a representative atom of an element relative to the weight of an atom of carbon ^{12}C, which has been assigned the value 12.

ATP: *See* adenosine triphosphate.

ATP synthase: An enzyme complex that forms ATP from ADP and phosphate during oxidative phosphorylation in the inner mitochondrial membrane.

auto- [Gk. *autos,* self, same]: Prefix meaning "same" or "self-same."

autoecious [Gk. *autos,* self, + *oikia,* dwelling]: In some rust fungi, completing the life cycle on a single species of host plant.

autopolyploid: A polyploid formed from the doubling of a single genome.

autoradiograph: A photographic print made by a radioactive substance acting on a sensitive photographic film.

autotroph [Gk. *autos,* self, + *trophos,* feeder]: An organism that is able to synthesize the nutritive substances it requires from inorganic substances in its environment. *See also* heterotroph.

auxins [Gk. *auxein,* to increase]: A class of plant hormones that control cell elongation, among other effects.

axial system: In secondary xylem and secondary phloem, the term applied collectively to cells derived from fusiform cambial initials. The long axes of these cells are oriented parallel with the main axis of the root or stem. Also called longitudinal system *and* vertical system.

axil [Gk. *axilla,* armpit]: The upper angle between a twig or leaf and the stem from which it grows.

axillary: Describing buds or branches arising in the axil of a leaf.

B

bacillus, *pl.* **bacilli** (ba•sĭl′ŭs) [L. *baculum,* rod]: A rod-shaped bacterium.

backcross: The crossing of a hybrid with one of its parents or with a genetically equivalent organism; a cross between an individual whose genes are to be tested and one that is homozygous for all of the recessive genes involved in the experiment.

Bacteria: The phylogenetic domain consisting of all prokaryotes that are not members of the domain Archaea.

bacteriophage [Gk. *bakterion,* little rod, + *phagein,* to eat]: A virus that parasitizes bacterial cells.

bacterium, *pl.* **bacteria:** A prokaryotic organism. *See also* Bacteria.

bacteroid: An enlarged, deformed *Rhizobium* or *Bradyrhizobium* cell found in root nodules; capable of nitrogen fixation.

bark: A nontechnical term applied to all tissues outside the vascular cambium in a woody stem. *See also* inner bark *and* outer bark.

basal body: A self-reproducing, cylinder-shaped cytoplasmic organelle from which cilia or flagella arise; identical in structure to the centriole, which is involved in mitosis and meiosis in most animals and protists.

base: A substance that dissociates in water, causing a decrease in the concentration of hydrogen ions (H^+), often by releasing hydroxyl ions (OH^-); having a pH, in solution, of more than 7; the opposite of acid. Also called an alkali.

basidioma, *pl.* **basidiomata:** A multicellular structure, characteristic of the basidiomycetes, within which basidia are formed.

basidiospore: A spore of the Basidiomycota, produced within and borne on a basidium following nuclear fusion and meiosis. Also called a basidiocarp.

basidium, *pl.* **basidia:** A specialized reproductive cell of the Basidiomycota, often club-shaped, in which nuclear fusion and meiosis occur.

basipetal development (or **differentiation**)**:** Development (or differentiation) that proceeds toward the base (away from the apex) of an organ; the opposite of acropetal development (or differentiation).

beneficial elements: Elements essential for only limited groups of plants or for plants grown under specific environmental conditions.

berry: A simple fleshy fruit that includes a fleshy ovary wall and one or more carpels and seeds; examples are the fruits of grapes, tomatoes, and bananas.

bi- [L. *bis,* double, two]: Prefix meaning "two," "twice," or "having two points."

biennial: A plant that normally requires two growing seasons to complete its life cycle, flowering and fruiting in its second year.

bilaterally symmetrical: *See* zygomorphic.

biofilm: An assemblage of bacterial cells attached to a surface and enclosed in a matrix of polysaccharides excreted by the bacteria.

biological clock [Gk. *bios,* life, + *logos,* discourse]: The internal timing mechanism that governs the innate biological rhythms of organisms.

biomass: Total dry weight of all organisms in a particular population, sample, or area.

biome: A complex of terrestrial communities of very wide extent, characterized by its climate and soil; the largest ecological unit.

biosphere: The zone of air, land, and water at the surface of the Earth that is occupied by organisms.

biotechnology: The practical application of advances in hormone research and DNA biochemistry to manipulate the genetics of organisms.

biotic: Relating to life.

bisexual flower: A flower that has at least one functional stamen and one functional carpel.

bivalent [L. *bis,* double, + *valere,* to be strong]: A pair of synapsed homologous chromosomes. Also called a tetrad.

blade: The broad, expanded part of a leaf; the lamina.

body cell: Vegetative or somatic cell. *See also* spermatogenous cell.

border cells: Rootcap cells programmed to separate from the rootcap and from each other; on their release, they may remain alive in the rhizosphere for several weeks and undergo changes in gene expression that enable them to produce and exude specific proteins completely different from those of the rootcap.

bordered pit: A pit in which the secondary wall arches over the pit membrane.

bract: A modified, usually reduced, leaflike structure.

branch root: *See* lateral root.

brassinosteroids: A group of growth-promoting steroid hormones in plants that play essential roles in a wide range of developmental processes, such as cell division and cell elongation in roots and stems, vascular differentiation, responses to light (photomorphogenesis), flower and fruit development, resistance to stresses, and senescence.

bryophytes (brī′o•fīts): Members of the phyla of nonvascular plants; the mosses, hornworts, and liverworts.

bud: (1) An embryonic shoot, often protected by modified scale-leaves; (2) a vegetative outgrowth of yeasts and some bacteria as a means of asexual reproduction.

bulb: A short underground stem covered by enlarged and fleshy leaf bases containing stored food.

bulk flow: The overall movement of water or some other liquid induced by gravity, pressure, or an interplay of both.

bulliform cells: Large epidermal cells present in longitudinal rows in grass leaves; believed to be involved in the mechanism of rolling and unrolling or folding and unfolding of the leaves. Also called motor cells.

bundle scar: Scar or mark left on a leaf scar by vascular bundles broken at the time of leaf fall, or abscission.

bundle sheath: Layer or layers of cells surrounding a vascular bundle; may consist of parenchyma or sclerenchyma cells, or both.

bundle-sheath extension: A plate of ground tissue extending from a bundle sheath of a vein in the leaf mesophyll to the upper or lower epidermis, or both; may consist of parenchyma, collenchyma, or sclerenchyma.

C

C_3 pathway: *See* Calvin cycle.

C_3 plants: Plants that employ only the Calvin cycle, or C_3 pathway, in the fixation of CO_2; the first stable product is the three-carbon compound 3-phosphoglycerate.

C_4 pathway: The set of reactions through which CO_2 is fixed to the compound phosphoenolpyruvate (PEP) to yield oxaloacetate, a four-carbon compound.

C_4 plants: Plants in which the first product of CO_2 fixation is a four-carbon compound (oxaloacetate); both the Calvin cycle (C_3 pathway) and the C_4 pathway are employed by C_4 plants.

callose: A complex polysaccharide, β1,3-glucan, synthesized in the plasma membrane and deposited between the plasma membrane and the cell wall; a common wall constituent in the sieve areas of sieve elements; also develops rapidly in reaction to injury in sieve elements and parenchyma cells.

callus [L. *callos,* hard skin]: Undifferentiated tissue; a term used in tissue culture, grafting, and wound healing.

calorie [L. *calor,* heat]: The amount of energy in the form of heat required to raise the temperature of one gram of water by 1°C. In making metabolic measurements, the kilocalorie (kcal), or Calorie—the amount of heat required to raise the temperature of one kilogram of water by 1°C—is generally used.

Calvin cycle: The series of enzymatically mediated photosynthetic reactions during which CO_2 is reduced to glyceraldehyde 3-phosphate (3-phosphoglyceraldehyde) and the CO_2 acceptor, ribulose 1,5-bisphosphate, is regenerated. For every three molecules of CO_2 entering the cycle, a net gain of one molecule of glyceraldehyde 3-phosphate results.

calyptra [Gk. *kalyptra,* covering for the head]: The hood or cap that partly or entirely covers the capsule of some species of mosses; formed from the expanded archegonial wall.

calyx (kā′lŭks) [Gk. *kalyx,* husk, cup]: The sepals collectively; the outermost flower whorl.

CAM: *See* crassulacean acid metabolism.

cambial zone: A region of thin-walled, undifferentiated meristematic cells between the secondary xylem and secondary phloem; consists of cambial initials and their recent derivatives.

cambium [L. *cambiare,* to exchange]: A meristem that gives rise to parallel rows of cells; a term commonly applied to the vascular cambium and the cork cambium, or phellogen.

capsid: The protein coat of a virus particle.

capsule: (1) In angiosperms, a dehiscent, dry fruit that develops from two or more carpels; (2) a slimy layer around the cells of certain bacteria; (3) the sporangium of bryophytes.

carbohydrate [L. *carbo,* ember, + *hydro,* water]: An organic compound consisting of a chain of carbon atoms to which hydrogen and oxygen are attached in a 2:1 ratio (CH_2O); examples are sugars, starch, glycogen, and cellulose.

carbon cycle: Worldwide circulation and utilization of carbon atoms.

carbon fixation: The conversion of CO_2 into organic compounds during photosynthesis.

carbon-fixation reactions: In photosynthetic cells, the light-independent enzymatic reactions concerned with the synthesis of glucose from CO_2, ATP, and NADPH. Also called light-independent reactions *and* dark reactions.

carnivorous: Feeding upon animals, as opposed to feeding upon plants (herbivorous); also refers to plants that are able to utilize proteins obtained from trapped animals, chiefly insects.

carotene (kăr′o•tēn) [L. *carota,* carrot]: A yellow or orange pigment belonging to the carotenoid group.

carotenoids (kărŏt′e•noids): A class of fat-soluble pigments that includes the carotenes (yellow and orange pigments) and the xanthophylls (yellow pigments); found in chloroplasts and chromoplasts of plants. Carotenoids act as accessory pigments in photosynthesis.

carpel [Gk. *karpos,* fruit]: One of the members of the gynoecium, or inner floral whorl; each carpel encloses one or more ovules. One or more carpels form a gynoecium.

carpellate: Pertaining to a flower with one or more carpels but no functional stamens. Also called pistillate.

carpogonium [Gk. *karpos,* fruit, + *gonos,* offspring]: In red algae, the female gametangium.

carposporangium [Gk. *karpos,* fruit, + *spora,* seed, + *angeion,* vessel]: In red algae, a carpospore-containing cell.

carpospore: In red algae, the single diploid protoplast found within a carposporangium.

carriers: Transport proteins that bind specific solutes and undergo conformational change in order to transport the solute across the membrane.

caryopsis [Gk. *karyon,* nut, + *opsis,* appearance]: Simple, dry, one-seeded indehiscent fruit with the pericarp firmly united all around the seed coat; a grain characteristic of the grasses (family Poaceae).

Casparian strip [after Robert Caspary, German botanist]: A bandlike region of primary wall containing suberin and lignin; found in anticlinal—radial and transverse—walls of endodermal and exodermal cells.

catabolism [Gk. *katabolē,* throwing down]: Collectively, the chemical reactions resulting in the breakdown of complex materials and the release of energy.

catalyst [Gk. *katalysis,* dissolution]: A substance that accelerates the rate of a chemical reaction but is not used up in the reaction; enzymes are catalysts.

category [Gk. *katēgoria,* category]: In a hierarchical classification system, the level at which a particular group is ranked.

cation [Gk. *katienai,* to go down]: A positively charged ion.

cation exchange: The replacement of mineral cations weakly bound to the surface of soil particles by other cations.

catkin: A spikelike inflorescence of unisexual flowers; found only in woody plants.

cavitation: Rupture of the water columns in the xylem by air bubbles.

cDNA: *See* complementary DNA.

cell [L. *cella,* small room]: The structural unit of organisms; in plants, cells consist of the cell wall and the protoplast.

cell division: The division of a cell and its contents, usually into two roughly equal parts.

cell plate: The partition that forms at the equator of the spindle in the dividing cells of plants and a few green algae during early telophase.

cell sap: The fluid contents of the vacuole.

cellular respiration: *See* respiration.

cellulase: An enzyme that hydrolyzes cellulose.

cellulose: A carbohydrate that is the chief component of the cell wall in plants and some protists; an insoluble complex carbohydrate formed of microfibrils of glucose molecules attached end to end.

cell wall: The rigid outermost layer of the cells found in plants, some protists, and most prokaryotes.

central mother cells: Relatively large, vacuolated cells in a subsurface position in apical meristems of shoots.

centriole [Gk. *kentron,* center, + L. *-olus,* little one]: A cytoplasmic organelle found outside the nuclear envelope and identical in structure to a basal body; found in the cells of most eukaryotes other than fungi, red algae, and the nonflagellated cells of plants. Centrioles divide and organize spindle fibers during mitosis and meiosis.

centromere [Gk. *kentron,* center, + *meros,* part]: Region of constriction of a chromosome that holds sister chromatids together.

chalaza [Gk. *chalaza,* small tubercle]: The region of an ovule or seed where the funiculus unites with the integuments and the nucellus.

channel proteins: Transport proteins that form water-filled pores that extend across cellular membranes; when open, channel proteins allow specific solutes to pass through them.

chemical potential: The activity or free energy of a substance; it depends on the rate of motion of the average molecule and the concentration of the molecules.

chemical reaction: The making or breaking of chemical bonds between atoms or molecules.

chemiosmotic coupling: Coupling of ATP synthesis to electron transport via an electrochemical H^+ gradient across a membrane.

chemoautotrophic: In prokaryotes, having the ability to manufacture their own basic foods by using the energy released by specific inorganic reactions. *See also* autotroph.

chiasma (kī • ăz′ma) [Gk. *chiasma,* cross]: The X-shaped figure formed by the meeting of two nonsister chromatids of homologous chromosomes; the site of crossing-over.

chitin (kī′tĭn) [Gk. *chiton,* tunic]: A tough, resistant, nitrogen-containing polysaccharide forming the cell walls of certain fungi, the exoskeleton of arthropods, and the epidermal cuticle of other surface structures of certain protists and animals.

chlor- [Gk. *chloros,* green]: Prefix meaning "green."

chlorenchyma: Parenchyma cells that contain chloroplasts.

chlorophyll [Gk. *chloros,* green, + *phyllon,* leaf]: The green pigment of plant cells; the receptor of light energy in photosynthesis; also found in algae and photosynthetic bacteria.

chloroplast: A plastid that contains chlorophylls; the site of photosynthesis. Chloroplasts occur in plants and algae.

chlorosis: Loss or reduced development of chlorophyll.

chroma- [Gk. *chroma,* color]: Prefix meaning "color."

chromatid [Gk. *chroma,* color, + L. *-id,* daughters of]: One of the two daughter strands of a duplicated chromosome, which are joined at the centromere.

chromatin: The deeply staining complex of DNA and proteins that forms eukaryotic chromosomes.

chromatophore [Gk. *chroma,* color, + *phorus,* bearer]: In some bacteria, a discrete vesicle delimited by a single membrane and containing photosynthetic pigments.

chromophore: The light-absorbing portion of a phytochrome molecule.

chromoplast: A plastid containing pigments other than chlorophyll, usually yellow and orange carotenoid pigments.

chromosome [Gk. *chroma,* color, + *soma,* body]: The structure that carries the genes. Eukaryotic chromosomes are visualized as threads or rods of chromatin, appearing in contracted form during mitosis and meiosis, and otherwise are enclosed in a nucleus; each eukaryotic chromosome contains a linear DNA molecule. Prokaryotes typically have a single chromosome consisting of a circular DNA molecule.

chrysolaminarin: The storage product of the chrysophytes and diatoms.

cilium, *pl.* **cilia** (sĭl′ē • ŭm) [L. *cilium,* eyelash]: A short, hairlike flagellum, usually numerous and arranged in rows.

circadian rhythms [L. *circa,* about, + *dies,* day]: Regular rhythms of growth and activity that occur in an approximately 24-hour cycle.

circinate vernation [L. *circinare,* to make round, + *vernare,* to flourish]: In ferns, the coiled arrangement of leaves and leaflets in the bud; such an arrangement uncoils gradually as the leaf develops.

cisterna, *pl.* **cisternae** [L. *cistern,* reservoir]: A flattened or saclike portion of the endoplasmic reticulum or a Golgi body (dictyosome).

citric acid cycle: The series of reactions that results in the oxidation of pyruvate to hydrogen atoms, electrons, and carbon dioxide. The electrons, passed along electron-carrier molecules, then go through the oxidative phosphorylation and terminal oxidation processes. Also called the Krebs cycle *or* tricarboxylic acid cycle.

clade: A monophyletic group, made up of an ancestor and all of its descendants.

cladistics: A system of arranging organisms following an analysis of their primitive and advanced features so that their phylogenetic relationships are accurately reflected.

cladogram: A line diagram that branches repeatedly and suggests phylogenetic relationships among organisms.

cladophyll [Gk. *klados,* shoot, + *phyllon,* leaf]: A branch resembling a foliage leaf.

clamp connection: In the Basidiomycota, a lateral connection between adjacent cells of a dikaryotic hypha; ensures that each cell of the hypha will contain two dissimilar nuclei.

class: A taxonomic category between phylum and order; contains one or more orders and belongs to a particular phylum.

cleistothecium [Gk. *kleistos,* closed, + *thekion,* small receptacle]: A closed, spherical ascoma.

climacteric: The large increase in cellular respiration evidenced by an increased uptake of oxygen and production of carbon dioxide during the ripening of many fruits, such as tomatoes, bananas, and peaches.

climax community: The final stage in a successional series; its nature is determined largely by the climate and soil of the region.

cline: A graded series of changes in some characteristics within a species, often correlated with a gradual change in climate or another geographic factor.

clone [Gk. *klon,* twig]: A population of cells or individuals derived by asexual division from a single cell or individual; one of the members of such a population.

cloning: Producing a cell line or culture, all of whose members are characterized by a specific DNA sequence; a key element in genetic engineering.

closed vascular bundle: A vascular bundle in which a cambium does not develop.

coalescence [L. *coalescere,* to grow together]: The union of floral parts of the same whorl, such as petals to petals.

coccus, *pl.* **cocci** (kŏk'ŭs) [Gk. *kokkos,* berry]: A spherical bacterium.

codon (kō'dŏn): Sequence of three adjacent nucleotides in a molecule of DNA or mRNA that form the code for a single amino acid or for the termination of a polypeptide chain.

coenocytic (se • nō • sĭ'tic) [Gk. *koinos,* shared in common, + *kytos,* hollow vessel]: Describing an organism or part of an organism that is multinucleate, with the nuclei not separated by walls or membranes. Also called siphonaceous, siphonous, *or* syncytial.

coenzyme: An organic molecule, or nonprotein organic cofactor, that plays an accessory role in enzyme-catalyzed processes, often by acting as a donor or acceptor of electrons; NAD^+ and FAD are common coenzymes.

coevolution [L. *co-,* together, + *e-,* out, + *volvere,* to roll]: The simultaneous evolution of adaptations in two or more populations interacting so closely that each is a strong selective force on the other.

cofactor: One or more nonprotein components required by enzymes in order to function; many are metal ions; others are coenzymes.

cohesion [L. *cohaerēre,* to stick together]: The mutual attraction of molecules of the same substance.

cohesion-tension theory: A model for the ascent of water in vascular plants. According to this theory, water is pulled up through the plant body in the xylem. This pull, or tension, is brought about by transpiration and/or use of water by the leaves, which results in a gradient of water potential from the leaves to the soil solution at the surface of the roots.

cold hardiness: The ability of a plant to survive the extreme cold and drying effects of winter weather.

coleoptile (kō'lē • op'till) [Gk. *koleos,* sheath, + *ptilon,* feather]: The sheath enclosing the apical meristem and leaf primordia of the grass embryo; often interpreted as the first leaf.

coleorhiza (kō'lē • o•rī'za) [Gk. *koleos,* sheath, + *rhiza,* root]: The sheath enclosing the radicle in the grass embryo.

collenchyma [Gk. *kolla,* glue]: A supporting tissue composed of collenchyma cells; common in regions of primary growth in stems and in some leaves.

collenchyma cell: Elongated living cell with unevenly thickened, nonlignified primary cell wall.

colloid (kŏl'oid): A permanent suspension of fine particles.

columella: The central column of cells of a rootcap.

community: All the organisms inhabiting a common environment and interacting with one another.

companion cell: A specialized parenchyma cell associated with a sieve-tube element in angiosperm phloem and arising from the same mother cell as the sieve-tube element.

competition: Interaction between members of the same population or of two or more populations to obtain a resource that both or all require and that is available in limited supply.

complementary DNA (cDNA): A single-stranded molecule of DNA synthesized from an mRNA template by reverse transcription.

complete flower: A flower having four whorls of floral parts—sepals, petals, stamens, and carpels.

complex tissue: A tissue consisting of two or more cell types; epidermis, periderm, xylem, and phloem are complex tissues.

compound: A combination of atoms in a definite ratio, held together by chemical bonds.

compound leaf: A leaf with a blade divided into several distinct leaflets.

compression wood: The reaction wood of conifers; develops on the lower sides of leaning trunks or limbs.

concentration gradient: The concentration difference of a substance per unit distance.

conducting phloem: The part of the inner bark that contains living, functional sieve elements and is actively engaged in the transport of food substances.

cone: *See* strobilus.

conidiophore: A hypha on which one or more conidia are produced.

conidium, *pl.* **conidia** [Gk. *konis,* dust]: An asexual fungal spore not contained within a sporangium; may be produced singly or in chains; most conidia are multinucleate.

conifer: A cone-bearing tree.

conjugation: The temporary fusion of pairs of bacteria, protozoa, and certain algae and fungi during which genetic material is transferred between the two individuals.

conjugation tube: A tube formed during the process of conjugation to facilitate the transfer of genetic material.

connate (kōn'āt): Referring to similar parts that are united or fused, such as petals fused in a corolla tube. *See also* adnate.

consumer: In ecology, an organism that derives its food from another organism.

continuous variation: Variation in traits to which a number of different genes contribute; the variation often exhibits a "normal" or bell-shaped distribution.

contractile vacuole: A clear, fluid-filled vacuole in some groups of protists that takes up water within the cell and then contracts, expelling its contents from the cell.

convergent evolution [L. *convergere,* to turn together]: The independent development of similar structures in organisms that are not directly related; often found in organisms living in similar environments.

cork: A secondary tissue produced by a cork cambium; made up of polygonal cells, nonliving at maturity, with suberized cell walls that are resistant to the passage of gases and water vapor; the outer part of the periderm. Also called phellem.

cork cambium: The lateral meristem that forms the periderm, producing cork (phellem) toward the surface (outside) of the plant and phelloderm toward the inside; common in stems and roots of gymnosperms and woody angiosperms. Also called phellogen.

corm: A thickened underground stem, upright in position, in which food is accumulated, usually in the form of starch.

corolla [L. *corona,* crown]: The petals collectively; usually, the conspicuously colored flower whorl.

corolla tube: A tubelike structure resulting from the fusion of petals along their edges.

cortex: (1) Ground-tissue region of a stem or root, bounded externally by the epidermis and internally by the vascular system; a primary-tissue region; (2) the peripheral region of a cell protoplast.

cotransport: Membrane transport in which the transfer of one solute depends on the simultaneous or sequential transfer of a second solute.

cotyledon (kŏt'ĭ•lē'dŭn) [Gk. *kotyledon,* cup-shaped hollow]: Seed leaf; generally absorbs food in monocotyledons and stores food in other angiosperms.

coupled reactions: Reactions in which energy-requiring chemical reactions are linked to energy-releasing reactions.

covalent bond: A chemical bond formed between atoms as a result of the sharing of two electrons.

crassulacean acid metabolism (CAM): A variant of the C_4 pathway; the enzyme PEP (phosphoenolpyruvate) carboxylase fixes CO_2 into C_4 compounds at night, then during the daytime, the fixed CO_2 is transferred to the ribulose bisphosphate of the Calvin cycle in the same cell; characteristic of most succulent plants, such as cacti.

cristae, *sing.* **crista:** Infoldings of the inner mitochondrial membrane that form a series of crests or ridges, containing the electron transport chains involved in ATP formation.

crop rotation: The practice of growing different crops in regular succession to aid in the control of insect pests and diseases, to increase soil fertility, and to decrease erosion.

cross-fertilization: The fusion of gametes formed by different individuals; the opposite of self-fertilization.

crossing-over: The exchange of corresponding segments of genetic material between the chromatids of homologous chromosomes at meiosis.

cross-pollination: The transfer of pollen from the anther of one plant to the stigma of a flower of another plant.

cross section: *See* transverse section.

crozier: A hooked tip formed by the apical cell of an ascogenous hypha that allows paired nuclei to divide simultaneously, one in the hypha and the other in the hook.

cryptogam: An archaic term for all organisms except the flowering plants (phanerogams), animals, and heterotrophic protists.

cultivar: A variety of plant found only under cultivation.

cuticle: Waxy or fatty layer on the outer wall of epidermal cells, formed of cutin and wax.

cutin [L. *cutis,* skin]: Fatty substance deposited in many plant cell walls and on the outer surface of epidermal cell walls, where it forms a layer known as the cuticle.

cyclic electron flow: In chloroplasts, the light-induced flow of electrons originating from and returning to Photosystem I.

cyclosis (sī•klō′sis) [Gk. *kyklosis,* circulation]: The streaming of cytoplasm within a cell.

-cyte, cyto- [Gk. *kytos,* hollow vessel, container]: Suffix or prefix meaning "pertaining to the cell."

cytochrome [Gk. *kytos,* container, + *chroma,* color]: Heme proteins serving as electron carriers in respiration and photosynthesis.

cytokinesis [Gk. *kytos,* hollow vessel, + *kinesis,* motion]: Division of the cytoplasm of a cell following nuclear division.

cytokinin [Gk. *kytos,* hollow vessel, + *kinesis,* motion]: A class of plant hormones that promotes cell division, among other effects.

cytology: The study of cell structure and function.

cytoplasm: The living matter of a cell, exclusive of the nucleus; the protoplasm.

cytoplasmic ground substance: *See* cytosol.

cytoplasmic inheritance: The inheritance of characteristics under the control of genes located in plastids and mitochondria.

cytosine: One of the four pyrimidine bases found in the nucleic acids DNA and RNA.

cytoskeleton: The flexible network within cells, composed of microtubules and actin filaments, or microfilaments.

cytosol: The cytoplasmic matrix of the cytoplasm in which the nucleus, various organelles, and membrane systems are suspended.

D

day-neutral plants: Plants that flower without regard to daylength.

de- [L. *de-,* away from, down, off]: Prefix meaning "away from," "down," or "off"; for example, dehydration means "removal of water."

deciduous [L. *decidere,* to fall off]: Shedding leaves at a certain season.

decomposers: Organisms (bacteria, fungi, heterotrophic protists) in an ecosystem that break down organic material into smaller molecules that are then recirculated.

dehiscence [L. *de-,* down, + *hiscere,* split open]: The opening of an anther, fruit, or other structure, which permits the escape of reproductive bodies contained within.

dehydration synthesis: The synthesis of a compound or molecule involving the removal of water. Also called a condensation reaction.

denitrification: The conversion of nitrate to gaseous nitrogen; carried out by a few genera of free-living soil bacteria.

deoxyribonucleic acid (DNA): Carrier of genetic information in cells; composed of chains of phosphate, sugar molecules (deoxyribose), and purines and pyrimidines; capable of self-replication and determines RNA synthesis.

deoxyribose [L. *deoxy,* loss of oxygen, + *ribose,* kind of sugar]: A five-carbon sugar with one fewer atom of oxygen than ribose; a component of deoxyribonucleic acid.

dermal tissue system: The outer covering tissue of the plant; the epidermis or the periderm.

desmotubule [Gk. *desmos,* to bind, + L. *tubulus,* small tube]: The tubule traversing a plasmodesmatal canal and uniting the endoplasmic reticulum of one cell with that of the adjacent cell.

determinate growth: Growth of limited duration, characteristic of floral meristems and of leaves.

deuterium: Heavy hydrogen, ^{2}H; a hydrogen atom with a nucleus containing one proton and one neutron. (The nucleus of most hydrogen atoms consists of only a proton.)

dichotomy: The division or forking of an axis into two branches.

dicotyledons: Obsolete term used to refer to all angiosperms other than monocotyledons; characterized by having two cotyledons. *See also* eudicotyledons *and* magnoliids.

dictyosome: *See* Golgi body.

differentiation: A developmental process by which a relatively unspecialized cell undergoes a progressive change to a more specialized cell; the specialization of cells and tissues for particular functions during development.

diffuse-porous wood: A wood in which the pores, or vessels, are fairly uniformly distributed throughout the growth layers or in which the size of pores changes only slightly from early wood to late wood.

diffusion [L. *diffundere,* to pour out]: The net movement of suspended or dissolved particles from a more concentrated region to a less concentrated region as a result of the random movement of individual molecules; the process tends to distribute such particles uniformly throughout a medium.

digestion: The conversion of complex, usually insoluble foods into simple, usually soluble forms by means of enzymatic action.

dikaryon [Gk. *di,* two, + *karyon,* nut]: In fungi, a mycelium with paired nuclei, each usually derived from a different parent.

dikaryotic: In fungi, having pairs of nuclei within cells or compartments.

dimorphism [Gk. *di,* two, + *morphē,* form]: The condition of having two distinct forms, such as sterile and fertile leaves in ferns, or sterile and fertile shoots in horsetails.

dioecious [Gk. *di,* two, + *oikos,* house]: Unisexual; having male and female (or staminate and ovulate) elements on different individuals of the same species.

diploid: Having two sets of chromosomes; the $2n$ (diploid) chromosome number is characteristic of the sporophyte generation.

disaccharide [Gk. *di,* two, + *sakcharon,* sugar]: A carbohydrate formed of two simple sugar molecules linked by a covalent bond; sucrose is an example.

disk flowers: The actinomorphic, tubular flowers of the Asteraceae; in contrast to flattened, zygomorphic ray flowers. In many Asteraceae, the disk flowers occur in the center of the inflorescence, the ray flowers around the margins.

distal: Situated away from or far from the point of reference (usually the main part of body); opposite of proximal.

DNA: *See* deoxyribonucleic acid.

DNA sequencing: Determination of the order of nucleotides in a DNA molecule.

domain: The taxonomic category above the kingdom level; the three domains are Archaea, Bacteria, and Eukarya.

dominant allele: One allele is said to be dominant with respect to an alternative allele if the homozygote for the dominant allele is indistinguishable phenotypically from the heterozygote; the other allele is said to be recessive.

dormancy [L. *dormire,* to sleep]: A special condition of arrested growth in which the plant and such plant parts as buds and seeds do not begin to grow without special environmental cues. The requirement for such cues, which include cold exposure and a suitable photoperiod, prevents the breaking of dormancy during superficially favorable growing conditions.

double fertilization: Broadly, two fertilization events in a single female gametophyte by two sperm cells from a single pollen tube. In angiosperms, the fusion of egg and sperm (resulting in a 2*n* fertilized egg, the zygote) and the simultaneous fusion of a second male gamete with the polar nuclei (typically resulting in a 3*n* primary endosperm nucleus); a unique characteristic of all angiosperms. By definition, double fertilization also occurs in gnetophytes, but the second fertilization event does not result in endosperm formation and instead forms an extra embryo that ultimately aborts.

doubling rate: The length of time required for a population of a given size to double in number.

drupe [Gk. *dryppa,* overripe olive]: A simple, fleshy fruit derived from a single carpel, usually one-seeded, in which the inner fruit coat is hard and may adhere to the seed.

druse: A compound, more or less spherical crystal with many component crystals projecting from its surface; composed of calcium oxalate.

E

early wood: The first-formed wood of a growth increment; contains larger cells and is less dense than the subsequently formed late wood; replaces the term "spring wood."

eco- [Gk. *oikos,* house]: Prefix meaning "house" or "home."

ecology: The study of the interactions of organisms with their physical environment and with one other.

ecosystem: A major interacting system that involves both living organisms and their physical environment.

ecotype [Gk. *oikos,* house, + L. *typus,* image]: A locally adapted variant of an organism, differing genetically from other ecotypes.

ectomycorrhizas: Mycorrhizas in which the fungus does not penetrate living cells in the roots. The hyphae grow between the cells of the root epidermis and cortex, forming a highly branched network called the Hartig net. The hyphae also form a mantle, or sheath, that covers the root surface.

edaphic [Gk. *edaphos,* ground, soil]: Pertaining to the soil.

egg: A nonmotile female gamete, usually larger than the male gamete of the same species.

egg apparatus: The egg cell and synergids located at the micropylar end of the female gametophyte, or embryo sac, of angiosperms.

elater [Gk. *elater,* driver]: (1) An elongated, spindle-shaped, sterile cell in the sporangium of a liverwort sporophyte that aids in spore dispersal; (2) clubbed, hygroscopic band attached to the spore of horsetails.

electrochemical gradient: The driving force that causes an ion to move across a membrane, due to the difference in electric charge across the membrane in combination with the difference in the ion's concentration on the two sides of the membrane.

electrolyte: A substance that dissociates into ions in aqueous solution and so makes possible the conduction of an electric current through the solution.

electromagnetic spectrum: The entire spectrum of radiation, which ranges in wavelength from less than a nanometer to more than a kilometer.

electron: A subatomic particle with a negative electric charge equal in magnitude to the positive charge of the proton, but with a mass of 1/1837 that of the proton. Electrons orbit the atom's positively charged nucleus and determine the atom's chemical properties.

electron-dense: In electron microscopy, not permitting the passage of electrons and so appearing dark.

electron transport: The movement of electrons down a series of electron carrier molecules that hold electrons at slightly different energy levels; as electrons move down the chain, the energy released is used to form ATP from ADP and phosphate. Electron transport plays an essential role in the final stage of cellular respiration and in the light-dependent reactions of photosynthesis.

element: A substance composed of only one kind of atom; one of more than 100 distinct natural or synthetic types of matter that, singly or in combination, compose virtually all materials of the universe.

embolism: The filling of vessels and/or tracheids with air or water vapor.

embryo [Gk. *en,* in, + *bryein,* to swell]: In plants, a young sporophyte before the start of a period of rapid growth (germination in seed plants).

embryogenesis: Development of an embryo from a fertilized egg, or zygote. Also called embryogeny.

embryophytes: The bryophytes and vascular plants, both of which produce embryos; a synonym for plants.

embryo sac: The female gametophyte of angiosperms, generally an eight-nucleate, seven-celled structure; the seven cells are the egg cell, two synergids, and three antipodals (each with a single nucleus), and the central cell (with two nuclei).

endergonic: Describing a chemical reaction that requires energy to proceed; the opposite of exergonic.

endo- [Gk. *endo,* within]: Prefix meaning "within."

endocarp [Gk. *endo,* within, + *karpos,* fruit]: The innermost layer of the mature ovary wall, or pericarp.

endocytosis [Gk. *endon,* within, + *kytos,* hollow vessel]: The uptake of material into cells by means of invagination of the plasma membrane; if solid material is involved, the process is called phagocytosis; if dissolved material is involved, it is called pinocytosis.

endodermis [Gk. *endon,* within, + *derma,* skin]: A single layer of cells forming a sheath around the vascular region in roots and some stems; endodermal cells are characterized by a Casparian strip within radial and transverse walls. In roots and stems of seed plants, the endodermis is the innermost layer of the cortex.

endogenous [Gk. *endon,* within, + *genos,* race, kind]: Arising from deep-seated tissues, as in the case of lateral roots.

endomembrane system: Collectively, the cellular membranes that form a continuum (plasma membrane, tonoplast, endoplasmic reticulum, Golgi bodies, and nuclear envelope).

endomycorrhizas: *See* arbuscular mycorrhizas.

endoplasmic reticulum: A complex, three-dimensional membrane system of indefinite extent present in eukaryotic cells, dividing the cytoplasm into compartments and channels. Those portions that are densely coated with ribosomes are called rough endoplasmic reticulum, and other portions with fewer or no ribosomes are called smooth endoplasmic reticulum.

endosperm [Gk. *endon,* within, + *sperma,* seed]: A tissue, containing stored food, that develops from the union of a male nucleus and the polar nuclei of the central cell; it is digested by the growing sporophyte before or after maturation of the seed; found only in angiosperms.

endosymbiosis: A symbiotic relationship in which one or more organisms live within the cells or body of a host without doing harm.

energy: The capacity to do work.

energy of activation: The energy that must be possessed by atoms or molecules in order to react.

energy-transduction reactions: *See* light reactions.

entrainment: The process by which a periodic repetition of light and dark, or some other external cycle, causes a circadian rhythm to remain synchronized with the same cycle as the modifying, or entraining, factor.

entropy: A measure of the randomness or disorder of a system.

enzyme: A protein that is capable of speeding up specific chemical reactions by lowering the required activation energy but is itself unaltered in the process; a biological catalyst.

epi- [Gk. *epi,* upon]: Prefix meaning "upon" or "above."

epicotyl: The upper portion of the axis of an embryo or seedling, above the point of insertion of the cotyledons (seed leaves) and below the next leaf or leaves.

epidermis: The outermost layer of cells of the leaf and young stems and roots; primary in origin.

epigeous [Gk. *epi,* upon, + *ge,* the Earth]: A type of seed germination in which the cotyledons are carried above ground level.

epigyny [Gk. *epi,* upon, + *gynē,* woman]: A pattern of floral organization in which the sepals, petals, and stamens apparently grow from the top of the ovary. *See also* hypogyny.

epiphyte (ĕp′ĭ•fīt): An organism that grows upon, but is not parasitic on, another organism.

epistatic [Gk. *epistasis,* stopping]: Describing a gene whose action modifies the phenotypic expression of a gene at another locus.

essential elements: Chemical elements essential for normal plant growth and development. Also called essential minerals *and* essential inorganic nutrients.

ethylene: A simple hydrocarbon, $H_2C{=}CH_2$; a plant hormone involved in the ripening of fruit.

etiolation (e′tĭ•o•lā′shŭn) [Fr. *etioler,* to blanch]: A condition involving increased stem elongation, poor leaf development, and lack of chlorophyll; found in plants growing in the dark or with a greatly reduced amount of light.

etioplast: Plastid of a plant grown in the dark and containing a prolamellar body.

euchromatin: Regions of chromosomes that undergo condensation and decondensation in the cell cycle; regions capable of gene transcripton.

eudicotyledons: One of two major classes of angiosperms, Eudicotyledonae; formerly grouped with the magnoliids, a diverse group of archaic flowering plants, as "dicots"; plants with an embryo having two cotyledons; abbreviated as "eudicot."

Eukarya: The phylogenetic domain containing all eukaryotic organisms.

eukaryote [Gk. *eu,* good, + *karyon,* kernel]: A cell that has a membrane-bounded nucleus, membrane-bounded organelles, and chromosomes in which the DNA is associated with proteins; an organism composed of such cells. Plants, animals, fungi, and protists are the four groups of eukaryotes.

eusporangium: A sporangium that arises from several initial cells and, before maturation, forms a wall with more than one layer of cells.

eustele [Gk. *eu-,* good, + *stēlē,* pillar]: A stele in which the primary vascular tissues are arranged in discrete strands around a pith; typical of gymnosperms and angiosperms.

evolution: The derivation of progressively better-adapted forms of life from simple ancestors; Darwin proposed that natural selection is the principal mechanism by which evolution takes place.

exergonic [L. *ex,* out, + Gk. *ergon,* work]: Energy-yielding, as in a chemical reaction; applied to a "downhill" process.

exine: The outer wall layer of a spore or pollen grain.

exocarp [Gk. *exo,* without, + *karpos,* fruit]: The outermost layer of the mature ovary wall, or pericarp.

exocytosis [Gk. *ex,* out of, + *kytos,* vessel]: A cellular process in which particulate matter or dissolved substances are enclosed in a vesicle and transported to the cell surface; there, the membrane of the vesicle fuses with the plasma membrane, expelling the vesicle's contents to the outside.

exodermis: The outer layer, one or more cells in depth, of the cortex in some roots; characterized by Casparian strips within the radial and transverse cell walls. Following development of Casparian strips, a suberin lamella is deposited on all walls of the exodermis.

exon [Gk. *exo,* outside]: A segment of DNA that is both transcribed into RNA and translated into protein; exons are characteristic of eukaryotes. *See also* intron.

expansins: A novel class of proteins involved in the loosening of cell wall structure.

eyespot: A small, pigmented structure in flagellated unicellular organisms that is sensitive to light. Also called a stigma.

F

F_1: First filial generation; the offspring resulting from a cross. F_2 and F_3 are the second and third generations resulting from such a cross.

facilitated diffusion: Passive transport with the assistance of carrier proteins.

family: A taxonomic group between order and genus; the ending of family names in animals and heterotrophic protists is *-idae;* in all other organisms, it is *-aceae.* A family contains one or more genera, and each family belongs to an order.

fascicle (făs′ĭ•kŭl) [L. *fasciculus,* small bundle]: A bundle of pine leaves or other needlelike leaves of gymnosperms; an obsolete term for a vascular bundle.

fascicular cambium: The vascular cambium originating within a vascular bundle, or fascicle.

fat: A molecule composed of glycerol and three fatty acid molecules; the proportion of oxygen to carbon is much lower in fats than in carbohydrates. Fats in the liquid state are called oils.

feedback inhibition: Control mechanism whereby an increase in the concentration of some molecule inhibits the further synthesis of that molecule.

fermentation: In the absence of oxygen, a process by which the NADH generated in glycolysis is reoxidized to NAD^+. This oxygenless, or anaerobic, process results in the formation of lactate (lactate fermentation) in bacteria, fungi, protists, and animal cells, and of ethanol, or ethyl alcohol, and carbon dioxide (alcohol fermentation) in yeast and most plant cells.

ferredoxin: An electron-transferring protein of high iron content; some ferredoxins are involved in photosynthesis.

fertilization: The fusion of two gamete nuclei to form a diploid zygote. Also called syngamy.

fiber: An elongated, tapering, generally thick-walled sclerenchyma cell of vascular plants; its walls may or may not be lignified; it may or may not have a living protoplast at maturity.

fibril: Submicroscopic threads composed of cellulose molecules, the form in which cellulose occurs in the cell wall.

field capacity: The percentage of water a particular soil can hold against the action of gravity. Also called field moisture capacity.

filament: (1) The stalk of a stamen; (2) one of the threadlike bodies or segments of certain algae or fungi.

fine roots: Roots actively engaged in the uptake of water and minerals from the upper 15 centimeters of soil. Also called feeder roots.

fission: (1) Asexual reproduction involving the division of a single-celled individual into two new single-celled individuals of equal size; (2) the division of plastids and mitochondria.

fitness: The genetic contribution of an organism to future generations, relative to the contributions of other organisms living in the same environment that have different genotypes.

flagellum, *pl.* **flagella** [L. *flagellum,* whip]: A long, threadlike organelle that protrudes from the surface of a cell. Bacterial flagella are capable of rotary motion and consist of a single protein fiber each. Eukaryotic flagella, which are used in locomotion and feeding, consist of an array of microtubules with a characteristic internal 9-plus-2 microtubule structure; they are capable of a vibratory, but not rotary, motion. A cilium is a small eukaryotic flagellum.

flavonoids: Phenolic compounds; water-soluble pigments present in the vacuoles of plant cells; those found in red wines and grape juice have been reported to lower cholesterol levels in the blood.

flavoprotein: A dehydrogenase that contains a flavin and often a metal and plays a major role in oxidation; abbreviated FP.

floral tube: A cup or tube formed by fusion of the basal parts of sepals, petals, and stamens; floral tubes are often found in plants that have a superior ovary.

floret: One of the small flowers that make up a composite inflorescence or the spike of grasses.

florigen [L. *flor-*, flower, + Gk. *-genes,* producer]: The plant hormone that promotes flowering.

flower: The reproductive structure of angiosperms; a complete flower includes calyx, corolla, androecium (stamens), and gynoecium (carpels); all flowers contain at least one stamen or one carpel.

fluid-mosaic model: Model of membrane structure, with the membrane composed of a lipid bilayer in which globular proteins are embedded.

follicle [L. *folliculus,* small ball]: A dry, dehiscent simple fruit derived from a single carpel and opening along one side.

food chain, food web: A chain or web of organisms existing in any natural community such that each link in a chain feeds on the one below and is eaten by the one above; there are seldom more than six links in a chain, with autotrophs on the bottom and the largest carnivores at the top.

fossil [L. *fossils,* dug up]: The remains, impressions, or traces of an organism that has been preserved in rocks in the Earth's crust.

fossil fuels: The altered remains of once-living organisms that are burned to release energy; oil, gas, and coal.

founder cells: The group of cells from which leaf primordia and root primordia are initiated.

founder effect: A type of genetic drift that results from the founding of a population by a small number of individuals.

FP: *See* flavoprotein.

free energy: Energy available to do work.

frond: The leaf of a fern; any large, divided leaf.

fruit: In angiosperms, a mature, ripened ovary (or group of ovaries) containing the seeds, together with any adjacent parts that may be fused with it at maturity; sometimes the term is applied informally, and misleadingly, as in "fruiting body," to the reproductive structures of other kinds of organisms.

frustule: The two-part cell wall of a diatom, made up of polymerized, opaline silica ($SiO_2 \cdot nH_2O$) and consisting of overlapping halves.

fucoxanthin (fū'kō•zăn'thĭn) [Gk. *phykos,* seaweed, + *xanthos,* yellowish-brown]: A brownish carotenoid found in brown algae and chrysophytes.

fundamental tissue system: *See* ground tissue system.

funiculus [L. *funiculus,* small rope or cord]: The stalk of an ovule.

fusiform initials [L. *fusus,* spindle]: The vertically elongated cells in the vascular cambium that give rise to the cells of the axial system in secondary xylem and secondary phloem.

G

gametangium, *pl.* **gametangia** [Gk. *gamein,* to marry, + L. *tangere,* to touch]: A cell or multicellular structure in which gametes are formed.

gamete [Gk. *gametē,* wife]: A haploid reproductive cell; gametes fuse in pairs to form zygotes, which are diploid.

gametic meiosis: Meiosis resulting in the formation of haploid gametes from a diploid individual; the gametes fuse to form a diploid zygote that divides to form another diploid individual.

gametophore [Gk. *gamein,* to marry, + *phoros,* bearing]: In bryophytes, a fertile stalk that bears gametangia.

gametophyte: In plants that have an alternation of generations, the haploid (n), gamete-producing generation, or phase.

gel: A mixture of substances having a semisolid or solid constitution.

gemma, *pl.* **gemmae** (jēm'ă) [L. *gemma,* bud]: A small mass of vegetative tissue; an outgrowth of the thallus, for example, in liverworts or certain fungi; it can develop into an entire new plant.

gene: A unit of heredity; a sequence of DNA nucleotides that codes for a protein, tRNA, or rRNA molecule, or regulates the transcription of such a sequence.

gene flow: The movement of alleles into and out of a population.

gene frequency: The relative occurrence of a particular allele in a population. Also called allele frequency.

gene pool: All the alleles of all the genes of all the individuals in a population.

generative cell: (1) In many gymnosperms, the cell of the male gametophyte that divides to form the sterile and spermatogenous cells; (2) in angiosperms, the cell of the male gametophyte that divides to form two sperm.

genetic code: The system of nucleotide triplets (codons) in DNA and RNA that dictate the amino acid sequence in proteins; except for three "stop" signals, each codon specifies one of 20 amino acids.

genetic drift: Evolution (change in allele frequencies) owing to chance processes.

genetic engineering: The manipulation of genetic material for practical purposes. Also called recombinant DNA technology.

genetic recombination: The occurrence of gene combinations in progeny that are different from the combinations present in their parents.

gene transfer: *See* transformation.

genome: The totality of genetic information contained in the nucleus, plastid, or mitochondrion.

genomic library: A library encompassing an entire genome in the nucleus or in the nucleoid of an organelle (mitochondrion, plastid) in eukaryotes or in the nucleoid of a prokaryote.

genomics: The field of genetics that studies the content, organization, and function of genetic information in whole genomes.

genotype: The genetic constitution, latent or expressed, of an organism, in contrast to the phenotype; the sum total of all the genes present in an individual.

genus, *pl.* **genera:** The taxonomic group between family and species; genera include one or more species.

geotropism: *See* gravitropism.

germination [L. *germinare,* to sprout]: The beginning or resumption of growth by a spore, seed, bud, or other structure.

gibberellic acid: The main gibberellin found in fungal cultures; occurs rarely in plants, but is commonly used in the malting of barley and the management of fruit crops.

gibberellins (jĭb•ĕ•rĕ'lĭns) [from *Gibberella,* a genus of fungi]: A class of plant hormones, the best known effect of which is to increase the elongation of plant stems.

gill: The strips of tissue on the underside of the cap in many hymenomycetes.

girdling: The removal from a woody stem of a ring of bark extending inward to the cambium. Also called ringing.

glucose: A common six-carbon sugar ($C_6H_{12}O_6$); the most common monosaccharide in most organisms.

glycerol: A three-carbon molecule with three hydroxyl groups; glycerol molecules combine with fatty acids to form fats or oils.

glycogen [Gk. *glykys,* sweet, + *-gen,* of a kind]: A carbohydrate similar to starch that serves as the reserve food in bacteria, fungi, and most organisms other than plants.

glycolysis: The anaerobic breakdown of glucose to form two molecules of pyruvate, resulting in the net formation of two molecules of ATP; catalyzed by enzymes in the cytosol.

glyoxylate cycle: A variant of the citric acid cycle occurring in bacteria and some plant cells; results in the net conversion of acetate into succinate and, eventually, new carbohydrate.

glyoxysome: A peroxisome containing the enzymes necessary for the conversion of fats into carbohydrates; glyoxysomes play an important role during the germination of seeds.

Golgi body (gôl'jē): In eukaryotes, a group of flat, disk-shaped sacs that are often branched into tubules at their margins. Also called a dictyosome. Golgi bodies serve as collecting and packaging centers for the cell and are concerned with secretory activities. The term "Golgi apparatus" (or "Golgi complex") refers collectively to all the Golgi bodies, or dictyosomes, of a given cell.

grafting: A union of different individuals in which a portion, called the scion, of one individual is inserted into a root or stem, called the stock, of the other individual.

grain: (1) *See* caryopsis; (2) the alignment of wood elements.

grana, *sing.* **granum:** Structures within chloroplasts, seen as green granules with a light microscope and as a series of stacked thylakoids with an electron microscope; contain the chlorophylls and carotenoids and are the sites of the light reactions of photosynthesis.

gravitropism [L. *gravis,* heavy, + Gk. *tropē*, turning]: The response of a shoot or root to the pull of the Earth's gravity. Also called geotropism.

ground meristem [Gk. *meristos,* divisible]: The primary meristem, or meristematic tissue, that gives rise to the ground tissues.

ground tissue: Tissue other than the vascular tissues, the epidermis, and the periderm. Also called fundamental tissue.

ground tissue system: All tissues other than the epidermis (or periderm) and the vascular tissues. Also called fundamental tissue system.

growth layer: A layer of growth in the secondary xylem or secondary phloem. *See also* annual ring.

growth ring: A growth layer in the secondary xylem or secondary phloem, as seen in transverse section; may be called a growth increment, especially where seen in other than transverse section.

guanine [Sp. from Quechua, *huanu,* dung]: A purine base found in DNA and RNA; name derived from "guano," because is abundant as a white crystalline base in guano and other kinds of animal excrement.

guard cells: Pairs of specialized epidermal cells surrounding a pore, or stoma; changes in the turgor of a pair of guard cells cause opening and closing of the pore.

guttation [L. *gutta,* drop]: The exudation of liquid water from leaves; caused by root pressure.

gymnosperm [Gk. *gymnos,* naked, + *sperma,* seed]: A seed plant with seeds not enclosed in an ovary; the conifers are the most familiar group; extant gymnosperms constitute a monophyletic group.

gynoecium [Gk. *gynē*, woman, + *oikos,* house]: The aggregate of carpels in the flower of a seed plant.

H

habit [L. *habitus,* condition, character]: Characteristic form or appearance of an organism.

habitat [L. *habitare,* to inhabit]: The environment of an organism; the place where it is usually found.

hadrom: The central strand of water-conducting cells found in the axes of some moss gametophytes and sporophytes.

haploid [Gk. *haploos,* single]: Having only one set of chromosomes (n), in contrast to diploid ($2n$).

hardwood: A name commonly applied to the wood of a magnoliid or eudicot tree.

Hardy-Weinberg law: The mathematical expression of the relationship between the relative frequencies of two or more alleles in a population. It demonstrates that the frequencies of alleles and genotypes will remain constant in a random-mating population in the absence of inbreeding, selection, or other evolutionary forces.

haustorium, *pl.* **haustoria** [L. *haustus,* from *haurire,* to drink, draw]: (1) A projection of a fungal hypha that functions as a penetrating and absorbing organ; (2) in parasitic angiosperms, a modified root capable of penetrating and absorbing materials from host tissues.

heartwood: Nonliving and commonly dark-colored wood in which no water transport occurs; it is surrounded by sapwood.

helical phyllotaxis: Leaf arrangement in which there is one leaf at a node, forming a helical pattern around the stem. Also called alternate phyllotaxis.

heliotropism [Gk. *helios,* sun]: *See* solar tracking.

hemicellulose (hĕm′ē•sĕl′ū•lōs): A polysaccharide resembling cellulose but more soluble and less ordered; found particularly in cell walls.

herb [L. *herba,* grass]: A nonwoody seed plant with a relatively short-lived aerial portion.

herbaceous: Referring to nonwoody plants.

herbarium: A collection of dried and pressed plant specimens.

herbivorous: Feeding on plants.

heredity [L. *heredis,* heir]: The transmission of characteristics from parent to offspring through the gametes.

hermaphrodite [Gk. for Hermes and Aphrodite]: An organism possessing both male and female reproductive organs.

hetero- [Gk. *heteros,* different]: Prefix meaning "other" or "different."

heterochromatin: Regions of chromosomes that remain highly condensed throughout the cell cycle and appear to be devoid of transcription; for example, the chromatin located in the centromere region of each chromosome and at the end sequences, called telomeres.

heterocyst [Gk. *heteros,* different, + *cystis,* bag]: A transparent, thick-walled, nitrogen-fixing cell that forms in the filaments of certain cyanobacteria.

heteroecious (hĕt′er•ē′shŭs) [Gk. *heteros,* different, + *oikos,* house]: In some rust fungi, requiring two different host species to complete the life cycle.

heterogamy [Gk. *heteros,* other, + *gamos,* union, reproduction]: Reproduction involving two types of gametes.

heterokaryotic [Gk. *heteros,* other, + *karyon,* kernel]: In fungi, having two or more genetically distinct types of nuclei within the same mycelium.

heterokonts: Organisms with one long, ornamented (tinsel) flagellum and one shorter, smooth (whiplash) flagellum; include oomycetes, chrysophytes, diatoms, brown algae, and certain other groups. Also called stramenopiles.

heteromorphic [Gk. *heteros,* different, + *morphē,* form]: Describing a life history in which the haploid and diploid generations are dissimilar in form.

heterosis [Gk. *heterosis,* alteration]: The superiority of the hybrid over either parent in any measurable character. Also called hybrid vigor.

heterosporous: Having two kinds of spores, designated as microspores and megaspores.

heterothallic [Gk. *heteros,* different, + *thallus,* sprout]: Describing a species with haploid individuals that are self-sterile or self-incompatible; two compatible strains or individuals are required for sexual reproduction.

heterotroph [Gk. *heteros,* other, + *trophos,* feeder]: An organism that cannot manufacture organic compounds and so must feed on organic materials that originated in other plants and animals. *See also* autotroph.

heterozygous: Having two different alleles at the same locus on homologous chromosomes.

Hill reaction: The evolution of oxygen and photoreduction of an artificial electron acceptor by a chloroplast preparation in the absence of carbon dioxide.

hilum [L. *hilum,* trifle]: (1) Scar left on a seed after its separation from the funiculus; (2) the part of a starch grain around which the starch is laid down in more or less concentric layers.

histones: The group of five basic proteins associated with the chromosomes of all eukaryotic cells.

holdfast: (1) Basal part of a multicellular alga that attaches it to a solid object; either unicellular or composed of a mass of tissue; (2) cuplike structure at the tips of some tendrils, by means of which they become attached.

homeo-, homo- [Gk. *homos,* same, similar]: Prefix meaning "similar" or "same."

homeostasis (hō′mē•ō•stā′sĭs) [Gk. *homos,* similar, + *stasis,* standing]: The maintaining of a relatively stable internal physiological environment within an organism or of a steady-state equilibrium in a population or ecosystem. Homeostasis usually involves feedback mechanisms.

homeotic genes: Genes affecting floral organ identity.

homeotic mutation: A mutation that changes organ identity so that the wrong structures appear in the wrong place or at the wrong time.

homokaryotic [Gk. *homos,* same, + *karyon,* kernel]: In fungi, having nuclei with the same genetic makeup within a mycelium.

homologous chromosomes: Chromosomes that associate in pairs in the first stage of meiosis; each member of the pair is derived from a different parent. Also called homologs.

homology [Gk. *homologia,* agreement]: A condition indicative of the same phylogenetic, or evolutionary, origin, but not necessarily the same in present structure and/or function.

homosporous: Having only one kind of spore.

homothallic [Gk. *homos,* same, + *thallus,* sprout]: Describing a species in which the individuals are self-fertile.

homozygous: Having identical alleles at the same locus on homologous chromosomes.

hormogonium, *pl.* **hormogonia:** A portion of a filament of a cyanobacterium that becomes detached and grows into a new filament.

hormone [Gk. *horman,* to stir up]: An organic substance produced, usually in minute amounts, in one part of an organism and transported to another part of that organism where it has a specific effect; hormones function as highly specific chemical signals between cells.

host: An organism on or in which a parasite lives.

humus: Decomposing organic matter in the soil.

hybrid: Offspring of two parents that differ in one or more heritable characteristics; offspring of two different varieties or of two different species.

hybridization: The formation of offspring from unlike parents.

hybrid vigor: *See* heterosis.

hydraulic redistribution: The passive movement of water from wet to dry soil via roots. It occurs at night or during periods of low transpiration (during the day in CAM plants) and is driven by water potential gradients in roots and soil.

hydrocarbon [Gk. *hydro,* water, + L. *carbo,* charcoal]: An organic compound that consists only of hydrogen and carbon atoms.

hydrogen bond: A weak bond between a hydrogen atom attached to an oxygen or nitrogen atom and another oxygen or nitrogen atom.

hydroids: Water-conducting cells of the moss hadrom; they resemble the tracheary elements of vascular plants, except for their lack of specialized wall thickenings.

hydrolysis [Gk. *hydro,* water, + *lysis,* loosening]: Splitting of one molecule into two by addition of the H^+ and OH^- ions of water.

hydrophyte [Gk. *hydro,* water, + *phyton,* plant]: A plant that depends on an abundant supply of moisture or that grows wholly or partly submerged in water.

hydrostatic pressure: The pressure required to stop the movement of water; measured in units called pascals (Pa) or, more commonly, megapascals (MPa).

hydrotropism: The directed growth of plant roots in response to a moisture gradient.

hydroxyl group: An OH^- group; a negatively charged ion formed by the dissociation of a water molecule.

hymenium [Gk. *hymen,* membrane]: The layer of asci on an ascoma, or of basidia on a basidioma, together with any associated sterile hyphae.

hyper- [Gk. *hyper,* above, over]: Prefix meaning "above" or "over."

hyperaccumulators: Plants that accumulate extraordinarily high amounts of heavy metals (concentrations 100- to 1000-fold higher than non-hyperaccumulators) in their shoots, principally the leaves, without suffering phytotoxic damage.

hypertonic: Referring to a solution that has a concentration of solute particles high enough to gain water across a selectively permeable membrane from another solution.

hypha, *pl.* **hyphae** [Gk. *hyphē,* web]: A single tubular filament of a fungus, oomycete, or chytrid; the hyphae together comprise the mycelium.

hypo- [Gk. *hypo,* less than]: Prefix meaning "under" or "less."

hypocotyl: The portion of an embryo or seedling situated between the point of attachment of the cotyledons and the radicle.

hypocotyl-root axis: The embryo axis below the point of attachment of the cotyledon or cotyledons, consisting of the hypocotyl and the apical meristem of the root or radicle.

hypodermis [Gk. *hypo,* under, + *derma,* skin]: One or more layers of cells beneath the epidermis that are distinct from the underlying cortical or mesophyll cells.

hypogeous [Gk. *hypo,* under, + *ge,* the Earth]: A type of seed germination in which the cotyledons remain under ground.

hypogyny [Gk. *hypo,* under, + *gynē,* woman]: Floral organization in which the sepals, petals, and stamens are attached to the receptacle below the ovary. *See also* epigyny.

hypothesis [Gk. *hypo,* under, + *tithenai,* to put]: A temporary working explanation or supposition based on accumulated facts and suggesting some general principle or relation of cause and effect; a postulated solution to a scientific problem that must be tested by experimentation and, if disproved or shown to be unlikely, is discarded.

hypotonic: Referring to a solution that has a concentration of solutes low enough to lose water across a selectively permeable membrane to another solution.

I

IAA: *See* indole-3-acetic acid.

imbibition (ĭm•bĭ•bĭshŭn): Absorption of water by and swelling of colloidal materials due to the adsorption of water molecules onto the internal surfaces of the materials.

imperfect flower: A flower lacking either stamens or carpels.

inbreeding: The breeding of closely related plants or animals; in plants, it is usually brought about by repeated self-pollination.

incomplete flower: A flower lacking one or more of the four kinds of floral parts, that is, lacking sepals, petals, stamens, or carpels.

indehiscent (ĭn′de•hĭs′ĕnt): Remaining closed at maturity, as are many fruits (samaras, for example).

independent assortment: *See* Mendel's second law.

indeterminate growth: Unrestricted or unlimited growth, such as by a vegetative apical meristem that produces an unrestricted number of lateral organs indefinitely.

indole-3-acetic acid (IAA): A naturally occurring auxin; a plant hormone.

indusium, *pl.* **indusia** (ĭn•dū′zi•ŭm) [L. *indusium,* woman's undergarment]: In a fern leaf, a membranous growth of the epidermis that covers a sorus.

inferior ovary: An ovary completely or partially attached to the calyx; the other floral whorls appear to arise from the top of the ovary.

inflorescence: A flower cluster, with a definite arrangement of flowers.

initial: A cell that remains within the meristem indefinitely and, at the same time, divides and adds cells to the plant body.

inner bark: In older trees, the living part of the bark; the bark inside the innermost periderm.

integral proteins: Transmembrane proteins and other proteins that are tightly bound to the membrane.

integument: The outermost layer or layers of tissue enveloping the nucellus of an ovule; develops into the seed coat.

inter- [L. *inter,* between]: Prefix meaning "between," or "in the midst of."

intercalary [L. *intercalare,* to insert]: Describing growth by cell division that occurs some distance from the meristem in which the cells originated.

interfascicular cambium: The vascular cambium arising between the fascicles, or vascular bundles, from interfascicular parenchyma.

interfascicular region: Tissue region between vascular bundles in a stem. Also called a pith ray.

intermediary cells: Specialized companion cells with abundant plasmodesmatal connections with bundle-sheath cells; involved in the polymer trapping mechanism of phloem loading.

internode: The region of a stem between two successive nodes.

interphase: The period between two mitotic or meiotic cycles; the cell grows and its DNA replicates during interphase.

intine: The inner wall layer of a spore or pollen grain.

intra- [L. *intra,* within]: Prefix meaning "within."

intron [L. *intra,* within]: In eukaryotes, a portion of mRNA, transcribed from DNA, that is removed by enzymes before the mRNA is translated into protein. *See also* exon.

ion: An atom or molecule that has lost or gained one or more electrons and thus is positively or negatively charged.

irregular flower: A flower in which one or more members of at least one whorl differ in form from other members of the same whorl.

iso- [Gk. *isos,* equal]: Prefix meaning "similar," "alike"; "is-" before a vowel.

isogamy: A type of sexual reproduction in which the gametes (or gametangia) are alike in size; found in some algae and fungi.

isomer [Gk. *isos,* equal, + *meros,* part]: One of two or more compounds identical in atomic composition but differing in structural arrangement; for example, glucose and fructose.

isomorphic [Gk. *isos,* equal, + *morphē,* form]: Identical in form.

isotonic: Having the same osmotic concentration.

isotope: One of several possible forms of a chemical element that differ from other forms in the number of neutrons in the atomic nucleus, but not in chemical properties.

K

karyogamy [Gk. *karyon,* kernel, + *gamos,* marriage]: The union of two nuclei following fertilization, or plasmogamy.

kelp: A common name for any of the larger members of the order Laminariales of the brown algae.

kinetin [Gk. *kinetikos,* causing motion]: A purine that probably does not occur in nature but that acts as a cytokinin in plants.

kinetochore [Gk. *kinetikos,* causing motion, + *chorus,* chorus]: Specialized protein complex that develops on each centromere and to which spindle fibers are attached during mitosis or meiosis.

kingdom: A taxonomic category, the broadest after domain; for example, Fungi or Plantae.

Kranz anatomy [Ger. *Kranz,* wreath]: The wreathlike arrangement of mesophyll cells around a layer of large bundle-sheath cells, forming two concentric layers around the vascular bundle; typically found in the leaves of C_4 plants.

Krebs cycle: *See* citric acid cycle.

L

L1, L2, L3 layers: The outer cell layers of angiosperm apical meristems with a tunica-corpus organization.

lamella (la•mĕl′a) [L. *lamella,* thin metal plate]: A layer of cellular membranes, particularly photosynthetic, chlorophyll-containing membranes. *See also* middle lamella.

lamina: The blade of a leaf.

laminarin: One of the principal storage products of the brown algae; a polymer of glucose.

lateral meristems: Meristems that give rise to secondary tissue; the vascular cambium and cork cambium.

lateral root: A root that arises from another, older root. Also called a branch root *or* a secondary root, if the older root is the primary root.

late wood: The last part of the growth increment formed in the growing season; it contains smaller cells and is denser than the early wood; replaces the term "summer wood."

leaching: The downward movement and drainage of minerals, or inorganic ions, from the soil by percolating water.

leaf: The principal lateral appendage of the stem; highly variable in both structure and function; the foliage leaf is specialized as a photosynthetic organ.

leaf buttress: A lateral protrusion below the apical meristem; represents the initial stage in the development of a leaf primordium.

leaf gap: In ferns, region of parenchyma tissue in the primary vascular cylinder above the point of departure of the leaf trace or traces.

leaflet: One of the parts of a compound leaf.

leaf primordium [L. *primordium,* beginning]: A lateral outgrowth from the apical meristem that will eventually become a leaf.

leaf scar: A scar left on a twig when a leaf falls.

leaf trace: That part of a vascular bundle extending from the base of the leaf to its connection with a vascular bundle in the stem.

leaf trace gap: In seed plants, region of parenchyma tissue in the primary vascular cylinder of a stem above the point of departure of the leaf trace or traces.

leghemoglobin: An oxygen-binding heme protein found at high concentrations in the cytosol of bacteria-infected nitrogen-fixing root nodule cells; helps regulate the concentration of oxygen, which is a potent irreversible inhibitor of nitrogenase, the enzyme that catalyzes the fixation of nitrogen.

legume [L. *legumen,* leguminous plant]: (1) A member of the Fabaceae, the pea or bean family; (2) a type of dry, simple fruit that is derived from one carpel and opens along both sides.

lenticels (lĕn′tĭ•sĕls) [L. *lenticella,* small window]: In vascular plants, spongy areas in the cork surfaces of stems, roots, and other plant parts that allow interchange of gases between internal tissues and the atmosphere through the periderm.

leptoids: Food-conducting cells associated with the hydroids of some moss gametophytes and sporophytes; they resemble the sieve elements of some seedless vascular plants.

leptom: Food-conducting tissue consisting of leptoids; surrounds the hadrom in the axes of some moss gametophytes and sporophytes.

leptosporangium: A sporangium that arises from a single initial cell and with a wall composed of a single layer of cells.

leucoplast (lū′kō•plăst) [Gk. *leuko,* white, + *plasein,* to form]: A colorless plastid; leucoplasts are commonly centers of starch formation.

liana [Fr. *liane,* from *lier,* to bind]: A large, woody vine that climbs on other plants.

lichen: A mutualistic symbiotic association between a fungus and a population of unicellular or filamentous algal or cyanobacterial cells. About 98 percent of the lichen-forming fungal species are ascomycetes, the remainder are basidiomycetes. Lichens are polyphyletic.

life cycle: The entire sequence of phases in the growth and development of any organism, from time of zygote formation to gamete formation.

ligase: An enzyme that joins together (ligates) two molecules in an energy-dependent process; DNA ligase, for example, is essential for DNA replication, catalyzing the covalent bonding of the 3′ end of a new DNA fragment to the 5′ end of a growing chain.

light reactions: The reactions of photosynthesis that require light and cannot occur in the dark. Also called light-dependent reactions *and* energy-transduction reactions.

lignin: One of the most important constituents of the secondary wall of vascular plants, although not all secondary walls contain lignin; after cellulose, lignin is the most abundant plant polymer.

ligule [L. *ligula,* small tongue]: A minute outgrowth or appendage at the base of the leaves of grasses and those of certain lycophytes.

linkage: The tendency for certain genes to be inherited together because they are located on the same chromosome.

lipid [Gk. *lipos,* fat]: One of a large variety of nonpolar organic molecules that are insoluble in water (which is polar) but dissolve readily in nonpolar organic solvents; lipids include fats, oils, steroids, phospholipids, and carotenoids.

loam soils: Soils containing sand, silt, and clay in proportions that result in ideal agricultural soils.

locule (lŏk′ūl) [L. *loculus,* small chamber]: A cavity within a sporangium or an ovary in which ovules occur.

locus, *pl.* **loci:** The position on a chromosome occupied by a particular gene.

long-day plants: Plants that must be exposed to light periods longer than some critical length for flowering to occur; they flower in spring or summer.

lumen [L. *lumen,* light, an opening for light]: (1) The space bounded by the plant cell wall; (2) the thylakoid space in chloroplasts; (3) the narrow, transparent space of endoplasmic reticulum.

lysis [Gk. *lysis,* loosening]: A process of disintegration or cell destruction.

lysogenic bacteria: Bacteria-carrying viruses (phages) that eventually break loose from the bacterial chromosome and set up an active cycle of infection, producing lysis of their bacterial hosts.

lysosome [Gk. *lysis,* loosening, + *soma,* body]: An organelle bounded by a single membrane and containing hydrolytic enzymes, which are released when the organelle ruptures and are capable of breaking down proteins and other complex macromolecules.

M

macrocyst: In cellular slime molds, a flattened, irregular structure, encircled by a thin membrane, in which zygotes are formed.

macroevolution: Evolutionary change on a grand scale, involving major evolutionary trends.

macrofibril: An aggregation of microfibrils, visible with the light microscope.

macromolecule [Gk. *makros,* large]: A molecule of very high molecular weight; refers specifically to proteins, nucleic acids, polysaccharides, and complexes of these.

macronutrients [Gk. *makros,* large, + L. *nutrire,* to nourish]: Inorganic chemical elements required in large amounts for plant growth, such as nitrogen, potassium, calcium, phosphorus, magnesium, and sulfur.

magnoliids: A clade, or evolutionary line, of angiosperms leading to the eudicots. The leaves of most magnoliids possess ether-containing oil cells.

major veins: The larger leaf vascular bundles, which are associated with ribs; they are largely involved in the transport of substances into and out of the leaf.

maltase: An enzyme that hydrolyzes maltose to glucose.

mannitol: One of the storage molecules of the brown algae; an alcohol.

mating type: A particular genetically defined strain of an organism that is incapable of sexual reproduction with another member of the same strain but capable of such reproduction with members of other strains of the same organism.

matrotrophy: Pertaining to a form of nutrition provided by the maternal gametophyte, such as a moss gametophyte providing nutrients to the zygote and developing sporophyte.

mega- [Gk. *megas,* large]: Prefix meaning "large."

megagametophyte [Gk. *megas,* large, + *gamos,* marriage, + *phyton,* plant]: In heterosporous plants, the female gametophyte; located in the ovule of seed plants.

megaphyll [Gk. *megas,* large, + *phyllon,* leaf]: A generally large leaf with several to many veins; its leaf trace (or traces) is (are) associated with a leaf gap in ferns and with a leaf trace gap in seed plants; in contrast to a microphyll. Also called a macrophyll.

megasporangium, *pl.* **megasporangia:** A sporangium in which megaspores are produced. *See also* nucellus.

megaspore: In heterosporous plants, a haploid (n) spore that develops into a female gametophyte; in most groups, megaspores are larger than microspores.

megaspore mother cell: A diploid cell in which meiosis produces four megaspores. Also called a megasporocyte.

megasporocyte: *See* megaspore mother cell.

megasporophyll: A leaf or leaflike structure bearing a megasporangium.

meiosis (mī•ō′sĭs) [Gk. *meioun,* to make smaller]: The two successive nuclear divisions in which the chromosome number is reduced from diploid ($2n$) to haploid (n) and segregation of the genes occurs; as a result, gametes or spores (in organisms with an alternation of generations) are produced.

meiospores: Spores that arise through meiosis and are therefore haploid.

membrane potential: The voltage difference across a membrane due to the differential distribution of ions.

Mendel's first law: States that the factors for a pair of alternative characteristics are separate, and only one may be carried in a particular gamete (genetic segregation).

Mendel's second law: States that the inheritance of one pair of characteristics is independent of the simultaneous inheritance of other traits; such characteristics are assorted independently, as though there were no others present (later modified by the discovery of linkage).

meristem [Gk. *merizein,* to divide]: Embryonic tissue regions, primarily concerned with formation of new cells.

meso- [Gk. *mesos,* middle]: Prefix meaning "middle."

mesocarp [Gk. *mesos,* middle, + *karpos,* fruit]: The middle layer of the mature ovary wall, or pericarp, between the exocarp and endocarp.

mesocotyl: The internode between the scutellar node and the coleoptile in the embryo and seedling of grasses (Poaceae).

mesophyll: The ground tissue (parenchyma) of a leaf, located between the layers of epidermis; mesophyll cells generally contain chloroplasts.

mesophyte [Gk. *mesos,* middle, + *phyton,* plant]: A plant that requires an environment that is neither too wet nor too dry.

messenger RNA (mRNA): The class of RNA that carries genetic information from the gene to the ribosomes, where it is translated into protein.

metabolism [Gk. *metabolē,* change]: The sum of all chemical processes occurring within a living cell or organism.

metaphase: The stage of mitosis or meiosis during which the chromosomes lie in the equatorial plane of the spindle.

metaphloem: Part of the primary phloem that differentiates after the protophloem and before the secondary phloem, if any secondary phloem is formed in a given taxon.

metaxylem [Gk. *meta,* after, + *xylon,* wood]: The part of the primary xylem that differentiates after the protoxylem and before the secondary xylem; the metaxylem reaches maturity after the portion of the plant part in which it is located has finished elongating.

microarray: A large array of short DNA molecules, each of known sequence, bound to a glass microscope slide or other solid support; used to monitor the expression of thousands of genes simultaneously.

microbody: *See* peroxisome.

microevolution: Evolutionary change within a population over a succession of generations.

microfibril: A threadlike component of the cell wall, composed of cellulose molecules, visible only with the electron microscope.

microfilament: *See* actin filament.

microgametophyte [Gk. *mikros,* small, + *gamos,* marriage, + *phyton,* plant]: In heterosporous plants, the male gametophyte.

micrometer: A unit of microscopic measurement convenient for describing cellular dimensions; 1/1000 of a millimeter; its symbol is μm.

micronutrients [Gk. *mikros,* small, + L. *nutrire,* to nourish]: Inorganic chemical elements required only in very small, or trace, amounts for plant growth, such as iron, chlorine, copper, manganese, zinc, molybdenum, nickel, and boron.

microphyll [Gk. *mikros,* small, + *phyllon,* leaf]: A small leaf with one vein and one leaf trace, not associated with either a leaf gap or a leaf trace gap; in contrast to a megaphyll. Microphylls are characteristic of the lycophytes.

micropyle: In the ovules of seed plants, the opening in the integuments through which the pollen tube usually enters.

microsporangium: A sporangium within which microspores are formed.

microspore: In heterosporous plants, a spore that develops into a male gametophyte.

microspore mother cell: A cell in which meiosis will occur, resulting in four microspores; in seed plants, often called a pollen mother cell. Also called a microsporocyte *or* pollen mother cell.

microsporocyte: *See* microspore mother cell.

microsporophyll: A leaflike organ bearing one or more microsporangia.

microtubule [Gk. *mikros,* small, + L. *tubulus,* little pipe]: Narrow (about 25 nanometers in diameter), elongated, nonmembranous tubule of indefinite length. Microtubules occur in the cells of eukaryotes; they move the chromosomes in cell division and provide the internal structure of cilia and flagella.

middle lamella: The layer of intercellular material, rich in pectic compounds, cementing together the primary walls of adjacent cells.

mimicry [Gk. *mimos,* mime]: The superficial resemblance in form, color, or behavior of certain organisms (mimics) to other, more powerful or more protected ones (models), resulting in protection, concealment, or some other advantage for the mimic.

mineral: A naturally occurring chemical element or inorganic compound.

minor veins: The small leaf vascular bundles, located in the mesophyll and enclosed by a bundle sheath; involved in distribution of the transpiration stream and uptake of the products of photosynthesis.

mitochondrion, *pl.* **mitochondria** [Gk. *mitos,* thread, + *chondrion,* small grain]: A double-membrane-bounded organelle found in eukaryotic cells; contains the enzymes of the citric acid cycle and the electron transport chain; the major source of ATP in nonphotosynthetic cells.

mitosis (mī•tō′sĭs) [Gk. *mitos,* thread]: A process during which the duplicated chromosomes divide longitudinally and the daughter chromosomes then separate to form two genetically identical daughter nuclei; usually accompanied by cytokinesis.

mitotic spindle: The array of microtubules that forms between the opposite poles of a eukaryotic cell during mitosis.

mixotrophy: The ability of an organism to utilize both organic and inorganic carbon sources.

mole: The amount of a substance with a weight, in grams, that is numerically equal to its atomic weight or molecular weight. The number of particles in 1 mole of any substance is always equal to Avogadro's number: 6.022×10^{23}.

molecular weight: The relative weight of a molecule when the weight of the most frequent kind of carbon atom is taken as 12; the sum of the relative weights of the atoms in a molecule.

molecule: Smallest possible unit of a compound, consisting of two or more atoms.

mono- [Gk. *monos,* single]: Prefix meaning "one" or "single."

monocotyledons: One of the two great classes of angiosperms, Monocotyledonae; plants having an embryo with one cotyledon; abbreviated as "monocots."

monoecious (mō•nē′shŭs) [Gk. *monos,* single, + *oikos,* house]: Having the anthers and carpels produced in separate flowers on the same individual.

monokaryotic [Gk. *monos,* one, + *karyon,* kernel]: In fungi, having a single haploid nucleus within one cell or compartment.

monomers [Gk. *monos,* single, + *meros,* part]: Small, repeating units that can be linked together to form polymers.

monophyletic: Pertaining to a taxon descended from a single ancestor.

monosaccharide [Gk. *monos,* single, + *sakcharon,* sugar]: A simple sugar, such as a five-carbon or six-carbon sugar, that cannot be dissociated into smaller sugar particles.

-morph, morph- [Gk. *morphē,* form]: Suffix or prefix meaning "form."

morphogenesis: The development of form.

morphology [Gk. *morphē,* form, + *logos,* discourse]: The study of form and its development.

movement proteins: Viral-encoded proteins that facilitate the movement of viruses from cell to cell through plasmodesmata.

mRNA: *See* messenger RNA.

mucilage: A highly hydrated polysaccharide. Root tips growing in soil are coated with large amounts of mucilage, which has its origin in the peripheral cells of the rootcap.

multigene family: A collection of related genes on a chromosome; most eukaryotic genes appear to be members of multigene families.

multiple epidermis: A tissue composed of several cell layers derived from the protoderm; only the outer layer assumes the characteristics of a typical epidermis.

multiple fruit: A cluster of mature ovaries produced by a cluster of flowers, as in the pineapple.

mutagen [L. *mutare,* to change, + Gk. *gen,* birth]: An agent that increases the mutation rate.

mutant: A mutated gene or an organism carrying a gene that has undergone a mutation.

mutation: Any change in the hereditary state of an organism; such changes may occur at the level of the gene (gene mutations, or point mutations) or at the level of the chromosome (chromosome mutations).

mutualism: The living together of two or more organisms in an association that is mutually advantageous.

myc-, myco- [Gk. *mykēs,* fungus]: Prefix meaning "pertaining to fungi."

mycelium [Gk. *mykēs,* fungus]: The mass of hyphae forming the body of a fungus, oomycete, or chytrid.

mycobiont [Gk. *mykēs,* fungus, + *bios,* life]: The fungal component of a lichen.

mycology: The study of fungi.

mycorrhizas (mycorrhizae): Literally meaning "fungus roots"; intimate and mutually beneficial symbiotic associations between fungi and roots; characteristic of most vascular plants. *See also* arbuscular mycorrhizas; ectomycorrhizas.

N

NAD⁺: *See* nicotinamide adenine dinucleotide.

NADP⁺: *See* nicotinamide adenine dinucleotide phosphate.

nanoplankton (năn′ō•plăngk•tŏn) [Gk. *nanos,* dwarf, + *planktos,* wandering]: Plankton with dimensions of less than 70 to 75 micrometers.

nastic movement: A plant movement that occurs in response to a stimulus, but in a direction independent of the direction of the stimulus.

natural selection: The differential reproduction of genotypes based on their genetic constitution.

nectary [Gk. *nektar,* drink of the gods]: In angiosperms, a gland that secretes nectar, a sugary fluid that attracts animals to plants.

netted venation: The arrangement of veins in the leaf blade that resembles a net; characteristic of the leaves of all angiosperms except for monocots. Also called reticulate venation.

neutron [L. *neuter,* neither]: An uncharged particle with a mass slightly greater than that of a proton, found in the atomic nucleus of all elements except hydrogen, in which the nucleus consists of a single proton.

niche: The role played by a particular species in its ecosystem.

nicotinamide adenine dinucleotide (NAD^+): A coenzyme that functions as an electron acceptor in many of the oxidation reactions of respiration.

nicotinamide adenine dinucleotide phosphate ($NADP^+$): A coenzyme that, in the form of NADPH, functions as an electron donor in many of the reduction reactions of biosynthesis; similar in structure to NAD^+, but contains an extra phosphate group.

nitrification: The oxidation of ammonium ions or ammonia to nitrate; a process carried out by a specific free-living soil bacterium.

nitrogen fixation: The incorporation of atmospheric nitrogen into nitrogen compounds; carried out by certain free-living and symbiotic bacteria.

nitrogen-fixing bacteria: Soil bacteria that convert atmospheric nitrogen into nitrogen compounds.

nitrogenous base: A nitrogen-containing molecule having basic properties (tendency to acquire an H^+ ion); a purine or pyrimidine; one of the building blocks of nucleic acids.

node [L. *nodus,* knot]: The part of a stem where one or more leaves are attached. *See also* internode.

nodules: Enlargements or swellings on the roots of legumes and certain other plants inhabited by symbiotic nitrogen-fixing bacteria.

nonclimacteric fruits: Fruits such as citrus fruits, grapes, and strawberries that do not exhibit the large increase in cellular respiration during ripening that is seen in climacteric fruits.

noncoding RNA: An RNA molecule that is not translated into protein.

nonconducting phloem: The part of the inner bark in which the sieve elements are dead. The phloem parenchyma cells and ray parenchyma cells of nonconducting phloem may remain alive and continue to function as storage cells for many years.

noncyclic electron flow: The light-induced flow of electrons from water to $NADP^+$ in oxygen-evolving photosynthesis; it involves both Photosystems I and II.

nonseptate: *See* aseptate.

nucellus (noo•sĕl′ŭs) [L. *nucella,* small nut]: Inner part of an ovule, in which the embryo sac develops; equivalent to a megasporangium.

nuclear envelope: The double membrane surrounding the nucleus of a cell.

nucleic acid: An organic acid consisting of joined nucleotide complexes; the two types are deoxyribonucleic acid (DNA) and ribonucleic acid (RNA).

nucleoid: A region of DNA in prokaryotic cells, mitochondria, and chloroplasts.

nucleolar organizer region: A special area on a certain chromosome that is associated with formation of the nucleolus.

nucleolus, *pl.* **nucleoli** (noo•klē′ō•lŭs) [L. *nucleolus,* small nucleus]: A small, spherical body found in the nucleus of eukaryotic cells; composed chiefly of rRNA that is in the process of being transcribed from copies of rRNA genes; the site of production of ribosomal subunits.

nucleoplasm: The ground substance of a nucleus.

nucleosome: A complex of DNA and histone proteins that forms the fundamental packaging unit of eukaryotic DNA; its structure resembles a bead on a string.

nucleotide: A single unit of nucleic acid, composed of a phosphate, a five-carbon sugar (either ribose or deoxyribose), and a purine or a pyrimidine.

nucleus, *pl.* **nuclei:** (1) A specialized body within the eukaryotic cell, bounded by a double membrane and containing the chromosomes; (2) the central part of an atom of a chemical element.

nut: A dry, indehiscent, hard, one-seeded simple fruit, generally produced from a gynoecium of more than one fused carpel.

O

obligate anaerobe: *See* strict anaerobe.

-oid [Gk. *oid,* like, resembling]: Suffix meaning "like" or "similar to."

Okazaki fragments [after R. Okazaki, Japanese geneticist]: In DNA replication, the discontinuous segments in which the 3′ to 5′ strand (the lagging strand) of the DNA double helix is synthesized; typically 1000 to 2000 nucleotides long in prokaryotes, and 100 to 200 nucleotides long in eukaryotes.

ontogeny [Gk. *on,* being, + *genesis,* origin]: The development, or life history, of all or part of an individual organism.

oo- [Gk. *oion,* egg]: Prefix meaning "egg."

oogamy: Sexual reproduction in which one of the gametes (the egg) is large and nonmotile, and the other gamete (the sperm) is smaller and motile.

oogonium (ō′o•gō′nē•ŭm): A unicellular female sex organ that contains one or several eggs.

oospore: The thick-walled zygote characteristic of the oomycetes.

open vascular bundle: A vascular bundle in which a vascular cambium develops.

operator: A segment of DNA that interacts with a repressor protein to regulate transcription of the structural genes of an operon.

operculum (o•pûr′ku•lŭm) [L. *operculum,* lid]: In mosses, the lid of the sporangium.

operon [L. *opus, operis,* work]: In the bacterial chromosome, a segment of DNA consisting of a promoter, an operator, and a group of adjacent structural genes; the structural genes, which code for products related to a particular biochemical pathway, are transcribed onto a single mRNA molecule, and their transcription is regulated by a single repressor protein.

opposite phyllotaxis: Leaf arrangement in which leaves occur in pairs at a node.

order: A category of classification between class and family; classes contain one or more orders, and orders are composed of one or more families.

organ: A distinct and visibly differentiated part of a plant, such as root, stem, leaf, or part of a flower.

organelle (or′găn•el) [Gk. *organella,* small tool]: A specialized, membrane-bounded part of a cell.

organic: Pertaining to living organisms in general, to compounds formed by living organisms, and to the chemistry of compounds containing carbon.

organism: Any individual living creature, either unicellular or multicellular.

osmosis (ŏs•mō′sĭs) [Gk. *osmos,* impulse or thrust]: The diffusion of water, or any solvent, across a selectively permeable membrane; in the absence of other forces, the movement of water during osmosis is always from a region of greater water potential to one of lesser water potential.

osmotic potential: The change in free energy or chemical potential of water produced by solutes; carries a negative (minus) sign. Also called solute potential.

osmotic pressure: The potential pressure that can be developed by a solution separated from pure water by a selectively permeable membrane; in the absence of other forces, the movement of water during osmosis is always from a region of greater water potential to one of lesser water potential.

outcrossing: Cross-pollination between individuals of the same species.

outer bark: In older trees, the dead part of the bark; the innermost periderm and all tissues outside it. Also called rhytidome.

outgroup: In a cladogram, a species or group of species that does not exhibit one or more shared derived characters found in the group under study, the ingroup.

ovary [L. *ovum,* egg]: The enlarged basal portion of a carpel or of a gynoecium composed of fused carpels; a mature ovary, sometimes with other adherent parts, is a fruit.

ovule (ō•vūl) [L. *ovulum,* little egg]: A structure in seed plants containing the female gametophyte with egg cell, all surrounded by the nucellus and one or two integuments; when mature, an ovule becomes a seed.

ovuliferous scale: In certain conifers, the appendage or scalelike shoot to which the ovule is attached.

oxidation: The loss of an electron by an atom or molecule. Oxidation and reduction (gain of an electron) take place simultaneously, because an electron that is lost by one atom is accepted by another. Oxidation-reduction reactions are an important means of energy transfer in living systems.

oxidative phosphorylation: The formation of ATP from ADP and inorganic phosphate; takes place in the electron transport chain of the mitochondrion.

P

pairing of chromosomes: Side-by-side association of homologous chromosomes.

paleobotany [Gk. *palaios,* old]: The study of fossil plants.

palisade parenchyma: A leaf tissue composed of columnar chloroplast-bearing parenchyma cells with their long axes at right angles to the leaf surface.

panicle (păn′ĭ•kûl) [L. *panicula,* tuft]: An inflorescence with a main axis that is branched and its branches bearing loose flower clusters.

para- [Gk. *para,* beside]: Prefix meaning "beside."

paradermal section [Gk. *para,* beside, + *derma,* skin]: Section cut parallel to the surface of a flat structure, such as a leaf.

parallel evolution: The development of similar structures having similar functions in two or more evolutionary lines as a result of the same kind of selective pressures.

parallel venation: The pattern of venation in which the principal veins of the leaf are parallel or nearly so; characteristic of monocots.

paramylon: The storage molecule of euglenoids.

paraphyletic: Pertaining to a taxon that excludes species that share a common ancestor with species included in the taxon.

paraphysis, *pl.* **paraphyses** [Gk. *para,* beside, + *physis,* growth]: In certain fungi, a sterile filament growing among the reproductive cells in the fruiting body; the term is also applied to the sterile filaments growing among the gametangia and sporangia of certain brown algae and among the antheridia and archegonia of mosses.

parasexual cycle: The fusion and segregation of heterokaryotic haploid nuclei in certain fungi to produce recombinant nuclei.

parasite: An organism that lives on or in an organism of a different species and derives nutrients from it, with the association being beneficial to the parasite and harmful to the host.

parenchyma (pa•rĕng′kĭ•ma) [Gk. *para,* beside, + *en,* in, + *chein,* to pour]: A tissue composed of parenchyma cells.

parenchyma cell: Living, generally thin-walled plant cell of variable size and form; the most abundant kind of cell in plants.

parthenocarpy [Gk. *parthenos,* virgin, + *karpos,* fruit]: The development of fruit without fertilization; parthenocarpic fruits are usually seedless.

passage cell: An endodermal cell of roots that retains a thin wall and a Casparian strip when other, associated endodermal cells develop thick secondary walls.

passive transport: Non-energy-requiring transport of a solute across a membrane down the concentration or electrochemical gradient by either simple diffusion or facilitated diffusion.

pathogen [Gk. *pathos,* suffering, + *genesis,* beginning]: An organism that causes a disease.

pathogenic: Disease-causing.

pathology: The study of plant or animal diseases, their effects on the organism, and their treatment.

PCR: *See* polymerase chain reaction.

pectin: A highly hydrophilic polysaccharide present in the intercellular layer and primary wall of plant cell walls; the basis of fruit jellies.

pedicel (ped′ĭ•sĕl): The stalk of an individual flower in an inflorescence.

peduncle (pe•dŭng•kûl): The stalk of an inflorescence or of a solitary flower.

pentose phosphate cycle: The pathway of oxidation of glucose 6-phosphate to yield pentose phosphates.

peptide: Two or more amino acids linked by peptide bonds.

peptide bond: The type of bond formed when two amino acid units are joined end to end by the removal of a molecule of water; the bonds always form between the carboxyl (—COOH) group of one amino acid and the amino (—NH_2) group of the next amino acid.

perennial [L. *per,* through, + *annuus,* year]: A plant in which the vegetative structures live year after year.

perfect flower: A flower having both stamens and carpels; hermaphroditic flower.

perfect stage: The phase of the life history of a fungus that includes sexual fusion and the spores associated with such fusions.

perforation plate: Part of the wall of a vessel element that is perforated.

peri- [Gk. *peri,* around]: Prefix meaning "around" or "about."

perianth (pĕr′ĭ•anth) [Gk. *peri,* around, + *anthos,* flower]: (1) The petals and sepals taken together; (2) in leafy liverworts, a tubular sheath surrounding an archegonium and, later, the developing sporophyte.

pericarp [Gk. *peri,* around, + *karpos,* fruit]: The fruit wall, which develops from the mature ovary wall.

periclinal: Parallel to the surface.

pericycle [Gk. *peri,* around, + *kykos,* circle]: A tissue characteristic of roots that is bounded externally by the endodermis and internally by the phloem.

periderm [Gk. *peri,* around, + *derma,* skin]: Outer protective tissue that replaces epidermis when it is destroyed during secondary growth; includes cork, cork cambium, and phelloderm.

perigyny [Gk. *peri,* around, + *gynē,* woman]: A form of floral organization in which the sepals, petals, and stamens are attached to the margin of a cup-shaped extension of the receptacle; superficially, the sepals, petals, and stamens appear to be attached to the ovary.

perisperm [Gk. *peri,* around, + *sperma,* seed]: Food-storing tissue derived from the nucellus that occurs in the seeds of some flowering plants.

peristome (pĕr′ĭ•stōm) [Gk. *peri,* around, + *stoma,* mouth]: In mosses, a fringe of teeth around the opening of the sporangium.

perithecium: A spherical or flask-shaped ascoma.

permanent wilting percentage: The percentage of water remaining in a soil when a plant fails to recover from wilting even if placed in a humid chamber.

permeable [L. *permeare,* to pass through]: Usually describing membranes through which liquid substances can diffuse.

peroxisome: A spherical, single-membrane-bounded organelle, ranging in diameter from 0.5 to 1.5 micrometers; some peroxisomes are involved in photorespiration, and others (called glyoxysomes) with the conversion of fats to sugars during seed germination. Also called a microbody.

petal: A flower part, usually conspicuously colored; one of the units of the corolla.

petiole: The stalk of a leaf.

pH: A symbol denoting the relative concentration of hydrogen ions in a solution; pH values run from 0 to 14; the lower the value, the more acidic the solution, that is, the more hydrogen ions it contains; pH 7 is neutral, less than 7 is acidic, and more than 7 is alkaline (or basic).

phage: *See* bacteriophage.

phagocytosis: *See* endocytosis.

phellem: *See* cork.

phelloderm (fĕl′o•durm) [Gk. *phellos,* cork, + *derma,* skin]: A tissue formed inwardly by the cork cambium, opposite the cork; inner part of the periderm.

phellogen (fĕl′o•jĕn): *See* cork cambium.

phenolics: A broad range of compounds, all of which have a hydroxyl group (—OH) attached to an aromatic ring (a ring of six carbons containing three double bonds); include flavonoids, tannins, lignins, and salicylic acid.

phenotype: The physical appearance of an organism; results from the interaction between the genetic constitution (genotype) of the organism and its environment.

phloem (flō′ĕm) [Gk. *phloos,* bark]: The food-conducting tissue of vascular plants; composed of sieve elements, various kinds of parenchyma cells, fibers, and sclereids.

phloem loading: The process by which substances (primarily sugars) are actively secreted into the sieve tubes.

phloem unloading: The process by which sugars and other assimilates transported in the phloem exit the sieve tubes of sink tissues.

phosphate: A compound formed from phosphoric acid by replacement of one or more hydrogen atoms.

phospholipid: A phosphorylated lipid; similar in structure to a fat, but with only two fatty acids attached to the glycerol backbone, with the third space occupied by a phosphorus-containing molecule; phospholipids are important components of cellular membranes.

phosphorylation (fŏs′fo•rĭl•ā′shŭn) [Gk. *phosphoros,* bringing light]: A reaction in which phosphate is added to a compound; for example, the formation of ATP from ADP and inorganic phosphate.

photo-, -photic [Gk. *photos,* light]: Prefix or suffix meaning "light."

photobiont [Gk. *photos,* light, + *bios,* life]: The photosynthetic component of a lichen.

photolysis: The light-dependent oxidative splitting of water molecules that takes place in Photosystem II of the light reactions of photosynthesis.

photon [Gk. *photos,* light]: The elementary particle of light.

photoperiodism: Response of organisms to duration and timing of day and night; a mechanism that has evolved for measuring seasonal time.

photophosphorylation [Gk. *photos,* light, + *phosphoros,* bringing light]: Formation of ATP in the chloroplast during photosynthesis.

photorespiration: The oxygenase activity of Rubisco combined with the salvage pathway, consuming O_2 and releasing CO_2; occurs when Rubisco binds O_2 instead of CO_2.

photosynthesis [Gk. *photos,* light, + *syn,* together + *tithenai,* to place]: Conversion of light energy to chemical energy; the production of carbohydrates from carbon dioxide and water in the presence of chlorophyll by using light energy.

photosystem: A discrete unit of organization of chlorophyll and other pigment molecules embedded in the thylakoids of chloroplasts and involved in the light-dependent reactions of photosynthesis.

phototropins 1 and 2: Two flavoproteins, localized in the plasma membrane, that are photoreceptors of blue light.

phototropism [Gk. *photos,* light, + *tropē,* turning]: Growth in which the direction of the light is the determining factor, such as the growth of a plant toward a light source; turning or bending in response to light.

phragmoplast: A spindle-shaped system of microtubules that arises between two daughter nuclei at telophase and within which the cell plate is formed during cell division, or cytokinesis. Phragmoplasts are found in all green algae except members of the class Chlorophyceae and in all plants.

phragmosome: The layer of cytoplasm that forms across the cell where the nucleus becomes located and divides.

phycobilins: A group of water-soluble accessory pigments, including phycocyanins and phycoerythrins, found in the red algae and cyanobacteria.

phycology [Gk. *phykos,* seaweed]: The study of algae.

phycoplast: A system of microtubules that develops between the two daughter nuclei, parallel to the plane of cell division. Phycoplasts occur only in green algae of the class Chlorophyceae.

phyllo-, phyll- [Gk. *phyllon,* leaf]: Prefix meaning "leaf."

phyllode (fĭl′ōd): A flat, expanded, photosynthetic petiole or stem; phyllodes occur in certain genera of vascular plants, where they replace leaf blades in photosynthetic function.

phyllotaxis: The arrangement of leaves on a stem. Also called phyllotaxy.

phylogeny [Gk. *phylon,* race, tribe]: Evolutionary relationships among organisms; the developmental history of a group of organisms.

phylum, *pl.* **phyla** [Gk. *phylon,* race, tribe]: A taxonomic category of related, similar classes; a higher-level category, below kingdom and above class; until recently, called "division" by botanists.

physiology: The study of the activities and processes of living organisms.

phyto-, -phyte [Gk. *phyton,* plant]: Prefix or suffix meaning "plant."

phytoalexin: [Gk. *phyton,* plant, + *alexein,* to ward off, protect]: A chemical substance produced by a plant to combat infection by a pathogen (fungi or bacteria).

phytochrome: A phycobilinlike pigment, found in the cytoplasm of plants and a few green algae, that is associated with the absorption of light; photoreceptor for red and far-red light; involved in a number of timing processes, such as flowering, dormancy, leaf formation, and seed germination.

phytomeres: A succession of repeated developmental units, consisting of a node with its attached leaf (or leaves), the internode below the leaf (or leaves), and the bud (or buds) at the base of the internode.

phytoplankton [Gk. *phyton,* plant, + *planktos,* wandering]: Aquatic, free-floating, microscopic, photosynthetic organisms.

pigment: A substance that absorbs light, often selectively.

pileus [L. *pileus,* cap]: The caplike part of mushroom basidiomata and of certain ascomata.

pinna, *pl.* **pinnae** (pĭn′a) [L. *pinna,* feather]: A primary division, or leaflet, of a compound leaf or frond; may be divided into pinnules.

pinocytosis: *See* endocytosis.

pistil [L. *pistillum,* pestle]: Sometimes used to refer to an individual carpel or a group of fused carpels.

pistillate: *See* carpellate.

pit: A recessed cavity in a cell wall where a secondary wall does not form.

pit membrane: The middle lamella and two primary cell walls between two pits.

pit-pair: Two opposite pits plus the pit membrane.

pith: The ground tissue occupying the center of the stem or root within the vascular cylinder; usually consists of parenchyma.

pith ray: *See* interfascicular region.

placenta, *pl.* **placentae** [L. *placenta,* cake]: The part of the ovary wall to which the ovules or seeds are attached.

placentation: The manner of ovule attachment within the ovary.

plankton [Gk. *planktos,* wandering]: Free-floating, mostly microscopic, aquatic organisms.

plaque: Clear area in a sheet of cells resulting from the killing or lysis of contiguous cells by viruses.

-plasma, plasmo-, -plast [Gk. *plasma,* form, mold]: Prefix or suffix meaning "formed" or "molded"; examples are protoplasm, "first-molded" (living matter), and chloroplast, "green-formed."

plasma membrane or **plasmalemma:** Outer boundary of the cytoplasm, next to the cell wall; consists of a single membrane. Also called cell membrane *or* ectoplast.

plasmid: A relatively small fragment of DNA that can exist free in the cytoplasm of a bacterium and can be integrated into and then replicated with a chromosome. Plasmids make up about 5 percent of the DNA of many bacteria, but are rare in eukaryotes.

plasmodesma, *pl.* **plasmodesmata** [Gk. *plasma,* form, + *desma,* bond]: The minute cytoplasmic threads that extend through openings in cell walls and connect the protoplasts of adjacent living cells.

plasmodium (plăz • mō′dĭ • ŭm): Stage in the life cycle of myxomycetes (plasmodial slime molds); a multinucleate mass of protoplasm surrounded by a membrane.

plasmogamy [Gk. *plasma,* form, + *gamos,* marriage]: Union of the protoplasts of gametes that is not accompanied by union of their nuclei.

plasmolysis (plăz • mŏl′ĭ • sĭs) [Gk. *plasma,* form, + *lysis,* loosening]: Separation of the protoplast from the cell wall because of the removal of water from the protoplast by osmosis.

plastid (plăs′tĭd): Organelle in the cells of certain groups of eukaryotes that is the site of such activities as food manufacture and storage; plastids are bounded by two membranes.

pleiotropy [Gk. *pleros,* more, + *tropē,* turning]: The capacity of a gene to affect more than one phenotypic characteristic.

plumule [L. *plumula,* small feather]: The first bud of an embryo; the portion of the young shoot above the cotyledons.

pneumatophores [Gk. *pneuma,* breath, + *-phoros,* bearing]: Negatively gravitropic extensions of the root systems of some trees growing in swampy habitats; they grow upward and out of the water and probably function to ensure adequate aeration.

point mutation: An alteration in one of the nucleotides in a chromosomal DNA molecule; an allele of a gene changes, becoming a different allele. Also called gene mutation.

polar molecule: A molecule with positively and negatively charged ends.

polar nuclei: Two nuclei (usually), one derived from each end (pole) of the embryo sac, which become centrally located; they fuse with a male nucleus to form the primary (typically $3n$) endosperm nucleus.

pollen [L. *pollen,* fine dust]: A collective term for pollen grains.

pollen grain: In seed plants, a microspore containing a mature or immature microgametophyte (male gametophyte).

pollen mother cell: *See* microspore mother cell.

pollen sac: A cavity in the anther that contains the pollen grains.

pollen tube: A tube formed after germination of the pollen grain; carries the male gametes into the ovule.

pollination: In angiosperms, the transfer of pollen from an anther to a stigma. In gymnosperms, the transfer of pollen from a pollen-producing cone directly to an ovule.

poly- [Gk. *polys,* many]: Prefix meaning "many."

polyembryony: Having more than one embryo within the developing seed.

polygenic inheritance: The inheritance of quantitative characteristics determined by the combined effects of multiple genes.

polymer (pŏl′ĭ • mer): A large molecule composed of many similar molecular subunits.

polymerase chain reaction (PCR): A technique for amplifying specific regions of DNA by multiple cycles of DNA polymerization, utilizing special primers, DNA polymerase molecules, and nucleotides; each cycle is followed by a brief heat treatment to separate complementary strands.

polymerization: The chemical union of monomers such as glucose or nucleotides to form polymers such as starch or nucleic acid.

polymer trapping: An active mechanism for phloem loading that does not involve symporters. This mechanism explains the accumulation in sieve elements of the polymers raffinose and stachyose, which are synthesized from sucrose in symplastically loading species.

polynucleotide: A single-stranded DNA or RNA molecule.

polypeptide: A molecule composed of amino acids linked together by peptide bonds; not as complex as a protein.

polyphyletic: Pertaining to a taxon with members derived from two or more ancestors not common to all members of the taxon.

polyploid (pŏl′ĭ • ploid): Referring to an organism, tissue, or cell with more than two complete sets of chromosomes.

polysaccharide: A polymer composed of many monosaccharide units joined in a long chain, such as glycogen, starch, and cellulose.

polysome or **polyribosome:** An aggregation of ribosomes actively involved in the translation of the same mRNA molecule, one after another.

pome (pōm) [Fr. *pomme,* apple]: A simple fleshy fruit, the outer portion of which is formed by the floral parts that surround the ovary and expand with the growing fruit; found only in one subfamily of the Rosaceae (including apples, pears, quince, and pyracantha).

population: Any group of individuals, usually of a single species, occupying a given area at the same time.

P-protein: Phloem protein; a proteinaceous substance found in cells of angiosperm phloem, especially in sieve-tube elements. Once called slime.

preprophase band: A ringlike band of microtubules, lying just beneath the plasma membrane, that delimits the equatorial plane of the future mitotic spindle of a cell preparing to divide.

pressure-flow hypothesis: The assertion that assimilates are transported from sources to sinks along a gradient of turgor pressure developed osmotically.

primary cell wall: The wall layer deposited during the period of cell expansion.

primary endosperm nucleus: The result of fusion of a sperm nucleus and the two polar nuclei in most angiosperms. The embryo sacs of the Nymphaeales and Austrobaileyales have only one polar nucleus.

primary growth: In plants, growth originating in the apical meristems of shoots and roots, in contrast to secondary growth.

primary meristem or **primary meristematic tissue** (mĕr′ĭ • ste • măt′ĭc): A tissue derived from the apical meristem; of three kinds: protoderm, procambium, and ground meristem.

primary metabolites: Molecules that are found in all plant cells and are necessary for the life of the plant; examples are simple sugars, amino acids, proteins, and nucleic acids.

primary pit-field: Thin area in a primary cell wall through which plasmodesmata pass, although plasmodesmata may also occur elsewhere in the wall.

primary plant body: The part of the plant body arising from the apical meristems and their derivative meristematic tissues; composed entirely of primary tissues.

primary root: The first root of the plant, developing as a continuation of the root tip or radicle of the embryo; the taproot.

primary tissues: Cells derived from the apical meristems and primary meristematic tissues of root and shoot; as opposed to secondary tissues derived from a cambium; primary growth results in an increase in length.

primordium, *pl.* **primordia** [L. *primus,* first, + *ordiri,* to begin to weave]: A cell or organ in its earliest stage of differentiation.

pro- [Gk. *pro,* before]: Prefix meaning "before" or "prior to."

procambium (prō • kăm′bē • ŭm) [L. *pro,* before, + *cambiare,* to exchange]: A primary meristematic tissue that gives rise to primary vascular tissues.

proembryo: An embryo in early stages of development, before the embryo proper and suspensor become distinct.

programmed cell death: The genetically controlled, or programmed, series of changes in a living cell or organism that lead to its death.

prokaryote [Gk. *pro,* before, + *karyon,* kernel]: A cell lacking a membrane-bounded nucleus and membrane-bounded organlles; Bacteria and Archaea.

prolamellar body: Semicrystalline body found in plastids arrested in development by the absence of light.

promeristem: The initiating cells (initials) and their most recent derivatives in an apical meristem; the least differentiated, or determined, part of an apical meristem.

promoter: A specific segment of DNA to which RNA polymerase attaches to initiate transcription of mRNA from an operon.

prophase [Gk. *pro,* before, + *phasis,* form]: An early stage in nuclear division, characterized by shortening and thickening of the chromosomes and their movement to the metaphase plate.

proplastid: A minute, self-reproducing body in the cytoplasm from which a plastid develops.

prop roots: Adventitious roots arising from the stem above soil level and helping to support the plant; common in many monocots, such as maize *(Zea mays).*

prosthetic group: A heat-stable metal ion or inorganic group (not an amino acid) that is bound to a protein and serves as its active group.

protease (prō′tē • ās): An enzyme that digests protein by the hydrolysis of peptide bonds. Also called a peptidase.

protein [Gk. *proteios,* primary]: A complex organic compound composed of many (100 or more) amino acids joined by peptide bonds.

prothallial cell [Gk. *pro,* before, + *thallos,* sprout]: The sterile cell or cells found in the male gametophytes, or microgametophytes, of vascular plants other than angiosperms; believed to be remnants of the vegetative tissue of the male gametophyte.

prothallus (prō • thăl′ŭs): In homosporous vascular plants, such as ferns, the more or less independent, photosynthetic gametophyte. Also called the prothallium.

protists: All organisms that do not have the distinct characteristics of fungi, animals, or plants.

proto- [Gk. *protos,* first]: Prefix meaning "first"; for example, Protozoa, "first animals."

protoderm [Gk. *protos,* first, + *derma,* skin]: Primary meristematic tissue that gives rise to epidermis.

proton: A subatomic particle with a single positive charge equal in magnitude to the charge of an electron and with a mass of 1; the basic component of every atomic nucleus; a common term for a hydrogen ion (H^+).

protonema, *pl.* **protonemata** [Gk. *protos,* first, + *nema,* thread]: The first stage in development of the gametophyte of mosses and certain liverworts; protonemata may be filamentous or platelike.

protophloem: The first-formed elements of the phloem in a plant organ; first part of the primary phloem.

protoplasm: A general term for the living substance of all cells.

protoplast: The protoplasm of an individual cell; in plants, the unit of protoplasm inside the cell wall.

protostele [Gk. *protos,* first, + *stēlē,* pillar]: The simplest type of stele, consisting of a solid column of vascular tissue.

protoxylem: The first part of the primary xylem, which matures during elongation of the plant part in which it is found.

protracheophyte: An organism with branched axes and multiple sporangia, but with water-conducting cells similar to the hydroids of modern mosses rather than to the tracheary elements of vascular plants; an intermediate stage in the evolution of vascular plants, or tracheophytes.

proximal (prŏk′sĭ • măl) [L. *proximus,* near]: Situated near the point of reference, usually the main part of a body or the point of attachment; the opposite of distal.

pseudo- [Gk. *pseudes,* false]: Prefix meaning "false."

pseudoplasmodium: A multicellular mass of individual amoeboid cells, representing the aggregate phase in the cellular slime molds.

pulvinus, *pl.* **pulvini:** Jointlike thickening at the base of the petiole of a leaf or petiolule of a leaflet, having a role in the movements of the leaf or leaflet.

pumps: Transport proteins driven by either chemical energy (ATP) or light energy; in plant and fungal cells, they typically are proton pumps.

punctuated equilibrium: A model of evolutionary change proposing that there are long periods of little or no change punctuated by brief intervals of rapid change.

purine (pū′rēn): The larger of the two kinds of nucleotide bases found in DNA and RNA; a nitrogenous base with a double-ring structure, such as adenine or guanine.

pyramid of energy: Energy relationships among various feeding levels involved in a particular food chain; autotrophs (at the base of the pyramid) represent the greatest amount of available energy; herbivores are next, then primary carnivores, secondary carnivores, and so forth. Similar pyramids of mass, size, and number also occur in natural communities.

pyrenoid [Gk. *pyren,* stone of a fruit, + *oides,* like]: A differentiated region of the chloroplast that is a center of starch formation in green algae and hornworts.

pyrimidine: The smaller of the two kinds of nucleotide bases found in DNA and RNA: a nitrogenous base with a single-ring structure, such as cytosine, thymine, or uracil.

Q

quantum: The ultimate unit of light energy.

quiescent center: The initial region in the apical meristem of roots that has reached a state of relative inactivity.

R

raceme [L. *racemus,* bunch of grapes]: An indeterminate inflorescence in which the main axis is elongated but the flowers are borne on pedicels of about equal length.

rachis (rā′kĭs) [Gk. *rachis,* backbone]: The main axis of a spike; in ferns, the axis of a leaf (frond), from which the pinnae arise; in compound leaves, the extension of the petiole corresponding to the midrib of an entire leaf.

radially symmetrical: *See* actinomorphic.

radial micellation: The radial orientation of cellulose microfibrils in guard cell walls; plays a role in the movement of guard cells.

radial section: A longitudinal section cut parallel to the radius of a cylindrical body, such as a root or stem; in the case of secondary xylem, or wood, and secondary phloem, a section parallel to the rays.

radial system: In secondary xylem and secondary phloem, includes all the rays, the cells of which are derived from ray initials. Also called the horizontal system *or* ray system.

radicle [L. *radix,* root]: The embryonic root.

radioisotope: An unstable isotope of an element that decays or disintegrates spontaneously, emitting radiation. Also called a radioactive isotope.

raphe [Gk. *raphē,* seam]: (1) A ridge on seeds, formed by the stalk of the ovule, in seeds in which the stalk is sharply bent at the base of the ovule; (2) a groove on the frustule of a diatom.

raphides (răf′ĭ • dēz) [Gk. *rhaphis,* needle]: Fine, sharp, needlelike crystals of calcium oxalate found in the vacuoles of many plant cells.

ray: A panel of tissue, variable in width and height, formed by the ray initials in the vascular cambium and extending radially in the secondary xylem and secondary phloem.

ray flowers: In Asteraceae, the flattened, zygomorphic flowers; in contrast to the actinomorphic, tubular disk flowers. In many Asteraceae, the ray flowers occur around the margins of the inflorescence, the disk flowers in the center.

ray initial: An initial in the vascular cambium that gives rise to the ray cells of secondary xylem and secondary phloem.

reaction center: The complex of proteins and chlorophyll molecules of a photosystem, capable of converting light energy to chemical energy in the photochemical reaction.

reaction wood: Abnormal wood that develops in leaning trunks and limbs. *See also* compression wood; tension wood.

receptacle: The part of the axis of a flower stalk that bears the floral organs.

recessive: Describing a gene whose phenotypic expression is masked in the heterozygote by a dominant allele; heterozygotes are phenotypically indistinguishable from dominant homozygotes.

recombinant DNA: DNA formed either naturally or in the laboratory by the joining of segments of DNA from different sources.

reduction [L. *reductio,* bringing back; originally "bringing back" a metal from its oxide]: Gain of an electron by an atom; takes place simultaneously with oxidation (the loss of an electron by an atom), because an electron that is lost by one atom is accepted by another.

regular: *See* actinomorphic.

regulator gene: A gene that prevents or represses the activity of the structural genes in an operon.

replicate: Produce a facsimile or a very close copy; term used to indicate the production of a second molecule of DNA exactly like the first molecule or the production of a sister chromatid.

replication fork: In DNA synthesis, the Y-shaped structure formed at the point where the two strands of the original molecule are being separated and the complementary strands are being synthesized.

repressor: A protein that regulates DNA transcription; acts by preventing RNA polymerase from attaching to the promoter and transcribing the gene. *See also* operator.

resin duct: A tubelike intercellular space lined with resin-secreting cells (epithelial cells) and containing resin.

resonance energy transfer: The transfer of light energy from an excited chlorophyll molecule to a neighboring chlorophyll molecule, exciting the second molecule and allowing the first one to return to its ground state.

respiration: An intracellular process in which molecules, particularly pyruvate entering the citric acid cycle, are oxidized with the release of energy. The complete breakdown of sugar or other organic compounds to carbon dioxide and water is termed aerobic respiration, although the first steps of the process are anaerobic. Also called cellular respiration.

restriction enzymes: Enzymes that cleave the DNA double helix at specific nucleotide sequences.

reticulate venation: *See* netted venation.

reverse transcription: The process by which an RNA molecule is used as a template to make a single-stranded copy of DNA.

rhizobia [Gk. *rhiza,* root, + *bios,* life]: Bacteria of the genera *Rhizobium* or *Bradyrhizobium,* which may be involved in a symbiotic relationship with leguminous plants that results in nitrogen fixation.

rhizoids [Gk. *rhiza,* root]: (1) Branched rootlike extensions of fungi and algae that absorb water, food, and nutrients; (2) root-hair-like structures in liverworts, mosses, and some vascular plants, occurring on free-living gametophytes.

rhizome: A more or less horizontal underground stem.

rhizosphere: The volume of soil around plant roots that is influenced by root activity.

ribonucleic acid (RNA): Type of nucleic acid formed on chromosomal DNA and involved in protein synthesis; composed of chains of phosphate, sugar molecules (ribose), and purines and pyrimidines. RNA is the genetic material of many kinds of viruses.

ribose: A five-carbon sugar; a component of RNA.

ribosomal RNA (rRNA): Any of a number of specific molecules that form part of the structure of a ribosome and participate in the synthesis of proteins.

ribosome: A small particle composed of protein and RNA; the site of protein synthesis.

ring-porous wood: A wood in which the pores, or vessels, of the early wood are distinctly larger than those of the late wood, forming a well-defined ring in cross sections of the wood.

RNA: *See* ribonucleic acid.

root: The usually descending axis of a plant, normally below ground, which serves to anchor the plant and to absorb and conduct water and minerals into it.

rootcap: A thimblelike mass of cells that covers and protects the growing tip of a root. It typically consists of a central column of cells, the columella, and a lateral portion, the lateral rootcap, which surrounds the columella.

root hairs: Tubular outgrowths of epidermal cells of the root; greatly increase the absorbing surface of the root.

root pressure: The pressure developed in roots as the result of osmosis, which causes guttation of water from leaves and exudation from cut stumps.

rRNA: *See* ribosomal RNA.

Rubisco: RuBP carboxylase/oxygenase, the enzyme that catalyzes the initial reaction of the Calvin cycle, involving the fixation of carbon dioxide to ribulose 1,5-bisphosphate (RuBP).

runner: *See* stolon.

S

samara: Simple, dry, one-seeded or two-seeded indehiscent fruit with pericarp-bearing, winglike outgrowths.

sap: (1) The fluid contents of the xylem or the sieve elements of the phloem; (2) the fluid contents of the vacuole, called cell sap.

saprophyte [Gk. *sapros,* rotten, + *phyton,* plant]: An organism that secures its food directly from nonliving organic matter. Also called a saprobe.

sapwood: Outer part of the wood of a stem or trunk, usually distinguished from the heartwood by its lighter color; part of woody stem in which conduction of water takes place.

satellite DNA: A short nucleotide sequence repeated in tandem fashion many thousands of times; this region of the chromosome has a distinctive base composition and is not transcribed.

savanna: Grassland containing scattered trees.

scarification: The process of cutting or softening a seed coat to hasten germination.

schizo- [Gk. *schizein,* to split]: Prefix meaning "split."

schizocarp (skĭz′ō • karp): Dry simple fruit with two or more united carpels that split apart at maturity.

sclereid [Gk. *skleros,* hard]:A sclerenchyma cell with a thick, lignified secondary wall having many pits. Sclereids are variable in form but typically not very long; they may or may not be living at maturity.

sclerenchyma (skle • rĕng′kĭ • ma) [Gk. *skleros,* hard, + L. *enchyma,* infusion]: A supporting tissue composed of sclerenchyma cells, including fibers and sclereids.

sclerenchyma cells: Cells of variable form and size with more or less thick, often lignified, secondary walls; may or may not be living at maturity; include fibers and sclereids.

scutellum (sku • tĕl′ŭm) [L. *scutella,* small shield]: The single cotyledon of a grass embryo, specialized for absorption of the endosperm.

secondary cell wall: Innermost layer of the cell wall, formed in certain cells after cell elongation has ceased; secondary walls have a highly organized microfibrillar structure.

secondary growth: In plants, growth derived from secondary or lateral meristems, the vascular cambium and cork cambium, that results in an increase in girth; in contrast to primary growth, which results in an increase in length.

secondary metabolites: Molecules that are restricted in their distribution, both within the plant and among different plants; important to the survival and propagation of the plants that produce them. There are three major classes: alkaloids, terpenoids, and phenolics. Also called secondary products.

secondary plant body: The part of the plant body produced by the vascular cambium and the cork cambium; consists of secondary xylem, secondary phloem, and periderm.

secondary root: *See* lateral root.

secondary tissues: Tissues produced by the vascular cambium and cork cambium.

second messenger: A small molecule formed in or released into the cytosol in response to an external signal; it relays the signal to the interior of the cell; examples are calcium ions and cyclic AMP.

seed: A structure formed by maturation of the ovule of seed plants following fertilization.

seed coat: The outer layer of the seed, developed from the integuments of the ovule.

seedling: A young sporophyte, which develops from a germinating seed.

segregation: Separation of the chromosomes (and genes) derived from different parents at meiosis. *See also* Mendel's first law.

selectively permeable [L. *seligere,* to gather apart, + *permeare,* to go through]: Describing membranes that permit the passage of water and some solutes but block the passage of others; as distinct from semipermeable membranes, which permit the passage of water but not solutes.

sepal (sē′păl) [L. *sepalum,* covering]: One of the outermost flower structures, a unit of the calyx; sepals usually enclose the other flower parts in the bud.

septate [L. *septum,* fence]: Divided by cross walls into cells or compartments.

septum, *pl.* **septa:** A partition or cross wall.

sessile (sĕs′ĭl) [L. *sessilis,* of or fit for sitting, low, dwarfed]: Attached directly by the base; referring to a leaf lacking a petiole or to a flower or fruit lacking a pedicel.

seta, *pl.* **setae** (sē′ta) [L. *seta,* bristle]: In bryophytes, the stalk that supports the capsule, if present; part of the sporophyte.

sexual reproduction: The fusion of gametes, followed by meiosis and recombination at some point in the life cycle.

sheath: (1) The base of a leaf that wraps around the stem, as in grasses; (2) a tissue layer surrounding another tissue, such as a bundle sheath.

shoot: The aboveground portions, such as the stem and leaves, of a vascular plant.

short-day plants: Plants that must be exposed to light periods shorter than some critical length for flowering to occur; they usually flower in autumn.

shrub: A perennial woody plant of relatively low stature, typically with several stems arising from or near the ground.

sieve area: A portion of the sieve-element wall containing clusters of pores through which the protoplasts of adjacent sieve elements are interconnected.

sieve cell: A long, slender sieve element with relatively unspecialized sieve areas and with tapering end walls that lack sieve plates; found in the phloem of gymnosperms.

sieve element: The cell of the phloem that is involved in the long-distance transport of food substances; sieve elements are further classified into sieve cells and sieve-tube elements.

sieve plate: The part of the wall of sieve-tube elements that bears one or more highly differentiated sieve areas.

sieve tube: A series of sieve-tube elements arranged end to end and interconnected by sieve plates.

sieve-tube element: One of the component cells of a sieve tube; found only in flowering plants and typically associated with a companion cell. Also called sieve-tube member.

signal transduction: The process by which a cell converts an extracellular signal into a response.

silique [L. *siliqua,* pod]: The fruit characteristic of the mustard family; two-celled, the valves splitting from the bottom and leaving the placentae with the false partition stretched between. A small, compressed silique is called a silicle.

simple fruit: A fruit derived from one carpel or several united carpels.

simple leaf: An undivided leaf; as opposed to a compound leaf.

simple pit: A pit not surrounded by an overarching border of secondary wall; as opposed to a bordered pit.

simple tissue: A tissue composed of a single cell type; parenchyma, collenchyma, and sclerenchyma are simple tissues.

siphonaceous [Gk. *siphon,* tube, pipe]: In algae, multinucleate cells without cross walls; coenocytic. Also called siphonous.

siphonostele [Gk. *siphon,* pipe, + *stēlē,* pillar]: A type of stele containing a hollow cylinder of vascular tissue surrounding a pith.

size exclusion limit: The effective pore size of a plasmodesma.

slime: *See* P-protein.

softwood: A name commonly applied to the wood of a conifer.

solar tracking: The ability of the leaves and flowers of many plants to move diurnally, orienting themselves either perpendicular to or parallel to the sun's direct rays. Also called heliotropism.

solute: A molecule dissolved in a solution.

solute potential: *See* osmotic potential.

solution: Usually a liquid, in which molecules of a dissolved substance, the solute (for example, sugar), are dispersed between molecules of the solvent (for example, water).

solvent: The substance present in the greatest amount—usually a liquid—in a solution; the substances present in lesser amounts are called solutes.

somatic cells [Gk. *soma,* body]: All cells except the gametes and the cells from which the gametes develop.

soredium, *pl.* **soredia** [Gk. *soros,* heap]: A specialized reproductive unit of lichens, consisting of a few cyanobacterial or green algal cells surrounded by fungal hyphae.

sorus, *pl.* **sori** (sō′rŭs) [Gk. *soros,* heap]: A group or cluster of sporangia or spores.

specialized: (1) Of organisms, having special adaptations to a particular habitat or mode of life; (2) of cells, having particular functions.

speciation: The origin of new species in evolution.

species, *pl.* **species** [L. *species,* kind, sort]: A kind of organism; species are designated by binomial names written in italics.

specific epithet: The second part of a species name; for example, *mays* of *Zea mays* (maize).

specificity: Uniqueness, as in proteins in given organisms or enzymes in given reactions.

sperm: A mature male gamete, usually motile and smaller than the female gamete.

spermatangium, *pl.* **spermatangia** (spûr′mă • tăn′jĭ • ŭm) [Gk. *sperma,* seed, sperm, + L. *tangere,* to touch]: In the red algae, the structure that produces spermatia.

spermatium, *pl.* **spermatia** [Gk. *sperma,* seed, sperm]:In the red algae and some fungi, a minute, nonmotile male gamete.

spermatogenous cell: The cell of the male gametophyte, or pollen grain, of gymnosperms, which divides mitotically to form two sperm.

spermatophyte [Gk. *sperma,* seed, + *phyton,* plant]: A seed plant.

spermogonium, *pl.* **spermogonia** (spûr′mă • gō′nĭ • ŭm) [Gk. *sperma,* seed, sperm, + *gonos,* offspring]: In the rust fungi, the structure that produces spermatia.

spike [L. *spica,* head of grain]: An indeterminate inflorescence in which the main axis is elongated and the flowers are sessile.

spikelet: The unit of inflorescence in grasses; a small group of grass flowers.

spindle fibers: Bundles of microtubules, some of which extend from the kinetochores of the chromosomes to the poles of the spindle.

spine: A hard, sharp-pointed structure; usually a modified leaf, or part of a leaf.

spirillum, *pl.* **spirilli** [L. *spira,* coil]: A long coiled or spiral bacterium.

spongy parenchyma: A leaf tissue composed of loosely arranged, chloroplast-bearing cells.

sporangiophore (spo • răn′jĭ • o • fōr′) [Gk. *spora,* seed, + *pherein,* to carry]: A branch bearing one or more sporangia.

sporangium, *pl.* **sporangia** (spo • răn′jĭ • ŭm) [Gk. *spora,* seed, + *angeion,* vessel]: A hollow unicellular or multicellular structure in which spores are produced.

spore: A reproductive cell, usually unicellular, capable of developing into an adult without fusion with another cell.

spore mother cell: A diploid ($2n$) cell that undergoes meiosis to produce (usually) four haploid cells (spores) or four haploid nuclei.

sporic meiosis: Meiosis resulting in the formation of haploid spores by a diploid individual, or sporophyte. The spores give rise to haploid individuals, or gametophytes, which eventually produce gametes that fuse to form diploid zygotes; the zygotes, in turn, develop into sporophytes. This kind of life cycle is known as an alternation of generations.

sporophyll (spō′rō • fĭl): A modified leaf or leaflike organ that bears sporangia; a term applied to the stamens and carpels of angiosperms, fertile fronds of ferns, and other, similar structures.

sporophyte (spō′rō • fīt): The spore-producing, diploid ($2n$) phase in a life cycle characterized by alternation of generations.

sporopollenin: The tough substance of which the exine, or outer wall, of spores and pollen grains is composed; a cyclic alcohol highly resistant to decay.

stalk cell: *See* sterile cell.

stamen (stā′mĕn) [L. *stamen,* thread]: The part of the flower producing the pollen, composed (usually) of anther and filament; collectively, the stamens make up the androecium.

staminate (stăm′ĭ • nat): Pertaining to a flower having stamens but no functional carpels.

starch [M.E. *sterchen,* to stiffen]: A complex, insoluble carbohydrate; the chief food storage substance of plants; composed of a thousand or more glucose units.

starch sheath: The innermost layer or layers of cells in the cortex of the stem or hypocotyl when this region is characterized by a conspicuous and rather stable accumulation of starch.

statocytes: Gravity-sensing cells; in the stem, they occur in the starch sheath; in the root, in the columella of the rootcap.

statoliths [Gk. *statos,* stationary, + *lithos,* stone]: Gravity sensors; starch-containing plastids (amyloplasts) or other bodies in the cytoplasm.

stele (stē′le) [Gk. *stēlē,* pillar]: The central cylinder, inside the cortex, of roots and stems of vascular plants.

stem: The part of the axis of vascular plants that is above ground, as well as anatomically similar portions below ground, such as rhizomes or corms.

stem bundle: Vascular bundle belonging to the stem.

sterigma, *pl.* **sterigmata** [Gk. *sterigma,* prop]: A small, slender protuberance of a basidium, bearing a basidiospore.

sterile cell: One of two cells produced by division of the generative cell in developing pollen grains of gymnosperms; it is not a gamete, and eventually degenerates.

stigma: (1) The region of a carpel that serves as a receptive surface for pollen grains and on which they germinate; (2) a light-sensitive, pigmented structure; *see also* eyespot.

stigmatic tissue: The pollen-receptive tissue of the stigma.

stipe: A supporting stalk, such as the stalk of a gill fungus or the leaf stalk of a fern.

stipule (stĭp′yūl): An appendage, often leaflike, on either side of the basal part of a leaf, or encircling the stem, in many kinds of flowering plants.

stolon (stō′lŏn) [L. *stolo,* shoot]: A stem that grows horizontally along the ground surface and may form roots, such as the runners of a strawberry plant. Also called a runner.

stoma, *pl.* **stomata** (stō′ma) [Gk. *stoma,* mouth]: A minute opening, bordered by guard cells, in the epidermis of leaves and stems through which gases pass; also used to refer to the entire stomatal apparatus: the guard cells plus their included pore.

stramenopiles: *See* heterokonts.

Strasburger cells: *See* albuminous cells.

stratification: The process of exposing seeds to low temperatures for an extended period before attempting to germinate them at warm temperatures.

strict anaerobe: An anaerobe that is killed by oxygen; can live only in the absence of oxygen.

strobilus, *pl.* **strobili** (strōb′ĭ • lŭs) [Gk. *strobilos,* cone]: A reproductive structure consisting of a number of modified leaves (sporophylls) or ovule-bearing scales grouped terminally on a stem; a cone. Strobili occur in many kinds of gymnosperms, lycophytes, and sphenophytes.

stroma [Gk. *stroma,* anything spread out]: The ground substance of plastids.

structural gene: Any gene that codes for a protein; as opposed to a regulatory gene.

style [Gk. *stylos,* column]: A slender column of tissue that arises from the top of the ovary and through which the pollen tube grows.

sub- [L. *sub,* under, below]: Prefix meaning "under" or "below"; for example, subepidermal, "underneath the epidermis."

suberin (sū′ber • ĭn) [L. *suber,* cork oak]: Fatty material found in the cell walls of cork tissue and in the Casparian strip of the endodermis.

subsidiary cell: An epidermal cell morphologically distinct from other epidermal cells and associated with a pair of guard cells. Also called an accessory cell.

subspecies: The primary taxonomic subdivision of a species. "Varieties" are used as equivalent to subspecies by some botanists, or subspecies may be divided into varieties.

substrate [L. *substratus,* strewn under]: (1) The foundation to which an organism is attached; (2) the substance acted on by an enzyme.

substrate phosphorylation: Phosphorylation—the formation of ATP from ADP and inorganic phosphate—that takes place during glycolysis.

succession: In ecology, the orderly progression of changes in community composition during the development of vegetation in any area, from initial colonization to attainment of the climax typical of a particular geographic area.

succulent: A plant with fleshy, water-storing stems or leaves.

sucker: A sprout produced by the roots of some plants that gives rise to a new plant; an erect sprout that arises from the base of stems.

sucrase (sū′krās): An enzyme that hydrolyzes sucrose into glucose and fructose. Also called invertase.

sucrose (sū′krōs): A disaccharide (glucose plus fructose) found in many plants; the primary form in which sugar produced by photosynthesis is translocated.

supergroup: One of seven subdivisions of the domain Eukarya.

superior ovary: An ovary that is free and separate from the calyx.

suspension: A heterogeneous dispersion in which the dispersed phase consists of solid particles sufficiently large that they will settle out of the fluid dispersion medium under the influence of gravity.

suspensor: A structure at the base of the embryo in many vascular plants. In some plants, it pushes the embryo into nutrient-rich tissue of the female gametophyte.

symbiosis (sĭm′bi • ō′sĭs) [Gk. *syn,* together with, + *bios,* life]: The living together in close association of two or more dissimilar organisms; includes parasitism (in which the association is harmful to one of the organisms) and mutualism (in which the association is advantageous to both).

sympatric speciation [Gk. *syn,* together with, + *patra,* fatherland, country]: Speciation that occurs without geographic isolation of a population of organisms; usually occurs as the result of hybridization accompanied by polyploidy; may occur in some cases as a result of disruptive selection.

symplast [Gk. *syn,* together with, + *plastos,* molded]: The interconnected protoplasts and their plasmodesmata; the movement of substances in the symplast is called symplastic movement, or symplastic transport.

sympodium, *pl.* **sympodia:** A stem bundle and its associated leaf traces.

syn-, sym- [Gk. *syn,* together with]: Prefix meaning "together."

synapomorphies: Character states (two or more forms of a character) that arose in the common ancestor of a group and are present in all of its members.

synapsis [Gk. *synapsis,* contract or union]: The pairing of homologous chromosomes that occurs prior to the first meiotic division; crossing-over occurs during synapsis.

synergids (sĭ • nûr′jĭds): In flowering plants, two short-lived cells lying close to the egg in the mature embryo sac of the ovule. *Amborella* is an exception, with three synergids.

syngamy [Gk. *syn,* together with, + *gamos,* marriage]: *See* fertilization.

synthesis: The formation of a more complex substance from simpler ones.

systematics: Scientific study of the kinds and diversity of organisms and of the relationships between them.

T

taiga: The northern coniferous forest.

tangential section: A longitudinal section cut at a right angle to the radius of a cylindrical structure, such as a root or stem; in the case of secondary xylem, or wood, and secondary phloem, at a right angle to the rays.

tannins: Phenolic compounds present in relatively high concentrations in the leaves of a wide variety of woody plants. The bitter taste of tannins is repellent to insects, reptiles, birds, and higher animals; used commercially in tanning, dyeing, and preparation of ink.

tapetum (ta • pē′tŭm) [Gk. *tapes,* carpet]: Nutritive tissue in the sporangium, particularly an anther.

taproot: The primary root of a plant, formed in direct continuation with the root tip or radicle of the embryo; forms a stout, tapering main root from which arise smaller, lateral roots.

taxon: General term for any one of the taxonomic categories, such as species, class, order, or phylum.

taxonomy [Gk. *taxis,* arrangement, + *nomos,* law]: The science of the classification of organisms.

teliospore (tē′lĭ • ō • spōr): In the rust fungi, a thick-walled spore in which karyogamy and meiosis occur and from which basidia develop.

telium, *pl.* **telia:** In the rust fungi, the structure that produces teliospores.

telomere: The end of a chromosome; has repetitive DNA sequences that help counteract the tendency of the chromosome, without telomeres, to shorten with each round of replication.

telophase: The last stage in mitosis and meiosis, during which the chromosomes become reorganized into two new nuclei.

template: A pattern or mold guiding the formation of a negative or complement; a term applied especially to DNA duplication, which is explained in terms of a template hypothesis.

tendril [L. *tendere,* to extend]: A modified leaf or part of a leaf or a modified stem forming a slender, coiling structure that aids in support of the plant's stems; found only in some angiosperms.

tension wood: The reaction wood of angiosperms; develops on the upper side of leaning trunks and limbs.

tepal: One of the units of a perianth that is not differentiated into sepals and petals.

testcross: A cross of an individual having a dominant phenotype with a homozygous recessive; used to determine whether the individual is homozygous or heterozygous.

tetrad (tĕt′răd): (1) A group of four spores formed from a spore mother cell by meiosis; (2) *see* bivalent.

tetraploid (tĕt′ra • ploid) [Gk. *tetra,* four, + *ploos,* fold]: Having twice the usual, or diploid ($2n$), number of chromosomes; $4n$.

tetrasporangium, *pl.* **tetrasporangia** [Gk. *tetra,* four, + *spora,* seed, + *angeion,* vessel]: In certain red algae, a sporangium in which meiosis occurs, resulting in the production of tetraspores.

tetraspore [Gk. *tetra,* four, + *spora,* seed]: In certain red algae, the four spores formed by meiotic division in the tetrasporangium of a spore mother cell.

tetrasporophyte [Gk. *tetra,* four, + *spora,* seed, + *phyton,* plant]: In certain red algae, a diploid individual that produces tetrasporangia.

texture: Of wood, refers to the relative size and amount of variation in size of elements within the growth rings.

thallophyte: Previously used to designate fungi and algae collectively, now largely abandoned.

thallus (thăl′ŭs) [Gk. *thallos,* sprout]: A type of body that is not differentiated into root, stem, or leaf. The term was commonly used when fungi and algae were considered to be plants, to distinguish their simple construction, and that of certain gametophytes, from the differentiated bodies of plant sporophytes and the elaborate gametophytes of the bryophytes.

theory [Gk. *theorein,* to look at]: A well-tested hypothesis, one unlikely to be rejected by further evidence.

thermodynamics [Gk. *thermē,* heat, + *dynamis,* power]: The study of energy exchanges, using heat as the most convenient form of measurement of energy. The first law of thermodynamics states that in all processes, the total energy of the universe remains constant. The second law of thermodynamics states that the entropy, or degree of randomness, of a system tends to decrease.

thermophile: An organism with an optimal growth temperature between 45° and 80°C.

thigmomorphogenesis: The alteration of plant growth patterns in response to mechanical stimuli.

thigmotropism [Gk. *thigma,* touch]: A response to contact with a solid object.

thorn: A hard, woody, pointed branch.

thylakoid [Gk. *thylakos,* sac, + *oides,* like]: A saclike membranous structure in cyanobacteria and in the chloroplasts of eukaryotic organisms; in chloroplasts, stacks of thylakoids form the grana; chlorophylls are found within the thylakoids.

thymine: A pyrimidine occurring in DNA but not in RNA. *See also* uracil.

Ti plasmid: A circular plasmid of *Agrobacterium tumifaciens* that enables the bacterium to infect plant cells and produce a tumor (crown gall tumor); a powerful tool in biotechnology in the transfer of foreign genes into plant genomes.

tissue: A group of cells organized into a structural and functional unit.

tissue culture: A technique for maintaining fragments of plant or animal tissue alive in a medium after removal from the organism.

tissue system: In plants, a tissue or group of tissues organized into a structural and functional unit in the plant or plant organ. There are three tissue systems: dermal, vascular, and ground, or fundamental.

tonoplast [Gk. *tonos,* stretching, tension, + *plastos,* formed, molded]: The cytoplasmic membrane surrounding the vacuole in plant cells. Also called the vacuolar membrane.

torus, *pl.* **tori:** The central thickened part of the pit membrane in the bordered pits of conifers and some other gymnosperms and some angiosperms.

totipotent: For a plant cell, having the potential to develop into an entire plant.

tracheary element: The general term for a water-conducting cell in vascular plants; tracheids and vessel elements.

tracheid (trā′kē • ĭd): An elongated, thick-walled conducting and supporting cell of xylem, with tapering ends and pitted walls without perforations, in contrast to a vessel element; found in nearly all vascular plants.

tracheophyte: A vascular plant.

transcellular pathway: A pathway from cell to cell, across the plasma membranes and tonoplasts.

transcription: The synthesis of messenger RNA, which is a copy of a portion of one strand of the double-stranded DNA helix.

transcription factors: Proteins that directly or indirectly affect the initiation of transcription.

transduction: The transfer of genes from one organism to another by a virus.

transfer cell: A specialized parenchyma cell with wall ingrowths that increase the surface area of the plasma membrane; apparently functions in the short-distance transfer of solutes.

transfer RNA (tRNA): Low-molecular-weight RNA that becomes attached to an amino acid and guides it to the correct position on the ribosome for protein synthesis; there is at least one tRNA molecule for each amino acid.

transformation: The transfer of naked DNA from one organism to another; transposons are often used as vectors in transformation when carried out in the laboratory. Also called gene transfer.

transgenic organism: An organism whose genome contains DNA, from the same or a different species, that has been modified by the methods of genetic engineering.

transition region: The region in the primary plant body showing transitional characteristics between structures of root and shoot.

translation: The synthesis of a polypeptide directed by the nucleotide sequence of messenger RNA.

translocation: (1) In plants, the long-distance transport of water, minerals, or food; most often used to refer to food transport; (2) in genetics, the interchange of chromosome segments between nonhomologous chromosomes.

transmembrane proteins: Globular proteins that traverse the lipid bilayer of cellular membranes; some extend across the lipid bilayer as a single alpha helix, others as multiple alpha helices.

transmitting tissue: A tissue similar to stigmatic tissue and serving as a path for the pollen tube in the style.

transpiration [Fr. *transpirer,* to perspire]: The loss of water vapor by plant parts; most transpiration occurs through stomata.

transpiration stream: The flow of water through the xylem, from the roots to the leaves.

transport protein: A specific membrane protein responsible for transferring solutes across membranes; transport proteins are grouped into three broad classes: pumps, carriers, and channels.

transposon [L. *transponere,* to change the position of something]: A DNA sequence that carries one or more genes and is flanked by sequences of nucleotides that confer the ability to move from one DNA molecule to another; an element capable of transposition, which is the changing of a chromosomal location.

transverse section: A section cut perpendicular, or at a right angle, to the longitudinal axis of a plant part.

tree: A perennial woody plant generally with a single stem (trunk).

tricarboxylic acid cycle (TCA cycle): *See* citric acid cycle.

trichogyne [Gk. *trichos,* hair, + *gynē,* woman]: In the red algae and certain ascomycetes and Basidiomycota, a receptive protuberance of the female gametangium for the conveyance of spermatia.

trichome [Gk. *trichos,* hair]: An outgrowth of the epidermis, such as a hair, scale, or water vesicle.

triglyceride: Glycerol ester of fatty acids; the main constituent of fats and oils.

triose [Gk. *tries,* three, + *-ose,* suffix indicating a carbohydrate]: Any three-carbon sugar.

triple fusion: In angiosperms, the fusion of the second male gamete, or sperm, with the polar nuclei, resulting in formation of a primary endosperm nucleus, which is triploid ($3n$) in most angiosperms.

triploid [Gk. *triploos,* triple]: Having three complete chromosome sets per cell; $3n$.

tritium: A radioactive isotope of hydrogen, ^3H. The nucleus of a tritium atom contains one proton and two neutrons, whereas the more common hydrogen nucleus (H) consists only of a proton.

tRNA: *See* transfer RNA.

-troph, tropho- [Gk. *trophos,* feeder]: Suffix or prefix meaning "feeder," "feeding," or "nourishing"; for example, autotrophic, "self-nourishing."

trophic level: A step in the movement of energy through an ecosystem, represented by a particular set of organisms.

tropism [Gk. *tropē,* turning]: A movement response to an external stimulus in which the direction of movement is usually determined by the direction from which the most intense stimulus comes.

tube cell: In male gametophytes, or pollen grains, of seed plants, the cell that develops into the pollen tube. Also called a vegetative cell.

tuber [L. *tuber,* swelling]: An enlarged, short, fleshy underground stem, such as that of the potato.

tundra: A treeless circumpolar region, best developed in the Northern Hemisphere and mostly found north of the Arctic Circle.

tunica-corpus: The organization of the shoot apex of most angiosperms and a few gymnosperms, consisting of one or more peripheral layers of cells (the tunica layers) and an interior (the corpus). The tunica layers undergo surface growth (by anticlinal divisions), and the corpus undergoes volume growth (by divisions in all planes).

turgid (tûr′jĭd) [L. *turgidus,* to be swollen]: Swollen, distended; referring to a cell that is firm due to water uptake.

turgor pressure [L. *turgor,* swelling]: Pressure within the cell resulting from the movement of water into the cell.

tylose [Gk. *tylos,* lump]: A balloonlike outgrowth from a ray or axial parenchyma cell through the pit in a vessel wall and into the lumen of the vessel.

type specimen: Usually a dried plant specimen housed in an herbarium; selected by a taxonomist to serve as the basis for comparison with other specimens in determining whether they are members of the same species.

U

umbel (ŭm′bŭl) [L. *umbella,* sunshade]: An inflorescence with individual pedicels arising from the apex of the peduncle.

unicellular: Composed of a single cell.

unisexual: Usually describing a flower lacking either stamens or carpels; a perianth may be present or absent.

unit membrane: A visually definable, three-layered membrane, consisting of two dark layers separated by a lighter layer, as seen with an electron microscope.

uracil (ū′ra • sĭl): A pyrimidine found in RNA but not in DNA. *See also* thymine.

urediniospore [L. *uredo,* blight, + *spora,* spore]: In rust fungi, a reddish, binucleate spore produced in summer.

uredinium, *pl.* **uredinia** [L. *uredo,* blight]: In rust fungi, the structure that produces urediniospores.

V

vacuolar membrane: *See* tonoplast.

vacuole [L. *vacuus,* empty]: A space or cavity within the cytoplasm filled with a watery fluid, the cell sap; a lysosomal compartment.

variation: The differences among the offspring of a particular species.

variety: A group of plants or animals below the species rank. Some botanists view varieties as equivalent to subspecies; others consider them divisions of subspecies.

vascular [L. *vasculum,* small vessel]: Pertains to any plant tissue or region consisting of or giving rise to conducting tissue; for example, xylem, phloem, or vascular cambium.

vascular bundle: A strand of tissue containing primary xylem and primary phloem (and procambium, if still present) and frequently enclosed by a bundle sheath of parenchyma or fibers.

vascular cambium: A cylindrical sheath of meristematic cells that divides to produce secondary phloem and secondary xylem.

vascular plant: A plant that has xylem and phloem. Also called a tracheophyte.

vascular rays: Ribbonlike sheets of parenchyma that extend radially through the wood, across the cambium, and into the secondary phloem; always produced by the vascular cambium.

vascular system: All the vascular tissues in their specific arrangement in a plant or plant organ.

vector [L. *vector,* bearer, carrier; from *vehere,* to carry]: (1) A pathogen that carries a disease from one organism to another; (2) in genetics, any virus or plasmid DNA into which a gene is integrated and subsequently transferred into a cell.

vegetative: Of, relating to, or involving propagation by asexual processes; also referring to nonreproductive plant parts.

vegetative cell: *See* tube cell.

vegetative reproduction: (1) In seed plants, reproduction by means other than producing seeds; apomixis; (2) in other organisms, reproduction by vegetative spores, fragmentation, or division of the somatic body. Unless a mutation occurs, each daughter cell or individual is genetically identical to its parent.

vein: A vascular bundle forming part of the framework of the conducting and supporting tissue of a leaf or other expanded organ.

velamen [L. *velumen,* fleece]: A multiple epidermis covering the aerial roots of some orchids and aroids; also occurs on some terrestrial roots.

venation: Arrangement of veins in the leaf blade.

venter [L. *venter,* belly]: The enlarged basal portion of an archegonium containing the egg.

vernalization [L. *vernalis,* spring]: The induction of flowering by cold treatment.

vessel [L. *vasculum,* small vessel]: A tubelike structure of the xylem composed of elongated cells (vessel elements) placed end to end and connected by perforations. Its function is to conduct water and minerals through the plant body; found in nearly all angiosperms and a few other vascular plants (for example, gnetophytes).

vessel element: One of the cells composing a vessel. Also called vessel member.

viable [L., *vita,* life]: Able to live.

volva [L. *volva,* wrapper]: A cuplike structure at the base of the stalk of certain mushrooms.

W

wall pressure: The pressure of the cell wall exerted against the turgid protoplast; opposite and equal in magnitude to the turgor pressure.

water potential: The algebraic sum of the solute potential and the pressure potential, or wall pressure; the potential energy of water.

water vesicle: An enlarged epidermal cell in which water is stored; a type of trichome.

weed [O.E. *weod,* used at least since the year 888 in its present meaning]: Generally an herbaceous plant not valued for use or beauty, growing wild, and regarded as using ground required by or hindering the growth of useful vegetation.

whorled phyllotaxis: Arrangement of three or more leaves or floral parts in a circle at a node.

wild type: In genetics, the phenotype or genotype that is characteristic of the majority of individuals of a species in a natural environment.

wood: Secondary xylem.

X

xanthophyll (zăn′thō • fĭll) [Gk. *xanthos,* yellowish-brown, + *phyllon,* leaf]: A yellow chloroplast pigment; a member of the carotenoid class.

xerophyte [Gk. *xeros,* dry, + *phyton,* plant]: A plant that has adapted to arid habitats.

xylem [Gk. *xylon,* wood]: A complex vascular tissue through which most of the water and minerals of a plant are conducted; characterized by the presence of tracheary elements.

Y

yeast: A unicellular fungus that reproduces primarily by budding; not a taxonomic group but merely a morphological growth form; encompasses zygomycetes, ascomycetes, and basidiomycetes.

Z

zeatin (zē′ă • tĭn): A plant hormone; a natural cytokinin isolated from maize.

zooplankton [Gk. *zōē,* life, + *plankton,* wanderer]: A collective term for the nonphotosynthetic organisms present in plankton.

zoosporangium: A sporangium bearing zoospores.

zoospore (zō′o • spōr): A motile spore, found among algae, oomycetes, and chytrids.

zygomorphic [Gk. *zygo,* pair, + *morphē,* form]: Describing a type of flower capable of being divided into two symmetrical halves only by a single longitudinal plane passing through the axis. Also called bilaterally symmetrical.

zygosporangium, *pl.* **zygosporangia:** A sporangium containing one or more zygospores.

zygospore: A thick-walled, resistant spore that develops from a zygote, resulting from the fusion of isogametes.

zygote (zī′gōt) [Gk. *zygotos,* paired together]: The diploid ($2n$) cell resulting from the fusion of male and female gametes.

zygotic meiosis: Meiosis in a zygote to form four haploid cells, which divide by mitosis to produce either more haploid cells or a multicellular individual that eventually gives rise to gametes.

All photographs not credited herein are by Ray F. Evert. All section openers and Chapters 2 and 25 openers are paintings by Rhonda Nass/Ampersand.

CHAPTER 1

Chapter Opener Joe McDonald/Animals Animals-Earth Scenes; **1.1** Courtesy of F. Hasler et al., the National Oceanic and Atmospheric Administration, and NASA; **1.2** © Stanley M. Awramik, University of California/Biological Photo Service; **1.3** NASA/JPL/JHUAPL/MSSS/Brown University; **1.4** © Stoelwinder/agefotostock; **1.5** © Carol B. Jones; **1. 6** A. E. Seaman Mineral Museum/Photo by Tom Waggoner; **1.7** © Inger Vandyke/OceanwideImages.com; **1.8** Dr. Jeremy Burgess/Science Photo Library/Photo Researchers, Inc.; **1.9** After W. Troll. 1937. *Vergleichende Morphologie der Hoheren Pflanzen,* vol. 1, pt. 1, Verlage von Gebrüder Borntraeger, Berlin; **1.10a** Dr. Anne La Bastille/Photo Researchers, Inc.; **1.10b** B. C. Alexander/Photo Researchers, Inc.; **1.10c** Martin Harvey/The Wildlife Collection; **1.10d** Jack Swenson/The Wildlife Collection; **1.10e** Fred Hirschmann; **1.10f** Stephen P. Parker/Photo Researchers, Inc.; **1.12 (left)** The Irish Image Collection/Superstock; **1.12 (center)** Ed Pritchard/GettyImages; **1.12 (right)** AP Photo/Charlie Riedel; **1.12 (line art)** Adapted from P. Ehrlich et al. 1977. *Ecoscience: Population, Resources, Environment.* W. H. Freeman and Company, New York; **1.13a** Michael J. Blaylock, Ph.D.; **1.13b** Photo by Ryan Somma; **1.13c** B. Moose Peterson/WRP; **1.14** José Martinez Zapater, University of Valencia, Spain; **1.15a** Richard Levine/Alamy; **1.15b** William Byrne Drumm/Drumm Photography.

CHAPTER 2

2.3 H. Curtis and N. Sue Barnes. 1994. *Invitation to Biology,* 5th ed. Worth Publishers, New York; **2.4a, b** Chemistry from H. Curtis and N. Sue Barnes. 1994. *Invitation to Biology,* 5th ed. Worth Publishers, New York; **2.4c** L. M. Beidler; **2.5b** Chemistry from H. Curtis and N. Sue Barnes. 1994. *Invitation to Biology,* 5th ed. Worth Publishers, New York; **2.6** M. Kruatrachue and R. F. Evert. 1977. *American Journal of Botany* 64, 310–325; **2.7** After H. Curtis and N. Sue Barnes. 1994. *Invitation to Biology,* 5th ed. Worth Publishers, New York; **2.8** Chemistry from H. Curtis and N. Sue Barnes. 1994. *Invitation to Biology,* 5th ed. Worth Publishers, New York; **2.10** B. E. Juniper; **Page 25** dwphotos/iStockphoto; **2.14–2.21** H. Curtis and N. Sue Barnes. 1994. *Invitation to Biology,* 5th ed. Worth Publishers, New York; **2.22a (photo)** Dr. Jeremy Burgess/Science Photo Library/Photo Researchers, Inc.; **2.22b (photo)** Dr. Morley Read/Science Photo Library/Photo Researchers, Inc.; **2.22c (photo)** Gerry Ellis/Ellis Nature Photography; **2.22d (photo)** © Bill Strode/Woodfin Camp & Associates; **2.23b** Carr Clifton/Getty Images; **2.24** Gerry Ellis/Ellis Nature Photography; **2.25a** Frans Lanting/Minden Pictures; **2.25b** Dwight Kuhn/Bruce Coleman, Inc.; **2.26** Albert F. W. Vick, Jr./National Wildlife Research Center; **2.27** Katherine Esau; **2.28b** Steve Solum/Bruce Coleman, Inc.

CHAPTER 3

Chapter Opener Russell Kightley Media; **3.1a** Michael W. Davidson at Florida State University; **3.1b** Courtesy of the National Library of Medicine; **3.2** A. Ryter; **3.3 (micrograph)** Michael A. Walsh; **3.6** Katherine Esau; **3.8** Susan E. Jordan Eichhorn; **3.9** A. Trojan and H. Gabryś. 1996. *Plant Physiology* 111, 419–425; **3.11** R. Bock and J. N. Timmis. 2008. *BioEssays* 30, 556–566; **3.12** Myron C. Ledbetter; **3.14** Roland R. Dute; **3.15** After W. W. Thomson and J. M. Whatley. 1980. *Annual Review of Plant Physiology* 31, 375–394; **3.16** David Stetler; **3.17–3.18** Katherine Esau; **3.19** M. Kruatrachue and R. F. Evert. 1977. *American Journal of Botany* 64, 310–325; **3.20a** Mary Alice Webb; **3.21** P. Boevink et al. 1996. *The Plant Journal* 10, 935–941; **3.22a, b** Roland R. Dute; **3.23** After B. Alberts et al. 1994. *Molecular Biology of the Cell,* 3rd ed. Garland Publishing Inc., New York; **3.24** Courtesy John E. Heuser; **3.25** After: H. Curtis and N. Sue Barnes. 1994. *Invitation to Biology,* 5th ed. Worth Publishers, New York; **3.27b** M. V. Parthasarathy et al. 1985. *American Journal of Botany* 72, 1318–1323; **Page 56** After Richard E. Williamson. 1986. *Plant Physiology* 82, 631–634, and B. Alberts et al. 1989. *Molecular Biology of the Cell,* 2nd ed. Garland Publishing Inc., New York; **3.28a** H. Curtis and N. Sue Barnes. 1994. *Invitation to Biology,* 5th ed. Worth Publishers, New York; **3.28b** Lewis Tilney; **3.29a** Brian Wells and Keith Roberts. In: B. Alberts et al. 1989. *Molecular Biology of the Cell,* 2nd ed. Garland Publishing, Inc., New York; **3.29b** After B. Alberts et al. 1989. *Molecular Biology of the Cell,* 2nd ed. Garland Publishing, Inc., New York; **3.31** After Katherine Esau. 1977. *Anatomy of Seed Plants,* 2nd ed., John Wiley & Sons, Inc., New York; **3.33** After R. D. Preston. 1974. In: A. W. Robards ed. *Dynamic Aspects of Plant Ultrastructure.* McGraw-Hill Book Company, New York; **3.34–3.35** After L. Taiz and E. Zeiger. 1991. *Plant Physiology.* Benjamin/Cummings Publishing Co., Redwood City, CA; **3.37a** W. A. Russin and R. F. Evert. 1985. *American Journal of Botany* 72, 1232–1247; **3.38** After B. Alberts et al. 1994. *Molecular Biology of the Cell,* 3rd ed. Garland Publishing, Inc., New York; **3.39** After Catharina J. Venverloo and K. R. Libbenga. 1987. *Journal of Plant Physiology* 131, 267–284; **3.41** W. T. Jackson. 1967. *Physiologia Plantarum* 20, 20–29; **3.45** J. Cronshaw; **3.46** P. K. Hepler. 1982. *Protoplasma* 111, 121–133; **3.47** Russell H. Goddard et al. 1994. *Plant Physiology* 104, 1–6.

CHAPTER 4

Chapter Opener Ottfried Schreiter/Photolibrary; **4.2** H. Curtis and N. Sue Barnes. 1994. *Invitation to Biology,* 5th ed. Worth Publishers, New York; **4.3** LatitudeStock/Alamy; **4.4** H. Curtis and N. Sue Barnes. 1994. *Invitation to Biology,* 5th ed. Worth Publishers, New York; **4.5** After H. Curtis and N. Sue Barnes. 1994. *Invitation to Biology,* 5th ed. Worth Publishers, New York; **4.6** After A. L. Lehninger. 1975. *Biochemistry,* 2nd ed. Worth Publishers, New York; **Page 80** Doug Wechsler/Earth Scenes; **4.9** After S. J. Singer and G. L. Nicolson. 1972. *Science* 175, 720–731; **4.10, 4.12** After B. Alberts et al. 1994. *Molecular Biology of the Cell,* 3rd ed. Garland Publishing, Inc., New York; **4.14** Brigit Satir; **4.15** After H. Curtis and N. Sue Barnes. 1994. *Invitation to Biology,* 5th ed. Worth Publishers, New York; **4.16** © David G. Robinson; **4.17** After N. A. Campbell. 1996. *Biology,* 4th ed. The Benjamin/Cummings Publishing Company, Inc., Menlo Park, CA; **4.18a, b** K. Robinson-Beers and R. F. Evert. 1991. *Planta* 184, 307–318; **4.19** E. B. Tucker. 1982. *Protoplasma* 113, 193–201.

CHAPTER 5

Chapter Opener Ed Reschke/PhotoLibrary; **5.2–5.4** H. Curtis and N. Sue Barnes. 1994. *Invitation to Biology,* 5th ed. Worth Publishers, New York; **5.5** After A. L. Lehninger. 1975. *Biochemistry,* 2nd ed. Worth Publishers, New York; **5.6** After H. Curtis and N. Sue Barnes. 1994. *Invitation to Biology,* 5th ed. Worth Publishers, New York; **5.7** Thomas A. Steitz; **5.8b** D. Sadava et al. 2008. *Life: The Science of Biology,* 8th ed. Sinauer Associates, Inc., Sunderland, MA; **5.9** H. Curtis and N. Sue Barnes. 1994. *Invitation to Biology,* 5th ed. Worth Publishers, New York; **5.10, 5.12** After H. Curtis and N. Sue Barnes. 1994. *Invitation to Biology,* 5th ed. Worth Publishers, New York; **5.13** H. Curtis and N. Sue Barnes. 1994. *Invitation to Biology,* 5th ed. Worth Publishers, New York.

CHAPTER 6

Chapter Opener V&A Images/Alamy; **6.1** Susan E. Jordan Eichhorn; **6.2** After H. Curtis and N. Sue Barnes. 1994. *Invitation to Biology,* 5th ed. Worth Publishers, New York; **6.8** H. Curtis and N. Sue Barnes. 1994. *Invitation to Biology,* 5th ed. Worth Publishers, New York; **6.10, 6.11a** After H. Curtis and N. Sue Barnes. 1994. *Invitation to Biology,* 5th ed. Worth Publishers, New York; **6.11b** John N. Telfold; **6.12** After H. Curtis and N. Sue Barnes. 1994. *Invitation to Biology,* 5th ed. Worth Publishers, New York; **Page 117** Ben Nottidge/Alamy; 6.13 After H. Curtis and N. Sue Barnes. 1994. *Invitation to Biology,* 5th ed. Worth Publishers, New York; **6.14b** © 1989 Egyptian Expedition of the Metropolitan Museum of Art, Rogers Fund, 1915 (15.5.19e); **Page 119** Henry Ausloos/Photolibrary; **6.15** H. Curtis and N. Sue Barnes. 1994. *Invitation to Biology,* 5th ed. Worth Publishers, New York.

CHAPTER 7

Chapter Opener NASA; **7.1** David G. Fisher; **7.2** Paul W. Johnson/Biological Photo Service; **7.3** Colin Milkins/Oxford Scientific Films; **7.4** Peter Gray. 1991. *Psychology.* Worth Publishers, New York; **7.5** Prepared by Govindjee; **7.6** Linda E. Graham; **7.10** After W. M. Becker et al. 1996. *The World of the Cell,* 3rd ed. Benjamin/Cummings Publishing Company, Inc., Menlo Park, CA; **7.12** After B. Alberts et al. 1989. *Molecular Biology of the Cell,* 2nd ed. Garland Publishing, Inc., New York; **7.13a** After L. Taiz and E. Zeiger. 2002. *Plant Physiology,* 3rd ed. Sinauer Associates, Inc., Publishers, Sunderland, MA; **7.13b** D. L. Nelson and M. M. Cox. 2008. *Lehninger Principles of Biochemistry,* 5th ed. Freeman, New York; **7.15** Dr. Jeremy Burgess/Science Photo Library/Photo Researchers, Inc.; **7.18** After H. Curtis and N. Sue Barnes. 1994. *Invitation to Biology,* 5th ed. Worth Publishers, New York; **7.20** After W. M. Becker et al. 1996 *The World of the Cell,* 3rd ed. Benjamin/Cummings Publishing Company, Inc., Menlo Park, CA; **7.25a** Courtesy Abdul Rahman; **7.25b, c** S. D. X. Chuong et al. 2006. *The Plant Cell* 18, 2207–2223; **7.26a (photo)** Leonard LaRue/Bruce Coleman, Inc.; **7.26b (photo)** Phil Degginger/Bruce Coleman, Inc.; **7.26a, b (line art)** After N. A. Campbell. 1996. *Biology,* 4th ed. Benjamin/Cummings Publishing Company, Inc., Menlo Park, CA.

CHAPTER 8

Chapter Opener International Flower Bulb Centre; **8.1** Moravian Museum, Brno, Czech Republic; **8.2a** John Bova/Photo Researchers, Inc.; **8.2b** Arnold Sparrow/Brookhaven National Laboratory; **8.3a** © 2008 Victoria Foe. Reproduced by permission of Garland Science/Taylor & Francis Books, Inc.; **8.3b** H. Curtis and N. Sue Barnes. 1994. *Invitation to Biology,* 5th ed. Worth Publishers, New York; **8.4** After B. Alberts et al. 1989. *Molecular Biology of the Cell,* 2nd ed. Garland Publishing, Inc., New York; **8.6** P. B. Moens; **8.8** B. John; **8.9** G. Ostergren; **8.11** H. Curtis and N. Sue Barnes. 1994. *Invitation to Biology,* 5th ed. Worth Publishers, New York; **8.12** After K. von Frisch. *Biology.* Translated by Jane Oppenheimer. 1964. Harper and Row Publishers, Inc., New York; **8.13–8.15** H. Curtis and N. Sue Barnes. 1994. *Invitation to Biology,* 5th ed. Worth Publishers, New York; **8.16** After A. J. F. Griffiths et al. 1996. *An Introduction to Genetic Analysis,* 6th ed. W. H. Freeman and Company, New York; **8.17a** Nik Kleinberg; **8.17b** Matt Meadows/Peter Arnold; **8.18** H. Curtis and N. Sue Barnes. 1994. *Invitation to Biology,* 5th ed. Worth Publishers, New York; **8.19** Anthony J. F. Griffiths; **8.21 (micrograph)** Dr. Max B. Schröder and Hannelore Oldenburg. 1990. *Flora* 184, 131–136; **8.22** Heather Angel; **8.23** After S. Ross-Craig. 1950. *Drawings of British Plants,* part IV. G. Bell and Sons, Ltd., London; **Page 171 (photo a)** G. I. Bennard/Oxford Scientific Films; **Page 171 (photo b)** Heather Angel/Biofotos; **Page 171 (photo c)** Wisconsin State Herbarium.

CHAPTER 9

Chapter Opener Kovalchuk Oleksandr/Shutterstock; **9.1** H. Curtis and N. Sue Barnes. 1994. *Invitation to Biology,* 5th ed. Worth Publishers, New York; **9.2a** A. Barington Brown/Science Source/Photo Researchers, Inc.; **9.2b** Will and Deni McIntyre/Photo Researchers, Inc.; **9.3–9.4** H. Curtis and N. Sue Barnes. 1994. *Invitation to Biology,* 5th ed. Worth Publishers, New York; **9.5** A. B. Blumenthal et al. 1973. *Cold Spring Harbor Symposia on Quantitative Biology* 38, 205–223; **9.6** H. Curtis and N. Sue Barnes. 1989. *Biology,* 5th ed. Worth Publishers, Inc., New York; **9.7–9.8** H. Curtis and N. Sue Barnes. 1994. *Invitation to Biology,* 5th ed. Worth Publishers, New York; **9.9** After H. Curtis and N. Sue Barnes. 1994. *Invitation to Biology,* 5th ed. Worth Publishers, New York; **9.10–9.11** H. Curtis and N. Sue Barnes. 1994. *Invitation to Biology,* 5th ed. Worth Publishers, New York; **9.12–9.13** After H. Curtis and N. Sue Barnes. 1994. *Invitation to Biology,* 5th ed. Worth Publishers, New York; **9.14** Hans Ris; **9.15** After W. M. Becker et al. 1996. *The World of the Cell,* 3rd ed. Benjamin/Cummings Publishing Company, Inc., Menlo Park, CA; **9.16** After B. A. Pierce. 2006. *Genetics. A Conceptual Approach,* 2nd ed. W. H. Freeman and Company, New York; **9.17a** Creative Commons; **9.17b** Carl Denton http://www.trilliums.co.uk; **9.17c** Courtesy Andreas S. Fleischmann; **9.17d** Joseph A. Marcus, Lady Bird Johnson Wildflower Center; **9.18 (micrograph)** James German; **9.18 (line art)** After P. Chambon. 1981. *Scientific American* 244 (May), 60–71; **9.19** After R. Lewin. 1981. *Science* 212, 28–32; **9.20** R. Allen, et al. 2007. *Proceedings of the National Academy of Sciences USA* 104, 16371–16376.

CHAPTER 10

Chapter Opener ARS/USDA, photo by Scott Bauer; **10.1** Courtesy of Golden Rice Humanitarian Board, www.goldenrice.org; **10.2** After B. Alberts et al. 1983. *Molecular Biology of the Cell.* Garland Publishing, Inc., New York; **10.3** After H. Curtis and N. Sue Barnes. 1994. *Invitation to Biology,* 5th ed. Worth Publishers, New York; **10.4** After N. A. Campbell. 1996. *Biology,* 4th ed. Benjamin/Cummings Publishing Company, Inc., Menlo Park, CA; **10.5** Courtesy of Jaideep Mathur, University of Toronto;

10.6 Keith Wood, University of California, San Diego; **10.7** After a photograph by John T. Fiddles and Howard M. Goodman; **10.8** H. Curtis and N. Sue Barnes. 1994. *Invitation to Biology,* 5th ed. Worth Publishers, New York; **Page 199 (photos a–d)** L. L. Hensel et al. 1993. *The Plant Cell* 5, 553–564; **Page 199 (part e)** © 2006. Juan Lazaro IV, International Rice Research Institute; **10.9** Courtesy of T. Erik Mirkov/Texas A&M University; **10.10** Roni Aloni et al. 1998. *Plant Physiology* 117, 841–849; **10.13** Stephen Ferreira, University of Hawaii at Manoa; **10.14a** Sara Patterson; **10.14b–d** Harry J. Klee; **10.14e** Richard M. Amasino; **10.15** Benjamin A. Pierce. 2003. *Genetics: A Conceptual Approach.* W. H. Freeman and Company, New York; **10.16** H. Lodish et al. 2008. *Molecular Cell Biology,* 6th ed. W. H. Freeman and Company, New York.

CHAPTER 11

Chapter Opener Paul van Gaalen/PhotoAsia; **11.1** By permission of Mr. G. P. Darwin, Courtesy of The Royal College of Surgeons of England; **11.2** After D. Lack. 1961. *Darwin's Finches.* Harper and Row, Publishers, Inc., New York; **11.3** H. Curtis and N. Sue Barnes. 1994. *Invitation to Biology,* 5th ed. Worth Publishers, New York; **11.4** Anthony J. F. Griffiths; **11.5** H. Curtis and N. Sue Barnes. 1994. *Invitation to Biology,* 5th ed. Worth Publishers, New York; **11.6a** M&C Photography/Peter Arnold, Inc.; **11.6b** Kent and Donna Dannen/Photo Researchers, Inc.; **11.7a** Heather Angel/Biofotos; **11.7b** Robert Ornduff, University of California, Berkeley; **11.8** J. Antonovics/Visuals Unlimited; **Page 217 a** imagebroker.net/ SuperStock; **Page 217 b** Lost in Montgomery http:// lostinmontgomery.wordpress.com/; **11.9** After J. Clausen and W. H. Hiesey. 1958. *Experimental Studies on the Nature of Species IV, Genetic Structure of Ecological Races.* Carnegie Institution of Washington Publication 615, Washington, D.C.; **11.10** Science Photo Library/Alamy; **11.12–11.13** H. Curtis and N. Sue Barnes. 1994. *Invitation to Biology,* 5th ed. Worth Publishers, New York; **11.14** Marion Ownbey. 1950. *American Journal of Botany* 37, 487–499; **Pages 224–225** Thomas Givnish, University of Wisconsin–Madison; **11.15a** Irvine Wilson/© DCR Natural Heritage; **11.15b** Heather Angel/Biofotos; **11.15c–e** C. J. Marchant; **11.16** Benjamin A. Pierce. 2003. *Genetics: a Conceptual Approach.* W. H. Freeman and Company, New York; **11.18a** Thase Daniel/Bruce Coleman, Inc.

CHAPTER 12

Section Opener acrylic painting by Rhonda Nass/Ampersand; Courtesy Priscilla and Michael Baldwin Foundation, *Vanishing Circles,* Arizona Sonora Desert Museum, Tucson, AZ; **Chapter 12 Opener** Creative Commons License; **12.1** Corbis/ Bettmann; **12.2a** Larry West; **12.2b** L. Campbell/NHPA; **12.2c** Imagery; **12.3** Wisconsin State Herbarium; **Page 237 (top photo)** G. R. Roberts; **Page 237 (bottom photo)** Courtesy Boleslaw Kuznik; **12.4** After N. A. Campbell et al. 2008. *Biology,* 8th ed. Pearson/ Benjamin Cummings, San Francisco; **Page 239** E. S. Ross; **Page 241 (photo a)** NASA image created by Jesse Allen; **Page 241 (photo b)** Thomas Timberlake/RBG Kew; **12.6** After M. Yukawa et al. 2005. *Plant Molecular Biology Reporter* 23, 359–365; **12.7 (photo)** Pete Gasson, Royal Botanic Gardens, Kew; **(barcode)** Sujeevan Ratnasingham, University of Guelph; **12.8** After N. A. Campbell et al. 2008. *Biology,* 8th ed. Pearson/Benjamin Cummings, San Francisco; **12.9a** L. V. Leak. 1967. *Journal of Ultrastructure Research* 21, 61–74; **12.9b** Helmut Konig and Karl Stetter; **12.9c** Katherine Esau; **12.10** L. E. Graham et al. 2009. *Algae,* 2nd ed. Benjamin Cummings, San Francisco, New York; **12.11** Adapted from Christian de Duve. 1996. *Scientific American* 274 (April), 50–57; **12.12a** L. Wingren; **12.12b** Linda E. Graham; **12.14a** Arthur Morris/ Visuals Unlimited; **12.14b** Larry West; **12.14c** K. B. Sandved; **12.14d** George Barron, University of Guelph; **12.15a** Matt Meadows/Peter Arnold, Inc.; **12.15b** Linda E. Graham; **12.15c** Kim Taylor/Bruce Coleman, Inc.; **12.15d** © James Watanabe, http:// seanet.stanford.edu; **12.15e** E. V. Gravé; **12.16a** L. Mellichamp/ Visuals Unlimited; **12.16b** James W. Perry; **12.16c** E. S. Ross; **12.16d** Robert Carr/Bruce Coleman, Inc.; **12.16e** Courtesy James Ellingboe; **12.16f** J. Dermid; **12.16g** Jeff Foott/Discovery Images/ Picturequest; **12.16h** J. Dermid; **12.16i** John Glover/Alamy; **12.16j** E. Beals.

CHAPTER 13

Chapter Opener Courtesy Mike Pearson/University of Auckland; **13.1** MSU Instructional Media Center/R. Hammerschmidt; **13.2** David Mencin, UNAVCO-United States; **13.3** D. Sadava et al. 2008. *Life: the Science of Biology,* 8th ed. Sinauer Associates, Inc., Sunderland, MA; **13.4** C. C. Brinton, Jr. and John Carnahan; **13.5** G. P. Dubey and S. Ben-Yehuda. 2011. *Cell* 144, 590–600; **13.6a** USDA; **13.6b** David Phillips/Visuals Unlimited; **13.6c** Richard Blakemore; **13.7** Hans Reichenbach; **13.9** P. Gerhardt; **13.10** © micrographia.com; **13.11 (micrograph)** M. Jost; **13.12a** E. V. Gravé; **13.12c** Winston Patnode/Photo Researchers, Inc.; **13.13 (photo)** Fred Bavendam/Peter Arnold, Inc.; **13.13 (line art)** After M. R. Walter. 1977. *American Scientist* 65, 563–571; **13.14a** Robert D. Warmbrodt; **13.14b** Paul W. Johnson/ Biological Photo Service; **13.15** Danita Delimont Creative/Alamy; **13.16** T. D. Pugh and E. H. Newcomb; **13.17** Germaine Cohen-Bazire; **13.18** J. F. Worley/USDA; **13.19a** M. V. Parthasarathy. **13.19b** Henry Donselman; **13.20** After G. N. Agrios. 1978. *Plant Pathology,* 2nd ed. Academic Press, Inc., New York; **13.21** Creative Commons; **13.22** R. Robinson/Visuals Unlimited; **13.23** Leonard Lessin/Peter Arnold, Inc.; **13.24** After sketch by Amine Noueiry; **13.25** Norm Thomas/Photo Researchers, Inc.; **13.26** M. Dollet and R. G. Milne, from M. Dollet et al. 1986. *Journal of General Virology* 67, 933–937; **13.27** After sketch by Amine Noueiry; **13.28a** Katherine Esau; **13.28b, 13.29** Jean-Yves Sgro; **13.30** Katherine Esau; **13.31** Th. Koller and J. M. Sogo, Swiss Federal Institute of Technology, Zürich.

CHAPTER 14

Chapter Opener Courtesy Tara AuBuchon www.flickr.com/photos/ taubuch/; **14.1a** G. W. Hudler, Cornell University; **14.1b** Photo © Biopix: JC Schou; **14.2** Charles M. Fitch/Taurus Photos; **14.3a** M. Powell; **14.3b, d** E. S. Ross; **14.3c** Thomas Volk; **14.4** R. J. Howard; **14.6** M. D. Coffey, B. A. Palevitz, and P. J. Allen. 1972. *The Canadian Journal of Botany* 50, 231–240; **14.7** E. C. Swann and C. W. Mims. 1991. *The Canadian Journal of Botany* 69, 1655–1665; **14.8a** SciMAT/Photo Researchers, Inc.; **Page 285 (photo b)** John Hogdin; **14.9** Creative Commons; **14.11** Eye of Science/Photo Researchers; **14.12** M. Powell; **14.13** John W. Taylor; **14.14** After R. Emerson. 1941. *Lloydia* 4, 77–144; **14.17** Glomus intraradices Schenk & Smith DAOM 234180 © Agriculture and Agri-Food Canada (1998); **14.18a** Thomas Volk; **14.18b** Gary J. Breckon; **14.20a** M. G. Roca et al. 2005. *Eukaryotic Cell* 4, 911–919; **14.20b** J. C. Pendland and D. G. Boucias; **14.21a** C. Bracker; **14.21b** Damian S. Neuberger; **14.21c** Bryce Kendrick; **14.23a** Andrew McClenaghan/Photo Researchers, Inc.; **14.23b** John Durham/Science Photo Library/Photo Researchers, Inc.;

14.24 G. L. Barron, University of Guelph; **14.26** D. J. McLaughlin and A. Beckett; **14.27, 14.28b** H. Lü and D. J. McLaughlin. 1991. *Mycologia* 83, 322–334; **14.29a** C. W. Perkins/Earth Scenes; **14.29b** Thomas Volk; **14.29c** Peter Katsaros/Photo Researches, Inc.; **14.29d** James W. Perry; **14.31** Courtesy Darrell D. Hensley, University of Tennessee, Entomology and Plant Pathology; **14.32a** Alan Rockefeller/Wikipedia Commons; **14.32b** R. Gordon Wasson, Botanical Museum of Harvard University; **14.33a** Jane Burton/Bruce Coleman, Inc.; **14.33b** © Doug Wechsler; **14.33c, d** Jeff Lepore/Photo Researchers Inc.; **14.34** E. S. Ross; **Page 303 (photos a, b)** G. L. Barron, University of Guelph; **Page 303 (photo c)** N. Allin and G. L. Barron, University of Guelph; **Page 307** John Webster, University of Exeter; **14.37a** E. S. Ross; **14.37b** Subhankar Banerjee/Courtesy Gerald Peters Gallery, Santa Fe/New York; **14.37c** Robert A. Ross; **14.38a, b** E. S. Ross; **14.38c** Stephen Sharnoff/National Geographic Society; **14.38d** Larry West; **14.39** E. Imre Friedmann; **14.41** V. Ahmadjian and J. B. Jacobs; **14.42** S. A. Wilde; **14.43** Bryce Kendrick; **14.44** D. J. Read; **14.45** B. Zak, U. S. Forest Service; **14.46a** Robert D. Warmbrodt; **14.46b** R. L. Peterson and M. L. Farquhar. 1994. *Mycologia* 86, 311–326; **14.47** Thomas N. Taylor, Ohio State University.

CHAPTER 15

Chapter Opener Photo by Jeanette Johnson. © In-Depth Images Kwajalein; **15.1** Courtesy James St. John; **15.2** Steve Lonhart/ SIMoN NOAA; **15.3** D.P. Wilson/FLPA/Getty Images; **Page 321 (photo a)** Bob Evans/Peter Arnold, Inc.; **Page 321 (photo b)** W. H. Hodge/Peter Arnold, Inc.; **Page 321 (photo c)** Kelco Communications; **15.4** *BioScience* 61, 421–422. © 2011 by American Institute of Biological Sciences. Photo courtesy Walter H. Adey; **Page 323 (photo a)** Florida Department of Natural Resources; **Page 323 (photo b)** Photo by Miriam Godfrey, Courtesy National Institute of Water & Atmospheric Research Ltd.; **15.5a** Biophoto Associates/Photo Researchers, Inc.; **15.6** CSIRO Marine Research/ Visuals Unlimited; **15.7 (micrograph)** Geoff McFadden and Paul Gilson; **15.8** J. Burkholder, North Carolina State University; **15.9a** Elizabeth Venrick, Scripps Institution of Oceanography, University of California, San Diego; **15.9b** Holly Kunz, Courtesy of Peggy Hughes and David Garrison; **15.9c** © PacificCoastNews.com; **15.10a, b** D. P. Wilson/Science Source/Photo Researchers, Inc.; **15.10c** Florida Department of Natural Resources; **15.12** Robert F. Sisson/National Geographic Society; **Page 329 (left photo)** Roger Steene/ImageQuestMarine.com; **(right photo)** Mark Spencer/Auscape International; **15.13** R. R. Powers; **15.14a** M. I. Walker/Science Source/Photo Researchers, Inc.; **15.14b** F. Rossi; **15.14c** Dr. Ann Smith/SPL/Photo Researchers, Inc.; **15.14e** Biophoto Associates/Science Source/Photo Researchers, Inc.; **15.16** C. Sandgren; **15.17** Damian S. Neuberger; **15.18a** Mauritius Images GmbH/Alamy; **15.18b, c** D. P. Wilson/Eric and David Hosking Photography; **15.19b** Coastal Ocean Association of Science and Technology, Dr. Brian Lapointe; **15.20** C. J. O'Kelly; **15.21** Specimen provided by John West; **15.24** Martha E. Cook; **15.25a** D. P. Wilson/Eric and David Hosking Photography; **15.25b** E. S. Ross; **15.25c** R. C. Carpenter; **15.25d** © Michael Guiry. AlgaeBase, Ryan Institute, NUI Galway, University Rd, Galway, Ireland; **15.26** M. Littler and D. Littler, Smithsonian Institution; **15.27a** Linda E. Graham; **15.27b** J. Waaland, University of Washington; **15.28 (micrograph)** C. Pueschel and K. M. Cole. 1982. *American Journal of Botany* 69, 703–720; **15.30** Ronald Hoham; **15.31** After G. L. Floyd; **15.32** After K. R. Mattox and K. D. Stewart. 1984. In: D. E. G. Irvine and D. M. John, eds. *Systematics of the Green Algae.* Academic Press, London, Orlando, FL; **15.33** Linda E. Graham; **15.34** W. L. Dentler/Biological Photo Service/University of Kansas; **15.36** D. L. Kirk and M. M. Dirk. 1986. *Science* 231, 51–54; **15.37** J. Robert Waaland/Biological PhotoService; **15.38a** Linda E. Graham; **15.38b** M. I. Walker/Science Source/Photo Researchers, Inc.; **15.39** Linda E. Graham; **15.40** After R. T. Skagel et al. 1966. *An Evolutionary Survey of the Plant Kingdom.* Wadsworth Publishing Co., Inc. Belmont, CA; **15.41a** James Graham; **15.41b** Courtesy Dr. Yuuji Tsukii, Hosei University, Tokyo; **15.41c** E. S. Ross; **15.42** D. P. Wilson/Eric and David Hosking Photography; **15.44a** Robert A. Ross; **15.44b** Grant Heilman Photography; **15.44c** L. R. Hoffman; **15.44d** V. Paul; **15.45** Linda E. Graham; **15.46** M. I. Walker/Science Source/Photo Researchers, Inc.; **15.47a** Lee W. Wilcox; **15.47b** © Wim van Egmond/Visuals Unlimited, Inc.; **15.48** After W. S. Judd et al. 2007. *Plant Systematics: A Phylogenetic Approach,* 3rd ed. Sinauer Associates, Inc., Sunderland, MA; **15.49** Linda E. Graham; **15.50** William H. Amos/Bruce Coleman, Inc.; **15.51a** A. W. Barksdale; **15.51b** A. W. Barksdale. 1963. *Mycologia* 55, 493–501; **15.53** Matteo Garbelotto; **15.54** After J. H. Niederhauser and W. C. Cobb. 1959. *Scientific American* 200 (May), 100–112; **15.55** Damian S. Neuberger; **15.57a** Victor Duran; **15.57b** Ed Reschke/Peter Arnold, Inc.; **15.58a, b** K. B. Raper; **15.58c, d** London Scientific Films/Oxford Scientific Films; **15.58e–g** Robert Kay.

CHAPTER 16

Chapter Opener Ian Cameron/Transient Light; **16.1** Brent Mishler; **16.2** D. R. Given; **16.3** After W. S. Judd et al. 2007. *Plant Systematics: A Phylogenetic Approach,* 3rd ed. Sinauer Associates, Inc., Sunderland, MA; **16.4a** After G. M. Smith. 1955. *Cryptogamic Botany,* vol. 2, *Bryophytes and Pteridophytes,* 2nd ed. McGraw-Hill Book Co., New York; **16.4b** R. E. Magill, Botanical Research Institute, Pretoria, South Africa; **16.5** Linda E. Graham; **16.6b** Damian S. Neuberger; **16.9** Karen S. Renzaglia; **16.10a** John Wheeler; **16.11a** Field Museum of Natural History; **16.11b** Dr. G. J. Chafaris/Dr. E. R. Degginger; **16.13a** Courtesy Walter Piorkowski; **16.15a, b** J. J. Engel. 1980. *Fieldiana: Botany* (new series) 3, 1–229; **16.15c** J. J. Engel; **16.16** K. B. Sandved; **16.17a** Larry West; **16.17b** Andrew Syred/Photo Researchers; **16.17c** After C. T. Ingold. 1939. *Spore Discharge in Land Plants.* Clarendon Press, Oxford; **16.18** Martha E. Cook; **16.19a** Courtesy Lawrence Jensen; **16.19b** Courtesy George Shepherd; **16.20a** Creative Commons, Reski Lab, University of Freiburg http://en.wikipedia.org/wiki/ File:Physcomitrella_Protonema.jpg; **16.20b** D. G. Schaefer. 2002. *Annual Review of Plant Biology* 53, 477–501. Photo courtesy of Dr. Didier Schaefer; **16.21** Courtesy Steve Reekie www.stevereekie.co. nz; **16.22** Charles Hébant. 1975. *Journal of the Hattori Botanical Laboratory* 39, 235–254; **16.23a** Damian S. Neuberger; **16.24** Rod Planck/Photo Researchers, Inc.; **16.25** Fred D. Sack and D. J. Paolillo, Jr. 1983. *American Journal of Botany* 70, 1019–1030; **16.26a** After C. T. Ingold. 1939. *Spore Discharge in Land Plants.* Clarendon Press, Oxford; **16.26b** R. E. Magill, Missouri Botanical Garden, St. Louis; **16.27a** A. D. Staffen; **16.27b** E. S. Ross; **16.29a** Andrew Drinnan; **16.29d** Damian S. Neuberger.

CHAPTER 17

Chapter Opener Courtesy Heather Sullivan and Alan Holditch www.mississippiferns.com; **17.1** Hans Steur/Visuals Unlimited; **17.2** M. K. Rasmussen and Stuart A. Naquin; **17.3** After A. S. Foster and E. M. Gifford, Jr. 1974. *Comparative Morphology of Vascular Plants,* 2nd ed. W. H. Freeman and Company, New York; **17.4** After Katherine Esau. 1965. *Plant Anatomy,* 2nd ed. John Wiley & Sons, New York; **17.5a, d** After K. K. Namboodiri and C. Beck. 1968. *American Journal of Botany* 55, 464–472; **17.5b, c** After Katherine Esau. 1965. *Plant Anatomy,* 2nd ed. John Wiley & Sons, New York; **17.7** After G. M. Smith. 1955. *Cryptogamic Botany,* vol. 2, *Bryophytes and Pteridophytes,* 2nd ed., McGraw-Hill Book Company, New York; **17.9** Claudio Pia; **17.10a** After David S. Edwards. 1980. *Review of Palaeobotany and Palynology* 29, 177–188; **17.10b** After J. Walton. 1964. *Phytomorphology* 14, 155–160; **17.10c** After F. M. Heuber. 1968. *International Symposium on the Devonian System,* vol. 2, D. H. Oswald, ed. Alberta Society of Petroleum Geologists, Calgary, Alberta, Canada; **17.11** Field Museum of Natural History; **Page 401 (part a)** Interpreted by M. K. Rasmussen and Stuart A. Naquin from T. L. Phillips and W. A. DiMichelle. 1992. *Annals of the Missouri Botanical Garden* 9, 560–588; **Page 401 (part b)** After W. N. Stewart and T. Delevoryas. 1956. *Botanical Review* 22, 45–80; **Page 401 (part c)** After H. P. Banks. 1970. *Evolution and Plants of the Past.* Wadsworth Publishing Company, Inc., Belmont, CA; **17.12** University of Aberdeen; **17.13** After David S. Edwards. 1986. *Botanical Journal of the Linnean Society* 93, 173–204; **17.14** After W. S. Judd et al. 2007. *Plant Systematics: A Phylogenetic Approach,* 3rd ed. Sinauer Associates, Inc., Sunderland, MA; **17.15** Specimen provided by Ripon Microslides, Ripon, WI; **17.17a** Damian S. Neuberger; **17.18a** David Johnson, Big Bend National Park; **17.18b, c** Damian S. Neuberger; **17.18d** Fletcher and Baylis/Photo Researchers, Inc.; **17.21** W. H. Wagner; **17.22** After M. K. Rasmussen and Stuart A. Naquin; **17.23a, e** W. H. Wagner; **17.23b** tororo reaction/Shutterstock; **17.23c** Stoelwinder/AgeFotostock; **17.23d** Nancy A. Murray; **17.23f** David Johnson; **17.23g** Geoff Bryant/Photo Researchers, Inc.; **17.24a** Creative Commons: Photo by Peggy Greb/ARS; **17.24b, c** U.S. Geological Survey - St. Petersburg Coastal and Marine Science Center; **17.26a** Bill Hilton Jr., www.hiltonpond.org; **17.26b** Creative Commons: Germany - Saarland - 05/2006; **17.27b** Courtesy Eric Guinther, AECOS Inc.; **17.28a** D. Cameron; **17.28b** R. Schmid; **17.29b** R. L. Peterson et al. 1981. *Canadian Journal of Botany* 59, 711–720; **17.31** Bill Ivy/Tony Stone Images, Inc.; **17.32a** Bill Beatty/Visuals Unlimited; **17.32b** C. Neidorf; **17.32c** Creative Commons; **17.32d** Stan Gilliam @ USDA-NRCS PLANTS Database; **17.34** D. Farrar; **17.36a** Damian S. Neuberger; **17.36b, c** W. H. Wagner; **17.37a** R. Carr; **17.37b** Gerry Ellis/Ellis Nature Photography; **17.39** Dr. Jeremy Burgess/Science Photo Library/Photo Researchers, Inc.

CHAPTER 18

Chapter Opener Creative Commons. Richard Sniezko/U.S. Forest Service; **18.2** After W. N. Stewart and G. W. Rothwell. 1993. *Paleobotany and the Evolution of Plants,* 2nd ed. Cambridge University Press, New York; **18.3** From H. N. Andrews. 1963. *Science* 142, 925–931; After A. G. Long. 1960. *Transactions of the Royal Society of Edinburgh* 64, 29–44; **18.4** Charles B. Beck; **18.5–18.6** M. K. Rasmussen and Stuart A. Naquin; **18.7** After P. R. Crane. 1985. *Annals of the Missouri Botanical Garden* 72, 716–793; **18.8** M. K. Rasmussen and Stuart A. Naquin; **18.9a** Field Museum of Natural History (transparency #B83046c); **18.9b** After P. R. Crane. 1985. *Annals of the Missouri Botanical Garden* 72, 716–793; **18.10** After W. S. Judd et al. 2007. *Plant Systematics: A Phylogenetic Approach,* 3rd ed. Sinauer Associates, Inc., Sunderland, MA; **18.11** After W. E. Friedman. 1993. *Trends in Ecology and Evolution* 8, 15–21; **18.12** J. Dermid; **18.13a** © 2008 Dennis Stevenson; **18.16** B. Haley; **18.18c** Gary J. Breckon; **18.22** N. Fox-Davies/Bruce Coleman Ltd.; **18.25a** W. H. Hodge/Peter Arnold, Inc.; **18.25b** E. S. Ross; **18.26** H. H. Iltis; **18.27** Grant Heilman Photography; **18.28a** Larry West; **18.28b** J. Burton/Bruce Coleman, Inc.; **18.29a** © Jaime Plaza Van Roon; **18.29b** Courtesy Natalie Tapson; **18.29c** Heather Angel/ Natural Visions; **18.30** Geoff Bryant/Photo Researchers, Inc.; **18.31 (photo)** Carolina Biological Supply Company; **18.32** Mark Wetter; **18.33** Gene Ahrens/Bruce Coleman, Inc.; **18.34** Sichuan Institute of Biology; **18.35** David A. Steingraeber; **18.36a** Knut Norstog; **18.36b** D. T. Hendricks and E. S. Ross; **18.37** Knut Norstog; **18.38a** James W. Perry; **18.38b** Runk and Schoenberger/Grant Heilman Photography; **18.39a** Gerald D. Carr; **18.39b** G. Davidse; **18.39c** © 2008 by Scott Mori; **18.40a** E. S. Ross; **18.40b, d** James W. Perry; **18.40c** K. J. Niklas; **18.41a** Chris H. Bornman; **18.41b, c** E. S. Ross.

CHAPTER 19

Chapter Opener Courtesy Jeffrey D. Karron; **19.1** Bjarke Ferchland/ Bruce Coleman, Inc.; **19.2** W. P. Armstrong; **19.3a** Ross Warner/ Alamy; **19.3b** Gary J. Breckon; **19.3c** © David G. Smith; **19.4a** E. S. Ross; **19.4b** E. R. Degginger/Earth Scenes; **19.4c** T. Davis/Photo Researchers, Inc.; **19.5a** E. S. Ross; **19.5b** COMPOST/VISAGE/Peter Arnold; **19.5c** G. Carr; **19.8a, c** Larry West; **19.8b, e** J. H. Gerard; **19.8d** Grant Heilman Photography; **19.10** E. S. Ross; **19.12a** Larry West; **19.12b** Specimen provided by Rudolf Schmid; **19.13a** Runk and Schoenberger/Grant Heilman Photography; **19.15a** Science Photo Library/SuperStock; **19.15b** Andrew Syred/Photo Researchers; **19.15c** Power and Syred/Photo Researchers; **19.15d** Cheryl Power/ Photo Researchers; **19.20** After E. M. Gifford and A. S. Foster 1989. *Comparative Morphology of Vascular Plants,* 2nd ed. W. H. Freeman and Company, New York, San Francisco.

CHAPTER 20

Chapter Opener © 2002–2011 Arapahoe County. All Rights Reserved; **20.1** Courtesy Nathan Cook; **20.2a** Taylor Feild, University of Toronto; **20.2b, c** Jaime Plaza, Botanic Gardens Trust, Sydney, Australia; **20.2d** Tammy Sage, University of Toronto; **20.3** Ross Frid/ Visuals Unlimited; **20.4** Joseph H. Williams; **20.5** James L. Castner; **20.6** Geoff Bryant/Photo Researchers, Inc.; **20.7** After M. G. Simpson. 2010. *Plant Systematics,* 2nd ed. Academic Press/Elsevier, Amsterdam, Boston; **20.8a** K. Simons and David L. Dilcher; **20.8b, c** David L. Dilcher and Ge Sun; **20.9a, c** E. S. Ross; **20.9b** James W. Perry; **20.10b, c** E. S. Ross; **20.10d** A. Sabarese; **20.11a** E. S. Ross; **20.12a** Larry West; **20.12b, 20.13** E. S. Ross; **20.14** Larry West; **20.15** E. S. Ross; **20.16** R. A. Tyrell; **20.17** D. J. Howell; **20.18a** Damian S. Neuberger; **20.18b, c** T. Hovland/Grant Heilman Photography; **20.20** T. Eisner; **20.21a** Imagebroker.net/ SuperStock; **20.22a** E. S. Ross; **20.22b** After R. T. Scagel et al. 1965. *An Evolutionary Survey of the Plant Kingdom.* Wadsworth Publishing Company, Inc., Belmont, CA; **20.22c** After L. D. Benson. 1957. *Plant Classification.* D. C. Heath and Company, Boston; **20.23a** E. S. Ross; **20.23b, c** K. B. Sandved; **20.24** After R. T. Scagel et al. 1965. *An Evolutionary Survey of the Plant Kingdom.* Wadsworth Publishing Company, Inc., Belmont, CA; **20.25, 20.27a** E. S. Ross; **20.27b** U. S. Forest Service; **20.28–20.29** E. S. Ross; **20.30** Robert and Linda Mitchell; **20.31a, c, d** T. Plowman; **20.31b** E. S. Ross.

CHAPTER 21

Chapter Opener Ms Fr 2810 fol.186 Pepper harvest and offering the fruits to a king, from the "Livre des Merveilles du Monde," c.1410–12 (vellum) by Boucicaut Master (fl.1390–1430) Bibliotheque Nationale, Paris, France/Archives Charmet/The Bridgeman Art Library Internat; **21.1** Anthony Bannister/ABPL/Earth Scenes; **21.2** E. S. Ross; **21.3** After M. Balter. 2007. *Science* 316, 1830–1835; **21.4a, b** Courtesy Brian Steffenson, University of Minnesota; **21.4c** Courtesy Luigi Riganese; **21.5** E. S. Ross; **21.6** W. H. Hodge/Peter Arnold, Inc.; **21.7** John Elk III/Bruce Coleman, Inc.; **21.8** E. S. Ross; **21.9** G. R. Roberts; **21.10a** A. Gentry; **21.10b** M. K. Arroyo; **21.11a** C. F. Jordan; **21.11b** M. J. Plotkin; **21.12** E. S. Ross; **Page 510** John Doebley; **21.13** M. J. Plotkin; **21.14** K. B. Sandved; **21.15** W. H. Hodge/Peter Arnold, Inc.; **21.16** Luiz C. Marigo/Photolibrary; **21.17** Courtesy Syngenta Seeds AB, Sweden; **21.18** Harvey Lloyd/Peter Arnold, Inc.; **21.19** Jes Aznar/AFP/Getty Images; **21.20** R. Abernathy; **21.21** Courtesy Lance Klessig; **21.22** © Micheline Pelletier/Sygma/Corbis; **21.23, 21.24a** Agricultural Research Service, U.S.D.A.; **21.24b** University of Wisconsin; **21.25** J. D. Glover et al. 2010. *Science* 328, 1638–1639. Photo courtesy Jerry Glover; **21.26** J. Aronson; **21.27** Robert and Linda Mitchell; **21.28** M. J. Plotkin; **21.29** Michael J. Balick.

CHAPTER 22

Section Opener acrylic painting by Rhonda Nass/Ampersand; Courtesy Priscilla and Michael Baldwin Foundation, *Vanishing Circles,* Arizona Sonora Desert Museum, Tucson, AZ; **Chapter Opener** FoodCollection/Superstock; **22.2** After A. S. Foster and E. M. Gifford, Jr. 1974. *Comparative Morphology of Vascular Plants,* 2nd ed. W. H. Freeman and Company, New York; **22.4** Daniel M. Vernon and David W. Meinke. 1974. *Developmental Biology* 165, 566–573; **22.5** Gerd Juergens; **Page 533** Werner H. Muller/Peter Arnold, Inc.; **22.8** Tom McHugh/Photo Researchers, Inc.; **22.9** James S. Busse.

CHAPTER 23

Chapter Opener Courtesy Joybilee Farm, 2010 www.fiberarts.ca; **23.13** H. A. Core et al. 1979. *Wood: Structure and Identification,* 2nd ed. Syracuse University Press, Syracuse, NY; **23.14** I. B. Sachs; **23.16** After Katherine Esau. 1977. *Anatomy of Seed Plants,* 2nd ed. John Wiley & Sons, Inc., New York; **23.18** D. S. Neuberger and R. F. Evert. 1975. *Protoplasma* 84, 109–125; **23.20a** Michael A. Walsh; **23.20c** R. F. Evert. 1990. Dicotyledons. In: *Sieve Elements. Comparative Structure, Induction and Development.* H.-D. Behnke and R. D. Sjolund, eds. Springer-Verlag, Berlin; **23.23** After K. Esau. 1977. *Anatomy of Seed Plants,* 2nd ed. John Wiley & Sons, Inc., New York; **23.24** R. F. Evert et al. 1971. *Planta* 100, 262–267; **23.25** J. S. Pereira; **23.27** T. Vogelman and G. Martin; **23.28** M. Daniel Marks and Kenneth A. Feldman.

CHAPTER 24

Chapter Opener Martin Gray/National Geographic Stock; **24.1–24.2** After J. E. Weaver. 1926. *Root Development of Field Crops.* McGraw-Hill Book Company, New York; **24.3** After W. Braune et al. 1967. *Pflanzenanatomisches Prakticum.* VEB Gustav Fischer Verlag, Jena; **24.4** Margaret E. McCully; **24.7** F. A. L. Clowes; **24.8** After Katherine Esau. 1965. *Plant Anatomy,* 2nd ed. John Wiley & Sons, Inc., New York; **24.9a** Robert Mitchell/Earth Scenes; **24.13** After W. Braune et al. 1967. *Pflanzenanatomisches Prakticum.* VEB Gustav Fischer Verlag, Jena; **24.14** Robert D. Warmbrodt; **24.18** E. R. Degginger/Bruce Coleman, Inc.; **24.19a** Dr. Morley Read/Shutterstock; **24.19b** © Floridata.com; **24.20** Robert and Linda Mitchell; **Page 575 (parts a, b)** After J. W. Schiefelbein, J. D. Masucci, and H. Wang. 1997. *Plant Cell* 9, 1089–1098; **Page 575 (parts c, d)** M. E. Galway et al. 1994. *Developmental Biology* 166, 740–754.

CHAPTER 25

25.5 After M. Lenhard and T. Laux. 1999. *Current Opinion in Plant Biology* 2, 44–50; **25.6** Mary Ellen Gerloff; **25.7b** Thomas Holton/Getty Images; **25.13c** W. Eschrich; **25.15–25.16** After K. Esau. 1977. *Anatomy of Seed Plants,* 2nd ed. John Wiley & Sons, Inc., New York; **25.17** Rhonda Nass/Ampersand; **25.19a** James W. Perry; **25.20** Rhonda Nass/Ampersand; **Page 596** After P. A. Deschamp and J. T. Cooke. 1983. *Science* 219, 505–507; **25.24a** M. Michele McCauley; **25.25** T. R. Pray. 1954. *American Journal of Botany* 41, 663–670; **25.31** Daniel J. Barta; **25.32a, b** J. Kang and N. Dengler. 2004. *International Journal of Plant Science* 165, 231–242; **25.32c, d** Julie Kang; **25.33** Joanne M. Dannenhoffer; **25.34** Raymon Donahue and Greg Martin; **Page 603** Daniel Berehulak/Getty Images AsiaPac; **25.36** After K. Esau. 1977. *Anatomy of Seed Plants,* 2nd ed. John Wiley & Sons, Inc., New York; **25.37** Shirley C. Tucker; **25.38** After L. Taiz and E. Zeiger. 2002. *Plant Physiology,* 3rd ed. Sinauer Associates, Inc., Publishers, Sunderland, MA; **25.39** After W. K. Purves et al. 2001. *Life: The Science of Biology,* 6th ed. Sinauer Associates, Inc., Publishers, Sunderland, MA; **25.41a** J. Bowman; **25.41b** Detlef Weigel; **25.42** James W. Perry; **25.43** Courtesy Paul Busselen/www.kuleuven-kortrijk.be/bioweb; **25.44** David A. Steingraeber; **25.45** James W. Perry; **25.46** G. R. Roberts.

CHAPTER 26

Chapter Opener Creative Commons; **26.18** Charles O'Rear/Corbis; **Page 627 (parts a, b)** After Peter M. Ray et al. 1983. *Botany.* Saunders College Publishing, Philadelphia; **26.22** I. B. Sachs, U.S.D.A., Forest Products Laboratory; **26.25** H. A. Core et al. 1979. *Wood: Structure and Identification,* 2nd ed. Syracuse University Press, Syracuse, NY; **26.27a** Galen Rowell, 1985/Peter Arnold, Inc.; **26.27b** C. W. Ferguson, Laboratory of Tree-Ring Research, University of Arizona; **26.28a** Qiang Sun; **26.29** Regis Miller.

CHAPTER 27

Chapter Opener David Nance/USDA-ARS; **27.4** After L. Taiz and E. Zeiger. 2002. *Plant Physiology,* 3rd ed. Sinauer Associates, Inc. Publishers, Sunderland, MA; **27.5** Roni Aloni, Tel Aviv University and Cornelia I. Ullrich, Darmstadt University of Technology; **27.6** Roni Aloni, Tel Aviv University; **27.7** After H. Fukaki and M. Tasaka. 2009. *Plant Molecular Biology* 69, 437–449; **27.8** Bruce Iverson; **27.10** F. Skoog and C. O. Miller. 1957. *Symposia of the Society for Experimental Biology* 11, 118–131; **27.12** J. D. Goeschl; **27.13** After L. Taiz and E. Zeiger. 2006. *Plant Physiology,* 4th ed. Sinauer Associates, Inc., Publishers, Sunderland, MA; **27.15** D. R. McCarty; **27.17** S. W. Wittwer; **27.18a** After M. B. Wilkins, ed. 1984. *Advanced Plant Physiology.* Pitman Publishing Ltd., London; **27.18b** J. E. Varner; **27.19** After J. van Overbeck. 1966. *Science* 152, 721–731; **27.20** Carolina Biological Supply Company; **27.21** Abbott Laboratories; **27.22** After L. Taiz and E. Zeiger. 2006. *Plant Physiology,* 4th ed. Sinauer Associates, Inc., Publishers, Sunderland, MA; **27.23** H. Fukuda. 2004. *Nature Reviews: Molecular Cell Biology* 5, 379–391. Courtesy H. Fukuda; **27.25** Jaideep Mathur, University of Toronto; **27.26** Kurt Stepnitz, MSU/DOE Plant Research Laboratory; **27.27** After Z. Lin et al. 2009. *Journal of*

Experimental Botany 60, 3311–3336; **27.28** L. Taiz and E. Zeiger. 2006. *Plant Physiology,* 4th ed. Sinauer Associates, Inc., Publishers, Sunderland, MA.

CHAPTER 28

Chapter Opener Rob Nelson; **28.4** F. D. Sack and J. Z. Kiss. 1989. *American Journal of Botany* 76, 454–464; **28.5** After B. E. Juniper. 1976. *Annual Review of Plant Physiology* 27, 385–406; **28.6** After I. O. Leyser. 1999. *Current Biology* 9, R8–R10; **28.7** R. Aloni et al. 2004. *Planta* 220, 177–182; **28.8** Stephen A. Parker/Photo Researchers, Inc.; **28.9** Jack Dermid; **28.10** After A. W. Galston. 1968. *The Green Plant.* Prentice-Hall, Inc., Upper Saddle River, NJ; **28.11a** Biophoto Associates/Science Source/Photo Researchers, Inc.; **28.11b, c** After B. Sweeney. 1969. *Rhythmic Phenomena in Plants.* Academic Press, Inc., New York; **28.12** After L. Taiz and E. Zeiger. 2006. *Plant Physiology,* 4th ed. Sinauer Associates, Inc., Publishers, Sunderland, MA; **28.13** Steve A. Kay and A. Millar; **28.14** After A. W. Naylor. 1952. *Scientific American* 186 (May), 49–56; **28.15** After P. M. Ray. 1963. *The Living Plant.* Holt, Rinehart, & Winston, Inc., New York; **28.17** U.S. Department of Agriculture; **28.21a** Richard M. Amasino; **28.21b** David Newman/Visuals Unlimited; **28.22** D. H. Keuskamp et al. 2010. *Plant Signaling & Behavior* 5, 655–662; **28.23** Richard M. Amasino; **Page 67** Courtesy Global Crop Diversity Trust; **28.25** After A. W. Naylor. 1952. *Scientific American* 186 (May), 49–56; **28.26** After K. Esau. 1977. *Anatomy of Seed Plants,* 2nd ed. John Wiley & Sons, Inc., New York; **28.27** Robert L. Dunne/Bruce Coleman, Inc.; **28.28** Runk and Schoenberger/ Grant Heilman Photography; **28.29** Janet Braam; **28.30a** J. Ehleringer and I. Forseth, University of Utah; **28.30b** Gene Ahrens/Bruce Coleman, Inc.

CHAPTER 29

Chapter Opener Courtesy Mark Brundrett; **29.1** Dana Richter, Michigan Technological University; **29.2a** Donald Specker/Earth Scenes; **29.2b** Biological Photo Service; **29.3** University of Wisconsin, Madison, Department of Soil Science; **29.4** E. Crichton/ Bruce Coleman, Inc.; **29.6** After B. Gibbons. 1984. *National Geographic* 166 (3), 350–388 (Ned M. Seidler, artist); **29.7** After F. B. Salisbury and C. W. Ross. 1992. *Plant Physiology,* 4th ed. Wadsworth Publishing Co., Belmont, CA; **29.8** H. Curtis and N. Sue Barnes. 1994. *Invitation to Biology,* 5th ed. Worth Publishers, New York; **Page 694 (photo a)** Dwight Kuhn/Bruce Coleman, Inc.; **Page 694 (photo b)** Bob Gibbons/ardea.com; **Page 694 (photo c)** Nuridsany et Perennou/Photo Researchers; **29.9** After D. E. Fosket. 1994. *Plant Growth and Development: A Molecular Approach.* Academic Press, San Diego, CA; **29.10a, b** B. F. Turgeon and W. D. Bauer 1982. *Canadian Journal of Botany* 60, 152–161; **29.10c** Ann Hirsch, University of California, Los Angeles; **29.10d** E. H. Newcomb and S. R. Tandon; **29.11a** Liphatech, Inc., Milwaukee, WI; **29.11b** J. M. L. Selker and E. H. Newcomb. 1985. *Planta* 165, 446–454; **29.12b** H. E. Calvert; **29.13a** Frank Greenaway © Dorling Kindersley; **29.13b** Courtesy of Carol A. Wilson and Clyde L. Calvin; **29.15** H. Curtis and N. Sue Barnes. 1994. *Invitation to Biology,* 5th ed. Worth Publishers, New York; **29.16** H. Lambers et al. 2008. *Trends in Ecology and Evolution* 23, 95–103; **Page 703 (photo a)** Grant Heilman Photography; **Page 703 (photo b)** J. H. Troughton and L. Donaldson. 1972. *Probing Plant Structures.* McGraw-Hill Book Company, New York; **Page 704** Foster/Bruce Coleman, Inc.

CHAPTER 30

Chapter Opener Teri Campbell/Corbis; **30.1** Stephen Hales; **30.2** photos courtesy of E. Raveh; **30.3** After L. Taiz and E. Zeiger. 2006. *Plant Physiology,* 4th ed. Sinauer Associates, Inc., Publishers, Sunderland, MA; **30.4** After D. E. Aylor et al. 1973. *American Journal of Botany* 60, 163–171; **30.5a** James L. Castner; **30.5b** Christi Carter/Grant Heilman Photography; **30.6** After M. Richardson. 1968. *Translocation in Plants.* Edward Arnold Publishers, Ltd., London; **30.7** After A. C. Leopold. 1964. *Growth and Development.* McGraw-Hill Book Company, New York; **30.8** After M. Richardson. 1968. *Translocation in Plants.* Edward Arnold Publishers, Ltd., London; **Page 714** Ottmar Bierwagen/Spectrum Photofile; **30.9** After L. Taiz and E. Zeiger. 1991. *Plant Physiology.* Benjamin/Cummings Publishing Company, Inc., Redwood City, CA; **30.12** After P. F. Scholander et al. 1965. *Science* 148, 239–246; **30.13–30.14** After M. H. Zimmermann. 1963. *Scientific American* 208 (March), 132–142; **30.15a** G. R. Roberts; **30.17** After M. Richardson. 1968. *Translocation in Plants.* Edward Arnold Publishers, Ltd., London; **30.18** H. Reinhard/Bruce Coleman, Inc.; **30.19** After E. Hausermann and A. Frey-Wyssling. 1963. *Protoplasma* 57, 37–80; **30.20** After T. E. Dawson. 1993. *Oecologia* 95, 565–574; **30.22** After J. S. Pate. 1975. *Transport in Plants,* I., *Phloem Transport,* M. H. Zimmerman and J. A. Milburn, eds. Springer-Verlag, Berlin; **30.25** Eberhard Fritz; **30.26** M. H. Zimmermann; **30.28** After N. A. Campbell. 1966. *Biology,* 4th ed. Benjamin/Cummings Publishing Company, Inc., Menlo Park, CA; **30.29** R. F. Evert and R. J. Mierzwa. 1986. In: *Plant Biology,* vol. 1, *Phloem Transport,* pp. 419–432. J. Cronshaw et al., eds. Alan R. Liss, New York; **30.30** A. Davidson et al. 2011. *Protoplasma* 248, 153–163.

INDEX

Note: Page numbers followed by f indicate figures; those followed by t indicate tables.

A

B

E

F

G

I

J

K

L

M

O

P

S

T

U

V

W

X

Y

Z

Metric Table

	Fundamental Unit	Quantity	Numerical Value	Symbol	English Equivalent
Area		hectare	10,000 m^2	ha	2.471 acres
Length	meter			m	39.37 inches
		kilometer	1000 (10^3) m	km	0.62137 mile
		centimeter	0.01 (10^{-2}) m	cm	0.3937 inch
		millimeter	0.001 (10^{-3}) m	mm	
		micrometer	0.000001 (10^{-6}) m	μm	
		nanometer	0.000000001 (10^{-9}) m	nm	
		angstrom	0.0000000001 (10^{-10}) m	Å	
Mass	gram			g	0.03527 ounce
		kilogram	1000 g	kg	2.2 pounds
		milligram	0.001 g	mg	
		microgram	0.000001 g	μg	
Time	second			sec	
		millisecond	0.001 sec	msec	
		microsecond	0.000001 sec	μsec	
Volume (solids)	cubic meter			m^3	35.314 cubic feet
		cubic centimeter	0.000001 m^3	cm^3	0.061 cubic inch
		cubic millimeter	0.000000001 m^3	mm^3	
Volume (liquids)	liter			l	1.06 quarts
		milliliter	0.001 liter	ml	
		microliter	0.000001 liter	μl	